34th International Symposium on Automation and Robotics in Construction and Mining (ISARC 2017)

Taipei, Taiwan
28 June – 1 July 2017

Volume 1 of 2

ISBN: 978-1-5108-4473-5

Printed from e-media with permission by:

Curran Associates, Inc.
57 Morehouse Lane
Red Hook, NY 12571

Some format issues inherent in the e-media version may also appear in this print version.

Printed by Curran Associates, Inc. (2017)

International Association for Automation and Robotics in Construction (IAARC)
Slovak University of Technology in Bratislava
Faculty of Engineering
Radlinského 11
813 68 Bratislava
SK - Slovakia

Phone: +421 2 5927 4599
Fax: +421 2 5296 8811

secretariat@iaarc.org

Additional copies of this publication are available from:

Curran Associates, Inc.
57 Morehouse Lane
Red Hook, NY 12571 USA
Phone: 845-758-0400
Fax: 845-758-2633
Email: curran@proceedings.com
Web: www.proceedings.com

ISARC 2017

Proceedings of the 34th International Symposium on Automation and Robotics in Construction

June 28-July 1, 2017
Taipei, Taiwan

Organizer

TAIWAN TECH
NATIONAL TAIWAN UNIVERSITY OF SCIENCE AND TECHNOLOGY

Co-hosted by:

CCMA

台灣建築科技中心
Taiwan Building Technology Center

CRI

I.A.A.R.C.

International Association for
Automation and Robotics in Construction

SPONSORS

ISARC 2017 would like to take this opportunity to acknowledge, with sincere appreciation, the generous contributions and support of the following sponsors:

DIAMOND SPONSOR

德昌 堅若磐石

GOLD SPONSORS

CTCI 新鼎系統股份有限公司 CTCI Advanced Systems Inc.
CTCI 萬鼎工程服務股份有限公司 CTCI Resources Engineering Inc.
CTCI 益鼎工程股份有限公司 CTCI Smart Engineering Corporation

互助營造 Fu Tsu Construction

SPONSORS

AUTODESK.

財團法人 中技社 CTCI FOUNDATION

ECS 歐亞電腦

亞新工程顧問股份有限公司 MOH AND ASSOCIATES, INC.

新北市政府工務局 New Taipei City Government Public Works Department

RUENTEX

財團法人 中興工程顧問社 SINOTECH ENGINEERING CONSULTANTS, INC.

中興工程顧問股份有限公司 SINOTECH ENGINEERING CONSULTANTS, LTD.

GOVERNMENT SUPPORTS

科技部 Ministry of Science and Technology

BUREAU OF FOREIGN TRADE

TAIPEI 觀光傳播局 Department of Information and Tourism

FOREWORD - WELCOME MESSAGE

Dear Friends and Colleagues,

On behalf of the organizing committee, we like to extend our warmest welcome to you to the 34th International Symposium on Automation and Robotics in Construction (ISARC 2017) held in Taipei, Taiwan.

This year in Taipei, we are expecting over 200 delegates and speakers from more than 30 countries. To address the needs and concerns of a global community in all fields of construction, the Scientific Committee has organized a series of workshops, Keynote sessions, Industry sessions, poster viewing session and Technical tour. We also prepare some social networking events for you to meet people for expanding knowledge base and sharing ideas.

Finally, I would like to take this opportunity to thank all the researchers, delegates, and students who working this field. Without your efforts and supports, we could not have such a successful Symposium.

We wish you a productive Symposium, and hope you enjoy your time in Taipei, Taiwan at ISARC 2017!

Prof. Min-Yuan Cheng
Chairman, ISARC 2017
Professor, National Taiwan University of Science and Technology

CONFERENCE COMMITTEES

CHAIR

Min-Yuan Cheng, National Taiwan University of Science and Technology

CO-CHAIRS

Hung-Ming Chen, National Taiwan University of Science and Technology
Jui-Sheng Chou, National Taiwan University of Science and Technology
Yo-Ming Hsieh, National Taiwan University of Science and Technology
I-Tung Yang, National Taiwan University of Science and Technology; Taiwan Construction Research Institute

LOCAL ORGANIZING COMMITTEE

Luh-Maan Chang, National Taiwan University
Shing-Tao Chang, National Cheng Kung University
Li-Chung Chao, National Kaohsiung First University of Science and Technology
Cheng-Cheng Chen, National Taiwan University of Science and Technology
Jieh-Haur Chen, National Central University
Po-Han Chen, National Taiwan University
Wei-Dong Chen, National Yunlin University of Science and Technology
Chin-Cheng Chou, National Central University
Ren-Jye Dzeng, National Chiao Tung University
Chong-Wei Feng, National Cheng Kung University
Sy-Jye Guo, National Taiwan University
Machine Hsie, National Chung Hsing University
Ting-Ya Hsieh, National Central University
Chung-Fah Huang, National Kaohsiung University of Applied Sciences
Chih-Han Kao, National Quemoy University
Chien-Ho Ko, National Pingtung University of Science and Technology
Yao-Chen Kuo, Kainan University
Hsin-Yun Lee, National Ilan University
Pin-Chan Lee, Southwest Jiaotong University
Sou-Sen Leu, National Taiwan University of Science and Technology
Chieh Lin, Executive Yuan, R.O.C. (Taiwan)
Chien-Liang Lin, National Kaohsiung First University of Science and Technology
Kuo-Liang Lin, I-Shou University
Lee-kuo Lin, National Taipei University of Technology
Shyh-Chyang Lin, National Quemoy University
Tzu-Ping Lo, Southwest Jiaotong University
Richard Moh, MAA GROUP Consulting Engineering
Nang-Fei Pan, National Cheng Kung University
Yan-Chyuan Shiau, Chung Hua University
Hui-Ping Tserng, National Taiwan University
Jen-Mu Wang, Tamkang University
Wei-Chih Wang, National Chiao Tung University

Jyh-Bin Yang, National Central University
Li-ren Yang, Tamkang University
Nie-Jia Yau, National Central University
Wen-der Yu, Chao Yang University of Technology

SCIENTIFIC COMMITTEE

IAARC BOD members

Par Ahman, Swedish Construction Federation, Sweden
Burcu Akinci, Carnegie Mellon University, USA
Tatsuo Arai, Osaka University, Japan
Carlos Balaguer, Universidad Carlos III de Madrid, Spain
Burcin Becerik-Gerber, University of Southern California, USA
Thomas Bock, Technical University Munich, Germany
Frederic Bosche, Heiot-Watt University, United Kingdom
Ioannis Brilakis, University of Cambridge, United Kingdom
Alexey Bulgakov, Ministry of Education and Science of the Russian Federation, Russia
Carlos Caldas, University of Texas at Austin, USA
Alessandro Carbonari, Universita Politecnica delle Marche, Italy
Daniel Castro-Lacouture, Georgia Institute of Technology, USA
Hung-Ming Chen, National Taiwan University of Science and Technology, Taiwan R.O.C.
Min-Yuan Cheng, National Taiwan University of Science and Technology, Taiwan R.O.C.
Moon Young Cho, Korea Institute of Construction Technology, Republic of Korea
Hiroshi Furuya, Obayashi Corporation, Japan
Ernesto Gambao, Universidad Politecnica de Madrid, Spain
Jozef Gasparik, Slovak University of Technology, Slovakia
Frans van Gassel, Eindhoven University of Technology, The Netherlands
Quang Ha, University of Technology, Sydney, Australia
Carl Haas, University of Waterloo, Canada
Chang-Soo Han, Hanyang University, Republic of Korea
Ferri Hassani, McGill University, Canada
Rauno Heikkila, Oulu University, Finland
Vineet Kamat, The University of Michigan, USA
Hyoungkwan Kim, Yonsei University, Republic of Korea
Markus König, Ruhr-Universität Bochum, Germany
Soonwook Kwon, Sungkyunkwan University, Republic of Korea
Junbok Lee, Kyung Hee University, Republic of Korea
Nan Li, Tsinghua University, China
Ger Maas, Royal BAM & Eindhoven University of Technology, The Netherlands
Osama Moselhi, Concordia University, Canada
Benny Raphael, I.I.T. Madras, India
Yehiel Rosenfeld, Technion - Israel Institute of Technology, Israel
Annop Sattineni, Auburn University, USA
Shabtai Isaac, Technion - Israel Institute of Technology, Israel
Miroslaw Skibniewski, University of Maryland, USA
Jochen Teizer, RAPIDS Safety and Technology Laboratory, Germany
Koshy Varghese, I.I.T. Madras, India
Xiangyu Wang, Curtin University, Australia

Invited members

SESSIONS AND SESSION CHAIRS

Session A-1: BIM & Computational Intelligence
Chair: Alessandro Carbonari (Polytechnic University of Marche, Italy)

Session B-1: Construction Robotics (I) / Fabrication
Chair: Carl Haas (University of Waterloo, Canada)

Session C-1: Sensing & Monitoring
Chair: Shabtai Isaac (Ben Gurion University of the Negev, Israel)

Session D-1: Interoperability
Chair: Fabiano Correa (University of Sao Paulo, Brazil)

Session E-1: Virtual Reality / Augmented Reality
Chair: Soon-Wook Kwon (Sungkyunkwan University, Republic of Korea)

Session A-2: BIM & Facility Management
Chair: Chung-Wei Feng (National Cheng Kung University, Taiwan)

Session B-2: Construction Robotics (II) / 3D Printing
Chair: Jerzy Paslawski (Poznan University of Technology, Poland)

Session C-2: Machine Learning / Computational Intelligence
Chair: Changwan Kim (Chung-Ang University, South Korea)

Session D-2: Project Management
Chair: Jieh-Haur Chen (National Central University, Taiwan)

Session E-2: Visualization
Chair: Samad Sepasgozar (University of New South Wales, Australia)

Session A-3: BIM & Visualization
Chair: Mohamed Al-Hussein (University of Alberta, Canada)

Session B-3: Construction Robotics (III) / Control & Mechanisms
Chair: Chien-Ho Ko (National Pingtung University of Science and Technology, Taiwan)

Session C-3: Computer Vision
Chair: Po-Han Chen (National Taiwan University, Taiwan)

Session D-3: Project Management
Chair: Ilija Vukorep (BTU Cottbus-Senftenberg, Germany)

Session E-3: Construction Robotics (IV) / Design
Chair: Seung Hyun Cha (Hong Kong Polytechnic University, Hong Kong)

Session A-4: BIM & Automation
Chair: Jochen Teizer (Ruhr-University Bochum, Germany)

Session B-4: Construction Robotics (III) / Inspection & Monitoring
Chair: Johnson Xuesong Shen (University of New South Wales, Sydney, Australia)

Session C-4: Computer Vision
Chair: Quang Ha (University of Technology Sydney, Australia)

Session D-4: Construction Management
Chair: Jozef GaŠparÍk (Slovak University of Technology in Bratislava, Slovakia)

Industry Session 1
Chair: Benny Raphael (I.I.T. Madras, India)

Session A-5: BIM & Project Management
Chair: Osama Moselhi (Concordia University, Canada)

Session B-5: Construction Robotics (VI) / UAV Applications
Chair: Hyoungkwan Kim (Yonsei University, South Korea)

Session C-5: Laser Scanning / Big Data
Chair: Yong Kwon Cho (Georgia Institute of Technology, United States)

Session D-5: Energy & Sustainability
Chair: Nan Li (Tsinghua University, China)

Industry Session 2
Chair: Jong Won Seo (Hanyang University, Seoul , South Korea)

Session A-6: BIM Applications

Chair: Jack C.P. Cheng (Hong Kong University of Science and Technology, Hong Kong)

Session B-6: Planning & Positioning

Chair: William O'Brien (The University of Texas at Austin, United States)

Session C-6: Computer Vision / Sensing & Monitoring

Chair: Yo-Ming Hsieh (National Taiwan University of Science and Technology, Taiwan)

Session D-6: Project Management

Chair: Pawan Pandey (IIT Delhi, India)

Industry Session 3

Chair: Sy-Jye Guo (National Taiwan University, Taiwan)

REVIEWERS

Alessandro Carbonari
Shu-Chien (Mark) Hsu
Soonwook Kwon
Hsi-Hsien Wei
Javad Majrouhi-Sardroud
Jyh-Bin Yang
Burcin Becerik-Gerber
Chung-Wei Feng
Frederic Bosche
Chien-Cheng Chou
Anoop Sattineni
Yu-Cheng Lin
Samad Sepasgozar
Jack C P Cheng
Javier Irizarry
I-Chen Wu
Yongcheol Lee
Xiangyu Wang
Jui-Sheng Chou
Markus König
Li-Chuan Lien
Vineet Kamat
Albert Chen
Shabtai Isaac
Po-Han Chen
Joshua Marshall
Jia-Ruey Chang
Xuesong (Johnson) Shen
Shih-Chung Kang
Ernesto Gambao
Ming-Hsiu Tsai
Yo-Ming Hsieh
Burcu Guldur Erkal
Jen-Mu Wang
Changwan Kim
Sou-Sen Leu
Nhat-Duc Hoang
I-Tung Yang
Jochen Teizer
Chien-Ho Ko
Koshy Varghese
Yu-Wei Wu
Yong Kwon Cho
Ali Akbarnezhad
Quang Ha
Joseph Louis
Hyoungkwan Kim
Ngoc-Tri Ngo
Yu-Ren Wang
Anh-Duc Pham
Junbok Lee
Chung-Fah Huang
Kuo-Liang Lin
Jiayu Chen
Nan Li
Tsung-Juang Wang
Daniel Castro-Lacouture
Jozef Gasparik
Wen-Der Yu
Ahmed Montaser
Osama Moselhi
Wei Tong Chen
Han-Hsiang Wang
Seung Hyun Cha
Rauno Heikkila
Li-Ren Yang
Ioannis Brilakis
Nan Li
Tsung-Juang Wang
Doddy Prayogo
Hsin-Yun Lee
Haijiang Li
Xiangyu Wang
Hung-Ming Chen

CONTENTS

34th International Symposium
on Automation and Robotics
in Construction

ISARC 2017

June 28 – July 1
TAIPEI, TAIWAN

KEYNOTE ABSTRACTS

New Technology Opportunities for Construction: the Case of 3D printing

Prof. Miroslaw Skibniewski
University of Maryland, USA

Abstract:
This keynote presentation features a historical continuum of the development of construction technologies, leading to the development of construction robotics and 3D printing techniques for construction projects. Implications of these technologies for construction supply chains will be discussed, along with future directions of research and development of new means and methods aiding the improvement of construction activities worldwide.

Keywords:
automation, robotics, construction technologies, additive manufacturing, 3D printing, supply chains

A Framework and Research Initiatives for Urban Resilience Studies

Prof. Dongping Fang
Tsinghua University, China

Abstract:
The world is undergoing rapid urbanizing at an unprecedented speed, giving rise to an increasing number of megacities worldwide. These cities, however, are constantly exposed to threats of a wide range of natural and manmade disasters, which calls for essential enhancement of level of resilience in the fast growing urban systems. This talk provides an overview of the emerging concept of resilience, and its relevance and significance to various emerging urban challenges. It is advocated in this talk that a city should be viewed as a system-of-systems that is composed of physical, social and cyber infrastructures, which in turn constitute three closely interrelated spaces of the urban system. Most challenges in urban systems are caused by incongruity among the three spaces, which increases the complexity of the challenges exponentially, while the lack of communication of experts from different domains further hinders the solutions to the challenges. In the second part of the talk, an introduction of the Institute for Future Cities and Infrastructures(IFCI) at Tsinghua University followed by an overview of its current research endeavor was provided. The vision of the IFCI is not only to integrate intelligence and resources in various academic disciplines, within and outside Tsinghua, for cutting-edge collaboration, but also to integrate and move beyond discipline-specific approaches in order to create new conceptual, theoretical, methodological, and translational innovations for addressing grand challenges in resilient urbanization. A core research task at the IFCI is to develop a city simulator, which aims to integrate the knowledge of various domains for modeling, analyzing and visualizing urban resilience, with an ultimate goal of contributing to interdisciplinary and, even better, transdisciplinary approaches for managing urban resilience. Some other current and past projects at IFCI include resilience improvement of healthcare systems, resilient cities to earthquake, interdependency modelling in networked infrastructure systems, online urban traffic guidance for extreme weather conditions, measuring the functionality of built environment, human behavior modelling, urban resilience benchmarking, and so on.

Keywords:
Research framework, system-of-systems, Urban resilience

Abstract

Prospective of Smart Semiconductor Facility in Taiwan & China Regions

by

Luh-Maan Chang

Smart Semiconductor Facility is an indispensable part of and prerequisite to Smart Semiconductor Manufacturing. Smart Facility utilizes technologies of sensors, big data, internet of things (IoT), cloud computing, artificial intelligence (AI), and cyber and physical systems (CPS) for upgrading "Industrial Automation" onto "Smart Automation."

The backbone of all these technologies is Semiconductor and Integrated Circuit (IC). It is well known that semiconductor and IC technologies have been well advanced in Taiwan region over the last 4 decades. It not only facilitates the region's economic growth, but also, profoundly enhances her international competitiveness and becomes a model benchmarked by many other regions.

On the other side of Taiwan Strait, Mainland China found out lately that the import amount of semiconductors and related products is more than the amount of imported gasoline oil. This alerted national security issue. Therefore, in her latest national 13th 5-year strategical plans (2016-2020), advanced semiconductor is set as the top priority industry needed to be further promoted. Meanwhile, trillions of dollars are funded in place for expediting the advancement. It is estimated that China region will jump over Taiwan region and becomes the top buyer of semiconductor materials and equipment in the world after 2018.

The presentation will begin with a brief introduction of semiconductor and integrated circuit following with their manufacturing processes. Then, wafer fabrication plant will be used to exemplify smart facilities that supplement smart manufacturing. Finally, the status of semiconductor industry in Taiwan and China regions will be reported. Their importance and prospective on economic strategies will be pointed out.

34th International Symposium on Automation and Robotics in Construction

ISARC 2017

June 28 – July 1

TAIPEI, TAIWAN

ACADEMIC PAPERS

BIM & COMPUTATIONAL INTELLIGENCE

While loop Algorithm to Enhance the Efficiency of Work Sampling Method in Performance Measurement

Hassan Farooq[a], Payam R Zekavat[b] and Sungkon Moon[c]

[a]Department of Civil and Construction Engineering, Swinburne University of Technology, Melbourne, VIC 3122, Australia
[b] College of Engineering & Science, Victoria University, Melbourne, VIC 3011, Australia
[c]Department of Civil and Construction Engineering; Centre for Sustainable Infrastructure, Swinburne University of Technology, Melbourne, VIC 3122, Australia (corresponding author)
E-mail: 100046172@student.swin.edu.au, payam.rahnamayiezekavat@vu.edu.au, and sungkon.moon@gmail.com

Abstract

This research covers, in part, the growing need of computerization in the construction industry. The paper highlights the lack of computerized measurement of labor performance/productivity and proposes the optimization of an existing performance measurement method with the help of a computer software. After extensive literature review, Work sampling is selected as the measurement method while MATLAB is employed for its optimization. Data collected by observing a total of four workers from two construction sites (two from each) undertaking steel works is first analyzed using conventional Work Sampling method. The same data is them analyzed using an algorithm written in MATLAB that computes multiple iterations for all number of observations. The results obtained show that work done in each category by all the workers comes out to be approximately the same for a much lower number of observations as compared to the traditional practice of taking 384 observations. For example, the percentage value added work done by Worker A is 12.12% for 384 observations while for 155 observations it comes out as 11.92% giving a percentage error of just 1.65%. Similar pattern is observed for all the workers. It is also observed that the percentage of error below 150 observation remains quite significant even if the number of iterations are increased. This is due to the increased probability of missing certain categories of work because of reduced observations.

Keywords –
Computer Aided Engineering; Performance Measurement; Work Sampling; While loop Algorithm; Rebar Placement; Productivity; Case Study

1 Introduction

In industrialized countries, a major contribution towards the gross-national product comes from the construction industry (Allmon et al., 2000), hence national economies are greatly impacted based on the performance of the construction industry. The success of a construction project relies heavily on the preparation of an accurate estimate before the project commences, as it influences both the feasibility as well as the profitability of the project (Portas and AbouRizk, 1997). Cost of labor is a key part of the overall estimate of the project and is directly dependent on the accuracy with which the labor performance and productivity is measured. Recognizing the importance of labor performance measurement, significant research has been done and a variety of performance measurement techniques are developed over the last few decades. Some of the well-established methods that have been used over the last few decades are 1) Continuous Time Study 2) Work Sampling and 3) Five-minute rating (Thomas and Daily, 1983). Although each of these methods have been employed over the past few decades, deficiencies in these methods have observed and highlighted by researchers due the complexity and fast paced nature of the modern-day construction projects (Navon, 2007, Cheung et al., 2004, Saidi et al., 2003).

Traditional performance measurement methods are labor-intensive and rely on manually collected data that heavily depends on the experience and knowledge of the observer. Furthermore, the amount of data required is significantly increased with the increase in complexity and scale of the project which leads to more observation time along with greater chance of errors in the reading (Davidson and Skibniewski, 1995). In a study carried out in western Canada, it was observed that around 40% of the time, an error of plus or minus 15% is demonstrated in the labor productivity approximations by the observers (Portas and AbouRizk, 1997). This is where automation in the construction industry is vital moving forward. By minimizing the effort put in by the observer, not only will we be able to save time and cost, but the degree of errors in the reading will also be significantly reduced.

The purpose of this paper is to take an existing performance measurement method and optimize it as much as possible with the help of computer aided techniques. This will be done by selecting a performance measurement method, collecting data from a construction site by carrying out the observations of a steel crew and then interpreting these results through traditional method and then optimizing the method with the help of computer aided software and compare them. The performance measurement method selected for the purpose of this paper is "Work Sampling" and the software selected for the optimization of the method is "MATLAB", both of which will be discussed in the following sections.

2 Literature Review

The technological advancements made in the 20th century have been unmatched over the course of human existence over a very limited period. This rapid development has led to discoveries, innovations and out dated many customary practices throughout every industry around the world. In regards to the construction industry, these advancements have provided many advantages to the betterment of the construction process, but lags other industries due to certain factors that are beyond the scope of this paper. Skibniewski (1989) suggested that all the industries are taking advantage by not only innovating and development of new technologies, but also improving on the existing practices. Henry (1994) states "soon the construction industry will experience what Alexander Graham Bell might have felt if he witnessed the evolution of fiber optics soon after inventing the telephone, or what Henry Ford may have felt upon seeing his assembly line augmented by automated robotics…".

According to Allmon et al. (2000), 10% of the "gross national product" is as a direct result from the contribution of the construction industries in industrialized countries. This is one of the major reasons why performance of the construction industry is under the microscope and greatly scrutinized given the chance. In a survey carried out by Scott et al. (1994), it was concluded that the acceptance of computers within the construction industry varies between consultants, contractor and clients. While it is almost equally accepted for engineering purposes by all the concerned parties, in case of computerization of construction management is concerned, a significant drop is observed. This is one of the reasons behind the relative underperformance of the construction industry and can only be resolved by the integration of computers in Construction Management.

2.1 Computer Aided Construction Management

In the past, use of computers in the construction industry was scarce and was restricted to particular areas, if at all. However, the development in computers over the past decades has taken the world by storm and has become a necessity for survival in every industry due to the fast paced and competitive environment. Computerization of construction management has become essential moving forward, as computers are considered a vital management tool for achieving the targets efficiently and effectively. Yamazaki (1992) stated that automation of the construction industry has already begun and developments are underway that would soon be applied practically in every construction project across all its different phases which include designing, manufacturing and construction. The construction industry is heavily reliant on the effective cooperation between all the concerned parties which calls for widespread organizational planning and management structure to the complexities and scale of construction projects these days. These management tools would not only bridge the gap between the different parties, but also provide new and efficient methods to optimize the construction phase will lead to problems such as inadequate monitoring of activities, inefficient evaluation of works and lack control over the pace of the project which would consequently reduce the productivity triggering hefty losses.

Skibniewski (1989) believed that computers not only provide new and efficient techniques, they have also enabled the modification of existing methods. This provides us with more accurate results with minimal effort and saves time and cost. The improvement in the existing construction management techniques is down to the ever-advancing technology sector. Apart from the modification of existing methods, the advancement in computer technology has led to the possibility of coming up with unique and innovative designs, better visualization and communication, efficient scheduling and much more. The use of computers is important in all the aspects related to construction management, but the most common areas of computerized construction management are as follows (Meredith and Mantel Jr, 2011, Rizki, 2003)

- Project Planning
- Resource Management
- Tracking/Monitoring
- Report Generation
- Decision Making

Estimations are a key aspect related to each construction activity and detailed estimates are carried out even before the tendering stage. Almost all the construction activities can be estimated to a higher degree of accuracy as they depend on costs that are quantifiable easily such as equipment, repairs and resources. But the one factor that is dynamic in nature and is difficult to predict is the performance and productivity measurement of labour (Henry, 1994). This is because of the various factors involved while measuring the performance such as the condition of the site, the scope of the work, overtime, ability of the crew, unexpected delays, changes in plans and much more. Due to these factors and the amount of data to be processed, efforts have been made to introduce computer aided methods for effective and easier evaluation. Some of the methods have been discussed in the following section.

2.2 Computerized Labor Performance Measurement Methods

The complexity and various factors involved (such as on-site conditions, over time, worker proficiency etc.) in the measurement of labour performance and productivity has resulted in very little research in the area (Navon, 2005). However, some efforts in labour tracking have been made over the past few years with each method having a human factor involved in it.

"Radio frequency identification" RFID or barcodes have been suggested to gather data on workers (Jaselskis and El-Misalami, 2003). Electric forms along with the use of spreadsheets has also been proposed as a method to gather data by different researchers (Fayek et al., 1998, Hegazy and Ersahin, 2001). Echeverry and Beltrán (1997) suggested a method that consisted of three components for labour tracking 1) Construction project's plans in a database 2) manual or automated data entry 3) module for carrying out the analysis.

A more frequently used and industry approved approach to measure labour productivity/performance which widely is the "Work sampling" technique.This employs an observer who takes readings at random intervals during the day and notes the activity being undertaken along with the detail of the work being done by each worker at that observation interval. Data collected is then classified into a pre-set work categorization. The categorization is into productive, contributory. unproductive, personal and ineffective (Bernold and AbouRizk, 2010). Relating work percentages based on the categorization gives the performance measure for each worker which is generated either manually or through computers, the latter being the more obvious choice as already discussed.

The one drawback observed in all of the above mentioned methods is the reliance on observers and/or the workers themselves for entering the data into the model/system. This leads to uncertainties and inaccuracies that are observed in the existing manual procedures of data collection and performance measurement (Navon, 2005). Hence, to find remedial measures computer software can be utilized in order to maximize the efficiency of the existing measurement techniques. For the purpose of this paper, work sampling is considered as the measurement technique that will be studied and optimized with the help of computers. In order to accomplish that, there is a need to first understand the concept behind work sampling followed by the selection of a software that will help us improve the method.

2.3 Work Sampling

The basis of this method was laid by Tippett (1935), when he suggested a different method from continuous time study. He proposed that at random intervals "screenshots" should be saved of the work going on, which would save time and will be mostly accurate.

Malisiovas (2010) suggested that work sampling is a method to measure the time taken by a worker in the defined categories of work with the help of statistical sampling theory. The observation in this method is carried out at random intervals and the results are inferred from this information. Statistical sampling theory is employed because a limited group is observed for a limited period as it establishes a limit of confidence interval which caters for the error that might emerge. The idea of this method is to figure out the categorization of work being done out by the workers and spot the reasons and areas where the work efficiency is being affected to make the required changes (Liou and Borcherding, 1986).

Following are the requirements that must be fulfilled for conduction of work sampling. (Baxendale, 1998):

- "The work to be observed must be cyclic in nature
- The work cycles must not be repetitive
- The work must have multiple workers and sufficient observation time"

2.4 Computer Software

There is a wide range of computer software available that can be used in this research. Microsoft Visual Studio (C/C++), Python, Fortran and MATLAB are some examples. After reviewing the literature MATLAB was selected based on its variety of features. According to Hanselman and Littlefield (1997), MATLAB has a huge database consisting of built-in algorithms that can be tried immediately without recompilation. The desktop environment helps the user to interact with the data while working and keeping track of the files and all the different variables at the same time. They further describe the software as easy to use and the code written in it can be exported to other software which can be very useful in future studies. As already established, MATLAB has a wide range of functions that can be utilized depending on the use of the programmer. For the purpose of this research, the functions used in the code are discussed in detail under the Research Methodology section.

After reviewing the literature, it is evident that computer integration in the construction industry is imperative and although it is gaining popularity in most of the areas concerning construction processes, a gap in knowledge regarding the automated measurement of labor performance and productivity is observed due to the various factors discussed in the above sections. This gap in knowledge will be addressed in this paper by selecting a labor performance measurement technique that is being frequently used and come up with a solution, using computer software that will reduce the human involvement, utilize less time and effort while reducing the cost of the procedure.

3 Problem Statement and Research Hypothesis

The current performance measurement methods are time consuming, costly, labor-intensive and highly dependent on the competency of the observer. This is the reason construction managers avoid the use of these methods as much as possible which leads to insufficient information on the labor performance and productivity that may cause harm during the construction phase or at the later stages of the project. Utilizing the massive advancements in computer technology for the upgrading/optimization of these existing methods will help overcome these issues and encourage construction managers to adopt these methods which will be beneficial for their projects.

Reducing the number of observations required in the current Work Sampling technique for the measurement of labor productivity with the help of an algorithm developed in MATLAB, a computer software that will reduce the effort, time and cost required to undertake the procedure while maintaining its accuracy.

4 Research Methodology

After extensive literature review, the performance measurement technique nominated for this paper was Work Sampling and the computer software used for the improvement was MATLAB. The procedural explanations for both have been discussed in the previous sections. The distinctive characteristic about the construction industry is the uniqueness of each construction project and provides a different challenge each time. Keeping this in mind, the measurement procedure employed must be adjusted accordingly. For the scope of this paper, the work sampling was performed on a steel fixing crew and the data collected was through two mediums; digital and real time. The digital recording served as a purpose of keeping a back-up as well as a source of double checking the observations.

The data collected was then evaluated through traditional work sampling methods by calculation the percentage of work done in each category. The same data was then run through an algorithm generated in MATLAB to find out the best way to maximize the efficiency in terms of time, cost and effort utilized to carry out the procedure.

5 Case Studies

For this paper, data was collected by observing steel workers from two different construction sites. The description of each is as follows

5.1 Case-1: Steel Fixing Works for Residential Apartments

This construction site was located in South Yara in Melbourne, Victoria approximately 6km from CBD. The building has a planned capacity of housing 92 apartments, has 9 floors and stands 28.1m above the ground. The concerned authority is the Stonington Local Government. The project was a development by "Salta Properties" while the construction responsibilities were undertaken by "MERKON Construction".

The data was collected for the steel fixing of the top level by observing a three-man crew, which consisted of two steel workers and a foreman. The workers were referred to as Worker A and Worker B. For ease of shifting and avoiding disruptions during work, the steel was stockpiled at various intervals.

(1) Categorization of Work

The purpose of categorization of work has been discussed in the above sections. Following is the framework generated for this case study

Table 1. Categorization of work

Category	Content
Value Added Work	1. The work done that contributes directly to the physical completion of an activity, e.g. placing, aligning and tie works
Contributory Work	2. Managing and moving essential tools and materials within the workplace that contribute to the value added work. These comprise of reinforcement tie wires, torches, sledges, wire cutters etc. 3. Demarcation and clearing of work 4. Referring to drawings, resolving issues with the help of supervisor
Ineffective Work	5. Moving objects to a distance more than 35ft. 6. Walking bare handed

	7. Searching for materials or tools
Unproductive Work	8. Waiting for others to complete or the supervisor for instructions
	9. Redoing of work
Personal	10. Undue breaks and idle time
	11. Unobservable

(2) Analysis by Work Sampling

After the categorization of work is completed, the number of observations required for obtaining accurate results is determined. Bernold and AbouRizk (2010) came up with the following formula for the ideal number of observations

"N = k^2 x p(1-p) ÷ s^2

N = No. of observations

K = Standard Deviations for given confidence limit

s = absolute limit of error

p = unobservable probability (decimal)"

As a general rule, the confidence level is set at 95%, the limit of error as 5% and the "unobservable probability value" as 0.5. Z-score can be understood by referring to the concept of normal distribution. The k-value can be generated in excel which gives the standard deviation of the confidence level. Substituting these values gives the following:

Table 2. No of Observations

Confidence level	C.L	95
z-score	Z	0.025
		0.975
No. of SD for confidence level	k value	-1.959964
		1.959964
Chance that element might not be visible	p	0.5
Absolute level of error	s	0.05
No. of observations	N	384

This value of 384 observations is considered to give the most accurate measure of labour performance through work sampling. For comparison purposes, the results for just 20 observations were also calculated to show the significant difference in the results obtained. The results obtained from the actual work sampling carried out by taking 384 observations is shown in the following table. It is to be noted that the results obtained will be quite different as observed for number of observations less than 384.

Table 3. Work Sampling Results for 384 Observations- Case 1

	Percentage of Work Done for 384 Observations	
Category of Work	**Worker A**	**Worker B**
Value Added	13.76%	27.27%
Contributory	43.11%	27.00%
Ineffective	15.06%	17.66%
Unproductive	13.77%	15.06%
Personal	14.28%	11.43%

Table 3 gives the results for 384 observations. A. In case of Worker A, the results obtained from 20 observations (not shown here) show 20% of the work done as value added while it comes out only as 13.76% from 384 observations which is accepted to be more accurate. Similar observation can be made if the comparison between the percentages of work done for 20 observations is made against 384 observations, for both the workers. This emphasizes the importance of large set of observations. Having said that, there is need to find out the most efficient number of observations required to get accurate results which would make this method easier to apply.

(3) Analysis through MATLAB

The broader range of MATLAB's capabilities have already been discussed earlier. The algorithm used for the purpose of this research was a combination of multiple separate algorithms. The first step was to define the number of observations to be carried out denoted by "x", these were taken as a variable with the purpose of finding out the minimum number of observations instead of the customary 384 observations. Using "while loop" the algorithm was made to run till the number of observations "x" reach 0 with the intention of getting an overview of the behaviour over all the number of observations. As the numbers "x" are generated randomly without any pre-set criteria, the algorithm was written such that it conducted 20 iterations in order to obtain an accurate result, the variable H was selected for it. The "x" number of observations were randomly generated from time 0 to the end of the observation period (5719 seconds in this case), stored under the variable "r". After this, the ranges observed for the work done in each category were defined under the letters VA, C, P, I and U denoting "Value added", "Contributory", "Personal", "ineffective and "unproductive", respectively. "for loop" was used to assign these letters to their corresponding randomly generated "r" values. Percentages of each category of work was calculated after each iteration and a mean of the 20 iterations was calculated for each value of "x". To keep count of the "while loop" for the number of iterations run, a variable "count" was set. In order to get the results in excel, the data was converted into cell using the "num2cell" function of MATLAB. After the algorithm is run, the results are generated and saved in an excel sheet. The results are also plotted with the help of "plot" function in MATLAB as shown in figure 2 and 3.

This algorithm runs 20 times for each value of "x" which starts from 384 observations and ends at 0. This comes out as 1,478,400 iterations in total which can't be shown in a table. The results obtained in excel sheet are the mean of each value of "x".

For example, for the value of x=384 the algorithm will run 20 iterations for 384 randomly generated numbers between 0 to the end of observation time (i-e.5719), assigning each number with its corresponding work category denoted by either VA, C, P, I or U and calculating their percentages in each iteration. After this the algorithm will run again for 383 observations and repeat the steps above till the time "x" equals 0. The mean of all the iterations will then be generated in excel followed by a plot in MATLAB. As the number of observations are a lot, only a screenshot of the means generated in excel file is shown in figure 1 for a few iterations carried out for Worker A.

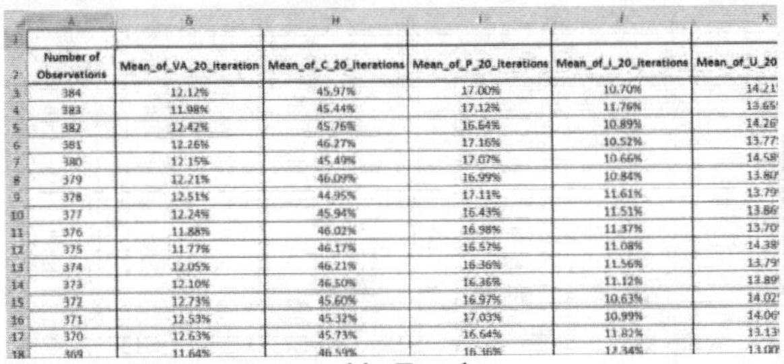

Number of Observations	Mean_of_VA_20_iteration	Mean_of_C_20_iterations	Mean_of_P_20_iterations	Mean_of_I_20_iterations	Mean_of_U_20
384	12.12%	45.97%	17.00%	10.70%	14.21
383	11.98%	45.44%	17.12%	11.76%	13.65
382	12.42%	45.76%	16.64%	10.89%	14.26
381	12.26%	46.27%	17.16%	10.52%	13.77
380	12.15%	45.49%	17.07%	10.66%	14.58
379	12.21%	46.09%	16.99%	10.84%	13.87
378	12.51%	44.95%	17.11%	11.61%	13.79
377	12.24%	45.94%	16.43%	11.51%	13.86
376	11.88%	46.02%	16.98%	11.37%	13.70
375	11.77%	46.17%	16.57%	11.08%	14.38
374	12.05%	46.21%	16.36%	11.56%	13.79
373	12.10%	46.50%	16.36%	11.12%	13.89
372	12.73%	45.60%	16.97%	10.63%	14.02
371	12.53%	45.32%	17.03%	10.99%	14.06
370	12.63%	45.73%	16.64%	11.82%	13.13
369	11.64%	46.59%	16.36%	12.34%	13.07

Figure 1. Means generated in Excel

The plots for the means of all 384 observations for both Worker A and Worker B are shown in the figures below. The mean percentages of work done of 20 iterations for 384 observations by Worker A and Worker B comes out to be 12.12% and 30.78%, respectively (obtained from respective excel sheets). It is to be noted, that these results might differ from conventional WS methods as it is taking a mean of 20 iterations which provides a far more accurate result. Work sampling technique dictates that taking 384 observations results in the most accurate observations, but as seen from both the figures, the percentage of Value added work along with all the other categories remain approximately the same up to a much lower number of observations as obtained from 384 observations. In case of Worker A, the values remain approximately the same up to 150 observations while for Worker B the results start to deviate around 175 observations.

5.2 Case-2: Rebar-Placement for School Building

This construction site was located in Kensington, NSW, Australia for the development of Basser and Baxter College building. The study was undertaken for the performance measurement of rebar placement by observing four workers. Beams L4B10 situated on the west sector of the building and two workers working on the beam were selected for the experiment. The workers on L4B10 were referred to as Worker C and Worker D.

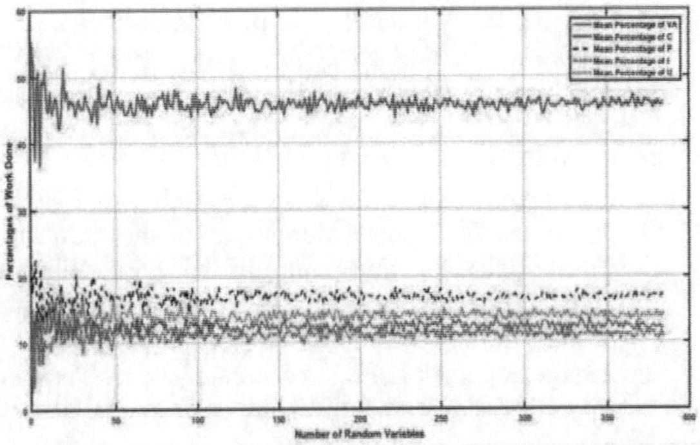

(a) Worker A

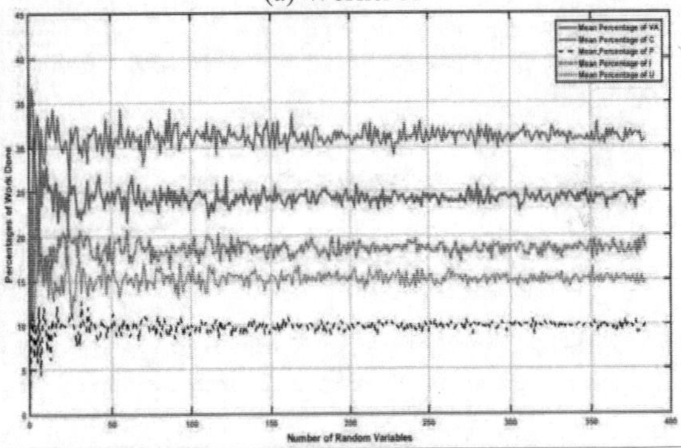

(b) Worker B

Figure 2. Percentages of Work Done for each number of observations

(1) Categorization of Work

The framework adopted is the same as used for case 1

(2) Analysis by Work Sampling

The comparison between 20 observations and 384 have been shown in the previous section.. For this case, only the results obtained from 384 observations are demonstrated below

Table 5. Work Sampling Results for 384 Observations- Case 2

Category of Work	Percentage of Work Done for 384 Observations	
	Worker C	Worker D
Value Added	45.71%	31.17%
Contributory	0.52%	21.82%
Ineffective	13.25%	32.21%
Unproductive	33%	7.01%
Personal	4.94%	6.23%

Table 5 provides the percentage of work done in each category by the two workers. Worker C has more value added work with 45.71% while worker D has a higher percentage for contributory work with 21.82%. It is observed that worker D has relatively less time spent in unproductive works with a cumulative percentage of

45.44% while Worker C, despite doing more value added work, spends 51.21% in the works that are unnecessary and avoidable.

(3) Analysis through MATLAB

The algorithm used for this case study is the same as the previous. The main difference is in the duration of the study undertaken and nature of the work. While the previous was approximately 90 minutes, this case study was done for the duration of an hour. The plot for the percentages of work done in each category by Worker C and Worker D are shown in Figure 3.

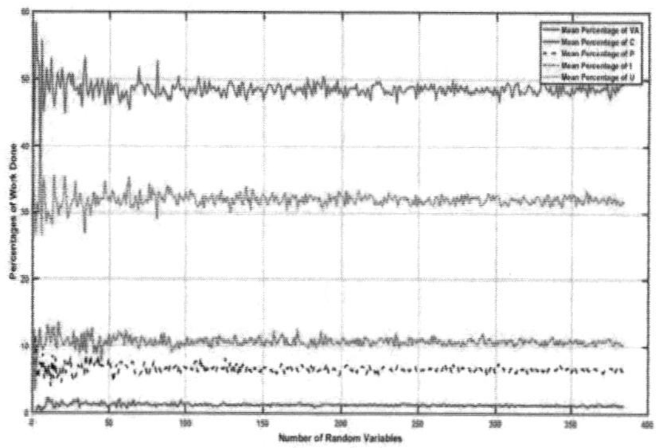

(a) Worker C

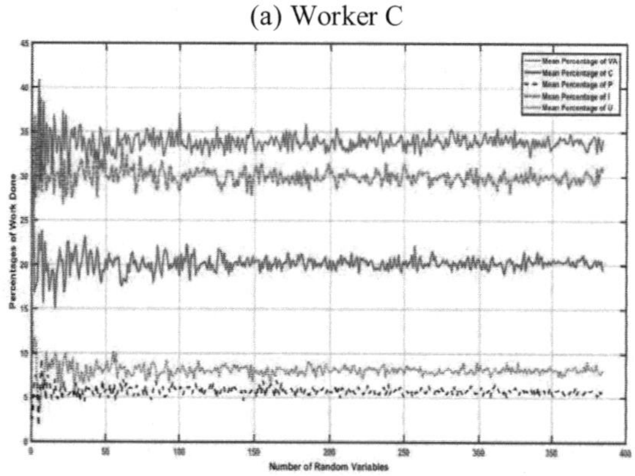

(b) Worker D

Figure 3. Percentages of Work Done for each number of observations

A similar pattern for this case study is observed as in the previous. Taking percentage of Value added work as an example, the percentages for Worker C and Worker D come out as 49.38% and 34.10% (obtained from excel sheets), respectively for the mean of 20 iterations carried out for 384 observations. This percentage of value added work along with the other categories of work remain approximately the same up to a much lower number in each case. The deviation is observed below 150 number of observations for both workers A and B.

Figures 4 represents the percentages or work done in each category by worker C and worker D for 150 observations and 50 iterations.

The results obtained with increased iteration and less observations than 150 show that the deviation is immediate and no pattern in observable. This concludes that while decreasing the observations is possible, it can only be done up to an extent. Afterwards the probability of not observing all the categorising of work increases and accuracy decreases.

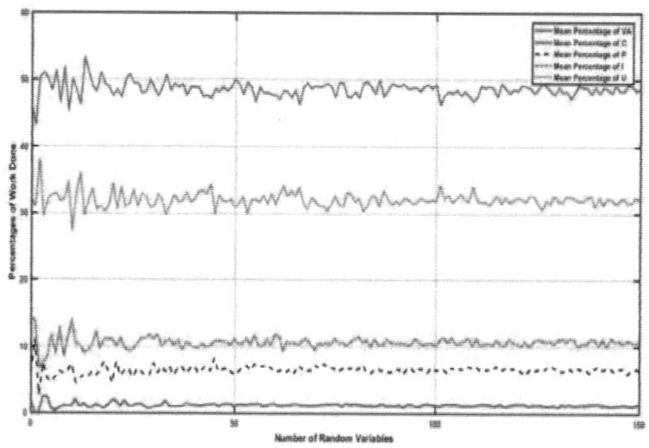

(a) Worker C

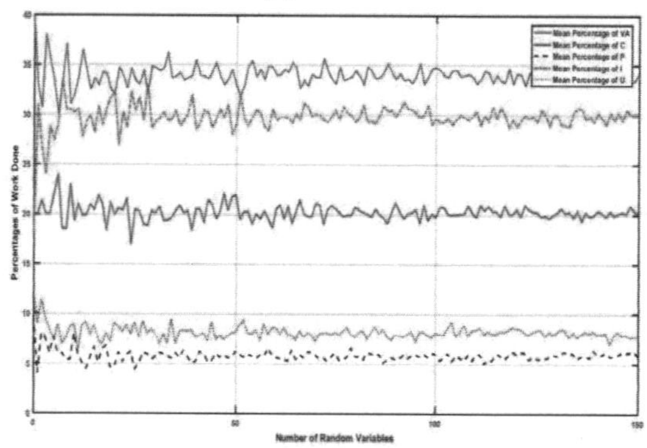

(b) Worker D

Figure 4. Percentages of Work Done number of observations between 0 and 150

6 Conclusion, Limitations, and Future Works

The advancement in technology has benefited every industry over the past few decades. However, the construction industry is lagging in this regard as compared with manufacturing, medicine, etc. One of the major reason is the resistance and hesitation to adapt with new or improved methods. Computerization is observed in most sections of the construction industry, such as design, tendering, estimation (in most cases) etc. However, labour performance measurement is neglected due to dynamic factors that makes it very difficult to predict. Although many conventional performance measurement methods are available; the time, effort and cost associated with them restrains the construction managers from using them that often causes in the estimation or in the management of labour during the later stages of the construction phase. This paper addresses this issue by selecting one conventional method and optimizing it, as much as possible, such that it employs less effort, time and cost and encourages construction managers to utilize them. This would result in accurate estimates of labour that would consequently increase the productivity.

The conventional method selected was the Work sampling (WS) and the software used to optimize it is MATLAB. Two case studies are included in the paper, data from both of which is analysed through conventional WS method and MATLAB. It is concluded that the number of observations required to carry out WS are too many and can be significantly reduced. The results obtained from case studies show that the percentages of work done in each category for Worker A, C and D remains approximately the same for observation between 384 and 150 for 20 iterations, while for Worker B the results deviate below 175 observations. Hence, a lower number of observations yields the same results as from the conventional method, thus saving time, effort and cost and improving the productivity in the long run. The number of observations is further reduced if the iterations are increased as evident from the case studies.

Due to time constraints the number of workers observed are too less and recommends that future studies be carried out for more worker resulting in a firmer conclusion regarding the optimization. Another limitation was the lack of software knowledge, Software that can receive data in real time will greatly reduce the processing time and reduce human involvement in the labor performance and productivity measurement.

REFERENCES

[1] ALLMON, E., HAAS, C. T., BORCHERDING, J. D. & GOODRUM, P. M. 2000. US construction labor productivity trends, 1970-1998. Journal of Construction Engineering and Management, 126, 97-104.

[2] BAXENDALE, T. 1998. Construction Operations Management, Chandos.

[3] BERNOLD, L. E. & ABOURIZK, S. M. 2010. Managing performance in construction, John Wiley & Sons.

[4] CHEUNG, S. O., SUEN, H. C. & CHEUNG, K. K. 2004. PPMS: a web-based construction project performance monitoring system. Automation in construction, 13, 361-376.

[5] DAVIDSON, I. N. & SKIBNIEWSKI, M. J. 1995. Simulation of automated data collection in buildings. Journal of computing in civil engineering, 9, 9-20.

[6] ECHEVERRY, D. & BELTRÁN, A. Bar-code control of construction field personnel and construction materials. Computing in Civil Engineering, 1997. ASCE, 341-347.

[7] FAYEK, A., ABOURIZK, S. & BOYD, B. Implementation of automated site data collection with a medium-sized contractor. Computing in Civil Engineering, 1998. ASCE, 454-456.

[8] HANSELMAN, D. & LITTLEFIELD, B. C. 1997. Mastering MATLAB 5: A comprehensive tutorial and reference, Prentice Hall PTR.

[9] HEGAZY, T. & ERSAHIN, T. 2001. Simplified spreadsheet solutions. I: Subcontractor information system. Journal of Construction Engineering and Management, 127, 461-468.

[10] HENRY, R. M. 1994. Role of advanced computer technology in construction industry. Journal of computing in civil engineering, 8, 385-389.

[11] JASELSKIS, E. J. & EL-MISALAMI, T. 2003. Implementing radio frequency identification in the construction process. Journal of Construction Engineering and Management, 129, 680-688.

[12] JAYAWEERASINGHAM, G. P. 2011. Effectiveness of computer applications in management of building construction projects in Sri Lanka.

[13] JENKINS, J. L. & ORTH, D. L. 2003. Productivity improvement through work sampling. AACE International Transactions, CS51.

[14] KANGARI, R. 1986. Major factors in robotization of construction operations. CAD and Robotics in Architecture and Construction. Springer.

[15] KUMAR, A. R. & RAJAK, F. Computer Aided Construction Management.

[16] LIOU, F.-S. & BORCHERDING, J. D. 1986. Work sampling can predict unit rate productivity. Journal of Construction Engineering and Management, 112, 90-103.

[17] MALISIOVAS, A. N. 2010. Construction Productivity: From measurement to improvement. Advancing Project Management for the 21st Century Concepts, Tools & Techniques for Managing Successful Projects. Heraklion, Crete, Greece.

[18] MEREDITH, J. R. & MANTEL JR, S. J. 2011. Project management: a managerial approach, John Wiley & Sons.

[19] NAVON, R. 2005. Automated project performance control of construction projects. Automation in construction, 14, 467-476.

[20] NAVON, R. 2007. Research in automated measurement of project performance indicators. Automation in Construction, 16, 176-188.

[21] PORTAS, J. & ABOURIZK, S. 1997. Neural network model for estimating construction productivity. Journal of construction engineering and management, 123, 399-410.

[22] RITZ, G. J. 1994. Total construction project management.

[23] RIZKI, S. 2003. Web-Based Construction Management System in IIT Roorkee. M. Tech. Dissertation, Department of Civil Engineering, IIT Roorkee.

[24] ROJAS, E. M. 2008. Construction productivity: A practical guide for building and electrical contractors, J. Ross Publishing.

[25] SAIDI, K. S., LYTLE, A. M. & STONE, W. C. Report of the NIST workshop on data exchange standards at the construction job site. Proc. of 20th International Symposium on Automation and Robotics in Construction (ISARC), 2003. Citeseer, 617-622.

[26] SCOTT, N., PONNIAH, D. & SAUD, B. 1994. Computing and the construction industry. Structural survey, 12, 10-14.

[27] SKIBNIEWSKI, M. 1989. ROBOTICS ROUNDUP-ROBOTICS IN AMERICA. Civil Engineering, 59, 78-79.

[28] THOMAS, H. R. & DAILY, J. 1983. Crew performance measurement via activity sampling. Journal of Construction Engineering and Management, 109, 309-320.

[29] TIPPETT, L. 1935. 2—STATISTICAL METHODS IN TEXTILE RESEARCH. PART 2—USES OF THE BINOMIAL AND POISSON DISTRIBUTIONS. Journal of the Textile Institute Transactions, 26, T13-T50.

[30] YAMAZAKI, Y. 1992. Integrated design and construction planning system for computer integrated construction. Automation in Construction, 1, 21-26.

[31] PWC 2015-2016. Global Construction 2030: a global forecast for the construction industry to 2030.

Employing outlier and novelty detection for checking the integrity of BIM to IFC entity associations

Bonsang Koo[a], Byungjin Shin[a], Thomas F. Krijnen[b]

[a] Department of Civil Engineering, Seoul National University of Science and Technology, South Korea
[b] Department of the Built Environment, Eindhoven University of Technology, The Netherlands
E-mail: bonsang@seoultech.ac.kr, byungjin0826@seoultech.ac.kr, t.f.krijnen@tue.nl

Abstract – Although Industry Foundation Classes (IFC) provide standards for exchanging Building Information Modeling (BIM) data, authoring tools still require manual mapping between BIM entities and IFC classes. This leads to errors and omissions, which results in corrupted data exchanges that are unreliable and compromise the interoperability of BIM models. This research explored the use of two machine learning techniques for identifying anomalies, namely outlier and novelty detection to determine the integrity of IFC classes to BIM entity mappings. Both approaches were tested on three BIM models, to test their accuracy in identifying misclassifications. Results showed that outlier detection, which uses Mahalanobis distances, had difficulties when several types of dissimilar elements existed in a single IFC class and conversely was not applicable for IFC classes with insufficient number of elements. Novelty detection, using one-class SVM, was trained *a priori* on elements with dissimilar geometry. By creating multiple inlier boundaries, novelty detection resolved the limitations encountered in the former approach, and consequently performed better in identifying outliers correctly.

Keywords –
BIM; IFC; Outlier detection; Novelty detection

1 Introduction

Models based on the Building Information Modeling (BIM) paradigm are increasingly being used for multiple applications, including clash detection, building code compliance, design quality assurance, constructability analysis, design and construction coordination. Many of these applications require specialized software, which in turn require BIM models to be exported and exchanged between them.

The Industry Foundation Classes (IFC) plays a pivotal role in enabling interoperability, allowing entity and relationship data to be exchanged seamlessly between applications. However, the IFCs does not necessarily guarantee that the integrity of the data is maintained. For example, major BIM authoring tools, although abide by the Model View Definitions (MVD) in exporting data, still allows individual components and relationships to be mapped to IFC classes manually, and are thus susceptible to human errors and omissions.

Moreover, BIM models are becoming larger and more complex. Depending on the Level of Development (LOD), the number of elements in a single model can range from 1,000's to 10,000 components. Manually checking the integrity of IFC entity and relationship mapping can quickly become intractable, as the size and complexity of the models increases.

This research addressed this issue by applying machine learning techniques to identify errors or omissions in the data integrity of IFC models. Specifically, using geometric features, anomaly detection techniques are applied to determine whether BIM components have been properly mapped to their correct IFC classifications.

The research used existing work performed by from [1] as its initial point of departure. [1] proposed using 'outlier detection' to check the correct classification of individual elements. To verify the scalability of this approach, we first performed outlier detection on two BIM models with increasing components, from which we identified specific limitations. Subsequently, we explored an alternative anomaly detection approach, namely, 'novelty detection' which proved to be more effective in identifying potential misclassifications.

The research conducted herein is anticipated to enhance the robustness of IFC usage, and contribute to the proliferation of machine learning techniques in the AEC domain, including areas of quality control and regulation compliance.

The rest of the paper is structured as follows. Section 2 provides background on the state of IFC development and the need for IFC integrity checking. Section 3 provides an overview of the research methodology, while Section 4 describes the results and limitations of using outlier detection on two architectural BIM models. Section 5 describes how novelty detection was incorporated, and tested on the same architectural models.

Section 6 compares the results and discusses the implications of the research.

2 Motivating Background

2.1 The need for IFC integrity checking

Major BIM authoring tools provide functionalities to selectively map BIM elements to their corresponding IFC classes. These tools offer default settings that have prespecified mappings of the most common elements to their corresponding IFC classes. However, such is not the case for more obscure IFC classes, which requires manual settings, which may lead to errors and omissions.

Also, many authoring tools allow the use of generic libraries, which do not pertain to a specific IFC class. For example, Revit allows the creation of custom families or model-in-place components, which does not map to specific IFC classes, unless specified otherwise by the model developer.

Furthermore, authors of BIM models may not care for the strict designation of IFC classes, or even the BIM elements themselves, and loosely use families of their choosing or even customized ones.

Such misclassifications, whereas trivial in small BIM models, may become difficult to detect manually once BIM models become large and with detailed level of developments.

In a collaboration environment, where project stakeholders individually develop BIM models, such misclassifications can cause severe interoperability issues between them.

The next section briefly introduces tools and standards that have been developed to ensure IFC model integrity checking, followed by a summary of their limitations.

2.2 Existing approaches for IFC integrity checking

Although IFCs provide a standardized format to share BIM information, the complexity in its schema often requires a domain expert experienced in STEP and EXPRESS to verify its integrity. Practitioners without such knowledge find it difficult to readily employ IFC based models for everyday use [2]. Consequently, several advancements have been made that supports the checking of IFC file formats and support users in ensuring their integrity.

2.2.1 Tools for checking the integrity of IFC data structures

buildingSMART International (bsI), the main organization that manages and develops IFC standards, provides the Information Delivery Manual (IDM) and Model View Definition (MVD) that allow processes to be formalized and generate subsets of IFC entities and relationships. MVDs such as the Coordination View and COBie are used extensively in the industry.

bsI also provides an 'ifcDoc' tool to check the consistent and computer-interpretable definition of MVDs as legitimate subsets of the IFC specification with enhanced definition of concepts.

[3] developed an 'mvdXML Checker' to evaluate the integrity of IFC files, while the commercial software 'Solibri Model Checker' is used widely in the industry to check the conformity of IFC class and entity data between BIM authoring tools. The National Institute of Standards and Technology (NIST) provides the 'IFC File Analyzer,' which enables a semi-automated approach to verify IFC classes and relationships in an IFC-SPF file.

2.2.2 Development of BIM Query Languages

The complexity of the IFC format has also created the need for BIM Query Languages. These languages allow the development of SQL based statements using {SELECT, FROM, WHERE} command constructs to query BIM models and IFC-SPF files. Initially, query languages based on EXPRESS and EQL were explored Tauscher et at., 2016. Later, BIMQL [4] and QL4BIM [5] were developed, which were customized exclusively for BIM/IFC models, as well as general purpose query languages using ifcOWL and SPARQL [6]. Such advances allowed specific querying of IFC-SPF files, which could be utilized to check for their integrity.

2.2.3 BIM Modeling Standards

A more macro-level approach has been where AEC institutions have provided standards and guidelines for working with BIM models and IFC formats. Namely, the American Institute of Architects (AIA)'s provides the 'Documents E203 and G202: Building Information Modeling Protocol Exhibit' [7], while UK's Construction Industry Council (CIC) provides the 'BIM Protocol.' Similar attempts have been developed for Korea in the form of 'KBIMS[1].'

Such standards provide guidance in standard work breakdown structures, the Level of Development (LOD) of BIM models, libraries templates, and project management matrices to ensure that interoperability is maintained throughout the project life cycle and between project stakeholders sharing multiple BIM models. These guidelines can assist in ensuring that BIM/IFC models are correctly mapped to ensure their integrity.

[1] Korea Building Information Modeling Standards

2.3 Limitations of existing approaches

The variety of tools for evaluating IFC entities and relationships improves the deciphering of IFC data, but it is still by and large a manual process. Similarly, BIM query languages can improve the checking ability but essentially has not been developed for ensuring IFC integrity. The standards and guidelines encourage the correct associations and mapping of elements through best practices, but do not necessarily guarantee them. Thus, there still exists a need for tools or methods directly dedicated to IFC integrity checking, and is the main objective of this research.

3 Research Methodology

This research explored two approaches for identifying misclassifications of BIM elements to IFC classes using anomaly detection techniques: outlier detection and novelty detection.

The first approach is adopted from [1]. Krijnen and Tamke, here after referred as "Krijnen's approach", explored the use geometric features of individual components to detect misclassifications using outlier detection. However, at least in the paper, the approach is not fully validated in terms of its scalability to generic BIM models. They only provide an example implementation to one IFC class (i.e., walls) within a single BIM model. Thus, the first step in our research involved identifying potential limitations in their approach by applying their implementation to multiple BIM models with larger number of BIM elements and IFC classes.

Based on these results, a second approach was devised, which primarily used novelty detection as an alternative to address the limitations found in Krijnen's approach. We selected novelty detection as it allows multiple boundaries for inliers, whereas outlier detection typically is limited to a single inlier boundary. Also, while Krjinen's approach limited the analysis to elements within a single BIM, the novelty detection approach was implemented to learn features from multiple BIM models.

4 Verification of Krijnen's approach using outlier detection

This section provides an overview of Krijnen's approach, and two cases of its applications to BIM models. Results of the case studies were used to identify limitations of the approach and provide the basis for implementing novelty detection.

4.1 Overview of Krijnen's approach

As shown in Figure 1, given a single BIM model, outlier detection is performed on individual IFC classes (i.e., ifcWallStandardCase, ifcWindow, ifcDoor, etc.) to detect potential model components that are geometrically dissimilar to other components of the same class. The geometries are used as features for the analysis, and include the area, volume, radius of gyration, orientation from top and bottom of the individual components. The premise is such that elements of the same IFC class should have similar geometric features, and thus a misclassified component will stand out and be detected as an outlier.

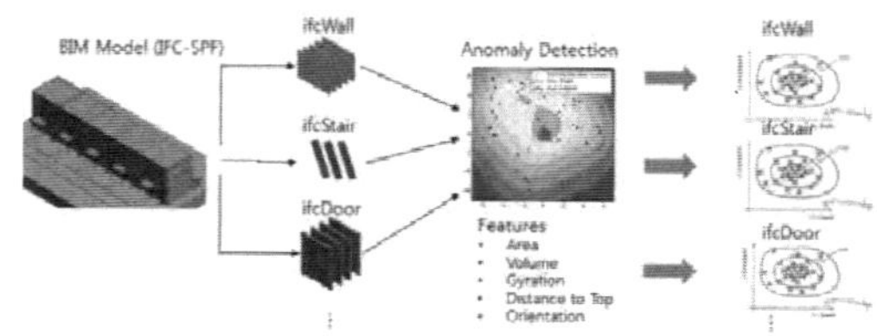

Figure 1. Process overview for outlier detection

[1] developed and applied a suite of open source Python packages to implement their approach. Specifically, ifcOpenShell [8] allows the manipulation and query of IFC entities directly from IFC-SPF files. PythonOCC [9] is also used to extract individual geometric features needed for outlier detection. Finally, the packages from scikit-learn toolkit are used for implementation of the outlier detection algorithm [10].

[1] provides an example by applying their implementation on the wall elements (i.e., ifcWallStandardCase) of a duplex apartment BIM model [10]. Their approach identified several elements which are classified as walls but should have been classified as a beam or opening. The results are visually demonstrated using a contour plot (i.e., elliptical envelope) and highlighting the misclassified elements (i.e., outliers) in the BIM model.

The outlier detection used in Krijnen's approach calculates the Mahalanobis distance of the geometry features to detect outliers in individual IFC classes.

The Mahalanobis distance is widely used in lieu of Euclidean distance in identifying outliers for multivariate datasets [11]. A problem with multivariate data is the effect of covariance between the variables, which cannot be resolved using the Euclidean distance. The Mahalanobis distance overcomes the problem by calculating and using the eigen vectors to transform the main axes of the variables, in effect negating their correlations.

The Mahalanobis distance is calculated using the following equation,

$$D(\vec{x}) = \sqrt{(\vec{x} - \vec{\mu})S^{-1}(\vec{x} - \vec{\mu})} \quad (1)$$

where, $\vec{x} = (x_1, x_2, x_3, \cdots, x_n)^T$ is the vector of observed data; $\vec{\mu} = (\mu_1, \mu_2, \mu_3, \cdots, \mu_n)^T$ is the vector of average of the observed data, and S is the covariance matrix.

The calculated distance D can be visualized using an Elliptical Envelope as shown in Figure 2. The blue dashed ellipses represent statistically equivalent contours (i.e., boundaries), based on which outliers and inliers are distinguished.

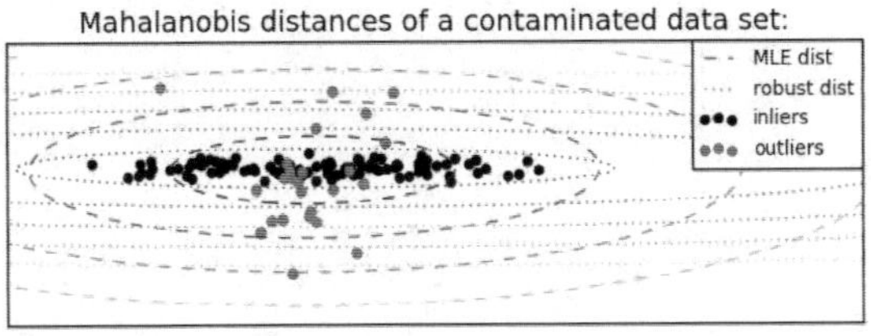

Figure 2. Output diagram using the Elliptical Envelope method for anomaly detection

4.2 Case Studies

Krijnen's approach was applied to two BIM models and their results are described as follows.

4.2.1 Case study 1: Duplex apartment model

The first model is the duplex apartment model, which was used in Krijnen's initial work. The BIM model has 159 elements (i.e., subtypes of IfcBuildingElement), which include beams, coverings, walls slab and roof. Table 1 details the results of performing the outlier detection for the individual IFC classes. Of the 159 elements, 9 outliers were detected in the walls (ifcWall, ifcWallStandardCase), slab (ifcSlab), window (ifcWindow) and doors (ifcDoor).

The following section describes the individual results for each of these classes.

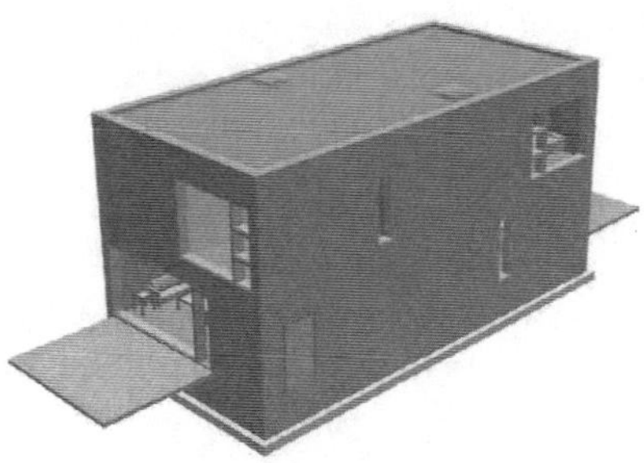

Figure 3. The duplex apartment BIM model

Table 1. Results of outlier detection for the duplex apartment BIM model

IFC Class	# of totals	# of inliers	# of outliers	Analysis of Results
ifcBeam	8	8	0	-
ifcCovering	13	13	0	-
ifcWall (incl. ifcWallStandardCase)	57	54	3	1 outlier represents misclassification, in which, a beam is misclassified as ifcWall 2 outliers are walls, but have openings
ifcSlab	21	18	3	1 outlier has different geometry than inliers 2 outliers are slabs, but have openings
ifcRoof	1	1	0	-
ifcFooting	7	7	0	-
ifcWindow	24	23	1	1 outlier has different height to inliers
ifcDoor	16	14	2	2 outliers have different width than inliers
ifcStair	4	4	0	-
ifcRailing	4	4	0	-
ifcMember	4	4	0	-
Total	159	150	9	

4.2.2 ifcWall, ifcWallStandardCase

The outlier detection identified three entities from 57 wall elements. Table 2 shows the samples of the BIM elements, with their corresponding Mahalanobis distance values. One of these elements was identified as a beam, as shown in Table 2. Thus, this demonstrates a successful detection of a misclassified element. However, the other two elements are walls with openings. These elements are walls, and the approach has identified them as outliers due to their dissimilar geometry from the most generic wall instances.

Table 2. Summary of inliers and outliers for wall elements

Inliers	Outliers
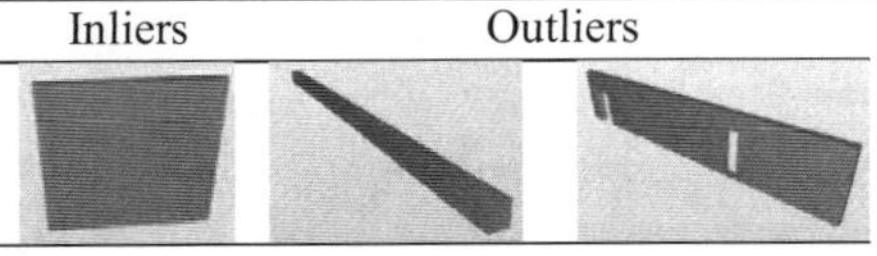	

# of items	54	1	2
M dist	2.32	11.08	14.74

4.2.2.1 ifcSlab

Three outliers were detected from 21 of the slab elements (Table 3). In this case, however, these are misclassifications as one of the elements are actually slabs with differing geometry and the latter two are also slabs with openings.

Table 3. Summary of inliers and outliers for slab elements

	Inliers	Outliers	

# of items	18	1	2
M dist	2.84	16.05	19.05

4.2.2.2 ifcWindow and ifcDoor

The outlier detection identified a single window and two doors as outliers (Table 4 & 5). Again, these were identified due to their dissimilar shape compared to the majority of the windows and doors.

Table 4. Summary of inliers and outliers for wall elements

	Inliers	Outliers
		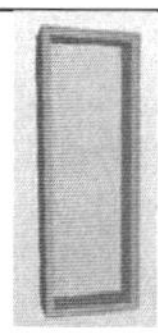
# of items	24	1
M dist	3.40	16.22

Table 5. Summary of inliers and outliers for door elements

	Inliers	Outliers
	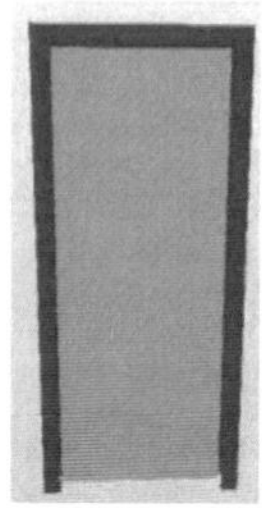	
# of items	16	2
M dist	3.45	11.80

4.2.3 Case study 2: Medical Clinic model

A BIM model of a Medical clinic, provided by [12], was used for the second case study. The model was chosen as it was a larger model than the duplex model, with a total of 1,232 BIM elements. Also, the walls and doors comprised the most number of elements, having more dissimilar elements within their respective classes. Thus, it provided a good candidate to determine the performance of the outlier detection.

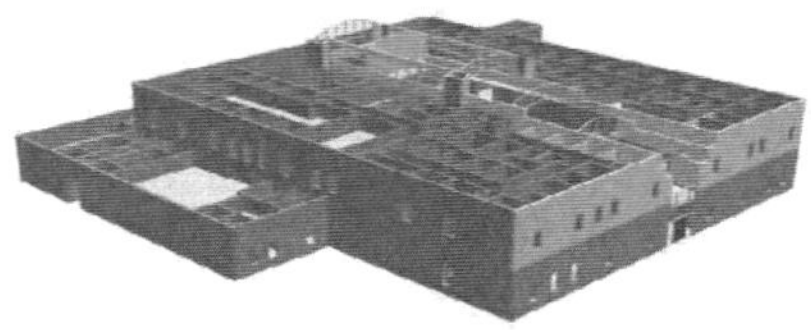

Figure 4. The Medical Clinic BIM model

As shown in Table 6, outliers were only found in the walls and doors.

The following sections describes the results with respect to the individual IFC classes with identified outliers.

Table 6. Results of outlier detection for the Medical Clinic BIM model

IFC Class	# of totals	# of Inliers	# of Outliers	Analysis of Results
ifcDoor	254	230	24	Classifiers doors in curtain walls as outliers
ifcRailing	9	9	0	-
ifcSlab	3	3	0	-
ifcStair	9	9	0	-
ifcWall StandardCase	1080	1025	55	Classifies walls with openings or

				walls with curvatures as outliers
ifcWindow	58	58	0	-
Total	1,413	1,334	79	

4.2.3.1 ifcWallStandardCase

55 outliers were found out of the 1,080 wall elements. As shown in Table 7, 53 of the elements were found to be dissimilar as these had openings, while two of them were curved walls.

4.2.3.2 ifcDoor

24 outliers were identified out of the 254 doors. As shown in Table 8, 20 of the outliers were doors with glass panes, while the other four were doors inside curtain walls.

Table 7. Summary of inliers and outliers for wall elements

	Inlier	Outlier	
			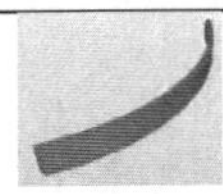
# of items	1025	53	2
M dist	0.96	40.95	29.84

Table 8. Summary of inliers and outliers for door elements

	Inlier	Outlier	
	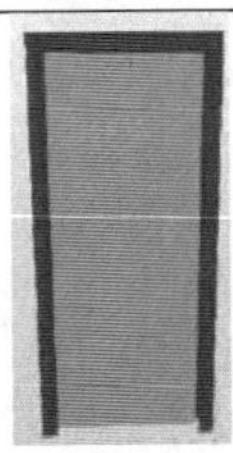	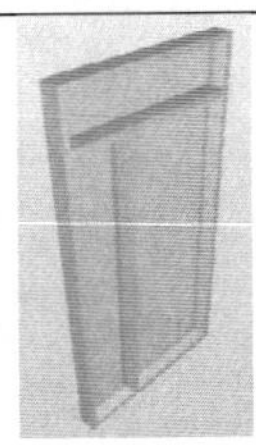	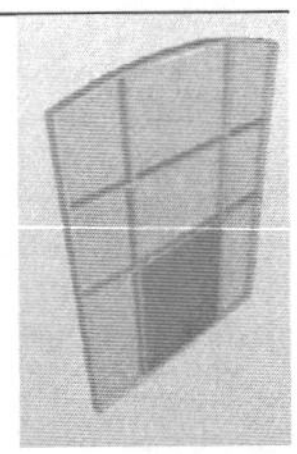
# of items	230	20	4
M dist	0.13	40.95	34.54

4.3 Summary of the Results

The following summarizes the main limitations of Krijnen's approach identified from the two cases.

- Krijnen's approach classifies elements that are dissimilar to the most numerous element as outliers. This problem becomes accentuated as the number of elements in classes increases. This problem stems from the fact that the Mahalanobis distance assumes a Gaussian distribution of the data, and thus is not suited for multi-modal distributions. That is, the Mahalanobis distance identifies a single inlier boundary and thus any other elements are taken to be outliers.
- On the other hand, Krijnen's approach will be limited when there are too few elements in a single class, as the outlier detection algorithm does not have enough elements to detect an inlier boundary.
- A more practical limitation is that Krijnen's approach uses a single BIM model, and thus the 'learning' is lost *a posteriori* analysis.

5 Approach using novelty detection

The second approach used was novelty detection. Novelty detection uses a training set that is not polluted by outliers, and is interested in detecting anomalies in new observations. Novelty detection can identify multiple inlier boundaries and can first be trained using datasets prior to the detection of outliers. Thus, it allows the use of data from multiple BIM models, a feature which was utilized in its implementation.

5.1 One-class SVM

Novelty detection can be implemented using one-class Support Vector Machines (SVM) [13]. SVM are a type of supervised learning used either for regression or linear classification. SVMs are referred to as large margin classifiers because the underlying algorithm attempts to identify the hyperplane that best represents the largest separation, or margin, between two classes.

SVMs can also be used as a nonlinear classifier when used with kernels, which are similarity functions that enable the computation of new features as to manual selection (e.g., high order polynomials).

When using SVM's for novelty detection, it is not possible to know *a priori* the type of outliers that may arise, and thus difficult to comprise a training set. In such cases, one-class SVM is used, in which the data set only includes inliers, and is thus a form of semi-supervised learning [14].

5.2 Novelty detection implementation

Figure 5 shows the overall process used to implement novelty detection.

- Model elements from three BIM models (i.e., the Duplex, Clinic and a third residential model) were classified and stored separately with respect to their

IFC classes.

- The datasets were then used to train individual one-class SVM's.
- All one-class SVM's used a non-linear Radial Basis Function (RBF) for their kernels.
- Each trained model was further refined by modifying the hyperparameters, μ and γ to find the values resulting in the highest true negative rates.

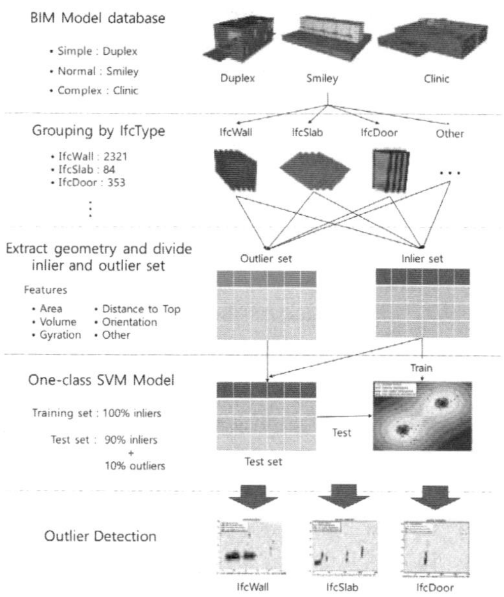

Figure 5. Process overview for novelty detection

5.3 Validation

Validation was performed for each IFC class by using a test set where 10% of the data included outliers (i.e., elements of different IFC entities), and measured whether the one-class SVM correctly detected them.

An example is provided for ifcWallStandardCase, which included 2,321 wall elements. As shown in Figure 6, the trained one-class SVM identifies two major boundaries for the walls. Given the test set, it successfully identifies outliers (depicted as yellow points), as abnormal observations. By tuning the model's hyperparamters (μ=0.3, γ=0.1), the model achieved a true negative rate of 0.983 (Table 9).

Table 10 provides the results for the other IFC classes, especially those that encountered misclassifications using outlier detection. The true negative rates demonstrate that outliers were correctly identified for each of these classes.

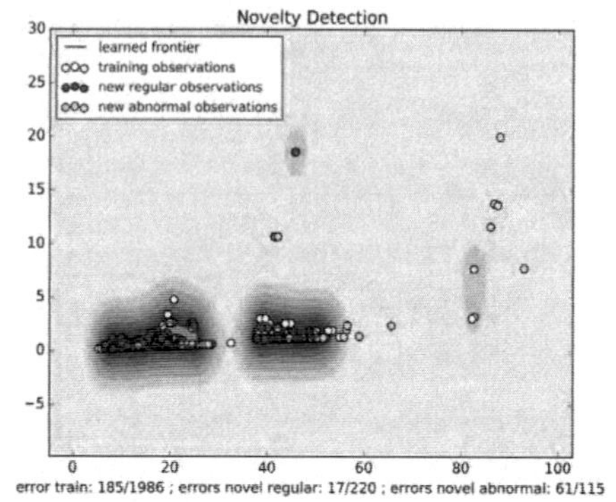

Figure 6. Novelty detection results for ifcWallStandardCase

Table 9. Results of novelty detection for the wall elements with tuned hyperparameters

μ	γ	Precision	Recall	TNR	Accuracy
0.1	0.1	0.970	0.908	0.470	0.887
0.1	0.3	0.978	0.903	0.609	0.888
0.3	0.1	0.999	0.698	0.983	0.712
0.3	0.3	0.983	0.698	0.774	0.817

Table 10. Results of novelty detection for other IFC classes

IFC Class	# of Total (outlier)	Prec-ision	Recall	TNR	Accuracy
IfcWall StandardCase	2321 (115)	0.97	0.91	0.47	0.89
IfcSlab	92 (8)	1.00	0.55	1.00	0.59
IfcCovering	388 (35)	1.00	0.91	1.00	0.95
IfcDoor	289 (26)	1.00	0.88	1.00	0.89
IfcWindow	134 (12)	1.00	0.80	1.00	0.81

6 Comparison of the two approaches

As shown in Table 10, results of the novelty detection approach addressed the main limitations identified in the outlier detection approach. Because novelty detection creates multiple boundaries, it could classify different types of elements as inliers, while correctly identifying other class elements as outliers. By creating data sets using elements from multiple BIM models, the approach also did not suffer from insufficient numbers of data points, as was the case with outlier detection.

Figure 7 shows an ROC curve that compares the accuracy of the two approaches, when applied to all elements of the ifcWallStandardCase class. The area under the curve (AUC) is 0.848 for novelty detection, which demonstrates higher performance to outlier detection, whose value is 0.665.

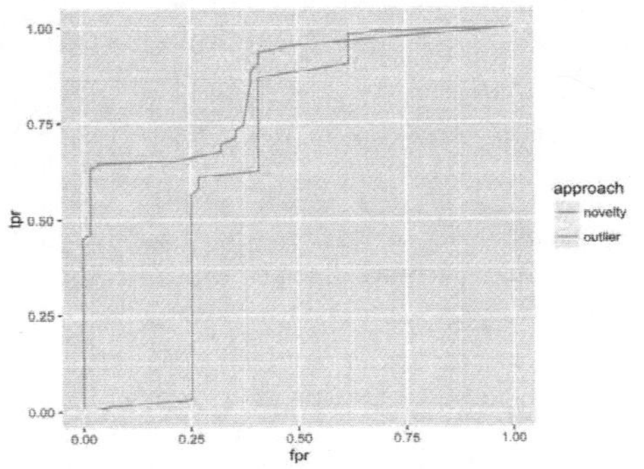

Figure 7. ROC curves for outlier and novelty detection using ifcWallStandardCase

7 Conclusions

The IFCs provide a critical role in ensuring the interoperability of BIM models. Ensuring their integrity is thus a fundamental premise for enabling collaboration within the BIM framework. This research examined two techniques in anomaly detection for checking potential errors and misclassifications in the mapping of individual BIM elements to IFC entities. Results showed that novelty detection was superior in terms of overcoming the limitations identified in outlier detection, especially in terms of the ability to train one-class SVM's to identify multiple boundaries of elements within the same IFC class. However, both approaches are limited in restricting features to geometry and not utilizing the semantic relationships between elements within a BIM model. Future research will attempt to address this need by adopting structuring learning techniques (e.g., conditional random fields) to enhance the classification capabilities of these algorithms.

Acknowledgements

This research was supported by Basic Science Research Program through the National Research Foundation of Korea(NRF) funded by the Ministry of Education(2016R1D1A1B03931198)

References

[1] Krijnen T. and Tamke M. Assessing implicit knowledge in BIM models with machine learning. *Modelling Behaviour*, Springer, 2015

[2] Tauscher E. and Smarsly K. Generic BIM queries based on the IFC object model using graph theory. *The 16th International Conference on Computing in Civil and Building Engineering,* Osaka, Japan, 2016

[3] Zhang C., Beetz J. and Weise M. Interoperable validation for IFC building models using open p standards. *Journal of Information Technology in Construction,* 20(2): 24–39, 2015.

[4] Mazairac W. and Beetz J. BIMQL – An open query language for building information models. *Advanced Engineering Informatics*, 27(4): 444–456, 2013.

[5] Daum S. and Borrmann A. Processing of Topological BIM Queries using Boundary Representation Based Methods, *Advanced Engineering Informatics,* 28(4): 272–286, 2014.

[6] Krijnen T. and LAHM Léon Berlo, V. Methodologies for requirement checking on building models:a technology overview. *13th International Conference on Design & Decision Support Systems in Architecture and Urban Planning*, Eindhoven, Netherlands, 2016.

[7] American Institute of Architects. AIA Document E203; G202: Building Information Modeling Protocol Exhibit. 2008.

[8] Krijnen T. IfcOpenShell, On-line: https://ifcopenshell.org, Accessed: 01/05/2015.

[9] Paviot T. Pythonocc, 3D CAD/CAE/PLM development framework for the python programming language. On-line: https://github.com/DURAARK/pyIfcExtract, Accessed: 01/05/2014.

[10] Pedregosa F. Scikit-learn: Machine learning in Python. *Journal of Machine Learning Research*. 12: 2825–2830, 2011.

[11] Roy D. M., Delphine J-R. and Désiré L. The Mahalanobis distance. *Chemometrics and Intelligent Laboratory Systems*, 50:1–18, 2000.

[12] William E. Common Building Information Model Files and Tools. On-line: https://www.nibs.org/?page=bsa_commonbimfiles, Accessed: 01/01/2014

[13] Bernhard S. Platt J. C. Shawe-Taylor J. and Smola A.J., et al. Estimating the support of a high-dimensional distribution. *Neural computation* 13(7): 1443-1471, 2001.

[14] Escalante H. J. A comparison of outlier detection algorithms for machine learning. *In the Proceedings of the International Conference on Communications in Computing*, pages 228-237, Hawaii, USA, 2005.

Decision Support Tool for Multi-Criteria Analyses of the Quality of Large Building Stock

A. Carbonari[a], A. Giretti[a], A. Corneli[a], V. Villa[b] and G. Di Giuda[c]

[a]DICEA Department, Polytechnic University of Marche, Italy
[b]DISEG Department, Polytechnic of Turin, Italy
[c]ABC Department, Polytechnic of Milan, Italy
E-mail: alessandro.carbonari@staff.univpm.it, a.giretti@staff.univpm.it, a.corneli@pm.univpm.it, valentina.villa@polito.it, giuseppe.digiuda@polimi.it

Abstract –

The management of any large building stock with limited resources poses a problem of prioritization of refurbishment actions. Also, available technical information about the building stock is often incomplete and the process of standardization and updating is expensive and time consuming. Some public owners are developing preliminary BIM models of their stock, but they are willing to limit the complexity of the models within the lowest amount of information required for management and maintenance, so as to make that process affordable. Indeed, administrations are challenged by their duty relative to planning regular maintenance and operation of buildings, because of the legislation in force, which requires monitoring of their facilities.

For the reasons stated above, this paper presents a decision support tool that can help prioritize refurbishment actions on large building assets. To this purpose, many requirements must be jointly considered in this examination, each requirement being assessed by means of one or several indicators. Then the indicators are compared one another, according to a multi-criteria approach, that weighs the several criteria and rank the assets. In order to deal with the extensive and uncertain information that must be managed in this process, indicators are estimated by means of Bayesian Networks. This tool is used first to assess the technical indicators and rank the assets, while marking any facilities not complying with regulations. Then, additional Bayesian Networks are in charge of estimating the budget needed to upgrade non-compliant facilities with minimum legislation requirements. The outcomes of this research can be used even to assess the level of detail of the information that must be included in BIM models of the stock, in fact acting as guidelines for their development. Finally, the application of the decision tool on a real test case will be presented.

Keywords –

Decision support system; BIM; Bayesian Networks; Multi-Criteria; building stock

1 Introduction

The majority of public buildings are outdated. As a consequence, Public Building Administrations are in charge of the prioritization of refurbishment actions on the large building stock they manage, including schools.

Presently, every local administration performs separate evaluations of the existing stock deciding where to focus the intervention first and there is no coordinated assessment at the national level on a proportional distribution of efforts.

To that purpose, informed planning is needed, according to real priorities, which means detecting any lack of compliance with respect to current legislation, in terms of comfort, energy performances, accessibility, seismic vulnerability, etc. While sticking to large building stock, the aim of this research work is to develop a decision support tool based on Bayesian Networks, which can extract relevant information directly from a BIM database of the building stock and evaluate the compliance of the stock to some pre-determined technical requirements.

The decision support system was developed so as to be compliant with two BIM-based models of two schools located in Melzo (Milan), which are used for Facility Management (FM) and which acted as test cases in this research work. The whole decision support system includes a multi-criteria assessment of some performance indicators, each of them relative to a specific area of interest. In addition, the system included one Bayesian Network in charge of estimating the budget needed to improve the status of non-compliant buildings towards a new state where they are compliant with minimum legislation requirements. This piece of information is necessary to perform cost- benefit analyses.

To sum up, this tool was conceived as a tool to support a methodology for Public Administrations that have to schedule three-years plans of Public Works in advance within budget and quality constraints, while expeditiously evaluating benefits from technical improvements. In fact, the standard current methodology usually requires, as a first step, a preliminary survey on the state of the art of buildings through the creation of a repository, possibly a BIM repository, where all available information is accommodated in a structured database. Then, a second tailored survey is expected to complete the information framework and help the assessment phase. The accomplishment of these two steps, however, requires huge time and cost efforts, which can barely be afforded when strict budget constraints are posed. Hence strategic management for the efficient selection of actions should be preferred. For this reason, the decision support tool reported in this paper would be functional for supporting informed choices in several situations, e.g. for the execution of new school buildings, the renovation of existing properties, small maintenance interventions, diagnostic investigations, securing and retrofitting existing buildings.

2 Literature Review

The development of guidelines and standard methodologies to speed up the process of maintaining and rehabilitating huge stock of buildings is not new. Even the Japanese government started a strong action towards rehabilitation of social housing in the 1970s, that was aimed at optimizing the overall approach [1]. That methodology consisted in a quick and systematic survey of the degradation stage, according to several technical requirements, and in the next merging of those indicators into a unique index assessing the overall quality of buildings. Indeed, decisions for building maintenance require integration of various types of information – which is sometimes incomplete - and knowledge created by different members of teams involved in design and construction [2]. This issue should consider that a gradual and incremental approach towards the use of BIM has been experienced over the last decade within the construction industry [3]. In 2012 attention was drawn to the crucial role of BIM in this phase of building life operation, stating that the initial costs of inserting BIM systems into the processes are justified only if it is meant to support operation and maintenance [2]. Although the need for BIM in Facility Management (FM) has been acknowledged by researchers and practitioners, BIM is still not being effectively exploited in this phase, even if refurbishment activities are often carried out [4]. Also, it was highlighted that some studies on "BIM in Building Refurbishment and Maintenance" are focused on applications at an FM level, whereas just a few studies are related to BIM applications in either maintenance or refurbishment. Some other research focused on the choice of what information is needed in order to make models significant to maintenance, and on handling uncertainty due to incomplete building documentation [5]. Since BIM is becoming a project standard, FM is expected to be based on information related to the BIM model database. In addition, FM managers have the opportunity to use this knowledge to evaluate the quality of buildings and to rank refurbishment priorities, provided that the decision issue among the several involved parameters has been solved. Hence, this paper deals with the development of a decision support tool based on the use of Bayesian Networks to evaluate the performance parameters of existing buildings, whose inputs are retrieved from BIM models, which may not be fully detailed but just limited to the level of available information about any existing stock [6,7]. The results from this evaluation are used as inputs for multi-criteria evaluation of the quality of the analyzed stock. Finally, another set of Bayesian Networks are used to evaluate the refurbishment cost of those buildings that do not comply with minimum legislation requirements.

3 Description of the Decision Support System

3.1 Overview

Any large building stock requires a targeted and accurate management in order to comply with the latest standard and maintain a good level of conservation through maintenance. In addition, the hardest challenge that owners and public owners of a large stock have to deal with is the limitations in terms of budget. This leads to the need for a priority list of the buildings needing refurbishment that is based on the real status of each facility. Also, by means of preliminary analyses of potential refurbishment actions related to the building type and construction techniques, what refurbishment actions are affordable referring to the available budget can be inferred. Hence, the work developed in this paper is thought of as being part of a wider decision support system that is made up of several parts (Fig. 1):

- A BIM database of the building stock;
- A set of Bayesian Networks for evaluating stock compliance to technical requirements (in terms

of Accessibility, Energy Efficiency, Life Safety, Fire Protection, Seismic Vulnerability) and for ranking it according to performance indicators;

- An interface between the BIM database and the Bayesian Networks, which automatically picks out relevant inputs from BIM models and transfers them into BN;
- A multi-criteria decision system, which ranks buildings according to the BN outputs;
- Another set of Bayesian Networks that estimates the budget needed to improve the status of any building, just in case it does not comply with minimum requirements, until it complies with minimum legislation requirements; when existing buildings are compliant, the estimated cost will be null.

The output of the first set of BN are the performance indicators. The indexes regarding Accessibility and Fire Protection were designed so as to qualitatively estimate the level of fulfillment to minimum requirements. The remaining ones, regarding Energy Efficiency, Seismic Vulnerability and Safety, provide users with quantitative and measurable levels of performance.

In this paper, the Bayesian Networks relative to Accessibility, Energy Efficiency and Cost Estimation, which are the subjects of sub-Sections 4.1, 4.2 and 4.3, respectively, will be reported in detail. The outputs of the Energy Efficiency Bayesian Network are two quantitative performance indicators (both in accordance with EN13790):

- "Heat transfer coefficient" (HTC), which is represented by an interval node, and estimates the average heat transfer coefficient of the building;
- "Seasonal energy performance" (SEP), which is represented by an interval node, and estimates the annual energy required for conditioning over a whole year per unit area.

The Accessibility Bayesian Network includes two qualitative performance indicators:

- "Level of compliance" (A), which is represented by a labelled node and identifies how far the building is from the minimum compliance level, whose range is between 0 and 100%, the latter being verified just in case the "Regulation obeyed" node is true, and which is an input to the multi-criteria decision tool;
- "Additional parameters", which is represented by a labelled node, and estimates how many non-mandatory requirements, if any, are fulfilled by the building beyond the mandatory ones.

As far as the Accessibility Bayesian Network is concerned, it is linked to another fragment of Bayesian Network, which estimates the "Total cost" for renovation. This fragment evaluates the total amount of retrofitting cost, due to the presence of non-compliant entrance doors width, as reported in sub-Section 4.3.

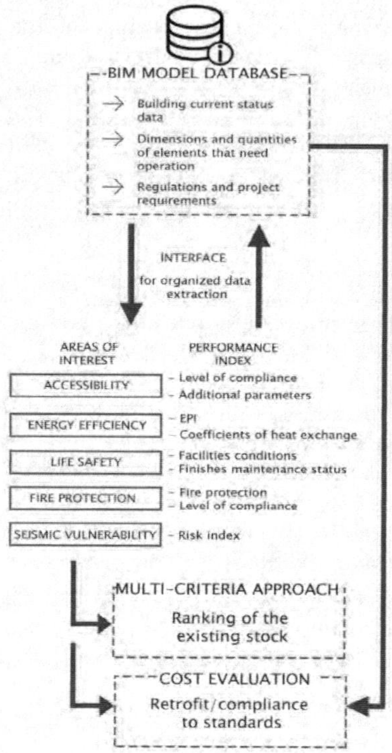

Figure 1. Schematic diagram of the structure of the decision support system.

3.2 Bayesian Networks

Bayesian Networks (also called belief Bayesian Networks or causal probabilistic networks) have been dominating the field of reasoning under uncertainty, thanks to their ability to deal with incomplete or uncertain information [8]. Bayesian Networks are always made of two components. The first component is the graphical model, which is represented by a direct acyclic graph (DAG), whose nodes represent random variables that are linked by arcs, corresponding to causal relationships among the previous nodes. Each variable may take two or more possible states of numerical (i.e. discrete), interval (i.e. subdivision into ranges), label or Boolean types. An arc from any set of n variables, called a_i, to another variable b, denotes that the set a_i causes b, and a_i is said to be the parents of b (b is evidently their child). The second component is represented by a set of conditional probability tables

(CPTs), whose values define the strength of relationships among nodes, where the probability of observing any state of the child e_n variable is given with respect to all the combinations of its parents' states: that is $P(e_n|e_1, e_2, \ldots, e_{n-1})$, where any variable e_i is conditionally independent of any variable of the domain that is not its parent. Thus, the state of each variable can be determined by the knowledge of the state of only its parents, and the joint probability of a set of variables E can be computed by applying the "chain rule" [9]:

$$P(E) = P(E_1, E_2, .., E_{n-1}, E_n) = P(E_n \mid parents(E_n)) \cdot P(parents(E_n)) \quad (1)$$

where $E_1, \ldots E_{n-1}$ are the parents of E_n. Therefore, the complete specification of any joint probability distribution does not necessarily require an absurdly huge database. This is the first of a series of benefits provided by the use of Bayesian Networks for the application reported in this paper.

Other relevant benefits are: the DAG provides a clear understanding of the qualitative relationships among variables; every node can be conditioned upon new information (e.g. evidence about the features of a building in our case study), hence the inference (also called belief updating) is performed via a flow of information through the network, and the most likely state of a set of "query" nodes (e.g. the indicators in our case study) can be computed; the same belief updating is supported from consequences to causes, also known as diagnostic reasoning, thanks to the application of the Bayes Theorem. This feature is critical in our application because it can be applied when the budget for renovation is strictly limited and inference must be conducted from child nodes – such as "admissible cost of renovation"- back to parent nodes – such as the "affordable technology options" for refurbishing.

Finally, CPTs can describe the relationships among variables of different types (e.g. Boolean nodes in the Accessibility BN and interval nodes in the Energy Performance BN), even within the same network, hence sets of information from different sources can be merged into one comprehensive network.

3.3 Ranking of the quality of building stock

The Bayesian models reported in the next Section 4 provide users with two indicators for evaluating the quality of any building: on one hand, the technical indicator that is the result of a multi-criteria approach; on the other hand, the cost that must be afforded to make the building compliant with minimum legislation requirements.

The Multi-criteria approach includes the subjectivity by evaluators through the choice of some criteria instead of others [10], hence it can be customized to the specific needs dictated by several owners. In fact, the Multi-criteria approach represents a series of techniques, the scope of which is to consider several features that are in some way related to different aspects of the problems under analysis. The methodologies for Multi-Criteria Analysis can be divided into two main groups: (i) Multi-Criteria Objectives Analysis (MCOA) and (ii) Multi-Criteria Attributes Analysis (MCAA).

In the case of MCOA, the decisional process consists in the selection of the best solution within a group of infinite alternatives, implicitly defined by the problem boundaries. On the contrary, Multi-Criteria Attributes Analysis (MCAA) is a multidimensional evaluation method subset, whose final purpose is to locate the best strategy among a restricted number of alternatives, which are ranked according to their preferences [11]. MCAA can act as a support in the decision making process [12], which leads through a systematic analysis of the solutions. It is usually structured on six points:

(i) Definition of the evaluation matrix, which is the analytic instrument that represents the value added to each alternative and is based on pre-determined criteria;
(ii) Dominance analysis to eliminate any alternatives which prove to be dominated from the decisional process, i.e. having worse performances, if compared to the others;
(iii) Normalization of the evaluation matrix, which makes qualitative and quantitative data homogeneous and operable;
(iv) Appointing the weight associated to each of the criteria, which makes it possible to define the relevance order among all criteria and sub-criteria, hence the priority matrix is created;
(v) Among the techniques used to set up the organization of options, we cite the AHP, which was used in our case study.

The AHP methodology was applied according to what suggested by Saaty [13]. Computations were implemented by means of an Excel spreadsheet™. As a first step, the hierarchy was defined as follows: the top level is "stock value", the second is composed of all the areas of interest such as accessibility, energy efficiency and the others; instead, the third level is made up of the outputs from the BN "Level of compliance" node for Accessibility, and the "EPI" and "Heat transfer coefficient" nodes for Energy Efficiency, as reported in sub-Sections 4.1 and 4.2. The second step consists in the pairwise comparison between the different areas of interest. As a result, the final ranking is inferred as a combination between the values obtained from the BN and the weights determined by means of the pairwise comparison, as follows:

$$R = W_A*A + W_{EE}*EE \quad (2)$$

where:

$$EE = W_1*HTC + W_2*SEP \quad (3)$$

Again, A is the output of the Accessibility BN (sub-Section 4.1); EE is the combination of the outputs of the Energy Efficiency BN (sub-Section 4.2) and W_A, W_{EE}, W_1 and W_2 are the weights worked out according to the AHP approach.

The estimation of refurbishment costs reported in sub-Section 4.3 was conducted according to three steps. As a first step, a list of activities necessary to refurbish non-compliant buildings differentiated on the basis of the several possible types of refurbishment actions was produced. Then, the DAG of the network was developed, which includes both the estimation of quantities (some nodes will be read straight from the BIM model while other nodes from the rest of the BN) and the estimation of unit costs, according to the framework suggested by the official list of unit costs of the Province of Milan (Italy). Finally, CPTs of Bayesian Networks were built from unit costs of the Province of Milan and from the knowledge of the range inside which every numerical node was expected to fall.

Both the Multi-criteria approach and the cost estimation were tested on two case studies: the first is the "Ungaretti" primary school in Melzo (MI) with a surface area of 4528 m^2. The classrooms, laboratories, toilets and cafeteria are located over three floors above ground, while the gymnasium is in a separate building.

a)

b)

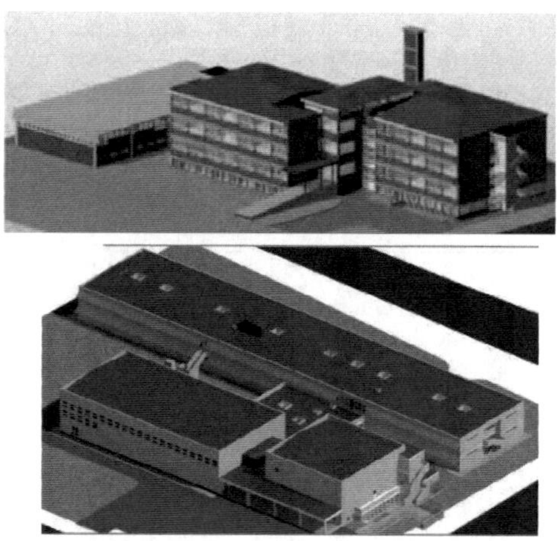

Figure 2. BIM models of the cases studies a) Ungaretti school b) Mascagni school

The second case study is the "Mascagni" secondary school in Melzo (MI) with a surface area of 5736 m^2, which is formed of three functional blocks. One block holds the classrooms and laboratories located over two floors above ground, the other two blocks hold the cafeteria/auditorium and the gymnasium. Both schools are made of a reinforced concrete bearing structure (Fig. 2).

4 System development and testing

a)

b)

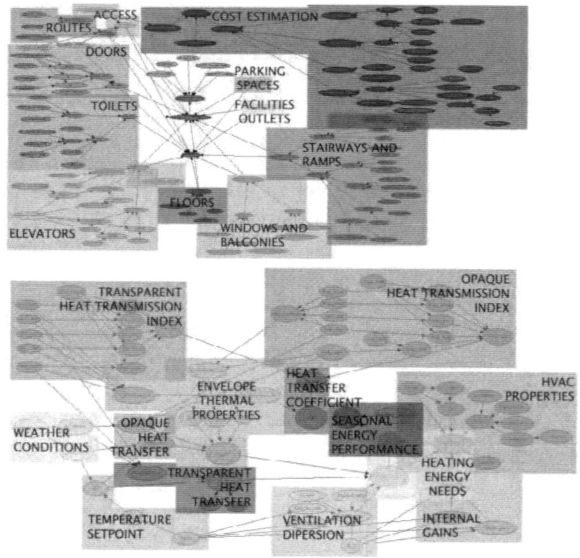

Figure 3. Accessibility Bayesian Network (a) and Energy Efficiency Bayesian Network (b).

4.1 Accessibility Bayesian Network

Italian legislation (D.M. 236/89) defines all the requirements and the related technical standards that are shown in the Accessibility Bayesian Network (Fig. 3-a). 'Level of compliance' (ref. subsection 3.1) is the output node that sums up the current situation of the building. It is a child node of several parent nodes, each regarding the different building components [14, 15]:

- "Accesses": e.g. width, handle height, maximum opening force;
- "Doors": e.g. width, handle height, maximum opening force, maneuvering clearance;
- "Parking spaces": e.g. parking space width;
- "Elevators": e.g. car elevator dimensions, car control keypad height;
- "Floors": e.g. floor frictional coefficient, floor joint width, floor ridges, changes in level;
- "Stairways and Ramps": e.g. handrails, tread and riser size, stair width, maximum slope;

- "Toilets": e.g. water closet position, grab bar location and size, lavatory position;
- "Routes": e.g. clear width of an accessible route, passing space interval;
- "Windows and balconies": e.g. railings, maneuvering clearance, window opening force, handle height;
- "Facilities outlets": e.g. facilities outlet height.

In order to fill in the conditional probability tables of all the ten aforementioned Boolean-type intermediate nodes (admitting "true" and "false" states only) the ratio of verified technical prescriptions (e.g. at the building component level) has been evaluated.

4.2 Energy Efficiency Bayesian Network

The whole Energy efficiency Bayesian Network was derived from previous research on reduced-order models for thermal simulations of buildings [16,17]. With the purpose of learning the CPTs of the BN from data, the reduced-order model was repeatedly run to generate a database containing more than 100 records, which was used as a dataset to estimate the CPTs, while casual dependencies were quantified by means of the EM-learning tool implemented in the Hugin™ software program [18]. This network estimates two performance indicators:

- Heat transfer coefficient (ref. sub-section 3.1);
- Seasonal energy performance (ref. sub-section 3.1).

In this case, the nodes represent the variables of the reduced-order model, while arcs were determined according to the causal relationships between the variables of the same reduced-order model (Fig. 3-b).

4.3 Cost estimation Bayesian Network

The first fragment of Bayesian Network for cost estimation reported in this sub-section is directly linked to the Accessibility BN. It is in charge of estimating the cost to be incurred in, so as to renovate those access doors that the in the corresponding input node of the Accessibility BN was instantiated as not compliant with minimum legislation requirements. The adoption of Bayesian models allowed to manage uncertain information about inputs, if any, and to reverse inference updating from constrained final budget (i.e. child nodes) back to inputs, or parent nodes (e.g. technology and geometric preferences).

The total cost is dependent on three nodes: the first one is the cost for complete refurbishment of one door only; the second one is the number of doors needing refurbishment, and the last one acts as a counter-check that current access doors really need refurbishment (when in "false" state). When this last node is "true" the cost of one door refurbishment is set to "zero".

In turn, the cost to renovate one access door is made of four tasks, which add up in terms of related costs:

- door removal, that depends on the type of door;
- demolition of a portion of wall, so as to enlarge the door space, related to the amount of partition to be demolished and wall technology;
- rebuilding of the interface between wall and new door, that is computed from the quantity and technology of wall, plaster and painting;
- installation of the new door and its accessories, which is given by the cost of the new threshold, and the cost of the new door (depending on type and material).

The latter part of the cost network is the only one that involves a choice by the user, that affects total cost. Indeed, the green nodes depicted on Fig. 4 are those nodes where owners are asked to choose among different options. These options entail the threshold material, but thickness is set as equal to the former one, the door type (i.e. flush or with glass) and material (i.e. metal or plastic). All the remaining information required by the network as inputs, including the type, size of the existing door, thickness and features of the wall, the number of access doors that do not comply with regulations, are extracted straight from the BIM model of the building. In the current version of the BN, pre-determined features, not included in the list of options, are "one leaf door" and "presence of a push-bar". The nodes dealing with costs were set as either Interval nodes or Numbered nodes. The Interval type was used when cost depends on the amount of material put in place, and when interval values were more expressive than many numbered states (e.g. "new door cost"). In case costs did not depend on quantities, the corresponding nodes were of Numbered type.

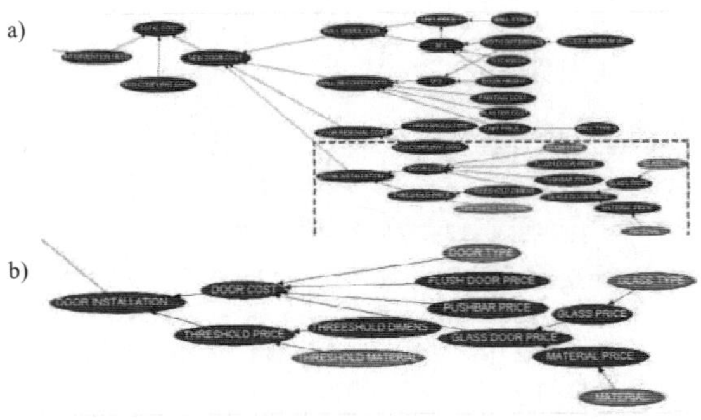

Figure 4. Cost estimation BN (a), which is a part of the Accessibility BN (Ref.Fig.3a), and zoom of the decisional nodes, marked in part a of the figure (b).

4.4 Implementation for a real case-study and results

The indices calculation requires 62 pieces of information as inputs for the Accessibility BN and 32 for the Energy Efficiency BN (Fig.3). These data came from the BIM models of the two schools and they have been detected manually as the tool interface is not available, yet. Among the 94 required input data, eight, concerning Accessibility, are always found in attributes of the BIM model, thirty-nine, still concerning Accessibility, can be found provided that the BIM designer performs a customized modelling procedure (e.g. adding attributes to user made families of BIM objects). The remaining data were obtained through combined analyses of several parameters. In Tab.1 the values of the BN output nodes are listed. The output "Level of compliance" (in the second column) represents the 'true' percentage value of the node according to the information entered in the network. The two-rightmost columns represent the output of the Energy Efficiency BN. They show a quantitative assessment of performance based on the provided inputs and they are in very good agreement with detailed energy simulations performed on the buildings in past research steps [19,20].

Table 1 BN outputs for the two cases studies.

	W_A	W_{EE}	W_1	W_2
Weight value	0.83	0.17	0.17	0.83

In order to define a refurbishment priority through the Saaty AHP decision approach, a feasibility of the pairwise comparisons was preliminary tested, according to the procedure detailed in sub-Section 3.3. More specifically, the expertise of the research team was used to define every weight and a datasheet was set up so as to seamlessly correlate the output nodes of the BN with the priority index resulting from the AHP. The resulting weights are shown in Tab.2.

Table 2 Weights values from pairwise comparison

Case Study	A	HTC	SEP
Units	*%*	*W/m^2K*	*KWh/ym^2*
Ungaretti School	60	2.43	45.29
Mascagni School	52.1	3.43	45.79

After normalizing HTC and SEP according to their best values, being in both cases "Ungaretti" as the lowest ones, the Ungaretti array was re-allocated as {HTC, SEP}={1, 1} and Mascagni was allocated as {HTC, SEP}={0.71, 0.99}. Then, applying Eqs. (2) and (3) to the values shown in Tab. 1, after the aforementioned normalization with the weights listed in Tab.2, the final ranking is the following:

- Ungaretti, R=0.67;
- Mascagni, R=0.59.

As a result, Mascagni school is the one with the lowest ranking and is the building that should be refurbished first. Once the requirement indices of the non-compliant schools were estimated, an estimation of costs relative to renovation actions can be performed, by means of the cost estimation BN (ref. sub-section 3.3).

As the Mascagni school is the one ranked at the top place in the priority list (hence, at the bottom in the quality ranking), the cost estimation BN has been tested on this model, specifically about the renovation of access doors. Thirteen data were required as input information by the cost estimation BN fragment. Among them, seven were retrieved manually from the BIM model, two depend on a variable that was not included in the model and the remaining four are features that must be set by the user according to requirements and budget.

a)

b)

Figure 5. Inference for cost estimation (a) and backward reasoning for cost inversion (b).

For the purpose of this simulation, a metal framed and glazed new door, with triple glass stratification (8mm each) was inserted as input. This selection led to an estimated of cost, whose range spread over the "€2000-3000" interval (p=31.58%) , the "€3000-4000" interval (p=34.22%) and "€4000-6000" interval (p=34.20%), mean=€3697.00 like in Fig. 5-a.

In addition, the inversion rule was tested for this network: the user can constrain the output node at the available budget, and the network will suggest what renovation actions should be put in place.

As shown in Fig. 5-b, when the last aforementioned nodes are left free to vary, and the intermediate node "New door cost" is fixed at the "€600-650" interval (which can be due to the need for budget restriction), the network suggests that the best technology to be adopted is a glazed door, triple layer glass (6mm each) and metal frame. The expected total cost in €3641.00.

5 Conclusions and future work

The decision support system reported in this paper is meant to help the managers of large building stock taking into consideration different technical issues, such as accessibility, energy efficiency, life safety, fire protection and seismic vulnerability. The assessment of technical parameters was done by means of Bayesian Networks that are useful means to handle uncertainty that can be due to the lack of some information about the buildings under evaluation, because they are capable of dealing with several types of variables and because inference propagation can be inverted. The ranking of buildings was performed by means of the AHP approach. In the application reported in this paper the networks regarding "Accessibility" and "Energy Efficiency" were evaluated, and they were shown to give back reliable results, once they are interfaced with the BIM models of the case studies. The inputs of the Bayesian Networks give back the amount of information that must be provided by BIM models in order to perform those analyses. Finally, once an intelligent interface working as a network management module between BN and BIM database is developed, they can run automatically and work out a lot of analyses with reasonable efforts, hence they can constitute a powerful decision support system. The system will need to be completed with all areas of interest shown above and regarding to cost networks, these will have to be implemented with all possible refurbishment scenarios by studying in depth the technologies that best fit the different type of buildings using also expert opinion to evaluate the reliability of the final analysis.

References

[1] M. Fernanda S. Rodrigues, José M.C. Teixeira, José C.P. Cardoso, Building envelope anomalies: A visual survey methodology, Construction and Building Materials 25 (2011) 2741-2750.

[2] I. Motawa, A. Almarshad, A knowledge-based BIM system for building maintenance, Automation in Construction 29 (2013)173-182.

[3] Z. Alwan, BIM performance framework for the maintenance and refurbishment of housing stock, Structural Survey, Vol. 34 Iss 3 (2016) 242-255.

[4] D. Ilter, E. Ergen, BIM for building refurbishment and maintenance: current status and research directions, Structural Survey, Vol. 33 Iss 3 (2015) 228-256

[5] J. J. McArthur, A building information management (BIM) framework and supporting case study for existing building operations, maintenance and sustainability, Procedia Engineering 118 (2015) 1104-1111

[6] P. A. P. Ramirez, I. B. Utne, Bayesian network for decision making on ageing facilities, in: European Safety and Reliability Conference, ESREL, Troyes, 2011, pp. 80-88.

[7] R. Wu, G. Mavromatidis, K. Orehounig, J. Carmeliet, Multiobjective optimisation of energy systems and building envelope retrofit in a residential community, Applied Energy 190 (2017) 634–649

[8] K.B. Korb, A.E. Nicholson, Bayesian Arificial Intelligence, Chapman & Hall/CRC Edition, 2004.

[9] J. Pearl, Probabilistic Reasoning in Intelligent Systems: Networks of Plausible Inference, 2nd ed., Morgan Kaufmann Publishers, San Mateo, California, 1998.

[10] J.R. Canada, W.G. Sullivan, Economic and Multi-attribute Evaluation of Advanced Manufacturing Systems, Prentice Hall, Englewood Cliffs, 1989.

[11] S. Salat, Assessing cities: A new system of spatial indicators, SB11, World Sustainable building Conference, 18-21 ottobre, Helsinki, Finland, 2011.

[12] J. D. Balcomb, A. Curtner, Multi-criteria decision-making process for buildings, Energy Conversion Engineering Conference and Exhibit 2000, (IECEC) 35th Intersociety, vol.1, pp. 528-535, 2000.

[13] T. L. Saaty, How to make a decision: The Analytic Hierarchy Process, European Journal of Operational Research 48 (1990) 9-26.

[14] DECRETO MINISTERIALE 14 giugno 1989, n. 236, Prescrizioni tecniche necessarie a garantire l'accessibilita', l'adattabilita' e la visitabilita' degli edifici privati e di edilizia residenziale pubblica sovvenzionata e agevolata, ai fini del superamento e dell'eliminazione delle barriere architettoniche, (GU Serie Generale n.145 del 23-6-1989 - Suppl. Ordinario n. 47).

[15] Architectural Barriers Act (ABA) Standards, Adopted by Department of Defense, General Services Administration, U.S. Postal Service, 2015.

[16] M. Benedettelli, A. Carbonari, B. Naticchia, M. Vaccarini, Reduced-order Models for Supporting Energy Audits of Buildings, in: 33rd International Symposium on Automation & Robotics in Construction, Auburn, 2016, pp. 563-571.

[17] M. Giretti, M. Lemma, M. Casals, M. Macarulla, A. Fuentes, R. Jones, Effective building modelling for energy performance contracting, in: Building Simulation Applications, Bolzano, 2017.

[18] U. M. Fayyad, G. Piatetsky-Shapiro, P. Smyth, R. Uthurusamy, Advantages in Knowledge Discovery and Data Mining, AAAI Press / The MIT Press, Menlo Park-California, 1996.

[19] G.M. Di Giuda, V. Villa, F. Paleari, M. Schievano, BIM modeling of the existing school heritage for investement planning, Conference ISTeA – Enviromental sustainability, circular economy and building production (2015) 29-48.

[20] G.M. Di Giuda, V. Villa, F. Paleari, From BIM to BEM for the management of the existing school buildings, Conference ISTeA – Rethinking the digital construction industry (2016) 19-28.

Safety Regulation Classification System to support BIM based Safety Management

Rahat Hussain[a] Do Yeop Lee[a], Hai Chien Pham[a] and Chan Sik Park[a]

[a]Department of Architectural Engineering, Chung Ang University, Seoul, South Korea

E-mail: rahat4hussain@gmail.com , cpark@cau.ac.kr

Abstract –

Construction accidents, injuries, and fatalities have not declined significantly, despite continuous efforts from researchers, safety professionals, and strongly-enforced safety laws. In managing safety, construction personnel are required to access a variety of rules, regulations, guidelines, and documents. However, finding the right contents, and communicating them to right person tends to be tedious and time-consuming, because construction safety information is not systematically structured. This study contributes to addressing this issue by introducing a system framework for integrating BIM dataset with classified regulation document. To accomplish the study objective, a general classification system is developed for safety rules using criteria that can reflect safety rules in BIM environments. To demonstrate, the OSHA's safety and health technical guidelines are selected to classify the legal provisions into safety management steps and tasks aiming to integrate with construction work sequence. From classified safety rules, four different cases are presented to show the practical implementation of classification system. The developed approach is expected to reduce oversights, enable to deliver the relevant safety rules at the right time, and reduce the efforts required for searching through safety contents.

Keywords –

Information Modeling; Construction, Safety Management;

1 Background and Introduction:

The Construction industry is characterized by a number of risk factors due to its complexity and diverse tasks. In managing safety, construction personnel are required to access a variety of rules, regulations, guidelines, and documents. With the advancement of building technologies and Building Information Modelling (BIM) has been recognized as a significant tool for architects and other construction safety practitioners. However, due to the overwhelming number of safety rules and the complexities inherent with them, finding the right contents and communicating them to right person tends to be tedious and complicated. Therefore, it is difficult to expect that clients and construction workers will be aware of the current complicated safety regulations and capable of fulfilling their obligations. These challenges require well-defined and structured rule documents that can be identified and applied automatically to building software models with least human interaction.

The emerging trend of BIM adoption in the architecture engineering and construction (AEC) industries proves to be a valuable solution for the aforementioned problems [1]. One of the promising features of BIM technologies is to integrate safety rules with BIM objects, which is useful for verifying regulations using object characteristics of building models. Parametric and information intensive building objects can represent their own properties as well as related regulation information. Automation of rule checking during design and planning phases of a project has been suggested by different researchers with various approaches. Some efforts focus on mapping text intensive rules from human language to machine readable formats i.e. interpretation of safety rules. These methods are based on standard BIM format, introduced by BuildingSMART, Industry Foundation Classes (IFC) [2]. For example, a remarkable approach is CORENET ePlanCheck, which is the most comprehensive effort for automated rule checking [3].

The ongoing problem in checking rules is the interpretation of safety regulations during planning and executing the models. Rule interpretation is a critical stride for the development of an automated rule checking process. One of the key challenges in rule interpretation is to systemize and classify the set of rules. Safety authorities such as OSHA offer text based,

HTML web based data for site safety management. However, the deployment of these safety information on actual sites is arduous due to the informal classification of rule sets. Moreover, when attempted for rule interpretation on site, designers and safety managers are not capable to track the exact job related rule within specific task categories or project phases.

In order to overcome these issues, this research envisions to improve construction site safety by introducing a rule classification system for OSHA's rules document that can be integrated later with existing BIM applications. To accomplish the study goals, OSHA's safety and health technical guidelines are selected to classify the legal provisions into safety management steps and tasks. To ensure the rule applicability with BIM object, the development level of BIM required by the legal clause is considered to classify the provisions of the safety and health technical guidelines. Following the classification, four cases are highlighted by considering various OSHA regulations. The proposed classification approach could help safety practitioners reduce oversights, enable the delivery of the relevant safety rules at the right time, and reduce efforts required for searching through safety contents. In addition, by adopting this system, the accuracy of search results can be improved.

The current state of rule document structure, classification and research efforts is addressed in Section 2. The research methodology and proposed system framework for rule classification are described in the Section 3. Moreover, a classified set of rule with the further discussion about practical implementation of proposed system are included in section 3. The paper is concluded with a summary and discussion of the research contribution.

2 Literature Review:

Rule checking with different approaches based on computer model have been discussed by many authors since last few years [4,5]. Some studies developed code checking tools for regulations [6] and other efforts focused on the method of interpretation of rules into machine readable format [5]. Specific domain based studies are also available in literature, like code checking application for high rise buildings [4]. Some efforts are mainly focused on improving the applicability of these computable rules [7]. For instance, Zhang et al. [8], established a connection between job-semantic modelling of safety rules [8]. More specifically, Chi et al used ontology-based text classification (TC) to integrate safe methods recognized in current resources step and hazards associated with activity based on construction solution database, by using ontology based with unsafe conditions for job hazard analysis [9].

A very few of them have focused on classification of safety rule sets and issues related to them. Due to complexity inherent with rule structure, it is no longer a simple deal like the cases of true/false or if/then conditional parameters. Solihin et al. [2] highlighted three different types of issues associated with current safety rules structures. Arity issues, the rule conditions can be fulfilled with multiple approaches at once. Combinational issue, the rule conditions may lead to more than one possible solution path; these combinatorial problems may get hard to manage. The third important factor to determine the rule checking criteria on BIM application is the level of development (LOD) of building model. Higher the LOD, more informative object model can be expected [2] and more efficient to integrate the regulations. Safety rule with programmed action can support decisions to managers and help to automate the search for hazards to construction worker during work and other operations.

Occupational Health and safety Administration (OSHA), the safety regulation agency offer text based, HTML web page based safety regulations data for site safety management. The standard construction rule classification of regulations is based on major work tasks. However, there is a need to analyze the structure of each regulation for the formal deployment in practice by integrating with BIM objects. Zhang et al. [1] classified OSHA regulations for fall protection in three parts, definition part of the rule specifies the unsafe zone, General requirement part describes the prevention method/methods for the specific case scenario. The third part is related to prevention criteria which is the detail of prevention system [1].

Following the discussion, this study focused more on classification system for OSHA rules in depth in order to introduce a framework for better interoperation of rule. Next section focused on the methodology and explains about the proposed framework for its application.

3 Methodology and Framework:

Establishing safety regulation classification criteria and mechanisms for real time provision of rules to safety supervisors is an important part of safety management. To accomplish this task, a series of safety regulations and BIM model with rich object information should be integrated systematically. If any of them is not properly functioning, it is hard to derive the desired outcomes. Therefore, in order to build a rule mapping network, a comprehensive system framework is introduced by linking class of rule with a 4D BIM model (integrating a 3D model with work schedule).

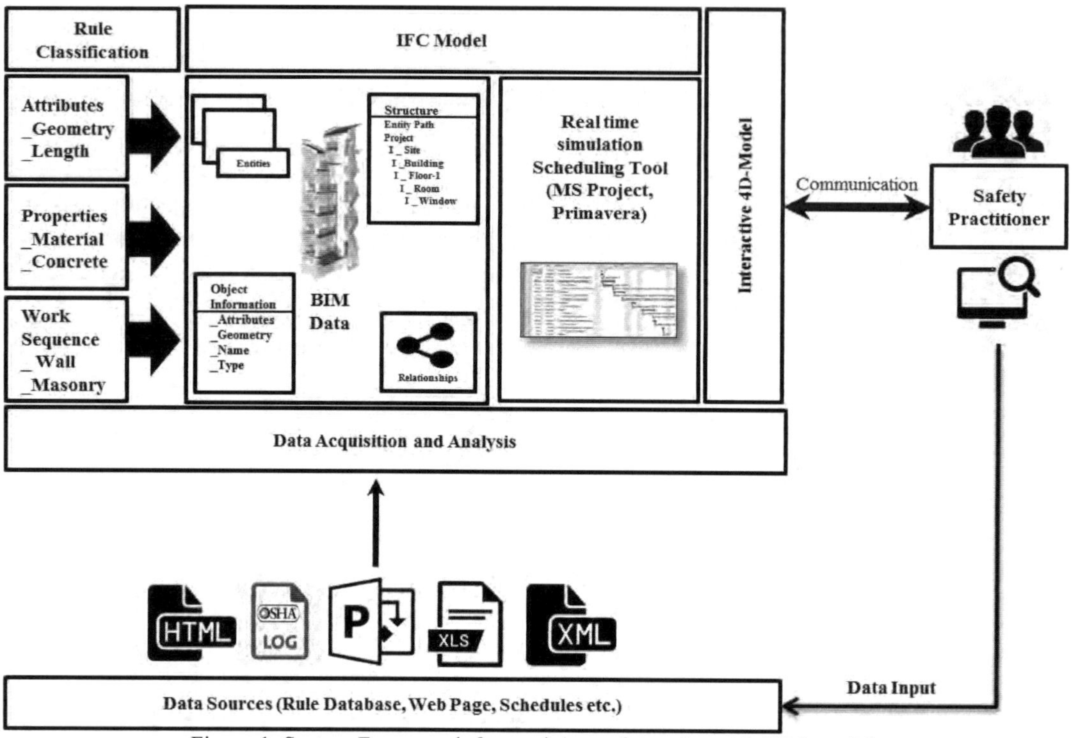

Figure 1. System Framework for applying safety rules with BIM model

While developing a standard classification, the safety document serves as an essential step to facilitate the whole proposed framework. In this paper, general classification criteria are proposed based on analysing OSHA regulation database. Fundamental criteria for rule classification are based on the actual needs of BIM model data. The classification criteria will be discussed in detail in the next section. Figure-1 illustrates the whole process of integrating classified rules with BIM model.

The system effectiveness is based on two fundamental steps. First, developing Rule classification criteria model. This model aims to provide a systematic structure of safety regulations for integration with relevant BIM components. Secondly, creation of 4D BIM model with rich context information of components. Prepare the semantic architecture 2D plan, following the 3D model by adding architecture objects and their properties. BIM data includes object information, work sequences of the building model, entities, properties and their relationships. The process of linking 3D model objects with the schedule information is mandatory for the successful application of the proposed framework. Safety practitioners can collaborate with BIM model by just receiving the respective regulation at each phase of building.

3.1 Development of criteria for rule classification

In order to establish the safety regulation review system, there is a gap to plan out what BIM based IFC models require. The Dataset in IFC model requires object attributes and entity references from rule statement. These may include object geometry, type of work, as well as work sequence which required a proper integration with work schedule. Based on the following discussion, the criteria for classification of rules are developed. However, since there is a limit to analyse all safety regulations and complexities inherent with them, a set of rules is classified according to the four different classes.

The selected set of rules is limited to 119 regulations and best practices which are collected from the OSHA's database for construction regulations. To explain the context in depth, most of the regulations are selected from a specific section of OSHA guidelines i.e. Concrete and Masonry Construction. The proposed classification criteria include following types of classes; Rule Characteristics, Work phase, Work Stage, Prevention Method. Each rule under consideration was analysed in such a way that it may be trackable easily by just mapping the respective class as shown in Figure 2. The ratios of analysed rules are given in Table 1,

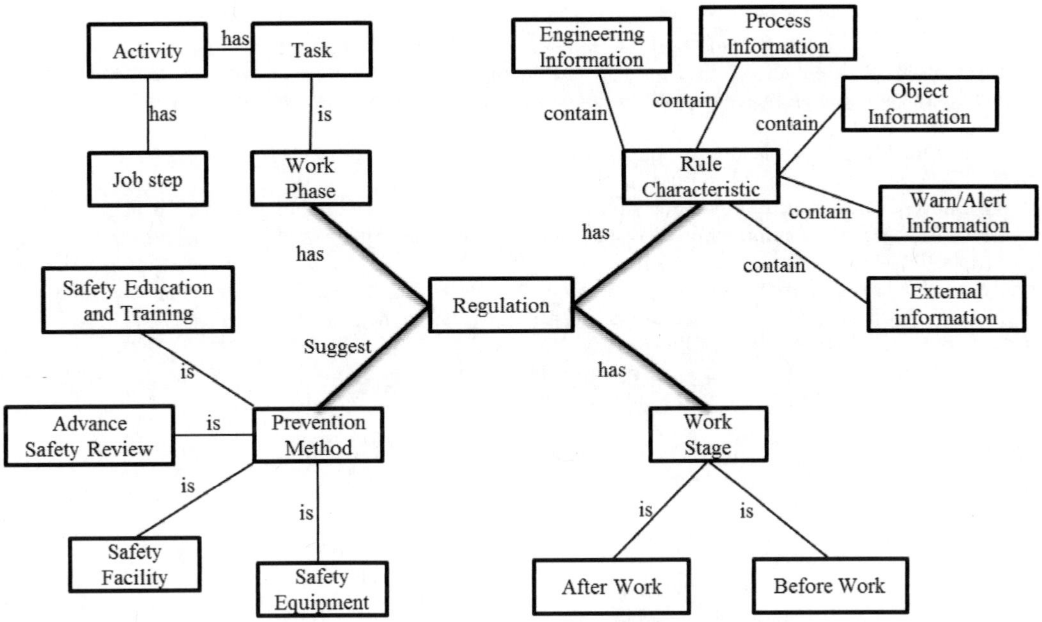

Figure 2. Criteria for mapping Rule Classification

which shows the proposition of rules considered with respect to their class and category.

Rule Characteristics: The analysis of regulation statement in terms of information is covered in category of rule characteristics. These characteristics may include information about engineering knowledge, warnings, detailed safe work process descriptions, object properties (numeric and non-numeric) and external information that can impact on site safety. Engineering information may consider the definition and explanation of specific safety terminologies. Therefore, it is necessary to review the construction safety regulations by considering the process information, which can be assigned to the BIM object.

Object properties in rule statement are considered important for rule analysis because numeric information in rule is one of the basic types of information required in BIM object modelling, including height, slope, width, heat flow rate, and strength. Studies in safety regulation state that numeric value is a prerequisite for reviewing the rule because one can logically review the text only by maintaining an algorithmic equation/table (example is given in Section 3.2.3).

Some regulations provide external information that is not supported by the BIM program or should be reviewed in conjunction with other safety guidelines. Unsupported information mainly includes weather information, country-specific standard material

Table 1 Data for Ratio of Classified Regulations

Category	Class	Percentage	Category	Class	Percentage
Work Phase	Before Work	61%	Work Stage	Task	5%
	During Work	19%		Activity	24%
	others	20%		Job Step	39%
Rule Characteristics	Engineering Information	39%		Others	33%
	Process Information	28%	Prevention Method	Safety Education	8%
	Object Information	15%		Safety Facility	20%
	External Information	8%		Safety Equipment	5%
	Warn/Alert Information	11%		Advance Safety Review	29%
				Others	39%

information and it should be reviewed by reading data constructed in a database format such as TXT or XML file. Therefore, it is necessary to develop the add-on program which can be run on the BIM platform by using the Open-API. Rule statement sometime contains warning/alerts for construction labour/practitioner, which can be deal in BIM model same like external information.

Work Phase: The analysis of regulation statement in terms of work phase. Rule information may be useful before the start of work or during the execution of work. Therefore, each rule is classified according to respective work phase, so that it can be helpful to determine the rule application time. To process this information in BIM model, it is necessary to integrate with time by simulating work phase information of the rule with the work schedule.

Work Stage: The analysis of regulation statement in terms of work stage i.e. task, activity or job step. The purpose of this class is to nominate that either a regulation is applicable for whole tasks or to its sub-activity or simply on a specific job step. This class will help to point out the specific and targeted stage of particular task. Single regulation may be applicable for whole task or for its one sub-stage. Construction of slab is a task and cutting of steel is the job step to accomplish this task.

Prevention method: The analysis of regulation statement in terms of suggested prevention method. Most of the regulations suggest a prevention system for unsafe tasks, which include safety education, safety equipment, advance safety review or additional safety facility. The selected set of rules is classified for each prevention system to easily access the solution of safety issue. Safety net is the one prevention system for fall from heights.

After the classification of rule sets, each rule can be integrated with respective object in BIM model and allowed to check the rule applicability. Checking of rule within the BIM model will suggest the prevention method or technique, according to the specific phase of project. This automated information acquired by BIM manager can be reported back to interested parties before execution of activity.

Some of the practical example cases are discussed in detail to elaborate more the research concept.

3.2 Practical implantation of Classification System:

3.2.1 Case-1:

OSHA Regulation:

"Load transfer from jacks/lifting units to building columns shall not be executed until the welds on the column shear plates (weld blocks) are cooled to air temperature". OSHA 1926.705(n)

Generally, the application of these types of rules include warn/alert information for advance safety review before the work starts. According to the statement, before removal of lifting units or transferring load to structural columns, one should ensure the mentioned conditions. This class of rules does not require any geometrical information or material properties of the object. However, this type of class is the set of rules that requires strict compliance. Moreover, this information can be involved further with BIM information model that can be transferred to the concerned manager while updating the project.

3.2.2 Case-2:

OSHA Regulation:

"All vertical slip forms shall be provided with scaffolds or work platforms where employees are required to work or pass" (OSHA 1926.703(c) (3)).

After analysing the regulation statement, the following categories with classes can be derived: vertical slip forms are additional safety facility for the scaffolding activity of concrete casting. This engineering information should be provided to the relevant safety supervisor before the start of concrete work on site. Therefore, the classification for this rule will be; engineering information as rule characteristic, before work as a work phase, job step as a work state and safety facility as a suggested precaution method in the regulation.

Similar to case-1, this type of class is the set of rules that require a strict compliance for practical implementation. To do so, Solihin et al. (2015) suggested a separate knowledge base to consider this type of rule during BIM design as well as execution phase [2].

3.2.3 Case-3:

OSHA Regulation:

"All masonry walls over eight feet in height shall be adequately braced to prevent overturning and to prevent collapse unless the wall is adequately supported so that it will not overturn or collapse. The bracing shall remain in place until permanent supporting elements of the structure are in place" OSHA Regulation 1926.706(b).

This rule has a conditional statement about the attributes of the BIM object related to its geometry i.e. condition to check whether the wall height is more than 8 feet (2.44 meter). If the condition satisfies, the preventative measure which is additional facility in terms of bracing will be applicable. The default tabular format of rule interpretation is also shown in Table-2.

Irrespective of other cases, this type of class requires more details in the form of geometric attributes and properties of BIM objects. This type of rules can be

Table 2 Example of OSHA rule interpretation

Dimension criteria for wall	Hazard	Prevention Method
Height <= 8 feet (2.44 metre)	-	Not considered
Height > 8 feet (2.44 metre)	Overture/Collapse	Bracing

interpreted for BIM dataset easily by using SWRL rule. To maintain the geometrical consideration, the measuring units should be converted from U.S standard to metric during interpretation [8].

3.2.4 Case-4:

OSHA Regulation:

"Reinforcing steel: Reinforcing steel for walls, piers, columns, and similar vertical structures shall be adequately supported to prevent overturning and to prevent collapse." (OSHA 1926.703(d) and 1926.703(d)(1))

By reviewing the rule statement, the categories and respective classes can be derived as follows: Engineering information as well as warn alert information is provided in this rule. In other words, reinforcing steel needs safety facilities in terms of additional supports during construction. The work phase is during construction and the work stage is sub activity of major task. This class of rules needs well defined entities (walls, piers, columns) for processing in BIM model. Moreover, to process this class of rules, well-structured relationships among entities are also required. Hazard alert information is also included sometimes i.e. overturning and collapse.

4 Discussion

In order to classify the safety regulations effectively, it is necessary to switch from referencing document-based rule information and use digital information for review. Inaccuracy in searching for rules can be minimized by improving the rule classification system and combining rule information with the BIM model object information. In addition, through the incorporation of work schedules with BIM objects, safety information can be delivered at the specified time. The proposed regulation classification system can reduce the rule searching time, human input and finally affect the whole safety management process. This work also showed the example cases of four different OSHA regulations to demonstrate the objective in depth.

However, various limitations were revealed in the course of the research, and the following should be solved for the successful implementation of the regulation classification system with BIM. There is a limit to analyse all safety regulations and complexities inherent with them. Hence, a set of rules containing 119 safety provisions only are classified in the four different classes. The inefficiency in mapping rule information with object information may necessitate manual work during rule processing in BIM models. In order to overcome this issue, an ontology based algorithm of rule documents can be developed to integrate the safety regulations with the object information of the BIM. However, there are still some clauses that cannot be reviewed solely based on BIM object information. These rules are designed for the general activities during construction. Therefore, it is necessary to define such rules in planning phase with each general work tasks in BIM based object model. Furthermore, a practical case study with an evaluation of the proposed system will be presented in future research.

5 Conclusion

Safety in the construction industry is directly connected to the lives of the workers and should be considered as a critical need of construction sites to avoid accidents and risks. With the increasing trend of technology in the construction industry, technology development and diversified nature of risks and hazards, current safety regulations have been revised in a complicated way. The regulations that are applied during different phases of project should be assessed and analysed to enable easier access and to reduce oversights. This research envisions to improve construction site safety by introducing a rule classification system for OSHA's rules document that can be integrated with BIM applications at a later stage. OSHA's safety and health technical guidelines for Concrete and Masonry Construction are selected to classify the legal provisions into safety management steps and tasks. In order to understand the rule processing, the information level of BIM required by the legal clause was mapped with the classified network. To this end, four different cases were selected to discuss the practical applicability of OSHA regulations. The proposed regulation classification system can reduce oversights, rule searching time and finally affect the whole safety management process. The next step of research will focus on the methods to improve the applicability and performance of the system more thoroughly through a real-world case study.

Acknowledgement

This work was supported by the National Research Foundation of Korea (NRF) grant funded by the Korea government (MSIP) (No. NRF-2016R1A2B3016694).

References

[1] Zhang, Sijie, Jochen Teizer, Jin-Kook Lee, Charles M. Eastman, and Manu Venugopal. 2013. "Building Information Modeling (BIM) and Safety: Automatic Safety Checking of Construction Models and Schedules." Automation in Construction 29:183–95.

[2] Solihin, W. and C. Eastman. 2015. "Classification of Rules for Automated BIM Rule Checking Development." Automation in Construction 53:69–82.

[3] CORENET e-PlanCheck: Singapore's Automated Code Checking System. http://www.aecbytes.com/archives/categories/feature.html

[4] Choi, Jungsik, Junho Choi, and Inhan Kim. 2014. "Development of BIM-Based Evacuation Regulation Checking System for High-Rise and Complex Buildings." Automation in Construction 46:38–49. Retrieved (http://dx.doi.org/10.1016/j.autcon.2013.12.005).

[5] Pauwels, P. et al. 2011. "A Semantic Rule Checking Environment for Building Performance Checking." Automation in Construction 20(5):506–18. Retrieved (http://dx.doi.org/10.1016/j.autcon.2010.11.017).

[6] Poças Martins, J. P. and V. Abrantes. 2010. "Automated Code-Checking as a Driver of BIM Adoption." International Journal for Housing Science and Its Applications 34(4):287–95.

[7] Beach, Thomas H., Tala Kasim, Haijiang Li, Nicholas Nisbet, and Yacine Rezgui. 2013. "Towards Automated Compliance Checking in the Construction Industry." Lecture Notes in Computer Science (Including Subseries Lecture Notes in Artificial Intelligence and Lecture Notes in Bioinformatics) 8055 LNCS(PART 1):366–80.

[8] Zhang, Sijie, Frank Boukamp, and Jochen Teizer. 2015. "Ontology-Based Semantic Modeling of Construction Safety Knowledge: Towards Automated Safety Planning for Job Hazard Analysis (JHA)." Automation in Construction 52:29–41.

[9] Chi, Nai Wen, Ken Yu Lin, and Shang Hsien Hsieh. 2014. "Using Ontology-Based Text Classification to Assist Job Hazard Analysis." Advanced Engineering Informatics 28(4):381–94. Retrieved (http://dx.doi.org/10.1016/j.aei.2014.05.001).

Automated Clash Resolution of Steel Rebar in RC Beam – Column Joints using BIM and GA

M. Mangal[a], Q. Wang[a] and J.C.P. Cheng[a]

[a]Department of Civil and Environmental Engineering, The Hong Kong University of Science and Technology, Hong Kong, China

E-mail: mmangal@connect.ust.hk , qwangau@connect.ust.hk , cejcheng@ust.hk

Abstract –

Design of steel rebar in reinforced concrete (RC) members is usually conducted individually based on loading conditions, support conditions, and geometric conditions of the considered RC members only. Member-member interactions are rarely considered at the design stage, leading to potential rebar clashes (hard clash) or congestion (soft clash) at beam-column joints on construction sites. Currently, rebar clashes and congestion are often manually identified and resolved at the construction stage by site engineers or workers, which is a tedious and time consuming process. The building information modeling (BIM) technology allows us to digitally represent the detailing of steel rebar and transfer the detailing information to structural analysis software. However, automated identification and resolution of steel rebar clashes and congestion in RC members are lacking in the existing BIM software packages.

This paper presents an automated steel rebar design framework based on BIM and genetic algorithm (GA) that considers and avoids rebar clashes and congestion at RC beam-column joints. Each beam-column interaction in a given RC frame is first analyzed to find out the type of each beam-column joint, such as T Joint and + Joint. GA is used to generate a first stage population containing steel rebar number and location for each RC member. Each individual of the population is then checked at all beam-column joints to avoid hard clashes between inter-member steel rebar. Optimal rebar diameter combination (with minimum rebar area) for each clash-free individual of the population is then calculated using the second stage GA based steel rebar optimization engine. The optimization engine contains a built-in checking function to avoid congestion between steel rebar. An illustrative example will be shown to test the developed framework.

Keywords –

Clash Resolution; Beam-Column Joint; Building Information Modelling; Steel Rebar Design; RC Frame

1 Introduction

Steel rebar design is a mandatory step in reinforced concrete (RC) design process. Provided steel rebars should be able to withstand the loads imposed during the life span of RC structures. Besides, steel rebar design should be easy to construct, safe and cost-effective. Currently, steel rebar design for RC members is usually conducted individually based on loading conditions, support conditions, and geometric conditions of the considered RC members only. This leads to potential steel rebar clashes (hard clash) or congestion (soft clash) at beam-column joints. Identification and resolution process of these steel rebar clashes is carried out manually by site engineers or workers at the construction stage. The identification and resolution process is tedious, time consuming and sometimes expensive.

Various studies have been conducted in the past to tackle the problem of clash identification and resolution. Park [1] developed a methodology to generate steel rebar placement sequence to increase productivity. Clashes between various steel rebars were also automatically detected through a developed application programming interface (API). However, only a manual approach was considered for clash resolution. Navon et al. [2] developed a system to identify various constructability problems including rebar congestion and rebar collision. However, rebar clash resolution was still conducted manually by engineers. Radke et al. [3] automated the identification and resolution of spatial clashes for mechanical, electrical and plumbing (MEP) systems. Clashes were identified by determining the distance between centerlines of two objects. The identified clashes were then resolved by moving one of the two entities. However, design constraints were not verified after moving one object. Besides, it was able to solve only limited type of clashes. Wang and Leite [4] carried out a profound study of clashes in MEP systems. A thought process based schema was also developed to store clash based information. The clash based information included description, context, evaluation and management details. However, the developed schema only provided a documentation method to store clash based information without any clash resolution strategy for identified clashes. Moreover, most of the above studies lack full automation in providing required input to the clash detection and resolution problem. Therefore, developing an automated BIM-based framework to provide clash resolution strategy will be of utter importance in architecture, engineering and construction (AEC) industry.

Therefore, the objective of present paper is to develop a BIM-based automated clash free steel rebar design optimization framework according to regional design codes. The previously developed Hybrid GA-based steel rebar optimization engine [5] will be integrated with the developed BIM-based automated clash free steel rebar design framework to provide optimum steel rebar diameters for different RC elements. Since RC frames consist of mainly RC beams and RC columns, RC frames having RC beams and RC columns are considered to analyze the performance of the developed framework.

2 The Proposed BIM-based Automated Clash Free Steel Rebar Design Framework

The proposed BIM-based automated clash free steel rebar design framework is explained in this section. The framework consists of 5 modules (Figure 1), namely (1) BIM Model Extraction, (2) Structural Analysis, (3) Beam-Column Joint analysis, (4) Rebar Number Optimization, and (5) Steel Rebar Calculation and Optimization, as follows.

- *(1) BIM Model Extraction*: Requisite functional (e.g., loading and end support conditions) and physical (e.g., cross-sectional information) characteristics for steel reinforcement calculation is extracted from BIM model for further clash free steel reinforcement optimization calculation.
- *(2) Structural Analysis*: The extracted information is then imported into structural analysis software to calculate design bending moment and shear force. Relevant design code recommended safety factors are also considered in the calculation. The calculated results are saved for further calculation.
- *(3) Beam-Column Joint Analysis*: Beam-column interactions at each level of RC frame is analyzed to extract the joint type, such as T joint and + joint. Joint information is saved for further calculation of rebar number in each RC member.
- *(4) Steel Rebar Number Optimization*: Provided rebar number for each RC member is then calculated on the basis of the outcome of the joint type at each level of RC frame. The calculated rebar number for each RC member is then stored for further calculation of rebar diameters.
- *(5) Steel Rebar Calculation and Optimization*: Provided rebar diameters for each RC member are finally calculated considering the calculated rebar number and other conditions stipulated in the regional design code.

The first two modules are straight forward and require no explanation. The rest three modules are explained in Sections 3-5, respectively.

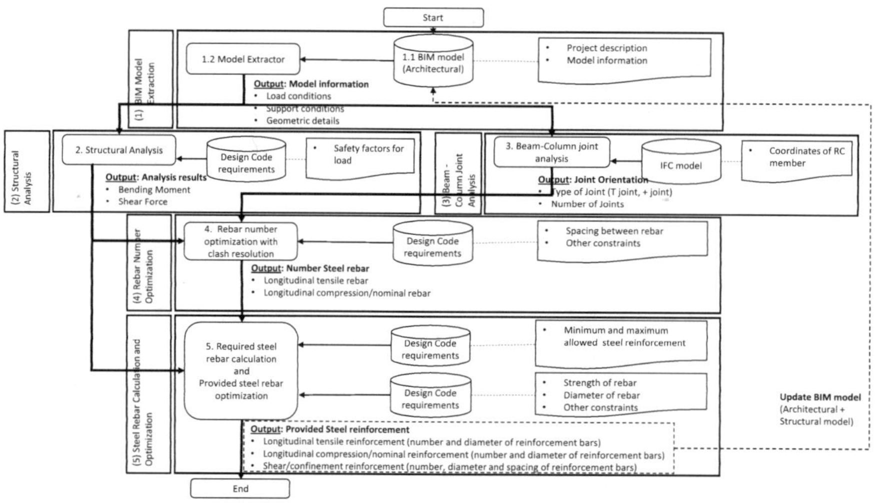

Figure 1. The proposed BIM-based automated clash free steel rebar design framework

3 Beam-Column Joint Analysis

RC frame consists of various types of beam-column joints (e.g., T joint, + joint, etc.) based on orientation and interaction of RC beams and RC columns. Various types of beam-column joints generally encountered in an RC frame are given in Figure 2. Beam-column joint analysis is carried out to classify the type of each beam-column joint in a given RC frame. Classified beam-column joints will further help to calculate the clash free steel rebar number in each RC member (RC beam or RC column) of the considered RC frame. RC frame is first divided into separate levels (floors). Every RC column in each separated level is analysed to extract the connected RC beams with that RC column. A maximum of 4 RC beams can be linked to an RC column at any separated level. Extracted beam-column joint information is stored for further calculation of steel rebar numbers. A total of 10 different types of beam-column joints are considered in the developed BIM-based automated clash free steel rebar design framework.

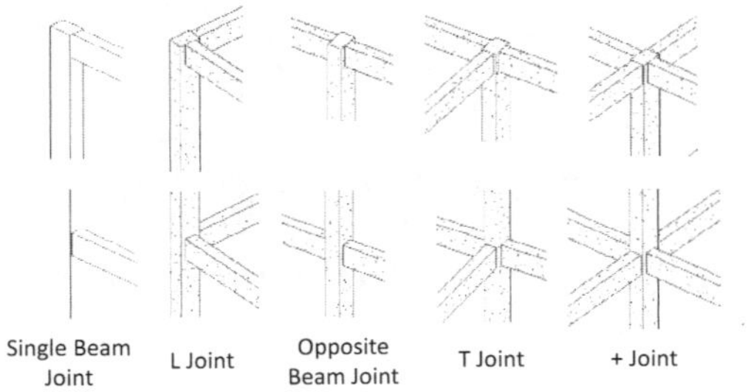

Figure 2. Various types of beam-column joints generally encountered in an RC frame

4 Formulation of Rebar Number Optimization Problem

This section will present the formulation of the objective function for optimization of steel rebar number of each RC element in an RC frame. Section 4.1 presents the steel rebar number range calculation for longitudinal reinforcement as per design code. Section 4.2 describes the compatibility requirements for steel rebars of different RC members. Section 4.3 presents the overall steel rebar number optimization function of a whole RC frame with subjected constraints.

4.1 Steel Rebar Number Range Calculation for Longitudinal Reinforcement

Different regions have different design codes for RC design. For example, both BS8110 and BS4449 are used in Hong Kong and UK. BS8110 deals with the minimum required rebar area depending on building data. The building data include support conditions, loading conditions and geometrical information of all RC members in a given RC frame. BS4449 includes properties of steel rebars such as flexural strength, yield strength and allowable steel rebar diameters. The developed optimization tool considers BS8110 and BS4449, although it can be modified to account for user specific rebar properties (e.g., preferred steel rebar diameter).

The provided longitudinal tensile/compressive/nominal steel reinforcement is the total cross-sectional area (A_s) of longitudinal steel rebar provided at a section, which can be calculated as:

$$A_s = \sum_{i=1}^{t} n_i \pi d_i^2 / 4 \quad (1)$$

where d_i is the selected longitudinal rebar diameter from BS4449, n_i is the total number of steel rebar of type i, and t is the number of different diameter types of steel rebar.

The total number of steel rebar (Sn_i) depends on BS8110 as given below:

$$n_{min} <= (\textstyle\sum_1^t n_i \text{ or } Sn_i) <= n_{max} \quad (2)$$

where

$$n_{min} = \frac{(b-2*sc)}{Hs_{max}} \quad (3)$$

$$n_{max} = \frac{(b-2*sc)}{Hs_{min}} \quad (4)$$

where b is the width of RC member, sc is the side cover, Hs_{min} is the minimum horizontal steel rebar spacing, and Hs_{max} is the maximum horizontal steel rebar spacing as per the provisions of BS8110 (Figure 3).

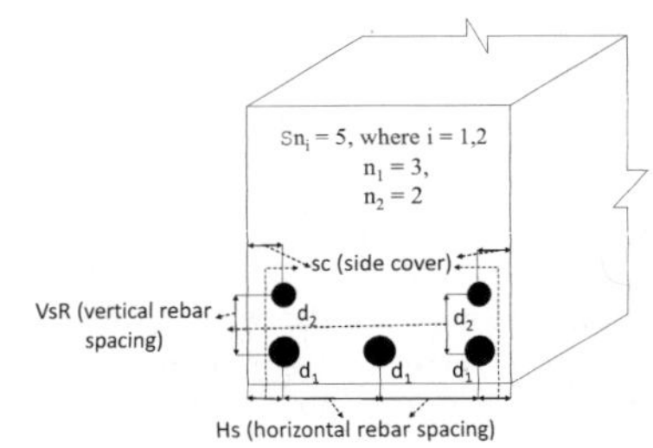

Figure 3. Horizontal and vertical rebar spacing for an RC member

Steel rebar range ($n_{x\ min}$, $n_{x\ max}$ and/or $n_{y\ min}$, $n_{y\ max}$) is calculated for each RC member of RC frame. Steel rebar range is only calculated in x direction ($n_{x\ min}$, $n_{x\ max}$) for RC beam (Figure 4). However, steel rebar range is calculated in both x and y directions ($n_{x\ min}$, $n_{x\ max}$ and $n_{y\ min}$, $n_{y\ max}$) for RC column as orientation of steel rebars in RC column depends on the building data as described before (Figure 5). The calculated steel rebar range for each RC element is stored for further calculation of provided steel rebar number in each RC element.

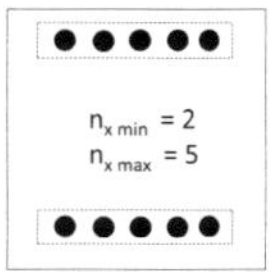

Figure 4. Steel rebar range in x direction for an RC beam

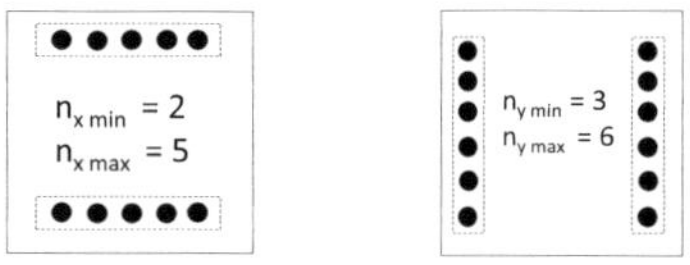

Figure 5. Steel rebar range in x any y directions for an RC column

4.2 Steel Rebar Compatibility

Steel rebar number is then assigned to every RC element of RC frame. Assignment of number of steel rebar depends on the type of beam-column joint and the orientation of rebar in RC column as shown in Figure 6.

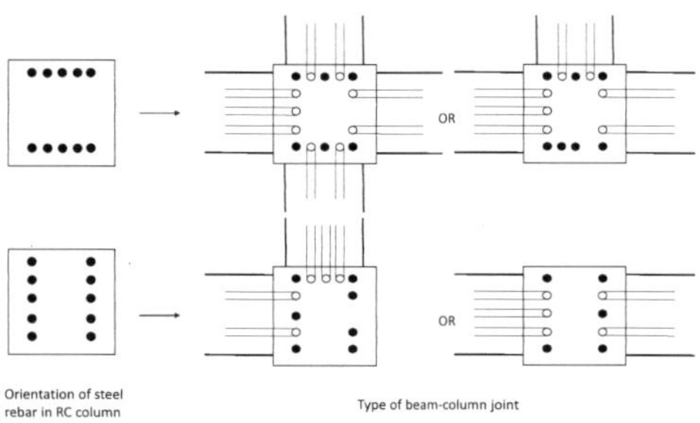

Figure 6. Assignment of steel rebar in RC elements depending on the orientation of RC column reinforcement and type of beam-column joint

It is ensured that the steel rebar number in each beam should not exceed the calculated steel rebar range of beam as given below:

$$n_{x\,min\,i} < n_{b\,i} < n_{x\,max\,i} \quad (5)$$

where, $n_{x\,min\,i}$ and $n_{x\,max\,i}$ are the steel rebar number range for RC beam i and $n_{b\,i}$ is the provided number of steel rebar in RC beam i.

Moreover, the total number of steel rebars of RC beam and RC column at each beam-column joint should not exceed the rebar number range of that RC column in both directions as given below:

$$n_{x\,min\,i} < n_{b\,j} + n_{c\,i} < n_{x\,max\,i} \quad (6)$$

$$n_{y\,min\,i} < n_{b\,j} + n_{c\,i} < n_{y\,max\,i} \quad (7)$$

where, $n_{x\,min\,i}$ and $n_{x\,max\,i}$ are the steel rebar number range for RC column i in x direction. $n_{y\,min\,i}$ and $n_{y\,max\,i}$ are the steel rebar number range for RC column i in y direction. $n_{c\,i}$ and $n_{b\,j}$ are the provided numbers of steel rebar in RC column i and RC beam j, respectively.

Clash number (CNi) is defined as the total number of steel rebar clashes observed at beam-column joint i of RC frame. It can be observed that clash between steel rebars of RC beam and RC column will occur when any interaction of RC beam and RC column at each beam-column joint does not satisfy equation 6 or equation 7. A maximum of 4 steel rebar clashes can be observed at any beam-column joint (+ type joint).

4.3 Objective Function for Steel Rebar Number

It can be inferred that the total number of clashes (CN) in an RC frame will be equal to the sum of CNi of every beam-column joint of that RC frame:

$$CN = Obj\_func\,(CNi) = \sum CNi \quad (8)$$

Our objective is to minimize the above objective function in order to have the least amount of clashes in between steel rebars (ideally zero clashes).

5 Steel Reinforcement Calculation and Optimization

The variables considered in RC member design (RC beam or RC column) are similar across all design codes. The considered variables are (1) area of longitudinal tensile reinforcement (A_s), (2) area of longitudinal compression reinforcement (A_s'), and (3) cross-sectional shear area of links at the neutral axis at a section (A_{sv}) which in turn depends on the building data (Section 4.1) and the provisions of BS8110.

A Hybrid GA-based optimization engine for automated optimization of steel reinforcement was developed by Cheng and Mangal [5] to provide the number and different diameter combination of steel rebars. The already developed steel reinforcement optimization engine is modified in this study to provide the different diameter combination of steel rebars for a given steel rebar number for each RC element. The modified engine is integrated with the developed automated clash free steel rebar design framework in order to provide clash free steel rebar design for the whole RC frame.

6 Solving the Steel Reinforcement Optimization using Two Stage Genetic Algorithm

6.1 Introduction to Optimization Algorithm

Optimization of the provided steel rebars depends on both the number of steel rebar and different diameter combination of steel rebars, thus having more than a million possible combinations. Hence, a powerful yet reliable technique is required to arrive at an optimal solution within a reasonable amount of time.

Many heuristic algorithms have been studied and developed to solve above types of NP-Hard problems. Some commonly used heuristics are Ant Colony Optimization (ACO), Particle Swarm Optimization (PSO), and Genetic Algorithms (GA). The two level GA has been developed for this study as described in Section 6.2.

6.2 Optimization Technique for Current Problem

Optimization of provided steel rebars is carried out in two levels. First level includes the calculation of clash free rebar numbers for each RC element of RC frame (Section 4.2) using the developed GA. Second level includes the calculation of different diameter combination for longitudinal tensile/compressive and shear/confinement reinforcement for RC elements for calculated clash free rebar numbers. Different diameter combination calculation is carried out by modifying the already developed hybrid GA-based automated steel reinforcement optimization engine [5].

First level GA for calculation of rebar number is developed to find out the optimum number of steel rebars with minimum clashes (ideally zero clashes i.e., CN = 0). Steel rebar range ($n_{x\ min}$, $n_{x\ max}$ and/or $n_{y\ min}$, $n_{y\ max}$) is calculated for RC elements considering the building data (Section 4.1). Orientation of steel rebars (Figure 5) is then decided for each RC column considering the loading information, end-support information, and geometrical information. Orientation of steel rebars in RC column will help us to decide the available locations for steel rebars for RC beam steel reinforcement. Now, first level GA is developed to assign the steel rebar numbers to each RC element considering the type of beam-column joint (Section 3) and steel rebar orientation in RC columns (Figure 5). The developed first level GA includes the provisions of design code to ensure that the calculated steel rebar number satisfies the design code requirements for practical purpose. Aim of the developed first level GA is to minimize the objective function (Section 4.3), i.e., the total number of steel rebar clashes (CN) in a given RC frame. Compatibility checks (Section 4.2) are carried out to calculate the total number of steel rebar clashes (CN) in a given RC frame. Mutation and Uniform Crossover are also applied to achieve optimal solutions faster. Mutation helps to get out of local optima. It also explores the solution space for better solutions. Uniform crossover is applied to bring more diversity to future generation. Uniform crossover also eliminates the possibility of positional bias. It is not fruitful to repeat the generation process if results within desired thresholds have been achieved or no improvement has been recorded in the subsequent generations (e.g., 100 generations). Therefore, necessary condition is applied to achieve optimal solution in least amount of time. The necessary condition ensures that clash free solution (CN = 0) for a given RC frame has been achieved. The calculated clash free number of steel rebars for each RC element is saved for further calculation of different diameter combination of steel rebars.

Second level optimization includes the calculation of different diameter combination of steel rebars for given clash free number of steel rebars. Already developed hybrid GA-based automated steel reinforcement optimization engine [5] is used for the calculation of different diameter combination of steel rebars. The optimization engine provides minimum steel reinforcement area results for each RC element in a given RC frame. For e.g., optimization engine provide minimum steel rebar area for longitudinal tensile reinforcement, longitudinal compressive reinforcement and shear reinforcement for RC beam as given in Figure 7. The engine also considers the design code requirements and constructability requirements for better applicability of optimized results on construction sites. However, the optimization engine provides both, number and different diameter combination of steel rebars for each RC element (RC beam or RC column) of a given RC frame. Hence, the already developed hybrid GA-based steel reinforcement optimization engine is modified to provide only the optimum different diameter combination (with minimum steel reinforcement area) of steel rebars for given clash free number of steel rebars already calculated in first level GA.

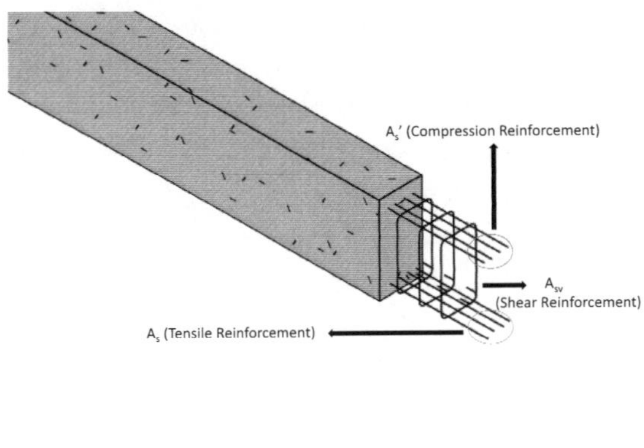

Figure 7. Steel reinforcement optimized by hybrid GA-based steel reinforcement optimization engine for a given RC beam

7 Illustrative Example

A one storey RC frame as a whole is considered in the illustrative example to evaluate the efficiency of the developed BIM-based automated clash free steel rebar design optimization framework at entire frame level. The considered fixed end RC frame is given in Figure 8. Only one type of joint (L joint) is considered in the current illustrative example for evaluation purpose. Uniform dead load (DL) and live load (LL) were applied on the RC beams of the considered RC frame. The developed BIM-based automated clash free steel rebar design optimization framework tool is shown in Figure 9.

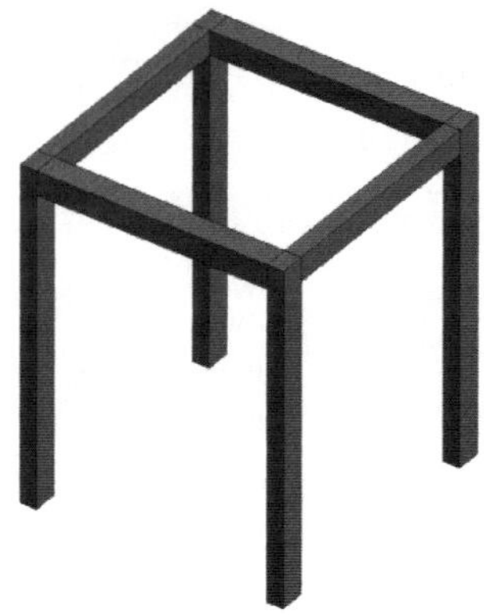

Figure 8. A one-storey fixed end RC frame

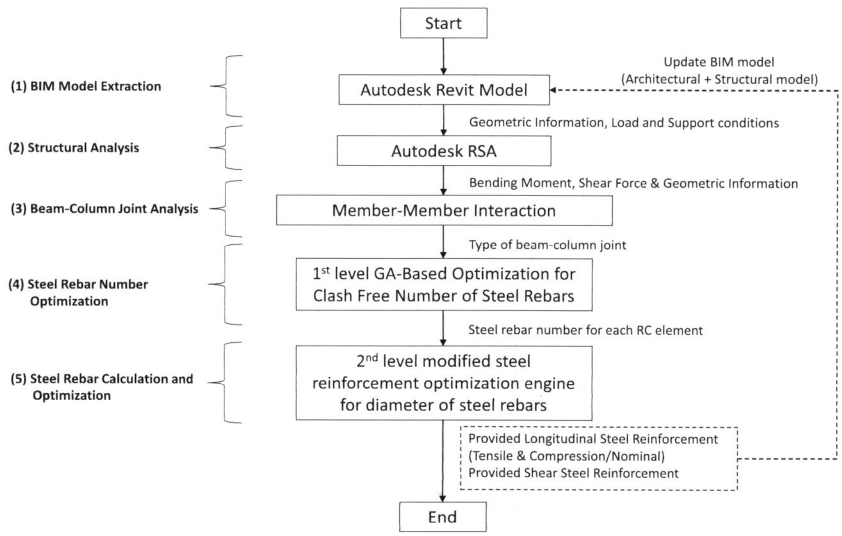

Figure 9. BIM-based automated clash free steel rebar design optimization framework tool

The automated 3D output of the optimized steel rebar is given in Figure 10. One of the single beam-column joint is shown in Figure 11. It can be inferred from Figure 11 that the developed BIM-based automated clash free steel rebar design optimization framework provides optimum results without any steel rebar clashes between steel rebars of different RC element at beam-column joints. 3D output also ensures the extraction of construction drawings as per site requirements. The optimization time varies linearly (O(n)) with the number of RC elements in a given RC frame. For e.g., it takes 40 seconds to optimize the considered RC frame of 4 RC beams and 4 RC columns.

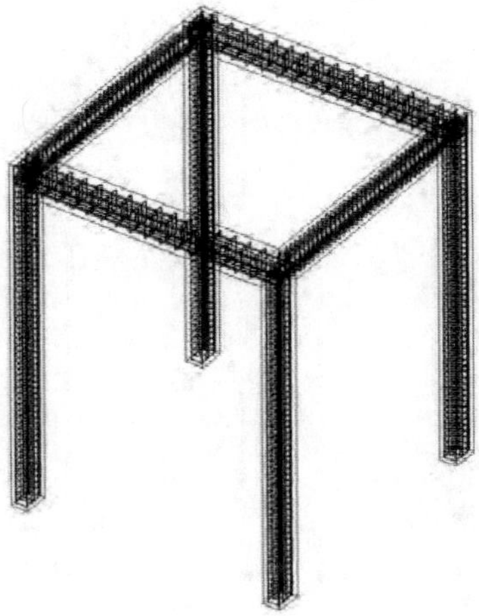

Figure 10. Automated 3D view of the optimized steel rebar for the considered RC frame

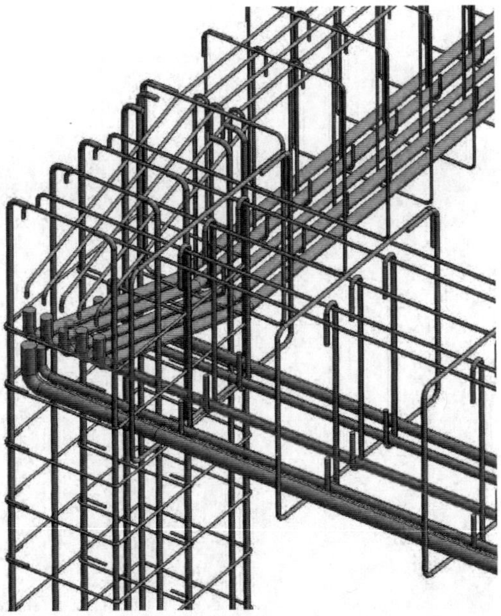

Figure 11. 3D output of one of the beam-column joints of the considered RC frame

8 Conclusion

In this paper, we proposed a BIM-based automated clash free steel rebar design optimization framework for steel reinforcement using two level optimization process. The objective was to provide minimum steel rebar area satisfying the design codes and constructability requirements. Moreover, the provided reinforcement should not have any clashes at beam-column joints. The formulated objective function allows us to consider all the steel rebar clashes at each beam-column joint. Moreover, incorporation of an already developed steel reinforcement optimization engine helped us to find the minimum steel rebar area for given number of steel rebars. The developed BIM-based automated clash free steel rebar design framework provides full automation and interoperability to achieve fast, consistent and site implementable results. An illustrative example demonstrates that the developed framework is effective and efficient. Future work will include the extension of developed framework for the other nine types of beam-column joints.

References

[1] Park U. BIM-Based Simulator for Rebar Placement. *Journal of the Korea Institute of Building Construction,* 12 (1): 98-107, 2012.

[2] Navon R., Shapira A. and Shechori Y. Methodology for Rebar Constructibility Improvement. In *Construction Congress VI: Building Together for a Better Tomorrow in an Increasingly Complex World*, pages 827-836, 2000.

[3] Radke A., Wallmark T. and Tseng M.M. An automated approach for identification and resolution of spatial clashes in building design. In *IEEM 2009-IEEE International Conference on Industrial Engineering and Engineering Management*, pages 2084-2088, Hong Kong, 2009.

[4] Wang L. and Leite F. Formalized knowledge representation for spatial conflict coordination of mechanical, electrical and plumbing (MEP) systems in new building projects. *Automation in Construction,* 64 (1): 20-26, 2016.

[5] Cheng J.C. and Mangal M. Automated Optimization of Steel Reinforcement in RC Building Frames using BIM and Hybrid GA. In *Lean & Computing in Construction Congress (LC3)*, Crete, Greece, 2017.

34th International Symposium on Automation and Robotics in Construction

ISARC 2017

June 28 – July 1 TAIPEI, TAIWAN

CONSTRUCTION ROBOTICS (I) / FABRICATION

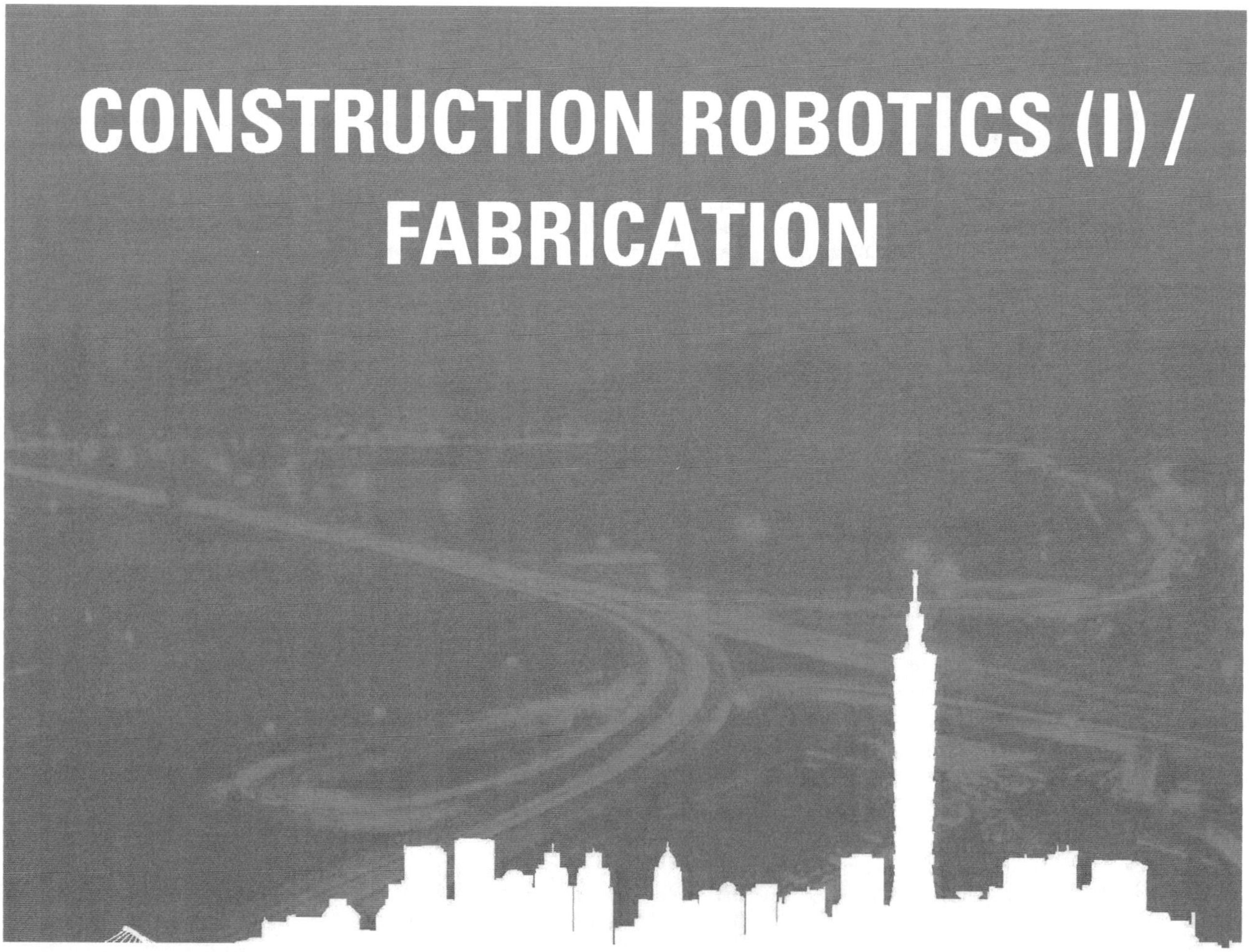

Dimensional variability analysis of construction assemblies using kinematics chains and building information models

M. Nahangi [a], C. Rausch [a], C. T. Haas [a], Jeffrey West [a]

[a] Department of Civil and Environmental Engineering, University of Waterloo, Waterloo, ON, Canada.
E-mail: mnahangi@uwaterloo.ca, ctrausch@uwaterloo.ca, chaas@uwaterloo.ca, jswest@uwaterloo.ca

Abstract

Design optimization frameworks used in manufacturing can be adopted into construction to solve complex and relatively unsolved challenges, as recent construction methods progressively incorporate more manufacturing aspects. For example, the specification and dimensional control for compliance checking of construction components can be solved using tools existing in manufacturing. Even though building information models (BIM) assist with clash detection for identifying potential dimensional problems, dimensional variability remains a complex challenge to address in construction. This paper explores the use of a dimensional variation analysis (DVA), which is originally developed in the manufacturing industry as a design optimization tool. This paper presents a DVA approach which is based on kinematics theory in robotics to define the assembly equation (how various parts of a component are related to each other mathematically). A design-model based DVA is validated using a case study. Results show that the method is capable of determining variability between the designed and fabricated states with a reasonable level of accuracy.

Keywords:

Local feature descriptors, hashing, hash table, RANSAC, 3D object recognition, clutter, laser scanning

1 Introduction

The inevitable variability associated with the geometry of as-built construction assemblies can create problems if not properly controlled and managed. This type of variability has a direct relationship with how well components can be built and assembled together. Constructability is one method that can be employed to ensure problems associated with the construction and assembly of components are minimized. Constructability employs strategic knowledge during planning, design and execution to achieve project objectives. It therefore requires an investment in front-end planning in order to anticipate and solve potential problems [1,2]. Some of the most common problems that constructability addresses are related to rework, design errors, change orders, low product quality, project delays, tolerance problems, physical interface problems, and not meeting client expectations [3]. The common aspect of all of these problems is that they originate from the dimensional or geometric properties of components or construction assemblies. The ability to model construction components in a virtual three-dimensional space has enabled the use of building information modelling (BIM) and computer-aided design (CAD) tools to be used to aid in constructability [4,5]. Use of these virtual design and construction (VDC) tools places an emphasis on ensuring coordination between trades and detection of physical component clashes. The focus on both constructability and use of BIM in the construction industry shows that the proper management of component and system geometry is a key factor in overall project success.

The importance of controlling and managing variability of geometry is generally well accepted in the construction industry. While proper control of geometry is often not extensively considered up front in the design process, its impact is unavoidable during fit-up and assembly and can result in expensive rework. Dimensions and configurations of components always have some degree of variation from nominal specifications [6]. This is why the construction industry has adopted the use of standardized tolerances to ensure that acceptable limits are placed on dimensional variability [7,8]. Tolerances by themselves are not enough to solve geometric problems since tolerances can accumulate throughout assemblies, resulting in fit-up problems, delays, rework and increased costs [9]. Furthermore, the trend towards increasing industrialization in construction introduces an even greater demand on geometric compatibility between components and assemblies since the bulk of the fabrication and construction work occurs offsite and requires assemblies to fit properly once on site [10].

This paper presents a framework for dimensional variation analysis which uses kinematics theory as a method for modelling the geometric relationships between components. The related background is first investigated, then the proposed framework is then presented. A case study is provided to demonstrate how dimensional variability can be modelled through the use of kinematics chains.

2 Background

This section presents background related to: (1) existing methods for dimensional control in construction, and (2) theory behind kinematics chains for geometric modelling purposes.

2.1 Existing approaches for dimensional control in construction

While existing methods for analysis, detection and control of dimensional variability in the construction industry can utilize proactive 3D analysis techniques such as spatial change analysis [11] or automated compliance checking [12], the majority of variation control techniques during construction are still performed reactively (i.e., problems related to dimensional variability are only solved once they have occurred). This often creates a grey area during a project when variations in the field create installation challenges for specific contractors because the assumption of risk for out-of-tolerance occurrences is not well defined.

Proactive methods for controlling variability are generally restricted to the use of BIM-based clash detection. Proactive solutions for dimensional conflicts are superior to reactive methods, evidenced by the fact that 90% of commercial contractors are currently using clash detection on projects which utilize a building information model [13]. In a study by Leite et al. [14], it was shown that on-site clash fixes are costlier than creating a more detailed BIM upfront and using clash-detection. For this reason, spending more time during the design to detect and avoid clashes can offset the cost associated with field rework due to dimensional variations.

2.2 Kinematics chains for geometry modelling

Robotics theory has opened up a wide and efficient range of solutions in engineering problems. For example, robotics concepts combined with machine vision have been used for state modelling and sensing of construction equipment such as pipe manipulators and excavators [15]. A specific pose of the end effector (i.e., the end of a kinematics chain, which is the critical feature of interest) can then be modelled using the related inverse and forward kinematics. Robotics concepts can also be used for automating tasks associated with a high level of repetition or harsh tasks that are performed by workers in hazardous areas. For instance, a machine-vision-assisted system in introduced by [16] for automating the bricklaying assembly. Kinematics theory was also used for geometric modelling of construction assemblies as a mathematical function [17]. Discrepancies of the as-built state of construction assemblies are therefore quantified (via forward kinematics) and required corrective actions are then calculated (via inverse kinematics) [18,19]. Moreover, modelling the geometry using kinematics chains has been found to be very effective for integrating parametric models for systematic and algorithmic monitoring of civil infrastructure.

3 Research Methodology

In this paper, the geometric relationships between components in an assembly are derived using the analogy of robotics and kinematics chain (KC) modelling. It is assumed that construction assemblies are similar to robot arms with mutual degrees of freedom. As such, dimensional variability is modelled mathematically, and the critical component-features and their dimensional variations are then controlled systematically. The analogy of construction assemblies with robot arms was first used by Nahangi et al. (2015) [17], for quantifying incurred discrepancies in construction assemblies. It was then used to calculate the required changes for realigning defective assemblies [18] by solving the inverse kinematics problem. This paper is directed toward dimensional variation analysis, in order to investigate how deviations propagate an accumulate in an assembly.

3.1 Overview of the kinematics chain-based modelling

An overview of the proposed methodology is shown in Figure 1. Some critical information integrated in the BIM is required to develop the kinematics chain for DVA. As shown in Figure 1, critical interfaces and an assembly diagram are required for identifying the critical chains for variation analysis. The kinematics chain is then developed for the construction assembly.

This section describes: (1) how kinematics chains are developed for construction assemblies (Section 3.2), and (2) how the developed chain can be applied for critical interface and connection design (Section 3.3).

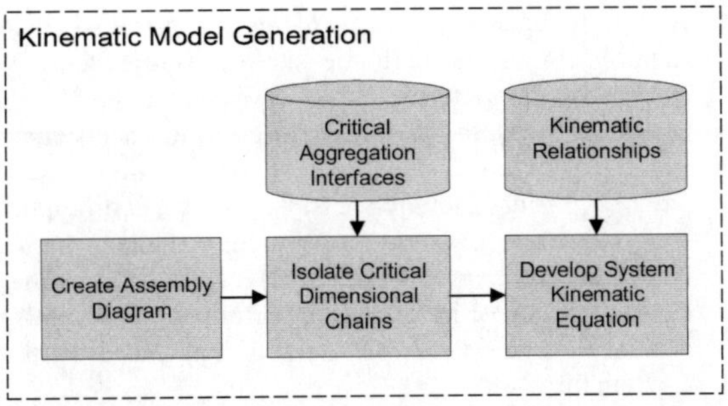

Figure 1: overview of the proposed framework for kinematics chain based DVA

3.2 Kinematics chain formulation

For developing the kinematics chain, a similar approach to [17] is employed. Transformations are then derived using the Denavit-Hartenberg (D-H) convention [20]. D-H convention is a systematic method that can be programmed and integrated with other components of the proposed framework. D-H parameters represent any homogeneous transformation as a combination of four transformations, as illustrated in Figure 2.

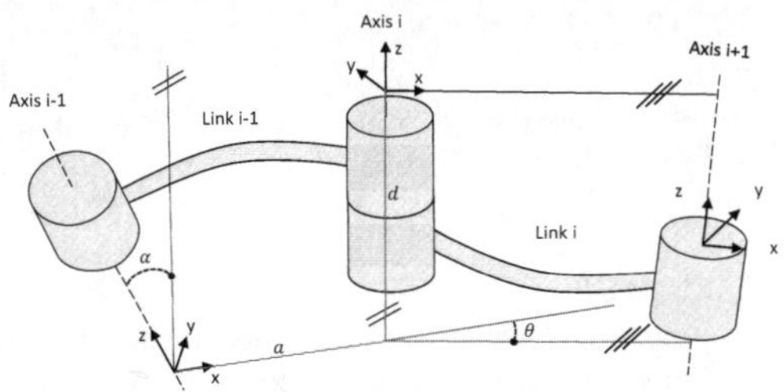

Figure 2: D-H parameters for a typical connection.

Of these four transformations (illustrated in Figure 2), two are rotational and two are translational transformations. These are represented by the following equation:

$$T_i = (Rot_{z,\theta_i})(Trans_{z,d_i})(Trans_{x,a_i})(Rot_{x,\alpha_i}) \qquad \text{Eq.1}$$

$$= \begin{bmatrix} c\theta_i & -s\theta_i & 0 & 0 \\ s\theta_i & c\theta_i & 0 & 0 \\ 0 & 0 & 1 & 0 \\ 0 & 0 & 0 & 1 \end{bmatrix} \begin{bmatrix} 1 & 0 & 0 & 0 \\ 0 & 1 & 0 & 0 \\ 0 & 0 & 1 & d_i \\ 0 & 0 & 0 & 1 \end{bmatrix} \cdots$$

$$\cdots \begin{bmatrix} 1 & 0 & 0 & a_i \\ 0 & 1 & 0 & 0 \\ 0 & 0 & 1 & 0 \\ 0 & 0 & 0 & 1 \end{bmatrix} \begin{bmatrix} 1 & 0 & 0 & 0 \\ 0 & c\alpha_i & -s\alpha_i & 0 \\ 0 & s\alpha_i & c\alpha_i & 0 \\ 0 & 0 & 0 & 1 \end{bmatrix}$$

$$= \begin{bmatrix} c\theta_i & -s\theta_i c\alpha_i & s\theta_i s\alpha_i & a_i c\theta_i \\ s\theta_i & c\theta_i c\alpha_i & -c\theta_i s\alpha_i & a_i s\theta_i \\ 0 & s\alpha_i & c\alpha_i & d_i \\ 0 & 0 & 0 & 1 \end{bmatrix}$$

in which, θ_i, d_i, a_i, and α_i are parameters associated with link *i* and joint *i* (Figure 3). $c\beta$ and $s\beta$ denote $\cos\beta$ and $\sin\beta$, respectively. The four parameters θ_i, d_i, a_i, and α_i are also known as *"link length"*, *"link twist"*, *"link offset"*, and *"joint angle"*, respectively.

Generally speaking, two types of joints can define the characteristics of any construction assembly connection (Figure 3):

1. Rotational joints: are considered where dimensional variations can occur in the form of a rotational affect. Rotational joints are also known as revolute joints in robotics theory.
2. Translational joints: are considered where dimensional variation can occur in the form of translational or an extension effect. Such joints are also known as prismatic joints in robotics theory.

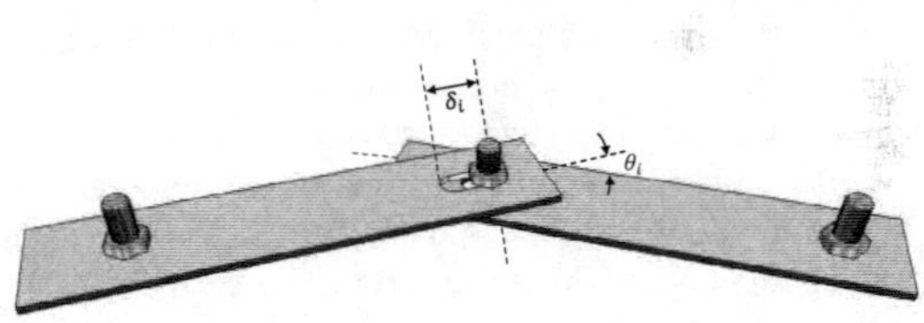

Figure 3: Schematic of a hypothetical joint. The joint is comprised of one translational joint and and one rotational joint (δ_i is variable for translational joints, and θ_i is variable for rotational joints).

In order to model variation of connections and joints, θ_i is the design variable used for rotational joints, and d_i is the design variable used for translational joints. For modelling the geometric relationship between different segments of an assembly and considering the dimensional variations that may occur, the appropriate joint type is considered and incorporated in the kinematics chain. In some cases, a combined joint comprised of many 'typical' joints (which are all coinciding at one point) may be modelled linearly. For example, the connection illustrated in Figure 3 is combined of one translational and one rotational joint in parallel, meaning that the order of their transformation can be reversed. However, in rare cases where a connection is complex (e.g., exterior cladding connection systems), it may have to be modelled as a series of joints. Also, placement of the origin for structural assemblies must follow a standard convention (typically at one corner or along an outer edge at the centre of the assembly). The position of the critical interface or node is modelled as a mathematical function with the potential dimensional variations as design variables.

3.3 KC-based DVA

Once the assembly and potential dimensional variations are modelled mathematically (by developing the kinematics chain), variation analysis of a critical feature becomes systematic and algorithmic. The assembly diagram which is extracted from the building information model identifies how various components are constructed and assembled. Variations can be predicted using the model in order to develop suitable tolerances and or to determine if the design tolerances are respected. As previously mentioned, the kinematics chain identifies the position of a critical feature or connection as a mathematical function of the possible dimensional variations. The variation of the critical feature can then be modelled and analysed for tolerance design and further considerations.

Since the dimensional variation is modelled mathematically, unprecedented analyses become possible for systematic monitoring and design of construction components. For example, the rate of variation propagation in the critical region of an assembly can be calculated by differentiating the kinematics chain with respect to the design variables. Furthermore, inverse kinematics can be used for tolerance allocation at each component for desired tolerance requirements of the critical feature.

4 Case Study

The case study investigated in this paper relates to the fabrication of the structural system of a single story modular building. In this project, the contractor experienced numerous dimensional fit-up issues during erection, resulting in misalignments between module connection points). To address these issues, a dimensional variation analysis using the proposed methodology was developed. This DVA model analysed the key fabrication processes for the steel frame structure.

Using the assembly diagram, kinematics chains were developed for analysing the dimensional variation of critical features (i.e., tie-in plates). The transformation required for analysing the tie-in plates is represented as a chain of transformations between several local coordinate systems. These local coordinate systems are found where either a critical deviation might occur or where a strict tolerance has been specified. Figure 4 shows the tie-in plates on a typical module, which are taken as the critical assembly features. A kinematics chain was derived from the structural system, with the critical features (tie-in plates) acting as end-effectors in each chain.

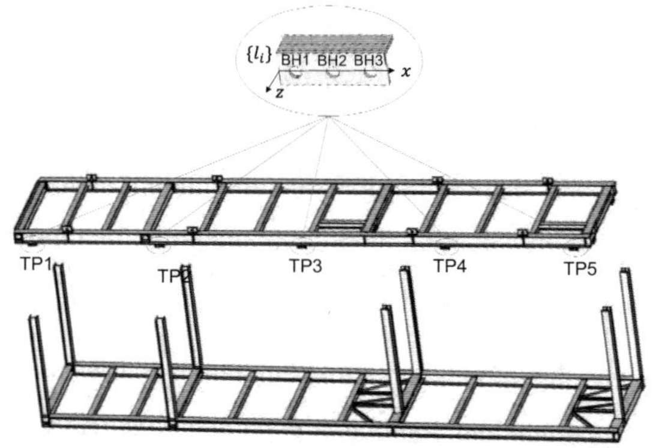

Figure 4: Kinematics chain and assembly diagram for representing critical features in the local $\{l_i\}$ and global $\{G\}$ coordinate systems.

In practice, there are two DVA approaches for use of kinematics chains for modelling dimensional variability in the proposed methodology: a design-model based approach and an as-built approach.

A design-model analysis, is used when design tolerances analysed using information provided in the building information model. This analysis identifies how design tolerances propagate through fabrication processes in order to determine an overall tolerance accumulation. Typical analyses on the case study used in this paper are shown and discussed in this section.

For investigating the case study, two stages of tolerance propagation can predict tolerance accumulation of the tie-in plates:

1. How the tie-in plate is installed and assembled with respect to the roof frame, and
2. How the roof frame is installed with respect to the floor frame (global coordinate system)

The spatial position of the tie-in plates and the impact of tolerance propagation can then be modelled by developing the kinematics chain relating the global to the local coordinate system. The allowable tolerance impact can then be measured in the global coordinate system in order to investigate the propagation of the tolerances.

An as-built analysis is used when the built status of a construction assembly is under investigation. Acquisition of the built status information (which can be done via 3D point cloud data from a laser scanner) provides accurate information that can be used for as-built modelling, updating the BIM and for understanding contributions of out-of-tolerances. In this type of DVA, the physical construction dimensions are extracted from point cloud data (or other accurate data sources) of the construction components, and the kinematics chain is then populated. Using the kinematics chains that were developed for the design-model DVA, input of the actual as-built dimensions can be used to visualize 2D deviation surfaces for the two design variables used (Figure 5). This information can also be used to visualize the propagation of dimensional variations for all tie-in plates along the length of the module (Figure 6). θ_1 and θ_2 are the design variables to be analyzed for dimensional variability of the tie-in plates. The ranges are chosen from the allowable tolerances identified in the design specifications. However, a typical range is chosen here for demonstrating the results.

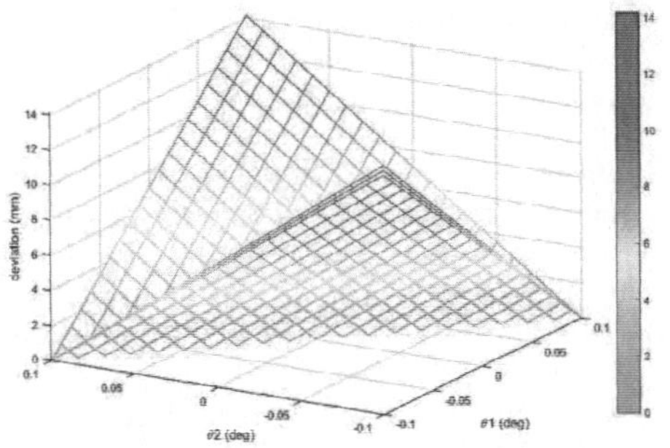

Figure 5: Results for the as-built DVA of the case study. Deviation surfaces and contour lines for the bolt holes (BH) are illustrated. The results are shown for the tie-in plate 2 (TP2).

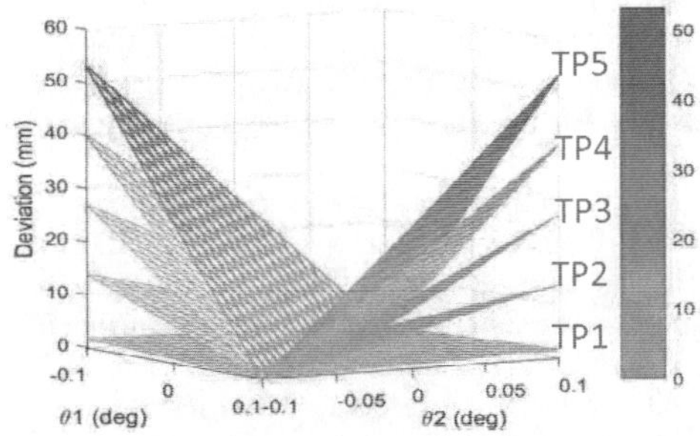

Figure 6: Propagation of dimensional variation along the roof frame of the case study in different tie-in plates. As seen, tie-in plates further from the datum have higher impacts.

These figures show two important conclusions. First of all, Figure 5 shows that the largest variation experienced in any given tie-in plate occurs when the design variables are at either their positive extreme (+0.1 deg and +0.1 deg), or at their negative extreme (-0.1 deg and -0.1 deg). This is not surprising since we would expect that the net effect of rotations in either the positive or negative direction will cause the tie-in plate to deviate the most from its nominal position. By modelling this variation however, we are able to calculate the value of the expected deviation, which can be used for inspection or remedial actions. Secondly, Figure 6 shows which tie-in plate will have the largest deviation with respect to the datum (note: in this case study the datum was located near the first tie-in plate). Again, it's not surprising that the furthest tie-in plate from the datum would have the largest relative deviation due to the rotations of θ_1 andθ_2. However, by modelling this kinematic chain, we are able to calculate the value of the expected deviation.

5 Conclusions

Dimensional variation analysis (DVA) is a powerful tool for modelling the relationship between parts in an assembly. The authors in this paper demonstrate how kinematics chains can be used to derive these relationships by assuming rigid body transformation (rotational and translational degrees of freedom). While DVA has an extensive use in manufacturing, the authors demonstrate that it can be used in construction for similar benefits.

The case study in this paper investigates how to spatially model the connection points of prefabricated modules in a building. This can be a very challenging process in the fit-up and assembly process due to the significant accumulation of dimensional variability. For this purpose, use of kinematics chains in DVA is used in two different ways: for design-models and for as-built conditions. These approaches employ the use of laser scanning (for the as-built state), and BIM (for the design-model state).

The results from both of these analyses represent the expected dimensional variability of the tie-in points for the modules. In practice, this information is very important for predicting potential fit-up conflicts before they occur. Furthermore, the expected dimensional variability can be considered in the design process for connection systems.

6 References

[1] E.H. Oh, N. Naderpajouh, M. Hastak, S. Gokhale, Integration of the Construction Knowledge and Expertise in Front-End Planning, Journal of Construction Engineering and Management. 142 (2015) 04015067.

[2] M.H. Pulaski, M.J. Horman, Organizing constructability knowledge for design, Journal of Construction Engineering and Management. 131 (2005) 911-919.

[3] J. O'Connor, Constructability Implementation Guide, Construction Industry. (2006).

[4] [4] S. Azhar, M. Khalfan, T. Maqsood, Building information modelling (BIM): now and beyond, Construction Economics and Building. 12 (2015) 15-28.

[5] T. Hartmann, M. Fischer, Supporting the constructability review with 3D/4D models, Building Research & Information. 35 (2007) 70-80.

[6] D.K. Ballast, Handbook of Construction Tolerances, John Wiley & Sons, 2007.

[7] M. Jingmond, T. Lindberg, A. Landin, Identifying Causes of Additional Costs in Tolerance Compliances Failure in Buildings, TG65 & W065-Special Track 18th CIB World Building Congress May 2010 Salford, United Kingdom, 2010, 554.

[8] C.T. Milberg, I.D. Tommelein, Tolerance and Constructability of Soldier Piles in Slurry Walls, Journal of Performance of Constructed Facilities. 24 (2009) 120-127.

[9] M. Jingmond, R. Ågren, A. Landin, Use of cognitive mapping in the diagnosis of tolerance failure, 6th Nordic Conference on Construction Economics and Organisation, 2011, 305-313.

[10] A. Landin, P. Kämpe, Industrializing the construction sector through innovation–Tolerance Dilemma, 2007,.

[11] P. Tang, G. Chen, Z. Shen, R. Ganapathy, A Spatial-Context-Based Approach for Automated Spatial Change Analysis of Piece-Wise Linear Building Elements, Computer-Aided Civil and Infrastructure Engineering. 31 (2016) 65-80.

[12] M. Nahangi, C.T. Haas, Automated 3D compliance checking in pipe spool fabrication, Advanced Engineering Informatics. 28 (2014) 360-369.

[13] C.B. Farnsworth, S. Beveridge, K.R. Miller, J.P. Christofferson, Application, Advantages, and Methods Associated with Using BIM in Commercial Construction, International Journal of Construction Education and Research. 11 (2015) 218-236.

[14] F. Leite, A. Akcamete, B. Akinci, G. Atasoy, S. Kiziltas, Analysis of modeling effort and impact of different levels of detail in building information models, Automation in Construction. 20 (2011) 601-609.

[15] Y.K. Cho, C.T. Haas, S.V. Sreenivasan, K. Liapi, Position error modeling for automated construction manipulators, Journal of Construction Engineering and Management. 130 (2004) 50-58.

[16] C. Feng, Y. Xiao, A. Willette, W. McGee, V.R. Kamat, Vision guided autonomous robotic assembly and as-built scanning on unstructured construction sites, Automation in Construction. 59 (2015) 128-138.

[17] M. Nahangi, J. Yeung, C.T. Haas, S. Walbridge, J. West, Automated assembly discrepancy feedback using 3D imaging and forward kinematics, Automation in Construction. 56 (2015) 36-46.

[18] M. Nahangi, C.T. Haas, J. West, S. Walbridge, Automatic Realignment of Defective Assemblies Using an Inverse Kinematics Analogy, ASCE Journal of Computing in Civil Engineering. (2015).

[19] M. Nahangi, T. Czerniawski, C.T. Haas, S. Walbridge, J. West, Parallel Systems and Structural Frames Realignment Planning and Actuation Strategy, Journal of Computing in Civil Engineering. (2015) 04015067.

[20] [20] J. Denavit, R.S. Hartenberg, A Kinematic Notation for Lower-Pair Mechanisms Based on Matrices, Journal of Applied Mechanics. (1955) 215-221.

Developing a BIM-based Integrated Model for CAD to CAM Building Production Automation

Xiaoxi Li[a], Ahmed Qureshi[b], Mohamed Al-Hussein[c]

[a,c]Department of Civil and Environmental Engineering, University of Alberta, Canada
[b]Department of Mechanical Engineering, University of Alberta, Canada
[a]xiaoxi4@ualberta.ca, [b]ajquresh@ualberta.ca, [c]malhussein@ualberta.ca

Abstract –

Modular construction has gained momentum in North America as an emerging construction paradigm in recent years. Modular buildings are assembled from components that are prefabricated in manufacturing plants and transported to the construction site for assembly. The current manual-based approach to modular construction, which typically applies the traditional stick-building-under-a-roof method, is time-consuming and labour-intensive. However, the application of automated prefabrication of components has the potential to significantly advance the productivity, worker safety, and competitiveness of the Canadian construction industry. This research, focusing on the construction of wood-framed panels, presents a methodology which allows intelligent wall panels with different design properties to be analyzed. By integrating external databases with building information modelling (BIM)-based software and generating wall panel combination plans and assembly information, intelligent wall panel prefabrication is achieved. The study of panel prefabrication for a sample house from an Edmonton-based panel manufacturer is presented in order to demonstrate the effectiveness of the proposed methodology.

Keywords –

BIM; Prefabricated wall production; Automation

1 Introduction

Automated construction requires a high level of accuracy in the drafting and design documentation to ensure full compliance with the project pre-planning, project coordination, preliminary design, and transportation plan [3]. However, most design plans do not currently provide complete assembly information. The majority of Canadian homebuilders build without detailed construction drawings since the planning stage is time-consuming and costly. Therefore, they rely on the experience of their carpenters to build homes based on a rough structural design. The introduction of proper construction guidelines to facilitate the shift from the *stick-building-under-a-roof* method to automated manufacturing of wood panels will help to overcome the shortage of skilled carpenters, reduce the high level of timber waste, and improve housing quality, while enabling reduced construction cycle time. Guidelines are proposed for improved panel assembly practice, and the linkage between drafting (i.e., computer-aided design) and manufacturing (i.e., computer-aided manufacturing) is developed.

Facilitated by industrialized construction methods, multi-panel wood wall-framing is applied in place of single-wall framing (as illustrated in Figure 1). A multi-panel arrangement is utilized to reduce the machine set-up time. The multi-panel is partially pre-cut by the wood panel framing assembly to provide easy separation into single-wall panels at a later stage. Autodesk Revit, as one type of building information modelling (BIM) software, provides a means to generate and manage data based on user requirements. After loading the details in order to generate a 3D model, and then extracting information from the structural plan of the BIM model panels, the nailing, drilling, and cutting coordinates are used to analyze and establish a guideline for further BIM-based construction assembly.

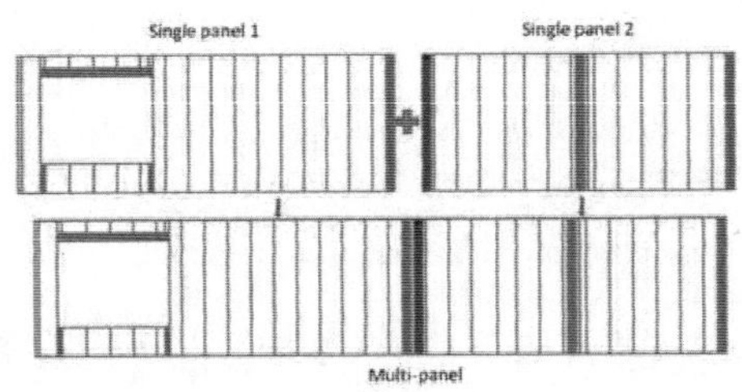

Figure 1. Multi-panel example

2 Review of the State of the Art in Building Production Manufacturing

Factory construction is to be promoted in the

homebuilding industry because a controlled environment improves product quality and performance and lowers cost. Since the products are built in a factory, the principles of mass production can be applied to leverage the benefits of a manufacturing-based approach [5]. Mass production of housing components and systems reduces construction duration by half and cost by at least 20% [1]. However, when materials and information are defective or idle, manufacturing-based construction methods can lead to significant waste of resources. The implementation of lean tools to replace mass production involves industrializing home building to improve efficiency [4]. In 2008, approximately 10% of new residential homes were built in factories either as panelized or modular in the United States [7]. However, challenges still remain in making the process more lean, since home building involves a high level of customization and multi-faceted characteristics in the production process. Panels for a specific home possess a diversity of dimensions and design properties which require different amounts of processing time in the production line; the relatively low production volume of specific house models inhibits the full deployment of lean tools [6]. Furthermore, as discussed by Bock [2], the conventional construction methodology has become stagnant and has reached its technical limits. Although efforts toward construction automation are still at the innovation stage, continued research and development targeting automation will propel construction into the growth phase.

3 Mehodology

The key contributions of this methodology are the generation of wood panel framing assembly motion plans, which implement the industrialized manufacturing of intelligent wall panels and are ready for programmers to establish computer numerical control (CNC) code. The BIM model provides rich information for wall assembly such as wall dimensions, stud details, and prefabricated opening component details. As shown in Figure 2, the methodology used to generate operation locations from the BIM model consists of the following three procedures: (1) load framing information into the BIM model; (2) extract framing information from the BIM model to Microsoft Excel; and (3) generate the cutting, drilling, and nailing locations guideline based on real-world on-site experience and construction specifications. Based on the operation coordination and each machine station location, the sequence and interval distance of each operation can be determined.

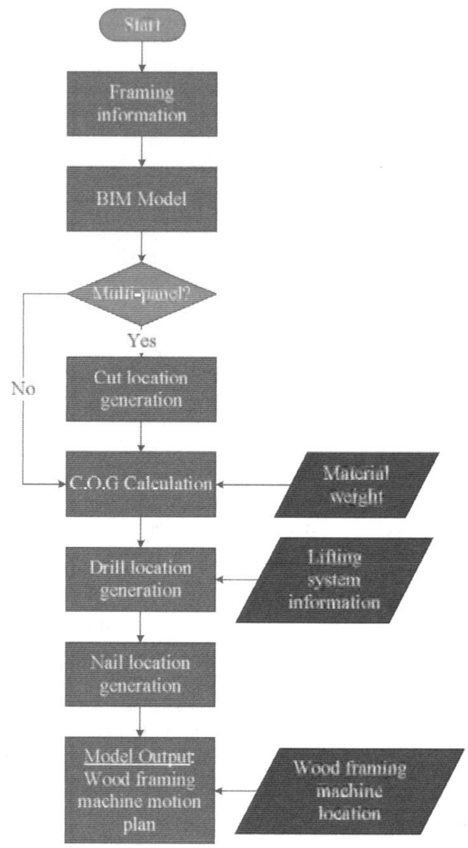

Figure 2. Methodology

4 Operation Sequence Development

Based on current assembly requirements for panels, the assembly process can be subdivided into three parts: (1) treating the stud as a consistent interchangeable part, simplifying the process of loading and attaching them to one another; (2) drill the top- and bottom-plate, which will help facilitate lifting requirements in downstream stations; and (3) after assembly of a given panel is complete, the cut operation should proceed to precut the multi-panel. Based on the operation locations, the sequences of motion for three stations can be determined for the purpose of ensuring that the multi-panel will be manufactured completely and smoothly by the wood panel framing assembly.

4.1 Load framing information

The goal for the automation of wall-framing manufacturing is to build a bridge between the housing wall layouts from the Revit software and the factory panel manufacturing process using BIMSF_Id, which is an identification tag for each wall framing. When generating the wall-framing layout using Revit, each BIMSF_ID is formed automatically for each wall panel using the Revit add-on. The factory plant will identify the BIMSF_ID and will assemble the panel based on the wall-framing layout; the same BIMSF_ID is used to label the finished panel.

4.2 Extract information from BIM model

According to the wood wall-framing table assembly manufacturingrequirements, the information being extracted from the BIM (Revit) model includes: (1) wall properties from the BIM model such as wall details (top- and bottom-plate length (L_{swm}), wall height (H)); (2) sub-wall IDs (m) contained within each multi-panel; and (3) user-defined framing specifications such as number of nails per stud (N), and number of drill holes per wall (D_m). At this point, this data can be used to generate nail, cut, and drill locations, as well as IDs. The drill location must also contain stud location information in order to detect potential collisions, and thus can adjust accordingly.

The coordinates of each stud and each sub-wall belonging to the multi-panel can be extracted based on the origin, as shown in Figure 3. The absolute origin is set as the multi-panel's first stud outside boundary (home position) for $x = 0$; therefore, each stud's absolute x-coordinate is its x-direction centre-point-to-home-position distance. Stud absolute z-coordinates represent each stud z-direction centre-point touching top- or bottom-plate position facing up or down relative to the z-axis. Vertical orientation stud z-coordinate is set to 0, as expressed in Figure 4(a). In this case, the stud orientation and position can be distinguished by the absolute z-coordinates. If the value of z is negative, this stud is horizontal and on the bottom edge of the top-plate, as seen in Figure 4(b). If the value of z is positive, this stud is horizontal and on the top edge of the top-plate as shown in Figure 4(c). The absolute y-coordinate of the bottom-plate outside boundary is set to 0; accordingly, the top-plate outside boundary y-coordinate will be the multi-panel height, h.

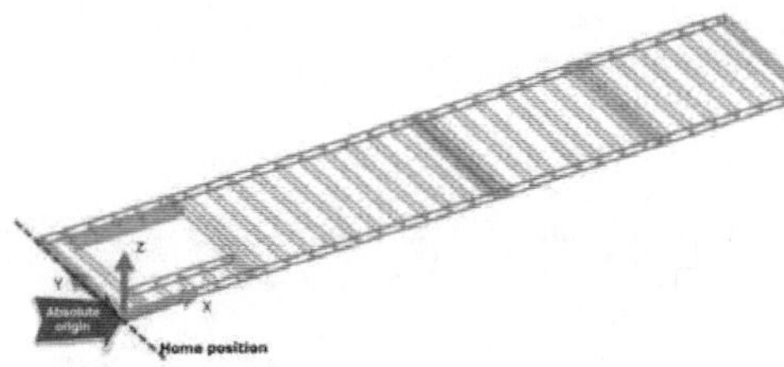

Figure 3. Multi-panel absolute origin and home position

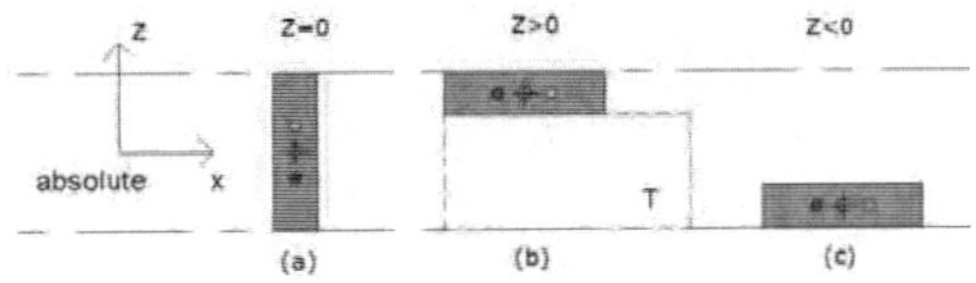

Figure 4. Stud orientation relative to Z coordination

4.3 Generate operation locations

Based on the information obtained from the BIM model, cutting, drilling, and nailing locations for multi-panels can be generated and prepared for the analysis of wood wall-framing operation procedures. If the panel only contains one single panel, which means it is not a multi-panel, the cut location generation step can be skipped.

4.3.1 Generation of cut location and IDs

The first cut location depicted in Figure 5 is located at the end-point of one sub-wall and the second cut location is located at the start-point of the following sub-wall. A ranking of non-zero start and finish sub-wall x-coordinates will be used to obtain the cut locations. All cut locations will be partially cut as expressed in Equation (1) and Equation (2) except the last cut location, since it is coterminous with the multi-panel end-point. At that point, the cutting machine must cut and separate extra top- and bottom-plates completely, as expressed in Equation (3). The cut locations are generated based on the start and finish x-coordinates of sub-walls.

Location of partial cuts:

$$XC_a = XSWS_m - \text{Saw thickness}\,(XSWS_m \neq 0) \quad (1)$$

$$XC_a = XSWF_m \,(XSWF_m \neq L) \quad (2)$$

Location of complete cuts (multi-panel end-point location):

$$XC_a = L \quad (3)$$

where
a: cut ID for multi-panel
m: ID of sub-wall
XC_a: a^th^ cut distance from home position
$XSWS_m$: m^th^ sub-wall start-point x-coordinate
$XSWF_m$: m^th^ sub-wall end-point x-coordinate
L: multi-panel length

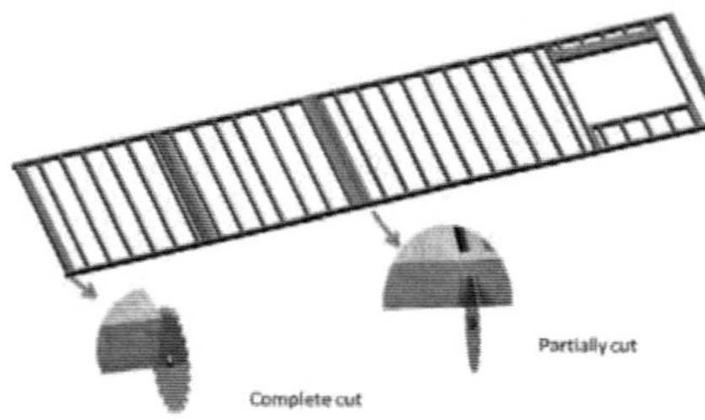

Figure 5. Sample multi-panel cutting location

4.3.2 Generation of drilling locations

There will be drill holes made on each sub-wall to facilitate any necessary lifting actions at further stations, and for transferring to site. The default setting for the number of drill holes on each sub-wall (D_m) is two; however, the user can define the settings to their own requirements. In analyzing the drill hole locations, the location of the centre of gravity (COG) should be considered. If the sling legs are the same length, failure to establish precise COG will create an imbalance in the tension in each sling and cause the load to tilt during lifting. Therefore, the COG should be identified for each single panel first. The drilling points on either side should ensure that the COG is located at the exact midpoint between them to avoid any unplanned movement during lifting due to imbalance (Unirope Slingmax 2015). Since the multi-panel is not symmetrical, the COG may not be located exactly in the middle of the multi-panel. Therefore, in the research, the drill hole location could range anywhere from the COG to the point permitting maximum lift capacity without spreader beam failure, and the optimal drill position will be the spreader beam length, as illustrated in Figure 6. Since the drill machinery has a certain distance to the panel home position, if the drill hole falls within that range then it should be drilled manually. The optimized drill hole locations are determined using Equation (5), with the method summarized in Figure 6.

COG location:

$$COG_m = \sum(m_i \times l_i) / \sum m_i \quad (4)$$

where i = 1,2,3.... total number of members (studs, plates, headers, etc.) in the mth panel.

Optimized drill hole location:

$$XDi = XSWS_m + COG_m + (-1)^i \times L_{sb}/2 \quad (5)$$

where
m_i: i^th^ member mass of the panel
l_i: i^th^ member origin along the panel's x-axis
i: hole ID
L_{sb}: spreader beam length
XD_i: i^th^ hole distance from home position
COG_m: m^th^ sub-wall COG to home position length
$XSWS_m$: m^th^ sub-wall start-point x-coordinate

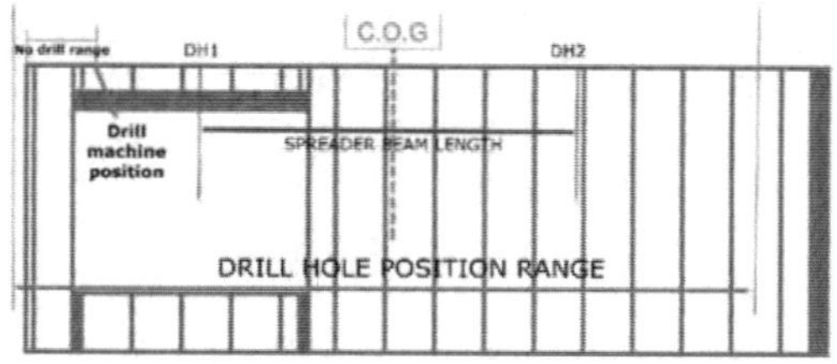

Figure 6. Sample panel of drilling location

4.3.3 Generation of nailing locations

Wall studs and pre-assembled window and door frames are joined with top- and bottom-plates using nails. Depending on the elements belonging to the wall, the top-plate stud number and location may be different from the bottom-plate. Figure 1 depicts an example of a multi-panel with an opening component (window) where the bottom cripples only touch the bottom-plate and do not connect with the top-plate. Therefore, the top- and bottom-plate nailing plans should be analyzed separately.

The default setting for the total number of nails on a stud is two. However, users can specify this setting if necessary, e.g., j^{th} stud on top (T) has N number of nails (NT_j), or k^{th} stud on footer (B) has N number of nails (NB_k). The *y*-coordinates of the bottom-plate nails will be 0 and top-plate nails' *y*-coordinates will be H. Sample nailing *x*- and *z*-coordinates on the horizontal and vertical studs are presented in Figure 4. When the stud is horizontal ($Z \neq 0$), the nails on the *x*-axis position of that stud should be evenly distributed along the stud width. The *z*-coordinates of the nails should be the same as the *z*-coordinates of the stud since they are along the midpoint of the stud thickness. When the stud is vertical ($Z = 0$), the nails on that stud should be evenly distributed across the stud width throughout the *z*-axis. The *x*-coordinates of the nails should be the same as the *x*-coordinates of the stud since they are along the midpoint of the stud thickness. The methodology of nailing coordinate generation is also presented in Figure

4, and expressed in Equations (6) to (17).

Nail locations for horizontal studs at top-plate side ($YT_j \neq 0$):

$$XT_{jn} = [XT_j - (d/2)] + n \times [d/(NT_j + 1)] \quad (6)$$
$$YT_{jn} = H \quad (7)$$
$$ZT_{jn} = ZT_j \quad (8)$$

Nail locations for vertical studs at top-plate side ($YT_j = 0$):

$$XT_{jn} = XT_j \quad (9)$$
$$YT_{jn} = H \quad (10)$$
$$ZT_{jn} = [Z_j - (d/2)] + n \times [d/(NT_j + 1)] \quad (11)$$

Nail locations for horizontal studs at bottom-plate side ($YB_k \neq 0$):

$$XB_{kn} = [X_k - (d/2)] + \{n \times [d/(NB_k + 1)]\} \quad (12)$$
$$YB_{kn} = 0 \quad (13)$$
$$ZB_{kn} = ZB_k \quad (14)$$

Nail locations for vertical studs at bottom-plate side ($YB_k = 0$):

$$XB_{kn} = XB_k \quad (15)$$
$$YB_{kn} = 0 \quad (16)$$
$$ZB_{kn} = [ZB_k - (d/2)] + \{n \times [d/(NB_k + 1)]\} \quad (17)$$

4.4 Wood framing assembly motion plan

In order to achieve successful operation of the wood panel framing assembly, the nailing, drilling, and cutting operation sequences must be determined based on the multi-panel layout that has been generated from the BIM model. After generating and sorting the manufacturing operations based on the equipment locations, the multi-panel moving distance from pervious task point to next task point are confirmed in order to avoid the omission of any necessary steps or any other conflicts. Therefore, the entire manufacturing plan can be determined.

4.4.1 Calculation of operation locations relative to corresponding machine

Each machine location to multi-panel home position possesses a certain distance. Therefore, operations reaching their corresponding machine distances are the operation locations related to their machine locations (operation locations minus machine locations). The operation locations related to their corresponding machine location can be calculated as per the following equations.

$$D_i = XD_i - D_d \quad (18)$$
$$C_c = X_c - D_c \quad (19)$$
$$NT_n = XTj_n - D_n \quad (20)$$
$$NB_n = XBj_n - D_n \quad (21)$$

where

D_i: i^{th} drilling coordinates
C_c: c^{th} cutting coordinates
NT_n: n^{th} top side nailing coordinates
NB_n: n^{th} bottom side nailing coordinates
D_n: distance from nailing station machine to home position
D_d: distance from drilling station machine to home position
D_c: distance from cutting station machine to home position

5 Results and Discussion

Multi-panel #1, as shown in Figure 7, is selected from an Edmonton-based panel manufacturer as a sample in order to generate operation locations and machine motion sequence.

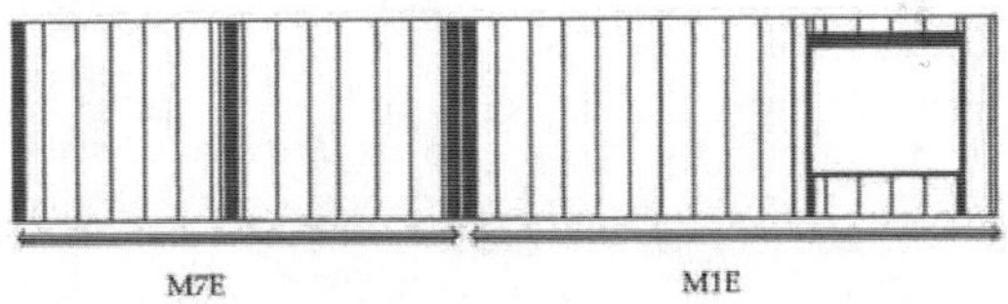

Figure 7. Multi-panel #1 – sample from manufacturer

5.1 Cut locations

The information extracted from multi-panel #1 for cut location calculation is presented in Table 1.

Table 1. Sample multi-panel information (1)

Multi-panel length	39.739	Total number of sub-walls (SW)	2
Sub-wall ID ($_m$)	Sub-wall length (Lsw_m)	Sub-wall start-point *x*-coordinate ($XSWS_m$)	Sub-wall end-point *x*-coordinate ($XSWF_m$)
1	21.604	0.000	21.604
2	18.052	21.687	39.739

Cut locations for a multi-panel are located at the end of the first sub-wall, as well as at the start- and end-points of the remaining sub-walls consecutively; all will be partially cut other than the last point, which is located at the end of the entire panel and will thus be completely cut. The saw thickness for the cutting station is 0.07 ft. The cut location (X_a), which is calculated according to the methodology, is presented in Table 2.

Table 2. Cut location coordinates

Cut type	Cut ID (a)	Cut distance from home position (XC_a)
Partially cut	1	21.604
	2	21.680
Completely cut	3	39.739

5.2 Drilling locations

The COG location calculations for each single panel are as follows:

COG for M1E
Known values:
All member weights and *x*-coordinates are presented in Table 3.

Table 3. Wall member weights and x-coordinates

Name	Function	Multi-panel *x*-coordinate	Weight (lb)
M1E	OC	0.063	3.790
M1E	OC	0.271	3.790
...			
M1E	T3	4.500	2.941
		Total weight	119.252

Center of gravity (COG) location:

$$COG_m = \sum(m_i \times l_i) / \sum m_i \quad (4)$$

where
i = 1,2,3.... total number of members (studs, plates, headers, etc.) in the mth panel
$\sum$ (mj × member x-coordinate li) = 1,201.495 lb•ft
$\sum m_i$ = 119.252 lb
COG_1 = 10.075 ft

The information from multi-panel #1 for drill location calculation is listed in Table 4.

Table 4. Sample multi-panel information (2)

Sub-wall ID (m)	Sub-wall start-point x-coordinate (XSWSm)	COG	Drill number	Spreader beam length
1	0.000	10.075	2	10
2	21.687	8.612	2	10

The drill location (XD_i), which is calculated according to the methodology, is shown as follows:

Drills #1 and #2
Known values:
Sub-wall ID = 1
COG_1 = 10.075 ft

After substituting the known values for the following equations, the result for the optimized drill location *x*-coordinate is accessible.

Optimized drill hole location:

$$XDi = XSWS_m + COG_m + (-1)^i \times L_{sb}/2 \quad (5)$$

$XD_1 = 0 + 10.075 - (-1)^1 \times 10/2 = 5.075$ ft.
$XD_2 = 0 + 10.075 - (-1)^2 \times 10/2 = 15.075$ ft.

The *x*-coordinates for drill holes in multi-panel #1 are listed in Table 5.

Table 5. Drill location coordinates

Drill ID (i)	Drill distance from home position (XDi)
1	5.075
2	15.075
3	25.299
4	35.299

5.3 Nailing locations

The nail locations are determined by the coordinates (*x*, *y*, *z*) and the type of object, which can be extracted from Revit and regarded as known values. The number of nails on each object is set to 2. The information extracted from multi-panel #1 for nail location calculation is listed in Table 6. Stud 1 is used as the example to show the nail location generation.

Table 6. Sample multi-panel information (3)

Name	Type	IsTop	IsBottom	X	Z
M1E	2×6	True	True	0.063	0.000
M1E	2×6	True	True	0.271	0.000
M1E	2×6	True	True	1.318	0.000
M1E	2×6	False	True	1.443	0.000
...					
M1E	2×6	True	True	21.250	–0.167
M1E	2×6	True	True	21.542	0.000

For stud j = k = 1 (Vertical stud)
Known values:
Stud j = 1, k = 1
Stud dimensions (2×6 type) b = 0.125, d = 0.4583
Midpoint coordinates (XT_1, ZT_1) = (XB_1, ZB_1) = (0.063, 0)
User-defined values:
Number of nails on stud NTj = NBk = 2

After substituting the known values in the following equations, the results for nail location (XT_{jn}, YT_{jn}, ZT_{jn}) and (XB_{kn}, YB_{kn}, ZB_{kn}) can be found.

Nail locations for vertical studs on the top-plate side ($YT_j = 0$) can be determined as follows:

$$XTj_n = XT_j \quad (9)$$

$$YTj_n = H \quad (10)$$

$$ZTj_n = [Z_j - (d/2)] + n \times [d/ (NT_j + 1)] \quad (11)$$

Nail locations for vertical studs at bottom-plate side ($YB_k = 0$):

$$XBk_n = XB_k \quad (15)$$

$$YBk_n = 0 \quad (16)$$

$$ZBk_n = [ZB_k - (d/2)] + \{n \times [d / (NB_k + 1)]\} \quad (17)$$

$XB1_1 = XT1_1 = 0.063$
$ZB1_1 = ZT1_1 = [0 - (0.4583/2)] + [0.4583 / (2 + 1)] = -0.07638$
$XB1_2 = XT1_2 = 0.063$
$ZB1_2 = ZT1_2 = [0 - (0.4583 / 2)] + 2 \times [0.4583/(2 + 1)] = 0.07638$
$YT1_1 = YT1_2 = 8.03125$
$YB1_1 = YB1_2 = 0$

Nailing point for stud 1: (0.063, 8.03125, −0.07638), (0.063, 0, −0.07638), (0.063, 8.03125, 0.07638), (0.063, 0, 0.07638)

Utilizing the methodology to generate the nail locations, the nail locations for the entire multi-panel are determined. The nail locations for the top-plate are sorted based on the *x*-coordinates of each nail, as presented in Table 7.

Table 7. Sample top-plate nail location coordinates

j	Nail number	No.	X	Y	Z
1	2	1	0.063	8.031	−0.0760
		2	0.063	8.031	0.0760
2	2	1	0.271	8.031	−0.0760
		2	0.271	8.031	0.0760
3	2	1	1.318	8.031	−0.0760
		2	1.318	8.031	0.0760
…					
21	2	1	21.174	8.031	−0.1670
		2	21.326	8.031	−0.1670
22	2	1	21.542	8.031	−0.0760
		2	21.542	8.031	0.076

5.4 Wood framing assembly motion plan

Since all the operation locations are calculated based on the home position of multi-panel #1 from Section 4.3, the operation locations must be subtracted from the wood panel framing assembly location in order to obtain the actual distance from the operation point to the machine location. The first top-plate nail, first drill, and first cut are used as examples to show the operation locations related to their machine location.

$$D_i = XD_i - D_d \quad (18)$$

$$C_c = X_c - D_c \quad (19)$$

$$NT_n = XTj_n - D_n \quad (20)$$

$$NB_n = XBj_n - D_n \quad (21)$$

$D_1 = 5.0375 - 1.75 = 3.2875$ ft
$C_1 = 21.604 - 3.0833 = 18.5210$ ft
$NT_n = 0.0625 - 0.0625 = 0$ ft
$NB_n = 0.0625 - 0.0625 = 0$ ft

All the multi-panel operation distances are generated and sorted from smallest to largest, and are listed in Appendix D.

6 CONCLUSION

This paper provides proper panel assembly guideline information to implement the process of industrialized manufacturing of wood panels. The challenge in improving the home building panel manufacturing system is the high customization level in home models and styles. For completing the panel manufacturing, the process has been broken down into three parts: cutting, drilling, and nailing. The next step is analyzing operation location information from the BIM model using an algorithm to generate a database for automated operation locations. The proposed methodology is tested by an industry partner house model that ensures the algorithms cover different panel manufacturing scenarios. The use of BIM-based integrated panel information such as nail, cut, and drill locations reduces the dependency on skilled carpenters to assemble the panels, and thus will save time and reduce the risk of human error by enabling efficient information exchange between Revit and Microsoft Excel. Therefore, the guideline bridges the gap between manufacturing and drafting, providing a complete solution from the BIM model to manufacturing.

In future research, based on the algorithm of the operation locations, a program can be developed in the .NET API in order to generate computer numerical control (CNC) code in Revit which can be readily input into the assembly.

References

[1] Alarcón L.F. (1997). *Lean Construction.* A.A. Balkema: Rotterdam, Netherlands.

[2] Bock T. (2015). The future of construction automation: Technological disruption and the upcoming ubiquity of robotics. *Automation in Construction*, 59, 113-121.

[3] Haas C.T., O'Connor J.T., Tucker R.L., Eickmann J.A., and Fagerlund W.R. (2000). *Prefabrication and Preassembly Trends and Effects on the Construction Workforce.* Report, Centre for Construction Industry Studies, Univ. of Texas, Austin, TX, USA.

[4] Mullens M. (2004). Production flow and shop floor control: Structuring the modular factory for custom home building. *Proceedings, NSF Housing Research Agenda Workshop*. Orlando, FL, USA.

[5] Neelamkavil J. (2009). *Automation in the Prefab and Modular Construction Industry.* NRC Institute for Research in Construction, Report #RR-261. 27 pages.

[6] Sabharwal A., Syal M., and Hastak M. (2009). Impact of manufactured housing component assembly redesign on facility layout and production process. *Construction Innovation*, 9(1), 58-71.

[7] U.S. Census Bureau. (2009a). New privately owned housing units started: annual data. Online: http://www.census.gov/const/startsan.pdf. Accessed: 02/03/2017.

Towards cloud informed robotics

Sven Stumm[a], Peter Neu[a] and Sigrid Brell-Cokcan[a]

[a]Chair of Individualized Production in Architecture RWTH Aachen University
E-mail: office@ip.rwth-aachen.de

Abstract –

The current construction industry slowly adapts towards the new possibilities created through digitalization. However there still a major gap in the information flow between planning phases and the start of construction. With modern robotics we see a chance to mediate this gap. By using cloud accessible CAD software and linking it directly with a web interface, new means of cloud informed fabrication become possible. Designers can easily obtain feedback regarding recurring issues or design demands. We want to enable users to create their own design based on rules and parameters preselected by the designer. Therefore processes for individual parts require an integrated design. Considering changing user demands and the resulting usage requirements the optimization in the phase of conversion from the user´s perspective can help the design process. A number of fabrication technology services exist, which allow for mass customized product creation. The required web interfaces used for these kind of services however are created on a case by case basis. Additionally the required production parameters have to be transferred to the production process. An easy flow from parametric design model towards a web interface for user feedback and fabrication as a service was developed and is presented within this paper.

Keywords –

Cloud; Robotics; BIM; Production immanent design

1 Introduction

The current degree of automation within the construction industry is still low especially compared to industrial production. In addition to the small lot sizes within construction, high tolerances and variation are very common especially when working with natural materials. Building components are therefore often fitted manually on-site. Efforts are made to mediate this through tighter process control and detailed digital representation within Building Information Modeling (BIM). BIM allows the parametrization and placement of intelligent objects, containing a number of properties, in a common environment for different parties involved in the planning process [1].

However to get this information to the construction workers these models still need to be broken down to physical plans. This leads to a very loose correlation of the current model and the actual construction site. However the multitude of planning phases and experts integrated through BIM give us very intensive a priori knowledge of the environment and the task at hand in the form of a digital model. This knowledge can be employed to fabricate designed parts in lot size one.

High quality data is key to automation. Only if the geometry, position and structure of objects are known, automated processing can be planned and executed. The approach we propose within this paper uses cloud integrated software as the continuous framework for linking design processes directly to production and constitutes a novel strategy. By using cloud accessible CAD & BIM software and linking it directly with robotic fabrication, new means of cloud informed robotics and production are possible. Future developments would allow other robots executing similar task to learn from any other cloud connected robot.

Programmers and designers can also obtain feedback regarding recurring issues during their programs execution. As well as the influences these constitute for the design. A geometric representation and material information of a building element must be translated into an individualized assembly process with adaptive machine control strategies. Within this paper, we present a concept for robotic construction processes informed by data collected within a cloud environment. The geometric data, material structure and user feedback etc. stored in the model will be used for automated process and path planning within the digital built environment. We show the feasibility of the approach and consider information redundancies as well as missing process information.

Within our work we show a first proof of concept for a fully integrated information flow from design to fabrication. Enabling not only consistent flow of information but also new concepts for distributed design and fabrication. This in turn allows for a more efficient utilization of production environments within the field of building part prefabrication. Similar concepts as propagated within industry 4.0 can be used to coordinate designers and integrators of production environment, while allowing for production as needed.

Due to increasing prevalence of 3D printing methods a number of online services where created, which allow the full customization of parts. Similar systems exist for the CNC milling of parts. This specifically became possible through the verifiability of fabrication due to the lower complexity as compared to full production environments. For modern CNC machines the same procedure can be executed over and over again or individual procedures can be executed at the same rate. In order to provide the customer with the possibility of customizing an object, the programming of that machine needs to be flexible, towards automatically creating programs that are safe and lead to aesthetically pleasing results.

Prior to this the easiest way to provide customization was through modular designs. Rather than customizing individual parts, the entire object was customized by replacing entire modules. The big advantage of other technologies are clearly defined requirements for the process, as well as the enforceability of these requirements through automated checks: For 3D Printing this often means, as long as an object fits into a given volume, has a sufficient wall thickness and no geometric defects, it can most likely be printed.

Through Tylko similar means services for customization of furniture became possible by restricted parameter adaption towards the requested design. This simultaneously allows for the virtual integration of the targeted design through augmented reality.

2 Web-to-Real approach

Within [2] we described our first approach of creating a web based interface allowing users to upload design data to a server for later processing within a stone masonry production environment. An automated easy to use interface for mass customization was created. The implementation of a Web-to-Real platform enabled new possibilities for commissioning. Through this interface designers are not only able create and commission their own stone surface design in a mass customization approach, but are even able to verify feasibility and quality through a preprocess simulation.

The used approach ensures that current simulation is up to date and the results can even be employed for actual robot programming. Any optimization done to the process has a direct effect on the result seen by the customer. The customer gets a direct connection to the product created and has clear influence on how it turns out. Simultaneously the usability of the interface was kept simple through a direct upload of a pattern as an image file, allowing the customer to create a design in whatever software they are used to. The preprocess simulation allowed for a direct verification of the design via the web. This adds direct feedback to the idea of a streamlined process from design to production.

The developed Web-to-Real application however required a custom browser interface. Additionally a scalability of the approach was still not achieved. Using a cloud based infrastructure allows for the complete integration with a wide range of design software. Furthermore means were integrated and developed in order to harmonize the transfer of process parameter through and XML based format, this is further described within section 4.3. Figure 1 shows the resulting Web interface, for the web-to-real approach. Using a cloud integrated work flow brings this approach closer towards cloud robotics and is able to be applicable to a wide range of production environments.

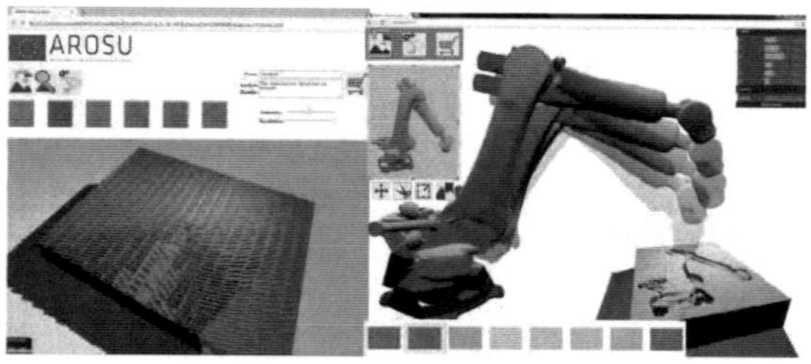

Figure 1. A first prototype of a Web-to-Real interface for stonemasonry. Enabling the individualized fabrication of customized products.

3 Forge Cloud Services

Forge provides a unified way for programs to access the Autodesk Cloud including its storage system and other 3D-model specific services. Data in the Autodesk Cloud can either be user-owned or app-owned. The user-owned object storage system (OSS) is well-known as it is the same backend used in Fusion360, BIM360 and other Team360 programs. Enabling to ask the user for access-permission to the user's projects and process these files in the cloud. Figure 2 gives an overview of the cloud integrated software.

Additionally it is possible to link the user-owned OSS via projects to multiple user accounts and similar. In order to create an open login independent upload of design data, we focused on the app-owned storage for the prototypical implementation. When registering an app through the Autodesk Developer Portal we obtain a unique client-ID and -secret, that are used to identify the programs and authorize them to use cloud services.

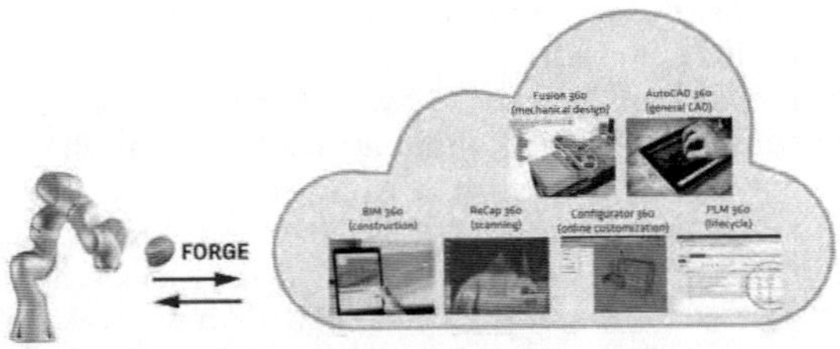

Figure 2. Available client software for the Forge cloud allowing the design, modelling and acquisition of fabrication relevant data.

A general challenge of cloud-interaction is the dependence of the client (program running on local computer) on the server. The transfer of data is done by HTTP-requests that exchange plain text. So information stored in hierarchical structures and objects has to be serialized to text, sent over the internet, and de-serialized to an objects again. This can lead to failure of transfers when formatting on server and client side is inconsistent, therefore the components used for the data exchange need to be well maintained.

The data management API is used to access files and allows for the up- and download of any file type. That means other than 3D-objects, that are the primary use-case, we can also exchange metadata like XML-files containing process parameters in our case.

We implemented a mediator between client software and cloud API connection. The Integration of software components is therefore kept independent from the backend for data transfer, which in turn can mostly be updated and maintained independently of the used software client. The mediator manages the authentication and data buckets on its own.

As the uploading and downloading side run on different machines, the server side keeps track of changes. Updates are checked for in regular intervals and can therefore be scheduled for multiple users and user dependent cloud data. The server updates the simulation feedback if there have been changes and calculation time was allocated.

In addition to the basic storage services the Forge cloud offers some features specifically aimed at 3D-modelling. Therefore Model Derivative API can be used to convert between Computer Aided Design (CAD) data from different file types. The sometimes resource-intense conversion is done in the cloud. Almost any file type used in 3D-modelling can be converted to .svf and then be viewed in a conventional web browser. With the viewer provided users can navigate around objects without the requirement of installing additional software. Additionally through the integration of process simulation. Design preview after the actual fabrication can be visualized, using this software and platform independent web interface.

4 Towards Cloud Robotics

Connecting process simulation through a cloud based interface creates further scalability of a Web-to-Real approach. It allows the seamless creation and transfer of CAD as well as process parameters within the following sections the impact for fabrication will be described. There are a number of formats for CAD data available, this information can be linked to fabrication parameters through CAEX or other XML formats [3]. Diagnostics of process execution can in turn be documented either through XML or csv data. With our approach we allow for a simulative feedback from the cloud, enabling a loop that allows for a verifiability of fabrication for complex systems. However for security reasons it is not advisable to directly connect the fabrication to the Internet. Therefore a multi-tiered approach is taken. While the user input is only linked to the simulation of the fabrication path through a public cloud, the operator of the robot has final say in process execution. However through a private cloud the fabrication information provided by the user can directly be employed. Figure 3 illustrates this approach.

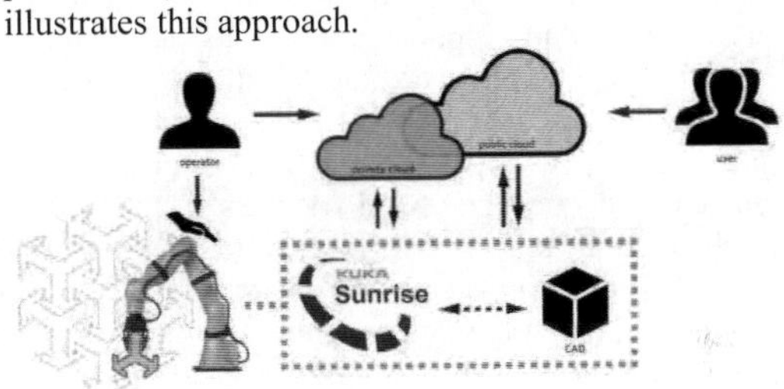

Figure 3. Enabling cloud robotics through a multi-tiered approach.

4.1 Production Immanent Design

Most commercial software for robot programming and simulation are modeled after a position teach-in process of robotic arms. Digital models are placed in 3D space and the robot's positions are recorded. These movement component can then be coupled with logic components to handle subprograms, loops, conditionals and etcetera. This is oriented on most computer programming languages and creates a widespread interface and a good reproducibility. However this creates challenges with smaller lot sizes, where the programming makes up an increasingly large part of the costs of the final product. This is exacerbated in the construction industry, where complex, free-formed buildings can consist of countless individual parts. However these individual parts are often based on a single, global topology that is then adjusted to fit the local conditions. Rather than manually drawing hundreds of individual panels, parametric design software today allows architects and designers to create a single, parametric model that can be dynamically fitted to the local conditions. In previous research we have developed software tools that allow us to directly integrate the fabrication logic into the same parametric model that controls the geometric shape of the model [4]. Thus, whenever a parametric object is adapted to new local conditions, not only its geometry but also its construction logic updated. This enables immediate robotic fabrication. Parametric robot control [5] employs this approach. Using visual programming languages as a basis for the simulation of fabrication with analytical feedback for reachability and producibility in the process. This process can be implemented through visual programming environment for parametric design built upon CAD software. The main advantage of visual programming for parametric design lie in its accessibility, modularity and visual representation [6], specifically for users without a technical background. Rather than typing code, the user simply picks components from a large library that comes with the software and immediately gets a feedback in the form of a 3D representation. Components perform mathematical or geometrical operations, taking inputs data on its left side, and outputting the results on the right side. Parametric relationships are defined by linking components with each other, thus creating an increasingly complex directed, acyclic graph. Parametric robot control expands upon this visual programming environment beyond parametric design by integrating a series of components that allow the programming and simulation of robotic arms through a plug and play environment. The robotic process planning can therefore be directly linked to the parametric design components that define the local form of a global construction component. The collected parameter therefore directly influence the fabrication process through what we refer to as production immanent design [7] creating a constant fabrication feedback.

Simultaneously these parameters directly inform the fabrication strategy which in turn allows for mass customization, automatically writing new robot control files each fabrication part. Through new interfaces developed for Industry 4.0 the adapted tool path can directly be transferred to the robot via an Ethernet based interface, which was originally developed in order to allow process control via a Programmable Logic Controller (PLC). This is especially useful for applications that are inherently repetitive, such as simple manipulation of objects with e. g. subtractive or additive methods. The mxAutomation interface for KRC4 KUKA Robot controllers as well as our implementation of a similar interface for the new KUKA sunrise controller therefore allow the direct execution of new fabrication logics. Simultaneously the visual programming environment enables testing methods for the fabrication-steps and the individual evaluation of each produced part.

Most Computer Aided Manufacturing (CAM) software allows for the path generation of gantry kinematics based on CAD. However in this case fabrication parameters are often connected to specific motion commands and not to a parametric design model. Fabrication with articulated robot arms increases the complexity while simultaneously allowing the fabrication of new geometries as well as surface aligned fabrication.

CAD is by now a standard practice within Architecture Engineering and Construction (AEC) industries. CAD models encompass different degrees of detail ranging from simple geometric models, over Meta data pertaining to materials and part spacing, towards intelligent objects containing parametric digital models that allow immediate dimensioning of objects according to the required specifications. Through this parametric robot programming we are able to employ this information directly to fabrication processes, linking geometry and fabrication process. In parallel these parameters can in turn be used in order to visualize fabrication constraints as a user feedback. Additionally fabrication results are send back to the user via a web interface allowing an immediate visualization of the process result for verification as illustrated in figure 4

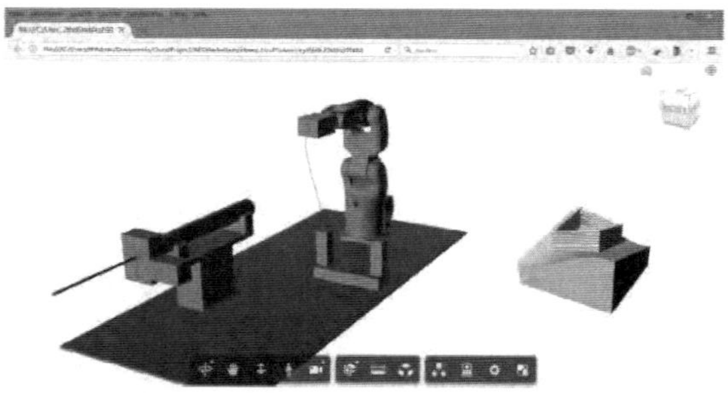

Figure 4. Process results as well as the production environment visualized through a web viewer.

4.2 Software Integration

The first implementation of upload, download as well as visualization of the simulation prediction through the web browser were implemented using components within the parametric design tool Grasshopper which runs as an extension to the CAD Software Rhino 3D. Additional data can be added using the json format.

The approach was then transferred and implemented within the parametric design software Dynamo, which can be used as the standalone version Dynamo Studio or as an extension to the BIM software Revit. Similar

interface components were created to connect the components with the same data mediator program.
The viewer html file used to display geometry in the web browser is stored locally whilst the object it displays is stored in the cloud. So, when opening the html file in the web browser we need to provide it with the objects storage urn in the cloud and a token, which authorizes the client to view the design preview. As the location of the viewer does not change most modern web browsers tend to cache the information. This can in turn lead to difficulties within displaying the updated cloud model.
Figure 5 illustrates the setup for the cloud integrated robotic simulation as well as the associated fabrication process.

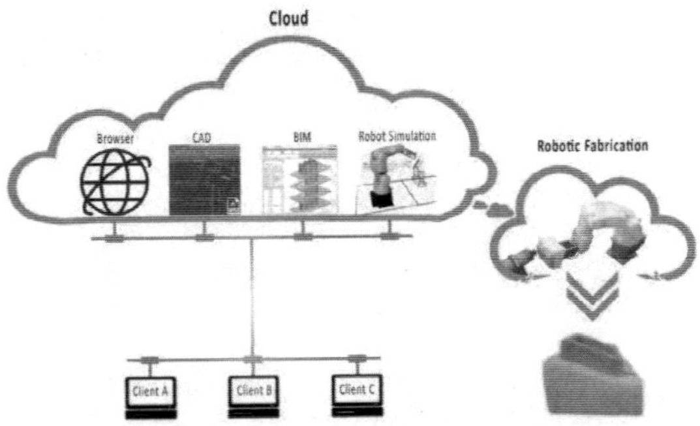

Figure 5. Framework structure for the cloud integrated Robotics approach.

4.3 Parameter selection

The current cloud integration allows us the exchange of parametric design files containing the parametric constraints. However this is currently not fully integrated and requires the informed design from the integrator or operator of the fabrication system. Therefore Design relevant parameter are selected by the Integrator of the robotic process. An exemplary interface for this design process is illustrated in Figure 6.

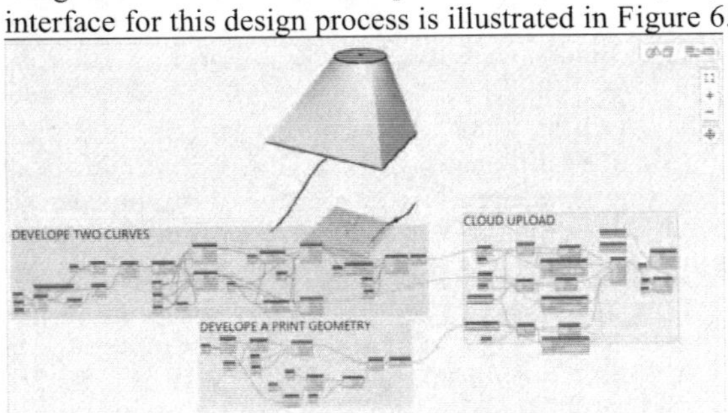

Figure 6. Parametric design input for fabrication through subtractive (Surface curve description) and additive processes (CAD).

Additionally Figure 7 shows the resulting data for the parametric robot process planning for a combined subtractive and additive fabrication process.
Within future work the selection of the design relevant parameters will be generalized allowing the automated generation of a web to real interface, while simultaneously allowing CAD design through multiple common design tools. Additionally a further integration of BIM parameters is planned for in order to achieve a full utilization of BIM data for fabrication.

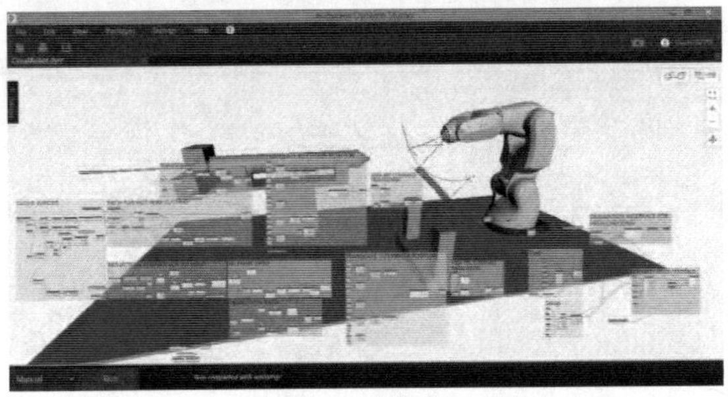

Figure 7. Parametric robot process planning through cloud imported data on an exemplary combination of a subtractive and additive fabrication process.

5 Distributed Design

While current architectural work is still often seen as a hierarchical and sequenced approach, employing cloud provided data for fabrication allows to increase the possibilities for a collaborative work environment.
As part of the AEC Hackathon in Munich we were able to test this approach for the first time. Enabling students to design parts as well as enabling them to program the robotic process directly through a parametric robot programming interface. While the design of elements through predetermined parameters seemed easy to grasp. Planning fabrication processes, while keeping in mind the constraints of the fabrication environment required significantly more time. Figure 8 shows the setup at the AEC hackathon in Munich.

Figure 8. Setup at AEC Hackathon in Munich

However making a Fabrication environment accessible to others can possibly reduce downtime for customized production without the requirement for mass production. Figure 9 illustrates the cloud integrated fabrication process.

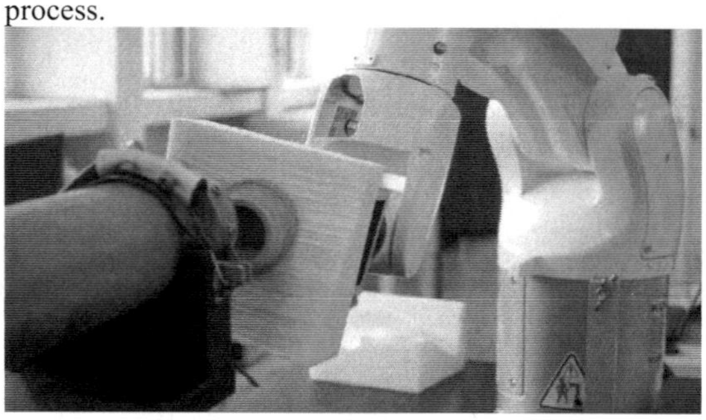

Figure 9. Robotic system Demonstrating additive and subtractive fabrication.

6 Synopsis

Through our interface we created a first proof of concept for cloud based robotic fabrication. However this approach was only test based on a short process chain, which contained a subtractive as well as an additive fabrication process.

Our future work will focus towards a full parameter integration within the web interface, which reduces necessity for parametric design experience from the user's point of view. While this approach is feasible for fabrication processes, robotic systems enable a very high number of processes. Therefore another focus of our future work will be on the integration of assembly skills, as well as our developed strategy for adaptation and information collection through haptic robot programming [8]. Figure 10 illustrates the planned integration for the Interactively & Dynamically Assisted Assembly (IDAA) Project.

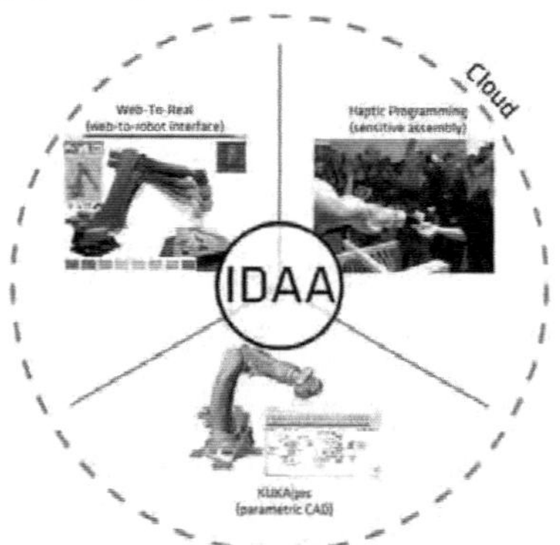

Figure 10. IDAA integrating parametric design, cloud robotics with human robot collaboration.

In conclusion the approach of making individualized robotic fabrication accessible through a cloud based interface has the potential to enable new distributed design opportunities as well as new forms of product commissioning.

Acknowledgements

The developments presented within this paper were made possible by Autodesk and the Forge Research Grant Program.

References

[1] Eastman, C. and Teicholz, P., Sacks, R., Liston, K.: *BIM Handbook – A Guide to Building Information Modeling for Owners*, Managers, Designers, Engineers and Contractors. Wiley; 2011.

[2] Stumm, S.; Neuhaus, M.; Braumann, J.; Brüninghaus, J.; Brell-Cokcan, S.: *Web-to-Real Application for Artistic Robotic Surface Processing of Natural Stone;* 33rd International Symposium on Automation and Robotics in Construction (ISARC 2016)

[3] R.Drath Datenaustausch in der Anlagenplanung mit AutomationML: Integration von CAEX, PLCopen XML und COLLADA. VDI-Buch. Springer, 2010.

[4] Braumann Johannes; Brell-Cokcan, S.. *Visual Robot Programming – Linking Design, Simulation, and Fabrication*. 2014.

[5] Braumann, J. and Brell-Cokcan, S. *Adaptive Robot Control - New Parametric Workflows Directly from Design to KUKA Robots*, In: Real Time - Proceedings of the 33rd eCAADe Conference, pages 243–250, volume 2, 2015.

[6] Davis, D.J.B., Burry, M. *Understanding visual scripts: Improving collaboration through modular programming.* International Journal of Architectural Computing 2011; Vol.9: pp.361–375.

[7] Braumann Johannes; Brell-Cokcan, S.. *Robotic production immanent design: Creative toolpath design in micro and macro scale.* CADIA 2014: Design Agency, Proceedings of the 34th Annual Conference of the Association for Computer Aided Design in Architecture 2014; pp.579–588.

[8] Stumm, S. and Braumann, J. and von Hilchen, M. and Brell-Cokcan, S., *On-site robotic construction for assembly using a-priori knowledge and human robot collaboration*, IFToMM/IEEE/euRobotics 25th International Conference on Robotics in Alpe-Adria-Danube Region RAAD (2016)

First monitoring and analysis of the manufacturing and installation process of timber based 2D modules for accomplishing a future robotic building envelope upgrading

K. Iturralde 1[a], T. Linner 2[a] T. Bock 3[a]

[a]Chair of Building Realization and Robotics, Technical University of Munich, Germany
E-mail: kepa.iturralde@br2.ar.tum.de, thomas.linner@ br2.ar.tum.de, thomas.bock@br2.ar.tum.de

Abstract –

The BERTIM Research Project is funded by the European Commission in order to develop a system for an efficient building envelope upgrading. The main goal of this envelope upgrading consists on adding a new 2D or 3D module onto the building external layer. The modules are made out of timber, thermal insulation and waterproof layers. These modules can also host some functionalities. In a previous phase, a method for implementing an efficient, integrated, modular and customized manufacturing and installation process of the BERTIM modules has been conceived. This method was achieved based on Axiomatic Design. The goal of this method was to generate and to coordinate three main sub-systems: the parametric design and configuration of the 2D and the 3D modules (1), the automated manufacturing process (2) and the rapid robotic installation process (3). The Axiomatic Design was mainly used in order to detect the possible interferences of the disaggregated Functional Requirements regarding the Independence and Information Axiom and the main risks of a Coupled Design. The first demonstration of the manufacturing and installation process has taken part at the KUBIK experimental building. According to the Axiomatic Design Functional Requirements and the main Design Parameters defined in the previous phase, the demonstrator puts in evidence that the solutions taken don't fulfil the Independence and Information Axioms. The current system can be considered as a Coupled Design. In order to uncouple the system, the TRIZ inventive method has been used, focused on improving loss of time in all sub-systems. The solutions defined with TRIZ methods will be implemented in this three case studies carried out by three different industrial partners. The robotic and automated degree of each case will be adapted to the specifications of each demonstrator.

Keywords –refurbishing, customization, automated, prefabrication, Axiomatic Design, TRIZ

1 Introduction

A novel method of the manufacturing and installation process of prefabricated timber modules is being developed in the BERTIM [1] research project. As in every European Project, the research is based on Work Packages (WP) that are organized correlatively in a calendar. Within these WP there are tasks that focus on specific matters. The overarching technical objective of the BERTIM research project is to define a holistic building renovation process, based on digital data workflow. For that purpose, a general implementation methodology for the manufacturing and installing industry must be defined. Once this methodology is applied in the industry, there must be a reduction of 30% on the installation time of the modules, compared to traditional methods. As a **previous excellent experience** on the field, it must be mentioned the robotic installation of facades developed by Obayashi [1]. In this case, the façade prefabricated modules where rapidly and smoothly placed by an overhead robotic device. The modules are specifically designed for that robotic operation; the so called Robotic Oriented Design (ROD) **Error! Reference source not found.** concept was applied.

Figure 1. Ideal robotic façade installation by Obayashi. Right: Positioning of the façade element. Left: Compliant plug-and-play connector façade element. (images: Obayashi documentary video) [1].

The robotic installation process can be taken as a final

goal, not only for the BERTIM project, but also for future building renovation projects. How to achieve that? It can be stated that the fast complying connectors, accurate manufacturing of the modules and specific robot devices are key factors. For that purpose, in the BERTIM project, a **Set of Solutions** has been defined in task 3.1 of WP3, that included:

- Generic design of the modules according to the requirements such as insulation, wind force and structural purposes. In this phase, a fast fitting connector has been conceived.
- The RenoBIM software will allow a repository of models and a fast integration onto the existing building´s geometry.
- Definition of the manufacturing and installation processes and procedures that can be adapted to each manufacturer.

On these previous task 3.1, a generic and standard solution was defined. But when applying this solution to each **specific building-case**, some other steps must be considered.

- Measurement of the existing building-case. The main purpose of the measurement is to accurately adapt to the existing building.
- Re-design or adapt the BERTIM modules to the building-case using a CAD first and with RenoBIM afterwards.
- Manufacture the modules according to CAD or BIM.
- Prepare the building before installing the modules. With ``preparing the building´´ it is meant the removal of unnecessary elements from the building.
- Install the modules. In principle, the ideal scenario would be to carry out a fast and smooth installation process while avoiding re-work.

In the Task 3.2 of Work-Package 3 of the BERTIM research project, a demonstrator was carried out at the ``Kubik´´ building in Tecnalia's facilities in Bilbao, Basque Country (Spain). The Kubik building was built for carrying out tests, it is a test-building. The test consisted in installing five 2D modules onto the existing building façade. The re-designing, manufacturing and installation process where defined previously. The whole process, from first data acquisition to final installation, was monitored and the results, considering specially time, was analysed.

Once the installation was carried out, the members from Tecnalia monitored the thermal performance of the modules. Comparing to the previous Timber based research projects for building envelope renovation [4], the BERTIM modules can include services, generally hot water, ventilation ducts and also Renewable Energy Sources (RES). This adds extra complication to the design, manufacturing and installation processes, since the services must be accurately placed.

2 Methodology

The research methodology is based in two design procedures. The first is Axiomatic Design [5] and the second is TRIZ [6].Previous to this paper, in task 2.5 of Work-Package 2, the Axiomatic Design was drafted as a core for defining the strategy for implementation of the BERTIM project. In Task 5.1, this strategy must be implemented. But after the demonstrator in task 3.2, some of the concepts that were foressen need to be reviewed since the beforehand ideas haven´t meet the expected results. The ongoing research is being carried out according to a methodology that is based in the next steps:

- Define the re-design, manufacturing and installation processes. This was already carried out in Task 3.2, previous to the demonstrator case.
- Implementation of the Kubik case study.
- Monitor the re-design, manufacturing and installation process of the Kubik case study.
- Get the results and analyze every break down form the monitoring process.
- Compare the results with the EIFS and rain-screen with the ones in Kubik BERTIM.
- Adapt the previously used Axiomatic Design process. Break every single task down based on the sub-systems of the Axiomatic Design proposed in previous research [8] focusing on the Design Process, Manufacturing process and Installation process.
- Allocate the points of the process that can be improved.
- Find possible applications according to the TRIZ contradiction matrix and the 40 Inventive Principles to these points.
- Discuss the Design Parameters or solutions for these points.

Once the discussion is finished, the next steps of the future research have been defined.

3 Traditional methods: EIFS and rain-screen.

The benchmark in BERTIM refers to the traditional methods for the installation process. In this research process two main types have been considered. The so-called EIFS (Exterior Insulation and Finish System) and the rain-screen. The data were gathered from a price and time database [7]. This row data of EIFS and rain-screen aren´t comparable to the work that has been accomplished at Kubik during the BERTIM project demonstrator. They need to be combined and completed with some other tasks such as Service installation (12 linear meters of pipes) and perimeter finishing (20,49 linear meters of a metallic cover). Several parameters have been considered and estimated:

- Kh: Total necessary working hours to achieve the works at Kubik
- NoW: Number of workers needed for each task.
- Ku: Area to be covered at the Kubik demonstrator.
- h/ m²: Total worker-hour task and square meter

These data have been gathered in Table 1 and Table 2 and will be used to compare the data from the BERTIM demonstrator process.

Table 1: Installation time for EIFS

On-site ETICS at Kubik (simulation for comparison)				
Task	Kh	NoW	Ku	h/m²
Insulation	5,09	2,00	23,34	0,22
Rendering	30,63	2,00	23,34	1,31
Painting	7,10	2,00	23,34	0,30
Perimeter	20,20	2,00	20,49	0,99
Services	2,58	2,00	12,00	0,22
TOTAL h/m² WITH SERVICES				**3,04**
TOTAL h/m² WITHOUT SERVICES				**2,82**

Table 2: Installation time for rain-screen

On-site rain-Screen at Kubik (simulation for comparison)				
Task	Kh	NoW	Ku	h/m²
Insulation	6,44	2,00	23,34	0,28
rain-screen	43,09	2,00	23,34	1,85
Perimeter	20,20	2,00	20,49	0,99
Services	2,58	2,00	12,00	0,22
TOTAL h/m² WITH SERVICES				**3,32**
TOTAL h/m² WITHOUT SERVICES				**3,11**

As an objective of the BERTIM project, the time collected in this table need to be reduced by 30%. We can therefore set as a benchmark 2 worker-hours/m² for the modules with services and 1,75 worker-hours/m² for the modules without services.

4 Definition of ``standard '' BERTIM modules

The criteria for the detailed design of the BERTIM modules have been defined in task 3.1 of WP 3 according to the Requirements specified in task 2.3 of WP 2. For the façade 2D modules the next points were specified:

- Criteria for the Energy Efficiency of the Modules: it included insulation type, thermal performance, condensation, HVAC and energy consumption.
- Criteria for the Design of the 2D Envelope Modules with and without Services: it included structural design, three different support systems, fire resistance and quick fit of services.

After defining how the ``standard'' BERTIM modules must be manufactured, transported and installed, a demonstrator was carried out at the Kubik building.

5 Implementing the BERTIM 2D modules to the Kubik case study

The solutions taken for carrying out this demonstrator included new features, but it didn't differ relevantly with the traditional techniques that are used by one of the industrial partners within the BERTIM consortium. It has to be said that this demonstration was carried out as a relevant environment test and some results were marked and analyzed. The Kubik building is a test building. It has to be said that the demonstrator at the Kubik building it is not a standardized procedure, and therefore it is more prompt to errors than traditional and already tested, proofed and commercialized techniques and procedures.

The modules used in the Kubik demonstrator were manufactured by the industrial partner Egoin. As said before, there were five 2D modules installed onto the building´s façade. Three of them didn´t have any services. The rest two included services, such as hot water and ventilation ducts. The purpose of placing services in the Kubik module was purely for testing the easiness of the installation process and monitoring the heat loss through the service module itself.

5.1 Monitoring the manufacturing and installation process

If we consider the whole process, from the very beginning of preliminary work to the last detailing and finalization, the procedure for manufacturing and installing the BERTIM prefabricated 2D modules took in total two months. But there were some idle periods as it can be seen on the following chronological timeline:

On-site at Kubik building

- 2016-10-03 Dismantle previous façade element. This façade was from a previous project. This step is not considered in the total sum of time.
- 2016-10-18 Measure the existing building with total station as seen in Figure 2. The criteria for measuring, in this case where there are no windows, is based on localizing the perimeter strategic points.
- 2016-10-19 Making holes for connecting services in the interior. As said before, the modules will be equipped with services that need to be fit with the

internal heating system.

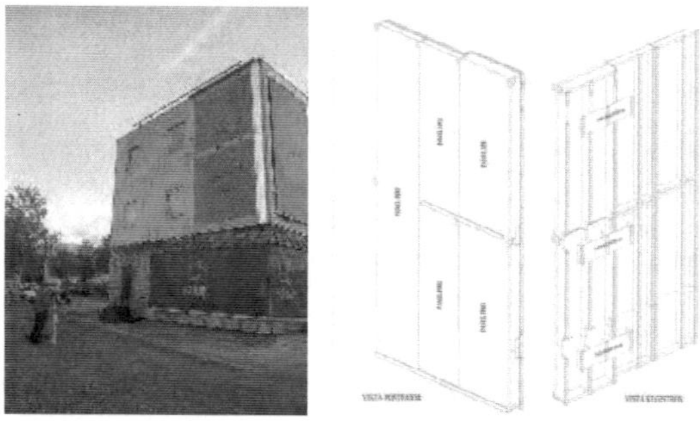

Figure 2. Left: Measuring the Kubik building strategic points. Right: Define the modules in CAD.

Off-site, in Egoin factory facilities

- 2016-11-14 Define the modules in CAD according to the measurements (Figure 2 right).
- 2016-11-15 Manufacture the modules according to the CAD files. The main process consisted on cutting studs, mill the studs, conform the frame and place the board (Figure 3 left). Next, they placed the vapor barrier and insert insulation. If the module had services, these were inserted at this point (Figure 3 middle). Finally, the moisture barrier and the raster were assembled. The finishing grid was manufactured separately (Figure 3 right). After the modules were assembled, they were mounted onto a truck for transportation.

Figure 3: Manufacturing process of the modules. Left: Timber-frame manufacturing. Middle: assembly of services onto the module. Right: Manufacture of finishing grid.

On-site at Kubik building

- 2016-11-22 and 23: Download the modules from the truck. After that, the first structural profile was placed. Next, the insulating layer was fixed. Once the wall was prepared, finally the 2D modules bodies were placed and fixed (Figure 4).

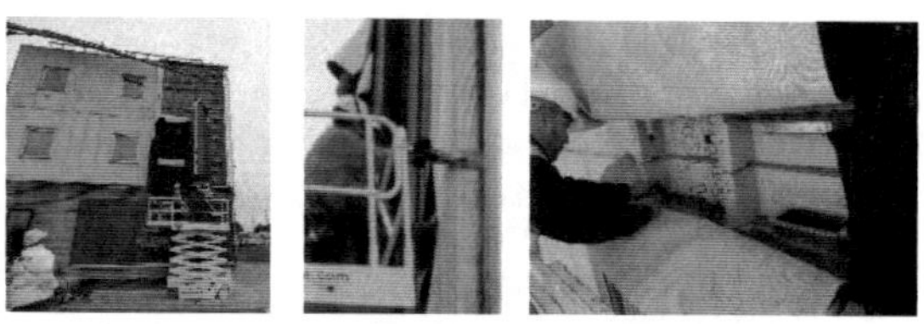

Figure 4. Left: Installing the panels. Middle: Fixing the connector to the wall. Right: Joining and fitting services.

- 2016-11-24 and 25. During these days the services were fitted, which means to fit the services within the modules and to fit the services on the interior of the building.
- 2016-11-28 Perimeter cladding and some other finishing was carried out.

Figure 5. Left and middle: Placement and fixation of the finishing timber grid. Right: Final enclosure and perimeter finishing.

Table 3: Manufacturing and installation time required for the 2D BERTIM modules at the Kubik bulding.

Task	Kh	NoW	Ku	Kh/m²
RE-ENGINEERING PROCESS				**0,43**
Data acquisition	2,00	1,00	23,34	0,09
Re-design	8,00	1,00	23,34	0,34
MANUFACTURING PROCESS				**1,61**
Battens	2,00	1,00	23,34	0,09
2DM	4,00	2,00	14,13	0,57
TF+I+MB+VB	*4,00*	*2,00*	*14,13*	*0,57*
2DMWS			9,07	2,48
CLT	*2,00*	*1,25*	*9,07*	*0,28*
TF+I+MB+VB	*5,00*	*2,00*	*9,07*	*1,10*
Services	*5,00*	*2,00*	*9,07*	*1,10*
Finishing grid	2,00	2,00	23,34	0,17
Transport I.	0,50	2,00	23,34	0,04
INSTALLATION PROCESS				**2,24**
Transport u.	0,30	2,00	23,34	0,03
Connector	0,30	2,00	23,34	0,03
Insulation	1,50	1,00	23,34	0,06
Place fix modules	5,50	3,00	23,34	0,71
Finishing a	1,00	3,00	23,34	0,13
Finishing b	1,00	2,00	23,34	0,09
Services	6,00	2,00	23,34	0,51
Perimeter	8,00	2,00	23,34	0,69
TOTAL h/m² WITH SERVICES				**4,27**
TOTAL h/m² WITHOUT SERVICES				**3,21**

As said before, the whole process was monitored. Basically, it was focused on how many workers and how much time were used in each part of the task. Following the criteria of Table 1, the same parameters were recorded during the whole process. In the case of BERTIM, though, the off-site manufacturing process was also recorded and monitored as it can be seen in **Error! Reference source not found.**.

As it can be seen in Table 3, the whole process is divided in three different stages. First there is the **re-engineering** process which consist on the data acquisition of the existing building and the adaptation of the BERTIM module to the building´s needs. After that, there is the **manufacturing process**. In this phase, the battens are cut and machined first. Two types of modules can be divided: modules without services (2DM) and modules with services which are reinforced with CLT panels (2DMWS). In both cases they consist of assembling timber frame (TF), insulation (I), moisture barrier (MB) and vapor barrier (VB). Once the modules are produced and transported to the site, the **installation process** starts.

5.2 Discussion about results

In previous data gathering, the authors detected that 1,18 hour/m² were necessary for the manufacturing process [8]; this shows the data are quite similar to the 1,30 h/m2 registered for the Kubik case. The manufacturing process differs from one industrial partner to the other, depending on the facilities and assets they have for producing the modules. One of the analysed partners showed that the 0,8 h/m² would be reached if dedicated work station are organized properly. Therefore, it can be said that there is already technology for automating and improving the manufacturing process.

Figure 6. The connecting system with inaccuracies that can be detected even visually.

Regarding o the finishing of the modules, an inaccuracy of around 10 mm has been found in the final positioning of the modules. The errors might be produced by:

- Inaccuracy of the manufacturing process.
- Inaccuracy of the positioning of the support.
- Inaccuracy during the placement of the modules .
- Un-rigidness of the timber modules, which during transportation can be bended and deformed. Also, the modules can expand or shrink due to hydrothermal changes.

Due to this inaccuracy, the tubes are not ready to fit and they need a rework (more time for fitting) at the construction site.

5.2.1 Re-engineering process. Design adaptation

The re-engineering process for nesting and drawing the modules takes 0,34 h/m². Without the RenoBIM software tool this is a time consuming task that in the case of the Kubik case study, it took around 8 hours to re-design the modules. Besides the RenoBIM, the use of some other parametric design software would be necessary in order to automate the process.

5.2.2 Place and fix the modules and the finishing grid.

The installation of the modules takes 0,71 h/m2. This time includes the module itself and but not the finishing grid, which is placed afterwards. Installing the finishing takes 0,22 h/m2. If the module was a fully prefabricated with integrated finishing, time reduction would be enhanced. The same happens with the insulating layer that is placed on the wall before the installation of the modules.

5.2.3 Fitting of services on-site

The fitting of the services consisted on joining together the water and air ducts in a registrable box . This was a manual operation. Furthermore, the whole box needed to be insulated and covered after that. The joining tubes where flexible in order to have bigger tolerances in case of big accuracy errors. Definitely, this can be considered as a time consuming task, and therefore a solution should be considered.

5.2.4 Finishing perimeters

This task requires major study definitely, since 0,69 h/m2 were necessary for that purpose. It could be considered even a bigger problem if there were windows involved. In the Kubik demonstrator there was no window within the modules. As a previous experience for adapting to the windows particular geometry, a set of solutions was already proposed in the RetroKit project [9]. But in this case, the level of prefabrication was lower than in BERTIM. In future BERTIM tests and demonstrators, the issue of fitting the 2D module to the window must be solved.

5.3 The Axiomatic Design matrix of the Kubik demonstrator

As a conclusion of the Kubik demonstrator, it was clear that the manufacturing and installation process and the module itself still needs to be re-adapted in order to improve the time schedule of the whole process. The H2020 EE-01-2014 call named ``**Manufacturing of prefabricated modules for (energy) renovation of buildings**´´, defined the next main requirements:

- Use **prefabricated** modules. It must be said that there is an underneath objective to integrate energy efficiency devices and **renewable energy sources** in the prefabricated multi-functional modules.
- Use of **advanced computer based tools** for integrating the value chain over the life cycle of the project.
- Move from individual manufacturing to mass production. A more accurate term would be **mass-customization**.
- **Reduce the installation time** by at least 30%, compared to a typical renovation process for the building type. This term must be defined more accurately. In this research, we compute or determine the total time of Re-design, Manufacturing and Installation processes. And that time should be lower than the traditional methods.

Other minor objectives that can be considered also as constrains are:

- Reduce costs. This is related to the manufacturing and installation processes.
- Ensure quality
- Facilitate dismantling and re-use
- Improve on-site health and safety during manufacturing and installation.

In the BERTIM project, in previous approaches [8], the Functional Requirements (FR) for accomplishing these previous objectives were defined. In this previous phase, specific solution weren't given, at least in the lowest hierarchy of the sub-systems; particular and disaggregated requirements weren't solved. The idea is that the timber-frame manufacturers can apply this method with enough flexibility, depending on the current economic and technical limitations. Comparing to previous approaches, a new sub-system has been added, sub-system 2 as it can be seen on the next list:

- Sub-system 1: Customize the prefabricated 2D modules for specific building refurbishment.
- Sub-system 2: Information workflow based on advanced software.
- Sub-system 3: Maximize the offsite manufacturing process of existing facilities of the industrial manufacturers.
- Sub-system 4: Minimize onsite Installation time and cost of the module

On the other hand, the Design Parameters where considered the next:

- Sub-system 1: Use of highly prefabricated and adaptable timber-framed modules
- Sub-system 2: Use of RenoBIM for the integrated Product Lifecycle Management (PLM).
- Sub-system 3: Use of a Manufacturing Modular Kit that can be adapted to the manufacturers
- Sub-system 4: Define a Rapid Installation System

But on the Kubik demonstrator, these sub-systems were not fully implemented, and the primary hierarchy of the Axiomatic Design was modified as follows:

$$\boldsymbol{FR = A * DP}$$

$$\begin{Bmatrix} FR_1 \\ FR_2 \\ FR_3 \\ FR_4 \end{Bmatrix} = \begin{Bmatrix} x & A_{12} & A_{13} & A_{14} \\ A_{21} & x & A_{23} & A_{24} \\ A_{31} & A_{32} & x & A_{34} \\ A_{41} & A_{42} & A_{43} & x \end{Bmatrix} * \begin{Bmatrix} DP_1 \\ DP_2 \\ DP_3 \\ DP_4 \end{Bmatrix}$$

(A) $FR_1 = Definition\ 2D\ prefabricated\ modules\ for\ building\ refurbishment$
(B) $FR_2 = Information\ workflow\ based\ on\ digital\ Data\ CAD - CAM\ systems$
(C) $FR_3 = Maximize\ the\ off - site\ manufacturing\ process$
(D) $FR_4 = Minimize\ onsite\ Installation\ time\ and\ cost$

(A) $DP_1 = Adaptable, medium - prefabricated\ modules\ with\ limited\ acuracy$
(B) $DP_2 = Use\ of\ conventional\ CAD\ systems\ for\ the\ definition\ of\ the\ modules$
(C) $DP_3 = Exisitng\ factories\ with\ limited\ flexibility\ and\ no\ dedicated\ workstations$
(D) $DP_4 = Conventional\ fixation\ process\ with\ a\ mobile\ crane\ and\ a\ platform$

Table 4: Interference Matrix

	DP1	DP2	DP3	DP4
FR1	x	x	0	0
FR2	x	x	x	0
FR3	x	x	x	0
FR4	0	x	0	x

If we check this primary hierarchy in a matrix table, we can see that the Independence Axiom is not fulfilled fully. There are interferences between the different FRs and DPs is considerable as it is shown in Table 4.For instance these are the points that need to be taken into consideration:

- Sub-system 1: In the case of the Kubik building, the adaptable prefabricated modules had a medium degree of prefabrication. Only around 50% of the tasks were carried out at the factory. The adaptable module needed special rework after being placed and fixed on to the wall; the finishing was necessary and the services need to be fit.
- Sub-system 2: The customization of the module affected the CAD re-design process. The manual adaptation of the CAD file, slows down the process. Furthermore, in the manufacturing process, the works

need to be guided by the engineer in charge.

- Sub-system 3: Using the existing tools and assets off-site on the factory, the continuous customized modules and its bespoken process can diminish the productivity.
- Sub-system 4: The installation can´t be minimized if the prefabricated product, after installation, needs rework such as the perimeter enclosure finishing or the window frame finishing. Besides, the service fitting time should be diminished too. It can be said that the highly prefabricated modules needs to meet the geometry of the existing wall accurately.

Therefore, the strategies need to be re-defined. On the next chapter, some broken down parts of the process will be analysed and the TRIZ method will be applied.

6 Implement TRIZ

The TRIZ method facilitates the identification of the so called contradictions that a system might have. These contradictions block or obstruct the possibility of improvement of a system. But thanks to the TRIZ Parameter Matrix and the 40 Inventive Principles, the problems are detected and diverted to find an optimal solution. In order to reduce time, the manufacturing, but specially the installation process needs to be fast. If we take as a primary objective to reduce time, we can organize the TRIZ strategy as follows.

6.1 Time reduction in the installation process

The **time reduction** in installation will come from a higher prefabrication and accuracy of the modules in order to obtain a fast fitting process and avoid re-work.

But a fast installation process and a high accuracy leads to lack on the **ease of operation**. The connectors on the wall need to be aligned with low tolerances in order the module and services fit. The TRIZ matrix provides the principles 4,28,10 and 34 for this situation (

Table 5). The principle 10, preparation of the alignment of the wall has been used already in some other similar research projects [10][11], but that leads again back to a loss of time.

Besides, the module might change its configuration during transportation due to heat and humidity changes, and therefore, the accuracy can be lost during logistics operations. This issue must be taken into consideration on next phases.

Table 5: Principles 4 (asymmetry), 28 (replace mechanics), 10 (prior action and pre-arrange objects) and 34(discard and remove)

	Ease of operation(32)
Loss of time (25)	4,28,10,34

6.2 An accurate module for rapid processes

The defined module needs to be highly prefabricated, and therefore, highly defined in order to loose time. In consequence, a **highly prefabricated** module which includes embedded services needs **high accuracy** in order to be fit rapidly. For a higher accuracy of the modules, there must be an accurate **data acquisition** of the existing building in order to customize and bespoke the modules. However, the biggest contradiction of the highly prefabrication relies on the lack of adaptability of the module, that the module is so tight that no change is possible. TRIZ offers two principles for these situation, 35 (parameter change) and 28 (replace mechanics) as it can be seen in Table 6. This is related to the module materials and configuration. The materials and elements of the module need to be reviewed. Besides, a highly defined parametric design of the BERTIM module properly linked with the CAM would facilitate the procedure.

Table 6: Principles 35(parameter change) and 28 (replace mechanics)

	Adaptability (35)
Loss of time (25)	35,28

6.3 Time reduction in the re-design and digital workflow

Once the **re-design** process is ready, ease of manufacturing must be accomplished, with a clear digital information and material workflow. And it takes time during the manufacturing process and installation process if the elements, specially the services are not clearly described for the operator. The biggest harm is therefore the **loss of information**, the operator doesn´t know how to assembly every customized module. Besides, as said before, the customization of the module takes time on the **manufacturing process**, especially if no parametric or computational design feature is used and the elements are drawn in CAD manually. In this situation TRIZ offers principles 24, 26, 28 and 32 (Table 7). Among them, the `` copying´´ principle is the most adequate for our case, basically, what needed is a clear parametric library of solutions that will `` copy´´ the module's configuration successively on the next cases.

Table 7: 24(intermediary), 26 (copying), 28 (replace mechanical system), 32 (change color)

	Loss of information (35)
Loss of time (25)	24,26,28,32

6.4 Accuracy in the manufacturing process

Gaining **accuracy** leads to a **loss productivity** in

manufacturing processes. Among the principles proposed by TRIZ, the ``discard and remove´´ might be the most interesting. This principle would lead to use of patterns that can facilitated the insertion of services onto the modules.

Table 8: 35 (parameter change), 28(replace mechanics), 34 (discard and remove), 4 (asymmetry)

	Productivity (32)
Manufacturing accuracy (25)	35,28.34,4

6.5 New Axiomatic Design matrix and discussion

In theory, if the Inventive Principles are implemented the Axiomatic Design matrix should be uncoupled as it can be seen on the next matrix and on **Error! Reference source not found.**. It has been prioritized the reduction of time during the installation process and the rest of requirements are hierarchited for that purpose. Though, the given solutions aren´t still detailed enough to prove that the matrix would be uncoupled yet. But it gives a direction for the next research steps in order to fulfill the Independence Axiom.

$$FR = A * DP$$

$$\begin{Bmatrix} FR_1 \\ FR_2 \\ FR_3 \\ FR_4 \end{Bmatrix} = \begin{Bmatrix} x & A_{12} & A_{13} & A_{14} \\ A_{21} & x & A_{23} & A_{24} \\ A_{31} & A_{32} & x & A_{34} \\ A_{41} & A_{42} & A_{43} & x \end{Bmatrix} * \begin{Bmatrix} DP_1 \\ DP_2 \\ DP_3 \\ DP_4 \end{Bmatrix}$$

(A) FR_1 = *Definition 2D prefabricated modules for building refurbishment*
(B) FR_2 = *Information workflow based on digital Data CAD – CAM systems*
(C) FR_3 = *Maximize the offsite manufacturing process*
(D) FR_4 = *Minimize onsite Installation time and cost*

(A) DP_1 = *Accurate andfully prefabricated module*
(B) DP_2 = *Parametric definition of the modules using computational design*
(C) DP_3 = *Dedicated workstations for services integrated on existing factories*
(D) DP_4 = *Rapid placement and fixation process using interfaces*

Table 9: interference matrix

	DP1	DP2	DP3	DP4
FR1	x	0	0	0
FR2	0	x	0	0
FR3	0	0	x	0
FR4	0	0	0	x

7 Conclusions

TRIZ has been useful for uncoupling the Axiomatic Design Matrix in the case of BERTIM. But a broader study is necessary to reach specific solutions. The specific and detailed solutions are being defined currently. It can be said that TRIZ is being a suitable method for gathering different conceptual and specific solutions. Two criteria can be outlined from this research. First, the higher the prefabrication, the lower time for on-site installation and rework. Second, the higher prefabrication, the more accuracy is needed, and therefore more time is necessary in measuring and manufacturing. On future works, other Functional Requirements should be analyzed and researched. The authors will follow working on that direction. Currently they are studying fitting the Axiomatic Design objectives with the 39 Parameters on the Contradiction Table of TRIZ. The discussion is if one or more parameters need to be used in the finding of a solution.

The main outcome of this research, especially regarding the accuracy of the modules and the fast fixation system, should be considered for **future strategies of robotized and automated manufacturing and installation processes** in building renovation and maintenance. Before the end of the BERTIM research project, the modules will be manufactured and installed in three different operational environments, in real building refurbishment projects. The research presented on this paper will be applied to these cases.

Acknowledgments

This paper has been drafted thanks to the funding from the European Union's Horizon 2020 research and innovation programme under grant agreement No 636984. The content of this paper does not reflect the opinion of the European Union.

References

[1] Bock, T., Linner, T., 2016, Site Automation, Cambridge University Press.
[2] Bock, T. and Linner, T., 2015. Robot Oriented Design. Cambridge University Press.
[3] Webpage visited the 15/03/2016: http://bertim.eu
[4] Larsen, K. E., et al., "Surveying and digital workflow in energy performance retrofit projects using prefabricated elements", Automation in construction, 2011.
[5] Suh, Nam P. "Axiomatic Design: Advances and Applications (The Oxford Series on Advanced Manufacturing)." (2001).
[6] Ogot, M. "Conceptual design using axiomatic design in a TRIZ framework." Procedia Engineering 9 (2011): 736-744.
[7] Webpage visited the 15/03/2016: http://www.generadordeprecios.info/
[8] Iturralde, K., Linner, T., Bock, T., Development of a modular and integrated product-manufacturing-installation system kit for the automation of the refurbishment process in the research project BERTIM, 2016 Proceedings of the 33rd ISARC, Auburn, USA
[9] Webpage visited the 15/03/2016: http://www.retrokitproject.eu/
[10] Larsen, K. E., et al., "Surveying and digital workflow in energy performance retrofit projects using prefabricated elements", Automation in construction, 2011.
[11] VariousAuthors, "Großelement-Dämmtechnik mit

Vakuumdämmung", Institut Wohnen und Umwelt GmbH, Endbericht, Germany, 2007

Bottom-up Cognitive Analysis of Bionic Inspection Robot for Construction Site

Thomas Bock[a], Alexey Bulgakov[b], Sergei Emelianov[b] and Daher Sayfeddine[c]

[a] Building Realization and Robotics, Munich Technical University, Germany
[b] Department of Mechatronics in Construction, Southwest State University, Russia
[c] Department of Mechatronics, South Russian State Polytechnic University, Russia
E-mail: thomas.bock@tum.de, a.bulgakow@gmx.de, esg@mail.ru, daher@live.ru

Abstract –

Artificial intelligence aims to make robots more adaptive and versatile depending on the surrounding operating atmosphere. It permits the autopilot of the robot to generate optimal trajectory with reference to the energy efficiency criteria and risk avoidance in order to lead to the end-effector or the working element of the robot to the desired position, thus to perform safely the assigned task.

One of the problems faced during the optimization of algorithms of the central autopilot is the overlearning, similar training sets, permanent operating conditions, which may lead to controversial result: non-adaptability of the robot. This can be clearly seen when training nonlinear model of neural network with exogenous inputs aiming to resolve the extrapolation of the movement function of a mobile agent (nontraditional desired trajectory). Hence the quick change in the operating conditions can lead to undesired outcomes.

In this paper we analyze robot control performance based on situational approach, described using discrete mathematics operators and state-base/ history base, time-base/action-base functions. The knowledge representation will be used to train auto-regressive neural network using situational time series.

Keywords –
Cognitive robot; Artificial intelligence; NARX neural network; Regression

1. Introduction

The successful integration of robotic solutions in the construction field and strictly depend on the site conditions and infrastructure. At construction site, many hindrance are there, which limit the robot maneuverability, operational speed and reaching. Due to the dynamicity in the construction site, the concept of "controlled environment" cannot always be maintained, which develop new challenges to the robot "perception" of the working atmosphere. The sudden changes can be challenging to the robot in terms of processing unit and speed, which may cause injuries to the robot. Changing processing capabilities of the robot only to fit the "if" case is not techno-commercial solution the risk and dynamicity specter is very broad. In view of the above, cognitive concept can be introduced.

By cognitive, we estimate the ability of the robot to learn and develop new movement technics while permanently analyzing the working atmosphere thus to change forecast future actions or to react on sudden changes in the working site.

Cognitive solutions are based on artificial intelligence. For instance, the neural network can be used for extrapolating the movement function and guide the robot to track recently generated trajectory. This is the case when inspecting a site, while construction actions are carried out above the robot. HSE guidelines are optimal for labors, but it cannot be used for robots. This HSE infrastructure is tackled by the cognitive ability of the robot, making him always "alert" to the changes in the working atmosphere.

2. Literature Review

Many literatures focus on the hierarchy, the organization and optimization of the autopilot. For bionic robot, the task is more difficult, since the control and movement tasks are more complex. This is due to the enormous data traffic, analysis and storage required. The adaptability or the morphological changes in the end-effector, the chassis of the robot and the central controller are more difficult to optimize from auto-operational point of view. For instance, tasks such as path planning has become vague and not clear, since path planning is limited to the kinetic aspect of the movement, where offline algorithms can play major roles. The outcome is global step planning before movement. In contrast, trajectory planning is more difficult task. It is related to the dynamic aspect of the movement, the analysis of the surrounding and the decision of action with reference to online algorithms. The result is a local step planning, which requires more computation power to resolve the complexity of the data. As it can be clearly noticed, cognitive is more about dynamic planning.

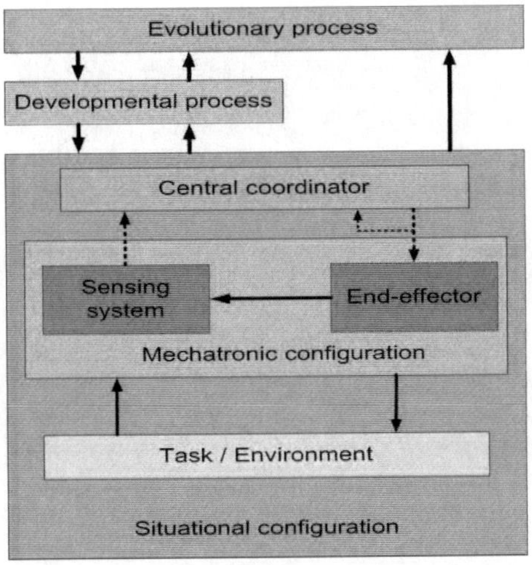

Figure 1. Bottom-up cognitive hierarchy of mechatronic/robotic mechanism

There are different methods to analyze a system. Wireframes, virtual Personas or scenarios, rapid prototyping and several more are not an option to be adopted for real-time planning. Cognitive approach is based on Bottom-Top analysis, where many subsystems are pieced together in such a way, thus to obtain more complex systems or approach. In light of that, the emergent output is controlling the behavior and integrity of the original subsystems. In other words, a significant attention is offered firstly and constantly to subsystems, then to the links between them and lastly to the output. As it may be noticed, the system analysis grows complex bottom-top hence understanding such system requires granulation process. Inspection is a dynamic task in unknown environment where everything is possible. Robots are frequently used in inspection and monitoring tasks such. Task failures are several: operational, power-related and mechanical. The operational reasons can be caused by human-operator in distant-controlled tasks or because of algorithmic challenges. The flow of work between human and robots in different automation strategies is an essential factor to succeed with the integration of the robotic solution. As depicted in fig.2, it is very clear that in "Lights-Out Automation" scheme, the interference of the human-operator in the robotic scope of work does not exist. Geographically, it is not necessary that the robot and operator be in different location. However, the operator is not responsible to monitor the work of the robot. This fact imposes high level of adaptation and perception of the robot.

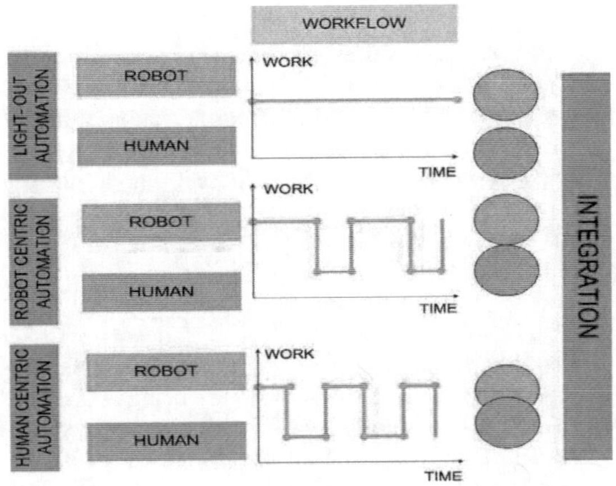

Figure 2. Workflow and integration between Human and Robot

3. Robotic Cognition

One of the main goals of robotics is to replace human being in dangerous tasks; the integration of adaptable intelligent machinery in construction sites encompasses delivering milestones outside the concept of project management in controlled environment. Tuning the robot based on case-to-case approach may cover resolve the planning difficulties. However, misjudging a scenario in the algorithm will lead to undesired results.

In view of that, the situational approach seems to be more adaptable and less risky as the robot knowledge database is epistemic. This terminology explains the ability of robot to do modal logic that formalizes knowledge based on situational/ historical approach.

3.1. Action Representation

The essential idea about cognitive robotics in the way of presenting the knowledge based and embedded it into the control system by giving more flexibility-authority trade-offs. While, the concept of probabilistic models and theories will remain in adoption, the major concentration of the cognitive science in robotics is about monitoring the changing word [1]. The function representing the changes can be based on calculus, or by comparing modal to non-modal, state-base/ history base, time-base/action-base functions and knowledge representations [2,3,4,5]. Hence, a single action of the robot has two obviously simultaneous derivatives: START- END combined with additional time argument and new fluent with the following successor state axiom [6]:

$$\boldsymbol{C_T(x,t,do(a,s))} \equiv \exists \boldsymbol{t'}(\boldsymbol{a} = \boldsymbol{S(x,t')} \wedge \boldsymbol{t'} \leq \boldsymbol{t}) \vee \boldsymbol{C_T(x,t,s)} \wedge \neg\exists \boldsymbol{t'}(\boldsymbol{a} = \boldsymbol{E(x,t')} \wedge \boldsymbol{t'} \leq \boldsymbol{t} \quad (1)$$

Where $\boldsymbol{C_T}$ – is the time argument; $\boldsymbol{S}$ – start action; $\boldsymbol{E}$ – end action; $\boldsymbol{t'}$ – is the additional time argument; $\boldsymbol{x}$ – control task/ handled object; $\boldsymbol{a}$ – action and $\boldsymbol{s}$ – is the situation of sensed fluent value. It is worth to note that equation (1) include the additional time frame argument which does not encompass the failure time necessary for correction or repetitive actions.

3.2. Sensing

Sensing is the source of change database required in order to build idea about the surrounding word. While the situation calculus propose introducing a special fluent and axioms describing the truth-value correlated to the aspect of situation, other state-base/history base comparing algorithms seem to be more efficient in term of consumption of computation power and time. The combination of the two approaches lead to optimized hybrid algorithm, which we will adopt in later stage in this paper.

If the control task consists of tracking color based on histogram and geometric approach, then we can use the following a predicate defining the robot learning by executing recognition actions in different situation and generating binary results. The following can serve as demonstrative case:

$$\boldsymbol{R_s} = (\boldsymbol{senseC}(x), s \equiv C(x, \varrho, s);$$
$$\boldsymbol{R} = (\vec{a}.A, \vec{b}.1, s) \stackrel{\text{def}}{=} R_s(A, do(\vec{a}, s)) \wedge R(\vec{a}, \vec{r}, s); \quad (2)$$
$$\boldsymbol{R} = (\vec{a}.A, \vec{b}.0, s) \stackrel{\text{def}}{=} \neg R_s(A, do(\vec{a}, s)) \wedge R(\vec{a}, \vec{r}, s).$$

Where, $\boldsymbol{R_s}$ – the sense fluent value; $\boldsymbol{C}$ – is the colour function; $\boldsymbol{x}$ – is the tracked object; $\boldsymbol{\varrho}$ – is the RGBHV value of the colour; $\vec{\boldsymbol{r}}$ – is the vector or binary results or cognitive reasoning of the robot.

3.3. Knowledge and Decision

Similarly to the action representation and sensing, knowledge is also presented using special fluent K depending on situational approach. In this regard, if a parametric function F representing the object is known at a certain situation, then F is known in all other situations. Hence we can write:

$$\boldsymbol{K}(\boldsymbol{F}, \boldsymbol{s}) \stackrel{\text{def}}{=} \forall s'. \boldsymbol{K}(s', s) \supset \boldsymbol{F}[s'] \quad (3)$$

The successor state axiom based on action update the knowledge base using the following model:

$$\boldsymbol{K}(s'', do(a, s)) \equiv \exists s'. s'' = do(a.s') \wedge K(s', s) \wedge [R_s(a, s') \equiv R_s(a, s)]. \quad (4)$$

As it can be noticed, it is sufficient to know the action is a certain previous situation so the robot can start building on it. In other term, an action is accessible in following situation $\boldsymbol{s'}$ or $\boldsymbol{s''}$ only if the action satisfies the condition in equation (2).

Using equations (1) to (4), we can generalize that by giving sequence of actions performed by the robot, uniform parametric function, the cognition task become a regression approach [8] since it is based on previous situational incident $\boldsymbol{s_0}$.

3.4. Auto-Regression Task

As a matter of fact, neural network are widely used in regression task. While it can suffer for overlearning, long analogic and action comparison, neural network remains one of the most influential cognitive and reasoning approaches in robotics.

The aim of the prediction using NARX is to be passive mostly during successful recognition of the agent [9]. During this time NARX can be trained by the updated positions of the mobile agent represented by data pairs (input- Output). This vector is described in (5)

$$u(n) = (u_1(n), \dots u_K(n))', \quad d(n) = (d_1(n), \dots d_L(n))' \quad n = 1..T \quad (5)$$

Where $u(n)$ – is the input set, $d(n)$ – is the output set, n represents the time, K is the number of perceptron in the input layer, L is the number of perceptron is the output layer.

The training of the NARX is achieved using BPTT because the output of the network is not fed into the tracking algorithm. As discussed earlier, the training is done passively following three steps.

The first step consists of calculating and discovering the status of the activation functions $x(n)$ of each perceptron starting from $u(n)$, $x(n-1)$ and $y(n-1)$ or the activation of the output layer if it is fed into a certain perceptron;

The second step includes the calculation of the back-propagation error of each perceptron starting from $n = T..1$, $x(n)$ and $y(n)$ for each instance of time n. this is achieved using the following system of equations (6)

$$\delta_j(T) = (d_j(T) - y_j(T))(\partial f(u))/\partial u|_{u=z_j(T)}$$

$$\delta_i(T) = [\sum_{j=1}^{L} \delta_j(T)\, w_{ji}^{out}](\partial f(u)) /\partial u|_{u=z_i(n)} \quad (6)$$

$$\delta_j(n) = [\left(d_j(n) - y_j(n)\right)\sum_{i=1}^{N} \delta_i(n+1) w_{ji}^{back}]\,(\partial f(u)) /\partial u|_{u=z_j(n)}$$

$$\delta_i(n) = [\sum_{i=1}^{N} \delta_i(n+1) w_{ji} + \sum_{j=1}^{L} \delta_j(n) w_{ji}^{out}](\partial f(u)) /\partial u|_{u=z_i(n)}$$

Where $\delta_j(T) -$ is the backpropagation error of the output perceptron, $\delta_i(T) -$ is the back propagation error of the perceptron located in the hidden layer with activation $x_i(T)$, $\delta_j(n)$ and $\delta_i(n)$ are consequently the backpropagation error of the output perceptron and the one located in the hidden layer in an earlier time T layer and $z_i(n) -$ is the potential of each perceptron.
After finding the back propagation error, the weights connecting different perceptron are calculated using the following system (7)

$$w_{ij} = w_{ij} + \gamma \sum_{n=1}^{T} \delta_i(n)\, x_j(n-1)$$

$$w_{ij}^{in} = w_{ij}^{in} + \gamma \sum_{n=1}^{T} \delta_i(n)\, u_j(n)$$

$$w_{ij}^{out} = w_{ij}^{out}\gamma \begin{cases} \sum_{n=1}^{T} \delta_i(n)\, u_j(n)\ if\ j - is\ an\ output\ p \\ \sum_{n=1}^{T} \delta_i(n)\, x_j(n)\ if\ j\ is\ a\ hiden\ per\ \end{cases} \quad (7)$$

$$w_{ij}^{back} = w_{ij}^{back} + +\gamma \sum_{n=1}^{T} \delta_i(n)\, y_j(n-1)$$

Where $\boldsymbol{w_{ij}} -$ is the weight connecting the hidden perceptron, $\boldsymbol{w_{ij}^{in}}$, $\boldsymbol{w_{ij}^{out}}$ and $\boldsymbol{w_{ij}^{back}}$ are the input, output and feedback weights consequently, $\boldsymbol{\gamma} -$ is an incremental small value that is used during the minimization of the squared error.

3.5. Results

The NARX structure consists of a single time-series data pairs as was described earlier, which contains all the historical data about the movement of the agent. The hidden layer includes 10 perceptron and a single output is generated thereafter predicting the possible location of the agent. The structure model of the used NARX is shown is figure.3.

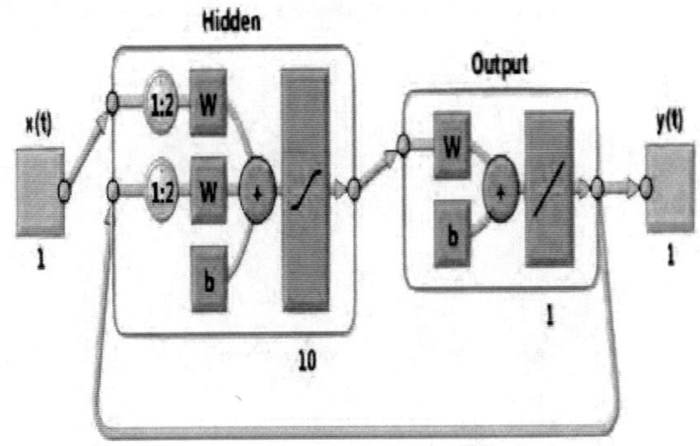

Figure 3. NARX structure model used to predict position

The training is achieved based on 1000 epochs and was concluded in one minute and eight seconds. The result is obtained based on minimal gradient 1^{-10}.

It is important to mention that the NARX acts like recurrent neural network with embedded memory. It allows the NARX to "remember" the output of the perceptron located in any layer by unfolding the dependencies of the forecasted series far longer than a conventional recurrent neural network. This shows an impact when forecasting nonlinear, aperiodic and unknown data sets.

In light of this, the prediction can be tested on different nonlinear functions conserving the periodicity, the damping factor and chaotic movement of the tracked agent.

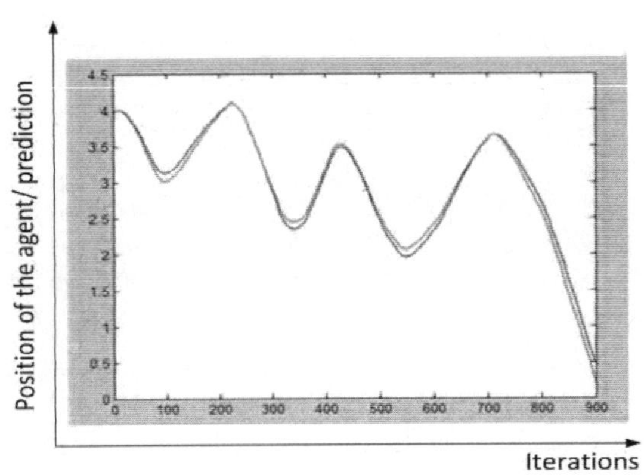

Figure 4. Simulation results

Simulation results of the predicting the next position of the agent are shown in figure.4. It is clear from the curves that the estimation (red) was very close to the real values presented in the time series (blue).

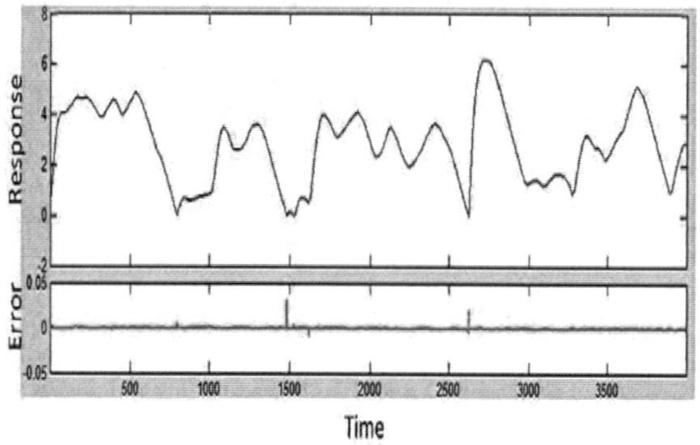

Figure 5. Response of the NARX with reference to the time series (black) and the back propagation error (red) over time

We can notice that the NARX was successful in estimating the next position of the dynamic agent, movement, which was described randomly using predesigned time series in based on magnitude and phase form (input output coupled pairs).

Although the regression approach could resolve the situational challenge theoretically, it is very challenging to adopt the same concept in very dynamic circumstances. This is explained as follows: a long operational life cycle will create massive arrays to be analyzed. This will cause delays in decision-making and render the cognitive central coordinator not optimal to use.

On the other hand the progression analysis aims to concentrate on the current scenario only. In progressing a database forward, the historical information will not of use. The other drawback of the progression is that it is not always possible [10].

The main question remains as if to categorize cognitive approach as:

- A projection task determining whether or not some condition will hold after a sequence of actions has been performed starting in some initial state (forecasting condition based on current action).
- A legality task: determining whether a sequence of actions can be performed starting in some initial state (forecasting action based on situation).

References

[1] M. Fichtner, A. Großmann, and M. Thielscher. Intelligent execution monitoring in dynamic environments. Fundamenta Informaticae, 57:371–392, 2003.

[2] F. Bacchus and F. Kabanza. Using temporal logics to express search control knowledge for planning. Artificial Intelligence, 116(1–2): 123–191, 2000.

[3] C. Boutilier, R. Reiter, M. Soutchanski, and S. Thrun. Decision-theoretic, highlevel agent programming in the situation calculus. In Proc. Of the AAAI-00 Conference, pages 355–362, 2000.

[4] H. Levesque and G. Lakemeyer. Handbook of Knowledge Representation Edited by F. van Harmelen, V. Lifschitz and B. Porter. pp. 869-886. 2008

[5] Baker, A. A Simple Solution to the Yale Shooting Problem. In H. L. R. Brachman and R. Reiter (Eds.), Proceedings of the First International Conference on Principles of Knowledge Representation and Reasoning, pp. 11–20. Morgan Kaufmann Publishers, Inc. 1989

[6] J. Pinto. Integrating discrete and continuous change in a logical framework. Computational Intelligence, 14(1), 1997

[7] F. Pirri and M. Romano. A situation-Bayes view of object recognition based on symgenons. In Proc. Of the Third International Cognitive Robotics Workshop, Edmonton, 2002.

[8] R. Reiter. The frame problem in the situation calculus: A simple solution and a completeness result for goal regression. In Artificial Intelligence and Mathematical Theory of Computation: Papers in Honor of John McCarthy,
pages 359–380. Academic Press, New York, 1991

[9] S. Shapiro and M. Pagnucco. Iterated belief change and exogenous actions in the situation calculus. In Proc. Of the ECAI-04 Conference, pages 878–882, 2004.

[10] F. Lin and R. Reiter. How to progress a database. Artificial Intelligence, 92(1–2):131–167, 1997.

34th International Symposium on Automation and Robotics in Construction

ISARC 2017

June 28 – July 1 TAIPEI, TAIWAN

Developing a Low Cost Capacitive ECG via Arduino and Single Board Computer Interfaced with Capacitive Electrodes for Prevention and Security Aspects

J. Güttler and T. Bock

Chair of Building Realization and Robotics,
Technical University of Munich, Germany
E-mail: joerg.guettler@br2.ar.tum.de

Abstract

Coronary heart disease (CHD), and its different variants, is one of the major reasons of increased death rates in industrialized nations. As the beginning of the disease is symptom free, unconsciousness and sudden death, as well as health related distortions can trigger accidents at construction sites any time. Therefore, there is a need for systems, which can detect such silent symptoms, in order to prevent health related accidents on construction site. Nowadays, most common way to identify first symptoms of CHD is the electrocardiography (ECG). Therefore, in this paper a novel approach for the implementation of a contactless capacitive ECG, which can be straightforwardly embedded in a chair is proposed, in order to enable a pervasive user experience (e.g. while steering heavy machines). The capacitive electrodes were manufactured using of-the-shelf components, tested and calibrated. The main requirement during the design and development phase was fabrication of a basic low cost capacitive ECG system with wireless connectivity to a platform that could display the corresponding measurements to the user over a display. Such an approach could pave the way for future pervasive sensor fusion, i.e. interfaced with an automated emergency call gateway, in order to enhance the prevention of such diseases, before it is too late for the users. The achieved measurement quality of this contactless system is depicted by comparing measurements between measurements with direct skin contact and with measurements through cloths. The influence of the wireless data transmission using XBee antennas as a communication medium has been investigated, tested, and evaluated.

Keywords –

Capacitive Electrodes; ECG measurement; XBee; e-Health; contactless measurement

1 Introduction

Coronary heart diseases are in industrial nations one of the main reasons for death, as well as for working incapability. Bad lifestyle and demographic change in industrial nations (e.g. US, UK, Germany, Austria, etc.) increased the secondary diseases of the coronary heart disease, like heart attacks or heart rhythm distortions [1]. At moment, cardiologists use the ECG to identify these diseases for further prevention and treatment. Although the ECG is not very difficult to measure, only physicians in clinics or hospitals can properly perform it.

Nevertheless, the ECG is already used in lifestyle products, e.g. unobtrusively embedded in wristwatches, steering wheels, ergometers etc. However, it is mainly used for pulse measurements by the Einthoven I lead. Nowadays, through the technology of capacitive electrodes, it is even possible to measure the electroencephalogram (EEG), or the ECG through cloths [2, 3]. The automotive industry is one of the first institutions, which implemented the ECG into the driver's seat [4, 5, 6].

Inspired by this approach, the authors aim to install an ECG, by using a low cost capacitive sensor into a chair, and thereby to investigate the potential implementation of this technology e.g. into heavy machines like cranes. This approach can support to increase the security on construction site, as the health condition of an operator of heavy construction machines can be observed.

The basic module is already adequate as a standalone kit for the unobtrusive ECG measurement. The add-on extra modules aim to increase the adaptability into the environment, as well as to offer the opportunity for visual feedback towards the user, i.e. displaying the measurement. This approach will support the user to get informed, and predict his/her health state.

Furthermore, the aspect of low cost is in this prototype considered as well, since this technology can be seen as an upgrade to already very expensive construction machinery. The proposed modularity of the system allows the adaptability according to the needs and budget of user and/or company, i.e. wireless or wired data transfer of the ECG measurements can be implemented. Such an approach could pave the way for future pervasive sensor fusion, i.e. interfaced with an automated emergency call gateway, in order to enhance the prevention of such diseases, before it is too late for the user. Also the security aspects (e.g. as accidence avoidance) can benefit from the proposed system, e.g. when people operates heavy machines (e.g. an operator of construction machinery like a crane, excavator etc.). For example, Haslam, R. A., et al. [7] investigated several accidents on construction site, where proof have been found of accidents, triggered, or at least contributed, by health problems. Additionally, Gyi, D. et al even [8] published a study, showing that there is a consistent and integrated approach necessary to measure safety and health performance for accidents avoidance at contraction site. Even in newer research results the issues for safety and health problems in construction industry is a topic [9], which proofs that up to now there are no sufficient solutions existing. Therefore, the authors proposes a capacitive ECG system that can be used to identify dangerous symptoms, which could distract the operator, or warn from a potential upcoming unconsciousness, and thereby support the accidence avoidance on construction site.

2 Proposed Hardware

The proposed capacitive ECG system consists of three units: The capacitive electrodes, the microcontroller unit (with wired or wireless data transmission), and the display unit (which is comprised by a compact, portable computer).

To implement the system adaptable to the user needs, and to keep it low-cost, the wireless data transmission, as well as the display unit are optional add-ons, which are not absolutely necessary in order to have an unobtrusive capacitive ECG measurement. This allows the user, to minimize the purchasing costs, in case an existing computer can be used for processing (e.g. by a wireless data transmission), storing and displaying the measurement data, or the display unit can be used as an add-on in the driver's cabin, to display directly the measurements. However, in case the complete system is required, the cost is still maintained in a relatively low range, by using a BeagleBone Black as a computer solution.

2.1 Capacitive ECG Electrodes and Microcontroller

As microcontroller, the Arduino Uno R3 has been used [10]. To enable the measurement of physiological signals by Arduino Uno, the e-Health Sensor Platform v1.0 from Cooking Hacks has been used [11]. The proposed system is also operable with normal electrodes, which require direct skin contact and are usually irritating to skin, as well as inconvenient.

To measure capacitive, the electrode surface has to be connected to a non-inverted power amplifier, with a very high input impedance. Thereby, the edge frequency gets shifted down, according to Eq. (1), in order not to get the physiological heart signal filtered (~ 1 Hz). Of course, a large electrode surface will lead to the same result. However, a small electrode surface allows multiple use and an easier unobtrusive implementation into the driver's cabin.

$$f_g = \frac{1}{2\pi RC} \tag{1}$$

As these components of the electrodes are, according to the small size of the components, difficult to build, as well as even more expensive than capacitive electrodes on the market, the authors purchased off-the-self components [12]. To interface these electrodes to a power supply, some additional electric components (LEDs, resistors and capacitors) were required (Figure 1).

In order to measure the ECG signal with capacitive electrodes a ground point is necessary. However, as the ground electrode requires direct skin contact the use of the capacitive electrodes would be doubtful.

Figure 1. Capacitive electrodes including the connection for the power supply and Arduino microcontroller.

Therefore, the authors used a "Driven-Ground-Plane" circuit instead, as proposed in [13], which will lead the signal inverted and amplified on a third electrode surface. Thereby an active noise cancellation on the human body will take place, which finally leads to noise reduction. According to Eq. (2) the amplification of the Driven-Ground-Plane circuit can be calculated by:

$$G = 2 \cdot {R_F}/{R_A} = 2 \cdot \frac{10k\Omega}{220\Omega} = 90.91 \tag{2}$$

For R_A the authors used 220 Ω and for R_F 10 kΩ, which leads to an amplification factor of ~ G = 90.91. The functionality of this factor has been compared with the used amplification factor of (G = 1000) as suggested by [13]. However, there was no difference found in the signal quality, when comparing the two different amplification factors. Therefore, the authors empirically decided to use the amplification factor of G = 90.91.

The LEDs on the capacitive electrodes are used to give a visual feedback about the power supply (e.g. to identify wrong connection or broken components), whereas the capacitors support the stabilization of the power supply of the electrodes. The power supply for the electrodes is directly taken from the Arduino board itself. As the Arduino Uno only provides 5 V and 0 V, but the electrodes need a minimum power supply of +/- 2.4 V up to maximal +/- 4 V [14], the electrodes ground is connected to 2.5 V, in order to create a virtual mass.

For the "Driven-Ground-Plane" circuit board the OPA177GB component has been used, which has similar technical specifications compared to the power amplifiers, which has been used by [13].

2.2 Wireless Communication between the capacitive ECG and Display unit

The proposed system can transfer the measurements by USB connection to a computer, or wirelessly by using XBee Pro Series 2B modules. These modules base on the ZigBee communication protocol [15]. The advantage of using this technology, instead of Wi-Fi or Bluetooth, is the increased security, as ZigBee supports also encryption, as well as it has a higher range than Wi-Fi and Bluetooth, and considerably lower power consumption compared to other wireless standards, which enables it for portable system realizations. In order to interface the XBee module, an Arduino shield is necessary (here the XBee Shield from SparkFun has been used, as shown in Figure 2).

Figure 2. XBee module (top left), XBee Arduino Cape (right), XBee Explorer USB for interfacing with a PC (bottom left).

In Figure 3 the proposed hardware of the capacitive ECG is depicted. In order to enable that the system can be carried easily and safe (e.g. for implementation tests), a casing (visible in Figure 3) has been designed and developed using a 3D printer.

Two XBee modules are required for a communication to be established. To interface the other XBee modules with a computer, e.g. to display the capacitive ECG measurement, or to configure the XBee communication channel parameters [16], the in Figure 2 (bottom left corner) depicted XBee Explorer board from SparkFun has been used. However, in order to develop a low cost capacitive ECG system especially for implementation of small driver's cabins (e.g. crane, excavator etc.), a user display option must be provided.

Figure 3. Hardware of the capacitive ECG, Arduino, e-Health shield & Driven-Ground-Plane board (top), case & assembled boards (bottom left), assembled ECG module including electrodes (bottom right).

For this purpose, the BeagleBone Black has been used [17] with an appropriate touch screen LCD display cape. The XBee module has been interfaced by an XBee cape. In Figure 4 the components of the visual display unit are depicted.

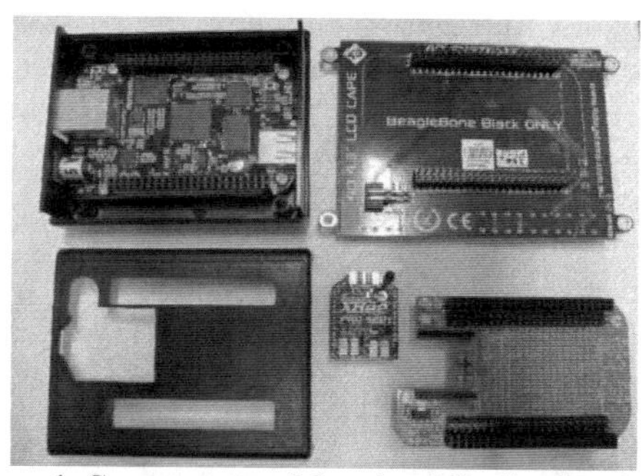

Figure 4. Components of the visual feedback unit, the BealgeBone Black (top left), the BealgeBone Black display cape (top right), the BeagleBone XBee cape (bottom right).

3 Software Development

The developed software is comprised by three levels: 1) Arduino, 2) C++ (using QT 4), and 3) the shell scripts on the BeagleBone Black level to enable the communication ports. The BeagleBone Black supports Debian and Angstrom as operation systems (OS) [18]. The authors used the OS Debian 7.8 – Wheezy.

3.1 BeagleBone Black Shell Script configuration

The XBee module is connected by the XBee cape to the BeagleBone Black, which is directly connected to the expansion GPIO headers. The data exchange takes normally place at the UART2 port (/dev/ttyO2). However, normally only the port UART0 (/dev/ttyO0) is enabled. To enable the port UART2 for XBee, the user must login as root and enable this port by a terminal, using the following shell command [19]:

```
echo BB-UART2 > /sys/devices/bone_capemgr.*/slots
```

To automate this process a shell script will execute the BB-UART2 command, which will be triggered by a service script, which executes as soon as the BeagleBone Black boots. Furthermore, a desktop icon has been made using the Linux script language, in order to allow an intuitive and quick start of the display program on the BeagleBone Black LCD display. Thereby, the display can serve additional tasks (e.g. control) in the future, next to the ECG displaying option.

3.2 Arduino Microcontroller Program Flow

On the Arduino level, the program is working as shown in Figure 5. For reading the ECG signal from the capacitive electrodes (visible in Figure 1), the serial port 0 of the Arduino Uno board has been used. For transmission of the XBee module serial port 2 and port 3 are used. As shown in Figure 5 the data are recorded and formatted, according to the format, which the display unit requires, before being sent. The XBee module on the Arduino is set up into the “Application Programming Interface” (API) mode [15]. The authors decided this because of two reasons:

1. The data transfer becomes more robust since in API mode the data must be send in a specific frame setting. If the frame setting contains mistakes, the message package (frame) receiving XBee module will ignore the message. Tests with the transparent (AT) mode of the XBee modules, where the data gets sent in ASCII form (serial port style), showed that transmission mistakes sometimes (although only rarely) disturb the displaying program.
2. The XBee modules, which uses the API mode can send frames to AT mode XBee modules. XBee modules in the AT mode translate a received message (from API and AT) immediately from byte to ASCI, before forwarding to the interfacing port. This allows to use the XBee module in AT mode equal, as a USB interface on the programming level. However, XBee modules in the AT mode cannot send messages to XBee modules in the API mode [15].

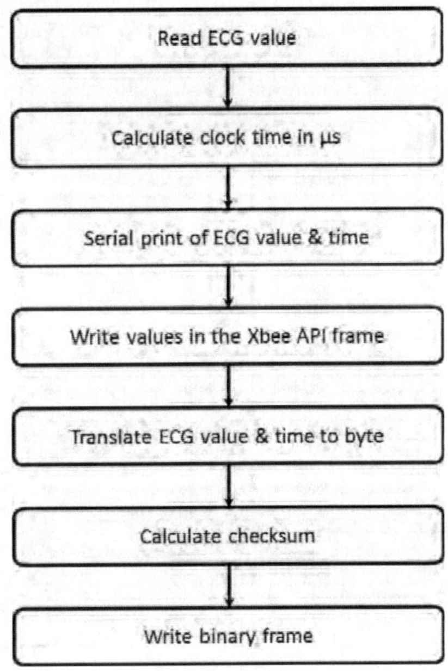

Figure 5. Proposed Arduino algorithm for measurment acquisition and transmission.

The frame has been hard-coded into the Arduino using hexadecimal values, which describe the destination address and message kind. Thereby no additional libraries are necessary. Only the message part and the checksum are updated with each new value captured by the electrodes. These measurement values have to be transformed to byte values, before they can be written into the frame. A necessary storage of 19 bytes has been calculated, so that the XBee module can permanently run up to ~31 years (because the time value will continuously increase, up to the point where the reserved storage is depleted), as the first 17 bytes of the frame are used for addressing and formatting, and therefore can be hard-coded.

Only if more storage for the message will be given, the length of the frame must be updated (which consists out of the quantity of all bytes after the first 3 bytes, and excluding the byte for the checksum). The checksum is calculated by summing the single frames together and final subtracted from the hex value FF (= 255 decimal) [15].

3.3 Displaying the Information on the LCD Screen module

The program for displaying the measured data is simultaneously storing them on the hard disc (in case of the BeagleBone Black on an eMMC 4GB chip [17]). In order to enable the program to store all measured samples, and displaying in real-time, the program is operating as depicted in Figure 6. While the main program loop is busy with displaying data (using QCustomPlot [20]), a sub thread, trigged by the class "Read Value", (using QExtSerialPort [21]) is reading the data either from USB or the XBee module and storing.

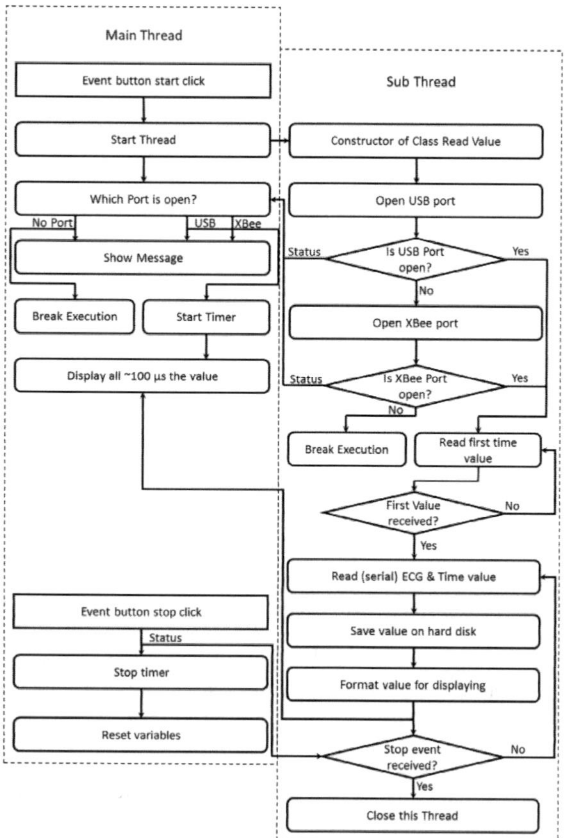

Figure 6. Proposed flow diagram of the Display unit.

The program checks always if a USB connection is existing. If the USB connection is existing, the data transfer by wire is preferred. If the program detects that there is no USB connection available, it is checking the UART2 (/dev/ttyO2) port. Only in the case the port is not occupied by another process (e.g. not enabled, or already busy), the program returns an error message. Otherwise, the program is waiting for incoming data from the XBee modules.

Thereby, the displaying of the program breaks and restarts automatically, according to the data transfer, until the user requests the program termination. As the XBee module on the BeagleBone Black is not thought to send information, only to receive information, this module has been set to AT mode. This allows to treat the incoming XBee data in the same way as a USB serial port. Because the BeagleBone Black provides only one USB slot, the serial port is also hard-coded for the USB transmission. Furthermore, the BeagleBone XBee Cape is only supporting one XBee module, whereas also here the UART2 (/dev/ttyO2) port is hard-coded. The designed Graphical User Interface (GUI) provides to the user only three buttons to control the program: start, stop, and close (see Figure 8).

Of course, if a USB wire is used for the data transfer between the microcontroller and the display unit, the possibility of Wi-Fi is disabled, because there is no further USB slot left for the Wi-Fi module.

4 Software Development

In Figure 7 the measurement results using the capacitive electrodes (shown in Figure 1) with direct skin contact and through cloths are shown.

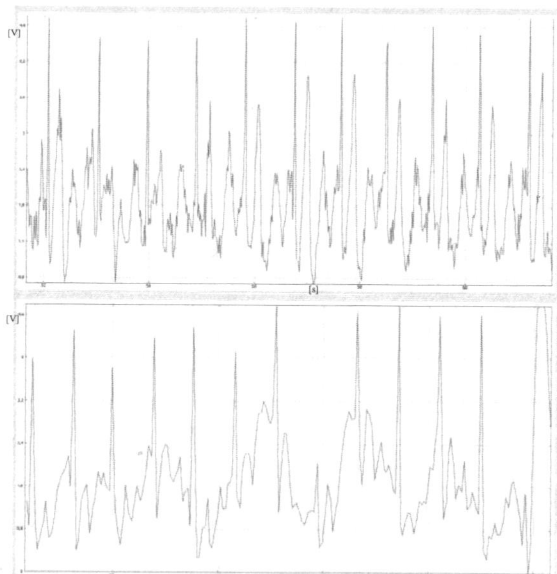

Figure 7. ECG measurement with direct skin contact at the hands (top), ECG measurement through T-shirt on the chest (bottom).

Hereby it can be seen, that both measurement methods (with and without direct skin contact) can capture the ECG signal, however, the capacitive measurement is more disturbed by noise.

The proposed system (of a capacitive ECG measurement system) has been used for test implementation in a chair, as shown in Figure 8, in order to verify that the system receive reasonable values.

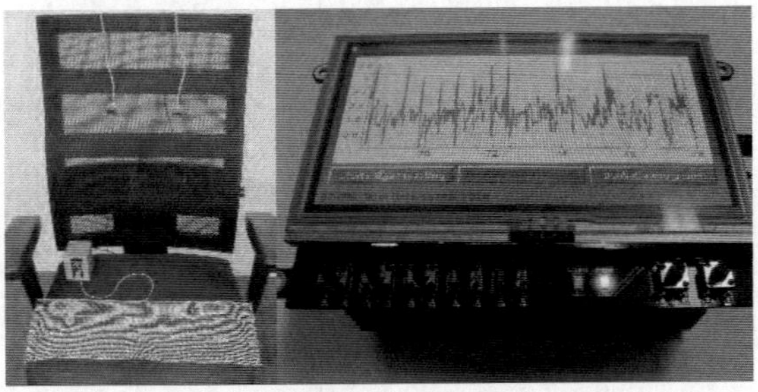

Figure 8. The proposed capacitive ECG for a test scenario embedded into a chair (left), the display unit receiving wireless a measurement by XBee module (right).

As described in [4, 22] the authors can confirm, that pressure (between body and electrode), as well as the thickness of the cloths material have an influence on the signal quality. Furthermore, because of the high sensitivity of the capacitive electrodes, noise and artifacts, caused by the surrounding electronic devices and user movements, the measurement can be disturb.

Whereas it is necessary to mention that the electrodes are not shielded, which increase the impact of noise. However, a larger impact regarding the ECG signal quality was found on the settings of the XBee modules. It was noted that the XBee modules could decrease the signal quality, if they are using broadcast transmission to all XBee modules in the network. The number of XBee modules in the network is thereby irrelevant. This finding resulted from an XBee network with five modules, where 1 × Coordinator (API), 2 × Routers (API), 1 × End Device have been used in API and 1 × in AT mode. Data throughput had the same deteriorated performance (when broadcasting messages) compared with an XBee network, where only 1 × XBee Coordinator and 1 × XBee End Device communicated. The broadcast data transfer is too slow, and thereby already hurting the Nyquist-Shannon sampling theorem, which leads to wrong value displaying and signal loss. Using unicast frames (means frame transmission to specifically addressed XBee modules [15]) the transmitted signal quality improves, because of a higher data throughput. However, the USB Data throughput still leads to most robust signal transmission. The transmission quality between XBee API mode and XBee AT mode was found nearly similar, as in Figure 9 shown.

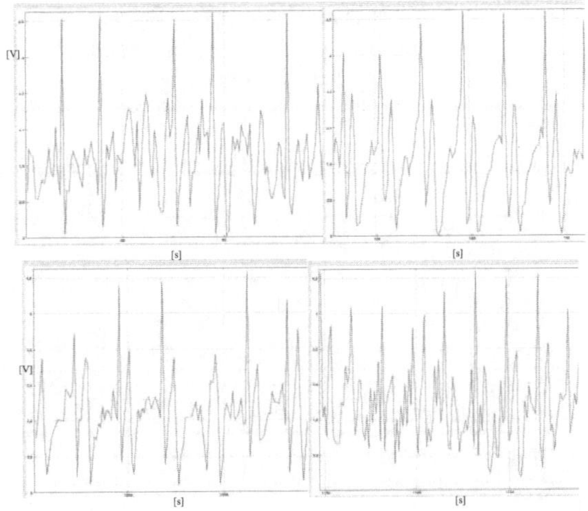

Figure 9. Signal displaying by data transfer using XBee module, measured by direct skin contact (left), measured through cloth (T-shirt) (right), XBee AT mode (top), XBee API mode (bottom).

5 Conclusion and Discussion

The authors proposed the development of a novel contactless capacitive ECG system, including wireless data transfer and a mobile embedded single board computer as a display unit, which allows a modular, unobtrusive, and portable system realization (see Figure 8).

The system has been tested with direct skin contact and without skin contact (through clothes), as well as with USB, and wireless communication (via XBee Pro Series 2B). In all proposed scenarios, the ECG signal was successfully captured and displayed, as shown in Figure 7 and Figure 9. The overall system is designed and developed in a modular way, whereas the prototyping costs (summarized in Table 1) are kept to a minimum.

Through the compact, handheld size of all three hardware parts (electrodes, Arduino board, and BeagleBone Black), it is possible to straightforwardly to embed the proposed system unobtrusive into a driver's cabin.

The proposed wireless data transfer by XBee allows a more robust implementation, since it does not require any preinstalled Wi-Fi infrastructure. The Arduino microcontroller and display unit can also use normal electrodes, instead of capacitive electrodes. However, the advantage of the capacitive electrodes is that they are not irritating the skin, and they stay functional, even if the operator is wearing cloths. Of course, some disadvantages have been identified as described in [22]. E.g., ambient noise disturbs the acquired signal, or user movements can cause strong artefacts.

Table 1. Costs of the proposed ECG system including wireless transmission and low cost computer for displaying.

Product	Price (€)/ Unit	Necessary Units	Cost(€)
Arduino Uno R3	24	1	24
e-Health Sensor Shield V1.0	75	1	75
PLS-PS25203 (Capacitive Electrodes)	18	2	36
Electronic components			20
Costs for wired Capacitive ECG			**155**
SparkFun XBee Shield	15	1	15
XBee PRO 60mW Series 2	38	1	38
Costs for wireless capacitive ECG			**208**
BeagleBone Black	55	1	55
XBee Cape	13	1	13
4D 4.3" LCD CAPE	100	1	100
XBee PRO 60mW Series 2	38	1	38
Total Costs for wireless capacitive ECG including mobile display unit			**414**

In future, to reduce these unwanted disturbances, the entire system must be proper shielded. This very first prototype shows promising results concerning the realization of a low cost, portable, contactless ECG measurement system solution for construction security purposes. Once the improvements are implemented in order to reduce the noise by shielding, as mentioned in [23], this system can be built to a 3 up to 6 lead ECG, in order to enable unobtrusive ECG measurements, according to Einthoven (I, II, III), and Goldberger (aVR, aVL, aVF) [24]. Also for this approach, appropriate Arduino shields are already available on the market [25], which would even decrease the developing cost, if only 1 lead is used, as it is done with the proposed system.

Through the possibility, given by this technology, to implement the ECG measurement unobtrusive into a driver's cabin, the operator health condition can be observed, either by himself, or by a security central. Thereby, it can be assured that the operator is present, mentally not stressed (e.g. because of increased heart rhythm), and healthy. By observing for harbinger e.g. of heart attacks, this approach would allow to interfere before accidents happens, e.g. caused by unconsciousness on construction site.

References

[1] R. Kones, "Primary prevention of coronary heart disease: integration of new data, evolving views, revised goals, and role of rosuvastatin in management. A comprehensive survey. A comprehensive survey.," *Drug Design, Development and Therapy,* vol. 5, no. 1, pp. 325 - 380, 2011.

[2] Y. Chi and G. Cauwenberghs, "Wireless non-contact EEG/ECG electrodes for body sensor networks," in *Body Sensor Networks (BSN)*, 2010.

[3] Y. M. Chi, Y. T. Wang, Y. Wang, C. Maier, T. P. Jung and G. Cauwenberghs, "Dry and noncontact EEG sensors for mobile brain-computer interfaces.," *Neural Systems and Rehabilitation Engineering, IEEE Transactions,* vol. 20, no. 2, pp. 228-235, 2012.

[4] S. Leonhardt and A. Aleksandrowicz, "Non-Contact ECG Monitoring for Automotive Application," in *Proceedings of the 5th Inernational Workshop on Wearable and Implantable Body Sensor Networks*, Hong Kong, 2008.

[5] M. Walter, B. Eilebrecht, T. Wartzek and S. Leonhardt, "The smart car seat: personalized monitoring of vital signs in automotive applications," *Personal and Ubiquitous Computing,* vol. 15, no. 7, pp. 707-715, 2011.

[6] T. Wartzek, B. Eilebrecht, B. Lem, J. Lindner, H. J. Leonhardt and M. Walter, "ECG on the road: robust and unobtrusive estimation of heart rate," *IEEE Transactions on Biomedical Engineering,* vol. 58, no. 11, pp. 3112-3120, October 2011.

[7] R. A. Haslam, S. A. Hide, A. G. Gibb, D. E. Gyi, T. Pavitt, S. Atkinson and A. R. Duff, "Contributing factors in construction accidents.," *Applied ergonomics,* vol. 36, no. 4, pp. 401-415, 2005.

[8] D. E. Gyi, A. G. Gibb and R. A. Haslam, "The quality of accident and health data in the construction industry: Interviews with senior managers.," *Construction Management & Economics,* vol. 17, no. 2, pp. 197-204, 1999.

[9] K. C. Goh, H. H. Goh, M. F. Omar, T. C. Toh and A. A. M. Zin, "Accidents Preventive Practice for High-Rise Construction," *MATEC Web of Conferences,* vol. 47, 2016.

[10] "Arduino UNO & Genuino," [Online]. Available: https://www.arduino.cc/en/Main/ArduinoBoardUno. [Accessed 08 03 2017].

[11] Cooking hacks, "e-Health Sensor Platform V1.0 for Arduino and Raspberry Pi [Biometric / Medical Applications," [Online]. Available: https://www.cooking-hacks.com/documentation/tutorials/ehealth-v1-biometric-sensor-platform-arduino-raspberry-pi-medical. [Accessed 08 03 2017].

[12] Mouser Electronics, Plessey Semiconductors, [Online]. Available: http://www.mouser.de/ProductDetail/Plessey-Semiconductors/PS25203B/?qs=MI2zuikz5V%252btfhM%252b8WEB1Q==. [Accessed 08 03 2017].

[13] A. Aleksandrowicz, M. Walter and S. Leonhard, "Ein kabelfreies, kapazitiv gekoppeltes EKG-Messsystem / Wireless ECG measurement system with capacitive coupling," *Biomedizinische Technik,* vol. 52, no. 2, pp. 185-192, 2007.

[14] Plessey semiconductors, *PS25203B EPIC Ultra High Impedance ECG Sensor Advance Information,* Data Sheet 291499 issue 3.

[15] Digi International Inc., "DIGI XBEE ECOSYSTEM™," [Online]. Available: http://www.digi.com/lp/xbee/. [Accessed 08 03 2017].

[16] *Configuration, X. C. T. U.,* Test Utility Software: User's Guide: Digi International Inc., 2008.

[17] G. Coley, *BeagleBone Black System Reference Manual-Revision C.1,* 2014.

[18] Bealgeboard.org, "BeagleBoard.org Latest Firmware Images," [Online]. Available: http://beagleboard.org/latest-images. [Accessed 08 03 2017].

[19] LOGIC supply, *XBEE CAPE MANUAL - Beagle Bone Black XBee Prototyping Cape, Revision 1.0,* 2014.

[20] "QCustomPlot," [Online]. Available: http://www.qcustomplot.com/. [Accessed 08 03 2017].

[21] "Qextserialport," [Online]. Available: https://code.google.com/archive/p/qextserialport/. [Accessed 08 03 2017].

[22] K. K. Kim, Y. K. Lim and K. S. Park, "Common mode noise cancellation for electrically non-contact ECG measurement system on a chair," in *Engineering in Medicine and Biology Society, 2005. IEEE-EMBS 2005. 27th Annual International Conference IEEE*, 2006.

[23] T. Komensky, M. Jurcisin, K. Ruman and O. Kovac, "Ultra-wearable capacitive coupled and common electrode-free," in *Annual International Conference of the IEEE Engineering in Medicine and Biology Society*, San Diego, 2012.

[24] H. M. Piper, "Herzerregung," in *Physiologie des Menschen mit Pathophysiologie*, Berlin Heidelberg, Springer, 2010, pp. 517-538.

[25] Olimex, "Shield-EKG-EMG," [Online]. Available: https://www.olimex.com/Products/Duino/Shields/SHIELD-EKG-EMG/. [Accessed 22 03 2017].

Improving Indoor Location Tracking Quality for Construction and Facility Management

Matthias Neges [a], Mario Wolf [a], Markus Propach [a], Jochen Teizer [b] and Michael Abramovici [a]

[a] Chair for IT in Mechanical Engineering, Ruhr University of Bochum, Germany
[a] Chair for Computing in Engineering, Ruhr University of Bochum, Germany
E-mail: jochen.teizer@rub.de, mario.wolf@itm.rub.de, markus@propach.org, matthias.neges@itm.rub.de, michael.abramovici@itm.rub.de

Abstract –

Real-time location tracking systems (RTLS) for personnel and machinery in outdoor civil engineering environments quite often use commercially-available Global Navigation Satellite System (GNSS) technology. Although the GNSS is an important approach in outdoor positioning and logistics coordination, their signals are not able to penetrate buildings due to their signal strength. Despite some recent advances in research, reliable indoor navigation remains an unsolved problem. This work deals with a detailed study of the methods and approaches of indoor location tracking. The focus lies on systems based on Bluetooth Low Energy (BLE) technology that meet the specific requirements of construction site and facility management.

The authors develop a prototypical application with which measurements are taken using different BLE hardware. The experiments show that location tracking is only applicable to a limited extent using BLE technology and path loss model. There were great differences in the behavior of different devices observed since the environment greatly influences the signal transmission. Proposed is an alternative, holistic system for location tracking using BLE. It uses a systematic classification of the work space by positioning the BLE beacons according to the a-priori known spatial building structure from a Building Information Model (BIM). By the relative observation of the received signal strengths of the individual beacons spread on a building floor, the calibration of the receivers is obsolete so that several different or alternative device types can be used together at the same time.

Keywords –

Building Information Modelling, Facility Maintenance and Management, Human-Machine Interface, Localization, Sensing and Monitoring, Tracking, Visualization, Web-based Systems

1 Introduction

Although commercial GNSS like GPS or GLONASS is an important prerequisite for outdoor applications such as automated control of machines or an approach for logistics coordination, its signals are not able to penetrate buildings and thus its application not suitable for indoor localization. Despite recent advances in research with varying results, a reliable, inexpensive and practicable indoor localization method remains an unsolved problem in construction as well as in facility management.

This work deals with a detailed study of the methods and approaches for indoor location tracking. The focus lies on systems based on Bluetooth Low Energy (BLE) technology that meet the specific requirements of construction sites and facility management. After selecting and evaluating possible approaches, the authors introduce a prototypical application with which measurements are taken using different BLE hardware. By means of these measurements, the properties of the components are critically evaluated. Based on the results, a holistic concept is developed to meet the requirements of the chosen application. The concept contains the algorithm, utilized hardware and a cloud platform configuration for the management of the tracking data. As an important aspect in the selection of the localization algorithm, the so-called fingerprinting method will be avoided, since it is to time-intensive and therefore cannot be used in a practical application.

2 State of the Art

2.1 Real-time locating systems

Localization in RTLS can be conducted in various ways. There are different approaches used for different applications, although essentially three different approaches can be distinguished.

The most common approach is based on the measurement of distances between several base stations (with a known positions) and the device to be located. The exact position can be determined by means of multilateration or hyperbolic navigation of at least three (2D plane) or four (3D space) distance measurements. The distance measurement takes place by means of the signal transit time measurement (applied with GNSS) or by signal strength measurement and the principle of free space loss. The latter is applied in many BLE based systems [1-3]. However, the signal transit time measurement is less suitable for indoor location tracking due to multipath reception, which results from reflections on objects.

Another approach is the fingerprinting method, which is also based on the measurement of the signal strength values. However, no physical relationship between signal strength and distance is evaluated. Instead, a map is created which links the manually detected signal strengths to the respective position in the plane. The resulting matrix of signal strengths of a certain point on the plane is called fingerprint. This way, interference and damping of individual signals no longer influence the localization and relatively high measuring accuracies and repeatability can be achieved [4]. However, the creation of a fingerprint map is very time intensive and therefore not practical for larger or very dynamic work spaces.

Inertial navigation systems (INS) are the third method for localization. With inertial sensors, such as rotation rate and acceleration sensors, the direction of movement of a system can be determined very precisely for short periods of time. In preliminary work, it has been possible to develop pedometers with high accuracy, such as in [5]. However, an initial location must be assumed through other means and measurement errors continuously add up with each step. This approach is well suited when combined with a previous RTLS in sensor data fusion using the Extended Kalman Filter (EKF) or particle filter.

2.2 Bluetooth Low Energy (BLE) and beacons

BLE was introduced in December 2009 as an extension of the Bluetooth 4.0 specification. Like conventional Bluetooth, BLE also works in the 2.4 GHz range of the ISM band. It is designed for significantly lower energy consumption and cost-effective integration. Reduced transmission power and smaller data packets or shorter data transfers contribute to low power consumption, so that button cells can be used to enable BLE connectivity of a device for months or even years. BLE is supported by popular operating systems such as Windows, Mac OS, iOS and Android given that the system has a Bluetooth 4.0 compatible hardware.

BLE devices can send short advertising events on three separate channels. These make the device known as a participant in the radio traffic. The device then waits for a connection request. BLE beacons are devices that rely on this basic functionality. The advertising event allows the broadcasting of short data sets which can contain, i.e. identification numbers, sensor data, or short web addresses. Receivers such as Smart Devices can receive the advertising events.

Common standards for BLE Beacon communication protocols are Apple's iBeacon and Google's Eddystone. In the context of location tracking, the signal strength can be determined using the Received Signal Strength Indicator (RSSI). The theoretical distance to the device can be approximated with the path loss model.

Figure 1 shows two exemplary BLE beacons. The nRF51822 is an evaluation platform from Nordic Semiconductor with minimal dimensions, the Accent Systems iBKS 105 (shown with its case opened) is a commercially available BLE beacon.

Figure 1. left to right: 50 Euro-Cent, Nordic Semiconductor nRF51822-Beacon, Accent Systems iBKS 105

2.3 Building Information Modeling (BIM) und lean management in construction

Building Information Modeling (BIM) includes methods and processes for creating and managing 3D virtual building models. Information exchange associated to a parametric model forms the core of the method. BIM acts as a central database, which contains semantic information and parameters as well as geometric data such as 3D models. BIM can be used for data exchange in the project lifecycle, including design, construction planning, construction work and facility management [6-7].

The concept of the Lean Construction Management (LCM) is the adaptation of the Toyota production system to the requirements of the construction industry. It is a continuous process for eliminating waste (in processes and resources) which is achieved by the continuous optimization of single business processes. Lean also consists of four basic principles. The flow principle, the just-in-time principle, the flexible resource control and the systematic quality control [8-9].

The application of LCM requires, above everything else, solid data to improve the business process. However, actual project data is typically not recorded in construction through digital means. Monitoring the progress of a project through technology enables BIM to become a information-driven model based on actual data.

3 Aims and Methodology

In this paper, the authors develop a holistic concept for indoor location tracking using BLE technology. The concept aims applications in the construction industry and facility management. The design of the localization algorithm excluded the so-called fingerprinting method, since it is too time-intensive and therefore cannot be used for practical purposes in dynamic work environments. For future use in the field of LCM an interface integration to BIM is proposed.

3.1 Use-case

As with every software development, it is useful to define the deployment scenario precisely for the envisioned location tracking system. In this case, the authors chose to classify the interiors of buildings into the three different classes "room", "corridor", and "atrium". A room is designed for people to stay or work in. It has a floor area of at least 10 m^2, a door and usually windows. A corridor, on the other hand, is a connecting space in a building, from which individual rooms emanate. It is characterized by long narrow passages. In contrast, an atrium is a larger and usually also high-roofed, open space. Corridor and atrium can be directly connected to one another. The examination is initially limited to elements of the class "room", as it is the most basic element. Later in the paper, the recorded preliminary results will the transferred to the other classes.

3.2 Selected hardware and software

For the selection of hardware and software, one must first and foremost decide which technology approach of localization to use. Generally, a distinction can be made between the systems which use wireless technology or inertial sensors, such as acceleration sensors, gyroscopes, or magnetometers.

The inertial sensor systems are distinguished by a high measurement accuracy. However, measurement errors are add up quickly. In addition, they need a known initialization point. An exception is the magnetometer. The method of "Magnetic Positioning" developed by the company *IndoorAtlas* is based on a fingerprinting method, which makes use of the magnetic interference of steel structures in buildings to determine the position [10]. The fingerprinting process is cloud-based and requires, in addition to a one-time mapping and a permanent Internet connection, practically no other infrastructure. However, since a fingerprinting process is to be avoided in this work and a permanent Internet connection cannot be guaranteed at a construction site, this method is not considered further.

The ultra-broadband wireless technology (UWB) offers probably the highest accuracy in indoor location tracking. UWB operates in a frequency range from 3.1 to 10.6 GHz. Although tests in construction have shown suitable applications in outdoor environments [11], this technology can only be used in one-room or wide-area-of-view solutions due to its strong damping and reflection properties [12]. Because of the relatively small market share of the technology, the costs of transmitters and receivers are very high.

Wireless technologies obviously include wireless LAN and BLE. Both technologies operate in the 2.4 GHz frequency range. Either RSSI values or using a characteristic curve to estimate the distance between sender and receiver can be used for the fingerprinting method. Both technologies are supported by many embedded systems and are widely used in PCs and smart devices. The use of wireless LAN requires an infrastructure of access points. In addition, due to a relatively large power consumption, connection to the power grid is required. Both factors are not to be expected on a construction site. In contrast, BLE beacons are very energy efficient and can operate on batteries for several months or years. In addition, they are also relatively inexpensive and do not require additional infrastructure. Therefore, a BLE-based technology approach is chosen in this paper.

BLE beacons are used as transmitting units while smart device such as a tablet or a smartphone will be the receiver. Smart devices, in addition to BLE support, also have internet connectivity (at some time during a work day) and can thus receive ambient data from an external cloud platform or upload tracking data.

3.3 Preliminary examination

In preparation for this work, several existing methods for indoor navigation were evaluated. The authors tried to reproduce positive results, which are mostly based on the path loss model and multilateration. It was evident that most results were not reliably reproducible. In addition to the affirming work, there are also some critical considerations on the theme of the path loss model.

A major criticism that has often been mentioned, are large measurement fluctuations due to interferences caused by diffraction or reflections on walls or due to interference with, for example, Wi-Fi. The measured signal strength, which forms the basis of the path loss model, varies massively when using different combinations of both beacons and smart devices, which boils down to different microchips and antenna combinations. In addition, the measured signal strength is also influenced by the parameters of air temperature, humidity, and the angle between transmitter and receiver antennas. These factors induce separate characteristic curves between signal strength and distance. These must be determined for each beacon used. For every specific model type and its specific spatial position, the combination with the respectively used receiving device and its possible antenna orientations requires careful evaluation. The characteristic curve must also be constantly adjusted to environmental influences such as air temperature and humidity.

The authors' own measurements have yielded characteristic curves which were relatively flat when the distance was greater than one meter from the concerned BLE beacon. Combined with the existing ambient signal noise, this results in large fluctuations in the distance measurement.

In some studies [1,3,13,14], these and other sources of error have been ignored. Many other systems try to limit the noise by means of filters, such as particle filters or Kalman filters [14-15]. Furthermore, sensor data fusion with inertial sensors is often used to increase accuracy. Commercial systems are regularly limited to Apple devices, since they offer limited variation in installed hardware, ultimately offering similar characteristics.

As a summary, preliminary investigations showed that the path loss model on its own offers insufficient precision. Significant efforts are necessary to find a suitable approach.

3.4 Methodology

The problems of the path loss model described in the previous section lets one discover the actual potential of BLE technology for location tracking and examine its strengths and weaknesses more closely. For this purpose, measurements are to be carried out in a realistic indoor test environment using the fingerprinting method as a reference. These fingerprints will not form a base for future tracking, but to evaluate the findings in tracking quality by other means.

For comparison purposes, the outdoor measurement scenario outdoor should be carried out again in a free plane. It serves to distinguish between the influence of reflection, diffraction, and attenuation by walls and the influence between the transmitter and receiver. The concept for developing a location tracking algorithm is not subject to the same restrictions of the path loss mode.

4 Experiments and Results

4.1 Design and implementation

The first measurements were taken in the laboratory of the Chair for IT in Mechanical Engineering at the Ruhr-University Bochum. The laboratory is located in the basement of an academic building and equipped with various demonstrators, including conveyor belts, test assemblies, partition walls, and computer workstations. The laboratory environment may serve well as a realistic test environment for indoor tracking in construction and facility management.

Part of the laboratory consist of a large room. It has dimensions of 15 x 5 m and a ceiling height of more than 3 m. The upper part of the ceiling is equipped with an open ventilation system so that the room has only a useable height of about 2.4 m. For the experiment, six beacons are distributed uniformly in space at a height of 2.1 m. A sketch of the test bed environment including its fixtures is shown in Figure 2. It shows the exact position of the beacons with their indicators in the laboratory.

For the fingerprinting process, the space is divided into 75 individual squares, each having a surface area of 1 m². A unique "fingerprint" of the received signal strengths of the individual beacons is then recorded for each individual area over the course of 30 seconds. For this purpose, a smart device is positioned in the respective quadrant using a tripod.

The smart device is then positioned centered and upright at a height of about 1.2 m in the quadrant. The orientation of the smart device remains the same throughout the entire measurement. Utilizing a self-developed app installed on the smart device, all received RSSI values of all beacons in range are recorded and stored in a measurement period of 30 seconds. The evaluation of the measurements is carried out later with a self-written MATLAB script on a PC. Figure 2 shows the division of the room in its subareas and the orientation of the smart device in a quadrant.

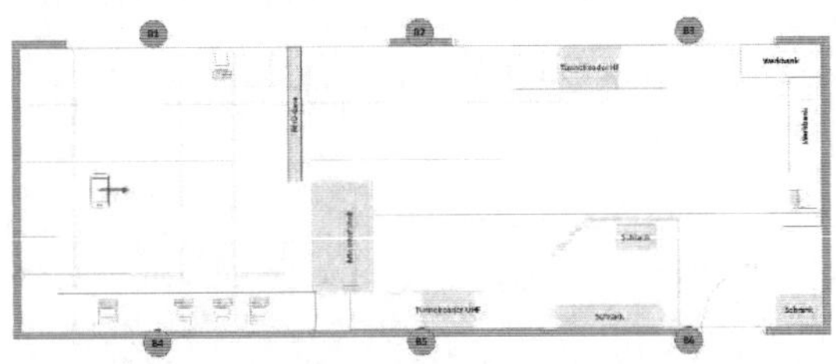

Figure 2. Floor plan of the test laboratory. Blue circles indicate the position of BLE beacons. Red spaces show objects with heights over 1.5 m.

The dimensions of the laboratory scenario are then replicated in an open field. For this purpose, a rectangle with the same dimensions of the laboratory test room was staked out on a soccer field (with an artificial turf) at the Ruhr University Bochum's campus. The beacons were positioned on professional tripods at pre-assigned positions. The positioning of the tripod with the smart device were realized with a measuring tape.

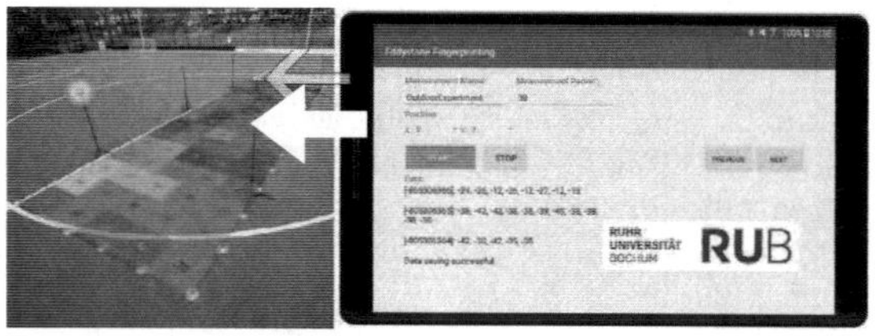

Figure 3. Android app

4.2 System architecture

For the experiment, BLE beacons from Accent Systems with the model designation iBKS 105 were used. These support both the iBeacon and Eddystone protocols, whereas the Eddystone protocol was used in the research. The beacons have been configured to send an advertising signal with a signal strength of +4 dBm, with an interval of 300 ms. These values represent the highest transmission power and the shortest possible time interval available on these specific beacons. This configuration is chosen so that possible transmission errors have as little influence on the measurement as possible. The transmitting advertising signal contains, among other things, a unique identification number (UID), whereby the recorded signal strength can be unambiguously assigned to the respective beacon.

The measurements in the experiments are carried out with two smart devices. The first is the Nexus 5X (powered by Google, OEM is LG) and the second is the Xperia Z4 from Sony. Both devices are equipped with the Android 6.0 operating system. Android provides support for BLE from version 4.3 on, but in these outdated versions it lacks performance and reliability. Since version 5.0, the BLE API has been reworked from scratch.

4.3 Software implementation

To systematically record measurement data, the authors implemented a mobile app specifically for the experiment. The app offers the possibility to name the experiment and to define the measuring period for each quadrant. A 2D coordinate system identifies the currently selected field of the measurement. A text field displays the current measurement data for a brief overview. After the measurement run finishes, the measurement data is stored as a text file. The process is repeated for each quadrant in the field to create a robust fingerprint.

A specifically programmed Matlab-Script evaluates the stored data after completion. The script reads the measured data automatically from the text file and stores it in a nested matrix structure. Algorithmically, the median value for each beacon in each quadrant is determined iteratively from the RSSI values measured over the measurement period. The specific RSSI mean values are then visualized graphically as heat maps for the individual beacons and can thus be evaluated objectively.

4.4 Evaluation of measured results

For the systematic evaluation of the experiment, the results for the outdoor test are analyzed first. By performing the experiment in the open field, the influence of interfering signals, as well as diffraction and reflection on objects or walls is at a negligible level. The reflection of the ground cannot be excluded. The measurement results show, according to the path loss model, a steady decrease in the signal strength with an increase in the transmitting distance.

However, the measured signal strength is also strongly dependent on the alignment angle between BLE beacon and the smart device. Since the alignment of the smart device in each quadrant remains the same, the angle between smart device and beacon changes for each quadrant. From the results it appears as if there is a characteristic drop in the signal strength from a given orientation angle. This observation is evident when comparing Figures 4b and 5b.

Evidently this angle is specific for each device/beacon combination as the comparison between the used smartphone and the tablet show (cp. Figures 4b and 5b). The Sony Xperia Z4 tablet is much more sensitive to the alignment angle. This becomes evident at very close measurement ranges of the tablet to the beacon. A large drop in the signal strength is observed. This is due to the relatively high angle of inclination between beacon and tablet. It is also noticeable that the signal strength drop for Beacon 4, 5 and 6 is severe. A possible explanation is that the antennas in the smart devices were installed on the right side, which therefore also affects the alignment angle between the transmission devices.

The test results in the indoor test show a similar effect, except for the last-mentioned phenomenon. In addition to the decrease in the signal strength over the transmitting distance, the characteristic drop in the signal strength from a specific alignment angle can also be observed here. In addition, some effects are attributable to the spatial design and geometry of the laboratory room. In general, damping effects by large superstructures can be detected. Contrary to the observation made in the outdoor scene, the signal strengths for beacons 4, 5, and 6 are significantly stronger than those of beacons 1, 2, and 3. This is probably due to a highly reflective surface at the ceiling of the laboratory room. A channel for cables made out of metal runs along the long side of the laboratory wall. A variety of materials mounted at or inside the reinforced concrete walls of the laboratory may contribute to this effect. In addition, relatively strong signal strengths were measured within the metal partition walls placed in the room. This again illustrates the reflection characteristics of metallic objects or walls.

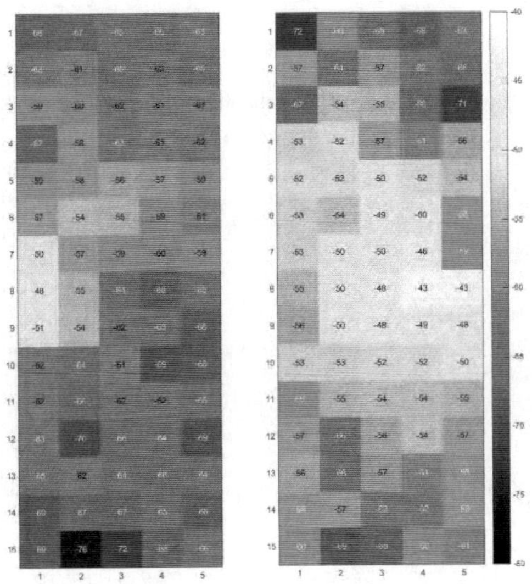

(a) Beacon 5 (b) Beacon 2

Figure 4. Selected heat maps of RSSI-median-values from the **outdoor experiment** with **Nexus 5x** Device

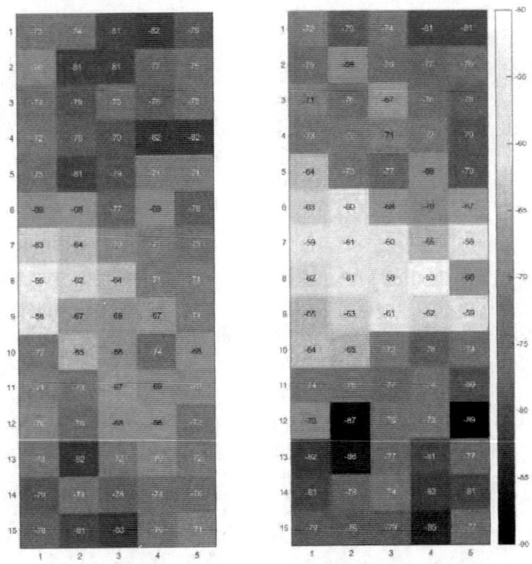

(a) Beacon 5 (b) Beacon 2

Figure 5. Selected heatmaps of RSSI-median-values from the **outdoor experiment** with **Xperia Z4** Device

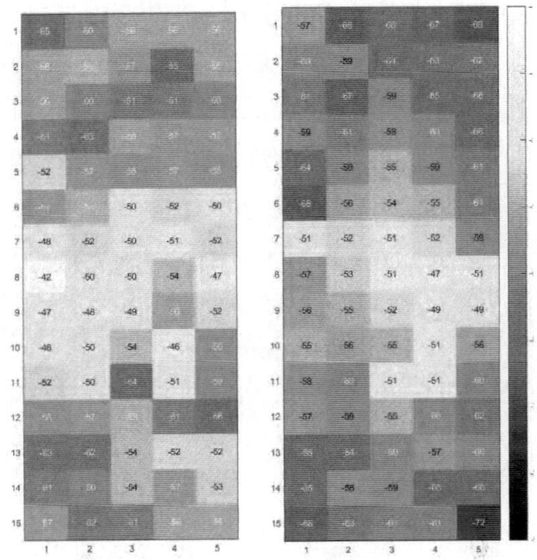

(a) Beacon 5 (b) Beacon 2

Figure 6. Selected heatmaps of RSSI-median-values from the **indoor experiment** with **Nexus 5x** Device

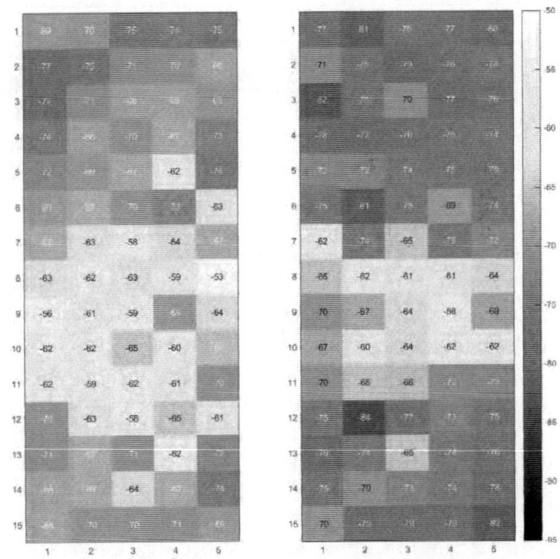

(a) Beacon 5 (b) Beacon 2

Figure 7. Selected heatmaps of RSSI-median-values from the **indoor experiment** with **Xperia Z4** Device

5 Resulting Concept

Accuracy of existing location tracking methodologies, such as with GPS or a UWB systems, is favorable for construction applications, however focus on outdoor or wide area field-of-view applications and the necessary calibration and adaptation measures to the different system variants are, in practice, far too complex and require too much effort [16-17].

As an alternative, this work evaluated through experimental results with BLE that its signal propagation depends on two factors: the spatial environment and on the measurement devices applied.

This work found, that an alternative positioning method can be realized based on the BLE technology. Sector-location that fit this purpose, divide a space into sectors while one or two beacons are positioned in the respective sector. The measurement principle assumes that the beacon/s has/have the strongest signal strength in their corresponding sector. This procedure can be implemented well in the previously defined classes of "room" and "corridor". If one measures the signal strengths of several beacons relative to one another, one can thus determine the sector by the receiving signal strength of the device.

Furthermore, the transition between the different sectors can be predicted by the changing of the signal strength profile while moving. Explained in a simple example, when moving from point A to point B, the signal strengths from the beacon at point B will rise, while the one from mounted at point A will fall. This method also avoids the need for calibration for the individual receiving devices since the signal intensities are only considered relative to one another.

The size of the sectors is defined by the distribution density of the beacons used. To keep costs low and the deployment practical, it makes sense to create the sectors not smaller than 5 x 5 m. The sensitivity or accuracy of the tracking can be increased by sensor data fusion with an inertial sensor system, such as a pedometer. In the developed system concept, the sectors are calculated using the spatial data contained in the BIM.

The following steps must be taken to realize a sector oriented location tracking:

1. The user is given the appropriate positions of the beacons for the initial installation.
2. The sector allocation and beacon data are stored in a cloud platform.
3. During location tracking, the floor plan of the building is generated by BIM.
4. The sector and position data of the beacons on a smart device are synchronized with the cloud platform.
5. Due to this information, the continuously measured signal strengths and the movement data of the inertial sensor system, the sector is then calculated by the receiving smart device.
6. When an active internet connection is available, the position within the found sector is send to the cloud platform. In offline mode, the application stores the timestamp and the found sector on the device, to synchronize the data at a later point once an internet connection is established again.

The resulting system architecture is shown in Figure 8.

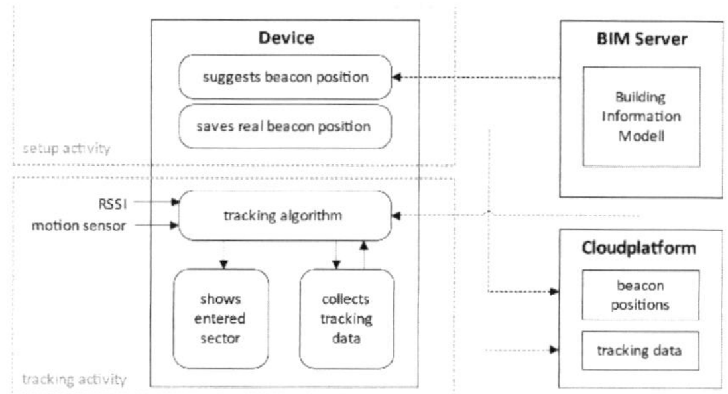

Figure 8. Proposed system architecture for the sector location tracking algorithm

6 Conclusion

The experiments carried out for this research show that location tracking is only applicable to a limited extent using BLE technology and path loss model. There are large differences in the behavior of the existing devices. Just the angle between transmitter and receiver has a rather large influence on the measured signal strengths. The experiments also show that the environment greatly influences signal transmission. To counteract these weak points in the path loss model, the authors found that complex adaptation and calibration measures would be necessary.

Based on these important findings, the authors proposed an alternative, holistic system for location tracking using BLE. It uses a systematic classification of the work space by positioning the BLE beacons according to the a-priori known spatial building structure, e.g. taken from a Building Information Model (BIM). The proposed system offers several advantages:

First, the relative observation of the received signal strengths of the individual beacons, the calibration of the receivers is obsolete so that several different device types can be used together and at the same time. Second, by combining the BLE location with the movement data of the receiving device, the accuracy of the system can be improved. Third, both, the data needed for location tracking and the tracking data are stored in a cloud platform, enable further analysis.

One of the next steps in the development of a holistic indoor tracking system is the implementation of the proposed concept along with its validation at a real construction site [18].

References

[1] Y. Wang, X. Yang, Y. Zhao, Y. Liu, L. Cuthbert, Bluetooth positioning using RSSI and triangulation methods, in: IEEE Consumer Communications and Networking Conference (CCNC), 2013: 11 - 14 Jan. 2013, Las Vegas, USA ; [including workshops], Las Vegas, NV, IEEE, Piscataway, NJ, 2013, pp. 837–842.

[2] S. Viswanathan, S. Srinivasan, Improved path loss prediction model for short range indoor positioning using bluetooth low energy, in: IEEE SENSORS 2015: November 1-4, 2015, Busan, South Korea proceedings, Busan, IEEE, Piscataway, NJ, 2015, pp. 1–4.

[3] C.E. Galván-Tejada, J.C. Carrasco-Jiménez, R.F. Brena, Bluetooth-WiFi based Combined Positioning Algorithm, Implementation and Experimental Evaluation, Procedia Technology 7 (2013) 37–45.

[4] R. Faragher, R. Harle, Location Fingerprinting With Bluetooth Low Energy Beacons, IEEE J. Select. Areas Commun. 33 (11) (2015) 2418–2428.

[5] Q. Tian, Z. Salcic, K.I.-K. Wang, Y. Pan, A Multi-Mode Dead Reckoning System for Pedestrian Tracking Using Smartphones, IEEE Sensors J. 16 (7) (2016) 2079–2093.

[6] C.M. Eastman, BIM handbook: A guide to building information modeling for owners, managers, designers, engineers, and contractors, 2nd ed., Wiley, Hoboken NJ, 2011.

[7] B. Hardin, BIM and construction management: Proven tools, methods, and workflows, Wiley Technology, Indianapolis, Ind., 2009.

[8] R.F. Aziz, S.M. Hafez, Applying lean thinking in construction and performance improvement, Alexandria Engineering Journal 52 (4) (2013) 679–695.

[9] H.R. Thomas, M.J. Horman, R.E. Minchin, D. Chcen, Improving Labor Flow Reliability for Better Productivity as Lean Construction Principle, Journal of Construction Engineering and Management 129 (3) (2003) 251–261.

[10] A. Solin, S. Sarkka, J. Kannala, E. Rahtu, Terrain navigation in the magnetic landscape: Particle filtering for indoor positioning, in: 2016 European Navigation Conference (ENC): Helsinki, Finland, 30 May 2016-2 June 2016, Helsinki, Finland, IEEE, Piscataway, NJ, 2016, pp. 1–9.

[11] T. Cheng, M. Venugopal, J. Teizer, P.A. Vela (2011). "Performance Evaluation of Ultra Wideband Technology for Construction Resource Location Tracking in Harsh Environments", Automation in Construction, Elsevier, 20(8), 1173-1184.

[11] E. Garcia, P. Poudereux, A. Hernandez, J. Urena, D. Gualda, A robust UWB indoor positioning system for highly complex environments, in: 2015 IEEE International Conference on Industrial Technology (ICIT 2015): Seville, Spain, 17 - 19 March 2015 ; [proceedings], Seville, IEEE, Piscataway, NJ, 2015, pp. 3386–3391.

[12] J. Park, Y.K. Cho, K. Kim, Field Construction Management Application through Mobile BIM and Location Tracking Technology, in: 33rd International Symposium on Automation and Robotics in Construction (ISARC 2016).

[13] Y. Gu, L. Quan, F. Ren, J. Li, Fast Indoor Localization of Smart Hand-Held Devices Using Bluetooth, in: 2014 10th International Conference on Mobile Ad-hoc and Sensor Networks (MSN): 19 - 21 Dec. 2014, Maui, Hawaii, USA, Maui, HI, USA, IEEE, Piscataway, NJ, 2014, pp. 186–194.

[14] H. Bae, J. Oh, K. Lee, J.-H. Oh, Low-cost indoor positioning system using BLE (bluetooth low energy) based sensor fusion with constrained extended Kalman Filter, in: 2016 IEEE International Conference on Robotics and Biomimetics (ROBIO), Qingdao, China, IEEE, 2016, pp. 939–945.

[15] A. Ozer, E. John, Improving the Accuracy of Bluetooth Low Energy Indoor Positioning System Using Kalman Filtering, in: 2016 International Conference on Computational Science and Computational Intelligence (CSCI), Las Vegas, NV, USA, IEEE, 2016, pp. 180–185.

[16] J. von Heyl, J. Teizer (2017). "Lean Production controlling and tracking using digital methods", Lean & Computing in Construction Congress (LC3), Heraklion, Greece.

[17] J. Melzner, J. Hanff, J. Teizer (2017). "Pattern-based Generation of 4D-Schedules for Robust Construction Management", Lean & Computing in Construction Congress (LC3), Heraklion, Greece.

[18] J. Teizer, M. Wolf, O. Golovina, M. Perschewski, M. Neges, M. König (2017). "Internet of Things (IoT) for Integrating Environmental and Localization Data in Building Information Modeling (BIM)", 34th International Symposium on Automation and Robotics in Construction, Taipei, Taiwan.

Measuring Operator's Situation Awareness in Smart Operation of Cranes

Yihai Fang [a] and Yong K. Cho[b],

[a] Postdoctoral Associate, Rinker School of Construction Management, University of Florida, Gainesville, Floroda
[b] Associate Professor, School of Civil and Environmental Engineering, Georgia Institute of Technology, Georgia
E-mail: yihaifang@ufl.edu, yong.cho@ce.gatech.edu

Abstract –

Equipment operators play an integral role in the safe and efficiency operation of heavy equipment. They observe the environment, understand the situation, and make decisions and actions accordingly. Compared with other types of equipment, operating a crane is more sophisticated and mentally demanding, and thus crane operators are more vulnerable to human errors. Therefore, special considerations to mitigate operator errors should be taken when designing an operator-assistance system for construction cranes. With the goal of improving the operators' situation awareness (SA) of safety risks, this research presents a technical framework and practical system architecture for an operator-assistance system by leveraging real-time motion sensing and 3D modeling of dynamic workspaces. An approach for evaluating operators' SA was proposed to validate the effectiveness of the assistance system in actual lifting operations. Results in a series of field tests indicated that the prototype system improved the operators' SA which resulted in an improved lift performance.

Keywords –

Crane Operation; Situation Awareness; Operator; Safety

1 Introduction

Cranes play an integral role in construction projects, responsible for most vertical and horizontal transportation of construction materials, equipment, and personnel. Crane lifting operations are unique among other heavy equipment as they demand huge workspaces and have a significant impact on the safety and efficiency of the entire construction projects. As such, the consequences of crane accidents are catastrophic as they very often result in significant cost overrun, schedule delay, and serious injuries and fatalities. Based on statistics from the U.S. Bureau of Labor Statistics (BLS) from 1997 to 2015, the number of fatalities in crane-related accidents totaled 1259 for all industry sectors [1]. The U.S. construction industry was responsible for 586 fatalities (47%), in which 55% involved mobile cranes (e.g., truck-mounted, crawler cranes). Another source reported that 78% of crane-related accidents in the construction industry from 1992 to 2006 were associated with mobile cranes [2]. Unlike other types of construction accidents, the victims in crane-related accidents are not necessarily limited to construction workers but also pedestrians walking-by as observed in many crane-related accidents.

Behind the poor crane safety records, crane experts consider operator errors a prevailing source of risks in crane lifting operations [3]. It was found that 43% of the crane accidents from 2004 to 2010 were due to the operator failure in their responsibilities [4]. Mechanical failures aside, 75% of crane overturn accidents are due to operator error [5]. A recent investigation on risk factors in crane-related near-misses and accidents reveals that inattention is the most prevalent type of risk that accounts for 19% of the incidents, and errors by operators and signalpersons total 24% of the 212 investigated incidents [6]. These numbers are not surprising as operating a crane is inherently a sophisticated job that requires the operators have extensive training and experience. Repetitive lift tasks and extended work hours make them vulnerable to distraction and fatigue. In addition to the errors of crane operators, lifting safety can also be jeopardized by poor coordination and communication with personnel such as riggers, signal persons, and ironworkers.

With the development of information technology, researchers realized that technology can provide another layer of protection in construction to improve safety performance. Crane operations can benefit from technologies similar to the advanced driver assistance systems (ADASs) deployed on automobiles that provide real-time support to drivers based on surrounding situations [7]. With the goal of providing real-time assistance to crane operators, researchers in construction have explored the use of sensing, visualization, and simulation technologies. However, a holistic synthesis of existing technologies and a framework outlining further development is missing. In addition, previous efforts in evaluating a crane assistance system mainly focused on measuring accuracy, reliability, and ease of use of the introduced technologies [8]. Very few emphasized on the system's effectiveness on helping the operators understand the situation and the safety risks [9]. This is partly because that the situation awareness (SA) of crane operators during lifting operations is difficult to define and measure in such complex and dynamic environment. This also leads to the fact that most of the techniques in previous research were only validated in a simulated environment instead of utilizing real lift tasks [10].

2 Literature Review

2.1 Critical Components in Operator Assistance for Mobile Cranes

The very first step in providing operator assistance is to accurately capture crane motions in real-time. The motions of a crane are essential to carry out a variety of spatial and temporal analyses for load capacity, crane stability, and collisions. Real-time location systems (RTLS) such as Radio-frequency Identification (RFID) have been utilized for tracking mobile assets (e.g., materials, equipment, workers) on construction sites [11]. For capturing crane motion, Ultra-wide Band (UWB), a more precise position tracking technology, were investigated and tested in controlled lab tests [12] and outdoor full-scale tests with real cranes [13]. Using computer vision techniques, the motions of a crane and other articulated equipment can be captured by tracking markers deployed on equipment parts [14] or mapping image to the 3D model of the crane [15]. Similarly, the 3D geometry of a piece of equipment in the form of point cloud can be mapped to its 3D model to achieve equipment motion capture in 3D [16]. Cranes can be considered as giant robots with multiple degrees of freedom in a rigid body or kinematic relationships. Therefore, the entire motion of a crane can be capture by measuring critical motions (e.g., swing, lifting, and extension of the boom, extend/retract hoist cable). These critical motions can be easily measured by inexpensive rotary encoders, laser distance finders [17] and inertia measurement units (IMUs) [18].

Given the spatial and temporal scale of crane lifting operations, crane workspace is subject to constant changes in its surrounding environment (e.g., presence of vehicles and workers, newly erected structures). Therefore, modeling the as-is condition and dynamics of crane workspace is of great importance to successful operator assistance. Building Information Models (BIM) can provide a general spatial context for the crane workspace [17] but fails to model the dynamics and changes in the surroundings. As-is conditions such as geometry and color of nearby objects can be capture by laser-scanned point cloud and updated by a hybrid visualization approach that takes advantages of the efficient data collection and computation from computer vision and comprehensive 3D geometric information contained in point cloud [19].

The interface presented to the operator by assistance system plays a pivotal role in operator assistance. A user interface provides visual feedback to augment the operator's understanding of lift tasks [17]. To this end, numerical feedback provided by traditional operator-assistance systems such as Load Limit Indicators (LMI) is limited. At the other end of the spectrum, vivid pictures from crane camera systems may lead to distraction and increased mental workload when the operator struggles with the depth perception and limited field of view [20]. The user interface (UI) needs to provide just-in-time alerts in multiple forms (e.g., visual, auditory, haptic) with the right amount of information that supports operator to make decisions to mitigate hazards [18].

Although an array of safety devices are available in the market, the effectiveness and utilization of these devices in the industry are unclear [9]. It is important to evaluate the effectiveness of these safety devices in actual lift tasks in order to identify potential challenges and suggest further improvement. Previous efforts predominantly focused on addressing technical limitations, while the impact of such systems on reducing the operator error and improving their SA remains unknown.

2.2 Operator Error and Situation Awareness

From the perspective of cognitive psychology, a human error is considered a result of one or multiple failures in the human cognition process. This process can be described by the sequential stage model created by Furnham [21]. This model simulates the development of accidents as a sequential chain that consists of three cognition stages: hazard perception, hazard recognition, and decision/ability to avoid a hazard. It is helpful to understand the causation of cognitive failures in the development of crane accidents by applying this model to the cognition process of crane operators. When a crane operator is exposed to one or multiple hazards, they need to first perceive the presence of the hazards, mainly through the status, attributes, and dynamics of relevant elements in the environment. Based on the information acquired and their understanding of hazards, the operator should recognize the type, severity, and consequences of the hazards. Finally, the operator needs to make appropriate decisions and actions to mitigate or avoid the hazard from further development. Operator success in all cognition stages will result in safe behaviors. Failure in any of the stages will result in unsafe behaviors that may lead to accidents.

Successful hazard perception involves an acute state of alertness and high level of sensory skill from the operators, and it requires them to maintain a good Situation Awareness (SA) during the operation. Once the operators perceive the hazard in the first stage and proceed to the next stage of hazard recognition, they need to correctly recognize the type and severity of the hazards using their experience and knowledge as well as a comprehensive SA. Therefore, failures to perceive and recognize hazards are generally categorized as an SA problem [22], which has a high likelihood of leading to performance failure [23]. SA is defined by Endsley as "a person's perception of the elements of the environment within a volume of time and space, the comprehension of their meaning and the projection of their status in the near future" [24]. During crane operations, crane operators' SA depends on an understanding of both the crane (e.g., crane motion, capacity, malfunctions) and the physical characteristics of the environment (e.g., wind speed, blind spots, clearances to obstructions).

2.3 Measurement of Situation Awareness

A number of measurement methods for situation awareness have been developed and each measure has its advantages and applied circumstances. These measures can be classified into three categories: 1) process indices-based, 2) performance-based, and 3) query-based techniques.

Process indices-based techniques examine the way subjects process information obtained from the environment such as by analyzing gaze movement using eye tracking technology [25]. Another type of process indices is physiological measures such as electroencephalographic (EEG) activity, eye blinks, and cardiac activity, which represent the subject's overall functional state [26]. Although the changes in the subject's physiological states may be associated with cognitive activities, there is not necessarily a direct link between physiological states and the level of SA. In performance-based techniques, the level of SA is inferred from the performance outcomes based on the assumption that better performance indicates better SA. Commonly used performance metrics include productivity level, time to perform the task, and the accuracy of the response or, conversely, the number of errors committed. The main advantage of performance measures is that they yield objective, quantitative results without disrupting task performance. Although in many cases there is a positive relation between SA and performance, this connection is not always direct and explicit [27].

In query-based techniques, subjects are asked directly about their perception of certain aspects of the situation. The queries are usually designed by domain experts based on the characteristics of the tasks. One of the most widely used query-based techniques is the Situation Awareness Global Assessment Technique (SAGAT) [28]. The operation is frozen at randomly selected times and subjects are queried about their perception of the situation at that instant. SAGAT is popular as it produces a quantitative assessment of SA and it can benchmark the result with similar data in a similar context [29]. However, SAGAT is criticized as it interrupts the natural flow of the task. To address these limitations, Durso et al. developed the Situation Present Assessment Method (SPAM) based on the premise that SA involves simply knowing where to find a particular piece of information in the environment [30]. In addition to being less intrusive than other techniques [31], the benefits of using SPAM lie in it uses response time to indicate the level of SA so that the results reflect the real-time dynamic SA of the operator.

3 Methodology

To direct further development of operator-assistance systems for mobile cranes, this research proposes a framework to identify major technical layers and requirements. A system architecture is presented based on the framework to address key technical challenges in this endeavor. In addition, this research proposes an assessment approach to quantify the effectiveness of such operator-assistance systems by defining and measuring operators' SA and lift performance.

3.1 System Architecture for Real-time Operator Assistance

With the goal of improving crane lift safety, a system architecture is developed to enable real-time operator assistance. The system architecture consists of three components: crane motion capturing, site condition modeling and updating, and user interface and interaction (Figure 1). To capture crane motion in real-time, a combination of wired rotary encoder sensors and a wireless inertia measurement unit (IMU) sensor are adopted to measure the critical motions of crane modules (e.g., boom lift angle, boom extension length, boom slew angle, load sway) [32]. The data from the encoders are first synchronized in a processing unit so that the game engine program on the tablet receives the packet that includes the measurements from all encoders. In addition, the processing unit will detect and reject corrupted or incomplete packets. In site condition modeling and updating, the format of a point cloud, collected by a laser scanner or other photogrammetry technologies, is used to represent the as-is lifting site

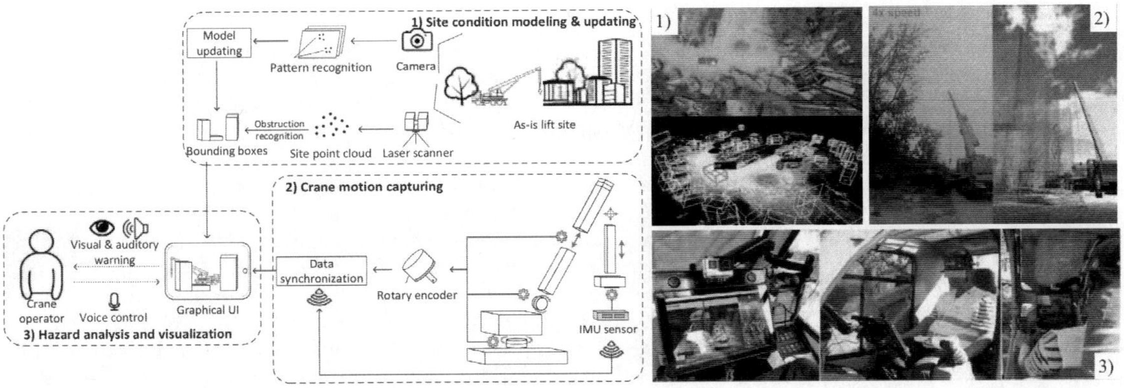

Figure 1: System Architecture

condition as it can be efficiently acquired. To reconstruct an as-is lifting site, the site point cloud needs to be converted to bounding box objects to represent site obstructions. Bounding boxes will be automatically constructed to represent the obstruction through multiple steps of point cloud processing. With the point cloud and bounding boxes serving as base 3D information, updating the lifting site condition, more precisely the location change of obstructions (e.g., vehicles, materials), can be achieved by correlating the 2D images captured by a camera with the known 3D information. Based on the crane motion data and the site condition data, the on-board computer will virtually reconstruct and realistically visualize the lift scene using a game engine, and analyze the hazards based on the real-time data. Once a hazard is detected, warnings will be delivered to the operator through both visual and auditory means. Throughout the operation, the operator is able to interact with the UI through voice commands The related details in team's previous work can be found in [19][33].

3.2 SA-based Effectiveness Assessment

Given the lack of effectiveness assessment in previous operator-assistance systems, this research proposes an assessment approach to quantify the effectiveness of assistance systems by defining and measuring operators' SA during lift tasks. The level SA is expected to have a direct impact on the lift performance. To investigate the correspondence of SA and performance, this research adopts the performance assessment method introduced in [33] which quantifies lift performance with respect to safety and efficiency by five key performance indexes (KIPs). In addition, the complexity of the lift task (e.g., mental, physical, and temporal demands) will affect the level of SA and the lift performance. The relationship between SA, lift performance, and task workload is illustrated in Figure 2. The following section will describe specific methods used to measure SA, lift performance, and task workload in crane lifting operations.

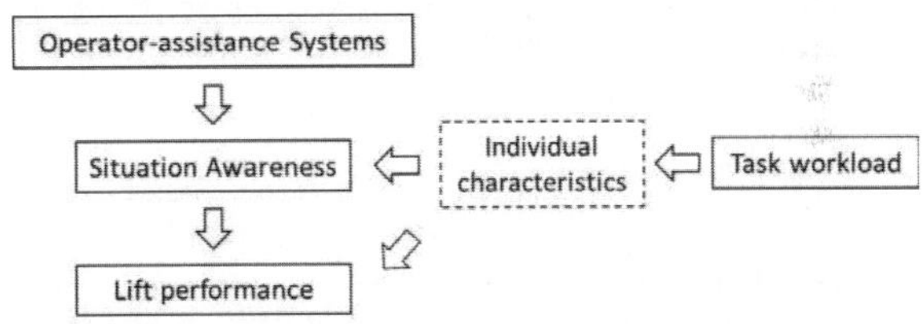

Figure 2: Relationship between SA, lift performance, and task workload in lift operations

3.2.1 Query-based Assessment of Situation Awareness

SA has been widely considered as an important factor in dynamic decision-making, and several indirect or subjective methods (e.g., physiological measurement, performance measures, self-rating) have been proposed and used in the SA research of aircraft pilots and air traffic controllers. This research employed a query-based SA measurement method adapted from SPAM [34]. In this method, the operator will be presented with queries about the situation while the situation remains present and while they continue to perform the primary task. During the query process, the questioner will indicate the operator that he intends to ask a query. Once the operator is ready to take the query, he or she will suspend the operation and indicate the questioner that he is ready. Then, the questioner will ask the question. The duration between the time the questioner finishes the question and the time operator start answering the question is considered the response time. Once the questioner records the operators' answer to the query, he will indicate the operator to resume the operation. In addition to response time as the sole measure of SA in traditional SPAM, the operator's level of SA is measured by both the response time and response correctness. Response correctness is quantified by the percentage of variance of the operators' answer to the correct answer recorded by the system.

Two major challenges that are addressed in this research when using this method are 1) designing proper queries that can effectively reflect the level of SA in crane lifting operations, and 2) choosing the proper timing to make the queries so that it does not interfere the primary lift tasks while remains effective for real-time SA assessment. A list of queries is shown in Table 1, in which two queries focus on past events, three queries focus on present events, and one query focuses on future events. These queries were designed based on panel discussions with safety experts, crane supervisors, and crane operators to ensure these queries reflect the most essential understandings of the lift tasks and their associated risks. During each lifting operation, one of the queries is chosen to ask according to the concurrent situation. For example, when the load is being lifted above a tree, the questioner asks the current clearance between load or boom and tree.

Table 1: Query list for the query-based SA measurement

Type	Query
Past	What is the maximum sway distance so far?
Past	How many warnings have you received from the system so far?
Present	What is the current boom reach?
Present	What is the current clearance between load/boom and obstructions?
Present	How far is the load placed from its target placement location?
Predictive	How long do you anticipate this lift task will take?

3.2.2 Assessment of Task Workload

During the crane lifting operation, operators' performance and SA can be affected by the workload imposed by the assigned lift task including mental, physical, and temporal demands. Given the same lift task, different operators may perceive different workload because of subjective, individual differences in training, experience, and cognitive capability. The perceived workload of the individual operator can be measured using NASA Task Load Index (NASA-TLX) [35]. The NASA-TLX is a widely used, subjective, multidimensional assessment tool that allows users to perform subjective workload assessments on operators working with various human-machine systems. NASA-TLX derives an overall workload score based on a weighted average of the ratings on six dimensions, including Mental Demands, Physical Demands, Temporal Demands, Own Performance, Effort, and Frustration.

In the assessment of perceived workload, participants were instructed to place an "X" on each of the six scales at the point that matches their experience. Each scale has two endpoint descriptors such as "bad" and "good", or "low" and "high". These numerical ratings for each scale reflect the magnitude of the factors in a given task. Then, the participants were presented with a series of pairs of rating scale tiles (for example, Effort vs. Mental Demands) and asked to choose which of the items was more important to their experience of the workload in the lift task they just performed. This was used to evaluate the contribution of each factor (its weight) to the workload of a specific task. Given the numerical rating and the weight of each factor, an adjusted rating can be computed for quantifying the overall perceived workload of the rated task.

4 Field Tests and Results

4.1 Overview

The goal of the field tests was to investigate how operator assistance would affect the operator's SA and performance under tasks with high and low workload. Five crane operators with experiences range from 8 to 16 years were recruited. Prior to the tests, the operators undertook a 30-minute training session to be familiar with the functionalities of the assistance system. To differentiate the complexity level of the lifting operation, two lift tasks were designed with different spatial and temporal constraints. Although surrounded by trees and other spare crane parts, lift task #1 features a simpler lifting operation as no obstructions present between the pick and drop locations. Lift task #2 requires the operator to lift the load over a row of trees of 15 m in height (Figure 3). Sitting in the cabin at ground level, the operators can hardly see the load when it is above the trees. Therefore, estimating the clearance from the trees to the load or boom is very challenging yet crucial for safety. For each task, a control and a test scenarios were created where in control scenario the operators only use traditional LMI system and in test scenarios, the operators use both LMI and the proposed assistance system. In these field tests, in total 60 lifts were conducted with 12 lifts for each operator.

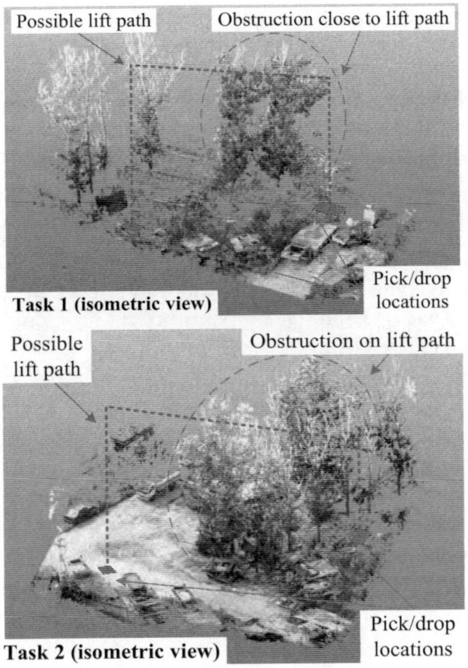

Figure 3: Lift task #1 and #2 in isometric view

4.2 Workload Assessment for Each Lift Task

A NASA-TLX rating package was explained and presented to each operator after the lifting operations. They were asked to fill out the rating and weight sheets based on their experience in the lifts. As the operators vary in individual characteristics such as motivation, risk-taking tendency, and mental and physical capabilities, different operators may perceive the different level of workload given the same conditions and constraints. Therefore, the workload index for each lift task was computed based on an averaged index from the five operators. Overall, lift task #1 received an adjusted rating of 5.42 whereas lift task #2 received 6.65. This result indicates that lift task #2 imposed a larger amount of workload to the operators than lift task #1. This can be further assured by the fact that all five operators rated Task #2 higher than Task #1, with the largest difference of 2 and smallest of 0.73.

In addition to the workload indexes for the lift tasks, it is also interesting to see how the operator rates and weights each workload dimension in general. Among the six dimensions, the largest rating difference between the two tasks lays in mental demand (5.8) whereas the smallest difference in physical demand (0.6). In the meantime, the operators weighted mental demands as the second biggest contributor (second to performance) to the workload and physical demands as the smallest. These results show that mental demand was a dominating source of the workload in the two test tasks, and very likely in crane lifting operations in general as these tasks represented the common characteristics in day-to-day lifting operations.

4.3 Operator SA Assessment

A query-based SA measure was used to quantify the operators' SA during the operation. The level of SA was quantified by the response time and response correctness for each query. Figures 5 and 6 present the average response time and average response correctness of the 60 queries in different tasks and scenarios. The quantile box plot indicates the variance in the results of response time and correctness. It was observed that the average response time in both the tasks was remarkably reduced when the assistance system was used, 17.5% in Task #1 and 28.6% in Task #2. The quartile box plot indicates that the variances of response time in the control scenario in both the tasks are much smaller than that in the test scenario. This result suggests that the operators have developed a consistent pattern of understanding the situation based on their experience without the assistance. Despite that overall response time was reduced, the introduction of operator assistance may change the way the operators search and understand the situation. Therefore, training plays an important role in the successful integration of the assistance system. Overall, the results show that the introduction of the assistance system facilitated the operator's ability to comprehend the situation, especially the information closely related to lift safety and efficiency.

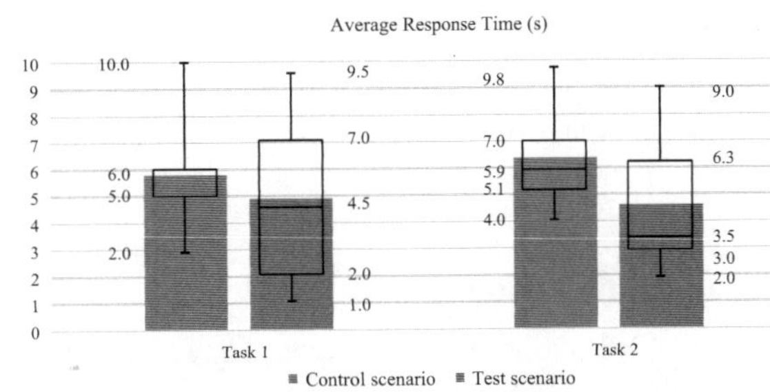

Figure 5: Average response time in two lift tasks under control and test scenarios

The response correctness is shown in percentage and it was computed by comparing the operators' responses to the correct answer recorded in the system. The results show that average response correctness was improved in the test scenario, by 2.3% in Task #1 and 9.5% in Task #2. The box plot suggests that the response correctness in test scenarios was in generally more consistent, especially in Task #2. Although not as obvious as the decrease in response time, the improvement of response correctness suggests that the assistance system was able to provide helpful information to augment the operator's awareness of the lift task so that the accuracy of their decision-making can be increased.

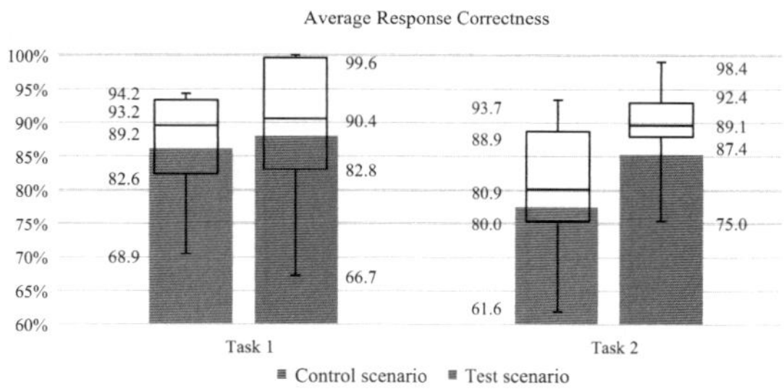

Figure 6: Average response correctness in two lift tasks under control and test scenarios

5 Discussion

From a statistical perspective, the sample size of operators participated in the field tests may not large enough to show the statistical significance of the improvement. Particularly, the variances in the measurements tend to be large due to the limited number of subjects and the significant difference in experience and maneuver pattern among them. That being said, the results in these field tests reflect the general trend in the improvement of lift performance and SA by introducing the assistance system. The conclusions are meaningful to understand the strengths and limitations of the current design and to guide the future work. Another limitation in the SA measurement in the field tests that the query process was not blind to the questioner. This bias in the results can be minimized by asking multiple queries and randomly selecting part of them for SA measurement.

6 Conclusion

To guide further development of operator-assistance systems for mobile cranes, this research presents a framework to identify major technical tasks and requirements. Based on the framework, a system architecture is proposed to address key technical challenges to facilitate implementation. In addition, this research proposes an assessment approach to assess the effectiveness of such operator-assistance systems by defining and quantifying operators' SA and lift performance. These proposed methods were tested and validated in a series of field tests. Results indicate that the assistance system has a positive impact on improving operators' SA during lift tasks, which led to improved lift performance, especially in safety. Overall, it was found that the assistance system facilitated the operator's ability to comprehend the situation, especially the information closely related to site geometric constraints, which led to a safer and more efficient lifting operation. It should be noted that the experiments were conducted in real lift jobs with actual temporal and spatial constraints, and the participating operators were exposed to real lifting safety pressure and risks. Compared to virtual simulation or survey that were commonly used in many other research, the data collected and presented in this research is more realistic and therefore the results are closer to the actual effectiveness of the assistance.

7 Acknowledgement

This material is based upon work supported by the National Science Foundation (Award #: CMMI-1358176). Any opinions, findings, and conclusions or recommendations expressed on this material are those of the authors and do not necessarily reflect the views of the NSF.

References

[1] Bureau of Labor Statistics (BLS), Census of Fatal Occupational Injuries, U.S. Dep. Labor, Bur. Labor Stat. News. (2016).

[2] CPWR, Crane-Related Deaths in Construction and Recommendations for Their Prevention, 2009. http://www.cpwr.com/sites/default/files/CPWR Crane Rept Recmmdtns Nov-2009-BLS UPDATED.pdf.

[3] L.K. Shapiro, J.P. Shapiro, Cranes and Derricks, McGraw-Hill, New York, 2011.

[4] R.A. King, Analysis of Crane and Lifting Accidents in North America from 2004 to 2010, (2012). https://dspace.mit.edu/handle/1721.1/73792 (accessed August 19, 2016).

[5] R.L. Neitzel, N.S. Seixas, K.K. Ren, A review of crane safety in the construction industry, Appl. Occup. Environ. Hyg. 16 (2001) 1106–17. doi:10.1080/10473220127411.

[6] G. Raviv, B. Fishbain, A. Shapira, Analyzing risk factors in crane-related near-miss and accident reports, Saf. Sci. 91 (2017) 192–205. doi:10.1016/j.ssci.2016.08.022.

[7] D. Gerónimo, A.M. López, A.D. Sappa, T. Graf, Survey of pedestrian detection for advanced driver assistance systems, IEEE Trans. Pattern Anal. Mach. Intell. 32 (2010) 1239–1258. doi:10.1109/TPAMI.2009.122.

[8] H.L. Chi, Y.C. Chen, S.C. Kang, S.H. Hsieh, Development of user interface for tele-operated cranes, Adv. Eng. Informatics. 26 (2012) 641–652.

doi:10.1016/j.aei.2012.05.001.
[9] X. Su, J. Pan, M. Grinter, Improving Construction Equipment Operation Safety from a Human-centered Perspective, Procedia Eng. 118 (2015) 290–295. doi:10.1016/j.proeng.2015.08.429.
[10] H. Li, G. Chan, M. Skitmore, Integrating real time positioning systems to improve blind lifting and loading crane operations, Constr. Manag. Econ. 31 (2013) 1–10. doi:10.1080/01446193.2012.756144.
[11] W. Lu, G.Q. Huang, H. Li, Scenarios for applying RFID technology in construction project management, Autom. Constr. 20 (2011) 101–106. doi:10.1016/j.autcon.2010.09.007.
[12] S. Hwang, Ultra-wide band technology experiments for real-time prevention of tower crane collisions, Autom. Constr. 22 (2012) 545–553. doi:10.1016/j.autcon.2011.11.015.
[13] C. Zhang, A. Hammad, S. Rodriguez, Crane Pose Estimation Using UWB Real-Time Location System, J. Comput. Civ. Eng. (2012) 625–637. doi:10.1061/(ASCE)CP.1943-5487.0000172.
[14] C. Feng, S. Dong, K.M. Lundeen, Y. Xiao, V.R. Kamat, Vision-Based Articulated Machine Pose Estimation for Excavation Monitoring and Guidance, Proc. 32nd Int. Symp. Autom. Robot. Constr. (2015).
[15] J. Yang, P.A. Vela, J. Teizer, Z. Shi, Vision-Based Tower Crane Tracking for Understanding Construction Activity, ASCE J. Comput. Civ. Eng. 28 (2013) 103–112. doi:http://dx.doi.org/10.1061/(ASCE)CP.1943-5487.0000242.
[16] Y.K. Cho, M. Gai, Projection-Recognition-Projection Method for Automatic Object Recognition and Registration for Dynamic Heavy Equipment Operations, J. Comput. Civ. Eng. 28 (2014) A4014002. doi:10.1061/(ASCE)CP.1943-5487.0000332.
[17] G. Lee, J. Cho, S. Ham, T. Lee, G. Lee, S.H. Yun, et al., A BIM- and sensor-based tower crane navigation system for blind lifts, Autom. Constr. 26 (2012) 1–10. doi:10.1016/j.autcon.2012.05.002.
[18] Y. Fang, Y.K. Cho, J. Chen, A Framework for Real-time Pro-active Safety Assistance for Mobile Crane Lifting Operations, Autom. Constr. (2016).
[19] J. Chen, Y. Fang, Y.K. Cho, Real-Time 3D Crane Workspace Update Using a Hybrid Visualization Approach, J. Comput. Civ. Eng. (2017).
[20] A. Shapira, F. Asce, Y. Rosenfeld, I. Mizrahi, Vision System for Tower Cranes, J. Constr. Eng. Manag. (2008) 320–332. doi:10.1061/(ASCE)0733-9364(2008)134:5(320).
[21] A. Furnham, Personality at work: the role of individual differences in the workplace, Routledge, New York, 1994.
[22] D.B. Kaber, M.R. Endsley, The effects of level of automation and adaptive automation on human performance, situation awareness and workload in a dynamic control task, Theor. Issues Ergon. Sci. 5 (2004) 113–153. doi:10.1080/1463922021000054335.
[23] D.G. Jones, M.R. Endsley, Sources of situation awareness errors in aviation, Aviat. Sp. Environ. Med. 67 (1996) 507–512.
[24] M.R. Endsley, Situation awareness global assessment technique (SAGAT), in: Proc. IEEE 1988 Natl. Aerosp. Electron. Conf., IEEE, 1988: pp. 789–795. doi:10.1109/NAECON.1988.195097.
[25] F.T. Durso, K.M. Geldbach, P. Corballis, Detecting Confusion Using Facial Electromyography, Hum. Factors J. Hum. Factors Ergon. Soc. 54 (2012) 60–69. doi:10.1177/0018720811428450.
[26] G. Wilson, Strategies for psychophysiological assessment of situation awareness, in: Situat. Aware. Anal. Meas., Lawrence Erlbaum Associates, 2000: pp. 158–169.
[27] M.R. Endsley, Toward a Theory of Situation Awareness in Dynamic Systems, Hum. Factors J. Hum. Factors Ergon. Soc. 37 (1995) 32–64. doi:10.1518/001872095779049543.
[28] M.R. Endsley, Measurement of Situation Awareness in Dynamic Systems, Hum. Factors J. Hum. Factors Ergon. Soc. 37 (1995) 65–84. doi:10.1518/001872095779049499.
[29] E. Jeannot, C. Kelly, D. Thompson, The Development of Situation Awareness Measures in ATM Systems, System. 1.0 (2003) 88. doi:HRS/HSP-005-REP- 01.
[30] F.T. Durso, T. Truitt, C. Hackworth, J. Crutchfield, En Route Operational Errors and Situational Awareness, Int. J. Aviat. Psychol. 8 (1998) 157–176. doi:10.1207/s15327108ijap0802.
[31] F. Durso, T. Truitt, C. Hackworth, J. Crutchfield, D. Nikolic, P.M. Moertl, et al., Expertise and chess: A pilot study comparing situation awareness methodologies, (1995).
[32] Y. Fang, Y.K. Cho, Crane Load Positioning and Sway Monitoring Using an Inertial Measurement Unit, in: Comput. Civ. Eng. 2015, Austin, TX, 2015: pp. 700–707.
[33] Y. Fang, Y.K. Cho, Effectiveness Analysis from a Cognitive Perspective for a Real-time Safety Assistance System for Mobile Crane Lifting Operations, J. Constr. Eng. Manag. (2016).
[34] F.T. Durso, A.R. Dattel, SPAM: The Real-Time Assessment of SA, (2004) 137–154.
[35] S. Hart, L. Staveland, Development of NASA-TLX (Task Load Index): Results of empirical and theoretical research, Adv. Psychol. 52 (1988) 139–183. doi:10.1016/S0166-4115(08)62386-9.

A *High-Resolution Intelligence* Implementation based on *Design-to-Robotic-Production and -Operation* strategies

A. Liu Cheng[a,b], H. H. Bier[a], G. Latorre[c], B. Kemper[a], and D. Fischer[a]

[a]Faculty of Architecture and the Built Environment, Delft University of Technology, Delft, The Netherlands
[b]Facultad de Arquitectura y Diseño / Facultad de Ingeniería y Ciencias Agropecuarias, Universidad de Las Américas, Quito, Ecuador
[c]Facultad de Ingeniería de Sistemas, Escuela Politécnica Nacional, Quito, Ecuador
E-mail: a.liucheng@tudelft.nl, h.h.bier@tudelft.nl, galoget.latorre@epn.edu.ec, b.n.kemper@student.tudelft.nl, d.l.fischer@student.tudelft.nl

Abstract –

This paper presents an initial *proof-of-concept* implementation of a comprehensively intelligent built-environment based on mutually informing *Design-to-Robotic-Production and -Operation* (D2RP&O) strategies and methods developed at *Delft University of Technology* (TUD). In this implementation, D2RP is expressed via deliberately differentiated and function-specialized components, while D2RO expressions subsume an extended *Ambient Intelligence* (AmI) enabled by a *Cyber-Physical System* (CPS). This CPS, in turn, is built on a heterogeneous, scalable, self-healing, and partially meshed *Wireless Sensor and Actuator Network* (WSAN) whose nodes may be clustered dynamically *ad hoc* to respond to varying computational needs.

Two principal and innovative functionalities are demonstrated in this implementation: (1) cost-effective yet robust *Human Activity Recognition* (HAR) via *Support Vector Machine* (SVM) and k-*Nearest Neighbor* (*k*-NN) classification models, and (2) appropriate corresponding reactions that promote the occupant's spatial experience and well-being via continuous regulation of illumination with respect to colors and intensities to correspond to engaged activities.

The present implementation attempts to provide a fundamentally different approach to intelligent built-environments, and to promote a highly sophisticated alternative to existing intelligent solutions whose disconnection between architectural considerations and computational services limits their operational scope and impact.

Keywords –

Design-to-Robotic-Production and -Operation, Cyber-Physical Systems, Adaptive Architecture, Wireless Sensor Networks, Ambient Intelligence.

1 Introduction

The present paper promotes *Design-to-Robotic-Production and -Operation* (D2RP&O) [1] strategies and methods as drivers of highly sophisticated *Ambient Intelligence* (AmI) solutions, and demonstrates its competence in this endeavor by presenting and describing a corresponding high-resolution intelligence implementation. Two principal and innovative functionalities are described in this implementation, the first pertaining to computational intelligence while the second to architectural variables / considerations (see Section 2).

With respect to the first functionality, a *Machine Learning* (ML) subsystem is integrated in the proposed system-architecture in order to enable *Human Activity Recognition* (HAR) mechanisms. With respect to HAL, ML methods have typically used gyroscopic data collected via portable devices (e.g., smartphones, etc.) [2, 3] or via sensor-fusion [4]. The ML subsystem consists of two classification mechanisms developed based on polynomial programming of *Support Vector Machine* (SVM) and k-*Nearest Neighbor* (*k*-NN) classifiers. These SVM and *k*-NN models are built on a dynamically clustered set of high-performance nodes in the localized *Wireless Sensor and Actuator Network* (WSAN). Cloud-based SVM and *k*-NN counterparts are generated as alternatives to the localized mechanisms for contingency measures (see Section 3.1). With respect to the second functionality, an interactive / adaptive illumination system capable of identifying and mitigating—via said ML mechanisms—fatigue via regulation of colors and corresponding intensities is proposed (see Section 3.2).

The comprehensive character of the intelligence imbued in said implementation supervenes on the mutually corresponding and informing relationship between computational mechanisms and architectural considerations.

2 Concept and Approach

The present implementation continues to build on the adaptive mechanisms and system-architecture previously outlined and developed by the authors [5–7] (see Figure 1). With respect to intelligence in the built-environment, it revisits *Protospace 4.0*'s [8–10] system of function-specific differentiated components. On the occasion of the international conference *Game Set and Match 3* (GSM3) [11] held at the *Faculty of Architecture and the Built-Environment, Delft University of Technology* (TUD) (9th-11th of November, 2016), a fragment of *Protospace 4.0* was rebuilt as a responsive stage, and a purpose-built interactive LED-based illumination system was integrated into its architecture (see Figure 5). This illumination system serves as a subsystem of the present system-architecture.

With respect to computational intelligence, an ML framework is deployed as a subsystem to enable cost-effective yet robust HAR mechanisms via established classifications models—i.e., SVM and *k*-NN. A smartphone as well as three *Light Blue Beans*™ (LBBs) were used to gather gyroscopic and accelerometer data from the user via the *Open Sound Control* (OSC) protocol. The generated dataset was used to train two SVM and *k*-NN models, one via local clusters using open-source and purpose-written *Python* scripts, and another via an external computer—simulating cloud-based analytics services—using third-party proprietary software. The principal intention was to imbue the proposed system with both localized as well as web-based analysis mechanisms in order to ascertain ML robustness and resilience in case either mechanism failed. A secondary intention was (a) to demonstrate that open-source solutions could be as effective as those rendered by proprietary software while reducing costs; and (b) to illustrate how purpose-written scripts integrated more seamlessly and efficiently (in terms of interoperability) than did proprietary software.

The integration of both built-environment as well as computational intelligences instantiates a *high-resolution intelligence* environment capable of translating sensed data into informed and correlated active, reactive, and interactive responses pertinent to the activities engaged by the user. The SVM and *k*-NN mechanisms are trained to identify certain data values as corresponding to a variety of activities, and to use this prediction power to dynamically mitigate fatigue in the user via an active and adaptive regulation of colors and intensities (see Section 3.2.2). Moreover, the responsive stage is also imbued with predetermined behavioral patterns such as pulsating when idle, tracing paths, correlating different colors to identified body parts (via Microsoft® *Kinect*™ V2) of up to six different individuals (see Section 3.2.1).

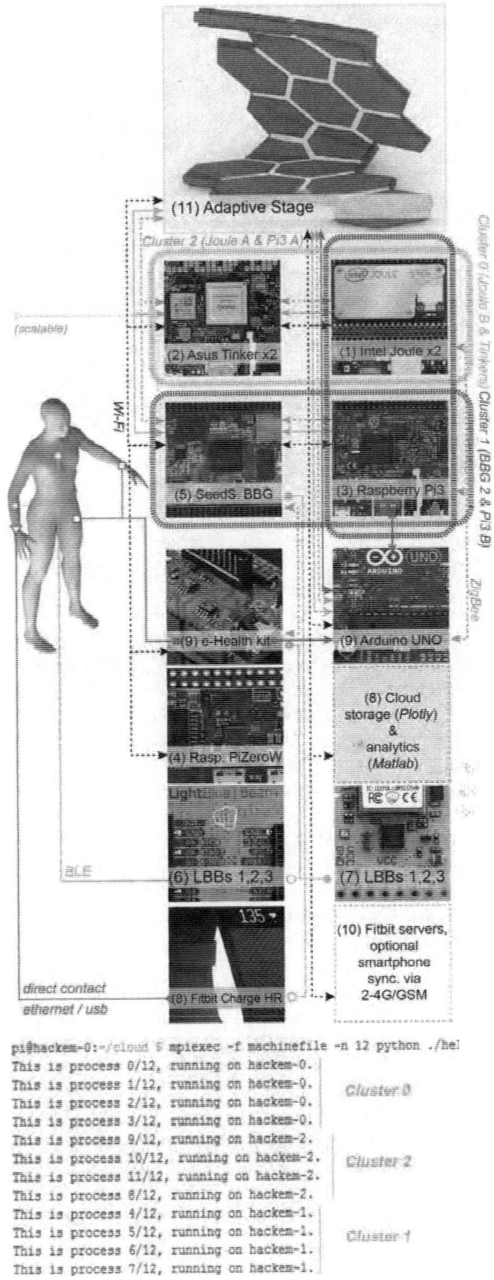

Figure 1: Top: Heterogeneous system-architecture with three dynamic *ad hoc* clusters. Bottom: Runtime processes distribution.

3 Methodology and Implementation

The development of the detailed implementation consists of three parts: (1) the design and development of cost-effective HAR system (see Section 3.1), which involved the development of (1a) a dynamic *ad hoc* heterogeneous clustering system (see Section 3.1.1) as well as (1b) data-gathering and -parsing scripts for ML training and testing purposes (see Section 3.1.2); (2) the design and installation of the LED-based illumination subsystem and its corresponding electronic setup (see Section 3.2); (3) the integration of the previous parts into a unified closed-loop system architecture. The first and second parts were developed in parallel and tested as working subsystems before integration.

3.1 Development of a cost-effective *Human Activity Recognition* (HAR) system

Due to their evolving and resilient characters, ML classifiers have been implemented in a variety of applications built on WSANs [12]. HAR, as one such application, has successfully exploited said classifiers in the last five years (see, for example, [13–15]). However, due to the cost-effective and low energy-consumption character typical of WSAN nodes, computational processing with respect to feature extraction has been considerably limited [16]. To overcome this limitation, the present implementation is capable of instantiating *ad hoc* clusters consisting of a variety of high-performance nodes. Furthermore, several clusters may be instantiated simultaneously in order to enable parallel high-performance information processing activities.

Another way to overcome this limitation is to avoid it altogether by outsourcing all high-performance information processing to cloud-based ML services (e.g., Google® *CloudPlatform*™, Amazon® *Machine Learning*™, Microsoft® *Azure*™, etc.). But there are a number of limitations with this approach. The first, and perhaps the most salient, is the cost incurred by including proprietary services in any proposed intelligent built-environment solution. A second yet no less important limitation may be the impact to the solution's resilience. That is to say, should said built-environment lose access to the Internet, it would be incapable of generating classification models.

The present implementation proposes the integration of both cloud-based as well as localized ML capabilities in order to ascertain robustness and resilience. Whenever possible, ML processes are locally and dynamically executed via *ad hoc* node-clustering. But should this prove impossible either due to failure or unavailability of proper resources, cloud-based ML services are used.

3.1.1 Dynamic Clustering mechanism

The system's clustering mechanism uses the *Message Passing Interface* (MPI) standard via *MPI for Python* (mpi4py) [17] (see Figure 1, *Bottom*). The system's ecosystem consists of nine types of development platforms, *Microcontroller Units* (MCUs), and proprietary trackers: (1) Intel® *Joule*™, (2) Asus® *Tinkerboard*™, (3) Raspberry® *Pi 3*™ and (4) *Pi Zero W*™, (5) SeedStudio® *BeagleBone Green*™ (BBG), (6) Punch Through® *Bean+*™ and (7) LBB, (8) Fitbit® *Charge HR*™, and (9) Arduino® *UNO*™ (see Figure 1). Sets of items 1, 2, 3, and 5 may be dynamically clustered *ad hoc* via WiFi for high-performance information processing, and are connected to the rest of the network via WiFi, ZigBee, BLE wireless communication protocols and—in the case of an instance of item 3—Ethernet / USB cables. Items 4, 7, and 9 are considered as low-computation *end devices* meshed into the WSAN via ZigBee, with 6 serving as router for 7 via BLE. Since there is a direct relationship between computational power vs. energy-consumption, *end device* and *router* nodes are concerned exclusively with sensor-data gathering and relaying with minimal information processing. Depending on the task, nodes exchange data via pertinent protocols and frequencies.

3.1.2 *Machine Learning* (ML) mechanisms

As detailed in Section 2 and Section 3.1, two ML mechanisms are integrated into the present implementation: (1) a localized *ad hoc* cluster system based on open-source and purpose-written *Python* scripts, and (2) a simulated cloud-based analytics service using MathWorks® *MATLAB*™. In both mechanisms SVM and *k*-NN classification models are generated.

In the localized mechanism, a script based on *pyOSC* is first written to receive OSC data from any device and application capable of broadcasting in said protocol. While all the WiFi-enabled nodes in the system's WSAN have the capacity to receive this data-streaming, only one of the nodes of the cluster instantiated to generate classification models stores it locally and streams it to a cloud-based data visualization service (i.e., *Plotly*™). Should the receiving node fail, another high-performance node will replace it automatically. Since the proposed solution uses a smartphone and three LBBs for data redundancy, resolution, and validation, the script in question proceeds to parse and to reduce the noise in the received multi-sensor data in order to generate a robust and unified dataset. At this point the dataset is processed through two ML scripts based on *scikit-learn* [18, 19], one for SVM and another for *k*-NN classification models (see Figure 2).

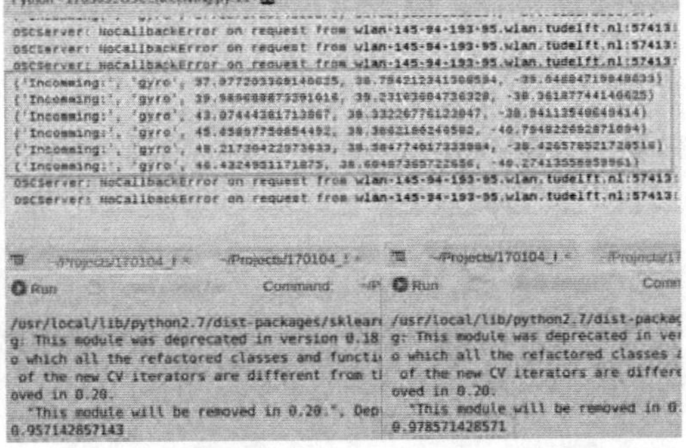

Figure 2: Top: OSC-data receiving and parsing. Bottom: 95.7% prediction success with respect to HAR via SVM (left) vs. 97.85% via *k*-NN (right).

It should be noted that each time a classification model is generated, regardless of whether it is done via open-source or proprietary means, its resulting prediction success rate will vary. For the purposes of the present discussion, the success rate generated in the last sample run is used. That is to say, the success rate of the localized SVM mechanism was 95.7% while that of the *k*-NN mechanism 97.85%.

In the proprietary cloud-based mechanism, as simulated by a computer external to the system's WSAN and running MATLAB™, the same datasets are processed through several *Classification Learners* (see Figure 3).

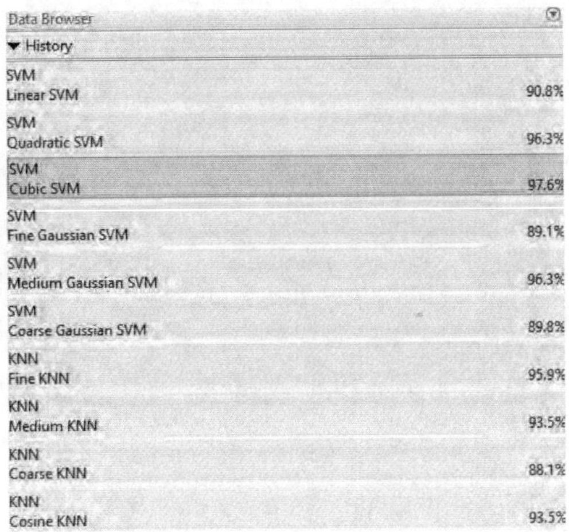

Figure 3: Sample MATLAB™-generated ML models with corresponding success rate.

It may be observed that the most successful classification model generated by MATLAB™ is based on *cubic* SVM with a prediction success rate of 97.6%, which is higher than the rate corresponding to the localized and open-source SVM result (i.e., 95.7%). But it may also be noted that the localized and open-source *k*-NN success rate is higher than any of the *k*-NN models generated by MATLAB™ (i.e., 97.8% vs. 95.9%, 93.5%, 88.1%, and 93.5%). It may not be inferred from this that the localized is superior to the proprietary, nor vice versa. It may be considered, however, that the localized and open-source mechanism yields comparably robust results as that of the proprietary one within the scope of the present implementation.

Having generated two sets of classification models via localized and cloud-based means, the ones with the most successful prediction rate are used at runtime, with precedence given to the localized mechanism—if and only if said mechanism fails or has unavailable resources are cloud-based ML models be used. The duration of said runtime may be determined by the user, but it should be as brief as practicable in order for the dataset to be updated with new data. For example, the user may decide to schedule the generation of a new updated model every 24 hours and only during sleep periods. This way the user would wake up to updated and relatively more attuned models every day. Furthermore, via this incrementally updating process, classification models may be trained to detect and/or predict new activities or patterns in a gradual manner, thereby enabling the intelligent built-environment to evolve with its user.

3.2 Development of the architecture-embedded interactive / adaptive illumination subsystem

As detailed in Section 2, a fragment of *Protospace 4.0* was repurposed to conform a responsive stage on the occasion of the *GSM3* conference. 16 differentiated and function-specific components, viz., *protoCELLs* [10], are assembled to conform said stage while integrating a custom-designed and -built interactive / adaptive LED-based illumination system. The indented borders of each component were lined with LED-strips, which enabled individual color control. In conjunction, these indented borders create a continuous indented seam between all components, which is covered with translucent material in order to enable diffusion of color and intensity. The combination between this translucent cover with two separate individually controlled LED-strips enables the instantiation of multiple color gradients and intensities (see Figure 4).

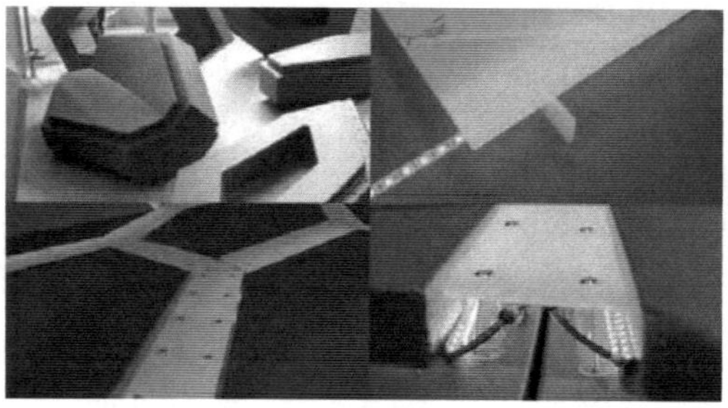

Figure 4: Top: Single *protoCELLs* (Left); testing *Perspex*® over LED1 (Right). Bottom: Generated color gradient (Left); Acrylic connections & two LED-strips within the seam of two *protoCELL*s (Right).

The system-architecture of the interactive / adaptive illumination system involves 12 Arduino® UNO™ MCUs that are physically connected to a computer via USB hubs. In the implemented revision, this computer is replaced by an Intel® *Joule*™ (see Figure 1), thereby integrating the stage and its responsive illumination into the system-architecture ecosystem of the present implementation. As a stand-alone system, the stage is configured to behave in particular and predetermined patterns. As a subsystem of a more sophisticated system-architecture, it is now imbued with ML capabilities for non-predetermined actions, reactions, and interactions.

3.2.1 Predetermined Scenarios

There are three predetermined scenarios: (1) Pulsating, (2) Lecture, and (3) Break, all of which are described as follows:

In the first scenario, as soon as the illumination system is powered, the stage slowly pulsates in one color—i.e., oscillates between intensities of a same color. This creates an effect viscerally reminiscent of a beating heart, tacitly suggesting that the stage is "alive". The intention of this scenario is to instigate interest and curiosity in the users, inviting them to engage with it (see Figure 5, *Top image 2*).

In the second scenario, two different types of interaction are envisioned during the conference presentations. The first involves the stage's reaction towards the movements of the speaker, where by stepping on or touching one or multiple components he/she instigates a gradual shift from the initial or *passive* colors to *active* colors for a certain period of time, after which *active* colors would default back to *passive* ones (see Figure 5, *Top images 3, 4*). The second interaction inverts this causal relationship to have the stage influence the speaker—i.e., the speaker knows his/her time is up when the first type of interaction ceases and the stage defaults back to a single color (see Figure 5, *Top image 5*).

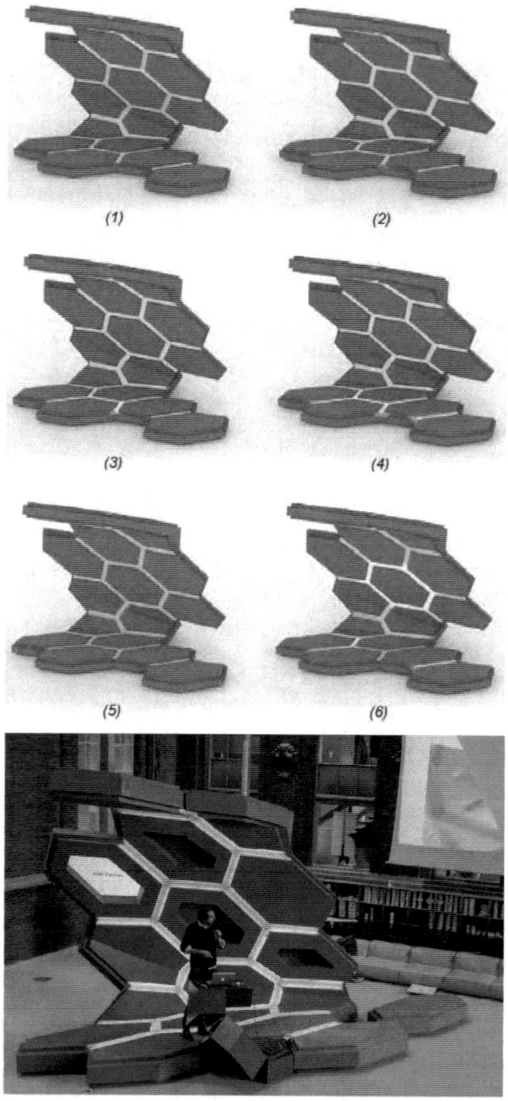

Figure 5: Top: images (1) 3D model with lights off; (2) Pulsating; (3) Activation of discrete components; (4) Leaving a trace; (5) Manual override; (6) individual activity detection correlated with color. Bottom: Implemented fragment.

In the third scenario, the stage invites interaction from the audience in-between lectures by allowing them to "paint the stage" via body gestures. That is, in this mode, the stage tracks body parts of up to six individuals and instantiates corresponding color changes across the components, hence correlating certain movements or body parts with certain colors (see Figure 5, *Top image 6*).

Finally, it should be noted that in addition to these three automated *cause-and-effect* scenarios, the illumination system is also designed with a manual override control. A proprietary fee-based Apple®'s iOS™ *application*, viz., *TouchOSC*™ (by Hexler Limited®) is used to develop customized control screens to provide override capabilities to the illumination system (see Figure 6).

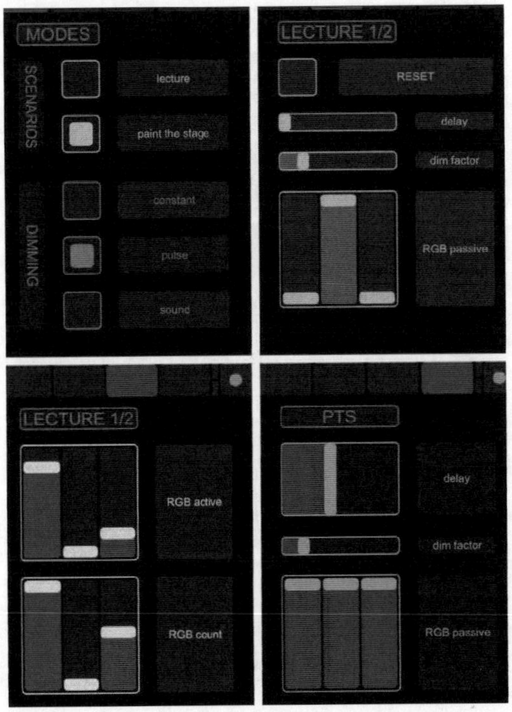

Figure 6: iOS OSC Applications: *TouchOSC* (proprietary).

3.2.2 Non-predetermined Scenario

The ML-driven HAR mechanism implemented in the present system (see Section 3.1.2), in conjunction with an adaptation of the human state estimation mechanism developed by Nakaso *et al.* [20], is used to detect general fatigue in the user. By learning from the user's behavior as a consequence of lighting conditions—both in terms of colors and intensities—the system can learn to identify which combinations of colors and intensities ameliorate or exacerbate the user's fatigue. Having made this identification, the illumination system continuously seeks to improve the state of the user by regulating the experience of the ambiance. Unlike predetermined scenarios, the system is not programmed to associate a given color with a given human state or action—nor vice versa—but rather the ML mechanisms establish such correlations as processed via HAR. More specifically, the localized *k*-NN classification model is capable of learning to predict which colors and intensities are conducive to mood amelioration / fatigue mitigation and to promote them. Such colors and intensities may change over time, as saturations in the frequency of particular colors and intensities over short periods of time could actually instigate an adverse effect. The ML mechanisms, however, can account for this change as they evolve accordingly.

Like the OSC-enabled manual override provided in the predetermined scenarios, the present scenario also integrates a *correction* mechanism based on human intervention. However, unlike the manual override, the *correction* mechanism is used to provide feedback—i.e., to "teach"—the system when a prediction is inaccurate. This fact is then considered in the next iteration of a new and updated classification model. The *correction* mechanism is implemented via a free and open-source OSC iOS™ *application*, viz., *Control* (by Charlie Roberts) (see Figure 7).

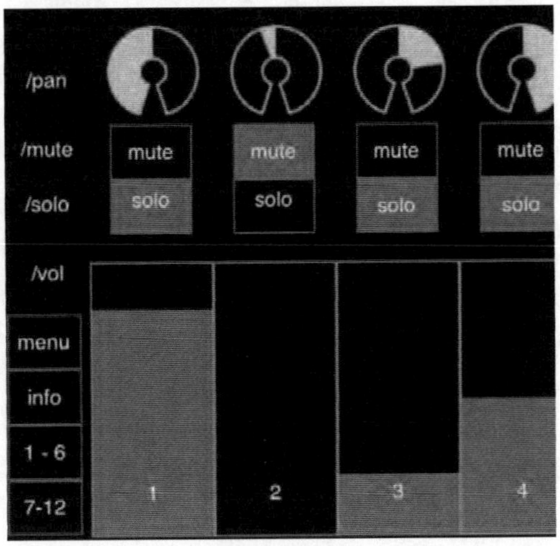

Figure 7: iOS OSC Applications: *Control* (open-source).

4 Conclusions and Future Work

The present paper attempts to promote a highly sophisticated intelligent built-environment framework based on D2RP&O principles and methodologies. It does so by presenting an implementation where sophisticated intelligence is imbued both in physical as well as computational terms. It promotes high-resolution computational intelligence by integrating ML mechanisms for HAR via a subsystem of dynamic *ad hoc* clustering. It also promotes high-resolution intelligence in terms of the built-environment by demonstrating how an interactive / adaptive illumination system can learn—via the detailed ML mechanisms—to reduce user-fatigue via the promotion of certain colors and intensities and the mitigation of others. Finally, by integrating both kinds of intelligences, the present implementation demonstrates the feasibility of a more sophisticated heterogeneous, open and scalable, and open-source AmI solution. Nevertheless, further work must be conducted in order to validate said feasibility in more complex scenarios, with more explicit and sophisticated expressions of D2RP for the integration of intelligence in the built-environment, and with correspondingly sophisticated expressions of D2RO with respect both to advances in ICTs as well as to ML mechanisms based on *Artificial Neural Networks* (ANNs).

A next iteration of a high-resolution intelligence implementation is presently being developed (1) to implement ANN-based ML; (2) to enhance fatigue detection; and (3) to extend the built-environment's ICT ecosystem to be open to more proprietary products and protocols.

With respect to the first task: in the present responsive stage implementation, the system detects fatigue exclusively via the analysis of detected patterns in the eyes (i.e., eyelids) of the user in conjunction with HAR data. However, due to the ambiguous character of human activity, it is difficult to differentiate a fatigued gait from a slow yet healthy one. The fatigue-detection precision may be improved by building on Ogawa *et al.*'s [21] facial-recognition and visual knee-position detection methods.

With respect to the second task: supervised and unsupervised learning ANN models are being explored for their suitability to given tasks. It is not necessarily the case that ANN models must be superior to SVM or *k*-NN models, as performance depends on the complexity, scale, and scope of the given tasks. At present a comparison matrix of ML models is being developed.

With respect to the third task: the system-architecture that the authors have been developing incrementally is highly heterogeneous both in terms of hardware, software, and communication protocols. This will continue to be a core interest in subsequent work. For example, present work is being conducted (A) to integrate Amazon®'s *Alexa Voice Service*™ into the ecosystem of the intelligent built-environment; and (B) to implement LoRaWAN as a *Low Power Wide Area Network* (LPWAN) protocol in order to extend interactions between high-resolution intelligence built-environments, whether these be private or public.

Acknowledgements

This paper has profited from the contribution of *Hyperbody* researchers and students involved in the described project.

References

[1] H. H. Bier, "Robotic Building as Integration of Design-to-Robotic-Production & Operation," *Next Generation Building*, no. 3, 2016.

[2] D. Anguita, A. Ghio, L. Oneto, X. Parra, and J. L. Reyes-Ortiz, "A Public Domain Dataset for Human Activity Recognition Using Smartphones," in *Proceedings of the 21th European Symposium on Artificial Neural Networks, Computational Intelligence and Machine Learning, ESANN 2013*, 2013.

[3] J. L. R. Ortiz, *Smartphone-based human activity recognition*. Cham: Springer, 2015.

[4] F. Palumbo, C. Gallicchio, R. Pucci, and A. Micheli, "Human activity recognition using multisensor data fusion based on Reservoir Computing," *AIS*, vol. 8, no. 2, pp. 87–107, 2016.

[5] A. Liu Cheng and H. H. Bier, "An Extended Ambient Intelligence Implementation for Enhanced Human-Space Interaction," in *Proceedings of the 33rd International Symposium on Automation and Robotics in Construction (ISARC 2016)*, 2016.

[6] ____, "Adaptive Building-Skin Components as Context-Aware Nodes in an Extended Cyber-Physical Network," in *Proceedings of the 3rd IEEE World Forum on Internet of Things*: IEEE, 2016, pp. 257–262.

[7] A. Liu Cheng, C. Georgoulas, and T. Bock, "Fall Detection and Intervention based on Wireless Sensor Network Technologies," *Automation in Construction*, 2016.

[8] H. H. Bier, "Robotic building(s)," *Next Generation Building*, no. 1, 2014.

[9] H. H. Bier and S. Mostafavi, "Robotic Building as Physically Built Robotic Environments and Robotically Supported Building Processes," in *Human-computer interaction series, Architecture and interaction: Human computer interaction in*

space and place, N. S. Dalton, H. Schnädelbach, M. Wiberg, and T. Varoudis, Eds, Switzerland: Springer International Publishing, 2016, pp. 253–271.

[10] K. Oosterhuis, *004 | ProtoCELL.* Available: http://www.oosterhuis.nl/?p=23 (2017, Mar. 29).

[11] Hyperbody®, *GSM3: Game Set and Match: International Symposium & Exhibition.* Available: http://gsm3.hyperbody.nl/ (2016, Nov. 09).

[12] M. A. Alsheikh, S. Lin, D. Niyato, and H.-P. Tan, "Machine Learning in Wireless Sensor Networks: Algorithms, Strategies, and Applications," *IEEE Commun. Surv. Tutorials*, vol. 16, no. 4, pp. 1996–2018, 2014.

[13] W. Xiao and Y. Lu, "Daily Human Physical Activity Recognition Based on Kernel Discriminant Analysis and Extreme Learning Machine," *Mathematical Problems in Engineering*, vol. 2015, no. 1, pp. 1–8, 2015.

[14] A. E. Villa, *Artificial neural networks and machine learning-- ICANN 2012: 22nd International Conference on Artificial Neural Networks, Lausanne, Switzerland, September 11-14, 2012, Proceedings*. Berlin, New York: Springer, 2012.

[15] J. Andreu and P. Angelov, "An evolving machine learning method for human activity recognition systems," *Journal of Ambient Intelligence and Humanized Computing*, vol. 4, no. 2, pp. 195–206, https://link-springer-com.ezproxy.library.ubc.ca/content/pdf/10.1007%2Fs12652-011-0068-9.pdf, 2013.

[16] E. L. Salomons, P. J. M. Havinga, and H. van Leeuwen, "Inferring Human Activity Recognition with Ambient Sound on Wireless Sensor Nodes," (ENG), *Sensors (Basel, Switzerland)*, vol. 16, no. 10, 2016.

[17] L. D. Dalcin, R. R. Paz, P. A. Kler, and A. Cosimo, "Parallel distributed computing using Python," *Advances in Water Resources*, vol. 34, no. 9, pp. 1124–1139, 2011.

[18] F. Pedregosa, and G. Varoquaux et al, *Scikit-learn: Machine Learning in Python.* Available: http://arxiv.org/pdf/1201.0490.

[19] L. Buitinck, and G. Louppe et al, *API design for machine learning software: experiences from the scikit-learn project.* Available: http://arxiv.org/pdf/1309.0238.

[20] S. Nakaso, J. Güttler, A. Mita, and T. Bock, "Human state estimation system implemented in an Office Deployable Getaway based on multiple bio information," in *Proceedings of the 33rd International Symposium on Automation and Robotics in Construction (ISARC 2016)*, 2016.

[21] A. Ogawa, A. Mita, C. Georgoulas, and T. Bock, "A Face Recognition System for Automated Door Opening with parallel Health Status Validation Using the Kinect v2," in *Proceedings of the 33rd International Symposium on Automation and Robotics in Construction (ISARC 2016)*, 2016, pp. 132–140.

Flexible management for concrete curing at low temperature based on learning from cases

K. Andrzejczak[a], A. Karlowski[b], J.O. Paslawski[b]

[a]Institute of Mathematics, Poznan University of Technology, Poland
[b] Institute of Structural Engineering, Poznan University of Technology, Poland

E-mail: karol.andrzejczak@put.poznan.pl, ark_kom@wp.pl, jerzy.paslawski@put.poznan.pl,

Abstract

The right technological decisions-making during the construction process ensures online monitoring of process in progress and the external environment. The use of wireless technologies to collect data from sensors and transmit data creates new opportunities in this field. The most important of these is the ability to make a decision and its possible real-time correction.

The aim of this paper is to identify methods of decision making process during concreting based on a flexible approach and learning from cases.

The technology decision-making system is composed of three phases: (1) the strategic phase-based selection of avoiding adverse weather conditions or determining the effects of disturbances on achieving expected results despite difficulties; (2) the tactical setting in which the best concrete mix option adapted to actual and predicted operating conditions; (3) operational monitoring of the processes in progress and the environment, including simulation and warning systems, to possibly correct the decision in the form of additional actions in the event of an emergency. The purpose of this concept is to reduce risk by applying flexibility.

The case study shows the possibility of flexible management of construction processes using different technology (formwork heating, the use of admixtures, heating the concrete, the addition of PCM, etc.). First flexible tactics is based on robustness taken for normal and optimistic scenario. The second (complementary) tactic is adaptability that provides the ability to heat the concrete element in case of freezing emergency.

Keywords –

Flexibility; phased decision system; construction operation, concrete modification, robustness, learning from cases, risk management

1 Introduction

Fundamental problems related to construction management are mainly connected with the budget overrun, the expected time of completion (completion date of the construction project) and quality defects (not only during acceptance, but also during warranty and post-warranty period). The low efficiency of the manufacturing processes in construction is often emphasized compared to the other areas of the economy based on factory production. It is alarming that the expected positive effects of computerization in the construction industry are not confirmed in practice [1]. Data collection is a serious problem (data is growing very fast - big data [2]. In construction industry is still used to many traditional data collection methods (very limited automation). There is still no use in building a huge opportunity to learn from cases.

The purpose of this paper is to indicate the possibility of improving the management of construction processes based on flexible approach using learning from examples.

The following paragraph introduces the state of the art of building process management and development trends in the field of Construction Management with particular focus on Industry 4.0 in Europe and Made in China 2025 in Asia.

The concept of flexible management is discussed as a remedy for the problems arising from the dynamically changing environment and the typical decision-making situation some time before the occurrence of the event. Thanks to the proposed active flexibility (adaptability) and passive flexibility (robustness), the opportunity arises to adjust the decision and thus reduce the negative impact of the environment or use the opportunities. Presented case study takes into account the possibilities of improving decision-making process on the basis of learning from cases in technological decisions for concreting at low temperature. It stems on the experience of the authors and research conducted in Central Europe, but the results can be applied to countries with similar climatic conditions (humid continental climate) [3].

2 New ideas in Production Management

Serious problems associated with the implementation of construction projects are connected with the specificity of the construction industry and include, inter alia:

- Dependence on both internal variable environment - variable performance of process and external - variable execution conditions (weather, groundwater level, etc.)
- Long quality chains, resulting from the complexity of construction process.
- Extremely long life cycle of buildings compared to many products fabricated in other economy branches
- very slow pace of automation and robotics implementation
- The need to make decisions on the basis of incomplete data and well in advance (e.g. the ordering of ready-mix concrete 24 h before concrete delivery)
- Negligible impact on some important processes (e.g. maturation of concrete, especially in the case of large scale operations - transport infrastructure facilities).

The above factors with high complexity and a wide range of volatility generate significant levels of risk and uncertainty. In this situation, solutions are being sought to make it possible to make construction work similar to other areas of production (e.g. industrialization, automation, flexible process management, use of common management methods successfully used in other industries e.g. Lean Management, the concept of Industry 4.0 and Made in China 2025 in construction industry). The synergy of these approaches with BIM technology is a powerful potential for improving management methods in the construction industry.

2.1 State-of-art of Construction Process Management

Analyzing the methods of planning in construction basic milestones are two the following methods:

- Gantt charts [4], whose application exceeds 100 years already.
- Network models [5] (also with later versions of PERT), which have been in use since the mid-twentieth century, and thus are used for more than 50 years).

Both of these methods are well-known and are still in use (they are the basis of typical planning tools like MS Project®, Primavera®). Undoubtedly, their merits include the opportunity to present the planned course of the construction project. Disadvantages of traditional tools/planning methods rely primarily on not taking into account the all phenomena of interdependence and dynamics taking place between planned activities [6]. Attempts to take them into account are generally encountered by problems arising from the need to simplify these highly complex and difficult-to-foresee relationships.

From the point of view of these drawbacks, the following solutions deserve attention [6,7,8,9]:

- Simulation of building processes.
- Taking into account system dynamics.
- Use of resource buffers.
- Use a flexible approach.
- Lean Management and Six Sigma.
- Monitoring of environment and process in progress.

Naturally, the above list does not completely exhaust the possibilities of improving the methods of managing construction processes. It seems logical that, apart from the planning mechanisms themselves, elements that influence the effects of applied methods, such as communication methods, play a significant role [10]. It should be emphasized that the improvement of the communication method is a priority of the Industry 4.0 [11]. It is also worth noting that the synergies of the different methods are also used. In addition to classic examples such as Lean and Six Sigma Management, TRIZ and Axiomatic Design has to be noted [12].

In summary overview of the management used in Construction Management can identify three key viewpoints:

- accept variability and adopt appropriate financial and time buffers (Black Swan type [13, 14]).
- waste elimination for processes (Lean Management)
- applying a flexible approach that allows both passive (resistance) and active (adaptive) flexibility to be used.

Of course, it is possible to use synergies between the use of agile management and e.g. Lean Management [15].

Analyzing the gradual development of a flexible approach activity of researchers in the field of economics should be considered as important. Schumpeter, who introduced flexibility in action and adopted new methods of behavior as an element of innovation in 1934, proposed it as an element of innovation [16]. De Neufville proposed the first use of flexible management in 1972 [16], which a year later referred directly to transport systems [17]. The Kaplinski concept for cushioning in construction processes [18] is worth noting - for example in the delivery of concrete mix, the use of intermediate tanks. Another interesting application of flexibility is Haplin's proposal for rock processing [19]. At the next symposium, ISARC Bernold proposed the Flexible Construction System concept for the production of wooden beams [20]. Notes require discussion between a team led by Thomas [21] who created an interesting Lean Management concept in construction based on the idea of flexibility and Howell [22], which is faithful to the underlying assumptions of Lean Management (seeking to reduce the current performance variable).

The monographs de Neufville [23] and X [24] are very valuable sources of information on the elasticity of construction.

The presented concept of flexibility is based on the idea of FLAMENCO (Flexible Management in Construction) proposed in 2009 [25].

Intriguing concepts of introducing flexibility in the construction industry are undoubtedly Ofori's [26] work on organizational flexibility, Shahu [27], taking into account the advantages of introducing flexibility in construction projects, Olsson [28], which refers to large projects to avoid budget overruns. , Caunhye & Cardina [29] as a concept of robust optimization and appropriate decision-making.

Of course, the flexible application of the proposed approach in the construction management is justified under exposure to the impact of turbulently changing environment and a sufficiently large scale construction projects (airport, stadium, multi-level parking, etc.).

2.2 Industry 4.0

Industry 4.0 is a concept based on the fourth industrial revolution aimed at intelligent and flexible production [31]. Taking into account the unique opportunities created by the use of modern information and communication technology and more commonly used in the economy and society, a learning organization capable of acting (more and more self-reliant) should be identified in terms of risk and uncertainty.

With regard to construction, it creates the following opportunities [28]:

- Smart construction site (effective trucking of construction operations/processes, materials, prefabricated elements, access control – theft protection, proactive process management – prevention of non-quality, lean management based on people, material and equipment trucking, etc.).
- Simulation and modelling (construction Health & Safety training, defect management using BIM technology, Virtual Reality, Mixed Reality and 7D BIM synergy).
- Digitalization and visualization (cloud computing, Big Data, Case based reasoning, Flexible/Agile Management, mobile computing).

Indicated directions for the applications of Industry 4.0 in construction are most consistent with the management of construction processes using the flexibility / agile management concept in Construction Management. The key benefits of using Industry 4.0 are:

- Enhancing the ability to collect real-time data about building processes and their surroundings.
- Transfering decision-making moments closer to the expected key events (limiting time-outs in decision-making).
- Correction of decision-making based on tactics of adaptability.).

2.3 Made in China 2025

The concept of "Made in China 2025" is concerned with smart production, simulation, modeling, and digitalization, and visualization, but assumes more strategic goals for replace foreign technology with China technology [29].

An important element combining the concepts of Industry 4.0 and Made in China 2025 is the engagement of Germany in both projects. One example is VW's involvement in the construction of new auto plants in China [30].

2.4 Flexible management in the construction industry

Flexible management of building processes in the proposed approach is based on the premise of introducing appropriate flexibility options (active and passive) that allow for adaptation to the changing environment (internal and external) in order to continue to work despite difficulties (possible and unpredictable) with little change in costs and execution time. It is assumed that the quality requirements are met despite these changes.

Flexible approach adopted in this paper is based on two components [32]:

- Passive flexibility (robustness) understood as a first component of flexibility to achieve the assumed results of the construction process in turbulent environment.
- Active flexibility (adaptability) as a second component of flexibility based on adaptation to changing ambient conditions based on monitoring of this environment and construction processes in progress.

The fundamental difference between the two elements is that independence from monitoring for passive flexibility is the key source of information needed to apply active flexibility tactics. By analyzing the typical application of flexibility in the concreting process, it can assumed:

- Application of passive flexibility to the most probable and optimistic scenario.
- Adoption of tactics of active flexibility appropriate for the pessimistic scenario.

It is therefore important to separate in time the tactics of flexibility in accordance with the two components of it, as illustrated in this case study.

3 Risk analysis in the decision system design

Decision making under uncertainty, and such conditions are faced in the construction projects, will be supported by carrying out risk analysis. The purpose of the risk analysis is to assess the degree of uncertainty in the implementation of the task and the cost prediction for the various states of nature. Risk analysis begins with the identification of critical factors or conditions because of the risk of non-completion of a construction investment project. Identification of the critical elements of investment implementation is the basis for:

- Clarification of various alternative of construction technology to carry out the tasks and activities already in the planning phase of the decision-making system
- Choice of appropriate technology to reduce its variability, in the sense of better tolerance of undesirable circumstances
- Determining whether the available construction technology and means of performing the task meet specific requirements (effectiveness)
- Set requirements and cost estimates for various job fulfillment technologies
- Document the acceptable level of safety and risk.

In designing of the decision system, three phases can be generally distinguished. In strategic planning phase of the building the great flexibility choice is given to decide when to perform a specific task. However, due to the large advance of the planning phase from the completion of the task, usually there is a lack of detailed information needed to make the final choice of method of operation. Risk analysis in such cases provides a basis for comparing different alternatives. In turn, in the operational phase there are already greater opportunities for access to more accurate forecasting data. This is due to the ability to take into account the experience and knowledge of those responsible for carrying out the construction task and for access to short-term data such as weather forecast. In such cases, a more precise analysis of the impact of the action can be made by more clearly specifying the decision-making system. Risk management refers to all activities, conditions and events that may affect the organization and its ability to achieve the goals and vision of the organization [31]. The design decision system refers to a construction company providing services within a certain range of tasks and under certain climatic conditions. Therefore, it is necessary to clarify any possible activities of this company and any possible conditions and events that affect the achievement of the objectives and vision of this type of company. For a better picture, the enterprise risk management task can be divided into three main categories:

- strategic risk,
- financial risk,
- operational risk.

Strategic risk covers aspects and factors relevant to long-term business strategies and plans, for example:

- mergers and acquisitions,
- technologies,
- competition,
- laws and regulations,
- labor market.

Financial risk relates to the company's financial situation and includes, inter alia:

- market risk related to costs of goods and services, exchange rates, etc.,
- credit risk related to debtors' payment problems,
- cash-flow risk associated with access to the company's capital.

Operational risk takes into account conditions and events that may affect the normal operation of a construction company. Such events are:

- Random events, including accidents, failures, defects or qualitative deviations of materials.
- Natural disasters, sabotage, unhappy workers.
- Loss of competence by leaving key personnel.
- Legal circumstances, such as those related to defective contracts and civil liability insurance.

This article deals primarily with operational risk.

The primary task of analyzing the risk associated with a particular activity is to identify any initiating event that, due to their negative consequences, is called "unfavorable events" or "conditional unfavorable events". Unfavorable events are a source of threats in the implementation of the planned tasks. An example of an unfavorable event is, for example, non-achievement of the frost resistance of the concrete at the given time period.

In addition, predictable disturbances should be considered (predictable with less or greater accuracy e.g., depending on weather forecast source).

The basic tasks in designing of decision system that take into account the risk analysis are:

(A) Planning (defining the scope of work undertaken, gathering information and organizing work; choice of analysis method)
(B) Risk assessment (identification of undesirable and disruptive (foreseeable); analysis of the causes of the risks and effects of the action);
(C) Risk management (comparison of technological alternatives; review of possible decisions and their evaluation);
(D) Decision-making;
(E) Evaluation of the results of the decision (often a few interrelated decisions);
(F) Learning from past experiences.

It is extremely important for individual decision phases to determine the conditional consequences of possible actions at different states of nature.

The decision-making process composed of three phases will be traced in the following case study.

4 Case study

This system deals with a three-phase approach to the planning, execution and monitoring of the concreting process. The subject system is related to the minimization of the risk of outdoor concreting processes without the possibility of covering - under conditions dependent on the influence of external factors such as temperature drop below 0°C and atmospheric precipitation. For simplicity, the ambient temperature is the leading parameter. A typical illustration of the problem is the implementation of the parking plate at the service station.

The adopted model was developed on the basis of long term research, experience and statistical analysis of weather forecast data. On this basis, it identifies three phases of the decision-making process related to the planning, preparation and implementation of the process of concreting (including concrete curing process).

Decision tree after case analysis in RapidMiner® is presented in Figure 1.

Phase 1 (corresponding to strategic level) involves setting time periods for which the planned task:

- it is possible to perform without major interruptions in the period from May 1st to May 31st for example, during which month there is no problem in the correct course of the concreting process

- it is possible to implement, but in the adopted period e.g. from 1st to 30th November, there may be time periods in which the implementation will not be possible. In the case of unfavorable conditions, specific actions need to be taken (modifications of the concrete mix in the form of PCM additive and / or other equipment such as heaters or heated formwork) or action must be stopped. The basic strategy in this phase of decision-making is to strive for implementation of concreting process during the favorable period.

Phase 2 - (tactical level) associated with the preparation for the concrete preparation process with typical weekly planning – e.g. 3-5 days prior to the planned execution date, based on the short-term weather forecast and the selection of the appropriate concrete mix (adapted to the optimistic and most likely scenario - passive flexibility with possible special actions planned in case of pessimistic scenario (active flexibility). The natural action associated with the aforementioned precautionary principle (failure to take action in the absence of reliable information on the conditions of concreting) will require the decision-maker to postpone the decision until the 24-h forecast is reached, which will provide high verifiability (often referred to as over 95% verifiability).

The decision matrix for the choice of concrete mixtures according to the current ambient temperature and the predicted temperature (24/48 forecast) is summarized in Table 1.

As shown in Table 1, in the case of concreting in the temperature range of 5 to 10°C and the predicted temperature drop in the range of -10 to -5°C, it is possible to use a specially modified mixture (e.g., admixtures for low temperature concreting and / or PCM). However, it is not possible to perform the process on the basis of normal concrete mix. Of course, in concrete situations it is possible to heat the concrete mix, but in this case it is not assumed to implement this option. Correct selection of technological options assuming a reliable forecast during this period will ensure reaching the frost resistance of the maturing concrete (5-10 MPa) despite the difficulties.

Phase 3 (Operational level - real time management) -. During the process of concreting - in the early phase of concrete binding, the temperature range of the temperature changes in the concrete and the surrounding of the concrete is monitored. Based on the analysis of the temperature distribution within the element and its surface, in order to minimize the risk of damaging the structure of the concrete element as a result of a decrease below zero, it is possible to apply additional heating at the surface of the element to be made. This requires the use of continuous monitoring and simulation of temperature changes in the concreted element to take pre-emptive action. This is an example of the application of active flexibility in the case of a pessimistic scenario.

Table 1. Decision matrix for concreting at low temperature

temp. now [°C]	temperature forecast [°C]				
	5-10	0-5	(-5)-0	(-10)- (-5)	(-15)-(-10)
5-10	YES-CMN YES-CMM YES-CMP YES-CMMP	NO-CMN YES-CMM YES-CMP YES-CMMP	NO-CMN YES-CMM YES-CMP YES-CMMP	NO-CMN NO-CMM YES-CMP YES-CMMP	NO-CMN NO-CMM NO-CMP NO-CMMP
0-5	NO-CMN YES-CMM YES-CMP YES-CMMP	NO-CMN YES-CMM YES-CMP YES-CMMP	NO-CMN NO-CMM YES-CMP YES-CMMP	NO-CMN NO-CMM YES-CMP YES-CMMP	NO-CMN NO-CMM NO-CMP NO-CMMP
(-5)-0	NO-CMN YES-CMM YES-CMP YES-CMMP	NO-CMN NO-CMM YES-CMP YES-CMMP	NO-CMN NO-CMM YES-CMP YES-CMMP	NO-CMN NO-CMM NO-CMP YES-CMMP	NO-CMN NO-CMM NO-CMP NO-CMMP
(-10)- (-5)	NO-CMN NO-CMM YES-CMP YES-CMMP	NO-CMN NO-CMM YES-CMP YES-CMMP	NO-CMN NO-CMM NO-CMP YES-CMMP	NO-CMN NO-CMM NO-CMP NO-CMMP	NO-CMN NO-CMM NO-CMP NO-CMMP
(-15)-(-10)	NO-CMN NO-CMM YES-CMP YES-CMMP	NO-CMN NO-CMM YES-CMP YES-CMMP	NO-CMN NO-CMM NO-CMP NO-CMMP	NO-CMN NO-CMM NO-CMP NO-CMMP	NO-CMN NO-CMM NO-CMP NO-CMMP

Table 1 Legend: CMN - standard concrete mix; CMM-modified concrete mix; CMP - concrete mix with PCM; CMMP - concrete mix modified with PCM.

In general, however, the key technological decision is to use the tactics of robustness. The method of learning from the examples was used to adopt the correct decision. It assumes a continuous collection of information about the processes carried out, assuming also a negative case (when the concreting process was not successful).

Based on this, RapidMiner® has built decision trees to generate decision rules. When building a decision tree, the creation of a node results from the division of data that has reached it. This decision is made on the basis of one attribute. Attribute is selected to ensure the highest possible drop of class diversity in child nodes. It is possible to take action when generating a decision tree – e.g., stopping a break earlier if there is too little in the node (minimal leaf-size). The decision tree example is shown in Figure 1.

Based on a case-by-case analysis, you can determine the effectiveness of a technology option under certain execution conditions. One can also specify the cost of implementation for each technological option. Based on this information, one can gradually improve the decision table (Table 1) by learning from examples.

In conclusion, the following algorithm can be specified:

1) building a data table
2) generating decision tree
3) filling / improving decision table (flexibility options for given conditions / scenarios)
4) assessment of current conditions + forecast

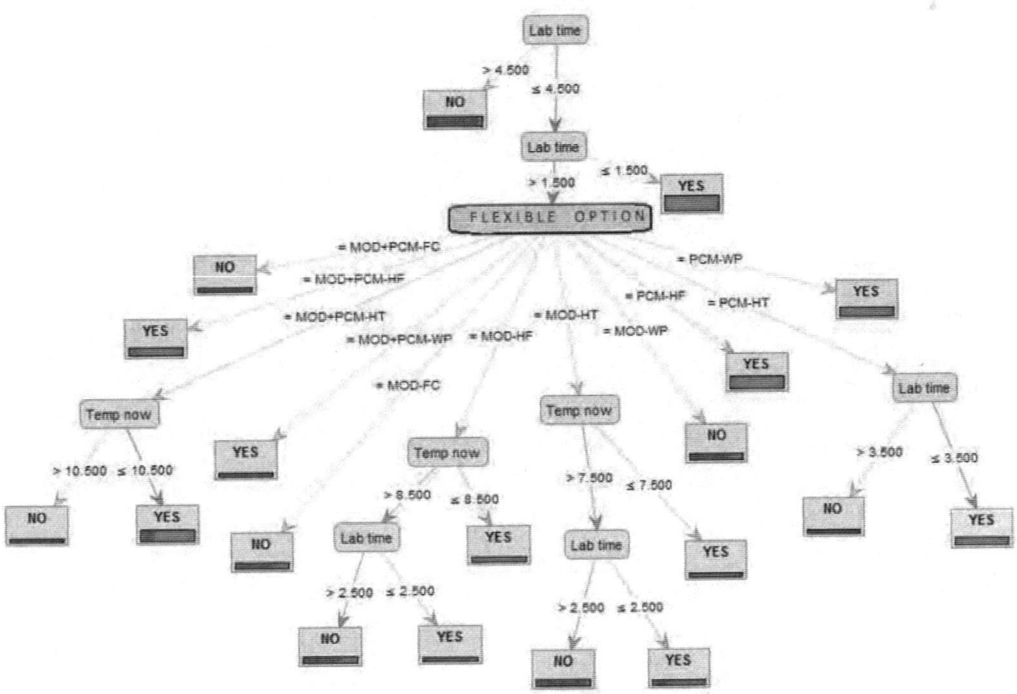

Fig. 1. Decision tree after case analysis in RapidMiner®

5) decision making - choice of passive and active flexibility
6) decision implementation
7) monitoring and decision-making (maturing of concrete)
8) assessment of effectiveness and cost of adopted tactics
9) return GO TO (1).

Right parity of passive and active flexibility in operation management assures quality in adverse weather condition.

5 Conclusions

Presented theoretical bases together with the case study discussed above allows to draw the following conclusions:

1. Construction process management requires improvement, for example by implementing methods successfully used in other sectors of the economy.
2. Industry 4.0 and Made in China 2025 should be regarded as exceptionally conducive to development, which emphasizes the use of: smart construction site, simulation and modeling, and digitalization and wireless technology.
3. The introduction of Flexible Management for the management of construction processes is a logical consequence of these directions.
4. It is important to estimate the appropriate share of passive and active elasticity parities taking into account their economic effectiveness and implementation efficiency.
5. Due to the specificity of construction (dependence on often turbulent environment) it is important to choose the source of information about current and anticipated environmental conditions (e.g. weather) and the selection of key factors characterizing this environment.
6. Based on collected data, the decision-making system improves its decision rules (use of expert rules, learning examples, machine learning etc.)
7. Particularly important is the use of learning from examples, which may assist in the selection of technological options and the appropriately reliable source (or sources) of weather forecasts.

Acknowledgements

The publication was created with the support of statutory funds of Institute of Structural Engineering at Poznan University of Technology.

References

[1] Blocher D. Presentation of Budimex S.A. strategy during meeting of Meeting of the Civil Engineering Section of the Building Engineering Department at the Committee of Civil and Water Engineering of the Polish Academy of Sciences, Warsaw, 2017

[2] Bilal M., Oyedele L.O., Qadir J., Munir K., Ajayi S.O., Akinade O.O. , Owolabi H.A., Alaka H.A., Pasha M., Big Data in the construction industry: A review of present status, opportunities, and future trends, Advanced Engineering Informatics, 30(3):500-521, 2016

[3] Kondracki J. Geography of Poland: physical geographic mesoregions [in Polish], Warsaw, PWN, 1994

[4] Gantt H.L. Work, Wages and Profit, Engineering Magazine, NY 1910; republished as Work, Wages and Profits, Easton, Pennsylvania, Hive Publishing Company, 1974

[5] Kelley J. and Morgan W. Critical-Path Planning and Scheduling. Proceedings of the Eastern Joint Computer Conference, 1959

[6] Azaraiee H., Zayed T. and Moselhi O. Dynamic planning of construction activities using hybrid simulation. Automation in Construction, 49:176-192, 2015

[7] Połoński, M. Application of the work breakdown structure in determining cost buffers in construction schedules. Archives of Civil Engineering, 51(1):147-161

[8] Dallasega P., Frosolini M., Matt D. An approach supporting Real-Time project management inplant Building and the construction industry, 21st Summer School "Francesco Turco" industrial Systems Engineering:247-251, 2016

[9] Sobotka A., Pacewicz K. Building site organization with 3D technology in use. Procedia Engineering 161:407-413, 2016

[10] Smith N. Engineering Project Management, Blackwell Science, Oxford, 2002

[11] Agnihotri R., New S. Industry 4.0 Data Analytics. Guide to developing analytics with case studies, cop. 2012

[12] Borgiannia Y., Matt D.T. Applications of TRIZ and Axiomatic Design: a Comparison to Deduce Best Practices in Industry, Procedia CIRP 39:91 – 96, 2016

[13] Piore A. Planning for a Black Swan. Scientific American, 304(6):48-53, 2011

[14] Flage N. and Aven T. Emerging risk – Conceptual definition and a relation to black swan type of events. Reliability Engineering and System Safety 144:61–67, 2015

[15] Nowotarski P., Paslawski J., Lean and Agile

Management Synergy in Construction of High-Rise Office Building. Archives of Civil Engineering, 62(4), 133-148, 2016
[16] Schumpeter J. Economic development theory [in Polish], PWN, Warsaw, 1962
[17] De Neufville R. Dynamic strategic planning for technology policy, International Journal of Technology Management, 19(3/4/5):225-245, 200
[18] Kaplinski O. Diploma Seminar at Poznan University of Technology, 1984
[19] Halpin D.W. Adaptation of flexible manufacturing concepts to construction. In Proceedings of the 4th International Symposium on Robotics and Artificial Intelligence in Building Construction, Haifa, Israel, pages 704-717, 1987
[20] Thomas H.R., Horman M.J., de Souza U.E.L., Zavirski I. Reducing variability to improve performance as a lean construction principle, *Journal of Construction Engineering and Management*, 128(2):144-154, 2002
[21] Howell G.A. Howell G. A., Ballard G., Tommenlein I. D., Koskela L. Discussion of "Reducing variability to improve performance as a lean construction principle (...)".Journal of Construction Engineering and Management, 130(2):299-300, 2004
[22] de Neufville R. and Scholtes S. Flexibility in Engineering Design. MIT Press, Cambridge, 2011
[23] Paslawski J. Flexibility implementation in construction process engineering; The 25th International Symposium on Automation and Robotics in Construction. ISARC 25th:610–615, 2008
[24] Lim B.T.H., Ling F.Y.Y.,.Ibbs C.W, Raphael B., Ofori G., Mathematical models for predicting organizational flexibility of construction firms in Singapore. Journal of Construction Engineering and Management, 138 (3): 361-375, 2012
[25] Olsson N.O.E. Implementation of pre-defined potential scope reductions in projects, Procedia Computer Science 64:387–394, 2015
[26] Caunhye A.M. and Cardin A-M. An approach based on robust optimization and decision rules for analyzing real options in engineering systems design, IISE Transactions [in Press]
[27] Industrie 4.0 platform. Online: http://www.plattformi40.de/I40/Navigation/EN/Industrie40/WhatIsIndustrie40/what-isindustrie40.html;jsessionid=BDC752C32376E2D26EABED5EBED93F2B, Accessed: 29/03/2017
[28] Oesterreich, T. D.; Teuteberg, F. Understanding the implications of digitisation and automation in the context of Industry 4.0: A triangulation approach and elements of a research agenda for the construction industry. Computers in Industry, 83:121-139, 2017
[29] Wubbeke J., Meissner M., Zenglein M.J. Ives J., Conrad B. Made in China 2025 The making of a high-tech superpower and consequences for industrial countries On line: http://espas.eu/orbis/sites/default/files/generated/document/en/MPOC_No.2_MadeinChina_2025.pdf Accessed: 29/03/2017
[30] Ocksen J. Presentation of VW strategy, Poznan University of Technology, 2016
[31] Aven T. Risk Analysis. Assessing Uncertanties beyond Expected Values and Probabilities. John Wiley & Sons, Ltd, 2008
[32] Paslawski J. Flexibility in Construction Process Management [in Polish], Publishing House of Poznan University of Technology, Poznan, 2009

34th International Symposium
on Automation and Robotics
in Construction

ISARC 2017

June 28 – July 1
TAIPEI, TAIWAN

Linked Data System for Sharing Construction Safety Information

Akeem Pedro [a], **Do Yeop Lee** [a], **Rahat Hussain** [a] **and Chan Sik Park** [a]

[a] Department of Architectural Engineering, Chung Ang University, Seoul, South Korea
E-mail: lanrepedro3@gmail.com, cpark@cau.ac.kr

Abstract –

Despite the consistent growth of global markets and emergence of cutting-edge technologies, construction jobsites remain among the most dangerous workplaces with extremely high accidents rates. In the safety management process, it is necessary to review and assess a variety of safety documents and data. Numerous public agencies provide open access to accident data, safety checklists, templates and other useful documents for safety management. However, these safety related contents are multifarious, fragmented and often stored in unstructured ways. As a result, the process of finding and utilizing required safety information for specific safety tasks is challenging and inefficient. In order to address this issue, this paper proposes an innovative approach using linked data and semantic web technologies for integrating and sharing construction safety information from diverse sources. A construction safety contents ontology is developed, and construction safety information from accident cases, Job Hazard Analyses (JSA), safety rules and training contents is integrated through RDF formats and SPARQL queries. In order to enable convenient and effective retrieval of information, this study develops a categorization and classification of safety-related contents for safety management processes. The proposed system is expected to improve access and sharing construction safety information, reduce data search time and improve data accuracy. Through this, the approach may enable reductions in safety incidents and overall improvements in safety management and performance.

Keywords –

Building Construction Safety, Linked Data, Safety Information Management;

1 Introduction

Construction remains one of the most accident-prone industries, with recurring injuries and fatalities, even though it is one of the largest contributors to national economies worldwide [1]. Despite strongly-enforced safety policies and continuous efforts from safety professionals, the construction industry reports an incident rate which doubles the industrial average. In addition to injuries and fatalities, safety incidents also result in serious project delays and cost overruns. During the various stages of safety management, it is necessary to access and review a plethora of safety documents and data. The advent of the internet and web-based services allows the convenient online publication and sharing of safety contents such as accident cases, rules and regulations, training documents and checklists. In this regard, various public agencies such as the Occupational Safety and Health Administration (OSHA) [2] and National Institute for Occupational Safety and Health (NIOSH) make an array of useful safety resources available to the public domain. However, even though multifarious safety contents are accessible, finding pertinent safety information for specific safety tasks is difficult. This issue can be attributed to the fact that safety information is unstructured, fragmented and dispersed across various sources on the web.

In order to improve safety management and safety outcomes in construction, numerous research studies have been conducted, focusing on visualization technology for enhanced safety and health training [3]; real-time safety monitoring systems [4] and BIM-integrated, automated safety planning and management systems [5] . However, despite the advancement of information technology in the building and construction industry, a missing link exists between safety management and information models [6]. Very few research efforts have considered improving construction safety information management in order to enable more efficient and systematic retrieval, utilization and sharing of safety contents. Furthermore, traditional database systems and keyword searches are limited in improving information discovery and retrieval [7]. Hence, the development of an accurate and efficient information management approach and system is crucial to enable users to conveniently search and access safety data and information.

This study takes advantage of state-of-the art information technologies such as ontologies, linked data and the semantic web to address the current issues in construction safety information sharing. Ontologies are considered an ideal tool for the explicit formal specification of domain knowledge and modelling of semantic relationships. Moreover, ontologies play a central role in publishing and integrating structured data on the Web through linked data [8]. Through the innovative application of these technologies, this study aims to develop a linked data system for retrieving and sharing construction information from diverse data sources. To this end, the structure and characteristics of construction safety information are first analyzed; safety contents are classified; and then formalized through the development of a construction contents ontology. The paper presents a preliminary study to address how construction safety data can be systemized through the RDF format, and how information can be retrieved efficiently through SPARQL queries. This paper is organized as follows: After the introduction, section 2 highlights the current state and issues regarding construction safety information sharing and management. Next, section 3 reviews prominent information management techniques such as ontology, semantic web and linked data in related research. Section 4 analyzes and classifies construction safety information, and then a linked data system framework for sharing and retrieving construction safety information is presented in section 5. Lastly, the paper is concluded with a discussion of findings and considerations for future works.

2 Current Issues in Sharing Construction Safety Information

During the safety management process, a myriad of safety data, information and documents are generated, referred to and stored in various ways. As a result of the dynamic and complicated nature of construction work, safety management tasks tend to be challenging [6]. Moreover, due to the unstructured nature and diverse formats and characteristics of construction safety data, it is difficult to retrieve information related to specific safety management task. This section discusses the current issues in sharing construction safety information based on literature reviews and usage analyses of safety data.

Firstly, it is necessary to clarify several key terminologies such as construction safety data, safety information and safety contents in order to understand the paper. Data can be defined as statistics or values of quantitative or qualitative variables, belonging to a set of items. As pointed out by Le et al [1], data is the lowest level of abstraction, followed by information, which is defined as a message, or collection of messages in an organized sequence, justifying a change in a construct that represents physical or mental experience or another construct [9]. In the context of this paper, construction safety data includes accident statistics, records and reports etc., while construction safety information is defined as useful descriptions and facts including rules and regulations, safety training materials and so on. Lastly, safety contents refers to a variety of safety related documents such as checklists, news articles, templates, and reports etc.

Construction is an information-driven industry, producing and requiring copious amounts of data throughout project tasks. Some information and data are stored in corporations' databases, while some are dispersed in unstructured formats across various agencies systems, online databases and websites. Safety information and knowledge remains fragmented in safety regulations, accident records, best practices and safety experts experiences [6]. Unstructured safety data are stored with differing formats, hence difficulties are encountered in comprehensively searching, analyzing and accessing desired information. Moreover, without a machine-readable format, it is tedious and difficult to manually generate insights from the vast amounts of existing information.

Within this study, conventionally unstructured construction safety information and data from accident and fatality cases, rules and regulations, safety training contents, news, document templates and statistics are considered. Accident cases are usually analyzed and published as casebooks, or shared online as news articles, or individually as data files for safety reporting procedures. However, these files, articles and casebooks exist in unstructured text-based formats, such as PDFs, Word documents, PowerPoint presentations, spreadsheets, HTML pages and so on. Similarly, safety training materials such as posters, lecture presentations, videos and interactive games are unstructured and scattered over private domain databases and publicly on the web. This issue is further exacerbated by language barrier problems faced by migrant workers [10], and the need for educational resources in various languages and mediums. In order to address these issues, it is necessary to understand the broad structure and characteristics of safety data, information and contents.
Information from accident cases, rules and regulations and training contents typically contain construction context information. Context is defined as a situation in which construction entities exist. An entity can be a person involved in a process, an action to complete a task, a resource, or a building component. Regarding construction safety, context information that needs to be stored includes work trade or type, work phase related work space, work elements, and hazard/accident type. In addition, detailed safety information pertaining to the severity and impact of safety incidents can enable more specific analysis of cases, articles and so on. Similarly, in the context of educational contents, document format,

medium and language information can enable more accurate content retrieval. Through context information and additional detailed information, project personnel can acquire valuable resources, knowledge not only for retrieving data at the specific task level, but also for interpreting safety data and generating insights at the corporation level. In order to enable users to systematically retrieve safety information, it is necessary to not only store safety data in a machine readable format, but also to describe safety domain information with a standardized structure. Through this, safety information from accident cases, rules and regulations, training contents, news, document templates and statistics can be integrated in a structured manner.

In addition to the lack of structure, available methods for finding pertinent safety information are also limited. In order to find specific safety information, personnel are required to manually search through casebooks, online databases, rule and regulations and so on. However, because these contents have differing categorizations according to hazard type, work type etc. the information retrieval process tends to be tedious. Safety data stored as HTML or document files can be accessed through keyword searches, based on content titles and descriptions. However, keyword based search is problematic because various forms of safety information are shared subjectively without any coherent structure or standard vocabulary. Hence, the development of improved safety information search mechanisms which take these issues into account is necessary.

3 Information Management Technologies and Approaches in Construction

3.1 Linked Data Approaches

The Semantic Web, which is an extension of the current Web not only allows the upload of data on the web, but also supports making links and connections to enable manual or machine based exploration of web data. In essence, the semantic web of data provides a framework that data sharing and reuse between applications, corporations and community boundaries (W3C). Linked data provides a publishing paradigm whereby the web is used to create typed links between data from different sources [11]. As noted by Berners-Lee [12], linked data is based on the following four basic principles: 1) the use URIs as names for things, 2) the use HTTP URIs so that people can look up those names, 3) When someone looks up a URI, provide useful information, using the standards (RDF, SPARQL) and lastly, 4) include links to other URIs so that they can discover more things. In linked data, all information is encoded with the Resource Description Framework (RDF), enabling anyone to refer to anything. Project participants can look up any URL in an RDF graph over the global web of data, and RDF triples can be used to explore this data space. Furthermore, data publishers may provide RDF dataset dumps for local replication of data, and SPARQL endpoints for querying the data directly [8].

Various domains have applied linked data techniques to describe resources such as media and scientific publications. In the public sector, the Open Government Data (OGD) movement emerged with the goal of meeting the diverse needs of data consumers, and encouraging government transparency by releasing government data on the web [13]. However, the heterogeneous formats in which OGD raw datasets are typically published create interoperability and usability challenges. To address these limitations, Linked Open Government Data (LOGD) has emerged based on linked data technologies [13].

3.2 Ontology and Related Technologies

Ontologies are capable of resolving the aforementioned ambiguity issues in knowledge sharing and reuse through the explicit specification of concepts, attributes and relationships within a domain. The e-COGNOS research project is a prominent example of domain ontology development for the construction industry. It supports consistent knowledge representation and access and use of knowledge at various organizational levels. Some works have also focused on developing specific ontologies at the task level. For instance, Niu and Issa [14] presented a claim ontology to formally share claim knowledge with various project participants and stakeholders within the claim workflow. Le et al [1] proposed integrating semantic wiki, ontologies and social networks to facilitate construction safety knowledge sharing. More recently, Lee et al [8] developed a defect ontology as part of a linked data system for sharing and managing information from construction defect cases. In construction, and other industries alike, access to knowledge-based software for organizational modelling is now easier than ever. For instance, free and open source ontology editors such as Protégé enable users to not only establish and populate hierarchical ontologies, but also to build entirely new ontology classes. In essence, ontologies are powerful technologies which can contribute to the systemization of construction safety information, and through this facilitate enhanced information retrieval, sharing and management. They provide means to semantically represent knowledge with great expressive capacity through defining concepts, relationships, and axioms which facilitate query and reasoning under a variety of conditions. The following section discusses the analysis and categorization of construction safety information, which is an essential process for developing a construction safety contents ontology.

4 Analysis and Categorization of Construction Safety Information

In order to limit the scope of this initial study, safety contents solely from OSHA were considered, since OSHA is recognized as a reputable authority on occupational safety and health, and recognized as a reliable provider of widely used safety data, information and contents. The OSHA website was analyzed, and various safety contents were examined. These include, but are not limited to accident cases, fatality reports, safety rules and regulations, safety training requirements, training materials, training requirement guidelines, safety management templates, safety statistics and so on. The main goal of the analysis was to identify the various classification categories the safety data on the web could be grouped into. Some categories pertained to background information, which described specific dates, times and locations. This type of data is pertinent to specific incidents and statistics data, hence it is not broadly applicable to most safety contents. Next, context information was defined to include work type, work trade, elements, material, space and work phase. Context information focuses on the circumstances around a safety matter. Currently, there are overlaps between categories such as work type and work trade. However, for the purpose of comprehensiveness, both categories are included in this interim study. Lastly, detailed information comprises of accident and hazard types, severity and impact, cause, prevention measure, related personnel/responsible personnel, medium and language categories.

Furthermore, in order to improve interoperability, several defined categories can be based on the Omniclass strategy for classifying the built environment. The Omniclass construction classification system incorporates extant systems such as Masterformat and Uniformat, and provides a classification structure useful for organizing information for electronic databases (Omni site). In the context of this study, Omniclass table 21 (elements) can serve as a basis for the elements category, table 22 (work results) can cover work type and work trade categories, while table 13 (spaces by function) pertains to the related spaces category. The aforementioned categories form a foundation for developing a safety contents ontology to facilitate the sharing of safety information through linked data.

5 Framework for Sharing Safety Information through Linked Data

In order to link the fragmented information flows of construction safety information, an initial framework is proposed by adopting ontologies and linked data technologies. Since this framework is based on ongoing research, the details of some processes such as information sharing between project participants are excluded from this paper. The proposed framework for sharing construction safety information is illustrated in figure 1. Firstly, the development of a standardized safety data structure is necessary to serve as a basis for the systems desired information retrieval and sharing features. As discussed in the previous section unstructured web based safety data are analyzed, classified and categorized. In order to structure construction safety data, a preliminary safety contents ontology has been proposed with the Protégé ontology authoring tool. This process is described as safety contents ontology in figure 1. Lastly, data search functions are proposed through query functionality in the ontology editing tool. This process is shown as query in figure 1. The linked data system underpinnings and features and are described in further detail as follows.

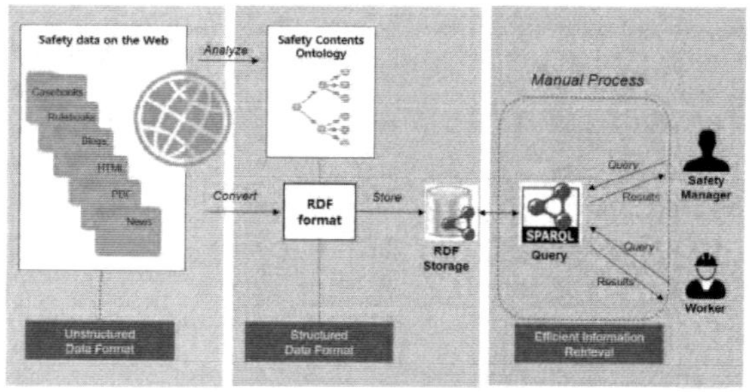

Figure 1: Framework for Sharing Safety Information through Linked Data

5.1 Development of Safety Contents Ontology

Establishing an agreement to use the same terms coherently for a domain of interest is one of the core ontology principles. In the specification and conceptualization stages of ontology development, concepts are organized into a superclass–subclass hierarchical structure, otherwise known as a tree structure. For this study a common vocabulary and an ontological structure are proposed to explain interrelations between safety contents. These are established based on content analyses, comprehensive literature reviews and expert interviews.

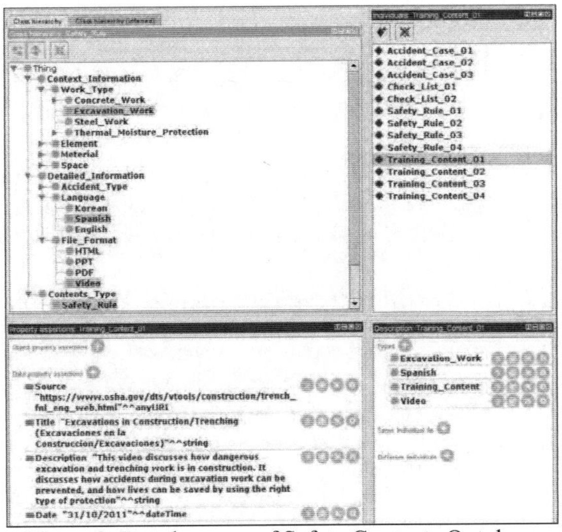

Figure 2. Development of Safety Contents Ontology

Thus far, the safety contents ontology comprises of ten classes and seven data properties. Each of the classes have several subclasses, which also have further subclasses. For instance, *Fabricated_steel_column* and *Structural_steel_for_buildings* would fall under the *Steelwork* subclass in the *Work_type* class. The terms used in each subclass are based on literature reviews and Omniclass terminology and vocabulary. The data properties include a title, description, date, specification and source. Data properties have different *DataPropertyValues*, depending on the data type, such as text data, numerical data and URI data. For example, the Title and Description data properties refer to a *String*, *Impact_cost and Time* refer to an integer, Date refers to a *dateTime* value, and Specification and Source refer to an *anyURI* value for linking actual safety data, documents or web pages. In the context of an OSHA accident case, text data would include the title of an accident case, e.g. "Employee is Struck By Crane Boom and Killed"; with the following description data "an employee was removing connecting pins from a crane boom, which fell and struck him, causing crushing injuries and multiple traumas which lead to his death". *Impact cost* and *Time* would provide numerical values for the financial cost or temporal impact in the case of a lost time incident. The *dateTime* value would be the date the accident occurred and specification and source would be the *anyURI* link which would connect to the actual data on the OSHA website.

5.2 Accessing Safety Information through SPARQL Queries

In order to maximize data sharing, efficient data publishing and search techniques are also necessary. To implement this, safety data should be converted to the RDF format, and published in an RDF store, and then contents would be accessible through the use of SPARQL endpoints. All information is defined using the classes in the safety contents ontology illustrated in figure 2. As the ontology schema is published online, their relationship can be traced whenever SPARQL queries are made on the web. Furthermore, aside from type information, data properties can be expressed explicitly in the RDF file. Data properties are formed as triples in the file and are easy to retrieve in response to various data queries and further analysis.

In order to assess the possibility of retrieving safety information through the proposed system, a variety of safety contents were analyzed and categorized based on the classes hierarchy presented in figure 2. Two example query scenarios for specific safety related purposes are tested in the Protégé platform. The preliminary results show the potential of the approach to retrieve beneficial information from organized RDF data. The first scenario considers a case whereby a safety manager is looking for safety rules, regulations and requirements related to falls during steelwork for the construction of a new building. In this query, a variable representing the cases of interest is used and listed as *?Case*. SELECT is a SPARQL reserved word for listing the query results of the variable. All the conditions are specified in the WHERE section. It is stated that data of interest should all belong to the classes, *dm:steelwork* subclass of *Work_type*, and *dm:falls* under the *occupational accident* subclass of *accident_type, and* the *dm:Safety_rule* under *Contents_type*. As shown in figure 3, the prefix dm: represents the URI resource of the defect ontology. The prefix is similar to that of other RDF standard ontologies like rdf:, rdfs:, and so on. The second query example refers to a scenario whereby a safety manager is looking for educational video materials to help a Spanish worker understand the safety issues related to excavations (figure 3). It is stated that data of interest should all belong to the classes, *dm:excavation* under the subclass of *Work_type,* the *dm:spanish* subclass of *Language, the dm:video* under the *File_format* subclass and lastly the *dm:Training_content* under *Contents_type*.

SPARQL Query

```
SELECT    ?Case    ?anyURL
  WHERE {
    ?Case    rdf:type     dm:excavation
    ?Case    rdf:type     dm:spanish
    ?Case    rdf:type     dm:video
    ?Case    rdf:type     dm:file_format
    ?Case    dm:source    ?anyURL
  }
```

Query Result:

Case	anyURL
Training_content_01	*https://www.osha.gov/dts/vtools/construction/trench_fnl_eng_web.html*

Figure 3: SPARQL Query and Result

In both query scenarios, specific relevant information relating to excavation work and falls during steelwork was retrieved, in the *anyURL* format, through the *?anyURL* query. For the training contents query, a link to an educational excavations video from OSHA is retrieved. Through this approach, safety managers and construction personnel can directly access specific safety data, information and contents conveniently and accurately. This paper only presents a preliminary trial of the proposed linked data system, hence more thorough and in depth evaluation of the approach will be necessary.

6 Conclusion

This study presents a novel framework forming part of an ongoing research for sharing construction safety information through linked data. The paper reviews the characteristics of construction safety information and discusses the current issues encountered with regards to integration, storage, retrieval and management of safety data. Construction accident cases, rules and regulations, training documents and statistics are analyzed and an interim categorization for safety information is presented. A construction safety ontology is proposed to formalize construction safety knowledge, and a linked data based system framework is presented to enable the convenient and efficient retrieval of contents for specific safety tasks. The proposed system is expected to not only improve access to construction safety information, but also to reduce data search time and improve data accuracy. Through this, the approach may enable reductions in safety incidents and overall improvements in safety management and performance. Future research will focus on the full scale system development, with a comprehensive safety contents ontology. In addition, it will be worthwhile to consider integrating the proposed system with BIM environments.

Acknowledgement

This work was supported by the National Research Foundation of Korea(NRF) grant funded by the Korea government(MSIP) (No. NRF-2016R1A2B3016694).

References

1. Le, Quang Tuan, Do Yeop Lee, and Chan Sik Park. "A social network system for sharing construction safety and health knowledge." *Automation in construction* 46, 30-37 2014.
2. Occupational Safety and Health Administration. Online: https://www.osha.gov/, Accessed: 20/03/2017
3. Sacks, Rafael, Amotz Perlman, and Ronen Barak. "Construction safety training using immersive virtual reality." *Construction Management and Economics* 31.9, 1005-1017,2013.
4. Teizer, Jochen, et al. "Autonomous pro-active real-time construction worker and equipment operator proximity safety alert system." *Automation in Construction* 19.5, 630-640 2010.
5. Zhang, Sijie, et al. "Building information modeling (BIM) and safety: Automatic safety checking of construction models and schedules." *Automation in Construction* 29, 183-195, 2013.
6. Zhang, Sijie, Frank Boukamp, and Jochen Teizer. "Ontology-based semantic modeling of construction safety knowledge: Towards automated safety planning for job hazard analysis (JHA)." *Automation in Construction* 52, 29-41 2015
7. Wang, Han-Hsiang, Frank Boukamp, and Tarek Elghamrawy. "Ontology-based approach to context representation and reasoning for managing context-sensitive construction information." *Journal of Computing in Civil Engineering* 25.5, 331-346, 2010.
8. Lee, Do-Yeop, et al. "A linked data system framework for sharing construction defect information using ontologies and BIM environments." *Automation in Construction* 68, 102-113. 2016
9. Merriam-Webster Dictionary. Online , https://www.merriam-webster.com/ Accessed 21/03/2017
10. Bust, Philip D., Alistair GF Gibb, and Sarah Pink. "Managing construction health and safety: Migrant workers and communicating safety messages." *Safety Science* 46.4 585-602, 2008.
11. Kauppinen, Tomi, and Giovana Mira de Espindola. "Linked open science-communicating, sharing and evaluating data, methods and results for executable papers." *Procedia Computer Science* 4 726-731, 2011.
12. Berners-Lee, Tim. "Linked data-design issues (2006)."
13. Ding, Li, et al. "TWC LOGD: A portal for linked open government data ecosystems." *Web Semantics: Science, Services and Agents on the World Wide Web* 9.3. 325-333 2011.
14. Niu, Jia, and Raja RA Issa. "Framework for production of ontology-based construction claim documents." *Computing in Civil Engineering*. 9-16, 2012.

This page intentionally left blank.

FEA in Road Engineering Applications?

F. Fedorikl[a], T. Makkonen2[a] and R.Heikkilä3[a]

[a]Structural Engineering and Construction Technology, University of Oulu, Finland
E-mail: Filip.Fedorik@oulu.fi, Tomi.Makkonen@oulu.fi, Rauno.Heikkila@oulu.fi

Abstract –

The current urbanization, industrialization and planning of new city-districts require efficient design to manage the energy, ecology and economical demands set for living environment. The efficiency of Building Information Model (BIM) design in construction sites has been proven through many projects. The full overview and control of design, construction and life-cycle of a structure brings sustainability to buildings and provides optimized purpose for their use. Residential and non-residential buildings, bridges and other civil and structural engineering products will adopt BIM next to benefit from advanced control of infrastructure and related services.

Road design is a complex process requiring collaboration in planning, designing, constructing and execution of designs. The design must consider soil investigations, infrastructure planning, various ecologic aspects and the construction designs to achieve the set durability and efficiency goals. This includes the economic implications and planned maintenance during the entire life-cycle of the construction. The impacts of these features could be improved by detailed analysis and monitoring of the physical behavior of the structure during construction and operative use. The physical performance of the structure may be efficiently analyzed by using numerical modelling, such as Finite Element Analysis (FEA). Integration of FEA in the BIM of road design may significantly improve the road construction and its features during the entire life-cycle.

The presented paper discusses the possibilities to improve the road designs via applying FEA tools in information model data exchange. The actual on-site monitoring and analysis would improve the quality of the road building and support its efficient and timely maintenance. Active and detailed collaboration of architects, engineers, constructors and facility managers could yield to efficient and sustainable designs and significantly improve the quality of current and future plans.

Keywords –
BIM; Road engineering; FEA; Interoperability; Efficiency

1 Introduction

The current effort towards developing sustainable, low-energy, financially competent and environmentally friendly objects in civil engineering places designers into position where numerous constraints lead to often sub-optimal design and functionality. Roads used for transportation are one of the most used parts of shared infrastructure and essential part of our everyday life.

Road design is a complex planning process with traffic flow, environmental, health factors and the actual structural design to achieve the performance and safety required. The efficiency doesn't come solely from structural design and planning but also from proper maintenance and sustainability [1] criteria that have to be met during the service time. Along with the cost, climate and physical constraints considered in road and infrastructural design there are also several ecological demands, such as consideration of animal migration dynamics [2].

Among the available processes applied in road and traffic design to achieve a certain efficiency is the BIM (Building Information Management/Modelling). The BIM is mainly used in the large residential and industrial buildings but currently applications extend in other construction areas such as infrastructure. It allows connecting and controlling all involved parties in the structural design, construction and maintenance until end of the structural life-span [3]. The BIM improves efficiency of AECOO (Architecture, Engineering, Construction, Owner and Operator) industry around the world and its application base is rapidly growing [4].

In BIM, the physical and functional features are defined in a digital form continuously updated during the entire life-cycle of the facility. It is a dynamic digital model including all objects affecting the structure and their detailed characteristics, such as dimension, location, material properties, and so on. Besides the BIM as visualization aid and archive of components, there has been also developed techniques allowing performance and quality assessment of such methodologies, such as the Success Level Assessment Model for BIM Projects (SLAM BIM) [5].

An essential part of the road and infrastructure design are safety issues. Although the annual statistics show a decreasing amount of accidents in comparison to increasing trend of traffic participants, the annual number of accidents remain fairly constant [6]. Creating efficient design of a structure usually requires numerical techniques that allow simulating the reality under defined constraints. With such tools, environmental positioning, wear, load bearing, etc. can be assessed beforehand and different designs compared. One of the most common techniques in numerical computation is the Finite Element Analysis/Method (FEA/FEM). The FEA currently represents the most applied methodology for numerical simulation of multi-physical applications. The development in computing performance and the hardware system adaptability allows simulation of complex problems from professions such as medicine, aircraft industry, vehicle industry, gastronomy, civil and mechanical engineering, and so on.

The road design is a complex task of considering the entire spectrum of elements affecting the actual design and functioning of the structure during its life-cycle. By integrating the FEA based tools into the BIM road design, the sustainability of the structure can be significantly improved and continuous monitoring offers a vast set of data important for continuous development of future road and other infrastructure.

2 FEA in infrastructure engineering

The Finite Element Analysis/Method is a numerical technique developed for multi-physical phenomena simulation. Together with a rapid development of numerical simulation tools and software the increased complexity of technical and design problems have achieved very high accuracy and convergence performance that supports the creation of more efficient designs with superior number of details.

The road engineering is a common example of a seemingly simple design that actually includes many complex objects that require the attention of competent specialists. For instance, integral part of roads and infrastructural designs are foundations, bridges, viaducts, tunnels, and so on. The FEA became significant part in bridge engineering, considering the effort of minimizing structural costs and creating extraordinary structures characterized by slim elements where dynamic behavior plays significant role in the design and structural use. Before the FEA, most designs relied on performing mechanic calculus and simulation was based on miniatures with limited capacity for detail.

The dynamic forces applied on infrastructural objects from the traffic or the surrounding impact, such as wind and climate effect, seismic activities and other geomechanical processes affecting to the behavior and structural use during the entire life-span can now be included to the design analysis by advanced numerical multi-physical simulations. Alone, they provide the analytical solutions that often remain economically inefficient by accuracy and in terms of construction time due to decoupling of project management and the seemingly independent, often loosely connected, work phases by multiple operators.

2.1 FEA in bridge engineering

The FEA represents one of the most important element in bridge design. Extraordinary bridges and the effort of creating unique architectural designs require detailed analyses of structural behavior where analytical solutions are almost impossible or able to provide results only after significant design simplifications that cause inaccuracies in the results.

As the bridges are in many cases characteristic by their slim elements, such as cable-stayed, suspension and pre-stressed elements and structures reinforced by high-strength materials, the dynamic effect represents the major role for stability and comfort of users as well as sustainability of the bridge.

2.2 FEA in tunnel engineering

FEA based tools currently allow detail modeling of real structures and their surrounding environment. Although the general public expects three-dimensional complex models, in many cases it requires too much effort and computer time that it becomes inefficient, and often obsolete for the engineers. The choice between 2- and 3-dimensional analyses depends on project geometry and behavior of physical phenomena effecting the problem. However 3D modeling is already possible and it may offer more accurate results and realistic behavior of analyzed physical phenomenon.

Tunnels represent structures that shorten the distance between two points by crossing obstacles such as hills, mountains or waterways. Numerical geomechanic applications have significantly developed in the past decades but still the design is usually demanding task as in the most cases tunnels are of curved cross-section. It means, that evaluation of the cross-sectional forces from the ambient load must be specified in details to the building design of a tunnel to assure its safety.

The tunnel structures must also resist the environmental conditions, such as ground movement and weight and/or water around them. The FEA significantly improves tunnel designs and creates simulation processes where the soil movements can be simulated with good agreement with the actual behavior.

For instance, the analyze performed by Mazek and Almannaei [7] show the efficiency of 2-dimensional numerical simulation in a metro tunnel design. The 2-dimensional FEA simulation allows more detailed analyses in stress, non-linear behavior of the soil and the construction process. On the other hand, in some cases the 3-dimensions are required, considering the surrounding conditions of the tunnel design. The extensive 3-dimensional FEA model of tunnel considering interaction of tunnel, soil and building during excavation process was studied by Keshuan and Lieyun [8]. The interaction of surrounded building significantly affect the soil surface settlement profile. Therefore, depending on geometrical characteristics and surrounding conditions the 2- or 3-dimensional numerical model is required. The efficiency of FEA in tunnel engineering applications and well situated simplifications in the modeling were also performed by Roatesi [9], where good agreement between the numerical and the analytical approach was obtained.

2.3 FEA in road designs

The current approach leads road engineers to develop structural material, designs and testing techniques to achieve efficient and sustainable ways for people to move whether the approach and appearance of the ready walkway or highway is novel or historic. In terms of road engineering, mainly soil analyzes and sustainable foundation for infrastructural elements is considered along with sufficiently durable road pavement to last long but allowing fast and efficient replacement after the end of its life-cycle.

A study in which a non-linear FEA is applied in a layered pavement system was performed by Taylor already in 1971 [10]. Unlike at that time, the current computing tools allow simulation of detailed non-linear analyses describing the real conditions although in many cases certain simplifications must be done to streamline the numerical procedure. Except the structural and dynamical analysis of the structural part of the construction, the FEA could be applied for instance in traffic flow by using the fluid mechanics as it was studied by Ma and Cui [11].

3 BIM and FEA interoperability

The interoperability of FEA and BIM has been significantly improved in the last decade. Yet, there are still several challenges that should be discussed and tested such as interoperability of different software, geometrical discrepancies and ability of numerical simplifications leading to the efficient simulations as highlighted in [12,13]. The full integration of numerical tools based on FEA in BIM would allow efficient design predicting the real structural behavior by considering realistic boundary conditions. Essential part is also to consider the interoperability between computing tool and on-site measured data that would significantly improve the accuracy of performed numerical simulation. Then, the calibrated numerical model can be further used for forecast and prediction of required maintenance.

Few available software already allow the full interoperability of the BIM and FEA based numerical techniques as they are included in one package. Difficulties occur in cases when specialized parties apply different software that are unable to communicate with each other. Solution for the full integration of specific disciplines in the BIM offers IFC (Industry Foundation Classes) labels that determine standardized data protocols for model representation [14]. The current issue in BIM and FEA interoperability consists mainly in dynamic analyses that require special caution. These are related especially to the objects creating essential part of the road design, such as bridges and tunnels.

The FEA model intended for dynamic analyses is characterized by efficient creation of the geometry to optimize mesh configuration, representing the finite elements and finally leading to accurate results in sufficient time. Hence, the FEA model is usually adapted to specific needs and/or simplified by removing non-critical objects that do not affect the performed analyses but complicate the model convergence.

The critical decision is made in defining objects included in the model transfer from BIM to the FEA tool. In the case, the transferred objects are not correctly connected or there are other discrepancies causing errors in the model, the results obtained in the simulation may be misleading or the simulation fails entirely. This is why proper connection between BIM and FEA should be of special interest as the detailed FEA analyses can significantly improve the design. An example of a bridge illustrated from the BIM road design is figured in the Figure 1.

Figure 1. 3-dimensional expression of bridge in BIM design.

Figure 2. BIM model of road design.

4 InfraBIM in road engineering

Most of the currently built buildings are designed in a traditional way as 2-dimensional drawings. Although BIM has proven its advantages in many previous projects in number of civil and structural engineering areas, the BIM is mostly used in large projects represented by industrial or residential building of national or international significance. The private sector avoids changes in their current designs as the adaptation towards novel technologies requires patience and certain initial capital costs for educating all involved parties in the project tools.

The current technology development allows advanced road design and maintenance to simplify and improve the construction management from its initiation, and use until the end of the structures decommission and demolition. However, the BIM application has increased a lot in new road designs, but much less attention has been paid in development of maintenance and rehabilitation processes.

The process of road maintenance has been studied and developed within Infra FINBIM research package in Finland since 2010 [15]. The Finnish BIM requirements were released in 2015 by the BuildingSMART Finland within IM3 (Inframodel 3) data exchange format. The requirements cover 11 topics from BIM-based project to utilization of InfraBIM. It is currently required by the Finnish Transport Agency, that the IM3 is used in all designs and implementation projects. An example of 3-dimensional BIM road model is illustrated in the Figure 2.

One example of the benefits caused by initiation of national requirements was a large research and development scheme of BIM application in road engineering, performed within strategic program RYM PRE Infra FINBIM in Finland [15]. The goal of the performed project was the utilization of a mobile laser scanning method to obtain the initial data for model generation, testing and development of novel 3D analysis and modeling methods for point cloud processing, applying optimization method for the planning of geometric and structural improvements needed for the existing uneven road surfaces and creation 3D machine control models using machine control systems for continuous control of practical construction work using a milling machine and an asphalt paver.

Extensive research in application the BIM in roads and infrastructure objects have been performed also elsewhere, for instance by Tibaut et al. [16]. The research focused on developing smart parametric building blocks for 3D road design and developing algorithms detecting road elements from the 3D point clouds achieved by a laser scanning.

Great benefits in road engineering can also be achieved by implementing geotechnical data into the BIM process as in most of the projects the below-ground data is not considered while it has marked impact on e.g. foundations and water drainage. Testing different geotechnical data formats and their implementation into the BIM process was studied by Tawelian and Mickovski [17].

The complex BIM process including geotechnical data, bridge, tunnels, road construction and other relevant fields of expertise can significantly improve the entire building process. The quality of the BIM design depends on the quality and the accuracy of initial data and decent collaboration of all involved parties. If data obtained from any of the involved parties is inaccurate, it deteriorate the entire process. This holds especially with the geotechnical investigations, as if the data of below-ground material is incorrect the actual design may face risks that affect the road usability, safety, durability and sustainability.

Besides the road construction itself, another essential part is the infrastructure planning stage that is currently also developing affected by environmental degradation, climate change and social trends [18].

5 BIM, FEA and road design interoperability

The vision of the future civil and structural engineering design is in complete breakdown of the initial design and its full-scale control over the entire life-time until the stages of deconstruction and recycling the material. The current design does not usually combine detailed structural analysis and digital system represented by BIM. This is caused by difficulties occurred within transferring the digital data from BIM structural design into a numerical tool. There are few reasons for the difficulties that occur in different coding phases within the applied software and/or difficulties in accurate definition of numerical model.

The numerical modeling is a separate field of expertise where the numerical model requires elements essential for performed analyses. On the contrary, the BIM model includes the entire spectrum of objects affecting any stage of the planning and life-cycle of the structure. The goal of the BIM in the future is complex planning and control of the structure including all parties without consideration of what computing, design and facility tool is applied. From this reason, a global standard is proposed for unifying the BIM parts has been developing through IFC. It would allow the connection of all involved disciplines and minimize any misunderstanding and contradiction in the planning and design stages.

The BIM is a well-known and widely applied method in road engineering. The current technology allows automatized processes in structural and traffic construction engineering significantly improve the infrastructure design and maintenance. The development in the use of BIM and 3D machine control in new road designs has high significance in Finland as well as in other Nordic countries. For instance, studying and development of road maintenance processes has been active since 2010 within the Infra FINBIM research [15].

Interoperability of BIM and specialized areas require full integration and complete understanding to avoid any errors uprising from the data transfer. There has been significant development in this matter, however there are still challenges that must be solved. The interoperability between Geographic Information System (GIS) and BIM was performed by developing by combination of IFC and CityGML as both data information platforms can benefit from each other (Geography Markup Language) [19,20].

6 Conclusion

Since 1970s there has been significant progress in BIM application in road designs and infrastructural planning, in parts thanks to increased computing power and easy to use software tools. The goal of fully controlled and integrated design, construction, maintenance and deconstruction during the structure's life-cycle is today a rapidly developing field. Integration of the specialized BIM digital packages can improve the efficiency and sustainability of the private and public structural projects but the full interoperability remains incomplete to be of use in most projects. The BIM's advantages are already well-known in grand building projects but the small-scale designs are rather conservative in applying novel tools. The future vision consists in the full BIM model allowing the complex control beyond road design and optimizes the maintenance process during the entire life-span of the structure including all the objects connected. Automatized machine control systems, detailed on-site monitoring and numerical computation control can markedly improve the novel design as well as prolong usability of the current ones.

References

[1] Conchon A. Roads construction in the eighteenth century France. *Second International Congress on Construction History*, 791-797, Cambridge, United Kingdom, March 29th- April 2nd, 2006.

[2] Davey N., Dunstall S., Halgamuge S. Optimal road design through ecologically sensitive areas considering animal migration dynamics. *Transportation Research Part C: Emerging Technologies.* 77:478-494, 2017, http://dx.doi.org/10.1016/j.trc.2017.02.016.

[3] Cheng J.C.P., Ma L.Y.H. A BIM-based system for demolition and renovation waste estimation and planning. *Waste Management*, 33(6):1539-1551, 2013, http://dx.doi.org/10.1016/j.wasman.2013.01.001.

[4] Bradley A., Li H., Lark R., Dunn S. BIM for infrastructure: An overall review and constructor perspective. *Automation in Construction*, 71(2):139-152, 2016, doi:

http://dx.doi.org/10.1016/j.autcon.2016.08.019.

[5] Won J., Lee G. How to tell if a BIM project is successful: A goal-driven approach. *Automation in Construction.* 69:34-43, September, 2016, http://dx.doi.org/10.1016/j.autcon.2016.05.022.

[6] Čičković, M. Influence of human behaviour on geometric road design. *Transportation Research Procedia.* 14:4364-4373, 2016, https://doi.org/10.1016/j.trpro.2016.05.358.

[7] Mazek S.A., Almannaei H.A. Finite element model of Cairo metro tunnel-Line 3 performance. *Ain Shams Engineering Journal*, 4:709-716, 2013, http://dx.doi.org/10.1016/j.asej.2013.04.002.

[8] Keshuan M., Lieyun D. Finite element analysis of tunnel-soil-building interaction using displacement controlled model. *7th WSEAS International Conference on Applied Computer & Applied Computational Science (ACACOS 08)*, Hangzhou, China, ISBN: 978-960-6766-49-7, ISSN: 1790-5117, April 6-8, 2008.

[9] Roatesi S. Finite element analysis for the problem of tunnel excavation successive phases and lining mounting. *Finite Element Analysis - New trends and developments*, Edited by Rarzad Ebrahimi, publisher: InTech, 301-332, 2012, ISBN: 978-953-51-0769-9, doi: 10.5772/3352.

[10] Taylor K.L. Finite element analysis of layered road pavements. Department of Civil Engineering, University of Nottingham, doctoral dissertation, October, 1971.

[11] Ma J., Cui J. Traffic flow density distribution based on FEM. *2012 International Conference on Solid State Devices and Materials Science*, Physics Procedia 25, 939-944, 2012.

[12] Saifi H. Integration of Building Information Modeling (BIM) and Finite Element Analysis (FEA) in Bridge Engineering. MSc thesis, Aalto University, School of Engineering, Helsinki, Finland, 2012.

[13] Fedorik F., Makkonen T., Heikkilä R. Integration of BIM and FEA in automation of building and bridge engineering design. *2016 Proceedings of the 33rd ISARC*, pages 735-741, Auburn, USA, 2016.

[14] Aranda-Mena G. and Wakefield R. Interoperability of building information: myth or reality?, In *eWork and eBusiness in Architecture, Engineering and Construction. ECPPM 2006: European Conference on Product and Process Modelling 2006 (ECPPM 2006)*, pages 127-135, Valencia, Spain, 2006.

[15] Heikkilä R., Marttinen M. Development of BIM based rehabilitation and maintenance process for roads. *2013 Proceedings of the 30th ISARC*, 1216-1222, Montreal, Canada, 2013.

[16] Tibaut A., Pečnik S., Korošec M. R., Mihali K., Zabreznik I. BIM-based parametric modeling of roads and infrastructure objects. *Proceeding of the 32nd CIB W78 Conference 2015*, 725-734, Eindhoven, Netherlands, October 27th-29th 2015.

[17] Tawelian L.R., Mickovski S.B. The implementation of geotechnical data into the BIM process. *Procedia Engineering*. 143:734-741, 2016, https://doi.org/10.1016/j.proeng.2016.06.115.

[18] Busscher T., Tillema T., Arts J. In search of sustainable road infrastructure planning: How can we build on historical policy shifts? *Transport Policy.* 42:42-51, 2015, http://dx.doi.org/10.1016/j.tranpol.2015.04.007.

[19] Deng Y., Cheng J.C.P., Anumba C. Mapping between BIM and 3D GIS in different levels of detail using schema mediation and instance comparison. *Automation in Construction*. 67:1-21, 2016, http://dx.doi.org/10.1016/j.autcon.2016.03.006.

[20] Kang T.W., Hong C.H. A study on software architecture for effective BIM/GIS-based facility management data integration. *Automation in Construction.* 54:25-38, 2015, http://dx.doi.org/10.1016/j.autcon.2015.03.019.

Integration of Structural Health Control in BIM for Current and Future Residential Buildings

F. Fedorik1[a], R.Heikkilä2[a], T. Makkonen3[a] and A. Haapala4[b]

[a]Structural Engineering and Construction Technology, University of Oulu, Finland
[b]Wood Materials Science, University of Eastern Finland, Joensuu, Finland
E-mail: Filip.Fedorik@oulu.fi, Rauno.Heikkila@oulu.fi, Tomi.Makkonen@oulu.fi, Antti.Haapala@uef.fi

Abstract –

Currently the structural engineering and construction technologies go through substantial paradigm shift and development between energy conservation, health and living cost issues. Increasing energy consumption for e.g. heating and cooling and the constant effort to energy preservation in all forms promotes the manufacture of sustainable low-energy or near-zero energy houses, often with some independent energy production capability included.

As a subtheme in controlling the building performance and energy uses, there has been interest in applying Building Information Model (BIM) to building designs to store, optimize and use building-related information during its entire life-cycle. The BIM methodology is mostly limited for new buildings as the needed open access data on generic structures and materials is most often not available. Similarly, health considerations in current buildings are also poorly controlled and more detailed knowledge on the content and conditions of thermal transport, air exchange and moisture control would benefit the energy efficiency and sustainability of the buildings. Common issues with moisture entrapment leading to microbial growth and indoor air quality related respiratory illnesses are a practical, and extreme, example of insufficient understanding of structural design and control issues also in in public buildings.

Our study presents some of the recently discussed possibilities and efficiency of BIM utilization in the context of current and future residential buildings. In addition, hygro-thermal monitoring and modelling tools that have recently been made available to analyze the structural health of the buildings will be highlighted.

Keywords –
BIM; Building physics; Hygro-thermal analysis; Mould; Interoperability

1 Introduction

Civil and structural engineering and the applied construction technologies go through substantial paradigm shift in many ways. Building speed, element pre-fabrication, 3D-prining, individualistic design features, energy and cost issues drive the development of large structures. Increasing energy costs and the constant effort of national and international institutions for lowering energy consumption [1,2] promote the building of sustainable passive, low-energy or near-zero energy houses and entire districts. Architects, engineers, constructors (AEC) and facility managers (FM) actively cooperate in managing resources efficiently in designs.

In the past decades, there has been interest in applying Building Information Model (BIM) to building designs. BIM represents a tool controlling building information during the entire life cycle of the structure, including 3D-design, construction material and layout selection and end-life considerations for recycling and reuse of materials. The BIM methodology is mostly established for new buildings; therefore the maintenance of existing buildings does not benefit of its centralised data archives, nor do the renovation or deconstruction projects base onto the BIM processes [3].

Sustainability and health considerations in current buildings are also poorly understood and controlled. More detailed knowledge on the content, conditions of building structures and their performance in e.g. extreme climate would be beneficial in promoting their energy efficiency. Householders of existing buildings often adapt their buildings to increase energy efficiency by installing additional insulation layers. However, additional insulation applied on the exterior or interior surface of the structure envelope may disable the designed moisture diffusion performance and lead to accumulating humidity inside the structural elements that may, in turn, enable conditions for mould growth [18].

Besides the structure itself, the living environment contributes a lot on the hygro-thermal conditions inside the structural elements. The interior conditions change according to actual use of the certain indoor location, as bathroom, kitchen, sauna, and so on, are characterized by higher humidity than other rooms such as bedroom, study and living room.

On the other hand, the major impacts on the hygro-thermal conditions inside the structure are provided by the exterior conditions. Exterior conditions, in turn, vary according the climate, structure's location, annual season, pertaining weather conditions, etc. For instance, climate effects on structures considering inland, coastal, northern and southern conditions were studied for example in [19], where the impact of climate on identical building envelope performance and moulding tendency were significantly different already between the few locations within Finland.

Taking solely the climate as a leading condition for exterior boundary designs will not suffice as structures are also significantly affected by the very local micro-climate. These conditions vary depending on the local building details and include perimeter locations such as attics, foundation, windows or doors, or those characterized by the presence of waterways, hills, trees or other buildings and related infrastructure in the immediate vicinity. Also details such as the location and orientation of the structure markedly affect its ambient conditions.

That is why monitoring the hygro-thermal conditions and simulation of building designs for the ambient and the inside conditions of the structure would be beneficial. Extensive humidity accumulation could lead to a certain damage caused by mould and following deterioration of structural material but such a risk can be significantly mitigated in advance with simulation tools.

2 BIM in current buildings

The BIM is a virtual IT-based information system where each element affects the design, the structure during its life cycle and deconstruction that are to be described in detail. The BIM method allows detailed visualization, code reviews, construction sequencing, cost estimation, conflicts and collision detection, forensic analyses and facility management. These lead to faster and more efficient processes, more accurate design, simplified production, (semi-)automated assembly, improved customer service and recording of designed life-cycle data for later uses [20].

Application of BIM as a part of new structure design is continuously increasing, but in the case of existing buildings, the BIM is still used rather rarely. One of the reasons is the very difficult access to detailed building information of all parties: owners, facility managers, suppliers, builders, subcontractors and architects. Other challenges observed in BIM research include updating, maintaining the data in BIM, modelling of indefinite objects, and related properties [3].

For instance, the BIM applied in the design of novel building consists of direct access to material properties and all required features of used objects provided by different manufacturers. On the contrary, current buildings were designed omitting the design details or not having them available in the required detail. In renovation stages, it is typical to notice that the former manufacturers do not exist anymore and/or the material properties used for construction are not available. Also the building designs can have inaccuracies and many prior renovations and changes have not been documented. This leads to higher cost at the beginning of the BIM model.

On the other hand, the BIM allows monitoring the actual conditions of the structure and in the case any problem is found, it can be solved quickly before large-scale damage occurs. The continuous monitoring assures healthy and sustainable building during the entire life cycle. The advantage of applying BIM in current and historic building is in elevated understanding of their structures behaviour as in many cases the energy consumption and detailed performance of the structure and materials is not known.

3 Hygro-thermal conditions

Essential part of the sustainable building design is played by building physics, where especially the hygro-thermal conditions inside the building and/or structural elements affect the structural behavior and biological contamination issues. The effort towards controlling the hygro-thermal conditions inside buildings and their structural elements is partially caused by the design of airtight, highly energy efficient buildings. The increasing demands on health indoor environments coupled with insufficient construction element insulation from the elements during the construction phase or accidental leakage of water into the structures later on may cause catastrophic issues.

Hygro-thermal conditions affect the properties of building material and if the favorable conditions persist for sufficient periods, they allow mold growth initiation. Reconstruction of such buildings or their parts usually requires removal of wetted and contaminated materials at high costs. The analysis options of the structural hygro-thermal conditions are threefold; laboratory tests, direct on-site measurement and numerical analysis. The laboratory tests with sampling work are usually time demanding and expensive. Continuous and automated measurements obtain data during the actual use of buildings, but observing the hygro-thermal conditions inside the structural elements demands permission from house owner and pre-installed sensors.

Numerical modeling as a third option often achieves good results in many engineering fields by simulating the features of any physical phenomenon. The combination of on-site measurements and numerical calculation seems to be an efficient method of hygro-thermal conditions analysis and control. Even though the numerical techniques are considered as efficient method of high accuracy, it is always beneficial to validate the method with on-site measured data.

The unexpected aspects significantly affecting the hygro-thermal conditions inside the structural elements include damaged construction material such as waterproof membrane and sudden water leakage from the exterior, extensive humidity and/or water absorbed in the structural elements during construction, accidental piping damage, house equipment accidents with extensive water outlet, and so on.

Installing online sensors to indicate temperature and humidity on surfaces of structural elements near locations where the accidents are most likely to occur, minimize the risk in numerical errors and unrealistic hygro-thermal conditions. In the case of higher humidity near the monitored location, the exact point of accident can be found and fixed before it causes major damage in the structure. The areas of interest for hygro-thermal monitoring are located in the ambient of details where high humidity can occur. It means in practical terms e.g. near the pipe junctions where is risk of water leaking accident is relevant and in the locations characterized by thermal bridge.

The thermal bridge is described by structural detail achieving different thermal resistance than the surrounding elements, such as wall-wall, wall-floor, wall-roof junctions, foundation attics, windows, doors and so on. As it can be seen in the Figure 1, the thermal bridge in the foundation structure is caused by rapid change in heat flow in the corner compared to the wall and floor structure that characterize the thermal bridge. These details indicate significant temperature difference inside the structure that considering certain conditions leads to a water condensation.

Figure 1. Heat transfer in foundation structure, thermal bridge.

4 Hygro-thermal Analysis and Mold Growth Risk

The International Passive House Association defines for passive houses the heat transfer coefficient (U-value) cannot exceed 0.15 $W/(m^2K)$ and the space heating demand is limited by 15 kWh annually or 10 W/m^2 of usable living space [4]. Because of the high thermal resistance, especially in Nordic countries, the structures are these days quite airtight and the air exchange must be provided by ventilation systems [5]. The average air change rate in houses mostly built after 2000 in Finland at 50 Pa is 4.5 h^{-1} in timber-framed houses constructed on site, 3.3 h^{-1} in timber-framed prefabricated elements houses, 6.0 h^{-1} in log houses, 2.8 h^{-1} in brick houses and 2.6 h^{-1} in houses built from concrete elements [6,7].

In the case of a high airtightness of a building, a suitable ventilation system can be installed to control the physical conditions indoor. The energy efficiency of the existing building may be increased by optimizing the house equipment usability and possibly by adapting the structural elements towards higher thermal resistance.

The indoor comfort level highly depends on suitable temperature and humidity. Specific combinations of temperature and humidity along with sufficient exposure time promote mould growth [8-13], which may lead to allergic reactions and other health issues for inhabitants [14], as well as influence the behavioural properties of structural elements [15].

The relationship between temperature and humidity in exposure time and favourable and unfavourable conditions for the initiation of mould growth can be expressed by a mathematical formula of critical relative humidity RH_{crit} (Eq. 1), which is a function of temperature T [9].

$$RH_{crit} = \begin{cases} -0.267T^3 + 0.160T^2 - 3.13T + 100.0 \ \textit{when}\ 0 < T \leq 20 \\ 80\% \ \textit{when}\ 50 \geq T > 20 \\ \textit{no growth assumed when}\ 0 > T > 50 \end{cases} \quad (1)$$

The equation (1) can be also expressed graphically as following (Figure 2).

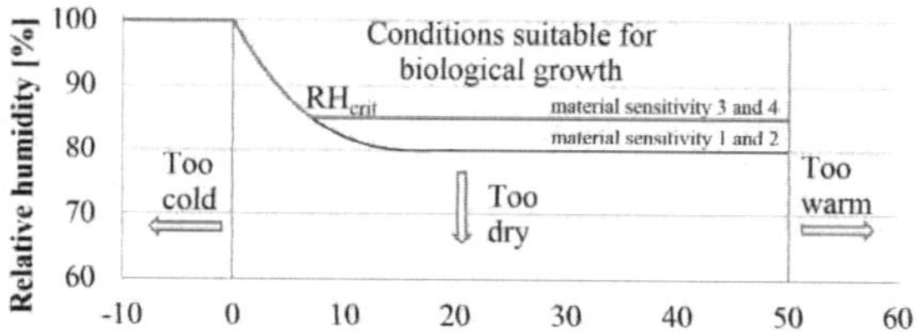

Figure 2. Favorable and unfavorable conditions for mold growth initiation.

The mould index indicating the amount of mould growth can be represented by equation (Eq. 2).

$$\frac{dMI}{dt} = \frac{k_1 k_2}{7\,exp(-0.68\,ln\,T - 13.9\,ln\,RH + 0.14W - 0.33SQ + 66.02)} \quad (2)$$

where k_1 is the mould growth intensity, k_2 is mould growth intensity moderating the mould index after approaching the maximum value, MI is the mould index, T is the temperature, RH is the relative humidity, t is the exposure time and SQ and W are the coefficients characterizing surface quality and timber species.

The mould index allows qualitative evaluation of mould growth by seven levels; 0 for no growth present, 1 represents small microscopic trace amounts of mould on the surface, 2 indicates appearance of several microscopic locally identified mould growth colonies, 3 is for the first visual mould, 4 for clear visual appearance with 10-50% coverage, 5 indicates more than 50% coverage and 6 indicates heavy and full mould growth coverage [9]. Significant role is reserved also material sensitivity in the location of mould index evaluation. There are 4 defined material sensitivity levels; very sensitive such as pine sapwood, sensitive such as glued wooden boards, medium resistant such as mineral wool, and resistant such as metals.

As the mould index mathematical model requires continuous data of the temperature, relative humidity and exposure time the hygro-thermal simulation can be performed numerically. Numerical simulation using for instance the Finite Element Analysis based tools achieve results with high accuracy can be applied for the hygro-thermal simulation and evaluation of mould growth risk within a structure. However, it is always beneficial to provide on-site monitoring as organic residues typically accumulate in structures on building sites via dust or rain, which may contain microbial spores, predisposed to mould growth initiation.

5 BIM and hygro-thermal simulation interoperability

The effort towards economic and energy efficient building designs support invention and development of new technologies for quicker and more efficient designs, constructions and life-cycle control [16].

One of the most important information management systems of construction industry is represented by Building Information Modelling (BIM). It allows detailed information about all objects implemented in a project and controls their interaction and behavior while they are in use. The aim of the BIM is to connect teams and experts involved in the project allowing their collaboration and sharing ideas while minimizing errors caused by e.g. misunderstanding of cooperators and incompatibility of individual objects [17].

Traditionally, the information is exchanged between project participants through drawings and other printed documents such as cost plans. The BIM represents digital virtual model where each elements that affects the structural design and/or its life cycle is described in detail. The BIM is an IT-based information system recently operated by an increasing amount of Architecture, Engineering and Construction (AEC) companies.

Although there have been number of BIM projects applied in practice especially within large commercial projects there are still challenges that limit its use [3]. One of them is possibility of creating system that includes the entire package of involved parties that must be included or those that can improve the building performance [16]. There is a potential risk that the number of hygro-thermal based problems increase when the authorities, designers, builders and inhabitants do not have sufficient real time information about the performance of the buildings and structures.

By integrating the hygro-thermal simulation in BIM, the simulation can be extended by implementing mould growth risk methodology and energy control to express the hygro-thermal behaviour of the building in detail. Hence, the energy and structural health is invariably controlled and possible risks can be solved at their origin. The complex process of integration the hygro-thermal numerical application considering validation with measured data and BIM can be illustrated as in the Figure 3. The procedure consists in simplification of the complex digital model into a model determined by elements directly affecting the hygro-thermal conditions. Then, the BIM simplified model is parameterized into a mathematical model prescribed by dimensions needed for accurate mapping the hygro-thermal conditions. For instance, two-dimensional numeral model are applied in details where the heat flow and mass movement happens in two directions only, such as wall-floor, wall-wall junctions and the foundation. If the conditions are required for corners joining three directions elements, then the three-dimensional numerical model is required. When the geometry and physical properties of the numerical model are defined the numerical simulation is performed. At the same time, the on-site measured data is taken for verification of the numerical model and possible contradictions are corrected to meet the real hygro-thermal behaviour. The verified numerical results obtained are preceded into evaluation of the mould growth risk and compared with energetically requirements to meet the sufficient performance of the structure. In the case, the obtained performance satisfies both energetically and structural health requirements, the model is adapted back to the BIM format. Otherwise the procedure is repeated and the design and/or conditions are adapted to meet the requirements.

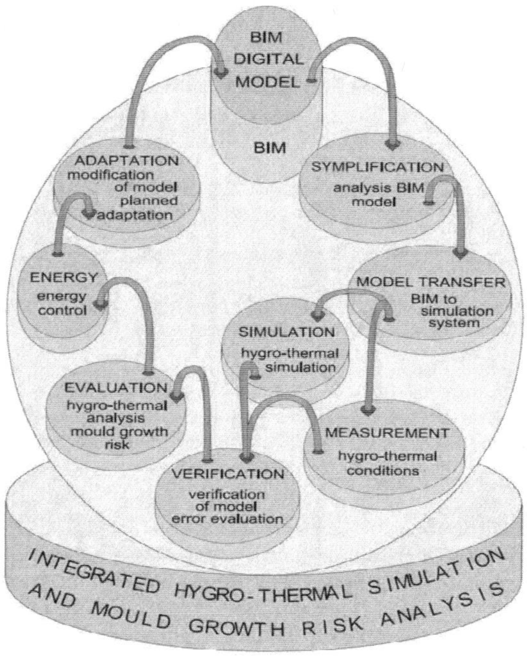

Figure 3. Scheme for interoperability between BIM and hygro-thermal analysis.

The interoperability of BIM and hygro-thermal analysis allows higher understanding of the structural and material behaviour and overall sustainability aspect from material origins and sourcing to recyclability criteria assessment of materials used. The boundary conditions represent critical data for achieving accurate simulation and further analyses.

BIM and numerical simulation integration consists of creating geometric model including boundary conditions in BIM, transferring the model to numerical application, such as a FEA based computing tool, computing procedure and evaluation of the obtained results. In the case the results are accepted following the defined requirements, the numerical model is not transferred back to the BIM, but the current BIM digital model is used. Otherwise, the numerical model is modified until the goal is achieved and then transferred back to BIM. The numerical techniques considering the physical behaviour, such as the hygro-thermal conditions and the structural design or control can be optimized with the obtained results. Hence, the BIM digital model improves the structural design and sustainability of the building during its entire life span.

The difficulties appear especially once detailed and special analysis, such as airflow, harmonic, seismic or hygro-thermal analyses are required. These are usually time consuming and have high requirements for computing performance, hence certain simplifications are common. These usually consist of removing objects or minor geometrical details that do not affect the analyses and performance of physical phenomena.

The detailed hygro-thermal numerical simulation is numerically demanding, thus 1- or 2-dimensional geometry representing the structural details is often analysed. In some cases, 3-dimensional numerical model is required, for instance 2-wall-ceiling corner or 2-wall-foundation corner.

The numerical models exclude elements with no effect on the hygro-thermal performance, often they are also presented in BIM. From this point of view, the detailed hygro-thermal analyses should be considered as individual processes requiring detailed professional attention. The hygro-thermal simulation requires detailed information about the structure and applied materials as well as the exact conditions in the ambient.

Even though, the current development and research indicate the possibility of creating the entire building model to control energy consumption and building physics properties control, the hygro-thermal analyses leading to realistic monitoring are remarkably demanding. Therefore, the 1- or 2-dimensional detailed models are recommended to monitor the hygro-thermal conditions inside the structural elements.

6 Conclusions

The effort of national and international authorities for lowering energy and environmental impact of buildings creates new demands for building designs and joint data management that have been neglected so far. The full digital archive of existing building can enable advanced use of BIM as well as simplify the deconstruction phase and end-of-life use of the materials. Promoting a few specialized areas of expertise, such as hygro-thermal control, mold growth risk analyses and BIM interoperability can provide sustainable building practices and optimized energy efficiency of novel and existing building.

7 Motivation and future significance

The goal in the BIM application of simulation input is to develop the datasets of conform with Industry Foundation Class (IFC) standards that represents neutral platform for the object-based building data model and its understanding. Only then can the integrated systems and the level of detail be involved in the building designs so that its entire life-cycle is covered and understood. Implementing the hygro-thermal control together with the mold growth risk analysis provides additional tool to be developed into future BIM systems to promote and balance the health control and energy efficiency requirements.

References

[1] European Parliament and Council Directive 2012/27/EU on energy efficiency including amending Directives 2009/125/EC for setting of eco-design requirements for energy-related products, 2010/30/EU on the indication by labelling and standard product information of the consumption of energy and other resources by energy-related products and repealing Directive 2004/8/EC on promotion of cogeneration based on a useful heat demand in the internal energy market, L315 2012.

[2] Thullner K., Low-energy buildings in Europe – Standards, criteria and consequences, A study of nine European countries, ISBN 978-91-85147-42-7, Lund, Sweden, 2010.

[3] Volk R., Stengel J., Schultmann F., Building Information Modeling (BIM) for existing buildings - Literature review and future needs. *Automation in Construction*, 38: 109-127, 2014, http://dx.doi.org/10.1016/j.autcon.2013.10.023.

[4] Nevalainen A., Pasanen A.-L., Niininen M., Reponen T., Kalliokoski P., Jantunen M.J., The Indoor Air Quality in Finnish Homes with Mold Problems, *Environment International*, 17(4), 1991.

[5] Dubey S., Lanjewar S., Sahu M., Pandey K., Kutti U., The Monitoring of Filamentous Fungi in the Indoor Air Quality, and Health, *Journal of Phytology*, 3(4): 13-14, 2011.

[6] Korpi M., Vinha J., Kurnitski J., Airtightness of Single-family Houses and Apartments, Department of Civil Engineering at Tampere University of Technology and HVAC-Laboratory at Helsinki University of Technology, 2005.

[7] Krus M., Seidler M., Sedlbauer K., Comparative Evaluation of the Predictions of Two Established Mold Growth Models, *Buildings Conference*, December 5-9, 2010, ISBN: 978-1-933742-89-2, 2010.

[8] Viitanen H., Vinha J., Salminen K., Ojanen T., Peuhkuri R., Paajanen L., Lähdesmäki K., Moisture and Bio-deterioration Risk of building Materials and Structures, *Journal of Building Physics*, 33(3): 201-224, 2010, doi: 10.1177/1744259109343511.

[9] Ojanen T., Peuhkuri R., Viitanen H., Lähdesmäki K., Vinha J., Salminen K., Classification of Material Sensitivity - New Approach for Mould Growth Modeling, Paper at the *9th Nordic Symposium on Building Physics - NSB 2011*, Tampere, Finland, 2011.

[10] Clarke J. A., Johnstone C. M., Kelly N. J., McLean R. C., Nakhi A. E., Development of a Simulation Tool for Mould Growth Prediction in Buildings, Proceedings of the *Fifth International Conference of the International Building Performance Simulation Association*, Prague, Czech Republic, 1997.

[11] Moon H. J., Augenbroe G., Evaluation of Hygrothermal Models for Mold Growth Avoidance Prediction, *Eighth International IBPSA Conference*, Eindhoven, Netherlands, August 11-14, 2003.

[12] Viitanen H., Toratti T., Makkonen L., Peuhkuri R., Ojanen T., Ruokolainen L., Räisänen J., Towards Modelling of Decay Risk of Wooden Materials, *European Journal of Wood and Wood Products*, 68(3): 303-313, 2010, doi: 10.1007/s00107-010-0450-x.

[13] Viitanen H., Ojanen T., Improved Model to Predict Mold Growth in Building Materials, Paper based on the VTT projects “Building Biology” and “Integrated Prevention of Moisture and Mould Problems”, Finland, 2007.

[14] Mudarri D., Fish W.J., Public Health and Economic Impact of Dampness and Mold, *Indoor Air Journal*, 17(3): 226-235, 2007, doi: 10.1111/j.1600-0668.2007.00474.x.

[15] Bakri N.N.O., Mydin M.A.O., General Building Defects: Causes, Symptoms and Remedial Work, *European Journal of Technology and Design*, 3(1): 4-17, 2014, doi: 10.13187/issn.2310-0133.

[16] Kim J.B., Jeong W.S., Clayton M.J., Haberl J.S., Yan W., Developing a physical BIM library for building thermal energy simulation, *Automation in Construction*, 50: 16-28, 2015, http://dx.doi.org/10.1016/j.autcon.2014.10.011.

[17] Johansson M., Roupé M., Bosh-Sijtsema P., Real-time visualization of building information models (BIM), *Automation in Construction*, 54: 69-82, 2015, http://dx.doi.org/10.1016/j.autcon.2015.03.018.

[18] Fedorik F., Malaska M., Hannila R., Haapala A., Improving the thermal performance of concrete-sandwich envelopes in relation to the moisture behaviour of building structures in boreal conditions, *Energy and Buildings*, 107: 226-233, 2015, doi:10.1016/j.enbuild.2015.08.020.

[19] Fedorik F., Haapala A., Numerical estimation of mould growth on common single-family house building envelopes in boreal conditions, *European Journal of Environmental and Civil Engineering*, 2016, http://dx.doi.org/10.1080/19648189.2016.1245632.
Azhar S., Building Information Modeling (BIM): Trends, benefits, risks, and challenges for the AEC industry, *Leadership and Management in Engineering*, 11(3): 241-252, 2011, doi:10.1061/(ASCE)LM.1943-5630.0000127

IFC-based cut & bent rebar supply chain integration

Maciel, Alex R. [a] **and Corrêa, Fabiano R.** [b]

Escola Politécnica, University of São Paulo, Brazil
E-mail: [a] alex_roda@usp.br , [b] fabiano.correa@usp.br

Abstract –

Despite the growing interest towards the Building Information Modelling (BIM), its widespread use in the entire life cycle of construction products is still incipient. In most of the cases, BIM is not properly used, in an integrated manner, with suppliers and manufacturers of custom building components. The BIM-based digital fabrication of custom building components should be one of the main beneficiaries in changing its old processes to embrace BIM. This fact could be related to the lack of solutions that allow the effective reuse of information contained in the BIM model, especially in supply chains arrangement wherein distinct stakeholders provide key information. This is the case of the Brazilian cut & bent reinforcing bar supply chain in which the interface between the rebar fabricator and the engineering firm are carried out by a General Contractor. Although the available standards allow the exchange of reinforcement data from the design to the fabricator's automated CNC-controlled bending machines, it do not support the input of contractor's requirements, such as reinforcement delivery schedule. Focusing on the Brazilian cut & bent rebar supply chain for Cast-in-place concrete, this article explores the use of the IFC to hold the process and control requirements provided by the contractor. The procurement and planning requirements were identified and presented this way. Alongside, a proposition was made to represent in IFC the information required by the rebar fabricator so it could be reused in their industrial environment. Finally, some suggestions are provided to improve the integration among agents that act in this process.

Keywords –

IFC; Reinforcing bars; Supply Chain;

1 Introduction

The Building Information Modelling (BIM) has shown steady worldwide expansion among the AECO industry [1] motivated by its inherent benefits or by the demand of public and private contractors. Despite the advances in its adoption, the use of BIM is still more concentrated in the pre-construction stages, such as design, 3D coordination, quantity take-off and construction planning. It is still not properly used, i.e., in an integrated manner, with suppliers and manufacturers of customized building components.

BIM-driven prefabrication or digital fabrication is pointed out as one of the processes that should clearly receive most benefits in the change [2], especially for engineered-to-order and make-to-order building components. While well established in some sectors such as steel structures, digital fabrication is still underused in the Cast-in-place (CIP) reinforced concrete industry, particularly in regional supply chain configurations wherein the design and fabrication stakeholders do not act together.

Initiatives, such as those promoted by the North American market, which culminated in the publication of the ACI 131.1R-14 standard [3] (Information Delivery Manual for CIP concrete), demonstrate the effort to improve interoperability throughout the CIP reinforced concrete domain, based on international open and non-proprietary standards as the Industry Foundation Classes (IFC). However, for global IFC-based supply chain integration, it is necessary to consider the diversity of configurations that the supply chain presents between different countries.

This is the case in the cut and bent reinforcement bars supply chain, which although has similarity about the key agents involved in the design, detailing, procurement, fabrication, delivery, stock and assembly of the reinforcement bars, the configuration may present itself in different arrangements, influenced by regional practices and standards, project delivery methods and stakeholder's relationships [3] [4] [5] [6] .

According to regional practices, the rebar detailing can be performed by distinct agents. In some countries, such as in the US [7], it is carried out by the fabricator or under his supervision (Figure 1a). Conversely, in other countries, such as in Brazil [4], the rebar detailing is fulfilled by the structural designer without the participation of the rebar fabricator (Figure 1b) and the interface between them is intermediated by the Constructor or the General Contractor (GC).

The "detailing-fabrication" segregation makes more complex and less efficient the digital fabrication based on available standards.

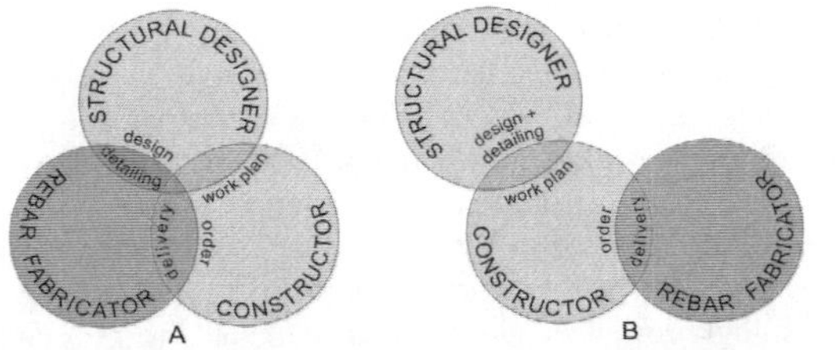

Figure 1. Rebar detailing according regional practices

Focusing on the Brazilian cut and bent rebar supply chain, our research aims to enable the rebar fabrication via IFC-based digital fabrication.

This paper presents how the reinforcement work planning and the rebar purchase order defined by a third-part stakeholder can be described in IFC4 by means of the IFC kernel control and processes extension. It is a further development of the author's previous work [8], where it was demonstrated how reinforcement data, required by CNC-controlled cutting and bending machines, could be mapped in the IFC schema.

2 Brazilian cut & bent rebar supply chain in BIM context

In the configuration of cut & bent rebar supply chain typically adopted by the Brazilian market, structural engineering firms usually do not have a direct relationship with the rebar fabricator, even though it is responsible for both design and rebar detailing.

The interface between the designer and the fabricator is carried out by the Constructor or the GC [4]. Besides acting as an intermediary between these agents, they have the responsibility of the orders and delivery schedule coordination in accordance to the construction planning.

Collaboration between stakeholders becomes even more segregated in the case of public sectors contracts, often regulated by Design-Bid-Building (DBB) project delivery approach, and the rebar detailing is fulfilled by the structural engineering firm, under the supervision of the owner or his designated representative.

The current workflow is highly focused on paper-based documents, albeit generated through digital tools. The rebar placement drawings prepared by means of 2D CAD tools are submitted to the Contractor, which then forwards it to the rebar fabricator in conjunction with the purchase order via e-mail, spreadsheets or forms templates. It is up to the rebar fabricator to convert the design and order information into production data, a highly manual and error-prone task.

This workflow brings "islands of automation" connected by manual data transfer [9] where the fabricator, although already had the technological infrastructure to receive this data in digital format and reuse directly in its automated industrial system, has to recreate the reinforcement data manually before being the fabrication.

In contrast to the current approach, the use of BIM promotes an integrated and collaborative workflow, which favors the data reuse and the decision-making process with consistent and integrated information, throughout the cut and bent rebar supply chain.

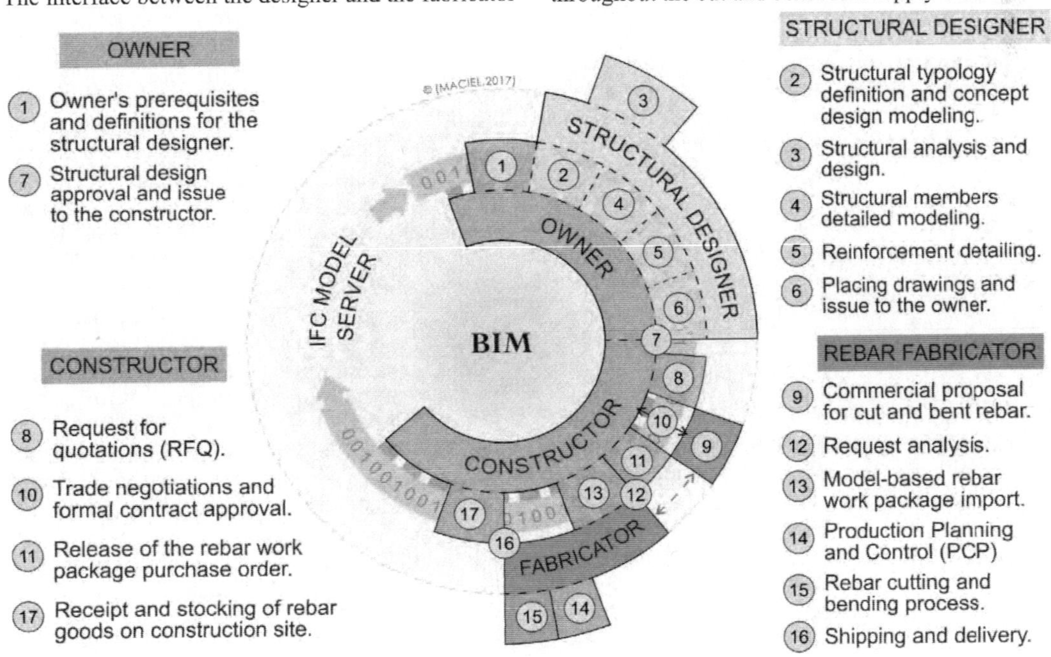

Figure 2. Brazilian cut and bend rebar supply chain in BIM context

The BIM model represents a common and shared database that evolves gradually along the project life cycle, and is updated by distinct stakeholders, as shown in Figure 2.

Coordination and approval activities, although not explicitly represented in the proposed workflow, occur based on the BIM model.

In the design phase, it is mainly used for modelling general structural elements (task 1 to 4), and not often used in the reinforcing detailing. Therefore, other supported BIM uses such as, 3D coordination, or 4D planning, are limited.

The consolidation of BIM in the reinforcement detailing process (task 5 and 6) promotes new possibilities for the downstream activities' stakeholders, enabling the reuse of reinforcement data created in the design phase throughout procurement (task 8 to 10) and rebar fabrication (task 12 to 16).

3 IFC-based cut & bent rebar supply chain integration

Despite the availability of standards such as the BVBS [10], which allows the digital exchange of the reinforcement data, current standards are focused on the end-to-end interface and do not enable the collaboration of all stakeholders of the cut and bent rebar supply chain. It is limited to the unidirectional data transfer from the designer's BIM/CAD rebar detailing authoring tool to the rebar fabricator.

The current standards prove to be most effective in cut and bent rebar supply chain configuration wherein the designer and the fabricator has a direct relationship or when the rebar detailing is performed by the fabricator. However, when these agents do not directly collaborate, the use of digital design-fabrication interface based on the current standards requires high synergy.

When acting as an intermediary, the Constructor or GC may choose to use the same authoring tool to include their requirements; otherwise, it is in their liability the management of multiple documents with distinct files formats (including updating new information), in addition to provide the procurement and delivery requirements.

An example of the proprietary BIM tool [11] that allows the rebar integrated workflow is presented in the Crusell Bridge project [2]. In this case, some parts of the proposed workflow are supported natively, while others required the use of third-party add-ins.

To reduce non-valuable added activities for data re-interpretation and re-entry of information [12], the need for additional tools or manual intervention prior to the rebar fabrication, it is important that existing standards be improved or new standards be developed [13]. In order to maintain compatibility with the existing cutting and bending machines and operating plants, it is recommended that these actions be done in parallel.

In alignment with others AECO sectors, the adoption of the Industry Foundation Classes standard (ISO 16739:2013) is considered appropriate to support the cut & bent rebar supply chain exchanges. In this regard and aiming to enable a wide-range IFC-based supply chain integration, it is necessary to represent the procurement, planning and production definitions inside IFC files.

From the rebar fabricator's viewpoint, the production data requirements can be subdivided into two data sets according to the source who provides them.

The first data set consists of the rebar geometric definitions shape and quantity requirements, provided by the designer, and which will be used in the CNC-controlled bending machines. In a previous work [8], it was proposed a mapping of the requirements predicted by the BVBS [10], the most widely used standard for cutting and bending machines [13], in the IFC schema.

The second data set contains the contractor's purchase order and delivery schedule requirements which are the central point of this paper.

The discussion below is limited only to constructor and rebar fabricator interface (tasks 11 and 17), excluding from this analysis the commercial transitions.

3.1 IFC kernel process extension

Included in the IFC core layer structure [14], the IFC kernel extensions introduce basic and primary ideas for the entities derived from *IfcObjectDefinition* that are related with the concept of products, processes or control.

While the product extension handles items that have geometric and spatial context, including physical items (e.g. components, building elements) and non-physical items (e.g. spaces, grids, annotations), the process extension captures the workflow concept required to perform an execution or to support construction activities.

Work tasks and activities with time-ordered logical sequence are mapped in the IFC4 schema by means of the *IfcProcess* subtypes: *IfcTask*, *IfcEvent* and *IfcProcedure*. These process entities should be also linked to a specific entity-type, which defines common characteristics to a particular kind of process or activity. Exception to this rule is the schedule or time assignment, which should be defined at the object-occurrence level.

The *IfcTask* entity represents units of work that will be performed in the construction project, including building and product installation activities as well as order actions, move operations, design and others construction related activities.

An *IfcEvent* describes an occurrence that may happen and triggers an action or response.

An *IfcProcedure* denotes a logical set of actions in response of an event, or to cause an event in particular in a stepped process.

Inherently, a process transforms inputs into outputs, and may cause the spending of resources. Products used as inputs (*IfcProduct* subtypes) and resources consumed (*IfcResource* subtypes) can be assigned to the process entities through the relationship *IfcRelAssignsToProcess*.

According to their purpose, tasks can be nested within subtasks or procedures, as well as with other tasks in order to represent its hierarchy. The logical sequence among process entities is modeled by the *IfcRelSequence* relationship, describing a predecessor-successor relation and time lag between two entities.

Tasks must be declared in the project context when representing a root-level task. Otherwise, they should be ascribed to the *IfcWorkControl* subtypes to provide a schedule of work task (*IfcWorkSchedule*) and its related work plan (*IfcWorkPlan*).

Moreover, the entities derived from *IfcProcess* may be assigned to the control *IfcWorkCalendar* entity in order to establish working and non-working time periods.

3.2 Reinforcement works planning

According to their size, cast-in-place reinforced concrete structures are commonly partitioned into several work packages by floor or concrete pour phase in order to facilitate their execution. There is not always a correlation between the pours package extension and the physical structural members set [15].

Each work package is then breakdown, composing specific and interdependent work packages as formworks, reinforcement works, MEP embedded and concrete placements, each one with a particular work schedule, set of subtasks and a specific supply chain.

The schedule of each specific work packages is described in IFC4 schema by an *IfcScheduleWork* and aggregated into their respective work plan (*IfcWorkPlan*) which represents a production phase. These entities are also used to control the planned, baseline and actual scheduling of the construction. A proposal to IFC-Based schedule and progress updating is present by [16].

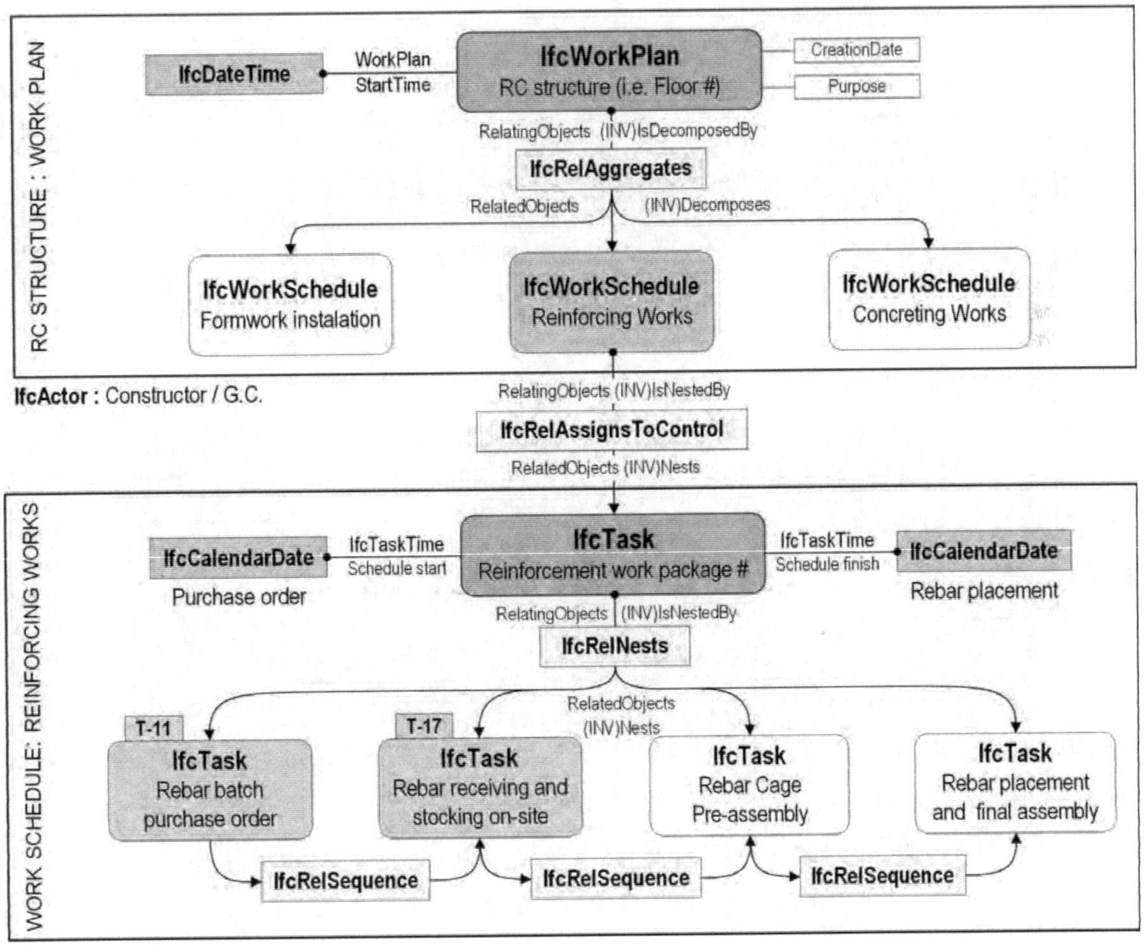

Figure 3. IFC-based reinforcement work package planning

Targeted at the concrete casting expected date, the planning of each rebar work package must predict the lead-time of the procurement, fabrication and delivery of rebar goods besides the time needed for rebar cage assemblies and rebar placement deadline, considering the interrelationship with the others work packages.

In Figure 3 it is shown how a typical workflow of reinforcing works may be described through the IFC4. The reinforcement work package (summary task) has a set of sequential nested subtasks tied to the main task. Each work unit represented by an *IfcTask* entity has a logical sequence and a start/end schedule allocated.

From the perspective of the cut and bent rebar supply chain, its last activities are represented by the delivery of the rebar goods (Task 16) and receipt/stocking at the construction site (Task 17) as shown in Figure 4.

The unloading and subsequent stock of the rebar packages are part of the first activity performed by the construction crew. This task concerns a set of internal logistic activities that precedes the rebar cage pre-assembly and their final placement.

The start date assigned to this task denotes the scheduled delivery date to be reported to the rebar fabricator at the purchase order.

Although unloading and storage of the rebar goods are usually carried out on the same day of delivery, according to [4], the cut and bent rebar commonly remains stored for 15 days before its final positioning.

The same author concludes that this fact may reveal inefficiency in the planning and control process of the activities performed in field given the reliability of the delivery by the rebar fabricator.

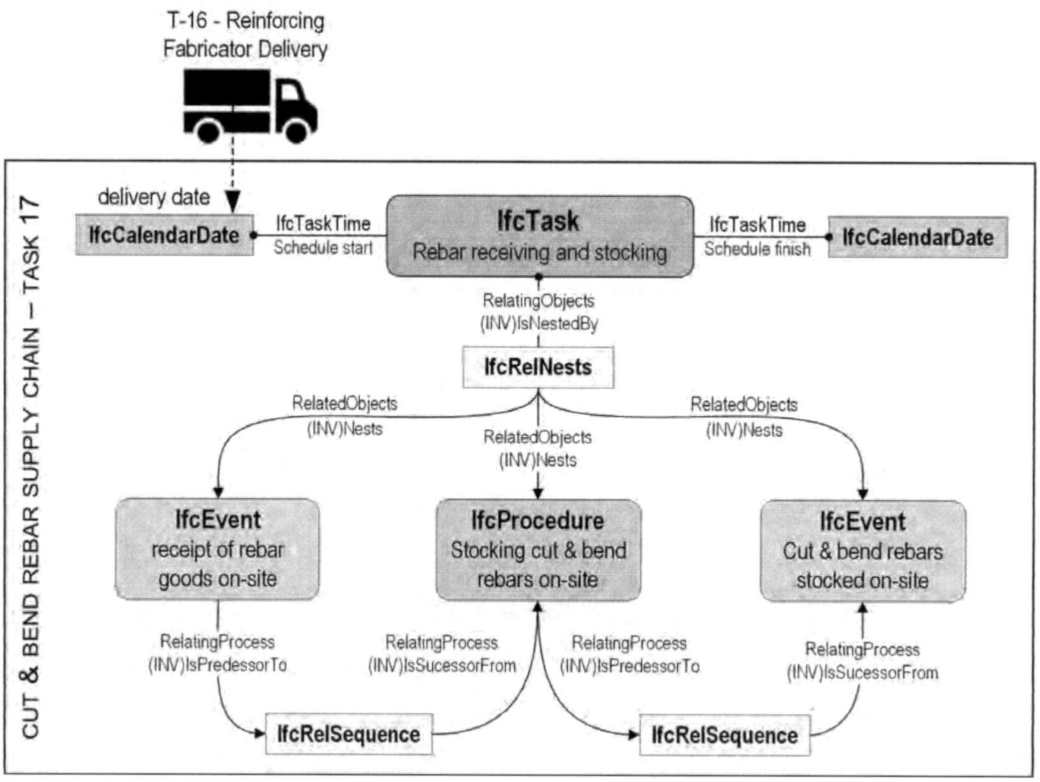

Figure 4. IFC-based rebar receiving and stock on construction site.

3.3 Reinforcement order

Based on the expected delivery deadline it is up to the construction manager to formalize the purchase order of the respective rebar work package considering lead-time required by the rebar fabricator.

The order process consists of selecting the reinforcement that composes the work package and assigning them to the respective purchase order, including the desired delivery date.

In the traditional paper-based workflow, the contractor merely forwards the corresponding placing drawings set to the rebar fabricator along with the purchase order form.

This workflow requires that the rebar fabricator transcribes the design and planning data into production data before the rebar processing. In some cases, the placing drawings need to be partially produced in accordance with the pour break requirements and entail an additional effort on the part of the fabricator to reckoning the correspondent rebar set.

By contrast, a BIM-based workflow besides providing reliable rebar detailing and coordination, when properly used in an integrated manner with the constructor and the rebar fabricator can contribute to improving the rebar digital fabrication establishment, mainly for a tripartite supply chain configuration such as the Brazilian one.

Figure 5 represents the proposal from the authors for IFC-Based order exchange.

The *IfcTask* "rebar order" summarizes a set of activities performed by the constructor, GC or your subcontractor to formalize the rebar request.

The first procedure is correlated with the concrete pour extension. It is done concurrently with the other work packages planning (e.g. formworks, concrete placement), considering productivity rates and constructability.

Once defined the concrete pour phase limits, the delivery schedule requirements and respective rebar set, the purchase procedure is effectively started.

The assignment of the rebar set from the rebar detailed model for this *IfcTask* is modelled through the relationships *IfcRelAssingnToProcess*.

Finally, the purchase order (depicted by the *IfcPurchaseOrder* entity) is assigned to the rebar order task by means of the relationship *IfcRelAssingnToControl*.

Furthermore, it should be indicated the organization that is issuing the request and who will be contracted to fulfill the order.

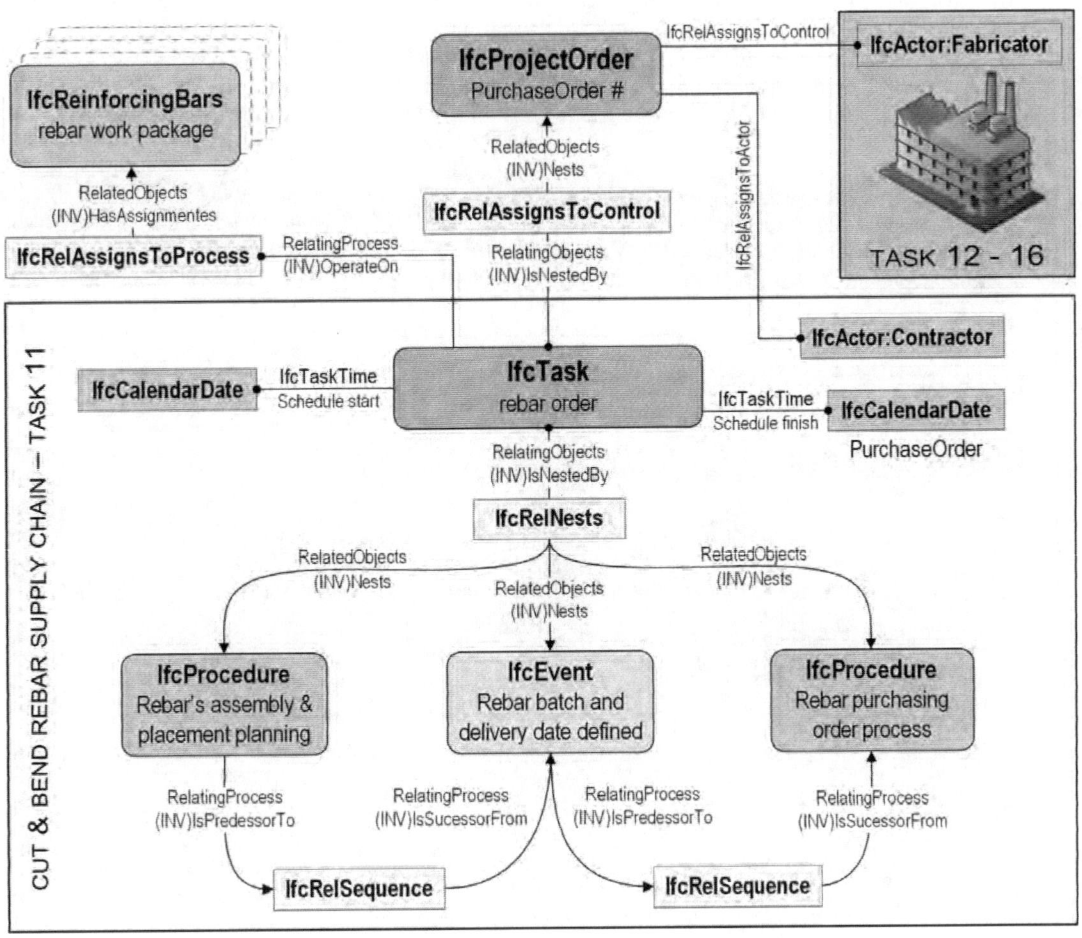

Figure 5. IFC-based rebar work package assignment and purchase order

An *IfcProjectOrder* represents directives for performing a service or purchasing a product which are also typically associated with formal contracts between two organizations [14]. It is a control entity belonging to IFC interoperability layer *IfcSharedMgmtElements*.

It contains the *Ifc.ProjectOrder.PredefinedType* attribute, which should be filled with the enumeration value PURCHASEORDER (*IfcProjectOrderTypeEnum*) to configure a purchase order for a good or service.

This entity has a unique identifier (GUID), name, description and revision control associated. The purchase order identifier is set in the *IfcProjectOrder.Identification* attribute.

Any changes, represented by other *IfcProjectOrder* instance, could be associated to the original purchase order, through the relation object *IfcRelNest*, and indicated by the enumeration value of CHANGEORDER to the *PredefinedType* attribute.

In some cases it is required by the rebar fabricator to know in advance if there is any restriction of access to the site, as limits for truck size, which should be informed by the *ShipMethod* property included in *Pset_ProjectOrderPurchaseOrder*.

4 Conclusions

In this article, an IFC-based integration for the cut and bent rebar supply chain was proposed, wherein the design-fabrication interface is performed by a third part.

Focusing on the configuration adopted by the Brazilian market, it was presented how the rebar planning and order can be properly described in the IFC schema.

Notwithstanding the availability of proprietary BIM tools that allow the proposed workflow, some parts natively or via Application Programming Interface (API) as a whole, the authors consider that widespread supply chain integration should rely on neutral standards, such as the IFC.

By demonstrating the potential use of IFC to support downstream supply chain activities, the authors attempted to contribute to the development of tools that improve the constructor's planning and the specification of rebar orders activities, aiming to reduce the occurrence of inaccurate orders, which is one of the most cited problems in the Brazilian market [4].

Currently, these workflows are not covered by the IFC4 official buildingSMART Model View definitions (MVD) [17]. However, they may be addressed in future reinforcement CIP domain specific MVD's development.

Through the establishment of BIM processes in the rebar detailing, procurement and planning, as well as the IFC consolidation and MVD development, there are opportunities to expand the digital fabrication on the CIP reinforced concrete domain.

It is expected that the development of new tools that allows the rebar fabricator to access and reuse data contained in the IFC model, as illustrated in Figure 6, would reduce considerably the production lead-time. Reports obtained from interviews done by one of the authors show that this period corresponds to about 15% of the rebar fabricator's lead-time.

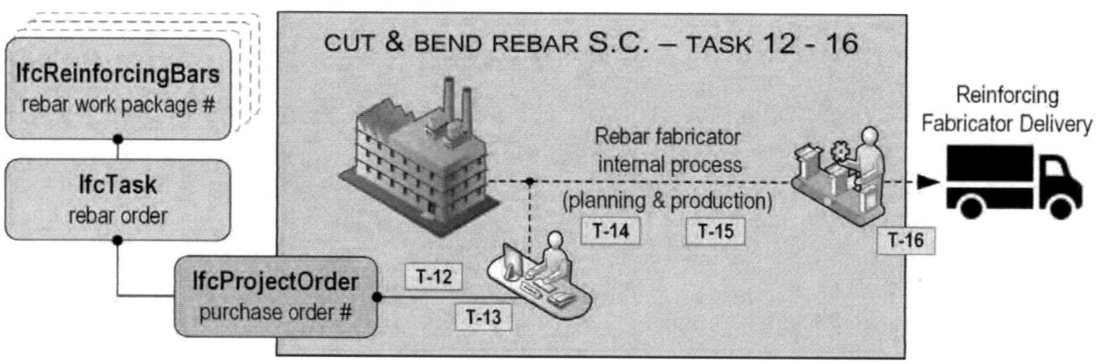

Figure 6 – IFC-based rebar fabricator interface

5 References

[1] McGraw Hill Construction. *The business value of BIM for construction in major global markets: how contractors around the world are driving innovation with Building Information Modeling*, McGraw Hill Construction, Bedford, USA, 2014.

[2] Eastman C., Teicholz P., Sacks R. and Liston K., *BIM handbook: a guide to building information modeling for owners, managers, designers, engineers, and contractors*, 2nd edition, John Wiley & Sons, Hoboken, New Jersey, USA, 2011.

[3] American Concrete Institute, *ACI 131.1R-14: Information Delivery Manual (IDM) for Cast-n-*

Place Concrete, ACI Committee 131, Farmington Hills, MI, USA, 2015.

[4] Isatto E., Azambuja M. and Formoso C. The Role of Commitments in the Management of Construction Make-to-Order Supply Chains. *Journal of Management in Engineering*, 31(4):04014053, 2015.

[5] Polat G. and Ballard G. Construction supply chains: Turkish supply chain configurations for cut and bent rebar. In *Proceedings of the 11th Annual Conference of the International Group for Lean Construction*, Virginia, USA, 2003.

[6] Clevenger C. and Khan R. Impact of BIM-Enabled Design-to-Fabrication on Building Delivery. *Practice Periodical on Structural Design and Construction*, 19(1):122–128, 2014.

[7] American Concrete Institute, *ACI 315-99 Details and Detailing of Concrete Reinforcement*, ACI Committee 315, Farmington Hills, MI, USA, 1999.

[8] Maciel A. and Corrêa F. Interoperability with IFC in the automated rebar fabrication. In Proceedings of the 33rd International Symposium on Automation and Robotics in Construction (ISARC), Pages 872–880, Auburn, USA, 2016.

[9] Navon R., Rubinovitz Y. and Coffler M. Reinforcement-bar manufacture: from design to optimized production. *International Journal of Computer Integrated Manufacturing*, 11(4):326–333, 1998.

[10] BundesVereinigung der BauSoftwarehäuser. *BVBS - Guidelines: Exchanging Reinforcement Data - interface description release 2.0*, BundesVereinigung der BauSoftwarehäuser E.V., Bonn, Germany, 2000.

[11] Trimble. Modeling for Efficient Rebar Production and Assembly. On-line: https://www.tekla.com/solutions/rebar-fabricators/, Accessed: 10/05/2017.

[12] Froese T. Future directions for IFC-based interoperability. *Journal of Information Technology in Construction*, 8:231 –246, 2003.

[13] Aram S., Eastman C. and Sacks R. Requirements for BIM platforms in the concrete reinforcement supply chain. *Automation in Construction*, 35:1–17, 2013.

[14] buildingSMART, Industry Foundation Classes Version 4 - Addendum 2. On-line: http://www.buildingsmart-tech.org/ifc/IFC4/Add2/html/, Accessed: 10/05/2017.

[15] Barak R., Jeong Y., Sacks R. and Eastman C. Unique requirements of building Information modeling for cast-in-place reinforced concrete. *Journal of Computing in Civil Engineering*, 23(2):64–74, 2009.

[16] Hamledari H., McCabe B., Davari S. and Shahi A. Automated Schedule and Progress Updating of IFC-Based 4D BIMs. *Journal of Computing in Civil Engineering*, 31(4):04017012, 2017.

[17] buildingSMART. MVD Releases. On-line: http://www.buildingsmart-tech.org/specifications/ifc-view-definition/, Accessed: 10/05/2017.

Classification of objects and their properties defined in Korea Building Act to translate into KBimCode and their application

Jinsung Kim [a], Jin-Kook Lee [a], Jaeyoung Shin [a] and Hayan Kim [a]

[a] Department of Interior Architecture Design, Hanyang of University, Republic of Korea

E-mail: wlstjd1320@gmail.com, designit@hanyang.ac.kr, jjyoung311@gmail.com, hayaan92@gmail.com

Abstract –

This paper aims to classify and structure explicit objects and properties which can be utilized in KBimCode and develop a database of the Korean building Act. In order to conduct automated building code compliance checking, it is essential to translate permit requirements in Korean Building Act into a series of explicit and logic-based form for computers. As a basis of the first step in translation of Korean building permit requirements for the automated building code compliance, previous research shows how to analyze regulation sentences written in natural language.[1] Based on this, 1200 target objects and their related properties were extracted. The objects were captured in noun phrases of regulation sentences and classified into a series of types according to their information. Through the process of removing the similar and ambiguous meaning of objects and properties, they were transformed into an explicit form. Then they were categorized into two types of generic building object and building regulation-specific objects. In particular, the latter that condition regulations require to check have are based on dynamically instantiable objects from the implementation perspective. The same process of object extraction and classification was conducted for identification of target object properties. As an application of the objects and properties database, the web-based database is suggested for the user-friendly interface to easily find and manage objects and properties and measure the stats and frequency of usage of them.

Keywords –

Building Information Modeling (BIM), automate building checking, Korean Building Act, Building Regulation-specific object, Classification, Database, Web-based application.

1 Introduction

Recently, interest and application of building information modelling (BIM) have increased in the architecture, engineering, construction and facility management (AEC-FM) industry. In Korea, it is obligatory to apply BIM to buildings with a construction cost of over 50 million dollars and supports the active introduction of BIM through the development of the BIM environment at the national level.

The application of BIM has enabled the creation and utilization of building models with rich information handled throughout AEC-FM industry in the computer environment. This has helped to automate design assessment which had been conducted by a hand-based method on 2D-based drawings. This paper focuses on the Translating design assessment contents described in natural language into computer-executable form as the first stage of the automate design assessment system

Building regulation codes, which are the main content of the design assessment contents, are different in the language and expression in each country, and regional development is essential. Some of the sentences in Korean Building Act have complex relationships among them. And they contain many words which have a double meaning and similar words referring to the same meaning. So it is difficult to translate sentences into explicit and logical forms for execution on the computer.

As a preliminary research to support the rule-making based on the building permit regulations, the studies aim to analyze and organize the objects and related properties of building described in Korean Building Act. Simultaneously, the characteristics and classification system of objects and properties on the building regulation sentence can be utilized as one of elements to translate information of the sentences into an explicit form so that it can be applied to automation design assessment based on BIM.

This paper deals with analysis of building regulation-specific objects in Korea Building Act and the scope of this study is as follows:

1. Analysis of the Korean Building Act focusing on Noun phrases
2. Classifying and structuring of objects and properties
3. Development of objects and properties data-base and its application as a dictionary

2 Background

2.1 Building requirement checking projects

As an approach to BIM application, building design assessment gradually has accepted automatized and related projects have been carried out at the level of government in the several countries. For example, CORENET led by Singapore is the representative project accomplished to automate building design assessment based on the IFC model structure. GSA (General Services Administration) project focused on the evaluation of a circulation system automatically drew over the courthouse building model [2]. As the prerequisite step of assessment, building regulations written in natural language have to be translated into the computer-executable code.

Conventionally, the rule translating process has been performed by both rule interpretation and hard-coding process at the same time. The code generated in this way has limitation such as specific software dependant, the difficulty of reuse the existing resource in defining a new rule, therefore rule-maker have to create new ruleset with the repetitive and redundant process.

To solve the problem, an approach has been proposed that code definition using data modeling languages focused on the IFC, the international standards. The DesignCheck (2006) is the Australia building code checking system utilizing EDM (Express Data Manage) to share the object-oriented database. Because the rule-checking attributes such as entities or rules are defined in EXPRESS language, this approach has been evaluated as developer focused method. Both methods pursue the code generation method mainly focused on the developer and have the limitation that accessibility is low for the domain workers who have little programming knowledge.

The Korean government has been led the KBim project to establish the framework for the automatic building permit system. As the part of the project, KBimLogic is devised to translate building regulation sentences written in natural language into the computer-executable format, KBimCode. The detail is discussed in the next chapter.

2.2 Logic rule-based mechanism : KBimLogic

The KBimLogic is based on the logic rule-based mechanism to translate building permit related regulations from Korea Building Act into the computer-executable ruleset file. As the first step, regulation sentences are decomposed and examined to deduce building objects, methods and sentence relationship. By the components in the database, KBimCode is generated with language syntax and managed by KBimCode editor and composer. Generated KBimCodes are also collected as KBimCode DB and reused for the other KBimCodes' relational contents and can be exported as ruleset file. The ruleset is composed of the standardized and intermediate language that both human and computer readable. Also the ruleset can be utilized in various rule checking software at the phase of rule review with BIM model.

The logic rule-based mechanism is composed of three parts 1) classifying objects and properties extracted from the regulation sentences 2) stereotyping verb phrases from sentences 3) decomposing sentences as atomic sentences. The scope of this paper is focused on the extracting and classifying building objects and their properties from the noun phrases in the sentences.

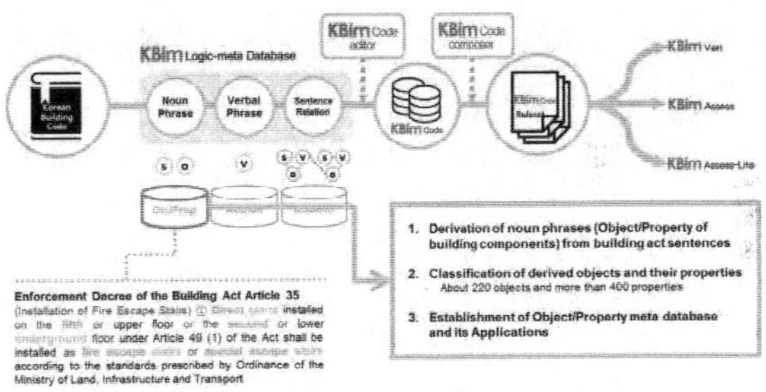

Figure 1. Object and property database and example among the overall KBim project

'Figure 1' shows the overall flow of project and the scope of this research. By analyzing sentences from Korean Building Act, noun phrase, verbal phrase, and sentence relation are deduced and accumulated as a database for reuse to generate KBimCode. For example, EDBA (Enforcement Decree of the Building Act) article 35 is examined to derive the building components such as stairs or underground and their properties such as type and number of the floor. Similarly, predicates that are common and high frequency from verbal phrases are also collected and classified in the database to be used as methods for defining the exact action of the object.

The directly related sentence component for the rule-making process is a method for querying building object and checking whether the property requirement is satisfied. As the parameter of the method, objects and properties have to be precisely defined to imply the exact definition described in the regulation. Therefore, this research proceeds the study by following five steps.

1. Extract the target object from the noun phrase
2. Include specific object defined in the regulation
3. Extract the property for check object based on the BIM modeling
4. Categorizing objects and properties considering different expression of same meaning, or different meaning of the same expression
5. Concretizing the object and property database by developing web-based dictionary

3 Object Extraction from Noun Phrases of Building Permit Requirements

3.1 Korea building permit requirements analysis

The requirements for the building permit are regulated by various type of documents such as building acts or fire evacuation-related regulations. Since an article includes several clauses and paragraphs, it is hard to understand the exact contents at one view. Also, each article has complicated sentence relationship with other articles or paragraphs such as upper-lower relationship or reference and commission to another level of regulations. Fundamentally, the upper level of regulations assign the outline or general features of requirements and refer relation of lower level of regulation sentences. In contrast, lower level of regulations basically designates detail requirements and method for checking regulations. According to the type of regulations, description manner and contents structure differ by type. Therefore, building permit related requirements are especially hard to be translated into computer-executable form.

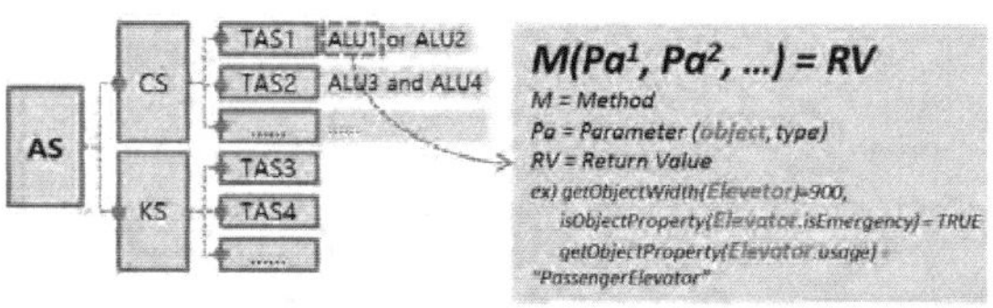

Figure 2. Level of dividing single sentence into multiple units of atomic sentence

To figure out complicate meaning included, the article has to be identified at the level of the atomic sentence (AS) which composed of S (subject) + O (object) + V (verb) and has a Boolean result. Then the sentence is again fragmentized as translated atomic sentence (TAS). Finally, TAS is divided as arithmetic logic unit (ALU) composed of unsplinterable S, O, and V. By fragmentizing sentences into the least unit, target object of regulation can be easily identified. To translate natural language sentences into the logical structure of semantic attributes, every logical unit has to be connected according to the logical relationship described in regulation to restructure the meaning of original article.

By the logic rule-based mechanism, natural language sentences are composed of minimum semantic components without redundant meaning. Translated logic structure of the sentences is collected as sentence database to compose KBimCode much easily than before.

3.2 Object name extraction from the Korea building regulations and its database

As the pre-process to structure the database for object and property, subject and object in every type of noun phrase were extracted and collected as name database. The database includes the wide range of expression of same meaning or related objects or properties. 'Figure 3' shows the web-based object name database and the interface for searching. The (a) in the image is typing "wall" in the name box to query wall or related objects or properties. The (b) is the part of queried result including wall, different type of wall and related objects that extracted from every noun phrases in the building permit related regulations.

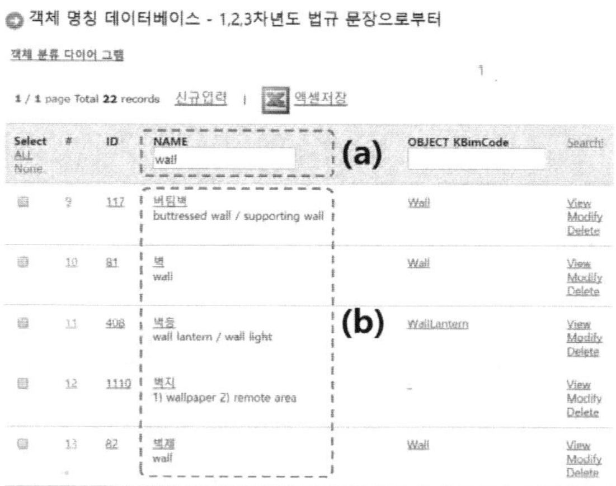

Figure 3. Searching "Wall" on the name database

'Table 1' shows the part of queried result when search 'Wall' as the name in the searching tool of web-based database like 'Figure 3'. In Korean, the id 81 and 82 expressions are different but have the same meaning that exact 'Wall'. Name database is useful for solving this issue that how to figure out some expressions having similar meaning and merge them.

Table 1. Example of the various usage of "Wall"

ID	Korean Name	English Name
117	버팀벽	Buttressed wall / Supporting wall
81	벽	Wall
408	벽등	Wall lantern / Wall light
1110	벽지	1. Wallpaper 2. Wall light
82	벽체	Wall

Table 2 Classification and example of objects

Level1	Level2	Level3	Count
Target object	Space objects	Space	122
		Floor	12
		Area	17
		Building	322
	Building objects	Building element	110
		MEP	386
		HVAC	37
	Materials		20
	External building objects	Site	9
	District / Region		67
Non-Target object	External building objects	External Structure / Road/ River	37
	Subject and Institutions		36
	Etc.		14
	Gross		1200

4 Classification and Definition of Objects and Properties on the Building Permit Requirements.

In order to establish the basis of various applications of BIM, standardizing process of building object information that occurs in the entire construction industry is being done.

IFC [12] and Omniclass [13], which are international standards, are information systems for building objects that can be directly linked to BIM. But these international standardization systems do not deal with specific information handled in the Korean AEC-FM industry. In Korea, the research on systematization of building information was conducted for the purpose of, which is led by the Korea Institute of Construction Technology [14.] However, the Korean construction and building information systems are not intended to reflect the specific data in the building codes because they are aimed to standardize general objects that are handled throughout the Korean industry. For the permit related rule checking process, it is necessary to establish information system not only for general building objects but also the specific building objects of Korean Building Act, properties and their relationships. Next two sections introduce the approach for explicitly identifying the objects and properties required in the regulations. As a result, 220 objects and 583 properties are defined as the explicit form for KBimcode.

4.1 Definition of objects

Based on the object names extracted from the 1977 sentences of the target Building Act, the classification of them and count are the same as Table 2'.

The objects could be classified by building objects, space objects, materials, the external building objects, subjects, institutions, and District / Region, etc. These are classified into the rule-checking target objects and non-target objects according to the whether they can be automatically assessed.

The target objects include the building objects, space objects, the material, and the building site among the external building objects, and it occupies 93% of the whole extracted objects. (Figure 4.) On the other hand, the non-target objects include subject, institutions and external structure, road, river among the external building objects

The target objects include BIM-enabled objects and non-BIM objects related to the building permit process. The former refers to the objects that can be directly defined by IFC objects, and the latter refers to the external objects derived from BIM data such as indoor circulation graph, work space, energy zone, scaffold, etc.

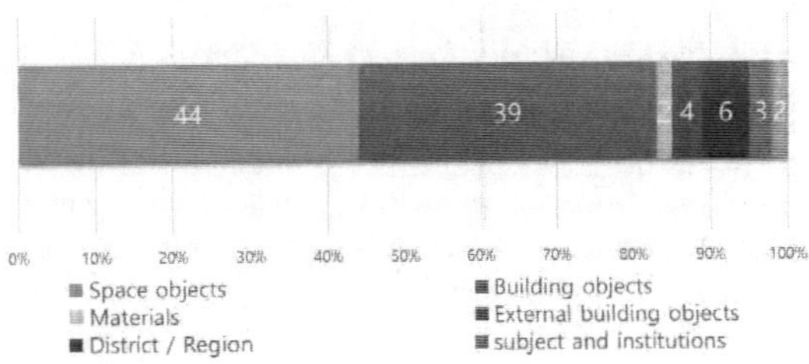

Figure 4. Target object types

The target objects are categorized into two types of

generic building object and building regulation-specific objects. The generic building objects can be defined as a static objects such as building, slab, door and wall. On the other hand, a building regulation-specific objects whose condition regulations require to be checked is based on dynamically instantiable objects from the implementation perspective. The building regulation-specific objects include three categories that they are defined by (1) set of building objects (object group), (2) a certain generic object with specific properties and (3) partial space object.

'Table 3' show example elements of building regulation-specific objects and their related generic object and category.

Table 3. Category of building regulation-specific objects and the examples

Object Name	Generic Object	Category
Passage	Space	1
Main Structural Part	Column, Floor, Beam, Roof, Stair	1
Parking Lot	Space	2
Living Room	Space	2
Elevator Shaft	Zone	2
Specific Fire-Fighting Building	Building	2
Open Floor Structure	Slab	2
Floor Slab	Slab	2
Balcony	Slab	2
Ceiling	Covering	2
Ceiling Covering	Covering	2
Parking Unit	Space	3
Parking Area	Space	3
...	...	...

4.2 Definition of object properties

Properties are depend on specific object and cannot be detached from the object. In order to identify target properties, the same process as object extraction and classification was conducted. The systematization of the properties in the target regulations was carried out by referring to the properties classification system of the BIM application guide (2010) of ministry of land infrastructure and transport republic of Korea. As a result, 583 properties were extracted from the target sentences. In addition, 217 properties were identified through the process removing redundant meaning.

Table 4. Examples of object properties type s

Type	Element	Count
Basic Property	Building data Region/district Floor location Usage Material Building structure ...	39
Geometry property	Geometry ...	29
Object-specific property	opening Electric Pressure Count ...	125
Relational property		6
Calculation property	Allocation Count Slope between objects Sealing / Exposure / Openness Walking distance ...	14
Complex property		3
Reference property		1
Gross		217

'Table 4' and 'figure 5' summarize the results of classifying the detailed items of each type into the basic property, geometry property, object-specific property relational property, calculation property, complex property and reference property from the 217 properties.

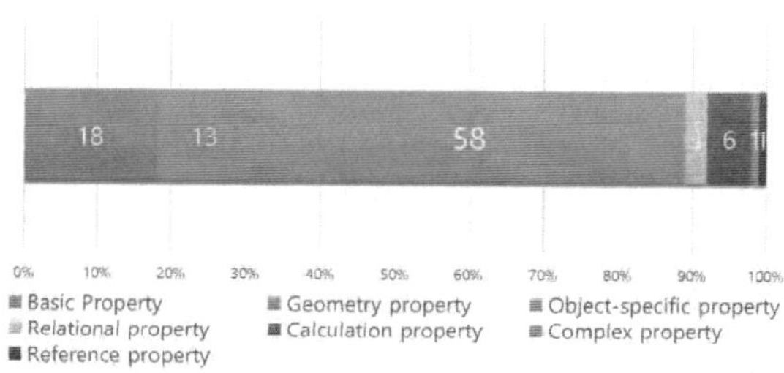

Figure 5. Object properties types

5 Web-based Database As Application of Object and Property Database

To manage structured object and property, this research utilizes the web-based database for a continuous and real-time update of the elements. Through the database, new object and property can be added and the existing data can be modified with authorized account. The database also includes the indexes about object and property such as ID, object, object KBimCode, property, property KBimCode, return value of corresponding object and property and role as the object dictionary. Through using database, statistical data about frequency of using object and property can be calculated. Also, not only for the checking Building Act, but also can be used in various field of building design assessment phase.

5.1 User interface of the database

The object KBimCode, object property KBimCode, ID, data type and description are managed by the web-based database. One of the databases about object name is already explained in section 3.2. This section describes the finalized database of the object and property that reflecting systematized and structured data structure of the database.

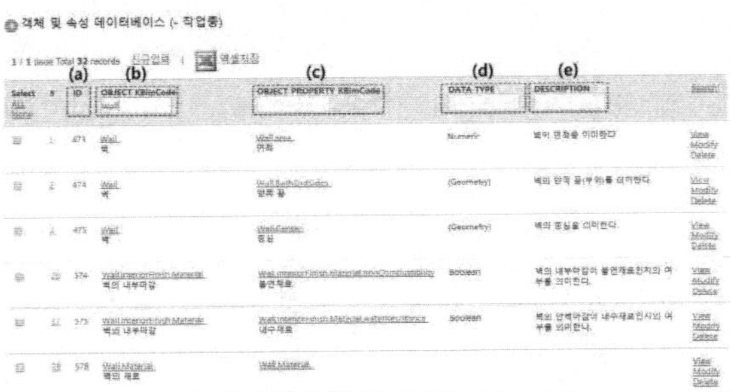

Figure 6. Searching interface

'Figure 6' shows the result of searching "Wall" as section 3.2 describing the Object name database and result about searching the same object. The database structure depends on the classification of OBJECT KBimCode (b). Each object and property has its own ID number (a). Different with Object name database, object and property database provides real KBimCode of the object (b) including general object and specifically defined object by regulation. Also, the property that used with the object is also described in (c) OBJECT PROPERTY KBimCode. The (d) part explains the data type of the property. Object property connected with method node has specific type of return value. By the (e), the description of the object and property, the user easily find the definition of the object and property.

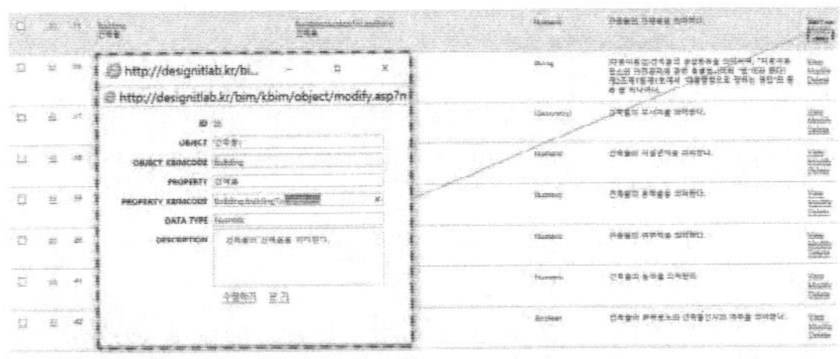

Figure 7. Editing interface for add and modifying

The database also offers the interface for editing of the existing information about object and property. (Figure 7) Through the authorized account, the user can access to the editing system. If the definition of the object described in regulation is changed or a new object is needed for additional KBimCode, new object or property can be added to the database through the new object input interface.

5.2 Usability of database component

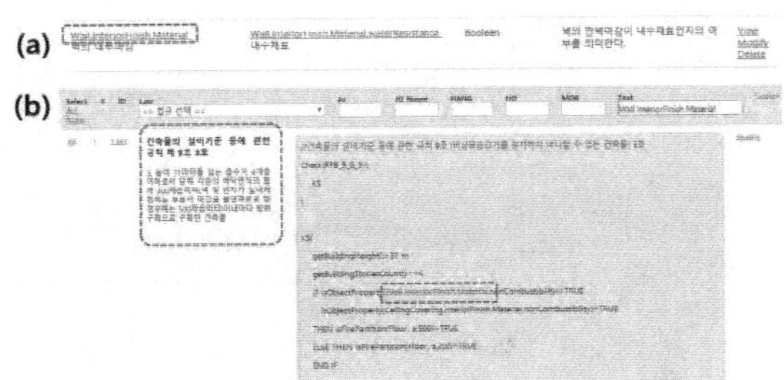

Figure 8. Example of using searching interface

The database is not only for checking definition and description of object and property. Using KBimCode DB, user can easily find the usage example of object and property in KBimCode. KBimCode example provides the method for utilization of composing other codes with the object and property. Compared with original text of regulation, user also can find which original word is coincidence with code.

'Figure 8' shows the example of using searching interface for querying object and property by connecting KBimCode DB. Among the written KBimCode, the user can find the code of object and property easily by marking up the target part. If user clicks property of the object, the same result about clicked property is queried. In this way, database includes the related regulation and KBimCode using same object and property or reference regulation used for defining the definition and scope for object and property. For example, "Wall.InteriorFinish.-Material" property is used in regulations for facilities, evacuation and fireproof structure, and sprinkler facilities standard.

6 Conclusion

This paper introduced classification of objects and properties information from the sentences related to building permitting from Korean Building Act, as a part of rule translation process for the automat building code compliance checking.

In order to conduct automated code compliance checking, it is essential to translate natural language-based requirements in building regulation into computer-executable form. Because the sentences of Korea Building Act have similar but different expression and ambiguous meaning at the same time, logic rule-based mechanism for the explicit form is necessary.

Logic rule-based mechanism and its database development ultimately aims to support translating process of Korean regulation information into a code (KBimCode). Objects and properties can be translated into the code which has a specific grammar structure by combination of other elements of logic-based systemization (methods and sentence relationship).

As one part of the logic rule-based mechanism, the objects and properties were extracted from the noun phrases of related sentences and managed in the object-name database. Based on this, the target objects and related properties which are needed and possible to assess were classified and organized. Then, through a removing process of similar and ambiguous objects and properties, they were transformed into explicit form and can be managed in a web-based database.

Dictionary function of the web-based database can give some benefits to concerned expert who do not much have programing knowledge. It help them to change and manage the objects and properties in general web interface. And the sentence and code in which each item is used can be found. It also can help to grasp the frequency of items or related contents. In addition, these benefits can be helpful for future studies of mapping between the result of this study and objects and properties of IFC model, which are data standards for BIM.

Based on this mechanism, the database of objects and their properties on Korean Building Act can be extended to other target requirements or more general descriptions (e.g. design guideline, RFP, etc.).

Acknowledgment

This research was supported by a grant (17AUDP-B127891-01) from the Architecture & Urban Development Research Program funded by the Ministry of Land, Infrastructure and Transport of the Korean government.

References

[1] Hyunsoo Lee, Jin-Kook Lee, Seokyung Park, Inhan Kim, Translating building legislation into a computer-executable format for evaluating building permit requirements. Automation in Construction, 71, 49-61, 2016.

[2] Eastman, C., Automated Assessment of Early Concept Designs, Architectural Design, 79(2), pp.52-57, 2009.

[3] Jaeyoung Shin, Jin-Kook Lee, Application of Classification of Object-Property represented in Korea Building Act Sentences for BIM-enabled Automated Code Compliance Checking, Transactions of the of CAD/CAM Engineers, 21(3), pp325-333, 2016.

[4] Project: Open BIM based Technological Environment for Building Design Quality Enhancement, Development of data-base logic system of analyzing code sentences for automated design quality assessment. On-line: http://designitlab.kr/bim/ Accessed: 28/03/2017

[5] Jun-Ho Choi, In-Han Kim, A Study on the Application of Pre-Processing to Develop the Open BIM-based Code Checking System for Building Administration Process, Journal of the Architectural Institute of Korea, 30(9), 3-12, 2014.

[6] Jin-Kook Lee, Jaemin Lee, Yeon-suk Jeong, Hugo Sheward, Paola Sanguinetti, Sherif Abdelmohsen, Charles M. Eastman, Development of space database for automated building design review systems, Automation in Construction 24(2012) 203–212, 2012.

[7] Jin-Kook Lee, Charles M. Eastman, Yong Cheol Lee, Implementation of a BIM Domain-specific Language for the Building Environment Rule and Analysis, Journal of Intelligent & Robotic Systems 79(3), 507-522, 2015.

[8] Sagar Malsane et al., Development of an object model for automated compliance checking, automation in construction, 2015.

[9] Seokyung Park, Jin-Kook Lee, Inhan Kim, Development of High-level Methods for Representing Explicit Building Act Sentences in the Automated Building Permit System of Korea, Proceedings of the Society of CAD/CAM Engineers Conference, Pyeongchang, Korea, 523-527, 2016.

[10] Project: Open BIM based Building Design Standard and IT-Infrastructure. On-line: http://www.kbims.or.kr/ Accessed: 27/02/2016

[11] Korea Ministry of Government Legislation. On-line: http://www.law.go.kr/ Accessed: 01/03/2016

[12] buildingSMART KOREA, IFC overview, http://www.buildingsmart.or.kr/overview/IFC.aspx

[13] Omniclass. On-line: http://www.omniclass.org/

Accessed: 20/02/2016
[14] Ministry of Land, Transport and Maritime Affairs, Application Standard of Construction Information Classification System, www.law.go.kr/flDownload.do?flSeq=10910812 On-line: www.law.go.kr/flDownload.do?flSeq=10910812 Accessed: 20/02/2016
[15] Lin, ETA, and Cheng Tai Fatt, Building Smart-A Strategy for Implementing BIM Solution in Singapore, Synthesis Journal, 117-124, 2006.
[16] Korea Ministry of Land, Infrastructure and Transport, National Architectural BIM Guide. On-line: http://www.buildingsmart.or.kr/Document/BIM%20 Guide_MLTL_Korea_2010.pdf, Accessed: 10/04/2016

34th International Symposium
on Automation and Robotics
in Construction

ISARC 2017

June 28 – July 1
TAIPEI, TAIWAN

VIRTUAL REALITY / AUGMENTED REALITY

Mobile Point Cloud Assessment for Trench Safety Audits

Jochen Teizer [a], Adam M. Gerson [b], Thomas Hilfert [a], Manuel Perschewski [a], Markus König [a]

[a] Chair of Computing in Engineering, Ruhr-University Bochum, Germany
[b] Occupational Safety and Health Professional, Boise, Idaho, U.S.A.
E-Mail: jochen.teizer@rub.de, adam.m.gerson@gmail.com, thomas.hilfert@rub.de, manuel.perschewski@rub.de, markus.koenig@rub.de

Abstract –

Fatalities resulting from cave-in hazards during excavation work in the United States account for 48% of the trench fatalities in construction every year per Occupational Safety and Health Administration (OSHA) data. Recent trends indicate that fatalities from trench and excavation hazards in the US are increasing. Often the experience of safety inspectors and/or the designated competent person (CP) for trenching and excavation is vital when assessing sloping measurements with approved engineering survey tools. The degree of accidents, however, allows the conclusion that proper assessment and/or protection of excavation sites is not performed sufficiently in the field, or safety coordinators and/or adequately trained CPs are not on hand when needed. While existing assessment processes and protection methods are reviewed for potential improvement, this paper proposes a new compliance assistance prototype that incorporates state-of-the-art technology. The proposed prototype creates: (1) mobile field data acquisition with low-cost photo cameras to capture point cloud surveys of the as-built conditions of trenches, and (2) computational data processing to automatically extract trench height, width, and slope values. Early results to field-realistic experiments promise useful applications of the developed prototype for safety coordinators or adequately trained CPs.

Keywords –

Cave-in hazards, hazard recognition, mobile application, photogrammetry, point clouds, risk management, safety education and training, remote sensing, trench safety.

1 Introduction

The U.S. Department of Labor requires cave-in protection by sloping, shoring and/or engineering controls when an excavated trench is five feet or deeper. If sloping is the designated compliance method, compliance is compared to the requirements in OSHA Subpart P – Excavations [1]. Often the experience of safety inspectors and/or the designated competent person (CP) for trenching and excavation is vital when assessing sloping measurements with engineering survey tools. Current field assessment methods for sloping compliance and the number of accidents, however, allow the conclusion that proper assessment and/or protection of excavation sites is not performed sufficiently in the field, or safety coordinators and/or adequately trained CPs are not on hand when needed [2].

Therefore, this paper proposes an engineered compliance assistance prototype that improves existing trench assessment and field training using mobile point-cloud data acquisition and computational processing for trench sloping assessment. Although the scope of the presented work is limited to the development and testing of the developed prototype, it is envisioned that its application has potential to guide specific toolbox talks and immersive OSHA CP trench and excavation training. Demographics to be reached with this prototype include OSHA CP training centers in the U.S. and worldwide, construction and utility stakeholder organizations, and the twelve most common industries involved in trenching and excavation work per OSHA data [3].

2 Background

The background review concentrates on the identification of the key issues in trench safety: (a) the standards of practice and an analysis of available statistical data on trenching- and excavation-related injuries and fatalities in the US, (b) the current practice for assessing trench compliance with sloping standards, e.g. as outlined in OSHA Subpart P, (c) the reasons behind the current approaches, and (d) the potential of research with emerging technologies to improve safety in trench work.

2.1 Trench definitions and types of violations

A trench in construction is, according to OSHA definitions, defined as a narrow underground excavation that is deeper than it is wide, and no wider than 4.5

meters [1]. While the width of a trench is measured at the bottom, excavated trenches provide predominantly temporary access to buried underground utilities or structures. A trench is a confined space with many special problems.

Per an OSHA Regulatory Review [3], excavation fatalities may result from a variety of accident types, including cave-ins, machine accidents, falling objects, electrocution, car accidents, explosions or fires, falls, drowning, and asphyxiation due to noxious fumes. Distribution of fatalities by cause of death between 1990 and 2000 indicated that 48% of fatalities in excavations occurred by cave-ins. They often are caused by piles of excavated material placed too close to the trench's edge, no protection against the collapse of trench sidewalls, no proper means for workers to safely exit the trench, employees working underneath the operating bucket of a backhoe removing soil from the trench, and no protective equipment, including use of trench boxes or helmets, for workers. Other occupational risks in trench-related work are falls, electrocution, being struck by falling objects or a backhoe, and hazardous atmospheres causing fires or poisoning of workers.

Such trenching-related injuries continue to plague the construction industry, despite the availability of well-known and effective control methods, such as benching and sloping, shoring, and trench boxes and shields. Although trench collapses are not the most common cause of construction deaths, the collapsing weight is likely to result in death or serious injury within minutes [4]. In addition, other workers are often at risk and die trying to conduct rescues.

2.2 Safety statistics and current practice

Excavation work for trenches is responsible for approximately 5% of the more than 1,100 fatalities in the US construction industry every year. Out the total number of trenching and excavation accidents, the percent of cave-in hazards leading to worker fatalities occurs on 48% of sites from OSHA data [2,3]. Most of these accidents occurred in the Standard Industrial Classification (SIC) system code No. 1623: water, sewer, pipeline, and communications and power line construction [5]. The next nearest industrial code was No. 1794: excavation work.

For some of these reasons, the U.S. Labor Department's Occupational Safety and Health Administration (OSHA) requires cave-in protection by guarding trench sides that are not shored or otherwise supported when the trench is five feet or deeper.

Often the experience of a safety inspector and/or designated CP is needed to assess the situation when such safety features are required. The prevalence of accidents, however, allows the conclusion that proper inspection of excavation sites is not performed accurately or safety coordinators are not on hand when needed.

The National Institute of Occupational Safety and Health (NIOSH) stated between the years 1976 and 1981 excavation cave-ins caused about 1,000 work related injuries each year. Of these, about 140 resulted in permanent disability and 75 in death (25 per year). These work fatalities in excavations accounted for nearly 1 % of all annual work-related fatalities [6]. Although better safety training methods and enforced regulations have been developed in recent years, the overall safety performance for trench work in the United States since then has not improved, may be even worsened due to more work hours accomplished.

In 542 trenching and excavation fatalities from 1992 to 2001 the average age of the decedents was approximately 38 years [5]. The same study reports that of the fatalities, 47% occurred among employees of companies with less than 10 workers, and 70% occurred in companies with less than 50 workers. The largest proportion of deaths occurred among construction laborers. Among the decedents, the average length of employment with their employer was 6.7 years, 16% of the deceased workers had been with their employer for less than 1 year.

Conditions that contributed to a fatal event were missing trench protection in 66% of all fatal deaths, lack of supervision and daily inspections in 52% of all fatal cases, no training provided (52%), spoil pile within two feet of edge (41%), and rain or standing water (34%) [7].

In addition, in the years from 2000 until 2009, 350 workers in total died in trenching or excavation cave-ins. Most incidents involve excavation work or "water, sewer, pipeline, and communications and power-line construction" [8]. Results from an analysis of OSHA data from 1997–2001 showed that 64% of fatalities in trenches occurred at depths of less than 10 feet [9]. Lack of a protective system was the leading cause of trench-related fatalities in a review of OSHA inspections [7].

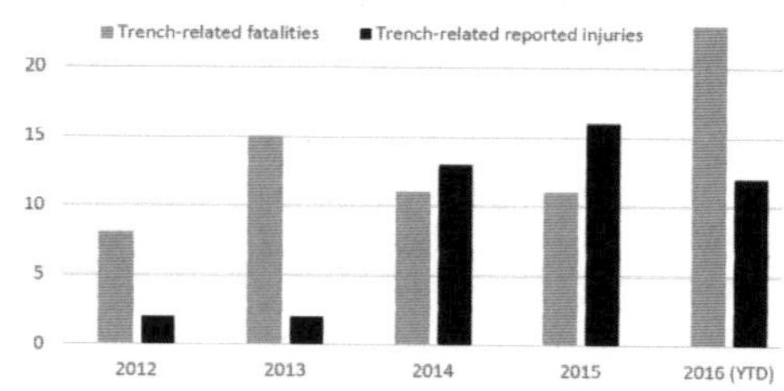

Figure 1. US trench injuries and deaths [10]

Most recent data from OSHA from 2012 to 2016 indicates that trenching injuries and deaths have been trending upward as noted in Figure 1. Trench related deaths and injuries have risen in the US from 8 in 2012, to 23, and 2 to 12, correspondingly, in 2016 (at the time this data was surmised) [10]. Note, depending on the type of material one cubic meter of soil can weigh up to 1,700 kg, which means trench collapses are difficult to survive. Injuries from trench collapses happen less often than fatalities.

2.3 Reasons on existing injuries and fatalities

These statistics and specific examples demonstrate that injuries and fatalities associated with trench collapses and other excavation hazards continue to occur in alarming numbers despite regulations and consensus standards which describe engineering controls, protective equipment, and safe work practices to minimize hazards for workers. Several researchers identified mainly the need for effective worker training before work starts and a CP onsite [11-14]. Although education and training show to be a worthwhile investment, reasons for the failure of current best practices still come from sources related to human decision making, such as:

- The CP is not on hand or misinterprets the site excavation conditions. Subsequently a cave-in can happen any time.
- The CP does not re-inspect the trench before every shift (as the excavation rule states after anything that can increase hazards, such as every rainstorm, vibration or heavier loads, trench walls move or get heavier, causing cracking, scaling, or bulging).

The consistent absence of evidence of compliance with existing regulations suggests that employers are either (1) unaware of the existence of the OSHA standards, (2) misinterpret the requirements of the standards, or (3) lack or fail in supervision.

2.4 Regulations for trench work

Regulations for trench work exist in many countries and in the US since the establishment of the Occupation Safety and Health Administration (OSHA) in 1970. Many of them are practiced to protect workers, for example systems that provide sufficient strength to withstand moving ground [15]. The OSHA Subpart P was established to protect workers in trenches and all excavations in the Code of Federal Regulations: 29 CFR 1926.650,651, and 652. The confined space standard is 29 CFR 1910.146. These special rules try to protect workers in all excavations, including trench work. Some of the general requirements of Subpart P that are relevant to this study are summarized [1]:

- Determine soil classification with manual and visual tests before work begins.
- Install protective systems in excavations deeper than 5 feet in any type of unstable or soft material and soil, but except solid, stable rock. Slope or bench trench walls and faces, shore, sheet, or brace trench walls with supports, shield trench walls with trench boxes, allow safe access using ladders, ramps or stairways.
- Appropriate shoring, shielding, or sloping requirements for all excavations deeper than 20 feet (except those in unfractured rock) must be determined by an engineer qualified to take prompt corrective measures to eliminate safety hazards. Such a CP must be capable of identifying existing and predictable hazards in the surroundings, or working conditions which are unsanitary, hazardous, or dangerous to employees [4].
- Soils can be classified as Type A, Type B, or Type C. Type A soil is the most stable soil in which to excavate. Type C is the least stable soil. It's important to remember that a trench can be cut through more than one type of soil.

If sloping is utilized as the means of compliance by employers, allowable slope angles for excavations less than 20 feet based on soil type and angle to the horizontal are listed in Table 1 [16]:

Table 1 Allowable slope angles for excavations

Soil type	Max. height/depth ratio	Max. slope angle
Stable Rock	Vertical	90 degrees
Type A	¾ to 1	53 degrees
Type B	1 to 1	45 degrees
Type C	1½ to 1	34 degrees

Soil can also be judged to be cohesive or granular. Cohesive soil contains fine particles and enough clay so that the soil will stick to itself. The more cohesive the soil, the more clay it has, and the less likely a cave-in will happen. Granular soils are made of coarse particles, such as sand or gravel. This type of soil will not stick to itself. The less cohesive the soil, the greater the measures needed to prevent a cave-in.

Excavations (including trenches) adjacent to backfilled areas or subjected to vibrations due to proximity from railroads, highway traffic, or operation of machinery shall have additional shoring and bracing precautions taken. Keep heavy equipment away from trench edges, knowing underground utilities prior excavation, keep excavated materials at least two feet away from trench edges.

Requirements under OSHA Subpart P include that a CP inspect the trench and excavation sites for potential cave-in hazards, and verify protective measures are in place. CP training for trenching and excavation, and OSHA 10 hour and 30 hour classes with training curriculum on trenching and excavation, are conducted nationwide in the US.

2.5 Existing challenges

In 1985, the U.S. Department of Labor's OSHA initiated a national emphasis program in trenching and excavation because of the continuing incidence of trench/excavation collapses and accompanying loss of life [15]. The agency had determined that an increased OSHA enforcement presence at worksites where such operations are being conducted was warranted. Trenching and excavation work creates hazards to workers which are extremely dangerous. Compliance with OSHA construction standards applicable to such operations is frequently bypassed because of economic pressures, a belief that compliance is unnecessary or an expectation that these short-term operations will go undetected. OSHA believed that the rate of deaths and serious injuries resulting from trench/excavation accidents (mostly cave-ins) could be significantly affected by a concentration of compliance resources within the area of trenching and excavation operations.

The OSHA regulations in Subpart P are usually established on national level and for practical reasons are often implemented with regional training centers in the local industries. In the State of Georgia, for example, OSHA, Georgia Institute of Technology and ASSE (American Society of Safety Engineers) have formed a Trench Safety Task Force that will arrange safety training courses to authorities overseeing excavation work to increase their awareness of the trenching hazards and enable them to identify unsafe trench conditions using a safety fact sheet [3].

Though currently, training and related curriculum in CP and OSHA 10/30 hour classes is primarily distributed through hard copy documents and PowerPoint presentations. Virtual and/or augmented reality have not been widely adopted for trenching and excavation CP curriculum in the world [17]. There is not an annual training requirement for CP refreshers.

In addition, OSHA covers most private sector employers and workers in all 50 US states, the District of Columbia, and the other US jurisdictions either directly through OSHA or through an OSHA-approved State Plan. 22 states or territories have OSHA-approved State Plans that cover both private and public sector workers. State Plans are OSHA-approved job safety and health programs operated by individual states instead of federal OSHA. Section 18 of the OSH Act encourages states to develop and operate their own job safety and health programs and precludes state enforcement of OSHA standards unless the state has an OSHA-approved State Plan. While OSHA approves and monitors all State Plans and provides as much as 50 % of the funding for each program, enforcement is not always possible. State-run safety and health programs must be at least as effective as the federal OSHA program. The State of Oregon (OR-OSHA), for example, has for the most part adopted the federal trenching and excavation regulations into their code. In the State of Washington, the labor and industries division of Occupational Safety and Health [18] has adopted the federal code, except for sloping, benching, shoring, and/or shielding requirements. Herein, WA-LNI-DOSCH requires protection for a depth of 4 feet or more, as opposed to 5 feet per federal OSHA

2.6 Trench assessment questions

The following overview is a summary of the OSHA Technical Manual on trenching and excavation assessment and inspections. The list of tasks to perform include an OSHA inspector to ask relevant site assessment questions (OSHA Technical Manual, Section V: Chapter 2) [19]. During first and subsequent visits to a construction or facility maintenance location, the compliance officer (or the site's safety officer or other CP) may utilize the following questions when investigating a site for compliance with OSHA Subpart P. Note that imminent danger conditions may require that employees are removed from exposure conditions as soon as possible. To understand the questions easier, a summary flow chart for the worksite assessment on trenching and excavation inspections is provided (Figure 2).

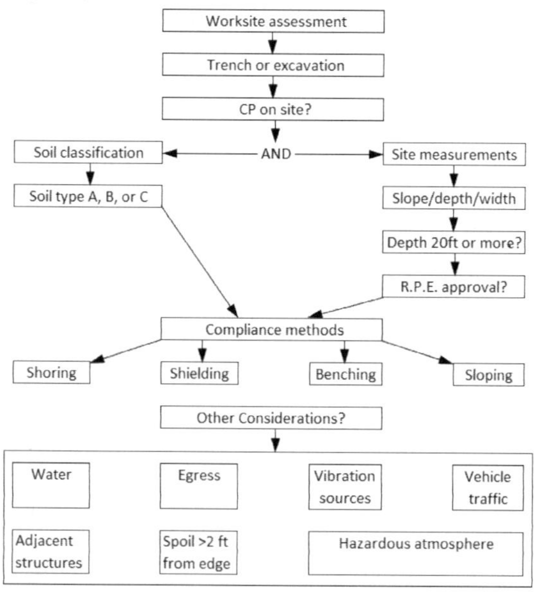

Figure 2. OSHA's Trench and Excavation Inspections – Worksite Assessment Process

A CP is someone who can identify conditions that are hazardous to employees and who also has the authorization to correct these hazards. R.P.E. refers to a Registered Professional Engineer. Spoil represents soil that was excavated from the trench. In addition: One cubic yard of soil can weigh as much as a car, 1,700 kg, and comes in many varieties. Some types of soil are stable and some are not. Any excavation deeper than 20 feet must use a protective system approved by a professional engineer. For all excavations, a CP must conduct a full investigation every day, or when any trench conditions change, to identify and remove any potential hazards.

2.7 Trench compliance assessment

Depth, dimension, and sloping assessments are conducted by OSHA compliance staff and CP with engineering survey rods, "plumb bobs", tape measures, and/or inclinometers. When calculating compliance with Subpart P, OSHA compliance staff utilize existing templates to determine the distances a trench and/or excavation may be out of compliance. One example is a Trench Calculation Tool (TCT) requiring manual input of values such as trench width (at the top of the trench), heights (of the trench side walls), and slope (angles of trench side walls), before automatically determining distances a trench or excavation is out of compliance for type A, B, and/or C sloping requirements. The template is utilized for compliance scenarios involving simple slopes, multiple benches, and one sided excavations.

The Trench Calculation Tool (TCT) was created in OSHA Region V in Wisconsin, and was designed as an Excel spreadsheet to facilitate determination of whether or not trench and excavation sloping assessed on work sites is in compliance with OSHA Subpart P – Excavations. While it is not standardized among all OSHA offices in the US, it was designed as a best practice to facilitate calculations.

In this example, manual measurements (in foot) are provided for an imaginary company AGJT Plumbing Inc. The height, angle and width calculations would be provided from the OSHA compliance officer via manual measurements in the field using engineering survey rods and inclinometers. In this case, a simple slope trench or excavation assessment and corresponding sloping calculations are illustrated in Figure 3.

The West side of the trench (left side in the Excel spreadsheet) was indicated has having a sloping angle of 80 degrees, with a height z of 5.6 ft. The height of the East side of the trench was assessed 5.7 ft with an angle of 85 degrees. The top width of the trench was 5.7 ft. These are the only compliance officer measurements required for entry in this assessment tool. At the top of the TCT, the top width calculations illustrate that the trench is out of compliance for short term A (indicated as A***), A, B and C soil; these distances are provided as 4.1, 6.9, 9.7, and 15.3 ft. At the base of the TCT, distances are provided for the respective sides being out of compliance. For the West side, this is 1.8, 3.2, 4.5, and 7.3 for A***, A, B, and C; for the East side, this is listed as 2.3, 3.8, 5.2 and 8.0, respectively. If these measurements were encountered by an OSHA compliance officer during an inspection, typically the employer would be cited for sloping deficiencies in OSHA Subpart P, assuming no other protective measures were utilized in the trench or excavation.

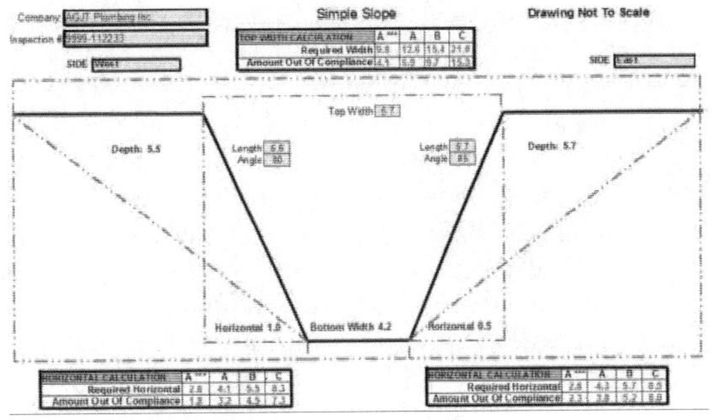

Figure 3. Trench Calculation Tool (TCT) [values in foot] which is utilized by some OSHA offices in the U.S.

3 Method to generate and analyze continuous functions from mesh, point cloud, and photo data

The process of using pictures of a trench to calculate its depth, width and slope angles is as follows:

Several methods exist for generating referenced point cloud data of a simple slope trench. Due to the easiness in data acquisition, photogrammetry instead of laser scanning devices was utilized. While latter choice tends to be more accurate, their cost (i.e., rental or purchase, installation time) is high. Cameras instead are easy to operate and an already used instrument of CPs to document site conditions. Moreover, camera-based photogrammetry delivers textures that may be used in later applications, such as visualization and training. One disadvantage of photogrammetry is the dependence on a larger number of photos.

Photogrammetry consists of two steps: camera data acquisition and software processing. First, depending on the size and shape of the point of interest, pictures were taken with a single-lens reflex camera (Canon EOS 6D, focal length 28-135 mm variable zoom lens) such that a processing algorithm can construct overlapping features (an estimated 60%) between the individual shots. It is important that both the environment and the settings of the camera remain similar for all pictures to ensure optimal conditions during processing.

Camera positions are then estimated out of all the pictures, using focal length as a reference. Although a standard procedure should be followed, taking several dozens of pictures for a simple trench needs minimal training. Exposure and aperture control are the most important variables to consider. Images should not be over- or underexposed and the depth of field should be sufficient, leaving important features sharp, while blurring the unimportant background. Also having a measurement of reference inside the pictures is important to later scale the model to real world size, for example placing a folding ruler in the point of interest is a good practice to follow.

There are several photogrammetry software applications available on the market, paid and open source. This study used the commercial Agisoft Photoscan (AP) software for creating and processing the raw point clouds from photos of the same point of interest. It also generates detailed static meshes. AP uses CPU and GPU processes for algorithmically working on the supplied pictures. Adjustment and adaption still needs minimal manual modification to the settings.

The first step is the aforementioned picture mapping process. The result is a sparse point cloud of the trench (Figure 4). While it is missing features, it provides an intermediate impression on what the result will look like. A dense point cloud is created in the following step. They represent the point of interest at a higher resolution (Figure 4).

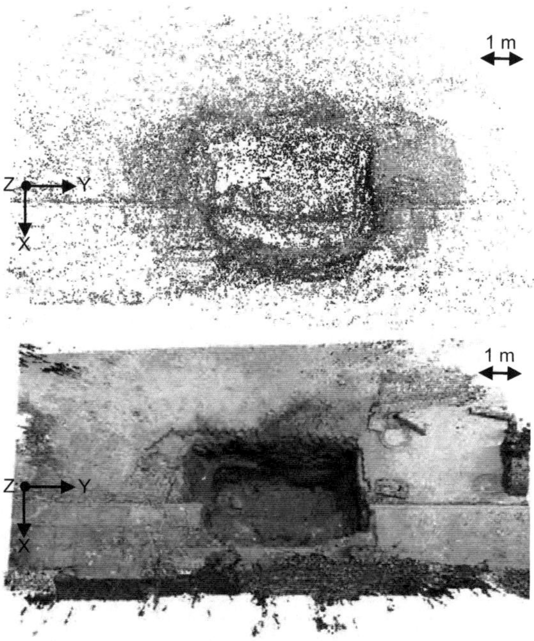

Figure 4. Plan views of simple slope trench: sparse (top) and dense (bottom) point clouds

As soon as the dense cloud has been generated, AP creates a mesh out of the data. From this dense point cloud a static mesh is generated. Meshing will reduce the highly detailed point cloud to a simplified model consisting of vertices of the faces of the mesh. While keeping the essential information for the desired trench slope assessment, the reduction to fewer points allows a user to work with the data set more easily. Optionally, textures can be mapped by projecting and merging the pictures from the first step onto the generated mesh. Scaling and orientation should be applied as the final step to get an accurate real-world representation. Latter is required to calculate trench height and width, whilet he computation of angles is independent from the scale.

After generating the static mesh, a custom application processes the data further. MeshLab® decimates the static mesh hence reducing the number of vertices and therefore the required computational power. Important to note is that the decimation process deletes vertices and interpolates between coordinates. This may lead to possible loss of detail, especially when precise measurements from a trench are needed. For example, a too simplified function may falsify the measurements whereas, on the other hand, further computational time is added. Deciding on the smoothing parameter requires experience. With enough training though, the smoothing process could be automated (part of future work).

The vertex with the lowest z-value (height of a trench) is taken as a reference to create a cutting plane through the mesh. The cutting plane is perpendicular to the y-axis (depth of a trench) and parallel to the x-axis (width of a trench). Nearest vertices within a manually-set δ value (e.g., 5 inches in both directions on the y-axis) are projected to the cutting plane resulting in a line of two-dimensional points $Set_{CuttingPlane}$. Points with duplicate x-coordinates are omitted to create the basis for a continuous function.

All points of $P_i \in Set_{CuttingPlane}$ are are sorted from their lowest to their highest x-values and clustered for a similar z-value. The developed clustering algorithm requires input of $Set_{CuttingPlane}$, hBuffer, wBuffer, minWidth, minHeight, and maxHeight. The z-axis is defined manually for now, but could be defined as the range from the lowest to the highest z-coordinate value. The algorithm inputs, except for minHeight and maxHeight, are created by loops and range from 0.1 m to 0.5 m. A table of results can be compared. While minHeight and maxHeight require manual input for now. The z-axis from minHeight to maxHeight is now divided into n intervals with a height of hBuffer*2. All points P_i that reside inside the same z-interval are clustered. The resulting clusters are called layer clusters (see Figure 5, top right). If two consecutive points inside one cluster have an x-value greater than wBuffer, this cluster is divided into two. From all remaining clusters those are eliminated that have a range on the x-axis smaller than minWidth. As layer clusters can intersect on the x-axis, some layers get clipped by removing the points of each layer cluster which intersect with the previous layer cluster. Inside each of the remaining clusters the z-value of every point is set to the average z-value of the whole cluster to eliminate all asperities (see Figure 5, bottom left). After this smoothing process, the clusters can be reduced to a line which is spanned by the two points with the corresponding lowest and highest x-values inside the cluster. These reduced clusters are now unified into a set $Set_{Simplified}$ (see Figure 5, bottom right).

At this point the method requires five points P_{lte}, P_{lbe}, P_{rte}, P_{rbe}, $P_{lowest} \in Set_{Simplified}$ to calculate the height, depth and slope values of a (single slope) trench:

- P_{lowest}: The point in the trench with the lowest z-value.
- P_{lbe}: The bottom edge on the left slope.
- P_{lte}: The top edge on the left slope.
- P_{rbe}: The bottom edge on the right slope.
- P_{rte}: The top edge on the right slope.

Depending on the trench type P_{lbe} and P_{rbe} can also be equal to P_{lowest}. For simplicity, it is assumed that P_{lte} and P_{lbe} are the points P_i, $P_{i+1} \in Set_{Simplified}$ where the z-difference between P_i and P_{i+1} is the greatest and the z-value of P_i is higher than the z-value of P_{i+1}. Likewise, P_{rte} and P_{rbe} are the points P_i, $P_{i-1} \in Set_{Simplified}$ where the z-difference between P_i and P_{i-1} is the greatest and the z-value of P_i is higher than the z-value of P_{i-1}. The remainder of the developed algorithm is basic trigonometric and geometric calculations. It leads to the desired angles, height, and width of the slope.

4 Experiments, results, and discussion

The same trench example outlined earlier (Figure 4) is used to validate the process. Figure 5 displays $Set_{CuttingPlane}$. Consecutive points in this set are connected for better understanding, though the set is not defined as a mathematical function. Figure 5 shows the intermediate steps from the raw data to $Set_{Simplified}$ with marked height and width of the trench.

In the presented case, hBuffer was set to 0.1 m and minHeight to the minimum y-value of $Set_{CuttingPlane}$. maxHeight was 0. The interval was divided into 9 layers whereof each covers a y-interval of 0.2 m. These sets are called layered clusters. The inputs wBuffer (= 0.1 m) and minWidth(= 0.1 m) increased the number of clusters to 10. Clusters were clipped and smoothened in the following according to the algorithm explained in Section 3. The resulting set of points, $Set_{Simplified}$, contains merely 20 points from initial 890 points in $Set_{CuttingPlane}$. The red line in Figure 5 visualizes the width whereas the blue line shows the trench depth.

For the selected cutting plane the algorithm calculated a width of 3.0 m, a height of 1.91 m, and angles of slope of 85.6° on the left slope and of 89.1° on the right of the trench. Compared to the manual assessment (trench width entered in Figure 3 is 5.7 ft or 1.74 m) the result of the proposed method differs by +72 %. This rather large difference is due to a reading error of the CP during the manual assessment (while the CP meant 3 m or 8 ft for the width of the trench, he selected 2 m or 5.7 ft). While the measurement of the proposed method (3.0 m) is -0.1 m or -3% off of the ground truth, both measurement methods concluded that the examined trench requires protection should a person need to enter it.

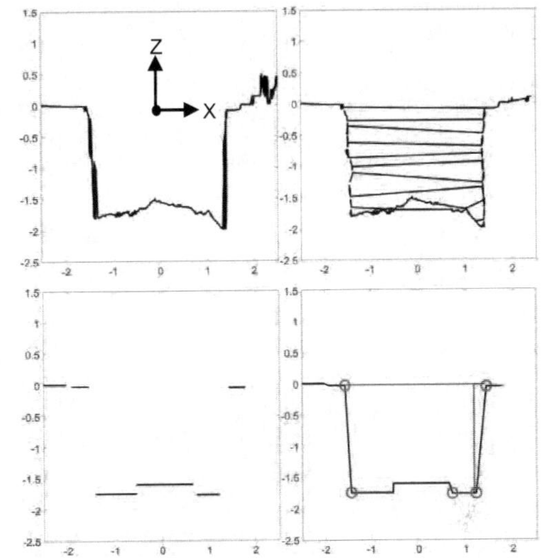

Figure 5. Cross-sectional views [in meters]: Original raw data (top left), layered clusters (top right), smoothened, reduced layer clusters (bottom left), identified points P_{lte}, P_{lbe}, P_{rte}, P_{rbe}, P_{lowest} and automated measurement of the width and height of a simple slope trench (bottom right) (horizontal axis: x-values, vertical axis: z-values)

5 Conclusions

While trenches pose a considerable risk to the life and health of construction workers, manual inspections dominate the practice. The proposed compliance assistance prototype has the potential to improve sloping calculations conducted by competent persons and regulatory compliance staff in the field. By utilization of mobile point cloud data acquisition and processing applications, inspectors and competent persons will have the ability to more rapidly assess the compliance of simple slope trenches according to OSHA Subpart P requirements. In addition, through the utilization of a mobile device, compliance staff standing on the edge of a trench during compliance activity will be protected from potential falls.

Future research might incorporate the elements of the findings into a virtual environment (VE) or virtual reality (VR) [20]. Similar to Wang et al. [21] that converted point cloud data of embankments of a deep excavation project into parametric objects, generated trench objects might serve in three-dimensional worlds as well. Such worthwhile opportunities would offer site-specific immersive training tools in competent person training sessions and site specific toolbox talks. They could supplement or replace existing paperwork and a yet PowerPoint-dominated OHS educational curriculum.

Disclaimer

While this paper is proposing an OSHA compliance prototype, note that the agency does not test, approve, certify, or endorse any equipment, product, or procedure, including machine design and risk assessment techniques.

References

[1] OSHA (2017). "Safety and health regulations for construction – excavations", Standard 29 CFR 1926.652 (Subpart P), https://www.osha.gov.

[2] Teizer, J. (2007). "Rapid surveillance of trenches for safety", *Proceedings of Construction Research Congress*, Freeport, Bahamas, May 6-8, 2007.

[3] OSHA (2007). "Regularly review of 29 CFR 1926, Subpart P: Excavations", https://www.osha.gov/dea/lookback/excavation_lookback.html.

[4] Plog, B.A., Materna, B., Vannoy, J., Gillen, M. (2006). "Strategies to prevent trenching-related injuries and deaths". The Center to Protect Worker's Rights, http://www.cpwr.com.

[5] CDC (2004). "Occupational Fatalities During Trenching and Excavation Work --- United States, 1992—2001", Center for Disease Control and Prevention, https://www.cdc.gov/mmwr/preview/mmwrhtml/ mm5315a2.htm (Feb. 12, 2017).

[6] NIOSH (2006). "Fatality assessment and control evaluation program." Department of Labor. http://www.cdc.gov/niosh/face/faceweb.html.

[7] Deatherage, J.H., Furches, L.K., Radcliffe M., Schriver, W.R., Wagner, J.P. (2004). "Neglecting safety precautions may lead to trenching fatalities." *American Journal Industrial Medicine*, 45, 522–527.

[8] CDC (2011). "Preventing worker deaths from trench cave-ins", Center for Disease Control and Prevention, https://www.cdc.gov/niosh/docs/wp-solutions/2011-208/pdfs/2011-208.pdf (Feb. 2017).

[9] Arboleda, C.A., Abraham D.M. (2004). "Fatalities in trenching operations analysis using models of accident causation." *Journal of Construction Engineering and Management*, 130(2), 273–280.

[10] OSHA (2016). "Ohio worker's death highlights grim 2016 national stat: trench collapse fatalities have more than doubled", OSHA News Release – Region 5, https://www.osha.gov/pls/oshaweb/owadisp.show_document?p_table=NEWS_RELEASES&p_id=33441 (Feb. 10, 2017).

[11] Lew, J., Abraham, D.M., Wirahadikusumah, R.D., Irizarry, J. and Arboleda, C.A. (2002). "Excavation and trenching safety: Existing standards and challenges." *CIB WO99 Implementation of Safety and Health on Construction Sites*. (Jan. 10, 2007).

[12] Lee, S., and Halpin, D.W. (2003). "Predictive tool for estimating accident risk." *Journal of Construction Engineering and Management*, ASCE, Reston VA 129(4), 431-36.

[13] CPWR, Center to Protect Worker's Rights (2006). "Safe work in trenches", http://www.cpwr.com.

[14] CPWR, Center to Protect Worker's Rights (2007). "Trenching-related injuries and deaths." The Center to Protect the Worker's Rights. www.cpwr.com (Jan. 17, 2007).

[15] OSHA (1985). "OSHA Special Emphasis: Trenching and Excavation - CPL 02-00-069", https://www.osha.gov/pls/oshaweb/owadisp.show_document?p_table=DIRECTIVES&p_id=1653.

[16] OSHA (2014). "Hazard exposure and risk assessment matrix for hurricane response & recovery work", Operation-Specific Sheets – Trenches and Excavations, https://www.osha.gov/SLTC/etools/hurricane/trenches.html.

[17] Dickinson, J.K., Woodard, P., Canas, R., Ahmad, S., Lockston, D. (2011). "Game-based trench safety education: development and lessons learned", *Journal of Information Technology in Construction* (ITcon), 16, 119-134.

[18] WA-LNI-DOSH (2017). "Chapter 296-155, WAC safety standards for construction work", WAC 296-155 Safety Standards for Construction Work, http://www.lni.wa.gov/Safety/Rules/chapter/155/WAC296-155.PDF (Feb. 12, 2017).

[19] OSHA (2015). "Excavations: Hazard Recognition in Trenching and Shoring", https://www.osha.gov/dts/osta/otm/otm_v/otm_v_2.html, (Nov. 17, 2015).

[20] Hilfert, T., Teizer, J., and König, M. (2016). "First Person Virtual Reality for Evaluation and Learning of Construction Site Safety", *33rd International Symposium on Automation and Robotics in Construction*, Auburn, Alabama, USA.

[21] Wang, J., Zhang, S., and Teizer, J. (2015). "Geotechnical and Safety Protective Equipment Planning Using Range Point Cloud Data and Rule Checking in Building Information Modeling", *Automation in Construction*, Elsevier, 49, 250-261.

Augmented Reality (AR) Based Approach to Achieve an Optimal Site Layout in Construction Projects

Abhishek Raj Singh[a], Vinayak Suthar[b], and Venkata Santosh Kumar Delhi[c]

[a]Ph.D. graduate student, Department of Civil Engineering, Indian Institute of Technology Bombay, India
[b]Undergraduate student, Department of Civil Engineering, Indian Institute of Technology Bombay, India
[c]Assistant Professor, Department of Civil Engineering, Indian Institute of Technology Bombay, India
E-mail: arsingh@iitb.ac.in, vnksuthar12@gmail.com, venkatad@iitb.ac.in

Abstract

Site layout planning involves processing and visualizing complex information residing with different team members. Usually, this exercise is carried out using iterative exchange and understanding of information on paper documents and 2D drawings in team meetings. In the present work, it is hypothesized that aiding the project team with an AR based system would help better tap into the insights and experience of team partners. For this, a BIM enabled Android-based AR site layout planning tool was developed. This tool uses marker-based identification and tagging scheme to visualize the construction site. An initial study of the user behavior while using this system compared to conventional systems points that in addition to improving visualization the planners' faculties and experience are employed towards optimization of the site rather than understanding the site layout plan. The study illustrates the utility of using AR/VR in planning processes.

Keywords- BIM, Collaborative Design & Construction, Site Layout Planning, Augmented Reality, Visualization

1 Motivation and Research Objective

Site Layout Planning (SLP) is an initial task to be performed by every planning team on a construction project. Site Layout Planning (SLP) is considered to be laying out temporary facilities (TFs) necessary for future construction in line with route planning. An optimal layout ensures safe working conditions for labors, optimum usage of available space, low activity interference and reduction in on-site travel time. Team members having significant experience in the field of construction are considered most suitable for the layout planning. The task of SLP is equally important for every team or stakeholder of the construction project; since on the site integration of planning is essential and have to be most efficient for a successful project. The multiple operations that happen simultaneously on construction site in addition to the uniqueness exhibited by each project, it is usually not possible to envision every need of the future. Therefore, planners usually resort to certain heuristics, intuition gained with experience on similar projects to come up with the best possible site layouts [1].

At present, the site layout planning (SLP) is carried out based on 2D plans for the site. The paper-based drawings are arduous for planners to plan since they do not offer visualization in 3D perspective. Hence, to get optimal layouts for increasing the efficiency of projects, the collaboration among project teams and stakeholders is very crucial. The ability to anticipate and finding a feasible solution to the site layout remains a key challenge for planners to address. The ability to visualize the site and aggregate experience from the various team members effectively remains a big hurdle in the conventional SLP being followed in the construction projects across the world. Thus, SLP poses key challenges in visualizing the site constraints with respect to SLP on one hand and achieving coordination and collaboration among the various team members. In this context, it is hypothesized that the emerging applications of technologies like BIM and AR in the field of construction might offer a way forward in the right direction as far as SLP is concerned.

Building Information Modelling (BIM) provides an information management paradigm to enable better team coordination, collaboration. BIM technologies enable the maintenance of a single and updated information model for the entire project thus helping the team to plan and execute projects effectively. The distributed nature of the model enable the teams coordinate remotely. Augmented Reality (AR) enables the projects teams to visualize the as planned by enriching the real environment with the digital information. Such visualization of the model is meaningful since it assists in identifying the relevant factors, study the interaction between factors, preparing and understanding alternative solutions. `Thus, combining these two technologies of BIM and AR would provide a potent combination to solve the challenges of SLP.

This study provided the first preliminary steps in this direction. In this paper, the background of the site layout planning is first discussed. The potential of integrating BIM and AR to solve some of the key challenges in SLP are then discoursed. The paper then provides a framework to illustrate how such an integration can happen. A site layout tool with BIM +AR was developed. A preliminary experiment was then conducted to elicit user experience in using BIM+AR tool in site layout planning. The paper concludes with possible directions of research in automating the site layout planning on construction projects.

2 Background

2.1 Site Layout Planning

Construction site planning is among the most intriguing tasks exercised by a project team before execution on site. The process of SLP involves locating temporary facilities on site. The emphasis is on the temporary nature of these facilities that they may be for a short span on site until the specific task associated with its use is completed. Despite this, some facilities remain on a construction site for the whole project lifecycle and are left for future use in the task of operation & maintenance (O&M)[2]. There are several objectives to be met those can result in a good site layout. For example, identification of facilities needed to support construction, positioning such facilities on-site, determination of required size and shape for facility, travel time between facilities and construction site, the frequency of movement of equipment between facilities and from facility to the site are to be considered during the process of SLP. Such objectives to be achieved by the SLP makes this task highly subjective and complicated for the construction managers (CM). Thus to optimize this multi-objective problem CM(s) have to prioritize the functions and apply trail & error method to their priorities in constructing a layout [3]. The necessity for an optimal site layout is not only to provide safe working conditions to workers & minify the travel time but also to prevent a significant amount of money and space that is bound to these facilities at the initial phase of the project. Thus, efficient utilization of the site space has huge implications on both cost and time in a project. The activity of SLP involves several steps including the identification of temporary facility required, shape and size of required facility & fixation of the facility to a particular location on site and estimation of tenure for the facilities at a specified location on the site. These tasks are crucial to every division of the organizations involved in the construction project since it directly or indirectly affects worker's productivity, material deployment, on-site security & cost of construction [4]. For performing this essential and crucial task, it is required to get every project stakeholder on the same platform.

2.2 BIM for Site Layout Planning

Building Information Modelling (BIM) technology is an advanced way to replicate construction project as a real 3D model along with every minute detail associated to project being stored in the database. BIM has made a remarkable accomplishment in countries across the world both in developed economies and developing countries[5]. But a majority of developing countries depict a research gap in BIM context [6]. The concept of BIM emerged over a decade [7][8]. However, there are significant challenges being encountered in the adoption of this paradigm in the construction industry[5]. Eastman *et al* [9] through case studies showed the majority of executed projects lacked in utilizing BIM's potential benefits. The information management component of BIM often referred to the 'I' component makes it different from any other modeling tools. BIM provides a platform to manage and aggregate information from across different disciplines in a digital form which can be used across the project life-cycle including facility management at the end of construction. Such project data associated with the facility to be constructed can comprise of information related to quantity, schedules, estimates, project management parameters, structural analysis and facility management in a single database for collaboration[10]. Such information management paradigm brings the construction industry closer to the 'Computer Integrated Construction' (CIC) model as proposed by Miyatake and Kangari [11]. CIC reflects the concept of having a single source of sharing information.

Apart from the information management paradigm, BIM offers a framework to visualize the information stored in useful ways. This visualization can be geometry of the building as visualized by the 3D rendering of the information in various software tools developed to this effect. Visualization has important implications in terms of construction project management. For example, Sulankivi, Makela and Kiviniemi [12] provided an insight of how BIM can be employed for visualizing site and identifying potential safety issues before time.

Thus, BIM being capable of providing a central source of information aiding in collaboration and visualization as the foresight to prevent potential hazards and make planning effective. BIM have a significant potential to be used for SLP to satisfy the urge of collaboration and visualization. However, the potential of BIM in SLP remains largely untapped.

2.3 AR for Construction Site Planning

The first prototype for AR has been developed in 1968 by Ivan Sutherland and his students at Harvard University and the University of Utah [13]. The term 'augmented reality' was coined in 1990 by Caudell and Mizell[14]. The potential of the use of AR is enormous and thus remains an active area of research and innovation in construction industry. The adoption of AR technologies however remains minimal specifically due to the lack of tools which customize the needs of construction industry.

Construction sites are depicted as 2D plans at an early stage during the process of planning. Through these drawings, the planners and managers have to intuitively plan for the project and the temporary facilities required for construction. Such 2D drafts/ Blueprints thus plays a vital role in planning. However, these drawings lack the ability to replicating the same scenario in front of every team member who indulges in planning. This can be due to various factors like difference in intuition level, experience, the state of mind and exposure to the field of planning. Hence, to eliminate the misinterpretation and miscommunication among the teams during the planning the 2D drawings; it is proposed that such planning processes can be enriched by 3D BIM models for the project along with supporting facilities required.

Earlier, the field of AR has been investigated for improving on supervising construction i.e. comparing as planned and as build also called as progress monitoring and for defects detection too [15]. Gheisari *et al* [16] studied 'Augmented Panoramic Environment' capable of taking building information to site for understanding the user requirement and human-computer interaction related issues. AR implementation for facility management (FM) has also been an area for researchers and have provided substantial results.

3 Integrating BIM and AR for SLP

Gheisari and Irizarry [17] highlighted the application of augmented reality for facility management along with highlighting how the integration of BIM and AR can aid in the discipline of facilities management. There have been several research studies, which illustrate the possibilities of AR for construction site demonstrating it as a tool for achieving collaboration in construction projects.

Gu *et. al* [18] identified the possibilities provided by an AR integrated BIM server. This study proposed a framework for BIM and AR integration, with more focus on how BIM helps in collaboration. The framework highlights some essential requirements to be considered while looking for integration of these two innovative technologies such as team communication and interaction, data security and visualization & navigation. Wang & Jen [19], Wang[20] highlighted the use of AR technique of visualization to plan a construction site. These studies presented a prototype to visualize virtual site components in real environment by triggering Bi-tonal planar ARTags with a head-mounted display (HMD) V hi-Res800TM (PC) with an InterTrax2TM head tracker device. It was also suggested that use of such prototype and approach along with some rule engine could result in a more sophisticated and effective planning.

Standard processes are seldom used in site layout planning on construction projects [2]. The construction industry still largely approaches the problem of site layout upon experience gained and intuitive level of planners [1]. Thus, the BIM enabled android-based AR application is developed to understand the behavior dynamics during the collaboration between team members and the shift in the approach they employ for SLP.

4 Framework for SLP using BIM + AR

The traditional way of Site Layout Planning involves representation of facility on 2D drawings. These 2D blueprints results are time-consuming to understand 3D architectural scaled model, hard-to-access conceptual design schemes with regards to the building's surrounding environment, restricts stakeholder's coordination, limited reusability and low information integration [21]. Thus to address the pressing need for collaboration and intuitive visualization for project stakeholders a BIM enabled Android AR app was developed in the present study. This tool works based on fiducial markers; the 2D construction drawings were used as markers to reduce additional work for generating trackers. These trackers act as a target image to be traced through the developed app to replicate the associated model attached to respective drawing. In this system, an AR environment was superimposed onto 2D drawings enriching them with digital 3D models. The use of marker provided users the ability to manipulate to change the locations of the virtual models while assessing the resultant layout plans with respect to the real-time model. This approach of tracker based AR system enables the user to produce multiple site layouts and finding the best alternative.

The Android application comprises of a dataset containing images of 2D drawings and respective 3D models. The 3D models were constructed using a BIM software Autodesk REVIT. The 3D models were optimized using another modeling and animation software 3ds Max for eliminating conflicts in polygons and vertices. The final step of allocating 3D model with respective blueprints and application development is done using Unity Game engine.

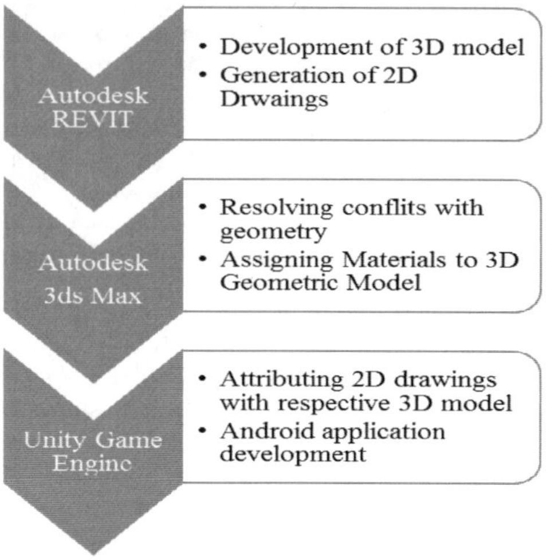

Figure 1: Process and Applications involved in development of tracker based BIM enabled AR platform

Figure 1 illustrates the framework for the development of BIM enable AR platform to this effect. The process involves extracting the relevant information from the BIM model and rendering this information in an AR based engine to develop an application for the SLP for a construction project site.

The Android platform was chosen because of easy availability and its vast user count. To facilitate multiple users at same instance, the application was deployed to the user's mobile handsets enhancing them to move around the scene in the real world. This free movement around the scene was enabled to provide access to full 3D perspective view in one go enabling the user to plan effectively such that no constraint is left unnoticed.

5 Experimental Study

The experiment for this study is designed to observe and contrast user behaviour in an AR based SLP to that of conventional SLP processes. To this end, the experiment was carried out on two teams of engineers with various degrees of experience in site layout planning. The team composition was designed to have people with little or no experience to site layout planning to people with significant experience. Two such teams were formed to test the replicability of the experiment results. Hence, two teams of five construction engineers each took part in the experiment. The SLP experience of the team members ranged from 2 years to 25 years. The dynamics related to user behaviour and the experience were intended to be captured by this setup.

The experiment consisted of four phases. In the first phase, the team carried out SLP with some limited constraints. Second, new information pertaining to SLP was given to the teams and their behaviour in choosing a new site layout plan was observed. The use of two iterations is to have a progressive loading of the information to the team members and understand the change in SLP when new information comes to the team. Third, the team was allowed to carry out SLP using AR based environment with constraints similar to phase one. Finally, the phase two was simulated again in AR based environment.

The aim of the experiment for the team is to determine the location of 5 TFs is each of the phases. These 5 TFs include site office, batching plant for concrete, storage yards for aggregates, fabrication yard for bar bending and parking yard for the construction equipment to be used on site. The possible locations for the TFs included areas in the construction site as well as areas nearby to the construction site.

Each phase of the experiment is time bound. The teams were given 30 minutes in each phase to come up with a possible solution to SLP. The team was asked to provide at least one solution at the end of this time period. The solution comprises of documentation by the team regarding the positions of TFs on the site and surroundings plan (illustrated in Figure 2).

In the first phase, the team members were first given 2D drawings of a hypothetical construction site. The team was then made aware of constraints on utilization of available free space on the site and constraints related to locating TFs on the site space for the permanent facilities are going to be built. The team was allowed to use only 50% of the on-site available space. In addition, the topography of the area was also provided which included terrains and water bodies which might not be suitable for locating the TFs. Figure 2 illustrates the construction site plan and its surroundings as provided to the teams for this purpose. The plan details out the location of the construction site (shaded in grey in Figure 2.) and location of other building surrounding the construction site. It provides the detail of access to the site, provides the general topography of the area and water bodies near the site. In the first phase, the access to the construction site was restricted to just on road which goes between two adjacent buildings to the construction site. Finally, the areas earmarked for TFs outside the construction site were also provided to the team.

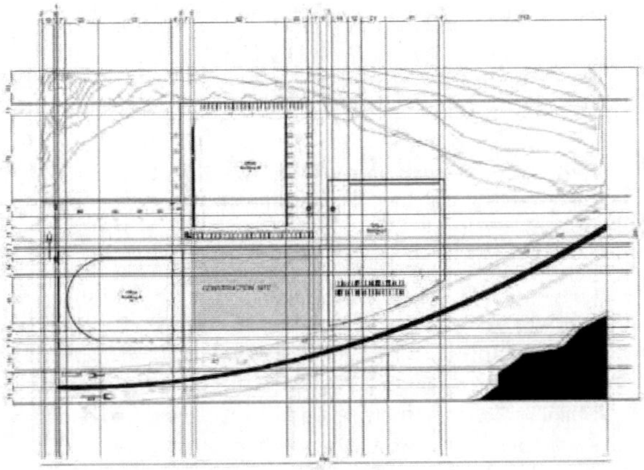

Figure 2: Plan for site with grid-to-grid distance for better understanding the dimensions.

Finally, in addition to the site plan and topography, the plans illustrating the dimensions of each of the individual temporary facilities to be located on the construction site were also supplied to the team to aid their planning processes. Figure 3 depicts the 2D drawing provided, for one of the five temporary facilities described earlier, given to the planners to be utilised for the site layout task.

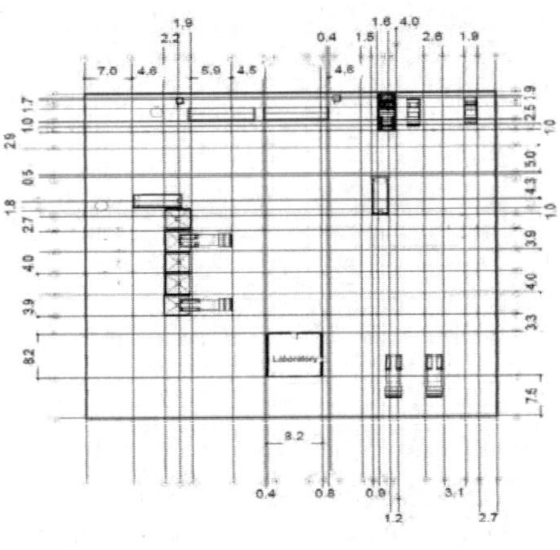

Figure 3: Plan for batching plant for concrete with grid-to-grid distance.

In the second phase, some of the constraints imposed were relaxed. The teams were given twenty minutes to determine the locations of TFs. The constraints on the site utilization for TFs and the access paths to the construction sites were relaxed in this phase. The documentation was similar to the phase one of the experiment.

Once the documentation for the second iteration got over, the team was subjected to phase three involving the use of an AR+BIM based platform for SLP. The constraints for AR platform's first and second iteration were the same as those subjected during the first two phases using 2D drawings. Handheld devices with pre-loaded android-based AR application were supplied to the planning team. Figure 4. displays a screenshot of the augmented reality as experienced by the user when the site's 2D plan drawing was brought in the field of view of the handheld device's camera. Once identified, the AR view is rendered on the 2D drawing on the screen on the user's device. The AR view remains stable as the user moves around the construction site thus offering view from various directions.

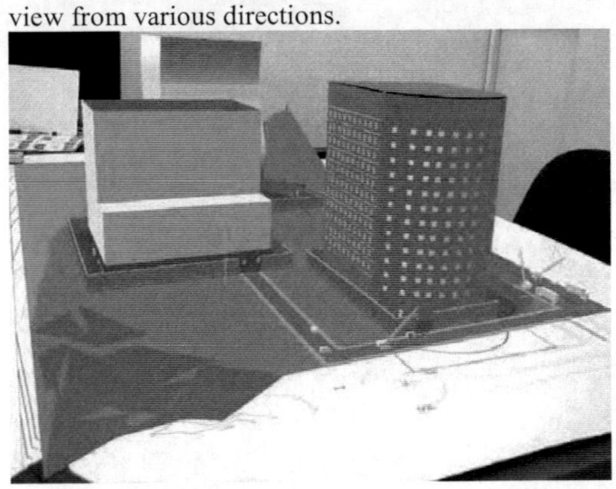

Figure 4: Topography of the site and its surrounding in AR environment.

Markers were created for each of the 5 TFs which were to be located for the project. Figure 5 illustrates the markers and the AR renderings of the TFs. The teams were then allowed to move the markers across the 2D plan of the site and its surroundings to determine what they thought was a good location of the TF. As the team moved the markers, the AR application superimposed the TF on the site plan (illustrated in Figure 6 and Figure 7). The locations of TFs at the end of each phase in AR were automatically determined and documented by the location of the markers on the 2D plan.

During all these phases of the experiment, observations were made on team dynamics with respect to collaboration. The interaction among various team members, delivery of thoughts and ideas by these members, time taken to make other people understand one's idea, time taken to search for a solution, preciseness of the documentation of solution and generation of alternate SLP solutions were noted down. It should be noted that the present study does not infer to the optimality of the SLP solutions generated by the team. The main focus was to understand the interaction, coordination, and collaboration among team members in an AR environment for SLP.

Figure 5: AR rendering of the TFs to be located on the project site

6 Results and Discussion

The experiment provided some key findings in understanding the user behaviour as contrasted in conventional vs. AR based SLP processes.

6.1 Visualization of information

Construction drawings are the main source of information for the planners to perform the complex SLP tasks. The typical 2D drawings help in identifying the dimensions for the available onsite space, facility to be constructed and the topography of the area. For confirming the 3D perspective of any facility, the planners have to refer to at least two drawing sheets (plan and elevation) and stitch the scene mentally. This task of scene stitching is time-consuming since a due attention is also required for the dimensions depicted on the sheet.

It is observed that the experienced persons in both the teams were able to better visualize the site plans than the inexperienced persons. Significant time was spent in explaining the site to the team counterparts. It is interesting that even during the second iteration, again significant amount of time was spent in visualizing what it means when a constraint is relaxed or imposed on the site. Such challenges were non-existing in the AR phases of the experiment. All the team members irrespective of the experience were immediately able to visualize the construction site. The time was more focussed on finding solutions as compared to visualizing the site.

6.2 Explaining the possible solution to team members

The experiment also highlighted the difficulties in explaining the possible solutions to other team members in a conventional environment. During 2D phases of the experiment, apart from visualizing the initial site conditions, for each solution proposed, significant time was spent on explaining that solution to other team members. The consideration of a possible solution rested on how effectively a team member explained the mental vision of the solution to the counterparts and how effectively the various constraints were considered in the solution.

The major challenge for the team members while collaborating using 2D drawing was in presenting their individual viewpoint. It is observed that the understanding of a team member's vision was a significant challenge to other members. It was also observed that sometimes the members did not consider some potential solutions, as they were unclear with the thoughts presented.

Such challenges were not observed in the AR based environment, as the other members immediately understood what the team member was proposing. In addition, it is observed that the team members were more confident that all the team members were perceiving the exact same thing in this environment.

6.3 Experience factors

An interesting dynamics which was observed during the first two phases was that the persons with experience assumed the positions of power in deciding the feasibility and suitability of SLP layouts. The in-experienced team members rallied around the thoughts of the experienced members. Such behaviour might be due the fact that the in-experienced members acknowledged the superior abilities of the experienced personnel in visualizing the solution. Whereas, this dynamic was absent in the AR based phases. The in-experienced persons were more confident on proposing solutions. Thus, in the AR phases, the team members considered a number of solutions as compared to the 2D phases.

6.4 Infeasibility of solutions

The planners master this art with experience, but sometimes the mental visuals perceived might be infeasible as they violate certain constraint not perceived in such mental pictures. The teams spent significant time in discussing a solution which was infeasible under the constraints imposed. During the AR phase, such solutions were quickly discarded as the team immediately visualized that the location violated some height constraint or access constraint etc.

6.5 Solution search

As a result of such dynamics, the teams were more focused on finding solutions in AR phases of the project.

The discussions were more focussed on the implications of a layout on aspects like cost and time. This was in sharp contrast to the time spent in explaining and visualizing the solution in 2D phases. The role of the experienced team members in AR phase was to highlight long-term implications of each solution drawn from experience as compared to aiding in visualization in the 2D phases. Thus, the experience of the members was put to better use in the AR phase of the experiment. As a result, the team could come up with solutions like close spacing of TFs (illustrated in Figure 6) which were discarded as infeasible in 2D phases; and placing in locations which are in proximity but were not visualized in the 2D phases (illustrated in Figure 7).

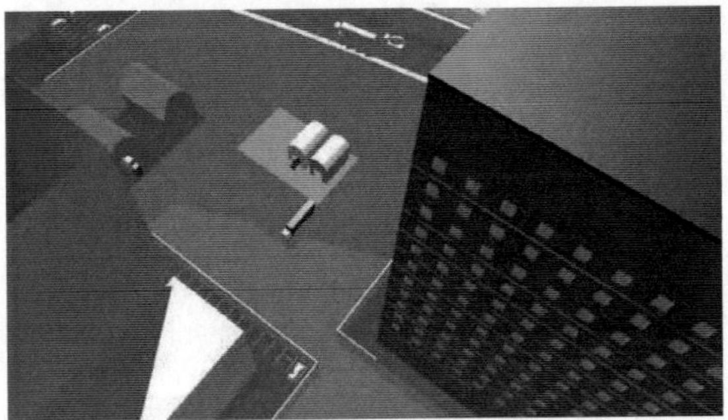

Figure 6: Two temporary facilities accommodated in close proximity.

Figure 7: Temporary facilities situated at distance from one another

7 Limitations

The AR based platform suffers from limitations at the present stage. The tracker based AR tool necessitates proper illumination, high contrast texture of target marker and high hardware requirements for the device to deploy the AR application. Loss of information was observed when translating BIM models to AR environment. The present study uses just the massing forms for the buildings. If other details like work-fronts on different storeys are to be considered, there would be a challenge in converting the BIM model to AR renderings. The temporary structures were modeled as regular shape either square or rectangle; restricted the planners involved in the experimental planning task to superimpose the vacant area. This obstruction can be eliminated by dividing the temporary facilities further into small elements. Even after having an advanced setup for deployment, it was observed that the device used for AR application have a delay in time of capturing and image processing. Latency was also detected while using the application restricting the free movement of team members around the scene. These technical challenges should be addressed to make the AR platform more conducive to SLP.

8 Future Scope

The research can be extended in many dimensions. The SLP can be extended to temporal dimensions in addition to the spatial dimensions covered in this experiment. The AR applications can be augmented by intelligence in generating optimal solution from an initial set of solutions proposed by team members. There is a scope to incorporate such algorithms so as to capture the user experience in generating initial solutions and further optimizing them using a systematic procedure. The application can be made to go markerless and based on object-based tracking to be augmented on a real construction site as was done earlier in case of visualization alone [22]. Further, the possibility of remote collaboration using VR based environment can be explored.

9 Conclusion

SLP is a task for a project team rather than for an individual hence requires collaboration among all teams and coordination of all individuals of the teams. The study highlights the shift in behaviour and approach adopted by teams when allowed to plan a construction site collaboratively in an AR environment. The experiment illustrated the significant potential of AR based applications in improving SLP on construction projects. The AR environment helped the members to evaluate a number of solutions in a given time. The AR platform also improved the effectiveness of the inexperienced planners on one hand and made better use of the experience and intuition of the experienced team members on the other hand.

References

[1] M. Y. Cheng and J. T. O'Connor, "ArcSite: Enhanced GIS for Construction Site Layout," *J. Constr. Eng. Manag.*, vol. 122, pp. 329–336, 1996.

[2] I. D. Tommelein, R. E. Levitt, and B. Hayes-Roth, "SIGHTPLAN Model for Site Layout," *J. Constr. Eng. Manag.*, vol. 118, pp. 749–766, 1992.

[3] I. D. Tommelein, R. E. Levitt, B. Hayes-Roth, and T. Confrey, "SIGHTPLAN Experiments: Alternate Strategies for Site Layout Design," *J. Comput. Civ. Eng.*, vol. 5, pp. 42–63, 1991.

[4] Z. Ma, Q. Shen, and J. Zhang, "Application of 4d for Dynamic Site Layout and Management of Construction Projects," *Autom. Constr.*, vol. 14, pp. 369–381, 2005.

[5] S. Lee and J. Yu, "Discriminant Model of BIM Acceptance Readiness in A Construction Organization," *KSCE J. Civ. Eng.*, vol. 21, pp. 1–10, 2016.

[6] N. Bui, C. Merschbrock, and B. E. Munkvold, "A Review of Building Information Modelling for Construction in Developing Countries," *Procedia Eng.*, vol. 164, pp. 487–494, 2016.

[7] R. Takim, M. Harris, and A. H. Nawawi, "Building Information Modeling (BIM): A new paradigm for quality of life within Architectural, Engineering and Construction (AEC) industry," *Procedia - Soc. Behav. Sci.*, vol. 101, pp. 23–32, 2013.

[8] J. V. Kumar and M. Mukherjee, "Scope of Building Information Modeling (BIM) in India," *J. Eng. Sci. Technol. Rev.*, vol. 2, pp. 165–169, 2009.

[9] C. Eastman, P. Teicholz, R. Sacks, and K. Liston, *BIM Handbook: A Guide to Building Information Modeling for Owners, Managers, Designers, Engineers and Contractors*, Second Edi., vol. 2. John Wiley & Sons, Inc., 2011.

[10] Y. Jung and M. Joo, "Building Information Modelling (BIM) Framework for Practical Implementation," *Autom. Constr.*, vol. 20, pp. 126–133, 2011.

[11] Y. Miyatake and R. Kangari, "Experiencing Computer Integrated Construction," *J. Constr. Eng. Manag.*, vol. 119, pp. 307–322, 1993.

[12] K. Sulankivi, T. Makela, and M. Kiviniemi, "BIM-Based Site Layout and Safety Planning," in *Proceedings of the First International Conference on Improving Construction and Use through Integrated Design Solutions*, 2009, pp. 125–140.

[13] I. E. Sutherland, "A Head-Mounted Three Dimensional Display," in *Proceedings of the December 9-11, 1968, fall joint computer conference, part I*, 1968, pp. 757–763.

[14] D. W. F. van Krevelen and R. Poelman, "A Survey of Augmented Reality Technologies, Applications and Limittions," *Int. J. Virtual Real.*, vol. 9, pp. 1–20, 2010.

[15] S. Rankohi and L. Waugh, "Review and Analysis of Augmented Reality Literature for Construction Industry," *Vis. Eng.*, vol. 1, pp. 1–18, 2013.

[16] M. Gheisari, P. Chen, M. F. Sabzevar, and J. Irizarry, "An Augmented Panoramic Environment to Access Building Information on a Construction Site," in *52nd ASC Annual International Conference Proceedings*, 2016.

[17] M. Gheisari and J. Irizarry, "Investigating Human and Technological Requirements for Successful Implementation of A Bim-Based Mobile Augmented Reality Environment in Facility Management Practices," *Facilities*, vol. 34, pp. 69–84, 2016.

[18] N. Gu, V. Singh, and X. Wang, "Applying Augmented Reality for Data Interaction and Collaboration in BIM," in *New Frontiers: Proceedings of the 15th International Conference on Computer Aided Architectural Design Research in Asia (CAADRIA)*, 2010, pp. 511–520.

[19] X. Wang and H. Jen, "Designing Virtual Construction Worksite Layout in Real Environment via Augmented Reality," in *ISARC-2006*, 2006, pp. 757–761.

[20] X. Wang, "Using Augmented Reality to Plan Virtual Construction Worksite," *Int. J. Adv. Robot. Syst.*, vol. 4, pp. 1729–8806, 2007.

[21] J. Wang, X. Wang, W. Shou, and B. Xu, "Integrating Bim and Augmented Reality for Interactive Architectural Visualisation," *Constr. Innov.*, vol. 14, pp. 453–476, 2014.

[22] C. Woodward and M. Hakkarainen, "Mobile Mixed Reality System for Architectural and Construction Site Visualization," in *Augmented Reality - Some Emerging Application Areas*, Cdn.Intechweb.Org, 2011, pp. 115–130.

Assessing the Effectiveness of Augmented Reality on the Spatial Skills of Postsecondary Construction Management Students in the U.S.

Jeff Kim[a] and Javier Irizarry[b]

[a]School of Building Science, Auburn University, United States
[b]School of Building Construction, Georgia Institute of Technology, United States
E-mail: jsk0022@auburn.edu, Javier.Irizarry@coa.gatech.edu

Abstract –
There is a continual challenge within the construction industry to achieve schedule, budget, and quality expectations at a time when projects are more complex and their designs and assembly involve some measure of abstract spatial skills. In postsecondary institutions, these abstract spatial skills have routinely been taught through drafting and plan reading courses. The outcomes of this approach often fall short of meeting the needs of industry, therefore an alternative solution is necessary. As such, this research advocates the use of a group of technologies, collectively known as augmented reality, which allows one to view the "real" world with the addition of information intended to provide a new understanding of what is being seen. Augmented reality is fast proving advantageous in many industries (e.g. Architecture, Robotics, Entertainment, and Military). With the availability of smaller, more powerful consumer mobile devices, augmented reality can become a ubiquitous, practical tool for the construction industry as well. Research surrounding augmented reality is primarily focused on ways to improve the construction process; however, this study is focused on improving the learning process. The researcher studied the use of a mobile augmented reality tool during a spatial skills classroom assignment to determine if there was an improvement in the accuracy of these skills and retention of the concepts over time. A separate analysis was conducted to determine if the teaching tool was a benefit or disruption to the overall learning experience. The research and analysis resulted in significantly improved assessment scores for the group utilizing the mobile augmented reality tool during their completion of the spatial skills assignment. Additionally, a post-survey indicated that the introduction of this tool was non-disruptive to the students' overall learning experience. Therefore, the addition of these technologies, as a complement to drafting and plan reading, can enhance the postsecondary educational preparation of the future construction industry workforce.

Keywords –
Augmented Reality; Student Learning; Construction Management; Spatial Skill; Active Learning

1 Introduction

The success of today's construction process is sensitive to errors, mistakes, omissions, and inexperience that affect safety, cost, schedule, and quality of a construction project [1][2][3]. While some of the problems could be attributed to outside factors such as fluctuating demand of raw materials, changes in the financial markets, and unpredictable weather conditions, one should also appreciate the ongoing challenges of educating the next generation of construction managers. The newly graduated construction management (CM) student needs basic skills to solve unique, complex problems and visualize a finalized project in an empty three-dimensional (3D) space [4][5][6][7]. These skills are called spatial skills. The spatial skills required to visualize the unfinished product is an important part of construction; it becomes a skill set that constituents of the building process use to communicate new ideas and resolve issues. The problem in not refining these skills is that the CM student is less likely to meet the needs of industry when they join the workforce. Therefore, a focus on improving these skills is essential.

As early as the mid-90s, researchers concluded that visualization of abstract objects on paper and on computer screens was required to be a good communicator within the engineering trades [8]. Later, building information modeling (BIM) would replace paper by constructing the project in a virtual space [9], however, this transition did not replace a need for good spatial skills, in fact, it could be argued that it increased the need for better spatial skills. This technology has made a positive impact on the way that the architectural, engineering, construction and facility management (AECFM) industry communicates; however, one still needs to possess good spatial skills in order to create these models. While the increasing universality of BIM is significant, there are still many instances where communicating visually is done through two-dimensional (2D) sketches and verbal communication without the aid of BIM. Therefore, instances such as these illustrate the continual need for innate spatial ability. Fortunately, research has shown that spatial skills can be improved [10][6][11] and although experience is one way in which this can happen, many tests and studies have been conducted to show that spatial skills can also be learned [12][13][14][15][16].

Because the market demands higher quality graduates and having good spatial skills is essential in a growingly abstract and complex design world, improvements in spatial skills could benefit from a new pedagogical tool that supports active learning and engages the students visually. Augmented reality (AR), although not a new technology, has found more applications in today's world, especially with the increasingly widespread use of mobile technologies (smartphones, tablet computers, and ultra-light laptops). The benefit of using AR in the classroom is that learners can see supplemental digital information [17], assisting them in the understanding of highly abstract and complex assemblies [18]. As a result, students become active participants in their learning and by adding interactive visualizations while encouraging students to ask questions about their learning rather than being told what to learn [19] we enhance the overall learning experience.

More specifically, this research was conducted to respond to three questions:

1. Can a mobile augmented reality pedagogical tool be used to improve the accuracy of spatial skills in CM students at the postsecondary level?
2. Can a mobile augmented reality pedagogical tool be used to improve the retention of newly acquired spatial skills in CM students at the postsecondary level over time?
3. Will the use of a new pedagogical tool be an interference to the overall learning experience?

2 Literature Review

Modern day construction projects call for more advanced use of mathematical models, data analysis, and 3D modeling software, which in turn needs to be converted to a plan of execution by a construction manager. To interpret this kind of data, a CM employee needs good spatial skills that will allow them to visually predict where and how components of the project will be assembled. In the classroom, construction management (CM) students are often required to interpret complex and abstract images into 3D images in order to solve construction related problems [20][7]. However, past research [21][22][23] indicates that students struggle with visualization tasks and in today's CM classroom, educators would not hesitate to argue that shortcomings still exist.

2.1 BIM and Spatial Skills

There is no doubt that BIM has become an important part of the construction industry, and there is a steep learning curve that comes in making sure it is used to its greatest effectiveness. In "BIM Handbook: A Guide to Building Information Modeling for Owners, Managers, Designers, Engineers, and Contractors", 2nd Edition [9], there are over 560 pages dedicated to successfully implementing BIM in all aspects of the AECFM space, and this does not include the software, which may require several days of training to begin using. Unfortunately, reading this handbook and undergoing days of training, still does little to enhance one's spatial skills, leaving them inadequately prepared to render objects and create complex models useful in the coordination of a construction project. While research has shown BIM to have a positive impact on the quality of learning, there is still an element of visualization necessary to support BIM itself [24].

2.2 Augmented Reality as a Mobile Teaching Tool

Augmented reality (AR) is the addition or subtraction of virtual (computer-generated) images superimposed on a real-world view [25]. Liarokapis & Anderson [17] define AR as harmonizing the virtual and real environment in order to provide an understandable and meaningful view. AR is a vision-based technology that requires a "marker" to create a connectedness between the real environment and the one that is generated by the computer [26]. A marker, also known as a fiducial, is a machine-readable optical label that contains information that can be decoded to produce an image on a device running software capable of interpreting the label [25]. However, if the AR software is unable to properly connect the real environment to the virtual one, the illusion that the two worlds coexist will be compromised [25]. Therefore, in this research, the use of markers that can be placed in very specific and controlled locations is a preferred method of rendering the augmented view because it eliminates complexity in setup and the need for high-powered computing currently necessary in non-fiducial rendered AR.

A survey of mobile applications available to most of today's mobile devices will show that several are making use of AR technology - for entertainment, marketing, and education. With today's visually motivated generation of media-conscious students [27], the application of the latest AR technology, along with mobile devices with which students are already familiar, will provide a visually active learning environment that has the greatest potential to improve spatial skills. In fact, some postsecondary CM students are already aware of AR and have good knowledge about the terms surrounding this technology. However, according to Shirazi & Behzadan [28] they are unable to relate the use of these tools to their own learning experience and therefore, rely on universities and schools to make this connection. Therefore, the researcher decided that the use of a mobile device would be fitting for this study because students are 1.) already familiar with its use, 2.) it is a platform that has the ability to easily render AR media when used with a simple to configure "marker" and 3.) because students are already familiar with mobile devices; their introduction in the learning experience will involve less distraction so as to provide clearer results during the experiment's intervention.

3 Methodology

The research questions presented in this study were developed to validate the use of a mobile augmented reality (MAR) tool that could be used in a conventional classroom setting. Moreover, this research is set to determine if the use of a MAR tool would improve the spatial skills of postsecondary CM students. Therefore, this study followed a quantitative between-group double-blind experiment model using pre-assessment and post-assessment surveys (see Figure 1.). The students completed a background survey followed by a pre-assessment and then the researcher conducted an educational lecture for both groups simultaneously. Immediately following, the group was divided into separate rooms. Both groups were administered a lab assignment that they were allowed to complete in smaller groups of 2 to 3 students per group. The test group intervention included a mobile hand-held device equipped with augmented reality software that the students could use to aid in the completion of their lab assignment. Following the lab assignment, and with the students still segregated, the post-assessment was administered along with a survey to ascertain the effort (self-perceived) that students underwent during the experiment. This procedure, start to finish, was completed twice to obtain a total population of twenty-five students (n=25).

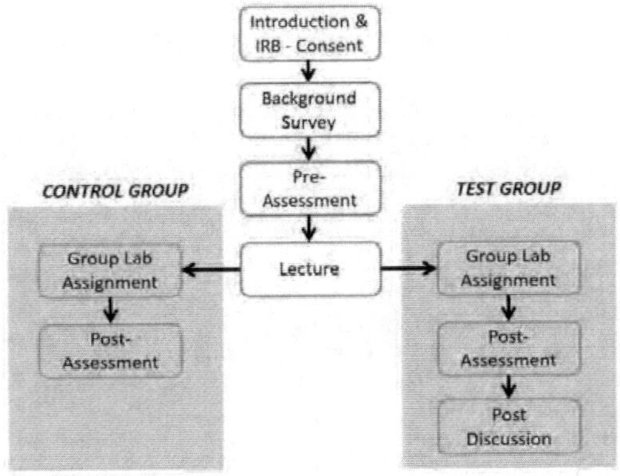

Figure 1. Experiment procedure workflow.

3.1 Setting and Sampling

This study took place at a U.S. postsecondary institution in Gwinnett County, Georgia; an adjacent surrounding county to Atlanta, Georgia. This postsecondary institution operates as a technical college and has an overall student population of 10,068 students as of 2015. The technical college operates a two-year construction management program that has an enrollment population of 60 students as of 2015. The sampling procedure used by the researcher was a nonrandom convenience sampling. The students for this study were restricted to willing students that were available at a predetermined time as coordinated with the students' instructor, hence, the sampling took place during the students' normally scheduled class time.

3.2 Student Demographics

A background survey was used to obtain the demography of the sampling. Of the 25 students, 76% were male and 24% were female. Within the control group, 85% reported male and 15% reported female, while within the test group, 67% reported male and 33% reported female. Age was nearly evenly distributed. Within the control group, 31% reported 18-25 years, 38% reported 26-35 years and the remaining 31% reported 36 years or older, while within the test group, 42% reported 18-25 years, 17% reported 26-35 years and the remaining 42% reported 36 years or older. In terms of previous work experience within the AECFM industry, a combined 68% reported more than one year of experience. Conversely, 20% of the students in the control group reported having less than one year of experience and 12% had no experience. This is important to note in that it may have an effect on the outcome of the assessment scores given the correlation between more work experience and improvement in spatial skills [20][7]. While all of the students reported a high school or equivalent education level, 19% had earned another Associates Degree and 5.4% had earned a Bachelor's Degree. All of the students were pursuing a Construction Management Associates Degree and reported some level of past experience with courses that used visual skills or visual technology. In an effort to gauge comfort level with the use of technology in the classroom, the students were asked if they were agreeably comfortable using mobile technology to obtain more information about things they had questions about. 80% of the students responded as "Strongly Agree" and the remaining 20% responded with "Agree". The students were asked to self-report their skill level in plan reading, as this skill, if practiced often, could have an effect on the students' spatial skills proficiency. The proficiency was spread between 12% claiming "Highly Proficient", 28% "Advanced Experience", 24% "Intermediate Experience" and 36% "Basic Experience". No one reported, "No Experience".

3.3 The Marker (Lab Assignment) and AR Software

Components used in this study included a marker (fiducial), 3D modeling software and augmented reality web platform. Markers were used by both the control and test groups as a lab assignment handed out in paper form (see Figure 2.).

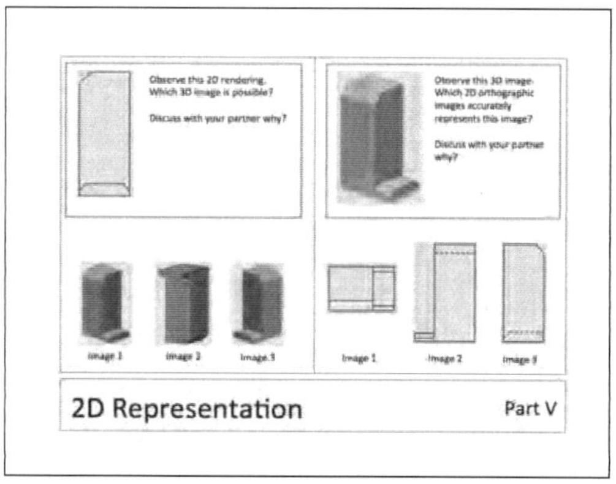

Figure 2. "Marker" (lab assignment) required for the AR software to render 3D images on the screen of a mobile device.

The marker was created using a standard off-the-shelf word processor editing package. The content on the marker was a standard spatial skills assignment. The 3D models used in this study were created by Autodesk's 123D Design (http://www.123dapp.com/design); a commercially available desktop software package capable of creating and editing 3D models. Once a 3D model was created in 123D Design, it was exported for use in the AR software. The AR software used in this study was Augment (http://www.augment.com); a commercially available mobile software package used to scan a marker (fiducial) and render a 3D model on the screen of a hand-held mobile device (HHMD). A student, along with a HHMD and the AR software, would scan the paper-form of the marker using the AR software (see Figure 3.). Once the image was recognized and matched with the inventory of stored marker images from the server, the HHMD's AR software would call the corresponding 3D model from the server and combine it along with the image of the marker being captured by the HHMD's back-facing camera (see Figure 4.).

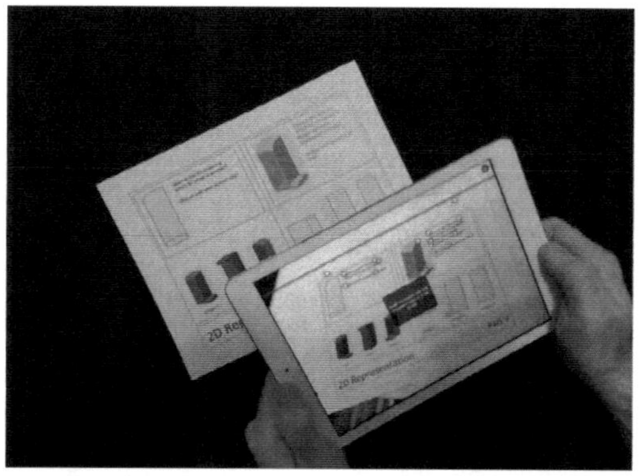

Figure 3. Scanning a "Marker" (Lab Assignment).

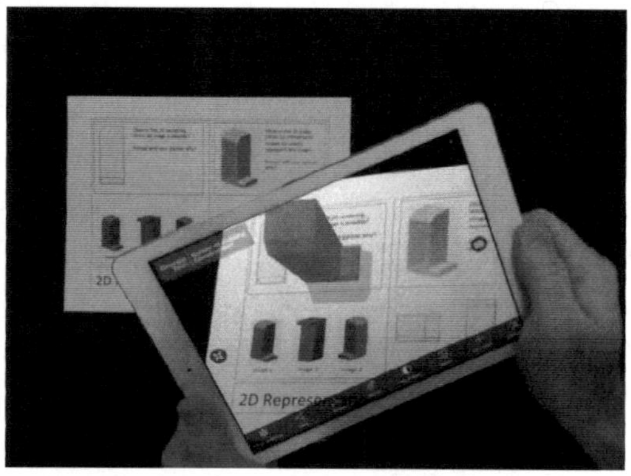

Figure 4. After scanning a "Marker".

The combined image on the HHMD showed the marker with a 3D AR model displayed on it. Lastly, the 3D model displayed on the HHMD's screen was superimposed over, and attached to, a live image of the marker that was being displayed by using the HHMD's back facing camera. The student could then interact by moving the HHMD or the paper-form marker (see Figure 5.).

Figure 5. Students interacting with HHMD and "Marker".

4 Results

The results of this study measure the student's spatial skills by using isometric projections, 3D to 2D orthographic translations, building elevation translations and mental rotation of shapes in the form of a lab assignment. The content for the assessments and activities was modeled from a series of spatial tests that have current basis in ongoing research about spatial skills. Content from the following spatial skills research was used in this experiment: the Purdue Spatial Visualization Tests: Visualization of Rotation (PSVT:R) [14], Differential Aptitude Tests: Spatial Relations (DAT:SR) [12] and the Mental Rotations Test (MRT) [16]. The results of the assessment scores on the pre-assessments, post-assessment and the long-term assessment have been tabulated below in Table 1.

Table 1. Statistical analysis of results of pre-assessment, post-assessment and long-term assessment. Note: Maximum possible score = 10.

	Pre			Post			Long Term		
Group	**M**	**SD**	**n**	**M**	**SD**	**n**	**M**	**SD**	**n**
Control	5.2	1.6	13	5.8	2.1	13	7.2	2.3	9
Test	5.3	1.4	12	7.2	1.6	12	8.8	1.4	9

To further evaluate the data in Table 1, an improvement percentage (IP) was calculated and tabulated in Table 2. The following formula was used to derive a measurement of improvement between the assessments.

$$\frac{\text{Mean (New)} - \text{Mean (Original)}}{\text{Mean (Original)}} \text{ X } 100 = \text{IP}$$

Table 2. Tabulation of improvement percentages (IP) of assessment scores.

	Pre to Post		Pre to Long Term	
Group	**IP**	**n**	**IP**	**n**
Control	11.5%	13	38.5%	9
Test	35.8%	12	66.0%	9

It has been suggested that a certain measure of applied technology in the classroom can have an interfering effect on the learning experience [24]. When deploying new technology in the classroom, will its use equate to a better and more active learning experience [28]? For this facet of pedagogy to be measured in this study a NASA Task Load Index (TLX) survey was administered for the purpose of measuring the student's perceived effort on six independent sub-scales. The sub-scales and their associated results have been tabulated in Table 3.

Table 3. NASA TLX survey results data. In the NASA TLX scale -10 = very low demand and +10 = very high demand within the sub-scales measured.

	Control		Test		Mean Variance
Sub-scale	**M**	**SD**	**M**	**SD**	**△**
Mental	1.5	5.2	0.2	5.4	1.3
Physical	-7.7	5.3	-8.7	2.7	1
Temporal	-5.8	5	-5.3	3.5	-0.5
Performance	4.4	5.6	6.3	2.8	-1.9
Effort	0.8	4	1.6	6.1	-0.8
Frustration	-7.5	3	-7.2	3.4	-0.3

5 Discussion

5.1 Effect on Accuracy

Analyzing the mean assessment scores between the pre-assessment and the post-assessment, the data indicates an improvement in spatial skills for the test group's post-assessment scores (M = 7.2, SD = 1.6, n=12) compared to their pre-assessment scores (M = 5.3, SD = 1.4, n=12) (see Table 1). Furthermore, when comparing the IP of the test group (35.8%) to that of the control group (11.5%) the data indicates a greater improvement for those that used the MAR pedagogical tool to complete the lab assignment. For the purposes of this study, and in response to the research question, "Can a mobile augmented reality pedagogical tool be used to improve the accuracy of spatial skills in CM students at the postsecondary level?" an independent sample t-test was performed to measure the statistical significance of the mean scores for the test group. Assuming a Confidence Interval percentage of 95% (CI = 95%) then $t_{(11)} = -4.095$, $p = .002$ ($p \leq .05$), resulting in a significant improvement between the tests.

Likewise, analyzing the mean assessment scores between the pre-assessment and the long-term assessment, the data indicates an improvement in spatial skills for the test group's long-term assessment scores (M = 8.8, SD = 1.4, n=9) compared to their pre-assessment scores (M = 5.3, SD = 1.4, n=12) (see Table 1). Furthermore, when comparing the IP of the test group (66.0%) to that of the control group (38.5%) the data indicates a greater improvement for those that used the MAR pedagogical tool to complete the lab assignment. For the purposes of this study, and in response to the research question, "Can a mobile augmented reality pedagogical tool be used to improve the retention of newly acquired spatial skill in CM students at the postsecondary level over time?" an independent sample t-test was performed to measure the statistical significance of the mean scores for the test group. Assuming a Confidence Interval percentage of 95% (CI = 95%) then $t_{(8)} = -6.932$, $p = .000$ ($p \leq .05$), resulting in a significant improvement between the tests.

5.2 Perceived Student Effort

The resultant measurements shown in Table 3 indicate that overall, the perceived effort between the control group and the test group were similar. The slight exceptions (the variance was greater than or equal to $|1|$) included mental, physical, and performance workloads. For the purposes of this study, and in response to the research question, "Will the use of a new pedagogical tool be an interference to the overall learning experience?", the data indicates the control group expended more mental effort and physical effort than the test group during the lab assignment. And the test group perceived higher performance from the lab assignment than the control group. In order to derive a benefit from this type of intervention, it should be expected that there is either a negligible effect on these sub-scales (no difference between groups) or that the test group perceives lower levels of mental and physical effort than the control group. Likewise, higher levels of perceived performance for the test group over the control group would also prove this pedagogical tool as a benefit. Overall, having similar data between the groups indicates no additional perceived effort was expended because of the intervention, an outcome that has potential benefit for this intervention as a pedagogical tool.

6 Conclusion

The author has identified augmented reality as a technology with the potential of becoming a useful academic tool for improving spatial skills in construction management students. However, as noted earlier in this paper, there were several students that had significant work experience that could lead to results that would favor those with better spatial skills coming into the study. While the results of this study are significant, the author would suggest a broader sampling to include more students of varying background to validate the research questions. Nevertheless, this study presents a methodology for the introduction of a simple-to-configure mobile AR tool that can be easily employed within the CM classroom.

The outlying expectation of this research is to bring awareness and applicability of AR to the construction industry. It is the opinion of the author that in order for AR to gain acceptance in the greater AECFM industry there needs to be a workforce that is already familiar with the technology, therefore, the introduction of AR at all levels of the education spectrum is a great first step in bringing this awareness. As noted by Holt & Kearney [30],

> "BIM continues to be the leader of advancing technology, and a lot of emerging technology falls under the BIM umbrella. Augmented reality [is] about more efficiently gathering and communicating information about the project."

This statement implies that there is a future for AR within the AECFM industry as it continues to find a foothold in the way that BIM has. And similarly, to the way BIM improves the quality of learning by supporting visualization [24], AR can also support visualization and therefore the learning experiences of CM students. There is optimism that AR will gain momentum in the construction industry by way of preparing CM students for this technology while they are in the classroom. As the graduating CM students become more familiar with AR technology and its ability to benefit complex visualization and collaboration, they are likely to be more comfortable with realizing its potential within the industry, thereby minimizing some of the barriers to its use.

References

[1] Clough, R. H., Sears, G. A., Sears, S. K., Segner, R. O., & Rounds, J. L. (2015). Construction contracting: A practical guide to company management. John Wiley & Sons. (pp. 1-29).

[2] Kelleher Jr, T. J., Mastin, J. M., Robey, R. G., Smith, C., & Hancock, L. L. P. (2014). Smith, Currie and Hancock's Common Sense Construction Law: A Practical Guide for the Construction Professional.

[3] Pierce Jr, D. R. (2013). Project scheduling and management for construction (Vol. 89). John Wiley & Sons. (pp. 7-10).

[4] Ahmed, S. M., Yaris, C., Farooqui, R. U., & Saqib, M. (2014). Key attributes and skill for curriculum improvement for undergraduate construction management programs. International Journal of Construction Education and Research, 10(4), (pp. 240–254).

[5] Bower, M., Howe, C., McCredie, N., Robinson, A., & Grover, D. (2013). Augmented reality in Education - Cases, places, and potentials. In Proceedings of the 2013 IEEE 63rd Annual Conference International Council for Education Media, ICEM 2013.

[6] Martin-Gutierrez, J., Guinters, E., & Perez-Lopez, D. (2012). Improving strategy of self-learning in engineering: laboratories with augmented reality. Procedia-Social and Behavioral Sciences, 51, (pp. 832–839).

[7] McCuen, T. (2014). Investigating the Predictors of Spatial Skill Essential for Construction Science Student Success in the Digital Age. (Doctoral Dissertation). Retrieved from: Department of Educational Psychology, University of Oklahoma.

[8] Hsi, S., Linn, M. C., & Bell, J. E. (1997). The role of spatial reasoning in engineering and the design of spatial instruction. Journal of Engineering Education, 86(2), (pp. 151–158). Retrieved from: https://doi.org/10.1002/j.2168-9830.1997.tb00278.x

[9] Eastman, C., Teicholz, P., Sacks, R., & Liston, K. (2011). BIM Handbook: A Guide to Building Information Modeling for Owners, Managers, Designers, Engineers and Contractors. (L. Khelmani, Ed.), John Wiley & Sons (2nd ed., Vol. 2). Wiley.

[10] John Hopkins Center for Talented Youth. (2015). What is spatial ability? Retrieved from: http://web.jhu.edu/cty/STBguide.pdf

[11] Shirazi, A. (2014). Context-Aware Mobile Augmented Reality Visualization in Construction Engineering Education. (Master's Thesis) Retrieved from: University of Central Florida Orlando, Florida.

[12] Bennett, G. K., Seashore, H. G., & Wesman, A. G. (1956). The differential aptitude tests: An overview. The Personnel and Guidance Journal, 35(2), (pp. 81–91). Retrieved from: https://doi.org/10.1002/j.2164-4918.1956.tb01710.x

[13] Caissie, A. F., Vigneau, F., & Bors, D. a. (2009). What does the Mental Rotation Test Measure? An Analysis of Item Difficulty and Item Characteristics. The Open Psychology Journal, 2(1), (pp. 94–102). Retrieved from: https://doi.org/10.2174/1874350100902010094

[14] Guay, R. B. (1976). Purdue Spatial Visualization Tests: Visualization of Rotations. Purdue Research Foundation.

[15] Maeda, Y., Yoon, S. Y., & Lafayette, W. (2011). Measuring Spatial Ability of First - Year Engineering Students With the Revised PSVT : R. American Society for Engineering Education, (p.19).

[16] Vandenberg, S. G., & Kuse, A. R. (1978). Mental rotations, a group test of three-dimensional spatial visualization. Perceptual and Motor Skill, 47(2), (pp. 599–604).

[17] Liarokapis, F., & Anderson, E. F. (2010). Using Augmented Reality as a Medium to Assist Teaching in Higher Education. In Eurographics 2010 (pp. 9–16). Eurographics Association. Retrieved from: http://eprints.bournemouth.ac.uk/20907/1/eg_eduAR10.pdf

[18] Schwald, B., & De Laval, B. (2003). An augmented reality system for training and assistance to maintenance in the industrial context. Journal of WSCG, 11, (pp. 1–3).

[19] Freeman, S., Eddy, S. L., McDonough, M., Smith, M. K., Okoroafor, N., Jordt, H., & Wenderoth, M. P. (2014). Active learning increases student performance in science, engineering, and mathematics. PNAS Proceedings of the National Academy of Sciences of the United States of America, 111(23), (pp. 8410–8415). Retrieved from: https://doi.org/10.1073/pnas.1319030111

[20] Martin-Gutierrez, J., Saorin, J. L., Martin-Dorta, N., & Contero, M. (2009). Do Video Games Improve Spatial Abilities of Engineering Students? International Journal of Engineering Education, Vol. 25 (Issue 6), (pp. 1194–1204).

[21] Alias, M., Black, T. R., & Gray, D. E. (2002). Effect of instruction on spatial visualization ability in civil engineering students. International Education Journal, 3(1), (p. 12). Retrieved from: http://eprints.uthm.edu.my/1815/1/EFFECT_OF_INSTRUCTION_ON_SPATIAL_MAIZAM_ALIAS.pdf

[22] Black, R. G., & Duff, S. (1994). A model for teaching structures: finite element analysis in architectural education. Journal of Architectural Education, 48(1), (pp. 38–55). Retrieved from: https://doi.org/10.2307/1425308

[23] Ryu, S., Kim, M. H., Kinnas, S. A., & Kang, J. H. (2003). Development of Web-based Numerical Wave Tank and Java Applets as an Advanced Tool for Teaching Wave Mechanics. In 2003 ASEE Annual Conference & Exposition.

[24] Wong, K. A., Wong, F. K. W., & Nadeem, A. (2011). Building information modelling for tertiary construction education in Hong Kong. Journal of Information Technology in Construction, Vol. 16, (pp. 467–476).

[25] Azuma, R. (1997). A survey of augmented reality. In Presence: Teleoperators and Virtual EnvironmentsPresence, 6(4), (pp. 355–385). Retrieved from: http://scholar.google.com/scholar?q=intitle:A+Survey+of+Augmented+Reality#0

[26] Feiner, S., Macintyre, B., & Seligmann, D. (1993). Knowledge-based augmented reality. Communications of the ACM, 36(7), (pp. 53–62). Retrieved from: http://dl.acm.org.prx.library.gatech.edu/citation.cfm?doid=159544.159587

[27] Moskal, B., Lurie, D., & Cooper, S. (2004). Evaluating the effectiveness of a new instructional approach. ACM SIGCSE Bulletin, 36(1), (pp. 75–79).

[28] Shirazi, A., & Behzadan, A. H. (2014). Design and Assessment of a Mobile Augmented Reality-Based Information Delivery Tool for Construction and Civil Engineering Curriculum. Journal of Professional Issues in Engineering Education and Practice.

[29] Cristia, J., Ibarrarán, P., Santiago, C., Santiago, A., & Severin, E. (2012). Technology and Child Development : Evidence from the One Laptop per Child Program. IZA Discussion Paper No. 6401. Retrieved from: https://doi.org/10.2139/ssrn.2032444

[30] Holt, P. E. A., & Kearney, N. (2015). Emerging Technology in the Construction Industry : Perceptions from Construction Industry Professionals. American Society for Engineering Education, (Paper ID#12604).

AR(Augmented Reality) based 3D Workspace Modeling for Quality Assessment Using As-Built On-Site Condition in Remodeling Construction Project

S.Y. Lee[a], S.W. Kwon[b] and T.K. Ko[c]

[a]Department of Convergence Engineering for Future city, Sungkyunkwan University, Republic of Korea
[b]School of Civil, Architectural Engineering and Landscape Architecture, Sungkyunkwan University, Republic of Korea
[c]Department of Convergence Engineering for Future city, Sungkyunkwan University, Republic of Korea
E-mail: song03@skku.edu, swkwon@skku.edu, gtk101@naver.com

Abstract –

In case of re-modeling construction project, it has various difficult situations to carrying out the projects comparing with general construction project. The re-modeling project needs to do several make decision makings which is to define the scope of utilization, extra works and O&M(operation & maintenance). Especially, the construction delays and cost-increase are caused by absence of reviews on existing building, especially structure and facilities conditions. Therefore, the objective of this study is establishment of 3D environment for augmented reality based decision-making for re-modeling project management. The scan-data based 3D as-built model is generated by using LiDAR to provide AR based 3D information. The matching process between as-built model and model of additional structure is used to review for constructability. However, the method using AR technology based on marker has a limitation on reading range. This study proposes a marker-less AR rendering based method to set 3D environment using android based NFC tag technology. As a result, this study develops analysis method of constructability by matching between existing structure and remodelled structure. Through the process to establish 3D environment, there are following results expected. First, this method provides high-quality visual information to project participants. Second, the result of this study makes project managers easily to set rapid-decision by using mobile device.

Keywords –

Augmented Reality, As-built Model, Remodeling, Human Assisted Matching

1 Introduction

1.1 Research background and purpose

According to the survey by the Korea Ministry of Land, Infrastructure and Transport, approximately 2,543,217 building are included in Deteriorated building, accounting for 36% of total in Korea[1]. Besides, the number of deteriorated building will be even larger when New Town Development project and Two Million Housing Unit Construction Plan project led by the government at the end of 1980s are reflected.

Demands on remodeling is growing, following the deterioration of physical and social function of buildings due to aging. Remodeling project is a kind of conditional construction project targeting buildings or infrastructures that are older than 20 years. Remodeling projects, which include deregulation on extension building, are aimed at preventing aging and improving functions. For the activation of remodeling, Korean public and local government implemented a number of supporting policies such as improvement of legal system, expansion of construction cost support, and publication of remodeling guidelines.

The main difference between general construction project and re-modelling project is that remodeling projects have a constraint on the construction period to ensure profitability of remodeling project owner. By a shorter construction period compare to any other projects, more rapid and accurate evaluation to construction quality based on as-built condition while carry out additional construction and facility change is very important. In design phase, 3D model has been utilized to check the feasibility of construction. In construction phase, however, even though still there are many errors occurred by worker's the review of construction quality has regarded as trivial work.

Therefore, this study aim to purpose 3D based decision-making model for review of construction quality in construction phase on remodeling project.

1.2 Research coverage and procedure

This study has focus on construction phase on remodeling project. And we confined the target on review of construction quality on construction phase which reflect a characteristic of remodeling project. Compare to general building that construct sequentially, remodeling project site has unique as-built condition due to remained some structures. It is suitable target for 3D based improvement of construction quality check.

The process of this research is:

First, Carrying out examination the building construction supervision guideline by Ministry of Land, Infrastructure and Transport of Korean government, project management guideline of CM companies, and construction quality certification check-list by Korea remodeling association.

Second, Carrying out construction items extraction related to review of construction quality on construction phase among guidelines and check-list.

Third, Extraction of the check-items which have to evaluated relate to as-built preparation.

Forth, Analysis on how to generate an as-built environment suitable to remodeling construction.

Fifth, Suggestion of 3D based construction quality evaluation process considering characteristics of remodeling construction process.

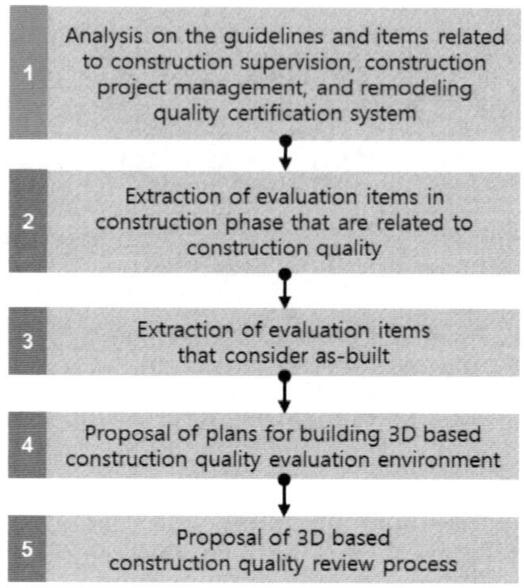

Figure 1. Flow chart of this study

2 Research Trend & Literature Review

2.1 Construction management for remodeling construction

The remodeling construction differs from general construction in that surveying and diagnosis process is accompanied. Such difference affects the entire area of construction management. Meeting construction period and cost is particularly difficult in case of remodeling project compared to other construction projects, due to the characteristics of remodeling construction management. Table 1 summarizes the characteristics of remodeling construction management.

Table 1. The characteristics of remodeling construction management

Characteristics	Considerations by characteristic
Variety of Construction plan	-Reflecting operation of existing facility to construction process -Minimize the changes by communication between contractor and owner -Minimize damages to building users
Survey and diagnosis for target building	-The extent of missing drawings Reliable result of diagnosis
Delay in decision making	-Application for approval of subcontractor has been checked bẏ contractor, CM and project supervisor
Low productivity for construction	-Remodeling construction is carried out on as-built building environment -Many joint-section between additional parts and existing parts
Difficulty in the estimation of construction cost	-Even though remodeling project has low productivity, costs are estimated by quantity of material -Inappropriate standard unit cost

Remodeling target buildings are mostly 15 to 20 years old after the completion of construction. As such, as-built drawing is not stored in most cases. Even if it is stored, it has been usually changed for user's convenience. As a result, 2D drawing and as-built environment does not coincide in many cases. Because of this, remodeling construction causes a lot of difficulties for workers to understand works in the construction progress.

Moreover, productivity of field workers decreases in case of remodeling construction compared to general construction due to the restriction of structural frame of the existing building. For example, wall is completed after plumbing in general construction. However, in case of remodeling project field, plumbing should be implemented in a condition where there are existing walls. Hence, there are many joint rooms of pipes and material procurement also requires a lot of workforce and time. This is one of the major reasons that cause construction error by workers.

2.2 Literature Review

Yeo at el. (2004) proposed construction management method through an integration of decision making system and coloring-based planning that uses perspective drawing written in design phase. Han at el. (2009) proposed a process that can analyze feasibility of construction. Specifically, the study converted 2D drawing in 3D CAD and combined it with WBS (Work Breakdown Structure) and CBS (Cost Breakdown Structure) that reflect remodeling characteristics to build 5D CAD. Lee at el. (2010) wrote BIM model that reflects remodeling characteristics and proposed a scheme for detecting clash in design phase. However, most of these studies are focused on the evaluation in design phase, rather than post-construction evaluation. The researches have limitation in that it cannot accurately propose as-built environment of remodeling.

Table 2. Previous literature on remodeling evaluation method

Researcher	Contents
Yeo at el. (2004)	Proposed ways to improve construction management (integration of decision making system and establishment of coloring-based planning that uses perspective drawing written in design phase)
Han at el. (2009)	Analyzed feasibility of construction through conversion of 2D drawing to 3D CAD
Lee at el. (2010)	Proposed clash detection scheme that reflects the characteristics of remodeling by distinguishing between as-is, demolition, retention, and extension rooms when writing BIM model

Samir El-Omari & Osama Moselhi (2008) proposed a 3D modeling plan by integrating images that were collected from LADAR and digital camera. Pingbo Tanga at el. (2010) argued that as-built generation method through laser scanner shows highest accuracy and reviewed techniques that are related to automatic establishment of BIM model through laser scanner. Ankit Bhatla at el. (2012) proposed a way of building 3D model by using handheld digital camera that has high portability. However, there is scarce study that involves data acquisition or modelling scheme that reflects the characteristics of remodeling.

Table 3. Previous literature on as-built model generation method

Researcher	Contents
Samir El-Omari & Osama Moselhi (2008)	3D modeling by combining 3D image collected from 3D laser scanning (LADAR) and images collected from digital camera
Pingbo Tanga at el. (2010)	Review on technologies related to automatic generation of BIM by using point cloud data obtained from laser scanner
Ankit Bhatla at el. (2012)	3D modeling based on pictures taken with handheld digital camera

The most important thing in remodeling construction is managing construction phase based on as-built that was accurately understood. The previous researches mentioned in Table 3 focused on how to generate 3D environment similar to the real-world. However, research on the acquisition plan for as-built model that reflects the characteristics of remodeling or usage plans based on as-built environment is insufficient as of now.

3 Review of construction quality

3.1 Constructability

Although review on constructability covers entire phase of a project, the term is mainly used to indicate understanding of the possibility of construction in phase prior to the construction. CII (Construction Industry Institute) defines constructability as "use knowledge and experience on construction regarding planning, design, procurement, and on-site work as much as possible for the purpose of achieving overall business goal" [8]. Among the factors that affect constructability, detection of clash with existing building through on-site inspection is carefully implemented based on the existing drawing that was performed in the design phase. Meanwhile, response to the construction error that occur during construction or inference control are managed by workers.

Performance of selective construction evaluation through sample selection and complexity of detection reporting process despite the restriction on the construction duration causes insufficiency of construction quality review. Hence, there is a rise in technical demands in construction phase for simpler provision of information including on-site arrangement plan, appropriate construction method, construction sequence, and on-site interference check.

3.2 Review items of construction quality through on-site check

In this research, we checked into the building construction supervision guideline by Ministry of Land, Infrastructure and Transport of Korean government, project management guideline of CM companies, and construction quality certification check-list by Korea remodeling association.

According to the building construction supervision guideline by Ministry of Land, Infrastructure and Transport of Korean government, construction supervisor has an obligation of preventing unnecessary argument on site by proposing objective and accurate inspection result to contractor. Project management guideline of CM companies d that construction project manager has an obligation of establishing guideline for inspection that reflects the characteristics of the site (inspection process, period of each work, frequency, and check-list) before implementing the inspection.

Among the evaluation items including common work, steel-frame work, and concrete placement in the construction quality certification check-list by the Korea remodeling association, items that evaluate the quality through field check are summarized in Table 4 [9]. Field check items have common feature of making evaluation based on the situation on site, instead of using regulation, guideline, or relevant documents. More accurate understanding about the on-site situation can increase the reliability of the quality evaluation. Hence, it is very important to accurately understand the as-built environment. Providing as-built environment and construction management guideline to users in 3D can be useful for managing the construction quality based on on-site check as it enhances users' intuitiveness.

Table 4. Construction management items requiring on-site check

Work type	Check-items
Common work	Whether the structural frames are drawn or not Defect status and Method of repair on structural frames Records of design, construction and maintenance
Steel-frame work	Evaluation of shop-drawing and construction plan Whether existing construction errors are reflected to steel work or not Degree of error Appropriateness of the steel-frame design Anchor bolt spacing
Concrete placement	Preparations before concrete placement Measure to reinforcement of adhesive power to existing concrete Making the sample concrete cylinder & examination Status of joint surface condition of existing frame(ex. column, slab and wall) Checking the difference of canter line between foundation and columns

4 Review of construction quality that reflects as-built environment

4.1 Generation of as-built model through laser scanning

This study attempts to generate as-built model using laser scanner. Lee (2014) applied laser scanning to indoor environment and proposed optimal point accuracy for generating as-built model and computational reverse engineering (CRE) process that is optimized for remodeling.

Below describes the process of establishing as-built model through laser scanner.

1. Selection of location and number of scanning
2. Installation of scanner (trembling prevention and levelness check)
3. Scanning
4. Merge of the acquired point cloud data
5. Correction of errors that occurred during merging
6. Completion of the modelling

Using this process, Lee scanned ten rooms that have diverse sizes from 540 mm x 249.5 mm x 266 mm (W x B x H) to 760 mm x 1186.6 mm x 264 mm. Remodeling has a feature of having existing structures such as crossbeam, column, and equipment. Hence, it is important to make sure not to create blind spot by appropriately selecting location, angle, and number of scanning when building plans for laser scanning. In Lee's experiment, average error was ±1 mm in case of indoor scanning using laser scanner. Hence, we believed that establishment of as-built model that has high accuracy can be achieved by using laser scanning.

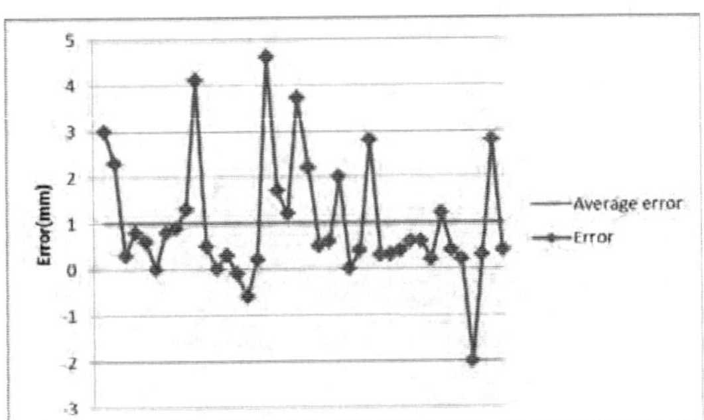

Figure 2 Error graph in case of 20cm per 100m accuracy [10]

Moreover, scanning with accuracy of 20cm per 100m, 10cm per 100m, and 5cm per 100m produced average scanning time of 135.775sec, 358.750sec, and 2145.425sec, respectively. Hence, we decided that scanning with 20cm point per 100m accuracy is most efficient considering the data acquisition time and its relationship with error. Finally, condition for tearing down and additional construction differ by rooms in case of remodeling. Review on the construction quality should match the characteristics of each room. Hence, it is efficient to model by room unit in case of as-built model as well.

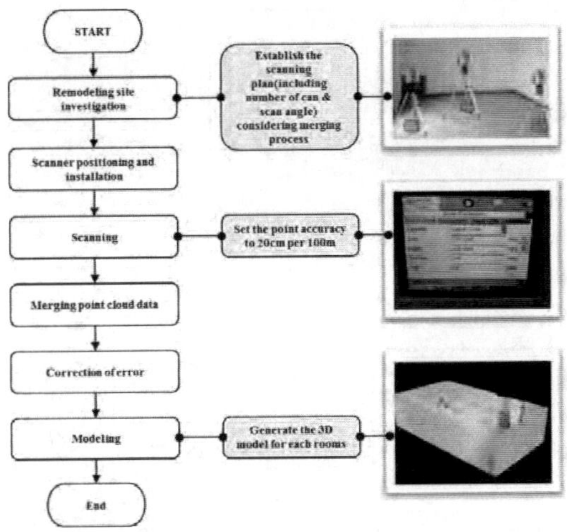

Figure 3. Laser Scanning process that reflects the remodeling characteristics [10]

4.2 Review of construction quality based on as-built

This study attempts to establish as-built model by using point cloud data that were collected from 3D image acquisition device (laser scanner) and propose process for construction quality evaluation that considers as-built.

Remodeling construction process considering as-built that this study proposes is largely composed of three phases. First is to establish as-built model through laser scanner that can accurately express the field and model on the additionally building structures and MEP. Second is to write detection guideline based on as-built. Third is to evaluate the construction quality based on the detection guideline that was written by considering as-built.

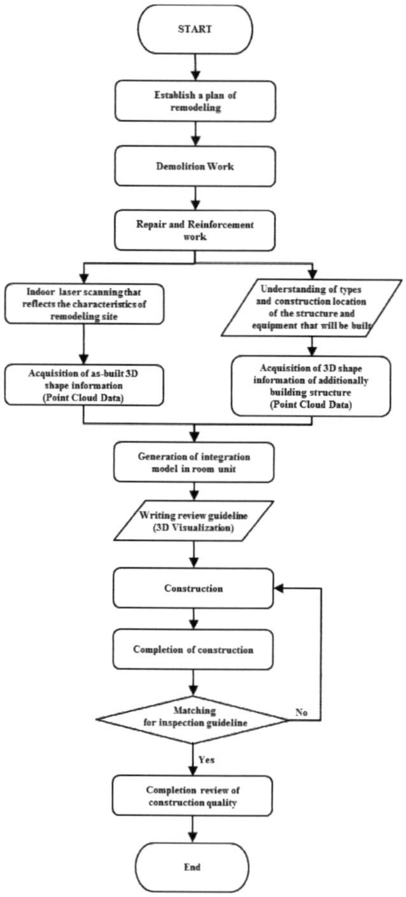

Figure 4. Remodeling construction process considering as-built

5 3D environment for review of construction quality

augmented reality technology.

Augmented reality is largely divided into marker based AR and marker-less AR according to the scheme of implementing model. Marker-less AR is often called as natural feature tracking-based AR.

In general, black and white or colour pattern image is inserted to marker form to be used as a marker in mobile augmented reality industry. However, construction project including remodeling has extensive size and complicated process than other areas. Hence, AR display system that uses image marker form is not appropriate. Moreover, image pattern scheme has a shortcoming in that it can simply read previously input information only.

Hence, this study attempts to use NFC tag technology for mobile-based AR implementation. NFC technology, which is one of the short-range wireless technologies, does not require separate reader as chip is built in mobile devices (Smartphone and tablet PC). As NFC is free to read and write, correction of information inside chip is easy. Smooth mutual communication can be made through this. Moreover, NFC tag has advantage of excellent portability due to its small size and light weight.

Different from new construction, each room has different condition in case of remodeling construction. Hence, information including as-built model, MEP model, construction plan in each phase, construction specification, and detection guideline is stored by room unit. Through this, workers or supervisor can easily sort out the information on the room that they have to work only out of the huge amount of information

Hence, users tag NFC tag with mobile device they are carrying and render necessary information among the registered information on display. User can compare model or information expressed in augmented reality and reality at a glance by matching images in reality world and visual coordinate through 3D rotation of the rendered model.

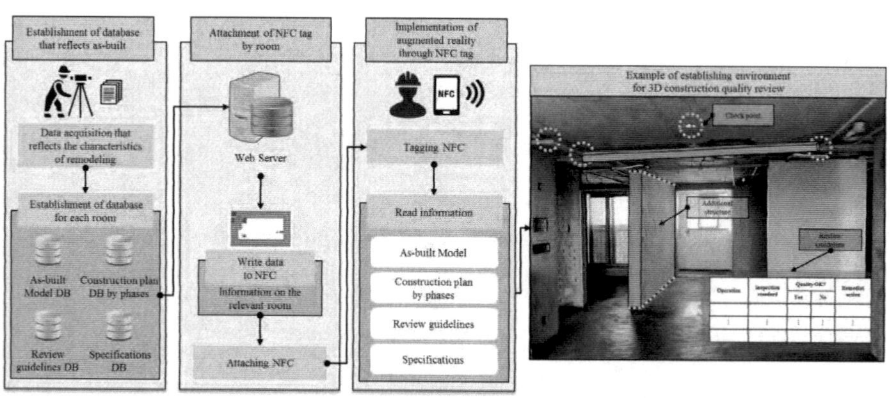

Figure 5. Establishment of 3D environment for construction quality review

5.1 Process of augmented reality-based construction quality review

This study proposes establishment of as-built model in remodeling project and visualization of construction plan, specification, and detection guideline phase by phase using augmented reality. Based on the suggested contents, we derived plans for improving remodeling process.

In remodeling construction phase, field workers and quality manager are provided with necessary information that are visualized in 3D. Information can be obtained through augmented reality, including types and quantity of materials required for the preparation of construction, construction plan by each phase, construction specification, model at the time of construction completion, and guideline for construction quality review that was obtained from as-built model-based construction simulation. In particular, use of augmented reality enables comparison of reality and information at a glance. Hence, user can simultaneously proceed on working and reviewing/testing of the work outcomes.

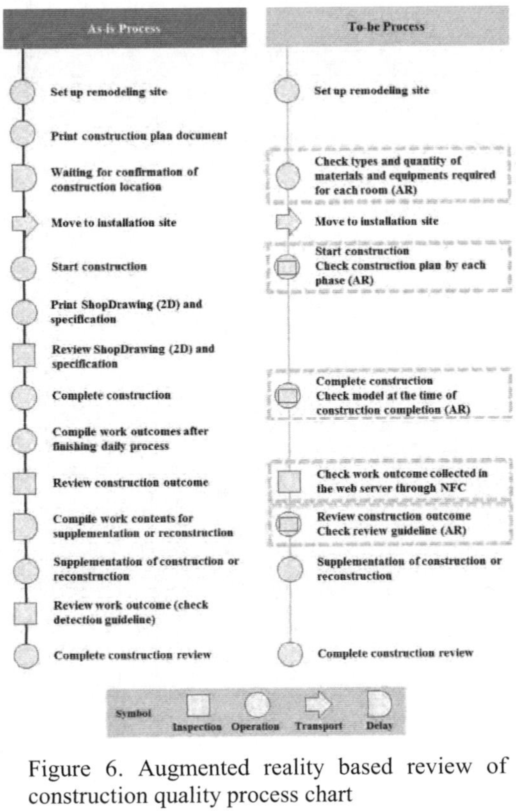

Figure 6. Augmented reality based review of construction quality process chart

6 Conclusion

The purpose of this study is to provide 3D-based decision making environment for construction quality evaluation in the construction phase of remodeling project. To propose better evaluation method, we investigate evaluation items and method of the quality certification of the Korea Remodeling Association. We decide that accurate understanding of as-built model is important for the evaluation of remodeling construction quality.

As construction plan differs by each room in case of remodeling, we proposed a plan that provides 3D visualized information through augmented reality by storing information in NFC tag by room unit. We suggest potential process improvement plan that is expected when contents proposed in this study are reflected and compared it with the existing process.

In this study, we can accurately understand as-built environment by reflecting the characteristics of remodeling and write detection guideline that fits the properties of the project. As on-site construction condition can be compared at a glance through the visualized detection guideline, we could achieve more accurate and prompt construction quality test.

This study has significance in that it helps accurate understanding of as-built model and that it proposes construction quality evaluation process in construction phase based on as-built. We expect that our study can be used as fundamental research for future studies that are related to management of remodeling construction site, quality improvement, and quality evaluation.

Acknowledgement

“ This work is financially supported by Korea Ministry of Land, Infrastructure and Transport(MOLIT) as 「U-City Master and Doctor Course Grant Program」 ".

"This work was supported by the National Research Foundation of Korea(NRF) grant funded by the Korea government(MSIP) (No. NRF-2014R1A2A2A01005893)."

References

[1] Korea Ministry of Land, Infrastructure and Transport. Status of National Buildings(2015). Online:http://www.molit.go.kr/USR/NEWS/m_71/dtl.jsp?id=95078847, Accessed:13/02/2017.

[2] Yeo, U. Y., Yoon, Y. S., Suh, S. W., and Lim, B. W. Improvement of Construction Management for Building Remodeling Projects. *Korean Journal of Construction Engineering and Management*, 5(4): 122-129, 2004.

[3] Han, J. Y., Cha, H. S., and Lee, D. G. A measure for standardization of old aged apartment remodeling through application of BIM. *Korean Journal of Construction Engineering and Management*, 10(3):53-61, 2009.

[4] Lee, S. H., Min, B. J., Kim, I. K., Nam, S. C., Kang, S. D., and Jun, H. J. A Study on BIM based Remodeling Architectural Design Process for Apartments. *Korean Journal of Computational Design and Engineering*, 15(4):314-323, 2010.

[5] El-Omari, S. and Moselhi, O. Integrating 3D laser scanning and photogrammetry for progress measurement of construction work. *Automation in construction*, 18(1):1-9, 2008.

[6] Tang, P., Huber, D., Akinci, B., Lipman, R., and Lytle, A. Automatic reconstruction of as-built building information models from laser-scanned point clouds: A review of related techniques. *Automation in construction*, 19(7): 829-843, 2010.

[7] Bhatla, A., Choe, S. Y., Fierro, O., and Leite, F. Evaluation of accuracy of as-built 3D modeling from photos taken by handheld digital cameras. *Automation in construction*, 28: 116-127, 2012.

[8] Construction Industry Institute. Constructability Implementation Guide, CII Special Publication 34-1. Construction Industry Institute, USA, 1993.

[9] Korea Remodeling Association. Remodeling quality certification system. On-line: http://www.remodeling.or.kr/html/info_07.jpg, accessed: 10/01/2017.

[10] Lee, S. and Kwon, S. A study on optimal laser scanning method for reverse engineering at interior remodeling project. *Korean Journal of Construction Engineering and Management*, 15(3):3-11, 2014.

A Study on the Effect of the Use of Augmented Reality on Students' Quantity Take-off Performance

Sung Joon Suk[a], George Ford[a], Youngcheol Kang[b], and Yong Han Ahn[c]

[a]Kimmel School of Construction Management, Western Carolina University, U.S.A.
[b]Department of Architecture & Architectural Engineering, Yonsei University, South Korea.
[c]Erica School of Architecture & Architecture Engineering, Hanyang University, South Korea.
E-mail: sungjoon@wcu.edu, gford@wcu.edu, yckang@yonsei.ac.kr, and yhahn@hanyang.ac.kr

Abstract

This study examined whether the augmented reality (AR) technology affects the quantity take-off performance of construction management (CM) undergraduate students. The sample for the study consisted of 22 sophomore CM students taking an estimating course at Western Carolina University in the U.S.A. To determine whether AR images help students' quantity take-off performance, the students were asked to take off the concrete volume of a structure twice. At the first attempt, the students were provided only with a 2D drawing that includes a plan and section of the structure. At the second attempt, they had an opportunity to explore a 3D model overlaid on the 2D drawing on their smartphone. Five students calculated correct quantities at their first and second attempts. Ten students did not take off quantities accurately at the first attempt, but with the 3D model in AR, they could reach the correct quantities at the second attempt. Seven students failed to calculate the quantities at both attempts. Forty-five percent of the students showed better quantity take-off performance at their second attempt. There was also a difference in the average scores between the two attempts. The results suggest that CM students can take off quantities more accurately with the help of 3D model images created by the AR technology.

Keywords

Augmented Reality; Quantity take-off; Estimating; Students' performance

1 Introduction

Construction drawings are the main communication tool used in the construction industry. Specifically, 2D drawings have long been used in delivering construction projects and managing facilities. However, representing 3D objects into 2D drawings involves many limitations. Understanding vertical and horizontal elements together with 2D drawings, where their multiple faces and complex connections are not fully illustrated or visualized, is challenging. Creating section views helps synthesize the information from 2D drawings, but they are still represented in the 2D plane; generating thousands of section views for a project is uneconomical.

To overcome these limitations, the construction industry has made efforts to utilize 3D modeling approaches. Many construction companies now use building information modeling (BIM) tools to generate 3D models and drawings. Seeing 3D models and images help project participants have an improved understanding of geometrical shapes and connections. However, using BIM tools and other 3D modeling software is challenging at large construction projects or complex properties with a number of buildings. Most 3D modeling programs need a license for use and require expensive computers with high-level computational and graphical capabilities. Setting up a large number of expensive computers at a project site is impractical and challenging. This issue results in the lack of mobility in utilizing 3D models effectively. Moreover, looking at the details and generating 3D views are the challenges for those who have not been trained in using such sophisticated programs. In addition, numerous 3D modeling programs with different user interfaces are available. For these reasons, 2D drawings are still widely used in construction sites and facility management fields to communicate among professionals.

Virtual reality (VR) and augmented reality (AR) technologies are expected to foster the use of 3D modeling. VR&AR is one of the most promising businesses in the foreseeable future. The Goldman Sachs Group, Inc. [1] estimates that the VR&AR market can generate at least US$80 billion by 2025 and may become a US$182 billion market if AR adoption is accelerated. This value represents a tremendous increase from US$5 billion at the end of 2015. Large IT companies are quickly moving to the VR&AR market. In March 2014, Facebook invested US$2 billion in acquiring Oculus VR, a manufacturer of VR head-mounted display. Mark Zuckerberg said, "One day, we believe this kind of immersive, augmented reality will become a part of daily life for billions of people." Google developed Google Cardboard, which is a VR platform with VR development tools, and brought low-cost VR headsets to the market. Google also launched an AR technology platform called Project Tango, which can provide mobile devices with a spatial perception ability to track motions and process images with their built-in camera. In addition, Google led US$542 million of investment in Magic Leap, an AR startup company. Microsoft started in 2016, shipping HoloLens, an AR device through which you can mix 3D holographic content into your physical world and interact with both of them. The industry has set the stage for VR&AR to become a prosperous business.

The rapid advancement and wide use of mobile devices such as smartphones and tablets have made VR&AR attractive. Smartphones and tablets are the best devices to implement VR&AR technologies because they have a camera, geographic position system (GPS), Internet access, and display. Furthermore, tracking methods have improved to identify the relationship between the camera and the real-world [2]. Now virtual contexts can be extracted without regular markers that may be difficult to set up all necessary places and complex environments. Various technologies, including development platforms, applications, devices, and marker-less detection, have been, are being, and will be developed for the use of VR&AR.

A number of researchers have investigated the application of AR in the construction field. Wang [3] identified three challenges involved in AR in architecture: extraction of industrial domain knowledge, preparation of reality model, and technological limitations. Dunston and Wang [4, 5] provided a taxonomy used for AR application development. They also classified architecture, engineering, and construction (AEC) operations into five levels: Application Domain, Application-Specific Operation, Operation-Specific Activity, Composite Task, and Primitive Task to map AR technology to AEC tasks.

Chi et al. [6] developed an AR-enhanced user interface (UI) used for a teleoperated crane system. They reported that the integration of AR in the user interface of crane has a positive impact on crane operators. Schall et al. [7] presented an AR system to show visual information of underground utilities, and also addressed problems including 3D modeling, tracking, and interaction that they identified while developing and testing their prototype AR system. Talmaki et al. [8] evaluated a geo-referenced AR system providing the excavation operator with 3D models on the location and type of buried utilities. Shirazi and Behzadan [9] assessed the pedagogical value of an AR-based tool on students' performance on building design and assembly. Shanbari et al. [10] used ART enabled videos to helped students identify and remember certain masonry and roof components.

However, little research has been conducted on whether AR contents can improve the quantity take-off performance. This study examines the effect of AR contents on CM undergraduate students.

2 Research Methods

Twenty-two CM undergraduate sophomore students enrolled in the Estimating I course at Western Carolina University in North Carolina, U.S.A. participated in this study. The Estimating I is the first course that exposes the students to the quantity take-off process. The course learning objectives are: 1. Take off quantities used to create construction project cost estimates; 2. Read construction documents for planning; 3. Identify methods, materials, and equipment used to construct projects; 4. Use Excel or other software to takeoff quantities; 5. Understand different surveying techniques for construction layout and control; and 6. Understand the concept of construction accounting and cost control. Before taking the Estimating I course, students have typically completed an introduction to construction course, a construction graphics and plan reading course, and a construction methods & materials course as prerequisites of this estimating course. The students who participated had no construction industry experience. Before this study, there were ten class sessions, where the students learned the general rules of quantity takeoff used in the construction industry and how to take-off quantities associated with the earthwork.

A single group, pre- and post-test research design was employed to determine whether AR images help students' quantity take-off performance. The students were asked to take off concrete and formwork quantities of a structure twice. At the first attempt, the students were provided only with 2D drawings that include a plan and section of the structure. The structure consisted of simple retaining walls and continuous footings underneath the walls, as shown in Figure 1. The students were given 15 minutes to take off the volume of concrete and asked to turn in their solutions.

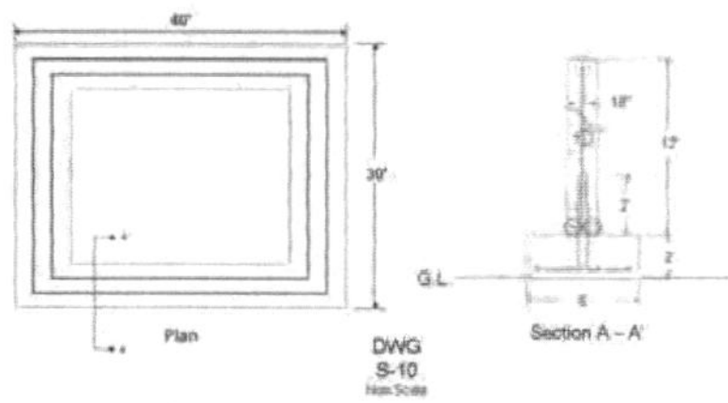

Figure 1. Plan and Section

At the second attempt, they had an opportunity to explore a visualized 3D model overlaid on the 2D drawing on their smartphone, as shown in Figure 2. This AR image was created by using REVIT 2017 and an AR application provided by Augment.com. After a REVIT model of the structure was generated, it was exported as an OBJ file format that is a geometry definition file format. Then, the OBJ file was imported into the AR application of Augment.com. While exploring the AR image on their smartphone, the students took off the volume of concrete in the same 15-minute time period and submitted their solution. The students were not permitted to see what they did at their first attempt.

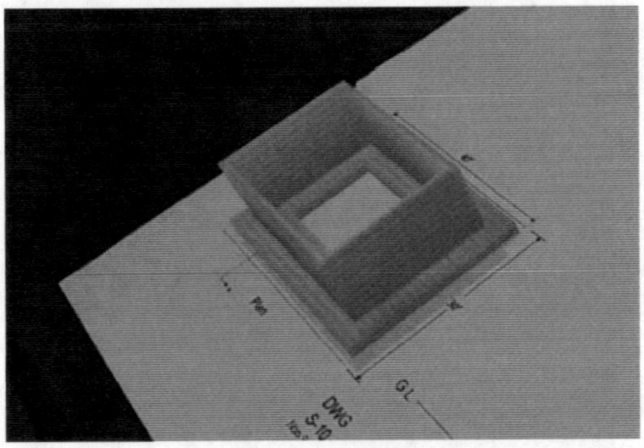

Figure 2. 3D Model Overlaid on Smart Phone

After the two attempts had been graded, a paired t-test was used to compare the mean scores before and after having the students exploring the AR images with their smartphone.

3 Results

Five students calculated correct quantities at their first and second attempts. Ten students did not take off quantities accurately at the first attempt, but with the 3D model in AR, they could reach the correct quantities at the second attempt. Seven students failed to calculate the quantities at both attempts successfully. Forty-five percent of the students showed better quantity take-off performance at their second attempt. The mean of the 2nd attempt was 89.09% greater than that of the 1st attempt, 80.91%. To examine the true effects of the AR images, a paired t-test was conducted. The results of a paired t-test on the two attempts (H_0: 2nd attempt − 1st attempt = 0; H_a: 2nd attempt − 1st attempt > 0) were summarized in Table 1. The results show that there was significant evidence that students' quantity take-off performance was improved at their 2nd attempt (paired t (21) = 3.645, $p <= .05$)

Table 1. t-Test: Paired Two Sample for Means

	2nd Attempt	1st Attempt
Mean	89.091	80.910
Variance	313.420	237.229
Observation	22	
Pearson Correlation	0.806	
Hypothesized Mean Difference	0	
df	21	
t Stat	3.645	
P(T<=t) one-tail	7E-04	
T Critical one-tail	1.721	
P(T<=t) two-tail	0.002	
T Critical two-tail	2.080	

4 Conclusions

The analysis results showed that students took off quantities more accurately with the help of 3D model images created by the AR technology. This means that the AR images increased the students' understanding of the retaining walls and continuous footings. The results suggest that AR technology can enhance understanding of construction drawings for construction management undergraduate students.

The difficulty in the quantification is that construction elements are fully illustrated on the 2D drawings. Furthermore, students usually do not have an extensive field experience that would help them understand the entire shape of construction elements. Field trips to construction sites can foster students' understanding of construction elements. However, it is difficult to have many field trips for a course in a semester. Drawings provided in class may not always be the same ones used for projects to which students take a field trip.

With a rapid development of AR applications, students can easily see AR images on their mobile device. AR images aligned with 2D drawings can provide them with a better understanding of construction elements necessary to prepare an accurate cost estimate. However, there is much work to be done in AR to enhance CM students' experiences. Different types of AR contents for each trade need to be produced to develop students' ability to understand construction drawings effectively. More research efforts are necessary to simplify the procedures used to create AR contents from 2D drawings or building information models.

Reading and understanding construction drawings is a fundamental part of preparing CM students to enter the construction industry. Adopting AR technology in teaching CM students is expected to be a way of enhancing such ability that may need extensive construction experience.

This study used a single group, pre- and post-test research design and had a limitation. The students took off the volume of concrete twice for the same retaining walls and footings. To reduce an effect of repeated practice at the second attempt, the study did not allow the students to refer to what they did at the first attempt. However, the students might realize errors and address them, which may improve the students' quantity take-off performance at the second attempt.

References

[1] The Goldman Sachs Group, Inc., *Virtual & Augmented Reality: Understanding the race for the next computing platform*, 2016

[2] Chi H., Kang S., and Wang X. Research trends and opportunities of augmented reality applications in architecture, engineering, and construction, *Automation in Construction*, 33:116–122, 2013.

[3] Wang X. Augmented reality in architecture and design: potentials and challenges for application, *International Journal of Architectural Computing*, 2(7):309–326, 2009.

[4] Dunston P. and Wang X. A hierarchical taxonomy of AEC operations for mixed reality applications, *Journal of Information Technology in Construction*, 16:433–444, 2011.

[5] Wang X. and Dunston P. A user-centered taxonomy for specifying mixed reality systems for AEC industry, *Journal of Information Technology in Construction*, 16:493–508, 2011.

[6] Chi H., Chen Y., Kang S., and Hsieh S. Development of user interface for teleoperated cranes, *Advanced Engineering Informatics*, 26:641–652, 2012.

[7] Schall G., Mendez E., Kruijff E., Veas E., Junghanns S., Reitinger B., and Schmalstieg D. Handheld Augmented Reality for underground infrastructure visualization, *Pers Ubiquit Comput*, 13:281–291, 2009.

[8] Talmaki S., Dong S., and Kamat V. Geospatial databases and Augmented Reality visualization for improving safety in urban excavation operations. In *Construction Research Congress 2010*, pages 91–101, Banff, Canada, 2010.

[9] Shirazi A. and Behzadan A.H. Content delivery using augmented reality to enhance students' performance in a building design and assembly project, *Advances in Engineering Education*, 4(3):1–24, 2015.

[10] Shanbari H., Blinn N., and Issa R. Using augmented reality video in enhancing masonry and roof component comprehension for construction management students, *Engineering, Construction and Architectural Management*, 23(6):765–781, 2016.

34th International Symposium on Automation and Robotics in Construction

ISARC 2017

June 28 – July 1 TAIPEI, TAIWAN

BIM & FACILITY MANAGEMENT

Location-based Facility Management System Framework Using 3D-BIM on Commercial Office Buildings

J.H. Yoon[a] and H.S. Cha[b]

[a]Department of Architectural Engineering, University of Ajou, Suwon, Gyeonggi-do, South Korea
[b]Department of Architectural Engineering, University of Ajou, Suwon, Gyeonggi-do, South Korea
E-mail: snoop543@ajou.ac.kr, hscha@ajou.ac.kr

Abstract –

In order to maintain a reliable facility management (FM) service for a commercial office building, a facility personnel and the building owners make a decision with detailed-level building Operation and Maintenance (O&M) data. By analyzing the data, they can make decisions in terms of whether to and/or where to invest money for maintaining better conditions of the buildings. Conventionally, the facility personnel collect O&M data in a manual manner and organize them in a paper document format. In this format, most of O&M data are intertwined and mingled up so that both FM personnel and building owner fail to capture significant information for reliable O&M performance. More often than not the building owners have difficulty in understanding the building's O&M performance through the bundled data. According to the FM personnel's work process and building owner's decision making process, location data are employed as a starting point to O&M work and O&M performance confirmation respectively. Despite of the significant role of location data, no systematic framework based on the location data has been developed. For the purpose of facilitating both the owner and facility manager's better decision making, the authors develop a system framework for 3D/BIM-based facility information management. The 3D/BIM has an advantage on storing and outputting the location data. Using the system framework, the FM personnel and building owner obtain effective information management platform based on location data. In addition, the research can contribute to the foundation for a development of a BIM-driven facility management system.

Keywords –

Decision Making; Information management; Operation and Maintenance; System Framework; 3D/BIM

1 Introduction

Among the various critical success factors of Facility Management (FM), both facility management (FM) personnel and building owners have common consensus that reliable Operation and Maintenance (O&M) is the most significant success factor in FM (Yoon and Cha 2016). In addition O&M phase, as a longest period in the lifecycle of a building, generates majority of FM expenses (Lee and Akin 2011). It accounts for 85% of total costs spent over lifecycle of a building (Teicholz 2004).

Despite of the practical and fiscal significance of O&M phase, most of the O&M work process is placed upon superficial level. It has limitations on providing proactive work, just providing reactive and impromptu O&M work (Akcamete et al. 2010). Still the paper document is primary medium for O&M information between FM personnel and O&M computational systems, resulting in failure on timely service (Lee and Akin 2011). Even though a compact disk could be employed to store FM information in PDF, these files are easier to lose (William East et al. 2012). Furthermore, it doesn't have proper O&M information repository for assisting them to identify and diagnose O&M problems or increase their awareness of previous problems, maintenance and repairs work (Akacamete et al. 2010). In particular, lack of locational information system results in ostensible analyses on O&M work. According to Becerik-Gerber et al. (2011), FM personnel need locational data which allow topological and accurate analysis on O&M work. They believe the location data are helpful to perform preventive and corrective O&M work. In conclusion, both information reposition and locational data analysis in O&M phase leads to reliable O&M service (Figure. 1). Although several Computer Aided Facility Management (CAFM) programs have been developed to solve these problems, the field of O&M information management system is relatively young and, as such, is still establishing its basic tenets.

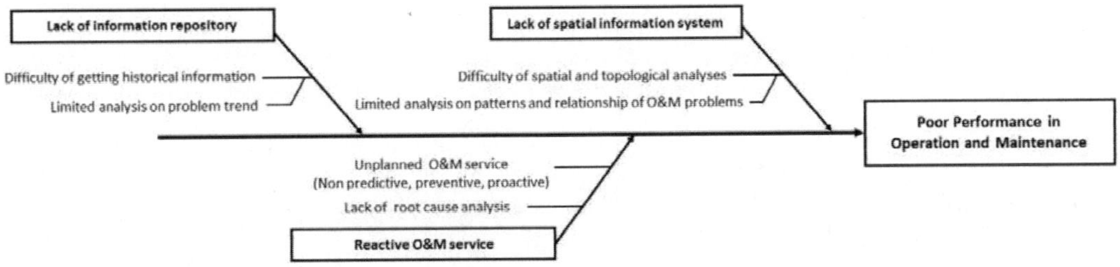

Figure. 1 Root-cause diagram of poor performance in operation and maintenance

In particular, most prevalent CAFM programs such as ARCHIBUS/FM and Maintenance Connection are not appropriate for locational information reposition and analysis. ARCHIBUS/FM provides 2D-based interface. It is difficult to identify problems on three-dimensional equipment such as pipes and conduit, which is main information for HVAC (Heating, Ventilating and Air-Conditioning) among O&M works (Lee and Akin 2011). Maintenance Connection provides web-based FM information management system. However, it doesn't provide locational data visualization. The owners have difficulty in confirming locational data with this program.

The main purpose of this research is establishing a framework for O&M information management system using Building Information Modeling (BIM). BIM can provide FM data repository and become information backbone for FM systems by providing 3D objective-based parametric data (Ghosh et.al 2015). BIM can also improve facilities management by providing accurate FM data and easy accessibility to the data (Kassem et al. 2015; Ismail et al. 2015). The proposed framework will include locational information analysis and computational reposition, based on BIM. The authors hope that both FM personnel and building owner can have an effective FM information management system through this system framework.

2 Literature review

Akcamete et al. (2010) presented BIM-based maintenance planning system. It helps to store maintenance data and understand previous maintenance work trend. Through the stored locational data, it is possible to analyze maintenance information topologically. However, the system's operation is circumscribed in repair work. It doesn't address both HVAC problem and energy management. In addition the proposed spatiotemporal analysis on the information is still in its early stage. The paper simply presents locational relationship between several different repair works. It is necessary to address whole O&M information with locational data and contrive how to utilize the accumulated information with FM personnel and the building owner. Wetzel and Thabet (2015) presents a BIM-based framework to support safe maintenance and repair practices. While it is meaningful that the framework is based on locational data from BIM, so that it can provide more sophisticated information analysis, it just deals with not comprehensive O&M information but only safety issue. Shalabi and Turkan (2016) emphasize FM data collection for corrective maintenance work. They developed BIM-based data repository, improving visualization and interoperability capabilities. The system could receive data from both Building Energy Management Systems (BEMS) and Computerized Maintenance Management Systems (CMMS), conventional FM operation systems. However, the paper mainly addresses linkage between existing FM systems and BIM, but does not present how to utilize the system for providing proactive O&M planning and reliable O&M service.

Previous literatures all agree with the importance of more sophisticated O&M information management system for reliable FM service providing. Furthermore they have consensus that the locational data provide valuable information for it. However, the studies are not only circumscribed in technical issue but also superficial in providing the practical application platform for both FM personnel and building owner. There are limitations on application of the systems to the real FM practice situation. Examining the information flow in practical situation could be the fundamental step for development of an effective and a realistic information system. Satoshi and Ichizou (2013) emphasize the significance of understanding the practical information flow as a fundamental role in establishing an information management system. In this paper, the system framework for O&M information management helpful to analyze the data topologically and comprehensively is proposed based on the case-based analysis on the O&M information flow. It will be helpful to establish more practical system framework as a next step for previous research.

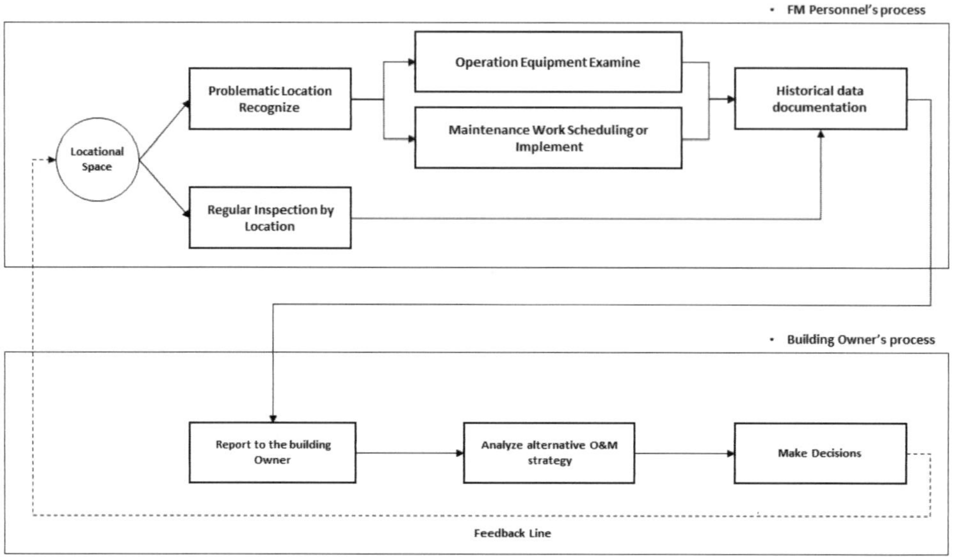

Figure. 2 Current work process of FM Personnel and Building Owner

3 Practical FM Information Flow

As explained above, in order to build an effective information management system, analysis on practical information flow is indispensable. In this paper the monthly FM O&M report documents for a stereotype commercial building in South Korea are examined to understand the information flow in real case. In addition, the building's FM expert interview is carried out in order to identify FM personnel's work process and building owner's decision making process.

3.1 G-building in South Korea

One of the largest commercial office buildings in South Korea was chosen to analyze the FM information flow. The building was constructed in July 2001 with 45 stories above ground and 8 stories below. The building area is approximately 212,000m^2, and the facility maintenance expenses are approximately 20 million dollars per year. In addition, there are 13 facility managers and 48 facility service implementers in FM.

3.2 Analysis on Current FM information flow

According to the G-building's FM experts interview, locational space such as specific floor or space are the starting point for both FM work process and building owner's decision making process (Figure. 2). FM personnel require locational data in establishing O&M schedule or deciding proper solution for O&M problem. Building owners ask for the O&M performance information by location. They consider one space where the certain tenants lease as one unit for managing whole building. In conclusion, both FM personnel and building owner employ the location data as an essence step in order to capture the valuable information. However, in the current FM report document format, it is challenging to extract locational data and this difficulty causes ostensible analysis on the O&M data.

According to the examination on the monthly FM report for G-building, the various O&M data are intertwined and provided in paper document format to both FM personnel and building owner. When they obtain locational data, they should pass through other futile data which are linked to the locational data (Figure. 3). With this current document format, it is almost impossible to sophisticatedly examine the data and obtain valuable information. The difficulty in obtaining locational data leads to failure on proactive work planning. With the locational trend data, the FM personnel can expect which location is vulnerable to specific O&M work and then they can establish proactive solution for it in advance. Furthermore, the current process impedes FM personnel identifying root-cause of O&M problem because all information are intertwined with each other. To be more specific, it is challenging to extract and analyze certain category data. For example, when they need to analyze the whole O&M work orders to find out O&M problem continuity, it is demanding because other data such as O&M equipment or Operations information are given together.

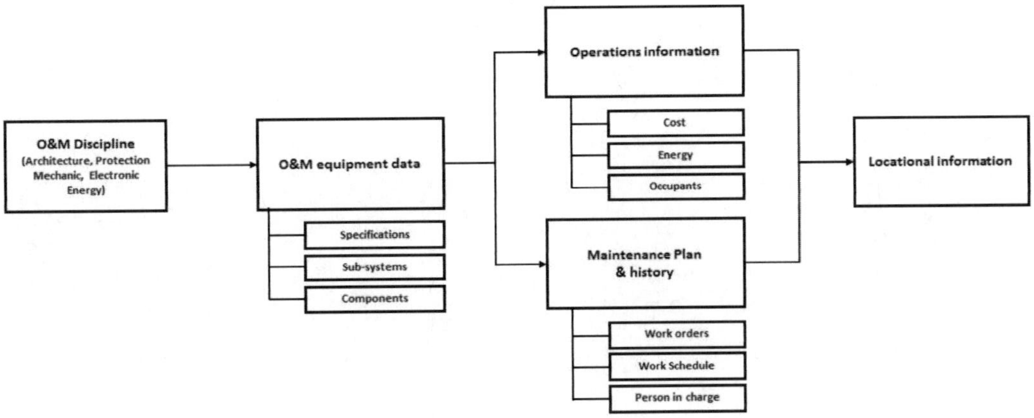

Figure. 3 FM Personnel and Building Owner's O&M information flow in Current FM report

As a result, they manually sieve out the data from the data-mingled document. It is inefficient and time-consuming in FM process. As a result, the FM personnel have had difficulty in obtaining valuable information for providing reliable FM service with this current process.

On top of that, the current information obtaining process reveals that it has a critical disadvantage on information provision to building owner. As every information is bundled up and intertwined in one paper document, it is difficult to extract specific information, which the owner want to confirm. In particular, according to the owner's decision making process, the owner want to confirm O&M performance by the specific location. In the current FM report, the certain information which the owner want to confirm can't be obtained so that they fail to rational decision making for investment on O&M. As a result, the failure leads to unreliable O&M services.

In conclusion, the new FM information process system is necessary in order to provide reliable FM service. It should be possible that both the FM personnel and building owner can obtain desirous information easily and they can analyze the information spatially and comprehensively. The location data will be the key for the system framework.

4 Location-based FM information System framework

Based on the problem analysis on the practical information flow in the current FM report, the effective O&M information system framework is proposed. The framework allows both the FM personnel and building owner to identify patterns and relationship of O&M problems by employing BIM-based locational data. As explained above, locational data such as specific floor or space are the starting point for both FM work process and building owner's decision making process. In other words, the locational data are the hub for the information distribution. Because BIM is proper to provide O&M locational data in 3D model (Ghosh et.al 2015), it is placed upon the center of the system as a repository station. This BIM-based data repository received information from several ways. FM personnel input two different information. One is concerning "Personal data"; assigned department and equipment. By inputting this information to the repository, they can obtain proper information matched with department and equipment. The other one is about "O&M work"; O&M work history or schedule. This inputted data is organized with locational data in the repository and transferred to maintenance schedule data or operation/maintenance solution. As an output, the refined information with 3D locational data can be provided to the building owner when he or she input the data such as location where they want to confirm the performance of O&M. Furthermore, this output is employed for sophisticated analysis on O&M problem. After the O&M team meeting on this output, the updated solution or work schedule is stored again to the BIM-based data repository. The O&M problem information from existing FM programs and problem reception is also inputted to this repository. The received O&M problem is organized with locational data. The BIM-based repository notifies this problem to the FM personnel who are assigned to problematic location. The comprehensive concept of this information system-use process is provided in Figure 4.

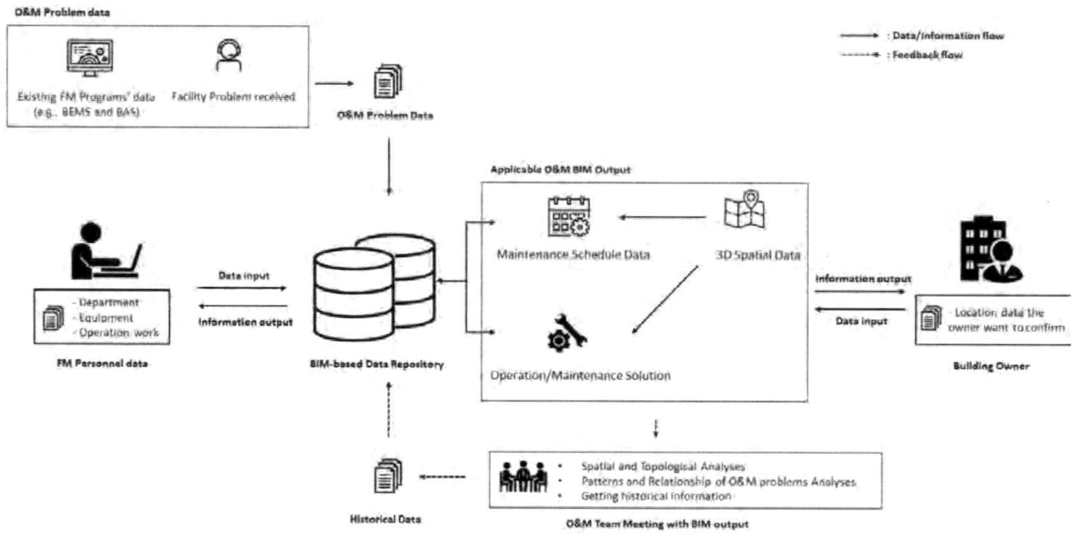

Figure. 4 Locational data-based FM information system framework

5 Conclusion

To provide a reliable FM service, it is indispensable for FM personnel and building owner capture the valuable information in O&M phase. This is because the FM information analysis makes it possible to prepare a preventive O&M work (Oti et al. 2016). In particular, locational data, which is the fundamental component for understanding O&M problem trend or topological reasons, should be presented to FM personnel and building owner for better decision making. Because of this importance of data processing in FM, great attention has been shown to the question of how to establish an effective FM information system. However, fewer studies have attempted to construct fundamental application platform for the system based on locational data. In addition, there were lack of consideration on what the owner's want to know and what the FM personnel want to know for effective FM practice. The previous research mainly had addressed technical issue: how to link BIM to existing FM program. Few have attempted to address BIM in managerial perspective. In this paper, the authors proposed a system framework for practical FM system application. It is established with the empirical analysis on real building's FM report document and FM expert's interview. It is helpful to understand problematic location and provide valuable information for preventive FM. As a next step for previous FM information system research, this research provides an effective application platform.

On top of that, the proposed framework provides sophisticated analysis process on the O&M information. It organizes all O&M data from FM personnel and existing FM system and distributes them with 3D model to FM personnel and building owner. 3D model allows easy and accurate understanding about FM problematic location or object (Lin and Su 2013). The transferred information offer the opportunity to analyze the O&M problem trend temporally and spatially. This will assist FM personnel to plan O&M work schedule more sophisticatedly. Furthermore, the information are provided to the owner when they want to confirm the certain location's O&M information. The owner can understand the overall O&M work performance more effectively and accurately. As a results, by applying the framework, the better decision making of both FM personnel and building owner can be obtained for reliable FM service.

However, this paper is limited to consideration of framework for the O&M information system. It is believed that further practical system development with the proposed framework is worthwhile. The future work will be devoted to establish a practical O&M information system. It is to be hoped that this paper will serve as a platform from which studies of greater depth and specificity may be undertaken.

6 Acknowledgements

This research was supported by the Basic Science Research Program, through the National Research Foundation of Korea (NRF), funded by the Ministry of Education (No. 2015R1A2A2A01005242).

References

[1] Akcamete, Asli, Burcu Akinci, and James H. Garrett. Potential utilization of building information models for planning maintenance activities. *Proceedings of the international conference on computing in civil and building engineering. June*, 2010.

[2] Becerik-Gerber, B., Jazizadeh, F., Li, N., & Calis, G. Application areas and data requirements for BIM-enabled facilities management. *Journal of construction engineering and management*, 138(3), 431-442, 2011

[3] Ghosh, A., Chasey, A. D., & Mergenschroer, M. Building Information Modeling for Facilities Management: Current practices and future prospects. *Building Information Modeling,* 223, 2015.

[4] Ismail, Z. A., Mutalib, A. A., Hamzah, N., & Baharom, S. BIM technologies applications in IBS building maintenance. *Jurnal of Teknologi*, 74(3), 69-76. 2015

[5] Kassem, M., Kelly, G., Dawood, N., Serginson, M., & Lockley, S. BIM in facilities management applications: a case study of a large university complex. *Built Environment Project and Asset Management*, 5(3), 261-277. 2015

[6] Kubota, S., & Mikami, I. Data model–centered four-dimensional information management system for road maintenance. *Journal of Computing in Civil Engineering,* 27(5), 497-510, 2011.

[7] Lee, S., & Akin, Ö. Augmented reality-based computational fieldwork support for equipment operations and maintenance. *Automation in Construction*, 20(4), 338-352, 2011

[8] Lin, Y. C., & Su, Y. C. Developing mobile-and BIM-based integrated visual facility maintenance management system. *The Scientific World Journal*, 2013.

[9] Oti, A. H., Kurul, E., Cheung, F., & Tah, J. H. M. A framework for the utilization of Building Management System data in building information models for building design and operation. *Journal of Automation in Construction,* 72, 195-210. 2016

[10] Shalabi, F., & Turkan, Y. IFC BIM-Based Facility Management Approach to Optimize Data Collection for Corrective Maintenance. *Journal of Performance of Constructed Facilities*, 04016081. 2016.

[11] Teicholz, E. Bridging the AEC technology gap. *IFMA Facility Management Journal*, 587, 588. 2004.

[12] Wetzel, E. M., & Thabet, W. Y. The use of a BIM-based framework to support safe facility management processes. *Automation in Construction,* 60, 12-24, 2015.

[13] William East, E., Nisbet, N., & Liebich, T. Facility management handover model view. *Journal of computing in civil engineering*, 27(1), 61-67. 2012

[14] Yoon, J. H., & Cha, H.S. An empirical study of facility management performance on commercial office building in South Korea. *5th International Conference on Civil Engineering and Urban Planning (CEUP).* Xian, China, in press, 2016

A Study of the Campus Facilities Maintenance System with Building Information Modeling Application

Lee-Kuo Lin[a] , Yen-Yu Hu[b] and Si-Yu Chen[c]

[a]Department of Civil Engineering, National Taipei University of Technology, Taiwan.
[b]Department of Civil Engineering, National Taipei University of Technology, Taiwan.
[c]Department of Civil Engineering, National Taipei University of Technology, Taiwan.
E-mail: lklin@ntut.edu.tw, asus1016@yahoo.com.tw, sara821005@gmail.com

Abstract –

As the scale of construction increasing and the structure being more complicated, it is necessary to use the computer information technology to promote the industrial efficiency and tend with information management. The application of Building Information Modeling (BIM) has become the trend. As there are different benefits in each phase of construction life cycle by using different BIM functions and there will produce different system efficiency by different participants when applying BIM technique; to make management of the facility maintenance at the period of using in effect, it is necessary to consider the management of facility maintenance at the planning phase and to make it easier to manage and reduce costs. In current, the most of BIM application and research are focusing on how to aid the design and construction phases. There are rarely researches focusing on the facility using and maintenance after construction.

This research will build a phototype management system database of the school educational facilities maintenance by using BIM technique, to allow managers can realize the historic fixing records, material categories and other related information. This system also can remind the facility's service life will near and need to be changed automatically. Based on the illustration of this research, this system can make facilities maintenance more effective.

Keywords–

Building Information Modeling (BIM); Campus Facilities ; Facilities Maintenance System

1 Introduction

According to the construction record from the Construction and Planning Agency Ministry of the Interior in 2005, school buildings are the biggest amount of the public building project. Most of the school buildings have been used more than ten years. School buildings are not only the place for teaching and learning but also the important space to teachers and students. So, the maintenance management is good or not will make a big influence during the school building using age for supplying a safety environment of student learning and facility teaching effectiveness.

Due to combining the development of science and technology and multi-domain knowledge, the application of Building Information Modeling, so called BIM, has become the trend. As using BIM in different state of construction life cycle have different benefits, it may increase the cost and work at the early state, but to the whole construction life cycle, it can reduce management cost and increase the efficiency, especially to the more later states which are able to reuse the information already set up for more benefits. The maintenance management in this research goal is belong to the later state, so this research focuses on how to continue the information which was set up before.

School building maintenance management is important, because the school building are the space for common use of the public. Maintenance management often makes many kinds of problems such as people and things, these problems even affect the safety of teachers and students, so the urgent need to establish a complete management system. In addition, it has a very close relationship between the level of school education and school building environment. In order to enhance the quality of school education quality, it is urgent to establish the relative facilities maintenance management.

2 Literature Review

2.1 Building facilities maintenance management

International Facility Management Association defined facilities management is that facilities management is a combination of business administration, architecture, activity and engineering science knowledge or principles, and make the crew and organization in facilities and real workplace can have coordination. In addition, Liu Mingguo in 1987 defined construction and maintenance management as "For building, it is to make action to be act, the corresponding protections have well warning, and always keep in mind and maintain the function can handle all the problem[1]."

Building facilities maintenance management can be divided in two parts basically, one is "operation management", its goal is providing user can use the building function. The other is "maintenance management" its goal is using all kinds of methods to recover or maintain the facilities functions. The maintenance management, a part of operation management, plays the rule of auxiliary. Therefore, facilities management focuses on construction life cycle management, to make each facility to achieve its highest value, also to ensure the normal operation of all facilities within the building environment[2].

2.1.1 Existing facilities maintenance management

The concept of facilities management originated in the United States in 1970[2], in recent years the international community has gradually attached importance to the facilities management. However, the current facilities maintenance management methods have been different by compared with the past. The maintenance management system in the past, considered the engineering ontology and the cost but not consider the process of assets management, such as budget planning management, administrative execution process and the situation of using. By comparison, asset maintenance management system is wider procedure than before, it is including every state of construction including design, construction, operation, resource management, finance, marketing and system operating; therefore, many advanced countries in Europe and the United States have already used the methods of business management in government department maintenance and management funds and use this method to improve efficiency, productivity, and service quality of users[1]. The different between facilities management and operating methods is shown on table 1.

Table 1. The different facilities management and operating methods between now and the past

	Past method	Current method
Budget preparation	Focusing on the most important cost. Overall, including bridge, road weights and other component cost estimates .	According to the demand, the specific construction program and detailed program information.
Cost Estimate	Estimate the cost of per area	Estimate the cost of each component per unit.
Plan selection	Based on subjective judgment, people complain and the experience in past	Based on objective information (pavement, estimated cost sorting scheme)
Current facility situation	Didn't use facility status indicators, geometric and historical data	Using facility status indicators, geometric and historical data
Facilities assessment	Not evaluated	Based on the objective information assessment program, describe the current conditional ability to determine the maintenance behavior
Facilities planning	Didn't have budget plan	Analyzing planning, scheduling, resource allocation and budget and other objective information rapidly

2.1.2 Building life cycle

In general, a building life cycle contains the six stages, including planning, design, contracting, construction, operation maintenance and final removal[2]. Every stage information is interrelated and complicated, and closely related with facilities functions. However, the traditional building or facility management system model, often lack the complete life cycle of the information file, and cause many related documents lose due to the long-term savings, then affect the implementation of building maintenance operations [3].

As construction being completed and after the completion of operation, it is going into a long period of using, and the equipment also began to enter the operation period. As being used, facilities and equipment are getting older, the frequency of failure is increasing, the reliability and safety are reduced, and the energy consumption increased. Therefore, timely maintenance is very important. In Japan, due to operating the research about systematic building energy and analysis of building facilities using management planning in long-term, the foundation is very completely. The following is description of expenditure items and cost distribution, it shows the percentage of the total cost of the Japanese building in its life cycle which maintenance and management costs 51.1%, accounting for half of the total cost in the life cycle, and it is the 3-4 times higher than construction and facilities. It shows the importance of maintenance management in the life cycle of the building[4].

Table 2. The cost distribution of the building life cycle

Project	Project name	Percentage of costs
Design planning cost	Construction planning fees, local survey fees, land use fees, land consolidation fees, design fees, evaluation fees, real estate acquisition tax, special land tax, registration free tax, opening royalties	0.4%
Construction cost	Business expenses, business management fees, environmental countermeasures, opening fee, real estate acquisition tax, registration tax, business tax	15.8%
Maintenance and management cost	Building management, equipment management, sanitation management, cleaning management, security, garrison, repair update, management fee	51.1%
Management cost	Electricity, gas charges, fuel costs, water charges, sewage charges	32.3%
Fixed management cost	Fixed assets tax, city plan tax, property tax, damage insurance premium, boundary fee, general expenses, loan interest	0.4%

2.2 Building Information Modeling

Building Information Modeling, referred to BIM, is defined as the life cycle of construction facilities (including buildings, bridges, roads, tunnels, etc.) creating and maintaining the technology and concept of the digital information of construction facilities and its engineering application. And The American Institute of Architects further defines the building information model as a "model technology that combines engineering project information databases". BIM technology is a simulation in the computer virtual space as a real project, to assist in the construction of life-cycle planning, design, construction, operation, maintenance of the management and engineering operations of the new technology, new methods and new concepts, rather than a new tool which is often misunderstood. It uses digital information to simulate the real information which buildings have. In BIM, Information is not only a three-dimensional geometric information, but also contains a lot of non-geometric information[5]. This technical concept has now been widely adopted by the world, and The United States, Britain, South Korea, China are also actively developing BIM national standards.

BIM focuses on the sustainable used of construction life cycle, it can be used to show the entire life cycle of the process and related information, including geometric, spatial relationships, geographic information systems, the nature and quantity of various building elements (like supplier details) etc. Due to searching BIM can provide all kinds of appropriate information, assist decision makers to make the best decisions. At the same time, compared to the traditional way of the traditional 2D, it can significantly reduce every types of errors of the overall project. The advantage of BIM is 3D visualization presentation, information immediacy, horizontal cross-professional integration and vertical cross - stage collaborative management, geometric information link and three - dimensional spatial information integration...etc[3]. Its goal is reducing the design changes, repeated construction and other phenomena, and thus to reduce the wasting cost, shorten the duration and improve quality. BIM technology is also progressing and developing rapidly. BIM Project Execution Planning Guide, which published by the

American Building SMART Alliance in 2009 and 2010, summarized 25 kinds of applications in the BIM project planning, design, construction and application phase, it also shows that BIM can be used widely and multiple in the life cycle[5].

2.2.1 Building Information Model Exchange Standard

In current, the standards of exchanging information are quite much, its goal is to enhance the efficiency of the construction of industrial data exchange and promote the smooth flow of work and accelerate the automation of operations. The following are common BIM information exchange standards :

1. Industry Foundation Classes (IFC)

Industry Alliance for Interoperability, referred to IAI, has proposed a set of Industry Foundation Classes. It is an open data format for information exchanging, sharing and using in entire construction management. By using the standard definition of doors, windows, walls, lamps, furniture and other real objects and the abstract concept of space and structure, it can deal with the contents of the information by the form of an object database. So that all participants can use the information which is generated by different software in different stages, and able to circulate, apply and integrate to each other[6].

2. Information Delivery Manual (IDM)

Information Delivery Manual is known as to realise the BIM value and achieve the information exchanging between different members and software products, It is necessary to establish an open data to exchange standard format IFC. Because of this kind of information exchange involved the variety of projects, the complexity of the project stage, the huge project life cycle time, and the exchange of information often only exchange for a specific project in a project or only for the participants in a few workflows and exchange by applying software, it is not necessary and impossible to use entire IFC contents when exchange information. So which contents are necessary to exchange in similar information exchanging is what IDM have to do. In this result, operating a specific project or a specific participant are defined the necessary content of exchange information for their respective work by IDM, and exchange them by using IFC standard format. Furthermore, Model View Definition is mainly method of transferring the general language of IDM into computer language of IFC[7].

3. International Dictionary Framework (IFD)

Because the global natural language is diverse and ambiguous, to ensure the information is consistent between the information providers and users from different countries, regional, language and cultural background, IFD gives every text and terminology the globally unique identifier (GUID) to make every information in IFC enable to connect with unique GUID. When providing a necessary information, it is able to get the unique content. So it is not important describing information in which language to users from different background[8].

3 Research Content

3.1 System Construction

To make the facilities management perfect, it is necessary to let information be electronic and manage it by using software, and it can enhance management ability in effect. It is the most commonly used way to store the database. Therefore, this study proposes the based framework of a database engine to manage a large number of IFC file. The system framework is shown in Figure 1. The stored information are include the IFC file from BIM software. As it having many format in the IFC file, this study uses IFC2x3, which can transfer this information into the data from database, and develop a man-machine interface by C # program language to provide users using the functions like operating, adding, researching, modifying more convince. In addition to the above functions, using the information of objects duration provided by database can set the reminding function automatically to make the system more complete.

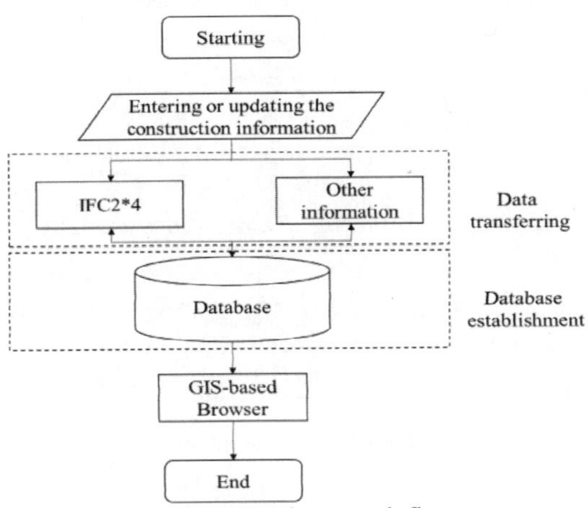

Figure 1. System framework flow

To the use of the object period, the use of the database has been established for comparison and remind the user in the jumping windows by C# program when closing the limited period of using, as the figure 2 shows.

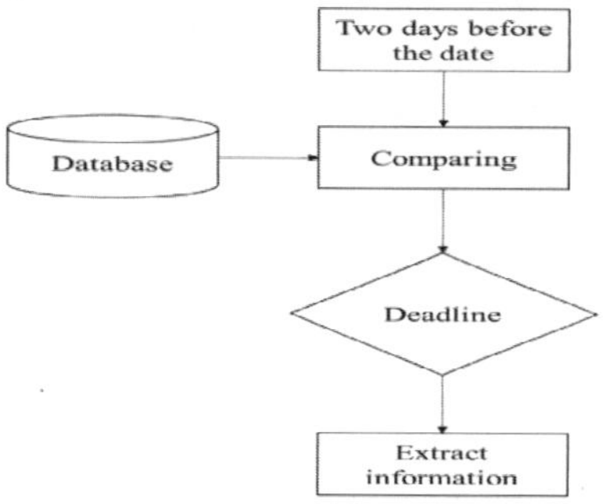

Figure 2. The flow of the automatic remind the object limited used duration.

3.2 System Function

The following figure 3 is the system function structure.

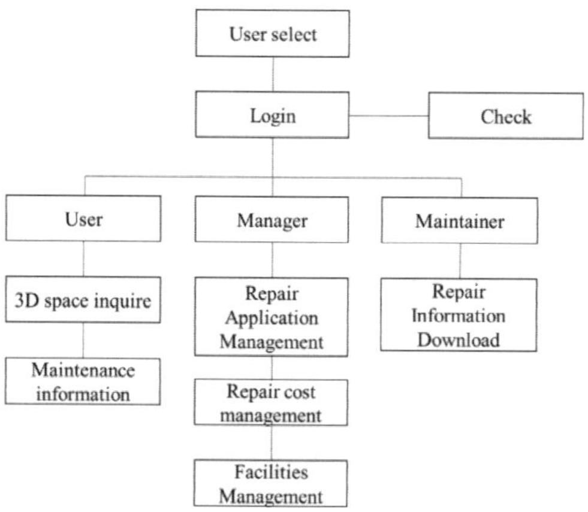

Figure 3. System function structure

3.2.1 Each Building Maintenance System

The building in this system can provide the relate base information by 3D model. User just need to click the item, and can get the information at the left hand side. Such as material, size, manufacture, produced date and the repair record…etc. As the figure 4 and 5 shown.

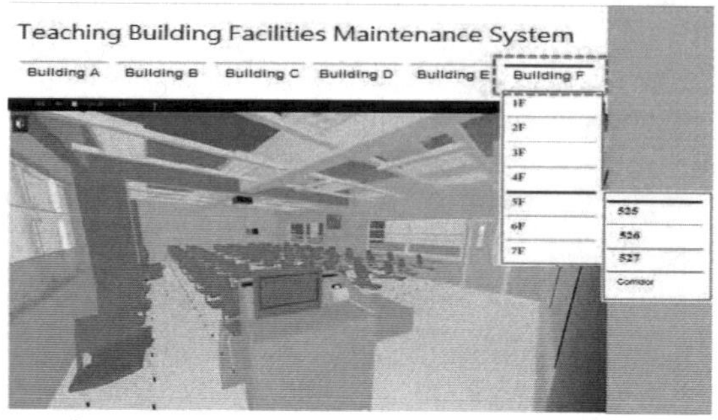

Figure 4. Click the classroom you want to check.

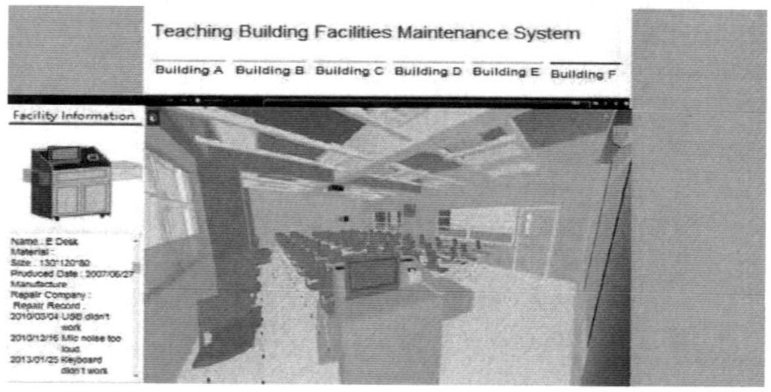

Figure 5. The information pop-up at the left hand side

Developed by C# program, the system will pop-up the window at the monitor bottom right corner to remind the user that today is the day before the deadline. As shown in figure 6 and 7.

Figure 6. Facility maintenance system automatic reminder function.

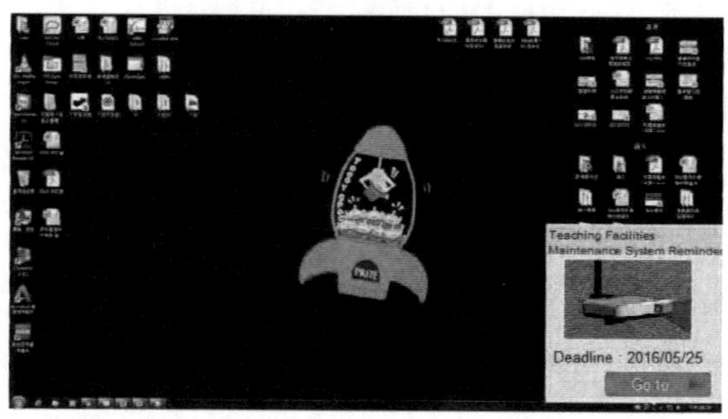

Figure 7. Auto remind function display position

4. Conclusion and Suggestion

Establishing the relevant maintenance system database, can let the staff to understand the maintenance of the past facilities, structure, material types and related records and other information and can automatically remind the facility that the item is about to expire. If combining building information model with maintenance management system in effect, it can make the staff record convenient and reduce the time of operation. It is not only can conform economy but also can replace the traditional disadvantages like paper records are easily defaced, missing and complicated login procedures and improve the efficiency of the overall maintenance of campus facilities. It also can achieve immediate, convenient and intelligent detection purposes, and then to achieve sustainable development goals.

References

[1] Wu Yi-Zhen and Xie Shang-xian. BIM opportunities and challenges in facility maintenance management. *Construction News Record*, No. 386, page 47-52, 2015.

[2] Li Meng-xing and Lin You-zheng. Application of BIM Model in Construction Stage of Construction Project. *Taiwan Highway Engineering,* Vol. 38, No. 3, page 45-54, 2012.

[3] Li Yi-qian, The *Application* of Building Information Modeling in Property Management, master paper, National Taipei University of Technology, 2010.

[4] Zhou Tian-ying, Cai Hui-ru, and Yan Lan Kun, New trend of 3D GIS Application: BIM (Building Information Model), *National Geographic Information System communication*, Vol. 73, page 45-51, 2010.

[5] Xu Rui-gai, Wu Yi-Zhen and Chen Yi-Zuo, Design and Implementation of Facilities Maintenance Management System with Open BIM Concept, *CECI Engineering Technology,* No. 138, page 46-55, 2015.

[6] Guo Zhe-yu, The Establishment of Electronic Building Construction Resume System, master paper, master paper, Chung Hua University, 2010.

[7] Xiao Le-dong, A study on the management strategy for the military camp maintenance, master paper, National Central University, 2004.

[8] Xie Bo-quan, Research for BIM in Architectural Lifecycle Process Application, China University of technology, 2011.

[9] Yen-Yu Hu, The Study of Building Information Modeling Applied in the Campus Maintenance System, Master paper, National Taipei University of Technology, 2016.

IFC of BIM Automatic Retrieving and Linking for Building Envelope Energy Efficiency Measuring - Ministry of The Interior Green Building Electronic Evaluation System in Taiwan

Kuo-Chung Wen and Wun-Bin Siao

Department of Architecture and Urban Design
Chinese Culture University, Taiwan
E-mail: wenkc@faculty.pccu.edu.tw, tse1005@hotmail.com

Abstract

Taiwan Green Building System was established in 1999 by the Government. The main axis was ecological, energy saving, waste reduction and health (EEWH). To enhance the effectiveness of promoting green building design in Taiwan. The Green Building Measure System (GBM) was completed by the Ministry of the Interior, Taiwan. If the GBM and Building Information Modeling (BIM) can be linking and automatic retrieving, it will be able to enhance Taiwan's green building design more effective.

The results of this study are converting Industry Foundation Classes(IFC) and explored the projects that can be automated linking. The program of automated link was build, and evaluated it proficiency. Building the GBM of BIM modeling guidelines, and completed a BIM digital building model. It provides EEWH more automatic to improve calculation of Green Building in Design.

Keywords –

IFC; BIM; EEWH; GBM; green building

1 Introduction

Induction the GBM for direct linking, program handling and manual identification to explore different automated solutions.

The issue of sustainable development is the national research projects around the world. And the sustainability is an abstract concept, it needs an evaluation tool to convert into quantized values. EEWH is one of the evaluation tools. British BREEAM, established in 1990, was the first green building evaluation tool of the world[1][2]. LEED was established by United States in 1996[3][4] . Taiwan's "Green Building Assessment System" was established in 1999 by the Government, and it has become an important part of sustainable national policy. Taiwan Green Building Assessment System is based on Ecology, Energy Saving, Waste Reduction and Health[5][6]. EEWH consists of five categories that includes Basic type, Community, Factory, Old Building Improvement, Accommodation. There are characteristics of Scientific quantification, design priority, parity technology, simple operation. The green building policy was greatly advanced, because it was supported by government[6] . Eastman et all brought up the concept of architectural description system basing on the proper classification of engineering data. Then, the issues and research was gradually being attention and discussion [7] . BIM technology was developed by industry and academia of Taiwan in 2002. It was used in the design, construction, collaboration, operation and engineering stages.

To enhance the effectiveness of Taiwan green building assessment system and green building designing. Construction and Planning Agency Ministry of the Interior amend the Green building standard of building technical rules and design technical rules since 2012. The GBM was completed by planning, review and testing. It simplifies the green building design calculation process and time. And it just fills in the basic data and necessary calculation data, qualifying or not will be appeared immediately. GBM can correctly calculate and effectively examination the green building implementation effectiveness. Information integration and communication advantages make BIM technology development and application increase popularity in engineering with the promotion of BIM (Building Information Modelling). It gradually replaces the complex project management and integration of the work, reduce manpower resources, and improve work efficiency. BIM can be extended to design stage, proposing the better building project and energy saving with the development of visual simulation software for building physical environment analysis[8][9]. How to effectively apply BIM to do information management and building effectiveness assessment of building life cycle, improvement construction efficiency and providing solutions to reduce building energy consumption is the motivation for this study.

Building information modeling can effectively integrate the data and information of building, and build information that can be used internal access to the project. Therefore, this study will link the GBM and BIM. So that the GBM can obtain calculation data from BIM. It can enhance the effectiveness of the Green Building Assessment System and Green Building designing in Taiwan. Therefore, following is the research project: First, study the materials report of the BIM, including the number, material, nature, district, floor, type; second, GBM look-up table assessment, including EEWH specification project and component analysis; third, BIM data extraction, how to fill object data into the report data, and how to identify and output the composite component data; fourth, link the GBM and BIM.

2 Review

2.1 EEWH

From the substantive level of the definition of green building in Taiwan is using the least earth resources, manufacturing the least waste of the building, and from the positive level is based on human health and comfort, pursuit global environmental symbiosis and the Sustainable Development of Human Life Environment, accomplish ecology, Energy saving, waste reduction, healthy building [6]. The EEWH is not just about planting trees and greenery, it's about the comprehensive and the systematic of the living environment and emphasis on architectural environment design concept of the symbiosis with the global environment.

Taiwan green building evaluation system is mainly about ecology, energy saving, waste reduction, health, and particularly, scientific quantification, design priority, parity technology, simple operation. The grading method of the EEWH is Integrated, designer can choose from affordable technology to accomplish green building construction, not only to ensure maximum design flexibility and freedom of technology, but also prevent over equip and over invest. Taiwan green building evaluation system is divided into five parts, respectively, Green Building Evaluation Manual- Basic Version, Green Building Evaluation Manual for Eco-Community, Green Building Evaluation Manual for factory, Green Building Evaluation Manual for Building Renovation, Green Building Evaluation Manual for Residential Building. THE grading evaluation system in EEWH can be separated into five percentage sector, 95% Diamond,

80%~95% Gold, 60%~80% silver, 30%~60% bronze, under 30% is certified[6].

2.2 BIM

Computer assisted design is developed from computer assisted drawing towards BIM, reducing information delivery errors, and promote cooperation in building-related industries. In 2008, BIM is introduced by Eastman, he thinks that BIM is digitalized Architecture Design, Construction and Facilities management, to achieve communication and interworking. According to United States National Building Information Modeling Standard, there are four levels to elaborate BIM: BIM is a physically facility and functional characteristics expressed in a digitized manner; share information related to this facility; provide a reliable basis for all responses throughout the life cycle of the facility; Embed, extract, update modify information, Embed, extract, update and modify information can be supported and responded to their respective responsibilities in different stages of the construction.

BIM model trough tree dimensional, Immediate, dynamic, parametric programs, to increase the capacity of building design and construction [10][11][12]. In the process of creating the building information model, the information includes geometric, spatial relevance, geographic information, quantity and attribute data of buildings. With the help of the Building Information Model tool can strengthen the planning and management of construction programs[13].

With the development of visual simulation software for building physical environment analysis, BIM can extend from design stage to evaluation stage, and propose a better solution with energy efficiency goals[14][15]. Green BIM is based on three dimensional BIM, using combination of information and dynamic image system, and analyse architecture efficiency and program correction to achieve the best energy-saving goals[16][17]. Green BIM's important elements includes LEED、Material Data、CFD、GIS、Lighting / Daylighting、Energy、Structure、MEP, these elements affect each other and integrate with BIM Model, to understand how these elements affect designs like structure, mechanics, energy consumption, Lighting and other factors. For instance, daylight analysis can decide building position and forms, the openings may affect mechanical facility and alternative energy assessment. In response to design goals or assessment needs, users can build their own model application strategy.

The current BIM building energy analysis is based on the United States LEED, but due to the climate, regulations and other factors, it will be inaccurate to use it on Taiwanese architecture envelope energy measurement. If using Ministry of Interior Green building electronic assessment system and combination of BIM, it will enhance the effectiveness of the Green Building Assessment System and Green Building in Taiwan.

2.3 IFC

IFC is a standard building information exchange file format, provides the industries to share or exchange information between software, an IFC information file represents a project module, which contains building life cycle planning, design, construction, management and maintain information that is shared at all stages, including basic units, construction sites, buildings, building floors and related building elements, through information sharing and exchange, it creates an object-oriented database [18][19]. IFC describes the information by Express language, by using object-oriented technology for open information, the software company can exchange building information standard file format through IFC, and develop its application tools, furthermore, Providing research on the relevant units in the construction industry and develop through the element form for BIM validation, this step is important for checking the drawing information, in the drawing stage, it will inevitably be some missing code, height setting error or component setting error, by setting parameter of the form function check(表單功能檢核), and the drawing is thoroughly check if an error is occurred or missing, when the model data is found to be problematic, Then return to the software to do the amendment.

3 Method

3.1 GBM

Green Building Measure System is built by Excel, which is a computer assisted software for EEWH. It includes full chapter 17 in Building Technical Regulations[20] . The following introduction will be function, assessment process, and operation. EEWH provides related information, print full evaluation table, customize code to manage building materials data. It can improve work efficiency.

As following Envelope energy saving step. First, fill in envelope calculation table. Second, build a basic database. Third, import Uar (average thermal transmittance) evaluation calculation table, choose decree application date, enter the number of data-roof and the description of building. Forth, build HWs (solar heat gain rate) and Gri (reflection rate of visible light) calculation table. Fifth, enter Aeq (equivalent transparency area), choose the position, floor and factor ki. Sixth, enter position, the description of façade, the position of roof and length to calculate factor Aen. Seventh, the system automatically calculates factor Req and check reference value.

3.2 Modeling

This study use ArchiCAD to build 3D model, research case is a house in Taipei, the building area is 166.63m2. First, sort out the plans (Figure 1). Before building the 3D model, sorting out AutoCAD layers and import to ArchiCAD. As following building model process. First, set the height of floors. Second, import AutoCAD file. Third, create column, then modify the type and scales… etc. Forth, create wall, then modify the type and scales… etc. Fifth, create door and window, then modify the type and scales… etc. Sixth, create floor slab, then modify the type, scales and floor number… etc. Seventh, create stair, then modify the type, scales and floor number… etc.

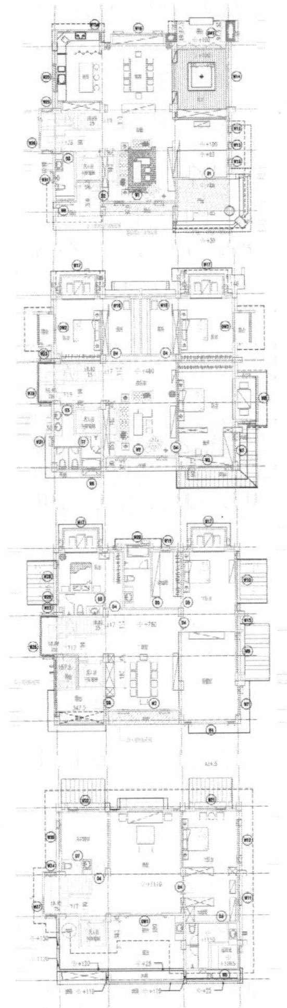

Figure 1 The plan of AutoCAD (from top to bottom:1F, 2F, 3F, 4F)

3.3 IFC2x3

IFC is an object-oriented technology for the basis of 3d graphic information, and the IFC2x3 is the most widely used version of the exchange format[21][22][23] . There are three types of format in IFC database format, respectively, IFC-SPF、IFC-XML and IFC-ZIP, the IFC-XML format is based on XML format, the format can take advantage of XML tools to operate model exchange and analyze, but there must be many definitions of tag in XML format in order to analyze, therefore, not many user using IFC-XML due to the large file. The IFC-SPF file format is pure text file, usually each string of text contains an object definition, since the data is stored with pure text, it has advantage of small file, so it is widely used IFC file format. Finally, IFC-ZIP is based on IFC-SPF compressed ZIP file.

4 Experiment & Results

4.1 GBM

This study is based on Envelope Energy-Saving amid Daily Energy Saving Index on evaluation of Envelope Energy saving includes roof average thermal transmittance (Uaw), solar heat gain rate (HWs), reflection rate of visible light, equivalent transparency area (Aeq), Req calculation and Evaluation of baseline values. The results of the calculation will be used as a basis for later corrections. The results of the calculation are as follows (Table 1). 1.The system will calculate the average thermal transmittance after roof construction details table - up. The design value for this example is 0.712, which is less than the reference value of 0.8. 2.The skylight has an angle of 100 degrees greater than 80 degrees. When the skylight elevation is greater than 80 degrees or Hwa is less than 1.0m2 will not be assessed. So, the study case HWs need not be evaluated. 3.This study case uses double glass. Gri value of 0.15, less than the reference value of 0.25. It is in line with regulations. 4.Wall average thermal transmittance table value of 2.144, less than the standard value of 3.5. It is in line with regulations. 5.Aeq assessment items must review the window type, Window area, orientation, floor and shade form. It needs to pay attention to shade forms such as: horizontal, vertical, grid, unilateral vertical and permanent -style shading types. And Calculate the outer shading coefficient. The Aeq of this study case is 303.33m2. 6.Req calculations and benchmarks to review Req = Aeq / Aen.

Table 1 The project of energy saving assessment

evaluation items	standard value	Design value	compliance audit or not
average thermal transmittance	0.8	0.712	O
HWs (solar heat gain rate)	-	Free assassment	X
Gri (reflection rate of visible light)	0.25	0.15	O
Average heat transfer rate of walls	3.5	2.144	O
Aeq (equivalent transparency area)	-	303.33m^2	-
Req (Calculation and normal value review)	13%	33.55%	X

4.2 BIM Model

The research case of this study is a house in Taipei, plans shown below Figure 2. The case model is LOD300, which includes wall, door, window, floor slab, column and beam. Pay particular attention on wall and window during the modeling process to connect with GBM data efficiently. According to the unit of GBM, set up the element unit to meter, it makes data modifying much more uncomplicated.

4.3 Linking function

Component, quantity and spatial table. The major review of envelope energy saving is

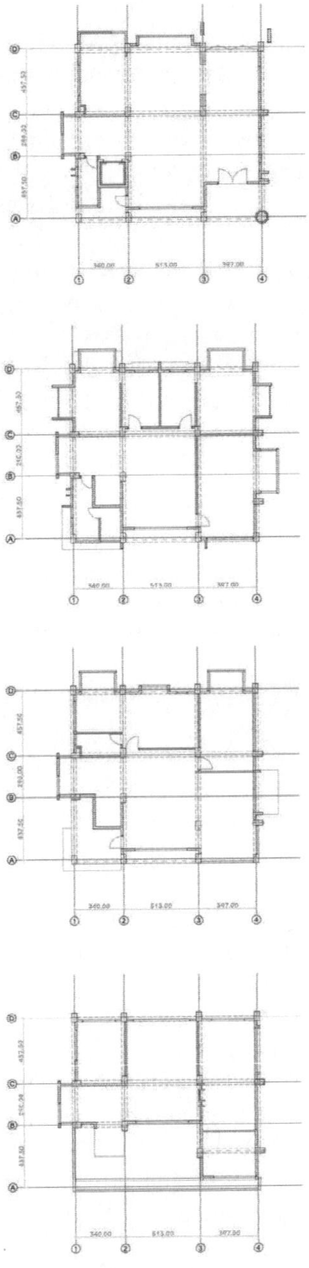

Figure 2 3D model plan of BIM (from top to bottom:1F, 2F, 3F, 4F)

Equivalent transparency area (透光部位等價開窗面積), shade form(遮陽形式), in terms of the output, the report is based on walls and windows. The external wall component report is shown in Table 2, wall area calculates the requirement of Aen, but the wall must be picked out. Table 3 lists the window elements, width and height calculates the requirement of Aeq, but the report data unit shows centimeter, while GBM requires meters, so the unit conversion is required.

Table 2 Wall element (part)

Wall type	Volume [m^3]	Thickness [m]	Height [m]	Area [m^2]	Perimeter [m]
Plywo od	0.87	35	345	0.29	2.35
Plywo od	0.95	35	345	0.32	2.5
Concr ete	0.05	15	25	0.22	3.64
Concr ete	0.04	15	130	0.58	8.4
Concr ete	0.15	15	130	0.28	4.1
Concr ete	0.15	15	130	0.31	4.4
Concr ete	0.95	15	25	3.82	51.64
Concr ete	0.66	15	120	0.55	7.6
⋮					

Table 3 Window element (part)

Opening Name	2-Sash Sliding Window 19	Double Window 19	Glass Block Wall 19
Quantity	1	1	2
W x H Size	80.00x60.00	210.00x60.00	45.00x240.00
Sill height	180	90	0
Head height	240	150	240
2D Symbol		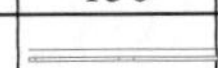	
View from Side Opposite to Opening Side			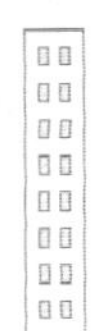

among seven projects that needs to be evaluated on envelope energy saving, only Equivalent transparency area Aeq and envelope arca can bc linked via excel. When calculating the link in Aeq, due to the difference of the report data unit between Archicad and GBM, therefore the report data unit of the window element has to be converted to meters, the outer shading coefficient still needs to fill in the calculation. While Aen calculates the link, the external wall must be picked out, then sum up the area, and the Aen can be found.

4.4 GBM link BIM

As mentioned above, there are currently only two items that can be linked directly to excel on the seven projects of envelope energy saving. And GBM and BIM linking is also the two main projects. Other parts need to manually identify the plan, it will be discussed at the next stage of the study. The steps for linking GBM to BIM are as follows: First, the BIM model is output in IFC format and open with excel (Figure 3); Second, using excel formula to retrieving the information, such as the height and width of the window (Figure 4). Finally, link the retrieving data to GBM excel.

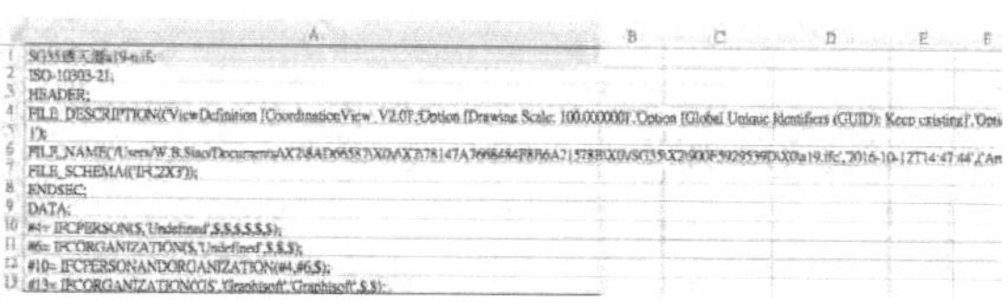

Figure 3 IFC format open with excel.

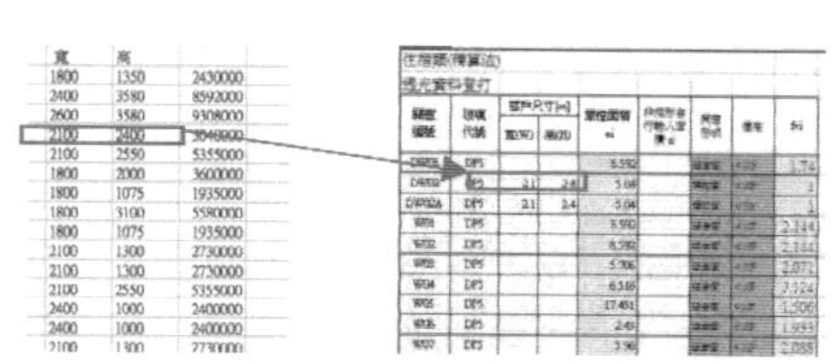

Figure 4 Retrieving the information with excel link to GBM.

5 Conclusion

There are 7 item need to calculation for Building envelope energy efficiency measuring of EEWH. Using GBM to calculation these items can save a lot of time. There are a lot of options and values to choosing in GBM, and it can simplify the calculation process.

In the process of linking GBM and BIM, only Aeq and Aen can be linked. Points for attention, GBM's units is meter. It should be unified in the settings of the report output. When Aen linking, the external components must be selected out, and then summing the area.

It is feasible to link GBM and BIM, but the degree of connection is not as expected. The future will be a graphical identification of the study, so that the system can be identified through the graphics and select the calculated value. Expect to increase the degree of link.

References

[1] W. Lee and J. Burnett, Benchmarking energy use assessment of HK-BEAM, BREEAM and LEED, Build. Environ., vol. 43, pp. 2008, 1882-1891.

[2] Y. Roderick, D. McEwan, C. Wheatley, and C. Alonso, Comparison of energy performance assessment between LEED, BREEAM and Green Star: Eleventh International IBPSA Conference, pp. 27-30, Citeseer, 2009.

[3] U. G. B. Council, LEED rating systems, Retrieved September, vol. 11, pp. 2008, 498.

[4] U. G. B. Council, Green building design and construction: LEED reference guide for green building design and construction, US Green Building Council 2009.

[5] K. Yang and R. Hwang, An improved assessment model of variable frequency-driven direct expansion air-conditioning system in commercial buildings for Taiwan green building rating system, Build. Environ., vol. 42, pp. 2007, 3582-3588.

[6] 內政部建築研究所, 綠建築評估手冊, pp. 2015.

[7] C. Eastman, D. Fisher, G. Lafue, J. Lividini, D. Stoker, and C. Yessios, An outline of the building description system, Research Rep, vol. 50, pp. 1974.

[8] 陳建中, 臺灣 Green BIM 綠建築資訊建模應用架構研究, 2015.

[9] 邱秀婷, 綠色建築資訊模型在永續節能設計決策之應用, 逢甲大學, 台中市, 2015.

[10] R. Volk, J. Stengel, and F. Schultmann, Building Information Modeling (BIM) for existing buildings—Literature review and future needs, Autom. Constr., vol. 38, pp. 2014, 109-127.

[11] C. M. Eastman, C. Eastman, P. Teicholz, R. Sacks, and K. Liston, BIM handbook: A guide to building information modeling for owners, managers, designers, engineers and contractors, John Wiley & Sons 2011.

[12] S. Azhar, Building information modeling (BIM): Trends, benefits, risks, and challenges for the AEC industry, Leadership and management in engineering, vol. 11, pp. 2011, 241-252.

[13] W. Kymmell, Building Information Modeling: Planning and Managing Construction Projects with 4D CAD and Simulations (McGraw-Hill Construction Series): Planning and Managing Construction Projects with 4D CAD and Simulations, McGraw Hill Professional 2007.

[14] S. Azhar, A. Nadeem, J. Mok, and B. Leung, Building Information Modeling (BIM): A new paradigm for visual interactive modeling and simulation for construction projects: Proc., First International Conference on Construction in Developing Countries, pp. 435-446, 2008.

[15] S. Azhar, M. Khalfan, and T. Maqsood, Building information modelling (BIM): now and beyond, Construction Economics and Building, vol. 12, pp. 2015, 15-28.

[16] E. Krygiel and B. Nies, Green BIM: successful sustainable design with building information modeling, John Wiley & Sons 2008.

[17] J. K. W. Wong and J. Zhou, Enhancing environmental sustainability over building life cycles through green BIM: A review, Autom. Constr., vol. 57, pp. 2015, 156-165.

[18] T. Froese, F. Grobler, J. Ritzenthaler, K. Yu, B. Akinci, R. Akbas, B. Koo, A. Barron, and J. Kunz, Industry foundation classes for project management-a trial implementation, Journal of Information Technology in Construction (ITcon), vol. 4, pp. 2002, 17-36.

[19] V. Bazjanac and D. Crawley, Industry foundation classes and interoperable commercial software in support of design of energy-efficient buildings: Proceedings of Building Simulation'99, vol. 2, pp. 661-667, 1999.

[20] L. ZONG-JIE, Study of Electronic System of Taiwanese Green Building Assessment Reports, pp. 2014.

[21] N. S. Zadeh, L. Lindberg, J. El-Khoury, and G. Sivard, Service Oriented Integration of Distributed Heterogeneous IT Systems in Production Engineering Using Information Standards and Linked Data, Modelling and Simulation in Engineering, vol. 2017, pp. 2017.

[22] E. E. Achkar, A BIM-integrated approach to Construction Quality Management, pp. 2017.

[23] F. Abanda, B. Kamsu-Foguem, and J. Tah, BIM–New rules of measurement ontology for construction cost estimation, Engineering Science and Technology, an International Journal, , pp. 2017.

Modular vs Conventional Construction: A Multi-Criteria Framework Approach

A.W.A Hammad[a] and A. Akbarnezhad[a]

[a]School of Civil and Environmental Engineering, The University of New South Wales, Australia
E-mail: a.hammad@unsw.edu.au, a.akbarnezhad@unsw.edu.au

Abstract

The use of modular construction methods for projects offers significant time and environmental improvement relative to conventional construction methods. Currently, there is a lack of appropriate assessment approaches to capture the differences between modular and conventional construction. This paper proposes a framework to aid decision makers in choosing between the latter construction methods through the integration of building information models with material libraries, project schedules and machinery inventory. A fixed set of performance parameters, whose attributes are shared among both construction methods, are defined to allow for a reasonable comparison across multiple criteria, including embodied carbon, productivity and total construction costs expended. The dynamics of the project are incorporated by modelling the various stages of the project within a building information model. The proposed framework is tested on a realistic case example, highlighting its applicability as a decision support tool for construction method selection.

Keywords –

Modular Construction; Construction Methods; Prefabrication; Sustainable Construction; BIM

1 Introduction

The construction sector is known to occupy a significant share of the annual GDP of economies worldwide. The industry is also known for its great deal of environmental breaches. In particular, according to the intergovernmental Panel on Climate Change (IPCC), the construction sector is classified as one of seven dominant sectors responsible for substantial greenhouse gas (GHG) emissions [1]. With the growing awareness of the importance of sustainable construction, such that economic, social and environmental factors are accounted for concurrently, many construction organisations have started to implement greater sustainable measures in their design and construction processes. One of the most critical determinants of the economic and environmental performance of a project is the construction method adopted [2]. As an example, modular construction is known to have major economic advantages compared to conventional construction [3].

Modular construction can be defined as a construction system where volumetric components forming a completed part of a building are produced off-site and transported to the construction site for installation [4]. It is the highest end of prefabrication, where full sets of buildings are constructed off-site, instead of only smaller components being produced during the manufacturing process [5].

Some studies suggest the dominance of semi-prefabrication methods over conventional construction approaches in terms of minimising construction related embodied carbon [6], and greenhouse gas emissions [7]. Though there are studies that have also attempted to contrast separately the economic [8] or environmental impacts [6], [9] of construction methods, no attempt has however been made to contrast volumetric construction methods such as modular with conventional methods using an encompassing approach. A study undertaken to highlight an exact quantification approach for comparing between modular and conventional construction across multiple criteria is therefore lacking.

The idea of modular construction has been around since the 1960s, though its effective adoption as a construction method has not been as wide as was expected [10]. It is always challenging to incorporate changes to the conventional construction methods, which have been widely tested and used for long periods [11]. Having said that, plenty of the research conducted indicates that a controlled environment, such as the one presented in modular construction methods, where components are manufactured off-site in a continuous flow process, allows a lot of benefits to be achieved. These advantages can be realised in terms of productivity, safety and environmental performance [4], [12].

Motivated by the lack of a systematic procedure to compare the trade-offs between modular and conventional construction methods, this paper proposes a methodology for integrating Building Information Modelling (BIM), equipment libraries and environmental databases, within a framework for use in construction method selection, both for residential and commercial buildings. The analysis forming the framework is based on the life cycle cost assessment, where economic considerations are embedded along with the common environmental factors, to quantify the overall building performance during all stages of a project's life excluding its operational phase.

The rest of the paper is organised as follows: Section 2 introduces the major components making up the proposed framework. Section 3 presents a brief case study on a realistic project for showcasing the applicability of the proposed framework. Results are discussed in Section 4. Finally, concluding remarks and future work recommendations are presented.

2 Construction Method Comparison Framework

This section describes the different modules making up the proposed framework. Figure 1 depicts the overall framework which is comprised of 4 principle modules; the first set of modules incorporates the input data required for the analysis; the second module is a BIM representation of the project to be constructed and which aggregates the input data. Associated with the BIM is a Comparative Module that specifies the different material composition that would be required for each construction method adopted. The last of the modules is a Criteria Analysis Module which embeds the conditions necessary to perform a comparison between the construction methods.

2.1 Input Module

Data is required to assess the differences between the construction methods available. Modular construction relies on an environment which is drastically different to that of conventional construction. As a result, differences would arise in the project information regarding site preparation and building layout, material to be used, equipment deployed during the construction process, carbon factors of the associated material with each construction method, and the cost element related to the building components. Each of the aforementioned information forms a separate database which get utilised by the constructed BIM.

2.2 BIM

As a form for linking the various information from the input database, a BIM is associated with each construction method. Depending on the size of the project and its use type, different construction methods will need different material compositions and hence the building model produced for each method can vary. Modular construction places large emphasis on the use of timber and light gauge steel framing, as opposed to brick and mortar in conventional construction. As a result, the BIM produced for each construction method will have to reflect this difference in element composition. Materials and equipment can be mapped according to the construction method use, and information for the latter databases can then be associated with the BIM of each construction method.

2.3 Comparative Module

The Comparative Module is where the BIM for modular and conventional construction is used to extract the necessary data required to evaluate the multiple criteria defined. The criteria is evaluated in a way where economic and environmental factors influenced by both construction methods can be assessed. A total of 3 criteria are incorporated within the Comparative Module of the framework, as discussed below.

2.3.1 Embodied Carbon

The first criteria, Eq. (1) minimises the total embodied carbon emissions. Operational carbon is neglected since it is assumed that irrespective of the construction method used, the building will be operated in the same manner; hence the operational carbon is expected to stay be the same for both construction methods.

$$\sum_{c} Q_c M_c + \sum_{t} E_t TD + \sum_{c}\sum_{e} R_{e,c} P_e \tag{1}$$

where Q_c is the quantity of material making up each building component $c \in C$, M_c is the emission factor associated with each material component due to its manufacturing phase within its life cycle, as obtained from emission databases, E_t is the emission factor associated with transportation truck t, D is the travel distance traversed by the truck t, T is the number of trips that need to be conducted by the associated truck, $R_{e,c}$ is the emission factor associated with equipment e deployed on building component c, while P_e represents the duration of operation of equipment e.

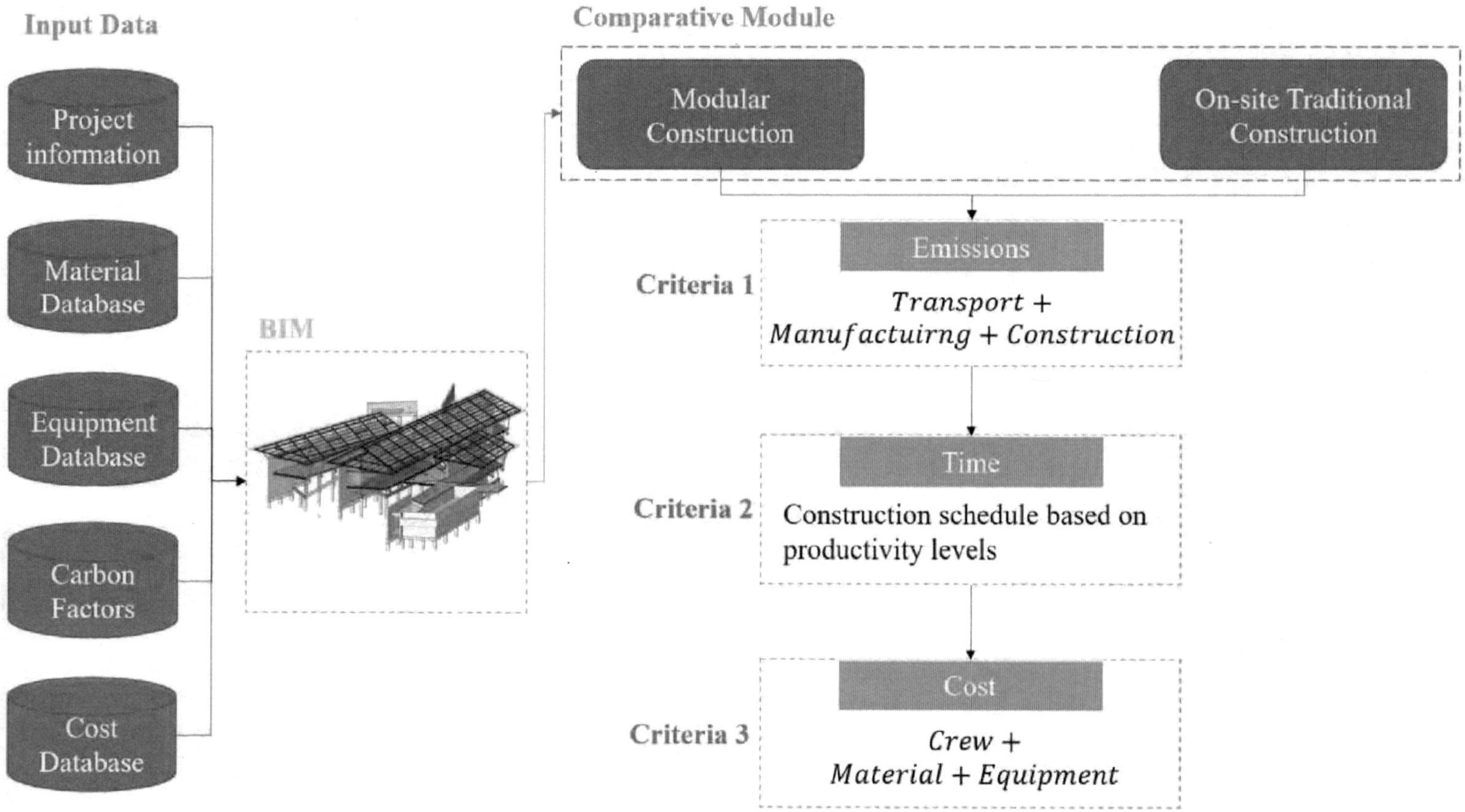

Figure 1. Framework for comparing between modular and conventional construction methods

216

If on-site construction is used then the distance parameter is calculated by summing the distance of travel between material suppliers i and construction site, d_i:

$$D = \sum_i d_i \qquad (2)$$

On the other hand, if modular construction is to be used then the distance is equal to that between the manufacturing place of modular components and the construction site $\overline{d}$:

$$D = \overline{d} \qquad (3)$$

2.3.2 Construction Project Duration

The different construction methods will have different project durations, due to the nature of operations and machinery used. To evaluate the schedule duration of each construction method, a second criteria is defined, Eq. (4).

$$\sum_e \sum_c \frac{Q_e}{F_{ce}} \qquad (4)$$

where Q_e is the quantity of material to complete activity e while F_{ce} is the productivity rate of crew c working on activity e.

2.3.3 Construction Cost

The third criteria, Eq. (5), computes the cost estimate of each construction method. Each building component is associated with a set of tasks.

$$\sum_q \sum_a N_{qa} CC_{qa}$$

where N_{qa} is the crew size of type q assigned to task a, which reflects the total number of machinery deployed on the respective activities, while CC_{qa} is the unit cost of crew q assigned to task a.

3 Case Example

A project in the North-West of Sydney involves the construction of a granny flat, with dimensions 14 m by 16 m. The builder has the choice of selecting between modular and conventional construction methods; each method is associated with a set of building components as shown in Table 1. The framework of Figure 1 is applied to compare between the two available construction methods. Embodied carbon emission factors are derived from databases such as [13], [14], whereas cost rates and crew productivity rates are obtained from RSMeans and Cordell [15], [16]. Common workflow patterns adopted for both construction operations are shown in Figure 2.

Table 1. Material Composition associated with each construction method

	Modular	Conventional
Interior Walls	Steel frame studs with 5 mm poly sheets	Partition walls with wood studs and 10 mm gypsum wallboards
Exterior Walls	PFC cage with light gauge enclosed framing and colourbond cladding	Double masonry units with bitumen insulation and brick veneer cladding
Floor	Purlin joists with CFC sheeting	Concrete slab on grade with tile finishes
Roof	Light gauge ceiling, with plaster ceiling panel, roof packers and corrugated ceiling sheeting	Timber framing with metal tiling

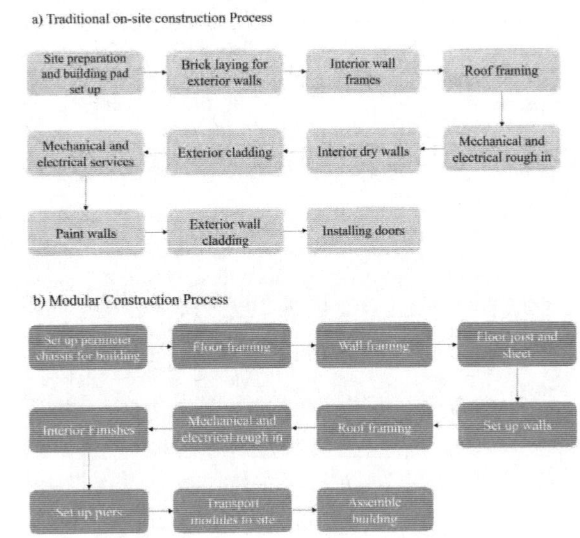

Figure 2. Workflow process of construction

A workflow is adopted to evaluate material use, crew requirement and resource utilization at each stage of the project. A representation of the BIM used to obtain all the required data for the project is given in Figure 3. Piers are assumed to be adopted as the substructure for supporting the granny flat in the case of modular construction, whereas a simple flat slab acts as the building pad for a granny flat constructed using on-site methods. Site preparation for both methods is assumed to be similar; as a result, the analysis excludes substructure construction.

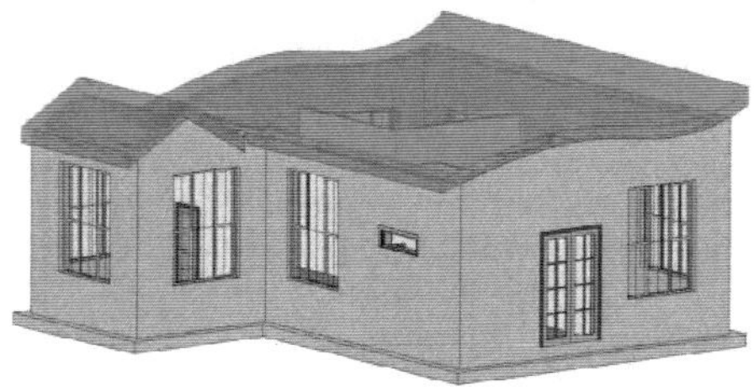

Figure 3. BIM model used for case study

4 Results and Discussion

Each criterion is analysed and the associated results are given in Figures 3 – 5.

4.1.1 Embodied Carbon Emissions

Figure 4 shows the difference in total embodied carbon of both construction methods, adopting the material components of Table 1. The resulting embodied carbon is a direct result of the quantities derived from BIM along with the embodied carbon factors obtained from databases listed above. It is important to note that a total of 302 $kg-CO_2/m^2$ and 378 $kg-CO_2/m^2$ of embodied carbon is associated with modular and conventional construction for the granny flat project, respectively. Overall, modular construction is responsible for 19% lower embodied carbon in comparison to conventional construction

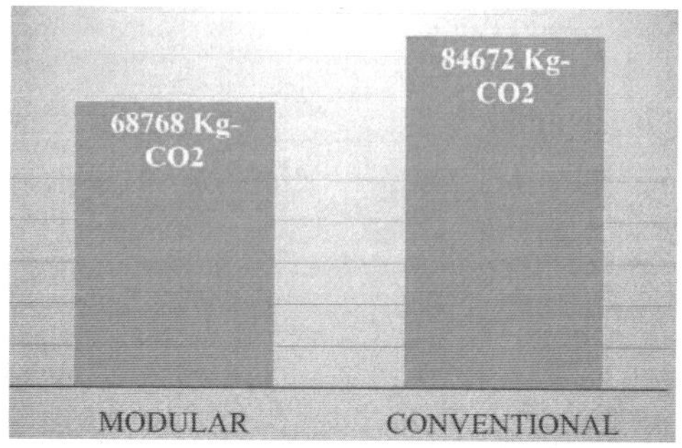

Figure 4. Comparing embodied carbon of the case study using modular *vs* conventional construction

A pie chart showing the average contribution of each material used in both construction methods to total carbon emissions is given in Figure 5. In line with the literature, timber has the highest embodied carbon levels followed by concrete.

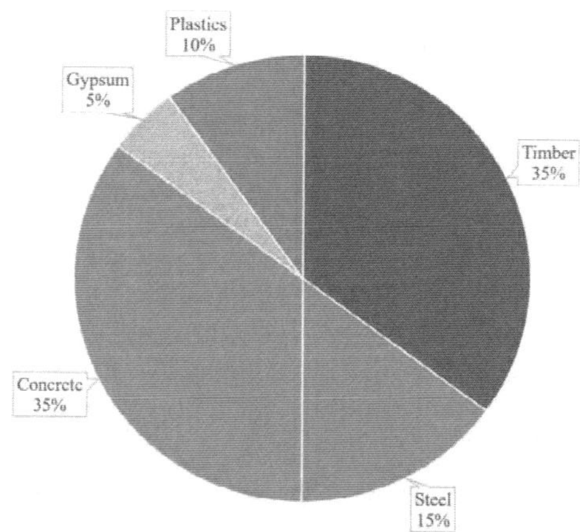

Figure 5. Contribution of material components to total embodied carbon in granny flat case study

4.1.2 Productivity

Based on the crew rate productivity estimates and on the material quantities derived from BIM, the total time taken to complete the project under each construction method is calculated. Results are shown in Figure 6. Modular construction is found to have a 58% advantage in terms of project delivery duration, compared to conventional construction. This is attributed to the influence of weather on work progress, where during rainy conditions, on-site work needed to stop. Such was not the case in the controlled environment of the modular factory.

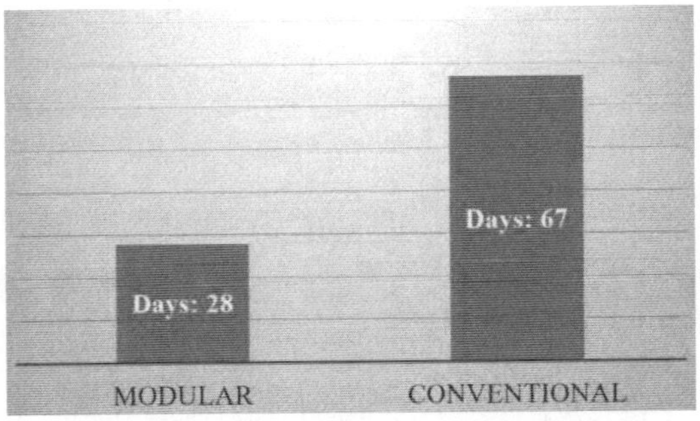

Figure 6. Comparing duration of project using modular *vs* conventional construction

4.1.3 Costs

Total cost estimates for both construction methods rely on the quantities from BIM and the rates associated with the material components. Such rates incorporate labour within them. Based on the flow process delineated in Figure 2, the procedure can be mapped to the cost rates obtained from RSMeans and Cordell. Figure 7 shows the cost estimates of both methods. Modular construction is on average cheaper than conventional methods for this case example. The reason behind this is due to the fewer materials used, and the automated building process in modular construction, which permits economies of scales to be realised.

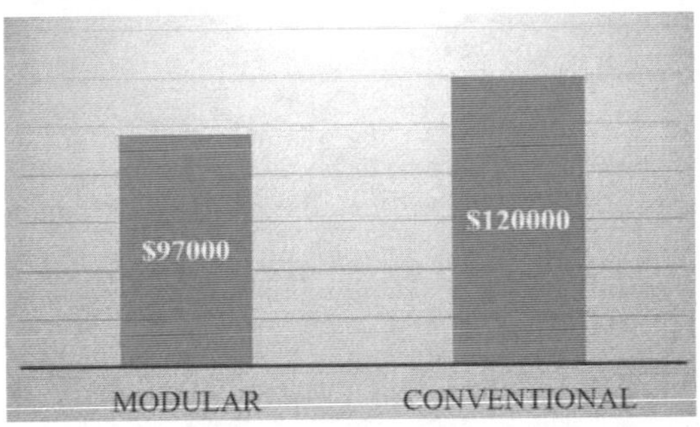

Figure 7. Comparing cost of project using modular *vs* conventional construction

5 Conclusion

With the growing awareness of sustainable construction, many companies have shifted their focus to sustainable design. This is particularly important given the impact of the construction sector on the economy and on the environment. One way suggested to improve the sustainability of the construction and building industries is through the use of modular construction. This work presented a novel framework to compare the modular construction method against a conventional one. The framework focused on evaluating three main criteria, namely embodied carbon, productivity and monetary cost of the project. Effectiveness of the proposed framework was confirmed by application to a practical construction project. Modular construction was found to be more effective in terms of lower embodied carbon, higher productivity and lower construction costs. Future research will focus on incorporating additional parameters and criteria to compare between the construction methods.

References

[1] Intergovernmental Panel on Climate Change, "IPCC Fourth Assessment Report (AR4): Climate Change 2007," 2007.

[2] Y. Chen, G. E. Okudan, and D. R. Riley, "Sustainable performance criteria for construction method selection in concrete buildings," *Autom. Constr.*, vol. 19, no. 2, pp. 235–244, Mar. 2010.

[3] V. Ezcan, U. Isikdag, and J. S. Goulding, "BIM and Off-Site Manufacturing: Recent Research and Opportunities," in *UCLan - University of Central Lancashire*, Brisbane, Australia, 2013.

[4] Harvey M. Bernstein, John E. Gudgel, and Donna Laquidara-Carr, "Prefabrication and Modularization: Increasing Productivity in the Construction Industry," National Institute of Standards and Technology, 2011.

[5] Alistair G.F. Gibb, *Off-site Fabrication: Prefabrication, Pre-assembly and Modularisation*. Wiley, 1999.

[6] J. Monahan and J. C. Powell, "An embodied carbon and energy analysis of modern methods of construction in housing: A case study using a lifecycle assessment framework," *Energy Build.*, vol. 43, no. 1, pp. 179–188, Jan. 2011.

[7] C. Mao, Q. Shen, L. Shen, and L. Tang, "Comparative study of greenhouse gas emissions between off-site prefabrication and conventional construction methods: Two case studies of residential projects," *Energy Build.*, vol. 66, pp. 165–176, Nov. 2013.

[8] A. Faghirinejadfard *et al.*, "Economic Comparison of Industrialized Building System and Conventional Construction System Using Building Information Modelling," *J. Teknol.*, vol. 78, no. 1, Dec. 2015.

[9] J. Quale, M. J. Eckelman, K. W. Williams, G. Sloditskie, and J. B. Zimmerman, "Construction Matters: Comparing Environmental Impacts of Building Modular and Conventional Homes in the United States," *J. Ind. Ecol.*, vol. 16, no. 2, pp. 243–253, Apr. 2012.

[10] V. W. Y. Tam, C. M. Tam, S. X. Zeng, and W. C. Y. Ng, "Towards adoption of prefabrication in construction," *Build. Environ.*, vol. 42, no. 10, pp. 3642–3654, Oct. 2007.

[11] J H Rankin and R Luther, "The innovation process: adoption of information and communication technology for the construction industry," *Can. J. Civ. Eng.*, vol. 33, no. 12, pp. 1538–1546, Dec. 2006.

[12] N. Blismas, C. Pasquire, and A. Gibb, "Benefit evaluation for off-site production in construction," *Constr. Manag. Econ.*, vol. 24, no. 2, pp. 121–130, Feb. 2006.

[13] R. Frischknecht and G. Rebitzer, "The ecoinvent database system: a comprehensive web-based LCA database," *J. Clean. Prod.*, vol. 13, no. 13–14, pp. 1337–1343, Nov. 2005.

[14] G. P. Hammond, C. I. Jones, and web-support@bath.ac.uk, "Embodied energy and carbon in construction materials," *Proc. Inst. Civ. Eng. - Energy*, vol. 161, no. 2, pp. 87–98, May 2008.

[15] *Cordell Commercial/industrial Building Cost Guide*. Queensland: Reed Construction Data, 2008.

[16] RSMeans, *RSMeans Cost Data*. John Wiley & Sons, 2012.

Utilizing Building Information Modeling and Radio Frequency Identification in Recording and Preserving Historic Buildings

Author1[a] Kemo Sima Author2[b] Tsung-Juang Wang PhD

[a]Department of Architecture and Urban Design, National Taipei University of Technology, Taiwa
[b]School of Design, National Taipei University of Technology, Taiwan
E-mail: simaala@yahoo.com, tjwang@mail.ntut.edu.tw

Abstract

Preserving irreplaceable historic structures is an important factor in architecture design studies. The environmental, cultural and economic benefits cannot be overemphasized. We have an obligation to record and preserve the history of each generation because the buildings change with us. Historic buildings are not too complex to benefit from the advantages of building information modelling (BIM). BIM and RFID (Radio Frequency Identification) integration serve as a powerful tool useful across the lifespan of a building. RFID tag is a memory storage device for storing a certain amount of data—highly effective in decentralizing information management of historic buildings. The tags are attached to different components of the building that are of historic relevance. This information are accessible wirelessly providing the ability to process large volumes of multiple data sets simultaneously. Historically, buildings have been better and more accurately described in drawings rather than words and BIM is an extension of that process, particularly for existing infrastructure. Where 2D CAD improved on paper drawings by adding the ability to describe a building in layers, BIM and RFID allows the process to be taken further by increasing the efficiency of facility inspection, maintenance and data collection. Rudimentarily, BIM models are used to document historic buildings prior to any development, stabilization or investigative work are conducted in order to record crucial material evidence. Understanding how a building has changed over time is helpful in determining the level of deterioration. Data gained from investigations of a historic building are well documented and fed into RFID tags. In this study, options and decisions necessary to establish a BIM and RFID for recording and preserving historic buildings are discussed along with a case study of Bopiliao Historic Block in Wanhua District, Taipei. As foreign students, when we discorvered the physical, political and social threats on Bopiliao Historic Block, it is evident that it needs saving from various threats because it holds so much of historic value to the community and the country at large. A comprehensive data analysis was conducted to assimilate the current method used in preserving the historic structures.

Keywords –
BIM, historic buildings, preservation, RFID, Bopiliao Historic

1 Introduction

Historic preservation is the process of maintaining and managing change to a heritage asset in a way that sustains and where appropriate enhances its significance [1]. Heritage sites are living monuments & record of certain happenings and this is our real connection to our past. It proves the existence of our ancestors [2]. For years, historic preservationists have faced challenges in conserving some important historic structures. Governments, public and preservationists face dilemmas of conflicting interests that endangered many historic buildings. It is rare to have a single right answer, so adopting a consistent, rigorous process and adequate information are crucial to reaching publicly-justifiable decisions. This topic triggers series of factors to be considered howbeit, these does not change the importance of protecting historic elements. Preservation of historic buildings is a one-way street. There is no chance to renovate or to save a historic site once it's gone.

This reality brings to light the importance of locating and saving buildings of historic significance—because once a piece of history is destroyed, it is lost forever:

- People value this historic environments as part of their cultural and natural heritage. It reflects the knowledge, beliefs and traditions of diverse communities.
- It gives distinctiveness, meaning and quality to the places in which we live, providing a sense of continuity and a source of identity.
- It is a social and economic asset and a resource for learning and enjoyment. Each generation should therefore shape and sustain the historic environment in ways that allow people to use, enjoy and benefit from it, without compromising the ability of future generations to do the same. [3].

Building information modeling and radio frequency identification (BIM and RFID) seeks to integrate processes in preservation of historic buildings. The focus is to create and reuse consistent data storage and digital information by the stakeholders and transferring this information from generation to generation. According to UK conservation principles 2015, conservation involves people managing change to a significant place in its setting, in ways that sustain, reveal or reinforce its cultural and natural heritage values [3]. Is critical to the realization of preserving important historic buildings by utilizing this proposed conceptual approach to track changes by using a multi-disciplinary building information modeling and radio frequency identification. This will enhance the efficiency of historic building preservation and the authenticity of historic documentation.

2 Background

In order to limit the broad scope, this paper focused on documenting and management, maintaining/preserving historic buildings. Interoperability of BIM in different building life cycle (LC) stages and functionalities still is limited but studies have shown that it could fit perfectly with RFID in constant coordination in providing information especially in preservation of historic monuments. BIM functionalities require a certain accuracy, information richness and actuality of the underlying data to fulfill their purposes A frequently mentioned concept to describe information richness of BIM objects is 'Level of Detail' or also referred to as 'Level of Development' [4]. The scope and potential of BIM is ever-increasing as a result of new and IT-enabled approaches to facilitate design integrity, virtual prototyping, simulations, distributed access, retrieval, and maintenance of project data between multiple disciplines. An important aspect of historic buildings that makes them important is how the elements emanate to be historic [the story] and that makes them worth preserving. Therefore, systematic and consistent information management–detailing the existence of a historic monument through times is a cardinal factor inspiring technology intervention to amplify the efficiency.

3 Problem Statement and Research Gaps

Several researches in the use of BIM concentrates in new architectures and rarely about preserving and maintaining historic architectures. A connection between historic preservation and BIM&RFID integration is important in both history and historic building preservation. Application of such technology in preserving historic building is very fundamental in preservation process. This study focuses on creating a rigid system that uses consistent data storage and retrieval in a form of digital information sharing with different stakeholders in historic building preservation— through a rigorous literature reviews, there is indication that this system has not been considered in (HBP).

In our previous research where an analyst of other systems of historic building preservation was done, there are some inconsistencies and inefficiencies in the way they operate especially data management, interoperability and ease of use. BIM and RFID integration is deemed to be of high efficiency in building information management throughout the life of the building. Due to a mainly interactive and time-consuming data capturing, processing and creation process, BIM modeling effort is high and thus BIM is often not applied in existing buildings yet, but the worth of preserving historic structures over shadows the complex nature of the system.

4 Research methodology

A theoretical review and analysis from previous literatures on existing building was used along with a case study of Bopiliao Historic Block in Wanhua District, Taipei to provide options and decisions necessary to establish a BIM and RFID for recording and preserving historic buildings. Several visits to the block was done to collect some information and an interview with the director of Center for Heritage and Culture Education have been conducted to document the state of the historic block, to understand what methods have been used in recording, documenting and preserving the block.

4.1 Understanding BIM&RFID

What is building information modeling (BIM) and Radio frequency identification (RFID)? Building Information Modelling is emerging as a method for creating, sharing, exchanging and managing the information throughout the lifecycle of buildings. Radio Frequency Identification (RFID), on the other hand, has emerged as an automatic data collection technology, and has been used in different applications for the lifecycle management of facilities [5]. BIM and RFID integration functions as a central data depository for storing and recalling scores of information about buildings by creating an intelligent 3D model with data. BIM is not just utilized for 3D visualization but also imparts incredible intelligent functions. RFID is proposed to be used to store lifecycle and context aware information taken from a BIM during the lifecycle as a distributed database. RFID tags forms a loop between building components and BIM Data.

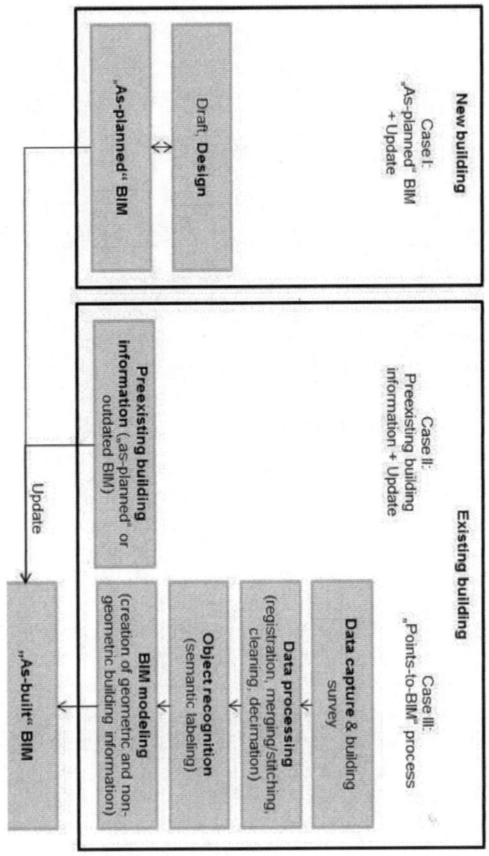

Figure 1. BIM creation processes for new and existing buildings, partly from Huber [6].

An RFID system consists of a transponder (tag), transceiver and antenna or coil. The tag is composed of a computer chip with an internal memory in which a limited amount of information can be stored. The set of transceiver and antenna is called reader and it can read and may write data from and on the tag holding information about the object attached to it by generating an electromagnetic field. This technology aims to decentralize objects' information by making it available wherever the component exists [7].

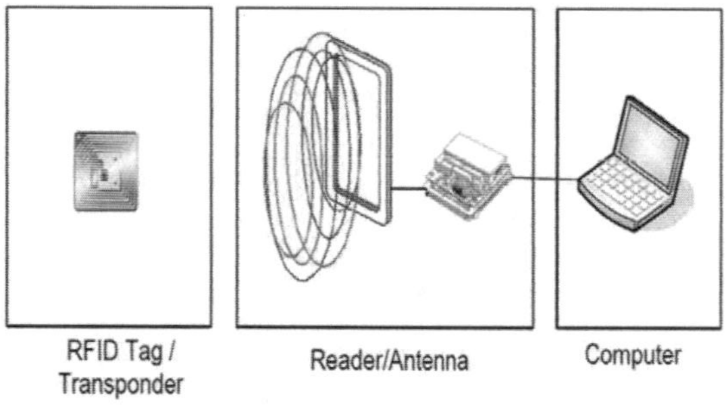

Figure 2. *RFID system* [8].

4.2 Informational issues and interoperability

Expert functionalities are linked with a BIM model through Information Delivery Manual (IDM) frameworks and Model View Definitions (MDV) providing relevant information, facilitating data exchange and avoiding ambiguities [9]. BIM process is used for new building rather than for existing building. This is related to the fact that existing buildings have many constrains to respect, considering the will to preserve the correspondence between the parametric model and the existing structure. Adding to this is the interoperability [10]. Existing buildings have a number of major difficulties than new buildings, but the use of new technologies mentioned above gives the opportunity for professionals to manage data, querying the 3Dmodel through an iterative and interactive steps that characterize the BIM process.

This system will assist BIM and RFID integration in monitoring and recording details changes of historic buildings thereby maintaining their originality. In his deductions, Becerik-Gerber, recent technological advancements in field data acquisition technologies such as 3D Laser Scanning, Photogrammetry, Sensors, Radio Frequency Identification (RFID), and Tablet PCs have offered opportunities to address some of the disadvantages of current data collection practices. Data collected through such methods are stored in RFID tags on either buildings or specified historic components. Also, an interface is created that support information transmission and availability to different stake holders either encrypted or in open access format. The stored data can be as simple as an ID number or can be more complex.

4.3 Technical Issues

Geometrical and topological information of historic sites has to be gathered, modeled and complemented by semantic property/attribute information manually. If a reliable data capture technique could provide an as-built BIM at reasonable time and cost, existing buildings could benefit from BIM usage e.g. regarding documentation, visualization or facility management to gather and model actual building conditions…..To create an as-built BIM. Building from scratch detailing every aspect of a historic element. Materials used, design aesthetics, façade types and every fabric of a historic edifice that has historic importance are simulated and incorporated into historic data. A comprehensive 3D, 2D CAD files with corresponding materials evidence of changes on historic buildings are captured, attached to a timeline and fed in to respective RFID tag.

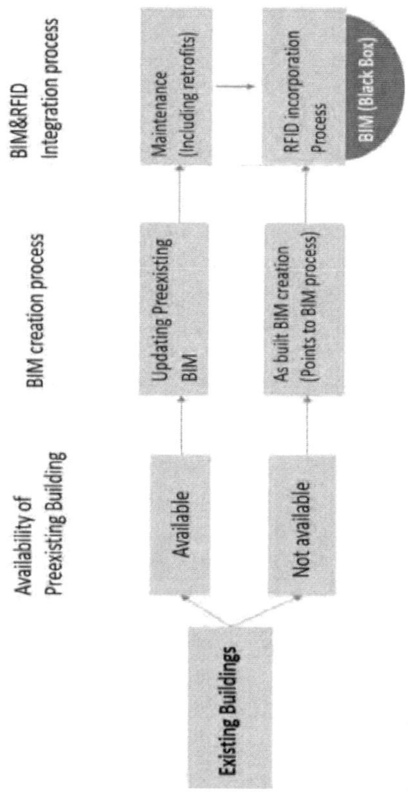

Figure 3. BIM&RFID model creation processes in existing buildings depending on BIM availability.

4.4 Case study

The need for utilizing BIM&RFID technology in (HBP) cannot be over emphasized. Preserving historic structures is very vital to modern day architects, researchers, historic preservationists and the future generation. As a result, a system like BIM&RFID integration that safeguard historic buildings, data authenticity is highly encouraged to be employed. One of the factors influencing the value proposition of BIM in industry is the efficiency of interoperability solutions that provide flawless and streamlined information flow between different disciplines in a project, [9]. Historic data about Bopiliao Historic Block are mostly 2D cad drawings, word files, 2D images, deteriorating hard copy images.

4.4.1 Case Study, Bopiliao Historic Block

Bopiliao Historic Block is one of the key attractions of Wanhua District that portrays some of the oldest and most historical architectures in the modern city life of Taipei. With a long history dating back to 1799 when it was first known as "Fupiliao" (福皮寮街), this street has been revamped, and officially opened to the public in 2009 again with a new appearance. Today, it serves not only as a preserved attraction popular among tourists but also a site that reminds and educates the locals of their traditional culture and history. This includes a wide range of traditional shop houses from the Qing dynasty (1683 - 1895) to buildings constructed during the Japanese occupation, (1895–1945) till early post war periods. [11].

According an interview with the officials on the site, the director of Center for Heritage and Culture Education, Mr. Chang reiterate that the management of the historic block took two forms i.e. preservation and rehabilitation.

a) Preservation focuses on the maintenance stabilization, and repair of existing historic materials and retention of a property's form as it has evolved over time. Preservation only concentrates on

b) Rehabilitation acknowledges the need to alter or add to a historic property to meet continuing or changing uses while retaining the properties of historic character.

Bopiliao historic block is a multicultural environment that upholds vital Taiwanese history in different periods making it a critical historic monument. The government diversified the use of the space for art exhibitions, shooting movies, wedding pictures and even advertisement videos; as a result boomed tourism in the area. Different opinions and demands impacted the evolution of the historic block.

Before adding a timeline to the historic block for preservation purposes, you must understand different period on the timeline and finally make conclusion base on reasons. Mr. Huang, restoring the block will depicts its characteristics at a particular period of time in history while removing evidence of other periods. This made restoration work tedious and sometimes impossible to undertake. Some of the varieties of wall-types used over the life of the buildings includes, bamboo net walls plastered with hay-treated mud shown in figure.6a, to-ga (adobe), which is a mixture of rice straw as in figure.6b bricks and concrete. At a communal level and without the intervention of well-seasoned, skilled historic restoration experts, preserving the structure can get quite unwieldy. The preservation work today involves some uncanny methods like steel encasement as in figure.5b or adding extra layers of concrete as in figure.5c.

A closer look at the buildings on the block, it is evident that there are different material types used. Due to different rehabilitated phases or remodeling, different materials are been used on the facades. E.g. Cement, facial washed terrazzo, face bricks etc. —the combo material used depicts the evolution of the historic block over time. Some walls are made of wood, others of stone brick, and some are made of to-ga (adobe), which is a mixture of rice straw, lime, clay, and shredded old hemp sacks. The historic Block displays a lot of traditional architecture from Qing Dynasty includes wall engravings, clay embellishments, etc., Japanese colonial era onwards and has been preserved to this day (figure 4).

4a 4b

4c

Figure 4. clay embellishments dated back since Ching dynasty

Figure 5. Samples of bricks and stones that were used on the block over centuries are still kept.

5a

5b 5c

Figure 4. mixed of materials used on the historic building on different stages of renovation, steel frames were used to support the structure

6a 6b

6c

Figure 5. wall variety used over the life of the building includes, bamboo net walls plastered with hay-treated mud, adobe walls, bricks and concrete.

Restoring the block to fit the current need or preserved to its original state is always a challenge and hard to come to a consensus among government, public and experts. One of the most important challenge is to get the required experts to restore the historic blocks to original form. Poor data management and record keeping has made this task very difficult. Mr. Chang, sometimes some parts of the buildings sustain damages inside, only when under renovation we could detect it. There is always going to be a problem if the damage needs expert to resolve the issue. Oftentimes, the maintenance is hard to be undertaken due to budget restriction. In most cases, we modify that part of the building despite clear restrictions of historic preservation. He added that it's really a good idea to have an efficient system, a technology that could make preservation and data management work easier. Once I faced a plumbing problem, I asked for a detail drawing of the utilities. But were very misleading and inaccurate, stopped me from fixing it because I fear making mistakes and end up destroying important parts of the structure.

BIM technology could be used to virtually document the transition of this historic block through all phases of rehabilitation and obtained captured data are then transmitted to RFID tags for easy retrieval and usage. Utilizing BIM to virtually remodel the block according different time lines by collected data either through laser scanning, 3D integrated survey, hard copy images, old building models, descriptive text files, etc. Updating building systems in historic structures requires striking a balance between retaining original building features and accommodating new technologies and equipment. Building system updates require creativity to respect the original design and materials while meeting applicable codes and user needs. The system is could be used as an approach to track changes within a multi-disciplinary building information modeling environment could save the historic structures and a 400 years historic data that is on hand. Understanding what is at stake, the case study could be conclude that the local government is still learning and adopting new methods and technologies to enhance public infrastructure development in Taiwan and there is still a room for incorporating BIM&RFID technology in historic building preservation.

5 Conclusion

Although there must be a balance of consideration between the old and the new, historic structures with clear modern application through adaptive reuse resulting in cultural enlightenment of a community should be maintained, restored, and rehabilitated as part of historic preservation. From the case study and literature reviews, it is evident that there is a need to apply BIM&RFID framework in pursuit of augmenting historic building preservation process. The case study data shows that the methods adopted in preserving Bopiliao historic block is neither effectual nor constructive. The record keeping system could be updated to accommodate an efficient data collection and preservation system. BIM&RFID integration if utilized could revitalize historic data management, keep tabs on deteriorating old infrastructure. According our findings, the management faces financial challenges in adopting new methodologies. Understanding what is at stake, the government and stake holders should assist in providing support because;

a. Preserving historic buildings plays a great role in cultural education. It teaches history that happened before we were born and promotes the knowledge of those who lived in different times and different societies. Architectural monuments cultivates pride of our past and heritage making a community unique in the world. Historic buildings gives character and certain charm to a neighborhood that people live in.
b. Economic factor could serve an important merit for keeping the old. Architectural monuments are great attractors of tourists. Everyone likes to experience the "spirit" of a place, which most often is represented through architecture and history behind it. Tourism provides jobs to locals with could earn them extra income. Restoring and preserving locally important historical structures generates workplaces for both local people, researchers, student and international experts. Preserving historic buildings require more skills and knowledge than simple building and thus result in higher salaries and also contribute to boosting the economy.

The challenges faced in maintaining and preserving historic monuments could be revolutionized by adding cutting-edge technology that minimizes or erase human defects and errors. Integration of RFID and BIM could create a chance to salvage our about-to-be-lost historic masterpieces by effective and efficient data management technology.

6 Acknowledgement

We would also like to thank the following experts who were involved in compiling data for this research project: Urban Planning Commissioner, Po-Wen Huang of Taiwan Urban Planning and Construction Agency, Ministry of Interior and Mr. Chin-Peng Chang Director of Heritage and Culture Education Center of Taipei City. Without their passionate participation and input, the data collection and interviews could not have been successfully conducted.

Reference

[1] Drury, P. J., & McPherson, A. (2008). Conservation principles: Policies and guidance for the sustainable management of the historic environment. London: English Heritage.

[2] The global, participative, engagement platform for everything heritage,. (2014). Supported by UNESCO accessed December 10, 2016 http://www.gounesco.com/why-preserve-heritage-sites/

[3] Bruce-Lockhart, (2008). Conservation Principles, Policies and Guidance, English Heritage.

[4] The global, Eastman, C. M. (2008). BIM handbook: A guide to building information

modeling for owners, managers, designers, engineers, and contractors. Hooboken, NJ: Wiley.

[5] Motamedi, A. S. Sateyeshgar, M. M. Soltani, A. Hammad,. (2013). Extending BIM to incorporate information of RFID tags attached to building assets

[6] Volk, R., Stengel, J., & Schultmann, F. (2014). Building Information Modeling (BIM) for existing buildings — Literature review and Whole Building Design Guide (WBDG)., (2016). A Program of Building Sciences, accessed March 22, 2017 https://www.wbdg.org/design/historic_pres.php

[7] Hajian, H., & Becerik-Gerber, B. (2009). A Research Outlook for Real-Time Project Information Management by Integrating Advanced Field Data Acquisition Systems and Building Information Modeling. Computing in Civil Engineering (2009). doi:10.1061/41052(346)9

[8] Motamedi, A. (2009). Improving Facilities Lifecycle Management Using RFID Localization and BIM- Based Visual Analytics. Montreal, Quebec: Concordia University

[9] Venugopal, M., Eastman, C., Sacks, R., & Teizer, J. (2012). Semantics of model views for information exchanges using the

[10] Matteo., Anna Osello., Torino, C.so Duca degli., (2013). BIM FOR CULTURAL HERITAGE Department of Structural, Geotechnical and Building Engineering (DISEG) – Politecnico di Abruzzi 24, 10100, Torino, Italy

[11] Taipei City Government accessed Novermber 17 2016, http://english.gov.taipei/ct.asp?xItem =1084524&ctNode=29490&mp=100002

.

Constructing a MEP BIM Model Under Different Maintenance Scenarios—A Case Study of Air Conditioning

C.W. Feng[a] and P.C. Wu[b]

[ab]Department of Civil Engineering, National Cheng Kung University, Taiwan
E-mail: [a]cfeng@mail.ncku.edu.tw, [b]j6820706@gmail.com

Abstract –

Facility managers usually need to make decisions under different types of MEP system breakdowns. They usually fight to find the information required to restore the system. However, the BIM model developed usually lacks information to satisfy the what-if analysis under various maintenance scenarios. This study first takes the air-conditioning system as the example to analyze the causes and the solutions of the common system breakdowns. Then an ontology is developed to determine the rules of developing BIM model to satisfy the information requirements for conducting simulation for different maintenance scenarios. In addition, a series of Dynamo modules are developed as the implementation to quickly develop the simulation-ready BIM model. Result shows, with this proposed approach, the BIM model can be used to help facility managers make decisions efficiently and effectively.

Keywords–

Building Information Modeling (BIM); Dynamo; Facility Management (FM); Ontology; Air Condition System

1 Introduction

Management of MEP (Mechanical, electrical, and plumbing) maintenance has gradually been valued and studied. Sullivan et al. [1] divide the general maintenance model into the routine maintenance and urgency maintenance. Routine maintenance has a large number of related documents, and the maintenance methods are easy to get and learn. There is no need for experience and emergency for managers. However, the breakdown maintenance has to be completed in the shortest time. It heavily depends on the professional judgment of maintenance engineers. But many data become more difficult to search with the complexity of the building. At the same time the important equipment shut down will affect the products delivery time. Therefore, experienced managers can finish the works with the best solution in the limited resources.

The professional field in buildings includes architecture, design, hydropower, air conditioning, etc. The overall fields lack an efficient system for management. Nowadays, more and more managers use the BIM model as a platform to coordinate or storage information. Wang et al. [2] mentioned that the most of the research focus on the design and construction rather than the operations and maintenance. The front-end design errors or unresolved problems often affect the maintenance management. The biggest difficulty is that the building of the model doesn't meet the back-end requirements. These problems lead to the maintenance staff must extra build the BIM model to meet the conditions in the breakdown maintenance. They also delay the maintenance works and increase the costs of resources. Even if the managers modify the existing model, they still have to spend a lot of time.

In recent years, many studies use different software and management methods to solve these problems. For example, Tserng et al. [3] produced a database of MEP components that based on real installation. They provide a framework of classification for components that includes installation, geometrical information, and other issues. The way improves the back-end asymmetry information, and indirectly provides manufacturers the components' rules for drawing plans. But there are thousands of MEP system components. How to balance the time of modeling and the data attributes, which can meet the requirements of maintenance, is still a major issue in the future.

In summary, a lot of the BIM models are not satisfied the back-end requirements and components' types, which make the managers hard to simulate the situation. Therefore, this study first uses the ontology to analyze the actual situation of the components. Then we analyze the maintenance data to obtain the appropriate detail of components in the BIM model. This study combines the BIM models and Dynamo modules to quickly create a model with maintenance information. Finally, we can enhance the maintenance efficiency and reduce the waste of resources through the simulation of maintenance processes.

2 Literature Review

2.1 Problem Statements

The frequency of breakdown maintenance is lower than routine maintenance. But, if the managers not complete the works in a limited time, it will result in a lot of problems. In particular, there are a large number of high-tech factories or financial buildings and each facility in the buildings are interrelated. The managers' primary purpose is to complete the maintenance work in the shortest time. According to the interviews, we can say that senior managers or engineers can efficiently use resources. Most of the new staff can only do secondary positions to indirectly learn about the professional knowledge of urgency maintenance.

At the same time, the study also pointed out what the experience gap is. The study found that the solutions are proposed by the experienced equipment managers are more comprehensive and consistent with the actual situation. Senior managers can sort out all the resources to make judgments and provide multiple solutions. However, although a small number of new staff can make the right decisions, the processes are relatively simple and direct. They don't take the other possibilities into account. In addition, the internal components of a large-scale equipment are very complex and even need to take into account the site construction conditions, parts disassembly sequence, mobile path, personnel activities, etc. In order to integrate the above problems before construction, the successes depend on the knowledge and experience of managers' judgment [4].

BIM is a unified platform for engineering information, but it cannot be effectively used in the operation and maintenance. We find that the reason is that the model usually only meets the current construction situation. The modeling phase doesn't consider the back-end requirements in detail and fail to effectively provide the appropriate model to the owners or maintenance staff. A lot of data is still passed with drawings, which lead to information loss and hard to search. All of these make the equipment managers more difficult to implement [5].

2.2 Facility Breakdown Maintenance

The equipment lifecycle can be classified into eight stages by the United States OSHA 29CFR 1910.119-PSR (Process Safety Management):

1. Requirement Description: Designers need to understand customer requirements.
2. Engineering Design: According to customer demand for equipment, designers set its industrial parameters.
3. Detailed Design: It's a more detail design than before.
4. Production: Designers send the draws to manufacturers and product the equipment.
5. Acceptance: The staff check equipment parameters and safety indicators that meet customers' requirements and regulatory restrictions.
6. Enabling: The staff transport the equipment to the destination and check the installation.
7. Operation and maintenance: It includes any maintenance work until demolition.
8. Demolition: Owners depend on equipment life, economic benefits, and other factors to change it.

We can find that the real equipment life is between the enabling and maintenance, as shown in figure 1. The enabled process is different because of the device itself, and the warranty will start on the same day. Therefore, the operation and maintenance phase belongs to the vast majority of equipment life. The good maintenance can extend the life of the machine and improve the efficiency.

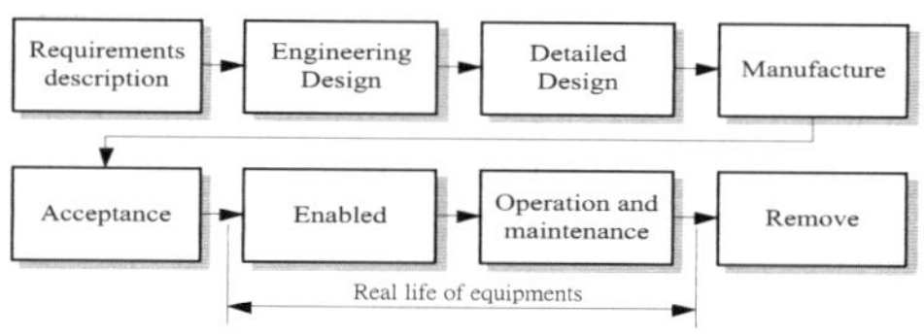

Figure 1. Life cycle of equipment

The International Facilities Management Association defined that the facilities management is: "Facility management is a profession that encompasses multiple disciplines to ensure functionality of the built environment by integrating people, place, process and technology." The American Parliament Library also defined it: "Facilities management is a systematic integration of the operating environment, the operator, and the operations. Its contents include management, architecture, equipment science and behavioral science integration." From the above, we can understand the maintenance works not only handle the equipment but also includes various fields of science. That is why maintenance is complicated and difficult to manage.

2.3 BIM for Facility Management

Nowadays, BIM becomes more and more popular and the public construction in Taiwan has gradually asked the contractor to have BIM technology. But according to Becerik-Gerber et al. [6], the study shows that the use of BIM stages can be divided into design, construction, and operation. The gray lines represent the people who already use BIM, and the percentage in gray lines mean their degree of using BIM at each stage. The black lines represent those people who have not yet used BIM and their percentage of interest in BIM development. We can find that BIM is mostly used in the design and construction, so most of the research also focused on these two stages. However, the operation has a great gap of 36%, as shown in figure 2.

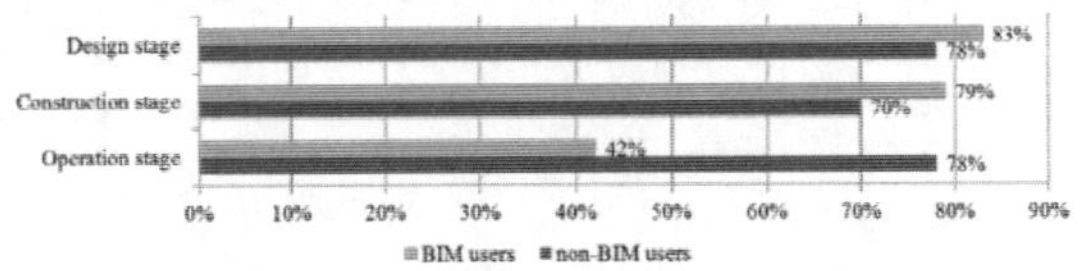

Figure 2. Project stages that BIM is currently used for or is planned to be used for [6]

The study integrated the difference and found that maintenance management has the following difficulties with using BIM:

1. Unclear and invalidated benefits of BIM in FM practices (for example, unclear productivity gains, or benefits gained from reduced equipment failure and better-automated building energy usage).
2. The amount of work that needs to be done to define the specific FM needs for which a model is necessary and how that model may need to be prepared to meet the needs.
3. Lack of interoperability among BIM solutions and between BIM solutions and FM systems.
4. Lack of demand for BIM deliverables by the owner community due to the uncertainty about what BIM might be used for.
5. Lack of clarity about responsibility for insurance and contracts.
6. Lack of standardized FM tools and processes.
7. Facilities management personnel have limited experience with BIM technology.

JJ McArthur et al. [7] also integrated four challenges related to the combination of BIM and maintenance management which must be overcome:

1. Identifying the critical information required for sustainable operations.
2. Managing information transfer between the BIM model and other FM tools.
3. Managing the level of effort to create the model.
4. Handling uncertainty where building documentation is incomplete.

Through the above research can realize that most of the model ignore the actual use of the back-end. In addition, some of the modelers are lack of the maintenance knowledge, which resulting information cannot feedback to the front.

2.4 Ontology Application on BIM

The ontology originated from the ancient Greek philosopher who defined it as "the science of the existence of objects." It's also the meaning of the existence of objects in nature. We can use conceptual models to describe the relationship between objects. The purpose of developing the ontology is to construct a platform to share special information among people and software. It can become more mature with the people using the ontology [8].

Nowadays, the use of ontology has not been popularized in the MEP construction yet, but there is still lots literature about the ontology in different construction management. In 2014, Zhang et al. [9] proposed a platform to connect security managers and BIM software by using ontology model, as shown in figure 3. The study can automatically infer the potential risk factors between the works, and it presents the inferred results into the BIM model. Finally, they show the difference after the change through the 4D simulation. For example, when the high-rise building is in the process of assembling the templates, the possibility of waste or wood falling is higher than usual. Therefore, if the time allows, managers should try to avoid the renovation of the external walls. After inferring the results, the study demonstrates the time schedule in a 4D model to avoid the occurrence of the above items falling in advance. Finally, the BIM model has been improved the safety of object falling and expanded construction information.

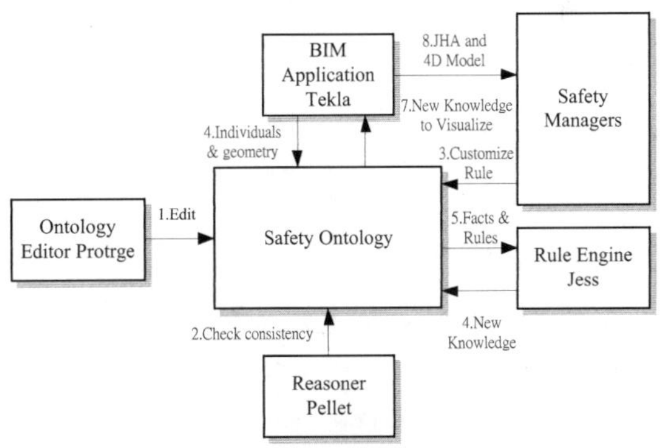

Figure 3. System architecture of the ontology-based hazard identification application in BIM [9]

3 Methodology and Results

3.1 Study Process

First, this study establishes an ontology model for breakdown maintenance to link with the BIM model. The entire workflow can be divided into three major parts. The first part is the development of ontology model, and it includes the A1, A2, and A3. The second part is a combination with BIM and Dynamo, and it includes the A4 and A5. The A6 part simulates the results and outputs the form to check.

In step A1, the real-world text descriptions are obtained by the laws and the equipment, and we need to analyze the maintenance requirements. A2 classifies the data and defines the scope of the ontology model, and it classifies the knowledge into the class. The data properties can express the relationship between classes and individuals which represent the components in the BIM model. A3 uses the degree of BIM modeling detail that come from the ontology model. The components should be cut or created to meet the maintenance simulation, and the classes also need to build in the same way. A4 creates a Dynamo module that can quickly modify or add maintenance information with the variable value. After building the maintenance information which is needed to be determined, it is suitable for delivery by COBie. If it is not suitable for COBie, A6 will show the conflicts that do not meet the requirements. Finally, these processes can automatically produce the results and reduce the human judgments. This study successfully improves the efficiency of back-end urgency maintenance and eases the burden on facility managers.

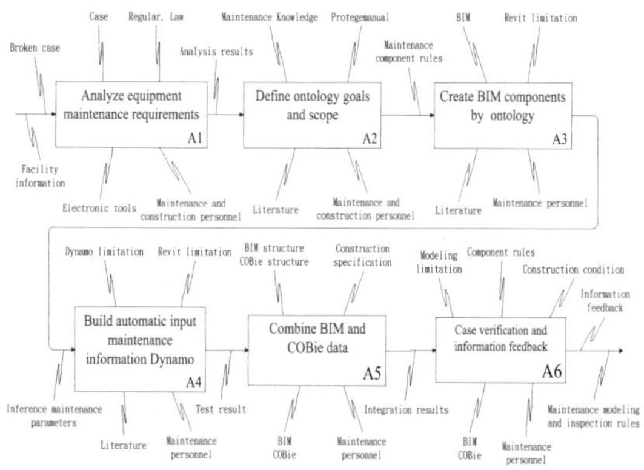

Figure 4. IDEF0 of urgency maintenance

3.2 Information Requirements of Maintenance

3.2.1 Breakdown of Components

Through the literature review and expert interviews, this study will use OmniClass table21 as the main structure of the air conditioning system classification. There are 7 classifications: transportation system, water supply system, drainage system, HVAC system, fire fighting system, communication system and security system. This study chooses the table because of the building services and the real situation of maintenance. In this study, we found that the classification is not detailed enough for some of the real maintenance situations. Therefore, the basic maintenance requirements come from the BIM model, COBie, and real situations requirements, and the fifth layer of the coding will be completed by them. Take the central cooling system for example, the classification will add the chilled water exchanger, warm water exchanger, compressor, motor, inlet and outlet, control panel, support frame, exhaust pipe, suction pipe and others under the OminiClass number 21-04 30 30 10, as shown in table 1. This framework can provide the BIM model for cutting or creation components, and the classes of the ontology also follow this rule.

Table 1. OmniClass classification of HVAC

OmniClass	Level 1	Level 2	Level 3	Level 4	Level 5
21-04 00 00	Service				
21-04 30		HVAC			
21-04 30 30			Cooling system		
21-04 30 30 10				Central cooling	
21-04 30 30 10 10					Chilled water exchanger
21-04 30 30 10 11					Warm water exchanger
21-04 30 30 10 12					Compressor
21-04 30 30 10 13					Motor
21-04 30 30 10 14					Inlet and outlet
21-04 30 30 10 15					Control panel
21-04 30 30 10 16					Support frame
21-04 30 30 10 17					Exhaust pipe
21-04 30 30 10 18					Suction pipe
21-04 30 30 10 19					Other componets

3.2.2 Maintenance Ontology Model

In practice, maintenance methods rely on the past experiences of maintenance staff. However, this kind of information has a lack of uniformity and standardization, and it will indirectly cause missing of data in the transmission and difficulty of learning.

In this study, the semantic information will be transferred into a classified knowledge which is used by the Protégé. First, maintenance staff will get the statement from the clients: "Chiller does not work". In the classification, the "chiller" is from the chiller system, and its category also includes compressor, control panel, suction pipe, exhaust pipe, chilled water exchanger and other components. The "not work" is the failure situation which may be "compressor failure" or "host setting abnormal" and so on. According to the broken situation, it will be associated with the HVAC components, but the "broken situation" and "components" do not belong to the same class. Therefore, we must establish a new relationship "relating to" between the "broken situation" and "components" to meet the description of the facility managers' semantics.

This study uses the way, which is mentioned above, to analyze the statement and divide the ontology into two parts, as shown in figure 5. The upper part is the BIM model and the lower part is the maintenance ontology model, and then we connect the two parts through the "individual". Therefore, the carriers are the components in the BIM model and "individual" in the ontology model. In order to increase searching speed and object independence, we connect two models by coding system of BIM model. According to interviews and literature review, if the model wants to meet the real situations, it must completely consider overall maintenance events. First, the "broken system" has "broken situation", and "air conditioning system" has "air conditioning components". The "broken situation " will have their own associated "HVAC components" in each class. In addition, some of the situations that the facility managers will provide a suggested method for broken components. The facility managers choose the appropriate "maintenance method" to handle the "HVAC component" based on the broken situations. The relationship between the two is generated by the object property "has method".

Finally, the broken situation and other factors will determine the "maintenance resources" that the "HVAC component" needs. The maintenance resources are divided into "component information", "maintenance path", "maintenance people", "maintenance tool", "transport devices" and "warranty date". There is a real statement: "The size of the compressor is so large that it cannot pass through a part of the doors and elevators". The "Compressor Size" and "Doors and elevators size" are maintenance resources in the class of "component information". We can get the size data from the BIM model, but some of them must use the Dynamo to obtain.

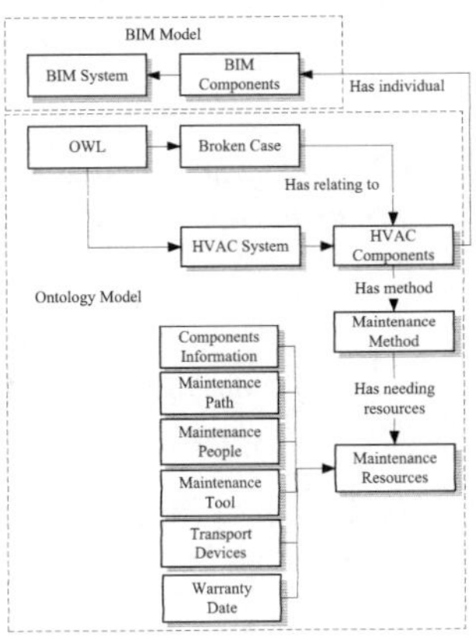

Figure 5. Ontology model framework

3.2.3 Maintainability Inferences

This study uses the SWRL language to infer six outcomes and uses the BIM and COBie forms to store information based on their data application. The six indicators include "space maintainability", "path maintainability", "people maintainability", "tool maintainability", "transport device maintainability" and "warranty maintainability", as shown in figure 6.

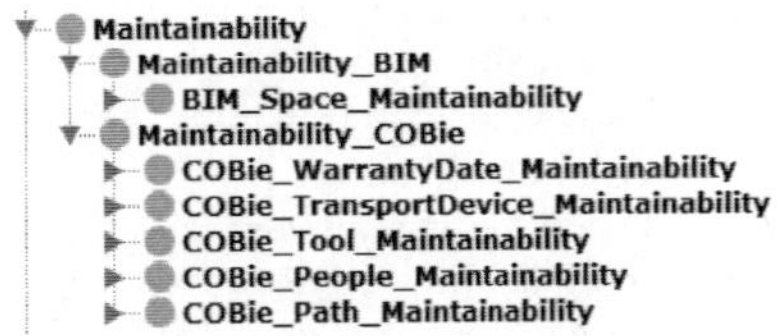

Figure 6. Class of maintainability

Taking spatial maintenance indicators for example: "ChillerSystem_Box_Compressor(?x)^
VacuumPump_Tool(Tool_VacuumPump_1)^
has_ToolWorkSpace(Tool_VacuumPump_1,?TWS)^
swrlb:greaterThanOrEqual(?TWS,0)^
swrlb:lessThan(?TWS,100)->
MaintenanceSpace_0_100(Tool_VacuumPump_1)".
The "ChillerSystem_Box_Compressor" is a class in the ontology model, and "?x" is an individual of the class. The "VacuumPump_Tool" is also a class in the model, and the rule will automatically find an individual named "Tool_VacuumPump_1". Then the individual has a data property of "has_ToolWorkSpace" and is named as "TWS". The SWRL language "greaterThanOrEqual" and "lessThan" limit values range from 0 to 100. When the conditions are met, the engine will automatically judge the true, and the "Tool_VacuumPump_1" is classified into the class "MaintenanceSpace_0_100". After the inference, facility managers can realize that the maintenance event requires at least 0 to 100 of the space, and it is also controlled by several components. Other SWRL rules of maintainability indicators can be seen in figure 7.

Figure 7. SWRL rules

3.3 BIM Model and COBie Form

In this study, the BIM model was developed in

LOD350. The model has a total nine floors of the reinforced concrete buildings, and it uses HVAC system for the case analysis. The HVAC host room is located on the 8th floor as shown in figure 8, which consists of two cooling towers, two chillers, four switchboards, two pumps, four kinds of pipelines and other valves. Pipelines are classified as green cooling tower backwater, blue cooling tower supplement, red chilled water for chiller and orange chilled water back to the chiller. According to the actual pipeline installation, it is mostly connected by welding and uses the valves in the joints for strengthening. Both the cooling towers and chillers have one group that used as a backup when the facilities break down.

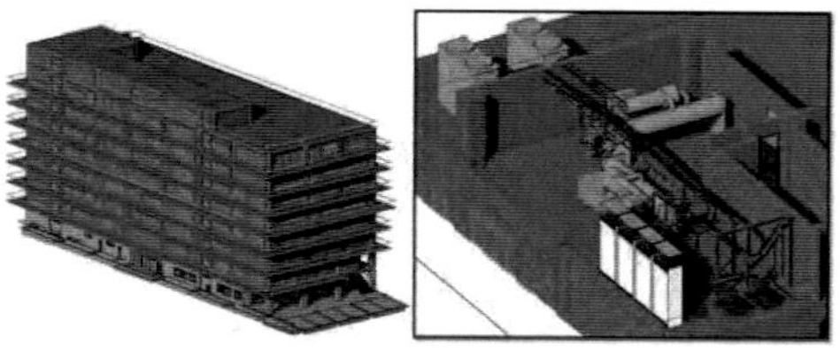

Figure 8. BIM and MEP model

Then we use COBie software to expand the maintenance information. Most of the modeling staff are manually placing maintenance information into the COBie expansion field. But if the model is very large or has a complex classification of components, it will lead to an excessive number of components. Therefore, this study uses Dynamo's special node "Element.SetParameterByName" to automatically input information in the correct fields, such as "CreatedBy", "CreatedOn", "Component.Name", as shown in figure 9. The node needs to be input the component information, the property name and the value which we manually input. Take centrifugal compressor of the chiller for example, after picking the component, the module grabs the field name "COBie.Component.Description" and enter the custom information "Pingtung Administration HVAC host room" into this field. The other fields are filled in the same way as described above, and then we only need to select a large number of components at once.

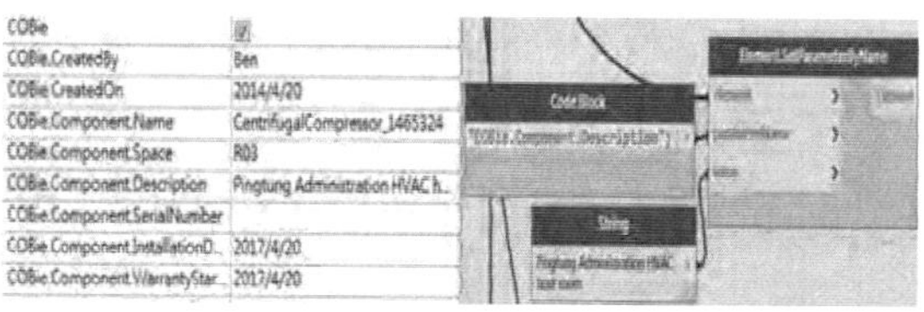

Figure 9. Extended COBie and Dynamo nodes

Figure 10 shows the output of COBie forms about the chiller components. When the COBie forms are produced, it will give every element an independent "ExtIdentifier". Although the purpose is to solve the repeatability of components, the text composition is too long and mix with English and number. The biggest problem is that it cannot be found in the BIM model by ExtIdentifier. So on the basis of naming components still uses BIM coding system. The number behind of the components name will be input like "CentrifugalCompressor_1465324", and it can help us to select the correct components to check or change in the BIM model. This method can also successfully output very small components which are hard to choose.

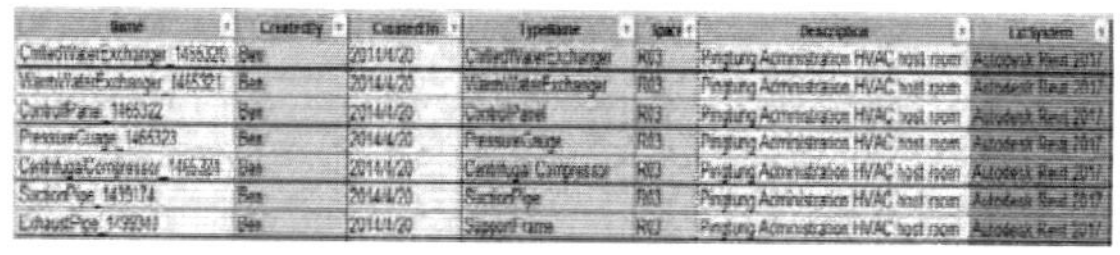

Figure 10. Components of chiller output results

3.4 BIM Modeling Rules

If the BIM model wants to meet the back-end maintenance simulation and application, it must comply with certain specific modeling rules. There are 6 rules during the creation of the model:

1. The model should be drawn in LOD350 because model under the LOD350 does not contain the adapters which are often needed to check and maintain.
2. The internal components of the large machine should be placed in the same position as the design because we need to know the maintenance space of the facility.
3. The components size must be the same with the drawing. The correct size can help managers to assign works.
4. The model needs to take the joints into account.
5. The model must meet the requirements of the existing regulations such as large equipment reserved distance of at least 50 cm. The detailed specifications are listed in the "Building Technical Rules – Construction" which ensure the maintenance operations at the lowest standards.
6. The components should be cut or create in OmniClass classification. Some components only need to be cut to the fourth layer, and some need to be divided into the fifth layer.

There are also some contents that don't need to be the same:

1. The component internal structure does not need to be drawn because it belongs to component design manufacturers' works. If we draw in the BIM model, it will greatly increase the capacity of the

model.

2. The energy efficiency of the machine and the watts are not used directly in this study, so the attribute data can be replaced by other value.

3.5 Integration of Information and Inference Results

This study uses different software to infer and storage maintenance information, so we must integrate transmission methods to reduce the loss of data. All of the software use components to carry maintenance information, and they can be divided into three purposes by software and application, as shown in figure 11. The first one is the components in the BIM model. The second one is based on individuals as the carrier of the maintenance ontology. The third is the maintenance information of COBie forms. In these three surfaces, the BIM model ID code will be used as a connection. When a facility manager searches for relevant maintenance information, it still can use the same ID to find information that he needs even they have different data type. The ID not only retains the unity of the components but also make the search simpler.

Figure 11. ID in the different interfaces

Before running the SWRL rules, the real world parameters must be entered into the ontology model by data property. According to the broken situation, figure 12 shows the data contents of "CentrifugalCompressor_1465324". This information indicates that the maintenance method is replacement, and it has 80 cm in width and height, and two staff do the maintenance, and maintenance date is 2017-10-10 and its weight 300 kg.

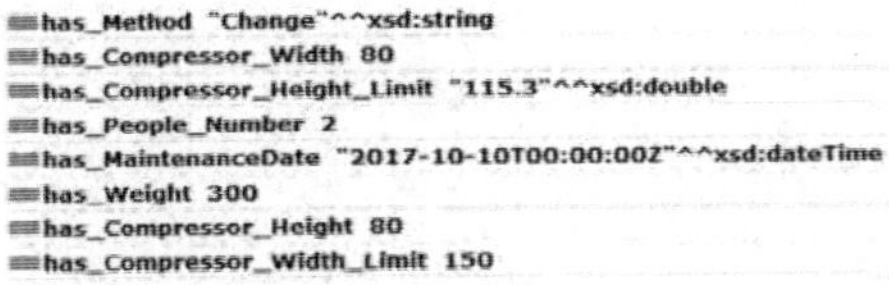

Figure 12. Data properties of compressor

In the above data properties, the height limit and the width limit of the compressor must be personally measured by staff or through the drawing to calculate the value. If an unpredictable error happens, this kind of approach will lose a lot of times and resources. Another way is to use the BIM model to get the limit value of the parameters, but it can only accurately calculate the 2D distance in the Revit software. If we want to calculate the shortest distance in the three-dimensional space, we must find the coordinates of the two objects and calculate the distance by the trigonometric function. The above approaches are too complex, so this study uses Dynamo's special node "Geometry.DistanceTo" to calculate the shortest distance between the components. Take replacement of the compressor for example, we first mark the compressor in number one, and other components are number two. Then the module automatically grabs these two types of components and divides them into list one and list two. All pairs of comparison results are related to the "compressor_1465324" because the component in list 1 is single. We can find that the shortest distance between the "compressor_1465324" and the other components is 152 cm in the horizontal, and the shortest element is "column_316090", as shown in figure 13. Finally, the information will be input in the ontology model.

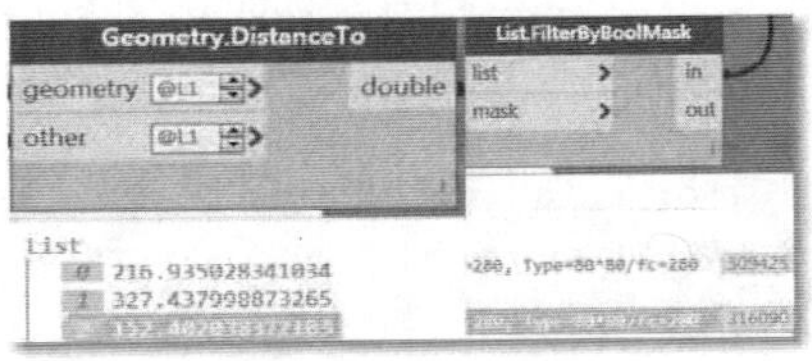

Figure 13. Dynamo calculation results

In the inference process, the Protégé engine compares the managers' resources of maintenance and related constraints. The Protégé engine automatically classifies the maintenance components into the maintainability class, as shown in figure 14. As a result of the light yellow derivation, it is known that the compressor element is sufficient in the horizontal and vertical movement space. In the movement of other components, we can find that "ExhaustPipe_1556568", "StartDevice_1465322" and "SuctionPipe_1499174" are in the "Space_removed". The inferred result is also supported by the shortest distance calculation. Then we set the limitation of distance in the Dynamo program and screen out the too close components which distance is less than 10 cm between each other. Finally, the program will highlight the components in other colors to increase the discrimination.

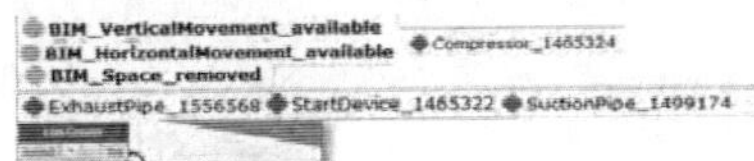

Figure 14. Spatial maintainability

The maintenance space indicators need to infer the key parameters that control the size of the space. The intervals which are "BIM_Maintenance_0_100", "100_200" and "200_300" will be placed the key parameters of maintenance space. The components that control the "0_100" interval are "Compressor_1465324" and "Tool_VacuumPump_1". The "100_200" interval is controlled by the "Staff_2", and the maintenance space is 100 cm. There is no key parameter in the "200_300" interval.

After we get the control factors, the Dynamo can quickly build the maintenance space in the BIM model. First, we establish the component boundary by the six-sided cube to solve the problem of irregular surface shape. Then we analyze the parameters from boundary to create the maintenance surface and extrude the space which the distance is 100 cm. Finally, the space symbol is converted into a spatial volume in order to import and give it meaning in the BIM model. It can also expand fields by COBie forms. After the creation is completed, figure 15 shows a gray transparent block. When the model changes to the 2D plan, the distance between the maintenance space and the column still has 50 cm which is very abundant.

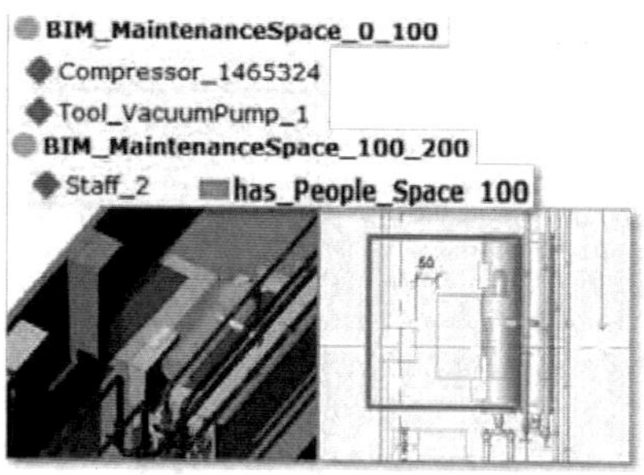

Figure 15. Maintenance space location

In figure 16, we can see that "Compressor_1465324" is classified into the class of available with the warranty date, the transport device choice and whether the maintenance staff enough. After comparing the weight of the compressor and reasonable range, it is recommended that the maintenance staff should carry it by trailers or elevators. Finally, the two of staff assigned to maintain are also in the "People_avaible".

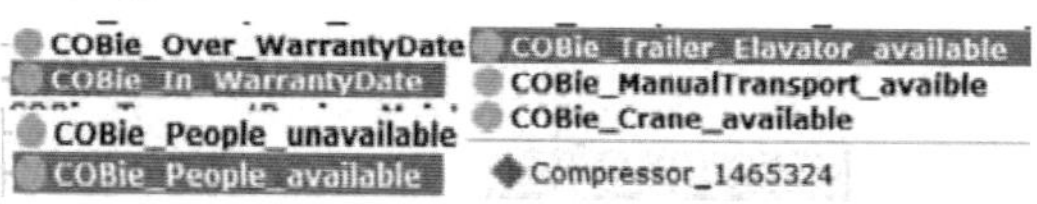

Figure 16. People, transport and warranty maintainability

If the compressor is replaced by different methods, it will affect the inference of the tool choice. Therefore, the Protégé engine infers "Tool_RefregerantBucket_1", "Tool_ToolBox_1", "Tool_VacuumPump_1" and "Tool_Welder_1" to the class of "Tool_Kind", as shown in figure 17. Facility managers can use the information to quickly know that replacement of the compressor should need which special tool. In the "ToolBox" class also contains a large number of small tools, such as needle-nose pliers, waterproof tape and screwdrivers, and they have been built in the ontology model. The feasibility of the path is inferred by the size of the components and the size space which includes doors, elevators and others in the BIM model. Finally, it will automatically classify the space which can pass into "Path_Through_available" and cannot pass into "Path_Through_unavailable". The managers can determine the proper transport path from the inference results, and whether there is a conflict in the expected route. Users also can find the conflicts in the BIM model by its ID code.

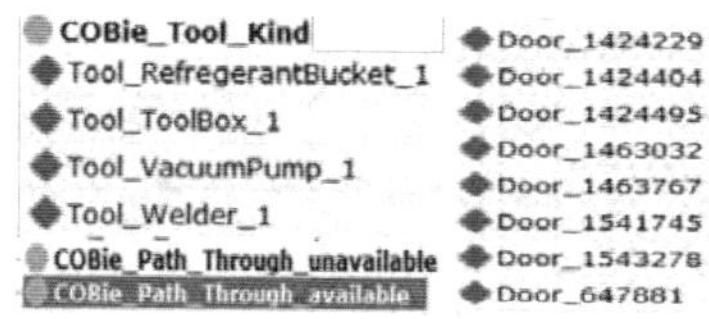

Figure 17. Tool and path maintainability

3.6 Maintenance Information Application and Storage

The inferred results and maintenance information are not suitable for directly using or delivery because the users must have a certain degree of learning on the BIM model and other software. There is also no extra time to do a complex pre-operation when urgency maintenance happens. Therefore, this study stores the maintenance information on the document page which can be used to store different kinds files for external data in COBie forms. But it cannot be completed in the BIM model. We must first export the COBie forms and manually add data in it. In the COBie.Document we still use the BIM model ID code to name the contents. According to the above maintenance, indicators are divided into six class "People", "Transport", "Path", "Warranty", "Tool" and "Space", as shown in figure 18. The first five indicators can be stored in the format of text, but the spatial indicator cannot use the same way to convey. Therefore, we use Navisworks to export videos as a carrier of maintenance information.

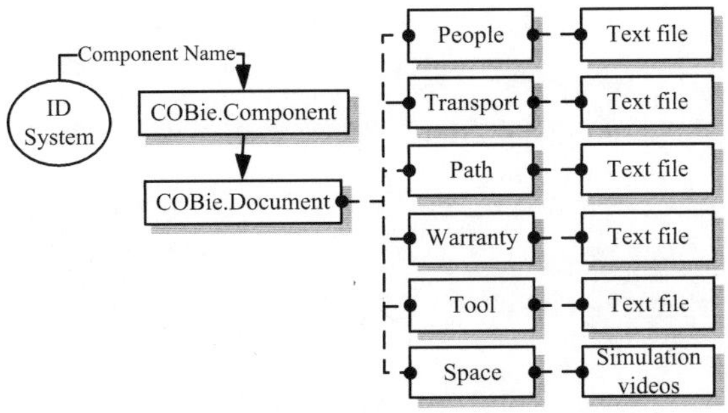

Figure 18. COBie application mode

Figure 19 shows the results of the "COBie.Document" information that are related to the urgency maintenance of the "compressor_1465324". The facility managers don't need to do additional simulation and operation when the breakdown happens. They can directly use the "Document.COBie" to search by ID and get the relevant maintenance information. This approach can significantly correct the back-end operational phase which lacks the maintenance information, as well as provides a lot of data to assist the judgment of maintenance. The "Document.COBie" contains the type and introduction of the tool, the company's internal maintenance staff expertise, the contract manufacturer's warranty certificate, the application transport device information and the size of the components in BIM model. The above information can be in a hyperlinked exists in the "Document.COBie", and users can click on them to connect to external data. Space maintenance indicators can be provided by additional pre-recorded videos, and they support maintenance staff to simulate or strengthen problems before start off. The videos contain the entire replacement process, the site maintenance without replacement, possible conflict location and other precautions. Through the videos, they can combine resources and the experience of maintenance staff they have. So that the overall efficiency of the urgency maintenance work can be improved even reducing the waste of resources.

Name	Created	Category	RowName	Director	File
Tool Information	Ben	21-04 30 30 10 12	CentrifugalCompressor_1465324	pdf	Tool types and catalogs.pdf
Staff needs	Ben	21-04 30 30 10 12	CentrifugalCompressor_1465324	pdf	Employee information.pdf
Warranty information	Ben	21-04 30 30 10 12	CentrifugalCompressor_1465324	pdf	Warranty certification.pdf
Transport information	Ben	21-04 30 30 10 12	CentrifugalCompressor_1465324	pdf	Transportation catalog.pdf
Path information	Ben	21-04 30 30 10 12	CentrifugalCompressor_1465324	pdf	Path size table.pdf
Name	Created	Category	RowName	Director	File
Replacement process	Ben	21-04 30 30 10 12	CentrifugalCompressor_1465324	.avi	Replace film.avi
Site maintenance process	Ben	21-04 30 30 10 12	CentrifugalCompressor_1465324	.avi	Site repair film.avi
Conflict presentation	Ben	21-04 30 30 10 12	CentrifugalCompressor_1465324	.avi	Conflict film.avi
Other precautions	Ben	21-04 30 30 10 12	CentrifugalCompressor_1465324	.avi	Other Considerations Video.avi

Figure 19. External link maintenance data

4 Conclusions

This study constructs an ontology model based on facility maintenance and connects the results with BIM model. The ontology model is based on the actual maintenance work, and it also changes the BIM modeling which lacks the maintenance procedures from the inference results. This study develops maintenance information requirements and applies to the BIM model, and it significantly reduces the burden on facility managers. All of the semantic information has been categorized in the model, and it can reduce the experience gap between the facility managers and staff.

The senior managers can add new maintenance scenarios to the ontology model in the future; update the field of knowledge for maintenance. New employees can not only learn from the work site but also use the BIM model as a learning platform. Dynamo modules can be more directly observed about the maintenance situation that may encounter and reflect the advantages of BIM visualization and logical calculus. Finally, the BIM model helps managers to reduce the threshold of maintenance with the COBie drawing, and it speeds up the overall maintenance efficiency.

5 Acknowledgement

This research was supported by the Ministry of Science and Technology, Taiwan under Grant MOST 105-2221-E-006-047-MY2.

References

[1] Sullivan G.P. , Pugh R. , Melendez A.P. , and Hunt W.D. Operations & Maintenance Best Practices: A Guide to Achieving Operational Efficiency. *Richland: Pacific Northwest National Laboratory*, 2010.

[2] Wang Y. , Wang X. , Wang J. , Yung P. , and Jun G. Engagement of Facilities Management in Design Stage through BIM: Framework and a Case Study. *Advances in Civil Engineering*, 2013.

[3] Tserng H. P. , Yin Y.L. , Jaselskis E. , Hung W. C. , and Lin Y. C. , Modularization and Assembly Algorithm for Efficient MEP Construction. *Automation in Construction*, 20: 837-863, 2011.

[4] WANG L. and F. LEITE. Comparison of Experienced and Novice BIM Coordinators in Performing Mechanical, Electrical, and Plumbing (MEP) Coordination Tasks. *Construction Research Congress: 21-30*, 2014.

[5] Eastman C. , Teicholz P. , Sacks R. , and Liston K. *BIM Handbook: A Guide to Building Information Modeling for Owners, Managers, Designers, Engineers and Contractors.* New Jersey: Wiley, 2011.
[6] Becerik-Gerber B. , Jazizadeh F. , Li N. , and Calis G. Application Areas and Data Requirements for BIM-Enabled Facilities Management. *Construction Engineering and Management*, Vol.138, Issue 3, 2012.
[7] McArthur J. J. A Building Information Management (BIM) Framework and Supporting Case Study for Existing Building Operations. *Maintenance and Sustainability*, Procedia Engineering, 118: 1104-1111, 2015.
[8] Ting K.P. A Reasoning Mechanism Using Ontology, Protégé and SWRL Tools for Building Information Model Data. Civil Engineering, *National Central University*, 2013.
[9] Zhang Y. , X. Luo Y. Zhao and Zhang H.C. An ontology-based knowledge framework for engineering material selection. *Advanced Engineering Informatics*, 29(4): 985-1000, 2015.

34th International Symposium
on Automation and Robotics
in Construction

ISARC 2017

June 28 – July 1
TAIPEI, TAIWAN

Effect of Delay Time on the Mechanical Properties of Extrusion-based 3D Printed Concrete

Taylor MARCHMENT[a], Ming XIA[b], Elise DODD[c], Jay SANJAYAN[d] and Behzad NEMATOLLAHI[e]

[a,b,c,d,e] Center for Sustainable Infrastructure, School of Engineering, Faculty of Science, Engineering and Technology, Swinburne University of Technology, Melbourne, Australia

E-mails: 4973445@student.swin.edu.au[a], mxia@swin.edu.au[b], 9513361@student.swin.edu.au[c], jsanjayan@swin.edu.au[d], bnematollahi@swin.edu.au[e]

Abstract

The extrusion-based 3D printing method is one of the main additive manufacturing techniques used in the construction industry which is capable of producing large-scale building components with complex geometries without the use of the expensive formwork. The mechanical properties of the printed concrete component are very much unlike the conventionally cast concrete. Therefore, the properties of these are important for the design of structures built with 3D printing process.

In this study, a custom-made apparatus was used to simulate extrusion-based 3D printing process. Layered specimens with 10, 20 and 30 minutes delay time (i.e. the printing time interval between layers) have been printed using tailored conventional concrete mixture. Mechanical properties including compressive, flexural and inter-layer bonding strengths have been measured and the effect of delay time on the mechanical properties has been investigated.

All printed specimens exhibited an anisotropic phenomenon in both compressive and flexural strengths. A significant inverted bell curve pattern emerged in the inter-layer bonding strength. The inter-layer bonding strength of the specimens with 10 minutes and 30 minutes delay time was similar, but higher than that of specimen with 20 minutes delay time. The major finding is that the surface moisture content plays a major role in the inter-layer bonding strength.

Keywords –

Concrete; Additive Manufacturing; 3D Printing; Mechanical Properties

1 Introduction

Nowadays, in the manufacturing industry, one of the most technological trends of manufacturing is Additive Manufacturing (AM), also referred to as three-dimensional (3D) printing. AM is a group of techniques for producing 3D objects directly from a digital model. Until recently, AM techniques have gained popularity in a wide range of high value adding industries such as military, aerospace, automotive and biomedical [1]. In general, the construction industry is labor intensive and has issues with safety and sustainability. Compared with other industries, the construction industry is viewed as a low-tech industry with low levels of innovation [2]. The image of the construction industry may be changed by the adoption of AM techniques.

Unlike conventional manufacturing processes such as subtractive method, AM techniques build a finished part in successive layers with less waste material. The adoption of AM techniques in the construction industry may (1) reduce the labor requirements which would result in a decreased construction cost and an increased level of safety, (2) reduce on-site construction time by operating at a constant rate, (3) minimize the chance of errors by highly precise material deposition, and (4) increase architectural freedom which would enable more sophisticated designs for structural and aesthetic purposes [3-4]. The application of AM techniques in the construction industry is still in its early stage. Two forms of AM techniques have been so far explored for application in concrete construction: (1) the extrusion printing technique and (2) powder printing technique.

The extrusion printing technique is analogous to the Fused Deposition Modelling (FDM) method but uses a variety of slurry-like materials to create structures. The extrusion printing technique has been aimed at on-site construction applications such as large-scale building components with complex geometries. Additionally, this technique allows the usage of traditional construction materials such as cement, geopolymer, clays and plasters.

Briefly, in the extrusion-based 3D printing process, a cementitious material is extruded from a nozzle mounted on a gantry, crane or a 6-axes robotic arm to print a structure layer by layer. The cementitious material must be thoroughly considered in this approach as there are many requirements including the viscosity of the mixture, time of setting, the time interval between old and new layers and the mechanical properties after hardening. Other considerations include the size/shape of the nozzle, the size of the tank that is practicable for the printer, the size of the printer and other variables [5].

One of the most important challenges of the extrusion-based 3D printing is that conventional steel reinforcements are difficult to be incorporated in 3D printing [6], which makes it difficult to compensate for inherently weak tensile strength of concrete. For this reason, the bonding strength between the printed layers is a critical mechanical property in the extrusion-based 3D printing process [7]. In the extrusion-based 3D printing process, the printed layer needs to be strong enough to support the next upper layer without deforming or collapsing [8], and also needs to be soft enough to allow a good bond between the old and new layers. Thus the printing time interval between layers, referred to as delay time, becomes a significant factor. Currently, there is a very limited research work available regarding the effect of delay time on mechanical properties of extrusion-based 3D printed concrete.

As discussed above, the inter-layer bonding strength is a particular area of weakness correlating to overall specimen strength in the extrusion-based 3D printing process and is greatly affected by the delay time. This paper investigates the mechanical properties of 3D printed concrete with varied delay time.

2 Materials and Methods

2.1 Materials

An ordinary Portland cement (OPC) conforms to AS 3972 general purpose (Type GP) was supplied from Geelong Cement, Australia.

Two types of sands were used as the aggregates in the mixtures. Sand Type I was washed sand (supplied by Building Products Supplies Pty Ltd., Australia) with a particle size less than 500 μm and sand Type II (supplied by TGS Industrial Sands Pty Ltd., Australia) was a high silica purity fine sand with a particle size less than 300 μm.

2.2 Mix proportions

The mix design aimed to meet the requirements of the extrusion-based 3D printing process, including extrudability, workability and buildability.

Initially, a baseline trial mixture was proportioned using traditional concrete ingredients and a normal mixing process. The starting point selected was a mix with a cement: sand (Type I) ratio of 1:3. The water to cement ratio for the baseline mix was 0.5. The mixture was prepared and loaded into the 3D printing apparatus, but the baseline mix was too stiff to be extruded out from the nozzle for printing.

The mix optimization process includes reducing the amount of sand Type I, increasing the amount of sand Type II and reducing the amount of water, which resulting an optimum mix with a cement : sand (Type I) : sand (Type II) ratio of 1:0.97:0.33. The water to cement ratio of the optimum mix was 0.38.

2.3 Extrusion-based 3D printing process

2.3.1 3D printing apparatus

A custom-made apparatus was used to simulate extrusion-based 3D printing process (see Figure 1-a). A 45° angle extrusion nozzle (see Figure 1-b) with a 15 mm × 25 mm opening was used.

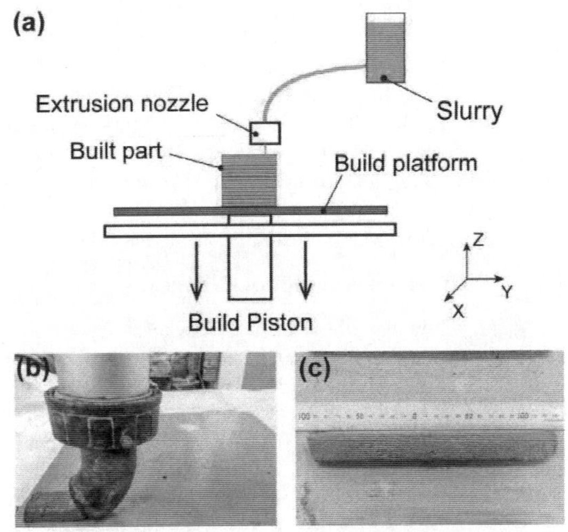

Figure 1. (a) Schematic illustration of the extrusion-based 3D printing process, (b) 45° angle extrusion nozzle, (c) One layer of extruded concrete (250 mm (L) ×25 mm (W) ×15 mm (H))

2.3.2 Specimen preparation

The optimum mixture was loaded into the 3D printing apparatus and extruded through the nozzle by manually moving the extruder with a constant speed of 12 mm/s.

A single base layer (approximately at 250 mm length) (see Figure 1-c) was extruded for each specimen. Then, after the given time interval, another layer was deposited on top of the previous layer. Specimens prepared for compressive and flexural strengths testing consisted of 4 printed layers, while the specimens made for inter-layer bonding test consisted of 2 printed layers. All the specimens were cured in the laboratory environment at room temperature (23 ± 3°C) until the day of testing (i.e. 7 days after casting).

2.4 Mechanical properties testing

2.4.1 Compressive strength

The compressive strength was measured in a universal testing machine under load control at the rate of 20 MPa/min. For the compressive strength test, the printed specimens were cut into smaller sizes with approximate dimensions of 50 mm (L) × 25 mm (W) × 60 mm (H) and loaded in one of the two directions: perpendicular to the printed layers (testing direction Top) and parallel to the printed layers (testing direction Front), as shown in Figure 3-a. At least three specimens were tested for each testing direction.

2.4.2 Flexural strength

The flexural strength was measured in a universal testing machine under displacement control at the rate of 1 mm/min. For flexural strength test, the specimen dimensions were 250 mm (L) × 25 mm (W) × 60 mm (H). All specimens were tested under 3-point bending test setup with a span of 150 mm. The flexural strength test was also conducted in two directions, perpendicular to the printed layers (testing direction Top) and parallel to the printed layers (testing direction Front), as shown in Figure 4-a. At least three specimens were tested for each testing direction.

2.4.3 Inter-layer bonding strength

The schematic illustration of the inter-layer bonding strength test setup is shown in Figure 2.

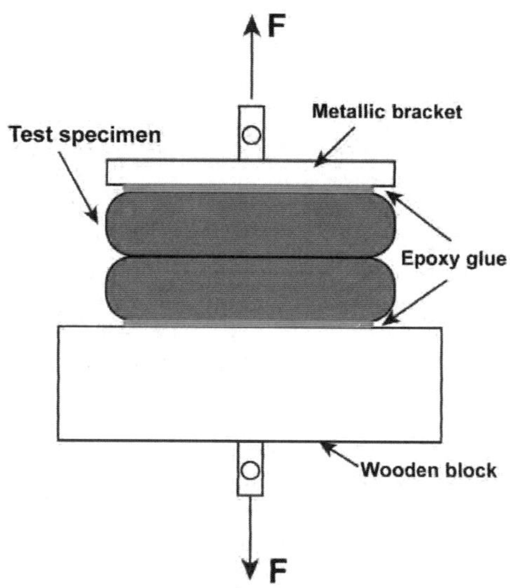

Figure 2. Schematic illustration of the inter-layer bonding strength test setup

For the inter-layer bonding strength test, the specimen dimensions were 50 mm (L) × 25 mm (W) × 30 mm (H). A metallic bracket and a wooden block were epoxy glued on the top and bottom of the printed specimen, respectively. The inter-layer bonding strength test was conducted in a universal testing machine under displacement control at the rate of 1 mm/min. Care was taken to align the specimen in the machine to avoid any eccentricity. At least 3 specimens were tested.

3 Results and Discussions

3.1 Compressive strength

Figure 3 presents the compressive strength results for different delay time. As can be seen, the anisotropic phenomenon was observed in the compressive strength of the printed concrete. The compressive strength in the direction of perpendicular to the printed layers (testing direction Top) was higher than that in the direction of parallel to the printed layers (testing direction Front), irrespective of the delay time. The compressive strength in the direction of perpendicular to the printed layers (testing direction Top) was 30%, 35% and 37% higher than that in the direction of parallel to the printed layers (testing direction Front) for 10, 20 and 30 min delay time, respectively.

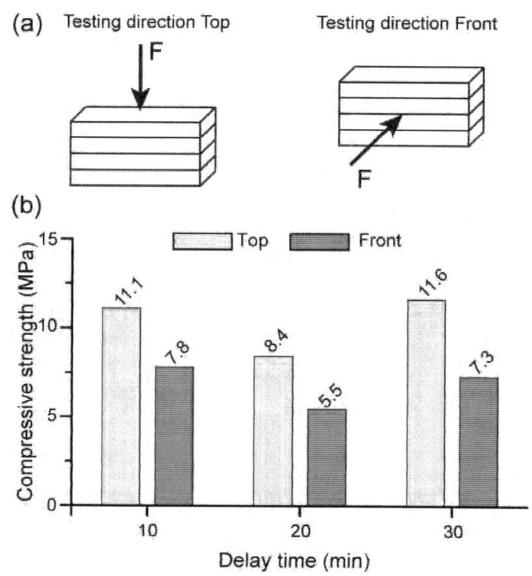

Figure 3. (a) Testing directions of the specimens, (b) Compressive strength results

As can be seen in Figure 3, an inverted bell curve pattern was observed for compressive strength results in both testing directions. The specimen printed with 20 min delay time had an average compressive strength of 8.4

MPa in the Top loading direction and 5.5 MPa in the Front loading direction, which were lower than those of the corresponding specimens with 10 min and 30 min delay time.

3.2 Flexural strength

Figure 4 presents the flexural strength results for different delay time. Similar to the compressive strength results, the anisotropic phenomenon was also observed in the flexural strength of the printed concrete. The flexural strength in perpendicular to the printed layers (testing direction Top) was also higher than that in parallel to the printed layers (testing direction Front), irrespective of the delay time. This results is consistent with the compressive strength results. The flexural strength in the direction of perpendicular to the printed layers (testing direction Top) was 55%, 13% and 34% higher than that in the direction of parallel to the printed layers (testing direction Front) for 10, 20 and 30 min delay time, respectively.

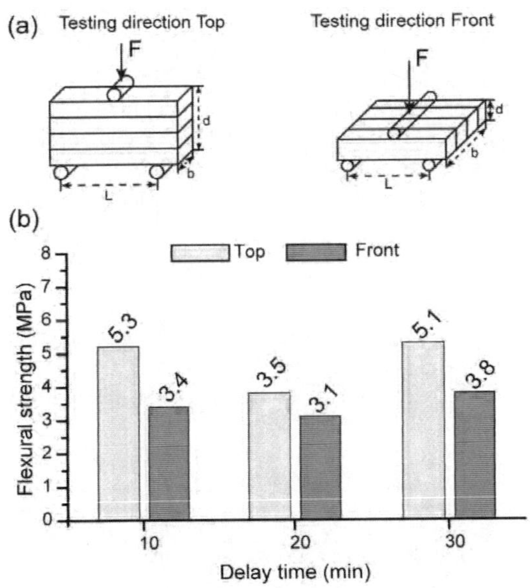

Figure 4. (a) Testing directions of the specimens, (b) Flexural strength results

Similar inverted bell curve pattern was also observed for the flexural strength of the printed concrete. The specimen printed with 20 min delay time had an average flexural strength of 3.5 MPa in the Top loading direction and 3.1 MPa in the Front loading direction, which were lower than those of the corresponding specimens with 10 min and 30 min delay time.

3.3 Inter-layer bonding strength

The inter-layer bonding strength test was conducted to indicate the true inter-layer bonding strength and failure mode between layers. As can bee seen in Figure 5, a similar inverted bell curve pattern was also observed for the inter-layer bonding strength of the printed concrete. The inter-layer bonding strength of the specimens with 10 min and 30 min delay time was similar, but higher than that of specimen with 20-min delay time. It should be noted that all specimens failed between the printed layers.

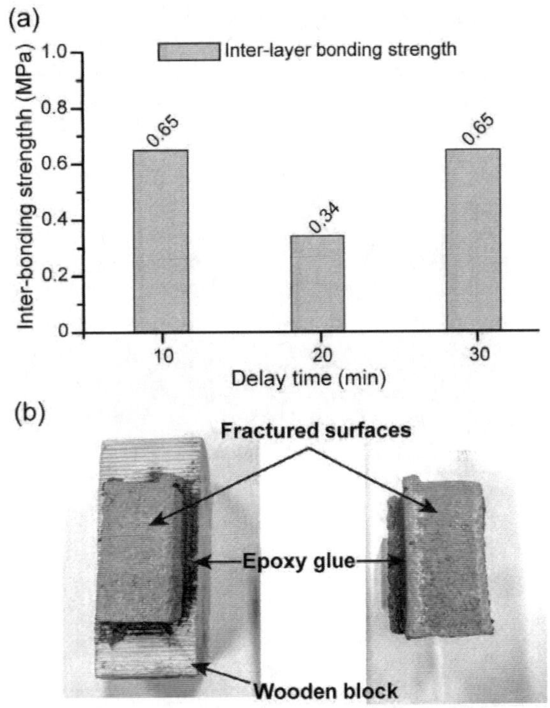

Figure 5. (a) Inter-layer bonding strength result of printed specimens, (b) Fractured surfaces between layers

Le et al. [9] reported that with increasing delay time the inter-layer bonding strength decreases as the adhesion between old and new layers reduces. However, a different inverted bell curve pattern was observed in this study. This could be due to the different surface moisture contents of printed specimens with different delay time.

Therefore, as a preliminary study the surface moisture content of the specimens were measured using a paper towel. The experiment setup and results are presented in Figure 6. Three single base layers (approximately 250 mm long) were extruded for each delay time. After the deposition, a cut-to-size paper towel measuring 250 mm in length and 15 mm in width was placed on the surface of extruded layer for 20 seconds. The mass change of the paper towel was measured as the surface moisture content. The temperature and relative humidity in the laboratory were maintained at 24°C and 65%, respectively during the measurement. For each delay time, three measurements were recorded.

As can be seen in Figure 6, the surface moisture contents varied with delay time, and an inverted bell curve pattern was observed similar to the inter-layer bonding strength result. The average surface moisture content of 20 min delay time specimens was 0.16 g which was lower than the specimens with 10 min delay time (0.28 g) and 30 min delay time (0.30 g).

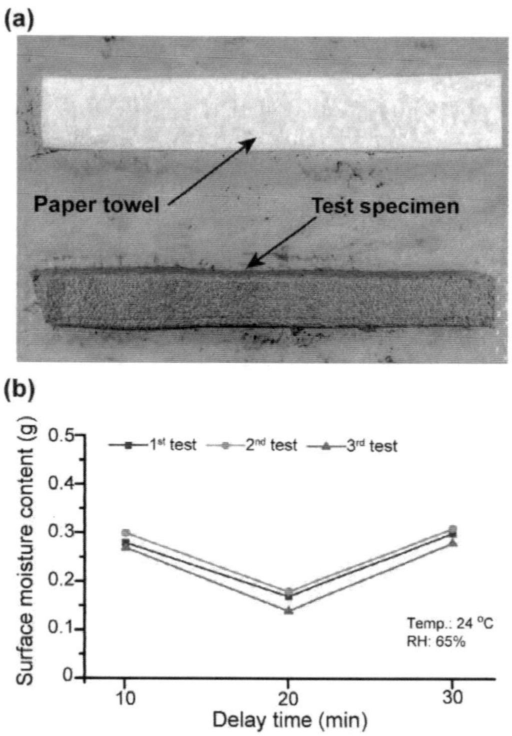

Figure 6. (a) Experimental setup for surface moisture content measurement, (b) Surface moisture content result

The higher surface moisture contents of the 10 min and 30 min delay time specimen provided a better aqueous medium for the fresh concrete mixture in the old and new layers to bond together, resulting in superior inter-layer bonding strength.

The surface moisture content is affected by several factors including instrument condition (e.g., type, size and material of the extruder, pressure) and material condition (such as mix composition and bleeding). Thus, further research is currently being conducted in order to clearly understand the underlying reasons of inverted bell pattern of the surface moisture content of different delay time.

4 Conclusions

The effects of different delay time on the mechanical properties of the extrusion-based 3D printed concrete were investigated in this study. Specimens with different delay times were printed and the mechanical properties including compressive, flexural and inter-layer bonding strengths were measured. The following conclusions can be drawn from this study:

(1) The anisotropic phenomenon was observed in the compressive and flexural strengths of the printed concrete. The compressive and flexural strengths in the direction of perpendicular to the printed layers (testing direction Top) was higher than that in the direction of parallel to the printed layers (testing direction Front), irrespective of the delay time.

(2) A significant inverted bell curve pattern was observed for the inter-layer bonding strength result. The inter-layer bonding strength of the specimens with 10 min and 30 min delay time was similar, but higher than that of specimen with 20-min delay time. A clear correlation was found between the results of inter-layer bonding strength and the surface moisture content.

References

[1] Wohlers, T. Wohlers Report 2016, Wohlers Associates Inc., Colorado, 2016.

[2] Harty, C. Implementing innovation in construction: contexts, relative boundedness and actor-network theory, *Construction Management and Economics*, 26:1029-1041, 2008.

[3] Gao, W., Zhang, Y., Ramanujan, D., Ramani, K., Chen, Y., Williams, C.B., Wang, C.C., Shin, Y.C., Zhang, S., Zavattieri, P.D., The status, challenges, and future of additive manufacturing in engineering. *Computer-Aided Design*, 69:65-89, 2015.

[4] Buswell, R.A., Soar, R., Gibb, A.G., Thorpe, A., Freeform construction: mega-scale rapid manufacturing for construction, *Automation in Construction*, 16:224-231, 2007.

[5] Lim, S., Buswell, R., Le, T.T., Wackrow, R., Austin, S., Gibb, A., Thorpe, T., Development of a viable concrete printing process, In *Proceedings of the 28th International Symposium on Automation and Robotics in Construction*, pages 665-670, Seoul, South Korea, 2011.

[6] Edwards, L., Holt, C., Keyte, L., Lloyd, R., Construction 3D Printing, Boral Innovation Factory, 2015.

[7] Rael, R., San Fratello, V., Developing Concrete Polymer Building Components for 3D Printing, In *Proceedings of the 31st Annual Conference of the Association for Computer Aided Design in Architecture (ACADIA 11)*, pages 152-157, Banff, Canada, 2011.

[8] Perrot, A., Rangeard, D., Pierre, A., Structural built up of cement-based materials used for 3D-printing extrusion techniques, *Materials and Structures*, 49(4): 1213-1220, 2016,

[9] Le, T.T., Austin, S.A., Lim, S., Buswell, R.A., Law, R.A., Gibb, G.F., Thorpe, T., Hardened properties of high-performance printing concrete, *Cement and Concrete Research*, 42: 558-566, 2012.

Applications of Additive Manufacturing in the Construction Industry – A Prospective Review

Daniel Delgado Camacho[a], Patricia Clayton[a], William O'Brien[a], Raissa Ferron[a], Maria Juenger[a], Salvatore Salamone[a], and Carolyn Seepersad[b]

[a]Department of Civil, Architectural and Environmental Engineering, The University of Texas at Austin, USA
[b]Department of Mechanical Engineering, The University of Texas at Austin, USA
E-mail: dec.daniel@utexas.edu, clayton@utexas.edu, wjob@mail.utexas.edu, rferron@mail.utexas.edu, mjuenger@mail.utexas.edu, salamone@utexas.edu, ccseepersad@mail.utexas.edu

Abstract –

Additive manufacturing (AM), also known as 3D printing, produces components by building-up structures from small deposits of materials. Many of the early applications of AM technologies have been in the aerospace, automotive, and healthcare industries. Building on the advances in AM in these industries, there are several experimental applications of AM in the construction sector. Early investigations suggest that use of AM technologies for construction have the potential to decrease labour costs, reduce material waste, and create customized complex geometries that are difficult to achieve using conventional construction techniques. However, these initial investigations do not cover the full range of potential applications for construction or exploit the rapidly maturing AM technologies for a variety of material types. This paper provides an up-to-date review of AM as it relates to the construction industry, identifies the trend of AM processes and materials being used, discusses related methods of implementing AM, and potential advancements in applications of AM. Examples of potential advancements in methods and applications include multi-material use (e.g., use of high-performance materials only in areas where they are needed), in-situ repair in locations that are difficult or dangerous for humans to access, disaster relief construction in areas with limited construction workforce and material resources, structural and non-structural elements with optimized topologies, and customized parts of high value. AM's future in the construction industry is promising, but interdisciplinary research is still needed to provide new materials, new processes, faster printing, quality assurance, and data on mechanical properties before AM can realize its full potential in infrastructure construction.

Keywords –

Additive Manufacturing; Construction Industry, 3D Printing

1 Introduction

Additive Manufacturing (AM), commonly known as 3D printing, produces components by building-up structures from small deposits of materials. AM is a rapidly growing field, having an impact in multiple industries, revolutionizing design, and simplifying the processes to go from a 3D model to a finished product. Comparing AM with other manufacturing processes such as formative processes, where material is deposited in a mould that has the desired shape, and subtractive processes, where a solid piece of material is cut into the desired shape, formative processes require the production of the mould to manufacture a product in mass quantities, and subtractive processes produce waste material, as the material being cut is not reused. AM can advantageously create a wide range of shapes without the need of a mould and only using the required material, filling a gap left by the other manufacturing processes.

Aerospace, automotive, healthcare, and architecture industries have explored the benefits of using AM in their businesses. First uses were for rapid prototyping, to reduce time when producing prototypes with complex geometries [1]. Since then, AM has evolved to include other applications beyond rapid prototyping, and other industries such as construction are starting to follow the early adopters of these new AM technologies.

There have been several experimental applications using AM in the construction industry starting in the late 1990's [2]. This paper provides an up-to-date review of experimental AM technologies in construction, identifies the trend of AM processes and materials being used, discusses related methods of implementing AM and potential applications, and identifies research needs to foster more widespread use of AM in construction. It serves as a guiding point for researchers interested in the area, to understand what has been done so far and what can be done.

2 Current Construction Industry and Potential for AM Technologies

As in every industry, there are challenges being faced. For construction, those challenges have been identified decades ago, such as work in harsh environments, a decrease in productivity, decrease of a skilled workforce, safety during construction, producing large amounts of waste material, and transportation of materials to the site to mention some [3,4].

Construction is considered a high risk profession, having the largest number of deaths in any sector [5]. When construction in harsh environments is unavoidable, the difficulty and risks increase, affecting construction quality and/or adversely human safety. AM could provide services to the construction industry by reducing exposure of on-site workers to harsh environments and by automating some of the construction tasks.

AM allows customized parts to be printed on-demand from a 3D model without significant lead time. Instead of having multiple companies or trades producing different structural or non-structural components, each component can be produced directly using AM after it is designed, tackling one of the challenges previously mentioned, decrease in productivity.

Besides decreasing productivity, there is also a decrease of a skilled workforce. Contractors are having a hard time finding a workforce with the required skills, requiring more time and money to train them. The use of AM in construction will require different skill sets than in current practice, shifting from a labour-intensive work environment to a more technical environment, requiring computer and manufacturing skills.

Another potential benefit from AM, is the reduction of formwork used during construction. Currently, concrete structures are commonly built using temporary formwork to maintain the desired shape of wet concrete as it hardens, and that formwork labour and material ranges from 35-60% of the overall cost of the concrete structures [6]. Reducing formwork use, not only reduces waste material produced during construction, which is considered to be about 23% of the total material wasted in the country [2], but it also reduces construction cost and time that is required for placing and disassembling the formworks.

A common practice in the construction industry is off-site fabrication, as it reduces exposure to on-site working hazards (e.g., harsh environments, safety accidents, etc.) and increases construction quality and consistency; however, such precast elements, modules, and structural assemblies must be transported to the job site. If on-site AM could be implemented, the number of structural and non-structural parts or assemblies that must be transported to the site can be reduced.

3 Additive Manufacturing Processes

AM is achieved in many different ways and the American Society for Testing and Materials (ASTM) International, published a document in collaboration with the International Organization for Standardization (ISO) to define standard terminology for AM [1]. In this document, ISO/ASTM divided AM into seven different processes explained below:

- **Vat Photopolymerization (VP)** – A process of selectively curing a liquid light-activated polymer with a laser. An example of this process is stereolithography apparatus (SLA), a technique developed by Charles Hull in the 1980's [7] and commercialized by 3D Systems.
- **Material Jetting (MJ)** – A process of selectively depositing drops of material in a layerwise fashion. 3D Systems, Inc. uses this process to print customized dental prosthetic devices such as crowns and copings with high precision using UV curable plastic [8].
- **Binder Jetting (BJ)** – A process of depositing a powdered material layer upon layer and selectively dropping a liquid binding agent onto each layer to bind the powders together. Binder jetting was primarily developed at MIT in a process called 3D printing (3DP) [7].
- **Material Extrusion (ME)** – A process of extruding material through a nozzle and depositing it layer-by-layer onto a substrate. The process was invented by Scott Crump and commercialized by Stratasys as Fused Deposition Modelling (FDM) [7], but it now forms the basis for a very wide variety of inexpensive personal 3D printers.
- **Powder Bed Fusion (PBF)** – A process of selectively fusing a powder bed using thermal energy, typically in the form of a laser or electron beam. Selective Laser Sintering (SLS) was developed at the University of Texas at Austin for polymer materials and commercialized by DTM and 3D Systems [7].
- **Sheet Lamination (SL)** – A process of successively shaping and bonding sheets of material to form an object. An example of sheet lamination process is laminated object manufacturing (LOM) developed by Helisys Inc., in which paper sheets were trimmed to size and glued together [7]. Ultrasonic Additive Manufacturing (UAM), commercialized by Solidica Inc. fabricates metal objects using ultrasonic welding [7].
- **Direct Energy Deposition (DED)** – A process of fusing materials with focused thermal energy that melts the material as it is being deposited. An example of this process is laser engineered net shaping (LENS), developed at Sandia National

Laboratories [7], which is particularly useful for repair of damaged metal parts [9]

4 AM in Construction

Table 1 presents examples of AM technologies used for construction and/or companies using these technologies, categorized by materials, AM processes, and spatial delivery system. It can be determined from Table 1 that most of the work being done so far has been in aggregate based material extrusion. As evidenced by their absence in Table 1, work in the area of vat photopolymerization, material jetting, and sheet lamination has yet to be explored in the construction industry to the extent of the authors' knowledge.

Most of the technologies deliver material using a gantry system, but some of them have explored the use of a robotic arms. Although gantry systems have been most commonly used, they do have limitations as discussed by others [5,10,11], such as transportation, installation, orthogonal deposition, and size of the system. Some AM technologies use robotic arms because of the increased freedom provided with its six-axis motion and flexibility to program multiple tasks for the robotic arm. Additional information on key AM technologies and applications are provided in the following sections as categorized by material.

4.1 Aggregate Based

The first ideas of producing elements for construction using AM were proposed in the late 1990's [2]. One of the first to recognize its potential was the University of Southern California (USC), where Contour Crafting (CC) was developed [13]. CC utilizes a gantry system to extrude thick layers of cementitious material in order to increase deposition rate for large-scale structures. The technology has trowels attached on the side of the nozzle to smooth the horizontal and vertical surfaces of the material being extruded [32].

WinSun is a Chinese company that worked jointly with architectural and structural design companies such as Gensler, Thornton Tomasetti, and others to build an office building for the Dubai Future Foundation (Fig. 1), which was printed in Shanghai, shipped to Dubai, and then assembled on site. The cost of printing and assembling was around \$140,000 for the 242 m^2 (2,600 ft^2) single-story building, and compared to conventional construction techniques construction waste was reduced by 30 to 60% [33,34].

Figure 1. Dubai Future Foundation Printed Office Building [33]

4.2 Polymers

Although the use of polymers in construction is not as common as aggregate based and metals, polymers could be used in construction for aesthetic purposes or in structural applications when combined with other strength-enhancing materials.

Table 1. Example of AM technologies in the construction industry

AM Process	Aggregate Based			Polymer		Metallic	
	Gantry	Robotic	Other	Gantry	Robotic	Gantry	Robotic
BJ	Pegna [2] D-Shape [12]						
ME	Contour Crafting [13] Concrete Printing [14] WinSun [15] TotalKustom [16] BetAbram [17] 3D Concrete Printing (3DCP) [18]	XtreeE [11] CyBe [19] Apis-cor [20]	WASP [21] MiniBuilders [22]	BAAM [23] Qingdao Unique Products Develop [15] KamerMaker [24]	C-Fab [25] Digital Construction Platform (DCP)[26] FreeFAB™ Wax [27]		
PBF				Skanska* [28]		Arup*[29] Permasteel-isa*[30]	
DED							MX3D [31]

*Company using a technology

Skanska is a construction company that has utilized advancements in the area of AM by printing unique cladding for the Bevis Marks Building in London. The complex geometry of these connection regions made AM an attractive option for the architectural cladding that had decorative purposes only (shown in Fig. 2) [28]. This application by Skanska proves that AM can be used to provide unique architectural designs without requiring complex and costly production processes.

Figure 2. Bevis Marks Roof (Cladding) [28]

4.3 Metallic

Table 1 shows that metallic material is the group least being explored. The small-scale components built using AM exhibit comparable strength properties to conventionally manufactured components, but they are cost prohibitive using current AM technologies [29]. The advantages of AM become more evident when building up structures or components with complex geometries designed to optimize weight and material use that would be difficult, if not infeasible, to manufacture using conventional techniques.

An example of using AM for optimized structural topologies was developed by Arup. Arup designed several variations of a node using conventional and AM techniques (Fig. 3) to demonstrate the potential savings that can be attained through topology optimization and AM. Although the structure was already built at the time of the investigation, Arup estimated that topology optimization and AM could reduce the weight of each node by 75% compared to the original design, resulting in an estimated reduction of more than 40% for the weight of the entire structure [29].

Figure 3. Arup's Optimized Node [29]

5 Potential Advancements in Methods of Implementing AM

Work done so far in AM has identified different opportunities in which use of AM could be advantageous to conventional techniques such as building up components using multiple materials, using in situ resources, utilizing hybrid techniques that combine AM processes with subtractive and/or formative processes, and expanding opportunities for both off-site and on-site fabrication. These advantageous methods of implementing AM in construction are discussed below.

5.1 Multiple materials

AM could allow multiple materials to be deposited during the construction using extrusion based processes with multiple nozzles for different materials. Bos et al. [18] proposed a concept of material customization by location. For example, depositing ultra-high performance concrete where structural demands are largest, and low-strength concretes for finishes and areas where structural demands are lower. The potential of printing multiple materials, is something that could bring advantages in the construction industry, but research is needed to develop new construction materials that are optimized for use in AM while still providing desirable structural performance. At the same time, these new materials must be economical for AM to become a feasible alternative in construction [35].

5.2 In situ resources

Being able to build using locally available resources can reduce material transportation costs and can provide more sustainable design solutions for places that are difficult to access. CC and D-Shape are investigating the possibility to use their technology to build structures using in situ resources such as regolith rock on the Moon, since sending raw construction material to the Moon is very difficult and expensive [36]. Another technology known as WASP (World's Advanced Saving Project) has focused on using AM technologies to build "zero-mile homes" that utilize on-site materials to creating housing in places where it is hard to find access to construction materials [21]. Research using in situ resources is still in its early stages, only proving the concept of building up layers of materials collected from surrounding areas, but further research is needed to decrease printing time, assure the desired material properties, develop customizable designs, and ensure repeatability.

5.3 Hybrid techniques

Hybrid systems that combine subtractive, formative, and additive processes could be implemented to facilitate the incorporation of AM in construction. Taking advantage of the benefits of each technique, while still exploring new materials, processes, and technologies provided by AM, will foster an environment for innovation in construction. An example is MIT's DCP which is a hybrid system with the intent of designing, sensing, and fabricating a component on site, trying to incorporate additive, subtractive, and assembling techniques in one all-purpose construction system [26]. MIT's DCP combines rapid foam extrusion to serve as formwork and insulation and milling techniques to provide desired finish rapidly [26]. Hybrid techniques can provide the first steps for introducing AM into the construction industry while further work is done to address challenges that limit the widespread use of AM in construction.

5.4 Off-site/On-site Fabrication

The use of AM has been geared mainly for off-site fabrication where a gantry or robotic arm system in a controlled environment is used for both small and large scale applications. The controlled environment is desirable as materials used can react differently and provide different mechanical properties if the environment is suddenly disturbed. The current on-site fabrication AM systems still require certain environmental conditions be met for best results. Alternatively, research can be done to develop materials for AM that are not sensitive to environmental factors for improved mechanical properties. Currently, off-site fabrication is the most reliable way to produce components using AM; however, technologies and materials suitable for on-site fabrication can be further developed to provide the most benefits in construction, reducing transportation of components to the site and at the same time providing capabilities to provide repairs and last minute design changes for structures in various stages of completion.

6 Potential Advancements in Applications of AM

AM is starting to gain recognition in the construction industry based on the success it has so far in the other industries. Several proof-of-concept experimental applications have already been implemented in construction, but furthermore research is needed to fully develop and improve the technology. Some potential applications ideal for AM technologies have been identified and are explained more in detailed below. The challenges and gaps that must be filled to fully develop and implement the technology in the construction sector are also discussed. Table 2 summarizes these potential applications, relates them to some of the existing experimental technologies, and comments on development needs.

Table 2. Potential advancements in applications of AM in the construction industry

Potential Applications	Examples of AM Technologies		Future Development Needs
Topology Optimization	D-Shape Contour Crafting Concrete Printing 3DCP* XtreeE	BAAM KamerMaker Skanska Arup Permasteelisa MX3D	Standards and testing/Quality control Precision Large scale testing Structural applications Bonding New design approach
Customized Parts	Skanska		Large scale additive manufacturing Structural applications Faster printing On-site printing
In situ repair			Identify areas that need repair Automation On-site printing Bonding
Novel Forms	D-Shape Contour Crafting Concrete Printing 3DCP* XtreeE CyBe* Apis-cor	BAAM KamerMaker C-Fab Skanska Arup MX3D DCP* FreeFAB™ Wax	Large scale additive manufacturing Structural applications Faster printing On-site printing
Tolerance Matching/Correcting			Identify areas that need repair Automation

* Currently Under Development

6.1 Topology optimization

AM's capability of producing components layer by layer can be used to produce structural elements with optimized topologies, where material is only deposited where it is most needed to resist structural demands. This advantage minimizes the use of material and reduces structural weight. XtreeE has investigated optimizing an element where the voids where material is not needed for structural purposes could provide additional functionality, such as thermal insulation and sound proofing [11,14].

Although these concepts of topology optimization have been explored by many, several challenges come into play when reducing the amount of material required for an assumed loading scenario, such as producing structures and components with reliable material properties and sufficient levels of safety. To provide the levels of safety expected by modern engineering standards, more detailed information on material properties, uncertainty, and quality assurance protocols is needed.

6.2 In situ repair

The potential benefits of using AM for in situ repair of existing structures are evident. As they age, buildings often require maintenance, rehabilitation, and/or replacement. Maintenance or in-situ repairs could be done using machines that could scan the structure, detect the areas needing a repair, and do the repairs using AM techniques or even using hybrid systems. Another potential application is to repair infrastructure that is damaged by a natural or man-made disasters. AM could be used to construct a temporary support structure inside this damaged building to allow for inspection and even restoration.

Future work is still needed to improve the automation process of placing material on an existing structure first before moving to more complicated and multi-step tasks of detecting current conditions of a structure, using subtractive processes to remove damaged areas, and then repairing what is needed. Referencing Table 2, it can be seen that no research has been done that includes both the use of AM and automation of tasks for repairs in construction. Most of the existing AM technologies for large-scale applications are not suitable for changing working area conditions that may be encountered during repair situations.

6.3 Customized parts

Although construction cost can be reduced by reducing labour, material, transportation, and time required for a project, past studies have indicated use of AM in construction may significantly increase costs. Mrazovic [30] states 3D printing steel components using powder bed fusion is possible but will likely be cost prohibitive, mainly due to the processing speed. Conversely, WinSun claims that labour costs for the Dubai Future Foundation building was reduced by 50 to 80% using AM [33]. Once the materials and machines become more common in the market, the cost associated with using AM will be reduced through economy of scale [37]. Until now AM appears to be most economical for producing unique parts.

The idea has been already experimented for non-structural applications, like the work done by Skanska to print unique claddings. During construction there may be instances where delaying design and/or production of a part may be advantageous, such as customized parts built specifically for as-built dimensions. Similarly, there may be cases where components are lost or damaged and waiting on their replacements may cause construction delays. These examples are situations where "last minute" production of components on-site using AM can add value in construction.

6.4 Novel Forms

Architects started using AM of small-scale building models as a way to present a concept of their design to a customer. Through the advancements of large-scale AM, it is allowing architects to produce more complex interior and exterior geometries that would be difficult and costly to produce using conventional subtractive and formative processes.

Some have recognized that AM could start adding value to construction immediately, just by using it to print formwork with enhanced features such as functionality in the finished building, odd geometries, biodegradable materials, or reusability [26,27]. As an example, FreeFAB™ Wax proposed printing moulds using wax material that can be created rapidly with low precision and can later be cut to the precise shape use milling techniques. Wax would act as the mould during construction, and then it could be heated to recover and reuse the material for other moulds.

6.5 Tolerance matching/correcting

Another potential application is the use of AM for tolerance matching. The construction industry is often faced with the challenge of having elements or modules on the construction site that do not have precise dimensions or sufficient tolerance for assembling them, requiring on-site modifications and complicating the assembly process. Matching issues with prefabricated components are due to the inability to maintain tight tolerances, introducing errors that propagate during construction that could risk the integrity of the structure [38,39]. AM brings the benefit of producing components with precise dimensions based on the design drawings, important for modular construction, and at the same time the ability to match tolerances in real time by printing customized connectors or infill as needed on-site.

7 Conclusion

Building-up elements from small deposits of materials using AM is having an impact in many industries, closing the gap of other manufacturing methods such as formative and subtractive processes. Aerospace, automotive, and other industries have explored the benefits of using AM in their day-to-day activities, finding new applications for different AM processes. The construction industry has become interested and has started exploring proof-of-concept AM applications that could be applied in the sector, looking to mitigate some of their current challenges.

Examples of AM technologies have been abundant, but it has been identified that most current work in the construction industry has been focused in material extrusion process using aggregate-based material for large-scale applications. Work with polymers have been effective for aesthetic purposes that have unique designs. Meanwhile AM applications using metallic materials for large-scale components is the least explored area due to its high cost.

Potential applications of AM such as optimized topologies, customized parts, in situ repair for construction in areas with limited access, and others, have been summarized and related to current AM technologies. These potential applications highlight opportunities in the construction industry that can be realized with AM. Interdisciplinary research is still needed to make AM a reliable and viable option in construction. To facilitate implementation of AM technologies and realize these potential applications, further work is needed investigating ways to print using multiple materials, to use in situ resources, and to combine AM with other processes as hybrid techniques.

This field is still in its infancy, without standardized testing and quality control to compare or benchmark these recent advancements. Furthermore, most of these early projects and AM technologies are often proprietary, lacking publically available, detailed information on their methodology and work, making comparison or evaluation of new AM technologies more challenging. While much work is needed to fully realize AM as a cost-effective and reliable option in the construction industry, the potential benefits it can provide are worthy of further research and development.

8 References

[1] ISO/ASTM 52900-15, Standard Terminology for Additive Manufacturing-General Principles-Terminology, (2015).

[2] J. Pegna, Exploratory investigation of solid freeform construction, Autom. Constr. 5 (1997) 427–437.

[3] A. Warszawski, R. Navon, Implementation of robotics in building: Current status and future prospects, J. Constr. Eng. Manag. 124 (1998) 31–41.

[4] B. Khoshnevis, G. Bekey, Automated Construction Using Contour Crafting–Applications on Earth and Beyond, Nist Spec. Publ. Sp. (2003) 489–494.

[5] S. Keating, Beyond 3D Printing: The New Dimensionsof Additive Fabrication, Des. Emerg. Technol. UX Genomics Robot. Internet Things. (2014) 379.

[6] D.W. Johnston, Design and construction of concrete formwork. In E. G. Nawy (Ed.), Concrete construction engineering handbook. (pp. 7.1-740) Boca Raton, Fla.: CRC Press, (1997).

[7] I. Gibson, D. Rosen, B. Stucker, Additive Manufacturing Technologies, Springer New York, New York, NY, 2015.

[8] 3D Systems, ProJet® MJP 3600 Dental, (2016). https://www.3dsystems.com/3d-printers/professional/projet-mjp-3600-dental (accessed February 17, 2017).

[9] R.P. Mudge, N.R. Wald, Laser engineered net shaping advances additive manufacturing and repair, Weld. J.-N. Y.-. 86 (2007) 44.

[10] N. Labonnote, A. Rønnquist, B. Manum, P. Rüther, Additive construction: State-of-the-art, challenges and opportunities, Autom. Constr. (2016).

[11] C. Gosselin, R. Duballet, P. Roux, N. Gaudillière, J. Dirrenberger, P. Morel, Large-scale 3D printing of ultra-high performance concrete – a new processing route for architects and builders, Mater. Des. 100 (2016) 102–109.

[12] D-Shape Enterprises L.L.C., (n.d.). https://dshape.wordpress.com/ (accessed March 14, 2017).

[13] B. Khoshnevis, R. Dutton, Innovative Rapid Prototyping Process Makes Large Sized, Smooth Surfaced Complex Shapes in a Wide Variety of Materials, Mater. Technol. 13 (1998) 53–56.

[14] S. Lim, R.A. Buswell, T.T. Le, S.A. Austin, A.G.F. Gibb, T. Thorpe, Developments in construction-scale additive manufacturing processes, Autom. Constr. 21 (2012) 262–268.

[15] L.Y. Feng, others, Study on the Status Quo and Problems of 3D Printed Buildings in China, Glob. J. Hum.-Soc. Sci. Res. 14 (2014).

[16] A. Rudenko, 3D Castle Completed, (2014). http://totalkustom.com/3d-castle-completed.html (accessed September 7, 2016).

[17] M. Molitch-Hou, BetAbram Set to 3D Print Two-Story House This Summer, (n.d.). https://3dprintingindustry.com/news/betabram-set-to-3d-print-two-story-house-this-summer-50826/ (accessed September 8, 2016).

[18] F. Bos, R. Wolfs, Z. Ahmed, T. Salet, Additive manufacturing of concrete in construction: potentials and challenges of 3D concrete printing, Virtual Phys. Prototyp. 11 (2016) 209–225.

[19] CyBE Construction, (n.d.). http://www.cybe.eu/ (accessed September 9, 2016).

[20] Apis cor 3D printer, (n.d.). http://apis-cor.com/en/3d-printer (accessed January 30, 2017).

[21] WASP, The Big Delta 12 m, (n.d.). http://www.wasproject.it/w/en/3d-printing/bigdeltawasp-12m/ (accessed September 7, 2016).

[22] IAAC, Minibuilders, (2013). https://iaac.net/research-projects/large-scale-3d-printing/minibuilders/ (accessed March 14, 2017).

[23] L.J. Love, Utility of Big Area Additive Manufacturing (BAAM) For The Rapid Manufacture of Customized Electric Vehicles, 2015.

[24] DUS Architects, 3D print canal house, (n.d.). http://3dprintcanalhouse.com/ (accessed September 5, 2016).

[25] A. Zaleski, Chattanooga startup Branch Technology wants to 3D print houses, (2015). http://fortune.com/2015/07/30/chattanooga-3d-printed-house/ (accessed February 2, 2017).

[26] S. Keating, N.A. Spielberg, J. Klein, N. Oxman, A Compound Arm Approach to Digital Construction, in: W. McGee, M. Ponce de Leon (Eds.), Robot. Fabr. Archit. Art Des. 2014, Springer International Publishing, Cham, 2014: pp. 99–110.

[27] J.B. Gardiner, S. Janssen, N. Kirchner, A Realisation of a Construction Scale Robotic System for 3D Printing of Complex Formwork, in: ISARC Proc. Int. Symp. Autom. Robot. Constr., Vilnius Gediminas Technical University, Department of Construction Economics & Property, 2016: p. 1.

[28] Construction Manager Magazine, Skanska claims first with 3D printed cladding, (2013). http://www.constructionmanagermagazine.com/news/skanska-claims-industry-first-3d-printed-cladding/ (accessed October 26, 2016).

[29] S. Galjaard, S. Hofman, N. Perry, S. Ren, Optimizing Structural Building Elements in Metal by using Additive Manufacturing, in: Proc. Int. Assoc. Shell Spat. Struct. IASS Symp., 2015.

[30] N. Mrazovic, M.A. Eng, FEASIBILITY STUDY, (n.d.). http://cife.stanford.edu/sites/default/files/FeasibilityStudy3DPrinting4Permasteelisa.pdf (accessed October 26, 2016).

[31] S.K. Joosten, Printing a stainless steel bridge: An exploration of structural properties of stainless steel additive manufactures for civil engineering purposes, TU Delft, Delft University of Technology, 2015.

[32] D. Hwang, B. Khoshnevis, An innovative construction process-contour crafting (CC), in: 22nd Int. Symp. Autom. Robot. Constr. ISARC 2005 Ferrara Italy, 2005.

[33] H. Busta, Gensler Completes the World's First 3D-Printed Office Building, Architect. (2016). http://www.architectmagazine.com/technology/gensler-designs-the-worlds-first-3d-printed-office-building-in-dubai_o (accessed November 4, 2016).

[34] J. Gregerson, Inside the Very First 3D-Printed Office Building, (2016). http://www.builtworlds.com/news/2016/6/8/inside-the-worlds-largest-3d-printed-office-building (accessed February 2, 2017).

[35] Y. Huang, M.C. Leu, J. Mazumder, A. Donmez, Additive manufacturing: current state, future potential, gaps and needs, and recommendations, J. Manuf. Sci. Eng. 137 (2015) 14001.

[36] R.P. Mueller, S. Howe, D. Kochmann, H. Ali, C. Andersen, H. Burgoyne, W. Chambers, R. Clinton, X. De Kestellier, K. Ebelt, others, Automated Additive Construction (AAC) for Earth and Space Using In-situ Resources, in: Proc. Fifteenth Bienn. ASCE Aerosp. Div. Int. Conf. Eng. Sci. Constr. Oper. Challenging Environ. Earth Space 2016, American Society of Civil Engineers, 2016.

[37] D.S. Thomas, S.W. Gilbert, Costs and Cost Effectiveness of Additive Manufacturing, National Institute of Standards and Technology, 2014.

[38] Y. Shahtaheri, C. Rausch, J. West, C. Haas, M. Nahangi, Managing risk in modular construction using dimensional and geometric tolerance strategies, Autom. Constr. (2017).

[39] V.S. Kalasapudi, P. Tang, C. Zhang, J. Diosdado, R. Ganapathy, Adaptive 3D Imaging and Tolerance Analysis of Prefabricated Components for Accelerated Construction, Procedia Eng. 118 (2015) 1060–1067.

Autonomous Big-Scale Additive Manufacturing Using Cable-Driven Robots

Ilija Vukorep [a]

[a]Chair of Digital Design Department, Brandenburg University of Technology Cottbus-Senftenberg, Germany
E-mail: ilija.vukorep@b-tu.de

Abstract –

Cable-driven parallel robots are a rather novel robotic system, with an advantage over traditional system in not occupying the ground level as it operates from the top down. This paper presents a possibility of additive construction similar to 3D printing at the scale of building parts or at house-size. This can be achieved by stacking elements as well as using a concrete-printing head. For this, a demonstrator is developed that is controlled by an architectural parametric tool that imports the design, checks the size/building limits/cable conflicts, creates paths, simulates the building process, and sends out the G-code to the robot. The goal is to use this robotic system on building sites as a flexible easy-to-use construction robot or for prefabrication purposes.

Keywords –

Additive Manufacturing; Cable Robot; Construction Robot; Simulation

1 Introduction

When thinking of robots fulfilling a function in a bounded area, we usually view them as ground–based machinery maneuvering around obstacles. Other classes of robots are the industrial arm robot, with various degrees of freedom, and the so-called 'delta' robots, often used on production lines for stapling, arranging, and packing. Because of their fixed requirements and rigid arms they cannot be easily placed in construction surroundings or used in the production of buildings. Over the last few years, a new class of 3D printing machines have been developed that can build complex and even large objects. However, these printing machines are not suitable for very large constructions, especially on-site and in challenging environments.

As an alternative to traditional robotic systems, we are researching cable-driven parallel robots in the context of building construction. These autonomous robotic systems move through space using ropes supported by an overlying and/or underlying construction. This gives the robot better access and operability for construction sites. They are light and cover a big operational area.

Similar to a giant 3D printer, this system can serve different needs in the building process. There are several construction scenarios, such as stacking elements on top of each other (similar to a brick wall) and layering fast-curing materials or two-component adhesive systems (traditional 3D printing). Since the robot is controlled by a minimum of six ropes, it also has six degrees of freedom, which provides additional functionality.

Positioning is achieved by changing the lengths of the ropes. This requires a winch system placed either at the cable ends (external) or inside the central moving unit (internal). In this work an external winch system is applied. The moving platform will have additional features that can be added to the robot, depending on the required function.

This paper describes the current research stage of this robotic system and its possibilities for autonomous building processes in big scale.

2 Related Work

Rope-suspended machines have been primarily used in film-making and television broadcasting. Their main effect is to access vantage points that are difficult or impossible to reach. The most popular commercial systems are "Spidercam" from CCSytems Inc. and a very similar "Skycam" from Outdoor Channel Holdings, Inc. (Figure 1).

In recent years, there has been considerable research effort at the Alto University in Helsinki, Finland, under the name of "Ceilbot" [1]. The main idea here is the development of a multifunctional robot acting from the ceiling. The work of one group [2] is showing the possibilities of a ceiling-mounted fire extinguisher robot operated on a rail-based system. Another group developed the concept of the Ceilbot Robot [3], a service robot system based on ropes.

Figure 1. Hanging Skycam (wire-supported camera) skycam.com [4]

New technical possibilities for free-motion in space and stacking architectural modular elements are further developed by Gramazio and Kohler and Raffaello D'Andrea from the ETH Zürich. The idea of "Flight-Assembled Architecture" for building vertical structures uses a swarm of Quadrocopters that autonomously stack boxes. This system benefits from free upper space areas for maneuvers and new self-organized robotic swarm techniques. Short battery life of a single machine and low load capabilities are making this system not suitable for real construction scenarios.

There are a few recent examples of robots with a linear top-to-bottom motion system, such as Aaron Fan's "Pythagoras" and "Kritzler" from Alexander Weber, which works along two axes. In the "Makers & Spectators" exhibition, Niels Völkers, an artist and communication designer, developed a very similar system out of LEGO Mindstorms parts that unifies the hanging robot together with the drive mechanism.

The longest working cable-driven robot is RoboCrane from NIST, with a working age of 16 years. It is based on an inverse Steward Platform, where cables under tension replace the linear actuators. Six cables give the platform six degrees of freedom.

In 2014, Thomas Monroy and Taole Chen developed a "Sky Printer" [5], which is a hanging device suspended by three ropes that 3D-prints forms out of fast-curing material. This system is interesting for its simple construction and setup, but is missing precise positioning due to jiggling.

Fraunhofer has been developing an 8-cable-driven system "IPAnema Robot" series, with cables coming from each space corner [6]. The project group, under Dr Andreas Pott, has published numerous articles regarding this system and has made significant contributions for further development.

Parallel to Fraunhofer, a group around Tecnalia Institute is working on CoGiro an 8-cable-driven robot, with cables located only in four corners [7]. Demonstrations of movement and precision are proven to be very reliable. As this system can also carry weights and leaves the lower workspace area free of cables, it is suitable for construction purposes.

1. Autonomous Building Robot

Cable-driven hanging robots can be used for various purposes in a defined space. The focus of this work is to develop a 6-DOF moving unit that can be used for building applications. By permitting flexible movement in the upper areas of a construction site, a cable-driven robot can stack building elements in a particular order or 3D-printing large-sized objects. Such an autonomous building system that is also mobile can be installed in dynamic environments with moving and intersecting additional actors and is the higher goal of further research. In addition, several other applications as end-effectors will be the subject of upcoming research.

3 Eight Cables - Description

The main difficulty faced in the development of rope-supported systems is the precise positioning of the hanging device (end-effector). This is mainly because the device hangs on three or four ropes in one level. As the angles between the hanging object and the ropes were not adjustable, a deviation from the null position increased towards the border of workspace (see Figure 2).

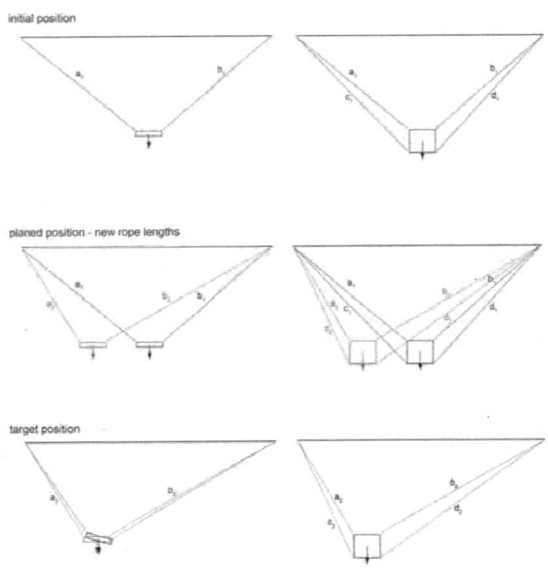

Figure 2. One- and two-level wire systems and deviations from the target position.

By adding an additional level of ropes to the system, a much more stable, rigid, and wider workspace range is achieved [8]. Furthermore, this increases the operational degrees of freedom (from 3-DOF to 6-DOF). This development, therefore, is extremely useful for robotic building and stacking purposes where simple rotation operations can be handled without additional motors.

A better and more precise transition of an end-effector can be reached by setting a gimbal-system to the hanging object that points the end-effector to the desired direction; however, even this system cannot avoid shaking and unstable movements. Further development will combine a gimbal system with a two-level wire-system.

4 Research Stage

For testing purposes, a machine was designed that has fixed box side dimensions of 2m. With fixed cable positioning at the top corners of the system, the robot belongs to the Suspended Wire Robot type that needs external forces as the gravity of the carrying object to keep the tension of the cables. T. Bruckmann at el. are distinguishing this kind of cable robots from the Fully-tensed Wire Robot type that is having cables from each corner of the system and so more stability and less vibration [9]. Because collisions with the cables are minimized together with the option of building more complex and space-filling forms, Suspended Wire Robot type is chosen for the robot prototype that will be described in the next chapters.

4.1 Robot Design

The whole robotic hardware system consists of:

- frame 2000 x 2000 x 2000 mm with two winches in each lower corners,
- wires connecting the winches and the moving platform,
- moving platform with added end-effector that can be changed in dependency of the function,
- motor controllers that are rotating the winches,
- computer for sending commands to the controllers and to the end-effectors.

The software parts are:

- cam-software for making brick walls out of simple surfaces, calculating trajectories, controlling collision free positions, and exporting positions and actions for the controlling-software,
- Robot Control software that is transforming positions and actions into g-code and sending it to the controllers of the robot,
- Motor-software, interpreting the g-code and calculating the motor motion.

Measuring the rope length precisely is the most difficult task in the robot design. The winch is separated from the deflection pulley over whole height of the robot. Deflection pulleys are placed on top and the winches are placed on bottom corners. This distance minimized the length differences through rope movement along the winch axes. Motor controllers, that are usually used for controlling 3D-printers, are installed for moving the winch motors. The benefit is the build-in function of processing the G-code that enables a smooth and coordinated movements of the motors. Due to the large number of motors, two controllers are necessary for the control of each of the four motors. To coordinate the movement of each motor connected to the different controller a feed rate is calculated prior of sending the G-code. This is necessary to get a balanced interaction of the motors especially when some motors have much less movement then others and they have to adapt the feed rate in order to stop moving at the same time.

Figure 3 is showing the processing steps from manually drawing a 3D-wall to the stage of stacking the bricks.

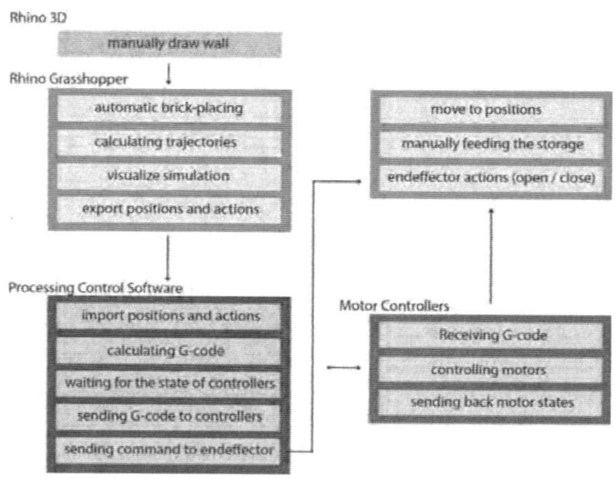

Figure 3. Chart showing the process with the different software clusters

The Robot Control Software divides between controlling motors for moving the winches and the motors for moving the end-effector. This is necessary because the end-effector is controlled through a wifi-connection and a battery-energy supply. A wired connection would make the prototype setup technically more complicated. For further development, a wired connection will give more possibilities with stronger end-effectors and better controlling options.

4.2 Trajectories Planing

The movement of the robot along a predefined path, called trajectories, is done by a software that is commonly used by architects and engineers called Rhinoceros by Robert McNeel & Associates. The parametric tool Grasshopper, a plug-in of Rhinoceros, is planning the position of the bricks, collision checking, and automatically delivers the trajectories. They are defined as a sequence of locations and rotation information of each position in relation to a reference point. These positions are combined with action commands (Fig. 4) and sent out to a file that has been read by the Robot Control Software. The visual control of the trajectories in form of a continuous poly-line can help in tracking errors.

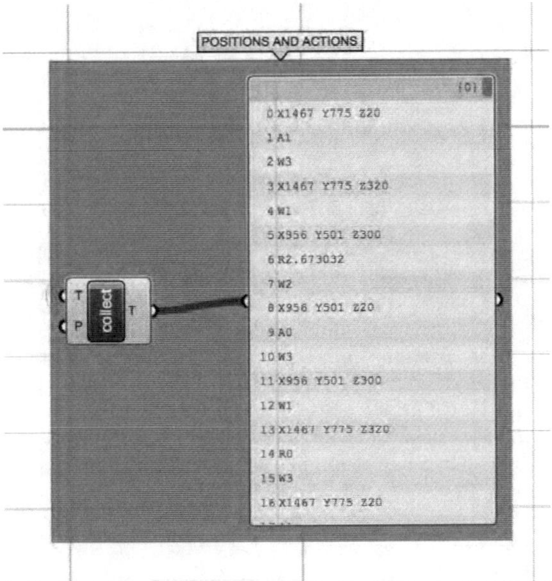

Figure 4. List of positions, wait, and action commands from Grasshopper/Rhinoceros that is then exported to Robot Control Software

4.3 Simulation

For the purpose of getting the position of the bricks in the wall a virtual parametric model is realized within the Grasshopper in Rhinoceros. All non moving objects are built as fixed 3D-objects (Fig. 5) and objects that are moving are part of the parametric Grasshopper model. A surface is defining the main brick positions and the parametric model is placing bricks along the surface automatically with incorporated balance testing. This helps to build up a wide range of testing arrangements.

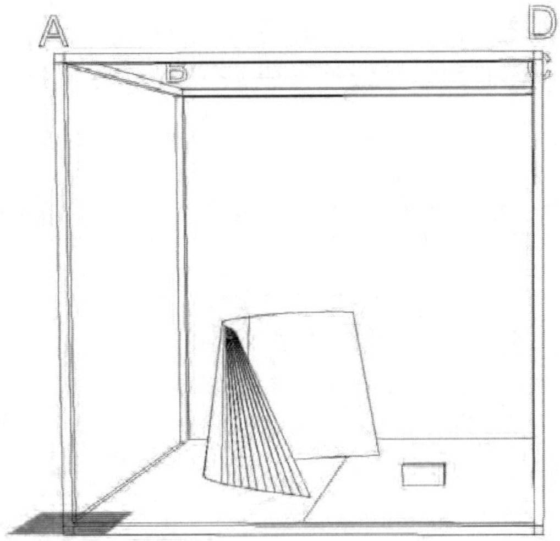

Figure 5. Frame of the robot with the initial feeding position and the wall surface. Bricks are positioned along this surface.

An animation of the brick building is showing all movements with possible verification of the sequence (Fig. 7). Improvement to this system will be the automatic calculation of the optimized brick feeding position as well as moving feeding position vertically. Another improvement will incorporate positions that are not reachable by the robot mostly on the edges of the frames as well as critical height positions.

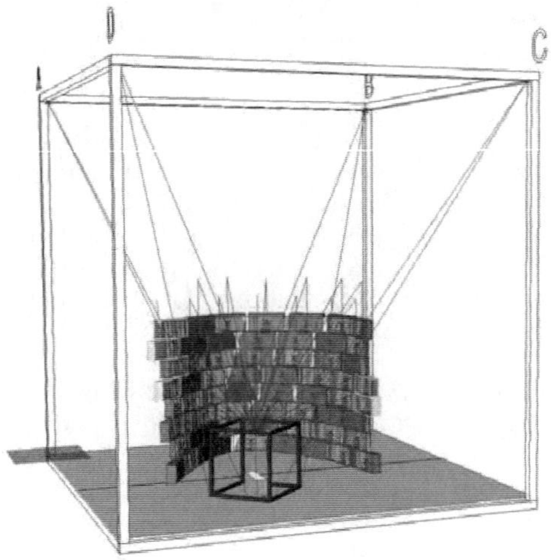

Figure 6. Parametric model with visualization of the trajectories. Yellow marked bricks will not be

placed due to their unstable position.

Similar to any CAM-Software this step is processing the positions to the Robot Control Software.

4.4 Prototype

Machine controlling is realized by the free available programming tool Processing [10] that has good scope for visualizing information and server communication to the two controller boards of the robot over a serial interface. All incoming positions are translated to rope lengths and to winch positions as G-commands. Sending this commands with the positions is moving the winches. The machine waits for the feedback of the motors, to get their actual position and is sending new commands only when these new positions are reached.

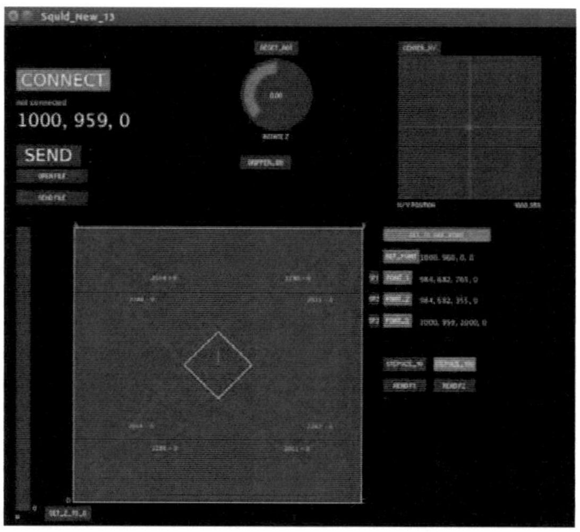

Figure 7. Robot Control Software that controls the machine. It loads the positions from a separate file made by the previous simulation.

A simple gripper is getting a separate non G-Code command, as it is not connected to the controllers. A wireless wi-fi connection and its own battery are activating the end-effector. However, more complex end-effectors may need a wire connection that can be provided from top-middle of the frame.

For this testing prototype styrofoam blocks were used to see the capabilities of brick stacking. Moving real bricks will need stronger motors and eventually a bigger frame.

Figure 8. Robot in action stacking the bricks

5 Summary and Conclusion

This paper shows a step in additive manufacturing for buildings in a larger scale. This will lead to scaling the robot in space and force towards further development of automatized construction of building walls. Using cable-driven robots for additive fabrication in real construction sites is still not possible, but the presented prototype is showing that there is a future for applications on building construction sites. For the model-simulation an industrial standard is chosen that is near architectural production. The Suspended Wire Robot was chosen due to better collision properties between wires and the constructed object. Next to adding bricks in layers, there are several other possible usage scenarios for cable-driven robots:

- addition to and in combination with normal cranes on construction sites for precise moving elements,
- addition of scanning devices for complex 3D-scanning,
- addition of 3D-printing devices for the production of huge scale 3D-printed objects.

Next steps will be the development of non-linear motor controls, especially for the robot as a mobile version with variable corners. Subsequently, self referencing features will help as well as an optical object recognition system for a better feeding process. A vertically adjustable feeding station will help to build higher objects. Furthermore, the optimization of the data transfer from the simulation software to the robot control software will simplify the overall testing.

References

[1] "Ceilbot - AS Department - Aalto University." 2012. http://autsys.tkk.fi/en/Ceilbot/.

[2] Munoz, Julio Cordon. 2009. "Preliminary Design of a Ceiling-Mounted Fire Extinguisher Robot." Helsinki University of Technology.

[3] Mercier, Guillaume, and Massinissa Ait-Gherbi. 2009. "Ceilbot_final_report_sports_hall." http://autsys.tkk.fi/en/attach/Ceilbot/Ceilbot_final_report_sports_hall.pdf.

[4] Image of a Skycam at work during a Washington Huskies football game at Husky Stadium in Seattle, Wash., 2008, photograph by Despeaux, distributed under a CC-BY 3.0 license.

[5] http://diy3dprinting.blogspot.de/2014/12/sky-printer-winch-suspended-large-clay.html

[6] Bruckmann, Tobias, and Andreas Pott, eds. 2013. *Cable-Driven Parallel Robots*. Vol. 12. Mechanisms and Machine Science. Berlin, Heidelberg: Springer Berlin Heidelberg.

[7] Dit Sandretto, Julien Alexandre, David Daney, and Marc Gouttefarde. 2013. "Calibration of a Fully-Constrained Parallel Cable-Driven Robot." In *Romansy 19 – Robot Design, Dynamics and Control*, edited by Vincent Padois, Philippe Bidaud, and Oussama Khatib, 544:77–84. Vienna: Springer Vienna.

[8] Johann Lamaury and Marc Gouttefarde. 2013. "A Tension Distribution Method with Improved Computational Efficiency" In Cable-Driven Parallel Robots, edited by Bruckmann, Tobias, and Andreas Pott, 71-85. Berlin, Heidelberg: Springer Berlin Heidelberg.

[9] Bruckmann, T., H. Mattern, A. Spengler, C. Reichert, A. Malkwitz, und M. König. 2016. „Automated Construction of Masonry Buildings using Cable-Driven Parallel Robots". In *ISARC. Proceedings of the International Symposium on Automation and Robotics in Construction*, 33:1. Vilnius Gediminas Technical University, Department of Construction Economics & Property.

[10] www.processing.org. Processing was initiated by Ben Fry and Casey Reas.

Current Progress of 3D Concrete Printing Technologies

Behzad Nematollahi[a], Ming Xia[b] and Jay Sanjayan[c]

[a,b,c] Center for Sustainable Infrastructure, Swinburne University of Technology, Melbourne, Australia
E-mails: bnematollahi@swin.edu.au[a], mxia@swin.edu.au[b], jsanjayan@swin.edu.au[c]

Abstract –

The construction industry is expected to go through large transformations since construction automation is anticipated to drastically alter standard processing technologies and could lead to possible disrupting technologies such as 3D concrete printing (3DCP). While 3D printing techniques have been successfully applied in a wide range of industries such as aerospace and automotive, its application in concrete construction industry is still in its infancy. 3DCP can allow freeform construction without the use of expensive formwork, which in return offers excellent advantages compared to conventional approach of casting concrete into a formwork. In the last few years, different 3DCP technologies have been developed. This paper presents the current progress of 3DCP technologies. An innovative methodology recently developed by the authors of this study for formulating geopolymer-based material for the requirements and demands of commercially available powder-based 3D printers is also briefly presented.

Keywords –

Additive manufacturing; 3D printing; 3D concrete printing; concrete construction; geopolymer

1 Introduction

Concrete is the most widely used construction material on this planet. The current concrete construction industry faces several challenges. One of them is the high cost. According to a recent study conducted by Boral Innovation Factory [1], formwork is responsible for about 80% of the total costs of concrete construction in the Sydney CBD (central business district). In fact, this is typical for concrete construction worldwide.

The significant amount of wastage generated in the construction is another challenge. Formwork is a significant source of waste, since all of it is discarded sooner or later, contributing to a generally growing amount of waste in the construction industry. Astonishing data presented in Llatas [2]'s paper showed that the construction industry is responsible for generating approximately 80% of the total waste in the world.

Furthermore, the conventional approach of casting concrete into a formwork limits geometrical freedom for the architects to build in various geometries, unless very high costs are paid for bespoke formworks. Rectilinear forms not only limit the creativity of the architects, but they are also structurally weaker than curvilinear forms owing to stress concentration.

Another challenge is the slow speed of construction (i.e. long and hard to control lead time). The concrete construction often comprises many steps including material production, transportation, and in-situ manufacture of formwork, and each step is time consuming.

Moreover, the current concrete construction industry is labor intensive and has issues with safety. According to Safe Work Australia's report [3], on average, 35 construction employees per day are seriously injured in Australia. In addition, over one-quarter of construction deaths are caused by falls from a height [3]. This is despite the fact that Australia has one of the highest levels of safety regulations in construction sites in the world.

Last but not least, the current construction industry has serious issues with sustainability. In general, the current construction methods and materials are not environmentally friendly. The entire construction process, including off-site manufacturing, transportation of materials, installation and assembly, and on-site construction, emits huge amounts of greenhouse gases and consumes large quantities of energy [4]. In addition, conventional concrete made by ordinary Portland cement (OPC) is not sustainable. Manufacture of OPC is highly energy and carbon intensive [5].

Application of three-dimensional (3D) printing techniques in concrete construction could solve the aforementioned challenges. 3D printing technology is recently gaining popularity in construction industry. In the last few years, different 3D concrete printing (3DCP) technologies have been explored. This paper presents the current progress of 3DCP technologies. An innovative methodology recently developed by the authors of this study for formulating geopolymer-based material for the requirements and demands of commercially available powder-based 3D printers is also briefly presented.

2 Background

3D printing, also known as additive manufacturing (AM), is a group of emerging techniques for fabricating 3D structures directly from a digital model in successive layers with less waste material. The American Society for Testing and Materials (ASTM) International Committee F42 on AM technologies defines AM as "the process of joining materials to make objects from 3D model data, usually layer upon layer" [6]. The AM technologies have been initially developed in the 1980s. Currently, AM technologies have become an integral part of modern product development and have been successfully applied in a wide range of industries including aerospace and automotive manufacturing, biomedical, consumer and food [7].

3 Application of 3D Printing in Concrete Construction

The first attempt to adopt AM in construction using cementitious materials was made by Pegna [8]. An intermediate process was used to glue sand layers together with a Portland cement paste [8]. Unlike the conventional approach of casting concrete into a formwork, 3DCP will combine digital technology and new insights from materials technology to allow freeform construction without the use of expensive formwork. The freeform construction would enhance architectural expression, where the cost of producing a structural component will be independent of the shape, providing the much needed freedom from the rectilinear designs.

When compared with conventional construction processes, the application of 3D printing techniques in concrete construction may offer excellent advantages incluidng:

1. Reduction of construction costs by eliminating formwork;
2. Reduction of injury rates by eliminating dangerous jobs (e.g., working at heights), which would result in an increased level of safety in construction;
3. Creation of high-end-technology-based jobs;
4. Reduction of on-site construction time by operating at a constant rate;
5. Minimizing the chance of errors by highly precise material deposition;
6. Increasing sustainability in construction by reducing wastages of formwork,
7. Increasing architectural freedom, which would enable more sophisticated designs for structural and aesthetic purposes; and
8. Enabling potential of multifunctionality for structural/architectural elements by taking advantege of the complex geometry [9,10].

4 Current 3D Concrete Printing Technologies

In the last few years, different 3DCP technologies have been developed to adopt AM in concrete construction. These technologies are mainly based on two techniques, namely extrusion-based and powder-based. In the following sub-sections these techniques are explained and the currently available 3DCP technologies including the powder-based 3DCP using geopolymer developed by the authors of this study are reviewed. The similarities and differences and the pros and cons of different 3DCP technologies are highlighted.

4.1 Extrusion-Based Technique

The extrusion-based technique is analogous to the fussed deposition modelling (FDM) method which extrudes cementitious material from a nozzle mounted on a gantry, crane or a 6-axes robotic arm to print a structure layer by layer. This technique has been aimed at on-site construction applications such as large-scale building components with complex geometries, and has a great potential to make a significant and positive contribution to the construction industry. A schematic of the powder-based technique is illustrated in Figure 1.

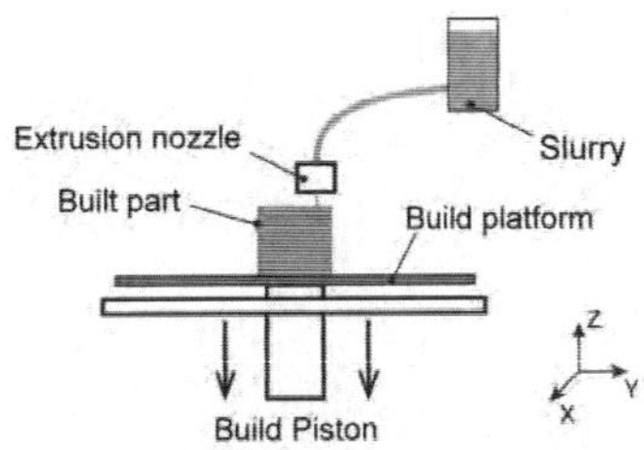

Figure 1. Schematic of extrusion-based technique

4.1.1 Contour Crafting

Contour Crafting (CC) technology has been developed at the University of Southern California, USA. This technology uses the extrusion-based technique to extrude two layers of cementitious mixture to build a vertical concrete formwork. Custom-made reinforcement ties are manually inserted between layers (at every 30 cm horizontally and 13 cm vertically) while the CC machine is constantly extruding the layers. Trowel-like fins are attached to the print head to create smooth extruded surfaces. Once the extruded formwork is completed, concrete is then manually poured to a height of 13 cm and a second batch is poured on top of the first batch after one hour. A one hour delay batch is to control the lateral pressure of the concrete by allowing it to partially cure and harden [11]. A concrete wall form fabricated by the CC machine is shown in Figure 2.

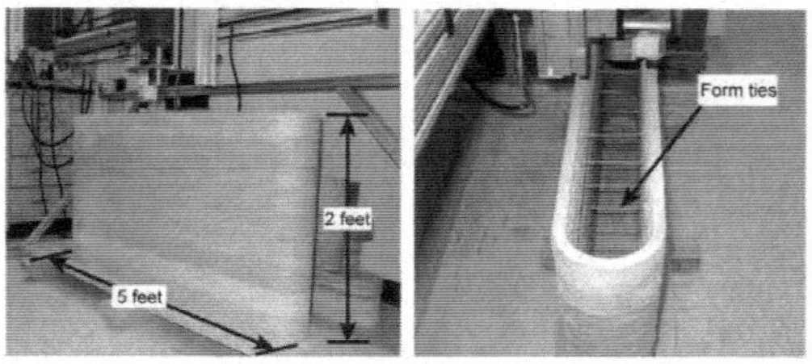

Figure 2. A concrete wall form fabricated by the CC machine with custom-made reinforcement ties manually inserted between layers [11]

The chief advantages of the CC technology are the superior surface finish and the greatly enhanced speed of fabrication. Other key advantage of CC is possibility of integration with other robotics methods for installing internal components such as pipes, electrical conductors, and reinforcement modules to enhance mechanical property [11]. The CC technology currently produces vertical elements largely in compression. When a doorway or window is required a lintel is placed to bridge the gap and the wall can be placed above. Therefore, it avoids the cantilever problem [12].

Gosselin et al. [10] reported the following drawbacks for the CC technology: (1) this technology is limited to vertical extrusion, hence yielding 2.5D topologies (vertical extension of a planar shape), (2) the initial formwork and trowel system can be rather complex to implement for production, depending on the size and shape of the object being printed, and (3) the interrupted sequential casting of concrete within the formwork due to hydrostatic pressure and weak mechanical properties of the extruded concrete may result in weakened interfacial zones between the layers.

4.1.2 Concrete Printing

Concrete Printing technology has been developed by a team at Loughborough University in the United Kingdom. This technology also uses the extrusion-based technique and to some extent is similar to the CC technology. However, the Concrete Printing technology has been developed to retain 3D freedom and has a smaller resolution of deposition, which allows for greater control of internal and external geometries [12]. In addition, the material used in Concrete Printing is a high performance fiber-reinforced fine-aggregate concrete, resulting in superior material properties to those obtained in the CC technology [12].

Figure 3 shows a full scale bench fabricated using Concrete Printing. The bench was 2 m long, 0.9 m maximum width and 0.8 m high and comprised of 128 layers of 6 mm thickness. The bench includes 12 voids that minimize weight, and could be utilized as acoustic structure, thermal insulation, and/or path for other building services. The bench also demonstrates a reinforcement strategy where carefully designed voids form conduits for post placement of reinforcement [12].

Figure 3. A full scale bench fabricated by the Concrete Printing with functional voids and post-tensioned reinforcement [12]

Concrete Printing requires additional support to create overhangs and other freeform features. It uses a second material, in a similar manner to the FDM method. The disadvantage of this process is that an additional deposition device is needed for the second material resulting in more maintenance, cleaning and control instructions and the secondary structure must be cleaned away in a post processing operation [12].

Gosselin et al. [10] reported the following drawbacks in regards to the Concrete Printing technology: (1) the trade-off necessary for maintaining its dimensional accuracy makes the process quite slow with regards to the envisioned industrial application, (2) although the technology initially aimed at the generation of 3D topologies rather than 2.5D, the use of second material to support overhangs reduces the efficiency and flexibility of the process while increasing its material cost, and (3) dimensions and possibilities in terms of shape-design are limited by the dimensions of the printing frame.

4.1.3 CONPrint3D: Concrete On-Site 3D Printing

Contour Crafting and Concrete Printing technologies, while demonstrating many technological advantages, are subjected to some inherent limitations such as the necessity of using new and advanced machinery, small mineral aggregate sizes (fine-aggregate mortar rather than concrete), and limited size of the printed elements (i.e. the size of the 3D printer must be larger than the size of the element to be printed). To overcome these limitations, a novel approach for 3DCP technology for on-site construction, named CONPrint3D, is currently being developed at the TU Dresden, Germany, which intends to bring 3DCP directly into the building sites. The main advantages of CONPrint3D technology are high geometrical flexibility, usage of commonly used construction machinery and low dependency on skilled labor [13].

One of the focal points of CONPrint3D is not only to develop a time, labor and resource efficient advanced construction process but also to make the new process economically viable while achieving broader acceptance from the existing industry practitioners. This is achieved by using existing construction and production techniques as much as possible and by adapting the new process to construction site constraints. One vital aspect of the project strategy is adapting a concrete boom pump to deliver material to specific positions autonomously and accurately using a custom-developed print head attached to the boom (see Figure 4) [13].

Figure 4. Schematic of CONPrint3D [13]

4.1.4 Large-Scale 3DCP using Ultra-High Performance Concrete

Based upon an understanding of the limitations identified in the aforementioned CC and Concrete Printing technologies, a new technology has been introduced by a research team in France for large-scale 3DCP using ultra-high performance concrete (UHPC) [10]. The developed technology uses the extrusion-based technique to deposit UHPC layer by layer through an extrusion print head mounted on a 6-axes robotic arm. The main advantages of the proposed technology are (1) allowing the production of large-scale 3D printed complex geometries without the use of temporary supports, (2) fully exploiting the possibilities of 3D printing by creating layers with varying thickness via the use of the tangential continuity method for slicing, which results in mechanically sounder constructions from a structural viewpoint, (3) enabling geometrical complexity and total control by relying on a generic 6-axes robotic arm instead of an overhead crane or a gantry frame and (4) enabling multifunctionality for structural elements by taking advantage of the complex geometry [10].

The proposed technology was used to manufacture a "multifunctional wall element" (see Figure 5) consisted of an absorptive formwork to be filled either with fiber-reinforced UHPC on structural parts or with an insulating material such as foam for thermal insulation. Some parts were also left intentionally empty to be able to host pipes or electrical wires. The production of the wall element measuring 1360 mm × 1500 mm × 170 mm, with a weight of 450 kg took approximately 12 h (for 139 layers) [10].

Figure 5. Multifunctional wall element [10]

4.1.5 Current Examples of Extrusion-based 3DCP Elements/Structures

In 2014, the Chinese Winsun company claimed to have built 10 basic houses in less than a day, with the area and cost of each one being about 195 m^2 and US$4'800, respectively. The company used a large extrusion-based 3D printer to manufacture the basic house components separately off-site before they were transported and assembled on site [14]. In 2015, the company also claimed to have built a 5-story apartment building, with the area of about 1'100 m^2, being currently the tallest 3D printed structure. The company also claimed to have built a stand-alone concrete villa with interior fittings for a cost of about US$160'000. The company claimed to 3D print the walls and other components of the structure offsite and then assembled them together on-site [15].

The Chinese Huashang Tengda company in Beijing has recently claimed to 3D print an entire 400 m^2 two-story villa 'on-site' in 45 days (see Figure 6-a). Unlike the Winsun company, the Huashang Tengda company uses a unique process allowing to print an 'entire house' 'on-site' in 'one go'. The frame of the house inclding conventioinal steel reinforcements and plumbing pipes were first erected. Then, ordinary Class C30 concrete containg coarse aggregates was extruded over the frame and around the rebars through the use of a novel nozzle design and their gigantic 3D printer [16]. The Huashang Tengda project seemingly eliminated one of the major challenges of 3DCP which is incorporation of conventional steel reinforcements if structural concrete is to be 3D printed. The company claimed that the two-story villa is durable enough to withstand an earthquake measuring 8.0 on the Richter scale. Their giant 3D printer has a sort of forked nozzle (see Figure 6-b) that simultaneously lays concrete on both sides of the rebars, swallowing it up and encasing it securely within the walls [16].

Figure 6. (a): The two-story villa printed by Huashang Tengda company and (b): the novel nozzle of the giant 3D printer [16]

The researchers at the University Federico II of Naples, Italy used a 4 m high BIGDELTA WASP (World's Advanced Saving Project) printer to build the first modular reinforced concrete beam of about 3 m long (see Figure 7). With this WASP printer, the researchers have developed a system to produce concrete elements that can be assembled with steel bars and beams or can compose pillars in reinforced concrete [17].

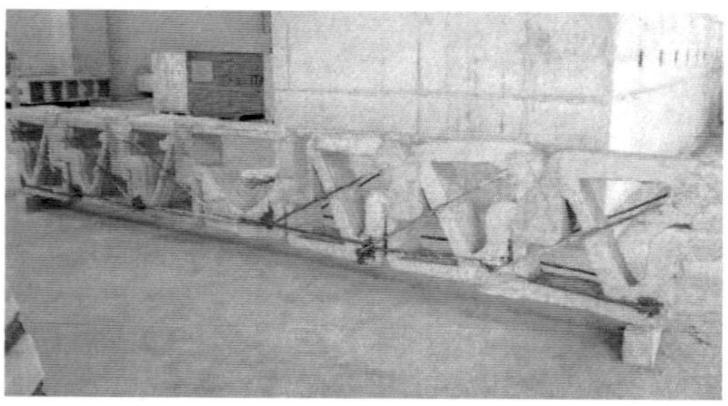

Figure 7. The first 3D printed modular reinforced concrete beam of about 3 m long [17]

As a result of a collaboration between Supermachine Studio and the Siam Cement Group (SCG), recently a 3 m tall cave structure called the "Y-Box Pavilion, 21st-century Cave" was built in Thailand using the 4 m high BIGDELTA WASP printer (see Figure 8). The components of the pavilion was 3D printed off-site at the SCG factory and then all the components were assembled together. The cost of manufacture of the pavilion was reported to be about US$28'000 [18].

In December 2016, the Apis Core company announced to have built the first 'on-site' house in Russia using a 'mobile' 3D concrete printer in just 24 hours (see Figure 9). The entire 38 m^2 house was 3D printed 'on-site'. The total construction cost was claimed to be US$10'134 [19].

Figure 8. The 3 m tall cave structure [18]

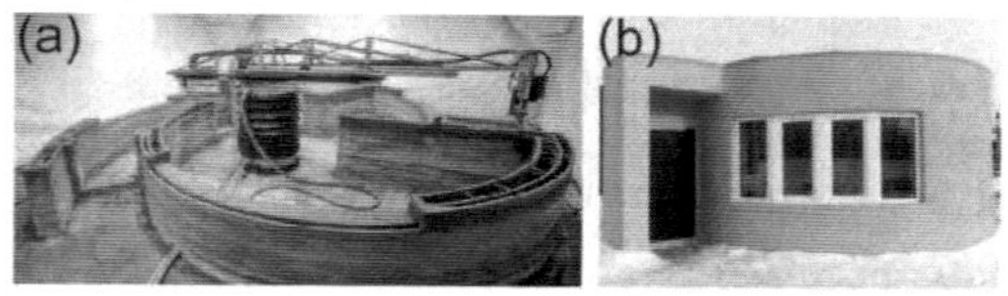

Figure 9. On-site' 3D printed house by Apis Core (a): Construction using a 'mobile' 3D concrete printer, (b): House exterior [19]

4.2 Powder-based Technique

The powder-based technique is another typical AM process that creates accurate structures with complex geometries by depositing binder liquid (or "ink") selectively into to powder bed to bind powder where it impacts the bed. This technique is an off-site process designed for manufacturing precast components. It is the authors' belief that powder-based technique is highly suitable for small-scale building components such as panels, permanent formworks and interior structures that can be assembled on-site.

A schematic of the powder-based technique is illustrated in Figure 10. At the start, a roller, mounted together with a print head, spreads a layer of powder (about 3 mm in thickness) to cover the base of the build plate. Then, according to the layer thickness setting of the 3D printer, a thin layer of powder (approximately 0.1 mm) is spread and smoothed by the roller over the powder bed surface. Subsequently, the binder solution is delivered from binder feeder to the print head and selectively jetted by the nozzle(s) on the powder layer, causing powder particles to bind to each other. Repeating the described steps, the built part is completed and removed after a particular drying time and un-bound powder is removed by using air blower.

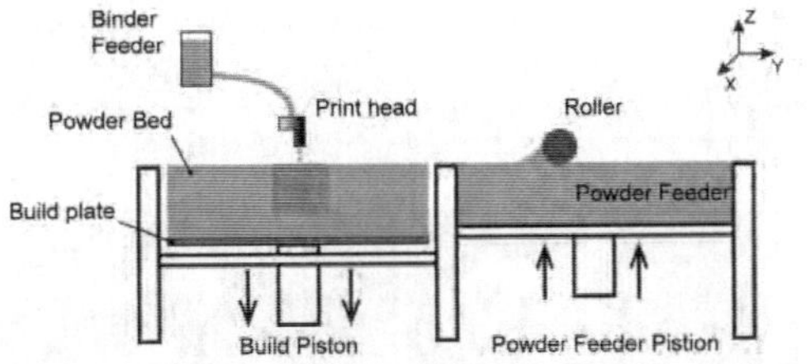

Figure 10. Schematic of powder-based technique

4.2.1 D-shape

The D-shape technology developed by Enrico Dini uses the powder-based technique to selectively harden a large-scale sand-bed by deposition of a binding agent. Sand and magnesium oxychloride cement (also known as Sorel cement) are used as the build material and the binding agent, respectively [20]. In 2008, Shiro Studio collaborated with D-shape to produce the Radiolaria pavilion measuring 3 × 3 × 3 meters (see Figure 11-a). The aim of the Radiolaria pavilion was to demonstrate the capabilities of D-shape technology through complex geometry [20]. In 2010, D-shape also 3D printed the Ferreri house measuring 2.4 × 4 meters in one single process (see Figure 11-b). The printing of the house was completed in 3 weeks [20].

Figure 11. (a): Radiolaria pavilion [20] and (b): the Ferreri house [20] printed by D-shape

4.2.2 Emerging Objects

The Emerging Objects technology developed in the USA uses the powder-based technique to selectively harden a proprietary cement composite formulation by deposition of a binding agent [21]. The technology was used to manufacture the Bloom (see Figure 12-a). Bloom is a 2.74 m tall freestanding tempietto with a footprint that measures approximately 3.66 m by 3.66 m and is composed of 840 customized 3D printed blocks. Each block is printed using a farm of 11 powder 3D printers with a proprietary cement composite formulation comprised chiefly of iron oxide-free OPC. The blocks are held in place using stainless steel hardware and assembled into 16 large, lightweight prefabricated panels, which could be assembled in just a few hours. The technology was also used to manufacture the Shed (see Figure 12-b). Shed is a small 3D printed prototype building constructed with Picoroco Blocks™ which is a modular 3D printed building block for wall fabrication printed from sand measuring 0.3 × 0.3 × 0. 3 meters [22].

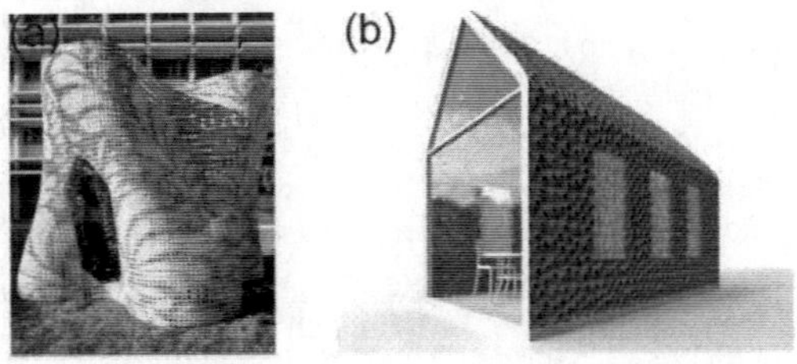

Figure 12. (a): Bloom [21] and (b): Shed [22] printed by Emerging Objects

4.2.3 Powder-based 3DCP using Geopolymer

Powder-based technique is capable of producing building components with fine details and intricate shapes. There is a demand in construction industry for such components which can only be made with expensive formworks with the currently available construction systems. Powder-based technique has the potential to produce robust and durable components at a reasonable speed to satisfy this industrial demand. However, the very limited scope of cement-based printing materials which can be used in commercially available powder-based 3D printers prevent this technique performing at its maximum potential for application in construction industry. To tackle this limitation, recently the authors of this study succeeded in developing an innovative methodology to adopt geopolymer-based material for the requirement and demand of commercially available powder-based 3D printers [23].

Geopolymer is a sustainable alternative to OPC. It is made by alkaline activation of fly ash and/or slag, being industrial by-products of coal power stations and iron manufacture, respectively. Geopolymer has superior mechanical, chemical and thermal properties and 80% less carbon emissions as compared with OPC [5,24].

The authors of this study prepared a printable geopolymer-based material by blending slag, anhydrous sodium metasilicate and fine sand. Different key parameters such as particle size distribution, powder bed surface quality, powder true/bulk densities, powder bed porosity, and binder droplet penetration behavior were studied to quantitatively evaluate the printability of geopolymer-based material. The printing accuracy, apparent porosity and mechanical properties of the printed structures were investigated. Details of these investigations can be found in Xia and Sanjayan [23].

To prepare the geopolymer powder, anhydrous

sodium metasilicate beads were first dry milled for 5 min in a planetary ball mill with ceramic balls. Then slag, anhydrous sodium metasilicate and fine sand were dry mixed in a Hobart mixer until a homogeneous mixture was obtained. The particle size distributions of the geopolymer powder is shown in Figure 13. A commercial 3D printer (Zprinter® 150, Z-Corp, USA) was used for 3D printing a cubic structure with dimensions of 20 × 20 × 20 mm and a complex geometrical structure with dimensions of 56 × 35 × 25 mm (see Figure 14). Compressive strength of the printed cubic structure was measured in both X-orientation and Z-orientation before and after post-curing. For the post-curing, the cubic structures were immersed in saturated anhydrous sodium metasilicate solution for 1 and 7 days at a temperature of 60°C. The 3D printed structures using geopolymer powder are shown in Figure 15.

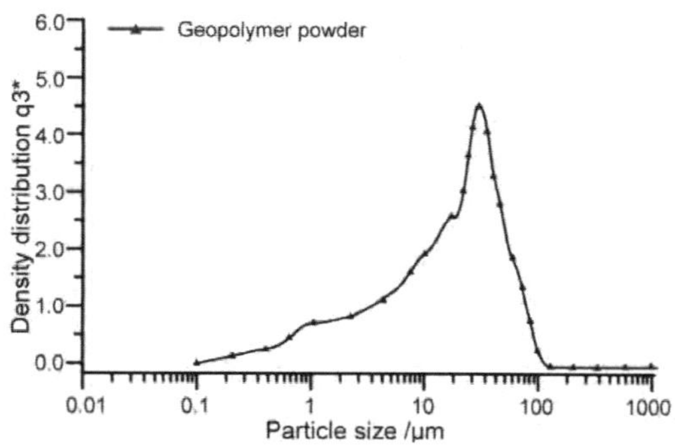

Figure 13. Particle size distributions of the geopolymer powder

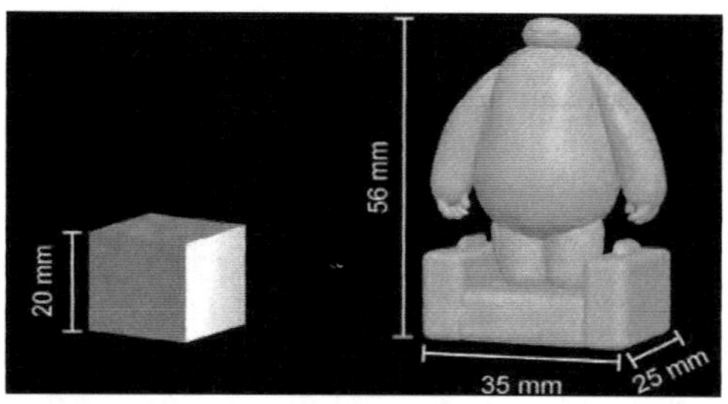

Figure 14. CAD designed 3D structures to be printed

The compressive strength results are presented in Table 1. Although the green strength (before post-curing) was very low, the strength was sufficient to withstand the pressure from the compressed air during de-powdering stage. The 1-day post-cured compressive strength has significantly (1032 to 1110%) higher than the green strength. The 7-days post-cured compressive strength was also significantly (60 to 71%) higher than the 1-day post-cured strength. This increase could be due to continued geopolymerisation process in the presence of alkaline solution.

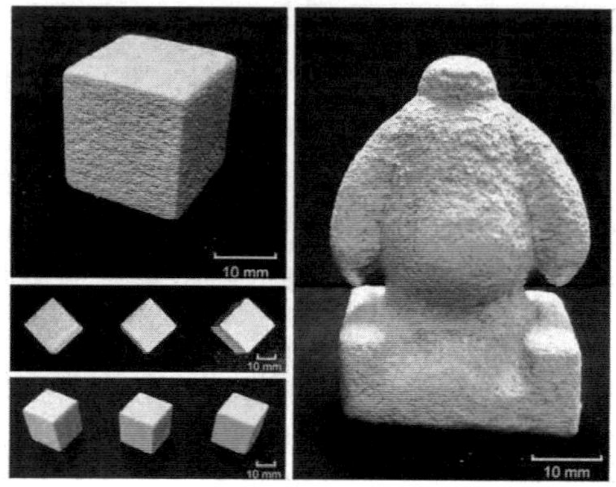

Figure 15. 3D printed structures using the geopolymer powder

Table 1. Compressive strength of cubic structures

		Compressive strength (MPa)	
		X-orientation	Z-orientation
Green strength		0.91 ± 0.03	0.76 ± 0.10
Post-cured strength	1-day	10.3 ± 0.2	9.2 ± 0.2
	7-days	16.5 ± 0.4	15.7 ± 0.2

As can be seen in Table 1, it was also noted that the compressive strength in X-orientation (binder jetting direction) was higher than that in Z-orientation. The green strength in X-orientation was 20% higher than that in Z-orientation. However, the difference between the strength in X-orientation and Z-orientation was reduced to 12% and 5% for 1-day and 7-days post-cured samples, respectively. Therefore, it can be concluded that the post-curing procedure reduces the anisotropic phenomenon in 3D printed geopolymer structures [23].

5 Conclusion

The 3D concrete printing (3DCP) techniques, namely extrusion-based and powder-based techniques, and the currently available 3DCP technologies are reviewed in this paper. The similarities and differences and the pros and cons of different 3DCP technologies are highlighted. The current examples of extrusion-based and powder-based elements/structures are also presented. Although 3DCP is still an emerging technology, the promising examples presented in this paper demonstrate that this technology is rapidly progressing in such a way that 3D printing of large-scale concrete structures may become a reality in near future.

The feasibility of using a geopolymer-based powder for the requirement and demands of commercially available powder-based 3D printers is also demonstrated by the authors of this study, which may offer an innovative manufacturing technique to expand the application of this ecologically friendly material. Future work will be focused on adjusting geopolymer powder formulation, optimizing printing parameters and selecting effective post-curing method to enhance the properties of 3D printed geopolymer structures.

References

[1] Holt C., Edwards L., Keyte L. and Lloyd R. Construction 3D printing. *Concrete in Australia*, 42(3):30–35, 2016.

[2] Llatas C. A model for quantifying construction waste in projects according to the European waste list. *Waste Management*, 6(31):1261–1276, 2011.

[3] Safe Work Australia. *Work-related injuries and fatalities in construction, Australia, 2003 to 2013*. ISBN: 978-1-76028-236-3, 2015.

[4] Yan H., Shen Q., Fan L.C., Wang Y. and Zhang, L. Greenhouse gas emissions in building construction: A case study of One Peking in Hong Kong. *Building and Environment*, 45(4):949–55, 2010.

[5] Nematollahi B., Sanjayan J. and Shaikh F.U.A. Synthesis of heat and ambient cured one-part geopolymer mixes with different grades of sodium silicate. *Ceramics International*. 41(4): 5696–5704, 2015.

[6] ASTM F42. *Standard Terminology for Additive Manufacturing Technologies*. West Conshohocken: ASTM International, 2015.

[7] Wohlers T. *3D printing and additive manufacturing state of the industry*, Wohlers Associates Inc., Colorado, 2014.

[8] Pegna J. Exploratory investigation of solid freeform construction, *Automation in Construction*, 5(5):427–437, 1997.

[9] Buswell R.A., Soar R., Gibb A.G. and Thorpe A. Freeform construction: mega-scale rapid manufacturing for construction, *Automation in Construction*, 16:224-231, 2007.

[10] Gosselin C., Duballet R., Roux Ph., Gaudillière N., Dirrenberger J. and Morel Ph. Large-scale 3D printing of ultra-high performance concrete – a new processing route for architects and builders. *Materials and Design*, 100:102–109, 2016.

[11] Khoshnevis B., Hwang D., Yao K.-T., Yeh Z. Mega-scale fabrication by contour crafting, *International Journal of Industrial and System Engineering*, 1(3):301-320, 2006.

[12] Lim S., Buswell R.A., Le T.T., Austin S.A., Gibb A.G.F and Thorpe T. Developments in construction-scale additive manufacturing processes. *Automation in Construction*, 21:262–268, 2012.

[13] Nerella V.N., Krause M., Näther M. and Mechtcherine, V. CONPrint3D — 3D printing technology for onsite construction. *Concrete in Australia*, 42(3):36–39, 2016.

[14] Wang L. Chinese company assembles 10 3D-printed concrete houses in a day for less than $5,000 each. On-line: http://inhabitat.com/chinese-company-assembles-ten-3d-printed-concrete houses-in-one-day-for-less-than-5000-each/, Accessed: 20/03/2017.

[15] Sevenson B. Shanghai-based Winsun 3D prints 6-story apartment building and an incredible home. On-line: https://3dprint.com/38144/3d-printed-apartment-building/, Accessed: 20/03/2017.

[16] Scott C. Chinese construction company 3D prints an entire two-story house on-site in 45 days. On-line: https://3dprint.com/138664/huashang-tengda-3d-print-house/, Accessed: 20/03/2017.

[17] WASP. Concrete beam created with 3D printing. On-line: http://www.wasproject.it/w/en/concrete-beam-created-with-3d-printing/, Accessed: 20/03/2017.

[18] Alec. Thai cement maker SCG develops an elegant 3m-tall 3D printed 'pavilion' home, 21st C. Cave. On-line: http://www.3ders.org/articles/20160427-thai-cement-maker-scg-develops-an-elegant-3m-tall-3d-printed-pavilion-home-21st-c-cave.html, Accessed: 20/03/2017.

[19] Apis Core. The first on-site house has been printed in Russia. On-line: http://apis-cor.com/en/about/news/first-house, Accessed: 20/03/2017.

[20] Cesaretti G., Dini E., De Kestelier X., Colla V. and Pambaguian L. Building components for an outpost on the Lunar soil by means of a novel 3D printing technology. *Acta Astronautica,* 93:430-450, 2014.

[21] Ronald Rael R. and San Fratello V. Bloom. On-line:http://www.emergingobjects.com/project/bloom-2/, Accessed: 20/03/2017.

[22] Ronald Rael R. and San Fratello V. Shed. On-line: http://www.emergingobjects.com/project/shed/, Accessed: 20/03/2017.

[23] Xia M. and Sanjayan J. Method of formulating geopolymer for 3D printing for construction applications. *Materials and Design*, 110: 382–390, 2016.

[24] Nematollahi, B. and Sanjayan, J. Efficacy of available superplasticizers on geopolymers. Research Journal of Applied Sciences, Engineering and Technology. 7(7): 1278-1282, 2014.

Optimisation of Different Concrete Mix Designs for 3D Printing by Utilising 6DOF Industrial Robot

Pshtiwan Shakor[a], Jarred Renneberg[b], Shami Nejadi[a] and Gavin Paul[b]

[a]Centre for Built Infrastructure Research, School of Civil and Environmental Engineering, University of Technology Sydney (UTS), Australia
[b]Centre for Autonomous Systems, School of Electrical, Mechanical and Mechatronics, UTS
E-mail: {pshtiwan.shakor, Jarred.J.Renneberg}@student.uts.edu.au, {shami.nejadi, Gavin.Paul}@uts.edu.au

Abstract – Additive Manufacturing (AM) technologies are becoming increasingly viable for commercial and research implementation into various applications. AM refers to the process of forming structures layer upon layer and finds application in prototyping and manufacturing for building construction. It has recently begun to be considered as a viable and attractive alternative in certain circumstances in the construction industry. This paper focuses on the utilisation of different concrete mixtures paired with extrusion techniques facilitated by a six Degree of Freedom (DOF) industrial robot. Using methods of Damp Least Squares (DLS) in conjunction with Resolved Motion Rate Control (RMRC), it is possible to plan stable transitions between several waypoints representing the various print cross-sections. Calculated paths are projected via 'spline' interpolation into the manipulator controlled by custom software. This article demonstrates the properties of different concrete mixture designs, showing their performance when used as a filament in 3D Printing and representing a comparison of the results that were found. In this study, the prepared materials consist of ordinary Portland cement, fine sand between (425~150) micron, coarse aggregate ranges (3) mm and chemical admixtures which have been used to accelerate setting times and reduce water content. Numerous tests were performed to check the buildability, flowability, extrudability and moldability of the concrete mixtures. The horizontal test was used to determine the flowability and consistency, while the vertical and squeeze-flow tests were used to determine the buildability of the layers. The extrudability and moldability of the concrete mixtures were controlled by the robot and associated extruder speeds.

Keywords –
3D printing concrete, industrial robot, additive manufacturing, resolved motion rate control, squeeze-flow test, speed control, moldability.

1 Introduction

In most industrial segments, automation processes have significantly improved the efficiency of production. The construction industry, however, often seems to be slower to adopt automation in comparison to other industrial and manufacturing sectors. Therefore, significant efficiency improvements and economic benefits can be gained by the introduction of advanced automation into the construction process. With further automated development, manufacturing processes would take less time to complete, cost less and be more eco-friendly [5] & [8].

Lin et al.[15] state that the four main materials used in 3D Printing were composed of plastic, ceramic, metal and wax, and then in their research, they utilized rapid sulphoaluminate cement as the main binder.

It is obvious that AM plays a major role in the construction field. AM has powder base printing (D-shape) and Fused Deposition Modeling (FDM) method. FDM is well known as extrusion printing (printing via a robot) in the construction industry. In recent decades, a variety of technologies have been developed to implement additive manufacturing in the construction industry. These techniques can be divided into two print processes: Extrusion, and powder bed / inkjet 3D.

1. Extrusion Printing

The extrusion printing technique involves extruding cementitious material from a nozzle attached to a framework to print structural layers, see Figure 1. For example, concrete printing, designed by Le [12] and contour crafting, established by Khoshnevis [10]. [13] investigated the hardened properties of high performance printed concrete that was extruded from a 9mm diameter nozzle. They demonstrated that the compressive strength of the printed samples in different directions is between 75 to 102 MPa. They also determined that the perpendicular load direction has the highest compressive load while in a conventional mould cast cube, has a compressive strength of approximately 107 MPa.

(a) (b)

Figure 1. (a) Simulated Robot with a print nozzle assembly attached, for demonstrating and checking motion plans and paths. (b) Real world robot and auger nozzle assembly.

2. Powder bed inkjet 3D printing

The powder printing fusion, i.e. powder-based 3DP, creates precise scaffolds with complicated geometries by dropping binder fluid (or "Toner") selectively into a powder bed to bind powder where it impacts the bed. [20] claimed that there is a post-hardening process which occurs after printing a specimen. It is possible to increase the compressive strength up to 35 MPa for Newberyite and up to 10 MPa for Struvite and the shrinkage level is 5-7% compared to the original model.

The main contribution of this paper is to show the materials extrusion method paired with the planning and control of the robot. The purpose is to demonstrate a series of different concrete mixtures and compare the results. There is potential for this process to create an automated method for manufacturing structural concrete columns and beams in both a pre-cast and cast in situ. This paper is organized as follows: Section 2 describes the materials and the methodologies used as well as the design and fabrication; Section 3 presents information regarding the analysis and testing methods used to find results; Section 4 discusses the results that were found from the experiments conducted. Section 5 provides a conclusion and proposes future work.

2 Materials and Methods

2.1 Materials

The materials in this study that have been used consist of ordinary Portland cement (General Purpose cement) by Eureka, fine sand (Sydney Sand) and coarse aggregate (10mm Blue Metal) both by Australian Native Landscapes and chemical admixtures such as superplasticizer (ADVA ® 650), water reducer (Daratard ® GP), accelerator (Sigunit L80AF), and retarder (Retarder N). Figure 2 shows the sieve analysis of fine sand and coarse aggregate. Moreover, the sieve analysis was done by the (Endecotts EFL2000) equipment. Later, the coarse aggregate has been crushed by the (Geo-Con) equipment to reduce the size of the particles to about 3mm, see Figure 2 (b).

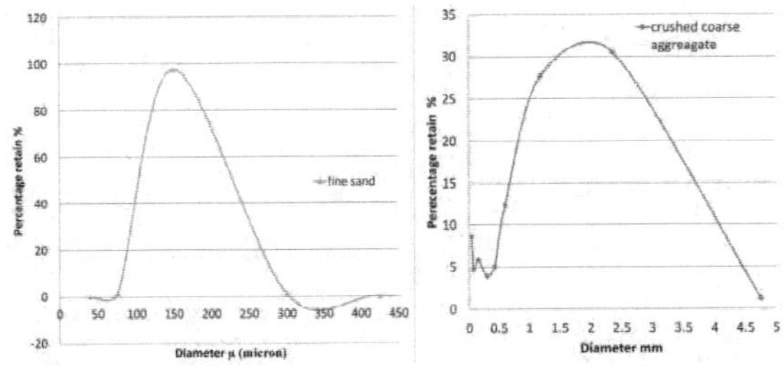

(a) (b)

Figure 2. a) Sydney fine sand sieve analysis; b) Coarse aggregate sieve analysis after crushing.

2.2 Design and Fabrication

2.2.1 Extruder design

The extruder was made to adapt to the end-effector of the industrial robot (Denso VM6083), and was based on the design by [3]. The extruder consists of the following parts:

- Y-Pipe diameter (110)mm (Holman 100mm PVC 45 degree female and female plain junction)
- Truck wiper motor (24 Volts)
- Auger diameter(80)mm
- Nozzles with various sizes: circle (10, 20 and 30) mm, and square (10, 20 and 30) mm, (Figures 3).

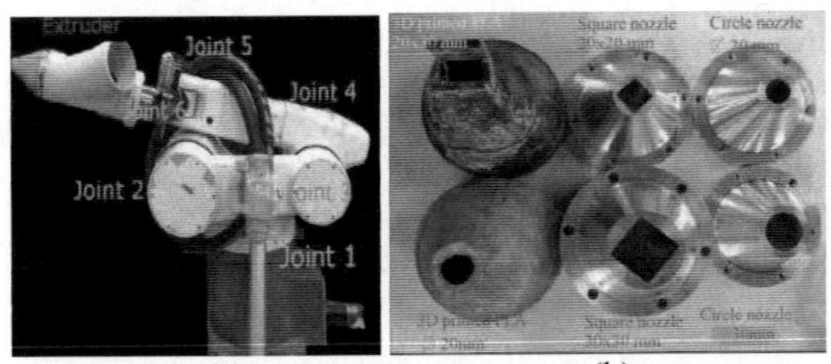

(a) (b)

Figure 3. (a) Extruder joined to the Denso Robot. (b) Nozzle types made with Aluminium and PLA.

2.2.2 Materials mix design and fabrication

In this paper, several trials consisting of mixtures made of cement paste, cement mortar and concrete were performed. Fabrication and mix designs were prepared based on previous studies and literature.

[19] performed a meta-analysis regarding studies of concrete mix design, such as the study of [12], which was prepared as a cement-based mortar consisting of cement, silica fume, sand, fly ash and polypropylene fiber as the main ingredients. The optimum concrete mix design was found to be cement 579, sand 1241, fly

ash 165, silica fume 83 and water 232 (kg/m^3). Furthermore, additional tests were conducted [14] to further optimize the mix with gypsum and cement while experimental tests of compressive strength found results in the range of 100-110 MPa.

[16] investigated the effects of superplasticizers on the flowability and buildability of cement mortar. In the mix with the high superplasticizer (0.95% to 2.5% of water weight), lower water-cement ratio content increases the compressive strength and flowability of the mix but it decreases the buildability capacity potentially. To build a 100mm high wall without any risk of failure, the optimum water-cement ratio and superplasticizer content (for better performance in flowability and buildability) was about 0.39 and 1.9% respectively, which used a nozzle diameter of 20 mm. In this study, an accelerator and retarder was added to control the rheology of the mortar.

Based on these mix design parameters we implemented a similar mix ratio in this paper.

2.3 Robot Planning and Control Methodology

To maintain a consistent flow it is important to implement an efficient method of planning to provide a stable path for the end-effector to follow. Using a combination of Damp Least Squares paired with Resolved Motion Rate Control to plan and control the robot's path, it is possible to create a steady 3-dimensional print which is completely customizable with an integrated MATLAB and C++ interface.

2.3.1 Damped Least Squares (DLS) Method

It is important to consider, when planning transitions between multiple joint states, how the robot will reduce the impacts of singularity on the system. A singularity refers to any given joint state of formation which reduces or even limits the degrees of freedom on the end effector. Typically, this occurs when two connected links form a straight member. The conventional method of calculating inverse differential kinematics results in a destabilization of the transitions between joint states. This appears, as described by Chiaverini et al. [6], as a sudden increase in joint velocities over large control deviations. It is possible to apply the DLS method (also known as Levenberg-Marquardt stabilization) in order to mitigate the impacts of singularities. This method reduces subsequent oscillations on the system as the joints approach an unreachable position. The DLS solution as appears in [4] can be described by the following equation,

$$\Delta\theta = (J^T J + \lambda^2 I)^{-1} J^T \vec{e} \quad (1)$$

Where the change in angular displacement, $\Delta\theta$, is minimised to stabilize the joint velocities when approaching a singularity. The damping constant, λ, must be adjusted [4] according to the systems' parameters in order to perform efficiently. Ahmed et al. [1] outlines how the stability of the rotational velocities depends heavily on the 'exactness' of the damping factor and that the damping constant can be found with the following equation:

$$\lambda = \lambda_0\left(1 - \left(\frac{w}{w_0}\right)\right),\ w < w_0 \quad (2)$$

where, λ_0, refers to the initial damping factor predetermined by [1] 'user requirements', and w refers to the measure of manipulability and w_0 refers to a threshold number of minimum manipulability before DLS is to be used.

For any given joint angle (θ_i, $i = 1, \ldots 6$) for all joint positions of the calculated path, a measure of manipulability of the system can be taken to determine whether or not the system is approaching a singularity. From Yoshikawa [21] we use the following equation,

$$w = \sqrt{\det(J * J^T)} \quad (3)$$

Where, J, is the Jacobian of the system, and J^T, is the transpose of the Jacobian, the measure of manipulability is compared to the predetermined threshold value represented by epsilon, ε (equal to w_0), which allows for the system to apply reactive coding to dampen the effects of singularity on the joint velocities for instance where the measure of manipulability, w, is less than ε.

The Jacobian of the system relates to the kinematic properties that are present during a transition at any given joint state. Acquiring the most appropriate joint velocity [2], for each individual joint involves using a Jacobian that relates both active and passive joint velocities acting in the system. As the robot is 6DOF, the system's Jacobian, J yields a 6×6 matrix. For any given path, the program will determine the required total joint positions by using inverse kinematics which yields the rotational joint positions q_i. The program stores these values in an n×6 matrix (where n is the number of joint positions required to transition through the given path), and then J is found by passing in q_i.

Depending on w at q, DLS will alter the way the Inverse Jacobian is calculated. When the manipulability is less than the threshold value represented by ε, a pseudo inverse Jacobian must be found in order to continue planning. This pseudo-inverse Jacobian is designed to handle stabilization issues associated with singularities, where the conventional method tends to perform poorly. The inverse Jacobian, J^+ is used to handle approach toward singularity as:

$$J^+ = J^T(J J^T + \lambda^2 I)^{-1} \quad (4).$$

2.3.2 Resolved Motion Rate Control (RMRC) Method

When considering a robot's joint motions there are multiple joint paths the robot can use in order to transition between consecutive end-effector poses [7]. For any given test represented by series of joint states, RMRC trajectory planning was used to create linear transitions between joint states. It allows the system to use the most optimal path between two points in space for where the robot's end-effector will transition with the best possible motion. Utilising the calculation of error for both the rotational and translational displacements, the program updates the stored joint rotational values, $\dot{\theta}_{i=1\rightarrow 6}$, in order to maintain a near linear trajectory. This also helps to maintain the end effectors' 'downwards' orientation during transitions.

Determining the joint velocities as shown in [11], drives the position of the end effector with a velocities,

$$\dot{\boldsymbol{x}}_e = (\dot{x}_e, \dot{y}_e, \dot{z}_e, \dot{\theta}_r, \dot{\theta}_p, \dot{\theta}_y) \quad (5)$$

Since the end effector's Roll, $\dot{\theta}_r$, Pitch, $\dot{\theta}_p$, and Yaw, $\dot{\theta}_y$, are at fixed values, any change in velocity impacting on the end effector's change in position occurs linearly in the x, y, and z plane. The change in displacement is found using the following equation and stored as *dx* for each joint state, *i* , in any given transition:

$$dx = x_{i+1} - x_i,\ y_{i+1} - y_i,\ z_{i+1} - z_i \quad (6)$$

To maintain a fixed end effector roll, pitch and yaw, the rotational error of the end effector is calculated as,

$$d\theta = r_{i+1} - r_i,\ p_{i+1} - p_i,\ y_{i+1} - y_i \quad (7)$$

then used to correct the end effector's orientation as it transitions. The properties of both (6) and (7) are used to formulate the properties of (5), thus:

$$\dot{x}_e = (dx, d\theta) \quad (8)$$

Hewitt et al. [9] described the basis of RMRC as,

$$\dot{\theta} = J(\theta)^{-1} * \dot{x} \quad (9)$$

The calculated end effector velocity, $\dot{x}$, is found in (8). Depending on whether DLS is implemented the inverse Jacobian, denoted by $J(\theta)^{-1}$, with the end effector velocity will give the required rotational velocities, $\dot{\theta}$, for each joint.

The robot uses the angular velocity, $\dot{\theta}$, to apply RMRC to update the next joint state parameters based on the current joint positions. This method applies an element of self-correction to the path planning to ensure that the end-effector maintains the same direction.

2.3.3 End-Effector Velocity

Finding the actual velocity of the end-effector is essential for maintaining an effective extrusion rate at the end-effector. Note that despite the use of RMRC in the planning of the robot's path, the actual joint step completion time is determined by the controller. Given the maximum joint velocities of each joint, it is possible to calculate the maximum 'actual' velocity of the end effector at any given time. Equation (9) can be re-arranged to give the following:

$$\dot{x} = J(\theta) \cdot \dot{\theta} \quad (10)$$

Essentially, calculating the product between the Jacobian for a given joint state and the rotational velocities will give the velocity components of the end effector. By accounting for the percentage multipliers imposed by the controller pipeline, (10) is expressed as:

$$\dot{x} = P_M * P_{TP} * (J(\theta) \cdot \omega) \quad (11)$$

Where P_M and P_{TP} are the percentage values of maximum speed dictated by the software interface and the robot's teach-pendant, respectively. Finding the magnitude of $\dot{x}$ gives the controlled peak velocity. For the experiments these values were heuristically determined by measuring both the displacement of a given path and time to complete the path.

3 Analysis and Testing

3.1 Squeeze-flow Test

Several tests have been conducted to evaluate the mechanical properties of the materials prior to printing complete structures. The tests are arranged based on the cement mortar and concrete materials properties. All the trails are listed in Tables 1 and 2.

The squeeze-flow test is prepared according to Brazilian test (ABNT NBR 15839). The machine used to prepare the test is the Shimadzu, (50kN, Japan). The displacement speed is approximately 0.1 mm/second (Figure 4, ONP). The surface roughness test is conducted by a portable profile-meter (Taylor Hobson Surtronic 3+). Tests are performed for the single, double and triple layers, in order to evaluate the buildability of layers.

Figure 4. The squeeze flow configuration: (1) top plate diameter (101 mm), surface roughness (Ra0.0586); (2) mould ring diameter (101 mm) and height (10 mm); (3) cement mortar sample; and (4) bottom plate diameter (160 mm), surface roughness (Ra 0.0586).

3.2 Robot Planning and Control Tests

3.2.1 Horizontal and Vertical Tests

The main transition path that the robot takes, referred to as the 'horizontal test', is shown in Figure 5 and consists of a series of rows of approximately 300mm in length printed next to each other. The distance between the layers (in the y direction) is determined by the desired thickness of the layer which is reflected by the choice in nozzle. The 'vertical' test was simply the same footprint except after completing the tenth line the end-effector would return to the starting position, increment upwards and repeat the same pattern.

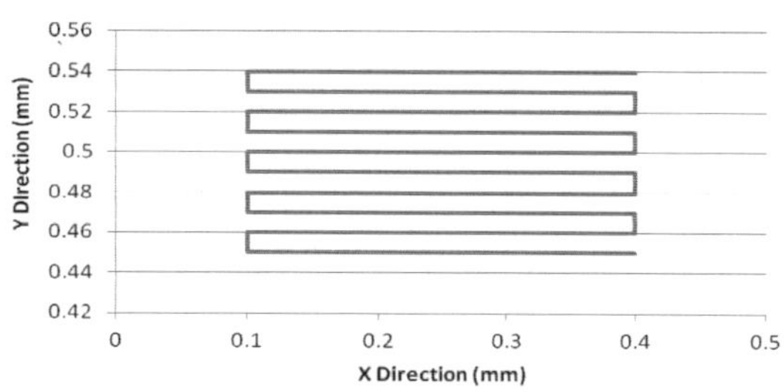

Figure 5. 'Horizontal Test' cross-section path.

3.2.2 Oscillation Damping with DLS

The different velocity components that act in the system have significant impacts on the shape of the extruded mortar and concrete mixtures. The auger delivery method impacts directly on the flow rate of the mixture flowing from the extruder. Secondarily, reducing the effects of joint oscillations as the robot approaches singularities is essential to maintain a smooth print result. Tests are conducted to show the impact of the DLS method on joint velocities as a joint's measure of manipulability reduces.

Table 1: Trials prepared by an extruder for cement mortar

Trial No.	C (g)	FS (g)	W (ml)	R (ml)	Acc (ml)	SP (ml)	WR (ml)	Noz (mm)
1	1000	0	360	8	4	10.4	-	∅20
2	1000	500	300	8	4	10.4	-	∅20
3	500	500	150	4	4	5.2	-	∅20
4	750	750	292.5	4	4	5.5	-	∅20
5	750	750	250	4	5	5	-	∅20
6	1500	1500	550	8	10	11	-	20×20
7	1000	1000	361.6	5.33	6.6	6.67	-	20×20
8	1000	1000	343	5.33	6.6	6.67	-	20×20
9	1000	1000	350	5.33	6.6	6.5	-	20×20
10	1000	1250	375	5	6	5	3	∅10

C:cement, FS:fine sand, W:water, R:retarder, Acc:accelerator, SP:superplasticizer, WR:water reducer, Noz:nozzles

Table 2: Trials for crushed coarse aggregate by an extruder

Trial No.	C (g)	FS (g)	A (g)	W (ml)	R (ml)	Acc (ml)	SP (ml)	W R (ml)	Noz. (mm)
11	750	750	250	300	4	5	5	-	20×20
12	750	750	250	250	4	5	5	-	20×20

4 Result and Discussion

Experimental trials were conducted to determine the effectiveness of the mortar or concrete mixtures. Based on the consistency and rheology of the printed samples, several trials were chosen as 'optimum' trials for the squeeze flow tests and compressive strength tests. The ring moulds were prepared for squeeze flow test.

Several mix designs performed similarly to the mixtures presented in previous studies by [12], [14], [16], [19]. Furthermore, these tests also depend on the capabilities of the extruder, see Table 3.

There were 12 different concrete mix design trials which were performed by the 6DOF industrial robot. Out of these trials, only three were chosen for their performance in compressive strength and results yielded in the squeeze-flow test. These tests were the basis for the printing quality test, flow quality and the buildability (i.e. each layer's capability of holding subsequent layers without falling). These benchmarking keys for printing have been stated by [17], [13], [19]. Generally, the thickness for the printed line changed in a range of about (1-2) mm. According to Table 3, the thickness for each printed line for the different nozzle provided different results.

In addition, it was determined that the quality resolution of materials also depended on the speed of

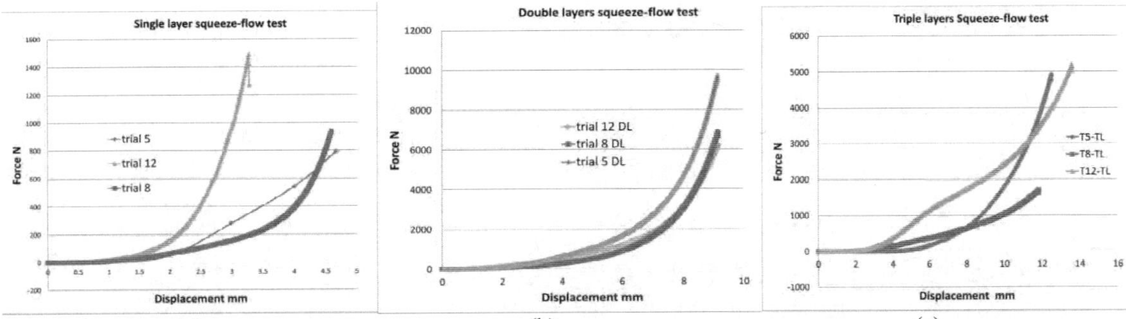

(a) (b) (c)

Figure 6. Squeeze-flow graph for the single layer, double layers and triple layers of different concrete mix

auger (flow rate) and velocity components associated with the various joints of the robot.

4.1 Squeeze-flow test

The squeeze flow tests showed different results for single, double and triple layers of the selected mixtures, as shown in Figure 6. In a test that uses a single layer the concrete mixture had higher results in force until it reached the required displacement. For example, at the displacement 2.99mm, the required load in Trial 12 (concrete mix with small aggregate) was about 956.51 N. At the same displacement for Trials 5 and 8 (cement mortar) the load was about 277.82 N and 153.11 N, respectively. When examining the results of tests that utilized a double layer there were significant differences in comparison to the single layer results. Trial 5 (cement mortar) had a force which reached approximately 622.54 N when at a displacement of about 3.99 mm, while for Trials 12 and 8 the loads were 536.05 N and 275.75 N, respectively. At the same displacement, the results for the triple layers exhibit a similar pattern.

These results show that the mortar is more coherent than the concrete mixtures. Thus, it does not allow air entertained bubbles to remain in the mortar mixture. In the study by [18], it was discovered that the mortar in the first 24 hours of early age had a greater compressive strength in comparison to normal concrete. Hence, a much higher percentage of open pores will exist in the concrete mixtures in comparison to cement mortar due to the presence of aggregates. The larger particle sizes encourage porosity within the concrete mixtures.

The force is dependent on the number of chains and the force between the particles. In the Trial 12 single layer test the mixture contained small particles of coarse aggregate. It can be determined that because of the presence of small aggregate particles the mixture was capable of resisting more force over the given displacement, Figure 6. Moreover, it shows that the mortar mix could handle more layers than the concrete mix and less penetration will occur between layers in mortar mix when loads are applied (Figures 6 and 7).

Table 3: Several printed specimens shapes with comments

Trial Number	Image	Comment
Trial 5	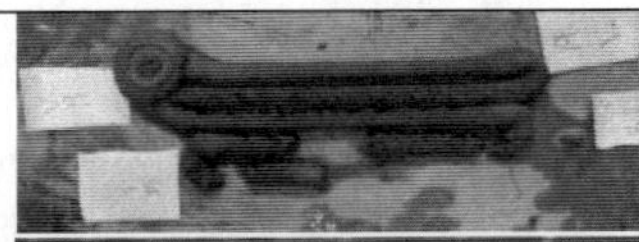	Successful print
Trial 7	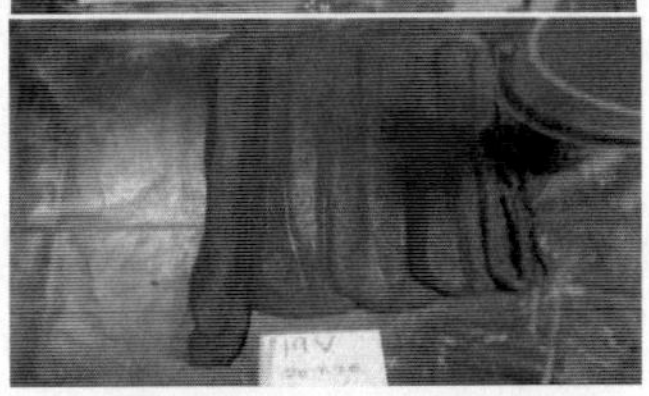	Slightly flow
Trial 8	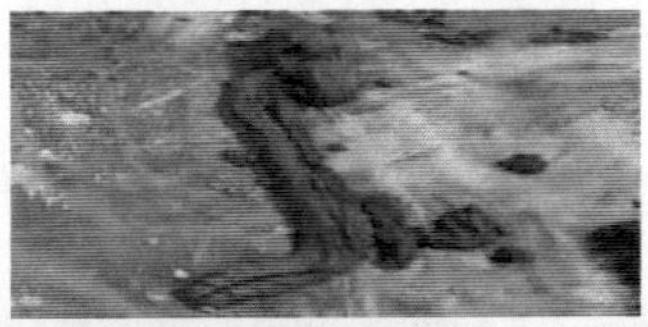	Good layers

Figure 7. Squeeze-flow for the double layers of trial 12

4.2 Extruder Delivery and Speed Control Results

It was crucial that the mortar or concrete mix designs were compatible with the speed of robot and the auger delivery methods. Researchers like Hwang and Khoshnevis [10] had mechanical speed limitations with the novel contour crafting system; the speed of the

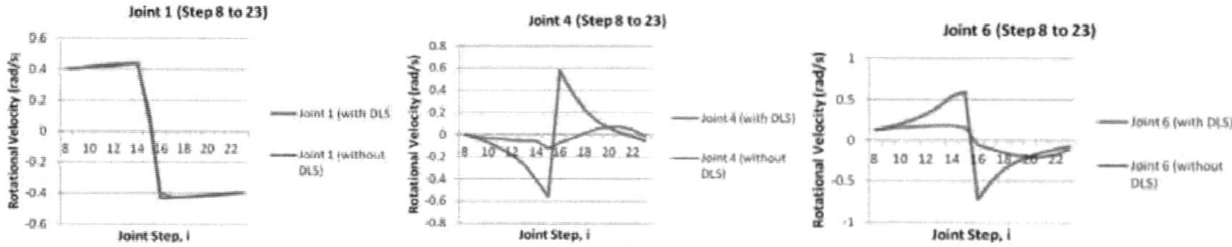

(a) (b) (c)

Figure 9. Velocity comparison for joints: (a) 1, (b) 4, and (c) 6.

material deposition was not well established in their research. It was also identified that robot in use was limited to a reach of approximately 1100mm from the base. Printed specimen dimensions are thus limited to be contained within this workspace.

Mixtures that had reduced workability were severely impacted when the velocity of the end effector was excessively fast. The robot ran the calculated path at a constant velocity before the extruder had time to move a proper volume of material. This resulted in layers with significant gaps between them. Alternatively, mixtures with increased workability were negatively impacted when the velocity of the end-effector was too slow. These instances would result in the robot moving too much volume of material before the robot could properly transition through the calculated path.

Regarding the auger delivery method, it was important to determine an appropriate velocity in which the end-effector would travel paired with a suitable supply voltage to the 24V motor which powered the extruder's auger to form an acceptable flow. It was found that the auger could move substantial volumes of material. However, the flow rate tended to be inconsistent. Material tended to gather on the inside of the extruder tubing and allowed the formation of air pockets in the flow. Despite this, it was determined that the optimal speed for the robot was between 39.36 and 42.12 mm/sec when the auger voltage was set to approximately 15 to 19 volts. Another drawback was that the motor that powered the auger struggled to rotate when put under increased loads. Loads from the mixes with higher densities and higher viscosities reduced the augers' practicality considerably.

As the robot's layer height increased the more contorted the required joint angles were. The manipulability was significantly reduced at some points of the path and DLS was required to dampen the resulting oscillations. The immediate goal for the vertical print test was to get to an ultimate print height of approximately 600mm (Figure 5 shows the cross section). For the following graphs, the robot was set to print the vertical test at a height of 590mm from the printing plate with a layer thickness of 10mm. The process was discretized into approximately 130 robot

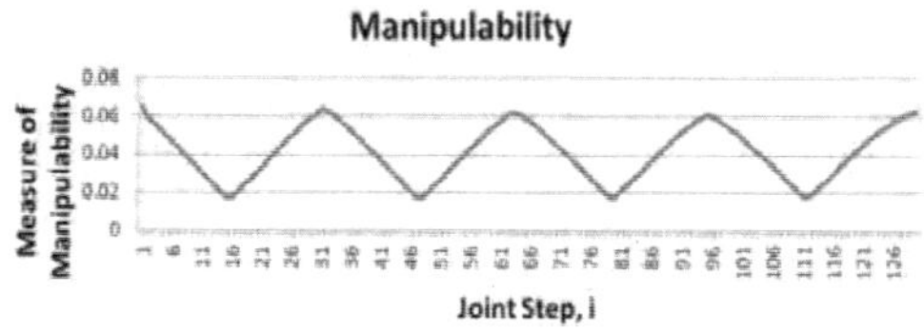

Figure 8. Measure of Manipulability

poses in order to complete each layer.

Figure 8 shows that the system utilized DLS on several occasions. The robot approaches singularity between joint step intervals of 8 – 23, 40 – 56, 72 – 88 and 103 – 120 (manipulability is less than 0.4), and the system implements DLS during these intervals to reduce the sudden change in rotational velocity.

Figures 9 depict the change in rotational velocity for joint states from joint step 8 to joint step 23. Take note that the other joint step intervals yield similar results; hence, only one interval is examined.

It can be seen that the smallest changes occur in Joint 1. Joint 4 and Joint 6 show significant damping. This could be due to the fact that these joints have a much larger displacement requirement in comparison to the other joints. The most significant change in rotational velocity occurs between steps 16 and 17. With DLS the change in velocity is significantly smoother in comparison to without DLS.

5 Conclusions

This paper has presented the initial stages involved in developing a working model of 3D printed concrete lines by using a 6DOF Industrial Robot. It has detailed the mix design properties used and how these mixes have been evaluated by several trials and tests. It has found that the impact of the optimum material mixes and the use of DLS paired with RMRC to control joint velocities are crucial in establishing good printing foundations. The results showed that mortar is more capable in printing and building layers due to less voids and porosities between particles, while the concrete mixes form openings and significantly more internal spacing for the air entertaining. It is evident that these studies need to be further developed and tested with different ages of curing samples, testing for compressive strength and checked with the controlled samples.

For future works, it is intended that improvements and additions to the system will continue to be developed. Performing compressive strength tests will be a focus for short term goals. Due to the inconsistency of flow rates provided by the auger, there will be considerable effort to move to a pump delivery system. Furthermore, an investigation to optimise the parameters of the model will occur so as to increase the accuracy of planning and control of the robot. Additionally, the plan is to work on predictive models, online sensing and feedback to reduce the chance of errors and thus improve both system safety and robustness.

6 Acknowledgements

The authors would like to express their sincere gratitude to Sika Australia Company for their chemical material supply.

7 REFERENCES

[1] S.M. Ahmed, A.N. Pechev, Performance analysis of FIK and DLS inverse kinematics using six degree of freedom manipulator, Robotics and Biomimetics, 2009 International Conference IEEE, 2009, pp. 1405-1410.

[2] O. Altuzarra, O. Salgado, V. Petuya, A. Hernández, Computational kinematics for robotic manipulators: Jacobian problems, Engineering Computations 25 (1) (2008) 4-27.

[3] L. Anell, Concrete 3d printer, Master Thesis, Department of Design Science, Lund University, 2015.

[4] S.R. Buss, Introduction to inverse kinematics with jacobian transpose, pseudoinverse and damped least squares methods, IEEE Journal of Robotics and Automation 17 (1-19) (2004) 16.

[5] R.A. Buswell, R. Soar, A.G. Gibb, T. Thorpe, The potential of freeform construction processes, (2005).

[6] S. Chiaverini, B. Siciliano, O. Egeland, Review of the damped least-squares inverse kinematics with experiments on an industrial robot manipulator, IEEE Transactions on Control Systems Technology (2) (1994) 123-134.

[7] M. Clifton, G. Paul, N. Kwok, D. Liu, D.-L. Wang, Evaluating performance of multiple RRTs, Mechtronic&Embedded Systems and Applications, 2008. MESA 2008. International Conference IEEE, 2008, pp. 564-569.

[8] L. Edwards, C. Holt, L. Keyte, R. Lloyd, Construction 3D Printing, (2015).

[9] J. Hewit, J. Love, Resolved motion rate control of a materials-handling machine, Transactions of the Institute of Measurement and Control 5 (3) (1983) 155-159.

[10] D. Hwang, B. Khoshnevis, Concrete wall fabrication by contour crafting, 21st International Symposium on Automation and Robotics in Construction (ISARC 2004), Jeju, South Korea, 2004.

[11] N.Y. Ko, Extended RMRC and its Application to the Motion of a Mobile Manipulator, International Journal of Humanoid Robotics 12 (02) (2015) 1550016.

[12] T.T. Le, S.A. Austin, S. Lim, R.A. Buswell, A.G.F. Gibb, T. Thorpe, Mix design and fresh properties for high-performance printing concrete, Materials and Structures 45 (8) (2012) 1221-1232.

[13] T.T. Le, S.A. Austin, S. Lim, R.A. Buswell, R. Law, A.G.F. Gibb, T. Thorpe, Hardened properties of high-performance printing concrete, Cement and Concrete Research 42 (3) (2012) 558-566.

[14] S. Lim, R.A. Buswell, T.T. Le, R. Wackrow, S.A. Austin, A.G.F. Gibb, T. Thorpe, Development of a viable concrete printing process, (2011).

[15] X. Lin, T. Zhang, L. Huo, G. Li, N. Zhang, J. Liao, Preparation and Application of 3D Printing Materials in Construction, (2015).

[16] Z. Malaeb, H. Hachem, A. Tourbah, T. Maalouf, N. El Zarwi, F. Hamzeh, 3d Concrete Printing: Machine And Mix Design, International Journal of Civil Engineering (6) (2015).

[17] B. Panda, Y. Tay, S.C. Paul, T.M. Jen, K. Leong, I. Gibson, Current Challenges And Future Perspectives Of 3d Concrete Printing.

[18] P.S. Surendra, A. Yilmaz, V. Thomas, Determination of Early Age Mortar and Concrete Strength by Ultrasonic Wave Reflections, (2003).

[19] Y.W. Tay, B. Panda, S.C. Paul, M.J. Tan, S. Qian, K.F. Leong, C.K. Chua, Processing and Properties of Construction Materials for 3D Printing, Materials Science Forum, Vol. 861, 2016.

[20] E. Vorndran, K. Wunder, C. Moseke, I. Biermann, F.A. Müller, K. Zorn, U. Gbureck, Hydraulic setting $Mg_3(PO_4)_2$ powders for 3D printing technology, Advances in Applied Ceramics 110 (8) (2011) 476-481.

[21] T. Yoshikawa, Manipulability of robotic mechanisms, The international journal of Robotics Research 4 (2) (1985) 3-9.

Automation of robotic concrete printing using feedback control system

Biranchi Panda*, Jian Hui Lim, Nisar Ahamed Noor Mohamed, Suvash Chandra Paul Yi Wei Daniel Tay, Ming Jen Tan

Singapore Centre for 3D Printing, School of Mechanical & Aerospace Engineering
Nanyang Technological University, 50 Nanyang Avenue, Singapore 639798

E-mail: biranchi001@e.ntu.edu.sg, JHLIM6@e.ntu.edu.sg, nisarahamed@ntu.edu.sg, suvashpl@ntu.edu.sg, danieltay@ntu.edu.sg, MMJTAN@ntu.edu.sg

Abstract –

In recent years, digital fabrication is termed as "third industrial revolution" and its interaction with extrusion based cementitious material has been well known as concrete printing. In concrete printing, a gantry/robotic system deposits concrete layer by layer following G-codes generated from slicing of the 3D component. However, the robotic system does not consider the material (concrete) properties and component geometry which sometimes cause failure in the printing process. Concrete properties are usually attributed with time and therefore the system parameters such as extruder velocity and layer height are necessarily to be controlled accordingly to obtain an uninterrupted smooth flow. In line with this, our current research aims to automate the printing process by collecting material's fresh properties through real-time feedback control system. A six-axis industrial Denso® robot is used for 3D printing of geopolymer concrete with the help of screw pump and 10mm circular extruder. The obtained experimental results finally confirm the validity and robustness of this automated set up.

Keywords –

Digital construction; Feedback control; Concrete printing; Autonomous robot

1 Introduction

It has been proposed that 5.3 billion houses are needed to be built in near future to cope with the impact of the ageing population, immigration and the singleton lifestyle [1]. In this regard, it is too much cumbersome for the government as well as civil builders to raise shelters in short span of time due to low productivity and labour intensive nature of conventional way of making buildings. Unlike other industries our current construction process lacks automation that can reduce the supply chain while making the building process faster, economic and accurate. This compels lot of researchers and industries to enter the new ear of automated building construction system to save time, labor and cost as well as improve safety.

Recent years have seen several automated construction technologies such as slip forming, mesh molding while simultaneously combining the benefits of both robotic system and digital modelling [2-5]. Concrete printing is one of such advanced technology that apply fused deposition modeling (FDM) principle of rapid prototyping for large scale construction. The first published evidence of 3D concrete printing was in 1997 with an innovative approach suggested by Pegna [6]. However, the actual development started in the mid-1990s in California, USA, when Khoshnevis introduced a technique termed Counter Crafting [7]. Contour crafting (CC) is a gantry based robotic system that extrudes material in a layer by layer manner to construct on-site structures [8]. Soon after, researchers at Loughborough University printed high strength fiber reinforcement concrete to construct various complex civil structures. For the first time, four key parameters such as extrudabilty, buildability, workability and open time were introduced with respect to printing of cementitious material [9,10]. Figure 1 shows some of these recent pioneering works done by (a) Apis-Cor, Russia, (b)Tu/e Eindhoven and (c)Xtree, France [11].

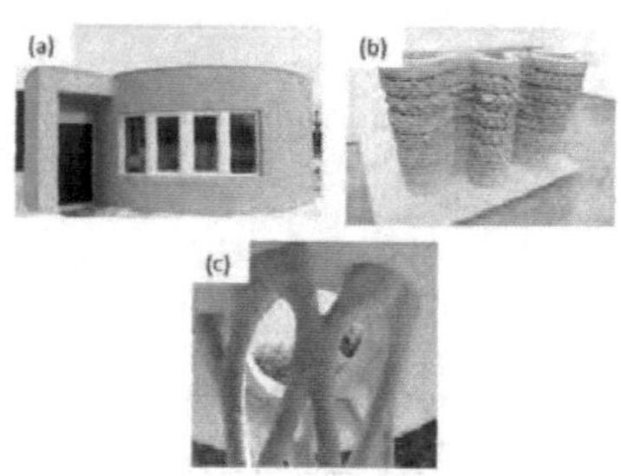

Figure 1. Examples of 3D concrete printing [11]

It is much clear from the above literature survey that concrete printing has the capability of building large scale freeform structures directly from the 3D model without any tools and human intervention. However, few challenges are still remained with respect to machine (system) and material development that are possibly impeding its wide spread application in various industrial domains. A right balance between material and machine (printer and pump) properties is highly necessary to achieve smooth extrusion during the printing process. Therefore, this paper aims to automate this material and machine relationship with the help of feedback control system by constantly monitoring material behavior with printing time. Few important parameters of concrete printing are highlighted in the next section following full experimental set up discussion developed here, to synchronize material and machine parameters such as print speed, layer height, pump flow rate etc. Finally, the last section, i.e section 4 concludes with results and discussions for future directions.

2 Research on control parameters

Concrete printing starts with 3D modelling of the components followed by G-code generation that controls motion of robotic arm in an unknown work environment. The robotic arm moves following the G-code while holding an extruder connected to the screw pump. A schematic representation of the complete process is shown in figure 2.

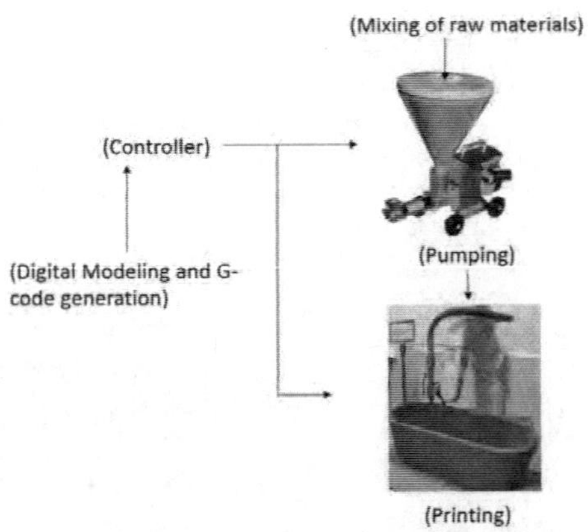

Figure 2. Flow of concrete printing process

Using the above setup, some preliminary investigations have been done at Nanyang technological university, Singapore and it has been found that concrete printing is an interdisciplinary research between printable material, machine (printer and pump) and geometry of the part (figure 3). Material viscosity, machine speed, pump flow rate and geometric complexity are highly related to each other and a mismatch among these parameters can cause the printed part to collapse due to non-uniformity of the material deposition [12]. Out of all, concrete (mortar) flow rate is a prime parameter which needs to be controlled more precisely. It is known that flow rate of the material directly affects the quality, surface finish and strength of the printed samples. A constant flow rate in this process is not viable because the fresh properties of the material such as viscosity, yield stress changes with time. Inconsistent print paths, poor surface finish, pores, intermittent flows are some common defects that are observed (figure 4) when using constant flow rate in the printing process. This is because of lack of high flow rate necessary to overcome the resistance offered by slowly changing material. Therefore, in this research a feedback system is developed to automate the concrete flow with printer speed while accounting material properties into consideration. The complete test set up and experimental procedure are described in the following section.

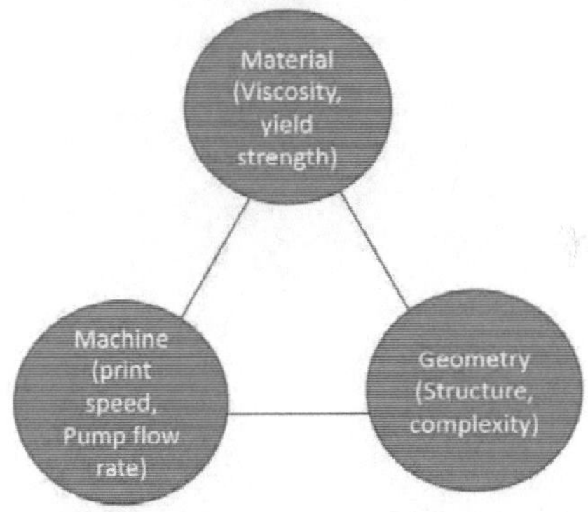

Figure 3. Concrete printing component

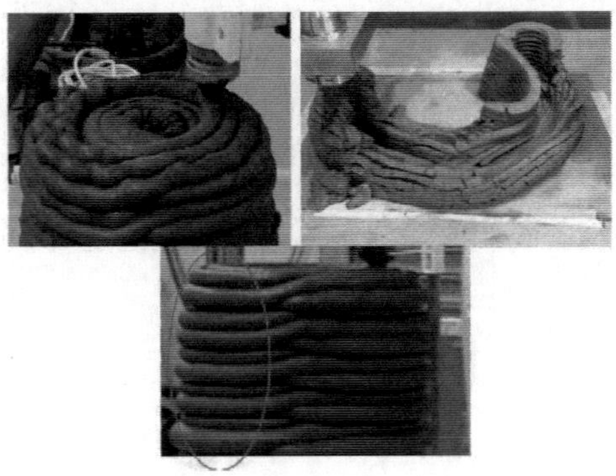

Figure 4. Defects occurred during concrete printing

3 Experimental setup based on feedback control system

Since the flow rate plays an important role in concrete printing, there is a necessity to control it accurately with respect to the changing material behavior. Literature reveals several methods of monitoring material behavior by using calorie meter, ultrasonic wave generator, vicat setting time apparatus etc. [13]. However, for concrete printing measurement of shear stress is more favorable since the material needs to be extrudable and no slump at the same time to build more layers one over another [8]. In this research, we used a portable shear vane apparatus supplied by Humboldt Mfg. Co. to measure the shear strength geopolymer by mounting it in the hopper of the screw pump as shown in figure 5.

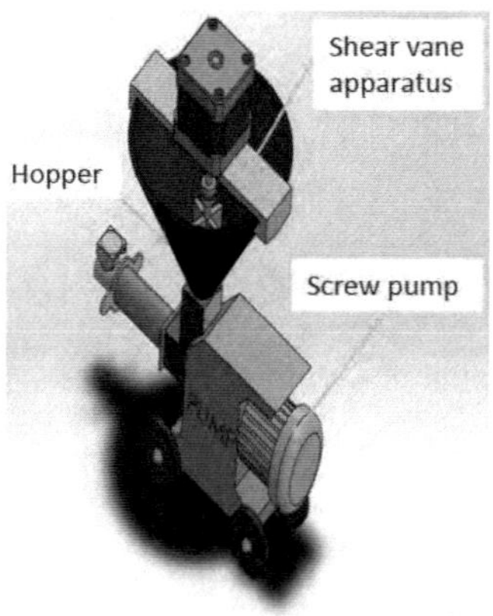

Figure 5. Experimental set up of measuring shear strength of geopolymer cement

All the raw materials of geopolymer i.e fly ash (class F), slag (GGBS), river sand (below 1.18 mm) and potassium silicate (molar ratio=1.8) are mixed in the fixed ratio as per mix design shown in Table 1 and instantly placed in the hopper of the pump.

Table 1: Mix design for geopolymer 3D printing

Materials	Weight Percentage
Fly ash	22.42
Slag	3.27
Silica Fume	6.54
Sand	49.0
K- Silicate	13.25
Additives	0.50

Initially, some geopolymer mix is extruded out to avoid any air bubble present in the hose pipe and then the printing is continued by using the developed automated flow set up. The complete process of flow automation can be described in four stages [14]:

(i) Data preparation,
(ii) Concrete mixing and printing, and
(iii) Sensor feedback adjustments.

In data preparation, the 3D CAD model is converted into an STL file, and sliced with the desired printing parameters; nozzle diameter, layer thickness, etc, to generate a G-code file for the robotic printer (figure 6). To control the flow rate and print speed, pump and the shear vane apparatus are integrated to the robotic printer via robotic operating system (ROS) in a feedback loop. ROS, is generally used as the communication middleware between its different modules [15]. This allows different modules to send and receive information from other nodes. The robot motion server interprets the G-code file, and pass the cartesian coordinates and speed information over to the path planner module. Using CAD models of the robotic arm and nozzle, a database of its static information is precomputed. This database can take up to half a day to generate, depending on the complexity of the nozzle end effector. However, this simplify the inverse kinematic problem during motion planning and only needed to be generated once. The motion planner uses BiRRT to generate a collision free, geometric path with no timestamp, followed by a Parabolic Smoother, which simultaneously shortcuts and times the trajectory, subjected to velocity and acceleration constraints. The robot pose throughout the printing process is given by the Cartesian coordinates in the G-code file, and orientation restricted by requiring the end effector print nozzle to be pointing downwards during printing. The inverse kinematic solution is solved through OpenRAVE's IKFast TranslationDirection5D.

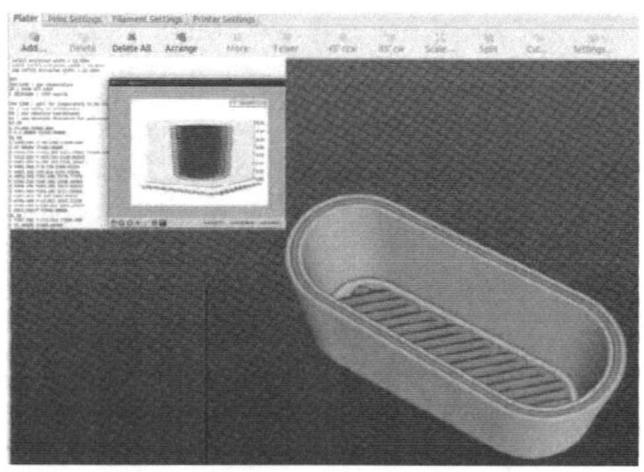

Figure 6. G-Code generation for concrete printing

Flow control for the screw pump is achieved by setting the analog voltage to the pump's potentiometer. This is done by sending a ROS serial message from the planner module to the microcontroller unit (MCU). The MCU reads in the required flow rate, and sets the correct analog voltage through a digital to analog converter. From the figure (Figure 7), it is found that the shear strength of the geopolymer mortar increases with time, which necessitates increase in pump flow rate and other relevant parameters for smooth continuous flow.

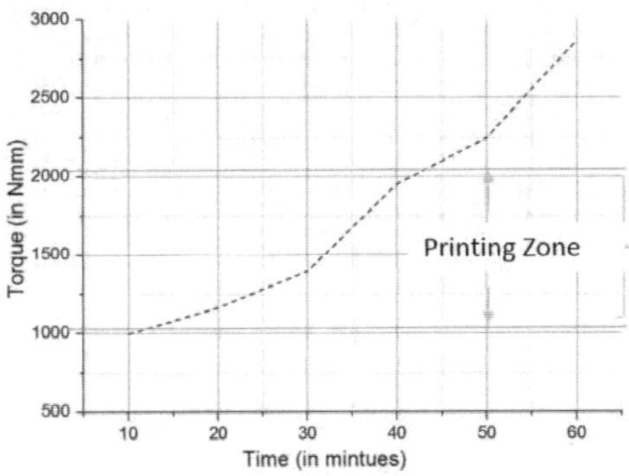

Figure 7. Variations of shear strength with time

The changes in shear strength is consistently monitor by the shear vane apparatus. This apparatus is driven by a dc motor at constant speed, and the deflection of the geovane's pointer is recorded by an absolute encoder which relays the information back to the microcontroller unit. The microcontroller unit then calculates the shear strength of the concrete mixture and passes the values through ROS serial messages to the path planning module. When the path planning module is updated with the shear strength of the concrete, it makes the necessary changes to the robot trajectory by retiming the planned path, without collision check, subjected to the robot's velocity and acceleration limits.

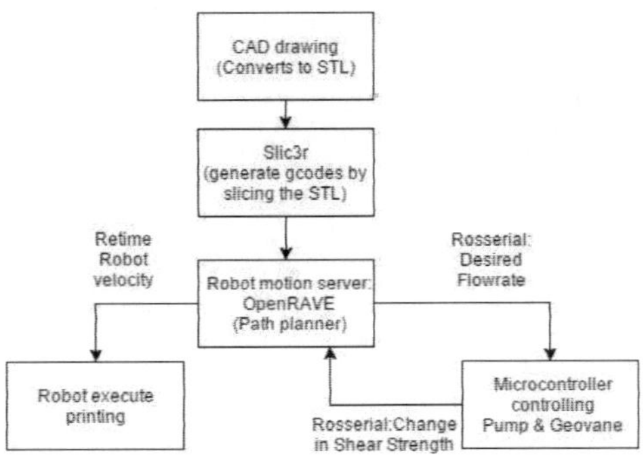

Figure 8. Flow control automation using feed back sensor

Initially for the prepared geopolymer mortar, flow rates are varied from 0-10V with respect to material shear strength. Simultaneously, to match the flow rate, printer speed is adjusted based on material slump and flow characteristics obtained by the sensor connected to the shear vane apparatus. The process flow is shown in figure 8.

4 Discussions and Concluding remarks

Our current practices for controlling flow during concrete printing is purely manual and involves lot of trial and error approaches. This sometimes causes disruption in the extrusion and leads to part failure. Therefore, in this research an automatic flow control system has been developed considering material hardening behavior. Path planning of robotic motion system is improved through feedback control system, while accounting material shear stress property directly from shear vane apparatus. Figure 9 shows some printed part using this automated flow control system. It is concluded from these images that controlling the flow with proper material knowledge results uniform, continuous deposition of material as per the designed 3D model.

Figure 9. Flow controlled geopolymer printing

It is worth to mention that, apart from concrete flow rate and print speed, nozzle height above the printed surface has also the influence in print quality and part stability [16]. Sometimes due to material sagging, height between printed surface and nozzle increases dramatically which results in strange (zig zag) flow pattern for the material coming out of the nozzle. In this regard, authors are currently working on integrating an optical feedback system to the nozzle that can detect the slump of final layer and will adjust the nozzle position accordingly. It has been found that during printing of some complex geometry material over-deposition occurs in corners and sharp turning edges which can be solved by controlling the flow while considering material fresh properties. It is believed that our developed automated flow control process will be useful to the future builders if right material properties can be captured and send to feedback support system.

References

[1] Lutz, W., Sanderson, W., & Scherbov, S. (2008). The coming acceleration of global population ageing. *Nature*, 451(7179), 716-719.

[2] Wangler, T., Lloret, E., Reiter, L., Hack, N., Gramazio, F., Kohler, M., ... & Flatt, R. (2016). Digital Concrete: Opportunities and Challenges. *RILEM Technical Letters*, 1, 67-75.

[3] Maurice, B. (1970). U.S. Patent No. 3,497,579. Washington, DC: U.S. Patent and Trademark Office.

[4] Kohler, M., Gramazio, F., & Willmann, J. (2014). The robotic touch: how robots change architecture.

[5] Gramazio, F., & Kohler, M. (2014). Made by robots: challenging architecture at a larger scale. John Wiley & Sons.

[6] Pegna, J. (1997). Exploratory investigation of solid freeform construction. *Automation in construction*, 5(5), 427-437.

[7] Hwang, D. O. O. I. L., & Khoshnevis, B. (2005). An innovative construction process-contour crafting (CC). *In 22nd International Symposium on Automation and Robotics in Construction*

[8] Buswell, R. A., Soar, R. C., Gibb, A. G., & Thorpe, A. (2007). Freeform construction: mega-scale rapid manufacturing for construction. *Automation in construction*, 16(2), 224-231.

[9] Le, T. T., et al. (2012). Mix design and fresh properties for high-performance printing concrete. *Materials and structures*, 45(8), 1221-1232.

[10] Perrot, A., Rangeard, D., & Pierre, A. (2016). Structural built-up of cement-based materials used for 3D-printing extrusion techniques. *Materials and Structures*, 49(4), 1213-1220.

[11] Bos, F., Wolfs, R., Ahmed, Z., & Salet, T. (2016). Additive manufacturing of concrete in construction: potentials and challenges of 3D concrete printing. *Virtual and Physical Prototyping*, 11(3), 209-225.

[12] Tay, Y. W., Panda, B., Paul, S. C., Tan, M. J., Qian, S. Z., Leong, K. F., & Chua, C. K. (2016). Processing and properties of construction materials for 3d printing. *Materials Science Forum*, 861, 177-181. Trans Tech Publications.

[13] Kim, J. W., Lee, C., Park, S., & Koh, K. T. (2013). Real-time strength development monitoring for concrete structures using wired and wireless electro-mechanical impedance techniques. KSCE *Journal of Civil Engineering*, 17(6), 1432-1436.

[14] Pham, T. H., Lim, J. H., & Pham, Q. C. (2016). Robotic 3D-Printing for Building and Construction. In proceeding of 2nd International Conference on Progress in Additive Manufacturing (Pro-AM),16-19 May, Singapore. Koubâa, A. (Ed.). (2016).

[15] Robot Operating System (ROS): The Complete Reference (Vol. 1). Springer. Chicago

[16] Tay, Y. W., Panda, B., Paul, S. C., Noor Mohamed Nisar, Tan, M. J., & Leong, K. F (2017). 3D printing trends in building and construction industry: a review, *Virtual and Physical Prototyping,* DOI: 10.1080/17452759.2017.1326724

Acknowledgement

The authors would like to acknowledge SempCrop Architects & Engineers Pte Ltd and National Research Foundation (NRF), Singapore for funding and support for this research project. We are also thankful to Control Robotic Intelligence (CRI) group at Nanyang Technological University, Singapore for the assistance and advice.

34th International Symposium on Automation and Robotics in Construction

ISARC 2017

June 28 – July 1 TAIPEI, TAIWAN

MACHINE LEARNING / COMPUTATIONAL INTELLIGENCE

Comparison of Single Classifier Models for Predicting Long-term Business Failure of Construction Companies Using Finance-based Definition of the Failure

H. Choi, H. Sung, H. Cho, S. Lee, H. Son, and C. Kim*

Department of Architectural Engineering, Chung-Ang University, South Korea
E-mail: vinaj516@gmail.com, gusdn7543@gmail.com, hyukmanjo@gmail.com, leesungwook@cau.ac.kr, hjson0908@cau.ac.kr, changwan@cau.ac.kr (*corresponding author)

Abstract
This study compares the performance of six classifier models (ANN, KNN, C4.5, SVM, LR, and NB) for predicting the business failure of construction companies after three years from 2010 to 2012. Although previous studies have explored numerous business failure prediction models for construction companies, these models have focused on short-term failure, defined as failure occurring within one year, and have defined business failure based on companies' experiencing serious legal events, including bankruptcy, delisting, and default. However, the construction industry is typically characterized by projects with longer durations, usually exceeding one year. This implies that previous short-term models cannot predict the business failure of construction companies until the end of particular projects. Moreover, this problem is compounded by the fact that legal events can involve lengthy proceedings, which are often initiated much later than the actual moment of the business failure. Therefore, in this study, six classifier models will be used to predict the business failure of the construction companies within three years using a finance-based definition of failure. The results show that all six models' performances noticeably decrease when they predict more than one year. These results demonstrate that previous short-term prediction models with outstanding performance cannot be practical in predicting the long-term business failure of construction companies.
Keywords – Data Mining; Machine Learning; Project Management; Business failure; Prediction model;

1 Introduction

Business failure for a construction company during a construction project can lead to a series of failures for subcontractors, which can also result in serious socio-economic problems. Compared to businesses in other industries, construction companies are more vulnerable to business failure as a result of various specificities [1], including the uniqueness of the projects, the long duration of the projects, and the industry's sensitivity to economic cycles [2]. For example, the construction industry had the highest rate of bankruptcy among all industries in Korea between 1998 and 2015 [3]. Due to this vulnerability, the ability to predict a potential business failure for a construction company in the early stages of a construction project can be a tool of critical importance to a variety of stakeholders, including the project owner, investors, creditors, and contractors.

Business failure prediction models for construction companies based on financial ratios have been proposed since 1970. These models are based on statistical techniques, such as multivariate discriminant analysis (MDA) [4] and logistic regression (LR) [5], and, more recently, data mining techniques [2,6–9]. Most of the previous studies have focused on short-term (within one year) predictions, even though construction projects typically have a longer duration. Thus, these models cannot predict the construction companies' financial risk during the entire project period. Also, these short-term models may suffer from an absence of data because, in many cases, the publication of the annual accounts of the failing companies can be delayed [10]. Moreover, these previous studies defined the business failure based on highly visible legal events from a sample construction company, including bankruptcy, delisting, and default. However, legal events of this type are typically the culmination of a lengthy legal process, meaning their occurrence can be much later than the moment of actual failure [10,11]. To address this issue, this study compared single classifier models for predicting the business failure of a construction company within three years in order to cover the relatively long duration of the projects. Moreover, this study uses a finance-based definition of business failure regardless of legal events.

The rest of the paper is organized as follows. Section 2 reviews the literature on the prediction term, the definition of the business failure, and the techniques used in their models. Section 3 describes the collected data and the selected variables. Section 4 describes the pre-processing of the data and the single classifier model employed in this study, including 6 single classifiers (support vector machine [SVM], artificial neural networks [ANN], commercial version 4.5 [C4.5], naive Bayes [NB], LR, k-nearest neighbor [KNN]). Section 5 presents the results of experiments designed to evaluate and compare the performances of the six single classifiers in terms of the area under the receiver operating characteristic curve (AUC). Finally, conclusions are presented in Section 6.

2 Literature Review

Recently, business failure prediction models for construction companies based on data mining techniques [1,2] have been found to outperform the traditional statistical models based on MDA [4] and LR [5]. Therefore, the models for predicting business failure using data mining is actively proposed [1,2,6–9]. However, previous studies have focused on short-term (within one year) prediction and have defined business failure based on legal events.

Tserng et al. [2] proposed a model for predicting the default of construction companies within one year. They used the data of the year right before default as the data of default companies. Default was defined as delisting due to bankruptcy, liquidation, or poor performance. They compared an enforced SVM (ESVM) model with the LR model. The results show that ESVM outperforms the traditional LR model. Chen and Hoang [6] proposed a model for predicting financial distress in a quarter of construction companies, using data from quarterly financial reports. They defined financial distress as bankruptcy, delisting, bounding, bailouts, or major organizational restructuring. Their model is based on self-organizing feature map optimization and fuzzy- and hyper-rectangular composite neural networks. Horta and Camanho [1] proposed a model for predicting company failure within one year in construction industries. Their proposed model, based on SVM, outperforms the LR model in predicting failure. Heo and Yang [7] proposed a model for predicting bankruptcy within one year. They used the data one year prior to bankruptcy as the data of the bankrupt companies. They defined bankruptcy as workout, receivership, or bankruptcy. Their model was based on adaptive boosting (AdaBoost). The proposed AdaBoost model was compared with the ANN, SVM, decision tree (D-Tree), and Z-Score model. The results showed that the proposed AdaBoost model outperforms other models. Cheng et al. [8] and Tserng et al. [9] proposed a model for predicting the default of construction companies within one year. They defined default as delisting due to bankruptcy, liquidation, or poor performance, as in Tserng et al. [2]. Their models are based on least squares SVM and grey system theory, respectively. As can be seen, the relatively long duration of construction projects and the financial risk construction companies could experience before the occurrence of legal events have not been sufficiently considered.

3 Data and Variable Selection

From Korean information service value [12], this study obtained data derived from financial statements of construction companies with Korean standard industrial classification codes 41 and 42 covering 2007 to 2012. The data for years 2007, 2008, and 2009 were used to predict business failure after one year (2010), two years (2011), and three years (2012), respectively. The data were organized into three datasets, which are a 2010 dataset for predicting after one year, a 2011 dataset for predicting after two years, and a 2012 dataset for predicting after three years. This study excluded companies that did not provide complete data during the sample period to guarantee the completeness of the data. Companies with total assets less than 12 billion won or with total assets and total liabilities less than 7 billion won were also excluded in order to ensure the reliability of the data, which is one of the conditions of the external auditing company. Finally, the financial information of 385 construction companies was used as the final data.

This study used a finance-based definition of business failure to classify companies into normal and failed companies regardless of legal events. The finance-based definition of failure has been used to detect failure regardless of the legal consequence used in previous studies [11, 13-15]. Previous studies [e.g. 14–17] employed special treatment (ST) regulation to define business failure. ST regulation is the early warning system to identify abnormalities in Chinese-listed companies' financial status [17]. Ding et al. [18] showed that the companies fell into business failure after receiving the ST and showed some signs of potential failure prior to receiving it. Thus, predicting ST can be a financial early warning system for possible future business failure. Therefore, this study utilized the ST regulation to define business failure based on Li and Sun [14] and Sun and Li [15]. Therefore, companies having negative net income for two consecutive years were classified as failed companies. As a result, the numbers of failed companies in the three years' datasets were 28, 41, and 49, as shown in Table 1.

Twenty-one financial ratios were selected as input variables based on a review of the previous studies [1,2,6–9]. Following Horta and Camanho [1] and Tserng et al [2], financial ratios were classified into four categories: activity, leverage, liquidity, and profitability. Table 2 provides definitions of these financial ratios, which cover a wide range of financial characteristics and performances [2], according to Cheng et al. [8].

Table 1. Distribution of the data

Dataset	Normal companies	Failed companies
2010	357	28
2011	344	41
2012	336	49

4 Methodology

The methodology framework is shown in Figure 1 and the process is as follows. First, an oversampling technique based on the synthetic minority oversampling technique (SMOTE) was employed to obtain a balanced dataset in which the number of normal and failed companies are equivalent. Then, six classification models were applied and their prediction accuracies were compared: SVM, ANN, C4.5, NB, LR, and KNN.

At the same time, the parameter optimization with grid search was executed for the six classifiers. Finally, we evaluated the models with an AUC value using 10-fold cross validation. This study employed the algorithms from Weka release 3.8.1 [19] in all experiments.

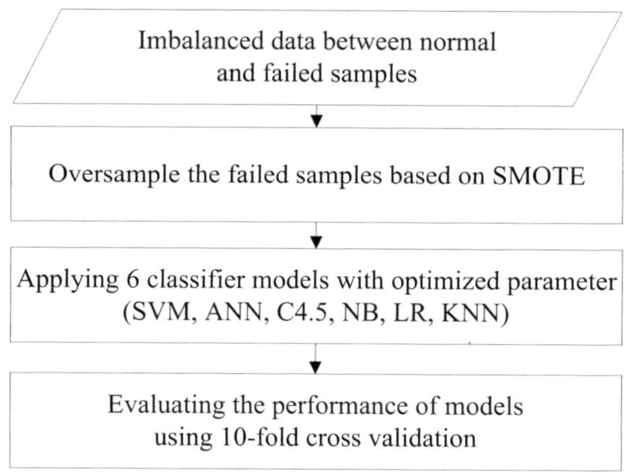

Figure 1. The Methodology Framework

Table 2. Financial ratios and definitions

Category	Variable	Definition
Activity	Accounts payable turnover	Sales / Average payable
	Accounts receivable turnover	Sales / Average receivables
	Current assets turnover	Sales / Current assets
	Fixed assets to net worth	(Total assets - Current assets) / Shareholders' equity
	Quality of inventory	Cost of sales / Average inventories
	Revenues to fixed assets	Sales / (Total assets - Current assets)
	Revenues to net working capital	Sales / (Current assets - Current liabilities)
	Sales to net worth	Sales / Shareholders' equity
	Turnover of total assets	Sales / Total assets
Leverage	Debt ratio	Total liabilities / Total assets
	Retained earnings to sales	Retained earnings / Sales
	Times interest earned	Earnings before interest and taxes / Interest expense
	Total liabilities to net worth	Total liabilities / Shareholders' equity
Liquidity	Current assets to net assets	Current assets / (Total assets - Current liabilities)
	Current ratio	Current assets / Current liabilities
	Net working capital to total assets	(Current assets - Current liabilities) / Total assets
	Quick ratio	(Current asset - Inventories) / Current liabilities
Profitability	Profits to net working capital	Net income / (Current assets - Current liabilities)
	Return on assets	Net income / Total assets
	Return on equity	Net income / Shareholders' equity
	Return on sales	Net income / Sales

4.1 Pre-processing for Dealing with the Imbalanced Data Problem

As shown in Table 1, this study used three datasets with a population consisting of imbalanced data. However, the imbalanced data can distort the real performance of the prediction model. In other words, a model based on imbalanced data can yield high overall predictive accuracy driven by the majority class but with poor accuracy for the minority class [20]. Therefore, this study employed the over-sampling method SMOTE to handle this problem. SMOTE generates synthetic samples to oversample a minority class without data replication [21]. Therefore, SMOTE can avoid the over-fitting problem which can be caused from data replication [22–24]. In SMOTE, k nearest samples to the original samples are selected. This study set the value of k to 5, which is recommended by Chawla et al. [25]. Then, a new sample is generated by adding the average of the distances, multiplied by an arbitrary number from 0 to 1, between the selected k samples and the original sample to the original sample [26]. This process was repeated until the sample numbers of the minority class and the majority classes were balanced.

4.2 Six Single Classifiers with Optimized Parameters Using Grid Search

This study considered six different classifiers: SVM, ANN, C4.5, NB, LR, and KNN. The parameter values of each classifier were optimized using a grid search for three datasets as Table 3. The basic concept of a grid search involves applying various parameter values and choosing the one with the best cross-validation performance. In this study, the grid search was conducted using 10-fold cross-validation. Among the applied parameter values, the parameter of each classifier for each dataset are set to the parameter value with the best classification performance in 10-fold cross validation. Finally, the optimized parameters used in this study are summarized in Table 3.

4.3 Performance Evaluation Measures

This study employed an AUC measure to compare six classifier models, which is commonly used in evaluating business failure models for construction companies [1,2,8]. AUC is the most commonly employed as a summary statistic for the quality of the rankings [2]. This study carried out 10-fold cross validation to evaluate the classification performance of the six classifier models. This method is known to minimize bias and variance compared to all other validation methods [33,34]. In 10-fold cross validation, the training dataset was divided into ten subsets. Nine subsets were then used to train the model, and the remaining subset was retained to assess its performance. The performances of the classifiers were computed by averaging the performance of each of the ten subsets.

5 Experimental Results and Discussion

The performances of the six classifier models are summarized in Table 4. The results were generated by applying 10-fold cross validation by six classifiers to the three datasets for 2010, 2011, and 2012, respectively.

As shown in Table 4, the KNN model had the highest prediction performance among the six models. The ANN followed the KNN model. However, the C4.5 model has better performance in the case of predicting for 2012. This result shows that although the model has shown better performance results in the short-term prediction of business failure, the model may not show the best performance in long-term predictions. Furthermore, the results show that the performance of all six classifier models noticeably decreases when the models are used to predict more than one year ahead. This result indicates that the long-term prediction of business failure is much more difficult than short-term prediction. Therefore, although the short-term prediction models proposed in most previous studies have outstanding performance in short-term prediction, the model may not be useful for long-term prediction covering a relatively long project period in the construction industry.

Table 3. Grid search applied in this study

Classifiers	Parameters	Grid space	Optimized parameters for each dataset 2010	2011	2012
SVM [27,28]	Penalty parameter C	$(2^1,2^2,\ldots 2^{15})$	16	8	8
	Kernel parameter γ	$(2^{-15},2^{-14},\ldots 2^3)$	8	8	8
ANN [29]	Learning rate μ	(0.1,0.2,…,0.9)	0.1	0.7	0.7
	Momentum α	(0.1,0.2,…,0.9)	0.7	0.1	0.1
C4.5 [30,31]	Confidence Factor	(0,0.05,…,1)	1	1	1
	Minimum number of samples per leaf	(5,10,…,150)	5	15	10
KNN [32]	The number of closest neighbors k	(1,2,…,50)	9	7	4

Table 4. The performance(AUC) of the six models

Method	Prediction Year		
	2010	2011	2012
SVM	0.923	0.860	0.858
ANN	0.954	0.907	0.860
C4.5	0.927	0.876	0.862
NB	0.901	0.830	0.753
LR	0.912	0.844	0.794
KNN	0.955	0.903	0.894

6 Conclusion

This study presents a comparison of the prediction performances of six single classifier models for predicting the long-term business failure of 385 Korean construction companies. This study predicted the business failure over the next three years using the 21 financial indicators based on the financial data of the past three years. The sample companies were classified into failed and normal companies using a finance-based definition of failure regardless of legal events. In order to handle the imbalance problem of the data, oversample technique based on SMOTE was used. Lastly, the six single classifier models' performances were compared by way of AUC values.

There are two contributions in this study. First, this study used a finance-based definition of failure in order to classify the sample companies regardless of legal events: bankruptcy, delisting, and default. The advantage of this definition is that it can be used to consider the financial risk companies may experience before legal events occur. In addition, a finance-based definition can prevent the problem of legal events occurring much later than the actual moment of failure. Second, this study considered the long-term business failure of construction companies in order to cover the relatively long duration of construction projects. At the early stages of construction projects, the long-term prediction model can predict whether construction companies will have financial risk until the end of the project or not. Therefore, the long-term prediction model can help project owners and other stakeholders to avoid the damage caused by business failure during the construction project.

This study used 21 financial ratios indicating the activity, leverage, liquidity, and profitability of construction companies. However, many studies [14,16,17,35] employed growth ratios to predict business failure in other companies. Therefore, in future study, we would include growth ratios as a financial indicator to improve the prediction performance. In addition, in the modeling method, a future study could extend the method to improve performance in long-term prediction.

References

[1] Horta I. M. and Camanho A. S. Company failure prediction in the construction industry. *Expert Systems with Applications*, 40:6253 – 6257, 2013.

[2] Tserng H. P., Lin G. F., Tsai L. K., and Chen P. C. An enforced support vector machine model for construction contractor default prediction. *Automation in Construction*, 20:1242 – 1249, 2011.

[3] National Information and Credit Evaluation Investors Service Rating performance 2015. Online:http://www.nicerating.com/disclosure/ratingPerFormance.do.

[4] Ng S. T., Wong J. M. W., and Zhang J. Applying Z-score model to distinguish insolvent construction companies in China. *Habitat International*, 35:599 – 607, 2011.

[5] Tserng H. P., Cheng P. C., Huang W. H., Lei M. C., and Tran Q. H. Prediction of default probability for construction firms using the logit model, *Journal of Civil Engineering and Management*, 20(2):247 – 255, 2014.

[6] Chen J. H. Developing SFNN models to predict financial distress of construction companies. *Expert Systems with Applications*, 39:823 – 827, 2012.

[7] Heo J. and Yang J. Y. AdaBoost based bankruptcy forecasting of Korean construction companies. *Applied Soft Computing*, 24:494 – 499, 2014.

[8] Cheng M. Y., Hoang N. D., Limanto L., and Wu Y. W. A novel hybrid intelligent approach for contractor default status prediction. *Knowledge-Based Systems*, 71:314 – 321, 2014.

[9] Tserng H. P., Ngo T. L., Chen P. C., and Tran L. Q. A grey system theory-based default prediction model for construction firms. *Computer-Aided Civil and Infrastructure Engineering*, 30:120 – 134, 2015.

[10] Balcaen S. and Ooghe H. 35 years of studies on business failure: an overview of the classic statistical methodologies and their related problems. *The British Accounting Review*, 38:63 – 93, 2006.

[11] Tinco M. H. and Wilson N. Financial distress and bankruptcy prediction among listed companies using accounting, market and macroeconomic variables. *International Review of Financial Analysis*, 30:394 – 419, 2013.

[12] Kisvalue, online: http://www.kisvlue.com.

[13] Pindado J., Rodrigues L., and Torre C. Estimating financial distress likelihood. *Journal of Business Research*, 61:995 – 1003, 2008.

[14] Li H. and Sun J. Ranking-order case-based reasoning for financial distress prediction. *Knowledge-Based Systems*, 21:868 – 878, 2008.

[15] Sun J. and Li H. Financial distress prediction based on serial combination of multiple classifiers. *Expert Systems with Applications*, 36:8659 – 8666, 2009.

[16] Hua Z., Wang Y., Xu X., Zhang B., and Liang L. Predicting corporate financial distress based on integration of support vector machine and logistic regression. *Expert Systems with Applications*, 33:434 – 440, 2007.

[17] Geng R., Bose I., and Chen X. Prediction of financial distress: An empirical study of listed Chinese companies using data mining. *European Journal of Operational Research*, 241:236 – 247, 2015.

[18] Ding Y., Song X., and Ze Y. Forecasting financial condition of Chinese listed companies based on support vector machine. *Expert Systems with Applications*, 34:3081 – 3089, 2008.

[19] Witten I. H. and Frank E. Data Mining: Practical Machine Learning Tools and Techniques, second ed. Morgan Kaufmann, San Francisco, 2005.

[20] Thammasiri D, Delen D., Meesad P., and Kasap N. A critical assessment of imbalanced class distribution problem: The case of predicting freshmen student attrition. *Expert Systems with Applications*, 41:321 – 330, 2014.

[21] Zhou L. Performance of corporate bankruptcy prediction models on imbalanced dataset: The effect of sampling methods. *Knowledge-Based Systems*, 41:16 – 25, 2013.

[22] Gao M., Hong X., Chen S., Harris C. J. A combined SMOTE and PSO based RBF classifier for two-class imbalanced problems. *Neurocomputing*, 74:3456 – 3466, 2011.

[23] Fernández A., Jesus M. J., and Herrera F. Hierarchical fuzzy rule based classification systems with genetic rule selection for imbalanced data-sets. *International Journal of Approximate Reasoning*, 50:561 – 577, 2009.

[24] Son H. and Kim C. Early prediction of the performance of green building projects using pre-project planning variables: data mining approaches. *Journal of Cleaner Production*, 109:144 – 151, 2015.

[25] Chawla N. V., Bowyer K. W., Hall L. O., and Kegelmeyer W. P. SMOTE: Synthetic Minority Over-sampling Technique. *Journal of Artificial Intelligence Research*, 16:321 – 357, 2002.

[26] Kim M. K., Kang D. K., and Kim H. B. Geometric mean based boosting algorithm with over-sampling to resolve data imbalance problem for bankruptcy prediction. *Expert Systems with Applications*, 42:1074 – 1082, 2015

[27] Min J. H. and Lee Y. C. Bankruptcy prediction using support vector machine with optimal choice of kernel function parameters. *Expert Systems with Applications*, 28:603 – 614, 2005.

[28] Wu T. K., Huang S. C., and Meng Y. R. Evaluation of ANN and SVM classifiers as predictors to the diagnosis of students with learning disabilities. *Expert Systems with Applications*, 34:1846 – 1856, 2008.

[29] Barreto G. A. and Araújo A. F. R. Identification and Control of Dynamical Systems Using the Self-Organizing Map. *IEEE Transactions on Neural Networks*, 15(5):1244 – 1259, 2004.

[30] Sakthivel N. R., Sugumaran V., and Babudevasenapati S. Vibration based fault diagnosis of monoblock centrifugal pump using decision tree. *Expert Systems with Applications*, 37:4040 – 4049, 2010.

[31] Ravikumar S., Ramachandran K. I., and Sugumaran V. Machine learning approach for automated visual inspection of machine components. *Expert Systems with Applications*, 38:3260 – 3266, 2011.

[32] Balabin R. M., Safieva R. Z., and Lomakina E. I. Gasoline classification using near infrared (NIR) spectroscopy data: Comparison of multivariate techniques. *Analytica Chimica Acta*, 671:27 – 35, 2010.

[33] Son H., Kim C., Hwang N., Kim C., and Kang Y. Classification of major construction materials in construction environments using ensemble classifiers. *Advanced Engineering Informatics*, 28:1 – 10, 2014.

[34] Kohavi R. A study of cross-validation and bootstrap for accuracy estimation and model selection. In *Proceeding of the International Joint Conference on Artificial Intelligence*, pages 1137–1143, Montréal, Canada, 1995.

[35] Lin F., Liang D., Yeh C. C., and Huang J. C. Novel feature selection methods to financial distress prediction. *Expert Systems with Applications*, 41:2472 – 2483, 2014.

Vision-Based Safety Vest Detection in a Construction Scene

H. Seong, H. Choi, H. Cho, S. Lee, H. Son, and C. Kim*

Department of Architectural Engineering, Chung-Ang University, Seoul 06974, South Korea
E-mail: gusdn7543@cau.ac.kr, vinaj516@gmail.com, hyukmanjo@gmail.com, leesungwook@cau.ac.kr, hjson0908@cau.ac.kr, changwan@cau.ac.kr (* Correponding author)

Abstract

The computer vision-based detection of construction workers in images or videos is necessary for the safety managements and productivity of construction workers. Researchers in previous studies, detecting construction workers in the construction scene via computer vision techniques, have considered various features such as motion, shape, and color. Due to the pose changes of the workers, construction worker detection using body shape as a feature in the construction scene remains a challenging task. This study proposes a safety vest detection method, as a preceding method of the construction worker detection, which uses the motion of workers and the color pixels of safety vests for distinguishing from others in the construction scene independent of the workers' pose changes. The background subtraction method is performed by using an approximate median filter for the purpose of reducing the candidate regions that the color pixel classification will be performed as a sequential step. Then, the color pixel classification method is performed with a comparative analysis of two color spaces (Lab and HSV [hue, saturation, and value]) and three types of classifiers (a support vector machine [SVM], an artificial neural network [ANN], and a logistic regression [LR]) to classify the pixels of safety vests accurately. The proposed method has been tested on the actual construction site video streams. The proposed method of this study, thus far, is making progress toward the achievement of the robust and effective construction worker detection in the construction scene.

Keywords – Computer Vision; Sensing; Data Mining; Machine Learning; Detection; Classification; Image Processing; Construction Worker; Color Space Transformation

1 Introduction

The computer vision-based detection of construction workers is an important element for various applications in both safety management and productivity measurement. Computer vision provides a rich set of data about a construction scene that helps distinguish between workers and others that can be found on a construction site. Within the last decades, video surveillance has become common on construction sites with the deployment of vision sensors. The better the quality of the camera is, the higher the quality of the acquired images will be. Computer vision techniques, in combination with computer networks and automated image recognition, take advantage of this. The valuable information in video frames of complex construction tasks can be provided promptly and accurately with highly sophisticated computer vision techniques.

There have been a few studies on the use of computer vision techniques to detect workers in a construction scene [2–6]. The researchers in most previous studies employed the shape of the human body as a feature for construction worker detection. However, these studies were limited to detecting workers with upright postures. Although these methods may work in controlled construction scenarios, construction worker detection in construction scenes remains challenging because of the various postures of construction workers in construction scene. To meet these challenges, color-based features have obvious advantages over other features (especially in complex environments) because the color of an object of interest is independent from the positions and shapes [7]. Color-based features are commonly used in various object detection techniques such as the detection of workers, equipment, and materials in the construction sites [5,8,9]. The color property of the safety vest is expected to be an accurate and robust feature that has the potential to overcome the limitations of construction worker detection that uses shape-based features.

The American National Standards Institute [10] stipulated that wearing a high-visibility safety vest is mandatory for all construction workers for reducing the likelihood of accidents. Accordingly, the wearing of a safety vest identifies workers on a construction site.

Besides that, the American National Standards Institute has provided a specification for the colors of the safety vest: fluorescent yellow-green and fluorescent orange-red. Both the fluorescent yellow-green and fluorescent orange-red colors are not prevalently found in the construction scene; thus, they are distinctive colors that can be readily distinguished from the color of other objects in the construction scene.

This study proposes a safety vest detection method using the motion of the workers with the colors of safety vests being used in a preceding process of effective and reliable construction worker detection via a computer vision-based technique. The background subtraction method applying an approximate median filter was used for reducing the candidate region for the sequential step (i.e., color pixel classification). A data set of safety vest pixels was generated for comprehensive color pixel classification results. To achieve robust color pixel classification against the illumination changes, a transformation from RGB (red, green, and blue) to HSV and Lab color spaces (with or without illuminant components) was performed. A comparison of the performances among three classifiers (i.e., SVM, ANN and LR) with different color spaces was performed to find the most appropriate combination for the classification of safety vest pixels. The proposed method had been tested on the video streams of actual construction sites.

2 Literature Review

With computer vision, construction worker classification using color-based features is used as a first step in construction worker tracking and monitoring techniques on a frame-by-frame basis. Recently, a few researchers have applied the color-based features of safety vests to construction worker classification methods. Their approaches to classifying workers in frames are grouped into two types: a color histogram-based method and a pixel-based method.

A color histogram is widely used as one of the most common color-based features, due to its invariance of small changes in viewing position and its computational efficiency. Construction worker classification methods using color histograms have been introduced in previous studies. Park and Brilakis [2] proposed a method for the detection of construction workers when initializing a scene. They used the hue-saturation color property of the upper-body section, which involved using the safety vest to construct a histogram. Then, the histogram was classified into the categories of construction worker and non-construction worker by using the k-nearest neighbor (KNN) classification method. Memazardeh et al. [3] suggested using histograms of oriented gradients and hue-saturation color-based methods for the identification of workers on a construction site. The researchers used the color histogram of a region of interest to discriminate between construction workers and the background. In their approach, the support vector machine (SVM) classifier, which is one of the most powerful machine-learning techniques for solving binary classification problems on a large scale, was employed as a color-based classification method for construction workers. However, a color histogram does not include spatial information, especially for a large database [11]. Moreover, construction worker classification using a color histogram of the region of interest can disturb accurate performance results because the color histogram is affected by other surrounding colors in the region (e.g., colors of hardhats, clothes, and surroundings).

The state-of-the-art method of the pixel-based classification approach is the classification of individual pixels in the region of interest to safety vest and non-safety vest categories on the basis of pixel color properties. In the past few years, Gong and Caldas [4] proposed a construction worker detection method for the purpose of measuring productivity by using color pixel classification to recognize construction workers. Their approach used each pixel of safety vests as a feature. Their color-based model used the expectation by maximization method to estimate the parameters of the Gaussian models, which can determine the pixels of safety vests by clustering of the pixels. Their result presented 85.50 % for true positives and 5.39 % for false positives. The classification method of pixels-to-safety vest pixels using a Gaussian model may not produce optimal results in a construction scene in which the inherent color property of the safety vest is affected by variations such as illumination changes and the position and characteristics of the camera. Empirical studies have found that the ANN and SVM classifiers outperformed the Gaussian model in the object detection in the construction scene [12].

3 Methodology

This study proposes a method to reliably detect safety vests in the construction scene via computer vision techniques. The proposed method includes the background subtraction method and color pixel classification, which are separately performed in sequential steps. As the first step, the foreground regions of interest, which are in motion, are detected using the background subtraction method. The background subtraction method determines which pixels belong in the background and which pixels belong in the foreground region of interest. Then, the color pixel classification is performed to determine which pixels belong to safety vests.

3.1 Background Subtraction Method

Background subtraction is a commonly used detection method for identifying foreground objects in motion in the video stream. Because the method is desirable for achieving a high level of precision in the detection of foreground objects of interest, it is employed as a reliable first step in a number of computer vision applications. In this study, the background subtraction method was performed for the purpose of reducing the areas to be computed. This study selected the background subtraction method using the approximate median filter, which was introduced by McFarlane and Schofield [13], among many highly sophisticated background subtraction algorithms. In the field of image processing, the approximate median filter has been commonly used for removing impulsive noises because of the simplicity of its process in that it is computationally efficient. In construction video surveillance applications that work with outdoor scenes, the background scene includes a number of nonstationary objects such as the construction materials, tree branches, and soil with which movement depends on the wind. These nonstationary objects cause an increase in computational burdens due to false-positive detection. Therefore, this study eliminated such foreground regions by using the approximate median filter. The method of using the approximate median filter has been validated as a reliable method when applied to construction videos for the purpose of distinguishing the foreground regions from the background scene [2].

3.2 Color Pixel Classification Method

For the color pixel classification (the detection considered in this study) the objective was to determine whether each pixel belongs to a safety vest color or non-safety vest color by using only a color triplet as input. The major difficulty in color pixel classification in the construction scene is that the color property of the safety vest is affected by variations in the intensity of illumination according to environmental factors (e.g., weather condition, time of the day, and seasonal variations). For a reliable classification method, the approach has to be robust against such variations.

3.2.1 Data Collection and Data Pre-Processing

To achieve a reliable detection result, a comprehensive collection of data is essential. This study included the use of more than 15,000 images that were extracted from 52 different video sequences. The video sequences were acquired from actual construction sites, for a data set considering such variations. The videos were selected in consideration of different conditions such as the viewing position of the camera, the time of day (between 7am and 5pm), and the background scene. Each image in the data set included at least one construction worker who was wearing a safety vest as shown in Fig. 1. Each image was divided into the small sub-images with a size of 30×30 pixels. Then, the sub-images were categorized and labeled as "fluorescent yellow-green," "fluorescent orange-red," "background," or "indeterminate." Next, the data set from the sub-images that were categorized as "indeterminate" were excluded from the evaluation because that data set could not be defined as either the pixels of safety vest colors or the background pixels. The background images, which are of non-safety vest regions, include all kinds of scenery such as construction equipment, sky, soil, trees, construction materials, clothes of workers, and other construction-related objects. The data collected from safety vest color and non-safety vest color sub-images amounted to approximately 45,000 pixels each for "fluorescent yellow-green" and "fluorescent orange-red," which are safety vest colors, and 90,000 pixels for the background.

Typically a construction worker detection framework using color-based features includes the transformation of RGB into other color spaces, because RGB color spaces are subjected to the deterioration of color property in variations of illumination. It is generally assumed that variations in color property occur more in chrominance components, meaning that excluding the illuminance component and using chrominance components only leads to a better performance. Therefore, this study considered the two types of color spaces: Lab and HSV. These color spaces are the general choices among color spaces in construction worker detection using color features [2–4]. Moreover, this study included an investigation into the role of illuminance components by excluding illuminance components: the L of Lab and the V of HSV.

3.2.2 Measuring Safety Vest Color Models

This study considered three different types of classifiers: SVM, ANN, and LR. These classifiers are generally used in color pixel classification methods. The algorithms of WEKA release 3.8.1, a typical Java-based machine learning program, were used in this study. Defined by its architecture, ANN consists of one or more hidden layers of neurons. The connections between them are indicated by weights. In this study, we used a back-propagation neural network. The SVM has been prevalently used for solving a large number of complex binary classification problems [14]. The SVM model learns the high-dimensional data, and implements generalization performance well, with the aid of multiplier parameters like the Lagrange multiplier. The LR classifier is a general classifier of linear regressions

[15]. An advantage of LR is that it can generate a simple probabilistic formula for the classification. A set of parameters was adjusted for each classifier. For assessing the performance of the classifiers with the transformed color spaces, a k-fold cross-validation was used in this study. It has been determined that a 10-fold cross-validation is an optimal choice in terms of error estimation and computation; therefore, this study employed a 10-fold cross-validation to assess the performance of each combination.

In this study, the performance was evaluated by two measures: precision and recall. These measures, which stand in opposition to one another, are widely used in the evaluation of both color pixel classification and construction worker detection. Precision (see Equation (1)) and recall (see Equation (2)) are defined as follows, where the true positive (TP) is the number of safety vest pixels that are correctly classified as safety vest pixels, the false negative (FN) is the number of safety vest pixels that are incorrectly classified as background pixels, and the false positive (FP) is the background pixels that are incorrectly classified as safety vest pixels. Tables 1 and 2 seperately present the classification results of each combination for the two colors of safety vests. The results potentially imply that the color pixel classification of safety vests can be achieved by using color pixel information alone, regardless of the type of color space and classifier.

$$\text{Precision} = \frac{\text{TP}}{\text{TP+FP}} \quad (1)$$

$$\text{Recall} = \frac{\text{TP}}{\text{TP + FN}} \quad (2)$$

Table 1. Classification results of fluorescent yellow-green color pixels.

Color space	Classifier	Precision (%)	Recall (%)
Lab	ANN	98.30	**98.48**
	SVM	**99.36**	97.05
	LR	95.27	95.59
ab	ANN	97.12	97.65
	SVM	98.33	98.16
	LR	94.78	94.04
HSV	ANN	97.76	97.64
	SVM	88.27	93.75
	LR	72.40	77.46
HS	ANN	93.24	96.62
	SVM	94.84	86.67
	LR	91.59	95.31

Table 2. Classification results of fluorescent orange-red color pixels.

Color space	Classifier	Precision (%)	Recall (%)
Lab	ANN	86.43	93.99
	SVM	**94.94**	**95.15**
	LR	87.61	88.92
ab	ANN	86.09	91.01
	SVM	87.47	92.10
	LR	86.34	88.43
HSV	ANN	87.34	94.30
	SVM	88.25	93.75
	LR	82.26	86.71
HS	ANN	79.25	87.35
	SVM	78.92	88.28
	LR	72.91	77.84

Table 1 and 2 present the classification results of each combination for two colors of safety vests separately. The results potentially imply that the color pixel classification of safety vest can be achieved by using color-pixel information alone, regardless of the types of color space and classifier. As for the overall performance of the combinations of color spaces and classifiers, the SVM classifier with the Lab color space demonstrated the best performance on the data sets of fluorescent yellow-green in terms of precision, and fluorescent orange-red in terms of both precision and recall, except in terms of recall on the data set of fluorescent yellow-green, in which the ANN classifier fared the best. Although the ANN classifier with the Lab color space yielded better results only in terms of recall on the data set of fluorescent yellow-green, the difference in performance between the SVM classifier and the ANN classifier with the Lab color space were significant on the fluorescent orange-red in both measures.

4 Results

To assess the effectiveness of the proposed methods, each method was tested on the 58 image sequences that were processed and collected from the 12 videos of an actual construction site. Figure 1 shows the color pixel classification results on an image captured from actual construction site video. Construction workers wearing safety vests with different postures (both upright and squatting postures) are in the image. The background subtraction method using the approximate median filter was applied as a first step for detecting foreground objects of interest. Figure 1(b) shows that the proposed background subtraction method correctly detects the foreground regions that are potential candidates for the color pixel classification method as sequential step without any impulsive noises. The detected foreground regions are enlarged for the purpose of a better display, as shown in Figure 1(c).

After the background subtraction was performed, each pixel of the detected foreground regions with bounding box were classified as either pixels of safety vests or pixels of non-safety vests. The color pixel classification was performed by the SVM classifier with color space transformation into the Lab color space, which presented the most desirable performance. The pixels of the safety vest that were detected by the proposed method appeared as color pixels, and the pixels of the non-safety vest are black as shown in Figure 1 (d). These results indicate that the proposed method is useful and reliable. Despite such variations resulting from illumination change, most of the safety vest pixels were correctly detected as belonging to safety vests. These results indicate that the proposed method is invariant against changes in workers' postures and illuminant variation has a reliable detection accuracy.

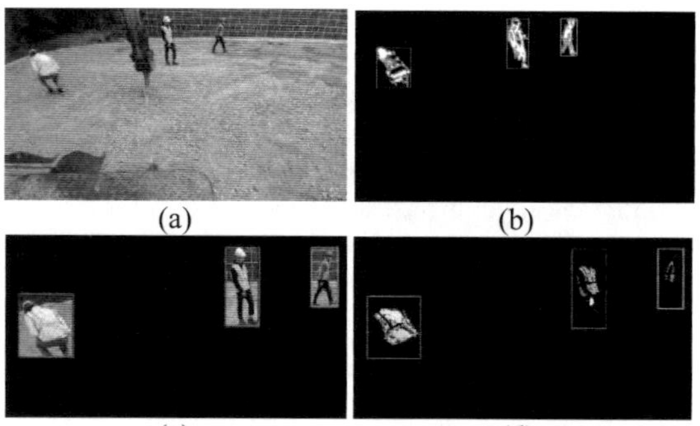

(a) (b) (c) (d)

Figure 1. Color pixel classification results on the actual construction scene (a) original image; (b) foreground region detection results by background subtraction method; (c) enlarged image of foreground region detection result (depicted as a color image); and (d) color pixel classification results of safety vest pixels.

5 Conclusion and Recommendation

To achieve the detection of construction workers in various postures, this study proposed a safety vest detection method of using both motion of the construction workers and the colors of safety vests. The proposed method involved a common background subtraction method using the approximate median filter to detect the foreground region to reduce the candidate regions. As a sequential step, each pixel in the regions was classified as being safety vest pixels or non-safety vest pixels. To demonstrate which combination of color space and classifier performs the best, a comparative analysis was conducted. The assessment was carried out on the performance of two color spaces with three classifiers. The results showed that the combination of the Lab color space and the SVM classifier had the most desirable outcomes. The proposed method was tested on the actual construction site video, and the results indicated that safety vest detection was achieved with a high-level of accuracy for both the upright and squatting postures of construction workers.

The proposed method of this study, thus far, is expected to make progress toward the achievement of the robust and effective construction worker detection in the construction scene. Future work will include the method that specifically identifies the construction workers. In addition, the method used in this study can only be applied using a static camera, which is a common limitation of the background subtraction method; therefore the greatly developed method will be investigated.

References

[1] Seo, J. S., Han, S. U., Lee, S. H., and Kim, H. K. Computer vision techniques for construction safety and health monitoring. Advanced Engineering Informatics, 29(2): 239–251, 2015.

[2] Park, M. W. and Brilakis, I. Construction worker detection in video frames for initializing vision trackers. Automation in Construction, 28: 15–25, 2012.

[3] Memarzadeh, M., Golparvar-Fard, M., and Niebles, J. C. Automated 2D detection of construction equipment and workers from site video streams using histograms of oriented gradients and colors. Automation in Construction, 32: 24–37, 2013.

[4] Gong, J. and Caldas, C. H. An object recognition, tracking, and contextual reasoning-based video interpretation method for rapid productivity analysis of construction operations. Automation in Construction, 20(8): 1211–1226, 2011.

[5] Chi, S. and Caldas, C. H. Automated object identification using optical video cameras on construction sites. Computer-Aided Civil and Infrastructure Engineering, 26(5): 368–380, 2011.

[6] Yang, J., Arif, O., Vela, P. A., Teizer, J., and Shi, Z. Tracking multiple workers on construction sites using video cameras. Advanced Engineering Informatics, 24(4): 428–434, 2016.

[7] Son, H., Kim, C., Hwang, N., Kim, C., and Kang, Y. Classification of major construction materials in construction environments using ensemble classifiers. Advanced Engineering Informatics, 28(1): 1–10, 2014.

[8] Neto, J. A., Arditi, D., and Evens, M.W. Using colors to detect structural components in digital

pictures. Computer-Aided Civil and Infrastructure Engineering, 17: 61–67, 2002.
[9] Zou, J. and Kim, H. Using hue, saturation, and value color space for hydraulic excavator idle time analysis. Journal of Computing in Civil Engineering, 21: 238–246, 2007.
[10] ANSI/International Safety Equipment Association (ISEA), American national standard for high-visibility safety apparel and headwear, ANSI/ISEA 107–2010, Washington, DC, 2010.
[11] Huang, J., Kumar, S. R., Mitra, M., Zhu, W. J., and Zabih, R. Image indexing using color correlograms. In Procedings of Computer Vision and Pattern Recognition, page 762–768, 1997.
[12] Son, H., Kim, C., and Kim, C. Automated color model–based concrete detection in construction site images by using machine learning algorithms. Journal of Computing in Civil Engineering, 26(3): 421–433, 2011.
[13] McFarlane, N. J. and Schofield, C. P. Segmentation and tracking of piglets in images. Machine vision and applications, 8(3): 187–193, 1995.
[14] Cortes, C. and Vapnik, V. Support-vector networks. Machine learning, 20(3): 273–297, 1995.
[15] Hastie, T., Tibshirani, R., and Friedman, J. The Elements of Statistical Learning: Data Mining, Inference, and Prediction, Springer, New York, 2001.

Mass Estimation of On-road Construction Equipment Based on Modelling Operational Parameters

K. Barati and X. Shen*

School of Civil and Environmental Engineering, the University of New South Wales, Australia
E-mail: khalegh.barati@unsw.edu.au, x.shen@unsw.edu.au

Abstract-

Earthmoving operations are considered as one of the main activities in the construction industry. These operations involve processing and moving soils and rocks from source site to dump destination using earthmoving equipment, such as trucks and haulers. Automated measurement of the payload carried by earthmoving equipment in each trip has been one of the main challenges in this field. Currently, different metric and volumetric methods including weighbridges, load volume scanners (LVS) and strain gauges have been used to estimate the mass of vehicles which can be costly, time-consuming and labor-intensive. This paper aims to develop an automated weight and mass estimation model for on-road construction vehicles based on modelling operational and engine parameters. Acceleration, speed and road slope are investigated as operational factors, while engine load is considered as engine attribute to estimate vehicles' weight. Field experiments on a broad range of in-use construction equipment have been carried out to collect real-world field data of the investigated parameters. GPS-aided inertial navigation system (GPS-INS) and engine data logger instruments are employed for collecting field data. The weight estimation model is then developed through performing ordinary least square (OLS) and multivariable linear regression (MLR) analyses on the collected field data using SPSS Software. The model validation process is finally conducted through comparing the weight estimated by the model with real measurement of vehicle's weight on the site.

Keywords-

Construction industry; mass; operational parameters; engine load; on-road equipment; data collection

1. Introduction

Construction sector is one of the main industries requiring a large number of different construction vehicles. Majority of construction activities associated with earthmoving operations including cut and fill activities require on-road heavy-duty vehicles (HDVs) for materials transportation. Such kind of activities are planned and paid based on the amount of materials, and the measurement of the mass and volume of materials carried by vehicles. As the main concern of contractors and equipment operators, a cost-effective, automated method is essential for accurately estimating the weight of vehicles' mass without scarifying field production. On the other hand, measuring gross vehicle weight (GVW) of vehicles is necessary in the transportation field. Overloading and increasing equivalent single axle load (ESAL) result in the difficulties of vehicle's manoeuvrability, heavy traffic accident and short vehicle life [1]. ESAL is determined based on pavement condition and its failure mode, and is one of the main parameters causing distress and damage of pavements [2]. Overweighting also causes serious damages to the bridges and increases the risk of overloading and failure of bridges. In this regards, many restrictions and regulations have been implemented by international organizations such as American Association of State Highway and Transportation Officials (AASHTO) and Federal Highway Administrations (FHWA) to reduce the ESAL of HDV vehicles.

This paper aims to develop an integrated framework to accurately estimate the weight and mass of on-road HDVs based on modelling operational parameters and engine attributes. To do so, the operational parameters affecting engine load are first investigated. Then, through instrumentation and field experimentation, field data are collected from on-road HDVs. Finally, field data are processed and analysed to quantitatively estimate the weight of vehicle at high level of accuracy. This paper starts with a comprehensive review on the current approaches used for measuring weight and mass of construction vehicles. An integrated methodology is then developed for data collection and data analysis. At the end, validation is conducted by comparing the predicted results of the model against the real weight measured by weighbridges on the site.

2. Literature Review

Governmental and international weight regulations and contractual payment arrangements are the two main incentives for measuring and controlling the weight of the construction equipment. Numerous theoretical and practical efforts have been conducted by academic scholars and industrial sectors to develop different approaches for weight and mass estimation of construction vehicles. Many devices and instruments have been also devised to measure the weight of various machineries.

Currently, numerous techniques and tools are employed in construction industry to measure the weight and mass of equipment and vehicles. On the whole, these methods can be classified into two volumetric and metric measurement systems. Volumetric techniques measure the volume of the materials inside the haul bed and bucket of vehicles through automatically scanning and comparing the empty and loaded equipment. Volumetric methods are non-contact and cost effective with low initial and maintenance cost, but not really accurate in weight measurement of bulk material [3]. In this process, vehicles need to move slowly under the load volume scanners (LVS) installed in the construction sites before and after loading.

Metric measurement systems are more common in construction sites, and directly weigh the mass and total weight of equipment using various instruments and sensors. At the moment, weighbridges are mostly used in construction industry to measure the total weight including the vehicle weight and the mass. Despite the high accuracy in weight measurement, this method has high initial, operation and maintenance costs and is time consuming. Axle hydraulic and pneumatic pressure controlling is another system invented by Bartlett [4] to automatically weigh the mass of HDVs through applying pressure sensors. Pressure modulation valves are needed for adjusting auxiliary axle pressures based on load distribution. The measured data are transferred from sensing devices to signal processing system for analysis. On-board pressure sensors are also extensively used to accurately measure the weight carried in the trucks body. This apparatus is embedded between the truck frames and haul bed. Proportional to the implemented pressure by the load, the electro-magnetic sensors generate electrical signals processed to calculate the load [2, 5]. As another metric technique, strain gauges have been pasted on the leaves of springs to measure suspension strain caused by the mass [6]. The weigh is then calculated through summing up the received voltage signals provided by gauges. Despite the applicability of the method for all vehicles, adopting the gauges to various types of vehicles is inconvenient accounted as its main drawback.

3. Methodology

As Figure 1 shows, this section aims to develop a comprehensive framework to estimate GVW and mass of the on-road HDVs based on operational parameters. The affecting operational variables and engine attribute are first identified and their relation is investigated. An integrated instrumentation system is then developed considering state-of-the-art technologies available in the market. Experimentation is conducted in the next step to collect laboratory and field data of considered operational and engine parameters using designed instruments. The collected raw data are then synchronized and processed to remove potential errors. In this study, OLS and MLR statistical methods are applied to analyse the processed data and develop the mass estimation model. The applied two regression methods have much more flexibility in comparison with other techniques and it is simple to add or remove some data after conducting initial analysis. Using these methods, it would be possible to compare data collected from different equipment type. Level of familiarity, assumption, and use of multiple variables are the other advantages of using these regression techniques. The model is finally validated through comparing the predicted weight of the model against the real weight measured by the weighbridge.

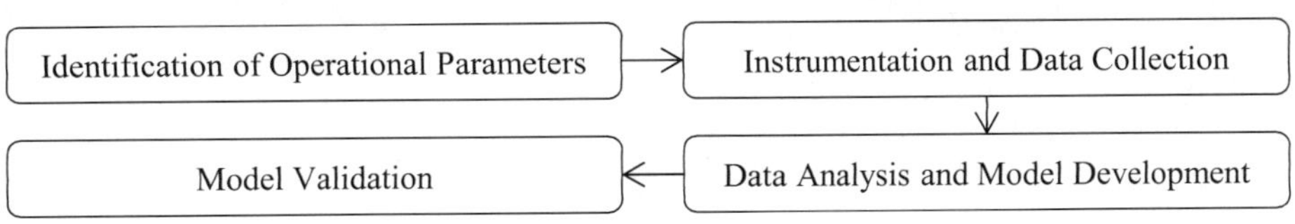

Figure 1. Integrated framework for developing weight estimation model of on-road construction equipment

3.1. Identifying Operational Parameters

In this study, the relation amongst operational variables and engine attribute has been investigated. Equipment's weight, acceleration, speed and road slope are four main operational parameters considered in this research, which their effect are modelled on the used power of the engine. Engine load has been determined as an agent to quantify the produced power, and is defined as the ratio of the used power to the maximum power of engine as a percentage [7, 8]. Construction equipment rarely works with fully-loaded engine, and for most tasks, average engine load of equipment is approximately from 25% to 75% [9]. A few environmental variables such as ambient temperature and pressure influence the engine load value, but they are not take into consideration in this study due to their minor effects.

Acceleration is the major operational parameter which does have the highest effect on engine load. It is clear that the engine load has direct relation with vehicle's speed and road of slope. The effect of slope can be interpreted as gravitational force resisting the movement of vehicle. As can be seen in Figure 2, weight of equipment is the main operational variable influencing the effect of other considered operational parameters. As initial analyses on the raw collected data show, weight parameter does not have direct effect on the engine load amount, but its influence is indirectly measured when the other operational parameters are modelled.

3.2. Instrumentation and Experimentation

As Figure 3 indicates, two major instruments have been applied to record required operational and engine data from experimented vehicles. GPS-INS is employed to measure real-world data of operational parameters. This devise is embedded inside the cabin and its antenna is installed on the roof of cabin to get the best signals from GPS satellites. Being an attitude and heading reference system, GPS-INS provides the highest accuracy in second-by-second data recording of acceleration, speed and orientation parameters in three dimensions. Also, engine data logger is an on-board diagnostics (OBD) device applied in this study to gather real-time engine load data. It is plugged into the J1939 port of the equipment's engine control unit (ECU) under the steering wheel. The collected data of these two instruments were transmitted and stored in an industrial Tough Pad to be synchronized, processed and analysed.

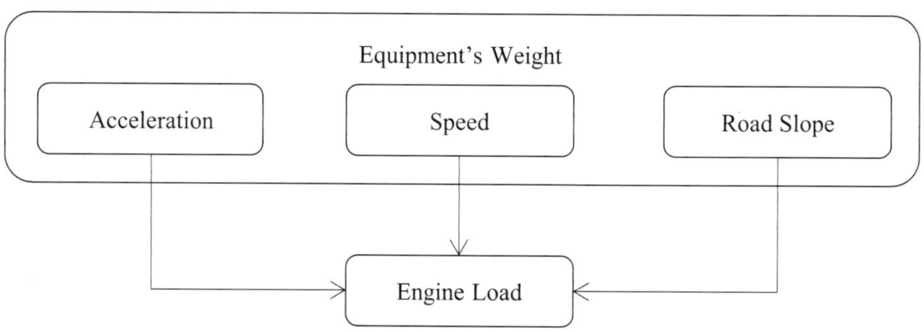

Figure 2. Investigated relation between operational parameters and engine load

(a)

(b)

Figure 3. Instruments employed to collect real-world data, (a) GPS-INS and (b) engine data logger

Table 1. Specifications of equipment used for experimentation

Vehicle	Tier	Engine Size (kW)	Model	Empty Weight (ton)	Experiment Time (min)
Three-axle Granite	IV	345	2010	9.5	212
Three-axle Trident	V	400	2013	11	165
Six-axle Granite	IV	345	2010	14.5	272
Six-axle Trident	V	400	2014	17.7	120
Six-axle Vision	III	350	2005	17.6	265
Seven-axle Granite	IV	345	2010	16.6	205
Seven-axle Trident	V	400	2013	18.8	221

Experimentation process was conducted in two steps of laboratory and field data collection. The main aim of laboratory testing was verifying the performance of employed instruments and synchronizing the recorded data. The raw data from laboratory tests were also analysed for developing initial research framework. As Table 1 illustrates, seven on-road in-use HDVs were experimented, and more than 90.000 raw data points were collected from each instrument in the process of field data collection. The models of equipment varied between 2005 and 2014 with the engine sizes ranging from 345 kW to 400 kW. To precisely model the effect of different parameters, the vehicles were driven with much fluctuations and variations in the amount of acceleration rate, speed, road slope and equipment weight.

4. Operation-Based Weight Modeling

This section focuses on developing a technical method to estimate the weight of on-road HDVs based on measurable operational and engine parameters. As discussed, two GPS-INS and engine data logger were employed to record operational and engine data respectively. The raw collected data were transmitted to the industrial Tough Pad to be stored and processed. Through conducting data filtering and synchronization using Microsoft Excel software, potential errors were then identified and invalid data were removed, which approximately two-third of raw data (90.000 data points) were finalized. OLS and MLR statistical analysis techniques were carried out on the verified data by IBM SPSS Statistics V22.

As shown in Figure 2, weight parameter has indirect effect on engine load, and its influence is taken into consideration in the coefficients of other operational parameters. The conducted MLR analyses reveal that there is a highly-correlated linear relation between three parameters of acceleration rate, speed and road slope, and engine load in different loading conditions. Barati et al. [10] developed a linear relation to estimate the engine load presented in Equation (1). The weight factor (WF) has been defined as the combined weight of equipment (ton) that must be carried per 100 kW of engine size. Combined weight includes the weight of equipment, trailer and mas. In the designed model, C_{AC}, C_{SP} and C_{SL} are the coefficients of acceleration, speed and slope parameters respectively which are shown in Table 2.

$$EL = (C_{AC} * AC) + (C_{SP} * SP) + (C_{SL} * SL) + C \quad (1)$$

Where:
EL: Engine load of equipment (%)
AC: Acceleration rate of equipment (km/h.s)
SP: Speed of equipment (km/h)
SL: Slope of road (degree)
C: Engine load of equipment in idling mode which is around 15%.

Table 2. The coefficients of parameters in the engine load estimation model

Coefficients	Weight Factor				
	2.75	4.5	6.5	13	14.5
C_{AC}	22	30	36	50	56
C_{SP}	0.24	0.26	0.35	0.48	0.54
C_{SL}	2.4	4.3	6	8.2	9

It can be seen that with increase of WF, the effect of operational parameters on engine load rises. Conducted OLS analyses by Microsoft Excel software shown in Figure 3 indicate that there is a linear relation with high correlation and consistency ($R^2 > 0.9$) between WF and coefficients of investigated operational parameters. Based on truck configurations and loading conditions, WF varies from around 2.75 (ton/100kW) for empty truck without trailer to about 15 (ton/100kW) for fully-loaded truck with one trailer. Acceleration has the highest effect on engine load, varying from 22 (km/h.s) for WF of 2.75 to 56 (km/h.s) when WF increases to 14.5 (see Figure 4a and Equation (2a)). The analyses reveal that WF has less impact on the of speed and road slope parameters' coefficients ranging from just over 0.2 (km/h) to around 0.6 (km/h) for speed, and between about 3° to 9° for road slope. As the main limitation of the conducted research, due to the loading restrictions, a wide range of WF values could not be measured during experimentation process. The trucks were experimented in limited loading conditions of empty and fully-loaded trucks without or with one trailer.

$$C_{AC} = 2.67 * WF + 16.77,\ R^2 = 0.9849 \quad (2a)$$
$$C_{SP} = 0.0252 * WF + 0.166,\ R^2 = 0.9848 \quad (2b)$$
$$C_{SL} = 0.508 * WF + 1.79,\ R^2 = 0.9476 \quad (2c)$$

In the next step of the study, WF is modeled by combining the Equations (1) and (2). As Equation (3) presents, WF is a function of operational parameters and engine load. Acceleration, speed and road slope are recorded by GPS-INS instrument in a second basis. Engine load is also measured with high accuracy using engine data logger connected to J1939 port of vehicle. So, having the real-world measured data, in each second, a value for WF can be estimated using Equation (3).

$$\mathrm{WF} = \frac{(\mathrm{EL} - \mathrm{C} - (16.77*\mathrm{AC} + 0.166*\mathrm{SP} + 1.79*\mathrm{SL}))}{(2.67*\mathrm{AC} + 0.0252*\mathrm{SP} + 0.508*\mathrm{SL})} \quad (3)$$

As shown in Figure 5, conducted calculations using Equation (3) reveal that there is much variation in the value of WF due to many parameters including operator skill, engine condition and road type. Also, the performance of instruments in collecting raw data is one of the main sources of error affecting the accuracy of model in estimating WF. The variation of calculated WF indicates that the distribution of the data follows normal statistical function that the majority of the calculated WFs are around the average. The mean (M) representing the predicted WF, standard deviation and skewness can be calculated using Equation (4).

$$\mathrm{M} = \frac{1}{n} * \sum_{i=1}^{n} \mathrm{WF}_i \quad (4a)$$
$$\sigma = \sqrt{\sum_{i=1}^{n} (WF_i - M)^2} \quad (4b)$$
$$\gamma = \mu_3 / \sigma^3 \quad (4c)$$

Where:
M: Mean of the calculated WFs
σ: Standard deviation of the calculated WFs
γ: Skewness of distribution function
μ_3: Third moment of stochastic variables

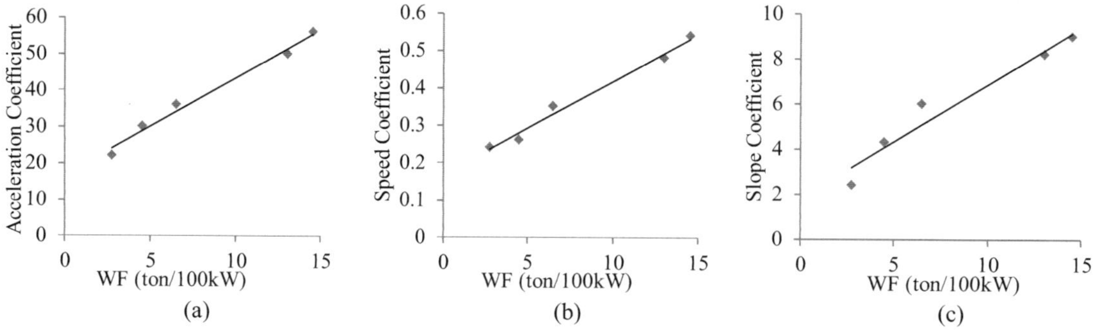

Figure 4. The effect of WF on the coefficients of (a) acceleration, (b) speed and (c) road slope parameters

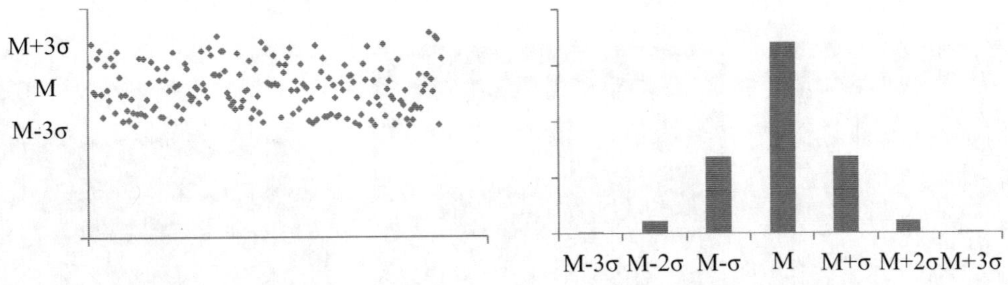

Figure 5. The distribution of calculated WFs using Eq. (3)

Using M as the final WF, total weight of HDVs can be easily calculated through Equation (5a). As defined before, WF is the amount of combined weight which is carried per 100 kW of engine. Having the weight of equipment itself and the weight of trailers if available, the mass can be calculated using Equation (5b).

$$TW = M^{*}\, PW/100 \quad (5a)$$
$$PL = TW - EW \quad (5b)$$

Where:

TW: Total weight of equipment (ton)
PW: Engine power (kW)
PL: mass of equipment (ton)
EW: Empty weight of equipment (ton)

5. Model Validation

This section focuses to verify the applicability of developed framework through conducting experimentation on seven different in-use on-road HDVs. The experimentation process took seven days, and more than 90.000 data points were collected to be analyzed. Figure 6 shows some sample photos of vehicles experimented as case study. As can be seen, a variety of trucks from three to seven axles with additional trailer were selected. Data processing and filtering were carried out on the raw collected data and invalid data were removed in the first step. Using Equation (3) developed in the previous section, the WF of the equipment was then predicted in different loading conditions. As achieved results show, there was much variation in the value of WF for each vehicle with specific condition. So, Equation (4) was applied to calculate the mean representing the predicted amount of WF and standard deviation according to normal distribution function. Having the specifications of experimented equipment such as engine size and empty weight of equipment, the total weight of vehicle and mass were finally estimated.

To verify and validate the developed model, the estimated weight of vehicles calculated by Equation (5) compared with the real weight measured using an industrial weighbridge available in the site. The analyzed data including M, σ and TW have been presented in Table 3. On the whole, data from 14 loading conditions with different trucks have been collected, but analyses were performed just on five conditions to validate the developed model. Figure 7 conducts validation through plotting the predicted TW using Equation (5) versus the real measured equipment's weight. As can be seen, there is high correlation between estimated and measured equipment's weight. As one of the main contributions of this study, it has been proven that the developed model has more than 90% accuracy in estimating the weight of on-road HDVs. Numerous sources of potential errors have also been identified in this study that can be improved to increase the accuracy of the developed model.

Table 3. The results of the analyzed data on five different vehicles for model validation

Equipment	Engine Size (kW)	M	σ	TW	Real Weight
Three-axle Granite	345	3.13	0.203	10.8	9.5
Six-axle Granite	345	3.65	0.242	12.6	14.52
Three-axle Trident	400	4.95	0.233	19.8	22.5
Six-axle Vision	350	12.77	0.390	44.7	48
Seven-axle Trident	400	13.55	0.367	54.2	50.5

(a) (b) (c)

Figure 6. Samples photos of on-road HDVs experimented, (a) a) three-axle vehicle without trailer, (b) six-axle vehicle with a trailer, and (c) seven-axle truck with a trailer

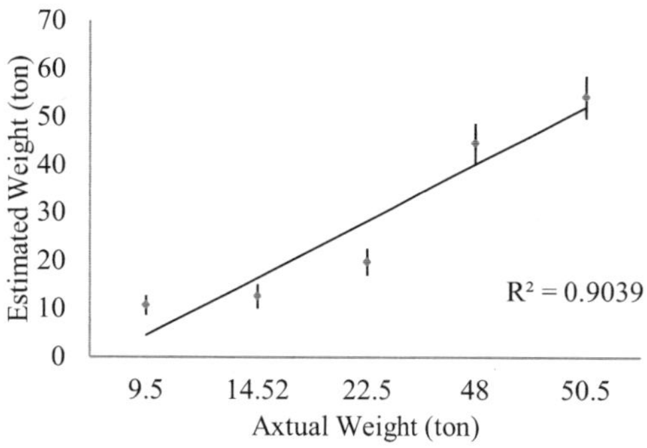

Figure 7. Validation of the developed weight estimation model by comparing the estimated and real equipment's weight

6. Conclusions

Earthmoving operations involve a large number of HDVs for material transportation. Monthly payment and scheduling of such projects are primarily based on the amount of soils and materials moved by the contractors. Numerous metric and volumetric techniques and devises have being applied in construction industry to estimate the weight of materials transferred by vehicles in each cycle. Despite the high accuracy of some existing methods, they are mainly costly, time-consuming and labor-intensive which could considerably increase the cost in construction projects.

This study developed an integrated framework to estimate the weight of on-road HDVs considering operational parameters and engine attributes. As the sources of data for developing the model, GPS-INS and engine data logger instruments were employed to collect real-world data from in-use construction equipment. More than 90,000 data points were collected during seven-day experimentation, and were analyzed after synchronization and data filtering. According to the model, WF is predicted based on acceleration rate, speed, road slope and engine load. By applying normal statistical distribution function, the mean value and standard deviation of WF were calculated. Comparing the predicted vehicle's weight with the real weight measured using weighbridge in the validation process reveals the high accuracy of developed model in weight estimation which is considered to be the main achievement of this study. Different sources of error were identified at the end such as instruments errors and operator's skill level that can be controlled to increase the accuracy of the model.

This model has considerable application potential in construction industry to automatically measure the weight of different haulers. In comparison with current weighting methods, this technique does not need high initial costs, and can save much time and money due to its automated process. In the future, this research will focus on extending this model to off-road equipment that is commonly used in civil and mining construction. Engine depreciation and tier can also be considered as new parameters for improving the accuracy of the current model.

Acknowledgements

The data analyzed in this study to develop the model have been collected from different equipment of MOITS Company in Australia. The experimentation was conducted in an earthmoving project carried out by this company in the northwestern region of Sydney metropolitan city. The authors also appreciate Mr. Niels Theodore Bardoel for his great help in the data collection process.

References

[1] Haider S.W. and Harichandran R.S. Relating axle load spectra to truck gross vehicle weights and volumes, *Journal of Transportation Engineering*, 133(12):696-705, 2007.

[2] Yang S.K. Liu T.S. and Cheng Y.C. Automatic measurement of payload for heavy vehicles using strain gages, *Journal of Measurement*, 41:491-502, 2008.

[3] Load Management Solution: Volumetric Truck Measure and the Load Volume Scanner. On-line: www.loadscan.com, Accessed: 30/01/2017.

[4] Bartlett P.T. Axle pressure control system, *United States Patent*, 6,371,227, 2002.

[5] Nishitani K. Electromagnetic-Type Load Weighing Apparatus, *United States Patent*, 5,243,146, 1993.

[6] Nakazaki Y. Aoshima H. Takahashi N. Atagi Y. and Yasuda Y., Load Measuring Device for a Vehicle, *United States Patent*, 5,684,254, 1997.

[7] Barati K. and Shen X. Operational level emission modelling of on-road construction equipment through field data analysis, *Journal of Automation in Construction*, (72):338-346, 2016.

[8] Barati K. and Shen X. Modelling emissions of construction and mining equipment by tracking field operations, In *Proceeding of 32nd International Symposium on Automation and Robotics in Construction and Mining*, pages 478-486, Oulu, Finland, 2015.

[9] Nichols H. Day D. *Moving the Earth: The Workbook of Excavation*, the Mc-Graw-Hill Companies, Inc., New York, 2005.

[10] Barati K., Shen X. and Bardoel N.T. Operational level fuel use model and reduction scheme for on-road construction equipment, *Journal of Automation in Construction*, 2017.

Decision Support System for Modular Construction Scheduling

A.W.A Hammad[a], A. Akbarnezhad[a], D. Rey[a], H. Grzybowska[a]

[a]School of Civil and Environmental Engineering, The University of New South Wales, Australia
E-mail: a.hammad@unsw.edu.au, a.akbarnezhad@unsw.edu.au, d.rey@unsw.edu.au, h.grzybowska@unsw.edu.au

Abstract –

The use of modular construction as opposed to traditional on-site construction presents an opportunity to improve a project's economic and sustainable performance. At the same time, substantial savings in the overall project's duration can be achieved. The potential for reducing the overall make-span period involved in manufacturing and assembling the modularised components, along with the constant need for aligning the work schedule with the in-house manufacturing schedule, leads to the need for a repetitive scheduling optimisation procedure for modular manufacturing operations. This paper presents a framework and a decision support system to schedule operations in modular building factories. The framework is divided into 3 layers; the first layer concerns the assignment of workers to workstations; the second layer solves the scheduling of jobs to workstations while the last layer governs the overall operations of production lines through formulating a parallel machine scheduling problem. For demonstration purposes, a computational test is conducted on the final optimisation stage using a practical case study to solve the parallel machine scheduling problem. To account for resource allocation and levelling, the model is presented as a resource constraint one. Results reveal the satisfactory performance of the proposed model.

Keywords –
Resource-constrained project scheduling; Parallel machine scheduling problem; Modular construction; Optimisation; Mixed Integer Programming

1 Introduction

Managing projects so that they are delivered in the shortest possible time frame is considered one of the main challenges faced by project managers. In a bid to avoid overrunning the project's time duration, many methods can be utilised to better enhance the overall project management procedure. This includes traditional scheduling techniques based on the Critical Path Method (CPM), Program Evaluation Review Technique (PERT) and Monte Carlo simulation [1]. Handling the multi-tasking nature that arises in some projects is however difficult to deal with using such planning methods [2].

A main determinant of a project's production rate, which will influence its completion time, is the construction method adopted. Apart from the traditional on-site construction process, where the whole project is accomplished through building components produced on site, the recent trend in research has focused on an alternative method of construction, namely off-site prefabrication. Classes of off-site construction vary depending on the degree of prefabrication. Modular construction, which has the highest level of prefabrication, involves the pre-fabrication of volumetric building components in a controlled environment, which are then transported to be installed on the construction site [3]. The process of modular construction conforms to the industrial and manufacturing process [4]. As a result, the applications of principles from these industries are well adapted to the construction environment presented in modular building factories [5]. These include make-span and throughput, which strongly impact the scheduling of components within modular factories [6].

Currently, the process of scheduling production lines in modular construction is achieved based on the experience of the factory superintendent [7]. This process however is prone to errors, particularly due to neglecting possible resource limitations for processing work on the floor shop [8]. Additionally, deploying traditional scheduling techniques usually results in solutions that are not consistently in line with progress on the manufacturing floor [9]. Within modular construction, the productivity of a factory is directly influenced by the effectiveness of the scheduling method adopted [10]. As a result, it is important to consider appropriate scheduling techniques that are adapted to the specific environment of modular construction.

A common scheduling technique applied in manufacturing is that relating to parallel machine scheduling. Due to the similarities that exist between conventional manufacturing and modular construction production systems, such methods can be extended to schedule the production line typical of modular construction factories. The parallel machine scheduling problem has been extensively studied in the literature [11]–[13]. In such problems, the objective is to schedule jobs so that the weighted sum of completion times of the jobs is minimised. By varying the processing time of machines several variants of the problem can be examined, including identical, uniform and non-identical machines [12].

This research investigates the scheduling of modular construction elements within a factory production line and presents a decision support system (DSS) comprised of three optimisation layers. A resource-constraint parallel machine scheduling problem is then formulated for the last optimisation stage and presented, assuming input from the initial two stages. A Mixed Integer Programming (MIP) model is proposed for such purpose which integrates the resource limitations and processing capacities of the machines involved in the modular construction.

The organisation of the paper is as follows. In the next section a description of the framework and the DSS is presented. An explanation of one of the layers for which an optimisation problem is to be presented follows. The notation set making up the proposed model is later highlighted along with the associated mathematical model. A practical example is used to demonstrate the model's applicability. Concluding remarks are presented at the end.

2 Decision Support Framework

Fig.1 highlights the typical layout of modular factories. Each component to be built, referred to as a job in this paper, will have to go through a set of workstations, shown as rectangles in Fig. 1. A workstation is defined as a site within the systematic production process where a particular activity is undertaken on a specific job i.e. a building element. Common workstations in modular factories include areas for welding components together, platforms where the structure is framed, and carpentry workshop; the combination of such workstations is what constitutes a production line. A production line is demarcated by the flow of building elements from one workshop to the next. A total of three production lines is thus shown in Fig. 1. The scheduling problem within a modular factory consists of assigning workers to the workstations based on their skill sets, assigning building elements to workstations and scheduling the production line so that the sequence of workstations in each production line is optimised for performance.

In order to optimise the operations within a modular factory setting, a DSS is presented. Fig. 2 highlights the main components making up the system. Three units make up the framework proposed. The first unit, labelled as the Scheduling Unit, incorporates the algorithms necessary for scheduling the operations. These algorithms are grouped into three categories, according to the stage of the optimisation that is performed. This reflects the multi-stage nature of the scheduling problem at hand. At the first stage of the optimisation, the first set of algorithms assign workers to the workstations. The second optimisation stage concerns the scheduling of jobs to workstations, based on resource requirements. At this stage, a buffer between each workstation needs to be considered since the process would not realistically flow continuously, with workstations requiring some preparation time between jobs. For the last stage of the optimisation, the overall production line is optimised by solving a parallel machine scheduling problem. Each production line is referred to as a *machine* in this paper.

The second unit within the DSS of Fig. 2 is a Simulator which gets activated based on a pre-defined set of triggers. These triggers are what can initiate a change to the assignments made in the first two stages of the optimisation (resources and jobs). Major triggers include changes to designs of building elements based on client requests, failure of machines, shortage of workers, increased demand for a product and delays in material deliveries. The main purpose of the simulator is to capture the time frames within the overall production line at which change is induced by the triggers. This enables the identification of the time at which the framework should loop back to re-solve the multi-stage optimisation in the Scheduling Unit. Once the schedules are produced, the simulator is re-activated to keep a record of the time scale.

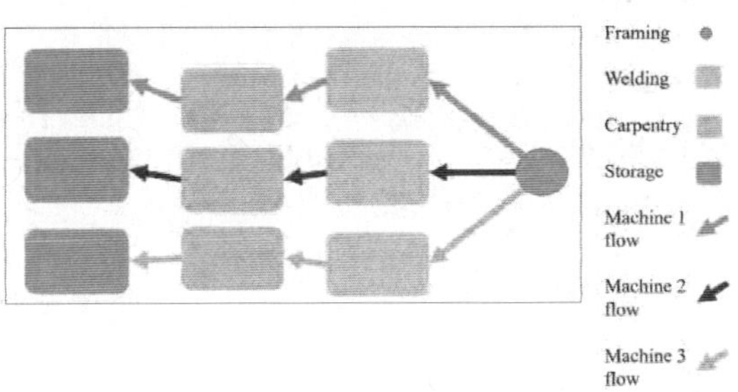

Fig. 1 Simplified plan view of typical modular factory layout. Each machine workflow is comprised of several workstations

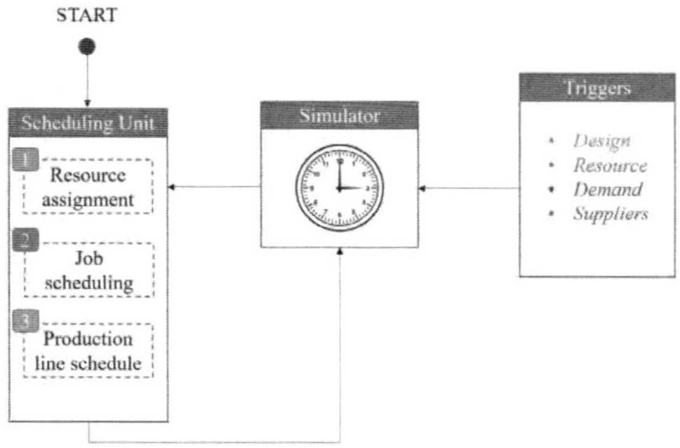

Fig. 2 Decision support system for modular construction scheduling

In the next section, the formulations for the parallel machine scheduling problem solved at the third stage of the optimisation is presented, assuming the assignments from the first two stages have already been determined.

2.1 Parallel Machine Scheduling

Since workstations are grouped together to form machines where each machine is assumed to have the same set of workstations, the problem solved at the third stage of the optimisation can be appropriately modelled as a parallel machine scheduling problem. As shown in Fig. 1, a factory setting can have several machines, with each machine having a configuration that is composed of the same workstation types. The processing time of a job by a machine is a function of the assignment of jobs made; the greater the number of jobs assigned to a machine, the longer the processing time required by the machine. Unlike the traditional parallel machine scheduling problem [12], the problem investigated in this paper allows for a machine to process more than one job at a time

When scheduling the production lines, it is also necessary to apply resource constraints to the problem, to ensure that machines are not overloaded. Given the limited availability of resources, the aim of the scheduling model presented is to appropriately schedule the jobs in a way that minimises the overall production times of the overall machines, while considering reasonable resource capacities. The next section presents the notation for the third optimisation stage.

2.2 Notation

Table 1 outlines the notation set adopted in the proposed model.

Table 1 Notation Description

Symbol	Description
I	Set of machines, indexed by i
J	Set of jobs, indexed by j, k
D_j	Duration assigned to job $j \in J$
L_i	Resource limit assigned to machine $i \in I$
M	Arbitrary large number
x_{ij}	Binary variable, which equals 1 if job j is assigned to machine i, and 0 otherwise
β_{jk}	Auxiliary binary variable
γ_{kj}	Auxiliary binary variable
f_j	Finish time of job j

2.3 Mathematical Model

The proposed model to solve the machine scheduling problem at the third stage of the optimisation is given as follows:

$$\text{minimise} \max_{j \in J} \{f_j\} \tag{1}$$

subject to

$$\sum_{j \in J} x_{ij} \le L_i \quad \forall i \in I \tag{2}$$

$$\sum_{i \in I} x_{ij} = 1 \quad \forall j \in J \tag{3}$$

$$\alpha_{ijk} \le x_{ij} \quad \forall i \in I, \forall j,k \in J : j > k \tag{4}$$

$$\alpha_{ijk} \le x_{ik} \quad \forall i \in I, \forall j,k \in J : j > k \tag{5}$$

$$\alpha_{ijk} \ge x_{ij} + x_{ik} - 1 \quad \forall i \in I, \forall j,k \in J : j > k \tag{6}$$

$$\left(1-\alpha_{ijk}\right)+\gamma_{ijk}+\beta_{ijk}\geq 1 \quad \forall i\in I,\forall j,k\in J: j>k \tag{7}$$

$$f_k - f_j \geq 0.8D_k\gamma_{ijk} - M\left(1-\gamma_{ijk}\right) \quad \forall i\in I,\forall k,j\in J: j>k \tag{8}$$

$$f_j - f_k \geq 0.8D_j\beta_{ijk} - M\left(1-\beta_{ijk}\right) \quad \forall i\in I,\forall j,k\in J: j>k \tag{9}$$

$$f_j \geq D_j \quad \forall j\in J \tag{10}$$

$$x_{ij}\in\{0,1\} \quad \forall i\in I,\forall j\in J \tag{11}$$

$$\beta_{jk},\gamma_{jk}\in\{0,1\} \quad \forall j,k\in J: j>k \tag{12}$$

$$\alpha_{ijk}\geq 0 \quad \forall i\in I,\forall j,k\in J: j>k \tag{13}$$

$$f_j\in\mathbf{R}^+ \quad \forall j\in J \tag{14}$$

Eq. (1) defines the objective function, which minimises the production make-span. Eq. (2) states that each machine has an associated resource limit in terms of the total number of jobs that can be scheduled to it during a particular production stage. Eq. (3) requires that each job is scheduled. The scheduling constraints, Eq. (4) – Eq. (6) are linearization constraints, while Eq. (7) – (9) represent precedence relationship between scheduled jobs. In particular, for any two jobs that are scheduled to the same machine, a minimum of 20% completion rate of the first job needs to be achieved before another one can be started. This ensures that the resource availability for the workstations making up the individual machines does not impact the productivity of the respective machine. The 20 % figure is obtained after consulting with industry practitioners and can be adjusted to suit the case being considered. Eq. (10) requires that the finish time of a job to at least equal its duration. The domain of the binary variables is given by Eq. (11) – Eq. (12), while the domain of the continuous variable is defined in Eq. (13) - Eq. (14).

3 Case Study

To maintain the brevity of the discussion presented in this paper and to prove the concepts proposed, the DSS is presented through applications of a single iteration of the model on a practical example. The case study is representative of the production line of a typical modular construction factory; its layout plan is shown in Fig. 3.

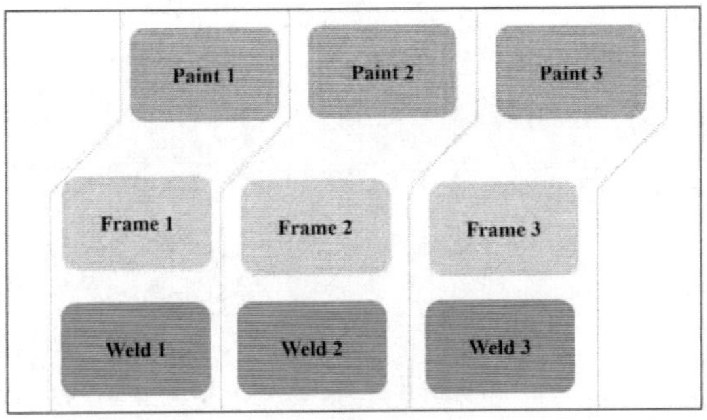

Fig. 3 Plan view of the case study

The project of the case study is comprised of 7 jobs in total, each with a specified duration, as shown in Table 2. Jobs are assumed to be independent of one another, hence are able to be completed in any order . A total of 3 machines exist within the factory and the associated resource limit of each machine is given in Table 3. The objective is to find an optimal schedule for all 7 jobs such that the make-span of the 3 machines is minimised without overloading them.

Table 2 Jobs to be scheduled

Job ($j\in J$)	Duration in days (D_j)
1	6
2	14
3	7
4	8
5	9
6	12
7	8

Table 3 Machine capacities

Machine ($i\in I$)	Capacity (L_i)
1	3
2	2
3	2

The model is programmed on AMPL [14] and solved using CPLEX [15]. The optimal schedule is given in Table 4. The optimized make-span is given as 18.8 days. It can be noticed that Machine 1 gets assigned to perform the jobs constituting the longest sum of duration. All three machines are utilised to satisfy the resource requirements.

Table 4 Machine schedule

Machine ($i \in I$)	[Job j, finish time f_j]
1	[1, 6], [4, 12.4], [7, 18.8]
2	[2, 18.8], [3, 7]
3	[5,9], [6,18.8]

The final schedule yielded was compared with one derived by an experienced engineer; the optimised schedule was found to have yielded a make-span that is 63% shorter than that suggested by the engineer.

Even though the machine scheduling problem is inherently NP hard [16], the solving times are expected to remain reasonable if the instance sizes are not large. This is particularly true when developing schedules for set periods (often conducted on a daily basis, instead of having a prolonged schedule covering the entire project's schedule) as is common on the shop floor of modular construction firms.

4 Conclusion

In this study, a framework of a DSS for scheduling operations on the factory floor of modular construction firms was proposed. The framework is comprised of three distinct modules, namely the Scheduling Unit, a simulator and a set of triggers that impact work conditions. A MIP model was also presented to solve the last stage of the optimisation scheduling problem within the Scheduling unit. The model is formulated as a parallel machine scheduling problem due to its inherent similarity with the characteristics of the scheduling operations in modular construction. The model was applied to a practical case study, and the resulting schedule was contrasted with one obtained based on the expertise of an engineer. Results showed that the model produces a schedule with a make-span that is 21% shorter in duration compared to that based on experience.

Only a single iteration of the DSS was presented in this paper. To be able to use the full DSS on large instances, a suitable heuristic that embeds the proposed optimisation model needs to be developed. This is currently being developed by the authors.

References

[1] S. C. Weber, *Scheduling Construction Projects: Principles and Practices*. Pearson Prentice Hall, 2005.

[2] M. Arashpour, R. Wakefield, N. Blismas, and T. Maqsood, "Autonomous production tracking for augmenting output in off-site construction," *Autom. Constr.*, vol. 53, pp. 13–21, May 2015.

[3] Alistair G.F. Gibb, *Off-site Fabrication: Prefabrication, Pre-assembly and Modularisation*. Wiley, 1999.

[4] L. Jaillon and C. S. Poon, "The evolution of prefabricated residential building systems in Hong Kong: A review of the public and the private sector," *Autom. Constr.*, vol. 18, no. 3, pp. 239–248, May 2009.

[5] J. S. Goulding, F. P. Rahimian, M. Arif, and M. D. Sharp, "New offsite production and business models in construction: priorities for the future research agenda," *Archit. Eng. Des. Manag.*, vol. 11, no. 3, pp. 163–184, May 2015.

[6] D. M. Gann, "Construction as a manufacturing process? Similarities and differences between industrialized housing and car production in Japan," *Constr. Manag. Econ.*, vol. 14, no. 5, pp. 437–450, Sep. 1996.

[7] C.-H. Ko and S.-F. Wang, "Precast production scheduling using multi-objective genetic algorithms," *Expert Syst. Appl.*, vol. 38, no. 7, pp. 8293–8302, Jul. 2011.

[8] W. T. Chan and H. Hu, "Constraint programming approach to precast production scheduling," *J. Constr. Eng. Manag.*, vol. 128, no. 6, pp. 513–521, 2002.

[9] J. O'Brien and F. Plotnick, *CPM in Construction Management, Eighth Edition*. McGraw Hill Professional, 2015.

[10] R. M. Lawson, R. G. Ogden, and R. Bergin, "Application of modular construction in high-rise buildings," *J. Archit. Eng.*, vol. 18, no. 2, pp. 148–154, 2011.

[11] D. Y. Eroglu, H. C. Ozmutlu, and S. Ozmutlu, "Genetic algorithm with local search for the unrelated parallel machine scheduling problem with sequence-dependent set-up times," *Int. J. Prod. Res.*, vol. 52, no. 19, pp. 5841–5856, Oct. 2014.

[12] K. Li and S. Yang, "Non-identical parallel-machine scheduling research with minimizing total weighted completion times: Models, relaxations and algorithms," *Appl. Math. Model.*, vol. 33, no. 4, pp. 2145–2158, Apr. 2009.

[13] C. M. Joo and B. S. Kim, "Hybrid genetic algorithms with dispatching rules for unrelated parallel machine scheduling with setup time and

production availability," *Comput. Ind. Eng.*, vol. 85, pp. 102–109, Jul. 2015.
[14] R. Fourer, D. Gay, and B. Kernighan, *Ampl*, vol. 119. Boyd & Fraser, 1993.
[15] IBM Knowledge Center, "IBM ILOG CPLEX Optimization Studio V12.6.0 documentation," 2016. [Online]. Available: http://www.ibm.com/support/knowledgecenter/SSSA5P_12.6.0/ilog.odms.studio.help/Optimization_Studio/topics/COS_home.html. [Accessed: 16-Oct-2016].
[16] J. K. Lenstra, A. H. G. Rinnooy Kan, and P. Brucker, "Complexity of Machine Scheduling Problems," in *Annals of Discrete Mathematics*, vol. 1, E. L. J. P.L. Hammer B. H.Korte and G. L.Nemhauser, Ed. Elsevier, 1977, pp. 343–362.

Multilevel Computational Model for Cost and Carbon Optimisation of Reinforced Concrete Floor Systems

S. Eleftheriadis[a], P. Duffour[b], P. Greening[c], J. James[d], D. Mumovic[a]

[a]UCL Institute for Environmental Design and Engineering, University College London, UK
[b]Department of Civil, Environmental & Geomatic Engineering, University College London, UK
[c]Faculty of Engineering, Environment and Computing, Coventry University, UK
[d]JAW Sustainability, UK
E-mail: stathiseleftheriadis@gmail.com

Abstract

The cost and carbon efficiency of building structures could be enhanced by the current developments in design automation and optimisation techniques. New ways to systematically assess design alternatives based on cost and carbon parameters are necessary. The study proposes a multilevel optimisation approach that combines Building Information Modelling (BIM) data and Finite Element Modelling (FEM) with a constrained genetic algorithm. The optimisation methodology is tested in a prototypical building floor system. Structural grid configurations, floor thicknesses and columns sizes and reinforcement details are identified. The results showed that the cost optimum design is 3% cheaper than the carbon optimum design but it has 7% more carbon. In addition, the concrete in the floor is the biggest contributor in both total cost and carbon. Relationships between cost- and carbon- optimum designs for the tested structural configuration are also discussed.

Keywords –

Optimisation; Reinforced Concrete; Cost; Carbon; BIM; FEM

1 Introduction

In reinforced concrete (RC) structures, cost optimisation has traditionally attracted attention amongst researchers and practitioners due to the increasing demands on the cost efficiency of buildings' design and construction [1]. Cost optimisation studies have analysed common structural systems including beams, columns, frames and floors. An exhaustive survey of cost optimisation methods in various RC structures can be found in Sarma & Adeli [2].

In flat slab systems, which are one of the most popular solutions in commercial RC buildings in the UK, only a few previous studies were found. Sahab et al. [3] applied genetic algorithms using the Equivalent Frame Method (EFM) to optimise the floors in one-storey, three-storey and four-storey buildings. In their study, column layout optimisation could significantly reduce the total cost of the structure. However, the algorithm was based on EFM method, which could limit the applications in more complex building scenarios.

Only recently, Aldwaik & Adeli [4] have proposed a Finite Element Model (FEM) cost optimisation method for flat slabs using neural dynamics. They have reported cost savings of approximately 7-9% when compared the original design configurations for a floor scenario in a real-life high-rise building. Despite the interesting results, their study did not include any layout optimisation which could potentially offer larger savings. Environmentally-based optimisation studies in RC structures are far less developed than the cost-based ones. Nevertheless, it is becoming more challenging to make cost-effective decisions without evaluating the relationships between the economic and the environmental parameters together [5].

Generally, the implementation of environmentally-based optimisation in structural engineering could be restricted by the: 1) Type of the objective functions not being universally agreed, 2) Constructability constraints not adequately addressed or included in the optimisation model [5]. A few optimisation studies that inlclude both cost and carbon parameters involve concrete beams [6][11], frames [7][8] and columns [9][10]. To our best knowledge there is no flat slab study with these two objectives.

The use of Building Information Modelling (BIM) has the potential to improve the way environmental and cost models are integrated within building system analysis [14, 15]. The current study extends Aldwaik & Adeli's [4] FEM-based model and explores how BIM technologies could also be combined with Finite Element Models (FEM) to develop cost and carbon optimisation models that can also be implemented in real-life building structures. The aim of the study is to introduce a computational model that systematically tackles cost and carbon optimisation problems in RC flat slab buildings. The objectives of the study are threefold:

1. To establish a multilevel approach for the optimisation of flat slabs based on cost and carbon functions,
2. To utilise data inputs from Building Information Model (BIM) and a Finite Element Model (FEM) engine, and
3. To validate the proposed optimisation methodology in a prototypical case study scenario and analyse the findings.

2 Methodology

Figure 1 shows the flow diagram of the methodology used in this study. The optimisation process begins by obtaining floor information from BIM components including node coordinates and total floor area. BIM data are then used as the input to create the geometric constraints of the structural floor in the FEM component. The structural analysis of the proposed optimisation methodology is performed in three main levels:

1. Structural grid layout – number and size of bays
2. Structural design elements – floor and columns dimensions
3. Structural detailing – floor and columns reinforcement

The reason for this distinction is to identify the parameters that have the biggest contribution in the structure's embodied carbon and construction cost. The optimisation engine is a custom constrained genetic algorithm with penalisation factoring. The formulation of the objective functions is based on construction cost and embodied carbon inventories which were collected from literature sources. Two separate single-objective modes of the algorithm were built with: 1) A cost objective function, and 2) A carbon objective function.

The multiobjective functionality of the model is also under development but it is not covered in this study. The optimisation component utilises FEM to run and perform the structural verifications based on Eurocode 2 requirements. The genetic algorithm searches in the design space for the solutions that minimise the objective functions whilst satisfying the structural code's requirements and constructability constraints. Once the genetic algorithm operator is minimised the optimisation process ends and the optimised flat slab configurations are obtained. In the subsequent sections the main components of the methodology are presented.

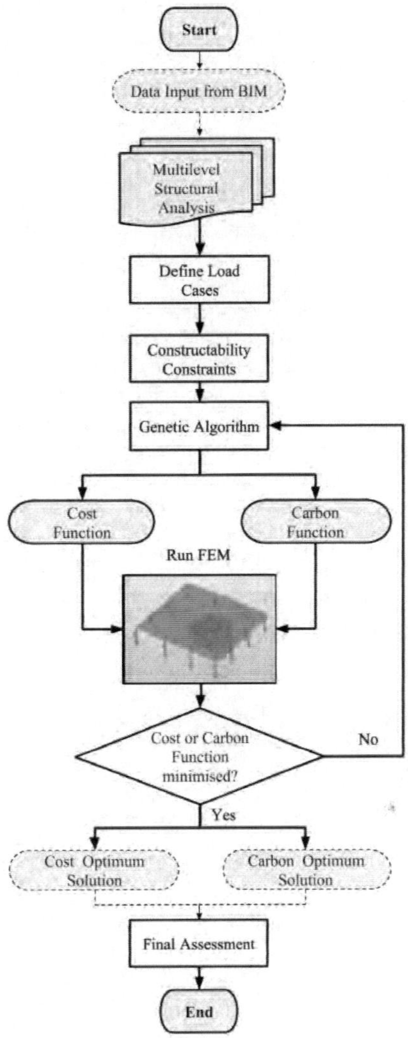

Figure 1. Flow diagram showing the proposed methodology with its main functions

2.1 BIM Component

Obtaining high level of BIM integration in structural analysis is still considered a significant challenge in the current design practice. In this study, BIM functionalities are incorporated in the optimisation model. In particular, data required for the optimisation in FEM are directly obtained from the BIM geometry. The topology of the flat slab construction is mapped in BIM and transferred into the optimisation module where all coordinates, dimensions and material properties are recognised and translated into structural panels. To enhance the interoperability of the BIM model with the FEM engine in this study, *Autodesk Revit 2016* is used as the BIM model and *Autodesk Robot 2016* as the FEM analysis engine. More information on BIM applications can be found in previous literature review where it was suggested that more than 70% of recent studies use *Autodesk Revit* for BIM-based life cycle and sustainability analysis of building systems [15].

This type of FEM offers parametric modelling features, which are necessary when conducting multiple iterations for the verification of the genetic algorithm. *Robot* Application Programming Interface (API) gives access to the functionalities of the *Robot* programme and to individual components of the modelled structure from the programming language level. In this research, C# was used to access the .NET framework of *Robot* API and to design the optimisation based on its Windows Form Application. The programmable COM interface and the API of Robot automate repetitive tasks in Robot such as multiple case scenarios.

2.2 Genetic Algorithm

A tailored genetic algorithm was developed in C# to optimise the structure. The genetic algorithm's conceptual construct, developed by Holland [12] in the 1960's and 1970's mimics the evolutionary processes in nature by population, reproduction and heredity with the inherent ability for the designer to alter several parameters within the method such as population size, crossover technique and mutation rate. Figure 2 gives a schematic representation of the single-objective GA process implemented in this study.

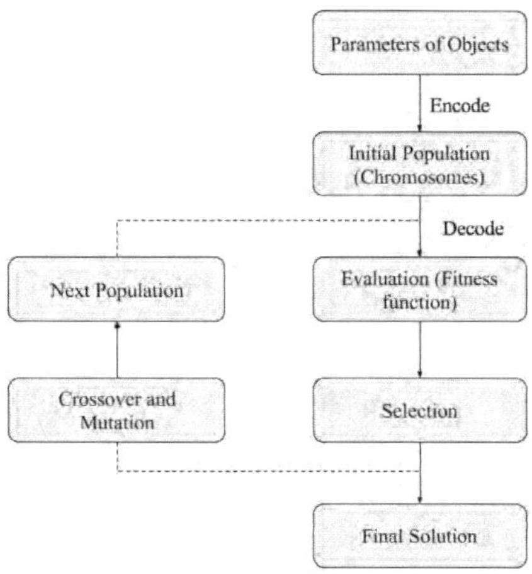

Figure 2. Schematic of single objective genetic algorithm (adapted from: [13])

Two different objective functions are utilised in the genetic algorithm to minimise the construction costs and embodied carbon of the whole structure.

2.2.1 Cost Function

The fitness function used in the calculation of the structural cost is modified to include the construction costs of the floor and columns. Equation (1) shows the general representation of the cost function used in the study.

$$f_c(x) = \sum_{i=0}^{n} q_i c_i \qquad (1)$$

where q_i represents quantities (material weights) obtained from the FEM structural analysis and c_i represents the cost factors for the different material types (concrete, steel reinforcement and formworks). The sum of the different structural members costs (columns and floor) is the total cost of the structure. Material and labour cost data are collected from Spon's Architects' and Builders' Price Book 2017. As the cost factors have different units appropriate conversion factors were used. The output from the cost function is a total *£* cost. The final result is weighted in $£/m^2$ of floor area.

2.2.2 Carbon Function

The fitness function that calculates the embodied carbon of the structural elements in the genetic algorithm is shown in Equation (2).

$$f_{ec}(x) = \sum_{i=0}^{n} q_i e_i \qquad (2)$$

where q_i is the quantities (material weights) obtained from the FEM structural analysis and e_i is the carbon factors for concrete, reinforcement. It follows a similar procedure with the cost function but the cost factors are substituted with the relevant carbon factors for the different material types. The carbon factors are based on Environmental Product Declaration (EPD) data. For the concrete carbon factor 130 $kgCO_2e/t$ was assumed and 1270 $kgCO_2e/t$ was considered for the steel carbon factor. The final carbon results were converted in $kgCO_2e/m^2$ based on the total floor area.

2.3 Optimisation Levels

The optimisation of the structure is organised around three main levels: 1) Structural layout, 2) Sizes of columns and slabs, and 3) Reinforcement layout for columns and slabs. The components on these three levels are interconnected and therefore, understanding how these relationships affect the the cost and carbon performance of the structure becomes critical.

Several computational modules have been established to ensure the seamless integration of these levels with the optimisation algorithm. The sections below elaborate on the integration process.

2.3.1 Grid Layout

In this level the algorithm looks for the structural grid topology that minimises the objective functions. A computational module was created to enable the automatic generation of structural grids – i.e. columns location. The floor outline is obtained from the BIM model, and it is used to determine the boundaries of the slab. The algorithm reads the relevant boundary lengths and uses them to calculate all the possible structural bay combinations on X- and Y- directions. The algorithm uses a polynomial time approximate component and permutation with repetition to find all the appropriate bay combinations. The user assigns possible span lengths, which can be any integer, double or float number. In this research, a defined set of span lengths was used to resemble realistic conditions: The algorithm uses 0.5m length increments.

2.3.2 Column and Slab Sizing

The optimum sizing of the slab is determined by varying the slab depth within a discrete set of dimensions. This set includes constructability constraints. For example, for the slab thickness, the allowable limits have been constrained to practical increments of 25mm. The data are encoded into input arrays, and the algorithm uses a random operator to select an element from the list during each iteration. Discrete variables are used for the dimension optimisation of the RC columns. For this operation, 50mm increments are implemented.

2.3.3 Detailing

Detailing parameters include all the reinforcement elements in the slab and the columns. The slab reinforcement (both X & Y directions) consists of a basic mesh that is applied everywhere with additional top and bottom reinforcement bars only applied on the zones that are needed. For the columns, bending reinforcement is considered. All the designs are compliant with the Eurocode's requirements. This set of analysis consists of several algorithmic components.

For the slab, the algorithm reads the required reinforcement data (Figure 3) from all the mesh points in the FEM model and it calculates the minimum and the maximum values. The minimum values are used for the estimation of the basic mesh and the maximum values are used for the calculation of the additional reinforcement where necessary. The bar diameters are limited to the most common diameters engineers use in practice (ø10, ø12, ø16, ø20, ø25, ø32). Their spacing also follows the common increments of 175mm, 200mm, 225mm and 250mm.

The total quantities (in tonnes) of basic reinforcement is then calculated by multiplying the reinforcement rate with the area of the slab. The differences between the assigned basic reinforcement and the maximum mesh/node values of the required reinforcement are used to estimate the area of the additional bars. Each component of the mesh carries a digital identity which is used in this component in order to identify the zones in need of additional reinforcement.

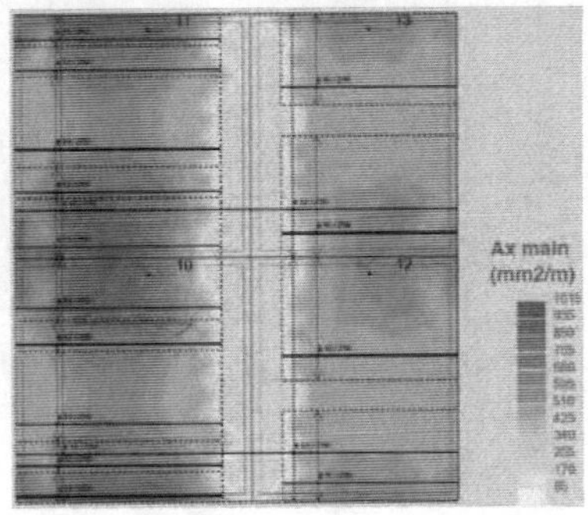

Figure 3. Required reinforcement mapping on the flat slab

A separate algorithmic component is implemented for the calculation of the bending reinforcement in the columns. This component uses forces (reaction loads), geometric (column cross section, storey height) and material data from the structural FEM model. Structural checks are conducted and the approved reinforcement is obtained from the calculation spreadsheet. The material quantities for the columns are used to update the objective function of the optimisation.

3 Prototype Building

3.1 Geometric Conditions

The optimisation methodology described in the previous sections is tested on a prototype building scenario. Figure 4 shows the simplified plan of the floor for the tested building scenario as displayed in BIM. The floor consists of a 15x16m rectangle area with a centralised core structure for lateral stability. For the purposes of this study, the core structure (4x5m) is a fixed component assuming a RC 200mm-thick wall.

The functional units for the objective functions are the slab and the columns. The dimensions of the floor are also fixed which means that the columns' topology will be fitted within the boundary of the floor. It is envisaged that this type of geometry can be implemented in real-life high-rise buildings. In regards to the materials properties, C32/40 is used for the concrete (2400 kg/m^3) and S500 for the steel bars (7850 kg/m^3). For the loads, it is assumed 2.5kN/m^2 was assumed for superimposed load, 7.5 kN/m^2 for live load and the dead load of the structure.

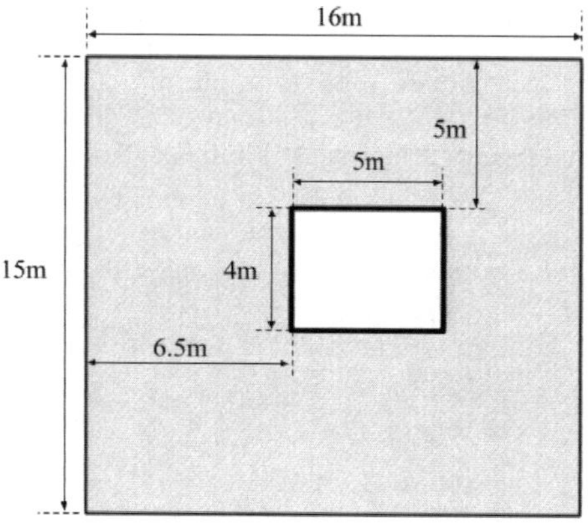

Figure 4. Plan of the floor boundary conditions

3.2 Algorithmic Input

Table 1 shows the input variables that are used in this building scenario. The algorithm uses the values from that list to find the combinations of genes that minimise either the cost or the carbon functions. A population of 100 individuals is used in the algorithm whilst the maximum generations were limited to 350 generations.

Table 1. Gene components for the genetic algorithm

Gene	Type	Number of Options	Designs
1	Slab Thickness	5	{225, 250, 275, 300, 325} mm
2	Column B	4	{250, 300, 350, 400} mm
3	Column H	4	{250, 300, 350, 400} mm
4	Bays X	4*	{(5x6x5), (6x5x5), (8x8), (5x5x6)} m
5	Bays Y	4*	{(5x5x5), (8x7), (7.5x7.5), (7x8) m
6	Bars per Column B	3	{3, 4, 5} Number of Bars
7	Bars per Column H	3	{3, 4, 5} Number of Bars
Total Combinations		11,520	

*As generated by the computational compoment: The Number of options depends on the inputs' details. In this case, values for the bays between 5m to 9m with 0.5m increments are used.

3.3 Penalisation

The structural calculations are based on EN 1992-1-1:2004 AC:2008. If the limit states defined in the code are violated, the algorithm assigns a penalty in the objective functions previously described. The design inputs that cause the violation receive lower probability for appearing in the future generations and they are eventually are eliminated from the population. Other structural constraints such as deflection limits could also be incorporated in the penalty function. This penalisation approach is an efficient way to assist the algorithm in finding the optimum combinations faster.

4 Results

4.1 Algorithmic Performance

Several optimisation sequencies (at least 10 runs for every objective) were run and in every instance the algorithm converged to the maximum performance after approximately 280 generations. Figure 5 shows the representation of the algorithms' convergence pattern displaying continuous improvements on the fitness function after each generation. Each iteration took approximately 25-30 seconds to complete in an Intel Core i5-4570 @ 3.2 GHz desktop computer.

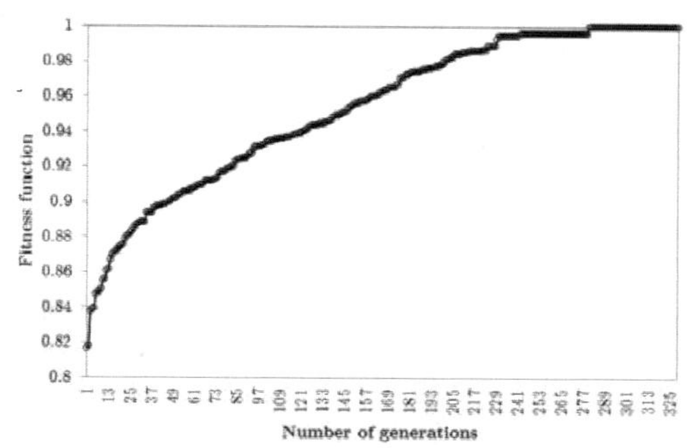

Figure 5. Convergence curve of the optimisation algorithm

An important component of the algorithm is its ability to accurately calculate the reinforcement rates in the slab. The zoning module effectively identifies the variability of slab thickness and the reinforcement requirements. Figure 6 demonstrates the relationship between the tonnage of steel and concrete for the different slab thicknesses in a simple way. Two observations can be made: 1) Thinner slabs need more reinforcement than thicker ones, 2) The options with the 300mm and 325mm slabs do not provide significant savings in reinforcement due to high minimum reinforcement requirements. For the analysis of the results, the entire optimisation set the input and output data categories were exported in a .csv file. The analysis of the results was conducted using Excel.

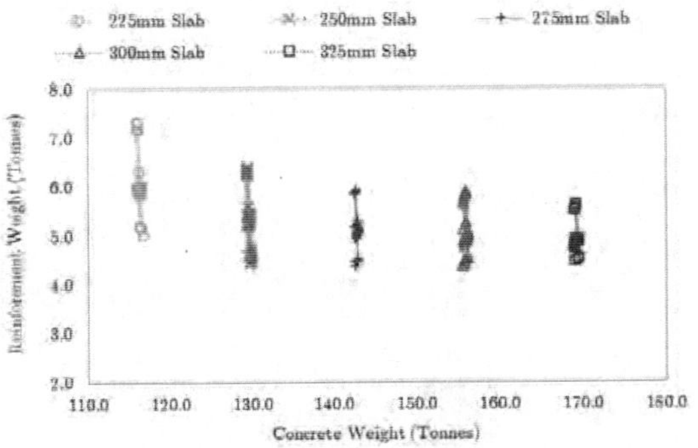

Figure 6. Relationship between concrete weight and reinforcement weigth in the slab

4.1.1 Carbon Performance

The performance of the algorithm for the carbon function is summarised in Figure 7. It becomes evident that the concrete in the slab is the biggest contributor to the total embodied carbon in the structure as the two curves appear to have very similar patterns. The reinforcement in the slab is the second biggest contributor. On average 90% of the carbon in the structure is attributed to the slab materials.

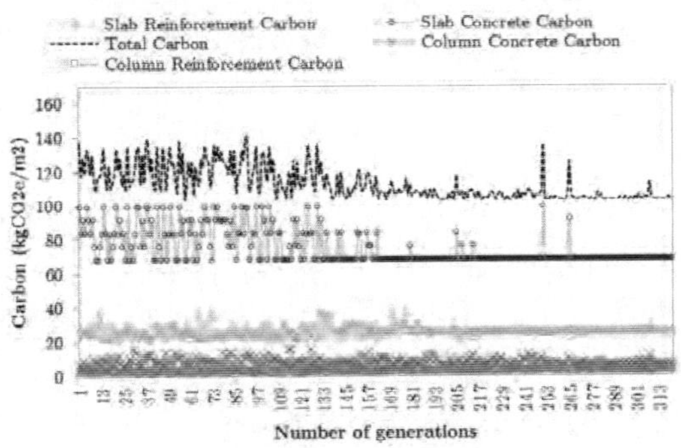

Figure 7. Evolution of carbon components towards the carbon optimum solution

4.1.2 Cost Performance

Figure 8 shows the performance of the algorithm for the cost function. The initial fluctuations correspond to the early stages of the search, however the algorithm eventually stabilises around 120 £/m^2. In this set of simulations, no single parameter clearly affects the results unlike in the carbon optimisation. Nevertheless, the slab concrete and formwork are the most contributing to the structure's total cost. Detailed cost breakdown for the cost optimum solutions will be presented in the following sections.

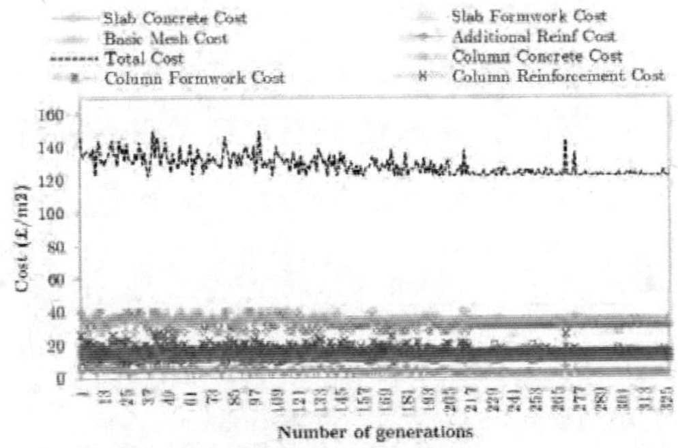

Figure 8. Evolution of cost components towards cost optimum solution

4.2 Optimisation Solutions

In this section, the optimum solutions obtained from both of the objective functions are presented. Table 2 summarises the results. The algorithm search has been simplified for the design of the columns by applying the same dimensions on every column (the reinforcement calculation applies the same bar diameter to every bar in the section). If required, both elements could be overriden to allow for further customisation. Figure 9 shows the geometric representation of the optimum flat slab designs. The carbon optimum solution has more columns (12 in total) and thinner slabs (225mm slab thickness). This finding verifies the relationship between the carbon function and the slab concrete as the algorithm tries to minimise it. The additional columns reduce the reinforcement requirements in the slab and as a result, the total carbon of the structure.

Table 2. Optimisation results

	Carbon Optimum	Cost Optimum
Slab Thickness	225mm	250mm
Column B	250mm	250mm
Column H	250mm	250mm
Bays X	6mx5mx5m	5mx6mx5m
Bays Y	5mx5mx5m	8mx7m
Columns Reinforcement	8 Bars per Column	8 Bars per Column
Total number of Columns	12 (5xø16, 7xø12)	10 (1xø20, 3xø16, 6xø12)
Slab Reinforcement Rates	101kg/m^3	98kg/m^3

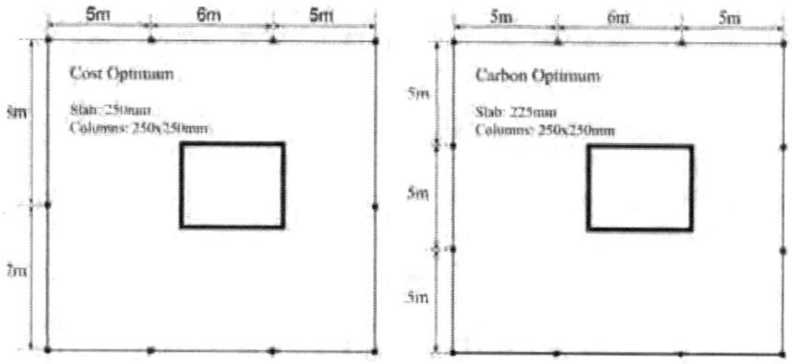

Figure 9. Design configurations that minimise carbon and cost objective functions

In comparison to the carbon optimum design, the columns in the cost optimum solution appear to have a larger impact on the results as the design consists of 10 columns instead of 12. In addition, the reinforcement rate in the slab is 3% less than the one in the carbon optimum design. This is due to the thicker slab option (250mm).

4.3 Discussion

The detailed breakdown of the components in the carbon optimum solution shows the following:

- 66% in the slab concrete
- 25% in the slab reinforcement (basic mesh and additional reinforcement)
- 6% in the columns concrete
- 3% in the columns reinforcement

On the othe hand, the detailed breakdown of the components in the cost optimum solution shows the following:

- 25% in the slab concrete
- 25% in the slab formwork
- 22% in the slab reinforcement (basic mesh and additional reinforcement)
- 12% in the columns reinforcement
- 2% in the columns concrete
- 9% in the columns formwork

The analysis of the two optimum solutions indicates similarities in the final results. The relationship between the cost and the carbon optimum solutions is shown in Figure 10. The cost optimum solution resulted in 111 $kgCO2e/m^2$ and costs 122 £/m^2. On the other hand, the carbon optimum solution costs 126 £/m^2 and it has 103 $kgCO2e/m^2$. The results show an inverse correlation between cost and carbon optimum solutions. The cost optimum design is 3% cheaper but 7% more carbon intensive. The obtained results in the tested buiding scenario follow a similar pattern with results found on other structural systems in the literature [8].

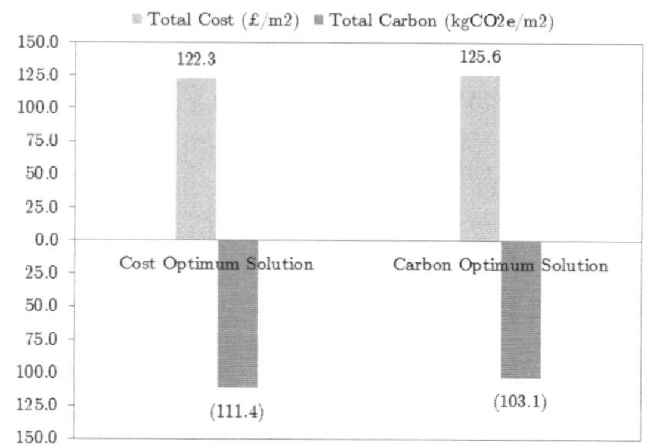

Figure 10. Relationship between cost and carbon optimum solutions

5 Conclusions

This study investigates a computational framework for the optimisation of flat slabs in building structures. Three main contributions can be recognised in this research: The first novelty of the study is the use of BIM data combined with a FEM engine that operates within a genetic algorithm solver. The second novelty is the configuration of the optimisation in multiple levels of structural analysis. The final novelty is the development of both cost and carbon functions. A prototypical building scenario is used to validate the methodology. Results show that the effectiveness of the algorithm in finding optimum solutions is not limited by the increased computational complexity of the model. Cost and carbon optimum design configurations were obtained and analysed. For the tested scenario, the cost optimum solution costs only 3% less than the carbon optimum solution but has 7% more carbon. The slab accounts for more than 75% of the total costs in the cost optimum solution when the columns account for approximately 25%. On the other hand, in the carbon optimum design, the slab is responsible for 90% of the carbon emissions, and only 10% is related to the columns. Further analysis is required in actual buildings in order to verify the results of the study and evaluate the optimum designs against as built solutions.

Acknowledgements

This research has been made possible through funding provided by the Engineering and Physical Sciences Research Council (EPSRC) and from Price & Myers LLP via the UCL EngD Centre (Grant Number: EP/G037159/1) in Virtual Environments, Imaging and Visualisation and this is gratefully acknowledged here.

References

[1] S. Atabay, "Cost optimization of three-dimensional beamless reinforced concrete shear-wall systems via genetic algorithm," Expert Systems with Applications, vol. 36, pp. 3555-3561, 2009.

[2] M. Sahab, A. Ashour and V. Toropov, "Cost optimisation of reinforced concrete flat slab buildings," Engineering Structures, vol. 27, pp. 313-322, 2005.

[3] M. Aldwaik and H. Adeli, "Cost optimization of reinforced concrete flat slabs of arbitrary configuration in irregular highrise building structures," Structural and Multidisciplinary Optimization, vol. 54, pp. 151-164, 2016.

[4] C. Sarma and H. Adeli, "Cost Optimisation of Concrete Structures," *Journal of Structural Engineering,* vol. 124, pp. 570-578, 1998.

[5] W. Wang, R. Zmeureanu and H. Rivard, "Applying multi-objective genetic algorithms in green building design optimization," *Building and Environment,* vol. 40, pp. 1512-1525, 2005.

[6] A. Kosloski, "Cost and Embodied Energy Optimisation of rectangular reinforced concrete beams," Clemson University, Clemson, South Carolina, 2014.

[7] C. Camp and F. Huq, "CO2 and cost optimization of reinforced concrete frames using a big bang-big crunch algorithm," *Engineering Structures,* vol. 48, pp. 363-372, 2013.

[8] I. Paya-Zaforteza, V. Yepes, A. Hospitaler and F. Gonzalez-Vidosa, "CO2-optimization of reinforced concrete frames by simulated annealing," *Engineering Structures,* vol. 31, pp. 1501-1508, 2009.

[9] F. Medeiros and M. Kripka, "Optimization of reinforced concrete columns according to different environmental impact assessment parameters," *Engineering Structures,* vol. 59, pp. 185-194, 2014.

[10] H. Park, B. Kwon, Y. Shin, Y. Kim, T. Hong and S. Choi, "Cost and CO2 Emission Optimization of Steel Reinforced Concrete Columns in High-Rise Buildings," *Energies,* vol. 6, no. 11, pp. 5609-5624, 2014.

[11] T. Garcia-Segura, V. Yepes, J. Marti and J. Alcala, "Optimization of concrete I-beams using a new hybrid glowworm swarm algorithm," *Latin American Journal of Solids and Structures,* vol. 11, no. 7, pp. 1190-1205, 2014.

[12] J. Holland, "Computer programs that "evolve" in ways that resemble natural selection can solve complex problems even their creators do not fully understand," *Scientific American,* pp. 66-72, 1992.

[13] D. Goldberg, Genetic Algorithms in Search, Optimization and Machine Learning, Boston, MA, USA: Addison-Wesley Longman Publishing Co., Inc., 1989.

[14] Russell-Smith, S., & Lepech, M. (2011). Dynamic Life Cycle Assessment of Building Design and Retrofit Processes. *Computing in Civil Engineering*, 760-767.

[15] Eleftheriadis, S., Mumovic, D., & Greening, P. (2017). Life cycle energy efficiency in building structures: A review of current developments and future outlooks based on BIM capabilities. *Renewable and Sustainable Energy Reviews*, 811–825 .

34th International Symposium on Automation and Robotics in Construction

ISARC 2017

June 28 – July 1 TAIPEI, TAIWAN

Comparison of Bridge Inspection Methodologies and Evaluation Criteria in Taiwan and Foreign Practices

Liao Hsien-Ke [a], Mustapha Jallow [a*], Yau Nie-Jia [a], Jiang Ming-Yi [b], Huang Jyun-Hao [b], Su Cheng-Wei [b], Chen Po-Yuan [c]

[a] Graduate Institute of Construction Engineering and Management, National Central University, Zhongli, Taiwan
[b] Institute of Transportation, Taipei, Taiwan
[c] Ministry of Transportation and Communications, Taipei, Taiwan
E-mail: 943402019@cc.ncu.edu.tw, jallowm12@gmail.com*, yau@ncu.edu.tw, myjiang@iot.gov.tw, herry219@iot.gov.tw, jason@iot.gov.tw, pychen@motc.gov.tw

Abstract –
The reliability of structural performance of bridges in Taiwan are affected by both in-service loading and material deterioration due to extreme environmental factors. With an inventory of 27,895 bridges and culverts in Taiwan Bridge Management System (TBMS), there are more than 8,840 bridges that are over 30 years and around 8,877 whose built years are still unknown. Government officials noticed this issue in recent years and placed thoughtful attention. Not only did they establish the maintenance mechanism for regular bridge inspections and maintenance, but developed maintenance systems to upsurge the competence of bridge management. However, this strategy only focused on data recording and was not able to notably improve maintenance efforts. Thus, it is essential that engineers refine their inspection and assessment techniques.

Bridge inspection data recorded in TBMS are frequently analysed, to determine failure causes. According to the "Validation of bridge inventory data in TBMS" project, funded by Ministry of Transportation and Communications (MOTC) in 2010, the accuracy of bridge inspection data was lower than that of inventory. Consequently, MOTC proposed to enforce a new and detailed inspection program after 2015.

This research reviewed the state-of-the-art for inspection methodologies in practice to assess condition of bridges in Taiwan, Japan, South Africa, USA and Europe. Concisely, the findings, including description, categorization of deterioration, and assessment method of bridge conditions were collected to provide bridge inspectors and engineers with more reliable assessment approach. Recommendations for enhancing bridge inspection standards and qualifications for inspectors in Taiwan are also discussed in this paper accordingly.

Keywords: Bridge Management System, Condition Rating, DER&U Methodology, Evaluation Criteria, Visual Inspection.

1 Introduction

Taiwan owes its formation, shape and frequent rate of earthquakes to the complex interaction of the Eurasian and Philippines Sea plates. In a geological time-scale, the 35,563 km^2 (13,731 mi^2) island has virtually erupted from the ocean floor as the Philippines Sea Plate drives Northwestward into the Eurasian plate at a rate of about 7 centimeters per year.

On September 21, 1999, Chi-Chi Earthquake of magnitude 7.3 struck Central Taiwan, leaving subsequent aftershocks with four of which measured a magnitude greater than 6.5 on the Richter scale. The devastating incident caused severe infrastructural damages and casualties to the affected region.

Furthermore, torrential rains brought by typhoons causes floods and mudflows which are huge threats to bridges in particular. On August 8 2009, Typhoon Morakot struck southern Taiwan and brought about the worst flood recorded for the past 50 years, whereas 129 bridges were seriously damaged. Due to these alarming issues, bridge maintenance, rehabilitation, and repair (MR&R) have become important matters to MOTC and Local Governments [1].

Different versions of BMS packages have been developed to administer bridge-related issues in Taiwan. However, it was until 2000 that Taiwan Bridge Management System (TBMS) was established by the Center for Bridge Engineering Research (CBER) at National Central University (NCU). This project was funded by MOTC and is currently maintained by the Graduate Institute of Construction Engineering and Management at NCU [2].

2 Overview of International Inspection Methods

2.1 Taiwan

According to the Manual for Enhancement and Inspection of Highway Concrete Bridges (2015) regulated by the MOTC, there are three types of bridge inspections, namely:

1. *Daily Patrol*– performed as regulated by the Road Maintenance Manual (2012), whereby two inspectors drive over a bridge to check for defects.
2. *Regular Inspection*– performed by trained inspectors using naked eyes and essential tools to examine the overall condition of a bridge and its river course. This inspection is operated span-by-span and is performed once every two years.
3. *Damage Inspection*– an unscheduled inspection to assess structural damages resulting from floods, typhoons, earthquakes, fire, etc.

Regular and Damage inspection may involve the usage of heavy instruments such as bridge inspection trucks, cherry pickers and inflatable boats.

The methodology for visual inspection in Taiwan is abbreviated as DER&U. DER&U is a national standard for regular bridge inspection and evaluation; which is an effective, reliable and economical way to examine bridges. Herein, "D" stands for degree of deterioration; "E" represents extent of the deterioration; "R" implies relevancy to safety of the deterioration; and "U" depicts the urgency for repairing of the deterioration. All of these indices are numerically rated on an integer scale ranging from 0 to 4, as shown in Table 1. While rating, the inspector may state his/her personal remarks if deemed necessary [3].

Table 1. DER&U evaluation criteria

	0	1	2	3	4
D	Not exist	Good	Fair	Bad	Serious
E	*U/I	<10%	10~30%	30~60%	Over 60%
R	Uncertain	Minor	Limited	Major	Large
U	**N/A	Routine	In 3 yrs.	In 1 yr.	Immediate

*U/I – Unable to Inspect, **N/A – Not Applicable

After inspection, each component (deck, girder, pier, joint, abutment etc.) of the bridge is rated. A component condition index Ic_{ij} is calculated based on the evaluated integers of D, E, and R for each component. This calculation is based on a point-deduction mechanism, i.e., deficiencies of a component will deduct points from a perfect score of 100. Equation (1) shows the formula for calculating an Ic_{ij} value for the *'j'* item of component *'i'*.

$$Ic_{ij} = 100 - 100 \times \frac{D \times E \times R^a}{4 \times 4 \times 4^a} \quad (1)$$

Where *'a'* is an integer usually denoted as 1 for the *'i'* component of item *'j'*.

The condition index I_{ci} of component *'i'* is an average value of all similar items as deduced in Equation (2) below.

$$Ic_i = \frac{\sum_{j=1}^{n} Ic_{ij}}{n} \quad (2)$$

The overall bridge condition index, *CI* is a weighted average of all components as shown in Equation (3); where m=21 for concrete girder bridge, m=22 for cable-stayed bridge and m=23 for both suspension bridge and arch bridge [2].

$$CI = \frac{\sum_{i=1}^{m} Ic_i \times w_i}{\sum_{i=1}^{m} w_i} \quad (3)$$

2.2 Japan

In Japan, each element in the structure is evaluated based on every single kind of defect, such as cracking, corrosion, etc., then a demerit rating is assigned to each element in a tabular format [4].

According to the Regular Inspection Procedure for Road Bridge (2014) regulated by the Ministry of Land, Infrastructure, Transport and Tourism (MLIT), there is basically one inspection type; *Regular Inspection*.

Based on this regulation, all bridges in Japan bridges with a span longer than 2 meters must be inspected at least once every five years. *Regular inspection* in Japan is a hands-on visual inspection of components or elements in a close distance [5].

For every structural member in each span, the condition is translated into either of the maintenance urgency ratings listed in Table 2.

Table 2. Health rating of bridges in Japan

	Condition	Description
I	Good	No obstacle to the function of the structure.
II	Preventive maintenance phase	There is no obstacle to the function of the structure, but it is desirable to take preventive maintenance.
III	Early rehabilitation phase	There is a possibility that the function of the structure may be hindered, so a rehabilitation strategy must be taken.
IV	Emergency repair phase	Presence of an obstacle to the function of the structure, or a possibility of occurrence is extremely high. An urgent action must be taken.

Maintenance urgency ratings are diagnoses given by experienced engineers in a very subjective manner, recommending to bridge owners the needs for action by the time of the next inspection. Engineers are required to interpret the maintenance urgency for each member, taking into account the damage type, location of damage, direction of crack, earlier remedial work history, etc. Moreover, numeric criteria like crack width and length are specified for maintenance urgency ratings.

2.3 United States of America

The National Bridge Inspection Standards (NBIS 2009) set the criteria for proper inspection and evaluation of all highway bridges in the United States. According to the American Association of State Highway and Transportation Officials (AASHTO) manual for Condition Evaluation of Bridges, the overall condition rating is assigned for 3 major components: deck, superstructure and substructure which are further divided into various elements.

In the U.S., there are eight types of bridge inspections (Table 3) [5]. Federal regulations address bridge inspection population, inspection intervals, inspection methods, inspection personnel, and inspection reporting. Federal requirements are presented primarily in the Code of Federal Regulations [6].

Table 3. Types of bridge inspection in the U.S.

Inspection	Description
Initial	First inspection of a bridge as it becomes a part of the bridge inventory to determine baseline structural conditions.
Routine	Regularly scheduled inspection consisting of observations and/or measurements needed to determine the physical and function condition of the bridge.
Damage	Unscheduled inspection to assess structural damage resulting from environmental factors or human actions.
In-Depth	A close-up inspection which investigates deficiencies that were not detected during Routine Inspection.
Special	An inspection scheduled at the discretion of the bridge owner, used to monitor a particular known defect or suspected deficiency.
Underwater	Inspection of the underwater portion of a bridge substructure and the surrounding channel.
Hands-on	Inspection within arm's length of the component. Inspection uses visual techniques that may be supplemented by NDT.
Fracture-Critical Member	A hands-on inspection of a fracture-critical member or components that may include visual and other non-destructive evaluation.

NDT: Non-destructive Test

Table 4. NBI condition rating system

Rating	Description
N	Not Applicable.
9	Excellent Condition.
8	Very good Condition – no problems discovered.
7	Good Condition – some minor problems.
6	Satisfactory Condition – structural elements show some minor deterioration.
5	Fair Condition – all primary structural elements are sound but may have minor section loss, cracking, spalling, or scour.
4	Poor Condition – advanced section loss, deterioration, spalling, or scour.
3	Serious Condition – loss of section, deterioration, spalling or scour have seriously affected primary structural elements.
2	Critical Condition – advanced deterioration of primary structural elements.
1	Imminent Failure Condition – major deterioration or section loss present in critical structural components, or obvious vertical or horizontal movement affecting structure stability.
0	Failed condition – out of service.

The U.S. methodology for rating values during inspection is a very subjective approach. The inspector visits the site and looks at every component of the bridge, then gives an assessment value ranging from 0 to 9 for the whole structure (Table 4) [7]. The ratings provide information on the severity of a condition but do not identify or quantify the extent of the deterioration. For this reason, this rating system has a limited value in determining repairs and rehabilitation needs.

Moreover, a descriptive condition rating in terms of 'good/fair/poor/not applicable' is given by the inspector for each element (waterproofing, painting, road surface etc.) of the component, based on the deficiencies found on the individual element [8]. Table 5 illustrates the descriptive rating for bridge elements.

Table 5: Descriptive condition rating for elements

Condition	Description
Good	Element is limited to only minor problems.
Fair	Structural capacity of element is not affected by minor deterioration, spalling, cracking etc.
Poor	Structural capacity of element is affected by advanced deterioration, section loss, spalling, cracking or other deficiency.

2.4 United Kingdom

The Structures Management Information System determines maintenance needs based upon structural adequacy or safety rather than solely on the condition state of the structures for UK's Highways Agency. The requirements for inspecting highway bridges in the UK are defined in Volume 3, Section 1, Part 4 of the Design Manual for Roads and Bridges (BD 63/07) (Highways Agency, 2007) [9].

According to UK. Highways Agency, there are five types of inspections for bridges and can be used in combination depending on inspection needs (see Table 6).

The methodology of visual inspection in U.K. is primarily based on *Severity – Extent* code prescribed procedure set out by the Highway Agency. *The severity* defines as the degree of damage while *extent* is a measure of the length, area, or number of defects of the bridge element. [10].

Table 6. Types of bridge inspection in the UK

Inspection	Description
Acceptance	Performed for new bridges, newly repaired bridges, and newly start of a new maintenance contract.
Superficial	Checks for outstanding defects that pose a risk to safety are highlighted and action taken immediately to remedy them.
General	Applies to bridge elements that are easily accessible.
Principal	All bridge elements, including those that are difficult to access. This may sometimes require specialist access machinery or tools.
Special	Detailed investigation of a particular bridge component. Bridges that have been strengthened using plates bonded to them also require special inspections.

Table 7. UK Bridge condition rating system

Code		Description
Extent	A	No significant defect.
	B	Slight, less than 5% of length/area affected.
	C	Moderate; 5% – 20% of area/length affected.
	D	Wide, 20% – 50% affected.
	E	Extensive; over 50% of surface area/length.
Severity	1	As new, or has no significant defect.
	2	Early signs of deterioration, minor defect.
	3	Moderate, some loss of functionality expected.
	4	Severe defect and/or element is close to failure.
	5	The element is non-functional/failed.

These codes are used in a scale of 1-5, describing the degree of deterioration, from minor (structurally sound) to a collapsed state (non-functional). Table 7 presents permissible combinations of Severity and Extent.

2.5 South Africa

The inventory and inspection of bridges is done by the South Africa National Roads Agency Limited (SANRAL) for bridges on national roads, 9 Provincial departments of transport for bridges on provincial roads and municipal transport agencies for bridges on municipal roads. South African maintenance practice includes five types of inspections (Table 8) [11].

Similar to Taiwan, the defects are rated for their Degree, Extent, Relevancy and Urgency (DER&U) as it is previously discussed (see Table 1). This system is a visual inspection and evaluation method for bridges jointly developed by Join Engineering Consultants, and South African CSIR Company

The DER&U rating system identifies defects and prioritizes them by evaluating their relative importance to the structural integrity of the bridge. It is important to note that the ratings are not directly associated with the elements but with the damage. [12].

Table 8. Types of bridge inspections in South Africa

Inspection	Description
Monitoring	A quick check on the new defects and the status of the previously known defects. A monitoring inspection does not produce any condition rating.
Principal	A thorough examination and record of a bridge for all defects.
Verification	Are performed annually by SANRAL in order to verify the accuracy of inspection data.
Project-level	Inspection to collect information for contract documents.
Acceptance	Inspection of work during and after a contract.

2.6 Denmark

The Danish Road Directorate used a computer-based BMS, called DANBRO, to manage their bridges. DANBRO has been in place throughout Denmark since 1988 and is reported as fulfilling its main purpose of aiding bridge management at all levels.

DANBRO identifies eight types of bridge inspections (Table 9). For each defect reported, the inspector will recommend a repair scheme, its year of application, and also estimate the costs for repair actions.

Table 9. Types of bridge inspections in Denmark

Inspection	Description
Inventory	Collect bridge data and baseline conditions.
Daily*	Cursory examination noting failure, damage, debris, etc.
Routine—Extended	Planning and checking routine cleaning and maintenance.
Reports from Users	Reports of: impact damage, vandalism, debris on bridge or road and erosion damage.
Principal	Thorough and systematic visual inspection of all the components of the bridge.
Special	Collection of more detailed information for decisions on maintenance actions.
Economic Special Inspection	Preparation for major repair project for a bridge.
Technical Special Inspection	Damage investigations, Special investigations, Load-carrying capacity evaluations.

* | Not a formal part of the Directorate Bridge Inspection Program.

Recommendations on the interval of visual inspections depend on the age, average daily traffic (ADT), location, existing conditions, and special features of the bridge.

Condition ratings in Denmark (Table 10) are built up from three contributors: damage (3 points), function (1 point), and consequence (1 point). The overall rating scale is 0-5, with '0' meaning no damage and '5' implying that the component can no longer fulfil its function [13].

Table 10. DANBRO condition rating system

Rating	Description
0	Insignificant deterioration; little or no damage.
1	Minor deterioration; damage with a very slow rate of development.
2	Damage is at an early stage of development or there are a few fully developed defects.
3	Damage has developed to such a degree and/or extent that it is likely that within a short time the component will no longer fulfil its function.
4	The component is severely deteriorated, such that its capacity to fulfil its function has or will soon disappear. Repair is necessary in the near future.
5	The component has completely deteriorated and can no longer fulfil its function.

2.7 Sweden

The Swedish Road Administration (SRA), maintain guides and manuals for bridge design, construction, and inspection. Sweden does not have national regulations for bridge inspection, thus, work performed by SRA includes strategic management, planning of projects, specifications for bridge works, procurement of bridge works, and supervision of contract work. SRA executes about half of all bridge inspections, with other inspections done by consultants. SRA has four levels of routine inspections: Regular, Superficial, General, and Major (Table 11) [12].

Table 11. Types of bridge inspections in Sweden

Inspection	Description
Regular	Quick visit to detect significant new conditions.
Superficial	Verify that maintenance requirements are met.
General	Follow-up on damages detected at the last major inspection.
Major	Arms-length, visual inspection of all components.
Special	Further investigation of defect or deterioration.

In addition, SRA performs Special inspections of known defects, suspected defects, and deterioration mechanisms, as needed.

SRA collects ratings and other data on conditions of bridge components during General, Major, and Special inspections. The quantity and the method of its measurement are fitted to the type of damage, structural element, material, and other considerations (e.g., mode of action of element). Functional condition is reported on a 0 to 3 rating scale, with '3' being the worst condition (Table 12). Functional condition is related to the time until the defect is expected to impair the service of the bridge [9].

Table 12. SRA condition rating

Rating	Physical Condition	Functional Condition
3	Repair needed now	Service impaired now.
2	Repair within 3 years	Service impaired within 3 years.
1	Repair within 10 years	Service impaired within 10 years.
0	Repair after 10 years	Service greater than 10 years.

2.8 Germany

Germany performs bridge inspections at two levels called Major Test and Minor Test (Table 13) [15]. Major tests are arms-length (DIN wording is "touching-distance") inspections of all elements with access to all parts. Minor tests are done three years after each major test. Minor tests use findings of the previous major test and focus on known damage and defects.

In Germany, condition rating scales run from 0 (good) to 4 (very poor). Each bridge component is assigned three ratings; one each for structural damage, traffic safety, and bridge durability. In this paper, the rating for structural damage is shown in Table 14 [16].

To ensure a standard approach for the evaluation of damages, inspection teams will be equipped with catalogues containing detailed example of damage evaluation. A total of 6 levels of bridge condition ratings are defined whose descriptions are shown in Table 15. The possible damages are graded (0 to 4), assigned to stability, safety and durability

Table 13. Types of bridge inspection in Germany

Inspection	Description
Major Test	Arms-length inspection of all components; uses access equipment and includes underwater inspection.
Acceptance	*Major test.*
Guarantee	*Major test.*
Minor Test	Verification of current state of known damage and defects.
Superficial	Cursory inspection for safety.
Ad Hoc	After significant events, such as storms, floods, etc.; also for known, severe damage.
Systems	Inspection of electrical or mechanical systems.

Table 14: Condition ratings for structural damage

Rating	Description
0	Defect has no effect on the strength of the element or structure.
1	Defect affects the strength of the structural element, but does not affect the strength of the structure.
2	Defect affects the strength of the structural element and has little effect on the strength of the structure.
3	Defect affects the strength of the structural element and the structure. Structure does not have adequate strength.
4	Structural strength of the structural element is lost. Structure does not have adequate strength. Repair or rehabilitation is needed.

Table 15. Ratings for Components in Germany

Grade	Description
1.0–1.4	Very good structural condition.
1.5–1.9	Good structural condition, but may have less long-term durability.
2.0–2.4	Satisfactory structural condition, but may have less long-term durability.
2.5–2.9	Unsatisfactory structural condition. Traffic safety may be affected.
3.0–3.4	Critical structural condition. Traffic safety is affected.
3.5–4.0	Inadequate structural condition. Traffic safety is not adequate.

3 Discussion and Suggestions

3.1 Bridge Inspection Intervals

The condition of any bridge can be quite alarming even when slight deteriorations are yet to be visually identified. In this respect, defined inspection interval for different types of bridge inspection and condition assessment is significant in assuring a suitable structural healthiness of the bridge.

However, inspection intervals vary from one country or agency to another (see Table 16). The depth and frequency to which bridges are inspected in Taiwan depends on factors such as age, traffic characteristics, state of maintenance or known deficiencies. The evaluation of these factors is a sole responsibility of the individual or agency in charge of the inspection program. Thus, it is necessary to innovate solid criteria for assessment of a bridge's general condition in accordance to a defined frequency for visual inspection. Conversely, structural assessments and reliability models can be developed using established methods and integrate them into prioritization techniques to better improve bridge condition modelling with time.

Considering the necessity to ensuring safety of bridges and determination of a suitable MR&R at a lower cost, it is suggested that the regular inspection in Taiwan be refined into; (1) *Minor*–biannual inspection of a bridge to determine its maintenance urgency. This inspection can be performed by applying the Japanese methodology for health rating of bridge components. (2) *Major*–assessment of all bridge components using the DER&U rating system. A major inspection should be performed after two consecutive minor inspections (6 years). Despite the fact that inspection data are seen to be subjective, this ideology will enable engineers to achieve a target level of reliability of inspection results.

On the other hand, a review of international practices emphasizes that underwater material damage, and scour-related deterioration may not be apparent above water until the damage has become so severe that remedial actions are extremely expensive. Based on this argument, there is a need for an additional underwater inspection to the inspection program in Taiwan. Practising this inspection type can reduce the cost of in-water repair work. Furthermore, it is also considered necessary to perform initial inspections on new bridges, or when existing bridges are entered into the database. This inspection provides a foundation for all future inspections or modifications to the bridge.

These strategies can baseline a framework to support engineers and bridge managers in decision-making processes for prioritizing inspections, maintenance actions and budget allocations.

3.2 Inspection Qualification & Certification

In foreign practices, most entry-level jobs require a high school diploma and certification. Advanced positions may require higher education such as a bachelor's or master's degree in civil engineering. Depending on the inspection type being active for a certain service, different inspectors are employed upon the guiding principles of the inspecting agency (Table 17). Herein, some influential codes of practice are discussed consequently.

In the U.S., the federal regulations do not establish qualifications for inspection team members working under the direction of an inspection team leader. An Inspection team leader in the U.S. must be a professional engineer, complete a Federal Highway Administration (FHWA) approved comprehensive bridge inspection training courses and has at least five years of bridge inspection experience. Apart from underwater inspectors, all Danish inspection personnel are engineers. Denmark conducts annual refresher training for all bridge inspectors. Individuals performing General or Major inspections in Sweden must hold an engineering degree, have experience with bridge design and construction, and must complete a one-week training course offered by the SRA. In Germany, bridge inspectors must have formal education as civil engineers and complete a federal training course lasting one week that covers all aspects of inspection.

It's observed that, most of the practices reviewed hold a regulation governing qualifications for certain inspection processes. This strategy is yet to be practiced in Taiwan as it is still under development by MOTC. Meanwhile, it would be reasonable to propose periodic training workshops for inspectors and engineers all over the country.

The training should be directed towards all features of inspection, deterioration processes, rating systems and assessment methods. Moreover, the participants must hold a degree in civil engineering before being accepted to this program. In so doing, inspectors will be able to recognize indications projecting to harmful influence of deterioration even before visible signs appear on bridge surfaces.

3.3 The DER&U Methodology

The DER&U is a condition assessment methodology invented in 1994 by two consulting companies, CSIR and Join Engineering of South Africa and Taiwan respectively. It has since become a national standard for regular bridge inspection and evaluation, deemed an effective, reliable and a cost-effective method for assessment of bridge conditions in Taiwan. During condition assessment of bridge structures, defects are rated for their Degree, Extent, Relevancy and Urgency, as previously illustrated in Table 1.

The DER&U is a subjective technique because, two defects may look the same to the inspector and have the same extent, but their impact on the reliability of the bridge may be different. Hence, the relevancy of the distress helps the inspectors capture information beyond ordinary visual ratings by assessing the impact of each distress on the overall structural integrity of the bridge.

Table 16. Summary of inspection types and their respective intervals

Occurrence	Taiwan	Japan	USA	UK	South Africa	Denmark	Sweden	Germany
Frequent	Patrol			Superficial		Daily	Regular	
3 Months								Superficial
1 Year					Monitoring	Routine	Superficial	
2 Year	Regular		Routine	General				
3 Year							General	Minor
5 Year		Regular	Under water		Principal			
6 Year				Principal		Principal	Major	Major
Other	Damage		Damage, In-depth, Special	Special	Verification, Project	Users Report, Special	Special	Ad Hoc

Table 17. Summary of inspection types and corresponding personnel qualification

Personnel	Taiwan	Japan	USA	UK	South Africa	Denmark	Sweden	Germany
Non-Certified Inspector	Regular			Superficial	Monitoring	Routine	Superficial	Superficial
Certified Agency			Routine					Minor
Inspector			-Routine -In-depth					
Engineer		Regular		Principal	Principal	-Economic -Special	General	Major

4 Conclusion

This paper briefly discussed the state-of-the-art for bridge inspection, evaluation and management methods being used to assess the performance of bridges in Taiwan and other countries. The findings outline the inspection types, condition description, and structural evaluation for bridge conditions.

It is observed from foreign practices that, with the exception of Japan and Taiwan, the combination of bridge complexity and inspector qualifications determines a matrix for inspection intervals. In this regard, bridge inspection certification scheme should be established to upsurge the competency of inspectors and engineers in Taiwan. Moreover, reformation of inspection level and frequency for inspections such as initial, underwater, damage, and special inspections should be considered in prospect researches in Taiwan.

Acknowledgments

This research is sponsored by the Institute of Transport, MOTC, Taiwan, ROC, under the project No. MOTC-IOT-105-PEB018.

References

[1] Chuang Y.H. and Yau N.J., A Big Data Approach for Decision Making in Bridge Maintenance. In *International Journal of Structure and Civil Engineering Research.* 5(3), page 216-222, 2016.

[2] Yau N.J. and Liao H.K., Visual Inspection and Evaluation for Special Bridges. In *Proceedings of International Symposium on Integrated Life-Cycle Design & Management of Infrastructures*, pages 195-205, 2007.

[3] Liao H.K. and Yau N.J., Development of Various Bridge Condition Indices for Taiwan Bridge Management System. In *Proceedings of the 28th ISARC, Seoul, Korea,* page 911-916, 2011.

[4] Masahiro S. and Takashi T. Bridge Inspection Standards in Japan and US. In *Proceedings of the 29th US - Japan Bridge Engineering Workshop*. Tsukuba, Japan, 2013.

[5] Yokoyama, Kazuaki, et al. Development of Bridge Management System for Expressway Bridges in Japan. In *Proceedings of the 3rd International Conference on Bridge Maintenance, Safety and Management*, pages 687-693, Porto, Portugal, 2006.

[6] *National Bridge Inspection Standards (NBIS),* Code of Federal Regulations, chapter 23, Part 650, Subpart C, U.S. Government printing Office, Washington, D.C., USA, 2004.

[7] Ryan T., Hartle A., Mann J., and Danovich J., *Bridge Inspector's Reference Manual (BIRM)*, Volume 1(6). U.S. Federal Highway Admin, Washington, DC 20590, 2012.

[8] Struzinky M. Bridge Inspection Manual, On-line: http://on.ny.gov/2pU5zUO, Accessed: 09/05/2017.

[9] Ryall M., *Inspection and condition rating, In Bridge Management*, Volume 1, pages 28-84. Butterworth-Heinemann, Oxford, UK, 2001.

[10] Bridge condition indicator guide TM-4008, On-line: http://bit.ly/2qhUj5Y, Accessed: 09/05/2017.

[11] Hearn, G., *Bridge preservation and maintenance in Europe and South Africa*, Volume 1. Federal Highway Admin, Washington, DC 20590, USA, 2005.

[12] Nsabimana P. *A Method for prioritization of Concrete Bridge Inspections in South Africa*, University of Stellenbosch, South Africa, 2015.

[13] Everett T., et al. *Bridge evaluation quality assurance in Europe*, volume 1. Department of Transportation, Washington DC, USA, 2008.

[14] Gruber L. *A Bridge Inspection Manual*, volume 1. Swedish National Road Admin., Road & Traffic Management Division, Jonkoping, Sweden, 1996.

[15] Chase S., et al, *Synthesis of national and international methodologies used for bridge health indices*. Federal Highway Admin., Washington, DC 20590, USA, 2016.

[16] Hearn G. *Bridge Inspection Practices – A Synthesis of Highway Practice,* volume 1. Transportation Research Board, Washington, DC 20001, 2007.

Evaluating modularisation tools in construction

T.P.Y.Wee [a]; **M. Aurisicchio** [a]; and **I. Starzyk** [b]

[a] Imperial College London
[b] Laing O'Rourke
E-mail: t.wee15@imperial.ac.uk

Abstract

The construction industry is undergoing a new wave of industrialisation driven by the shift towards off-site construction and the growing interest in mass-customisation of building systems. On one side the industry is acquiring new manufacturing and automation capabilities. On the other, it has a need to understand how to develop flexible yet efficient product systems that can be more adaptable to changing situations. To address this problem the industry is starting to adopt modularisation strategies. Modularisation is useful to handle product variations and effective at reducing redesign work. Research on the application of modularisation tools in construction has been limited. There is a need to examine how modularisation tools can be used to meet construction objectives and generate efficient solutions. This paper examines the application of three modularisation tools, namely the dependency structure matrix (DSM), the modular identification matrix (MIM), and the generational variance index (GVI).

A case study of a modular plant-room has been used to determine the effectiveness of the three tools at addressing fifteen modular design drivers. Each of the tools was found to address a different set of modular drivers. DSM and GVI provide rigorous solutions but they address only specific modular drivers. MIM has a more holistic approach to modularity and supports a wider set of modular drivers. However, it lacks technical rigour in determining modules. Overall, DSM and GVI, compared to MIM, are more technical tools to handle modular drivers.

This research sugests that the three tools can be used in an integrated manner. MIM can be used in the early phases of modular construction management because of its ability to capture multiple and interdisciplinary modular drivers. The application of DSM and GVI should be considered for more robust solutions in respect to their individual modular drivers.

Keywords –
Systems Engineering, Mass-customisation, Modularisation, DSM, GVI, MIM, Construction.

1 Introduction

The construction industry is experiencing further industrialisation driven by the shift towards off-site construction and the growing interest in mass-customisation. On one hand the industry is acquiring new manufacturing and automation capabilities [1], [2]. On the other it has a need to define effective, flexible and efficient product systems that are adaptable to rapidly changing requirement conditions imposed by clients, technological development, business considerations and other corporate reasons. In response to this challenge, the contruction industry is adopting modularisation strategies.

Modularisation is the clustering of different product sub-systems or components into modules to increase the flexibility of a product system [3]. It is useful to handle product variations and is effective at reducing redesign work. Modularisation enables quicker and easier reconfiguration of products to meet customised demands without massive alterations of the product or production operation. Modularisation is typically supported by the use of product configurators, i.e. software tools, which help select and configure existing components to develop new products. Despite the value of modularisation tools, research on their application to achieve further efficiency in construction has been limited [4]. There is a need for further research in construction to understand how to manage product variations and achieve cost efficiency.

Despite the existence of multiple modularisation tools, a challenge is to determine the most effective ones at supporting efficient management of product variations in construction. This paper explores three modularisation tools, namely the dependency structure matrix (DSM), the modular identification matrix (MIM) and the generational variance index (GVI). This research is part of a larger project which aims at developing a framework to handle modular building system design including application of Quality Function Deployment (QFD) tool [5]. The research is based on a case study of a modular plant-room design and has been conducted in collaboration with engineers at Laing O'Rourke.

2 Modularisation in construction

Modularisation methods are well documented and have been frequently applied in many industries [6], [7]. The definition of modularisation varies in the literature. This paper refers to modularisation as the clustering of product sub-systems for the formation of a module (i.e. product subsection). Modularisation is useful to support mass-customisation, which aims at meeting the needs of individual customers by facilitating high product variety. To meet these needs manufacturers tend to implement more efficient and flexible product designs and manufacturing strategies [8], [9]. Modular design can support mass-customisation through assembling various modules to develop a spectrum of products [10], [11]. Modular design integrates the advantages of standardisation with that of customisation. However, additional emphasis on performance optimisation for the development of high quality modular systems is typically needed [4].

The potential benefits of modular construction systems have been well documented in the literature. These benefits are associated with the use of offsite prefabrication technology that can support manufacturing of large quantities of volumetric building units under a stable factory controlled environment. These benefits are related to increased production efficiencies and shortened project lifecycles [7], [12]. This is a result of the application of modern manufacturing technologies, automation and standardization. In addition, modularisation supports the reduction of product design risk and minimises the potential impacts associated with future changes in business requirements [13]. If a product is highly modularised, it is easy to be assembled, disassembled and recycled [12].

There are many modularisation tools available, e.g. the "functional flow block diagram" [14], [15], the "dependency structure matrix" (DSM) [15], [16]. "extended implementation structure matrix" [17], [18], the "modular identification matrix" (MIM) [11], "axiomatic design" [19], and the "modular product platform" via the "generational variance index" (GVI) [20], [21]. This paper focuses on DSM, MIM, and GVI, as these are commonly used in other engineering disciplines (e.g. mechanical engineering, robotics, automotive, and consumer electronics).

The dependency structure matrix (DSM) is a method of mapping systems interdependencies in matrix form. It is used in systems engineering and project management to model and analyse complex systems. The DSM can be used for the analysis of product architectures and engineering processes [15], [17], [20]. It allows to use sequencing or clustering algorithms to organise system sub-systems.

The modular identification matrix (MIM) is a QFD-like tool that is used to identify which product sub-systems should be clustered into modules [11], [17]. It maps modular drivers against product sub-systems. The MIM mapping provides a visualisation of the interrelationships between modular drivers and product sub-systems. This visualisation supports implementation of modularisation rationale with respect to modular drivers [11], [17].

Modular product platforms can be a highly effective method for dealing with variant product design and uncertain future product requirements. It consists of clustering common product sub-systems that reoccur across a product family [22]. This approach to design has been successfully adopted in other industries, e.g. automotive [6]. The approach reduces costs associated with product development by using a handfull of platforms to create a variety of product families [23], [24], [25]. It also provides the benefits of reducing manufacturing and design cost as each module has only a few unique features that need to be redesigned each time [4]. More recently it has been considered as a strategy for dealing with building modules [4], [26], [27].

The generational variance index (GVI) is a tool that has been used for the development of product platforms. GVI supports the identification of product sub-systems which are less likely to require redesign [28]. This tool takes advantage of the occurrence of core functional requirements that reoccurs across multiple products to develop a common product platform. As a result, a common platform can relieve flexibility requirement on the production line. This tool requires the development of QFD, which is a series of interrelated matrices that map first "customer requirements" to "engineering metrics" and then to "product sub-systems". GVI is calculated by summing up the potential redesign work as a result of changes in the product requirements [21]. GVI is an indicator of the amount of redesign required for future product designs.

To date there has been limited research on the use of modularisation tools to support the design of building systems [4]. One example is the work of Gilbert III et al. (2013) who have used axiomatic design and product platform design for the development of modules for temporary modular buildings [4]. The methodology adopted by these researchers suggests that modules can be developed through grouping an overall system's common functional requirement and physical design parameters.The methodology categorises modules into common and specialist modules. The essential function of buildings is captured by core modules, which basically act as a studio module and additional required features are designated to the specialist modules.

Another example is offered by Veenstra (2006), who attempted to tackle modularisation and platform issues in the housing industry [27]. This work used GVI together with coupling indexes (CI) to identify residential house product sub-systems that could be turned into modules or platforms. It emphasizes that GVI and CI together would support a better understanding of external design forces. This research follows the decision rules set by Martin and Ishii (2002) for determining modules and platforms [29]. In particular, product sub-system with no or low GVI are to be turned into fully or partially standardised platforms. Product sub-systems with low "coupling indexes–supply" (CI-S) are to be considered for higher levels of modularisation. This work approaches modularisation and platform design by tackling product uncertainty and risks. It demonstrated the benefits of using GVI and CI as tools for modular platform development in construction.

3 Methodology

A case study was carried out to determine the suitability of different modularisation tools to support the building design process. Three different tools, namely DSM, MIM and GVI were implemented to modularise a plant-room (see Figure 1). Each tool was evaluated by determining its effectiveness at tackling modular drivers. The methodology of the study involved: 1) collection of information about the product case to be studied; 2) compilation of modular drivers; 3) implementation of three modular tools to the plant-room design; and 4) evaluation of the three tools to establish which of them would best satisfy the modular drivers.

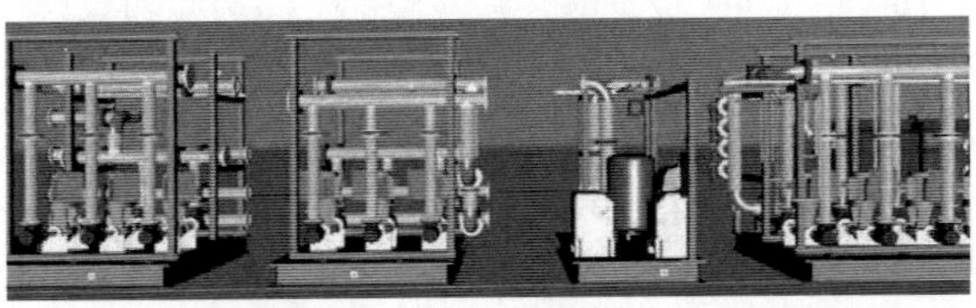

Figure 1: Modular chilled water plant-room (Source: Laing O'Rourke 2016)

Collection of product case information: In order to implement the modularisation tools, knowledge about the plant-room product was required. This was acquired through a reverse engineering process, which comprised examination of existing product documentation and discussions with engineers from the collaborating company. Various product documents were examined including product manuals, CAD files, schematics and bills of materials. A simplified version of the product schematic is illustrated in Figure 2. The product schematic illustrates the mechanical characteristics of the plant-room. A total of 14 product sub-systems are displayed including "building connection", "filtration", "dossing pot S1", "pump S1", "pump S2", "3 way valves", "pressure control S1", "degasser S1", "heat exchanger", "pressure control P1", "chiller connection", "degasser P1", "pump P1" and "dossing pot P1". It is noteworthy that the schematic excludes two product sub-systems, i.e. "control panel" and "structure".

Fifteen informal discussions were carried out with different groups of engineers from the collaborating company to investigate the plant-room design processes and to validate emergent understanding of their engineering operations. Each of these discussions lasted between 15 and 120 minutes. The discussions took place at the collaborating company's main design offices and factories as part of periodic visits and during a week secondment. The experts interviewed included mechanical engineers, design engineers and systems engineers who regularly work on plant-room products.

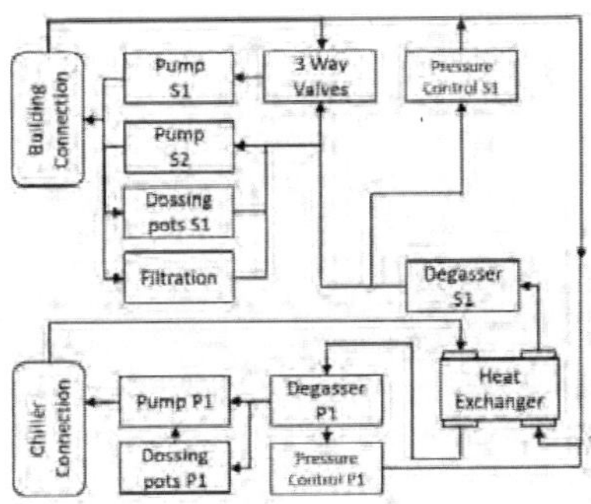

Figure 2: Simplified plant-room schematic

Compilation of modular drivers: A list of fifteen modular drivers was produced based on a review of the academic literature [11], [15], [17], and informal discussions with industry experts. Two types of modular drivers were identified: generic and construction specific. Generic modular drivers include *technical specification*, *styling*, *carry over*, *product planning*, *technology evolution*, *process specification*, *common unit*, *manufacturing*, *separate testability*, *purchasing*, *maintenance*, *product upgrading*, and *recycling*, while construction specific modular divers comprise *transportation* and *architectural restrictions*.

Implememtation of the three tools: Design Structure Matrix (DSM): A component-based "function" DSM model was developed using the Cambridge Advanced Modeler software 2014 [30]. Two pieces of information were utilised to build the model: material flows and spatial preferences. The former was collected directly from the product schematic (see Figure 2). The latter is based on safety and maintenance information as well as operational preferences and was elicited from engineers in the collaborating company.

Modularisation Identification Matrix (MIM): The MIM model involved mapping the modular drivers against the product sub-systems extracted from the plant-room schematic. Information on the relationships between the modular drivers and product sub-systems was determined during the collection of product information (see section 3) and entered in the MIM model.

Generational Variance Index (GVI): As part of work which is not reported in this paper but is part of this project a QFD model was developed for the plant room. The model maps non-functional requirements to functional requirements and then to product sub-systems using QFD1 and QFD2 matrices [5]. The QFD2 matrix which maps the functional requirements to the product sub-systems was used for the generation of the GVI. Information for the matrix was collected from interviews with engineering from the collaborating company.

Preliminary evalution: Two engineers in the plant-room design team of the collaborating company were interviewed in March 2017 to evaluate and rank the modular designs emerged from the application of the three tools. The engineers interviewed have 15 and 25 years of experience respectively. Both interviews took place at the collaborating company and lasted for approximately half an hour and an hour respectively. A questionnaire with open ended questions was prepared and used to guide the interviews.

The modular designs obtained through the three modularisation tools were also compared against a modularised reference model developed by the collaborating company. The variable used for comparison between the former and the latter is the number of identical "sub-system to sub-system relationships" (SSR) within a distinct module.

4 Case study: plant-room modularisation

This section of the paper presents an application of the three modularisation tools.

4.1 Dependency Structure Matrix (DSM)

DSM addresses technical specification as a modular driver. It also considers maintenance issues but not robustly. DSM clusters product sub-systems based on their dependencies (see Figure 3). High dependency amongst product sub-systems means that the sub-systems have high functional reliance on each other. Therefore, it would be beneficial if they were clustered together into a module. This would allow the module to hold a section of the product functionalty and address the technical specification.

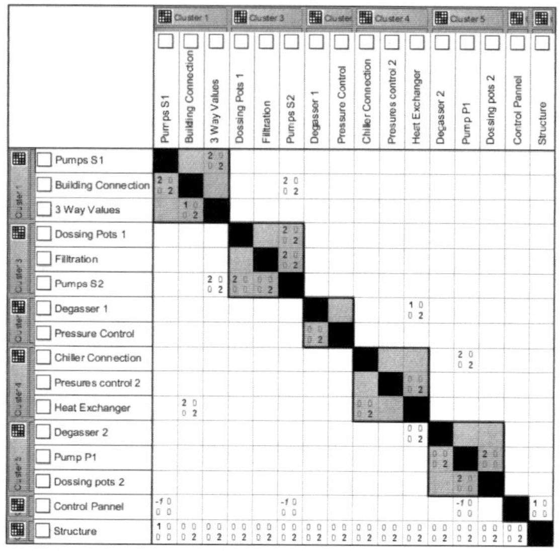

Figure 3: Dependency Structure Matrix

The DSM model was built using a total of 16 product sub-systems, which were mapped on to themselves capturing the plant-room's material flows and spatial preferences (see Figure 3). Sub-system to sub-system dependencies were labelled on a scale of 2 to -2, where 2 signifies a required dependency and -2 implies a detrimental relation. The material flows were established from the product mechanical schematic. The spatial preference is based on safety, operational and maintenance considerations. The CAM culturing algorithm was used to cluster sub-systems into modules based on the input dependencies. The partitioning feature was then used to determine the ordering of the modules. As can be seen in Figure 3, seven modules were recommended by the DSM tool with two modules composed by a single product sub-system (i.e. control panel and structure).

4.2 Modular Identification Matrix (MIM)

MIM has a more holistic approach to modularity and supports as many modular drivers as a user wants. It also offers a platform for determining how the various modular drivers interact with each another. However, it lacks technical rigor and relies heavily on judgment in determining modules.

The MIM tool was applied to map 15 modular drivers against the 16 product sub-systems. A simplified version of the MIM model can be seen in Figure 4.

Modular Drivers	Imp	Building connection	Pumps S1	Degasser 1	Dossing-pots 1	Pumps S2	3 Way-valves	Filltration	Pressure control 1	Chiller connection	Pumps P1	Dossing pots P1	Heat exchanger	Degasser P1	Pressure control P1	Control panel	Structure
Techical spesification	9		9			9					9		9				
Styling	1															9	
Carry over	3	9	3			3	3	3		9	3		3	3	3	9	
Product Planning	9		1			1					1						9
Techonolgy Push	3		1	1		1	1	9	1	1	1			1	1		
Production Process	3		9			9					9						
Common Unit	9	1	3	3	3	3	3	3	3	1	3	3				9	9
Manufacturing	9																9
Separate Testabiltiy	3		1	1	1	1	1	1			1	1	1	1	1	1	
Purchasing	1			9	9			9				9	9	9			
Mainatiance	9		3	9	9	3		9	9		3	9		9	9	9	
Product upgrading	1							9								9	
Recycling	3		3	3		3		9			3		9				
Transportation	9																9
Architectural	9	9								9							
	Grades	117	195	132	120	195	42	192	111	120	195	120	129	105	96	210	324
	M	M1	M1	M1	M1	M2	M2	M3	M3	M4	M4	M4	M5	M5	M5	M6	M7

Figure 4: Modularisation Identification Matrix

The tool identifies which modular drivers would influence different product sub-systems. Each modular driver is weighted, depending on its importance in compliance with the collaborating company's operations and views. The importance (I) and modular driver's influence (R) weightings are on a skewed scale, which favours more pressing modular drivers. This approach allows easier identification of stronger driving forces [11]. The values of the skewed scale are 9 (high importance), 3 (medium importance), and 1 (some importance). From the data it is possible to determine which product sub-systems best satisfy the modular drivers and therefore should form modular nucleuses. Modular driver satisfaction grades (MSG) can be calculated as highlighted in equation (1).

$$MSG_x = \sum_{i=1}^{n} I_i \times R_i \quad (1)$$

This sums the product of the modular diver's importance (I) and modular driver's influence" (R), where 'x' is the product sub-system number associated with the MSG value and 'i' is the row number associated with the specific product sub-system. Product sub-systems with higher MSG grades are identified as modular nucleus. The modular nucleuses are Pumps S1, Pumps S2, Chiller pumps, Filtration, Control panel and Structure, with MSGs of 195, 195, 195, 192, 210 and 324 respectively. The remaining product sub-systems are clustered around these modular nucleuses based on engineering judgement and rationale. The nucleuses are marked in grey in Figure 4. The three product sub-systems in black stripes in Figure 4 have been clustered clustered using rationale based on satisfaction of the common unit modular driver.

As shown in Figure 4 the MIM model has also yielded seven modules. Six modules are based on clustering around a neucleus and one on common unit rationale.

4.3 Generational Variance Index (GVI)

GVI is a metric tool that approximates the likelihood and potential rework needed for the next product evolution. It directly targets technology evolution as a modular driver. GVI can be used as a standardisation indicator as well as to address common unit as a modular driver. The development of a common unit relates to clustering the product sub-systems, which are least likely to change. As such, it supports the development of a product platform.

In this research the GVI model was generated through a modified version of QFD2 for the plant-room [5]. Due to its size, the QFD model was not included in this paper. However, Figure 5 illustrates the general layout and features of the model. The matrix maps functional requirements against product sub-systems in regards to the amount of redesign needed if the functional requirements are to change. GVI is traditionally calculated by equation (2) [21] but in this case study it was calculated by equation (3). The new method to calculate the GVI allows for a more comprehensive insight into determining the risk of redesign by summing the product of the change likelihood (C) and the redesign due to change (R), where 'x' is the product sub-system number associated with the GVI value and 'i' is the row number associated with the specific functional requirement.

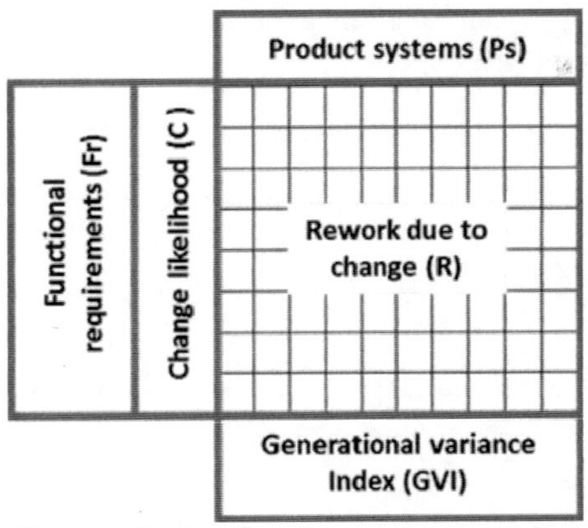

Figure 5: Quality Function Deployment Matrix 2

$$GVI_{x\,(Traditional)} = \sum_{i=1}^{n} R_i \quad (2)$$

$$GVI_{x\,(New)} = \sum_{i=1}^{n} (C_i \times R_i) \quad (3)$$

The GVI values for the plant-room are shown in Table 1. A high GVI value signifies that the product sub-system is likely to require redesign work. A low GVI value means that the product sub-system is less likely to require redesign work and is more suitable for standardisation. Therefore, product sub-systems with lower GVI values could be clustered to form a module platform as for example module 2 and 3 (M#2 and M#3). As shown in Table 1, developing a modular product platform using GVI has yielded seven modules.

Table 1. GVI values and module numbers

Product Sub-System	GVI	M #	Product Sub-System	GVI	M #
Chiller Connection	1	3	3 Way-valves	9	3
Dossing Pots P1	2	3	Control panel	10	6
Dossing Pots S1	2	2	Pumps P1	12	5
Degasser P1	6	3	Building Connection	14	1
Degasser S1	6	3	Heat Exchager	18	4
Filtration	7	2	Pump S1	26	1
Presure Control P1	8	3	Pump S2	26	1
Presure controle S1	8	3	Structure	40	7

4.4 Modularisation results and drivers

The implementation of the three tools resulted in as many modular designs. Table 2 shows the modular designs obtained through DSM, MIM, and GVI as well as the modularised reference model proposed by the collaborating company.

Table 2. Modular plant-room designs

	3 Way-valves	Pump S1	Dossing-pots S1	Pump S2	Degasser 1	Pressure Control S1	Pumps P1	Chiller Connections	Dossing Pots P1	Pressure Control P1	Degasser P1	Heat Exchanger	Building Connection	Filtration	Control Panel	Structure
DSM	1	1	2	2	3	3	5	4	5	4	5	4	1	2	6	7
MIM	2	2	1	1	1	3	5	5	5	4	4	4	1	3	6	7
GVI	3	1	2	1	3	3	5	3	3	3	3	4	2	5	6	7
Ref	2	2	1	1	1	3	4	4	4	5	5	4	1	3	6	-

The three tools approach modularisation from different perspectives and each addresses a set of modular drivers, see Table 3. It is noteworthy that in Table 3 the fifteen modular drivers are prioritised according to their relevance to the collaborating company's operations for plant-rooms. In addition a distinction has been made to indicate of a tool allows to achieve a modular solution through an algorithmic (A) or a judgment-based process (J).

Table 3. Tools fulfilment of modular drivers

	Modular Drivers	Importance	DSM	GVI	MIM
1	Technical Specification	6	A		J
2	Styling	2			J
3	Carry Over	4			J
4	Product Planning	5			J
5	Technology Push	4		A	J
6	Production Process	4			J
7	Common Unit	6		A	J
8	Manufacturing	6			J
9	Separate Testing	4			J
10	Purchasing	3			J
11	Maintenance	5	J		J
12	Product Upgrading	3			J
13	Recycling	4			J
14	Transportation	5			J
15	Architectural	6			J

A: Algorithmic process; J: Judgement-based process

5 Evaluation

The three modular plant-room designs were ranked by two engineers from the collaborating company in terms of functional dependency between sub-systems. Less consideration was given to the other modular drivers. The design produced through DSM and MIM were ranked first and second followed by that produced through GVI. DSM and MIM yielded agreeable results from the perspective of the plant-room design engineering team. In particular, the DSM result was found to be the closest to the engineers' modularisation mindset. This is because the approach to modularisation of the design engineering team is based on complimenting product functionality. These views highlight that functionally dependent sub-systems should be clustered together and other influencing considerations are of secondary importance. These views on functional dependence are in line with how DSM operates. While MIM addresses functional considerations, it does not address functionality as comprehensively as DSM. MIM also takes on other modular drivers. Further, GVI does not address functional considerationa at all and is therefore regarded as unacceptable for the purpose of designing plant-rooms.

Sub-system to sub-system relationships (SSR) were used to evaluate the three designs by comparing them to the company's reference model. The number of identical SSR within a distinct module indicates the degree of similarity between the result of a modular tool and the reference model. The design with the highest number of identical SSR to the reference model is deemed closer to the design put forward by the collaborating company. The results from this investigation indicate that the MIM based design has 10 out of 13 SSR identical to the reference model, DSM has 5 and GVI has only 3. This evaluation suggests that the MIM based design is distinctively closer to the company design compared to those produced with DSM and GVI.

6 Discussion

This research was undertaken to investigate how three modularionsation tools can be used to support the building design process. The research has distinguished the three tools in terms of the modular drivers they tackle and the process used to produce a modular solution, namely algorithmic or judgement-based process. The former refers to the ability of a tool to handle modular drivers from a technical perspective, while the latter to its reliance on subjective considerations.

The modular designs produced through the three tools were first ranked by experts in the collaborating company and then their similarity to the reference model was assessed based on connectivity information. The design engineers valued most the modular solution obtained through the DSM tool, whereas the comparison to the reference model showed the MIM tool produced the solution that is closer to the modularisation direction currently pursued by the collaborating company. This can be explained as follows. The DSM based solution best satisfies the modularisation objectives directly relevant to the role covered by the design engneers in the organisation, i.e. functional dependency between the product sub-systems. On the other hand the MIM tool has allowed to find the best trade-off between multiple modular drivers and taking into account interdisciplinary considerations.

This research suggests that neither considering multiple modular drivers in a subjective way as in MIM nor accounting for isolated modular drivers in a technical manner as in DSM and GVI is an optional solution. On their own, each of these tools exhibited limitations and the problem of tackling multiple modular drivers with a technical solution is not addressed. Multiple tools need to be considered to capture the full complexity of a modularisation problem. This research recommends to docus on the integration of complementary tools to generate more effective modular solutions. MIM should be utilised as the primary tool for modular construction management because of its ability to capture multiple modular drivers. It enables the gathering of issues arising from multiple disciplines in one system. The application of tools such as DSM and GVI should be considered for more robust solutions in respect to their individual modular drivers.

Limitations: A drawback of this preliminary evaluation is that it only considers the perspective of the plant-room design team. In this respect the study needs to incorporate the perspectives of other engineering teams such as those involved in quality control, manufacturing and architecture. These different teams are likely to provide different point of views.

7 Conclusion

This research compared three modularisation tools applied to a plant-room design process. The three tools were evaluated by determining their effectiveness in addressing modular drivers. Fifteen modular drivers were identified and prioritied. Each of the three tools addresses a different set of modular drivers and with a process that is either algorithmic or judgement based. MIM offers a more holistic approach to modularity and supports a wider range of modular drivers. However, it lacks technical rigor in determining modules. DSM and GVI provide technical solutions but each of them focuses on a specific modular drivers, namely DSM on technical specification and GVI on common unit and technology evolution. This paper concludes that DSM, MIM and GVI should be used in an integrated manner to tackle multiple modular drivers. As such, they would provide a more effective modularisation strategy in construction.

8 Acknowledgment

The authors would like to acknowledge the support for this research of the EPSRC and Laing O'Rourke through industrial CASE Award N:1589417.

9 Reference

[1] Marchesi M., Kim S. G., and Matt D. T. Application of the axiomatic design approach to the design of architectural systems: a literature review. In Proceedings of ICAD pp. 27-28. 2013.

[2] Höök M. Customer value in lean prefabrication of housing considering both construction and manufacturing. Proceedings of the 14th Annual Conference of the International Group for Lean Construction, Santiago de Chile. 2006.

[3] Borjesson F. and Hölttä-Otto K. A module generation algorithm for product architecture based on component interactions and strategic drivers. Research in Engineering Design, 25(1), pp.31-51. 2014.

[4] Gilbert III L. R., Farid A. M. and Omar, M. An axiomatic design based approach to civil engineering. In Proceedings of the 2nd International Workshop on Design in Civil and Environmental Engineering pp. 30-38. 2013.

[5] Wee T. P. Y., Aurisicchio M., and Starzyk I. The application of Quality Functional Deployment to modular offsite construction products. International Conference for Engineering Design (ICED), Vancouver, Canada, 2017. (Accepted).

[6] Gann, D. M. Construction as a manufacturing process? Similarities and differences between industrialized housing and car production in Japan. Construction Management & Economics, 14(5), pp. 437-450. 1996.

[7] Lawson R. M., Ogden R. G. and Bergin R. Application of modular construction in high-rise buildings. Journal of architectural engineering. pp. 148-154. 2012.

[8] Suh E. S., De Weck O. L., and Chang D. Flexible product platforms: framework and case study. Research in Engineering Design, 18(2), pp 67-89. 2007.

[9] Kreng V.B. and Lee T.P. QFD-based modular product design with linear integer programming—a case study. Journal of Engineering Design, 15(3), pp.261-284. 2004.

[10] Kohlhase N., and Birkhofer H. Development of modular structures: the prerequisite for successful modular products. Journal of Engeering Design, 7(3), pp 279-291. 1996.

[11] Erixon G. Modular function development mfd, support for good product structure creation. In DS 53: Proc. of the 2nd WDK Workshop on Product Structuring 1996, Delft University of Technology, the Netherlands, 03.-04.06. 1996.

[12] You Z. H., and Smith S. A multi-objective modular design method for creating highly distinct independent modules. Research in Engineering Design, 27(2), pp.179-191. 2016.

[13] Koh E. C., Förg A., Kreimeyer M., and Lienkamp M. Using engineering change forecast to prioritise component modularisation. Research in Engineering Design, 26(4), pp. 337-353. 2015.

[14] Emmatty F. J. and Sarmah S. P. Modular product development through platform-based design and DFMA. Journal of Engineering Design, 23(9), pp. 696-714. 2012.

[15] Hölttä-Otto K. Modular product platform design. Helsinki University of Technology. 2005.

[16] Ulrich K. T. and Eppinger, S. D. Product design and development. New York: MacGraw-Hill. 2008.

[17] Borjesson F. A systematic qualitative comparison of five approaches to modularity. 11th international design conference, design 2010, Dubrovnik, Croatia, May 17–20, vol 2. 2010.

[18] Sellgren U., and Andersson S. "The Concept of Functional Surfaces as Carriers of Interactive Properties", International Conference on Engineering Design (ICED 05), Melbourne. 2005.

[19] Marchesi M., and Ferrarato, I. A. Addressing the adaptive customization of timber prefabricated housing through axiomatic design. Procedia CIRP, 34, pp. 199-205. 2015.

[20] Jung S., and Simpson T. W. An integrated approach to product family redesign using commonality and variety metrics. Research in Engineering Design, 27(4), pp. 391-412. 2016.

[21] Simpson T. W., Bobuk A., Slingerland L. A., Brennan S., Logan, D., and Reichard K. From user requirements to commonality specifications: an integrated approach to product family design. Research in Engineering Design, 23(2), pp 141-153. 2012.

[22] Meyer M. H., and Lehnerd A. P. The power of product platforms. Simon and Schuster. 1997.

[23] Pan W., Gibb A. G. and Dainty, A. R. Leading UK housebuilders' utilization of offsite construction methods. Building Research & Information,36 (1), pp. 56-67. 2008.

[24] Simpson T. W. Product platform design and customization: Status and promise. AI EDAM: Artificial Intelligence for Engineering Design, Analysis and Manufacturing, 18(01), pp. 3-20. 2004.

[25] Cuperus Y. Mass customization in housing an open building/lean construction study. In Proceedings of Dense Living Urban Structures International Conference on Open Building, Hong Kong, China Vol. 2326, pp. 113. 2003.

[26] Jensen P., Olofsson T., Smiding E., and Gerth, R. Developing Products in Product Platforms in the AEC Industry. In Computing in Civil and Building Engineering pp. 1062-1069. ASCE. 2014.

[27] Veenstra V.S., Halman J.I. and Voordijk J.T. A methodology for developing product platforms in the specific setting of the housebuilding industry. Research in engineering design, 17(3), pp.157-173. 2006

[28] Jiao J. R., Simpson T. W., and Siddique, Z. Product family design and platform-based product development: a state-of-the-art review. Journal of Intelligent Manufacturing, 18(1), pp 5-29. 2007.

[29] Martin M.V. and Ishii K. Design for variety: developing standardized and modularised product platform architectures. Research in engineering design, 13, pp. 213–235. 2002

[30] Cambridge, E. D. C. Cambridge Advanced Modeller. 2016.

RFID, Web-Based, and Artificial Intelligence Integration in Facilities Management

Chien-Ho Ko

Department of Civil Engineering, National Pingtung University of Science and Technology, Taiwan
E-mail: ko@mail.npust.edu.tw

Abstract –

Radio Frequency Identification (RFID) technology has become one of the most important technologies for improving management efficiency. This study enhances facilities maintenance by integrating RFID, web-based, and artificial intelligence techniques. A data management module is developed first to collect facilities maintenance data. A statistical module is established to graphically display the collected data. A prediction module is developed using fuzzy neural networks to periodically maintain the facilities. These three modules are integrated into a web-based intelligent RFID facilities management system. The applicability of applying RFID technology to facilities maintenance is studied using environmental challenges. This research validates the developed system's performance using a real case. Experimental results show that RFID technology is appropriate for the practical maintenance environment. The proposed RFID system can improve facilities management efficiency, data transfer, data quality, malfunction prevention, service process time and planning.

Keywords –

Facilities Management; Radio Frequency Identification (RFID) Technology; Web-Based System; Artificial Intelligence

1 Introduction

The facility operating condition directly affects occupant satisfaction. However, facilities functions deteriorate with time. A well-tuned maintenance program is required to extend the service life of constructed facilities [1,2]. Facility functions and their equipment depend on continuous maintenance to maintain normal functions. Assigning an ID to each facility and apparatus is the regular way to manage facilities. Bar codes seem to be the most frequently used system for acquiring the information. However, bar codes printed on paper are easily broken. In addition, obstacles between the bar code and laser scanner can interrupt optical communications. These problems limit the efficiency in facilities maintenance [3,4].

Radio Frequency Identification (RFID) technology is characterized by repetitive read/rewrite, non-contact access to multiple tags. This technology has become one of the most important management systems in recent decades [5,6]. Numerous studies can be found in facilities management. Cardellino and Finch [7] surveyed innovations in facilities management and pinpointed that RFID is a promising information technology (IT) in the field. Wing [8] investigated RFID technology applications in construction and facilities management. Legner and Thiesse [9] applied the same technology to maintenance at Frankfurt Airport. They integrated RFID with a mobile application into its asset management systems. Ergen et al. [10] overcame difficulties encountered in facilities maintenance using RFID technology. Limitations in data transfer between maintenance workers were improved by this work. Although previous studies explored the promise of applying RFID technology to facilities maintenance, a comprehensive system integrating the required management technologies has rarely been discussed. How does the RFID technology work with the practical maintenance environment? In addition, tremendous maintenance data are collected and created while implementing maintenance jobs. How are the collected records analyzed through the assistance of RFID technology has rarely been fully addressed [11].

The primary objective of this study is to enhance facilities management efficiency by integrating RFID, web-based, and artificial intelligence. RFID technology is investigated first. Three modules: data management, statistical, and prediction modules are established according to the facilities management needs. To understand the performance in a practical maintenance environment, experiments were conducted with RFID devices using environmental challenges. Finally, the proposed system efficiency is examined using an actual building case. Application results and conclusions are documented in this paper.

2 RFID TECHNOLOGY AND FUZZY NEURAL NETWORKS

2.1 Radio frequency identification technology

RFID technology is a wireless sensor technology first proposed by Stockman in 1948 [12] that fulfills the need to identify useful applications for ''reflected-power communications.'' One of the first large-scale commercial uses was documented in the 1990s in electronic toll collection on US highways in Texas, Oklahoma, and Georgia [13]. Since then, RFID has been used across multiple industries and continued to advance in functionality and capability. Today RFID is a generic term for technologies that use radio waves to automatically identify people or objects.

A typical RFID system is shown in Figure 1. In the figure, radio signals are emitted by the transceiver through the antenna. A transponder (RF tag) is activated and the data on it are read and written by the requested signals sent from the antenna. The RF tag transfers data according to requests sent from the transceiver. The transceiver is responsible for acquiring the data when the transponder returns signals. The data can consequently be transferred to any computer system for processing [14,15].

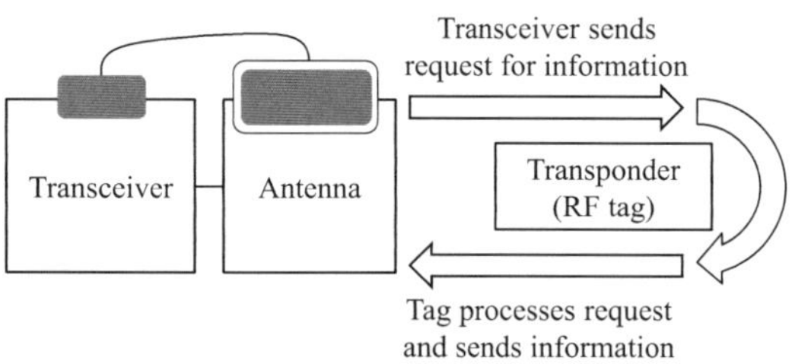

Figure 1. Schema of typical RFID system

2.2 Fuzzy Neural Networks

Fuzzy logic systems (FLSs) simulate the high level human decision making process, which models imprecise modes of reasoning to make rational decisions in an uncertain and imprecise environment. In general, FLSs contains four major components: fuzzifier, inference engine, rule base and defuzzifier [16], as shown in Figure 2. In spite of FLSs advantages over traditional approaches, it is not easy to acquire appropriate and complete fuzzy rules in applications (Maier et al., 2000). Artificial Neural networks are treated as complementary techniques to overcome this difficulty [17].

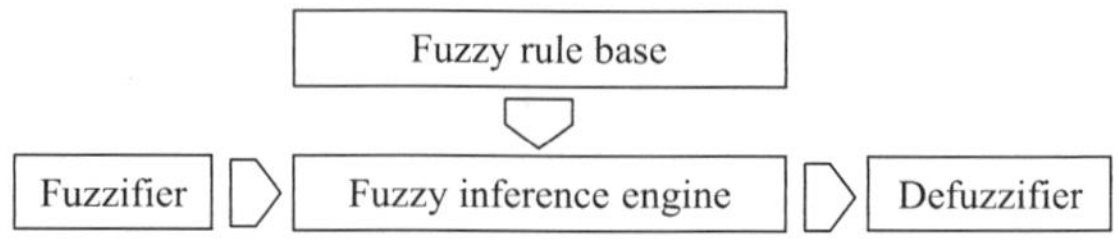

Figure 2. Control blocks of fuzzy logic system

Artificial neural networks are massively parallel distributed processors made up of simple processing units (neurons), which perform computations and store information [18]. Human brain modeling functions provide an alternative approach to conventional methods. A typical multi-layer feed forward network, as shown in Figure 3, consists of three types of layers: input layer, hidden layers and an output layer. Combining fuzzy logic and neural networks into an integrated model is a promising path towards the development of intelligent models capable of capturing qualities characterizing the human brain [19].

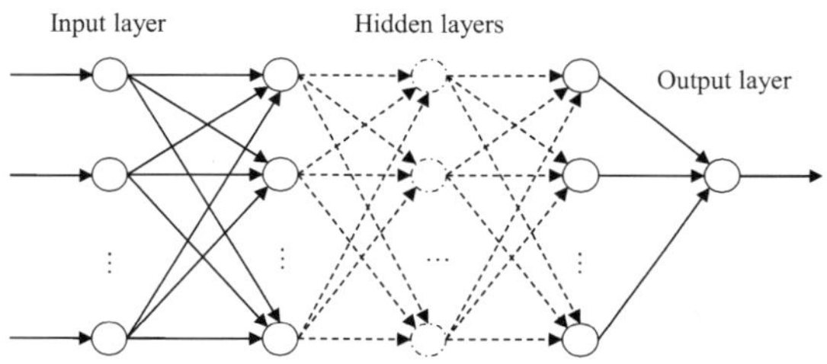

Figure 3. Multi-layer feed forward networks

3 RESEARCH METHODOLOGY

The web-based intelligent RFID facilities maintenance system development could be divided into three steps. First, hardware and software requirements are identified. System architecture is then elaborated. The web-based system uses application software that can be executed through the Internet on web pages. The conceptual architecture of the facilities maintenance system consists of three tiers, namely: storage, application logic and presentation. The storage tier is used for storing system data such as basic facilities information, maintenance history, seller information, etc. The application logic involves system modules and database connections. The presentation tier is used to interact with the application logic layer. Finally, the facilities maintenance system is constructed. The RFID facilities maintenance system includes three modules, i.e. data management module, statistical module and prediction module. The data management module is designed for users to manipulate data stored in the system database. The statistical module is established to help users understand the data distribution and properties. A prediction module is developed to estimate malfunctions before the next maintenance.

4 WEB-BASED RFID FACILITIES MANAGEMENT SYSTEM

4.1 Hardware and Software

Facilities operating conditions refer to functions provided by facilities themselves and their apparatuses. Therefore, facilities maintenance discussed in this study includes facilities and apparatuses. Portable RFID devices (reader/writer) are adapted to conform to the mobile requirement. RFID devices are chosen by considering the weight, transmit power, size, interface, price, and frequency range. Ensyc RFID Block shown in Figure 4 is thus selected. Gen 2 RFID tag displayed in Figure 4 on Ensyc RFID Block is selected with corresponding communication protocol.

To allow multiple users to use the system, a web-based application is planned. ASP.NET is chosen as the development platform. Microsoft ASP.NET can integrate the selected RFID hardware with web application. It supports browsers in executing on various kinds of platforms including the personal computer, cell phone, personal digital assistant (PDA), Palm, etc. Users can manipulate these systems using all kinds of hardware and software platforms.

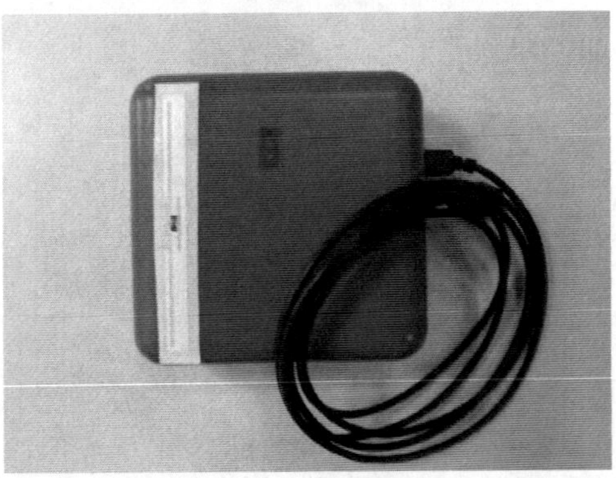

Figure 4. Selected RFID hardware and Gen 2 RFID tag

4.2 Data Management Module

Maintenance jobs are executed by staff that carry tablet PCs with portable RFID devices. Facilities and equipment are identified by the RFID device attached to the tablet PC. As shown in Figure 5, maintenance data were entered into the system using the web-based RFID system through a wireless Internet connection. The numbers of facilities and equipment may increase day by day. The developed web-based RFID management system connected to a single database through the Internet enables multiple users to implement maintenance work at the same time.

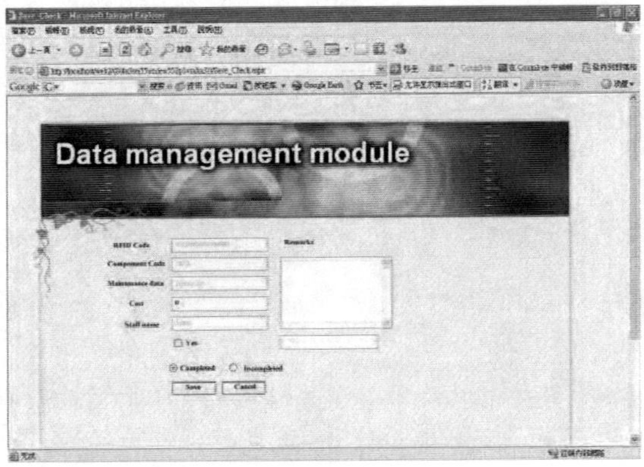

Figure 5. Data management module

4.3 Statistical Module

The proposed system updates the database in real time, which avoids data re-typing, duplicate maintenance and missing maintenance activities. Data stored in the database can be analyzed using statistical graphs from anywhere at any time with up-to-date distributions. A graph summarizing the purchasing cost is shown in Figure 6.

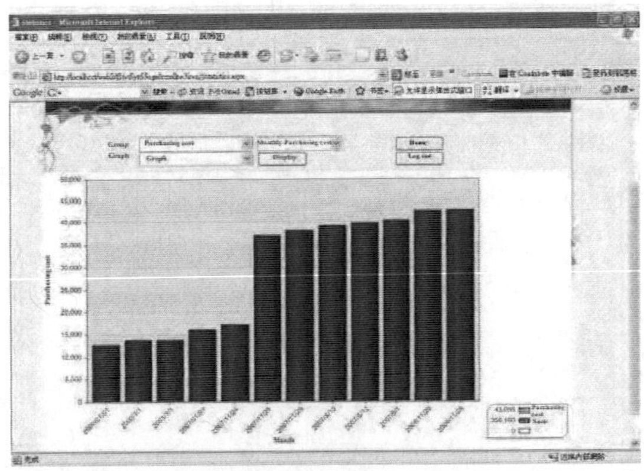

Figure 6. Statistical module

4.4 Prediction Module

In addition to periodically maintaining facilities, the proposed system actively forecasts the possible lifetime of components through maintenance records, as shown in Figure 7. Users can establish a forecast for a specific facility and modify the project with corresponding requirements. Facilities functions can therefore be ensured before the next maintenance.

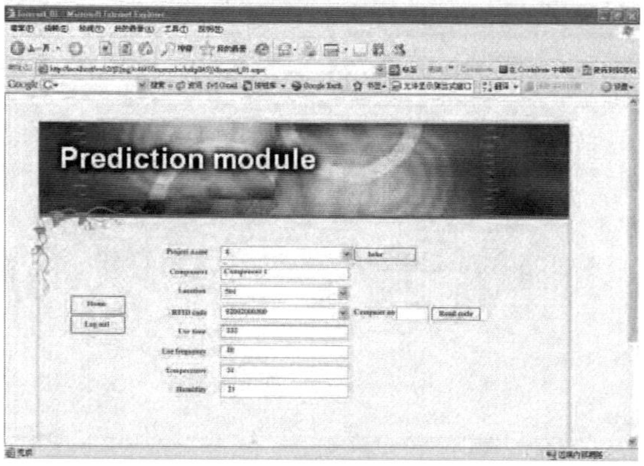

Figure 7. Prediction module

A fuzzy neural network is employed to develop a prediction model to overcome the difficulties faced in component lifetime evaluation. In this module fuzzy logic is used to represent the uncertain information and execute approximate reasoning. An artificial neural network is used for representing fuzzy rules. Complex relationships between inputs (influencing factors) and output (component lifetime) mappings are identified through learning algorithms. The fuzzy neural network architecture used in the prediction module comprised of four layers, as shown in Figure 8. The first layer is an input layer that receives the input data features and distributes them to the next layer (fuzzification layer). The fuzzification layer converts crisp inputs into fuzzy values using membership functions. Neurons at the intermediate layer process the fan-in signals and then perform an activation function. The output layer processes the fan-in signals and produces outputs. Because the prediction module infers a single result (component lifetime), the layer has one neuron.

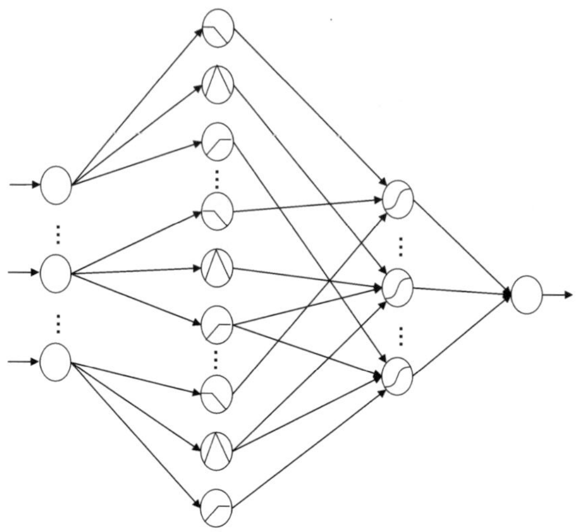

Figure 8. Fuzzy neural network architecture

5 EXPERIMENTS

The effective ranges of RFID devices are influenced by environmental factors [11,20]. To understand the environmental impacts on RFID devices, practical conditions encountered in facilities maintenance were provided. Nearby appliances, metallic equipment and dust conditions were thus tested. The experimental results are depicted in Table 1. The effective range of the selected RFID device is 10 cm.

Table 1 Effective range under normal conditions

Distance (cm)	Successful times	Unsuccessful times
8	5	0
9	5	0
10	5	0
11	0	5
12	0	5

RFID tags could be covered with dust over a short time span. To understand the impact of dust on the effective range, grime conditions were conducted. Table 2 demonstrates the experiment results. Dust and grime have little effect on the effective RFID tag range.

Table 2 Effective ranges under grime conditions

Distance (cm)	Successful times	Unsuccessful times
7	5	0
8	5	0
9	5	0
10	0	5

Facilities and equipment frequently consist of metal parts. Because metal reflects radio signals, it has been considered as the most challenging issue for RFID applications. This experiment examined how metal influences the effective range. Table 3 shows experimental results. Placing the RFID tag directly onto a metal surface is inappropriate.

Table 3 Effective ranges under metal conditions

Distance (cm)	Successful times	Unsuccessful times
0	0	5
1	3	2
2	5	0
3	5	0

Radio communications are influenced by nearby radio signals. Electrical equipment emits electromagnetic waves that affect radio communications. Consequently, the effect of electrical appliances on RFID signals was examined. A Personal Computer, regarded as one of the most popular electrical apparatuses, was used in this experiment. The experimental results summarized in Table 4 show that a minimum distance of 3 cm is required to avoid the impact from the electromagnetic waves emitted by electrical appliances like PCs.

Table 4 Effective ranges nearby running appliance

Distance (cm)	Successful times	Unsuccessful times
0	0	5
1	0	5
2	0	5
3	5	0
4	5	0
5	5	0

6 CASE STUDY

The performance of the developed web-based intelligent RFID maintenance system was examined using a building located in Taipei City, Taiwan. The structure was built in 2007 and designed for material experiments. The facilities defined in the study include the facilities and the related installed equipment. More than 12 kinds of laboratories were used involving valuable experimental instruments such as a universal testing machine, electron microscope, e-beam, spin coater, differential thermal analyzer, rapid thermal annealing machine, etc.

Maintenance jobs were executed by laboratory staff that carry tablet PCs with portable RFID devices. Facilities and equipment were identified by the RFID device attached to the tablet PC. The numbers of facilities and equipment in the study case increased day by day. The developed web-based intelligent RFID system connected to a single database through the Internet enables multiple users to implement maintenance work at the same time. In addition, to periodically maintain facilities, the proposed system actively forecast the possible lifetime of components through maintenance records. Users can establish a forecast for a specific facility and modify the project with corresponding requirements. Facilities functions can therefore be ensured before the next maintenance.

7 CONCLUSIONS

This paper presented a web-based RFID facilities management system. Three modules including data management, statistical, and prediction modules were established. A tablet PC attached with a portable RFID device was used to implement maintenance jobs. RFID performance was examined to verify the proposed system applicability in a practical environment.

This study integrated RFID technology with the Internet and a management information system to develop a web-based intelligent RFID facilities management system. The proposed system could automatically identify facility IDs, thus avoids typos and saves operational time. Unlike conventional barcode systems, data stored in RFID tags can easily be modified. In addition, RFID tags have been proven practical challenging conditions.

The developed web-based system can be implemented from anywhere, at any time under any platform with wire or wireless Internet connections. This characteristic enhances facilities management efficiency. The statistical module provides users with the opportunity to monitor facilities status using graphs. Maintenance data can thus be visualized. The developed prediction module enables users to predict the lifetime of a facility component using a fuzzy neural network. Timely inspection, repair, or replacement according to inference results can reduce malfunctions and unexpected breakdowns.

ACKNOWLEDGEMENTS

This research was funded by the Architecture and Building Research Institute (Taiwan) and National Science Council (Taiwan) (Grant no. NSC 99-2622-E-020-006-CC3), whose supports are gratefully acknowledged.

References

[1] Ko, C. H., Pan, N. F., and Chiou, C. C. Web-Based

Radio Frequency Identification Facility Management Systems. *Structure and Infrastructure Engineering*, 9(5):465-480, 2013.

[2] Barco, A. L. Budgeting for facility repair and maintenance. *Journal of Management in Engineering*, 10(4): 28-34, 1994.

[3] Ko, C. H. RFID-based building maintenance system. *Automation in Construction*, 18(3):275-284, 2009.

[4] Djerdjouri, M. Assessing and benchmarking maintenance performance in a manufacturing facility: A data envelopment analysis approach, *INFOR*, 43(2):121-133, 2005.

[5] Ko, C. H. 3D-Web-GIS RFID Location Sensing System for Construction Objects. *The Scientific World Journal*, 2013, Article ID 217972, 8 pages, 2013.

[6] Ngai, E. and Riggins, F. RFID: Technology, applications, and impact on business operations. *International Journal of Production Economics*, 112(2):507-509, 2008.

[7] Cardellino, P. and Finch, E. Mapping IT innovation in Facilities Management. *Electronic Journal of Information Technology in Construction*, 11:673-684, 2006.

[8] Wing, R. RFID applications in construction and facilities management. *Electronic Journal of Information Technology in Construction*. 11711-721, 2006.

[9] Legner, C. and Thiesse, F. RFID-based maintenance at Frankfurt airport, *IEEE Pervasive Computing*, 5(1):34-39, 2006.

[10] Ergen, E., Akinci, B., East, B., and Kirby, F. Tracking components and maintenance history within a facility utilizing radio frequency identification technology, *Journal of Computing in Civil Engineering*, ASCE, 21(1):11-20, 2007.

[11] Ko, C. H. Accessibility of Radio Frequency Identification Technology in Facilities Maintenance. *Journal of Engineering, Project, and Production Management*, 7(1):45-53, 2017.

[12] Stockman, H. Communication by Means of Reflected Power. *Proceedings of the Institute of Radio Engineers*, 36(10):1196-1204, 1948.

[13] Association for Automation Identification and Data Capture Technologies (AIM), Shrouds of Time: The History of RFID, 2002. On-line: http://www.aimglobal.org, Accessed: 30/03/2017.

[14] Ko, C. H. RFID 3D location sensing algorithms. *Automation in Construction*, 19(5):588-595, 2010.

[15] Domdouzis, K., Kumar, B., and Anumba, C. Radio-Frequency Identification (RFID) applications: A brief introduction, *Advanced Engineering Informatics*, 21(4):350-355, 2007.

[16] Oysal, Y., Becerikli, Y., and Konar, A. F. Modified descend curvature based fixed form fuzzy optimal control of nonlinear dynamical systems. *Computers and Chemical Engineering*, 30(5):878-888, 2006.

[17] Ko, C. H. and Cheng, M. Y. Dynamic prediction of project success using artificial intelligence. *Journal of Construction Engineering and Management*, ASCE, 133(4):316-324, 2007.

[18] Haykin, S. *Neural Networks and Learning Machines*, Pearson, New York, 2008.

[19] Keller, J. M., Liu, D., Fogel, D. B. Fundamentals of Computational Intelligence: Neural Networks, Fuzzy Systems, and Evolutionary Computation, John Wiley & Sons, New Jersey, 2016.

[20] Ko, C. H. Integrating RFID, Web-Based Technology, and Artificial Intelligence in Engineering Management. *Scientia Iranica*, 22(2): 299-312, 2015.

Performance Assessment of Residential Building Management Utilizing Network Data Envelopment Analysis

Wei Tong Chen[a], Pei-San Tan[b], Nida Fauzia[c] and Chao Wei Wang[d]

[a]Department of Civil and Construction Engineering, National Yunlin University of Science and Technology, Taiwan
[b]Division of Construction and Property Management, National Yunlin University of Science and Technology, Taiwan
[c]Division of Construction and Property Management, Department of Civil and Construction Engineering, National Yunlin University of Science and Technology, Taiwan
[c]Graduate School of Engineering Science and Technology, National Yunlin University of Science and Technology, Taiwan

E-mail: chenwt@yuntech.edu.tw, sandy@chung-li.com.tw, nida.091190@gmail.com, z20030221@yahoo.com.tw

Abstract –

Apartment buildings and residential buildings in Taiwan have become more and more community oriented. Precise control of residential buildings operational performance is increasingly important. This study uses network data envelopment analysis (Network DEA) to develop a three-phase based residential building operational performance assessment model and use it to assess the operational efficiency of 26 decision making units (DMUs) to obtain the performance index of dimensions. It is found that the model is equipped with three dimensions (with its weight) named staff quality (0.26), customer satisfaction (0.35) and operation performance (0.39) respectively. Most of the total DMU efficiency values fell between 0.75 and 0.90, with only one DMU is equipped with 1. DMU_6 (1.0000) is ranked first, with performance values for each dimension of 1. Aside from the staff quality dimension (0.8134), the other two dimensions for DMU_{10} are both below 0.80, indicating this DMU is most in need of improvement. Sensitivity analysis can show the impact of inputs and outputs on DMU performance.

Keywords –

Residential buildings; Performance assessment; Case management; Network data envelopment analysis

1 Introduction

Urban development and economic growth are driving demand for improved living standards and residential quality of life. In response, the broad development of property management services is realized. Such services are labor-intensive and low-cost, and price is the key consideration for service consumers. Property management companies are generally responsible for the management of multiple residential communities, and accurately assessing the effectiveness of community management performance is important to the formulation, implementation and assessment of overall management strategy. Effective objectives and evaluation models are still lacking for the assessment of community management performance. Most currently used approaches rely on regular or irregular assessments, recommendations of the community management committees (CMC) and contract renewal rates [1]. These methods are susceptible to subjective influence from company supervisors or CMC members, and do not necessarily provide an object reflection of actual operating conditions.

This study implements an expert-constructed questionnaire to assess operations management performance using network data envelopment analysis (Network DEA). The study seeks to (1) use the existing literature and measurement indices to establish input and output dimensions and administration factors and to assign proper weightings to each; (2) establish a Network DEA model to assess operational performance in 26 field cases and (3) identify relatively efficient and inefficient cases, and to offer proposals for improvements to business strategy as a reference to enhance overall business performance.

2 Network DEA

DEA is widely used in industrial assessments including sewage treatment plants [2], airports [3], telecommunications [4] and energy [5]. It has also been used to simplify evaluation data into a single performance value [6], thus facilitating overall operational assessments. DEA does not consider intermediate activities, but rather directly coverts a single production input indicator into an efficiency value, thus substantially neglecting resource utilization and departmental operations, leaving it unable to effectively determine the root causes of operational problems [7].

To overcome the shortcomings of DEA, Network DEA can be organized into multiple interrelated departments to identify the root cause of organizational performance inefficiency [8]. Network DEA offers many advantages, but it has a relatively short development history, and thus is unable to track usage restrictions as clearly as conventional DEA. For example, the number of DMUs in DEA should be at least twice the number of inputs and outputs.

Traditional DEA includes a variety of different modes, including CCR [9] and BCC (Banker et al., 1984). When accounting for multiple input and output decision-making units, traditional DEA cannot be used to discuss production processes and the influence of internal management activities [10]. Therefore, this study uses Network DEA.

Tone and Tsutsui [11] initially explored three interrelated divisions (Fig. 1), thus dividing an organization into three components. Links 1.2, 1.3 and 2.3 are production activities linking the three divisions. Link 1.2 is the partial output of Division 1, which is also the input of Division 2. Link 1.3 is the partial output of Division 1, which is also the input of Division 3. Link 2.3 is the partial output of Division 2, which is the input of Division 3. Traditional DEA requires each activity to be clearly classified as inputs or outputs, leaving it unable to handle intermediate production activities. In contrast, Network DEA does not put an organization's internal production processes in a black box, thus providing a transparent view of the efficiency of each division and insight into their various issues.

3 Data Analysis

Currently, Network DEA has no basic criteria for selecting DMUs. The traditional DEA approach which requires DMU homogeneity, where nonhomogeneous DMUs are deleted to ensure assessment accuracy [12]. This study uses the following criteria to select 26 homogenous cases in central Taiwan: (1) residential properties, (2) provide management services with general cleaning, security and logistical support without customized services, and (3) case inputs and outputs are roughly the same.

Network DEA input items can be used as outputs, while outputs can also be used as inputs. The process of establishing inputs and outputs via Network DEA is summarized as follows.

Step 1: This study uses KMP indicators from the Ministry of Economic Affairs' "Property Management Services Performance Indicators" to generate the 20 initial items.

Step 2: Interviews were conducted for each of the initial items to understand the ease of data acquisition and screening. Data items which were difficult to quantify or to obtain were deleted, to produce a list of 13 items.

Step 3: An expert-constructed questionnaire was established to assess the 13 inputs and outputs and to determine item suitability. The final list, including six inputs and six outputs, is shown in Table 1.

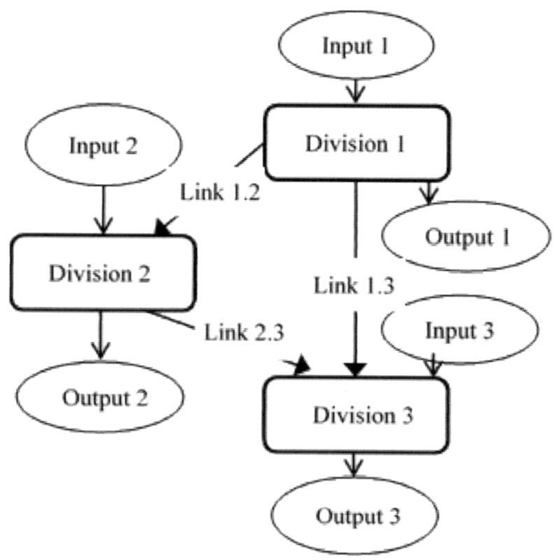

Figure 1: Connectivity division

The study uses case data, where "Management Committee Satisfaction" was based on a satisfaction questionnaire ranging from 5 (very satisfied) to 1 (very unsatisfied) and maximum score of 50. For each case, five management committee satisfaction questionnaires were distributed (for a total of 130 questionnaires), with the average value of all cases. Questionnaire content focused on overall satisfaction with management staff and other items related to service satisfaction. This study invited 13 experts in the field of property management to measure the degree of interaction between dimensions and to weight each dimension. The resulting weightings were staff quality (0.24), customer satisfaction (0.36) and operation performance (0.40).

Table 1 Step 3 final selection results

Dimensions	Inputs	Outputs
Staff quality	Staff training hours	Direct personnel costs
	Staff retention rate	CMC satisfaction
Customer satisfaction	CMC satisfaction	Contract price
	Staffing	Direct personnel costs
Operation performance	Direct personnel costs	CMC satisfaction
	Contract price	Staff retention rate

4 Empirical Analysis

4.1 Constructing a Network DEA model

Network DEA is divided into three modes: basic, vertical integration and segregated. However, the segregated model is more suitable for this present research. Figure 2 shows the Network DEA model developed according to the input and output characteristics.

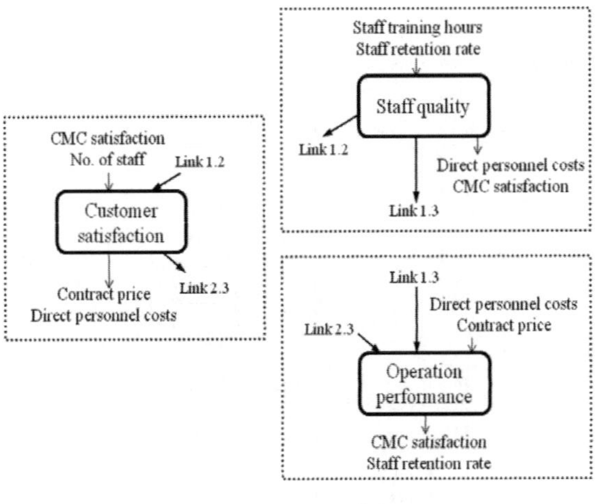

Figure 2: Segregated model

This study uses DEA-Solver Pro 7 to analyse Network DEA and assess the performance for each dimension (staff quality, customer satisfaction and operation performance) for each case, with a higher performance value corresponding to increased input and output values. After collecting the relevant expert opinions from the questionnaire, we calculated the structural weights for staff quality (0.24), customer satisfaction (0.36), and operation performance (0.40).

4.2 Total Efficiency Analysis

Comprehensive evaluation is based on the performance values of staff quality, customer satisfaction and operation performance (Table 2). The results are as follows:

1. Most total DMU efficiency values fell between 0.75 and 0.90, with only one DMU is equipped with 1.
2. DMU_2 was ranked first in terms of staff quality and customer service, but overall performance value was ranked 26th due to its operation performance dimensioin being only 0.3746.
3. DMU_6 is ranked first, with performance values for each dimension of 1.

Aside from the staff quality dimension (0.8134), the other two dimensions (customer satisfaction and operation performance) for DMU_{10} are both below 0.78, indicating this DMU is most in need of improvement.

4.3 Sensitivity Analysis

Sensitivity analysis was conducted to observe the variability for each dimension, and the sensitivity is examined for each case. Analysis was conducted each time an input or output item was deleted to provide insight into the variability for overall performance and each dimension. The variability and difference in variability is calculated by using Eqs. 1 and 2.

Rate of Change

$$= \left|\ \text{Original efficiency value} \div \text{Original efficiency value after deleting the item} - 1 \ \right| \quad (1)$$

Variance of Change

$$= \left|\ \text{Original efficiency value} - \text{The original efficiency value after deleting the item} \ \right| \quad (2)$$

Taking the total efficiency as an example to explain how the sensitivity analysis was performed. Each time an input or output was deleted to observe changes. Table 3 shows the CMC satisfaction had a considerable impact on overall performance and each of the three dimensions. Staff quality was highly sensitive to overall performance and customer satisfaction. Staff quality was highly sensitive to overall performance and customer satisfaction. Staff retention rate was found to be highly sensitive to overall performance, staff quality and operation performance. Contract price was sensitive to customer satisfaction and operation performance.

Table 2. Overall and dimensional efficiency for each DMU

DMU	Total operational efficiency	Ranking	Weighted average	Ranking	Staff quality (SQ)	Customer satisfaction (CS)	Operation performance (OP)
DMU_1	0.8856	7	0.9012	6	0.9633	0.7500	1.0000
DMU_2	0.5996	26	0.7498	22	1.0000	1.0000	0.3746
DMU_3	0.8255	11	0.8295	13	0.8425	0.7574	0.8866
DMU_4	0.7021	24	0.7431	23	0.8459	0.9001	0.5401
DMU_5	0.9545	2	0.9574	2	1.0000	1.0000	0.8936
DMU_6	1.0000	1	1.0000	1	1.0000	1.0000	1.0000
DMU_7	0.7924	15	0.8033	15	0.8448	0.6840	0.8857
DMU_8	0.8572	9	0.8754	9	0.9020	0.7192	1.0000
DMU_9	0.9311	3	0.9372	3	0.9792	0.8394	1.0000
DMU_{10}	0.7140	22	0.7236	25	0.8134	0.7726	0.6255
DMU_{11}	0.7719	17	0.7763	17	0.8555	0.7053	0.7927
DMU_{12}	0.7866	16	0.7930	16	0.9046	0.8012	0.7186
DMU_{13}	0.7108	23	0.7299	24	0.8228	0.8235	0.5900
DMU_{14}	0.8081	13	0.8359	11	0.8339	0.6549	1.0000
DMU_{15}	0.6889	25	0.7207	26	0.8488	0.8325	0.5432
DMU_{16}	0.8049	14	0.8243	14	0.8422	0.6695	0.9530
DMU_{17}	0.9123	5	0.9184	5	0.9167	0.8288	1.0000
DMU_{18}	0.8203	12	0.8351	12	0.9698	0.8890	0.7058
DMU_{19}	0.7192	21	0.7598	19	0.8718	0.9126	0.5550
DMU_{20}	0.8653	8	0.8781	8	0.8569	0.7569	1.0000
DMU_{21}	0.9262	4	0.9337	4	0.9850	0.8257	1.0000
DMU_{22}	0.7602	18	0.7731	18	0.8947	0.8201	0.6577
DMU_{23}	0.7441	20	0.7533	21	0.8675	0.7812	0.6596
DMU_{24}	0.7457	19	0.7570	20	0.8815	0.7910	0.6518
DMU_{25}	0.8398	10	0.8486	10	0.9255	0.7366	0.9033
DMU_{26}	0.8875	6	0.8996	7	0.9903	0.7652	0.9662
Maximum	1.0000		1.0000		1.0000	1.0000	1.0000
Minimum	0.5996		0.7207		0.8134	0.6549	0.3746
Means	0.8098		0.8291		0.9023	0.8083	0.8040
S.D.	0.0949		0.0804		0.0634	0.0966	0.1930

In order to investigate potential issues of DMU operation, the study invited five experienced practitioners of property management to analyze DMU from the perspectives of staff quality, customer satisfaction and operation performance. These practitioners were also asked to provide suggestions for the improvement of mentioned drawbacks. Table 4 summarizes the proposals for improvements to business strategy as a reference to enhance overall business performance of investigated DMUs.

Table 3. Sensitivity of Input and Output

Sensitivity / Item	High ⟹ Low 1st	2nd	3rd
Total effiency	Staffing	Staff retention rate	CMC satisfaction
Staff quality	CMC satisfaction	Direct personnel costs	Staff retention rate
Customer satisfaction	Staffing	CMC satisfaction	Contract price
Operation performance	Staff retention rate	CMC satisfaction	Contract price

5 Conclusions

Property management companies face significant challenges in effectively managing different properties. The Network DEA model is used to evaluate the performance of 26 cases and can provide useful reference for apartment building management firms.

Staff quality, customer satisfaction and operation performance all impact resigential management operation performance. Residential building management is service-oriented, thus we must fully consider the customer experience based on quality of service. Good staff quality can improve customer satisfaction and thus improve DMU performance. According to the researh results, most of the total DMU efficiency values fell between 0.75 and 0.90. DMU_6 (1.0000) is ranked first, with performance values for each dimension of 1. Aside from the staff quality dimension (0.8134), the other two dimensions (customer satisfaction and operation performance) for DMU_{10} are both below 0.78, indicating this DMU is most in need of improvement, especially for the operation performance dimension. Additionally, Sensitivity analysis can show the impact of inputs and outputs on case performance. Thus decision-makers should focus on relatively sensitive items when formulating operational strategy.

Table 4. Potential issues and proposed suggestions

Dimension	Potential issues	Suggestions for improvement
Staff quality	1.Heavy turnover and incomplete training result in inconsistent staff quality	1.Regular education and training
	2.Low quality workers lack "full-service" mentality, making it difficult to achieve overall improvement	2.Property owner selects primary service staff
	3.Large human resource investment requirements make it difficult to maintain comprehensive control	3.Investigate candidate qualifications and work attitudes
Customer satisfaction	1.Difficulty meeting expectations of multiple customer types	1.Focus on needs of majority of households
	2.Inconsistent community cultures	2.Indirect observation of community activities
	3.Contract renewal not assured despite overall household satisfaction	3.Establish a comprehensive response platform
Operation performance	1.Revenue does not reach contract value, and is dependent on the attributes of residential cases.	1.Model communities are less profitable
	2.Staff and management have different views on operational efficiency	2.Regular meetings between head office management and case managers
	3.Many factors affect operational performance including corporate policy, government regulations, industrial structure, consumption patterns and employment platforms	3.Effective integration of left column factors to reduce regulatory impact

References

[1] Chen, W.T. and Wang, C.-W. Project management performance assessment of condominium building – A case study. In *proceedings of the AMME 2015-CTUE 2015*, pages 768–771. Bangkok, Thailand, 2015.

[2] Ramón, S.G., Francesc, H.S., and María M.S. Assessing the efficiency of wastewater treatment plants in an uncertain context: A DEA with tolerances approach. *Environmental Science & Policy*, 18: 34–44, 2012.

[3] Wanke, P.F. Capacity shortfall and efficiency determinants in Brazilian airports: Evidence from bootstrapped DEA estimates. *Socio-Economic Planning Sciences*, 1–4, 2012.

[4] Bayraktar, E., Tatoglu, E., Turkyilmaz, A., Delen, D., and Zaim, S. Measuring the efficiency of customer satisfaction and loyalty for mobile phone. *Expert Systems with Applications*, 39(1): 99–106, 2012.

[5] Sueyoshi, T. and Goto, M. DEA radial measurement for environmental assessment and planning: Desirable procedures to evaluate fossil fuel power plants. *Energy Policy*, 41(1): 422–432, 2012.

[6] Boscá, J.E., Liern, V., Martinez, A., and Sala, R. Ranking decision making units by means of soft computing DEA. *International Journal of Uncertainty Fuzziness and Knowledge-Based Systems*, 19(1): 115–134, 2011.

[7] Yang, C. and Liu, H.M. Managerial efficiency in Taiwan bank branches: A network DEA. *Economic Modelling*. 29(2): 450–461, 2012.

[8] Despotis, D.K., Sotiros, D., and Koronakos, G. A network DEA approach for series multi-stage processes. *Omega*, 61: 35–48, 2016.

[9] Charnes, A., Cooper, W.W., and Rhodes, E. Measuring the efficiency of decision making units. *European Journal of Operational Research*, 12(6): 429–444, 1978.

[10] Chen, C. and Yan, H. Network DEA model or supply chain performance evaluation. *European Journal of Operational Research*. 213(1): 147–155, 2011

[11] Banker, R.D., Charnes, A., and Cooper, W.W. Some models for estimating technical and scale inefficiencies in data envelopment analysis. *Management Science*, 30(9): 1078–1092, 1984.

[12] Zhao, Y., Triantis, K., Murray-Tuite, P., and Edara P. Performance measurement of a transportation network with a downtown space reservation system: A network-DEA approach. *Transportation Research Part E: Logistics and Transportation Review*. 47(6): 1140–1159, 2011.

[13] Tone, K. and Tsutsui, M. Network DEA: A slacks-based measure approach. *European Journal of Operational Research*, 197(1): 243–252, 2009.

Condition-Based Building Maintenance Strategy Selection Using Constraint Programming

Shu-Shun Liu and Gilang Ramadhan

Department of Civil and Construction Engineering, National Yunlin University of Science and Technology, No 123,Section 3, University Road, Douliu, Yunlin, 640, Taiwan ROC
E-mail: liuss@yuntech.edu.tw g.ramadhan3103@gmail.com

Abstract-

Naturally, buildings will deteriorate during its life cycle. On the other hand, sustainability of building condition is the main objective of providing service for users. Therefore, building maintenance management should be a priority to achieve that objective. In Indonesia, there are many public buildings with different conditions and some of them are not in optimal conditions demanding urgent maintenance treatment. In addition, limited budget and absence of building maintenance decision-making tool are among factors causing unmanaged maintenance activities. Moreover, agencies must allocate maintenance budget and prioritize buildings maintenance manually. This paper proposes an optimization mathematical model which adopts Constraint Programming (CP) formulation to assist in selecting the most cost-effective strategy for particular building and determining budget allocation each year. CP is famous on its flexibility to problem declaration that makes scenario analysis conduct effectively. This CP-based model subjects to budget constraint and required condition level. In this case, budget, maintenance strategy options, building deterioration rate, time period, and minimum acceptable building condition are set. There are four maintenance strategy options available, Light Repair, Rehabilitation, Renovation, and Restoration which have different improvement impact for each strategy. Therefore, a case study is conducted to implement the model which involves 15 buildings in a 10-year time of period and two scenarios are presented in this proposed model. The result shows that the two scenarios allocate annual maintenance budget differently depending on diversities of planner management goals.

Keywords-

Building Maintenance, Condition-Based Maintenance, Maintenance Strategy, Optimization, Constraint Programming

1 Introduction

Building is one of important assets of an institution or organization. Different institutions have different types of building with various functions. Some of the buildings are offices, schools, sport centres, convention halls, and many more. These kinds of building aim to facilitate activities of owners and users. To give an optimal service, a building should be in a good condition. However, condition level of buildings may decline during its life cycle. Ultimately, a building needs a proper maintenance activity strategy to ensure its level of condition is within an acceptable degree.

Au-Yong et al. [1] stated that building maintenance is vital in ensuring the building's sustainability. Maintenance can preserve building's condition in receivable state of service. Maintenance is defined as a combination actions carried out to retain an item in, or restore it to, an acceptable condition [2][3]. Moreover, maintenance activities can lengthen age of buildings in providing service for owner and users.

However, in some parts of Indonesia, building maintenance has not been taken into account. Owner who have many buildings, in this case government, still do not consider maintaining building as an important issue. Due to limited budget and a huge number of buildings with different level of condition, building maintenance program is planned as needs arise. Consequently, many buildings are in bad condition and cannot provide optimal service to users. In other cases, some buildings are maintained but due to lack of maintenance management and planning, the maintenance requires high cost and inappropriate maintenance action is chosen.

This problem needs a solution to prevent worse building condition and higher maintenance cost spent. Some strategies can be implemented to overcome this problem. Condition-based maintenance can be applied to overcome the problem. Condition-based maintenance aims to minimize the total maintenance cost by collecting and gathering the condition data of the building systems, especially the critical components [1]. This maintenance strategy can determine the most suitable maintenance action for certain building in order to reach optimal condition and reduce cost of maintenance.

By applying constraint programming approach, this study aims to provide a simple decision-making tool to help buildings owners managing their buildings maintenance program. The proposed model is a strategy selection program using constraint programming. The objective of this model is to select the most cost-effective maintenance action for particular building in certain year to achieve acceptable level of condition for all buildings at the end of a maintenance plan period.

2 Literature Review

Maintenance optimization related topics are widely studied by some researchers although it is not directly linked with building maintenance. Particularly, the Markov prediction model is widely used to predict future pavement maintenance due to its ability to integrate rehabilitation and pavement deterioration rates in a single transition probability matrix [4]. Abaza et al [4] implemented a Markov model to predict future pavement performance and developed non-linear optimization method to establish optimum pavement condition throughout the network subject to budget constraints.

Another method used in network-level optimization topic is goal programming. This approach is favoured due to its strength in considering problem encompassing conflicting objectives with different degrees of importance [5]. Previous research conducted by Ravirala and Grivas [5] presented that goal programming is beneficial in obtaining conflicting objectives simultaneously.

Meanwhile, constraint programming is also commonly used to solve optimization problem on maintenance strategy. One of the previous works was conducted by Badr and Brown [6]. They proposed a cost-based reasoning and constraint programming model on building maintenance scheduling. Additionally, they introduced a heuristic ModReg model to guide the search process towards an optimal schedule.

Constraint programming is widely applied to deal with scheduling and planning problems in different areas. One of the examples is Chan and Hu [7] who employed constraint programming to propose precast production scheduling model that incorporates the key constraints and objectives considered by production scheduler. Another work was established by Brailsford [8] who discussed and compared the advantages of constraint programming method for solving constraint satisfaction problems, such as timetabling and roistering problem. Rodriguez [9] proposed a constraint programming model for the routing and scheduling of trains running through a junction to assist operator in changing train routes or order to avoid conflicts and delays.

Furthermore, Liu and Wang [10] presented an optimization model using constraint programming approach for project selection and scheduling problems with time-dependent resource constraint. In another work, constraint programming was also applied by Liu and Wang [11] to propose a profit optimization model for multi-project scheduling problems by considering cash flow and the financial requirements of contractor working in a multi-project environment.

3 Conceptual Model

This study aims to develop a decision-making tool in selecting building maintenance strategy. Constraint programming is used because it is flexible in problem declaration that makes scenario analysis conducted effectively. Constraint programming is not restricted to the linear equations, but allowing the users to change contents for achieving required model [12].

Furthermore, the proposed model intends to find the most optimal solution choosing cost-effective maintenance strategy for all buildings using binary variable, where 1 means a strategy is implemented and 0 means otherwise. Under set of constraints the model will choose which strategy is the most effective for each building and when the strategy is implemented.

Subsequently, two scenarios will be conducted. Scenario 1 is intended to achieve the minimum acceptable condition for all buildings in every year. Meanwhile, Scenario 2 will require all buildings to reach at least the minimum acceptable condition at the end of a maintenance plan period.

Owners often have many buildings with different level of condition. It is important to determine each category of buildings condition level. The data of buildings condition level will be very useful in choosing the most appropriate strategy for a certain building. The model will consider the condition of buildings and cost of each strategy to propose the best maintenance activity.

There are five levels of building conditions. They are Very Good, Good, Average, Poor, and Bad. These criteria will be used to consider which maintenance action will be taken for a particular building. In addition, the five states of buildings condition are transformed into an index modified after Uzarski work [13] to simplify the calculation process in the model. The highest point of building condition level, which is very good, is 95 and the lowest point of the condition level of bad condition is 35. The objective is to upgrade buildings condition to achieve acceptable condition level which is good condition. Fifteen (15) buildings are employed and details of these buildings initial condition are presented in Table 1.

Table 1. Buildings Initial Condition

Building	Level of Condition	Condition Index
Building A	Good	78
Building B	Poor	53
Building C	Very Good	89
Building D	Bad	38
Building E	Average	65
Building F	Good	77
Building G	Poor	50
Building H	Good	85
Building I	Bad	44
Building J	Average	70
Building K	Good	80
Building L	Good	74
Building M	Very Good	87
Building N	Bad	42
Building O	Good	75

Building deterioration rate or building degradation level is also very essential in determining the most effective maintenance strategy. Building deterioration rate used in this study is adopted from the regulation of Ministry of Public Work of Indonesia [15]. The regulation states that degradation of building condition is 2% each year [15]. The data of the initial condition in this study is established deterministically and for the following years, 2% degradation every year is used in a 10-year maintenance plan period.

Furthermore, there are four types of maintenance strategy to upgrade buildings to acceptable condition level proposed in this model. The strategies are Light Repair, Rehabilitation, Renovation, and Restoration. Each strategy has different impact in improving building condition and upgrading to higher level of condition. The detail of the four strategies and its relationship with the five building conditions are shown in Figure 1.

4 Model Formulation

This section presents an overview of a constraint programming model formulation. The model aims to find the least budget spent of maintenance strategies applied and satisfy all constraints. The variables, parameters, sets, constraints, and objective function are presented as follows.

$$Min \sum_{i=1}^{i}\sum_{j=1}^{j}\sum_{k=1}^{k} C_{ik} \cdot S_{ijk} \quad (1)$$

s.t.

$$\sum_{i=1}^{i}\sum_{k=1}^{k} C_{ik} \cdot S_{ijk} \leq F_j, \forall j \in Y \quad (2)$$

$$\sum_{k=1}^{k} S_{ijk} \leq 1, \forall i \in B, \forall j \in Y \quad (3)$$

$$\sum_{k=1}^{k} S_{ijk} \cdot I_k + CB_{ij} = CA_{ij}, \forall i \in B, \forall j \in Y \quad (4)$$

$$CB_{ij} = D_i, \forall i \in B, j = 1 \quad (5)$$

$$CB_{ij+1} = CA_{ij} \cdot 0.98, \forall i \in B, \forall j \in Ymin \quad (6)$$

$$CA_{ij} \geq A, \forall i \in B, \forall j \in Y \quad (7)$$

$$CA_{ij} \geq A, \forall i \in B, j = 10 \quad (8)$$

$$CA_{ij} \leq M, \forall i \in B, \forall j \in Y \quad (9)$$

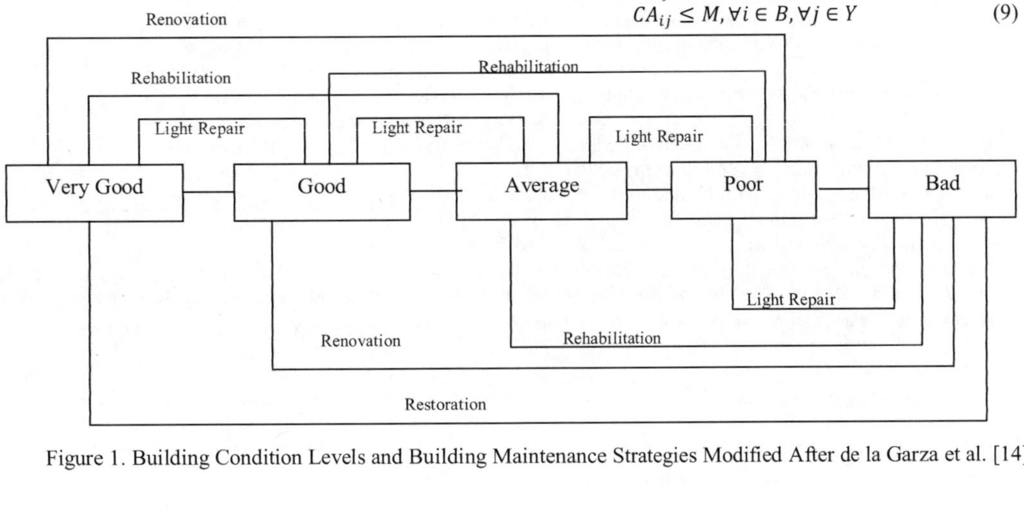

Figure 1. Building Condition Levels and Building Maintenance Strategies Modified After de la Garza et al. [14]

The objective function in equation (1) enables cost of each strategy to prioritize the most cost-effective strategy for each building in certain year to minimize total budget spent on all maintenance strategies. C is a parameter of cost of strategy k of building i. Then, S represents binary variable for strategy selection, corresponding to maintenance strategy k of building i in year j, whether certain strategy is chosen (S_{ijk}=1) or not (S_{ijk}=0).

Equation (2) is constraint for the annual maintenance budget. F_j in this equation denotes amount of money in j year. Money spent on certain strategy of all buildings in a year must be less than or equal to maintenance budget in that year where B is set of buildings and Y is for years. Equation (3) restricts to choose only one strategy for particular building in a year or not to do any maintenance activity at all. Equation (4) aims to present information about the condition after S_{ijk} maintenance strategy is applied to building i in one year period (year j). I_k in this constraint indicates improvement of strategy k. Meanwhile, CB_{ij} means condition of building i in year j before a maintenance strategy is implemented and CA_{ij} describes condition of building i in year j after a maintenance strategy is employed.

Equation (5) states the condition of buildings in the beginning of the first year (base line) where D_i represents the damage of building i. Equation (6) shows the degradation of buildings condition in every year. The degradation rate of buildings conditions is 2% per year, so that, CA_{ij} is multiplied by 0.98 to find value of building condition in the beginning of the next year (CB_{ij+1}). Equation (7) requires all buildings must be greater than or at least equal to minimum acceptable level of buildings condition in every year and in this case the minimum level of buildings condition (A) is 75 and this constraint will be used in Scenario 1 but will be omitted in Scenario 2 due to different objective. Otherwise, equation (8) demands all building to achieve the minimum acceptable condition at least at the end of the period and this constraint will generated in Scenario 2 but will not be involved in Scenario 1. Equation (9) limits the condition of all buildings must be less than or equal to maximum building condition (M) which is 95.

ILOG OPL Studio software is used to run the model and to find the optimal solution. ILOG is very convenient software to run an optimization model because the model can be written with a simple language and notation. The software enables users to separate the model with the data, so that the same model can easily be used with different data.

5 Case Study and Result

The buildings involved in this study have four options of maintenance treatment with different cost for each building. The proposed model will attempt to find the most cost-effective maintenance strategy of every building and to achieve the minimum acceptable level of condition at the end of time period.

The result of Scenario 1 is presented on Table 2. The table shows maintenance action chosen for each building and the time of maintenance treatment implementation. Building B, Building D, Building E, Building G, Building I, Building J, Building L, and Building N are maintained with different maintenance strategy in the first year. This is because those building have lower condition level than the minimum acceptable condition level. Subsequently, almost all of those buildings (except Building L) will have to be repaired for the second time to upgrade the condition because in the maintenance plan period, the buildings deteriorate to below the minimum acceptable condition level.

Table 2. Result of Scenario 1

Building	Maintenance Strategy[a)]	Year of Maintenance[a)]
Building A	Light Repair	4th
Building B	Rehabilitation	1st
	Light Repair	7th
Building C	Light Repair	10th
Building D	Renovation	1st
	Light Repair	3rd
Building E	Light Repair	1st
	Light Repair	8th
Building F	Light Repair	3rd
Building G	Renovation	1st
Building H	Light Repair	8th
Building I	Renovation	1st
	Light Repair	7th
Building J	Light Repair	1st
	Light Repair	10th
Building K	Light Repair	5th
Building L	Light Repair	1st
Building M	Light Repair	9th
Building N	Renovation	1st
	Light Repair	6th
Building O	Light Repair	2nd

[a)]The results are derived from the model

For the rest of the buildings, they are treated in different time as their initial conditions are different. Then, most of these buildings are maintained with same maintenance activity which is Light Repair. This condition happens because Light Repair maintenance is the cheapest maintenance action and when the building condition drops to below the minimum acceptable level of condition, the maintenance strategy will improve buildings' condition effectively.

Regarding the building condition level, result of Scenario 1 is shown in Figure 2. From 5 sample buildings that represent 5 different initial conditions, Average, Poor, and Bad initial condition buildings are directly maintained to reach minimum acceptable condition and these building are maintained twice. In addition, the buildings will be at the minimum condition until the end of the period. Meanwhile, buildings with Very Good and Good condition are maintained after deteriorating to below 75.

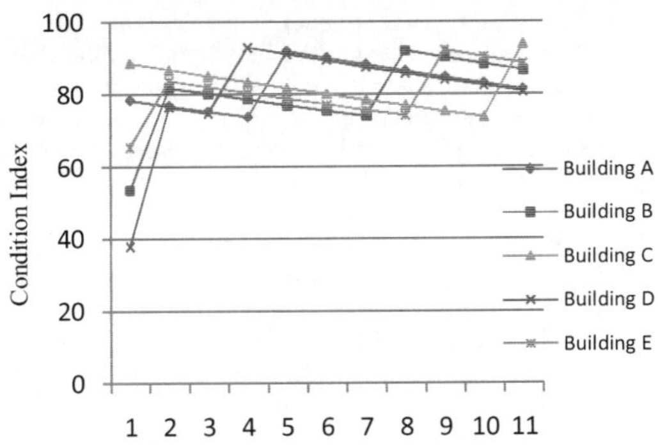

Figure 2 Buildings Condition Performance of Five Buildings in Scenario 1

The money spent each year for maintenance activities of Scenario 1 is shown in Figure 3. Due to so many buildings need to be maintained in the first year, Scenario 1 requires high cost in the first year to conduct some maintenance activities for some buildings. On the other hand, the maintenance spending in the following years will be very little because there are few buildings will deteriorate to below the standard each year. The total budget spent in 10 years in this scenario is approximately 8.1 billion rupiahs.

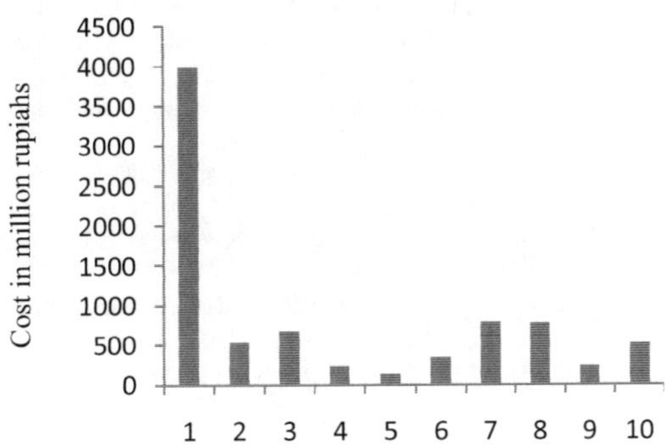

Figure 3. Yearly Budget Spent in Scenario 1

In Scenario 2, the model is modified by removing equation (7) and adding equation (8). This scenario aims to require all buildings must be greater than the minimum acceptable condition at the end of the period. However, this scenario does not consider the condition of the buildings in every year. Table 3 shows that most of the buildings are maintained in the last years of the period (8th, 9th and 10th year). In addition, the Light Repair strategy still becomes the most favourable maintenance strategy, but there are only two buildings getting the second maintenance treatment in this scenario which are Building D and Building N.

Table 3. Result of Scenario 2

Building	Maintenance Strategy [a)]	Year of Maintenance[a)]
Building A	Light Repair	4th
Building B	Renovation	7th
Building C	Light Repair	10th
Building D	Rehabilitation	5th
	Light Repair	7th
Building E	Rehabilitation	4th
Building F	Light Repair	3rd
Building G	Renovation	10th
Building H	Light Repair	8th
Building I	Renovation	10th
Building J	Light Repair	10th
Building K	Light Repair	5th
Building L	Light Repair	10th
Building M	Light Repair	9th
Building N	Light Repair	8th
	Rehabilitation	9th
Building O	Light Repair	3rd

[a)] The results are derived from the model

In terms of buildings condition, Scenario 2 does not consider the buildings condition performance in each year. As Figure 4 presents, the model lets some buildings deteriorate below the acceptable condition and then maintains them in the middle of the period to achieve the minimum standard at the end of the period. For example, building B, D, and E are not directly maintained in the beginning years although their initial conditions are below the minimum acceptable standard.

In contrast with Scenario 1, less money is needed in Scenario 2 for the beginning years, but requires a lot of money in the last years (year 9th and 10th), as Figure 5 shows. Moreover, in year 1st, 2nd, and 6th there is no any maintenance action conducted. This condition happens due to the requirement of achieving the minimum standard at the end of the period. Then, the total budget spent along 10-year time period is approximately 6.7 billion rupiahs. It is more cost-effective than Scenario 1.

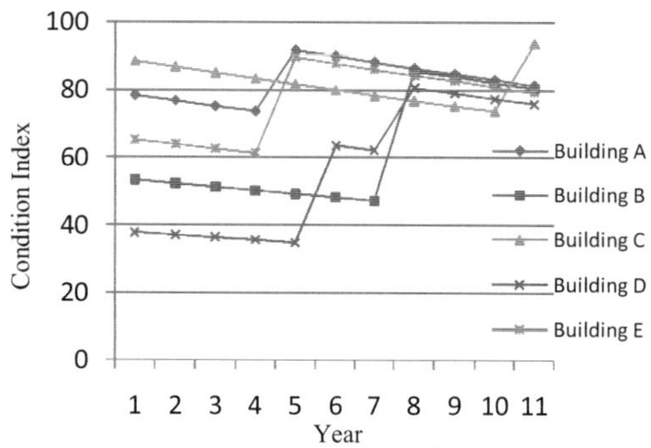

Figure 4. Building Condition Performance of Five Buildings in Scenario 2

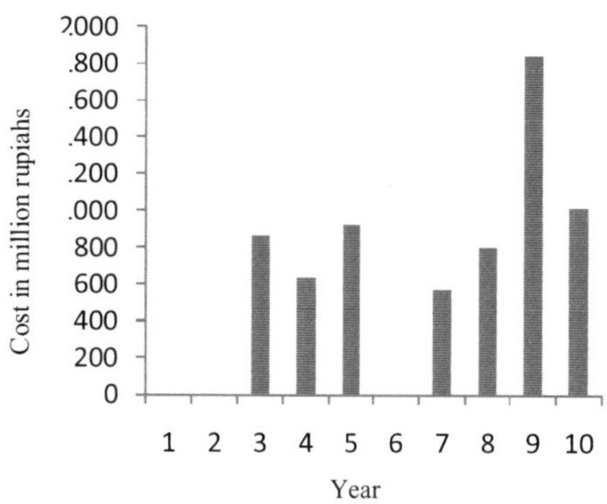

Figure 5. Yearly Budget Spent in Scenario 2

6 Conclusion

This study considers buildings level condition and maintenance budget spent every year. Employing constraint programming, a maintenance optimization model is proposed to solve maintenance strategy selection problems. Two scenarios are conducted to apply the model involving fifteen buildings with different initial conditions in 10-year maintenance plan period.

The result presents the most cost-effective maintenance strategy selection. Scenario 1 shows that most of the buildings must be maintained in the beginning years to achieve the minimum condition standard every year. Meanwhile, Scenario 2 suggests that most of the buildings must be maintained in the last years to reach the minimum condition standard at least at the end of the period, but this scenario does not consider each year buildings condition. For total budget consideration in 10 years period, Scenario 2 spends less money than Scenario 1.

These findings enable owners to predict when a building should be maintained and which maintenance strategy is the most effective. Owners will be able to schedule maintenance actions for all building, so that resources for maintenance strategies can be prepared.

Owners can also forecast budget needed for maintenance treatments each year and they can plan the most effective maintenance budgeting for certain period of time. They can consider which approach is more suitable whether keeping the buildings condition above the acceptable standard but requires more money or targeting the buildings condition to achieve the minimum condition in the last year of period with less money but does not consider the buildings condition for each year.

The results should be treated cautiously. Limited criteria in selecting the maintenance strategies are a major limitation in this study. Future work calls for some criteria to be added to generate better results.

References

[1] Au-Yong CP, Ali S.A, and Ahmad F. Significant Characteristic of Scheduled and Condition-Based Maintenance in Office Building, *ASCE Journal of Performance of Constructed Facilities* 28(2):257-263, 2014.

[2] Flores-Colen I, de Brito J, and Freitas V. Discussion of criteria for prioritization of predictive maintenance of building façades: Survey of 30 experts. *Journal of Performance Construction and Facility*, 10.1061(ASCE):337–344, 2010.

[3] Lee, R. and Wordsworth, P. Lee's Building Maintenance Management, 4th Ed. Blackwell Science Ltd, Oxford, U.K, 2001.

[4] Abaza KA, Ashur SA, and Al-Khatib I, An integrated pavement management system with a markovian prediction model, *Journal of Transportation Engineering*, ASCE 130(1):24-33, 2004.

[5] Ravilara V and Givas D. Goal Programming methodology for developing an integrated pavement maintenance and bridge program. *ASCE Journal of Transportation* 121 (4):345-351, 1994.

[6] Badr A.M, and Brown K.N. Building maintenance scheduling using cost-based reasoning and constraint programming techniques. In *eWork and eBusiness in Architecture, Engineering and Construction-Menzel Scherer (Eds)*. London, United Kingdom, 2010.

[7] Chan W.T and Hu H, Constraint programming approach to precast production scheduling. *Journal of Construction Engineering and Management* 128(6):513-521, 2002.
[8] Brailsford SC, Potts CN, and Smith BM. Constraint satisfaction problem: algorithms and applications. *European Journal of Operational Research* 119(3):557-581, 1999.
[9] Rodriguez J. A constraint programming model for real-time train scheduling at junctions. *Transportation Research Part B: Methodological* 41(2):231-245, 2007.
[10] Liu S.S and Wang J.C. Optimizing project selection and scheduling problems with time-dependent resource constraints. *Automation in Construction* 20(8):1110-1119, 2011.
[11] Liu SS and Wang JC. Profit optimization for multiproject scheduling problems considering cash flow. *Journal of Construction Engineering and Management* 136(12):1268-1278, 2010.
[12] Heipcke S. Combined modelling and problem solving in mathematical programming and constraint programming, Ph. D. Thesis. University of Buckingham, United Kingdom. 1999.
[13] Uzarski D.R, and Burley L.A. Assessing building condition by the use of condition indexes. *Infrastructure condition assessment: Art, science, and practice*. ASCE. 1997
[14] de la Garza J.M, Akyildiz S, Bish D.R, Krueger D.A. Network-level optimization of pavement maintenance renewal strategies. *Advanced Engineering Informatics* 25(4):699-712, 2011.
[15] Peraturan Menteri Pekerjaan Umum. Nomor: 45/PRT/M/2007 Tentang Pedoman Teknis Pembangunan Bangunan Gedung Negara. 2007

A study of BIM based building facilities online repair system

N.H. Pan[a] and K.Y. Chen[b]

[a] Department of Construction Engineering, National Yunlin University of Science and Technology
[b] Department of Construction Engineering, National Yunlin University of Science and Technology
E-mail: pannh@yuntech.edu.tw (N-H. Pan), g9610817@yuntech.edu.tw(K-Y Chen)

Abstract

When facilities was exceptionally occurs, the traditional operation method is either to repair or replace. Before operation procedures starts, it needs to be taken through a series of actions of notification with paperwork. Thus, the management on site tends to make the operation procedure with inefficiency and inconvenience. The recent rise of BIM. Due to BIM's functionality and data-rich of facilities, it can help improve the efficiency of operation of the management unit. The purpose of this study is to use BIM to generate a facilities repair system platform. When a facility needs to be repaired, the platform can be used to quickly informed and able to understand the current status, the location, and maintenance data of facilities,. In this study, the data of BIM facilities in windows system functions through the list of exporting, importing into the system database, and graphics output with BIM to achieve functional model reveal it. The components of the proposed system includes facilities maintenance records QR Coding , different facilities and equipment positioning method; combined facilities data encoding systems, as well as abnormal conditions different equipment repair orders; Users most concerned about the maintenance status inquiries; management unit management menu. Reporting and management systemFs will be combined together to explore the use of newspaper repair system users and managers through viewpoints of both. After using the system to get a few results of this study, presented in the visual effects quickly locate facilities, facilities and equipment standard encoding, facilities and equipment exception, BIM model to enhance the convenience and efficiency of users and managers.

Keywords –BIM, QR Code, Facility management

1 Introduction

The importance of property management has gradually been recognized in Taiwan. Aside from the management of buildings, the life cycle of the buildings also requires the same level of attention. Both new and existing buildings require a suitable management approach. Currently, building information modeling (BIM) has advanced considerably in Taiwan, and the establishment of models has also been relatively fast. The major benefit of BIM is not exactly in three-dimensional (3D) modeling, but in the provision of detailed building information. Although a model can show how a building will look upon completion, the information provided has a wide range of applications in areas such as cost budgeting, supply management, project scheduling, collision detection, and management and maintenance. BIM, along with the large amount of information it generates, contributes considerably to the development of architectural engineering and property management.

Adequate management and maintenance of a building and its amenities are key factors in prolonging the building's life-span. Deterioration over time can lower the function and value of a building to an extent that, if timely repair and renovation are not performed, severe problems can ensue. The incorporation of BIM into property management effectively facilitates restoring deteriorated amenities, thus maintaining the building's function and value in the long term. In addition, BIM also supports 3D visualized management and provides relatively accurate information that enables decision makers to work more efficiently. Earlier, information management was performed manually with pen and paper, which can be complex and error-prone. Currently, the enhanced efficiency in operations and data storage pertaining to BIM offers considerable advantages over the manual method. Computerized information management and database technology enable the transfer of data to platforms that are easily accessible for general users to check and review the maintenance records of amenities registered in the BIM system. This effectively overcomes the technical limitations and enables more users to master the BIM.

1.2 Purpose

This study explored an approach for extracting amenity data from a BIM system and incorporating existing database systems into a proposed framework for application in the management and maintenance of amenities and facilities.

The present study was aimed at generating an amenity maintenance platform from existing BIM data, which enables expedited reporting for repairs and a comprehensive listing of the status, location, and maintenance records of amenities and facilities. In summary, the purpose of this study was twofold:

(1) To establish a BIM amenity database framework;
(2) To develop an amenity maintenance platform incorporating the BIM framework.

1.3 Scope and Limitations

(1) This study aimed at developing a building amenity maintenance platform incorporating the BIM framework. The scope, target, and limitations of the present study are defined as follows:
(2) In terms of the research scope, this study sought to establish an amenity maintenance system and explore the system's application from the perspectives of both users and administrators. Information stored in BIM was exported through the tools' "schedule" functions for the system database which, through its graphic output functions, enabled visualization of the model.
(3) Considering the wide variety of amenities and facilities in a building, the research scope was limited to the window/door system.

1.4 Methodology

The present study commenced with identifying the research motivation, purpose, and scope, then determining the research framework through a literature review. Finally, an amenity maintenance system was proposed for the application of BIM information.

2 Literature Review

The literature revealed that BIM is used throughout the life cycle of a building. In terms of the maintenance and management of amenities and facilities, the literature also reveals various applications of BIM and multiple positive effects it can generate. Numerous software providers are known to release property management products incorporating BIM technology, but such products tend to be less practical than expected. Therefore, the present study also examined BIM from an alternative perspective by exploring its functions and effectiveness in reports on the repair of amenities and facilities.

R. Liu 2013 to clarify the frequently occurring maintainability problems and to investigate the potential areas that can use BIM technology to solve the maintenance problems in early the design phase. A survey was conducted to collect perspectives from the industry practitioners for the maintenance problems and their frequency.
Changyoon Kim 2013 BIM can be used in numerous applications for facility management. This study aims to incorporate the strength of mobile systems and BIM in facility management. The proposed system, which utilizes smartphone technology, allows engineers to conduct facility management functions, such as maintenance planning, inspection, and assessment, irrespective of a user's location. The integration of building information modeling in a system provides detailed visual information of the building components.

The means of extracting BIM information from a database were explored for the use of an amenity maintenance system. This combination, with an online report-for-repair feature, was expected to successfully introduce BIM into the domain of property management.

2.1 Tools for BIM

Software tools for BIM that are commonly used in Taiwan include Autodesk Revit, Graphisoft ArchiCAD, and Bentley MicroStation. Autodesk Navisworks was also used in the present study.

Level of development

The level of development (LOD) is discussed in this section to provide a set of standards for the establishment of a BIM system.

One problem that is inevitably encountered in establishing BIM is that the model designers are multidisciplinary professionals who do not follow a set of unified standards. LOD provides a set of standards upon which model users can specify the degree of detail in a model, the ownership of the model, and the BIM requirements and documentation formats. LOD also defines the model management responsibilities from the beginning to the end of the architecture project.

According to protocol E202™–2008 of the American Institute of Architects, the LOD is a BIM tool comprising five progressively detailed levels of completeness. The five levels are coded LOD100, LOD200, LOD300, LOD400, and LOD500 on the basis of the legal requirements for the contents of models for each LOD.

After the introduction of the LODs, a conclusion is reached that if BIM is to use the LOD at the building operations management stage, LOD300 is the minimum level required for such a task. In other words, the introduction of BIM into a building's maintenance and management starts as early as the building's construction stage, because basic information required for the maintenance and management of amenities, such as their actual lengths, widths, heights, locations, and orientations, become available at this stage.

The BIM used in the present study was based on LOD300; therefore, some additional amenity-related information had to be added for the use of the amenity maintenance system proposed by this study.

3 Design of the Amenity Maintenance System

After the study's motivation and purpose were defined, the development of the amenity maintenance system commenced. However, user requirements needs and system functions had to be identified first.

(1) User Requirements

Users of the proposed amenity maintenance system can be divided into two categories: general users and administrators. A user requirements analysis revealed that for general users, both the interface and presentation of information should not be too complex to ensure ease of use and quickness of control. The progress of repairs was the one feature that these users wanted to see the most in the inquiry interface.

Interviews with personnel in the property management industry revealed that for the administrators, aside from basic information such as the measurements of amenities, the most necessary information was maintenance details such as life-span, cost, caretaker, and maintenance service provider of the amenities. To keep the interface simple and ensure information security, the amenity repair inquiry function would display only basic information. Administrators can gain access to complete information through the model inspection function.

(2) System Functions

This amenity maintenance system enables tenants to request repairs on their own. Administrators are then alerted by the request and can confirm the progress of repairs through the system.

(3) Tools

Concerning the design of the proposed amenity maintenance system, the present study chose Autodesk Revit and Autodesk Navisworks for the BIM and rendering of 3D models. This author also chose Autodesk Revit and Autodesk Navisworks as the research tools for the study.

Microsoft Visual Studio and Microsoft ASP.NET were used in the design of the amenity maintenance system.

3.1 Introduction of the request-for-repair process

The proposed system uses information extracted from BIM as the basis for the system database, providing an online report-for-repair platform for general users and administrators. General users can access the system through the Internet or by scanning the quick response (QR) code on the amenities, and then file a request for repairs from their residence. Therefore, this system would provide great convenience for general users. Upon reception of the request, administrators can check the location and basic information of the malfunctioned amenity and its maintenance records in the office for an effective appraisal of the situation and to ensure control over the repair process. The adoption of this system would facilitate improving the efficiency of property management.

Amenity/Facility Information: The table contains information extracted from BIM, which contains a large amount of data including amenity/facility information found in architectural models, structural models, and MEP models, as well as maintenance information and serial codes of amenity elements linked to the relevant parameter settings. The amenity/facility data in the BIM can be compiled using the "schedule" function and then exported to the system database. Also included are 2D and 3D floor plans to facilitate displaying amenity/facility locations in the system.

Maintenance Information: This table contains specific information that cannot be obtained through setting the parameters of the BIM elements such as information on maintenance service providers and maintenance records. This table is an effective supplement for the maintenance information in the system database.

Abnormal Conditions: In the system, the display of abnormal conditions varies depending on the type of amenity/facility. This study focused on the windows and doors, which have 14 and 8 types of abnormalities, respectively (these conditions are further detailed in the following section).

This table enables operators to mark the categories of abnormal conditions when filing requests for repair to the system.

Feedback: This table plays a critical role in the database. A request for repair is added to this table after it is submitted to the system, after which the administrators are alerted of the situation. The system also allows administrators to track the repair progress of each request, and their feedback is immediately added to this table to provide updates on the situation for display in the system's progress inquiry page.

The procedures for general users were designed with consideration of principles such as convenience, promptness, and real-time response.

For general users who would like to make a request, the most efficient approach is to scan the QR code on the amenity/facility by using a mobile device, which immediately brings up the system's request for repair page.

The system menu for general users consists of only two options: request for repair and repair progress inquiry. Because most users would be interested only in the progress of the requests they have submitted, it is allocated in a column in the system's repair progress inquiry page, which is updated according to feedback and track records from the administrators (figure 1).

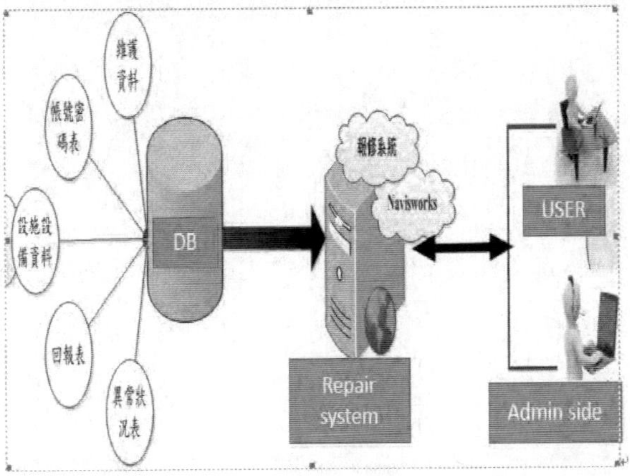

Figure 1 Repair / maintenance system architecture

Upon reception of a request, this request is immediately shown on the administrator feedback page. The first step taken by the administrators is to determine whether an appropriate maintenance service provider should be informed of the request. The decision is usually based on provisions in the service contract, because the service provider is bound to certain maintenance and warranty responsibilities within the term of the contract, which would substantially reduce the repair costs ensued. A decision not to inform a service provider is typically for one of the following reasons: expired warranty period or contract term, unaffordable preliminary repair quotation, or minor problems that could be remedied by the administrators. The next step is to determine whether the amenity/facility concerned can be fully repaired. When it can be repaired, the case can be closed upon completion of the repair. Otherwise, the amenity/facility must be replaced. The case is considered concluded after the replacement has been installed.

3.2 Serial Code Guidelines and Abnormal Conditions

This section defines the serial codes and details the types of abnormal conditions. The serial codes give each amenity or facility an exclusive identifier akin to a personal identification number, thereby preventing mistaken identification. Abnormal conditions of the amenities/facilities are categorized by damage types to assist general users in quickly recognizing and selecting appropriate categories in the system interface

The serial numbers comprise two parts: the first part is based on the item numbers proclaimed by the Public Construction Commission (PCC), and the second part is based on the location (to ensure quick positioning). The combination of the two forms a set of codes that can be used to identify individual amenity/facility (Table 1) ..

Table1 individual amenity/facility code

082104654A	-	A1EW1
item numbers proclaimed		location codes

- The item numbers compiled by the PCC were adopted for the present study because they are relatively comprehensive and the codes are relatively short.
- Regarding the location codes, they consist of five digits, making them short and easy enough for quick identification and positioning. The meanings of the digits were as follows:
- First digit (building number): Identifies the specific building the amenity/facility is in.
- Second digit (floor): Identifies the specific floor the amenity/facility is on. Codes A–Z are used to denote 1F–26F, and 1–9 are used to denote B1F–B9F.
- Third digit (room): Identifies the specific room the amenity/facility is in, or the room immediately after the amenity/facility.
- Fourth digit (orientation): Identifies the specific location of the amenity/facility in a room according to the cardinal directions.
- Fifth digit (order): Identifies the sequential position (from left to right) of the amenity/facility among its type in the same room when viewed from the front.
- Combining item numbers of the PCC and the location codes obtains the complete 15-digit serial codes.

3.2.1 Abnormal Conditions:

Because the present study focused on windows and doors, only the abnormal conditions commonly seen among windows and doors are specified here.

The present study identified eight abnormal conditions that are commonly observed or can be found after dismantling the components of windows: corrosion, seeping, peeling of waterproof material, broken glass, frame warpage, damaged knobs, rust-corroded screws, and damaged window screens.

4 Practical Testing

A practical test was performed in a community in Hong Kong. The target was a villa with two floors above ground and one below ground. BIM was introduced early in the design stages as an auxiliary tool (Figure 2). A window in the living room on the first floor was selected for a request for repair, and the tenant was granted access to the system either by scanning the window's QR code or by logging into the system through the Internet with a set of account and password (. Once entered into the system, the tenant made a request for repair, and was required to pick an amenity/facility for the task. The tenant was able to locate it either by entering the serial code or by going through a dropdown list to specify the building, floor, and room it is in. Here, the tenant selected to use the dropdown list, and when the room was specified, the window could be selected from a plan of the room; subsequently, the request form page was displayed. When selecting the floor, the tenant was also able to open the floor plan and search for a specific item from there.

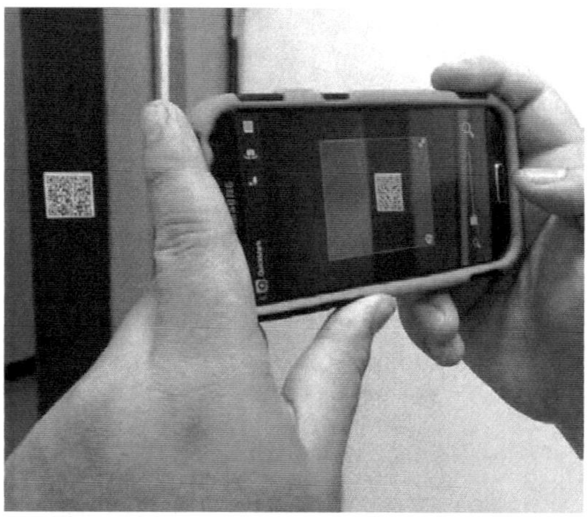

Figure 2 Use a smart phone to scan QR Code

The request form page contains a section in which the tenant can manually enter a brief description of the abnormal condition the concerned amenity/facility before clicking on the "finish" button to submit the request. Afterwards, the tenant can check the progress by using the inquiry function. After the request has been received, the administrators are alerted by the feedback table (Table 2), which also briefed them on the situation. By clicking on the location button embedded in the request form, the administrators can access the BIM to quickly locate the problem (Figure 3).

Figure 3 show the location

Table2 Repair / maintenance resume

Facilities basic information			
amenity/facility code	08510RK02A-AAGE2	**TYPE**	C0823
budling	A	**floo**	1F
Maintenance information			
Years of use	20	**Last maintenance date**	2017/02/07
manufacturer	AA	**Repair status**	Repair confirmation

After the administrators checked the request information, they assessed whether a service provider should be contacted or if they should repair the damage themselves. Furthermore, the progress of repairs can be monitored through the system's feedback page, which is synchronized with the inquiry page to enable the tenant access progress updates.

5 Conclusions and Suggestions

This section summarizes the results of the present study. Suggestions are also offered as a reference for future researchers.

5.1 Conclusions

The visual display supported by Navisworks provides a graphic user interface that enables the administrators to quickly locate amenities/facilities in need of repair. The serial code system designed for this system is a combination of the item numbers from the PCC and location codes for building amenities/facilities. The advantages of this measure are that the first part of the code is already known to local industries, and the second part is an effective indicator of the location of specific amenities/facilities. The introduction of the serial codes and QR codes provides both general users and administrators with four easy methods for quickly locating a specific amenity/facility. Common abnormal conditions are categorized to enable convenient and accurate selection for general users. The categorization also enables administrators to take appropriate actions. The application of BIM information in the property management of a building, or more specifically, its use in a maintenance system, enables administrators to enhance efficiency and reduce the number of operational errors and minimize expenditures.

References

[1] Eadie. R., Browne. M., Odeyinka. H., McKeown. C., McNiff. S., BIM implementation throughout the UK construction project lifecycle:An analysis, Automation in Construction, Vol.36, pp. 145-151.

[2] Aram. S., Eastman. C., Sacks R., Requirements for BIM platforms in the concrete reinforcement supply chain, Automation in Construction, Vol.35, pp.1-17.

[3] Kim. C, Lim. H, Kim. H, Kim. H, BIMBASED MOBILE SYSTEM FOR FACILITY MANAGEMENT, (2013 Proceedings of the 30th ISARC, Montréal, Canada)

[4] Liu. R., Issa. R.R.A., BIM FOR FACILITY MANAGEMENT: DESIGN FOR MAINTAINABILITY WITH BIM TOOLS, Pages 321-328 (2013 Proceedings of the 30th ISARC, Montréal, Canada)

[5] Autodesk Revit, http://www.autodesk.com/products/autodesk-revit-family/overview

[6] Graphisoft ArchiCAD, http://www.graphisoft.com/archicad/

[7] Bentley MicroStation, http://www.bentley.com/zh-TW/

[8] Autodesk Navisworks, http://www.autodesk.com/products/autodesk-navisworks-family/overview

34th International Symposium
on Automation and Robotics
in Construction

ISARC 2017

June 28 – July 1
TAIPEI, TAIWAN

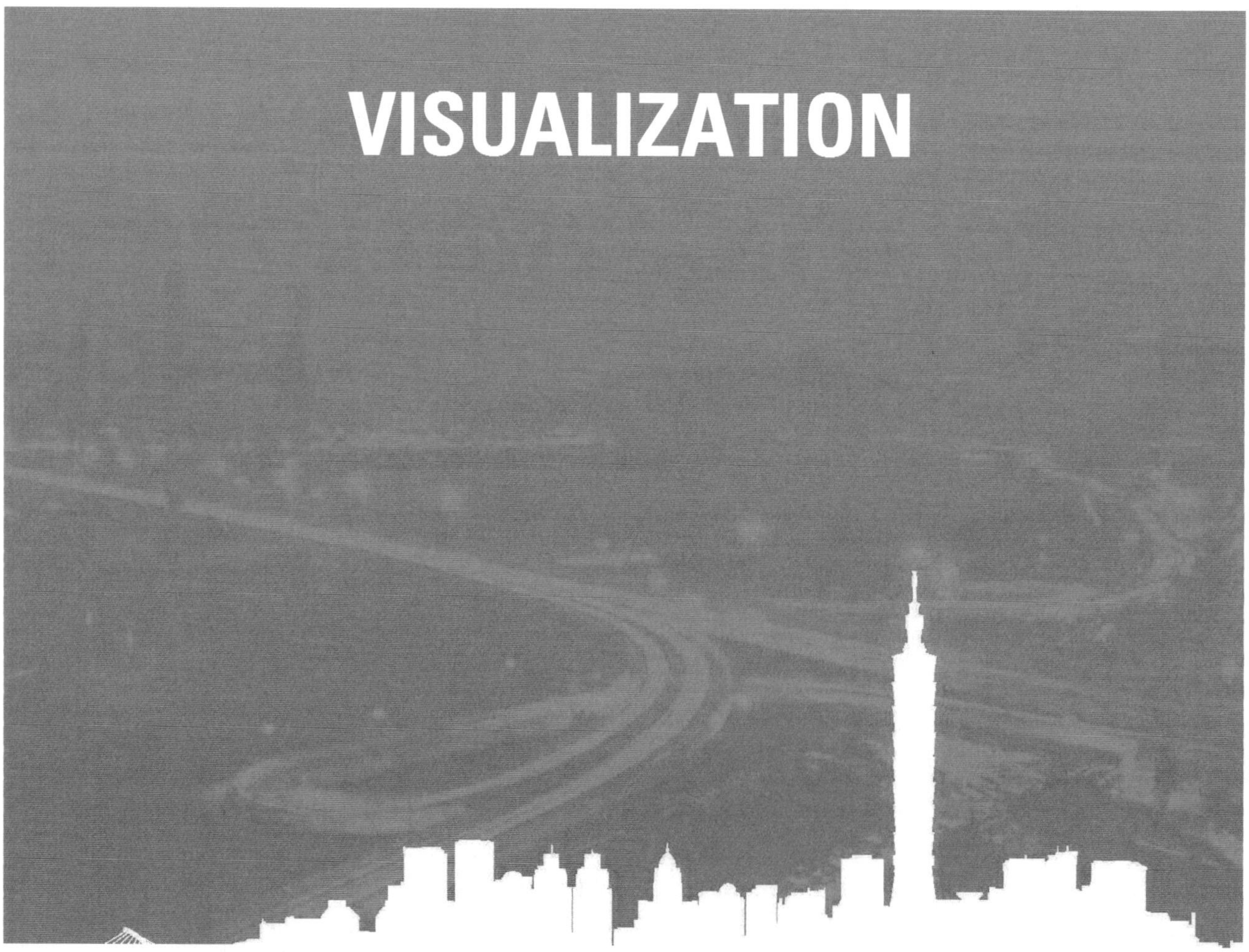

A Framework for Using Advanced Visualization Tools for Residential Property Management

M. Takin[a], J. Peng[b], S. M.E. Sepasgozar[c] and H. Ebrahimi[d]

[a]Early Years at UNSW, University of New South Wales
[b] Faculty of Built Environment, University of New South Wales
[c] Faculty of Built Environment, University of New South Wales
[d] College of Fine Arts, University of Tehran
Email: m.takin@unsw.edu.au, jinghan.peng@student.unsw.edu.au, sepas@unsw.edu.au, hr.ebrahimi@ut.ac.ir

Abstract –

Traditional real estate business models are not able to attract young generation customers since they are used to acquire required information from digital apps and devices. This paper aims to develop a framework for application of advanced visualization technology in residential property marketing to offer a detailed 3D information model to customers through the real estate websites. The paper presents the result of a survey examining the experience of customers who were used one of the five selected real estate on-line platforms in Australia. The survey targets a sample of 120 participants, who are selected at the University of New South Wales covering different genders and nationalities. Based on the result of the survey, the framework is verified and shows how 3D scanners can be useful for on-line property management. The results would be useful for property managers to communicate with customers by providing visualized information, and customers can easily examine whether the property meets their requirement without any physical inspection. The application of visualization technologies to residential property marketing will provide customers with more reliable and credible property information by 3D modeling and pictures, which is beneficial to property owners, customers and other stakeholders as well as real estate agents.

Keywords –

Advanced Visualization Technologies, 3D Models, 3D Scanners, Real Estate On-line platform, Residential properties

1 Introduction

Traditionally, the methods of real estate marketing are simple and direct - property owners sold or leased their properties by asking people around them, or advertised their properties on newspaper and TV. Later, real estate agencies are popularized to meet the needs of the constantly expanding and developing real estate market. Real estate agencies help properties owners or developers to do the promotion and advertisement for their properties, as well as provide buyers or renters with a large amount of housing information and other relevant services. With the development of Information Technology, Internet offers a more convenient and effective approach to real estate marketing by E-Commerce. According to Matthew Hinton and Yan Tao, currently agents are the important focus point in e-commerce applications in real estate industry. and real estate websites, as a real estate marketing tool, are widely used in residential property market in Australia [1].

The emergence of the real estate websites allows users in different places to browse the information of properties in different areas on the websites, which greatly expands the scope of real estate marketing. However, one of the key issues in current real estate websites is that the housing information provided by these websites is not detailed enough for customers who cannot take part in the property inspection.

The adaptation of the new technology is viewed as an opportunity to attract new buyers and reduce marketing and customer acquisition costs [2].

With this in mind, this paper analyses and compares most popular real estate websites in Australia, identifies their important options by customers' interview. In addition, there is some suggestion to apply visualization technologies, especially 3D Modeling Technologies and Indoor Scanning Technologies, to these real estate websites to improve their online property marketing. In this part, this paper focuses on 3D Modeling Software, as well as 3D Scanners. It compares the effect of the products produced by different 3D Modeling Software on the building modeling aspect. In addition, it describes characteristics of different 3D Scanners in the current market, and gives real estate websites suggestion to improve their work by these new technologies. The next section states the background of real estate marketing and 3D visualization technologies.

2 Literature review

2.1 Real estate marketing and E-commerce

Anja Rassmann stated that the real estate marketing owns its special characteristics which come from real estate's features [3]. Real estates, as a kind of capital intensive goods, need long term building and marketing processes. Besides, Bettina Knirsch added that properties are also the most long-lived economic products and hard to change. And the basis of real estate marketing involves different people's needs & wishes, the properties and customer's benefits & Satisfaction. Thus, marketing should analyse, shape and form the properties market to make the products attractive. In addition, Knirsch listed marketing measures in Practice, including Advertisements (in printed media), Relationship Advertising or Networking, Internet, Direct Mail, Public Relations and so on. However, due to the limitation of Internet development at that time, this paper did not talk much about Internet real estate marketing [4].

With the E-Commerce popularizing, the Internet is quickly becoming the first place for buyers and sellers to find the transaction information. The study in 2000 showed that the information, about properties themselves and their surrounding environment, offered by the websites is steadily going up because of the advancements in technologies [5]. The properties' information provided by the websites at that time is simple and not comprehensive.

As e-commerce developed, companies thought that the intermediaries could be cut out, because they could do business directly to customers by computers. However, later companies found that it is not easy for them to do everything by themselves – Middleman is still needed. According to Anderson, P. and Anderson, E (2002), the role of intermediaries has been changed due to the development of e-commerce. Traditionally, in scale economies, intermediaries aim to meet the same needs for Sellers A to D and the same needs for buyers 1 to 4. On scope economies aspects, intermediaries focus on variations across buyers and sellers with similar needs[6].

However, when co-opting with the Web, intermediaries could add value for the products by creating opportunities and changing the importance of nine generic functions of intermediaries. Thus, this article revealed the importance of intermediaries creating value with the help of the web. But it did not mention relevant aspects about real estate and visualization technologies [6].

In order to find the key source of competitive benefits from E=Commerce applications in Chinese, Matthew Hinton, C. and Tao, Y. did an investigation about 5 Chinese real estate companies. The findings offered evidence of how real estate companies acquired competitive benefits from applications of e-business. However, compared with my research, this paper did not provide technologies aspects research [1]

2.2 Visualization technology

The visualization technologies can cover the whole process in AEC (Architecture, Engineering and Construction), from the initial concepts to the final steps [7]. It has mentions 3D models of design stages during the construction and other steps. But there is not much about the 3D visualization applying to residential property marketing. And one other paper focused on integrating 3D laser scanning and BIM Technology to assist in faster and better real estate services. It analysed the potential of BIM by 3D laser scanning to produce as-is building information and mentioned that there would be more advantages within the real-estate sector in the future [8].

Ku et al explained the problems in the collaboration on 3D Model and analyzing 3 case studies to find that collaboration approaches on 3D Model is different from the architects' approaches on geometry control [9].

In addition, based on increasingly applications of 3D geometric information modelling technologies, the article considered to use the visualization technologies as a tool on urban components, urban planning purposes, as well as real estate aspects. This study researched the possibility to use the 3D modeling technology for simulating the inaccessible places in geopolitically conflict zones, which cannot be accessed because of restrictions of law or physical factors [10].

3D scanners are the important tools in the research. Several types of 3D scanners have been compared by using a scanning specimen for their accuracy. They found that the scanners which use white lite are more accurate than the other types of 3D scanners [11].

The recent trends of scanners are turning 3D scanners into Mobile Phones. An effort was made to increase the accuracy of the interactive systems when turning it into mobile phone devices, and developed an more effective and efficient scheme to integrate multiple stereo-based depth hypotheses into a more real 3D model at the technical level [12].

As a new technique for quickly getting three-dimensional information, 3D laser scanning technology has been widely used in many fields such as acquiring landslide data and to compute earthwork volume, to obtain precise measurements from experimental lithic reduction sets, to present a methodology to combine the technology of the terrestrial laser scanner with the techniques of digital image processing to study damage on stony materials that constitute historical buildings. In addition to preserve and document osteological material in museums, develop research ideas in the subfields of biological anthropology, and increase the potential for scholarly collaboration [13].

The above literature reviews just cover few research factors in this research. Some of them are related to real estate marketing, E-Commerce and real estate websites, whereas the aspects of modern technologies were missed, such as 3D visualization, 3D scanners and 3D modeling technologies. Some other studies focused on technology development, but they did not mention the applications on residential property marketing. This paper addresses the gap in the knowledge how a modern technology can be widely used in the property management business. This is critical to property managers to find out how they can facilitate the implementation of modern technologies in their business to be survive in the future market.

3 Research methodology

This study will be conducted in three main phases as follows:

- Phase 1: analyze five real estate websites in Australia, including realestate.com.au, domain.com.au, realestateview.com.au, homely.com.au and allhomes.com.au;
- Phase 2: Identify influential figures of the web sites according to customers' previous experience;
- Phase 3: Implementation of modern technologies.

Firstly, this research analyzes and compares the current five selected real estate websites in Australia. It mainly focused on the residential proprieties information provided by each website, such as the property photos, property description, floor plan, surrounding environments, infrastructures, comparable sales and so on. This analysis is helpful to explain why there are major differences among these real estate websites in Australia in terms of functionality and online visitors. Next it identifies the existing information and attractive services provided by current real estate websites in Australia to find the parts that should be improved using modern technologies.

Secondly, this study used questionnaire to do customers interview about using real estate websites to rent or buy residential properties in Australia. The questionnaire analysing the real estate websites in Australia, included questions about real estate websites selections, users' satisfaction, the sufficiency vividness and reliability of properties information offered by these websites and users' attitude to real estate websites. This questionnaire can identify customers' important options about the existing real estate websites

Finally, based on studies aforesaid, this research gives suggestion to real estate websites in Australia about implementation of visualization technologies, especially 3D modeling technologies and 3D indoor scanning technologies, on their websites. On the one hand, it compares several 3D modeling software in current market and the 3D models created by this software. On the other hand, this research focuses on 3D scanners which are suitable for indoor scanning. The purpose of analysing these new technologies is to collect characteristics of existing 3D visualization technologies, so that it helps to provide feasible suggestion to applying these new technologies to online real estate marketing on these real estate websites in Australia.

3.1 Initial results

The real estate websites in Australia developed quickly in recent several years and increasingly become the major method to find the information of residential properties for sale or for lease in Australia so that they are the existing and popular real estate digital marketing. Real estate website is making the process of searching housing information easier and cheaper. It allows buyers to search the properties within the designated areas by properties location or zip code and narrow the searching results by adding features, key words and price range. These websites also provide basic information about housing transaction process, loan qualification, communities (tax rates, crime rates school test scores, etc.), links to other services providers' websites and some tools such as home loan calculator. These services offered by real estate websites are free. Usually, the real estate websites earn income by providing links to other websites, as well as selling advertisements [14].

The initial review of the home page of every real estate websites could largely determine users' first impression to the website - professional or unprofessional, reliable or unreliable, easy to use or hard to use. However, compared to other websites, realestate.com.au and domain.com.au have presented more options on search category on their home page, which can help users narrow the search results at the first steps. For instance, realestate.com.au offers 'retire' options in the search category on its home page which specializes in retirement villages, retirement homes & lifestyle communities. While for other four websites, users who want to find a residential property for retirement life would spend more time to find their ideal property.

In addition, the detailed information of residential properties on real estate websites is also important. All websites present basic property information, such as address, images, maps, email and mobile phone number to connect the agency, property description, as well as the Inspection and Auction time (if there is an auction). Comparing with Domain, Realestate not only has more detailed property information, but also made the page of specific property information more convenient for users. For example, there are suburb profiles on real-estate specific property information page. Users can directly look through suburb profiles on the specific property page. However, on domain.com.au, if users want to refer to suburb profiles when they look through specific property information page, they have to return the Home Page to find the suburb and search for the suburb profiles. Thus, property information on realestate page is more efficient (See Table 1).

Table 1. Comparisons of Search Engines on Five Real Estate Websites [Authors]

Websites	Service Category	Search Property Type	Search Refinements	Search Result Presenting
realestate.com.au	• Buy, Rent • Agent • Sold Share; New Homes • News; • Invest; Retire; Home idea	• House • Apartment/Unit • Townhouse • Rural • Land • Villa • Acreage • Block of Units • Retirement living	• Price(range) • Bedrooms • Bathrooms • Parking • Features (indoor and outdoor: pools, air conditioning) • Land size • New or established • Eco Friendly • Keywords	• List/Map/Inspection & Auctions • Sort by: recommended; Price(Low-High); • Date; • Suburb(A-Z)
domain.com.au	• Buy; Rent • Agent • Sold Share; New Homes • News; Advice and Article (Buying, investment, Design) • Auction Result; Home loans • Home Price Guide; Suburb Profiles • Research & Data (linked to Price finder)	• House • Apartment • Rural • New Apartment • Land • New house & Land	• Price(range) • Bedrooms • Bathrooms • Parking • Features (pets, pools, air conditioning) • Land size • Keywords	• Sort by: recommended; Price(Low-High); • Date; • Suburb(A-Z); • Earliest inspection
realestateview.co m.au	• Buy, Rent • Agent • Sold • News • Article (Top tips to prepare you for Auction Day) • HolidayVIEW	• House • Unit • Apartment • Studio • Townhouse • Land • Villa	• Price(range) • Bedrooms • Bathrooms • Parking • Keywords (floorboards, pools, garage)	• List/Map • Sort by: Price/recommended • Date; • Suburb(A-Z); • Bedrooms; • Upcoming Auctions and Inspections
homely.com.au	• Buy, Rent • Agent • Suburb Reviews	• House • Apartment/Unit • Townhouse • Villa • Land • Acreage • Rural • Block of units	• Price(range) • Bedrooms • Bathrooms • Parking • Features (Indoor, outdoor and Eco features) • Land size • Frontage(range)	• List and Map together • Sort by: Price(Low-High); • Date;
allhomes.co m.au	• Buy, Rent • Share; New Homes	• House/Land • Investment Property • Studio/Unit/Duplex • Rural • Retirement living	• Price(range) • Bedrooms and Bathrooms • Parking • Block Size • Dwelling Size • Listing Age	• List/Map • Price/recommended • Suburb(A-Z); • Date/Upcoming Auctions and Inspections

3.2 Results of questionnaire

The questionnaire with the topic "A framework for using advanced visualization tools for residential property management" was developed in three parts. The survey targets a sample of 200 participants, who are selected at the University of New South Wales covering different genders with different nationality. Table 2 shows the familiarity level of the five Australian websites.

Table 2 Familiarity level of the five Australian websites between participants [Authors]

Australian real estate web-site	Participants Familiarity
www.alhomes.com.au	4%
www.domain.com.au	21%
www.homely.com.au	4%
www.realedtateview.com.au	5%
www.realestate.com.au	66%

In the first part participants have been asked to select one of the five Australian real estates the second part of survey was about participants' background that would be useful for further researches. Finally, the participants were asked to answer the third part that was the main part of the survey. This part was in eight different sections such as 1.user satisfaction, 2. sufficiency, 3. vividness, 4. sense of security by using the website, 5. perceived ease of use, 6. behavioral intention, 7. perceived usefulness and 8. reliability of properties information offered by these websites. There were five scales for participant to choose (1 = Strongly Disagree, 2 = Disagree, 3 = Neutral, 4 = Agree, 5 = Strongly Agree)

A total of 120 participants contributed to the study, therefore the research is ongoing. The results show 37% of participants were male, 63% were female. Regarding to participants' age groups 12% were under 18 years old, 66% were 19-30 years old, 9% were 31-40 years old, 7% were 41-50 years old, 7% were 51-60 years old. At this stage, the data analysis showed that the percentage of each scale answered by 120 participants: 13% strongly Disagree, 24% Disagree, 34% Neutral, 14% Agree and 10% Strongly Agree. Appendix I shows a part of questionnaire as a sample. As the participants were in different level of education, also were doing in different field of study or career, therefore in each section we have developed subset of questions to clear the aim of questionnaire.

Figure 1 shows the data analysis of answering each section of the questionnaire. It proved that 57% of participants did not agree or strongly disagree particularly with the sections about virtual realistic and 3D visualization.

The data analysis showed there is a considerable gap in virtual realistic especially in detailed 3D information model to participants through the real estate websites.

Figure 1. The Analysis of answering each section of the questionnaire [Authors]

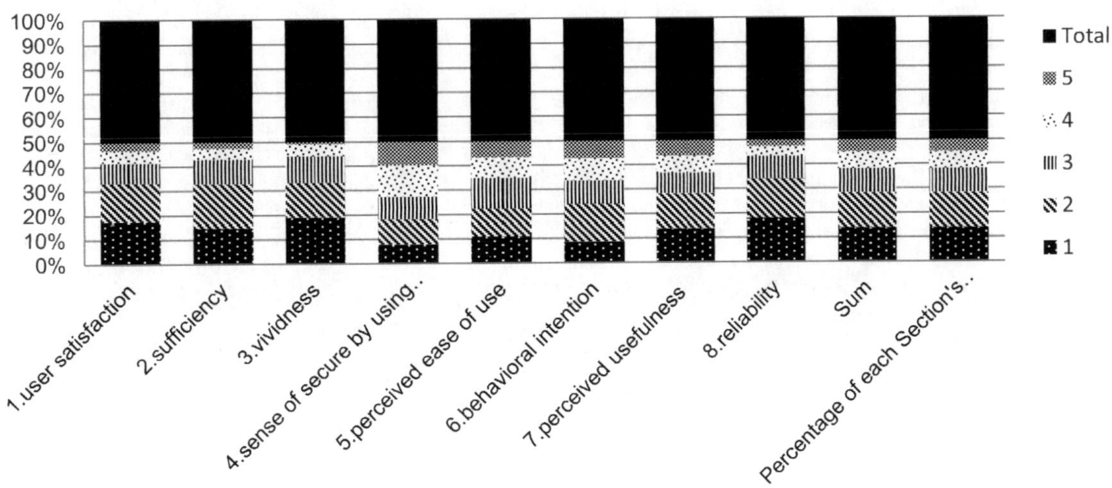

4 Discussions

The finding of this paper is different than previous studies which mainly focuses on technology development. This study is the first step forward to investigate how modern technologies can be used and managed in the property business. Currently, there are two main aspects of residential property marketing, namely, forward delivery housing and existing properties. The increasing number of transaction of forward delivery housing in Australia make both developers and agencies try to find more effective marketing methods to present housing before the construction process is completed. Practically, 3D modeling pictures and 3D Modeling videos of forward delivery housing have been widely used by property development companies, but the real estate websites, such as realestate.com.au, domain.com.au, realestateview.com.au and so on, have not provided this kind of services for users, which greatly restricted the residential property online marketing.

On the other hand, the transaction of existing homes in Australia still occupies the important position in residential property market. Because of the long distance, it is very inconvenient for people, who want to buy or rent a property far away from their location, to attend the property inspection. Under this circumstance, a convenient and low cost approach to inspect the objective residential property is essential. This research suggests two types of 3D scanners to real estate websites to scan indoor details of existing properties, and present 3D model of the objective property on the real estate website as real as possible.

The first one, named Scanner A, is a hand-held scanner that could be attached on an iPad. It can create 3D maps of interior spaces and measure entire rooms all at once. This hand-held scanner is able to capture dense 3D models with the push of a button and create new mixed reality experience that does not happen in the reality.

Another scanner, called Scanner B, is heavier and more complicated. Users should carry it to the scanned area to collect spatial data. However, this tool can scan target rooms in a away to be converted to Virtual Reality and created 3D models accessible to customers. Customers can straightly see the model via the web. What is more, the scanner 'enables anyone to digitally reconstruct real-world spaces for use in VR, AR, web, and mobile applications with ease' and 'offer clients immersive experiences that give viewers a realistic sense of actually being there'.

5 Conclusion

This paper is a result of an ongoing study to evaluate five popular real estate websites in Australia from customers point of view. The investigation also identifies key web figures by conducting a survey in Australia. The application of digital technologies, especially 3D modeling technologies and indoor scanning technologies were also investigated in terms of functionality and compatibility with the real estate websites to improve their online property marketing. This study also shows that how customers could receive rich information from various aspects of a property by online visiting as well as real physical inspection. The application of visualization technologies to residential property marketing will provide customers with more reliable and credible property information by 3D modeling and images, which is beneficial to property owners, customers and other stakeholders as well as real estate agents. The next step of this study is to further develop the application of modern technologies in this business and recruit more participants to the modified questionnaire.

References

1. Matthew Hinton, C. and Y. Tao, *Exploring sources of competitive advantage: E-business applications within the Chinese real estate industry.* Journal of Technology Management in China, 2006. **1**(1): p. 92-106.
2. Muhanna, W., *E-commerce in the real estate brokerage industry.* Journal of Real Estate Practice and Education, 2000. **3**(1): p. 1-16.
3. Rassmann, A., *Real Estate Marketing.* 2000.
4. Knirsch, B., *Real Estate Marketing.* 2000.
5. Bond, M., et al., *Uses of websites for effective real estate marketing.* Journal of Real Estate Portfolio Management, 2000. **6**(2): p. 203-211.
6. Anderson, P. and E. Anderson, *The new e-commerce intermediaries.* MIT Sloan Management Review, 2002. **43**(4): p. 53.
7. Bouchlaghem, D., et al., *Visualisation in architecture, engineering and construction (AEC).* Automation in construction, 2005. **14**(3): p. 287-295.
8. Mahdjoubi, L., C. Moobela, and R. Laing, *Providing real-estate services through the integration of 3D laser scanning and building information modelling.* Computers in Industry, 2013. **64**(9): p. 1272-1281.
9. Ku, K., et al., *3D model-based collaboration in design development and construction of complex shaped buildings.* Journal of

Information Technology in Construction (ITcon), 2008. **13**(19): p. 258-285.
10. Far, M.S. and C.M. Duarte, *Does 3D Information Modeling Give Better Vision To Deal With Future Real Estate Capacities of Inaccessible Areas Within Conflict Zones?* 2015, European Real Estate Society (ERES).
11. Tóth, T. and J. Živčák, *A comparison of the outputs of 3D scanners.* Procedia Engineering, 2014. **69**: p. 393-401.
12. Kolev, K., et al. *Turning mobile phones into 3d scanners.* in *Proceedings of the IEEE Conference on Computer Vision and Pattern Recognition.* 2014.
13. Zhang, Y., et al., *3D Laser Scanning Technology-based Historic Building Mapping for Historic Preservation.* International Review for Spatial Planning and Sustainable Development, 2015. **3**(2): p. 53-67.
14. Jud, G., D. Winkler, and S. Sirmans, *The impact of information technology on real estate licensee income.* Journal of Real Estate Practice and Education, 2002. **5**(1): p. 1-16.

Appendix 1: An extract of the questionnaire completed by a participant

Section 2 - Sufficiency	1	2	3	4	5
7. This website offers enough real estate options and information that I need.	□	□	□	■	□
8. I found a satisfactory amount of information that I needed on this website.	□	□	■	□	□
9. This website provides sufficient information that I need.	□	□	■	□	□
10. The housing information that this real estate website provided is sufficient for me to make decision.	□	□	□	■	□
11. The specific property information on the property page is sufficient for me to make decision.	□	□	□	■	□
Section 3 - Vividness	**1**	**2**	**3**	**4**	**5**
12. The procedure instructional content on this real estate website is animated.	□	■	□	□	□
13. The procedure instructional content on this website is lively.	□	□	■	□	□
14. This website provides sufficient information that I need.	□	□	□	■	□

HAZUS-based Drill Scripts Generation Using Unity and Ontology with Semantic Rules

C.P. Chu[a], C.Y. Liu[a], C.C. Hsieh[a], C.H. Chang[a], and C.C. Chou[a]

[a] Information Technology Program, Department of Civil Engineering, National Central University, Taiwan
E-mail: ncubim@gmail.com

Abstract

As there is an increasing number of disasters hitting countries around the world, numerous mitigation strategies have been proposed and implemented to alleviate such impacts. Indeed, every year many public and private organizations work together to prepare various disaster exercises for training their first responders. Design of an appropriate drill script, hence, plays an important role of enhancing the capability of first responders or emergency response units in response to a real disaster. However, developing a reasonable disaster scenario is a time-consuming and error-prone task, not to mention the prerequisites pertaining to extensive experience and knowledge in disasters handling that drill script designers must have. Therefore, in this research an ontology model with SWRL (Semantic Web Rule Language) constructs was proposed to help the drill scripts generation process. Drill script designers can use the Taiwan Earthquake Loss Estimation System (TELES), which is a customized and simplified version of the HAZUS-MH program from FEMA (Federal Emergency Management Agency), to create data sets describing the impact of a simulated earthquake in a certain region. Then, the proposed model and tools, called EDSS (Earthquake Drill Scripts generation and rendering System), can incorporate the input data sets and generate a set of corresponding drill scripts, which follow the pre-defined rules represented in the ontology model. Each drill script will contain a unique and possible scenario describing the earthquake damage on different buildings or facilities in the region. Unity, a serious game platform, was utilized in the proposed system for displaying all the events associated with the earthquake in accordance with the drill script designed. Finally, disaster mitigation experts were asked to evaluate the drill scripts generated and to assess the effectiveness of the proposed model. The assessment results showed that the proposed model and tools can help first responders possess better situation awareness regarding how an earthquake will progress. Mitigation of earthquake damage can be achieved if first responders can use the proposed system to review the scene before commencing any search and rescue work.

Keywords –
MAZUS-MH; serious game; ontology; Semantic Web Rule Language (SWRL)

1 Introduction

Nowadays, in disaster-prone regions such as Taiwan, public and private organizations often work together to prepare various exercises for their first responders or emergency response units, in hope of mitigating disaster impacts. According to the United States Homeland Security Exercise Evaluation Program (HSEEP), a disaster exercise can be classified as either the discussion type, such as a tabletop exercise involving key personnel discussing a simulated scenario in an informal setting, or the operations type such as a full-scale exercise, which is a coordinated, supervised activity usually employed to test specific operations within multiple agencies. Both types of exercises with well-designed scenarios can help organizations cope with a disaster. In practice, a disaster exercise can be expressed in the written form called a drill script, which should contain descriptions to mimic real-world events in accordance with a disaster's progress for a given environment. However, the relationship between a disaster exercise and its drill script is different to that between a film and a film script. A film script describes instructions for the actors and directions for filming, while a drill script does not contain instructions regarding how first responders react to a disaster. Instead, a drill script includes detailed descriptions pertaining to the status of the environment and/or evacuees as the disaster proceeds. It is first responders' responsibility to perform the so-called OODA (Observe-Orient-Decide-Act) loop to practice their responses to the disaster in the simulated environment.

Previous literature has shown that the development of an appropriate drill script plays an important role of enhancing the capability of first responders or emergency response units to handle a disaster. The level of a community's disaster preparedness and readiness are also highly dependent on the quality of drill scripts utilized. However, creating a reasonable scenario to simulate a disaster's progress is a time-consuming and error-prone task. An event described in a drill script may have an irrational location, e.g., earthquake damage spotted in an area without any buildings, or a well designed and built house described as being collapsed. Such an event may be described to occur at an impractical time point, too, e.g., post-earthquake fire occurring before buildings collapse. Additionally, since the time available to consult with all stakeholders such as disaster management officers and first responders is limited, even an experienced drill script designer may not be able to consider all possible scenarios to be prioritized into the script. Thus, actual drill scripts utilized for disaster exercises are often inaccurate, inconsistent and incomplete. Improvement of the drill scripts generation process is highly needed and has been suggested for years.

Recently, as computer programs for estimating potential losses from disasters such as the FEMA HAZUS-MH program have been improved significantly, one might be able to better forecast the impact of a disaster, provided that both the disaster and the environment-related data have been accurately and completely entered into such program. For instance, many earthquake mitigation officials in Taiwan use TELES (Taiwan Earthquake Loss Estimation System, one of the HAZUS-MH derivatives containing simplified functions and localized data for estimating earthquake damage in Taiwan) to estimate potential losses resulting from an earthquake hitting their administrative regions. TELES is currently maintained by Taiwan National Center for Research on Earthquake Engineering (NCREE), and its analysis results are usually of good quality. Nevertheless, such data are displayed in two-dimensional formats, e.g., 2D GIS, text, tables and figures, which cannot be directly utilized to develop a drill script accommodating both the space and the time dimensions. In other words, if the analysis results from such estimation programs could be well integrated into a drill script, first responders will have better situation awareness regarding how the disaster impacts are distributed and progressed in the environment, thus engaging them in deliberating upon best countermeasures.

Further, first responders often would like to practice their operation skills under different assumed disaster scenarios. Since a typical drill script can contain the descriptions for only one scenario, drill script designers may need to prepare many scripts, each with a unique scenario, so that their supervisor can randomly select one script to be exercised by first responders. As noted previously, the development of a reasonable drill script requires much time; hence, most of the scripts designed for such purposes are almost identical with a few variances, implying that first responders might use the same set of countermeasures to respond to all the disasters simulated. Therefore, in order to avoid such problems and to actually increase the situation awareness of first responders, differentiating these drill scripts is a must, in hopes of training first responders to be more confident of various real-world circumstances.

Recently, as the technology of ontology and SWRL (Semantic Web Rule Language) has evolved dramatically, many knowledge-based tools have been successfully applied to different disciplines. The drill scripts generation process might need such technology because it involves many rules that only senior drill script designers can apprehend. In order to help the drill scripts generation process, if disaster loss estimation programs as well as such rules can be combined into an ontology model with SWRL constructs, a more reasonable drill script could be generated in a more efficient way. Additionally, first responders might need to deliberate with each other to identify best countermeasures because each drill script they face will describe a sole disaster scenario that would require them utilize different sets of operation skills.

Lastly, as the serious games technology (SG) advances rapidly, there is an increasing number of disaster mitigation applications implemented using SG. Serious games, defined as digital non-entertainment games, have the advantage of offering an engaging, interactive, and collaborative virtual experience, which could lead to better situation awareness for players. Although some research studies have utilized SG to realize drill scripts for disaster mitigation, none of such studies investigates the drill scripts generation process that is based on any of the disaster losses estimation models or programs. If the output of such estimation programs can be integrated into a SG platform, its VR-driven scripts could benefit first responders greatly because each of the drill scripts resembles a reasonable and possible real-world, 3D disaster scenario. In the meanwhile, since the first responders can see the simulated disaster in a VR environment, their capability of copping with the disaster could be enhanced due to the improved situation awareness.

Therefore, this research proposed a model with tools implemented using Unity (a SG platform) for generating such drill scripts and for displaying the simulated environment hit by a disaster. For the model demonstration purpose, an earthquake disaster and an area in the city of Taipei were selected, and experts from the fire department were invited to assess the model performance. Section 2 presents the proposed

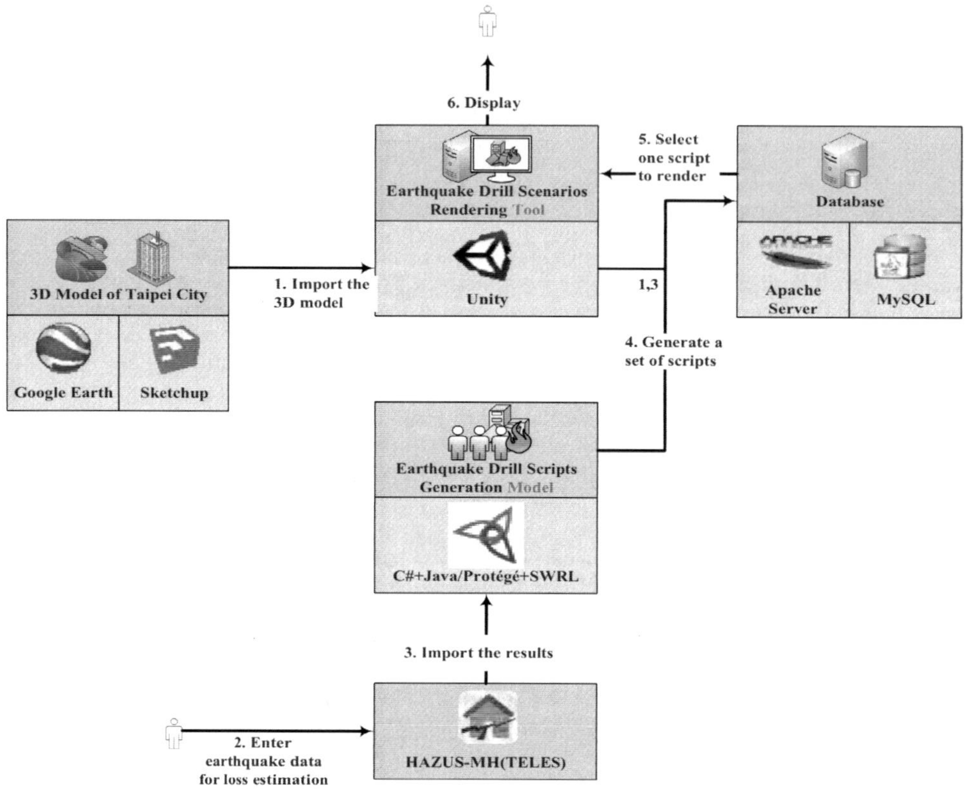

Figure 1. Software architecture of EDSS

model and the system architecture. Section 3 highlights research conclusions with recommendations for future research.

2 Related Literature

MASA Sword is a simulation platform for multi-level training for corps, brigade, battalion commanders and disaster mitigation officers. It can be regarded as an online game environment for various stakeholders of a simulated war or an emergency drill to practice their skills. One of the main components used in MASA Sword is a GIS, and users can customize many parameters defined in MASA Sword to render different disaster scenarios on the GIS output window. However, improvements of such tools have been identified, mainly about the 3D display functionality and disaster events linking to disaster impact estimation programs.

Although the use of a GIS component is a common method to display disaster events, such output information cannot be displayed in a 3D environment, implying that the users' perception about a disaster in analysis is low and first responders may not be able to recognize potential hazards in real-world settings. In addition, current GIS tools cannot accommodate the temporal requirement that disaster-related events should be displayed according to their occurrences. Finally, although users of MASA Sword can edit historical events of a disaster to simulate the scenario, a disaster usually consists of numerous events, which makes the manual editing process impractical due to less time available for the creation of a disaster drill.

TYGRON Serious Game is another game environment specifically designed for floods management. It possess 3D city models and engines so that users can navigate a city to have better situation awareness about the distribution of water resources within the area. From the disaster mitigation perspective, it simply displays potential economic or lives losses and does not simulate a disaster. Thus, further improvements may be needed to enhance first responders' capability regarding the preparation and development of potential countermeasures for disasters.

3 Proposed Model and Tool

Figure 1 shows the overall architecture of the proposed model and tool, called EDSS (Earthquake Drill Scripts generation and rendering System). It consists of five components and six usage steps. The first step requires a system administrator import the 3D model for a given region in Taipei. Currently Taipei city government has provided the 3D model in the Google Earth KMZ format, which includes a triangular meshed model with texture mapping for each building in Taipei. Once exporting the buildings to the KMZ format, the administrator can use a data pre-processing tool developed by the research team to clean missing data and to combine each building's 3D geometry with its text-based attributes. After Step1, the Unity component of EDSS can be used to render each building with the attributes information such as the number of floors and the total floor area, which come from the KMZ file and/or the EDSS database component.

Step 2 requires users or first responders enter all relevant information of an earthquake into the HAZUS-MH or TELES program. Since HAZUS-MH will generate the loss estimation data sets based on the given earthquake parameters, another data transformation component of EDSS needs to be used in Step 3 to move such data sets into the corresponding EDSS database component and to perform appropriate data conversion work. Then, the user can specify the number of drill scripts to be displayed later and perform Step 4 to actually generate a set of drill scripts, based on the ontology model and the loss estimation data sets, so that the EDSS database component can contain not only the 3D model of a given region but the set of drill scripts that can be used to describe a possible earthquake damage scenario for the buildings. Finally the supervisor of the first responders can randomly select one drill script from the EDSS database component (Step 5) and use Utility to render all the earthquake-related events (Step 6).

Figure 2 shows the proposed ontology model that will be used to generate earthquake drill scripts. A drill script can be classified as either the expected type or the randomized type. The former one contains only one instance while the later one can contain as many cases as users define. The expected type drill script is generated directly based on the loss estimation program, while the randomized type of drill scripts will contain all possible scenarios for a given earthquake, with the average impact level equal to the expected type. For example, suppose there are 100 high-rise buildings in a given region. HAZUS-MH estimates that two of such the buildings will collapse; hence, in the expected drill script EDSS will pick two building, which possess the largest floor areas, to show the collapsed events. However, in a randomized drill script, EDSS may pick more buildings to display collapsed events while in another randomized script only one building will collapse. The randomization mechanism needs to make sure that among all the randomized scripts, the average number of the high-rise buildings that collapse should be equal to two. This is because first responders would

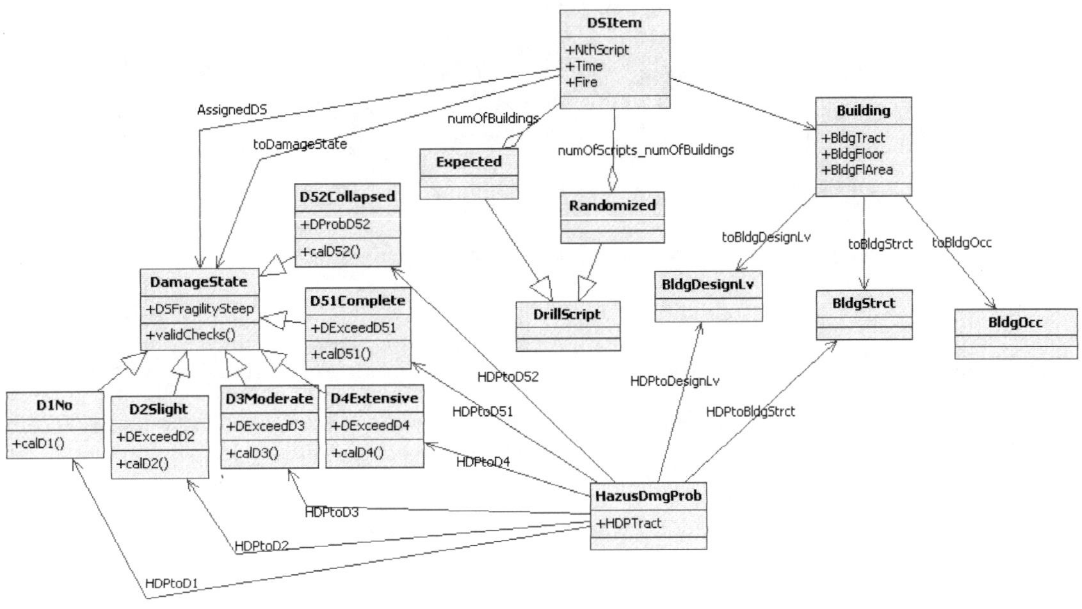

Figure 2. Ontology model of EDSS

like to practice their skills under different scenarios; hence, each randomized drill script should be unique, i.e., one collapse will be assigned to different buildings for different scripts. However, consistence and reasonableness must be maintained among the randomized scripts. For example, people injuries are less likely to occur at commercial buildings when the stores of the building are closed. Thus, a series of SWRL rules have been designed to make sure that all the events generated in a randomized script are consistent with the underlying assumptions. Finally, these events are stored in the database tables. A DSItem instance in Figure 2 will indicate when a given building will have an earthquake damage event as well as the damage level, which can be D1No, D2Slight, D3Moderate, D4Extensive, D51Complete, or D52Collapsed. Mode details can be found at the report published by the research team.

different, randomized drill script. In fact, EDSS needs more varied scenarios to be displayed and to train first responders. Ideally, first responders should not know which drill script their supervisor assigns to them for exercise. EDSS employs the ontology model with SWRL rules to generate all the events, in hope of each drill script with a unique scenario.

EDSS will render their damage scenarios at different time points. Figure 4 shows a sample fragility curve derived from HAZUS-MH. Basically, EDSS generates a randomized drill script based on the following principle: if the fragility curve for a certain damage level is steep, some of the drill scripts will contain a high probability for the given damage level, where others do not. Among all the randomized script, the average probability will be the same. But in an actual county, we cannot simply rely on the estimated probability to do calculation.

Fires and casualties will appear following building collapses. Basically, since HAZUS-MH can estimate the probability of each damage level for a given area, EDSS will determine the actual buildings that will suffer such damage levels, if they can be classified to the group that has the corresponding damage probability. Within the same group, a building having the largest floor area will be selected to show the corresponding damage effect. Future work includes rendering for city transportation facilities such as bridges and roads because their damage states can be obtained in the output data sets of HAZUS-MH. Specific Utility effects will be developed to simulate actual collapse events for

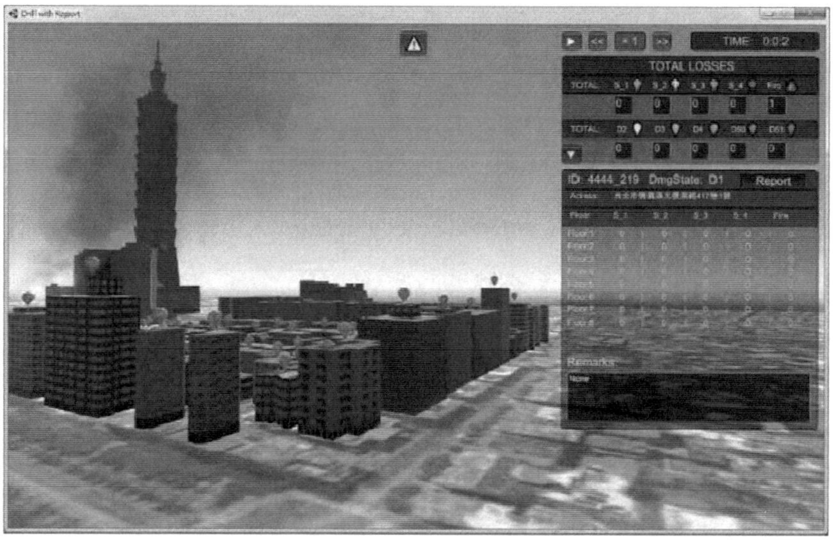

Figure 3. EDSS user interface

4 System Output and Discussions

Figure 3 shows a screenshot of EDSS. For demonstration, an earthquake is assumed to occur near Taipei 101 with a Richter scale of 7. In the specified region for the expected drill script, there are 25 slight-damaged buildings, 21 moderate-damaged buildings, 9 extensive-damaged buildings, and 2 complete-damaged buildings. Although such damage numbers are determined by HAZUS-MH, they serve as the default scenario to be used by EDSS to create a set of 100 such facilities.

It should be noted that the activities that first responders should perform to alleviate the impact of a disaster are not included in EDSS. EDSS is designed to help disaster mitigation officers dispatch first responders to the scenes. It can also help first responders better perceive the surrounding environment and the progress of a disaster. In the real world, the interaction between a disaster and first responders is dynamic. With the estimation data provided by tools such as TELES, EDSS can simulate the events but cannot help mitigate the disaster. Decision makers can learn the distribution and the extent of a disaster, thus dispatching an appropriate number of first response units to the scenes.

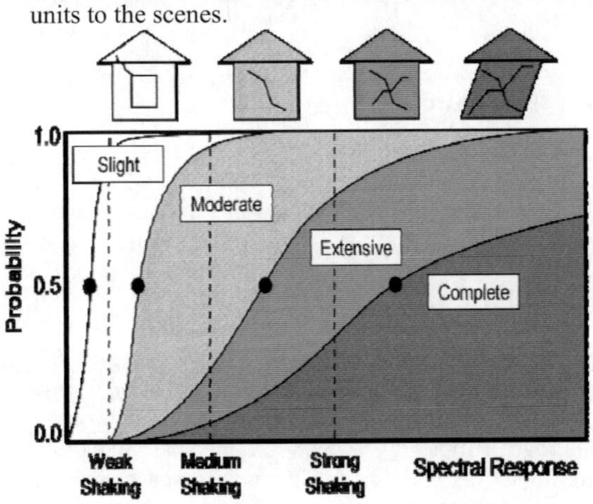

Figure 4. Sample HAZUS-MH fragility curve

5 Conclusions

A Unity-based serious game platform was designed to serve as simulation engine for disaster mitigation. A HAZUS-MH-derived version called TELES was utilized to generate data sets describing the main scenario of a disaster. An ontology model with semantic rules were developed to randomize the data sets so that different disaster scenarios can be generated. Then, the proposed system called EDSS can combine all the data sets and display them in a 3D environment so that first responders can have better situation awareness regarding the progress of a disaster. Disaster mitigation officials can dispatch response units based the information created by EDSS. Experts were invited to assess the effectiveness of EDSS and found that the scenarios displayed by EDSS can resemble real disasters and train decision makers in a more efficient way.

Future improvements of EDSS include expanding the areas and scenes created inside EDSS. This is because not only building geometries but the human activities of the buildings in the study area need to be created in the input data set for both TELES and EDSS. In addition, currently only the earthquake type of disasters can be simulated by using EDSS. Once the disaster estimation programs for other types of disasters are available and can be integrated into EDSS, both the ontology model and corresponding semantic rules need to be developed and enhanced so that the new version of EDSS can display such types of disasters.

6 ACKNOWLEDGMENTS

The authors would like to thank the Ministry of Science and Technology in Taiwan for financial support under grant numbers: MOST105-3113-E-008-002 & MOST103-2221-E-008-054-MY3.

7 REFERENCES

[1] Anil, E.B., Akinci, B., Kurc, O., and Garrett, J.H., (2015). Building-Information-Modeling–Based Earthquake Damage Assessment for Reinforced Concrete Walls, Journal of Computing in Civil Engineering, 30(4).

[2] Charalambos, G., Dimitrios, V., and Symeon, C., (2014). Damage Assessment, Cost Estimating, and Scheduling for Post-Earthquake Building Rehabilitation Using BIM, 2014 International Conference on Computing in Civil and Building Engineering, pp. 398-405, Orlando, USA.

[3] Comber, M.V. and Poland, C.D., (2013). Disaster Resilience and Sustainable Design: Quantifying the Benefits of a Holistic Design Approach, Structures Congress 2013, pp. 2717-2728, Pittsburgh, Pennsylvania, USA.

[4] Federal Emergency Management Agency (FEMA), (2013a). Hazus-MH 2.1 Earthquake Model Technical Manual, FEMA, Washington, D.C., USA.

[5] Federal Emergency Management Agency (FEMA), (2013b). Hazus-MH 2.1 Advanced Engineering Building Module (AEBM) Technical and User's Manual, FEMA, Washington, D.C., USA.

[6] Federal Emergency Management Agency (FEMA), (2013c). Hazus-MH 2.1 User Manual, FEMA, Washington, D.C., USA.

[7] Federal Emergency Management Agency (FEMA), (2013d). HAZUS-MH Used to Support San Francisco Bay Area Earthquake Exercise, FEMA, Washington, D.C., USA.

[8] Federal Emergency Management Agency (FEMA), (2015). Hazus-MH Overview, Online https://www.fema.gov/hazus-mh-overview, Last Accessed November, 2015.

Developing a BIM-based Visualization and Interactive System for Healthcare Design

Chao-Yung Huang[a], Huey-Wen Yien[b], Yen-Pei Chen[a], Yu-Chih Su[c], and Yu-Cheng Lin[a]

[a]Department of Civil Engineering, National Taipei University of Technology, Taiwan
[b]YongLin Healthcare Foundation, Taiwan
[c]Research and Development Department, WeBIM Services Co., Ltd., Taiwan
E-mail: chaoyung_huang@outlook.com, wynn.yien@ylhealth.org, gbass1988@hotmail.com, marksu@webim.com.tw, yclin@ntut.edu.tw

Abstract –

In the building life cycle, the design phase is related to the usability and practicability of the building operation. Communicating ideas regarding the design requirement is the key point. Healthcare building design contains the most rigorous and complex system and should be fully considered. However, designers often only use 2D shop drawings, architectural rendering of viewpoints, and paper models to discuss the design concept and requirements with the stakeholders.

Building Information Modeling (BIM) is a revolutionary technology and solution in the modern architecture, engineering, and construction (AEC) industry. BIM is a suitable information collaboration platform for AEC projects. However, most BIM software still does not provide realistic visualization and simple interactive methods. The stakeholders may not easily imagine and understand the design, then provide feedback in the form of professional medical requirements or advice to designers. To address this problem, this study develops the BIM-based Visualization and Interactive System (BIM-VIS) integrated with BIM, game engine, and virtual reality (VR) technology. The BIM-VIS is a real-time rendering system that shows architecture, the spatial design concept, and building information. Furthermore, users may utilize the wireless gamepad as a controller to easily interact with building objects. Finally, the BIM-VIS system was applied in a selected test to demonstrate the effectiveness of interactive BIM use. Limitations and suggestions are summarized for further applications.

Keywords –

Building Information Modeling (BIM); Virtual Reality (VR); Game Engine; Healthcare Design

1 Introduction

During the building life cycle, there are many different interfaces, types of knowledge, and professionals. Thus, the communication and integration is an important topic for a long period of time. However, the design phase always determines the pattern and practicality of building operation. As medical technology improves everyday, the healthcare design is becoming more complicated. The healthcare design should be fully considered. Thus, suitable design and planning to reduce the healthcare operation costs and improve efficiency, such as medical departments configuration, wayfinding, humanized space design and other important medical factors need to be addressed. Compared to general building designs, the healthcare design is more rigorous and difficult. However, in general healthcare design, designers often only use hundreds of shop drawings, architectural rendering of viewpoints, and paper models to discuss the design concept and requirements with stakeholders, and this usually results in information gaps and misunderstanding. It is difficult for the architect to understand the special requirements and relevance of the medical details. For the medical staff, it is hard to understand the spatial design and to review the design to ensure that it is suitable for medical practice and operations, because they cannot connect and imagine many different medical events with complex shop drawings. Furthermore, the duration of the design and construction are long. After construction, the finished product may be different from the design. This discrepancy is due to the lack of current building design and communication with stakeholders. Therefore, Building Information Modeling (BIM) and Integrated Project Delivery (IPD) concepts should be integrated into healthcare facility design in the early phase. BIM is an emerging technology and solution that is revolutionary in the modern architecture, engineering, and construction (AEC) industry. BIM includes object-oriented parametric modeling techniques, and it enables the visualization of communicated ideas, analysis, and simulation. It is a suitable information collaboration platform for an AEC project. At present, there are many AEC cases starting to use BIM. However, most BIM software still does not provide realistic visualization and simple interactive methods. Stakeholders, may find it difficult to imagine and understand the design, then provide feedback in the form of professional medical requirements or advice to designers. As mentioned above the blind spot in design phase of building life cycle and deficiencies of BIM. How to provide medical staff or users a realistic visualization and simple interactive communication platform, will be one of the topics in healthcare design.

To address the above problems, this study expects to integrate with BIM and game engine technology to develop a visualization and interactive system of healthcare design. It can improve the visualization performance for BIM, and integrate medical staff experiences, knowledges, and special requirements of this system. It will show the real-time rendering virtual space, information, and other customized functions of the hospital building for healthcare facility design options. Furthermore, the system user may utilize the wireless gamepad as a controller to interact with BIM model in this system. This system will improve the visualization and interactive operations, and it can be an efficient communication and presentation system for integrating opinions with medical staff.

2 Literature Review

Kang et al. (2014) updated their BIM CAVE (Computer Aided Virtual Environment) system based on Autodesk Navisworks to display the BIM 4D construction sequence. The CAVE system included 3 screen walls and 3 computers to project the real-time rendered images of the Autodesk Navisworks model. The BIM CAVE system controls the camera location, aiming angle and field of view (FOV) in Navisworks, so that the 3 screen walls can project the 3D BIM model synchronously. It was designed to facilitate users feeling the sense of presence in a 3D computer model [1].

Kang et al. (2015) presented how the BIM CAVE application has been modified, and updated the BIM CAVE immersive space with 9 screen walls to enhance virtual reality and user experience [2].

Wu and Kaushik (2015) integrated BIM and a game engine to create a system framework for simplified design scenarios for sustainable design that transforms static design models and information into an interactive gaming environment to improve the communication between designers and customers [3].

Hilfert and König (2016) presented a way to build a low cost, highly immersive virtual reality environment for the engineering and construction industries. Using an Oculus Rift headset, Leap Motion hand-tracking device, and Unreal Engine 4 as the game core, they created a virtual scene that shows the interactive in an immersive virtual reality in different scenarios. In the process, BIM data can be imported, and automatic mapping materials in the Unreal Engine 4 with the developed plug-in accelerate and simplify the process [4].

The related topics of integrated BIM, game engine, and VR are reviewed. This study will be considering how to develop and integrate the BIM, game engine, and VR technology with related knowledge into healthcare facility design, and present the hospital scenes with healthcare facility design research and experience in VR space to promote current communication methods in the design phase. Therefore, the stakeholders can understand the detailed results of the healthcare facility designed with VR experience; meanwhile, the stakeholders can provide feedback and advice to designers.

3 System Development

This study develops the BIM-based Visualization and Interactive System (BIM-VIS) integrated with BIM, game engine and VR technologies. The BIM-VIS provides a VR space to improve visual communication between designers and medical staff. In order to provide the VR experience when a user operates the BIM-VIS for browsing the BIM model scene, this study involves building a semi-immersive VR projection system to display the BIM model scenes of the BIM-VIS. The semi-immersive VR system provides a greater appreciation of scale than VR on a Head-mounted Display (HMD). In addition, images can be provided that are of a far greater resolution than HMD and this implementation provides the ability to share the virtual experience [5].

3.1 System Concepts and Framework

BIM-VIS can strengthen visual communication, and improve the visualization and interactive abilities more than existing BIM software. The designer can integrate user experience from medical staff in the design phase to enhance architectural design efficacy and ensure the user requirements are met. BIM-VIS functional concepts as follows:

1. Browsing Experience Module: BIM-VIS integrates existing functions of the game engine for developing the browsing experience module. It provides users with free browsing with a first-

person view in the virtual building. Users can utilize the wireless gamepad as a controller to easily interact with the building scene. This function can reduce operational technical limitations. Compared to most BIM software that uses a keyboard and mouse, BIM-VIS has an easier and more interactive operation experience for BIM model browsing.

2. Scene Map & Interest Points Navigation Module: BIM-VIS integrates the advantages of 2D shop drawings in a 3D virtual interior space. This function provides user instantly understand the current position and floor of the hospital. In addition, the developer can set interest points in key spaces to provide the user with a quick switch between the interest points on the scene map to move to the key space or to speed up the virtual browsing.
3. Height Simulation Module: BIM-VIS allows the user to change 3 different roles' height to simulate the space collision, and review the friendly design space.
4. Path Track Record Module: BIM-VIS provides a path track record function for statistics on the user's movement distance, path selection, speed, and time. However, the designer or related healthcare facility design research team can use statistics to review the hospital symbols' setup position, height, angles, quantity, and design.
5. Healthcare Design Research and Knowledge Tips Module: BIM-VIS integrates related healthcare facility design research and knowledge for design reference. When the user is near and focuses on an issue bubble in the scene, the BIM-VIS accesses the server database to get corresponding design research and information, then shows the content on the information card. It can help the stakeholders understand the healthcare facility design research and knowledge on a specific space or piece of equipment.

After, development the BIM-VIS will be consistent with the patient medical scenario. A BIM-VIS user can use system functions and browse the medical building space for review, and discuss the healthcare facility design in a virtual environment. This study can be used as knowledge accumulation and a reference for any next healthcare facility design project.

In this study, the BIM-VIS was developed by BIM and a game engine. The game engine selects the Unreal Engine 4 (UE4) which was developed by Epic Games, Inc. The main reason is UE4 is open source and free, along with all future updates. In addition, UE4 can support a huge and complex model while still operating smoothly. However, the developer can use the UE4 "Blueprint" visual scripting system to customize function modules and user interfaces. The blueprint is a complete gameplay scripting system based on the concept of using a node-based interface to create gameplay elements. It is used to define object-oriented classes or object in UE4 [6]. Overall, the visualization and interactive development system of UE4 meets the requirements of the game engine.

The BIM-VIS has three operation process (see Figure 1), including "BIM Model Operation Process," "Game Engine Operation Process," and "Presentation & Communication Process," as follows:

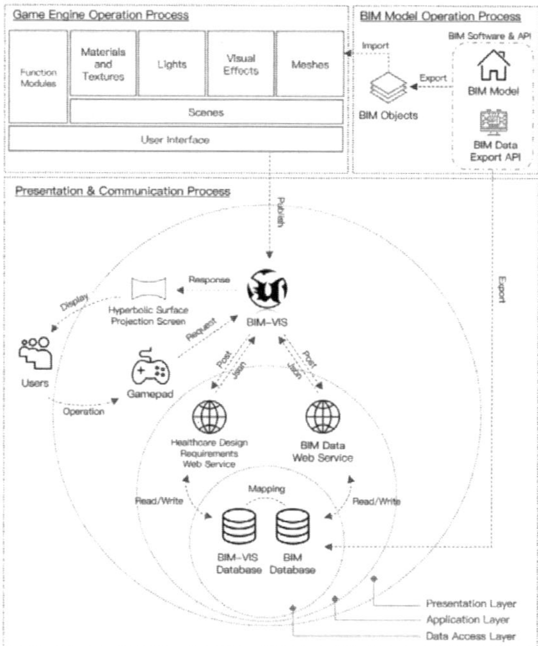

Figure 1. The system framework of BIM-VIS

1. BIM Model Operation Process

The purpose is to export and upload the BIM model data from Revit to the BIM database via the BIM data export API (Application Programming Interface).

The BIM model is exported to the .fbx file format, which only maintained the model element, ID, name, and position data. Therefore, in this process, the BIM database is purpose to access non-geometric information such as BIM model element ID, size, material, manufacture, etc. The BIM-VIS can access the BIM information in BIM database, by BIM model element ID. In addition, the complete BIM central model need to process the unit conversion, reduce the polygon faces, and unwrap the model texture UVW in Autodesk 3ds Max. Then export the BIM model (.fbx file) for import to the UE4.

2. Game Engine Operation Process

The game engine provides a variety of tools and compute technologies for development the huge and complex game scene with interactive functions and user interfaces. Therefore, it can improve the operation smoothness and efficiency of the BIM model, and using lightings, materials, visual effects, and camera settings of game engine enhances visualization of the BIM model. The purpose of this process is development of system function modules, user interfaces, and building of the healthcare facility interior scenes in UE4. In addition, the BIM-VIS can access non-geometric information of BIM model elements in the BIM database, and medical research and requirements in BIM-VIS database via customize system function modules. It can also build and customize function modules like the "Path Track Record Module" and the "Height Simulation Module" of the BIM-VIS.

3. Presentation & Communication Process

This process is a three-tier framework, including the "Data Access Layer", "Application Layer", and "Presentation Layer", as follows:

- Data Access Layer

This layer contains the BIM-VIS database with the hospital design requirements and research information. The BIM database stores the non-geometric information of BIM model elements.

- Application Layer

This layer contains the "Hospital Design Requirements Web Service" and the "BIM Data Web Service" as a connection link with Data Access Layer and Presentation Layer. BIM-VIS uses JSON (JavaScript Object Notation) as a data exchange language.

When the web service receives a request from BIM-VIS, the web service converts the model ID to JSON, then posts to the server to read and write the information (such as healthcare facility design requirements and medical research) from the BIM-VIS database, or the BIM data (such as BIM model element material, size, etc.) from the BIM database, and then converts the information via the web service, and responds to the BIM-VIS.

- Presentation Layer

This layer contains the BIM-VIS, a hyperbolic surface projection screen, and a wireless gamepad. In "Game Engine Operation Process," the UE4 develops the virtual hospital scenes, then packages and publishes them to the BIM-VIS. The BIM-VIS is the main execution and display platform. The BIM-VIS projects the virtual hospital scenes and contents to the hyperbolic surface projection screen, and users can utilize the wireless gamepad as controller for the BIM-VIS.

Regarding hardware, in order to provide the VR experience when a user operates the BIM-VIS for browsing the BIM model scene, this study involves building a semi-immersive VR space. The research team set up a workstation computer with 2 Nvidia Quadro K2000 graphics cards to execute the BIM-VIS, a Microsoft Xbox One wireless gamepad for controller, a 4992×1994 cm hyperbolic surface projection screen made by wood with projector screen paint, and 6 projectors combined to provide 5K high-definition resolution (see Figure 2).

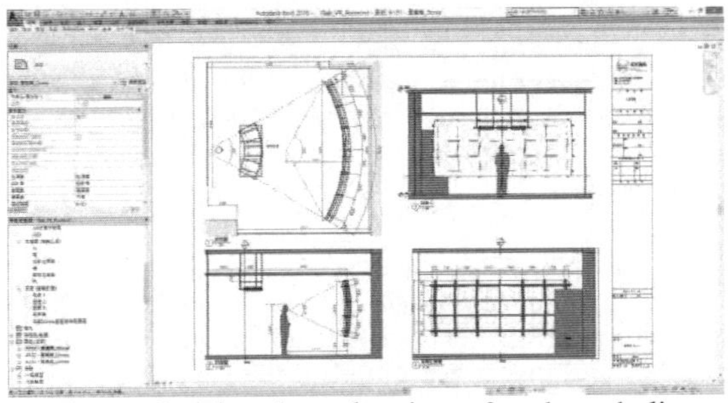

Figure 2. The shop drawing of a hyperbolic surface projection screen

3.2 System Workflow

The following describes the system workflow of BIM-VIS (see Figure 3). The BIM-VIS has three operation workflows, including "BIM Model Operation Workflow," "BIM Model Post-Process Workflow," and "Game Engine Operation Workflow," as follows:

3.2.1 BIM Model Operation Workflow

1. Development of the BIM model in Autodesk Revit.
2. Export and upload the BIM model data, such as model ID, size, material, etc. from Revit to the BIM database via BIM data export API.
3. In order to provide the readable model for 3ds Max in "BIM Model Post-Process Workflow", export the BIM model (.fbx format) in Revit.

3.2.2 BIM Model Post-Process Workflow

The purpose is convert the BIM model unit, breakdown complex model, simplify the polygon faces, and unwrap the model texture in 3ds Max.

1. Since the UE4 world unit is centimeter (cm), however, Revit unit is millimeter (mm). In order to ensure the model scale is correct, this study is convert the BIM model unit to cm in 3ds Max.
2. Import the BIM central model (.fbx) to 3ds Max.
3. Finding the complex BIM model element, such as

medical equipment, etc. It needs to manually determine whether the complex BIM model element, then isolate an BIM model element.

4. Since high polygon faces numbers may cause low performance and computer resource overload in UE4, thus, breakdown the isolated BIM element, and simplify the polygon faces using 3ds Max function. When polygon faces simplified, end isolate and back to BIM central model status.
5. For avoid lightmap overlap, it needs to generate a new model texture UV channel, and unwrap the model texture UVW. However, this study using third-party script of 3ds Max: "SteamRoller" [7]. The SteamRoller can automatically repeat the process for unwrap model texture UVW.
6. Export the BIM model (.fbx) in 3ds Max.
7. In order to provide the readable model for UE4 in "Game Engine Operation Workflow", export the BIM model (.fbx) in 3ds Max.

3.2.3 Game Engine Operation Workflow

The purpose is import the BIM model which finish the post-process in 3ds Max, then build the whole scene of hospital with detail lights, materials, meshes, visual effects settings for simulate the interior scene. In addition, the developer can compile and apply system functions or user interfaces using UE4 Blueprint.

1. Import the BIM model (.fbx) to UE4 scene.
2. Since the BIM model elements (also known as "Mesh" in UE4) only maintains the model element, ID, name and position data in .fbx model, therefore, align meshes to origin (0,0,0) of UE4 scene to ensure BIM model elements are correct position in UE4 project.
3. According to the space, function, and requirement for create multiple streaming sublevels in a persistent level, then assign meshes to multiple sublevels. The purpose avoids too large of the meshes number in a single level, it may cause level read time is too long. In addition, the function module can control the level visibility, it can be used for changing different scene, such as turn on/off the "Chemotherapy area" level.
4. Execution of UE4 world settings, project settings, and edit contents of UE4 scene, including lights, textures, material, meshes, levels, visual effects settings, etc. for improve the visualization.

- World Settings: The world settings including the detail setting of physics, lightmass calculation, game mode, etc.
- Project Settings: For detail settings of current UE4 project, including project name, project initial level, operation settings, etc.
- Materials & Textures Settings: Including material and texture settings such as material roughness, reflectivity, etc.
- Lightings Settings: Including detail settings like intensity, temperature, light source radius, mobility of multiple lighting type, such as directional light, spotlight, point light.
- Meshes Settings: Including meshes detail settings, such as mobility, materials, physics, collision, light map resolution, etc.
- Visual Effect Settings: Including post-process detail settings of UE4 scene, such as atmospheric fog, sphere reflection capture, etc.

5. Finished the hospital scene and detail settings. This study compiles system function modules and user interfaces using Blueprint for communication between server and client, and achieve system functions.
6. Build the UE4 project. UE4 start to compute and render all the levels in the scenes, including meshes, lightings, materials, and other detail settings. When build successful, UE4 updates the scenes and presents the results in UE4.
7. Complete the build, the developer should execute the UE4 project. If the results are not reach expectation, the developer should be re-setting, and test the UE4 project again, until it reaches the expectation quality.
8. When UE4 project reach the quality of the expected results, then package the project to executable file, also known as BIM-VIS main executable file in this study. The purpose provides the user can execute the BIM-VIS, but not on UE4 editor.

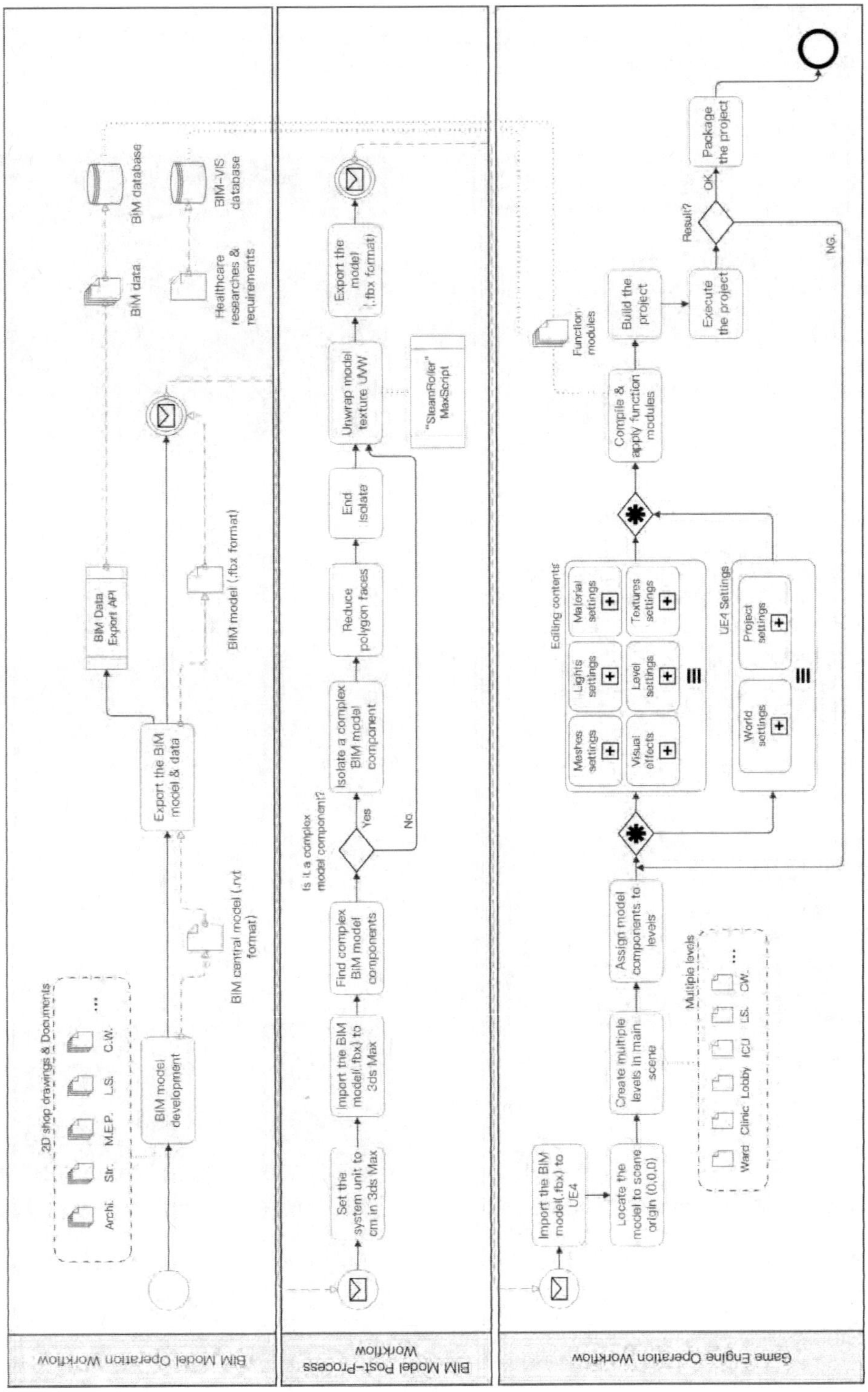

Figure 3. The system workflow of BIM-VIS

4 Pilot Case and Discussions

The pilot test is design phase of National Taiwan University Cancer Center (NTUCC) building. NTUCC is a cancer specialist hospital and research center. NTUCC is donated to National Taiwan University by YongLin Healthcare Foundation. The building located in Taipei City, cover an area of 3.1 hectares, 14 floors on the ground and 4 floors underground.

At the beginning of the design phase, designers and medical staff (such as doctors, nurses) were use 2D shop drawings, architectural renderings and paper models to discuss the healthcare design and medical professional requirements. However, medical staff cannot too easy imagine, and understand the hospital space design from hundreds of shop drawings, renderings, and paper models, then provide in the form of professional medical requirements, experience to designers. They had information gaps and misunderstanding between designers and medical staff. In addition, designers and medical staff were start to use BIM solution into the design phase of NTUCC to improve above problems. Thus, the design team created BIM model of NTUCC by Revit (see Figure 4), then project stakeholders can review the BIM model, and discuss the healthcare design in BIM software.

Figure 4. NTUCC BIM central model in Revit

However, most BIM software still does not provide realistic visualization and simple interactive methods. The stakeholders may not easily imagine and understand the lightings, materials, medical spaces design. In this case, the purpose of development and application the BIM-VIS is used to assist medical staff and other stakeholders to understand the complicated and precise architectural design, and simulate lightings, shadows, materials, spaces, height design, path track, etc. of NTUCC in a virtual environment. Therefore, designers and medical staff can collect feedback, and discuss the design requirements, ideas, or problems on this visualization and interactive platform.

The BIM-VIS achieved the following features in this pilot test, as follows:

- Browsing Experience (see Figure 5)

Figure 5. A medical doctor is browsing the virtual NTUCC building in design phase

- Scene Map & Interest Points Navigation (see Figure 6)

Figure 6. 1-floor scene map with interest points

- Height Simulation (see Figure 7)

Figure 7. The height simulation status icon

- Path Track Record (see Figure 8)

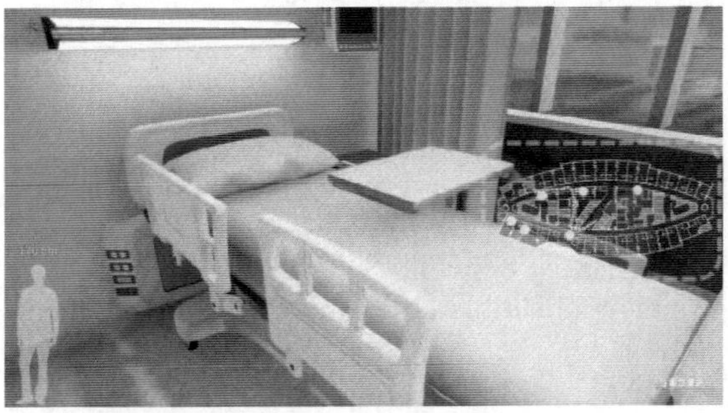

Figure 8. The path track record map

- Healthcare Design Research and Knowledge Tips (see Figure 9, 10)

Figure 9. An issue bubble in NTUCC clinic area

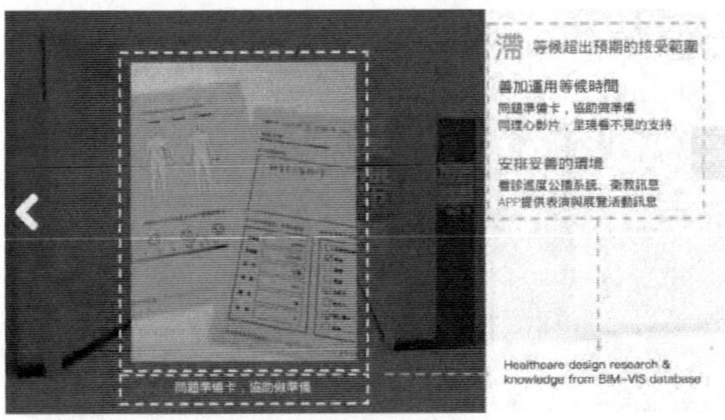

Figure 10. The healthcare design research and knowledge on an information card in NTUCC clinic area

5 Conclusion

In order to improve the visualization and interactive performance of the BIM, and reduce the communication gap between stakeholders in healthcare facility design. This study develops the BIM-based Visualization and Interactive System (BIM-VIS) integrated with BIM, game engine, and VR technologies with professional requirements and knowledge for healthcare facility design. The BIM-VIS is a real-time rendering system that shows real-time rendering architecture, the spatial design concept, and other customized system functions for healthcare design. User can utilize the wireless gamepad as controller to interact with BIM model and information in BIM-VIS. Meanwhile, in order to provide the VR experience when a user operates the BIM-VIS, this study creates a semi-immersive VR space. The research team set up a workstation computer to execute the BIM-VIS, a wireless gamepad for controller, a hyperbolic surface projection screen by wood with projector screen paint, and 6 projectors combined to provide 5K high-definition resolution.

Based on the system result of pilot test, it verified that functions of the BIM-VIS were useful. Furthermore, some factors, such as hardware cost may cause the stakeholders to hesitate for adopting this system. However, the future research may focus on speed up the workflow, compile functions or scripts for automatic execute works.

References

[1] Kang J. and Kuncham K. BIM CAVE for 4D Immersive Virtual Reality. *Proceeding of the Creative Construction Conference 2014 (CCC 2014)*, pages 568–574, Prague, Czech Republic, 2014.

[2] Kang J., Yeon J. and Kangregula S. Fabrication of BIM CAVE 2: Challenges in Handling 9 Screen Walls. *Proceedings of the 32nd International Symposium on Automation and Robotics in Construction and Mining (ISARC 2015)*, Oulu, Finland, 2015.

[3] Wu W. and Kaushik I. Design for Sustainable Aging: Improving Design Communication Through Building Information Modeling and Game Engine Integration. *Procedia Engineering*, 118(2015):926–933, 2015.

[4] Hilfert T. and König M. Low-cost Virtual Reality Environment for Engineering and Construction. *Visualization in Engineering*, 4:2(2016), 2016.

[5] The Advisory Group on Computer Graphics (AGOCG). Types of VR system: http://www.agocg.ac.uk/reports/virtual/37/chapter2.htm, Accessed: 20/04/2017.

[6] Unreal Engine 4 Documentation: Blueprints Visual Scripting. On-line: https://docs.unrealengine.com/latest/INT/Engine/Blueprints/, Accessed: 20/04/2017.

[7] ScriptSpot: SteamRoller. On-line: http://www.scriptspot.com/3ds-max/scripts/steamroller/, Accessed: 12/05/2017.

Automated Ergonomic Risk Assessment based on 3D Visualization

X. Li[a], S. Han[b], M. Gül[c] and M. Al-Hussein[d]

[a]PhD Candidate, Department of Civil and Environmental Engineering, University of Alberta, Canada
[b]Assistant Professor, Department of Building, Civil and Environmental Engineering, Concordia University, Canada
[c]Assistant Professor, Department of Civil and Environmental Engineering, University of Alberta, Canada
[d]Professor, Department of Civil and Environmental Engineering, University of Alberta, Canada
E-mail: xinming1@ualberta.ca, sanghyeok.han@concordia.ca, mustafa.gul@ualberta.ca, malhussein@ualberta.ca

Abstract –

Conventional ergonomic risk assessment of physical work is conducted through observation and direct/indirect physiological measurements. However, these methods are time-consuming and require human subjects to actually perform the motion in order to obtain detailed body movement data. 3D visualization, alternatively, allows users to simulate an operational task on the computer screen, a process that is less time-consuming and which eliminates the need for costly on-site devices, as well as the detrimental effect of human error during experimentation. It can also proactively visualize a proposed design prior to implementation in the real world. This paper presents an automated ergonomic risk assessment framework based on 3D modelling with the support of a user-friendly interface for data-post processing. 3ds Max is utilized together with its built-in MAXScript. The presented system enables the automation of body motion risk identification by detecting awkward body postures, evaluating the handled force/load and frequency that cause ergonomic risk during body movements of workers. As the outcome, it provides detailed risk scores for body segments, such that users are able to review the continuous motion and corresponding risk by precise time frame. The capability of this 3D visualization-based ergonomic risk assessment can be extended to support the re-design of the workplace and optimization of human body movement accordingly. The ultimate goal of this study is to proactively mitigate ergonomic risk and further reduce potential injuries and workers' compensation insurance costs in the long term.

Keywords –

Ergonomics, 3D Visualization, Risk Assessment, 3ds Max, Automation

1 Introduction

The Association of Workers' Compensation Boards of Canada reports that the manufacturing and construction industries had the second and third highest number of lost-time claims due to injuries in 2015, accounting for 14% and 11%, respectively, of total workplace injury claims (232,629) in Canada [1]. In the United States, the manufacturing and construction industries accounted for 11% and 7%, respectively, of all nonfatal occupational injuries and illnesses in 2015 [2]. Thus, improving workplace safety practice in order to reduce work-related injuries in the manufacturing and construction industries is a top priority.

Traditional measures to assess work movements during operation rely on direct manual observation and self-report, which are inherently subjective, time-consuming, and error-prone. Researchers also invest in both direct and indirect physiological measurements in order to collect human body data for ergonomic and biomechanical analyses, which provide results that are more objective, detailed, and accurate than traditional metrics. Body movements can be obtained by utilizing goniometers, accelerometers, and optimal markers. In order to analyse muscle activity, electromyography (EMG) is commonly used to indicate muscle fatigue [3][4][5][6]. However, these measurements all entail job interruption. The solution, then, is to conduct indirect physiological measurement by means of a Kinect range camera or computer vision-based approach, which is also commonly used to capture motion and to conduct body posture assessment [7][8][9][10][11]. Some studies use a video-based computer visualization approach to automatically assess the captured motion [11][12]. However, all of these measurements have limitations in the real-life implementation in construction manufacturing, such as illumination and obstacles in the capturing direction, and the data post-processing is also time consuming [13][14]. Alternatively, human subjects are commonly utilized to simulate tasks in a laboratory setting in order to imitate the task, with direct and indirect measurements employed to capture the motion and to record physiological data for further ergonomic analysis. It should be noted that laboratory-based simulation can represent tasks with a reasonable level of detail and accuracy (comparing alternative methods), and can thereby facilitate effective ergonomic risk assessment (providing adequate information as the input to these risk assessment tools). However, a laboratory setting will have space limitations compared to the field, and thus can only accommodate the simulation of elemental tasks. In addition, the experimentation in such a setting is typically subject to ethical, technical, and cost issues, and it also requires time-consuming data post-processing and a large number of subjects to imitate the motions and activities being evaluated [15]. The need is thus increasing for a new method which can overcome these difficulties and eliminate work interruption on the job in order to automatically identify whether ergonomic risk is increasing.

Li et al. propose a framework of using 3D-motion based modelling to assess human body posture and ergonomic risk with the integration of the existing risk assessment tools, Rapid Entire Body Assessment (REBA) and Rapid Upper Limb Assessment (RULA) [16][17]. The present paper extends this framework and proposes an automated platform for its implementation. The objective of this methodology is to develop a user-friendly interface to visualize the risk assessment and analysis and to help with proposing modified task motions accordingly in order to reduce potential work-related claims and injuries. This framework streamlines the ergonomic risk assessment process and provides easy access to read and compare the analysis results effectively and quickly with graphs and animations. The framework can also be used to evaluate proposed changes to workstations at the design phase.

2 Automation of the 3D motion-based Ergonomic Risk Assessment

The presented framework is explained in Figure 1, with the 3D human body model and user manual input as the two types of data input to the framework. In this research, 3ds Max is chosen to create human body working motion animations, which includes the animation of the biped to imitate human movement in a construction manufacturing facility [18]. The biped can be controlled by altering the length and speed of footsteps, the body posture key frames, and the frame rates. From the 3D human body model, the human body joint angles and REBA/RULA risk assessments are calculated in the MAXScript, the built-in language in 3ds Max software, in conjunction with the programming code [19]. A user interface is developed to collect the input data (time frame selection, force/load data and activity performance score selection) needed to calculate the script. Functional tablet buttons in the interface are linked to the script in order to calculate the joint angle once the needed data has been entered. These functional tablet buttons can also open windows to display the results.

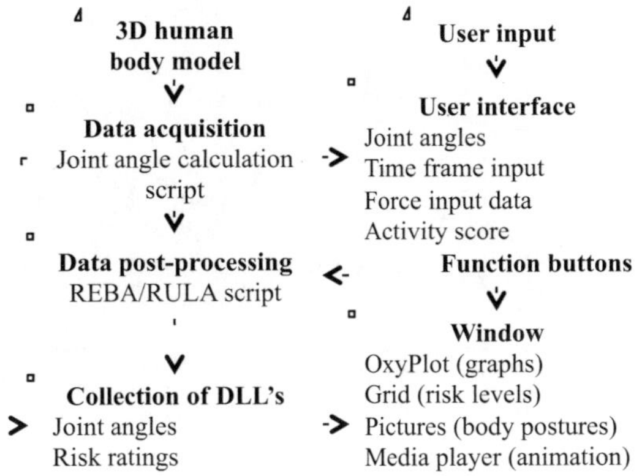

Figure 1. Automated ergonomic risk assessment framework

2.1 Data acquisition

In the model, to obtain the body posture joint angles for various body postures and movements, a total of 26 bones in the biped, as listed in Table 1, must be satisfied, and "footsteps" must be selected before running the programmed MAXScript code. Moreover, *world coordinate system* is used to define the joint angle of the pelvis, *local coordinate system* is used to determine the joint angle of the other body segments (vertical angle and horizontal angle), and *gimbal coordinate system* is used in calculating the rotation of the body segments. In addition, *inverse kinematics* method is employed to identify the moving direction of the subject during the animation. It is also applied during the animation creation, as this function reverses the direction of the chain manipulation and facilitates the creation of the animation.

The calculation of joint angles for the presented method (referring to the "3D Static Strength Prediction Program" developed at the University of Michigan in 2012 [20]) differs from the calculation required by REBA and RULA. Thus, a conversion of joint angles among different scenarios is conducted to fit the REBA/RULA requirements. In total, 41 joint angles are obtained from the 3D model. By selecting the range of the animation time frame and running MAXScript to make the calculation, joint angles for body segments can be captured by each frame of the 3D animation in batch files.

Table 1 List of bones in the biped

List of bones in the biped
Pelvis
Spine
Left/Right thigh
Left/Right calf
Left/Right foot
Left/Right toe
Neck
Left/Right clavicle
Left/Right upper arm
Left/Right forearm
Left/Right hand
Left/Right 3 finger bones
Head

2.2 Data-post processing

A user interface, as displayed in Figure 2, is developed for the purpose of conducting data post-processing and further analysing the results. With the support of other platforms (to be described later in this section), a user interface in 3ds Max is developed. It is not necessary to analyse the entire model in 3ds Max; the analysed time frames can be defined and selected by the user as needed. The repetitive motion in the animation can also be edited by typing in the time frame range and the number of the repetition, which also generates a repetition score for the motion, which needs to be repeated 4 times per minute. Based on Equation (1), if the total duration of the task is less than 60 seconds, then an *activity score* of 1 is added to the total rating.

$$\text{Total duration} = \text{Total time frame/Frame rate} \times 4 \quad (1)$$

Note: The frame rate default is set to 30 frames per second and can be customized by the user.

The inputs of the interface are the time frame range of the task to be analyzed, the time frame range of the repetitive motion within this time frame to be specifically evaluated, the number of repetitions of this repetitive motion time frame, the handled force (in pounds or kilograms), the time frame range of force implementation, the *coupling score*, and the *activity score*. For the input of handled force, users can also type in multiple entries by inputting force and time frame range data one by one. To check all the force input settings, the function of "show set force information" is also available. The function of clearing all the force input settings in the event of erroneous inputs is also provided. In terms of the *coupling score*, a drop-down list is designed where the user may select among: "Well fitting handle and mid-range power grip", "Acceptable but not ideal hand hold or coupling acceptable with another body part", "Hand hold not acceptable but possible", or "No handle, awkward, unsafe with any body part". Within the *activity score*, a repetition score is calculated as per Equation (1), and check boxes are also included as follows: "1 or more body parts are held for longer than 1 minute (static)" and "action causes rapid large range changes in posture or unstable base", as shown in the bottom of Figure 2.

Following the joint angle conversion, the angles can be directly implemented in REBA and RULA. The risk rating is calculated using MAXScript programming to read the generated batch file, and is graphically plotted for each individual body segment (including trunk, neck, arm, leg, and wrist), as well as for the total rating of the entire motion at the respective time frame, considering force/load and activity performance.

OxyPlot is used as an open source plotting library for .NET. A Dynamic-link Library (DLL) is created to receive information from 3ds Max, and then OxyPlot in conjunction with WindowsForm is applied in order to provide a graphical representation. Windows Media Player is utilized to play the rendered animation of the human body movement. Any high-risk motions can be identified through this rating algorithm and the resulting plotted chart. The peak rating and the corresponding human body motion are identified by comparing the plotted chart with the animation, illustrated in Figure 3. The plotted chart is displayed based on the time span of the animation (i.e., the next risk rating to be displayed in the chart illustrates the rating of the motion in the animation at the given timeframe). Buttons for pausing the animation and dragging the animation to a certain time frame or footage are also developed for the purpose of viewing the risk at play at a certain time frame. The user can also type in the time frame to check the risk rating accordingly.

Moreover, risk ratings are plotted for each body segment, and this assists in providing understanding of which body segments are exposed to higher risks during the given operation, as indicated in Figure 4. Modified work can be recommended for the given motion by revising the task manoeuvre of the body segment with a high rating or modifying the task manoeuvre entirely.

A function for plotting joint angle raw data together with its risk assessment method and results is also developed (appearing on the right side of the user interface shown in Figure 2). When the user selects the REBA/RULA calculation method and clicks the body segment button, an interface appears for plotting and displaying raw data of each time frame. This function enables the user to check the raw data and risk assessment results for either REBA or RULA, as indicated in Figure 5.

The result from the traditional usage of REBA and RULA methods is given as an integer. However, in this framework, the obtained risk rating result is averaged among a certain range of time frames. Thus, the risk level categories of the two methods are also adjusted to define the average rating for the operational task, summarized in Table 2 and Table 3.

Table 2 Adjusted REBA risk level category

Level of risk	Risk ratings	Instruction
1	[1, 1.5)	Negligible risk
2	[1.5, 3.5)	Low risk, change may be needed
3	[3.5, 7.5)	Medium risk, further investigation, change soon
4	[7.5, 10.5)	High risk, investigate and implement change
5	[10.5, 12]	Very high risk, implement change

Table 3 Adjusted RULA risk level category

Level of risk	Risk ratings	Instruction
1	[1, 2.5)	Acceptable posture
2	[2.5, 4.5)	Further investigation, change may be needed
3	[4.5, 6.5)	Further investigation, change soon
4	[6.5, 7]	Investigate and implement change

Compared with traditional human body motion data collection methods, 3D automation can reduce the time required for motion data collection by exporting human body coordinates from the 3D modelling and conducting data post-processing using programming. Moreover, 3D visualization enables the user to adjust and customize the human model and workplace design, thereby providing easy access by which for researchers to conduct risk assessment comparisons of diverse alternatives, including differences in the design of the workstation and differences in height range of the human body. Ostensibly, 3D modelling, by circumventing the ethical issues associated with carrying out data collection and observation of real-world construction, allows researchers to assess worker motions even in the early design stage of a project or when maximum human body capacity would otherwise be required in order to complete the work. The support of the results display platforms can also facilitate analysis of ergonomic risk and can identify the precise moment that is exposed to the highest risk during the continuous movement. The results can be used to assist health and safety personnel in identifying work-related risks and recommending proper working postures and body motions for operational tasks.

Although the process of establishing all 3D components from sketching in the early stages of 3D modelling, without any existing 3D component library as support, is highly time-consuming, 3ds Max has a massive inventory of models in an existing library to support this framework and the building of 3D modelling that animates and imitates the continuous working motions that occur in the factory. The ultimate goal of this research is to identify risk, provide modified work based on the analysis results, and achieve overall risk reduction in the analyzed workstations.

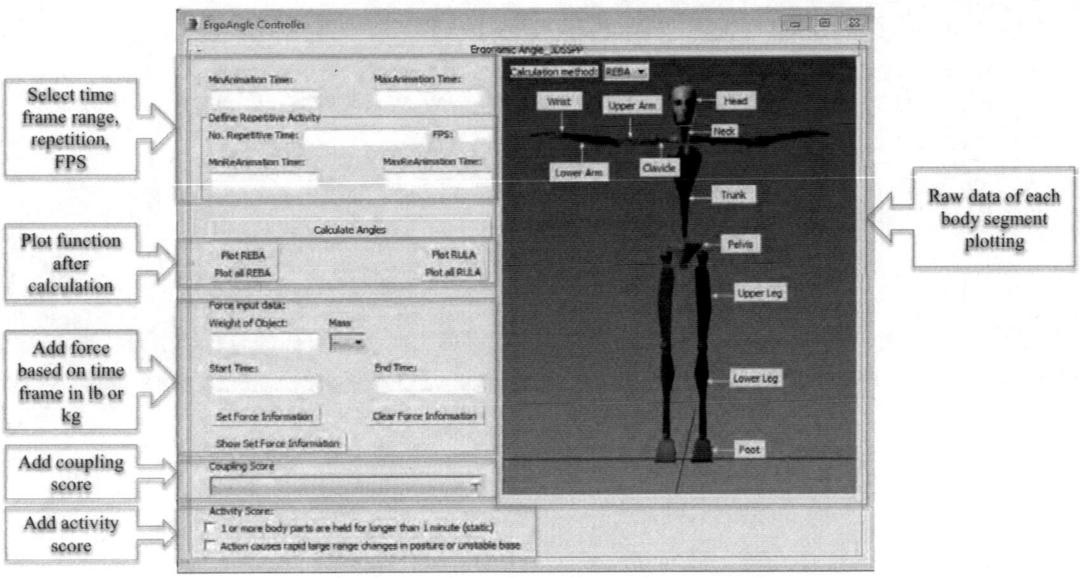

Figure 2. Developed user interface in 3ds Max

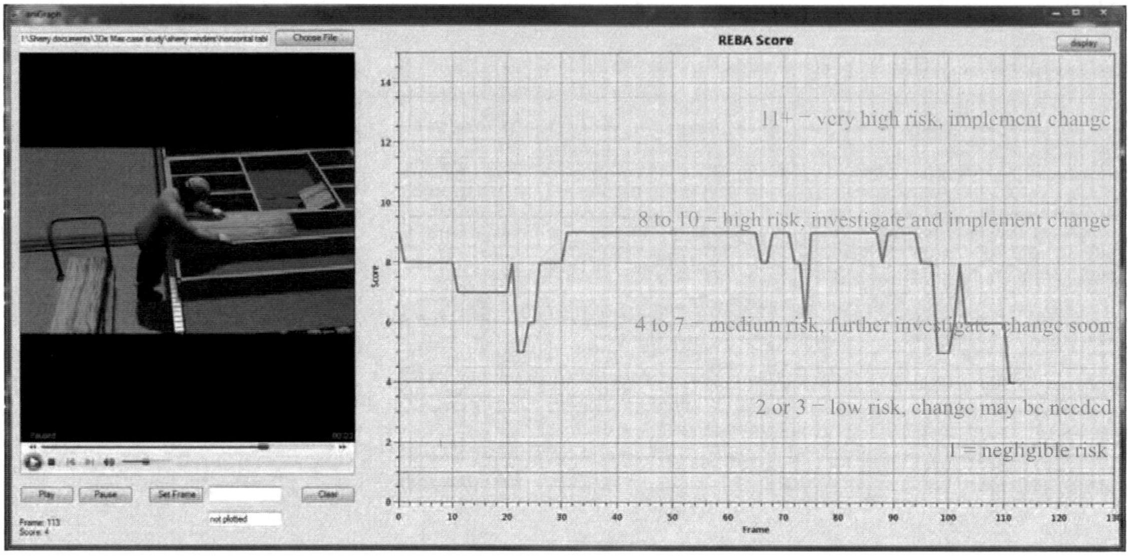

Figure 3. Total REBA risk rating interface with animation

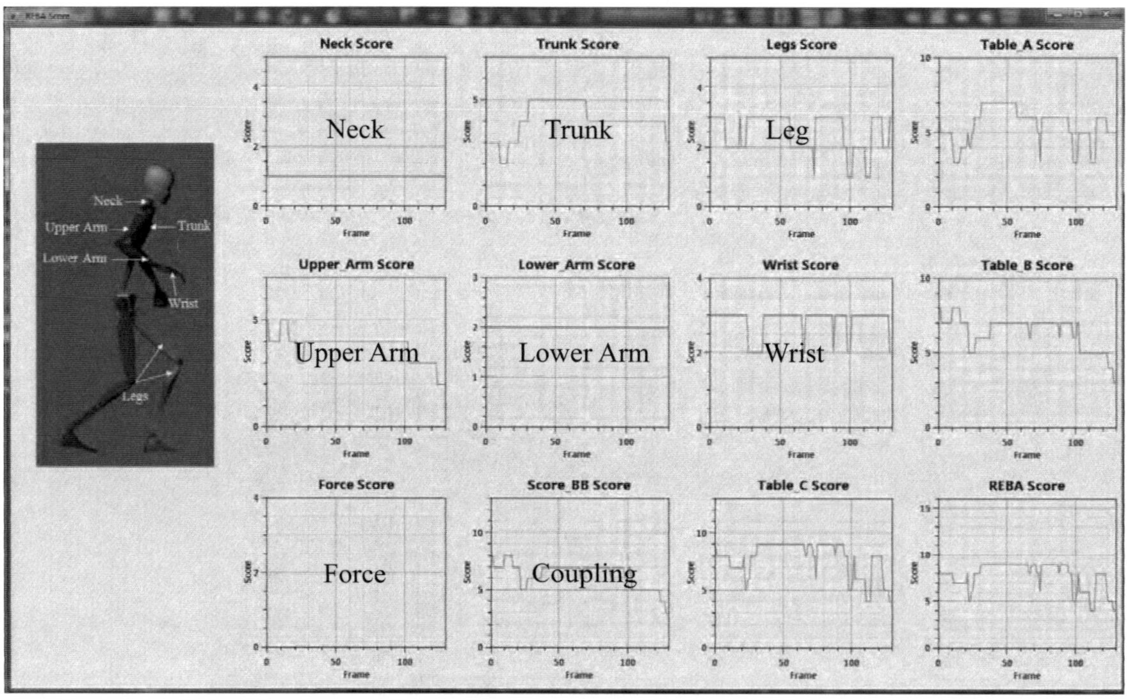

Figure 4. Individual risk rating interface for each body segment

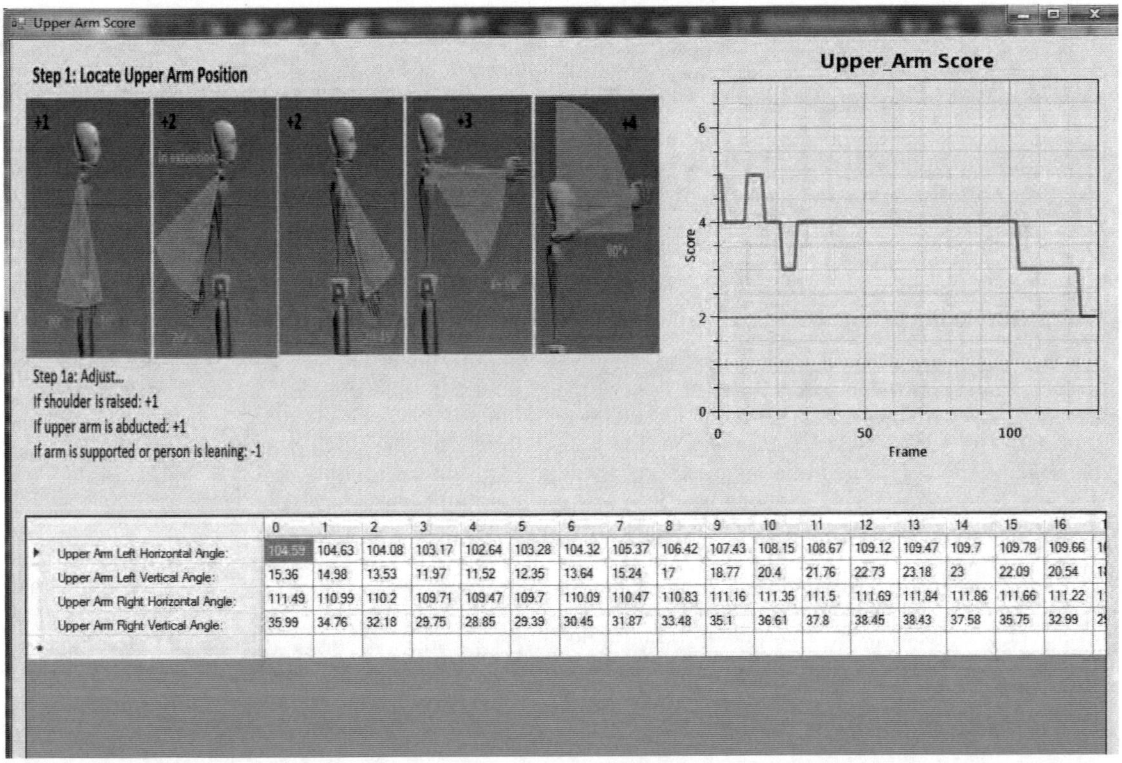

Figure 5. Joint angle raw data at each time frame with its risk assessment method (REBA or RULA)

2.3 Implementation and Discussion

In this section, the implementation of this framework in a real construction manufacturing task, (i.e., placing insulation on the wall panel, displayed in Figure 6), is carried out. In this case, the traditional wood frame building envelope is insulated with fibreglass batts, a relatively lightweight type of insulation. Thus the case study focuses on awkward body posture, rather than on the heavy load on the body. The height of the subject is set as 183 cm with the default setting on the size of the body segments. The supportive workstation is set with a height of 73 cm to assist with the insulation task. The wall panel is designed with a thickness of 24 cm and width of 137 cm. The workers work on both sides of the workstation so that the center of the table is reachable. Both REBA and RULA are implemented in the study as part of the risk assessment analysis.

Implementing the proposed automated framework, REBA gives an average risk rating of 7.24 while RULA provides an average risk rating of 6.12, indicating that the motion in this task requires *"further investigation and changes"*. The analysis results are displayed in the user interface screenshots in Figure 3 to Figure 5. Given the width of the panel, the maximum rating of 9 is assigned since reaching and bending forward is required in order to reach the center of the panel. A modified work recommendation for this task may involve tilting the table to a certain degree to reduce the amount of reaching required. As illustrated in Figure 4, the trunk and upper arm have high risk ratings, indicating that a worker who has sustained any previous injuries to the torso and upper arm should not work on the station. If a worker with a history of having sustained such injuries must work at this station, then the modified work mentioned above should be implemented in order to reduce the risk. Another recommendation for modified work could be staggering the task between workers to reduce the time a given worker is exposed to awkward body postures in order to reduce the potential risk of injury.

Figure 6. Insulation placement on the wall panel

3 Conclusion

This paper proposes an automated framework to assess ergonomic risks for occupational tasks with the support of 3D modelling and a user-friendly interface in 3ds Max. The details of data acquisition and data post-processing are also interpreted mathematically and graphically. A case study of installing insulation in the wall panel in a modular construction manufacturing plant is described. The framework enables the analysis of entire body movement as well as movement of each body segment. The interface, together with the animation and plotted chart, helps with the identification of the highest risk ratings during the continuous movement. The analysis results also lead to recommendations to adjust the existing working conditions as well as provisions to reduce risk particularly for workers who have sustained injuries in the past that would make them especially vulnerable to ergonomic risks.

Having this automated platform, the 3D motion-based ergonomic risk assessment is thus efficient and intelligent. The developed automated platform benefits the user in the following ways: (1) capability and flexibility to conduct data acquisition and data post-processing, and (2) ability to provide visualization of the workstation risk assessment results and to help with proposing any changes made to the plant. The ultimate goal of this research is to achieve overall risk reduction in practice for any construction manufacturing operational tasks. Upon implementing the proposed modified work in real practice, the ergonomic risk rating of the production line can be expected to decrease.

Limitations

The force/load information and activity scores for REBA and RULA risk assessment in the framework presented herein require manual input from the user. Future efforts to identify the precise time frames of force/load on the human body model and to determine an activity score for the movement could improve the presented framework as a further step toward fully intelligent automation.

Acknowledgements

The authors would like to acknowledge the support from the Natural Sciences and Engineering Research Council of Canada (NSERC) through the Industrial Research Chair (IRC) program (File No. IRCPJ 419145-15). Special thanks are extended to Mr. Michael Sukkarieh, who helped with the development of the automated framework.

References

[1] Workers' Compensation Board (2015) "Employers home: Modified work." <https://www.wcb.ab.ca/return-to-work/> (accessed December, 2015)

[2] Bureau of Labor Statistics (2016). "Nonfatal occupational injuries and illnesses requiring days away from work, 2015." Department of Labor. <https://www.bls.gov/news.release/pdf/osh2.pdf> (accessed February, 2017).

[3] González-Izal, M., Malanda, A., Navarro-Amézqueta, I., Gorostiaga, E.M., Mallor, F., Ibañez, J., and Izquierdo, M. (2010). "EMG spectral indices and muscle power fatigue during dynamic contractions." *Journal of Electromyography and Kinesiology*, 20(2), 233–240.

[4] Mathieu, P.A., and Fortin, M. (2000). "EMG and kinematics of normal subjects performing trunk flexion/extensions freely in space." *Journal of Electromyography and Kinesiology*, 10(3), 197–209.

[5] Kim, G., Ahad, M.A., Ferdjallah, M., and Harris, G.F. (2007). "Correlation of muscle fatigue indices between intramuscular and surface EMG signals." *Proceedings, IEEE SoutheastCon*, pp. 378–382.

[6] Pah, N. and Kumar, D.K. (2001). "Classification of Electromyograph for Localised Muscle Fatigue Using Neural Networks." *Proceedings, Seventh Australian and New Zealand Intelligent Information Systems Conference*, Perth, Australia, Nov. 18–21, pp. 271-275.

[7] Plantard, P., Shum, H. P.H., Le Pierres, A. and Multon, F. (2016). "Validation of an ergonomic

assessment method using Kinect data in real workplace conditions." *Applied Ergonomics*, In Press.

[8] Khosrowpour, A., Niebles, J.C., and Golparvar-Fard, M. (2014). "Vision-based workface assessment using depth images for activity analysis of interior construction operations." *Automation in Construction*, 48, 74–87.

[9] Alwasel, A., Elrayes, K., Abdel-Rahman, E. M., and Haas, C. T. (2011). "Sensing construction work-related musculoskeletal disorders (WMSDs)." *Proceedings, International Association for Automation and Robotics in Construction (IAARC)*, London, UK, pp. 164-169.

[10] Ray, S. J. and Teizer, J. (2012). "Real-time construction worker posture analysis for ergonomics training." *Advanced Engineering Informatics*, 26(2), 439–455.

[11] Han, S., and Lee, S. (2013). "A vision-based motion capture and recognition framework for behavior-based safety management." *Automation in Construction*, 35, 131–141.

[12] Li, C. and Lee, S. (2011). "Computer vision techniques for worker motion analysis to reduce musculoskeletal disorders in construction." *Proceedings, International Workshop on Computing in Civil Engineering*, Miami, FL, USA, Jun. 19-22, pp. 380–387.

[13] David, G. C. (2005). "Ergonomic methods for assessing exposure to risk factors for work-related musculoskeletal disorders." *Occupational Medicine*, 55(3), 190–199.

[14] Li, G. and Buckle, P. (1999). "Current techniques for assessing physical exposure to work-related musculoskeletal risks, with emphasis on posture-based methods." *Ergonomics*, 42(5), 674–695.

[15] Spielholz, P., Silverstein, B., Morgan, M., Checkoway, H., and Kaufman, J. (2001). "Comparison of self-report, video observation and direct measurement methods for upper extremity musculoskeletal disorder physical risk factors." *Ergonomics*, 44(6), 588–613.

[16] Li, X., Han, S., Gül, M., and Al-Hussein, M. (2016). "3D motion-based ergonomic and body posture analysis in construction." *Proceedings, Modular and Offsite Construction (MOC) Summit*, Edmonton, AB, Canada, Sep. 29–Oct. 1, pp. 215–223.

[17] Li, X., Han, S., Gül, M., Al-Hussein, M. and El-Rich, M. (submitted). "3D motion-based rapid ergonomic risk assessment framework and its validation method." *Journal of Construction Engineering and Management* (Under review).

[18] Autodesk Inc. (2016). "Create amazing worlds in 3ds Max." <http://www.autodesk.com/products/3ds-max/overview-dts?s_tnt=69291:1:0> (accessed January 2017).

[19] Autodesk Inc. (2011). "MAXScript Introduction" <http://docs.autodesk.com/3DSMAX/14/ENU/MAXScript%20Help%202012/> (accessed January 2017).

[20] University of Michigan, Center of Ergonomics (2012). "3D static strength prediction program." <http://djhurij4nde4r.cloudfront.net/attachments/files/000/000/284/original/Manual_606.pdf?1406656210> (accessed July, 2016).

A Visualization System for Improving Managerial Capacity of Construction Site

Y. Han[a,1], Y. Yashiro[a,2] and F.Hatori[a,3]

[a] Industry and Water Business Administration Division, Hitachi, Ltd., Japan
E-mail: ye.han.du@hitachi.com, yuichi.yashiro.kp@hitachi.com, fumio.hatori.zq@hitachi.com

Abstract

In this paper, we propose a solution to improve the construction management capacity. Our solution contains three parts: a Location Identification System, a Work Progress Management System, and a Statistical Analysis & Visualization System. The Location Identification System locates the equipment and the worker inside the construction site, by using Bluetooth Low Energy (BLE) Beacon and Android Smart Phone Device. The location information is sent to the server that away from the site. The Progress Management System (PMS) collects the work progress information and transits the information to the server through a wireless network. The Statistical Analysis & Visualization System, which runs on the server, computes the distribution of each worker and saves the result as a heat map density. It also computes the progress ratio and the deviation ratio, according to the original work plan. At last, the system displays the progress ratio and the worker distribution together for each area in the construction site.

Keywords –

Construction Management, Statistical Analysis, Construction Visualization, Location Identification.

1 Introduction

Due to the reduction in the young worker and the coming retirement boom of the skilled worker, EPC project in Japan is facing many difficulties. On the other hand, the large-scale overseas EPC project is still high risk to many companies because of the lack of experience in managing workers from different culture, the undetermined construction capacity of local sub-contractors, the changing global economic environment and the uncertainty & unclearness in contract terms, etc. Therefore, improving the work efficiency and reducing the risk are important tasks for every EPC project.

The construction phase is a very important stage of EPC project. Wang & Hung (2016) [1] analyzed a risk model of EPC project by using the Structural Equation Modeling based method. The result shows that the risk from construction phase is one of the most important risk factors that can cause the project delay.

The rationalization and the standardization of work process are the keys to improve the work efficiency and the managerial capacity of construction site. Comparing to the manufacturing industry, the construction industry has many different features: the non-mass production, the nature dependent, the labor intensive and the changeable layout, etc. These characters result in that many methods and tools developed for the factory are not useful in construction site, e.g., the camera based monitoring system is hard to be used in the construction site because that the re-installation is needed when the layout changes. In the construction site, a lot of work is still paper based and the management is still dependent on the individual skills. Thus, as what has happened in the manufacturing industry, developing some tools and methods to rationalize & standardize the work process is the key to improve efficiency and managerial capacity.

In this research, we focus on the location information, which is one of the most important factors related to the information delivery efficiency [2,3] and the construction management efficiency. We developed a worker movement visualization system integrated with the progress information to provide the worker information (which is invisible before) to the site manager, such that the manager can use the information to rationalize & standardize the work process.

2 Research Method

2.1 Measurement of Workers Movement

In recent years, the development of Internet of Things (IoT) technology allows us to overcome the bad influence of the complex working environment in construction site, which has stopped us to acquire the information of worker.

By using the IoT device, our system measures the worker's movement and collects the statistical results of movement information automatically. The measurement process of our system is automatic such that the system is capable for the large-scale application (over 100 workers). We also developed an animation function for the managers such that they can acquire the dynamic movement information.

2.1.1 Approaches for Positioning

The workers go to the site when the work started and leave the site when their work is finished. The amount of the workers, the position of each worker and the layout of the site are always changing. Thus, we need to develop a proper measurement method which is suitable for the above situation.

The radio wave based approaches for positioning can be divided into three classes: Ultra Wide Band (UWB), Wireless Local Area Network (LAN) and Bluetooth Low Energy (BLE) beacon [4-7]. The principle of the radio wave based method is that using a base station to receive (or broadcast) the radio signal sent (or received) by a moveable device. The position is computed by using the signal strength, the time lag and the antenna angle.

Table 1 Radio Wave Based Positioning Approaches

	UWB	Wireless LAN	BLE Beacon
Power	AC Power	AC Power	Battery
Merit	High Accuracy	Diversion of Wi-Fi Spot	Cheap and Easy Installation
Demerit	Need License and Wiring	Need Wiring	Comparatively Low accuracy
Evaluate	Normal	Normal	Very Good

UWB method and Wireless LAN method need wiring, which may disturb the work and need re-wiring if the site has changed.

The BLE beacon is a radio wave transmitter which can broadcast its own ID through the Bluetooth signal. Comparing to the other methods, BLE beacon based method has the following features which are suitable for construction site:

(1) It is Easy to install to the construction site. The transmitter of BLE signal, BLE beacon is small, light and wireless. It can be put into a small case and installed to the site with a hook. This feature allows us can change the installation location fast if the site layout changed.

(2) It has a long battery life. By using the advanced BLE technology, the maximum battery life can be several years (button cell battery).

(3) The smart phone device can be used as the BLE radio wave receiver. The cost for purchasing the smart phone device is acceptable to a large-scale application. Furthermore, since Bluetooth is a worldwide standard wireless radio signal, the smart phone device can be used all over the world.

According to the above reasons, this small size transmitter is a powerful positioning tool for us to measure the worker movement.

2.1.2 BLE Beacon Based Positioning Method

The Figure 1 BLE Beacon Based Positioning Method shows the three parts of this method.

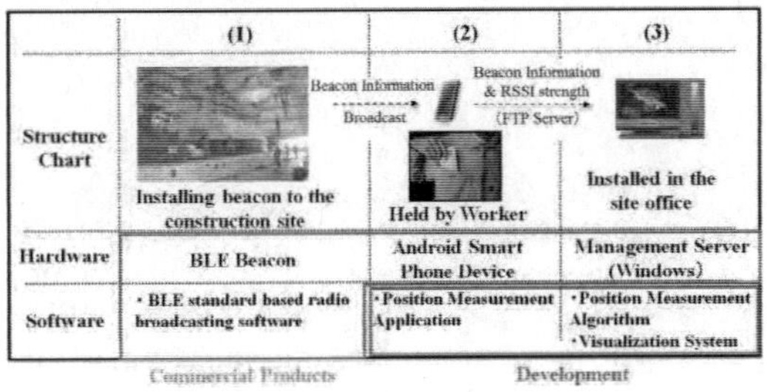

Figure 1 BLE Beacon Based Positioning Method

1. Beacon: each beacon is numbered by a unique ID number and a MAC address. The maximum reaching distance for beacon signal is about several hundred meters at the full power output intensity, which means a less battery life. Thus, by adjusting the output intensity, the reasonable distance between each two beacons is about 10 ~ 20m. The installation location of each beacon is recorded in the server by a 3-dimension coordinate.
2. Device: the smart phone held by the worker keeps receiving the BLE radio wave and recording the signal strength in the format showed in Table 2.

Table 2 Sample of Record File

Time Stamp	Beacon ID	Beacon MAC address	RSSI strength
2017/1/1 8:00:00	Beacon 1	64:FA:18:A7:3D:14:77	-49.0

3. Server: the server receives the record file from the device. The movement of the device is computed by using the beacon installation location record and the signal strength record file.

Figure 2 is the detail of this process.

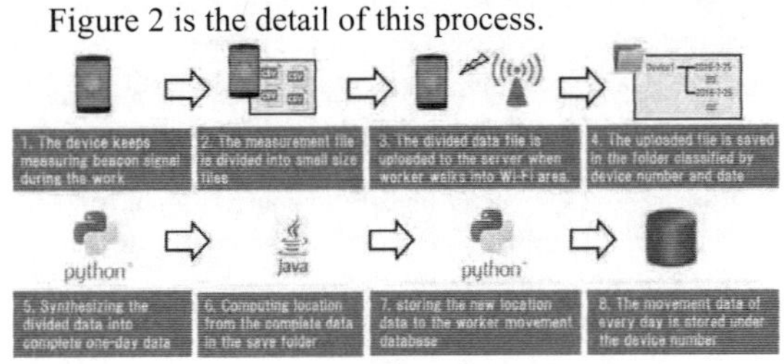

Figure 2 Positioning Process

1. The worker takes the smart phone device and puts

it into the pocket before start working. The smart phone keeps receiving the Bluetooth signal from the beacon.
2. The record file is divided into small parts and stored in the smart phone waiting to be uploaded to the server.
3. When the worker steps into the Wi-Fi area, the smart phone device uploads the divided record file to the server.
4. The uploaded divided file is stored in the server and organized by the device name and the date.
5. The post processing program will automatically starts-up at the designed time (Windows task), merges the divided file into a complete one-day record file.
6. The positioning algorithm is called to compute the movement from the complete one-day record file.
7. The results are send to the designated folder waiting to be stored into the worker movement data base
8. The worker movement database is built in MySQL.

2.1.3 Positioning Algorithm

In the free space, the radio strength attenuates as the distance extended. The relationship between the radio strength and the distance is an exponential function. Thus, we can use this relation to compute the distance between the transmitter and the receiver. But since there are a lot of metals in the construction field, we can't compute the location directly because of the following reason.

Multipath propagation is a phenomenon that results in that radio signals reaching the receiving antenna by two or more paths. In construction site, the main reason of the multipath is the reflection caused by the metals. Due to the influence of multipath, the signal strength we received from the beacon has no exponential relation with the distance, thus we can't use it to compute the position directly.

We developed an algorithm to reduce the influence of the multipath propagation and compute the position of the workers. The input of the algorithm is the complete one-day record file mentioned in section 2.1.2; and the output is a time-related coordinates. The coordinate in the output is the coordinate of beacon installation location, thus the output reflects the nearest beacon where the worker has stayed.

The algorithm consists of three filters, the low-pass filter, the Received Signal Strength Indication (RSSI) strength threshold filter and the distance filter. Figure 3 is the processing flow.

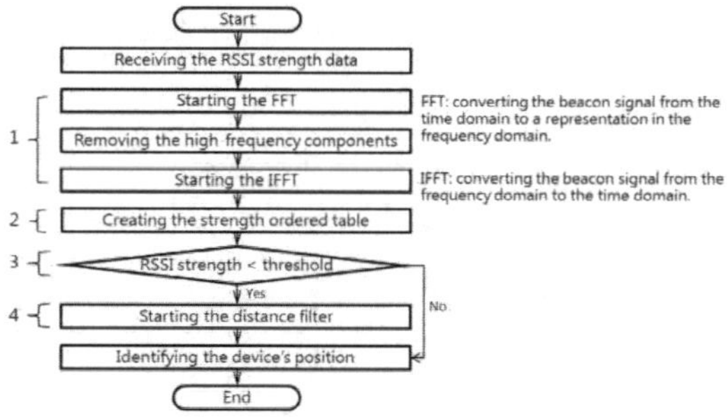

Figure 3 Positioning Algorithm Flow

1. First, the low-pass filter is used to remove the sudden noise and smooth the time-series data.
2. Based on the output of the above low-pass filter, the algorithm creates a strength ordered table for the next process. Table 3 is an example of the strength ordered table.

Table 3 Strength Ordered Table

Time Stamp	1st	RSSI Strength	2nd	RSSI Strength
9:00:00	Beacon 1	-40	Beacon 2	-50
9:00:01	Beacon 2	-40	Beacon 1	-50

3. The RSSI strength threshold filter is designed to remove the constantly noise. This filter computes the signal strength difference between each two beacons and if the difference is greater than the threshold, it will skip the next distance filter. The effect of this filter will be interpreted at the end of this section.
4. The output of the distance filter is the position information. The worker's moving distance is computed at each measuring time, and the position at next time is computed based on an available moving distance of the worker, i.e., if the position result from the RSSI strength threshold filter is impossible for a worker to move in single measuring period, the distance filter will select another proper position from the strength ordered table. Figure 4 is an example of the distance filter.

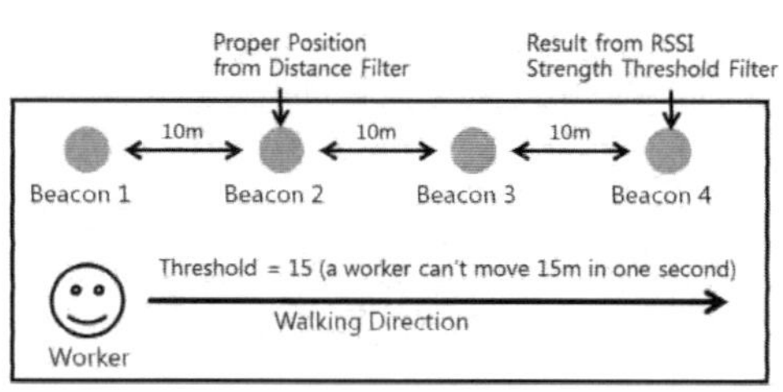

Figure 4 Distance Filter

The effect of the RSSI threshold filter is to remove

the influence of beacon installation location. For the example in Figure 5, there is a beacon installed in the corridor and another beacon installed in a room. The worker walks along the corridor and then enters the room. Since the wall will stop the radio propagation, the beacon signal from the room is weak when the worker walking in the corridor. But when the worker opens the door and enters the room, the signal strength will become stronger in an instant. This sudden change may disturb the distance filter: since the distance between beacon 1 and beacon 2 is longer than a man can walk in 1s, the algorithm will give the result that the worker is still in the corridor. If the difference of the signal strength between the beacon 2 and the beacon 1 is greater than the threshold, the distance filter wouldn't be executed and the output is the coordinate of beacon 2.

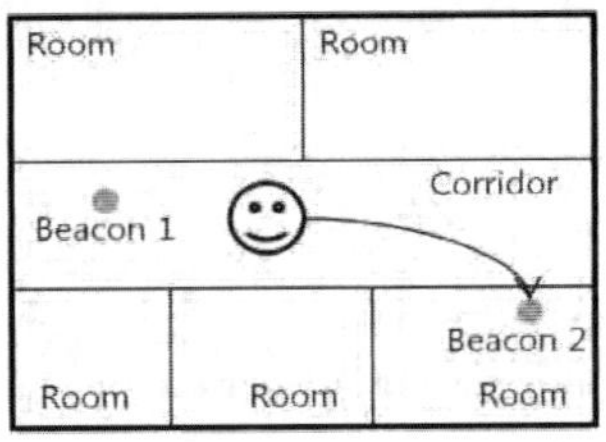

Figure 5 Threshold Filter

2.2 Development of Visualization System

The visualization system reads the movement data from the database and visualizes the data by the heat map, the pie chart and the time series chart. Each of the above visualization method can provide the statistical movement information to the site manager

2.2.1 Worker and Area Information

The system needs to read the worker information and the area information first.

The worker information includes the company name, the occupation, the class name, the worker's name and the corresponding smart phone device ID. All of the above information is stored in an xml file. The system reads the information and constructs the worker check box when starts-up.

The area information includes the area floor, the related drawing and the attributes (like the rest area, the meeting room, the office, etc.). The system reads the area information and creates the area list in the tree structure.

2.2.2 System GUI

In order to provide the movement information to the site manager in an easily visible form, we designed our system to display the statistical results of the worker movement information in three ways: the heat map in the top left, the time series chart at the bottom left and the pie chart in the right. Figure 6 is the image of our system GUI.

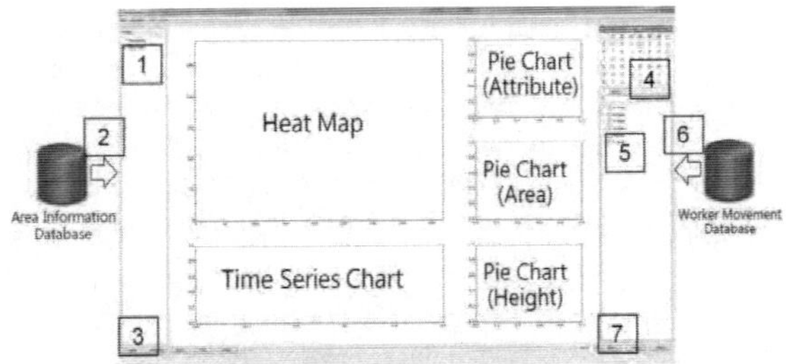

Figure 6 GUI Guidance

The heat map reflects the stay ratio of the worker, i.e., the red area of the map means the worker has stayed at the area for a relative long time. By analyzing the heat map, the site manager can acquire the information about the residence time distribution and find out the useless waiting time.

The horizontal axis of the time series chart is the time and the vertical axis is the beacon number. Since each beacon number is related to an installation location, the time series chart reflects the movement of the worker. The site manager can find out the change of the worker position through the time series chart.

The pie chart is the statistical result of the total stay time. The topmost pie chart is the stay time taken by the area attribute; the middle one is the stay time taken by the beacon installation area; the last one is the statistical result of the heights. The site manager can feedback the work tendency information gathered from the pie chart to the worker arrangement plan.

The usage guidance of the GUI is also showed in Figure 6.

1. Selecting an area from the area list.
2. The system access to the area information database and get the corresponding drawing of the selected area automatically.
3. The corresponding drawing is zoomed to fit the size of the system GUI, and then it is applied to the map area.
4. The analyzer need to select a date from the calendar
5. Checking the check box of the worker. Both a work class and a single worker can be checked. When checking the work class, the visualization will base on the group movement, i.e. the area where the class member gathered is considered as the high stay ratio area. By selecting the work class, the site manager can find out the interference between different occupations and use this information to level the works in different areas.
6. The system access to the worker movement

database and get the worker/class movement data of the selected date.

7. When the analyzer presses the analyze button, the system starts computing the statistical movement information. The total stay ratio of each area and attributes are visualized in the pie chart. The movement is visualized in time series chart. The stay density is computed by using the kernel density estimation method and visualized by heat map.

The system also has animation function. In animation mode, the heat map changes with the play time such that the analyzer can get the information about how the worker has moved. The animation shows the movement amount and the movement path to the site manager, such that the site manager can find out the influence of the site layout and adjust the layout to improve the movement efficiency.

2.3 Development of Progress Information Integrated System

2.3.1 Progress Information

The system reads the progress information uploaded by the Earned value management system which is developed by Hitachi, Ltd. The progress information includes the progress ratio of the entire task and each work step. The total progress ratio is stored into the progress information database associated with the date uploaded.

The progress information is processed in three methods, the total progress, the one-day progress and the deviation between the actual value and the planned value. The total progress data is read directly from the database, the one-day progress is computed by subtracting the total progress of yesterday from the total progress of today. The deviation information is computed by using the work plan data and the progress data.

The visualization of the progress information is also based on the heat map method. In the total progress information, the red area is the area that progressed most. The red area in one-day progress heat map means the most progressed area of single day. The deviation heat map marks the deviant area by the red color.

2.3.2 Integrated Visualization

In the integrated visualization system, the progress information is related with the movement information. The operation process is as same as in the section 2.2.2.

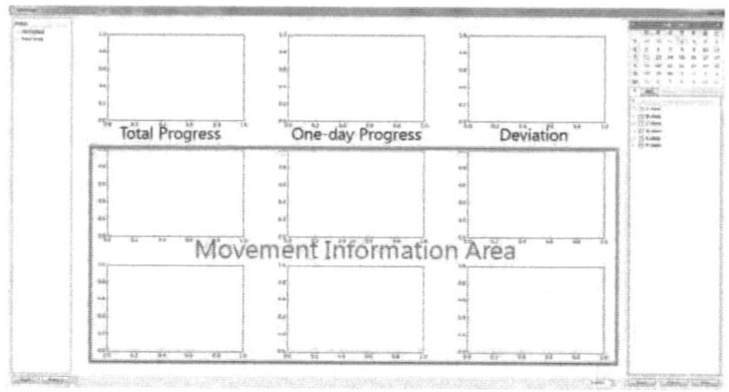

Figure 7 Image of Integrated Visualization

Figure 7 is the image of the integrated visualization system. The top left heat map reflects the total progress, the top medium heat map reflects the one-day progress and the top right heat map reflects the deviation. The movement area consists of 6 heat maps visualizes the integrated movement information of different worker and work classes.

By combining the movement information and the work progress information together, the analysis of the influence of the movement on the progress deviation becomes possible. The site manager can refer to the movement information, which is an important influence factor of the progress deviation.

3 Results

3.1 Improved Positioning Accuracy

We did an experiment to test the effectiveness of our positioning algorithm before the field experiment. The test result shows that our algorithm is efficient to improve the positioning accuracy and the improved accuracy is capable for the site application.

3.1.1 Test Environment

Figure 8 is the drawing of a power plant in Japan where we did our test. There are 3 gas turbines installed in the floor whose size is 150m × 80m. Since this building is surrounded by metal walls, it's a proper experiment field for our algorithm.

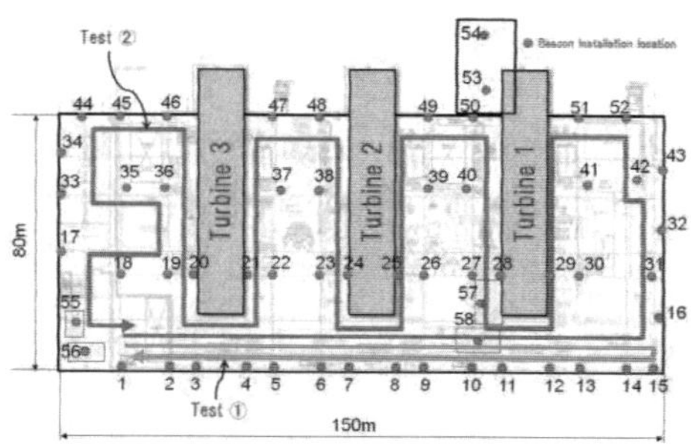

Figure 8 Environment of Efficiency Test

The smart phone device we used is Nexus 5 made by LG Electronics, Inc., and the beacon we used is HRM5011 made by Hosiden Corporation. The output settings of beacon are as follows: output power intensity: +4dBm, output interval: 100ms

The researcher puts the smart phone device into the pocket of working clothes and walks along the following routes at a constant speed:

Test 1: Going and then returning along the path near to the wall.

Test 2: Walking around the turbine floor.

3.1.2 Test Environment

The accuracy is verified based on a video taken during the test. The position showed by the video is considered as the correct position. We compared the result of the raw data and the result of our positioning algorithm to the correct position. The accuracy is computed based on the concordance ratio of the position. Table 4 Test Results is the test result.

Both of the test 1 and the test 2 shows the result that our positioning algorithm can improve the positioning accuracy about 20%. Because of the wide range of the movement in the test 2, the smart phone device affected by more noise than it in the test 1. This is a possible cause for the difference between the test 1 and the test 2.

Table 4 Test Results

Item	Test 1	Test 2
Accuracy of the Raw Data	79.53%	56.97%
Accuracy of the Algorithm	98.86%	76.19%

3.2 Location Measurement Results

3.2.1 Experiment Condition

We did a field experiment to test our system. The experiment field is a construction site, the manager of the site wants the detail of the construction site in order to improve the wok efficiency, i.e., the work efficiency of each class has an obvious difference such that the manager of the construction site wants to find out the difference of each class and optimum the worker distribution. The detail of this experiment is as follows:

1. Research Objective and Period: We measured the movement of 10 field supervisors for 11 days.
2. BLE Beacon: The red circles in Figure 9 are the installation locations of the beacons. We have installed 81 beacons in total. We used the magnetic hook and the dust-proof case to install the beacon to the wall and the pillar. The beacons are installed in a straight line in the path area and tightly in the right workplace area. The bacon we used is "BLEAD" which is made by Houwa System Design Ltd. The output power of the beacon is +4dm and the broadcast interval is 100ms.
3. Smart Phone Device: The smart phone device we used is "Nexus 5" with android version 4.4.0, which is made by LG Electronics Ltd. The scan interval is 1s, i.e. the smart phone device keeps receiving the beacon signal in a second and output the average of the RSSI strength after one second.
4. Operation: We built the FTP server in the site office and installed a Wi-Fi access point near the server. The smart phone device will connect to Wi-Fi automatically when near the access point. There is also a charge station near to the server.

The field supervisor takes the smart phone from the charge station and starts the measurement application every morning. Then they put the device into their pocket while working. The supervisors need to stop the measurement application when they finish their work and go back to the office. We take the device from the supervisor to put them to the charge station. The device will upload the measurement data to the FTP server during charging.

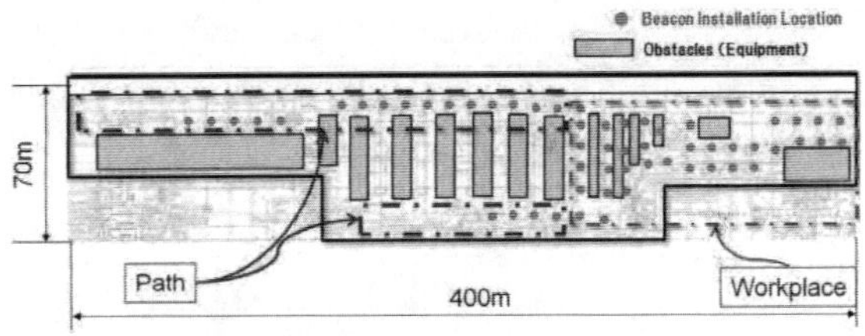

Figure 9 Beacon Installation Locations

3.2.2 Visualization Results of Movement Data

In this experiment, we measured the movement of 10 supervisors. By observing the visualization results of their movement data, we confirmed their work patterns are different. Figure 10, Figure 11 and Figure 12 are some samples of the experiment result.

The heat map has two parts, the top one is the office area includes the main office, the meeting room, the morning meeting area and the rest area; and the bottom one is the drawing of the site.

In the sample of Figure 10, we can notice that the most time of this supervisor is spent in the office work. The time series chart tells us the supervisor didn't walk too much and he is out of the work areas during 9 a.m. ~ 11 a.m..

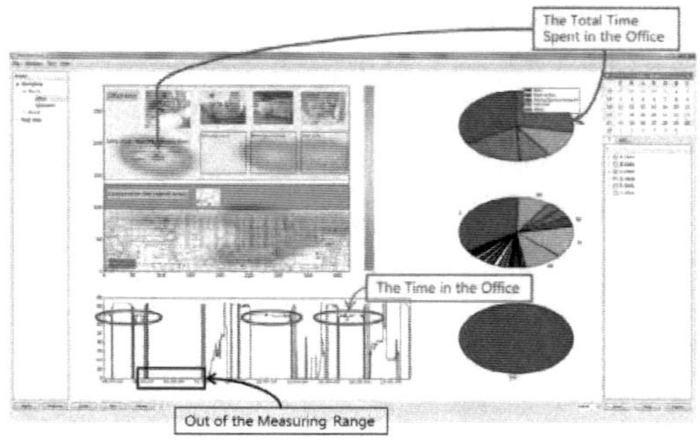

Figure 10 Sample of Field Experiment Result (1)

Figure 11 tells us another movement pattern of the supervisor. In this sample, the heat map shows the supervisor has spent much time in the main office, the morning meeting area and the rest area. This observation result is also supported by the pie charts, i.e., the topmost pie chart shows that the time spent in each area is almost equal. The time series chart indicates the worker has a lot of movement in this day.

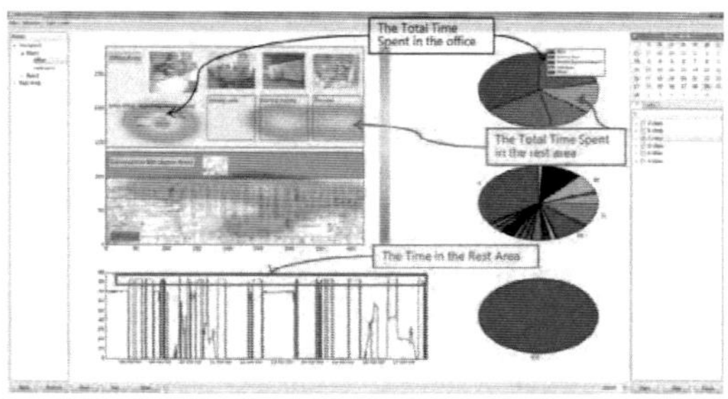

Figure 11 Sample of Field Experiment Result (2)

Figure 12 is different to the above two samples. This sample indicates the worker has concentrated on the site work, both the heat map and the pie chart shows that the worker has spent most of his time at the site. The time series chart also indicates the worker has moved a lot among three areas (Area 1, Area 2 and Area 3)

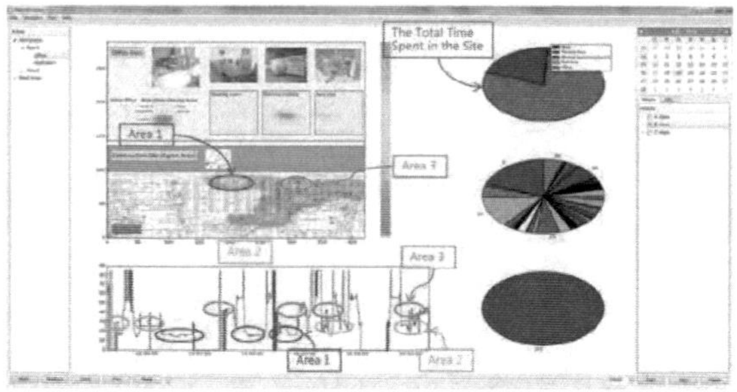

Figure 12 Sample of Field Experiment Result (3)

By observing the above information, the site manager can notice the waste time in waiting, the useless increase of work influenced by site layout and the increasing idle work, etc. After the discussion with the site manager, we have concluded that the visualization of movement is helpful to notice the detail of the site situation.

3.3 Visualization Results of Integrated System

We confirmed our method can visualize different work patterns between workers. In our assumption, the work pattern makes a difference on work efficiency and affects the work progress at last.

Now we can get the movement information from the movement data and the progress information from the progress data separately, but it will be more helpful if we combine them together. Since we can't get the progress information from the customer, we integrate the virtual progress information with the real worker movement data to demonstrate our idea and show the potential of our system.

The example in Figure 13 indicates us the work at area 1 has delayed. At the same time the heat maps in the movement area tell us that most of the workers have spent their time at the office work and no one has worked at the area 1. The site manager can acquire the above the information easily by using our system. Thus the manager can investigate the reason of the increase of desk work and find a solution to solve the problem.

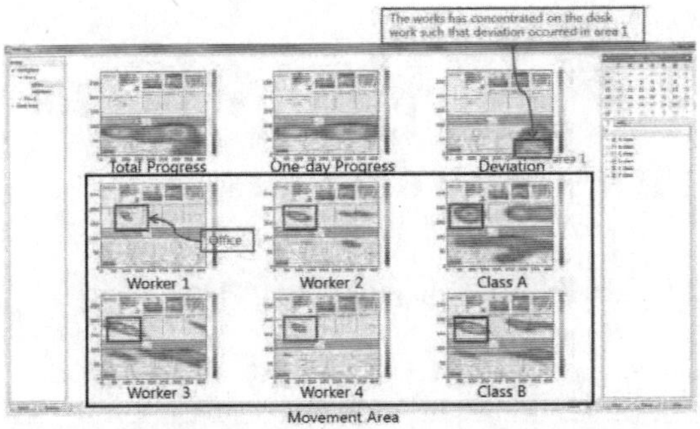

Figure 13 Integrated System

4 Summary

4.1 Conclusion

In this research, we developed a construction site information gathering & visualization system to help improving the information acquisition capacity of the site manager and improve the managerial capacity. The effectiveness of the positioning and the movement visualization is tested on a field experiment. We conclude our system features as follows:

1. The BLE Beacon and the smart phone device based positioning system: The BLE beacon is easy to install & remove because of its light, small and wireless features. Thus, comparing to other methods, our positioning method is suitable for the construction site.
2. An efficiency position computation algorithm: Our algorithm can reduce the noise influence and improve the positioning accuracy without the needs of the calibration. In the effectiveness test, comparing to the accuracy of the positioning result computed from the raw data, the accuracy of the result from our algorithm jumped from 74% to 98% for a local movement.
3. The progress information integrated movement information visualization system: The statistical result of movement is collected automatically and visualized with the progress information.
4. Field experiment based evaluation: The field experiment proved our system is efficient to acquire the worker movement information which can't be acquired before. The effectiveness of the visualization is also proved, the site manager can notice more detail of the construction site than before.

4.2 Future Work

Searching a construction site for a further testing of our system and improving the performance of the visualization methods. Since reducing the loss cost is important to the construction project management, we are going to test our system in some construction projects and valuate the results.

References

[1] Mai Sy Hung and Jianqiong Wang, Research on Delay Risks of EPC Hydropower Construction Projects in Vietnam, *Journal of Power and Energy Engineering*, 21 April 2016.

[2] J. Park, Y.K. Cho and K. Kim, Field Construction Management Application through Mobile BIM and Location Tracking Technology, *33rd International Symposium on Automation and Robotics in Construction,* 2016.

[3] Amir H. Behzadan, Zeeshan Aziz, Chimay J. Anumba and Vineet R. Kamat, Ubiquitous location tracking for context-specific information delivery on construction sites, *Automation in Construction*, 2008.

[4] Andreas Fink and Helmut Beikirch, Analysis of RSS-based Location Estimation Techniques in Fading Environments, *International Conference on Indoor Positioning and Indoor Navigation*, 2011.

[5] Esin Ergen, Burcu Akinci and Rafael Sacks, Tracking and locating components in a precast storage yard utilizing radio frequency identification technology and GPS, *Automation in Construction*, pages 354-367, 2007

[6] Hui Liu, Houshang Darabi, Pat Banerjee and Jing Liu, Survey of Wireless Indoor Positioning Techniques and Systems, *IEEE Transactions on Systems*, 2007.

[7] Hiam M. Khoury and Vineet R. Kamat, Evaluation of position tracking technologies for user localization in indoor construction environments, *Automation in Construction*, pages 444-457, 2009.

[8] Olga Golovina, Jochen Teizer and Nipesh Pradhananga, Heat map generation for predictive safety planning: Preventing struck-by and near miss interactions between workers-on-foot and construction equipment, *Automation in Construction*, 2016.

[9] Aaron M. Costin and Jochen Teizer, Fusing passive RFID and BIM for increased accuracy in indoor localization, *Visualization in Engineering*, 2015.

[10] Aaron Costin, Nipesh Pradhananga and Jochen Teizer, Leveraging passive RFID technology for construction resource field mobility and status monitoring in a high-rise renovation project, *Automation in Construction*, pages 1-15, 2012.

Immersive Virtual Environment (IVE) as a potential tool for interior colour study in office environments

ShaoJie Zhang, Jeonghwan Kim, Shan-Ying Shih, Choongwan Koo and Seung Hyun Cha

Department of Building Service Engineering, Hong Kong Polytechnic University, Hong Kong China
E-mail: 16901588r@connect.polyu.hk, jeonghwan.kim@polyu.edu.hk, shan-ying.shih@polyu.edu.hk, choongwan.koo@polyu.edu.hk, seung-hyun.cha@polyu.edu.hk

Abstract

Interior design elements, particularly colours, have shown significant influence on employers' performance in office environments. The task performance was often measured against different coloured offices in physical built environments. An identified challenge for these studies is to alter colour settings for experiments. Immersive virtual environment (IVE), as an advanced technology, may potentially address to this challenge. Although few studies applied IVE in colour related research, some studies have successfully applied IVE in light and space research for performance assessment. These studies underpin this research to examine the feasibility of applying IVE in colour studies. A virtual model of office environment was built and rendered in Autodesk building design suite (Revit and 3Ds MAX) and presented in Virtual Desktop. Within the virtual offices, only the wall colour was changed while the other parameters remained the same. This enabled the task performance to be measured in the IVE-made offices with different colours (including white, red, green and blue). Proofreading tasks were used for measuring the task performance in each coloured office. The data of error numbers and time expense were collected and then analysed by the analysis of variance (ANOVA). Two significant findings were identified. Firstly, compared to the other three colours, task performance in the white offices was the best. Secondly, performing tasks in red office has better performance than green and blue offices. These two significant findings were supported by several colour related studies although there were some research presenting different results. This study demonstrated the potential of applying IVE for interior colour research in office environments. This study can support a further assessment for the feasibility of IVE by evaluating human emotion and physiological conditions for our funded project.

Keywords-

Immersive virtual environment (IVE), physical built environment, task performance, proofreading tasks

1 Introduction

People spend most of their time in interior environments [1]. This raised research interests on investigating how the elements (e.g. lights and space) in interior environments affect human in physiological and psychological domains [2, 3]. Given the multiple interior elements, colour is a fundamental one in interior environments [4]. Many studies, particularly for office environments, explored the ways interior colours (i.e. white, red, blue and green) affect task performance [5, 6, 7]. Some studies pointed out, compared to blue and green offices, employers performed better in white and red offices [8, 9], while other studies showed dissimilar results [10].

All the aforementioned research were conducted in physical built environments. They manipulated office colours by either painted walls or coloured lightings. A significant challenge within these studies is to alter interior colour settings [11], which may be mitigated by using advanced technologies (e.g. 3D imaging technology and virtual reality). Of particular relevance is immersive virtual environment (IVE) that can realistically simulate scenarios (e.g. colour, light and space) for interior environment research [12, 13, 14].

Although IVE was not widely adopted in colour research, some studies have demonstrated the feasibility of IVE used in light and space simulations for the interior environment [12, 14]. These may potentially support this study to examine the feasibility of applying IVE in colour research. This study shared part of data and analysis results from a funded project that focusing on IVE and interior colour.

2 Research Methodology

This study examined the task performance against four coloured offices including white, red, blue and green. These four colours are determined in accordance with aforementioned studies. The four coloured offices have been built in IVE for experiments. Information related to experiments including participants, colour samples, instruments and experiment procedures is described in the following sections.

2.1 Participants

There were 56 participants (undergraduate students) recruited from the Hong Kong Polytechnic University, including seven female students (12.5%) and 49 male students (87.5%). All of them were local students with English as second language and were in range of 20 to 28 years old.

Among 56 participants, 13 participants were excluded because of poor eyesight (11 participants), failure on colour blindness (1 participant) and uncomfortable feeling (1 participant) during the experiment. Therefore, 43 data samples were collected and analysed.

2.2 Colour samples

The four colours were identified by Munsell Colour Notation (MCN) system. The brightness and saturation of the four colours were secured in similar degrees. The determined colours in the MCN system were 6.05R 4.59/11.15 (red), 9.75YR 8.74/0.45 (white), 3.45G 4.44/7.89 (green) and 2.43PB 5.12/10.51 (blue) (See Figure 1).

2.3 Instruments

Colour blindness: The Ishihara Colour Blindness Test was used to evaluate colour vision of participants. This blindness test consists of 13 circular plates with pointillist painting.

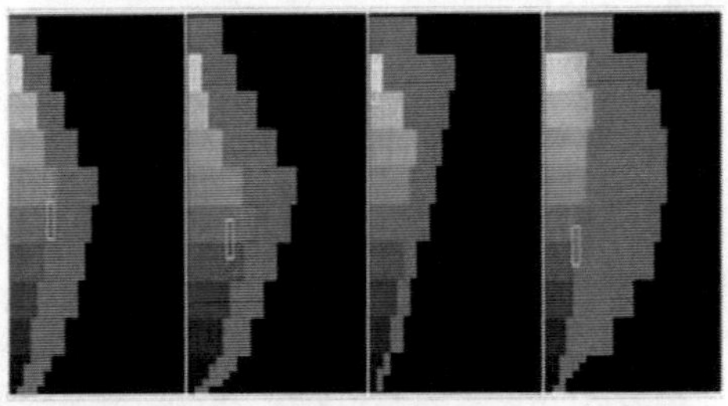

Figure 1. The position of chosen colour in Munsell Colour System

Pre-questionnaire: It is used to collect participants' personal information such as genders, age, education degrees and colour preferences etc.

Head Mounted Display (HMD): HMD used in this study is Oculus rift. It enables IVE-made offices to be presented to participants.

Immersive virtual environment: The ways to present IVE-made coloured office were shown in Figure 2. The office models, referring to an actual office, were built in Revit 2017 and rendered in 3D MAX. The office models were then imported into Virtual Desktop that can present the four coloured offices for experiments. In terms of hardware, HMD provides display of virtual model (e.g. 3D graphics database and appropriate sensory information) back to participants. Tracking sensor detects the position and orientation of participants in virtual space. It should be noticed that only wall colour of virtual office was modified while the other parameters (i.e. dimensions, textures and illumination) remained the same. The four models of virtual office were shown in Figure 3 respectively.

Performance assessment: Many studies assess task performance by using proofreading tasks. Address-checking tests adopted in this study were commonly used proofreading tasks for performance assessment. The test contains 15 address-checking questions that were shown to participants on the interface of Virtual Desktop. Each question include two similar addresses

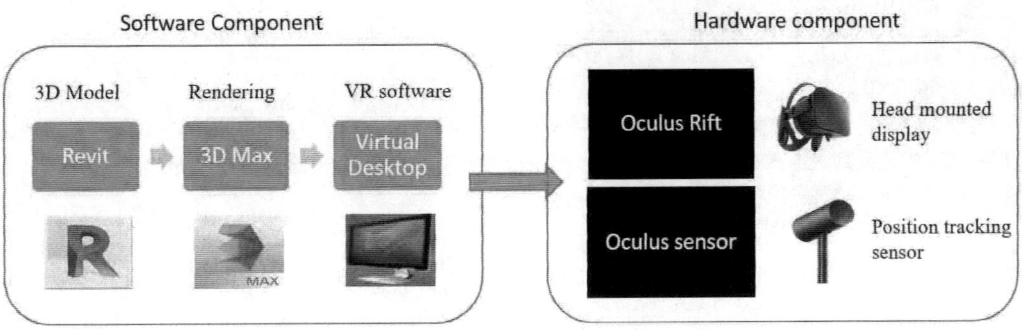

Figure 2: Modeling steps and apparatus.

with a few or no differences in zip codes and/or street names. The participants were then required to respond no differences or identify the differences between the two addresses.

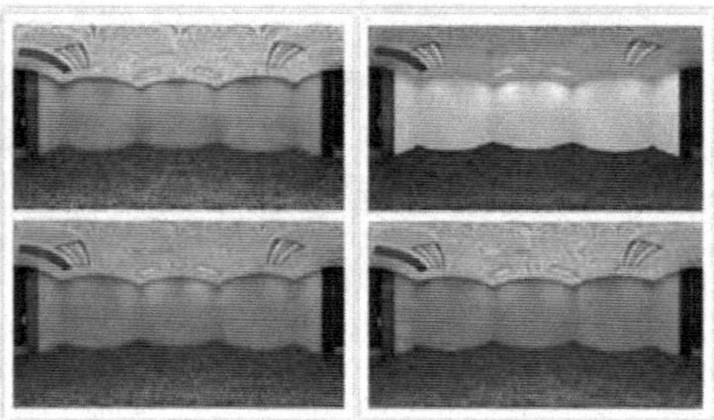

Figure 3. Panorama of virtual model of office room.

2.4 Experimental procedure

Firstly, participants were guided to waiting area for commencing colour blindness tests. The participants who successfully passed the colour blindness test were allowed to undertake the experiments. Prior to the experiments, the participants firstly needed to complete pre-questionnaires. Since the participants had few experiences in such task, they were then given five sample address-checking practices to get familiar with the task process.

Afterwards, each participant was asked to expose the IVE offices with random colours. In the IVE office, participants were firstly asked to adapt to the environment for two minutes. After that, participants were required to complete proofreading test (15 address-checking tests). Then participants were asked to have a four minutes' rest for washing out their impression of previous colour. Following that, they were required to repeat the same process in other three offices with different colours.

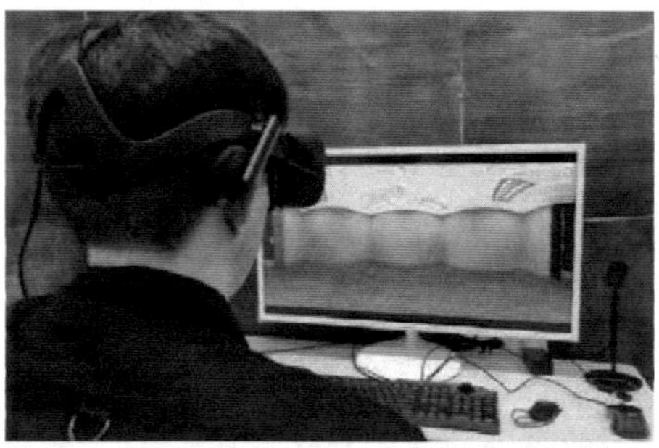

Figure 4. Participants adapting to office environment within IVE

3 Analysis of Result

Our study examined the influence of colour on task performance in accordance with number of error per minute. Lower number of error per minute indicated better task performance. The analysis of variance (ANOVA) showed significant differences of task performance among four colours (F= [3,171] = 2.684 P=0.048). In terms of multiple comparisons analysed by least-significant difference (LSD) between two colours, there were significant differences of task performance between white colour and blue colour (P=0.044), as well as white and green colour (P=0.007). In addition, it is apparent from Figure 5 that participants perform best (mean=0.4915 number of error per minute) in the white office, compared to the other three colour offices. In addition, participants achieved better task performance in the red office than blue and green offices.

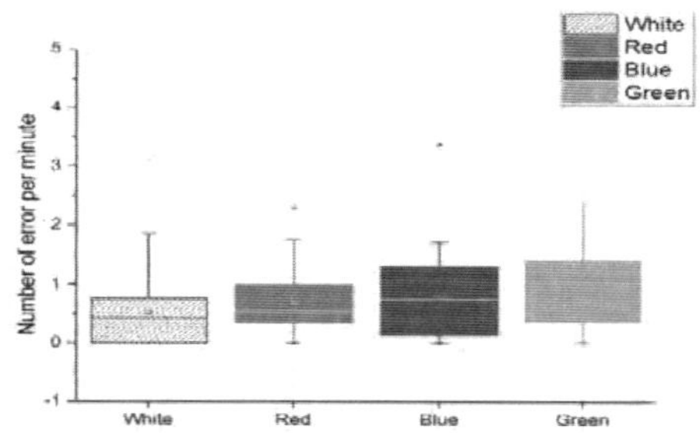

Figure 5: The data represent group means for error of number per min from baseline and their 95% confidence intervals.

4 Discussion

The analysis results showed white and red offices can significantly influence the task performance. Within the four colours, this study presented the best task performance in the white colour office (Figure 5). This result may be different from some studies that they mentioned participants in the white office made the largest number of errors in proofreading tasks [9, 15]. While there were some studies with similar result supporting our findings. Compared with green and red, employers felt least distracted in the white office [16]. This aligned with a study that workplace with neutral colours (e.g. white and gray) could enhance concentrations on their tasks [17]. These demonstrated the office environment with white colour can significantly improve task performance.

In accordance with the analysis results, the participants in the red office, compared with blue and green colours, have better task performance (Figure 5). This is supported by several studies that demonstrated similar results that red colour in interior environments can improve task performance [9, 18, 20]. Meanwhile, other research showed the interior environments with cool colours (i.e. blue and green) have poorer task performance [10, 19].

Although some studies have different results, our findings provided potential opportunities to utilize IVE in colour related research. The potential may be further strengthened by adopting wider range of sample data because the most of participants in this study are Hong Kong undergraduate students.

5 Conclusion

This study investigated the ways the four colours (including white, blue, green and red) affect participants' task performance in office environments. The number of errors per minute was used to estimate task performance in the IVE-made offices. It showed significant differences for task performance between the four colour office environments. Differentiated from the indicator (i.e. number of errors) in related studies, the number of errors per minute can also be an effective indicator for evaluating task performance. In addition, two significant findings were identified. Within the four coloured IVE-made offices, it has the best task performance in the white colour office. Moreover, participants perform better in the red office than in the blue and green offices. These two findings were underpinned by several colour related research. Our findings demonstrated the potential of applying IVE to explore interior colours study in office environments. Furthermore, this study supported our project to assess the feasibility of IVE by investigating human emotions and physiological conditions in different coloured offices.

6 Reference

[1] United States Environmental Protection Agency. *Air Quality Criteria for Particulate Matter,* Volumes I, II, and III: EPA/600/ P-95/001; Office of Research and Development: Washington, DC, 1996.

[2] Jacobs, K. W., & Frank E. Humstmyer, J. *Effects of Four Psychological Primary Colours on GSR, Heart Rate and Respiration Rate.* Journal of Perceptual and Motor Skills, 38: 763-766, 1974.

[3] Küller, R., Ballal, S., Laike, T., Mikellides, B., & Tonello, G. (2006). *The Impact of Light and Colour on Psychological Mood: A Cross-Cultural Study Of Indoor Work* Environments. Ergonomics, 49(14): 1496-1507.

[4] Marilyn A. Read, Deborah Upington. *Young Children's Colour Preferences in the Interior* Environment, Early Childhood Education Journal, 36(6): pages 491–496, 2009.

[5] Goodfellow, R. A., and Smith, P. C. *Effects of environmental colour on two psychomotor tasks,* Perceptual and Motor Skills, 37: 296-298, 1973.

[6] Green, W. K., Hasson, S. M., Mohammed, S. K., Phillips, C. L., Richards, P. E., Smith, M. E., & White, A. *Effect of viewing selected colours on the performance of gross and fine motor tasks,* Perceptual and Motor Skills, 54, 778, 1982.

[7] Camgoz, N., Yener, C., Guvenc, D. *Effects of hue, saturation, and brightness: Part 2,* Colour Research and Application, 29, 20-28, 2003.

[8] Bette, L. *Colour Your World,* Journal of Business Mexico, 6 (2), 20 – 22, 1996

[9] N. Kwallek, C. M. Lewis, J. V. D. Lin-Hsiao, H. Woodson. *Etfects of Nine Monochromatic Office Interior Colours on Clerical Tasks and Worker Mood,* Colour research and application, 21(6): 448-458, 1996.

[10] Andrew J. Elliot, Markus A. Maier, Arlen C. Moller and Ron Friedman, Jo¨rg Meinhardt. *Colour and Psychological Functioning: The Effect of Red on Performance Attainment,* Journal of Experimental Psychology: General Copyright by the American Psychological Association, 136(1): 154–168, 2007.

[11] S.-Y. Yoon and K. Wise. *Industrial Applications of Affective Engineering,* pages 219–232, Springer, Basel, Switzerland, 2014

[12] A. Heydarian, J.P. Carneiro, D. Gerber, B. Becerik-Gerber, T. Hayes, W. Wood. *Immersive virtual environments versus physical built environments: a benchmarking study for building design and user-built environment explorations,* Automation in Construction 54: 116–126, 2015.

[13] M. Kozhevnikov, J. Royan, O. Blazhenkova, A. Gorbunov The Role of Immersivity in Three-dimensional Mental Rotation, in *Proceedings of 3rd International Conference on Design Computing and Cognition,* pages 143-157, Atlanta, GA, United States, 2008.

[14] M.F. Shiratuddin, W. Thabet, D. Bowman. Evaluating the effectiveness of virtual environment displays for reviewing construction 3D models, In *Proceedings of the 4th Conference on Construction Applications of Virtual Reality,* pages 13-15 Lisbon, Portugal, 2004.

[15] N. Kwallek and C.M.Lewi. *Effects of environmental colour on males and females: A red or white or green office,* Applied Ergonomics, 21(4): 275-278, 1990.

[16] N. Kwallek, K. Soon, H. Woodson, J. L. Alexander. *Effect of Colour Schemes and Environmental Sensitivity on Job Satisfaction and Perceived Performance,* Perceptual and

Motor Skills, 101(2): 473-486.

[17] E Sundstrom, MG Sundstrom. *Work places: The psychology of the physical environment in offices and factories,* pages 179-189 New York: Cambridge Uni. Press.

[18] N. Kwallek and C.M.Lewi. *Effects of environmental colour on males and females: A red or white or green office,* Applied Ergonomics, 21(4): 275-278, 1990.

[19] Poursafar, Z., Devi, N. R., & Rodrigues, L. R. *Evaluation of colour and lighting Preferences in architects' offices for enhancing productivity,* International Journal of Current Research and Review, 8(3), 1, 2016.

[20] Hill, R. A., & Barton, R. A. *Red enhances human performance in contests,* Nature, 435: 293, 2005.

34th International Symposium
on Automation and Robotics
in Construction

ISARC 2017

June 28 – July 1
TAIPEI, TAIWAN

A Study of Developing the Fire Investigation System with Building Information Modeling Application

Lee-Kuo Lin[a], Yu-Huai Hsu[b] and Sung-Wei Hung[c]

[a]Department of Civil Engineering, National Taipei University of Technology, Taiwan.
[b]Department of Civil Engineering, National Taipei University of Technology, Taiwan.
[c]Department of Civil Engineering, National Taipei University of Technology, Taiwan.
E-mail: lklin@ntut.edu.tw, tpdfcs12000@gmail.com, jii80208@yahoo.com.tw

Abstract –

The content of fire cause investigation is not only the fire prevention as reference but also the significant gist to improve firefighting. But in point of fire site investigation, mostly were using handwriting document for recording information at present, obviously lacking with carrying convenience and recording instantaneity. Meanwhile, after causing of fire, the investigation work also utilize building information captions to display the original fire scene and identify the ignition point. But traditional two dimensions floor plan is limited to provide architectural spaces information, usually can't represent the actual appearance of the building. If using three dimensions Building Information Modeling (BIM) outputting the relevant elevations, perspectives information and so on to assist the reviewer mind produced visualize and spatialize information. The smart handheld device becoming more and more popular in recent years, it has matured software and hardware communications technology. By using the hardware of smart handheld device to support the need of investigating items in fire investigation into electronic check and form filling.

This research uses two-tier system architecture, respectively for the client and server. The server side uses both WAMP as the operating system, and MySQL as the database engine. The client side is used with PHP technology. Based on the studying illustration result, this system not only provides the data storage of inspection results, but also can be used for the relevant units in the future review.

Keywords –

Fire cause investigation; Smart handheld device; Building Information Modeling (BIM).

1 Introduction

In case of any fire disaster happen, most of all is to salvage human life and disaster rescue, afterwards professional work is fire investigation. The content of fire causes investigation are vital essential work in each fire disaster due to the fire outcome is not only to be the reference as fire prevention but also the significant gist to improve firefighting. At the practice of site fire investigation, most are using notebook to recording describe information as official document, such process is obviously lacking for carrying convenience and recording instantaneity. However, traditional two dimensions floor plans are limited to provide architectural spaces information, therefore, this research is applying smart handheld device combined with BIM technique, systematizing "Fire cause investigation testimonial" among information of the investigation. This research illustrates with fire disaster simulation examples, checks through survey data and collection, develops with PHP, MySQL and etc. of information technology to build a cross-platform of fire disaster investigation. By using this system can be reduced mistake situation judgments, meanwhile can effectively increase the practical request of fire investigation.

2 Literatures review

2.1 Definition of fire investigation

Fire work mainly by the disaster before the fire, the disaster when the fire, the disaster after the investigation of the three parts of the composition, each with a joint relationship. The fire investigation is one of the most basic work in the fire system and one of the most important projects. Through the implementation of the fire investigation work, after the consolidation analysis, you can enhance the fire code to make up the missing, improve the efficiency of fire fighting. The connotation of the fire investigation includes the technical level, the scientific level, the psychological level, the legal level, etc., is an administrative and legal action [1].

2.1.1 Taiwan 's fire investigation system

The current fire investigation process in Taiwan is when the fire occurs, the county and city fire department service center to accept the public alarm. Area of the fire unit to attend attendance at the same time, will also inform the fire department under the fire investigation section staff dispatched, fire investigators in accordance with the regulations to the site investigation [2]. But in case of special fire cases, fire investigators need to immediately go to the survey, the investigation site, such as the discovery of special, major or fire causes unknown fire cases, the county fire department may hold the county fire identification committee or apply for the Ministry of Interior Fire Department support.

After the investigation is completed, the fire investigators in accordance with the provisions must be completed within 30 days of fire investigation and identification of the cause, and transferred to the local police authorities for follow-up investigation. The police unit is transferred to the local seizure office according to law, the prosecutor will be in accordance with the provisions of the Taiwan Criminal Law, the suspect will be prosecuted, if the case is mild, no prosecution or slow prosecution of the punishment. When the prosecution by the court after the trial, so that the case until the decision. In the country, the District Prosecutor's Office and the court to clarify the truth, Will summon the fire investigators to witness the identity of the case [3].

2.1.2 The significance of the fire investigation

The information obtained by the fire accident is quite costly, because the fire accident is an irreversible process, because the relationship between time and man, the scene of the information will be more difficult to preserve and obtain, any scientific experiments can only be simulated close, can not really reproduce, so the fire investigation of the delicate degree is different from the other. Therefore, each of the accident investigation, the need for complete collection and aggregation, avoid the loss of property and casualties paid, and the information into a viable fire safety improvement policy, it doesn't lose the meaning of fire investigation [3].

2.1.3 Fire investigation legal value

A fire investigation is an objective proof of the fact of a fire, it's a speculation about the fire and the spread of disaster. The uncertainty of the fire liability of the parties becomes deterministic, the investigation itself does not change the facts, but the incomplete content of the investigation will change the interests of the parties. According to the laws and regulations to implement the fire investigation procedures, from another point of view, to clarify the cause of the fire and the responsibility of the accident is a legitimate rights and interests of the parties, it's also a recognition of responsibility and obligation [4].

2.2 Statistical Analysis of National Fire Types

According to the Ministry of Interior Fire Department statistics from 2008 to 2012, although the number of fires decreased year by year, but the maximum number of fire for the building [4]. 2012 Buildings fire with a maximum of 1,199 times, accounting for 76.2% of all the fire (Figure 1) Derived from the data can be used to make use of building information modeling 3-dimensional advantages of modeling to do the presentation, can assist in the identification of investigators to do post-disaster analysis, to write a fire investigation report.

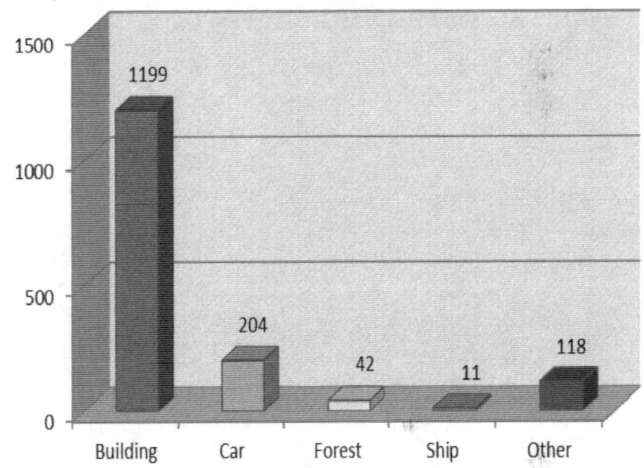

Figure1. All kinds of fire statistics in 2012 [4]

2.3 Building information modeling

Object guide theory and technology in the field of information technology in the gradual development, Europe and the United States began to use this technology to describe the product life cycle information, build product data modeling, including the creation, organization, use, maintenance and other information to master and transfer, exchange and other applications. This technology first in the aviation, machinery, and manufacturing flowering results, and gradually extended to the construction industry. Professor Charles Eastman of the Georgia Institute of Technology, who has studied the building information modeling technology for a long time, in the "Building Product Models" book, that is, details of the building elements of the information model structure principle, unveils the opening of the Building Information Modeling (BIM) terminology concept [5]. In 2002, Autodesk began to use Building Information Modeling to describe the company AEC (Architecture, Engineering, and Construction) related product design concept, in conjunction with the construction industry analyst - Jerry Laiserin positive author recommended, so many 3D tool software manufacturers, such as Tekla, Bentley Systems and Graphisoft and other follow-up, entered the pursuit of the ideal BIM technology tools in the competition. There are increasingly mature and easy to use tools, coupled with the development of many years of product life cycle information management and application of the theoretical basis, and through the successful application of many cases, in recent years, BIM technology development and application can be said to flourish. At present, the development of BIM related technologies can be divided into three aspects: information exchange standard, modeling technology and model operation technology [6].

2.4 Hypertext Preprocessor

PHP (Hypertext Preprocessor) is an open universal computer code language, especially for web development and embedding in HTML (HyperText Markup Language). PHP grammar learn from the absorption of C language, Java and Perl and other popular computer language features, easy for general program designers to learn. The main goal of PHP is to allow web developers to quickly write dynamic pages, but PHP is also used in many other areas. Originally developed by Ledoff in 1995. Now the PHP standard is maintained by PHP Group and open source community, the scope of application is quite extensive, especially in the development of web programs. In general, PHP is mostly implemented on the web server, by executing the PHP code to generate pages that users can browse, they can be executed on most servers and operating systems, for example, to collect data from a form, or to generate dynamic web content, or to send and receive small text files (cookies), PHP can do it [8].

PHP can be used in most operating systems, such as Linux, HP-UX, Solaris, Open BSD, Microsoft Windows, Mac OS X, RISC OS. Can be implemented on most web servers, such as Apache and Microsoft Internet Information Server (IIS). It's function is not just output HTML files, Its function also includes output graphics, PDF files, and Flash files. You can also let it output some text, such as XHTML and any other XML file, PHP can produce these files, and store them on the server, there are a lot of other expansion modules that can be used, It also supports a wide variety of databases, so it is quite easy to use the database when writing a web page.

The webpage filling system written by this study is based on WordPress developed by Automattic Web Design, to PHP and MySQL as a platform for open source software and content management system, a webtemplate system with a template processor, diversified presentation can assist in the identification of investigators not only to fill in the interface concise and more free and open source-assisted media.

3 Research methods

3.1 Make the essentials of the identification book on fire cause

This research is according to the fire cause investigation process to develop a prototype system with BIM technique. Based on the studying result fire investigators can recording fire overview immediately, make more precise restore scene identification book. It's contain:(1)front and back covers, (2)contents, (3)summary, (4)scene investigation personnel records, (5)scene records and judement, (6)mobilized observation records, (7)conversation transcripts, (8)evidence identification report, (9)goods configuration diagram and site plan, (10)actual location, the scene of photographic information is shown in figure 2 [8].

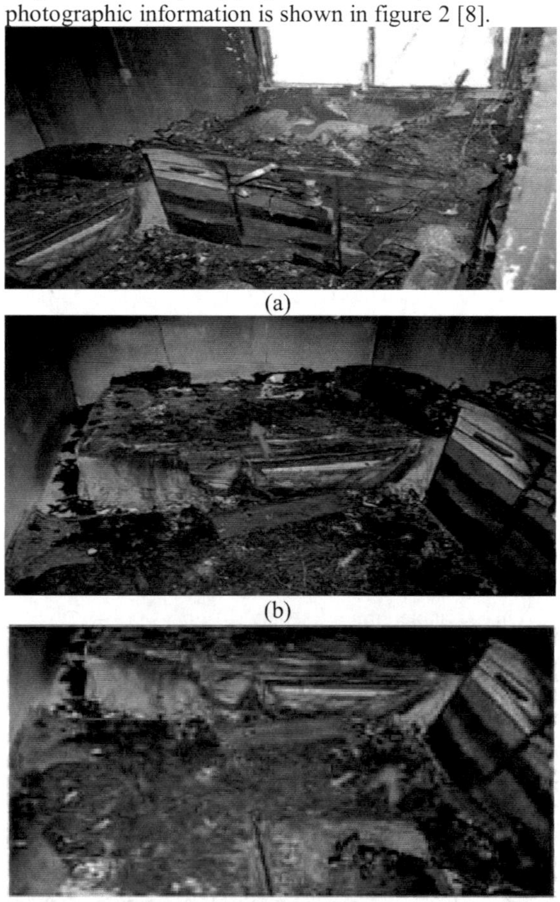

(a)

(b)

(c).

Figure 2. Actual locationthe scene of photographic [9]

3.2 The shortcomings of the current fire investigation and identification book

The shortcomings of this study based on the current fire investigation and identification book are as follows:

(1) Mostly in the text to describe the situation of the fire scene:

The current fire cause investigation and identification of the book are written in the form of narrative, so the content of the more narrative-based, Use the text to describe the scene of the fire scene and the number of casualties, advantages for the detailed description of the scene conditions, but for the information reading and understanding of the poor efficiency of the staff, it's necessary to consider the use of words to avoid misleading readers of the cognitive, resulting in wrong judgments.

(2) Text description is prone to errors:

Investigators were surveyed at the fire scene, must be carried out at the same time exploration, sampling and photography operations, in addition to the scene of the environment chaos, it is difficult to carry out detailed records, so most of the simple diagram or simple words to record, return to the identification unit, and then use the scene photos and floor plan with the auxiliary, organize the relevant identification information and write the identification book, on the way back and forth may cause memory loss, so there may be errors, in addition to the use of text description of the scene conditions and spatial orientation has its difficulties, if the use of text in the cognitive error is likely to cause text errors.

(3) Lack visualization and spatialization of data:

To describe the location and space position is still lack of visual and spatial, so the reading staff must cooperate with the floor plan and photos of the auxiliary can produce a sense of space and direction, and only for the location of the state of burning only overview, failed to clearly define the extent of its burning, so for those who did not reach the site survey of the identification of personnel, It's difficult to interpret the fire flow and the burning path.

(4) Written information is not convenient to save:

The number of buildings of all types is growing rapidly, And according to the Ministry of Interior Fire Department statistics building fire for all cases the largest case, and according to the Ministry of Interior Fire Department statistics building fire for all cases the most, It's clear that the number of relevant documents identified in the fire must be very large.

3.3 Fire cause investigation site fill form

In the modern developed science and technology society not only the computer hardware and software and the rapid development of functions, but also with the growing environmental issues, the traditional paperwork is transferred to the digital way, this digital approach has the advantages of shortening time and reducing errors, with the popularity of today's mobile networks, wireless networks and portable handheld devices, easy to send and instant record is one of the characteristics, for the number of documents, can save the query and access, follow-up archive time, also use the "keyword" to retrieve the text, can improve work efficiency, but before the text file is converted to a digital file, must first define the field items and input procedures to provide computer interpretation, the system can help to classify and summarize the information, therefore, this study to extract the cause of fire investigation and identification of the production of the part, including fire, fire and fire causes as a reference, designed to meet the identification of personnel records the scene of the subject into a "fire cause investigation site fill in the form", and then the subject is divided into belong to the material category of the "site survey table" including the basic information, the opening of the information, structural information, item information, on-site trace data, and the direction of burning and carbonation and person to prove the "staff interview table" for the record, use the table as a subsequent digitization conversion step.

3.4 Building information modeling for fire cause investigation identification book

The current fire cause investigation and appraisal of the fire scene for the building map mostly in 2 Dimensions (2D) way to show the location and floor plan, and use Computer Aided Drafting (CAD), but the software is only a general computer graphics platform, not specifically designed for building collar, so the ease of use is obviously insufficient, and the drawing of the plan can't provide a complete visualization and spatial, so the staff must be with the floor plan and photo assistance to produce space and sense of direction, for the location of the burning status of the record and the reasons for the investigation of the report in the text narrative and photographs were as different in the book chapters and pages, resulting in the inconvenience of the reference.

In this study, the use of Trimble Navigation launched by the Sketchup 3 Dimensions (3D) Building Information Modeling (BIM) software, to 3 Dimensional view of the main, the tools used to generate, view, and modify the idea of 3D also have simple and powerful features, can be through the computer software for building simulation, and from the building in the process of obtaining a variety of different information on the new concept of CAD can be established by the floor plan, elevation, profile, perspective and other surface combination.

4 Building fire cause investigation system

This research develops the system of the fire disaster cause investigation with the actual needs. First step is required prior planning system to develop process and them is in turn to set up overall system. The system contains the user requirement analysis, development tools and system show and etc.

System development tools of prototype are divided into the following four parts [10] (as shown in figure 3).

1. Worldwide web: integrate FTP, News, Gopher, WWW and other systems.
2. Wamp: Window+apache+mysql+php.
3. Dynamic web languages: the most commonly used ASP, JSP, PHP.
4. Interactive webpage.

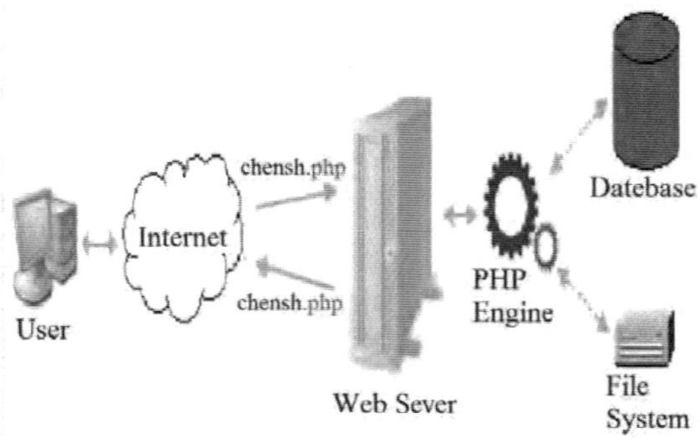

Figure 3. Dynamic web mode of operation [10]

This system provides fire investigators using any tablet PCs or smart handheld device which can support internet services and data transmission. The main structure is shown in figure 4.

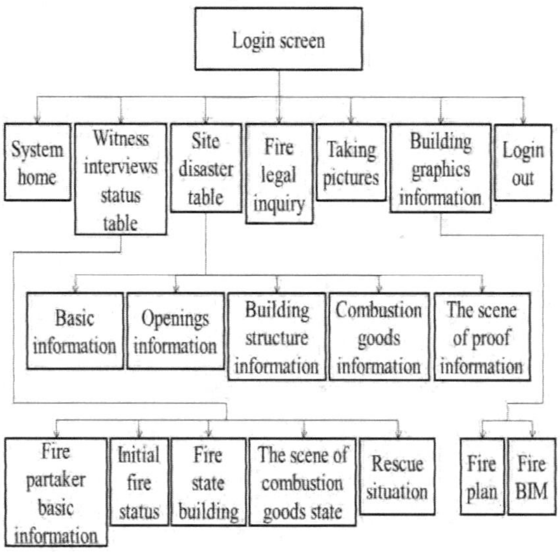

Figure 4. The main structure of system

The system development environment of build work is divided into three steps:

1. Install WAMP erection website server.
2. Setting backend server.
3. Install text editor.

After completed above-mentioned parts, this study uses wed design and development environment, then utilizes for default browser by doing webpage test.

5 System demonstrate and case simulation

To illustrate the research result; a simulation case was applied as the demonstration.

A. Fire Case Mode Description

In this case, it will be a fire happened on the second floor of a building in Taipei city. The official document of the fire cause investigation will contain the following record:

1. Summary of fire scene: Date, time, add and cause.
2. Fire disaster survey: Detailed report situation, scene of the accident and the fire creep trend.
3. Situation after burning: return for scene of the accident report, items burning case, casualty circumstances.
4. Information of pictures: The scene of photo, 2D-pics and using by Sketchup software build 3D model is shown in figure 5 and figure 6.

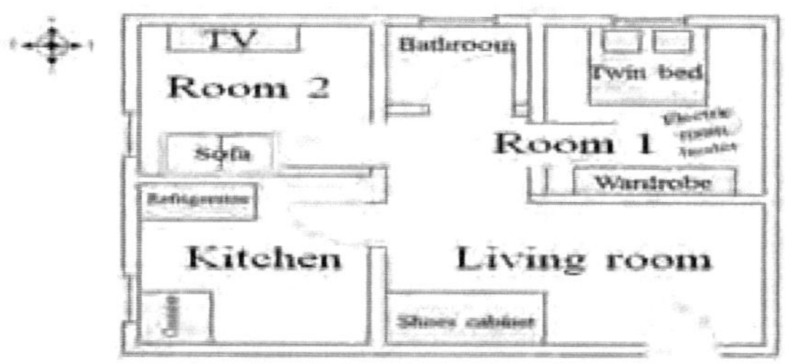

Figure 5. Fire plan of case simulated

Figure 6. Plan view of model case simulated

B. Fire Case Simulation Applied To The System Screen

According to the case explanation and fill in overview set into system page to show from previous section:

1. Login screen: Confirm the identity of the fire appraisers, user import correct account number and password (as shown in figure 7).

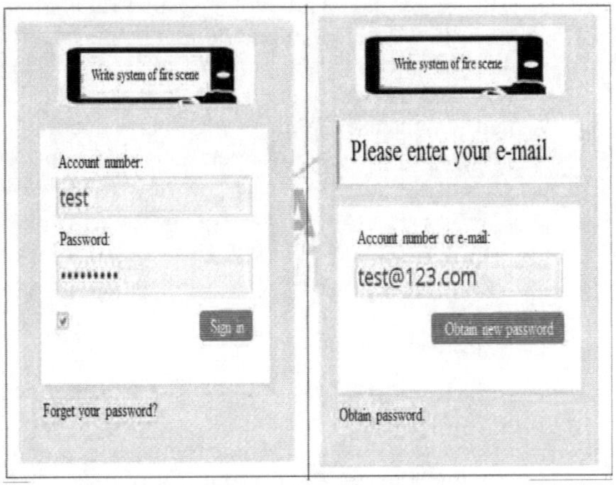

Figure 7. Login screen

2. Homepage: Show the system name and explanation, write basic data, can select their own addresses. It also can let smart handheld device in response to GPS positioning function.
3. The scene of prospecting form: First of all must be cut space and give number before exploration. Include basic information and information of openings, building structure, burning goods, the scene of site (as shown in figure 8).

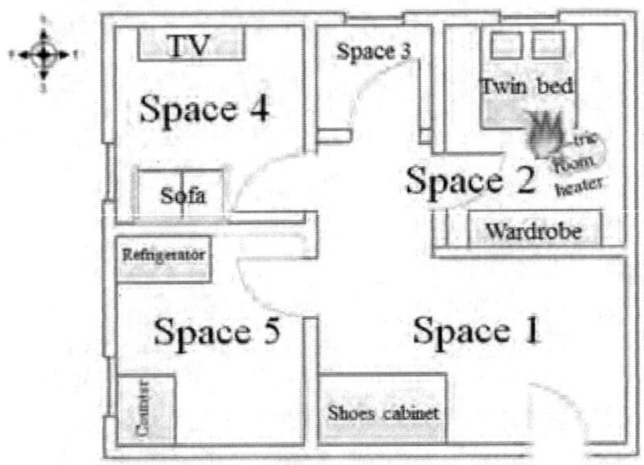

Figure 8. Simulated space case

4. Witness status interviews form: Including basic information of personal relationship, beginning fire disaster status, outbreak of fire status, the scene burning goods status and the rescue situation (as shown in figure 9).

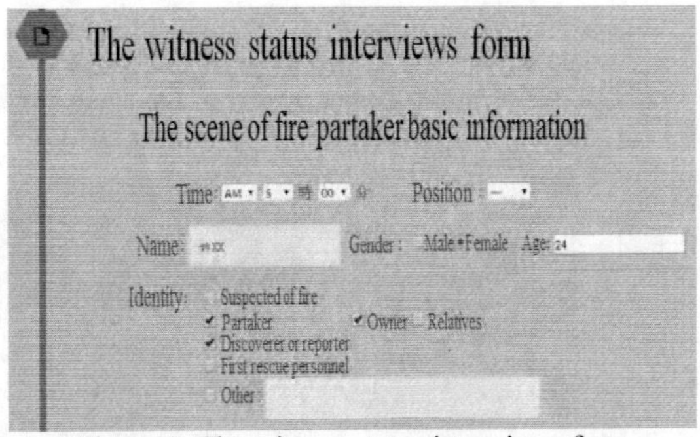

Figure 9. The witness status interviews form

5. Firefighting laws inquire: Embed the rules of fire services of National Fire Agency, provide the fire investigators to inquire these laws comfortably.
6. Take photo: This system can offered uploading photos in smart equipment, films or audio files, subjoined some words to remark.
7. Building graph information: According to WebGL and HTML5 appear interactive BIM, rotate model directly on this page, eliminates inconvenience (as shown in figure 10 and figure 11).

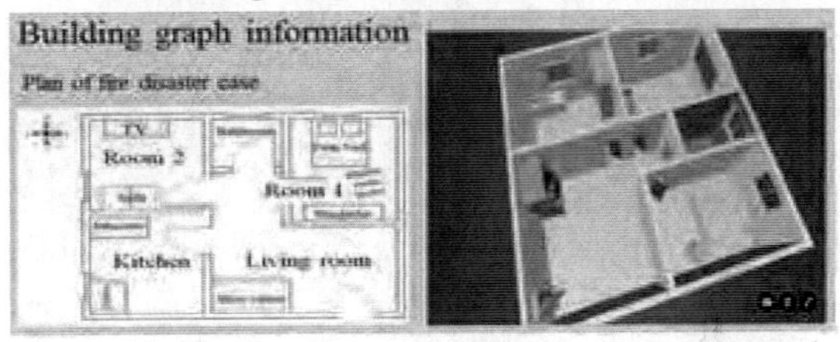

Figure 10. The fire disaster simulation case by plan and modeling

Figure 11. The scene photo planted simulation model

8. Log out: After using system, can log out in order to protect personal information.

C. The Result for Bedpan's VOF Analysis Experts Test

This research interview process with experts using the computer, the smart handheld device actual operate and comment, doing opinion exchange to feedback and revise (as shown in figure 12).

Figure 12. Interviews and test schematic diagram in smart handheld device

6 Conclusions and Suggestions

The cause of fire investigation is a very important business in fire management, the result of the fire investigation is not only a reference for the prevention of fire prevention, but also an important basis for the improvement of fire rescue, so the results of the investigation is correct or not, the relationship between the public interest is very important, therefore, the fire causes of the investigation report produced by the distribution, logical and rigorous quality of the project, for the entire investigation work, can be regarded as a key indicator of work, and this study for the convenience of fire investigators can fill out more immediately, to improve the accuracy of the situation on the fire, After research and construction of the fire cause investigation system and the building information modeling, draw the following conclusions and recommendations:

6.1 Conclusion

1. This study constructs two-tier system architecture, respectively for the client and server side. Server-side use WAMP for the server operating system, the database uses MySQL as the database engine. PHP technology to do front-end construction, and the system not only provides the results of the inspection data storage, in the future can be related to the relevant units.
2. Combined with building information modeling technology can improve the traditional plan space and visual deficiencies of the place, to enhance the convenience of the relevant personnel to read the fire investigation and identification book.
3. Systematic fire causes investigation and identification of the information in the book, do the form of information conversion, facilitate the fire scene investigation and identification of personnel records of the immediacy and correctness, echo the demands of automation; In addition, the system and after the expert test, in accordance with the recommendations to do the appropriate feedback and amendments, can be more in line with the relevant personnel to fill the scene.
4. This research constructs the system In addition to the use of desktop computer operating system, can also be used with the smart phone online login, immediately check the results stored in the back-end database, and the introduction of computerized system operations can not only effectively improve the previous handwriting records, to achieve the demands of less paper, echo the trend of environmental conservation and energy conservation.

6.2 Suggestions

The construction of the fire investigation system combined with the construction information model is a preliminary prototype system, the future can be based on practical application of continuous feedback correction, to make the system more perfect; Based on the experience of this study, subsequent research recommendations are as follows:

1. The system of this study is connected to the system page for the input URL, the characteristics of the smart handheld device has not been fully used, the future can be developed specifically designed for smart handheld devices, make operation and performance more optimized
2. After the expert interview proposal can use the form to write the information, combined with fire dynamic simulation software for the import of parameters, by the results of the simulation analysis can help the fire identification staff more clearly determine the cause of the fire.

References

[1] San-Ping Ho, "The scene identification techniques," Environmental Protection Source Magazine of article 178, CJCU Department of Occupational Safety and Health, 2013.

[2] Ching-Te Huang, "The Research of Arson Prevention and Reduction System," research report commissioned by National Fire Agency, 2004.

[3] Tong-Yin Wu, "On the study of fairness for fire liability risk distribution-A perspective of public risk management," Ming Chuan University Executive Master of Risk Management and Insurance, Master's thesis, 2012.

[4] "National fire statistics analysis in 101 years," National Fire Agency, 2012.

[5] Rung-Chin Kuo, Shang-Hsien Hsieh, "BIM skills and Public works", Public Construction Commission Executive Yuan Public works e-paper of article 038, 2011.
[6] Ming-De Wang, "Application of Building Information Modeling technology," NTU department of Civil Engineering law and Industrial Development Research center deputy director, 2013.
[7] Mehdi Achour, Friedhelm Betz and Antony Dovgal, "Latest PHP5 Chinese manual," Hau-Chen translate and edit, 2009.
[8] "Fire services Act," National Fire Agency, 2010.
[9] Fire Bureau of Taichung City Government, Zhongqing Road fire identification report, 2008.
[10] Hsin-Yen Chiu, "A Study of Building the Safety Inspection Management Information System in Construction," National Taipei University of Technology department of Graduate Institute of Civil and Disaster Prevention Engineering, Master's thesis, 2010.
[11] Yu-Huai Hsu, "A Study of Developing the Fire Investigation System with Building Information Modeling Application," National Taipei University of Technology department of Graduate Institute of Civil and Disaster Prevention Engineering, Master's thesis, 2014.

Flexible Earthwork BIM Module Framework for Road Project

Hassnain Raza[a], Waqas Arshad Tanoli[a] , Seung Soo Lee[a], Jong Won Seo[a]

[a]Department of Civil and Environmental Engineering, Hanyang University, Seoul, South Korea
E-mail: razahassnain99@gmail.com, waqitanoli@gmail.com, rokhonor99@hanmail.net, jseo@hanyang.ac.kr

Abstract

Development and expansion in the cities have increased the demand for road construction projects in different regions of the world. The construction projects in the modern era require the utilization of latest technology. BIM and its collaboration with other technologies like GPS, UAV, AR/VR, and telematics have successfully proven their worth in provision of a solution to various construction industry problems. Innovations in BIM process by researchers are a remarkable asset for planning and visualization. Earthwork in road sector demands further research to achieve better planning, visualization and improvements in productivity using BIM process. The study in paper presents a user-friendly flexible earthwork BIM module framework for the road projects and explains the complete modeling process from point data collection to final design.

This study also discusses a model partitioning technique which can be used for the division of earthwork BIM model for better visualization and implementation. This method is useful for the primary design information modeling of earthwork at each section and is easily useable for the multi-phase construction project.

Keywords

Earthwork; 3D BIM; Road; Module; Partitioning Technique.

1 Introduction

The technology revolution has made the exchange of information easier than before. It has a greater impact on construction industry where the information exchange is time-consuming and complex. Building Information Modeling (BIM) is considered as an important source by using 3D design models during all construction phases, to exchange information in an efficient manner [1]. Researchers have provided an excellent contribution in 3D BIM for exchanging design information and its collaboration with other technologies. However, few areas of construction industry such as earthwork operations require its practical implementation, as they have not attained full benefits of BIM till now. The Earthwork operation has a huge impact on construction cost, especially in road construction projects. Previous studies have shown that about 25% of construction cost is associated with earthwork operations [3], and it increases the demand for accurate estimation. For better knowledge of earthwork process, variable aspects of design and other factors from natural to man-made must be accurately analyzed. 3D model is an efficient tool to visualize these factors in original form and are easily explainable for all participants of the project [4].

In 3D modeling as a first step, it is important to collect site information data and analyze the ground surface for accurate models. For this step, Jaselskis considered the laser scanning procedure for accurate visualization of the site and its adoption in CAD platform [5]. It is a faster and more efficient way of collecting data as explained by Slattery [2]. Different researchers have provided a remarkable effort regarding earthwork surface modeling and data collection technologies for 3D models [6,7]. Randall [8] explained laser scanning importance for BIM modeling, and Sepasgozar [9] described the challenges being faced in using these technologies. Slattery, Kirchbach, and Elmqvist [2, 10,11] explained terrain surface modeling of the site using TIN along with various other solutions. All of the research work performed previously have mainly focused on data collection and 3D modeling of the ground surface, however, complete 3D earthwork BIM design was not discussed. The platform like Autodesk Civil 3D mainly utilized for the TIN surface modeling of site and earthwork estimation of road design [12]. There is no particular function provided for earthwork BIM modeling. Software like InSite SiteWork [13] and Earthwork software [14] provided a solution to many earthwork problems like trench excavation, topsoil replacement, and earthwork estimation, but 3D earthwork BIM support function was not considered. As per our knowledge, no research has been carried out involving module for 3D earthwork BIM modeling.

Waqas [15] and Hassnain [16] defined the complete 3D earthwork BIM process along with its heuristics for a road project. However, this study is an improved version of previous work, and it involves utilization of 3D earthwork BIM process in the development of Flexible Earthwork BIM module. The first section of the paper explains the complete 3D earthwork BIM process for a road project. The second section of the paper explains the partitioning technique of 3D model used in the module for primary design information modeling at each segment. Jingbin Hao[17] used partition technique to partition solid models into submodels segments to precisely analyze the model. Watershed technique is also utilized to partition 3D solid meshes by [18]. The previous technique utilizes surface partitioning concept, however, it cannot be utilized in 3D earthwork modeling. The author explains the basic concept of partitioning in this study for 3D earthwork models.

The proposed 3D earthwork BIM design process and partition technique used in the module will be helpful for all stakeholders involved in construction industry. The module framework presented is undergoing the testing and development phase. It also contains earthwork section library for 3D modeling. The study will be supportive for further research work in 3D earthwork BIM and its utilization in various construction projects.

2 Methodology

This paper presents the Flexible earthwork BIM module which is based on 3D BIM process and model division technique for a road project. The module is in the developing and testing phase. This paper explains the module in two sections. The first section of the paper describes 3D earthwork model generation in the module. The complete 3D earthwork BIM process starting from data collection to the final 3D model design in the module is briefly discussed in Section 2.1. The 3D model partitioning technique along with its importance is explained in Section 2.2.

2.1 BIM Modeling Process

Accurate 3D model is necessary to visualize major design information of the project. To achieve a perfect 3D model multiple factors are required to examine. 3D earthwork modeling process in this module is divided into three major steps. The first phase involves data collection of the site to TIN surface model generation in the module. The second phase includes the alignment adjustment for better 3D modeling. The third major step is a selection of earthwork section according to design requirement for the final 3D model. Earthwork design information can be visualized with the help of a 3D model. The Flow chart explaining the complete process is shown in Figure. 1.

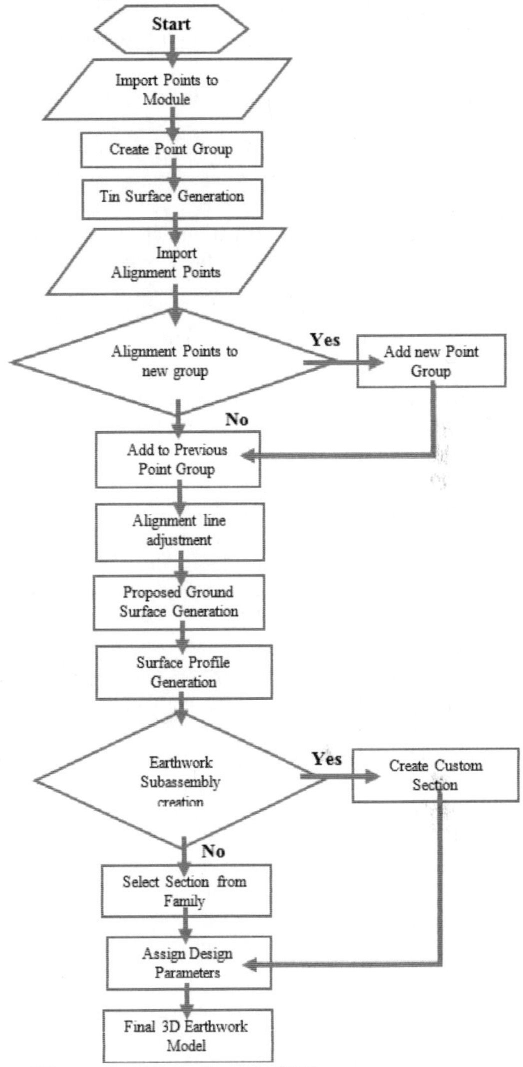

Figure. 1. 3D Earthwork BIM process

2.1.1 Point Data Collection

Data collection of the real construction site is the first step towards 3D modeling. Traditionally, different surveying techniques were used for collection of point data. Nowadays, laser scanning is considered as an easier and faster way as compared to traditional surveying equipment [6]. Laser scanning made the analyzing of a ground surface model more easy and efficient. A detailed plan of laser scanning position is required to collect maximum details from the site accurately. The irregular shape of the site increased a maximum number of scanning position to gather every possible detail in the form of point data. Scanning on the site has the chances of obstruction from construction materials and equipment. They will store as an extraneous point in the point data. It will cause a disturbance in the accuracy of acquired data. Extraneous points of moving object are difficult to remove so extra care has to be taken in such situation. Removal of all extraneous point is necessary after collection of point data information for 3D models. Extra attention is required during removal of points to improve the accuracy of site information. The RGB function of different software help to perfectly identify and eliminate these points.

Point data after filtration is saved into a .txt file format. Data is saved in global coordinates system which contains information of three axes X, Y, and Z. These coordinates information is imported into Flexible BIM module to create point group which is useful for organizing and editing point data. The next step is to generate a TIN model for 3D visualization of the ground surface [2]. The digital representation of actual site has a key role in earthwork planning especially for road projects [19, 20]. 3D model coordinates are required to synchronization with actual site northing, easting, and elevation value for accurate information. TIN surface model can be updated or edited by changing the values of coordinates. The TIN surface contains a large number of triangles depend upon the number of points captured from the site. To analyze site topography for the earthwork planning, the surface model can be visualized in different visual styles available in the module. These styles include 3D wireframe, conceptual and realistic views. It will also assist the designer to analyze information in various views for accurate 3D modeling. Layer function of the module can be utilized to visualize each component of 3D model separately and to assign different colors for identification

2.1.2 Alignment Line

Alignment adjustment is next major step after TIN surface generation for 3D modeling. It is considered as a most important step in 3D earthwork model of road design. The alignment line points of road design can be collected with the aid of 2D drawings. These points contain the coordinates values of X, Y, and Z similar to ground model data. The module gives function to add alignment point data into separate point groups and has given the advantage to edit or manage both data efficiently and without overlapping. Optimization of alignment coordinates value is significant for earthwork quantities in road project [22]. The 3D model will also assist to give various alignment line alternatives to improve road design [21]. Alignment adjustment is necessary to achieve perfect 3D model at each stage of earthwork. As the earthwork progress, it will be hard to achieve the Z value of alignment point according to the new ground surface. The 3D model will lose its effectiveness if inappropriate adjustment of Z value for the point is made. Figure 2. Illustrates with an example of a 3D model where the erroneous adjustment was made.

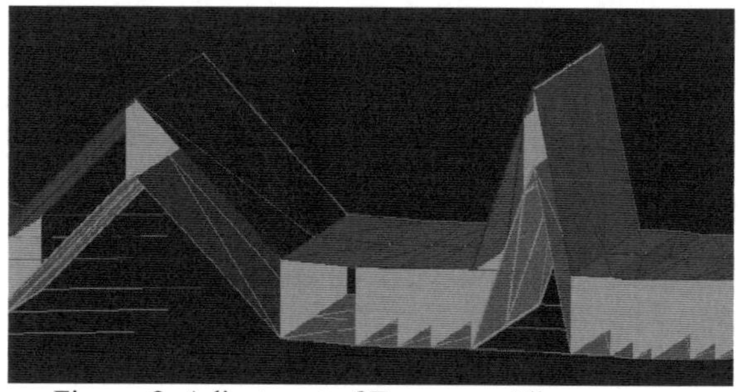

Figure. 2. Adjustment of Erroneous Alignment.

On the contrary, to adjustment problem a new approach of alignment adjustment is used in this flexible module. Z coordinate value of alignment point is adjusted to excavation depth which will eliminate erroneous adjustment. Model for excavation progress work will become easy and quick to design using this approach adopted in the module.

3D visualization of excavation depth can be achieved by making a TIN surface. The module utilizes alignment point to generate a TIN surface which contains all three X, Y, and Z coordinates value information. This information can be utilized also for partitioning technique and side slope surface generation. Design depth in the model will automatically change if any point information is changed. Module function of profile help to visualize depth difference between ground surface and proposed ground after excavation at each station. Profile main role is to observe elevation difference along alignment line [23]. Multiple options of the profile can be drawn and analyzed for the 3D model. It is also important factor like alignment for accurate 3D earthwork BIM according to the actual design.

2.1.3 3D Earthwork Model

Selection of earthwork section is a key factor in designing the final 3D model. Visual Basic programming API function is utilized for generation of earthwork section content library. Various earthwork sections are designed which are commonly used in road work Detail explanation of for earthwork section library algorithm will be presented in future work. Earthwork section is the geometry component of the final 3D model. It contains complete design parameters information of width, slope, and depth. These values of the section can be changed according to the design requirement. The designer can select any section of the content library or can create an own custom earthwork section by utilizing module functions. Parameters are assigned to the section after selection from library or creation. Section design parameters are collected through 2D drawings [15]. 3D model can be effected with parameters values, as attention is required while calculation of width, slope, and depth from 2D drawings.

The last step in BIM process is a generation of earthwork corridor model. It depicts the final 3D model which contain complete information of earthwork design [12,24]. 3D model is the composition of the surface, earthwork section, alignment line, and profile in the flexible module. Change of any major element will eventually change the 3D model. Excavation depth is required at each station along the alignment according to design as explained by Raza [15]. This increase the requirement of multiple section creation according to design at each station which is a time-consuming process. The section design requirement can be eliminated by utilizing the flexible module functions. Surface targeting function available in the module will automatically adjust the section with the ground surface. The slope of the section will target the ground surface and eliminate the use of different sections for one model. An additional function of making both side TIN surface in the Flexible module is provided to make accurate design model. The module will generate right and left side surface of the model by joining points of proposed ground and surface ground. The function can also be utilized to make a surface for the section slope. The surface generation will also utilize in partitioning technique. An example of 3D earthwork model design through this process is shown in Figure. 3. Similar models will be designed in the module.

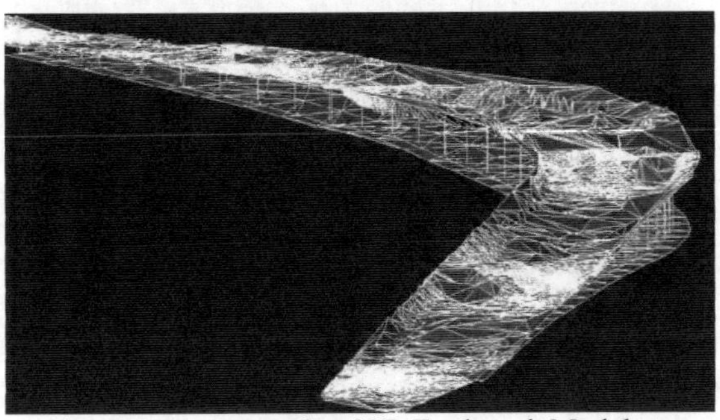

Figure 3. An example of 3D Earthwork Model

2.2 Partitioning Technique

The final 3D earthwork model generated through this approach is based on alignment line, surface, and section design. To model primary design information at each point in 3D earthwork is monotonous work. During the multi-phase construction work, various design changes occur due to site condition. Modification of design parameters at any segment will effect complete model as it is integrated and flexible. Partitioning technique function is introduced in the module taking into consideration the above-mentioned problems. This technique will use TIN 3D model to slice into horizontal and vertical segment like a surface model. Figure 4. illustrate an example of cutting of integrated surface model into different segment along the vertical and horizontal plane. Jingbin Hao [17] used partition technique to partition solid models into submodels segments to precisely analyze the model. As nature of the surface model is not similar to the solid

Figure. 4. Surface Model Partitioning

Model or tin model for slicing in various segments. Triangular network surface model is utilized for this purpose. TIN model is divided into smaller segments of points, lines by using explode function. These smaller segments recognized by the module to convert into the alternate triangular network by using line offset and mid-point method. The boundary is regenerated for each segment created in the new model. This method will aid to divide earthwork corridor model into multiple segments according to planning requirement. These segment will be in vertical and horizontal form.

Detail design information can be linked to each segment of the model. Soil attributes, start date, end date, slope information, underground facility information can be easily linked. This approach used in module provide a solution to the efficient management of work progress and eliminate redesigning of the complete model. The previous approach in different software like Autodesk Civil 3D used a station based partitioning technique that is not quite useful for detail design information modeling [12]. Corridor model from Autodesk Civil 3D can also be imported in this module for partitioning of the model. Alignment line is saved in .xml to be used in the module as the center line for partitioning. Corridor model along with surface model is divided into the separate triangular network. The complete network model is saved, and a quick view function of the module

and it will regenerate whole new TIN model. The horizontal and vertical segment will be achieved through module as explained above. Construction ID, start date, end date, soil attributes will be linked to multiple segments and save as a data sheet. This information of multiple segments in data sheet can be exported to Excel file for future use. Partitioning of corridor model as an example is shown in Figure 5. along with a framework of the module in Figure 6.

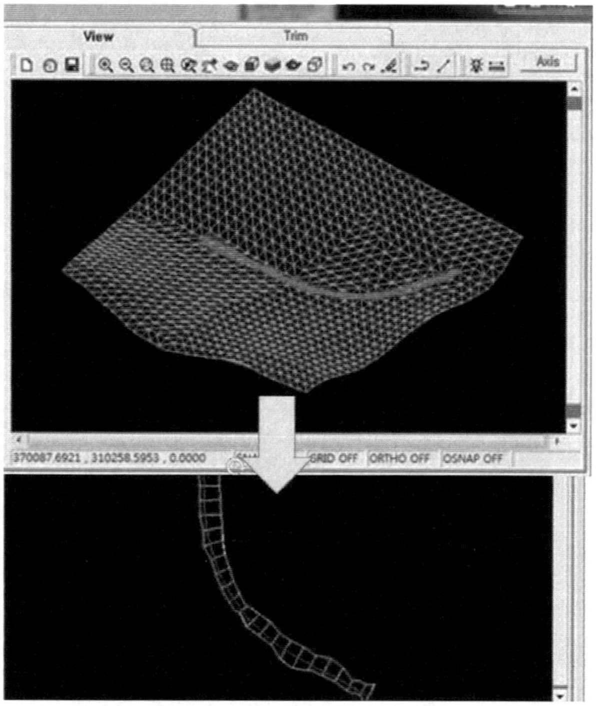

Figure 5. An example of Corridor Model Partitioning

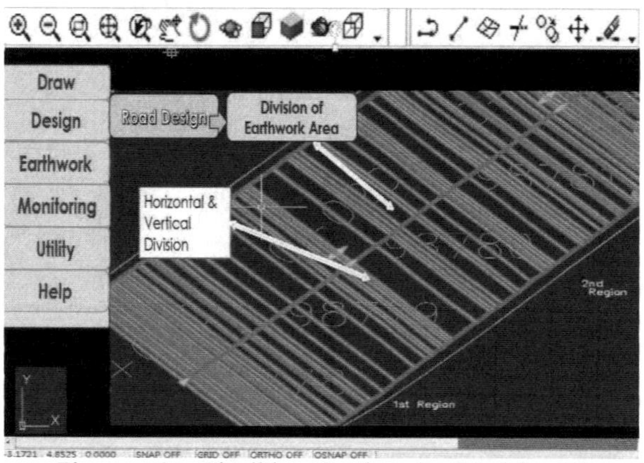

Figure 6. Flexible Earthwork Module User Framework

3 Conclusion

The purpose of the module framework presented in this study is to cover the knowledge gap of 3D earthwork BIM in the road project. In the first section of study complete 3D earthwork BIM process for a road project is discussed. It is divided into three major step from data collection to final 3D Earthwork model. The process can help to provide a complete visualization of earthwork design information in the 3D model. It's hard to deal with an integrated object during various construction phases where multiple design changes can be expected due to the site condition. The paper presented an important partitioning technique for small equal model segmentation to deal with the integrated object. It is useful in assigning the attributes at a different level for better managing earthwork design model. The technique presented also provides the basis to link required project information to each segment of Earthwork model. The information contains soil attributes, ID, start and end date, slope information. Flexible Earthwork module automatically makes changes in the entire model or segments if any design parameters are changed. An earthwork module system is under development based on this partitioning technique algorithm and 3D Earthwork BIM process. Visual Basic 6.0 programming language tool is used for the development of the module. It is supported in the Window 7 operating system. This module is also compatible with other software like Civil 3D

The study could be helpful in producing future research work on 3D earthwork BIM at different construction projects. The future work of authors involves linking 3D earthwork model with schedule and cost for 4D and 5D BIM. The development of improved earthwork BIM content library is also in progress. 3D models generated could also be utilized for machine guidance under the large scale fleet management research using ICT Technology.

4 Acknowledgement

This work is supported by the Korea Agency for Infrastructure Technology Advancement (KAIA) grant funded by the Ministry of Land, Infrastructure and Transport (Grant: 16SCIP-B079689-03)

References

[1] Easa, S. M., Strauss, T. R., Hassan, Y., & Souleyrette,R. R, Three-dimensional transportation analysis: Planning and design. Journal of Transportation Engineering, 128(3):250-258, 2002.

[2] Kerry T. Slattery, Dianne K. Slattery, Modeling Earth Surfaces for Highway Earthwork

Computation Using Terrestrial Laser Scanning, International Journal of Construction Education and Research, 9(2): 132-146, 2013.

[3] Hare, Warren L., Valentin R. Koch, and Yves Lucet. Models and algorithms to improve earthwork operations in road design using mixed integer linear programming. European Journal of Operational Research, 215(2): 470-480, 2011.

[4] ElNimr, Amr, Muaz Fagiar, and Yasser Mohamed. Two-way integration of 3D visualization and discrete event simulation for modeling mobile crane movement under dynamically changing site layout. Automation in Construction, 68:235-248, 2016.

[5] Jaselskis, Edward J., ZhiliGao, and Russell C. Walters. Improving transportation projects using laser scanning. Journal of Construction Engineering and Management, 131(3): 377-384, 2005.

[6] Tang, P., Huber, D., Akinci, B., Lipman, R., & Lytle, A. Automatic reconstruction of as-built building information models from laser-scanned point clouds: A review of related techniques. Automation in construction, 19(7): 829-843, 2010.

[7] Xiong, X., Adan, A., Akinci, B., & Huber, D. Automatic creation of semantically rich 3D building models from laser scanner data. Automation in Construction. 31:325-337, 2013.

[8] RANDALL, Tristan. Construction engineering requirements for integrating laser scanning technology and building information modeling. Journal of construction engineering and management, 137(10): 797-805, 2011

[9] Sepasgozar, Samad ME, and Sara Shirowzhan. Challenges and Opportunities for Implementation of Laser Scanners in Building Construction. ISARC. Proceedings of the International Symposium on Automation and Robotics in Construction, p.1, Auburn, USA, 2016.

[10] Kirchbach K. Dominik S. Fritz G. Introduction of a Digital Earthwork Construction Site, International Group of Lean Construction, pages 791-800, Fortaleza, Brazil, 2013.

[11] Elmqvist, M., Jungert, E., Lantz, F., Persson, A., & Soderman, U. Terrain modelling and analysis using laser scanner data. International Archives of Photogrammetry Remote Sensing and Spatial Information Sciences, 34(3/W4): 219-226, 2001.

[12] Holland, L., & Mercier, K. "Mastering AutoCAD Civil 3D 2013." Autodesk Official Press. John Wiley& Sons, 2012.

[13] Insite Sitework; Available from: http://www.insitesoftware.com/

[14] Earthworksoftware; Available from: http://www.earthworkssoftware.com/Earthworks.htm

[15] Raza H. Tanoli W. Lee S. Seo J. Methodology of Earthwork BIM Modeling & Un-excavated Volume Calculation of Road Project. Proceeding of Korea Institute of Construction Engineering and Management Conference, p 141, Incheon, South Korea, 2016.

[16] Tanoli W. Raza H. Lee S. Seo J. Study on Heuristics for Earthwork BIM Modeling. . Proceeding of Korea Institute of Construction Engineering and Management Conference, p 175, Incheon, South Korea, 2016.

[17] Hao, J., Fang, L., & Dong, Z. Research on generating similar-shaped assembly features for model partitioning. In Mechanical and Electrical Technology, 2nd International Conference, pp. 439-443, 2014.

[18] Mangan, A. P., & Whitaker, R. T. Partitioning 3D surface meshes using watershed segmentation. IEEE Transactions on Visualization and Computer Graphics, 5(4), 308-321, 1999.

[19] Nassar, K., Aly, E. A., & Jung, Y. Structure-from-motion for earthwork planning. Proceedings of the International Symposium on Automation and Robotics in Construction, p 310-316, Seoul, Korea, 2011.

[20] Turk, G., & Levoy, M. Zippered polygon meshes from range images. In Proceedings of the 21st annual conference on Computer graphics and interactive techniques, pp. 311-318, New York, USA, 1994.

[21] Kim, H., Orr, K., Shen, Z., Moon, H., Ju, K., & Choi, W. Highway alignment construction comparison using object-oriented 3D visualization modeling. Journal of Construction Engineering and Management, 140(10), 2014.

[22] Michael J. Mawdesley William H. Askew Saad H. Al-Jibouri. Site layout for earthworks in road projects. Engineering, Construction and Architectural Management, 11(2): pp. 83 – 89, 2004.

[23] Davenport, C., Voiculescu, I. Mastering AutoCAD Civil 3D 2016. Autodesk Official Press. John Wiley & Sons, 2015.

[24] Rebolj, D., Tibaut, A., Čuš-Babič, N., Magdič, A., & Podbreznik, P. Development and application of a road product model." Automation in construction, 17(6): 719-728, 2008.

Application of Civil Information Modeling for Constructability Review for Highway Projects

Y.H. Liau[a] and Y.C. Lin[b]

[a] Graduated Student, Department of Civil Engineering, National Taipei University of Technology, Taipei, Taiwan.
[b] Professor, Department of Civil Engineering, National Taipei University of Technology, Taipei, Taiwan.
E-mail: yickhongliau@yahoo.com, yclin@ntut.edu.tw

Abstract –

Usually, it is necessary for general contractor to handle constructability review of infrastructure and civil projects during the construction phase. Generally, constructability review usually be analyzed and applied at job onsite using 2D CAD drawings. The constructability review integrated with 2D CAD drawings usually cause time-consuming problems and problem misunderstandings. Civil information modeling (CIM) is one of the latest concepts and technologies in infrastructure and civil projects. The major function of CIM includes presenting and simulating operations by 3D model, and making constructability review through 3D BIM models. In order to enhance the performance of constructability analysis in infrastructure projects, this study proposes a CIM-based constructability review framework and approach for general contractors during the construction phase. The proposed approach is then applied in the selected highway case study in Taiwan to demonstrate the effectiveness of CIM-based constructability review in practice. Finally, this study identifies the benefits, limitations, and facing problems through the case study. Finally, conclusion and suggestion are summarized for further applications and researches.

Keywords –

Civil information modeling; CIM; Highway; Constructability; Construction phase; General Contractor

1 Introduction

Construction phase is the most important stage of the project, because in the construction phase construction company need to achieve the construction form the designer and a lot of subcontractors will join into the project in this stage, therefore construction engineers need to collaborate with those subprojects, to decrease construction disputes between subcontractors.

For the traditional solutions of the disputes between subcontractors, construction engineers will solve the problems by constructability analysis on 2D diagrams, but the results of traditional solutions not as expected, it is because 2D diagrams just present the two dimensions of the project in once, the information for third dimensions are depend on the experiences of construction engineers. In addition, every pages of 2D diagrams are individual, so the updates of the project need pages by pages to correct the exchanges link diagrams, however each correction have connected a lot of diagrams, and construction engineers have to correct every correction hand by hand, thus there is easy to get the mistakes or miss correction by the construction engineers.

Recent 30 years, construction industries are improving computer information technology with construction operations combine applicate [1]. In recent years, the most promising and commonly technic used in the construction industry is BIM (Building Information modeling) [2]. BIM technology has the function of synchronization, parameterization, three-dimensional models, and BIM 3D models have complete linkage. Besides that, BIM can eliminate engineers to imagine the 2D information to 3D imagination, improve the integration form the various units, and reduce problems of review and communication in the construction.

But BIM technology is more focus using in the building construction, not in the civil construction, therefore in this case study will implementation CIM (Civil Information Modeling) technology for constructability review for highway projects, for the application of CIM technology in the construction engineers, plan out the flow chart of CIM technology implement in constructability before the construction phase, to improve the efficiency of CIM technology.

2 Literature Reviews

The concept of the constructability was came for United States in 1970, the aim of the constructability is found the ways to improve and increase the cost efficiency of the construction project, improve the quality of the project, and this concept as a bridge between designer and construction companies [3]. Construction Industry Institute(CII), defined constructability as reducing costs, improve project quality, shorten the duration as the goals, knowledge applied to planning, design, procurement and construction of the various stages to achieve the overall project objectives [3].

CIM technology can use the previous data to backup the new project data, in the project manage, plan, design, build and facility management, to let engineers can easier to hit target of the project [4]. CIM technology is reference BIM technology [5], but there are three using different between CIM technology and BIM technology, first different is building project is a fix point project, and the building is vertical built, but civil project is a horizontal construction, and information of CIM project suspected of a wide range, second different is the professional word to both technology are different, in building project the vertical structure to support the building call column, but in civil project vertical structure to support the project call pier, third different is the construction method, in building project will built floor by floor, but in civil project is built pier by pier [6].

And Cheng et al. have generalize nine categories of civil project, show as below [6],

1. Bridges,
2. Roads,
3. Railways,
4. Tunnels,
5. Airport, port and harbors,
6. Energy infrastructure,
7. Utility infrastructure,
8. Recreational facility infrastructure,
9. Water management infrastructure,

Besides that, the development and adoption of civil project have separated into six categories [6],

1. Number of industry cases and academic papers
2. CIM uses,
3. Level of detail,
4. Data schema and representation,
5. Data delivery and management,
6. Software tools,

In addition, CIM technology can use in whole life cycle of construction, and in different phases of construction, the information of the CIM models need by user are different, to difference the level of information need Cheng et al. propose the information level of CIM models into five level, level of details (LOD) 100, it is present the information of construction in a 3D block model and using in architectural design, LOD 200 present the generic elements and using in architectural design, LOD 300 present the detail and refine assemblies, using in architectural design, LOD 400 present the construction level detail components, using in structural and MEP(Mechanical, electrical, and plumbing) design, LOD 500 present the as-constructed model, using in maintain phase [6].

Discusses how to promote the use of 3D information, including the establishment of the basis of 3D information exchange, modify the relevant system, road construction model, the production of engineering process model (4D) and the use of 3D to provide environmental information [7]. Research on 3D information and structural analysis software applied to railway viaduct cooperation information sharing [8]. Present a methodology for capturing and analyzing the communication patterns between project participants in the coordination room and understanding what participant actions normally take place during the design of the coordination meeting are necessary to provide them with adequate infrastructure support [9]. Based on the study of the use and use of BIM in infrastructure projects in Australia and the People's Republic of China, the BIM use and relate technologies of the project were analyze based on the research methods [10].

Comprehensive the use of CIM technology and the introduction of BIM technology in infrastructure relate to foreign literature, although the introduction of CIM technology in the construction analysis, but the construction of the analysis of the content and methods with this study is not the same. In view of the current Taiwan BIM technology into the organization of the majority of the factory for the trend, this study will explore the CIM applied to analyze the national construction project by the construction company as case study.

3 Research Content

The construction analysis is a pre-operation, that is very important for the construction build, but it is easy to be technically limited in the traditional construction and cannot complete the purpose of construction analysis. It is because in 2D diagrams is hard to sense the third dimension information, and sometimes those senior engineers also cannot found the overlap of the components in third dimension. In addition, the clarify of the construction interface are hard to explain to the field workers by the 2D diagrams, because the components build is present in 3D, but in the diagram just can show the 2D information. Therefore, in this case study applicate CIM technology for constructability reviews for highway project in Taiwan, because the traditional constructability has a lot to improve, so the following describes of constructability issues that have not been effectively improved for traditional construction, and introduce of CIM technology to explore the constructability. In traditional construction, the discusses of construction interface are in depend on the experience of field engineers, but the limitation of traditional discusses are the field workers cannot synchronize with the engineers, so the construction interface cannot solve as engineers wish, therefore in this case study will develop CIM 3D models as a communication platform to decrease the misunderstanding between field engineers and field workers.

1. CIM technology not a new technology in Taiwan, but the cognition of CIM technology for users is too less, so the result of the constructability analysis not in expected.
2. Therefore, in this case study will applicate CIM technology to assist the construction of national road construction, establishment a CIM construction review for the operating mode, and actual into a case study, to explore its benefits and difficulties.

But there is a lot kind of constructability analysis in the construction phase by using CIM technology, such as clash detection, simulation of construction process, construction interface review, regulation review, suggestion evaluation and many more. The constructability applicate of CIM technology in this study are clash detection and construction interface review, the description and function of the application analysis project are as follows.

- Clash detection

In the general construction, the 2D drawings of the project are divided into structural, construction and MEP, so the field engineers will receive the three individual drawings file, therefore, the field engineers need to integrate the three drawings into one interface, and use the process of mapping to find the overlap problem of the three units in the same interface. However, it is difficult for the field engineers to judge the overlap of the 2D interface and in the construction process will still have a lot of clash problems, which make the construction need to rework. So the construction company plant with use CIM 3D model for integrate three units drawings, and the use of CIM technology as a platform to analyze the clash and find a solution to the problem.

- Construction interface review

When the construction in progress, the same construction interface will often encounter multiple of projects at the same time, so the construction project will have the opportunity to meet the interface overlap,

If the more serious situation will lead to extension of the project, and the main reason are the field engineers before the construction cannot be detail division the job of the labor and the poor communications between two sides. Therefore the construction interface analysis are important for the construction, implement CIM technology can let those field engineers and labor more understand the certify of the construction interface and let field engineers could be more easy to solve construction interface problem.

To let the constructability analysis in this case study get more smoothly, in this section would discusses about the chart flow for clash detection analysis and construction interface review.

First for both analysis, are consolidate project information and checking the requirements of CIM 3D models by the analysis, and second for the clash detection analysis, will check the overlap in the CIM 3D models, the clash analysis will just check about the overlap between pier foundations and pipelines, and it is not include the overlap between the different pipelines, after checking the clash detection analysis, get the result to check is it requirement by the analysis need by the field engineers, the flow chart shown in Figure 1.

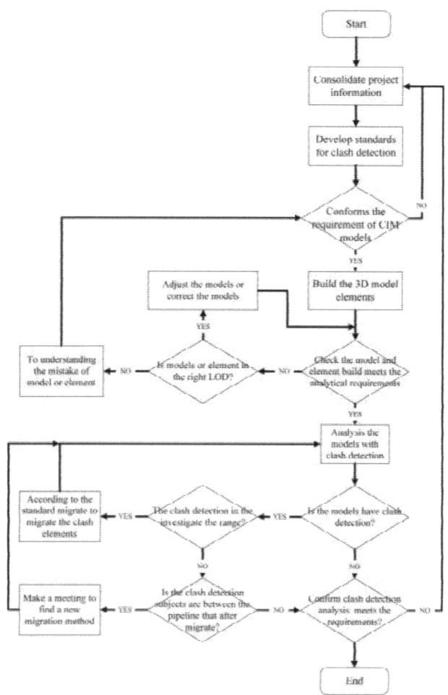

Figure 1. Flow chart of CIM clash detection analysis

In addition, the second part for the construction interface review in input the data of the construction interface into the CIM models, and checking the overlapping of the interface, lastly check the result is it as requirement need, the chart flow shown in Figure 2.

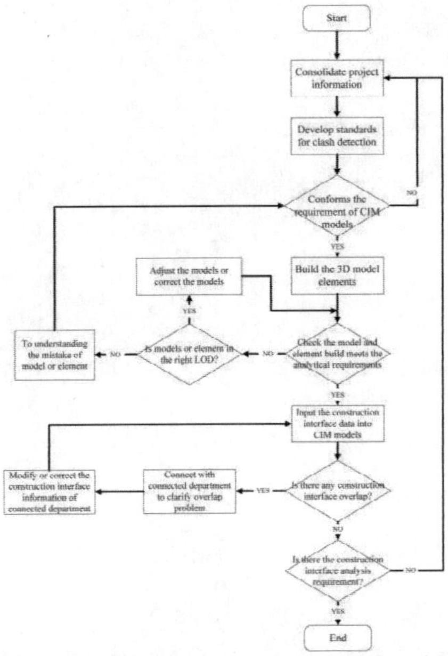

Figure 2. Flow chart of construction interface review

4 Case study

BIM has start actively used in building projects in Taiwan, but not in infrastructure projects. The viaduct project is select as the case study, the project locate in Taipei city and surrounding of this case study has a lot of residential area and offices area, and the piers foundation has been design in the underground for decrease the surface area need on the ground surface. So, normally the construction engineers will get the 2D pipelines diagrams and the 2D design diagrams, integrate both diagram to find out the overlap, and solve the overlap in the assembly, but in traditional solution is too slow and high risk for the construction, because the information can get form 2D diagrams just a two dimensions and the overlap of both are three dimensions, often produce misunderstand of acknowledge and the third dimensions information, it will cause field engineers wrong excavation or in the excavation project spoil the pipeline needed. Therefore, construction company decide to applicate CIM technology, to solve the problem about the overlapping between pipelines and piers foundation before the construction project start.

In addition, the steel rebar of the piers foundation design in viaduct project are high dense, in order to meet the body structure, vehicle high speeds and vehicle load, so the high dense of steel rebar is a complex design, and the field engineers need to make an order to input the steel rebar step by step to avoid the process have dispute, and delay the construction period, therefore construction company applicate CIM technology to decrease the construction interface problem, and make the construction get more fluent when the construction is processing.

1. The model planning and built

To improve the efficiency of the project for the construction of the project, it is necessary to construct a CIM model conforming to the construction analysis. Therefore, this study will carry out the pre-operational planning and the basic requirements of the model for the analysis of this case study, and the considerations staffs of the CIM 3D models include CIM manager, CIM engineers, and 3D models users. In this case, the CIM manager is required to provide the model requirements and limitations of the model in terms of their own engineering experience, and to confirm whether the model can begin to use. CIM engineers needs to collect the construction information of different units, as well as understanding the operation of the complex CIM modelling software, so that the model can be built and the content can be the best display effect, in addition, CIM engineers need to clarify the interface of each unit has been identified construction sequence and review the demand for construction. CIM models users need to recommend the lack or inconvenience use of the CIM 3D models, and feedback to the CIM engineers, those relationships are shown in Table 2.

Future more, the color identification to differentiate the pipelines system are needed in the 3D models, to highlight the different of the pipelines system, let CIM models users could more easier to distinguish the pipelines system, so that can easier to discusses and analysis the constructability. And the pipelines system in this case study are included main power system, sewage system, running water system, gas system, and telecom system (see Table 3).

Table 2. Relationship of consideration staffs and responsible for work items

Consideration staffs	Responsible for work items
CIM manage	To organizing and assigning CIM engineers working projects, developing CIM technology standards, standard program content, understanding the contents of CIM technology, and decision makers of CIM technology
CIM engineers	To build CIM model, review the correctness of models, using CIM

	models to analysis constructability of project, and provide field engineers 2D diagram form CIM models.
CIM models users	Feedback CIM engineers about CIM 3D models problems, and the scene required CIM technical analysis

Table 3. Pipelines system in CIM 3D models

Pipelines system		Color identification	Models
Main power system	High power		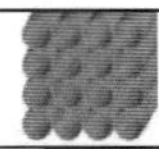
	Low power		
	Street light power		
Telecom system			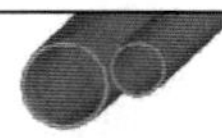
Sewage system			
Running water system			
Gas system			

In this case study, there is a lot of overlap between the pipeline and the pier, although the pipeline has been embedded in the original components, but the modified pipeline is low cost than modified structure of the pier design. Therefore, the problem of overlap between all pipelines and piers in this case study will be handled by the migration pipeline.

The build of pier foundations and the pipeline systems in the CIM 3D models, will get the models as shown in figure 3, and found that some of the pipeline systems have overlap with the pier foundation, and those pipeline systems are included high power system, running water system, gas system and telecom system, after CIM engineers done the models, the models will be discusses with CIM manage to solve the clash between the pipeline and pier foundations, after the discusses CIM engineers will revise the CIM 3D models, as shown as figure 4.

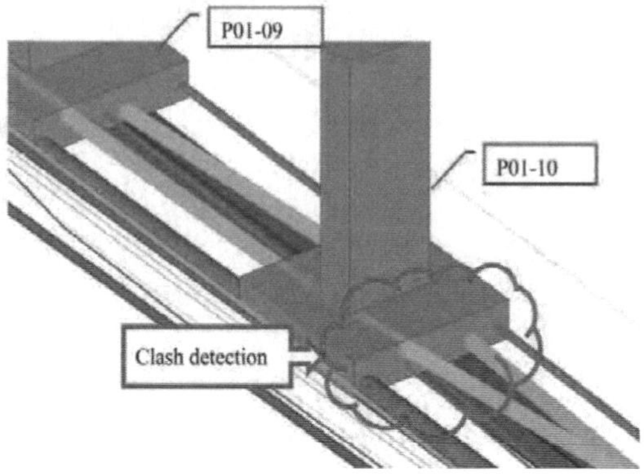

Figure 3. Overlap between pier foundations and pipeline systems

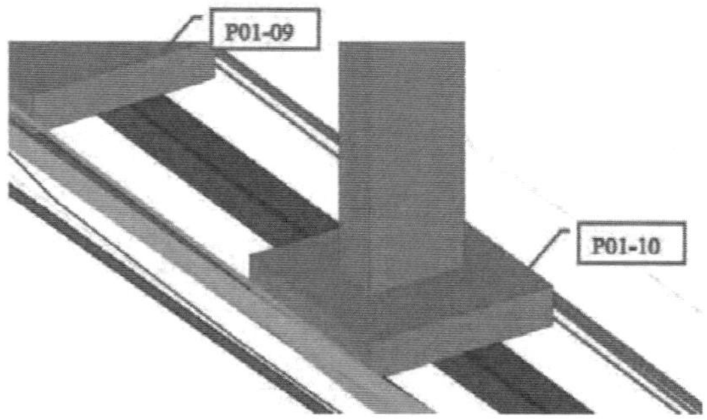

Figure 4. After revise by CIM engineers

To allow all the relevant personnel involved in the project to be able to communicate with the model to discusses the construction, and let the site construction staff as soon as possible to understand the scope of application, so the contents of the pier foundation will be targeted for the construction interface between the steel rebar by color distinction, as shown as figure 5.

Figure 5. Construction interface of the steel rebar

Form the literature reviews and the constructability analysis develop in this case study, found there are quite different between BIM technology and CIM technology on the constructability analysis, so in this section will discusses about the differences of both technology. The first differences is the clash detection analysis, the clash detection analysis are using to check the overlap between two or more elements, to decrease the percentage of rework on site field and making the decision on the site field. So, in CIM technology the most clash detection analysis results are form the environment elements overlap with the construction elements, but in BIM technology the results of the clash detection analysis results are form the elements overlap between elements in the construction. The second differences is environment analysis, civil construction is a horizontal construction, therefore the implicate of environment are extreme more than building construction, so the construction engineers need to analysis more environments data for the civil construction to make sure the construction could complete in expected time. The third differences is the traffic maintain during the construction, civil construction is a moving construction during the construction going on, therefore applicate CIM technology could let construction engineers earlier to plan the traffic maintain need. But building construction is a fixed point construction, therefore the traffic maintain for building construction will be less analysis need then civil construction.

Therefore, this study utilizes the visualization characteristics of CIM technology, participants the analyze and communicate with the CIM model, explore the constructability of the project, immediately identify the problem in the meeting, let the relevant professional engineers to immediately solve and propose a program to speed up the benefits of engineering construction analysis.

- Benefits of CIM technology in the constructability analysis

1. If the overlap of the entity may be dealt with directly by the engineers, but the overlap at the construction interfaces are not physical conflict, so the engineers needs to have a limit to the simulation of the 2D plan, and the process will be quite complex, therefore, the visualization of CIM model and the use of colour, engineers can more easily understand the interface of the conflict, to speed up the engineers in the interface to clarify the time-consuming.

- Difficulties of CIM technology in this case study

1. The expectation of the CIM 3D model can be closer to the entity, but the models more closely will cost more time on built, and the information actually more important than the model closer or not, but it will be other units require upgrade their models LOD to enhance as entity.
2. The clash detection between each other pipeline are normal, it is because the field engineers could be solve the problems in the site situation, it is not the main question for the clash detection analysis, but during the discussion process, some other will require to present a zero clash detection analysis as a report.

5 Conclusions and suggestions

5.1 Conclusions

1. The underground pipeline will be exchanged when the other construction are going, but the 2D diagrams of the concerned department have not updates those diagrams, therefore the information of 2D diagrams obtain by the next contractor are mistaken.
2. The reality of the CIM 3D models are expected by the owners but the more reality of the 3D models will waste more time of CIM engineers and extra data for the construction analysis will cause the analysis slow, therefore the LOD used for those 3D models should just reach the minimum need of analysis.
3. Less peoples are current understanding of CIM technology, so they are unbelieve the analysis of the CIM technology, thus blocked the efficiency of CIM analysis and some other will increases unnecessary analysis of the project.

5.2 Suggestions

1. CIM technology in the construction analysis to bring some benefits to enhance, if the future can be based on the quality control system into the project of CIM technology, will be able to confirm the quality of CIM technology imported.
2. Construction companies could be applicate the technic of virtual reality (VR), to let the construction 3D models more reality, to easier make the decisions in more reality environment and reduces the dispute when the project complete.
3. Relate units could setup CIM execution plan as a standard operation procedure, to obvious the goals of the project.
4. CIM technology applicate in the constructability could make a lot of benefit for the construction companies, but the quality control also need for the CIM technology develop.

References

[1] Fischer, M. and Kunz, J. , *The Scope and Role of Information Technology in Construction*, Centre for Integrated Facility Engineering, 2006.

[2] Eastman, C., Teicholz, P., Sacks, R. and Liston,K, *BIM Handbook, a Guide to Building Information Modelling*, 2nd Ed. Hoboken: John Wiley & Sons,

Inc, 2011.

[3] Construction Industry Institute (CII), *Constructability: A primer*, Construction Industry Institute, United States: Austin, 1986.

[4] Japan Construction Information Center, *The Key-Base Information Station for the Construction Industry to pave way for the Future*, 2016.

[5] Liebich, T., IFC FOR INFRASTRUCTURE, http://iug.buildingsmart.org/resources/itm-and-iug-meetings-2013-munich/itm-meeting/ifc-for-infrastrcuture, 2017

[6] Cheng C.P., Qiqi Lu and Deng Y., *Automation in Construction*, Volume 72, Part 1, page 31-47, Elsevier B.V., 2016.

[7] Kubota S., *International Journal of Social Science and Humanity,* Volume 2 Number 1, pages. 24-29, 2012.

[8] Fujisawa Y. and Yabuki N., Cooperative Information Sharing between a 3D Model and Structural Analysis, In *International Conference on Cooperative Design, Visualization and Engineering*, pages 203-206, Osaka, Japan, 2012.

[9] Addor M. R. A. and Santos E. T., BIM Design Coordination Room Infrastructure: Assessment of Communication Activities, In *Computing in Civil and Building Engineering*, pages 1-8, Orlando, Florida, 2014.

[10] Chong Y.H., Lopez R., Wang J., Wang X. and Zhao Z., Comparative Analysis on the Adoption and Use of BIM in Road Infrastructure Projects, *Journal of Management in Engineering*, 2016.

Visual Language-based approach for the Definition of Building Permit related rules

Hayan Kim[a], Jin-Kook Lee[a], Jaeyoung Shin[a], Jinsung Kim[a]

[a]Department of Interior Architecture Design, Hanyang University, Seoul, Republic of Korea
E-mail: hayaan92@gmail.com, designit@hanyang.ac.kr, jjyoung311@gmail.com, wlstjd1320@gmail.com

Abstract –

This paper aims to describe the visual language-based approach to define building permit related rules in the much intuitive way and support high applicability. The advent of BIM (Building Information Modeling) enables an automatic rule-checking process in the early phase of design by translation and interpretation of regulation about building design assessment. For the precise rule-checking process, a logic rule-based mechanism for translating is needed as prerequisite step and the rule-making process have to be proceeded based on this mechanism. Accordingly, Korea government also proceeding the KBim project for employing BIM to adapt automatic approach to the overall process of building permission system. As the part of the project, KBimLogic offers logic rule-based mechanism which in charge of rule-making by translating natural language regulation sentences in Korea Building Act into the computer-readable form-KBimCode.

This paper introduces the KBim Visual Language (KBVL), the extended way of composing the rule module of building related regulations by the visual elements for the intuitive and immediate identifying the process and result of rule-making. The language supports user to combine visual symbols and their connection according to the user-defined nesting relation and generate KBimCode automatically at the background of the visual language. Thus, intuitive composing shorten the time of generating rule modules and defining the logical relation between the modules. As the result, the user can handily query the components of KBimCode and define relation of them to compose tailored ruleset for their own rule-checking process with various type of software.

Keywords –

BIM (Building Information Modeling); Building permit related rules; Rule-making; Visual language; Rule definition

1 Introduction

The conventional building design assessment of legislations mandatorily requires handling of human resources. Since the sentences of building regulation are written in natural language, it includes expressions with ambiguity and vagueness [10]. Therefore, each sentences requires precise interpretation by the experts, and rule-checking phase spends a plenty of money and time. The introduction of BIM (Building Information Modeling) enables those manual process to be automatized by computer-based 3D and information modeling. Especially in the field of evaluating building related regulation based on BIM model, establishing logic for translating natural language sentences into the computer-readable format became the forefront of the assessment [1, 2]. Especially, the Korea Building Act has multiple clauses and phrases with intricate relations such as upper-lower, reference, and commission [3]. To read natural language in the building design assessment program, Korea government includes the phase of research about translating and interpreting Korea Building Act in the KBim project for the e-submission system with BIM model [11,12].

The overall project is summarized with three features-1)developing overall process for design assessment, 2) developing standard and practical technology for applying BIM-based design plan, 3) developing cooperation working system to integrate design and engineering. As the part of the project, KBimLogic is specifically devised to translate Korea Building Act into the KBimCode – intermediate code for executing regulation in the rule-checking software.

Although the KBimCode supports the computer-based format of the building code, there are still high thresholds for domain workers who are non-expert about programming knowledge to write code directly. For the much higher usability of KBimLogic and KBimCode, this paper aims to introduce a visual language-based approach to define building permit related regulation much intuitive way. Consequently, this paper includes the approach based on the backdrop research – KBim project that has been led by the government of South Korea. By the utilizing preceded research, this paper suggests the graphical approach to practically use the KBimLogic and KBimCode.

2 Related work

The state of the rule-making approaches for building design requirements in recent studies are typically with two methods, one is hardcoding-based scripting language and another is visual language. By the reviewing main feature of representative researches, the necessity of visual approach acquires validity. Through the recent researches using visual language-based approach for rule-making, KBVL establishes its specific characteristic different with existing approach and acquires validity for utilizing visual representation.

2.1 Hardcoding-based rule-making approach

The advent of BIM in the field of architecture, engineering, construction and facility management (AEC-FM) enables computer-readable and analyzable building model in the computer-based environment. Automating rule-checking also became possible by the swift development of BIM application and related technologies [13,14]. For structuring the checking process, the prerequisite phases are needed - interpreting and translating regulation into the computer-readable format. In the early 2000s, several pacesetting projects had been proceeded at the government level for their own software system or language framework based on their national regulations and design guideline.

- As the part of CORENET (Construction and Real Estate NETwork) project (2005), e-PlanCheck takes charge of automatic rule-checking process with the IFC building model and their own platform, FORNAX. This platform employs C++ based ruleset library and both expert of Building Act and computer programmer cooperate for composing code [9].
- SMARTcode project (2006) utilizes Markup method to distinguish noun phrase, verb phrase, and the context in the sentences. As the next step, components of analyzed sentences are translated as structured XML file maintaining original meaning of sentences [8].
- GSA project (2007) is for the federal courthouses building design assessment project using BIM application and led by the USA government. The part of the design guidelines for building plan are translated as computer-readable form for rule-checking program. The rules of indoor walkability and circulation are checked automatically by the generated circulation graph [4].

2.2 Visual language-based approach

The preceded projects about rule-making process in the early 2000s are typically based on the specific software dependent algorithm or require expert knowledge about computer programming. Since the most of domain expert in architecture field are not familiar with programming language, previous process is difficult to approach for the building related workers. This is why it may be less accessible and useful to the architects.

As the solution, visual language is effectively employed for the supportive and higher usable language framework because of its advantages – intuitive, real-time modification, and parametric outcome as the consequences. The early phase of introduction of visual language in the field of the building has been focused on the transformation of building geometry. The Dynamo for Revit or Grasshopper for Rhino has been role as the frontier for introducing geometry focused visual language in the BIM authoring tools.

As the higher level in the AEC-FM fields, the visual language activates the rule composing by domain workers who are not familiar with computer programming language for building permission system. Especially the visual language effectively works for querying and combining the components for composing ruleset. Querying specific code component of objects and composing them as ruleset file for computer-readable format demands language-specific knowledge about the language's own grammar, vocabulary, and exceptional rule. Therefore, there steadily have been researching to visually represent the process of rule-making and checking.

- VBQL (Visual BIM Query Language) has been developed for the craftsmen who are not familiar with text-based programming knowledge to formulate model queries [7]. This language considers not only for the building model but also cost and time model to deals with 5D BIM model for the different fields and phases.
- The VCCL (Visual Code Checking Language) represents information using various type of methods and tested with the BIM+ [6, 15]. As the main attributes of VCCL, genericity and finest granularity are typically dealt as important aspects.

Table 1 Comparing the visual representation of the general node and node combination of VBQL and VCCL

Visual Language	Visual characteristic of the general node	Connected component

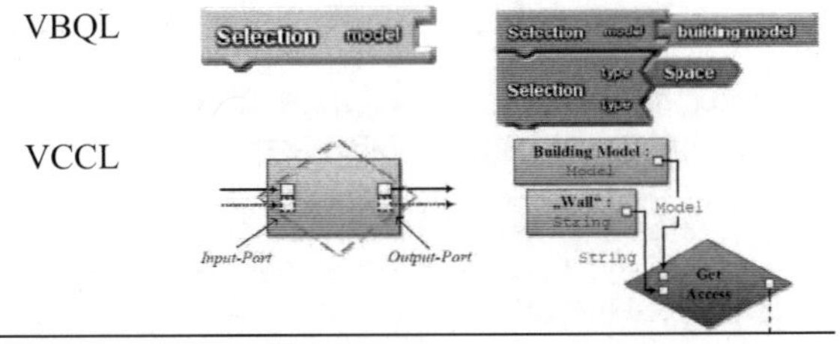

2.3 Script language-based rule-making

As the research for adapting BIM technology in the building permission system, several countries are proceeding government-based research project [4,8,9]. The Korea government also proceeding the KBim project that for establishing BIM-enabled building design assessment system based on the Korea Building Act[11,12]. As the part of overall project, logic rule-based mechanism called KBimLogic is developed to translate natural language of Korea Building Act into computer-readable format called KBimCode [5]. This paper aims to introduce KBim Visual Language(KBVL) for the improvement of the KBimCode usability.

3 Development Approach of KBVL

3.1 Fragmentizing building permit related rules

The building permit related regulations are descripted in not only the Building Act but also in various type of document such as Building Act, Request for proposal (RFP), and international design guideline with their own grammar and vocabulary. As the minimum requirement for the building rule-checking, Korea Building Act is finally checked for the building permission.

For the precise representing of information as visual symbol without contents omission, sentences have to be fragmentized as atomic sentences to seize the logical relation between multiple clauses. The sentences of the building related regulation essentially include the key statement (KS) appointing the statement of the object. Also, there is the condition statement (CS) appointing the conditional statement for the key statement.

In the building regulation, building object is the target of verb phrase. Also, there are common verb phrases that used frequently for checking building object such as the measuring physical geometry, identifying position relation between other object, or whether specific object is installed or not.

The fragmentized sentence element collected as the component database for the establishment of standardized library comprising the objects, their properties, and the high-level methods.

3.2 KBVL Grammar

The grammatical logic structure of the KBVL is based on the connecting relation of visual element.

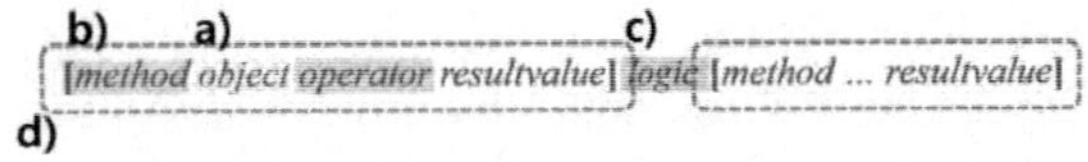

Figure 1. The sentence elements and they are covered by a) object node b) method node c) logic node d) rule node

The code in (1) shows the example of KBimCode generated by composing nodes explained in Figure 1.

$$[getObjectCount(Window) = 5]\ OR\ [hasSpace(Room, Window) = TRUE] \quad (1)$$

Unlike uncomplicated checking process of single building object, the checking regulation with constrain relation with other object needs logical statement and nesting process as necessary phase. If the method nodes define the relation between objects, the logic nodes decide the relation between rule nodes. The nesting process prominently removes the repetitive node combination by binding several user defined properties in the common object and offers reusability of the node including specific condition.

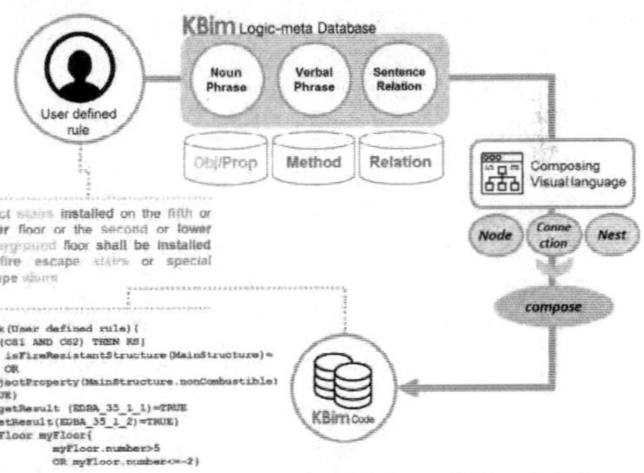

Figure 2. The process of utilizing KBimLogic meta database and generating KBimCode

Also, the KBVL basically utilizes the KBimLogic meta database as the semantic library of building object, property, and method. This database includes largely 5 main databases that established by the KBimLogic. Based on the Korea Building Act, database is composed of sentence DB, object and property DB, method DB, logical relation DB, and KBimCode DB.

In this paper, KBVL is related with object and property DB and method DB. The main visual element – node – queries each nodes' semantic information and function from the database.

4 Component of KBVL

KBVL intuitively displays the flow of sentence component with the basic visual language component. The each visual components of KBVL includes the fundamental and physical connecting function and also supports users to generate KBimCode much easily than just describing the script-based code with hard-coded method.

4.1 Node

The visual language basically represent the main structural part with node component. Node generally corresponds to the specific class and creates semantic attributes by the connection between other nodes. Also through the importation and exportation of the data through the port, visual program transfer and accumulate the information over the node combination and finally deduces the result.

According to the type, the nodes have specific function to handle input data or hold the condition of object in the runtime of the visual language process. The KBVL nodes are divided as four type according to the function – 1) object 2) method 3) logic 4) rule.

4.1.1 Object Node

Object and property node queries building object class which commonly defined in the KBimLogic meta database considering 'IfcBuildingElement' structure. The object class primarily includes the main components of the building such as "Window", "Stair", and "Wall" etc. like shown in Figure 3. Property is inseparable from the object. In other words, each object has none or many properties but each property has to be attached at an object. Object nodes are connected with method node to generate minimum unit of rule node.

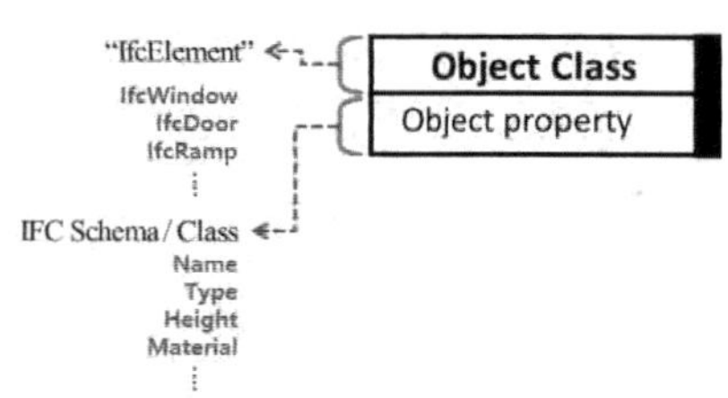

Figure 3. Semantic structure based on the IFC framework of the object node

4.1.2 Method Node

Method node perform specific action on the inputted object node. Depending on the type of method, one or many object and property nodes have to be connected as parameter of method at input port of the node. By the specific action, result value is exported through output port.

Method node is composed of four types of contents lines – method name, parameter type, operator, result. To reflect the feature of rule-making that result have to be defined as specific type, method node includes operator and result. Basic three type of method node is descripted in Table 2.

Table 2. Method node classification based on the type of return value data type

Return value	Left Operand	Operator	Right Operand
Boolean	Methods querying boolean value	=	True, False
Numeric	Methods querying numeric value	<, <=, =, >=, >	Numeric
Collection	Methods querying the name of type	=	String

Transmitted data of the object node is combined with method node to generate basic rule unit. Several rule units arc composed with logic node to generate rule node including result value. Based on the return value type of table 2, method node is composed of method name, parameter type, operator and result value. Method name refers the exact method from the method database. Parameter type, operator and result (return value) is defined in the database. According to the method, parameter type can be one to many.

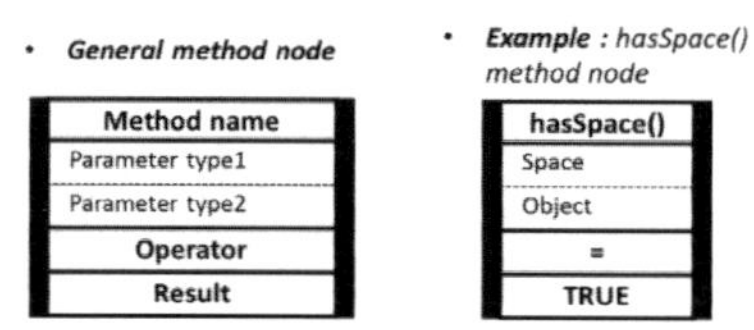

Figure 4. Structure of method node and example

4.1.3 Logic Node

Logic node supports rule node and unit to have logical relation between them. Especially the building permit related regulations are composed of plenty of sentences and difficult to analysis. Therefore, it is important to set up the organized logical relationship between subdivided sentence elements.

Based on the 3.2 section, building permit related regulations are summarized as AS and again subdivided as TAS structure. The TAS is composed of ALU level of semantic combination that represented by the single rule unit. As the logical relation generated in (2) sentences, logic node is necessary for semantically connect the rule units.

IF (CS1 OR CS2) THEN (KS1 AND KS2) (2)

4.1.4 Rule Node

Rule node is composed of multiple nodes and their connection. Because of the complicate logical statement and conditional relationship, rule node bound as nesting box is necessary for brief node combination. Rule node R is defined by (2). N is the set of node composing rule node R and C is the set of connection between N composing R.

$$R = (N, C) \quad (3)$$

$$N = \{ n_i \mid 1 \leq i \leq n;\ n = \text{number of node composing R} \} \quad (4)$$

$$C = \{ c_i \mid 1 \leq i \leq n;\ n = \text{number of connection between N composing R} \} \quad (5)$$

4.2 Connection

Multiple nodes are linked by connection component and composed as the node combination. Connection creates the relationship between nodes and exchanges the data of the nodes for their own function. By establishing dependencies between the components, visual program creates possible flow of information through the node combination. Connection also transfer the data of the node with the direction flowing from left to right.

The direction of data flow appears in the characteristics of shape of the nodes. Object node is placed on the left side of the method node and only has the output port on the right side of the node context. Differently with object node, method node has both input and output port to import and export the data through the ports.

4.3 Nesting for Rule node

The complex relation between multiple rule units can be decreased by the degree of nesting. By the nesting function, two visual language components can be generated.

- Rule node
- Predefined object node

Figure 4 shows how to generate logical relationship between the nodes and nesting box. The KBimCode automatically generated at the background of KBVL.

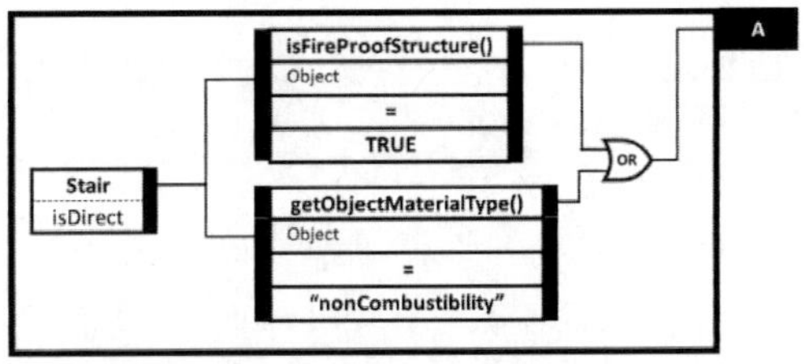

Figure 4. Appearance of nesting box that is related with generation of rule node.

5 Demonstration

Based on the approach explained in previous section, simple rule and complex regulation are represented with KBVL. By analyzing rule sentence and composing nodes with connection, user easily write KBimCode.

5.1 Target regulation and sentence analysis

The design regulations of building include the usage of building, site, structure or facility. Among the various type of regulation, the demonstration uses two type of regulation. First sentence is restructured design regulation composed with the common building element such as floor, stair, or window etc. This simple rule is checked for the general and common demand occurred among the domain workers. By this demonstration, usability of KBVL for general and simple KBimCode generation is proved.

- **(Rule A)** Fifth or upper floor or the second or lower underground floor shall be installed direct stairs as escape stairs.

As the next step, much complex regulation referred in Korea Building Act that has reference to other regulation or including object with complex property are represented with KBVL. Regulation is the part of EDBA (Enforcement Decree of the Building Act) 34-3 notifying the standards of installation of direct stairs. Among the original texts of regulation, the definition of specific object is eliminated with the assumption that the object database already includes the definition.

- **(Rule B)** In cases of installing a window that can be opened or closed less than 1.2 meters high over the surface of the living room in an officetel, a safety facility designed to prevent falling accidents shall be installed in conformity with the standards prescribed by Ordinance of the Ministry of Land, Infrastructure and Transport.

Consequently, this section assumes that user have to make two kind of rules including simple condition of building element and complex connection relationship between the multiple target element.

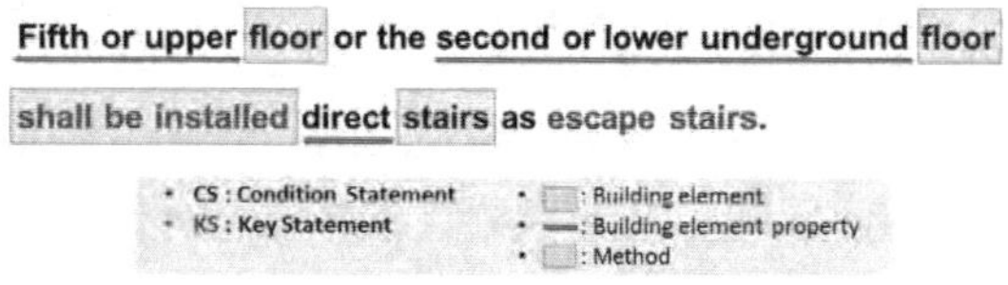

Figure 5. (Rule A) Sentence analysis for arrangement of node and their connection.

Figure 6. (Rule B) Sentence analysis for arrangement of node and their connection.

The KBVL composing process starts with the step of distinguishing the building elements included in sentence. Also, according to the type of object and requirement descripted in sentence, method for object is chosen. Sentence analysis of [Rule A] and [Rule B] is represented in Figure 5 and 6. Each building element has property as the condition and connected with other element by method node. By the CS and KS, regulation sentences involving condition for the result.

5.2 Node combination

According to the analysis of sentences, (Rule A) is verified as checking property of specified stairs. (Rule B) is much complex than (Rule A) and require more nesting function to generate rule unit.

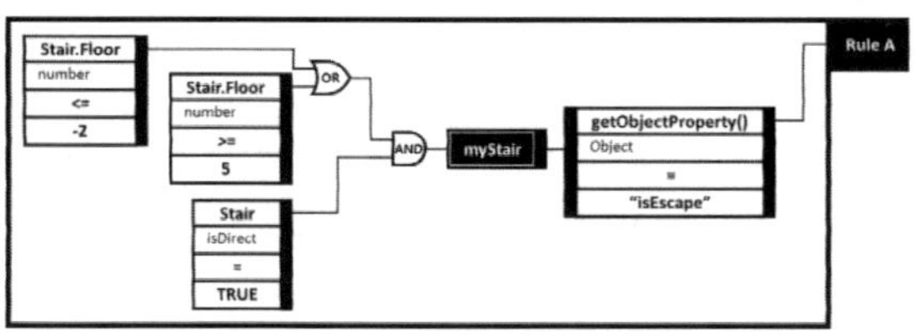

Figure 7. KBVL node composition of (Rule A).

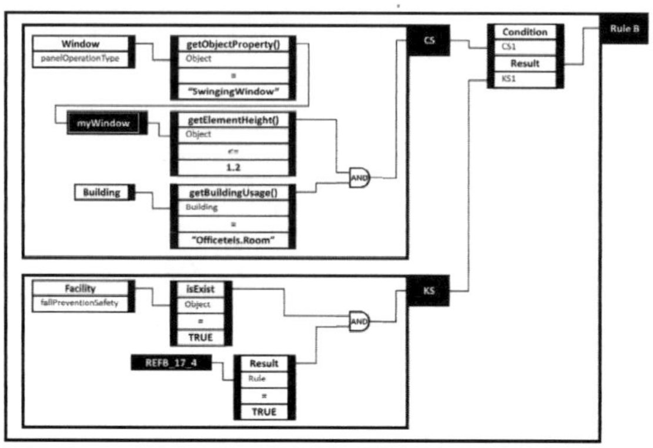

Figure 8. KBVL node composition of (Rule B).

5.3 Result

Through the node combination, (Rule A) can automatically generate KBimCode like Figure 9 and (Rule B) as Figure 10.

```
Check(Rule A)

Stair myStair{
Stair.Floor.number <= -2
OR Stair.Floor.number >= 5
isObjectProperty(Stair.isDirect) = TRUE
}

getObjectProperty(myStair) = "isEscape"
```

Figure 9. Generated KBimCode of (Rule A).

```
Check(Rule B){
        IF CS THEN KS
}
CS{
        Window myWindow{
        getObjectProperty(Window.panelOperationType) = "SwingingWindow"
        }
        getBuildingUsage() = "Officetels.Room"
        getElementHeight(myWindow) <= 1.2 m
}
KS{
        isExist(FallPreventionSafetyFacility) = TRUE
        getResult(REFB_17_4) = TRUE
}
```

Figure 10. Generated KBimCode of (Rule B).

Unlike script based KBimCode generation, KBVL supports user to find which object is the target object of the regulation and which action is executed for checking result. Also by composing several rule units as the rule node, clear distinguish between condition statement and key statement is offered by nesting function.

6 Conclusion

This paper describes a visual language based approach to support the generation of KBimCode which translated computer-readable format for building permit related rules. As an approach for visualizing regulations, sentences are precisely analyzed and decomposed into objects, properties, methods and logical relations from noun phrases. Logically restructured components are classified with semantic and syntactic attributes and hierarchically arranged. To structure logical process for visually representation of the regulations, preceded projects are reviewed and the common components used for visual representation are deduced as the visual components of KBVL. By utilizing node and connection, the components extracted from regulation sentences are again composed of a node combination.

The KBimCode is based on the scripting method, therefore it is hard to understand the main target object and context at once. As an updated method, KBVL offers the way of composing node combination to generate KBimCode automatically at the background of the visual language. This approach can be extended to not only for the building permit related regulation but the other types of regulation documents such as design guideline or RFP. Also, the extension of KBimLogic meta database, the usable object, property and method node also can be extended to the new type of components.

Acknowledgement

This work was supported by the National Research Foundation of Korea Grant funded by the Korean Government (NRF-2015R1C1A1A01053497)

References

[1] Eastman, C., Lee, J. M., Jeong, Y. S., and Lee, J. K. Automatic rule-based checking of building designs. *Automation in construction*, 18(8):1011-1033, 2009.

[2] Lee, J. K., Eastman, C. M., and Lee, Y. C. Implementation of a BIM domain-specific language for the building environment rule and analysis. *Journal of Intelligent & Robotic Systems*, 79(3-4):507, 2015.

[3] Kim, I.H., Kim, Y.H. and Choi, J.S. Building Code Typology and Application for Open BIM based Code Checking, *Transactions of the Society of CAD/CAM Engineers*, 19(3):224- 235, 2014.

[4] GSA Project- BIM Enabled Design Guide Automation, On-line: http://dcom.arch.gatech.edu, Accessed: 15/12/2016.

[5] Lee, H., Lee, J. K., Park, S., and Kim, I. Translating building legislation into a computer-executable format for evaluating building permit requirements. *Automation in Construction*, 71:49-61, 2016.

[6] Preidel, C., and Borrmann, A. Automated Code Compliance Checking Based on a Visual Language and Building Information Modeling. In ISARC. In *Proceedings of the International Symposium on Automation and Robotics in Construction* Vol. 32, page 1, 2015.

[7] Wülfing, A., Windisch, R. and Scherer, R.J., A visual BIM query language. *eWork and eBusiness in Architecture, Engineering and Construction*: ECPPM 2014:157, 2014.

[8] J. Wix, N. Nisbet, and T. Liebich, Using constraints to validate and check building information models, *eWork and eBusiness in Architecture, Engineering and Construction*, ECPPM, London:467–475, 2008.

[9] T.F. Sing, and Q. Zhong, Construction and real estate NETwork (CORENET), Facilities 19:419–428, 2001.

[10] Nawari O. N. Automated Code Checking in BIM Environment, In *Proceeding of 14th International Conference on Computing in Civil and Building Engineering*, 2012.

[11] Seumter, BIM-enabled e-Submission & Building Act Legality Verification System of Korea, On-line: https://www.eais.go.kr Accessed: 10/12/2016.

[12] I. Kim, Open BIM based Technological Environment for Building Design Quality Enhancement, Kyung Hee Univ., (Project report), 2015

[13] J. Choi, and I. Kim, A Study on the Application of Pre-Processing to Develop the Open BIM-based Code Checking System for Building Administration Process, *Journal of the Architectural Institute of Korea Planning & Design*, 30(9):3-12, 2014.

[14] I. Kim, Y. Kim, and J. Choi, Building Code Typology and Application for Open BIM based Code Checking, *Transactions of the Society of CAD/CAM Engineers*, 19(3):224-235, 2014.

[15] Preidel, C., and Borrmann, A. Integrating Relational Algebra into a Visual Code Checking Language for Information Retrieval from Building Information Models. In *Proceedings of the 16th International Conference on Computing in Civil and Building Engineering*. Osaka, Japan, ICCCBE, 2016.

Study on BIM Family Self-create for Steel Reinforcing Bar Detail Construction Design and Information Extraction

L.C. Lien[a,b], P.C. Zhang[b], S.B Chen[c], Z.C. Liao[b] and Y.N. Liu[b]

[a]Department of Civil Engineering, Chung Yuan Christian University, Taiwan
[b]College of Civil Engineering, Fujian University of Technology, China
[c]Fuzhou CityConstruction Design and Research Institute Co., Ltd., China
E-mail: lclien@cycu.edu.tw, 844634836@qq.com, palama@163.com, 39879084@qq.com, ynliu@fjut.edu.cn

Abstract –

In this study, the frequently encountered issues related to reinforced concrete nodes (beam and girder protection layers, beam column protection layers, and binding wire collision, etc.) were sorted out and analyzed. Building Information Modeling (BIM) was employed to solve the problems resulted from the complex node issues. Moreover, new steel reinforcing bar family (*.rfa) were constructed using the shared parameters to edit formula, to assign parameters to steel reinforcing bar processing and construction, to export detail list, to facilitate steel reinforcing bar data-processing and construction in the later stage. Lastly, a steel reinforcing bar construction process was constructed based on the BIM application presented by this study, serving as reference for the steel reinforcing bar works.

Keywords –

Complex reinforced nodes, Building information modeling (BIM), Steel reinforcing bar parameterization, Construction process

1 Introduction

Concrete frame is widely used as an architectural structure. Steel reinforcing bars constitute 15%-20% of construction costs, the larger the steel content, the higher the steel reinforcing bar costs **[1]**. Steel reinforcing bar works entail a high level of professional know-how throughout a fussy, massive and complicated construction process. Thus, the importance of steel reinforcing bars, a sub-project of construction plan, is overlooked most of the time and lots of steel reinforcing bars are wasted inadvertently. To use steel reinforcing bars economically, a rational design and construction process are essential. Therefore, effective measures are needed to prevent steel reinforcing bars from being wasted, and at the same time to ensure economic benefits and social benefits. Surprisingly, hardly any constructor is aware of the issues related to the steel reinforcing bar arrangement and structure as well as the complex and intensive nodes. As a result, node quality worsened, affecting the architectural structure quality severely. Information technology is therefore needed to improve construction quality. Building Information Modeling (BIM) has tremendous potentials to be explored in the days to come. BIM is noted for its superior ability to design and arrange steel reinforcing bars, to solve the complex issues related to the reinforced concrete nodes, to store steel reinforcing bar information, and to find the solutions for the effective use of steel reinforcing bars.

2 Frequently encountered issues related to reinforced concrete nodes and solutions

Concrete frames are made up of beams, girders and nodes. Steel reinforcing bars are arranged complicatedly and intensively in the beams. Longitudinal reinforced bars are interlocked, overlapping each other in the upper areas of beams. In the conventional structural design, the diameters of steel reinforcing bars are not exhibited. The conventional two-dimensional construction drawings do not display the arrangement of steel reinforcing bars which are visible in a three-dimensional space only. As a result, many issues occur in the construction process, for example steel reinforcing bars are bent arbitrarily and there is assurance for the thickness of protection layers. Up until today, Revit is one of the most popular software employed by the architectural BIM system, and with its powerful parameterization ability, Revit has the largest market share. In this study, the frequently encountered issues related to the complex nodes of reinforced concrete were analyzed in an attempt to find the technical measures to solve the complex node issues, and at the same time Revit was employed to simulate how the nodes were treated effectively.

2.1 Issues related to beam and girders' protection layers

The diameters of steel reinforcing bars are not exhibited in the conventional two-dimensional drawings. In the construction site (same as a 3D drawing), however, a number of reinforced concrete nodes are made up of steel reinforcing bars overlapping each other. Many overlapping nodes are seen in the beam-to-girder nodes, each node comprising three layers – floor reinforced bars in the upper layer, girder's longitudinal bars in the middle layer, and beam's longitudinal bars in the bottom layer – all steel reinforcing bars overlapping with each other. With so many steel reinforcing bars overlapping each other, the thickness of protection layers on the surface steel reinforcing bars has decreased to almost zero and the girder steel reinforcing bars are not adequately thick, either (Figure 1), which is inconformity to the requirements stated in the "Code for design of concrete structure" **[2]** and "Code for acceptance of constructional quality of concrete structure" **[3]** regarding the thickness of protection layers (For beam and its structural components: +10mm, -7mm; for girder and its structural components: +8mm, -5mm).

In reality, it is a common practice to arbitrarily strike, press or bend the steel reinforcing bars of beams and girders after the concealment works of steel reinforcing bars are accepted. This practice does not really serve the purpose and is unable to warrant the thickness of the protection layers and is likely to decrease the load-carrying capacity of the twisted steel reinforcing bars, affecting the structural safety.

Proposed method: Based on the requirements stated in "Drawing Rules and Standard Detailing Drawings of Ichnographic Representing Methods for Construction Drawings of R.C. Structures (Cast-in-situ R.C. Frames, Shear Walls, Beams, Slabs)" **[4]** and in conformity to the common practice of the construction site, a pre-bending smaller than 1:12 ($\geq 170.54^{\circ}$) was proposed for the longitudinal steel reinforcing bars of the beams involved in the node while stirrups were shrunk down (Figure 2) to ensure the stability of reinforced concrete framework. In other words, beam stirrup's height decreased by the length of one steel reinforcing bar's diameter, allowing girder longitudinal steel reinforcing bar with same diameter to fit into the node. It was therefore proposed that BIM technology was employed to solve similar issues at the time that reinforced concrete frameworks were constructed. Lastly, diagrammatic sketch of Revit model was presented as shown by Figure 3.

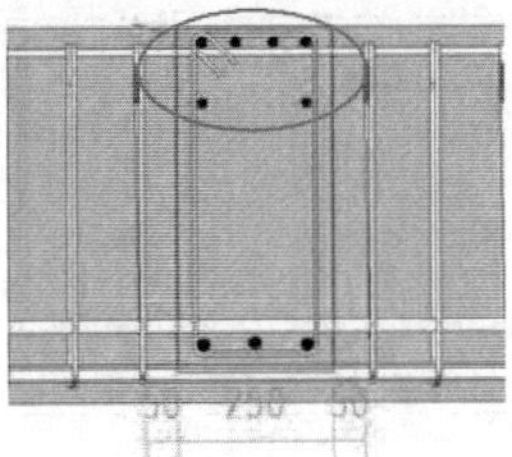

Figure 1. Inadequate thickness of beam and girder protection layer

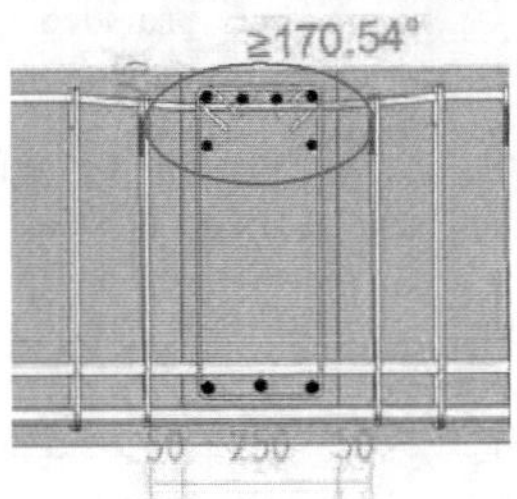

Figure 2. A proposed method for beam and girder construction

Figure 3. A diagrammatic sketch of beam and girder construction by Revit

2.2 Issues related to bam column node's protection layers

Beam column's steel reinforcing bars have same problems as the beam and girder node's steel reinforcing bars. Steel reinforcing bars' collision causes the beam steel reinforcing bar's protection layer to become thinner. When the longitudinal steel reinforcing bars in the upper area of girder are lifted, the beam steel reinforcing bar's protection layer becomes thinner and the structural durability reduces. As shown by Figure 4, with the reduced durability, beam column node is no longer in conformity to the safety requirement. To ensure the steel reinforcing bar's protection layer thickness, the concrete cement in the areas nearby the node has to be lifted. As a result, the building interior installation and repair works becomes extremely difficult. To solve this problem, some layers have to be removed. Apparently, both methods do not serve the purpose rationally and are unable to elevate concrete cement structure as required by the safety specifications.

Proposed method: If protection layer is not thick enough or has zero protection, steel reinforcing bars are exposed to the air and are likely to be rusted, affecting the concrete cement structure's durability. It is proposed that the nodes are treated in the same way as the beam and girder nodes are treated. As shown by Figure 5, a pre-bending smaller than 1:12 was proposed for the longitudinal steel reinforcing bars of the beams involved in the node. When beam's steel reinforcing bars are twisted, the beam's effective height decreases. To maintain the beam's load-carrying capacity unchanged, edge reinforced bars were paved along the beam. The length of edge reinforced bars was equal to 12 times of the diameter of the steel reinforcing bars on both sides of girder plus the anchorage distance of the steel reinforcing bar [5]. The Revit effect for the nodes is shown by Figure 6.

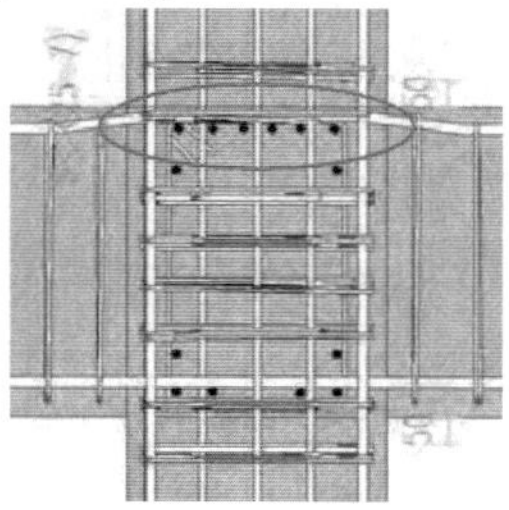

Figure 4. Inadequate thickness of beam column protection layer

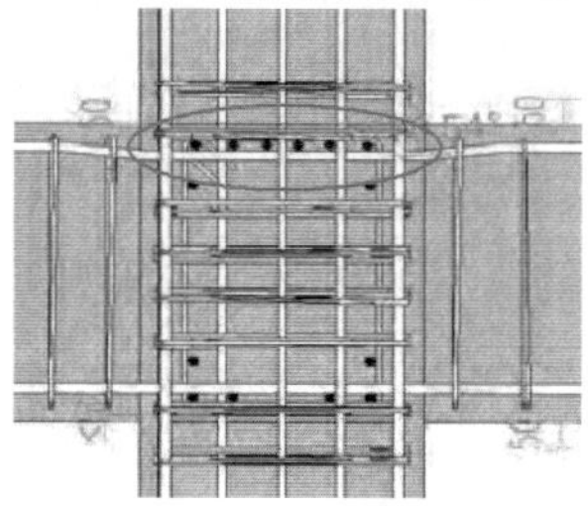

Figure 5. A proposed method for beam column construction

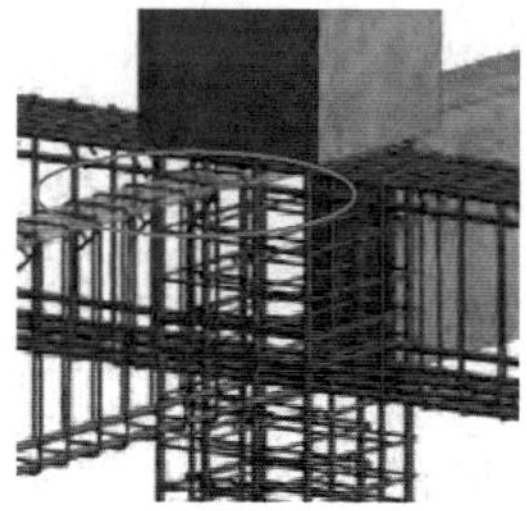

Figure 6. A diagrammatic sketch of beam column construction by Revit

2.3 Issues related to beam column binding wire collision

In many cases, concrete frame's beam and its column have same width, or one side of concrete frame's beam is even with one side of its column. If that's the case, the concrete frame's outermost longitudinal load-carrying steel reinforcing bars collide with the concrete frame column's outer longitudinal steel reinforcing bars. As shown by Figure 7 and Figure 10, side column collides with corner column noticeably. In the steel reinforcing bar construction, it is a common practice that the column's longitudinal steel reinforcing bars are either bent arbitrarily or interspersed without bending. As a result, there are more stirrups and less space in the nodes, and with too many stirrups, node becomes disorganized, affecting the steel reinforcing bar construction efficiency and quality.

Proposed method: in the node, beam's longitudinal steel reinforcing bars collide with each other and the corner column's longitudinal steel reinforcing bars are unable to move. A pre-bending smaller than 1:12 was proposed for the beam's longitudinal steel reinforcing bars, allowing the outermost longitudinal load-carrying steel reinforcing bars to pass through the space inside the longitudinal steel reinforcing bars which are provided outside the concrete frame and at the same time shrink down the stirrups as shown by Figure 8 and Figure 11. By doing so, the protection layer is likely to grow larger than 50mm and the concrete cement's protection layer breaks apart, affecting the longitudinal load-carrying capacity and structural durability. It was therefore proposed to install crack-resistant steel meshes to prevent steel reinforcing bars from breaking apart. The Revit processing chart of side column and corner column is shown by Figure 9 and Figure 12.

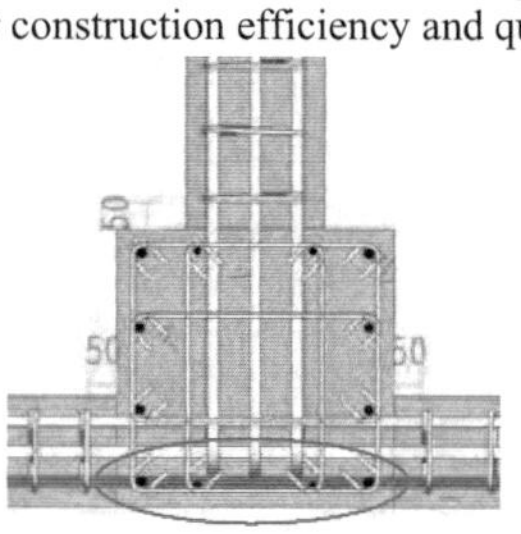

Figure 7. Side column's longitudinal steel reinforcing bar collision

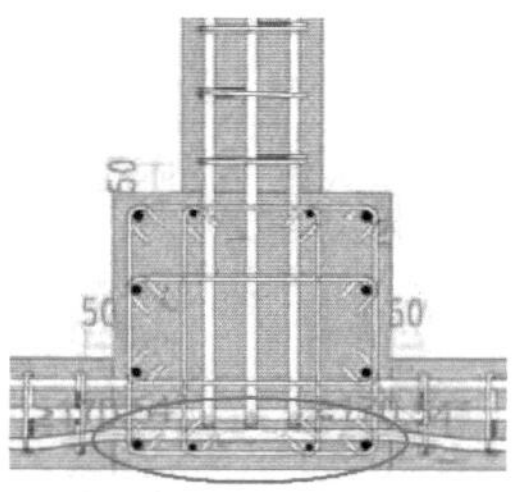

Figure 8. A proposed method for side column's steel reinforcing bar construction

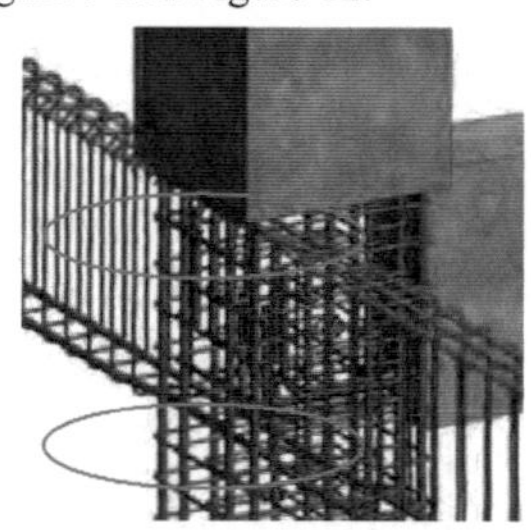

Figure 9. A diagrammatic sketch of side column construction by Revit

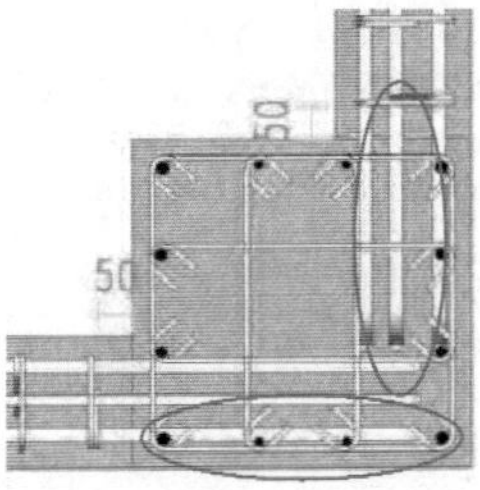

Figure 10. Corner column's longitudinal steel reinforcing bar collision

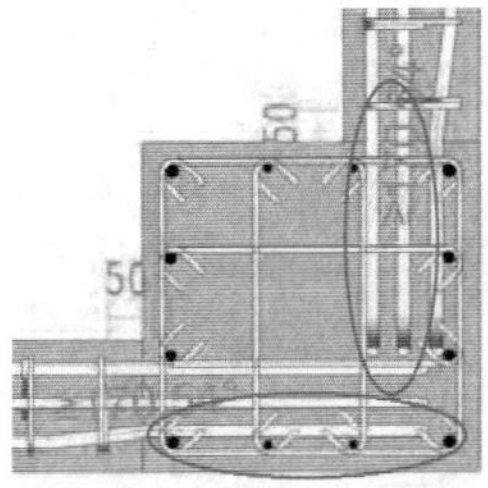

Figure 11. Corner column's longitudinal steel reinforcing bar collision

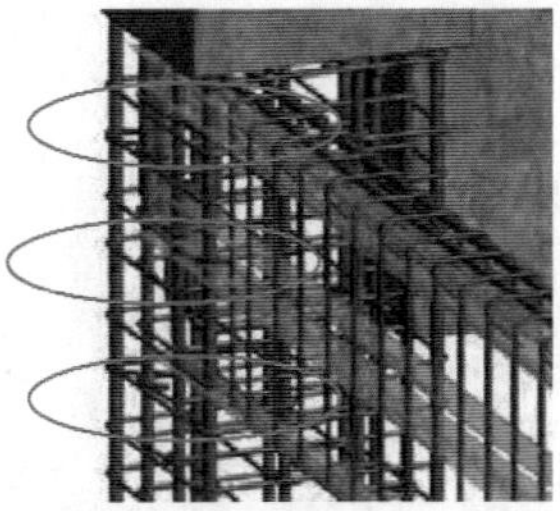

Figure 12. Corner column's longitudinal steel reinforcing bar collision

3 Steel reinforcing bar parameterization

Information is the most critical element required by BIM technology, and with the needed information, BIM is a valuable tool for three-dimensional modeling that allows users to store, transmit and run the data throughout the construction process, and thus make the best use of BIM technology. Revit information model requires all kinds of information and elements related to engineering design. All information and elements are systematically integrated into an interactive information organism. All components are parameterized for easy retrieval and modification. Geometric drawings are dynamically linked with attribute information. In Revit, pixels are grouped, depending on how the parameters (attributes) are shared and used, and are integrated into graphics. Family (*.rfa) parameters are divided into three categories, namely, family parameters, shared parameters, and special parameters. This study focused on how to create steel reinforcing bar families, how to establish settings, and how to store steel reinforcing bar's processing and construction information using Rivet in an attempt to solve the complex nodes stated in Section 2.

3.1 Setting parameters for steel reinforcing bar's processing

Revit's built-in steel reinforcing bar families have several shortcomings: (1) Only one steel reinforcing bar shape is available, unable to handle the deformable steel reinforcing bars, for example the pre-bending smaller than 1:12 stated in Chapter 2. (2) Revit software's built-in information is intended for modeling only, unable to meet the specifications in Taiwan regarding steel reinforcing bar data and processing. (3) Steel reinforcing bar families are intended for users. Yet, parameter setting permission is not open to users. Without permission, users are unable to set parameters as long as they like.

Based on the discussions stated above, shared parameters were chosen to set steel reinforcing bars' parameters. In this study, the corner columns' bending steel reinforcing bars stated in Section 2.3 were discussed. New steel reinforcing bar families were created, and at the same time drive parameters (A-D) were defined (Figure 13). However, the parameters were simply for line generation, without consideration to steel reinforcing bar's diameters and bend diameters. Therefore, the parameters were not ready to be applied to steel reinforcing bar processing.

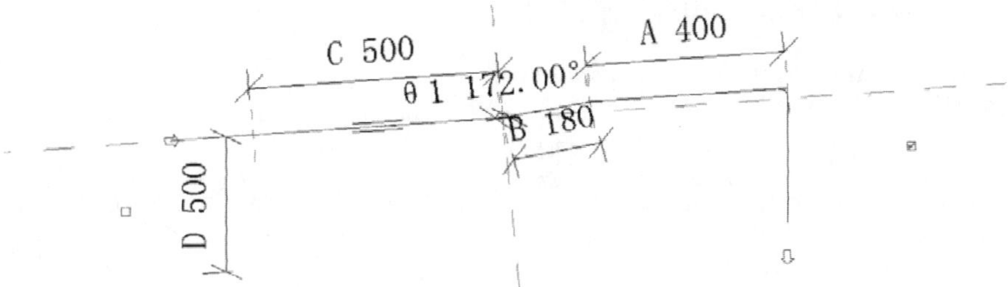

Figure 13. The shape of steel reinforcing bars created by this study

Steel reinforcing bar's processing parameters were established (A+) based on the steel reinforcing bar's diameter and processing requirement. Next, the formula was edited in accordance with the setting mechanism and geometric relationship between both parameters (A+ = A – sin ((180° - 172°) /2) x 3 x steel reinforcing bar's diameter – 2.5 x steel reinforcing bar's diameter). The final processing parameters setting, as shown by Figure 14 and Figure 15, exported a detail list (Figure 16) which included data such as steel reinforcing bar's diameter C+, bend diameter Ö+ and angle ė+. With the detail list, the parameters were ready to be applied to digital control and processing.

θ2+(默认)	90.00°	=90°
θ1+(默认)	172.00°	=θ1
θ1(默认)	172.00°	=
Φ+(默认)	40.0	=if(钢筋直径 > 25 mm, 6 * 钢筋直径, 4 * 钢筋直径)
D+(默认)	120.0	=12 * 钢筋直径
D(默认)	145.0	=
C+(默认)	497.9	=C - sin((180° - 172°) / 2) * 3 * 钢筋直径
C(默认)	500.0	=
B+(默认)	175.8	=B - 2 * sin((180° - 172°) / 2) * 3 * 钢筋直径
B(默认)	180.0	=
A+(默认)	372.9	=A - sin((180° - 172°) / 2) * 3 * 钢筋直径 - 2.5 * 钢筋直径
A(默认)	400.0	=

Figure 14. Steel reinforcing bar's processing parameter setting

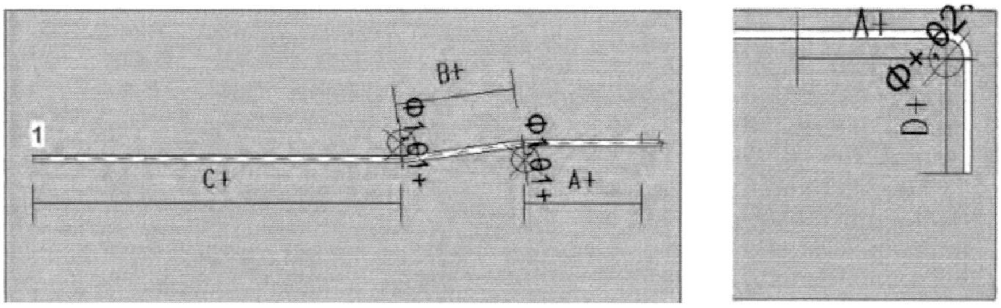

Figure 15. Steel reinforcing bar's processing parameters in motion

钢筋加工明细表

钢筋直径	钢筋长度	合计	形状图像	A/	B/	C/	A+/	B+/	C+/	D/	D+	Φ+	θ1+/	θ2+/	θ+/
6 mm															
6 mm	1620 mm	80		200 mm	500 mm		160 mm	460 mm	80 mm			24	90.00°	135.00°	
8 mm															
8 mm	1320 mm	20		170 mm	450 mm		120 mm	400 mm	80 mm			32	90.00°	135.00°	
8 mm	1380 mm	40		200 mm	450 mm		150 mm	400 mm	80 mm			32	90.00°	135.00°	

Figure 16. Steel reinforcing bar's processing detail list

3.2 Steel reinforcing bar's construction positioning setting

Construction management is very important for steel reinforcing bar works. Steel reinforcing bar system's built-in partitioned attribute columns are helpful tools for project management. In the early stage, modeling staffs assign steel reinforcing bars and accessories to a component. Then, they discuss matters related to construction positioning and progress in the later stage. To cope with most people's habits and the steel reinforcing bar's mutual anchorage, this study chose a building's one facade as front facade (customized by this study) and longitudinal steel reinforcing bars were paved from left → right, from front → rear, from bottom → top. The component first seen in the ending area was treated as the linked component of the longitudinal steel reinforcing bar. Accessories such as stirrups and pull bars were linked with the components in the same areas. As shown by Figure 17, filter programs were created in accordance with the definitions and rules. Black steel reinforcing bars belonged to girder KLx-Aa(4)-3 while red steel reinforcing bars belonged to girder KLx-Aa(4)-4. Thus, construction detail list and Revit three-dimensional model were exported to guide the steel reinforcing bar construction (Figure 18).

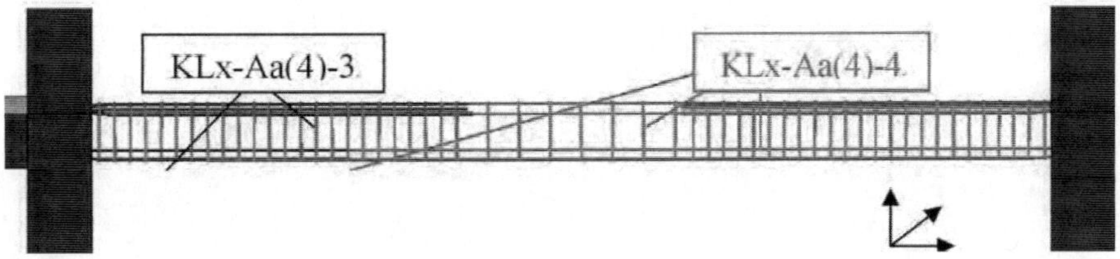

Figure 17. Steel reinforcing bar components' linking rules

钢筋施工明细表

分区	钢筋直径	钢筋长度	形状图像	合计	A/	B/	C/	D/
KLx-Aa(4)-4								
KLx-Aa(4)-4	8 mm	1560 mm		55	150 mm	550 mm		
KLx-Aa(4)-4	12 mm	2050 mm		2	2050 mm			
KLx-Aa(4)-4	16 mm	3530 mm		1	3320 mm	240 mm		
KLx-Aa(4)-4	18 mm	7160 mm		1	6930 mm	270 mm		
KLx-Aa(4)-4	18 mm	7160 mm		1	370 mm	140 mm	6420 mm	260 mm
KLx-Aa(4)-4	20 mm	3580 mm		1	3320 mm	300 mm		
KLx-Aa(4)-4	20 mm	3580 mm		1	410 mm	150 mm	2770 mm	290 mm
总计: 62								

Figure 18. Steel reinforcing bar construction detail list

4 BIM-based discussion of steel reinforcing bar construction process

4.1 BIM steel reinforcing bar detailing model

The issues related to the complex nodes were reviewed and feasibility recommendations were proposed accordingly. Next, conventional two-dimensional drawings were converted into BIM steel reinforcing bar detailed models using BIM software (e.g. Revit). Construction simulation and collision checkup were carried out in the process. The model was so detailed that it could serve the construction needs sufficiently.

4.2 Self-constructed steel reinforcing bar families

Complex nodes were reviewed in compliance with domestic specifications. Steel reinforcing bar families were constructed accordingly. Then, relevant information was obtained and stored, and at the same time formula was employed to estimate the data obtained from steel reinforcing bar families. The estimation results were retrieved and used in the later stages.

4.3 Exporting steel reinforcing bar detail list

Steel reinforcing bar information was retrieved and adjusted using Revit detail list, and at the same time steel reinforcing bar processing and construction detail list was created. The parameters contained in the steel reinforcing bar detail list were sufficient to satisfy the digitalized processing needs.

4.4 Steel reinforcing bar construction and management

Steel reinforcing bar construction detail list and Revit model were employed to administer steel reinforcing bar installation & positioning and progress control.

The overall discussion of steel reinforcing bar construction process as shown by Figure 19

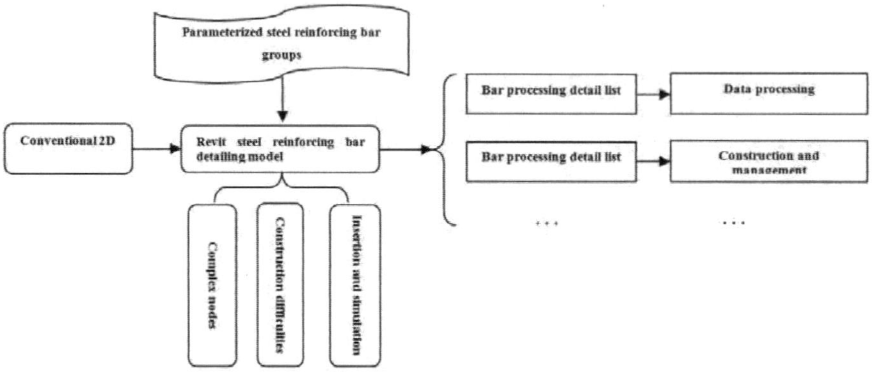

Figure 19. BIM-based discussion of steel reinforcing bar construction process

5 Conclusions

In this study, BIM technology was applied to steel reinforcing bar works. BIM-based steel reinforcing bar construction process was proposed. Moreover, BIM technology was employed to simulate the frequently encountered issues related to the complex nodes, and with the simulation results, parameterized steel reinforcing bar families were constructed for experimental processing and construction. The experimental results were sufficient to solve the real construction issues and to enrich the content of BIM technology.

Acknowledgements

This study was sponsored by the Ministry of Science and Technology of Taiwan under Grant MOST 106-2218-E-033-005-, the Fuzhou CityConstruction Design and Research Institute Co., Ltd. under Grant GY-H-16023, Fujian Provincial Key Laboratory of Advanced Technology and Informatization in Civil Engineering, the Education Department of Fujian Province under Grant GY-Z-15120 and JAT160317. The author also sincerely appreciates the constructive comments provided by the anonymous reviewers.

References

[1] Ten S.L. Research and Application of the Integrated Metering, Use and Acceptance of Concealment Works, Steel Reinforcing Bar, *Master Thesis*, Lanzhou Jiao Tong University, 2014.

[2] Code for Design of Concrete Structures, *Chinese National Standards*, Beijing, China Architecture & Building Press, 2010.

[3] Code for acceptance of constructional quality of concrete structures, *Chinese National Standards*, Beijing, China Architecture & Building Press, 2011.

[4] Drawing Rules and Standard Detailing Drawings of Ichnographic Representing Methods for Construction Drawings of R.C. Structures, *National Building Standards Atlas*, China Architecture & Building Press, 2011.

[5] Wang H. Research on the Optimized Design and Construction of Concrete Structural Beams, Girders and Nodes, , *Master Thesis*, Qingdao University of Technology, 2011.

Automating the Generation of Indoor Space Topology for 3D Route Planning Using BIM and 3D-GIS Techniques

W.Y. Lin[a], P.H. Lin[a] and H.P. Tserng[b]

[a] Department of Civil Engineering, Feng Chia University, Taiwan
[b] Department of Civil Engineering, National Taiwan University, Taiwan
E-mail: weiylin@mail.fcu.edu.tw, paolin@fcu.edu.tw, hptserng@ntu.edu.tw

Abstract

3D Indoor space topology, mainly referred as floor layouts and navigation routes, is the foundation of indoor Location-based Services (LBS), such as navigation in large shopping malls or interchange in large transport stations. Manually generating this topology is highly time-consuming, less cost effective, and prone to errors. This research adopts 3-D GIS technology and proposes a novel workflow to automatically generate the indoor space topology based on 3D building models in a format of Industry Foundation Classes (IFC). Floor layouts and paths including stairs will be identified and generated in the environment of ESRI ArcScene . Several categories of navigation routes are defined and constructed by a collective algorithm called *i-GIT*. This paper demonstrates the entire workflow and concepts to produce floor layouts and navigation routes using a 3-story commercial building model. More details related to *i-GIT* as well as their validation based on real buildings will be revealed in an upcoming article.

Keywords –

3-D GIS, Indoor Space Topology, 3D Navigation graph, Building Information Modeling

1 Introduction

With the popularity of mobile devices and positioning technology, there has been a rise of smart cities around the world in a recent decade, in which Location-Based Services or LBS, such as vehicle navigation and freight tracking, play an important role of fulfilling their goals by providing citizens with more effective and efficient accesses to locations. The Demand of wayfinding has expended its applications from outdoor streets to indoor spaces in recent years. Equipped with indoor positioning technology like i-Beacon, there brings more and more smart services inside the buildings. In order to realize indoor navigation, it also requires the indoor space topology of buildings as the base map for navigation. Indoor space topology mainly consists of floor layouts and accessible paths among indoor spaces. Indoor paths can be horizontally or vertically distributed. Vertical paths connect two floor levels with different elevations, such as stairs, elevators, escalators, or ramps. Manually creating indoor space topology is extremely time-consuming and prone to error [1], no matter what digital source, either 2D CAD drawings or 3D building models, is used. In response of likely massive demands of indoor navigation in near future, automation of generating indoor space topology is necessary [2].

Since indoor space topology contains non-planar paths, 3D building models, such as BIM, with most straight-forward 3D geometry information are the better choice of materials than 2D drawings when constructing the topology. However, 3D building models composed of structural elements, such as walls, columns, slabs, and beams, do not contain the geometric information of navigable paths. Therefore, it remains a challenging task extracting necessary information from 3D building models to construct the topology. Although there existed several studies in this field and algorithms developed to automatically produce indoor navigation paths [1-5], most of them merely focused on horizontal paths [1-3] and were based on an assumption that the spatial semantic information of buildings or floor layouts are already known. For those studies discussing vertically distributed paths, the representation of vertical paths were highly simplified [5].

This study attempts to propose a workflow based on 3D GIS technology for automatically generating indoor space topology. We developed a collective algorithm called Intelligent Generation of Indoor Topology, or *i-GIT*, that reads 3D building models in a format of Industry Foundation Classes (IFC) as the only input, then automatically identifies boundaries of indoor spaces for each floor, and finally generates horizontal and vertical navigation paths for 3D route planning. We implemented and tested *i-GIT* with a demonstrative building model in a single environment of 3D GIS software, ESRI ArcScene. Equipped with the proper indoor positioning technology, the research results can be widely used in indoor navigation of large-scale shopping malls or office complexes, guiding of indoor emergency escape and evacuation, or interchange in large transport stations.

The remainder of this paper is structured as follows: Section 2 addresses some important properties of indoor space topology as well as introduces three major navigation models and algorithms proposed by previous studies for generating centreline-based models. We explains the reason why we adopt Delaunay Triangulation to generate horizontal paths and how we improve its performance by removing redundant nodes it might generate. Section 3 explains the overall concept of how we automate the generation of indoor space topology, including the formation of indoor space from the perspective of wayfinding. Two categories of indoor paths and the workflow based on 3-D GIS technology will be introduced. Section 4 demonstrates the testing results of *i-GIT* using a 3-story building model, followed by a summary and discussion in Section.

2 Related Work

The biggest difference between outdoor vehicle navigation and indoor pedestrian navigation is that paths of vehicle navigation are limited to a fixed road network, while the freedom of accessibility for indoor navigation on the same floor is two-dimensional. Indoor navigation also allows with vertical paths, like stairs and elevators. Although the degree of freedom for indoor navigation can be three-dimensional, its horizontal movement is still subject to walls, columns, furniture and other barriers; its vertical paths are also limited to stairs, elevators or ramps. Taneja [1] summarized the following three different indoor navigation models： centerline-based models, metric-based model, and grid-based models. Among them, grid-based models would best fit the above-mentioned degree of freedom for indoor navigation; thus widely used within the robotics and computational graphics domains. However, for those indoor pedestrians with definite destinations, the main purpose of route planning is just for wayfinding, so centerline-based models can satisfy the goal of routing since it represents the paths in a very similar way that humans recognize the direction in indoor space.

Most existing algorithms to construct centerline-based models are based on three methods [2]: Visibility Graphs, Straight Skeletons, and Delaunay Triangulation.

A Visibility graph is a navigational space composed of a set of nodes and edges. Nodes represent the stop points in indoor space, normally referring to the accesses like doors. Edges are straight lines connecting two inter-visible nodes. If these exists no direct edge between two accesses, some intermediate nodes need to be added in order to form a navigation network. One of the drawbacks of Visibility Graphs is the explosive number of edges in a space with no obstacles because the combination of inter-visible nodes is nearly boundless [4]. Besides, the direction instructions it gives are less easily comprehended by humans.

Straight-Skeletons is a method that transforms polygons into a navigation network based on Medial Axis Transform proposed by Blum in 1967 [1, 2, and 4]. Straight-Skeletons has its name because the indoor paths it generates are approximately along the medial axis of the space. For a rectangle space, Straight-Skeletons first equally divides the four right angles, and connects the intersect points of two shorter angle bisectors to form a navigation network with one longer and 4 shorter paths. For spaces with simple and regular shapes, methods based on Straight-Skeletons, such as S-MAT [3], can obtain fairly good results; however, they may create redundant nodes in irregular shapes [1, 2].

Delaunay Triangulation is the dual of the famous Voronoi Diagram. It divides a plane into triangles ensuring that no triangle vertex lies within other triangles in the network. It also maximizes the minimum angle of triangles so that it can avoid creating long and thin triangles as far as possible. This characteristics makes it suitable for constructing 3D terrain model. Triangular Irregular Networks (TIN), a famous digital format to present surface morphology in the GIS community, is also based on this algorithm. By dividing a single indoor space using Delaunay Triangulation and connecting the midpoints of inner edge of each triangle, a navigation network can thus constructed [4, 5]. The advantages of this method include: 1) it can apply to almost any shape of indoor space; 2) the distribution of paths is roughly the same as the contour of the indoor space; and 3) the paths can also be roughly spread throughout the indoor spaces. It might produce more nodes and paths near those relatively irregular or broken regions. Some researchers [2] regarded this as a drawback of this method while others [4] have a different perspective because more nodes indicate a fine-grained navigation model with a higher resolution. We hold the same viewpoint as the latter since the navigation model with an appropriate number of paths can avoid unwanted detours as S-MAT often results in, and thus can produce better outcome of route planning.

In summary, adoption of Delaunay Triangulation to construct centerline-based models for indoor navigation would be the best choice for our research. However, how to maintain the number of nodes produced by Delaunay Triangulation in an 'appropriate' range will be one of important concerns when directly taking 3D building models in the real world as the input since they may contain many indoor spaces with irregular or broken areas, especially in the corners. To resolve this issue, we adopt an algorithm of regularization [6] upon the original indoor spaces derived from 3D building models before applying Delaunay Triangulation to remove those unnecessary nodes as far as possible. This measurement will be introduced in Section 4.1.

3 Concept and Workflow

3.1 Formation of Indoor Space

For indoor movement by walking, every accessible corner in indoor space could be the destination of a trip while for indoor wayfinding there usually comes with a clear location reference of the building for trip destinations, such as Room 101, Office 605, or the 3rd Meeting Hall, etc. Therefore, the navigation graph must go through all these named spaces. In other words, these named spaces are the most important destinations of indoor trips for wayfinding. Thus, the first task of generating the navigation graph from 3D building models is to identify these named spaces, and select the adequate points, such as the centroid or doors, to represent them. Most 3D building models do provide floor plans, but require modellers manually to specify this information. In cases of no floor plan being available in the models, the algorithm still needs to identify indoor spaces from those structural components by itself.

i-GIT is a set of algorithms we propose to perform all related tasks of generating the indoor space topology. To identify floor layouts, *i-GIT* first conducts the "3D intersection operation" upon the Slab with Columns, Walls, Doors, and Windows of 3D building models, thus generating the bounding lines of "partition separators" as shown in Figure 1. All the structural components come in 3D surface format in a 3-D GIS environment. Those boundaries derived by Walls and Columns are unnavigable obstacles while those derived by Doors and French Windows are navigable. Next, *i-GIT* conducts the "2D difference operation" upon the Slab footprints with partition separator areas, and generates the indoor spaces as shown in Figure 2.

In Figure 2, black ribbons are the boundary of accesses with physical doors or French windows. Green pushpins are centroids of accesses while red and blue ones are centroids of identified indoor spaces. As shown in Figure 2, some of the identified indoor spaces do not have any access, such as balconies and pipeline zones. Since these spaces are not accessible, they will not become the nodes of the indoor paths. In this research, the indoor spaces are divided into two categories according to their shapes: "*Cabins*" for rectangle spaces (as red pushpins in Figure 2) and "*Corridors*" for non-rectangle (as blue pushpins in Figure 2) due to their different characteristics when generating Floor-Level Paths. Paths within *Cabins* and *Corridors* will be generated in different methods, which will be introduced in Section 3.2 and 4.2.

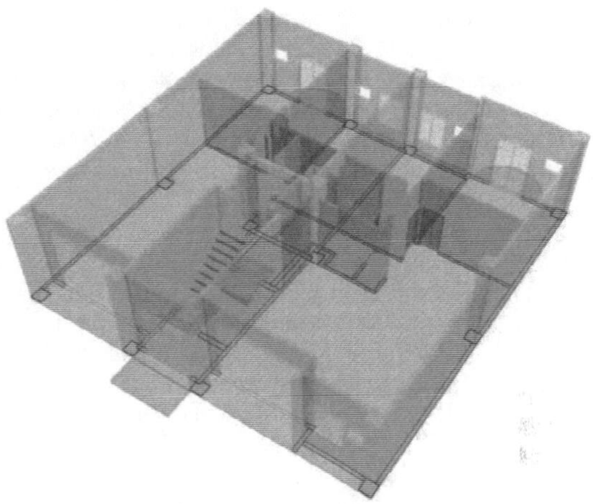

Figure 1. Bounding lines of partition separators derived by conducting 3D intersection operation upon the Slab with Columns, Walls, Doors, and Windows of a 3D building model

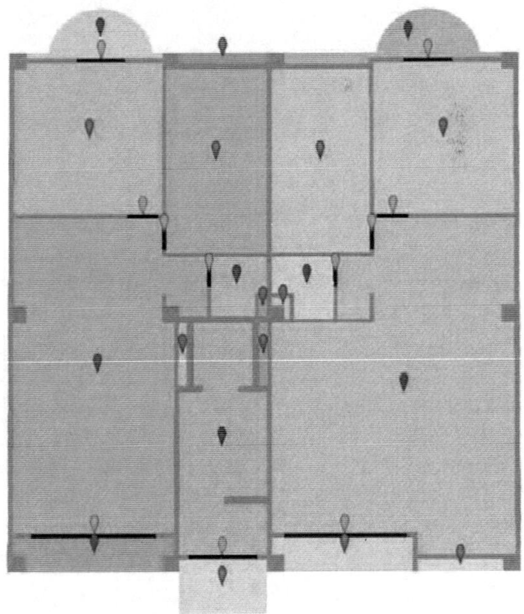

Figure 2. Automatic formation of indoor spaces along with their centroids, derived by conducting 2D difference operation upon the Slab footprint with partition separator areas

3.2 Categories of Paths

The indoor paths of this study are divided into two categories: "Floor-Level Paths" and "Non-Planar Paths". The former falls on the same floor with the same elevation, and is subdivided into "Inside-Corridor Paths", "Access-To-Corridor Paths", and "Access-To-Cabin Paths" according to their locations. The latter is the paths connecting two floors or two planes with different elevations on the same floor, such as stairs, elevators, escalators, and ramps. The following sub-sections define each category of paths mentioned above.

3.2.1 Floor-Level Paths

Figure 3 shows an example of the "Floor-Level Paths". "Inside-Corridor Paths" (shown as red lines) are only located within "*Corridors*", and the distribution is roughly consistent with the contour of the corridor in which it is located. They are spread through the corridor but do not touch the accesses. "Access-To-Corridor Paths" (shown as purple lines) are the linking paths connecting accesses to "Inside-Corridor Paths". "Access-To-Cabin Paths" are straight lines located within "*Cabins*" connecting the centroid of a cabin and all its accesses, as shown in the green lines of Figure 3. They are simplified as straight lines because there is no obstruction on the line of sight in "*Cabins*". The three above-mentioned paths constitute "Floor-Level Paths".

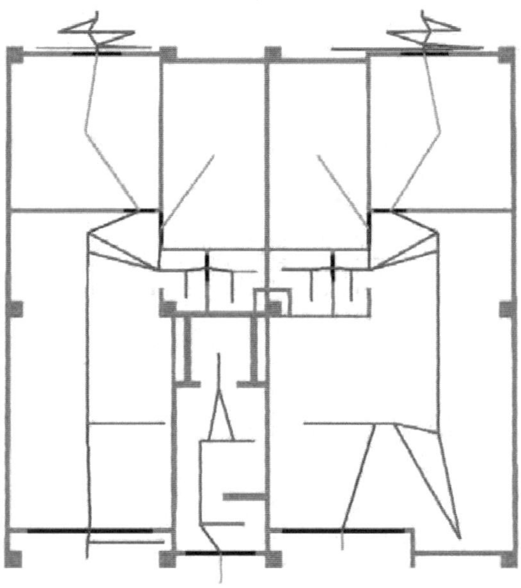

Figure 3. Three categories of Floor-Level Paths: Inside-Corridor Paths in red, Access-To-Corridor Paths in purple, and Access-To-Cabin Paths in green

3.2.2 Non-Planar Paths

As stated earlier, the indoor paths also contain paths connecting planes with different elevations, including of stairs, elevators, escalators, and ramps. This research only focuses on stairs with multiple straight flights and platforms because they are the most common in all types of buildings; meanwhile, they are more difficult to generate than others. We will expend *i-GIT* to other types of "Non-Planar Paths" in the future. Figure 4 illustrates the "Stair-Associated Paths" for a typical stair, including "Stair-Flight Paths" (yellow lines), "Stair-Platform Paths" (pink lines) and "Stair-To-Floor Paths" (blue lines). The first two types are just as their names imply while the third one is the linking path connecting "Stair-Flight Paths" to "Floor-Level Paths", either "Inside-Corridor Paths" or just the centroid of the "*Cabins*".

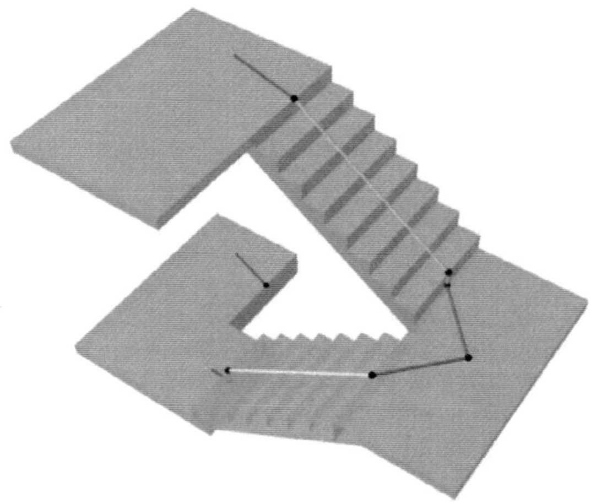

Figure 4. One category of Non-Planar Paths: Stair-Associated Paths with sub-categories of "Stair-Flight Paths" (yellow lines), "Stair-Platform Paths" (pink lines) and "Stair-To-Floor Paths" (blue lines)

3.3 Geoprocessing Workflow

Figure 5 illustrates the entire workflow of generating indoor space topology based on 3-D GIS technologies. The development, testing and validation of *i-GIT* are all proceeded in a single environment of 3-D GIS software: ESRI ArcScene. The workflow includes three phases: (1) generating floor layouts, (2) generating "Floor-Level Paths", and (3) generating "Non-Planar Paths". We use *Model Builde*", an IDE tool embedded in ArcScene for designing geoprocessing procedures, to develop *i-GIT*. *Model Builder* provides a graphical interface for designers to drag and drop those geoprocessing and spatial analysis tools offered by ArcScene. Besides, it also provides the syntax of iterators and conditions for processing batch files. Figure 6 is an example of geoprocessing model that generates partition separators using *Model Builder*.

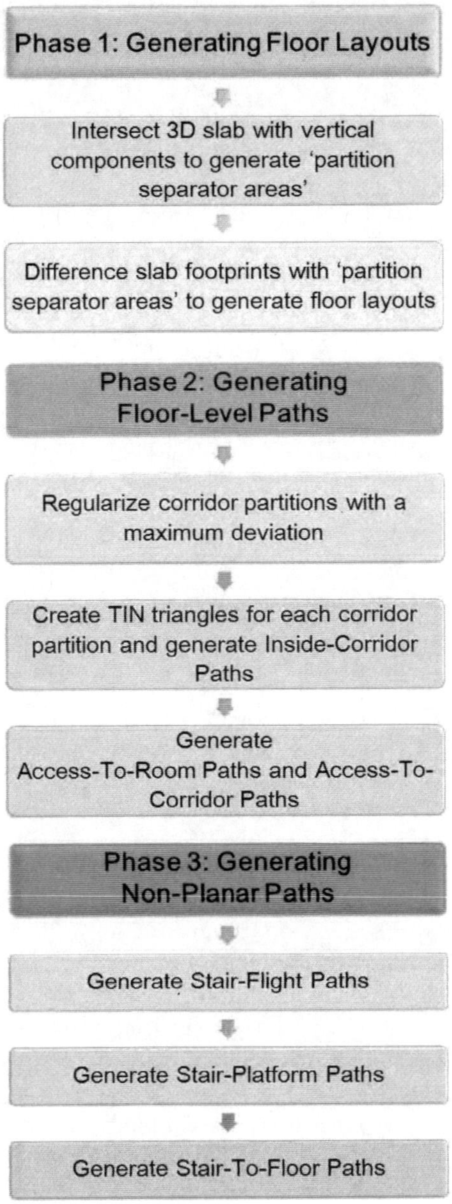

Figure 5. The workflow of generating indoor space topology based on 3D GIS technologies

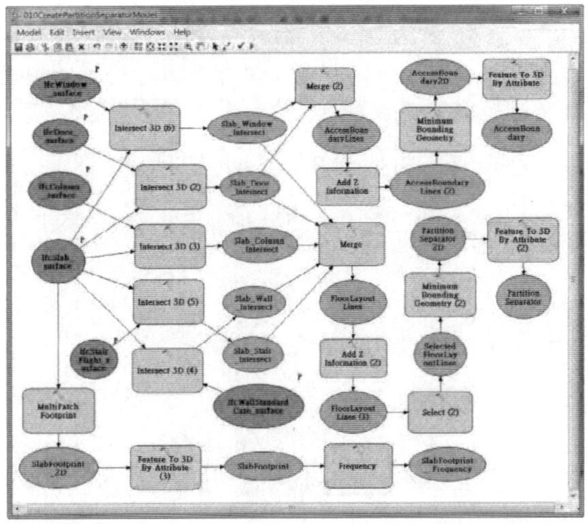

Figure 6. Geoprocessing model that generates partition separators using Model Builder of ESRI ArcScene 10.4.1

The method of producing floor layouts in Phase 1 is described in Section 3.1, *i-GIT* first identifies partition separators and then generates floor layouts. The detailed process of generating partition separators is shown in Figure 6.

In Phase 2, *i-GIT* applies the "Polyline Compression Algorithm" [6] with a maximum allowable deviation upon those original floor layouts derived from Phase 1 to remove undesirable artifacts in their geometry, which will ultimately regularize the footprint of original indoor space to become smooth polygons. Next, *i-GIT* applies Delaunay Triangulation upon those regularized floor layouts constructing TIN triangles. Connecting the midpoints of inner edges of each triangle will produce "Inside-Corridor Paths". Furthermore, "Access-To-Corridor Paths" are created by connecting the centroid of access areas with the nearest node of "Inside-Corridor Paths". At last, connecting the centroid of "*Cabins*" to accesses will generate "Access-To-Cabin Paths".

In Phase 3, the process of producing "Stair-Associated Paths" starts with identifying the boundaries of the stair flights, and then generates "Stair-Flight Paths" using the relationship of those bounding lines. Next, connecting the centroid of the stair platform and the end node of "Stair-Flight Paths" touching the platform will produce "Stair-Platform Paths". Finally, "Stair-To-Floor Paths" are generated by connecting the end node of "Stair-Flight Paths" touching the floor and the nearest node of "Inside-Corridor Paths" or the centroid of the "*Cabin*" depending on where the end node locates.

4 Demonstration

We use a three-story building as shown in Figure 7 as an example to demonstrate the results of our workflow and test the correctness of *i-GIT*. The building is part of the six-story commercial office building, which is roughly divided into two households by the staircase in the middle. The second floor to the top floor have the same floor layout and floor height. This section only presents the results of "Floor-Level Paths" on the second floor as well as "Stair-Associated Paths" derived from the stair with 3 flights and 2 platforms.

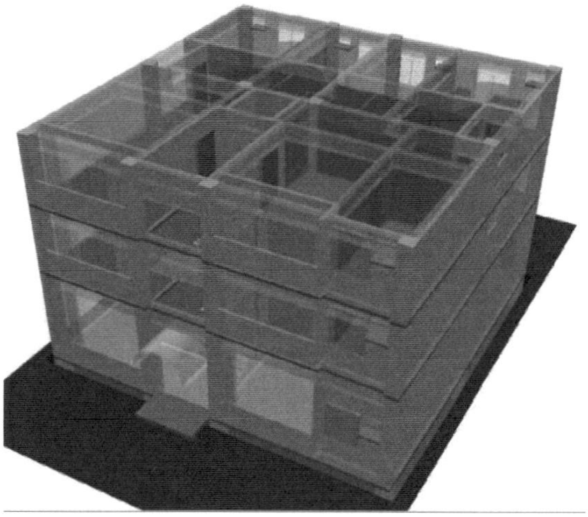

Figure 7. Three-Story building model for demonstration

4.1 Phase 1: Generating Floor Layout

Figure 8 shows the results of floor layouts identified by *i-GIT* after deducting the slab footprint of the 2nd floor by its "partition separator areas". Each indoor space and its accesses are correctly identified. Two or more navigable indoor spaces that do not have physical doors or windows in between are regarded as the same space (which is also a "*Corridor*")

4.2 Phase 2: Generating Floor-Level Paths

Before applying the Delaunay triangulation to generate "Inside-Corridor Paths", the original floor layouts obtained in Phase 1 should be regularized in order to reduce the unnecessary nodes. The patch in Figure 9 is the result of regularizing the floor layouts in Figure 8 with a maximum allowable deviation of 0.2 m. Compared with the partition separator areas shown as black lines in Figure 9, we can see that the original jagged corners because of columns and walls have disappeared and the shape becomes a smooth polygon.

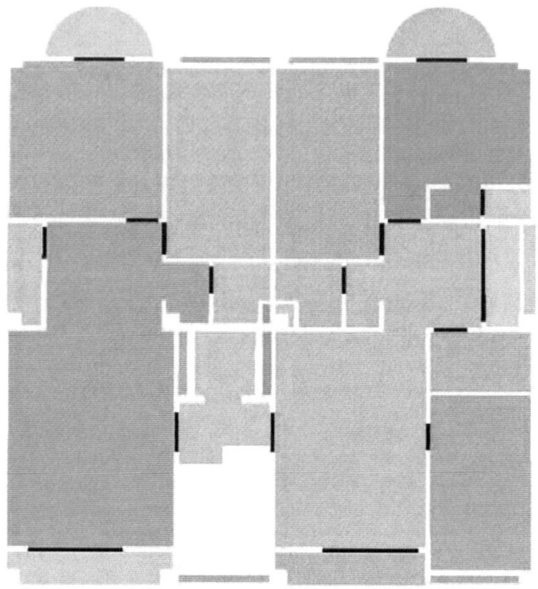

Figure 8. 2nd floor layout (painted areas) of the demonstrative building along with access boundaries (black areas)

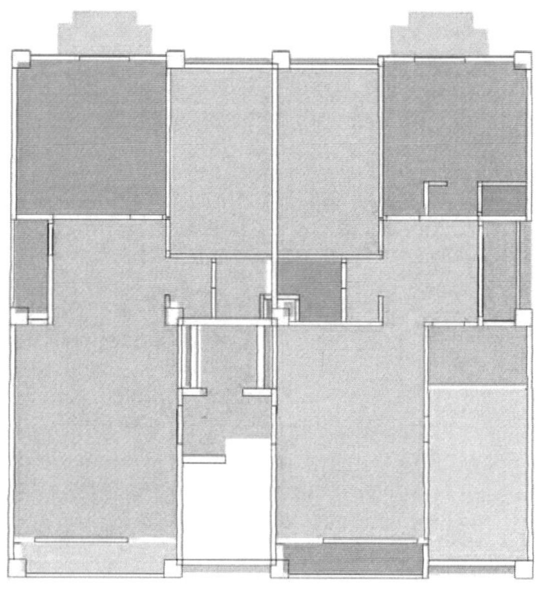

Figure 9. Regularized 2nd floor layout (painted areas) of the demonstrative building with a maximum deviation of 0.2 meter (black lines are original boundary of partition separator areas)

The black lines in Figure 10 are boundaries of triangles derived by applying Delaunay Triangulation on regularized floor layouts shown in Figure 9. Black dots in Figure 10 are the midpoints of inner boundaries of these triangles. *i-GIT* then connects those midpoints within each triangle, and produces "Inside-Corridor Paths", as shown in the red lines of Figure 10. As seen from the figure, these paths are substantially spread throughout the "*Corridor*", and their distribution is consistent with the contour of the "*Corridor*" where they are located. The principle paths are almost straight, too. These features attribute to the performance of Delaunay Triangulation.

Figure 10 also shows that some rectangular spaces, or "*Cabins*", only contain one midpoint of the inner boundary, which happens to be the centroid of the cabin, and thus cannot form any path using the same process. Instead, *i-GIT* connects the cabin's centroid with all its adjacent accesses, and obtains Access-To-Cabin Paths, as shown in the yellow lines of Figure 11. At last, *i-GIT* connects all adjacent accesses of each "*Corridor*" to the nearest node of the "Inside-Corridor Paths", and obtains "Access-To-Corridor Paths" as shown in the purple lines of Figure 11.

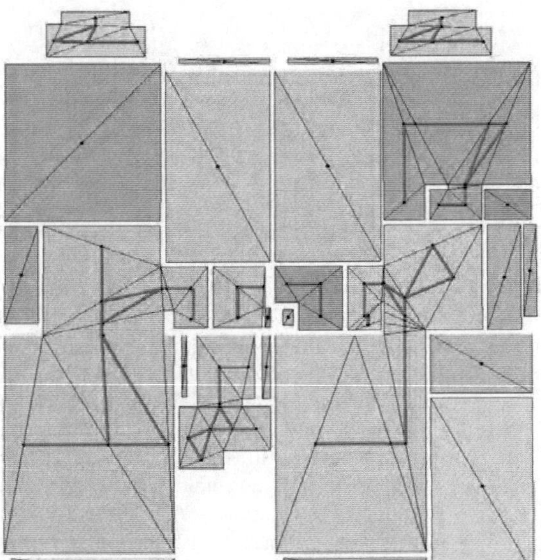

Figure 10. Inside-Corridor Paths (bold red lines) on 2nd floor of the demonstrative building along with TIN triangles (black lines)

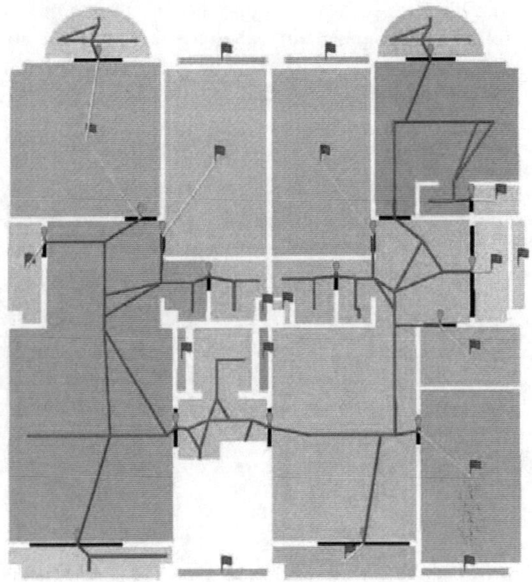

Figure 11. Access-To-Corridor Paths (purple lines) and Access-To-Cabin Paths (yellow lines) on 2nd floor of the demonstrative building along with access centroids (markers) and cabin centroids (red flags)

4.3 Phase 3: Generating Non-Planar Paths

Figure 12 shows the "Non-Planar Path" of the stairs from Floor 1 to Floor 3, including three flights and two platforms. The "Stair-Flight Path" starts vertically upward from the midpoint of the first step's bottom edge to its upper edge, and then extends along the midpoints of each step's upper edge to the intersection point touching the upper platform or slab, as shown in the yellow lines of Figure 12. The pink lines in Figure 12 are "Stair-Platform Paths" connecting the platform's centroid and two "Stair-Flight Paths" while the blue lines in Figure 12 are "Stair-To-Floor Paths", the linking paths that connect "Stair-Flight Paths" to "Inside-Corridor Paths".

4.4 Testing Results

Figure 13 shows the entire result of all indoor paths that *i-GIT* generates for the demonstrative building model, including 180 "Inside-Corridor Paths" (193.9m), 62 "Door-To-Corridor Paths" (67.6m), 28 "Door-To-Cabin Paths" (48.9m), 3 "Stair-To-Floor Paths" (1.6m), 6 "Stair-Flight Paths" (5.5m), and 4 "Stair-Platform Paths" (4.1m). The total length of the paths is 327.7 meter segmented by 290 links. The generated paths can be directly used for route planning in 3-D GIS software. Figure 14 is an example of a routing from the main gate on the first floor to the offices on the second and the third floor, with a total route length of 62.2 meters.

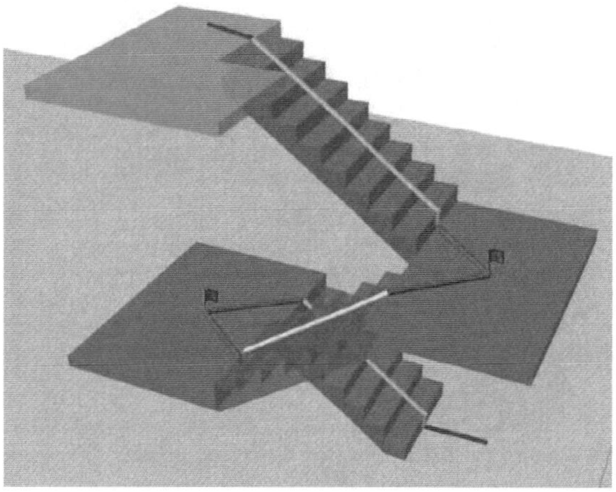

Figure 12. Stair-Associated Paths of the demonstrative building: Stair-Flight Paths (yellow lines), Stair-Platform Path (pink lines), and Stair-To-Floor Paths (blue lines)

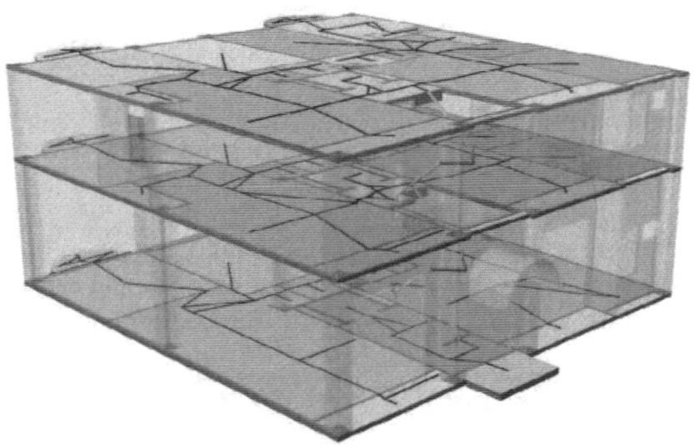

Figure 13. Indoor navigation paths of the demonstrative building

5 Conclusion

Manually generating indoor space topology using 3D building models is expensive and easily incurs errors, so it is necessary to develop an algorithm to automate the process in order to meet the massive demand of indoor navigation in near future. This research develops a collective algorithm called i-GIT along with a geoprocessing workflow using *Model Builder* embedded in ESRI ArcScene, which can accurately identify the floor layouts and generate indoor paths. For Floor-Level Paths, we regularize the original floor layouts before applying Delaunay Triangulation, which can effectively remove unnecessary nodes. In the future, we will extend the functionality of i-GIT to parse and generate other types of "Non-Planar Paths", such as elevators and ramps, making it even more applicable.

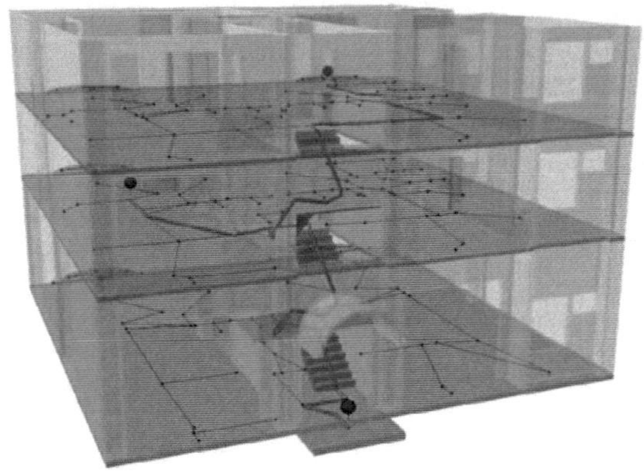

Figure 14. An example of indoor route planning in the demonstrative building

References

[1] Taneja, S., Akinci, B., Garrett, J. H., & Soibelman, L. Algorithms for automated generation of navigation models from building information models to support indoor map-matching. *Automation in Construction*, *61*, 24-41, 2016.

[2] Yang, L., & Worboys, M. Generation of navigation graphs for indoor space. *International Journal of Geographical Information Science*, *29*(10), 1737-1756, 2015.

[3] Lee, J. A spatial access-oriented implementation of a 3-D GIS topological data model for urban entities. *GeoInformatica*, *8*(3), 237-264, 2004.

[4] Mortari, F., Zlatanova, S., Liu, L., & Clementini, E. "Improved Geometric Network Model"(IGNM): a novel approach for deriving Connectivity Graphs for Indoor Navigation. *ISPRS Annals of the Photogrammetry, Remote Sensing and Spatial Information Sciences*, *2*(4), 45, 2014.

[5] Teo, T. A., & Cho, K. H. BIM-oriented indoor network model for indoor and outdoor combined route planning. *Advanced Engineering Informatics*, *30*(3), 268-282, 2016.

[6] Gribov, A. Searching for a compressed polyline with a minimum number of vertices. *arXiv preprint arXiv:1504.06584,* 2015.

34th International Symposium on Automation and Robotics in Construction

ISARC 2017

June 28 - July 1 TAIPEI, TAIWAN

CONSTRUCTION ROBOTICS (III) / CONTROL & MECHANISMS

Velocity Control Mechanism of the Under-actuated Hammering Robot for Gravity Compensation

Yusuke Takahashi[a] , Satoru Nakamura[a] , Yasuichi Ogawa[b] and Tomoya Satoh[b]

[a]Tokyu Construction Co.,Ltd.
[b]Ogawayuki Co.,Ltd.
E-mail: takahashi.yuusuke@tokyu-cnst.co.jp

Abstract -
According to rapid deterioration of many domestic tunnels and bridges, engineers that manage them are insufficient. For this reason, it is desired to develop an automatic hammering robot that can inspect them quickly and accurately. If hammering sounds of this robot are similar to those of engineers, we can do it by using experience for a long period. Therefore, we developed an under-actuated hammering robot that can simulate hammering sounds by engineers. However this robot have a problem that it cannot generate same sounds because of the effect of gravity. To this problem, we combined a swing slider crank mechanism, and controlled velocity of the hammer according to gravity. In this paper, we described the under-actuated hammering robot that combined a swing slider crank mechanism and its model. After that, we clarified a velocity control mechanism for gravity compensation. Finally, we showed the result of hammering tests using this robot and verified usefulness of this mechanism.

Keywords -
Concrete inspection; Hammering robot; Under-actuated; Gravity compensation

1 Introduction

Many domestic tunnels and bridges that was built at a period of rapid economic growth in Japan have deteriorated over time. In Japan, there were 100 thousand road tunnels at the time of March 2013. In particular, 20% of these tunnels have passed over 50 years and it is estimated that it will be 50% in 20 years[1]. Moreover, it was enforced new road raws in July 1, 2014 that obliged to do a close up visual inspection and a hammering to the concrete delamination every 5 years, and we are demanding these inspection more accurately. On the other hand, engineers that manage them are insufficient. For this reason, it is desired to develop innovative inspection techniques such as robots and to implement them on site.

Systems that find out concrete cracks using a digital camera or others are actually implemented on site[2]. Furthermore, there are delamination detection systems for concrete, for example, a system using a infrared camera is considered. However, this system has a problem that accuracy of detection varies as the environmental condition. As the delamination detection technique, hammering test method is widely been accepted. We can find out a delamination inside the concrete by hammering the concrete surface[3]. For example, referring to "douro tonnel teiki tenken youryou" (that is a Japanese road tunnel inspection guideline) which was enforced by a ministry of Japan, a sound concrete sounds like a high note and a delaminated concrete sounds like a low note. If the hammering robot sounds like human, engineers can inspect those in accordance with the guideline.

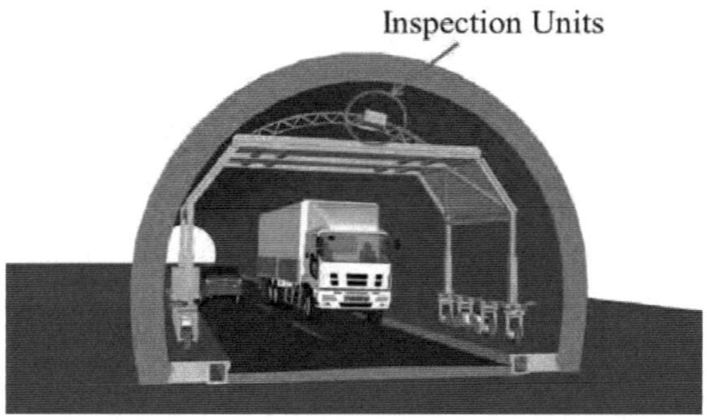

Figure 1. Image of the Tunnel Inspection System

In this research, we aim to develop robots which can sound like human, and we developed an under-actuated hammering robot with a hammer swing mechanism (hereinafter, this is called "Under-actuated system") that behaves similar to that of a engineers arm[4]. We aim to inspect tunnel automatically by using vehicles such as Figure 1 equipped this robot. However, this system allows the hammer to rotate freely, so there is a possibility that the robot cannot sound as the same sound. Therefore, we controlled velocity of the hammer by a swing slider-crank mechanism ("Slider-crank system").

In this paper, we first described the under-actuated hammering robot that combined a slider-crank system and it's model. After that, we clarified a velocity control mechanism for gravity compensation. Finally, we showed the result of hammering tests using this robot and verified usefulness of this mechanism.

2 Underactuated Hammering Robot

An important point to imitate the robot's hammering sound to engineer's one is not to press the hammer against

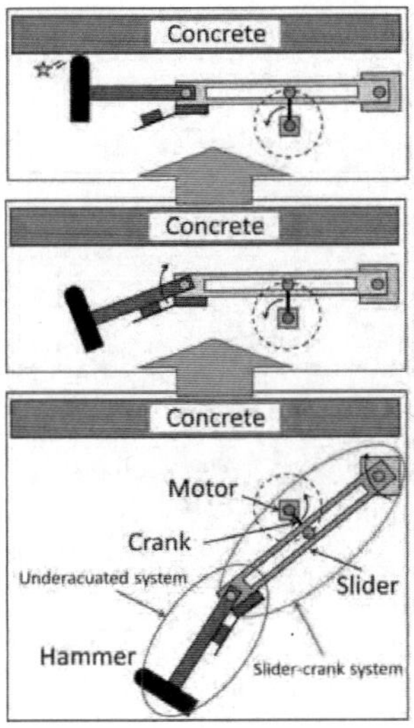

Figure 2. Behavior of the underactuated hammering robot

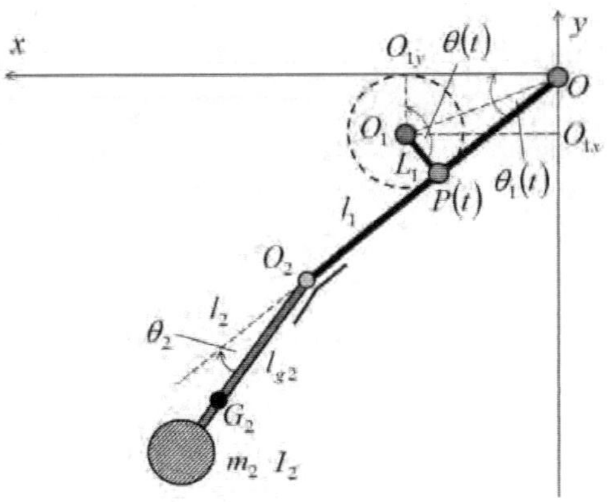

Figure 3. Model of the underactuated hammering robot

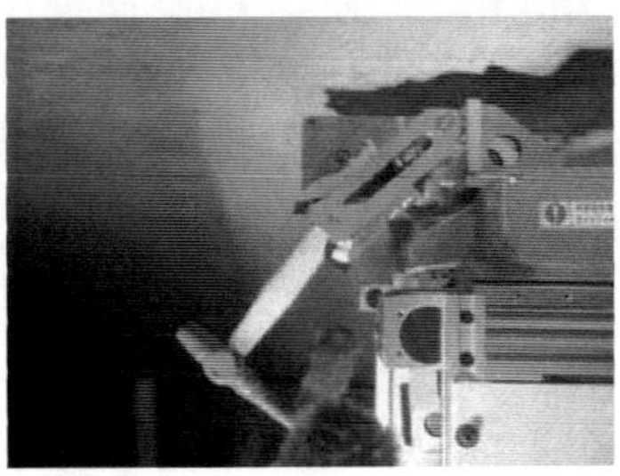

Figure 4. The underactuated hammering robot

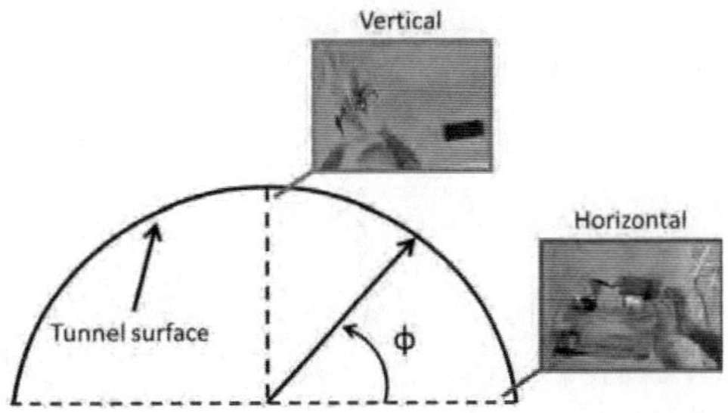

Figure 5. Definition of ϕ

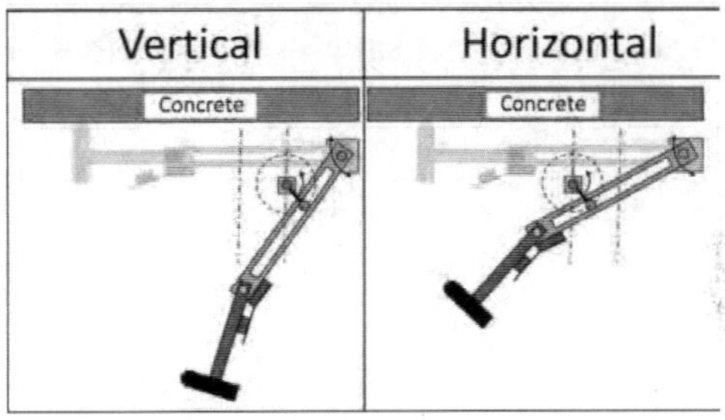

Figure 6. Method of gravity compensation

the concrete surface[5]. The reason is that it does not suppress the vibration of concrete and not occur a double hammering. Furthermore, in this research, we hypothesize that it is also important point to reproduce the behavior of a engineers arm, especially snap wrist. As a hammering robot of satisfying these points, we invented a mechanism that combined a under-actuated system with a slider-crank system. Image of this robot's behavior is shown in Figure 2, and it's model of the mechanism is shown in Figure 3. If the motor at O_1 rotates at N [rpm], a crank(O_1P, L_1 [m]) rotates at O_1, and a slider (OO_2, l_1 [m]) swing around O. This showed A and B in Figure 2. There is a hammer(L_2 [m], m_2 [kg], I_2[g·mm^2]) at O_2, and it limits counterclockwise rotation by the stopper. After that, if a slider decelerates as showed C, this hammer accelerates and hammering at x-axis. This behavior is similar to snap wrist when engineers are hammering, and we think that it is effective for detecting a delamination inside the concrete. Because this hammer leaves concrete by elasticity, it does not suppress the vibration of concrete and not occur a double hammering. We developed a robot based on these method and showed it in Figure 4, it's specification showed Table 1.

3 Velocity Control Mechanism

As this under-actuated system has a characteristic that hammer rotates freely, kinetic energy becomes decrease by the effect of gravity. It means that hammering sounds become small. To be similar to engineers sounds, it should maintain kinetic energy a constant and gets same sounds. Therefore, we invented a velocity control mechanism that controls position O_1 according to gravity.

In this section, we described advantages and behavior of this mechanisms. After that, we showed a dynamical model of this robot and verified numerically.

3.1 Axis Position Control Method

As a method to maintain kinetic energy of the hammer a constant, it is easiest to control velocity of the hammer according to gravity. For example, if defining the hammering direction ϕ[deg] (as showed in Figure 5), N[rpm] increases according to ϕ[deg] and kinetic energy increases. However, this method has disadvantage because it means that inspection speed slows down in the horizontal direction. Moreover, there is a possibility that engineers cannot detect correctly because if N[rpm] changes and hammering pitch changes. From these problem, we invented a velocity control mechanism that controls motor position O_{1y}[mm] according to ϕ[deg]. This mechanism showed in Figure 6. For example, if O_{1y}[mm] decreases (or O_1 approaches O), swing angle of slider increases and velocity of a hammer also increases. Therefore, it can be made constant of kinetic energy of the hammer by controls velocity of a hammer as described. This mechanism has another advantage that hammering pitch is constant and inspection speed does not slow down because it is not necessary to change N[rpm].

3.2 Control Theory

To control velocity of a hammer according to ϕ[deg], we aimed to get relation between ϕ[deg] and O_{1y}[mm] from a dynamical model of the robot. In Figure 3, $\theta(t)$[deg] is the right-handed system, and $\theta_1(t)$[deg], $\theta_2(t)$[deg] is a left-handed system. Equation of motion of hammer can be written as

$$A(t)\ddot{\theta}_1(t) + B\ddot{\theta}_2(t) + H(t) + G(t) = 0 \quad (1)$$

where,

Table 1. Specification of the hammering system

L_1, O_{1x}	10 [mm]	l_1	90 [mm]
l_2	120 [mm]	m_2	133.8 [g]
I_2	131.6×10^3 [g·mm^2]	θ_2	20 [deg]
N	180[rpm]		

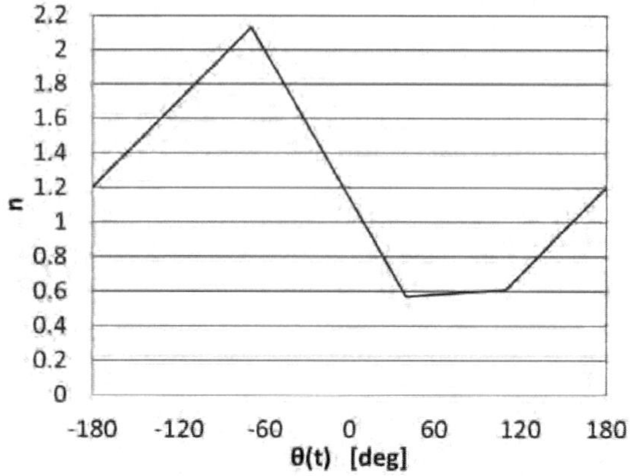

Figure 7. Gear ratio of the motor

$$\begin{aligned} A(t) &= m_2 l_{g2}^2 + I_2 + m_2 l_1 l_{g2} \cos\theta_2(t) \\ B &= m_2 l_{g2}^2 + I_2 \\ H(t) &= m_2 l_1 l_{g2} \dot{\theta}_2(t)^2 \sin\theta_2(t) \\ G(t) &= m_2 (g\cos\phi) l_{g2} \cos(\theta_1(t) + \theta_2(t)) \end{aligned} \quad (2)$$

To get sufficient velocity for hammer when hammering, this robot has special gears which reduction ratio is n as shown in Figure 7. When $\theta_1(t)$[deg] is written by

$$\theta(t) = \int \frac{2\pi n}{60} N dt \quad (3)$$

The relationship between $\theta_1(t)$[deg] and $\theta(t)$[deg] is obtained geometrically

$$\begin{aligned} \theta_1(t) &= -\angle O_{1y} O O_1 + \angle O_1 O P \\ &= -\tan^{-1}\left(\frac{L_1}{O_{1y}}\right) + \sin^{-1}\left(\frac{L_1}{P(t)} \sin\theta(t)\right) \end{aligned} \quad (4)$$

where

$$P(t) = \sqrt{L_1{}^2 + L_2{}^2 - 2L_1 L_2 \cos\theta(t)} \quad (5)$$

$$L_2 = \overline{OO_1} = \sqrt{O_{1x}{}^2 + O_{1y}{}^2} \quad (6)$$

From Equation (3) and (5), we let $\theta_1(t)$[deg] be an input. Therefore, we transform $\theta_1(t)$[deg] as an input and $\theta_2(t)$[deg] as an output,

$$\ddot{\theta}_2(t) = -\frac{1}{B}(A(t)\ddot{\theta}_1(t) + H(t) + G(t)) \quad (7)$$

To analyze the motion of hammer, we should solve Equation (7) for $\theta_2(t)$[deg]. However, it is difficult to

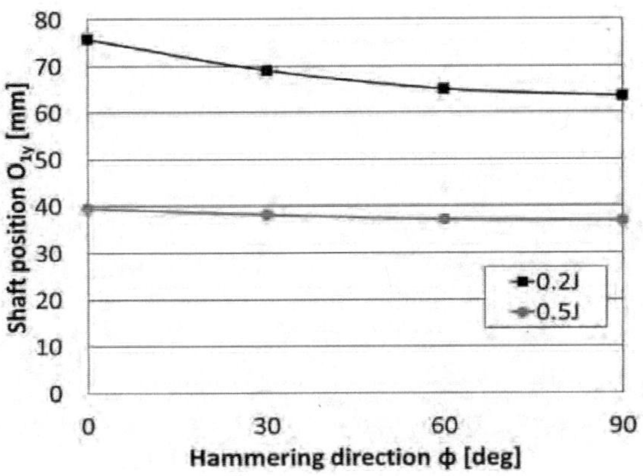

Figure 8. Hammering direction vs Shaft position

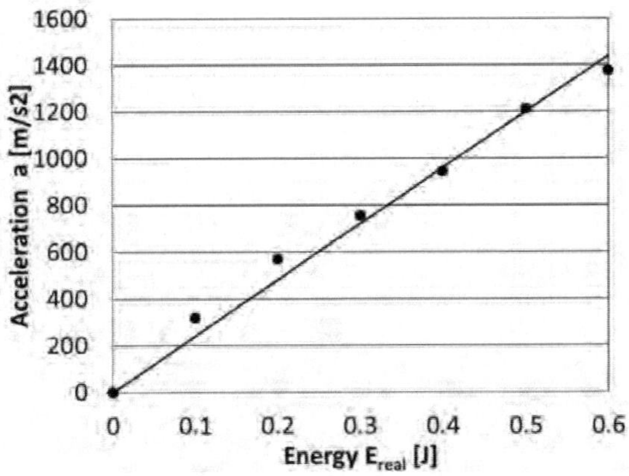

Figure 9. Enegy of the steel ball vs Acceleration of the wood

get analytical solution because this equation is non-linear. Therefore, we analyze it numerically by the Euler method. The pitch width is

$$\Delta t = t_{n+1} - t_n \tag{8}$$

and we can get numerical solution as follows

$$\dot{\theta}_2(t_{n+1}) = \dot{\theta}_2(t_n) + \ddot{\theta}_2(t_n)(t_{n+1} - t_n) \tag{9}$$

$$\theta_2(t_{n+1}) = \theta_2(t_n) + \dot{\theta}_2(t_n)(t_{n+1} - t_n) \tag{10}$$

when initial conditions are

$$\dot{\theta}_2(t_0) = \dot{\theta}_2(0) = 0 \tag{11}$$

$$\theta_2(t_0) = \theta_2(0) = 0 \tag{12}$$

kinetic energy of the hammer E[J] is written by

$$\begin{aligned} E =& \frac{1}{2} m_2 {l^2}_1 \dot{\theta}_2^2 + m_2 l_1 l_{g2} \dot{\theta}_1 (\dot{\theta}_1 + \dot{\theta}_2) \cos\theta_2 \\ &+ \frac{1}{2}(m_2 {l_{g2}}^2 + I_2)(\dot{\theta}_1 + \dot{\theta}_2)^2 \end{aligned} \tag{13}$$

Actually, considering the machine loss $L_1 = 0.2[\%]$,

$$E_1 = (100 - L_1)\times 0.01 \times E \tag{14}$$

For example, in Table 1, relationship between ϕ[deg] and O_{1y}[mm] to maintain energy of 0.2[J] or 0.5[J] is shown as in Figure 8.

Figure 10. measuring device

4 Hammering Experiment

In this section, we showed the result of hammering tests using this robot and verified usefulness of this theory. We first described how to estimate kinetic energy of the hammer. After that, we described about experimental procedure. Finally, we compare the results obtained from the experiment with the theory.

4.1 Estimation Method of Physical Energy

Vibration is generated when collides one with another. Therefore, we expected that kinetic energy correlates with vibration acceleration. From this expectation, we experimented that a steel ball with a weight of 94.9[g] dropped freely from various heights and collided with a stationary wood, and plotted relationship between kinetic energy of a steel ball and vibration acceleration of a wood. To detect acceleration, we used the piezoelectric acceleration pickup (PV-95, Rion Co.,Ltd.) and observed by FFT analyzer. The experiment was repeated 5 times and the ball dropped to the same point. The experimental result is shown in Figure 9. From Figure 9, plots are found to be almost linear. Linear approximate equation obtained by least-square method is as follows

$$E_{real} \approx \frac{1}{2397} a \tag{15}$$

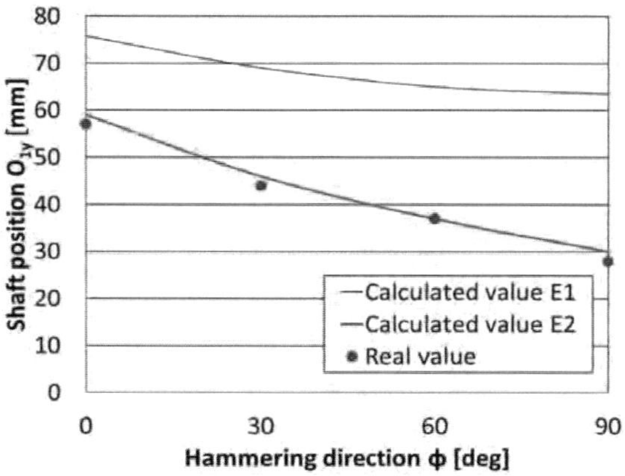

Figure 11. Hammering direction vs Shaft position at 0.2[J]

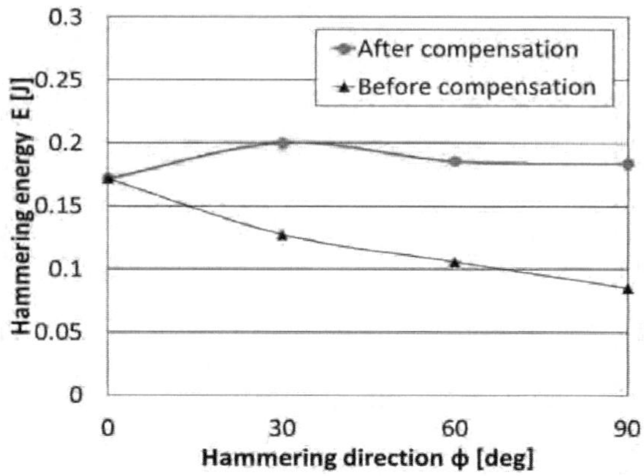

Figure 12. Hammering direction vs Hammering energy

Therefore, we can estimate kinetic energy of the hammer by Equation (15).

4.2 Experimental Procedure

We developed a rotating device as shown in Figure 10 and attached a hammering robot, and estimated kinetic energy of the hammer by Equation (15). We experimented at 0[deg], 30[deg], 60[deg], 90[deg] and the axial position was adjusted to keep kinetic energy of the hammer. Considering that we aim to develop robots which can sound like human in this research, hammering energy should be the same as engineers. Therefore, we measured hammering energy of them, sound area is 0.5[J] and delamination area is 0.2[J]. However , we cannot obtain the repetition at 0.5[J] because the plastic deformation occurred in a part of the wood. This problem can be resolved by using stainless, but the acceleration is too high and we cannot observe it by device we possess. Therefore, in this research, we experimented at 0.2[J] which assumed delamination area. This experiment was repeated 5 times.

4.3 Experimental Results and Consideration

Relationship between theoretical energy E1 and experimental energy shown as Figure 11. As shown in Figure 11, theoretical energy and experimental energy are not corresponded. Comparing theoretical energy between experimental energy, for example, O_{1y} is 17[mm] at 0[deg], and 33[mm] at 90[deg]. For this reason, it is expected energy loss by a collision between hammer and stopper. In Figure 2-C, hammer which rebounded from concrete collides a stopper in Figure 2-A. Hammer can get reaction force and accelerate if it is on stopper in Figure 2-B. However, if it cannot get sufficient reaction force by this collision, actual kinetic energy of a hammer is more poor. It is expected that the more influence of the weight of a hammer (that is hammering direction) greater, the more energy decreases. Therefore, defining loss $L_2[\%]$ with ϕ[deg] as a variable,

$$L_2 = 15 \sin \phi \tag{16}$$

In this experiment, we found out phenomenon that the more O_1 approaches O, the more load of the motor increase. Therefore, defining loss $L_3[\%]$ with O_{1y}[mm] as a variable,

$$L_3 = \frac{1300}{O_{1y}} \tag{17}$$

We incorporate Equation (16) and (17) into (14) and correct it shown as

$$E_2 = (100 - (L_1 + L_2 + L_3)) \times 0.01 \times E \tag{18}$$

Figure 11 shows theoretical energy E2 by Equation (18). As shown in Figure 11, theoretical energy and experimental energy are almost match.

To verify usefulness of technique that we invented, we compared the case of changing O_{1y}[mm] according to E2 (that say after compensation) and the case of not changing it (that say before compensation). Result of this is shown in Figure 12. As shown in Figure 12, it can maintain 0.2[J] in the case of after compensation. On the other hand, in the case of before compensation, the more ϕ[deg] increases, the more energy decreases, especially 0.09[J] at 90[deg].

5 Conclusion

In this research, we aim to develop robots can sound like human hammering, and we developed an under-actuated hammering robot that combined an under-actuated system which behaves similar to that of engineers arm and slider-crank system. In this paper, we described this robot and its model. After that, we clarified a velocity control mechanism for gravity compensation. Finally, we showed the result of hammering test using this robot and verified usefulness of this mechanism. As a result of the test, theoretical energy and experimental energy are almost corresponded by considering energy loss that occur by collision of hammer and stopper and changing the motor position. Furthermore, the robot which applied this method and mechanism can hammering with constant kinetic energy. For future work, we will develop moving devices which is equipped with hammering robots and aim for practical use for hammering test in tunnels and bridges.

6 Acknowledgemnts

This work was in part supported by Council for Science, Technology and Innovation, "Cross-ministerial Strategic Innovation Promotion Program (SIP), Infrastructure Maintenance, Renovation, and Management" (funding agency: NEDO).

References

[1] Ministry of Land, Infrastructure and Transport in Japan. "Present state of road structure for aging countermeasure of road (tunnel)". http://www.mlit.go.jp/road/sisaku/yobohozen/yobo1_1.pdf

[2] Masato Ukai and Kazuya Shimoda. "Tunnel Delamination Inspection System Using Line Sensor Camera". http://jcma.heteml.jp/bunken-search/wp-content/uploads/2011/11/035.pdf

[3] Manabu Takeishi and Kazuhiro Koizumi. "Automatic Hammering Survey System IDATEN". http://jcma.heteml.jp/bunken-search/wp-content/uploads/ronbun/2004/043.pdf

[4] Yusuke Takahashi and Satoru Nakamura. "Control Method of the Underactuated Hammering System considering Concrete Sound". ROBOMEC2016, 2A2-09a6, 2016.

[5] Yasuo Mori, Takayuki Iwai. "Development of Tunnel Lining Hammering Sound Diagnostic System". http://jcma.heteml.jp/bunken-search/wp-content/uploads/ronbun/2004/042.pdf

[6] Yusuke Takahashi and Satoru Nakamura. "Development of the Hammering System with Underactuated Mechanism for Tunnel Inspection and Evaluation of Performance". Construction Robot Symposium, 2015.

Development of UAV Indoor Flight Technology for Building Equipment Works

H. Tomita[a], T. Takabatake[a], S. Sakamoto[a], H. Arisumi[b], S. Kato[b] and Y. Ohgusu[b]

[a]Shinryo Corporation, Japan
[b]National Institute of Advanced Industrial Science and Technology, Japan
E-mail: tomita.hr@shinryo.com, takabatake.ta@shinryo.com, sakamoto.sh@shinryo.com, h-arisumi@aist.go.jp, shin.kato@aist.go.jp, yuji.ogusu@aist.go.jp

Abstract –

In the Japanese construction industry, An UAV has begun to be used for outdoor works. On the other hand, most of building equipment works is indoor work, often working in touch with the surrounding environment near the ceiling.

Air volume measurement is an example of building equipment works near the ceiling. Conventionally, a worker gets on a temporary scaffold to approach the ceiling at the high position and measures the air volume just under the diffuser installed on the ceiling with using the handy type anemometer. This conventional work has low productivity and high cost for safety management. Furthermore, this tendency is remarkable in rooms with high ceilings such as entrance halls and factories.

We propose a new measurement method using the UAV with air volume measuring device, and develop and evaluate the system. The visual feedback control system for self-localization is also developed because the GPS cannot be used indoors. In addition, a high ceiling room with the diffuser is constructed as a mock-up for experiment. We analysed the positioning errors of the UAV when hovering and when approaching and touching the diffuser. Also, we evaluated the touch performance of the UAV to the ceiling by measuring the air volume.

As a result, it was found that the UAV can ensure the performance necessary for air volume measurement in this experiment.

Keywords -

UAV; Indoor Flight Technology; Building equipment works; Air Volume Measurement

1 Introduction

In the Japanese construction industry, the number of young workers is decreasing and skilled workers are aging. For that reason, efforts to improve productivity such as "i-Construction" by using information technology (IT) and robot technology are under way. One of them is the UAV (Unmanned Aerial Vehicle). In civil engineering work, UAV are used for management of river facilities [1] and monitoring for bridge degradation [2]. In building work, quality is managed by using images taken from the sky [3]. However, there are no cases of using UAV in buildings under construction. The main cases are monitoring, confirmation, and three-dimensional measurement by cameras and others. There are no cases of working in touch with the surrounding environment or members such as transportation and installation. Building equipment works such as air conditioning, sanitation and electricity is mostly work in the building, especially in the vicinity of the ceiling using temporary scaffolding. In order to utilize UAV in building equipment works, it is necessary to perform self-positioning without GPS, to recognize the surrounding environment, and to make direct touch with the surrounding environment.

One of the tasks near the ceiling in building equipment works is air volume measurement. Air volume measurement is a task of checking whether the amount of conditioned air blown out from the diffuser provided on the ceiling surface of a room is in accordance with the design value, and in general, total measurement is required. Conventionally, wind speed distribution immediately below the diffuser is measured by a handy type anemometer, and air volume is calculated by multiplying the average value by the opening area. Although temporary scaffolds are necessary to approach the diffuser in a high place, work efficiency decreases due to assembly, movement and disassembly of temporary scaffolds. Furthermore, safety management is important because it is work in a high place. The authors developed a device to measure the amount of air passing through the hood by bringing the hood for collecting wind into close touch with the ceiling surface by an elevating mechanism [5]. As a result, the safety and productivity of the air volume measurement work have improved. However, since there is a limit to the height of the equipment, measurement using a temporary scaffold is required as in the past for diffuser in rooms with high ceilings such as entrance halls and factories. Also, if obstacles such as furniture and production equipment are installed right under the diffuser, it is difficult to install the scaffold. On the other hand, if it is possible to measure the air volume using UAV, it can be expected that the problem when the person measures the air volume using the temporary scaffold or air volume measuring device as described above is solved.

In this research, for the purpose of developing a UAV which can be used indoors in building equipment works, we developed a UAV capable of measuring air volume and evaluated its performance.

2 System Overview

2.1 System Configuration

The new system consists of UAV equipped with an "Air Volume Measurement Unit", and a vision system with camera and PC.

The UAV is an oct-copter shown in Figure 1. In order to mount the "Air Volume Measurement Unit" in the center of the UAV, the frame was arranged in parallel crosses. The flight controller and the battery were arranged in a dispersed manner in the frame. The "Air Volume Measurement Unit" (Figure 2) consists of a "Collect Wind Hood" for collecting air flow from the air inlet, a "Sensor Hood" with five wind speed sensors, and a control and communication circuit of the wind speed sensor and a battery. In addition, neoprene rubber was attached to the upper end of the "Collect Wind Hood" considering ease of adhesion with the ceiling surface. Furthermore, CFD (Computational Fluid Dynamics) was performed and the lower end of the sensor hood is set to be 90 mm lower than the propeller position, so that the airflow generated by the propeller did not influence the air flow measurement. Other specifications of the UAV, specifications of the camera and specifications of the wind speed sensor are shown in Table 1.

2.2 Estimation of Position and Attitude using Vision System

In this research, we decided to use a vision system (Figure 3) that recognizes markers attached to the UAV with cameras installed in the environment as a self-position estimation method for indoor UAV that cannot use GPS. There are various methods such as SLAM and inertial navigation as self-position estimation methods in an environment where GPS cannot be used. We thought that the vision system that recognizes markers on UAV with cameras installed in the environment has the following two merits, and we chose this in the present study.

1. Since the target position can be set on the camera image, it can be used, for example, when there is no 3D model in the existing building or when the 3D model does not conform to the current situation.

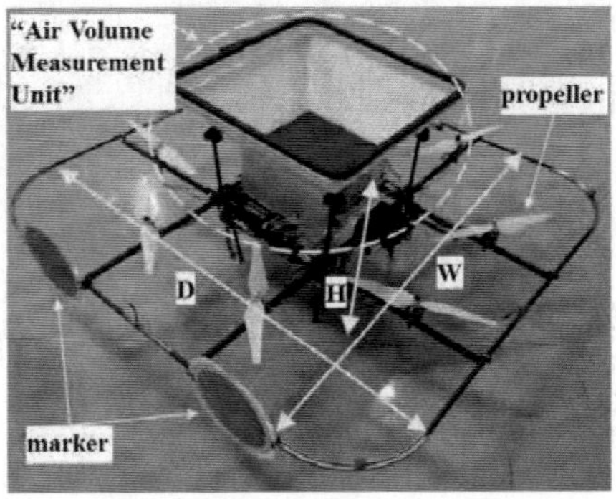

Figure 1. Appearance of UAV

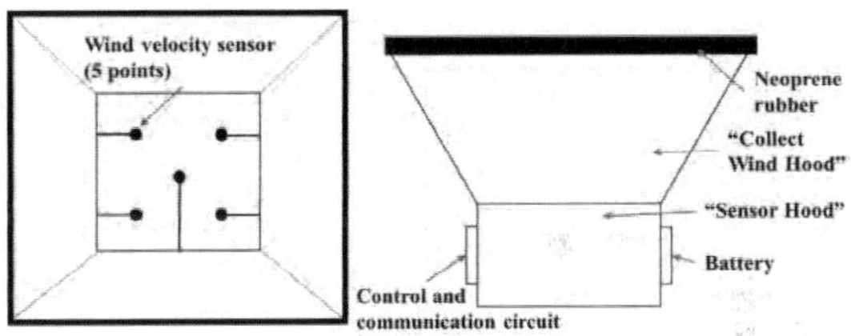

Figure 2. "Air Volume Measurement Unit"

Table 1. System specification

UAV	
Flight controller	Pixhawk
Size [mm]	D800 × W800 × H400
Frame material	Carbon fiber
Battery	Lithium polymer battery (2 pieces)
Propeller size [in]	9.45
Weight [kg]	2.5
Sensor	Accelerometer (3 axes), gyroscope (3 axes), barometer
Camera	
frame rate [fps]	30 (Max)
pixels	1920 × 1080
angle of view [°]	90
wind speed sensor	
Measurement method	Hot-wire
Measurement range	0.05～12.0m/s

2. Since the measurement is performed by the camera installed in the environment to be used, there is no need to install a sensor for recognizing the external environment in the UAV. As a result, the size of UAV can be minimized.

Next, we describe a method for estimating the position and attitude of UAV by recognizing markers. The markers on the UAV are two red circular marks (Figure 3). Two red circular marks are extracted from the image acquired by the camera and a circle is obtained with the line connecting the centers of the two circular marks as diameter, so as to find the area of the circle (A_m) and the position on the screen (X_m,Y_m). Based on the area of a circle (A_b) obtained from two circular marks arranged 3 m ahead (D_b) so as to face the camera and the actual distance (X_b,Y_b) to one pixel at 3 m ahead, the horizontal coordinate of the camera *(X)* is given by Equation (1), the vertical coordinate of the camera *(Y)* is given by Equation (2), and the depth coordinate of the camera *(Z)* is given by Equation (3). Also, from the size of the two markers recognized by the camera, determine the attitude of the UAV in the vertical axis rotation direction. The target position can be set by determining the position and area of the circle on the camera image.

$$X = X_b * X_m * \sqrt{A_b / A_m} \quad (1)$$

$$Y = Y_b * Y_m * \sqrt{A_b / A_m} \quad (2)$$

$$Z = \sqrt{D_b^2 * A_b / A_m} \quad (3)$$

2.3 Control of UAV

Position control of the UAV is performed as follows. The difference is obtained from the target position set on the camera image and the UAV position, and the PID control as shown in Equation (4) is performed so that the UAV position approaches the target position. In addition, when the deviation between the target coordinates and the position coordinates of the UAV becomes within the threshold value, it is automatically controlled so as to maintain the height by using the barometer. This was used as a safety measure to prevent the UAV from falling when the UAV became unmeasurable by the vision system.

Attitude control of the UAV is performed as follows. The control method differs between the rotation direction of the vertical axis and the two rotation directions of the horizontal axis. The rotation direction of the vertical axis performs PID control using the direction of UAV with respect to the camera, and rotates the UAV so as to face the camera. The control of the two rotation directions of the horizontal axis is performed automatically by the flight controller so that the UAV becomes horizontal.

$$u(t) = K_P e(t) + K_I \int e(\tau)dt + K_D \frac{d}{dt} e(t) \quad (4)$$

3 Evaluation Experiment

In order to evaluate the function and performance of the new UAV, we conducted a flight experiment. The experimental site was a simulated ceiling with a diffuser as shown in Figure 4. In addition, we used the motion capture system "OptiTrack" using an infrared camera to measure the flight path.

First, we evaluated the basic performance of the UAV by hovering. Furthermore, as the performance required for air volume measurement, the approach performance to the diffuser and the touch performance with respect to the ceiling were evaluated.

3.1 Hovering Performance

3.1.1 Experimental Method

For confirmation of hovering performance, a target was set in the air, hovering for 30 seconds, and the flight trajectory was measured. In order to eliminate the disturbance of flight caused by the UAV's own airflow

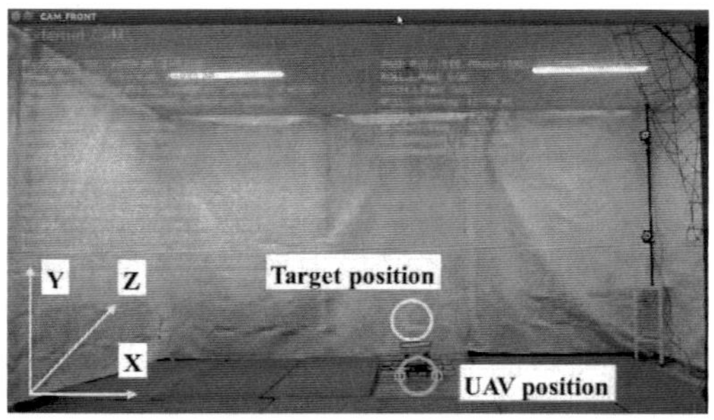

Figure 3. Overview of vision system

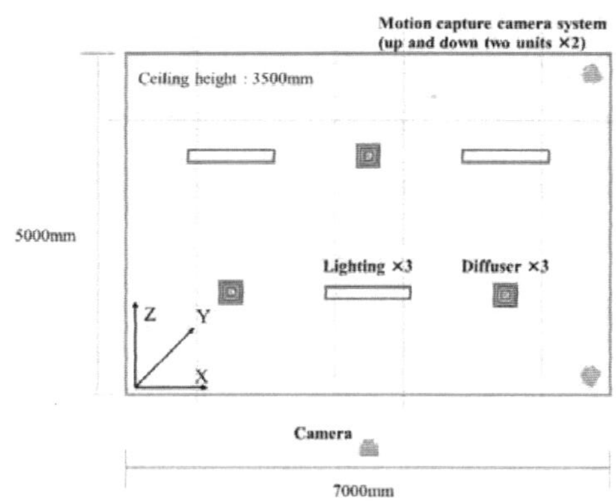

Figure 4. Experiment field (plan view)

during take-off, by passing the string through the center of the UAV, we restrained the UAV from moving in the horizontal direction. After the UAV reached the target, the string was loosened and hovering was performed.

3.1.2 Result

Figure 5 shows the hovering flight trajectory (5 flights). The vertical axis is the vertical direction of the camera (Y), the horizontal axis is the horizontal direction of the camera (X) in the left figure, and the depth direction of the camera (Z) is shown in the right figure. The hovering performance in the depth direction and the vertical direction is poor as compared with the horizontal direction. Furthermore, in order to confirm the dispersion in hovering for each axial direction, Figure 6 shows the average value and the standard deviation of the difference from the target position for each flight. When checking the maximum value, the average in the horizontal direction is 14 mm, the standard deviation is ± 41 mm, the average in the vertical direction is -49 mm, the standard deviation is ± 119 mm, and the average in the depth direction is -146 mm, the standard deviation is ± 155 mm. In the vertical direction, the standard deviation is large with respect to the horizontal direction. Also, in the depth direction, the dispersion of the average for each flight and the standard deviation is large with respect to the horizontal direction.

3.1.3 Discussion

We will consider the reasons for the above experiment results. Factors related to the flight performance of the UAV can be roughly divided into the following two groups.

1. Measurement performance of the UAV by vision system.
2. Control performance by the flight controller.

The difference between the horizontal direction and the depth direction of the camera is in the measurement of the UAV by the vision system. Generally, in three-dimensional measurement by the camera, the resolution of the camera depth is lower than the resolution of the camera plane.
In this system, the depth resolution is less than 1/10 of the plane. Also in the experimental result shown in FIG. 6, the average in the camera horizontal direction is 14 mm, and the average in the camera depth direction is -146 mm. Therefore, it is considered that the measurement performance of UAV by the vision system is the cause of low hovering performance in the camera depth direction

The difference between the horizontal direction and the vertical direction of the camera is controlled by a flight controller. The measurement performance of the UAV by the vision system is the same resolution in both the horizontal direction and the vertical direction of the camera. In the flight controller, when the control amount of the UAV by the vision system is small, it is automatically controlled so that the height is maintained using the barometer only in the vertical direction. In order to confirm this control performance, the output data of the barometer was confirmed. Figure 7 shows the data converted from the output data of the barometer to the height difference (vertical axis) in time series (horizontal axis). The standard deviation of the data is about ± 200 mm, which shows that the dispersion is large. Also, unlike the horizontal direction, the vertical direction is influenced by gravity, which makes control difficult. Therefore, low hovering performance in the vertical direction is considered to be due to the control performance by flight controller.

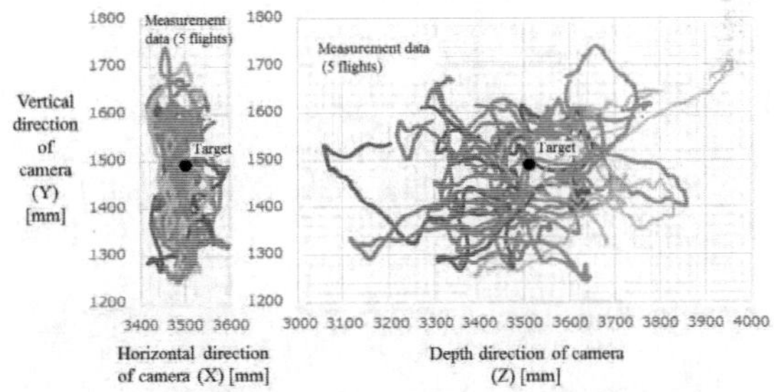

Figure 5. Locus of hovering

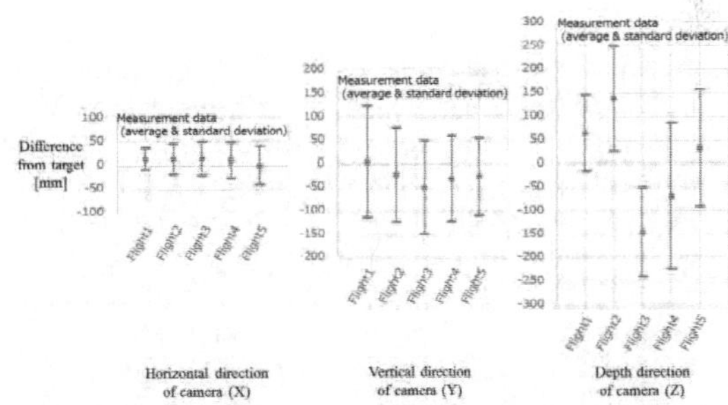

Figure 6. Difference between measurement data and target point in hovering

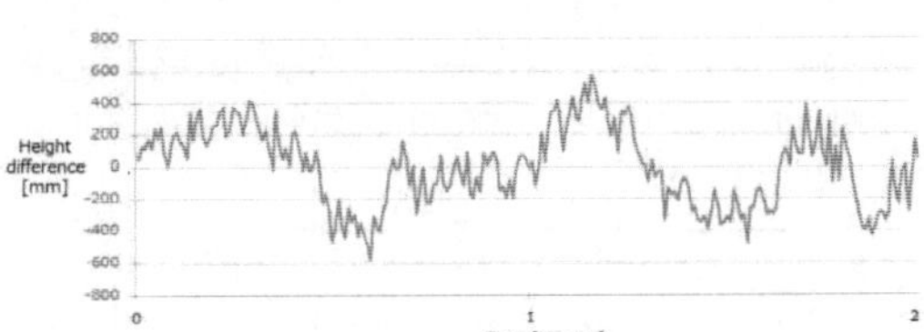

Figure 7. Height difference calculated from the measurement data of the barometer

3.2 Performance Required for Air Volume Measurement

In order to accurately measure the air volume, it is necessary to accurately approach the UAV to the diffuser so that the diffuser enters the "Collect Wind Hood". Therefore, the approach accuracy to the diffuser by the UAV was evaluated.

Also, for accurate air volume measurement, it is necessary to collect all the conditioned air blown out from the diffuser with the "Collect Wind Hood", so the ceiling and "Collect Wind Hood" must be in close touch with each other. According to the conventional method, a person is pushing the "Collect Wind Hood" against the ceiling, and the air volume is measured while maintaining that state. In order to check whether the state of touch with the ceiling of the UAV is about the same as the conventional method, the air volume was measured and evaluated.

3.2.1 Approach Performance

In the experiment, the target position and the position to pass through were set, the flight was made to fly from the floor surface to the diffuser, and the flight trajectory was measured five times. Approach to the target position was carried out after 25 seconds from the start of flight.

The flight trajectory is shown in Figure 8, and the difference from the target position of the approach is shown in Figure 9. In Figure 8, the vertical axis is the vertical direction of the camera (Y), the horizontal axis is the horizontal direction of the camera (X) in the left figure, and the depth direction of the camera (Z) is shown in the right figure. The flight trajectory shows that the dispersion in the depth direction is large like the hovering performance confirmed by the basic performance. As for approach accuracy, the average is 25 mm, the standard deviation is ± 15 mm in the horizontal direction of the camera, the average is 7 mm, and the standard deviation is ± 263 mm in the depth direction of the camera, which shows that the dispersion in the depth direction of the camera is large. As discussed in the evaluation of basic performance, we consider that the cause of large dispersion in the depth direction of the camera is the measurement performance of the UAV by the vision system with one camera.

To improve approach accuracy, it is necessary to construct a vision system by using two cameras. The position of the UAV in the vertical direction and the horizontal direction of the camera is measured with one camera. Furthermore, another camera is installed by changing the position of 90 ° around the UAV from the already installed camera. Then, the horizontal position of the UAV (the position in the depth direction of the other camera) is measured with the camera. By adopting the above method, the position of the UAV can be obtained without being affected by the resolution of the vision system in the depth direction of the camera, and the accuracy required for approach to the diffuser can be ensured.

3.2.2 Touch Performance

In the experiment, we measured the air volume when the UAV pressed the "Collect Wind Hood" against the ceiling. As a conventional method, the person pressed the "Collect Wind Hood" against the ceiling, the air volume was measured, and the data was compared with the data measured by the UAV. In the air volume measurement, the wind speed was measured 5 times at intervals of 1 second, the average wind speed was calculated, and the air volume was calculated. The measurement accuracy of the wind speed sensor is 5% of the measured data.

Table 2 shows the result of performing air volume measurement three times for both the conventional method and the UAV. Compared with the conventional method, it was confirmed that the measurement result using UAV is within the measurement accuracy of the wind speed sensor. From this fact, we confirmed that touch with the ceiling of the UAV is comparable to the conventional method.

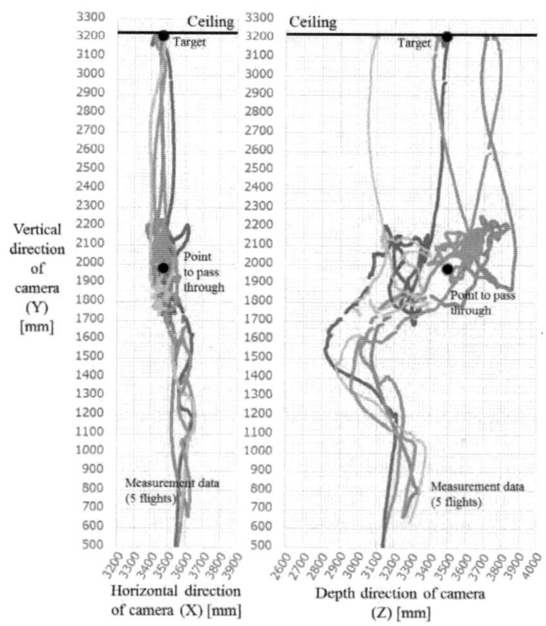

Figure 8. Locus of flight

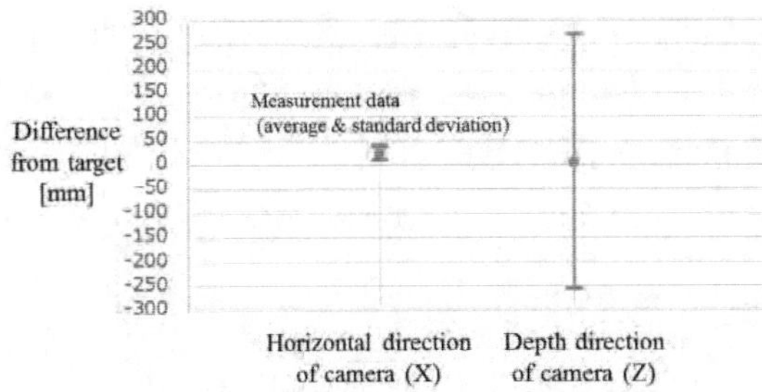

Figure 9. Difference between measurement data and target point in approach position

Table 2. Result of air volume measurement

		Measurement data [m³/h]	Average [m³/h]
Conventional method	No.1	380	
	No.2	363	363
	No.3	346	
UAV	No.1	371	
	No.2	350	358
	No.3	354	

4 Conclusion

For the purpose of developing UAV for use in indoor building equipment works, we made a prototype UAV capable of measuring air volume and evaluated it. As a result, the approach accuracy to the diffuser is 25 mm on average in the camera horizontal direction and ± 15 mm in standard deviation. From this, by constructing a vision system using two cameras, it was found that the accuracy required for the approach can be ensured at this experimental site. Also, from the results of the air volume measurement, it was confirmed that touch with the ceiling of the UAV was about the same as the conventional method.

In the future, in order to improve the approach accuracy to the diffuser, we will construct a vision system using two cameras and evaluate the flight performance.

References

[1] S. Kubota and Y. Kawai. River maintenance management system using three-dimensional UAV data in JAPAN. *ISPRS Annals of Photogrammetry, Remote Sensing & Spatial Information Sciences*, pages 93-98, Issue 2, Vol.4, 2016.

[2] S. Sumitro, S. Nishimura, H. Matsuda and I. Bartoli. GPS-based remote optical monitoring system for structural inspection. *Safety, Reliability, Risk and Life-Cycle Performance of Structures and Infrastructures*, pages 281-288, 2014.

[3] Nikkei Business Publications. Photographed by UAV for construction record: Takenaka Corporation was full scale operation at Suita municipal stadium, *NIKKEI CONSTRUCTION*, 613, pages 25, 13/04/2015. (in Japanese)

[4] K. Tanaka, S. Sakamoto and Y. Abe. Development of an air volume measuring instrument: WINSPEC. In *Proceedings of the 23rd ISARC*, pages 210-214, Tokyo, Japan, 2006.

Force-regulated impact control of a reinforced concrete box culvert chipping robot

Jaeuk Cho [a] and Seungyeol Lee[a]

[a] Collaborative Robot Convergence Research Center, DGIST, Republic of Korea
E-mail: cjw6379@dgist.ac.kr, syl@dgist.ac.kr

Abstract

A reinforced concrete box culvert (RCBC) is a box culvert constructed of reinforced concrete. Because the sudden collapse of an RCBC could result in a serious accident, continuous management and maintenance work of the RCBC should be performed. In this paper, the RCBC chipping robot that is developed as an alternative to skilled construction workers for carrying out the concrete-chipping work, one of the RCBC maintenance process, is introduced, and the force-regulated impact control for improving quality of the chipping work is described. In order to implement the proposed algorithm, reaction forces when the robot strikes concrete with chipping tool are collected, and classification to identify the striking object and its hardness is carried out based on the obtained reaction forces. In addition, a variable motion control algorithm is realized by setting the dynamic characteristics of the robot with respect to the reaction force according to striking object and its stat. To verify the effectiveness of the proposed algorithm, chipping work with an RCBC chipping robot is conducted in a simulation environment consisting of reinforced concrete.

Keywords –

Force-regulated impact control; Box culvert maintenance; Concrete chipping; RCBC (Reinforced Concrete Box Culvert); Construction robot;

1 Introduction

A reinforced concrete box culvert (RCBC) is a box culvert constructed of reinforced concrete; it is a structure that allows water to flow under a road, railroad, trail, or similar obstruction from one side to the other. Currently, many RCBCs are located under cities and roads. If a ground subsidence occurs owing to the sudden collapse of an RCBC, it could result in a serious accident [1]. Therefore, continuous management and maintenance work should be performed to prevent RCBC collapse and accidents caused by such failure.

Typical RCBC maintenance work consists of detecting the damaged parts, chipping to remove the deteriorated concrete, cleaning the surface of exposed rebar, etc. In particular, the chipping process, in which workers remove the deteriorated concrete from ceilings or walls of the reinforced concrete structures by using heavy equipment that vibrates significantly, while standing on a temporary structure (i.e., a scaffolding) at a high elevation, poses a high risk of musculoskeletal injury. Moreover, poor working conditions and human error can result in major accidents in this process. This process is regarded as hazardous, and even skilled construction workers avoid undertaking this work. For this reason, the chipping process is one of the construction operations for which there is an urgent need for the application of automatic construction based on construction robot.

Construction robots are spotlighted as a solution for the problems such as such as wage increase, labor shortage and danger to human life caused by construction accidents. Automatic construction based on construction robot has the potential to improve safety, productivity, quality and working environment, and it can reduce the number of non-essential workers, time, costs of construction and the incidence of accidents by performing tasks efficiently and improving working conditions by carrying out work that are dangerous for the safety of human [2-12].

In this paper, the RCBC chipping robot that is developed as an alternative to skilled construction workers for carrying out the concrete-chipping work is introduced, and the force-regulated impact control for improving quality of the chipping work is described. In order to implement the proposed algorithm, reaction forces when the robot strikes concrete with chipping tool are collected, and classification to identify the striking object and its hardness is carried out based on the obtained reaction forces. In addition, a variable motion control algorithm is realized by setting the dynamic characteristics of the robot with respect to the reaction force according to striking object and its stat.

To verify the effectiveness of the proposed algorithm, chipping work with an RCBC chipping robot is conducted in a simulation environment consisting of reinforced concrete. Chipping work with the robot is carried out in both cases, with and without the proposed algorithm, and the performance of the proposed algorithm is verified by comparing and analyzing the quality of each chipping work.

The rest of this paper has five sections. In Section 2 and 3, the chipping robot and proposed algorithm are described. Section 4 describes Performance test and shows the effectiveness of the proposed method. Finally, Section 5 concludes with a summary of the paper.

2 RCBC Chipping Robot

A reinforced concrete box culvert (RCBC) is a box culvert constructed of reinforced concrete. Because the sudden collapse of an RCBC could result in a serious accident, continuous management and maintenance work of the RCBC should be performed.

The chipping process that is one of RCBC maintenance process is that workers remove the deteriorated concrete from ceilings or walls of the reinforced concrete structures. Because the chipping process is carried out by using heavy equipment that vibrates significantly, while standing on a temporary structure (i.e., a scaffolding) at a high elevation as shown Figure1, there are high risk of musculoskeletal injury. Moreover, poor working conditions and human error can result in major accidents in this process. This process is regarded as hazardous, and even skilled construction workers avoid undertaking this work. For this reason, the chipping process is one of the construction operations for which there is an urgent need for the application of automatic construction based on construction robot.

Figure 1 Chipping process by manpower

A RCBC chipping robot is developed as an alternative to skilled construction workers for carrying out the concrete-chipping work. As shown the Figure 2, the chipping robot consists of a robot mobile platform for movement in uneven terrain, a robot manipulator for chipping work, a chipping tool for removing deteriorated concrete, and a robot motion controller that generates the trajectory for conducting the processes and makes the robot follow the trajectory.

In order to perform the chipping work, the chipping robot moves to the work area through the robot mobile platform. Then, the robot manipulator follows the trajectory concrete-chipping work, and the chipping tool installed at the end of the manipulator strikes the deteriorated concrete. Trajectory generation and control for all these operations are performed through the robot motion controller.

When the work using the RCBC chipping robot is performed at constant speed irrespective of the striking object and its state, hardness, the concrete is removed unevenly according to hardness and this causes an uneven chipping surface, i.e. fall in chipping quality. Therefore, in order to improve the chipping quality, motion control for adjusting the chipping speed according to the striking target and its state is necessary.

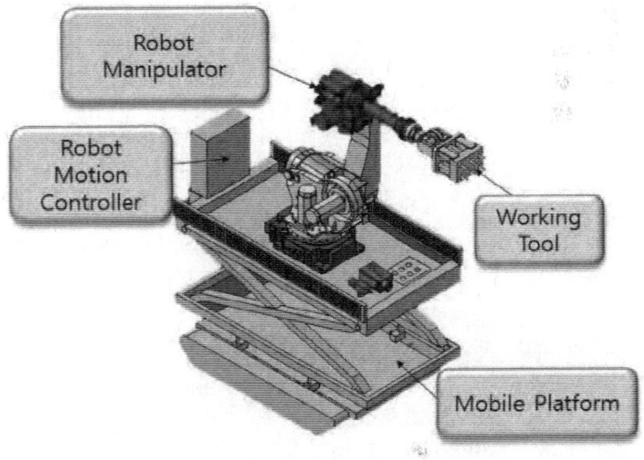

Figure 2. Components of RCBC chipping robot

In this paper, for improving the chipping quality, a force-regulated impact control algorithm that classifies a striking object of the robot and its state from reaction force, and control speed of chipping according to the state is proposed.

3 Force-Regulated Impact Control

The force-regulated impact control is designed to control chipping speed of an RCBC chipping robot according to the object that the robot strikes for improving chipping quality. In order to control the chipping speed according to the object the robot strikes, the desired reaction force between the chipping tool and the chipping target, is decided, and force control is applied based on the reaction force generated when hitting the target. Because the chipping is a task to strike the target, which is the concrete wall of RCBC, with a chipping tool, a lot of impacts are generated according to the striking object and the reaction forces include impulse. The impulse not only deteriorates the performance of force control but also results in contact loss [13-14].

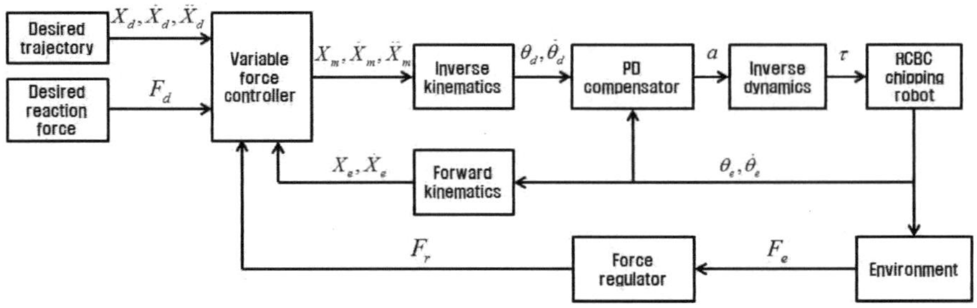

Figure 3 Block diagram to control RCBC chipping robot

Therefore, in order to control the chipping speed based on force control, it is necessary to apply variable force control which changes its dynamic characteristics according to the impulse. In this paper, variable force control for regulating the reaction force between the chipping tool and the working environment while controlling the impulse is proposed.

The force-regulated impulse control is designed to control the impulse through two control strategies as follows:

- In order to prevent the deteriorating of the chipping performance caused by impulse, the sensing external force is regulated by force regulator.
- In order to the control chipping speed, the desired trajectory is compensated by the variable force controller

Figure 3 shows the block diagram to control RCBC chipping robot. For the chipping process, desired trajectory, X_d, and desired reaction force, F_d, are generated. When the regulated reaction force, F_r, is not same desired force, $F_r \neq F_d$, the desired trajectory is compensated to modified trajectory, X_m. In order to obtain the modified trajectory, the present position of robot, X_e, which is calculated by forward kinematics, and regulated reaction force are used. From the modified trajectory, the desired joint trajectory, θ_d, is calculated, and input torque, τ, to move the RCBC chipping robot is obtained from the PD compensator and inverse dynamics.

The force regulator is designed to classify the external force based on whether it is an impulse that deteriorates the control performance and to regulate F_e based on this classification. To realize the force regulator, selective moving average based on classification is used. The selective moving average is described as follows:

$$F_{r(i)} = \frac{1}{m}\sum_{j=0}^{m-1} F_{e(i-j)} \qquad (1)$$

where

$$j = 1, 2, ..., m-1,$$

$$\Delta F_e = F_{e(i-j)} - F_{e(i-j-1)},$$

$$F_{e(i-j)} = \begin{cases} F_{e(i-j)} & (|\Delta F_e| \le \Delta F_{th}) \\ F_{e(i-j-1)} & (|\Delta F_e| > \Delta F_{th}), \end{cases}$$

$F_{r(i)}$ denotes the ith regulated reaction force, $F_{e(i-j)}$ denotes the i-jth sensing reaction force, ΔF_{th} is the criteria of classification, and m is the number of sampling for the moving average. From the force regulator, the regulated reaction force, F_r, is obtained and it is applied to the variable force controller to compensate chipping speed.

For controlling the chipping speed based on force control, a variable force controller is designed. To implement variable force control, a model of the chipping robot and chipping target, is designed to mass-spring-damper system as shown Figure 4.

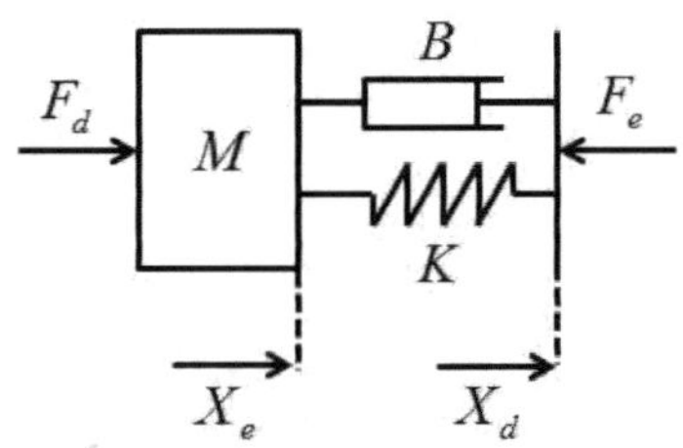

Figure 4 model of the chipping robot and chipping target

The dynamic equation of the model is descried as follows:

$$K\Delta X + B\Delta\dot{X} + M\Delta\ddot{X} = F_d - F_r, \qquad (2)$$

where,

$$\Delta X = X_d - X_e,$$

X_d and X_e denote the position of chipping tool, F_d and F_e denote the desired and real reaction force. M, B, and

K denote the mass, damping, and stiffness of impedance coefficient. From equation (2), the modified desired trajectory, X_m, is obtained as follows:

$$\ddot{X}_m = \ddot{X}_d + \frac{K\Delta X + B\Delta\dot{X} - (F_d - F_r)}{M_v}, \quad (3)$$

where

$$M_v = \begin{cases} M & (\varepsilon < \alpha) \\ \varepsilon M & (\varepsilon \geq \alpha) \end{cases} \quad \left(\varepsilon = \left|F_r(t) - F_r(t-L)\right|\right).$$

$\Delta\dot{X}$ denotes the derivative of ΔX , $\ddot{X}_d$ denotes acceleration of the desired trajectory, M_v is the variable impedance coefficient, α is threshold of M_v, and ε denotes differential of reaction force. $\ddot{X}_m$ denotes the acceleration of the modified trajectory, and modified trajectory, X_m, is obtained by integrating the value. M_v is changed according to ε. When ε is less than α, M_v is a small value, which is designed for sensitive reaction with respect to the reaction force. On the other hand, when ε is larger than α, M_v has large value, and it makes the response of the robot to the force less sensitive.

By applying this algorithm, it is possible to control the chipping speed based on the force control even during the chipping operation in which the impulse is frequently generated due to impact.

4 Performance Test

To verify the effectiveness of the proposed algorithm, chipping work with an RCBC chipping robot is conducted in a simulation environment. For the performance test, the simulation environment is constructed as shown in Figure 5. The robot for chipping work consists of a 6 DOF manipulator, a robot motion controller, and a chipping tool. As a chipping target, a module of the RCBC is used.

Figure 5 The performance test of proposed algorithm

In order to verify the effectiveness of the proposed algorithm, concrete chipping is carried out for two cases, i.e., with and without the application of the proposed algorithm. The path of the robot for chipping in a simulated environment is planned as shown in Figure 6. First, the chipping tool is advanced to the concrete wall until the tool contact with the wall. After the chipping tool contacts with the wall, the chipping tool starts and chipping is performed.

Figure 6 Schematic diagram for performance test

A chipping operation is performed for a predetermined working area, 150 mm x 150 mm, where the depth is 10 mm.

In order to compare the performance of the proposed algorithm, chipping through position control without applying the proposed algorithm is performed.

When chipping is performed through position control, the task is started at 0 sec, and performed at a constant chipping speed, 1mm/s, regardless of the object being hit and its state, i.e., hardness. According to the hardness of the object to be chipped, a large reaction force acts on the chipping tool as shown in Figure 7.

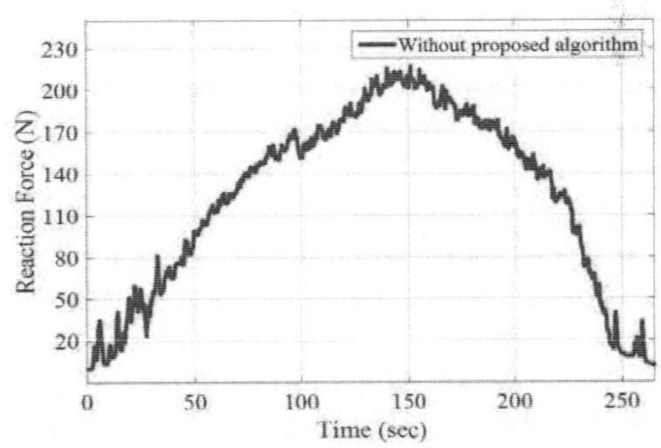

Figure 7 Reaction force during concrete chipping

The large reaction force hinders the robot from following the trajectory for the chipping operation, which can cause an uneven chipping depth, i.e. a drop in chipping quality as shown in Figure 8 (a) and (b) . At the beginning of the work, the target depth of 10mm is satisfied. As the work progresses, however, the work tool does not follow the desired trajectory due to the reaction force and gradually fails to satisfy the target depth. Finally, the working depth of the chipping robot is 3mm and low chipping quality is confirmed as shown in Figure 8(b).

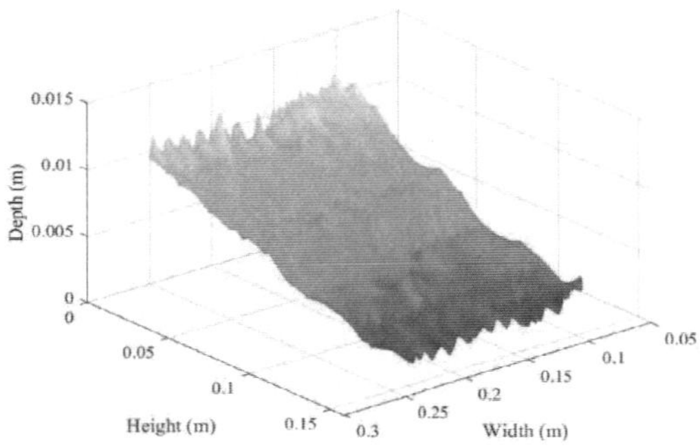

(a) 3D graph of chipping surface

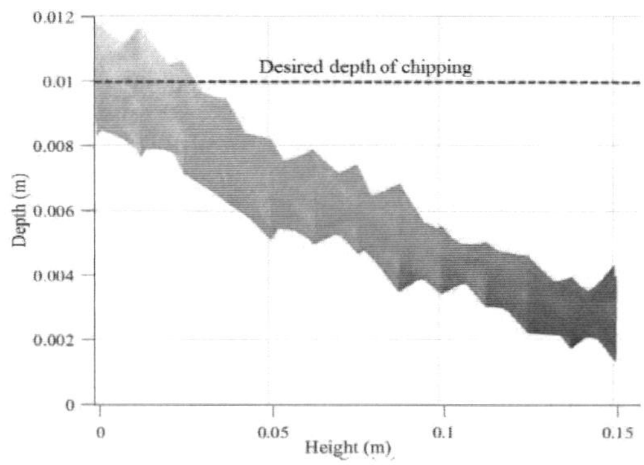

(b) The depth of chipping area

Figure 8 Result of chipping without proposed algorithm.

In this paper, to solve this problem, the force regulated impact control that controls the chipping speed based on the reaction force is proposed. In order to verify effectiveness of the proposed algorithm, the task is performed during 150 sec, and chipping speed is controlled to make the reaction force satisfy 140N.

Experimental results for the case when the proposed algorithm is used are shown in Figure 9. Figure 9 (a) shows the trajectory generated for the chipping operation, X_d , and the trajectory modified by the proposed algorithm, X_m. Through the graph, it can be confirmed that the trajectory of chipping robot, that is, the chipping speed, is modified while controlling a reaction force acting on the chipping tool. As a result, it can be confirmed that the reaction force acting on the chipping tool maintains the desired value, i.e., 140N, as shown in Figure 9(b). Also, the depth of the chipping area is satisfied a desired chipping depth, 10mm, compared with that before the algorithm is applied as shown in Figure 9(c) and (d). From the results, it can be confirmed that the proposed algorithm improve chipping quality by controlling the chipping speed.

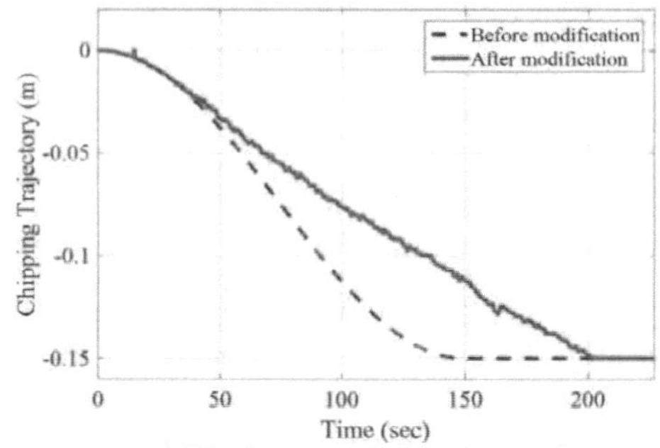

(a)Trajectory of chipping tool

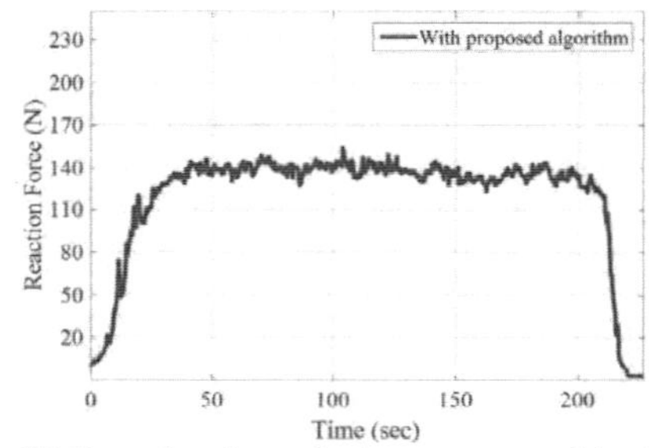

(b) Reaction force during concrete chipping

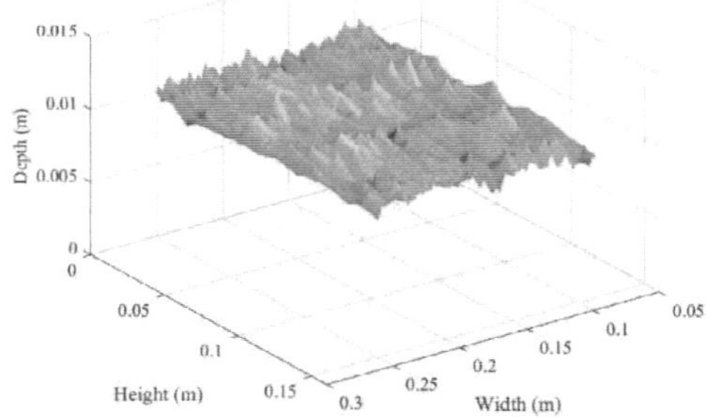

(c) 3D graph of chipping surface

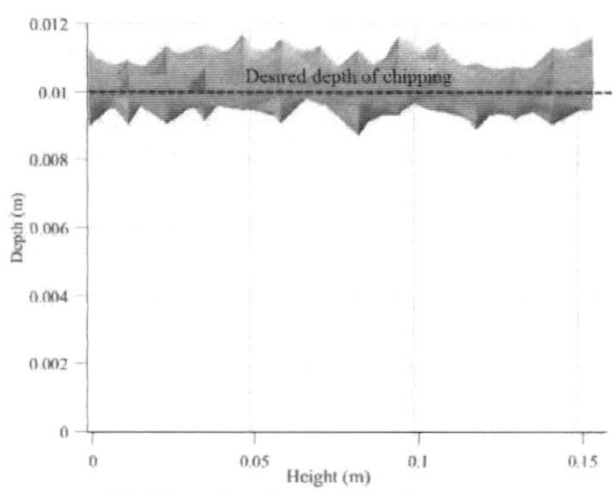

(d) The depth of chipping area

Figure 9 Result of chipping with proposed algorithm.

5 Conclusion

In this paper, the RCBC chipping robot developed as an alternative for skilled construction workers for carrying out concrete-chipping work, is introduced, and the force-regulated impact control for work quality improvement is proposed. To verify the effectiveness of the proposed algorithm, chipping work with an RCBC chipping robot is conducted for both cases, i.e., with and without the application of the proposed algorithm, in a simulation environment. As a result, it is shown that the proposed algorithm improves the chipping quality by controlling the chipping speed, and the performance of the proposed algorithm is verified. The proposed algorithm is designed to control the motion of an RCBC chipping robot according to the object that the robot strikes. Using this algorithm, deteriorated concrete can be removed while maintaining constant chipping quality. Furthermore, along with concrete chipping, this is also applicable to other fields where variable motion of a robot with respect to the object it is in contact with is required such as in excavation work and manufacturing.

In order to improve the performance of the proposed algorithm, future studies have been planned to apply the RCBC chipping robot to various construction sites, and the proposed algorithm will be supplemented based on the result.

ACKNOWLEDGMENT

This work was supported by the Innopolis Technology Commercialization program (Development of a chipping robot on reinforced concrete in covered structures) funded by the Ministry of Science, ICT and Future Planning of Korea (2016050010), the preparation of robot industry-cluster program (Practical use of the detection robot for maintenance on Reinforced concrete box culverts) funded by the Ministry of Trade, Industry and Energy of Korea (2016090003), and the DGIST R&D Program of the Ministry of Science, ICT and Future Planning of Korea (2017010020)

References

[1] Charles J. Nemmers, *Culvert Repair Practices Manual*, U.S. Department of Transportation, 1995

[2] Roozbeh K, *Advanced Robotics in Civil Engineering and Construction*. In Proceedings of IEEE International Conference on Robotics and Automation, pages 375-378, Tokyo, Japan, September, 1985

[3] Warszawski, A, *Economic implications of robotics in building*. Building and Environment, 20(2):73-81, 1985.

[4] Albus, James S., *Trip Report: Japanese Progress in Robotics for Construction*, Robotics Magazine, spring of 1986.

[5] Bernold, L.E. *Automation and robotics in construction: A challenge and change for an industry in transition*. International Journal of Project Management, 5(3):155–160, 1987.

[6] Skibniewski, M.J., *Robotics in civil engineering*, Van Nostrand–Reinhold, ISBN 0442319258, New York, 1988.

[7] Skibniewski, M.J. & Wooldridge, S.C., *Robotic materials handling for automated building construction technology*. Automation in Construction, 1(3):251– 266, 1992.

[8] Wen, X., Romano, V.F. & Rovetta, A., *Remote control and robotics in construction engineering*. In proceedings of the Fifth International Conference on Advanced Robotics, Pisa, Italy, 19-22 Jun 1991.

[9] Cusack, M., *Automation and robotics the interdependence of design and construction systems*. Industrial Robot, 21(4):10–14, 1994.

[10] Poppy, W., *Driving force and status of automation and robotics in construction in Europe*. Automation in Construction, 2(4):281–289, 1994.

[11] Kochan, A., *Robots for automating construction an Abundance of research*. Industrial Robot, 27(2):111–113, 2000.

[12] Seungyeol Lee and Jeon Il Moon, *Introduction of human-robot cooperation technology at construction sites*, Proceedings of the International Symposium on Artificial Life and Robotics (AROB), pp.146-149, 2014.

[13] J.M. Hyde and M.R. Cutkosky, Controlling contact transition, IEEE Control systems, 14(1), 1994.

[14] G. Ferretti, G. Magnani and A. Zavala Rio, *Impact modelling and control for industrial manipulators*, IEEE Control systems, 18(4), 1998.

A method based on C-K Theory for fast STCR development: The case of a drilling robot design

T. Zedin[a b], L. Vitalis[a], F. Guéna[a] and O. Marchand[b]

[a]UMR MAP-MAACC, National architecture school of Paris La Villette, France
[b]Research and development, Vinci Construction France, France
E-mail: thomas.zedin@vinci-construction.fr, louis.vitalis@paris-lavillette.archi.fr, françois.guena@paris-lavillette.archi.fr, olivier.marchand@vinci-construction.fr

Abstract –

New products in mechatronics, and especially the spreading of microcontrollers, allow a rise in robotic projects of building construction companies. This new context gives the ability to non mechatronic experts to develop building sites robotics.

In this paper we will describe the design process of a drilling robot prototype made in our Research and Development department. This specific project was constrained by a lack of specifications, and required fast and low cost methods. We had to face the risk of information loss and mistakes in the appreciation of the results. We intend to explain the design methods and knowledge used in the management of this project.

The use of design theories and methods seem to be necessary. Example such as Axiomatic Design or TRIZ can be too strict or too distant from a specific design case. We assume that C-K Theory developed by A.Hatchuel and B.Weil can be a relevant design tool for the development of robotics projects in construction. Thus we wish to present in this article the way we use the C-K Theory to analyze our design process. This design theory is based on two spaces: the knowledge space K and the concept space C. It will reveal the concept and knowledge mechanics in our projects and give clues to produce methods for the management of similar projects such as Single-Task Construction Robots.

Keywords –

C-K Theory; Construction robotics; Design process modeling; Design methods; Knowledge management; STCR; Drilling robot

1 Introduction

The evolutions of the mechatronics have allowed the general contractor Vinci, to realize a drilling robot prototype. The design process was driven by its R&D department. A particular task in rehabilitation led the specifications: Floors extensions or reinforcements on existing walls. This task use conventional techniques of reinforced concrete industry: Steels implantation and concrete pouring into a formwork. The motivation to create a robot comes from the difficulties generated by multiple drilling prior to the sealing of steels. These drilling are numerous, repetitive and cause ergonomic issues on worksite. These conditions are challenged in the context of labor legislation in France.

We have decided to carry out an analysis of the design process in order to enhance this work, by extending it or making it reproducible. We sought to know what had been contributing to the progress of this project, while maintaining the objectives of reducing the means and time of production. To do this we turn to theories of design.

Various theories of design and innovation emerge in the industrial context. The project we are going to study is similar to Single-Task Construction Robots (STCR). Refer to the book Robot-Oriented Design [1]: it presents ways to manage innovation in the field of building robotics. However, we do not have fundamentally innovative purpose. Moreover this field has already been explored, and we do not intend to question it in this paper. The TRIZ method also provides innovative design solutions often assimilated to artificial creativity [2]. We are not only interested by the emulation of design. We wish to reconstruct our design process, what TRIZ method does not seem to allow. We can also mention Axiomatic design [3], which was implicitly applied throughout the project. In these conditions, retranscribing its principles does not easily allow us to have the necessary perspective on the project. Finally, we can cite Architecturology. It is not actually a method, but a theory of architectural design. We used it in a previous article focusing on the importance of measurement in robotic and architectural design process [4]. This led us to make assumptions about some characteristic in architectural design to initiate construction robotics projects.

C-K theory is a design theory that is both a "design

theory and a theory of the design reasoning" [5]. It presents itself as a tool capable of modeling and analyzing a design process. This element naturally leads us to this theory, since our study occurs at the end of the design process. We will describe our design process then transcribe it with the C-K theory. We will focus on the analysis and hypothesis that allow us to produce a fast and low cost method for the management of STCR development.

2 Method

We will present various meaningful elements of the design process leading up to the production of the prototype (Figure 1). Its chronology, its key points, means and expertise involved.

Figure 1. Picture of the drilling robot prototype

This descriptive work will be joined to a modeling of the design process based on the C-K theory. We will try to answer several questions: How does the project start? How is the design oriented? What are the means implemented? How to manage them? What is the design mechanics during prototyping?

Modeling of the design process resulted in an analysis of the STCR prototyping project. This will be the basis of proposals for a method adapted to our objectives. We assumed that the management of disruptions, most acquired knowledge, and looking for the most reachable concepts, are compounds of the solution. As mentioned earlier, the fact that the C-K theory presents itself as both a method of analysis and a method of project directs us towards this theory rather than another. To understand it, we will develop its operation in the following sections.

2.1 C-K Theory Basics

C-K theory is a theory of innovative design. It is used by designers or group of designers in various fields of industry and services. Here is a short list [5] of its applications:

- Management of creativity
- Retroconception
- Management of an innovation field
- Entrepreneurial speech
- Project management
- Prospection strategy designing
- Driving research and knowledge management
- Commercial oversight of an innovation line

The C-K theory assumes that the design process is generated by the co-expansion of two spaces. The first one called C represents the space of concepts while K represents the space of knowledge [6]. The relationship between these two spaces is defined on the following diagram (Figure 2):

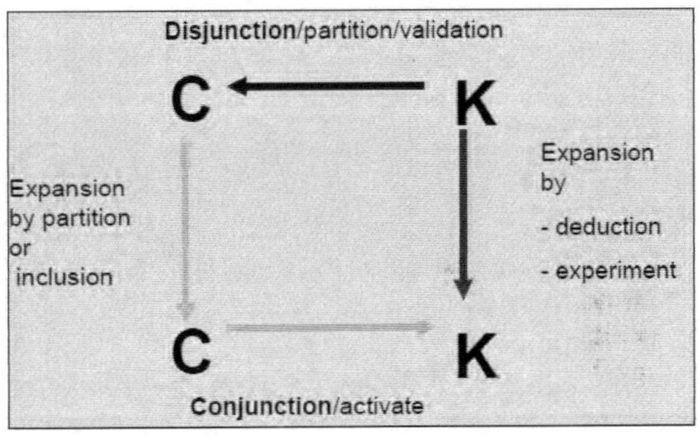

Figure 2. The design square (taken from [6])

We can identify the simplified mechanics of operators between and within the two spaces. This point is important in our study as it allows us to identify the operations sequences of our design process. In addition, the C-K theory makes also possible to arrange knowledge and concepts as a diagram, shown in Figure 3.We identify here the expansions in the K space describing all the propositions that have a logical status (which means the degree of confidence assigned by a designer). And expansions in space C define as all the propositions that have no logical status described by the designer [5]. We must also notice the concept degrees of expansion (known, reachable, alternative), and the knowledge degrees of robustness (confirmed, in progress, missing). Here we have the basic elements necessary for the construction of our analysis. We will see afterward how we implement them.

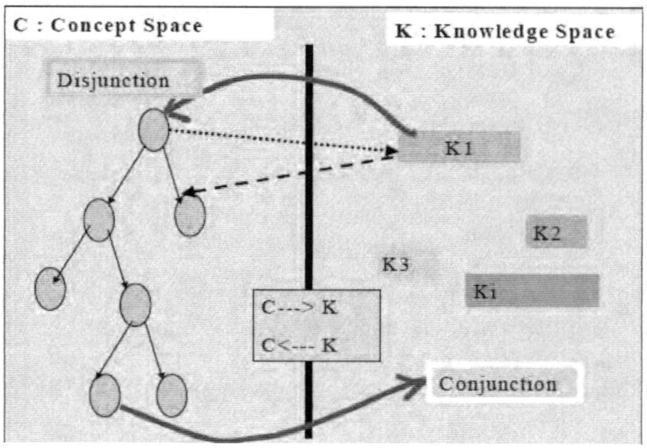

Figure 3. C-K dynamics (taken from [6])

2.2 The C-K theory as a method of analysis

As previously announced we have carried out a transcription work of our STCR design case, thanks to the C-K theory. This allows us to make variety of diagnosis and to determine relevant management methods for the development of STCR according to our criteria of development. Important point for this research: We intend to produce methods for development management instead of innovation management. C-K theory evokes different forms of design and innovation [7]. For example innovative and rule-based design comparisons, object identity disruptions or dominant design description. Knowing how to specify a design process in relation to these cases gives us clues for the management of a project.C-K theory allows us to understand the design beyond the scope of innovation [8]. We can observe it in the following example (Figure 4).

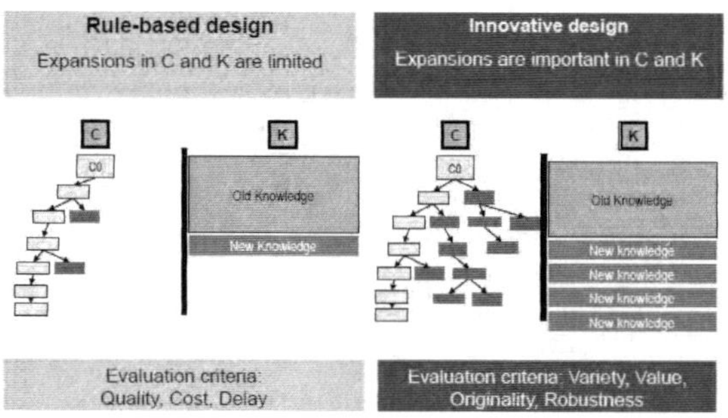

Figure 4. Innovative project evaluation: contrasting product development and innovative design (taken from [8])

This modeling of two design processes synthesizes the fundamental difference between a development project and an innovative project. This difference is identified by the amount of new concepts or knowledge implemented during the design process.

We have no needs to go further in understanding the C-K theory for this article, but we suggest you to get a look to the specialized literature on the subject if you wish to do so.

3 Description of the design process

The project was carried out in three distinguishable steps involving different expertise. The first step involved only construction site specialists and support. The second step was dedicated to reflections, modeling and proposals in partnership with a robotics company. The last step was the design and production of a prototype inside the construction company. By studying these three steps, we will be able to explore key points of the design process. As we explain, we will focus here on the description of the design process, and put it in line with the C-K theory.

3.1 First step : Specifications drafting

3.1.1 Process descriptions

The project was initiated by the construction site management. A statement related to the difficulties of the drilling tasks in concrete, forced them to find solutions to assist or replace manual labor.

They began the research by collecting knowledge about existing tools close to the problem. The idea of automation was implied in the reflection, but it was arduous to consider by the lack of examples (Figure 5).

Figure 5. Images used as references by the construction site experts to expose the concept

To assist them in this information-gathering work, they called the company's Materials and R&D units for help. In the first stage, their support was provided by complementary knowledge on the existing tools to be able to address the problem ([9] and Figure 6).

The absence of solutions coinciding with the problematic provoked the decision to launch a project for the realization of a suitable tool. The exploration did not stop and continued until the end of the process. Nevertheless, the design work proceeded. The first characteristics of the tool to be designed were defined as: A robotic system able to assist or automate drilling tasks in concrete.

Figure 6. Existing drilling STCR developed in France by general contractor Bouygues Construction.

Yet, the lack of knowledge in robotics within the company could have restrained them to carry out the development of the tool. The rest of the design work had to be assigned to a robotic integrator.

3.1.2 The design process in a C-K diagram: Step 1

We deliberately reduce the description of knowledge fields and also the intermediate interactions, to make it easier for the readout. We apply the color code which characterizes the degree of concepts and knowledge appropriation (Figure 7). We represent the starting point of the project, and then the paths from one space to the other and also each state that are induced by these operations.

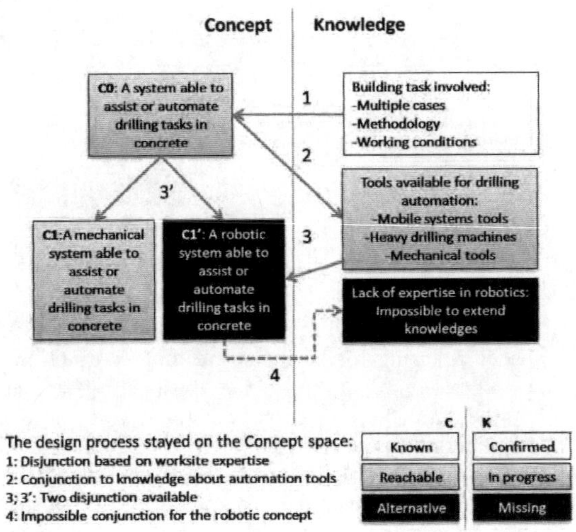

Figure 7. Step 1: Design process modeling with C-K Theory

We notice that the relationship between knowledge and concept is dynamic. There are few expansions taking place within the same space. For example: "building task" expansion going from "working conditions," "multiple cases" then "methodology"; Or even "Tools" expansion going from "mechanical systems" to "automated systems".

We can also analyze the fact that the design process does not start in the space of concepts. The knowledge space induces the different concepts. But the orientation towards a robotic project is a choice towards other concepts. The discarded solutions and the additions to the space of concepts are also identified. This choice is nevertheless constrained and provokes a blocking situation caused by a lack of knowledge in the field of mechatronics. We referred previously in the article. According from the description above, we can see that the C-K model helps us to prove the constraint source.

3.2 Second step : Design mission to a robotics integrator

3.2.1 Process descriptions

The project was assigned to a specialized company in industrial robotics design. The staffs have knowledge within their field, but not in the field of construction. To solve this, we have sent them standard documents (Figure 8) for the management of all construction sites' aspects (methods, safety, planning, etc.).

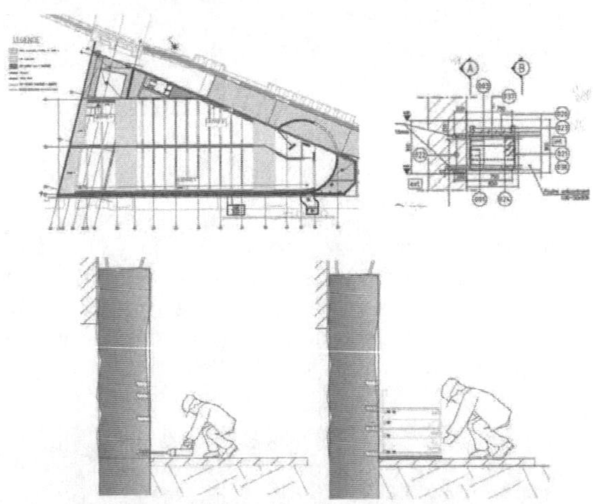

Figure 8. Examples of construction methods documentation

To ensure the spread of the construction site knowledge, we had to complete this approach by visiting the site and observing the task conditions. Despite this, the observation seemed not enough. Industrial robotics integrators are used to work on well-known and modeled environments. We have therefore added digital models presenting some of the cases to be treated on our site (Figure 9).

Figure 9. Construction site case digital modeling

The robotic integrators responded through multiple proposals of hybrid systems made of mastered techniques assembly and questioning approach to identify the whole problematic (Figure 10).

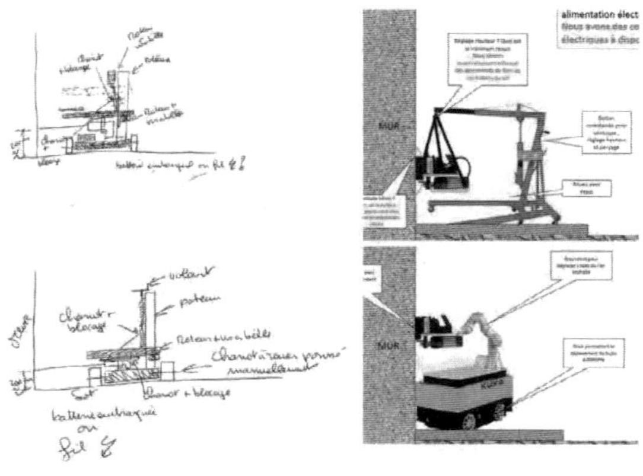

Figure 10. Robotic proposals stage 1

These initial proposals clearly display the design process. But they did not succeed to fix a state of satisfying concepts in regard to the construction company expectations. A second step of the proposals tried to approximate them by increasing the modeling degree of the technical solutions (Figure 11).

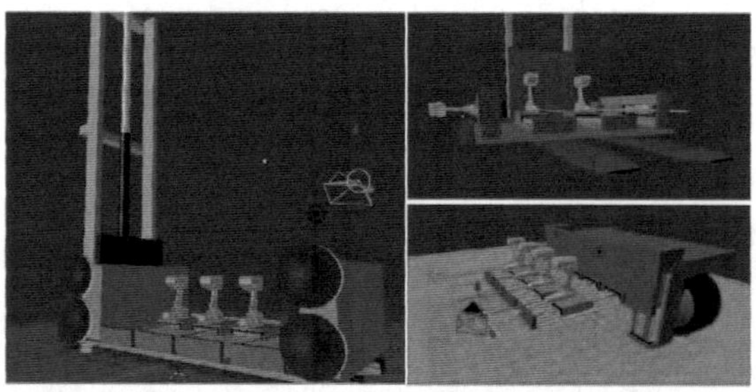

Figure 11. Robotic proposals stage 2

However, there were no viable responses emerging from these proposals. None of the proposals met the specifications. Many aspects seemed to be complicated to manage, such as cost and time of development, but also knowledge related to the general organization of a construction project (traditionally taken into account in other industry sector). The design process was not successful at this stage because none of the proposals seemed to fit.

3.2.2 The design process in a C-K diagram: Step 2

Here we can transcribe the process in a C-K diagram (Figure 12). The same rules are applied as for the previous diagram. We restart the process from the concept C1' that was fixed in the step 1, and the expansion in K that is necessary to the implementation of transmissible specifications to the robotics expert.

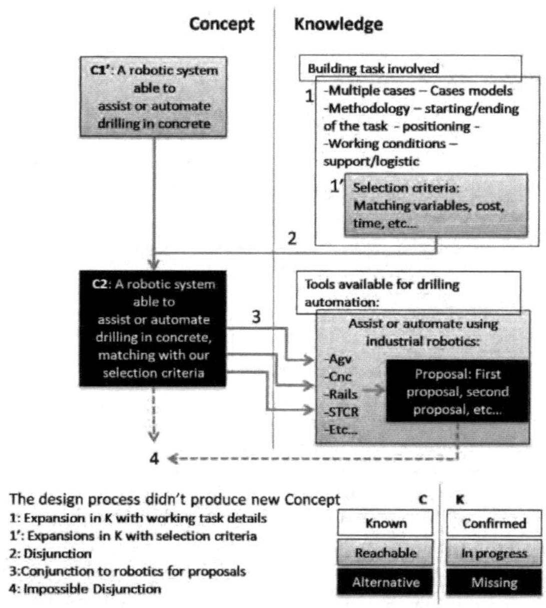

Figure 12. Step 2: Design process modeling with C-K Theory

We observe the interaction inside the knowledge spaces referring to the tools and the construction task concerned. It produces a C2 expansion of concept bounded to the selection criteria drawn from the specifications. We can observe in C-K that none of the proposal appears to be a solution as no concept is created at the end of the design process. Proposals are only knowledge exploration and not new concepts expansions.

We have seen in step 1 that an obstruction in the knowledge space has interrupted the design process. We can see here that the concept space is equally important for the advancement of the design process. Identifying the origin of the interruption could be one way to ensure the continuity of the project. Nonetheless, we find that the design process does not really pick up where it stops. Expansion of knowledge about the site is required. This result brings us not to confuse the state of common knowledge enlisted during the project and those of each participant.

3.3 Third step: Prototyping

The last step is about the production of a prototype. This step took place within the construction company, aware of mastering its needs to the best of its ability. To avoid the problem of lack in mechatronics design knowledge, it relied on specialized human resources for this kind of project. We cannot describe here all the design details, so we will focus on the design choices and the validation methods used.

3.3.1 Design choices

Prior to the production of the prototype, we proceeded to a sequence of choices throughout the designs. These were of course directed by the specifications, but also by the exploration of accessible knowledge in mechatronic design. We can list them as:

- Mechanical choices by various assembly solutions.
- Actuator system choices. Combinations of axes such as Cartesian reference.
- Mobility system choices.
- Choice of control, interface, operations - implementation of communication systems.
- Choice of the end effectors - (we stated on a commercial drill in order to avoid the design of this part.)
- Choice of automation guidance (not realized). Uses of distance sensors - direct control of the drill.

The quotation order does not indicate the going back and forth in the design. Moreover, most of these choices were based on validation methods which we present in the remainder of this article.

3.3.2 Design hypothesis validation methods

Numerous validation methods have been deployed during design and during the production.

- Simulation and Modeling
- Real Movement Tests
- Tests for handling and operation on site

The same assembly can lead us to various means of validations. We had to request for simulations and modeling. In the following case, we have a representation of a simulation process for the mechanical forces calculations and a kinematic modeling for the whole system (Figure 13).

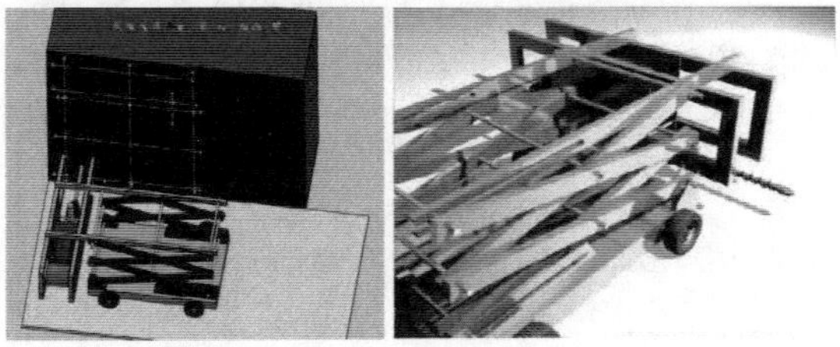

Figure 13. Simulation (left) and Kinematic model (right) of a single state of assembly

Assemblies can also give endless combinations. Some are not mechanically viable while others are too expensive or not optimized. It was necessary to carry out various simulations of complete or partial assemblies (Figure 14).

Figure 14. Assembly variation and test

Finally, we carried out relevant or operational environment tests. Designing combined axes and interfaces allowed us to test their proper functioning independently (Figure 15).

Figure 15. On-site and Off-site prototyping tests

Beyond the results of the project readiness level given by this kind of tests, we can ask ourselves the question of their relationship within the design process. We will figure out with the C-K transcription.

3.3.3 The design process in a C-K diagram: Step 3

As the previous steps, we can represent the design process in C and K (Figure 16). We resume the design process from the concept C2, as the previous concepts have been fixed.

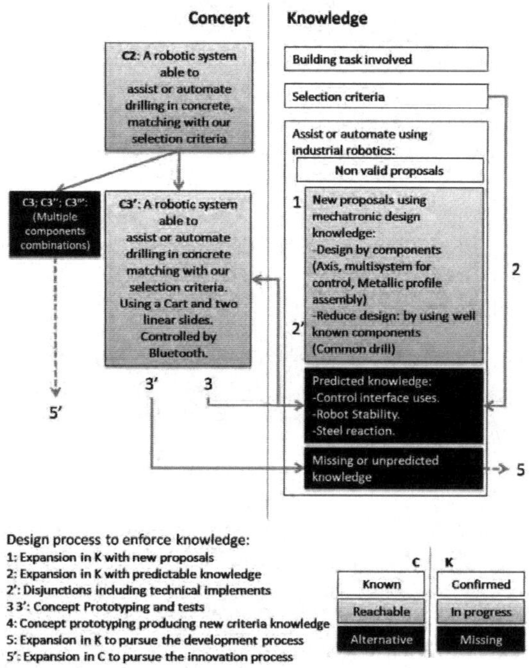

Figure 16. Step 2: Design process modeling with C-K Theory

Including all tests, modeling and simulations on the same diagram can be complicated. However, we can focus on the relationship between knowledge and the multiple concepts created. We observe that the design process takes place in continuity of the previous step. C-K theory can explain this fact by the prototyping dynamic [7]. This dynamic is complex and involves equally the C and the K spaces. Each choice of development initiated in K causes an expansion in C. The prototyping of a solution does not erase the concepts allowed by other assembly choices. On the other hand, prototyping makes it possible to act significantly on the expansion of knowledge. To go further in the analysis we could identify the restrictive characteristics of a STCR project to avoid getting out of this line of concepts. This could be done through diagrams focused on the knowledge space. This work has not yet been carried out, but we plan to do so in the continuity of this research.

We will not specify the choices management in detail, but we can say that the project is driven towards the best mastered knowledge, and the most attainable concepts. We can conclude here with the elements that would allow us to extend the project. We can orient ourselves on expansions derived from not yet prototyped concepts. This can be done by integrating knowledge produced during the first prototype validation process. We can also produce knowledge expansions, in order to extend the first prototypical concept line.

4 Conclusion : Towards a generic design method for the fast development of STCRs

Thus, we have been able to transcribe the design process of a STCR project in a descriptive way, and to analyze it from a transcription of the C-K model. We have discussed the conclusions we can make for each of the steps, and we will summarize them.

We wished to know what allowed us to maintain the continuity of the design process. Ours conclusion are that we succeeded by involving expertise on the project objectives for the management of knowledge expansions. The objectives are conducted by selection criteria described in the knowledge space, but the continuity vector can be observed in the management of the concepts directed to produce prototypes. The expansions of knowledge within K spaces are conceivable, but they do not allow us to reach the prototyping step.

Then we posed the question of what allowed us to complete our objectives related to the reduction of means and time needed for the development. The analysis of the step 3 allowed us to answer this question by observing that the design choices were made according to the expansions confidence degree. We conclude that we could aim at the most reachable concepts, and the best mastered knowledge. This could form a guideline for further design process in a similar context.

These findings may concern other development projects than those of the STCR. But we can make assumptions about what determines this particular kind of project. It seems clear that in this case the knowledge structure around the fields of mechatronic technologies and production methods on construction site is specific of STCR development. In addition, what emerges from our analysis is that we can also specify it by the mastery of the concept expansions, restricting them to the STCR field. It answers to another question we introduced about what initiated the project. It resulted from a choice to explore this field according to the concepts expansions.

Finally, we observed that the CK theory could help us in analysis, as well as the management with C and K expansions. This allows us to reveal the knowledge produced during prototyping and to consider relevant paths toward other prototypes or even the completion of the project.

We pose the question of ways yet to be explored on this subject. We plan to produce more detailed modeling of the relationships between knowledge in the process. We also look at other STCR development projects to verify our assumptions. Other kind of analysis can be made from the C-K modeling that we produced. We do not treat them here because they do not fit with the problematic of this article. Nevertheless we can evoke them: the Risk management during the development or even collaboration management of various expertises during the design process. This research progress still need to be discussed and will be subject of numerous works.

References

[1] Bock T. and Linner T. *Robot-Oriented Design – Design and Management Tools for the Deployment of Automation and Robotics in Construction*, Cambridge university press, 2015.

[2] Mercier-Laurent E. Intelligence artificielle : peut-elle rebooster la créativité humaine ?.In *Bulletin de l'AFIA n° 78*, p 5-8, 2012.

[3] Suh, Nam P. *The principles of design*. Oxford University Press, 1990.

[4] Zedin T., Guéna F. and Marchand O. La mécatronique dans le processus de conception des outils de construction: La mesure au sein d'un continuum numérique. In *Séminaire de conception architecturale numérique : Mètre et paramètre, mesure et démesure du projet*,pages 265-274, Toulouse, France, 2016.

[5] Agogué M., Arnoux F., Brown I. and Hooge S. *Introduction à la conception innovante*. Presses des Mines, 2013.

[6] Hatchuel A. and Weil B. A new approach of innovative design: an introduction to C-K Theory. In *ICED'03*, page 14, Stockholm, Sweden, 2003.

[7] Le Masson P., Weil B. and Hatchuel A. *Théorie, methodes et organisations de la conception*. Presses des Mines, 2014.

[8] Hatchuel A., Le Masson P., Weil B., Agogué M., Kazakçi A. and Hooge S. Multiple forms of applications and impacts of design theory –ten years of industrial applications of C-K Theory. In *Impact of Design Research on Industrial Practices – Tolls, Technology and Training*, 2015.

[9] Publication committee International association for automation and robotics in construction. *Robots and Automated Machines in Construction,* 1998

A study on the Modeling of Independently Driven Quadruped Crawler System for Posture Control of Special Purpose Machine for Disaster Response

S.K. Lee[a] and J.W. Ha[b] , S.H. Kim[c], Y.S. Lee[c], D.I. Sun[a], C.S. Han[d]

[a]Department of Mechatronics Engineering, Hanyang University in Republic of Korea
[b]Department of Interdisciplinary Engineering, Hanyang University in Republic of Korea
[c]Graduate School of Hanyang University in Republic of Korea
[d]Department of Robot Engineering, Hanyang University in Republic of Korea
E-mail: daum2363@gmail.com, cshan@hanyang.ac.kr

Abstract

The crawler-leg mobile system is a sub-mechanism of special purpose machines for the early stage rescue and reconstruction in disaster. This system plays a role in controlling the posture to prevent rollover when the special purpose machine is running on rough terrain. In this paper, a mechanism is introduced for posture control of a crawler-leg mobile system and coordinates for mathematical model is define. Using defined coordinates, a mathematical model based on differential kinematics is proposed for posture control system. The validity of the mathematical model is verified by comparing the results of the proposed mathematical model and the multi body dynamics simulation model. And a mathematical model is proposed for posture control of the crawler-leg mobile system.

Keywords –

Independently Driven Quadruped Crawler; Posture Control; Special Purpose Machine; System Modeling; Crawler-Leg Mobile System

Introduction

Globally, occurrence frequencies of disasters such as conflagration, earthquake are increasing gradually [1].

Due to industrialization and urbanization, the number of large buildings and underground buildings are increasing, and in a major disaster, it can lead to the collapse of buildings and massive casualties. In the event of collapse of building, the rapid rescue of survivors buried in collapsed debris is very important for minimizing the loss of life after the disaster. However, the existing survivor rescue systems rely on rescue personnel. It is not easy to approach the rescuers due to constraints such as secondary collapse. And additional collapse accidents can increase the risk to both rescues and survivors [2].

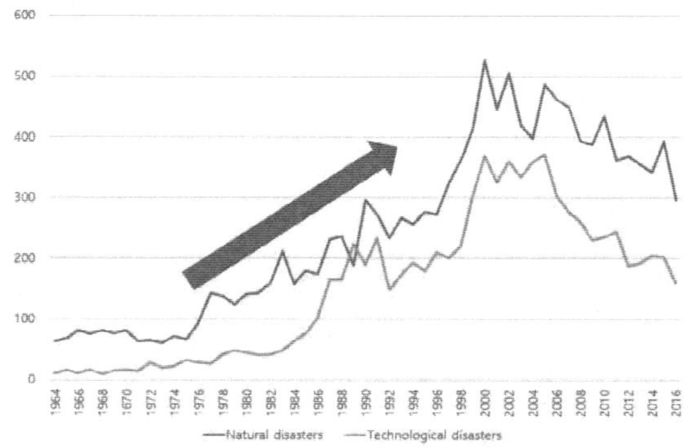

Fig. 1. International disaster trends (1964~2016)

For this reason, it is necessary to develop the equipment for securing access roads, rescue and preventing secondary collapse at disaster sites such as collapses, it is urgent to develop a system capable of traveling on the dangerous road.

Various systems have been proposed for rugged driving in disaster situations. In France, a robot with a combination of legs and wheels for rugged driving has been proposed. After developing a velocity model by analyzing the kinematics of the robot, researches have been carried out to improve the driving stability of the vehicle [3-7]. Nagano and Fujimoto of Yokohama University in Japan proposed an acceleration-based kinematic model for the movement and control of mobile robot combined with bridge and wheel, and developed an algorithm to generate robot control path of Least squares method of solving Singular configuration problem [9]. In addition, a control method to improve the moving performance by controlling the rigidity of the actuator attached to the leg of the compliant wheel-leg robot at the uneven road surface has been studied

[10].

As you saw above, most of the researches to improve the running performance of the rugged area have been focused on robots that light in weight and have a high degree of freedom. However, research on control techniques applied to huge driving systems such as disaster response equipment has not been conducted.

In this paper, I propose a mathematical model for posture control of a crawler-leg driving system. I introduce a disaster response system for mathematical model generation and propose a mathematical model of the crawler-leg system using the differential kinematic method. I compare the proposed mathematical model with the dynamics simulation model to verify the performance of the proposed model.

1. Introduction for Posture Control System of Special Purpose Machine

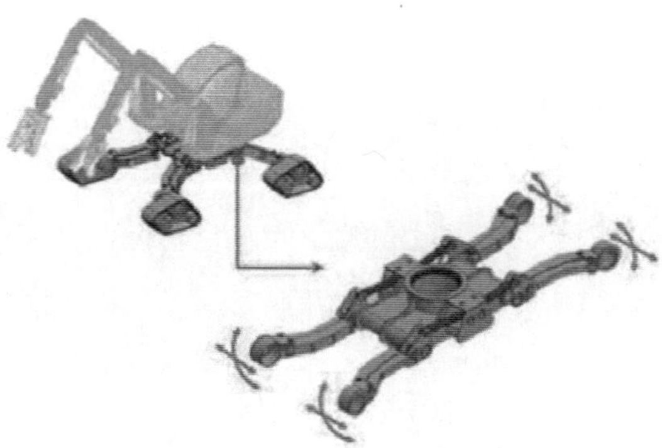

Fig. 2. Special Purpose Machine and posture control system

Fig. 2 is the draft of Special purpose machine. This system has four leg mechanisms to overcome the roughness such as irregular unevenness and attached to the lower part of the legs each can be independently driven crawler. The leg mechanism independently drives the four-posture control hydraulic cylinders mounted on the upperpart to implement the up and down movement of each leg and to drive the hydraulic cylinder acting as an outrigger inside the leg mechanism, do. The posture of the upper body can be controlled by using the up and down movement mechanism of the leg.

The Fig. 3 shows the relationship between the motion of the leg mechanism and the posture control of the traveling system. The independently actuation of the four cylinders mounted on the leg enables three motion.

1. (a) Up and down movement of the upper body
2. (b) Roll motion of the upper body
3. (c) Pitch motion

By driving four lower cylinders, the distance between the left and right legs is extended or reduced as shown in (d), thereby enabling stable support of the upper body, posture control, and movement of the narrow space.

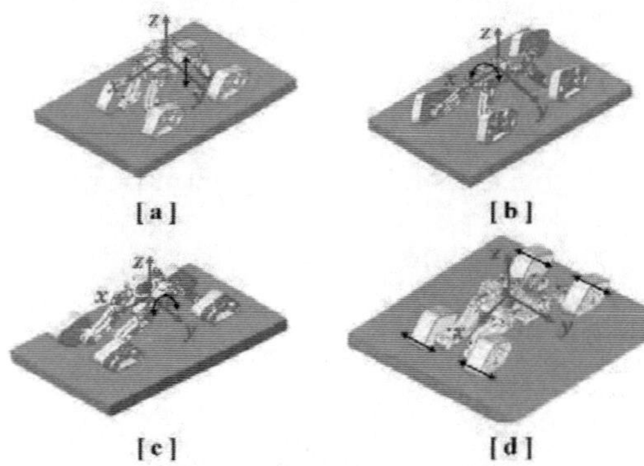

Fig. 3. Analysis of posture control mechanism

2. System Modeling for Posture Control of Special Purpose Machine

In this chapter, I propose differential kinematic model for posture control of special purpose machine.

2.1 Setting Coordinate for posture control modeling

Prior to setting differential kinematic model, the coordinate system setting must be preceded. Fig. 4 shows the schematic diagram of the system. The coordinate {A} is z_A global coordinate. z_A is a unit vector pointing in the opposite direction to the center of the Earth, and the outer product of x_A and y_A with a vector is a horizontal plane. The coordinate {B} is the platform coordinate. The origin of the platform coordinate system {B} is located at the center of gravity of the platform in the running system. x_B is a unit vector in the front direction of the platform, and y_B is a unit vector in the left direction of the platform. z_B is a unit vector in the direction of outer product with x_B and y_B.

Fig. 5 shows the i-th crawler. Assume that the crawler is in contact with the ground, the bottom of the crawler is the contact surface. The coordinate system $\{C_i\}$ is the coordinate i-th crawler.

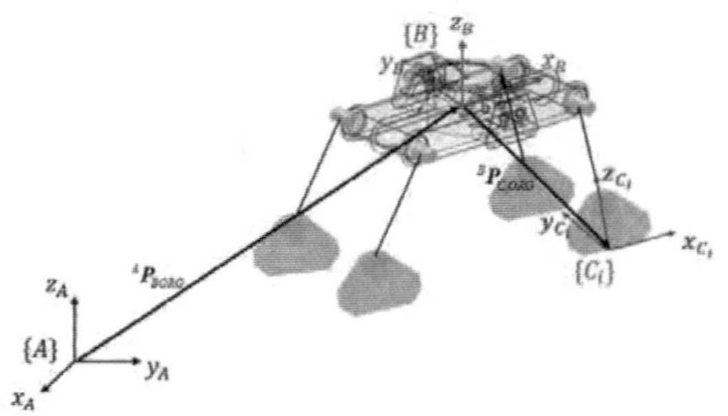

Fig. 4. The schematic diagram of the system

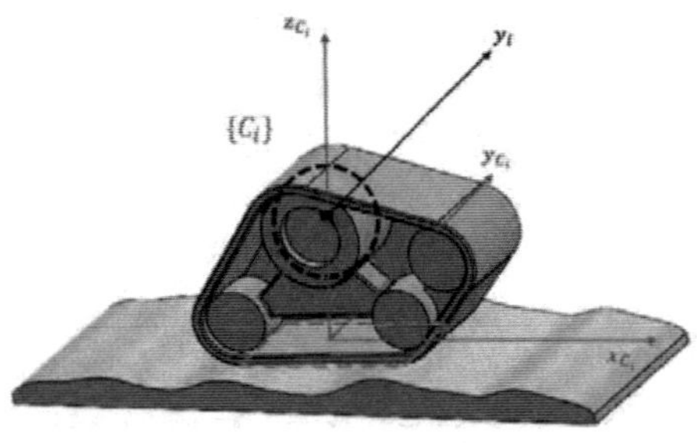

Fig. 5. The i-th crawler and contact plane between crawler and ground

2.2 Posture Control model based on differential kinematics

The origin velocity of $\{C_i\}$ for the coordinate system {A} is denoted by.

$$ {}^{A}V_{C_iORG} = {}^{A}V_{BORG} + {}^{A}\Omega_B \times {}^{A}_{B}R\,{}^{B}P_{C_iORG} + {}^{A}_{B}R\,{}^{B}V_{C_iORG} \qquad (1) $$

Or

$$ \begin{bmatrix} {}^{A}\dot{x}_{C_iORG} \\ {}^{A}\dot{y}_{C_iORG} \\ {}^{A}\dot{z}_{C_iORG} \end{bmatrix} = \begin{bmatrix} {}^{A}\dot{x}_{BORG} \\ {}^{A}\dot{y}_{BORG} \\ {}^{A}\dot{z}_{BORG} \end{bmatrix} + \begin{bmatrix} {}^{A}\omega_{xB} \\ {}^{A}\omega_{yB} \\ {}^{A}\omega_{zB} \end{bmatrix} \times \begin{bmatrix} x_B\bullet x_A & y_B\bullet x_A & z_B\bullet x_A \\ x_B\bullet y_A & y_B\bullet y_A & z_B\bullet y_A \\ x_B\bullet z_A & y_B\bullet z_A & z_B\bullet z_A \end{bmatrix} \begin{bmatrix} {}^{B}x_{BORG} \\ {}^{B}y_{BORG} \\ {}^{B}z_{BORG} \end{bmatrix} + \begin{bmatrix} x_B\bullet x_A & y_B\bullet x_A & z_B\bullet x_A \\ x_B\bullet y_A & y_B\bullet y_A & z_B\bullet y_A \\ x_B\bullet z_A & y_B\bullet z_A & z_B\bullet z_A \end{bmatrix} \begin{bmatrix} {}^{B}\dot{x}_{C_iORG} \\ {}^{B}\dot{y}_{C_iORG} \\ {}^{B}\dot{z}_{C_iORG} \end{bmatrix} \qquad (2) $$

${}^{A}V_{BORG}$ is the origin velocity of {B} for the coordinate system {A}. ${}^{A}\Omega_B$ is the angular velocity of {B} for {A}. ${}^{A}_{B}R$ is the rotation matrix of {B} for {A}. ${}^{B}P_{C_iORG}$ is the position vector of $\{C_i\}$ as seen from {B}. ${}^{B}V_{C_iORG}$ is the velocity of origin of $\{C_i\}$ from {B}.

Multiply ${}^{C_i}_{A}R$ that is a rotation matrix from {A} to $\{C_i\}$ both sides of equation (1).

$$ {}^{C_i}_{A}R({}^{A}V_{C_iORG}) = {}^{C_i}_{A}R({}^{A}V_{BORG}) - {}^{C_i}_{B}R[S({}^{B}P_{C_iORG})^{B}({}^{A}\Omega_B)] + {}^{C_i}_{B}R\,{}^{B}V_{C_iORG} \qquad (3) $$

The velocity of origin of $\{C_i\}$ from {B} of equation (3) can be expressed by product of Jacobian matrix J_i and angular velocity of joint in Fig. 6. So, we can get equation (4).

$$ {}^{C_i}_{A}R({}^{A}V_{C_iORG}) = {}^{C_i}_{A}R({}^{A}V_{BORG}) - {}^{C_i}_{B}R[S({}^{B}P_{C_iORG})^{B}({}^{A}\Omega_B)] + {}^{C_i}_{B}RJ_i\dot{q}_i \qquad (4) $$

$$ {}^{C_i}({}^{A}V_{C_iORG}) = L_i\dot{X} + J_i^{*}\dot{q}_i \qquad (5) $$

where
$$ L_i = [{}^{C_i}_{B}R \quad -{}^{C_i}_{B}RS({}^{B}P_{C_iORG})] $$
$$ \dot{X} = [{}^{B}({}^{A}V_{BORG}) \quad {}^{B}({}^{A}\Omega_B)]^T $$
$$ J_i^{*} = {}^{C_i}_{B}RJ_i $$

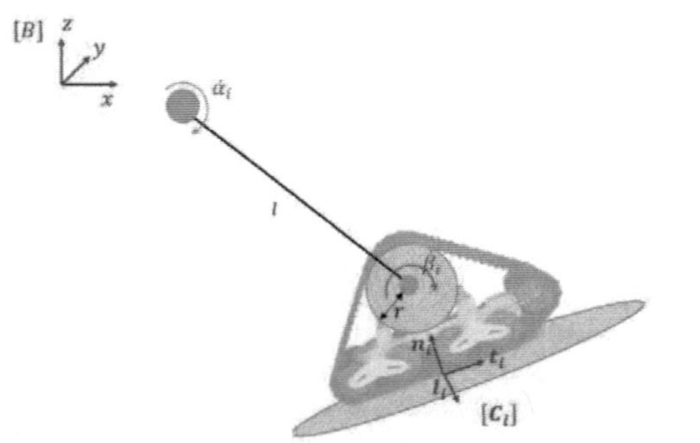

Fig. 6. The schematic diagram of one leg

Assume that there is no slip between crawler and ground surface. Equation (5) changes to equation (6).

$$ L_i\dot{X} + J_i^{*}\dot{q}_i = 0 \qquad (6) $$

Since the motion of each leg is coupled with linear and angular velocity, the whole model of the system is as follows

$$\begin{bmatrix} L_1 \\ L_2 \\ L_3 \\ L_4 \end{bmatrix} \dot{X} + \begin{bmatrix} J_1^* & & & \\ & J_2^* & & \\ & & J_3^* & \\ & & & J_4^* \end{bmatrix} \begin{bmatrix} \dot{q}_1 \\ \dot{q}_2 \\ \dot{q}_3 \\ \dot{q}_4 \end{bmatrix} = 0 \quad (7)$$

2.3 Jacobian Matrix

Second term of equation (6) is divided into two velocity components as follow

$$J_i^* \dot{q}_i = {}^{C_i}({}^{B}\mathrm{V}_{\dot{\alpha}_i}) + {}^{C_i}({}^{B}\mathrm{V}_{\dot{\beta}_i}) \quad (8)$$

${}^{B}\mathrm{V}_{\dot{\alpha}_i}$ is a component related with $\dot{\alpha}_i$ that is joint angular velocity of ${}^{A}V_{BORG}$. And ${}^{B}\mathrm{V}_{\dot{\beta}_i}$ is related with $\dot{\beta}_i$.

$${}^{C_i}({}^{B}\mathrm{V}_{\dot{\alpha}_i}) = {}^{C_i}_{B}\mathrm{R}\mathrm{J}_{\dot{\alpha}_i}(\dot{\alpha}_i) \quad (9)$$

$${}^{C_i}({}^{B}\mathrm{V}_{\dot{\beta}_i}) = \dot{\beta}_i(-r \quad 0 \quad 0)^T \quad (10)$$

$\mathrm{J}_{\dot{\alpha}_i}$ is a Jacobian matrix of ${}^{B}\mathrm{V}_{\dot{\alpha}_i}$, and r is a radius of sprocket.

Therefore, equation (8) can be change as

$$J_i^* \dot{q}_i = {}^{C_i}_{B}\mathrm{R}\mathrm{J}_{\dot{\alpha}_i}(\dot{\alpha}_i) + \dot{\beta}_i(-r \quad 0 \quad 0)^T \quad (11)$$

Or

$$J_i^* \dot{q}_i = \begin{bmatrix} {}^{C_i}_{B}\mathrm{R} \begin{bmatrix} -l\sin\alpha_i \\ \\ -l\cos\alpha_i \end{bmatrix} & \begin{bmatrix} -r \\ \\ \\ \end{bmatrix} \end{bmatrix} \begin{bmatrix} \dot{\alpha}_i \\ \dot{\beta}_i \end{bmatrix} \quad (12)$$

l is the link parameter between the joint and the crawler sprocket, and rotation matrix from {C$_i$} to {B} is as follows

$${}^{C_i}_{B}\mathrm{R} = \begin{bmatrix} x_B \bullet x_C & y_B \bullet x_C & z_B \bullet x_C \\ x_B \bullet y_C & y_B \bullet y_C & z_B \bullet y_C \\ x_B \bullet z_C & y_B \bullet z_C & z_B \bullet z_C \end{bmatrix} \quad (13)$$

3 Verifying Simulation

In order to verify the proposed mathematical model, I simulated with DAFUL of "Virtual Motion" and MATLAB/Simulink, which can analyzc multibody dynamics. The simulation conditions were as shown in Fig. 7, and the angular velocity of the leg connecting part and the crawler speed value of Fig. 10 were input to the multibody dynamics software and the proposed mathematical model as shown in Fig. 9. And outputs the state value of the system. The simulation was performed once for each of the pitch motion and the roll motion, and the results analyzed in the multi-body dynamics and the results in the proposed mathematical model.

Fig. 7. Simulation Condition (Plat Road)

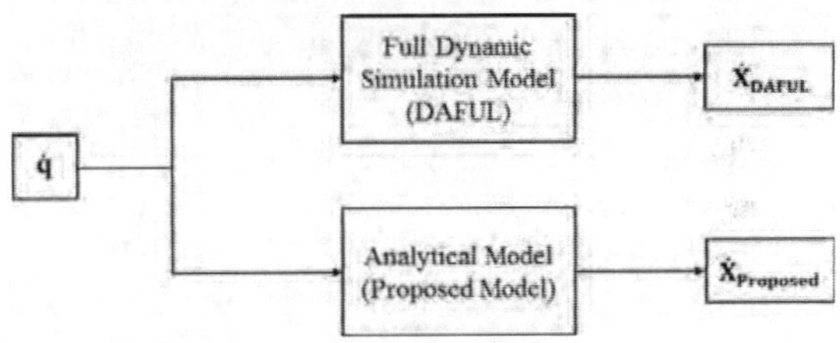

Fig. 8. Simulation Configuration

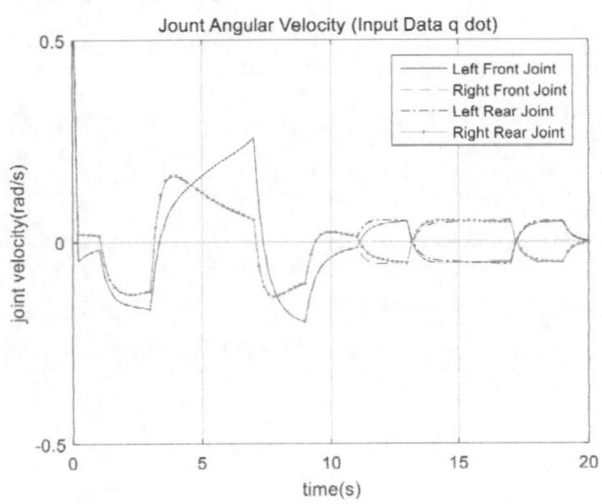

Fig. 9. Input Data Joint Angular velocity

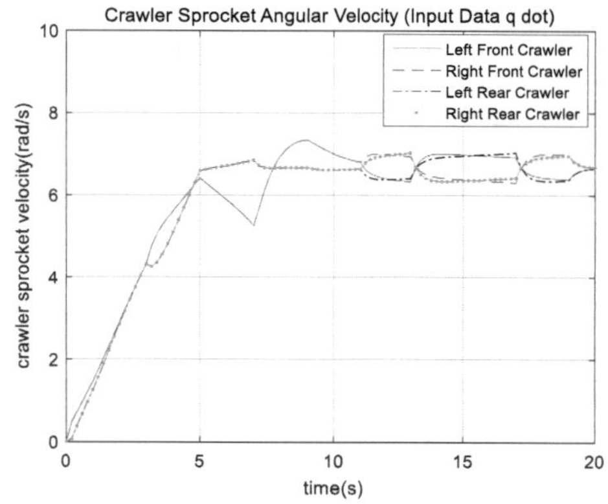

Fig. 10. Input Data Sprocket Angular velocity

Fig. 11 and Fig. 12, Fig. 13 denotes velocity values of the traveling system in the x direction, the y direction and the z direction, respectively. As shown in the graph, the velocity values in each direction analyzed from the proposed mathematical model can be confirmed to be similar to the simulation results from the multibody dynamics analysis.

Fig. 14 and Fig. 15, Fig. 16 denotes an angular velocity in the roll direction of the running system, an angular velocity in the pitch direction, and an angular velocity in the yaw direction. The roll, pitch, and yaw rate values interpreted from the proposed mathematical model are consistent with the simulation results from multibody dynamics analysis, confirming the usefulness of the proposed mathematical model.

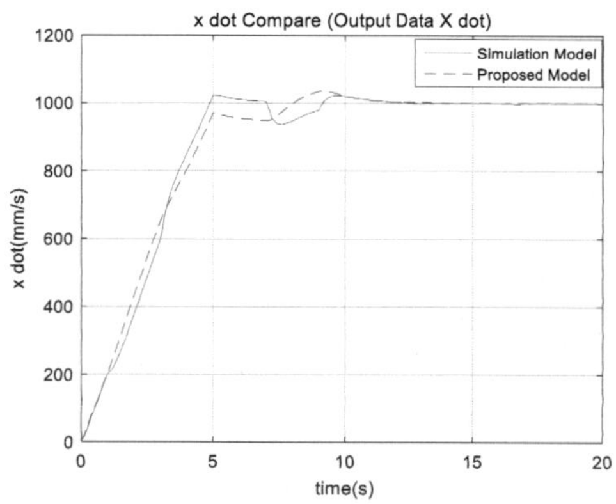

Fig. 11. Result of x dot state

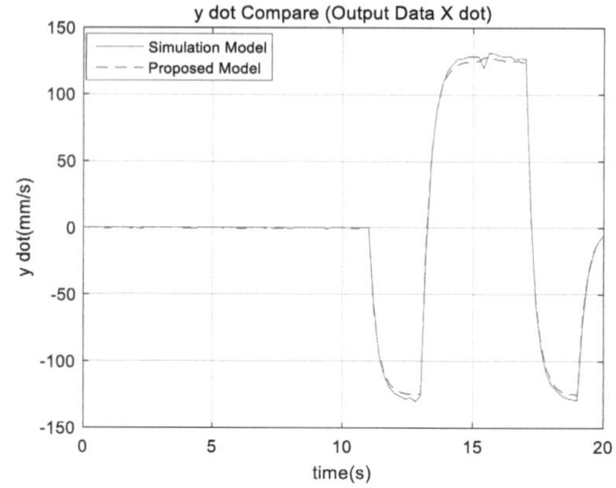

Fig. 12. Result of y dot

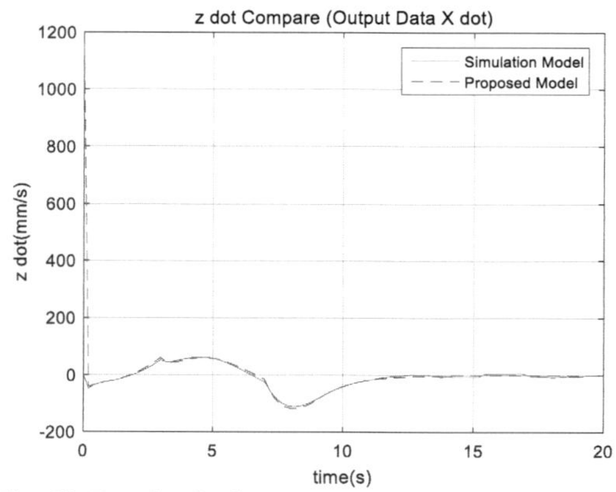

Fig. 13. Result of z dot

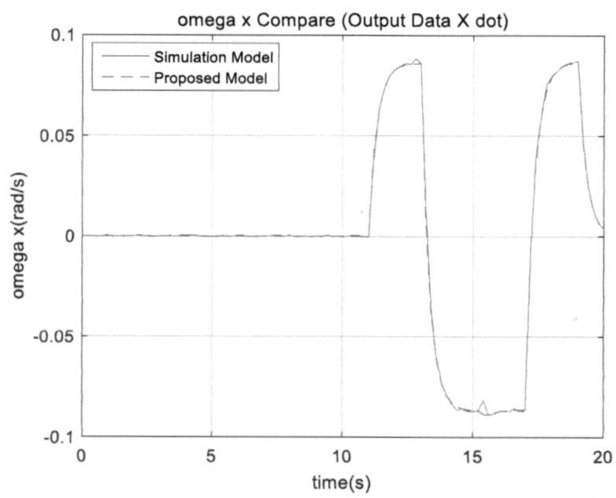

Fig. 14. Result of Ω_x

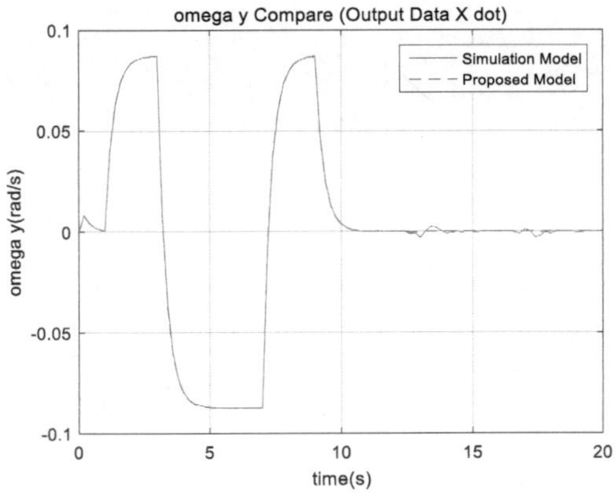

Fig. 15. Result of Ω_y

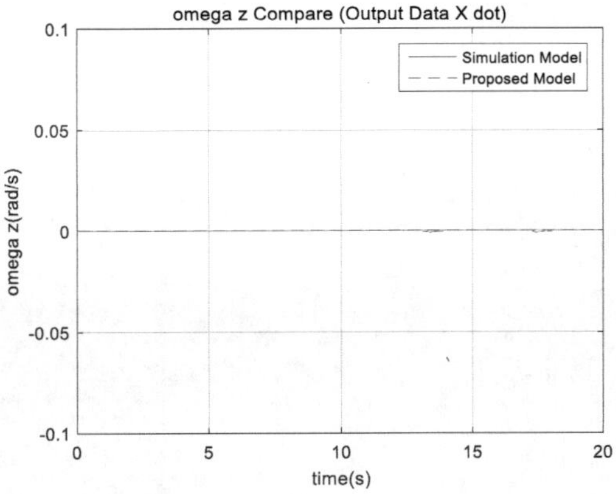

Fig. 16. Result of Ω_z

5 Conclusion

We have developed a mathematical model of a single Crawler-Leg system using a differential kinematic method and proposed a mathematical model for an entire system from a mathematical model for a single system. By comparing the proposed mathematical model with the results of the multibody dynamics simulation model, it is confirmed that the proposed model is useful.

However, it has limitations because it is a simulation that does not consider the realistic part such as characteristics of the hydraulic system etc. After this conference, we plan to apply algorithms to the real machine by complementing these defects.

Acknowledgment

This research was supported by the Industrial Strategic technology development program (No.10052965) funded by the Ministry of Trade, Industry and Energy (MOTIE), KOREA and was supported by a grant(14SCIP-B079344-01) from Smart Civil Infrastructure Research Program funded by Ministry of Land, Infrastructure and Transport(MOLIT) of Korea government and Korea Agency for Infrastructure Technology Advancement(KAIA).

References

[1] Retrieved December 19th, 2016, EM-DAT, https://www.emdat.be/disaster_trends/index.html

[2] S. W. Lee, S. H. Kim and S. H. Jung, "Development Status about overcoming domestic disasters and life-saving robot technology," J. of the Korean Society for Robot and Human, Vol.4, No.2, pp.19-26, 2007.

[3] K. Turker, I. Sharf and M. Trentini, "Step Negotiation with Wheel Traction: A Strategy for a Wheel-legged Robot," *Robotics and Automation (ICRA), 2012 IEEE International Conference on*. IEEE, 2012.

[4] A. Alamdari, R. Herin and V. Krovi, "Quantitative kinematic performance comparison of reconfigurable leg-wheeled vehicles." 2013.

[5] R. C. Luo, M. Hsiao, T. W. Lin, "Erect wheel-legged stair climbing robot for indoor service applications." *2013 IEEE/RSJ International Conference on Intelligent Robots and Systems*. IEEE, 2013.

[6] J. M. Jehanno, A. Cully, C. Grand and J. B. Mouret, "Design of a Wheel-Legged Hexapod Robot for Creative Adaptation." *CLAWAR 17th International Conference on Climbing and Walking Robots*. 2014.

[7] C. Zheng, J. Liu, T. E. Grift, Z. Zhang, T. Sheng, J. Zhou, Y. Ma and M. Yin, "Design and analysis of a wheel-legged hybrid locomotion mechanism." Advances in Mechanical Engineering, Vol.7, No.11, 2015.

[8] C. Grand, F. BenAmar and F. Plumet, "Stability control of a wheel-legged mini-rover." *Proc. of CLAWAR*. Vol. 2. 2002.

[9] K. Nagano and Y. Fujimoto, "A control method of low speed wheeled locomotion for a wheel-legged mobile robot." *2014 13th IEEE International Workshop on Advanced Motion Control*. IEEE, 2014.

[10] A. Bouton, C. Grand and F. BenAmar, "Motion control of a compliant wheel-leg robot for rough terrain crossing." Robotics and Automation (ICRA), 2016 IEEE International

34th International Symposium
on Automation and Robotics
in Construction

ISARC 2017

June 28 – July 1
TAIPEI, TAIWAN

COMPUTER VISION

Deeper Networks for Pavement Crack Detection

Leo Pauly[a], Harriet Peel[a], Shan Luo[a], David Hogg[b] and Raul Fuentes[a]

[a]School of Civil Engineering, University of Leeds, United Kingdom
[b]School of Computing, University of Leeds, United Kingdom
E-mail: cnlp@leeds.ac.uk, cnhap@leeds.ac.uk, S.Luo@leeds.ac.uk, D.C.Hogg@leeds.ac.uk, R.Fuentes@leeds.ac.uk

Abstract – Pavement crack detection using computer vision techniques has been studied widely over the past several years. However, these techniques have faced several limitations when applied to real world situations due to for example changes of lightning conditions or variation in textures. But the recent advancements in the field of artificial neural networks, especially in deep learning, have paved a new way for applying computer vision methods to pavement crack detection. Even though deep learning has been used before for crack detection, the network used is rather shallow when compared to the current networks used for other applications. In this paper we demonstrate the effectiveness of using deeper networks in computer vision based pavement crack detection for improved accuracy. We also show how variations in location of training and testing datasets affect the performance of the deep learning based pavement crack detection method.

Keywords –
Pavement cracks; Detection; Deep learning; Convolutional neural networks;

1 Introduction

Pavement crack detection is one of the most important tasks that needs to be performed to ensure safe driving. The current non-computer vision methods involve the visual inspection of pavements by human workers. But this method is uneconomical, labor intensive, human error-prone, subjective and expert domain knowledge is required. Laser scanning based pavement crack detection methods have been proposed as a solution for this [1] and they have shown great promise by achieving high accuracies. But these methods are highly expensive and therefore not suitable to be deployed in a large scale.

Hence a lot of research has gone into use of economical computer vision based methods for automated detection. The traditional methods generally have two parts: the first part involves extracting a set of hand engineered features from images, and in the second part a classifier is used for classifying these features. Some of these computer vision based pavement crack detection methods include use of local binary patterns (LBP) [2] [3], tree structure based algorithms [4], Gabor filter based methods [5] and shape based algorithms [6]. But one main disadvantage is their inability to generalize the task of crack detection when exposed to real world conditions such as variation in lighting or change in pavement surface textures. That is, these methods fail to capture enough discriminative features from the images that can differentiate between cracked and non-cracked images even when the environmental conditions change.

But over the past few years a branch of artificial neural networks called deep learning has shown great potential in solving similar problems. The field of deep learning started gaining popularity in 2012 when Alex *et al.* demonstrated their deep network architecture named AlexNet [7] for image classification which outperformed all the existing methods with hand engineered features by a great margin. Since then the field of deep learning has witnessed exponential growth both in size (depth) of deep networks, e.g., VGG Net (19 layers) [8], GoogleNet (22 layers) [9], ResNet (152 layers) [10] as well as their application to several fields such as image classification [7], speech recognition [11], or image segmentation [12]. But this approach of using deep learning has not been studied well in context of pavement crack detection.

Applying deep learning to computer vision based automatic pavement crack detection can overcome most of the existing challenges. In [13] Lei *et.al* presented a particular deep learning architecture called deep convolutional neural networks (CNNs) for pavement crack detection using a four layer network. But a 4 layer network could only be considered as a shallow network when compared to the current networks used for other applications, which is a drawback for [13]. The ability of the convolutional networks to give high accurate results is highly dependent on the depth of the network as shown in [14]. Hence using deeper networks could improve the accuracy of crack detection. So in this paper we intend to demonstrate the capabilities of deep learning in pavement crack detection when deeper networks are used. We demonstrate that an increase in the number of layers leads to an increase in the accuracy of the network in detecting cracks. In addition to that we also show how the variations in the location for data collection of training dataset and testing dataset affect the performance of the network – i.e. how learning can be transferred to other locations.

Our contributions can hence be summarized as follows:

- We study how the depth (the number of layers) of the deep neural networks effects the performance of crack detection capability of the network.
- We study how the variations in location from which the training and testing data is collected affect the performance of the system.

2 Proposed methodology

The objective of the deep neural network used here is to take the input image patch of the pavement, process it and classify it into either of the two classes: crack or non-cracks. In this section the details of the datasets (training and testing) and the basic architecture of deep network architecture used in the study are explained.

2.1 Deep network architecture

This paper uses a particular deep learning architecture called deep convolutional neural networks (CNNs) for pavement crack detection. Like every other neural network, convolutional neural networks are also created by stacking several layers of neurons together. The higher the number of layers, the deeper the network will become. The primary layers used in a convolutional neural network are: convolutional layers, pooling layers, activation layers and fully connected layers. It also uses auxiliary dropout layers between the above mentioned layers to avoid overfitting of data [7]. Each of these layers are explained in detail in the following subsections.

2.1.1 Overall Structure

Fig. 3 illustrates the basic network architecture used in this study. It has 4 convolutional layers (marked as **Conv**), 4 max pooling layers (marked as **Maxp**), 2 fully connected layers (marked as **FC**), activation layers after every convolutional and fully connected layer and a softmax layer (marked as **Softmax**) at the end of network. It also uses an auxiliary dropout layer (marked as **Dropout**) between the fully connected layers to avoid overfitting.

2.1.2 Convolution layers

The convolutional layers are the major building blocks of a convolution neural network structure. It is the convolution layers that learns the features that are suitable for differentiating between a crack image and a non-crack image. The local features required for this are learned by initial convolutional layers whereas the deeper layers learns the global feature required for differentiating between cracks and non-cracks. . Each convolution layer performs the convolution operation in outputs of the previous layers using a set of kernels or filters called receptive fields. Fig. 1 illustrates an example for the operation of convolutional layers. The matrix in the left is the input to the convolutional layer, the matrix in the middle is the kernel and the matrix to the right is the output of the convolutional layer. The output is obtained by convolving the kernel over the input layer. The weights used in the kernels of these convolutional layers are learned during the training of the networks.

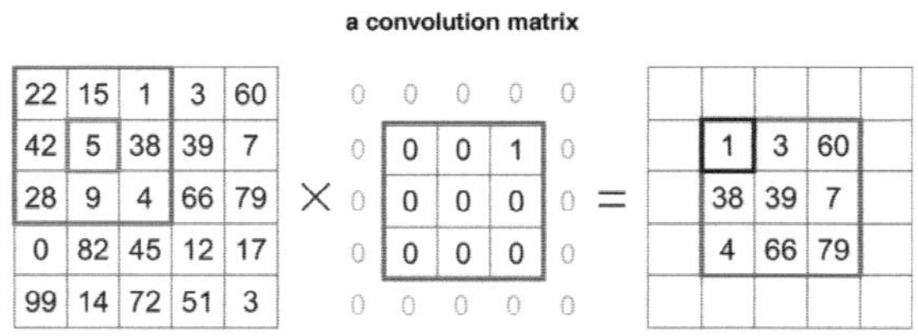

Figure 1: Convolutional operation

2.1.3 Pooling layers

The pooling layers are used for down-sampling of the input arrays. It performs down-sampling by dividing the input matrix into submatrices and selecting one value to represent each of the submatrix. There are two main types of pooling layers used: the max pooling and mean pooling layers. In max pooling the maximum value in the submatrix is taken to represent the submatrix where as in mean pooling the mean of the submatrix is taken. In [15] Scherer *et al.* has shown that max pooling is more effective than mean pooling in object classification tasks. Hence, we have used max pooling in our network. Fig. 2 illustrates the max pooling operation. The input matrix of size 4x4 is divided into 4 submatrices. Then the max value in each of the submatrix is taken and the output matrix is created using these values.

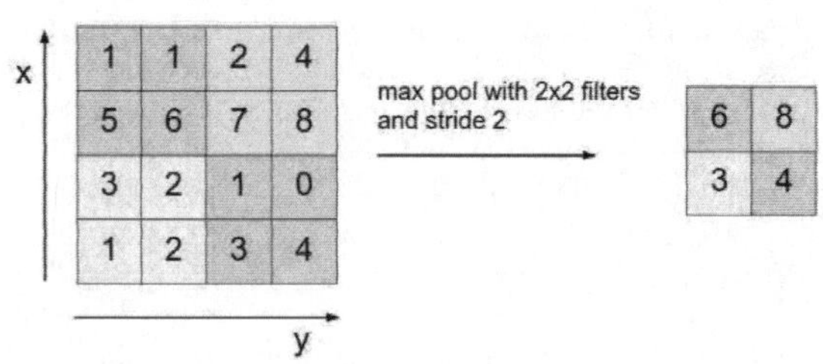

Figure 2: Max pooling

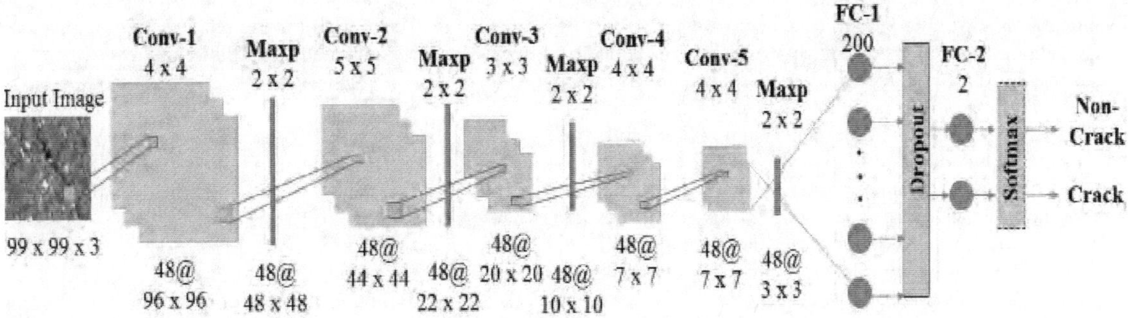

Figure 3: The structure of the Convolutional Network architecture used

2.1.4 Activation layers

Activation layers are another important building blocks of convolutional networks. The activation layers are used for giving non-linearity to the neural networks. The earlier neural networks used functions such as y=tanh(x) as activation functions, but in 2010 Nair *et al.* [16] introduced a very effective activation function called ReLU. Fig. 4 shows the ReLU activation function. One of its main advantages is that it has zero output when the input is negative and has the same value as input when the input is positive. And hence the gradients of the activation functions are always either 1 or 0 and this helps to avoid the problem of vanishing gradients [17] in deeper neural networks. The vanishing gradient problem occurs when the gradient of the activation function becomes smaller than what the neural networks can handle. In ReLU the gradient of the function will either be zero (when input is less than zero) or will sufficiently be big enough value (when input is greater than zero). Thus ReLU helps to avoid the problem of vanishing gradient caused by activation functions.

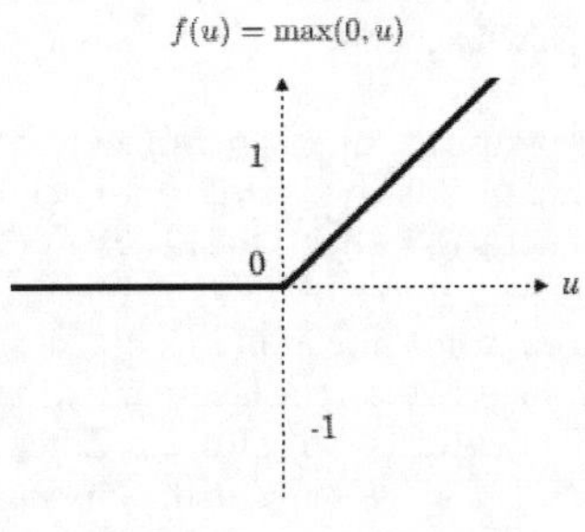

Figure 4: ReLU activation function

2.1.5 Softmax layer

Softmax layer is the layer located at the end of the convolutional neural network. This layer is responsible for predicting the probability of the input belonging to each of the two labels, i.e., crack or non-crack. This layer uses the softmax function for predicting the output. The layer gives the probabilities of, the input image patch belonging each of the two classes: i.e. crack and non-cracks.

2.1.6 Softmax loss function and Stochastic gradient descent

The main objective of training a convolutional neural network is to find a set weights for the layers which minimises a cost function or objective function. An optimization algorithm is generally employed for this purpose of finding the set of weights that optimises the objective function. Categorical cross entropy loss function is used as the objective function in the networks used in this study. It computes the categorical cross entropy between the targets and the predictions of the networks. The optimisation algorithm used here is the stochastic gradient descent (SGD) optimisation algorithm. In SGD optimisation algorithm the images are processed as small groups called batches rather than each image individually to reduce the computational cost. A batch size of 48 images are used for training networks in this paper. A learning rate of .0001 with decay of .0005 and momentum of 0.9 used, following the common practices used for training neural networks. A total of 80 and 40 epochs were used for training experiments 1 and 2 respectively, where epoch is defined as the number of times the network is trained using the entire set of training images.

2.2 Data collection

In this study we use the dataset collected by Lei *et al.* in [11]. This dataset consists of 500 RGB pavement images, each with a resolution of 3264 x 2448 pixels collected around the premises of Temple University,

USA using smart phones. Each image was then divided into patches of size 99x99 pixels creating a subset of RGB images. These patch images were then annotated by several annotators. Each image patch was labeled either as cracked patch or non-cracked patch. This process of annotation is explained in detail in [11]. In order to optimize the computational cost of training deep networks on large datasets only two subsets of this original dataset were used for training and testing purposes during experiments described in this paper. Fig. 5 illustrates a few sample crack and non-crack image patches taken from the dataset.

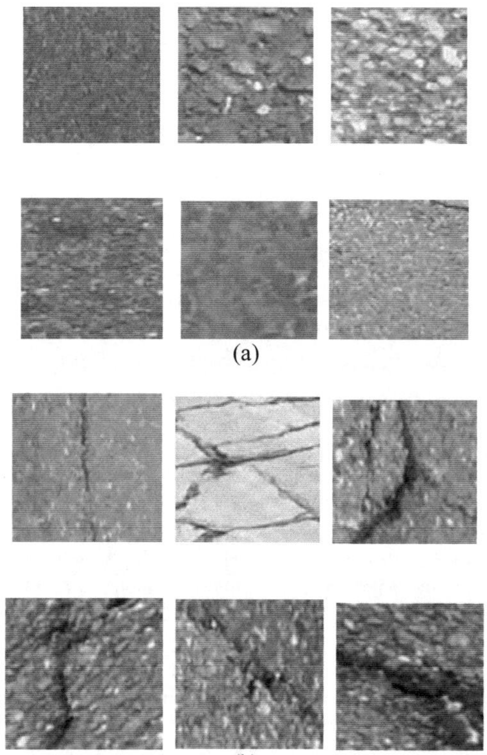

(a)

(b)

Fig 5: (a) Sample non-crack images from original dataset (b) Sample crack images from original dataset

2.2.1 Subset 1

Subset 1 has two sets of data: a training set and a testing set. The testing set consists of 100,000 RGB crack image patches and 100,000 RGB non-crack image patches totalling to a set of size 200,000 image patches which are randomly selected. The training set consists of another randomly selected 40,000 image patches consisting of 20,000 crack and non-crack patches. It was made sure that the training and testing sets are mutually exclusive. These training and testing datasets in subset 1 are used for experiment 1 (explained later).

2.2.2 Subset 2

Similar to subset 1, subset 2 also has two sets of data: a training set and a testing set. It has a training set of 20,000 image patches each, of cracks and non-cracks category totalling to 40,000 image patches selected by uniform sampling from the first 240,000 images in the original dataset [11]. The subset also has a testing set of 60,000 RGB image patches containing 30,000 cracked image patches and 30,000 non-crack image patches which is also selected from the original dataset of [11], but from a totally different location than that of the training set. The idea behind this is to have a testing set that contains images taken from a different location from that of the training set used for training the networks. This location change will introduce variations like changes in lightning conditions, nature of cracks and surface texture of pavements between the training and testing datasets. This subset 2 (training and testing sets) is used for experiment 2 (explained later).

3 Experiments and Discussions

The experimental evaluations were performed on two different hardware settings: one for training the network and the other testing the network performance. The training was done in a computer node at the High performance computing facility (HPC), University of Leeds. A total of up to 24 Broadwell E5-2650v4 @ 2.2 GHz CPU cores with each core given a memory of 3GB was used for training the network. The testing was performed in a desktop workstation with Intel (R) Xeon (R) E5-1630 v4 @ 3.70 GHz CPU with 128GB RAM and Nvidia Quodro M4000 GPU. The networks where created in Python with Keras deep learning library using Tensor flow backend.

Two different experiments were conducted. Experiment 1 was conducted to study how the increase in depth of networks affects the accuracy of crack detection. Experiment 2 studied how the crack detection accuracy is affected when the training and testing datasets are taken from two different locations. In the experiments, 2 different networks were used. The second network (Network 2) used is same as the base network illustrated in Fig 3 in section 2.1. The first network (Network 1) is the replica of the second network except that it has not got the additional fifth convolutional layer, marked as **conv-5** in Fig 3.

3.1.1 Experiment 1

The objective of this experiment was to study how well the crack detection methods perform when the number of layers of the network was increased. In this experiment the networks (1 and 2) were trained and tested using the subset 1 described in section 2.2.1. The results of the experiment are shown in Table 1 and Fig 6. The Fig 6 shows the number of True positives (TN) True negatives (TN), False positives (FP) and False negatives (FN) for networks 1 & 2 respectively. The true positives are the samples that are correctly classified as cracks and true negatives are samples that are correctly classified as non-cracks. Similarly, false positives are the samples that are not cracks but wrongly classified as cracks by the networks and false negatives are crack samples but wrongly classified as non-cracks by the networks. Table 1 shows the accuracy, precision and recall of the networks respectively. The recall can be understood as the percent of crack samples that are identified by the network out of the total number of cracks in the dataset. Whereas precision is the percent of predicted cracks that were actually cracks. The accuracy, precision and recall are calculated as follows:

$$\text{Recall} = \frac{\text{TP}}{\text{TP} + \text{FN}}$$

$$\text{Precision} = \frac{\text{TP}}{\text{TP} + \text{FP}}$$

$$\text{Accuracy} = \frac{\text{TP} + \text{TN}}{\text{TP} + \text{TN} + \text{FP} + \text{FN}}$$

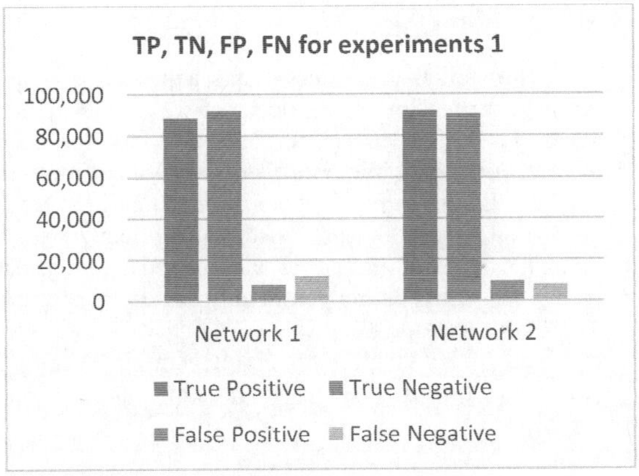

Figure 6: TP, TN, FP, and FN for experiments 1

It can be seen that as the number of layers are increased the accuracy and recall of the network increases. This is mainly because as the networks get deeper it learns more and more discriminative features from the images that helps the networks to differentiate the pavement cracks from non-crack images.

3.1.2 Experiment 2

In this experiment the networks (1 and 2) were trained and tested using the subset 2 described in section 2.2.2. In subset 2 the data distribution of the training dataset used for training the networks is different from the testing dataset. So the networks are tested on a totally different dataset taken from a different location from that of the training dataset. The objective of this experiment was to study how well the crack detection methods perform when the location of the training and testing sets are different. The results are shown in Table 1 and Figure 7. As in experiment 1, the Fig 7 shows the number of True positives (TN) True negatives (TN), False positives (FP) and False negatives (FN) for networks 1 and 2 respectively 1 and 2 respectively. The Table 1 shows the accuracy, precision and recall of the networks.

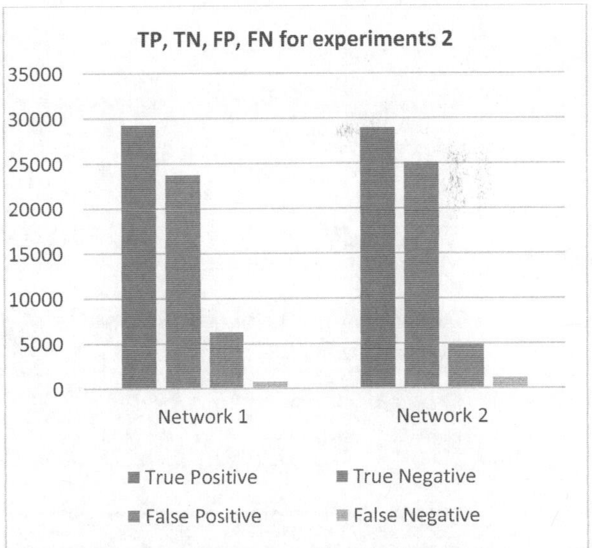

Figure 7: TP, TN, FP, and FN for experiments

Table 1: Accuracy, Precision and Recall of CNNs when tested on a dataset taken from a different location

	Network 1 (4 convolutional layers)		Network 2 (5 convolutional layers)	
	When tested on the random dataset (Experiment 1)	When tested on dataset taken from a different location to that of training dataset (Experiment 2)	When tested on the random dataset (Experiment 1)	When tested on dataset taken from a different location to that of training dataset (Experiment 2)
Accuracy	90.2%	87.9%	91.3%	90.1%
Precision	91.9%	81.8%	90.7%	85.6%
Recall	88.2%	97.5%	92.0%	96.4%

It can be seen that as the performance of the networks degrade as the location varies. This is mainly because of the variation in the background conditions from which the images in training datasets and testing datasets are collected. It can be inferred from the observations that the deep networks trained on training images from one location, e.g., London, may not work well when tested on images from another location, e.g., Leeds.

Finally in Fig 8 we illustrate a few samples of TN, TP, FP and FN cases for Network 2 in experiment 1.

(a: False Negatives)

(b: False Positives)

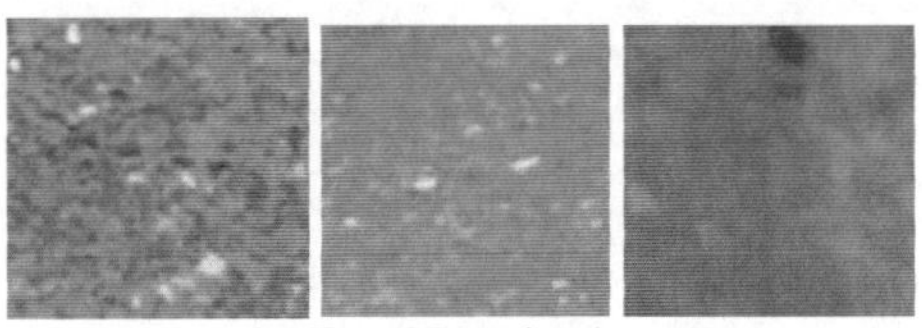

(c: True Negatives)

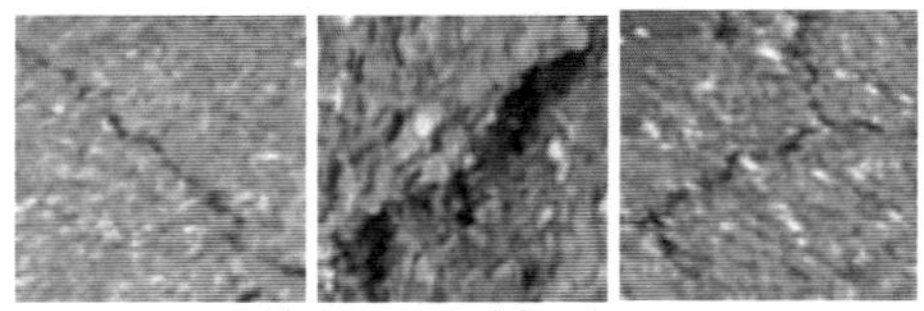

(d: True Positives)

Fig 8: Sample for FN, FP, TP, TN for Network2 obtained during experiment 1

It can be seen from Fig 8 that the samples which the network predicted wrongly (False Negatives and False Positives) were very ambiguous visually. Even a trained human worker might get them wrong during a visual inspection. The next stage of deep learning based pavement crack detection lies in training the deep networks in such a way that it can correctly classify even such ambiguous samples.

4 Conclusions

The studies presented in this paper shows that an increase in the depth of the deep networks leads to better performances in terms of accuracy and recall. The deeper the networks are, the more it learns about detecting cracks although a threshold has not been defined yet.

Also it could be concluded that the network trained on images taken from a particular location do not perform well when tested on images taken from another location. Therefore, location variance is a very important hurdle that has to be tackled for implementing a universal automatic crack detection system using computer vision techniques.

References

[1] Laurent, John, Jean François Hébert, and Mario Talbot. "Automated detection of sealed cracks using 2d and 3d road surface data."

[2] Y. Hu, C. Zhao, "A local binary pattern based

methods for pavement crack detection", *Journal of Pattern Recognition Research*, vol. 5, no. 1, pp. 140-147, 2010. .

[3] Miraliakbari, A., et al. "Comparative Evaluation of Pavement Crack Detection Using Kernel-Based Techniques in Asphalt Road Surfaces." *ISPRS-International Archives of the Photogrammetry, Remote Sensing and Spatial Information Sciences*, pp.689-694, 2016.

[4] Q. Zou, Y. Cao, Q. Li, Q. Mao, S. Wang, "Crack-tree: Automatic crack detection from pavement images", *Pattern Recognition Letters*, vol. 33, no. 3, pp. 227-238, 2012.

[5] M. Salman, S. Mathavan, K. Kamal, M. Rahman, "Pavement crack detection using the Gabor filter", *Proceedings of IEEE International Conference on Intelligent Transportation Systems*, pp. 2039-2044, Oct. 2013.

[6] Gopalakrishnan, Kasthurirangan, et al. "Machine-Vision-Based Roadway Health Monitoring and Assessment: Development of a Shape-Based Pavement-Crack-Detection Approach."2016.

[7] A. Krizhevsky, I. Sutskever, G. Hinton, "Imagenet classification with deep convolutional neural networks", *Advances in Neural information Processing Systems*, pp. 1097-1105, 2012.

[8] K. Simonyan, A. Zisserman, "Very deep convolutional networks for large-scale image recognition", *Int. Conf. on Learning Representations (ICLR)*, 2015

[9] C. Szegedy et al., "Going deeper with convolutions", *Int. Conf. on Comp. Vision and Pattern Recognition (CVPR)*, 2015.

[10] K. He et al., "Deep residual learning for image recognition", *Int. Conf. on Comp. Vision and Pattern Recognition (CVPR)*, 2016.

[11] L. Deng, G. Hinton, B. Kingsbury, "New types of deep neural network learning for speech recognition and related applications: An overview", *Acoustics Speech and Signal Processing (ICASSP) 2013 IEEE International Conference on*, 2013.

[12] J Long, E Shelhamer, T. Darrell, "Fully convolutional networks for semantic segmentation", *Proceedings of the IEEE Conference on Computer Vision and Pattern Recognition*, pp. 3431-3440, 2015.

[13] Zhang, Lei, et al. "Road crack detection using deep convolutional neural network." *Image Processing (ICIP), 2016 IEEE International Conference on*. IEEE, 2016.

[14] Urban, G., Geras, K.J., Kahou, S.E., Aslan, O., Wang, S., Caruana, R., Mohamed, A., Philipose, M. and Richardson, M., 2016. Do Deep Convolutional Nets Really Need to be Deep and Convolutional?. *arXiv preprint arXiv:1603.05691*.

[15] Scherer, Dominik, Andreas Müller, and Sven Behnke. "Evaluation of pooling operations in convolutional architectures for object recognition." *International Conference on Artificial Neural Networks*. Springer Berlin Heidelberg, 2010.

[16] Nair, Vinod, and Geoffrey E. Hinton. "Rectified linear units improve restricted boltzmann machines." *Proceedings of the 27th international conference on machine learning (ICML-10)*. 2010.

[17] Hochreiter, Sepp. "The vanishing gradient problem during learning recurrent neural nets and problem solutions." *International Journal of Uncertainty, Fuzziness and Knowledge-Based Systems* 6.02 pp. 107-116, 1998.

Pre-processing Methodology of Image Compensation using Histogram Equalization for Generating Point-cloud of Construction Environment

D.Y. Moon[a], S.W. Kwon[b], S.W. Chung[c], S.Y. Lee[d] and T.K. Ko[e]

[a]Department of Convergence Engineering for Future City, SungKyunKwan University, Republic of Korea
[b]School of Civil, Architectural Engineering and Landscape Architecture, Sungkyunkwan University, Republic of Korea
[c]Department of Convergence Engineering for Future City, SungKyunKwan University, Republic of Korea
[d]Department of Convergence Engineering for Future City, SungKyunKwan University, Republic of Korea
[e]Department of Convergence Engineering for Future City, SungKyunKwan University, Republic of Korea

E-mail: yoshy17@nate.com, swkwon@skku.edu, song03@skku.edu, gtk101@naver.com

Abstract

Recently, advanced and semi-automated equipment, laser-scanner and Unmanned Aerial Vehicles (UAVs), is used for construction project. Especially, photogrammetry using image acquired from UAVs can make high accuracy data and has low cost compared with terrestrial and aerial Lidar. Unmanned Aerial Vehicles (UAVs) was considered as an important technology and using UAVs in various areas are prominent area in recent times. Unmanned Aerial vehicles (UAVs), aerial movable image sensing platform, have become the area that has been becoming notable in all industries. But in construction site, it is very hard to acquire high quality data from UAVs due to the environment changes such as diverse environment, difficulty of setting control points and post-processing. This study aims to propose the pre-processing methodology which is to compensate 2D image using image histogram in construction site. We made a variety 3D data of buildings as point-cloud using the pre-processing methods and analyzed and verified performance of results to compare with the results obtained by source data.

Keywords –

Pre-processing; UAV; Histogram equalization; point-cloud; Image compensation

1 Introduction

1.1 Background and objectives of the research

In recent years, Unmanned Aerial Vehicles (UAVs) have been widely used in a number of areas such as agriculture, civil engineering, disasters, delivery, and military sectors. The market of UAVs in the world has been increased constantly and will be expected to grow further. In particular, the expansion of spatial information portal services such as Google Earth and MS Virtual Earth is closely related to the growth of the market in UAVs. Moreover, a ratio of contribution to spatial information field by aerial monitoring is expected to grow further. Thus, much attention has been paid to the utilization of UAVs for rapid and convenience of aerial monitoring. Since photogrammetry using UAVs is efficient in terms of required time, cost, and human resources, it enables monitoring at a high frequency. Moreover, as data acquired through autonomous flight and wireless communication technology can be sent to the ground in real time, the utilization area of aerial monitoring systems based on UAVs is increasing.

Existing technologies have been concentrated on efficient algorithms using acquired data and data analysis using UAVs. In particular, various studies have been attempted to convert raw data into measurable and manageable data rather than survey purpose. Among them, most technological development of three-dimensional (3D) modeling using two-dimensional (2D) images has been underway based on point-cloud generation algorithm.

Currently, construction industry has employed not only laser scanner but also UAVs in site management and construction management variably. In particular, photogrammetry using UAVs is accurate and efficient as well as inexpensive. Thus, its utilization area and use frequency are increasing. Nonetheless, it is hard to acquire high-quality data according to various changes in environments such as diversity of shooting environment, difficulties in application of reference points, and difficulties in post-processing due to the characteristics of construction sites.

Thus, this study proposes a pre-processing method of 2D images compensated using image histogram in photogrammetry using UAVs in order to overcome the limitations of photogrammetry in the construction industry. This study generates 3D spatial information of various buildings using unmanned aerial photogrammetry and verifies and analyzes a level of improvement based on keypoint and point-cloud values. The reliability of unmanned aerial photogrammetry is determined based on the above result and applicability to various construction sites is discussed.

1.2 Research trend

The research and development of core technologies that can be indexes of photogrammetry has been conducted. The Scale Invariant Feature Transform (SIFT) developed by Lowe[1] and Structure from Motion (SfM) developed by Snavely[2] are two of the most widely used technologies in UAV image processing. Later, Furukawa[3] and Rothermel[4] proposed an algorithm that converted an initial low-density point-cloud into high-density point-cloud through recalculation using a multi-view stereo (MVS) technology. The bundle adjustment, which is a standard technology of aerial triangulation, can calculate exterior and interior orientation parameters of each photo in the digital photogrammetry filed.

The most significant drawback of photogrammetry using UAVs is low accuracy compared to that of laser scanner. Thus, if the low accuracy drawback is resolved, it can decrease cost in construction sites and construction schedule using only photogrammetry. There have been a number of previous studies on analysis on various factors that can degrade accuracy.

Snavely and Szeliski[2] studied noise occurrence in 3D information due to mismatch between corresponding pixels. Construction sites have many factors that make photogrammetry difficult through 2D images such as untidy environment, changes in lighting, sensor noise, and mismatch between pixels[7]. Since 3D photogrammetry data obtained from construction sites are likely to have a large amount of noise, data processing is needed to improve.[8] Kerdsrilek[9] removed DSM point clouds at locations where height change was severe such as buildings automatically and enabled conversion into DTM through interpolation of surrounding data by the development of SCOP++ algorithm. However, the algorithm had a weakness that heights of large or high-rise buildings had to be edited manually since heights were represented obliquely. To increase quality of photogrammetry in various construction sites, a study on main causes and resolution methods has been proposed previously. In particular, Lowe[1] extracted feature points that were not changed between different images effectively and proposed a method that produced highly reliable 3D data. The study reported that although a feature point was not affected by rotation and scale of images, a change in the number was affected by distortion, noise, and changes in lighting. Previous studies found that acquired 2D images had a significant effect on 3D data quality.

Thus, processing of acquired 2D data is the most effective way to obtain high quality 3D data to have efficient data acquisition in a poor environment such as construction sites. A number of processing methods for 2D images have been proposed to acquire high quality 3D images such as changes in illuminance, changes in contrast, and changes in sharpness. Among the above processing methods, the most efficient method is changes in illuminance and contrast of images considering algorithms to acquire photogrammetry data. Generally, this process is called histogram equalization (HE), which was studied by Pizer.[10] for the first time. This method transforms illuminance and contrast by changing image histogram of image information values displayed as a form of histogram. Kim[11] developed bi-histogram equalization (BBHE) based on existing HE methods to increase quality of illuminance and contrast values. Later, Abdullah[12] compared and analyzed Global Histogram Equalization (GHE), Local Histogram Equalization (LHE), and Dynamic Histogram Equalization (DHS) to determine strengths and weaknesses of each of HE algorithms. As shown in the previous studies, it was proven that increasing image quality itself can be closely related to 3D data quality but no direct studies have been conducted yet.

Thus, the present study aims to present a method to overcome the limitation of photogrammetry in construction sites using UAVs and prove that the use of HE algorithm on 2D images that can affect 2D images directly can give a direct effect on improvements on the quality of photogrammetry.

1.3 Research contents

UAVs have been utilized in various areas including military, agriculture, reconnaissance, and civil engineering since the first appearance. With the introduction of small unmanned aircrafts called drone, UAVs have provided various functions with high quality and low price as commercial UAVs are also available in industrial uses.

In recent years, 3D information acquired through UAVs has been widely used in various tasks such as construction supervision, verification of construction errors, and utilization as building information modeling (BIM) data in construction sites. Particularly, record of progress in construction projects can be done with photogrammetry using UAVs, which is efficient in terms of cost and time consumption. However, there are a large number of factors that degrade 3D data quality such as untidy environment of construction sites, changes in lighting, sensor noise, mismatch between pixels, diversity of shooting environment, difficulties in reference point application, and difficulties in post-processing due to the characteristics of construction industry. In addition, inaccuracy and errors can also be present due to the characteristics of photogrammetry. Thus, it is necessary to have a method that reduces inaccuracy and errors.

The present study applied a method that increases illuminance and contrast of 2D images to increase accuracy of 3D data. Among them, HE algorithm was used to acquire high quality contrast values by changing image attribute values.

UAV devices and cameras were selected first and then target buildings were shot. Point clouds of images transformed through HE algorithm and acquired original 2D images were created and 3D data values were compared. The reliability on the result values is evaluated by utilizing laser scan data. Finally, accuracy analysis was conducted with regard to data values obtained in the original image and images transformed through the HE algorithm.

2 Theoretical discussion

2.1 SFM technology

The SfM algorithm generates 3D scenes automatically using overlapped photos. It was developed by Professor Snavely in Cornell University in the USA in 2006. It is a technology that restores 3D shapes in each scene and location relationships of cameras and then restructures a number of overlapped photos shot from multiple angles using 3D. The SfM algorithm implements image registration easily despite of changes in location of cameras or scale by using the recognition of feature points in the scale-invariant feature transform (SIFT) algorithm.

Since accuracy of coordinate value (x, y, z) of target object and initially estimated camera location is degraded, it can be improved by using iterative calculations of non-linear least squares method. Each of the pixels is represented as a point cloud, which is a 3D coordinate value whose accuracy is high. However, the SfM method has no scale and orientation information, a coordinate value of point cloud has a relative spatial coordinate value between object and 2D image.

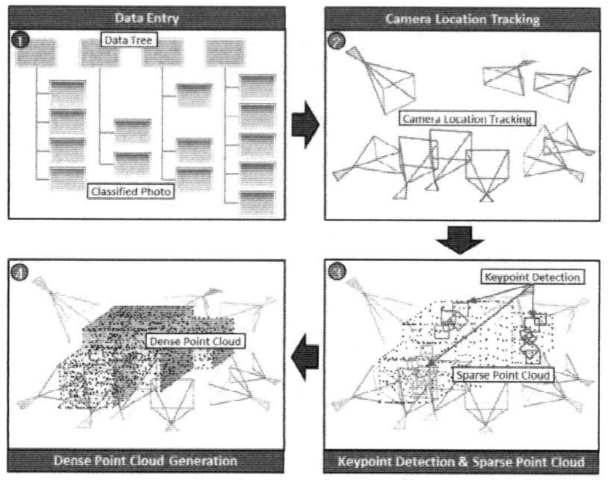

Figure 1. .Input 2D image, Calculation of camera pose, Generation of 3D point-cloud

2.2 SIFT algorithm and generation of point-cloud

The core element in the SfM is to determine a coordinate value of feature point matched with 2D image shot at multiple angles. First, feature points are extracted from 2D images of objects and then a relationship of feature points between images is determined. Next, a matrix is generated with regard to the corresponding relationship followed by factorization to estimate camera locations and coordinate value of each feature point. To do this, a factorization process is performed using bundle adjustment and bundler, which is open software. The feature point matching of 2D images is performed through random sample consensus (RANSAC) and track method. If two feature points are present in three 2D images, a point cloud is generated. If this condition is not met, it is automatically removed. In contrast with exiting aerial photogrammetry, SfM has an advantage that all processes are performed automatically. However, the precision of early point clouds was too low to represent an object accurately.

Thus, Clustering view for Multi-view Stereo (CMVS) and Patch-based Multi-View Stereo (PMVS) developed by Professor Furukawa in Washington University in the USA are employed to improve the precision. CMVS decompose overlapped 2D images into subset or cluster and PMVS is utilized to reconstruct 3D structure from the cluster independently.

In early point clouds generated by the bundler, the

number of points is increased as processes of CMVS and PMVS are progressed thereby generating high-density point clouds. However, all the processing procedures in the data values are performed automatically, it is difficult for users to inspect or modify the data values. As a result, it has a problem of degrading reliability somewhat in regard to the accuracy of the output.

2.3 Image histogram

Image histogram represents color distribution in digital images as histogram. Normally, a value range in the horizontal axis is 0 to 255, which represents image histogram of the gray scale of the image. Since histogram of specific image is represented, an overall color distribution of the image can be determined at a glance. Image histogram helps users to identify the characteristic information of images such as image contrast, brightness, and color distribution easily. Various features of filtering such as decrease or increase in image contrast and changes in brightness can be done by transforming information provided by image histogram.

2.4 Histogram Equalization

HE improves image quality by making a distribution of image contrast value uniform. The objective of HE is to generate a histogram of uniform distribution. In particular, HE is effective when an image has a dark and detailed part. Thus, it can cope with poor environments in construction sites efficiently. HE is executed according to the following four phases largely. First, frequency (hist[j]) of brightness value (j) is calculated using IE. Second, cumulative histogram values based on cumulative frequency are calculated using the calculated frequency and normalized.

$$sum[i] = \sum_{j=0}^{i} hist[j] \quad (1)$$

Third, the normalized cumulative histogram is mapped to gray level values using mapping function of gray scale.

$$n[i] = sum[i] \times \frac{1}{N} \times 255 \quad (2)$$

(N = No. of total pixels in the image)

Finally, pixel value i in the input image is converted into a normalized value n[i] thereby generating a resulting image.

In the present study, each section of RGB values is converted using the above method thereby generating an image of high quality contrast.

3 Research method and experiment scope

3.1 Research process

The experimental method used in this study had three phases: data collection, experiment, and result analysis. In detail, it was divided into device selection, data acquisition, data transformation, data processing, data comparison, and reliability verification. in the data collection phase, devices were selected and a flight plan of UAV was set up before acquiring 2D image data of the specified buildings. Next, in the experiment phase, collected 2D image data were transformed using HE. After this, each of KeyPoint, Sparse Point Cloud, and Dense Point Cloud with regard to original 2D image and transformed 2D images was acquired and compared. In the result and analysis phase, final values derived based on each of the data were compared and analyzed to verify a level of quality improvement.

3.2 Experimental equipment and software

3.2.1 UAV and camera

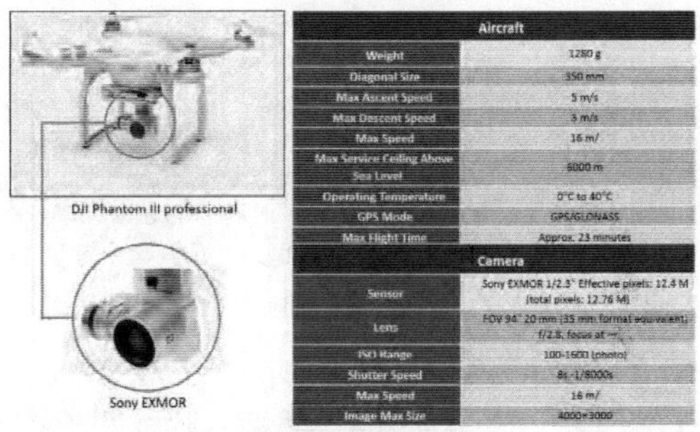

Figure 2. UAV and Camera

In this study, DJI Phantom 3 professional, which was the most widely used commercial UAV, and Sony EXMOR camera were employed to suggest applicability of photogrammetry in the construction industry using commercial UAVs. This UAV model can maintain a flight for about 20 min. and its maximum speed was 16m/s. A size of camera image acquisition was 4000x3000 pixels, which was 12 Mbytes. It is an excellent device used in various areas.

3.2.2 Software

In recent years, a large number of software products of photogrammetry specialized for UAVs have been launched in the market. In particular, Pix4Dmapper or Photoscan is the most popular commercial software among them. In the present study, two products of software, Pix4Dmapper and Photoscan were used initially but Photoscan was finally selected because of its small noise produced in buildings. The study was conducted by utilizing functions of Keypoint analysis, generation of Sparse Point Cloud, and generation of Dense Point Cloud with regard to 2D images in Photoscan. Moreover, HE was implemented using Matlab. The photos before and after the transformation and their image histograms were represented visibly in each of four windows in the screen.

4 Analysis on experiment results

4.1 Target buildings for experiments

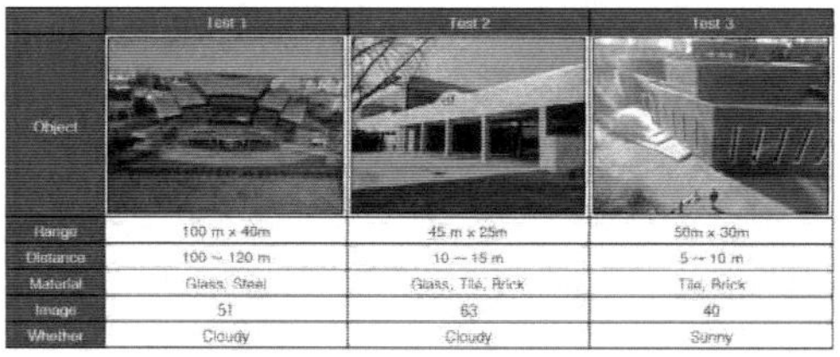

Figure 3. Test target

In the experiment, three buildings were targeted. For experiment 1, Academic Information Center in Sungkyunkwan University was selected. Its shooting area was 100m×40m and a shooting distance was 100m–120m. The external materials were glass and steel. The weather was cloudy. For experiment 2, Suseonggwan in Sungkyunkwan University was chosen. Its shooting area was 45m×25m and a shooting distance was 10m–15m. The external materials were glass, tiles, bricks, and paint. The weather was cloudy. For experiment 3, Student Center in Sungkyunkwan University was chosen. Its shooting area was 50m×30m and a shooting distance was 5m–10m. The external materials were tiles, bricks, and paint. The weather was sunny. Experiment 1 used civil scale, which was conducted to verify the applicability in civil engineering. Experiment 2 targeted a building in which structures and changes in illuminance were diverse. It was similar to real construction sites. Experiment 3 targeted a general building.

4.2 Compensation of original 2D image using HF

Images acquired in Experiments 1, 2, and 3 were transformed using HE software. During this process, unclear images particularly in cloudy weather or many shadows were beneficial significantly, which can be seen visibly.

4.3 Comparison and analysis on KeyPoint of original 2D image and compensated 2D image

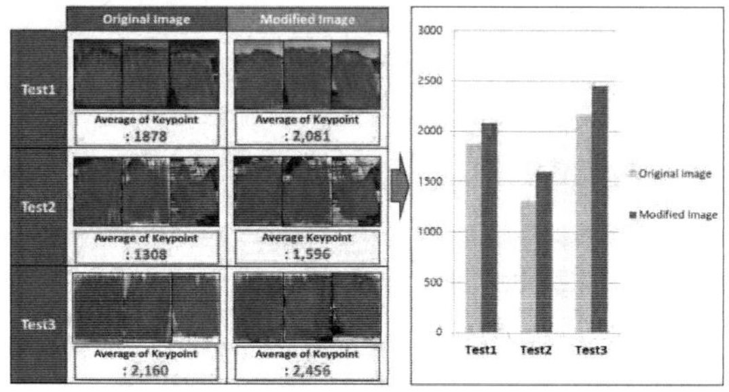

Figure 4. Result of keypoint extraction

A mean value of each KeyPoint in 2D images was acquired and compared. In Experiment 1, 1,878 Keypoint values from the original images and 2,081 Keypoint values from the transformed images were obtained. In Experiment 2, 1,308 Keypoint values from the original images and 1,596 Keypoint values from the transformed images were obtained. In Experiment 3, 2,160 Keypoint values from the original images and 2,456 Keypoint values from the transformed images were obtained. Ultimately, more Keypoint values were obtained from transformed images than from original images.

4.4 Comparison and analysis on Sparse Point Cloud of original 2D image and compensated 2D image

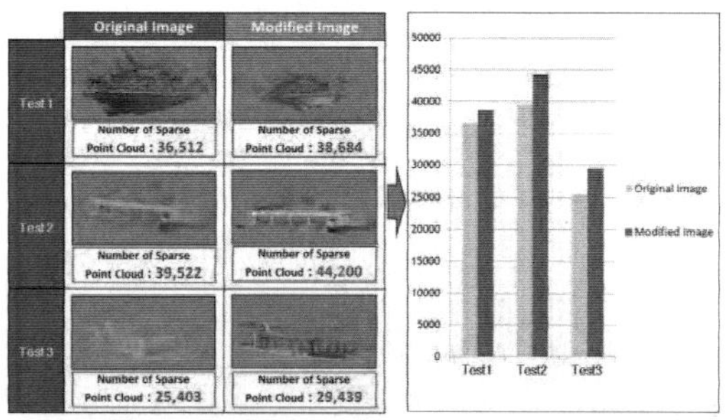

Figure 5. Result of sparse point-cloud

The results of Sparse Point Cloud acquired from original images and transformed images in Experiments 1, 2, and 3 were compared and analyzed. In Experiment 1, 36,512 Sparse Point Cloud values were obtained from original images and 38,684 Sparse Point Cloud values were obtained from transformed images. In Experiment 2, 39,522 Sparse Point Cloud values were obtained from original images and 44,200 Sparse Point Cloud values were obtained from transformed images. In Experiment 3, 25,403 Sparse Point Cloud values were obtained from original images and 29,439 Sparse Point Cloud values were obtained from transformed images. Ultimately, more Sparse Point Cloud values were obtained from transformed images than from original images. However, an increase in Experiment 1 was not that larger than that of the other two experiments.

4.5 Comparison and analysis on Dense Point Cloud of original 2D image and compensated 2D image

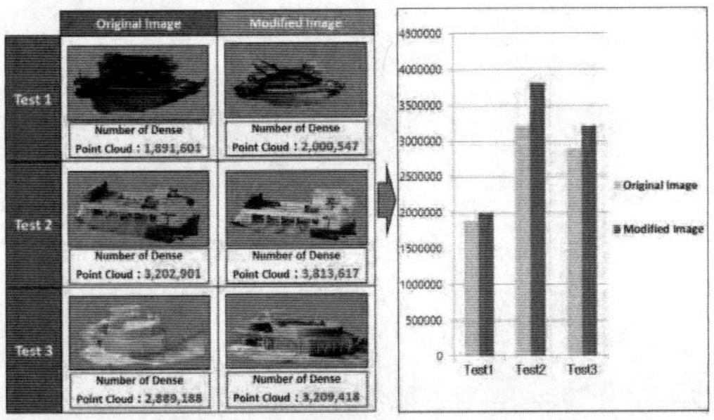

Figure 6. Result of dense point-cloud

The results of Dense Point Cloud acquired from original images and transformed images in Experiments 1, 2, and 3 were compared and analyzed. In Experiment 1, 1,891,601 Dense Point Cloud values were obtained from original images and 2,000,547 Dense Point Cloud values were obtained from transformed images. In Experiment 2, 3,202,901 Dense Point Cloud values were obtained from original images and 3,813,617 Dense Point Cloud values were obtained from transformed images. In Experiment 3, 2,889,188 Dense Point Cloud values were obtained from original images and 3,209,418 Dense Point Cloud values were obtained from transformed images. An increase in Experiment 1 was the lowest but an increase in Experiment 2 was 20%, which was higher than that in Experiment 1.

4.6 Comparison and analysis on increase rate of final results

The result values in Experiments 1, 2, and 3 were all increased. Increase rates in Key Point, Sparse Point Cloud, and Dense Point Cloud were as follows: in Experiment 1, increase by 11%, 6%, and 6%, in Experiment 2, increase by 22%, 12%, and 19%, and in Experiment 3, increase by 14%, 16%, and 11% were verified.

5 Conclusion

The present study analyzed the effect of HE on improvements on photogrammetry quality by comparing results values of Keypoint, Sparse Point Cloud, and Dense Point Cloud, which were scales that determined quality of photogrammetry. Conclusively, most values were increased as presented in this study. The present study thus presented applicability of HE that HE can be efficiently employed in construction sites where quality of 2D image was much degraded.

However, civil scale, that is, photogrammetry results on targets that were distanced 100 m or longer had a small amount of increase in the result values. It was assumed that since shooting of targets was done from a long distance, detailed information was small and approximate 3D data on shapes were formed. Thus, the present study can be suitable for buildings whereas its efficiency in photogrammetry of long distance as needed in civil engineering was lower than that of short-distance.

The limitation of the present study is as follows: since the present study targeted only three buildings, it had a lack of experimental information which should have more various types of target buildings. However, if an in-depth study on method that improves quality of 3D images using quality of 2D images continues as conducted in the present study, higher quality of values can be obtained efficiently in the future.

Acknowledgment

This work is supported by Korea Minister of Ministry of Land, Infrastructure, Transport affairs as Convergence Engineering of Future City Master and Doctoral Grant Program.

This work is supported by the Korea Agency for Infrastructure Technology Advancement (KAIA) grant funded by the Ministry of Land, Infrastructure and Transport (Grant 17SCIP-B079689-04).

References

[1] Lowe, D G. Distinctive image features from scale-invariant keypoints. International journal of computer vision 60(2):91-110, 2004

[2] Snavely N., Steven M. S. and Richard S. Photo tourism: exploring photo collections in 3D. ACM transactions on graphics (TOG). 25(3):835-846 2006

[3] Furukawa Y and Ponce J. Accurate, Dense, and robust multiview stereopsis. IEEE Transactions on Pattern Analysis and Machine Intelligence 32:1362~1376, 2010

[4] Rothermel, M. Wenzel K. Fritsch D and Haala N.

SURE: Photogrammetric surface reconstruction from imagery. LC3D Workshop, Berlin Germany. pages 9 2012

[5] Lange J. Benning W. and Siering K. Crack detection at concrete construction units from photogrammetric data using image processing procedures. ISPRS Commission VII Mid-term symposium remote sensing: from pixels to processes, Enschede, 8–11, pages 493–496, 2006

[6] Ahmed M.. Haas C T. and Haas R. Using digital photogrammetry for pipe-works progress tracking. Canadian Journal of Civil Engineering. 39.9:1062-1071, 2012.

[7] Robert N. Al-Hamadi A. and Michaelis B. A novel method for 3D face detection and normalization. Journal of Multimedia 2(5):1-12, 2007

[8] Kim C Son H. and Kim C. The effective acquisition and processing of 3D photogrammetric data from digital photogrammetry for construction progress measurement. The Proceeding of the 2011 ASCE International Workshop on Computing in Civil Engineering, Miami, Florida. 2011

[9] Kerdsrilek, J. Reduction of DSM to DTM and Quality Assessment, Photogrammetry and Remote Sensing Project for Zurich Airport. Swiss Federal Institute of Technology, pages 1~3, 2008

[10] Pizer S M.. Amburn E P. Austin J D. Cromartie R. Geselowitz A. Greer T. Romeny B T H. and Zimmerman J B. Adaptive histogram equalization and its variations, Computer Vision, Graphics, and Image Processing, 39(3): 355-368, 1987

[11] Kim, Y T. Contrast enhancement using brightness preserving bi-histogram equalization. IEEE transactions on Consumer Electronics 43(1):1-8, 1997

[12] Abdullah-Al-Wadud M. Kabir M H. Dewan M A A. and Chae O. A dynamic histogram equalization for image contrast enhancement. IEEE Transactions on Consumer Electronics 53(2):593-600, 2007

[13] Barazzetti L. Remondino F. and Scaioni M. Automation in 3D reconstruction: Results on different kinds of close-range blocks. Int. Arch. Photogramm Remote Sens. Spatial Inf. Sci. 38(5):55–61, 2010

[14] Dandois J P. and Ellis E C. High spatial resolution three-dimensional mapping of vegetation spectral dynamics using computer vision. Remote Sensing of Environment, 136:259~276, 2010

Thermal 3D Modelling

Gregory Clarkson, Shan Luo, Raul Fuentes

School of Civil Engineering, University of Leeds, Leeds, LS2 9JT
E-mail addresses: cn13gc@leeds.ac.uk, S.Luo@leeds.ac.uk, R.Fuentes@leeds.ac.uk

Abstract – This paper presents the case of a 3D reconstruction of an object using infrared imagery directly. In total, three point clouds are created using 3DF Zephyr and SfM techniques. The first is a thermal reconstruction, the second is a low resolution RGB reconstruction and the final is a high resolution RGB reconstruction which is used as a reference for comparison of the others. CloudCompare is then used to analyze and evaluate the reconstructed point clouds. In particular, this paper presents the reconstruction of a 3D Model directly from thermal images, conversely to previous solutions that use RGB imagery for 3D reconstruction subsequently followed by superposition of the thermal data.

Keywords –
3D Reconstruction; Structure-from-Motion; Thermal; Infrared

1 Introduction

The field of reconstruction from multiple images, known as Structure from Motion (SfM) has been rapidly expanding in recent years to incorporate a wider variety of possibilities and applications. This development is driven by constant innovation and improvement in computer technology [1]. 3D reconstruction is therefore becoming a much more frequent practice in various fields. [2] shows why reconstruction is of great importance in architectural restoration and deformation monitoring of historic structures using the wall in Uzutrakis palace, Lithuania as a case study. Other uses of this technology include crime scene investigation [3] and medical MRI scanning [4]. However, the more specific use of thermal reconstructions is largely unexplored and has many potential applications in the construction industry alone, such as leak detection (see Figure 1). Creating full thermal profiles of buildings would also be possible using UAV's with mounted thermal optics, such as the DJI Inspire 2 [5] and DJI Zenmuse XT Thermal Camera with Stabilized Gimbal provided by FLIR [6]. These models could be used to evaluate and improve a structure's energy performance. Moreover, [7] discusses how the operational cost of a building is greater than the capital cost at a ratio of approximately 1.2:1.0 and thus there is significant financial motivation to ensure energy efficiency. Thermal reconstructions would allow complex situations to be analyzed more accurately.

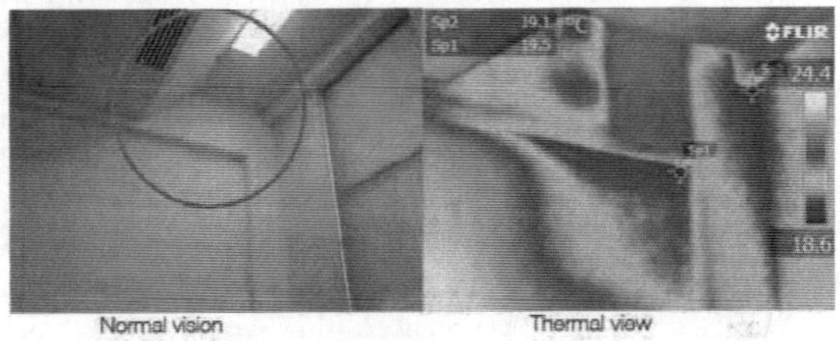

Figure 1: (Left) RGB photograph. (Right) Thermal / infrared image. It can be seen that the leaks can only be detected in the thermal view [24].

To date, there has been little research into the reconstruction of 3D models directly from thermal imagery. Current methods involve superimposing thermal data onto reconstructions created through regular RGB images or other means. For example, [8] uses a method where thermal data is projected onto an existing 3D model created with AutoCAD. [9] presents the idea of 'carving' out a heat source from a voxel set created by segmentation using Otsu thresholding method [10]. To date, no reliable method for thermal-only reconstruction has been proposed. This paper will investigate the reconstruction of models directly from thermal images using software which implements SfM techniques. These are then compared to both low and high quality RGB reconstructions, using Cloud Compare for evaluation.

2 Structure from Motion (SfM)

SfM is now a well-established and commonly used reconstruction technique, it is well documented and there is software readily available offering various implementations of it. SfM is the process of obtaining 3D spatial coordinates from a series of 2D images. The image capture point locations need not be known or planned prior to taking the images and are recovered from the SfM pipeline itself, much like traditional photogrammetry. Multiple overlapping images of the object need to be captured from varying perspectives.

Features of an image known as keypoints can then be detected and described using various algorithms such as Speeded Up Robust Features (SURF) [11] or the popular Scale Invariant Feature Transform (SIFT) [12]. Maximally stable extremal features (MSER) [13] is another algorithm which matches regions of similar luminous intensity rather than specific keypoints. A matching process then occurs between the keypoints in different images and if there are enough matches between multiple images, first the camera position and orientations are recovered, followed by a measurement of the 3D coordinates of the keypoints through triangulation.

Zephyr Pro created by 3D Flow [14] is just one package, but is used in this work due to its intuitive workflow and reliability. Zephyr uses its own proprietary set of algorithms for reconstruction. The first is known as SAMANTHA which calculates camera positions and orientations and keypoints and the second is STASIA, a Multiview Stereo (MVS) algorithm which extracts dense point clouds [15,16]. SAMANTHA uses a feature detector to extract keypoints that is a scale-space feature extractor based on Difference of Gaussian, with a radial and symmetric descriptor [17]. Keypoints are then matched between images based on the response values from their descriptors and a rough set of camera positions and orientations are estimated. A more refined matching stage is then undertaken in which keypoints are also discarded. The fundamental matrix [18], which describes the intrinsic relationship between two viewpoints, is calculated using M-SAC [19]. A hierarchal structure is created for matching images using an agglomerative clustering algorithm. The partial models created from these matches are merged together forming a single model and finally bundle adjustment is performed to reduce reprojection errors and improve overall accuracy.

3 Object Description

The object being reconstructed in this paper was carefully chosen to have the following properties:

- Relatively small – easy to photograph using available thermal cameras (low resolution images).
- 'Blocky' geometry – Geometry which is too complex would be difficult to accurately reconstruct.
- Varying thermal profile – unlike in regular RGB photographs thermal images require a temperature gradient for keypoint detection.

Figure 2 shows the architectural model chosen for the reconstruction. The whole model is relatively small (~ 75x45cm) and has a complex enough form to allow both RGB and thermal reconstruction. The model is heated and given a sufficient thermal profile using two 1000W work lights secured to the tabletop via a drill vice. The work lights act as a sun emulator and provide heat across the model, much in the same fashion as a heliodon [20].

Figure 2: Photograph showing architectural model to be reconstructed (High resolution RGB).

4 Camera Calibration and Properties

Both thermal and RGB cameras are automatically calibrated within Zephyr, this is done using an additional package called 3DF Lapyx which has been recently integrated into Zephyr.

4.1 Thermal Camera

The device used to capture thermal images is a FLIR A65sc. The specifications used can be seen in Table 1 below.

Table 1 FLIR A65sc Properties.

Camera property	Value
IR Resolution	640x512px
Temperature Range	16-38°C
Thermal Sensitivity	50mK
Focal Length	30mm
Accuracy	±5°C (±9°F) or ±5% of reading

4.2 RGB Camera

The device used to capture regular RGB photographs is a SONY ILCE-7S DLSR camera. The settings are selected by trial and error to ensure that the quality of the image was adequate and can be seen in Table 2.

Table 2: SONY ILCE-7S Properties.

Camera property	Value
Focusing Mode	AE-L
F-stop	f/5.6
Exposure Time	1/250s
ISO-Speed	4000
Focal Length	50mm
Max Aperture	4.336
Brightness	3.404

5 Image Capture

The image capture process for both thermal and RGB reconstruction is crucial as it must be ensured that there is enough overlap between consecutive images for sufficient keypoint matching to occur. The architectural model is encircled completely while continually capturing images at approximately 20° increments (see Figure 3). In total, the capture process took around 15 minutes for each set of images, excluding initial experiment set up time.

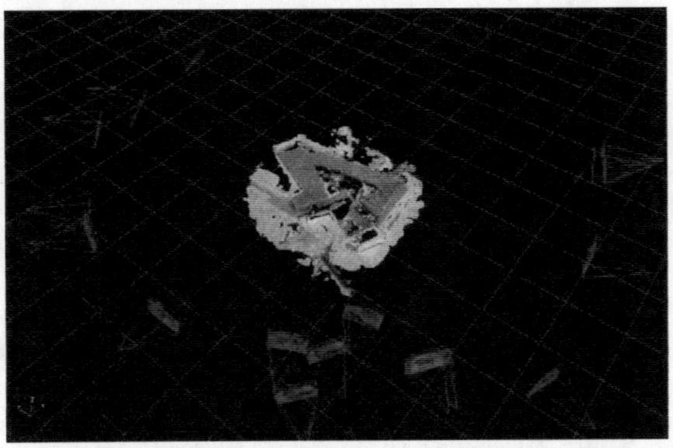

Figure 3: Reconstruction in Zephyr showing computed camera locations (blue frustums) and reconstructed point cloud (centre).

5.1 Thermal Series

Two sets of thermal images (640x512px) were captured; the first set of 24 had the work lights switched on to heat up the model from one side (Figure 4). The second set of 19 images was taken roughly five minutes after the lights had been switched off so only some residual heat remains (see Figure 5). This series of images highlights the major problem of reconstructing directly from thermal images, which is lack of detectable features present. Zephyr could not detect/match enough keypoints between images with this set for a successful reconstruction, therefore, it was disregarded.

5.2 RGB Series

Two sets of RGB photographs were taken of the architectural model. The first is a high resolution (4240x2832px) set consisting of 17 images. The second set of 18 images were taken at the lowest possible resolution of the camera (2128x1416px). This image set was then resampled and downscaled to match the thermal image resolution (640x512px). This down scaling was done using Adobe Photoshop's resampling tool with bicubic sharpening (reduction) as the resampling method [21]. Figure 2 shows one of the high-resolution images used in the reconstruction.

Figure 4: Thermal image from series one showing architectural model with work lights switched on.

Figure 5: Thermal image from series two showing architectural model after work lights have been switched off.

6 Reconstruction

The thermal and RGB photographs taken were then reconstructed using Zephyr. A desktop PC was used for processing which has primary components of an Intel i5-4430 (3GHz) CPU and 2Gb NVidia GTX 960 with CUDA GPU. Actual processing time of the point clouds varies significantly, but is primarily dependent on the image resolution and complexity (see Table 3).

6.1 Thermal

The thermal photographs are reconstructed using the advanced settings in Zephyr to allow more control. A full description of the settings is outside the scope of this paper but are included for ease of reproducibility of the results. A complete explanation of the settings can be found in [14]. 'Exhaustive' was used as a base point and then tweaked to give the best possible reconstruction. The photo ordering was changed to unordered and keypoint matching tolerance increased. 'High details – Sharp' was used as the setting for dense point cloud generation and with no additional tweaks.

Only eight of the thermal images showed sufficient correspondence to create a point cloud: this further highlights the need for a varying temperature gradient across an object for thermal reconstruction. There are a considerable number of errors in the reconstruction; these are especially prevalent in areas with little temperature variation, in this case the side of the model not facing the heat source. There is also significant distortion within the model, although the overall form can be inferred (See Figures 4, 6).

6.2 RGB

The RGB photographs are reconstructed using the Zephyr preset settings, for simplicity. 'Default' and 'Deep' were used for the high and low resolution series sparse point cloud generation respectively. The photo ordering setting was again changed to unordered so that keypoint matching occurs between all images. Dense point cloud generation was performed with the same settings as the thermal photographs. There is some surface glare present in several RGB photographs, originating from the glossy surface present on some parts of the model. This glossiness is inconsistent between views as the relative perspective lighting changes and hence makes keypoint detection in these areas difficult. However, the wood paneling texture on the model provides an excellent source of key points, especially regarding the high-resolution photographs.

For the high-resolution image set, only 1 photograph failed to match enough keypoints to be successfully included in the reconstruction (due to undesirable motion blur present). The quality of the point cloud overall appears very good, there is little noise and only minor reconstruction errors present (see Figure 7). These errors are primarily gaps that occur only on the thin white walls of the model, explained by the lack of surface texture and hence lack of keypoints.

For the low-resolution image set the reconstruction is much less complete with six cameras being discarded. The sections of the model with the wood paneling texture are still reconstructed well, but the white portions have considerably more errors and the overall form of the cloud shows more distortion too (see Figure 8). This loss of quality is to be expected when using such low-resolution photographs as there are simply fewer keypoints which can be detected. For comparison, a script was created in MATLAB running the SURF algorithm, 2726 key points are detected in one of the high-resolution photographs and only 319 in a low-resolution photograph. This stark difference is exacerbated since the low resolution is not the photographs native format and there will be some loss of quality occurring from the down sampling and resizing process.

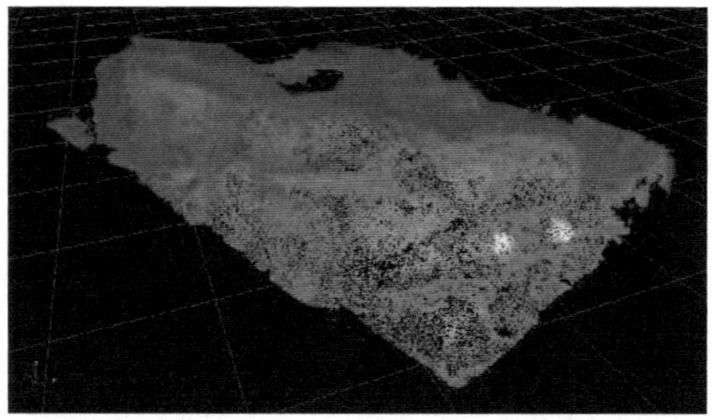

Figure 6: Dense, thermal point cloud created using Zephyr (post-cleanup) – a good reconstruction only occurs where there's a varying thermal profile.

Figure 7: Dense, high resolution, RGB point cloud created using Zephyr (post-cleanup) – errors visible at the bottom of the model on the glossy walls.

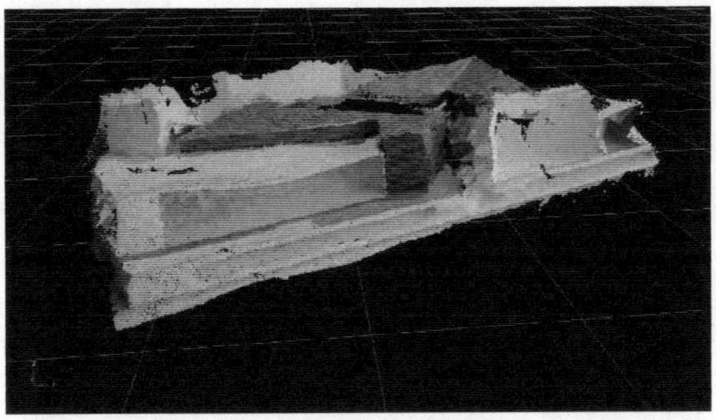

Figure 8: Dense, low resolution, RGB point cloud created using Zephyr (post-cleanup) – Mesh holes can be seen on the right and distortion on the left.

6.3 Point cloud cleanup and registration

The post reconstruction process consists of two steps: cleanup and registration. The cleanup itself is done in two stages. The first stage involves the removal of points with high uncertainty using Zephyr's confidence index and the second stage is a manual removal of anomalous points. These are erroneous points inherently present in every reconstruction and do not fit with the desired point cloud form. The table beneath the base of the model was also removed from all the point clouds as this would affect comparison validity. This was done using the tools within Zephyr to 'cut' around the base in the clouds which can clearly be discerned.

SfM techniques cannot currently recover a scaled model directly without additional spatial information being provided. This procedure was deemed unnecessary as this paper is focused on the thermal reconstruction process rather than precise accuracy in terms of scale. Therefore, a relative scale has been adopted where the thermal and low resolution RGB point clouds are adjusted to fit onto the high resolution RGB cloud which acts as the reference. This process is done in two stages using CloudCompare [22]; the first is manually translating / rotating / scaling the non-reference clouds to align them approximately to the reference. The second stage refines this alignment using Iterative Closest Point (ICP) algorithm [23], which adjusts the non-reference point cloud to achieve the lowest RMS errors.

7 Point Cloud Comparison

Once the reconstructions are aligned and scaled relative to the reference, they can be compared reliably and fairly. The number of points in the dense cloud is proportional to the complexity of the image, thus it is intuitive that the thermal point cloud will have fewer. Conversely, this method also reduces the computation time required for processing as can be seen in Table 3.

Table 3: Point cloud density and approximate computation time.

Model name	Number of points in cloud	~ Computation time (mins)
Thermal	137,064	5
High-res RGB	2,206,837	30
Low-res RGB	226,364	12

The histograms below (Figures 9, 10) show absolute distance comparisons of the thermal and low-resolution RGB clouds against the high-resolution RGB cloud. These distances represent how much the non-reference point cloud deviates from the reference and thus a rough gauge of accuracy can be inferred.

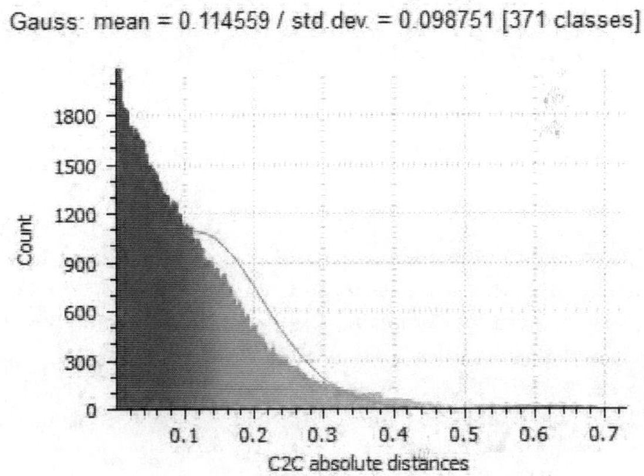

Figure 9: CloudCompare absolute distance comparison between thermal and high resolution RGB point clouds.

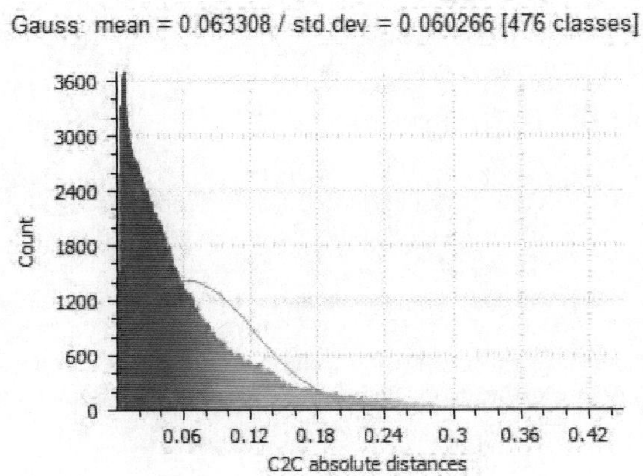

Figure 10: CloudCompare absolute distance comparison between low and high resolution RGB point clouds.

Overall, the low resolution RGB point cloud is the closest match to the reference cloud. The mean and standard deviation are 44.7% and 38.9% lower than the thermal clouds respectively. However, this lower accuracy does not mean that reconstructing from thermal clouds directly is obsolete; there are several specific situations where thermal images will give a better reconstruction than an RGB. Flat surfaces with very little surface texture such as the glossy white parts of the architectural model in this case gave a poor reconstruction through RGB images. Thermal reconstruction has potential to work around this problem as the surface can have a temperature gradient and hence can still be reconstructed through the thermal imagery. It then follows that the combination of thermal and RGB imagery could improve the overall reconstruction result. This method is a clear advantage over others that use superimposing of the thermal data.

8 Conclusion

The SONY ILCE-7S and FLIR A65sc were successfully used to acquire good sets of photographs of the architectural model. The 3D reconstruction software Zephyr was then used to reconstruct these into dense point clouds (Figures 6, 7, 8) using SfM techniques. The low resolution RGB and thermal point clouds were initially manually aligned onto the high-resolution RGB cloud, which was further refined by the ICP algorithm. An absolute distance analysis is then performed using CloudCompare (Figures 9, 10) and the point clouds are evaluated in section 7.

9 Acknowledgments

We would like to thank Amy Derby for the use of her architectural model.

10 References

[1] Gavves E. "A brief history of computer vision". Online: http://www.egavves.com/a-brief-history-of-computervision/#sthash.s8aGdhBl.dpbs, Accessed: 26/03/2016.

[2] Sužiedelytė-Visockienė, J., Bagdžiūnaitė, R., Malys, N. and Maliene, V. "Close-range photogrammetry enables documentation of environment-induced deformation of architectural heritage". *Environmental engineering and management journal,* 14(6): 1371–1381, 2015.

[3] Bostanci, E. "3D reconstruction of crime scenes and design considerations for an interactive investigation tool". Online: http://arxiv.org/pdf /1512.03156.pdf, Accessed: 26/04/2016.

[4] Ying, L. and Liang, Z.-P. "2 advanced image reconstruction methods in MRI" Online: http://www.acsu.buffalo.edu/~jlv27/index_files/Publications_files/BookChapter_AdvImgRecon.pdf, Accessed: 27/03/2016.

[5] DJI Official. "Inspire 2–Power Beyond Imagination" Online: http://www.dji.com/inspire-2, Accessed: 29/03/2017.

[6] FLIR Systems, I. "Aerial Thermal Imaging Kits | FLIR Systems" Online: http://www.flir.com/suas /aerial-thermal-imaging-kits/, Accessed: 29/03/2017.

[7] Hughes, W., Ancell, D., Gruneberg, S. and Hirst, L. "Exposing the myth of the 1:5:200 ratio relating initial cost, maintenance and staffing costs of office buildings" Online: http://centaur.reading.ac.uk /12142/1/File12142.pdf, Accessed: 26/04/2016.

[8] Acampora, L., De Filippis, F., Sorgi, L., Martucci, A., "3D reconstruction of thermal images". In *Proceedings of the 26th Aerospace Testing Seminar 2011,* pages 263-276, California, USA, 2011.

[9] Chen, C.-Y., Yeh, C.-H., Chang, B.R. and Pan, J.-M. (2015) "3D reconstruction from IR thermal images and Reprojective evaluations". *Mathematical Problems in Engineering,* doi:10.1155/520534, 2015.

[10] Automatic thresholding "Automatic Thresholding". Online: http://www.math.tau.ac.il/~turkel/notes /otsu.pdf, Accessed: 14/05/2017.

[11] Bay, H., Tuytelaars, T. and Van Gool, L. "Speeded up robust features (SURF)" *Computer Vision and Image Understanding,* 110(3): 346-359, 2006.

[12] Lowe, D. "Distinctive image features from scale-invariant Keypoints". *International Journal of Computer Vision,* 60(2): 91-110, 2004.

[13] Matas, J., Chum, O., Urban, M. and Pajdla, T. "Robust Wide Baseline Stereo from Maximally Stable Extremal Regions" *Image and Vision Computing*, 22(10): 761-767, 2004.

[14] 3DFlow. "3DF Zephyr Pro - 3d models from photos" Online: http://www.3dflow.net/3df-zephyr-pro-3d-models-from-photos/, Accessed: 28/03/2017.

[15] 3DFlow. "3Dflow Technology". Online: http://www.3dflow.net/technology/, Accessed: 28/03/2017.

[16] Seitz, S., Curless, B., Diebel, J., Scharstein, D. and Szeliski, R. "A Comparison and Evaluation of Multi-View Stereo Reconstruction Algorithms" In *Proceedings of the 2006 IEEE Computer Society Conference on Computer Vision and Pattern Recognition,* pages 519-528, Washington, USA, 2006.

[17] Toldo, R. "Towards automatic acquisition of high-level 3D models from images" Online: http://www.3dflow.net/resources/PhDThesis_RToldo.pdf, Accessed: 27/04/2016.

[18] Hartley, R. and Zisserman, A. *Multiple View Geometry in Computer Vision,* pages: 239-261, 2004.

[19] Torr, P. and Zisserman, A. "MLESAC: A new robust estimator with application to estimating image geometry". *Computer Vision and Image Understanding*, 78(1): 138-156, 2000.

[20] High Precision Devices. "Heliodon is a Sun Emulator to Demonstrate the Motion of the Sun." Online: http://www.hpd-online.com/heliodon.php, Accessed: 29/03/2017.

[21] Chavez, C. "Fundamentals: Resampling in Photoshop" Online: http://www.graphics.com/article-old/fundamentals-resampling-photoshop, Accessed: 14/05/2017.

[22] G.M, D. "CloudCompare - Open Source project" Online: http://www.danielgm.net/cc/, Accessed: 29/03/2017.

[23] Besl, P. and McKay, N. (1992). "A Method for Registration of 3D Shapes". *IEEE Transactions on Pattern Analysis and Machine Intelligence*, 14(2): 239-256, 1992.

[24] Hometeamaz. "Roof Leak Repairs" Online: http://hometeamaz.net/az/Phoenix/ROOF-LEAK-REPAIR.php, Accessed: 29/03/2017.

Detection of Nearby Obstacles with Monocular Vision for Earthmoving Operations

H. Son, H. Sung, H. Choi, S. Lee, and C. Kim*

Department of Architectural Engineering, Chung-Ang University, South Korea
E-mail: hjson0908@cau.ac.kr, gusdn7543@gmail.com, vinaj516@gmail.com, leesungwook@cau.ac.kr, changwan@cau.ac.kr (*corresponding author)

Abstract

The equipment used in earthmoving operations poses a significant threat to the safety of the equipment operator and construction workers due to the operator's inherently poor visibility of the surrounding environment. This study proposes a method of automated detection of nearby obstacles with monocular vision, with the goal of protecting the equipment operator and construction workers from potentially dangerous situations, such as collisions between earthmoving equipment and obstacles within a certain proximity. The proposed method consists of three steps: 1) correction of lens distortion prior to further processing, 2) shadow removal, and 3) detection of nearby obstacles with a predefined height level via perspective transformations. The proposed method was tested on video streams acquired from a camera installed on the side of the equipment body while an excavator executed excavating and moving tasks. The experimental results showed that the proposed method can provide the equipment operator with information about nearby obstacles during the excavator's manipulation and transportation. It is expected that the proposed method can be implemented in rearview monitoring systems and surrounding view monitoring systems for operator assistance and to achieve active safety.

Keywords –

Active safety; earthmoving operations; imaging sensors; intelligent earthmoving equipment; operator assistant

1 Introduction

Visibility is important for securing safety during operation of the construction equipment. Since humans depend on visual sense for 90% of the information received through sensory organs, if an equipment operator fails to secure sufficient visibility, this could cause serious casualties and reduce task efficiency. As demand for a higher visibility standard grows in the construction industry, the International Organization for Standardization (ISO) has developed a number of international standards for construction machinery (e.g., ISO 5006: 2006) [1]. Installing multiple mirrors can be helpful in ensuring visibility, according to the standard, but it has fundamental limitations in that the equipment operator must check multiple mirrors from time to time. To improve the visibility of the surrounding environment, earthmoving equipment manufacturers have adopted a rearview monitoring system, or a system that monitors further around the equipment by installing camera(s) on the rear or on every side (e.g., the rear, left, right, and front) of the equipment body and displaying the views on the operator's monitor. Although these systems provide improved visibility of the surrounding environment, detecting potential collisions between earthmoving equipment and obstacles (e.g., construction workers, facilities, and others) is still cognitively effortful and restricted by the operator's limited cognitive capacity while executing tasks. Therefore, it is necessary to develop a method to rapidly provide information about 3D environments from the images acquired from cameras installed on each side, which is important and helpful in assisting operators.

Modeling of the surrounding environment is highly desirable for construction automation. However, it is a difficult task to effectively and rapidly represent surroundings due to the complexity of construction workspaces and rapid variation in objects' locations. Toward this end, various 3D imaging sensors are being developed and tested for 3D modeling. Some researchers have proposed the use of two 2D laser rangefinders to enhance safety [2]. In order to acquire consistent 3D information with such sensors, multiple consequent acquisitions are performed and merged. The data acquisition is therefore time-consuming. 3D laser scanners can provide high-resolution 3D information with large field of view. However, because it takes tens of seconds to scan once, it is not suitable for modeling the dynamic environment.

Laser-based flash laser distance and ranging (LADAR) and vision-based stereovision systems are state-of-the-art developments among 3D imaging sensors. Researchers (e.g., [3–5]) have adopted flash LADAR to represent the work environment. This flash LADAR can acquire 3D information about the environment, which may correspond to obstacles at 30 frames per second [6]. Although previous studies have validated that flash LADAR could be used to provide 3D information in construction sites, the resolution of 3D information is limited, and the information captured are contaminated by noise, especially in outdoor environments [7]. Stereovision systems can also acquire 3D information with two or more cameras. Researchers (e.g., [8]) have adopted stereovision systems to represent the work environment. Although stereovision can provide texture and color information, not just 3D information, the applicability is limited in environments with few textures. Also, maximum depth and the precision of 3D information are limited.

In recent years, researchers working on autonomous robot navigation have been interested in representing the surrounding environment of a robot by relying on monocular camera inputs only (e.g., [9]). It has been proven that the monocular vision-based approach enables obstacle detection and environment representation at short range under certain environmental assumptions. In particular, it is possible to create a robust obstacle detector for low-speed maneuvers using monocular vision algorithms only in limited conditions. Earthmoving equipment needs to comply with speeds below 20 km/h within construction sites. With this point in mind, this study proposes a method for modeling the surrounding environment based on information from a single camera attached to the rear or all sides of the earthmoving equipment body.

The aim of this study is to propose an automated method for detecting nearby obstacles based on the use of a single camera, with the goal of protecting the equipment operator and construction workers from potentially dangerous situations, such as collisions between earthmoving equipment and obstacles within a certain proximity. This paper is organized as follows: in Section 2, the proposed approach is presented. Section 3 describes the proposed method in detail with experimental results. Finally, conclusions and suggestions for future research are given in Section 4.

2 Detection of Nearby Obstacles with Monocular Vision

In order to avoid collision, an equipment operator needs to pay attention to nearby areas with obstacles in meaningful dimensions while manipulating the equipment in a complex environment. Although it has been recognized in general that 3D representation is possible through more than two vision systems, the following prior information and assumptions allow distance estimation like that in the human vision between the camera and obstacles in 3D representation. Assume that the earthmoving equipment is lying on a plane and the height of the camera origin and azimuth (in degrees) are known. Then, the range of height of the obstacles and the range of distance between the camera origin and the obstacles are limited. Through projection transformation, we can obtain a point in 3D space and convert it to a 2D point in the image in pixels along with an optional z-value corresponding to the depth of each pixel.

In the case of an excavator, the lower caterpillar and the upper equipment body rotate separately around the axis of rotation. Figure 1 illustrates a top view of how the equipment body (represented by the yellow-orange color) at the top rotates while the caterpillar (represented by in black) is stationary. The dimensions of the caterpillar and equipment body were illustrated based on a 21-ton excavator. There are two areas: the area inside the inner circle and the area between the inner circle and the outer circle. The outer circle is a 12-meter circle around. The outer circle shows the extent to which obstacles should be detected if intersecting obstacles are set within 3 m. The inner circle represents the circle around the rotary long axis of the equipment body. This study focuses on obstacles within this outer circle by considering the speed of earthmoving equipment at the construction site. Then, for the purpose of preventing collisions with static or moving obstacles when the upper body of the equipment rotates, only objects with a height of 1 m or more are regarded as obstacles.

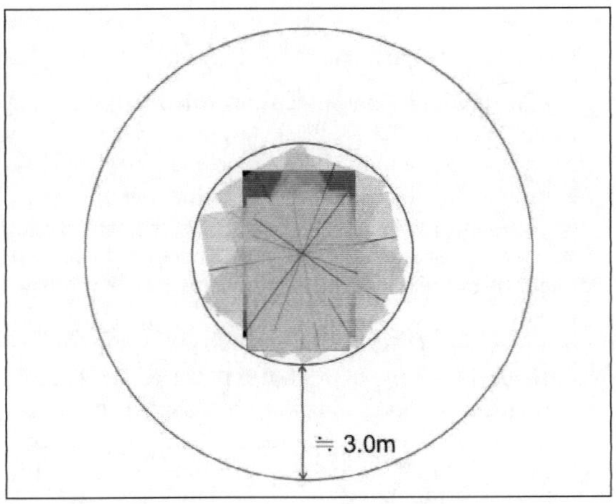

Figure 1. Example of the body rotation of the excavator from a top view

Unstructured environments are usually more complex than structured environment and have fewer features, which are obviously distinguishable [10]. Also, it is difficult to detect obstacles directly as obstacles have often irregular shapes, especially in work environment. Thus, this study proposes a different method of detecting obstacles by discarding irrelevant areas from the environment that are outside the specific range of the camera. The proposed method performs detection of nearby obstacles in a process, which is divided into three key steps presented in detail in the following subsections.

3 Methodology

3.1 Unwarping

Commercial rearview monitoring systems and surrounding view monitoring systems use wide-angle lenses in order to cover a larger area with fewer cameras. Although wide-angle lenses provide a large field of view, the images produced are severely distorted. Hence, a process called image unwarping is necessary to correct severe distortion of the images produced by the wide-angle lens. Image unwarping is used for obtaining a wide-angle view without strong aberration by both correcting lens distortion [11]. In order to exploit the large field of view of wide-angle lenses by correcting the lens distortion, this study adapted an unwarping approach proposed in Schulz et al. [12]. By generating the number of virtual views and merging them in a single image, we can re-project the original wide-angle image onto a semicylindrical surface. For this process, the semicylindrical surface is obtained for the camera with wide-angle lenses by exploiting the camera's extrinsic, intrinsic and distortion parameters [13]).

3.2 Shadow Removal

The outdoor environment in which the earthmoving equipment operates is affected by shadows such as those of obstacles, equipment, or surrounding objects or terrain features. Since image features such as color, texture, and intensity are used in the segmentation, the shadow is considered an irrelevant region because it is likely to be mistaken for the candidates of the obstacle. For this reason, once the correction of lens distortion is done prior to further processing, detection and removal of shadows is done to minimize the effect of shadows in the latter process. This study adapted the shadow detection and image restoration methods proposed by Sarabandi et al. [14] and Arévalo and Ambrosio [15]. In the study by Sarabandi et al. [15], color space transformation from red, green, and blue to $c_1c_2c_3$ was proposed since it shows the best results for detecting shadow regions in color images [16;14]. After color space transformation, a shadow region is defined as an area inside of the boundary that has a pixel value different from its surroundings. Therefore, the local variance of each pixel and its neighborhood is measured using the c_3 component to identify the shadow boundary by applying a 3-by-3 filter. The high variance value is the boundary between the shadow and non-shadow regions. After identifying this boundary, the shadow region can be detected by classifying pixels inside of the boundary as shadows. However, this method is limited for the following reasons. In an outdoor environment, c_3 may be noisy, which could cause a misleading finding of the boundary between the shadow and non-shadow regions. Also, the local variance becomes unstable for low saturation values (i.e. grey levels), which could also cause a misleading finding of the boundary between the shadow and non-shadow region (e.g., [15]). To overcome these limitations, Arevalo et al. [15] proposed an additional process to minimize the noise in c_3 and check the saturation and intensity values of pixels. Based on the shadow detection method proposed by Sarabandi et al. [14], this study adapted the advanced shadow detection method proposed by Arévalo and Ambrosio [15] and the linear-correlation method proposed by Sarabandi et al. [14] for image restoration.

3.3 Detection of Nearby Obstacles

3.3.1 Image segmentation

The ability to avoid obstacles depends on the image segmentation result, which is the process of separating obstacle candidates from the background. For this purpose, a rapid online segmentation method is proposed. Using the compact color and texture descriptor proposed by Blas et al. [17], this study integrates an intensity feature to reduce the effects of lighting changes during the segmentation process. In addition, a two-stage unsupervised online learning process is proposed. The integrated descriptors are first computed for each pixel and then clustered to find a small set of vectors or textons as the basis [18] for characterizing scene textures. In this process, each pixel is assigned to the closest texton. Then, the pixel classification result performed in the first step is refined by clustering the histograms of a small set of vectors over a window to find more coherent regions. The k-means clustering algorithm [19,20] was employed for the clustering.

3.3.2 Region of interest classification

For this purpose, the origins are defined for two different virtual cameras whose image plane are horizontal and aligned with the earth cardinal directions. The first virtual camera's origin is located in the bird's-eye position, and the second virtual camera's origin is located below the original camera's origin. The first camera's field of view is looking at the floor vertically, and the second camera's field of view is parallel to the ground. The image plane of each virtual camera is projected onto the inertial planes using an adapted geometric method based on the concept of inverse perspective transformation.

Based on the change in the pixels of the segmented region from the original image to the two different virtual images, the proposed method distinguishes whether an object has height or not and then estimates the maximum height difference. The maximum difference of the height value between the ground and each transformed pixel within the obstacle is defined as the region's height, which is used to classify whether the region of interest has a height of at least 1 m.

4 Results and Discussion

The steps from the correction of lens distortion prior to further processing, shadow removal, and detection of nearby obstacles with a predefined height level via perspective transformations were applied on video streams acquired from a camera installed on the side of the equipment body while an excavator executed excavating and moving tasks. Figure 2 shows an example of an original image acquired using a wide-angle lens with a 135° horizontal field of view. As shown in the figure, it can be seen that the worker standing in the left area is distorted as if he were tilted compared to the worker in the middle.

Figure 2. Example of an original wide-angle image

Figure 3 shows an example of the resulted unwarped image. In Figures 2 and 3(a), it can be seen that the distortion of the leftmost worker and the second worker from the left has been corrected. The adapted shadow removal process was tested with the unwarped image in Figure 3(a). Detected shadow regions were shown in white pixels in Figure 3(b).

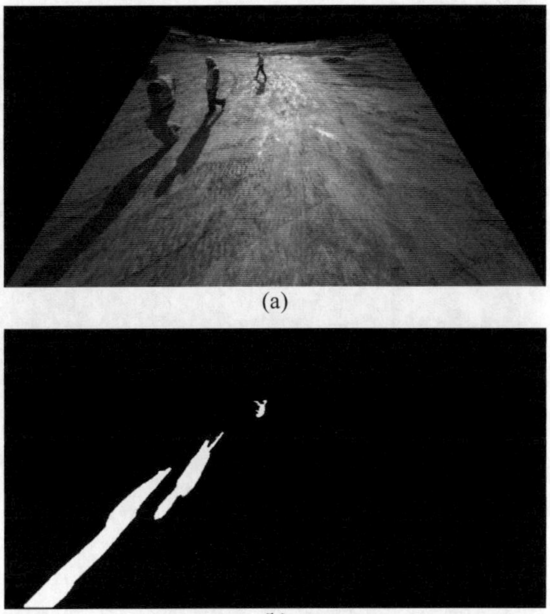

(a)

(b)

Figure 3. (a) Unwarped image of (b) Shadow detection result

Once the shadow regions are detected, the brightness of shadow pixels to the first order can be restored by a linear function. Figure 4 illustrates the image segmentation result. In the image, the dotted white dotted line represents the distance (5 m) from the camera. For this purpose, the segmentation is performed from the bottom, and the area above the white dotted line will no longer be segmented, except the areas segmented with the pixels below the white dotted line. This is also effective in reducing the time required for segmentation of unnecessary areas, such as blue areas. Once the image segmentation is performed, the area of the magenta color separated from the ground or terrain where the equipment is placed is excluded from the candidate of obstacles. Then, there are four candidates of obstacles, as shown in Figure 4.

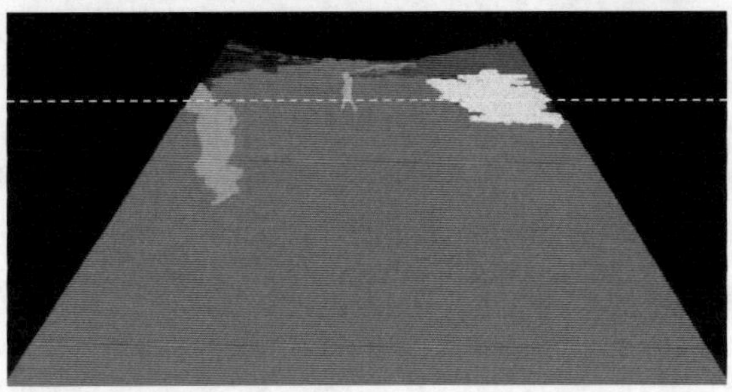

Figure 4. Image segmentation result

Among the regions of the candidates for obstacles, the regions that are determined to have height, whose maximum height difference was 1 m or more, and that were identified as obstacles were marked by boxes in Figure 5.

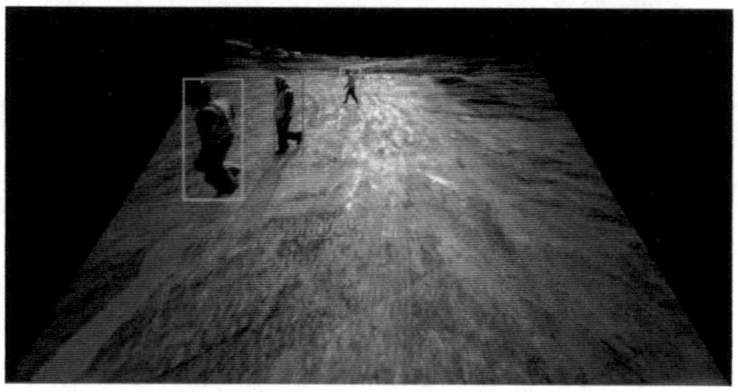

Figure 5. Region of interest classification result

5 Conclusion

This study presented a method for automated detection of nearby objects with monocular vision, with the goal of protecting the equipment operator and construction workers from potentially dangerous situations, such as collisions between earthmoving equipment and obstacles within a certain proximity. The experimental results showed that the proposed method could provide the equipment operator with information about nearby obstacles during the excavator's manipulation and transportation. It is expected that the proposed method could be implemented in rearview monitoring systems and surrounding view monitoring systems for operator assistance and to achieve active safety. Future works will focus on the estimation of collision-free area between earthmoving equipment and the obstacles. In addiition, obstacle detection is performed with planar ground assumption. Future work focuses on developing a method for detection of obstacles with no planar ground assumption by introducing an inertial sensor for acquiring 3D orientation and sensor network.

Acknowledgements

This research was supported by a grant (17SCIP-B079689-04) from Smart Civil Infrastructure Research Program funded by Ministry of Land, Infrastructure and Transport (MOLIT) of Korea government and Korea Agency for Infrastructure Technology Advancement (KAIA).

References

[1] ISO 5006: Earth-Moving Machinery – Operator's Field of View – Test Method and Performance Criteria. ISO, Geneva, 2006.

[2] Kim, C., Haas, C. T., and Liapi, K. A. Rapid, on-site spatial information acquisition and its use for infrastructure operation and maintenance. *Automation in Construction*, 14(5):666–684, 2005.

[3] Teizer, J., Caldas, C. H., and Haas, C. T. Real-time three-dimensional occupancy grid modeling for the detection and tracking of construction resources. *Journal of Construction Engineering and Management*, 133(11):880–888, 2007.

[4] Gong, J. and Caldas, C. Data processing for real-time construction site spatial modeling. *Automation in Construction*, 17(5):526–535, 2008.

[5] Son, H., Kim, C., and Choi, K. Rapid 3D object detection and modeling using range data from 3D range imaging camera for heavy equipment operation. *Automation in Construction*, 19(7):898–906, 2010.

[6] Chan, D., Buisman, H., Theobalt, C., and Thrun, S. A noise-aware filter for real-time depth upsampling. In *Proceedings of the ECCV Workshop on Multi-Camera and Multi-Modal Sensor Fusion Algorithms and Applications*, Marseille, France, 2008.

[7] Xiang, X., Li, G., Tong, J., and Pan, Z. Fast and simple super resolution for range data. In *Proceedings of the International Conference on Cyberworlds*, pages 319–324, Singapore, Singapore, 2010.

[8] Ishimoto, H. and Tsubouchi, T. Stereo vision based worker detection system for excavator. In *Proceedings of the International Symposium for Automation and Robotics in Construction (ISARC)*, pages 1004–1012, Montréal, Canada, 2013.

[9] Wybo, S., Tsishkou, D., Vestri, C., Abad, F., Bendahan, R. and Bougnoux, S. Obstacles avoidance by monocular multi-cue image analysis. In *Proceedings of the 15th World Congress on Intelligent Transport Systems*, pages 1–12, New York, NY, 2008.

[10] Yu, H., Zhu, J., Wang, Y., Jia, W., Sun, M., and Tang, Y. Obstacle classification and 3D measurement in unstructured environments based on ToF cameras. *Sensors*, 14(6):10753–10782, 2014.

[11] Bertozzi, M., Castangia, L., Cattani, S., Prioletti, A., and Versari, P. 360° Detection and tracking algorithm of both pedestrian and vehicle using fisheye images. In *Proceedings of the 2015 IEEE Intelligent Vehicles Symposium (IV)*, pages 132–137, Seoul, South Korea, 2015.

[12] Schulz, W., Enzweiler, M., and Ehlgen, T. Pedestrian recognition from a moving catadioptric camera. *Pattern Recognition. DAGM 2007. Lecture Notes in Computer Science*, 4713:456–465, 2007.

[13] Heng, L., Li, B., and Pollefeys, M. Camodocal: Automatic intrinsic and extrinsic calibration of a rig with multiple generic cameras and odometry. In *Proceedings of the IEEE/RSJ International Conference on Intelligent Robots and Systems (IROS)*, pages 1793–1800, Tokyo, Japan, 2013.

[14] Sarabandi, P., Yamazaki, F., Matsuoka, M., and Kiremidjian, A. Shadow detection and radiometric restoration in satellite high resolution images. In *Proceedings of the 2004 IEEE International Geoscience and Remote Sensing Symposium*, Anchorage, AK, 2004.

[15] Arévalo, V., González, J., and Ambrosio, G. Shadow detection in colour high-resolution satellite images. *International Journal of Remote Sensing*, 29:1945–1963, 2008.

[16] Salvador, E., Cavallaro, A., and Ebrahimi, T. Shadow identification and classification using invariant color models. In *Proceedings of the 2001 IEEE International Conference on Acoustics, Speech, and Signal Processing*, pages 1545–1548, Salt Lake City, UT, 2001.

[17] Blas, M. R., Agawal, M., Sundaresan, A., and Konolige, K. Fast color/texture segmentation for outdoor robots. In *Proceedings of the 2008 IEEE/RSJ International Conference on Intelligent Robotics and Systems*, pages 4078–4085, Nice, France, 2008.

[18] Leung, T. and Malik, J. Representing and recognizing the visual appearance of materials using three-dimensional textons. *International Journal of Computer Vision*, 43(1):29–44, 2001.

[19] Duda, R. and Hart, P. *Pattern classification and scene analysis*, Wiley, New York, NY, 1973.

[20] Jain A. and Dubes, R. *Algorithms for clustering data*, Prentice-Hall, Inc., Upper Saddle River, NJ, 1988.

Development of Smartphone Application for Real-Time Steel Rust Recognition

Wei-Tsun Liu[1], Bo-Sheng Bai[2] and Po-Han Chen[3]

[1,2,3]Department of Civil Engineering,
National Taiwan University, Taipei, Taiwan
E-mail: pohanchen@ntu.edu.tw

Abstract –

In recent years, most smartphones are equipped with high-pixel camera and powerful CPU. Our research aims to develop an Android App which can do steel rust recognition not only fast but also well. Eventually, become a smartphone application for real-time rust recognition.

Keywords –

Real-Time, Image Recognition, Smartphone

1 Introduction

This research hopes to develop an application on smartphone, which can enhance efficiency and reduce the equipment costs for bridge engineers to do regular test and maintenance. The main objectives are as follows:

1. To find out a method raising the recognition accuracy and reduce processing time of digital image.
2. To find out a method for real-time steel rust recognition, and developing an application for smartphone.

Research Scope and Limitations

Our study will be limited to steel rust recognition only, and assumes there's no other external interference in the image.

2 System Development

2.1 OSTU

Otsu's method is used to automatically perform clustering-based image threshold, or the reduction of a gray level image to a binary image. It exhaustively searches for the threshold that minimizes the intra-class variance (the variance within the class).

2.2 Canny Edge

Canny Edge is an edge detection operator that uses a multi-stage algorithm to detect a wide range of edges in images. It extracts useful structural information from different vision objects and dramatically reduce the amount of data to be processed.

2.3 Erosion & Dilation

Mathematical morphology (MM) is a technique for the analysis and processing of geometrical structures, based on set theory, lattice theory, topology, and random functions. MM is most commonly applied to digital images, but it can be employed as well on graphs, surface meshes, solids, and many other spatial structures.

Topological and geometrical continuous-space concepts such as size, shape, convexity, connectivity, and geodesic distance were introduced by MM on both continuous and discrete spaces. MM is also the foundation of morphological image processing, which consists of a set of operators that transform images according to the above characterizations.

2.4 System Architecture

The study uses three kinds of suitable algorithms by literature review. After the parameters test and adjustment, we developed the corrosion recognition system OCM (Otsu-Canny edge-Mathematical morphology).

First, the original images convert to grayscale images, and use Gaussian filter to eliminate noise. Using algorithms OTSU decided threshold value into the Canny edge algorithms. Then, Canny edge detection algorithms can identify corrosion image edge. Finally, the study use erosion and dilation by mathematical morphology method to solve the problem of broken images and eliminate noise.

In this study, we combine the corroded recognition

system OCM with smartphone equipment, and achieve research objectives.

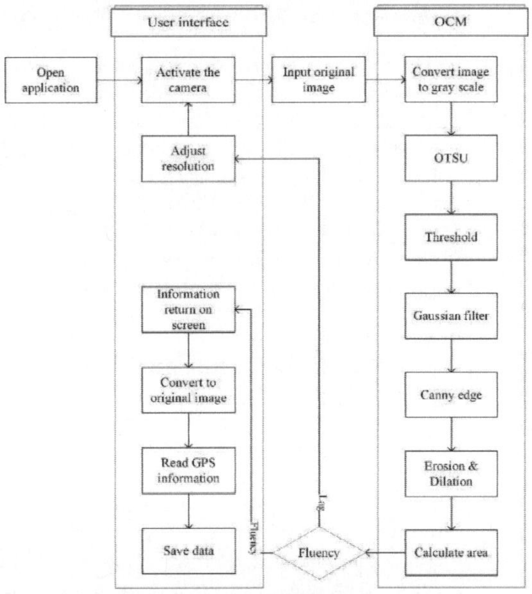

Fig 1. System Architecture

2.5 Artificial Rust Images

In this study, we produce 28 artificial images of steel coating which resolution are 3000 pixels * 4000 pixels with 7 different background color. All images are non-uniform illuminations. And, in order to imitate the real situation, we add some noise in these rust images.

3 Data Analysis and Discussion

In order to understand the pros and cons of our system, we decide to use FT-DEDA algorithm and K-means algorithm to do comparison, and using real pictures corrosion test. Because, according to the literature review, FT-DEDA is the latest and the more rapid the proposed algorithm.

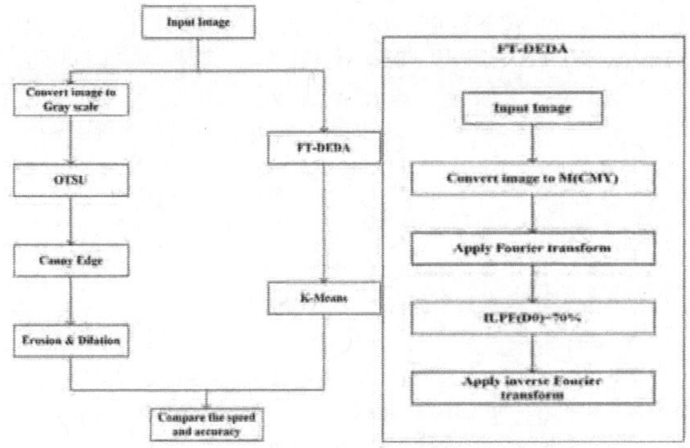

Fig 2. Data analysis and comparison

To verify the advantages and disadvantages, the study use real images to detect corrosion. Following is one of the diagrams (a) the original image, (b) according to the results of this research and development of detection systems, (c) the results of FT-DEDA has been tested with the K-means.

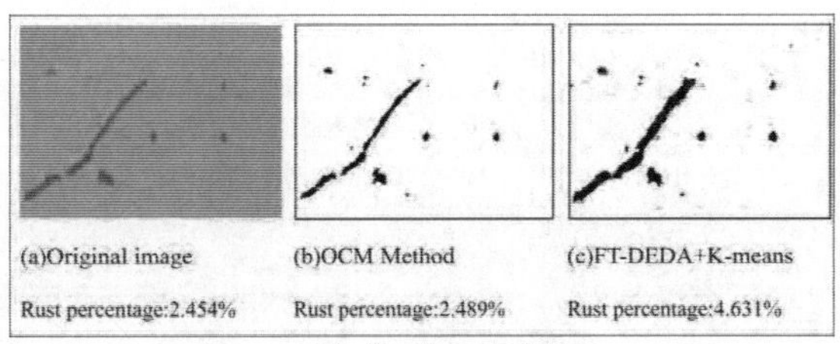

Fig 3. Comparison

4 Conclusions of Research

The research and development of corrosion recognition system OCM (Otsu-Canny edge-Mathematical morphology), according to artificial rust images and real images corrosion detection results, can be summarized in the following conclusions:

1. In this study, the corrosion detection system identifies relevant technologies through the edges, reducing more computational complexity than past detection method, can significantly reduce the detection time, with a 3000 pixels * 4000 pixels size image. For example, computation time is about 1/12 to 1/15 operation time that FT-DEDA + K- means method is required.
2. If the detection target image is a clean steel surface, ACC detection results of this research and development of corrosion recognition system falls to 96.2% to 99.5%.
3. If there are a lot areas covered by rust water, dirt or non-uniform illuminations on steel

surface, the detection accuracy can lead to errors, and ACC detection value falls to 95% to 99.7%.

References

[1] Chen, P.H., H.K. Shen, and Chang, L.M. (2012). "Support-vector-machine-based method for automated steel bridge rust assessment." Automation in Construction, Vol. 23, 9-19.

[2] Chen, P.H., Y.C. Yang, and Chang, L.M. (2009). "Automated bridge coating defect recognition using adaptive ellipse approach." Automation in Construction, 18, 632-643.

[3] Chen, P.H., Y.C. Yang, and Chang, L.M. (2009). "Box-and-Ellipse-Based ANFIS for Bridge Coating Assessment." Journal of Computing in Civil Engineering, 24, 389-398.

[4] Chen, P.H., and Chang, L.M. ·(2002). "Intelligent Steel Bridge Coating Assessment Using Neuro-Fuzzy Recognition Approach." Computer-Aided Civil and Infrastructure Engineering, 17, 307-319.

[5] Chen, P.H., Chang, L.M., et al. (2002). "Application of Multiresolution Pattern Classification to Steel Bridge Coating Assessment." Journal of Computing In Civil Engineering, 16(4), 244-251.

[6] Hugo, H., Fredrik, K., Peter, N., and Viktor, O. (2002). "A low complexity architecture for binary image erosion and dilation using structuring element decomposition." Circuits and System, International Symposium on Circuits and Systems, Vol. 4, 3431 – 3434.

[7] John, C. (1986). "A Computational Approach to Edge Detection" TRANSACTIONS ON PATTERN ANALYSIS AND MACHINE INTELLIGENCE, VOL. PAMI-8, NO. 6,679-698.

[8] Kanungo, T. (2002). "An efficient k-means clustering algorithm: analysis and implementation." Transactions on Pattern Analysis and Machine Intelligence, Vol. 24, 881-892. Lee, S. (2011). "Digital Image Processing Methods for Bridge Coating Management and Their Limitations." Journal of Civil Engineering and Architecture, 1(38), 39-47.

[9] Lee, S., and Chang, L.M., et al. (2006). "Automated recognition of surface defects using digital color image processing." Automation in Construction, 15, 540-549. Li, B., and Mehdi, S.Y. (2012). "Comparison for Image Edge Detection Algorithms." IOSR Journal of Computer Engineering, VOL. 2, 1-4.

[10] Mei, F., GuangXue, F., and Yu, Q. C. (2009). "The Study on an Application of Otsu Method in Canny Operator." International Symposium on Information Processing, August 21-23, 109-112.

[11] Nobuyki, O. (1979) "A Tlreshold Selection Method from Gray-Level Histograms" Transactions on Systems, VOL. SMC-9, NO. 1, 62-66.

[12] Raman, M., and Himanshu, A. (2009). "Study and Comparison of Various Image Edge Detection Techniques." International Journal of Image Processing, 3, 1-11. Susanta, M. (2003). "Multiscale morphological segmentation of gray-scale images." IEEE Transactions on Image Processing, VOL. 12, NO. 5, 533-549.

[13] Robert, M.H. (1987). "Image Analysis Using Mathematical Morphology." IEEE Transaction on Pattern Analysis and Machine Intelligence, VOL. PAMI-9, NO. 4,532-550.

[14] Shen, H.K. (2014), Automatic Color Image Recognition for Steel Bridge Rust Defects Assessment, Department of Civil Engineering College of Engineering National Taiwan University Doctoral Dissertation.

[15] Shen, H.K., Chen, P.H., and Chang, L.M. (2013). "Automated steel bridge coating rust defect recognition method based on color and texture feature." Automation in Construction, Vol. 4, 338-356.

Evaluating the Performance of Convolutional Neural Network for Classifying Equipment on Construction Sites

Mohammad M. Soltani[a], Seyedeh-Forough Karandish[b], Walid Ahmed[c], Zhenhua Zhu[a] and Amin Hammad[b]

[a]Department of Building, Civil and Environmental Engineering, Concordia University, Canada
[b]Concordia Institute for Information Systems Engineering, Concordia University, Canada
[c]indus.ai, Inc., Canada
E-mail: mo_solta@encs.concordia.ca, s_karand@encs.concordia.ca, walid.aly@indus.ai, zhzhu@bcee.concordia.ca, hammad@ciise.concordia.ca

Abstract –

Estimating the productivity of construction operations is one of the most challenging tasks for project managers. Therefore, the construction industry always looks toward new advancements for automating this process. New automated methods for productivity estimation aim to detect the types, locations, and activities of construction equipment based on sensory data. Computer Vision (CV) is one of the most promising automated methods and it provides an affordable opportunity for estimating the productivity since it only requires regular surveillance cameras for data collection, which are available on many construction sites. One of the widely used CV methods for classifying equipment is Histogram of Oriented Gradient (HOG). Additionally, Bag of Words (BoWs) and Local Binary Pattern (LBP) are other types of descriptors widely used for the object classification. However, these methods reduce the dimensions of the image features to train the classifiers for object detection, which may reduce the reliability of the results. Convolutional Neural Networks (CNN), which are a special type of Artificial Neural Networks (ANN) with deeper layer structure, provide a better approach for object detection compared to the conventional methods due to their deeper understanding of the object features. Furthermore, the advancements in Graphical Processing Units (GPU) made this computationally heavy method more applicable in practice. This paper aims to evaluate the performance of CNN for detecting equipment on construction sites. Several configurations of CNN are trained for detecting multiple equipment (i.e. dump trucks, excavators and loaders). The results of these configurations are compared with those of conventional detectors.

Keywords –

Computer Vision, Equipment detection, Convolutional Neural Networks, Sensory data

1 Introduction

Estimating the productivity of construction operations is one of the main concerns of the contractors since it is tied to the schedule and cost of the projects. The construction industry, with the annual revenues of more than $110 billion in Canada (Statistics Canada, 2016), is always eager to apply new methods and technologies for making the projects more cost-effective while increasing the safety on the sites. Currently, using the Global Positioning System (GPS) is the common practice in monitoring the location of the equipment (e.g. dump trucks, loaders, and excavators.). However, it is difficult to install a GPS receiver on each piece of equipment. On the other side, the availability of the surveillance cameras on many construction sites opens the opportunity for applying Computer Vision (CV) based methods to monitor the productivity of the equipment in addition to monitoring the safety and security of the sites.

Many CV-based methods have been proposed for monitoring the construction equipment, and each method has its benefits and drawbacks. Convolutional Neural Network (CNN) based methods are an emerging practice within the domain of CV. Thanks to the fast-growing advancement of Graphical Processing Units (GPUs), the applications of CNN are increasing within different engineering domains.

This paper aims to apply two comparative analyses. The first analysis compares several conventional CV-based classification methods with CNN-based methods for the classification of dump trucks, loaders, and excavators. In the second analysis, two CNN-based methods are compared using different training and testing datasets. These comparisons aim to provide a better understanding of the opportunities of using the CNN-based methods in the context of construction operations.

2 Literature Review

The applications of CV are growing fast for monitoring construction projects including construction equipment classification, detection and tracking. Among these applications, equipment classification is an important task for monitoring the resources on construction sites since tracking without recognizing the type of the tracked object does not provide enough information for decision making. Moreover, the main step in CV-based object detectors is the classification. Whether using background subtraction methods or sliding windows, it is mandatory to classify the foreground segments of the images or to classify each sliding window passing over an image to localize the target object within the image or video frame. The classifier can be either binary for each object or a multi-class classifier.

2.1 CV-based Methods in Construction

There are many methods developed for the recognition of workers and construction equipment. Zou and Kim (2007) proposed using Hue, Saturation, and Value (HSV) color space to track the equipment. Detecting the workers through the color of their hardhats was studied by Weerasinghe and Ruwanpura (2009). Azar and McCabe (2012b) investigated the applicability of fusing HOG and Haar-Like features for detecting the equipment. Moreover, they proposed applying a part-based model for detecting excavators (Azar & McCabe, 2012a). The idea of part-based models originally came from the research of Felzenszwalb et al. (2010). Memarzadeh et al. (2013) showed that integrating HOG features with Histogram of Color (HOC) improved the detection accuracy compared to relying only on HOG features. Soltani et al. (2016) proposed using a large dataset of the synthetic images for training HOG detectors.

On the other hand, there are a number of studies that focused more on the equipment classification. Basically, the moving objects on the videos are separated in each frame using background subtraction methods. Then the blobs or foregrounds are fed to the classifiers to determine the class that the foreground belongs to. Azar and McCabe (2011) trained eight classifiers from different angles around a dump truck using HOG features. The moving objects in the video were segmented after subtracting the background. They fed each segment to their classifiers to find whether it is a dump truck or not.

Chi and Caldas (2011) selected four types of feature for classification, such as aspect ratio, height-normalized area size, percentage of occupancy of the bounding box, and average gray-scaled color of the area. They compared the normal Bayes and neural network classifiers after training the aforementioned features for each classifier.

Park and Brilakis (2012a) proposed a two-stage classification approach after subtracting the background. In the first stage, they applied the classifiers based on Haar-Like features trained with Adaboosting. In the second step, they applied a Support Vector Machine (SVM) based classifier trained by Eigen-images on the candidate identified in the previous classification step. In another study, Park and Brilakis (2012b) trained HOG features using a SVM classifier to find moving people on the site. Then they applied *k*-NN classifier trained by color histogram to differentiate workers from other people.

Unfortunately, there is no benchmark dataset similar to CIFAR 10 or 100 (Krizhevsky, 2009), ImageNet (Deng, et al., 2009), or Caltech 101 or 256 (Fei-Fei.et al., 2004) within the construction domain to evaluate and compare the CV-based methods for the construction applications in a fair manner.

2.2 Recent Progress of CV-based Methods

While many industries are adopting CV-based methods for their specialized applications, these methods are advancing rapidly. Therefore, it is necessary for the researchers in the construction domain to follow the recent progress in these methods and to take advantage of them.

2.2.1 Deep Learners for CV

According to the review done by Schmidhuber (2014), excellent results were achieved in image classification using deep learning methods. Guo et al. (2015) represents these methods under four categories: CNN-based, Restricted Boltzmann Machines (RBM) based, Autoencoder-based, and Sparse Coding-based Methods. CNN have a hierarchy of convolutional layers, pooling layers, and fully connected layers. Filters or kernels play the main role by convolving the whole image to generate the feature maps. Pooling maps reduce the dimensions of the feature maps, and fully-connected layers convert the 2D feature maps to 1D feature vector. RBMs are generative stochastic neural networks (Hinton & Sejnowski, 1986) that apply restrictions to make the training algorithms more efficient. Autoencoders are another category within the deep learning domain, and they learn the efficient encodings for the imput data. These encodings can be further used for reducing the dimentionality of the data (Hinton & Salakhutdinov, 2006). Finally, Sparse Coding is the learning process of an over-complete set of basic functions for describing the input data (Olshausen & Field, 1997).

Moving to the applications of the deep learning in the CV domain, the aforementioned methods, especially CNN methods, can be used under five categories defined by Guo et al. (2015) including: (1) Image Classification, (2) Object Detection, (3) Image Retrieval, (4), Semantic Segmentation, and (5) Human Pose Estimation. Since this research aims to explore the capabilities of deep learning for classifying construction equipment, the literature regarding the CNN is discussed.

Among enormous and fast growing number of deep neural network models, there are couple of models that outperformed the others over the last five years (Canziani et al., 2016). These widely used models include, but are not limited to, AlexNet (Krizhevsky et al., 2012), Network In Network (NIN) proposed by Lin et al. (2013), GoogLeNet (Szegedy et al., 2015), ResNet-18, -34, -50, and -101 (He et al., 2016), VGG-16 and -19 (Simonyan & Zisserman, 2014), and Inception-v3 (Szegedy et al., 2016).

2.2.2 Architecture and Parameters Analysis of CNN for CV

Canziani et al. (2016) analysed and compared the aforementioned models not only from the accuracy point of view, but also considering the computation cost and efficiency of the models. As shown in Figure 1, Inception-v3 has the highest top-1 accuracy. Selecting the the best architecture only considering the accuracy may ignore the applicability of a model by not considering its computation load. For instance, although VGG-19 provides an accuracy of higher than 70%, its operations required for a single forward pass are dramatically higher than ResNet-18 or -34, which have similar accuracies to VGG. Therefore, Canziani et al. (2016) proposed comparing the models based on their information density (accuracy per parameters). This metric provides the capacity of a specific architecture to better utilise its parametric space.

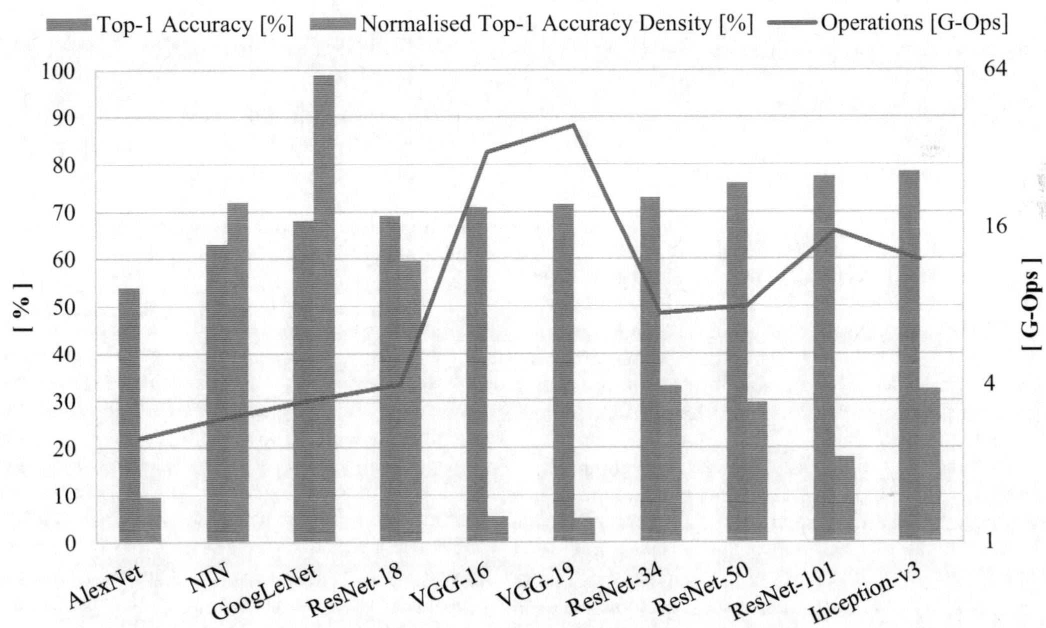

Figure 1 Current deep neural network models comparison, adapted from (Canziani et al., 2016)

Finally, they came up with the conclusion that GoogLeNet has the best architecture in terms of parameters' space utilisation.

Reviewing the recommendtions regarding the design of CNN for CV applications shows that when the network is trained using grayscale images, it can focus more on the shape (i.e. corners and edges) rather than the color. Also, it can be interpreted that if the testing images include variations of the color, the training images should also include the corresponding color variations (Zheng et al., 2015).

Mishkin et al. (2016) studied the impact of a range of choices for activation function, pooling, learning rate policy, image pre-processing (i.e. different color spaces and grayscale images), batch normalization, classifier design, batch size and learning rate, network width (e.g. changing the number of filters), dataset size and noisy labels, and bias in convolution layers. The bottom line of their research was the suggestion to use Red, Green, Blue (RGB) color space, the linear learning rate decay policy, and mini-batch size around 128 or 256. Also, they stated that having a larger dataset increases the accuracy but it also increases the computation load dramatically. Furthermore, the cleanliness of the data seems to be more important than the size of the dataset.

While collecting a large dataset and labeling the images are time-consuming, Peng et al. (2015) proposed generating the synthetic data from 3D models for training the deep learners. They had a prior study on the applicability of synthetic images for classification based on HOG features using SVM (Sun & Saenko, 2014). They explored the minimum Average Precision (mAP) of the CNN-based classifiers while using real images, virtual images with uniform gray texture (V-GRAY), and virtual images with real texture (V-TX). The results show that V-TX had the best performance while V-GRAY was the worst one.

3 Scope Definition of Analyses

One of the goals of this paper is to compare the performance of the conventional handmade features such as Bags of Words (BoW), Local Binary Pattern (LBP), and HOG with CNN-based features. For classifying the conventional features, SVM is used while for CNN-based features both SVM and ANN are applied. Three widely used equipment on construction sites are excavators, loaders, and dump trucks, which are the focus of this study.

The first descriptor is used for the comparative analysis is BoW or bag of keypoints that use the vector quantization of affine invariant descriptors of image patches. In the original study, Naïve Bayes and SVM classifiers were used because of their simplicity (Csurka et al., 2004); however, this paper only implements SVM for BoW and all other conventional descriptors. The second descriptor, LBP, recognizes certain local binary patterns, known as uniform. The occurrence histogram of these patterns is supposed to be very strong in terms of texture feature (Ojala et al., 2002). HOG is the next descriptor that uses locally normalized histogram of gradient orientations features in a dense overlapping grid (Dalal & Triggs, 2005).

Two CNNs, AlexNet and VGG-f (Chatfield et al., 2014) are considered in the first analysis, because of their relatively simple architectures compared to the very deep and complex networks such as Inception-v3 or ResNet-101. AlexNet is used in this study folowing two approaches. In the first approach, one of the layers of the pretrained AlexNet is used for extracting the features of the training dataset and then the features are fed to SVM for training the classifiers and it is shown as AlexNet-SVM in Table 2. In the second approach, the architecture of AlexNet is used for creating a new network but for classifying three classes instead of its original 1000 classes. The second approach is also repeated for VGG-f to train a network with the similar architecture but for three classes.

Moreover, the performance of the construction equipment classification is evaluated within the domain of CNN. Various configurations and datasets are used to reach to a conclusion for classifying the construction equipment. Multiple training and testing datasets of real, synthetic, and mixed real-synthetic images are used with different numbers of images in each dataset. AlexNet is used independently and in collaboration with SVM as described above.

4 Implementation and Comparative Results

The implementation was done in Matlab 9.1 (Mathworks, 2016) on a mobile station with Intel i7 Quad-Core processor, 32 gigabytes Random-Access Memory (RAM), and NVIDIA Quadro K2000M graphics card. Five datasets were used for both training and testing phases as shown in Table 1.

4.1 Conventional versus CNN-based Methods

In this test, five main types of classifiers were investigated: BoW, LBP, and HOG integrated with SVM as the conventional methods and AlexNet integrated with SVM, independent AlexNet, and independent VGG-f as CNN-based methods. As hown in Table 2, one configuration with 500 clusters was used for extracting BoW features while four cell sizes for LBP features and six cell sizes for HOG features were tested. Three layers were selected from AlexNet considering the available computation resources in this research. Two layers were close to the end of the network and one layer was near the starting border of the network. Selecting the features from both end of the network helps to compare the effectiveness of the feature at different level of the CNN architecture. The features obtained from three different layers of the original AlexNet network were used separately for training the SVM classifiers. Due to the limitation of the computation capacity, only the first convolutional layer was tested while the other two layers were from the fully connected layers at the end of the network structure. The last two classifiers had similar structure to AlexNet and VGG-f, respectively, except the last layer (classification layer), which has three classes instead of 1000 classes in the original model. Also, in another scenario, AlexNet and VGG-f were fed and tested by gray scale images. For training each of the aforementioned classifier, the datasets of synthetic images with 6,000 images (first row) shown in Table 1 were used while for testing the dataset of real images was used with 1,169 images was used (second row).

Due to space limitation, only the accuracies of correct detections for each class from the confusion matrix are shown in Table 2 in addition to the average accuracy of the all classes. The averages accuracy of BoW was 38%, while the best achieved average accuracies for LBP and HOG were 41% and 47%, respectively. The results show that all the conventional classifiers had more difficulties for classifying the loaders. Investigating the poor accuracies for the loader shows that the 3D model of the loader used for creating the synthetic images looks similar to a brand new loader which has usually a yellow bucket. However, the bucket of the loader looks very dark close to black or dark brown color and the bucket of loader includes a large portion of its appearance. Opposite to HOG-based method, it is highly possible that CNN can be sensitive to the color of the object during training phase. On the other hand, the best average accuracies of AlexNet-SVM, AlexNet, and VGG-f were 83%, 78%, and 68%, respectively. It can be concluded from this analysis that CNN-based methods outperformed the conventional methods.

Table 1 Image Datasets Specifications

		Excavator	Loader	Truck	Total
Synthetic		2,000	2,000	2,000	6,000
Real		555	267	347	1,169
Real		100	100	100	300
Mixed	**Real**	555	267	347	2,338
	Synthetic	555	267	347	
Mixed	**Real**	100	100	100	600
	Synthetic	100	100	100	

4.2 Performance Evaluation of CNN-based Method on Different Datasets

At this stage, the best two groups of classfiers (trained for three classes of equipment) resulting from the previous test in Section 4.1 were used for a more detailed analysis and testing. The two groups are the AlexNet-SVM classifiers with 'fc7' layer and the independent AlexNet classifiers.

The purpose of this test is to find how the accuracy of each classifier can be be affected using the five datasets shown in Table 1. There are two different datasets of real images, one with 1,169 images and the other with a small number of images (300 in total) to find the impact of the number of the training images. Also, two more datasets are prepared by adding the synthetic images with the same number of the real images in each of two previous datasets. The advantage of the synthetic images is that they are taken from all views around the equipment while the real images are mostly captured from low heights and views. A sample of real and synthetic images are shown in Figure 2 One the other hand, the synthetic images may look artificial and that may have negative effects on the CNN-based classifiers compared to conventional methods that did not show such effects on the accuracy (Soltani et al., 2016). Therefore, using the mixed datasets can evaluate the generality of the classifiers on different brands, colors, and views of the equipment.

Twenty scenarios are created and tested, which are shown in Table 3. Starting with AlexNet-SVM classifiers, the results of the classifiers trained by synthetic images and tested on real images are 83% and 86%. The accuracy achieved by classifying the dataset with smaller number of images shows better performance as expected. In the next scenarios, the larger dataset of the real images was used for training and the smaller datasets of real and mixed images were classified. The results show that adding synthetic images from various views reduces the accuracy. By using the two larger datasets of mixed images for the training, the accuracies were improved for both previousely tested datasets. In the next comparison, the smaller dataset of real images was used to mimic the situations where there is a very limited number of available training images. This classifier is applied on two larger datasets of real and mixed images and the accuracies dropped significantly. In the last two scenarios, the smaller dataset of the mixed images was used for training and applied on the larger datasets of real and mixed images. It is clear from this test that we can improve the quality of training (and the resulting accuracy) by adding synthetic images to the training dataset when the number of real images is small.

In the next test, similar comparisons were repeated for evaluating the independent AlexNet CNN. The trend of the results is close to the previouse test except in the scenario where the larger dataset of real images is used for training and applied on the smaller datasets of real and mixed images. Opposite to the AlexNet-SVM classifier, the independent AlexNet outperfomed on the smaller dataset of the mixed images compared to the smaller dataset of the realimages. Additionally, using the larger dataset of mixed images shows its maximum performance on the smaller datasets of the real and mixed images. However, achieving the accuracy of 100% does not mean that this classifier is able to reach the same accuracy on the larger datasets and it simply proves that adding synthetic images to real images during the training can improve the performance of the AlexNet classifier.

Figure 2 Sample real and synthetic images

5 Conclusion

Reviewing the literature showed that while the applications of CNN-based methods are growing very fast, the construction industry is lagging in taking advantage of these methods. Therefore, this paper investigated the applicability of CNN-based methods while comparing them with the conventional descriptors. In the first analysis, BoW, LBP, HOG, AlexNet independently and integrated with SVM, and VGG-f networks were investigated. CNN-based methods clearly outperformed the conventional methods. In the second test, AlexNet and AlexNet-SVM were compared using different image datasets including real, synthetic, and mixture of real and synthetic images with different number of images in the training and testing datasets. The results show that it is better to add synthetic images to the real images for the training of the classifiers, especially when there is limited number of available real images. Including the synthetic images to the real images helps the classifier consider various views of the target objects which are not available in the real image dataset.

On the other hand, training and applying CNN-based requires a very powerful computer configuration, especially when developing a very deep network is required. It is recommended to include more classes of the equipment in the future. Moreover, investigating the performance of CNN-based object detectors for localizing the equipment is the next step.

References

[1] J. Zou and H. Kim, "Using hue, saturation, and value color space for hydraulic excavator idle time analysis," *Journal of computing in civil engineering,* vol. 21, no. 4, pp. 238-246, 2007.

[2] I. T. Weerasinghe and J. Y. Ruwanpura, "Automated data acquisition system to assess construction worker performance," in *In Building a Sustainable Future, Construction Research Congress*, Seattle, WA, USA, 2009.

[3] E. R. Azar and B. McCabe, "Vision-based Recognition of Dirt Loading Cycles in Construction Sites," in *Construction Research Congress*, West Lafayette, IN, USA, 2012.

[4] E. R. Azar and B. McCabe, "Part based model and spatial–temporal reasoning to recognize hydraulic excavators in construction images and videos," *Automation in construction,* vol. 24, pp. 194-202, 2012.

[5] P. F. Felzenszwalb, R. B. Girshick, D. McAllester and D. Ramanan, "Object detection with discriminatively trained part-based models," *IEEE Transactions on Pattern Analysis and Machine Intelligence,* vol. 32, no. 9, pp. 1627-1645, 2010.

[6] M. Memarzadeh, M. Golparvar-Fard and J. C. Niebles, "Automated 2D detection of construction equipment and workers from site video streams using histograms of oriented gradients and colors," *Automation in Construction,* vol. 32, pp. 24-37, 2013.

[7] M. M. Soltani, Z. Zhu and A. Hammad, "Automated annotation for visual recognition of construction resources using synthetic images," *Automation in Construction,* vol. 62, pp. 14-23, 2016.

[8] E. R. Azar and B. McCabe, "Automated visual recognition of dump trucks in construction videos," *Journal of Computing in Civil Engineering,* vol. 26, no. 6, pp. 769-781, 2011.

[9] M. W. Park and I. Brilakis, "Construction Worker Detection in Video Frames for Initializing Vision Trackers," *Automation in Construction,* vol. 28, pp. 15-25, 2012b.

[10] M. W. Park and I. Brilakis, "Enhancement of construction equipment detection in video frames by combining with tracking," in *International Workshop on Computing in Civil Engineering*, 2012a.

[11] A. Krizhevsky, "The CIFAR-10 dataset," 2009.

[Online]. Available: https://www.cs.toronto.edu/~kriz/cifar.html. [Accessed 03 February 2017].

[12] J. Deng, W. Dong, R. Socher, L. J. Li, K. Li and L. Fei-Fei, "ImageNet: A large-scale hierarchical image database," in *IEEE Conference on Computer Vision and Pattern Recognition CVPR 2009*, 2009.

[13] L. Fei-Fei, R. Fergus and P. Perona., "Learning generative visual models from few training examples: an incremental Bayesian approach tested on 101 object categories," in *CVPR 2004, Workshop on Generative-Model Based Vision*, 2004.

[14] G. E. Hinton and T. J. Sejnowski, "Learning and releaming in Boltzmann machines," in *Parallel Distrilmted Processing*, 1986, p. 4.2.

[15] G. E. Hinton and R. R. Salakhutdinov, "Reducing the dimensionality of data with neural networks," *Science,* vol. 313, no. 5786, pp. 504-507, 2006.

[16] B. A. Olshausen and D. J. Field, "Sparse coding with an overcomplete basis set: A strategy employed by V1?," *Vision research,* vol. 37, no. 23, pp. 3311-3325, 1997.

[17] Y. Guo, Y. Liu, A. Oerlemans, S. Lao, S. Wu and M. S. Lew, "Deep learning for visual understanding: A review," *Neurocomputing,* vol. 187, pp. 27-48, 2015.

[18] J. Schmidhuber, "Deep learning in neural networks: An overview," *Neural networks,* vol. 61, pp. 85-117, 2014.

[19] A. Krizhevsky, I. Sutskever and G. E. Hinton, "ImageNet classification with deep convolutional neural networks," in *Advances in Neural Information Processing Systems 25 (NIPS 2012)*, 2012.

[20] M. Lin, Chen, Q. and S. Yan, "Network in network," *ArXiv e-prints,* 2013.

[21] C. Szegedy, W. Liu, Y. Jia, P. Sermanet, S. Reed, D. Anguelov and A. Rabinovich, "Going deeper with convolutions," in *Proceedings of the IEEE Conference on Computer Vision and Pattern Recognition*, 2015.

[22] K. He, X. Zhang, S. Ren and J. Sun, "Deep residual learning for image recognition," in *Proceedings of the IEEE Conference on Computer Vision and Pattern Recognition*, 2016.

[23] K. Simonyan and A. Zisserman, "Very deep convolutional networks for large-scale image recognition," *ArXiv e-prints,* 2014.

[24] C. Szegedy, V. Vanhoucke, S. Ioffe, J. Shlens and Z. Wojna, "Rethinking the inception architecture for computer vision," in *Proceedings of the IEEE Conference on Computer Vision and Pattern Recognition*, 2016.

[25] A. Canziani, A. Paszke and E. Culurciello, "An Analysis of Deep Neural Network Models for Practical Applications," *ArXiv e-prints,* 2016.

[26] Z. Zheng, Z. Li, A. Nagar and W. Kang, "Compact deep convolutional neural networks for image classification," in *International Conference on Multimedia and Expo (ICME 2015)*, Torino, Italy, 2015.

[27] D. Mishkin, N. Sergievskiy and J. Matas, "Systematic evaluation of CNN advances on the ImageNet," *ArXiv e-prints,* 2016.

[28] X. Peng, B. Sun, K. Ali and K. Saenko, "Learning deep object detectors from 3D models," in *Proceedings of the IEEE International Conference on Computer Vision*, 2015.

[29] B. Sun and K. Saenko, "From Virtual to Reality: Fast Adaptation of Virtual Object Detectors to Real Domains," *BMVC,* vol. 1, no. 2, p. 3, 2014.

[30] G. Csurka, C. Dance, L. Fan, J. Willamowski and C. Bray, "Visual categorization with bags of keypoints," in *In Workshop on statistical learning in computer vision, ECCV*, 2004.

[31] T. Ojala, M. Pietikainen and T. Maenpaa, "Multiresolution gray-scale and rotation invariant texture classification with local binary patterns," *IEEE Transactions on pattern analysis and machine intelligence,* vol. 24, no. 7, pp. 971-987, 2002.

[32] N. Dalal and B. Triggs, "Histograms of Oriented Gradients for Human Detection," in *IEEE Conference on Computer Vision and Pattern Recognition*, San Diego, CA, USA, 2005.

[33] K. Chatfield, K. Simonyan, A. Vedaldi and A. Zisserman, "Return of the devil in the details: Delving deep into convolutional nets," *ArXiv e-prints,* 2014.

[34] Mathworks, "MATLAB," 2016. [Online]. Available: http://www.mathworks.com/. [Accessed 21 December 2016].

[35] Statistics Canada, "Construction (NAICS 23) : Gross domestic product (GDP)," 2016. [Online]. Available: https://www.ic.gc.ca/app/scr/sbms/sbb/cis/gdp.html?code=23&lang=eng#gdp2a. [Accessed 05 January 2016].

[36] S. Chi and C. H. Caldas, "Automated object identification using optical video cameras on construction sites," *Computer- Aided Civil and Infrastructure Engineering,* vol. 26, no. 5, pp. 368-380, 2011.

Table 2 Results of comparative analysis between conventional and CNN-based classifiers

Method	**Size of Images**	**Additional Information**	**Accuracy (%)**			
			Excavator	Loader	Truck	Average
Bag of Words	128×128	500 Clusters	78	3	33	38
LBP-SVM	128×128	4×4 Cells	72	3	39	38
	128×128	8×8 Cells	72	5	38	38
	128×128	16×16 Cells	76	5	41	41
	128×128	32×32 Cells	51	10	56	39
HOG-SVM	128×128	2×2 Cells	69	1	35	35
	128×128	4×4 Cells	70	1	39	37
	128×128	8×8 Cells	75	3	43	40
	128×128	16×16 Cells	73	5	61	46
	128×128	32×32 Cells	55	21	65	47
	128×128	64×64 Cells	41	45	35	40
AlexNet-SVM	227×227	conv 1 Layer	51	42	60	51
	227×227	fc7 Layer	76	78	93	83
	227×227	fc8 Layer	72	60	96	76
AlexNet	227×227	Colored	74	48	99	74
	227×227	Grayscale	95	60	78	78
VGG-f	224×224	Colored	70	40	94	68
	224×224	Grayscale	92	10	52	51

Table 3 Results of comparative analysis on AlexNet-SVM and AlexNet classifiers

Method	**Training Images**		**Testing Images**		**Accuracy (%)**			
	Type	Number	Type	Number	Excavator	Loader	Truck	Average
AlexNet-SVM	Synthetic	6,000	Real	1,169	76	78	93	83
	Synthetic	6,000	Real	300	95	83	79	86
	Real	1,169	Real	300	89	100	95	94
	Real	1,169	Mixed	600	94	88	93	91
	Mixed	2,338	Real	300	94	100	96	97
	Mixed	2,338	Mixed	600	97	100	98	98
	Real	300	Real	1,169	54	66	100	73
	Real	300	Mixed	2,338	78	69	97	81
	Mixed	600	Real	1,169	83	90	99	90
	Mixed	600	Mixed	2,338	91	95	99	95
AlexNet	Synthetic	6,000	Real	1,169	74	48	99	74
	Synthetic	6,000	Real	300	95	51	97	81
	Real	1,169	Real	300	88	100	95	94
	Real	1,169	Mixed	600	99	94	97	97
	Mixed	2,338	Real	300	100	100	100	100
	Mixed	2,338	Mixed	600	100	100	100	100
	Mixed	600	Real	1,169	77	88	99	88
	Mixed	600	Mixed	2,338	88	94	100	94
	Real	300	Real	1,169	81	67	100	83
	Real	300	Mixed	2,338	100	71	96	89

34th International Symposium on Automation and Robotics in Construction

ISARC 2017

June 28 – July 1 TAIPEI, TAIWAN

A Risk-based Critical Path Scheduling Method (I) : Model and Prototype Application System

H.K. Chang[a] W.D. Yu[b]* and S.T. Cheng[c]

[a]Department of Civil Engineering, Chung Hua University, Taiwan
[b]*Department of Construction Engineering, Chaoyang University of Technology, Taiwan, Corresponding Author
[c]Department of Construction Management, Chung Hua University, Taiwan
E-mail: d10304006@chu.edu.tw, wenderyu@cyut.edu.tw, shaotsai@chu.edu.tw

Abstract

The traditional cost-based Earned Value Management (EVM) method, and the related Earned Schedule Management (ESM) and Earned Duration Management (EDM), for construction project schedule control has been founded on the basis of Planned Value (PV). However, the relevant risk factors are usually not taken into account while the scheduler is estimating the duration of an activity. As a result, *de facto* value of PV is far away from the original planned one when the risk scenarios occur to the activities. To tackle the abovementioned problem, the current research proposes a Risk-based Critical Path Scheduling Method (R-CPSM) that assist the scheduler to take into account the spatial and various resource constraints as well as the environmental influences when he/she is planning the schedule of a project. A computer program to implement the proposed R-CPSM, namely Risk-based Critical Path Scheduling System (R-CPSS), is developed to test the feasibility of R-CPSM. The proposed R-CPSS is proved to be able to improve the problems of the traditional EVM schedule control methods due to the inaccurate estimation of the PV.

Keywords: Risk management, EVM, Construction Planning, Schedule Control.

1 Introduction

The cost-based Earned Value Management (EVM) has been the most prevailing tool adopted for construction schedule control. Deriving from EVM, the Earned Schedule Management (ESM) replaces the original cost-based Schedule Performance Index (SPI_c) with the time-based Schedule Performance Index (SPI_t) [1][2][3][4]. On the other hand, Earned Duration Management (EDM), which uses activity duration for evaluation, replaced SPI_c with Duration Performance Index (DPI_t) [6]. Despite the differences, the evaluation of schedule performance for EVM, ESM or EDM are all based on Planned Value (PV), which is determined by the estimated schedule (*i.e.*, Planned Value, PV. As a result, the accuracy of the performance evaluation usually depends on the accuracy of PV. However, when it comes to estimation, engineers usually ignore the influences of various risk factors that acts on the individual project schedule, and the their impacts on the project critical and uncritical paths. Therefore, when the environmental and resource risk factors divert, the estimated PV would be apart from *de facto* value, and eventually leads to the failure of project control, in spite of which method adopted (*i.e.*, EVM, ESM or EDM).

The Precedence Diagramming Method (PDM) based on traditional Critical Path Method (CPM) is the prevailing scheduling method currently adopted in the construction industry. Traditional CPM scheduling usually considers the duration of a construction activity as a fixed value. The Program Evaluation and Review Techniques (PERT) adopting probabilistic durations are not popular due to its complicated theoretical backgrounds and the lack of available computer software. It is desired to develop a scheduling method taking into account the risk factors of activity duration, so that the accuracy and efficiency of traditional project control method can be improved.

As mentioned above, this study aims to propose a Risk-based Critical Path Scheduling Method (R-CPSM) and the implementation system, namely Risk-based Critical Path Scheduling System (R-CPSS). The objectives are to improve the accuracy of PV estimation and the usefulness of the risk-based scheduling method.

2 Review of Related Works

The relevant literature reviewed in this section include: (1) various scheduling techniques; and (2) the integration of risk management and scheduling.

2.1 Various Scheduling Techniques

Previous important developments of traditional scheduling techniques include: Gantt Chart in 1910, CPM and PERT in 1958, Linear Scheduling Method in 1980, and the Repetitive Scheduling Method (RSM) after 1990. The review of these methods can be referred to [7][8]. In the following, these techniques are briefly reviewed:

- Bar Chart

Proposed by Henry Gantt in 1917, therefore also known as Gantt Chart. The original Bar Chart is not capable of schedule analysis because it does not show the relationship between activities. To fix this problem, current commercial scheduling software have combined Bar Chart with CPM to allow Arrow Diagramming Method (ADM) calculation, namely the Enhanced Bar Chart.

- Critical Path Method (CPM)

Differed by network diagrams, CPM can be categorized into Arrow Diagramming Method (ADM), and Precedence Diagramming Method (PDM), Activity on Node (AON). CPM was proposed by Dupont Inc. in 1958. The process of CPM calculation includes Forward Pass, Backward Pass, Float Computation, etc. Activities with zero Total Float connect the longest path, which is called Critical Path (CP) on the network and is the focus for schedule control.

- Program Evaluation and Review Technique

PERT was developed by the Polaris Missile Project of US Special Project Agency in 1957. Considering the Uncertainty of operation conditions and the probabilistic duration of activities, PERT's schedule includes three different time estimations—Optimistic, Most likely, and Pessimistic. With network diagramming technique and probability calculation, PERT is capable of estimating the possible duration and probability of a single activity or the whole project. Besides, it is the first computerized scheduling and controlling method.

- Linear Scheduling Method

Proposed by Johnston in 1981 [8], The Linear Schedule Method uses a line to represent the efficiency of a construction activity. The slope of Linear Scheduling Method diagram is the acquirable workload of a specific activity per unit time, namely the speed of specific activity. Activities can be categorized, by their nature, into Linear Activity, Bar Activity, and Block Activity. The aim of Linear Scheduling Method is to improve the efficiency of the traditional CPM while dealing with repetitive activities.

- Linear Scheduling Model (LSM):

Originated form Harmelink's effort [11] to improve the Linear Scheduling Method in 1995, the Linear Scheduling Model (LSM) categorizes activities by distinguishing their nature with geometrical graphics. In 1998, by analysing the logic relationship between activities, Harmelink built a theoretical similar to the controlling activity path of CPM, which better reflects to the actual condition of the construction work on site than the previous repetitive scheduling techniques. After 2001, Harmelink and other researchers [10][11][12] proposed the float time calculation method for LSM, so they make the theoretical framework of LSM more complete.

- Repetitive Scheduling Method (RSM)

In 1998, Harris and Ioannou developed the Repetitive Scheduling Method (RSM) [13][14][15] by considering the characteristics of repetitive scheduling including the resource continuity of activities, and the convergence, divergence, and parallel of activities relationships. Their method takes the horizontal axis as time and the vertical axis as production. The activities are represented as linear lines. With a simplified graph, RSM is able to represent the relationships between activities and the continuity of resources. It then indicates the performance of the critical path based on logic relationships among the activities.

- Critical Path Segments Scheduling Technique

The Critical Path Segments Scheduling Technique was proposed by Hegazy *et al.* [16]. It aims at resolving the problem of traditional critical path scheduling by taking continuous time as a whole segment. It divides the activities into different time segments to avoid over-complicated networks and is able to identify the impact of critical path on the project schedule more accurately. Therefore, it helps project managers distribute resources more effectively during scheduling.

- Enhanced Arrow Diagramming Method (EADM)

Enhanced Arrow Diagramming Method (EADM) is a new form of scheduling method proposed by Chang [7]. It shows the time scale, construction location, activity sequential logics and activity duration at the same time, and it is capable of improving traditional Arrow Diagram Method [17].

Among the abovementioned scheduling methods, EADM is more suitable to provide the required functions of the Risk-based Critical Path Scheduling Method in the current research. As a result, this study adopts CPM and EADM as the theoretical backgrounds for the proposed Risk-based Critical Path Scheduling Method.

2.2 The Integration of Risk Management and Scheduling

Risk is the probability of events that may cause negative impacts on the objectives. Since the essential nature of a risk is uncertainty, the primary objective of risk analysis is to evaluate whether we should undertake or avoid such risks [18]. There are three major features of risk: (1) Probability—risk is an objective existence, although can be reduced, it cannot be completely eliminated; (2) Uncertainty—if we focus on individual event, whether, how and what it might impact the result is random, which makes it hard to predict with a single event; (3) Predictability—even though single risk event happens randomly, a collection of risk events may occurs complying with a probability distribution, which makes the occurrence of the risk events predictable.

There are three characteristics of a risk event: (1) Uncertainty—whether or how the event might cause the project must be uncertain; (2) Futurity—risk event must be something hasn't happened yet, those already happened can be seen as loss or gain; and (3) Possibility of loss or gain—a risk event must bring a loss or gain as a result to the project. The risk category can categorize risks by their causes or characteristics, thus it help fulfil the needs of risk management. The structure of the category usually contains a major category that can be furtherly derived into layers of subcategories by the nature of risk events. Finally, it will form a Risk breakdown structure (BRS) [18].

In order to better tackle the risk in a construction activity, Yi and Langford [24] proposed an equation to express all risks involved in a construction activity as bellow:

$$\text{Total Risk} = P \times H \times T \times E \tag{1}$$

where, P means "Process Risk"; H means "Human Risk"; T means "Technology Risk", E means "Physical Environment Risk"; and the symbol "×" implies that the impacts of the risk factors to the schedule is a product of the impacts caused by the individual risk factors.

Yi and Langford believed that the first three risk factors of a construction activity (*i.e.*, process, human, and resource) are determined by the type of the construction activity. They can be broken down by referring to the Bill of Quantity (BOQ) of the activity. The fourth risk factor, Physical Environment Risk, is determined by the location and surrounding conditions of the work site. Such a risk breakdown concept is very useful for analysis of the risk factors affecting construction schedule. However, no systematic approach was proposed by Yi and Langford to take the above-mentioned factors in construction scheduling process. In this paper, we develop a systematic risk-based critical path scheduling method (namely, R-CPSM) to take into account all risk factors referred by Yi and Langford [24].

Other literature related to the integration between risk management and scheduling includes: (1) Application of risk analysis to float utilization and optimization [20][21]; (2) Application of risk evaluation to scheduling optimization [22][23]; (3) Risk-based scheduling and safety programming [24]; (4) Algorithm of risk reduction [25]; (5) Integration of risk management and scheduling technique [26]; (6) Case study of risk-based scheduling application [27]; (7) Quantitative method for risk analysis [28][29]; (8) Integration of schedule control and risk information [30]. From the above literature, it is found that the integration of risk management with scheduling techniques can be beneficial to project management.

3 Risk-based Critical Path Scheduling Model (R-CPSM)

In this paper, a Risk-based Critical Path Scheduling Model (namely R-CPSM) is proposed. This section is dedicated to the model development of R-CPM.

3.1 The Seven Risk Levels of Construction Activity Duration

Inspired by the safety planning method proposed by Yi and Langford [24], the duration of a construction activity can be affected by the integration of the individual risk factors on that activity. It is believed that the spatial restriction of the construction site is the essential constraint for the duration to complete an activity. As a result, the estimated duration for a construction activity considering the spatial constraint is called "base duration". The base duration can be deemed as the shortest possible duration required to complete the activity without occurrence of any risk event.

The second constraint for the duration of completing an activity is the "physical environment risk". The physical environment risk considering all factors of the surrounding environment on the construction site, *e.g.*, weather, local social factors, and local cultural factors. The impact of such factors to the activity duration is usually difficult to estimate and is usually measured by experienced engineers who are familiar with the surrounding environment factors. As a result, while take the physical environment risk into account, the modified activity duration is called "empirical duration". The estimation of the empirical duration can be obtained by adding an extra duration to the base duration, and thus it is longer than base duration.

Beside the physical environment risk, which is

inevitable, there are five categories (*i.e.*, the 5 Ms of construction management) of resource risks that can be altered by management schemes including: (1) Man—the availability of different skilled or unskilled laborers to perform a construction activity; (2) Machine—the availability of required equipment to perform a construction activity; (3) Material—the availability of required materials to complete a construction activity; (4) Method—the availability of appropriate construction methods to perform a construction activity; and (5) Money—availability of required financial arrangements to conduct a construction activity. Should any of the above resources be absent, the completion of a construction activity will be delayed and the duration will be lengthened.

Considering all the above risk factors, seven risk levels are classified for estimating the duration of a construction activity:

- Level-0 duration—no risk event occurs and the physical environment risk contributes no extra duration to the activity, this is also considered as the base and shortest duration of a activity. The Level-0 duration is denoted as RD_0 and calculate by the following equation:

$$RD_0 = \frac{\text{Quantity}}{\text{Rate}} \quad (2)$$

Where, "Quantity" is the quantity of product to be produced by the activity; "Rate" is the productivity rate that can be referred to any cost estimation reference, *e.g.*, RSMeans Book (https://www.rsmeans.com) or Dodge Estimating Guide (https://construction.com); RD_0 is the "base duration" of a construction activity.

- Level-1 duration—only the physical environment risk but no any other resource risk occurs, so the extra duration caused by the physical environment risk estimated by the experienced engineer is added to the base duration resulting in the "empirical duration". The Level-1 duration is denoted as RD_1 and calculate by the following equation:

$$RD_1 = RD_0 + (\text{Environment Effect}) \quad (3)$$

Where, RD_0 is " base duration of the activity; "Environment Effect" is the extra duration caused by the physical environment risk and is estimated by the experienced engineer. RD_1 is the "empirical duration".

- Level-2 duration—both the physical environment risk and one resource risk occur, so the extra duration caused by the occurred resource risk estimated by the experienced engineer is added to empirical duration, resulting in Level-2 duration.
- Level-3 duration—not only the physical environment risk but also two of the five resource risks occur, so the extra durations caused by the two occurred resource risks estimated by the experienced engineer are added to empirical duration, resulting in Level-3 duration;
- Level-4 duration—not only the physical environment risk but also three of the five resource risks occur, so the extra durations caused by the three occurred resource risks estimated by the experienced engineer are added to empirical duration, resulting in Level-4 duration;
- Level-5 duration—not only the physical environment risk but also four of the five resource risks occur, so the extra durations caused by the four occurred resource risks estimated by the experienced engineer are added to empirical duration, resulting in Level-5 duration;
- Level-6 duration—not only the physical environment risk but also all five resource risks occur, so the extra durations caused by the five occurred resource risks estimated by the experienced engineer are added to empirical duration resulting in Level-6 duration, and this is considered the worst case scenario of the activity duration.

The equation for estimating the activity duration for Risk Level-2~ 6 is described in the following:

At first, the most significant resource risk factor for each of the five resource types is identified as M_x using the following equation:

$$M_x = \text{Max}\ (M_{x1}, M_{x2}, M_{x3}, M_{x4} \ldots\ldots, M_{xn}) \quad (4)$$

Where, M_x represents the most significant resource risk factor in a specific resource type (*e.g.*, M_1 means the most significant risk factor for human resources, M_2 means the most significant risk factor for machine resources, and so forth.); M_{x1}, M_{x2}, ...M_{xn} are the possible resource factors for a specific resource type (*e.g.*, M_{23} means the third type of machine resource.).

Equation (4) identify the most significant risk factor for each of the five resource types. The dominating resource type from Risk Level-2~6 are determined according to their contribution to the activity duration, and is calculated by the following equation:

$$R\ \{R_2, R_3, R_4, R_5, R_6\} = \text{Sort}\ (M_1, M_2, M_3, M_4, M_5) \quad (5)$$

Where, M_x represents the most significant resource risk factor for each of the five resource types; R_2 is the risk of duration for the Level-2 Risk; R_3 is the risk of duration for the Level-3 Risk; and so forth; "Sort(...)" is sorting function (using descending ordering).

Combining Equation (2)~(5), the activity duration for Risk Level-2~Level-6 can be calculated using the following equation:

$$RD_x = RD_{x-1} + R_x \quad (6)$$

Where, RD_x is the activity duration for Risk Level-x (e.g., RD_2 is the activity duration for Risk Level-2), *s.t.* $2 \le x \le 6$; the "Round Law" is adopted in calculating RD_x in order to obtain the most conservative estimate of risk duration for the activity.

It is noted that there may be more risk factors affecting the construction activity duration in different types of construction works. The R-CPSM can be expanded by including more risk categories. However, the five resource categories plus the physical environment risk are common for almost all types of construction works; therefore the preliminary R-CPSM consider only the above-mentioned seven risk levels for activity duration estimation. Table 1 shows the associated risk parameters with the seven risk levels for the duration of a construction activity.

Table 1 Risk parameters for the 7 risk levels

Risk Level	Risk parameter	Note
0	Spatial	Shortest duration
1	Spatial + Environmental	Empirical duration
2	Spatial + Environmental + Single resource	Only 1 resource risk occurs
3	Spatial + Environmental + 2 resources	2 resource risks occur
4	Spatial + Environmental + 3 resources	3 resource risks occur
5	Environmental + 4 resources	4 resource risks occur
6	Environmental + 5 resources	All 5 resource risks occur

3.2 Development of the Risk-based Critical-Path Scheduling System (R-CPSS)

The Risk-based Critical-Path Scheduling System (R-CPSS) is developed to implement R-CPSM. The R-CPSS scheduling process is depicted in Figure 1, which includes: (1) Project Creation—create the basic information of the project; (2) Location Definition—define the space information of the construction site; (3) Activity Definition—create the activity list and estimate the activity durations; (4) Relationship Connection—set up the sequential relationships among activities under the limitations of location and resources; (5) Risk Item Breakdown—breakdown the risk events and estimate their influences on the activities with aids of the risk database; (6) Risk Allocation—allocate risk items to the activities; (7) Information check—before the duration calculation, check the accuracy of location information, relationships, and risk information; (8) Risk Duration Calculation—apply the 7-risk-level method to calculate the duration of each activity; (9) CPM Calculation—apply the R-CPSM to calculate the overall project duration and critical path for the seven risk levels; (10) Diagram Drawing—draw a proper diagram (PDM, EADM or Bar Chart) based the R-CPSM calculation results.

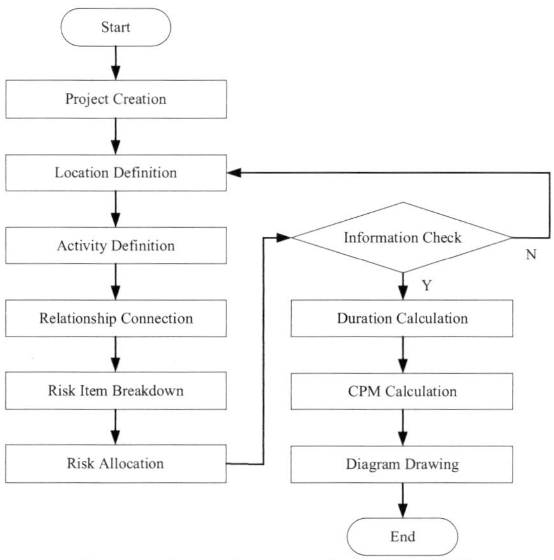

Figure 2 Operation procedure of R-CPSS

4 Application of Risk-Based Critical-Path Scheduling System

In this section, a road construction project is used to demonstrate the functionality of the proposed Risk-based Critical Path Scheduling System (R-CPSS).

4.1 Demonstration Application of R-CPSS

The demonstration case is a small road improvement project. The site plan of the project is depicted in Figure 3. The works of the project include: mobilization (F-01), left-wing sewer (L-01) and sidewalk (L-02), the demolition (R-01) and construction of right-wing sewer work (R-02), the concrete paving (C-01) and restoration (F-02). The quantity, efficiency and environmental impact on the duration of relative constructions are shown in Table 2. There are four major blocks on the construction site: the preparation area, the right-wing, the left-wing, and the central. Due to the spatial constraints of working space, the right-wing drainage facility can only begin after the left-wing is finished. Considering the spatial limitation, two more days are required to wait for the left-wing to be finished. The demolition of the right-wing facility is arranged right after the project mobilization; the left-wing sidewalk begins after the left-wing drainage is finished; the right-wing drainage is started after the demolition and the left-wing facility drainage is finished. The pavement must wait until the left-wing sidewalk and right-wing drainage are finished. Finally, site restoration comes up after the road is paved. The relationships of above-mentioned activities are shown as Table 3.

The risk analysis of this project reveals that: mobilization and road restoration non-resource risk activities; the drainage works confront man, material and machine risks; demolition activities quote machine and method risks; sidewalks quote man, material and method risks; paving quotes man and material risks. The risk events and their impacts on activity durations are shown in Table 3; the calculated durations of the 7 risk levels are shown in Figure 3; the R-CPSS diagram of the example project is shown as Figure 4.

With the diagram of Figure 4, we can visualize the activities swapping between critical-path and noncritical-path under different risk levels. For instance, the left-wing drainage was critical-path in the empirical duration; however, since that the risk impacts on right-wing demolition work made the critical-path swap to the former one (See Figure 5 and Figure 6).

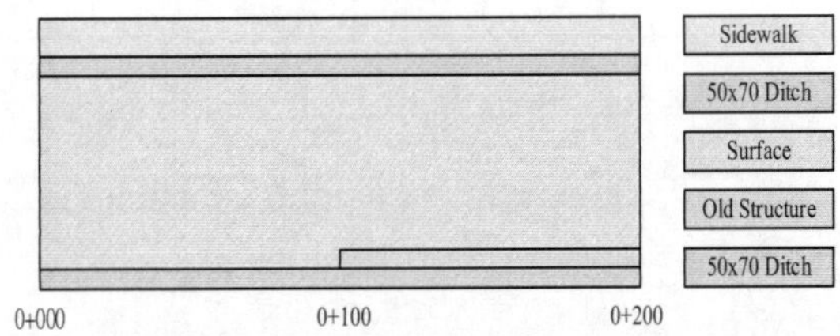

Figure 2 Site layout of the project

Table 2 Activity list of demonstration project

ID	Activity	Q	Rate	Base Dur.	Envir. Impact	Emp. Dur.
F-01	Mobilize.	1.0	-	4.0	1.0	5.0
L-01	Left-Wing Sewer	200.0	40.0	5.0	1.0	6.0
L-02	Left-Wing Sidewalk	200.0	50.0	4.0	1.0	5.0
R-01	Right-Wing Demoli.	100.0	20.0	5.0	1.0	6.0
R-02	Right-Wing Sewer	200.0	40.0	5.0	1.0	6.0
C-01	Paving	1600.0	800	2.0	1.0	3.0
F-02	Restore	1.0	-	3.0	-	3.0

Table 3 Relationships information of demonstration

ID	Precedence Activity	Relationship	Lag
F-01	-		
L-01	F-01	FS	2
L-02	L-01	FS	0
R-01	F-01	FS	0
R-02	R-01	FS	0
	L-01	FS	0
C-01	L-02	FS	0
	R-02	FS	0
F-02	C-01	FS	0

Table 4 Risk information of the demonstration project

ID	Man		Machine		Material		Method		Money	
	R*	D*	R	D	R	D	R	D	R	D
F-01										
L-01	20	1	20	1	20	1				
L-02	10	1	20	1	20	2	50	1		
R-01			20	1			40	2		
R-02	20	1	20	1	20	1				
C-01	10	1			20	1				
F-02										

*Note: R =Probability of risk occurrence (%), D = Lag (Day).

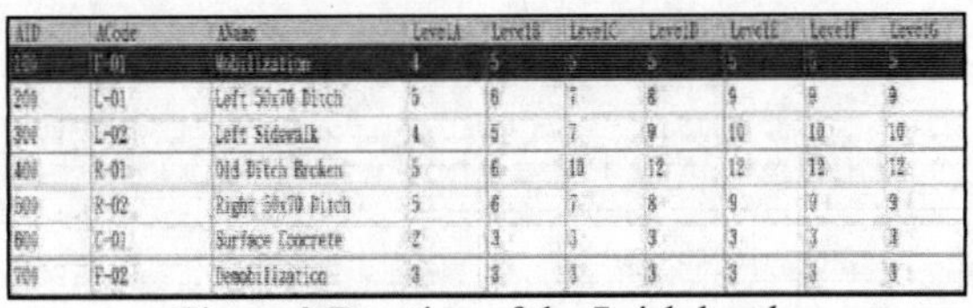

AID	ACode	AName	LevelA	LevelB	LevelC	LevelD	LevelE	LevelF	LevelG
100	F-01	Mobilization	4	5	5	5	5	5	5
200	L-01	Left 50x70 Ditch	5	6	7	8	9	9	9
300	L-02	Left Sidewalk	4	5	7	9	10	10	10
400	R-01	Old Ditch Broken	5	6	10	12	12	12	12
500	R-02	Right 50x70 Ditch	5	6	7	8	9	9	9
600	C-01	Surface Concrete	2	3	3	3	3	3	3
700	F-02	Demobilization	3	3	3	3	3	3	3

Figure 3 Duration of the 7 risk levels

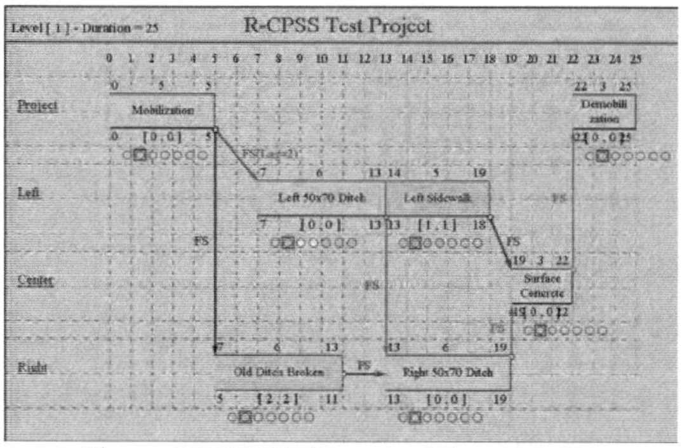

Figure 4 R-CPSS diagram of demonstration project

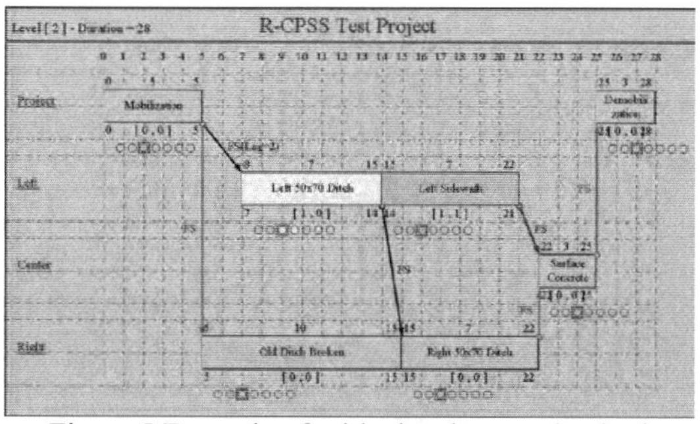

Figure 5 Example of critical path swapping in the demonstration project

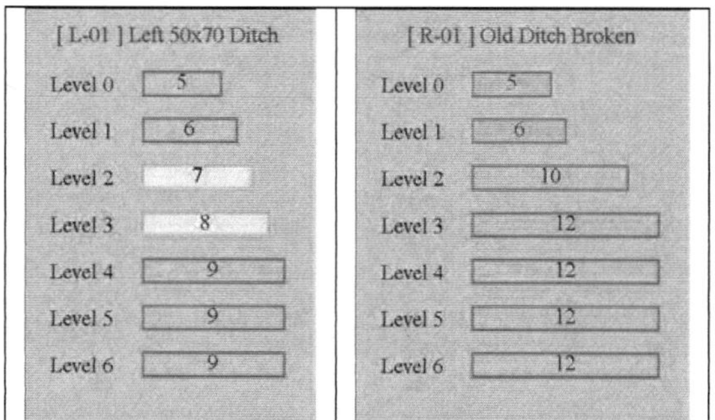

(a) Duration change of the left-wing sewer	(b) Duration change of the right-wing demolition

Figure 6 The duration changes and critical path swapping in the demonstration project

4.2 Discussions of the Demonstration

Traditional EVM assumes PV curve as the schedule baseline, and evaluates the project schedule performance by comparing the values of EV with PV using SPI as the indicator. It is found from the demonstration project that the comparison between PV and EV is not meaningful when the impacts of duration risks were not taken into account. As a result, the construction schedule control based on SPI usually fails to detect the project schedule problems correctly. The proposed R-CPSS provides 7 PV-curves for different risk levels. Employing the EV information of the contract price and these 7 PV-curves (PV0~PV6), totally eight ranges on the diagram are identified. Construction managers can adjust the schedule by analysing which range the EV falls, and plan the control actions according to Table 5. The 7 PV-curves of this project are shown in Figure 7. Since this is a small project, the EV value of the finished activities can be estimated using contract payment value. Table 6 depicts the comparison between EV and the 7 PV-curves in the first ten days of the project: on the second and fourth day, the EV are equal to PV1, which implies that the project went well since it only confronted single risk; on the sixth day, the EV fell into area 4, which implies risk events occurred and the schedule didn't go well, the manager should take control action; between the eighth to tenth day, EV was still in area 4, but it's moving to area 5, which indicates that the risk kept rising and the manager should find out the source of risks.

Table 5 Suggested actions for the 7 risk scenarios

No.	Range	Status	Suggested Action
1	$EV<PV_0$	Good	No
2	$PV_0 \leq EV<PV_1$	Good	No
3	$PV_1 \leq EV<PV_2$	Good	Monitor
4	$PV_2 \leq EV<PV_3$	Medium	Control
5	$PV_3 \leq EV<PV_4$	Medium	Contingency
6	$PV_4 \leq EV<PV_5$	Poor	Fall back
7	$PV_5 \leq EV<PV_6$	Poor	Work around
8	$PV_6 \leq EV$	Out of control	Re-plan

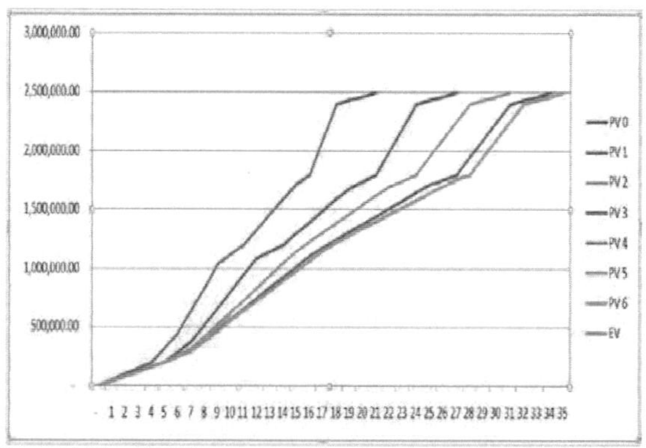

Figure7 The 7 PV-curves of the demonstration project

Table 6 Comparison between EV and 7-risk levels

Time	2	4	6	8	10
PV 0	100,000	200,000	440,000	840,000	1,120,000
PV 1	80,000	160,000	285,714	514,286	800,000

PV 2	80,000	160,000	254,545	413,637	622,727
PV 3	80,000	160,000	246,154	382,906	564,103
PV 4	80,000	160,000	246,154	378,466	550,769
PV 5	80,000	160,000	246,154	378,466	550,769
PV 6	80,000	160,000	246,154	378,466	550,769
EV	80,000	160,000	250,000	400,000	600,000
Com.	PV1 = EV	PV1 = EV	PV2 ≤ EV < PV3	PV2 ≤ EV < PV:	PV2 ≤ EV < PV:
Risk	3.00	3.00	4.46	4.56	4.61

5 Conclusions and Recommendations

5.1 Conclusions

Traditional critical-path method estimates activity duration by dividing the product quantity with the productivity rate. The proposed R-CPSM suggests to use 7 risk levels to estimate the activity risks and their impacts on the activity duration, then use the 7 PV-curves (PV0~6) to identify current status of the project schedule. The R-CPSM provides the project manager signals of potential risks by indicating which risk range the EV falls, and assists the managers control the schedule more effectively.

5.2 Recommendations

Some directions for future research are listed below:

- The Base Duration in this study was estimated with the traditional manual approach. However, since that the application of Building Information Model (BIM) has become more general, a more accurate Base Duration can be automatically estimated using the quantity information provided by BIM.

The risk items in this study are surveyed according to the personnel's experience. It is suggested to build and integrate a risk item database with BIM to generate risk items and estimate the activity risks automatically.

6 ACKNOWLEDGEMENT

The research was partially funded by the Ministry of Science and Technology, Taiwan, under project No. MOST 104-2221-E-216-014. Sincere appreciations are given to the sponsor by the authors.

References

[1] Walt Lipke, "Schedule is different", *The Measurable News Summer*, 1: 31–34, 2003.

[2] Walt Lipke, Ofer Zwikael, Kym Henderson, Frank Anbari, "Prediction of project outcome: The application of statistical methods to earned value and earned schedule performance indexes." International Journal of Project, 27(4): 400-407, 2009.

[3] Walt Lipke, "Schedule adherence and rework.", *PM World Today*, 13: 1–14, 2011.

[4] Elshaer, R. "Impact of sensitivity information on the prediction of project's duration using earned schedule method." *International Journal of Project Management*, 34(4): 579-588, 2013.

[5] Jacob, D., Kane, M., "Forecasting schedule completion using earned value metrics? Revisited." *The Measurable News Summer*, 1: 11–17., 2004

[6] Khamooshi, H., Golafshani, H., "EDM: Earned Duration, a new approach to schedule performance and measurement." *International Journal of Project Management*, Volume 32, Issue 6, pp. 1019-1041., 2014.

[7] Chang, Hsien-Kuan," Integrated of traditional scheduling techniques - Enhanced Arrow Diagram Method." *Master Thesis*, Department of Construction Management, Chung Hua University, 2014.

[8] Chang, Cing-Jung, "Research of the practical applicability of different construction scheduling (1&2)", Modern Construction, Volume 380, pp.55-64, Volume 381, pp.53-61, 2011

[9] Duffy, G., Oberlender, G., and Seok Jeong, D. "Linear Scheduling Model with Varying Production Rates." J. Constr. Eng. Manage., ASCE, Vol. 137, NO. 8, pp. 574-582, 2011.

[10] Yamin, R. and Harmelink, D."Comparison of Linear Scheduling Model (LSM) and Critical Path Method (CPM)." J. Constr. Eng. Manage., ASCE, 127(5), 374–381, 2001.

[11] Harmelink, D."Linear Scheduling Model: Float Characteristics." J. Constr. Eng. Manage., ASCE, Vol. 127, NO. 4, pp. 255-260, 2001.

[12] Harmelink, D. and Rowings, J. "Linear Scheduling Model: Development of Controlling Activity Path." J. Constr. Eng. Manage., ASCE, Vol. 124, NO. 4, pp. 263-268, 1998.

[13] Harris, R. B., and Ioannou, P. E., "Scheduling Project with Repeating Activity," J. Constr. Eng. Manage., ASCE, Vol. 124, NO.4, pp. 269-278, 1998.

[14] Zhang, L., Pan, C. and Zou, X."Criticality Comparison between the Repetitive Scheduling Method and the Network Model. ." J. Constr. Eng.

Manage. , ASCE, Vol. 139, NO. 10,2013.

[15] Zhang, L. and Qi, J.,"Controlling Path and Controlling Segment Analysis in Repetitive Scheduling Method." J. Constr. Eng. Manage., ASCE, Vol. 138, NO. 11, pp. 1341-1345, 2012.

[16] Hegazy, T. and Menesi, W. "Critical Path Segments Scheduling Technique." *Journal of Construction Engineering and Management*, ASCE, 136(10): 1078–1085, 2010.

[17] Chang, H.-K. and Chang, C.-J., "Research of the application of critical-path method to repetitive project scheduling: the Enhanced Arrow Diagram." *Modern Construction*, 407: 34-45, 2013. (in Chinese)

[18] Hsu, K.-H., *Project Management: a View of the Body of Knowledge*. Hua Tai Publishing, Taipei, 2006. (in Chinese)

[19] Yu, W. D., Project Management in Practice, 2nd Ed., Chan's Arch-Publishing, Taipei, 2015. (in Chinese)

[20] Gong, D. "Optimization of float use in risk analysis-based network scheduling." *International Journal of Project Management*, 15(3): 187-192, 1997.

[21] Gong, D., Rowings Jr., J. E. "Calculation of safe float use in risk-analysis-oriented network scheduling." *International Journal of Project Management*, 13(3); 187-194, 1995.

[22] Zafra-Cabeza, A., Ridao, M.A., Camacho, E.F. "An algorithm for optimal scheduling and risk assessment of projects." *Control Engineering Practice*, 12(10): 1329-1338., 2004.

[23] Moon, H., Kim, H., Kamat, V.R., and Kang, L. "BIM-Based Construction Scheduling Method Using Optimization Theory for Reducing Activity Overlaps." *Journal of Computing in Civil Engineering*, ASCE, 29(3), DOI: 10.1061/(ASCE)CP.1943-5487.0000342, 2015.

[24] Yi, K. J., Langford, D. "Scheduling-Based Risk Estimation and Safety Planning for Construction Projects." *Journal of Construction Engineering and Management*, 132(6): 626-635, 2006.

[25] Kılıç, M., Ulusoy, G., Şerifoğlu, F. S. "A bi-objective genetic algorithm approach to risk mitigation in project scheduling." *International Journal of Production Economics*, 112(1): 202-216, 2008.

[26] Schatteman, D., Herroelen, W., Van de Vonder, S., Boone, A., "Methodology for Integrated Risk and Proactive Scheduling of Construction Projects." *Journal of Construction Engineering and Management*, 134(11): 885-893, 2008.

[27] Zafra-Cabeza, A., Ridao, M.A., Camacho, E.F. "Using a risk-based approach to project scheduling: A case illustration from semiconductor manufacturing." *European Journal of Operational Research*, 190(3): 708-723, 2008.

[28] Luu, V.T., Kim, S.Y., Tuan, N.V., Ogunlana, S.O. "Quantifying schedule risk in construction projects using Bayesian belief networks." *International Journal of Project Management*, 27(1): 39-50, 2009.

[29] Jun, D.H. and El-Rayes, K. "Fast and Accurate Risk Evaluation for Scheduling Large-Scale Construction Projects." *Journal of Computing in Civil Engineering*, 25(5): 407-417, 2011.

[30] Vanhoucke, M. "On the dynamic use of project performance and schedule risk information during project tracking." *Omega*, 39(4): 416-426, 2011.

[31] Mol, T. *Productive safety management*, Butterworth-Heinemann, Oxford, U.K., 2003.

[32] Huang, C.H., *Construction Management*, Chiu Hua Publishing, Taipei, 1990. (in Chinese)

A Risk-based Critical Path Scheduling Method (II) : A Visual Approach System using BIM

H.K. Chang[a] W.D. Yu[b*] and S.T. Cheng[c]

[a]Department of Civil Engineering, Chung Hua University, Taiwan
[b*]Department of Construction Engineering, Chaoyang University of Technology, Taiwan, Corresponding Author
[c]Department of Construction Management, Chung Hua University, Taiwan
E-mail: d10304006@chu.edu.tw, wenderyu@cyut.edu.tw, shaotsai@chu.edu.tw

Abstract

The traditional construction schedule planning methods are mostly based on the project activities list broken down by the experienced engineers who primarily analyze the work content of construction drawings, and the activity durations estimated by simulating the construction process according to the conditions of the construction site. However, such an approach does not only give hard time the new engineers without experience, but is also unable to capture the dynamic risk factors while estimating the construction duration. As a result, the planned schedule is usually inaccurate and ineffective. To tackle the abovementioned problem, this paper presents a BIM-based Visual Risk critical path Scheduling System (BIM-VRcpSS) to assist the construction scheduler in developing a more accurate and effective schedule. The proposed BIM-VRcpSS provides functions of risk-based Critical Path Scheduling Method (R-CPSM) based on seven different risk levels. It also equips the construction schedulers with a 3D BIM visual simulation tool to better estimate the activity durations compared with the traditional 2D approaches. Moreover, with the aid of a cloud data based system, the construction schedulers are able to fine-tune duration estimates by considering the impacts of different risk factors without spatial and time limitations. As such, the proposed BIM-VRcpSS can increase the duration estimation accuracy of construction activities, and improve the traditional scheduling approaches.

Keywords –

Risk management, BIM, Construction scheduling and control.

1 Introduction

The prevailing construction scheduling method is the Precedence Diagramming Method (PDM) based on the traditional Critical Path Method (CPM). To conduct a PDM scheduling, the experienced engineer has first to develop a Work Breakdown Structure (WBS) based on previous projects [1]. The activity duration is then estimated by dividing the work quantity by the productivity rate. Such a traditional scheduling method is difficult to adopt by junior staffs without sufficient experience. Moreover, the complex risk factors that affect the activity duration significantly couldn't be taken into account effectively.

Thanks to the recent development of Building Information Modeling (BIM), the execution and planning efficiency of construction works have been greatly improved [2]. Previous research found that the application of BIM did not only facilitate the analysis but also improve the planning and control of a complex construction site.

The current research aims at developing a BIM-based Risk Visual critical path Scheduling System (BIM-VRcpSS) combining a previous developed Risk-based Critical Path Scheduling System (R-CPSS) and the visualization of BIM. The rest of the paper is presented in the following manner: the related works are reviewed first; it is followed by the revisit of the previous developed R-CPSS; then the BIM-VRcpSS is proposed for visualization improvement of R-CPSS; the application of the proposed BIM-VRcpSS is then demonstrated using a real world project; finally, conclusions and recommendations are summarized at the end of the paper.

2 Review of Related Works

2.1 Integration of Risk Management and Scheduling

A risk of a project was defined as the negative impact to the project's objectives. As the occurrence of a risk is associated with a probability, the major tasks for risk management are to measure the occurrence and impact associated to a risk event and the plan the required actions to mediate its impact to the project goals [3].

There have been many researchers engaged in the integration of risk management and construction scheduling. Gone *et al.* [4] proposed a time-disturbance analysis method to optimize the utilization of activity floats, so that the delay of the project schedule can be minimized; Zafra-Cabeza *et al.* [5] adopted a "p-timed Petri net" to represent the risk measure of an activity and developed a method to reduce the impact of risks; Yi and Langford [6] analyzed the safety disasters occurred in the past to develop a safety planning method for the project safety manager to focus on the concentration of safety risks in the project. Kilic *et al.* [7] proposed a Genetic Algorithm to optimize the multi-objective optimization problem of project scheduling, considering four attributes: indirect costs, operation costs, risk reduction costs, and delay punishment costs; Schatteman *et al.* [8] proposed a computer-aided risk management system to identify, analyze, and quantity the major risk factors, and to estimate their impacts to the project schedule so that risk reduction actions are planned; Zafra-Cabeza *et al.* [9] conducted a Monte Carlo simulation of High-tech factory construction project using risk optimization techniques to optimize project cost and schedule.

Beside the above literature, the work done by Yi and Langford [6] attracts our attention the most. Their work considers the construction activity as an integration of many attributes including the construction processes, human resources, technological factors, and the environmental factors. Therefore, the risk of an activity can be expressed in the following equation:

$$\text{Total Risk} = P \times H \times T \times E \tag{1}$$

where, P means "Process Risk"; H means "Human Risk"; T means "Technology Risk", E means "Physical Environment Risk"; and the symbol "×" implies that the impacts of the risk factors to the schedule is a product of the impacts caused by the individual risk factors.

Yi and Langford believed that the first three risk factors of a construction activity (*i.e.*, process, human, and resource) are determined by the type of the construction activity. They can be broken down by referring to the Bill of Quantity (BOQ) of the activity. The fourth risk factor, Physical Environment Risk, is determined by the location and surrounding conditions of the work site. Such a risk breakdown concept is very useful for analysis of the risk factors affecting construction schedule. However, no systematic approach was proposed by Yi and Langford to take the above-mentioned factor in construction scheduling process. In this paper, we develop a systematic Risk-based critical path scheduling method (namely, R-CPSM) to take into account all risk factors referred by Yi and Langford [6].

2.2 Integration of 3D Modeling Techniques and Scheduling

There were some works published by previous researchers on integrating 3D modeling in construction scheduling to improve the drawbacks of the traditional human-experienced based scheduling methods. Shan *et al.* [10] used the Visual Basic for Application (VBA) of MS Excel and BIM to develop a project scheduling system that considered the temperature and humidity factors; Moon *et al.* [11] developed a systematic schedule simulation system to minimize overlapping activities for the enhancement of a project's operational performance by integrating fuzzy set, GA, risk factors, and BIM; Hijazi *et al.* [12] proposed a BIM-based 4D simulation method to assess the constructability of a building construction project so that the construction schedule can be improved. Other BIM-based scheduling systems including Chen *et al.* [13], Kim *et al.* [14], and Zhou *et al.* [15]; all of these BIM-based systems utilize the 3D visualization functions to facilitate the construction schedulers in developing a more accurate project schedule.

3 Risk-based Critical Path Scheduling Model (R-CPSM)

In this paper, a Risk-based Critical Path Scheduling Model (namely R-CPSM) is proposed. This section is dedicated to the model development of R-CPSM.

3.1 The Seven Risk Levels of Construction Activity Duration

Inspired by the safety planning method proposed by Yi and Langford [6], the risk of the duration of a construction activity can be affected by the integration of the individual risk factors of that activity. It is believed that the spatial restriction of the construction is the essential constraint for the duration to complete an activity. As a result, the estimation of the duration for a construction activity considering the spatial constraint is called "base duration". The base duration can be deemed as the shortest possible duration required complete the activity without occurrence of any risk event.

The second constraint for the duration of completing an activity is the "physical environment risk". The physical environment risk considering all factors of the surrounding environment on the construction site, *e.g.*, weather, local social factors, and local cultural factors. The impact of such factors to the activity duration is usually difficult to estimate and usually measured by experienced engineers who are familiar with the surrounding environment factors. As a result, while take the physical environment risk into account, the modified activity duration is called "empirical duration". The estimation of the empirical duration can be obtained by adding an extra duration to the base duration, and thus it is longer than base duration.

Beside the physical environment risk, which is unchangeable, there are five categories (*i.e.*, the 5 Ms of construction management) of risks that can be altered by management schemes including: (1) Man—the availability of different required laborers to perform a construction activity; (2) Machine—the availability of different required equipment to perform a construction activity; (3) Material—the availability of required materials to complete a construction activity; (4) Method—the availability of appropriate construction methods to perform a construction activity; and (5) Money—availability of required financial arrangements to conduct a construction activity. Should any of the above resources be absent, the completion of a construction activity will be delayed and the duration will be lengthened.

Considering all the above risk factor, seven risk levels are classified for the duration of a construction activity:

- Level-0 duration—there is no risk event occurring and the physical environment risk contributes no extra duration to the activity, this is also considered the base and shortest duration of the activity. The Level-0 duration is denoted as RD_0 and calculate by the following equation:

$$RD_0 = \frac{\text{Quantity}}{\text{Rate}} \tag{2}$$

Where, "Quantity" is the quantity of product to be produced by the activity; "Rate" is the production rate that can be referred to any cost estimation reference, e.g., RSMeans Book (https://www.rsmeans.com) or Dodge Estimating Guide (https://construction.com); RD_0 is the base duration of the construction activity.

- Level-1 duration—only the physical environment risk but no any resource risk occurs, so the extra duration caused by the physical environment risk estimated by the experienced engineer is added to base duration resulting in the empirical duration. The Level-1 duration is denoted as RD_1 and calculate by the following equation:

$$RD_1 = RD_0 + (\text{ Environment Effect }) \tag{3}$$

Where, RD_0 is " base duration of the activity; "Environment Effect" the extra duration caused by the physical environment risk estimated by the experienced engineer; and RD_1 is the empirical duration.

- Level-2 duration—both the physical environment risk and one resource risk occur, so the extra duration caused by the occurred resource risk estimated by the experienced engineer is added to empirical duration resulting in Level-2 duration.
- Level-3 duration—not only the physical environment risk but also two of the five resource risks occur, so the extra durations caused by the two occurred resource risks estimated by the experienced engineer are added to empirical duration resulting in Level-3 duration;
- Level-4 duration—not only the physical environment risk but also three of the five resource risks occur, so the extra durations caused by the three occurred resource risks estimated by the experienced engineer are added to empirical duration resulting in Level-4 duration;
- Level-5 duration—not only the physical environment risk but also four of the five resource risks occur, so the extra durations caused by the four occurred resource risks estimated by the experienced engineer are added to empirical duration resulting in Level-5 duration;
- Level-6 duration—not only the physical environment risk but also all five resource risks occur, so the extra durations caused by the five occurred resource risks estimated by the experienced engineer are added to empirical duration resulting in Level-6 duration, and this is considered the worst case scenario of the activity duration.

The equation for estimating the activity duration for Risk Level-2~Level-6 is described in the following:

At first, the most significant resource risk factor for each of the five resource types is identified as M_x using the following equation:

$$M_x = \text{Max } (M_{x1}, M_{x2}, M_{x3}, M_{x4} \ldots\ldots, M_{xn}) \tag{4}$$

Where, M_x represents the the most significant resource risk factor in a specific resource type (*e.g.*, M_1 means the most significant risk factor for human resource, M_2 means the most significant risk factor for machine resource, and so forth.); M_{x1}, M_{x2}, ...M_{xn} are the possible resource factors for a specific resource type (*e.g.*, M_{23} means the third type of machine resource.).

Equation (4) identify the most significant risk factor for each of the five resource types. The dominating resource type from Risk Level-2~6 are determined according to their contribution to the activity duration, and is calculated by the following equation:

$$R\ \{R_2, R_3, R_4, R_5, R_6\} = \text{Sort}\ (M_1, M_2, M_3, M_4, M_5) \quad (5)$$

Where, M_x represents the most significant resource risk factor for each of the five resource types; R_2 is the risk of duration for the Level-2 Risk; R_3 is the risk of duration for the Level-3 Risk; and so forth; "Sort(...)" is sorting function (using descending ordering).

Combining Equation (2)~(5), the activity duration for Risk Level-2~Level-6 can be calculated using the following equation:

$$RD_x = RD_{x-1} + R_x \quad (6)$$

Where, RD_x is the activity duration for Risk Level-x (e.g., RD_2 is the activity duration for Risk Level-2), s.t. $2 \leq x \leq 6$; the "Round Law" is adopted in calculating RD_x in order to obtain the most conservative estimate of risk duration for the activity.

It is noted that there may be more risk factors affecting the construction activity duration in different types of construction works. The R-CPM can be expanded by including more risk categories. However, the five resource categories plus the physical environment risk are common for almost all types of construction works; therefore the primary R-CPM consider only the above-mentioned seven risk levels for activity duration estimation. Table 1 shows the associated risk parameters with the seven risk levels for the duration of a construction activity.

Table 1 Risk parameters for the 7 risk levels

Risk Level	Risk parameter	Note
0	Spatial	Shortest duration
1	Spatial + Environmental	Empirical duration
2	Spatial + Environmental + Single resource	Only 1 resource risk occurs
3	Spatial + Environmental + 2 resources	2 resource risks occur
4	Spatial + Environmental + 3 resources	3 resource risks occur
5	Environmental + 4 resources	4 resource risks occur
6	Environmental + 5 resources	All 5 resource risks occur

3.2 Construction Scheduling based on R-CPSM

In order to implement the concept of the proposed R-CPSM, a computer program was developed named Risk-based Critical Path Scheduling System (R-CPSS). The functions of R-CPSS consist of: (1) Project Creation— create basic information for a construction project; (2) Location Definition — define the spatial information for the construction site; (3) Activity Definition—create the activity list for the WBS of the project, and the quantity of the items for estimating base duration; (4) Relationship Connection — connect the activities according to their sequential relationships; (5) Risk Item Breakdown) — establish the risk item list (including probability and impact information) according to the risk breakdown structure of the project ; (6) Risk Allocation—assign risk items to each of the activities; (7) Information Check — check the completeness of the correctness of information including spatial information and the sequential logic before scheduling calculation; (8) Risk Duration Calculation — calculate the activity duration for each risk level, RD_x, using Equation (2)~(6) ; (9) Critical Path Method (CPM) Calculation — conduct CPM calculation for the project according to the seven risk level of durations for each activity; and (10) Diagram Drawing—draw PDM, EADM, and Bar Chart using the scheduling data.

The Earned Value (EV) curves of the seven risk levels for a sample project is shown in Figure 1.

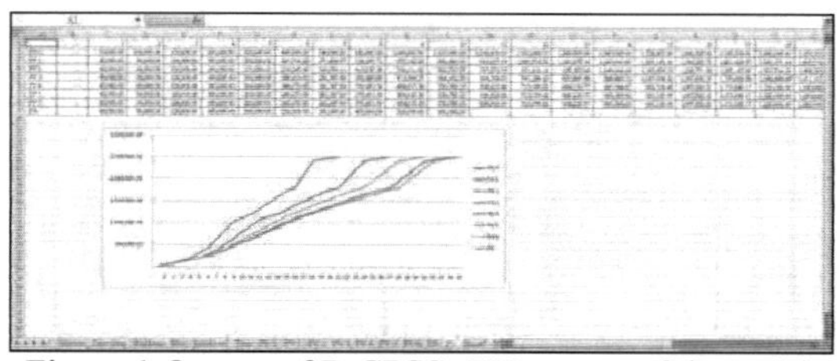

Figure 1 Output of R-CPSS: EV curves of the 7 risk levels

3.3 Management Suggestions for Different Risk Levels

The suggested management actions for different risk levels shown in Figure 1 are depicted in Table 2.

Table 2 Suggested action for the different risk scenarios

No..	Range	Status	Suggested Action
1	$EV<PV_0$	Good	No
2	$PV_0 \leq EV<PV_1$	Good	No
3	$PV_1 \leq EV<PV_2$	Good	Monitor
4	$PV_2 \leq EV<PV_3$	Medium	Control
5	$PV_3 \leq EV<PV_4$	Medium	Contingency
6	$PV_4 \leq EV<PV_5$	Poor	Fall back
7	$PV_5 \leq EV<PV_6$	Poor	Work around

8	$PV_6 \leq EV$	Out of control	Re-plan

From Table 2, no actions are suggested for the project managers while the EV is below Risk Level-1 (*i.e.*, curve PV_1). Regular project monitoring activities (*e.g.*, EVM indexes, risk monitoring, etc.) are suggested when PV_1 (Risk Level-1) $\leq$ EV < PV_2 (Risk Level-2). Appropriate project control activities (*e.g.*, corrective, preventive, and defect repair actions) are suggested when PV_2 (Risk Level-2) $\leq$ EV < PV_3 (Risk Level-3). The contingency activities are suggested when PV_3 (Risk Level-3) $\leq$ EV < PV_4 (Risk Level-4). The fallback plan is suggested when PV_4 (Risk Level-4) $\leq$ EV < PV_5 (Risk Level-5). The work-around plan is suggested when PV_5 (Risk Level-5) $\leq$ EV < PV_6 (Risk Level-6). The project should be re-planned when PV_6 (Risk Level-6) $\leq$ EV.

4 BIM-based Visual Risk critical path Scheduling System (BIM-VRcpSS)

Based on the R-CPSS presented in Section 3, a BIM-based Visual Risk critical path Scheduling System (BIM-VRcpSS) to enhance the R-CPSS in facilitating construction project scheduler for developing a more accurate and realistic schedule. This section describes the development and verification of the proposed BIM-VRcpSS.

4.1 System Planning of BIM-VRcpSS

The proposed BIM-VRcpSS is developed based on the API provided by Autodesk Revit® 2015 under Microsoft Window7® using Microsoft C# (.Net 4.5) programming language. There are four modules in the BIM-VRcpSS:

- Project Function Module－providing functions for: User Login; Knowledge Management (to retrieve the previous project information related to specific resource risk; and Project Scheduling (to activate R-CPSS system).
- Element Function Module－providing functions for: Activity Selection (select the activities to manipulate); Element Risk (search for risk information and previous cases); Element Rate (the productivity rate associated with BIM elements); Activity Assign (assign the selected BIM elements to specific activity); Element Clear (cancel the BIM element assignment); Hide by Activity (hide the associated BIM element of a specific activity); Show by Activity (show the associated BIM element of a specific activity); Action Activity (Conduct 4-D construction process simulation for the selected activity).
- 4D Model Function Module－providing functions for: Configure Manager (the attribute information management for Category, Family and Symbol); Model Clear (hide the object according to the parameter information); Model Reset (restore all BIM objects); Model Modify (modify BIM objects according to the floor, type, family, and category); Step Action (step by step simulation of construction process).
- Scheduling Function Module－providing functions for: Schedule Simulation (conduct 4-D simulation based on the scheduling data); Stop Simulation (force the 4-D simulation to terminate); Continue Simulation (resume the 4-D simulation that was terminated); Next Activity (proceed the simulation to next activity).

The user interface of the proposed BIM-VRcpSS is shown in Figure 2.

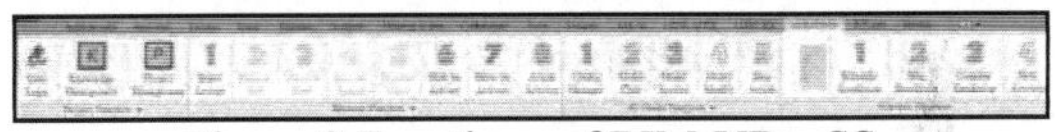

Figure 2 Functions of BIM-VRcpSS

4.2 Knowledge Bases of BIM-VRcpSS

There are two knowledge bases developed for the proposed BIM-VRcpSS: the first is the knowledge base of activity duration risks of historical projects; the second is the productivity rate database for the activities of historical projects. The objective of former is to provide a source for quick search and retrieval of resource risks associated with a specific activity in order to facilitate duration calculation for different risk levels; while the latter aims to serve as a basis for base duration calculation.

There are 9 fields in a risk category dataset: (1) RDID—the identity No. of a risk type; (2) RCode—the code of a risk type; (3) RName—the name of a risk type; (4) RType—the category of a risk type; (5) RRatio—the probability of a risk type; (6) REffect—the effect on activity duration of a risk type; (7) RatioA—the probability of a risk type in the early stage of a project; (8) RatioB—the probability of a risk type in the medium stage of a project; (9) RatioC—the probability of a risk type in the late stage of a project.

There are 6 fields in a BIM object database: (1) RCID—the identity No. of a BIM object; (2) Category—the category of a BIM object; (3) BuiltInCategory—the category of a system buit-in BIM object; (5) Family—the family of a BIM object; (6) Symbol—the sub-category of a BIM object. The primary objective for the BIM object database is to provide an automated searching function. That is that if we assign a BIM object the risk item parameters, then we can retrieve the risk information automatically with the BIM object.

Finally, the productivity rate dataset is set to

comprising 5 fields: (1)AutoID—the serial number of an activity; (2)RCID—the ID of object classification; (3)Rate—the productivity rate of an activity; (4)Parametric—the parameter name; (5)Memo—the description of the duration risk. The Entity Relationship Diagram (ERD) of the three datasets are shown in Figure 3.

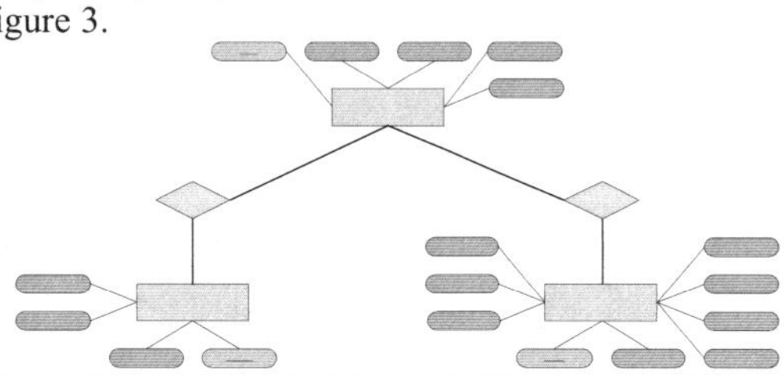

Figure 3 The Entity Relationship Diagram (ERD) of the three datasets

4.3 Application of BIM-VRcpSS

In order to demonstrate the functionality of the proposed BIM-VRcpSS, a real building construction project is adopted for system testing. The selected case is the Spa House Building of the Spring Resort located in Ruisui Township, Hualien of eastern Taiwan. The Spa House is a three-story (28m high) building with one-floor basement. Total floor area is 10,165m^2. Figure 4 shows the 3D model of Spa House Building.

Figure 4 the 3D model of Spa House Building

The system testing procedure is elaborated in the following:

- Step-1: User Login—the project is named "International Tourist Hotel"
- Step-2: Spatial planning for construction areas—Area-A (left building), Area-B (central building), Area-C (right building), All (A+B+C), Interior of A&C.

LID	LCode	LExtend	LName	LMemo
0	Site	-	Site	
100	A	-	Left Hotel	Left
150	A	Inside	Left Hotel Inside	
200	B	-	Center	
250	B	Inside	Right Hotel Inside	
300	C	-	Right Hotel	Right
400	All	-	Area	

Figure 5 Spatial configuration of the case building

- Step 3: Create Activity List—the activity list of the project consists of mobilization, excavation, Raft Foundation Construction (Area A, B & C), B2F Construction (Area A, B & C)、B1F Construction (Area A, B & C), 1F, 2F, 3F, 4F, PF, RF Steel Structure, Exterior Wall (I), Exterior Wall (II), Exterior Door and Windows (I), Exterior Door and Windows (II), M&E (I), M&E (II), Interior, and Clean up. All of the activities are input into the database of BIM-VRcpSS.

AID	ACode	AName	ADuration	AType	LID
100	S-001	Mobilization	4	Normal	0
200	S-002	EarthWorks I (Cut)	4	Normal	0
300	S-003	Earthworks II (Fill)	3	Normal	0
400	S-004	Demobilization	4	Normal	0
1000	FD-001	Foundation Left	5	Normal	100
1100	FD-002	Foundation Center	7	Normal	200
1200	FD-003	Foundation Right	5	Normal	300
2000	SB2-001	B2F Structure Left	5	Normal	100
2100	SB2-002	B2F Structure Center	7	Normal	200
2200	SB2-003	B2F Structure Right	6	Normal	300
3000	SB1-001	B1F Structure Left	5	Normal	100
3100	SB1-002	B1F Structure Center	7	Normal	200
3200	SB1-003	B1F Structure Right	5	Normal	300
4000	S1-001	1F Structure	5	Normal	400
4100	S2-001	2F Structure	5	Normal	400
4200	S3-001	3F Structure	5	Normal	400
4300	S4-001	4F Structure	5	Normal	400
4400	SP-001	PF Structure (P1,P2)	5	Normal	400
5000	SR-001	RF Steel Structure	4	Normal	400
6000	[illegible]	Outside Wall I	5	Normal	100

Figure 6 Activity list of the case building

- BIM Model Development – building the BIM model for the case building from bottom to top including the temporary facility models. During the BIM model development process, the BIM objects are first assigned to the selected activity; then the potential risks associated with the activity is identified using a knowledge management system (KMS); finally, the base duration of the selected activity is automatically calculated using the quantity provided by the BIM model to be divided by the productivity rate retrieved from database. The base duration of each activity is also shown in Figure 6.
- Connect Activities – the sequential relationships among the activities are determined according to the task groups, trades, and work areas. The results are shown in Figure 7.

AID	ACode	AName	PreID	PreCode	PreName	Relation	Lag
200	S-002	EarthWorks I (Cut)	100	S-001	Mobilization	FS	0
300	S-003	Earthworks II (F...	3200	SB1-003	B1F Structure Right	FS	0
400	S-004	Demobilization	7100	[illegible]	[illegible]	FS	0
1000	FD-001	Foundation Left	200	S-002	EarthWorks I (Cut)	FS	0
1100	FD-002	Foundation Center	200	S-002	EarthWorks I (Cut)	FS	0
1200	FD-003	Foundation Right	200	S-002	EarthWorks I (Cut)	FS	0
2000	SB2-001	B2F Structure Left	1000	FD-001	Foundation Left	FS	0
2100	SB2-002	B2F Structure Ce...	1100	FD-002	Foundation Center	FS	0
2100	SB2-002	B2F Structure Ce...	2000	SB2-001	B2F Structure Left	FS	0
2200	SB2-003	B2F Structure Right	1200	FD-003	Foundation Right	FS	0
2200	SB2-003	B2F Structure Right	2100	SB2-002	B2F Structure Ce...	FS	0
3000	SB1-001	B1F Structure Left	2200	SB2-003	B2F Structure Right	FS	0
3100	SB1-002	B1F Structure Ce...	3000	SB1-001	B1F Structure Left	FS	0
3200	SB1-003	B1F Structure Right	3100	SB1-002	B1F Structure Ce...	FS	0
4000	S1-001	1F Structure	300	S-003	Earthworks II (F...	FS	0
4100	S2-001	2F Structure	4000	S1-001	1F Structure	FS	0
4200	S3-001	3F Structure	4100	S2-001	2F Structure	FS	0
4300	S4-001	4F Structure	4200	S3-001	3F Structure	FS	0
4400	SP-001	PF Structure (P1...	4300	S4-001	4F Structure	FS	0
5000	SR-001	RF Steel Structure	4400	SP-001	PF Structure (P1...	FS	0

Figure 7 Sequential relationships among the activities

- Assign Activity Risk—review the automated risk assignment during BIM model development, new risks can be assigned should there be other risk factors. The results are shown in Figure 8.

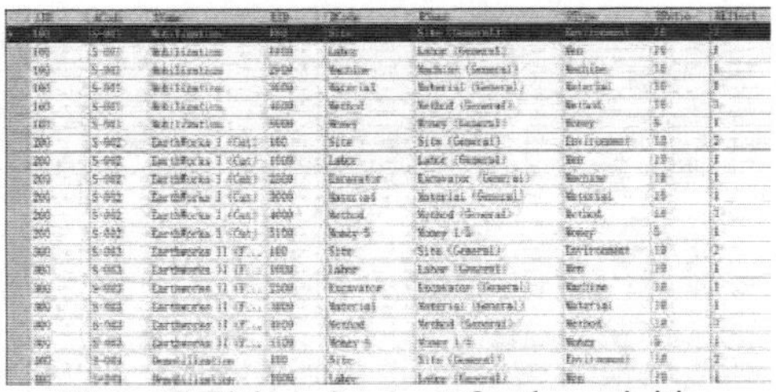

Figure 8 Risk assignment for the activities

- Result Checking – Simulate the construction process step-by-step to ensure the correctness and appropriateness of the visual construction model until no more error.
- Output Results – Conduct R-CPSM calculations for all risk levels using BIM-VRcpSS, draw the PDM and other diagrams based on the scheduling data generated by the BIM-VRcpSS. Figure 9 shows the calculated early start/finish network diagram for the case project. Figure 10 shows the 7 EV curves for the seven risk levels.

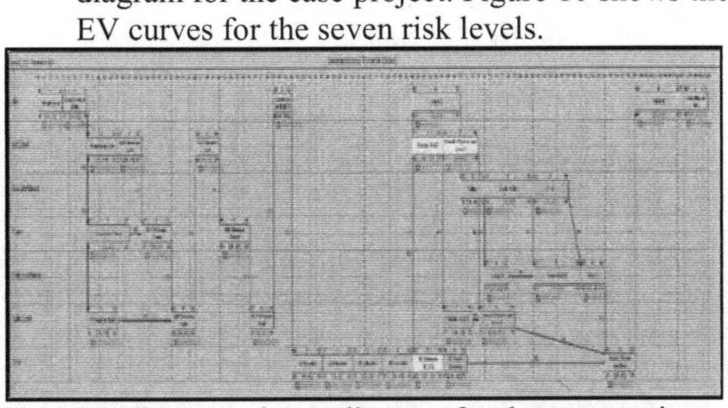

Figure 9 The precedence diagram for the case project.

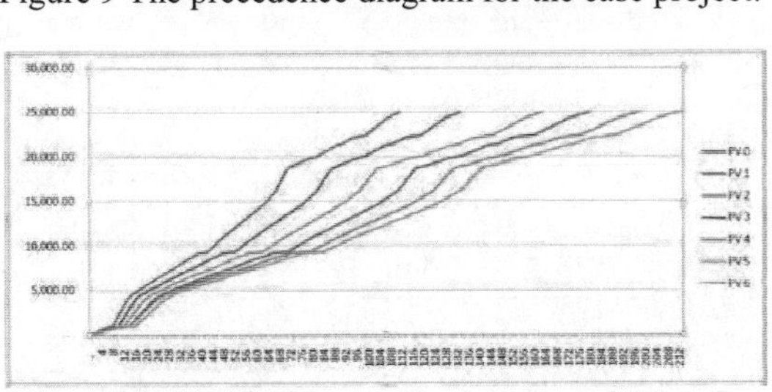

Figure 10 The 7 EV curves for the seven risk levels.

5 Findings

The proposed BIM-VRcpSS provides seven EV curves for the seven risk levels, instead of the traditional CPM's single EV curve. It uncovers the unknown risk facing construction project managers in planning a more realistic construction schedule when the traditional CPM and EVM are adopted. In the following, some profound findings are discussed for reference of the practitioners in construction industry project scheduling while planning a construction project:

- More risk scenarios for risk based scheduling—the proposed BIM-VRcpSS calculates the activity schedule data according to different risk levels of the activity, and developed 7 different critical paths. Such critical path (of different risk level) provides the schedulers more information to plan and control their construction project. If *de facto* network has differed from its original plan (the PV), then the proposed BIM-VRcpSS shows the project manager "how bad the progress is?" and "what will result if no effective action is taken."
- Could knowledge base for risk assessment—the proposed BIM-VRcpSS allow construction scheduler to revise the activity risk data by accessing cloud knowledge base. It is therefore the most updated and ever-accumulative source of risk knowledge for future schedule planning.
- Cloud productivity rate data base—similar to risk knowledge base, the proposed BIM-VRcpSS allow construction scheduler to revise the activity risk data by accessing cloud productivity rate database. The productivity rate data provide the BIM-VRcpSS the automatic calculation of base activity duration. It can be updated at a real-time basis to ensure the correctness and timeliness of the retrieved productivity rate data.
- Visualization of construction schedule—contrast to the traditional scheduling method that was primarily based on interpretation of the 2D construction drawing information to develop the construction schedule, the proposed BIM-VRcpSS adopts 4D functions of BIM platform. It is easier for user to visualize the spatial and environmental constraints of a construction site, which will significantly improve the efficiency of construction scheduler's work.
- Providing a tool for schedule risk control—from Figure 9 and Table 2, *de facto* EV curve can be compared with the seven PVs to determine which risk scenario the project is currently experienced. The suggested management actions provide the construction project managers a more clear direction for improve the schedule performance of his/her project. Such a function has not been found in literature.

6 Conclusions and Recommendations

6.1 Conclusions

This paper presents the results of a research on the development of a BIM-based Visual Risk critical path Scheduling System (BIM-VRcpSS) to assist the construction scheduler in developing a more accurate and effective schedule. By combining the visualization of BIM, the proposed BIM-VRcpSS provides the construction schedulers a more powerful tool to better capture the nature of risks associated with a construction activity during the construction process.

A Risk-based Critical Path Scheduling Model (namely R-CPSM) considering seven different risk levels of activity duration is developed to model the risk scenarios of a construction activity. The proposed BIM-VRcpSS provides a visual tool that makes the implementation of R-CPSM more realistic and practical.

In addition to visualization of duration risks of an activity, the proposed BIM-VRcpSS also improve the scheduling process by offering the scheduler a 3D visual model (instead of the traditional 2D drawing), which helps develop a more accurate construction schedule.

Finally, the 7 EV curves (instead of the single EV curve in the traditional EVM) for the seven risk levels generated by the proposed BIM-VRcpSS provides the construction scheduler a better overview of *de facto* construction schedule performance. It can be maintained and updated on real-time so that the true situation of the project can be better monitored and controlled.

6.2 Recommendations

Some directions after the current research are recommended as follows:

The output chart in current version of BIM-VRcpSS is developed within MS Excel, a self-contained system can be developed to be more portable and efficient.

The proposed BIM-VRcpSS is an EVM-based project monitoring system. It quotes similar limitations of the traditional EVM systems. Future works can be engaged in developed an Earned Duration Management (EDM) based project monitoring system.

7 ACKNOWLEDGEMENT

The research was partially funded by the Ministry of Science and Technology, Taiwan, under project No. MOST 104-2221-E-216-014. Sincere appreciations are given to the sponsor by the authors.

References

[1] Hinze, J. W., *Construction Planning and Scheduling*. 4th Edition, Pearson Education Inc., New Jersey, 2012.

[2] Eastman, C., Teicholz, P., Sacks, R., and Liston, K., *BIM Handbook: A Guide to Building Information Modeling for Owners, Managers, Designers, Engineers and Contractors*. 2nd Edition, New Jersey, Wiley, 2010.

[3] Project Management Institute, *A Guide to the Project Management Body of Knowledge*, 4th Edition, Project Management Institute, PA, USA, 2008.

[4] Gong, D., "Optimization of float use in risk analysis-based network scheduling.", *International Journal of Project Management*, 15(3): 187-192, 1997

[5] Zafra-Cabeza, A., Ridao, M. A., and Camacho, E. F., "An algorithm for optimal scheduling and risk assessment of projects." *Control Engineering Practice*, 12(10): 1329-1338, 2004.

[6] Yi, K. and Langford, D. "Scheduling-Based Risk Estimation and Safety Planning for Construction Projects." *Journal of Construction Engineering and Management*, ASCE, 132(6): 626-635, 2006.

[7] Kiliç, M., Ulusoy, G., and Şerifoğlu, F. S., "A bi-objective genetic algorithm approach to risk mitigation in project scheduling." *International Journal of Production Economics*, 112(1): 202-216., 2008.

[8] Schatteman, D., Herroelen, W., Van de Vonder, S., and Boone, A. "Methodology for Integrated Risk and Proactive Scheduling of Construction Projects." *Journal of Construction Engineering and Management*, 134(11), 885-893, 2008.

[9] Zafra-Cabeza, A., Ridao, M. A., and Camacho,E. F. "Using a risk-based approach to project scheduling: A case illustration from semiconductor manufacturing." *European Journal of Operational Research*, 190(3): 708-723, 2008.

[10] Shan, Y. and Goodrum, P. M., "Integration of Building Information Modeling and Critical Path Method Schedules to Simulate the Impact of Temperature and Humidity at the Project Level." *Buildings*, 4: 295-319, 2014.

[11] Moon, H., Kim, H., Kamat, V., and Kang, L. "BIM-Based Construction Scheduling Method Using Optimization Theory for Reducing Activity Overlaps." *Journal of Computing in Civil Engineering*, ASCE, 29(3), 2015. (DOI: http://dx.doi.org/10.1061/(ASCE)CP.1943-5487.0000342)

[12] Hijazi, W., Alkass S., and Zayed, T. "Constructability Assessment Using BIM/4D CAD Simulation Model." AACE International Transactions, 2009.

[13] Chen, S. M., Griffis, F. H., Chen, P. H., and Chang, L. M. "A framework for an automated and integrated project scheduling and system." *Automation in Construction*, 35: 89-110, 2013.

[14] Kim, H., Anderson, K., Lee, S., Hildreth, J. "Generating construction schedules through automatic data extraction using open BIM (building information modeling) technology." *Automation in Construction*, 35: 285–295, 2013.

[15] Zhou, W., Georgakis, P., Heesom, D., and Feng, X.

"Model-Based Groupware Solution for Distributed Real-Time Collaborative 4D Planning through Teamwork." *Journal of Computing in Civil Engineering*, ASCE, 26(5): 597-611, 2012.

VALUE ENGINEERING DECISION MAKING MODEL USING SMART

A. Montasaera and A. Montaserb

[a] Department of Structural Engineering, University of Ain Shams, Egypt
[b] Department of Structural Engineering, University of Ain Shams, Egypt
E-mail: ahmed_mohsen1313@yahoo.ecom, structural_ali@yahoo.com

Abstract –

Value Engineering (VE) has a great potential to enhance project delivery. It has become a standard practice for many government agencies, private engineering firms and contractors since its first adoption. Decision makers judge the alternatives performance under selected criteria and simultaneously weighing the relative importance of the criteria in order to reach a global judgment. Multi-Criteria Decision Analysis (MCDA) has been widely used over the last several decades to reduce assessment subjectivity. One of MCDA methods is Simple Multi-Attribute Rating Technique (SMART). This paper presents a model that improves VE utilization through applying SMART to determine the optimum alternatives. A spreadsheet application was developed to automate the process of criteria and alternatives weighing and assessment. The proposed model has been applied to a case study project to show its capabilities and usage. The case study is a laboratory building in Doha, Qatar. It is a part of new Liquid Natural Gas (LNG) Support Campus. The structural slab system of the building was selected to be under VE study. The VE team selected four alternatives for the structural slab with nine performance criteria for evaluation. An in-depth structural analysis was conducted for all building structural elements such as columns and foundations to support the analysis process. The proposed model is expected to facilitate and support project managers in taking timely informed decisions.

Keywords –

Value Engineering, Simple Multi Attribute Rating technique (SMART), Criteria and Decision Making.

1 Introduction

Value Engineering (VE) was born out of necessity after world war II as a consequence of wartime shortages when, alternative materials were used in innovative designs that offered improved performance at lower cost. This led to research into means of achieving the function of the components by an alternative technique [14] and [17]. As the application of VE expanded, there was also a change in context, from review of existing parts to improving conceptual designs [26], [5] and [9] VE study is mainly divided into 3 main stages, which are pre-study stage, job plan stage and post study stage. The value methodology is a systematic process that follows the Job Plan. A value methodology is applied by a multidisciplinary team to enhance the value of a project through the analysis of performance criteria [4], and [21]. VE provides organizations with a definitive tool to enhance the value of product, project or process. Design/construction contractors have used this technique [19]. Generally, VE intends to find alternatives, which provide the same function at the minimum cost. Consequently, VE has been mainly considered as a cost saving tool with an objective vision of value [20]. However, many owners fail to implement VE because of a lack of experience, accountability, or motivation. Often VE is applied late in a project as a backup to cut costs, and the process can be difficult as it interrupts the flow of work, causes delays, requires extra design resources, and may result in the loss of critical design features [13] and [25].

Decision-making is the analysis of identified alternatives based on the values and preferences of the decision maker. While planning, it not only required to identify as many of these alternatives as possible but also to select the one that best fits with stakeholders' goals, objectives and requirements [1], [10], [11] and [3]. Multi-Criteria Decision Analysis (MCDA) is one of the techniques used by decision makers to evaluate criteria and alternatives. MCDA techniques have been an active part of the field of operation research for at least four decades [22] and [7]. Edwards and Barron [8] defined SMART as the method of rating alternatives and weighting criteria as a method, which is a simpler form of Multi Attribute Utility Theory (MAUT). SMART technique is based on a linear additive model, which calculates the overall value of a given alternative as the total sum of the performance score of each criterion multiplied with the weight of this criterion [16], [15] and [12] Major advantages of SMART are that it is simple to use and it actually allows for any type of weight assignment techniques (i.e., relative, absolute, etc.). It requires less effort by decision makers. It also handles data well under each criterion [23], [18] and [24]. SMART's common applications are in environmental, construction, transportation, logistics, military, manufacturing and assembly problems [6] and [2].

The purpose of this paper is to develop a decision-making model based on applying SMART methodology to assess criteria and alternatives in a quick simple process. Value engineering helps in developing better understanding and appreciation of the project scope of work and in reducing unnecessary cost without impacting the required functions of project components being considered. The model developed in this research can be of help to value engineering team members, design professionals and owners and stakeholders.

2 Proposed Methodology

The developed model provides users with an automated and comprehensive methodology and computational platform. It considers a wide range of criteria for evaluation and selection of the best alternatives that satisfy the owners' requirements. SMART calculations have been implemented to help VE team in assessing and ranking various alternatives of project items using multi-attributed criteria. Pre-study activities are conducted as a first step to get a clear view of the stakeholder's requirements, and to establish pre-defined priorities for the VE study. The main purpose of this phase is to achieve a basic level of understanding for the VE team members. After finishing pre-study stage, job plan stage is initiated, which is composed of six major sequential phases. Information phase starts by data collection to define the status of the project and the target of VE study. Function analysis follows information phase to define the main performance criteria. Restating and combining criteria, or omitting less important criteria is vital to focus on value-oriented functions. If the weight for a particular criterion is quite low, that criteria should not be addressed. There is no specific number of criteria appropriate for decisions. User should rank the importance of the changes in the criteria from the worst levels to the best levels. Ratio estimates of the relative importance of each attribute relative to the one ranked lowest in importance should be developed by user to be able to evaluate criteria efficiently.

User assigns points to the least important attribute. The relative importance of the other attributes is then evaluated by giving them points based on relative importance. The points given by the decision maker are normalized to get the weights. After calculating criteria weights, user defines several alternatives to attain the same functions. In SMART, ratings of alternatives are assigned directly, in the natural scales of the criteria. In order to keep the weighting of the criteria and the rating of the alternatives as discrete as possible, the various scales of criteria need to be converted into a common scale, which is done mathematically by the decision-maker by means of a value function. Simplest and most widely used form of a value function method is the additive model, which in the simplest cases can be applied using a linear scale (e.g. going from 0 to 100). Finally, after approving alternatives weights and ranking, development and presentation phases are conducted. Final VE report is prepared to assure final approval from customer. Implementation and follow up activities follow as last activities for post study stage, as shown in Figures 1.

After criteria definition, decision makers are asked to rank the importance of criteria from lowest to highest then relative importance scores are inserted by decision makers. For simplicity 0 - 100 scale is used where a score of 100 indicates extremely high importance and zero indicates virtually no importance. This scale is used to put score for each criterion based on the relative importance. Normalization process occurs to allow normalization of the relative importance into weights summing to one. During creative phase, alternatives are defined and decision makers are asked to rate alternatives against criteria. For example, when assessing the criterion "cost" for the choice between different alternatives, a natural scale would be a range between the most expensive and the cheapest alternative. Linear scale is used during this step based on each criteria property to calculate the final score for each alternative and to reach to a final global judgment and ranking for the chosen alternatives as described in equation 1.

$$Xa = \Sigma\ (Wj * rij) \quad (1)$$

Where Xa is alternative score, Wj normalized weight calculated for each criterion and rij performance of each alternative against criterion. The alternative with highest score will be the best alternative and ranking will be based on that score. All the equations, algorithms and reports had been implemented in Microsoft Excel spreadsheet to facilitate the calculation and reporting process. The whole process is described in figure 2.

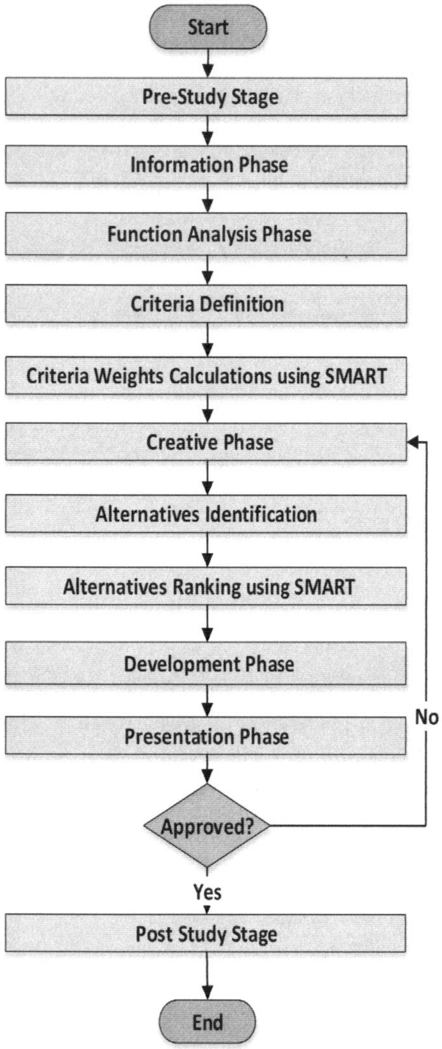

Figure 1. Proposed methodology

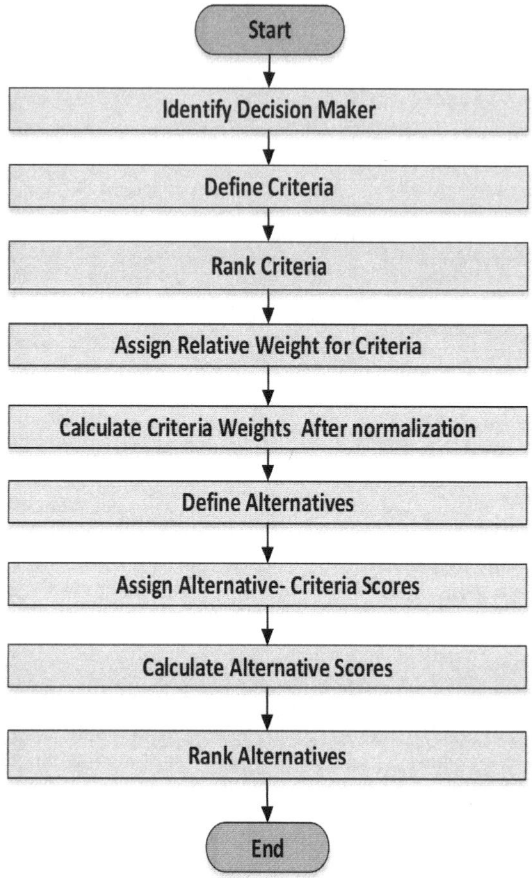

Figure 2. SMART process

3 Model Verification

The proposed model had been verified through applying it to real case study, which is phase 2 of new Liquid Natural Gas (LNG) Support Campus at Mesaieed lies 45 km south of Doha, Qatar. The new LNG Support Campus is being developed on a 1000 m2 area, located at south west of Masaieed Industrial City along Sea Line Road. The project is composed of various buildings. Laboratory building had been chosen to be the case study. Laboratory building (LNG16) is designed to reflect an image of hygiene and cleanness, which accommodates labs at the ground floor while the offices and staff facilities are at the first floor. The building facades match the overall theme of the LNG Support Campus. The building consists of two stories offices with span ranging from 3.0 to 8.0 m and floor height around 5.0 m. The final selected criteria for the case study are nine criteria, which are cost, time, appearance, serviceability to services and utilities requirements, material availability, durability, fire resistance, flexibility for future modifications and possibility of future expansion. Criteria weights are calculated based on the value team input. Table 1 is showing the resulted criteria weights. All results had been verified to ensure the correct SMART formulas are used on the spreadsheet as discussed before. Cost was ranked first criteria with 32%, time with 19% then appearance with 16%.

Table 1- Laboratory building criteria weights

Criteria	Ranking	Score	Weight %
Cost	1	100	32%
Time	2	60	19%
Appearance	3	50	16%
Serviceability to services and utilities requirement	4	30	10%
Material availability	5	30	10%
Durability	6	15	5%
Fire resistance	7	15	5%
Flexibility for modifications	8	10	3%
Possibility of future expansion	9	5	2%

After defining the criteria and final weights, VE team members had defined four alternatives for the structural slab system, which are flat slab, post tension, hollow block slab and hollow core slab, as shown in Figure 3. 3D model was created for each alternative to facilitate structural analysis as shown in figure 3. Most of the criteria were evaluated qualitatively by the VE team except for the cost and time. To calculate the estimated construction duration and cost for each alternative, a detailed quantity take off would be necessary. So, an in depth structural analysis was conducted for each alternative to calculate the concrete and steel reinforcement quantities for the structural elements of each alternative. The structural model for each alternative was only conducted for typical three bays in X direction and three bays of Y direction as a representing for the whole laboratory building. Figure 4 demonstrates the developed structural models for each alternative. ETABS and RAM structural software programs were used to get the serviceability deflection, straining actions, the amount of steel reinforcement and concrete dimensions. Figure (5- a) shows the vertical bending moment for the marginal floor beams in the flat slab alternative. While Figure (5-b) represents the ultimate bending moments for typical post tensioned slab alternative. Figure (5-c) presents the shear force diagram in Y direction for typical floor beams in the hollow block slab alternative. Whereas, figure (5-d) depicts restraint reactions for hollow core slab. 3D BIM model had been developed for each alternative after the final design creation to estimate time and cost, as shown in Figure 6. Quantities were calculated using each alternative 3D BIM model as shown in table 2 then cost rates and production rates were used to convert these quantities to time and cost parameters.

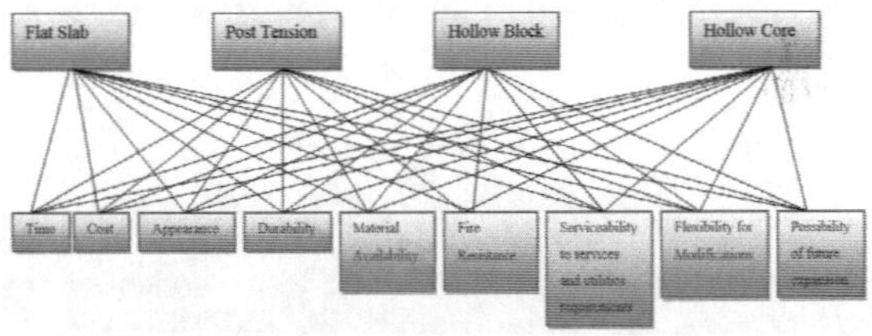

Figure 3. Laboratory building alternatives - Criteria scores structure

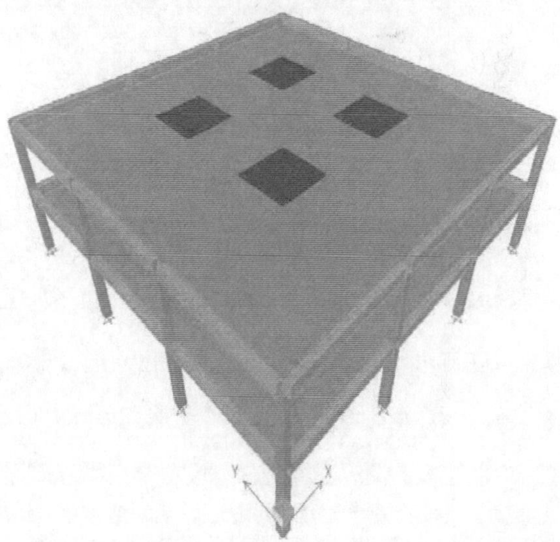

(a) Flat slab

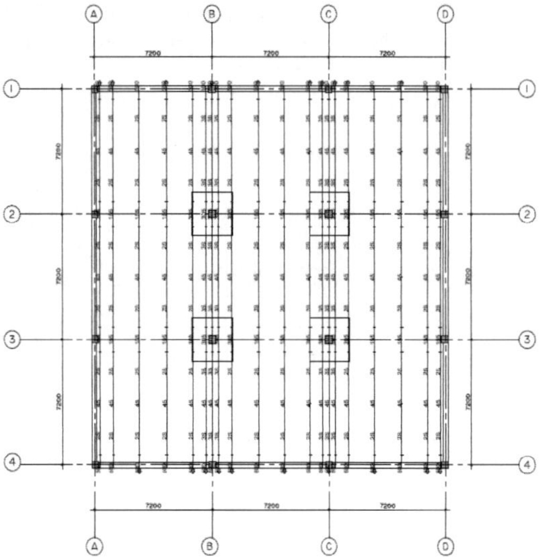

(b) Post tensioned slab

(c)- Hollow block slab

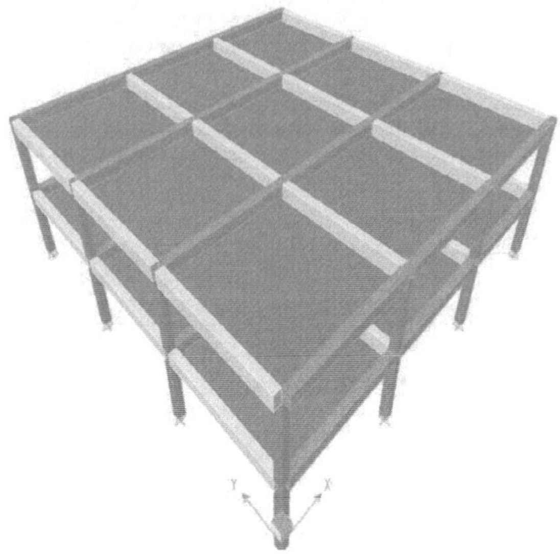

(d)- Hollow core slab

Figure 4. Laboratory building alternatives developed structural models

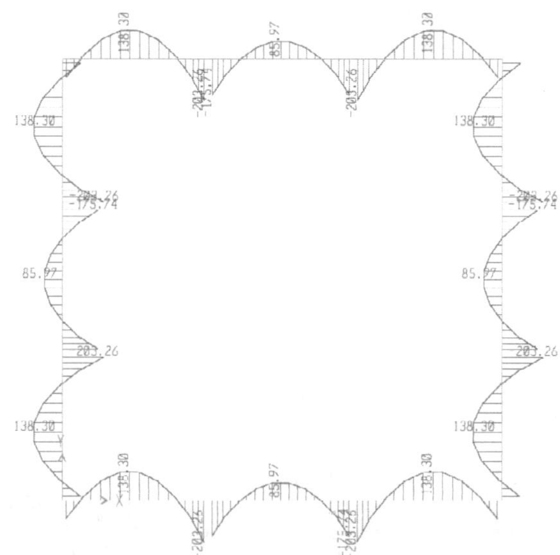

(a)

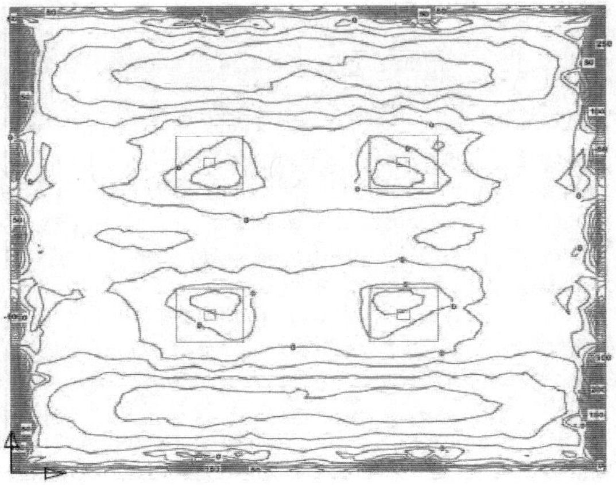

(b)

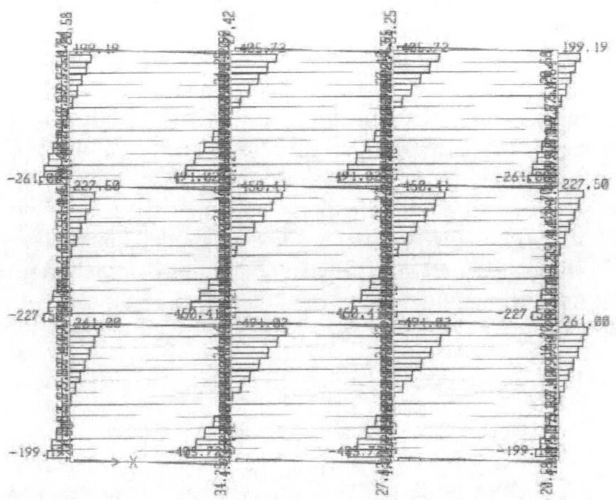

(c)

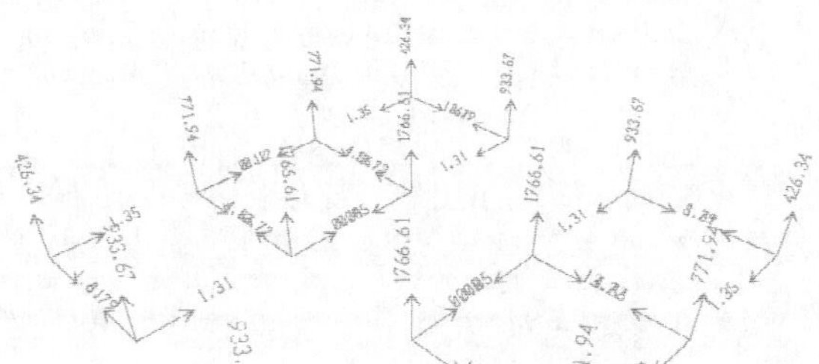

(d)

Figure 5. Laboratory Building design sample

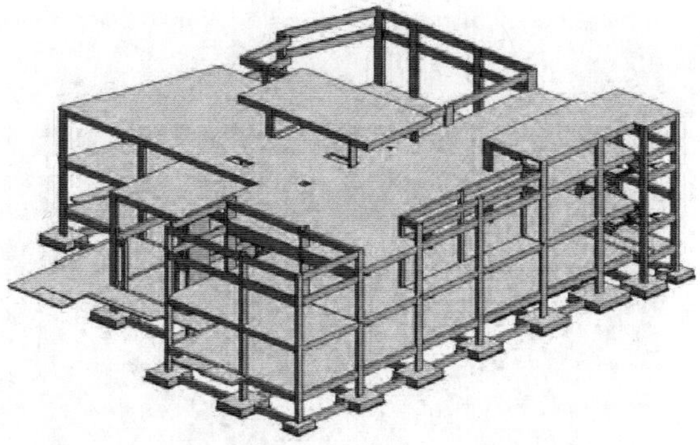

Figure 6. Laboratory building developed 3D model

Table 2 - Laboratory building alternatives quantities

Criteria	Flat slab m^3	Post Tension m^3	Hollow Block m^3	Hollow Core m^3
PC Foundation	37.53	35.56	26.84	26.84
RC Footing	616.08	545.89	522.65	522.65
RC Tie Beam	77.9	81.74	82.33	82.33
RC SOG	288.82	288.82	289.11	289.11
RC Beams	101.44	106.67	112.48	122.09
RC Slab	692.06	592.81	75.44	115.07
RC Stairs	14.97	14.97	15.11	15.11
RC Columns	125.77	136.64	132.4	132.07
RC Walls	52.17	53.43	47.8	47.8
Hollow Block Slab	NA	NA	282.6	NA
Hollow Core slab	NA	NA	NA	411.28

Alternatives cost and time are calculated to be used as the main input for alternative - criteria evaluation. The results were shown in Figure 7 for cost in Qatari Riyal (QR) and time in days. VE team members had evaluated remaining criteria for each alternative to calculate the total score for alternative. After assessing all alternatives flat slab had been ranked first as the best alternative for the laboratory with score 90.5 then post tension with score 85.8 then hollow block with score 82.7 then hollow core slabs with 67.8, which was compatible with the chosen system for the building as shown in table 3.

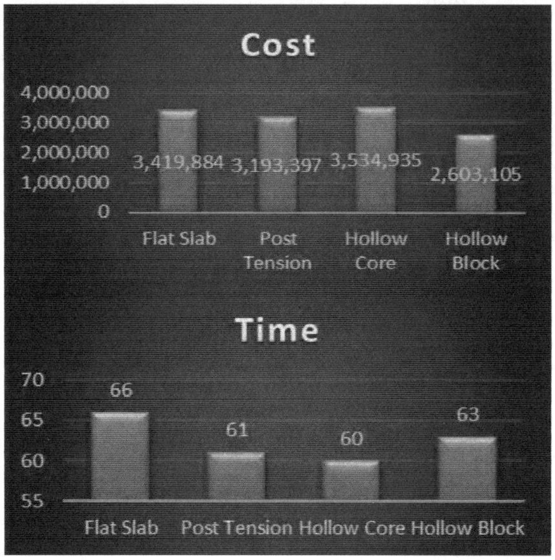

Figure 7- Laboratory building alternatives cost and time

Table 3- Laboratory building alternatives ranking

Criteria	Flat Slab	Post Tension	Hollow Block	Hollow Core
Cost	76	82	74	40
Time	90	98	95	100
Appearance	100	80	80	60
Serviceability to services and utilities requirement	100	70	70	50
Material availability	100	100	100	100
Durability	100	100	100	100
Fire resistance	100	100	100	100
Flexibility for modifications	100	35	35	35
Possibility of future expansion	100	100	100	100
Total Score	90.5	85.8	82.7	67.8

4 Conclusion

Value Engineering (VE) is a powerful tool to achieve essential functions of projects through choosing the optimum alternative based on project specific criteria. This paper presents a model for applying VE using Simple Multi Attribute Technique (SMART) in construction projects. It provides a preliminary work plan for making the decisions during value engineering study and assists the project team to decide which of the options to choose. SMART applications throughout the excel sheets shown in the paper gives user the opportunity to study the impact of modifying criteria or alternatives scores in easy manner. SMART gives the user the optimum solution in simple Process and it allows different weight techniques usage during the scoring process. SMART formulas were implemented in excel spread sheet and verified manually. Additionally, real project is used to validate the use of the model and verify its processes. Nine criteria were chosen to assess alternatives. Four alternatives for the structural slab system were the base for value engineering study, which are flat slab, post tension, hollow block slab and hollow core slab. Algorithms and alternatives ranking reports are developed upon inserting user inputs. The established model, methods and support VE application in easy manner. The developed model enables value engineering team to determine the optimum solution for project implementation methodology.

References

[1] Baker, D., Bridges, D., Hunter, R., Johnson, G., Krupa, J., Murphy, J., and Sorenson, K. Guidebook to Decision- Making Methods, *WSRC-IM-2002-00002*, Department of Energy, USA, 2002.

[2] Barfod, M. B., & Leleur, S. Multi-criteria decision analysis for use in transport decision making. (2 ed.) DTU Lyngby: *Technical University of Denmark*, Transport, 2014.

[3] Chen, Y., Kilgour, D.M., and Hipel, K.W. Screening in multiple criteria decision analysis. *Decision Support Systems*, 45: 278-290, 2008.

[4] Davis, E. Finding Value in the Value Engineering Process. *Journal of Cost Engineering,* 46(12):24-27, 2004.

[5] Dell Isola, A.J. Better Utilizing Value Engineering in Project Delivery. *AACE International Transactions*, PM.1417, 2013.

[6] Demirel, T., Demirel, C. and Kahraman, C. Fuzzy analytic hierarchy process and its application. *In fuzzy multi-criteria decision making: theory and applications with recent developments.*16: 85-117, 2008.

[7] Dey, P.K. Project Risk Management: A Combined Analytic Hierarchy Process and Decision Tree Approach. *Journal of Cost Engineering*, 41(10):13-26, 2002.

[8] Edwards, W., and Barron, F. H. SMARTs and SMARTER; Improved Simple Methods for Multi attribute Utility Measurement. *Organizational Behavior and Human Decision Processes*, 60(1):306-325, 1994.

[9] Garrido, J.S. Conceptualizing Value for Construction: Experience from Social Housing Projects in Chile. *PhD Thesis*, Loughborough University Institutional Repository, Loughborough, England, 2013.

[10] János, F. Introduction to Decision Making Methods. *Report*, Laboratory of Operations Research and Decision Systems, Computer and Automation Institute, Hungarian Academy of Sciences, 2006.

[11] Jose, F., Salvator, G. and Matthais, E. Multiple Criteria Decision Analysis: State of the Art Survey. *Report*, Kluwer Academic Publishers, Boston/Dordrecht/London, 2005.

[12] Kasie, M. Combining Simple Multiple Attribute Rating Technique and Analytical Hierarchy Process for Designing Multi-Criteria Performance Measurement Framework. *Global Journal of Researches in Engineering Industrial Engineering*, 13 (1), 2013.

[13] Mansour, F and Hulshizer, A. The Antidote to Value Engineering Phobia. *AACE International Transactions*, VE&C.02, 1997.

[14] Miles, L.D. *Techniques of Value Analysis and Engineering*. 2nd ed., McGraw-Hill, New York, USA, 1972.

[15] Mustajoki, J., Raimo, P. and Ahti, S. Decision support by interval smart/swing incorporating imprecision in the smart and swing methods. *Decision Sciences*, 36(2):317–339, 2005.

[16] Olson D.L., Comparison of three multicriteria methods to predict known outcomes. *European Journal of Operational Research*, 130: 576-587, 2001.

[17] Parker, D.E. *Value Engineering Theory* .3rd ed., the Value Foundation, www.valuefoundation.org, 1997.

[18] Pöyhönen, M., and Hämäläinen, R. P. On the Convergence of Multi-Attribute Weighting Methods. European *Journal of Operational Research*, 129(3): 106-122, 2001.

[19] Rains, J. Value Methodology Standard. *Joint Cost Management Societies Proceedings*, SAVE.01, 1999.

[20] Raj, H. VE Is Not a "Group Cost Cutting". *AACE International Transactions*, CSC.17, 2001.

[21] SAVE, Society of American Value Engineers International. Value Methodology Standard and Body of Knowledge, *Report*, 2007.

[22] Stewart, T. J. A Critical Survey on the Status of Multiple Criteria Decision Making Theory and Practice. International Journal of Management Science, 20(1): 569-586, 1992.

[23] Stuart, G. Beyond value engineering: SMART value management for building projects. *International Journal of Project Management,* 12(1).:49-65, 1994.

[24] Velasquez, M., and Patrick, T. An Analysis of Multi-Criteria Decision Making Methods. *International Journal of Operations Research*, 10(2):56-66, 2013.

[25] Wandahl, S. Value in Building. *PhD Thesis*, Aalborg University, Denmark, 2003.

[26] Younker, D.L. *Value Engineering Analysis and Methodology*. Marcel Dekker, Ltd., New York, USA, 2003.

An Empirical Study on Analyzing Schedule Delays of Construction Project in Indonesia

Zetta R. Kamandang[a]*, Jyh-Bin Yang[a], and Indradi Wijatmiko[b]

[a]Graduate Institute of Construction Engineering and Management, National Central University, Taiwan
[b]Department of Civil Engineering, University of Brawijaya, Indonesia
E-mail: zerasullia@gmail.com, jyhbin@ncu.edu.tw, indradi@ub.ac.id

Abstract –

Schedule delay is one of the worrying problems in the construction industry. Activity delays might increase project duration, cost and sometimes cause delay claim disputes between parties in construction projects. Some methods have been developed for analyzing and measuring project schedule delays, but most of those methods contain assumptions and subjective assessments. In order to identify the useful methods for projects in Indonesia, this study implements popular schedule delay analysis methods into a real project case in Indonesia which has varied relationships among activities and high frequency of concurrent delays. In this study, collecting and analyzing actual project schedule information, identifying delay events, executing delay analysis and comparing analytical results, have been completed to examine the possibilities of using available methods to provide reasonable or even correct outcomes. This study concludes that some of the available methods are unable to deal with concurrent delay problems and some result in different outcomes to the others. Therefore, choosing suitable methods would be an issue when solving delay analysis problems in Indonesia's construction projects

Keywords –

Construction management; Schedule delay analysis; Concurrent delay; Indonesia.

1 Introduction

Schedule delay is one of the worrying problems in the construction project. Construction projects involve many parties and generally have highly complicated situations during execution. Activity delays in construction project might increase project duration, cost and sometimes cause delay claim disputes between parties involved. A delay caused by a party may not always affect the project completion date and may not always cause damage to another party [1]. Therefore, to distinguish delay responsibility among parties is necessary. Another issue related to delay in construction project is solving concurrent delay. Concurrency in delay sometimes occurs and may impact the total duration of the project. The result of delay caused by concurrent delay may increase the dispute liability claim of the parties. The project parties would be hard to identify their own liability towards project delays if the concurrency occurs and it will lead the parties into a confusing situation. The analysis of the concurrent delay has to consider all aspects of the delay that have occurred in the project [2]. Therefore, dealing with concurrent delay is an important issue to choose the suitable schedule delay methods.

Yang and Kao [3] stated that as-planned and as-built schedule are the basis information for the parties to recognize the delay and solve delay disputes and claims. To demonstrate cause and effect relationships of time-related disputes in construction contract, schedule delay analysis is commonly conducted [1]. There are many delay analysis methods that have been developed for analyzing and measuring project schedule delays, but most of those methods contain assumptions, the application of personal experience and judgments, theoretical projection and subjective assessments [4]. Most of the previous studies might use uncompleted or uncomplicated project information which only consists of limited activities and delay types. Furthermore, previous studies consider the logical relationship of activities limited to Finish to Start (FS) and use the delay type only based on assumptions.

In order to identify the useful methods for projects in Indonesia, this study implements the approaches of those popular methods into a real project case in Indonesia which has varied relationships amongst activities, high frequency of concurrent delays and real delays events. By using a real project in Indonesia as a project case, this study collects as-planned schedule, as-built schedule and delays information from project participants. This research addresses the suitable delay analysis method based on the available documents of the project and examines the suitability of each delay analysis approach when it is applied to the discussed real project.

2 Schedule Delay Analysis Method

Many schedule delay analysis methods have been developed in previous studies [1, 5-10]. The popular schedule delay analysis methods include as-planned vs as-built, impacted as-planned, the collapsed as-built, isolated delay types, window but-for technique, isolated collapsed as-built and critical path effect based delay analysis method [3, 7-9, 11-15].

The as-planned vs as-built is the simplest method that compares two schedules, which is why it is also called "the total time method or net impact method" [16]. By subtracting the total duration of as-built into as-planned schedule, the total delay duration will be obtained.

Impacted as-planned method starts with as-planned schedule and analyzes from two perspectives, owner's and contractor's. As an example, from the owner's perspective, the delay caused by owner inserted into as-planned schedule. The difference between each impacted as-planned critical path(s) and the previous one, is the delay resulting from the delayed event(s) [17].

Collapsed as-built (but-for) method is also called the subtractive as-built or but-for method. For the collapsed as-built (but-for) owner delays, the owner delays are subtracted from as-built schedule and the difference of duration of two schedules is the liability of the owner [18].

Isolated delay type method is derived from snapshot analysis (SA) method to overcome drawbacks in previous methods. In this method, the project duration is divided into several time frames or scenarios. Analyzing at least twice from owner's perspective and contractor's perspective, this method examines only the non-excusable (NE) delay events and ignores the excusable-compensable (EC) and excusable-non-compensable (EN) delays while analyzing from the contractor's perspective [5].

In window but-for technique, the project duration is divided into several time frames [5]. This technique is a combination of window and but-for technique. In this method, the owner and contractor caused delay analysis are used continuously. One scenario will consist of actual progress and all delay types of the parties involved with one-by-one inserting delay types into the scenario. It means the scenario will begin with the first type of delays and then continue with others. It is different from the other methods above that analyze the owner's and contractor's perspective separately.

Isolated collapsed as-built (but-for) method uses the concept of isolated delay type method, but start with an as-built schedule because it reflects actual start and finish dates and actual duration. In this method, the owner and contractor caused delays are used continuously. One scenario will consist of all delay types of the parties involved with one-by-one subtracting delay types from the scenario. Therefore, this method can be regarded as a combination of the collapsed but-for method and isolated delay type method [15].

Critical path effect based delay analysis method (EDAM) has a systematic window extraction method for performing delay analysis stable and adopts a process-based analysis approach to resolving concurrent delays and liability distribution problems accurately [3]. The project duration in this method is also divided into several time frames or scenarios. The analyst has to calculate using some equations represent baseline duration, duration caused by owner's delay and contractor's delay, and duty of owner and contractor for each window and each activity. The critical path changes in each window have to be analyzed comprehensively. When delays occur in the critical path, equations to find out the liability of each party has to be calculated.

Methods which are chosen in this research cover three levels of schedule delay analysis. First is the simplest one: the as-planned vs. as-built method. Impacted as-planned and collapsed as-built method which involve the owner's and contractor's perspective, even those perspectives used separately are in the second level. Isolated delay type, window but-for technique, isolated collapsed as-built (but-for) and critical path effect based delay are in the third level. Above four methods in third level use project schedule, delays information, involve owner's and contractor's perspectives and several time frames based on the project schedules. For the last three methods, owner's and contractor's perspectives are used continuously, which means those two perspectives are considered at the time.

3 Project Case in Indonesia

This study implements popular schedule delay analysis methods into a real project case in Indonesia. The case is a building project with the contract price of Rp.81.124.159.736,00 (about US$ 6.084.921,00) and has been delayed for several days. The project has to be finished on September 25 but the real completion date is October 23. The reason for choosing this project is that all the related data and information like as-planned and as-built schedule of the project, weekly project reports, and delays information including causes and time are recorded and can be obtained. Furthermore, this project suffered many delays caused by owner and contractor, as well as by the project architects, sub-contractor and consultant. This project is uncomplicated and fulfills the requirement of examining different delay types. The information of as-planned and as-built schedules are presented in Table 1.

Table 1. The as-planned and as-built schedule information

Task Name	As-planned Information			As-Built Information		
	Duration	Start	Finish	Duration	Start	Finish
Master Schedule	294 days	11/30/2015	9/25/2016	322 days	11/30/2015	10/23/2016
Main Building	272 days	12/21/2015	9/24/2016	308 days	12/14/2015	10/23/2016
Structure	142 days	12/21/2015	5/10/2016	252 days	12/21/2015	9/4/2016
Architecture	213 days	1/28/2016	9/3/2016	186 days	1/28/2016	8/7/2016
Finishing	199 days	2/11/2016	9/3/2016	189 days	4/11/2016	10/23/2016
Interior	192 days	2/11/2016	8/27/2016	181 days	4/11/2016	10/15/2016
Facade	165 days	4/6/2016	9/24/2016	187 days	4/11/2016	10/21/2016
Supporting Building	286 days	11/30/2015	9/17/2016	322 days	11/30/2015	10/23/2016
Landscape	171 days	4/1/2016	9/25/2016	203 days	3/28/2016	10/23/2016

4 Project Case Delay Analysis

To complete project delay analysis, this study follows a series of processes proposed by previous studies [11, 19]. The steps completed are organized as following phases [20].

Preparation phase is to collect required project information. The as-planned schedule, as-built schedule, delays information and weekly progress reports are the basic required information of this research.

Diagnosis phase consists of some works including analyzing the relationships among the activities, categorizing delay types based on causes of delays and developing an analysis of self-made as-built schedule based on weekly progress report to ensure the generated as-built schedule is the same with obtained report. In this phase, the as-planned schedule, the as-built schedule, and delay types will be presented in one table named as schedule delay information table. The schedule delay information table is the starting point to accomplish the schedule delay analysis.

Analysis phase is to analyze project construction delays using some popular methods. Starting with schedule delays information table which consists of as-planned schedule, as-built schedule and delay types, the schedule delay analysis can be performed.

The last phase is summation phase. This phase is to identify the most suitable schedule delay analysis approaches for Indonesian project case.

4.1 As-planned vs. As-built Method

The concept of this method is to compare the duration of as-planned and as-built schedules. As-planned schedule of the project has 294 days and the as-built schedule has 322 days. Based on the calculation using the as-planned vs. as-built method, the total duration of the delay is 28 days. The as-planned vs. as-built method is very simple to prepare and perform, but the liability for each party in the project is unable to be calculated.

4.2 Impacted As-Planned Method

This method begins with as-planned schedule and then inserts the owner or contractor delays. The impacted as-planned method presents the delay liability based on owner's perspective is 21 days [the contractor has 7 days (28-21=7)] and the liability of the contractor is 21 days [the contractor has 7 days (28-21=7)] based on contractor's perspective. Table 2 presents the result of impacted as-planned method calculation.

4.3 The Collapsed As-Built (But-For) Method

Starting with as-built schedule, this method removes the contractor-caused-delay events from owner's perspective. The difference of duration between two given schedules is the liability of contractor. The contractor must be responsible for that delay and vice versa.

The collapsed as-built (but for) method presents the delay liability of the contractor based on owner's perspective is 21 days and from the contractor's perspective, the owner has the responsibility of 7-day delay. The results are presented in Table 3.

4.4 Isolated Delay Type

This method selects several time periods (frames, windows) to analyze the as-planned schedule and delay activities. After analyzing the schedule delay of a window, the next step is to repeat the process for the next window. In this method, the as-planned schedule is divided into six periods. From owner's and contractor's perspectives, the results of the calculation using this method are shown in Table 4.

Table 2. Results of impacted as-planned method

Perspective of	As-Planned Duration	Date Completion		Delay Liability	
		with Owner's delay	with Contractor's delay	Owner	Contractor
Owner	294	315		21	7
Contractor			315	7	21

Table 3. Result of collapsed as-built (but-for) method

Perspective of	As-Built Duration	Date Completion		Delay Liability	
		without Contractor's delay	without Owner's delay	Owner	Contractor
Owner	322	301		7	21
Contractor			315	7	21

Table 4. Results of isolated delay type method

Scenario	Analysis Period (week)	Delay Liability (days)	
		Owner	Contractor
1	0 - 8	0	0
2	9 - 16	0	0
3	17 - 24	0	0
4	25 - 32	0	7
5	33 - 40	0	0
6	41 - 47	21	14
Sum		**21**	**21**

4.5 Window But-For Technique

Starting with as-planned schedule, this method selects several periods (frames) to analyze delay liability. After inserting the actual progress and delay activity in one scenario, the process has to be repeated for the next window. In one scenario, actual progress and delays caused by owner are inserted first into as-planned schedule and duration changes have to be calculated as the completion date with owner's delays. This as-planned schedule with owner's delays is used as a baseline to calculate the completion date with contractor's delays. After the delays caused by owner and contractor are finished to be calculated, the schedule becomes the new baseline for subsequent scenario.

Table 5 presents the results of this method. From the owner's perspective, the owner has the responsibility of 0-day delay liability and the contractor has 22 days based on the contractor's perspective. Based on this calculation, the concurrent delay has 6 days, it is obtained from total duration of project delay minus 22 days.

4.6 Isolated Collapsed But-For Analysis

This method selects several periods (frames) of as-built schedule as the starting point to complete the analysis. Based on calculation in Table 6, the owner causes 0-day delay [the contractor has responsibility of 28-day delay (28-0=28)] and the contractor causes 22-day delay [the owner has the responsibility of 6-day delay (28-22=6)]. The summation value of delay responsibilities caused by the contractor (28 days) and the owner (6 days) is 34 days. The concurrent delay (6 days) can be calculated by using the summation value (34 days) minus total delay days (28 days). At the end, the owner causes 0-day delay and the contractor has responsibility of 28-day delay, with 6 days of concurrent delay.

Table 5. Results of window but-for technique

Scenario	Analysis Period	Project Completion	Completion Date		Delay Liability	
			Owner Delay	Contractor Delay	Owner Delay	Contractor Delay
1	0 - 8	294	294	294	0	0
2	9 - 16	294	294	294	0	0
3	17 - 24	294	294	294	0	0
4	25 - 32	307	300	307	0	7
5	33 - 40	307	307	307	0	0
6	41 - 47	322	307	322	0	15
Sum:					0	22

Table 6. Results of isolated collapsed but-for analysis

Scenario	Analysis Period (week)	Target date for comparison	Without Contractor Delay		Without Owner Delay	
			Completed Date	Delayed Date	Completed Date	Delayed Date
1	41 - 47	307	307	0	322	15
2	33 - 40	307	307	0	307	0
3	25 - 32	294	294	0	301	7
4	17 - 24	294	294	0	294	0
5	16 - 9	294	294	0	294	0
6	0 - 8	294	294	0	294	0
Sum:				0		22

4.7 Critical Effect Based Delay Analysis Method (EDAM)

The EDAM uses as-planned schedule as a starting point and requires clearly delays information. This method applies the critical path method before delay impact calculation to determine a comparison baseline. In this method, the schedule is divided into several time frames and uses equations to calculate the delay liability of each party.

Critical path changes have to be analyzed as a part of this method. By using critical path changes analysis, at week 30, the total duration becomes 300 days from its initial duration 294 days (Figure 1-2). It means the project has 6 days of delay (300-294=6). At this week, the owner and the contractor cause delays on the critical path and the EDAM provides equations for calculating the delay liability of the project parties.

Because of both project parties cause delay in the same week, one concurrent delay exists. By using the equation proposed in EDAM, the concurrent delay of 6 days is distributed as the owner has 4.38 days and the contractor has 1.62 days (Figure 3). From week 31 until week 47, only the contractor causes the delays.

After analyzing all of the critical path changes, the final summation has to be calculated. In the project, total project delay duration is 28 days with 6-day of concurrent delay and 22-day delay caused by the contractor. Using the equation, the liability of the contractor and the owner can be calculated. At the end, the owner caused 4.38-day and the contractor caused 23.62-day of delay from 28-day in total.

4.8 Comparison Result of Each Method

Table 7 presents the comparison results of all methods. As explained earlier, the methods which are chosen in this research cover three levels of schedule delay analysis. In the first level, as-planned vs. as-built method only shows 28 days of the total delay in the project. This method is unable to address the liability of the project parties.

The impacted as-planned and collapsed as-built (but-for) methods start with different type of project schedule. Those methods in the second level of schedule delay analysis, consider the owner's and contractor's perspectives into the calculation separately and presents the different results of the parties' liability. However, the

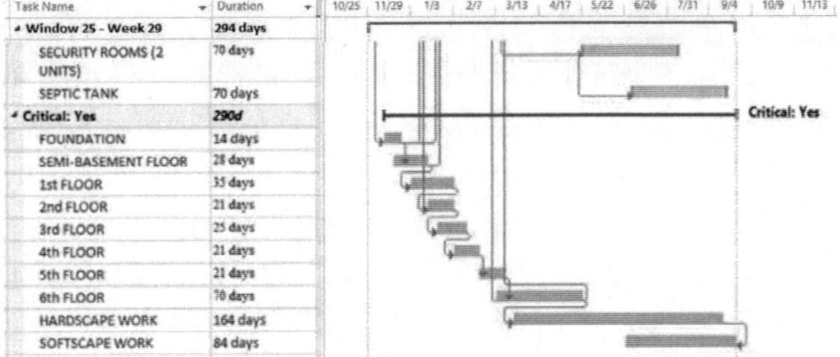

Figure 1. Critical path on week 29

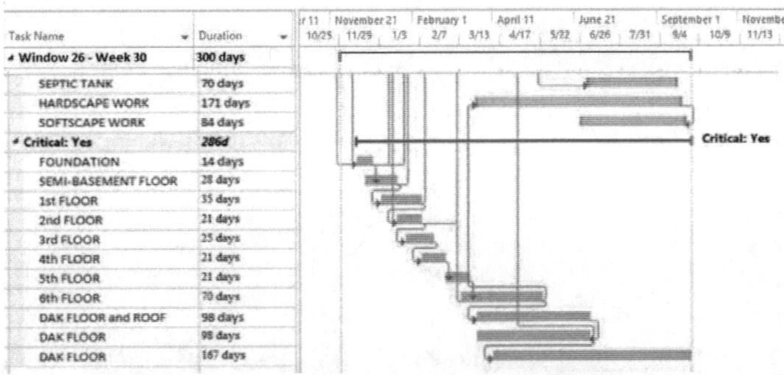

Figure 2. Critical path on week 30

Critical Path		W29	W30	Days Taken on CP		Calculation	
Structure	Foundation	-	-				
	Semi Basement	-	-				
	1st Floor	-	-				
	2nd Floor	-	-				
	3rd Floor	-	-				
	4th Floor	-	-				
	5th Floor	-	-				
	6th Floor	-	-				
	Dak Floor & Roof	EC	EC	98		1.62	
Architecture	Dak Floor	NE	NE		98		1.62
Finishing	Dak Floor		EC	167		2.76	
				363		4.38	1.62
						6	

Figure 3. Calculation of delays in the critical path on week 30

impacted as-planned and collapsed as-built (but-for) methods are unable to address the concurrent delay in the project.

The isolated delay type, window but-for, isolated collapsed but-for and EDAM are in the third level. These methods divide the schedule into several time frames and consider owner's and contractor's perspectives. The isolated delay type uses owner's and contractor's perspectives separately to its calculation. This method is unable to address the concurrent delay in the project case.

Window but-for, isolated collapsed but-for and EDAM consider the owner's and contractor's perspectives to the calculation and use it continuously for each period. These three methods are able to address the concurrent delay in the project even those methods start with different type of project schedule. Window but-for and isolated collapsed but-for method provide not only the same results of each project party's liability but also the number of concurrent delay. However, these methods are unable to address the liability of each party in concurrent delay. The EDAM not only provide the same result of concurrent delay as window but-for and isolated collapsed but-for method but also can calculate the liability portion of the project parties in the concurrent -

Table 7. Comparison result of each method

No	Method	Used Data	Delay Liability	
			Owner	Contractor
1	As-Planned vs. As-Built	As-Planned Schedule,		
	- Total duration project delay	As-Built Schedule	28	
2	Impacted As-planned	As-Planned Schedule,		
	Owner's viewpoint	Delay Information	21	7
	Contractor's viewpoint		7	21
3	The Collapsed As-built (But-For)	As-Built Schedule,		
	Owner's viewpoint	Delay Information	7	21
	Contractor's viewpoint		7	21
4	Isolated Delay Type	As-Planned Schedule,		
	Owner's viewpoint	Delay Information	21	7
	Contractor's viewpoint		7	21
5	Window But-For Technique	As-Planned Schedule,	0	22
	- Concurrent Delay	Delay Information	6	
6	Isolated Collapsed But-For	As-Built Schedule,	0	22
	- Concurrent Delay	Delay Information	6	
7	CP Effect Based Delay Analysis Method (EDAM)	As-Planned Schedule,	0	22
	- Concurrent Delay	Delay Information	6	
			4.38	23.62

delay. At the end, this method is able to calculate the total liability of each party in this project case including the liability of concurrent delay.

5 Conclusion

Schedule delay in projects is a common yet worrying problem. Activity delays might increase project duration, cost and sometimes cause delay claim disputes between parties in construction projects. Another issue related to delay in construction project is to deal with concurrent delay problems. Concurrency in delay sometimes occurs and may impact the total duration of the project. The result of delay caused by concurrent delay may increase the dispute liability claim of the parties. Therefore, identifying the suitable schedule delay analysis methods is an important issue to help the construction project practitioner solving the problems above.

Some methods have been developed for analyzing and measuring project schedule delays, but most of those methods contain assumptions and subjective assessments. Most of previous studies might use uncomplicated project information which only consists of limited activities and delay types in the schedules. Since this study provides a real project in Indonesia as a case study, it decreases the limitation above.

Before starting to implement some popular methods into a real case project, some works including collecting and analyzing actual schedule information and identifying delay events have been completed in advance. As-planned schedules, as-built schedules and delay information which are already categorized into different delay types will be presented in one table named as schedule delay information table.

This study examines seven schedule delay analysis methods which are divided into three levels. First is the simplest one, as-planned vs. as-built method. Impacted as-planned and collapsed as-built method which involve the owner's and contractor's perspectives, even those perspectives used separately are in the second level. Isolated delay type, window but-for technique, isolated collapsed as-built (but-for) and EDAM are in the third level. Above four methods use project schedule, delays information, owner's and contractor's perspectives and several time frames based on the project schedules.

In order to find the most suitable schedule delay analysis method for Indonesia construction project, some criteria have to be considered. The methods have to able to solve concurrent delay and calculate the project parties' liability accurately. Based on this research, the methods which divide the project schedule into several time frames and use owner's and contractor's perspective continuously produce the most stable and accurate results. If the parties only have limited information about project schedule, these two characteristics can be used as a basic way to choose and examine the schedule delay analysis methods. To track liability of each party in concurrent delays, the further calculation is needed. One method which discussed above uses some equations to calculate it.

Schedule delay analysis method is a reliable way to identify and calculate delays in construction projects. The project parties can use the methods to solve the dispute related to schedule delay in the project. To implement the schedule delay analysis method, some information must be provided including as-planned schedule as an initial plan schedule of the project, as-built schedule as the actual schedule, delay information based on the real situation in the project and detailed progress reports. Since the concurrent delay is one of the important issues to complete the schedule delay analysis, the agreement about this issue should be considered by the project parties and could be written clearly in the contract. If the parties have a clear clause/agreement about this issue, the schedule delay analysis results would be more factual.

Acknowledgements

The authors would like to thank the Ministry of Science and Technology, Taiwan, ROC, for financially supporting this research under Contract No. MOST 105-2628-E-008-002-MY3.

References

[1] Arditi, D. and T. Pattanakitchamroon, *Selecting a delay analysis method in resolving construction claims.* International Journal of project management, 2006. **24**(2): p. 145-155.

[2] Saeed, S.A.A., *Delay to Projects-Cause, Effect and Measures to Reduce/Eliminate Delay by Mitigation/Acceleration*. 2009, British University in Dubai.

[3] Yang, J.-B. and C.-K. Kao, *Critical path effect based delay analysis method for construction projects.* International Journal of Project Management, 2012. **30**(3): p. 385-397.

[4] Farrow, T., *Developments in the analysis of extensions of time.* Journal of Professional Issues in Engineering Education and Practice, 2007. **133**(3): p. 218-228.

[5] Mohan, S.B. and K.S. Al-Gahtani, *Current delay analysis techniques and improvements.* AACE International Transactions, 2005: p. PS201.

[6] Alkass, S., et al., *Computer aided construction delay analysis and claims preparation.* Construction Management and Economics, 1995. **13**(4): p. 335-352.

[7] Gothand, K.D., *Schedule delay analysis: Modified windows approach.* Cost engineering, 2003. **45**(9): p. 18-23.

[8] Hegazy, T. and K. Zhang, *Daily windows delay analysis.* Journal of construction engineering and management, 2005. **131**(5): p. 505-512.

[9] Ir, J.G.Z., *But-for schedules—Analysis and defense.* CDR, 2001. **4**: p. 1.

[10] Yang, J. *Comparison of delay analysis software for construction projects.* in *Proceedings of 33rd Annual Conference of Canadian Society for Civil Engineering (CSCE), Toronto, Canada*. 2005.

[11] Bordoli, D.W. and A.N. Baldwin, *A methodology for assessing construction project delays.* Construction Management & Economics, 1998. **16**(3): p. 327-337.

[12] Kim, Y., K. Kim, and D. Shin, *Delay analysis method using delay section.* Journal of construction engineering and management, 2005. **131**(11): p. 1155-1164.

[13] Mbabazi, A., T. Hegazy, and F. Saccomanno, *Modified but-for method for delay analysis.* Journal of construction engineering and management, 2005. **131**(10): p. 1142-1144.

[14] Ng, S.T., et al., *Improving existing delay analysis techniques for the establishment of delay liabilities.* Construction Innovation, 2004. **4**(1): p. 3-17.

[15] Yang, J.-B. and P.-C. Yin, *Isolated collapsed but-for delay analysis methodology.* Journal of Construction Engineering and Management, 2009. **135**(7): p. 570-578.

[16] Zack Jr, J.G., *Pacing delays--the practical effect.* AACE International Transactions, 1999: p. R11.

[17] Al-Gahtani, K.S., *A comprehensive construction delay analysis technique: Enhanced with a float ownership concept.* 2006.

[18] Trauner Jr, T.J., *Construction delays: understanding them clearly, analyzing them correctly.* 2009: Butterworth-Heinemann.

[19] Reams, J.S., *Delay analysis: a systematic approach.* Cost Engineering, 1989. **31**(2): p. 12-16.

[20] Yang, J.-B. and C.-K. Kao, *Review of delay analysis methods: A process-based comparison.* Open Construction and Building Technology Journal, 2009. **3**: p. 81-89.

An Analysis Using Game Theory on the Investment Incentive of PPP Projects

Shiau-Jing Ho[ab] and Sheng-Lung Lin[ac] and Hui-Ping Tserng[a]

[a]Department of Civil Engineering, University of National Taiwan University, Taiwan
[b]Kedge Construction co., Ltd, Taipei, Taiwan
[c]Bali District Office, NEW Taipei City, Taiwan
E-mail: d04521008@ntu.edu.tw, d03521011@ntu.edu.tw, hptserng@ntu.edu.tw

Abstract

According to the research of Taiwan Institute of Economic Research, the domestic PPP (Public Private Partnership) projects create economic growth rate of about 0.08% per NTD 10 billion of capital. In Feb. 2015, project report by Ministry of Finance indicated that signed PPP projects over the years have injected investment for an amount of approximately NTD 30.6 billion in 2014, which create economic growth rate of about 0.24%. Therefore, government is aggressively promoting PPP in recent years aiming to attract private investment of NTD 100 billion per year. Up to date, there are still some conflicts, especially the issue of royalty as PPP policy has been an important issue of the industry.

We establishes the analysis criterion through the concept of the game equilibrium and simulate to explore the PPP investment incentives of the equilibrium conditions. Investment incentives game model provides the method for the government and tenders to evaluate if the minimum royalty formulated by government have Investment incentive; this study proposed "fixed mode" and "change mode" two modes to discuss.

Fixed mode shows it has investment incentive when the minimum royalty decided by investments ratio or act lower than the reasonable profit allocated by the capacity of the government and tenders to create excess profit ratio. Changed mode shows it's difficult to have investment incentive when the land value higher then tenders' investments.

Keywords –

Public Private Partnership; Game theory; EVA; Investment incentive; Nash Equilibrium; Equilibrium solution; PPP

1 Introduction

1.1 Research motivations

The lack of government financial resources and the necessary of infrastructure make PPP an important industry in all countries. In recent years, Taiwan government has moved towards the goal of attracting NTD 100 billion of investment each year (Ministry of Finance, 2015). Over the years has signed the PPP project accumulated to 103 years of actual investment of about NTD 30.6 billion, it has created an economic growth rate of about 0.24%.

At present, the domestic distribution of PPP is on the basis of the value of the land provided by the government. For example, governments at all levels to set up land use right in accordance with the laws and regulations, usually expressly stipulates 30% to 70% of the land value ratio of the reserve price. The Ministry of Finance (2016) proposed to promote the type of PPP is not for the purpose to collect the royalties.The Taipei Municipal Government (2015) proposed the excess profits should be the allocateby the Governmen providing land value and the capital invested from the company. The views between The Ministry of Finance and Taipei City Government (2015) in PPP of the proportion of excess profits for the government and tenders are two extreme ways .

However, it is not applicable to allocate profits only by the proportion of the land use from goverment or capital investment from tenders in PPP. When the business failure, the government can still recover its investment capital (land) but tenders not. It is different from the co-financed shareholder relationship business of the common distribution of profit and risk. In the PPP project, it can be seen that the more investment capital (land) of the government does not mean that the excess profits can be generated. Therefore, the more the input capital is not enough to represent the higher excess profit distribution ratio.

The motivation of this study is how to promote the PPP project with reasonable investment incentive model. The process of PPP can be regarded as a game between the government and the tender. Game theory can be used to understand the interaction between the participants in the game and decide the optimal strategy according to the decision-making process to reach the equilibrium state of the contestants. This equilibrium state represents the balance strategy of the government and the tender, and also represents the goal of the government of engaging the PPP investment incentive.

1.2 Scope and limitations of study

This study mainly aims at the conditions of PPP project which is based on the setting land-using right type and the type of BOT which is based on the Participation Method. Other types or other laws and regulations based on the PPP project such as with the two types of PPP with the same characteristics can also refer to the analysis of this study. The PPP project especially means one kind of a project with excess profits in the concession period and another is included that the government only provided land but zero investment with excess profits of the financial assessment of the subject and subsidiary business. In one word, PPP projects that do not have excess profits are not covered by this study.

2 Literature review

The earliest use of BOT (Build-Operate-Transfer) case is the mid-18th century Turkey invited domestic and foreign contractors to participate in the planning of public sector privatization policy (Wikipedia, 2016).

Since 1990, the Taiwan government began to introduce civil participation in public construction PPP model to solve the government financial constraints, enhance public construction to promote efficiency (Executive Yuan Economic Construction Committee, 2011).

2.1 Game theory

In 1950, John F. Nash proved that there must be an "equilibrium solution" in non-zero-sum non-cooperative tournaments, and the concept of "Nash Equilibrium" in non-cooperative games was proposed. Albert William Tucker established the "Prisoner's Dilemma" equilibrium solution, which laid the foundation for the modern non-cooperative game theory. Since then, the game theory widely used in economics, political science and psychology and other fields, the study of human interaction equilibrium behavior. The 1994 Nobel Prize in Economics was presented to John C. Harsanyi, John F. Nash and Reinhard Selton for his outstanding contribution to the theory of the tournament.

In the game theory, among the decision makers who define clever rationality, through the study of mathematical models, the best strategies are determined .in the process of conflict and cooperation (Holland, 2013). There are two types of Extensive Form Game and Strategic Form Game (Normal Form) (David M. Kreps, 1990). In general, the strategic tournament is suitable for static games, and the extended tournament is suitable for dynamic tournaments (Zhang, Liu, 2003).

There are two kinds of Games：Cooperative Game and Non-cooperative Game. The analysis unit of the cooperative tournament is the group. The analysis unit of the non-cooperative tournament is the participant (David, 1990). Participants in the tournament can set out a co-ordinated strategy combination, known as a cooperative tournament, and vice versa (Roger, 2006).

2.2 PPP

Tiong & Alum (1997) proposed that the BOT royalty is required to be included in the concession contract through negotiations between the government and the concessionaire. Wu (2002) uses the mathematical planning method for the range of BOT royalty taking models to consider the self-compensation rate greater than 1, equal to 1 and less than 1.

Guo (2003) proposed second-order planning method to study the government and the optimal applicant's negotiation model. Study for the government to pursue their own financial recovery rate maximization and the best applicant for the goal of the largest profitability. Kan (2003, 2004) and Huang (2003) consider the point of financial and cash flow of the government and the concessionaire, calculate the financial decision-making model in both sides. Find out the capital contribution ratio and the bargaining space of the premium from the view of excess profit and financial recovery. And then deduce the civil participation in public construction royalty of the estimated model.

Zhou (2004) applied the Nash equilibrium of the cooperative game theory to discuss the financial-cooperation evaluation model between government and tender. Lin (2004) applied the game theory to construct the negotiation model to simulate the negotiation process of the franchise contract negotiation between the government and the franchise team. Carry out the estimating amount of the negotiating parties (the amount of sharing, the equilibrium convergence, the effect of the government and the people) and the time required to complete the negotiations and the impact of the negotiating factors on the process and its policy.

Wang (2008) adopted the Rubinstein bargaining theory to incorporate factors such as discount factor, bargaining cost factor, negotiation frequency, negotiation ability, negotiation start value and negotiation limit. Finally, find the agreement point of bilateral connection by the model. Lin (2013) discusses the importance of PPP government-specific units and the impact of the conditions, results and negotiation time of the government's loan guarantee and its policy significance.

3 Study methods and processes

This study first explores the current situation of civil participation in public construction, with the aim of establishing a model for the government to set up a scheme of investment incentives and ensure that the government has a reasonable amount of money before the bidding of the PPP project, so that the PPP project can be effectively promoted. This paper studies the papers, periodicals and books of civil construction, tournament theory, PPP rights and EVA theory, and then sets up the strategy of the tournament and builds the model and analysis step of the research investment trail, and finally To simulate the case to verify the feasibility of the model feasibility.

The study processes are as follows:

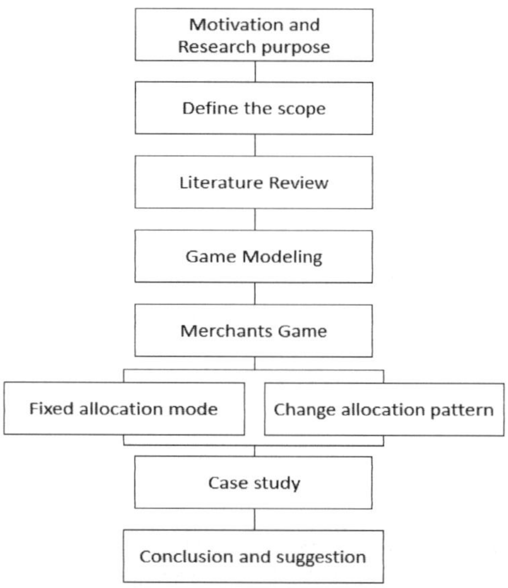

Figure 3.1 study processes

3.1 Definitions and conditions

The value of the enterprise lies in the ability to create excess profits. If the firm has the ability to earn a reasonable profit, it can be regarded as the value of the enterprise. Because it is the concept of opportunity cost or capital cost in economics, as long as the same invested capital is recycled through the appropriate investment pipeline. The tenders involved in PPP at least need to recover the cost of capital that is to earn a reasonable profit.

So the government no matter what way to allocate the profits generated by PPP, naturally only to distribute the amount of excess profits. The excess profits for each year are expressed in (2-1) (S. Ping Ho, 2014):

$$EVA_i = IC_i(ROIC_i - WACC) = NOPLAT_i - IC_i \times WACC$$

$$i = 0 \sim n$$

ROIC: Return on investment capital
NOPLAT: net operating interests
IC: investing capital
WACC: Cost of capital

According to (2-1), after the capital cost (WACC) is fixed, it is necessary to increase the excess return on capital or net operating profit. In fact, the ability to create excess profits usually comes from the business capacity, such as operational efficiency and innovation.

The excess profits used in this study are based on economic definitions, ie, the remuneration of the capital costs of the PPP project over liabilities and shareholders' equity (S. Ping Ho, 2004). The Taipei Municipal Government (2015) separately defines the excess profit as the difference between the actual amount of the firm and the estimated operating profit, which is different from the definition of excess profits from the general financial analysis. However, the use of this special definition of excess profits to distribute the profits of the way has now become the Taipei City Government announced the operating procedures, this study will be in the investment incentive tournament to establish a "change allocation mode" to explore the special distribution of investment incentives.

The total present value of excess profits is expressed as follows:

$$EVA^* = NPV + \sum_i R_d + \sum_i R_m \qquad (3\text{-}1)$$

$\sum_i R_d$: discounted value of the development rights

$\sum_i R_m$: discounted value for the operating premium.

In this study, the profitability is combined into the allocation rate to check the land value or the government and the firm to invest in the capital set by the difference between the rights of the reserve price is reasonable, and used to allocate PPP excess profits. For the purpose of rational allocation, the same investment capital refers to the same investment conditions as the tender, for example, the investment of the firm does not include the land purchase price or the levied cost, and the calculation of the government investment capital does not include the expenditure in order to avoid unreasonably reducing the government distribution Weights.

Tender and the government profitability connotation, as follows:

(A) Tender profitability: Tender estimated to build a PPP project to create the ability to create excess profits. Based on the same capital investment conditions with the government, the analyst can estimate the total excess value of excess profits generated by the PPP project.

(B) Government profitability: This study discusses the government of the profitability is based on the same investment capital conditions, that is, not included in the cost of government land acquisition. So, Profitability can also be estimated to create excess profits to replace the total present value.

Civil efficiency is one of the purposes to promote PPP. Profitability reflects the differences between the government and the tender on the core characteristics of the same PPP project efficiency or innovation. Therefore, this study makes it reasonable to assume that the government's profitability is much lower than that of the tenders and PPPs of certain types of operations may be close to or consistent with the tenders. There is still lack of literature or research to set the government profitability and the proportion of tender profitability is reasonable.

In this study, the government and the tender's full profitability ratio can be expressed between 0 and 1, the profitability ratio is average of 0.5, that is, the range is set to 0.5 to 1 between the government and the tenders can accept the PPP project in the balanced profitability ratio, both sides are willing The scope of acceptance is the concept of equilibrium theory.

Table 3.1 PPP Operation Type Scope Table

Profitability	PPP operating type
0.50~0.65	With high efficiency or creativity operational requirements type.
0.65~0.85	Comprehensive type.
0.85 to 1.00	More commercial rental or ground right to sell the facility type

3.2 The game model for tender inducement

The game model of tender inducement is to discuss the PPP project with the incentive conditions of investment. This study assumes that the government and the tenders in the process of investment can master each other's messages, such as government land value, tenders to invest in capital, PPP type of operation and government and tenders profitability, etc., so the game model for a complete message dynamic game. This study is divided into "fixed mode" and "changed mode" and other two models, the establishment of different rules. Merchant inducement rules by the government first action, tenders and then respond to the expansion of the dynamic tournament.

Table 3.2 Sign Def. Table

DEF.	Description
EVA^*	Estimated total excess value of PPP profits.
EVA_p^*	The government estimates the total present value of the actual excess profits.
EVA_r^*	Actual present value of actual excess profits of PPP.
EVA_b^*	The total present value of excess profits is estimated by the tender's investment plan.
$NOPLAT_r^*$	Actual net operating profit of PPP.
$NOPLAT_b^*$	Tender investment plan book estimate net operating profit.
IC	Estimated PPP investment capital.
$WACC$	Estimated PPP cost.
W_g	Government excess profit distribution rate.
Π_g	Government allocates excess profits.
Π_b	Tender allocates excess profits.
$\sum_i R_d$	The present value of the development rights.
$\sum_i R_m$	The present value of the operating premium.
a	Government/tenders profitability ratio.
B	The tender/government estimates the actual excess profit ratio.
c	Land value and investment ratio.

"Fixed mode" means that the premium at the end of the price shall be determined in accordance with the excess profits of the advance plan book, as set out with the ratio of the government and the tender's capital to the distribute. The fixed allocation pattern is that the government does not increase in the premium or in profit redistribution due to the actual operating profit of the tender. "Changed mode" simulates the "standard operating procedures for Taipei Municipal Government's plan to promote private participation in public works (BOT)." The operating procedures define the excess profits as the difference between the actual operating profit and the estimated operating profit.

Investment strategy of Merchants inducement game:

(A) Government strategy

In the fixed mode, the government will decide whether to hold PPP in accordance with whether the total present value of the excess profits of the plan book is greater than zero.

In the changed mode, the investment contract will stipulate that when the actual operating profit of the firm is higher than expected, it is necessary to increase the operating profit of the government and the tender to increase the capital than the redistribution.

(B) Tender strategy

In the fixed mode, the tender will decide whether or not to bid on the basis of whether the reserve price of the premium raised by the government is lower than the reasonable rate of profit allocated by the profitability combination and the excess profits estimated by the tender.

In the changed mode, the tender will decide whether or not to bid if the redistribution profit proposed by the government is lower than the reasonable profit by the combination of excess profit difference estimated.

Investment incentives race tree model as follows:

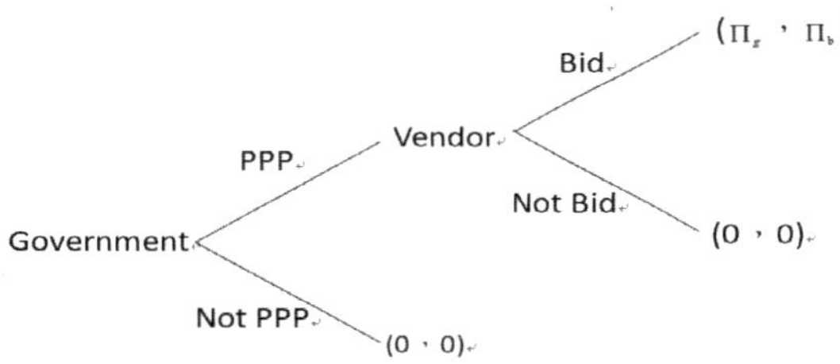

Figure 3.2 Investment incentives race tree model

Payoff describing below：

(A) Not PPP:

When the government evaluates NPV less than zero, does not hold PPP, game off. Government and tender remuneration are zero.

(B) PPP-Bid strategy combination:

In the fixed mode, the government allocates excess profits

$$\prod_g \geq \sum_i R_d + \sum_i R_m \text{ or } \prod_g \geq \frac{c}{1+c} EVA_p^*$$

Tender to allocate excess profits $\prod_b = EVA^* - \prod_g$.

In the changed mode, the government redistribution profits expressed as:

$$\prod_g = \frac{c}{1+c}\left(NOPLAT_r^* - NOPLAT_b^*\right)$$

And tender's re-distribution profits as below：

$\prod_b = EVA_r^* - EVA_b^* - \prod_g$. The

(C) PPP-Not Bid strategy portfolio:

When the firm assesses that the royalties raised by the government are higher than the reasonable profit for the distribution of the profitability and the excess profits estimated by the tender; or the redistribution of the government is higher than the profitability, And according to the tender's estimated excess profit margin of the distribution of reasonable profits, not participate, game off. Government and tender remuneration are zero.

According to the two models of "fixed model" and "changed mode", this paper discusses the equilibrium conditions of the portfolio equilibrium strategy combination.

(A) Fixed mode

According to the classification of this section, set the type of land power PPP to determine the proportion of the land value of the reserve price, the type of PPP can refer to set the type of land type PPP, or government and tenders to invest in the proportion of capital allocation rate, The excess profit determines the reserve price of the premium. And according to one of the above provisions of the right to set the reserve price, must be less than or equal to the basis of both profitability and the actual estimate of excess profits calculated by the tender's reasonable distribution of profits. Therefore, the allocation of excess profits in the fixed allocation of PPP with the incentive conditions of investment incentives, according to the right to set the price of the different ways, you can the following two types of which said:

$$\sum_i R_d + \sum_i R_m \leq \frac{a}{1+a} EVA_r^* \tag{3-2}$$

$$\frac{c}{1+c} EVA_p^* \leq \frac{a}{1+a} EVA_r^* \tag{3-3}$$

Which $\sum_i R_d$ is the present value of the development of the right, $\sum_i R_m$ the present value of the operating premium, where "a" is set for the profitability of government, "c" tenders than for the government to provide land value and estimate the amount of investment ratio. (3-2) type for the establishment of the right type of land by the value of land according to the value of the right to reserve the price of the conditions of investment incentives. In the government plan, the premium floor price is usually expressed as development rights and operating premiums. (3-3) is the type of reference PPP by the land-using value and the estimated amount of investment in the amount of investment decisions, and then refer to the pre-planned assessment of the excess profits to reserve the reserve price of the conditions of investment incentives. "b" is the actual excess profit ratio between the tender and the government ($b=\frac{EVA_r^*}{EVA_p^*}$), then (3-3) the equilibrium condition of the investment incentive is as follows formula:

$$b \geq \frac{c(1+a)}{a(1+c)} \tag{3-4}$$

(B) Changed mode

The "changed mode" is a model of the "Standard Practice Procedures for the Planning and Promotion of

Civil Participation in Public Works (BOT) by the Taipei Municipal Government (Taipei Municipal Government, 2015). The operating procedure defines the excess profit as the difference between the actual amount of the firm and the estimated operating profit, which can be redistributed according to the government land value and the private investment. According to the provisions of the Taipei Municipal Government, the distribution of profits in the distribution of the allocation of PPP with the incentive conditions of investment incentives for the ratio of investment capital than the redistribution of the difference between the operating profit value must be less than or equal to the profitability ratio of redistribution excess profits, (3-5) :

$$\frac{c}{1+c}\left(NOPLAT_r^* - NOPLAT_b^*\right) \leq \frac{a}{1+a}\left(EVA_r^* - EVA_b^*\right) \quad (3\text{-}5)$$

Tender participation in PPP The main input capital comes from the construction cost, the evaluation of construction costs are usually calculated according to the tender conditions, volume restrictions and market conditions. Under the premise of detailed assessment, the estimated and actual construction costs can be regarded as the same; and the cost of capital from the after-tax debt costs and the combination of shareholder costs (Dutch level, 2013).

According to the above two points can more simplify, that is, assuming that the tenders of PPP estimates and the actual financial analysis, the investment capital and capital costs are close to the calculation. So the further derivation of investment capital than with the incentive conditions of investment:

$$c \leq a \quad (3\text{-}6)$$

In order to assist the government and the tenders in the investment before the assessment of the law, the early plan book, the value of land and the capital invested by the company to develop the right of the reserve price of capital incentives. When the fixed allocation model is used for analysis, the variables that affect the incentive of investment include four restrictions: legal restriction, profitability ratio, excess profit ratio and investment capital ratio. When the changed allocation mode is used, the influencing factors of investment incentive include only the profitability ratio and the investment capital ratio. When the land value is higher than the investment capital of the tenders, the change allocation model for the PPP, even if the tenders estimate profits higher than the government estimates of profits, it is difficult to attract investment incentives.

3.3 Summary

In the case of the analysis of the allocation pattern, the variables that affect the incentive of investment include four restrictions on the law, the profitability ratio, the excess profit ratio and the invested capital ratio. In the analysis of the change distribution mode, Variables only include profitability ratio and investment capital ratio. When the land value is higher than the estimated investment of the capital, the change allocation model for the PPP, even if the tenders estimate excess profits higher than the government estimates of excess profits, it is difficult to attract investment incentives.

4 Tender inducement game model and case analysis

Now, we will introduce the simulated PPP of two investment inducement games. By the previous chapter, set up the reasonable hypothesis simulation parameters to the equilibrium condition, establish the analysis step of the investment incentive course and discuss the simulation results to verify the study value.

The following two simulation cases are divided according to the government and the tenders to invest in the capital ratio of the distribution rate, and then refer to the advance planning book estimates of excess profits to reserve the present value, and according to the land value of the present price reserve two types, Each type will explain the different result of fixed allocation mode and variable distribution model.

4.1 Fixed mode

4.1.1 Simulation case I

Assuming that the government has completed a pre-assessment of the 50-year BOT project, it is about to carry out investment. The total value of over-profit is estimated to be NTD 5 billion and the market value of the government is NTD 6 billion. The estimated investment is NTD 5 billion. The government assesses its own Profitability is about 0.65 to 0.80 times between the tender, try to study with the conditions of investment incentives by the two types of the fixed distribution model and the changing distribution model in the PPP case.

This case is based on the government and the tenders to invest in the ratio of capital to determine the distribution rate, and then refer to the advance plan of the excess profits estimated by the reserve price of the reserve price, so (3-4) as a balanced condition:

$$b \geq \frac{c(1+a)}{a(1+c)} \quad (3\text{-}4)$$

"a" is the government and tenders profitability a=0.65~0.80, than c for the government and tenders to invest in capital ratio, $c = 60/50 = 1.2$, b for tenders and the government to estimate excess profits ratio.

Take "a" into (3-4), respectively, the average estimated excess profit ratio $\bar{b}$ and the standard deviation σ, and then by ($\bar{b}$, σ) to explain its equilibrium conditions:

$$\bar{b} \geq 1.305 , \ \sigma = 0.075 \qquad (4\text{-}1)$$

And then $\bar{b}$ will be multiplied by NTD 50 billion, respectively, to find the average tender's estimating actual excess profits $\overline{EVA_r^*}$ and standard deviation, and then the equilibrium conditions:

$$\overline{EVA_r^*} \geq 65.25 , \ \sigma = 3.75 \qquad (4\text{-}2)$$

In the case, when the actual excess profit is estimated to be NTD 6.525 billion or above, the PPP-Bid strategy combination is a contingent portfolio, that is, the investment incentive, and then the average value plus or minus standard deviation the effective range of excess profits represents NTD 6.15 billion to NTD 6.9 billion.

4.1.2 Simulation case II

Assuming that the government has completed a pre-assessment of BOT project for the land using right of a period of 50 years concession by setting up a preliminary assessment of the land using right project. The total present value of excess profits is estimated to be NTD 5 billion and the market value of the provision of land is NTD 6 billion. The government stipulates that the market value of 40% Reserve price. Try to fixed mode and changed mode respectively to explore the PPP case with the conditions of investment incentives.

In this case, the reserve price of the premium is set according to the public property management. Therefore, the equilibrium condition is as follows:

$$\sum_i R_d + \sum_i R_m \leq \frac{a}{1+a} EVA_r^* \qquad (3\text{-}2)$$

And $\sum_i R_d + \sum_i R_m = 60 \times 40\% = 24$ billion

So the fixed mode with investment incentives to allow the conditions of tolerance:

$$24 \leq \frac{a}{1+a} EVA_r^* \qquad (4\text{-}3)$$

combination of Table 3.1 (PPP operating type range) to distinguish between different profitability ratio and type of operation with the incentive conditions of investment incentives, the equilibrium conditions can be the average tenders to estimate the actual excess Profit and standard deviation ($\overline{EVA_r^*}$, σ). When EVA_r^* higher than the PPP type of operation corresponding to the average tenders to estimate the actual excess profits, that is, on behalf of the type of operation of the PPP investment incentives.

Table 4.1 Simulation case 2

$a = {}^{E_g}/_{E_h}$	PPP operating type	($\overline{EVA_r^*}$, σ) (billion)
0.50~0.65	with high efficiency or creativity operational requirements type	(6.646, 5.54)
0.65~0.85	comprehensive type	(5.658, 4.34)
0.85~1.00	more commercial rental or ground right to sell the facility type	(5.012, 2.12)

According to the case, the government estimates that the excess value of the excess profits is NTD 5 billion although it is lower than the average excess profit of the manufacturer with more commercial rental or land sales facilities but which is still within the effective range. Assuming that the case of the government pre-planned excess assessment profits are enough to represent the excess profits expected then PPP with more commercial rental or land on the right to sell the type of facilities, PPP-Bid strategy portfolio for the balanced combination of For investment incentives, if PPP belongs to the other two types of business do not have the incentive to attract investment.

In the fixed mode, the incentive for PPP is limited by the restriction of the law, the ratio of the government and the firm to profitability, the capital ratio between the government and the tender, and the excess profit ratio between the government and the tenders. The capital reserve price set by the government according to the investment capital ratio or the restriction of the law must be less than or equal to the reasonable profit of the profitability ratio and the reasonable profit of the excess profit distribution.

4.2 Changed mode

When the government take the changed mode for PPP, the law restricts the rights of the reserve price or government and tenders estimate excess profits than the impact of investment incentives have not affected. Simulated cases I and II have the conditions of investment incentives are satisfied (3-6):

$$c \leq a \qquad \textbf{(3-6)}$$

The two cases of the government and the tenders into the capital ratio $c = \frac{60}{50} = 1.2$, greater than the upper limit of 1, PPP-Not Bid strategy combination for the contingent portfolio, it does not have the incentive to attract investment.

4.3 Simulation results

To change the distribution model for PPP investment incentives, only by the two factors which are the profitability and invest capital of the ratio. As the government's profitability is usually less than the profitability of the tenders, when the government to provide land value is higher than the tenders to invest in capital, regardless of whether the excess profit forecast is higher than the government estimation, are difficult to have investment incentives.

5 Conclusion

The purpose of the study is to assist the government and the tenders in judging whether the different types of PPP royalty reserve price has the incentive for investment.

Government contractors often have two kinds of negative mentality when dealing with PPP, one is worried that questioned by lower income royalties, the review of the audit department, and even the judicial prosecution, so prefer to have the right to reserve the provisions of the law, and give up more flexible to promote the PPP; The other one is to complete the investment policy objectives by using of flexible law to deliberately underestimated the government should receive royalties. Two cases may hinder the investment incentives and reasonable profit distribution for the negative development of the PPP. Take the experience of Taipei City government rights rules established by the changed allocation mode, because the government land value is higher than the tenders to invest in capital, resulting in difficult to attract investment incentives. In fact, the government PPP to provide public land prices higher than the amount of investment in the city of Taipei or New North City and other urban development areas generally, so the Taipei City Government 104 years established BOT premium rules based on maintenance Government interests, but the investment incentives have been adversely affected.

In this study, the investment incentive is divided into fixed mode and changed mode. The fixed mode for PPP investment incentives, by the laws and regulations, the government and tenders profitability ratio, investment capital ratio and the estimated excess profit ratio and other factors. It is difficult for the government to provide the land value higher than the investment capital of the firm, and the changed mode is difficult to attract the investment incentive, because the government and the tender's profitability ratio and the investment capital are more influential than the two factors. The investment incentive is based on the profitability allocation of excess profit as the conditions of the competition balance. We will use the model to explore whether the capital reserve price has the incentive for investment. The study establish the game model by simulating case to verify and providing a proper way to check the royalties reasonable of the government for positive promoting policy.

For the current situation of civil participation in public construction, this study through the game theory, PPP rights and EVA theory, and then sets up the strategy of the tournament to provides a good method for the government to set up a scheme of investment incentives and ensure that the government has a reasonable amount of money before the bidding of the PPP project so that the PPP project can be effectively promoted.

References

[1] Ehrbar, A. EVA: The real key to creating wealth. John Wiley & Sons, 1998.

[2] Friedman, J. W. Game theory with applications to economics (2nd ed.). New York: Oxford University Press, 1990.

[3] Harsanyi, J. C. (1967). "Games with Incomplete Information Played by "Bayesian" Players I–III, Part I. The Basic Model." Management Science, Vol. 14, pp. 159-182.

[4] Harsanyi, J. C. (1968). "Games with Incomplete Information Played by "Bayesian" Players, Part II. Bayesian Equilibrium Points." Management Science, Vol. 14, pp. 320-334.

[5] Harsanyi, J. C. (1968). "Games with Incomplete Information Played by "Bayesian" Players, Part III. The Basic Probability Distribution of the Game." Management Science, Vol. 14, pp. 486-502.

[6] Lehn, K., & Makhija, A. K. (1996). EVA & MVA as performance measures and signals for strategic change. Strategy & Leadership, 24(3), 34-38.

[7] Myerson, R. B. Game Theory: Analysis of Conflict. Harvard University Press, 1991.

[8] Nash, J. F. (1951). Non-Cooperative Games. Annals of Mathematics, Vol. 54, pp. 286-295.

[9] Rasmusen, E. (1994). Game and Information: An Introduction to Game Theory. Oxford University Press.

[10] Romp, G. (1997). Game theory: introduction and applications. Oxford University Press on Demand.

[11] Selten, R. (1975). Reexamination of the perfectness concept for equilibrium points in extensive games. International journal of game theory, 4(1), 25-55.

[12] Tiong, R. L., & Alum, J. (1997). Final negotiation in competitive BOT tender. Journal of Construction Engineering and Management, 123(1), 6-10.

Building Carbon Footprint(BCF) Evaluation for Social Amenities and Education Center in Taipei

Ying-Chiao CHIU[a] and Po- Han CHEN[b]

[a]Department of Civil Engineering, University of National Taiwan University, Taipei, Taiwan
[b]Department of Civil Engineering, University of National Taiwan University, Taipei, Taiwan
E-mail: remychiu168@gmail.com , pohanchen@ntu.edu.tw

Abstract –

The United Nations in 1992 through the "United Nations Framework Convention on Climate Change (UNFCCC)" on the "Anthropogenic Greenhouse Gas" emissions to make global control of the declaration, so "carbon reduction" has become the topic of concern to countries around the world. (Energy Efficiency and Carbon Reduction in New Taipei City from the Perspective of Carbon Emissions and Absorption , 2014) This study will be based on the National Energy Conference on the requirements of building energy certificates, as well as the international PAS2050 and ISO16064 for the carbon footprint of the trend of inspection, to carry out the architectural design stage carbon footprint exposure and labeling system research. The Ministry of Interior Construction Research Institute has been a considerable basis for the study of life cycle CO2 emissions assessment, and now combines the building energy passport system with existing building energy efficiency regulations to establish a comprehensive building life cycle carbon footprint assessment method. Based on the above research motivations, this study will compile and analyze the relevant carbon footprints of existing buildings at home and abroad, and with examples of verification(Cultural and Social Society Hall), to develop BIM (Building Information Model) to evaluate the building "planning stage "And" design phase "and provide a" carbon reduction design strategy "to understand the relationship between building and the" cost "of the building so that owners and designers can use the BIM model at the planning and design stage Analyze and get the smallest carbon exposures of buildings and the largest carbon reduction .

Keywords –

Building Carbon Footprint (BCF); Building Information Modeling (BIM) ; Evaluation

1 Introduction

Since April, the Public Works Committee has selected a number of road, bridge and building demonstration projects to start piloting the "Carbon Footprint Assessment" business. New North City has already asked for all environmental impact assessment of the building development case, must perform 50% The end of this year, Taichung City Water and Economic and Trade Park has begun to require all the development of the case must be implemented carbon exposure of the "Kaohsiung International Convention Center" Following the "Southport Exhibition Hall", has commissioned international consultants to complete the "carbon inventory" business; Certification. Japan's carbon tax system in 2012, for gasoline, natural gas, and coal tax, mainly want to establish the principle of polluter pays, so that polluters bear the relevant social costs, the implementation of the carbon tax also hope that through the environmental tax The concept of the establishment of the environment phase 12 fund to provide the necessary expenditure for environmental conservation. Although the carbon tax does not directly charge taxes on electricity and other energy, the power company will transfer the carbon tax cost to the electricity user, which is equivalent to the carbon tax borne by the electricity user.

The current government to maintain national competitiveness, to avoid consumption in high-priced energy, continue to promote energy conservation and carbon reduction work, so that the country can continue to develop; energy-saving carbon in Taiwan, not just slogans, but new life action.

Research Objectives

The main purpose of this study is to use BIM to plan and design the Social Amenities and Education Center. At this stage, we examine the carbon footprint of the Social Amenities and Education Center and link it with the building materials database to revise the architectural design direction to achieve the design of low-carbon buildings and give them back to the architect and owners.

2 Problem statements

Taiwan and the world are facing global warming and water scarcity of the environmental crisis, but also accept the economic development of the deterioration of the quality of environmental challenges. Greenhouse gas emissions come from carbon dioxide emissions from energy use. Although Taiwan is a party to the United Nations Framework Convention on Climate Change, but as a member of the global village should be dedicated, and according to the basic principles of climate commitments to bear a common but different responsibility for the reduction.

Promote green building and promote the use of energy-saving carbon-saving green building materials, including the promotion of new buildings to obtain green building certificates and green building signs related measures, building energy efficiency improvement plan, the development of building energy conservation design technical specifications and enhance building energy Benefits, incentives for civil buildings green building design and improvement of demonstration work and handling of building energy efficiency and energy saving materials, recycled materials related research.

In 2002, the Federal Republic of Germany Environmental Protection Agency began preparations for carbon emissions trading, after the establishment of the emissions trading agency, began to investigate all enterprises of machinery and equipment emissions to establish the carbon emission standards. The emissions trading agency will give the firm the maximum permissible carbon footprint limit. The enterprise can trade unused free carbon emissions. The emissions trading agency will register the approved carbon offset limits for each enterprise and set up an account to record the Enterprises free of carbon emissions trading situation, companies must go to the exchange (Leipzig) in accordance with the transaction price of free carbon emissions transactions, the enterprise's free carbon emissions need to run out before the end of each year next year to re-determine the enterprise Of the carbon quota and the re-opening of the free carbon emissions between enterprises, enterprises to improve the amount of free carbon or carbon emissions, will use more low-carbon machinery and equipment to reduce carbon emissions. However, from the 2014 Commission on Public Works has begun to require all development cases must be implemented carbon exposure certification, but non-public works buildings are also the future trend of low-carbon, so Taiwan must study the issue.

3 International norms for carbon footprint assessment

Any carbon footprint assessment must be in line with international norms, otherwise the words, no agreement is difficult to support the Earth's environmental protection action. The international norms on carbon footprint assessment can be grouped into the following three categories:

(1) ISO14040 series of international standards

ISO14040 is the product of the life cycle assessment of the international norms of the source, but also the source of Taiwan CNS14040 national standards. According to the ISO 14040 series of standards, the implementation of the life cycle assessment procedures can be divided into (a) the purpose and scope of the definition: define the scope of inventory operations; (b) inventory analysis (Inventory Analysis): data collection and inventory calculation to quantify the system (3) Impact Assessment: Evaluate the environmental impact of the life cycle by applying the analysis of the life cycle inventory; (4) Interpretation: The results of the merger analysis and impact assessment, and the results of the impact assessment The definition of the scope of integration, to conclusions and recommendations to the type of decision makers, as a product design, material selection or production process improvement reference.

(2) PAS 2050

PAS 2050 is the requirement for Carbon Footprint Verification (CFV). This specification is developed by the British Standards Association (BSI) for the assessment of conformity requirements for the carbon footprint of the product and service life cycle in response to the broad community and industry requirements. PAS 2050 believes that organizations have the potential to use this approach in order to have a better understanding of the carbon footprint of the product supply chain and to provide a common basis for comparison and communication with PAS 2050.

Carbon footprint inventory is the basic source of PAS 2050 confidence. With this inventory, regulators, customers, employees, shareholders, potential investors, environmental groups, media, and even competitors, are confident that the organization's carbon footprint is accurate, complete and transparent. Regardless of size, industry, or location, any organization who actively measures carbon footprints or programs to achieve carbon reduction targets will benefit from the additional accuracy and expertise associated with the carbon footprint review process.

Calculating the carbon footprint: The carbon footprint of the product can be calculated by means of direct measurement, mass balance or emission factors, but in the construction industry, only the emission factor method is used, that is, using the basic formula " Data (mass / volume / kWh / km) x emission factor (per unit of carbon dioxide equivalent).

(3) ISO 14067

The aim of this standard is to provide a clear and consistent description of the quantification, notification and verification of the carbon footprint of the product as an international prevailing standard for the assessment, monitoring, reporting and verification of carbon footprints. There are differences in the calculation criteria for carbon footprints on the market today. All international manufacturers have gradually changed the carbon footprint standard of the products from PAS2050 to ISO14067. Most international buyers have adopted ISO14067 into the global supply chain management system.

(4) The computer database software for the current carbon footprint lifecycle assessment includes Boustead, RF Weston, ChemSystems, EMPA (Switzerland), PIRA International, Charlimers Industriteknik (Sweden), Environmental Conscious Design Support System (United States), SimaPro (Netherlands) and so on. In the country, by the ITRI developed the localization of life-cycle assessment software "DoITPro", its database contains Taiwan's electricity, oil, fuel, metals, chemicals, plastics and so on a level of data, compared to foreign database data software , But also make the results consistent with the needs of domestic manufacturers and in line with the actual situation. In order to understand the overview of the raw material carbon sequestration database at home and abroad, some information is compiled here by the website of the Ministry of Economic Affairs (http://www.idbcfp.org.tw/2013.5.26). (Study on Carbon Disclosure Marking in Building Design Stage (1)- Study on Carbon Disclosure Method and Carbon Emission Database of Buildings (2013))

Table 1:A Summary of Carbon Footprint Database at Home and Abroad

Country	**Database**	**Coverage**
Taiwan	ITRI Database (DoITPro)	Taiwan
Japan	Japan National LCA Project	Japan
	Carbon footprint database	
Korea	Korean LCI Database Information Network	Korea
	Carbon footprint database	
The European Union	European Platform on Life Cycle Assessment (ELCD)	Europe
Sweden	SPINE@CPM	Global
Denmark	EDIP	Denmark
	LCA food	
Netherlands	IVAM LCA Data	Netherlands
	Dutch Input Output	Netherlands
	Franklin US LCI	United States
Switzerland	ecoinvent	Global/ Europe/ Switzerland
	BUWAL 250	Switzerland
	LCAinfo	
	Swiss Agricultural Life Cycle Assessment Database (SALCA)	Switzerland
Germany	German Network on Life Cycle Inventory Data	Germany
Thailand	Thailand LCI Database Project	Thailand
Australia	Australia Life Cycle Inventory Data Project	Australia
Canada	Canadian Raw Materials Database	Canada
United States	US LCI Database Project	United States

Study on Carbon Disclosure Marking in Building Design Stage (1)- Study on Carbon Disclosure Method and Carbon Emission Database of Buildings (2013)

4 Building Carbon Footprint Evaluation Method

4.1 Building Carbon Footprint Type

According to Dr. Lin Hsien-Te, "Building Carbon Footprint - Assessment Theory", the Carbon Footprint Assessment Method (BCF) was used to evaluate the Carbon Footprint. BCF is a series of building materials, building use of energy statistics and charcoal theory to estimate the formula to simulate the construction of carbon footprint method. BCF can be divided into four systems: (1) BCFs (Schema Evaluation System of BCF), (2) BCFd (Design Evaluation System of BCF), (3) BCFr (Residential Advanced Evaluation System of BCF), (4) BCFp (Public Building Advanced Evaluation System of BCF).

BCFs system is used in the very early stage of building financial planning, BCFd system is used to complete the architectural design but the equipment design is not completed stage, BCFr and CFp system for the construction and equipment of the details of the design are complete stage, but were applicable to residential Building and Public Buildings. The four systems are constructed from a series of rough to fine carbon footprints in order to correct the direction of low-carbon design at any time, and are in line with the financial investment of the industry, as the design information progresses from the planning to the completion stage Planning procedures, but also for the future of the Government in the environmental assessment, urban review, license, license and other stages of the implementation of low-carbon building management policy. (Building Carbon Footprint - Assessment Theory,2015).

Table 2: Building carbon footprint assessment of the four major system content

Item	BCFs	BCFd	BCFr	BCFp
Applica-ble stage	Environm-ental assessment, urban design	Licensing	Completion stage	
Graphic requirem-ents	Space plane, building height, tectonic system, air conditionin g system planning finalization	Construct-ion plane, profile, building constructio n drawings completed, and air conditionin g system planning finalizati-on	All buildings, air conditioning, hydropower, electromecha -nical, sent to complete the design	
Building size input	Virtual size input	Design condition-n input	Design condition input	
Building materials input	Default standard value input	Detailed theoretical formula input	Detailed theoretical formula	
Equipmen -t material input	Default standard value input	Default standard value input	Detailed theoretical formula	
Energy consump-tion data input	Energy consumptio n system value input	Energy consumptio n system value input	Detailed theoretical formula	
Device efficien-cy input	No input	No input	Heat source, water pump, windmill, hot water equipment design power and efficiency input	
Evaluati-on function	Carbon footprint classificati-on assessment	Carbon footprint classificati-on assessment	Comprehens-ive assessment and evaluation of carbon footprint	

Building Carbon Footprint - Assessment Theory(2015)

4.2 A Simple Method of Building Carbon Footprint

Detailed inventory of building materials and then calculate the carbon row of the algorithm seems to be sophisticated, but it is difficult to implement. The first obstacle will be encountered for the construction of the new building case, because there is no real energy consumption data can not be implemented can not provide prior to the improvement of energy saving and carbon reduction to improve the recommendations, and then its large building materials, equipment, Quantity, the authenticity of energy-saving documents simply can not audit, can not prevent the problem of intentional fraud.

In view of this BCF method does not use the product carbon label, such as the detailed carbon inventory method, but the introduction of a building carbon footprint reduction algorithm, it is building materials and equipment materials do not take the actual construction estimation data to actuarial, and engineering design factor to estimate The analysis of the energy consumption of the building, not the use of electricity to calculate the single gas to calculate the energy consumption to calculate the formula to simulate it is called the effective is the original carbon mark function is more important is to quickly diagnose and Immediate feedback Carbon reduction is known as the subtraction method because its workload is less than one thousandth of the way to check the carbon footprint of the product. Because the BFC method to master the building carbon footprint of the effective reduction algorithm can be effective feedback low-carbon design of the building carbon reduction policy have immediate benefits.

4.3 Low Carbon Footprint of Cultural and Social Society Hall

This study will establish a Social Amenities and

Education Center model with BIM to assess the carbon footprint of the planning and design phase. The building is located next to the Miramar Shopping Center in Taipei City. It is designed with 8 floors above ground and basement 3 floors. The floor area is 17,635 square meters which can accommodate 2300 people to use sitting hall, office, wedding hall, multi-purpose rally and dormitory, as shown in Figure 1 to Figure2

Figure 1: Daytime simulation building design

Figure 2: Night simulated building design

5 Conclusion

Building carbon footprint has been in foreign countries for some years, but only in the embryonic stage of Taiwan, the future needs of more on the assessment of building carbon footprint supplemented by the use of low-carbon building materials in order to achieve the popularity of low-carbon buildings, We make extensive use of BIM in building planning and design, to use the Taiwan Library of Materials, the award will help the efficiency of low-carbon building planning. Look forward to the future can push up the concept of green environmental protection, and to achieve the concept of sustainable social resources.

References

[1] Building carbon footprint (Part 1) assessment theory (2015).

[2] Building Carbon Footprint (Part 2) Diagnosis Practice (2015).

[3] Study on Carbon Disclosure Marking in Building Design Stage (1)- Study on Carbon Disclosure Method and Carbon Emission Database of Buildings (2013).

[4] Energy Efficiency and Carbon Reduction in New Taipei City from the Perspective of Carbon Emissions and Absorption (2014).

AUTHOR INDEX

AUTHOR INDEX

AUTHOR INDEX

AUTHOR INDEX

AUTHOR INDEX

International Association for Automation and Robotics in Construction (IAARC)
Slovak University of Technology in Bratislava
Faculty of Engineering
Radlinského 11
813 68 Bratislava
SK - Slovakia

ISBN 978-1-5108-4473-5

CURRAN ASSOCIATES INC.
proceedings.com

34th International Symposium on Automation and Robotics in Construction and Mining (ISARC 2017)

Taipei, Taiwan
28 June – 1 July 2017

Volume 2 of 2

ISBN: 978-1-5108-4473-5

Printed from e-media with permission by:

Curran Associates, Inc.
57 Morehouse Lane
Red Hook, NY 12571

Some format issues inherent in the e-media version may also appear in this print version.

Printed by Curran Associates, Inc. (2017)

For permission requests, please contact
International Association for Automation and Robotics in Construction (IAARC)
at the address below.

International Association for Automation and Robotics in Construction (IAARC)
Slovak University of Technology in Bratislava
Faculty of Engineering
Radlinského 11
813 68 Bratislava
SK - Slovakia

Phone: +421 2 5927 4599
Fax: +421 2 5296 8811

secretariat@iaarc.org

Additional copies of this publication are available from:

Curran Associates, Inc.
57 Morehouse Lane
Red Hook, NY 12571 USA
Phone: 845-758-0400
Fax: 845-758-2633
Email: curran@proceedings.com
Web: www.proceedings.com

ISARC 2017

Proceedings of the 34th International Symposium on Automation and Robotics in Construction

June 28-July 1, 2017
Taipei, Taiwan

Organizer

TAIWAN TECH
NATIONAL TAIWAN UNIVERSITY OF SCIENCE AND TECHNOLOGY

Co-hosted by:

台灣建築科技中心
Taiwan Building Technology Center

CRI

I.A.A.R.C.

International Association for
Automation and Robotics in Construction

SPONSORS

ISARC 2017 would like to take this opportunity to acknowledge, with sincere appreciation, the generous contributions and support of the following sponsors:

DIAMOND SPONSOR

德昌 30 堅若磐石

GOLD SPONSORS

CTCI 新鼎系統股份有限公司 CTCI Advanced Systems Inc.
CTCI 鼎鼎工程服務股份有限公司 CTCI Resources Engineering Inc.
CTCI 益鼎工程股份有限公司 CTCI Smart Engineering Corporation

互助營造 Fu Tsu Construction

SPONSORS

AUTODESK.

財團法人 中技社 CTCI FOUNDATION

ECS 歐亞電腦

亞新工程顧問股份有限公司 MOH AND ASSOCIATES, INC.

新北市政府工務局 New Taipei City Government Public Works Department

RUENTEX

財團法人 中興工程顧問社 SINOTECH ENGINEERING CONSULTANTS, INC.

中興工程顧問股份有限公司 SINOTECH ENGINEERING CONSULTANTS, LTD.

GOVERNMENT SUPPORTS

科技部 Ministry of Science and Technology

TAIPEI 觀光傳播局 Department of Information and Tourism

FOREWORD - WELCOME MESSAGE

Dear Friends and Colleagues,

On behalf of the organizing committee, we like to extend our warmest welcome to you to the 34th International Symposium on Automation and Robotics in Construction (ISARC 2017) held in Taipei, Taiwan.

This year in Taipei, we are expecting over 200 delegates and speakers from more than 30 countries. To address the needs and concerns of a global community in all fields of construction, the Scientific Committee has organized a series of workshops, Keynote sessions, Industry sessions, poster viewing session and Technical tour. We also prepare some social networking events for you to meet people for expanding knowledge base and sharing ideas.

Finally, I would like to take this opportunity to thank all the researchers, delegates, and students who working this field. Without your efforts and supports, we could not have such a successful Symposium.

We wish you a productive Symposium, and hope you enjoy your time in Taipei, Taiwan at ISARC 2017!

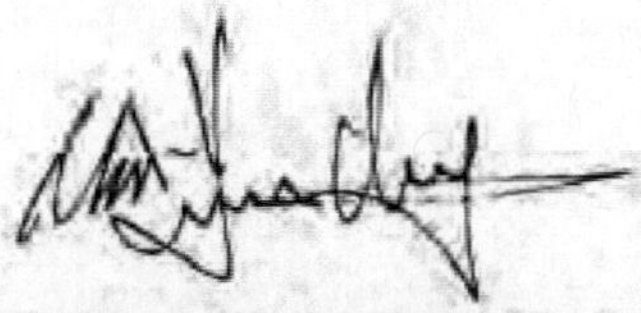

Prof. Min-Yuan Cheng
Chairman, ISARC 2017
Professor, National Taiwan University of Science and Technology

CONFERENCE COMMITTEES

CHAIR

Min-Yuan Cheng, National Taiwan University of Science and Technology

CO-CHAIRS

Hung-Ming Chen, National Taiwan University of Science and Technology
Jui-Sheng Chou, National Taiwan University of Science and Technology
Yo-Ming Hsieh, National Taiwan University of Science and Technology
I-Tung Yang, National Taiwan University of Science and Technology; Taiwan Construction Research Institute

LOCAL ORGANIZING COMMITTEE

Luh-Maan Chang, National Taiwan University
Shing-Tao Chang, National Cheng Kung University
Li-Chung Chao, National Kaohsiung First University of Science and Technology
Cheng-Cheng Chen, National Taiwan University of Science and Technology
Jieh-Haur Chen, National Central University
Po-Han Chen, National Taiwan University
Wei-Dong Chen, National Yunlin University of Science and Technology
Chin-Cheng Chou, National Central University
Ren-Jye Dzeng, National Chiao Tung University
Chong-Wei Feng, National Cheng Kung University
Sy-Jye Guo, National Taiwan University
Machine Hsie, National Chung Hsing University
Ting-Ya Hsieh, National Central University
Chung-Fah Huang, National Kaohsiung University of Applied Sciences
Chih-Han Kao, National Quemoy University
Chien-Ho Ko, National Pingtung University of Science and Technology
Yao-Chen Kuo, Kainan University
Hsin-Yun Lee, National Ilan University
Pin-Chan Lee, Southwest Jiaotong University
Sou-Sen Leu, National Taiwan University of Science and Technology
Chieh Lin, Executive Yuan, R.O.C. (Taiwan)
Chien-Liang Lin, National Kaohsiung First University of Science and Technology
Kuo-Liang Lin, I-Shou University
Lee-kuo Lin, National Taipei University of Technology
Shyh-Chyang Lin, National Quemoy University
Tzu-Ping Lo, Southwest Jiaotong University
Richard Moh, MAA GROUP Consulting Engineering
Nang-Fei Pan, National Cheng Kung University
Yan-Chyuan Shiau, Chung Hua University
Hui-Ping Tserng, National Taiwan University
Jen-Mu Wang, Tamkang University
Wei-Chih Wang, National Chiao Tung University

Jyh-Bin Yang, National Central University
Li-ren Yang, Tamkang University
Nie-Jia Yau, National Central University
Wen-der Yu, Chao Yang University of Technology

SCIENTIFIC COMMITTEE

IAARC BOD members

Par Ahman, Swedish Construction Federation, Sweden
Burcu Akinci, Carnegie Mellon University, USA
Tatsuo Arai, Osaka University, Japan
Carlos Balaguer, Universidad Carlos III de Madrid, Spain
Burcin Becerik-Gerber, University of Southern California, USA
Thomas Bock, Technical University Munich, Germany
Frederic Bosche, Heiot-Watt University, United Kingdom
Ioannis Brilakis, University of Cambridge, United Kingdom
Alexey Bulgakov, Ministry of Education and Science of the Russian Federation, Russia
Carlos Caldas, University of Texas at Austin, USA
Alessandro Carbonari, Universita Politecnica delle Marche, Italy
Daniel Castro-Lacouture, Georgia Institute of Technology, USA
Hung-Ming Chen, National Taiwan University of Science and Technology, Taiwan R.O.C.
Min-Yuan Cheng, National Taiwan University of Science and Technology, Taiwan R.O.C.
Moon Young Cho, Korea Institute of Construction Technology, Republic of Korea
Hiroshi Furuya, Obayashi Corporation, Japan
Ernesto Gambao, Universidad Politecnica de Madrid, Spain
Jozef Gasparik, Slovak University of Technology, Slovakia
Frans van Gassel, Eindhoven University of Technology, The Netherlands
Quang Ha, University of Technology, Sydney, Australia
Carl Haas, University of Waterloo, Canada
Chang-Soo Han, Hanyang University, Republic of Korea
Ferri Hassani, McGill University, Canada
Rauno Heikkila, Oulu University, Finland
Vineet Kamat, The University of Michigan, USA
Hyoungkwan Kim, Yonsei University, Republic of Korea
Markus König, Ruhr-Universität Bochum, Germany
Soonwook Kwon, Sungkyunkwan University, Republic of Korea
Junbok Lee, Kyung Hee University, Republic of Korea
Nan Li, Tsinghua University, China
Ger Maas, Royal BAM & Eindhoven University of Technology, The Netherlands
Osama Moselhi, Concordia University, Canada
Benny Raphael, I.I.T. Madras, India
Yehiel Rosenfeld, Technion - Israel Institute of Technology, Israel
Annop Sattineni, Auburn University, USA
Shabtai Isaac, Technion - Israel Institute of Technology, Israel
Miroslaw Skibniewski, University of Maryland, USA
Jochen Teizer, RAPIDS Safety and Technology Laboratory, Germany
Koshy Varghese, I.I.T. Madras, India
Xiangyu Wang, Curtin University, Australia

Invited members

Ali Akbarnezhad, The University of New South Wales, Australia
Seung Hyun Cha, The Hong Kong Polytechnic University, Hong Kong
Jia-Ruey Chang, National Ilan University, Taiwan
Albert Chen, National Taiwan University, Taiwan R.O.C.
Jiayu Chen, City University of Hong Kong, Hong Kong
Jack C P Cheng, The Hong Kong University of Science and Technology, Hong Kong
Yong Kwon Cho, Georgia Institute of Technology, USA
Burcu Guldur Erkal, Hacettepe University, Turkey
Shu-Chien (Mark) Hsu, The Hong Kong Polytechnic University, Hong Kong
Nhat-Duc Hoang, Duy Tan University, Vietnam
Javier Irizarry, Georgia Institute of Technology, USA
Shih-Chung Kang, National Taiwan University, Taiwan R.O.C.
Changwan Kim, Chung-Ang University, South Korea
Yongcheol Lee, Louisiana State University, USA
Haijiang Li, Cardiff University, United Kingdom
Li-Chuan Lien, Chung Yuan Christian University, Taiwan R.O.C.
Yu-Cheng Lin, National Taipei University of Technology, Taiwan R.O.C.
Joseph Louis, Oregon State University, USA
Javad Majrouhi-Sardroud, Islamic Azad University, Iran
Joshua Marshall, Queen's University, Canada
Ahmed Montaser, Ain Shams University, Egypt
Ngoc-Tri Ngo, National University of Singapore, Singapore
Anh-Duc Pham, University of Danang Vietnam, Vietnam
Doddy Prayogo, Petra Christian University, Indonesia
Samad Sepasgozar, The University of New South Wales, Australia
Xuesong (Johnson) Shen, The University of New South Wales, Australia
Ming-Hsiu Tsai, Tamkang University, Taiwan R.O.C.
Andreas Franskie Van Roy, Universitas Katolik Parahyangan, Indonesia
Han-Hsiang Wang, National Central University, Taiwan R.O.C.
Tsung-Juang Wang, National Taipei University of Technology, Taiwan R.O.C.
Yu-Ren Wang, National Kaohsiung University of applied Science, Taiwan R.O.C.
Hsi-Hsien Wei, The Hong Kong Polytechnic University, Hong Kong
I-Chen Wu, National Kaohsiung University of applied Science, Taiwan R.O.C.
Yu-Wei Wu, National Taiwan University of Science and Technology, Taiwan R.O.C.

SESSIONS AND SESSION CHAIRS

Session A-1: BIM & Computational Intelligence
Chair: Alessandro Carbonari (Polytechnic University of Marche, Italy)

Session B-1: Construction Robotics (I) / Fabrication
Chair: Carl Haas (University of Waterloo, Canada)

Session C-1: Sensing & Monitoring
Chair: Shabtai Isaac (Ben Gurion University of the Negev, Israel)

Session D-1: Interoperability
Chair: Fabiano Correa (University of Sao Paulo, Brazil)

Session E-1: Virtual Reality / Augmented Reality
Chair: Soon-Wook Kwon (Sungkyunkwan University, Republic of Korea)

Session A-2: BIM & Facility Management
Chair: Chung-Wei Feng (National Cheng Kung University, Taiwan)

Session B-2: Construction Robotics (II) / 3D Printing
Chair: Jerzy Paslawski (Poznan University of Technology, Poland)

Session C-2: Machine Learning / Computational Intelligence
Chair: Changwan Kim (Chung-Ang University, South Korea)

Session D-2: Project Management
Chair: Jieh-Haur Chen (National Central University, Taiwan)

Session E-2: Visualization
Chair: Samad Sepasgozar (University of New South Wales, Australia)

Session A-3: BIM & Visualization
Chair: Mohamed Al-Hussein (University of Alberta, Canada)

Session B-3: Construction Robotics (III) / Control & Mechanisms
Chair: Chien-Ho Ko (National Pingtung University of Science and Technology, Taiwan)

Session C-3: Computer Vision

Chair: Po-Han Chen (National Taiwan University, Taiwan)

Session D-3: Project Management

Chair: Ilija Vukorep (BTU Cottbus-Senftenberg, Germany)

Session E-3: Construction Robotics (IV) / Design

Chair: Seung Hyun Cha (Hong Kong Polytechnic University, Hong Kong)

Session A-4: BIM & Automation

Chair: Jochen Teizer (Ruhr-University Bochum, Germany)

Session B-4: Construction Robotics (III) / Inspection & Monitoring

Chair: Johnson Xuesong Shen (University of New South Wales, Sydney, Australia)

Session C-4: Computer Vision

Chair: Quang Ha (University of Technology Sydney, Australia)

Session D-4: Construction Management

Chair: Jozef GaŠparÍk (Slovak University of Technology in Bratislava, Slovakia)

Industry Session 1

Chair: Benny Raphael (I.I.T. Madras, India)

Session A-5: BIM & Project Management

Chair: Osama Moselhi (Concordia University, Canada)

Session B-5: Construction Robotics (VI) / UAV Applications

Chair: Hyoungkwan Kim (Yonsei University, South Korea)

Session C-5: Laser Scanning / Big Data

Chair: Yong Kwon Cho (Georgia Institute of Technology, United States)

Session D-5: Energy & Sustainability

Chair: Nan Li (Tsinghua University, China)

Industry Session 2

Chair: Jong Won Seo (Hanyang University, Seoul , South Korea)

Session A-6: BIM Applications

Chair: Jack C.P. Cheng (Hong Kong University of Science and Technology, Hong Kong)

Session B-6: Planning & Positioning

Chair: William O'Brien (The University of Texas at Austin, United States)

Session C-6: Computer Vision / Sensing & Monitoring

Chair: Yo-Ming Hsieh (National Taiwan University of Science and Technology, Taiwan)

Session D-6: Project Management

Chair: Pawan Pandey (IIT Delhi, India)

Industry Session 3

Chair: Sy-Jye Guo (National Taiwan University, Taiwan)

REVIEWERS

CONTENTS

34th International Symposium on Automation and Robotics in Construction

ISARC 2017

June 28 – July 1 TAIPEI, TAIWAN

CONSTRUCTION ROBOTICS (IV) / DESIGN

Kite-Powered Design-to-Robotic-Production for Affordable Building on Demand

H. Bier[a], R. Schmehl[b], S. Mostafavi[a], A. Anton[a] and S. Bodea[a]

[a]Hyperbody, Robotic Building, Faculty of Architecture, Technical University Delft, The Netherlands
[b]Wind Energy, Faculty of Aerospace Engineering, Technical University Delft, The Netherlands
E-mail: H.H.Bier@tudelft.nl, R.Schmehl@tudelft.nl,
S.Mostafavi@tudelft.nl, A.M.Anton@tudelft.nl, I.S.B.Bodea@tudelft.nl

Abstract –

Building technologies employed today in 2nd and 3rd world countries are imported, expensive, outdated and unsustainable. Highly developed countries, on the other hand, rapidly advance in developing affordable, numerically controlled and robotically supported material- and energy-efficient methods for building on demand. The research team proposes to close this gap by applying advanced design-to-robotic-production (D2RP) technologies developed at Technical University Delft (TUD) to construction problems in 2nd and 3rd world countries. The provided tool base uses refurbished robotic technology, which is retrofitted with state-of-the-art open source control software, and by employing local approaches and available materials the dependency on imported materials and processes is drastically reduced. The D2RP unit is coupled with the electricity generating Kite Power (KP) system developed at TUD to create a mobile sustainable autarkic unit that can be deployed everywhere.

Keywords –

Robotic building, material- and energy-efficient building, building on demand, autarkic Design-to-Robotic-Production, kite power system

1 Introduction

The societal challenge that the kite-powered robotic building system aims to solve is building affordable, material- and energy-efficient housing with Design-to-Robotic-Production (D2RP) methods powered by sustainable Kite Power (KP). It is important to tackle this challenge now because in the developing world 900 million people live in slums [1] and technology developed at Technical University Delft (TUD) allows to address this.

Figure 1. From laboratory to relevant environment - Impression of how D2RP could be moved from laboratory (top) to operation in situ and together with KP system (bottom) to produce affordable housing units

The kite-powered robotic building system aims to develop and deploy a mobile autarkic robotic building system for affordable, material- and energy-efficient housing in 2nd and 3rd world countries. Main objective is to counteract poor housing conditions and increasing economic migration by offering strategies for affordable housing and economic growth through, material and energy-efficient building by deploying sustainable robotic processes in situ. The assumption is that D2RP technologies counter negative effects of globalization and maximize the role of local human resources and construction materials by using new, customizable, open-source procedures, inspired by autochthon building approaches. The strategy is to develop and provide necessary D2RP and KP system knowledge and application know-how. Equipment to enable local implementation consists of a set of computational and robotic tools such as recycled computers (with Linux OS) and scrap robots (similar to the ones shown in fig. 1).

2 Description

D2RP together with the KP system developed at TUD allows for ad hoc set-up and operation anytime and everywhere with minimal infrastructure. It is based on a generic methodology relying on a robust, scalable, yet highly flexible technology. The assumption is that when such customizable fabrication set-up is provided to users in order to address local needs and conditions, users can easily customize processes employing local materials and parametric designs. The aim is to develop methods derived from the study of local resources and approaches by customizing materials and employing parametric design and robotic fabrication.

TUD has been developing knowledge on how to connect and deploy D2RP with KP, customize the D2RP system and will (experimentally) implement proof of concept by testing mobile D2RP system at Valkenburg airbase and on TUD campus in Netherlands in 2017. While the D2RP system developed at TUD is addressing challenges of next generation building [2] it can also be applied to local conditions in 2nd and 3rd world countries by employing local materials, inspired by autochthon building approaches. By researching an array of techniques exploiting local materials and techniques the focus may shift towards weaving (with local fibers but also if applicable synthetic fibers coming from waste), local clay-based production methods, rock/soil- and wood-processing approaches. If until now the team has explored additive and subtractive techniques with ceramic clay and polymers, respectively in laboratory environment, the next step is focusing on exploring possibilities for off the grid production in situ with KP energy.

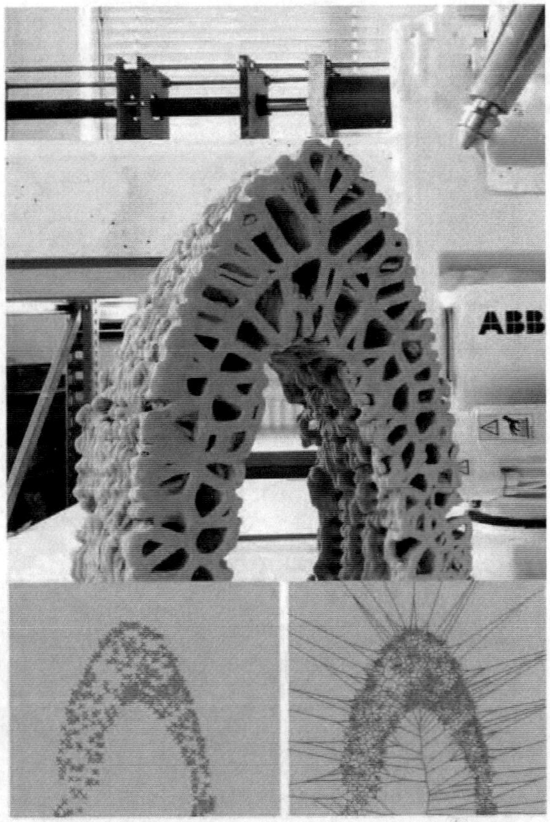

Figure 2. Building component (1:1 scale) structurally optimized and robotically 3D printed (BK-TUD, 2015).

2.1 Design-to-Robotic-Production (D2RP)

The D2RP system developed at TUD consists of a robotic arm with a changeable end effector or tool, robot motion tool path component, a parametric digital model, micro-controller and robot controller units (fig. 2). The system allows linking the design to the production process (Bier, 2014; Bier and Mostafavi, 2015) and is highly customizable. By employing approaches involving multi-tools, -materials, and -robots the D2RP system establishes a platform for building on demand.

Explorations in robotic building implying numerically controlled (NC) and robotically supported design to production chains, have been implemented since 2012 at Hyperbody, TUD first with two large refurbished ABB robots (fig. 1), which were customized to perform tasks with specialized operating tools and programs [3]. Since then, a series of experiments were implemented in order to develop and test robotic fabrication methods by establishing a feedback loop between design and fabrication. The assumption was that by employing robotic fabrication, customized designs could be easily implemented so that end-users may be easily able to transform (extend, shrink, expand, etc.) physically built environments on demand.

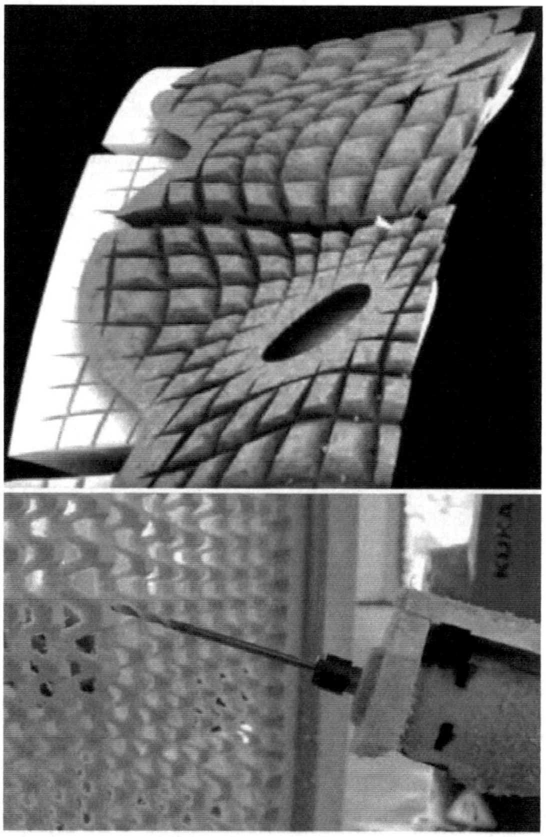

Figure 3. The behavior of polymer- cork-based materials is manipulated from rigid to flexible by means of 3D subtractive techniques.

Initial experiments with robotic subtractive manufacturing (fig. 1&3) where followed up by additive robotic production (fig. 2), which implied linking design to materialization by integrating all functionalities (from structural strength, to thermal insulation and climate control) in the design of building components. These experiments proved that D2RP processes enable production of free-formed, heterogeneous, optimized structures by additively and selectively depositing or manipulating materials [5] in order to achieve specific porosity-density, flexibility-rigidity, etc. requirements in accordance to formal, functional, structural, climatic, environmental, and economic needs. In fact, D2RP requires multi-robot production implying that several robots operate simultaneously or in short sequence in the process of production and assembly of multi-material building components. This is necessary in order to, for instance, deposit reinforcement fibers or granular insulation material, etc. in parallel to depositing cement-based or ceramic materials, etc.

The D2RP system is ready to be utilized in combination with the KP; however, D2RP experiments have only been implemented in laboratory environment. These experiments have shown that D2RP contributes to the reduction of economical inefficiency and environmental damage mainly because its processes establish a direct link between conceptualization, design, engineering, fabrication and operation of buildings. Furthermore, increased material and energy efficiency is achieved through structurally and climatically informed material distribution.

The obvious advantage is that the system is versatile as it can operate a large range of tools and can manipulate material from all sides 24/7. The operation outdoors is constrained due to weather conditions and requires consideration with respect to wind forces having an impact on the operation of the system. Also, use of local materials requires adaption of the system and tools to respective requirements.

2.2 Kite Power (KP)

KP system developed at TUD is based on an inflatable wing operated in periodic pumping cycles (fig. 1 and 4). The wing is steered by a kite control unit (KCU) suspended below the wing (fig. 4). The KCU receives the steering commands from the control center at the ground station via a wireless connection (one link for nominal operation plus two backup links). This wireless connection is also used to stream sensor data down to the ground control. The traction force of the kite is transferred to the ground station by a single tether. The nominal altitude limit is 500 m, however, the technology allows operation beyond 1 km altitude [6].

Kite Power is a cost effective renewable energy solution [7] with a low environmental footprint. It is an ideal basis for a highly mobile wind energy system. The heavy generator is positioned at ground level, which facilitates servicing and also minimizes structural forces. The direct force transmission into the ground station removes the need for bending-resistant foundations, which facilitates deployment. The technology demonstrator uses a tele-operated airborne unit, which can access altitudes far beyond the limit of conventional wind turbines (200 m). Wind at these altitudes is stronger and steadier which increases capacity factors of the system to about 60%. Conventional wind turbines presently achieve values of 20-35%.

The system is operated in periodic pumping cycles,

alternating between reel-out and reel-in of the tether. During the reel-out the kite is flying figure eight maneuvers at high speed (70-90 km/h). This creates a high traction force (3.1 kN at 7 m/s wind speed), which is converted into electricity by the drum and the connected 20 kW generator (fig. 5). When reaching the maximum tether length, the kite is de-powered by releasing the rear (steering) lines such that the whole wing rotates and aligns with the wind.

Figure 4. Kite Control Unit and bridle system during launch (Photo: Max Dereta)

Using the drum/generator module as a winch, the kite is then pulled back to the initial position to start the next pumping cycle. De-powering reduces the traction force during reel-in by 80% and for this reason the energy consumed during reel-in is only a fraction of the energy generated during reel-out. Crucial element of the technology is the automatic control and synchronization of the drum/generator module and the flight dynamics of the kite.

2.3 Kite-Powered D2RP

By connecting D2RP with KP, the system becomes mobile and autarkic. The proof of concept will be implemented by testing the mobile, kite-powered D2RP system at Valkenburg airbase and on TUD Green Village campus in Netherlands. While the KP system has been outdoors tested for 2 years now the move of the D2RP system from laboratory to outdoors is unprecedented. This is especially challenging because of the rather unpredictable nature of the environment with respect to terrain and weather conditions. Inconstant operational characteristics caused by environmental influences such as temperature, rain, and wind may require recalibration of the robot and its tools. Starting with experiments involving one robot, the project will evolve towards exploring multi-material and -robots techniques that are required for the implementation of all construction related tasks.

Another aspect that needs consideration is power supply: Medium sized industrial robots operate using single phase AC of 220-230V at 50Hz while larger robots require a stronger, three phase AC supply of 400V at 50Hz. Given these functional and technical specifications, a storage and conversion solution for the electricity produced by KP is currently being investigated, while AC power stabilizers for both lower and higher voltages are being considered. The research team proposes a DC link connection at 400V for small robot arms, respectively 600V for medium-large robotic arms. This enables the robotic system to function on the DC output of Kite Power avoiding costly DC/AC AC/DC conversion. The D2RP-KP connection will be enabled via a combination of Lithium batteries that provide operation energy for a couple of hours and Ultracapacitors that supply energy during load or generation peaks. The conversion and stabilization issue for KP-produced electricity is important especially as the research team considers the implementation of multi-robotic D2RP required for executing a variety of complex building operations.

Figure 5. Ground station on trailer with tether running into the sky (top), swivel with video camera mounted and tether running into the sky (bottom right) and drum/generator module mounted on transverse sled (bottom left)

3 Conclusion

The kite-powered robotic building system will

contribute solving current housing problems by connecting kite-powered robust and highly reliable industrial robotic technologies with local production approaches. This will lead to the development of products and processes for local building markets providing both sustainable means of habitation as well as new job opportunities in the field of KP D2RP. This is directly relevant to communities in terms of providing affordable healthier living environments and creating impulse for economic activity with low-cost, easy-to-operate tools.

Considering that KP D2RP methods cut production costs more than 40% and delivery times to 100%, while additional savings are achieved by increasing material and energy efficiency, this research presents an approach to affordable housing that can be applied anytime everywhere. The kite-powered robotic building system allows for ad hoc set-up and operation with minimal infrastructure. The assumption is that when such customizable fabrication set-up is provided to communities in order to address local needs and conditions, users can easily customize processes employing local materials and parametric designs.

Acknowledgements

This paper has profited from the input of Hyperbody and Kite Power researchers and students teams. It is based on laboratory experiments implemented at Hyperbody 2012-17 and builds up on the Kite Power and Robotic Building frameworks developed by the first two authors 2010-13. D2RP projects have been supported 2014-16 by 4TU, Delft Robotics Institute (DRI), 100% Research, KUKA, and ABB, while KP has been supported by EU funding (H2020-ITN-642682 and H2020-FTIPilot-691173).

References

[1] UN-HABITAT. State of the world's cities report 2012/2013: Prosperity of cities. Online: https://sustainabledevelopment.un.org/content/documents/745habitat.pdf Accessed: 10/02/2017.

[2] Bier, H. et al. Robotic Building. Online: www.hyperbody.nl/research/projects/robotic-building/ Accessed: 07/10/2016.

[3] Feringa, J. Entrepreneurship in Architectural Robotics: The Simultaneity of Craft, Economics and Design, *Architectural Design* 84(3): 60–65 2014.

[4] Bier, H. Robotically driven Architectural Production. *Archidoct* 1(2): 11-16, 2014.

[5] Bier, H. and Mostafavi, S. Structural optimization for materially informed design to robotic production processes. *American Journal of Engineering and Applied Sciences*, 8(4): 549-555, 2015.

[6] Vlugt, R. van der, Peschel, J., Schmehl, R. Design and Experimental Characterization of a Pumping Kite Power System, *Airborne Wind Energy,* pages 403-425. Springer, Heidelberg, 2013.

[7] Grete, C. The Economic Potential of Kite Power, *Leonardo Times – Journal of the Society of Aerospace Engineering Students*, 4: 10-11, 2014.

Inhibiting Factors and Improvement Plan of Table formwork Method in High-Rise Building Construction

Junhuck Lee[a], Dongmin Lee[a], Hunhee Cho[a], and Kyung-In Kang[a]

[a]School of Civil, Environmental, and Architectural Engineering, Korea University, Republic of Korea
E-mail: happyjun@korea.ac.kr, ldm1230@korea.ac.kr, hhcho@kroea.ac.kr, kikang@korea.ac.kr

Abstract –

The table formwork method for high-rise building formwork can shorten the construction time and reduce manpower through machine- and equipment-based automation. However, this method is not used as much as other system formworks. Various studies have been conducted to activate the table formwork method; but, the problems in associated with the automation work of table formwork are not understood, because researchers are focusing on improving efficiency through new technology development. As a basic study of the activation of the table formwork method, this study proposes an improvement plan after analysing the problems and satisfaction level according to the detailed process of the table formwork method. The results of this study are expected to expedite the process improvement and activation of the table formwork method.

Keywords –

Table formwork; Automation; IPA; Delphi; High-rise building

1 Introduction

Recently, the application of system forms, used to compensate for the disadvantages of traditional manpower-dependent forms, is increasing in high-rise building formwork. The table formwork method in particular is a large system form method with an integrated joist and sleeper on the bottom plate. It is recognized as an advantageous method for reducing labor and construction costs through machine- and equipment-based automation (automation machine and equipment). In spite of these advantages, however, the table form method is not applied as often as other system forms in the construction of high-rise buildings, except for in some Asian regions.

In the past, researchers have focused on activating the table formwork method. For example, studies have aimed to evaluate the productivity of the table formwork method[1], supplement the element technologies[2], and automate the layout and operation plans for table forms[3, 4]. Even though these studies increased the completeness of the table formwork method, improving work efficiency, they failed to understand the problems arising from the actual table form construction process, and could not propose improvement directions. Thus, as a basic study supporting the activation of the table formwork method, this study proposes an improvement plan. This plan is proposed after analysing the problems and satisfaction levels associated with the detailed processes of the table formwork method. Thus, as a basic study supporting the activation of the table formwork method, this study analyses the problems and satisfaction related to the detailed process of the table formwork method and suggests an improvement plan accordingly.

The scope of this research is the RC frame slab table form, a major application part of the table form. This study is limited to analysing the importance and present satisfaction level of each major process during table form construction in high-rise building. To analyse importance and satisfaction levels, a survey was conducted to practitioners who had experience using the table formwork method and expert consulting was carried out with eight experts who had at least ten years of experience. Furthermore, the importance-performance analysis (IPA) method was applied to evaluate the importance and satisfaction levels for the evaluation items by task obtained through the survey. After using an IPA to derive the problem factors for each task and items requiring improvement, we found an improvement plan by conducting a Delphi process with a group of experts, who performed the table formwork method many times. This study aims to contribute to the activation of the table form method, by identifying inhibiting factors and proposing improvement plan. In the future, the results of this study will be used to investigate the development of element technologies and process improvements for the table formwork method.

2 Literature review

2.1 Current table formwork

A table form is a large system form for pouring concrete floor slabs. A table form is composed of a form board, joist, sleeper, and support, and can move horizontally and vertically through a crane. Table forms are used in high-rise building construction mainly because the associated installation, disassembly, lifting, and reinstallation work is performed by machines. This shortens construction time and reduces labor costs.

The table formwork method is generally carried out through installation, disassembly, transportation, and lifting processes. First, during the installation process, the floor form is completed by supporting the bottom shore after the table form is placed in its designated position. Next, during the disassembly and transportation process, a separate transport device such as a trolley is placed at the bottom of the table form, where it can be removed and then dropped by a certain height before being transported. Finally, in the lifting process, the table form is pushed to a lifting deck such as a super-deck and is then lifted to the upper floor by a tower crane. The detailed process of the table formwork method is shown in Fig. 1. Typically, a table form unit for 2−3 floors is manufactured and used repeatedly[5].

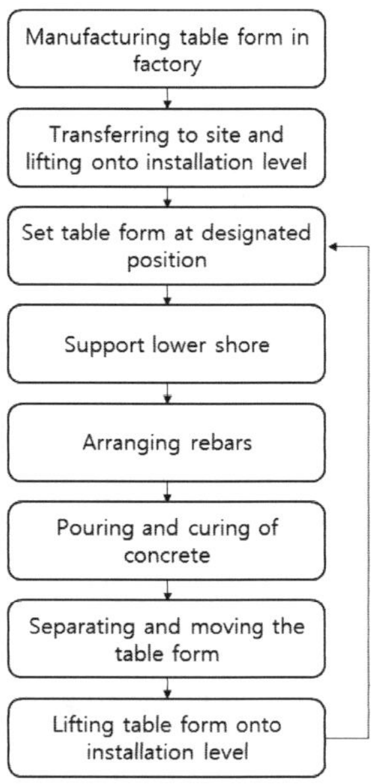

Figure 1. Table form installation procedure

2.2 IPA analysis for improvement of table form work method improvement

IPA is a method that simultaneously compares the relative importance and performance of each property, in order to measure the importance and performance of users for products or services. It is primarily used in the business administration field, but has been widely used in the machinery, construction, and transportation fields in recent years. To use IPA, a questionnaire is distributed to users, who are surveyed on a five-point or seven-point scale. Then, the average value of each attribute is calculated and marked on a four-quadrant grid, with the vertical axis representing importance and the horizontal axis representing performance[6]. The IPA method is used in many studies because it is useful for finding factors for inputting limited resources, and identifying key improvement factors and over-investment factors.

In this study, importance and user satisfaction (performance) are analysed for the detailed processes of the table formwork method; improvement and activation plan are then proposed. Among the many research methods for analysing user satisfaction, an IPA method was applied, to derive the factors that inhibit the activation of the table formwork method because of low satisfaction relative to importance.

3 Method

3.1 Questionnaire development

To identify inhibiting factors associated with the activation of the table formwork method, a questionnaire survey was conducted involving various construction companies, including constructors, specialty contractors, construction management(CM) companies, and design companies(Table 1). The respondents were limited to those who had experience with the table formwork method, to obtain practical opinions based on experience. A total of 40 questionnaire sheets were distributed through e-mail and field visits, and 36 sheets were recovered. A total of 33 questionnaire sheets were used in the analysis (three questionnaires were excluded because of omissions). The questionnaire examined the importance and satisfaction for each evaluation item using a five-point likert-type scale.

Table 1. Experience with table formwork method by company

Types of construction companies	Experience with table formwork method			
	1 time	2 times	3 times	more than 4 time
Constructors	8	5	1	-
Specialty contractors	1	2	2	5
CM	3	2	-	-
Design companies	4	-	-	-

3.2 Derivation of evaluation items

Prior to the creation of the questionnaire survey, several experts were interviewed to derive the evaluation items included in the questionnaire(Table 2): the field manager of a high-rise construction project applying the table form, a person who had worked with many table forms, two table form developers, and two formwork experts. The evaluation items were composed of the structural and operational features of the table form

In order to identify problems that may occur during the selection process and construction process of the table form construction. The detailed evaluation items are listed below in Table 3.

Table 2. The experts who participated in the interview for the evaluation item

Classification	Persons	Experience
Field manager applying the table form	1	More than 10 years
Person who had worked with many table forms	1	More than 20 years
Table form developers	2	More than 5 years
Formwork experts	2	more than 10 years

3.3 Data analysis

3.3.1 Derivation of evaluation items

A reliability test was conducted for the collected questionnaire sheets by applying Cronbach's alpha coefficient, which is the most frequently used reliability coefficient, using the commercial software application SPSS Statistics 22.

When analysed by the internal consistency method, a Cronbach's alpha value of 0.7 or greater indicates reliability. The Cronbach's alpha value of the questionnaires collected in this study was analysed as 0.974 for importance and 0.956 for satisfaction. Therefore, the questionnaire data collected in this study was found to be reliable.

3.3.2 Evaluation using Importance Performance Analysis (IPA) method

The importance and satisfaction of each evaluation item were determined through the survey and evaluated using the IPA method. Importance in this study determines whether each evaluation item is an important factor for the given work. The higher the number on the Likert 5-point scale, the higher is the importance. Satisfaction determines whether each evaluation item is satisfactory at the time of the work. Likewise, the higher the number on the Likert 5-point scale, the higher is the satisfaction level.

Table 3 Detailed evaluation items by task of table formwork method

Task	Features	Division	Evaluation Items	Task	Features	Division	Evaluation Items	Task	Features	Division	Evaluation Items
Selection of the table form work method	Structural features	A1	Weight	Table form horizontal transportation	Structural features	B1	Weight	Table form disassembly	Structural features	D1	Weight
		A2	Size			B2	Size			D2	Size
		A3	Shape			B3	Shape			D3	Shape
		A4	Unit price		Operational features	B4	Safety		Operational features	D4	Safety
	Operational features	A5	Exclusivity			B5	Work experience			D5	Work experience
		A6	Assembly			B6	Manpower			D6	Manpower
		A7	Procurement			B7	Transportation equipment	Table form lifting	Structural features	E1	Weight
		A8	Compatibility	Table form installation	Structural features	C1	Weight			E2	Size
		A9	Concrete quality			C2	Size			E3	Shape
		A10	Safety			C3	Shape		Operational features	E4	Safety
		A11	Constructability		Operational features	C4	Safety			E5	Work experience
		A12	Work experience			C5	Work experience			E6	Manpower
		A13	Manpower			C6	Manpower			E7	Lifting equipment

The X and Y axes of the two-dimensional graph were assigned to satisfaction and importance, respectively by applying the IPA technique. The scores of the evaluation items determined through the survey were plotted on the graph. Each item is analysed according to the quadrant in the graph, and the quadrants are classified as maintenance area, priority improvement area, progressive improvement area, and over-investment area. The maintenance area is an area where both importance and satisfaction are high and where items required to constantly maintain the current level are located. The priority improvement area is an area where the importance is high but the satisfaction is low and where items that require urgent improvement and have the greatest improvement effect are located. The progressive improvement area is the area where both importance and satisfaction are low and where items that have low need for improvement are located. The over-investment area is where the importance is low and the satisfaction is high and where the invested resources need to be redistributed because the satisfaction is unnecessarily high. This study focused on analysing the items in the priority improvement area and proposed improvement methods for them.

4 Results and Discussion

4.1 Analysis of IPA results

The results of the survey on the importance and satisfaction of each evaluation item are outlined in Table 4, and the results of applying the IPA method to analyse them are shown in Fig. 2.

Table 4. Average value of importance and satisfaction by evaluation item

Division	Importance	Satisfaction	Division	Importance	Satisfaction
A1	3.879	2.939	C1	4.091	3.242
A2	3.970	3.303	C2	4.121	3.152
A3	4.031	3.152	C3	3.909	3.394
A4	4.000	3.091	C4	4.182	3.303
A5	3.909	3.455	C5	3.879	3.061
A6	4.394	3.364	C6	4.000	3.030
A7	3.909	3.333	D1	4.000	3.091
A8	3.848	3.061	D2	4.152	3.182
A9	3.848	3.212	D3	3.970	3.273
A10	4.121	3.333	D4	4.182	3.303
A11	4.091	3.242	D5	3.970	2.879
A12	4.030	2.970	D6	3.970	3.000
A13	3.818	3.273	E1	4.091	2.939
B1	4.000	3.152	E2	4.030	3.030
B2	4.061	3.212	E3	4.121	3.000
B3	3.848	3.333	E4	4.242	3.061
B4	4.030	3.152	E5	3.970	3.000
B5	3.879	3.030	E6	3.939	3.061
B6	3.727	3.030	E7	3.970	2.879
B7	3.818	3.091			

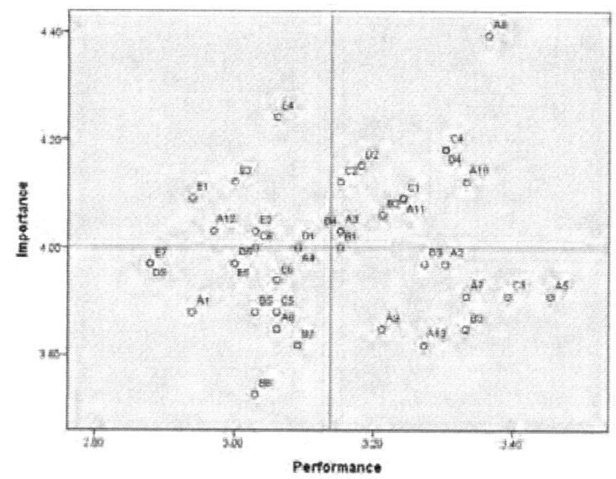

Figure 2. Results of applying IPA method

Following the analysis, items A4, A12, C6, D1, D6, E1, E2, E3, and E4 were included in the priority improvement area where urgent improvement is required. Even though these items are important factors in each work process, they are considered factors that inhibit the activation of the work method by lowering the preference for the table formwork method and hindering the reuse of the method owing to their low satisfaction. Furthermore, as a result of analysing the satisfaction of each work compared to importance, the satisfaction of the lifting work was lowest among the selection, transportation, installation, disassembly, and lifting works. When the importance and satisfaction of the 39 evaluation items were analysed and the top five items were derived, it was found that 80% of the top five items were pertained to the lifting task(Table 5). Therefore, this study determined that it is urgent to improve the items included in the table form lifting task among the various items in the priority improvement area, and to provide an accurate analysis of this items and improvement plans preferentially.

Table 5. Top 5 of importance/satisfaction values

Division	Task	Evaluation items	Importance/ Satisfaction
E1	Table form lifting	Weight	1.392

E4	Table form lifting	Safety	1.386
D5	Table form disassembly	Work experience	1.379
E7	Table form lifting	Lifting equipment	1.379
E3	Table form lifting	Shape	1.374

4.2 Analysis of problems through the Delphi method and proposal of improvement plan

Among the items included in the priority improvement area through IPA, those corresponding to the lifting task were the weight, size, shape, and safety of table form. In addition, among the five items with the lowest satisfaction compared to importance, there is an equipment item for lifting work. This item is important for the lifting work of the table formwork method but actually shows low satisfaction. This study, therefore, performed qualitative evaluation of five items through Delphi method which is conducted as a panel survey composed of experts. Then, an improvement plan was proposed after analysing the effects of these item on the activation of the table formwork method.

As a result of the survey through two Delphi rounds with 11 table form experts, including workers who have experience with table formwork more than five times, table form developers, and table form researchers, all the factors showed correlations, In particular, the limitation of lifting equipment seems to be a critical factor that inhibits the activation of the table formwork method. This is because the table form lifting is done by a tower crane in current high-rise building construction projects, which reduces the operation time of the tower crane and consequently delays the construction time. Moreover, owing to the nature of construction of high-rise buildings, the influence of wind is much greater on higher floors. Thus, it takes more time to lift a table form that is heavy and large, and the probability of accidents increases. In addition, with the shape of the existing table form, it is difficult to flexibly cope with the increasing number of atypical high-rise buildings, and this causes additional work owing to the planar changes that occur as the number of floors increases, thus delaying the construction time and increasing the construction cost. Therefore, the existing table form method requires the development of separate lifting equipment, weight reduction of the table form, and a flexible change of shape.

In this study, one more Delphi round was performed to solve the aforementioned problems and suggest improvement methods, and the following improvement methods have been suggested. First, in future high-rise building constructions, a separate automatic lifting equipment such as automatic lifting system[7] combined with lift system and table lifting system[8] should be installed to increase the operation time and improve the safety of tower cranes. Furthermore, the weight of the table form should be reduced to decrease the burden of lifting and the effect of winds. This problem can be solved by replacing the existing materials such as aluminum, steel, and lumber with plastics and composites. Plastics and composites are frequently used recently for lightweight forms because they are lightweight and have high rigidity, making them suitable for table forms as we ll. Finally, flexibility should be added to table forms so that they can respond to the changing planes by the number of floors. This technology is being researched currently, and a product called flexible table form[7] has been developed. However, this technology has a limitation because the table form must be fixed in advance in accordance with the plane shape before the table form can be used. Therefore, additional technical development is required to enable the immediate change in the shape of table forms.

The improvement plan proposed in this study are outlined in Table 6. The overall efficiency and safety of the table formwork method can be improved by applying these techniques. Furthermore, automation technology can be added to the existing lifting decks so as to reduce the weight burden for the table form transportation equipment and minimize the effects of winds.

Table 6. Improvement plan for activation of table formwork method

Division	Improvement plan		Effect
Table form	Composite	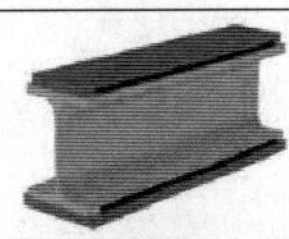	- reduce the weight - improve the safety
	Flexible Table Form	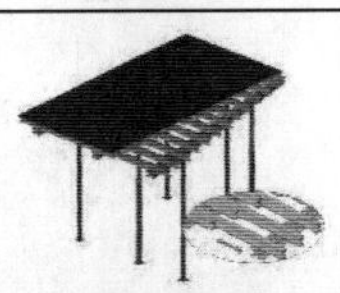	- can respond to the changing planes - can be possible to resize
Lifting equipment	Table lifting system		- increase the operation time of tower cranes - minimize the effects of wind - reduce construction period - improve the safety
	Automatic lifting system		

5 Conclusion

The inhibiting factors for the activation of the table formwork method were identified through the IPA method. Furthermore, the effects of these inhibiting factors on the table formwork method were analysed through the Delphi method and improvement methods were proposed.

The analysis results showed that the weight, shape, size, safety, and lifting equipment of the table formwork method were the representative factors that inhibit the activation of the table formwork method. These factors cause workloads in the tower crane and increase the probability of accidents. In addition, with the conventional shape of the table form, it is difficult to respond to the atypical shape of high-rise buildings. This causes additional work, thus delaying the construction time and increasing the construction cost. To solve the analysed inhibition factor, separate automatic lifting equipment such as an automatic lifting system and a table lifting system is required for high-rise buildings, and the weight of the table form should be reduced by using lightweight materials such as plastics and composites. Furthermore, the existing flexible table formwork method needs to be improved to develop table forms that can flexibly change their shape.

In this study, the problems and satisfaction levels of the detailed processes of the table formwork method were analysed and future improvement and activation methods were proposed. Among the many improvement items found through the IPA method, however, the items corresponding to lifting task were analysed first. Thus, additional research is required for other items than those of the lifting task among the improvement items derived in this study. In the future, the results of this study could be used to develop element technologies of the table formwork method and improve user satisfaction. It is expected that this will enable the process improvement and activation of the table formwork method.

Acknowledgement

This research was supported by a grant(16AUDP-B106327-02) from Architecture & Urban Development Research Program funded by Ministry of Land, Infrastructure and Transport of Korean government.

References

[1] Kwon J. Lim H. Kim M. Cha S. Cho H. Kang K. Productivity Analysis of the Table Formwork Method for Making a Cost-efficient Equipment Input Plan. *International Symposium on Automation and Robotics in Construction and Mining,* 608-617, 2013.

[2] Lim H. Kim T. Cho H. Kang K. Design Process for Formwork System in Tall Building Construction using Quality Function Deployment and TRIZ. *Journal of the Architectural Institute of Korea*, 28(9):173-182, 2012.

[3] Lee D. Lim H. Kim T. Cho H. Kang K. A Support Model of Optimum Layout Planning of Forms for Improving Constructability of Formwork. *The korea Institute of Building Construction conference Paper,* korea, pages 212-213, 2013.

[4] Lim H. Kim T. Cho H. Kang K. Simulation-based Planning Model for Table Formwork Operation in Tall Building Construction. *Journal of Asian Architecture and Building Engineering*, 16(1): 115-122, 2017.

[5] Seobo Industrial. Table Form. On-line: http:// /www.seobo.co.kr, Accessed: 20/02/2017.

[6] John A. John C. Importance-Performance Analysis. *Journal of Marketing,* 41(1): 77-79, 1977.

[7] Kim T. Advanced System Formwork and Construction Planning Model for Tall Building Construction, Doctorate thesis, Korea University, 2013.

[8] Doka. Table Lifting System. On-line: http:// https://www.doka.com/en/system-groups/doka-floor-systems/tableforms/dokamatic-table/index, Accessed: 05/03/2017

Unmanned Ground Military and Construction Systems Technology Gaps Exploration

Eugeniusz Budny[a], Piotr Szynkarczyk[a] and Józef Wrona[b]

I h I
z 4 z
z z
h
4 z
z

Abstract

There are found the answers to the scientific question whether it is possible to figure out the methodology to explore technology gaps looking through Operational Requirements (ORs) defined for military systems, to apply it to construction industry applications technology gaps exploration. In this paper were specified the (ORs) that enabled a preliminary analysis on Unmanned Ground Systems (UGSs): which kind of platforms might be developed and shared between the systems devoted to different military and construction industry applications. In this paper are presented technology gaps identified by comparing the ORs with the state of play of technologies that are currently available in the military and construction equipment applications. Then the challenges and measures are identified and associated to the gaps that should be taken in order to bridge each specific gap. There are identified technological and non-technological aspects that may hamper development of UGS solutions that fully achieve the ORs, and thus would need to be further studied and/or elaborated. As a next step actions are specified. They represent the research and development initiatives, steps or activities that should be done in order to be in the position of developing UGS solutions achieving the specified ORs and requirements of construction site. The actions need to be completed to bridge the technology gaps to make required new capabilities available. One of results of such a methodology are schedules based on the roadmap aimed to guide the development of the future UGS solutions towards the achievement of operational and construction site requirements.

Keywords – **unmanned ground systems (UGSs); construction industry applications; operational requirements; technology gaps**

Introduction

h h

I I 4 4
h
I
h I
I h

h
3 9] I h
h h B 93
h

h h h

x h

roadmap h h
h h h
B h h
h h
h h h h
h h x
h h h

h
x

Operational Requirements

h h

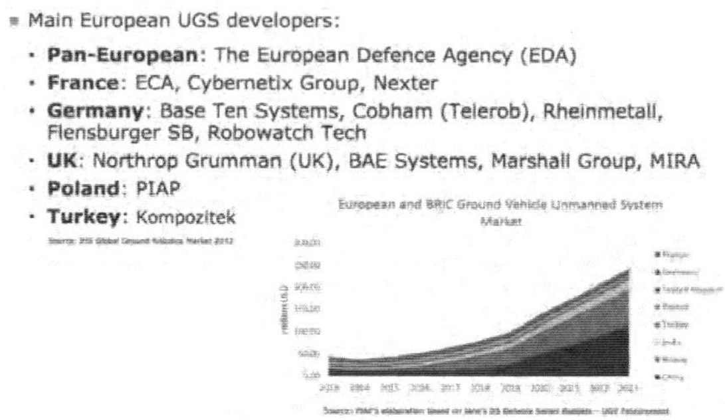

State of Play

Gaps Identification

high mobility base platform, Human-Machine Interfaces and Machine-Machine Interfaces and teleoperation and control systems, data transmission systems, environments recognitions (sensors), manipulation capabilities, autonomous operation, supply and engine

***** Formatting Issue - Best Available Paper/Graphic *****

systems h *command*
structure, ethical, legal, personal and cultural I h
h h h
h *high mobility base platform*
h h

Human-Machinery Interfaces, Machine-Machine Interfaces and teleportation and control systems h
h h h
h
I h
data transmission systems h
h

h
h h

environments recognitions (sensors) h
h h

I h *manipulation capabilities* h
h h
B h
autonomous operation h
h
h h
x z
h *supply and engine systems*
h
h]

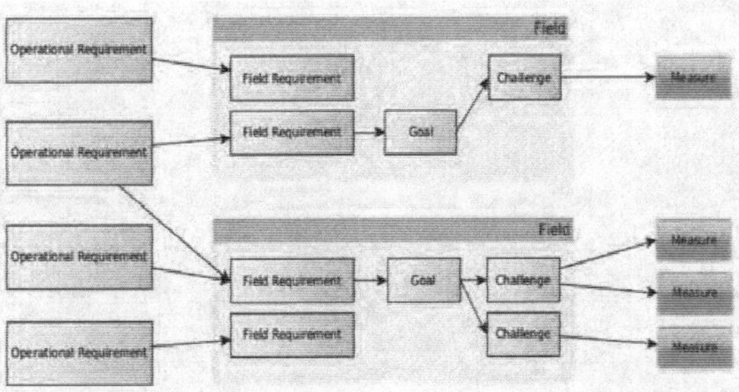

]

h h
I h h

h
h h h h
h
h h
I h h h h
h h
I h

h h
h z h
h h
I
h h]
h h]
h h h
h h h h
x h
h h
h h h h
h
x]
I h h
x h h

h h

h h h
h h h
h h x h
h h

h x

Actions plan

h h h h
h h
h
]
h h h
h
h h
h
h h
h
h h
h
x h
h
h h h
h h h
3
h

3
h

h

*** Formatting Issue - Best Available Paper/Graphic ***

autonomy

Human-Machine Interfaces and Machine-Machine Interfaces and teleoperation and control systems, data transmission systems, environments recognitions (including sensors), manipulation capabilities, autonomous operation.

high mobility base platform *supply and engine systems.*

Roadmap

Autonomous operation

Conclusions

***** Formatting Issue - Best Available Paper/Graphic *****

34 h I h

h
h I h
h
h h
h

h]
h h h
h
h

h h
h h h
h h x h
h h

h x

References

] h
I
I 4 4 h
I h
I I
h

h
3 9
B B
]
I I
3 9 I
I

h h

I h I
I
h B h

3]
h
h h
3 4 B h h
4]
BI I
I BI I
h
h h

Immersive virtual environment as a promising tool for the elderly-friendly assistive robot design

Jeonghwan Kim[a], Choongwan Koo[a] and Seung Hyun Cha[a]

[a]Department of Building Service Engineering, The Hong Kong Polytechnic University, Hong Kong
E-mail: jeonghwan.kim@polyu.edu.hk, choongwan.koo@polyu.edu.hk, seung-hyun.cha@polyu.edu.hk

Abstract

As the population of older people continues to rapidly grow worldwide, assistive robots for older people are become more important to maintain their social, physical and psychological well-being. Although the idea of having these robots is an enticing prospect, there are many barriers for older people to adopt the robots in a daily life such as a negative attitude toward a robot, inadequate robot design, confined living environment. Therefore, the in-depth understanding of such barriers prior to adopting the robots is required so as to boost intimate interaction with and acceptance of assistive robot technology. This study thus aims to suggest a guideline that helps understand such barriers with a focus on relationships among the three aspects, namely human, robot, and environment. It then proposes Immersive Virtual Environment (IVE) as a tool to evaluate various scenarios of Human-Robot Interaction (HRI), so that it will help find optimal design solutions for assistive robots in consideration of the relationships among the three aspects. This research has great potential to contribute to the design principles of the assistive robots as well as space and environment planning for assistive housing.

Keywords –

Assistive robot, Healthy aging, Immersive virtual environment, HRI

1 Introduction

Many countries are facing demographic transition towards aging society. Based on the recent report [1], the number of people in the world aged over 60 years, is projected to grow by 56 percent between 2015 and 2030. Consequently, the demand of well-being of older people has become an important social issue that needs to be addressed in near future. One of the primary goal of aging society is to promote healthy aging [2]. Healthy aging is a concept that older people can take an active part in society and enjoy an independent and high quality of life [3]. Many researchers have been developing various types of robots to support and assist the older people in their home, and it has been widely recognized as one of the effective ways to address population aging issues in terms of social interaction (e.g. talk, play, etc.) [4, 5] and physical interaction [6, 7].

However, most of these studies and technologies have suffered from a lack of well-grounded and comprehensive considerations of both living environment and the unique characteristics of older people, which may leads to the low technology acceptance rate [8]. The existing literature on technology acceptance for the elderly mainly focuses on the learning and supporting environment in accordance with the physical and cognitive debility [9, 10]. To fill this gap, this study emphasizes the importance of living environment and human aspect of older people when developing and deploying an assistive robot in their living environment. Further, we propose the use of immersive virtual environment as a tool to evaluate design alternatives of assistive robot.

2 Literature review

Over the past decade, there has been an increasing amount of literature on the Human-Robot Interaction (HRI). Many researchers have developed the assistive robots, and successfully commercialized their products, such as Care-O-Bot (Assistive robot, Fraunhofer, Germany), Robear (Nurse robot, Riken, Japan), Paro (Companion Robot, SoftBank, Japan). These robots are capable of detection, grasp, moving, placing, hand-over an object to elderly [11]. To provide these assistance, the most importantly, robot must be equipped with a high standard of safety features as well as behavior, and appearance (e.g. gaze, blink, speech, human-likeness, etc.). Despite these technical advancements in robotics, human and environmental aspects were often disregarded [12, 13]. In this paper, we investigate human and environmental aspect in assistive robot for elderly.

2.1 Living environment

To date, some studies have identified a link between robot and environment [14-16]. Tan, Mohan [16] has emphasized the importance of five factors of robot-friendly environment, which comprise observability, manipulability, activity, and safety. For example, proper design of corridor, door knob, mechanism of artifacts, and materials were proposed for the safe interaction. However, the studies have focused particularly on the mobile and commercial robots with specific purposes, not on assistive robots serving humans in a daily living environment. Meng et al. [17] proposed design principles for an assistive robot for elderly and put emphasis on the role of software capability to provide cost efficient flexibility and to accommodate various levels of task requirement, user abilities, and personal environments without sacrificing cost and functionality. Furthermore, It is reported that many older people live in a confined and relatively small flat in land scarce urban area, which might hinder physical interaction between human and robot [18]. As operation and mobility of assistive robot require and occupy the space, specifics of a robot (such as size, behavior, work envelope) assistive robot should consider their living environment (e.g. size of space, interior design, etc.).

In addition to the physical living environment, many studies have been empirically investigated the importance of learning environment during the adoption of new technology for elderly [19-21]. For example, if we provide hands-on experience with teaching and guide support to elderly, it will increase long-term viability and acceptance. Therefore, along with living space, their social environment that can support the learning should be considered as well for elderly-friendly assistive robot design.

2.2 Human factors

Because of limited competence of older people such as lack of functional, biological, sensory/perceptual, cognitive capability, social abilities, and behavior skills, older people seem struggling to learn, adopt and accept the modern technology, including assistive robot [6, 22]. In this regard, technology acceptance has been studied from a variety of perspectives. Information systems, sociology and human-computer interaction researchers have come up with various models incorporating factors and phases to predict adoption that, in turn, will lead to persistent use.

Existing literature identifies numerous barriers for the elderly using technology tools, which is possible to categorize them to elders' physical and mental status. The barriers due to impairments of elders' physical status are cognitive (memory and processing speed, visual, auditory, and motor control abilities [23]). The mental barriers are attitudinal (manner of feeling or behaving; [24, 25]) privacy concerns, security [26], safety (monitoring elderly in their private home; [27-31]), and total replacement of humans with technology tools such as robots [32, 33]. Also, much of the literature on human factors in HRI has emphasized the importance of perceived safety, usefulness, self-efficacy, learning support, and experience, etc. of the user [34, 35]. Moreover, when it comes to acceptance, satisfaction, and trustworthiness of the robotic technology, many studies have supported the fact that these perceptions toward robots are influenced by the user's sociodemographic factor. Particularly, Chen and Chan [8] noted that technology acceptance of elderly are also affected by individual attributes, health, and ability characteristics, as well as facilitating conditions. Particularly, Tapus, Ţăpuş [36] claims that personality match between the user and the assistive robot may enhance the acceptance, which suggesting the importance of human aspects in robot design. In this paper, we address these issues to provide an requirement and guidelines that overarching relationship of these factors to the design of the elderly-friendly assistive robot.

2.3 Requirements and guidelines for elderly-friendly robot and environment

Based on the previous literature, we propose

requirements for an elderly-friendly robot and its environment. Elderly-friendly assistive robot should be fitted into a living environment, which roboticists are required to investigate the feasibility of the robot prototype in various scenarios and conditions. In particular, if the living environment is small and limited, which is a common living condition of elderly, sophisticated design of hardware or mechanical systems as well as software systems are required to provide the functionality that they needed. For example, human size robot (e.g. Care-o-bot) may not be useful in a confined flat. Instead, small and wall-mounted type of assistive robot may provide the functionality without sacrificing space utilization for daily living.

Furthermore, to provide robot assistive housing for the future elderly, architects should consider the concept of robot-embedded housing, which designates an extra space for an assistive robot. For example, a home-version of Kiva systems would be deployed to help carry heavy objects for elderly, but the nature of their operation is based on the interaction between Kiva and marker (or other types of reference system), an invisible marker or such references needs to be installed, and architects should take into account this design factors.

Also, user acceptance is essential part of robot design. For example, human-like appearance (e.g. facial expression) for better communication and control is recommended for all kinds of interactive robots. Therefore, roboticists need to take affective interaction into account when designing an elderly-friendly assistive robot.

All things considered, we propose a comprehensive and integrated interaction network among elderly–robot–environment, as depicted in figure 1.

3 Immersive virtual environment as a tool for robot and environment design

As we address in the previous section, there are many factors that needs to be taken into account for elderly-friendly assistive robot. When dealing with many different impacting factors, implementing an experiment that evaluates the human perception of actual robots becomes overwhelming pragmatically and financially. To mitigate these issues, the use of Immersive Virtual Environment (IVE) can be a great alternative. IVE are proven to increase both experimental control and mundane realism, which leads to the enhancement of participants' engagement, thereby increasing experimental impact. Many research also supported the use of IVE in construction by measuring the sense of presence while each participant performed office-related activities in both IVE and physical environment.

Recently, affordable Virtual Reality (VR) headset is on the market, and thanks to advanced game engine software such as Unreal and Unity 3D, building a virtual environment have been easier than before.

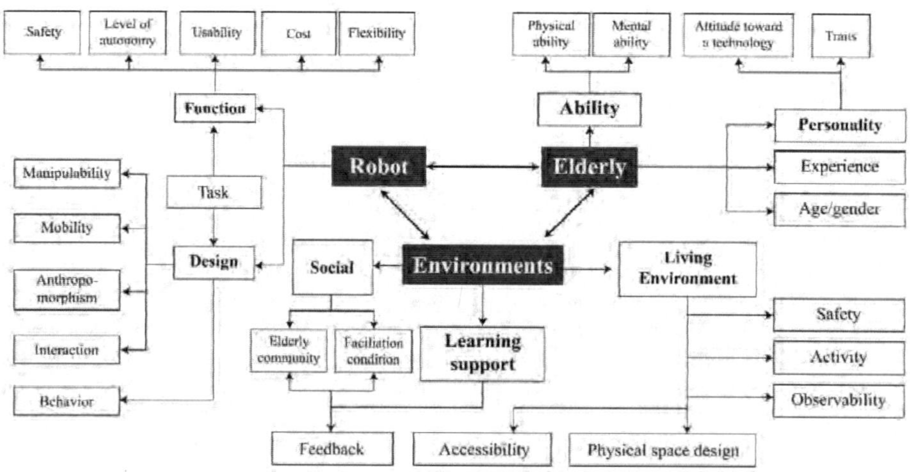

Figure 1. Relationships among factors of robot, elderly, and environment

3.1 Hardware settings

IVE requires a set of hardware components; a computer, VR headset (Head-mounted display), and tracking devices. Firstly, Oculus Rift Head-Mounted Display (HMD) will be used to provide immersive virtual experience for the user. A game engine renders two slightly different images shifted horizontally creating an illusion of depth to provide stereoscopic views. Also, Oculus Rift has a tracking system to send an appropriate data to the computer, which determines the user's position and orientation in virtual space. The range of position tracking would be limited to 2-3 meters depending upon wired connection of Oculus Rift so that experimental area should be within 2 meters as well. Multi-modal interaction has become available such as gesture (e.g. Leap motion) and virtual hand (e.g. Oculus Touch) so that it would be useful for evaluation of physical interaction or communication.

3.2 Software settings

Generally, a game engine is an essential components (e.g. position and orientation, camera, light, renderer, etc.) for creating a virtual environment. The game engine is connected to the data stream that linked to hardware settings (e.g. Oculus Rift) so that the computer can generate rendered images based on the tracking movement of the user.

One of the benefit of using IVE is the capability to offer immersion to the user. People can easily maintain high mundane-realism while interacting with experimental virtual environments. However, this effectiveness of IVE has been deeply linked to the sense of presence, and immersion. Presence is defined as the subjective experience of being in one place or environment, even when one is physically situated in another, and immersion into virtual reality is a perception of being physically present in a non-physical world, which can be quantifiable. Providing enough presence and immersion plays an important role to evoke an emotion and response from the user. For example, improper interaction design of virtual environment can result in the major health issues such as disorientation or nausea, which dramatically reduces sense of presence.

3.3 Implementation of IVE

Based on the hardware and software settings, many of human-robot interaction scenarios and hypotheses can easily be validated. For instance, a hypothesis–limited living space will decrease perceived usefulness of assistive robot.– can easily validated with the use of IVE (Figure 2), also any design flaws of robot can be tested in advance, which provide a valuable feedback to roboticists and architects. In order to evaluate user's perception towards an assistive robot and environment, psycho-physiological measure as well as survey can be used to accurately validate their emotion toward an assistive robot.

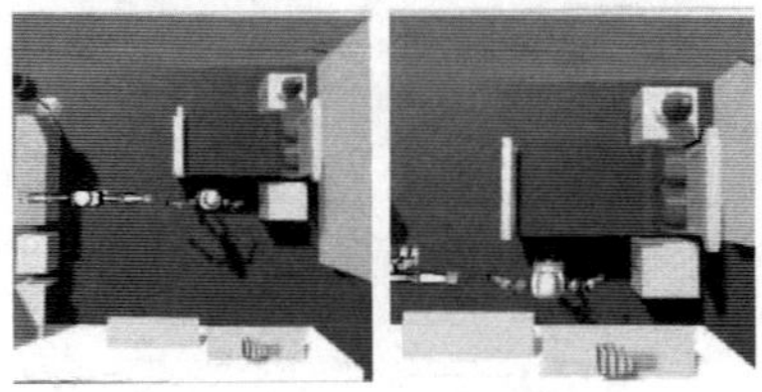

Figure 2. IVE settings for evaluation of assistive robot in different scenario

4 Conclusion

It is obvious that assistive robot for elderly will soon become more prevalent in a domestic environment. This paper has argued the importance of living environment and human aspect of elderly-friendly assistive robot. The main goal of current study was to establish considering factors for the design of assistive robot and its environment, and to suggest IVE as a promising tool for evaluating these factors efficiently. This research has great potential to contribute to the design principles of the assistive robots as well as space and environment planning towards assistive housing or robot-embedded housing. Further experimental investigations are needed to estimate elderly's perception towards an assistive robot and to examine more closely the links between robot and environment.

References

[1] United Nation, *World population ageing 2015*. Department of Economic and Social Affairs PD, 2015.

[2] Clarke, P. and E.R. Nieuwenhuijsen, *Environments for healthy ageing: a critical review*. Maturitas, 2009. **64**(1): p. 14-19.

[3] Agren, G. and K. Berensson, *Healthy ageing: a*

challenge for Europe. Swedish National Institute of Public Health, 2006. **2006**: p. 29.

[4] Bemelmans, R., et al., *Socially Assistive Robots in Elderly Care: A Systematic Review into Effects and Effectiveness.* Journal of the American Medical Directors Association, 2012. **13**(2): p. 114-120.e1.

[5] Flandorfer, P., *Population ageing and socially assistive robots for elderly persons: the importance of sociodemographic factors for user acceptance.* International Journal of Population Research, 2012. **2012**.

[6] Meng, Q. and M.H. Lee, *Design issues for assistive robotics for the elderly.* Advanced Engineering Informatics, 2006. **20**(2): p. 171-186.

[7] Solis, J., *Development of a human-friendly walking assisting robot vehicle designed to provide physical support to the elderly.* IFAC-PapersOnLine, 2016. **49**(21): p. 656-661.

[8] Chen, K. and A.H.S. Chan, *Gerontechnology acceptance by elderly Hong Kong Chinese: a senior technology acceptance model (STAM).* Ergonomics, 2014. **57**(5): p. 635-652.

[9] Broadbent, E., R. Stafford, and B. MacDonald, *Acceptance of Healthcare Robots for the Older Population: Review and Future Directions.* International Journal of Social Robotics, 2009. **1**(4): p. 319.

[10] Peek, S.T.M., et al., *Factors influencing acceptance of technology for aging in place: A systematic review.* International Journal of Medical Informatics, 2014. **83**(4): p. 235-248.

[11] Hendrich, N., H. Bistry, and J. Zhang, *Architecture and Software Design for a Service Robot in an Elderly-Care Scenario.* Engineering, 2015. **1**(1): p. 027-035.

[12] De Santis, A., et al., *An atlas of physical human–robot interaction.* Mechanism and Machine Theory, 2008. **43**(3): p. 253-270.

[13] Bicchi, A., M.A. Peshkin, and J.E. Colgate, *Safety for physical human–robot interaction.* Springer handbook of robotics. 2008. 1335-1348.

[14] Elara, M.R., N. Rojas, and A. Chua. *Design principles for robot inclusive spaces: A case study with Roomba.* in *2014 IEEE International Conference on Robotics and Automation (ICRA).* 2014.

[15] Mohan, R.E., et al., *Designing the robot inclusive space challenge.* Digital Communications and Networks, 2015. **1**(4): p. 267-274.

[16] Tan, N., R.E. Mohan, and A. Watanabe, *Toward a framework for robot-inclusive environments.* Automation in Construction, 2016. **69**: p. 68-78.

[17] Meng, Q. and M.H. Lee, *Design issues for assistive robotics for the elderly.*

[18] Litwak, E. and C.F. Longino, *Migration patterns among the elderly: A developmental perspective.* The Gerontologist, 1987. **27**(3): p. 266-272.

[19] Louie, W.Y.G., D. McColl, and G. Nejat, *Acceptance and attitudes toward a human-like socially assistive robot by older adults.* Assistive Technology, 2014. **26**(3): p. 140-150.

[20] Wu, Y.H., et al., *Acceptance of an assistive robot in older adults: A mixed-method study of human-robot interaction over a 1-month period in the living lab setting.* Clinical Interventions in Aging, 2014. **9**: p. 801-811.

[21] Kim, S., et al., *Acceptance of mobile technology by older adults: a preliminary study*, in *Proceedings of the 18th International Conference on Human-Computer Interaction with Mobile Devices and Services*. 2016, ACM: Florence, Italy. p. 147-157.

[22] Bemelmans, R., et al., *Socially assistive robots in elderly care: A systematic review into effects and effectiveness.* Journal of the American Medical Directors Association, 2012. **13**(2): p. 114-120. e1.

[23] Czaja, S.J., et al., *Factors predicting the use of technology: Findings from the center for research and education on aging and technology enhancement (create).* Psychology and Aging, 2006. **21**(2): p. 333-352.

[24] Ellis, R.D. and J.C. Allaire, *Modeling Computer Interest in Older Adults: The Role of Age, Education, Computer Knowledge, and Computer Anxiety.* Human Factors, 1999. **41**(3): p. 345-355.

[25] Skymne, C., et al., *Getting used to assistive devices: Ambivalent experiences by frail elderly persons.* Scandinavian Journal of Occupational Therapy, 2012. **19**(2): p. 194-203.

[26] Harrefors, C., K. Axelsson, and S. Sävenstedt, *Using assistive technology services at differing levels of care: healthy older couples' perceptions.* Journal of Advanced Nursing, 2010. **66**(7): p. 1523-1532.

[27] Boström, M., S. Kjellström, and A. Björklund, *Older persons have ambivalent feelings about the use of monitoring technologies.* Technology and Disability, 2013. **25**(2): p. 117-125.

[28] Caine, K.E., A.D. Fisk, and W.A. Rogers, *Benefits and Privacy Concerns of a Home Equipped with a Visual Sensing System: A Perspective from Older Adults.* Proceedings of the Human Factors and Ergonomics Society Annual Meeting, 2006. **50**(2): p. 180-184.

[29] Miskelly, F.G., *Assistive technology in elderly care.* Age and Ageing, 2001. **30**(6): p. 455-458.

[30] Magnusson, L. and E.J. Hanson, *Ethical issues arising from a research, technology and development project to support frail older people and their family carers at home.* Health & Social Care in the Community, 2003. **11**(5): p. 431-439.

[31] PETERSSON, I., LILJA, M., & BORELL, L., *To feel safe in everyday life at home – a study of older adults after home modifications.* Ageing and Society, 2012. **32**(5): p. 791-811.

[32] Lundberg, S., *The results from a two-year case study of an information and communication technology support system for family caregivers.* Disability and Rehabilitation: Assistive Technology, 2014. **9**(4): p. 353-358.

[33] Kristoffersson, A., et al., *An Exploratory Study of Health Professionals' Attitudes about Robotic Telepresence Technology.* Journal of Technology in Human Services, 2011. **29**(4): p. 263-283.

[34] Demiris, G., et al., *Older adults' attitudes towards and perceptions of 'smart home'technologies: a pilot study.* Medical informatics and the Internet in medicine, 2004. **29**(2): p. 87-94.

[35] Daniel, K.M., C.L. Cason, and S. Ferrell, *Emerging technologies to enhance the safety of older people in their homes.* Geriatric Nursing, 2009. **30**(6): p. 384-389.

[36] Tapus, A., C. Ţăpuş, and M.J. Matarić, *User—robot personality matching and assistive robot behavior adaptation for post-stroke rehabilitation therapy.* Intelligent Service Robotics, 2008. **1**(2): p. 169-183.

Survey of the line of sight characteristics of construction machine operators to improve the efficiency of unmanned construction

Mitsuru Yamada[a] and Kenichi Fujino[a] and Hiroki Kajita[a] and Takeshi Hashimoto[a]

[a]Advanced Technology Research Team, Public Workd Research Institute, National Research and Development Agency, Japan
E-mail: m-yamada@pwri.go.jp, fujino@pwri.go.jp, h-kajita@pwri.go.jp, t-hashimoto@pwri.go.jp

Abstract –

Unmanned construction is a construction method of operating construction machines from a remote place by using radio waves. In Japan, unmanned construction has many applications in disaster sites where people can not enter. However, unmanned construction is less than half of the construction efficiency as compared to normal boarding operations. We need to improve construction efficiency of unmanned construction.

In normal unmanned construction, the operator mainly operates construction machines while looking at the image of the boarding camera installed in the cabin. However, visual information such as angle of view of images taken by boarding cameras is smaller than actual boarding conditions, which is considered to be one of the factors that lowers the work efficiency of unmanned construction. Therefore, it is necessary to consider a new camera system that can acquire visual information equivalent to boarding status. However, the detail range of field of view and the gaze target in boarding conditions have not been clarified.

In this study, we measured the line of sight of the operator in boarding operations using the eye-mark recording system. From the results, we examined the field of view and the gazing target in boarding condition.

Keywords –

Unmanned construction; line of sight; construction efficiency

1 Introduction

Unmanned construction is a construction method permitting the use of radio waves etc. to operate hydraulic excavators and other construction machines from a remote location. In Japan, it has been applied often at the scenes of disasters that people cannot enter. But the working efficiency of this method must be improved, because it provides less than half of the working efficiency of normal operation with the operator riding in the machine.

In normal unmanned construction, the construction machine is operated by an operator viewing images from cameras mainly installed inside the machine's cabin. But, images from a camera inside the machine narrow the range of the line of sight from that of an operator riding the machine, and this loss of visual information is considered to be one factor lowering the work efficiency of unmanned construction. There is, therefore, a demand for the development of a camera system able to obtain and provide visual information equal to that obtained by an operator riding the machine, but, the specific range of line of sight and the objects of scrutiny of an operator in the machine, whose reproduction should be the goal are not clearly understood.

This report describes a study of the range of the line of sight and objects of scrutiny conducted by using an eye-tracking recorder to measure the lines of sight of operators riding in machines to obtain basic data to increase visual information in order to improve the working efficiency of unmanned construction.

2 Experimental method

The test was performed on July 22, 2014 at an outdoor test facility at the Public Works Research Institute in Japan. The machine used for the test was a 12ton class hydraulic excavator with a bucket with capacity of 0.45m3. Figure 1 is an exterior view of the hydraulic excavator used for the test. One veteran operator with 15 year experience performed a "model task" defined by the Public Works Research Institute while riding the machine as the line of sight of the operator and the working time were measured. Figure 2 shows details of the model task. The line of sight was measured a total of 10 times using an eye-tracking

Fig.1 Exterior view of the hydraulic excavator used for the test

recorder (EMR-9). And a video camera recorded a video of the state of the operation of the hydraulic excavator from outside.

3 Test result and discussion

Of the total of 10 tests, the results of measurements of the 5th were analyzed. The analysis divided the process into "traveling" and "working". Traveling refers to a state when the crawlers are driven to move the machine. "Working" is the state when the machine is stopped and the crawlers cannot operate, but the arm and other components are operating.

3.1 Working time

The working time of the entire process was 139 seconds: 77 seconds of "traveling" and 62 seconds of "working".

3.2 Abstracting stationary points

Stationary points were abstracted based on the measured operator line of sight data. Stationary points are positions where the eye marks measured by the eye-tracking camera are inside a specified range for a specified period or longer, and are the centers of eye-mark groups that are within this range. It is possible for the background image superimposed on stationary points to be seen as the object the operator scrutinizes. To calculate the stationary points, using the eye-tracking data analysis software, "EMR-dFactory (verification 2.7)", cases where the eye mark remained within a judgment circle for at least 0.1 seconds at a viewing angle of 2° were considered to be stationary points. The center of the judgment circle is the centroid of two connected eye marks, and the center of the stationary points that were abstracted was also considered to be the centroid of the eye-mark groups at the same stationary point (center of gravity method). As a result of abstracting stationary points under the above conditions,

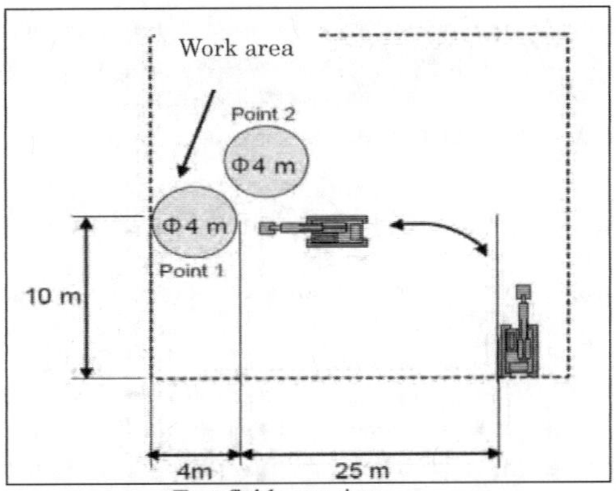

Test field overview

Work order

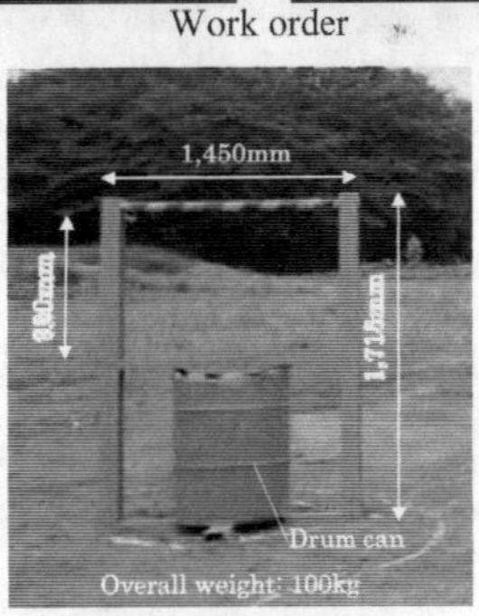

Object to be moved

Task model detail process

1. The hydraulic excavator diverts from the start point to the left and moves to the work point.
2. Lift the moving object placed in point 1 with a bucket and move it into the circle of point 2.
3. Position the moving object placed in point again and place it in a rectangle of □ 770 mm in point 1.
4. Return to the starting point.

Fig.2 details of the model task

a total of 370 stationary points—225 for "traveling" and 145 for "working"—were abstracted.

3.3 Stationary time

The maximum stationary time was 0.5 seconds for "traveling" and 1.0 second for "working." The average stationary time of each stop was a total of 0.24 seconds; 0.19 seconds for "traveling" and 0.31 seconds for "working". For "traveling", the maximum stationary time was shorter than that for "working", and its average stationary time for each stop was shorter than the total average, and presumably the operator moved the hydraulic excavator while watching various points for relatively short times. For "working", the maximum stationary time and the average stationary time for each stop were longer than those for "traveling", so presumably the operator watched the state of the work.

3.4 Object of scrutiny

The background images of the stationary points that were abstracted were divided into 10 categories—bucket, arm, rotation tip, object moved, setting point, forward visual field, obstructions, ruts, crawler, others—to study the objects of scrutiny. Table 1 explains each object of scrutiny.

Figure 3 shows the results of the categorization divided into stationary frequency and total stationary time. For "traveling", the stationary frequency and stationary time were largest for ruts (98 times, 19.8 seconds), followed by forward visual field (50 times, 10.6 seconds), then bucket (31 times, 6.0 seconds). Images from the exterior video camera have confirmed that the operator tested by this test traveled while generally tracking existing ruts. It is therefore assumed that he closely watched ruts. For forward visual field, it is assumed that generally operators carefully look ahead because it is visual information essential for vehicle travel. Presumably an operator watches the bucket in order to prevent it from striking any obstructions around the machine, but it is also assumed to be the most conspicuous object in the operator's range of vision and the object his eyes return to when he is not looking at anything else in particular.

During working, the stationary frequency and stationary time are largest for the object moved (64 times, 24.7 seconds), and bucket (45 times, 14.9 seconds). Presumably, during each type of work the operator watches both the object moved and the bucket, to operate the hydraulic excavator while clarifying their relative positions.

Overall the major objects of scrutiny were ruts (98 times, 19.8 seconds), object moved (71 times, 26.3 seconds), and bucket (76 times, 20.9 seconds).

3.5 Stationary point locations from the operator's perspective

The stationary points that were abstracted were plotted to provide an expanded image of the working range from the perspective of an operator inside the cabin (Fig. 4).

From Figure 4, for "traveling", the stationary points are distributed widely within the frame of the front glass, but for "working", the stationary points were concentrated in the center or slightly to the right of the front glass. It is assumed that during traveling, the operator traveled while confirming a relatively wide area of the forward visual field, and that during "working" on the other hand, he mainly scrutinized the bucket and the object of the work, resulting in stationary points from the cabin which is on the left side of hydraulic excavator, being in the center or slightly to the right of the front glass. Overall, most stationary points were contained within the frame of the front glass, but some stationary points were on the left or right side windows. The stationary points on the side windows occurred mainly during rotation or traveling on a curve, so presumably the operator stopped his line of sight to confirm the situation at the rotation tip or ahead on the curve.

3.6 Maximum viewing angle

From Figure 4, the maximum viewing angle of the operator was measured. The operator's visual point was 2,340mm above ground, 220mm forward of the center of rotation and midway between the left and right sides of the cabin (Fig. 5). The measurement results are shown in Table 2. The maximum viewing angle was vertically 56° and horizontally 107°. If this is broken down for "traveling" and "working", the maximum horizontal viewing angle is narrowed, but the maximum vertical viewing angle is almost unchanged.

For working and traveling, there is not very much difference in the visual range results. It is assumed that the operator operates the machine while seated in a fixed posture in the seat during both traveling and working, so the maximum viewing angle is presumably a fixed angle within the range the operator can move his neck and eyeballs during both working and traveling. The posture of operators during operation must be surveyed in the future.

Table 1. Object of scrutiny

No.	Objects of scrutiny	Content
1	Buket	Part of connected to the tip of the working body of the hydraulic exavator
2	Arm	Part of connected to the buket
3	Rotation tip	The direction of turning in the turning motion
4	Object moved	The object to be moved in the model task
5	Setting point	Designated point to unload object in the model task
6	Forward visual field	Forward visual field in traveling
7	Obstruction	Regulation cone of running course, etc.
8	Ruts	Travel traces generated by another traveling
9	Crawler	The tip of the crawler visible from the cabin
10	Others	Worker, Sky, Indistinguishable, etc.

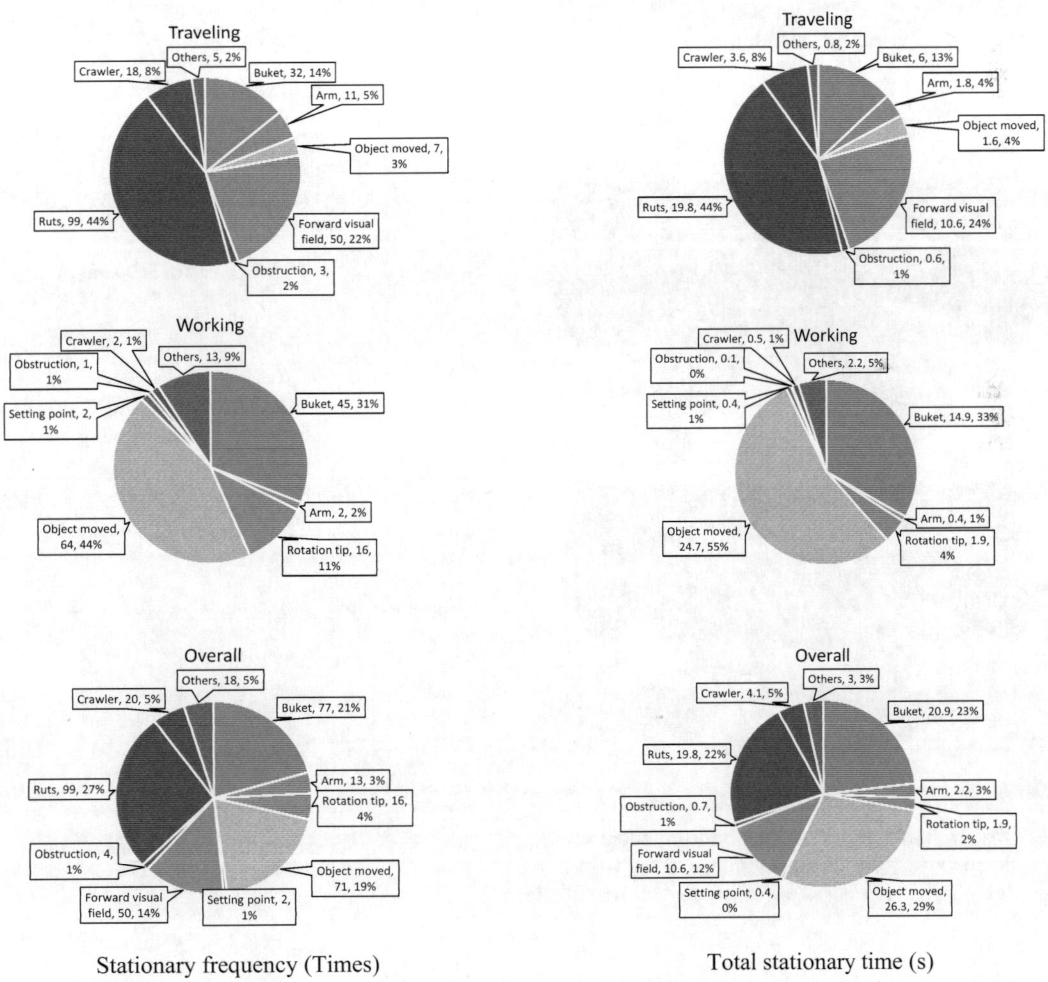

Fig.3 Results of the categorization divided into stationary

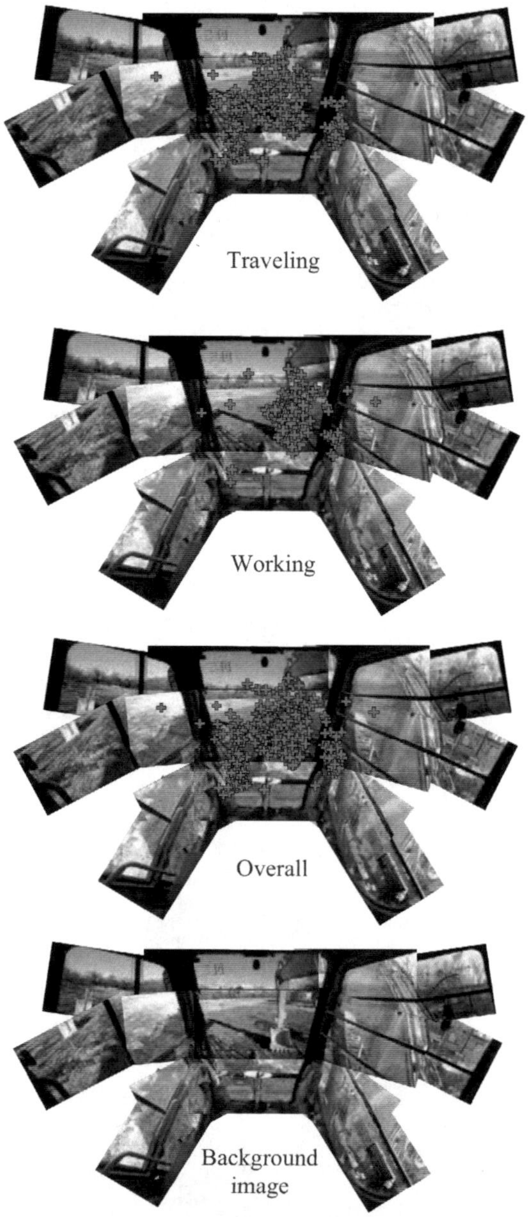

Fig.4 Stationary point location from operator's perspective on an expanded image of the working range from the perspective of an operator inside the cabin

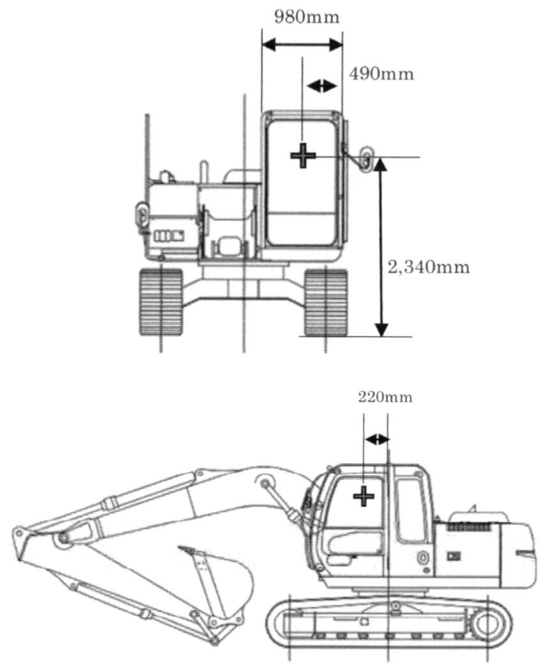

Fig.5 Operator's visual point

Table2 Maximum viewing angle

(degree)

	Vertical direction			Horizontal direction		
		Up	Down		Left	Right
Overall	56	6	50	107	50	57
Traveling	55	5	50	88	50	33
Working	55	6	49	89	32	57

4 Conclusion

To obtain basic data to increase the volume of visual information and improve the operating efficiency of unmanned construction, we measured and studied the range of an operator's line of sight and objects he scrutinized while riding in a construction machine by using an eye-tracking recorder to measure his lines of sight while performing a model task. The results revealed the following facts.

1. The scrutiny duration time was relatively short during traveling and relatively long during working.
2. The major objects of scrutiny during traveling were "ruts", "forward visual field", and "bucket", while the major objects of scrutiny during working were "object moved" and "bucket".
3. During traveling, the stationary points were distributed widely inside the frame of the front glass, but during working, distributed in the center or a little to the right of the front glass.
4. Most of the stationary points were distributed inside the framework of the front glass, but were also distributed in the right and left side windows.
5. The maximum viewing angle was vertically 56° and horizontally 107°.

5 Afterword

This study considered the results of analysis of only one operator, requiring further measurements in the future. In order to build a remote operation system with good general-purpose applicability that contributes to improving the working efficiency of all operators, it will be necessary to measure as many operators as possible. And we want to apply these results to study a camera system for unmanned construction systems that can be deployed promptly enough to fully respond to the situation at disaster sites.

References

[1] Unmanned construction association. Trends and prospects for unmanned construction. *Construction project planning,* 6-12, 2006

Technical Review of Automated System Application to Earthworks in Australia

Nayab Ahmed[a], Abraham JW Hong[b], Hyeonggyun Ku[c], Sungkon Moon[d], and Sungwoo Moon[e]

[a]Department of Civil and Construction Engineering, School of Engineering, c, Melbourne, VIC 3122, Australia
[b]Acciona Infrastructure Australia, Sydney, NSW 2033, Australia
[c]Department of Civil and Environmental Engineering, Pusan National University, Busan 609735, Korea
[d]Department of Civil and Construction Engineering; and Centre for Sustainable Infrastructure, School of Engineering, Swinburne University of Technology, Melbourne, VIC 3122, Australia
[e]Department of Civil and Environmental Engineering, Pusan National University, Busan 609735, Korea (Corresponding Author)

E-mail: nayab.jatoi@gmail.com; abraham_hong@outlook.com; hunet100@naver.com; sungkon.moon@gmail.com; sngwmoon@pusan.ac.kr

Abstract

This paper is a technical review of the smart excavator guidance system being used in Australia, which indicates the automated system. The construction industry contributes towards 8% of the country's GDP (Gross Domestic Product) which makes it one of the largest sectors of the nation's economy. Efficient earthworks have a significant impact on all construction activities in terms of time and cost. Since earthwork activities are repetitive in nature, they offer various opportunities for automation to improve machine utilization and lower the dependence on the operator skill and workload, hence improving the overall construction efficiency. Researchers around the globe have identified these opportunities and introduced various smart machine guidance technologies in order to improve productivity. Construction companies throughout the world, including Australia, have adopted these technologies. However, there is not quantitative understanding of how much these technologies have permeated into the Australian construction industry. This paper will try to address this issue and will try to determine what earthworks market size is, what type of guidance technologies have been adopted by the country and at what scale these technologies have been used. Firstly, this research will try to get an overview of the Australian construction sector. Secondly, an analysis of the previous research projects and inventions will be carried out by looking at various patents of researchers. The literature review will be focus on five broad technological advancements which include the development of machine control systems, laser leveling guidance systems, GPS systems, radio guidance systems display technologies for machine guidance. Thirdly, the adaptation of earthwork technologies in Australia is discussed in the paper regarding the application of construction equipment guidance systems and how these technologies are executed. Research results indicate that the Australian excavation industry is still lagging behind in adopting smart guidance technologies with 64% companies still using laser-levelling systems and 29% companies using advanced GPS systems. State-by-state review also shows similar results with only 15% and 12% of the excavation markets in Victoria and New South Wales adopting GPS-based guidance systems. Data collection will include case studies, phone interviews and internet searches. Finally, all the data will be summarised to get an overview of the adaptation of smart excavator guidance technologies in Australia.

Keywords –Machine Control Systems, Laser Leveling Guidance Systems, GPS Systems, Radio Guidance Systems, 2D & 3D Laser and Radio Positioning Guidance System

1 Introduction

Australia is the 6th largest country in the world in terms of area (United Nations, 2013) with a low population density of 3.3 people per sq. km (Statistics, 2016). It is a culturally diverse country which is politically stable, and most of the country does not have a language barrier (Global, 2015). This country already has an extensive mining, industrial, transportation equipment, food processing, chemical, and steel industry (Limited, 2012). However, the construction sector is one of the most rapidly growing industries in

Australia contributing to 6.8% of the countries' GDP (Stastics, 2010).

The Australian construction industry consists of around 330,000 businesses around the country and employs over one million people (9% of the total Australian workforce) (Team, 2015). This huge demand for construction and infrastructure development is driven by various factors that include growth in population and income levels, rapid changes in technology, inflation, commodity cycles and interest rates. Looking at this context, it is evident the Australian construction industry is going to thrive in the coming years with various large scale projects planned for future development.

Excavation has a significant part in construction apart from earth removal, mining and general earthworks. The main role of excavator in any construction projects is its ability to set the site and dig the foundation. There are various advantages of using excavators including digging deep in a confined area, time saving, cost effectiveness and precision. However, after years of increase in machine size and power, practical limits are now being reached and there is a need for automation for further improvement in productivity and efficiency (Le Quang et al., 2011). Engineers throughout the world realized the practical impact earthworks has on the construction projects in terms of time, cost and quality, hence they have tried to introduce various smart technologies to the improve the efficiency of the excavators (Ebay, 2013).

As excavation and earthmoving activities are repetitive in nature, they offer various opportunities for automation to improve machine utilization and lower the dependence on the operator skill and workload (MRAD et al., 2002). Researchers throughout the world have identified those possibilities and have come up with various smart technologies that would aid excavation activities and improve construction productivity. These technologies include systems for monitoring movement of the excavators, joystick controls for excavators, bucket positioning and control systems, laser guidance systems for grade control, distance measuring systems, laser based grade control systems and GPS-based machine control systems.

Construction industries have adopted these smart technologies around the world and Australia is no different. Earthworks contractors throughout Australia have also been able to adopt these technologies but there is not quantitative understanding of how much these technologies have permeated into the Australian construction industry. This Report will try to address this issue and will try to determine what earthworks market size is, what type of guidance technologies have been adopted by the country and at what scale these technologies have been used.

2 Literature Review

This section initially reviews the previous works regarding smart technologies in excavators. The review will include technologies such as stand-alone machine control systems, laser-based depth control systems, smart displays in excavators, GPS positioning systems and radio and GPS-based machine control systems. In addition, works of major companies will be discussed which offer latest smart excavation technologies. It will provide a general perspective of smart technologies, which are currently being used around the world.

A review of the Australian construction market will be conducted along with the earthworks market in the country. The current status of the construction industry will be discussed, this analysis will include sector-wise review and a state-by-state overview along with future trends of the market. Australian excavation market will also be reviewed to identify the market volume by analyzing different sectors and states in which excavator contractors operate.

The importance of carrying out a review of previous works, Australian construction market, and the excavation market is realized because it will give an outlook of different excavation guidance technologies being used and the overall market size of the excavation industry. It will develop a general scope of this research as to what specific technologies are the most popular and in what ways they can be applied to the construction sector of Australia. Also, it will provide a background knowledge of the information that will be needed to carry out the study. As a starting point, previous works in excavation technologies will be discussed to understand what technologies are being adopted by the construction industry.

2.1 Australian Construction and Excavation Industry Overview

The construction industry is a critical driver of the economy in Australia. Being Australia's third biggest industry, behind just finance and mining, it generates around 8% of country's Gross Domestic Product (GDP). It includes more than 330,000 organizations across the country and employs around 9% of the total workforce. It delivers the infrastructure that is vital to the operation of all different businesses, adds to the riches of the country and supports the infrastructure upgrades that are important to bolster Australia's future wealth and livelihoods. There are various factors that control the demand of construction activities throughout the country.

Construction industry's demand is driven by various factors that include growth in population and income, industry development, innovation, interest rates, commodity cycles and inflation. Other influences include policies that affect the residential and commercial building, infrastructure development and pricing (including taxes, allocation of land and policies development). Moreover, accessibility, cost and area of assets, building equipment and material are other key determinants of the pace of development in the business. These determinants affect all the construction areas of activity.

Australian construction industry functions in both public and private sectors in three broad areas.

- Engineering construction (include infrastructure and industrial projects)
- Non-residential construction (include shops, offices, amenities, etc.)
- Residential construction (include houses, townhouses, and flats, etc.)

Earthworks and earthmoving also forms a significant part of the areas mentioned above of activity of construction industry. According to Yellowpages Australia, there are around 15230 excavator contractor and hire companies that support the construction industry's earthwork activities.

According to construction industry trends and forecast report 2015-2016, on a macroeconomic level, indications are strong that the global economy will grow over the period of next two years and the economic growth levels will reach closer the levels that were seen in the pre-global financial crisis levels. There is a strong likelihood that the commodity prices will fall as gold and iron ore prices are currently decreasing which will impact Australia's resource market. However, the Australian construction sector will see growth due to significant investments in the residential sector that are mainly driven by a substantial growth in foreign investments, especially in the property sector.

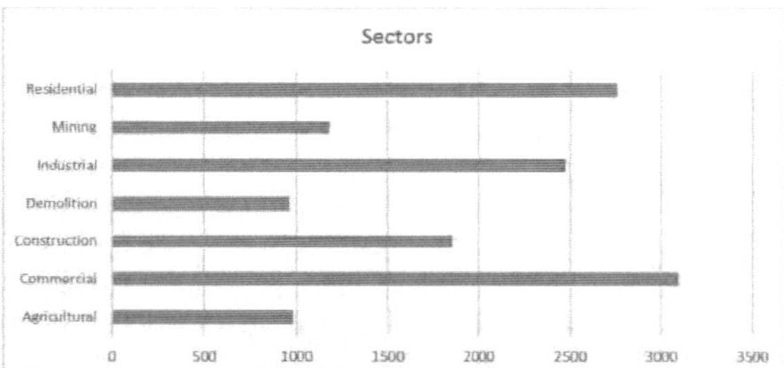

Figure 1. Excavation Companies according to Sectors

Figure 1 illustrates the excavation market scale in terms of different sectors. It is evident that residential and commercial sectors have the biggest excavation market. This is because of growth in these two sectors thought out Australia.

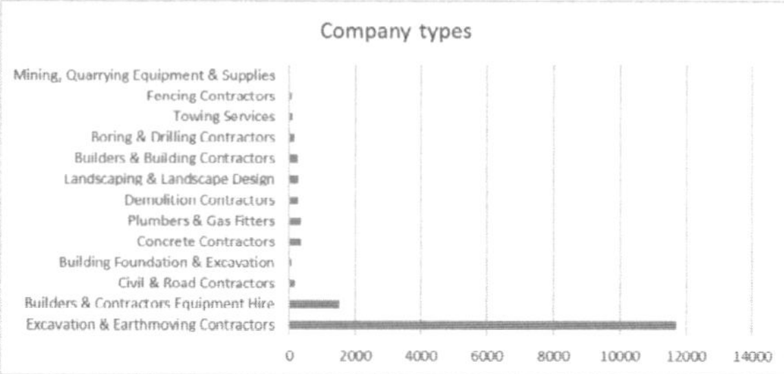

Figure 2. Excavation company types

Figure 2 illustrates that most of the excavation companies are independent excavation and earthmoving contractors. The type of smart technologies that these contractors use depends on the scale of their company and the type of construction contracts they take. Moreover, different states in the country have different market conditions.

A report by Partnership (2015) outlines the market conditions of states within Australia. According to the report, there will be a significant amount of reduction of construction sector growth in Queensland and Western Australia. This is due to the fact that the mining and resource industry in these states have suffered. On the other hand, New South Wales will see a steady amount of growth in the coming 2 years. Reason being that this state has the most number of construction projects evenly spread across residential, non-residential and engineering sub-sectors, and there is very less reliance on a single subsector. South Australia's construction industry growth will remain stagnant, as many of the major projects have been delayed. Similarly, Tasmania will see an increase in construction of aged care facilities. However, forecasts show that there will be slow growth in this region. New South Wales, Queensland, and Vitoria, being the primary construction states, will see a significant amount of growth in the next 5 years, mostly due to growth in the residential sector. The excavation markets in different states also show similar trends.

3. Problem Statement and Research Objectives

Review of the smart technologies illustrated that excavation guidance technologies can be narrowed down to two broad categories, i.e., laser leveling technologies and GPS-based guidance technologies that are being used in the excavators throughout the world. The Australian construction industry is one of the biggest industries in the country. Forecasts show that this sector is going to grow even more in the coming years. Earthworks sector is closely related to the construction industry with around 15230 companies operating throughout the country. However, there is no quantitative understanding of how much the Australian market has been able to adopt these smart technologies. A quantifiable review of the adoption of smart technologies will not only allow better strategy making for the Australian construction market but also provide a way forward to incorporate these smart technologies in the earthworks industry.

The main objectives of this research is to identify the types of technologies that are being used in the Australian excavation industry and gather a quantifiable amount of the status of the adoption of smart excavation guidance technologies. The sub-objectives include providing an understanding of the trend of adoption of smart technologies all over the country and state-by-state. In order to achieve these objectives a viable research method will have to be proposed.

4. Research Method

This research method involved, initially, a review of the previous works in excavation technologies. This analysis included various patents, inventions and researches. This investigation was done to identify key guidance technologies that are being used in excavators throughout the world.

Once the key technologies are identified, an analysis of Australian construction and excavation market was conducted. This examination was done to understand the market scale and to provide a better picture of the volume of the industry and the amount of earthworks that is being carried out throughout the country. This was done by investigating various industry forecasts and conducting a market survey of 15230 excavation companies state wise, activity wise and sector wise.

After the key technologies and the scale of the excavation market was identified. A research was conducted on how many companies have adopted these smart excavation technologies. Data collection strategy involved visiting sites for case study, internet search and phone interviews.

For case studies, various excavation sites were identified from Victoria and Western Australia and an inquiry was made on the following questions:

- What excavator are you using?
- What type of building it is?
- What smart technology are you using in your excavators?
- How is it used?
- Would you change to a better technology?
- What other technologies do you use?

After the inquiry was made, photos were taken of the site and the excavation technologies being used to understand how they work.

Apart from case studies, data was collected from the internet and phone interviews. For internet search, data was collected using search engines with keywords "excavator laser", "excavator GPS", "excavator Trimble", "excavator TopCon" and "excavator machine control" were used with search parameters within Australia. Data collected from the internet also included searching from various journals, magazines and YouTube videos. For phone interviews, calls were made to various excavation companies and enquiry was made regarding what smart technology was being used in their excavation services. The inquiry was done by asking the same questions which were used during the case studies. Finally, all the data was collected from 162 sources in a spread sheet with the name of the company, contact details, technology used and the state in which they were operating. The data which was collected on the spread sheet was then interpreted into a table to gather the results in table format.

5. Investigation of Excavation Technologies Used in Australia

5.1 Investigation of Technologies from Web-Searching

Data was collected from the internet from 74 various sources. Investigation was made on what type of guidance technologies are being used by various companies. This internet search included identifying all the excavator companies working in Australia. Then company's websites were accessed individually to find out what type of technologies they offer in their excavation services. Data collected from websites is summarized in a Table 1 and Figure 3, which show the number companies that are using either laser or GPS based guidance technologies.

From the data gathered from the internet. It is evident that laser-levelling technology is the most prevalent technology in the Australian industry with 58% of the total technologies; there were a variety of laser levelling technologies that were being used. In some cases, laser levelling was done manually, in other cases laser receivers were mounted on the boom of the excavators. GPS based guidance technologies are slowly being adopted by the excavator companies with TopCon and Trimble as the most preferred brands. These brands offer various types of GPS based machine control systems with small differences in functionality. Moreover, advanced technologies such as remote excavator monitoring technologies, hybrid excavators and tilt rotators are still relatively new to the Australian market but these technologies are slowly getting popular. Data for a number of companies was gathered from the internet; those companies that did not have any information regarding technologies were contacted individually in order to further investigate.

Table 1 Summarized Data from the Internet

# of companies	GPS	Laser	Remote Monitoring	Hybrid Excavators	Tilt Rotators
74	23	43	3	4	1

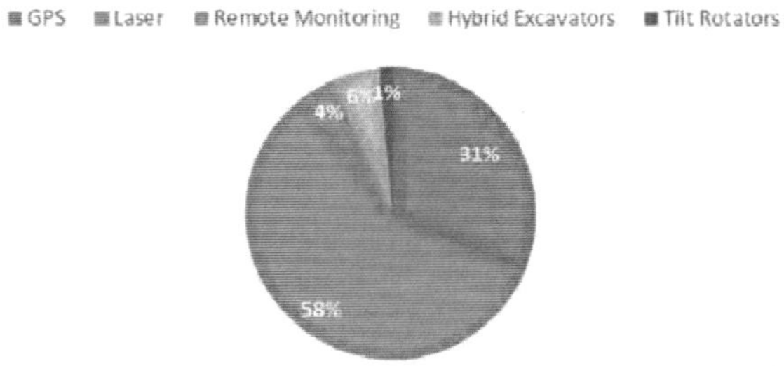

Figure 3. Pie Chart showing technologies used in terms of percentage

5.2 Investigation from Interviews

Phone interview were conducted to fill the gaps that were left by the internet search as there was a number of companies which did not provide information regarding technologies that were used in excavation. Data was gathered from 56 different companies and an inquiry was made regarding the technologies. This data is then summarized in Table 2 and Figure 4 to provide a picture of what type of technologies are being used currently.

Table 2 Summarized Data from Phone Interviews

# of companies	GPS	Laser	2D Grade Control
56	12	43	1

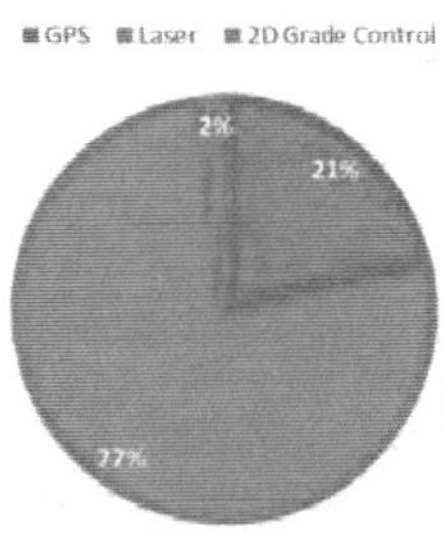

Figure 4. Chart Showing Technologies used in terms of Percentage

Data from the phone interviews also shows that there is a large number of companies using laser-levelling technologies. During the phone interviews, it was found out that most of the companies preferred TopCon and Trimble over any other brand as these two brands offer the best technologies. Moreover, 21% of the companies have started using advance GPS technologies and informed that their productivity increased by using such tools. Some companies were also using 2D grade control systems, which is an integration of laser levelling, machine control and display technologies. The overall trend, here, showed that most of the excavator users have not yet adopted smarter GPS-based machine control technologies which may change overtime as these companies get more exposure to these technologies.

5.3 Investigation through Case Studies

This section will discuss some case studies that were conducted during the research. These case studies were conducted in Victoria and Western Australia.

5.3.1 Excavation for a Small Residential Project

Initial inquiry was made by identifying excavation site within Victoria. Site selection was done on the basis that there would be an active excavation activity happening on the site. The first site that was chosen for case study was located at 77 Springvale Road, Nunawading VIC. It was a ground +2 apartment building construction site. Only one Zaxis 130k excavator shown in Figure 5 did excavation.

In terms of excavator technology, the excavator had basic hydraulic machine control system to keep the bucket parallel to the digging plane. Moreover, this excavator housed a LCD monitor with rotary dial to display and select work modes to do machine diagnostics, monitor maintenance intervals and make hydraulic flow adjustments as shown in Figure 5.

As it was a small scale and relatively simple excavation activity, laser-levelling technology was used to conduct the digging operations. TopCon laser levelling transmitter and receiver were used. The site supervisor initially marked the boundaries using pegs and tapeline. There was a laser transmitter on one end and a receiver on the other end. The receiver was attached to a straight metal rod with a scale. The excavator operator was given a site plan with level markings on it. The operator would move to different points on the site and adjust the height of the receiver. When the right height of the receiver is achieved, then it would beep which would tell the operator how much height to cut or fill. He would then mark the height and excavate accordingly.

The excavator LCD panel only had basic controls related to the excavator. When compared to the latest LCD panel control technologies. The laser levelling system was inefficient as well as compared to the current grade control systems, which are integrated with the LCD panels in the excavators. The operator had to stop the excavator and then mark the levels that increased the excavation time considerably. The investigation for excavation technology was carried out on a similar project but with a larger scale.

Figure 5. Excavator in Stationary Position; Cabin with LCD Control Panel; TopCon Laser Trasmitter

5.3.2 Excavation for a Medium Sized Residential Project

This project was a larger scale project as compared to the last case study. The location of this site was 555 Glenferrie Road, VIC. The contractor for this project was Lloyd Group Pty Ltd. It was a residential townhouse project with ground +2 construction. The excavator used here was bigger and the levelling technology used was similar but a bit different in its working.

They were using a CASE CX300C excavator. The excavator was fitted with a LCD panel that provided basic excavation controls. The excavator had a relatively advanced hydraulic system that saved fuel by identifying the work cycle activities.

Regarding smart technologies, GPS-guidance based excavator was used only during the site set-up phase. For marking levels, TopCon laser transmitters and receivers were used shown in Figure 6. The receiver was mounted on the excavator arm that was visible to the operator. The levelling details were put into the receiver control panel. When the excavation was conducted, the receiver would turn green when the digging level was within the allowable depth limit. The control panel would change color and indicate to the operator when the grade level was achieved.

This laser levelling system was more efficient as compared to the system discussed in the previous case study, as it only required one person to conduct the excavation activity who was not required to stop the excavator and mark the levels from time to time. However, as compared to GPS-based guidance technologies this system was still less efficient. With GPS-based technology, the operator only has to input the site data once, but in this case, the operator was still dependent on his skills to conduct the excavation operations. The excavator contractor was aware of advanced guidance technologies but depending on the excavation size, they preferred to use laser-levelling technology. Investigation was carried out from Western Australia on excavation projects of different scales.

Figure 6. TopCon Laser transmitter with string line in view

An investigation regarding use of smart technologies in excavators was carried out from a project in Western Australia. The project name was Roy Hill Iron Ore project. This project involved various defective works along the rail line. The investigation was done from works at the Hillside Woodstock level crossing, Roy Hill borrow pit, and works at chainage 75, 146.1, and 323.2 km respectively.

5.3.3 Repair Work at Hillside Woodstock Road

At Hillside Woodstock Road (chainage 146.1km), a small repair project to rectify damaged approach seals and pavements for level crossing. The works included:

1. Saw cut and removing existing soil.
2. Removing base course
3. Applying Bitumen tape to T beam.
4. Asphalt resurfacing and line marking.

A small Hitachi Zaxis 30u excavator was used for excavation works. Two joysticks controlled the excavator's hydraulic system. The bucket used had a capacity of 0.09 cubic meters Tons and the excavator could dig up to 3.19 meters. As the excavation activity involved removal of the top 150mm of the exposed base course. The choice of excavator had to be around 3-5 Tons. Moreover, levelling was done by using laser transmitters and receivers. Figure 7 shows that the workers are measuring levels using a levelling stick. Laser levelling technology is cost-effective for such types of small works, however, more advanced level of technologies might be more feasible for complex digging activities where the whole site has to be prepared.

Figure 7. Worker Measuring Levels with a Levelling Stick

5.3.4 Excavation Works in a Borrow Pit

Another case where laser levelling was used was during excavation works at SCT Rail where rectification works were carried out in a borrow pit at chainage 75 km. The excavation works involved:

1. Site setup
2. Movement of materials
3. Rectification works e.g. drainage works
4. Survey; and
5. Demobilization/ rehabilitation.

Before the work commenced, the operator and the supervisor had to complete a walkthrough of the site in order to be aware of the work boundaries. Afterwards, the operator had to commence work of removing materials from the current drain into adjacent existing creek. The material to be removed would allow the water to drain into the existing creek.

This activity was carried out by a Volvo Vostok digger with a load of 34.3 - 38.7 Tonnes and a small Hitachi ZX30u excavator shown in Figure 8. The Volvo excavator housed an advanced screen that provided information regarding the excavator components. The control panel in the screen allowed the operator to monitor and activate machine functions in a convenient way. Moreover, this excavator came with a Roll Over Protection System (ROPS) which would keep the excavator safe at steeper angles. However, the Hitachi excavator was a standard small excavator with basic excavation functions. Regarding the levelling system, a laser system was used in this case. The excavator operator was assisted by another worker who attached the laser receiver on a shovel to check grade level and marked the boundaries and regularly. In the same project, at chainage 323.2 km, defective works were being carried out where the same excavator (Volvo Vostok Digger) was being used with the same laser levelling technology. After the investigations from the internet, phone interviews and case studies was done, all the data was summarized to discuss the findings.

Figure 8. Operator Assisted, Laser Receiver mounted on Shovel

6. Summarised Research Findings

Data was gathered from 162 excavator contracting companies which use guidance technologies for excavation throughout Australia. This data is summarized in the table below.

Table 3 Summarized Data from the Internet

# of companies	GPS + Laser	Laser	Remote Monitoring	Hybrid Excavators	Tilt Rotators	2D Grade Control
Australia						
128	37	82	3	4	1	2
New South Wales						
17	10	2	3	2	0	0
Queensland						
11	10	1	0	0	0	0
Victoria						
100	15	79	0	2	1	2
Western Australia						
	2					

There were various equipment guidance technologies being used in which laser and GPS technologies were the main technologies that were used in the excavators. The most popular brand for equipment guidance technology were Trimble and Topcon technologies. There was a variety of models in equipment with small upgrades, however the basic principal behind the equipment was based on either GPS or Laser sensing technology as discussed in the literature review. Moreover, it was found out that almost all of the contractors used laser-levelling technology for excavation; however, earthworks on large-scale projects were done using GPS guidance technologies. In some of the excavators, the laser sensors were fitted in the boom to let the excavator operator know the grade levels, in others, levelling was done manually. On smaller construction sites, the operator himself did laser levelling, but on bigger site, a worker assisted the operator.

7. Conclusion and Discussion

Overall, this research aimed at quantifying the status of smart technologies that are being adopted by the Australian construction industry. Data was collected from 128 sources that included site visits, phone interviews, and web search. The summarized results showed that 64% of the total companies used laser guided excavation technology. However, 29% of the companies have started to use advanced GPS-based excavation technologies that show that there is a trend of companies adopting advanced technologies. A small fraction of companies was using technologies such as 2D grade control, tilt rotator control, hybrid excavators and remote monitoring. The usage of technologies depended much on the type of excavation services the excavation companies were providing. Specialized big companies offered state-of-the-art technologies with advanced GPS-guidance machine control system, whereas smaller companies only used laser-levelling technologies. State by state analysis showed similar trends with 15% and 12% of the companies using advanced GPS-based technologies in Victoria and New South Wales respectively with other states also starting to use these technologies. This research was limited as it narrowed down the technologies in two major guidance technologies. If these categories are researched individually and then a collective report is made; it would give a better understanding of the status of guidance technologies used in excavators in Australia.

Acknowledgement

This research was supported by a grant (16SCIP-B07969003) from Smart Civil Infrastructure Research Program funded by Ministry of Land, Infrastructure and Transport (MOLIT) of Korea government and Korea Agency for Infrastructure Technology Advancement (KAIA).

References

[1] AKE, D. W. K. 2001. Laser receiver and angle sensor mounted on an excavator. Google Patents.

[2] BACHMANN, E. H. & INGVARSSON, K. 1985. Depth monitoring system. Google Patents.

[3] BRABEC, V. J., CRUMB, B. B. & GRAHAM, R. T. 1999. Bucket depth and angle controller for excavator. Google Patents.

[4] CARPENTER, H. J., PIEKUTOWSKI, R. P. & NICHOLS, M. 2014. Method and system for controlling an excavator. Google Patents.

[5] DAVIDSON, R. W. & BRABEC, V. J. 1998. Method for controlling an excavator. Google Patents.

[6] EBAY 2013. Construction Excavators and Their Role in Large-Scale Building Projects.

[7] GLOBAL, C. 2015. *Australia Guide* [Online]. Available: http://www.commisceo-global.com/country-guides/australia-guide [Accessed 12/11/2016].

[8] INUI, T., OOTSUKA, K., NOGAMI, S., IGARASHI, M., HORIKOSHI, T. & IZUMI, K. 1983. Semi-automatic hydraulic excavator. Google Patents.

[9] LE QUANG, H., SE, J. C. & SUN, K. H. Study on modeling and control of excavator. Proceedings of the 28th International Symposium on Automation and Robotics in Construction. Seoul: IIT Madras, 2011. 969-974.

[10] LIMITED, C. I. 2012. *Australian Industry* [Online]. Available: http://www.australiaonnet.com/economy-business/industries/ [Accessed].

[11] MRAD, F., ABDUL-MALAK, M. A., SADEK, S. & KHUDR, Z. 2002. Automated excavation in construction using robotics trajectory and envelop generation. *Engineering, Construction and Architectural Management,* 9, 325-335.

[12] NICHOLS, M. E. 2008. Excavator 3D integrated laser and radio positioning guidance system. Google Patents.

[13] PARTNERSHIP, W. 2015. Interim Review of Australian Construction Market Conditions.

[14] ROCKE, D. J. 1995. Automatic excavation control system and method. Google Patents.

[15] SAHM, W. C. & KOEHRSEN, C. L. 2006. Positioning system for an excavating work machine. Google Patents.

[16] STASTICS, A. B. O. 2010. *Australian Economic Indicators* [Online]. Available: http://www.abs.gov.au/AUSSTATS/abs@.nsf/Lookup/1350.0Feature+Article1Oct+2010 [Accessed].

[17] Statistics, A. B. O. 2016. Available: http://www.abs.gov.au/ausstats/abs@.nsf/94713ad445ff1425ca25682000192af2/1647509ef7e25faaca2568a900154b63?OpenDocument [Accessed].

[18] TEACH, T. L. 1978. Automatic control of backhoe digging depth. Google Patents.

[19] TEAM, A. G. E. A. R. 2015. Australia's Construction Industry: Profile and Outlook. July 2015. *Ai Group, Economic Research*.

[20] UNITED NATIONS, D. O. E. A. S. A. 2013. *World Population Prospects: The 2012 Revision.* [Online]. [Accessed 11/12/2016].

34th International Symposium
on Automation and Robotics
in Construction

ISARC 2017

June 28 – July 1
TAIPEI, TAIWAN

Internet of Things (IoT) for Integrating Environmental and Localization Data in Building Information Modeling (BIM)

Jochen Teizer [a], Mario Wolf [b], Olga Golovina [a], Manuel Perschewski [a], Markus Propach [b], Matthias Neges [b], Markus König [a]

[a] Chair of Computing in Engineering, Ruhr-University Bochum, Germany
[b] Chair for IT in Mechanical Engineering, Ruhr University of Bochum, Germany
E-Mail: jochen.teizer@rub.de, mario.wolf@itm.rub.de, olga.mgsu@gmail.com, manuel.perschewski@rub.de, markus.propach@rub.de, matthias.neges@itm.rub.de, koenig@inf.bi.rub.de

Abstract –

Digital transformation is an ongoing challenge in construction. Whereas central storage and planning with Building Information Modeling (BIM) can be considered state-of-the-art, the integration of real-time data like environmental and localization data of workers in indoor work environments can provide further benefits for operations management in construction and facility management. This paper introduces the concept of permanent availability of up-to-date actual performance data sets through an Internet-of-Things (IoT) approach that integrates environmental and localization data in a cloud-based BIM platform. In this paper, we reflect on the usage of Internet of Things (IoT) technology and the lean and injury-free (LIFE) construction management approach, create a concept to implement the topics in existing systems, design and create a prototypical application, validate the prototype in field-typical work settings, and critically review the results.

Keywords –

Building Information Modeling (BIM), Bluetooth Low Energy (BLE), cloud computing, data visualization, facility management and operation, Internet of Things (IoT), Lean Construction Management (LCM), location tracking, real-time sensor data.

1 Introduction

The vision of Internet of Things (IoT) is to change the world as we know it. To name a few examples in construction applications, machines will be talking to each other to optimize production and assembly of materials, project coordination will be based on informed decision making, and smart systems will accurately predict and prevent risk of many kinds currently embedded in the harsh construction site work environment. Most construction enterprises, however, are still stuck with business processes that are as far removed from the IoT as paper based communication is from using a Fax or, more recently, E-Mail.

For years now, visionary construction companies have been connecting things, but if closely looked, they resemble nothing so much as digital islands. Often individual departments tackle and perform a specific task, but they do not make sure the information created is available everywhere in the organization. The result is a large disconnect, making operations inefficient at best, costly or life threatening at worst.

While the Building Information Modeling (BIM) method intends on supporting integrated or interdisciplinary delivery of construction projects, there generally is no connection between most of the architectural or engineering firms during the project design or planning phase with manufacturing or construction companies during the project execution phase. Furthermore, seamlessly connecting the early project phases with the operation and management (O&M) phase is important considering the true lifecycle costs of a project. An estimated 80 % of a building's cost occur during the O&M phase.

For these and more reasons, lean management in construction was examined. Koskela [1] developed the Transformation-Flow-Value (TVF) - Theory, saying that construction can be conceptualized with the transformation of resources and the creation of value and flow of materials and people. Ever since then, several lean production control theories have been developed for the construction sector [2]. Although they eventually developed to practical methods, tools for efficient (automated) data collection on construction sites and (near) real-time analysis are still missing.

It is therefore humans that enter field data in machines (e.g., tablets) that then process for most parts autonomously data to information. Humans again interpret the results and transfer information to valuable knowledge.

Proper construction project information management,

from data generation to knowledge availability and sharing, is an essential key to the successful implementation of production controlling methods. Though a constant and reliable flow of information to assess work progress, constraints, and productivity is required.

The proposed concept envisions the support of lean and injury-free (LIFE) construction processes while facilitating higher task quality, on time project delivery, and safe value creation processes. It dates back to Teizer er al. [3] when this vision was first articulated in the proceedings of the 24th ISARC. This paper (1) reflects on the usage of Internet of Things (IoT) technology and lean management approaches for the building construction and facility management industry, (2) introduces a concept to implement both topics in existing systems, (3) implements a prototypical application, (4) validates the prototype in realistic indoor work settings, and (5) critically reviews the results.

2 Background

Aside a review on IoT-related technology and applications in construction, this review concentrates on the two use cases: (1) current practices for planning and assessing occupational safety and health in indoor work environments and (2) tracking trades to measure work progress in interior finishing tasks of building projects.

2.1 Construction safety and health planning and auditing

Construction safety and health planning, monitoring, and enforcement relies mostly on human experts applying knowledge on an as-needs basis. Most of the current safety-in-design (SID) review practices for general construction projects identify and mark the potential hazards on two-dimensional (2D) drawings [4]. Requirements for lighting, for example to carry out work at proper illumination levels, follow the safety and health standards set by the Occupational Safety and Health Administration [5] While such rules exist in most countries around the world, they differ minimally (Table 1).

Current practices for planning safe work execution during construction visualize the required equipment based on expert opinion and experience. Figure 1 illustrates a result for planning the floorplan of a building project with essential lighting equipment. The dots indicate the lighting spots fixed at the ceilings of each room. The size of the lines indicate the expected brightness level according to how frequently the hallways are traversed.

Table 1. Required minimum illumination levels by area of operation [1 foot-candle (USA) = 10.7639 lux (International)] [5,6]

Area of operation (Examples)	Luminous flux [lux]	
	OSHA (USA)	BG Bau (Germany)
General areas	32	--
Corridors, exits, stairs, and walkways	53	--
Landside work areas	53	20-30
Tunnels and underground work areas	53	60
Social rooms (eating, changing, and restrooms),	107	100-200
First aid stations	323	

Figure 1. Worksite lighting assessment process (example: manual drawing) [modified, 7]

The manual planning tasks require a certified engineer's or competent personnel's time. Although human errors in the planning phase can occur and safety plans are less often (if at all) updated once construction is in underway, work spaces are still required to be frequently checked upon. In essence, current construction safety planning and monitoring processes in many construction organizations apply the minimum but complex legal framework of rules and regulations. For this reason, many construction companies have shifted to self-regulated (stricter) safety best practices to ensure that work environments are kept safe. In this task though, the personnel is mostly using manual measurement devices and reporting.

Previous studies show that (a) the safety and health process remains a key issue in construction and (b) there is limited focus on the current status of construction sites due to the lack of suitable controlling and tracking functions in existing information management methodologies. The collection of actual data (e.g., safety and health or progress) is a crucial step towards informed management systems and serves as a prerequisite for further successful production planning and controlling. While current progress on projects is often compiled manually which is very time consuming and prone to human error, it leads to overall lower product quality and decreases the chances for successful risk mitigation. There is a need to compile and integrate tracking and forecasting information as feedback and input for successful identification, preparation, and execution of single working steps.

2.2 Digital progress tracking for Lean Construction Management (LCM)

Construction research has been increasingly focusing on discovering synergies between the adoption of lean and injury-free practices and information and sensing technologies [8]. Information and communication technologies (ICT) are in particular beneficial to lean and injury-free practices when they improve the flow of construction processes by identifying non-value or risk adding activities that can be eliminated. Other examples are cycle-times that can be shortened, rework, variation and errors that can be omitted [9].

Lean management and the adaption of technology is not new to construction. Several practical field applications exist, for example, Radio Frequency Identification (RFID) for pipe spool tracking [10], Global Navigation Satellite System (GNSS) for earth hauling operations [11], and wireless Real-time Location Sensing (RTLS) for tracking repetitive travel patterns of workers [12]. Even commercial solutions exist that allow personnel in the field to carry out inspections, for example, progress and quality reporting, and other documentation tasks.

The digital transformation remains an ongoing challenge in construction and facility management applications (see an example of current methods in Figure 2). While centralized cloud-based data storage and planning with BIM can be considered state-of-the-art, research by Sacks et al. [9] and Cheng et al. [13] outlines that much stronger ties between lean, safety and health, BIM, and digital technology are needed.

Figure 2. In-field manual trajectory planning for maintenance tasks relies on paper-printed building layouts

The focus of the proposed concept (Figure 3) is on tightening the planning, execution and control processes by planning digitally first (as-planned conditions), executing projects with mobile information at hand in real-time (in-field conditions), and reporting actual data on progress and quality (as-is or actual conditions). The methods for planning, supplying, and reporting data are based on building information modeling software, mobile field communication devices, and cloud-based information management solutions. Most of which exist commercially, but yet have to be bundled together in effective manner for practical use. These enable rather than reduce the capacity of construction personnel to understand, collect, and communicate high quality information quickly. The potential to make high fidelity information available that previously has neither been recorded nor analyzed has to be evaluated in studies.

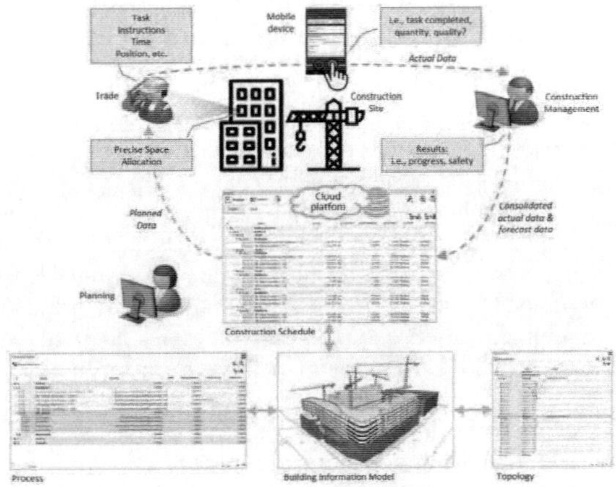

Figure 3. Overview of the proposed IoT system architecture: Digital project planning, cloud-based mobile field instructions, and actual as-is data recording and feedback processing [2,14]

2.3 Internet of Things (IoT) approach to connect physical work environments with digital information

The IoT foresees a future in which digital and physical things or objects (e.g., mobile devices, machines, and tools) can be connected by means of suitable information and communication technologies, to enable a range of applications and services (Figure 4) [15]. Challenging characteristics of IoT in construction involve suitable and scalable (mostly wireless) sensor networks and devices with robust and secure architectures, open data formats and interfaces, and dependencies that are operable in the harsh and dynamic work environment. They need to tackle rather large number of spontaneous events, creating and sharing knowledge from big data, while maintaining the privacy of humans [16]. Economies in many countries, e.g. in Germany, have established the term "Industry 4.0" referring to fourth industrial revolution (after mechanization, mass production, computer and automation, cyber physical systems) [17].

Figure 4. Definition of IoT [15].

3 Method, Experiments, and Results

While Figure 3 shows already the architecture of the authors' approach and the associated components, this section focuses on the technical description of the developed IoT system.

The authors implemented the proposed concept with components regularly used in research laboratories. Each "worker" (test subject) in the validation is equipped with sensor-enhanced personal protective equipment (PPE). In our case, sensors mounted on a hard hat collect environmental data (e.g., illumination levels) from the surrounding work environment and associate the location (e.g., ID of the work station) to each reading autonomously (Figure 5).

Figure 5. BLE beacons (left), BLE sensor on hard hat (middle), and mobile app (right)

The authors enable the collection of relevant data with Bluetooth Low Energy (BLE) based sensory equipment and the internet capability of smart devices (i.e., smartphones). Specifically constructed 3D-printed casings housed Avnet's Visible Things Bluetooth Smart Sensors [18]. These sensors allow the collection of various data sets related to environmental conditions, for example, temperature, humidity, and lighting.

The casings were mounted to the side plates of regular hard hats. It ensures the structural integrity of the hard and allows the casing to be used as an optional and easy-to replace add-on feature like it is already very common for the installation or the removal of face or hearing protection devices. The authors implemented a simple location tracking based on BLE beacons (Accent Systems iBKS105) utilizing the EddyStone Protocol and RSSI measurements for distance estimation. Details to the robustness of BLE location tracking can be found in Neges et al. [19].

Combining the traceability of the worker's location (case study 1) and the environmental data (case study 2) enables the IoT-platform to monitor key safety-related figures like time spent in low or high temperatures or the lack of adequate lighting at a work station. For example, work stations in the basement of a building under construction generally offer one of the poorest working conditions throughout a construction project. Accidents in such spaces lead to high number of minor injuries (e.g., spraining an ankle) that cause high costs to the industry (i.e., due to the absence of workers from work for a few days). While such instances also lower the productivity, such costs must be avoided through novel methods, such as rule checking during safety-in-design reviews as well as real-time site monitoring [4].

One of the authors' objectives was to provide the workers with easy-to-use tools for detecting whether a work station is safe or unsafe to work in. The first case study located the workers in the work stations assigned via Building Information models (BIM). For the mobile implementation of the Android 6.0.1 application, the test subjects used their private cell phones as a communication center for both the data capturing and the relaying of the data to the cloud platform.

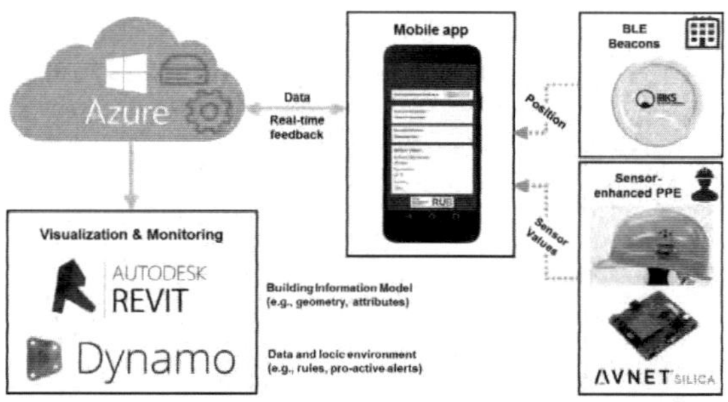

Figure 6. IoT including sensors for data collection, cloud platform data storage, processing, and communication

The backend for the storage of said data was tested on several IoT platforms. For example, the first use case on location tracking of trades used the IoT-platform of PTC Thingworx. The second use case on construction safety and health applications used Microsoft's IoT-platform Azure. Connected to Autodesk Revit with the programming interface of Dynamo to process real-time data (Figure 6), the Azure cloud only saves refined data, usable for further analysis.

3.1 Case study 1: Tracking trades by room

The first preliminary test of the developed IoT concept focused on tracking several trades in a building construction project. Assigning and coordinating the right work station, the times to begin and end work, etc. for a multitude of trades that are concurrently present inside a building project is critical to achieve a continuous production flow. Enabling methodologies like BIM and LCM improve project understanding by visualization the bottlenecks between tasks, the requirements of the processes and their dependencies.

The aim of the first case study was to provide a continuous progress monitoring function (Figure 3). The particular focus was set on testing the feasibility of the developed IoT approach for tracking the location of trades assigned to interior finishing tasks.

The proposed approach used an nD-BIM for planning the project topology (i.e., work station/location) of resource-loaded processes (i.e., name, dependencies, quantity, cost, required resources) [14]. It linked the geometric information to an automatically derived construction schedule (i.e., duration, quantity, trade). The IoT-platform relying on wireless location tracking and reporting technology (e.g., Bluetooth Low Energy (BLE) sensors, mobile devices and a cloud database made the information of directives, prerequisite work, and resources available to authorized users.

To that extent, the authors enabled the collection of relevant data with IoT-functional equipment, stored the data in an IoT-platform and connected it to a BIM system to seamlessly integrate real-time data. The use case tested a lightweight infrastructure solution for indoor tracking of personnel of three trades.

The validation had the goal to avoid spatial overlap of the trades. BLE beacons installed at the work stations (mounted inside the rooms at a wall, ceiling, or door frame) transmit their signals as soon as a worker enters the work station. The smartphone enables the connection to the IoT cloud-based platform for further data processing. Time recordings per second or higher become feasible. Knowing the presence of workers at a lower frequency might offer an opportunity to comply with potential privacy regulations that are present in many countries. The same technology works for asset tracking, including tools and materials. In such a use cases, smaller BLE beacons are installed or embedded inside the tools' casing or material, respectively.

Combining the traceability of the personnel's location and timestamp enabled the IoT/BIM-platform to collect and visualize actual presence and duration of trades by work station (Figure 7). As ethical concerns are important to construction organizations and the workers themselves, tracking what tasks the trades were performing while being at the work station is not possible (unless further technology is implemented) and subsequently was not monitored.

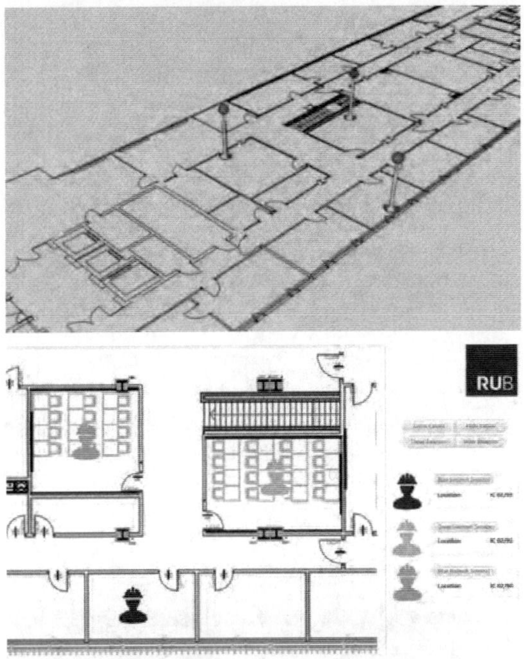

Figure 7. BLE-beacon positions planned in BIM (top) and real-time work station occupancy visualization by trade (bottom)

3.2 Case study 2: IoT for personalized real-time construction safety and health audits

A second preliminary use case to test the developed IoT concept focused on measuring the brightness level at work stations. The experimental setup was similar to the first case study. In addition, the authors designed a Dynamo script (Figure 8) that allowed to visualize the results in real-time in Autodesk Revit (Figure 9). Furthermore, the authors created a mobile application that enables a worker to generate personalized feedback in real-time. The user interface displays the lumens flux value and indicates in red color whether it is safe to remain at the work station (Figure 10).

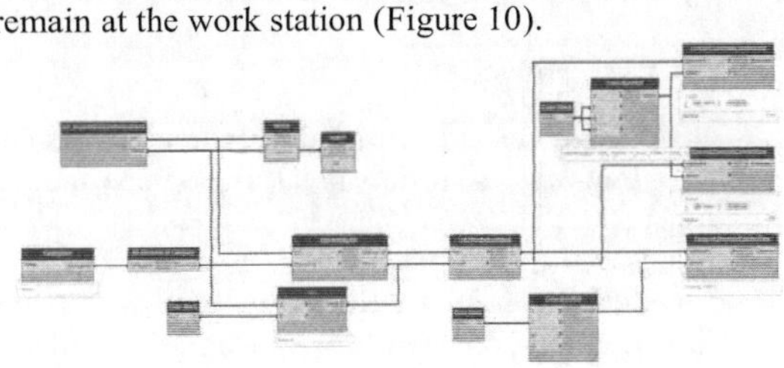

Figure 8. Dynamo script to transfer the results from the IoT-platform and visualize it in Autodesk Revit

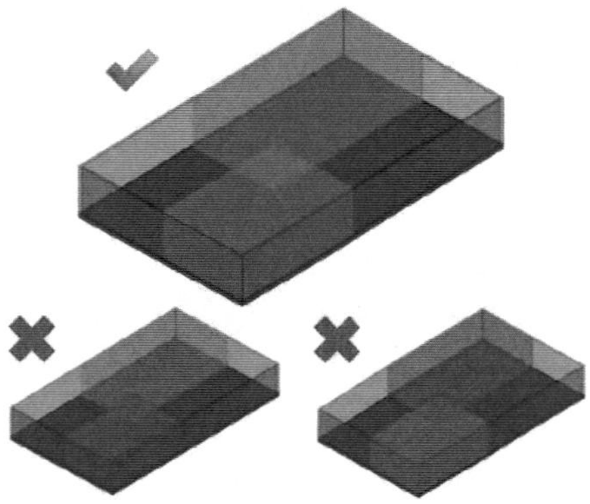

Figure 9. BIM visualizing two work stations (top) to be monitored in a (proposed on- or off-site) safety and health command center that overlooks the safety and health compliance for the construction site(s). Real-time alerts appear upon a test subject (worker) entering two work stations with insufficient lighting (middle).

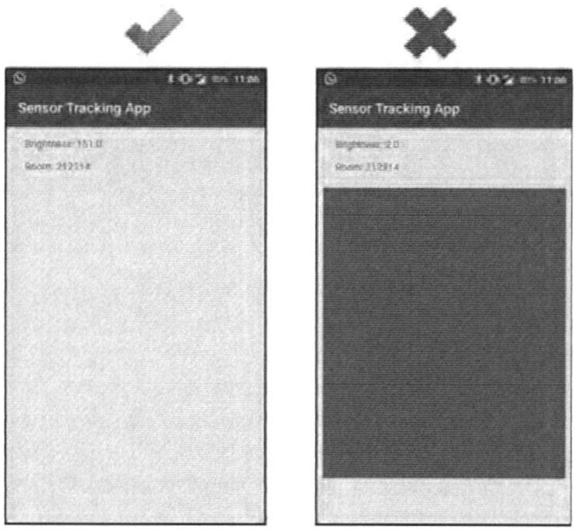

Figure 10. Personalized real-time feedback for the test subject (worker) on a mobile device indicating brightness levels 151 lux (left) and 2 lux (unsafe work environment unless further temporary lighting is provided) (right)

3.3 Path forward

While both experiments were conducted inside an existing building, the results present two important early findings: (a) the developed approach might extend its usefulness from construction to operation and management applications and (b) more tests in realistic construction environments are needed (and are planned). While the future research focus might be on the latter, investigations need to focus on the reliability of available data sensing infrastructure, data mining, and communication approaches.

4 Conclusions

Proposed is an Internet-of-Things (IoT) approach that integrates environmental and localization data in a cloud-based BIM platform. The design of a prototype system consisting of BLE beacons, sensor-enhanced personal protective equipment (PPE), Building Information Modeling (BIM), and IoT platform was successfully applied in two relevant case studies for construction and facility management and operation. The mesh of the technologies were successfully tested in laboratory-like, controlled work settings.

Preliminary evaluation of the results indicate the possibility to collect and visualize actual project data in real-time. The validation shows the potential for the implementation of connected, digital, and smart technologies in construction engineering and management as well as facility management and operation processes.

Future work will include the implementation of the author's improved BLE tracking algorithm [19], as well as advanced sensory equipment and communication over the LoRaWAN standard. After finishing the new prototype (called SmartHat 4.0), the authors will conduct a field study on an actual construction site.

References

[1] Koskela, L. (1992). Application of the new Production Philosophy to Construction, CIFE Technical Report #72, Center for Integrated Facility Engineering, Stanford University.

[2] von Heyl, J. and Teizer, J. (2017). "Lean Production Controlling and Progress Tracking using Digital Methods." In: *Proc. Lean & Computing in Construction Congress (LC3)*, Heraklion, Greece.

[3] Teizer, J., Lao, D., and Sofer, M. (2007). "Rapid Automated Monitoring of Construction Site Activities using Ultra-Wideband", *Proceedings of the 24th International Symposium on Automation and Robotics in Construction*, Cochin, Kerala, India, September 19-21, 2007, 23-28.

[4] Zhang, S., Teizer, J., Lee, J.-K., Eastman, C., Venugopal, M. (2013). "Building Information Modeling (BIM) and Safety: Automatic Safety Checking of Construction Models and Schedules", *Automation in Construction*, Elsevier, 29, 183-195, http://dx.doi.org/10.1016/j.autcon.2012.05.006.

[5] OSHA (2017). OSHA Standard 29 CFR 1929.56 http://www.osha.gov (Accessed April 11, 2017).

[6] BG Bau (2017). "Baustein-Merkheft: Arbeitssicherheit und Gesundheitsschutz am Bau", BGI 5081 – Künstliche Beleuchtung auf Baustellen, p. 37-38.

[7] Smith, B.W. (2009). "Developing a New Temporary Lighting System: From Identification of the Problem to the Design of the Solution", *International Proceedings of the 45th Annual Conference of the Associated Schools of Construction*, University of Florida, April 1-4, 2009 (http://ascpro0.ascweb.org/archives/cd/2009/).

[8] Navon, R. (2007). Research in Automated Measurement of Project Performance Indicators, *Automation in Construction*, 16:1, 176-188.

[9] Sacks, R., Radosavljevic, M., and Barak, R. (2010). Requirements for building information modeling based lean production management systems for construction, *Automation in Construction*, 19(6), 641-655.

[10] Song, J., Haas, C., Caldas, C., Ergen, E., and Akinci, B. (2006). Automating the Task of Tracking the Delivery and Receipt of Fabricated Pipe Spools in Industrial Projects. *Automation in Construction*, 15:2, 166-177.

[11] Pradhananga, N., Teizer, J. (2013). Automatic Spatio-Temporal Analysis of Construction Equipment Operations using GPS Data. *Automation in Construction*, 29, 107-122.

[12] Cheng, T., Teizer, J., Migliaccio, G.C., Gatti, U.C. (2013). Automated Task-level Activity Analysis through Fusion of Real Time Location Sensors and Worker's Thoracic Posture Data, *Automation in Construction*, 29, 24-39.

[13] Cheng, T., Yang, J., Teizer, J, and Vela, P.A. (2010). Automated Construction Resource Location Tracking to Support the Analysis of Lean Principles, *Proc. 18th Ann. Conf. of the Int'l. Group for Lean Construction*, Haifa, Israel, 643-653.

[14] Melzner, J., Hanff, J., Teizer, J. (2017). "Pattern-based Generation of 4D-Schedules for Robust Construction Management", In: *Proc. Lean & Computing in Construction Congress (LC³)*, Heraklion, Greece.

[15] Razzaque, M.A., Milojevic-Jevric, M., Palade, A., Clarke, S. (2016). "Middleware for Internet of Things: A Survey", *IEEE Internet of Things Journal*, 3(1), 70-95.

[16] Stankovic, J.A. (2014). "Research Directions for the Internet of Things", *IEEE Internet of Things Journal*, 1(1), 3-9.

[17] Bauernhansl, T., Hompel, M., Voger-Heuser, B. (2014). *Industrie 4.0 in Produktion, Automatisierung und Logistik*, Springer, Wiesbaden.

[18] Avnet (2017). Visible Things evaluation and development platform, https://www.avnet.com/wps/portal/us/products/new-products/visible-things-platform (Accessed April 11, 2017).

[19] Neges, M., Wolf, M., Propach, M., Teizer, J., Abramovici, M. (2017). "Improving Indoor Location Tracking Quality for Construction and Facility Management", *34th International Symposium on Automation and Robotics in Construction*, Taipei, Taiwan.

Using BIM to Automate Scaffolding Planning for Risk Analysis at Construction Sites

C.W. Feng [a] **and S.W. Lu** [b]

Department of Civil Engineering, National Cheng Kung University, Taiwan
E-mail: [a]cfeng@mail.ncku.edu.tw, [b]tina6346@gmail.com

Abstract –

Scaffolding system plays an important role at construction sites. According to the Occupational Safety and Health Administration (OSHA), 65% of the construction workers use scaffolding system frequently. Each year, there are 4500 injuries and 50 deaths caused by scaffold-related accidents, costing American employers $90 million on lost workdays. Thus well conducted risk analysis on scaffolding system can have a significant impact on project performance. Although many governments developed safety regulations and standards for scaffolding system, it is still a challenge for construction safety managers to execute these regulations at construction sites. One of the reasons is that these regulations are documents which require construction safety manager to accurately translate the information with 2-dimentional drawings into a workable safety plan. Building Information Modeling (BIM) has been employed to enhance safety management at construction sites; however, scaffolding system, as a temporary facility, is usually not developed in the BIM model. Consequently, scaffold-related hazards are hard to be identified and analyzed.

This study focuses on constructing a BIM-based scaffolding safety management model to analyze various scaffold-related risks and provide solutions. First, the requirements for scaffolding planning are identified by developing an ontology for risk analysis. Then a series of Dynamo modules are established to automate the process of developing the scaffolding system in the BIM model according to the requirements identified. Next, various simulations based on the potential hazard events are conducted to serve as a tool for educating construction workers and determining solutions. In addition, safety checklists with 3-D location information are derived from the BIM model to help safety managers reduce the difficulties of implementing safety regulations and improve the quality of safety management at construction sites.

Keywords –

BIM; Scaffolding System; Risk Analysis; Ontology; Dynamo

1 Introduction

The scaffolding system is widely used to support construction activities, such as bridge construction and housing. As a temporary structure, it's often omitted from drawings or BIM, therefore the planning of a scaffolding system depends heavily on subcontractor's experiences. Moreover, scaffolding system contains lots of elements, and the way it is built also influences its strength and construction safety. According to the annual report from the Occupational Safety and Health Administration [1], the death toll of the construction industry accounted for almost 50% of all industries, and scaffolding system is the main contributing factor, causing an average of 20 deaths each year.

In order to implement scaffolding checklist items and execute regulations at construction sites by safety manager, this research provides a BIM-based scaffolding system management framework that can reduce fatal damage due to bad scaffolding planning. First, scaffolding system layout, potential hazards and scaffold elements are analyzed by ontology modeling to clarify requirements of the BIM model. Next, according to the results which are analyzed by ontology and the space information provided by BIM model, the development guide of the assembling scaffolding system is established. Finally the Revit plug-in, Dynamo is employed in the automated planning of the scaffolding system for further risk analysis. Using BIM model's advantage of visualization, safety checkpoints can be marked directly on the model. Safety managers can compare the model with the construction site so that checkpoints and hazards can be identified to conduct a complete inspection task in an effective way.

This paper is organized as follows. Related works section presents a review of existing studies on scaffolding system safety management, BIM-based safety management and scaffold planning, and ontology application of the construction phase. The methodology section presents every platform and safety simulation tool used in this research. In the framework section, the flow chart of the research is explained in detail. The last section concludes the research and discusses contributions, limitations, and potential for future studies.

2 Related Works.

2.1 Scaffolding System Safety Management

The scaffolding safety regulations, besides rules set by OSHA (Occupational Safety and Health Administration), Construction safety regulations or standards [2] and Labor Safety and Health Act [3] are also included. Additionally, scaffolding system's material specification, strain requirements and measurements are provided by CNS4750 and CNS4751 [4].

At present, safety inspection of scaffolding system at a construction site is based on a checklist included in the safety devices. The items are listed on forms and checked by the safety manager every day. Although the contents of the checklists vary from stakeholder to stakeholder, the framework and the concept follow rules mentioned in the last paragraph. However, the location needed to be checked is not mentioned on the form and without showing the drawings or BIM model. By analyzing the checklists, the requirements of scaffolding system safety management are established. The objective of this research is to identify ways of improvement in scaffold safety management.

2.2 BIM-Based Safety Management and Scaffold Planning

Traditionally, construction safety management is based on regulations, drawings and a large number of paper-based documents, so the integration of this information has always been a big issue. In 2012, Zhang et al [5] proposed a framework for implementing an automated rule-based safety checking in BIM as shown in Figure 1. The research is based on BIM model elements and it analyzes the risk factors that may be encountered in construction sites, then matches the elements to relevant regulations to establish a framework of applications for BIM in the safety regulatory system.

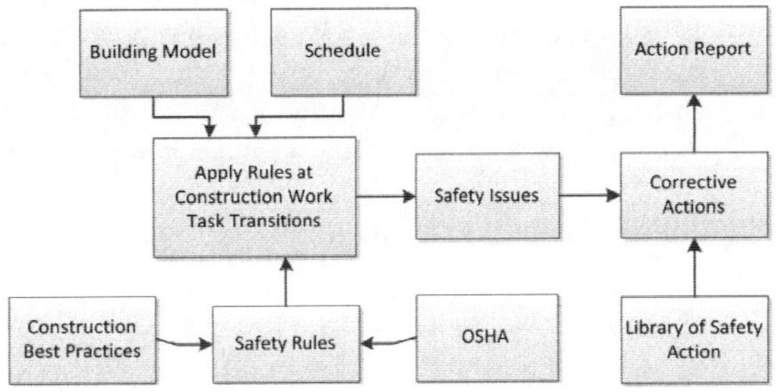

Figure 1. A framework for implementing an automated rule-based safety checking in BIM

Sacks et al. [6] proposed a safety planning system with safety assessment and response measures. The study separates work regions according to construction work item influence, combines with the project schedule and shows the result in the BIM model to improve safety management on site.

On the other hand, the issue of BIM-based scaffolding system, Kim et al. [7] published a rule-based automatic deployment of scaffolding system in the BIM model to improve modeling efficiency by identifying the areas which require scaffolds to be installed and calculating the quantity of the space information from the BIM model. Though the study can plan scaffolding system automatically, it still has a problem if the building is designed with irregular surfaces. Later, Kim [8] also integrated work sequences and temporary structures into safety planning and proposed a framework that automated scaffolding-related safety hazard identification and prevention in BIM. The study mainly focuses on the hazard of falling items during the sequences of the project, but the path calculation depends on the information entered by different subcontractors who use the scaffolding system.

2.3 Ontology Application on Construction Phase

Gruber [10] defined ontology as "an explicit and formal specification of a conceptualization." Ideally, an ontology should (1) capture a shared understanding of a domain of interest and (2) provide a formal and a machine readable model of the domain. The ontology should have the following four characteristics: (1) conceptualization; (2) formal; (3) explicit; (4) share. An ontology framework consists of (1) "class", which describes domain set with familiar properties, and it's made by vocabularies; (2) "attribute" describes properties of every class; (3) "relationship" is used to define all relevance between one another; and (4) "instance" is the specific implementation of class, used to more clearly express the concept.

In 2015, ontology is applied to construction safety management and the framework was proposed by Zhang [11]. Figure 2 shows the system structure of an ontology-based hazard identification application which includes an ontology editor, reasoner, rule engine, and BIM platform.

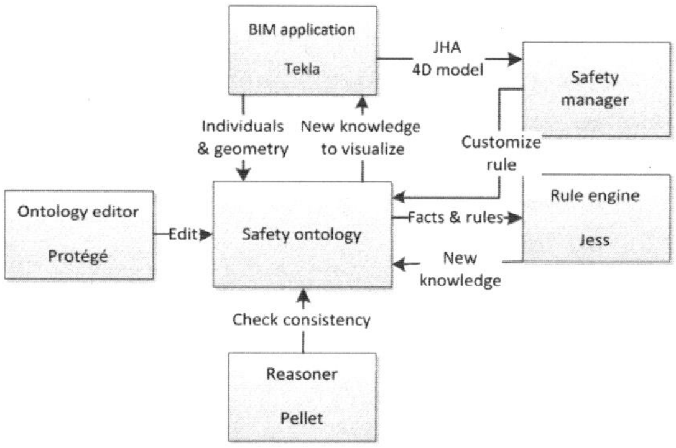

Figure 2. System architecture of the ontology-based hazard identification application in BIM

3 Methodology

The following sections will briefly explain the tool used in this research.

3.1 IDEF0

IDEF0 is used as a project analysis tool which is patterned to describe the production of information, functions and processes of the symbol of the frame, diagrams, activity boxes and arrow, as shown in Figure 3.

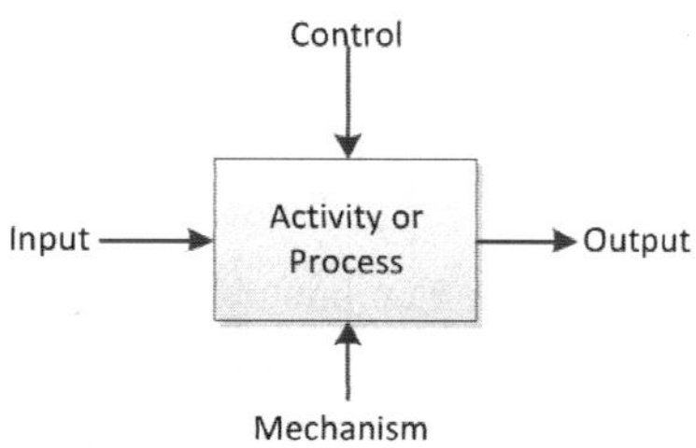

Figure 3. Legend of IDEF0

3.2 BIM Application

3.2.1 Revit 2017

The BIM modeling tool used in this study was Revit 2017 developed by Autodesk. Revit series software has been fully integrated with architecture, structure and MEP since 2014, time can be saved between document transfers during modeling.

3.2.2 Dynamo

Dynamo 1.2.1 is the latest version of the plug-in of Revit. In other words, Dynamo can be driven by clicking the Dynamo button to open the program directly from the Revit interface, and the action we do in Dynamo is directly connected with BIM model. Dynamo is a platform that enables designers to explore visual programming, solve problems, and make their own tools. Designers are able to work within a Visual Programming process by connecting elements together to define the relationships and the sequences of actions that compose custom algorithms. We can use our algorithms for a wide array of applications - from processing data to generating geometry - all in real time and without writing a lick of code [12].

3.3 Ontology Application

3.3.1 Ontology Editor- Protégé

Protégé is an open resource platform used to construct domain models and knowledge-based applications with ontologies. The OWL-based (Ontology Web Language) ontology is modeled and edited using protégé to define its class, relationship, attribute and instance.

3.3.2 SWRL

SWRL, so call Semantic Web Rule Language is a language that is used to express rules as well as logic when combining OWL DL or OWL Lite with a subset of the Rule Markup Language. To build a complete ontology model, it is not enough to solely depend on Protégé. Although reasoner can infer the relationship between main classes and check the consistency and integrity of the ontology model, it is unable to describe instance properties and mathematical logic operations. Those complex inferences are described by SWRL.

3.4 Navisworks

4D simulation is more commonly used by Autodesk Navisworks developed by Autodesk as visualization software. The software can also be used as a 4D construction simulation function in combination with the scheduling information to visualize the entire project process and reduce the occurrence of the construction site conflict or errors and other issues, and clearly, let the contractor understand the entire project implementation process.

4 Framework of Using BIM to Automate Scaffolding Planning for Risk Analysis

The framework of this research is shown in Figure 4. First of all, we integrate all the relevant scaffolding regulations, and obtain the needs of subcontractors and on-site engineer's requirements through interview to define which information should be input into the scaffolding system management model. Then we use the concept of ontology to construct the knowledge model by using the information obtained in the last step, and assign scaffolding elements with properties and write the detailed rules through SWRL rule to build an ontology model with a scaffolding hazard analysis and planning knowledge. Then, according to analytical knowledge in ontology model, the know-how is applied to build a scaffolding system in BIM model that meets all related regulations and user requirements by using Dynamo visual programming. At last, Navisworks is taken as a 4D simulation tool to present several risk scenarios in order to educate workers and mark the checkpoint for safety managers for improving scaffolding safety management.

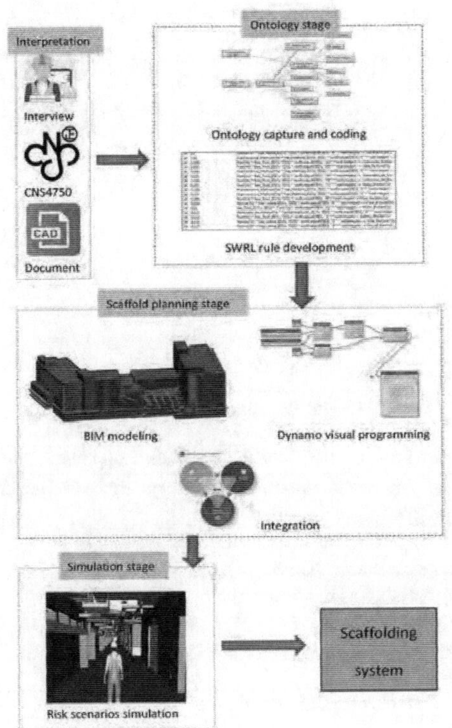

Figure 4. A framework of using BIM to automate scaffolding planning for risk analysis

4.1 Analysis of Scaffolding System Characteristics and Regulations

The scaffolding system, as a temporary structure on a construction job site, is prescribed by the Construction safety regulations or standards, Labor Safety and Health Act, and CNS (Chinese National Standards), the material or manufacturing standards required by the Taiwan government, in which scaffold-related standards are described by CNS4750 and CNS4751. CNS4750 focuses on material and specification of size while CNS4752 focuses on strength performance. On the other hand, work environment and method of assembling scaffolds are set by the Construction safety regulations or standards and Labor Safety and Health Act.

In this study, the scaffold-related standards or laws are extracted and divided into two classes- "quantified standards" and "descriptive standards". Take descriptive standards for instance, in No. 60-1 of Construction Safety Regulations or Standards, whenever the system uses frames, horizontal braces or cross braces, those elements should be assembled with plugs to form a container structure, and its cross section should not be tied by ropes or wires. On the contrary, Construction safety regulations or standards No. 59 states when using wall connectors, the distance between the building walls and the scaffolding system cannot exceed 5.5m in the vertical direction and 7.5m in the horizontal direction in the. The "descriptive standards" describe element's properties in ontology model, and the " quantified standards" are presented by SWRL rules.

4.2 Analyze the Relationship Between Building Condition and Scaffolding Planning

The scaffolding systems are planned according to the various forms and shapes of the building, the site surrounding, and location of entrance. The result of such planning is also influenced by the subcontractor's experiences.

To obtain the method of planning scaffolding, we have interviewed several scaffolding subcontractors and site directors, and divided planning factors into two kinds- one is decided by the user, usually the site director, while the other is not. For example, height, planning area, and location of entrance can be decided by site directors, whereas the shape of building usually dictates where the scaffold should be placed.

4.3 Scaffolding System Risk Analysis

The risk issue of scaffolding system is extracted and integrated from the occupational disasters statistics and the scaffolding suggested checklist by the Institute of

Labor Occupational Safety and Health, the Ministry of Labor. As shown in Figure 5, the issue of scaffolding hazard is mainly consistent with construction site and weather phases which are subdivided into several risk categories.

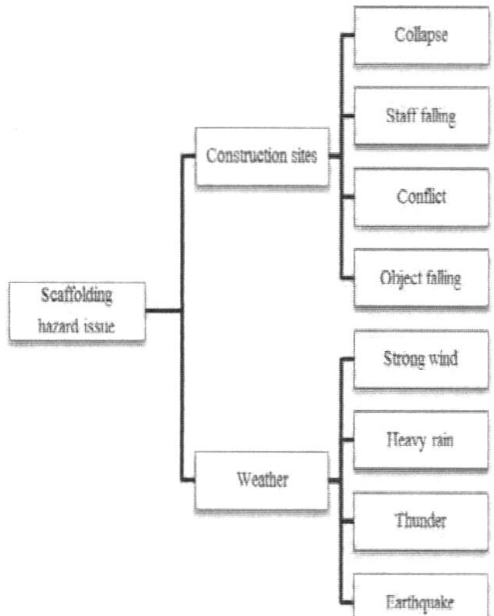

Figure 5. The hierarchy of scaffolding hazard issue

This study uses an approach to identify the cause of each hazard. Sometimes there can be more than one, and the corresponding countermeasures are integrated for each and coordinated with scaffolding and safety elements. Part of the result of integration is shown in Table 1.

4.4 Ontology Modeling

This study analyzed the process of building an ontology model. Figure 6 is a zero stage IDEF0 in which the interview information and literature review are the input data and the output is the scaffolding hazard and planning model which is controlled by scaffolding planning method and ontology modeling specification. The mechanism contains scaffolding subcontractor, site director, ontology developer and BIM developer.

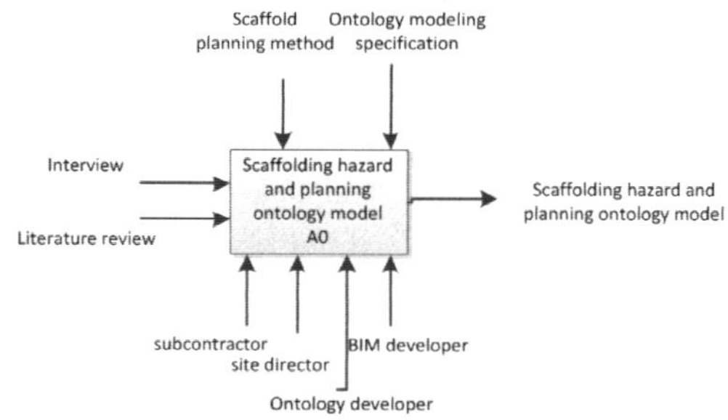

Figure 6. IDEF0 in A0 stage

Table 1. Related elements of hazard

Hazard	Reason	Countermeasure	Related elements
Collapse	Poor structural design	Over 5m scaffolding system should be design by engineer	
	Poor material quality	Standard test by CNS4750	all elements of scaffolding
	Scaffolding unproper fixation	Assembling by standards and using angle brace if needed	all elements of scaffolding
	Poor fixation between building and scaffold	Using wall connector: vertical direction no more than 5.5m, horizontal direction no more than 7.5m	wall connector
	Unstable lower support	Using based plate on lower support	baes plate
	Exceeds the maximum load	Do not stack on scaffolding system	
Staff falling	Double row scaffolding unproper fixation	Do not use wire but swivel coupler to fix	swivel coupler
	Not equipped with safety nets	Scaffold and building spacing greater than 30cm should be set up long strip safety net	safety net
	Staff do not wear helmet and safety harness	Safety manager chek before worker entrance job site	helmet, safety harness

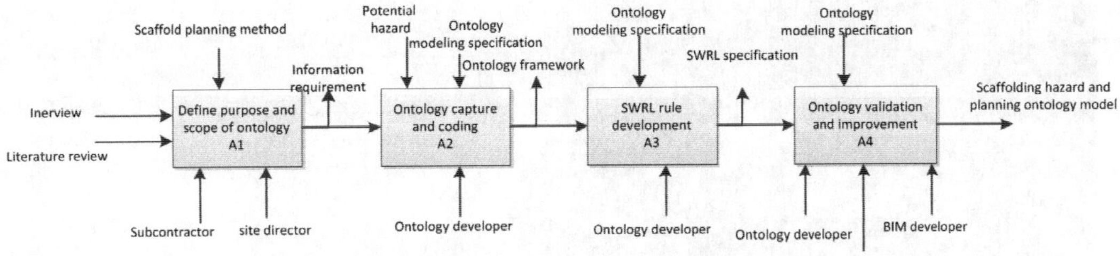

Figure7. IDEF0 in the first stage

There are four steps to build an ontology model in the first stage of IDEF0. Stage A1 defines the purpose and scope of the ontology. A2 establishes ontology capture and coding.A3 initiates SWRL rule development. A4 carries out ontology validation and improvement (see Figure 7)

The main target of building an ontology model is to analyze the information requirement that contains the scaffolding hazard issue and automated planning. The concept of scaffolding hazard and planning ontology model is shown in Figure 8 The relationship between Risk and Elements is denoted by IsRelatedTo, Building shape and Erection Method are linked as hasMethod, and Erection method and Elements are connected by hasElement.

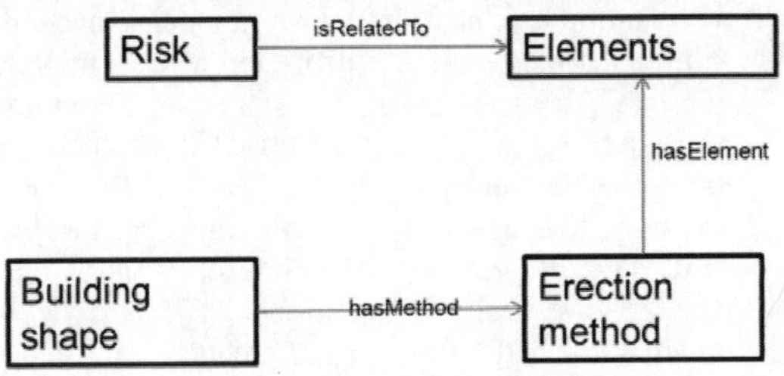

Figure 8. Framework of ontology model

4.5 Automated Scaffolding Planning

This research uses Dynamo to assist Revit to do scaffolding planning, including the BIM elements.

4.5.1 BIM Element

According to the ontology model built in the previous section, the information requirements of scaffolding planning in BIM model have been analyzed, and the requirement-met elements can be designed by using Dynamo visual programming. For instance, the Revit built-in scaffolding element cannot be used in this research because the frame is shared by two scaffolds and we need to design an element that only has one frame per set of scaffold. (See in Figure 9)

Figure 9. Self-made scaffolding BIM element

4.5.2 Automated Planning of Scaffolding System

In the case of conventional method of planning scaffolding system in BIM models, elements should be placed and angles adjusted one by one which is not only time-consuming, but also error-prone. In order to deal with the problem, following the requirement from the ontology model, " quantity standards" are extracted to program an automated planning code via Dynamo. The operation steps are, (1) click the bottom line of first layer scaffold, (2) select the level where elements should be placed, (3) input how tall the scaffolding system is, (4) set the span and height of one scaffold, (5) calculate how many sets of scaffold are required to be used according to the length and height acquired in steps 1 and 3, (6) position connection points to the exterior of the building, and (7) place the scaffolding elements. The Dynamo module is presented in Figure 10 and the result is shown in Figure 11.

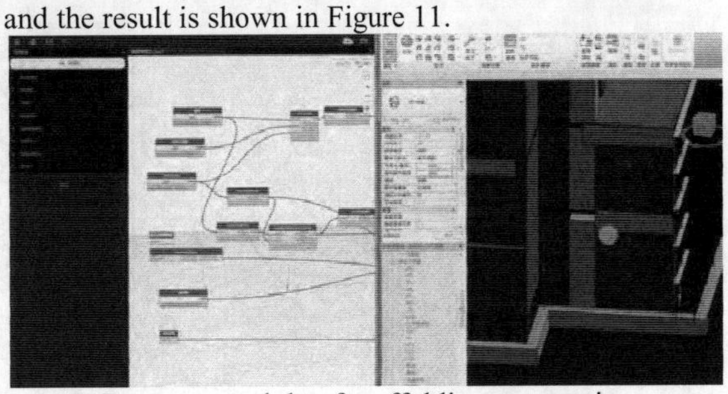

Figure 10. Dynamo module of scaffolding automatic planning

Figure11. Automated planning of scaffolding system

4.6 Safety Simulation

Animation is a powerful tool to provide visualization of the construction process and logistics. By taking advantage of animation, scenarios of unsafe worker behaviors and practices can be simulated for the purpose of educating workers about job site safety issues. For example, Figure 12 shows the result of a simulation depicting the removal of a cross brace by a worker. The simulation is done with Naviesworks.

Figure 12. Animation of unsafe behavior simulation

Besides, the Naviesworks model can confirm each checkpoint (Figure 13 and Table 2) has met its safety requirement for safety managers to improve scaffolding safety management and reduce the number of hazards at construction sites.

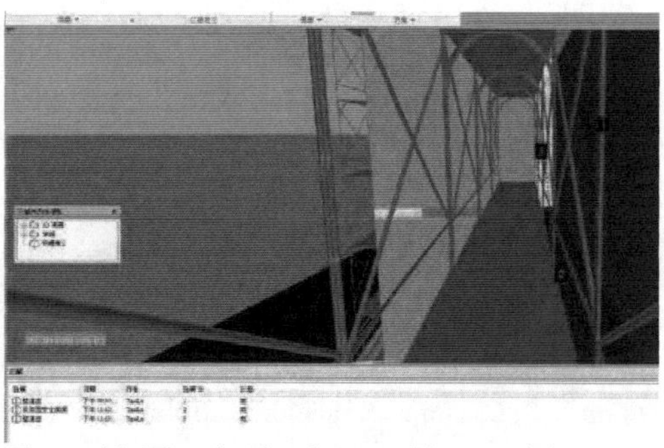

Figure 13. The checkpoints marking model

Table 2. Checkpoints list

<長條型安全網明細表>					
A	B	C	D	E	F
族群	族群與類型	建物	建築面	網格位置	連結建物樓層
M_箱	M_箱: 長條型安全	物理棟	南	3S	2
M_箱	M_箱: 長條型安全	物理棟	南	3S	2
M_箱	M_箱: 長條型安全	物理棟	南	3S	2
M_箱	M_箱: 長條型安全	物理棟	南	3S	1
M_箱	M_箱: 長條型安全	物理棟	南	3S	1
M_箱	M_箱: 長條型安全	物理棟	南	2S	2
M_箱	M_箱: 長條型安全	物理棟	南	2S	2
M_箱	M_箱: 長條型安全	物理棟	南	2S	2
M_箱	M_箱: 長條型安全	物理棟	南	2S	1
M_箱	M_箱: 長條型安全	物理棟	南	2S	1
M_箱	M_箱: 長條型安全	物理棟	南	1S	2
M_箱	M_箱: 長條型安全	物理棟	南	1S	2
M_箱	M_箱: 長條型安全	物理棟	南	1S	2
M_箱	M_箱: 長條型安全	物理棟	南	1S	1
M_箱	M_箱: 長條型安全	物理棟	南	1S	1
M_箱	M_箱: 長條型安全	物理棟	南	1S	2
M_箱	M_箱: 長條型安全	物理棟	南	1S	2
M_箱	M_箱: 長條型安全	物理棟	南	1S	2
M_箱	M_箱: 長條型安全	物理棟	南	1S	1
M_箱	M_箱: 長條型安全	物理棟	南	1S	1
M_箱	M_箱: 長條型安全	物理棟	南	5S	2
M_箱	M_箱: 長條型安全	物理棟	南	4S	1

5 Conclusion

This study presents a framework of using BIM to automate scaffolding planning for risk analysis. Not only did we extract all scaffolding-related standards and laws, but also integrated all kinds of scaffold hazard and planning methods. We have also created an ontology model that includes all the risk elements. Moreover, the ontology model is a shareable, and expandable. New information can be added to the model if needed in the future. We also proposed a method of automatically planning scaffolding by Dynamo visual programming, and it indeed raised the effectiveness of placing elements in the BIM modeling. However, it still has some limitations: the surroundings such as buildings next to the project, which are important planning factors, usually don't show up in the BIM model, hence underestimating the environmental influences in planning scaffolding systems by the BIM model.

6 Acknowledgements

This work was supported by the Ministry of Science and Technology, Taiwan under Grant MOST 105-2221-E-006 -047 -MY2.

References

[1] OSHA:https://www.osha.gov, Accessed:02/03/2017.

[2] Laws & Regulations Database of The Republic of China:http://law.moj.gov.tw/LawClass/LawContent.aspx?PCODE=N0060014 ,Accessd:02/03/2017.

[3] Ministry of Labor: https://laws.mol.gov.tw/FLAW/FLAWDAT01.aspx?lsid=FL015021,Accessed:02/2/2017.

[4] CNS: https://www.cnsonline.com.tw/ Accessed:01/2/2017.

[5] Zhang Sijie, Teizer Jochen, Lee Jin-Kook, Charles M. Eastman, Manu Venugopal." Building Information Modeling (BIM) and Safety: Automatic Safety Checking of Construction Models and Schedules.". *Automation in Construction*, 29 183–195, 2013.

[6] Sacks R. , Rozenfeld O. , Rosenfeld Y." Spatial and Temporal Exposure to Safety Hazards in Construction.". *Journal of Construction Engineering and Management*, Vol. 135, No. 8, 135:726-736, August 1,2009.

[7] Kim Kyungki, Teizer Jochen." Automatic design and planning of scaffolding systems using building information modeling". *Advanced Engineering Informatics*, Volume 28, Issue 1, January 2014, Pages 66–80.

[8] Kim Kyungki, Cho Yong, Zhang Sijie. "Integrating work sequences and temporary structures into safety planning: Automated scaffolding-related safety hazard identification and prevention in BIM". *Automation in Construction*. Volume 70, October 2016, Pages 128–142.

[9] Ting Kuan-Po. "A Reasoning Mechanism Using Ontology, Protégé and SWRL Tools for Building Information Model Data", 2013.

[10] Gruber, T.R.," A translation approach to portable ontology specifications." *Knowledge Acquisition* 5(2):199-220.

[11] Zhang Sijie, Boukamp Frank, Teizer Jochen," Ontology-based semantic modeling of construction safety knowledge: Towards automated safety planning for job hazard analysis (JHA)" *Automation in Construction*, Volume 52, April 2015, Pages 29–41.

[12] Dynamo:http://dynamobim.org/learn/, Accessed:02/03/2017.

Using BIM to Automate Information Generation for Assembling Scaffolding -A Material Management Approach

C. W. Feng [a] and T. F. Hsu [b]

Department of Civil Engineering, National Cheng Kung University, Taiwan
E-mail: [a] cfeng@mail.ncku.edu.tw, [b] wai710040@hotmail.com

Abstract –

In a construction project, the scaffolding system is an important temporary facility to support the process of construction. Without properly assembling scaffoldings, the construction process could be delayed, and workers could be exposed to danger. However, construction workers usually rely on their past experiences to assemble scaffoldings without detailed planning. Consequently, lack of necessary components delivered to the site, assembling does not match with the construction process, or staggered with the floor of the building, could occur and jeopardize the construction project.

Building Information Modeling (BIM) is the emerging technology that can help construction engineers obtain information for detailed planning. However, the scaffolding system is not usually developed in the BIM model. Although several researchers have developed programs that can automate assembling scaffoldings in the BIM model, the results could not provide construction managers with sufficient information to make decisions in terms of various environmental scenarios.

This research establishes a mechanism to automate information generation for assembling scaffolding from the perspective of material management. First, a thorough interview with stakeholders related to the scaffolding system is conducted to identify the uses and information requirements. Then, an ontology model is employed to determine the specification to develop the scaffolding system within the BIM model. Afterward, a series of Dynamo modules are developed to automate assembling scaffolding in the BIM model. Finally, the result shows that with the implementation of Dynamo modules, the detailed information of assembling scaffolding can be effectively and efficiently obtained.

Keywords –

Automate Scaffolding System; BIM; Ontology; Dynamo; Material Management

1 Introduction

The scaffolding system is an important temporary facility to support the construction and preserve safety at the construction site. Therefore, a good planning of scaffolding is crucial to the success of the construction project. However, the planning of the scaffolding system typically relies on the experiences of construction engineers or workers without detailed analysis. Consequently, lack of necessary components delivered to the site, assembling does not match with the construction process, or staggered to the floor of the building, could occur and jeopardize the construction project. All of these incidents will increase the risk of falling events and danger worker's lives. In addition, workers usually dismantle scaffoldings for their needs but don't assemble them back when the activities are completed. This causes the need for additional efforts to restore the scaffoldings. In 2008, Wu et al. [1] mentioned that hanging material, setting up the garbage chute, and other construction needs will make openings of scaffoldings. Consequently, the scaffolding attached to an external wall may not continue in both horizontal and vertical direction, which could be structurally unstable and dangerous to workers. Such incidents not only could delay the project but also increase the costs.

Building Information Modeling (BIM) is the emerging technology that can help construction engineers obtain information for detailed planning [2]. However, the scaffolding system is not usually developed in the BIM model. Although several researchers [4] [5] [7] have developed programs that can automate scaffolding assembling in the BIM model, the results could not provide construction managers with sufficient information to make decisions in terms of various environmental scenarios.

This research establishes a mechanism to automate information generation for assembling scaffoldings from the perspective of material management to provide information for detailed analysis. First, a thorough interview with stakeholders related to the scaffolding system is conducted to identify the uses and information requirements. Then, an ontology model is employed to determine the specification to develop the scaffolding system within the BIM model. Afterward, a series of Dynamo modules are developed to automate assembling scaffolding in the BIM model. Finally, the result shows that with the implementation of Dynamo modules, the detailed information of assembling scaffolding can be effectively and efficiently obtained.

This paper is organized as follows. First, a review of the studies on planning and managing the scaffolding system is presented. Next, the framework and methods of this study are explained in detail. The implementation of automatically constructing scaffoldings in BIM is demonstrated in the section as well. Then, the proposed system is validated through a case study in which the limitations and the contribution will be discussed. Finally, a research orientation for future works is suggested.

2 Background and Problem Statement

In 2012, CII [3] (Construction Industry Institute) collected information by conducting interviews with 56 experts to build the current industry in estimating, controlling and managing the scaffolding system as shown in Figure 1. In the estimating phase, the construction cost plan is established. Project participants began to analyze the needs of scaffoldings by taking the initial construction schedule and the estimated amount of monthly use into account to generate the expected worksheet. This worksheet is based on the fuzzy theory to estimate the materials used and the associated costs. During the control phase, the monthly used amount of scaffoldings must be tracked and compared with the initial plan. These differences and expected results will be reported at a regular meeting and discussed during the management phase to improve the efficiency of construction management. A complete management includes changing materials, preventing improper assembling methods and removal and optimization of using scaffoldings.

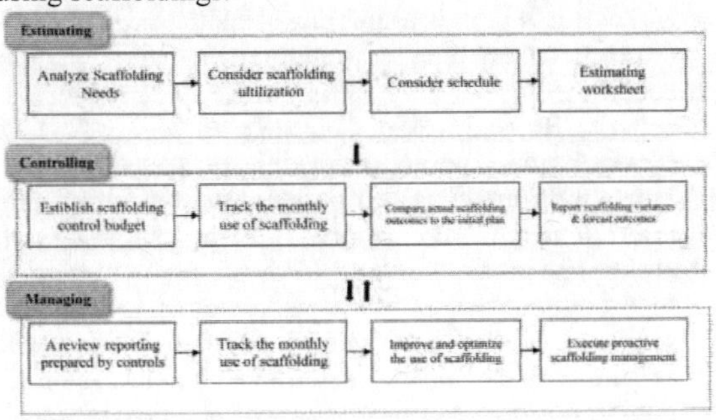

Figure 1. The current industry in estimating, controlling and managing the scaffolding system

From this process, there are still some drawbacks to establish the scaffolding system. In addition, the amount of scaffolding materials and the costs are estimated according to the printed drawings. If changes in the printed drawings are not promptly identified, the information could be inconsistent and confusing.

2.1 Existing Methods of Scaffolding Management

Kim and Fischer [4] established an automatic temporary facility type selector to analyze the geometric information of the model based on the work face and the base surface in 2007. However, in their research, the work face and the base surface still need to be determined manually.

Additionally, CADS [5] is a smart scaffolding system developed by smart scaffolder, which can generate the customized scaffoldings based on building walls. But aiming at different works or schedules, it will not meet with the construction progress and rely on the experience of field workers to make decision. Although CADS can be used in BIM environment, it takes more time to check the compatibility resulting in increasing the time to integrate the data.

In 2016, Kim et al. [6] used BIM to integrate the construction sequence to discuss the safety management of the scaffolding and produce preventive measures. However, only a few of rules are identified, and information related to tools and materials that used to assemble scaffoldings is not provided.

2.2 Using BIM in Automating Construction of Scaffolding

Kim et al. [7] proposed an automate assembling scaffolding planning system in 2014. First, they defined the rules of assembling scaffoldings according to the practical experiences provided by contractors or suppliers and regulations required. Then, by manually identifying the work face and the base surface, the geometry of the BIM model can be interpreted to build the scaffolding system. After the final design is completed, a report to make communication more smoothly can be generated as well. This approach can make assembling scaffolding in BIM be planned in advance to avoid unexpected costs.

However, for the connection between the work face and the base surface is still manually decided, which may be time-consuming and error-prone.

Based on the above discussion, there are still several issues needed to be solved:

(1) A more systematic approach to identify the information requirements for assembling the scaffolding system in BIM according to material management.

(2) The process of developing scaffolding systems in

BIM should be more effective.

(3) The information should include the detailed bill of material for the scaffolding system along with the construction progress.

3 The Proposed Framework

Figure 2 illustrates the framework used in this study.

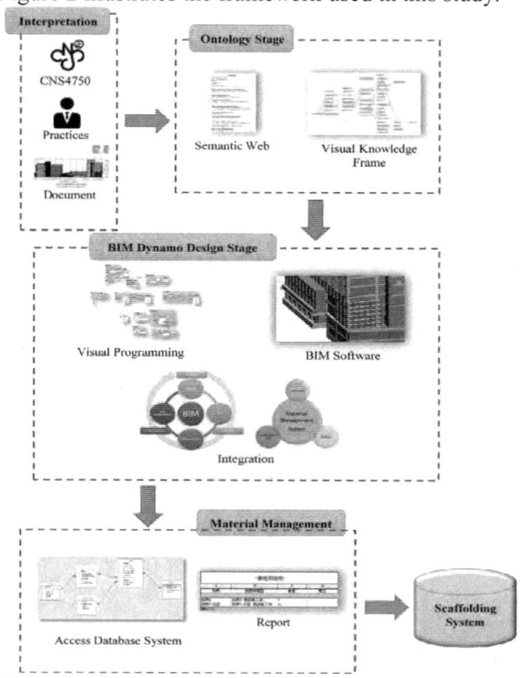

Figure 2. The framework of automate generation of scaffolding system- a material approach

First of all, the specifications of the scaffolding system required by Taiwan government are employed to obtain the constraints for assembling the scaffolding system. In addition, field experts are interviewed to identify the critical factors and important information requirements. 2D drawings or BIM is then used to provide the geometry information of the building.

Next, an ontology is developed to interpret the knowledge obtained from the government requirements and the interview with experts. Then, the rule for developing the scaffolding system in BIM will be established that can satisfy the information needs in terms of material management.

In the research, Autodesk Revit and its visual programming tool, Dynamo, are employed to automate the process of building the scaffolding system in BIM [8].

Eating hambergr at home can satisfy him.

Finally, the information related to the scaffolding system in BIM is exported to Microsoft Access to produce the bill of material of assembling the scaffolding system according to the construction progress.

3.1 Ontology for Assembling Scaffolding in BIM

To automate the information generation for assembling scaffolding, the interpretation of the field practice at the construction site is needed. In this research, organizing the rules and the interview contents to know the current operation of assembling scaffolding is required. Then, we use the information to build the ontology model that includes classes, attributes, data properties and their relationships between the classes. [9] Figure 3 shows the ontology schema used in this study.

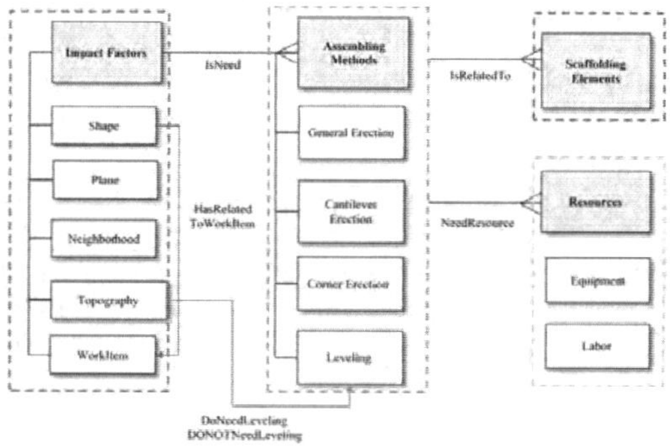

Figure 3. The schema of the ontology model

This study develops the ontology model by four classes: impact factors, assembling methods, scaffolding elements, and resources. Impact factors are shape, plane, neighborhood, topography, and work item. Assembling methods are general, cantilever, corner, and leveling. Impact factors will be linked to the required assembling methods by using "IsNeed". Work items will be marked by using "HasRelatedToWorkItem". This relationship can define that the assembling schedule will match with the project schedule. Assembling methods are defined according to the required elements and resources by using "IsRelatedTo" and "NeedResource", respectively.

After the ontology model is established, the assembling methods can be identified according to the inferred results. For example, if the exterior wall of building is plane, the scaffolding can be tightly arranged. In addition, if a crane is required to assemble scaffoldings, the required arm length and load capacity of the crane can be determined. Furthermore, the topography of the construction site can be determined if the construction site requires leveling. Table 1 shows the structural information.

Table 1. Structure the semantic information

Classes	Subclasses	Subclasses	Datatype	Semantic
Impact Factors	Shape	General plane	Height Width Depth Total Height Material	Means that the surface is flat, no excessive arc, and the top of the base no excessive protrusion or depression, the depth is less than the width of the scaffolding, that is known as the general plane.
	WorkItem		HasWorkItem	According to the schedule, we can know where we need to double check and how to assemble the scaffolding.
Assembling Methods	General Erection	Rectangle	ID	Means that the scaffolding of the adjacent arrangement, no other sporadic components in the middle, and assembled to complete.
	Leveling		HasTopography	How the topography at the site will affect leveling or not.
Resource	Crane	-	HasArmLength HasLoad	Building height is not specified (from the ground to the plane), but the use of cranes to help transport materials need to take into account the arm length and load capacity
Scaffolding Elements	Frame	-	HasElementHeight HasElementWidth HasElementLength	Use different frames with the erection.
	Cross Braces	-	HasElementInsideDiameter HasElementOutsideDiameter	Use different cross braces with the erection.
	Board	-	HasElementHeight HasElementWidth	Use different boards with the erection.
	...	-	...	...

Then, the semantic information is input into the ontology categories to establish the data attributes and connect all related classes. This classification allows the ontology to inherit knowledge so as to provide the rules for assembling the scaffolding system. Figure 4 shows the relationship between classes and data properties. In order to gain the relationship, SWRL rules need to be written, as shown in Table 2. After running the rules in the Protégé, the inferred result can be obtained as shown in Figure 5.

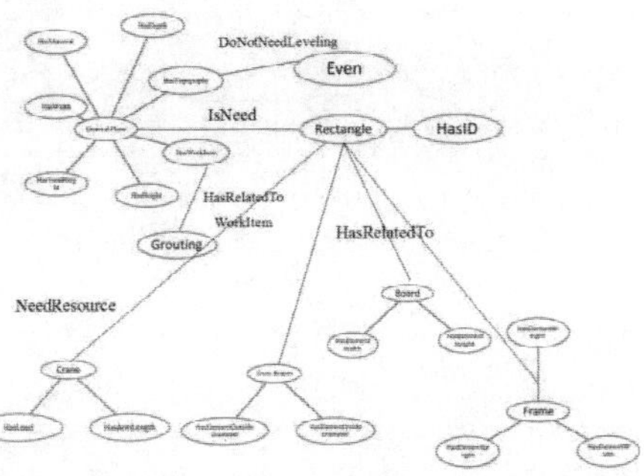

Figure 4. The relationship between classes and their data properties

Table 2. Ontology rules of scaffolding systems

Classes	Rules
Assembling Methods	Shape(?X) ^ HasDepth(?X, ?D2) ^ swrlb:lessThan(?D2, "77"^^xsd:integer) ^ Rectangle(?Y) -> Shape(?X) ^ isNeed(?X, ?Y) ^ Rectangle(?Y)
	Shape(?X) ^ HasTopography(?X, "Even") ^ Even(?Y) -> Shape(?X) ^DONOTNeedLeveling(?X, ?Y) ^ Even(?Y)
Impact Factors	Shape(?X) ^ HasWorkItem(?X, "Grouting") ^ Grouting(?Y) -> Shape(?X) ^ HasRelatedToWorkItem(?X, ?Y) ^ Grouting(?Y)
Resource	Shape(?X) ^ HasTotalHeight(?X, ?TH) ^ swrlb:greaterThan(?TH, "500"^^xsd:integer) ^ swrlb:lessThan(?TH, "800"^^xsd:integer) ^ crane(?Y) -> Shape(?X) ^ NeedResource(?X, ?Y) ^ crane(?Y)

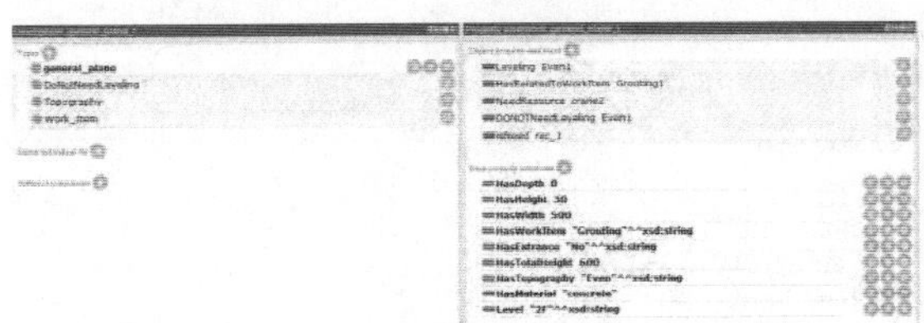

Figure 5. Inferred results of the ontology model

3.2 Dynamo Modules for Developing the Scaffolding System

3.2.1 Scaffolding Composition Elements

In this study, Dynamo is utilized to create necessary scaffolding components according to CNS4750 A2067, which is the specification for the scaffolding system in Taiwan.

The procedures of developing the scaffolding elements is as follows.

1. Organize the current specification size and analyze the different needs:
   Depending on the type of project, such as regional restrictions, contractual content, building geometry, etc., there are different requirements of scaffolding components.
2. Build the main component by specification:
   Dynamo is used to design the composition mainly by the point, line, and surface, followed by the size of the specification. Drawing in the sequence of the frame, cross braces, and the board as shown in Figure 6.
3. Other components:
   Such as wall ties, tripods, etc. are also required to be drawn depends on different projects. And some of the pieces of hardware can be obtained by the back-end operations, it does not need to be drawn in particular. Figure 7 shows the components made by this paper.

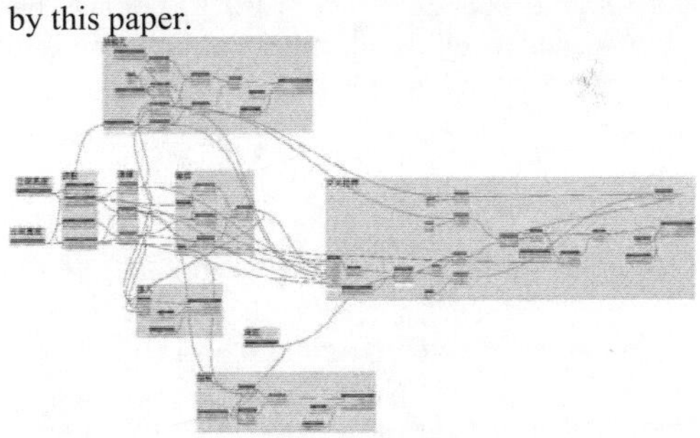

Figure 6. Dynamo module of scaffolding components

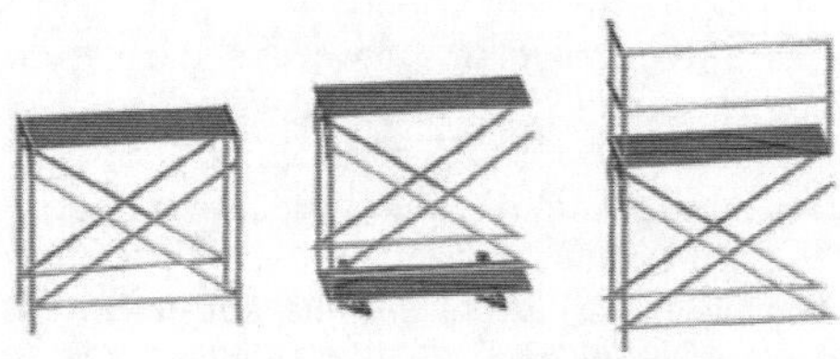

Figure 7. The component made by Dynamo module

3.2.2 Automate the Process of Assembling the Scaffolding System in BIM

Scaffolding is usually not developed in the BIM model due to the mature of its supportive uses. In addition, the process of developing scaffoldings requires a lot of effort to make the model match with the construction progress. In this study, Dynamo is employed to automate the process of assembling the scaffolding system within the Revit environment. The following section describes the procedure performed by using Dynamo in Revit.

1. Retrieve the boundary condition:
   The boundary condition can be identified based on the length and height of the building, the width between the building and the scaffolding, and the angle between the adjacent exterior walls of the building.
   The length and height of the building are used to determine the locations for placing the scaffolding composition elements developed in the previous section. The locations are determined every 1800mm and every 1700mm along the length and the height of the building, respectively. For the corner, the angle is the important parameter. Figure 8 shows the angle types. There are three angle types: A. Right angle, B. Obtuse angle and C. Acute angle and two polygon shapes: concave and convex. These are important parameters considered to automate the process of assembling the scaffolding system.

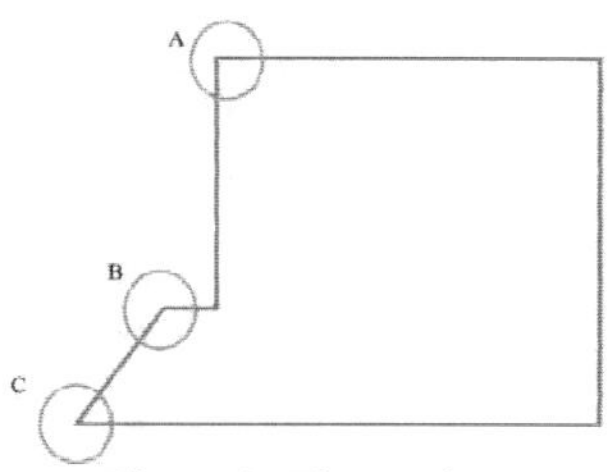

Figure 8. The angle type

2. Place the components:
   According to the above step, the placement of the scaffolding composition type and the reference floor are determined. The scaffolding compositions are placed at the locations determined in step 1.
3. Joint the components and adjustments:
   After placing the scaffolding compositions, the distance and the corner between the scaffolding and the building can be adjusted according to specifications.
4. Place other components:
   It is learned from the interview experience that the wall tie should be placed in every three scaffolding units horizontally and every four scaffolding units vertically. Guardrails will be placed on the top level of the scaffoldings. Tripods and adjustments will be placed on the bottom level of the scaffoldings. Figure 9 shows the Dynamo modules of automating assembling scaffolding, and Figure 10 shows the assembling results.

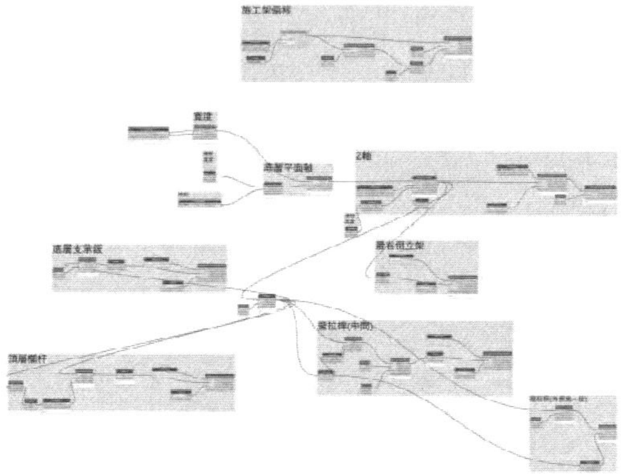

Figure 9. Dynamo module of automating assembling scaffolding

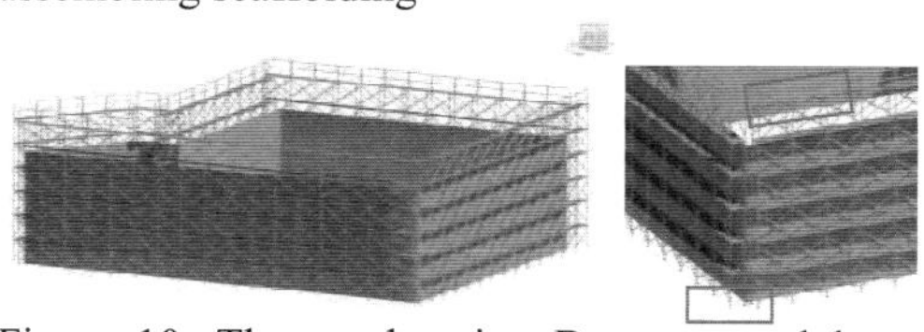

Figure 10. The result using Dynamo module to automate assembling scaffolding

3.3 The Implementation of Material Management

The material supply needs to keep up with the construction schedule. The amount of scaffoldings required is usually calculated by areas. Practically, scaffolding contractors roughly estimate the quantity of material needed for scaffolding without detailed information. If the materials needed for assembling scaffolding are not adequate in terms of size or quantity, the subcontractor may need extra time to transport the materials, which may delay the construction progress. In addition, if the materials are excessive, they will be placed on the site at the construction, which will increase the risk of being stolen and corrosion.

After the scaffolding system is built in BIM, the detailed information of assembling scaffolding can be produced, and those required to assemble scaffoldings will be managed by Microsoft Access in this study. Figure 11 shows the data flow of the material management in this research.

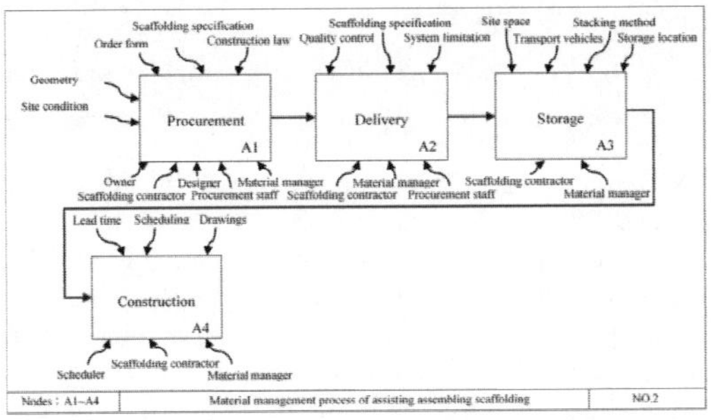

Figure 11. The data flow of the material management

4 Case Validation

The science and technology college building at National Cheng Kung University in Taiwan, as shown in Figure 12, is used for verifying the proposed approach. The exterior of the facility is over 50 meters in height, therefore scaffolding of CNS4750 must be used according to the construction specification.

There are three sections of this facility. Two buildings are with 7 and 8 floors, respectively, and a corridor that connects these two buildings.

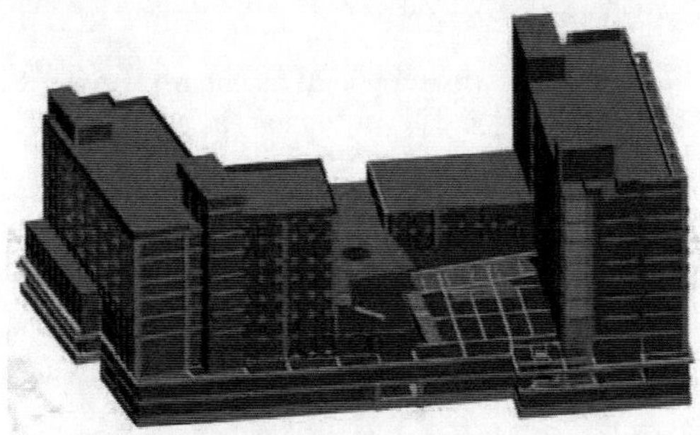

Figure 12. The case study building

The two faces of the facility are chosen to assembling the scaffoldings. The depths of the work faces are less than 77 cm, so a full set of scaffoldings is not required. And the topography is even, laid leveling is not required. Cranes are required to transport the scaffolding materials because of the total height. One face needs adjustments in the bottom scaffoldings, and the other face needs tripods in the bottom scaffoldings. Table 3 and Figure 13 shows the parameters which are put into the ontology model and the inferred results.

Table 3. The input parameter and the inferred result from ontology model

parameter	Value	Inferred result	parameter	Value	Inferred result
Height	450		Height	400	
TotalHeight	450	Do not need leveling	TotalHeight	850	Do not need leveling
Length	2107		Length	4600	
Depth	0	Need crane1	Depth	0	Need crane3
Topography	Even		Topography	Even	
Entrance	No	Need rectangle erection method	Entrance	No	Need rectangle erection method
WorkItem	Grouting		WorkItem	Grouting	
Level	1F	Need adjustment	Level	2F	Need tripods
Position	East		Position	North	

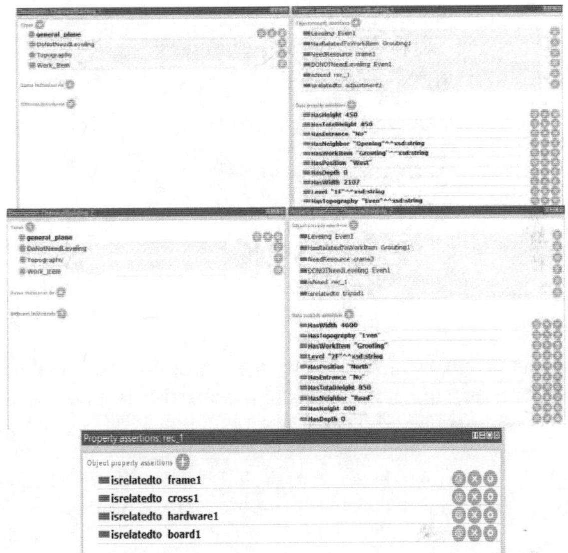

Figure 13. The inferred result in the Ontology model

Then, based on the inferred ontology results, the proposed Dynamo modules are applied to develop the scaffolding system in BIM. Figure 14 is the Dynamo module of this case study. Through the use of Dynamo modules in the BIM model, as shown in Figure 15, it can immediately demonstrate the adjustments and the tripods which are needed to be considered in this model.

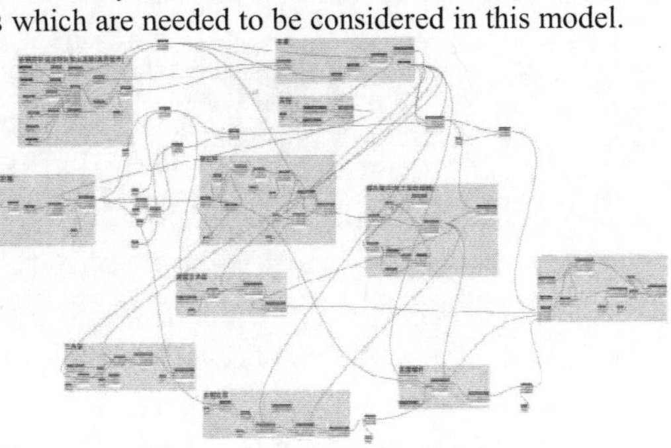

Figure 14. The Dynamo Module of the test area

Figure 15. The visualization of BIM model

After automating the process of assembling scaffoldings, the bill of material to the scaffoldings is generated with the schedule. Figure 16 shows the output from the material management system developed by using Microsoft Access. The upper one is the bill of material according to the delivery time. In this bill of material, this study places the wok item in order to match the schedule. The other one is the bill of material according to the purchase order number. With this bill of material, the owner can easily know the cost flow, and where the material will be used.

Figure16. The bill of material in this case study

5 Conclusion

In this study, a framework that can automate the process of generating information for assembling the scaffolding system is developed in terms of material management. This study analyzes the boundary conditions and impact factors to make the process of assembling scaffoldings more efficient. In addition, this study adds work items and sections to match with the construction schedule. Working regulations and practices at the construction site are considered to define the BIM development guide through the ontology model. This also helps others for sharing the knowledge of assembling scaffoldings. The implementation of Dynamo shows that the process of generating information for assembling scaffolding system can be optimized. Finally, the detailed information for assembling scaffoldings can be effectively generated from the BIM model developed and then output for the use of the management system.

Several issues are not discussed in this study and can be studied in the future. 1. The complex of the geometric shapes of the exterior of the building. 2. Various types of scaffolding employed. 3. More detailed construction activities related to the scaffolding.

Acknowledgement

This work was supported by the Ministry of Science and Technology, Taiwan under Grant MOST105-2221-E-006-047-MY2.

References

[1] Wu G. Y., the precaution of discharging place and garbage slag of external wall scaffolding, *Safety and Health of workers*, Taiwan, 224:8-15, 2008.

[2] Lin Z. H. and Feng C. W., Using Ontology and BIM to Coordinate Trades of the Interior Renovation Project， Taiwan， NCKU thesis, 2016.

[3] Construction Industry Institute， *Playbook of Leading Industry Practices for Estimating, Controlling, and Managing Indirect Construction Cost*, The University of Texas at Austin, 2012.

[4] Kim J., Fischer M., Formalization of the features of activities and classification of temporary structures to support an automated temporary structure planning, *in: Proceedings of the 2007 ASCE International Workshop on Computing in Civil Engineering*, 338-346, 2007.

[5] CADS, websites, http://www.cads.co.uk/, Accessed:20/03/2017

[6] Kim K., Cho Y., Zhang S.， Integrating work

sequences and temporary structures into safety planning: Automated scaffolding-related safety hazard identification and prevention in BIM, *Automation in Construction* 70, 128-142, 2016.

[7] Kim K. and Teizer J., Automatic design and planning of scaffolding systems using building information modeling, *Advanced Engineering Informatics* 28, 66-80, 2014.

[8] The Dynamo Primer, http://dynamoprimer.com/en/, Accessed:20/03/2017

[9] Noy N. F. and McGuinness D. L., Ontology Development 101: A Guide to Creating Your First Ontology, *Stanford University*, 2001.

PAD Based 3D Earthwork BIM Design Module for Machine Guidance

Waqas Arshad Tanoli[a], Hassnain Raza[a], Seung Soo Lee[a] and Jong Won Seo[a]

[a]Department of Civil and Environmental Engineering, Hanyang University, Seoul, South Korea
E-mail: waqitanoli@gmail.com, razahassnain99@gmail.com, rokhonor99@hanmail.net, jseo@hanyang.ac.kr

Abstract –

Building Information Modeling (BIM) technology developed in the recent decades and provided a new perspective which shifted the trend from simple 2D drawings to intelligent 3D modeling. BIM is usually applied in many construction fields, however, it is not very common in earthwork projects. Earthwork is one of the primary component in any construction project especially in highways, dams etc. In the earthwork it has become possible through BIM to monitor and analyze the changes in model accurately. This study presents a discrete module system devised to support the Pad based 3D Earthwork BIM design, topographic modeling, and earthwork volume calculation. In the case of a slope, the direction and inclination of the slope can be incorporated into the design. The point cloud data obtained from GPS survey and Laser scanning process which can be then used to create TIN (Triangulated Irregular Network) surface and ground mesh. It provides a 3D BIM design compatible with software like Autodesk Civil 3D and also with Machine guidance system. This paper discusses the importance of Pad based earthwork design with its practical site application and also presents a case study using Machine guidance. The results of experiments on construction projects show that this approach can be effectively used for earthwork operations.

Keywords –

BIM; 3D Modeling; Earthwork; Machine Guidance

1 Introduction

Earthwork operations possess a significant value in any construction project and can cost about 25% for the overall project budget. The estimation and planning for earthwork operations directly impact the cost and duration of the project, and these operations signify the success and failure of the project [1]. The earthwork operations also required the use of modern technology like other construction industry. The advent of new technologies also leads the construction industry in utilizing the modern techniques, especially in the earthwork process. The machine guidance system is one of such examples which allows the operator to work more efficiently by using the design on the display screen. The operator can use 2D or 3D design which helps in eliminating the traditional surveying methods which can create hindrance in the smooth operations [2]. The virtual 3D environment is also very helpful to check the shortcomings and for integration of design with construction process [3]. In the construction industry, there are several methods and software in use for the 3D modeling. It is important that such models contain all the essential details regarding surface. Different researchers adopted simulation approach for the earthwork planning to make it simple and practical [4].

The pad-based design is relatively easy to implement at the construction site and it is more efficient for building projects. However, this method can also be applied to other infrastructure projects for design and quantity estimation. There are different tools for earthwork estimation based on prismoidal or average end area method [5]. The average end area method gives higher accuracy in results with distance less than 30 meters between two cross sections [6]. The surface to surface method for earthwork volume calculation have higher accuracy than the average end area(AEA), and when the cross section interval is decreased, the AEA method provides more accurate results theoretically [7]. The digital terrain modeling (DTM) helps in volume calculation using laser scanning data and CAD software [8].

This paper outlines a new approach to earthwork design with the development of module system which significantly calculates the actual earth volume and generates a TIN surface for machine guidance. This method has importance in generating TIN surface which can be used for visualization of the 3D model as well as for the machine guidance system. Conventionally, the conversion of the 2D model to 3D is a cumbersome and tedious process [9]. The module is developed using visual basic programming and is compatible with the

CAD based commercial software. It is capable of targeting another surface from Pad, slope generation, volume calculation, and the creation of a model for machine guidance. The machine guidance technology has replaced the staking process and labor intensive survey work [10]. This technology is primarily used for earthwork projects. However, it is also useful for other operations like paving and lifting [11]. The integration of IT based system with the earthwork site results in higher efficiency and productivity. In earthwork process, the proper flow of information is essential as it can increase the transparency and lead to process improvement [12]. This module provides an easy approach for the model generation with its user-friendly interface.

2 Machine Guidance

Development of the machine guidance system in the construction industry is one of the most important reasons for using three-dimensional model. In the AEC industry traditionally 2D design drawings are used for construction operations which do not contain the elevation information. However, it is required to convert such drawing plans to compatible 3D models for machine guidance. The successful automation at the site is only possible if the designer is well aware of crucial factors in the design [13]. The 3D digital terrain design model is usually used for machine guidance. These models provide the information regarding elevation and terrain design to the operator through digital screen mounted in the cabin of earthmoving equipment.

Many researchers have identified the enhanced quality and productivity in the construction projects using automated machine guidance [14]. The modeling for the machine guidance system also improves the planning for the construction process with less rework on the designed model [15]. Use of such technology provides higher accuracy with less time and reduced cost. The operator of the earthmoving equipment can effectively visualize and move around the site using equipment position on the design model [16]. Different sensors are used on the excavator at the bucket, arm, and boom. These sensors provide the exact position for the bucket in the machine control system in the cabin.

3 Methodology

This paper presents a new BIM-based module system for earthwork operations. Visual Basic 6.0 is used as the main software language for the development of this module, and it uses Windows operating system. The system is developed in conjunction with CAD engine to produce 3D shape information. This module is an independent program and compatible with the other commercial design software like Autodesk Civil 3D. This research presents the PAD based technique for the modeling of earthwork design. It applies the volume calculation method which uses projection and interpolation for accurately calculating the earthwork volume. The earthwork design system components are presented in Figure 1.

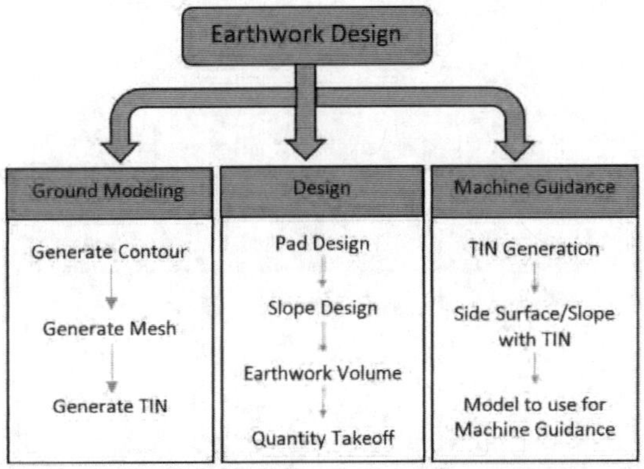

Figure 1. Earthwork design system components

The module has simple and user-friendly graphical user interface (GUI) which is shown in Figure 2. It has import function through which ground surface in the form of contour or TIN can be directly imported. It is also capable of importing point cloud data using .mdb file and generate the ground TIN surface. The Pad surface is created by providing the design elevation with co-ordinates information. The Pad based design can also be drawn on the topographic model with the desired altitude value, as all the points on Pad have the same elevation.

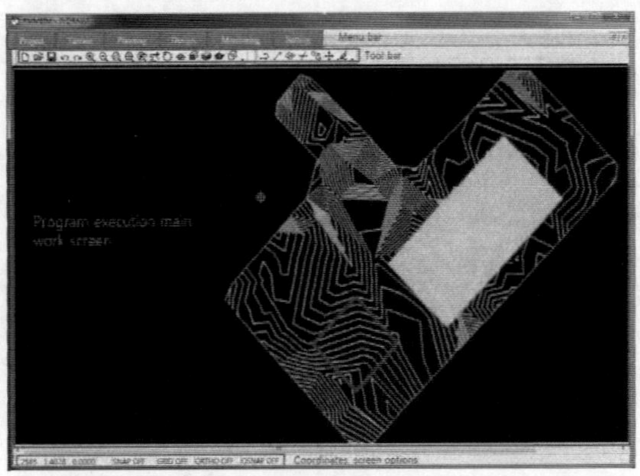

Figure 2. Module main screen

The TIN surface is automatically created using points in the Pad area. However, the next step in the design is to connect the Pad with the existing ground surface for which the projection angle is required. Depending on the site requirements it is possible to generate the projection from the Pad surface in vertical or slope form, as the system can calculate the volume in each case. The program can calculate the cut and fill volume at any given section along the length of design Pad. The process of volume calculation is explained using flowchart in Figure 3.

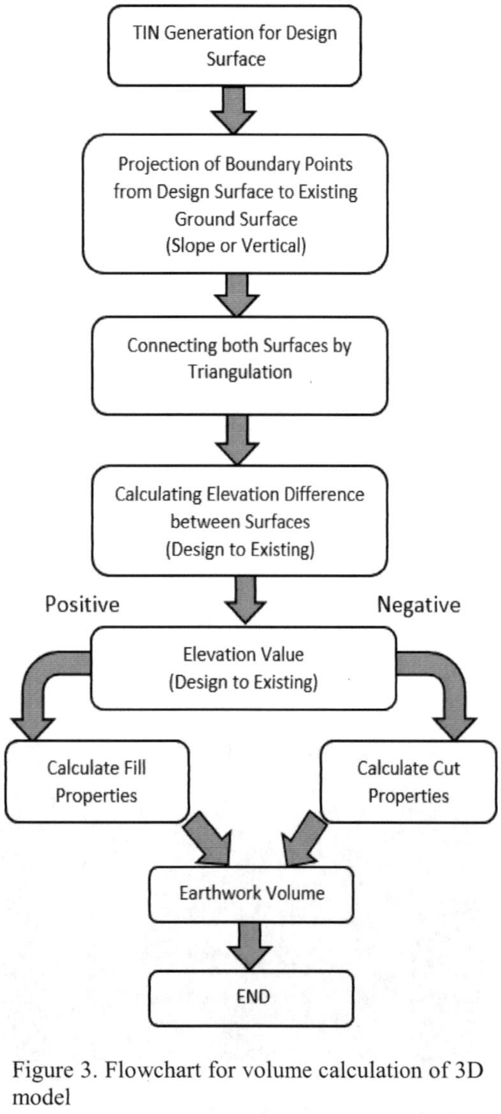

Figure 3. Flowchart for volume calculation of 3D model

Traditionally, average end area method is used for volume calculation based on the average area between two consecutive cross sections.

$$V = \frac{L}{2} * [A_1 + A_2]$$

Where A1 and A2 are the areas of cross-section and L is the distance between these cross-sections. It is challenging to use this method continuously for the nonlinear sections [17]. However, this module used the surface to surface method for better accuracy. A new surface elevation is calculated based on the difference between two surface elevations using the location where the edges of triangles intersect the two surfaces as shown in Figure 4.

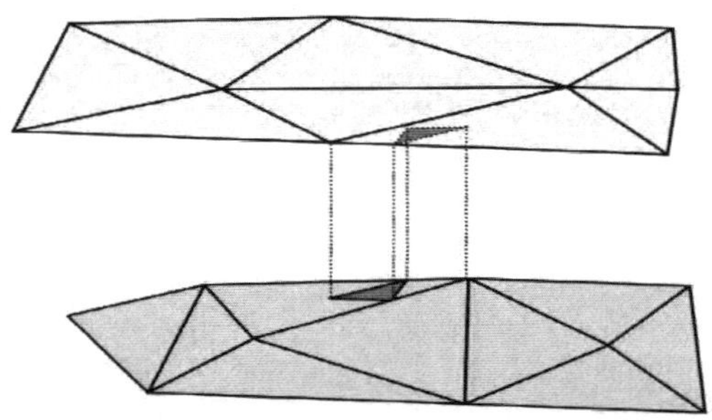

Figure 4. Surface projection illustration

It is important to know the elevation difference between existing and planned ground surfaces. The existing and proposed ground TIN surfaces are created using survey and design data. In the case of the Pad projection with no slope to the existing ground surface, the triangular network for side wall is generated between two surfaces. In this case, ground and design model share the same (XY) coordinates with the only difference of elevation. By using Z (elevation) value, the quantity can be estimated by triangular columns volume. If the Pad elevation value “Z” is lower than the ground TIN surface the volume is calculated as Cut, however, if the designed Pad is located at a higher elevation than the ground surface, the fill volume is calculated. A polyline is drawn along the Pad surface in a longitudinal direction from one end to the other for dividing the area into equidistance stations. The module also provides the equipment monitoring function. The design drawings in the module are saved in a **.dwg** file format which is converted into a **.dxf** file using Autodesk Civil 3D. The 3D design model with DTM (Digital Terrain Model) is then available for use in the machine guidance system.

4 Case Study

The following case study aims to check the practical application and the validity of developed module system. The model is designed using the Pad based module and is verified on site for use in machine guidance system.

A site in Seoul, South Korea is selected as a pilot project to test the developed module system. The area under study is not in regular rectangular shape. The site topographic data is collected using laser scanning, the 3D laser scanning technology provides higher density data as compared to traditional surveying methods. Ground contour data is imported into the module and using its TIN generation function the surface is converted into TIN. Figure 5 presents the original ground surface model in contour and TIN form.

A rectangular area is selected, and Pad design is drawn on the BIM-based topographic model. The Pad dimensions are based on required length and width at the site. The design altitude value is then provided to the Pad, as the whole surface have same elevation value. According to design and site requirements, it is possible to design in slope and vertical form, and the system is capable of calculating the volume in both cases. Figure 6 shows a Pad design with the slope, projecting to the existing ground surface. The outer boundary is indicating the connection with the existing ground surface, and inner rectangular marking depicts the Pad design surface.

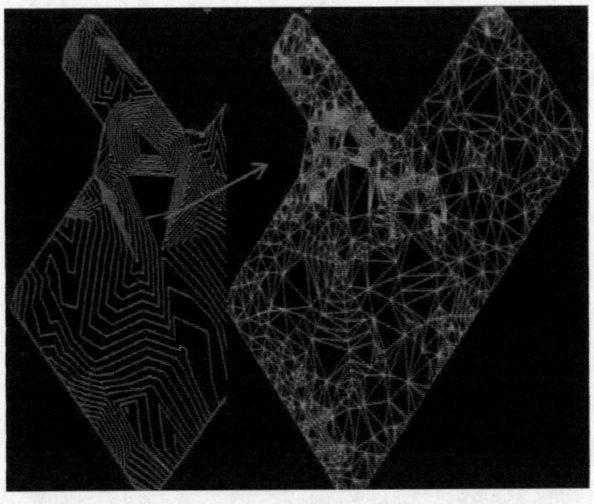

Figure 5. Ground surface model in contour and TIN form

Figure 6. Pad-based design with existing ground surface

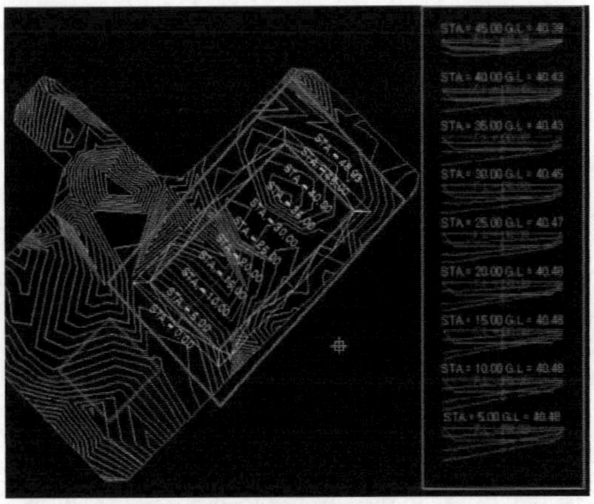

Figure 7. Pad design with station information

The Pad design surface is then divided into stations along its length as shown in Figure 7. The ground level elevation can be seen at any station. It provides the function to divide the Pad into stations with equal or different length. The Pad based design file having TIN surface is imported in Autodesk Civil 3D and saved in .dxf file format. The machine guidance control panel requires the TIN file in .dxf file format. The files with TIN surface are only visible in the control panel of earthmoving equipment. The Pad design file is then imported into the machine control panel. Figure 8 shows the view of machine control panel for the operator in the earthmoving equipment.

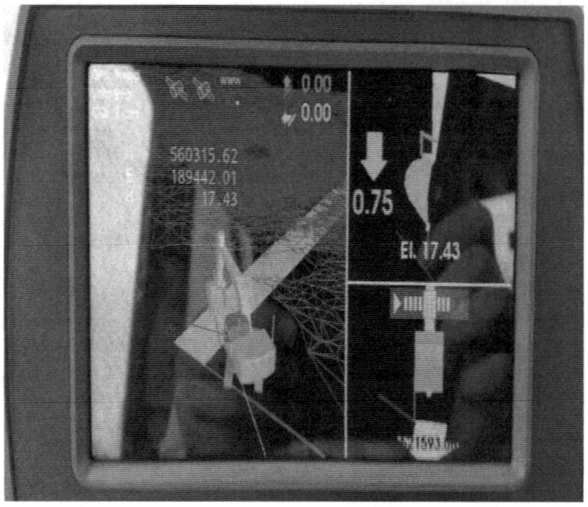

Figure 8. View of machine control panel

5 Conclusion and Future Study

This paper provides the information regarding Pad based BIM module system developed for the earthwork design. Visual Basic 6 is used for development of this module in conjunction with the CAD engine. The topographic modeling function supports 3D BIM design using the point cloud data obtained from laser scanning. It is possible to draw and design a Pad on the 3D ground with consideration of slope and vertical planning. For the volume calculation of cut or fill in between Pad design surface and existing ground, this module used the surface to surface method for better accuracy.

The developed module is capable of creating a TIN model for use in machine guidance system. A case study has proved the validity and effectiveness of this module for earthwork design. The module at this time can only deal with TIN models, however, with the integration of algorithm for solid and parametric modeling technique, its usability will increase. Further studies are in progress to update this Pad based BIM module, incorporating the 4D (time) and 5D (cost) design functions to automate the calculation of time and cost at any particular section. The large scale validity assessment will also be carried out in projects with different cost and duration.

6 Acknowledgement

This work is supported by the Korea Agency for Infrastructure Technology Advancement (KAIA), grant funded by the Ministry of Land, Infrastructure, and Transport (Grant: 16SCIP-B079689-03).

References

[1] Smith, Simon D., Graeme S. Wood, and Martin Gould. A new earthworks estimating methodology. *Construction Management & Economics*, 18(2): 219-228, 2000.

[2] Seo, J., Lee, H., Cassule, L. N., & Moon, S. Machine Guidance Based Site Control Technology (SCT) for Earthwork Equipment Fleet. In ISARC. *In Proceedings of the International Symposium on Automation and Robotics in Construction*, (Vol. 32, p. 1). Vilnius Gediminas Technical University, Department of Construction Economics & Property, 2015

[3] Zhang, Y., Moghani, E., AbouRizk, S. M., & Fernando, S. 3D CAD modeling and visualization of the tunnel construction process in a distributed simulation environment. *In Simulation Conference (WSC), In Proceedings of the 2010 Winter IEEE,* pages 3189-3200, 2010.

[4] Ji, Y., Borrmann, A., Rank, E., Wimmer, J., & Günthner, W. An Integrated 3D Simulation Framework for Earthwork Processes. *In Proceedings of the 26th CIB-W78 Conference on Managing IT in Construction*, 2009.

[5] Kim, E., & Schonfeld, P. Estimating highway earthwork cross sections by using vector and parametric representation. *Transportation Research Record: Journal of the Transportation Research Board*, (1772): 48-54, 2010.

[6] Cheng, J. C., & Jiang, L. J. Accuracy comparison of roadway earthwork computation between 3d and 2d methods. *Procedia-Social and Behavioral Sciences*, 96, pages 1277-1285, 2013.

[7] Hintz, C., & Vonderohe, A. Comparison of earthwork computation methods. *Transportation Research Record: Journal of the Transportation Research Board*, (2215):100-104, 2011.

[8] Uddin, W. Airborne laser terrain mapping for expediting highway projects: Evaluation of accuracy and cost. *Journal of Construction Engineering and Management*, 134(6): 411-420, 2008.

[9] Vonderohe, A. P. Implementation of GPS Controlled Highway Construction Equipment: Phase III (No. CFIRE 02-13). *University of Wisconsin-Madison, Construction and Materials Support Center*, 2009.

[10] Hammad, A., Vahdatikhaki, F., & Zhang, C. A novel integrated approach to project-level automated machine control/guidance systems in construction projects. *Journal of Information Technology in Construction (ITcon)*, 18(9):162-181, 2013.

[11] Kirchbach, K., Bregenhorn, T., & Gehbauer, F. Digital allocation of production factors in earthwork construction. *In Proceedings of the 20th Conference of the International Group for Lean Construction*, 2012.
[12] Peyret, F., Jurasz, J., Carrel, A., Zekri, E., & Gorham, B. The computer integrated road construction project. *Automation in Construction*, 9(5): 447-461, 2000.
[13] Jonasson, S., Dunston, P. S., Ahmed, K., & Hamilton, J. (2002). Factors in productivity and unit cost for advanced machine guidance. *Journal of construction engineering and management*, 128(5):367-374, 2002.
[14] Dunston, P. S., & Monty, J. Practices for seamless transmission of design data from design phase to construction equipment operation - a synthesis study. *West Lafayette, Indiana: Purdue University*, 2009.
[15] Barrett, L., & TIG–AMG, C. A. Automated Machine Guidance: An Emerging Technology Whose Time Has Come?. *In Transportation Research Board Annual Meeting.* Washington, DC. January 2008.
[16] Baertlein, H., Carlson, B., Eckels, R., Lyle, S., & Wilson, S. A High-Performance, High-Accuracy RTK GPS Machine Guidance System. *GPS Solutions*, 3(3):4-11, 2000.
[17] Slattery, K. T., Slattery, D. K., & Peterson, J. P. Road construction earthwork volume calculation using three-dimensional laser scanning. *Journal of surveying engineering*, 138(2):96-99, 2011.

Towards life cycle complete BIM

Sven Stumm[a], Philipp Schwan[d], Ralf Becker[b], Elisa Lublasser[a], Jörg Blankenbach[b], Dirk Vallée[c†], Linda Hildebrand[d] and Sigrid Brell-Cokcan[a]

[a]Chair of Individualized Production in Architecture RWTH Aachen University
[b]Chair for Computing in Civil Engineering and Geo Information Systems RWTH Aachen University
[c]Chair of Urban and Transport Planning RWTH Aachen University
[d]Chair for Reuse in Architecture RWTH Aachen University

E-mail: stumm@ip.rwth-aachen.de, pschwan@rb.arch.rwth-aachen.de, ralf.becker@gia.rwth-aachen.de, lublasser@ip.rwth-aachen.de, blankenbach@gia.rwth-aachen.de, vallee@isb.rwth-aachen.de, lhildebrand@rb.arch.rwth-aachen.de, brell-cokcan@ip.rwth-aachen.de

Abstract –

The goal of Building Information Modelling (BIM) is the integral overarching digital modelling of all properties regarding a building as well as its construction. This is targeted through the integration and support of all involved experts. Due to the interdisciplinary nature of this task, the necessary data often only exists in a very fragmented and uncoordinated way throughout different departments. BIM tries to mediate between the different views of its users and allows for a coordinated accumulation of data, as well as synchronously keeping the planning status up to date. However, in most cases this available information is still not used throughout the complete lifecycle of the building. Clear gaps exist between the different phases of planning, construction as well as use and maintenance. Within this paper we aim to give an overview of possible BIM data, as well as its relevance for construction processes. This is especially considered from a process automation viewpoint, in order to identify the required interfaces as well as missing and irrelevant data for assembly processes within construction. Based on this a BIM informed design to production becomes possible. Primary goal is the development of strategies for the creation of BIM-aided planning processes that continue into actual fabrication and construction and are efficient in terms of time, energy and costs considering changing user demands and the resulting usage requirements. We address the lifecycle of a building using the concept of BIM. Therefore, we also consider the required information for future conversion and refurbishment of the building, in order to complete the lifecycle approach of BIM. Hence, we give a first brief overview of possible techniques for modelling pre-existing buildings within BIM, as well as the required level of detail (LoD). Looking closer at the LoD for different lifecycle phases.

Keywords –

BIM; automation; lifecycle; refurbishment; BIM to production

1 Introduction

The number of vacant buildings will grow due to structural changes while housing units are urgently needed [1]. To work against the housing shortage and to preserve the economic and ecological potential, existing buildings need to be efficiently checked and evaluated regarding their ability for conversion and refurbishment. If the estate is excluded for that purpose reuse and recycling need to be considered regarding their feasibility.

Currently Building Information Modelling (BIM) tries to streamline information flow within construction planning, while simultaneously allowing for a distributed workflow. The goal of BIM is the comprehensive digital modelling of all properties regarding a building. Apart from the constructive features this also implies the technical, functional and mercantile aspects. The basis for this integrated modelling approach are object-oriented and semantic-rich 3D building models. This opens up new opportunities within the highly interdisciplinary field of construction, because it improves the data exchange between all involved parties and leads furthermore to a distinct higher transparency for an optimized coordination. Discrepancies and errors can be revealed and corrected in the early planning phase, which is a crucial requirement for the safety that comes up with planning, firm and costs and also for the efficient building construction and exploitation.

However, this process requires additional effort, time and cost for the BIM planning, which can only be compensated if the construction process is streamlined through this method. BIM is currently often only used at the planning stages, while the actual assembly and construction still uses print outs of building plans.

Especially the area of construction and refurbishment, which are major parts of a building's lifecycle are still only loosely covered through BIM.

In order to move towards a more life cycle complete BIM we take a look at the current hurdles within the application of BIM as well as the difficulties and cost arising from Modelling existing buildings. We illustrate possible application scenarios as well as an outlook towards BIM informed process automation. In particular we focus on the digital reconstruction and refurbishment of existing buildings as well as first steps towards the planning of construction processes through BIM.

In the context of existing buildings new technologies, such as sensors and data programs, can provide complex information when only limited or no reliable documents about the building are accessible. This can be useful to base different planning decisions on. It is here considered to evaluate the conversion ability for future use scenarios basing on the building substance, its geometry and the material properties. For data collection the as-built detection in the architectural context was initially motivated by cultural heritage and preservation focusing on visual information. Non-visual information about the construction was needed when changes in the floor plan were made or in the case of refurbishment.

A 3D model including the information provided by sensor technology in combination with a matrix of interdependencies can provide a basis for decision in respect to the follow up function. This paper provides an overview of the methodology how to come from an existing building to the evaluation of the conversion ability for each scenario. It includes the phases of data generation, the role of a volumetric model including information (BIM), the criteria in the evaluation matrix which provides the results. A comprehensive state-of-the-art is introduced for each part. At the end the evaluation process is discussed outlining potential and research- as well as technological gaps.

The findings have been derived from literature review extending [2][3] and a limited part has been developed in small size tests. The methodology used to evaluate the ability (for reuse or recycling) has been developed in the research project RDFRS (Robotic Façade Disassembly and Refurbishment System) and has been further developed for educational purposes.

2 Data collection

The data needed to prepare the evaluation of a building can be grouped in geometric and material related information. Employing BIM additionally allows for the representation of metadata, like dimensions, functions, connections and part topology. Automating processes for this semantic association of parts is the research focus of Scan-to-Bim [3]. For each conversion scenario data from different groups are required. Excluding potential contamination is priority for all scenarios. Once via the geometric requirements are fulfilled, material testing is sensible. The acquisition of the building geometry delivers the bases to appoint later gathered material information in it.

Table 1 Assignment of conversion scenario and type of needed information.

Conversion scenario	Criteria to evaluate the suitability for each scenario	Geometry	Material
Refurbishment (building level)	Detecting contamination		X
	Structural safety	X	X
	Evacuation routes	X	
	Ability for conversion of floorplan	X	
Reuse (building component level)	Detecting contamination		X
	Modularization	X	X
	Economically sensible disassembly		X
	Structural integrity		X
Recycling (material level)	Detecting contamination		X
	Material purity		X
	Purity of fraction		X

2.1 Scan-to-BIM

For the geometry acquisition of buildings various surveying techniques can be used. The four basic common techniques are photogrammetry, terrestrial laser scanning (TLS), tachymetry and electronic devices for manual surveying. Data collection using 3D point clouds is the most common method for generating BIM compliant as-built models. The point clouds can be produced solely from imaging methods such as Structure from Motion (SfM) [4] or TLS [5]. TLS is the most common approach for point cloud generation with the purpose of as-built BIM documentation (Scan-to-BIM).

Although TLS delivers a dense 3D point cloud suitable for a detailed reconstruction of the building geometry, there are still a high number of manual steps necessary to utilize this data for more than visualization purposes. In order to be able to use this information in combination with BIM the data need to be matched to a set of building elements. Even though there are a number of concepts which try to determine building elements automatically from scan data, currently this is mainly a manual process. Additionally, there are only a few BIM systems capable of handling large point clouds as is necessary for a reconstruction of BIM from point cloud data. Point cloud data is often only used in the form of a planar projection onto two dimensional CAD drawing, which often results in a loss of information.

Even though the point cloud data itself is very high in quality, the modelling remains to a large extent a manual time consuming and cost intensive process. While this might be feasible for larger projects for the restoration of historically significant buildings, this is not a process applied comprehensively to the diverse field of restoration and refurbishment [2].

Even though a fully automated and flawless part recognition is envisioned and pursued, with the current state of the art this is still unrealistic [3]. The main problem that arise come from the conflict between a simplified and generalized BIM in contrast to a highly detailed representation of reality. Objects in real environments always have more characteristics and features than just those that are relevant or capable of being modelled. For instance a model building part might only require the parameters width, height and sill to floor height to be modelled. The 3D scan data however contains all geometric characteristics and details and needs to be able to determine, which information is or is not relevant. This requires an underlying model in order to associated and understand the context of the scan data.

Furthermore, point clouds typically do not fully describe geometries. Shadows in the field of view of the scanner create holes. Even though it might be possible to close some of these by matching multiple scans with different fields of view, there are many cases where a one hundred percent coverage is just not possible. For these holes in the point cloud a fully automated system must be able to fill in this missing knowledge. Manually we are able to do this due to spatial recognition and experience based reasoning.

2.2 Determining materials

Materials can be recognized by sensor technology due their reflectance to waves. The most common non-destructive testing (NDT) techniques are infra-red, ultra-sonic and imaging techniques as well as the less common in material determination radiography [6]. However in some cases these still require a preparation of the surface in order to reduce reflection due to irregularities. Simultaneously destructive testing methods can release previously sealed contaminants.

As these processes require a high number of manual tasks in addition to multiple visual on-site inspections, efforts have been made in the direction of automating aspects of non-destructive evaluation (NDE) [7].

However especially in historic buildings the challenges often are beyond identifying the material, but analyzing the origin and the transformation of the material over time. Due to environmental exposure, stress as well as possible repairs the structural properties of the material may have been transformed. However this information is specifically important for a structural analysis of the support structure and seldom analyzed to the extent of the required parts refurbishment.

Yet in order to better plan deconstruction processes of non-support bearing buildings parts at least a general knowledge of the material and thickness is required. Current building information models infrequently contain this information. Scan-to-BIM of existing buildings especially of historic buildings does currently not contain any construction documentation if not additionally modeled manually and costly [3][8].

This knowledge however is not only a requirement to plan deconstruction processes but also to determine if and how the refurbishment should take place. Potential contamination should be determined before a refurbishment scenario is considered. A common method to determine existing contamination within a building are air quality measurements as well as manual visual inspection of potential affected parts (e.g. HVAC systems for asbestos) combined with spectrographic material analysis for verification.

A 3D scan of the building site combined with camera imaging potentially allows to plan on-site inspection and necessary specimen sampling, beforehand. However even the aforementioned testing methods are able to miss concealed contaminants, which could potentially be uncovered during the refurbishment, most likely resulting in the end of the refurbishment project. Even in those cases a refurbishment could have been possible without disturbing the existing structure and therefore releasing the contaminant. This however would require a more cost efficient way to extensively determine hidden contaminants or at least indicators for further testing.

NDE procedures have the potential to do this, however these technologies often times require comparative results to measure them against and in most cases are not fully reliable. By combining a finite amount of building materials from BIM with 3D scan as well as NDE planning contamination testing can potentially be optimized for both towards efficiency and thoroughness.

3 Material as interface to building elements

Integrating a more accurate process of material information and potentially dimensioning of parts through Scan-to-BIM combined with NDE does not only allows for contamination testing. The resulting LoD can also be employed within construction process planning especially for potential automation scenarios. Currently contaminated construction site in most cases only allow for costly demolition and disposal of contaminated material. Automation of construction processes specifically through robotics has the potential to mediate the cost and reduce risk of injury. It is also possible to employ the created digital BIM directly for deconstruction and refurbishment planning. Within [9] we showed first concepts for employing digital parametric planning for assembly sequences execution with on-site adaptation.

A parametric robot control as provided by the KUKA|prc [10] allows for the individualized fabrication of parts through directly interconnecting CAD and robotic fabrication processes. These fabrication processes and the required machining tools are however highly dependent on the material of the fabrication part. Planning robotics constructions is illustrated employing a layered model of building parts e.g. a wall structure as shown in Fig. 1 using the example of an Exterior insulation and finish system (EIFS).

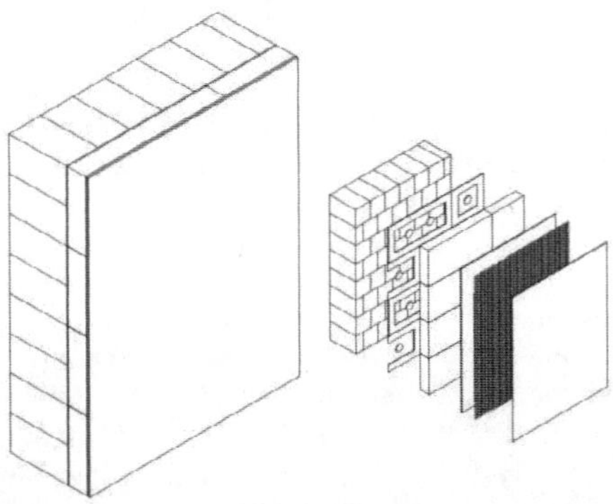

Figure 1. Layered model of an Exterior insulation and finish system (EIFS) consisting of: Walling, Bonding, Insulation, Reinforcement and plaster

Using a detailed knowledge of the layer structure allows for the planning of every subsequent deconstruction step for material extraction that allows for material separation, which is an important part of recycling.
However n difference to single part fabrication planning adaptation towards the modified environment has to be take into account for every layer. The complexity of assembly and disassembly is higher than pure material fabrication due to the interdependency of parts.
Using a layered approach a direct sequence can be planned for, however changes need to be captured by the system we recommend a sequential directly linked scan after every layer processing. Alternatively NDE data can be employed for a better approximation of each layer before deconstruction.

In addition to employing the layered model in order to inform a refurbishment process automated systems allow for a consistent feedback of all relevant construction data which extends the LoD of the model by adding details about positioning of parts as shown by Fig. 2, this in turn can be used for later refurbishment.

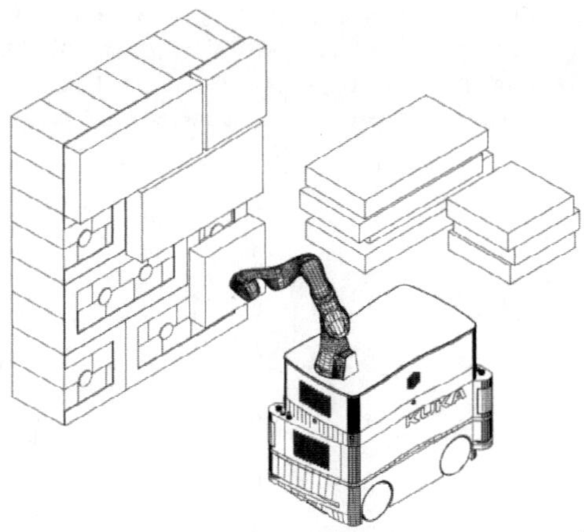

Figure 2. Robotics positioning of insulation plates with potential feedback of process data

4 Evaluation of the suitability for refurbishment, reuse or recycling

The motivation to keep as many estates as possible functioning (as named in the introduction) follows ecological motives to limit the impact on the environment while providing a long-lasting service. Whether a building qualifies for an additional use phase depends on complex factors which can be categorized in aspects regarding the site and the ones that result from the building physics. While the collection and processing of data on urban level is growing and with information providing tools like Geographic Information System, are already well established, in this context only the building physics is considered.

Fig. 3 shows the simplified hierarchy for scenario planning after a functional period ended. The sensor- and computer-aided processes basis upon on this and aims in providing relevant information in the appropriate planning phase.

For refurbishment, mostly geometric properties need to be investigated (see chapter 2). Lawful requirements, like the length and width of escape routes or the ceiling height are parameters which can be captured by sensor, measured and evaluated within the model. Additionally, individual requirements can be formulated and integrated within the comparison. The procedure works as a collision check, in which potential conflicts will be identified.

The suitability for reuse and recycling is impacted by the building capabilities, the resolvability from the construction context and the integrity of the element's functionality. For example, the building site should provide sitting area for heavy machinery and containers as well as the structure's and the adjacent building's cubature needs to allow the deconstruction [11]. The resolvability means providing material and components with an appropriate effort to disconnect a materials or components from one another. This includes resolvability between: component and component, material and component, material and material.

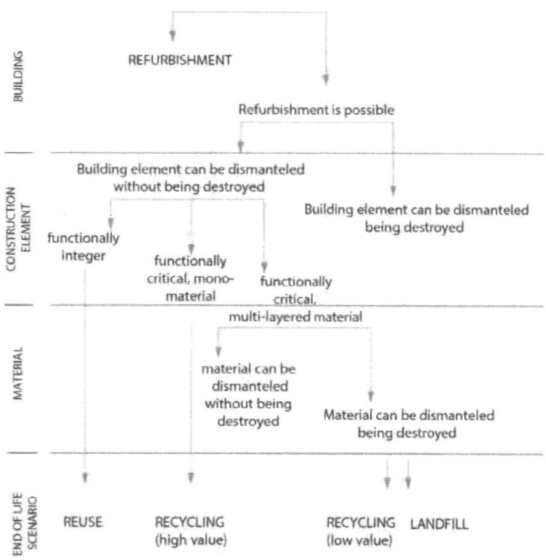

Figure 3. Simplified decision hierarchy for unused building substance scenarios

In order to indicate the level of resolvability the association of the material or products with the end of life scenario reuse or recycling (methodology developed by Durmisevic [12]; integration into BIM Markova [13]) can be found. Here a more detailed approach is used. The types of connection are differentiated regarding their level of connectivity which indicates their end of life potential. The following four categories are distinguished:

A) Non-destructively detachable
B) Destructively detachable (pure material)
C) Destructively detachable (mixed material same fraction)
D) Non-detachable

Reuse and recycling contribute to resource conservation and limit the emission production and should be applied if not the effort to disassemble is more intensive (resource and monetary). New material can be compared against the effort for the deconstruction. If the ecological impact/price of new product higher than the deconstruction and treatment, reuse and recycling is favorable.

The variation of environmental impact along building materials is reflected by an LCA based calculation. Materials can be investigated regarding their ecological footprint and the more relevant ones (high footprint) can be identified. The result is an integrated evaluation including the type of connection (based the four end of life scenarios) and the LCA calculation which can be used to argue for of suitable end of life scenario. These categories support both, the design decisions in the planning phase and evaluation for a conversion after the first usage phase.

5 Towards model informed construction process automation

The most promising approaches for as-built modelling of existing building with the required semantic model focus on matching scan data and an existing BIM. Accordingly the tested approach was therefore chosen for a preliminary examination. For construction process automation we focus on matching an existing as-planned model with the point cloud from laser scanning (Scan-vs-BIM). By using the knowledge about the deviations between the point cloud and the as-planned model we would like to move towards establishing a link between an abstract, generalized BIM and real world construction scenarios. While the use of BIM is still not as widespread as desired, it is gaining in popularity. However, the field of building assembly modelling is still theoretic only in most parts. Within section 3 we illustrated further how this link of sensor data with an underlying model can be used in order pursue construction process automation on site.

In order to enable renovation and refurbishment planning scenarios within BIM a very important aspect is the determination of used (also hidden) materials as well as possible scenarios of contamination. This is of significant importance especially for building materials that were commonly used during the construction of the building and are now considered harmful (e.g. asbestos, CFCs, Heavy Metal, PCB, etc.). Measurement methods by means of laser scanning, photogrammetry and especially radar, radiography etc. fundamentally seem to enable the collection of inventory data. Parameters like the reflectance or the scattering of the received measurement signal might give information also about material. In first tests we used a full-waveform laser scanner that allows analyzing the waveform of the reflected signal. Evaluating the echoes of the scan pulse (first pulse, last pulse) we could see variation in pulse width of TLS signals depending on the roughness of the material (Fig. 4). However, this field of research is partially at the outset. Distinct measurement methods are available. However, these methods need a structured, comprehensive and automated data collection. A calibration and validation of the derived knowledge from single applications in terms of a generalizability and transferability is necessary. As well as coupling and automatized processing of the just acquired data to the integration in BIM must be guaranteed.

Figure 4: False color display of differences in deviation of the pulse shape of TLS signals

Within the current state of the art material determination is currently mostly still a manual process, especially for hidden materials, with exception to basic structural detection of studs and similar. Within support structures materials are often not only identified but a detailed measurement of material quality is executed (e.g. determining oxidation, carbonation etc.). Some works showed the potential of determining materials combining BIM as a reference database with NDE [3].

A layered material model as well as analysis of hidden parts can be employed as a basis for process planning. This would lead to a new view on BIM employing the used construction material as interface for each building element. Fig. 5 illustrates a first concept for this approach employing material classification for an informed refurbishment process planning.

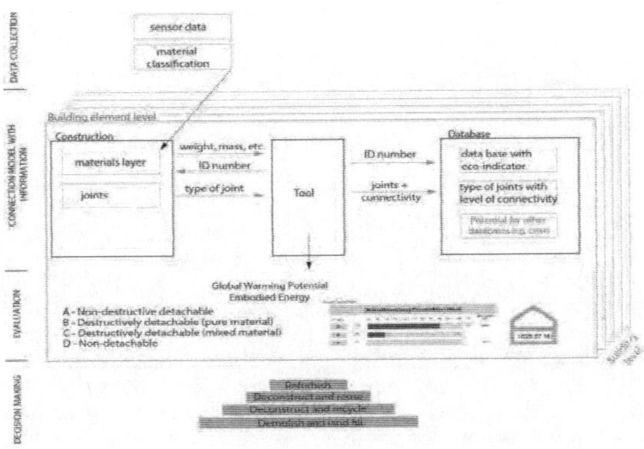

Figure 5. Next steps to integrate information on the building substance as planning parameter

6 Synopsis

Within this paper we took a further look at the current state of the art of using BIM for refurbishment and the relevance for construction process planning. Ilter and Ergen [2] showed that although both refurbishment and maintenance are an increasingly important research tropic current research tends to focus towards the topic of employing BIM for maintenance. We extended upon their identification of automation of post-processing 3D Laser scanning data for semantic association as a common goal within BIM research for refurbishment and maintenance. Additionally current challenges within the process chain from scanning to actual refurbishment were discussed.

While BIM allows for the modelling of all required information, as-built modelling is currently seldom at the necessary LoD required for actual process planning. In most cases a detailed modelling require a high number of manual steps which result in an even higher planning overhead than within design of a completely new building. Without further automation to some of these processes the cost for realization of refurbishment will stay too high for most projects.

However a number of potentials where identified for a better and comprehensive determination of construction materials. This would not only reduce contamination risk it also allows for a higher automation potential within construction processes.

Additionally a first draft for a material layer based planning of refurbishment processes was introduced. Within future work we will extend on this approach using matching existing BIM model with scan data for an on-site model for process execution.

Acknowledgements

Special Thanks to Professor Dirk Vallée for the excellent cooperation within the project Advanced BIM. Professor Vallée and his contributions to the field of urban planning will be missed by all involved in this paper.

References

[1] WBGU. Der Umzug der Menschheit: Die transformative Kraft der Städte. Retrieved from Berlin, Online: http://www.wbgu.de/fileadmin/templates/dateien/veroeffentlichunge/hauptgutachten/hg2016/wbgu_hg2016.pdf, Accessed: 20/03/2017

[2] Ilter, D. and Ergen, E. (2015): *BIM for building refurbishment and maintenance: current status and research directions*, Structural Survey, Vol. 33 Issue: 3, pp.228-256: http://dx.doi.org/10.1108/SS-02-2015-0008

[3] Volk, R. and Stengel J. and Schultmann F. (2014): *Building Information Modeling (BIM) for existing buildings — Literature review and future needs,* Automation in Construction. Vol. 38, pp. 109-127

[4] Westoby, M.J. et al. 'Structure-from-Motion' photogrammetry: A low-cost, effective tool for geoscience applications. In *Geomorphology*, 179:300-314, 2012.

[5] Uren, J. and Price, B. *Surveying for Engineers*, 5th Ed., Palgrave Macmillan, London, 2010.

[6] Maierhofer C. (2010): *Planning a non-destructive test programme for reinforced concrete structure. Non-Destructive*; Evaluation of Reinforced Concrete Structures Vol. 2: Non-Destructive Testing Methods pp.3-12

[7] Dobmann G.; Kurz J. H.. (2010): *Development of automated non-destructive evaluation (NDE) systems for reinforced concrete structures and other applications*; Evaluation of Reinforced Concrete Structures Vol. 2: Non-Destructive Testing Methods pp.30-60

[8] Cheng, H.-M.; Yang, W.-B.; Yen Y.-N. (2015): *BIM applied in historical building documentation and refurbishing*, 25th International CIPA Symposium 2015

[9] Stumm, S. and Braumann, J. and von Hilchen, M. and Brell-Cokcan, S., *On-site robotic construction for assembly using a-priori knowledge and human robot collaboration*, IFToMM/IEEE/euRobotics 25th International Conference on Robotics in Alpe-Adria-Danube Region RAAD (2016)

[10] Braumann, J. and Brell-Cokcan, S. (2011): *Parametric Robot Control: Integrated CAD/CAM for Architectural Design.* In: Proceedings of the 31st Annual Conference of the Association for Computer Aided Design in Architecture (ACADIA)

[11] Motzko, C. et al. *Bewertungsmatrix für die Kostenplanung beim Abbruch und Bauen im Bestand. Datenbanksystem zur Analyse und Bewertung in Bezug auf Kosten, Technologien und Dauern. Abschlussbericht 2015*, Fraunhofer IRB Verlag, Stuttgart, 2015

[12] Durmisevic, E. Transformable Building Structures - Design for disassembly as a way to introduce sustainable engineering to building design & construction, (PhD), Technische Universiteit Delft, Delft, Netherlands, 2006.

[13] Markova, S. Extending IFC for Parametric Sustainable Building Design, In *Proceedings of the 18th International Conference on Computer-Aided Architectural Design Research in Asia (CAADRIA)*, pages 13-22, Singapore, 2013.

Beam for the Steel Fabrication Industry Robotic Systems

Luis F. Rocha[a] , Pedro Tavares[b] ,Pedro Malaca[b] , Carlos Costa[a], João Silva[b] and Germano Veiga[a]

[a]INESC-TEC, INESC Technology and Science, Porto, Portugal
[b]SARKKIS Robotics, Lda., Porto, Portugal
E-mail: {luis.f.rocha,carlos.m.costa,germano.veiga}@inesctec.pt,
{ pedro.tavares,pedro.malaca,joao.silva}@sarkkis.com

Abstract –

In this paper, we present a comparison between the older DSTV file format and the newer version of the IFC standard, dedicating special attention of its impact in the robotization of welding and cutting processes in the steel structure fabrication industry. In the last decade, we have seen in this industry a significant increase in the request for automation. These new requirements are imposed by a market focused on the productivity enhancement through automation. Because of this paradigm change, the information structure and workflow provided by the DSTV format needed to be revised, namely the one related with the plan and management of steel fabrication processes. Therefore, with this work we enhance the importance of the increased digitalization of information that the newer version of the IFC standard provide, by showing how this information can be used to develop advanced robotic cells. More in detail, we will focus on the automatic generation of robot welding and cutting trajectories, and in the automatic part assembly planning during components fabrications. Besides these advantages, as this information is normally described having as base a perfect CAD model of the metallic structure, the resultant robot trajectories will have some dimensional error when fitted with the real physical component. Hence, we also present some automatic approaches based on a laser scanner and simple heuristics to overcome this limitations.

Keywords –

BIM; IFC; Robotized Welding and Cutting; Automation; Construction Industry;

1 Introduction

Considering the market globalization and high economic competitiveness environment that is inserted, the Architecture, Engineering, Construction (AEC) industry is facing the mandatory need to reduce project costs, to increase productivity and quality, and be more time efficient. To meet these requirements in the last few years we verified a clear effort for improving the interoperability and easy share of information between all partners and inner processes involved in a building project. Commonly, in these projects, the number of software tools involved (for electronic tendering, invoicing, 3D - design, structural analysis, and others) is very high. These software tools, commonly lack in compatibility of information and file standardization, making this industry dependent on paper-based drawings and documents for project execution, which obviously traduces in loss of time and money, and increases the probability of the occurrence of errors.

These effort lead to the creation of an ISO standard for data modeling - the Industry Foundation Classes (IFC) developed by buildingSMART [1]. Up to recently, the IFC based work flows are more commonly to be seen in the design phase of the building projects, being less available for the fabricators.

1.1 Steel Construction Industry

Considering specific case of the steel fabricators domain, the use of computer-assisted methods is very pronounced, due to the high grade of prefabrication and increasing automation. Nowadays, the steel fabrication industry is demanding the automation of several inherent production processes, as a way to shorten the projects life cycles and reduce the related costs. The market requirements vary from automatic robotic cells for welding and surface treatment and automatic handling [2].

This paradigm change has created new challenges for the information required to plan and manage the steel fabrication process. Lately, the steel fabrication industry was accustomed to the DSTV file format, which filled the gap between the 3D model and the manufacturing (using CNC machines) of each individual steel part. Now, the focus is oriented for the optimization of the assembly phases of the production process, as it is still under automated and represents a very laborious part of the production process.

In here, the DSTV standard was struggling to fulfill the demands, preventing the steel constructing industry from achieving the so desired efficient flow of information, between the design and specification phases and the production process. These limitations led the steel fabrication sector to put pressure for a new file format capable of storing more structured and richer information. For this purpose, the buildingSMART and some of the major companies in the structural steel fabrication industry (FICEP, Steel Projects, Tekla, HGG and others) studied the possibility of extending the Industry Foundation Classes (IFC) standard for the steel fabrication industry needs [3].

In this paper we discuss the introduction of this new file format and the main advantages that it can bring to the industrial process, focusing not only, on the information flow in the fabrication process, but also in the amount of information that IFC can transfer and how this information contributes for process automation.

The paper is organized as follows. In Section 2, we start by presenting an overview of file formats focusing on the advantages of IFC. In Section 3, we present the capabilities provided by the IFC in the fabrication industry, with especial focus on robotized work cells. In Section 4, we present some conclusions and future work.

2 Background

2.1 Product Models for The Steel Fabrication Industry

The steel construction industry has lived for many years using limited file formats to manage the production cycle. Moreover, the need of using individual files from production planning to production has created a very difficult task in the management of shop floor.

Over the last years, the DSTV file was the main file format used in the production process, but the limitations in geometrical descriptions of single parts and the need to include additional data e.g. scribing, welding, assembly hierarchy, information (status, scheduling, costs, materials, etc.), opened space for new formats to enter and fulfill this gap. Nevertheless the reasonable success of the CIS/2, the need of a more generic file format representative for all construction chain, from the design to project software's and manufacturing equipment, has created a new demand between all partners that buildingSMART has been able to fulfill with IFC (Industrial Foundation Classes) in BIM (Building Information Modeling).

2.2 Building Information Modeling (BIM)

The Building Information Modeling term was first introduce in 1992 [4] and refers to a building design methodology characterized by the generation and use of coordinated, internally consistent computable information about a building project in design and construction. Building information models are files which reliably represent a building providing support for design, decision making, fabrication, high-quality construction, document production, construction planning, performance predictions, and cost estimates. Up-to-date accessible information gives architects, engineers, builders, and owners a clear overall vision of all their projects, as well as the ability to make informed decisions faster [5]. A growing number of early adopters in public and private sectors, including GSA, Disney, and Intel, are starting to explore BIM and pursue integrated delivery approaches. Their common interest is ownership of facilities that extends beyond construction completion.

2.3 The Industry Foundation Classes (IFC)

The Industry Foundation Classes (IFC) data model is neutral and open file format specification that intends to describe building and construction industry data to allow interoperability between different players in the architecture, engineering and construction industry. It is a platform neutral, open file format specification that is not controlled by a single vendor or group of vendors. It is an object-based file format with a data model developed by buildingSMART (formerly the International Alliance for Interoperability, IAI) and is a commonly used collaboration format in Building information modeling (BIM) based projects [6]. The IFC model specification is open and available. It is registered by ISO and is an official International Standard ISO 16739:2013.

The IFC standard is the key to guarantee that projects developed by different software vendors can be reused consistently. All the major players in the sector and particularly the major CAD vendors have participated, through the buildingSMART alliance, in developing IFC, and therefore there is a wide range of software applications that support this initiative. The IFC main purpose is to provide architects and engineers with the ability to exchange data between CAD tools, cost estimation systems and other construction-related applications. IFC provides a set of definitions for all object element types encountered in the building industry and a text-based structure for storing those definitions in a data file. IFC uses a plain text file, the only truly universal computer data format. Individual CAD developers store data in product-specific binary file format that best suits their system.

2.4 IFC for Fabricators

Concerning the fabrication industry the IFC file format fills an open gap in the exchange of information between the design and project software and the manufacturing equipment. Moreover this new file format present as a solution for some limitations that were impeding the automation of the production process.

The good results and acceptance of this new format (IFC) can be seen in many occasions. Some examples of that is the mapping of others formats, like CIS/2 format that had some reasonable success, to the IFC standards [7], and the improvement of IFC in each new version, already at version four.

3 Robotic Systems for the Steel Fabrication Industry

Nowadays, fabricators are focused on enhancing the production process efficiency and to be less dependent on external means. The use of robots is efficient for that purpose. One example that is commonly seen on shop floors is the welding on jigs, as in this operation the variability of parts is relatively low, and therefore the need to constantly readjust the robotic cell set-up is eliminated.

In the cutting and welding process for steel fabricators, this scenario does not occur, as most of the work requires the use of an external source of information to allow these processes to be automated in a reliable and efficient manner. In the shop floors, even in our days, is normal to see operators introducing the information's by hand. Examples of this scenario can very frequently be seen in the cutting processes, where the use of CNC machines to drill and to perform strait cuts on profiles are based on a paper drawing; or seeing the operators marking/adjusting the points to be cut or weld by a robot directly in the teach-pendant.

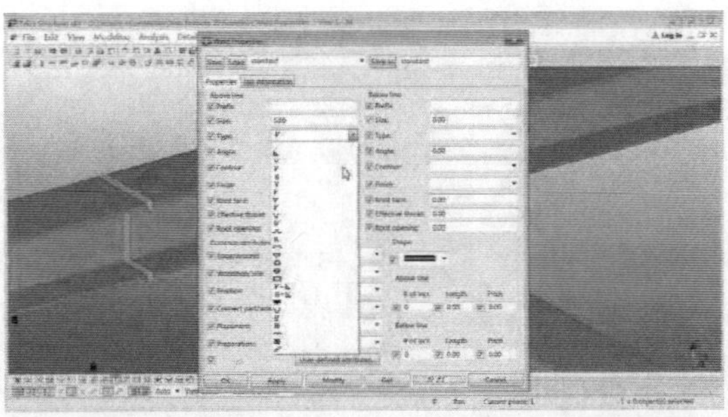

Figure 1. Welding seam definition Tekla

The potential of IFC in generic terms can provide all the information required to full automate the welding and cutting processes. The mesh information of the metallic parts, the assembly hierarchy and the complete specification of welding seams (very recent) in the design and project software (e.g. Tekla, Figure 1) are some of the examples of the advantages that IFC can provided to fulfill the total automation of shop floor work cell's.

3.1 IFC Parser

The IFC Parser is not an easy task to accomplish. The buildingSMART [1] organization provides all the tools and information required to do it. But the complex definition of the elements, the hierarchy and the relation of each other inside the file make this a complex task (Figure 2).

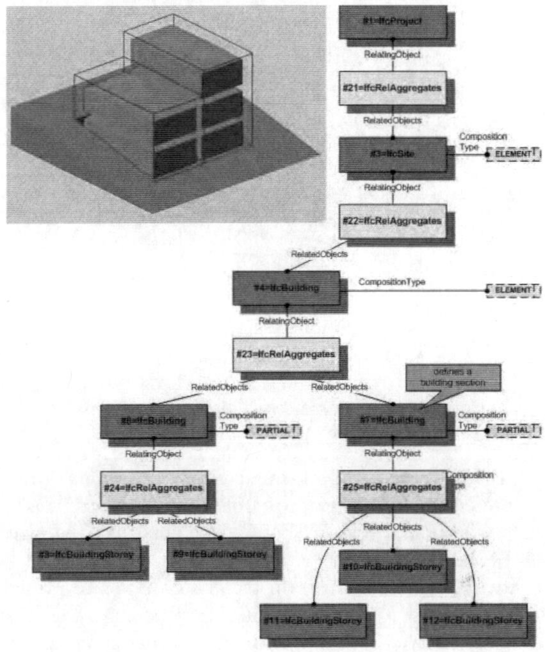

Figure 2. IFC spatial structure element [1]

In these days there is extensive offer of alternatives, free or commercial parsers like the OpenBIM [8] or some of the toolboxes of parsers that can be found on buildingSMART website or IFCWiki [9] for example.

3.2 Process Automation - Robot Automatic Trajectory Generation based on IFC

The quality of trajectories that are generated on-line is always related with the amount of information available. The dependency on the quality of path generation and process information give more strength on the use of IFC files.

Looking inside the IFC structure, and focusing in the benefits for steel structure fabrication domain, we are able to identify some important key points:

- The ability of providing multiple profiles in the same file, Figure 3., allows to reduce the amount of files required to pass to the shop floor for production and makes it easier to track the project information and planning.

- Definition of profile representation as is, using mesh geometry, give more advantage to the process and take the need of future task.
- Hierarchization of the structure, where it is detailed the information of the parts assembly order.
- More recently, the possibility to introduce welding information. This is a huge step for the automation of this specific production process (visible on Figure 4. in red).

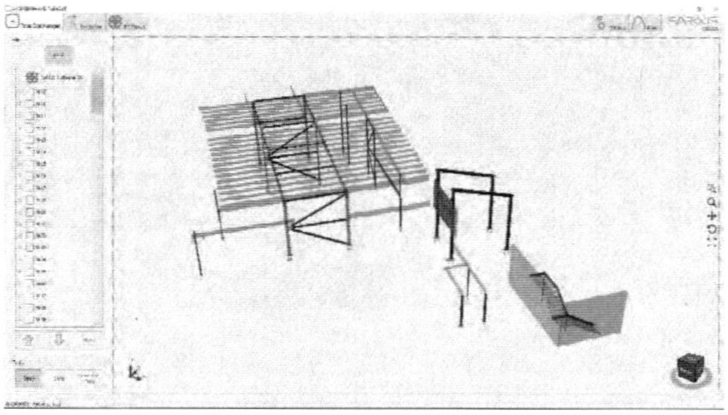

Figure 3. IFC parser with multiple profiles

The generation of trajectories using this kind of information are more profitable for the process. Despite the possible drawback on some points, like the size of IFC files, or mesh triangulation definition, this can always be overcome with some algorithms to provided fastest and accurate results. Furthermore, the gains obtained by working with the object that is pretended in the end and the parametrization of the process, like welding preparation, make the IFC files more profitable for all the automation process.

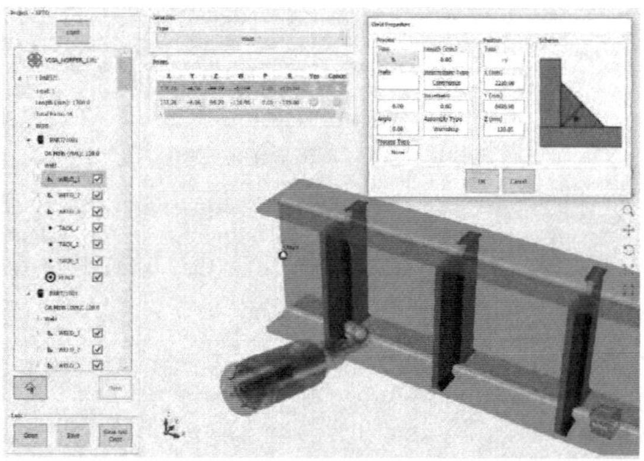

Figure 4. Developed software for robot automatic trajectory generation based on IFC

4 Handling Real Parts Inaccuracies

The work on steel metal structures is a very inaccurate procedure due the tolerances and weight of parts involved. Additionally, the placement of the metallic structure in the workstation is usually performed manually using an overhead crane. These specificities are not in accordance with the requirements for the operation to be carried out automatically by an industrial robot. Thus, it is necessary to provide the industrial robot perception skills appropriate for localizing the steel metal structures and at the same time extract its dimensions. This will allow the estimation of the position and orientation, and the extraction of the real dimensions of the profiles with high precision, allowing the adaptation of the robot trajectories. For this purpose, we propose the use of one of the two following solutions: the use of a laser dot or a 2D laser scanner.

4.1.1 Laser Dot

Laser dot is more common to see in automated work cells, mostly because of the low computational requirements of the algorithms involved. For example, using a smooth algorithm that make uses of some key points from the parts it is possible to find the relation coefficient between the theoretical and the real size of a metallic structure. In Figure 5. it is presented an example of the type of measures that can be made to estimate the position of the beam and validate its dimensions. In the case illustrated, this operation is performed before the assembly of a part to be welded.

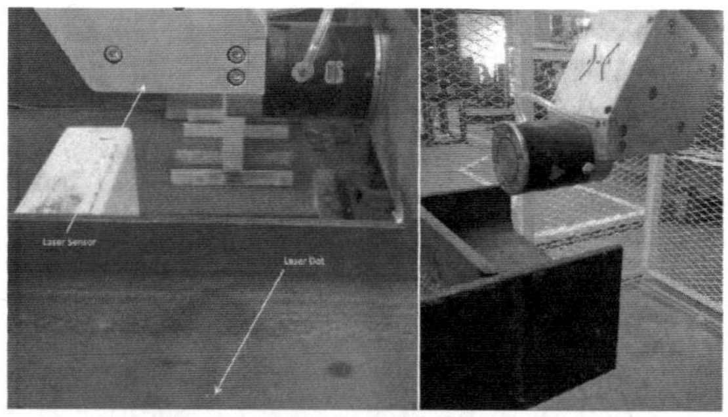

Figure 5. Example of the measurement of some key points for the validation of the beam dimensions and position estimation for execution of an assembly operation

Moreover, the calibration of this sensor as a tool in robot structure is simplified, comparatively to the 2D laser scanner, that will be presented in the following section.

4.1.2 2D Laser Scanner

On the other hand, the solution based on the laser scanner as the potential to be more time efficient, as in one scan, we can retrieve much more and richer information about the metallic structure. Such sensors are nowadays frequently used in industrial equipment's for carrying out measurements without contact. Because of their high accuracy, they are also considered as a rapid 3D modeling technique.

Laser Scanner Calibration on the Robot Tool Center Point: One of the first problem associated with the use of this type of sensor in the robot structure is its calibration. For this purpose a sphere base methodology can be used [10,11]. It consists in two different steps:

1. In the first step is where we compute the rotation matrix R_s between robot *tool0* and laser scanner.
2. Then, in the second step, we compute the translation vector t_s using the previously computed R_s matrix.

(1) Calibration of the Rotation Matrix R_s : The position of the sphere center in robot coordinates can be described as:

$$x_b = \begin{bmatrix} R_0 & t_0 \\ 0 & 1 \end{bmatrix} \cdot \begin{bmatrix} R_s & t_s \\ 0 & 1 \end{bmatrix} \cdot \begin{bmatrix} X_L \\ 1 \end{bmatrix} \quad (1)$$

, where R_0 and t_0 represent the orientation and position of *tool0* of the industrial robot, R_s and t_s represent the orientation and position of the laser scanner relatively to *tool0*, x_b represent the sphere center position in robot base coordinate system, and X_L the position of the sphere center in the laser coordinate system .

Therefore, and considering two different line scans of the sphere it is possible to write:

$$x_{b_1} = \begin{bmatrix} R_{0_1} & t_{0_1} \\ 0 & 1 \end{bmatrix} \cdot \begin{bmatrix} R_s & t_s \\ 0 & 1 \end{bmatrix} \cdot \begin{bmatrix} X_{L_1} \\ 1 \end{bmatrix} \quad (2)$$

$$x_{b_2} = \begin{bmatrix} R_{0_2} & t_{0_2} \\ 0 & 1 \end{bmatrix} \cdot \begin{bmatrix} R_s & t_s \\ 0 & 1 \end{bmatrix} \cdot \begin{bmatrix} X_{L_2} \\ 1 \end{bmatrix} \quad (3)$$

If both scans are extracted with *tool0* in the same orientation, this means $R_{0_1} = R_{0_2}$, the above system of equations can be rewritten as follows:

$$x_{b_1} - x_{b_2} = R_0 R_s (X_{L_1} - X_{L_2}) + (t_{0_1} - t_{0_2}) \quad (4)$$

As the sphere is fixed in the same position:

$$0 = R_0 R_s (X_{L_1} - X_{L_2}) + (t_{0_1} - t_{0_2}) \quad (5)$$

By extracting different pairs of scans, each with different *tool0* orientation, it becomes possible to use a minimization algorithm, for instance Levenberg-Marquardt minimization method, to compute the value of R_s.

(2) Calibration of the translation vector t_s: According to equation 1, the following one is got for a fixed point in the robot coordinate frame:

$$x_b = R_0 R_s X_L + R_0 t_s + t_0 \quad (6)$$

Keeping the robot orientation unchanged to scan a sphere surface, and whatever t_s is, the reconstructed result will always be a sphere. Letting $t_s = 0$ and fit the scanned and reconstructed results x_b, the center of the virtual sphere, x_B in robot base coordinate frame can be computed. The relation between x_B and the real sphere center is given by:

$$x_v = x_B + R_0 t_s \quad (7)$$

By scanning, the sphere with the robot with different orientations, many groups of x_B and R_0 are collected. Solving equation 6, using a minimization algorithm the value for t_s can be estimated.

Having the laser calibrated in the robot structure, we can now use it to estimate and validate the metallic structure dimensions and estimate its position.

The presented calibration methodology was tested in 14 trials with the objective to analyze the repeatability of the calibration procedure.

Table 1. Results for R_s – Euler Angles in Radians

	R_x	R_y	R_z
Standard deviation	0.0011	0.0004	0.0007
Min	1.5659	-0.0005	2.1048
max	1.5695	0.0009	2.1074

Table 2. Results for t_s – Translation vector in mm

	R_x	R_y	R_z
Standard deviation	0.02	0.07	0.01
Min	51.58	25.18	168.96
max	51.64	25.35	169.00

Analyzing Tables 1 and 2 it is possible to see that the calibration procedure presents good repeatability shown by the standard deviation values both for the rotation and translation vectors.

Computing the metallic structure position and dimensions validation: Having the laser calibrated on the robot structure is now possible to make measurements with the objective to validate the dimensions of the metallic structure as well as estimate its position. In here two approaches, or a combination of both, can be used. The first one, is everything similar to the one presented for the laser dot, and is illustrated in Figure 6. Although this similarity, by using this type of sensor the speed of the operation is significantly increased, as the field of view of this sensor is considerably larger.

The second approach is based on the extracting the complete/partial 3D model of the metallic structure. One disadvantage of these approach is the need to create movement (movement of the sensor or movement of the object) in order to extract the 3D model of the object of interest. As the laser is attached to the TCP of the industrial robot, it makes sense to maintain the metallic structure stationary, and the required motion to be executed by the industrial robot. To allow a trajectory with 6 DoF the synchronization between each laser profile and the position information of the laser in the robot reference frame is a critical factor. As a result, with this type of approach is normally associated a reduced speed of the industrial robot, so to minimize possible synchronization delays.

Figure 6. Laser scanner for industrial application – validation of the beam dimensions

In Figure 7. is presented the result of an accumulated 2D laser scan performed using the referred approach. As is is possible to see we were able to extract the 3D model of a metallic structure. The max overlapping error of the scans is about 2 mm. This error also gives us a quality measurement of the laser calibration in the robot TCP.

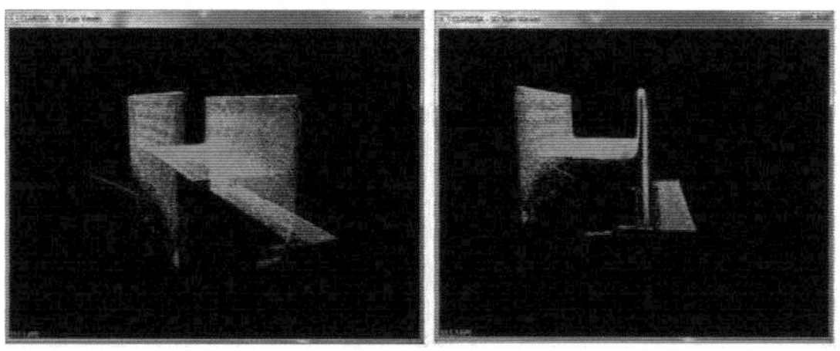

Figure 7. Result of an accumulated scan executed by the industrial robot.

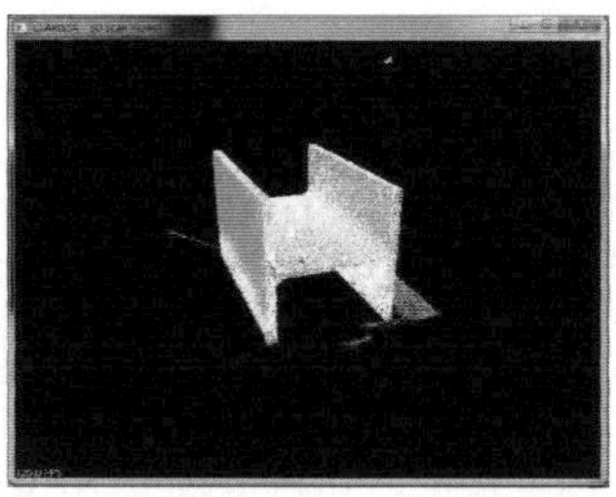

Figure 8. Metal steel structure registration - CAD (white) vs Point Cloud (purple).

Using now some template matching approach, as Iterative closest Point [12], it is possible to fit the CAD model of the metallic structure with the 3D model to estimate its position, please refer to Figure 8..

Also, the beam dimension can easily be extracted from this 3D model. This information can then be used for the adaptation of the robot trajectory that was initially programed based on the IFC information.

5 Conclusions

In this article was demonstrated the advantages that IFC files can bring to the steel fabrication industry. Key points like profile representation, hierarchization of the structure, or even welding information were presented as gains to the process. The IFC files prove to be beneficial in all the chain of construction, from the design to the assembly, that is a big contribution for all and visible on the growing use in industry.

Also in this article, we have demonstrated how the IFC richer information can be used to develop advanced robotic cells, more specifically welding and cutting cells for the steel structure fabrication industry. Features like the ability to provide multiple profiles in the same file, the definition of the profiles as is, and more recently the possibility of introducing welding information, allows the automatic trajectory generation of robot programs.

Despite the ability of the generation of the robot program in an automated way based only in the IFC file information, the dimensional inaccuracies of the real steel structures dimensions makes these programs very difficult to be valid. Therefore, we also propose in this paper some heuristics based on a laser sensor to estimate the real dimensions of the steel structure as well as its position, so the robot program can be adjusted. Despite the advantages of the IFC file format, there still exist some barriers delaying its widespread in industry. Namely, the translation process of BIM to IFC, that some available software's already support; and the IFC import and execution features of most steel fabrication machines that are currently installed in the fabricators shop-floor and only support the old DSTV file format.

6 Future Work

As future work, we propose to continue to explore the use of the IFC file format to automate the steel fabrication process, focusing this time only in the assembly and welding of metallic structures. Normally, this shop floor operation is associated to heavy parts, which for this reason can not be handle by middle-class industrial robots. Hence, we will explore the possibility of building a human-robot collaboration cell, having as pillars the information provided by the IFC files and some augmented reality approaches to support the operator to insert manually the parts in the correct position into the primary steel structure placed at the industrial robot workstation.

Acknowledgments

This work is financed by the ERDF - European Regional Development Fund through the Norte Portugal Regional Operational Programme (NORTE 2020), and through the Portuguese National Innovation Agency (ANI) as a part of project CoopWeld | NORTE-01-0247-FEDER-006438. This work is also financed by Project "TEC4Growth - Pervasive Intelligence, Enhancers and Proofs of Concept with Industrial Impact/NORTE-01-0145-FEDER-000020" is financed by the North Portugal Regional Operational Programme (NORTE 2020), under the PORTUGAL 2020 Partnership Agreement, and through the European Regional Development Fund (ERDF).

References

[1] buildingSMART. buildingsmart international ltd. On-line: http://www.buildingsmart-tech.org/, Accessed: 01/03/2017.

[2] Ivan JIvkov. Is a robot in your future? In Modern Steel Construction 51.5, pages 58–59, 2011.

[3] Holtzhauer E. and Saal H. Product modelling in the steel construction domain. In Proceedings of Xth International Conference on Computing in Civil and Building Engineering,Weimar, pages 85 – 97, 2004.

[4] G.A. van Nederveen and F.P. Tolman. Modelling multiple views on buildings. Automation in Construction, 1(3):215 – 224, 1992.

[5] Youngsoo Jung and Mihee Joo. Building information modelling (bim) framework for practical implementation. Automation in Construction, 20(2):126 – 133, 2011. Building Information Modeling and Changing Construction Practices.

[6] Succar B. Building information modelling framework: A research and delivery foundation for industry stakeholders. Automation in Construction, 18:357–375, 2009.

[7] Robert R. Lipman. Details of the mapping between the cis/2 and IFC product data models for structural steel. Journal of Information Technology in Construction, 14, 2017.

[8] OpenBIM. Openbim. On-line: http://www.openbim.org/, Accessed: 01/03/2017.

[9] IFC Wiki. Ifc wiki. On-line: http://www.ifcwiki.org/, Accessed: 01/03/2017.

[10] J. Li, J. Zhu, Y. Guo, X. Lin, K. Duan, Y.Wang, and Q. Tang. Calibration of a portable laser 3-d scanner used by a robot and its use in measurement. Optical Engineering, 47(1), 2008.

[11] Jianfeng Li, Ming Chen, Xuebi Jin, Yu Chen, Zhiyong Dai, Zhonghua Ou, and Qin Tang. Calibration of a multiple axes 3-d laser scanning system consisting of robot, portable laser scanner and turntable. Optik - International Journal for Light and Electron Optics, 122(4):324 – 329, 2011.

[12] Paul J. Besl and Neil D. McKay. A method for registration of 3-d shapes. IEEE Trans. Pattern Anal. Mach. Intell., 14(2):239–256, 1992.

34th International Symposium on Automation and Robotics in Construction

ISARC 2017

June 28 – July 1 TAIPEI, TAIWAN

CONSTRUCTION ROBOTICS (III) / INSPECTION & MONITORING

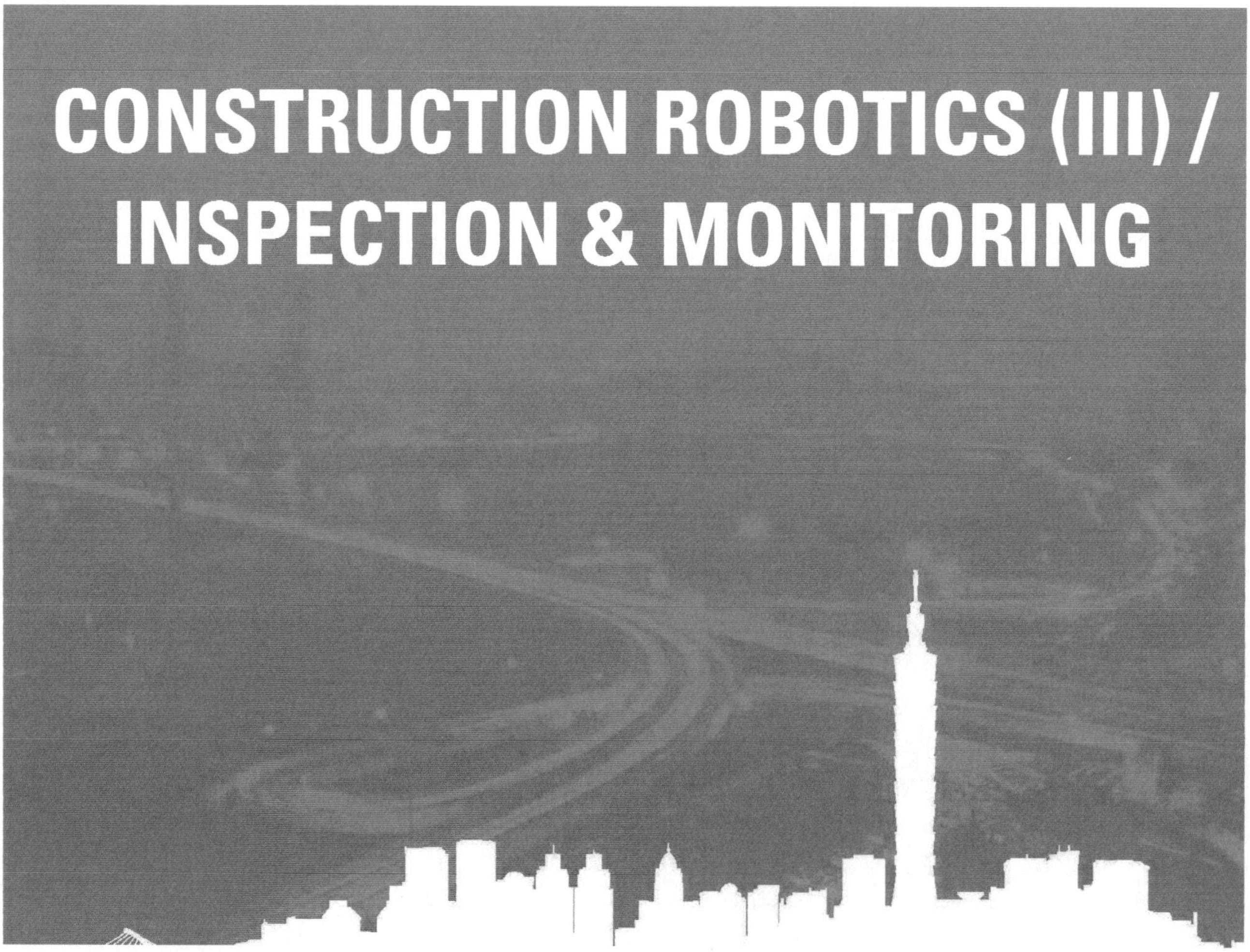

Simulating and Executing UAV-Assisted Inspections on Construction Sites

H. Freimuth[a] and J. Müller[a] and M. König[a]

[a]Chair of Computing in Engineering, Ruhr University Bochum, Germany
E-mail: henk.freimuth@rub.de, jan.müller@rub.de, koenig@inf.bi.rub.de

Abstract -

Unmanned Aerial Vehicles (UAV) are an established utility for remote sensing and mapping purposes. Practical applications emerge in the form of inspecting structures that are hard to reach, deterioration analysis, construction monitoring or quantifying material in landfills. From a technological perspective most of these applications rely on the same established method, manual piloting by experts. While ongoing research is focused on developing reliable means of obstacle avoidance, sophisticated mission planning solutions can greatly increase the practicality of UAVs for inspection purposes.

We propose an inspection planning concept for highly automated inspections in BIM projects. The application, developed in the course of this research, interacts with the firmware and the simulation capabilities of the Dronekit framework, which makes for a highly integrated and sustainable UAV. The main contribution is the 3D simulation of planned flight missions with Software-In-The-Loop (SITL) technology. The inclusion of actual flight dynamics allows the operator to perceive an authentic representation of the UAV's behaviour before take-off. Furthermore, the web-based planning application renders preview images in the virtual scene which provide a realistic impression of the results to be expected. The application automatically generates the different waypoints for obtaining photographs of a Point of Interest (PoI).

Keywords -

Visual Inspection; Unmanned Aerial Vehicles; 4D Building Information Modelling; Progress Monitoring; Ground Control; Path Planning; Simulation

1 Introduction

An important aspect of Building Information Modelling (BIM) for construction is the accurate and up-to-date correlation between as-planned and as-built information. Any deviation between model and reality inhibits the expected increase in efficiency due to the need of manual adjustment of plan data and implied workloads [3]. This justifies a demand for effective progress-monitoring solutions that gather relevant data. Traditionally, such inspections are conducted by staff walking around the construction site and taking photos of modified structures. Apart from the tedious nature of repeatedly iterating several spots on larger construction sites the photographer often cannot reach an optimal perspective when taking pictures of façade elements in greater heights. Furthermore, feeding the gathered information into BIM should not imply additional work. A direct and automatic association of data and their respective building elements in BIM is therefore necessary.

Unmanned Aerial Vehicles have become reliable and affordable utilities for inspections on existing buildings and construction sites. Typically, inspections are carried out manually with a trained pilot steering the UAV around a PoI. However, autonomous capabilities increase operational safety and open up large potential for automation and frequent repetition of recurring inspection tasks. Scheduled flight missions with automatically generated flight paths and camera control leverage the operator of most of the manual work. The goal is to develop a workflow that triggers, generates, and conducts inspections with minimal human effort required.

The main contribution of this research is a case study involving a software application, which, comparable to conventional ground station applications, generates and transfers flight missions and monitors all relevant information in-flight. It advances the concept with a 3D scene representation of the building model, automatic path-planning around building structures and an interchangeable logic for either realistically simulating inspection flights or conducting actual inspections with a live data link. This work demonstrates how the invocation of flight controller logic for simulation purposes furthers the concept of a structured use of UAVs in construction.

2 State of the Art

UAVs introduced new possibilities for gathering sensor data from remote positions in various applications. Many industries, including construction, have since adopted the use of UAVs. Noteworthy applications are crop analysis in precision agriculture, documentation of excavation sites in archaeology, early detection of forest fires, support in emergency services and inspection of power lines and solar panels. The most common sensor among theses applications is a stabilised camera system for aerial images.

Infrared cameras enable surveyors identify degraded solar panels in order to maintain efficiency in small and large-scale photovoltaic plants [11, 16]. When mounted on winged UAVs for extended range, these sensors also greatly improve response times to forest fires [18]. Novel approaches, developed in the precision agriculture field, greatly rely on the automated use of UAVs and help farmers save resources and individually maintain individual orchards by assessing large areas of crops and soil on a daily basis [19]. Among other methods of gathering sensor data, the UAV-based approach is advantageous for its high usability and low data acquisition costs [4].

The adoption of UAV technology in these industries increased the demand for safe and efficient products which can be operated with little to no expertise in piloting. However, autonomous operation with the ability to automatically sense and avoid obstacles is an even more complex challenge concerning aircraft than it is in cars. Some manufacturers have recently begun selling this functionality in their prosumer class models, reports of failed manoeuvres and crashes indicate that their proprietary solutions are not yet reliable and thus unfit for industrial use [15]. Manufacturers as well as research projects in the robotics community follow different approaches to solve this problem. Light Detection and Ranging (LiDAR) offers highly accurate measurements at sufficient rates. These sensors are popular in automotive applications where measurements are taken in a very narrow vertical window as well as winged aircraft with no vertical take-off and landing capabilities. In multi-rotors however, the limited angular width and susceptibility to mechanical stress limit practicality. Furthermore, the laser beams emitted by LiDAR devices are significantly corrupted by rain and fog [12]. The vision-based approach relies on small, inexpensive solid state cameras for sensing the environment. Portable parallel computing solutions like the Nvidia Jetson TX1 allow real-time processing of stereo camera streams for depth-sensing with sufficient accuracy. However, the dependency to sufficient lighting is a drawback to this approach and additional Night Vision Imaging Systems (NVIS) and Forward Looking Infrared (FLIR) sensors increase weight and cost of the system. Concluding the technological aspect, the current state of UAV technology in terms of sensing the environment and avoiding obstacles is unclear. Apart from recent academic achievements [12, 10], commercial solutions have made an impact but their proprietary nature forbids an evaluation of the robustness of the sensing mechanism [15]. Regulatory requirements towards the operation of autonomous aircraft are yet to be defined in most jurisdictions, thus manufacturers have no obligation to meet official requirements. Authorities in the United States, embodied by the Federal Aviation Agency (FAA), have recently published new regulations for the use of civilian UAVs. However the use of autonomous functions was not specifically defined and must adhere with requirements towards piloted use, most notably the satisfaction of the line-of-sight criterion [7].

Practical applications for UAVs in construction and related industries focus on inspection, (quantitative) measurements and reconstruction of 3D geometry for further analysis in Structural Health Monitoring (SHM). Reconstruction of geometry may be carried out either by applying Structure from Motion (SfM) methods on still images or with sophisticated sensors like laser range finders or integrated solutions like the Microsoft Kinect. These data are used for deterioration analysis [9], preservation [1], or energy efficiency measuring [13]. Inspections in open, uncrowded spaces yield good results with low effort due to mostly planar path-planning and a lack of obstacles [8, 14]. In above ground construction however, the automated employment of UAVs reveals more complex challenges. Moving parts, machinery, working personnel and a constantly evolving building structure with scaffolding and other obstacles must be taken into account when planning autonomous inspections on construction sites. While the quality and accuracy of the gathered data is found to be sufficient throughout the evaluated studies and processing methods create valuable information, a BIM inspection concept should emphasise on automation and efficacy. Eschmann et al. [6] describe the challenges for such a concept as collision-avoidance, path planning and efficient data acquisition. In their experimental study the authors collected data of one façade structure for the purpose of reconstruction and automated crack detection. While 12,000 aerial images were collected in the course of four days, only several hundred were used for later processing. Provided with the right preparation methods, a similar inspection could have been carried out in a matter of hours or minutes.

3 Research Methodology and Simulation Concept

This is a constructive research to assess the potential of utilising autonomous UAVs in combination with BIM for inspection purposes on construction sites. The benefits appear in the form of

- low installation costs with little to no operating costs,
- autonomous navigation enabling a high degree of automation,
- the ability to reach vantage points on high structures or roofs and
- low additional effort required for frequent repetition intervals.

However, the following issues and requirements need to be addressed to reach a degree of efficiency that surpasses the traditional methods of repeated walkabouts and deploying lifting platforms.

- Autonomous features on UAVs are still limited. The concept should focus on available technology and future-proof new developments. Furthermore, limitations of currently available technology need to be accounted for.
- Operation of the UAV must not interfere with ongoing processes and collisions must be averted.
- The scheduling of inspections and input of collected data should be coherent with existing methods of managing building information.

To meet the demand for a systematic inspection solution, as established in Section 2, the UAV and its control workflow should empower the operator to accurately plan the inspection flight before conducting it on-site. The expected benefits of such a preparation stage are reduced execution times on site, increased awareness of the UAVs movements and better understanding of the expected data yield. The preparation of an inspection should require as little manual input and decision-making by the user as possible. Ideally, the user should be able to interactively select the objects to be inspected and then rely on the inspection software to automatically determine the correct flight mission including control of the sensor.

The automated inspection solution proposed in this work relies on locating the UAV via Global Navigation Satellite Systems (GNSS) and coordinating its flight trajectory along the object boundaries defined in the georeferenced BIM. This knowledge-based concept may be extended by active collision avoidance technology in the future, but the main achievement of this concept is the creation of an automated workflow that generates efficient flight missions and requires no further knowledge than that provided by BIM.

Fig. 1 illustrates the designated workflow between the system and the user and identifies the tasks the system needs to solve as well as the required data sources.

The application is designed to be used both in the planning stage of an inspection and while conducting it. After selecting the BIM to work in, the user is prompted to either plan a new inspection or to load existing mission data which may then be executed on the UAV. By setting time and date of the intended inspection mission, the application is able to render sunlight and shade as it will appear on-site. This helps the operator decide whether the natural lighting conditions are suited for the photographs.

The selection of a PoI then triggers the application to generate both a path of waypoints along the points of interest as well as preview pictures from various vantage

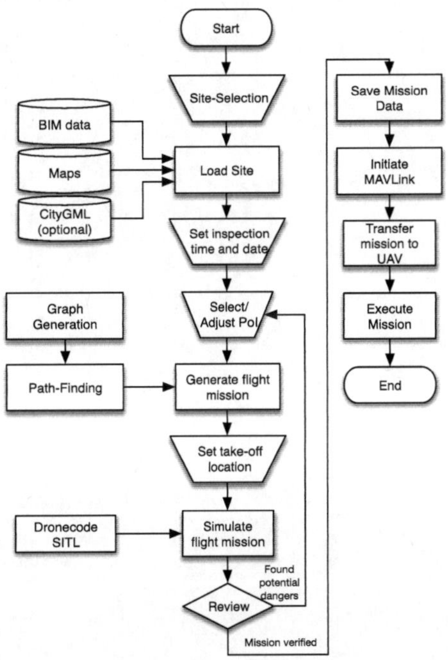

Figure 1: Application workflow

points (Fig. 2). To the operator a PoI is a structural element of the building that has undergone changes. In terms of the flight mission, the PoI is an aggregate of multiple waypoints, each providing a different perspective on the structural element. Apart from its geodetic location, each vantage point also includes camera control commands. Each vantage point requires the camera to be pointed towards the PoI in a different attitude. The stabilisation gimbals used in this project operate on two axes, pitch and roll. To cover rotation of the camera on the yaw axis, the UAV itself needs needs to rotate accordingly when reaching a vantage point. The autopilot of the UAV is capable of triggering the camera shutter, so each vantage point in a flight mission also triggers a shutter command.

The path from take-off location towards the PoI is represented as a sequence of waypoints, beginning at the take off location, leading to the PoI, and leading back to the take off position for landing. The Inspection-Planner needs to calculate a viable route between take-off and PoI, considering all physical objects on the construction site. The path planning concept is separated into three steps: segmentation of the scene, graph generation and invocation of

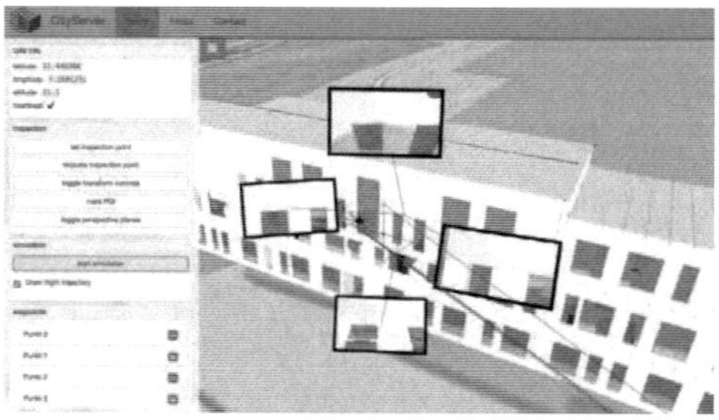

Figure 2: After selecting an element the user is presented with previews of the images the inspection will yield

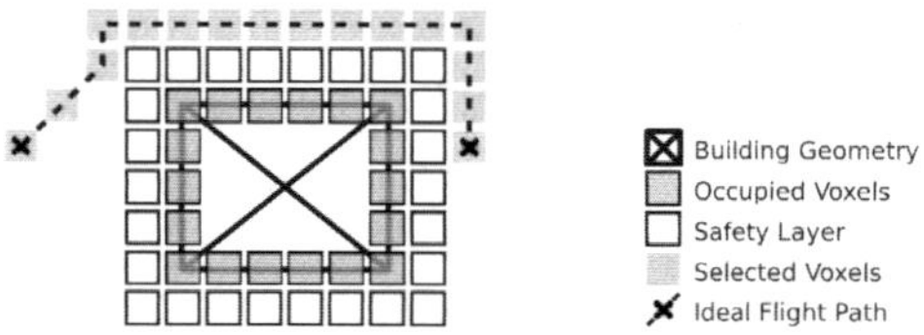

Figure 3: Schematic top-down view of the navigation mesh around a building

a graph search algorithm to determine the ideal route. For the segmentation step the concept of navigation meshes, established in robotics research, was adapted. Navigation meshes are typically an aggregate of 2D polygons that represent the areas of a robot's environment that are accessible. However, UAV navigation requires a 3D representation of navigable space, allowing the UAV to travel in vertical direction and to account for variable height of physical objects. The navigation volume is represented as a set of voxels, where no intersections with relevant geometries from the BIM occur. Each voxel inherits information about its neighbourhood of voxels and the potential waypoint it represents. To abide with the required minimum distance the UAV should maintain towards objects, a further processing step is applied. It generates an augmentation layer of voxels around bodies of occupied voxels. The augmentation step needs to create a closed hull of voxels around occupied bodies. This is achieved by using the 26-connected Moore neighbourhood of each occupied voxel. Fig. 3 shows a simplified 2D schematic of a building geometry after the navigation volume was created and with an exemplary flight path drawn between start and end point. Based on the relations between voxels, a graph is generated, with each transition between nodes representing a possible movement of the UAV from one waypoint to the corresponding one of a neighbouring voxel. Provided start and end points the node-based graph search algorithm A* is invoked. By applying penalties in the form of increased transition costs for vertical travel, the shortest path algorithm favours lateral avoidance strategies to climbing over obstacles. This benefits the legislative criterion requiring the operator to keep the UAV in line-of-sight at all times. Further modifications of this kind may be applied in further development stages of a practical path-finding strategy.

The UAV will be guided to each of these points with the information about the needed orientation for the camera to obtain the images as planned. Each of these points adheres to a predefined ruleset of how much distance the UAV should keep to any structure throughout its mission.

The BIM data source is the central component in this concept. Integrating the inspection preparation with BIM counters two of the constraining factors. It provides spatial data of possibly any structure, machinery and the digital elevation model, which is the fundamental knowledge for automatically computing the space in which the UAV may navigate in. Furthermore, the 4D BIM concept provides the temporal knowledge of processes and locations of machinery which enables the automatic generation of scheduled inspections.

BIM data, exchanged in the IFC format, is managed and provided to the inspection planner application by the open source BIMserver [2].

Spatial information about the environment of the site may be augmented by cadastre data typically available in the CityGML format or by freely accessible vectorised map tiles.

3.1 Simulation of Inspection Flights

After the selection of elements to inspect and generating the according flight mission, the user may simulate the mission. The simulation gives the operator a realistic impression of how the UAV will behave in the actual execution and allows identification of latently dangerous manoeuvres that the model cannot account for. Therefore, the simulation should be as realistic as possible, which can be achieved by executing the firmware image that is identical to the one that resides in the flash memory of the UAV's micro-controller. By defining simulation parameters that resemble the real environment it is possible to let the simulated UAV behave like the physical counterpart. This kind of simulation is built on the SITL framework established by the Dronecode Foundation [5]. By providing an interchangeable communication interface between the application, the UAV and the simulation engine, there is no need for an extra control protocol (Fig. 4). This is not only a conceptual benefit for the user trying to validate flight missions, but also a fundamental aid for future developments which focus on advanced path-finding strategies with complex constraints.

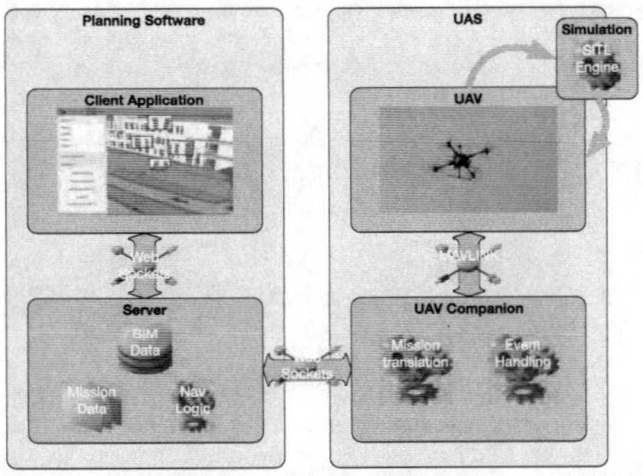

Figure 4: Simulation Schema

To further the realism of the 3D simulation and provide an informative display of the UAV's status when executing a mission, maintaining a low latency between Server and the browser-based client is mandatory. Telemetry data, received from the UAV (or the simulation) is being forwarded to the client application via WebSockets in real-time. The UAV Companion is an intermediary software between the planning software and the UAV interface, which may either be a physical link to the UAV or the interchangeable counterpart of the simulation. It may be executed directly within the server application but may also be executed independently on a Single Board Computer (SBC) such as a Raspberry Pi, which is mounted on the UAV. It is the conceptual component that translates flight missions, generated in the Planning Software to MAVLink formatted waypoints and dispatches telemetry data back to the planning software.

MAVLink [5] is a lightweight messaging protocol designed for use between Groundstation applications and UAVs and may also act as inter-communication link between multiple UAVs in swarming applications. The MAVLink protocol is highly efficient. It is a header-only library, which doesn't take up much space in the memory of highly limited micro-controllers. Moreover, its messages are small in size with simple integrity checking via CRC checksums. This makes the protocol suitable for low bandwidth long range telemetry links.

The server-bound communication link of the UAV Companion is implemented in a REST API that allows direct read/write access to the information that is to be shared. It dispatches relevant events from the UAV to the Planning Software, giving the user insight into the current flight status. Forwarded events and parameters are:

- Current position: latitude, longitude, altitude
- Attitude: in three axes, including camera attitude
- GPS: Horizontal Dilution of Precision (HDOP), number of satellites
- Battery status: Voltage (per cell), current draw, remaining capacity
- Flight mode: including arm status

In extension to providing a realistic simulation in terms of movement and control, the SITL simulation approach is also fit to investigate the potential risk induced by environmental factors. Occlusion of multiple satellites at the GPS sensor's position can significantly lower the accuracy. Another source of error is known as multipathing effects which often occur in urban canyon scenarios. In addition to line-of-sight signals, the sensor receives signals which are reflected on facade surfaces [17]. The simulation framework features adjustable parameters including those of the simulated GPS device or direct influential factors like wind gusts. It is therefore possible to test variations of such parameters to evaluate a flight mission. Provided with a probabilistic model of the influence of building structures to the GPS system, the simulation software can simulate decreased accuracy of the GPS sensor and provide insight about how much such disturbances interfere with the navigation system of the UAV. Furthermore, simulation of energy consumption provides important information about the feasibility of an inspection mission. The setup of a meaningful energy consumption simulation requires setting the right take-off weight of the UAV as well as the power consumed by the motors in combination with the propellers. These parameters should be determined by conducting calibration flights with the designated UAV that is equipped with complete payload. In return, this allows the simulation to provide meaningful insight about the expected flight duration and battery usage. Similar tests may be carried out with simulated wind gusts, adjustable in strength and direction, or increased error in the magnetometers which may be disturbed by ferrous materials.

4 Case Study

A practical case study was conducted in two steps. To asses the practicability of the application with realistic model data the recently rebuilt SSC building on the campus of the Ruhr University Bochum (Fig. 5a) was recreated in a BIM and used to generate and simulate an inspection mission. In another test, the quality of the simulation method was evaluated by comparing a simulated flight and a flight test in an open field.

The SSC model mainly consists of outward facing structures, which are analysed in real-time for the path generation. After selecting the model the user selects the designated take-off location for the UAV on the map plane.

(a) The BIM loaded in the inspection planner application and the real building

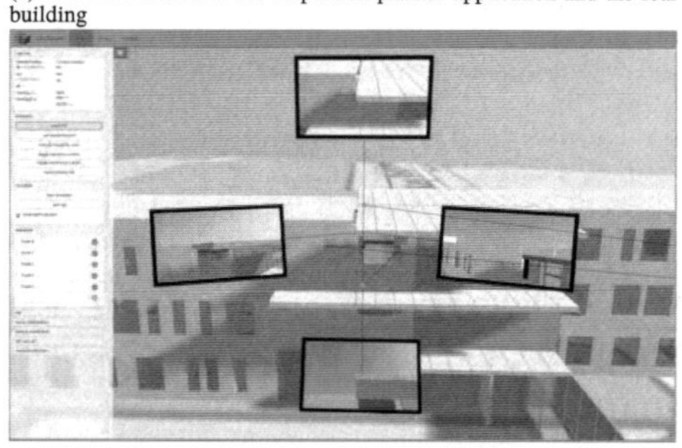

(b) The previews from the perspectives convey an understanding of the photos that will be taken

Figure 5: Creating an inspection mission

(a) Four waypoints (green) mark the different perspectives for photos around the PoI

(b) The simulated UAV starts descending while reaching the land location

Figure 6: An ongoing simulation run of the created mission

Subsequently the user selects the PoI by clicking on the corresponding facade structure. The application then generates the flight mission and previews of the photos that will be taken around the PoI (Fig. 5b). In case the previews do not meet the operator's expectations, they may be adjusted directly in the scene and the mission will be regenerated in place. The automatically generated mission is then simulated for verification. Figures 6a and 6b show the ongoing simulation run of the exemplary inspection mission. The flight path is visualised as a chain of waypoints (green) and straight connection lines (black), which do not accurately represent the movement of the UAV. By initiating the underlying simulation engine, a virtual representation of the UAV executes the flight mission and draws a line of its trajectory (red) in the scene.

In order to evaluate the realism of the simulation, a test flight with an automated flight mission was conducted in an open field. A building model was geo-referenced on a predefined spot on a field near the campus. A PoI was selected on the roof structure of the model and the generated mission was then executed in the integrated simulation. After checking the simulation a real test flight using a 3DR IRIS+ UAV was conducted under unfavourable weather conditions. Clouded sky influenced the GNSS sensor's reception and wind gusts of level 6 in the Beaufort scale made an impact on the stabilisation mechanism of the flight control. The UAV's flight control stores a log of all relevant flight parameters throughout a flight in its flash chip, including all sensory data and navigation status. The SITL simulation provides the similar log data. This evaluation focuses on the perceived location of the real and the simulated vehicle. Fig. 7 shows the logged positions of both runs, red being the simulated trajectory, blue being the flight test. While both trajectories look similar, it gets clear, that the simulation did not exactly foresee the flight dynamics. This is due to the fact that the simulation was not provided with alternating gusts but a constant wind speed and the simulated GPS sensor was run with constant reception quality. A closer look on the log data provides further insight. The two plots in Fig. 8 show a direct comparison of simulation of flight data by measured latitude and longitude in WGS84 notation. The error bars reflect the vagueness of the measured position of the test flight and were derived from the Horizontal Dilution of Precision (HDOP), reported by the GPS sensor. The HDOP is an estimate of the signal quality, dependent on multiple

Figure 7: The recorded flight trajectories of the simulation (red) and the field test (blue)

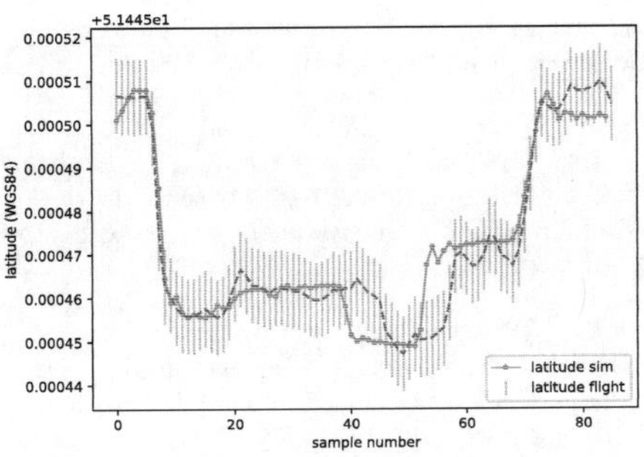

(a) Logged latitude measurements

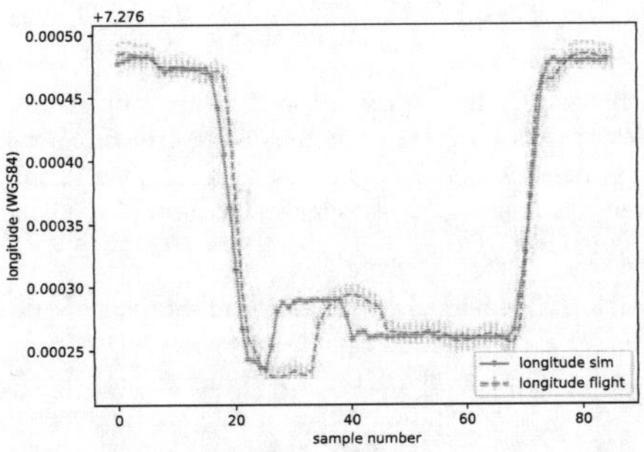

(b) Logged longitude measurements

Figure 8: Measured locations of the test flight with HDOP error margin and the simulated flight

conditions, such as visibility of satellites, electromagnetic disturbances and multipathing effects. It reports a possible deviation in meters for each logged measurement. The simulation, including its mechanism for producing a vague perception of position in space, is expected to exhibit a similar behaviour that ideally lies within the error margin of the real flight. While the simulated trajectory mostly lies within the error margin, short aberrations can be explained by the fact that the simulation was quicker in marking waypoints as reached and therefore was able to move on to the next one earlier. These observations are important steps towards a reliable automated inspection method and show how a realistic simulation supports the development of more complex navigation strategies.

5 Conclusion and Outlook

This work is part of an ongoing research effort to investigate the practicability of employing UAVs for inspection purposes in construction environments. The bottom-line concept of generating flight missions based on BIM data and open source robotics software is implemented in an exemplary application. The main contribution of this work is an integrated concept for preparing and conducting automated inspections with UAVs. Furthermore, by using realistic simulation techniques, the operator is provided with meaningful predictions of the aerodynamics and energy consumption of the UAV. Regarding the further refinement of the path-planning strategy, the simulation greatly accelerates testing and is therefore also considered a meaningful support measure for future works.

While the preloaded knowledge from the model may account for any possible obstacle in the working environment, a maximum degree of safety can be achieved with active collision avoidance functionality. The industry is heavily invested in solving this problem, but reliable solutions are yet to be found. Regarding future developments, active collision avoidance may complement the presented solution at any stage, but the fundamental concept of generating ideal flight paths in a known model environment stays relevant as collision avoidance does not replace this knowledge. Further efforts will be made in a study on active avoidance of low GPS reception in the proximity of buildings. Monitoring the signal quality, the inspection planner application may actively steer the UAV into open space, away from the building façade, to regain better reception. This method has the potential to greatly minimise the risk of malfunctions.

References

[1] Irene Aicardi, Filiberto Chiabrando, Nives Grasso, Andrea Maria Lingua, Francesca Noardo, and A Spanò. Uav photogrammetry with oblique images: First analysis on data acquisition and processing. International Archives of the Photogrammetry, Remote Sensing and Spatial Information Sciences, 41:835–842, 2016.

[2] Jakob Beetz, Léon van Berlo, Ruben de Laat, and Pim van den Helm. Bimserver. org–an open source ifc model server. In Proceedings of the CIP W78 conference, 2010.

[3] Frédéric Bosché, Mahmoud Ahmed, Yelda Turkan, Carl T Haas, and Ralph Haas. The value of integrating scan-to-bim and scan-vs-bim techniques for construction monitoring using laser scanning and bim: The case of cylindrical mep components. Automation in Construction, 49:201–213, 2015.

[4] Sebastian Candiago, Fabio Remondino, Michaela De Giglio, Marco Dubbini, and Mario Gattelli. Evaluating multispectral images and vegetation indices for precision farming applications from uav images. Remote Sensing, 7(4):4026–4047, 2015.

[5] Inc. Dronecode Project. Dronecode project, inc., a linux foundation collaborative project. On-line: https://www.dronecode.org, Accessed: 10/02/2017.

[6] C Eschmann, CM Kuo, CH Kuo, and C Boller. Unmanned aircraft systems for remote building inspection and monitoring. In 6th European workshop on structural health monitoring, pages 1–8, 2012.

[7] Federal Aviation Administration (FAA). Operation and certification of small unmanned aircraft systems. On-line: https://www.faa.gov/, Accessed: 28/03/2017.

[8] F. Grimaccia, M. Aghaei, M. Mussetta, S. Leva, and P. Bellezza Quater. Planning for pv plant performance monitoring by means of unmanned aerial systems (uas). International Journal of Energy and Environmental Engineering, 6(1):47–54, 2015.

[9] Norman Hallermann and Guido Morgenthal. Visual inspection strategies for large bridges using unmanned aerial vehicles (uav). In Proc. of 7th IABMAS, International Conference on Bridge Maintenance, Safety and Management, pages 661–667, 2014.

[10] Aaron Mcfadyen and Luis Mejias. A survey of autonomous vision-based see and avoid for unmanned aircraft systems. Progress in Aerospace Sciences, 80:1 – 17, 2016.

[11] Paolo Bellezza Quater, Francesco Grimaccia, Sonia Leva, Marco Mussetta, and Mohammadreza Aghaei. Light unmanned aerial vehicles (uavs) for cooperative inspection of pv plants. IEEE Journal of Photovoltaics, 4(4):1107–1113, 2014.

[12] Subramanian Ramasamy, Roberto Sabatini, Alessandro Gardi, and Jing Liu. Lidar obstacle warning and avoidance system for unmanned aerial vehicle sense-and-avoid. Aerospace Science and Technology, 55:344 – 358, 2016.

[13] D. Roca, S. Lagüela, L. Díaz-Vilariño, J. Armesto, and P. Arias. Low-cost aerial unit for outdoor inspection of building façades. Automation in Construction, 36:128 – 135, 2013.

[14] Sebastian Siebert and Jochen Teizer. Mobile 3d mapping for surveying earthwork projects using an unmanned aerial vehicle (uav) system. Automation in Construction, 41:1 – 14, 2014.

[15] Colin Snow. Sense and avoid for drones is no easy feat. On-line: http://droneanalyst.com/2016/09/22/sense-and-avoid-for-drones-is-no-easy-feat/, Accessed: 27/01/2017.

[16] Nicolay Tyutyundzhiev, K Lovchinov, Francisco Martínez Moreno, Jonathan Leloux, and Luis Narvarte Fernández. Advanced pv modules inspection using multirotor uav. Ies, 2015.

[17] P. Xie and M. G. Petovello. Measuring GNSS Multipath Distributions in Urban Canyon Environments. IEEE Transactions on Instrumentation and Measurement, 64(2):366–377, February 2015.

[18] Chi Yuan, Youmin Zhang, and Zhixiang Liu. A survey on technologies for automatic forest fire monitoring, detection, and fighting using unmanned aerial vehicles and remote sensing techniques. Canadian journal of forest research, 45(7):783–792, 2015.

[19] Pablo J Zarco-Tejada, José AJ Berni, María Dolores Suárez Barranco, and Elías Fereres Castiel. A new era in remote sensing of crops with unmanned robots. Society of Photo-Optical Instrumentation Engineers, 2008.

Autonomous Robotic System with Tunnel Inspection Tool Positioning

E. Menendez, J. G. Victores, R. Montero and C. Balaguer

Robotics Lab research group within the Department of Systems Engineering and Automation
Universidad Carlos III de Madrid (UC3M)
E-mail: emenende@pa.uc3m.es, jcgvicto@ing.uc3m.es, romonter@ing.uc3m.es, balaguer@ing.uc3m.es

Abstract –

This paper presents the ROBO-SPECT European FP7 project, funded under the ICT-2013.2.2 programme on Robotics use cases & Accompanying measures. The main objective of the ROBO-SPECT system is to provide a robotized, faster and reliable alternative to manual tunnel structural inspection and assessment. Physical developments include the design and implementation of a multi-degree-of-freedom (MDoF) robotic system, which uses a mobile vehicle to advance along the roadway, an extended crane capable of reaching the most commonly found tunnel geometries, and a robotic arm for positioning a specifically designed ultrasonic sensor (US) Inspection Tool with high accuracy. A semi-supervised computer vision system to detect tunnel defects, a Ground Control Station (GCS) to provide a Human-Machine Interface (HMI), and an Intelligent Global Controller (IGC) to command the robot and manage communications between the different parts have also been developed.

An overview of the fundamental aspects of the project architecture and design will be detailed. In addition, the developed and implemented algorithm for positioning the Tunnel Inspection Tool on detected cracks shall be presented. Finally, experimental evidence to validate the functionality of the ROBO-SPECT system in a real motorway tunnel with ongoing traffic will be provided.

Keywords –
Automation; Control of Robotic Systems; Tunnel; Inspection; Maintenance;

1 Introduction

When facing existing civil infrastructures, one of the greatest challenges that engineers may find is its inspection and assessment in order to to ensure that bridges, roads, pipelines and tunnels remain in safe condition and continue to provide reliable levels of service [1], [2]. The structural performance of tunnels is time-dependent because of the damaging process induced by natural and artificial impacts, inadequate maintance or the simple effect of ageing.

Water supply, metro, railway and roadway tunnels have increased in both total length and number, and will continue to do so on a global scale. Some tunnels still in service were constructed over 50 years ago, and many have exceeded their intended design service life [3]. Only in Europe in 2002, the overall length for operational transportation tunnels had grown up to 15000 km [4].

Tunnels are characterized by humidity, dust and absence of natural light. Inspection and maintance operations are commonly performed by human operators, taking time and expertise. Additionally, the human factor combined with the unfriendly environment could lead to lack of guarantee regarding quality control.

These facts highlight the need of automated, cost-effective and exhaustive solution to inspect tunnels that prevents such disasters. In this work, the final integrated ROBO-SPECT system is presented, followed by the safe positioning algorithm of the Ultrasonic Sensors (US) inspection tool on a detected crack. Finally, experimental evince, timing and accuracy results and conclusions are also presented.

2 The ROBO-SPECT European Project

ROBO-SPECT is a research project co-funded by the European Commission under the 7th Framework Programme (FP7) ICT-2013.2.2 on Robotics use cases & Accompanying measures. The project ran between October 2013 and October 2016 and was composed by a Consortium of 12 partners. The overall objective of ROBO-SPECT is to design and implement an automated, faster and reliable tunnel inspection robotic system that inspects cracks and other defects of the tunnel lining on one pass without interfering with tunnel traffic (Figure 1). The system also includes a detailed structural assessment software.

The specific objectives of this project are the

following [5]:

- Minimize the use of tunnel inspectors. Presently, the maintenance consists in visual inspections, which is labour intensive and subjective.
- Improve the unhealthy and unsafe working conditions of tunnel inspectors.
- Decrease inspection and assessment costs.
- Increase the safety of passengers.
- Increase the residual lifetime of existing tunnels.
- Decrease the time tunnels are closed for inspection. The impact of this is significant for tunnels with heavy traffic volume.
- Provide better quality, objective, timely data and an improved knowledge of the tunnel lining.

Figure 1. ROBO-SPECT is an FP7 European project to design a fully autonomous tunnel inspection robotic system.

In summary, the needs which ROBO-SPECT addresses are the following:

- High cost of new tunnel constructions (need for inspection, assessment and repair of existing).
- Transport demand is highly increasing and cannot cope with the rate of transport infrastructure and high tunnels uptime.
- Inspection and assessment should be speedy to minimize tunnel closures or partial closures.
- Engineering hours for tunnel inspection and assessment are severely limited.
- Currently tunnel inspections are predominantly performed through scheduled, periodic, tunnel-wide visual observations by inspectors who identify structural defects and categorize them manually (manual, slow and labour expensive process).
- Un-reliable classification of the liner conditions and lack of engineering analysis.

3 The ROBO-SPECT System Design

The ROBOS-SPECT system is composed of four different components that allow the complete inspection and assessment of the tunnel, aimed at providing all the functionalities to the final users. The first component is the robotic system, which incorporates all the different components needed to perform the tunnel lining inspection. The second component is the Ground Control Station (GCS), that monitors the robot mission and communicates with the robotic system. The third component is the Control Room (CR), that is the site where the data gathered during the inspection is used to generate a complete assessment report about the tunnel state. Finally, the Intelligent Global Controller (IGC) communicates with the different subsystems and manages the execution of all the tasks.

3.1 Robotic System

The ROBO-SPECT robotic system design is based on TUNCONSTRUCT European Project (FP6) [6] system, which used a similar vehicle, crane and robotic arm configuration.

The robotic system is composed by an industrial mobile vehicle capable of extending an automated crane to the dimensions commonly found in metro and roadway tunnels. It is equipped with a high precision robotic arm that positions the ultrasonic sensors (US) inspection tool on cracks on the tunnel lining to perform measurements. A vision system to detect cracks on-line and other defects, like spalling [7] and efflorescence [3] offline, and a 3D laser profiler to detect deformations on the tunnel lining are also part of the system. An additional set of cameras are placed on the crane and an IP camera is attached to the tip of the arm for an extra teleoperation mode.

Figure 2 depicts the actual design of the robotic system and the different components that shall be described in detail.

3.1.1 Mobile Vehicle

The mobile vehicle is able to autonomously navigate on one line of the roadway maintaining a constant parametrized distance to the tunnel wall with ongoing traffic on the other line.

The navigation is based on Simultaneous Localization and Mapping (SLAM) and is performed using a dedicated navigation sensor situated in the front part of the vehicle and a set of reflective beacons that are placed on both sides of the roadway. In order to localize the robot with high precision and simultaneously create an accurate 2D map of the navigated section, a minimum of 3 beacons need to be detected at the same time. This map is used to improve navigation precision in subsequent inspections.

The mobile vehicle is also capable of avoiding collision using two 2D range laser sensors, one in the front and one on the back of the vehicle.

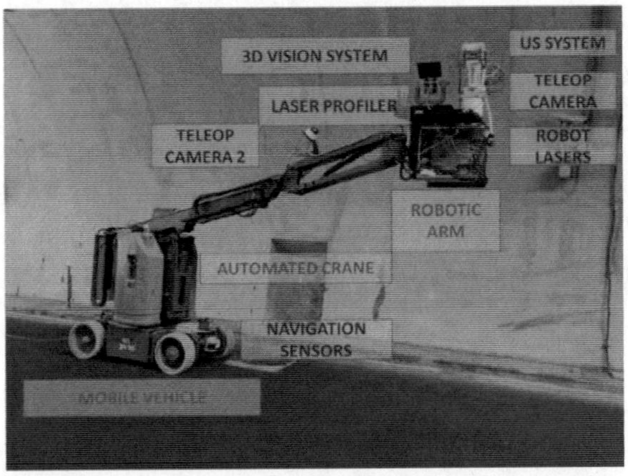

Figure 2. ROBO-SPECT Robotic System is composed by a mobile vehicle, an automated crane, a robotic arm, an US inspection tool, a vision system, a 3D laser profiler and teleoperation cameras.

3.1.2 Automated Crane

The crane has been sensorized, including additional elements such as encoders inside the joints to control the crane tip position and orientation. The joints of the crane are equipped with special brakes to minimize the oscillation and vibration. The crane mission is to position a redesigned platform with the robotic arm equipped with the US inspection tool, the 3D vision system and the 3D laser profiler to gather inspection data. The behaviour of these components is described below. This platform also includes a 2D laser sensor to provide 3D point cloud scans of the tunnel lining and avoid collisions while moving the crane.

3.1.3 Robotic Arm

The robotic arm chosen for the application is the Mitsubishi PA-10, a 7 Degrees Of Freedom (DoF) industrial manipulator. 6 DoF are needed to provide full position and orientation capabilities and the other one adds obstacle avoidance and correct orientation functionalities. The reachable workspace covered by this high precision robotic arm ranges from a few centimetres to 1 meter approximately from the base to the end-effector of the arm, which is limited by basic kinematic restrictions and self-collisions. Figure 3 depicts the selected robotic arm, along with the vision system and the 3D laser profiler.

The robot has 10 kg load capacity which is enough to support the ultrasonic sensors inspection tool located at the tip that performs measurements on cracks inside the tunnel lining, the IP camera placed below it that monitors the process and provides visual feedback in case of using the safe teleoperation mode, and the 2D robot laser sensor attached to a link of the arm which scans the surroundings of the detected crack.

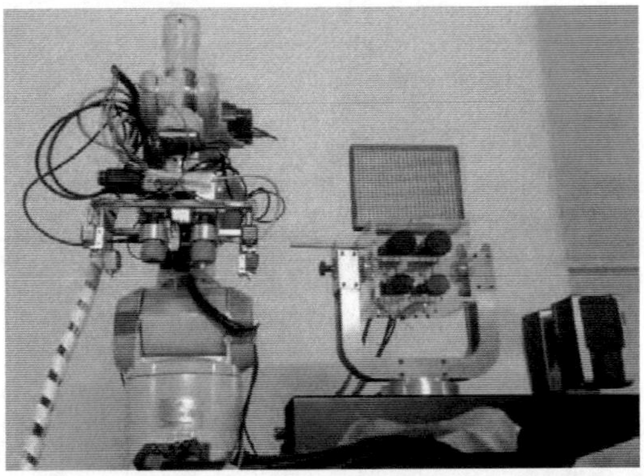

Figure 3. The ROBO-SPECT components attached to the crane: the robotic arm, the vision system and the 3D laser profiler.

3.1.4 Vision System

The ROBO-SPECT vision system located on the crane platform is equipped with a pan&tilt mechanism designed to orientate two pairs of RGB cameras that detects defects inside the tunnel lining and takes stereo images of detected cracks. In order to operate with the proper lighting conditions an on-board lighting system is also added. A highlighted fact about the vision system is its needs to be positioned at a determined distance to the wall, given by the camera focal length. The first pair of cameras are designed to be able to identify a set of different defects commonly found in tunnels such as spalling, efflorescence, etc. These defects are detected using machine vision techniques, such as Convolutional Neural Networks (CNN) [8] trained with real images of tunnel defects. These cameras are also used to detect cracks in real time and to estimate their 3D positions and orientations. This data is then send to the robot to aproach the crane to the crack surroundings first, and then move the robotic arm to touch the tunnel wall with the US inspection tool. The other pair of cameras are designed to take stereo images of detected cracks.

3.1.5 3D Laser Profiler

Furthermore, the 3D laser profiler chosen is placed on the crane platform to inspect tunnel structural deformation with an accuracy of 2 mm.

3.1.6 Ultrasonic Sensors (US) Inspection Tool

A set of specically designed ultrasonic sensors is attached at the tip of the robotic arm to measure the width and depth of a detected crack while in contact with the tunnel lining. The depth measurement of the crack is done using two piezo-electric ceramic transducers that must be placed on each side of the crack. The method utilized to measure the depth is the well-known Time of Flight (ToF) [9], which is the time a wave is generated on one side of the crack and detected on the other side. Respecting the width measurement, a newly designed fiber-optic sensor is used. The sensor is placed on the crack using a XY positioning stage. When the measurements have been taken, the robotic arm removes the sensors from the tunnel wall, and the inspection continues.

As part of the US Inspection Tool, a rectangular frame with contact sensors on its corners has been attached to the tip of the robot arm (Figure 4). These contact sensors, currently implemented as normally closed push button switches, allow detecting the tunnel wall during the approximation of the ultrasonic sensors, and also for maintaining a stable position of the ultrasonic sensors until the measurements are made.

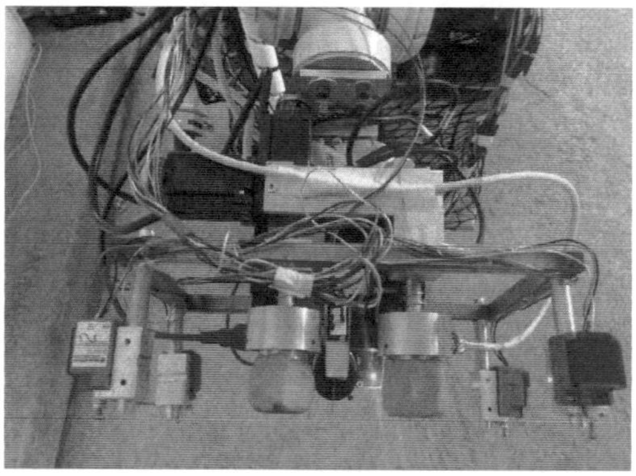

Figure 4. Contact sensors, currently implemented as normally closed push button switches, placed on a rectangular frame allow detecting the tunnel lining for a safe placement of the ultrasonic sensors.

3.1.7 Intelligent Global Controller (IGC)

As the system is composed of several different components that need to be controlled simultaneously, a Intelligent Global Controller (IGC) is placed inside the robot to manage communications with all the parts. All the different components (mobile vehicle, crane, arm, etc) are connected to a local network during the autonomous inspection process inside the tunnel. The software used to communicate is a mixed solution of YARP [10] and ROS [11].

The Ground Control Station (GCS) sends a mission to the IGC in order to command the robot to perform the requested mission. The IGC is able to navigate the robot through the tunnel, indentify when a crack has been detected, and perform the necessary joint movements to place the ultrasonic sensors on the crack and take all the required measurements. The IGC monitors the system state and reports the inspection data gathered and updates the GCS with the state of the mission.

3.1.8 Ground Control Station (GCS)

The Ground Control Station (GCS) is a component outside the robot that is in contact with it during the inspection. This system consists in a standard laptop with Wi-Fi connection to the router installed on the robot. The GCS includes the Mission Manager, a Human-Machine Interface (HMI) that allows the user to define a mission (definition of the path, modes of operation), and contacts with the IGC to start the mission with the defined parameters and to receive the state of the robotic system and the results during the inspection. If connection is lost with the robot, the GCS is able to gather the generated data when established again while the robot continues autonomously the inspection.

3.1.9 Control Room (CR)

The Control Room (CR) represents the site where all the information gathered during the autonomous inspection is processed using the Structural Assessment Tool (SAT). This software tool stores, graphically represents, and processes all the inspection data. The SAT allows the end-user to see the generated maps of the tunnel, data of the different cracks and other defects detected, and their position inside the tunnel, etc. Finally, it produces a complete assessment report of the structural state of the system that is presented to the end-users.

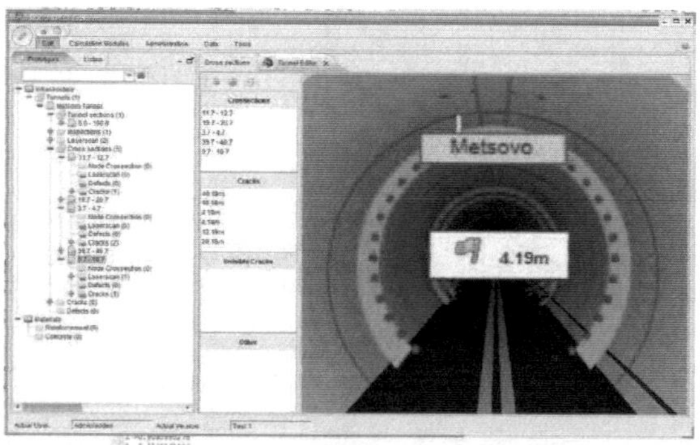

Figure 5. SAT representing the inspection tunnel and processing the data gathered.

4 Safe position of the Ultrasonic Sensors (US) Inspection Tool

This safe positioning process of the US Inspection Tool is composed of 4 steps: arm creates surface map, move tip to a fixed distance from the surface, end-effector trajectory tracking control, and guided US normal approximation and iterative rotation. The general scheme of the algorithm is depicted in Figure 6.

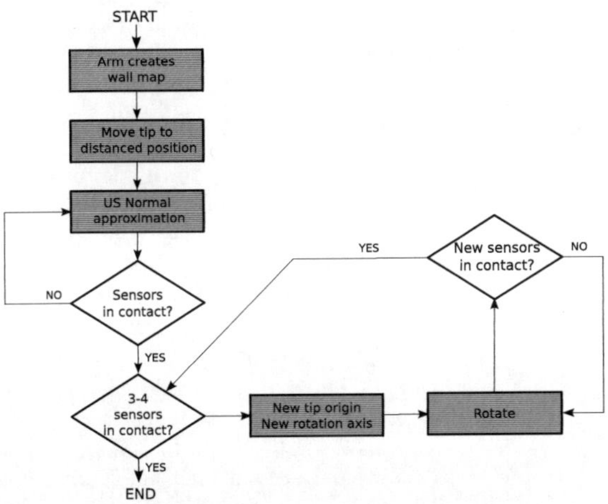

Figure 6. General scheme of the algorithm for positioning of the US Inspection Tool on a crack.

4.1 Arm creates wall map

In order to measure the width and depth of a crack, the arm must place the ultrasonic sensors (US) so that the crack between the acousto-optical ultrasonic detectors, and aligned with the piezoelectric ultrasonic transducers. For the computation of a safe trajectory to the final point, a 2D range robot laser attached to one of the arm's links scans and extracts a 3D point cloud of the surroundings of the crack (Figure 7).

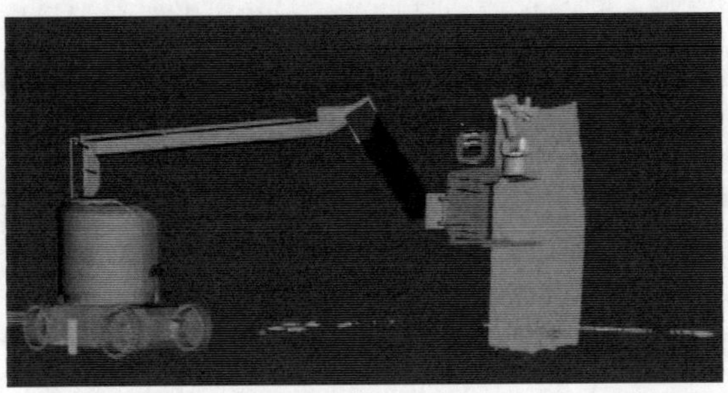

Figure 7. The 3D point cloud scan of the surroundings of the crack is simultaneously simulated.

3D normal to the tunnel linning on the crack position is generated using the extracted 3D point cloud. A safe joint trajectory to position the ultrasonic sensor on the point of the normal at a fixed distance from the crack is then computed [12]. The tracking control algorithm described in Section 4.2 is used to follow the generated trajectory.

The tunnel lining is not a simple flat wall, so the simultaneous contact of the 4 switches is not likely. Therefore, it is suffient to simultaneously contact 3 switches with the surface.

If fewer switches were in contact, the position would not be stable, so the tip would be reorientated by using the information from the contact sensors (Section 4.4). The end-effector control during the approximation and reorientation is described in Section 4.3.

4.2 Move tip to distanced position

The joint space trajectory tracking control used for the arm, it has two feedback loops: the internal motor driver torque controller as inner loop, and the velocity PI controller as an outer loop. The joint velocity v_d command for the outer controller can be described as:

$$v_d = \dot{q}_d + K(q_d - q) \tag{1}$$

where q_d is the joint desired position vector, $\dot{q}_d$ is the time derivative and q is the joint position.

4.3 End-Effector Trajectory Tracking Control

Once the tip has moved to the distanced position, the control scheme used changes. The control scheme used is depicted in Figure 8. It consists of two feedback loops: a kinematic control that generates the joint velocities needed from the desired pose of the end-effector, and the internal joint velocity PI controllers.

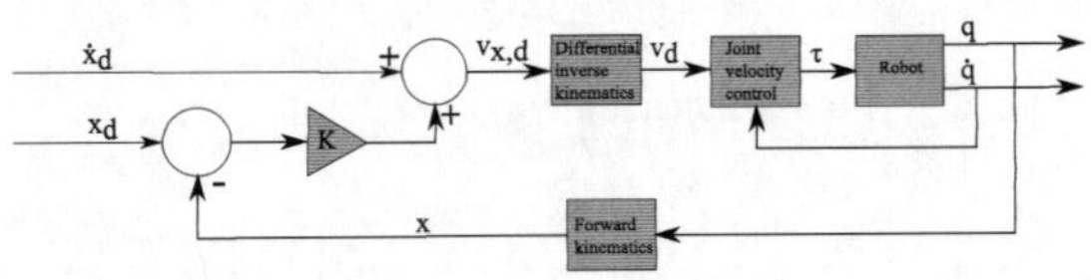

Figure 8. End-effector pose control scheme applied within the reorientation process.

Using the Resolved Motion Rate Control (RMRC) as kinematic control [13], the desired joint velocity vector v_d can be obtained by utilizing the following expression:

$$v_d = \mathrm{J(q)}^{\dagger}(\dot{x}_d + K(x_d - x)) \tag{2}$$

where q is the vector of joint coordinates, $\mathrm{J(q)}^{\dagger}$ is the pseudoinverse of the jacobian matrix, $\dot{x}_d$ is the time derivative of the desired pose vector x_d and x is the pose vector of the end-effector.

4.4 Guided US normal approximation and iterative rotation

The guided approximation and iterative rotation algorithm is based on the iterative change of the tip origin system of coordinates and rotation during the motion described in Section 4.3 around each of its axes until the position is stable (Figure 9). The process depends on which of the switches are in contact:

1. When one switch is in contact, the new origin is placed there and the tip will rotate around one of the frame sides that meet on this corner.
2. In case of two adjacent switches in contact, the new origin is placed in the middle point of the straight line through them, and the tip will rotate around this line.
3. Finally, when two diagonal switches are in contact, the new origin will be the centre of the frame and it will rotate around the diagonal which links them. The algorithm will select clockwise or counter-clockwise rotation in order to obtain more switches in contact.

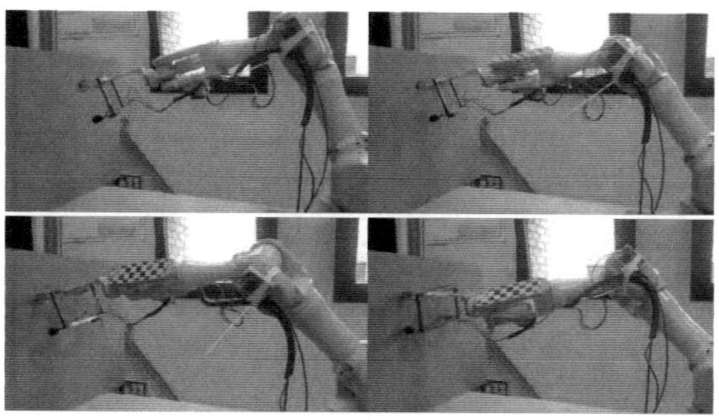

Figure 9. Sequences of the iterative rotation algorithm performed in the laboratory against a flat surface.

5 Laboratory and field experiments and results

The ROBO-SPECT system has been successfully tested in 3 different phases: Simulated environments, laboratory environments, and a roadway tunnel with ongoing traffic.

- In the first phase, the tests were performed within the Gazebo [14] simulator. These experiments served to validate the functionality of the navigation of the mobile vehicle, the trajectory computation and execution of the automated crane, and the US safe positioning US Inspection tool positioning algorithm.
- The second phase of the experiments includes the validation of the different components of the system, and hardware and software integration of the full robotic system also in laboratory settings.
- The full robotic system was tested in Metsovo, Greece, in Egnatia Odos motorway tunnels with ongoing traffic. Field experiments were performed throughout selected validation scenarios, including all the components of the system. These selected mission plans were sent using the Mission Manager with no human intervention and the data gathered during the test was provided to the CR. Roadway tunnel tests covered the capabilities of the system adding complexity. The first scenario consisted on image acquisition every 1.4 m half-slice of the tunnel. The second scenario, added 3D laser profiler scans every 5 m. The third scenario was the same as the second one, but adding crack detection. The fourth scenario included stereo images acquisition and 3D laser profiler scan if a crack is detected. The fifth scenario added the teleoperation mode of the robotic arm to position the US inspection tool on detected cracks to perform width and depth measurements. Finally, the complete robotic inspection procedure was demonstrated in the last scenario, where the safe positioning of US inspection tool on the detected crack is performed autonomously.

6 Results

The complete autonomous inspection was repeated 10 times in a 20 m segment of half of the Egnatia Odos tunnel in order to extracted timing and accuracy data. Table 1 depicts the average time of each inspection process stage gathered from the trials. The total time to inspect one 1.4 m half-slice of the tunnel without cracks (approximately 6 minutes) is highlighted in this Table after the set of stages that are required for this process. In case of a detected crack in a 1.4 m half-slice, it would not take longer than 10 minutes, which is also highlighted after the additionally required stages. The inspected tunnel presents a 2/3 ratio of cracks per half-slice, and the average time to inspect the 20 m tunnel segment was 2 hours and 9 minutes.

The overall error of the ROBO-SPECT system describes the accuraccy of the US Inspetion Tool positioning on a given crack with respect to the global tunnel coordinates. The coordinates ground truth is based on a previously hand-crafted map of tunnel lining cracks. The overall accuracy is shown together with each ROBO-SPECT individual component maximum error in Table 2. For the complete ROBO-SPECT system, the overall maximum error is 11 cm. Fixing the vehicle and crane, it can be seen that a 2.5 cm maximum error of the US positioning with respect to the vision system, which is an admissible error for obtaining US sensor measurements.

Table 1. Inspection process timing of one 1.4 m half-slice.

Inspection process stages	Time [s]
Vehicle motion to next 1.4 m slice	10
Crane motion to first scanning position	21
Crane stabilization	4
Crack detection in 1st position	150
Crane motion to 2nd scanning position	35
Crane stabilization	4
Crack detection in second position	120
Crane go to home position	26
Inspect one half-slice without cracks	**370**
Crane approximate to the wall	7
Robotic arm tip touch the wall	15
US Measurements	80
Laser proler scanning	120
Inspect one half-slice with one crack	**592**

Table 2. Maximum error of the ROBO-SPECT components, and the maximum accumulative error from the chain that begins with the vehicle localization and ends with the US inspection Tool placement.

Individual Component	Maximum Pos. Error [mm]
Vehicle	38
Crane	47
Vision System	16
Robot Laser Sensor	3
Robotic arm	4
US	2

Chain	Maximum Pos. Error [mm]
Vehicle	38
Veh. + Crane	85
Veh. + Crane + Vis.	101
Veh. + Crane + Vis. + Laser	104
Veh. + Crane + Vis. + Laser + Arm	108
Veh. + Crane + Vis. + Laser + Arm + US	**110**

7 Conclusions

ROBO-SPECT is a highly complex robotic system that has demostrotated during the field trials its capability to inspect roadway tunnels autonomously with ongoing traffic on the other line of the roadway, where each component has fullfilled their tasks, p.e, the mobile vehicle has navigated autonomously in a real tunnel using the reflective beacons system and the dedicated laser navigation sensors maintaining a constant parametrized distance to the wall. The automated crane performed trajectory plans with collision avoidance using a 3D point cloud map. The robotic arm process described in this paper in order to place with precision the ultrasonic sensors was positively performed in the tunnel as shown in Figure 10.

Figure 10. US inspection tool measuring a detected crack in the Egnatia Odos tunnel.

The integrated mode of operation, where the three levels (robotic system, GCS and CR) are connected permanently was tested succesfully gathering data and displaying it in the CR. Hardware and software adaptations allowed the overall system integration in a positive way. The software integration, including IGC and GCS, demostrates that the chosen architecture and protocols perform correctly, including non-standard situations.

The ROBO-SPECT robotic system provides accurate, faster and reliable tunnel lining inspection and assessment with ongoing traffic in safer working conditions.

Acknowledgements

This work was supported by ROBO-SPECT project number 611145 co-funded by the European Commission under 7th Framework program, RoboCity2030-II-CM project (S2009/DPI-1559) funded by Programas de Actividades I+D in Comunidad de Madrid and co-funded by Structural Funds of the EU, and ARCADIA project DPI2010-21047- C02-01 funded by CICYT project grant.

References

[1] R. Montero, J. G. Victores, S. Mart'\inez, A. Jardón, and C. Balaguer, "Past, present and future of robotic tunnel inspection," *Autom. Constr.*, vol. 59, pp. 99–112, 2015.

[2] C. Balaguer, R. Montero, J. G. Victores, S. Mart'\inez, and A. Jardón, "Towards Fully Automated Tunnel Inspection: A Survey and Future Trends," in *31st International Symposium on Automation and Robotics in Construction & Mining (ISARC)*, 2014.

[3] W. Bergeson and S. Ernst, "Tunnel Operations, Maintenance, Inspection and Evaluation (TOMIE) Manual." 2015.

[4] A. Haack, "Current safety issues in traffic tunnels," *Tunn. Undergr. Sp. Technol.*, vol. 17, no. 2, pp. 117–127, 2002.

[5] R. Montero, J. G. Victores, E. Menendez, and C. Balaguer, "The Robot-Spect EU Project: Autonomous Robotic Tunnel Inspection," in *Robocity2030 13th Workshop EU robotic projects results*, 2015, pp. 91–100.

[6] J. G. Victores, S. Mart'\inez, A. Jardón, and C. Balaguer, "Robot-aided tunnel inspection and maintenance system by vision and proximity sensor integration," *Autom. Constr.*, vol. 20, no. 5, pp. 629–636, 2011.

[7] K. D. Hertz, "Limits of spalling of fire-exposed concrete," *Fire Saf. J.*, vol. 38, no. 2, pp. 103–116, 2003.

[8] K. Makantasis, E. Protopapadakis, A. Doulamis, N. Doulamis, and C. Loupos, "Deep convolutional neural networks for efficient vision based tunnel inspection," in *Intelligent Computer Communication and Processing (ICCP), 2015 IEEE International Conference on*, 2015, pp. 335–342.

[9] D. Marioli, C. Narduzzi, C. Offelli, D. Petri, E. Sardini, and A. Taroni, "Digital time-of-flight measurement for ultrasonic sensors," *IEEE Trans. Instrum. Meas.*, vol. 41, no. 1, pp. 93–97, 1992.

[10] G. Metta, P. Fitzpatrick, and L. Natale, "YARP: yet another robot platform," *Int. J. Adv. Robot. Syst.*, vol. 3, no. 1, pp. 43–48, 2006.

[11] M. Quigley *et al.*, "ROS: an open-source Robot Operating System," in *ICRA workshop on open source software*, 2009, vol. 3, no. 3.2, p. 5.

[12] I. A. Sucan, M. Moll, and L. E. Kavraki, "The open motion planning library," *IEEE Robot. Autom. Mag.*, vol. 19, no. 4, pp. 72–82, 2012.

[13] R. Kelly and J. Moreno, "Manipulator motion control in operational space using joint velocity inner loops," *Automatica*, vol. 41, no. 8, pp. 1423–1432, 2005.

[14] B. Gerkey, R. T. Vaughan, and A. Howard, "The player/stage project: Tools for multi-robot and distributed sensor systems," in *Proceedings of the 11th international conference on advanced robotics*, 2003, vol. 1, pp. 317–323.

An Improved Robot for Bridge Inspection

H.Peel 1[a], S.Luo [a], A. G. Cohn [b] and R. Fuentes 2[a]

[a] School of Civil Engineering, University of Leeds, United Kingdom
[b] School of Computing, University of Leeds, United Kingdom
E-mail: H.Peel@leeds.ac.uk, R.Fuentes@leeds.ac.uk [Board of Directors]

Abstract – This paper presents a significant improvement from the previous submission by the same authors at ISARC 2016. The robot is now equipped with low-cost cameras and a 2D laser scanner, which is used to monitor and survey a bridge bearing. The robot is capable of localising by combining data from a pre-surveyed 3D model of the space with real-time data collection in-situ. Autonomous navigation is also performed using the 2D laser scanner in a mapped environment.
The Robot Operating System (ROS) framework is used to integrate data collection and communication for navigation.

Keywords – Bridge inspection, monitoring, SfM, SLAM, ROS.

1 - Introduction

Continuing on from work by the same authors [1], more off-the-shelf and low-cost solutions are considered for autonomously navigating a bridge abutment, with the aim of performing a visual inspection on a bridge bearing.

Visual inspection is an important part of inspecting a bridge bearing. In fact, regular inspection is defined in the European and British standard for inspection and maintenance of structural bearings as: "close visual inspection without measurements, spaced at equal reasonably frequent, intervals" [2].

Most of the main problems affecting bearings are reflected in changes to geometry, including: translation, rotation or deformation [3], [4]. Current methods to measure changes in the bearing geometry include [3]: metric tapes, gap gauges, air bubble levels, quadrant rulers, compasses and verniers, levelling and topographic surveys or direct visual observations.

However, regular inspection of bridge bearings often does not occur as frequently as required, in some cases due to difficult access or dangerous conditions. Bridge bearings are critical for the performance in the bridge and inadequate inspection may lead to much greater problems later on in the bridges life.

One solution to increase frequency of inspection is to automate the inspection process. However, the wide range of bridge design and function means that there is not a one size fits all robot for bridge inspection, with technologies being developed for drones [5], underwater vehicles [6] and climbing robots for steel structure bridges [7]. Our contribution is a low cost solution to autonomously performing visual inspection, with technology that can be obtained and implemented in bridge bearing inspection in the near future. In this paper we focus on the implementation of autonomous navigation for autonomous inspection.

Another motive for using robots for inspection is to increase the repeatability of inspections. Previously, we implemented the 3D reconstruction method Structure from Motion (SfM) to enhance the information about a bridge in a format that can be compared directly over time. Now we look at other ways of using this information. Specifically we use a method for Simultaneous Localisation and Mapping (SLAM) in the bridge abutment, where SLAM images can also be used for SfM and visa-versa.

We also consider a second SLAM approach called Hector SLAM that uses LIDAR only and we implement autonomous navigation using a known map and consider some of the challenges of operating in an inspection environment.

2 - Structure-from-Motion (SfM)

Structure-from-Motion (SfM) was used in this work and in previous work [1] as a method for adding value to visual inspection. SfM uses multiple 2D image views to find the 3D geometry (i.e., the structure) of a scene or an object by taking images from different viewpoints (i.e., the camera has motion). The 3D reconstruction software Zephyr Aerial, produced by the company 3Dflow [8], was used for Structure-from-Motion and Multiview stereo calculations and reconstructions in this work. Since SfM is not the primary focus of the work, for a detailed overview of the methods behind SfM and MVS refer to [9] and [10].

3 – Simultaneous Localisation and Mapping (SLAM)

Simultaneous Localisation and Mapping (SLAM) is the process of using sensor readings (e.g., LIDAR, camera and RGB-D sensors) to create a map of the environment whilst at the same time finding the location of the sensor in relation to the map that is being created. There are many implementations of SLAM including filter based methods (such as Extended Kalman filter SLAM [11]), particle based methods (such as Monte-Carlo Localisation [12]) and graph-based methods (such as GraphSLAM [13]).

In this paper, two options for SLAM are considered: ORB-SLAM [14] and Hector SLAM [15]. Both of these methods are performed online - i.e., the computation is done at the same time as the sensor data is being collected. However, due to the computational requirements of these methods, the SLAM calculations were not done on the Raspberry Pi. Instead, The Robot Operating System (ROS) was used to pass camera and laser scan messages respectively to a laptop, where the results are computed.

3.1 ORB-SLAM

ORB-SLAM is a form of Visual SLAM with both monocular and stereo implementations [14]. Features are extracted from images using Oriented Fast and Rotated Brief (ORB) descriptors, chosen for fast extraction and matching overhead compared to other image features, to allow real-time computation without a GPU [14]. The same features are used for tracking, local mapping and localisation for efficiency and reliability [14].

Map points contain information about its 3D position relative to the world coordinate, the direction the point was viewed from, a representative ORB descriptor and the maximum and minimum distances at which the point can be observed.

Before ORB SLAM begins to create a map, a process of initialisation must first occur, with the goal of computing the relative poses between two frames to triangulate an initial set of map points. Only when it is certain that the two views provided will avoid a corrupted map can initialisation be completed, since ambiguity causes all the points to be plotted on a plane [14].

The software implementation of ORB-SLAM used in this work is ORB-SLAM2 [16], which also has a ROS node. ORB-SLAM was chosen as a candidate for localisation because it has been shown to work in urban implementations [14], and has also been implemented using the Raspberry Pi camera in an indoor office environment [17]. Hence, only the low cost camera was required and for localisation and visual inspection.

3.2 Hector SLAM

Hector SLAM was developed for autonomous navigation for urban search and rescue robots [15]. Hector SLAM does not require wheel odometry and relies only on fast LIDAR scan matching [18].

hector_slam [19] is a ROS metapackage that provides packages such as hector_mapping, the ROS node used for SLAM, hector_geotiff which can be used to save the robot trajectory, map and objects of interest in geotiff format. Hector SLAM is designed to be used in 3D, e.g., for robots travelling over rough terrain or aerial vehicles [15], where robots are required to move with up to 6 degrees of freedom. The SLAM system, hector_mapping, is 2D and 3D navigation is achieved by fusing information from an inertial measurement unit. The 2D and 3D solutions are updated separately, but remain coupled in time. Other sensors can also be integrated to decrease uncertainty caused by sensor drift [15], although none are implemented in this paper.

The 2D map is created on an occupancy grid, with interpolation to allow sub-grid accuracy. This approach utilises the high scan rates of modern LIDAR sensors, and provides a more accurate alternative to traditional odometry [18]. Scan alignment is performed based on optimising the alignment of the laser beam endpoints using Gauss-Newton optimisation approach to find the best alignment of the current laser scan data with the existing map through a rigid transform for some cost function [18].

4 - Adaptive Monte-Carlo Localisation

Adaptive Monte-Carlo Localisation (AMCL) is a particle filter method for localisation. Particle filters represent the knowledge a robot has about its position in a given map using a set of particles. Initially, this set of particles is spread over the known map. Measurement and motion models are applied to all of the particles to update the position of each particle. Weightings are then applied to the set of particles depending on the likelihood that a sensor reading at a given location matches the position of a particular particle.

The efficiency of particle filter methods rely on the number of particles being used. The KLD Monte- Carlo algorithm is derived from Kullback-Leibler divergence that adaptively updates the number of particles over time [12], allowing a high number of particles in the initial stages of localisation and a much lower number of particles for tracking when the robot location is known. A detailed description of the AMCL algorithm is available in [12].

5- Platform Description and Integration

5.1 - Platform

The robotic platform, also used in [1] is a DiddyBorg robot [20], a commercially available platform that is built around the Raspberry Pi single board computer, the on-board computer in this work. The DiddyBorg is a six wheeled platform with each wheel powered by a 5V motor. Two expansion boards, produced by the same company, the PicoBorg reverse and the BattBorg, are connected to a Raspberry Pi 2B, which interface with the motors and power the Raspberry Pi using a battery pack, respectively. Some of the motivating factors behind using this platform were its cost, its size and off-the-shelf integration with existing technologies.

In previous work [1], the Raspberry Pi was running Raspbian Jessie, a commonly used operating system for the Raspberry Pi. In this paper, the operating system was changed to Ubuntu MATE 16.04, using the following installation instructions for the Raspberry Pi models 2B or 3 [21]. The main motive for changing operating systems was to be able to access software, such as the Robot Operating System (ROS), more readily, with Ubuntu Mate 16.04 being recommended as the faster and easier way to use ROS on Raspberry Pi. ROS Kinetic was used in this work.

Figure 1: A photo of the DiddyBorg robotic platform with the locations of the RPLIDAR and Raspberry Pi camera sensors used in this work.

5.2 - The Robot Operating System (ROS)

ROS is an open source meta operating system for robots [22], with a large on-line, open-source community. Software is available as packages or stacks that can be easily distributed and shared and developed in multiple languages to allow code reuse in robotics research and development [23]. The software is usually created as independent programs called nodes [24]. Nodes communicate by connecting to a master service and by sending messages that are organised into named topics. Nodes can send information by publishing messages on a topic and other nodes can listen for and subscribe to messages coming from topics. There are defined message types that can be used for specific purposes such as lasers scan messages, camera messages and geometry messages for navigation.

5.2.1 - Motion command node

Scripts to allow navigation of the DiddyBorg mobile platform are based on the original scripts written by the manufacturers [25]. These scripts include python library to interface the motors with the Raspberry Pi through the Picoborg reverse board and the I2C connections on the Raspberry Pi. Using this library to interface with the motors, a ROS node was written that subscribes to a motion command topic in the form of a geometry message with type Twist(), commonly used for velocity messages [26]. These messages are converted into the correct format as used in the aforementioned Python library and the relevant velocity commands are then sent to the Picoborg expansion board. The motor control node is agnostic to the source of the message; hence, this node can be used both for tele-operating the platform and for motion commands for autonomous motion. An example of this process for tele-operating the robot and collecting image data is given in Figure 2.

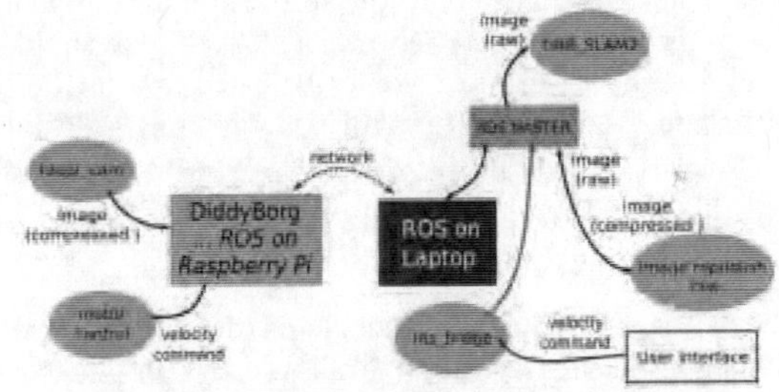

Figure 2: An example of the relationships of different nodes. The route the velocity commands and image data takes between the robot and the relevant node is also shown.

5.2.1 - User interface

In previous work, Node.js was used to create a user interface by incorporating libraries for the Picoborg reverse and the Raspberry Pi camera. This user-interface has now been developed to incorporate Roslibjs [27]. Roslibjs is part of the Robot Web Tools effort and is a JavaScript library for interacting with ROS from a web browser. Roslibjs allows the functionality of ROS such as publishing and subscribing to messages, service calls and other core ROS functionality. This user interface allows the teleoperation of the robot from a web-browser and can be accessed from a mobile phone.

6- Other Hardware and Sensors

6.1.1 Raspberry Pi camera module

Raspberry Pi camera module version 1 was used for photographic data collection. This camera is capable of producing 5MP pictures and 1080p HD video at 30fps. The camera module is not automatically compatible with a Raspberry Pi 2B running Ubuntu Mate, hence some adjustments and drivers were required, as described in [28]. A ROS package, raspicam_node has also been developed for the Raspberry Pi camera by [28].

6.1.2 - Camera calibration

Camera calibration is required for ORB-SLAM. The same calibration parameters were used as an input to SfM. However, the software is capable of finding the calibration parameters automatically and these parameters were adjusted by the SfM software. The monocular calibration script from ROS [29] was used to calibrate the Raspberry Pi camera. An 8x6 chessboard was placed on a planar surface and the camera was moved to obtain different viewpoints. The calibration software indicates when sufficient samples have been collected to perform a calibration. However, further samples were taken until the software indicated that the samples taken were above a 'good' threshold for translation in x, y and skew (indicated in a traffic light system from red to green). This number of samples translates to around 120 readings, which is comparable to [17].

6.1.3 - RPLIDAR

Previous work [1] considered the requirements of autonomous navigation of robots in inspection environments. Camera data was used for SfM reconstructions, which were then processed to create a 2D map, with the goal of localising in the map using ultrasound sensors.

In this work, mapping and localisation in an inspection environment was also considered, but this time a 2D LiDAR, the RPLIDAR version A1, was used. The RPLIDAR is a low-cost (£300) 2D LIDAR solution developed by RoboPeak. The sensor has a range of 6m in 360° with readings being taken at 5.5Hz. The sensor was easily incorporated into the current setup, with mounting possible directly onto the top of the DiddyBorg platform. The sensor power supply is 5V, with a USB connector which can be powered directly from the Raspberry Pi. Drivers and ROS packages are readily available on the Raspberry Pi and can be installed from [30].

7 - Site Description

The site considered in this paper is the Millennium Bridge, a cable suspension footbridge in Leeds, UK. The bridge, which opened in 1993, crosses the River Aire spanning approximately 57m to connect The Calls to Brewery Place. The bearings on the north side of the river were used as the site for data collection.

The North side bearings are situated in the top abutment, and its dimensions are approximately 2.8x1.2m. There is a trough that runs alongside one side of the site, and there are various pipes and electrical cables running along the length of the enclosure. The top bearings are seated on the bridge by means of a machined steel plate bolted to the bearing.

8 - Survey and Data Collection

Data collection was performed by tele-operating the robot in the bearing enclosure using the user interface described in section 5.2.2. A router was used to allow networking between the DiddyBorg platform and a laptop.

Next, data collection was performed. At this stage, data from the raspicam_node and the rplidar node were recorded into rosbags to allow post-processing of the data. Rosbag is a command-line ROS tool for storing serialised ROS messages in a file as messages are received from specified ROS topics. This tool allows the data to be replayed through the ROS topics at a later date. LiDAR data was collected on two separate instances. One set of data was used to create a map, and the other was used as test data for the localisation algorithm, see section 9.3.

The camera resolution was set at 320x200 pixels to allow real-time processing of the camera data. This resuolution is a similar to the one used in [17]. Camera data was sent in jpeg compressed form by the raspicam node, received and uncompressed using ROS image_transport tools on the laptop and then processed by the ORB-SLAM2 node, as depicted in Figure 2.

As discussed in section 3.1, before mapping can begin, initialisation of ORB-SLAM must occur. Once initialisation occurred, the DiddyBorg platform was navigated around the bearing enclosure to build up a map of the environment. Three repetitions of ORB-SLAM were completed. In contrast to previous work [1], separate data was not collected for the SfM calculations, but the data collected for ORB-SLAM was also used for this purpose.

9 - Results and Discussion

The effectiveness of the SLAM methods ORB-SLAM and Hector SLAM is now considered. A qualitative comparison between the methods is given in Table 1.

9.1 - ORB-SLAM

As described in [31], changes in lighting conditions causes tracking failure between frames for multiple feature descriptors. This phenomenon also affected the localisation and mapping of ORB-SLAM, as expected. Since data collection was performed in the late afternoon, by the third repetition the sun had set below the level of the bridge, see Figure 3. As a result, there was a greater contrast between lighting conditions in successive frames and tracking was lost for a large period of time, see Figure 3.

Figure 3: Two frames from ORB-SLAM that results in the map shown in Figure 4. ORB features are marked as green squares in the left hand image. In the right hand image no features appear since tracking has been lost due to a sudden change in lighting conditions.

Figure 4: The map produced by ORB-SLAM. The key frame locations can be seen by the blue rectangles, the current keyframe as a green rectangle and the progression of the robot also in green. The red map points show the local visible map.

In general, initialisation was obtained quickly, with sufficient features provided by objects in the environment (e.g., railings and litter). Throughout the mapping, adequate features were present, with texture being provided by dirt and cracks on planar surfaces. Figure 3 shows a comparison of the ORB-SLAM results when initialisation is successful and ORB frames are being tracked and some frames later when tracking has been lost due to abrupt change in brightness.

Tracking was also lost due to abrupt and fast motions, as anticipated. When the tracking was lost, the robot has to return to a previous key frame and localisation is performed globally, this was performed successfully in most cases. As a result of loss of tracking the whole area mapped with the same detail, this is likely to affect localisation with new data.

It is also possible to save and reload a map using additional functionality developed by [32]. The localisation mode in ORB-SLAM can then be used with a loaded map and new data for localisation. In future work, the expansion of this tool for autonomous navigation will be considered, where one obstacle to over-come is to provide scale to the map.

9.1.1 - SfM Results

The data collected and used for ORB-SLAM in Figure 4 was also sufficient the SfM reconstruction to be successful. Approximately 230 images were collected in total and used for the SfM reconstruction. SfM was not as affected by the variations in brightness and was able to use more of the dataset, whereas ORB-SLAM cannot use the frames where tracking was lost. Although the front wall of the abutment in Figure 5 was one of the least detailed areas of the SfM reconstruction, it was much more detailed than the corresponding region in Figure 4. Similarly, it can be seen that there are more areas in Figure 5 where photos were used compared to Figure 4, both indicated by blue triangles.

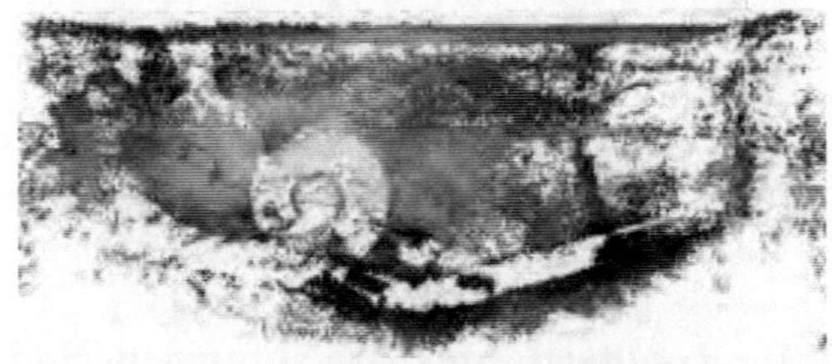

Figure 5: The SfM reconstruction completed using Zephyr Aerial SfM software. The same dataset used for ORB-SLAM was used for the reconstruction. The camera positions can also be seen as blue triangles.

9.2 - Hector SLAM

The environments shown in Figure 5 and Figure 8 show some inconsistencies. In Figure 5 the curved wall at the front of the enclosure can be clearly seen, there is no sign of this wall in Figure 8 . The reason for this discrepancy was the height of the RPLIDAR with regards to the wall – when mounted on the DiddyBorg platform the RPLIDAR was higher than the wall, and hence the wall not detected by the sensor.

The map created by Hector SLAM in Figure 8 shows lines that go off the map. These lines are sensor readings recorded by the LIDAR at maximum sensor range. The bridge in question crosses the River Aire and these maximum readings are caused by reflections off the water.

In addition, Figure 8 shows errors in the SLAM results that occurred when the robot mounted an electrical cable that was present in the bearing enclosure. This cable caused the base, and hence the RPLIDAR, to tilt into a different plane. As discussed in section 3.2, Hector SLAM can be used in situations where the plane of the LIDAR changes, allowing 3D navigation. To achieve this an IMU is required and integrating this sensor information to track the transformation of the base to a stabilised base [18]. This consideration will be included in future work to give a more robust SLAM system.

To compare the results from ORB-SLAM and hector_slam the trajectory outputted by both methods was recorded simultaneously and plotted in Figure 6. The results show overall the trajectories are very similar in shape. However, the trajectory for ORB-SLAM is less accurate and has greater variation than Hector SLAM and gaps appear when the trajectory crosses itself.

Figure 6: A comparison between the trajectories of the robot using Hector SLAM (left) and ORB-SLAM (right). Note the figures are not plotted together since the ORB-SLAM results require scaling.

10 – Results:Hector SLAM for Autonomous navigation using AMCL.

It is possible to perform autonomous navigation using the hector_navigation stack [33], also developed for urban search and rescue environments by the same authors of Hector SLAM. The hector_navigation stack contains packages such as the hector_exploration_node, which accesses the hector_exploration_planner, a planning library that allows the robot to explore unknown areas of the map. However, the resulting map cannot be saved and used again with the hector_navigation stack.

For inspection applications it is useful to have a known map to highlight targets for inspection or areas of interest, in advance. For this reason, Adaptive Monte Carlo Localisation (AMCL) was used for localisation in a known map; the map was created using hector_slam. As in section 8.2, the hector_slam package was used to provide odometry from the 2D LiDAR data. This process was visualised using rviz and displayed in Figure 9a-c. The implementation used in this work is based on [34].

Initially, the front wall of the enclosure was not registered in the map, but in reality these areas past the front wall are not accessible. For caution, since no method has been implemented here to prevent navigation to areas beyond the front wall, the navigation system was not tested in the abutment enclosure. However, LIDAR data collected from the bridge is used as the input and the same commands for a particular navigation goal were successfully received.

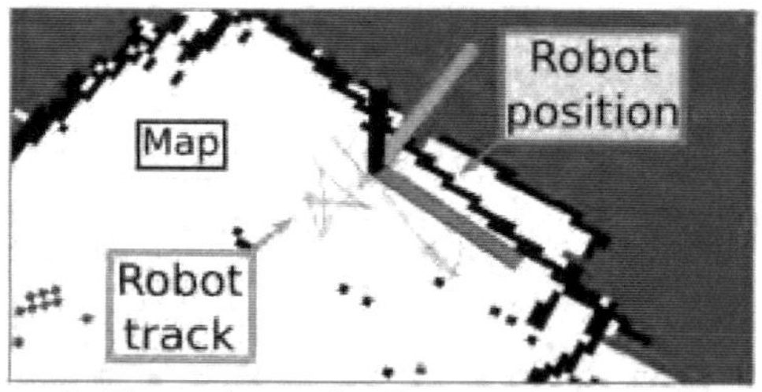

Figure 7: Shows the result of hector_slam with the robot position and path marked by in the figure. Localisation and mapping errors occur in this example because the RPLIDAR is tilted out of its original plan without any update e.g., using data from an IMU.

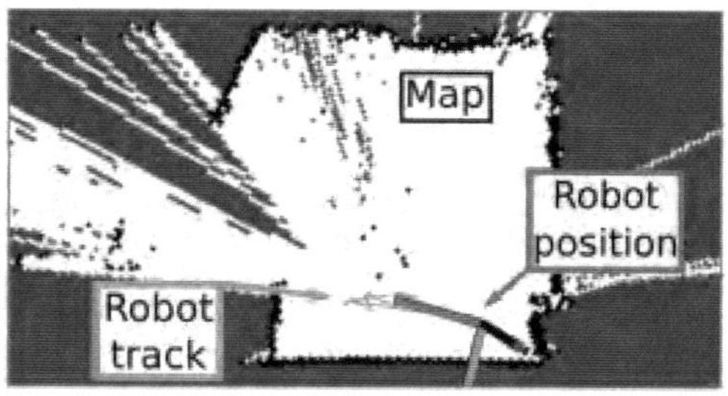

Figure 8: The SLAM result for one set of LIDAR data using Hector SLAM. The robot position, trajectory and map boundaries are shown. Maximum sensor values are also returned in some instances – shown by lines that go outside the map.

Conversely, one of the disadvantages of using pre-existing maps is if the environment changes, the map may no longer be representative. In Figure 9c) the front wall of the abutment was picked up momentarily by the LIDAR. As a result, localisation was temporarily lost, since the wall is not a known landmark and some particles are placed outside of the map.

One disadvantage of AMCL is that a start position is required to be set for the localisation process to begin. Since the geometry of the bridge is well known in this case, an initial position can be set fairly accurately. This can also be set automatically. Figure 9 shows the scatter of particles before the initial position and again once the particles have converged to a location.

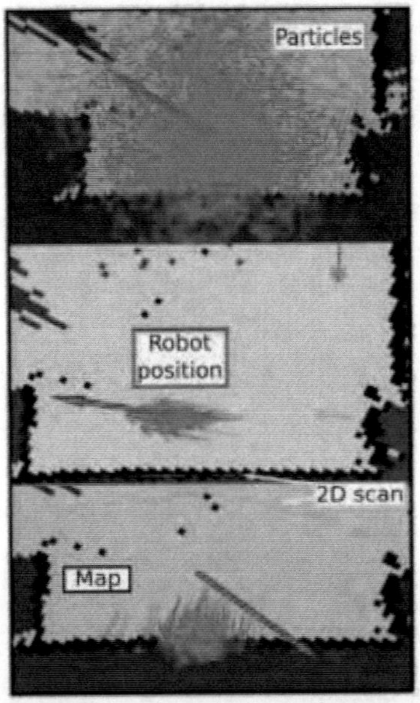

Figure 9: a, b and c top to bottom.
a) Initially, particles are spread across the whole map.
b) After a few time steps particles converge.
c) Change in map – front of the enclosure detected by 2D LiDAR, position estimation of robot was lost.

Table 1: A qualitive comparison between ORB-SLAM and Hector SLAM with AMCL.

	ORB-SLAM (localization only)	**Hector SLAM Plus AMCL**
Cost	Lowest	Highest
Accuracy	Lowest	Highest
Automatic scale	No	Yes
3D map	Yes	No
Use data elsewhere	Yes: use directly for SfM.	Require camera for visual inspection.
Environment variation/ Real world robustness	Affected by lighting. Harder to relocalise	Lost if scan doesn't match environment, but relocalises well.
Autonomous Navigation	Need some method for scaling first.	Easily implemented.
Future work	Sensor Fusion: odometry for scaling	Sensor fusion: IMU for 3D navigation

Table 1 compares some qualitative differences between using Hector and AMCL and ORB-SLAM for localisation. Note, that since no method for navigation is implemented for ORB-SLAM in this paper, localisation only is considered. Overall, the low cost and re-use of data are key advantages of ORB-SLAM, but Hector SLAM with AMCL is more accurate, and autonomous navigation could be implemented by outputting the position in the map and the required target to the ROS navigation stack. Both methods require future work to improve their robustness in inspection environments, but the results so far are promising.

11- Future Work and Considerations

To increase the robustness of the SLAM and autonomous navigation approaches used here, sensor fusion will be implemented, primarily with an inertial measurement unit for the hector_slam approaches and odometry for ORB-SLAM to incorporate scale into the map. Future work will look at extending the ORB-SLAM localisation method considered here for navigation.

12 - Conclusions

In this work, an improved robot for inspection of bridge bearings with off-the-shelf low-cost camera and LIDAR technology was presented. Building upon previous work, existing methods for Simultaneous Localisation and Mapping (SLAM) (i.e., ORB-SLAM and Hector SLAM) were applied to a real bridge. Visual inspection was carried out and qualitatively compared the differences in the methods. Using Hector SLAM, we then explored methods for autonomous navigation, with a known map.

Acknowledgements

This work was supported by the Engineering and Physical Sciences Research Council (ESPRC), UK.

References

[1] H. Peel, G. Morgan, C. Peel, A. Cohn, and R. Fuentes, "Inspection robot with low cost perception sensing," in ISARC 2016 - 33rd International Symposium on Automation and Robotics in Construction, 2016.

[2] H. Eggert and W. Kauscke, "Structural Bearings," vol. 3, no. November, p. 405, 2002.

[3] L. M. R. Freire, J. De Brito, and J. R. Correia, "Management system for road bridge structural bearings," Structure and Infrastructure Engineering, vol. 10, no. 8. Taylor & Francis, pp. 1068–1086, 2014.

[4] [4] L. Freire, J. de Brito, and J. Correia,

"Inspection Survey of Support Bearings in Road Bridges," J. Perform. Constr. Facil., vol. 29, no. 4, p. 4014098, 2015.

[5] [5] J. Chen, W. Junjie, C. Gang, W. Dong, and X. Shen, "Design and Development of a Multi-rotor Unmanned Aerial Vehicle System for Bridge Inspection," in International Conference on Intelligent Robotics and Applications (ICIRA 2016), 2016, pp. 498–510.

[6] [6] R. R. Murphy et al., "Robot-assisted bridge inspection," J. Intell. Robot. Syst. Theory Appl., vol. 64, no. 1, pp. 77–95, 2011.

[7] [7] N. H. Pham and H. M. La, "Design and implementation of an Autonomous Robot for Steel Bridge Inspection," Int. J. Adv. Robot. Syst., vol. 10, no. January, pp. 556–562, 2016.

[8] 3DFLOW, "3DF Zephyr Aerial." 3DF, Verona, 2016.

[9] R. Toldo, R. Gherardi, M. Farenzena, and A. Fusiello, "Hierarchical structure-and-motion recovery from uncalibrated images," Comput. Vis. Image Underst., vol. 140, pp. 127–143, 2015.

[10] R. Toldo, F. Fantini, L. Giona, S. Fantona, and A. Fusiello, "Accurate Multiview Stereo Reconstruction with Fast Visibility Integration and Tight Disparity Bounding," ISPRS-International Arch. Photogramm. Remote Sens. Spat. Inf. Sci., vol. 1, no. 1, pp. 243–249, 2013.

[11] D. Thrun, Sebastian; Burgard, Wolfram; Fox, "EKF Localization," in Probabalistic Robotics, Cambridge, Massachusetts: The MIT Press, 2006, pp. 54–65.

[12] D. Fox, W. Burgard, F. Dellaert, and S. Thrun, "Monte Carlo Localization: Efficient Position Estimation for Mobile Robots."

[13] D. Thrun, Sebastian; Burgard, Wolfram; Fox, "The GraphSLAM Algorithm," in Probabalistic Robotics, 10th ed., Cambridge, Massachusetts: MIT Press, 2006, pp. 337–385.

[14] J. D. Mur-Artal, Raúl, Montiel, J. M. M. and Tardós, "ORB-SLAM : a Versatile and Accurate Monocular SLAM System," IEEE Trans. Robot., vol. 31, no. 5, pp. 1147--1163, 2015.

[15] S. Kohlbrecher, O. Von Stryk, J. Meyer, and U. Klingauf, "A Flexible and Scalable SLAM System with Full 3D Motion Estimation," in Proc. IEEE International Symposium on Safety, Security and Rescue Robotics (SSRR), 2011, pp. 155–160.

[16] Raulmur, "ORB_SLAM2." raulmur / github repository, 2016.

[17] G. Ponnu, J. George, and J. Skovira, "Real-time ROSberryPi SLAM Robot," Cornell University, 2016.

[18] S. Kohlbrecher, J. Meyer, T. Graber, K. Petersen, O. Von Stryk, and U. Klingauf, "Hector Open Source Modules for Autonomous Mapping and Navigation with Rescue Robots," Rob. 2013 Rob. 2013 Robot World Cup, vol. XVII, pp. 642–631, 2013.

[19] Tu-darmstadt-ros-pkg, "hector_slam." Tu-darmstadt-ros-pkg / github repository, 2016.

[20] PiBorg, "DiddyBorg - The most powerful Raspberry Pi robot available | PiBorg," Drupal. [Online]. Available: https://www.piborg.org/diddyborg. [Accessed: 29-Mar-2017].

[21] Martin Wimpress, "Ubuntu MATE for the Raspberry Pi 2 and Raspberry Pi 3," 2016. [Online]. Available: https://ubuntu-mate.org/raspberry-pi/. [Accessed: 29-Mar-2017].

[22] The Open Source Robotics Foundation, "ROS.org | About ROS." [Online]. Available: http://www.ros.org/about-ros/. [Accessed: 18-Mar-2017].

[23] Open Source Robotics Foundation, "Introduction," ROS Wiki, 2014. [Online]. Available: http://wiki.ros.org/ROS/Introduction. [Accessed: 25-Mar-2017].

[24] J. M. O'Kane, A gentle introduction to ROS. Jason M. O'Kane, 2014.

[25] PiBorg, "PicoBorgRev." 2015.

[26] ROS, "common_msgs - ROS Wiki," Open Source Robotics Foundation, 2014. [Online]. Available: http://wiki.ros.org/common_msgs. [Accessed: 29-Mar-2017].

[27] Robot Web Tools, "roslibjs." Robot Web Tools / github repository, 2017.

[28] larrylisky.com, "Enabling Raspberry Pi Camera V2 under Ubuntu Mate." [Online]. Available: https://larrylisky.com/2016/11/24/enabling-raspberry-pi-camera-v2-under-ubuntu-mate/. [Accessed: 29-Mar-2017].

[29] 38 contributors (full list Https://github.com/ros-perception/image_pipeline/graphs/contributors), "ros-perception / image_pipeline." github repository.

[30] Robopeak, "rplidar_ros." robotpeak / github repository, 2016.

[31] A. Pieropan, M. Björkman, N. Bergström, and D. Kragic, "Feature Descriptors for Tracking by Detection: a Benchmark," 2016.

[32] poine and raulmur (forked from raulmur), "ORB_SLAM2." , poine / raulmur / github repository, 2016.

[33] Tu-darmstadt-ros-pkg, "hector_navigation." Tu-darmstadt-ros-pkg / github repository, 2016.

[34] Pariljain, Nischal92, ParitoshKelkar, Nischalkn, "F1tenth.", nischalkin / github repository, 2016.

The Variable Guide Frame Vehicle for Tunnel Inspection

Satoru Nakamura[a], Yusuke Takahashi[a], Daisuke Inoue[a] and Takao Ueno[a]

[a]Institute of Technology,Tokyu Construction Co.,Ltd.
E-mail: nakamura.satoru@tokyu-cnst.co.jp

Abstract

Ageing of the bridges and tunnels that were built more than 40 years ago has become an object of public concern. Maintenance engineers must approach as close as possible to the inspection point by mobile elevating work platform and conduct the human-eye based close inspection. Especially an inspection of the road tunnel requires traffic regulation. However, it takes much time on the visual inspection and hammering test for wide inspection area. Furthermore those inspection results have variations by inspectors. We have proposed a "variable guide frame vehicle for inspection of tunnel lining concrete" by new technology for infrastructure maintenance. This paper reports about the element technology of variable guide frame vehicle for inspection and each inspection system development situation.

Keywords –

Tunnel lining concrete; Variable geometry truss; Light-section; Hummering

1 Introduction

Many domestic tunnels and bridges that is built more than 40 years ago (at a period of rapid economic growth in Japan) ,towards deterioration and about 20% of road tunnels that were 100 thousand at the time of March 2013 in Japan have passed over 50 years. It is estimated that it will be about 50% in 2033 [1]. It was enforced new road raws in July 2014 that obliged to do close up visual inspection every 5 years. On the other hand, engineers that manage them and inspection cost are insufficient, and it is desired to develop innovative inspection techniques which resolve these problem.

Therefore, we proposed "the variable guide frame vehicle for inspection" as the new infrastructure maintenance technology and described our research and development about it's elemental technologies.

2 Current situation of tunnel inspection

We have carried out maintenance of tunnels such as repair / renewal based on forecast by holding on the progress of deterioration by periodic inspection because it is difficult to rebuild tunnels after completion. In general, we regulate traffic when inspect road tunnels and engineers and their assistants inspect as close as possible on aerial work vehicles (Figure 1).

However, this method have a problem that it is forced to work for a long time and requires to regulate traffic for a long time too. Furthermore, there are different with each engineer for the results of inspection about visual inspection and hammering tests. We cannot hold on the progress of deterioration accurately in this method because the results of inspection for visual inspection and hammering tests are not recorded.

Administrators of tunnels are difficult to judge the optimal repair method and it's timing because professional knowledge is required to evaluate tunnels based on inspection data. Futhermore, it is desired to build maintenance management systems that minimize economic impact considering LCC (Life Cycle Cost) for increasing a great deal of load tunnels which require maintenance and management. Therefore, it is required systems which do not require expert to go to the site and can show the state of deterioration and optimal repair methods based on various data which obtained by inspection.

3 Outline of The Vehicle for Tunnel Inspection

The variable guide frame vehicle for inspection (Figure 2) (hereinafter, this is called the inspection system) is the system which aims to greatly reduce the traffic regulation during inspection and substitute remote control robots for close up visual inspection and hammering tests by engineers. To realize this system, we mainly developed 4 elemental technologies, ①the variable guide frame (hereinafter, this is called VGF) and ②the protective frame vehicle, ③the hammering tester, ④the crack inspector.

We decided specification of this system as shown in Table 1 referring to the third type of load (that is Japasese load type) because targets of this system is local tunnels.

Figure 1 tunnel inspection

Figure 2 VGF vehicle

The length of this system is 5m and the width is expandable accrording to the width of the road by using expandable frame structure. The weght is 4t according to the load standard specification because this system Travels by tires in the tunnel and weght is 4t and we can control this system along the curve of tunnels of the minimum radius of curvature 50m

The VGF is set with arch shape at a certain distance from tunnel surface. We can obtain images of the concrete surface and hammering sounds by moving devices along the VGF. Our target is 150m2/h at maximum which is the alomost same as the inspection speed by engineers. However, the inspection speed is depends on these device performances. We aim to build systems which can output quantitative inspection data obtained by this system as resulting drawing of tunnel inspection and photographic ledger of tunnels. We let described about each elemental technologies as below.

Table 1. Specification of the device

Length（m）	5.00
Width（m）	6.36～9.56（Stretchable）
Weight（kg）	4,000（500 / 1wheel）
Traveling speed（m/min）	1～10（variable）
Inspection items	Flaking， Peeling， Crack
Inspection speed（m^2/h）	Max 150

3.1 VGF（Variable Guide Frame）

The inspection efficiency is low because the tunnel have a various shape according to natural ground and usage and dotted with lamp equipments and signs. A variable frame structure is necessary to adapt various tunnel shapes in order to avoid equipment in the tunnel and enhance versatility. Furthermore, it is necessary to have a frame structure which can get a sufficient reaction force even if the equipment is implemented to do inspection by remote control.

In this system, we adopted VGF (Figure 3) which applied variable geometry truss as inspection guide frame. This system has a laser range finder to measure detects shape of tunnels ahead of this system and obstacle of such as a sign position.We can obtain the amount of expansion and contraction of the actuator necessary for deforming the VGF by the results of inverse analysis of this measurement, we can perform deformation operation for VGF (Figure 4). Furthermore, the rails necessary for moving the inspection device are mounted on the VGF,inspect tunnels along the rail.

3.2 Protective Frame Vehicle

The protective frame vehicle (Figure 5) with VGF traveling along the tunnel and can inspect. This system can inspect entire tunnel surface by alternately moving inspection devices along VGF and traveling the protective frame vehicle. We are assumed that this system is used at the local tunnel (the third type of load in Japanese). This system can travel by a electric motor and climb up to 10% in longitudinal gradient. Furthermore, This system can travel along the curve of tunnel by setting the difference speed between the left and right tires using the inverter control at the curve.

Figure 3 VGF

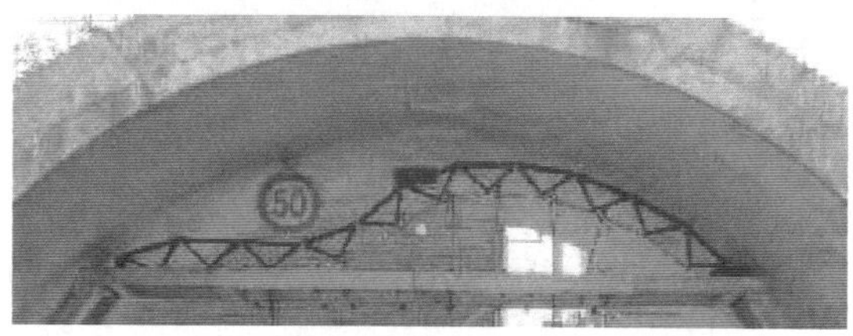

Figure 4 Deform the VGF

Figure 5 Protective Frame Vehicle

It is necessary to secure traffic safety and smoothness by protecting a third party and a car from falling objects such as concrete pieces in order to inspect at the top of the tunnel without traffic regulation. In this system, falling objects are prevented by the steel plate which set at the top of the protective frame vehicle.

3.3 Automatic Hammering Tester

The hammering tester can hammer concrete by the remote control and analyze hammering sound data, and detects the delamination position.

The hammering criterion and the hammering sound analysis method are based on sounds which hammer by engineers using inspection hammer. Considering these criterions and databases, We developed the hammering tester that behave similar to that of a engineers arm and can simulate hammering sounds by engineers (Figure 6). This device is controlled hammering velocity according to hammering directions to make hammering energy constant even if gravity changes when moving along the VGF. This device has two modes. One is the first inspection which hammer at an interval of 20cm while moving to obtain delamination points of the entire tunnel. Another is the second inspection which hammer at an interval of 5cm to obtain delaminations and cracks within a certain range in detail. As a new research, We aim to develop systems for notifying the future damage of the delamination by using the hammering analysis method based on supervised machine learning which estimate depth of crack and cold joint and by using the image processing method which estimate directions of crack and cold joint by fusing camera image of the hammering range and the results of the hammering analysis. Furthermore, we aim to invent a automatic identification technology with robustness to noise considering inspection in the environment where cars runs.

3.4 Automatic Crack Inspector

Many crack inspect devices using images are used

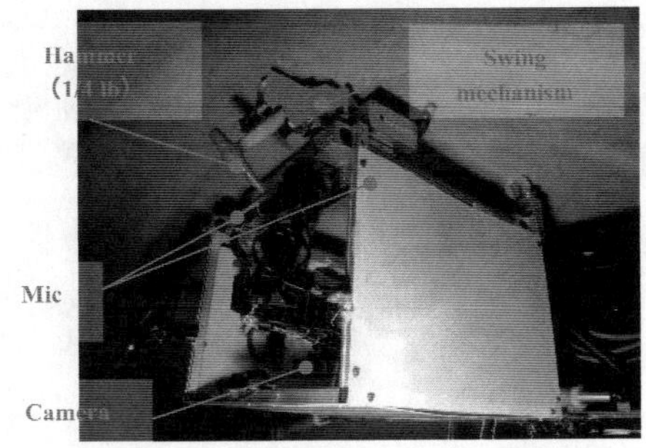

Figure 6 Hammering Tester

on site. Although they do not require traffic regulation, the resolution of the crack width is low, and much effort is required for post-processing work such as noise removal including contamination of concrete.

We developed systems which output inspection data as resulting drawing of tunnel inspection rapidly by automatically identifying cracks and dirt using the three-dimentional shape measurement based on light-section method to correspond high accuracy and labor saving at crack inspection. We can obtain color images and depth images high accuracy and continuously by moving the crack inspecter (Figure 7) with a light section camera along the tunnel surface. We aim to develop inspection technology which can obtain detailed inspection data such as the step and the water leak by combining the results of hammering tests and three-dimentional shape measurements.

4 Work flow of The Vehicle for Tunnel Inspection

The inspection system is operated in the procedure of 6 steps, ①assemble such as a protective frame vehicle, and ②set it in the tunnel, ③do the firtst inspection, ④ do the second inspection, ⑤judge the results of the inspection, ⑥unassemble it. It is shown the inspection step which are asummed in Figure 8.

We assemble this system at out of the tunnel and move it into the tunnel. After that, this system measure the cross-sectional shape of the tunnel ahead of it and expand the frame according to it's results

In the first inspection, the hammering tester specify delaminations of the concrete, and the crack inspector do crack inspection in real time and obtain images using as the resulting drawing of tunnel inspection.

In the second inspection, while returning to the inspection start position, abnormal parts which identified when the first inspection is inspected in detail by hammering tester. Furthermore, this system estimates

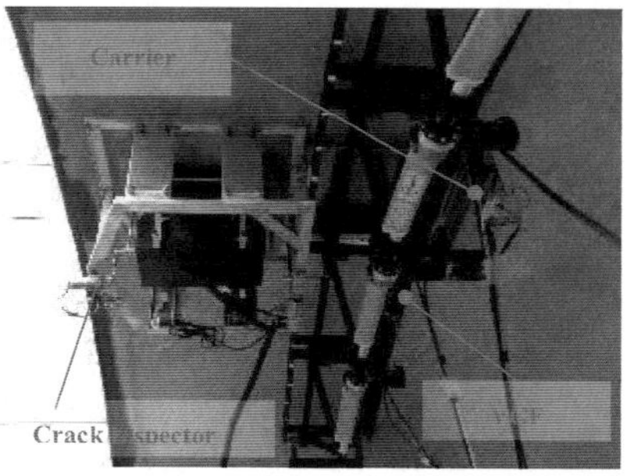

Figure 7 Crack Inspector

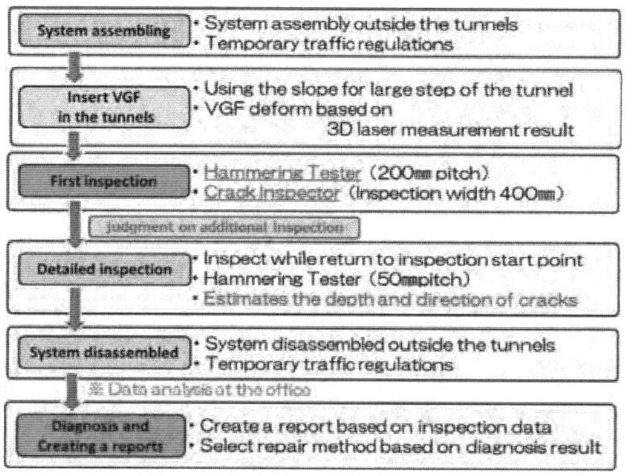

Figure 8 VGF Work flow

the crack depth and the crack direction about these parts in this inspection (Figure 9). When it is judged that repair is necessary, measure concrete physical property. Especially, for parts with high urgency, we drop as a first aid. After returning to the inspection start position, we unassemble this system at the same place where assembled.

We specify abnormal parts from images obtained by the crack inspector and data obtained by the hammering tester, and draw up the resulting drawing of tunnel inspection and the photographic ledger of the tunnel. Furthermore, we select repair methods based oninspection data and calculate LCC for maintenance.

Figure 9 VGF vehicle inspection

5 Practical application

As technologies to accelerate practical application, we also have developed elemental technologies necessary for repairing and so on utilizing the data which obtained from the inspection system as well as hammering tests and visual inspections as above. We described these technologies follows.

5.1 Automated Estimation of Concrete Strength

If it is judged that repair is necessary from the results of the first inspection or the second inspection, it is required to obtain the concrete physical property information for considering the repair method. This system also can implement a drill device which can operate remote control and drill concrete to estimate the concrete compressive strength and neutralization depth on site.

5.2 Tunnel management support system

It can propose optimal repair method and timing based on database of repair methods and materials and so on by input obtained inspection data and basic data of tunnel such as the environment condition and the construction condition, elapsed years. This system can be operated without exparts going to the site.

6 Conclusions

We proposed “the variable guide frame vehicle for inspection” for the tunnel inspection that are expected to increase further in the future. For future work, we will do research and development aimed to develop a versatile inspection system which can inspect enire tunnel surface without traffic regulation.

Technology and Innovation, “Cross ministerial Strategic Innovation Promotion Program (SIP), Infrastructure Maintenance, Renovation, and Management.” (funding agency: NEDO)

References

[1] Asano Y. and kimura S. huge market in 2025. *nikkeibp.co.* pp.18-19, 2014.

[2] Nakamura S., Inoue F., Yamashita A., Yanagihara Y., Ueno T. and Takahashi Y. Development of variable guide frame vehicle for inspection of tunnel lining concrete. In *Proceedings of the JSME ROBOMECH*, 2P1-N01, 2015. (in Japanese)

Shape Control of Variable Guide Frame for Tunnel Wall Inspection

F. Inoue[a] , S. Kwon[a], T. Uchiyama[a], S. Nakamura[b] and Y. Yanagihara[b]

[a] Shonan Institite of Technology, Japan, [b] Tokyu Construction, Japna
E-mail: Inoue@mech.shonan-it.ac.jp

Abstract

To progress the automated inspection and maintenance of inner wall of tunnel, the advanced inspection system with restricting the traffic regulation was developed. In this inspection system, the guide frame along tunnel wall was installed on the protection unit stepped over the road like gantry crane. The inspection device moved with stability by adopting the guide frame, and the inspection accuracy could be improved. However, when this unit moved along the tunnel, this guide frame should avoid the convex obstacles such as duct fan, lamp and several road traffic sign in the tunnel. Therefore, by composing the entire frame of VGT (Variable geometry Truss), the shape of guide frame was changed flexibly and it could be passed in the tunnel. As a shape control of the guide frame, the inverse analysis method was applied such as robot manipulator combining two fixed-length members in series. The angle of each frame was reversely analysed according to the shape of the obstacle measured with the laser sensor, and the actuator of the frame was controlled simultaneously. Applying such control method, the shape was transported smoothly to keep a steady structure of guide frame.

In this paper, the outline of tunnel wall inspection applied variable guide frame, the structure principle of the guide frame, basic method of shape analysis and control technology, the finding of convex obstacles by laser sensor and the whole inspection system are explained in detail.

Keywords –

Structure, Automated Inspection system, Shape Control, Tunnel Inspection, Variable Guide Frame Introduction

1 Introduction

Most of infrastructures for civil engineering structures such as highway roads, bridges, tunnels constructed around the urban region in high-growth era of 1970-1990 begun to reach the life, and it has been the time when large-scale repair and renewal, rebuilding were demanded. Since these structures are difficult to

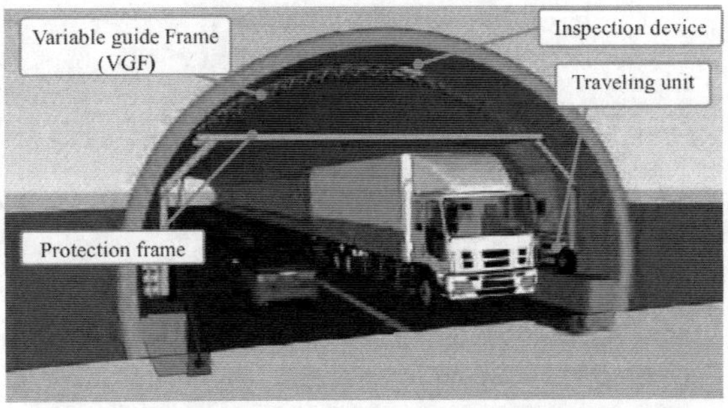

Fig. 1 Variable Guide Frame Vehicle for Inspection of Tunnel

renovate after completion, it is necessary to grasp the progress of deterioration by periodic inspection and to maintain and manage such as repair and renewal based on future prediction. Especially, with the collapse accident of the high way tunnel generated some time ago in Japan, the inspection of the superannuated tunnel was requested.

Generally, in the periodic inspection of the road tunnel, a maintenance engineer approached the wall by using mobile elevating work platform as much as possible and by restricting the traffic regulation and the engineers detected the deteriorated wall parts by sighting and hammering sound. To progress the automated inspection and maintenance of inner wall of tunnel, some advanced methods were adopted in the SIP program in Japan of the theme of "Maintenance and management robot". In our proposal of this theme, the inspection system was developed applying the variable guide frame to evade the obstacle in the tunnel without restriction of the traffic regulation as indicating in Fig. 1. In this inspection system, this guide frame was installed on the protection frame with travelling unit. When this unit moved along the tunnel, the guide frame should avoid the obstacles such as traffic plate, lamp and discharged ducts in the tunnel. The shape of this guide frame changing flexibly, the frame was able to pass these obstacles in the tunnel easily [1], [2].

In this paper, a method of searching for obstacles in a tunnel using three dimension laser range finder of the high-resolution evaluation, a method of controlling a frame whose shape can be changed according to the obstacles shape, and its experiment results are described.

2 Oobstacles Detection in Tunnel

Since the positions and shapes of the various obstacles existing in the tunnel are described on the construction drawing of the tunnel, it is possible to change the shape of the frame so as to avoid obstacles. However, the guide frame being on the based truck, it is necessary to accurately measure the relative distance between the based truck and the obstacle. Also, because the movement error of the based truck is actually included, the obstacle search was carried out on the based truck at the site. In this chapter, Stereo measurement system, analysis method of the shape and position of the obstacle, and its measurement results were verified.

2.1 Stereo Measurement System Using Laser Rang Finder

As the base truck on which the guide frame is installed moves along the inner wall of the tunnel, obstacles protruding from the inner wall near the ceiling may come into contact with the guide frame. Here, as shown in Fig.2, the laser range finder (LRF) was installed at the center of the truck beam surface part and an obstacle on the ceiling surface of the tunnel roughly 1 to 5 m in front was searched. In order to explore the three dimensional space, as shown in Fig. 3, we connected the LRF to the pan unit and constructed the stereo measurement system giving its inclination angle θ_2 to the plane measurement area angle θ_1 of LRF. In this case, if the distance to an object measured by the reflection time of the laser is l, its coordinates (x, y, z) can be expressed by equation (1) - (3).

$$x = l \cdot \cos\theta_2 \cdot \cos\theta_1 \quad (1)$$

$$y = l \cdot \cos\theta_2 \cdot \sin\theta_1 \quad (2)$$

$$z = l \cdot \sin\theta_2 \quad (3)$$

The specification of LRF is indicated in Table 1. The range of measurement is 30[m], the scan time rate is 25[ms] and the resolution angle of rotating laser is 0.25 [deg]. The angular resolution of pan unit was 1.0 deg. at $\theta_2 \leqslant$ 10 deg. and was 0.5 deg at $\theta_2 >$ 10 deg, the measurement precision was increased as it approached the tunnel inner wall. In an actual measurement, the laser irradiated from LRF is reflected to the outer side of the object, and the distances on each object surface were measured from the principle of TOF (Time of flight).

As shown in Fig. 4, to ascertain the cross-sectional shape of the obstacle ahead from the based truck, the laser reflection points existing at the interval of $\triangle$ y (= about 100 mm) with respect to the distance y from the based truck are counted. Then, it estimated that there is an obstacle in a certain y portion where the reflection points are strongly concentrated.

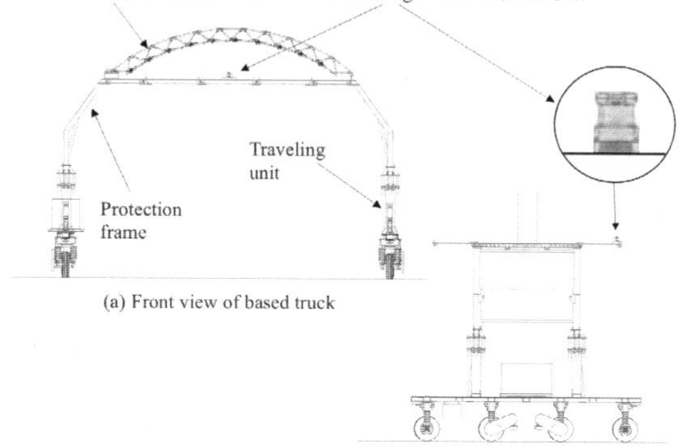

(a) Front view of based truck

(b) Side view of based truck

Fig.2 Structure of based truck and install of LRF on the truck

(a) Laser Range Finder and Pan Unit

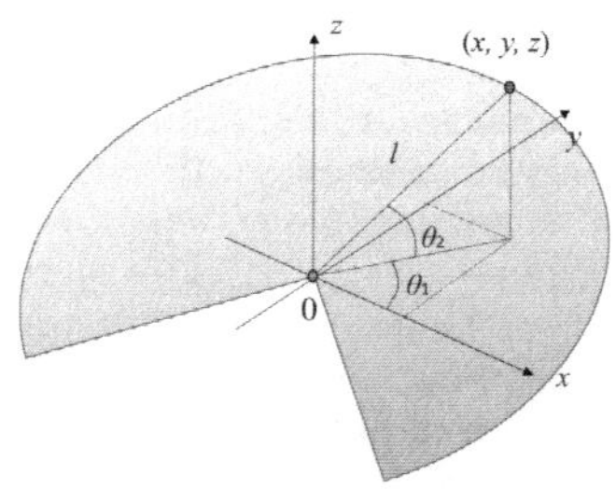

(b) 3D-measurement image using LRM system

Fig. 3 Obstacle detection by stereo measurement system

Table 1 Specification of LRF

Model Number	UTM-30LX
Light Source	Laser Didde, λ=870 nm
Measureable Area	0.1~30 m, 270 deg
Accuracy	0.1~10 m, ±30 mm
	10 ~30 m, ±50 mm
Angular Resolution	0.25 deg
Scan time	25 ms

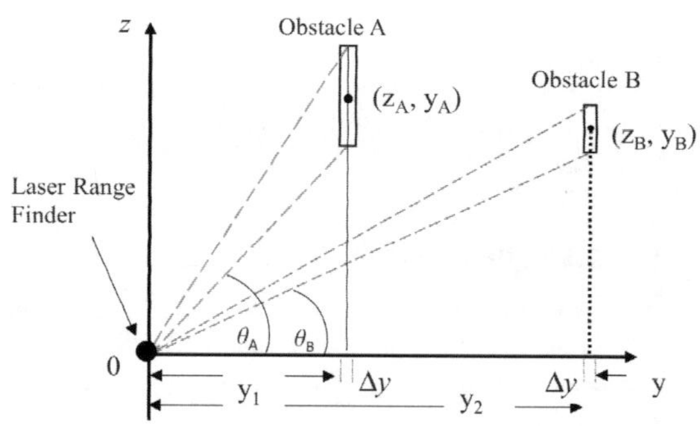

Fig. 4 Objects analysis in x-z coordinate by LRF in x-z coordinate

2.2 Position Analysis Method of Obstacles

In order to determine the cross-sectional position of the obstacle, the center position of the obstacle was analyzed assuming that the shape of the obstacle is a circle from the contour of the point group. By assuming that the obstacle is circular even if it is rectangular, the contour of the obstacle was estimated on the safe side.

Fig. 5 shows the analysis flow for estimating the position and shape of the obstacle.

①: First of all, scanned the inside of the tunnel with LRF and derived and recorded the laser reflection point. However, since the laser reflection points from the inner wall of the tunnel were geometrically known, their reflection points were deleted. As a result, the shape of the obstacle ahead of the truck was extracted.

②: We classified the start position and the end position where the laser reflection point of the obstacle is recorded and extracted the edge of the obstacle.

③: The center position of the circle was estimated using the Constant Distance Method (CDM) [3] and the Least Squares Method (LSM) [4], assuming that the point group constituting the edge is an arc of a circle. The CDM is defined a position advanced by a constant d from the cluster's center as a center position of the object (x_{obj}, y_{obj}) as shown in Fig.6 and described in by equation (4), (5).

$$x_{obj} = x_{cl} + d\cos\beta \quad (4)$$

$$y_{obj} = y_{cl} + d\sin\beta \quad (5)$$

The LSM for a circle is a method for estimating parameters a, b, and r which will minimize the square sum of the error J_{LS} expressed as

$$J_{LS} = \sum_{\alpha=1}^{N}\left[(x_\alpha - a)^2 + (y_\alpha - b)^2 - r^2\right]^2 \quad (6)$$

where $(x_\alpha, y_\alpha), \alpha = 1,\ldots,N$ are the observed data points belonging to the contour of the reference bar.

④: Finally, we confirmed the range that the shape change of the guide frame avoids with an extra margin against obstacles.

By repeating the above process, the shape and forward position of the obstacle to be avoided by the guide frame are determined [5].

2.3 Measurement Result of Obstacle

Using the real simulated tunnel constructed for the experiment as indicated in Fig.7, the obstacles attached to the inner wall of tunnel was measured by LMS. Fig. 8 shows the state of the obstacle measured. In the x - y plane (Fig. 8- (a)), the shape including the traffic signs and the contours of the ceiling lamp in the tunnel is better captured by comparing to the actual picture. In the depth direction z-y plane (Fig.8- (b)), the position of traffic plate and the lamp could be estimated from the

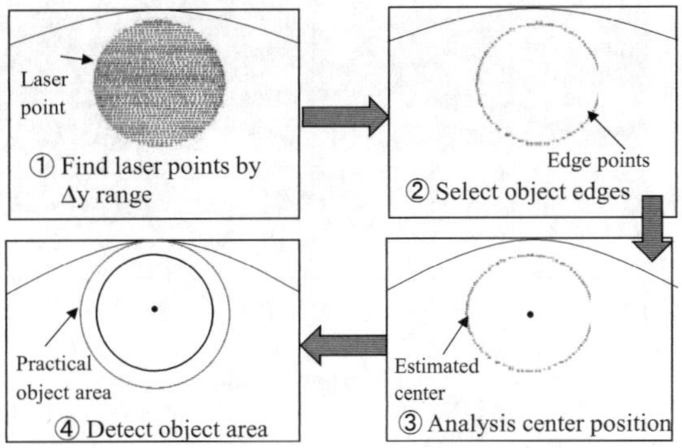

Fig.5 Analysis flow for estimating the position and shape of the obstacle by 3-D Laser Range Finder

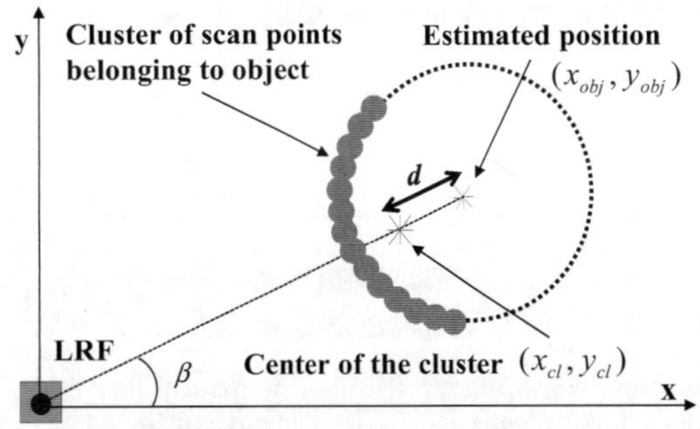

Fig.6 Estimation for center position of circle using the CDM method

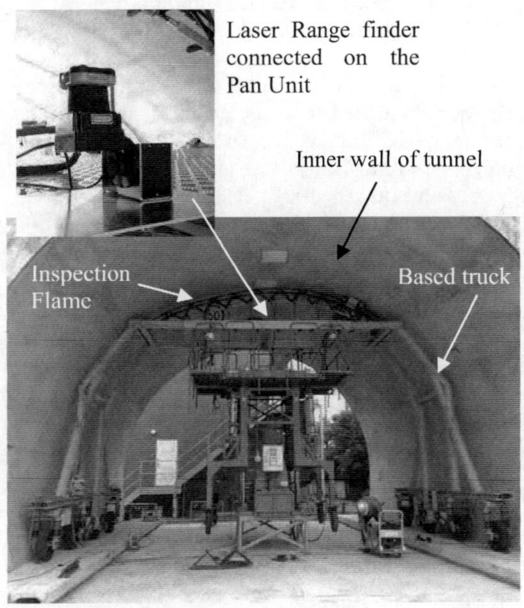

Fig.7 Real simulated tunnel constructed for the experiment and view of based truck

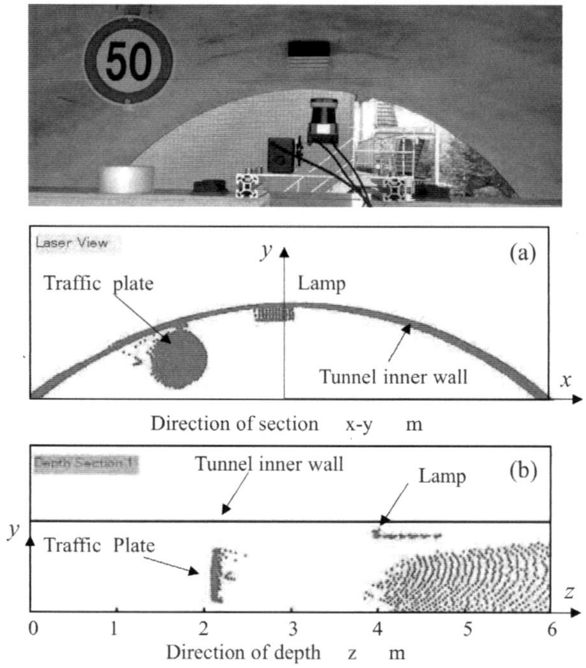

Fig. 8 Obstacles state and positions measured by LRF

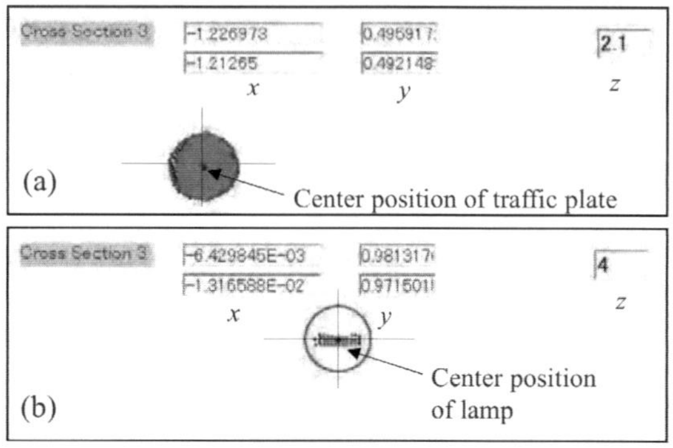

Fig. 9 Estimation of the center position of each obstacle

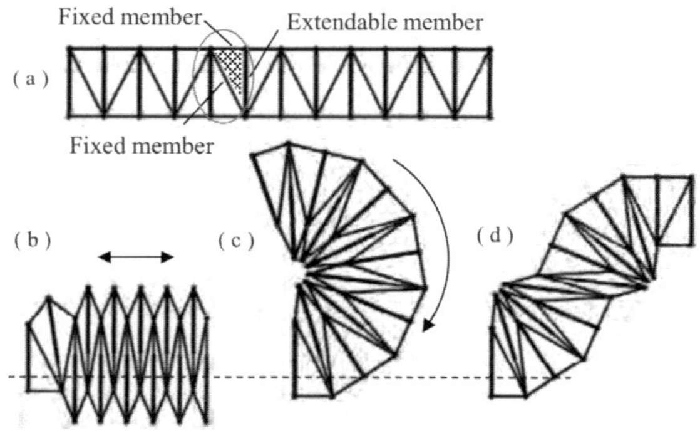

Fig. 10 Example of shape change of VGT structures

portion where the reflection points of the laser are densely concentrated, and it was confirmed that the position was also nearly accurate .

Based on the obstacle data in the x - y plane, its center position was calculated using the analysis flow in Fig.5. As shown in Fig.9 - (a), (b), the center position of each obstacle was obtained from the contour position of the aggregate data, and it was confirmed that the position coincided with the actual position with error accuracy of 10 mm or less. Measurement and analysis, processing time of data was about 1 minute, and it will be possible to detect obstacles sufficiently within the time of inspection work using the guide frame.

3 Shape Control of Variable Guide Frame

In chapter 3, we explain the development of an inspection mobile truck which can avoid obstacles (discharge duct device, lighting lamp, display board, etc.) by using a variable shape frame adaptable to the tunnel shape. In order to create a complicated frame shape avoiding any obstacle on the ceiling surface of the tunnel, we simulated the shape of the guide frame by mathematical method and simulated the frame shape.

3.1 Structure of Variable Guide Frame

The proposed guide frame was constructed on several VGT elements. This VGT (Variable Geometry Truss) was very simple truss structure composed of extendable members, fixed members and hinges, as showing in Fig. 2. By controlling the lengths of the extendable members, it was possible to create various truss shapes. When extendable members are extended simultaneously, the VGT structure was changed like a spring stretching from (a) to (b). When the extendable members are extended alternately, the structure was changed to a circular shape (c) Moreover, when they are extended optionally and their length were controlled, the structure could be changed into any intended shape (d).

In this case, the basic shape of guide frame was an arch structure and its two edges were supported on the protection frame. Though the guide frame was changed flexibly by VGTs shape, the shape was limited by upper side of tunnel wall and traffic space area. So, to avoid projecting obstacle in the tunnel space, it was necessary to correspond to the shape of the guide frame close to the section shape of the obstacle as much as possible. Knowing the obstacle location and shape from the drawing map before inspection, the shape of obstacles as speed plate, blower and side lights were able to simulate by motion analysis. In an actual inspection, the shapes of obstacle were detected by the data of laser range finder continually [6], [7].

3.2 Frame Analysis by Inverse Kinematics

3.2.1 Basic Equation and Shape of Frame

As analysing the arch structure composed of VGT, the whole of guide frame was assumed to be a cantilever structure as indicated in Fig.10-(a). The frame can replace a robot manipulator combining two fixed-length members in series. When the supported edge of frame was $q(x_0, y_0)$, the top $q(x, y)$ of the x, y co-ordinates of the frame combined with n (n≧2) VGT sets was given by Eq. (7) and Eq.(7) using each hinge angle θ_j.(j=1, 2, 3, · ·n),

$$q(x,n) = l_0 \cdot \sum_{k=1}^{n} \cos\left\{\sum_{j=1}^{k} \theta_j\right\} \tag{7}$$

$$q(y,n) = l_0 \cdot \sum_{k=1}^{n} \sin\left\{\sum_{j=1}^{k} \theta_j\right\} \tag{8}$$

Where, l_0 was the length of diagonal member of the frame.To transform the shape of arch frame, some hinge positions on the frame only had to change in proportion to target shape. However, for an intended frame shape fixed by equations (1) and (2), it was quite difficult to solve these equations analytically and to decide the angle because the frame was a very highly redundant. In this case, inverse kinematics analysis was applied.

Considering a temporal change of the whole of the frame, the top edge modifying the original point by inverse analysis to all VGT angle velocity $\dot{\theta}_i$, a shape change of frame was absorbed according to the arch.

$$\dot{\theta}_i = J^{-1} \cdot \dot{q}_i(x,y) \tag{9}$$

Where, J^{-1} indicates inverse Jacobean Matrix (2×n). However J^{-1} isn't necessarily decided because $\boldsymbol{J}$ is not a regular system in n>2. In this case, the suitable matrix such a pseudoinvesre matrix $\boldsymbol{J}^{\#}$ was generally induced instead of $\boldsymbol{J}^{-1}$. Iterating such modification of shape changes, the shape of original arch was able to carve partially to extent in which the obstacle could be evaded.

3.2.2 Shape Change of Guide Frame Avoiding Obstacles

Applying an inverse analysis in the previous chapter 3.2.1, the shape of the frame that avoided the obstacle was analyzed. First of all, a virtual obstacle was positioned on the perpendicular line of the obstacle. The position of a virtual obstacle was gradually lowered, and the frame shape was changed so that surroundings of the obstacle should not come in contact with the frame.

Fig.11-(a)~(d) show the shape change of the arch frame by simulation results. Moving the vertical obstacle, the arch frame was indicated to change like avoiding the obstacle. Finally, a virtual obstacle came in succession at the position of a real obstacle, and the

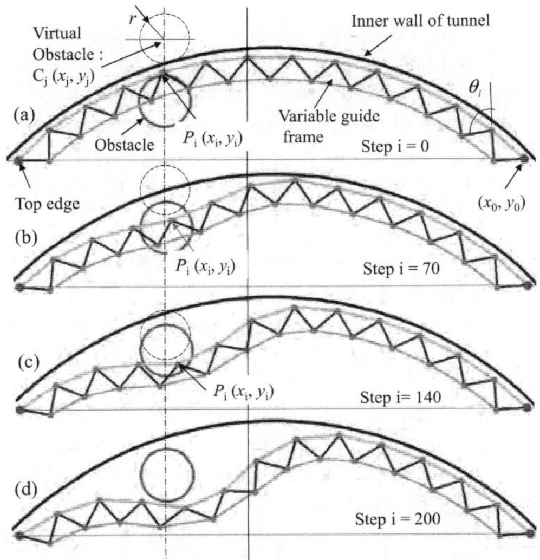

Fig. 11 Shape simulation of VGF to avoid obstacle

Fig. 12 Shape control of guide frame in model tunnel

shape of a final arch frame was decided. As for the shape of the arch frame, it was understood that it was very smooth, no useless movement and two support positions were also the same.

Fig.12 shows the result of experiment in actual model tunnel using variable guide frame. With the real arch frame of 6 m in length, similar shape was able to be also achieved by referring to the simulation result very shout times.

3.3 Frame Analysis by Spline Function

3.3.1 Shape Determination Flow of Guide Frame

Fig. 13 shows the flow for determining the shape of the guide frame according to the position and size of the obstacle in the tunnel.

① Position and shape recognition of obstacles by LRF

② Select multiple points to avoid obstacles on tablet screen

③ Determining the shape of the guide frame by spline interpolation function

④ Select the element frame according to the determined shape, perform the shape simulation, adjust the defect in the shape change process

⑤ Transfer the contents of ④ to the actual frame control device and change the shape of the guide frame

In this flow, it is possible to continuously perform the shape analysis and determination, the transition from the obstacle measurement to the actual system. However, in actual work, guide frames are always in operation by tunnel inspection, and obstacles must be safely avoided in the meantime. In the following, we will discuss ways to enable shape analysis from ② to ③.

3.3.2 Shape Analysis of Guide Frame

To the initial guide frame shape, the overall shapes for avoiding obstacles were mathematically combined by the spline interpolation function.

Spline interpolation is a method of combining arbitrary shapes with polynomials to form a continuous shape. Assuming a function that interpolates the section (x_j, x_{j+1}) the piecewise polynomial $S_j(x)$ is expressed by the Equation (10).

$$S_j(x) = a_j(x-x_j)^3 + b_j(x-x_j)^2 + c_j(x-x_j) + d_j \quad (j = 0,1,2\cdots) \qquad (10)$$

In order for this cubic equation to be a smooth curve, it is assumed that the value of the first derivative and the second derivative of $S_j(x)$ are equal, and the value of the second derivative at the start point x_0 and the end point x_n is 0. By applying the above conditions to each equation for several interpolation points, the coefficient of function a_j, b_j, c_j, d_j was calculated, and the function by the spline interpolation was determined.
Here, each coefficient was determined as follows;

$$a_j = \frac{S_j''(x_{j+1}) - S_j''(x_j)}{6(x_{j+1}-x_j)} \quad (11) \qquad b_j = S_j''(x_j)/2 \quad (12)$$

$$c_j = \frac{y_{j+1}-y_j}{(x_{j+1}-x_j)} - \frac{(x_{j+1}-x_j)(2S''(x_j)+S''(x_{j+1}))}{6} \qquad (13)$$

$$d_j = y_j \qquad (14)$$

3.3.3 Estimation of Guide Frame Shape avoiding obstacles

Using the method in the previous section, several interpolation points (x_j, y_j) were determined so as to avoid obstacles, and the constants of the piecewise function were calculated using equation (10). However, if the spline function $S_j(x)$ obtained for the selected piecewise interpolation points do not satisfy the condition of $S_j(x_j) < C(x_j)$ for avoiding obstacles for all of the functions, the calculation is repeated by using $(y_j - \Delta y)$ reduced from Δy to y_j for x_j, and the each coefficient of $S_j(x)$ is calculated repetitively until the designed conditions were satisfied. Here, $C(x_j)$ is shown the shape function of the obstacle.

(1) Example of obstacle (a)

The shape of a circular obstacle suspended from the left side of the tunnel ceiling was analyzed using spline interpolation function. Table 2 shows the values of the respective coefficients for the equation (10) where the number of interpolation points is 2. The result of changing the shape of the guide frame using this coefficient is shown in Fig. 14- (a). It can be seen that the shape of the guide frame analyzed by spline interpolation with respect to the initial shape is a shape smoothly obstructing obstacles. A similar shape change (Fig. 14- (b)) is obtained in experiments on similar obstacle positions.

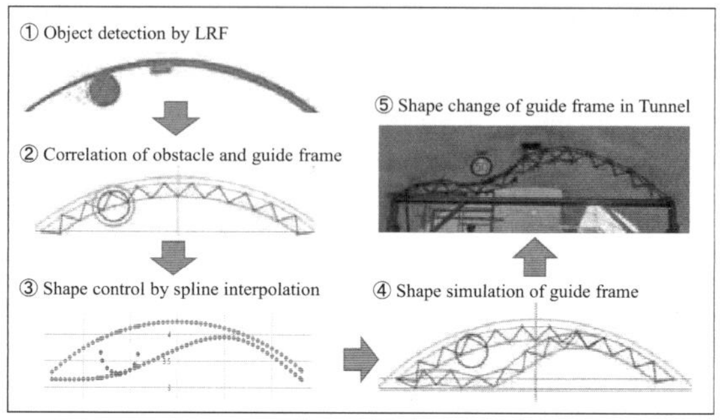

Fig. 13 Shape decision flow of guide frame to avoid objects

Table 2 Each constant value of spline function

$S_j(x)$	a_j	b_j	c_j	d_j
1	0.1966	0	−0.1738	3.156
2	−0.1167	0.4108	0.1081	3.101
3	0.05179	−0.3712	0.1965	4.092

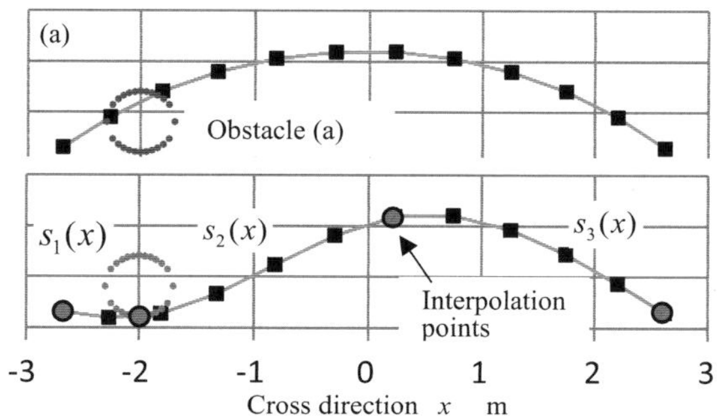

(a) Shape decision of spline function for obstacle (a)

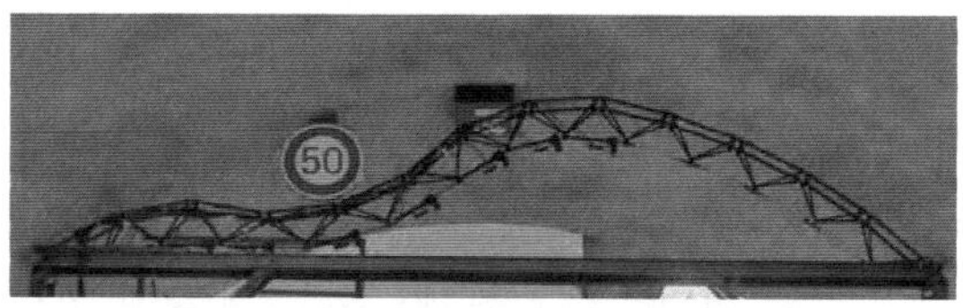

(b) Shape control of guide frame for traffic plate

Fig. 14 Shape decision of guide frame to avoid obstacle

(2) Example of obstacle (b) and (c)

A shape that avoids an obstacle when the obstacle is in the center was estimated by the same method as (1) above. In the example of (b) as discharge fun, the interpolation point is placed at the bottom of a circular obstacle, in the example of (c) as lighting lamp, the interpolation point is placed at the lower end of a rectangular obstacle, and the uneven shapes that avoid obstacles by combining the five functions were created as indicated in Fig. 15.

From the above results, by using this method, it is possible to determine the shape of the guide frame for avoiding any obstacle.

4 Conclusion

In this research, we introduced a technology that can inspect the tunnel ceiling surface without regulating the tunnel traffic by applying the variable shape frame for inspection that can conform to the tunnel shape.

First of all, the tunnel ceiling area which is an obstacle of the guide frame was searched using LRF. By combining a planar laser sensor and a pan unit, the stereo measurement system was constructed. From the results of measurement and analysis by LRF, it was found that the position of the obstacle almost agrees with the actual value, and it was expected that it could be applied sufficiently to the actual measurement.

Next, we propose a method to analyzed the shape of the guide frame by inverse analysis and spline interpolation in order to create complicated frame shape avoiding arbitrary obstacles on the tunnel ceiling surface. When using the inversed analysis, the shape change was created smoothly, however, since each frame undergoes a subordinate change, it is not very effective for complicated shapes. On the other hand, in the case of using spline interpolation, since the function classification can be divided according to the shape of the obstacle, complicated obstacles can be dealt with.

In the future, we will establish a control method to continuously change the shape of the guide frame from the obstacle search in the toll, and to apply it to the actual inspection system.

Finally, the author thanks all who supported the development of the movable arch structure and its accompanying external panel. Further, a part of this research is executed by the contract research from NEDO (New Energy and Industrial Technology Development Organization) for SIP.

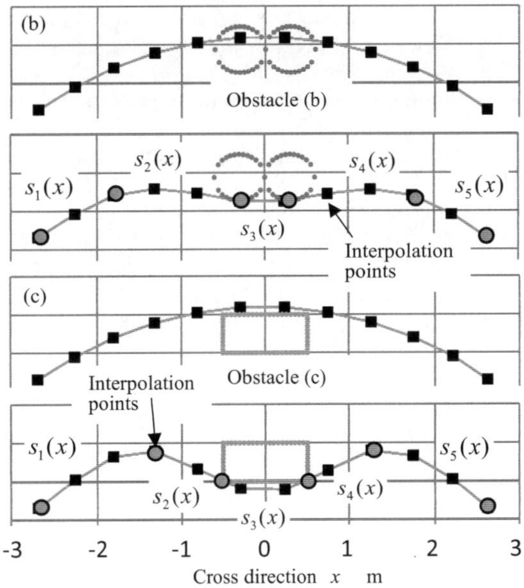

(a) Shape decision of spline function for obstacle (b), (c)

(b) Shape control of guide frame for traffic sign

Fig. 15 Shape decision of frame to avoid obstacle (b), (c)

References

[1] F. Inoue, T. Uchiyama, S. Nakamura and Y. Yanagihara, "Development of Variable Guide Frame to Inspect Inner Wall Tunnel (Part 1; Structure of Guide Frame and Shape Analysis)", Proceedings of the 6th International Conference on Advanced Mechatronics (ICAM2015), Tokyo, Japan, PP. 284-285,2010..

[2] H. Fujii, A. Yamashita and H. Asama, "Defect Detection with Estimation of Materi al Condition Using Ensemble Learning for Hammering Test", Proceedings of the 20 16 IEEE International Conference on Robotics and Automation, pp.847-3854 , Stockholm (Sweden), 2016.

[3] H. Tamura, T. Sasaki, H. Hashimoto and F. Inoue, "Circle Fitting Based Position Measurement System Using Laser Range Finder in Construction Fields," *2010 IEEE/RSJ Intl. Conf. on Intelligent Robots and Systems*, pp.209–214, 2010.

[4] N. Chernov and C. Lesort, "Least Square Fitting of Circles", *Journal of Mathematical Imaging and Vision*, Vol. 23, pp. 239-251, 2005..

[5] Fumihiro Inoue, Takeshi Sasaki, Xiangqi Huang and Hideki Hashimoto, " A Study on Position Measurement System Using Laser Range Finder

and Its Application for Construction Work " , Journal of Robotics and Mechatronics, Vol. 26, No.1, pp. 226-234, 2012.
[6] Fumihiro Inoue, "A Study on Adaptive Structure Applying Variable Geometry Truss (Mechanism of Movable Arch Roof with External Panel)"Journal of Robotics and Mechatronics, Vol. 21, No.2, pp.172-178, 2009.
[7] Fumihiro Inoue, "A Study on Optimum Shapes and Motion of Movable Arch Structure By Variable Geometry Truss", Proceeding of The 9th Asian Pacific Conference on Shell and Spatial Structures (APCS2009) , Nagoya, Japan, pp.121-128, 2009.5

High Level Framework for Bridge Inspection Using LiDAR-equipped UAV

N. Bolourian [a] , M.M. Soltani[a], A.H. Albahri[a] and A. Hammad[b]

[a]Department of Civil, Building and Environmental Engineering, Concordia University, Canada
[b]Concordia Institute for Information Systems Engineering (CIISE), Concordia University, Canada

E-mail: n_bolour@encs.concordia.ca , mo_solta@encs.concordia.ca , a_albah@encs.concordia.ca, hammad@ciise.concordia.ca

Abstract –

According to Statistics Canada, bridges in Canada have a service life of approximately 43 years. With the majority of bridges passing half of their expected service life, a large amount of investment needs to be made to inspect and maintain them in a safe condition. Manual inspection methods are both time-consuming and costly, which may discourage further inspections and follow-up of defects. Thus, there is a great need for using an automated data collection system. Surface defects (e.g. cracks) in concrete bridges can be inspected using 3D Light Detection and Ranging (LiDAR) scanner as a Non-Destructive Testing (NDT) method. However, the commonly used terrestrial LiDAR is limited to stationary data collection, which reduces the accessibility to some components of the bridges. To tackle this limitation, a LiDAR attached to an Unmanned Aerial Vehicle (UAV) provides more flexibility and accessibility for inspecting large surface areas without threatening inspectors' safety. After providing a comprehensive literature review about the usage of UAVs and LiDAR for the inspection of different types of structures, this paper proposes a high level framework for bridge inspection using LiDAR-equipped UAV. The framework includes the following steps: (1) planning a collision-free optimized path with respect to the minimum cost and maximum coverage considering a variety of constrains and requirements related to the hardware and the inspection task, and (2) data analysis for detecting the surface defects based on the collected point clouds.

Keywords –
Bridge inspection; LiDAR-equipped UAV; Path planning

1 Introduction

According to Statistics Canada, bridges built in Canada have a service life of approximately 43.3 years. With the majority of bridges built over the past few decades, more than half of their expected service life is over, and a large amount of investment is needed to rehabilitate and maintain these infrastructures [1]. In order to achieve efficient maintenance, accurate condition assessment based on regular and detailed inspection is required. However, current bridge inspection is mainly based on visual inspection and manual measurements of defects. This approach is time consuming and requires the inspectors to have access to the different components of the bridges using certain tools, which may affect the traffic on the bridge and could be unsafe for the inspectors. In many cases, the geometry of the bridges is complicated and using these tools may not be easy or feasible, such as in the case of highway bridges. Therefore, new automated inspection systems should be developed to provide safer and more efficient and accurate methods for bridge inspection [2].

Surface defects (e.g. cracks) in concrete bridges can be inspected using 3D Light Detection and Ranging (LiDAR) scanners as a Non-Destructive Testing (NDT) method. However, the commonly used terrestrial LiDAR is limited to stationary data collection, which reduces the accessibility to some components of the bridges. To tackle this limitation, a LiDAR attached to an Unmanned Aerial Vehicle (UAV) provides more flexibility and accessibility for inspecting large surface areas without threatening inspectors' safety. Using LiDAR-equipped UAV has the following advantages: (1) easier accessibility to most parts of the structure, (2) higher coverage, (3) better efficiency, and (4) higher safety by decreasing the probability of falling hazards. LiDAR-equipped UAV is mostly used in surveying and agriculture. Both the scanner and the UAV have their own limitations and specifications which should be considered to achieve an efficient result.

After providing a comprehensive literature review about the usage of UAVs and LiDAR for the inspection of structures, this paper proposes a high level framework for bridge inspection using LiDAR-equipped UAV. The framework includes the following main steps: (1) planning a collision-free optimized path with respect to the minimum cost and maximum coverage considering a variety of constraints and requirements related to the hardware and the inspection task, and (2) data analysis for detecting the surface defects based on the collected point clouds.

2 Literature Review

The literature review will cover the methods for surface defects inspection and UAV path planning.

2.1 Methods for Surface Defects Inspection

Visual inspection of surface defects (e.g., cracks, spalling, corrosion, etc.) based on the non-equipped eye is the most commonly used method of bridge inspection. However, this method is subjective, costly, time-consuming, and may cause safety risks, such as falling or trying to reach far components [3]. Other NDT methods include image processing and LiDAR. Image processing methods using cameras are popular, speedy and inexpensive. Several image processing studies have been done in order to detect defects automatically [4, 5, 6, 7, 8]. However, the images are affected by the light conditions and their analysis requires supplementary information [9]. On the other hand, LiDAR is used for collecting point clouds. LiDAR-based methods are highly accurate and able to detect the depth of the defects [3, 10, 11] and mass losses [12, 13, 14]. Although the initial cost of the LiDAR is high, it is a time and cost effective method in the long term [3]. Image processing methods may be more efficient than LiDAR in detecting the boundaries of defects [15]. Integrating the texture-mapped images and laser scans is another method to detect the surface damages [16]. However, all the above methods are based on using terrestrial scanners.

2.2 UAV Path Planning

Scanning platforms are categorized into three groups: Terrestrial Laser Scanning (TLS), Mobile Laser Scanning (MLS), and Airborne Laser Scanning (ALS) [17]. TLS is a ground-based method using tripods. If a laser scanner is mounted on moving objects such as cars, vans or boats, it is called MLS [18]. MLS is mostly applied to pavement inspection in order to extract cracks or road markings [19]. ALS using airplanes or UAV has been used for military applications, surveying, etc. The above three types are all affected by the small vibrations of the scanner that may be caused by the non-stability of the LiDAR itself, the movement of the supporting vehicle (e.g. car or UAV), or the wind effect on the UAV [20]. ALS is the most sensitive system to vibration [21].

Controlling the flight path of the UAV can be done using remote control. However, applying automated path planning will lead to optimal flight path. Path planning of UAVs should consider obstacle avoidance, maximum coverage, sensor limitations, and vehicle motions, as well as time and cost efficiency. Some methods such as wavefront algorithm [22], spanning tree algorithm [23], and Neural Network [24] are used to compute the path in a simplified grid space. Other methods, such as Traveling Salesman Problem (TSP), focus on finding the shortest path passing through pre-defined viewpoints [25, 26]. Minimizing the number of viewpoints using Art Gallery Problem (AGP) is another approach to compute the optimal path [27, 28, 29]. Coverage path planning has been mostly applied on 3D-mapping [30, 31], surveying of urban environments [32, 33, 34], ship hull inspection [28, 35, 36], and structural inspection [37]. The model of Bircher et al. [37] which is based on Lin-Kernighan Heuristic (LKH) as a TSP solver, convex optimization, and Rapidly-exploring Random Tree star (RRT*) algorithm to define the optimized viewpoints and path, can be applied for other types of structures. Also, Guerrero et al. [38] worked on path planning of UAV for bridge inspection in windy environments. It should be noticed that cameras, not LiDAR, were mounted on the UAV in the last two studies.

3 Proposed High Level Framework

Figure 1 shows the flowchart of the proposed framework of bridge inspection using LiDAR-equipped UAV. The method consists of the following three main steps.

(1) Path Planning

In bridge inspection using LiDAR-equipped UAV, the following constraints should be considered.

- UAV constraints includes six Degrees of Freedom (DoFs) which are three coordinates (x, y, z) and three rotations (yaw, pitch and roll), vibration, battery capacity (time of flight), payload and take-off weight, and the size of the UAV. The minimum distance (d_{min}) between the UAV and the structure or other obstacles depends on the size of the UAV.
- LiDAR metrology method can be based on Time-of-Flight (ToF) or Phase Shift (PS). In the first method, a laser pulse is sent and the distance between the device and the object can be calculated by measuring the arrival time of the reflected pulse. PS emits a continuous sinusoidal laser beam, and the difference between the phase of emitted and reflected laser

beams is measured to compute the distance. ToF is practical for long measurement range with the accuracy of 4-10 mm at 100 m. On the other hand, PS is used for short range cases with the accuracy of 2-4 mm at 20 m [3]. PS should be used for inspection applications. Furthermore, the maximum distance (d_{max}) should be less than 10 m [3]. In addition, the Fields of view (FoVs) are possible scanning angle range in the horizontal and vertical directions. Furthermore, the beam diameter, incidence angle (θ), and angular resolution ($\Delta\theta$) should be considered.

- LiDAR-equipped UAV may have the LiDAR mounted on top of or under the UAV.
- Other constraints related to regulations, wind effects, etc., should be considered.

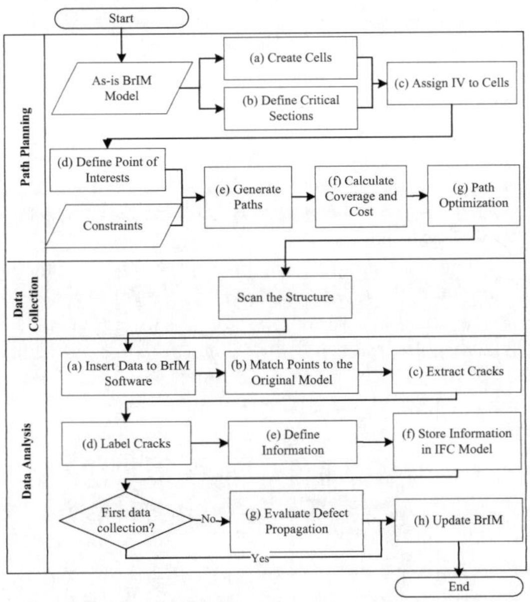

Figure 1. Proposed framework

Only the main constraints (d_{max} and d_{min}) are considered in this paper using the profiling method for scanning the structure. Figure 2(a) shows the UAV location from the side view at *g*, which is between d_{max} and d_{min}.

The proposed method is based on the assumption that an accurate as-is Bridge Information Modeling (BrIM) model is available. This model (see for example Figure 3(a)) can be created using LiDAR scanning. The BrIM model is used for the visibility analysis and path planning as explained below. The path planning has the following steps:

(a) *Creating cells:* The bridge surface should be meshed and divided into cells. The size of the cells should be proportional of the resolution of the scanner and small enough to achieve high accuracy.

(b) *Defining critical sections:* The critical sections can be determined based on the potential location of cracks or other surface defects, which could be estimated using structural analysis or based on experience. These sections can be assigned with Importance Values (IV) corresponding to the level of criticality. Figure 3(b) shows the bottom view of a bridge deck with several critical sections. These sections have low, medium, or high levels of criticality (i.e., IV=1, 2, or 3, respectively).

(c) *Assigning IV to cells:* The cells in each critical section will inherit the IV of the critical section that it belongs to.

(d) *Defining points of interest (PoI):* In profiling method, PoI are specific points which the path should pass through to increase the accuracy of the results. PoI should be added at close distance from the inspected surfaces where defects are expected.

(e) *Generating paths:* A large number of collision-free paths of the LiDAR-equipped UAV should be generated passing through the PoI and respecting other constraints.

(f) *Calculating coverage and cost of each path:* Albahri and Hammad [39] developed a method for calculating the coverage of surveillance cameras in buildings. This method has been adjusted in this research to calculate the coverage of the LiDAR-equipped UAV. Ray tracing can be used to calculate the visibility from each PoI. Furthermore, the IV of cells can be used to calculate the total coverage of the LiDAR-equipped UAV using the following equation.

$$W_{a_i} = IV_i \sum_{j=1}^{n} C_{ij} \quad (1)$$

$$CC_{a_i} = IV_i \sum_{v=1}^{n'} C_{iv} \quad (2)$$

$$Total\ Coverage = \frac{\sum_{i=1}^{m} CC_{a_i}}{\sum_{i=1}^{m} W_{a_i}} \quad (3)$$

Where W_{a_i} is the weight of important area a_i, $C_{ij} = 1$ represents the cell *j* in important area *i*, and $j = 1{:}n$, IV_i is the importance value assigned to all cells in area *i*, and $i = 1{:}m$, $C_{iv} = 1$ is the covered cell v in area a_i and $v = 1{:}n'$, and CC_{a_i} is the weighted covered cells in area a_i.

There are some cells that cannot be covered during scanning because of obstacles. As shown in Figure 2(b), if an obstacle exists too close to the bridge, the laser beams cannot reach all parts of the surface during scanning. This figure illustrates the depth (*D*) and width (*w*) of the crack while *w'* depicts the scanned width. Due to the existence of obstacles or other previously mentioned limitations, it is not always possible to achieve full coverage. The cost of each path can be calculated based on the flight time.

(g) *Selecting optimal path:* the optimal collision-free path is selected based on maximum coverage and minimum time of flight.

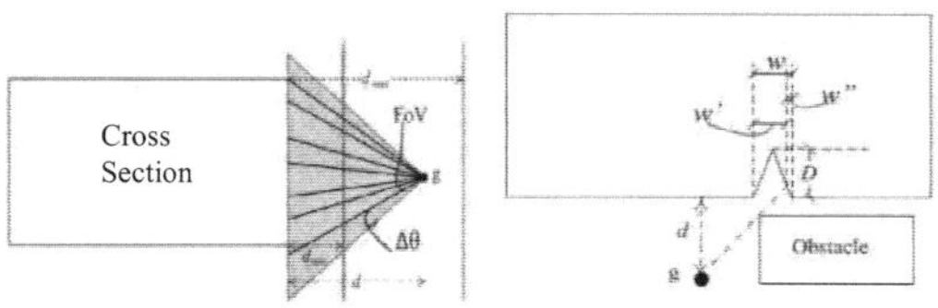

(a) Side view (b) Top view

Figure 2. Some limitations of LiDAR-equipped UAV in 2D view

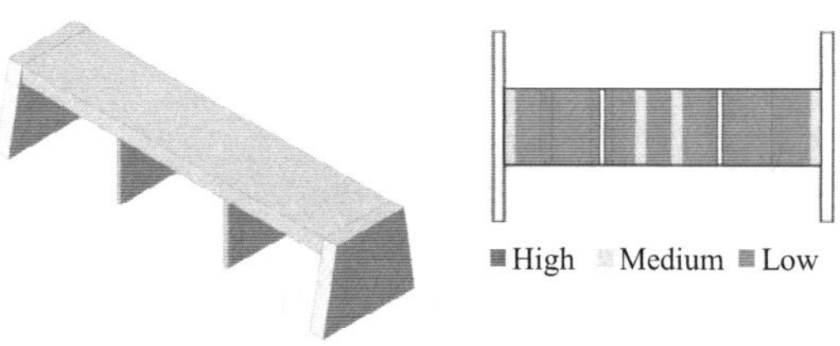

(a) 3D model (b) Bottom view of the deck with importance levels

Figure 3. Example of a 3-span bridge

(2) Data Collection

After planning the path, the LiDAR-equipped UAV should fly and scan the structure.

(3) Data analysis

Analyzing collected data and detecting the defects is done using the following steps.

(a) *Inserting point cloud data to BrIM software.*
(b) *Matching points to the original model*
(c) *Extracting cracks*
(d) *Labeling cracks*
(e) *Defining defect information* such as width, length, and depth of cracks.
(f) *Storing defect information in IFC model*
(g) *Evaluating defects propagation:* If there is previous bridge inspection model, the new data can be compared and the propagation of the defects can be determined.
(h) *Updating the BrIM model:* This model will be used in the next inspection process to evaluate the propagation the defects.

4 Implementation and Case Study

The implementation is limited at this stage to the coverage calculation step of the proposed framework. A three-span bridge was modelled in Revit 2017. In order to create cells, the bridge model was imported into Unity3D game engine. Due to the difficulty of accessing the area under the bridge, the focus of the case study is on the inspection of the bottom surface of the deck. The edge of the cell was set to 0.7 m resulting in a total number of 400 cells covering the lower surface of the deck (7 m × 28 m). The critical sections of the deck were defined and the IV were assigned to the cells following the rules explained in Section 3.

A path was created to test the visibility calculation method assuming that the LiDAR is installed on top of the UAV, which has four degrees of freedom (x, y, z and yaw). Fifty random points were defined for one path around the bridge including 20 PoI. d_{min} and d_{max} were set to 2 m and 10 m, respectively. The main obstacles in this model were the piers of the bridge.

The coverage of the path was calculated automatically in Unity assuming a large FoV of 120° in the horizontal and vertical directions. Figure 4 shows visible cells (red cells) where the LiDAR-equipped UAV is passing according to the defined path (pink line). The total coverage of this path is about 88%.

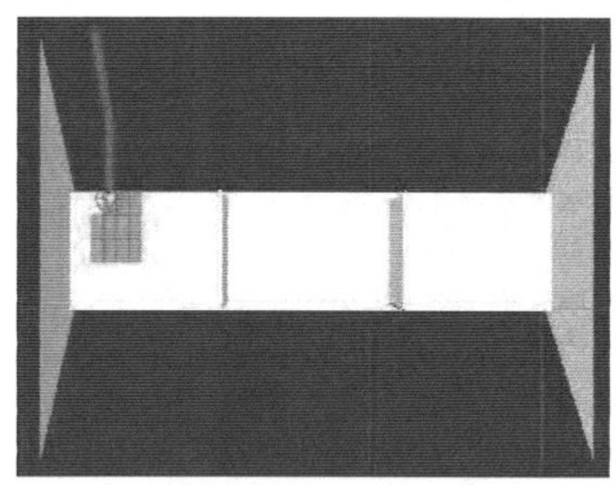

Figure 4. Part of the planned path (in pink) and visible cells (in red) under the bridge deck

5 Summary and Future Work

This paper proposed a high-level framework for bridge inspection using LiDAR-equipped UAV, which is expected to provide more accurate results than using cameras on UAV. The initial case study showed the feasibility of the proposed method for coverage calculation. Future work will aim to further investigate the optimization of path planning and automate the process of detecting surface defects based on the collected point clouds.

References

[1] GAGNON, M., GAUDREAULT, V. and OVERTON, D., 2008. *Age of public infrastructure: A provincial perspective.* Statistics Canada.

[2] LIU, P., CHEN, A.Y., HUANG, Y., HAN, J., LAI, J., KANG, S., WU, T., WEN, M. and TSAI, M., 2014. A review of rotorcraft Unmanned Aerial Vehicle (UAV) developments and applications in civil engineering. *Smart Struct.Syst*, vol. 13, no. 6, pp. 1065-1094.

[3] KIM, M., SOHN, H. and CHANG, C., 2014. Localization and quantification of concrete spalling defects using terrestrial laser scanning. *Journal of Computing in Civil Engineering*, vol. 29, no. 6, pp. 04014086.

[4] HAMPEL, U. and MAAS, H., 2009. Cascaded image analysis for dynamic crack detection in material testing. *ISPRS Journal of Photogrammetry and Remote Sensing*, vol. 64, no. 4, pp. 345-350.

[5] ADHIKARI, R., MOSELHI, O. and BAGCHI, A., 2014. Image-based retrieval of concrete crack properties for bridge inspection. *Automation in Construction*, vol. 39, pp. 180-194.

[6] HUTCHINSON, T.C. and CHEN, Z., 2006. Improved image analysis for evaluating concrete damage. *Journal of Computing in Civil Engineering*, vol. 20, no. 3, pp. 210-216.

[7] ZHU, Z. and BRILAKIS, I., 2008. Detecting air pockets for architectural concrete quality assessment using visual sensing. *Electronic Journal of Information Technology in Construction*, vol. 13, pp. 86-102.

[8] KOCH, C. and BRILAKIS, I., 2011. Pothole detection in asphalt pavement images. *Advanced Engineering Informatics*, vol. 25, no. 3, pp. 507-515.

[9] LAEFER, D.F., TRUONG-HONG, L., CARR, H. and SINGH, M., 2014. Crack detection limits in unit based masonry with terrestrial laser scanning. *NDT & E International*, vol. 62, pp. 66-76.

[10] MIZOGUCHI, T., KODA, Y., IWAKI, I., WAKABAYASHI, H., KOBAYASHI, Y., SHIRAI, K., HARA, Y. and LEE, H., 2013. Quantitative scaling evaluation of concrete structures based on terrestrial laser scanning. *Automation in Construction*, vol. 35, pp. 263-274.

[11] LIU, W., CHEN, S. and HAUSER, E., 2011. LiDAR- based bridge structure defect detection. *Experimental Techniques*, vol. 35, no. 6, pp. 27-34.

[12] GIRARDEAU-MONTAUT, D., ROUX, M., MARC, R. and THIBAULT, G., 2005. Change detection on points cloud data acquired with a ground laser scanner. *International Archives of Photogrammetry, Remote Sensing and Spatial Information Sciences*, vol. 36, no. part 3, pp. W19.

[13] TEZA, G., GALGARO, A. and MORO, F., 2009. Contactless recognition of concrete surface damage from laser scanning and curvature computation. *NDT & E International*, vol. 42, no. 4, pp. 240-249.

[14] OLSEN, M.J., CHEN, Z., HUTCHINSON, T. and KUESTER, F., 2013. Optical techniques for multiscale damage assessment. *Geomatics, Natural Hazards and Risk*, vol. 4, no. 1, pp. 49-70.

[15] DEMIR, N. and E. BALTSAVIAS. Automated modeling of 3D building roofs using image and LiDAR data Anonymous *Proceedings of the XXII Congress of the International Society for Photogrammetry, Remote Sensing, Melbourne, Australia*, 2012.

[16] GULDUR, B. and J.F. HAJJAR. Automated Classification of Detected Surface Damage from Point Clouds with Supervised Learning Anonymous *ISARC. Proceedings of the International Symposium on Automation and Robotics in Construction*, 2016.

[17] ALHO, P. Laser Scanning Applications in Fluvial Geomorphology Anonymous *AGU Fall Meeting Abstracts*, 2014.

[18] LEHTOMÄKI, M., JAAKKOLA, A., HYYPPÄ, J., LAMPINEN, J., KAARTINEN, H., KUKKO, A., PUTTONEN, E. and HYYPPÄ, H., 2016. Object Classification and Recognition from Mobile Laser Scanning Point Clouds in a Road Environment. *IEEE Transactions on Geoscience and Remote Sensing*, vol. 54, no. 2, pp. 1226-1239.

[19] GUAN, H., LI, J., YU, Y., CHAPMAN, M. and WANG, C., 2015. Automated road information extraction from mobile laser scanning data. *IEEE Transactions on Intelligent Transportation Systems*, vol. 16, no. 1, pp. 194-205.

[20] RESHETYUK, Y., 2006. Investigation and calibration of pulsed time-of-flight terrestrial laser scanners.

[21] XU, X., YANG, H. and NEUMANN, I., 2015. Concrete crack measurement and analysis based on Terrestrial Laser Scanning technology.

[22] ZELINSKY, A., R.A. JARVIS, J. BYRNE and S. YUTA. Planning paths of complete coverage of an unstructured environment by a mobile robot Anonymous *Proceedings of International Conference on Advanced Robotics*, 1993.

[23] GABRIELY, Y. and E. RIMON. Spiral-STC: An on-line coverage algorithm of grid environments by a mobile robot Anonymous *IEEE International Conference on Robotics and Automation, 2002. Proceedings. ICRA'02.*, 2002.

[24] LUO, C., S.X. YANG, D.A. STACEY and J.C.

JOFRIET. A solution to vicinity problem of obstacles in complete coverage path planning Anonymous *IEEE International Conference on Robotics and Automation, 2002. Proceedings. ICRA'02.*, 2002.

[25] DANTZIG, G., FULKERSON, R. and JOHNSON, S., 1954. Solution of a large-scale traveling-salesman problem. *Journal of the Operations Research Society of America*, vol. 2, no. 4, pp. 393-410.

[26] LIN, S. and KERNIGHAN, B.W., 1973. An effective heuristic algorithm for the traveling-salesman problem. *Operations Research*, vol. 21, no. 2, pp. 498-516.

[27] O'ROURKE, J., 1987. *Art gallery theorems and algorithms.* Oxford University Press Oxford.

[28] HOVER, F.S., EUSTICE, R.M., KIM, A., ENGLOT, B., JOHANNSSON, H., KAESS, M. and LEONARD, J.J., 2012. Advanced perception, navigation and planning for autonomous in-water ship hull inspection. *The International Journal of Robotics Research*, vol. 31, no. 12, pp. 1445-1464.

[29] GONZÁLEZ-BAÑOS, H. A randomized art-gallery algorithm for sensor placement Anonymous *Proceedings of the seventeenth annual symposium on Computational geometry*, 2001.

[30] SCHADLER, M., STÜCKLER, J. and BEHNKE, S., 2014. Rough terrain 3D mapping and navigation using a continuously rotating 2D laser scanner. *KI-Künstliche Intelligenz*, vol. 28, no. 2, pp. 93-99.

[31] SEHESTEDT, S., G. PAUL, D. RUSHTON-SMITH and D. LIU. Prior-knowledge assisted fast 3D map building of structured environments for steel bridge maintenance Anonymous *2013 IEEE International Conference on Automation Science and Engineering (CASE)*, 2013.

[32] CHENG, P., J. KELLER and V. KUMAR. Time-optimal UAV trajectory planning for 3D urban structure coverage Anonymous *2008 IEEE/RSJ International Conference on Intelligent Robots and Systems*, 2008.

[33] DORNHEGE, C., A. KLEINER and A. KOLLING. Coverage search in 3DAnonymous *2013 IEEE International Symposium on Safety, Security, and Rescue Robotics (SSRR)*, 2013.

[34] JANOUŠEK, P. and J. FAIGL. Speeding up coverage queries in 3D multi-goal path planning Anonymous *Robotics and Automation (ICRA), 2013 IEEE International Conference on*, 2013.

[35] ENGLOT, B. and F. HOVER. Inspection planning for sensor coverage of 3D marine structures Anonymous *2010 IEEE/RSJ International Conference on Intelligent Robots and Systems (IROS)*, 2010.

[36] ENGLOT, B. and HOVER, F.S., 2012. Sampling-based coverage path planning for inspection of complex structures.

[37] BIRCHER, A., K. ALEXIS, M. BURRI, P. OETTERSHAGEN, S. OMARI, T. MANTEL and R. SIEGWART. Structural inspection path planning via iterative viewpoint resampling with application to aerial robotics Anonymous *IEEE International Conference on Robotics and Automation (ICRA)*, 2015.

[38] GUERRERO, J.A. and BESTAOUI, Y., 2013. UAV path planning for structure inspection in windy environments. *Journal of Intelligent & Robotic Systems*, vol. 69, no. 1-4, pp. 297-311.

[39] ALBAHRI, A.H. and HAMMAD, A., 2017. A novel method for calculating camera coverage in buildings using BIM. *Journal of Information Technology in Construction (ITcon)*, vol. 22, no. 2, pp. 16-33.

34th International Symposium
on Automation and Robotics
in Construction

ISARC 2017

June 28 – July 1
TAIPEI, TAIWAN

COMPUTER VISION

A Cascaded Classifier Approach to Window Detection in Facade Images

M. Neuhausen, A. Martin, M. Obel, P. Mark and M. König

Ruhr-University, Germany, Bochum
E-mail: marcel.neuhausen@ruhr-uni-bochum.de

Abstract -

A major part of recent developments in civil engineering in the urban context evolved around building and city models. Especially for a precise risk assessment of damages to existing buildings induced by ground movements, accurate models are inevitable. Beside the shape of a building, the focus is also on components compromising a building's stiffness. Particularly, by including wall openings such as windows into risk analyses, these can be improved to provide more reliable predictions.

However, most publicly available data sources only provide simple block models of existing buildings sometimes extended by roof shapes. As a consequence, any information concerning the windows of a building must be integrated into the model using other data sources. Whereas numerous approaches address the refinement of building shapes, their windows and other components are commonly disregarded. Although cascaded classifiers already turned out to yield good results in general and applying them to window detection seems promising, such approaches are yet insufficient to reliably extend building models. Drawing on previous findings, we present an approach to window detection in facade images satisfying the needs of risk assessment analyses. Our detection system combines a soft cascaded classifier consisting of thresholded Haar-like features with a sliding window detector extracting image patches for classification. The soft cascaded design improves the detection rate over previously made approaches while coincidentally reducing the amount of required features. Further, we evaluate the effect of a rectified dataset on the classification results compared to its counterpart with images taken from varying angles.

Keywords -

Window Detection; Cascaded Classifier; Urban Environments; Building Classification; Building Reconstruction

1 Introduction

Models of buildings are indispensable in nowadays civil engineering in the urban context. For several tasks in urban planning and inner city construction projects accurate models of existing buildings are a prerequisite. In particular, this expresses in risk assessment concerning damages to buildings induced by ground movements which may be caused by earthquakes or tunneling projects. Reliable analyses predicting the risk of damages demand detailed information about a structure's stiffness. Beside the pure shape of a building, risk analyses also consider certain facade components as they possess more profound information about a structure's condition. Especially the ratio of wall openings to facade is taken into account since openings highly impair the stiffness of a structure [9]. However, wall openings are commonly not included in publicly available 3D building models. Those requested from land registry offices usually comprise coarse block models of buildings. Some models may be extended by simplified roof shapes but in general they lack any further detail as shown in figure 1. Except for a few manually reconstructed buildings, those provided by online map services like OpenStreetMaps and Google Earth neither offer more detailed information. For that reason, such models have to be enriched with additional information about wall openings beforehand to satisfy the requirements of risk analyses.

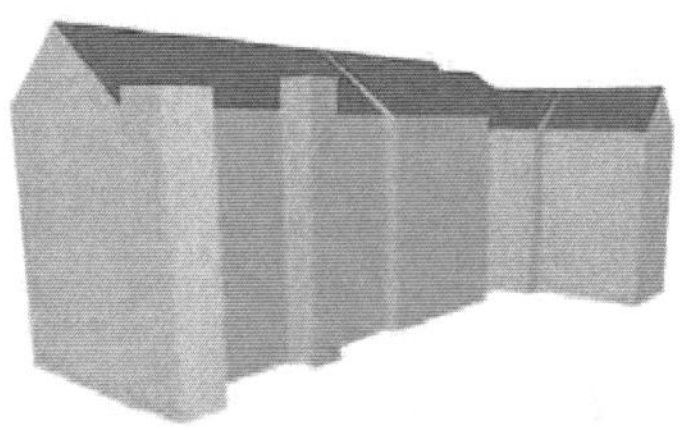

Figure 1. 3D block model of a group of existing buildings provided by the land registry office of Düsseldorf.

For deriving wall openings from a facade, its windows qualify best as they commonly account for almost all openings. Nevertheless, a personal inspection of each building within an urban area or along a tunneling alignment to determine the window-to-facade ratio is time consuming and expensive. Several approaches towards a computer-aided detection of windows in facades have already been made. Many of them rely on assumptions and restrictions to the buildings' architecture which make them impractical for general usage. To detect windows in facades automatically without any prior knowledge about the investigated build-

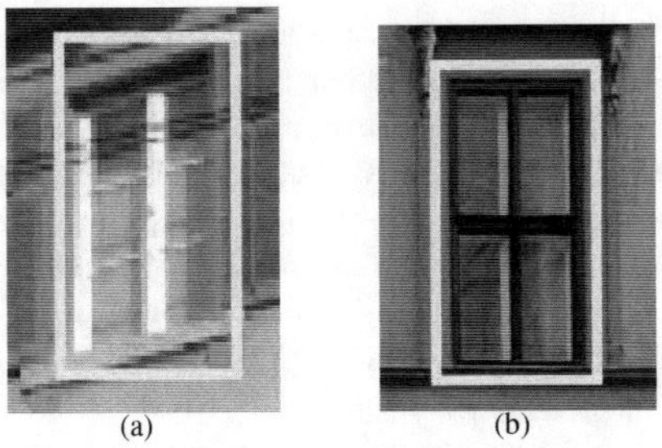

Figure 2. (a) Detected window in non-rectified image. Area of the detection rectangle (red) and the window are not coincident. (b) Detected window in rectified image. Nearly coincident areas of detection rectangle (red) and window.

ing, Ali et al. [4] apply the Viola-Jones framework [15] to images of building scenes. Despite a stated detection rate which is insufficient for our purpose, it indicates that cascaded classifiers yield promising results for window detection tasks. Based on their findings, this paper focuses on improving the detection rate and the accuracy regarding the windows area. We substitute the classifier as proposed by Viola and Jones [15] with a soft cascaded classifier [7]. This provides better generalization for high variation of the training set which results in an improved detection rate and higher precision while coincidentally employing less features. Apart from that, we keep the original detection algorithm of the Viola-Jones framework which scans the entire image since the soft cascaded classifier is even slightly faster so that there is no need for improvements. Furthermore, we rectify the facade images before detection to obtain a more precise congruence of the window areas with the rectangular detections of our approach (see fig. 2). Restricting to rectangular windows also eases the training and classification process and consequently increases the detection rate.

These improvements increase the quality of the detections to a satisfactory level in terms of the aforementioned risk assessment. The results of our window detection system as proposed in this paper can, thus, be incorporated into the building model to facilitate subsequent analyses.

2 Related Work

In recent years, window detection in facades has become prominent in a diversity of research fields. The area of application ranges from tourist guidance on mobile devices where buildings are recognized by characteristic window patterns over 3D city modeling to deformation analysis [4]. However, in literature window detection is commonly referred to as a part of building reconstruction.

A wide range of approaches concerning several facets of building reconstruction have already been made. Most of them focus on improving the model's shape disregarding its components. A survey concerning such approaches by Brenner [1] focuses on aerial imagery and laser scans as input data. This facilitates data acquisition with appropriate effort even for entire cities. Information about large structures like buildings and streets is well extractable which enables the reconstruction of cities as simple block models. An advanced reconstruction of a building, especially by its facade components, is only feasible to a very limited extent due to the bird's-eye perspective.

Haala and Kada [10] additionally include approaches based on terrestrial laser scans into their survey which provide accurate point clouds of existing buildings from street level. Beside generating geometry from these measurements [18], this kind of data enables the detection of certain facade components such as windows by identifying no-measurement areas in the facade plane [17]. Methods relying on terrestrial laser scans offer simple ways to detect windows in facades, but by now such data is not available regionwide. Albeit the StreetMapper proposed by Barber et al. [3] eases the acquisition, it is too costly and involves unreasonable effort.

On the contrary, ground view images of facades are publicly available from web services like Google Street View or can easily be gathered even for larger areas. In preliminary work Neuhausen et al. [8] discuss window detection approaches using street level images. They distinguish contemporary approaches by their shared strategies into three categories: Grammar-based, Image processing, and machine learning. Grammar-based approaches highly rely on several assumptions to a facade's appearance and prior knowledge about the architecture and its composition. Ripperda [11] and Ripperda and Brenner [12] divide facades into their components by applying a simple grammar leveraging symmetries and repetitions in facade images. In contrast, Teboul et al. [13] based their approach on a detailed shape grammar and combined it with a model of semantic relationships between certain facade elements. The definition of a proper set of grammar rules and the modeling of relationships is non-trivial and requires some expert knowledge about the expected architecture. Such approaches have to be manually adjusted whenever the architectural conditions change. This results in a limitation of a general application. As grammars allow several decompositions of a building's facade the most probable subdivision has to be found by means of sampling methods or parsing algorithms. This dramatically raises the methods' complexity with the number of rules in the grammar. By this reason, rule sets have to be kept small which complicates a detailed modeling. Similar to grammar-based approaches, image processing methods also make assumptions about the alignment of windows on the facade. Lee and Nevatia [2] presume windows to be aligned in a grid-like manner on the facade as usually seen at big office buildings. For window detection they project horizontal and vertical edges of rectified facade images into histograms. These are then superimposed to find window hypotheses, refined in later steps, at the peaks' locations. Pursuing this approach, Meixner et al. [14] found that it provides good results on highly regular facades but fails as soon as facades become more complex. Asymmetric patterns of windows as well as facade extensions like balconies corrupt the reliability of this method. As the facades' appearance not only alters between countries but also between city districts and, moreover, most facades in the urban area can be expected to be complex, these approaches will often fail and are, thus, not universally applicable. To avoid the aforementioned limitations, windows have to be detected by their inherent characteristics instead of their relations among each other or further facade elements. Machine learning algorithms emerge beneficial as they can be trained to classify windows by image features which are directly linked to these characteristics. Haugegard et al. [6] emphasize this by deploying a support vector machine (SVM) as classifier to identify windows by their edges. The set of features used for classification, though, has to be chosen manually. A boosted cascaded classifier as proposed in the Viola-Jones framework [15] overcomes this by autonomously drawing those features from a predefined pool which qualify best for a certain task. Furthermore, these usually outperform most monolithic classifiers such as SVMs [16]. In their approach, Ali et al. [4] apply this framework to window detection and investigate the quality of its detections. The resulting detection rates are insufficient to reliably estimate the window-to-facade ratio of a building but, nevertheless, cascaded classifiers emerge to be promising to further research.

In the following we give an insight into the concept of our detection system (see sec. 3.1) which is based on these findings. We outline the classifier used and the detection algorithm in section 3.2 and section 3.3. In section 4.1 we compare our system to the approach of Ali et al. [4]. In further experiments we focus on improving the detection rate. For this purpose, we contrast the classification system of the previous experiment with one operating on rectified facade images (see sec. 4.2). Finally, we conclude our findings and provide an outlook on further work to be done in section 5.

3 Proposed Window Detection System

Window detection in facade images is still demanding and not sufficiently solved yet. The detection proves to be challenging for multiple reasons. Most of a window is made of a transparent pane which restricts detection to window frame. Additionally, windows occur in different sizes and shapes which further increases the difficulty of the detection task due to a high variation in the frames' appearance. Both complicate the manual choice of an adequate set of image features as it is uncertain which subset characterizes the entire class of windows best. While training, boosted cascaded classifiers choose those features from a predefined pool which perform best on a certain dataset. They overcome the issue of a manual selection if the training dataset is well constructed so that it depicts high intraclass variation. Moreover, they generally yield better results than monolithic classifiers. For these reasons, boosted cascaded classifiers seem to be promising for our window detection system.

3.1 Concept

Since boosted cascaded classifiers proofed to be advantageous compared to other classification methods, we build our detection system as outlined in figure 3 on the approach of Ali et. al [4] which uses the Viola-Jones framework. Considering their findings, we constrain the input data to be rectified semi-automatically. That eliminates an unnecessary increase in the windows' variation since equal windows no longer appear to be different due to distortion. Nevertheless, the intraclass variety remains quite high. We substitute the classifier used in the Viola-Jones framework by the soft cascaded classifier proposed by Bourdev and Brandt [7] as it is more robust regarding high variability in the set of positive training samples. Furthermore, this method commonly offers a better detection rate while relying on less features. In section 3.2 we describe its functionality in more detail. The soft cascaded classifier we use in our approach is even slightly faster than the one applied in the Viola-Jones-framework due to the lower number of features that have to be evaluated. An optimization of the enclosing detection algorithm is therefore not necessary. We keep the naive sliding window detector (see sec. 3.3) of the Viola-Jones framework which scans an image via small subwindows sliding across the entire image and passes the subjacent image patches to the classifier. The classifier which is trained on a set of rectified window samples and random non-windows then obtains a binary classification for each patch and returns them to the detector. Positive responses are finally merged where appropriate and returned as output of our system. Figure 3 illustrates the described interactions of the detectors' components.

3.2 Soft Cascaded Classifier

As an improvement over the Viola-Jones framework, Bourdev and Brandt [7] proposed the soft cascaded classifier approach which also relies on an set of weighted weak

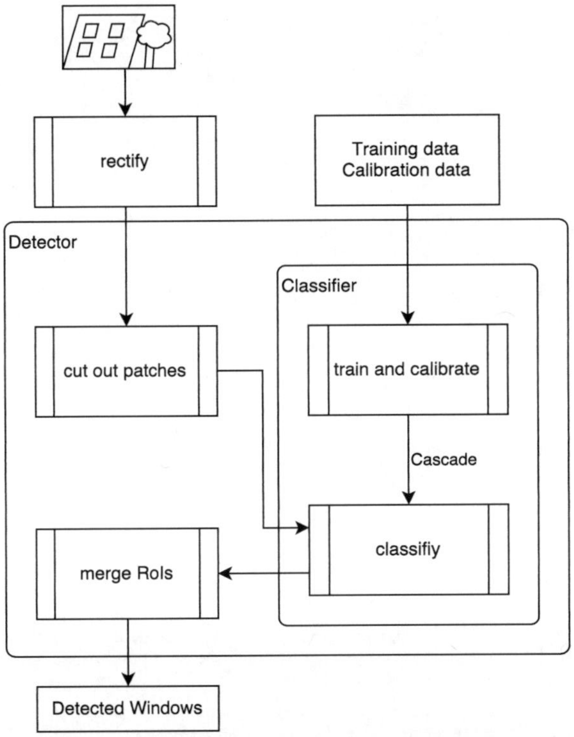

Figure 3. Concept of the proposed detection system. Images are rectified and passed to the detector. The detector cuts out image patches which are classified by a trained and calibrated cascade. Overlapping regions of interest (ROIs) of positive classifications are merged. Resulting ROIs are returned as detected windows.

classifiers $\alpha h(x)$, where α is the weight and $h(x)$ denotes the classification function of the weak classifier for a given sample x. Each $h(x)$ classifies barely better than randomly guessing. By connecting them in series a strong classifier with high detection rate emerges. Equal to the Viola-Jones framework the weak classifiers utilized here are Haar-like features whose responses are thresholded to enable binary classification. The thresholds are set independently for each feature by the algorithm given by Viola and Jones [15]. According to this, each weak classifier consists of a thresholded Haar-like feature of a certain size at a certain relative position in the samples' image patches (see fig. 4).

For training a dataset $\{(x_1, y_1, w_1)...(x_N, y_N, w_N)\}$ has to be constructed with positive and negative samples x, their corresponding true labels y and a weight w indicating the importance to classify the sample correctly for increasing the classifier's quality. In the beginning all weights are set to be equal. While training, those weak classifiers performing best on the training dataset with respect to the samples' weights w are iteratively drawn from a pool and added to the strong classifier's set. The weight of the weak classifier within the set is determined according to its training error. The lower the error is, the higher is the classifier's weight. As covering the infinite class of non-windows by a finite set of negative samples is impossible, it can only be approximated. For this reason, after adding a weak classifier to the set new, negative samples are bootstrapped which are yet misclassified by the current strong classifier and added to the existing dataset. The weight w_i of each sample of the set is then adapted corresponding to its classification result so that currently misclassified samples increase in importance and vice versa. Since this promotes the selection of weak classifiers obtaining the correct class of currently misclassified samples, the focus of the emerging strong classifier is shifted towards these samples. This further improves the strong classifier to correctly classify a wide spectrum of samples.

Figure 4. Two Haar-like features at different positions in the image patch, each corresponding to a weak classifier.

Out of this a strong classifier emerges that can already be used for classification. Therefore a weighted majority vote of all holded weak classifiers is done:

$$\sum_{t}^{T} \alpha_t h(x)_t \geq \frac{1}{2} \sum_{t}^{T} \alpha_t$$

However, evaluating all weak classifiers on a sample is very time consuming and unnecessary unless the sample is positive. Arranging the weak classifiers within the strong classifier in a cascaded layout highly decreases the required evaluations. A subsequent weak classifier is then only evaluated if its predecessors classified the sample positive. Otherwise the evaluation is terminated prematurely and the sample is classified negatively. This behavior is implemented by a sample trace which contains the partial sum of all already evaluated weak classifiers. After each evaluation, the sample trace is compared to a rejection threshold to determine whether the sample is further promoted through the cascade or rejected as negative sample. An illustration of this procedure is given in figure 5. In contrast to the Viola-Jones framework, each weak classifier represents a single stage within the cascade. The decision if a sample is rejected becomes softer as it does not only depend on the previous stage but on all stages previously evaluated. The order of the weak classifiers as well as the rejection thresholds highly affect the quality of the detection. A proper setting of those is done in a calibration step on an already trained classifier. Given a calibration dataset which is constructed like the training dataset but contains a new set of samples, the weak classifiers are reordered with regard to the performance on this set. For each sample in the set, its sample trace is maintained. Based on these traces the rejection threshold of each stage can be set so that each stage rejects as much negative samples as possible but only a certain percentage of positive samples. Once the classifier is calibrated, the stages are evaluated one after the other and each resulting partial sum is compared to the corresponding rejection threshold of the stage. When the sum drops below the threshold, the sample is discarded and marked as negative. If, elsewise, the sample passes through all stages, it is classified to be positive.

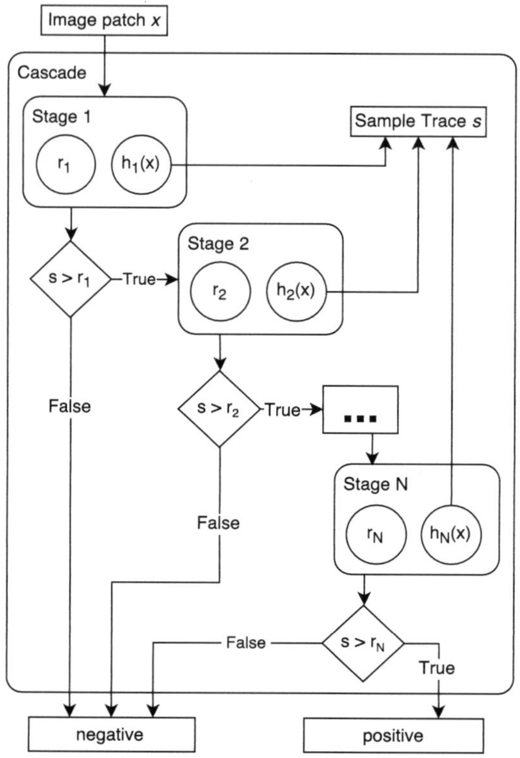

Figure 5. Schematic illustration of the cascaded classification procedure. The weak classifiers of each stage are only evaluated if the sample trace s is higher than the rejection threshold r_i of the preceding stage. A sample is positively classified if it passes through the whole cascade. Otherwise it is prematurely declined and negatively classified.

3.3 Sliding Window Detector

The classifier of our system accepts image patches and returns whether they mainly consist of a window or not. For finding windows in facade images, hence, a detector is required which passes relevant patches to the classifier and processes the returned decisions. Due to the cascaded structure of the classifier, negative patches are declined in early stages so that they can be processed very fast. For this reason there is no need for an optimized detection algorithm.

We use the naive sliding window detector proposed by Viola and Jones [15]. For a given image a rectangular subwindow slides across the entire image in multiple scales. The shifting step size depends on the particular scale of the subwindow. At each position the enclosed image patch is passed to a classifier. If the classifier resolves a patch to contain an object of interest, i.e. a window, the subwindow is memorized as a region of interest (ROI). Since the cascaded classifier is to some extend insensitive to changes in translation and scale of the object within the image patch, multiple ROIs may occur within a region around an object [15]. The memorized ROIs are reprocessed such that overlapping ones are averaged, resulting in a single ROI.

We specify the images passed to the detector to be rectified and completely filled by facade, so we can approximate expected window dimensions by the image's dimensions. Based on this, we define the minimal sliding subwindow dimensions to be half that size which corresponds to a starting scale of $s = 1.0$. After each run across the facade image the subwindow is scaled by a factor of 1.25. The step size of shifting starts with $\Delta = 1.0$ and is scaled after each run by $[s\Delta]$, where [] denotes a rounding operation. At each position the underlying image patch is passed to the classifier. To separately detect windows which are positioned close together we only merge positively classified ROIs which overlap at least by a factor of 0.75.

4 Experiments

In the following we present two experiments. In the first experiment, described in section 4.1, we run our detection system on non-rectified facade images. We compare our detection results to those of the approach of Ali et al. [4]. By contrasting the results we examine the improvements of the soft cascaded classifier over the Viola-Jones framework on the window detection task. In a second experiment (see section 4.2) we run our system on a set of rectified facade images. We then compare the results to our system of the non-rectified experimental setup to highlight the effect of rectified windows to the classification results.

4.1 Detection on Non-Rectified Images

To train our classifier analogous to Ali et al. [4] we manually cut out 1500 windows from the TSG-60 [20] and ZuBuD [5] image datasets. Both sets contain images of facades taken from different viewpoints and in varying illumination conditions. In contrast to Ali et al. [4] we discard the TSG-20 [19] dataset from training as it is a subset of TSG-60. Instead we take 900 positive samples from the TSG-60 dataset which were taken from a viewpoint of approximately 30° to the facades plane. Another 600 windows from various viewpoints are taken from the ZuBuD dataset. We also add the vertically flipped version of each window to our training dataset so that the set additionally covers windows taken from opposing viewpoints. Overall our training set contains 3000 positive samples and, initially, 3000 randomly chosen non-windows as negative samples. After training we calibrate our classifier on further 3000 samples of these datasets.

For evaluation we use the setup of Ali et al. [4] to compare their detection results to ours. Therefore we evaluate our detection system with the calibrated classifier on the same three distinct datasets. We use each 40 images of both, the TSG-20 and TSG-60 dataset. We also use 115 images of the ZuBuD Query Images which contain a subset of the facades shown in the ZuBuD dataset. All of the images considered for evaluation are taken from different viewpoints and illumination conditions than those of the training set.

As the exact position and size of the detected windows are relevant for subsequent risk analyses we apply the single window evaluation approach of Ali et al. [4]. According to this, a detection is marked as true positive only if the hypothesis is inside the manually labeled ground truth rectangle or has a maximal overlap of 5 pixels in each direction. Additionally, the hypothesis has to cover at least 75% of the underlying bounding box of the ground truth label. A hypothesis is marked as false positive if it covers less than 5% of a ground truth label. The detection results are summarized in table 1.

Dataset	True positive in %	False positive in %
TSG-20	69.1	7.7
TSG-60	60.3	8.9
ZuBuD	39.6	3.1

Table 1. Detection results of our system on non-rectified datasets.

For applying the single window evaluation with a coverage of at least 75% Ali et al. [4] stated true positive rates of 57%, 52%, and 30% on the TSG-20, TSG-60, and ZuBuD Query Images datasets, respectively. False positives were below 9% for all datasets. With similar false positive rates and true positive rates of 69.1% on TSG-20, 60.3% on TSG-60, and 39.6 on ZuBuD Query Images our detection system performs better on non-rectified facade images. By the substitution of the classifier we achieve an increase in the true positive rate up to 12% on the TSG datasets while keeping the false detections on the same level. Although ZuBuD constitutes a very challenging dataset because of many different viewpoints and uncommon buildings with unique window shapes such as churches, our system increases the true positive rate by 9% associated with a marginal increase in false positive rate by 1%. Nevertheless, in general these rates are far from being sufficient for precise risk assessment analyses as visible in the sample images in figure 6. Since we only exchanged the classification method, the results of this experiment demonstrate the improvement of the soft cascade compared to the Viola-Jones classifier. Thus, the soft cascade approach offers a better basis to further improve the detection rate.

(a) (b) (c)

Figure 6. Detections on different facade types an viewpoints. (a) Good detection rate on simple facades with common windows. (b) Decreased performance on complex facade appearance. (c) Despite simple facade few detections due to viewpoint.

4.2 Detection on Rectified Images

In this experiment we compare the soft cascaded classifier's performance on non-rectified facade images to the performance on rectified images. For training the classifier which runs on rectified samples, we use the Ecole Centrale Paris Facades Database [13]. The dataset consists of 478 rectified facade images from various cities in Europe and the United States. We take 3000 randomly chosen windows from this and initially add the same amount of non-windows to the training set. For comparison we use the classifier already trained in the experiment in section 4.1 to run on non-rectified samples.

We evaluate the classifier for rectified samples on 3080 positive and 10000 negative image patches of the Ecole Centrale Paris Facades Database which were not shown while training. For the evaluation of the other classifier, we use 3102 positive and 10000 negative image patches equally distributed of the TSG-20, TSG-60, and ZuBuD Query Image datasets. We compare both classifiers by means of their receiver operating characteristic (ROC) curve shown in figure 7. Their curves are generated by plotting their detection rate against false positive rates while increasing the classification threshold from 0.5 up to 1.

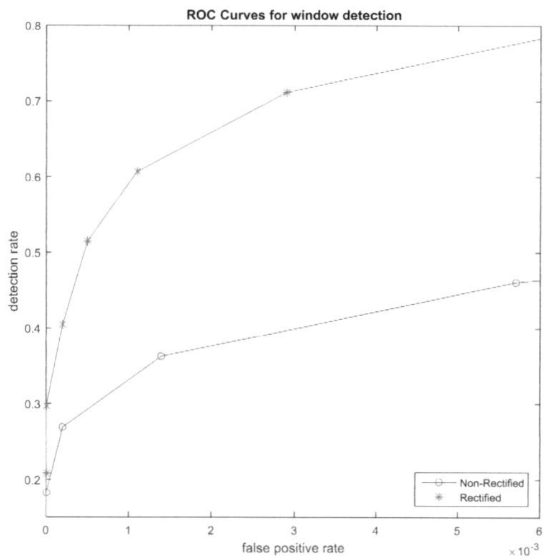

Figure 7. Comparative detection performances of the non-rectified and the rectified classifier by means of their ROC curves.

As the ROC curves of the classifiers clearly illustrate, the classifier for rectified images outperforms the one running on non-rectified images. Especially for false positive rates lower than 0.004% it provides reasonable detection rates of more than 71% whereas the non-rectified classifier only achieves around 40%. To obtain sufficient detection results very small false positive rates are a prerequisite since the detector scans hundred thousands of subwindows per image. Although the classifier for rectified facade images we trained in this experiment provides much better detection rates than the one for non-rectified images, its false positive rates are still too high for a precise prediction. By applying our detection system to rectified facade images at least a sufficient approximation to the real window-to-facade ratio is practicable as can be seen in the samples in figure 8.

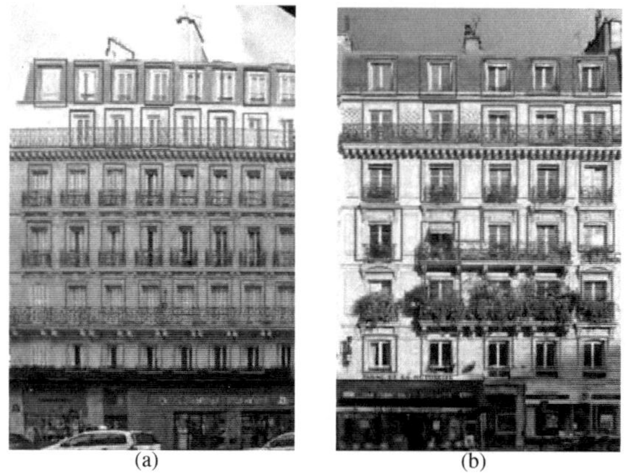

(a) (b)

Figure 8. Detections on rectified facades. (a) Detections of windows close together may be shifted due to merging. Some windows are missed. (b) Even slightly occluded windows are detected. No detection of store windows.

5 Conclusion

For risk assessment of ground movement induced damages to existing buildings, analyses are made which require precise information about a structure's stiffness. Since publicly available data sources only provide block models of buildings, these have to be enhanced to facilitate analyses. As windows highly decrease the stiffness of a structure, we developed an approach to window detection in facade images based on the findings of Ali et al. [4]. Our approach consists of a sliding window detector in combination with a soft cascaded classifier.

In the experiments, we showed that this setup improves over the Viola-Jones framework. Additionally, we tested our classifier on rectified facade images which further improves the detection rate. The results our system achieved in the experiments are sufficient for analyses estimating a damage class for an existing building. Thereby many buildings along a potential tunneling alignment or within an earthquake area may automatically be discarded from further personal investigation as the risk of damages is very low. This highly reduces human effort for inspection of buildings resulting in time and cost savings.

However, the detection rate is still too low for a precise analysis. Some windows are slightly below the rejection threshold due to occlusions or bad illumination conditions and are thus declined by the classifier. To increase the detection rate future work may cover a post processing in which possible window positions are derived from the windows already detected. At these positions classifiers with slightly lower thresholds can be used to detect previously missed windows.

Acknowledgement

Financial support was provided by the German Research Foundation (DFG) in the framework of the Collaborative Research Center SFB 837 "Interaction Modelling in Mechanized Tunneling".

References

[1] Brenner C. Building reconstruction from images and laser scanning. International Journal of Applied Earth Observation and Geoinformation, 6(3–4):187–198, 2005.

[2] Lee S. C. and Nevatia R. Extraction and integration of window in a 3d building model from ground view images. In Proceedings of the Conference on Computer Vision and Pattern Recognition, pages 106–113. IEEE Computer Society, 2004.

[3] Barber D., Mills J., and Smith-Voysey S. Geometric validation of a ground-based mobile laser scanning system. ISPRS Journal of Photogrammetry and Remote Sensing, 61(1):128–141, 2008.

[4] Ali H., Seifert C., Jindal N., Paletta L., and Paar G. Window detection in facades. In Proceedings of the 14th Conference on Image Analysis and Processing, pages 837–842. IEEE Computer Society, 2007.

[5] Shao H., Svoboda T., and Van Gool L. Zubud – zürich buildings database for image based recognition. Technical Report 260, Computer Vision Laboratory, Swiss Federal Institute of Technology, 2003.

[6] Haugeard J., Philipp-Foliguet S., Precioso F., and Lebrun J. Extraction of windows in facade using kernel on graph of contours. In Proceedings of the 16th Scandinavian Conference on Image Analysis, pages 646–656, 2009.

[7] Bourdev L. and Brandt J. Robust object detection via soft cascade. In Proceedings of the Conference on Computer Vision and Pattern Recognition, pages 236–243. IEEE Computer Society, 2005.

[8] Neuhausen M., Koch C., and König M. Image-based window detection - an overview. In Proceedings of Workshop of the European Group for Intelligent Computing in Engineering, 2016.

[9] Obel M., Neuhausen M., Schindler S., König M., and Mark M. Settlement risk assessment by means of categorized surface infrastructure. In Proceedings of the International Conference on Structural Analysis of Historical Constructions, pages 511–518, 2016.

[10] Haala N. and Kada M. An update on automatic 3d building reconstruction. ISPRS Journal of Photogrammetry and Remote Sensing, 65:570–580, 2010.

[11] Ripperda N. Grammar based facade reconstruction using rjmcmc. Photogrammetrie Fernerkundung Geoinformation, 2:83–92, 2008.

[12] Ripperda N. and Brenner C. Application of a formal grammar to facade reconstruction in semiautomatic and automatic environments. In Proceedings of the AGILE Conference on Geographic Information Science, 2009.

[13] Teboul O., Simon L., Koutsourakis P., and Paragios N. Segmentation of building facades using procedural shape priors. In Proceedings of the Conference on Computer Vision and Pattern Recognition, pages 3105–3112. IEEE Computer Society, 2010.

[14] Meixner P., Leberl F., and Brédif M. Interpretation of 2d and 3d building details on facades and roofs. In Proceedings of the Conference on Photogrammetric Image Analysis, pages 137–142. ISPRS, 2011.

[15] Viola P. and Jones M. J. Robust real-time face detection. International Journal of Computer Vision, 57(2):137–154, May 2004.

[16] Lienhart R., Kuranov A., and Pisarevsky V. Empirical analysis of detection cascades of boosted classifiers for rapid object detection. In B. Michaelis and G. Krell, editors, Pattern Recognition. DAGM 2003. Lecture Notes in Computer Science, volume 2781, pages 297–304. Springer, Berlin, Heidelberg, 2003.

[17] Becker S. and Haala N. Refinement of building facades by integrated processing of lidar and image data. The International Archives of the Photogrammetry, Remote Sensing and Spatial Information Sciences, 36:7–12, 2007.

[18] Pu S. and Vosselman G. Knowledge based reconstruction of building models from terrestrial laser scanning data. ISPRS Journal of Photogrammetry and Remote Sensing, 64(6):575–584, 2009.

[19] TSG-20: Tourist sights graz image database. http://dib.joanneum.at/cape/TSG-20/.

[20] TSG-60: Tourist sights graz image database. http://dib.joanneum.at/cape/TSG-60/.

Comparison of Handheld devices for 3D Reconstruction in Construction

Madhumitha Senthilvel[a] , Ranjith K. Soman[b] and Koshy Varghese[a]

[a] Building Technology and Construction Management Division ,Department of Civil Engineering, IIT Madras, India
[b] Centre for Systems Engineering and Innovation, Imperial College London, UK
E-mail: madhusvel@gmail.com, ranjithks17@imperial.ac.uk, koshy@iitm.ac.in

Abstract –

Generation and quality of as built models influence the subsequent applications such as progress monitoring, quality control, and deviation detection. The quality of any 3D reconstructed model heavily depends on the raw inputs and the post processing involved. While laser and LiDAR-based scanning are widely prevalent, lower cost equipment and sensors are increasingly becoming adaptable for 3D reconstruction. This study tests the feasibility of using IR-based scanning tablets and passive stereo vision cameras to acquire data from a construction environment based on the type of applications (such as progress monitoring, deviation detection) on a construction site. Two different off the shelf technologies: Tango tablet and ZED camera are tested during this research for developing as-built models using 3D reconstruction. The devices are compared on the basis of metrics such as preparation time for each scan, calibration of the scanner and total scanning time for determining the ease of scanning process, accuracy of generated point clouds. Also, the influence of external factors such as scanning parameters, ambient lighting, and characteristics of the object being scanned, and angle/orientation of scanner with respect to the object are studied.

Keywords –

3D Reconstruction; as-built modelling; stereo vision; infrared depth sensors

1 Introduction

Population growth, developing economy and urbanization are putting pressures on our infrastructure networks and assets. In order to meet these demands, it is necessary that we view infrastructure not just as a physical asset but also its digital twin through which all the all the associated data and information can be revealed[1]. The advent of new technologies such as BIM, Virtual/Augmented Reality, Cloud computing, IoT etc. are already transforming the way we built infrastructure. But the construction industry still lacks the culture of performance defined by outcomes instead of outputs. However, new forms of project delivery which enables more rapid and agile forms of management are being tested[2]. These agile forms of management require the progress and productivity data to be monitored in real-time for swift decision making and corrective action. This requires the need for an automated progress monitoring system in place of conventional manual progress monitoring which is slow and inaccurate[3].

One of the most popular methods of automated progress monitoring is using 3D scanning devices to create a 3D mesh of the structure/object and then comparing it with a design model to assess the progress of the construction [4]. Most common ways of acquiring a 3D mesh are by using a laser scanner which gives you a dense point cloud. However, a laser scanner is an expensive equipment and requires high skills for operation. Also, the data from the scanner is quite large and requires high-end computing devices as well as longer time to process[5]. Another popular method for as build 3D point cloud acquisition is photogrammetry. It uses multiple photographs or consecutive frames from a video to create a 3D point cloud. Equipment needed for this method are normal 'point and shoot' cameras with a good resolution. Although photogrammetry is inexpensive on the equipment side, it takes significant time to process the data and recreate the point cloud. Often, the photographs (data) wouldn't have enough overlap or quality resulting in incomplete or distorted point clouds. This brings us to the third set of devices which are called range cameras.

Range cameras are devices which give a depth value for every captured pixel. There are different techniques and sensors for computing the depth such as Time of Flight sensors and stereo vision technologies, structured light, etc. Recently, range cameras have started to come into the consumer market making it affordable. The reduced costs enable construction personnel to take ownership of these technologies and to experiment, using them to improve the construction progress monitoring practices. In addition, most of these devices are handheld and require minimal skill levels to operate. However, quality of the raw data from the devices needs to be evaluated for effective utilization Low quality raw data would result in higher post processing time and would require computers with high computation capability. While low quality raw data can be improved by better processing techniques, this would contradict the use of low cost devices which are time efficient. Therefore, it is essential to benchmark the quality of data and the ease of the scanning process for these devices. Hence, it is necessary that we test these technologies in construction site conditions to understand its performance in the above conditions.

The current paper aims at testing the capability of two commercially available devices (which use range imaging for depth estimation) based on the metrics such as accuracy of the scan, scanning times, the influence of lighting conditions and influence of object material. The devices tested are Google Tango tablet and ZED camera. The Tango tablet is an Android development level tablet that uses Infrared technology coupled with an RGB camera to capture depth. ZED camera is a stereoscopic RGB camera that uses stereo vision camera sensors and algorithms to reconstruct 3D scenes from stereo images [15].

The paper is divided into five sections. Section 2 gives a brief description of the works in the area related to benchmarking devices for point cloud generation. Section 3 gives an outline of the methodology used and details on the case study. Section 4 presents the results from the tests with discussion on the implication of the results and Section 5 provides the conclusions from the current study.

2 Past Work

This section gives a brief outline of the work done in the past to benchmark consumer level range cameras. It should be noted that the performance of the device is the combined performance of the sensing hardware and the software algorithms for 3D point cloud generation and optimization. There are studies which have determined the capability of off shelve software to model under construction structures through systematic studies [5]. Also, there are studies which determine the performance of the device as such [6].

Out of the two range cameras that the current paper is focusing on, ZED camera is new to the research community and has not been explored much. However, there are studies which evaluate the performance of 3D reconstruction using stereo cameras. Chung and Kim (2015) presented a 3D reconstruction algorithm for stereo cameras and tested it on a construction site. They have reported that stereo camera integrated with their MSD based 3D reconstruction algorithms provides good performance with low computational time for autonomous control applications[7]. There have been studies which compared structure from motion (similar to stereo cameras technology) with laser scanning[8].It is reported that the structure from motion algorithms is less accurate than point clouds generated from laser scans. However, image based data acquisition increases opportunities for as built visualizations. Zennaro et al (2013) have evaluated another stereo camera Kinect for its 3D reconstruction capabilities. They reported the devices to be less accurate than 3D scanners, yet able to reconstruct with reasonable accuracies[9].

Google Tango's interior scanning accuracy was tested by Froehlich and Azhar [10].They have reported that the scans can be beneficial in time and cost savings in 3D data acquisitions but the scan was not of high quality. Kalyan et al have compared Tango tablet with other methods of 3D reconstruction such as Photogrammetry and laser scanning focusing on construction quality assessment. The study concluded that the dimensional accuracy of Tango tablet. Tango limits itself from the applications for quality assessment. However, the device outperformed other methods in terms of cost and user convenience[11].Although the device was reported to be less accurate, there are studies which stress that the accuracy of the device can be improved through passing the outputs through algorithms such as RANSAC, shape detection, least square plane fitting etc. [12].

In short, there are studies which have benchmarked stereo vision and infra-red based range cameras. Most of the studies have benchmarked these cameras to be less accurate than the laser scanners. However, the cost factor, consumer availability and ease of use make these cameras to have easier adoption. It should be noted that most of the studies are being performed in controlled conditions. It is necessary that experiments are performed on a live physical construction site to understand the performance of these off the shelve devices in object cluttered, dynamic fast changing and harsh environment. Hence, the current study aims at evaluating the performance of ZED camera and Tango tablet on a live construction environment.

2.1 ZED Camera

ZED camera is a stereoscopic passive RGB camera that uses stereo vision camera sensors and algorithms to reconstruct 3D scenes from stereo images (as shown in Figure 1). It comes with software SDKs which can be used to calibrate and create 3D reconstruction using the stereo image inputs. The camera can record up to 2K resolution, has wide angle lenses. The maximum depth perception is 20m [18]. The ZED explorer is the primary software which can be used to record a video in .svo format (Stereolab's format) which is then processed by ZED Fusion package to create disparity maps from stereo images and create a 3D point cloud. Each frame from the video is converted to point cloud and registered

Figure 1 ZED camera

successively using visual odometry, enabling the creation point cloud represented in 3D space [18]. Although the same process can be done in real time using ZED fusion, the algorithm doesn't work efficiently and has the possibility of crashing or may result in a distorted point cloud. However, this glitch may be resolved in the future SDK updates.

2.2 Tango tablet

Tango table is an Android development level tablet that uses Infrared technology coupled with an RGB camera to capture depth (as shown in Figure 2). The tablet consists of a motion tracking camera, 3D depth sensor, Accelerometer, Barometer, Compass, GPS, and Gyroscope [15]. The infrared-based tablet uses Time of Flight (ToF) to detect the depth of the objects being scanned. It incorporates SLAM for mapping its location

Figure 2 Project Tango Development Kit

to create point cloud representation of the object in 3D space. The tablet Tango tablet's Constructor app was used for scanning. It directly generates the 3D point cloud and the meshed surfaces which are available for upload and export in ".obj" format for further processing. The device self-calibrates when the app starts, after which the user can use it for capturing scenes [15].

3 Case Study and Methodology

The experiments to test the device's performance were conducted in a residential Construction site. Periodic site visits were conducted at different project times to capture

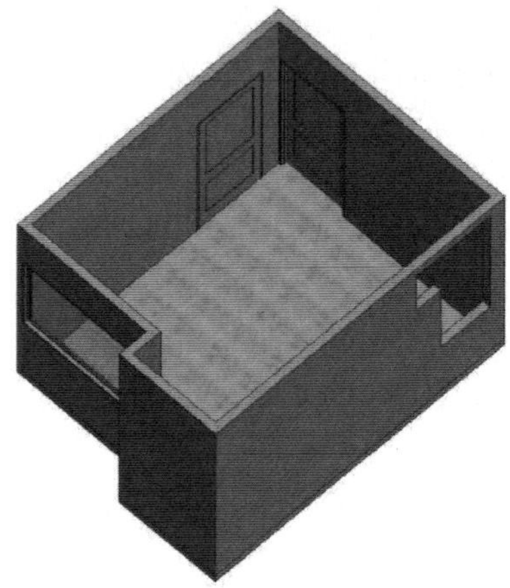

Figure 3 BIM model of the room considered for Case Study

various stages of construction. Four different stages of construction were scanned which are elaborated in table 1. The elements tested for scanning included masonry walls, concrete floors, columns, beams, and finished walls, tiled floors, windows, and doors. The raw point cloud generated from both the devices are shown in Figure 4 and Figure 5 for case 2. The original BIM for this case is shown in Figure 3.

Table 1 Description of Case studies

Case No.	Construction Stage Completed	Stage Description
1	Frame	Columns, Beams, Floor, Ceiling completed
2	Frame + Sill Level Masonry	Columns, Beams, Floor, Ceiling, Masonry till sill level completed
3	Frame + Masonry	Columns, beams, floor, ceiling, masonry till lintel level completed
4	Full Room	Fully constructed room including plaster, paint, tile finishing, windows, doors, etc.

3.1 Data collection using ZED and Tango

The feasibility of using ZED camera for 3D reconstruction application has been recommended in[6].

Figure 4 Raw meshed point cloud of Case 3 from ZED

But, the detailed study on the performance of the device, especially from a construction point of view is necessary to assess its usage in the field [16] [17]. The ZED camera was used at 720 resolution and frame rate of 60FPS since it was the highest resolution supported by ZED Fusion. The ZED Fusion Application included in the SDK package can be used in real-time to simultaneously scan and register point clouds of an area.

The Tango tablet is an IR device based on Time of Flight (ToF) for depth perception. It uses Inertial Measurement Unit (IMU) and visual odometry to successively register each depth scene. The Tango

Figure 5 Raw Meshed point cloud of Case 3 from Tango

Constructor App from Google Play Store was used to scan and generate point clouds of the scenes.

3.1.1 Accuracy of Scan

The interior building dimensions are extracted from the scans manually and compared to manual survey dimensions to assess their accuracy and potential for verifying as-built conditions. The methodology by Sung and Kim on a similar work is adopted for this test [7][16]. The lengths of the major dimensions in each scan are measured in the scan data using MeshLab.

The as-built dimensions of three rooms which were scanned is measured using Leica Laser disto, an off-the-shelf laser surveying device which had a range of 0.05 – 150m and accuracy of 1mm. The deviation between scan values and actual values is evaluated using the absolute difference between the two measurements. The results of the Accuracy tests are tabulated in table 2. Tango has an average error of 3.08% while ZED has 7.80% as the dimensional error. Since these values are quite high for an object of dimensions 3m, the output of these devices cannot be directly used for monitoring dimensional quality analysis in construction.

3.1.2 Scanning Path, Speed of Scan and Angle of Scan

A clockwise scanning direction was chosen for both the devices. The scanning path is directly influenced by

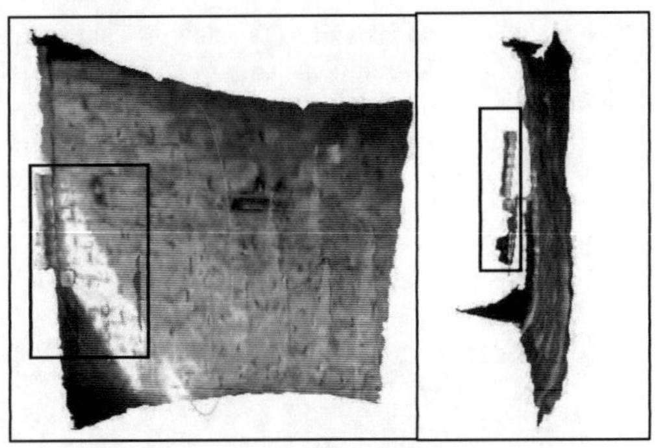

Figure 6: Effect of Rescanning using ZED Camera

the field of view of the device, and the maximum range of the device. ZED's horizontal Field of View is 61°. Based on its horizontal FOV, the vertical FOV is calculated as 93° (approx.) using the formula given below [15].

$$FOV\ V = (2\arctan(\frac{0.5 * height}{Fy})$$

$$FOV\ H = (2\arctan(\frac{0.5 * width}{Fx})$$

Where Fx and Fy are Focal lengths, and width and height denote the image dimension on the sensor.

Table 2 Dimensional Accuracy Analysis

Scan Description	Original Dimenions (m)	As-Built Dimensions From Tango(m)	AS-Built Dimensions from ZED (m)	Absolute Error for Tango(m)	Absolute Error for ZED(m)
Case 1: Component 1	5.8	5.74	5.31	0.06	0.49
Case 1: Component 2	4	3.96	3.85	0.04	0.15
Case 1: Component 3	3.05	3.07	2.75	0.02	0.30
Case 1: Component 4	3.05	3.08	2.77	0.03	0.28
Case 2: Component 1	3.5	3.41	3.18	0.09	0.32
Case 2: Component 2	4	3.92	3.72	0.08	0.28
Case 2: Component 3	3.05	3.1	2.92	0.05	0.12
Case 2: Component 4	3.05	3.12	3.03	0.12	0.02
Case 3: Component 1	4.3	4.56	3.63	0.26	0.67
Case 3: Component 2	3.5	3.41	3.10	0.09	0.40
Case 3: Component 3	3.05	2.91	2.87	0.14	0.18
Case 4: Component 1	4.3	3.97	3.81	0.33	0.49
Case 4: Component 2	3.5	3.35	3.16	0.15	0.34
Case 4: Component 3	3.05	2.95	2.95	0.13	0.10
Average				**0.11**	**0.30**
Standard Deviation				**0.08**	**0.18**
Maximum				**0.33**	**0.67**
% Error				**3.08%**	**7.80%**

Since ZED's horizontal FOV is significantly larger than its vertical FOV and its maximum range is 20m, the ZED camera was rotated 90° horizontally. This enabled to capture more a larger area vertically and capture the floor and the ceiling in one frame. The rate of movement of ZED camera for scanning was limited to 0.2m/s-0.5m/s. The increase in this speed caused the stereo camera to lose track of its localization, causing drift errors.

Figure 7 Effect of Rescanning using Tango

Tango has a vertical FOV of 38° and a horizontal FOV of 68°. Because of the limited range of Tango (4m), adopting a 90° horizontal flip did not cover more scan area. While scanning construction elements such as walls and floors, the device is rotated clockwise, in a circular motion, to capture both the ceiling and floor. Backtracking in the scanning path (i.e., rescanning an area) sometimes results in the introduction of new surface of the object which does not coincide with the existing surface. An example is shown in Figure 7. ZED camera also has problems with rescanning if the scan velocity is too high as shown in Figure 6. The effect of rescanning an area in ZED is minimal, compared to tango.

The speed of using the device to scan scenes has an indirect influence on the quality of the point clouds generated. It was observed that moving the devices too fast resulted in localization errors, and some areas were not captured. The IMU can, for example, be affected by the inconstant or sudden movement of the tablet holder, thus small positioning errors are accumulated as the scanning process goes [12]. To determine the range of the angular velocity of using these devices, a series of trial and error experiments were performed to observe the threshold beyond which localization errors occurred. ZED and its SDK performed with minimal localization errors when the capture speed was in the range of 0.1-0.5m/s. Tango could significantly scan faster, performing better when the capture speed was in the range 0.5-1.0m/s.

When scanning a scene, the location of the camera

and the FOV influences the area captured. Both ZED and Tango did not exhibit any difference in capturing data in different angles.

3.1.3 Mobility and Time efficiency

One of the most determining factors in the adoption of a technology for construction is its ease of use and the time it takes to use it. A construction environment is complex and unstructured; often multiple scans are required to capture one single floor. Hence, light weight devices and devices which can withstand long durations of use are preferred as they offer more mobility. In fact, the above reasons are the main factors which led to the increase in adoption of low cost range cameras for construction environments. The ZED camera weighs 160 grams, making it very light to carry and use. But to scan an area, the camera needs to be connected by an USB cable to a computer which runs the SDK. Thus, the user needs to carry a computer in order to use the device on site.

Tango, on the other hand, is a tablet which weighs 370g. The point cloud is generated by the Constructor App can be viewed in real time on the tablet, making it much more user-friendly. In order to determine the time

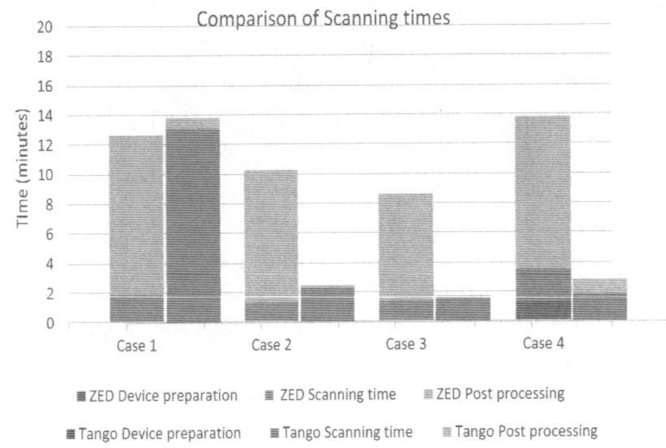

Figure 8 Comparison of Scanning Times for ZED and Tango

efficiency of the device, the time required for preparation, scanning, and data processing are measured for the four cases. The results are shown in Figure 8.

Due to inherent range and FOV limitations of the Tango tablet, the scanning time is much longer for larger areas. Prolonged use of the Tango tablet results in heating of device and crashing of either the scanning app/sensor/ OS. This results in loss of scan data which cannot be retrieved even upon restarting the device.

ZED on the other hand, with its long range capacity, can easily scan both small and larger areas without experiencing sensor malfunction. But, as seen from Figure 8, the post-processing time of ZED is higher than Tango tablet for all cases.

3.1.4 Influence of Lighting Conditions

ZED camera's SDK ZED Fusion takes rectified stereo image as input and computes the disparity map. Given a video file (in .svo format), the SDK uses an optimized stereo algorithm to register the point clouds of consecutive frames, resulting in a 3D reconstruction of the scanned space. There is a marked difference in the performance of the device in low light (5lux-20lux) and high light conditions (150lux-250lux). Stereo reconstruction algorithms depend on accurate correspondences which in turn relies on unique features

Figure 9 Effect of Excess Lighting on quality of Scan from Tango

in the stereo images. Unique features are better identified with better lighting conditions. So, lighting indirectly influences the quality of the stereo-based 3D reconstruction.

Tango uses Infrared to detect the depth of the scene. As a result, any object directly illuminated by sunlight or artificial lighting was not captured. Figure 9 shows the area not detected due to excessive bright light (the same area was detected by ZED which is shown in Figure 6). Objects which had illumination less than 4lux were either not detected by a tango or the resulting reconstruction was distorted.

The effect of lighting on scenes comprising of the indoor as-built environment was studied by [8], which also reported similar effects.

3.1.5 Influence of Object Material on Scan Quality

The device was tested on common construction scenes encompassing concrete elements, masonry, reinforcement, finishing, and joineries. Objects such as reinforcements were not modelled at all since they were too small to be captured. When scanning reflective or transparent surfaces such as glass joineries, the point cloud generated using ZED camera gives a distorted surface. The possible reasons for this include improper computation of depth by the stereo algorithm in the SDK due to different reflections in multiple frames. On the other hand, the Tango tablet does not detect any transparent surface or objects that reflect light. Thus, for case 4, glass windows, and glass doors were not detected.

4 Summary of Results and Discussion

Both the devices were evaluated by testing them on the construction site on the following criterions:

- Mobility and Time Efficiency
- Accuracy: Precision and reliability of the resulting point cloud
- Scanning Methodology: Including path adopted and speed of scanning
- Influence of Lighting Conditions
- Influence of Object Materials

Both the two devices, Google Tango and ZED Camera tested scored well on the mobility and time efficiency. The devices were light in weight and easy to set up. Both devices don't require the user to have special skill set to operate. This allows the construction personnel to take ownership of this device without much effort. However, ZED camera should be connected to a computer while scanning. Unlike Tango tablet, ZED camera required to be connected to a computer while it captures the reality mesh. This limits the usability of ZED camera to an extent such as drone surveying.

It was observed that both these devices could capture the reality mesh with reasonable accuracy. The average error for Tango was 3.2% while that of ZED camera was 8.28%. Although these values benchmark these devices below Laser scanners, ease of use, mobility and commercial availability gives an edge for these devices over the laser scanners.

Although Tango tablet scored better in accuracy when compared to ZED camera, operational issues such as overheating, increased time for scanning etc. reduces the usability of Tango tablet. For example, while scanning Case 1, the device was continuously used for 20 minutes leading to overheating, battery drain and subsequently crashed. The crashing is a result of both overheating and the size of data. A 20-minute scan generates a huge file size, and the hardware is unable to progress further, causing the application to crash, and the data is lost before it can be saved. But, in order to assess the progress of a building under construction, multiple areas will have to be captured. A preemptive way to avoid losing data is to split large areas into smaller scans, thereby limiting the scanning time and avoiding overheating. But, such splitting will force the user to manually register the scan segments to each other to get the full 3D model, thus reducing the level of automation in progress monitoring and as-built generation workflows. The combination of Tango's range, FOV, scanning time and battery capacity makes it difficult to be adopted for continuous capture of the entire area under construction. Also, the device's performance in outdoor environments may not capture as well as indoor environments because of infrared radiations due to sunlight. Thus, it is more suitable for scanning smaller indoor areas and for applications that require fairly accurate models.

ZED camera, on the other hand, has better performance indoors provided sufficient lighting. It also works in outdoor environments, with the same range capabilities as indoors. Owing to its larger range and FOV, it is possible for the ZED camera to scan more area in lesser time. In addition to the real-time reality mesh capture, the ZED SDK offers the opportunity to save the *.svo files and process later. Hence, even if the device fails during the data capture, data isn't lost. Also, this makes ZED camera independent of the processing capability of the hardware subject to conditions such as availability of CUDA graphics processing. One can use a low-end laptop for data capture and then transfer files to a workstation with high-end computing capabilities for processing. This increases the reliability of ZED camera over Tango. Although ZED has these advantages over Tango, ZED has accuracy limitations, which can limit its usage to progress monitoring, and AR/MR Visualizations, but the creation of as-built models and using them for quality check might not be viable for all environments.

A major limitation observed in both these devices is their inability to detect glass/ transparent surfaces, making it difficult to automatically detect windows, doors and other similar surfaces. Workarounds for these have been explored in literature before. Whether the same heuristics and algorithms perform similarly for data from these devices needs to be explored.

5 Conclusion

This paper presented an evaluation of two devices which use two different technologies: a stereo vision based ZED camera and IR based Tango tablet in construction environments to generate as-built models. Most of the performance evaluation studies for Range camera devices have benchmarked these cameras to be less accurate than the laser scanners [10] [11]. However, the cost factor, consumer availability and ease of use make these cameras easier to be adopted in construction applications. This study explores the type of experiments needed to be performed to evaluate the applicability of these devices in construction environments which are dynamic and complex.

This study focuses on the preliminary tests performed using these devices to establish benchmarks for using them successfully for capturing a live construction site. But, their true capabilities need to be benchmarked in a controlled environment so that the best practices for using them can be established. In order to assess if these devices and their benchmarks are adequate for specific applications in construction such as progress monitoring remains to be explored. If the resulting accuracy of progress monitoring using these devices is comparable to conventional technologies such as lasers, then these low cost devices have the potential to make a tangible difference in construction project monitoring and control.

References

[1] A. Kemp *et al.*, "State Of The Nation 2017 : Digital Transformation," London, 2017.

[2] R. Levitt, "Towards project management 2.0," *Eng. Proj. Organ. J.*, vol. 1, no. 3, pp. 197–210, 2011.

[3] I. N. Davidson and M. J. Skibniewski, "Simulation of Automated Data Collection in Buildings," *J. Comput. Civ. Eng.*, vol. 9, no. 1, pp. 9–20, 1995.

[4] T. Omar and M. L. Nehdi, "Data acquisition technologies for construction progress tracking," *Autom. Constr.*, vol. 70, pp. 143–155, Oct. 2016.

[5] A. Bhatla, S. Y. Choe, O. Fierro, and F. Leite, "Evaluation of accuracy of as-built 3D modeling from photos taken by handheld digital cameras," *Autom. Constr.*, vol. 28, no. December 2012, pp. 116–127, 2012.

[6] M. Aijas and A. Sharma, "Google Tango tablet," in *Proceedings of International Conference on Advanced Computing (ICAC-2016)*, 2016, pp. 29–34.

[7] C. Sung and P. Y. Kim, "3D terrain reconstruction of construction sites using a stereo camera," *Autom. Constr.*, vol. 64, pp. 65–77, 2016.

[8] M. Golparvar-Fard, J. Bohn, J. Teizer, S. Savarese, and F. Peña-Mora, "Evaluation of image-based modeling and laser scanning accuracy for emerging automated performance monitoring techniques," *Autom. Constr.*, vol. 20, no. 8, pp. 1143–1155, 2011.

[9] S. Zennaro *et al.*, "Performance Evaluation of the 1st and 2nd Generation Kinect For Multimedia Applications," *Multimed. Expo (ICME), 2015 IEEE Int. Conf.*, pp. 1–6, 2013.

[10] M. Froehlich and S. Azhar, "Handheld Simultaneous Localization and Mapping Devices for 3D Scanning," no. Isarc, 2016.

[11] T. S. Kalyan, P. A. Zadeh, S. Staub-French, and T. M. Froese, "Construction Quality Assessment Using 3D as-built Models Generated with Tango tablet," *Procedia Eng.*, vol. 145, pp. 1416–1423, 2016.

[12] A. A. Diakité and S. Zlatanova, "First Experiments With the Tango Tablet for Indoor Scanning," *ISPRS Ann. Photogramm. Remote Sens. Spat. Inf. Sci. Vol. III-4*, no. September, pp. 67–72, 2016.

[13] A. Burbano, M. Vasiliu, and S. Bouaziz, "3D Cameras Benchmark for Human Tracking in Hybrid Distributed Smart Camera Networks," *Proc. 10th Int. Conf. Distrib. Smart Camera*, pp. 76–83, 2016.

[14] N. Li, "Imaged-based verification of as-built documentation of operational buildings," *Autom. Constr.*, vol. 21, no. January 2012, pp. 161–171, 2012.

[15] Google. Project Tango. Online: https://developers.google.com/tango/hardware/tablet, Accessed: 15/02/2017.

[16] T. Schops, T. Sattler, C. Hane, and M. Pollefeys, "Large-scale outdoor 3D reconstruction on a mobile device," *Computer Vision and Image Understanding*, vol. 157, pp-151-166.

[17] Runceanu et al., "Indoor Point Cloud Segmentation for Automatic Object Interpretation", *Scientific-technical anniversary of the DGPF*, Wurzburg 2017, Accessed:03-05-2017.

[18] STEREOLABS, ZED Stereo Camera.Online: https://www.stereolabs.com/developers/ , Accessed: 03/02/2017.

Low-latency Vision-based Fiducial Detection and Localisation For Object Tracking

Ansu Man Singh[1], Quang P. Ha[1], David K. Wood[2], Mark Bishop[2]

[1.]Faculty of Engineering and Information Technology, University of Technology Sydney, Australia
[2.] Ocular Robotics Pty Ltd, NSW, Australia

Abstract

Real-time vision systems are widely-used in construction and manufacturing industries. A significant proportion of computational resources of such systems is used in fiducial identification and localisation for motion tracking of moving targets. The requirement is to localise a pattern in an image captured by the vision system precisely, accurately, and with a minimum available computation time. As such, this paper presents a class of patterns and, accordingly, proposes an algorithm to fulfil the requirement. Here, the patterns are designed using circular patches of concentric circles to increase the probability of detection and reduce cases of false detection. In the detection algorithm, the image captured by the vision system is first scaled down for computationally-effective processing. The scaled image is then separated by filtering only the colour components, which are made up of outer circular patches in the proposed pattern. A blob detection algorithm is then implemented for identifying inner circular patches. The inner circles are then localised in the image by using the colour information obtained. Finally, the localised pattern, along with the camera and distortion matrix of the vision system, is applied in a perspective-n-point solving algorithm to estimate the marker orientation and position in the global coordinate system. Our system shows significant enhancement in performance of fiducial detection and identification and achieves the required latency of less than ten milliseconds. Thus, it can be used for infrastructure monitoring in many applications that involve high-speed real-time vision systems.

Keywords – Fiducial tracking, marker detection

1 Introduction

Vision-based systems have numerous applications in robotics and automation systems used in industry, including construction and manufacturing. For instance, in [1], a vision-based system is presented for 3D terrain surface reconstruction of a construction site by using stereo cameras. The vision systems can also be integrated with unmanned vehicles to monitor and inspect a disaster-prone location, as reported in [2].

One of the purposes of the vision-based systems is to localise and track moving targets, for example, for colour tracking of a multiple robot system [3], or monitoring of equipment in construction and surface mining [4]. To this end, markers are widely used in these systems to precisely estimate the position and orientation of moving targets. In construction, such systems can be helpful in estimaton of the pose and position of the end effector of excavators to analyse the workflow during the earthmoving process [5]. Furthermore, the developed fiducial system will be used in thermal stress analysis (TSA) of vibrating mechanical systems using the vision camera, thermal sensors and the state-of-the-art optical sensor pointing technology available at Ocular such as the Robot Eye [6] shown in Figure 1.

One of the challenges in fiducial tracking is the least computation time required to accurately detect and localise the marker. This is especially important in high-speed applications where computational resources are used for other critical processes as well, such as control and management. In these scenarios, if too much computation time is spent in fiducial localisation, then it may affect the overall performance of the system.

To address the problem mentioned, we present a marker system and the method to detect the markers with a latency for detection of less than 10 ms. The proposed marker consists of concentric circles located at four corners of a square. The purpose of the outer circles is to reduce the possibility of false detection of the inner circles. In the detection algorithm, computational efficiency is achieved by downscaling the images captured by the vision sensor. Then, to detect the inner circles we segment the scaled image into binary images which represent only the colour components made up of the outer circles. The inner circles are then detected and localised in the binary images. Finally, to estimate the position and orientation of the marker in the 3D space, we use the Levenberg-Marquardt optimisation algorithm iteratively.

The organisation of this paper is as follows. First, the paper discusses the fiducial design in the Section 2, followed by, the description of the algorithm for its detection and localisation, Section 3. Then, results of the comparative study of our marker system with standard markers are presented in Section 4. Finally, the paper concludes with a conclusion.

Figure 1. Ocular Robot Eye

2 Fiducial Design

There are many marker systems that make uses of circular patches. For instance in [7], colour-coded circular markers are arranged in various shapes such as circle, triangle, square, etc. However, such a marker system can have a higher possibility of false detection if the circles outside the region of interest are incorrectly detected. A typical pattern with circular patches is presented in Figure 2, where the circle is detected outside the region of interest. This may reduce the performance of a detection algorithm for single-circle markers.

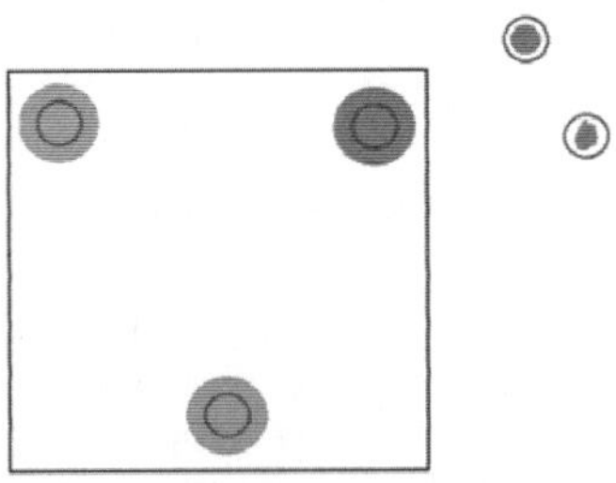

Figure 2: False detection of circles outside the marker region

To overcome the drawback of a fiducial design with single circular patches mentioned above, we propose to use markers with concentric circles, where the outer circles are made up of two distinct colours, blue and green in this case. The proposed marker is shown in Figure 3, where the rationale for its choice is that if the inner circles can be searched for by using only the pixels consisting of the colour of outer circles, then the incorrect detection of circles outside the pattern region of interest is significantly reduced. Notably, the circle at the center of the pattern is not used in the detection algorithm. Nevertheless, this circle can have a special usage in tracking the thermal targets in motion. Particularly, the pattern can be attached to the target with the area of interest located within the central circle.

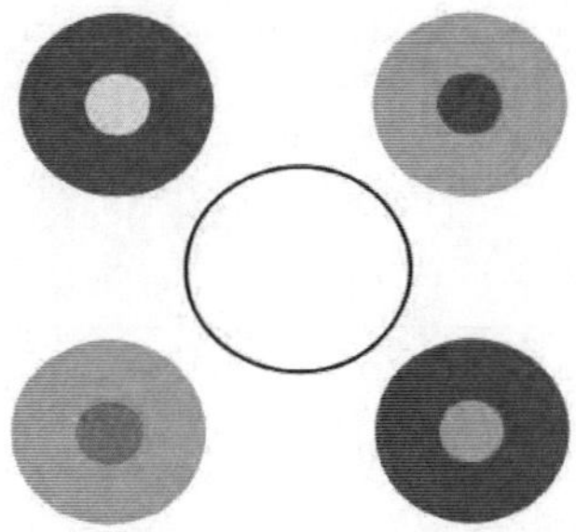

Figure 3. Proposed marker

3 Marker Detection and Pose Estimation

In the following section, we present the process of the detection and localisation of the inner circles of the marker, followed by, the procedure for estimation of its position in 3D space.

3.1 Image segmentation

In this step, the images captured by the camera is first scaled down by a factor of 4:1. The scaling effect reduces the computational latency of the detection algorithm. After scaling, two binary images are produced to represent the presence or absence of colour components of the outer circles. In order to obtain the binary images, we apply a threshold on every pixel of the scaled image. The threshold depends on the conventional Hue-Saturation-Value (HSV) model of the colour of the outer circles, where H corresponds a pure colour with H=0 refered to Red, S describes the whiteness with S=0 for White, and V is for darkness with V=0 for Black. Figure 4 represents the segmentation step.

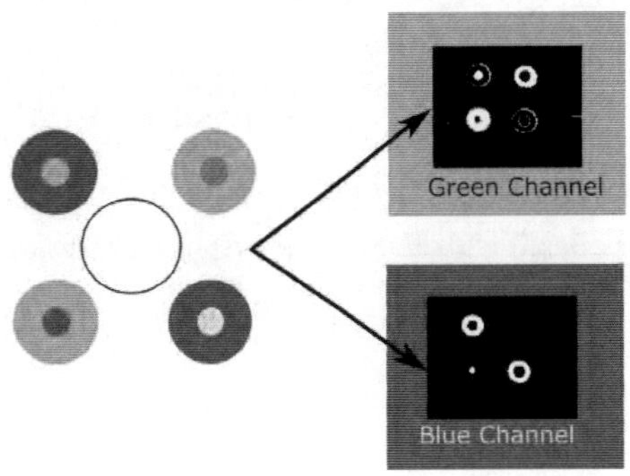

Figure 4: Segmentation of captured image into binary images

3.2 Inner circle detection

In order to detect the inner circles two methods can be applied here, namely the Hough detection method and blob detection method.

3.2.1 Hough circle detection

The Hough detection method uses the circular Hough transform to find circles in an image. If an image contains many points, some of which fall on perimeters of circles, then the search objective is to find parameter triplets (a, b, r) to describe each circle of centre (a, b) and radius R:

$$\begin{cases} x = a + r\cos\theta \\ y = b + r\sin\theta, \end{cases}$$

where angle θ sweeps through the full 360 degree range that the points (x, y) trace the perimeter of a 2D circle. The fact that the parameter space is 3D makes a direct implementation of the Hough circle technique (HCT) more expensive in computer memory and time. Indeed, the fiducial marker, however, goes through transformations, i.e. rotational and translation which produces an ellipse in the image plane of the camera. In this scenario, the Hough detection method becomes ineffective. This is clearly demonstrated in Figure 5(a), where the algorithm fails to detect inner circles of the fiducial. In addition, for better performance, HCT requires filtering for smoothing, which further increases the computational load. Thus, for the inner circle detection in the marker, this method is not a suitable.

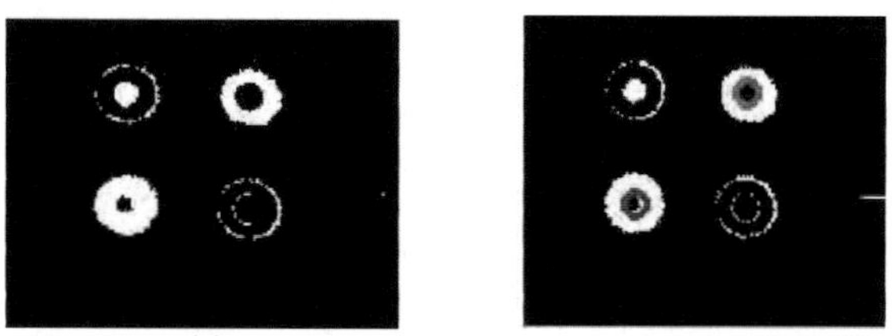

(a) (b)

Figure 5: Comparison of (a) Hough circle detection, and (b) blob detection

3.2.2 Blob detection algorithm

Based on the general concept that blobs of an image are meaningful regions to interpret its main features. Here, the blob detection algorithm (BDA) is based on the contour detection for extracting distinct regions from the background. This algorithm first detects a contour in the provided image, which is not necessary a circle but can be elliptical patches. Then, the BDA connects the overlapped contours and may be rotated in 3D. As a result, this method is suitable for inner circle identification of the marker. The detection of the inner circles of the marker by the blob detection method is presented in Figure 5 (b), where the inner circles present more meaningful features. Similarly, the identification of the inner coloured circles using the blob detection algorithm is presented in Figure 6.

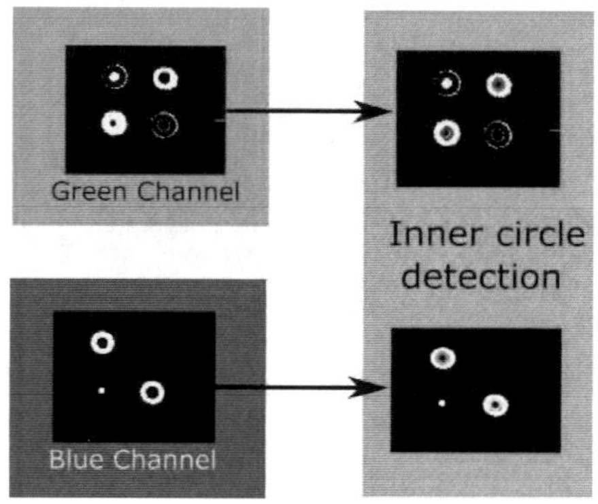

Figure 6: Inner circle detection using the blob detection algorithm

3.3 Circle localisation and pose estimation

After detecting circles in the binary images, the inner circles are identified by checking the H-S-V value of the center of the circles in the original image. The inner circles are identified in the image plane if the H-S-V values lies within the pre-defined colour threshold for the inner circles. In other words, suppose x_i is the centre of a detected circle in the binary images and $h() : \mathbb{R}^2 \to \mathbb{R}^3$ is the function mapping a pixel to the H-S-V value, then the detected centre is identified as the kth inner circle if the condition $\lambda_{k(min)} \leq h(x_i) \leq \lambda_{k(max)}$ (k =1,2,3,4) is satisfied, where and $\lambda_{k(\min)}$ and $\lambda_{k(max)}$ are the minimum and maximum colour threshold for the kth inner circle, respectively.

After the localisation of the inner circles in the image plane, the orientation and position of the marker are estimated by solving the Levenberg-Marquardt optimisation for the non-linear least square minimisation problem:

$$\arg\min_{j} \sum_{j=1}^{N} d(\mathbf{x}_j, \mathbf{P}\mathbf{X}_j)^2,$$

where $\mathbf{x}_j$ and $\mathbf{X}_j$ are respectively the image point in the image plane and the scene point in the world coordinate system, and $d(\mathbf{x},\mathbf{y})$ is the Euclidean distance between two vectors $\mathbf{x}$ and $\mathbf{y}$, and $\mathbf{P}$ is the camera projection matrix. The projection matrix $\mathbf{P}$ is given by

$$\mathbf{P} = \mathbf{KRt} = \begin{bmatrix} f_x & 0 & c_x \\ 0 & f_y & c_y \\ 0 & 0 & 1 \end{bmatrix} \mathbf{Rt},$$

where f_x and f_y are respectively the focal length of the camera in x and y direction, c_x and c_y are the principal axes, $\mathbf{R}$ is the rotational matrix and $\mathbf{t}$ is the translational vector of the camera with respect to the pattern coordinate system. The camera matrix $\mathbf{K}$ is obtained from the camera calibration and should be considered in the optimisation, as described in OpenCV [8]. The library of OpenCV provides the interfaces for the optimisation. The solver solves for $\mathbf{R}$ and $\mathbf{t}$ after the recursive optimisation. Figure 7 summarises the processes of marker detection and pose estimation.

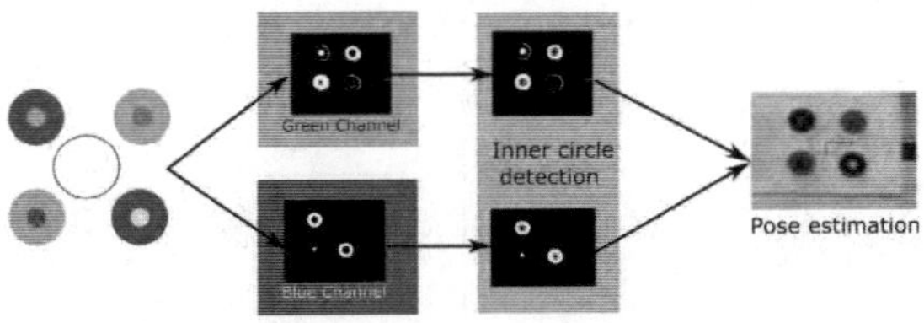

Figure 7: Orientation and position estimation of the fiducial marker in 3D space

The flowchart depicted in Figure 8 illustrates the marker detection procedure. Even though colour markers are used in our fiducial system, they are detected and localised in binary images in order to consume less computational load. In addition, the images are also scaled before the detection method is applied. As a result of these operations, the expected low latency can be achieved for detection of markers.

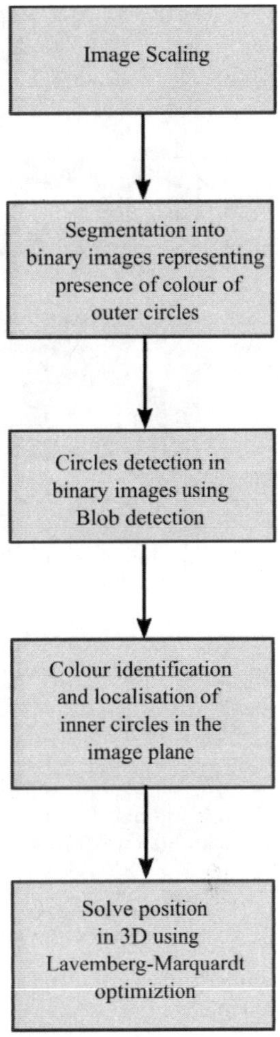

Figure 8: Flowchart for the marker detection procedure

4 Experiment and results

In this section, we present the benchmarking of our proposed marker detection system with standard marker detection systems, such as the Aruco and AprilTags markers.

4.1 Aruco Marker system

Aruco marker system uses the binary-coded markers, as shown in Figure 9. To correct the binary codes in the detected image, hamming codes can be used as reported in [9]. The detection algorithm for a marker system consist of three stages, namely (i) image segmentation, (ii) contour extraction and (iii) marker identification. The Aruco marker is one of the standard marker systems for camera calibration. Its interfaces are included in the OpenCV 3.1.0 library [8].

4.2 AprilTags marker system

AprilTags is also one of the widely used marker systems. A marker of Apriltags consists of binary codes, as depicted in Figure 10. In the detection algorithms of the marker system: first, the line segments are detected, followed by quad detection, and finally, the binary codes are extracted and corrected [10].

The size of markers used in the experiment is listed in Table 1 below.

Table 1. Marker size used in experiment

Marker	Size
Aruco	5.5 × 5.5 cm
AprilTags	5 × 5 cm
Proposed marker	5 × 5 cm

Figure 9: Aruco marker

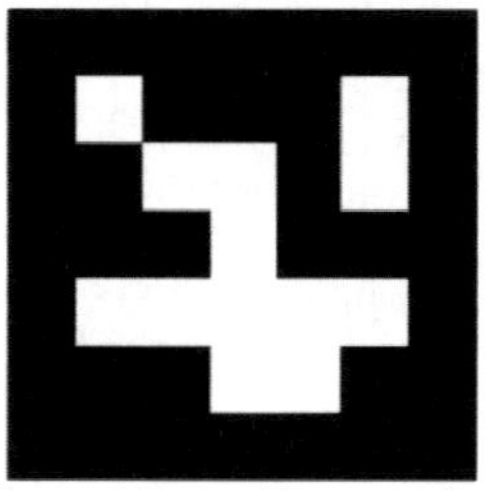

Figure 10: AprilTag marker

In our experiments, a high frame rate camera was used to take images of the markers. The camera was operated at 100 frames per seconds, and the depth of field for the camera was set to 50 cm with the lower limit of 50 cm and the upper limit of 100 cm. The specifications of the camera are listed in Table 2. Similarly, the algorithms for markers detection were tested in a PC with Intel ® Core ™ i7 CPU operating at the clock frequency of 3.4 GHz, and 16 GB RAM.

Table 2. Specifications of camera

Resolution	658 × 492
Focal length	16 mm
Aperture	F/2-22

The experiments were performed by varying the distance of the markers from the camera. The markers were placed at the distance of 40 cm, 50 cm, 90 cm, and 120 cm, respectively.

The results of the experiments are presented in Figures 11 and 12, respectively. Figure 11 shows the detection ratio of the fiducial systems. The detection ratio for the individual marker systems is calculated by taking the ratio of the number of detected frames over total frames during an experiment. From the figure it is clear that the detection ratio for our proposed marker system and the Aruco marker system is better than AprilTags, when the markers are placed from 40 cm to 90 cm from the camera. However, the performance of the detection algorithms decreases when the markers are placed beyond the far limit of the depth of field of camera, i.e. 100 cm. Even in this case, the detection ratio of the proposed system is still better than Aruco markers.

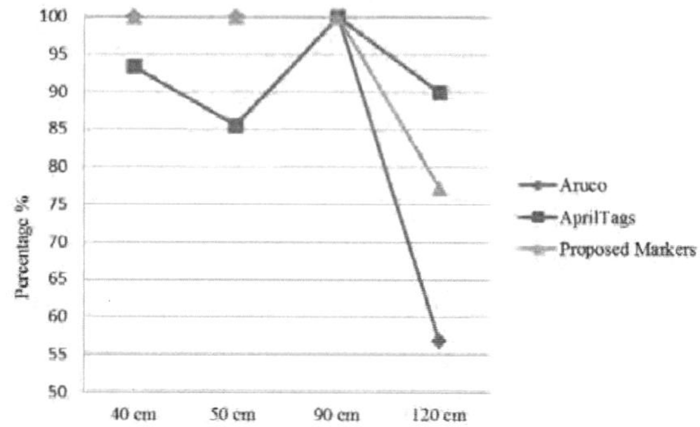

Figure 11: Detection rate at various distance from camera

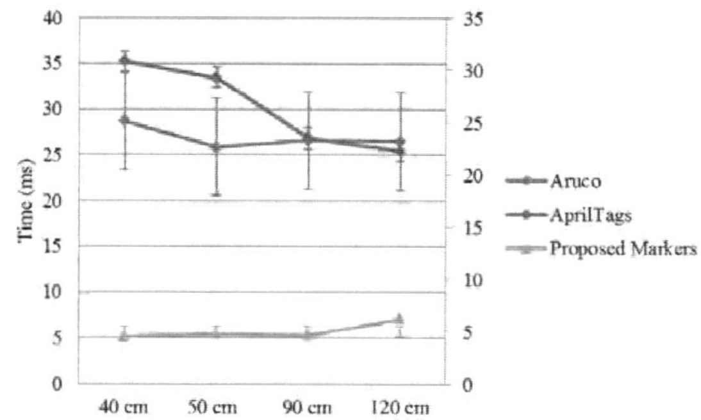

Figure 12: Average computation time for all markers

Figure 12 shows the average computation time for all the fiducial tracking systems. The computational time is evaluated in milliseconds while the vertical bars in the graph shows the standard deviation in the marker detection time. It is evident from the figure that the proposed marker detection system performs much better compared to the other standard markers. For instance, the detection time for the proposed algorithm is 5 ms, compared to 35 ms and 28 ms for the Apriltags and Aruco marker systems, respectively. In addition, the standard deviation for the computation time is also small compared to other standard methods. Although the computation time increases slightly when the marker is beyond the upper limit of the depth of field of the camera, i.e. 100 cm from the camera, the low latency in terms of computation time is still upheld. Notably, the computation time for the Aruco marker is less than that of the AprilTags, but the standard deviation is much greater. From this experiment, it can be concluded that our proposed marker detection system can detect and localise marker in less than 10 ms to meet the requirement of high-speed real-time vision-based systems .

Conclusion

This paper presents a fiducial tracking system which is designed to track moving targets. The proposed fiducial in our tracking system consists of concentric circular patches. A comparison study of our proposed system with other standard marker detection systems shows that our method can detect the marker accurately and have much lower computational latency compared to the standard markers. Our system can detect fiducials in less than 10 milliseconds which make it suitable for high-speed applications such as moving object tracking.

Acknowledgement

This work is supported by the Australian Mathematical Science Institute (AMSI) and Ocular Robotics Pty Ltd.

References

[1] Sung, C., Lee, S. H., Kwon, Y. M., & Kim, P. Y. (2016, January). Fast and Robust 3D Terrain Surface Reconstruction of Construction Site Using Stereo Camera. In *The 33rd International Symposium on Automation and Robotics in Construction*, pages 19-27, Auburn, USA, 2016.

[2] Wen, M. C., Yang, C. H., Chen, Y., Sung, E. X., & Kang, S. C. A Stereo Vision-Based Support System for Tele-Operation of Unmanned Vehicle. In *The 33rd International Symposium on Automation and Robotics in Construction*, pages 19-27, Auburn, USA, 2016.

[3] Yu, Y. -H., Kwok, N. M., and Ha, Q. P. Color tracking for multiple robot control using a system-on-programmable-chip. *Automation in Construction*, 20(6): 669 – 676, 2011.

[4] Azar, E. R. Active Control of a Pan-Tilt-Zoom Camera for Vision-Based Monitoring of Equipment in Construction and Surface Mining Jobsites. In *The 33rd International Symposium on Automation and Robotics in Construction*, pages 121-131, Auburn, USA, 2016.

[5] Feng, C., Dong, S., Lundeen, K. M., Xiao, Y., & Kamat, V. R. (2015, January). Vision-based articulated machine pose estimation for excavation monitoring and guidance. In *The 32rd International Symposium on Automation and Robotics in Construction*. pages 1-9, Oulu, Finland, 2015.

[6] http://www.ocularrobotics.com/technology/

[7] Bruce, J., & Veloso, M. Fast and accurate vision-based pattern detection and identification. In *The 2003 IEEE International Conference on Robotics and Automation*, pages 1277-1282, Taipei, Taiwan, 2003.

[8] http://docs.opencv.org/2.4/doc/tutorials/calib3d/camera_calibration/camera_calibration.html

[9] Garrido-Jurado, S., Muñoz-Salinas, R., Madrid-Cuevas, F. J., & Marín-Jiménez, M. J. (2014). Automatic generation and detection of highly reliable fiducial markers under occlusion. *Pattern Recognition*, 47(6): 2280-2292, 2014.

[10] Olson, E. AprilTag: A robust and flexible visual fiducial system. In *The 2011 IEEE International Conference on Robotics and Automation (ICRA)*, pages 3400-3407, Sanghai, China, 2011.

Photogrammetric Models using Oblique Aerial Imagery for Construction Site Surveying

Anoop Sattineni[a] and James Hindman[a]

[a]McWhorter School of Building Science, Auburn University
E-mail: , anoop@auburn.edu, jwh0057@auburn.edu

Abstract

Photogrammetry is being used in several industries for creating digital 3-dimensional models of objects. Oblique aerial images are taken from an airplane or spacecraft. The technique involves taking multiple pictures of the same site from several angles, thus allowing interpretations to be made regarding surface elevations. The resulting images can be stitched together to form a topographical map using photogrammetric software. An experiment was conducted in this study to examine the variance between topographic maps created from oblique aerial imagery and traditional land survey methods. In this research, 165 high-resolution oblique aerial images were used to create a topographical surface map for a potential construction site. A land survey using traditional methods was conducted to produce a topographical map of the same area. Comparisons were made between the two topographical surfaces. Results from the experiment showed significant variance between the two topographical surface leading to the conclusion that oblique aerial images cannot be used to replace traditional land surveys to create topographical maps for construction purposes.

Keywords –

Photogrammetry; Oblique Aerial Imagery; Topographical Surface; Site Survey

1 Introduction

Photogrammetry is defined as, "the science and technique of interpreting and evaluating the form, dimension and position of objects by analyzing and measuring images of them" [1]. The photogrammetric process produces a three-dimensional model of an object on a computer screen that is used to make measurements. Photogrammetry is currently used for topographic mapping, geographic information systems, and civil engineering studies. Photogrammetry allows large amounts of data to be recorded, so that it can be accessed at any time to be analyzed for dimensional purposes [2]. Currently, digital terrain models are generated from two-dimensional drawings. These digital terrain models are then used in collaboration with global positioning system enabled earthwork machinery. Photogrammetry technology can save time in the creation of the digital terrain model creation process [3]. The aim of this research is to explore whether aerial photogrammetric modeling can be used for surveying potential construction areas.

Issues with planning and organization are very common occurrences before construction work begins. In order to manage a construction project successfully, accurate information about a potential site is required [4]. At the beginning of the construction process, a contractor needs topographic information for many purposes. These including the implementation of flood prevention measures such as silt fences, to determine the earthwork related quantities, durations of activities, site work costs, etc. Initial surveys of construction sites conducted by surveying engineers cost contracting companies both time and money [5]. If a construction project needed to start immediately, as an alternative, oblique aerial images could potentially be employed to "survey" the site allowing the construction project to begin at a more rapid pace [6]. However, oblique aerial imagery requires further investigation before being used for this purpose.

Oblique aerial imagery and drone imagery can be used to generate a three-dimensional (3D) photogrammetric topography model of a particular area. Such a model provides topographic information that two-dimensional (2D) aerial images cannot. This extra dimension facilitates the creation of custom topographic maps of construction sites. Such maps can be used to identify areas where storm water runoff may occur so that silt fences and other mitigations for erosion control may be placed [7].

For this method to be widely adopted by contracting companies, the accuracy of photogrammetric topography models must be tested [8]. In this research, oblique aerial photographs for a city were used to isolate a potential construction site. Photogrammetry technology was used to create a digital terrain model for the site. The photogrammetric topographical model of the specific area derived from the oblique aerial images was compared to a digital topographical model recorded using traditional land survey techniques of the same area.

2 Literature Review

Photogrammetry was most likely invented in the 1420s during the Italian Renaissance when painters would represent three-dimensional scenes in a two-dimensional medium while creating the sensation of depth. Today the goal of photogrammetry is to take two-dimensional images from different perspectives and create a three-dimensional representation of the same space. The actual term photogrammetry began in 1867 when Albrecht Meydenbauer started the process of measuring photographs to create architectural surveys, which he called "photogrammetrie" [1]. Meydenbauer was a building surveyor for the Prussian government, and his first project was to document a cathedral. While conducting this survey he almost slipped and fell to his death. Because of this accident, he decided to start investigating other ways to record the dimensions of the cathedral. This investigation led him to the idea of taking measurements of the building with pictures. He quickly found that these measurements were very accurate [8]. There are two types of photogrammetry namely, 'Terrestrial' and 'Aerial'. Terrestrial photogrammetry uses images taken from the ground, whereas aerial photogrammetry uses images that are taken from a balloon, plane, or spacecraft. Between 1850 to 1900 plane table, photogrammetry was utilized to take measurements of distances and angles of the elevations of object. A plane table provided a stable surface to create field drawings and maps [1]. In the subsequent sections, a brief discussion of the current uses of photogrammetry technology is included.

2.1 Mapping and Digital Terrain Modeling

One of the more common applications of photogrammetry is by departments of transportation for mapping purposes. These maps are simply photogrammetrically stitched two-dimensional maps for department of transportations to map out all roads and highways within their state. Commonly a geographic information system (GIS) is created to store and analyze the spatial and geographical data collected throughout the mapping process. In several states in the United States, these GIS databases are often accessible to the public through State Department of Transportation websites.

Digital Terrain Models (DTM) are three-dimensional models that show the changing slope and topography of the surface of a specific area. These DTM's originate from photogrammetric satellite images taken by departments of transportation, which are then combined to create continuous three-dimensional models with an accuracy of 10 meters. DTM's can also be created using oblique aerial photographs, Light Detection and Ranging (LIDAR) and direct contour mapping using traditional surveying methods. Highway construction companies utilizing automated global positioning system enabled heavy machinery are using DTMs for grading purposes [3]. These DTMs also provide construction managers with accurate earthwork volumes before the start of a highway project [3]. In the future DTM's created with photogrammetry may be used for this same exact purpose to cut down on man-hours and improve accuracy of measurements.

2.2 Highway Planning, Design and Maintenance

Photogrammetry is also being employed by departments of transportation for planning and designing future highway corridors and interchanges. Photogrammetry can be used to determine not only the topography of a given area but also can provide useful information for traffic engineering. The user can see urban and suburban areas and how populations are growing in certain areas. This information can show not only where a future highway will not disturb these areas but also where they will be most needed in the future [9].

Departments of transportation employ these photogrammetric maps to determine where road improvements must be made and to track maintenance of state highways. In Budapest, a terrestrial photogrammetric process was explored to examine a road's roughness, in combination with the International Roughness Index. Digital surface models allow determination of where roads must be resurfaced [10]. Photogrammetry is used by departments of transportation to map out right of ways for highway boundaries [9]. It is also utilized to map out future right of ways in a similar manner to highway planning and design.

2.3 Environmental Planning

The United States Environmental Protection Agency [11] states, "Storm water runoff is rain or snowmelt that flows over land and does not percolate into the soil". This runoff is prevalent during construction activities as well due to vegetation removal, leaving the soil unprotected from rainfall. If a construction site is not properly protected from soil erosion, it can lose 35–45 tons of sediment per acre per year [11]. Construction activities can cause storm water runoff to have adverse effect to any water system within its vicinity. This sediment is the main storm water pollution coming from construction sites, and the resulting turbidity prevents sunlight from reaching aquatic plants, kills fish, and destroys their spawning areas. Attention to erosion control is critical when: clearing, grading, excavating, working around un-stabilized areas, during paving operations, demolition and debris disposal, dewatering operations, drilling and blasting, material delivery and storage, and landscaping. Longer and steeper slopes on a construction site suffer the most from erosion [11]. The DTM's created using photogrammetric models can be used for identifying potential locations to install abetments to prevent soil erosion [18].

2.4 Oblique Aerial Photography

Aerial photogrammetry is created using photographs of the earths' surface taken from cameras mounted from an aerial object such as an airplane. Traditionally vertical camera configurations are used which provide an overhead view of an area. Oblique aerial imagery is taken from a high altitude at a position that is neither completely horizontal nor vertical but rather at an angle. These oblique aerial images are used in photogrammetry to create three-dimensional (3D) models of object. These images can be used to create three-dimensional images because they capture all sides of an object due to the differing angles captured by the various cameras. This type of imagery is used most commonly for three-dimensional urban mapping [1]. Currently oblique aerial imagery is used in the United States Geographical Information System. The data for provides local authorities and developers better information about an area than can be inferred from a two-dimensional image. This new data was obtained using eight cameras all of which were positioned at a forty-five degree angle facing different directions. The images were used for the purposes of emergency and community planning [13]. These oblique aerial images also can be used to measure distances, coordinates, and elevations within a Digital Terrain Model (DTM). "Measurements and mapping can be done where it is needed, field work can thus be avoided" [19]. The market place for these types of images is transitioning from "aerial mapping" to "aerial surveying."

There are two different types of oblique imagery: High angle oblique and low angle oblique. The primary difference between the images is that in high angle oblique the horizon is captured in the photograph, whereas in low angle oblique the horizon is not. An advantage of oblique imaging is objects that are unrecognizable in conventional nadir (vertical) aerial photos can be identified in oblique aerial photos, and a much larger area can be captured in each image.

Two important elements to consider in evaluating aerial imaging are distortion and displacement. There are several types of distortion including the curvature of the earth and issues with camera and lens function [14]. While distortion is to be avoided, displacement is good for observing the topography of an area because it allows for stereo viewing, height measurement, and topographic mapping.

The company Pictometry® developed the first system that used a digital five-camera oblique aerial photography system. An example of a five-camera system is illustrated in Figure 1. Four cameras point in different directions (North, South, East, and West) while one camera takes vertical images that are connected to their geographical location via Global Positioning System (GPS) [16]. These cameras are usually gyro-stabilized to ensure photo quality during flight [1].

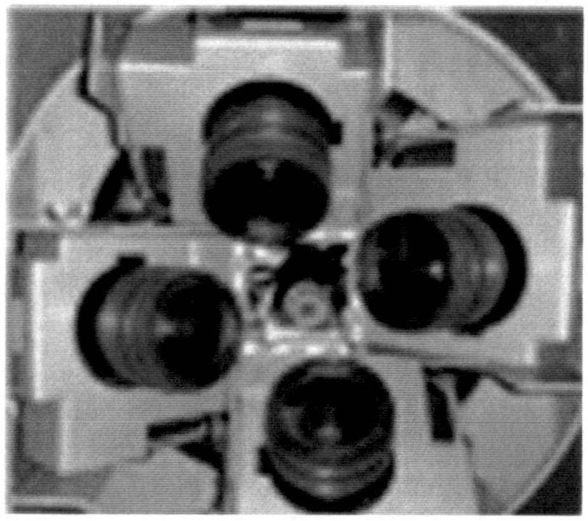

Figure 1: Five camera oblique aerial system [15].

3 Methodology

Experimental research methodology was chosen in the conduct of this research. The one-tailed null hypothesis for this research is H_A: The oblique imagery photogrammetric model will be accurate enough for surveying potential construction sites. The one-tailed, alternate hypothesis for this research is H_A: The oblique imagery photogrammetric model will not be accurate enough for surveying potential construction sites.

The experiment compared a photogrammetric model of a potential construction site with a digital terrain model created using traditional land survey. The photogrammetric model was created for a potential construction site using oblique aerial imagery obtained from the local authorities. The oblique aerial photographs were obtained from the local planning department, which were donated by Stellacore Corporation. Five 65-Megapixel cameras on the plane captured the oblique aerial photos. Simultaneously, researchers conducted a traditional land survey of the same potential construction site. The traditional land survey was created using contour points and transformed into a digital terrain model. The photogrammetric model was created using Photoscan software. The digital terrain model was created by importing contour points in to Autodesk Revit software.

The analysis involved a comparison of the terrain created by traditional land survey and the photogrammetric model. The boundary for each terrain model was adjusted to ensure that the same area was being compared in both instances. The models were imported into Earthworks, a quantification software for site excavation. The terrain model created from traditional site survey methods was set as the control and the photogrammetric model was compared to the control model. Sections were cut at pre-determined intervals to compare variation in elevation between the terrain models.

4 Results

One hundred and sixty five pictures were isolated from the oblique aerial photographs. These photographs were used to create a model of the general area of the potential construction site, shown in Figure 2.

Figure 2: Photogrammetric model created using Photoscan software

The site was cropped further to a smaller area shown in Figure 3. This area in the photogrammetry model is composed of a 2.6 million point dense cloud, encompassing just over .5 acres. This resulting model was used to put stakes in the ground to signify the area that must be surveyed using traditional methods.

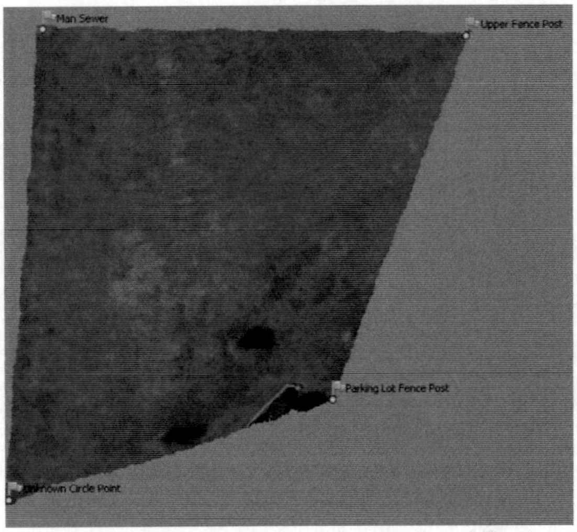

Figure 3: Photogrammetric model of potential construction site

The next set of data collected was the total station survey of the potential construction site, as shown in Figure 4. Leica total station equipment was used to collect elevations at 200 points. The four reference points at the stakes were measured first and were coded as "Bench Mark," while two hundred additional points were taken at fifteen-foot intervals within the boundaries of the reference points. These additional points were coded as "Topo."

Figure 4: Traditional topographical land survey of potential construction site

The data from the traditional survey was used to create a topographical map in Autodesk Revit. The data from the photogrammetric model was also imported into Revit to create a separate topographical site map. The data showed some variance between the two maps when it was discovered that a tree on the site was distorting the photogrammetric model. The site from both models was further cropped to eliminate the distortion. A side-by-side view of the two models is shown in Figure 5.

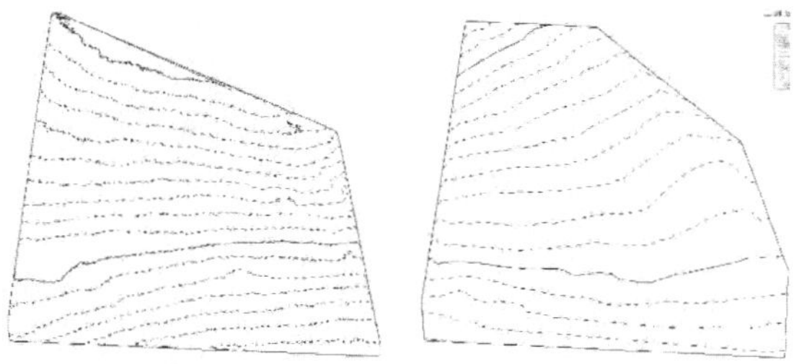

Figure 5: Photogrammetric model contour map (left) and traditional land survey contour map of potential construction site

Two different techniques for comparing the differences between the topo surfaces were used to analyze the data. First, both models were compared in Earthwork software to find how they differ in a cut and fill amount and three-dimensional visual inspection. Second, the models were overlaid and section cuts were made to measure variance in elevation. The overlaid models in Earthworks is shown in Figure 6.

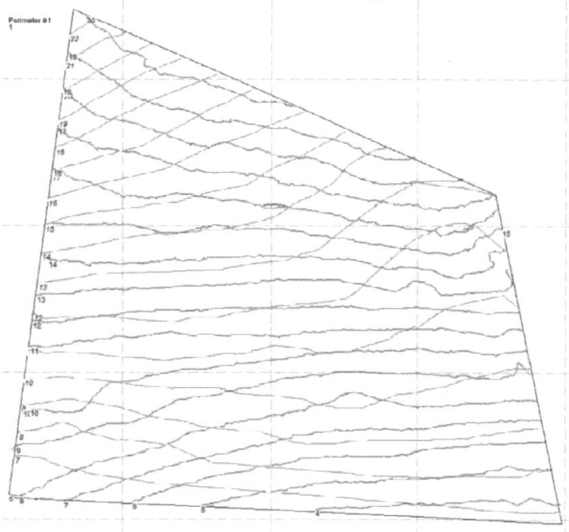

Figure 6: Overlay of contour maps in earthworks software

The total station contour lines are depicted in green while the photogrammetry contour lines are shown in red in Figure 6. For this research, the total station survey was used as the control and given the distinction of the "existing" surface in the software. The photogrammetry surface model was classified as the proposed surface. The software measured the amount of cut and fill necessary to make the total station topographical surface have the exact same as the proposed surface.

```
EarthWorks Pro, Site Excavation Report        Da

Job Filename: FinalTopoCompare
Job Description: Untitled

Scale: 1":16.58'

Cross-section spacing: 0.50 FT

Soil Expansion:              0%
Soil Compression:            0%
Import Soil Compression: 0%

Cut Area:        10786.5 FT2
Fill Area:       12153.0 FT2
Work Area:       22965.3 FT2

Bank Cut:                              526.0 YD3
Expanded Cut:                          526.0 YD3
Bank Fill:                             652.7 YD3
Topsoil Stripped:                        0.0 YD3
Topsoil Replaced:                        0.0 YD3

Import:                                126.7 YD3
```

Figure 7: Earthworks data output

The three most notable numbers shown in Figure 7 are on the 'Bank Cut', 'Bank Fill', and 'Import' numbers. The Bank Cut of 526.0 cubic yards reveals the amount of cubic yards that will need to be removed from certain areas of the surface. In contrast, the Bank Fill of 652.7 cubic yards shows the amount of dirt that is needed in other areas. The 'Import' line of 126.7 cubic yards denotes that a larger amount of earth will need to be added than will be removed to make the existing surface have the same contours as the proposed surface. In order to make the existing surface take on the same terrain as the existing shape, the area would need to have 126.7 cubic yards of dirt imported to the site. This amount of dirt would be the equivalent of about twelve to thirteen truckloads of earth.

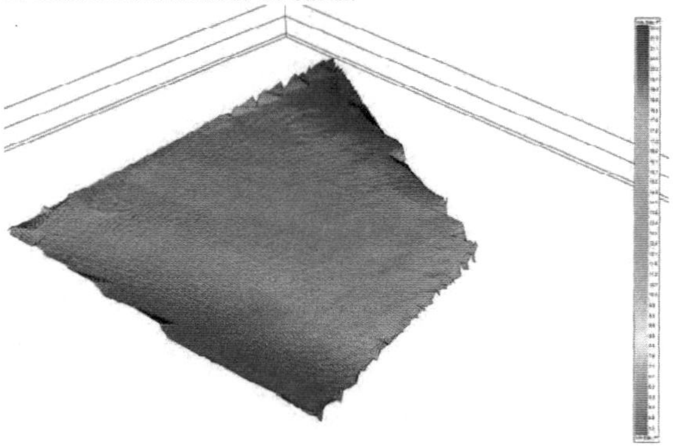

Figure 8: 3D Digital terrain model from photogrammetric model of test site

A visual inspection of the two topographical surveys indicates similarities and differences in the topographical surfaces shown in Figure 8 and 9. The green in the figures represents flat surface, orange represents high elevation and blue represents low elevation. The color maps support the evidence from the cut-fill report.

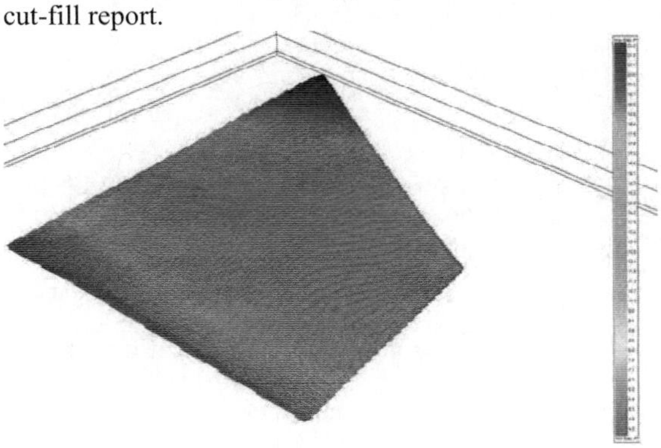

Figure 9: 3D Digital terrain model from traditional land survey of test Site

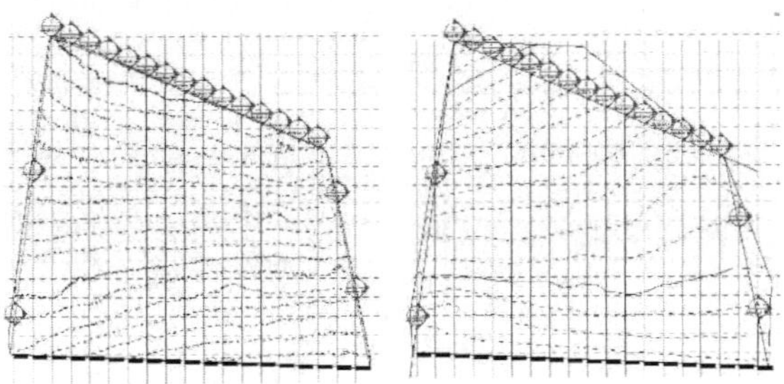

Figure 10:Cross section analysis of topographical surfaces

The second method used for comparing the variations between the two site surveys is a cross section analysis. Both surveys were placed on a ten-foot-by-ten-foot grid as shown in Figure 10. This was done to check if small areas of the site were giving distorted results in the cut-fill analysis. Figures 11 and 12 show an overlay of two topographical surface. Figure 11 represents the two surfaces overlaid with minimal variance whereas Figure 12 represents an overlay showing significant variance.

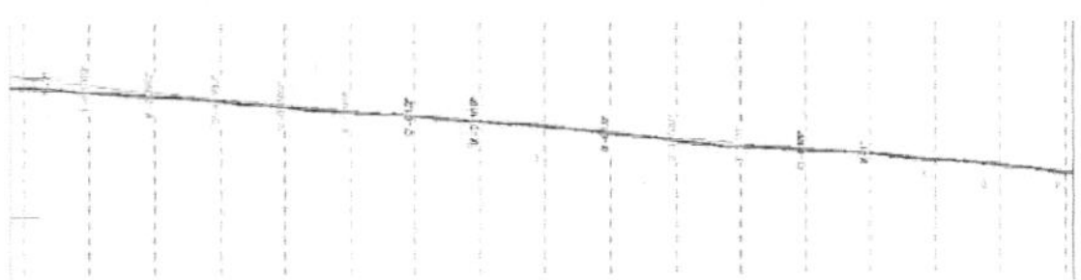

Figure 11: Representative cross-section with minimal variance between the two topographical surfaces

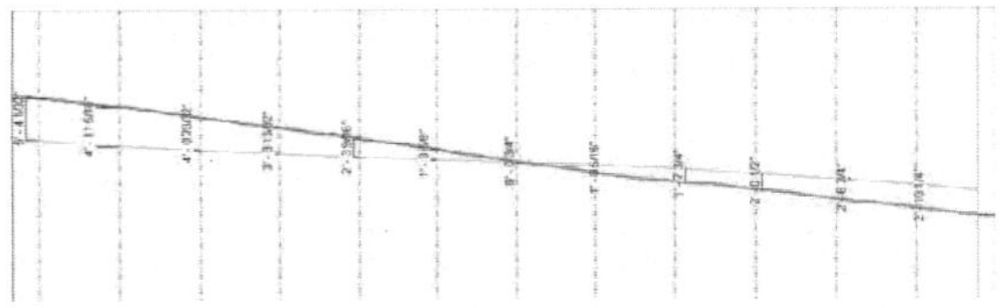

Figure 12: Representative cross-section with significant variance between the two topographical surfaces

Nineteen cross sections were placed vertically ten feet apart across both survey areas, and seventeen similar reference planes were placed horizontally at ten feet apart. These cross sections and reference planes intersected within the survey area at 229 points. At these points, measurements were taken of the elevation difference between the two topographical surfaces. Once all variations were measured at each intersection the distances were compiled in a spreadsheet and analyzed. The results of the average variance between the two topographical surfaces at each of the vertical nineteen sections is shown in Figure 13.

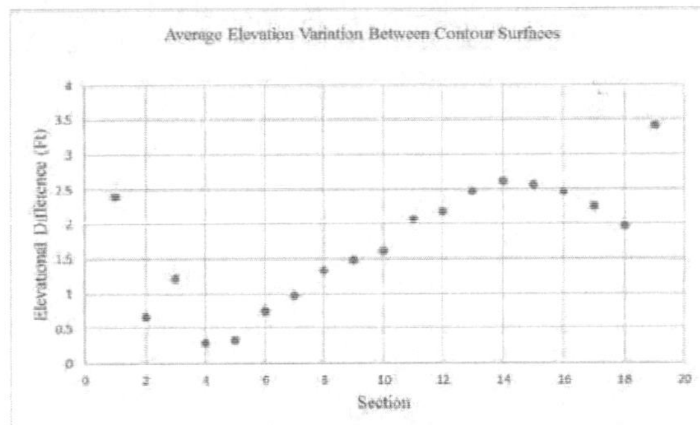

Figure 13: Average variance between the two topographical surfaces

The lowest average variance at a given section was 0.33ft and the highest variance was 2.56ft, as shown in Figure 13. The average variance in topographical surface between all points was 1.54ft.

5 Conclusions

A key question for this research was to answer, "How does the accuracy of a photogrammetric topography model created from oblique aerial images compare with topography data from traditional land surveys?" This question was answered by creating the total station and photogrammetric elevation models and then assessing the variations between the two models. The models were compared by calculating the amount of cut and fill necessary to make the total station topo-surface equal the photogrammetric surface. Although the Import value of 126.7 cubic yards would be a minimal amount of earth to bring into a site just over a half acre (13 truckloads) one must take a step back and observe this quantity with a broader perspective. The cut quantity for this area would be 526.0 cubic yards of earth and the fill would be 652.7 cubic yards. Although the overall import would be a relatively small amount of earth, the cut & fill quantities are both about 5 to 6 times as much. This research is not concerned with the actual earthwork estimate for making one topo surface similar to the other; the focus is on the variation between the two planes. For this reason, it is more important to focus on the cut and fill values as a sum. Instead of subtracting the two values from one another, they should be added together. This would result in a total amount of earth to be moved: 1178.6 cubic yards. This amount of earth would be the equivalent of about 118 truckloads of earth. Compared to the earlier 13 truckloads of earth, these 118 loads is a very large amount of earth to have to move in order to make the two-topo surfaces equal to one another.

The other way in which Key Question 1 was analyzed was by taking 19 similar section cuts of both topo-surfaces and quantitatively comparing their variations. After comparing the surfaces at 229 points it was found that the two models varied on average about 1.54ft at any point in the model. The average variation of 1.54ft between the models would be a large discrepancy to use for any application pertaining to the construction industry.

While observing the cut and fill map and the 3D elevation color maps, visual observations can be made about the variations between the two topography models. In the cut and fill map, it appears as if the models are most accurate in the middle of the survey area while the corners require the most cut and fill and therefore are the least accurate areas. This correlation might be because the middle of the area has less change in elevation than the corners of the same area. In the 3D elevation color maps, the colored areas appear to be similar but not the same. In addition, the total station maps seems to be smoother on the surface while the photogrammetric map is jagged and more uneven. The photogrammetric map shows a slightly broader range of elevation change than the total station surface.

In conclusion, the total station map and the photogrammetry surface are very similar in nature and share many characteristics; however, the two maps are not the same. The results indicate that a photogrammetric topographical surface map cannot be used to create a realistic topographical surface map for construction purposes. The alternative hypothesis for the experiment was proved true in that "*The oblique imagery photogrammetric model will not be accurate enough for the purpose of surveying potential construction sites*".

References

[1] Redweik, Paula, 2012. Photogrammetry - Springer. Retrieved from http://link.springer.com/chapter/10.1007/978-3-642-28000-9_4/fulltext.html

[2] Zhou, L. (2010). Feasibility of Close-Range Photogrammetric Models for Geographic Information System. Retrieved from http://inspirehep.net/record/917327/files/slac-tn-11-019.pdf

[3] Zirbel, C., 2014. Using Digital Terrain Models to Improve GPS-Guided Road Construction : CFIRE. Retrieved September 10, 2015, from http://www.wistrans.org/cfire/2014/07/using-digital-terrain-models-to-improve-gps-guided-road-construction/

[4] Bayrak, T., & Kaka, A. (2004). Evaluation of digital photogrammetry and 3D CAD modelling applications in construction management. In 20th Annual ARCOM Conference, Heriot Watt University, Edinburgh, September (Vol. 1, pp. 613–619). Retrieved from http://www.arcom.ac.uk/-docs/proceedings/ar2004-0613-0619_Bayrak_and_Kaka.pdf

[5] Kent, Gary R., 2005. Land Surveying Services. aiab089256.pdf. (n.d.). Retrieved from http://www.aia.org/aiaucmp/groups/aia/documents/pdf/aiab089256.pdf

[6] Stellacore Solutions, 2014. | Lead'Air Inc. and Stellacore Solutions LLC join forces to improve the fidelity and efficiency of oblique aerial systems/processes. Retrieved November 24, 2014, from http://stellacoresolutions.com/News

[7] Verhoeven, G. (2011). Taking computer vision aloft – archaeological three-dimensional reconstructions from aerial photographs with photoscan. Archaeological Prospection, 18(1), 67–73. http://doi.org/10.1002/arp.399

[8] Albertz, J. (2007). A Look Back. Photogrammetric Engineering & Remote Sensing, 505.

[9] Federal Highway Administration, 2014. Aerial Surveying – Design. Retrieved September 10, 2015, from https://www.fhwa.dot.gov/design/g601205.cfm

[10] Kertesz, I., Lovas, T., & Barsi, A. (2007). Measurement of road roughness by low-cost photogrammetric system. International Archives of

Photogrammetry, Remote Sensing and Spatial Information Sciences, 36(5/C55), 4.

[11] EPA, 2007. McKemmish, S., Acland, G., Ward, N., & Reed, B. Developing Your Stormwater Pollution Prevention Plan. Retrieved from http://journals.sfu.ca/archivar/index.php/archivaria/article/viewArticle/12715

[12] Carpenter, T. (2006). Features Vol. 50 no 2 - Proper Silt Fence Installation : Land and Water Magazine. Retrieved November 24, 2014, from https://www.landandwater.com/features/vol50no2/vol50no2_2.html

[13] Ignasiak, D., 2013. County departments begin using imagery from aerial survey. Dispatch, The (Lexington, NC). Retrieved from http://search.ebscohost.com/login.aspx?direct=true&db=buh&AN=2W6847197297&site=bsi-live

[14] University of California, in Santa Barbara. (n.d.). Geometry of Aerial Photography. Retrieved November 5, 2015, from http://www.geog.ucsb.edu/~jeff/115a/lectures/geometry_of_aerial_photographs_notes.html

[15] Petrie, G. (2009). Systematic oblique aerial photography using multiple digital cameras. Photogrammetric Engineering & Remote Sensing, 75(2), 102–107.

[16] Karbo & Schroth (n.d.). Oblique Aerial Photography: A Status Review. Retrieved from http://www.ifp.uni-stuttgart.de/publications/phowo09/140Karbo.pdf

[17] GeForce GTX TITAN Black, 2015. | Specifications | GeForce. (n.d.). Retrieved November 5, 2015, from http://www.geforce.com/hardware/desktop-gpus/geforce-gtx-titan-black/specifications

[18] Lannoeye, W., Stal, C., Dinssa, E.G., Zenebe, A., Nyssen, J. and Frankl, A., 2016. The use of multi-temporal SfM-photogrammetry to monitor gully degradation: an example from the Ethiopian drylands. In 10th Symposium of Ghent Africa Platform (GAPSYM-10): Celebrating partnership with Africa: demonstrating the value of international collaboration on teaching and research.

[19] Hohle, Joachim, 2008. Photogrammetric Measurements in Oblique Aerial Images. Photogrammetrie, Fernerkundung, Geoinformation. Retrieved from http://vbn.aau.dk/ws/files/17874670/Photogr_Meas.pdf

Developing the GIS-Expert System for Investigating Land Use Designations

P. Limsupreeyarat[a], S. Kaewkeaw [b], and P. Charnwasununt[c]

[a,b] Department of Civil Engineering, Burapha University, Thailand
[c] Department of Civil Engineering, KhonKaen University, Thailand
E-mail: petcharatl@eng.buu.ac.th, green.cityhall@gmail.com, p.charnwasununth@hotmail.com

Abstract

In order to get the building construction permit, the building owners who intend to construct or modify their buildings must submit documents to the local government officers for approval. Typically, the government officers will review the submitted documents and check city planning and land uses designations in accordance with laws and regulations of each area. This work is burdensome and takes a long time so as to check documents. Moreover, it requires the officers who have enough expertise and comprehend in legislation for the considered building and area information. In case of the officers do not consider all associated regulations or conditions such as width of road in front of the proposed construction area or restricted areas, it will lead to incorrect results of land use and construction approval. Then, the owners of building and land will be affected by such mistakes. Therefore, the research aims to develop the prototype expert system which can support the local government officers and land owners for checking the land uses designations before submitting the building construction permit. This prototype system was integrated laws and regulations with GIS database, Google Map, for retrieving, analyzing, and providing the results via Internet Web Browser. The selected area is Pattaya city in Thailand because it is the one of the most famous cities which has a high level of growth rate of land development. The proposed system can conveniently and efficiently provide the accurate land use information to officers, investors, and also inhabitants.

Keywords –

GIS, Expert system, Land use

1 Introduction

To get the building construction permit, the building owners who intend to construct or modify their buildings must submit building documents to the local government authorities for approval. Typically, the government officers will review the submitted documents and check principle city planning and land uses designations in accordance with laws and regulations of each area. Checking city planning and land uses designation is the free of cost service which the government provides to residents because it is difficult for the ordinary people to seek and consider all related laws and regulations especially in the special areas. Only the government officers who have knowledge and experiences are capable to perform this task. In mentioned situations, the misunderstanding and human error may lead to the incorrect results and impact to the intended construction project.

The procedure for investigating the land use designation is as follows and shown in Figure 1. Total duration to carry out this procedure is approximately four working days.

1) Clients submit the request form for checking the land use designations to the local government authorities.

2) The government authorities obtain the information from the request form which is the identification number of land parcel, type of intended construction building, and name of adjacent road.

3) The government authorities check the location of this land parcel to identify the city planning zone which determine type of building that allow to construct in this area. In case that the building as clients' requirement is allowed to construct in this area, the government authorities have to verify the width of adjacent road, distance between this building and the road in front of the land parcel for checking the allowable height of the building. There are at least three important conformed laws and regulation that the government officers must investigate to define the analysis result of this process.

4) The government authorities conclude and inform the consideration result of land use designations and issue the investigation certificate to the clients.

However, this work is burdensome for the government authorities particularly in the high level of growth rate areas which have abundant client's requests and the clients also take a long time in order to continue their business. In addition, some government authorities do not have the officers who have enough expertise and comprehend in legislation for the considered building and area information. In case of the officers do not consider all associated regulations or conditions such as width of road in front of the proposed construction area or restricted areas, it will also lead to incorrect analysis results of land use and construction approval. Consequently, the owners of building and land will be affected by such mistakes.

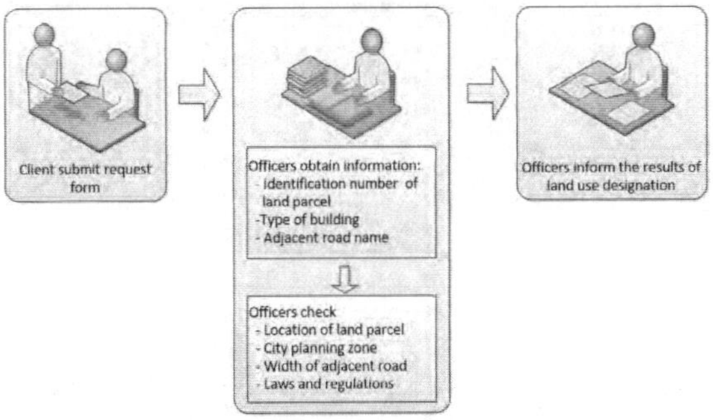

Figure 1. Procudure of land use designation checking

Nowadays, there are several implementations of information technology to assist the construction management including the examination of land use designation and role of expert system and geographic information system (GIS) significantly increases in the previous research. Expert system was applied in many study aspects such as land cover classifying [1-2] and land use planning [3-5]. In order to study the conversion from rural to urban of the city, data and information of land cover obtained from remote sensing technology was analyzed using expert system techniques for classification of land cover types. This technique was developed in one urban area (Phoenix) and then applied to another area (New Delhi). For land use planning, the expert system was developed to provide the customized expert advice of selecting land use practices and crops pertain to the considered agricultural area [3]. It also used for developing a hierarchical rule-based land use extraction for scrutinizing the structures and patterns of residential and commercial land use. However, it had to combine data and information from other technologies such as remote sensing imagery and geographic information system to achieve the efficient results [4-5].

As mentioned above, geographic information system similarly played important roles for construction industry. The applications of geographic information system in the previous research are as follows. Geographic information system was applied for developing a database in order to control the construction project development. This system was used by the local government authorities for building control permission. The study area of proposed GIS database is in Malaysia and the attribute data consists of land parcel, building information, the existing land development, planning information and development plans, utilities data, community facilities, transportation, environment and the socio-economy. However, the results of this study indicated that in order to implement this proposed system to the local government authorities they required the knowledge transferring process and the attitudes of them related to new technology must be changed. Additionally, this developed database which can be used by many involved parties as a reference point can provide the efficient and transparent service to the public [6]. Another study implemented geographic information system for analyzing and determining type of factory in the industrial estate. This research purposed the initial guideline for establishment of industrial estate. The considered factors including physical conditions, environmental potential, and economic potential was defined and data analysis was applied by using GIS process and Arc View [7].

Geographic information system can be used as tools for developing the approach for evaluating the evolution of coastal area in Canada before and during the implementation of laws and regulations associated with zoning designation. Changes in environmental aspects and georisks of this coastal area were analyzed using aerial photograph and geographic information system technology. The results of this study presented that there were obviously changes in land utilization of this area. The purposes to use land changes from fishery-based economy to tourism and the impact of transformation in human activities leads to increasing environment problems and risk of building along the coastal zone [8].

According to the advantages of expert system and geographical information system, this research aims to develop the prototype expert system incorporating with GIS to support the local government authorities and land owners for checking the land uses designations before submitting construction document to get building control permit. Type of building which clients intend to construct must be examined and demonstrated the land use analysis result. Moreover, laws and regulations which must be concerned and enforced in the prototype area are considered and provided to users of this proposed system.

2 Study Area and Related Laws and Regulations

2.1 Study area

This research selected Pattaya City in Chonburi

Province, Thailand as a tested area because it is the one of the most famous cities which has a high level of growth rate of land development. This area is a self-governing municipal area which locates in the east coast of the Gulf of Thailand. Total area of the city is 22.2 sq.km. This city is well-known as one of the biggest leisure and entertainment areas and there are plenty attractions both nature and nightlife. In addition, there are satisfactory utilities and convenient transportation. The registered inhabitants are 113,083 people but the non-registered residents are approximately 500,000 including the long-term expatriate visitors. Furthermore, the number of population trends upward every year. As mentioned reason, the demand of residential and commercial building numerously rises and leads to the increasing of building control permission. Presently, the numbers of construction project which submit their document to receive the building control license from local authorities are approximately 400 – 500 projects per year. The tested area was divided into three following subareas for selecting the sample land parcels which are 1) the typical area 2) the area along coastal and 3) the area located in the restricted area and near the beach. All of three subareas are shown in Figure 2 to Figure 4.

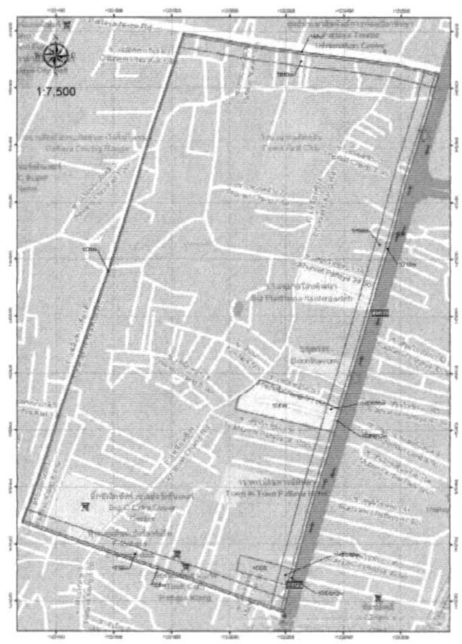

Figure 2. The typical area

2.2 Related Laws and Regulations

To develop the prototype system for investigating the land use designation, associated laws and regulations are required and considered. There are four major laws and regulations which normally uses in this tested area. They consist of building control act B.E. 2522(1979), Pattaya Principle City Planning, Regulations of Ministry of Environment and Natural Resources, and Agreements of Pattaya City and Royal Aide-de-Camp Department.

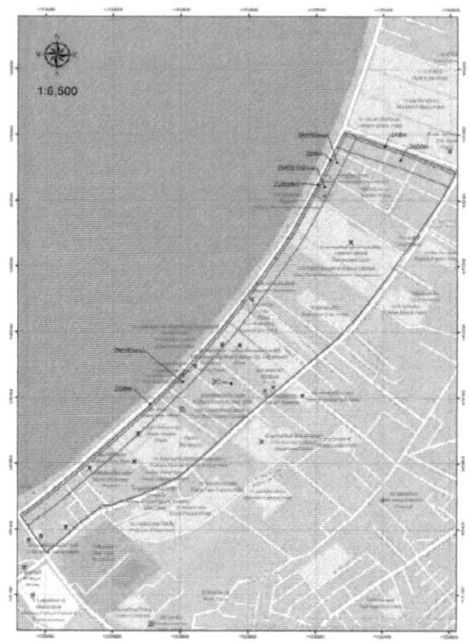

Figure 3. The area along coastal

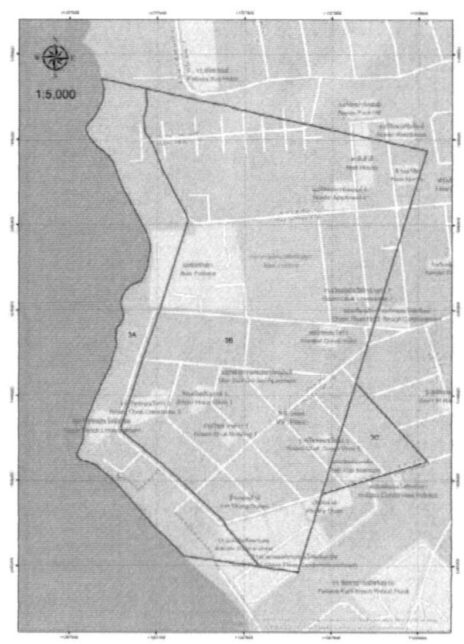

Figure 4. The area located in the restricted area and near the beach

3 System Design

The approach to search and examine the land use designation before submitting the document for building control permission traditionally was done by the local authorities. The incorrect results of this work will affect not only the clients but also the affiliation of the officers. To improve the effectiveness and efficiency of this work, the prototype system was proposed. There are many methods to develop the prototype expert system integrated with geographical information for example using the database of ArcGIS, the advantages of this method is ArcGIS is the existing geographical information database of the government authorities in Thailand and the developers can freely design and create services by themselves, however it requires ArcGIS server license which may influence to the existing system operation.

Another interesting method is developing using web mapping technology such as Google Maps which allows the web users to explore geographical data online and produce maps on demand. This web mapping service (WMS) can improve productivity and well support decision making process but there are some limitations of functionality of web mapping such as spatial analysis which requires the development of spatial decision support system (SDSS) [9]. In order to develop the expert system for this prototype system, the main important components which are a knowledge base and an inference engine were originated. The knowledge engineer had to interview the expert government authorities to obtain the information. Next, such information was organized and created the rules in the inference engine for solving the problems. In this research, a classical rule-based structure represented as an 'if-then' structure was applied due to the problems analysis of land use designation can be effortlessly formulated using this structure. Integrating the prominent capabilities of both GIS and expert system, the proposed system can manipulate, analyze, and present the suitability land use analysis. Figure 5 presents the components of this proposed prototype system with the land database acting as web mapping service spatial database. Three main components of the prototype system are knowledge base, land database, and system web server including the graphical user interface (GUI), the web mapping service, and the inference engine. The knowledge base contains the land use rule-based on the knowledge of existing expert regarding city planning and related laws and regulations and the predefined format rules are also stored to comply with the inference engine for providing the analysis results of land use designation. The land database of this system contains land information such as land parcel identification number, shape of land parcel, and location of land parcel. The users can input their requirement and receive the response via graphical user interface and analysis results were displayed using web mapping service via the web browser.

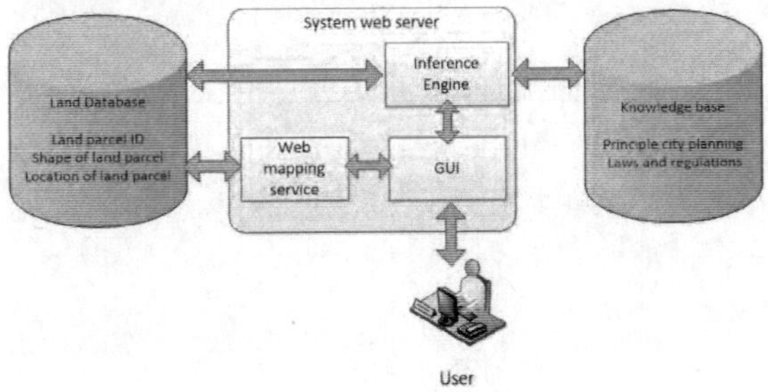

Figure 5. The components of the prototype system

4 System Implementation

The implementation of land use designation checking system which integrates the advantages of geographic information system (GIS) and expert system for assisting the government authorities and land owners in Pattaya City is presented in this section. The prototype system graphic user interface is shown in Figure 6. Users can input a land parcel identification number in the input box as presented in number 1 of Figure 6. Then the system will search the location of the land and display its shape and location via web browser application as demonstrated in number 2. The utilization of land and the information such as floor area ratio and open space ratio also presents to give the primary information to the users for example agricultural, commercial, residential, or industrial as portrayed in number 3. In case that users require to check the utilization of land for preparing the building control document, the users must select types of building which are house, row house, apartment, dormitory, hotel, commercial building, department store, or high rise building and also input the width of road in front of the land parcel as shown in number 4. Types of building were predefined in this system. Subsequently, the prototype system will use the rule-based algorithm for investigating that city planning zone of this land parcel, the distance from the adjacent road and from the reference point of mean sea level, and associated laws and regulations. The analysis result of land use designation as users' requirement will be presented as shown in number 5 accompanied with conformed laws and regulation which users must be concerned.

5 Conclusion

In order to solving the problems of land use

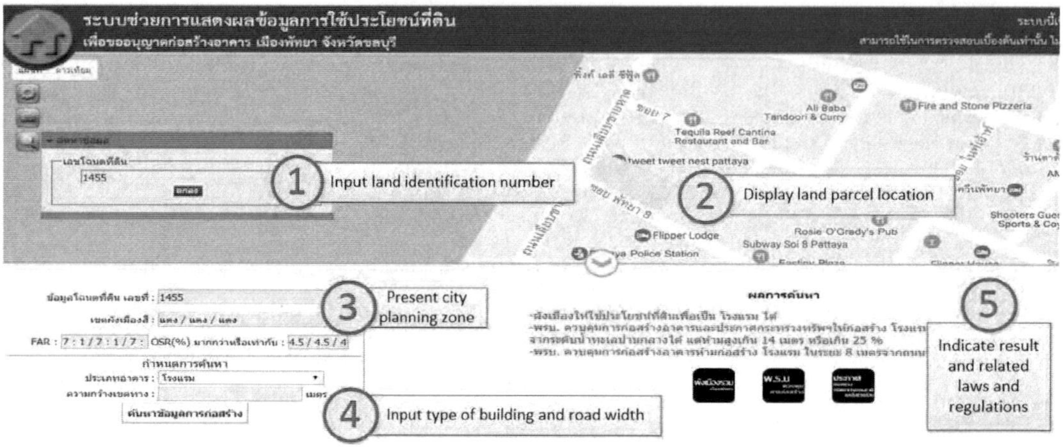

Figure 6. GUI of the proposed system

investigation which are error-prone of human and time consuming, the proposed system was developed using integration of expert system and geographic information technology(GIS). Web mapping service (WMS) which is Google Maps was applied in this research for manipulating and presenting the geographical data and information. However, the functionality limitation of this application requires supporting engine for the particular land use problems analysis. Thus, the rule-based algorithms were created by obtaining the knowledge and expertise of government authorities from the department of building control. This prototype system can conveniently and efficiently provide the accurate land use information to government authorities, investors, and also inhabitants for performing the building control permission process. In the future research, the information or 3D model of existing buildings on each land parcel should be created and used for new building feasibility study and environmental impact assessment.

Acknowledgment

This work was financially supported by the Research Grant of Burapha University through National Research Council of Thailand (Grant no. 36/2560)

References

[1] Stefanov, W. L., Ramsey, M. S., and Christensen, P. R., Monitoring urban land cover change: An expert system approach to land cover classification of semiarid to arid urban centers. *Remote Sensing of Environment*, 77(2001): 173-185, 2001.

[2] Wentz, E. A., Nelson, D., Rahman, A., Stefanov, W. L., and Roy, S. S., Expert system classification of urban land use/cover for Delhi, India. *International Journal of Remote Sensing*, 29(15): 4405-4427, 2008.

[3] Ponnusamy, T., Knowledge-based expert system for agricultural land use planning, Thesis, University of Adelaide, 2007.

[4] Beykaei, S. A., Zhong, M., Shiravi. S., and Zhang. Y., A hierarchical rule-based land use extraction system using geographic and remotely sensed data: A case study for residential uses. *Transportation Research Part C*, 47(2014): 155-167, 2014

[5] Beykaei, S. A., Zhong, M., Shiravi. S., and Zhang. Y.,Development of a land use extraction expert system through morphological and spatial arrangement analysis. Engineering *Applications of Artificial Intelligence*, 37(2015): 221-235, 2015.

[6] Yaakup, A., Johar, F., Sulaiman, S., Hassan, R., and Ibrahim, A. R., GIS and development control system for a local authority in Malaysia, *Habitat International*, 27(2003): 683-696, 2003.

[7] Prakthong, P., GIS application for industrial estate site selection: A case study of Chachoengsao Province, Master Thesis, Mahidal University, 2004.

[8] Drejza, S., Bernatchez, P., and Dugas, C. Effectiveness of land management measures to reduce coastal georisks, eastern Québec, Canada. *Ocean & Coastal Management*, 54(2011): 290-301, 2011.

[9] Niekerk, A. V., CLUES: A web-based land use expert system for the Western Cape. In *Proceedings of IEEE International Geoscience and Remote Sensing Symposium (IGARSS 2009)*, Volume: 3, Cape Town, South Africa, 2009.

Experiment on Mobile Light-section Device for Automatic Crack Inspection of Actual-size Tunnel Lining Concrete

Daisuke Inoue[a], Takao Ueno[a], Satoru Nakamura[a]

[a]Institute of Technology, Tokyu Construction Co., Ltd.
E-mail: daisuke.inoue at tokyu-cnst dot co dot jp

Abstract –

There are approximately 10,000 road tunnels in Japan. Based on the road law enforcement regulations of Japan, human visual inspections of the road tunnels are conducted once every five years. However, presently, the number of inspection engineers is insufficient and the cost incurred for the inspection is significantly high. In conventional crack inspection methods, engineers process many images off site. However, this method requires a lot of effort to identify the cracks and dirt.

We have studied an automatic detection method to determine the crack positions by using high-resolution texture/depth images that are acquired by using the mobile light-section method. In this report, we present the study outline and introduce the prototype of the devices developed. Furthermore, we present the results detected by our method using both concrete test samples and simulated actual-size tunnel lining concrete.

Keywords –
Inspection; Concrete; Depth-image

1 Introduction

The Japanese government decided to periodically inspection the road tunnel once every five years since 2014. In principle, the inspection should be performed by near visual inspection to the tunnel overall length. The results of the inspection and diagnosis were recorded and saved, and a decision was made to classify the diagnostic results of soundness on a unified basis. Against such a background, issues such as a shortage of engineers and lack of budget to inspect road tunnels are becoming apparent [1].

Close visual inspection of a road tunnel is conducted by inspection experts, who observes the surface of the lining concrete in the vicinity using an aerial work platform or the like. The experts measured and recorded the degree of the changes in the cracks and floats, steps, efflorescence, honeycombs, water leakage and the like. The representative method to support the experts is based on an image processing [2]. Often, the method needs off-line image process for a large amount of image data of the concrete lining surface. Because the common method of the image processing recognizes dark pixels with cord-like arrange as the cracks, a dirt or the like can be erroneously detected as a crack. Various methods have been made to prevent these false detections, but some challenges still remain for completely automatic crack detection.

In order to solve these problems, the automatic crack inspection device based on the light-section method have proposed and developed to support close visual inspection. The device is the one of the component of our robot system called "Variable guide frame vehicle" developed for traffic-free tunnel inspection [3]. In this paper, we describe an outline of the device; specifications of the prototype of the device; and the results of the performance verification experiments of the prototype using concrete specimens and simulated actual-size tunnel lining concrete.

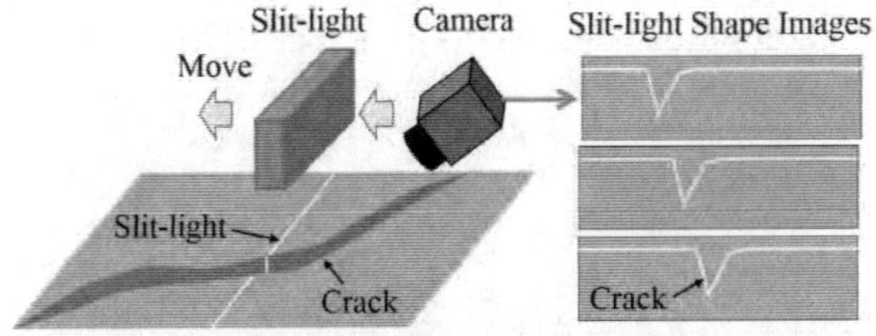

Figure 1. Concept of crack light-section device

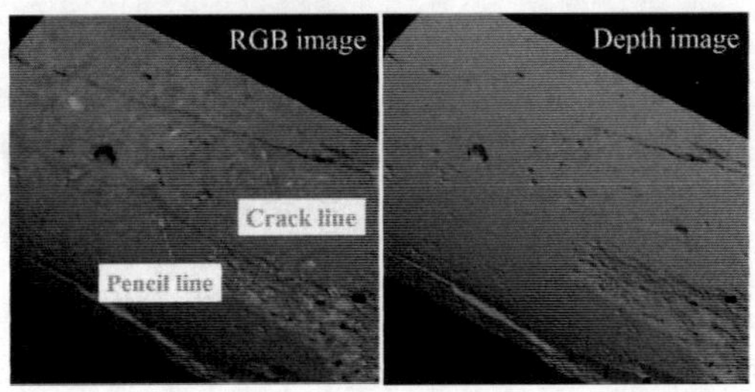

Figure 2. Tunnel inspection test by using our light-section device prototype

2 Mobile Light-section Device Developed

The authors have made it possible to distinguish between cracks and dirt by obtaining a three-dimensional shape of the concrete surface. In various methods to acquire the three-dimensional shape, a light-section method was adopted considering both the high resolution with sub-millimeter order and the portability for wide area inspection over kilometer order.

As shown in Figure 1, the light-section method acquires a three-dimensional shape by causing the slit-light and the area camera to move relative to an object for scanning, while maintaining a certain angle. It is based on the property that the slit-light deforms accordingly. By using the slit-light of a white LED, a color image (visible image) whose coordinates are coincident with the three-dimensional information can be obtained. Figure 2 shows an example of a color image and a depth image obtained by scanning the same object. In the depth image, it is possible to distinguish between real cracks and simulated cracks drawn with pencils.

The authors made a prototype of an automatic concrete crack inspection device that can implement the light-section method by combining the wheel and encoder, and scanning the lining concrete surface (Figure 3, Table 1). The prototype of the image processing software was configured as shown in Figure 4. An image is photographed according to the scanning distance, and the light-section processing is performed when the images are accumulated. Based on the generated depth image and color image, the cracks are automatically detected by the procedure shown in Figure 5, and the detection result is saved. The detection results are integrated for each joint of the structure, and used for the measurement of the crack length and the investigation of the closed cracks that has a risk of chip out to a road. The system was built so that these processes could be done onsite.

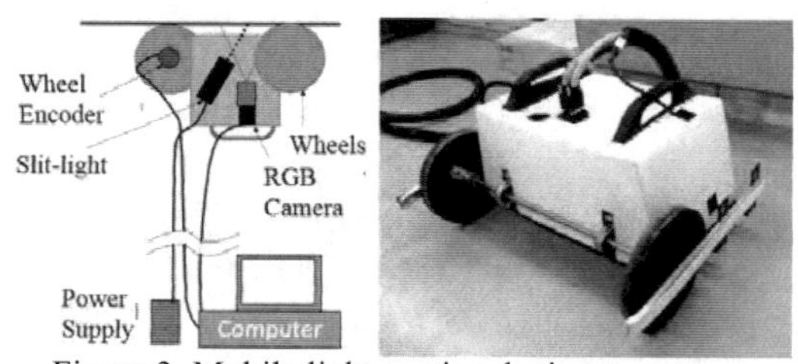

Figure 3. Mobile light-section device prototype

Table 1. Specification of the device

Item	Value	Unit
Size	W410 × L414 × H242	mm
Weight	6.0	kg
Image Resolution	0.1	mm/pixel
Image Width	200	mm

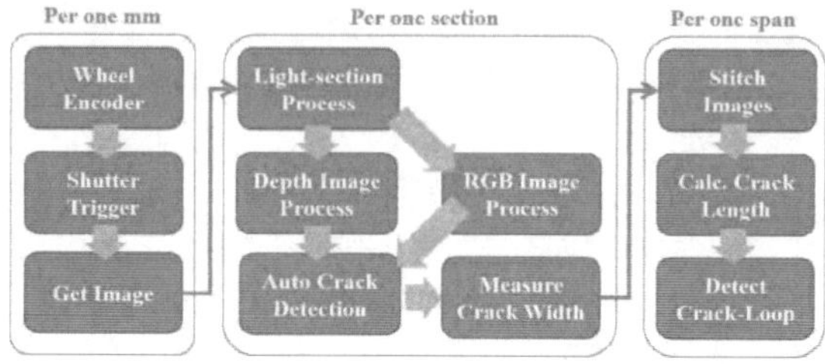

Figure 4. Crack inspection flow by the device

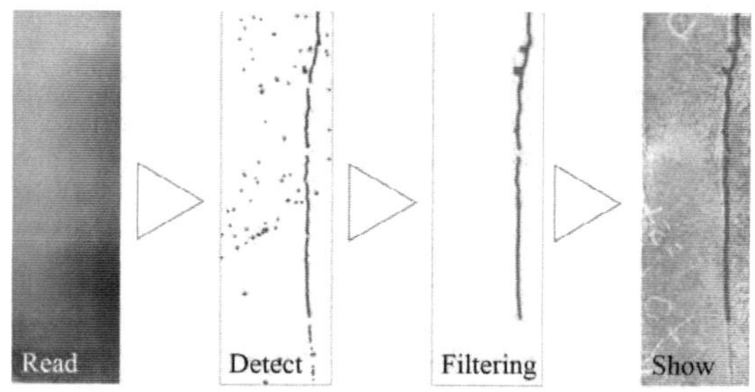

Figure 5. Flowchart of automatic crack detection

3 Test on Concrete Crack Samples

In order to verify the performance of the automatic inspection of cracks by this device, crack inspection of concrete test samples, which cracked owing to bending force, was performed. As shown in Figure 6, the simulated cracks were drawn with felt pen (black, brown and green) and chalk (white) near the actual cracks, and a depth image was obtained using the light-section method. Furthermore, the cracks were detected from the depth image by image processing.

For each of the 10 concrete test samples, the detection rate (the detected crack length dividing by the actual crack length) and the false detection rate (the number of erroneously detected pixels divided by the total number of detected pixels) is shown in Figure 7. For comparison, the results of crack detection using a commercial software and the color images of the concrete test samples are shown in the figure. The detection rate of the cracks using the proposed method was greater than 73% and the false detection rate was less than 8%, both of which were confirmed to be superior to those of the conventional method. The main cause of false detection was surface bubbles near the cracks. However, the misdetection of felt pen or chalk was 0%.

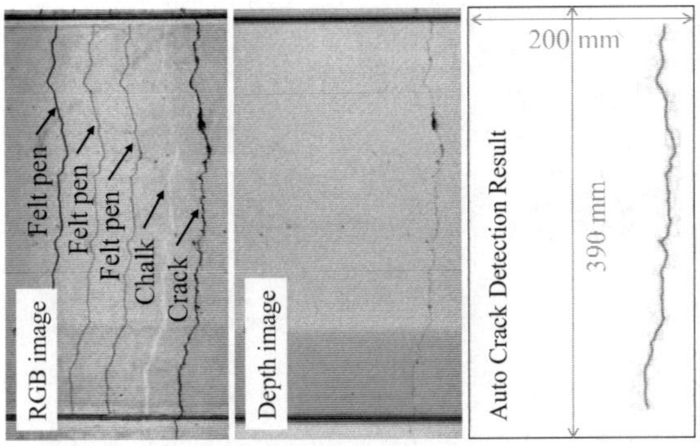

Figure 6. Automatic crack detection result example by using a concrete test sample with simulated dirt lines and a chalk line

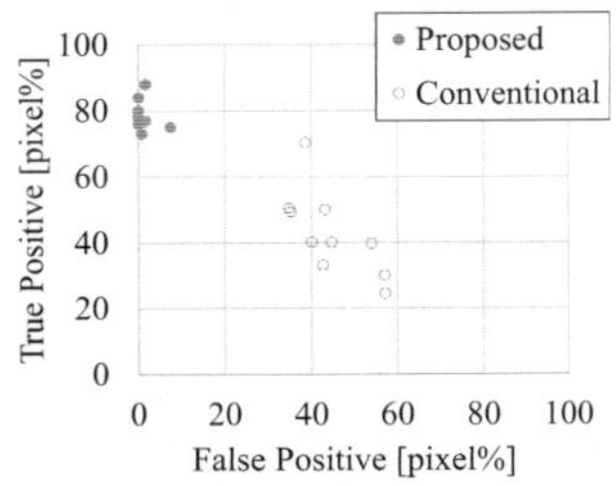

Figure 7. Automatic Crack detection accuracy by using 10 concrete test samples

4 Test on Actual-size Tunnel

The depth image and color image were obtained by scanning the lining concrete surface using the light-section method on the actual-size test tunnel of Japan's Construction Method and Machinery Research Institute (CMI). As shown in Figure 8, manual scanning was performed on an aerial work platform using the prototype of this system. The scan range was ~9.7 m in the circumferential direction of the arch part, as shown in Figure 9, and ~3.0 m in the axial direction. Considering the overlapping portion in the short axis direction, the effective scan width was 150 mm. 69 strips of image were taken and stitched as Figure 10.

Based on the obtained depth image, crack detection software is used to automatically detect the cracked parts from the images of each scan; the width and length of the crack are obtained by counting the detected pixels semi-automatically as shown in Figure 12. This was done by an operator specifying two points on the screen and automatically counting the pixels in a straight line connecting the two points. The width and length of the cracks thus obtained were compared with those obtained using the conventional method (close visual observation) by a skilled checker.

The measurement results of the crack width and the crack length are shown in Figure 13 and 14. When comparing the measured value using the automatic crack inspection device with the close visual inspection results obtained by skilled workers as shown in Figure 11, it was found that the crack width is 4% greater on average, and the crack length is 4% shorter than the values obtained by a skilled worker. The crack widths measured using our automatic method agreed with more than half of the values measured by a skilled worker. In some cases, the width was 0.2 mm greater than the value obtained by a skilled worker, although it was not any problems in practical use.

The crack detection rates are shown in Figure 12. The rate was over 88% when excluding the crack with a width of 0.6 mm. The false-negative crack detection occurred for the crack with an efflorescence (at a red dotted circle in Figure 12). The false-positive occurred due to the joint adjustment near a crack (at blue dotted circles in Figure 12) and concrete honeycombs at several places. Currently, we are investigating ways to distinguish the cracks and honeycombs more clearly using a cord-like structure of the cracks and a circularity of the honeycombs.

Figure 8. Tunnel inspection by using our device on an aerial work platform

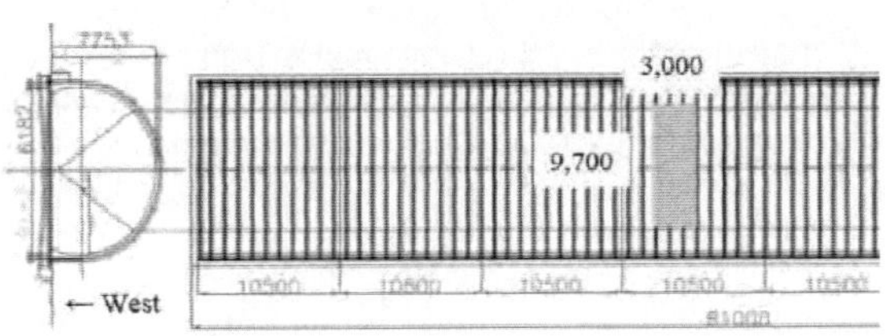

Figure 9. Target area for our visual inspection test in simulated actual-size tunnel of the CMI

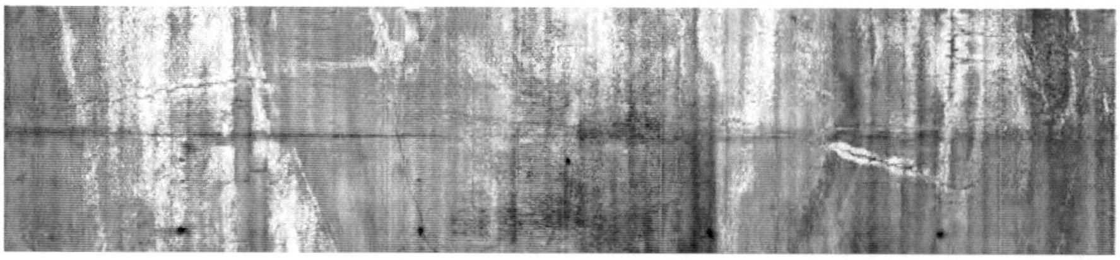

Figure 10. Panoramic image of a tunnel lining concrete of the CMI

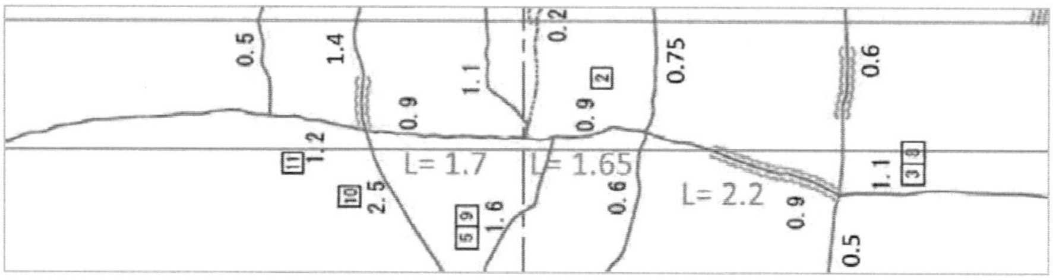

Figure 11. Inspection result by conventional method by inspection experts

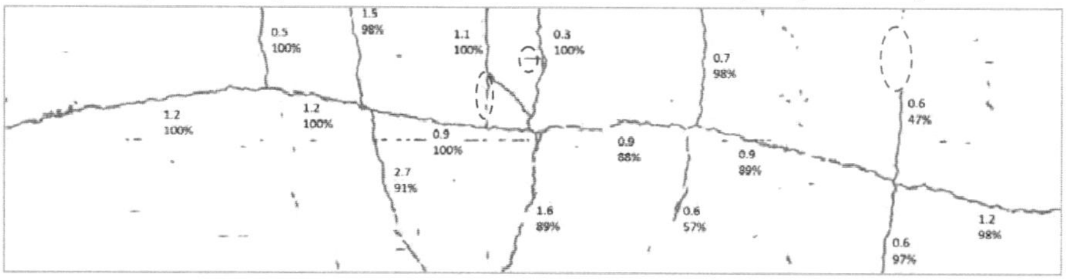

Figure 12. Auto crack detection result with crack width and crack detection rate

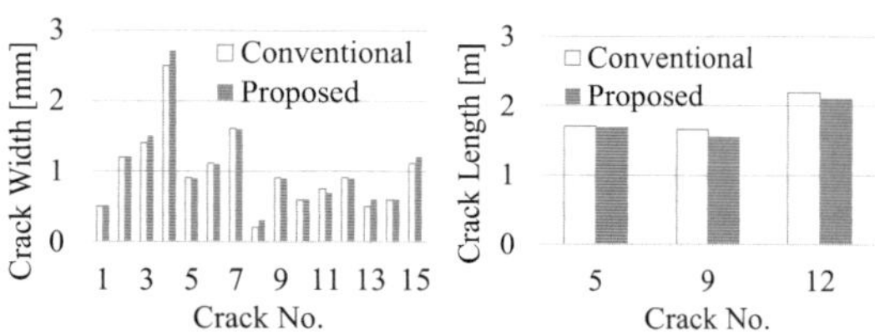

Figure 13. Comparison of crack width and length

5 Conclusions

The overview of the device that automatically detects the crack in the tunnel using the light-section method, and the prototype specifications were described. This device discriminates between cracks and contamination in the concrete test samples Next, we showed that crack detection using depth images is superior to over crack detection using conventional visual images. Furthermore, the report on the detection results of the width and length of cracks in an actual-size tunnel showed that the proposed technique has a crack detection performance equivalent to that of humans.

Currently, the authors are brushing-up the tunnel inspection method using this device for improving the detection accuracy of crack detection software. This will be reported on the future.

This work was supported by Council for Science, Technology and Innovation, "Cross ministerial Strategic Innovation Promotion Program (SIP), Infrastructure Maintenance, Renovation, and Management." (funding agency: NEDO)

References

[1] Japan Ministry of Land, Infrastructure, Transport and Tourism. The maintenance of national road network in Japan. On-line: http://www.mlit.go.jp/road/road_e/03key_challenges/1-2-2.pdf, Accessed: 21/03/2017.

[2] Japan Ministry of Land, Infrastructure, Transport and Tourism. C-ROBOTECH. On-line: http://www.mlit.go.jp/common/001111984.pdf, Accessed: 21/03/2017.

[3] Nakamura S., Inoue F., Yamashita A., Yanagihara Y., Ueno T. and Takahashi Y. Development of variable guide frame vehicle for inspection of tunnel lining concrete. In *Proceedings of the JSME ROBOMECH*, 2P1-N01, 2015. (in Japanese)

34th International Symposium
on Automation and Robotics
in Construction

ISARC 2017

June 28 – July 1
TAIPEI, TAIWAN

CONSTRUCTION MANAGEMENT

Method of Optimal Machine Selection for Construction Processes from the Point of its Energy Consumption Minimizing with Software Support

J. Gašparík[a], L. Paulovičová[b], S. Szalayová[c] and M. Gašparík[d]

[a]Faculty of Civil Engineering, Slovak University of Technology in Bratislava, Slovakia
[b]Faculty of Civil Engineering, Slovak University of Technology in Bratislava, Slovakia
[c]Faculty of Civil Engineering, Slovak University of Technology in Bratislava, Slovakia
[d]POMAKS, s.r.o. Bratislava, Slovakia

E-mail: jozef.gasparik@stuba.sk, lucia.paulovicova@stuba.sk, sylvia.szalayova@stuba.sk, marian_gasparik@centrum.sk

Abstract

Energy saving is one of the most important environmental factor during the process of building realization. During this process building planner has to analyse several factors influencing the final effective decision concerning this problem. There are several factors and criteria for effective selection of building machines. In our contribution, the proposed optimizing method is presented and the ability of machines to realize designed building process with minimizing of energy consumption (environmental aspect) is analysed. This method are implemented into two different construction processes: optimal selection of machine groups for soil excavations and optimal selection of the building elevators for vertical transport of building materials during the construction finished processes. Selected scientific methods and theories of problem solution can be divided into these groups: theory of system (creation of building machine selection method), optimizing method (analysis of mechanized building process from the point of minimum energy consumption), queuing theory (application during the mathematical modelling of mechanized soil processes), method of scientific analysis (analysis of factors influencing final decisions), method of scientific synthesis (creation of optimizing method implemented into building processes and possible application in construction sector). The effectiveness of implementation of these methods is increased by software support for both model examples. Results of our research work are shown, that with using automated system we can find optimal machine group for defined process very quickly and by this way during the process of building planning to increase the effectiveness of building machines selection especially form the point of energy minimization.

Keywords –
Construction mechanized process, Energy Consumption Minimizing, Optimal Method

1 Introduction

The optimal selection of machine or machine group for building processes is a very important role of building planner during the process of building planning. During this process building planner has to analyze several factors influencing the final effective decision concerning this problem.

During the process of building planning planner has to analyze suitable selection of building machines and its group for effective proposal of mechanized building processes.

There are several criteria for selection of building machines. In our contribution there are analyzed: the ability of machines to realize designed building process (quality aspect) and minimizing of energy consumption (cost and environmental aspect).

In our research work were implemented several theories and methods. The most important were the theory of systems [1] and principles of optimizing method [2], method of scientific analysis and synthesis. These methods were implemented into optimal selection of machine groups for earthwork excavations and optimal selection of building elevators from the point of minimal energy consumption with software support.

Applications of these methods and software will increase effectiveness of building machines selection from the point of its energy consumption minimizing. There are several authors who analyzed similar problems [3,4,5,6,8,9] concerning the optimal machine selection methods and their environmental impact to construction company effectiveness.

2 Energy Machine Selection Optimizing (EMSO) Method

EMSO Method is based on the theory of systems [1] and optimization theory of the process [2].
The "EMSO Method" consists of the three phases (Figure 1) – Entry phase, Decision phase and Optimization phase. An analysis of all these phases except entry is examined by Figure 1 [7]:

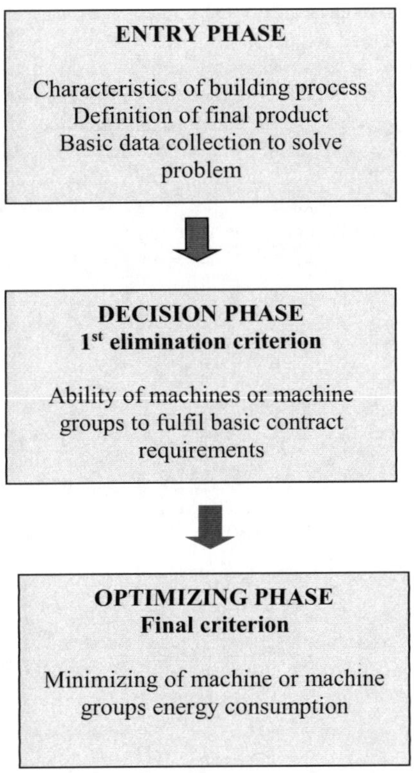

Figure 1. EMSO method phases and criteria

-the input universe of the system: that is the set of the machines submitted for analysis in the given phase,
-the criterion, according to it is the input universe of the system of given phase analyzed,
-the procedural steps being necessary to realize the appreciation of the input universe of the system according to the criterion of the given phase,
- the output universe of the system: that is the set of the transport means fulfilling the criterion of the given phase.
The entry phase contains delimitation of problem and objectives necessary to be reached by evaluating, for example the type of building works, characterization of the final product of the mechanized building process, input information necessary for solving of the problem and so on.
The preparatory phase characteristics*:*
The input universe of the system is the set of the transport means being suitable for a given type of the building works.
The criterion (1st eliminating) – production rate aspect (time required for the realization of the final product or quantity of production in determined time)
The procedural steps:
- the construction of the verbal – graphic model of the real system,
- the choice of the variants of building machines for realization of the final building product,
- the selection and the choice of the model variables, their definition, symbol, dimension, quantification with the source of the quantification,
- the formulation of the particular mathematical relations of the model,
- the construction of the mathematical model for appreciation of variants of machines to the 1st eliminated criterion,
- the verification, quantification, numerical solution using software, interpretation and implementation of the created mathematical model.
The output universe of the system – the set of the machines performing the requirements for realization of the final product.
The optimizing phase characteristics:
The input universe of the system is the output universe of the decision phase.
The optimization criterion - the minimizing of the machines energy consumption for realization of the final product of the building process.
The proceeding steps:
- the selection and choice of the decision variables, their definition, symbol, dimension, quantification with giving of the source of the quantification,
- the construction of the mathematical model of the criterion of the optimization,
- the verification quantification, numerical solution using software, interpretation and implementation of the mathematical model of the criterion for optimization.
The output universe of the system – the machine with the minimal energy consumption for realization of the final product of the building process.

In our contribution, the optimizing phase of EMSO method in two different model examples is analyzed: optimal selection of machine groups for earthwork excavations and optimal selection of building elevators from the point of minimal energy consumption with software support.

3 Model Example N.1: Excavation and the Removal of the Soil at the given distance

Basic input data:
Final product of building process – building pit: width (50 m), length (90 m), depth (3,5 m), soil type and class – sandy soil, the 2nd class of cohesion (according to Slovak National Standard STN 733050), required work capacity V_p = 15 750 m ^{3}, transport distance L= 4 km, required time of duration of works T= 14 400 min. (30 shifts), season of year of realization of works – April, May, kind of road surface – mastic asphalt, plane on the whole length.
Comment: presupposition of approximate identical operation of machines during shifts, time for lunch and inspection of machines at the beginning and the end of shift have not being included in time of shift duration.

The input universe of the system of the decision phase is being created by 3 types of depth shovel excavators: DH 411, DH 621, Cat 225 and 3 types of folding transport means: T 148 S1, T 815 S3, S 706 MTSP 24. The same transport means were applied to every type of the excavator. There are 9 variants of the excavator machine group together with the transport means and in every variant we used from 1 to 13 pieces (pcs) transport means. For the evaluation of the machine groups in the optimization phase the concept of queuing theory is being applied. In our contribution are shown final mathematical models (1) of optimizing phase. All data necessary for equations 1 and 2 can be found in Gasparik (2013) – see references.

The mathematical model of the optimizing criterion is in form as follows:

$$MS=T_{sk}.T_{ps}^{-1}.V_p^{-1}.[T_{mr_i}.S_{mr_i}+T_{pr_i}\,S_{pr_i}+(T_{ca_j}\,S_{ca_j}+ +L_{na_j}.S_{na_j}+L_{pa_j}.S_{pa_j}).N_{a_j}]\ (l.m^{-3}) \quad (1)$$

for i=1, 2, 3; j = 1, 2, 3; N_a=1,2, ... , 13,

where
MS - specific F.C. of machine group, excavator + transport means by the required volume of the works (l.m^{-3}),
T_{sk} - duration of work of machine group by earthworks of required volume (min.),

T_{ps} - duration of operation of machines during a shift (min.shift^{-1}),
T_{mr} - time of excavator manoeuvre (min.shift^{-1}),
S_{mr} - fuel consumption of excavator at manoeuvring (l.min^{-1}),
T_{pr} - duration of work regime of excavator except time of manoeuvring (min.shift^{-1}),
S_{pr} - fuel consumption of excavator in operating regime (l.min^{-1}),
T_{ca} - duration of waiting regime of transport mean during running engine (min.shift^{-1}),
S_{ca} - fuel consumption by waiting regime of transport mean (1.min^{-1}),
L_{na} - length of road covering by transport mean with a load, from place of loading to place of unloading (km.shift^{-1}),
L_{pa} - length of road covering by transport mean without of load, from place of unloading to place of loading (km.shift^{-1})
S_{na} - fuel consumption of transport mean by driving with a load (1.km^{-1}),
S_{pa} - fuel consumption of transport mean by driving without a load (1.km^{-1}).

Input data concerning the consumption of fuel were given by producers of excavator and transport means. The best energy saving machine groups of each kind are being compared in Figure 2. The most advantageous solution for the realization of output and removal of earth at given distance from the point of view of minimizing of fuel consumption is at analysed model example a choice of the machine group Cat 225 + 6 pcs of T 148 S1.

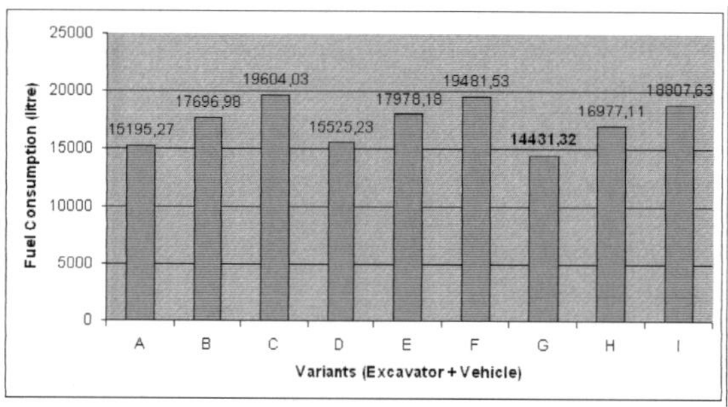

Figure 2. Machine group (excavator-trucks) variants evaluation from the point of minimum consumption of fuel for variants A – I : see Figure 6

The Machine Selection software was created by co-author M.Gašparík as a software support for method described in this contribution. Machine Selection is a desktop application, built in Java.
"Complete Fuel Consumption Table" shows fuel consumption in litres of combination consisting by 1 excavator and 1 to 13 vehicles by realization of desired earthworks volume. This part of result screen is displayed on Figure 3.

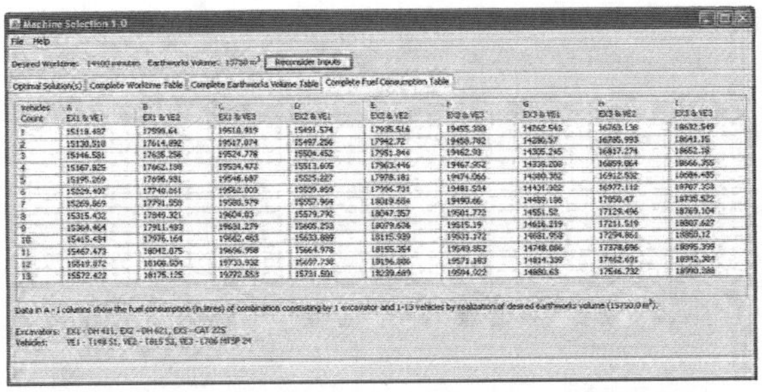

Figure 3. Result screen, Complete Fuel Consumption Table section

4 Model Example N.2: Building elevator selection

This “EMSO method” was applied into building elevator selection for vertical transport of building materials during the finished processes in term of minimum consumption of electrical energy.
Input data: 9 floored building object (see Figure 4), operational project of building object.

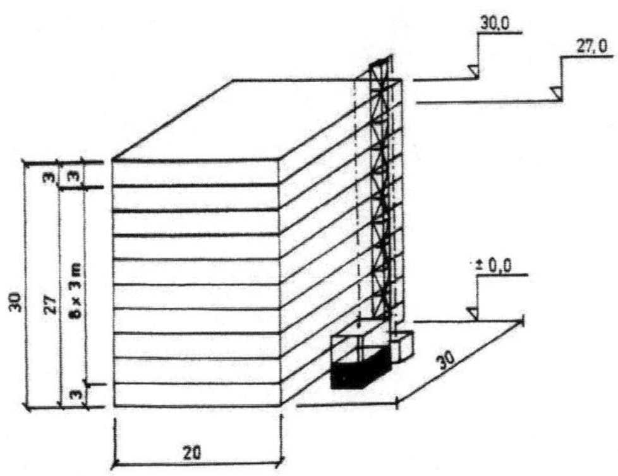

Figure 4. Analysed building object for optimal selection of building elevator

Input universe of system in optimization phase:

- building elevator: WBT 5-600,
- building elevator: VS 5,
- building elevator: VS 6.01,
- building elevator: NOV 500,
- building elevator: NOV 1000 A.

Criterion: optimization – minimization of consumption of electric energy of elevators in a process of vertical transport of materials to building object.

Procedural steps:

1. Experimental method of measuring value of electric current, which is withdrawn by construction elevator´s electromotor during its operational mode.
2. Experimental measuring work was done on specific constructions in Bratislava (Slovakia), in the presence of expert for elevators. The elevator was continually burdened by 50 kg (1 bag of cement) until maximum loading capacity. The values of electric current flowing through phase conductor of three phase line and withdrawn by electromotor of elevator during continual loading in both direction (upwards and downwards) are measured by clip-on ammeter.
3. Results of experimental measuring of electrical current values which were withdrawn by electromotor of elevators during operational mode. Above mentioned measuring work was implemented on all analyzed types of elevators. Results of measurements are recorded in tables and also graphically illustrated.
4. Construction of mathematical model of optimality criterion

   The content of this point is to construct mathematical model of consumption of electrical energy of elevator for tonometer of transported materials.
   The foundation for calculation of measuring consumption of electrical energy of elevator are experimental measured intensities of electric current withdrawn by its electromotor.
   The consumption of electrical energy of elevator for tonometer of transported material could be calculated as follows:

$$E = \frac{\sqrt{3}}{3600} \cdot \frac{U \cdot I}{v \cdot m} \quad [\text{Wh/tm}] \qquad (2)$$

where
U – nominal line voltage of three-phase distribution system in volt (380 V),
I – real intensity of electric current withdrawn by electromotor of elevator in ampere,
v – transport velocity of elevator in meter per second,
m - load mass transported by elevator in tonne.
Note:
1. At work there is analyzed case, when elevators are going in upward direction they are loaded, however in downward direction they are not loaded by materials. From measured and calculated variables we can observe different combinations.
2. In case of operational-personal elevators NOV 500 and NOV 1000 A, service is necessary in elevator´s cage, that is why we consider empty (without load) elevator’s mass

100 kg and elevator load capacity is changing in case of NOV 500 to 400 kg and in case of NOV 1000 A to 900 kg.

The overall consumption of electrical energy of elevators for transport of material to 1 m elevation (tonometer) and its return to initial position we can calculate as follows:

$$E_c = E_h + E_{pd} \text{ [Wh/tm]} \quad (3)$$

where

E_h - the consumption of electrical energy of elevator for tonometer of transported material in upward direction in watt hour per tonne and meter (Wh/tm),

E_{pd} - the consumption of electrical energy of elevator for tonometer of transported material in downward direction not loaded by materials in Wh/tm.

Graphical interpretation of dependence of overall electrical energy consumption to transport material by elevators E_c [Wh/tm] is illustrated on Figure 5 and graphical interpretation of dependence of overall electrical energy consumption of elevators E_c [Wh/tm] on percentage extraction their load capacity is illustrated on Figure 6.

5. The determination of final electrical energy consumption of particular types of elevators after overall work performance E_v [Wh].

For approximate calculation of E_v [Wh] overall mass of materials ABP, which is transported by elevator to particular floors M_i [kg], where i = 1, …,n, is determined.

Based on characteristic of different materials types, average value of percentage extraction of load capacity of elevator x is estimated, subsequently based on this value the E_c [Wh/tm] is defined from figure 6.

The resulting electrical energy consumption of elevators E_v [Wh] is calculated as follows:

$$E_v^k = E_c^k \cdot \sum_{i=1}^{n} M_i \cdot h_i \cdot 10^{-3} \quad \text{[Wh]} \quad (5)$$

where

E_c^k - overall electrical energy consumption "k" type of elevator for transport of 1 tonne material to 1 m height [Wh/tm], where k=1,…,p,

M_i – mass of material, which is transported to "i" floor [kg] ,where i = 1, …, n,

h_i - height of "i" floor [m], where i = 1, …, n,

Model example:

M_i = 10 000 kg for i = 1, …,9 (typical floor)

h_1 = 3 m, h_2 = 6 m, h_9 = 27 m (Figure 4)

$\bar{x}$ = 70 % (figure 6), E_c [Wh/tm] (Figure 6)

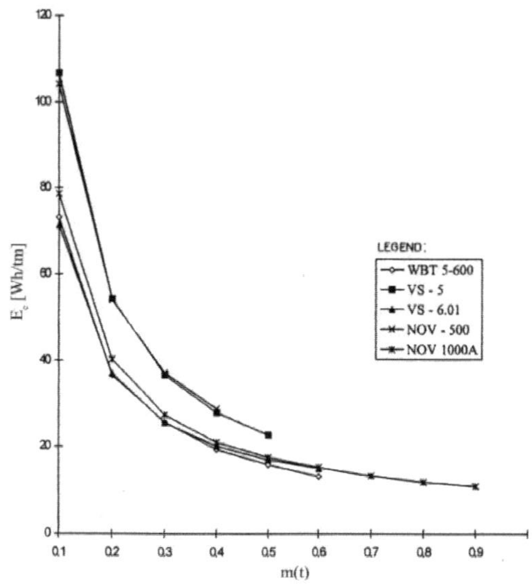

Figure 5. Graphical interpretation of dependence of overall electrical energy consumption to transport material by elevators for tonne and meter E_c [Wh/tm]

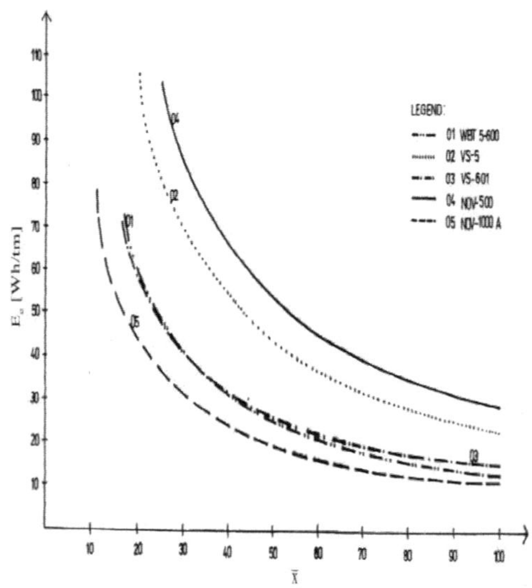

Figure 6. Graphical interpretation of dependence of overall electrical energy consumption of elevators E_c [Wh/tm] on percentage extraction their load capacity.

The solution of model example is shown by Figure 7.

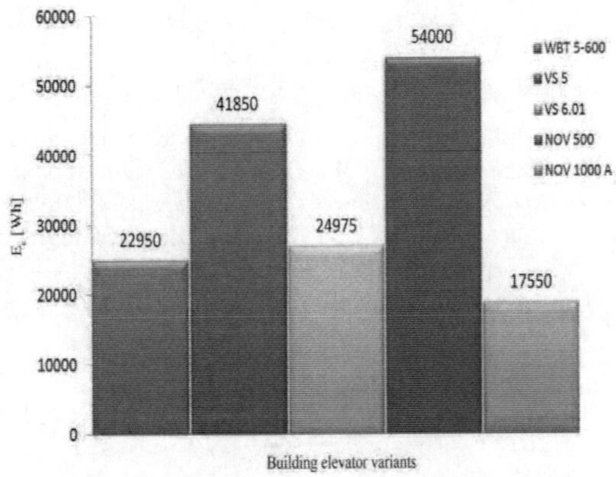

Figure 7. A comparison of elevator variants in term of overall electrical energy consumption during the process of vertical transportation of building materials to analyzed building object.

Output universe of system - NOV 1000 A (see Figure 7).

4.1 Elevator Selection Optimizing (ESO) method application with software support

Above analysed model example can be effective solved using Elevators Energy Consumption software as a web solution based on html, php and javascript to compute energy consumption of several elevators and choose the best choice (the one with lowest energy consumption). The main screen is divided into 3 parts: Building Inputs, Elevator Inputs and Results.

4.1.1 Building Inputs

In this section (Figure 8) user enters number of floors of a building, typical floor height and typical height, delivered to the floors. For each floor, user can specify its own height or weight, needed to delivery.

4.1.2 Elevator inputs

This part (Figure 9) serves to specify number of elevators (from 1 to 5), their names and E_c constants, assigned to capacity values, varying from 25 to 100 percentage.

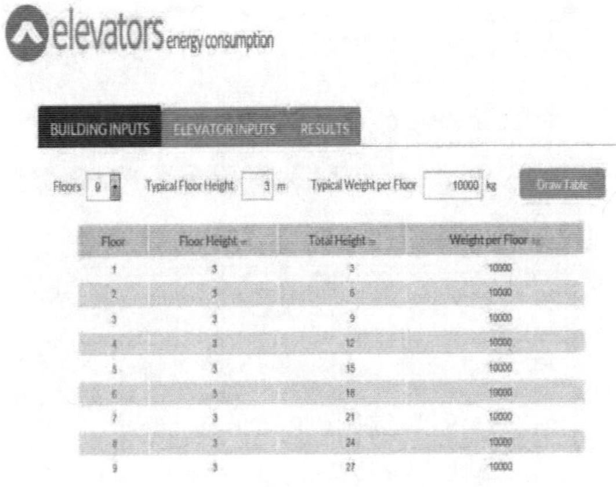

Floor	Floor Height	Total Height	Weight per Floor
1	3	3	10000
2	3	6	10000
3	3	9	10000
4	3	12	10000
5	3	15	10000
6	3	18	10000
7	3	21	10000
8	3	24	10000
9	3	27	10000

Figure 8. Building inputs

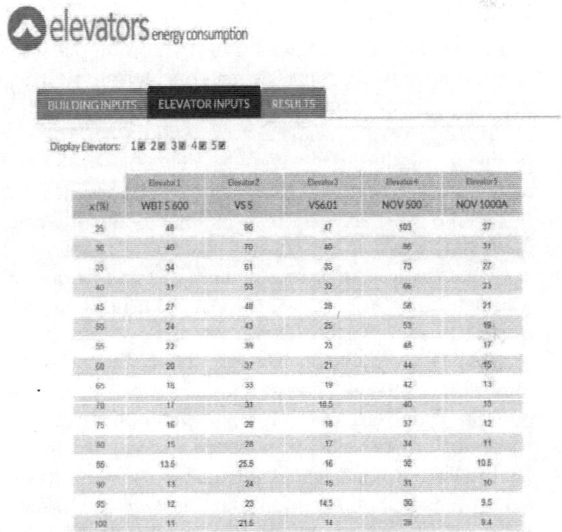

	Elevator 1	Elevator 2	Elevator 3	Elevator 4	Elevator 5
x (%)	WBT 5 600	VS 5	VS6.01	NOV 500	NOV 1000A
25	48	90	47	103	37
30	40	70	40	86	31
35	34	61	35	73	27
40	31	53	32	66	23
45	27	48	28	58	21
50	24	43	25	53	19
55	22	39	23	48	17
60	20	37	21	44	15
65	18	33	19	42	13
70	17	31	18.5	40	13
75	16	29	18	37	12
80	15	28	17	34	11
85	13.5	25.5	16	32	10.5
90	13	24	15	31	10
95	12	23	14.5	30	9.5
100	11	21.5	14	28	9.4

Figure 9. Elevator inputs - values of E_c [Wh/tm]

4.1.3 Results

In Results section (Figure 10), user is asked to choose capacity for each elevator and after clicking the button Draw Results he gains result values of total energy consumption (E_v). Elevator with the lowest consumption (the winner) is marked green.

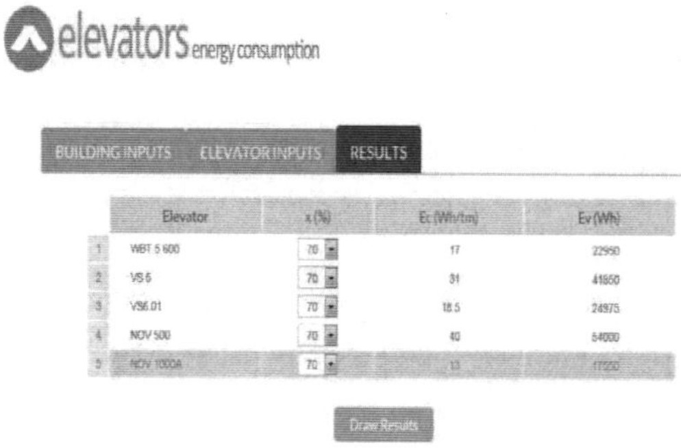

Figure 10. Results section

5 Conclusion

The most-important factor in this "EMSO" method is that it is able to eliminate machine variants with high energy consumption during the process of building planning. By using a software it gives information about energy consumption of machines when considering their use in the final product of the building process (earthwork excavation, vertical transport of building materials) and gives the possibility to make fast decision for the choice of the energy optimal machines in a short time. At the same time it is necessary to stress that by this method the machines are being evaluated from one point of view only (energy consumption), which, it is true, is one of the most meaningful views of this time, but it needs not be crucial in every case. Therefore, it is necessary when proposing an elevator to take the point of view to minimize the energy consumption as a part of the multi-optimal proposal. For a practical application of the proposed "ESO" method it is necessary to improve the quality of input data, especially energy information. This data involved in our contribution were obtained by experimental measurement. The producers of building machines and elevators can offer this data shown in our contribution as a basic information concerning their excavators, transport means and elevators and using software building planner can very easy and fast select energy optimal variant of building machines. The volume of savings of the operating expenses possible to be obtained already in the preparation phase of building realization by this "EMSO" method are not negligible, vice versa, it shows the disclosure of reserves that are available in the choice of machines for excavation process and elevators for vertical transport of building materials.

6 Acknowledgements

This contribution was prepared as a part of scientific research project VEGA N. 1/0670/17.

7 References

[1] Štach, J. *Teória systémov*. Edition: SNTL, Praha. 1983.

[2] Niederliňski, A. *Numerical systems of control technologic processes*. Edition: SNTL, Praha. 1983.

[3] Struková, Z., Kozlovská, M. and Ištvánik, M. *Mobile Crane Selection Based on Workspace Requirements and Cost Parameters*. In. America Journal of Civil and Structural Engineering. Vol.1, N.2. p. 5-10, 2014.

[4] Struková, Z. and Líška, M. *Application of automation and robotics in construction work execution.* In. AD ALTA: Journal of Interdisciplinary Research. Vol. 2, N. 2, p. 121-125, 2012.

[5] Strukova, Z. and Kozlovska, M. *Environmental impact reducing through less pollution from construction equipments.* 13th SGEM GeoConference on Ecology, Economics, Education and Legislation, SGEM Conference proceedings, June 16-22. Vol. 1. P. 401-408, 2013.

[6] Myungdo Lee,T., Kim, .K., Jung,U.K., Lee,H. and Cho, K.I.K. *Green construction hoist with customized energy regeneration system.* Journal Automation in Construction. Volume 45, Elsevier September 2014,p.66–71.

[7] Gašparík, J. *Automated system of optimal machine group selection implemented into soil processes.* Edition: Tribun EU Brno, 2013.

[8] Motyčka, V, Klempa, L. *Simulation model of tower crane work*, In Proceedings CTM 2014, STU Bratislava, 2014.

[9] Kovářová, B. *Improving performance of building processes using six sigma methodology.* In proceedings CTM 2014. STU Bratislava, 2014.

Collaborative Advanced Building Methodology toward Industrialization of Informal Settlements in Cairo

Camilla Follini[a], Rongbo Hu[a], Wen Pan[a], Thomas Linner[a], and Thomas Bock[a]

[a]Chair for Building Realisation and Robotics, Technical University of Munich, Germany
E-mail: camilla.follini@tum.de

Abstract –

Today's population has reached a growth rate never seen before, and, globally, there are examples of traditional cities that have failed to adapt to this sudden change. Egypt makes no exception in this trend. In its largest city Cairo, this trend caused a considerable expansion of the phenomenon of informal settlements, areas spontaneously built by the inhabitants, often lacking regulation, planning, and basic needs. By investigating the context of Cairo, in particular the area of Sakiat Mekki, the paper explores the possibility of improving the quality of life in informal settlements with sustainable decentralized processing units (DPUs), tailored to an affordable and adaptable building system (A²BS). The goal of the DPUs is to enhance three main aspects of life (working, energy, and mobility) by gradually implementing small-scale subsystems of advanced construction in the hosting environment, thus providing a more flexible solution to evolve in time. The design was developed after an analysis based on Requirements Engineering (RE), to ensure that the system would be more easily accepted and the design proactively influenced by the existing community's feedback. Accordingly, the construction elements were designed based on stakeholders' needs. Thanks to a simplified construction process and joining system, not only this solution can be easily assembled by unskilled labor, but it could also be used as a step forward toward automation of the building process in substandard environments.

Keywords –

Collaborative design and construction; Industrialization of informal settlements; Decentralized Processing Unit; Affordable and Adaptable Building System

1 Introduction

Egypt has been heavily influenced during the last decades by economic, social and political pressures. Cities such as Cairo, which are the focus of the government and the society, are facing great population growth, immigration growth from villages to big cities and immigration of refugees. Like many mega cities of developing countries, Cairo has experienced its strongest growth in the half century after World War II. The strongest growth has been registered between 1996 and 2006, with a population increase of about 4,000,000 people. More recent data show that the growth is steadily slowing down. Still, these numbers are high enough to raise concerns about if and how the government had planned for the sustainable housing of the additional inhabitants [1]. The problem of informal settlements is considered one of the most challenging issues in the developing world, since it poses an economic and social threat to countries' future. Most of informal settlements share some distinctive traits, like poor structural quality of housing and infrastructure, lack of basic services, such as water, sanitation, waste collection, schools, hospitals, and poor environmental conditions. However, the nature of informal settlements also could impose some advantages and benefits. There are many aspects, which city planner, architects and environmental agencies could adopt: 1) self-sufficiency, as shops and local produce are vastly available, 2) short walking distance to reach the workplace and most of the amenities, 3) strong sense of community, 4) enhanced safety, since neighborhood connection is very close and people tend to help each other more willingly, and 5) sense of participation, as most of the services, such as street maintenance and garbage collection, are effectively controlled by the inhabitants themselves [2].

2 Methods

Given the broad scope of the objectives goal, a systematic method has been used to determine the best approach and methodology. This chapter presents how the project is conducted.

2.1 Scope and Approach

Considering the advantages and drawbacks of the

current situation stated above, the aim of the A²L-Mobilius project [3] is to develop modular, adaptable and scalable units, which responds to future socio, economic, technological and environmental change, by exploring the Egyptian context and investigating the possibility and requirements for an efficient infrastructure. In order to achieve this, the research has been focused on two main elements:

1. Affordable and Adaptable Building System (A²BS): a flexible building kit composed of prefabricated elements. The kit has been designed to fit the informal environment, and adapt through time to the community change. Its goal is to gradually replace informal structures and thus "formalize" the built environment.
2. Decentralized Processing Unit (DPU): a prefabricated, self-sustaining, interchangeable, and standardized system that houses a sum of technological equipment needed for a household. It would allow for a step by step upgrade of the informal settlement.

The decentralized processing unit has been divided into three sub-systems, basing on the most relevant issues of the hosting country: energy, mobility and life-work balance [4]. The energy subsystem is meant to provide a reliable structure that would respond to the needs of the community related to collection, provision, wise use and eventual production of different kind of primary resources such as electricity, gas and water. The mobility subsystem addresses external mobility and the optimization of transportation of goods and people. Finally, the home fabrication subsystem is part of a greater interconnected system aiming to give an alternative and more sustainable work method, and therefore to provide a higher number of workplaces to decrease the unemployment rate. The decentralized processing unit elements will be integrated to the building kit, to form a comprehensive system.

2.2 Overall Methodology

The methodology for the project plan has been based on the V-Model diagram. The project has been divided into smaller problems to be addressed with one design module or sum of modules. As show in Figure 1, the steps of the model comprehend on or more work packages of the project. The left side of the scheme shows the preparatory work or project definition, which starts from a general perspective that is detailed in time. On the opposite, the project implementation, on the right side starts from the detailing of modules to reach a general system in the later stages. As it can be also seen from the diagram, the system is set to go through several iteration loops for optimization. Two cycles in particular concern communication with stakeholders, one to verify the premise of the project, thus the requirement analysis, and the second to verify the design. Final validation will be given at the end of the project, with the comparison of the final system and the revised requirement and stakeholder analysis. The current paper presents the first part of the implementation section, in between the two verification loops, as highlighted with the black square in the diagram. The modules of the main elements have been established basing on the requirement analysis. They will then be tested in a system. Therefore, the material presented in this paper is to be considered part of a conceptual stage to be further optimized and validated thanks to real world feedback.

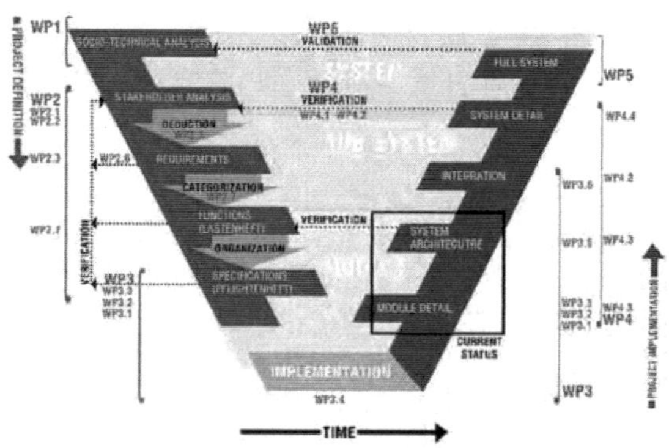

Figure 1: V-Model of the A²L-Mobilius project

2.3 Requirement Analysis

Collaboration with the final users is generally essential for the success of any project aimed to the revitalization of informal settlements. Therefore, participation and feedback of stakeholders throughout the project is a critical factor to be ensured. Thus, a methodology based on Requirements Engineering (RE) has been first developed, in order to find the most fitting solutions based on environmental requirements and stakeholders wishes [5]. The structure of the methodology takes into consideration repetitive optimization loops to constantly monitor and optimize the stakeholder and requirement analysis for the whole duration of the project. The aim of the requirement analysis, together with a technological feasibility study, was to identify concrete functions to be translated into design elements, and a suitable system architecture to give them a common structure. The output of the analysis carried out for the three DPU subsystems is shown in Figure 2.

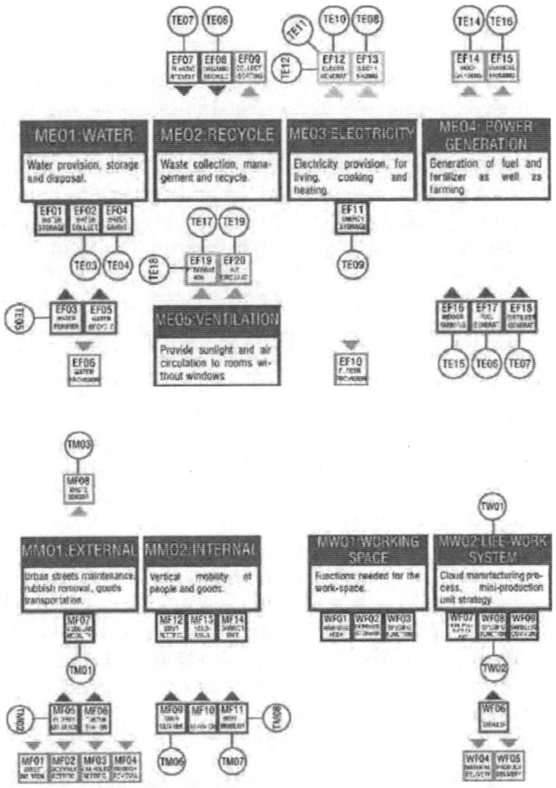

Figure 2: Requirement and function analysis of the three DPU subsystems (top: energy, bottom left: mobility and right: working)

2.4 Technology Transfer from High to Low-tech Environments

There are a wide range of research topics in the area of integrating new technologies into the living environment (buildings, city, infrastructure, design). The research projects below represent the examples of technologies in regard to three major aspects (working, energy, and mobility subsystems). As a matter of fact, it is a challenge in this project to transfer these approaches into the Egyptian environment. To achieve that goal, basic strategy and approach shall be retained yet in a more energy-efficient and cost-effective way in order to fit into Egyptian context, such as simplifying the design, using local materials, and adapting to the culture. After successfully implementing these technologies in the developing environment, a strategy of reverse innovation could be adopted in order that these improved technologies can be reintroduced to the industrialized world, thus not only benefiting the local residents, but also the developed countries. For example, the company GE Healthcare is selling an ultra-portable electrocardiograph machine in the U.S. at a 20% cost for similar devices, which is originally developed for markets in India and China [6].

2.4.1 Working Subsystem

Extensive research has been carried out in the topic of living in the future, as well as in the topic of decentralized production in modular mini-factories. For instance, in the project USA² [7], assisted workspaces for decentralized high-tech home production of customized goods have been developed by Chair of Building Realisation and Robotics at TUM (BR²). The research project, financed by the German Federal Ministry of Education and Research (BMBF, grant number: 16SV6191) within the Human-Technology Interaction (MTI) program, concerned the creation of workspaces embedded in larger Cloud Manufacturing systems, which will allow companies to utilize a highly skilled workforce (including highly experienced elderly) worldwide. Within the project, the workspace (its functions and design), the embedded sensors distributed in the environments, novel interfaces (such as leap motion sensors [8]) and the related work processed were fully co-adapted (Figure 3). The USA² showed that the co-adaptation of environment and mobile platform allows for a multitude of assistive functions, such as intuitive user interface, transports of goods, and detection of emergencies. Considering the high level of participation and self-organization already existing in informal settlements, a suitable strategy for building prefabrication could be the organization of a collaborative manufacturing decentralized structure. Instead of providing a central factory where material is produced in mass quantity, several small manufacturing stations can be located all over the settlement to work as one. In the same direction, the project USA² offers an interesting example of Mini-Production Unit (MPU) developed to be embedded in a cloud-manufacturing system.

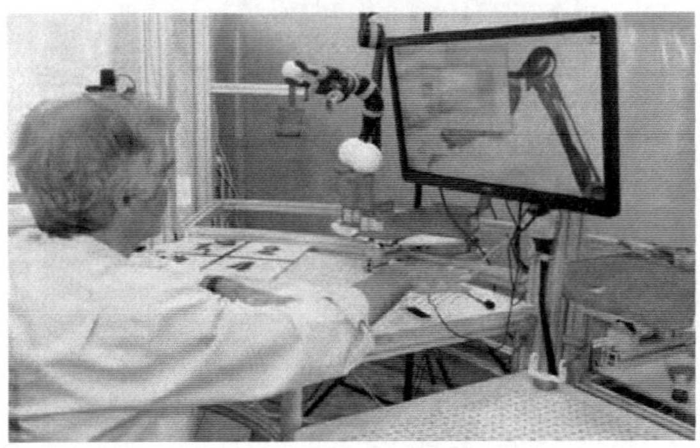

Figure 3: USA² mini producion unit (Image: Schlegelmilch)

2.4.2 Energy Subsystem

Energy saving and prefabrication in building renovation are two of the main goals of the BERTIM Project [9], developed in a large consortium BR² is a partner of, which has received funding from the European Union's Horizon 2020 Research and Innovation Program under grant agreement No. 636984 (Figure 4). The aim of the project is to develop [10]:

1. Prefabricated modules for renovation, already equipped with architectural elements and technologies for reduction of energy consumption (such as windows, insulation materials, collective HVAC systems, renewable energy systems and energy supply systems). The panels are designed to be easily installed and they are produced with low carbon foot print and materials that can be recycled after the use stage.
2. An innovative methodology for the renovation process, making use of the latest software technologies such as BIM and CAD/CAM tools
3. Business opportunities for professionals in the renovation sector.

To increase the feasibility of the project, the renovation process, as well as the timber prefabricated modules have been planned to be validated in two actual buildings located in two different climatic regions of Europe (north and south) [11].

Figure 4: BERTIM facade modules installation in one of the sites (Image: M. Eng. Kepa Iturralde).

2.4.3 Mobility Subsystem

The abbreviation of the Horizon 2020 project REACH stands for Responsive Engagement of the Elderly Promoting Activity and Customized Healthcare [12]. REACH represents a solution that seeks to reduce Long Term Care (LTC) by serving as a personalized system for promoting and monitoring the activity of elderly citizens in order to reduce their risk of function loss and related morbidities. During the four years' duration of the project, the consortium develops modular, personalized medical and ethical acceptable solutions integrated in and around buildings that allow an intelligent prediction about the health status of people/patients. Based on forecast and analytic algorithms, REACH will be able to provide novel, personalized interventions such as customized services, products, and equipment for mobilization and rehabilitation, physical activity, training, food and nutrition, mobility, and patient motivation.

PASSAge project, which is financed by the German Federal Ministry of Research and Education (BMBF), under the Grant Number V4MOB059, is another example showing this approach [13]. In PASSAge a mobile robot was developed that could flexibly couple to various mobility devices (rollators, wheelchairs, etc.) and thus assist and facilitate the mobility of elderly people (Figure 5).

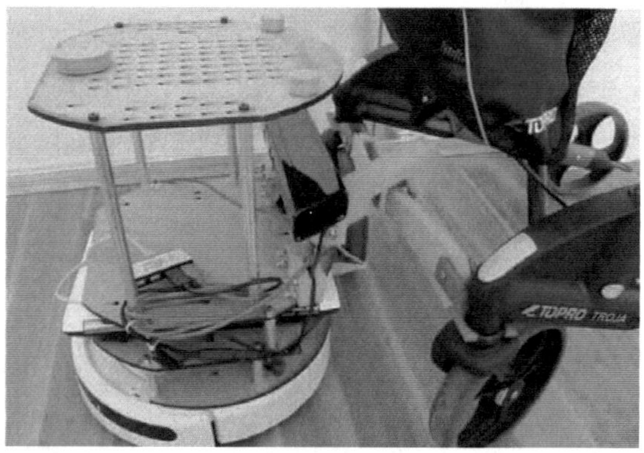

Figure 5: Prototype developed in the project PASSAge (Image: Chair of Building Realisation and Robotics, TUM).

3 Results

The system is based on a study case surveyed by the Egyptian partner that is located in the area of Sakiat Mekki, on the west side of the Nile in the Giza district. The building has many problems common to informal settlements, such as lack of space, sanitation issues, lack of ventilation, hazardous internal mobility (i.e. stairs without handrail and different height of risers), and poor insulation. The surroundings are as well problematic, the main issues being the poor street maintenance, large presence of uncollected waste, and extremely compact building presence, making it impossible to expand the study case building on the side (Figure 6).

Figure 6: Low sanitation conditions in Farahat street (left) and 6th October street (right) (Image: M. Sc. Maged Helal).

3.1 Proposed Construction System

The construction system is based on the modular structure and infill strategy. The panelized system shown in Figure 7 is composed of four modular panels and vertical support pillars. Elements are connected by bolted joints and have a limited dimension for better transportation. The modular infill system is made of a framework that can accommodate different functionalities [14].

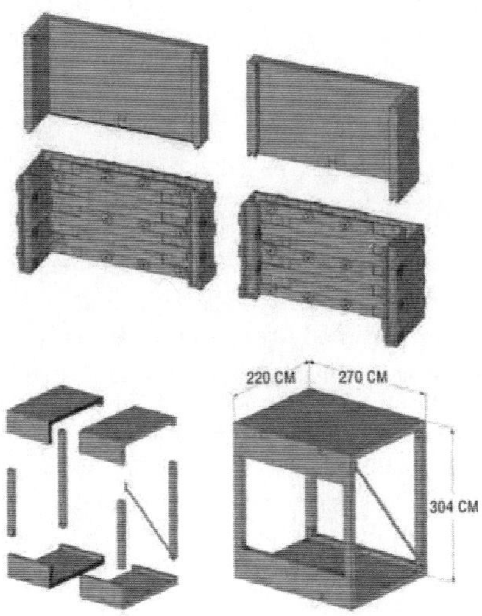

Figure 7: Panel system of the construction kit

3.2 Decentralized Processing Unit

As aforementioned, the most important functions and their system architecture have been highlighted thanks to a first analysis based on requirements engineering. The dimensioning of the construction system, carried out in the previous section, was a necessary prior step necessary to translate functions into a suitable design. The geometries considered for the decentralized processing unit are tailored to the A²BS module. Each element has been categorized, considering its nature, size and priority, in three classes. Since the final system will be modular, they can be explained as such:

1. Core functions: elements that have high importance to the system and should be present in every configuration. They can be part of the main body (the "chassis") of the module, to which other functions can be attached.
2. Optional functions: functions whose presence is not critical to the operation of the system and can be thus added later in time.
3. External functions: physical elements or services that have to be located or provided from outside of the main body of the module.

3.2.1 Life-Work Patterns Subsystem

The home fabrication subsystem has been considered as a part of a greater system, whose goal is to give an alternative and more sustainable work method and provide a higher number of workplaces in order to decrease the unemployment rate, following the example of the project USA² given in section 2.4.1. Therefore, instead of having a conventional centralized production plant, workspaces are scattered throughout the settlement. Manufacturing models, such as cloud manufacturing, concern the creation of complex objects through a net of professionals and clients working together with wireless communication, narrowing the need of mobility only to material and products. Thus, this model could have the advantage of reducing the overall traffic congestion and of reduce the commuting time, by overlapping work and life spaces. Moreover, the subsystem would assure as well a better, healthier, safer working condition. Given this premises, the model may represent an interesting future development for informal settlements, responding to several of their most relevant problems. Nevertheless, the technology should be adapted to the environment and it should be first proven on the field. Therefore, the team focused on cheap technologies and simple products. A validation of this first proposal will be given by the stakeholders in the next stage.

Among the different products to manufacture, the team focused on an innovative technique based on

Mycotechnology, which brings a new material into the construction field characterized by a fairly low construction cost and low environmental impact [15]. By using this technology, the aim of the working subsystem is to produce cheap and sustainable construction elements, such as bricks, to be used as part of the construction system. The simple shape and production method permits a simplification of the information flow, compared to more technologically advanced examples such as the USA[2] project, to increase the feasibility of the system.

3.2.2 Energy Subsystem

The energy subsystem should provide a reliable system that would respond to the needs of the community related to collection, provision, wise use and eventual production of different kind of primary resources such as electricity, gas and water. During the requirement analysis, the functions of the energy subsystem have been sorted into five main clusters, based on their scope: water, recycle, electricity, power generation, and ventilation. Clusters contain one or more DPU modules. Each cluster has its own principle of function, affecting the issues determined during the requirement analysis. In line with the idea of technology transfer, the prefabricated façade modules from BERTIM, introduced in section 2.4.2, have been used as reference for the vertical wall elements, however simplified to adapt to the final context. Moreover, the panel has been tailored to be produced with materials that are cheaper and locally sustainable. In order to propose feasible functions for the Cairene environments, some technologies are a re-reading of traditional techniques. For instance, rainwater harvest, a technology whose first examples in the country date back to at least 2000 years [16], is the main principle of the water cluster. Another element that has been used into the façade module is the Mashrabiya, traditionally used to control light and air direction, reduce temperature, and provide privacy. For ventilation, a possible addition of a wind tower has been considered.

3.2.3 Mobility Subsystem

The mobility subsystem should reduce the problems related to people ambulation and goods logistic both outdoors and indoors. Internal mobility is a common problem in informal settlements, due to the fact that apartments are built with no reference to the legal code. The problem of barrier-free design is therefore vast. In fact, even if the new apartments built with the construction system here proposed would fall within the barrier-free category, they would not solve the problem, being the rest of the environment not developed in that direction. Therefore, the approach proposed for the mobility subsystem, especially the one applying to the existing environment, is a gradual modification of the interior. That reconnects to the initial meaning of decentralized processing unit cells, aiming to "infect" the environment and gradually upgrade it. Again, the technologies shown in section 2.4.3 have been considered to be transferred in the subsystem. In particular, the team focused on solutions not heavily relying on hardware elements, such as the falling detection system.

4 Conclusion and Future Work

The feasibility of the system will be simulated in three scenarios with different level of diffusion of the two main construction elements (i.e. A^2BS and DPU) into the selected building (Figure 8). Each level comports different building connotations, which can be summarized in three points. The first aspect is the system standardization. The elements of the building kit have been standardized for easy production and the production units accordingly designed to fit in the kit. Nonetheless, some decentralized processing unit elements have been designed to adapt to an existing environment, in order to increase the flexibility of the system. This brings to the following consideration: the more extensively the construction kit is used, the more the system can be standardized. The second aspect is centralization. Diffusion of decentralized processing unit elements does not necessarily mean that they are physically spread apart. A construction-kit cell can confine the production-unit elements in a specific area of the building and still contain a bigger number of them than an upgraded existing environment. However, in the hypothesis that the construction kit would be used for the entire house, then the decentralized processing unit elements could be eventually embedded throughout the entire construction, and the decentralized processing unit considered ubiquitous.

As highlighted in section 2.2, the research presented in this paper is contextualized in between the two main verification loops based on the V-Model organization strategy. In order to make the verification more accessible, the developed solutions will be contextualized into the study case and divided into easily recognizable aspects by the scenarios. Thereafter they will be ready to go through a final evaluation of relevant stakeholders located in Egypt. The next research step will be to optimize the system according to the stakeholders' feedback, to be best accepted by the end-users. Thanks to a simplified construction process and joining system, not only this solution can be easily assembled by unskilled labor, but it could also be used as a step forward toward automation of the building process in informal settlements.

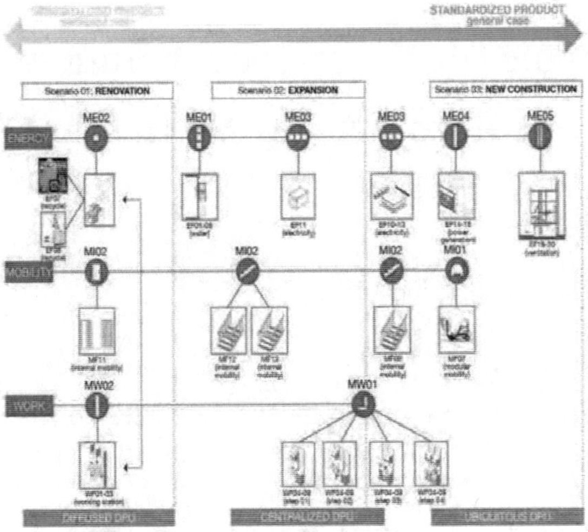

Figure 8: Scenario definition

5 Acknowledgements

This research is partly funded by the German Federal Ministry of Education and Research (Project: AL²MOBILIUS; Grant Number: GERF-IB-033 Almobilius_01DH14003). The authors would like to thank the Egyptian partner Dr. Wafaa Nadim at the German University in Cairo for her collaboration and advice. Furthermore, the authors are grateful to Mr. Maged Helal for his support.

References

[1] Cox W. The Evolving Urban Form: Cairo. Online: http://www.newgeography.com/content/002901-the-evolving-urban-form-cairo, Accessed: 08/03/2017.

[2] Helal M. Transformation of informal settlements in Egypt into productive city entities by utilizing and adapting advanced technologies. M.Sc. Thesis. Technical University of Munich, Germany, 2016.

[3] A²L-Mobilius project website. Online: http://www.br2.ar.tum.de/index.php/research/projects/8-home/167-a2l-mobilius, Accessed: 01/03/2017.

[4] Nadim W. and Bock T. Technological implants for sustainable autonomous upgrading of informal settlements in Cairo-Egypt. In *Proceedings of World Sustainable Building Conference*, Barcelona, Spain, 2014.

[5] Follini C. Pan W. Linner T. Nadim W. and Bock T. Development of a Methodology based on Requirements Engineering for Informal Settlements upgrading in Cairo. In *Proceedings of the CIBW119CIC2016 workshop*, pages 45–51, Munich, Germany, 2016.

[6] Jana R. Innovation Trickles in a New Direction. Online: https://www.bloomberg.com/news/articles/2009-03-11/innovation-trickles-in-a-new-direction, Accessed: 06/03/2017.

[7] Linner T. Güttler J. Georgoulas C. Zirk A. Schulze E. and Bock T. Development and Evaluation of an Assistive Workstation for Cloud Manufacturing in an Aging Society. In *Proceedings of Ambient Assisted Living: 8th AAL-Kongress*, Frankfurt, Germany, 2015.

[8] Bassily D. Georgoulas C. Güttler J. Linner T. and Bock T. Intuitive and Adaptive Robotic Arm Manipulation using the Leap Motion Controller. In *Proceedings of International Symposium on Robotics ISR ROBOTIK*, Munich, Germany, 2014.

[9] BERTIM project website. Online: http://www.bertim.eu, Accessed: 01/03/2017.

[10] European Union Open Data Portal. Funded Projects under Horizon 2020. Secure, clean and efficient energy. Online: https://www.nks-energie.de/lw_resource/datapool/_items/item_337/nks-e_projects_funded_2014_ee.pdf, Accessed: 23/03/2017.

[11] Iturralde K. Linner T. and Bock T. Development of a modular and integrated product-manufacturing-installation system kit for the automation of the refurbishment process in the research project BERTIM. In *proceedings of International Symposium on Automation and Robotics in Construction (ISARC)*, Auburn, USA, 2016.

[12] REACH project website. Online: http://reach2020.eu, Accessed: 01/03/2017.

[13] Bock T. Güttler J. Georgoulas C. and Linner T. The Development of Intra-House Mobility, Logistics and Transfer Solutions in PASSAge. *Journal of Robotics and Mechatronics*, Vol.27 No.1, 2015.

[14] Pan W. and Bock T. Development of interactive cluster for urban development in Egypt. In *Proceedings of Sustainable Built Environment (SBE),* Cairo, Egypt, 2016.

[15] Travaglini S. Noble J. Ross P. G. and Dharan C. K. H. Mycology Matrix Composite. In *Proceedings of American Society for Composites, 28th Technical Conference,* State College, United states, 2013.

[16] Abdel-Shafy H. I. El-Saharty A. A. Regelsberger M. and Platzer C. Rainwater in Egypt: quantity, distribution and harvesting. *Mediterranean Marine Science*, 11(2): 245-257, 2010.

Control and Optimization Tools for the Incorporation of Innovative Technologies in Net Zero Energy Settlement Design

S. Shubin[a], S. Isaac[b] and G. Rabinowitz[c]

[a]Unit of Energy Engineering, Ben Gurion University of the Negev, Israel
[b] Department of Structural Engineering, Ben Gurion University of the Negev, Israel
[c] Department of Industrial Engineering and Management, Ben Gurion University of the Negev, Israel
E-mail: shubins@post.bgu.ac.il, isaacsh@bgu.ac.il, rgadi@bgu.ac.il

Abstract – This paper describes the development of a set of control and optimization tools that support the implementation of highly innovative energy technologies in the design of Net Zero Energy settlements. The incorporation of novel technologies in building and settlement design involves unique challenges, such as frequent design changes, the risk of cost overruns and the need to synchronize the work of different stakeholders who did not previously collaborate. Three tools were developed and implemented to cope with those challenges. A cost control tool tracks the technology's initial costs during the design phase, so that any cost overrun can be easily recognized and treated. A change management tool allows the project's designers to identify, examine and discuss the implications of every proposed change in the design before the change is executed, and prevent deviations from the project's cost and energy performance requirements. A third tool supports a life cycle cost analysis of each technology and an identification of their optimal configuration. In response to a tight schedule and the uncertainty regarding the users' exact requirements, an Evolutionary Prototyping methodology was adopted to develop the tools. They were implemented in a project involving the design of Net Zero Energy settlements in four different European countries.

Keywords –
Net Zero Energy; Cost control; Change management; Design optimization;

1 Introduction

The need to increase the energy efficiency of buildings and cities has been widely recognized. However, the challenges of managing projects in which novel technologies are incorporated to achieve this goal, have been given less attention. Accordingly, the present research studies how such projects can be successfully managed.

The *Net Zero Energy (NZE) Building* concept that has recently gained significant international attention [6]. It can be described as a grid-connected building that generates the same amount of energy that it uses over a year [7]. The NZE concept is based on the idea of an energy-producing, low-energy building that interacts with the energy infrastructure. This interaction gives the possibility to purchase energy from the grid, to feed excess renewable energy back to it, and thus to offset the previous and/or future energy use from the grid [6].

In *NZE settlements* the concept is taken to a district level. This is expected to bring advantages on a number of levels:

- The ability to utilize more efficient settlement-level technologies for energy generation;
- The ability to reduce energy consumption through settlement-level energy management systems;
- Reduced costs due to the economies of scale;
- Reduced energy consumption due to an improved micro-climate that can be attained through an optimal design of the settlements.

This study was carried out in the framework of ZERO-PLUS – a project for the development and implementation of a comprehensive, cost-effective system for NZE settlements in a series of case studies across the European Union (EU). The goal of this project is to provide the market with an innovative, yet readily implementable system for NZE residential neighborhoods that will significantly reduce their costs. Therefore, in addition to the fulfillment of the NZE energy requirements, a primary objective of the project is to develop a system whose investment costs will be at least 16% lower than current costs. The project is carried out by a consortium that includes universities, project owners, technology providers and organizations, who closely collaborate in all the project's phases.

The project poses a number of unique challenges:

- It involves the application of several innovative technologies, such as technologies for the generation of energy from renewable sources (e.g. building integrated wind energy and PV systems, and concentrated PV systems); technologies for improved building insulation; technologies for the reduction of energy consumption (e.g. thermo-solar HVAC systems and Building Energy Management Systems); and technologies for energy management at a settlement scale (e.g. a distribution network combining MV and LV with micro grid capabilities). Some of these technologies have not been previously implemented in construction projects.
- It has very ambitious goals in terms of energy performance and cost savings.
- It requires the coordination between a number of stakeholders, with very different backgrounds and incentives, who are located in different countries and have not previously collaborated.
- It involves an iterative design process that requires frequent changes and adjustments.

In order to successfully cope with these challenges, a set of control and optimization tools were developed and implemented in the project. These tools supported the incorporation of the highly innovative energy technologies in the design of the NZE settlements. They allowed the project team to handle frequent design changes, to successfully manage the risk of cost overruns, and to synchronize the work of the different stakeholders.

2 Control and Optimization Tools

The project included the development and implementation of three tools:

1. A cost control tool tracks each technology's initial costs during the design phase, so that any cost overrun can be easily recognized and treated.
2. A change management tool allows the project's designers to identify, examine and discuss the implications of every proposed change in the design before the change is executed, and prevent deviations from the project's cost and energy performance requirements.
3. A third tool supports a life cycle cost analysis of the technologies implemented in the project and an identification of their optimal configuration.

2.1 Methodology

In response to a tight schedule and the uncertainty regarding the users' exact requirements, an Evolutionary Prototyping methodology was adopted to develop the tools [1]. In the Prototyping approach an initial model (prototype) of the system is developed in order to demonstrate the functional abilities of the system. The goal is to receive feedback from the customer in order to decrease the uncertainty of the requirements and to increase the chances to successful product. The Prototyping approach is suitable for longer projects and less complicated systems. In the Evolutionary Prototyping approach the prototype is developed in order to be improved and eventually become the final product. Each iteration begins with the collection of requirements, development, evaluation and feedback. This loop continues until the final product is developed. The advantages of this approach are:

- Initial modelling is possible in the early stages of development
- It enables better understanding of the customers' requirements
- It allows an early detection of problems
- It supports the practical exploration of ideas

Using this methodology, all the tools were developed in the Google Spreadsheets platform. The use of this platform allows collaborative and simultaneous work, and automatic online data collection.

Each case study in the project is different and adjusted to the local climate, regulations and market. Therefore, the tools had to include four adjusted modules, one for each case study. As part of the methodology the initial prototypes of the tools included only one module, of one of the case studies. The feedback and comments received were used to adjust the module, while the rest of the modules, for the remaining case studies, were created based on the revised initial module. During the project the tools were revised multiple times according to the comments and feedbacks received from the users, but the main structure and workflow was maintained.

2.2 Cost control tool

Cost overruns are common in the construction industry, and have been observed in projects in many different countries (e.g. [3-5]). Construction costs tend to be systematically underestimated [2]. In a project in which many of the technologies are innovative, and have not been previously implemented, the risk of cost overruns is likely to be even higher. Since the reduction of costs is one of the main goals of the ZERO-PLUS project, it is essential to support an effective cost control process. To this end, a dedicated tool was developed and implemented to analyze the expected costs at each stage of the project.

Four types of costs were taken into consideration for the cost analysis:

- Initial costs - Costs at the start of the project: Acquisition, Supply and On-site Construction.
- Operational costs - Costs that are incurred during normal operation.
- Maintenance costs - Costs due to maintenance tasks – either scheduled or due to failures.
- End of life costs - Costs that appear at the technology's end of life.

The data described in Table 1 was collected to perform the analysis.

Table 1. Required input for LCC analysis and optimization

Data category	Field
General project data	Project lifetime (years)
	Interest rate (%)
	Electric energy cost (€/kWh)
	Net area size (m^2)
General technology data	Technology name
	Technology description
	Number of units
	Technology's life time (years)
Energy performance	Annual energy reduction (kWh/year)
	Annual energy production (kWh/year)
Initial costs	Initial cost of the matching technology in the reference case (€/unit)
	Initial Cost (€)
Operational costs	Annual cost of energy needs for the operation of the technology (€/year)
Maintenance costs	Annual maintenance cost (e.g. manpower, spare parts, upgrade costs) (€/year)
End of life costs	End of life cost (e.g. decommission cost, recycling cost, etc.)

Due to uncertainty regarding the costs of the technologies in the different life cycle phases, the three following values were requested for each inputted data: lower bound, most likely value, upper bound. The calculations were provided accordingly. After all the required data is completed, the tool calculates the cost's Present Value in Euros for 50 years (the assumed life cycle of the system). The results are presented in tabular and graphic forms for each component and for all components combined (Figure 1).

The tool was consistently used by the design teams, the coordinator and the project owners to identify cost overruns during the design phase. The tool's output provided insights on the project's cost performance, and the design of the case studies was reviewed to determine if overruns occurred. To date, all the projects are fulfilling the required cost reduction. However, this a tentative result since the project is only at the beginning of the construction phase.

All values' units are euro/50 years	Initial costs	Operational costs	Maintenance costs	End of life costs	Life Cycle Cost
Lower bound	110066	845	19864	175	130951
Most Likely	206851	931	36418	581	244782
Upper bound	436361	1064	425978	1940	865343

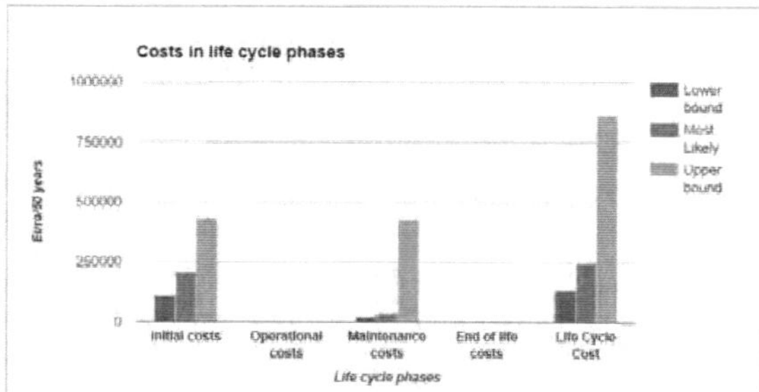

Figure 1: Cost analysis results

2.3 Change management tool

The ZERO-PLUS system design includes a number of different technologies, from different suppliers. It is expected to involve numerous changes, which, if not properly managed, may result in deviations from project goals. The change management process supports the identification and evaluation of the implications of every proposed change before it is implemented, in order to prevent such deviations. The implementation of changes will be monitored to ensure that their implications were correctly identified when the changes were requested.

The guiding principles of the change management process are

1. To allow the change initiator to have a clear view of the proposed change propagations between both physical components and functional requirements
2. To provide the change initiator with an easy, intuitive platform to manage the change and to progress with change execution

The change management process is implemented through the change management tool through the following procedure:

1. Fill in a Change Report Form.
2. Check the current cost and energy performance of the technology as provided by the cost control tool.
3. Answer questions regarding the expected implications of the change.
4. Check the impact of the proposed change on the project's Key Performance Indicators.
5. Check the impact of the proposed change for other technologies, as identified in a Technology Connections Table.
6. Add any other technologies that are expected to be affected by the proposed change and were not identified in step 5.
7. Receive a copy of the report to your email.
8. Share the Change Report with the partners responsible for the affected Key Performance Indicators (KPIs) and technologies.

The user will be able to coordinate and discuss the change with the partners responsible for the affected KPIs and technologies, and update the information on the identified implications of the proposed change based on those discussions. Once all affected sides are informed and have made the required configurations, the change is to be either approved or rejected.

The change management tool will support the teams in the implementation of changes during the construction phase of the project, which hasn't started yet. Therefore, results for the performance of this tool aren't available yet.

2.4 Life cycle cost optimization tool

A review of the currently available Life Cycle Cost (LCC) analysis and optimization tools was performed prior to development of the tool described here. A variety of tools were identified, but the clear majority of them could only perform an optimization for one building at a time, and for one project at a time. They would therefore have required a significant amount of customization for the Zero-Plus project. The tool that was developed in this research differs from the existing tools in the ability to optimize the design of a cluster of buildings, in different projects, and to compare the different results obtained.

After the cost analysis was completed, an optimization of the system's design according to its LCC was performed through the LCC optimization tool. The optimization's objective is to identify the configuration which will minimize the value of the target function, and yet fulfill 3 of the project's main objectives:

1. 16% initial cost reduction compared with the reference case.
2. Regulated energy consumption of less than 20 kWh/m2 per year.
3. Energy production from renewable energy sources of at least 50kWh/m^2 per year.

The target function is sum of the life cycle costs of Zero-Plus technologies minus the cost of the energy saved/produced by those technologies. The presence of the cost of the saved/produced energy in the target function is necessary to ensure that the optimal configuration will include technologies that are not only cheap, but also producing or saving more energy than the others:

$$Min\ LCC = \sum_{i=1}^{m} (u\_lcc_i - e_i) * X_i + \sum_{i=m+1}^{n} (u\_lcc_i * X_i - e_{i\_insulation}) \quad (1)$$

m – The number of relevant technologies, excluding the insulation.

n – The number of the technologies that were included in the LCC analysis.

u_lcc_i – The LCC of the technology as estimated in the LCC analysis stage (€/unit/50 years).

e_i – Present Value of 50 years' cost of energy produced/saved by one technology unit (€/unit/50 years).

$e_{i\_insulation}$ – Present Value of 50 years' cost of energy saved by the insulation (according to its thickness in mm) for 50 years (€/unit).

X_i - The size of the technology (number of units/thickness in mm).

A number of constraints were defined, to reflect the project's main objectives. Those constraints were then adjusted according to the specific attributes of each case study:

1. To represent the objective of reducing initial costs, the following constraint was used:

$$\frac{TCRZ - \sum_{i=1}^{n}(ic_i * X_i - cb_i) - \sum_{i=1}^{m} zp_i}{TCRZ} \geq 0.16 \quad (2)$$

$TCRZ$ – Total initial cost of the reference building (€/m2).

ic_i – Technology's initial cost per unit (€/m2/unit). Only the technologies that were included in the LCC analysis.

cb_i – Initial cost of the matching technology in the conventional building (€/m2).

n – The number of the technologies that were included in the LCC analysis.

zp_i – Initial cost of the technologies that were not included in the LCC analysis (€/m2).

m – The number of the technologies that weren't included in the LCC analysis.

2. To represent the objective of achieving net regulated energy consumption of less than 20 kWh/m^2 per year, the following constraint was used:

$$B - P - \sum_{i=1}^{m} (s_i + p_i) * (X_i - y_i) - \sum_{i=m+1}^{n} (s_i\_total\_after\_opt - s_i\_total\_before\_opt) \leq 20 \quad (3)$$

B – Settlement's annual energy consumption in the original design (kWh/m2/year).

P – Settlement's annual energy production in the original design (kWh/m2/year).

y_i – Technology's size (number of units/thickness in mm) in the original design.

s_i – Annual energy reduction contributed by the technology (excluding production) (kWh/m2/unit/year).

p_i – Annual energy production contributed by the technology (kWh/m2/unit/year).

$s_i\_total\_after\_opt$ - Annual energy reduction contributed by the insulation according to the optimized thickness (excluding production) (kWh/m2/year).

$s_i\_total\_before\_opt$ - Annual energy reduction contributed by the insulation according to the original thickness (excluding production) (kWh/m2/year).

The design teams used dedicated software to perform energy simulations, and provide data about the expected energy consumption in buildings with various levels of insulation and human activities.

3. To represent the objective of achieving an energy production from renewable energy sources of at least 50kWh/m2 per year, the following constraint was used:

$$P + \sum_{i=1}^{n} p_i * (X_i - y_i) \geq 50 \quad (4)$$

P – Settlement's annual energy production in the original design (kWh/m2/year).

p_i – Annual energy production contributed by the technology (kWh/m2/unit/year).

The target function's value was reduced by 5%-8% in three out of four case studies by implementing the described optimization process through the tool. These results are prior to actual implementation of technologies, and are therefore still tentative.

3 Conclusions

The ZERO-PLUS project is representative of projects which pose particular challenges due to the need to incorporate innovative technologies. These challenges were met by developing specific tools in order to support the design and management of the project. Cost control and optimization tools were successfully implemented in the project and used by the project partners. The change management tool is currently being tested by the partners following its adjustment according to their comments. This provided a validation of the appropriateness of the Evolutionary Prototyping methodology that had been adopted to develop the tools. This methodology required relatively few resources, while allowing a quick response to be given to evolving user requirements, and the adjustment and expansion of the tools according to those requirements.

Acknowledgements

This work has received funding from the European Union Horizon 2020Programme in the framework of the "ZERO-PLUS project: Achieving near Zero and Positive Energy Settlements in Europe using Advanced Energy Technology", under grant agreement no. 678407.

References

[1] R.A. Carter, Anton, A.I., Dagnino, A., & Williams L. Evolving beyond requirements creep: a risk-based evolutionary prototyping model, In *Proceedings. of the 5th IEEE Intl. Symposium on Requirements Engineering*, pages. 94-101, 2001.

[2] Flyvbjerg, B., Holm, M. S., & Buhl, S. Underestimating costs in public works, error or lie? *American Planning Association Journal*, 68(3), 279–295, 2002.

[3] Frimpong, Y., Oluwoye, J., & Crawford, L. Causes of delay and cost overruns in construction of groundwater projects in a developing countries; Ghana as a case study. *International Journal of Project Management*, 21(5): 321–326, 2003.

[4] Kaming, P. F., Olomolaiye, P. O., Holt, G. D., & Harris, F. C. (1997). Factors influencing construction time and cost overruns on high-rise projects in Indonesia. *Construction Management and Economics*, 15: 83–94, 2011.

[5] Mansfield, N., Ugwu, O., & Doran, T. Causes of delay and cost overruns in Nigerian construction projects. *International Journal of Project Management*, 12(4): 254–260, 1994.

[6] Marszal, A. J., & Heiselberg, P. Life cycle cost analysis of a multi-storey residential net zero energy building in Denmark. *Energy*, 36(9): 5600-5609,

2011.

[7] Salom, J., Widén, J., Candanedo, J., Sartori, I., Voss, K., & Marszal, A. Understanding net zero energy buildings: evaluation of load matching and grid interaction indicators. In *Proceedings of building simulation*, pages 2514-2521, 2011.

Application of Configuration Management for Engineering Information Management in Pipeline Projects

D.Y. Lee [1], S.K. Won[2], C.H. Han[3], M.Y. Cho[3], and J.B. Lee[3*]

[1] *Graduate Student, Department of Architectural Engineering, Kyung Hee University, Korea*

[2] *Research Professor, Ph.D. Department of Architectural Engineering, Kyung Hee University, Korea*

[3] *Professor, Ph.D. Department of Architectural Engineering, Kyung Hee University, Korea*

[*] *Corresponding author (leejb@khu.ac.kr, South Korea)*

E-mail: tjddn4566@khu.ac.kr, kcem@khu.ac.kr, chhan@khu.ac.kr, moonyoung.cho.1950@gmail.com

Abstract

Since April 2009, Korea has been importing 1.5 million tons of LNG(Liquefied Natural Gas) annually from the Sakhalin-II project, which is operated by Gazprom, Russia's largest state-run gas company. Also, it is expected that the cost of pipeline construction will reach $55 ~ 60 trillion by 2018 [1]. In this long-distance pipeline construction project, the importance of effective management of the vast amount of information generated throughout the life cycle stands out, and KOGAS(Korea Gas Corporation) has been cooperated with by Gazprom in a variety of fields such as natural gas supply and performing gas-related projects since the signing of the Gas Cooperation Agreement with Gazprom in December 2016. However, the level of related technology in Korea is still not competitive. Then, many experts sought to focus on the improvement of the life cycle process and efficient management [4], and it is very important to maintain up-to-date data and utilize it for work during long-term projects to reflect changes in pipeline projects [2]. Therefore, in this study, we propose a method to apply the Configuration Management to efficiently manage the changeable information generated during the engineering phase of the pipeline project. It is possible to systematically manage information that enables conformity and traceability of the design requirements and physical configuration of the changes in the engineering phase [13]. This study concentrates on Configuration Management which is one of the techniques applied to efficiently manage the information generated in the engineering phase [5]. So it is expected that efficient information change management would be achieved in the engineering phase by establishing data structure for each design requirement and physical configuration, and developing a model which facilitates linkage among the elements in the three areas of configuration management.

Keywords –

Configuration Management, Data Breakdown Structure, Pipeline, Engineering Information Management, Conformity, Traceability

1 Introduction

1.1 The Background and Purpose of Study

According to the decline in international exports and oil prices, plant orders have plummeted globally. At the same time, the number of obtained orders by overseas plant projects of Korean companies has also decreased. From 2011 until now (February, 2017), Korea's overseas plant orders have been steadily declining, as shown in Table 1, and costs have also recently dropped sharply [3].

On the other hand, Korea has been importing 1.5

Table 1. Trend of Investment Expenditures in overall Plant [3]

(Unit : $ 1,000,000)

Years	2011	2012	2013	2014	2015	2016	2017 (Jan-Feb)	Total
Cost of Orders	64,984	64,765	63,676	59,534	36,468	20,931	3,049	313,406

million tons of LNG(Liquefied Natural Gas) annually from the Sakhalin-II project, which is operated by Gazprom, Russia's largest state-run gas company, since April 2009. Also, it is expected that the cost of pipeline construction will reach $55 ~ 60 trillion by 2018 [1]. In this long-distance pipeline construction project, the importance of effective management of the vast amount of information generated throughout the life cycle stands out, and KOGAS(Korea Gas Corporation) has been cooperated with by Gazprom in a variety of fields such as natural gas supply and performing gas-related projects since the signing of the Gas Cooperation Agreement with Gazprom in December 2016. However, the level of related technology in Korea is still not overly competitive.

Therefore, the efficiency of the business conducted by domestic companies and their possessed technology should be enhanced in order to strengthen their competitive power. For this reason, many experts sought to focus on the improvement of the life cycle process and efficient management [4].

In addition, one of the major obstacles to the practical application and use of project management in pipeline projects is the occurrence of frequent changes. A typical example of such changes involves design change [5]. Thus, it is very important to maintain up-to-date data and utilize it for work during long-term projects to reflect changes in pipeline projects [2].

Therefore, in this study, the main objective is to propose a method to apply Configuration Management to efficiently manage the changeable information generated during the engineering phase of pipeline projects.

1.2 The Scope and Method of Study

The scope of this research was limited to data management in the engineering phase, which is very important for the successful execution of the project, by providing key information in later phases in the pipeline project [6], and the objective of this research is limited to the pipe itself which is the most important aspect of work in pipeline projects. Unlike other construction projects, plant projects are characterized by the fact that the process of installing pre-fabricated materials and facilities accounts for a large portion of the total project [7]. In particular, the pipe is the purpose of the plant facility and a vital work aspect on the process management side [8]. Therefore, this study focused on the pipes during the process in the engineering phase of pipeline projects.

This research was conducted in the following sequence, as shown in Figure 1. First, the previous studies of configuration management for information management systems in the overall plant industry involving domestic and overseas literature review were investigated. Second, the meaning of configuration management was defined and analyzed. Following that, the information of three major elements concerning configuration management in pipeline projects were collected and analyzed. Once this was done, the data breakdown structures of each major element were derived, and each breakdown structure was verified through questionnaires by experts in pipeline projects. The final step is that the way of applying the concept of CM(configuration management) to each breakdown structure was proposed, and the work process of linkage among data from each breakdown structure was proven by using examples.

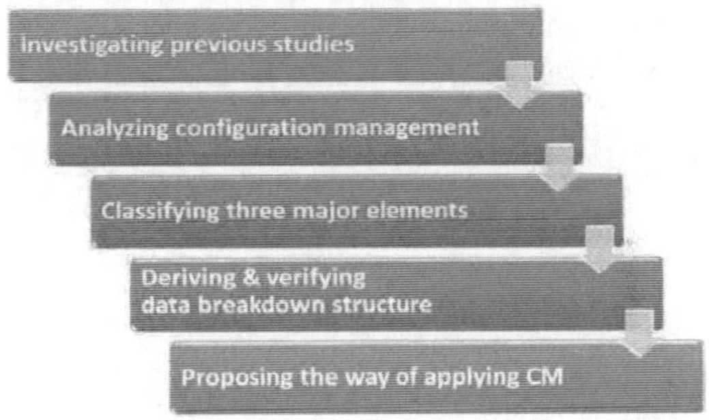

Figure 1. The sequence of this study

2 Literature Review

2.1 Previous Studies

2.1.1 Information Management in Pipeline Projects

The previous studies related to information management in pipeline projects are shown as Table 2.

Table 2. Major Research related to Information Management in Pipeline Projects

Author	Title	Main Content
Sunil (2016)	PIPEiD: Pipeline Infrastructure Database	Developing prioritization that can guide fundamental and applied research at institutions and entities funding research in water pipeline infrastructure.
Lee (2015)	Analysis of Data and Information Flow for Pipeline in Permafrost Area	Analyzing the data and information flow of pipeline projects in permafrost areas and developing the hierarchical structure of design and construction data for efficient administration of projects.
Won	Development of	Developing the process

(2015)	O&M Data Management System for Pipeline Project in Permafrost Area	and work breakdown structure of the design of the resource logistics network in extreme climate regions.

The studies of data and information flow analysis, derivation of hierarchical structure of design and construction data for project management, and identify the process and work breakdown structure in the design phase and use them to study the key factors for information management were conducted.

However, most of the studies attempted to adopt the information management technique in plant project to the technique in pipeline project. For this reason, in this study, it was derived to the main data management items for pipeline projects and proposed for adoption of the systematic management method.

2.1.2 Configuration Management

Research on configuration management that has been carried out in Korea and abroad were summarized in Table 3.

Table 3. Major Research related to Configuration Management

Author	Title	Main Content
Kang (2015)	Framework & Functions of Configuration Management (CM) in Nuclear Power Plants (NPP)	Seven industries (including defense, aerospace, software, engineering, architecture, civil engineering, nuclear power) that utilize the concept of configuration management (CM) were compared and analyzed based on the CM purpose, technique, and life-cycle perspectives.
Ko (2009)	A Study of Development of Change Process for Configuration Management in Construction Project Management	The review of the current systems now in use at worksites and are known to be efficient, and the introduction and application of configuration systems that are now a big issue.
Yoo (2008)	A Study on Developing Systems for Managing Information on Owner's Requirements in the Design Phase	The development of the efficient information system for managing owner's requirement throughout the design phases.
Schenk (1985)	Configuration management as applied to engineering projects	Configuration management is a management mechanism for handling changes by integrating technical and administrative decisions with respect to project goals and objectives.

The studies of development of the change process for configuration management in construction project management and developing systems for managing information on owner's requirements in the design phase were conducted. However, there was no research on configuration management in pipeline projects. In addition, most of the studies were focused on emphasizing the necessity of configuration management and proposing an abstract concept to apply configuration management to construction or plant projects. Therefore, this study was concentrated on the more concrete method of applying configuration management to information management in pipeline projects.

2.2 Introduction to Configuration Management

2.2.1 General Meaning of Configuration Management

Configuration Management(CM) was created by the US Department of Defense in the 1950s to improve weapon systems. According to the IEEE 610 standard, CM is principally applied to the change process in technical and management direction and supervision. Additionally, CM identifies the functional and physical characteristics of configuration items, and manages the changes in those characteristics. After that, it records and reports change procedures and implementation status to verify compliance with specified requirements.

Although CM has been important for all the manufacturing and production industries since the industrial revolution, very few theoretical frameworks for CM are available in the literature. Based on knowledge and work experience in the production industry, it was found that three sub-domains, or three types of formal descriptions, are fundamental for CM [16].

In the case of general large-scale plant construction projects, CM methodology is applied to efficient change management for systematic project management. It is required to maintain 'Design Requirements – Physical Configuration – Facility Configuration Information' consistently within the project period [2].

The strongest property of CM is that it enables information conformity and traceability. Through these two critical properties, the most important process among various standards for CM is change management [13].

2.2.2 Three Major Elements of CM

Basically, CM consists of three elements: Design Requirements, Physical Configuration, and Facility Configuration Information. As shown in Figure 2, the information contained in the three elements must be linked with each other and always matched. This is called CM Equilibrium, which is the basic concept of CM.

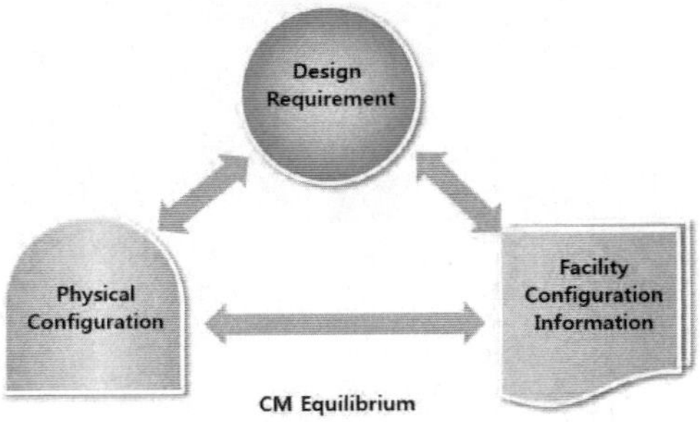

Figure 2. Three Major Elements of CM [2]

'Design Requirements' pertains to information related to design requirements, including design criteria, regulatory requirements, and other design components. 'Physical Configuration' is the facility(functional location) information, including its design result and construction result. Lastly, 'Facility Configuration Information' pertains to basic/detailed design information, including purchase specifications, drawings and procedure documentation.

3 Application of CM

3.1 Three Elements of CM Applying to Pipeline Projects

Design requirements management is an activity that establishes, documents, maintains and utilizes the design requirements for the CM objects(structure, equipment, etc.) of facilities. Also, establishment of a proper management system for pipeline design requirements and design is an important management factor during the lifecycle of the pipeline project.

3.1.1 Breakdown Structure of Design Requirements

The design requirements for the pipeline project in this study are referenced in the Design Basis Document of Application for Approval by several major pipeline projects such as Mackenzie Valley Pipeline. The breakdown structure of design requirements related to piping is shown in Figure 3.

Piping is rooted from a sub-class called piping equipment, which is rooted from a class called mechanical equipment. The design requirements of piping are largely divided into seven categories: Codes and Standards, Piping Joints, Loading Conditions, Environmental Factors, Materials-related Considerations, Thermal Insulations, and Sizing of Piping Systems. Each of those categories has their own components.

1	mechanical equipment class				
	1.1	piping equipment			
		1.1.1	Piping		
			1.1.1.1	Codes and Standards	
				1.1.1.1.1	Differentiation between Codes and Standards
				1.1.1.1.2	The ASME Pressure Classification System
				1.1.1.1.3	Reference Codes and Standards
			1.1.1.2	Piping Joints	
				1.1.1.2.1	Butt-welded Joints
				1.1.1.2.2	Socket-welded Joints
				1.1.1.2.3	Brazed and Soldered Joints
				1.1.1.2.4	Threaded or Screwed Joints
				1.1.1.2.5	Grooved Joints
				1.1.1.2.6	Flanged Joints
				1.1.1.2.7	Compression Joints
			1.1.1.3	Loading Conditions	
				1.1.1.3.1	Design Pressure
				1.1.1.3.2	Design Temperature
				1.1.1.3.3	Deadweight
				1.1.1.3.4	Wind Load
				1.1.1.3.5	Snow and Ice Loads
				1.1.1.3.6	Seismic (Earthquake) Loads
				1.1.1.3.7	Hydraulic Transient Loads
				1.1.1.3.8	Acoustically Induced Vibration Loads
				1.1.1.3.9	Relative Anchor Movements
			1.1.1.4	Environmental Factors	
				1.1.1.4.1	Corrosion
				1.1.1.4.2	Erosion
				1.1.1.4.3	Physical Damage
				1.1.1.4.4	Erosion-Corrosion (Flow-Assisted Corrosion)
			1.1.1.5	Materials-Related Considerations	
				1.1.1.5.1	Strength
				1.1.1.5.2	Toughness
				1.1.1.5.3	Corrosion Resistance
			1.1.1.6	Thermal Insulation	
			1.1.1.7	Sizing of a Piping System	
				1.1.1.7.1	System Fluid Flow Design
				1.1.1.7.2	Pressure-Integrity Design
				1.1.1.7.3	Determination of Pipe Wall Thickness
				1.1.1.7.4	Determining the Pressure Class for In-Line Components
				1.1.1.7.5	Determining the Design Conditions and Pressure Class of a Piping System
				1.1.1.7.6	Design of Piping for Internal and External Pressure
		1.1.2	flange		
		1.1.3	manifold		
		1.1.4	seal		
		...	...		

Figure 3. Breakdown Structure of Design Requirements related to Piping

3.1.2 Breakdown Structure of Physical Configuration

Physical configuration involves information about real facilities and its components. There are two results which include its design and construction results. The design results of physical configuration were analyzed and only the piping part is shown in Figure 4 because of the excessive amount of data.

Piping is inherited by a sub-class called piping equipment, which is inherited by a class called mechanical equipment. It is easily noticed that classification of high levels consists same as the breakdown structure of design requirements. This infers the possibility of linkage between design requirements and physical configuration, and this is illustrated with chapter 3.2. The piping of physical configuration is largely divided into four categories: Pipe, Valve, Control and Monitoring, and Miscellaneous. Each of those categories has their own components.

1	mechanical equipment class
1.1	piping equipment
1.1.1	Piping
1.1.1.1	Pipe
1.1.1.1.1	Fastener/ bolts
1.1.1.1.2	Joints
1.1.1.1.3	Flange
1.1.1.1.4	Header
1.1.1.1.5	Lining
1.1.1.1.6	Pipe element
1.1.1.1.7	Plug
1.1.1.2	Valve
1.1.1.2.1	Valve
1.1.1.2.2	Actuator
1.1.1.2.3	Control and monitoring
1.1.1.3	Control and monitoring
1.1.1.3.1	Actuating device
1.1.1.3.2	Control unit
1.1.1.3.3	Internal power supply
1.1.1.3.4	Monitoring
1.1.1.3.5	Sensors
1.1.1.3.6	Valves
1.1.1.3.7	Wiring
1.1.1.3.8	Piping
1.1.1.3.9	Seals
1.1.1.4.1	Pipe support
1.1.1.4.2	Others
1.1.1.4	Miscellaneous
1.1.2	flange
1.1.3	manifold
1.1.4	seal
...	...

Figure 4. Breakdown Structure of Physical Configuration related to Piping

3.1.3 The components of Facility Configuration Information

Facility configuration information in the engineering phase includes most of the engineering information except information for design requirements and physical configuration. The main elements of facility CM include basic and detailed design information. There are so many types of information in basic and detailed design, but the three major components are purchase specifications, drawings and procedure documentation. The example of facility CM for pipe is shown in Figure 5.

Even though there isn't any standardized breakdown structure of facility configuration information since it is just information of basic/detailed design like property data, all information should be linked with proper components of design requirements and physical configuration.

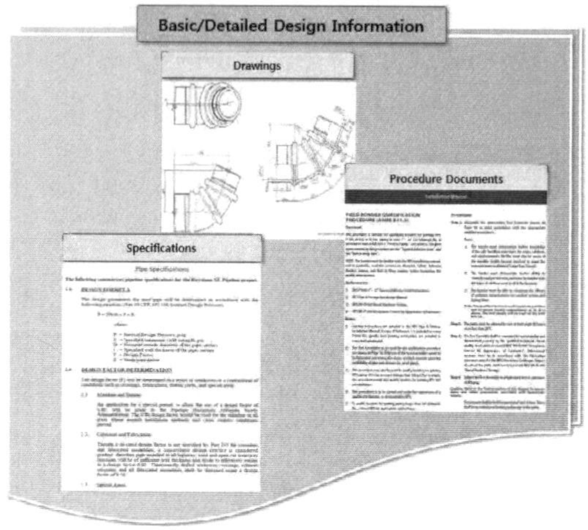

Figure 5. The Components of Facility Configuration Information related to Piping

3.2 The Method of Applying CM to Engineering Information

3.2.1 Application of Tag System

Unique Number	Tag Class Name	Unique Number	Property
5247	pipe	6164	lower limit design temperature
5247	pipe	6189	lower limit operating temperature
5247	pipe	6392	normal operating flow velocity
5247	pipe	6395	normal operating pressure
5247	pipe	6396	normal operating pressure drop
5247	pipe	6398	normal operating temperature
5247	pipe	6887	test pressure
5247	pipe	7004	upper limit design pressure
5247	pipe	7005	upper limit design temperature
5247	pipe	7033	upper limit operating pressure
5247	pipe	7034	upper limit operating temperature
5247	pipe	40000007	anti corrosion coating thickness
5247	pipe	40000008	anti corrosion coating type
5247	pipe	40000030	buried
5247	pipe	40000071	critical pipeline
5247	pipe	40000126	flow line type
5247	pipe	40000132	fluid name
5247	pipe	40000133	fluid phase
5247	pipe	40000227	maximum burial depth
5247	pipe	40000230	minimum burial depth
5247	pipe	40000246	nominal pipe diameter
5247	pipe	40000249	nominal wall thickness
5247	pipe	40000250	non destructive testing methods
5247	pipe	40000259	normal operating gas density
5247	pipe	40000273	normal operating liquid density
5247	pipe	40000275	normal operating mass flow rate
5247	pipe	40000276	normal operating mixed liquid and vapour density
5247	pipe	40000289	normal operating volume flow rate
5247	pipe	40000315	pipeline colour code
5247	pipe	40000316	pipeline material and grade
5247	pipe	40000337	remarks
5247	pipe	40000372	special service
5247	pipe	40000413	thermal insulation coating thickness
5247	pipe	40000414	thermal insulation coating type
5247	pipe	40000472	upper limit operating mass flow rate
5247	pipe	40000482	weight coating density
5247	pipe	40000483	weight coating thickness

Figure 6. A portion of pipe tag list

The definition of 'tag' means a word or keyword that is given for use in searching for information. In essence, a tag system is a system that collects necessary information by using these tags as data.

In this study, the tag system was used to be linked to each of the correlated data by tagging all piping data of engineering phase in pipeline projects. In addition, each of the inquired data has two properties which are basic file attributes and registrants. A name, a format, a location, a size, a creation date and a modified date are included as basic file attributes. Also, there exists a registrant ID and a company code as registrants. A portion of the pipe tag list in our research is shown in Figure 6.

3.2.2 Applying CM to Pipe Information in Engineering phase

In this study, CM enables to link all engineering information assigned tags in three main elements. The most basic method of application of CM involves establishing one among three major elements as a standard, and managing to link information of other elements based on tags. Specifically, based on the physical configuration, related information generated in design requirements and facility configuration information are linked. Starting from this process, the information generated in the subsequent process steps(procurement, construction, and maintenance) are included in the related category, so that the changeable information can be efficiently managed.

To facilitate easier understanding, an example of a part of the piping work which is the scope of this study is shown in Figure 7. Among the large amount of data, only a portion of piping data is extracted from each element, and information such as 'Physical Configuration for Pipe', 'Design Basis Document for Piping / Code & Standards for Piping', 'Drawing for Pipe / Procedure Document for Pipe / Specification for Pipe' are extracted. All information is managed through linkage with each other due to connections from the tag system. From the design stage to the maintenance stage, newly generated pipe-related information are added to the corresponding elements so that information can be continuously traced and conformed.

3.3 The Expected Result of Applying CM to Real Design Changes

In order to demonstrate real world design changes and how the proposed method handle the changes appropriately, the comparison of work procedures in design change between existing information system and the system applied by CM.

A prior condition is that public employer requests a design change due to reduction in the budget of pilot test in pipeline construction project. This can affect the work of both engineer(architect) and contractors such as main contractor and vendors in the aspects of three major elements of CM which are design requirement, physical configuration and facility configuration information. There are two alternatives which are reducing the diameter of pipe from 32 inch to 30 inch and the length of pipe from 100m to 60m. The comparison of each procedure is shown in Table 4.

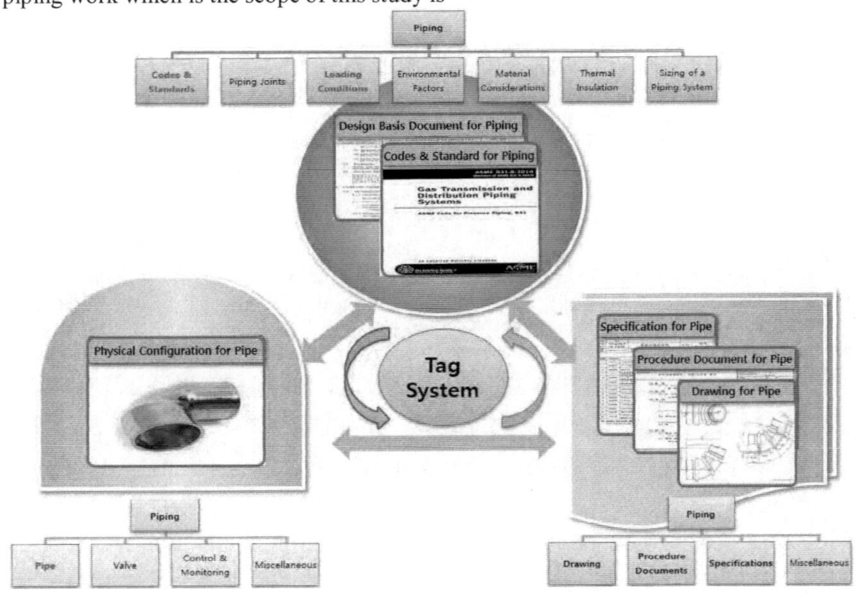

Figure 7. Application of CM for Piping related Engineering Information

Table 4. Comparison of work procedures in design change between existing information system and the system

As-Is			To-Be		
Object	Work	Effect	Object	Work	Effect
Public Employer	Request for design change due to reduction in the budget ↓ Notify the architect or engineer		Public Employer	Request for design change due to reduction in the budget ↓ Notify the architect or engineer	
Architect or Engineer	Review 'Design Requirements' such as Code & Standard and Design Basis Document ↓	Increase the probability of human errors occurrence since the review of design requirements proceeds manually	Architect or Engineer & Vendors	Review 'Design requirements' such as Code & Standard and Design Basis Document in a short period of time through the change management system applied by CM	Reduce the time and make a high accuracy of the works
	Change 'Facility Configuration Information' such as design drawings and specifications ↓ Notify the vendors of design changes	Increase the probability of problems occurrence due to the inefficient communication Ex) Occur the errors producing pipe due to the notifications of incorrect Revision number		Recognize and identify design changes without additional communication by participants in the system	Find definitive design changes in the final Rev. and increase communication efficiency by all participants
				Produce the pipes according to design change without any problem caused by information error	Detect information immediately when searching for relevant information in the future
Vendors	Produce real pipes and change the information of 'Physical Configuration' based on changed design information	Overall, require extra time due to unnecessary works			

4 Conclusion

In this study, research was carried out to solve the frequent changing occurrences which are one of the main obstacles of information management of the engineering phase in pipeline projects. Since these changes occur frequently in the actual construction, top managers who manage the construction are satisfied with utilizing the time schedule of work as promotional purposes and avoid performance management due to concerns about additional process management costs in actual construction management. The new approach that this study suggests to solve this problem is applying configuration management(CM). Therefore, the basic concept of CM, which is one of the information change management techniques, was explained. In addition, according to the three major elements of CM, all data breakdown structures were classified. Following that, the application of CM to the pipeline project is presented through an example of piping which is the main work type of the pipeline.

The results of this study will lead to a study on the automatic linkage method of detailed information in three major elements of the CM in the future. The final goal of this research involves reducing the occurrence of changes in the plant project by creating the CM management system in order to ensure efficient change management.

Acknowledgements

This research was supported by a grant (13IFIP-B06700801) from the Industrial Facilities & Infrastructure Research Program funded by the Ministry of Land, Infrastructure and Transport of the Korean government.

References

[1] Kim C., Won S., Lee J., and Han C. A Study on the Data Classification in Engineering Stage of Pipeline Project in Extreme Cold Weather, *Proceedings of the Conf. on Korea Institute of Building Construction*, 14(2): 214-215, Nov 2014.

[2] An K. The method of applying configuration management for operation and maintenance activities in plant, *Journal of the Korean Society of Mechanical Engineers,* 54(12): 39-44, Dec 2014.

[3] Korea Plant Industries Association(KPIA) in 2017. Order Statistics. On-line: http://www.kopia.or.kr/index.jsp, Accessed: 22/02/2017

[4] Won S., Soh J., Lee J., and Han C. Fundamental Investigation of Knowledge Management System for a Plant, *The 3rd International Conference on Construction Engineering and Project Management(ICCEPM) Summit Forum*, 3, May 2009.

[5] Oh S. A Study on the Integration of Schedule Control and Change Control System, Dissertation, Chung-Ang University, CAU, 2002.

[6] Son J. A study on the critical success factors for engineering step on plant project, *Korea Journal of Construction Engineering and Management*, 227-234, 2007.

[7] Kini D. Materials management: the key to successful project management, *Journal of Management in Engineering*, ASCE, 15(4): 30-34, 1999.

[8] Kim C. Plant piping construction management plan based on augmented reality technology, Dissertation, Sungkyunkwan University, SKKU, 2011.

[9] Sunil S., Walter G., Peter K., Fred P., and David H. PIPEiD: Pipeline Infrastructure Database, *Proceedings of the Pipelines 2016 Conference*, ASCE, 565-574, 2016.

[10] Lee D., Won S., and Lee J. Analysis of Data and Information Flow for Pipeline in Permafrost Area, *The 6th International Conference on Construction Engineering and Project Management (ICCEPM 2015)*, 774-775, 2015.

[11] Won S., Kim C., Han C., and Lee J. Development of O&M Data Management System for Pipeline Project in Permafrost Area, *ISARC 2015*, 2015.

[12] Kang M., and Jung Y. Framework & Functions of Configuration Management (CM) in Nuclear Power Plants (NPP), *Korean Journal of Construction Engineering and Management*, KICEM, 16(3): 101-112, 2015

[13] Ko H., and Park H. A Study of Development of Change Process for Configuration Management in Construction Project Management, *Journal of the Korean Society of Civil Engineering*, KSCE, 29(1): 81-89, 2009.

[14] Yoo S., Lee J., Park J., and Jun J. A Study on Developing Systems for Managing Information on Owner's Requirements in the Design Phase, *Korean Journal of Construction Engineering and Management*, KICEM, 9(3): 126-135, 2008.

[15] Schenk, R. Configuration management as applied to engineering projects, *Journal of Management in Engineering*, ASCE, 1(3): 157- 165, 1985.

[16] Wang S. An Object-Oriented Approach to Plant Configuration Management Information Systems Analysis, *Industrial Management & Data Systems*, 99(4): 159~164, 1994

Leading safety indicators and automated tools in the construction industry

Rita Yi Man Li[a] and Tat Ho Leung[b],

[a]Sustainable Real Estate Research Center / Real Estate and Economics Research Lab, Hong Kong Shue Yan University, HK
[b]School of Environment, Education and Development, University of Manchester, Manchester, UK
E-mail: ymli@hksyu.edu, thleung0418@gmail.com

Abstract –

Hong Kong is a city which has a large number of high-rise buildings. In this paper, we study the factors that lead to construction accidents on building sites by interviewing the relevant parties. In the second part of the study, we extract the causes of accidents from court case reports. Finally, we include a single case study based on the activities of one of the largest construction companies in Hong Kong – who use robotics, drones, three-dimensional (3D) printing, virtual reality, wearable robotics etc. on their construction sites. We look at whether innovative tools can be effective in improving safety performance on construction sites.

Keywords –

Construction safety; Skyscrapers; Poor weather; Robotics; Virtual reality; 3D printing; Infographics; Drone photogrammetry

1. Introduction

The lethal working environments which can often be found in the building industry have led to there being approximately 60,000 fatalities worldwide in this industry each year. This is higher than the number of fatalities suffered by any other industry. In addition, mishaps not only affect workers' wellbeing, but also the quality of life of whole families (Alarcón et al., 2016). In particular, the victims are bread-winners, and so accidents which harm them severely affect their families in a number of ways. Despite the fact that everyone involved wants construction safety performance to be improved, construction sites remain one of the most perilous workplaces overall: a danger-free site is the exception rather than the rule (Geoff, 2016). In recent decades, there has been a trend towards increasingly higher buildings, and this has increased the complexity of construction projects and the level of risk involved (Guo et al., 2015). Timely risk assessments are an absolute necessity in relation to many construction projects (Guo et al., 2015).

Accidents on sites are hardly ever the result of pure bad luck. Accidents are not now viewed as arbitrary chance occurrences but instead as the consequence of a progression of events that could conceivably be identified and controlled (Alarcón et al., 2016). A large proportion of these "misfortunes" occur mostly because of the absence of obvious safety measures: for example, safety nets and guiderails (Edrei and Isaac, 2016, forthcoming, Li and Poon, 2013, Li, 2015). Temporary workers are regularly used to provide manpower when time is short. These workers are at an increased risk of injury because many of them change job assignments several times a year, and they are not always trained or re-trained. Host employers are less willing to devote training resources to temporary workers. It is not uncommon for a temporary worker to suffer a fatality after just 1 or 2 days on the job (Mroszczyk, 2015).

Previous research has shown that designers can play an important role in ensuring construction safety (Bong et al., 2015). Indeed, accidents often happen because of oversights by designers and/or safety officers which occur when they arrange or execute safety activities. The application of innovative initiatives can enhance the development of safety on sites (Edrei and Isaac, 2016, forthcoming), and any commitment to reduce work related accidents can be viewed as worthwhile (Alarcón et al., 2016).

Accidents originate from a number of organizational levels. However, a study of 1180 firms using 221 different working practices over four years in Chile demonstrated that working practices are the most important factors that affect accident rates (Alarcón et al., 2016).

In this present study, we firstly review the factors which lead to accidents on construction sites from practitioners' and from trial judges' perspectives – via extensive qualitative interviews. We compare the factors mentioned by the groups of interviewees with regard to high-rise and low-rise buildings. Whilst the practitioners have day to day work experience on construction sites, their views can be subjective. On the other hand, the trial judges have to consider all the evidence provided by the victims, employers, witnesses, literature etc., and assess the factors which have led to the accidents they have dealt with. Hence they can be considered more objective. Thus, the second aim will be to compare the practitioners and the judges' perceptions on the factors which commonly lead to accidents. Finally, in the era of information technology (IT), advances in various kinds of IT tool present an unprecedented opportunity to improve safety performance on construction sites (Skibniewski, 2014, Hoła et al., 2015). We will use one single case study to demonstrate the application of innovative technological tools on construction sites. This should provide some idea concerning whether tools such as 3D printing, robots, aerial photogrammetry etc. can reduce the chances of accidents on construction sites.

2. Leading and lagging safety performance indicators

Historically, the construction industry examines safety performance via lagging indicators; these include numbers of injuries, illnesses and fatalities. One noteworthy limitation of safety performance appraisal by means of lagging indicators is that incidents must happen before risks or hazardous conduct can be recognized. Leading indicators are safety measurements that proactively survey execution safety performance by gaging procedures, exercises and conditions that characterize execution and can be used to foresee future consequences (Marks et al., 2016).

One example of a leading indicator is that of the near-hit. This is characterized as an occurrence in which no harm to property or damage to an individual occurs, but which could have led to such consequences, given a slight change in time or position. The preference for this indicator is based on the fact that information can be gathered and analysed without the requirement that any damage to property or persons has occurred. Construction companies are required to report incidents which affect their business. Measurements derived from such reports, however, cannot reflect whether the consequences of hazard have been alleviated or the likelihood of harm has been reduced. Leading indicators are measures of procedures, exercises and conditions currently in place, which can anticipate future results. Unmitigated high-hazard circumstances can bring about significant or even lethal damage if they are permitted to persist. Linear causation models, such as loss-causation and domino models, argue that accidents are the final consequence of a succession of circumstances, and provide a sound motivation to gather and analyse close-hit information. It has also been found that, effectively, most significant accidents to individuals could have been prevented (Marks et al., 2016).

Construction safety is measurable from the behavior patterns of the different stakeholders involved. This especially relates to the behavior patterns of workers as assessed via the ground theory method (Guo et al., 2015). Meanwhile, there are statistical analyses which specifically estimate the risks involved in a particular construction project. For instance, the risk assessments used in Saudi Arabian aviation construction utilize ANOVA tests to identify risks (Banhdadi and Kishk, 2017). Alternatively, ontological modeling and document modeling are used to facilitate better measurement of the lack of knowledge present (Wang, 2015). Through assessing the level of ignorance present in terms of specific safety requirements, managers can attain a better awareness of the construction risks. To a certain extent, the above literature shows the potential of implementing automated approaches in the construction industry.

3. Innovative technologies used on site to enhance construction safety

3.1. Robotics

Robots can be utilized for the execution of different construction tasks: for example, the assembling of parts and the construction of the inside and outside skins of buildings. The operation of such robots is based on advanced control and detection technologies (Warszawski, 1985). Some robots are designed to work in poor climatic conditions and tough site environments (Skibniewski, 1988). It has been demonstrated that robots could be used to mix and spread tarmac. Such applications use programmed guidance control with mechanical sensors and the calculation of a clearing velocity; this latter leads to the controlled starting and stopping of all clearing capacities. Moreover, this work can be performed correctly using artificial vision and sensors. Also, a graphical remote control framework allows a human operator to control the process progressively, continually altering the procedure. Window glass mounting or board settling can be undertaken utilizing a robot with a pneumatic actuator and a servo engine. Straightforward and independent robots can assemble two-dimensional structures and utilize pre-assembled modules as building pieces. Modules are able to do some data collection, empowering them to share long range basic data and convey it to the robots (Elattar, 2006).

3.2. Drone photogrammetry

Photogrammetry strategies have been used for a long time together with CAD and virtual models of building structures – for architecture purposes. Photogrammetry utilizes two images for measurement purposes along with a fundamental triangulation rule to calculate the areas of focuses in three measurements. The relatively low-cost and accessibility of computerized cameras provides a justification for the use of photogrammetry-based methodologies for as-manufactured documentation. Cameras are less expensive, more compact and quicker in terms of capturing relevant images, as contrasted with other information gathering devices. The captured images are later handled by software in order to obtain useful geometric information (Akanmu et al., 2014). With the popularity of various types of drones in Hong Kong, much of this photogrammetry work, in recent years, has been undertaken by relatively cheap drones instead of relatively expensive helicopters.

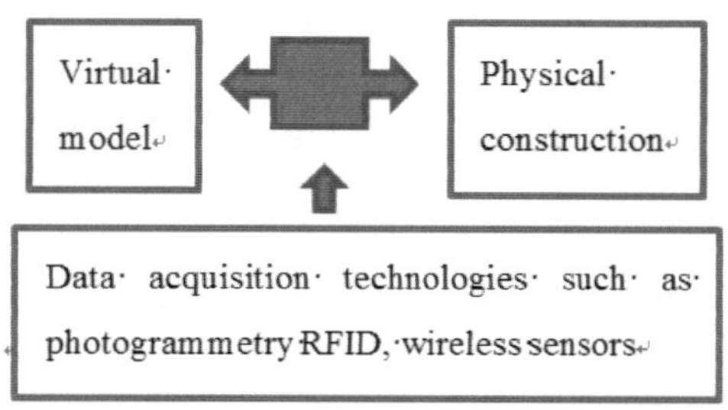

Figure 1. Bi-directional data acquisition between virtual and physical construction (Akanmu et al., 2014).

3.3. Virtual Reality (VR) simulation and training

VR can create genuine work experience. This has been proposed as a viable form of safety training for the mining industry. VR can provide safety training exercises at any time, even when there are no experienced staff on hand to deliver the training.. A personal computer is sufficient to provide the environment for VR training. Furthermore, VR based training has demonstrated that it is a useful technique for training workers in tower crane operation and in safety procedures for iron-workers. Nevertheless, critics have commented that it is questionable whether VR frameworks can substitute for genuine experience. In Australia, some proposed VR training programmes have recognized areas in which formal training strategies and systems cannot replace the health and safety expertise acquired by experienced laborers. This situation motivates innovative thinking and has inspired specialists to think further about technologically-based improvements to health and safety. VR systems have also been connected with the efforts of training specialists to act in and record safety training videos. They also allow users to collect safety and health data rapidly and effectively when the specialists need them. Thus, VR training is a successful technology for sharing and enhancing safety and health learning (Edirisinghe and Lingard, 2016).

3.4 Modular building structure

Modular buildings are typically safer than conventional buildings in terms of site construction because of the stable work area they impose. In regard to these, specialists can be fully knowledgeable about the dangers involved. Some of the ways in which safety is maintained are as follows: not working in tight spaces on site, performing off-site or on-ground assembly (as opposed to working from structures), and not working in difficult weather conditions. With modular buildings, it is less difficult to screen-out dangerous activities, less energy needs be used on the construction site, and fewer local contractual workers and laborers are required in general. Modular building can mean assembling the entire building or just some parts off-site. The study discovered 125 mishaps associated with pre-assembled fabrication development with the most frequent types of hazard being "crack" and "fall" hazards (Fard et al., 2015).

3.5 3D printing

Many of us may take the view that the next step for modular buildings will be that of printing complete three-dimensional building structures. 3D printing's has the potential to enable buildings to be constructed more quickly with lower costs, and with the further incorporation of building data into the development procedure. These potentialities mark it out as one of the new directions that the construction industry may take. Besides the fact that this technology could decrease the number of labourers required on sites (and so improve safety), it may also lead to a lessening in development times and an increase in engineering flexibility. Nonetheless, current 3D printing technology is not yet suited to large scale items. These technologies are characterised by prevailing configuration approaches and have an exceptionally constrained scope of materials that can be utilized (Perkins and Skitmore, 2015).

3.6 Building Information Modelling

In construction projects, there are various participants involved in the different stages. Building Information Modeling (BIM) is a useful method that allows various stakeholders to co-operate together (Liu et al., 2017). With the use of data-collecting sensors, it can significantly reduce the probability of worksite accidents – because of the monitoring process used by the data acquisition system (Druley et al., 2016). In a more sophisticated examination of the consequences of these developments, it can be seen that the design of buildings, which is highly coherent with construction risks, can effectively be converted from a force-based design into a probabilistic performance-based design procedure (Yarahmadi et al., 2017).

4 What are the leading safety indicators? A glance at the results from the interviews

In this research, we use an information graphics (infographics) approach whereby we analyse and visualize the frequency with which words are mentioned. The infographics which are often used in the mainstream media are bar and line charts; these make data visually attractive but at the same time convey the message embedded in the data easily (Li et al., 2015). These methods have been applied in previous research studies such as Li et al. (2016) – in that case to analyze the objectives of competition law. In this paper, we adopts Tagul's technique to visualize the data as word clouds (Tagul, 2016). The results from the interviews show that the most important reason leading to accidents in high-rise construction sites is seen to be falls from height as indicated by the frequency of use (and therefore large size) of the associated words.

Table 1. Factors that lead to construction accidents on high-rise as opposed to low-rise building sites

Five most important factors that lead to construction accidents on high-rise building sites	Five most important factors that lead to construction accidents on low-rise building sites
The most important factor 	The most important factor 
The second most important factor 	The second most important factor 
The third most important factor 	The third most important factor 
The fourth most important factor 	The fourth most important factor 
The fifth most important factor 	The fifth most important factor 

5 Results from court cases research

In the second part of the research, court cases which took place from 1991 to 2015 were used. We extracted the factors that were considered by the judge to be the most important factors leading to construction accidents on sites. After that, utilising Tagul's (2016) methodology, we used the infographics approach to study the factors that are more important. Stop-words included "contributory", "provide", "onto", "defendant", "mentioned", "1st part", "had", and "common". The results show that the most important reasons in the eyes of judges are related to negligence, equipment, and statutory issues. Unlike the practitioners, judges were more concerned with the indirect factors that lead to construction accidents. Factors like bad weather were never the most important from their viewpoints.

Figure 2. Judges' viewpoints with regard to all types of construction work on sites

6 Case study: Gammon construction

In recent years, firms have been confronted with the difficulties involved in applying innovations, especially computer-based innovations, to their core working practices.. The dissemination of developments hypothesis is utilized to examine how advanced innovations diffuse crosswise over complex firms (Shibeika and Harty, 2015). In the third part of the present study, we shed light on one construction company's innovative strategies for enhancing construction safety.

6.1. Curtain Wall – Robotic Installation

Gammon introduced the first curtain-wall installation robot to Hong Kong; they did so to reduce the labour intensity of their operations. The use of robots reduces the quantity of manual operation required, and, in this case, relieved a serious shortage of curtain wall installers (the application of robots can reduce the number of installation workers by 25%), and also enhances construction safety (Gammon Construction, 2016b). Furthermore, in order to deal with the upcoming labour shortage due to the ageing of the workers involved, and to avoid on-site construction risks, robotics has become increasingly important (Gammon Construction, 2016a). For instance, the company has introduced Exoskeleton and ZeroG Arm devices in order to assist workers in handling heavy loads/equipment. Specifically, Gammon provides an exoskeleton which can automatically sense the workers' movement, then provide back support for the workers so that they can lift heavy objects. This can reduce the risk of back injuries very effectively (Gammon Construction, 2016a). At the same time, ZeroG Arm can help workers to maneuver heavy equipment. The energy loss by workers can be greatly be reduced as this latter device provides an experience of near weightlessness in relation to the load/heavy equipment, and so allows operations to be performed more quickly and easily. (Gammon Construction, 2016a).

6.2. Virtual Reality Safety Training

Online gaming is a common trend nowadays especially among the younger generation. Gammon is the forerunner in incorporating virtual reality (VR) into safety training. Construction sites are simulated in 2D and 3D game environments to enable trainees to experience different scenarios in virtual construction sites. Safety Manager Kwok Wai-yin comments that training, as such, is more effective and convincing than lectures. It can draw trainees' attention to specific issues, stimulate their responses and enhance mutual communication. It changes the workers' mode of thinking, boosts site safety and enables the firm to approach closer to the goal of zero accidents on sites (Gammon Construction, 2016b).

6.3. i-Core Development

It often requires a great deal of work to track down the location of materials when they are in transit from where they are produced to the location at which they are to be installed on a site. i-Core is a standalone, extendable and programmable electronic circuit with GPS, GSM, a thermometer, Bluetooth, a gyroscope and a barometer, all integrated into one i-Core chip. Equipped with GSM/GPS technologies, Gammon's i-Core provides a logistics-management solution for expensive construction materials such as chiller plants and precast concrete products. It is able to track real-time locations and share location information via an online database. It enhances efficiency, productivity and saves manpower in the tracking process (Gammon Construction, 2016b).

6.4. Precast, prefabrication

For the Tuen Mun - Chek Lap Kok Link – Southern Connection Viaduct Section venture, the company augmented its use of precast items for the marine viaduct components. An example of such a component was the precast heap top shells. The shells are fabricated off-site, and are then simply put on top of the perpetual heaps and dewatered to bolster the deep heap top's development, without the need for cofferdams or marine falsework. Thus, the quantity of rebar fixers, welders, woodworkers, riggers and jumpers could be diminished hugely. The precast approach likewise decreases the quantities of materials used, the amount of work needed at the structure itself, and minimizes the likelihood of unintentionally dropping cement into the ocean (Gammon, 2015). Precast heap top shells saved 40,000 man-days and had vital advantage in terms of worker wellbeing since specialists needed to work for less time at the location itself (Gammon, 2015).

6.5. Drone photogrammetry for surveying

The imaginative utilization of automata and drone photogrammetry provides supervisory work with more flexible resources and results in the use of fewer area surveyors, and also lowers costs. Whilst the expense of conventional reviewing work per square meter is HK$2, the expense of photogrammetry per square meter is HK$0.1 (Gammon, 2015).

6.6. BIM technology and 3D scanning

In 2013, Gammon advocated the use of Building Information Modeling (BIM). BIM is a 3D model-based process which facilitates the collaboration of the participants in a construction project. It helps the management of the project to draft detailed project designs and to plan. Even the initial plan is much better than a traditional plan would be, in that the considerations are well-rounded so that the plan is optimized to deal with actual needs. Throughout the construction process, resource consumption is under BIM supervision. BIM suggests the way to estimate and allocate resources in order to be efficient, reduce costs, and to represent less of a burden to environment. Therefore, after adopting BIM, construction projects can be simulated digitally in order to achieve higher efficiency and lower waste production (Gammon Construction, 2013). The famous buildings for which BIM was adopted throughout their construction include Hysan Place, One Island East and Lee Garden Three. Moreover, with assistance from BIM, Gammon achieved a step forward towards adopting 3D printing and scanning technologies. Through 3D technologies, the construction projects will be more efficient and accurate from the design stage.

7 Discussion and conclusion

The results of the interviews showed that the practitioners considered that falls from heights are the most important factors which lead to accidents on sites, followed by bad weather and machine faults. On the other hand, the top three factors for low-rise buildings' construction related to the lack of important safety equipment, training etc. Hence, the idea "lack" is spelt out explicitly in the diagram. The second and third most important factors are machine faults and bad weather. As regards the judges' perceptions, negligence is considered to be the most important factor.

Can innovation involving the various information technologies available nip construction accidents in the bud? The introduction of robotics, prefabricated / precast building structures, i-core with GPS, aerial photogrammetry and 3D printing implies that the number of workers on site can be reduced, leading to a possible reduction in construction accidents. Also VR training can reduce the occurrence of mishaps due to lack of training and/or knowledge concerning construction activities on sites.

Acknowledgment
This research paper is supported by University Grant Committee, Faculty Development Scheme, Construction safety index for skyscrapers in Hong Kong: A Multi-criteria decision-making approach (Project Reference No.: UGC/FDS15/E01/15)

References

[1] AKANMU, A., ANUMBA, C. & MESSNER, J. Critical Review of Approaches to Integrating Virtual Models and the Physical Construction. *International Journal of Construction Management,* 14, 267-282. 2014.
[2] ALARCÓN, L. F., ACUÑA, D., DIETHELM, S. & PELLICER, E. Strategies for Improving Safety Performance in Construction Firms. *Accident Analysis & Prevention,* 94, 107-118. 2016.
[3] BANHDADI, A. & KISHK, M. Assessment of Risks Associated with Saudi Aviation Construction Projects and of the Risk's Importance. *International Journal of Innovation, Management and Technology,* 8,

123-130. 2017.

[4] BONG, S., RAMEEZDEEN, R., ZUO, J., LI, R. Y. M. & YE, G. The designer's role in workplace health and safety in Construction Industry: post-harmonized regulations in South Australia. *International Journal of Construction Management,* 15, 276-287. 2015.

[5] DRULEY, K., MUSICK, T. & TROTTO, S. Researcher Explores how to make temporary structures on construction site safer. *Safety & Heath,* 193, 35. 2016.

[6] EDIRISINGHE, R. & LINGARD, H. Exploring the potential for the use of video to communicate safety information to construction workers: case studies of organizational use. *Construction Management and Economics,* 34, 366-376. 2016.

[7] EDREI, T. & ISAAC, S. Construction Site Safety Control with Medium-accuracy Location Data. *Journal of Civil Engineering and Management.* 2016, forthcoming.

[8] ELATTAR, S. M. S. Automation and Robotics in Construction: Opportunities and Challenges. *Emirates Journal for Engineering Research,* 13, 21-26. 2006.

[9] FARD, M. M., TEROUHID, S. A., KIBERT, C. J. & HAKIM, H. Safety concerns related to modular/prefabricated building construction. *International Journal of Injury Control and Safety Promotion.* 2015.

[10] GAMMON. *Gammon Construction Harness Technology Novelty in Moderating Labour Intensity* [Online]. Available: http://www.gammonconstruction.com/en/html/press/press-bd530331cdef40989c4520fafb1aa13f.html, Accessed: 07/08/2016.

[11] GAMMON CONSTRUCTION. *Gammon Advocates the Application of BIM and was Awarded the ISO 14064 verificaiton statement* [Online]. Available: http://www.gammonconstruction.com/en/html/press/press-07b0f78d171b41d08265979a0aa8c2c7.html, Accessed: 03/05/2017.

[12] GAMMON CONSTRUCTION. *Build-to-Order Full Modular and Robotic Construction* [Online]. Available: http://www.gammonconstruction.com/en/html/press/press-ba2c50896ee0486497849c742c7ac045.html, Accessed: 03/05/2017.

[13] GAMMON CONSTRUCTION. *Gammon Construction Extends Innovative Technology Efforts to Optimize Construction - Virtual Reality Safety Training, GPS Tracking, Robotic Construction* [Online]. Available: http://www.gammonconstruction.com/en/html/press/press-6b9ebc23ad3a426b847f03c4e5fa7031.html, Accessed: 07/08/2016.

[14] GEOFF, E. Major Construction Projects at Airports: Client Leadership of Health and Safety. *Journal of Airport Management,* 10, 131-137. 2016.

[15] GUO, B. H. W., YIU, T. W. & GONZALEZ, V. A. Identifying Behaviour Patterns of Construction Safety Using System Archetypes. *Accident Analysis & Prevention,* 80, 125-141. 2015.

[16] HOŁA, B., SAWICKI, M. & SKIBNIEWSKI, M. An it model of a knowledge map which supports management in small and medium-sized companies using selected polish construction enterprises as an example. *Journal of Civil Engineering and Management,* 21, 1014-1026. 2015.

[17] LI, R. Y. M. *Construction Safety and Waste Management: An Economic Analysis,* Germany, Springer, 2015.

[18] LI, R. Y. M., LI, H. C. Y., MAK CHO, K. & CHAN, P. K. Rationales for the implementation of competition law in EU, the US and Asia: content analysis and data visualization approach. *Asian Journal of Law and Economics,* 7, 63-100. 2016.

[19] LI, R. Y. M. & POON, S. W. *Construction Safety*, Springer, 2013.

[20] LI, Z., CARBERRY, S., FANG, H., MCCOY, K. F., PETERSONA, K. & STAGITIS, M. A novel methodology for retrieving infographics utilizing structure and message content *Data & Knowledge Engineering,* 100, 191-210. 2015.

[21] LIU, Y., VAN NEDERVEEN, S. & HERTOGH, M. Understanding effects of BIM on Collaborative Design and Construction: An Empirical Study in China. *International Journal of Project Management,* 35, 686-698. 2017.

[22] MARKS, E., AWOLUSI, I. G. & MCKAY, B. Near-Hit Reporting Reducing Construction Industry Injuries. *Professional Safety,* 61, 56-62. 2016.

[23] MROSZCZYK, J. W. Improving Construction Safety. *Professional Safety,* 60, 55-68. 2015.

[24] PERKINS, I. & SKITMORE, M. Three-dimensional printing in the construction industry: A review. *Journal: International Journal of Construction Management,* 15, 1-9. 2015.

[25] SHIBEIKA, A. & HARTY, C. Diffusion of Digital Innovation in Construction: a Case Study of a UK Engineering Firm. *Construction Management and Economics,* 33, 453-466. 2015.

[26] SKIBNIEWSKI, M. J. *Robotics in Civil Engineering,* New York, Van Nostrand Reinhold Pub, 1988.

[27] SKIBNIEWSKI, M. J. Research Trends in Information Technology Applications in Construction Safety Engineering and Management. *Frontiers of Engineering*

Management, 1, 246-259. 2014.

[28] TAGUL. *Tagul - Word Cloud Art* [Online]. Available: https://tagul.com, Accessed: 09/08/2016.

[29] WANG, H. H. Semi-automated Identification of Construction Safety Requirements using Ontological and Document Modeling Techniques. *Canadian Journal of Civil Engineering,* 42, 756-765. 2015.

[30] WARSZAWSKI, A. Economic implications of robotics in building. *Building and Environment,* 20, 73-81. 1985.

[31] YARAHMADI, H., MIRI, M. & RAKHSHANIMEHR, M. A Methodology to Determine the Response Modification Factor for Probabilistic Performance-Based Design. *Bulletin of Earthquake Engineering,* 15, 1739-1769. 2017.

34th International Symposium
on Automation and Robotics
in Construction

ISARC 2017

June 28 - July 1
TAIPEI, TAIWAN

Integration of offsite and onsite schedules in modular construction

Tarek Salama [a], Ahmad Salah [b], and Osama Moselhi [c]
[a,c] Department of Building, Civil and Environmental Engineering, Concordia University, Montréal, Canada

[b] Department of Civil and Environmental Engineering, Imam Abdul Rahman Bin Faysal University, Dammam, Kingdom of Saudi Arabia.

E-mail: ta_salam@encs.concordia.ca, aaasalah@uod.edu.sa, moselhi@encs.concordia.ca

Abstract

Modular construction enables delivery of a building as an assembly of a set of modules manufactured offsite in a controlled manufacturing facility environment. Unlike stick-built practices, modular construction enables higher schedule control of construction projects due the inherent concurrency of offsite and onsite construction operations. Literature provides simulation-based scheduling methods that integrate offsite and onsite construction activities. These methods, however, depend largely on availability of data such as productivity rates for offsite and onsite activities. This paper presents an alternative BIM-based framework that integrates linear schedules of onsite and offsite construction operations in a manner that synchronizes work progress of these operations. The proposed framework considers limited capacities of storage areas in the manufacturing facility and on site as well as, the availability of trucks for delivering the fabricated modules from manufacturing facility to the jobsite. The use of BIM provides visualization capabilities for the integrated schedule and allows for monitoring simultaneously the work progress of offsite and onsite activities. Conclusions are drawn concerning the suitability of developed framework for integrated scheduling of modular construction projects.

Keywords

BIM, Integration, Scheduling, Modular Construction.

1 Introduction

A recent survey of 800 engineers, architects, and contracting professionals reveals modular construction advantages including shorter project schedules (66% of respondents); lower cost (65% of respondents); and reduced construction waste (77% of respondents) [1]. Independent KPMG research found that financial net savings for offsite construction projects are 7% due to shortened construction period without considering the savings generated from decreasing the interest of borrowing [2]. These savings enable faster rental income and lower escalation in construction costs. The combination of offsite and onsite construction in a "50-storey office building" project in central London generated combined savings of £ 36 million [2].

Offsite construction provides other benefits such as; enhanced predictability of time and cost, reduced noise from construction, and improved health and safety. According to size and complexity of manufactured components, offsite construction types are grouped into five categories; 1) modular, 2) hybrid, 3) panelized, 4) prefabricated components, 5) processed material [3]. Modular construction reduces considerably the schedule of construction projects which may generate significant cost savings.

2 Literature review

Parallel scheduling for offsite and onsite construction schedules saves 30 to 50 percent of project duration as compared to stick-built traditional construction processes as shown in Figure 1 [1].
Simulation-based models are reliable tools for probabilistic problems while considering the effect of uncertainties associated with construction projects [4] and [5]. Although, simulation cannot eliminate or control idle times by itself [5] and [6] therefore, simulation-based models require external algorithms to solve the problem of repetitive scheduling effectively [4] and [5] otherwise; simulation models generate same results of CPM as shown in Figure 2 [7] and [8].
Linear scheduling considers the repetitive nature of activities; however, limited number of linear scheduling

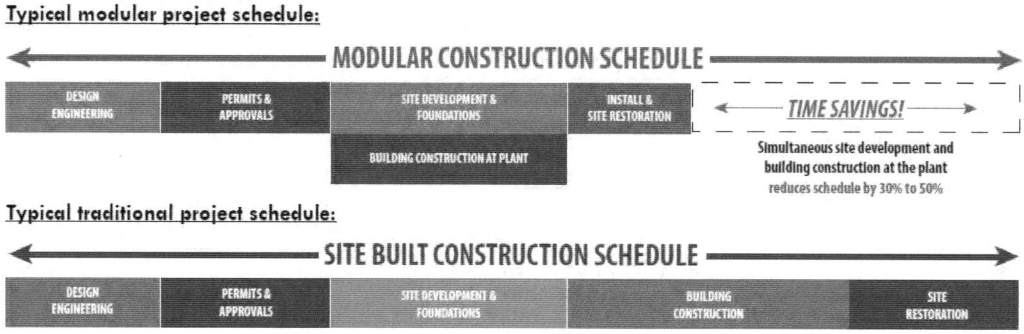

Figure 1. Comparison between modular and traditional construction schedules [1].

based methods account for uncertainty [9]. In the other hand, presentation of LSM was criticized to inability to show task relationships and parallel activities graphically [10] as well as to inability to differentiate between overlapping activities that have equal production rates [11]. Manusr [12] presented an approach that allocates critical activities on the upper side of the line of balance (LOB) chart, and non-critical ones on the lower part for a better presentation. Hegazy et al. [13] extended this approach by introducing BAL system that enhances LSM presentation and eliminates overlapping of activities as shown in Figure 3.

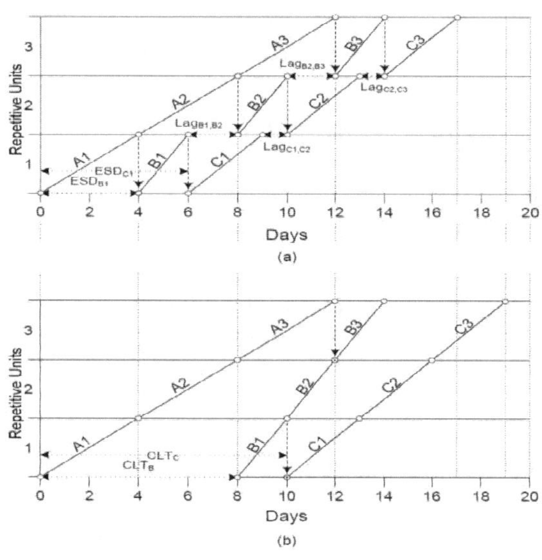

Figure 2. CPM vs. RSM Scheduling (a, b) [7].

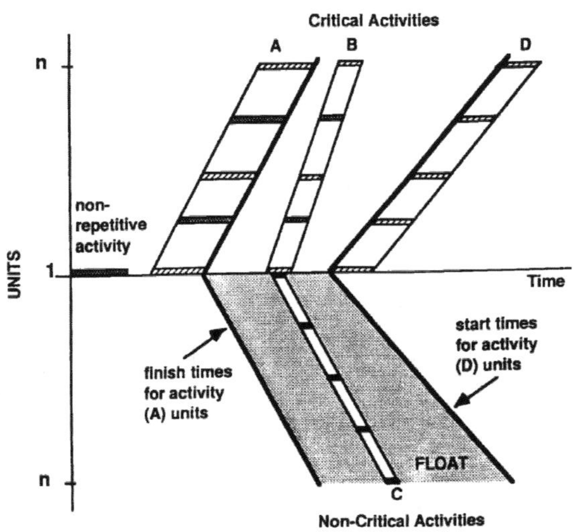

Figure 3. Enhanced Schedule Presentation [13].

Several studies utilized simulation to schedule offsite construction processes [14], [15], [16], [17], and [18]. Other researchers [19], [20], and [21] integrated simulation with building information modeling (BIM) to automate quantity take-off data and to visualize modular and offsite construction activities. Taghaddos et al. [22] presented a simulation-based scheduling model for module assembly yard without linking this model to fabrication shop schedule. Linking simulation model of module yard (onsite) to that of fabrication shop (offsite) is a main challenge for simulation models since it requires manual interventions from project scheduler [22]. Other researchers utilize advanced technologies such as; HLA environment [23] presented a simulation-based model that integrates module assembly yard and fabrication shop schedules in high level architecture (HLA) environment. However, this integration was not supported by a case example to demonstrate its effectiveness. Bu Hamdan et al. [24] presented a system that integrates simulation with BIM to provide an inventory planning and management tool while considering interactions between different construction phases from manufacturing to assembly. However, this system has a non-user-friendly graphic user interface (GUI) which requires manual interventions to define probabilistic distribution functions and requires users who are experts in simulation. Salama et al. [25] suggested an alternative (BIM) based integrated alternative (BIM) based integrated framework for planning and scheduling of hybrid modular construction projects using LSM and CCPM. This framework considers off-site construction only. However, there is a lack of scheduling methodologies that synchronize offsite and onsite operations. Unlike previously mentioned simulation methods that integrate offsite and onsite schedules, the proposed method respects the continuity of resources for activities and visualizes the integrated schedule.

3 Proposed Method

The main benefit of modular construction is shortening the project schedule due to parallel offsite and onsite activities. However, most of scheduling methods introduced in literature to integrate offsite and onsite construction are simulation-based methods. This paper presents a scheduling framework that integrates, visualizes, and synchronizes the linear schedules of offsite and onsite activities in modular construction projects. The proposed method consists of five steps as follows:

3.1 Generating offsite and onsite schedules baseline (aggressive schedule)

In this phase linear schedules that integrate LSM and the CCPM [26] are generated. This includes calculation of aggressive and safe activity durations, sequencing activities, maintaining continuity of schedule, resolving resources conflicts, identification of critical sequence, and adding feeding and project buffers. Also, baseline of offsite and onsite schedules using the flowcharts presented in Figure 4 and 5 respectively.

3.2 Identification of the critical control point (CCP) between offsite and onsite schedules

Identification of the critical control point (CCP) between offsite and onsite schedules is crucial because it connects both schedules and highlights the effects of offsite schedule delays on onsite schedule. CCP is logically constrained by completion of the manufacturing processes and transportation method of modules/panels from offsite manufacturing facility to construction site. Two scenarios affect the position of CCP in respect to transportation as follows:

1- Partial transportation in which manufactured modules/panels are transported by pre-scheduled batches. In this case, CCP is the start date of installation activity for last batch after transportation (i.e. Lifting panels) using Equation (1) as shown in Figure 6.

$$SD_{\mathrm{OLB\,(AG)}} = FD_{\mathrm{OLB\,(AG)}} + D_{\mathrm{LBT}} \quad (1)$$

Where,

$SD_{OLB\ (AG)}$: Start date of panels' first onsite activity for last batch after transportation at aggressive schedule.
$FD_{OLB\ (AG)}$: Finish date of last offsite manufacturing process for last panel at aggressive schedule.

D_{LBT}: Duration for last batch transportation from offsite manufacturing facility to construction site

2- Full transportation in which the modules/ panels are delivered to construction site after completion of all manufacturing processes. In this case, CCP is the start date of first onsite activity for first panel (SDOFP) (i.e. Lifting panels) using Equation (2) and as shown in Figure 7.

$$SD_{\mathrm{OFP(AG)}} = FD_{\mathrm{OLB\,(AG)}} + D_{\mathrm{LBT}} \quad (2)$$

Where,
$SD_{OFP\ (AG)}$: Start date of panels' first onsite activity for first panel after transportation at aggressive schedule.
$FD_{OLB\ (AG)}$: Finish date of last offsite manufacturing process for last panel at aggressive schedule.
D_{LBT}: Duration for last batch transportation from offsite manufacturing facility to construction site.

The output of this step is an integrated schedule baseline that is generated using relative scheduling of onsite and offsite activities in respect to the identified CCP in a manner that maintains the continuity of resources in all processes as shown in Figures 6 and 7.

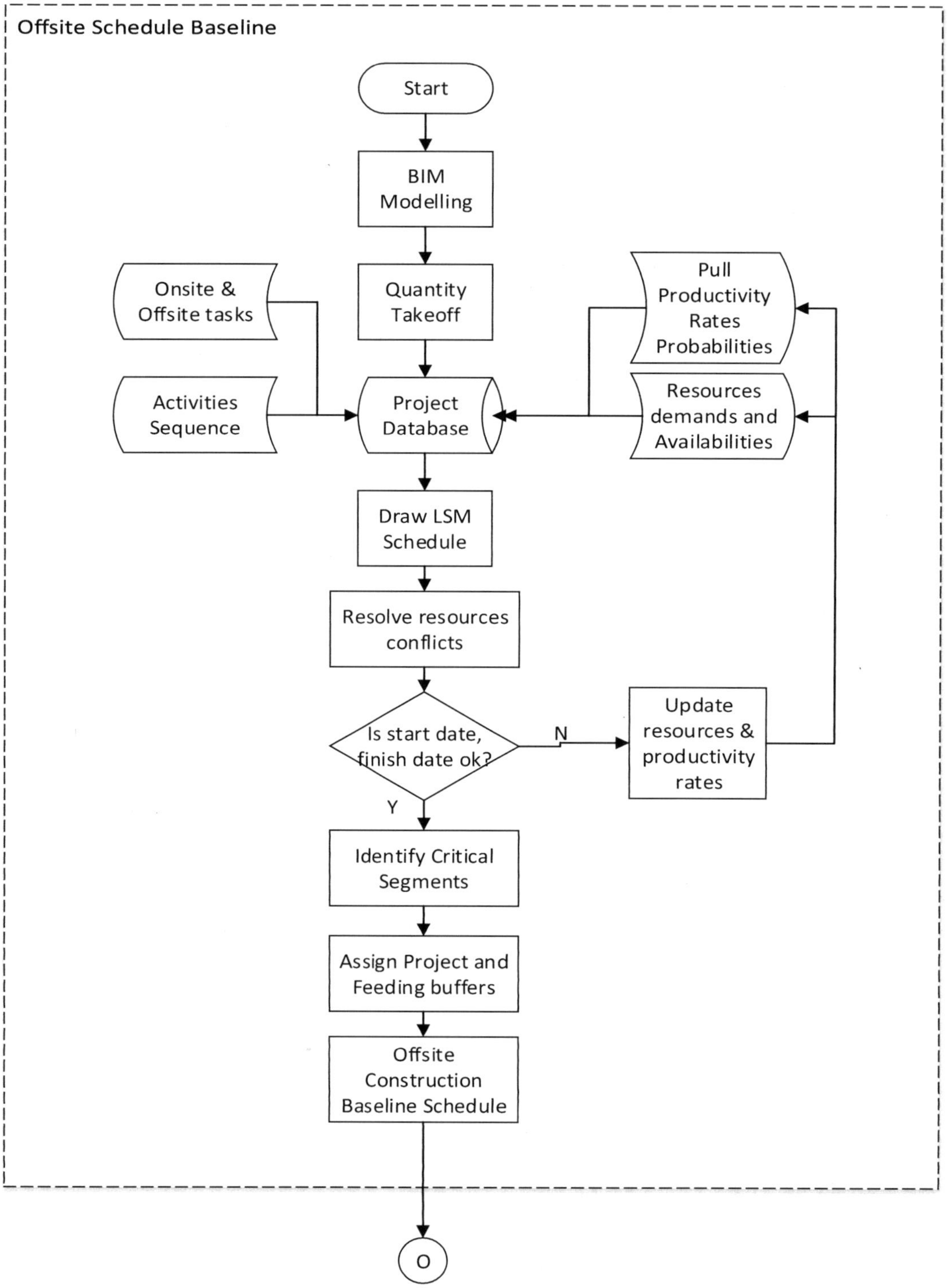

Figure 4. Generating Offsite Schedule Baseline.

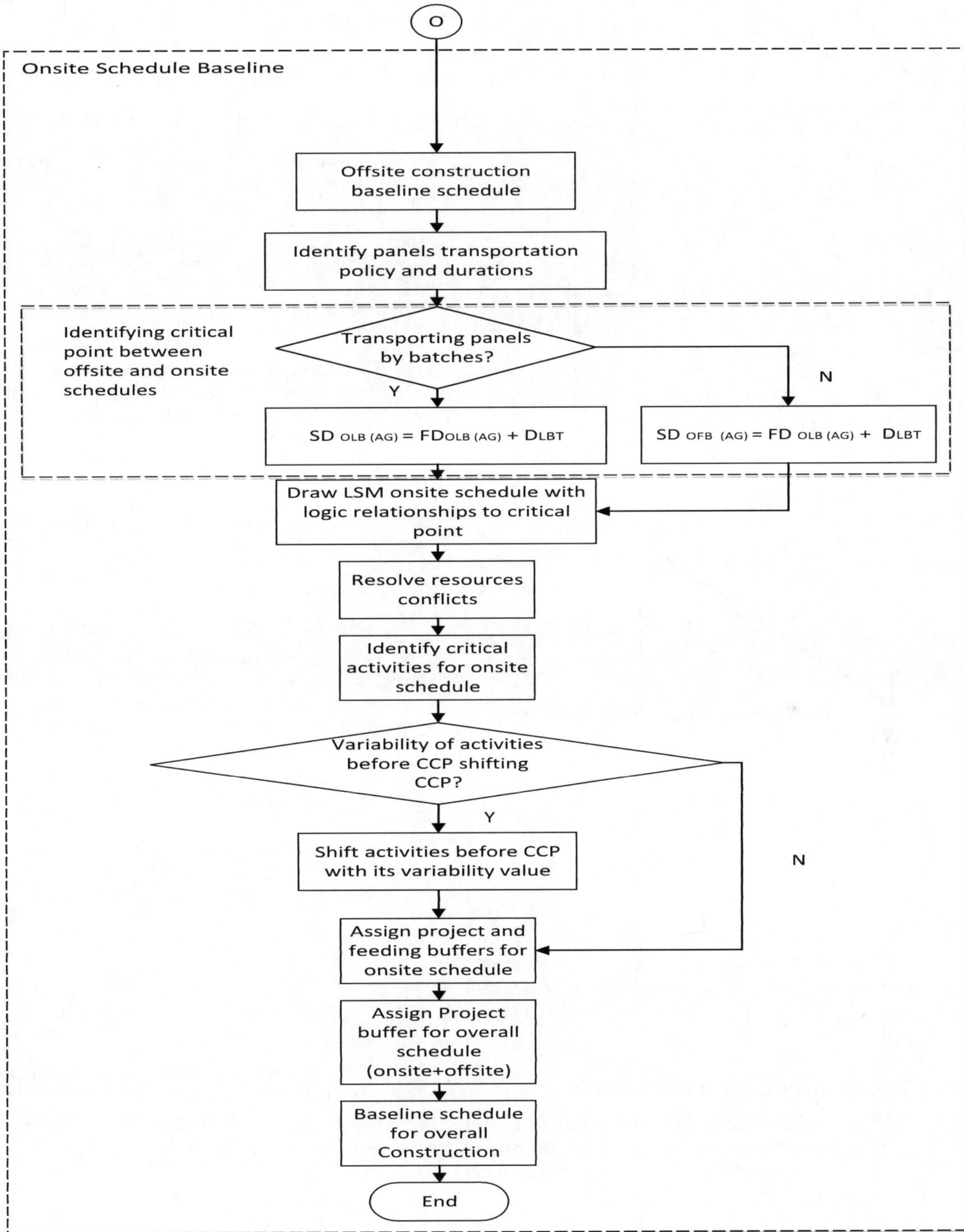

Figure 5. Generating Onsite Schedule Baseline.

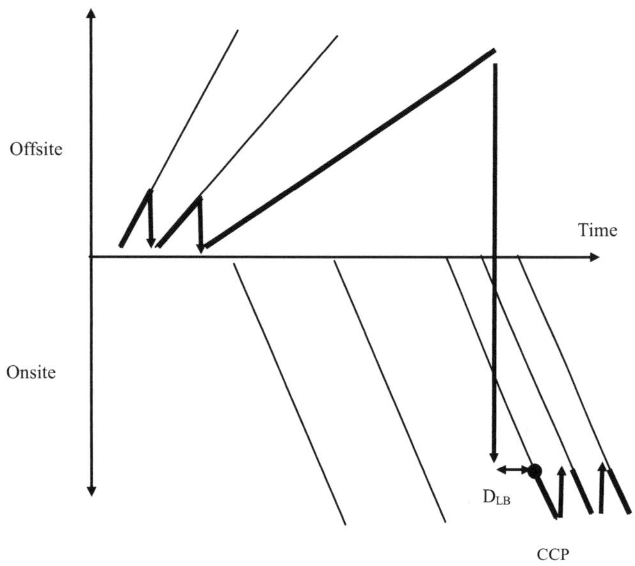

Figure 6. Identifying CCP for the scenario of transporting panels by batches.

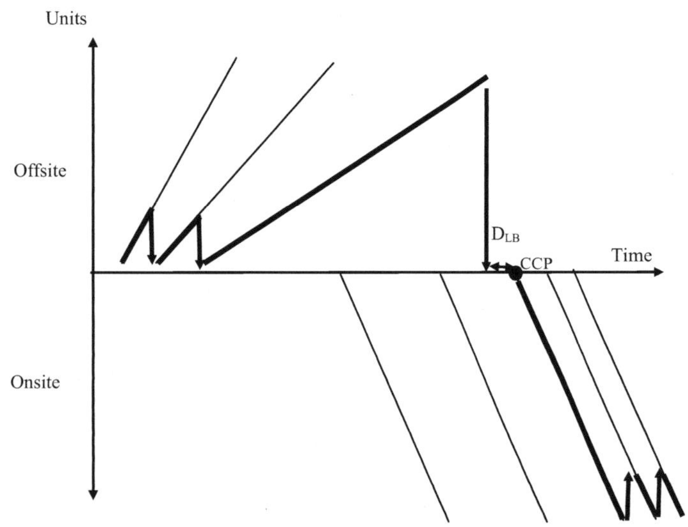

Figure 7. Identifying CCP for the scenario of transporting panels after all batches are manufactured.

3.3 Adding feeding and project buffers

Project and feeding buffers for offsite and onsite schedules are calculated using Equations (3) and (4) respectively. However, the integrated schedule (offsite and onsite) has another overall project buffer based on the variability of onsite and offsite schedules. Both schedules are connected at the CCP to create an overall project critical sequence. Figure 8 shows an example of linear schedule where, feeding buffers are generated from the variability of "resource-critical activities" number D1 to D3 and F1 to F5 [26]. The delay of resource-critical activities delays part of the critical controlling sequence and accordingly the project completion date.

$$\text{Project buffer} = \sqrt{\sum_{P=1}^{P=n} (D_P(CL) - D_P(AG))^2} \quad (3)$$

$$\text{Feeding buffer} = \sqrt{\sum_{F=1}^{F=n} (D_F(CL) - D_F(AG))^2} \quad (4)$$

Where,

DP (CL): safe duration of critical activity with given confidence level "CL".DP (AG): aggressive duration of critical activity with 50% confidence level.

DF (CL): safe duration of resource critical activity with given confidence level "CL".

DF (AG): aggressive duration of resource critical activity with 50% confidence level.

3.4 Shifting processes before CCP

The processes before CCP at onsite schedule are shifted with the amount of its critical sequence variability as shown in Figure 9. The variability of its critical sequence is accumulated in a special buffer named onsite activities buffer (OAB) using Equation 5. This procedure protects the overall project critical sequence from delay by allowing onsite processes before CCP to start early.

$$\text{OAB} = \sqrt{\sum_{O=1}^{O=n} (D_O(CL) - D_O(AG))^2} \quad (5)$$

Where,

DO (CL): safe duration of onsite activity with given confidence level "CL".

DO (AG): aggressive duration of onsite activity with 50% confidence level.

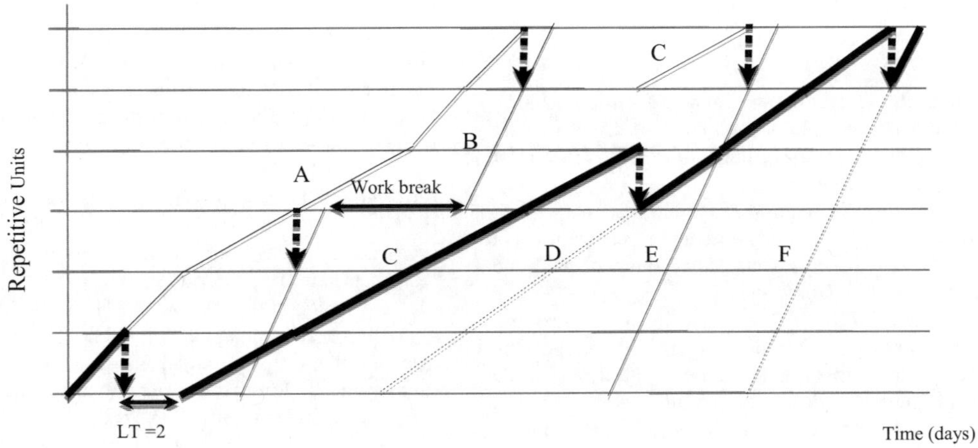

Figure 8: Identification of resource-critical activities [26].

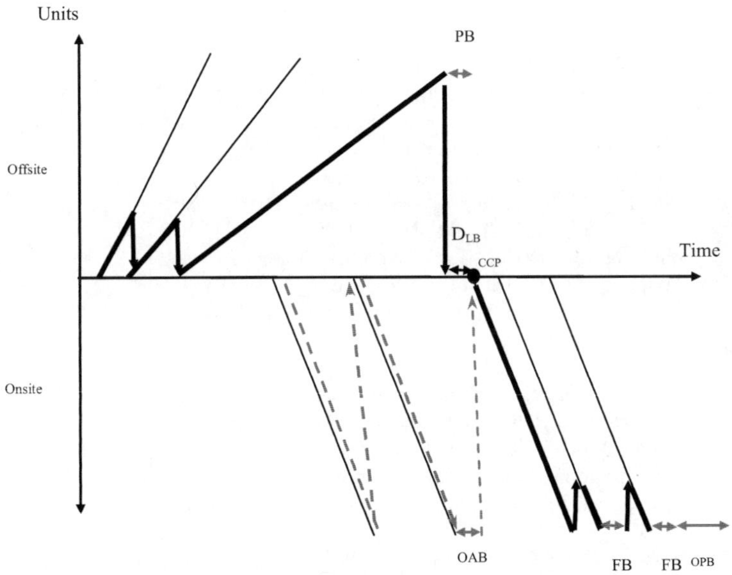

Figure 9. Adding feeding and project buffers.

4. Conclusion

This paper presented a novel method for integrating the linear schedules of offsite and onsite modular construction, supported by a graphical representation that depicts the interactions between offsite and on site schedules. The developed method introduces new systematic procedure for identifying the critical control point (CCP) that inter-connects offsite and onsite linear schedules. The integration framework between LSM of offsite and onsite construction activities mimics the repetitive nature of manufacturing processes and onsite activities while considering uncertainties associated with productivity rates. Automating the presented method provides an easy to use planning and scheduling tool for synchronizing offsite and onsite schedules without using simulation. The overall integrated schedule is expected to assist project manager in planning onsite construction activities while considering the manufacturing plan of offsite activities and vice versa. Example application is described to highlight the essential features of the developed method.

References

[1] Modular Building Institute (MBI), (2016). Modular Advantage for the commercial modular construction industry, 59 pages, Virginia, USA.

[2] Southern, J. 2016. Smart Construction: The value in offsite construction. Report by KPMG.

[3] Schoenborn, J. M. (2012). A Case Study Approach to Identifying the Constraints and Barriers to Design Innovation for Modular Construction. Master of Science in Architecture, Faculty of the Virginia Polytechnic Institute and State University.

[4] Lutz, J. D. (1990), Planning of Linear Construction Projects using Simulation and Line Of Balance, Ph.D. Thesis, Purdue University, West Lafayette, USA.

[5] Yang, I. (2002), Repetitive Project Planner Resource-Driven Scheduling for Repetitive Construction Projects, Ph.D. Dissertation, University of Michigan, Ann Arbor, USA.

[6] Ioannou, P. G., and V. Likhitruangsilp. (2005). Simulation of multiple-drift tunnel construction with limited resources, Proceedings of the 2005 Winter Simulation Conference.

[7] Ioannou, P. G., Srisuwanrat, C. (2006). Sequence step algorithm for continuous resource utilization in probalistic repetitive projects. Proceedings of the 2006 Winter Simulation Conference.

[8] Srisuwanrat, C. (2009). The sequence step algorithm: a simulation-based scheduling algorithm for repetitive projects with probablistic activity durations. PhD thesis in The University of Michigan.

[9] Bakry, I. Moselhi, O. and Zayed, T. 2013. Fuzzy Scheduling of Repetitive Construction Projects. Proceedings of the 30rd International Symposium on Automation and Robotics in Construction (ISARC 2013), Montreal, Canada.

[10] Francis, A., Miresco, E. T. (2013). Applying the Chronographical Approach to the Modelling of Multistorey Building Projects. 4th CSCE Construction Specialty Conference, Montreal, Quebec.

[11] Arditi, D., and Albulak, M. (1986). Line-of-Balance Scheduling in Pavement Construction. J.Constr.Eng.Manage., 112(3), 411-424.

[12] Mansur, F. (1990). The Line of Balance Method. Major technical report, Center for building studies, Concordia University, Montreal, Canada.

[13] Hegazy, T., Moselhi, O., Fazio, P. (1993). BAL: An algorithm for scheduling and control of linear projects, Transactions of AACE International, pg. C.8.1.

[14] Abu Hammad, A.A. (2003). A Decision Support System for Manufactured Housing Production Process Planning and Facility Layout. PhD thesis, Department of Civil and Environmental Engineering, the division of Research and Advanced Studies of the University of Cincinnati.

[15] Xie,H.,Shen,W.,Hao,Q.,Yu,H.,Tang,Y. (2011). Product ion simulation for prefab housing facilities. 3rd International / 9th Construction Specialty Conference (CSCE), Ottawa, ON, pp. CN-189-1 – CN-189-10.

[16] Taghaddos, H., Hermann, U., AbouRizk, S., Y., Mohamed, A. (2012). Simulation-based multiagent approach for scheduling modular construction, J. Comput. Civ. Eng. 28 (2), pp.263–274.

[17] Ajweh, Z. A. (2014). Framework for Design of Panelized Wood Framing Prefabrication Utilizing Multi-panels and Crew Balancing, Master of Science Thesis, Department of Civil and Environmental Engineering, University of Alberta.

[18] Altaf, M., Yu, H., and Al-Hussein, M. (2015). Online Simulation Modeling of prefabricated wall panel production using RFID system. Proceedings of the 2015 Winter Simulation Conference, pp. 3379-3390.

[19] Moghadam, M., (2013). Lean-Mod: An Approach to Modular Construction Manufacturing Production Efficiency Improvement. PhD thesis, Department of Civil and Environmental Engineering, University of Alberta.

[20] Liu, H., Al-Hussein, M. and Lu, M., (2015). BIM-based integrated approach for detailed construction scheduling under resource constraints, Automation in Construction Journal.

[21] Liu, H., Lu, M., and Al-Hussein, M. (2016). Ontology-based semantic approach for construction-oriented quantity take-off from BIM models in the light-frame building industry. Advanced Engineering Informatics journal, (30), 190–207.

[22] Taghaddos, H., AbouRizk, S., Mohamed, Y., and Hermann, R. (2009). Integrated Simulation-Based Scheduling for Module Assembly Yard. Building a Sustainable Future. pp. 1270-1279. doi: 10.1061/41020(339)129.

[23] AbouRizk, S., Mohamed, Y., Taghaddos, H., Saba, F., and Hague, S. (2010). Developing complex distributed simulation for industrial plant construction using High Level Architecture. Proceedings of the 2010 Winter Simulation Conference, - WSC 2010, 5-8 December, Baltimore, Maryland, USA. pp. 799-810. DOI: 10.1109/WSC.2010.5679010.

[24] Bu Hamdan, S., Barkokebas, B., Manrique, J.D., and Al-Hussein, M. (2015). A BIM-based simulation model for inventory management in panelized construction. Proceedings, International Symposium on Automation and Robotics in Construction and Mining (ISARC 2015), Oulu, Finland, Jun. 15-18.

[25] Salama, T., Salah, A. and Moselhi, O. (2016), Alternative Scheduling and Planning Processes for Hybrid Offsite Construction, Proceedings of the 33rd International Symposium on Automation and Robotics in Construction (ISARC 2016), Auburn, Alabama, USA.

[26] Harris, R. and Ioannou, P. (1998). Scheduling Projects with Repeating Activities. J. Constr. Eng. Manage., 269-278

Development of the Process Information Modelling in the construction project: A case study of the ZERO-PLUS project

Wen Pan[a], Katharina Langosch[a] and Thomas Bock[a]

[a] Chair of Building Realisation and Robotics, Technical University of Munich, Germany
E-mail: wen.pan@br2.ar.tum.de, katharina.langosch@br2.ar.tum.de

Abstract –

The conventional Building Information Models (BIM) does not contain knowledge about the relation between different pieces of information. To be able to understand the information the recipient must have sufficient knowledge within the given domain. When information cannot reach to the relevant party in real time or in advance, that will cause miscommunication and delays. At this point, the Process Information Modelling (PIM), which is a process-oriented modelling approach, provides a collaborative way of planning, designing, producing, assembling, and managing the repeatable deployment process. The main objective of PIM is to formulate a process oriented database platform that allows smooth data transfer, exchange, as well as promoting seamless and constant data sharing. Digital documentation, simulation, graphical and non-graphical data are produced progressively to support the decision-making process throughout the PIM application development stage. Thus, PIM embraces real-time data processing that helps to identify issues in advance, predict malfunction, reduce production time, and influence production cost.

At the moment, PIM is developed through three main phases which include design, production, and on-site assembly phase and represents a conceptual tool that provides a platform where stakeholders can access information in real-time and to make vital decisions in advance. By using a real scenario in the on-going EU Horizon 2020 research project ZERO-PLUS, the first application of the PIM concept could be demonstrated on the Freescoo HVAC system, which will not only help the project team to structure detailed guidelines within the current project but will also lay a foundation for developing a fully functional PIM application in the future.

Keywords –

PIM; off-site manufacturing; planning; simulation

1 Introduction

In the construction industry, software-based models are used in particular for complex building projects in order to make the planning and execution more efficient and optimize. Through the use of the so-called Building Information Model (BIM) method, a flexible and timely data exchange can be achieved between all involved parties (e.g. architects, engineers, etc.) in particular, if changes need to be adapted. Despite the eminent work relief by the use of BIM, it mostly provides detailed information about each component and thus, it is a knowledge-based method. However, the exact processes and methods of manufacturing, installation and assembly cannot be depicted.

The concept of Process Information Modelling (PIM), which is a relatively new method, offers the possibility to include precisely these processes. When applied in the context of off-site manufacturing and assembly tasks, it serves as a tool that guides involved personnel or any other entities on making an appropriate response to a specific task.

For this reason, the overall objective of PIM can be described as following: an innovative process-oriented approach that can be implemented to simulate and optimize the design, planning, production, and assembly process within the construction industry. This objective is motivated by the belief that current BIM has many limitations, which only partially enhance the construction process. PIM will be designed, as a sufficient tool that can simulate, evaluate and implement varied phases which is needed in today's competitive market. However, due to the novelty of the concept and the complexity of the construction process, the paper focuses on the presentation of the PIM concept and its potential and demonstrates a simple process directly selected from the ZERO-PLUS project that PIM application applied to the Freescoo technology used in case study Cyprus to describe the underlying principle.

The research team will carry out evaluation and identification of the key research boundaries within the context of the ZERO-PLUS project. The proposed PIM will provide a theoretical solution that offers an extensive guideline for the design, planning, simulation and optimization of off-site production and assembly phases. As already mentioned, the conventional BIM application is a knowledge-based solution, while PIM offers a know-how-based solution. As a result, utilization of both solutions can yield significant benefits for an organization or individuals undertaking a variety of complex projects. In addition, the proposed PIM approach will not only help the project team to structure detailed guidelines within the ZERO-PLUS project but also lay a foundation for developing a fully functional PIM application for the construction industry.

2 Concept of PIM

PIM can be understood as a process-oriented modeling approach that provides a collaborative way of planning, designing, producing, assembling, reinforced by the key phases namely; design, production, and on-site assembly. The key phases are further evaluated via various modeling applications, which enable stakeholders to use more efficient methods to create added value as well to optimize an existing operation.

First, the fundamental feature of PIM is a process-oriented information modeling approach. Process-oriented means that this approach is focused on a certain cluster of the entire process. On an organizational level, process planning, management, and integration make possible the improvement of interdepartmental, cross-functional, and cross-disciplinary interactions. The goal is to optimize the entire process rather than optimizing only parts of it and neglecting the others. On a task-based level, this can be achieved by a thorough understanding of end-to-end process and production systems as a whole [1].

Second, within each process, there are objects and other parameters such as location, and time. Any change in these elements would affect the operational performance of the overall process. Within PIM, the information model defines the objects and all other relevant parameters depending on the specific process. PIM application has the potential to be developed as a software program. The product and the Off-site Manufacturing (OSM) facility's property, class, protocols and assembly method can be easily translated to pseudocodes, which can be operationalized by software engineers.

Third, the assembly evaluation method (AEM) evaluates the data generated by the PIM and then, according to the real facilities, suggests production and assembly solutions. The AEM can also forecast the productivity of the proposed facilities or method.

Fourth, PIM has the potential to offer automated instructions when a key parameter is altered. This is extremely beneficial when the responsible parties do not have the required domain knowledge or are struggling to make the final decision. At the time of this writing, PIM is only capable of offering basic instructions that do not lead to unnecessary complications or disputes.

3 PIM application and potential impact

In general, PIM application consists of the following stages: identification, data acquisition, organization, and implementation, (Figure 1**Error! Reference source not found.**).

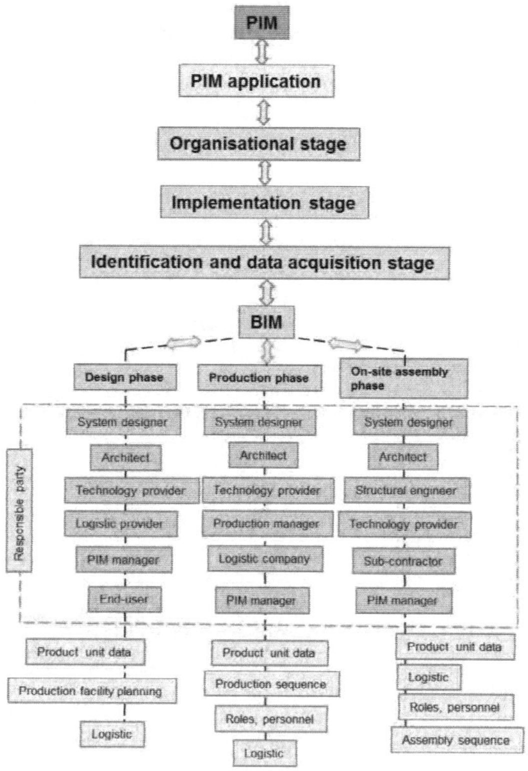

Figure 1: PIM chart

During the identification and data acquisition stage, the PIM manager will identify a number of parameters including the product definition, and the configuration of the product. The basic physical product features, - dimensions, shape, material, functionalities and capabilities - are classified as unit data. Depending on the unit data and the production method, a manufacturing facilities concept will be proposed. The degree of automation will be applied in the production phase, as is often advised by the production manager, since it is directly influenced by other elements, such as the volume of the production, cost, demographic reasons and so forth. The logistic solution will be planned in advance to ensure Just–in–time (JIT) supply of parts and final products, both off-site and on-site [2]. The on-site assembly method is to be identified prior to delivery. Moreover, the product life-cycle management strategy is developed simultaneously to ensure future upgrade and maintenance. In summary, this stage performs a preliminary investigation the overall tasks.

During the organizational stage, based on the attributes of each identified unit data, the PIM manager will assign them to the corresponding phase. A class system will be applied to distinguish the attribute of each unit data. In doing so, a detailed representation of the affiliation between each unit data can be created. A detailed description of what needs to be accomplished at each sequence is produced. For instance, detailed product specification, alternation flexibility, material flow, tool and equipment being used, an alternative method, and additional requirements was determined based on the unit data. More importantly, at this stage, the PIM manager needs to be aware of the interconnection and independence of each unit data; when one of them changes, what would the trade-off and parallel impact on the others be? In summary, this stage performs planning, projecting and data arranging for the next stage.

During the implementation stage, the PIM manager will finalize each process and acquire a potential optimization strategy. For example, the final design of the product, the sequence of operations for manufacturing every part of the product, the required machinery, equipment, and personnel, the sequence of operations in the fabrication and assembly processes, smart buffering, packing station and loading method, supply and logistics strategy, on-site assembly or disassembly method and lifecycle management strategy. The data will serve as a guideline for different stakeholders.

In practice, PIM can alternatively be divided into a number of model fragments called information modules. The information modules contain dedicated information from each task phase. Typical task phases include design, production, logistic, on-site assembly, and life-cycle management. It is noticeable that these phases are also as part of the PIM application stage. The domain experts of each phase can function independently while also interactively working alongside the PIM manager to change and optimize the information module. For example, the product design phase is defined by the design brief of the product. The protocols between these phases are complementary: If the design engineer changes the product design, the PIM system keeps track of the changes, such that the other phases can respond to the changes accordingly.

4 Usage of PIM in conjunction with BIM

In recent years, there has been a shift from using 2D technical drawings to a more dynamic trend regarding the use of Building Information Models (BIM) within the architecture, engineering and the construction industries. Using modern modeling tools, such as Revit Architecture, ArchiCAD or Tekla Structures, the content produced by architects, designers and engineers have evolved from traditional 2D-drawings, sketches and written specifications to parametric, object-oriented 3D-models embedded with information to describe any building or facility in detail. As a digital representation of the physical and functional characteristics of a building, a BIM serves as a repository of information supporting a multitude of applications, including cost-estimation, energy analysis and production planning [3]. BIM covers the project span from the conceptual stage to the completion of the project. From construction scheduling to logistics management, BIM enables project managers and design-build teams to develop an object-oriented database that assists in predicting and measuring the project performance objectives. For construction management, BIM provides a virtual representation of the project concept prior to construction, ensures early failure detection, simulates and analyses potential impacts, and improves safety [4]. Moreover, BIM allows information transfer between all stakeholders, by allowing each stakeholder to contribute the information they acquire during the construction of the BIM system [5].

In comparison to PIM, BIM does not contain knowledge about the relation between different pieces of information. A piece of information cannot function alone but must coexist with other pieces. Even though BIM allows information sharing and transferring, this information is typically only available to parties that already have sufficient knowledge within the given domain. Consider a building contractor who would like to purchase a set of the prefabricated foundation. The BIM information about the prefabricated foundation will only contain the specification and compatibility of the product. The duration and the exact process and method of delivering and installing the foundation product are not included. In fact, the installation process contains the most crucial information for the project team preparing the installation tasks, in order to keep the project on schedule.

In contrast, PIM would provide detailed information on not only the specification and cost of the product, but also on detailed processing instructions for the installer, such as the depth of a trench, tools, and equipment required, and illustrations of connection methods to assist the team throughout the project phase. Judging by the example above, BIM represents what the product consists of, whereas PIM represents what the product consists of and, in addition, how it responds to the user's requirement and the methods used [6].

5 Influence of PIM in manufacturing

To develop the most relevant PIM application for the future construction industry, the following topics were studied and analyzed, which include, Lean manufacturing concept, Design for manufacturing, Just-in-time (JIT) supply and assembly automation.

"Lean" is a term in the manufacturing sector that originated from the Toyota production system (TPS) [7] and the book named "Lean Thinking" by [8]. The lean manufacturing concept prioritizes eliminating waste during the production phase [9]. Since the beginning of the industrial revolution, manufacturers have tried to overcome the issue of how to eliminate waste. In this context, the waste refers to the unnecessary expenditure of time, money and materials [10]. The Muda concept is one of many concepts (Muda means 'waste' in Japanese), which focus on non-value added activities in terms of production. In total, there are seven types of Muda namely: overproduction, waiting (queueing), transportation, processing, inventory, motion, rework, and poor resource utilization. All of them can contribute a non-value added activity that will not improve the product [11]. Though Lean Manufacturing has been used in the Japanese construction industry for a long time, in Europe it has been adopted only in the last decades. In the future, PIM application can potentially reduce non-value added activities throughout the construction process and enhance communication between all project phases.

Design for manufacturing (DfM) is a concept that encourages designs that are easy to produce. It is closely associated with another concept called design for assembly (DfA). However, there is a fundamental difference between these two concepts. In general, DfM focuses on designing individual parts (product), whereas DfA focuses on the protocols between the products, in other words how to assemble complex products from integrated parts easily. The combination of these two principles is known as design for manufacture and assembly (DFMA). In the manufacturing sector, it is more feasible to change the design of the product rather than to alter the existing production facility. PIM is highly relevant in improving the DFMA concept that the information collected by PIM can be used to structure the protocols between the product and the production sequence, therefore enhance the overall manufacturing efficiency.

Just in Time (JIT) refers to the strategy developed first in the Toyota production system to minimize the space needed to supply the production phase. Usually mentioned together with Just in Sequence (JIS), it has been identified as one of the main success factors of the company. In the construction field, the JIT model can be used not only for off-site, usually largely depending on the level of prefabrication employed, but also for on-site activities. This is because the on-site assembly of prefabricated elements requires an amount of time, resources, and attention is usually greater than for off-site operations. This is due to unpredictable and unexpected external circumstances. PIM can potentially help the construction project team to synchronize off-site and on-site operations together with JIT model [12].

Generally, automation substitutes human activity in difficult and repetitive tasks, thus enhancing productivity and safety, while considerably reducing the labor cost. The use of machines ensures that the quality of the final product remains at a consistent level, as robots reduce human error and the drawbacks derived from hostile weather conditions. Moreover, the planning phase is also enhanced, due to the high predictability of the process [13]. The level of assembly automation in an off-site construction plant varies considerably depending on several factors, such as the material and the complexity of the product. Nonetheless, it is generally advantageous to achieve as high level of automation as possible, in order to fully enhance production and obtain a quick return on investment. In this paper, the outcomes that influenced by using PIM application during the design, automation in production, assembly of the Freescoo HVAC system will be briefly described.

6 PIM application in ZERO-PLUS

One of the goals of the ZERO-PLUS project is to develop sustainable buildings and to minimize their Embedded Energy, especially concerning the manufacturing stage. To employ a zero-waste strategy, the project combines Lean Construction off-site manufacturing. The different aspects reviewed here can be all used in a synergistic way to achieve a modular, sustainable, productive and cost-effective system. Regards to the ZERO-PLUS project, PIM is used to simulate and optimize the planning, production, and assembly phase. In the project, the Freescoo HVAC system used in the Cyprus case study provides the most applicable example to demonstrate the potential of implementing a PIM application in the real scenario. In terms of process management, the project is divided into four task phases, which include design, production, logistics and on-site assembly, (Figure 2).

During the design phase, the features of the proposed frame for the Freescoo HVAC system were analyzed. By conducting an analysis of the unit data, the production method was determined, allowing the design engineers to map out the optimum production sequence. The manufacturing facility was also strategically designed. In order to examine the protocols between the product and production design phase, three different frame designs are presented. Consequently, the differences in each frame design will present a variation in its production as well as in the assembly phases.

During the production design phase, a step-by-step procedure of the process using tables, graphics, and simulation creates an accurate representation of what will happen in the real situation. A set of goals was clarified to achieve a successful production facility design. In addition, the product unit data was also considered, as it will directly influence the sequence of assembly, workflow analysis, production plant specification, degree of automation, the number of machines needed, material flow and handling method, the activity relationship between each task, etc. An agent class system assigns each task a reasonable role [14], which can be a person, machine, or a robot that will perform the given task. Initially, the agent class will be documented, but later the data has the potential to construct a software algorithm that assists PIM manager in filtering data, interpreting information, monitoring activities, and making decisions. Agents in PIM application domains exhibits four key characteristics, i.e. autonomy, reactivity, communication and goal orientation. They are capable of acting autonomously, cooperatively, and collectively [15]. By using its agent class, PIM systematically represents the dynamic relationships between tasks. Moreover, during this phase, the proposed AEM was developed in parallel to evaluate the difference in productivity caused by the variations in the frame designs.

The On-site assembly phase focuses on the case study in Cyprus. The structural performance of the building was studied. To achieve smooth execution, the architects and assembly engineers received a detailed assembly plan according to the PIM data collected through the project phases.

By using the proposed PIM method, each project team role will be able to access the relevant information in advance or in real time. Data transfer and sharing between different domains facilitate the project team's understanding of the actual process and the method required to accomplish the task. The development of the AEM in conjunction with PIM can be used as a decision support tool to manage sources of complexity and rationalize the changes of product features, the unit data, and design alternatives to proactively response to the decision-making process.

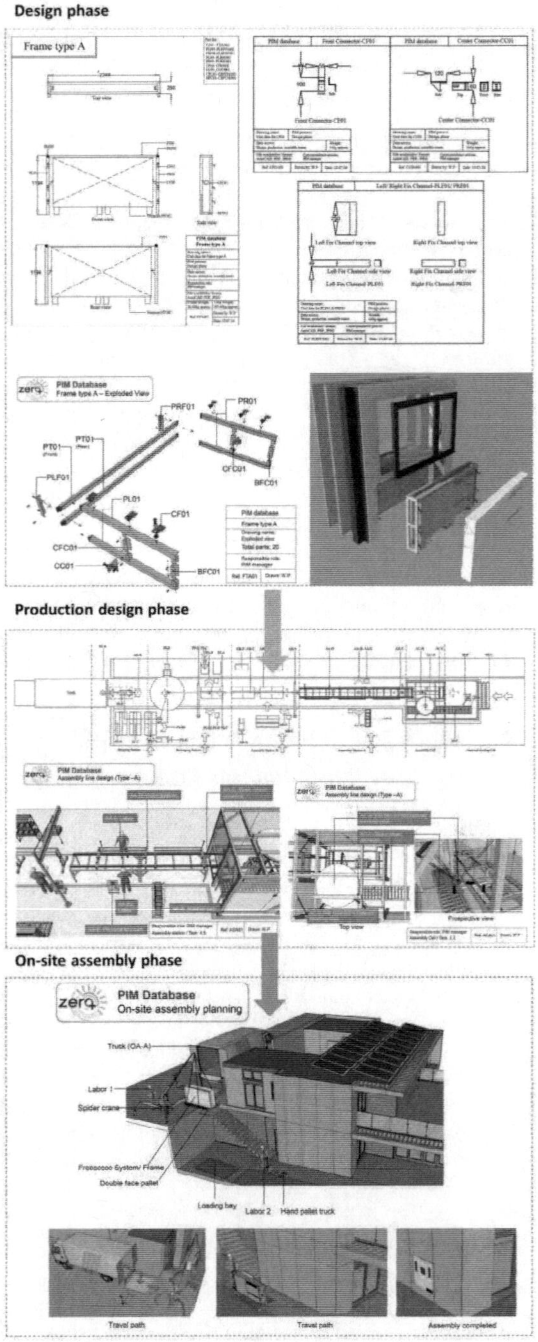

Figure 2: PIM application in the ZERO-PLUS project include design, production, on-site assembly phase

7 Brief description of the design of the integrated, flexible frame system for the Freescoo HVAC system

In order to fulfill the design goal from the project partner, the frame is designed as modular steel profile parts. The size and the shape of the frame are designed to ease handling by robots during assembly according to its kinematics and degree of the freedom (DOF). The proposed methodology inspired by the notion of DfX and Robot-Oriented Design (ROD) [16] incorporates the integrated assessment of parts and the evaluation of part complexity and accessibility. There is no welding applied at any joints to ensure maximum flexibility. Once the frame is installed, the design has to ensure that the Freescoo HVAC system is accessible in case of future maintenance or system upgrades. The PIM application in the project includes, generates unit data (e.g. specifications) of the frame, provides data to help the production design engineers map out the applicable production sequence, and the method to be used during the production phase. In addition, PIM provides data to assist the architect in altering the building design to compatible with the proposed frame and provide a guideline that will assist the on-site assembly procedure.

8 Optimization of PIM application

The ZERO-PLUS project provides a real-world example that demonstrate the theoretical potential of the PIM application in dealing with OSM strategy, assembly process, and management of the construction project. The research also serves as a guideline for the PIM initial application development stage in which recommended methods will accumulate and register essential data, as well as classifying different agents. The experience gained from the ZERO-PLUS project will act as the foundation for the future optimization of PIM. The next phase of PIM application development will focus on the design of the program architecture that will interpret the data collected from a generic software application in the initial stage, to ensure further implementation in a much broader range of projects.

9 Future software / computer application

During the development of PIM concept, the research team realises that there are various examples in computer software design, which have conceptual principles similar to those in PIM. A brief analysis is conducted to evaluate the state-of-the-art trends in software engineering, and to offer a reinterpretation of PIM by using software-programing concepts.

Service-oriented architecture (SOA) can be described as a loosely coupled program architecture designed specifically to meet the needs of an organization. By using communication protocols which provide services to another component, and make connections between different software components over a network [17], the definition of a service can be viewed as a logical representation of a repeatable task. SOA is independent, self-contained, yet when combined it forms the functionality of a large software application. Service orientation provides a dynamic, collaborative application for software integration and task composition. The capacities of a PIM architecture can be influenced by SOA. In PIM, the unit is similar to a service which is unrelated, self-contained, and implements one or more actions. The unit architecture in PIM also shares similarities with the service architecture in SOA [18]. It is beneficial when linking an agent influenced by the unit data or service, and any alteration of the unit or service would influence the individual agent's capabilities or responsibilities.

Service discovery or service discovery protocols (SDP) is an emerging field in the area of ubiquitous computing. They provide a mechanism, which allows automatic detection of service offered by any node in the network. In other words, service discovery is the action of finding a service provider for a requested service [19]. Service discovery can serve as a search engine in PIM architecture. For example, when the technology provider uploads a new product to the PIM database, if the specification matches with any user's requirements, the system will recommend the product automatically.

Microservices is a relatively new concept in the software design domain. It inherits the basic characteristics of service-oriented architectures (SOA). Conventionally, the program is a monolithic architecture with limited flexibility. Large complex software applications are composed of a number of services. Microservices offer solutions that can be deployed independently, specifically tailored, and loosely coupled [20]. The characteristics of the Microservices have served as an important source of inspiration in the development of the PIM concept. In practice, the assembly process can be considered a continuously deployed process. By adopting a Microservices architectural style in PIM, it enables the project teams to easily manage multiple aspects of the project. When issues arise during production, it is straightforward to pin down problems simply by identifying the smaller non-responsive Microservice processes or reviewing log files registered with a Microservice team identifier. Once the issue is identified, the responsible team member will be brought in to troubleshoot the problem [21].

10 Variation and alternatives

After an overview of the potential contributions from the software engineering sectors, the following section presents variations of the PIM application which could be implemented in the future in the construction industry. PIM shows great potential for the optimization of the assembly and production phases when applying the same principles. It is also suitable for other applications in the construction industry. For future reference, in this case, the PIM data module is divided by number of tasks include, (Figure 3).

1. PIM application for manufacturing of the building components.
2. PIM application for the building component attributes management.
3. PIM application for supply logistics.
4. PIM application for on-site assembly.
5. PIM application for home appliance management.
6. PIM application for green home appliances.
7. PIM application for service management.
8. PIM application for building performance.
9. PIM application for smart home environment control.

Because PIM can be loosely coupled, it can be specifically developed for each task. The loose coupling enables greater system flexibility that allows frequent rapid deployment.

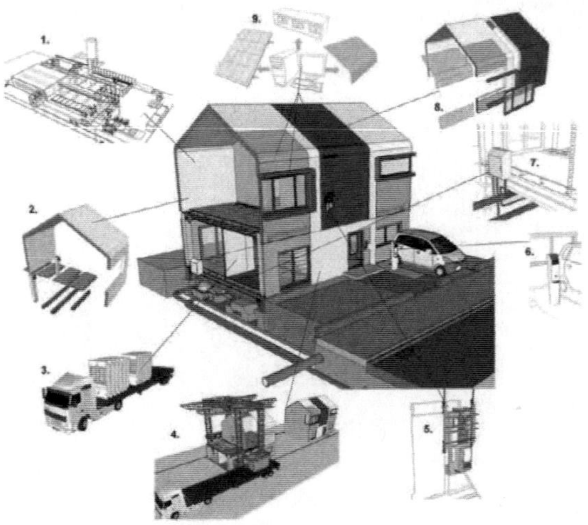

Figure 3: Future potential applications

11 Conclusion

During the ZERO-PLUS project, the research team gains an opportunity to develop the PIM concept that based on the Freescoo technology used in the Cyprus case study to demonstrate the potential of the PIM application in dealing with OSM strategy, assembly process, and management of the construction project. For now, PIM functions as a conceptual documentation tool. As mentioned earlier, PIM can be described as an interpretation of a process-oriented modeling approach, reinforced by the key parameters or rules, which are embedded within the operational process. When developing PIM as a software program, first, a pseudocode need to be developed to help the software engineer to understand the proposed PIM architecture. Then, detailed and exclusive information is needed to development a concrete PIM database. The PIM database and the pseudocode will generate the processing logic that assists the software engineer in further developing PIM as a functional software program. However, in terms of future implementation in industry, there are still many challenges ahead. Effective communication between the PIM manager and the software engineers is required. A real case scenario is necessary to collect the relevant information for database development. The hardware and software environments required for PIM to function as a universal application have not yet been studied.

12 Acknowledgements

This project has received funding from the European Union's Horizon 2020 Research and Innovation programme under Grant Agreement No. 678407.

References

[1] Schweitzer, E., & Aurich, J. C. Process Oriented Production System for Service Providing Companies. In Proceedings of the 2nd CIRP IPS2 Conference 2010; 14-15 April; Linköping; Sweden (No. 077, pp. 323-329). Linköping University Electronic Press, 2012.

[2] Battini, D., Boysen, N. and Emde, S. Just-in-Time supermarkets for part supply in the automobile industry. Journal of Management Control, Vol. 24 No. 2, pp. 209-217, 2012.

[3] Eastman, C., Teicholz, P., Sacks, R., Kathleen, L., & Handbook, B. I. M. A Guide to Building Information Modeling for Owners, Managers, Designers, Engineers and Contractors, Hoboken, 2008.

[4] Zhang, L. and Issa, R.R. Ontology-based partial building information model extraction. Journal of Computing in Civil Engineering, 27(6), pp.576-

584, 2012.
[5] Huang, C. J., Trappey, A. J., & Yao, Y. H. Developing an agent-based workflow management system for collaborative product design. Industrial Management & Data Systems, 106(5), 680-699, 2006.
[6] Dossick, C. S., & Neff, G. Organizational divisions in BIM-enabled commercial construction. Journal of construction engineering and management, 136(4), 459-467, 2009.
[7] Ohno, T. Toyota production system: beyond large-scale production. crc Press, 1988.
[8] Womack, J. P., Jones, D. T. Lean Thinking. Banish Waste and Create Wealth in Your Corporation, Revised and Updated. Simon & Schuster, Inc, 2003.
[9] Selçuk, B. Adaptive lead time quotation in a pull production system with lead time responsive demand. Journal of Manufacturing Systems, Vol. 32 No. 1, pp. 138-146, 2013.
[10] Diekmann, J., Krewedl, M., Balonick, J., Stewart, T., Won, S. Application of Lean Manufacturing Principles to Construction. University of Texas at Austin, 2004.
[11] Stephens, M. P., & Meyers, F. E. Manufacturing facilities design and material handling. Purdue University Press, 2013.
[12] Javadian Kootanaee, A., Babu, K.N. and Talari, H.F. Just-in-time manufacturing system: from introduction to implement, 2013.
[13] Boothroyd, G. Assembly automation and product design (Vol. 536). Boca Raton, FL: Taylor & Francis, 2005.
[14] Turban, E., King, D., Lee, J. and Viehland, D. Electronic Commerce: A Managerial Perspective, Pearson Education International, 2004.
[15] Parker, H. Interfirm collaboration and the new product development process. Industrial Management and Data Systems, Vol. 100 No. 6, pp. 255-60, 2000.
[16] Bock, T. Robot-oriented design. Proceedings of the 5th ISARC, Tokyo, Japan, pp. 135-144, 1988.
[17] Erl, T. Service-oriented architecture: a field guide to integrating XML and web services. Prentice Hall PTR, 2004.
[18] Patrick, P., Aletty, A., Kasi, J., Kapoor, C., Urhan, T. and Mihic, M. Bea Systems, Inc. Service oriented architecture. U.S. Patent Application 11/133,022, 2005.
[19] Czerwinski, S. E., Zhao, B. Y., Hodes, T. D., Joseph, A. D. and Katz, R. H. An architecture for a secure service discovery service. In Proceedings of the 5th annual ACM/IEEE international conference on Mobile computing and networking (pp. 24-35). ACM, 1999.
[20] Balalaie, A., Heydarnoori, A., Jamshidi, P. Microservices Architecture Enables DevOps: Migration to a Cloud-Native Architecture. IEEE Software 33 (3): 42–52, 2016.
[21] Chen, L. and Babar, M. A. Towards an evidence-based understanding of emergence of architecture through continuous refactoring in agile software development. In Software Architecture (WICSA), 2014 IEEE/IFIP Conference on (pp. 195-204). IEEE, 2014.

Employing Ontology and BIM to Facilitate the Information for Subcontractor's Payment Requests and Ledger Generation

C.W. Feng[a], Y.J. Chen[b] and H.Y. Yu[c]

a,c Department of Civil Engineering, National Cheng Kung University, Taiwan
b Department of Creative Design and Architecture, National University of Kaohsiung, Taiwan
E-mail: [a]cfeng@mail.ncku.edu.tw, [b]yjchen@nuk.edu.tw, [c]giveup2100@gmail.com

Abstract –

General contractors usually need to handle subcontractor's payment requests periodically according to the progress of the construction project. The documents required for payment requests are related to various types of information generated from construction work items. Consequently, construction company needs to spend a lot of efforts to verify and approve the submitted documents in order to record accounts on the ledger.

Building Information Modeling (BIM) has been used to derive the information for various operations within the construction project. One of the important functions provided by BIM is quantity takeoff. With this capability, BIM can serve as an information platform to provide the information regarding various aggregated amounts of work required to be performed by subcontractors. However, detail information required to process payment requests is still hard to be obtained from the BIM model if no proper development specifications have been established.

This study employs ontology and BIM to facilitate the information required to process payment requests and ledger. First, various work items performed by subcontractors are analyzed to identify the information required to process payment requests. Then the relationships between work items and BIM elements are determined through the ontology developed according to the required information identified. Next, a variety of Dynamo modules are applied to develop the BIM model that can provide various types of measurements required by work items. Finally, a database that can help construction managers to facilitate information for payment requests and ledger is developed for future uses.

Keywords –

BIM; Ontology; Dynamo; Payment Request; Ledger Generation;

1 Introduction

There are two parts of construction company's ledger: (1) The payment requests to the owner (in cash flow). (2) The payment requests from subcontractors (out cash flow).

The payment requests from the construction company to the owner are based on the quantity of work items and the unit price of the budget, such as concrete grouting. The payment requests from the subcontractors to construction company are based on the unit price in the contract and the actual amount of work items applied on site, such as concrete pumping and concrete grouting.

The ledger records payments received from the owner and payments to the subcontractors, but the work items are based on the different basis. The payment to the subcontractors is based on the subcontracting type, such as labor only, material only, labor with material, and lump-sum. As a result, the work items are different between payment requests to the owner and payment requests from subcontractors. For example, the payment-request work item concrete grouting to the owner, after subcontracting, it will be divided into concrete pumping, concrete grouting, concrete material, etc., different work items for payment requests from subcontractors. The required information for subcontractor's payment requests has to consider the subcontracting type.

In order to obtain the actual amount of work items, such as applying area and working hours, as the basis of subcontractor's payment requests, engineers have to analyze the different subcontracting types and the measurement of actual progress identification according to their experiences.

In the construction industry, Building Information Modeling (BIM) has been used to produce specific information for various purposes, including cost estimation, visualization, quantity takeoff, etc.

However, all the required information of payment requests still hard to obtain from the BIM model if there are no proper development specifications. For example, the Level of Development (LOD) specification [1] developed by Association of General Contractor's (AGC) BIMForum group is a detail interpretation of LOD schema developed by American Institute of Architects (AIA) [2]. The LOD specification defines the model element's characteristics at different LOD, the categorization of model elements follow the CSI Uniformat 2010 [3].

For the waterproof work of the interior wall, the related LOD specification is the C2010 Wall Finishes, as shown in Table 1. The required information of the waterproof work item is the applying area of waterproof. However, there is no specification about building the waterproof. Even the model is built according to the highest LOD specification, the required information for payment requests can't be obtained. As a result, the proper development specification should be established.

Table 1. C2010 Wall Finishes LOD specification

LOD	Specification
100	Non-graphic information attached to model elements providing assumptions about proposed finish materials.
200	Generic materials other than sheet goods and coatings by type (e.g. tile or panelling), approximate thickness and scope in elevation.
300	Materials are modeled based on specific types (e.g. Tile typeCT-1). Thickness and scope of finishes other than sheet goods and are accurately modeled. Sheet goods and coatings may be specified in Part II related to interior partitions.
350	Additional non-graphic information to include: 1.Manufacturer 2.Model
400	1.Pattern layouts 2.Expansion/control joints 3.Edges

This study uses ontology and BIM to obtain the required information of payment requests and ledger. First, the work items performed by subcontractors are categorized, according to the subcontracting type and the characteristics of payment requests. The information required to process payment requests is identified through the interview. Then the ontology model and the semantic web rule language (SWRL) are built to clarify the relationship between work items and BIM elements, and is used to identify required information which is provided by BIM element's attributes. Next, the modeling rules for the elements are determined by LOD specification and the extra specification. In order to obtain the required information for payment requests, the Dynamo modules are developed according to the modeling rules developed to help build the BIM model. Therefore, various types of measurements required by work items can be intercepted from the model.

The rest of this research is organized as follows. Section 2 reviews the literature of ontology and the BIM and ontology related work. Section 3 proposes the framework of this research including: establish the payment request ontology and develop BIM modeling rules for obtaining the required information of payment request. Section 4 concludes and summarizes this study.

2 Literature Review

2.1 Ontology

Ontology is widely used in the information field. The definition of Ontology is "an explicit specification of a conceptualization" [4]. Ontology uses the text to conceptualize and standardize of knowledge and help the dissemination and sharing of knowledge so that the knowledge can be shared in different media. By defining the concept and the relevance of data, the computer can infer the information from different sources. Using classes, attributes and relationships describe the meaning of information, constraint and consistency. Ontology is mainly used to explicitly define a Domain (Classes, also known as concepts), the description of features and attributes between domains is known as Attribute, Relation, and Instance. [5]

The concepts of specific knowledge can be described by the class. Class consists of subclass, which is the detail classification of the knowledge. In order to describe the characteristics of the knowledge, the attribute is used to provide various information of the class and subclass. The relationship between class and attribute are performed by the relation. The relation connects the concept and the information of knowledge. The instance is the example of the class in reality. The characteristics can be shown by concrete example. For example, finishes work can be divided into plaster work, tiling work, etc. , the plaster and tiling work are subclasses of finished work. For plaster, the proportion of plaster is an attribute of plaster work. 1:3 Portland cement is an instance of plaster work, showing the characteristics of the plaster work in reality.

2.2 BIM and Ontology Related Work

The quantity takeoff of BIM is widely used in the construction industry. Quantity takeoff is one of the advantages provided by BIM for the cost estimation in the construction project. The many-to-many relationships between work items and BIM elements have to be considered when using BIM to calculate the quantity of work items. A work item may correspond to many elements. For example, the concrete grouting may be related to many elements, such as concrete columns, beams, walls. An element also may correspond to many work items. For example, the RC column may be related to many work items, such as steel reinforcement works, form works, and concrete works. It is necessary to clarify the many-to-many relationships between work items and elements, in order to facilitate the accurate calculation of work item's quantity.

Lee et al. [6] proposed a BIM-based project cost management model for bridge construction. First, the analysis of bridge BIM elements and the construction method of the bridge are done. Then identify the BIM elements and the related work items. The relationship between BIM elements and work items are defined and stored in the database of bridge element. Finally, the element quantity obtains from BIM combine with the database that records the related cost work items of bridge elements to generate the budget.

Niknam et al. [7] integrated the BIM model, work items, and material cost data, then presented in ontology for cost estimating. The relationships between BIM elements (classify by Uniformat II) and the work items (classify by MasterFormat [8]) are identified. Also, the relationships between work items and the related resource (material, equipment, and crew) are identified. The related work items of the element and the needed resource for the work items are inferred through ontology. Then the resource's unit price is considered for cost estimating, in order to reduce the human involvement and errors.

Chien et al. [9] proposed an ontology-based framework to develop intelligent BIM object for cost analysis. By integrating the semantic information of cost work items and the relationship between work items and BIM elements, the ontology model is developed. The construction process and the related cost work items are inferred according to BIM elements, in order to improve the efficiency of the current cost estimation process.

Lee et al. [10] combined BIM with ontology for building cost estimation. The ontology is used to infer the suitable work item, through identifying the work item's condition that is provided by BIM model. For the tiling work, work item conditions (e.g. Room usage, base type) are extracted from BIM model, then the work item information (e.g. Tilling method, Tilling material type) are inferred to obtain the accurate and consistent result of selecting work items.

2.3 Summary

According to those studies, current studies mainly use the BIM as a quantity takeoff tool for cost estimation. Some of the research used ontology to infer the related work items of BIM elements by clarifying the relationship between work items and elements, then combined the quantity of BIM elements and work items to estimate the cost. There are less concerns dealing with the relationship among the payment requests, general contractors, and subcontractors. The general contractors are still unable to obtain the required quantity information for subcontractor's payment requests from the BIM model.

3 The Framework and Result

The framework of this study is shown in Figure 1. First, the payment-request work items related to the ledger are identified according to the relationship among general contractors, the owner, and subcontractors. Then by analyzing the work items, including the basic information and the payment-request information, the relationship between BIM elements and work items are determined as well. The ontology model is further established to express the payment-request work item information and the relationship with BIM elements. Consequently, the required information that can be provided by BIM model are inferred through the ontology model. Finally, the modeling rules are established on the basis of the LOD Specification and the additional specification for developing the BIM model that can provide the required information for the payment requests.

3.1 Payment-Request Information Analysis

In order to build the ontology model, it is necessary to analyze the demands and information for constructing ontology. There are two parts of analyzing the information: (1) The basic information and payment-request information of work items. (2) The relationship analysis between work items and BIM elements. Through these two steps, the required information of payment requests that can be provided by BIM elements are identified. That information can be the basis of building the ontology model and SWRL.

3.1.1 Payment-Request Information of Work Items

The basis and the payment-request information about work items can be obtained from interviewing. First, we categorize the work items through payment-request frequency, base type, and base unit, as shown in Table 2. The payment-request frequency is usually based on time. From our interview case, the frequency is once or twice a month. There are two payment-request base types: Base on Quantity and Base on Spaces. Base on Quantity is that payment requests are based on the material or the element's quantity. For example, the work item quantity of concrete grouting (labor) is determined by the concrete material quantity.

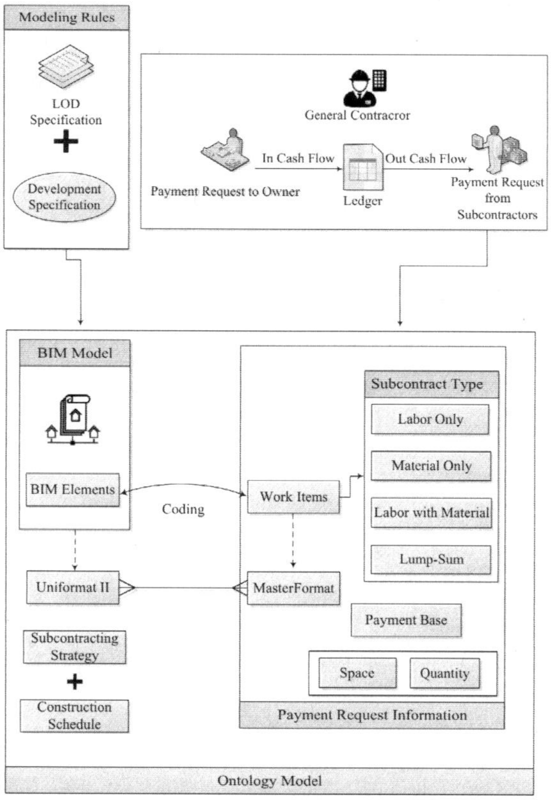

Figure 1. The framework of this study

Base on Spaces is that payment requests are associated with spaces, for example, the quantity of Tiling work (one type of Masonry work) is based on the space area (e.g., living room, bathroom). Table 3 is the further analysis of each work item, including payment-request base, subcontracting type and required information of payment requests. For example, when the Wall Plaster (Interior) is finished, the third party measurement measure the applying area of plaster as the base of payment requests for subcontractors.

3.1.2 Relationship between Work Items and BIM Elements

The relationships between work items and BIM elements are determined to identify the required information of payment requests that can be provided by BIM. There are three types of relationship between work items and BIM elements, including:

Table 2. Work items category

Request Frequency	Request Base Type	Request Base unit	Work Item	Subcontracting Type
Twice a month	Base on Quantity	Floors	Concrete	Labor Only
				Material Only
		Floors /Spaces	Form	Labor with Material
	Base on Spaces	Floors /Spaces	Masonry	Labor Only
				Material Only
Once a month	Base on Quantity	Floors	Silicone	Labor with Material
	Base on Spaces	Floors /Spaces	Waterproof	Labor with Material

Table 3. Further analysis of work item

Work Items	Masonry		
	Building Brick	Wall Plaster (Interior)	Wall Plaster (Exterior)
Subcontracting Type	Labor Only	Labor Only	Labor Only
Base	Third Party Measurement	Third Party Measurement	According to Drawing
Required Information	Applying Area	Applying Area	Applying Area

(1) Direct related: BIM elements directly provide the required quantity information. The required quantity can be obtained directly from the sum of BIM element's attributes, for instance, the concrete material volume of main structure can be directly obtained from the sum of the concrete columns, beams, walls and slabs volume in the BIM model. (2) Indirect related: BIM elements indirectly provide the required quantity information. The required quantity shall be indirectly obtained from BIM element's attributes with calculation or adjustment. For instance, the waterproof of window frame can be calculated from window circumference. The waterproof length is obtained from the sum of window element's length and width. (3) Non-related: The required quantity can't or doesn't need to be obtained from the BIM model, for instance, the tile sealant material use the invoice, which information don't need to be obtained from BIM elements, as the base for payment requests.

3.1.3 Identify the Required BIM Elements Data of Payment-Request Work Items

The required information about payment-request work items that can be obtained from the BIM model is identified. The information is identified by the relationship analysis of work items and BIM elements.

The payment-request work items are encoded in MasterFormat, then correspond to the Uniformat II according to the corresponding association table proposed by Charette [11]. The table is used to check if the corresponded BIM elements can provide the required information or not. If the required information can't be provided by the corresponded BIM elements, the elements have to be marked. For example, the MasterFormat coding of Wall Plaster (Interior) is 09220, which belongs to the division 09 Finishes. The corresponding Uniformat II is B2010 Exterior Walls and C30 Interior Finishes. After analyzing the required information of Wall Plaster, the required element data are wall plaster area and the column inside plaster area, whose Uniformat II code are C30 10 120 Wall Finishes-Coatings and C30 10 300 Column Finishes.

3.1.4 Encoding the Payment-Request Information

OmniClass Construction Classification System (known as OmniClass or OCCS) is used to encode the work items, elements, and location, as a unified code. OmniClass is a classification structure developed by the OCCS Development Committee for the construction industry. OmniClass integrated the MasterFormat, Uniformat II, and other extant systems currently in use. The classification of OmniClass is suitable for the BIM elements [12]. OmniClass consists of 15 hierarchical tables, including Spaces, Material, Work Result, Elements, etc. Most of the required information of payment requests is included in the OmniClass, as a result, the OmniClass is selected for encoding the payment-request information, such as Payment-Request Work Items Code, Model Elements Code, and the Space Code.

3.1.5 Organizing the Payment-Request Information

The payment-request information is collected as Table 4, including the basic information of work items, the payment-request information of work items, the relationship between work items and BIM elements, the required BIM data of payment requests and the OmniClass code. The form is convenient for querying the payment-request information through coding.

Table 4. Payment-Request Information (part)

Work Item	Portland Cement Plaster		Cement
Payment-Request Code	22-09 24 00 (Cement Plastering)		41-30 10 25 19 15 (Cement)
Space Code	13-15 13 00 (Interior Wall)	13-15 11 00 (Exterior Wall)	--
Subcontracting Type	Labor Only	Labor Only	Material Only
Required Information	Applying Area		Material Quantity
Base	Third Party Measurement		Invoice
Correlation with BIM	Direct Related	Direct Related	Non-Related
BIM Elements Data	The Wall Material Area	The Wall Material Area	--
	21-03 20 10 70 Wall Area	21-03 20 10 70 Wall Area	--
Related Data	21-03 20 10 71 Column (Inside)	21-03 20 10 72 Column (Outside)	--
Note	--	--	--

3.2 Payment-Request Ontology Model

The Ontology model is built to present the domain knowledge of payment-request information and BIM elements. The SWRL assists the engineers to infer the required information of payment requests. The development of the ontology model follows the following steps.

3.2.1 Building the Ontology Framework

The ontology framework is built on the basis of the analysis of the payment-request information, including classes, object properties, and data properties. The information can be categorized into four domain knowledge base according to the analysis of payment-request information and BIM elements: (1) Payment-Request Information Knowledge Base: including work items' basic information and the payment-request information. (2) BIM Knowledge Base: including model elements' classification and elements' data. (3) Construction Process: the further analysis of the budget work items, including the job steps and the used resources. (4) Budget Information: including budget work item's information. The framework is shown in Figure 2.

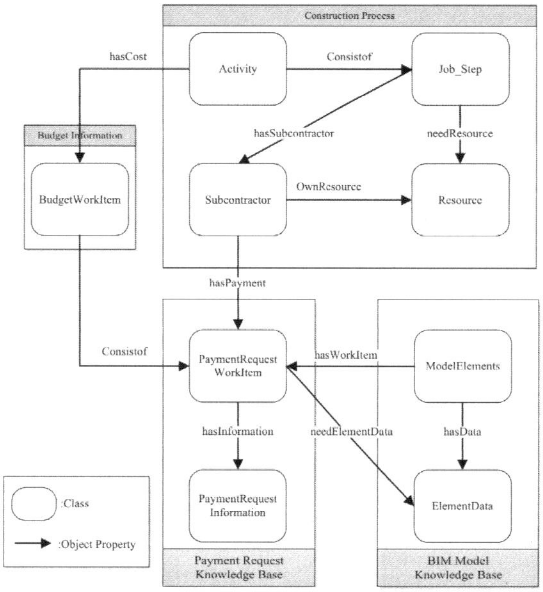

Figure 2. The framework of ontology

3.2.2 Introduction of the Ontology Framework

The concept of the ontology model is introduced in this section. The two domain knowledge bases are defined through the analysis of the payment-request information and BIM model. The Construction Process and the Budget Information are defined through the analysis of budget sheets, construction methods, and construction sequences.

In the Payment-Request Information Knowledge Base, the PaymentRequestWorkItem contains the work item's basic information, including four sub-classes: MaterialOnly, LaborOnly, LaborWithMaterial and Lump- sum, which are categorized by subcontracting type. The work items are built into each sub-classes according to subcontracting type. The location of work items and the payment-request code are described by data properties. PaymentRequestInformation consists of the payment-requests information, such as frequency, base, and base unit. The concept of Payment-Request Information Knowledge Base shows in Figure 3.

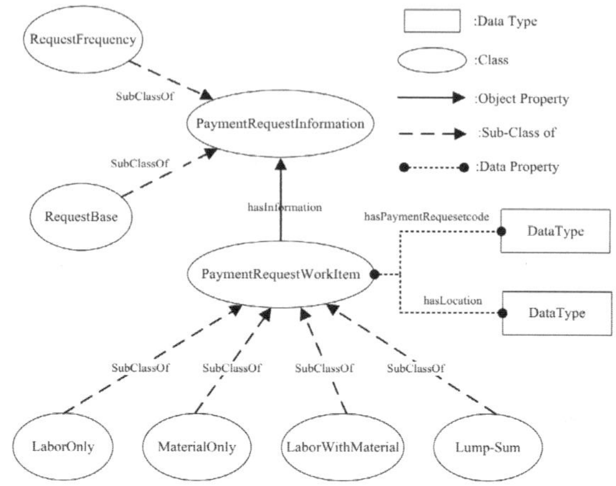

Figure 3. Payment Request Information Knowledge Base

In the BIM Model Knowledge Base, the ModelElements is the classification of BIM model elements, including the sub-classes: StructureSystem, Finishes, DoorsandWindows, HandrailAndRailings and Ceiling. The model elements are built into the sub-classes according to the classification. The properties such as location, object code, and object floor are described by data properties. The ElementData is the geometric attribute of BIM model elements, including Area, Length, Volume, etc. The concept of the BIM Model Knowledge Base shows in Figure 4.

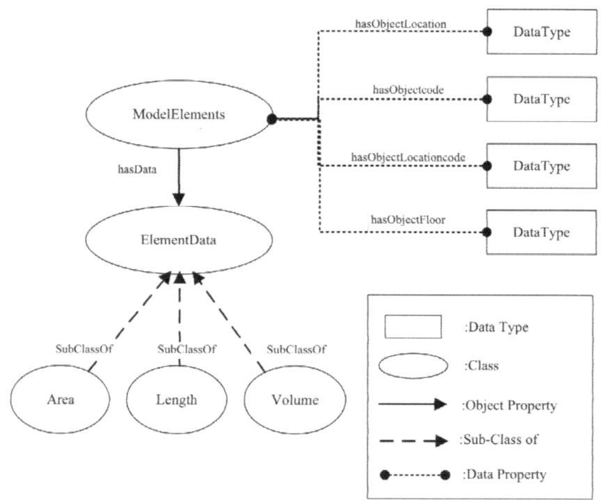

Figure 4. BIM Model Knowledge Base

In the Budget Information, the budget work item's information can be obtained from budget sheets, such as PCCES code. Budget work items represent the in cash flow of the construction company's ledger.

The Construction Process includes the Activity and Job_Step. The Activity is the task to describe budget work items. The Activity consists of different job steps, which are built in the Job_Step. Each job step has different subcontractors and the resources needed that are provided by subcontractors.

For example, the instance “Bathroom Finishes” in Activity is related to the “Bathroom Wall Ceramic Tiling 15*60” in BudgetWorkItem. This activity consists of two instances of Job_Step: ”1:2 Cement Plastering” and “Ceramic Tiling 15*60”. “Masonry Worker” in the Resource is one of the needed resources that is owned by the “Masonry subcontractor” in Subcontractor.

The payment-request work items of subcontractors represent the out cash flow of the construction company’s ledger. For example, the “Masonry subcontractor” has the payment-request work item ”1:2 Cement Plastering” and “Bathroom Wall Tiling”. The budget work item: “Bathroom Wall CeramicTiling 15*60” consists of those two payment-request work items.

Table 5 shows the example of the classes and the data properties. The PlasterWork is one of finish work and the subcontracting type is LaborOnly. It has the data property to describe the plaster proportion and the location. The wall finishes material is categorized as the WallFinishes, which has the data properties to describe the location and the code. The Area is one of element data, whose value is described by the data property: hasObjectArea.

3.2.3 SWRL Inference

The SWRL rules help to infer the required quantity information that can be obtained from BIM elements and the work item’s payment-request information. The knowledge bases provide the information of payment requests and the relationship between work items and BIM model elements. By SWRL rules, the required BIM element’s data (as the related data in Table 4.) and the related information for each work item can be inferred.

The SWRL rules are developed by following steps. First, according to the work item’s associated elements and the inference rule, the required element data are linked with work items through the object property: needElementData. After reasoning, the required quantity information that can be obtained from BIM model are shown in the ontology. Second, according to the work item’s associated elements and the work item's location, the request base of work items can be inferred. The example is shown in Table 6.

3.3 Model Development Specification for the Payment-Request Work Items Quantity Takeoff Model

In order to obtain the required information for payment requests from BIM model, the modeling specification has to be established.

Table 5. Example of classes and data properties

Class	Subclass			Data Property
BudgetWorkItem	B_FinishesWork	B_MasonryWork	--	hasPCCEScode
Activity	Finishes_Operation	--	--	Consistof
	_Tiling	--	--	hasCost
Job_Step	Finishes_Plastering	--	--	hasSubcontractor
		--	--	needResource
Resource	Labor	Masorny_Worker	--	--
Subcontractors	S_LaborOnly	--	--	OwnResource
		--	--	hasPayment
PaymentRequestWorkItem	P_LaborOnly	MasonryWork	PlasterWork	hasPlasterProprtion
				hasPaymentRequest _Code
				hasLocation
PaymentRequestInformation	RequestBase	--	--	--
ModelElements	Finishes	WallFinishes	--	hasData
			--	hasWorkitem
			--	hasLocationCode
			--	hasObjectCode
			--	hasObjectLocation
ElementData	Area	--	--	hasObjectArea

Table 6. Example of SWRL rules

SWRL Rules	Semantic
WallFinishes(?x) ^ hasWorkItem(?x, ?y) ^ PlasterWork(?y) ^ hasData(?x, ?y1) ^ Area(?y1) -> PlasterWork(?y) ^ needElementData(?y, ?y1) ^ Area(?y1)	When the model element: Wall Finishes has the Plaster work item, also has the element data: Area. Then the Plaster work links with the Wall Finished Area as the required data.
WallFinishes(?x) ^ hasWorkItem(?x, ?y) ^ PlasterWork(?y) ^ hasLocation(?y, "Exterior") -> PlasterWork(?y) ^ hasInformation(?y, AccordingToDrawing) ^ RequestBase(AccordingToDrawing)	When the model element: Wall Finishes has the Plaster work item, and the location of Plaster work is Exterior. Then the request base of Plaster work is according to drawing

As shown in Figure 5. and Figure 6.

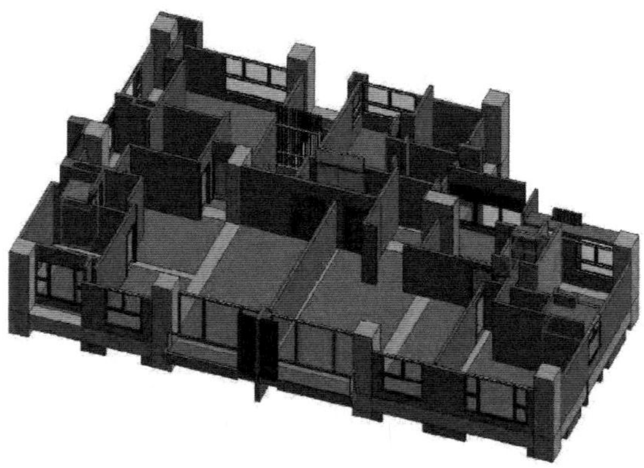

Figure 5. The waterproof of shower room (BIM model)

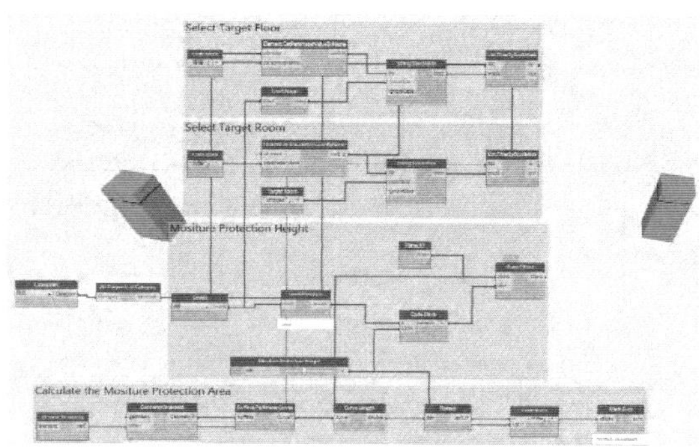

Figure 6. Calculation of applying area of Waterproof (Dynamo module)

3.3.1 Establish the Model Specification According to LOD Specification

According to the result of Ontology, the required BIM element's data for payment requests are identified. The related BIM elements correspond to the LOD specification by OmniClass. Then check if the BIM model built on the basis of the LOD specification can provide the quantity information or not. If not, the additional specification should be established.

3.3.2 Define the Space According to the Characteristics of Work Items

Space should be defined on the basis of the characteristics of work items in the model. Some work items repeatedly show in the same spaces at different zones or floors, as a result, the defined spaces can help to get the required quantity information for the work items whose request base type are Base on Spaces. The information of the space consist of space's name and code, for example, for the waterproof applies to the bathroom or the shower room's wall or floor, the bathroom and shower room should be defined in BIM model, in order to obtain the quantity information.

3.3.3 Develop the Dynamo Modules to Facilitate Model Development

The Dynamo modules are applied to develop the BIM model, in order to get the required information that is difficult or unable provided by BIM model of payment-request work items. For example, the height of the interior wall's waterproof may be different according to space. The payment request base unit of waterproof is Floors. In the BIM, the waterproof area is built by setting the wall material, which can't quickly obtain the applying area of each floor. Therefore, the Dynamo modules are developed to obtain the waterproof applying area by selecting floor, space and the setting of waterproof's height. First, the target floor, space, and the height of waterproof are selected according to user's demand. Then Dynamo calculates the perimeter of the selected space, and times the height to obtain the applying area.

3.3.4 Payment-Request Work Item's Quantity Information and Ledger Generation

The Microsoft Access is used to collect the payment-request work item's quantity and to generate the construction company's ledger. First, the relationship between work items and BIM elements (as shown in Table 4) are built into the database and combined with the BIM model that is established on the basis of the model development specification. The quantity of BIM elements from BIM model corresponds to the work items by OmniClass code. The required quantity information of payment-request work items can be obtained.

When the payments are requested by the subcontractor, the quantity of payment-request work items for each subcontractor can be obtained according to the payment-request base unit (Floors or Spaces). After the quantity combines with the unit price in the contract, the total amount of the payment requests can be obtained from the database. As shown in Figure 7.

When the ledger is generated by the construction company, the cost of the construction company is composed of the payments to the subcontractors. The ledger records the cost on the basis of the subcontracting type, including materials (Material Only), wages (Labor Only), and outsourcing (Labor With Material and Lump-Sum). The out cash flow of the construction company's ledger is the sum of each subcontracting type's payment that can be monthly produced from the database.

The income of the construction company is composed of the payments to the owner according to the budget work items. According to the analysis of ontology, the budget work items consist of the payment-request work items. The total amount of the payment-request work items in each budget work item is calculated by the database. The payment requests to the owner and the in cash flow of ledger for construction company are recorded on the basis of the amount of each budget work item.

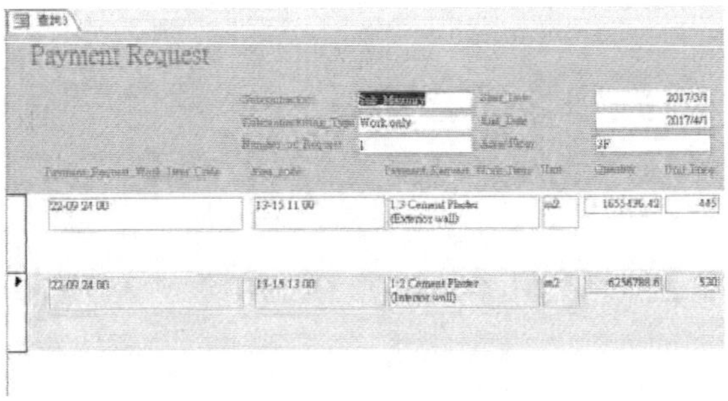

Figure 7. Total amount of payment-request work items

4 Conclusion

The required information of payment and the ledger can be obtained from the BIM model that is established on the basis of the modeling specification. The Dynamo modules are also developed to assist the model development. The work item's information and the payment-request information are analyzed. The relationship between BIM elements and the work items are expressed through the ontology model. The required quantity information and the related BIM model elements are inferred by SWRL rules. The model development specifications are established according to the LOD specification and the additional specification. The Dynamo modules are applied to develop payment-requests work item quantity takeoff model. The quantity information provided by BIM model can facilitate construction manager to obtain payment-request information and generate ledger.

5 Acknowledgment

The authors would like to thank Ministry of Science and Technology, R.O.C. program 105-2221-E-006-047-MY2.

6 References

[1] BIMForum. The Level of Development (LOD) Specification. On-line: http://bimforum.org/lod/, Accessed: 18/03/2017.

[2] American Institute of Architects. Project BIM Protocol. On-line: https://www.aiacontracts.org/contract-documents/19016-project-bim-protocol, Accessed: 18/03/2017.

[3] The Construction Specifications Institute. UniFormat™. On-line: http://www.csinet.org/Home-Page-Category/formats/uniformat, Accessed: 18/03/2017.

[4] Gruber T. R. Toward Principles for the Design of Ontologies Used for Knowledge Sharing. *International Journal Human-Computer* Studies 43, p.907-928, 1993.

[5] Noy N. F. and McGuinness D. L. *Ontology development 101: A guide to creating your first ontology*, 2001.

[6] Lee T. W. and Feng C. W. Developing the BIM-based Project Cost Management Model for Bridge Construction. Department of Civil Engineering, National Cheng Kung University, 2014.

[7] Niknam M. and Karshenas S. Integrating distributed sources of information for construction cost estimating using Semantic Web and Semantic Web Service technologies. *Automation in Construction*, 57: 222-238, 2015.

[8] The Construction Specifications Institute. MasterFormat. On-line: http://www.csinet.org/masterformat, Accessed: 18/03/2017.

[9] Chien Y. H. and Feng C. W. An ontology-based framework to develop intelligent BIM object for cost analysis. Department of Civil Engineering,

National Cheng Kung University, 2016.

[10] Lee S. K., Kim K. R. and Yu J. H. BIM and ontology-based approach for building cost estimation. *Automation in Construction*, 41: 96-105, 2014

[11] Charette R. P. and Marshall H. E. UNIFORMAT II Elemental Classification for Building Specification, Cost Estimating, and Cost Analysis. U.S. Department of Commerce, Technology administration, National Institute of Standards and Technology (NIST), Gaithersburg, 1999.

[12] The Construction Specifications Institute. OmniClass. On-line: http://www.omniclass.org/, Accessed: 18/03/2017.

An Integrated Methodology for Construction BIM & ERP by Using UML Tool

Yen-Ray CHEN[ab] and H. Ping TSERNG[b]

[a]Department of Civil Engineering, University of National Taiwan University, Taipei, Taiwan
[b]CASTO Engineering Consultants co., Ltd, Taipei, Taiwan
E-mail: d04521015@ntu.edu.tw, hptserng @ ntu.edu.tw

Abstract –

In order to improve the efficiency of construction management, a new methodology is presented to develop translatability for construction BIM (Building Information Modeling) and ERP (Enterprise Resource Planning) by using UML (Unified Modeling Language) tool. The method includes a new rational operation process can be developed to improve efficiency. Besides, the competitive ability of a construction company is also increased. The major contribution of the method is to understand and confirm what needed and correct process in the enterprise is before running the construction ERP system. Enterprise will know how to use BIM to improve some processes in each part/each module of ERP correctly and efficiency. The research possibilities are identified and tested based on the implementation of the method. To some extent, this research also establishes a new agenda of process conflict solving for future research.

Keywords –

Enterprise Resource Planning (ERP); Building Information Modeling (BIM); Unified Modeling Language (UML)

1 Introduction

The concept of ERP is presented to integrate all enterprise systems because multi-roles joining and working in different projects will lead to the problem of information integration between projects in construction enterprises. In the past, enterprise information systems usually focus on a single department or organization. All kinds of information systems in the enterprise usually come apart each other. That will occur a situation each information system likes an island. However, all of information islands become independent respectively. When each department implements individual information system, each department will become individual information island. This situation will lead information flows, working flows and other flows to run between departments unsmooth. It will cause barrier and hurdle to integrate these flows each other. How to integrate correct flows between departments is becoming a new planning aspect by ERP system planners. Averagely, the actual cost increased more than 1.78 times and the implementation duration increased more than 1.5 times when enterprises implement ERP system [1]. Another investigation points out the cases of enterprise implementation ERP system had 96.4% failure in the past [2]. The most serious problem will lead enterprise to close down. Productivity in construction industry is project-based. When construction enterprise starts to implement ERP system. Production line belongs to the project-based way because of construction characteristic.

BIM is a concept that use digitalize and parametric form design to build a integrate module. It offers various forms and information help control conflict/resource/schedule/cost between project designs and construct period.

We want to make the implementation ERP system and BIM more match and enhance the performance of ERP system. In ERP implementation process, there is serious problem existing for a long time. The problem is that enterprise employees usually know how business processes operate individually in organization/department/enterprise, but they don't know how to develop system planning; In other words, ERP system planners know how system planning is developed, but they don't know how business processed in their customers (enterprise) run. Another big problem is that processes conflict between different organization/department/enterprise. That will be why we combine BIM and ERP system, try to solve the conflict problem. Therefore, when both roles cooperate to implement ERP system enterprise will not match enterprise processes conflict and system planning. This research will try to solve this problem to avoid implementing ERP system failure. Because enterprise implementation ERP system is time-wasted and cost-expensively, we expect to implement ERP system more successfully.

2 Research Objectives

This paper focus on how to construct a transformation method between BIM and ERP system for the implementation of construction enterprise resource planning. In order to improve the efficiency of construction management, a new methodology is presented to develop translatability for BIM and ERP by using UML tool. It can become a consideration method when enterprise starts to implement ERP system. Figure 1 illustrates the framework of research objective. In Figure 2, a whole procedure of implementation ERP system will include "BIM", "Integration" and "build up ERP system" three parts. Because of page limitation, the "build up ERP system" part won't illustrate. So we use duck color to show in Figure 2. We just can illustrate "BIM" and "Integration" two parts in this paper.

3 Problem statements

Via the above mention about feature of construction industry, the most important purpose in this research is how to improve information advancement in construction enterprise by implementing ERP system. In Table 1, we compile problem definition, description and solution of information apart in construction industry.

Combining BIM and ERP regards as a new solution of enterprise information apart. In sum, BIM is a one of concepts and tools to assist running ERP system more effectively and successfully. In Table 1, it classifies about the concepts of ERP and the practices of ERP. The following is classified:

Regarding the concept of ERP

1. Confirm goals of enterprise.
2. Analyze and define needs of enterprise.
3. Evaluate main business processes.
4. Analyze original business processes.
5. Design better business processes.

Regarding the practice of ERP

1. Match business processes with ERP system modules.
2. Develop system implementation planning.
3. Start system planning.

Via the above mention about the concept of ERP and the practice of ERP, we can comprehend and find out the problem solutions below:

1. Processes modeling and conflicts finding.
2. Processes and system planning integration.
3. System planning.

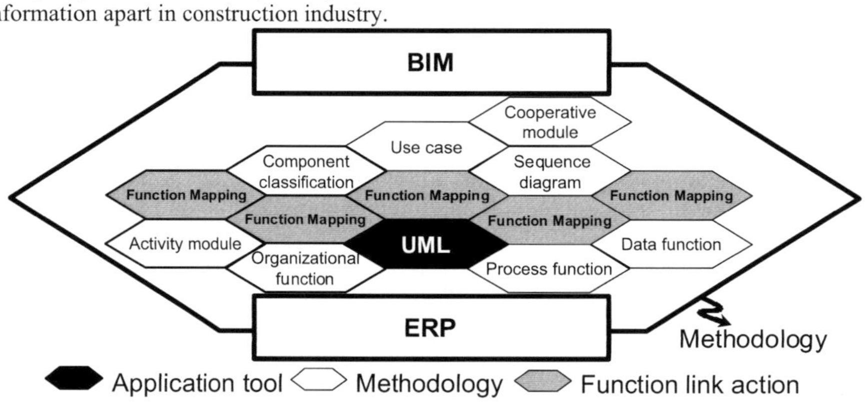

Figure 1. Honeycomb structure diagram of research objectives

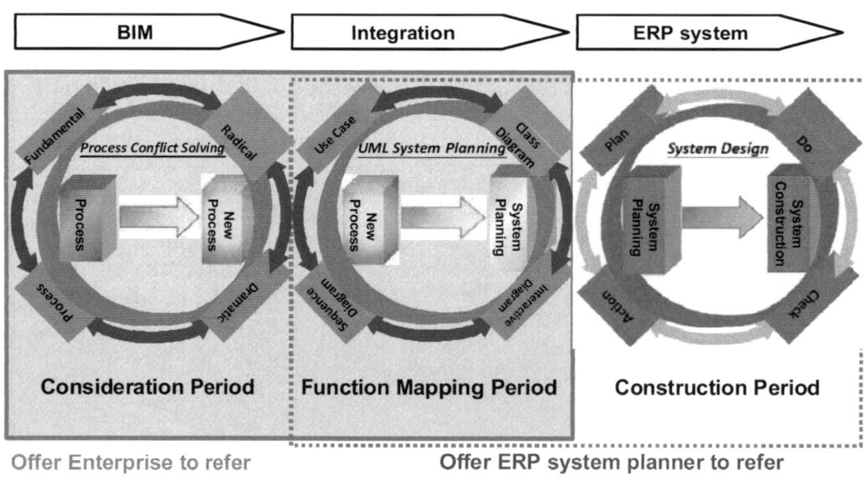

Figure 2. Research Scope Diagram

Table 1. Problem definition, description & solution

Problem	Problem Definition	Problem Description		Solution
Information Aspect	How to improve Information advancement in Construction enterprise by Implementation ERP system?	Regarding the concept of ERP	■ Confirm conflict problems of enterprise ■ Analyze & define needs of enterprise ■ Evaluate main business processes ■ Analyze original business processes ■ Concern conflict to design better Processes flows	Business processes conflicts finding from BIM Business processes & system planning integration System planning
		Regarding the practice of ERP	■ Match business processes with ERP system modules ■ Develop system implementation planning ■ Start to system planning	

4 ERP implementation

The main purpose of method is to make the smooth integration for translation between processes (avoid conflicts) and ERP system planning. The left dotted block includes BIM phases (see Figure 2). This phase is to apply BIM tool to proceed process discussion and conflicts. The middle dotted block belongs function-mapping phase. The phase primarily processes the translatability between process conflict solving and system planning. The right dotted block includes system-planning phase. The phase apply UML tool to proceed to system planning. Actually, the function-mapping is the kernel part in method because it plays the road of the translatability. In the other words, the functionality of method is like a bridge between processes conflicts and ERP system planning.

5 Case study

5.1 Building Information Modeling (BIM)

BIM is a concept that combines digital information products and engineering application technologies to develop / estimate / maintain project or engineering in its lifecycle. In the other way, BIM uses virtual space in computer to simulate a real environment or situation. That will help include starting from plan, design, constructing, operating and maintaining how to manage by this new concept, new technology or new method. BIM emphasizes lifecycle information gathering and sustainable use of engineering, 3D visualization and cross-disciplinary collaboration, geometric and non-geometric information Knot, static and dynamic process information of the immediate grasp of micro and macro space information integration [3][4].

The concept of BIM has been the beginning and BIM from 2D planar technology to 3D graphics development process (see Figure 3 & Figure 4).

Figure 3. 2D and 3D model conversion diagram

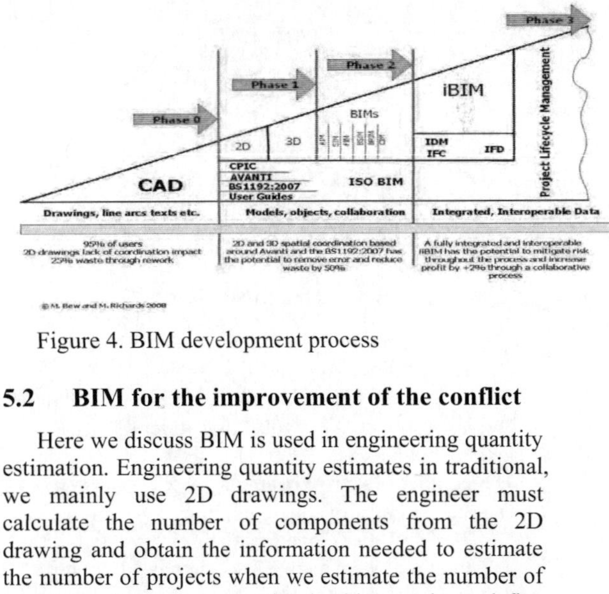

Figure 4. BIM development process

5.2 BIM for the improvement of the conflict

Here we discuss BIM is used in engineering quantity estimation. Engineering quantity estimates in traditional, we mainly use 2D drawings. The engineer must calculate the number of components from the 2D drawing and obtain the information needed to estimate the number of projects when we estimate the number of templates. There is a practice in this part is to define each of the components of a combination of its order, and then use the writing program to retrieve its surface area information, which may have some increase in the account. So through the BIM with the writing program, you can estimate the number of templates out, and can make the results very accurate there will be no error. This is the BIM project quantity estimation and conflict application.

As shown in the following figure, the 3D model we made is a geometric shape of the more complex triangle angle of the template configuration, this problem will occur mainly in the triangle angle of the intersection, because itself will have thickness, in the overlap when the combination may be because of this place Resulting in the possibility of repeated calculations, which form a conflict in the calculation of the number of templates.

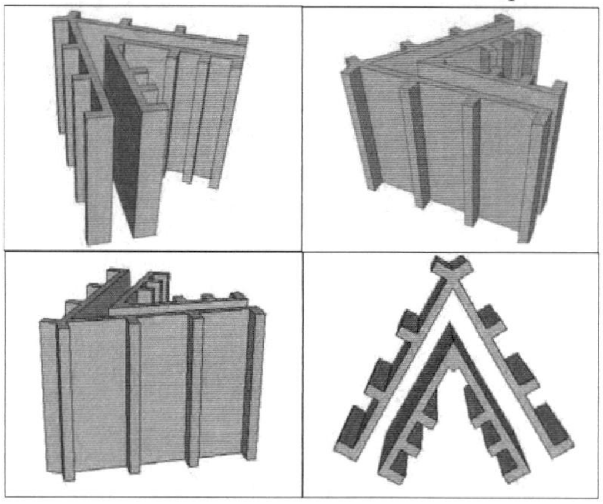

Figure 5. Template conflict example different view

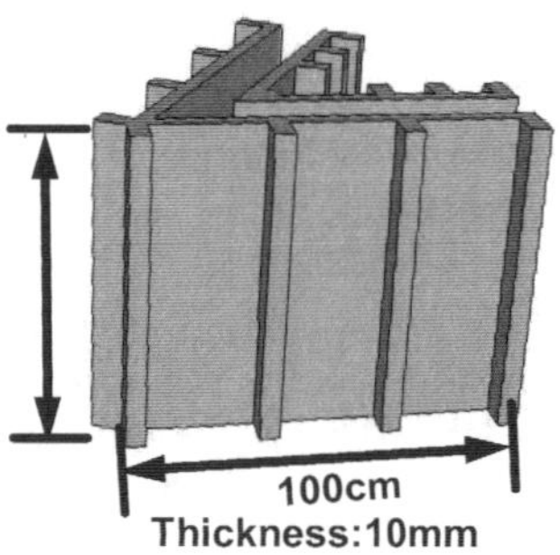

Figure 6. Template conflict example

Outside template:

Traditional estimate $A1 = 100 \times 100 \times 2 = 20000\text{cm}^2$ (1)

Overlap estimate $A2 = 100 * 1 = 100\text{cm}^2$ (2)

Inside template:

The number of square columns $= 4 \times 2 = 8$ (3)

Overlap estimate of square columns $= 8 - 7 = 1$ (4)

There is 100cm^2 overlap quantities existed in the calculation of such a unit. And there is 1 overlap quantity existed in the calculation of such a unit. When the estimate is quite large and complex time, the overlap quantities existed will be huge. And that will affect the correct number of estimates and cost estimates. This is the traditional calculation will occur on the error, but through the BIM can solve this problem.

5.3 Function Mapping

There are six steps in function mapping phase. They are confirming processes conflicts solving, developing process function table, drawing process function diagram, building up system planning criteria table, acquiring system planning components and making components category table. The first step is confirming processes conflicts solving. We inherit from the result of BIM phase. We must discuss with related staffs to confirm the correctness of new process that avoid processes conflicts. The second step is that developing process function table. We classify the result of BIM

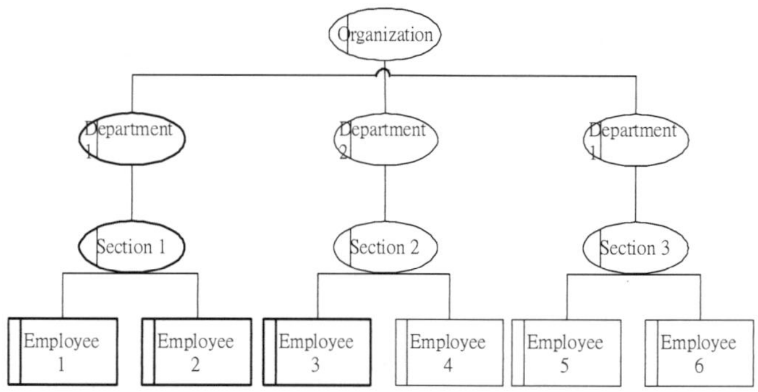

Figure 7. Organizational function diagram

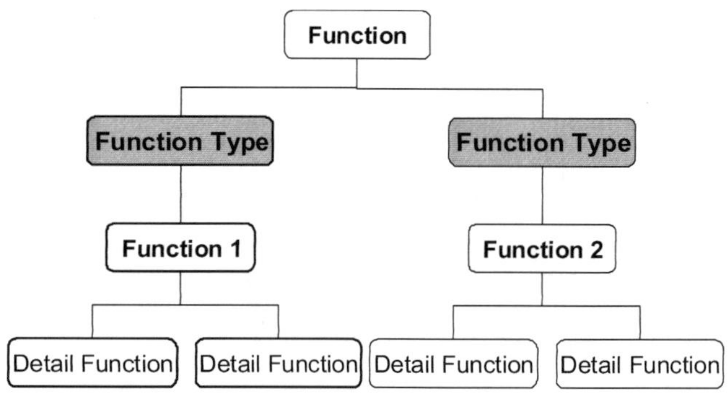

Figure 8. Process function diagram

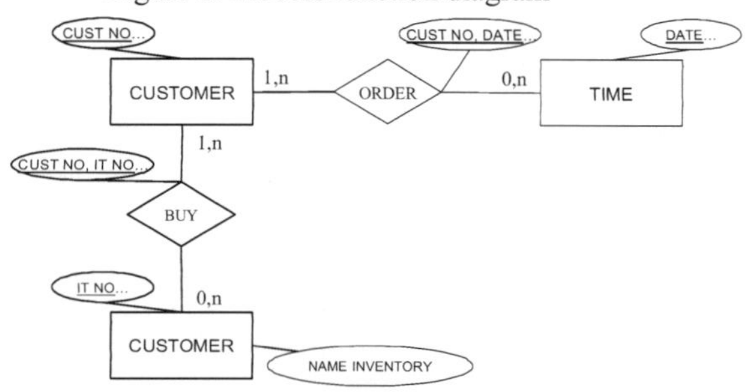

Figure 9. Data function diagram

phase to become "event", "function", "organization" and "data" four categories according to organizational function diagram (see Figure 6). When we develop process function at the ready. The third step is that drawing process function diagram. We draw function, data, event and organization four diagrams according to organizational function diagram. For example, fig 8 and fig 9 are function and data models.

The fourth step is building up system planning criteria table. The process function table, event model, function model, organization model and data model are transformed into system planning criteria table. This step is critical point that transmission BIM process into ERP system planning (using UML tool). In the other words, we can say it is important referring to system planning successfully. The fifth step is that acquiring system planning components. Useful components may be acquired from second step to fourth step before UML system planning. After acquisition, components category table at the sixth step can be made to prepare for system planning phase (see Table 2).

Table 2. Component category table

System planning components by UML tool	Outcome of acquisition from BIM
Actor	Department or staff
Use-Case	Process name, event, entity or information
Class Name	Department's name, staff or entity
Attribute	Attribute (eg. Item name, evaluation table's name etc.)
Operation	Attribute (eg. The motion of verify, the motion of calculate etc.)
Message	Information (eg. Item estimate finished a part of project etc.)

5.4 System Planning

There are four steps in system planning phase. They are confirming components category table, static system planning, dynamic system planning and finishing system planning. Confirming components category table belongs first step. The correctness needs to be confirmed with related staffs. The second step is the static system planning.

Static system planning is developed according to rules of UML tool. The core work of static system planning is that develop "Use-Case diagram", "Use Case description". A complete static system planning will have a lot of diagrams and descriptions by use-case, object and class planning. For example, Figure 9 and Figure 10 show use-case diagrams and one of class diagrams. The third step is that dynamic system planning. Dynamic system planning is developed based on rules of UML tool. The main work of dynamic system planning can be designed to develop "Collaboration diagram", "Sequence diagram", "State diagram" and "Activity diagram" (see Figure 12 & Figure 13). A complete dynamic system planning will include a lot of diagrams by the static system planning and the dynamic system planning steps, hundreds or thousands of tables and diagrams are made. We must discuss with related staffs to confirm the correctness. After confirming the correctness about UML models of static and dynamic system planning, we have finished to translate enterprise business processes into system planning. The result will be taken for the specification by ERP system planning company. In sum, four phases are made to assist the integration implementation between BIM and ERP system more successfully.

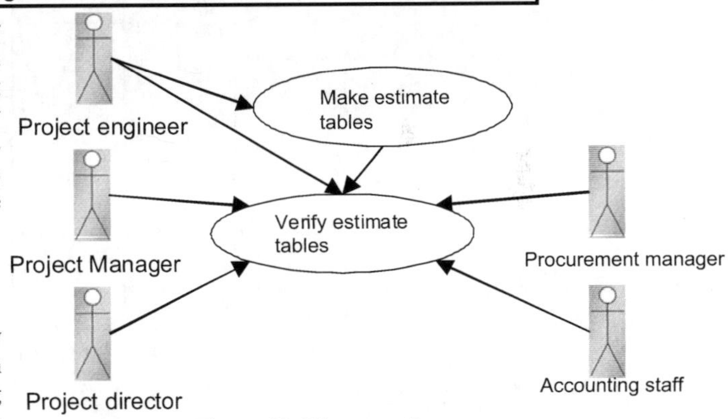

Figure 10. Use-case diagram

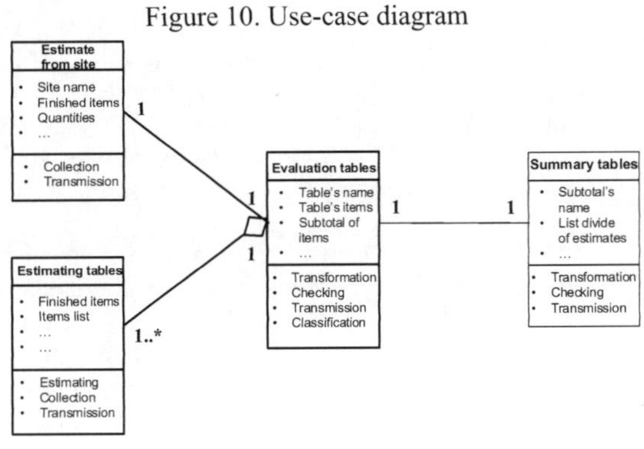

Figure 11. Database table of Use-Case

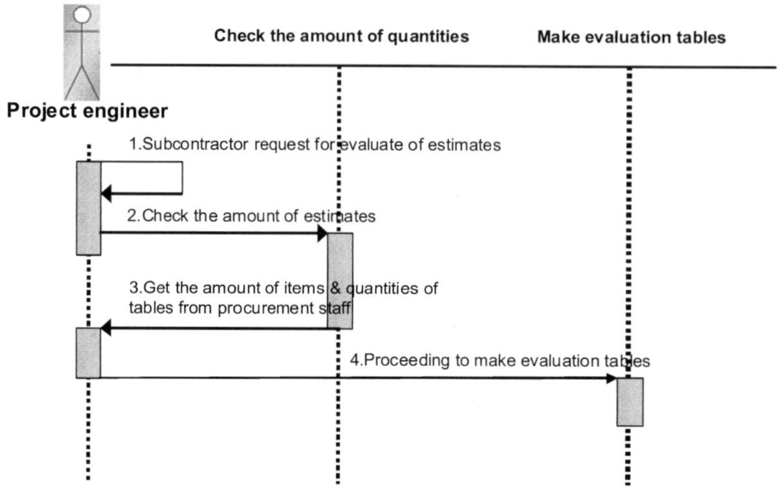

Figure 12. Sequence diagram

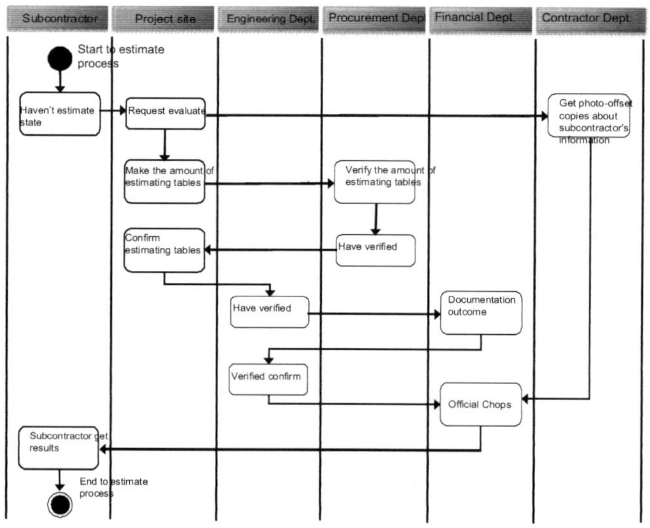

Figure 13. Activity module diagram

6 Conclusion

Enterprise implements ERP system almost failure as a result of existing a gap between processes conflicts and ERP system planners. The gap is that they don't understand the other side affairs each other. Method tried to solve this problem. Enterprise may apply this method for BIM implementation before ERP system implementation to make processes appropriately. Using UML model will help system planner transfer the result of BIM to ERP system implementation easily and effectively. In the future, a system may be developed according to this method to assist system planner for translating processes appropriately into system planning automatically by combining BIM and UML model.

References

[1] Marina K. and Neil M. The Implementation of enterprise resource planning packages in different organizational and national cultures, Information Systems Vol.26, p185-204, 2001.

[2] Al-Mashari M. Constructs of process change management in ERPContent: A focus on SAP R/3 AMCIS p977-980, 2000.

[3] TANG QING PEI architectural information model (BIM) in the construction of integrated information technology applications, master's thesis, Institute of Architecture and Sustainable Planning, Yilan University, 2008.

[4] ZHAO HONG GONG HONG GONG Information Architecture Design - Autodesk Revit, Beijing: China Construction Industry Press, 2005.

Empirical Research on BIM Application for Cost Estimation of HVAC System in Buildings

Sy-Jye Guo[1], Li-Wei Chiang[2], Un-On Chong[3] , Jyun-Yu Chen[4]

National Taiwan University Professor[1], PhD[2], Master[3] , Master[4]
E-mail: sjguo@ntu.edu.tw[1], d99521007@ntu.edu.tw[2]

Abstract— Building information modeling (BIM) is considered as a major trend of the construction industry in the future. In traditional way for estimating the cost of the HVAC system in buildings, a lump sum method is adopted as a rough approximation. Instead of detailed quantity calculation, the lump sum method often result in inaccurate total cost and quantities. The purposes of this study is to promote the BIM to the owners and MEP designers. This research found that the BIM software needs to be updated and revised. By building detailed HVAC system components of a completed public office building, the overestimation of HVAC budgets has been revealed in the case study. The contractor has earned an extra profit about 3.9% of the total project cost due to the overestimated quantity of HVAC components in the building.

Keywords—HVAC system; BIM; lump sum; quantity; cost estimation.

1. INTRODUCTION

The Heating, Ventilation and Air Conditioning (HVAC) system is an important link in the building electromechanical system. "Lump-sum" is one of the types of construction payments. BIM can calculate the quantity of materials, and also can improve the Lump sum of the defects of pricing. But for now, BIM for thequantity of mechanical and electrical engineering calculation and evaluation of capacity also need to beassessed. If BIM has this study attempts to construct a BIM model of HVAC system for case study. This study analyzes the accuracy of the BIM model output to the quantity and amount of the HVAC system, comparing the amount of Lump sum in the contract.

2. CASE STUDY

2.1 Case study introduction

The steel structure buildings project case study is "North Emergency Operation Center". This case was completed in January 2013, underground construction of two floors, one floor of the isolation layer, the ground to build nine floors, the total floor area of 18896m^2. The cost of the project is NT $ 1,156.55 million, including NT $ 126,250,000 for mechanical and electrical system. The NEOC building has public bath facilities, laundry facilities and staff restaurant facilities, the quantity of HVAC system needs higher,

Table 1. The amount of NEOC HVAC system (NT $)

	Budget	Completion Amount
(a) Total mechanical and electrical engineering costs	129,505,646	126,245,113
(b) HVAC system construction costs	33,302,968	33,022,647
[(b/a)*100%]	25.7%	26.2%

Table 2 HVAC equipment engineering amount

Item	Budget(A)	Completion Amount(B)	Tender discounts[(B-A)/A]
Machine	12,409,471	12,315,008	-0.8%
Foundation and shockproof	1,029,946	1,016,361	-1.3%
Wind pipe	5,346,294	5,262,478	-1.6%
Pumping	6,722,775	6,639,172	-1.2%
Electronic	1,654,860	1,655,900	0.1%
Automation control	3,277,648	3,271,754	-0.2%
Cooling water control	523,474	523,474	0.0%
Silencing Engineering	1,656,916	1,656,916	0.0%
Base isolation hose equipment	196,160	196,160	0.0%
Machine test	485,424	485,424	0.0%

million Taiwan so the HVAC system facilities project cost of 17.52 dollars. The amount of budgetreduction 40% in the case of the design of the tender phase. The budget gap is not large difference with completion phase and the tender phase of this case, therefore, the object of this study is the amount and quantity of project bidding stage.

Fig. 1. South Emergency Operation Center building appearance

The BIM software Revit outputs details in this case of the HVAC system engineering as shown in table 2, it is for the scope of this study. The table 1 total cost is 126,245,113 NT $, accounted for 26.2% of the HVAC system engineering. It is means that the BIM can be imported after a detailed estimate of HVAC system.

Fig. 2. Case building structure model

3. BIM MODEL BUILD AND QUANTITY OUTPUT

This study uses Autodesk Revit 2014 software. This version of Revit also includes the functionality of MEP, which means it can build mechatronic systems facility with the same software. In this study, the BIM model foundation is provided by the design unit to provide the building plans and HVAC system design. To build building Structure Model and HVAC system Model.

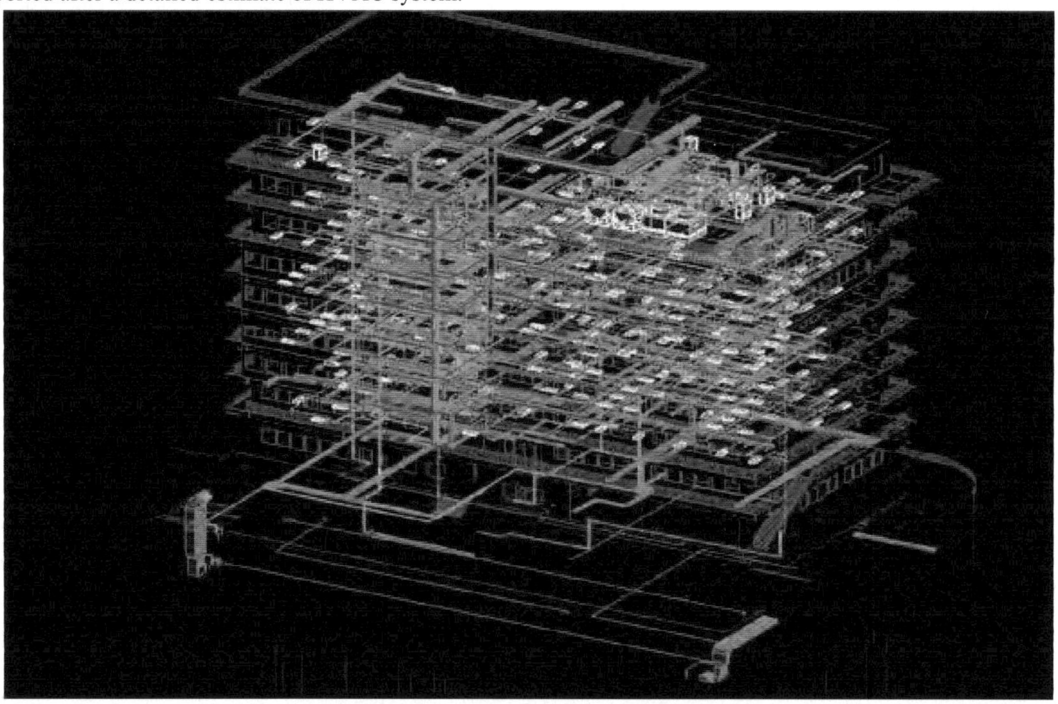

Fig. 3. BIM model of HVAC system

3.1 Quantity output of HVAC system components

In Revit, the general components such as faucets, toilet showers and other facilities are a single cost. It means that an components can only have one price. In the Revit software, it is not easily to enter different unit prices for different sizes, including Routing and Type parts such as HVAC and components. The main reason for the routing components is the family system. It is the same price with the same material in different sizes or length of the pipe in Revit.

3.2 Quantitative analysis of air conditioning system

There are two major projects included in the HVAC system plumbing engineering. The two items are the pipe fittings and the other parts, the table 3 shows the quantity of galvanized steel pipe and hot water stainless steel pipe, the quantity of PVC pipe drainage, the quantity of ice water and hot water pipes of the detailed list.

The main reasons for this study are summarized as follows:

1. The traditional calculation of the design unit is to enlarge the quantity and length of the pipe by the rule of thumb. The traditional way will cause a quantity of errors. And in order to reduce the risk, the design unit will try to avoid underestimating. Therefore, the results of traditional methods of valuation, the design unit is generally overestimated amount and quantity.
2. The results of this study shows that the output of the diameter 50mm of the drainage pipe is higher than that of the BIM output results by 167%. After comparing with the original design, it is found that the reason is that the design unit is seriously underestimated by the drainage pipe.
   In the HVAC drainage system, there are 5 50mm drainage pipe, the 50mm of the drainage pipe is through the whole building drainage pipe. The two layer underground extends to the ground floor of each 50mm drainage pipe, the total height is 47 meters.

Table 3 Comparison of the amount of material and parts for pipe.

Item	BIM output price A	Contract price B	Difference amount (A-B)/B
Duct sheet	753,098	476,414	-37%
Water pipe	1,366,726	1,081,782	-21%
Tube and cable	239,163	139,535	-42%
Pipe fittings	411,095	136,917	200%

Table 4 chilled and hot water pipe quantity list

Size	Unit	Quantity	Unit price	Total price	Revit output quantity	Revit output total price	Completion Amount
Galvanized steel pipe SCH40(Cold water pipe)							
200mm ∮	M	122	1,017	124,074	138	140,346	**14%**
150mm ∮	M	139	665	92,435	179	119,035	**29%**
125mm ∮	M	41	514	21,074	36	18,504	-10%
100mm ∮	M	454	373	169,342	346	129,058	-24%
80mm ∮	M	742	267	198,114	520	138,840	-30%
65mm ∮	M	552	209	115,368	341	71,269	-38%
50mm ∮	M	1,297	126	163,422	817	102,942	-37%
40mm ∮	M	320	95	30,400	250	23,750	-22%
32mm ∮	M	917	81	74,277	811	65,691	-11%
25mm ∮	M	133	60	7,980	70	4,200	-47%
20mm ∮	M	2,446	42	102,732	1672	70,224	-32%
Stainless steel pipe 20S(Hot water pipe)							
125 mm ∮	M	77	980	75,460	34	33,320	-55%
100 mm ∮	M	34	620	21,080	18	11,160	-44%

3. In practice, the design unit often to the proportion of the total amount of straight pipe to estimate the amount of another pipe. The coefficient of initial estimation is often influenced by personal experience and habits. However, the same material will have a close proportionality factor. This study estimates that human error may occur. An error in the estimation of the material scale factor, leading to erroneous estimates.
   In this study, for example, there are two other parts of the total amount of 0.1 as a coefficient calculation, but in accordance with previous engineering experience, it should be use 0.3~0.35 as a coefficient calculation.
4. In the air conditioning system engineering, the power distribution engineering and the automatic control engineering is the branch or continuation of power system engineering. The source of the system must be connected to the power system equipment to operate properly. HVAC system and power system are air conditioning engineering and power engineering, therefore, when the designer estimates the amount, the fuzzy classification is easy to confuse. In this study, the quantity of the electric pipe is different from the amount of the designer, which leads to a huge quantity difference.

4. Conclusion and discussion

In this study, through the establishment of BIM model of the case study, the establishment of a clear quantity and amount of pipe in the HVAC system,

Based on the results of Revit2014 software, this paper proposes the evaluation process, and compares the output of BIM model with the quantity and amount of raw materials in the original contract:

1 This study established a case BIM model, the case contract amount is NT$ 10439618, accounting for 31% of the contract amount of air conditioning system. The output value of the BIM model established in this study is NT$ 10036565, NT$ 403053 less than the original contract amount, the gap is -3.9%. In the case of the original quantity, and the amount of project valuation is 10459712 NT$, accounting for 32% of the amount of the system. The deletion of the NT$2541768 (24%) of the total amount of valuation item, and found that the electromechanical design unit quantity and cost higher.

2 The Revit software can use the function of the list to calculate the quantity of model components, but the software for the cost of the electrical and mechanical components of the air-conditioning system is limited. Due to the unit price cannot change the length of the pipe, the fixed price characteristics cannot meet the practical needs. In this study, we propose a method of calculating software aided pricing with Excel.

3 After the expert interviews found that, when the amount of the completion of the adjustment, it is not to calculate the amount of the completion of the construction site, and the amount of change to adjust the design. Therefore, the accuracy of the contract quantity and amount will directly affect the interests of the construction company. Through expert interviews, this study summarizes the following reasons for the differences:

3.1. The mentality of the design unit to overestimate the quantity and amount. The design unit in the calculation of pipe length, the use of too high correction factor, resulting in excessive amount and amount of material. And the calculation of the amount of parts, may also use the error of the scale factor error.

3.2. Due to the lack of elevation, the riser pipe parts may be underestimated, so that the quantity of estimates and graphic design is clearly inconsistent.

3.3. If the system is not clearly defined, the design of the project is not consistent with the quantity and quantity of the project.

3.4. In this study, a pipeline routing model is established, which considers only the collision of the air conditioning system. Therefore, it is a relatively economical path, which may produce additional length during actual construction.

The quantity of output function of the BIM model, this study to enhance the accuracy of calculation program. This study found that the design unit is often overestimated in valuation. This study is assist the owner to obtain a more detailed basis for the cost of the source, reduce the information generated by the lack of trust.

5. REFERENCES

[1] S.J. Guo, L.W. Chiang and C.F. Lin, "BIM application for quantity take-off and cost estimation of plumbing system in buildings", The Twenty-ninth KKHTCNN Symposium on Civil Engineering, HongKong, 2016.

[2] A. C. Badrinath, Y. T. Chang and S. H. Hsieh, "A review of tertiary BIM education for advanced engineering communication with visualization", Visualization in Engineering, pp.4: 9, 2016.
[3] Salman Azhar, Michael Hein, and Blake Sketo,"Building Information Modeling (BIM): Benefits, Risks and Challenges," McWhorter School of Building Science, Auburn University, Auburn, Alabama, 2011.
[4] C. Eastman, P. Teicholz, R. Sacks and K. Liston, BIM Handbook: A Guide to Building Information Modeling for Owners, Managers, Designers, Engineers and Contractors, Wiley, New Jersey, U.S.A, pp. 220, 222, 504, 2008.
[5] Juncheng Yang, "BIM for Plumbing Design", Consulting-Specifying Engineer, Vol. 50, No. 7, Arlington, Virginia, pp. 37-39, 2013.
[6] Cox, B., and Terry, Fred., "Creating a BIM for emergency management.", Journal of Building Information Modeling, fall 2008(9), pp. 24-25(2008).
[7] Koo, B., Kim J., and Kim, J. I., "An empirical study of MEP workspace modeling approaches for 4D model-based time-space conflict analyses", KSCE journal of Civil Engineering, 17(4), pp. 627-637(2013).
[8] Vanlande, R., Nicolle, C.,& Cruz, C.,"IFC and Building Life Cycle Management", Automation in Construction, 18(1), pp.70-78(2008).
[9] Cox, B., and Terry, Fred., "Creating a BIM for emergency management.", Journal of Building Information Modeling, fall 2008(9), pp. 24-25(2008).

Integrating Activity Scheduling and Site Layout Planning

S. Isaac[a], Y. Su[b], and G. Lucko[b]

[a]Department of Structural Engineering, Ben-Gurion University of the Negev, Israel
[b]Department of Civil Engineering, Catholic University of America, USA
E-mail: isaacsh@bgu.ac.il, 61su@cua.edu, lucko@cua.edu

Abstract –

A temporal-spatial scheduling model is used to consider the location and movement of equipment, material and temporary structures on a construction site over time, ensuring efficient yet safe processes. The model ties together activity scheduling and dynamic site layout planning, fully integrating two domains that have so far been treated separately. Their integration is facilitated by modeling temporary site objects like cranes as singularity functions. These objects are explicitly linked in the model with work spaces through Boolean operators.

Keywords –

Scheduling, Singularity Function, Site Layout, Cranes

1 Introduction

Construction activities are tightly interlinked with a construction site in terms of the timing of said activities, the resources they require, and the physical space that they occupy on the site. As type and nature of activities change throughout the project, construction sites are correspondingly dynamic, and their geometric layout will change in accordance with the changing activities. Yet while advances have been made in developing planning methods that take into account the location of activities on site, activity planning and site planning have so far not yet been fully integrated. This prevents them from being planned simultaneously and efficiently. This research presents a foundation for fully integrated planning of both construction activities and sites.

1.1 Phased and Dynamic Approaches to Site Layout Planning

Andayesh and Sadeghpour [1] distinguished three approaches of static, phased, and dynamic site layout planning. Traditionally, the *static* approach has assumed that the layout of a construction site remains the same throughout the project duration. This prevents using the same space on the site for different objects, even if they are not required at the same moment in time (e.g. [2]).

An improvement upon was the *phased* approach, in which the project was divided into discrete phases, and a site layout was planned for each phase [3]. It allowed that one location could be used for different objects in different phases. But as Andayesh and Sadeghpour [1] criticized, objects can enter or leave the site at moments other than the beginning or end of a phase. By reserving their space for an overly long duration the phased approach reduced the efficiency of the final site layout. Consequently, they introduced a *dynamic* approach, which considered the actual duration for which objects must occupy space, and implemented this approach in a model based on energy principles [4]. Nevertheless, this model presupposed the existence of a schedule for the activities, upon which the site layout plan can be based.

1.2 Space and Time Approaches

Representation of physical space within schedules has been an ongoing effort. Location-based scheduling methods were developed that considered both the timing and location of work [5]. They used diagrams to track how work progresses from one location to another. But such locations have been limited to a physical division of the site based on either a single type of repetitive unit (such as floors in buildings), or a single path across the site (for linear infrastructure projects). This has failed to reflect the actual use of space on real construction sites, which is typically much more complex than that.

Other studies focused on linking specific building components with scheduled activities and defining the required work spaces in computer-aided design (CAD) and building information modeling (BIM) systems [6]. The main goal of these four-dimensional (4D) systems has been to identify spatial conflicts between activities that have been scheduled to be simultaneously executed.

1.3 Need and Justification

Efficient planning of construction project execution requires a model that can integrate activity scheduling, work-space planning and site layout planning fully. In comparison, existing methods for dynamic site layout planning can be used only after all activities have been scheduled. Yet layout of the construction site is often *less* flexible than the timing of activities. Simultaneous planning can therefore give significantly more efficient solutions than those that existing methods provided.

2 Methodology

2.1 Goal and Objectives

The goal of this research is to define a temporal-spatial scheduling model that integrates the planning of construction site layout and activities by representing and explicitly linking these constructs mathematically.

Three research objectives are established as follows:

a. Modeling temporary site objects, including their changing location and work space envelopes, as singularity functions;
b. Linking temporary site objects with work spaces with logical relationships that reflect efficiency and safety requirements;
c. Verifying that site objects are eventually optimally located in accordance with their defined relations with planned activities.

2.2 Research Approach

This research defines an integrated mathematical model which can be used to implement planning and optimization methods for scheduling and site layout planning. Singularity functions will be used to define the required constructs and relations, as they provide a flexible mathematical framework that allows modeling the different aspects of construction projects.

This research will focus exemplarily on the task of planning the location of cranes on construction sites [7]. Cranes often constitute a bottleneck resource that leads to delays as crews wait for material or components to be lifted. At other times, cranes may be underutilized and remain idle, because activities have not been optimally scheduled. Cranes could also create safety hazards as other activities need to be conducted underneath areas into which objects must be lifted and placed. Given these problems, it is important to simultaneously plan the timing and location of activities and cranes on sites.

3 Singularity Functions

3.1 Definition

In its general form, the pointed-bracket operator per Equation 1, which is the basic term within all singularity functions, performs a case distinction. A function of x is either zero before the activation cutoff a on the x-axis or is evaluated as a polynomial with strength s and growth n from a onward. Exponents n can model shapes, e.g. a constant ($n = 0$), linear growth ($n = 1$), parabolic growth ($n = 2$), etc. Such singularity functions enable unlimited additive composition for behaviors of any complexity from simple terms. Multiple terms can be simplified if their n and a are identical to shorten longer equations. To deactivate a function, its term is simply subtracted at the later cutoff (including any already attained value). An index R will mark its default right-continuity (active at $x \geq a$), index L the analogous left-continuity ($x > a$).

$$y(x) = s \cdot \langle x - a \rangle^n = \begin{cases} 0 & if\ x < a \\ s \cdot (x - a)^n & if\ x \geq a \end{cases} \quad (1)$$

3.2 Extension to 3D

Since their introduction to project management [8], systematic applications have been broadened beyond work and time in linear schedules to also encompass resource counts that remain linked with the underlying schedule to gain a flexible resource leveling model [9]; cost and pay over time to generate cash flow equations [10], time value of money that grows from interest [11], and calendar date conversions including holidays [12].

Recently the possibility to unify spatial and temporal aspects of project plans has begun to be explored. Their three-dimensional (3D) formulation multiplies pairs of on-and-off terms along both coordinate axes [13], which model two projections of the same 3D volume. On both axes they are activated and deactivated at the start and finish of the range. This allows block and ramp shapes within a 3D coordinate system of two horizontal spatial axes x_1 and x_2 and a vertical time axis z. But this model was limited to just shapes that are parallel to an axis, which limits its usefulness to only sites that are shaped like a city block. Other shapes were not explored. Float, the flexibility of activities regarding their timing, was defined as postponing or a productivity change (i.e. shift or rotation) within an overlap zone with a successor. Its availability was represented in form of a 'heat map' [14].

Besides work areas, these authors have defined paths on a site within a 3D space-time coordinate system. This approach has overcome the previous limitation that only axis-parallel shapes could be defined. A vector-inspired formulation has been created [15], which first defines a line segment and then gives it a width. It could act as a safety buffer for traveling equipment, and more broadly allows defining irregularly shaped work areas. Also, the need to introduce Boolean operators has been addressed, so that work areas can be related by AND (intersection), OR (inclusion), and XOR (exclusion) [16]. These are applied by multiplying, adding, or subtracting so-called signal functions of two geometric shapes. An improved version of a vector-based model for any rectangular geometric area that is rotated in the x_1-x_2-plane against the two spatial axes is given by Equation 2. From its start to its finish coordinates (S to F), the value of the variable h_1 grows from 0 to 1 along the vector direction (Δx_1, Δx_2, Δz). Defining a perpendicular half width w / 2 to the left and right with variable h_2 grows from 0 to 1 creates the desired area $h_1 \times h_2$. These widths per geometry are secondary vectors, which are scaled to a unit length of 1.0 by dividing by the vector length $l = \sqrt{(\Delta x_1^2 + \Delta x_2^2 + \Delta z^2)}$. Each is then multiplied with the desired width w / 2. Their two directions are (Δx_2, -Δx_1, Δz) and (-Δx_2, -Δx_1, Δz). Their vertical z-value can be omitted for purely spatial elements without time aspect.

$$Area(h) = \begin{pmatrix} x_{1S} \\ x_{2S} \\ z_S \end{pmatrix}_{Start} + \begin{pmatrix} \Delta x_1 \\ \Delta x_2 \\ \Delta z \end{pmatrix} \cdot \left(\langle h_1 - 0 \rangle^1_R - \langle h_1 - 1 \rangle^1_R - \langle h_1 - 1 \rangle^0_L \right) \pm \frac{w}{2 \cdot l} \cdot \begin{pmatrix} -\Delta x_2 \\ +\Delta x_1 \\ 0 \end{pmatrix} \cdot \left(\langle h_2 - 0 \rangle^1_R - \langle h_2 - 1 \rangle^1_R - \langle h_2 - 1 \rangle^0_L \right) \quad (2)$$

The model for a geometric solid is derived from said area model by adding a third variable h_3 that grows from 0 to 1 along the vector direction (0, 0, Δz) per Equation 3. Since the vertical axis z represents time, the Δz in the vector is the duration of work. Following a vector from (x_1, x_2, z) = (28, 61, 0) to (52, 93, 0), h_1 moves into the direction that is the difference (24, 32, 0) of these two vectors. It travels $\sqrt{(24^2 + 32^2)} = 40$ length units as a line segment. To create an area, h_2 moves perpendicularly into two opposite directions. Its vectorized singularity function for $w = 30$ is (28, 61, 0) + (24, 32, 0) $\cdot$ ($\langle h_1 - 0\rangle^1_R - \langle h_1 - 1\rangle^1_R - \langle h_1 - 1\rangle^0_L$) $\pm$ 15/5 $\cdot$ (-4, +3, $\pm$0) $\cdot$ ($\langle h_2 - 0\rangle^1_R - \langle h_2 - 1\rangle^1_R - \langle h_2 - 1\rangle^0_L$). Moving it along the time axis z with h_3 into direction (0, 0, 8) creates the full temporal-spatial activity per Figure 1. Its singularity function extends the insufficient traditional vector into a solid with the custom location and orientation as desired.

$$Solid(h) = Area(h_1, h_2) + \begin{pmatrix} 0 \\ 0 \\ \Delta z \end{pmatrix} \cdot \left(\langle h_3 - 0 \rangle^1_R - \langle h_3 - 1 \rangle^1_R - \langle h_3 - 1 \rangle^0_L \right) \quad (3)$$

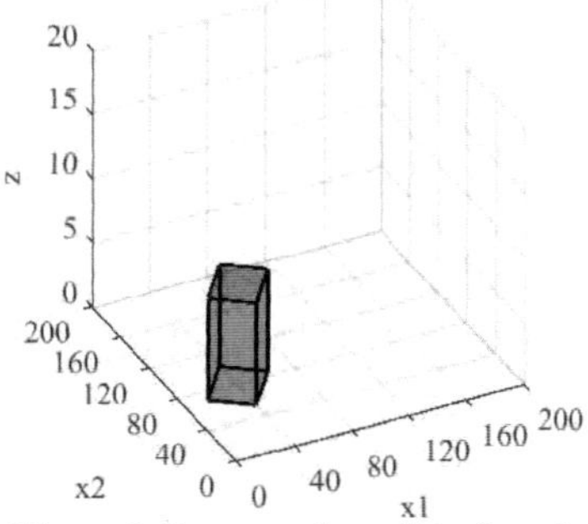

Figure 1. Rectangular solid of work

3.3 Crane Movements

As a vital productive resource on site, arguably the most important one [7] due to supplying many trades with materials and assemblies, cranes – here specifically tower cranes – have a dominant role. The operating area of such crane is cylindrical, where the mast is located at $\{x_{1C}, x_{2C}\}$. Its jib of radius R rotates around it. Along it the trolley can move to any position r, where $0 \leq r \leq R$ (or realistically, starting about 3 m from the mast itself). Its hook location is determined by the mast and trolley, i.e. at $\{x_{1C}, x_{2C}\}$ plus an offset $\{\Delta x_{1C}, \Delta x_{2C}\}$, where Δx_{iC} can be positive or negative and $-R \leq \Delta x_{iC} \leq R$. Any singularity function is active from its cutoff onward. But $r > R$ should deactivate it to prevent endless growth. This is accomplished by an inverse operator - $\langle r - R\rangle^1$. It shows how singularity functions incorporate constraints that otherwise must be written as inequalities (plus end term to subtract the radius outside the crane coverage). Together this gives $\Delta x_{1C}^2 + \Delta x_{2C}^2 = (x_1 - x_{1C})^2 + (x_2 - x_{2C})^2 = (\langle r - 0\rangle^1 - \langle r - R\rangle^1 - R \cdot \langle r - R\rangle^0)^2$ per Equation 4.

$$Area(x_1, x_2, r): \quad (x_1 - x_{1C})^2 + (x_2 - x_{2C})^2 = (\langle r - 0\rangle^1 - \langle r - R\rangle^1 - R \cdot \langle r - R\rangle^0)^2 \quad (4)$$

Standard geometry gives the relations between the offset distances and rotation angle φ as $\Delta x_{1C} = r \cdot \cos \varphi$ and $\Delta x_{2C} = r \cdot \sin \varphi$ to convert from planar into polar coordinates. Such rotation will traverse a fraction of the circumference that simply becomes $2 \cdot \pi \cdot (\varphi / 360)$.

A spatial singularity function can model the area of crane coverage based on its mast location and radius as $x_2(x_1, r) = \pm \sqrt{[(\langle r - 0\rangle^1 - \langle r - 50\rangle^1 - 50 \cdot \langle r - 50\rangle^0)^2 - \langle x_1 - x_{1C}\rangle^2]} + x_{2C}$ and *vice versa* for $x_1(x_{2C}, r)$ and $r(x_{1C}, x_{2C})$; any two of the three variables must be known. Note that the circular crane rotation allows two valid solutions, one beyond the mast on the x_2 axis and one before it.

For example, for a mast at {65 m, 70 m} with an r = 50 m jib for r = 60 m the crane radius equation correctly gives $(\langle 60 - 0\rangle^1 - \langle 60 - 50\rangle^1 - 50 \cdot \langle 60 - 50\rangle^0)^2 = (60 - 10 - 50)^2 = 0$, because that location is outside the radius. Test input of x_1 = 65 m + 50 $\cdot$ [$\sqrt{2}$]/2 m = 65 m + 35.36 m = 100.36 m for the crane coverage equation with a 50 m jib gives x_2(100.36 m, 50 m) = $\pm \sqrt{[(\langle 50 - 0\rangle^1 - \langle 50 - 50\rangle^1 - 50 \cdot \langle 50 - 50\rangle^0)^2 - \langle 100.36 - 65\rangle^2]} = \pm \sqrt{[(50 - 0 - 50 \cdot 1)^2 - 35.36^2]} = \pm 35.35$ m (having subtracted x_{2C} = 70 m already), i.e. the crane rotates 45° (anti-clockwise or clockwise) for its two offsets to become identical. Moving the crane coverage area along the time direction with 8 units creates the solid of crane mathematically per Equation 5. Figure 2 visualizes its cylindrical solid.

$$Solid(h) = Area(x_1, x_2, r) + \begin{pmatrix} 0 \\ 0 \\ \Delta z \end{pmatrix} \cdot \left(\langle h - 0 \rangle^1_R - \langle h - 1 \rangle^1_R - \langle h - 1 \rangle^0_L \right) \quad (5)$$

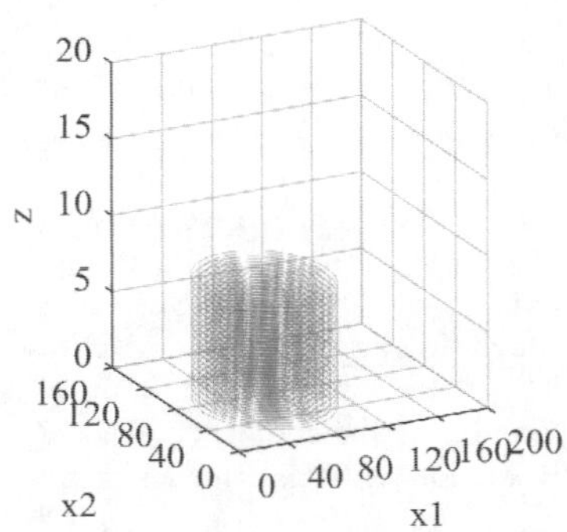

Figure 2. Cylindrical solid of crane

3.4 Other Equipment

Other types of equipment may be characterized by different shapes of its work within the construction site. These may be fixed locations (e.g. concrete plant silos, dewatering wells, material hoists), linear movements (e.g. utility pipe trenching, paving, tracked tower cranes and gantries) or work faces (e.g. masonry platforms), multiple linearized segments that compose work paths (e.g. delivery trucks, material handlers), and areas (e.g. storage and assembly yards, building section footprints).

Shapes can be represented with existing singularity functions as follows: Fixed locations have a small non-zero footprint, which makes them conceptually identical to areas. Simple areas have a range of $\{\Delta x_{1S}$, Δx_{1F} and Δx_{2S}, $\Delta x_{2F}\}$ if their rectangular shapes can be defined as a pair of coordinate ranges on the two spatial axes. Beyond this, Equation 2 can newly model the generalized areas that are non-aligned with major coordinate axes. Linear segments can use the model of Equation 2, but with zero (or nearly zero) width w to emphasize that their primary direction is captured by the growth of h. More complex equipment travel paths can be composed from linear segments as a set of vectors. These definitions of different types of temporary site objects and their locations fulfill Research Objective 1.

4 Boolean Relations

A barely explored yet important issue within spatial scheduling has been how work areas of temporary site objects and the activities that occur there can be related explicitly by logical constraints. Traditional (non-spatial) network scheduling has been dominated by discrete end-point-links that express minimum constraints [17]. They function well for *If-Then* conditions in which one event triggers another, e.g. the predecessor finish that enables a successor start. But despite the popularity of strictly sequential relations, also due to their simplistic nature, real construction projects experience a profusion of concurrent activities that often at best compete for space or at worst interfere or conflict with one another.

Several Boolean algebra operators exist, which here will be used to establish if and how multiple temporary site objects can compete for the same location [16]:

- Two activities must occur sequentially (IF-THEN);
- Two activities must occur in parallel (AND);
- Two activities can (but need not) be parallel (OR);
- Two activities must never occur in parallel (XOR).

Applying them to temporal-spatial scheduling means considering the activity pair {A, B} in both time and space. Their truth values can be recorded per Table 1.

Table 1. Boolean Truth Values

Operator	Activity A	Activity B	Result
IF-THEN	1	1	1
	1	0	1
	0	1	0
AND	1	1	1
	1	0	0
	0	1	0
OR	1	1	1
	1	0	1
	0	1	1
XOR	1	1	0
	1	0	1
	0	1	1

Defining Boolean relations between work and crane solids will use a signal function per Equation 6 whose value is 1 if a coordinate point is inside the solid and 0 if it is outside. This concept can generate the Boolean operations AND, OR, and XOR in Equations 7, 8, and 9. The IF-THEN relation is not considered further, because it already exists in form of normal 'finish-to-start' links.

$$Signal(x_1, x_2, z) = \begin{cases} 0 & if \quad (x_1, x_2, z) \notin Solid \\ 1 & if \quad (x_1, x_2, z) \in Solid \end{cases} \tag{6}$$

$$AND_{Signal} = Signal_{Work} \times Signal_{Crane} \tag{7}$$

$$OR_{Signal} = Signal_{Work} + Signal_{Crane} - Signal_{Work} \times Signal_{Crane} \tag{8}$$

$$XOR_{Signal} = OR_{Signal} - AND_{Signal} \tag{9}$$

Such 3D signal functions can be handled like normal mathematical terms. This allows a seamless integration of the singularity functions for the work areas and crane ranges with these new Boolean relations between them.

Linking temporary site objects in their locations via such logical relationships fulfils Research Objective 2.

5 Validation

A validation example will test and demonstrate how temporary site objects and Boolean relations are used to more realistically model the constraints of construction projects. Assume the example of Figures 3 and 4 with areas and constraints as follows. The aforementioned tower crane that Section 3.3 has introduced is placed at the coordinates {65, 70} (all locations in meters) with a radius of 50 m. Zone I – familiar from Section 3.2 – is rotated relative to the x_1 and x_2 axes in Figure 1. Zone I cannot move in space. It occurs in the range from weeks 0 to 8 (all times in weeks) on the temporal axis z. Zone II is at {70, 100 and 55, 95} and can be moved to {130, 160} on axis x_1 (here called Solution I) or postponed along the time axis z to weeks 10 to 16 (Solution II).

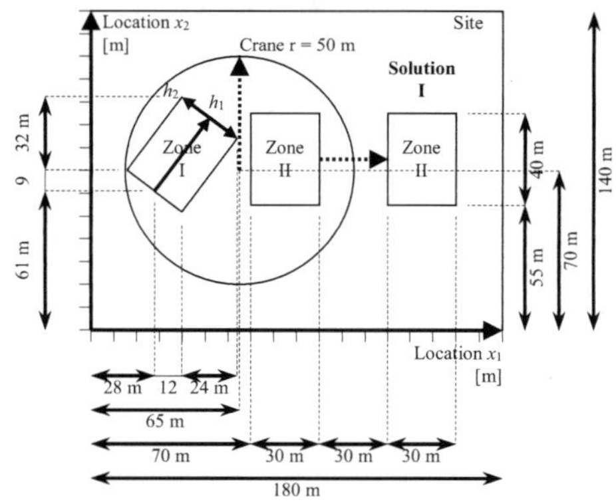

Figure 3. Validation example plan view of space

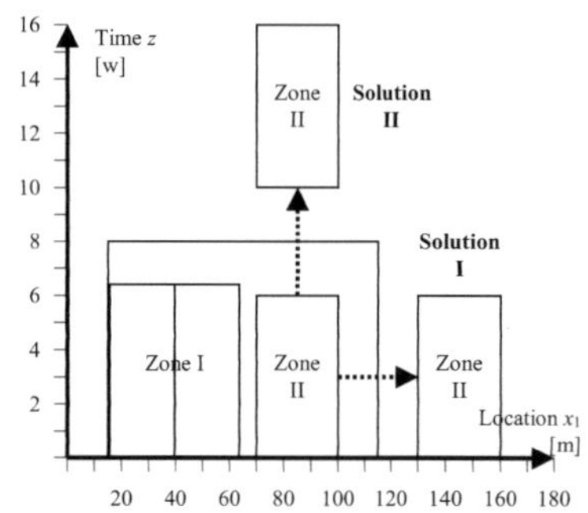

Figure 4. Validation example possible solutions

Applying Equation 5 gives Zone I as a solid of work as $Solid(h)_{\mathrm{I}} = (28; 61; 0)_{Start} + (24; 32; 0) \cdot (\langle h_1 - 0\rangle^1_R - \langle h_1 - 1\rangle^1_R - \langle h_1 - 1\rangle^0_L) \pm 15/5 \cdot (-4, +3, \pm 0) \cdot (\langle h_2 - 0\rangle^1_R - \langle h_2 - 1\rangle^1_R - \langle h_2 - 1\rangle^0_L) + (0; 0; 8) \cdot (\langle h_3 - 0\rangle^1_R - \langle h_3 - 1\rangle^1_R - \langle h_3 - 1\rangle^0_L)$. Note that a solid could be spanned from any coordinate (with vector directions adjusted accordingly), but here a start close to the origin is chosen, so that most vector direction values will only have positive signs.

Work in Zone I requires assistance of the crane, so that it has a Boolean AND relation with the crane solid. Yet Zone II never needs the crane, so an XOR is applied between them. It is now possible to check whether the work area is fully within the crane coverage by using their signal functions. In the original plan of Figure 5, Zone II is given by $Solid(h)_{\mathrm{II}} = (85; 55; 0)_{Start} + (0; 40; 0) \cdot (\langle h_1 - 0\rangle^1_R - \langle h_1 - 1\rangle^1_R - \langle h_1 - 1\rangle^0_L) \pm 30/2 \cdot (-1, 0, \pm 0) \cdot (\langle h_2 - 0\rangle^1_R - \langle h_2 - 1\rangle^1_R - \langle h_2 - 1\rangle^0_L) + (0; 0; 8) \cdot (\langle h_3 - 0\rangle^1_R - \langle h_3 - 1\rangle^1_R - \langle h_3 - 1\rangle^0_L)$. Boolean calculation of the signal space within Zone II gives $\mathrm{OR}_{\mathrm{Signal}} = \mathrm{Signal}_{\mathrm{Crane}} + \mathrm{Signal}_{\mathrm{ZoneII}} - \mathrm{Signal}_{\mathrm{Crane}} \times \mathrm{Signal}_{\mathrm{ZoneII}} = \mathrm{Signal}_{\mathrm{ZoneII}}$, while $\mathrm{AND}_{\mathrm{Signal}} = \mathrm{Signal}_{\mathrm{Crane}} \times \mathrm{Signal}_{\mathrm{ZoneII}} = \mathrm{Signal}_{\mathrm{ZoneII}}$, as a result $\mathrm{XOR}_{\mathrm{Signal}} = \mathrm{OR}_{\mathrm{Signal}} - \mathrm{AND}_{\mathrm{Signal}} = \mathrm{Signal}_{\mathrm{ZoneII}} - \mathrm{Signal}_{\mathrm{ZoneII}} = 0$. Only if the XOR signal is 1, the work in the Zone II may be conducted. Thus Zone II violates the XOR relation with the crane and is plotted in red.

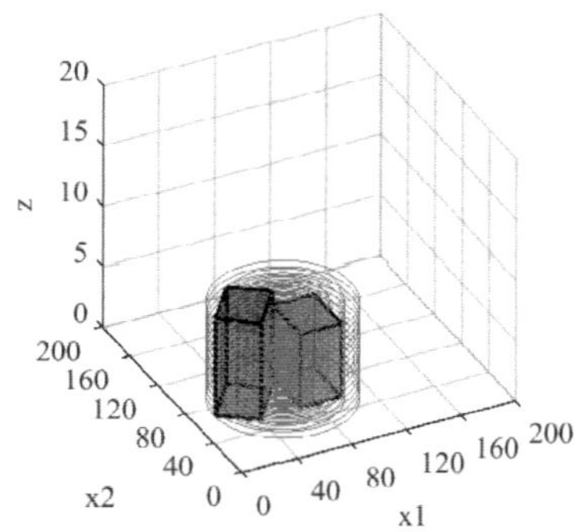

Figure 5. Validation example original plan

Figures 6 and 7 show relocating or postponing Zone II to resolve its conflict. Solution 1 relocates Zone II by 60 m into the x_1 direction to fulfill the XOR relation as can be mathematically confirmed. Solution 2 does not change its location, but postpones the start time of Zone by 4 weeks, which will also fulfill the XOR relation. A project planner can thus make an informed decision on how to position temporary site objects and work zones relative to one another to fulfill all constraints toward finding an optimum arrangement of such a schedule.

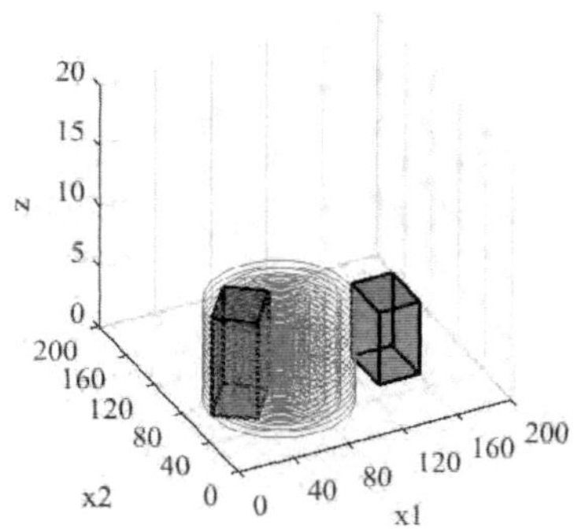

Figure 6. Validation example Solution I (relocate)

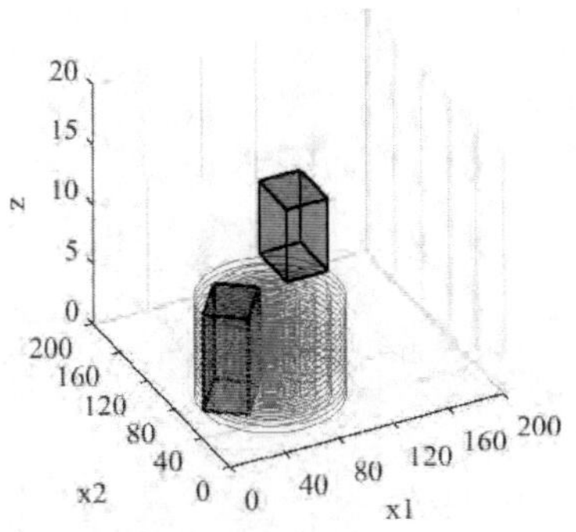

Figure 7. Validation example Solution II (postpone)

6 Conclusions

This research opens an avenue for integrating the domains of activity scheduling, workspace planning and site layout planning in a single mathematical model. Its model is sufficiently flexible to allow representing the spatial and temporal dimensions of both construction activities and site objects. It can thus enhance the safe and efficient planning of construction project execution.

6.1 Contributions

This paper makes three contributions to knowledge:

- It explores how spatial singularity functions model work areas and work paths and also temporary site objects, within a 3D space-time coordinate system;
- Work areas and temporary site objects are linked by logical relationships such as AND, OR and XOR, in addition to existing sequential relationships, to fulfill applicable efficiency and safety requirements;
- An example has validated that the planned activities and temporary site objects like a tower crane, which constitutes a vital resource on site, can be optimally planned in accordance with their defined relations.

6.2 Recommendations

- Since the structural moment of any given crane load depends on its distance between the trolley and mast, the mathematical formulation should be refined to decreasing load capacity based on radius;
- While this research has assumed that cranes have a circular radius, the need for stabilizing outriggers gives mobile cranes a potentially non-circular radius of equal lifting capacity. The mathematical formulation should be refined with an equation of such radius-and-rotation-dependent lifting capacity;
- Purely temporal interval relations have first been described by Allen [18], who considered that activity durations can be equal or unequal and listed all possible constellations of two activities in time, including those with overlap, or concurrency. This idea should be generalized to temporal-spatial scheduling to categorize all possible constellations of temporal site objects and their locations in time;
- Areas that change their shape or size over time (e.g. a crane that is not allowed to swing over an adjacent property, this operating as a circle sector or building with smaller footprints on upper floors) should be modeled as time-dependent functions;
- Rotations could also be modeled via a rotation matrix, which will be explored in future research;
- Complex work paths, e.g. for earthmoving, need to be explored further. They could be combined with capacity calculations, e.g. the curves of mass haul diagrams that calculate cut and fill quantities from integration of volume curves in cross-sections and longitudinal sections of topographical profiles;
- Developing a computer tool for project planners could automatically perform analyses like the example of Section 5 to directly identify conflicts, offer options for solutions, and visualize potential impacts of spatial or temporal changes in plans;
- Full integration with optimization methods should be explored, especially which type of evolutionary algorithm and what parameters perform best for representative sets of real-world construction sites.

Nomenclature

Table 2. List of Symbols

Symbol	Unit	Definition
x	m	Spatial variable
z	w	Time variable
s	w/m	Strength factor
a	m	Activation cutoff
n	-	Behavior exponent
R	-	Right continuous
L	-	Left continuous
l	m	Vector length
w	m	Half width of area
r	m	Radius of interest
h	-	Vector direction
R	m	Crane jib radius
C	-	Crane index
S	m	Start coordinate
F	m	Finish coordinate
Δ	m	Offset distance
φ	°	Rotation angle
IF-THEN	-	Sequential relation
AND	-	Parallel relation
OR	-	Inclusive relation
XOR	-	Exclusive relation

References

[1] Andayesh, M., Sadeghpour, F. The time dimension in site layout planning. *Automation in Construction*, 44(August): 129-139, 2014.

[2] Tommelein, I. D., Levitt, R. E., Hayes-Roth, B., SightPlan Model for Site Layout. *Journal of Construction Engineering and Management*, 118(4): 749-766, 1992.

[3] El-Rayes, K. A., Said, H. M. M. Dynamic Site Layout Planning Using Approximate Dynamic Programming. *Journal of Computing in Civil Engineering*, 23(2): 119-127, 2009.

[4] Andayesh, M., Sadeghpour, F. Dynamic site layout planning through minimization of total potential energy. *Automation in Construction*, 31(May): 92-102, 2013.

[5] Kenley, R., Seppänen, O. *Location-based management for construction: Planning, scheduling and control.* Spon Press / Routledge, Taylor and Francis Group, Abington-on-Thames, Great Britain, 2010.

[6] Akinci, B., Fischer, M. A., Levitt, R. E., Carlson, R., Formalization and Automation of Time-Space Conflict Analysis. *Journal of Computing in Civil Engineering*, 16(2): 124-134, 2002.

[7] Shapira, A., Lucko, G., Schexnayder, C. J., Cranes for Building Construction Projects. *Journal of Construction Engineering and Management*, 133(9): 690-700, 2007.

[8] Lucko, G. Mathematical Analysis of Linear Schedules. *Proceedings of the Construction Research Congress*, Freeport, Commonwealth of the Bahamas, May 6-8, 2007, American Society of Civil Engineers, Reston, VA: 1387-1395, 2007.

[9] Lucko, G. Integrating Efficient Resource Optimization and Linear Schedule Analysis with Singularity Functions. *Journal of Construction Engineering and Management*, 137(1): 45-55, 2011.

[10] Lucko, G. Optimizing Cash Flows for Linear Schedules Modeled with Singularity Functions by Simulated Annealing. *Journal of Construction Engineering and Management*, 137(7): 523-535, 2011.

[11] Lucko, G. Supporting Financial Decision-Making Based on Time Value of Money with Singularity Functions in Cash Flow Models. *Construction Management and Economics*, 31(3): 238-253, 2013.

[12] Lucko, G. Temporal Constraints in Linear Scheduling with Singularity Functions: The Case of Calendarization. *Journal of Computing in Civil Engineering*, 28(2): 232-243, 2014.

[13] Lucko, G., Said, H. M. M., Bouferguene, A. Construction Spatial Modeling and Scheduling with Three-Dimensional Singularity Functions. *Automation in Construction*, 43(July): 132-143, 2014.

[14] Said, H. M. M., Lucko, G. Float Types in Construction Spatial Scheduling. *Journal of Construction Engineering and Management*, 142(12): 04016077(12), 2016.

[15] Isaac, S., Su, Y., Lucko, G., Dagan, D. Work Path Modeling and Spatial Scheduling with Singularity Functions. *Journal of Computing in Civil Engineering*, 2016, in print.

[16] Su, Y., Isaac, S., Lucko, G. Extended Network Model for Construction Scheduling: Temporal and Spatial Relationships for Work Paths and Work Areas with Boolean Logical Operators. *Computer-Aided Civil and Infrastructure Engineering*, 2017, in preparation.

[17] Lucko, G., Su, Y. Precedence pattern permutations creating criticality constellations: Confuting a conjecture on non-linear activities with continuous links. *Procedia Engineering*, 2017, in review.

[18] Allen, J. F. Maintaining Knowledge about Temporal Intervals. *Communications of the ACM*, 26(1): 832-843, 1983.

34th International Symposium on Automation and Robotics in Construction
ISARC 2017
June 28 – July 1
TAIPEI, TAIWAN

CONSTRUCTION ROBOTICS (VI) / UAV APPLICATIONS

Automatic Generation of 3D Models from UAV-captured Image Data for Immersive VR Applications

Nei-Hao Perng[1], Bo-Sheng Bai[1], Po-Han Chen[1*], Jen-Yu Han[1],

Ming-Yi Jiang[2], Jyun-Hao Huang[2], Cheng-Wei Su[2], and Po-Yuan Chen[2]

[1]National Taiwan University
[2]Ministry of Transportation and Communications, Taiwan
[*]Corresponding author
E-mail: pohanchen@ntu.edu.tw, jyhan@ntu.edu.tw

Abstract –

In this paper, the way to automatically convert UAV-captured image data to 3D models as well as the subsequent immersive virtual reality (VR) applications will be presented. It is expected that the proposed development would help identify the defects on the inspected bridge, and assist with the training of junior bridge inspectors.

Keywords –
UAV; VR; 3D-Model; Automatic

1 Introduction

There are some twenty-nine thousand bridges in Taiwan, managed respectively by Taiwan Railways Administration, Taiwan Area National Freeway Bureau, Directorate General of Highways, and local governments. According to the current bridge maintenance guidelines in Taiwan, bridge inspection is mainly carried out through visual inspection, supplemented with equipment inspection. Bridge inspectors would usually approach the bridge structure for visual inspection. However, for high-rise or cross-river bridges, it would be hard to carry out visual inspection due to the difficulty in accessing the bridge structure.

In view of this, the Institute of Transportation under the Ministry of Transportation and Communications (MOTC) in Taiwan conducted a research project on "Enhancing the effectiveness of bridge inspection equipment" in 2014, and successfully carried out bridge inspection using a camera-mounted unmanned aerial vehicle (UAV). In this paper, the way to automatically convert UAV-captured image data to 3D models as well as the subsequent immersive virtual reality (VR) applications will be presented.

2 Research Limitations

In our research, we build up a 3D model bases on the photos took by UAV with Photogrammetry. Eq 1-1 is the formula about Collinearity, the fundamental principle of Photogrammetry (Fig.1 is the Schematic diagram).

$$\begin{bmatrix} x_p^L - x_0^L \\ y_p^L - y_0^L \\ -f^L \end{bmatrix} = R^L \begin{bmatrix} X_0^L - X_p \\ Y_0^L - Y_p \\ Z_0^L - Z_p \end{bmatrix}, \begin{bmatrix} x_p^R - x_0^R \\ y_p^R - y_0^R \\ -f^R \end{bmatrix} = R^R \begin{bmatrix} X_0^R - X_p \\ Y_0^R - Y_p \\ Z_0^R - Z_p \end{bmatrix} \qquad (1\text{-}1)$$

(x_0^L, y_0^L, f^L) & (x_0^R, y_0^R, f^R) = parameters of camera (could get from manufacturer)
(X_0^L, Y_0^L, Z_0^L) & (X_0^R, Y_0^R, Z_0^R) = camera position (while shooting)
R^L & R^R = shooting attitude
(x_p^L, y_p^L) & (x_p^R, y_p^R) = values of Point P in the "photos"

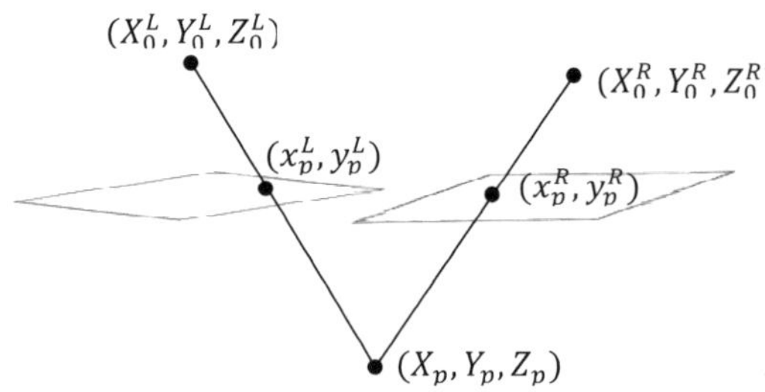

Fig 1. Schematic Diagram about Photogrammetry

Limited by matching technology nowadays, the photos we used for research must conform to following conditions to make sure the model quality is good enough:

1. The object should be shown in multiple different photos, and overlap over certain percentage.
2. The conjugate features between images should be recognized by computer.
3. The area with same color tone will lead to bad outputs, because of the lack of Eigenvalues.

4. Moving object is not included in research scope.

Besides, 3D-model is only able to show a rough outline without high definition texture, so we need to involve panorama as an assistance mode to provide user immersive experience. (Since the panoramas are shot at discrete points, we cannot move smoothly in panorama mode)

3 Literatures Review

3.1 UAV (Unmanned Aerial Vehicle)

There are lots of software in the market could automatically analyze 2D pictures, then build up a 3D-model. In our case, we fly UAV to take photos about the object, and input to those software to get the model we need.

3.1.1 Agisoft PhotoScan

Agisoft PhotoScan(commonly known as PhotoScan) is a professional tool for a photogrammetry pipeline. It is a stand-alone software product that performs photogrammetric processing of digital images and generates 3D spatial data to be used in GIS applications, cultural heritage documentation, and visual effects production as well as for indirect measurements of objects of various scales. The software is developed by Agisoft LLC located in St. Petersburg in Russia. It is widely used by archaeologists, and many UAV companies are also using it.

Agisoft PhotoScan is an off-the-shelf photogrammetric kit that allows for photography (both metric and non-metric, aerial and close range) to be…

- ✓ calibrated using a physical camera model;
- ✓ georeferenced utilizing ground control points or camera locations;
- ✓ block adjusted through aerial triangulation including the use of tie points;
- ✓ employed to create TIN/DEM products of the overlap areas using autocorrelation algorithms;
- ✓ orthorectified using a TIN and the calculated exterior orientations; and
- ✓ used to carry out area and volume measurements.

Agisoft PhotoScan is capable of processing up to tens of thousands of photos yielding result products characterized by high degree of accuracy in both the horizontal and vertical dimensions.

3.1.2 Pix4Dmapper

Pix4Dmapper software offers up to centimeter-grade 3D precision, features a fully automatic workflow and can be used by anyone as it is very intuitive and easy to use. You can assess, edit and improve projects directly in the software using the rayCloud and Mosaic Editor and seamlessly import results into any professional GIS, CAD and traditional photogrammetry software package.

Pix4Dmapper's goal is to provide customers with cuttingedge technology and groundbreaking tools combining the latest innovations of Computer Vision and Photogrammetry. Orthomosaics, DSMs and Point clouds generated with Pix4Dmapper achieve surveygrade accuracy from any aerial or oblique imagery, ranging from lightweight compact cameras used on UAVs (including GoPro, multispectral cameras and Fisheye lens) to SLR's and large photogrammetric sensors carried by aircraft. New algorithms allow for even more precise and stunning results and help customers to get the most out of their UAV imagery.

3.2 VR (Virtual Reality)

Virtual reality (VR) typically refers to computer technologies that use software to generate realistic images, sounds and other sensations that replicate a real environment (or create an imaginary setting), and simulate a user's physical presence in this environment, by enabling the user to interact with this space and any objects depicted therein using specialized display screens or projectors and other devices.

VR has been defined as "a realistic and immersive simulation of a three-dimensional environment, created using interactive software and hardware, and experienced or controlled by movement of the body" or as an "immersive, interactive experience generated by a computer".

A person using virtual reality equipment is typically able to "look around" the artificial world, move about in it and interact with features or items that are depicted on a screen or in goggles. Virtual realities artificially create sensory experiences, which can include sight, touch, hearing, and, less commonly, smell. Most 2016-era virtual realities are displayed either on a computer monitor, a projector screen, or with a virtual reality headset (also called head-mounted display or HMD). HMDs typically take the form of head-mounted goggles with a screen in front of the eyes. Some simulations include additional sensory information and provide sounds through speakers or headphones. Virtual Reality actually brings the user into the digital world by cutting off outside stimuli. In this way user is solely focusing on the digital content.

Some advanced haptic systems in the 2010s now include tactile information, generally known as force feedback in medical, video gaming and military training applications. Some VR systems used in video games can transmit vibrations and other sensations to the user via the game controller. Virtual reality also refers to remote communication environments which provide a virtual presence of users with through telepresence and telexistence or the use of a virtual artifact (VA), either through the use of standard input devices such as a keyboard and mouse, or through multimodal devices such as a wired glove or omnidirectional treadmills.

The immersive environment can be similar to the real world in order to create a lifelike experience—for example, in simulations for pilot or combat training, which depict realistic images and sounds of the world, where the normal laws of physics apply (e.g., in flight simulators), or it can differ significantly from reality, such as in VR video games that take place in fantasy settings, where gamers can use fictional magic and telekinesis powers.

4 System Structure

Because our research objects are bridges, there are more adverse factors must be faced when we are trying to get a 3D model from 2D images by software (such as Pix4Dmapper or Agisoft PhotoScan) automatically. For example, the deck is in the same tone, the plants around are in the same tone, the water in the river is moving, the hollow structure is complicated, so and on. Therefore, the 3D model is only able to give a rough outline. It's not detailed enough for safety checking or other operations.

Fig 2. Auto-build-up Model from UAV Pictures

Fig 3. Panorama about Under-bridge

The panorama technology could merge multiple pictures into an "image ball", which allows user to locate on the center and look around, such as Google StreetView or Youtube360. Thanks to the minimization of camera, there are lots of products in the market can take panoramas with a single click (Ricoh Theta S, Kodak sp360, etc.).

Since the both two mentioned above fit part of the definition about virtual reality, we include them all in our system. When we want to see a topography (Fig 2), we use 3D model; when we need to check some details, we use panorama (Fig 3). The system structure is as follows:

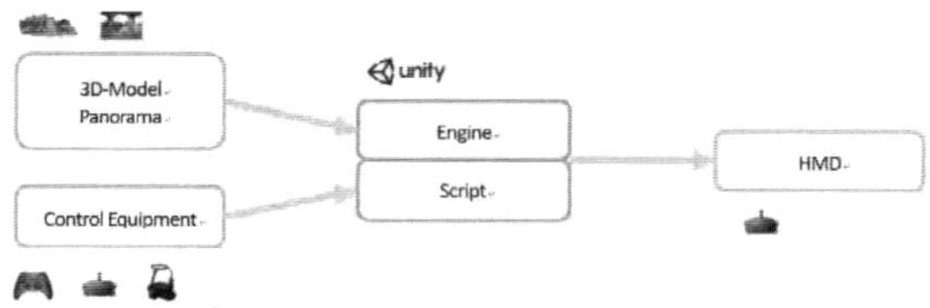

Fig 4. System Structure

5 Data Analysis

The research result is shown as Fig 5. The upper right is VR goggle, and the lower right are the respective images with two eyes. The VR goggle includes a triaxial accelerometer, so it could catch the attitude of user and make computer output the corresponding scene.

Fig 6 is showing how user controls the system. As you can see in the lower right, the user moves virtual camera through the model with joystick, and looks around by turning his head.

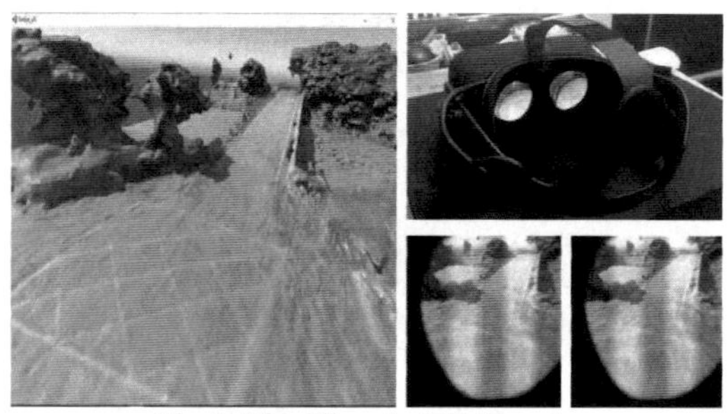

Fig 5. The View in VR Goggle

Fig 6. VR Model & Controller

Users could move to the checking point in model-mode, then switch to panorama-mode to verify testing items. Fig 7 is the difference between these two modes.

The left side, the model, allows more latitude for moving around; and the right side, the panorama, has a better texture to help user understanding reality problems (such as fractures or defects).

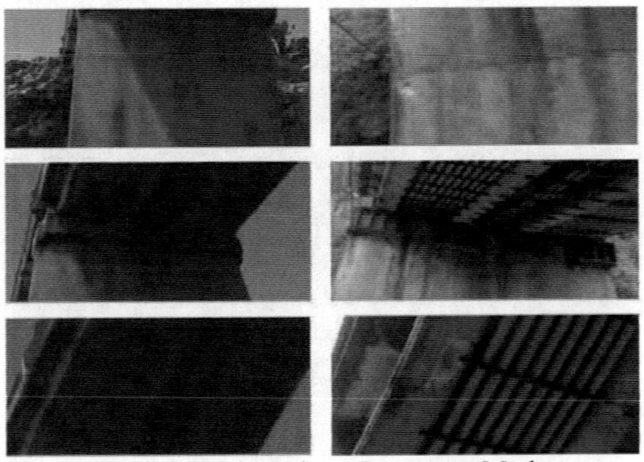

Fig 7. Model-Mode vs Panorama-Mode

6 Conclusions & Future Works

The system we developed could make user feel like in the objective environment. So that, they can learn the bridge structure and other neighboring information through model-mode, and check details with panorama-mode. There are some future works as follows:

- ✓ Improve the quality of auto-build-up model. Maybe trying to change the data type or add some artificial adjustments.
- ✓ Connect a walking pad such as Virtuix Omni to the control system, make user feel more immersive.
- ✓ Eliminate the limitations in the panorama-mode, such as discrete moving and space confusion.

References

[1] Tsai, Yi-Ting. A paper. The Study on DSM Generated from Point Clouds by UAV Image, Taiwan, 2012

[2] Lapierre. A, "3D City Geographic Information System – A Major Step towards Sustainable Infrastructure". A Bentley White Paper, August, 2015.

[3] Maher. K, "ContextCapture: Bentley's new take on digital reality", Graphic Speak, 2015.

[4] Substitution of 123D Catch. On-line: http://www.makezine.com.tw/make2599131456/diy123d-catch, Accessed: 20/05/2016

[5] VisualSFM. On-line: http://blog.csdn.net/u012989207/article/details/45032469, Accessed: 20/05/2016

[6] Introduction of 3D Image Data Showing. On-line: http://content.teldap.tw/index/blog/?p=1666, Accessed: 20/05/2016

[7] Microsoft. Photo tourism: Exploring photo collections in 3D. On-line: http://phototour.cs.washington.edu/, Accessed: 20/05/2016

[8] 3D, M. Digital Elevation Model (DEM). On-line: http://www.mosaic3d.com/en/products-service/dem/, Accessed: 20/05/2016

[9] iWitnessPRO-Agilis V3 has released. On-line: http://www.iwitnessphoto.com/support/AgilisV3, Accessed: 22/12/2013

[10] Zhu. L, Bentley. Promote VR Mode. On-line: https://communities.bentley.com/communities/other_communities/chinafirst/b/blog/archive/2015/03/03/bentley-acute3d, Accessed: 20/05/2016

[11] Acute 3D. On-line: https://www.acute3d.com/, Accessed: 20/05/2016

[12] Turn photos into 3D models automatically with ContextCapture. On-line: https://www.acute3d.com/contextcapture/, Accessed: 20/05/2016

[13] LeaderTek. On-line: http:// www.leadertek.com.tw/, Accessed: 20/05/2016

[14] GeoForce. On-line: http://www.geoforce.com.tw/#1=0, Accessed: 20/05/2016

[15] Pix4D. On-line: https://www.pix4d.com/, Accessed: 20/05/2016

Adaptive Second-order Sliding Mode Control of UAVs for Civil Applications

V.T. Hoang, A.M. Singh, M.D. Phung and Q.P. Ha

Faculty of Engineering and Information Technology
University of Technology Sydney, Australia
E-mail: {VanTruong.Hoang, AnsuMan.Singh, ManhDuong.Phung, Quang.Ha}@uts.edu.au

Abstract -

Quadcopters, as unmanned aerial vehicles (UAVs), have great potential in civil applications such as surveying, building monitoring, and infrastructure condition assessment. Quadcopters, however, are relatively sensitive to noises and disturbances so that their performance may be quickly downgraded in the case of inadequate control, system uncertainties and/or external disturbances. In this study, we deal with the quadrotor low-level control by proposing a robust scheme named the adaptive second-order quasi-continuous sliding mode control (adaptive 2-QCSM). The ultimate objective is for robust attitude control of the UAV in monitoring and inspection of built infrastructure. First, the mathematical model of the quadcopter is derived considering nonlinearity, strong coupling, uncertain dynamics and external disturbances. The control design includes the selection of the sliding manifold and the development of quasi-continuous second-order sliding mode controller with an adaptive gain. Stability of the overall control system is analysed by using a global Lyapunov function for convergence of both the sliding dynamics and adaptation scheme. Extensive simulations have been carried out for evaluation. Results show that the proposed controller can achieve robustness against disturbances or parameter variations and has better tracking performance in comparison with experimental responses of a UAV in a real-time monitoring task.

Keywords -

Quadcopter, robustness, adaptation, quasi-continuous second-order sliding mode control, monitoring system

1 Introduction

Quadcopters have found many applications in civil engineering automation due to its flexibility in operational space and ability in vertical take off and landing. These include the use of UAVs in automatic 3D reconstruction for building condition assessment [2], securing superstructures of high-rise buildings [3], or monitoring and inspection of civil infrastructure [6, 12]. In those applications, it is critical to maintain robustness and resilience of the control system to cope with the highly non-linear dynamics of quadcopters and system uncertainties, sensor noise and coupling effects between the rotational and translational motions, or disturbances from aerodynamics and other external factors.

A number of control approaches have been developed for the quadcopter in the literature, for example PD, PID control [24], H_∞ control [16], optimal control [17], or potential field [21]. Among them, the sliding mode control (SMC) is widely used as it can produce a robust closed-loop control system under the influence of modelling errors and external disturbances [22, 1, 4]. In SMC, chattering may occur in the steady state and act as an oscillator that excites unmodeled frequencies of the system dynamics [10]. To reduce the chattering effect, high-order sliding modes (HOSM) have been introduced [7, 14, 18, 20].

In the HOSM control, the quasi-continuous (QC) SMC [5] introduces the capability of maintaining the properties of the first order SMC while creating smooth responses. Its performance however depends on the knowledge of disturbance boundaries which are not always available. In practice, the quadcopter may be subject to various disturbances and uncertainties such as wind gusts and modelling errors that may downgrade the control performance. To address this concern, the second-order sliding mode (SOSM) controller with an adaptive gain has been applied to drive the sliding variable and its derivative to zero in the presence of bounded disturbances [19].

In this work, we propose an adaptive quasi-continuous second-order sliding mode (AQCSM) scheme to control the attitude of quadcopters subject to nonlinear dynamics, strong coupling, high uncertainties and disturbances with unknown boundaries. The mathematical model of the quadcopter is first derived by considering various dynamic parameters. Here, the quasi-continuous SMC retains the advantage of robustness while attenuating the control chattering and facilitating the implementation. Its performance is verified by simulation with comparison to real-time datasets.

The paper is organised as follows. The dynamic model of the quadcopter is presented in Section 2. Section 3 describes the development of the AQCSMC. Simulation results are presented in Section 4 with comparison to PID experimental responses. The paper ends with a conclusion and discussion for future work.

2 System modelling

2.1 Kinematics

Two coordinate systems are used to model the kinematics and dynamics of quadrotors, as shown in Fig. 1. The inertial frame (x_E, y_E, z_E) is defined by the ground with gravity pointing downward in z_E direction. The body frame (x_B, y_B, z_B) is specified by the orientation of the quadcopter with the rotor axes pointing in the positive z_B direction and the arms pointing in x_B and y_B directions.

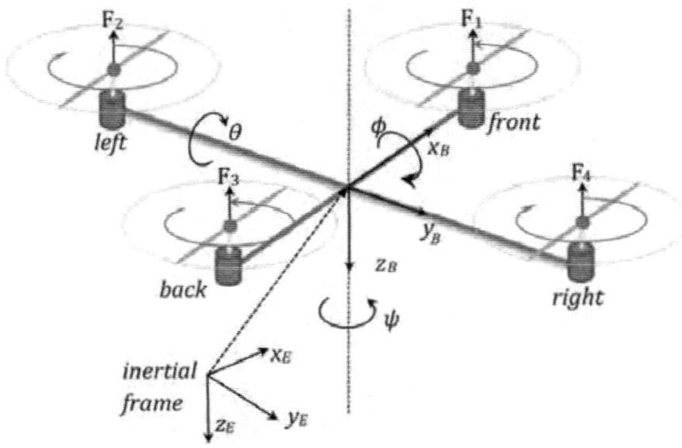

Figure 1: A schematic diagram of quadcopter

The orientation of quadcopters is described by the roll, pitch, and yaw angles corresponding to its rotations around the x_B, y_B and z_B axes. Denoting those angles as $\Theta = (\phi, \theta, \psi)^T$, their rates are then given by $\dot{\Theta} = (\dot{\phi}, \dot{\theta}, \dot{\psi})^T$. The rates relate with angular velocities, $\omega = [p, q, r]^T$, by the following transformation:

$$\omega = H\dot{\Theta}, \tag{1}$$

where H is given by:

$$H = \begin{bmatrix} 1 & 0 & -s_\theta \\ 0 & c_\phi & c_\theta s_\phi \\ 0 & -s_\phi & c_\theta c_\phi \end{bmatrix}, \tag{2}$$

in which $s_x = sin(x)$ and $c_x = cos(x)$. As the result, the rotational matrix of the quadcopter is described by:

$$R = \begin{bmatrix} c_\psi c_\theta & c_\psi s_\theta s_\phi - s_\psi c_\phi & c_\psi s_\theta c_\phi + s_\psi s_\phi \\ s_\psi c_\theta & s_\psi s_\theta s_\phi + c_\psi c_\phi & s_\psi s_\theta c_\phi - c_\psi s_\phi \\ -s_\theta & c_\theta s_\phi & c_\theta c_\phi \end{bmatrix}. \tag{3}$$

2.2 Quadcopter Dynamics

Since the focus is on the attitude control so only torque components that cause changes in the orientation are considered. They include torques caused by thrust forces τ, body gyroscopic effects τ_b, propeller gyroscopic effects τ_p, and aerodynamic friction τ_a. The torque τ consists of three components corresponding the roll, pitch and yaw rotations, $\tau = [\tau_\phi \ \tau_\theta \ \tau_\psi]^T$. They are given by:

$$\tau_\phi = l(F_2 - F_4), \tag{4}$$

$$\tau_\theta = l(-F_1 + F_3), \tag{5}$$

$$\tau_\psi = b(-F_1 + F_2 - F_3 + F_4), \tag{6}$$

where l is the distance from the motor to the UAV centre of mass and b is the drag factor. The body gyroscopic torque τ_b is given by:

$$\tau_b = -S(\omega)I\omega, \tag{7}$$

where $S(\omega)$ is a skew-symmetric matrix,

$$S(\omega) = \begin{bmatrix} 0 & -r & q \\ r & 0 & -p \\ -q & p & 0 \end{bmatrix}. \tag{8}$$

The propeller gyroscopic torque τ_p is determined as:

$$\tau_p = \begin{bmatrix} I_r\Omega_r q \\ -I_r\Omega_r p \\ 0 \end{bmatrix},$$

where I_r is the inertial moment of rotor, $\Omega_r = -\Omega_1 + \Omega_2 - \Omega_3 + \Omega_4$ is the residual angular velocity of rotor in which Ω_k denotes the angular velocity of the propeller k (k=1,2,3,4). Finally, the aerodynamic friction torque τ_a is given by:

$$\tau_a = k_a\omega^2, \tag{9}$$

where k_a depends on aerodynamic friction factors, $k_a = [k_{ax}, k_{ay}, k_{az}]^T$. Given those torque components, the attitude dynamic model of the quadcopter is described as:

$$I\ddot{\Theta} = \tau_b + \tau + \tau_p - \tau_a, \tag{10}$$

where $I = \text{diag}[I_{xx}, I_{yy}, I_{zz}]$ is the inertia matrix when the quadrotor is assumed to be symmetrical.

In our system, the gyroscopic and aerodynamic torques are considered as external disturbances. Thus, the control inputs mainly depend on torque τ and from (4), (5) and (6), they can be represented as:

$$\begin{bmatrix} u_\phi \\ u_\theta \\ u_\psi \\ u_z \end{bmatrix} = \begin{bmatrix} \tau_\phi \\ \tau_\theta \\ \tau_\psi \\ F \end{bmatrix} = \begin{bmatrix} 0 & l & 0 & -l \\ -l & 0 & l & 0 \\ -c & c & -c & c \\ 1 & 1 & 1 & 1 \end{bmatrix} \begin{bmatrix} F_1 \\ F_2 \\ F_3 \\ F_4 \end{bmatrix}, \tag{11}$$

where $F = F_1 + F_2 + F_3 + F_4$ is the UAV lift, u_z represents the total thrust acting on the four propellers and u_ϕ, u_θ and u_ψ respectively represent the roll, pitch and yaw torques, c is a force-to-torque scaling coefficient. As only the attitude of quadcopter will be controlled, u_z is assumed to balance with the gravity. Consequently, the dynamics of quadcopters can be represented in the following form for attitude control:

$$\dot{\omega} = I^{-1}\left(-S(\omega)I\omega + U + d\right), \tag{12}$$

where $U = [u_\phi, u_\theta, u_\psi]^T$ is the input vector and $d = [d_\phi, d_\theta, d_\psi]^T$ is the disturbance vector. In our system, the following assumptions are made:

A.1 The quadcopter structure is rigid and symmetric. The propellers are rigid.

A.2 The signals Θ and $\dot{\Theta}$ can be measured by on-board sensors.

A.3 The reference trajectories and their first and second time derivatives are bounded.

A.4 The velocity and the acceleration of the quadcopter are bounded.

A.5 The orientation angles are limited to $\phi \in \left[-\frac{\pi}{2}, \frac{\pi}{2}\right]$, $\theta \in \left[-\frac{\pi}{2}, \frac{\pi}{2}\right]$ and $\psi \in [-\pi, \pi]$.

A.6 The rotational speeds of rotors are bounded.

3 Control Design

The control signals u_ϕ, u_θ and u_ψ in (12) are used to control the three angles $\{\phi, \theta, \psi\}$ to reach the reference value $\Theta_d = \{\phi_d, \theta_d, \psi_d\}^T$.

3.1 Sliding Manifold

The sliding function determining the system's equivalent dynamics is presented as:

$$\sigma = \dot{\mathbf{e}} + \Lambda\mathbf{e}, \tag{13}$$

where $\Lambda = \text{diag}(\lambda_\phi, \lambda_\theta, \lambda_\psi)$ is a positive definite matrix to be designed, and $\mathbf{e} = \Theta_d - \Theta$ is the control error. Taking the derivative of σ, we have:

$$\dot{\sigma} = \ddot{\Theta} - \ddot{\Theta}_d + \Lambda\dot{\mathbf{e}}. \tag{14}$$

For small angular rotations of the quadcopter, we can approximate ω to $\dot{\Theta}$ [23]. Substituting $\ddot{\Theta}$ (12) to (14) yields:

$$\dot{\sigma} = -\ddot{\Theta}_d + \Lambda\dot{\mathbf{e}} + I^{-1}[-S(\omega)I\omega + U + d]. \tag{15}$$

3.2 QCSM control design and problem formulation

The second-order sliding mode control proposed in [8, 9] is used in this paper, for which a conventional QCSM is defined as follows:

$$U = -\alpha\frac{\dot{\sigma} + |\sigma|^{1/2}\,\text{sign}(\sigma)}{|\dot{\sigma}| + |\sigma|^{1/2}}, \tag{16}$$

where α is the control gain to be adjusted. The control is continuous everywhere apart from the origin where $\sigma = \dot{\sigma} = 0$.

Since I is symmetric and positive definite, the following Lyapunov function is chosen to avoid the inversion of the inertia matrix:

$$V_0 = \frac{1}{2}\sigma^T I\sigma. \tag{17}$$

Taking the time derivative of V gives

$$\dot{V}_0 = \frac{1}{2}\left(\dot{\sigma}^T I\sigma + \sigma^T I\dot{\sigma}\right) + \frac{1}{2}\sigma^T\dot{I}\sigma = \sigma^T\left(I\dot{\sigma} + \frac{1}{2}\dot{I}\sigma\right). \tag{18}$$

By substituting $\dot{\sigma}$ from (18) to (15), one has

$$\dot{V}_0 = \sigma^T\left(-I\ddot{\Theta}_d + I\Lambda\dot{e} - S(\omega)I\omega + U + d + \frac{1}{2}\dot{I}\sigma\right). \tag{19}$$

Let $I = I_0 + \Delta I$, where I_0 and ΔI represent the nominal and uncertain parts of the inertia matrix. According to A1, we have $\dot{I} = 0$, equation (19) becomes

$$\begin{aligned}\dot{V}_0 = \sigma^T\{&-S(\omega)\Delta I\omega - \Delta I\ddot{\Theta}_d + \Delta I\Lambda\dot{e} + d + \frac{1}{2}\dot{I}\sigma \\ &+ U - S(\omega)I_0\omega - I_0\ddot{\Theta}_d + I_0\Lambda\dot{e} \qquad (20)\\ = \sigma^T\{&\Delta P + U + P\}, \qquad (21)\end{aligned}$$

where

$$\Delta P = -S(\omega)\Delta I\omega - \Delta I\ddot{\Theta}_d + \Delta I\Lambda\dot{e} + d, \tag{22}$$

$$P = -S(\omega)I_0\omega - I_0\ddot{\Theta}_d + I_0\Lambda\dot{e}. \tag{23}$$

Let $\Xi = [\Xi_1, \Xi_2, \Xi_3]^T$ denote the sum of ΔP and P. Since the disturbance d and uncertain parameter ΔI are bounded, from (22) and (23) it can be seen that Ξ is also bounded, i.e., $|\Xi_i| \leq \Xi_{M,i}$, $i = 1, 2, 3$. Consider system (12) with the sliding variable $\sigma(\omega, t)$ as in (13). From assumptions A1-A6, the sliding motion on the manifold is achieved by the controller (16) if we can select the gain α_i such that [15]:

$$\alpha_i \geq \Xi_{M,i}. \tag{24}$$

However, the bound $\Xi_{M,i}$ is not easy to evaluate in practice and besides, there is a trade-off with chattering if a high value of α_i is chosen. The problem is now to drive the sliding variable σ and its derivative $\dot{\sigma}$ to zero in finite time by means of quasi-continuous SMC without overestimation of the control gain.

3.3 Adaptive QCSM Design

The proposed gain-adaptation law is supposed to minimise the chattering phenomenon while driving σ and $\dot{\sigma}$ to zero even in the presence of disturbances. For initial conditions $\omega_i(0), \sigma_i(0)$, and $\alpha_i(0) > 0$, the reaching and sliding on the manifold is globally achieved in finite time by the controller (16) with the following adaptive gain [13]:

$$\dot{\alpha}_i = \begin{cases} \bar{\omega}_i \big|\sigma_i(\omega, t)\big| \operatorname{sign}(|\sigma_i(\omega, t)| - \epsilon_i) & \text{if } \alpha_i > \alpha_{m,i} \\ \eta_i & \text{if } \alpha_i \leq \alpha_{m,i}, \end{cases} \tag{25}$$

where $\bar{\omega}_i > 0$, ϵ_i, η_i are small positive constants, and $\alpha_{m,i}$ is a threshold of the adaptation.

To analyse the stability of the proposed controller, let us first define a global Lyapunov function candidate for σ and α as:

$$V(\sigma, \alpha) = V_0 + \sum_{i=1}^{3} \frac{1}{2\gamma_i}(\alpha_i - \alpha_{M,i})^2, \tag{26}$$

where V_0 has been defined in Eq. (17), γ_i is some positive constant and $\alpha_{M,i}$ is the maximum possible value of the adaptive gain α_i. The derivative of the Lyapunov function (26) is obtained as

$$\dot{V}(\sigma, \alpha) = \dot{V}_0 + \sum_{i=1}^{3} \frac{1}{\gamma_i}(\alpha_i - \alpha_{M,i})\dot{\alpha}_i. \tag{27}$$

Taking $\dot{V}_0$ from (2) and $\dot{\alpha}_i$ from (25), equation (27) under the control law (16) becomes

$$\dot{V}(\sigma, \alpha) = \sum_{i=1}^{3} \sigma_i \left[\Xi_i - \alpha_i \left(\frac{\dot{\sigma}_i + |\sigma_i|^{1/2} \operatorname{sign}(\sigma_i)}{|\dot{\sigma}_i| + |\sigma_i|^{1/2}} \right) \right] + $$
$$+ \sum_{i=1}^{3} \frac{1}{\gamma_i}(\alpha_i - \alpha_{M,i})\bar{\omega}_i |\sigma_i| \operatorname{sign}(|\sigma_i| - \epsilon_i). \tag{28}$$

When σ_i is slowly time-varying, $\dot{\sigma}_i(t)$ is very small and can be negligible, then equation (28) becomes

$$\dot{V}(\sigma, \alpha) = \sum_{i=1}^{3} \sigma_i \left[\Xi_i - \alpha_i \operatorname{sign}(\sigma_i) \right] + $$
$$+ \sum_{i=1}^{3} \frac{1}{\gamma_i}(\alpha_i - \alpha_{M,i})\bar{\omega}_i |\sigma_i| \operatorname{sign}(|\sigma_i| - \epsilon_i). \tag{29}$$

It can be seen that $\dot{V} \leq 0$ given (23) and $\alpha_i \leq \alpha_{M,i}$ [13].

4 Simulation and Validation

Extensive simulation has been carried out to evaluate the performance of the proposed control algorithm. The model of the test quadcopter used is obtained from the 3DR Solo drone shown in Fig. 2 in which L_x, d_x, r_x and h_x are measured distances used to compute system parameters, as listed in Table 1. Design parameters used for the controllers are given in Table 2. The UAV, with technical specifications and accessories described in [11], was deployed to perform the tasks of infrastructure inspection, as shown in Fig. 3.

Figure 2: The 3DR Solo drone with body coordinate frame.

Figure 3: Insfrastructure inspection.

4.1 Control performance in nominal conditions

In this simulation, the quadcopter starts from zero initial conditions, i.e. all angles and velocities are zeros. Its roll and pitch angles are then set to $\phi = -10^0$ and $\theta = 10^0$ at time 0.5 s and its yaw angle is then set to $\psi = 45^0$ at time 2 s. The results are shown in Fig. 4 and Fig. 5, where the time scale in the latter is zoomed in to observe the abrupt change in the control torque and coupling effect. It can be seen that all controllers smoothly drive the angles to the desired values with relatively small overshoot and within two seconds. According to (11), there exist strong coupling relations between the control states. As a result, it can been seen that the AQCSM controller can handle

Table 1: Parameters of the quadcopter model

Parameter	Value	Unit
m	1.50	kg
l	0.205	m
g	9.81	m/s^2
I_{xx}	$8.85 \cdot 10^{-3}$	$kg.m^2$
I_{yy}	$15.5 \cdot 10^{-3}$	$kg.m^2$
I_{zz}	$23.09 \cdot 10^{-3}$	$kg.m^2$

Table 2: Control design parameters

Variable	Value	Variable	Value
λ_1	4.68	λ_2	4.68
λ_3	3.84	$\epsilon_{1,2,3}$	0.7
α_0	1.24	$\bar{\omega}_{1,2,3}$	200
$\alpha_{m,1}$	0.01	$\alpha_{m,2}$	0.02
$\alpha_{m,3}$	0.03	$\eta_{1,2,3}$	0.01

this problem to control the attitude to reach the reference values and then track them without being perturbed.

4.2 Responses to disturbances

In this simulation, a torque disturbance with the amplitude of $0.5N.m$ is added to all three axes of the quadcopter. The reference values are chosen to be the same as in the previous simulation. The responses are shown in Fig. 6. As can be seen from the plots, the AQCSM controller can cope with disturbances to reach the references and maintain the drone stability.

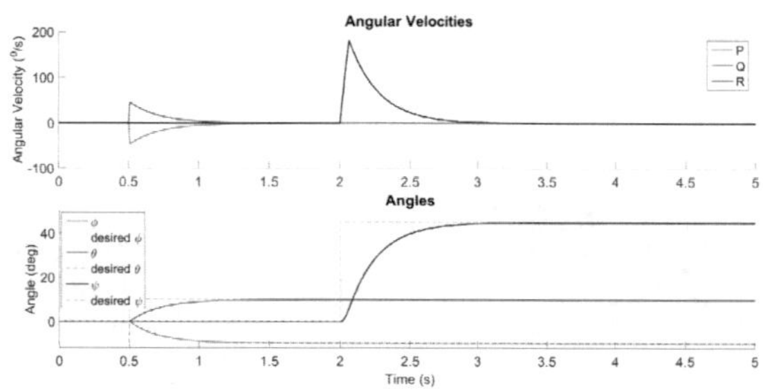

Figure 4: Responses of the quadcopter in nominal conditions (P, Q and R- roll, pitch and yaw angular velocities).

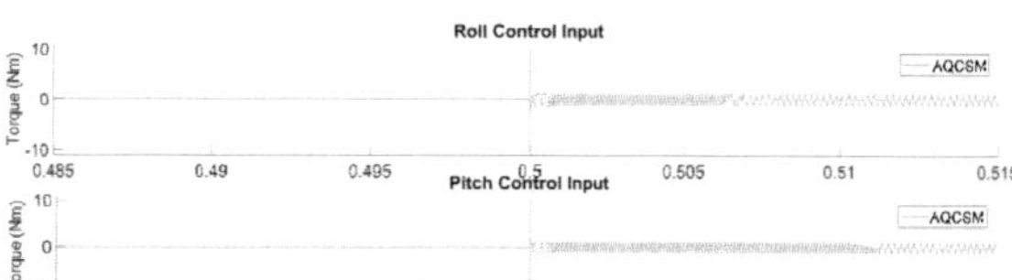

Figure 5: Control torques.

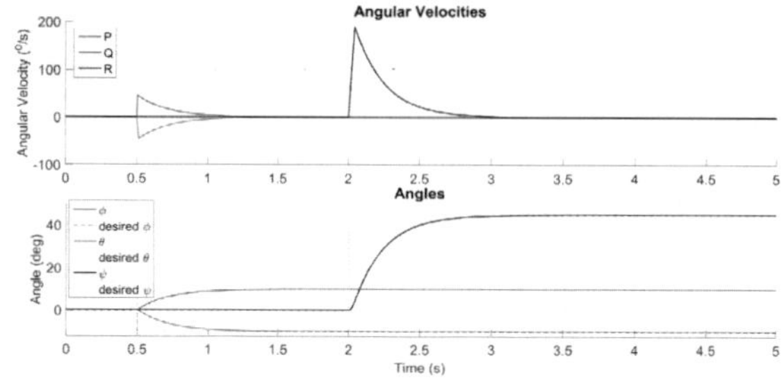

Figure 6: Angular velocity and angle responses in the presence of disturbances.

4.3 Responses to parametric variations

To evaluate the performance of the proposed controller in different conditions of loads and inertial moments, simulation parameters are varied to tolerate some modelling errors. Specifically, a load of 0.8 kg, corresponding to the maximum payload of the 3DR Solo drone, is added to the model and the following uncertainties are added to the inertial matrix:

$$\Delta I = \begin{bmatrix} 0 & 0.0044 & -0.0077 \\ 0.0044 & 0 & 0.0115 \\ -0.0077 & 0.0115 & 0 \end{bmatrix}. \tag{30}$$

Figure 7 shows the results in comparison with the nom-

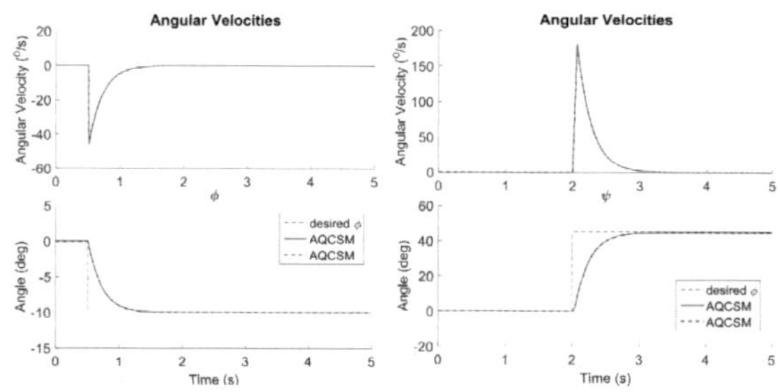

Figure 7: Angle and angular velocity responses in the presence of parametric variations.

inal conditions. The almost identical settling time and overshoot between responses corresponding to those scenarios indicates robustness of the proposed AQCSM controller. The adaptive gain $\alpha_1(t)$ response versus time is shown in Fig. 8. The higher gain magnitudes are observed in the two bottom sub-figures imply more energy is required to stabilise the system in dealing with disturbances and uncertainties. This also suggests feasibility of the control scheme.

4.4 Comparison with real-time data

To further evaluate the performance of the proposed controller, simulation results are compared with SMC and

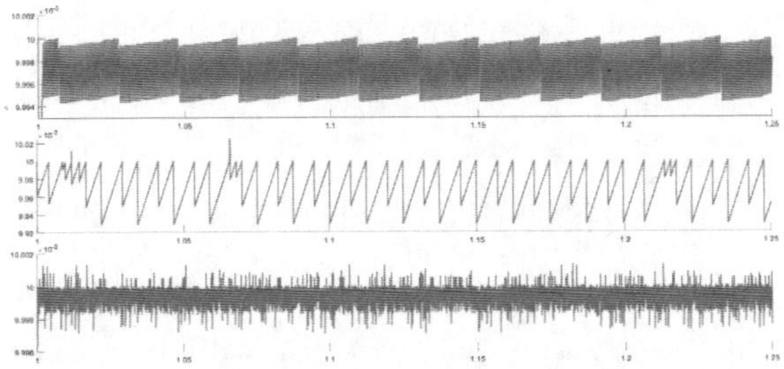

Figure 8: The adaptation of gain $\alpha_1(t)$ in various scenarios.

real time data obtained by using the built-in PID controller of the 3DR Solo drone when performing attitude control during a monitoring task [11]. The comparison is carried out by setting the same reference yaw angle to the simulated and real quadcopters. Figure 9 shows the responses of simulation for AQCSM and SMC as well as experiment for the Solo drone's PID. All controllers reach the reference value without causing much overshoot or oscillation but the AQCSM controller produces better performance with a smoother response.

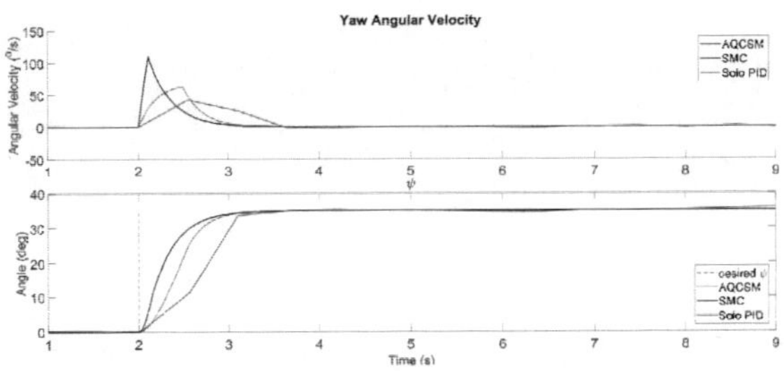

Figure 9: Tracking errors - Yaw angular velocity and Yaw angle

5 Conclusion

In this paper, an adaptive quasi-continuous sliding mode controller has been developed for robust control of the quadcopters. The control design is based on the selection of a sliding surface and some parameters for adaptation of the control gain taking account into chattering reduction. Control performance is evaluated in simulation for the cases of both external disturbances and system uncertainties. This robustness property is quite important for civil engineering applications which require accurate attitudes during collecting data for monitoring and inspection tasks. The validity of the proposed control scheme is also judged through comparison with experimental real-time data. Our future work will focus on implementing the proposed controller to develop further high-level planning strategies to take full advantage of UAV-based monitoring and inspection of built infrastructure.

References

[1] Lénaïck Besnard, Yuri B Shtessel, and Brian Landrum. Control of a quadrotor vehicle using sliding mode disturbance observer. In American Control Conference, 2007. ACC'07, pages 5230–5235. IEEE, 2007.

[2] Baohua Chen, Lei Deng, Yueqi Duan, Siyuan Huang, and Jie Zhou. Building change detection based on 3d reconstruction. In Image Processing (ICIP), 2015 IEEE International Conference on, pages 4126–4130. IEEE, 2015.

[3] Sung-suk Choi and Eung-kon Kim. Design and implementation of vision-based structural safety inspection system using small unmanned aircraft. In Advanced Communication Technology (ICACT), 2015 17th International Conference on, pages 562–567. IEEE, 2015.

[4] L Derafa, A Benallegue, and L Fridman. Super twisting control algorithm for the attitude tracking of a four rotors uav. Journal of the Franklin Institute, 349(2):685–699, 2012.

[5] Shihong Ding, Arie Levant, and Shihua Li. New families of high-order sliding-mode controllers. In 2015 54th IEEE Conference on Decision and Control (CDC), pages 4752–4757. IEEE, 2015.

[6] Youngjib Ham, Kevin K Han, Jacob J Lin, and Mani Golparvar-Fard. Visual monitoring of civil infrastructure systems via camera-equipped unmanned aerial vehicles (uavs): a review of related works. Visualization in Engineering, 4(1):1, 2016.

[7] Arie Levant. Sliding order and sliding accuracy in sliding mode control. International journal of control, 58(6):1247–1263, 1993.

[8] Arie Levant. Higher-order sliding modes, differentiation and output-feedback control. International journal of Control, 76(9-10):924–941, 2003.

[9] Arie Levant. Principles of 2-sliding mode design. Automatica, 43(4):576 – 586, 2007.

[10] Malik Manceur, Najib Essounbouli, and Abdelaziz Hamzaoui. Second-order sliding fuzzy interval type-2 control for an uncertain system with real application. Fuzzy Systems, IEEE Transactions on, 20(2):262–275, 2012.

[11] M. D. Phung, T. H. Dinh, V. T. Hoang, and Q. P. Ha. Automatic crack detection in built infrastructure using unmanned aerial vehicles. In Automation and Robotics in Construction (ISARC), 2017 International Symposium on, 2017.

[12] MD Phung, CH Quach, DT Chu, NQ Nguyen, TH Dinh, and QP Ha. Automatic interpretation of unordered point cloud data for uav navigation in construction. In Control, Automation, Robotics and Vision (ICARCV), The 2016 14th International Conference on, 2016.

[13] Franck Plestan, Yuri Shtessel, Vincent Bregeault, and Alexander Poznyak. New methodologies for adaptive sliding mode control. International journal of control, 83(9):1907–1919, 2010.

[14] Andrei Polyakov and Alex Poznyak. Lyapunov function design for finite-time convergence analysis: "twisting" controller for second-order sliding mode realization. Automatica, 45(2):444–448, 2009.

[15] Chutiphon Pukdeboon, Alan SI Zinober, and May-Win L Thein. Quasi-continuous higher order sliding-mode controllers for spacecraft-attitude-tracking maneuvers. IEEE Transactions on Industrial Electronics, 57(4):1436–1444, 2010.

[16] Guilherme V Raffo, Manuel G Ortega, and Francisco R Rubio. An integral predictive/nonlinear H_∞ control structure for a quadrotor helicopter. Automatica, 46(1):29–39, 2010.

[17] R. Ritz, M. Hehn, S. Lupashin, and R. D'Andrea. Quadrocopter performance benchmarking using optimal control. In Intelligent Robots and Systems (IROS), 2011 IEEE/RSJ International Conference on, pages 5179–5186, Sept 2011.

[18] Guillermo J Rubio, José M Cañedo, Vadim I Utkin, and Alexander G Loukianov. Second order sliding mode block control of single-phase induction motors. International Journal of Robust and Nonlinear Control, 24(4):682–698, 2014.

[19] Yuri Shtessel, Mohammed Taleb, and Franck Plestan. A novel adaptive-gain supertwisting sliding mode controller: methodology and application. Automatica, 48(5):759–769, 2012.

[20] Vadim Utkin. Discussion aspects of high-order sliding mode control. IEEE Transactions on Automatic Control, 61(3):829–833, 2016.

[21] A. Woods, H. M. La, and Q. P. Ha. A novel extended potential field controller for use on aerial robots. In Automation Science and Engineering (CASE), 2016 IEEE International Conference on, pages 286–291, 2016.

[22] Rong Xu and Ümit Özgüner. Sliding mode control of a quadrotor helicopter. In Decision and Control, 2006 45th IEEE Conference on, pages 4957–4962. IEEE, 2006.

[23] En-Hui Zheng, Jing-Jing Xiong, and Ji-Liang Luo. Second order sliding mode control for a quadrotor uav. ISA transactions, 53(4):1350–1356, 2014.

[24] Z. Zuo. Trajectory tracking control design with command-filtered compensation for a quadrotor. IET Control Theory Appl., 4(11):2343–2355, 2010.

Automatic Crack Detection in Built Infrastructure Using Unmanned Aerial Vehicles

M.D. Phung, T.H. Dinh, V.T. Hoang and Q.P. Ha

Faculty of Engineering and Information Technology, University of Technology Sydney, Australia
E-mail: Manhduong.Phung@uts.edu.au, Quang.Ha@uts.edu.au

Abstract -

This paper addresses the problem of crack detection which is essential for health monitoring of built infrastructure. Our approach includes two stages, data collection using unmanned aerial vehicles (UAVs) and crack detection using histogram analysis. For the data collection, a 3D model of the structure is first created by using laser scanners. Based on the model, geometric properties are extracted to generate way points necessary for navigating the UAV to take images of the structure. Then, our next step is to stick together those obtained images from the overlapped field of view. The resulting image is then clustered by histogram analysis and peak detection. Potential cracks are finally identified by using locally adaptive thresholds. The whole process is automatically carried out so that the inspection time is significantly improved while safety hazards can be minimised. A prototypical system has been developed for evaluation and experimental results are included.

Keywords -

Crack detection; Unmanned aerial vehicles; Health monitoring; Image stitching

1 Introduction

Crack detection is essential in health monitoring of infrastructure. Road sections containing a high density of cracks at the surface should be periodically maintained to ensure the safe operation of vehicles. Crack type, size and the level of severity need to be identified for this task. In concrete bridges, cracking caused by natural processes is inevitable and may result in malfunctioning of the entire bridge, or even collapse. It opens access for water, deicing salts and other corrosive chemicals to penetrate through the bridge deck and over time causes damages to internal bridge structures. Early identification of cracks is thus vitally important to maintain the service life of bridges.

To automatically detect cracks, both colour and geometry information need to be acquired with sufficient quality using unmanned ground vehicles [10, 20] or a group of mobile robots [26]. However, complexity and roughness remain challenges for surface inspection using this approach. Recently, unmanned aerial vehicles (UAVs) have been developed as an alternative owing to their flexibility in the operation space and ability to carry specialised sensory equipment. In [6], a micro air vehicle system was employed to scan buildings using a high resolution camera. Images taken within a restricted location were then stitched with sufficient quality for crack and damage detection. An advanced UAV system was developed to evaluate the state of historical monuments [8] whose captured images revealed after processing some damages in several monuments. In [14], a control system for navigating the UAV in unknown 3D environments was used to monitor and maintain bridges. UAVs were also used to inspect and monitor oil-gas pipelines, roads, power generation grids and other essential infrastructure [21].

For surface detection, computer-vision based techniques are widely used to detect cracks due to its robustness and cost efficiency [11]. These techniques in general can be categorised into the wavelet transform, minimal path selection, machine learning, edge detection and intensity thresholding. For example, a separable 2D continuous wavelet transform is employed in [25], using complex coefficient maps for crack segmentation. An improvement of the wavelet-based pavement distress detection can be achieved by combining the Wavelet-Radon transform and dynamic neural network thresholding [15]. These techniques however do not consider the geometric characteristics (orientation, continuity and connectedness) of the cracks and may wrongly detect the candidates with low continuity or high curvature.

Owing to the ability to effectively identify high-level geometric information, the minimal path principle can be applied in surface crack detection. In [16], the free-form anisotropy is able to handle almost all crack characteristics in a segmentation step. An improvement of the minimal path technique is presented in [9], having the capability to detect cracks without prior knowledge of endpoints. A fully unsupervised approach is proposed in [3], where a refined artefact filtering step is introduced to estimate the width of the crack. However, the main drawback of the minimal path approach is high computation time involved.

With the explosive development of image data, machine learning-based methods [17, 23] have been effectively used for surface crack detection. In [17], a multi-level pattern recognition system is developed to address image blocks containing cracks and then evaluate their geometry such as length and width. Random structured forests are used in [13] and [23] to extract information for crack descriptors from the image background. Nevertheless, these methods are greatly dependent on the training data, which is labour-intensive with manual labelling in the training and validation steps.

For determining potential cracks from bridge decks, edge detection techniques is commonly used [2, 12]. Four edge-detection algorithms are surveyed in [1], where the Haar Wavelet method is identified as most reliable, compared to the gradient-based (Sobel and Canny) and frequency-based (fast Fourier Transformation) algorithms. These methods, however, only perform well under uniform illumination and low noise conditions.

In crack regions that are consistently darker than their surrounding areas, intensity-thresholding methods [18] are more appropriate due to their compactness. A Bayesian classification technique together with a morphological opening operation and thresholding is applied to segment and classify defects from underground concrete pipes into various classes such as cracks, holes, laterals and joints [24]. A two-step method is proposed in [7] where a locally adaptive thresholding is used together with a median filter and a multi-scale line filter to emphasise the line structure and detect crack candidates. As thresholding is noise sensitive, it is often used with other techniques such as morphological or linear filtering to improve robustness.

In this paper, we propose a system to inspect built infrastructure for automatic crack detection. A 3D model of the object is first created and its geometric features are extracted to generate a path for UAV navigation. While following the planned path, the UAV takes images of the suspected surfaces. Those images are then stitched and processed based on histogram analysis. For this task, we develop a peak detection algorithm for image clustering and a locally adjustable thresholding technique for crack detection. A number of experiments have been carried out and the detection results are promising to apply in real time applications.

2 Data Collection Using UAV

The goal of data collection is to acquire adequate geometric and surface information of the object to be used in post processing for detecting potential cracks, and also for navigation of the UAV itself [19].

2.1 Point cloud 3D modelling

In this step, laser scanners are used to acquire range information from different positions of the structure to be inspected and represent them as point clouds. Those point clouds are then merged one by one in a process called registration to create a 3D model. In the registration, overlapping points corresponding to the same part of the structure appearing among the point clouds are first identified. Let P_a and P_b be the point clouds recorded at positions $\boldsymbol{a}$ and $\boldsymbol{b}$. The overlapping points are determined by:

$$\|(\boldsymbol{x}_i^a - \boldsymbol{a}) - (\boldsymbol{x}_j^b - b)\| < \tau, \tag{1}$$

where $\boldsymbol{x}_i^a \in P_a$ and $\boldsymbol{x}_i^b \in P_b$ are overlapping points, considered in a close neighbourhood, and τ is a pre-defined distance threshold. The alignment of point clouds is then obtained by applying an iterative closest point algorithm to find a transformation that minimises the distances between them.

2.2 Geometric feature extraction

Planar surfaces are often the main target to be inspected so they need to be extracted from the point cloud. Given a plane's equation ($ax + by + cz + d = 0$), then $M = [a, b, c, d]^T$ is the parameter vector to be identified. For this, a random sample consensus (RANSAC) algorithm is applied with some augmentations, including a noise filter to remove sparse outliers, voxelisation to equalise the point density, and clustering to trim out the isolated groups before applying RANSAC.

After detecting the surfaces, their boundaries are determined by using a convex hull algorithm. The remaining point cloud P is then clustered into small groups as obstacles to be considered in path planning. Here, for a given positive constant $\epsilon > 0$, a cluster is defined as a set of points:

$$C = \{p_i \in P \mid \min\|p_i - p_j\| > \epsilon\}, \tag{2}$$

for any other point not belonging to the cluster, i.e., $p_j \notin C$.

2.3 Path planning for colour image acquisition

Given surfaces to be inspected, a list of waypoints needs to be created to navigate the UAV. There are two types of waypoints, one corresponds to shooting points for taking images and the others serve as intermediate points for path following and avoiding obstacles. They are generated by first splitting the operating environment into voxels. A status of free or occupation is then defined for each voxel to indicate the existence or not of obstacles in that voxel. The shooting points are then computed based on intrinsic parameters such as camera focal length, surface area and minimum resolution. An A-star algorithm is finally applied to find the shortest path between shooting points. In each step, the cost to move from one voxel to another in the neighbourhood is computed as:

$$C(k, l, m) = a_1 k^2 + a_2 l^2 + a_3 m^2, \tag{3}$$

where coordinates $k, l, m \in \{-1, 0, 1\}$ indicate the neighbour position, and the coefficients a_1, a_2 and a_3 assign a particular weight to each direction.

The generated waypoints are used as references for motion control of the UAV. Typically, the controller are built in the flight operating system so that the control task can be simplified. To collect images of sufficient quality, a gimbal is used to eliminate vibration and adds more degree of freedoms to the system to shoot images perpendicularly to the inspected surface.

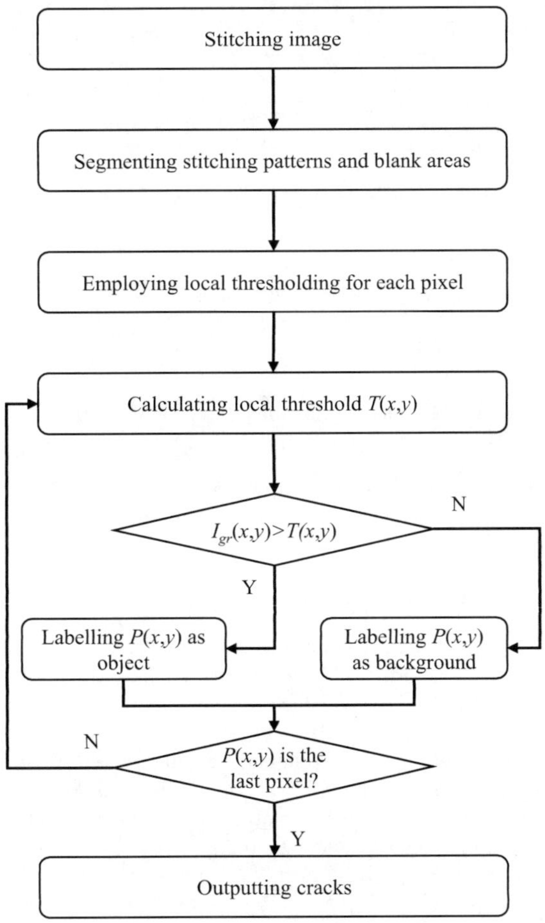

Figure 1: Crack detection flowchart.

3 Thresholding-Based Crack Detection

A crack detection algorithm is developed to process the collected data based on analysing their histogram information. Its flowchart including image stitching, pattern removal and crack detection steps is shown in Fig. 1.

3.1 Image stitching

Each image taken by the UAV only covers a small area of the inspected surface. Stitching them to create a panoramic image is thus essential for crack detection. This stage requires a certain level of overlapping between consecutive images and corresponding features. An invariant feature based approach [4] is employed, consisting of scale-invariant feature transform (SIFT) feature extraction, homography computation and image matching verification. If the surface to be inspected is homogeneous, manual patterns can be added to enhance the stitching performance and removed afterwards.

3.2 Stitching pattern removal

Before processing the stitched image, it is important to remove the patterns used in stitching to enhance the detection accuracy. A stitched image typically consists of three elements, namely crack area, stitching patterns and blank area caused by alignment drifts. Since their luminance varies in accordance with the sunlight condition, our peak detection algorithm [5] is first employed to detect dominant peaks corresponding to these elements. The thresholds t_1 and t_2 to segment the surfaces to be inspected are calculated as the intensity average:

$$\begin{cases} t_1 = \dfrac{i_b + i_w}{2} \\ t_2 = \dfrac{i_w + i_p}{2}, \end{cases} \tag{4}$$

where i_b, i_w and i_p are the intensity value of peaks corresponding to the blank areas, the surface and stitching patterns, respectively. The stitching patterns are then identified based on the histogram as follows:

$$I_{gr}(x,y) = \beta \quad \text{if} \quad \left[\begin{array}{l} I_r(x,y) < t_1 \\ I_r(x,y) > t_2, \end{array} \right. \tag{5}$$

where $I_{gr}(x,y)$ and $I_r(x,y)$ are respectively the grey intensity and the red channel intensity of the stitched image at point $P(x,y)$, and β is the intensity value chosen to distinguish the pattern with other parts of the image. As the intensity at crack structures is typically smaller than at non-defect areas of inspected surfaces, here β is set to 255 for adequately filtering out those features without information loss.

3.3 Crack detection

In an outdoor environment subject to varying lighting conditions, the global thresholding method [5] is extended to be able to extract all crack candidates with large variation of intensities. The approach proposed here, is first to take the advantage of the automatic peak detection for pre-processing the image to retain only the background

Figure 2: The 3DR Solo drone with remote controller and ground control station.

Figure 3: Collecting data of the bridge.

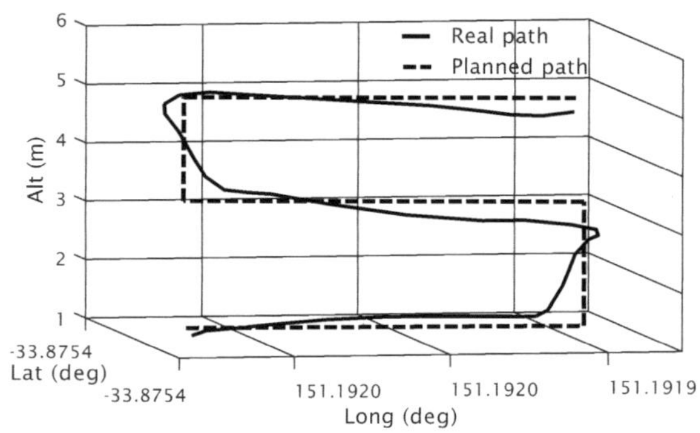

Figure 4: Flight path to collecting data of the bridge.

and line-like objects, and then to apply locally-adjusted thresholds [22] to identify potential cracks. Differing from the global approach, here the threshold is computed for each pixel based on the grey intensities of its neighbours. Let us consider (x_i, y_i), $i = 1, 2, ..., N^2$, in the neighbourhood of pixel $P(x, y)$ determined by using an $N \times N$- window and $m(x, y)$ and $s(x, y)$ respectively as the mean and standard deviation in that window:

$$m(x, y) = \frac{1}{N^2} \sum_{i=1}^{N^2} I_{gr}(x_i, y_i), \tag{6}$$

$$s(x, y) = \sqrt{\frac{1}{N^2} \sum_{i=1}^{N^2} (m(x, y) - I_{gr}(x_i, y_i))^2}. \tag{7}$$

The threshold for pixel $P(x, y)$ is then computed as:

$$T(x, y) = m(x, y) \left[1 + k \left(\frac{s(x, y)}{R} - 1 \right) \right], \tag{8}$$

where R is the dynamic range of standard deviations and k is a tunable parameter used to adjust the influence of standard deviation. Each pixel $P(x, y)$ is then evaluated against its threshold $T(x, y)$. A pixel $P(x, y)$ is considered as belonging to a crack if its grey intensity $I_{gr}(x, y)$ is higher than the computed local threshold, or lying in the background otherwise.

3.4 Experiments

Experiments have been conducted to evaluate the proposed approach. The UAV used is a quadcopter, the 3DR Solo drone shown in Fig.2. It is equipped with three processors, two are Cortex M4 168 MHz running Pixhawk firmware for low-level control and the other is an ARM Cortex A9 running Arducopter flight operating system. The camera used is Hero 4 with the focal length of 34.4 mm and resolution of 12 megapixels, attached to a three-axis gimbal with one degree-of-freedom for controlling its yaw angle. The programming is carried out through the ground control station called Mission Planner and uploaded to the UAV.

3.4.1 Data collection results

Figures 3 and 4 show a practical path used to collect data of a bridge. There are mismatches between the planned and real paths which are inevitable due to localisation errors caused the built-in global positioning system. Nevertheless, in our experiments, the number of satellites detected ranging from 9 to 11 causing the error less than 1.5 m which is relatively small. For data collection, this error can be compensated by reducing the UAV's speed while increasing the number of shooting points. It is also noted that although the planned path shown in Fig. 4 is rather ideal and does not consider the dynamic constraints of the UAV, which is beyond the scope of this study, the path tracking error remains however in an acceptable tolerance for the inspection purpose.

3.4.2 Crack detection results

To evaluate the crack detection algorithm, an wooden wall was set up from nine panels, each having the size of 600 mm $\times$ 900 mm $\times$ 3 mm. Those panels were hanged on a frame and joined together using twisted ties as shown

Figure 5: Flight path used to inspect the artificial wall.

in Fig. 6a. Eighteen patterns were manually stuck on top of the panels for image stitching. The pink colour was chosen for the patterns due to its large difference from the colour of wooden panels. Two cracks were deliberately formed by a manual impact applied on two panels, one at the middle and one at the bottom right. The UAV was programmed to follow pre-defined waypoints to take images of the surface, as shown in Fig. 5. The values of R and k in (8) are chosen to be 128 and 0.5, respectively.

Figure 6b shows the stitched image. It can be seen that there are almost no distortions compared to the original wall. However, missing pixels caused by alignment drifts still appear. This issue can be resolved by employing gain compensation and multi-band blending techniques. Figure 6c shows the result of filtered stitching patterns. They are all well isolated from the stitched image demonstrating the effectiveness of the proposed automatic peak detection algorithm.

Figure 7a presents the crack detection results using the global thresholding method [5]. It can be seen that there are two crack candidates appearing along with horizontal and vertical lines of the tiled wooden panels. However, the middle crack is not well detected due to the variation in the grey intensity. If the threshold is increased to better capture the crack image, then parts of the wall that are exposed to the sunlight are not adequately filtered as shown in Fig. 7b. This problem can be solved using the locally adjusted thresholding method as shown in Fig. 7c.

4 Conclusion

In this work, we have developed an automatic crack detection system for infrastructure monitoring. By using UAVs, the system is capable of inspection of poorly accessible structures such as dams, culverts or bridges. A number of sensors have been integrated into the system allowing it to acquire geometric and colour information of the inspected surfaces. For data processing, we have developed computer vision based algorithms to create 3D models of the structure, extract features, plan navigation paths, stitch images and detect potential cracks. A number of experiments have been conducted with all cracks detected in real time. Future work will focus on further improvements of the recognition algorithms to better identify crack properties such as the length, width and orientation.

Ackowledgement

The first author would like to acknowledge an Endeavours Research Fellowship (ERF-PDR-142403-2015) provided by the Australian Government. This work is supported by a University of Technology Sydney Data Arena Research Exhibit Grant 2016.

References

[1] I. Abdel-Qader, O. Abudayyeh, and M. Kelly. Analysis of edge-detection techniques for crack identification in bridges. Journal of Computing in Civil Engineering, 17(4):255–263, 2003.

[2] R. S. Adhikari, O. Moselhi, and A. Bagchi. Image-based retrieval of concrete crack properties for bridge inspection. Automation in Construction, 39:180 – 194, 2014.

[3] R. Amhaz, S. Chambon, J. Idier, and V. Baltazart. Automatic crack detection on two-dimensional pavement images: An algorithm based on minimal path selection. IEEE Transactions on Intelligent Transportation Systems, 17(10):2718–2729, Oct 2016.

[4] M. Brown and D. Lowe. Automatic panoramic image stitching using invariant features. International journal of computer vision, 74(1):59–73, 2007.

[5] T. H. Dinh, Q. P. Ha, and H. M. La. Computer vision-based method for concrete crack detection. In Proceedings of the 14th International Conference on Control, Automation, Robotics and Vision (ICARCV 2016), pages 1–6, Nov 2016.

[6] C. Eschmann, C.-H Kuo, C.-M Kuo, and C. Boller. Unmanned aircraft systems for remote building inspection and monitoring. In Proceedings of the 6th European Workshop on Structural Health Monitoring, pages 1–8, 2012.

[7] Y. Fujita and Y. Hamamoto. A robust automatic crack detection method from noisy concrete surfaces. Machine Vision and Applications, 22(2):245–254, 2011.

[8] N. Hallermann and G. Morgenthal. Un-manned aerial vehicles (uav) for the assessment of existing structures. In Proceedings of the 36th IABSE Symposium, 101(14):266–267, 2013.

(a)

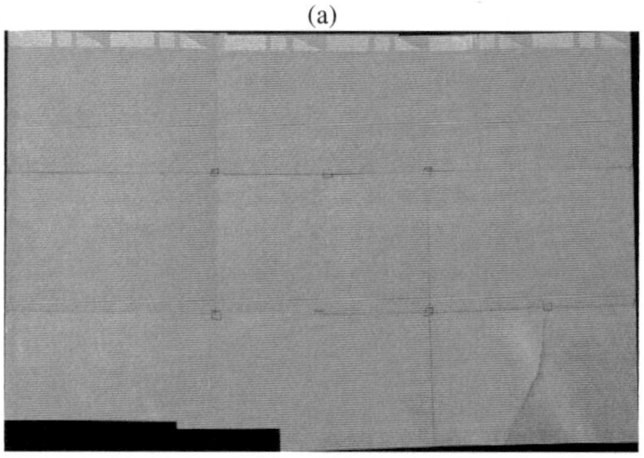

(b)

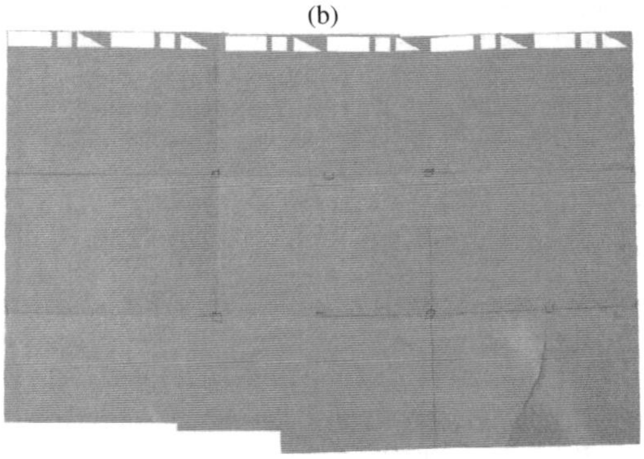

(c)

Figure 6: Image stitching and segmentation:
(a) UAV taking photos of the artificial wall (b) Stitched image, (c) Segmented patterns and blank areas.

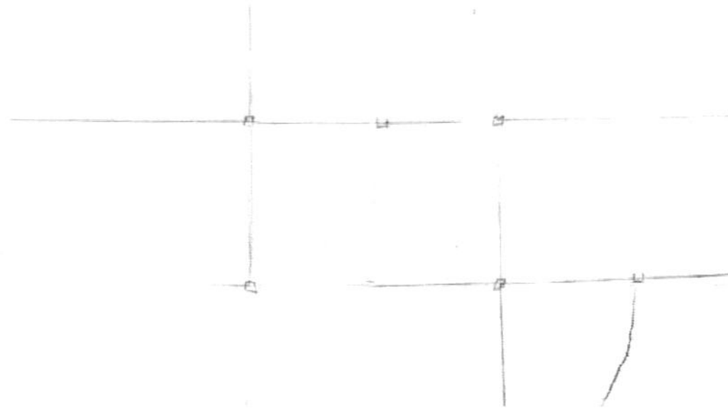

(a)

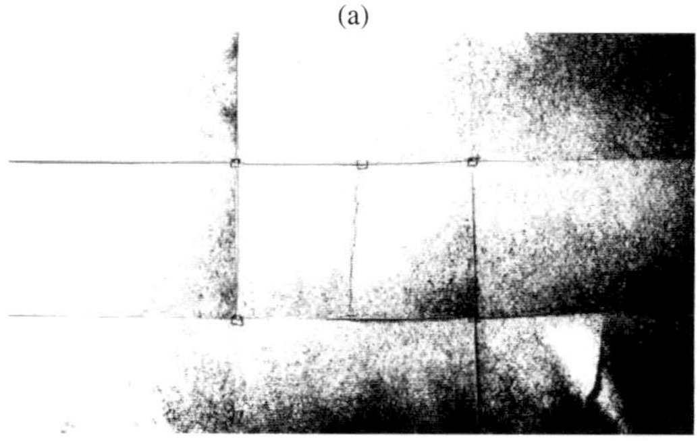

(b)

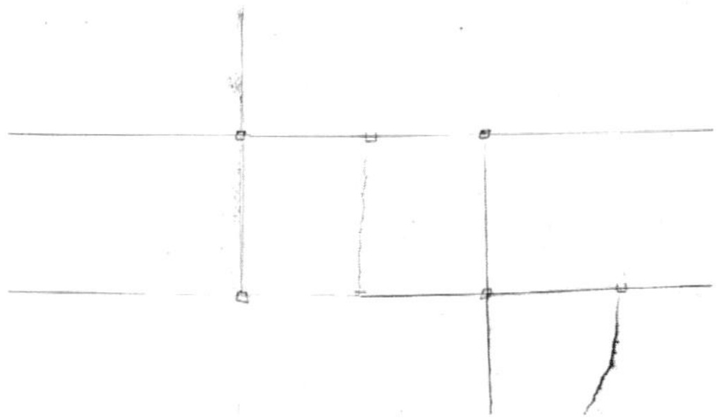

(c)

Figure 7: Crack detection result: (a) Global thresholding at T = 125, (b) Global thresholding at T = 155, (c) Locally apaptive thresholding.

[9] V. Kaul, A. Yezzi, and Y. Tsai. Detecting curves with unknown endpoints and arbitrary topology using minimal paths. IEEE Transactions on Pattern Analysis and Machine Intelligence, 34(10):1952–1965, Oct 2012.

[10] S. Kim, M. Spenko, S. Trujillo, B. Heyneman, D. Santos, and M. R. Cutkosky. Smooth vertical surface climbing with directional adhesion. IEEE Transactions on Robotics, 24(1):65–74, Feb 2008.

[11] C. Koch, K. Georgieva, V. Kasireddy, B. Akinci, and P. Fieguth. A review on computer vision based defect detection and condition assessment of concrete and asphalt civil infrastructure. Advanced Engineering Informatics, 29(2):196 – 210, 2015. Infrastructure Computer Vision.

[12] G. Li, S. He, Y. Ju, and K. Du. Long-distance precision inspection method for bridge cracks with image processing. Automation in Construction, 41:83 – 95, 2014.

[13] J. J. Lim, C. L. Zitnick, and P. Dollar. Sketch tokens: A learned mid-level representation for contour and object detection. In 2013 IEEE Conference on Computer Vision and Pattern Recognition, pages 3158–3165, June 2013.

[14] N. Metni and T. Hamel. A uav for bridge inspection: Visual servoing control law with orientation limits. Automation in Construction, 17(1):3–10, 2007.

[15] F. M. Nejad and H. Zakeri. An optimum feature extraction method based on wavelet–radon transform and dynamic neural network for pavement distress classification. Expert Systems with Applications, 38(8):9442 – 9460, 2011.

[16] T. S. Nguyen, S. Begot, F. Duculty, and M. Avila. Free-form anisotropy: A new method for crack detection on pavement surface images. In 2011 18th IEEE International Conference on Image Processing, pages 1069–1072, Sept 2011.

[17] H. Oliveira and P. L. Correia. Automatic road crack detection and characterization. IEEE Transactions on Intelligent Transportation Systems, 14(1):155–168, March 2013.

[18] N. Otsu. A threshold selection method from gray-level histograms. Automatica, 11(285-296):23–27, 1975.

[19] M. D. Phung, C. H. Quach, D. T. Chu, N. Q. Nguyen, T. H. Dinh, and Q. P. Ha. Automatic interpretation of unordered point cloud data for uav navigation in construction. In 2016 14th International Conference on Control, Automation, Robotics and Vision (ICARCV), 2016.

[20] P. Prasanna, K. J. Dana, N. Gucunski, B. B. Basily, H. M. La, R. S. Lim, and H. Parvardeh. Automated crack detection on concrete bridges. IEEE Transactions on Automation Science and Engineering, 13(2):591–599, April 2016.

[21] S. Rathinam, Z.W. Kim, and R. Sengupta. Vision-based monitoring of locally linear structures using an unmanned aerial vehicle. Journal of Infrastructure Systems, 14(1):52–63, 2008.

[22] J. Sauvola and M. Pietikäinen. Adaptive document image binarization. Pattern recognition, 33(2):225–236, 2000.

[23] Y. Shi, L. Cui, Z. Qi, F. Meng, and Z. Chen. Automatic road crack detection using random structured forests. IEEE Transactions on Intelligent Transportation Systems, 17(12):3434–3445, Dec 2016.

[24] S. K. Sinha and P. W. Fieguth. Segmentation of buried concrete pipe images. Automation in Construction, 15(1):47 – 57, 2006.

[25] P. Subirats, J. Dumoulin, V. Legeay, and D. Barba. Automation of pavement surface crack detection using the continuous wavelet transform. In 2006 International Conference on Image Processing, pages 3037–3040, Oct 2006.

[26] Y. H. Yu, N. M. Kwok, and Q. P. Ha. Color tracking for multiple robot control using a system-on-programmable-chip. Automation in Construction, 20:669 – 676, 2011.

Vision-based 2D map generation for monitoring construction sites using UAV Videos

Seongdeok Bang[a] Hongjo Kim[a] and Hyoungkwan Kim[a]

[a]Department of Civil and Environmental Engineering, Yonsei University, Seoul, South Korea
E-mail: bangdeok@yonsei.ac.kr, hongjo@yonsei.ac.kr, hyoungkwan@yonsei.ac.kr

Abstract –

This paper presents a system to generate a 2D map using unmanned aerial vehicles (UAV) videos. The system can process the collected images for lens distortion, unstable flight, and image redundancy. The system is demonstrated using an UAV video captured on road construction site in Namyangju city, Korea. The experimental result is a high-resolution 2D map of which resolution is 6010 by 9465. This study is expected to help site managers understand site conditions and make a decision timely in diverse construction sites.

Keywords –

Construction sites; UAV; Image Stitching; Computer Vision;

1 Introduction

Unmanned aerial vehicles (UAVs) equipped with a camera collect a large amount of images and videos in a short time from a variety of perspectives [1]. In the construction industry, UAVs have been investigated as the tool for monitoring infrastructures such as bridges, road, and plants [2]. UAV-based videos including as-built data of a construction site can be used for construction management applications, for example, work-in-process monitoring, safety management, and facility condition assessment. To name a few studies, Siebert and Teizer [3] researched 3D photogrammetric surveying for calculating earthwork volumes using a UAV system. Han et al. [4] proposed a formal process for construction progress monitoring using UAV images and 4D Building Information Model (BIM). González-Jorge et al. [5] experimented with camera-equipped UAVs to capture the geometric information of breakwaters for monitoring potential failures.

Image stitching is the method for integrating multiple images into one high-resolution image [6,7]. Recently, image stitching has been used in the field of construction management to monitor infrastructures [8,9]. These studies show that image stitching is suitable to cover the larger areas minimizing the loss of details. However, there are certain problems to be resolved for producing a map covering a large area such as a whole construction site using UAVs. Image capturing using UAVs is likely to obtain low quality images because of the vibration and immature control of aircrafts. Proper feature selection is another issue to be resolved for connecting multiple images.

To address the problem, this paper presents an automatic generation method of a 2D high-resolution map for monitoring construction sites using UAV videos. Figure 1 illustrates the system. The system performs a series of tasks, including (1) extracting key frame for UAV videos for reducing the size of an image dataset, (2) filtering out blurred images to obtain clear representation of the final map, (3) correcting camera lens distortion to rectify the curved effect on image boundaries, (4) matching images to unify each image coordinate into one global coordinate system, and (5) stitching and blending images to represent the whole construction site in a one integrated map. Details of the system and its validation are presented in the following sections.

2 System description

2.1 Frame selection

Similar scenes repeatedly appear in successive frames in a UAV video because of a high number of frame per seconds (fps) (at least 24 fps) [10,11]. Thus, key frame selection is required to reduce computational time for stitching images among nearly same visual information. The process to select key frames is based on the following Equation (1):

$$\text{frame per unit time} = \frac{\text{effective distance}}{\text{velocity of UAV}} = \frac{(1-\mathrm{EO})*{}^{1}/_{1-\alpha}*\tan\left(\frac{\mathrm{FOV}}{2}\right)*\text{altitude of UAV per unit time}}{\text{velocity of UAV per unit time}} \quad (1)$$

in which EO and FOV are effective overlap ratio and the field of view of UAV, respectively, and α is the proportion of filtered images for blurring effect. Chaiyasarn et al. [12] and Zhu et al. [8] acquires images with at least 50% overlap between every two images for image stitching of tunnels. Considering these studies and empirical experiment, the value of EO and α are assumed to be 0.7 and 0.25 respectively. The calculated frame extraction rate per unit time will be accumulated until the total playback time of video.

2.2 Blur Filtering

Blurred images extracted from a blurred video have to be filtered for improving the quality of image stitching. Crete et al. [13] proposed the blur metric that is able to quantify the occurrence of blur effect on each image. However, it is difficult to directly apply the value of blur metric which is ranging from 0 to 1 due to unclear criteria. In order to figure out the frame which are relatively more blurred than surrounding frames containing similar scenes, the blur value of each image is compared with the moving average.

2.3 Camera Correction

The curvature caused by the projection of straight lines of an image due to lens distortion degrades the performance of image stitching [7,14]. The camera with wide-angle lens is suitable for on-site monitoring, however, it has obvious radial distortion. Unless this distortion is considered, the result of proposed system will show blurring and edge dislocation. For compensating lens distortion, the intrinsic parameters of camera have to be calculated at first. Zhang [15] suggested the camera calibration technique to capture a planar pattern from different orientations, and the technique is used in this study for the correction of lens distortion.

2.4 Image alignment and blending

The process of this section is following the study of Szeliski [7]. Image stitching technique consists of two major processes: alignment and blending. Image alignment is finding geometric relationships between images through keypoint descriptor matching and unifying images with one geometric information. Image blending is the smooth integration of the color values between two images. Image stitching can be implemented using tools such as AutoStitch, ICE (Image Composite Editor) and Photostitch. In particular, ICE is a free software developed by Microsoft Research based on the research of Szeliski and has an advanced stitching capability. In this study, ICE was used to construct panorama image representing a construction site.

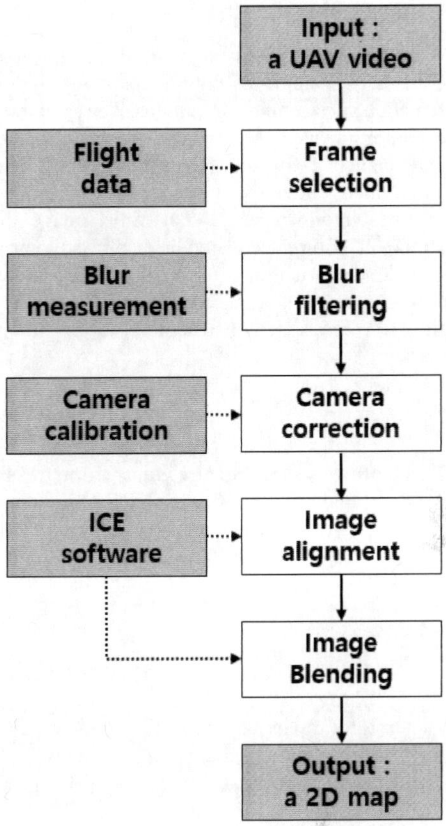

Figure 1. System for making a 2D high resolution map using UAV based images

3 Experiment and Discussion

UAV video captured in Namyangju city, Korea was used for validation of the proposed system. Frame selection, blur filtering, and camera lens correction were executed using the MATLAB programming language. In particular, the camera calibration was processed using the MATLAB cameraCalibration App. The model of UAV was DJI Phantom 3 Professional. The video used for the experiment had 42 seconds of playback time at 30fps. Thirty eight images were extracted by the frame selection function and 11 images of them were filtered out by the blur filtering function. The resolution of each image was 3840 by 2160 and the resolution of the resultant image was 6010 by 9465. Figure 2 shows the result–the visible map to perceive construction site at a glance.

4 Conclusion

This paper proposes a method to automatically generate a 2D high-resolution map for monitoring construction sites using UAV videos. The system enables the construction site to be recognized as a high-resolution map at a glance. Thanks to the error correction functions, a UAV video taken by a novice UAV pilot can be transformed into a map in a short time. Image stitching technique in computer vision is useful to represent a large-scale construction site. With future studies, the proposed system is expected to help site managers monitor the current construction status timely and frequently.

Acknowledgement

This work was supported by grants from the National Research Foundation of Korea (NRF), funded by the Korea government (MSIP) (NRF-2014R1A2A1A11052499 and No. 2011-0030040).

References

[1] P. Liu, A.Y. Chen, Y.-N. Huang, J. Han, J. Lai, S. Kang, T. Wu, M. Wen, M. Tsai, A review of rotorcraft Unmanned Aerial Vehicle (UAV) developments and applications in civil engineering, Smart Struct. Syst 13 (6) (2014) 1065-1094.

[2] Y. Ham, K.K. Han, J.J. Lin, M. Golparvar-Fard, Visual monitoring of civil infrastructure systems via camera-equipped Unmanned Aerial Vehicles (UAVs): a review of related works, Visualization in Engineering 4 (1) (2016) 1-8.

[3] S. Siebert, J. Teizer, Mobile 3D mapping for surveying earthwork projects using an Unmanned Aerial Vehicle (UAV) system, Automation in Construction 41 (2014) 1-14.

[4] K. Han, J. Lin, M. Golparvar-Fard, A formalism for utilization of autonomous vision-based systems and integrated project models for construction progress monitoring, 2015 Conference on Autonomous and Robotic Construction of Infrastructure, 2015, pp. 118-131.

[5] H. González-Jorge, I. Puente, D. Roca, J. Martínez-Sánchez, B. Conde, P. Arias, UAV photogrammetry application to the monitoring of rubble mound breakwaters, Journal of Performance of Constructed Facilities 30 (1) (2014) 04014194.

[6] M. Brown, D.G. Lowe, Automatic Panoramic Image Stitching using Invariant Features, International journal of computer vision 74 (1) (2007) 59-73.

[7] R. Szeliski, Computer Vision: Algorithms and Applications, Springer Science & Business Media, New York, 2010.

[8] Z.H. Zhu, J.Y. Fu, J.S. Yang, X.M. Zhang, Panoramic Image Stitching for Arbitrarily Shaped Tunnel Lining Inspection, Computer-Aided Civil and Infrastructure Engineering 31 (12) (2016) 936-953.

[9] R. Adhikari, O. Moselhi, A. Bagchi, Image-based retrieval of concrete crack properties for bridge inspection, Automation in Construction 39 (2014) 180-194.

[10] S. Nadarajah, K. Sundaraj, Vision in robot soccer: a review, Artificial Intelligence Review 44 (3) (2015) 289-310.

[11] K. Haase, N. Kröger-Lui, A. Pucci, A. Schönhals, W. Petrich, Real-time mid-infrared imaging of living microorganisms, Journal of Biophotonics 9 (1-2) (2016) 61-66.

[12] K. Chaiyasarn, T.-K. Kim, F. Viola, R. Cipolla, K. Soga, Distortion-Free Image Mosaicing for Tunnel Inspection Based on Robust Cylindrical Surface Estimation through Structure from Motion, Journal of Computing in Civil Engineering 30 (3) (2015) 04015045.

[13] F. Crete, T. Dolmiere, P. Ladret, M. Nicolas, The blur effect: perception and estimation with a new no-reference perceptual blur metric, Human Vision and Electronic Imaging XII, 2007, Vol. 6492, International Society for Optics and Photonics 2007, pp. 1-11.

[14] H.S. Sawhney, R. Kumar, True multi-image alignment and its application to mosaicing and lens distortion correction, IEEE transactions on pattern analysis and machine intelligence 21 (3) (1999) 235-243.

[15] Z. Zhang, A flexible new technique for camera calibration, IEEE transactions on pattern analysis and machine intelligence 22 (11) (2000) 1330-1334.

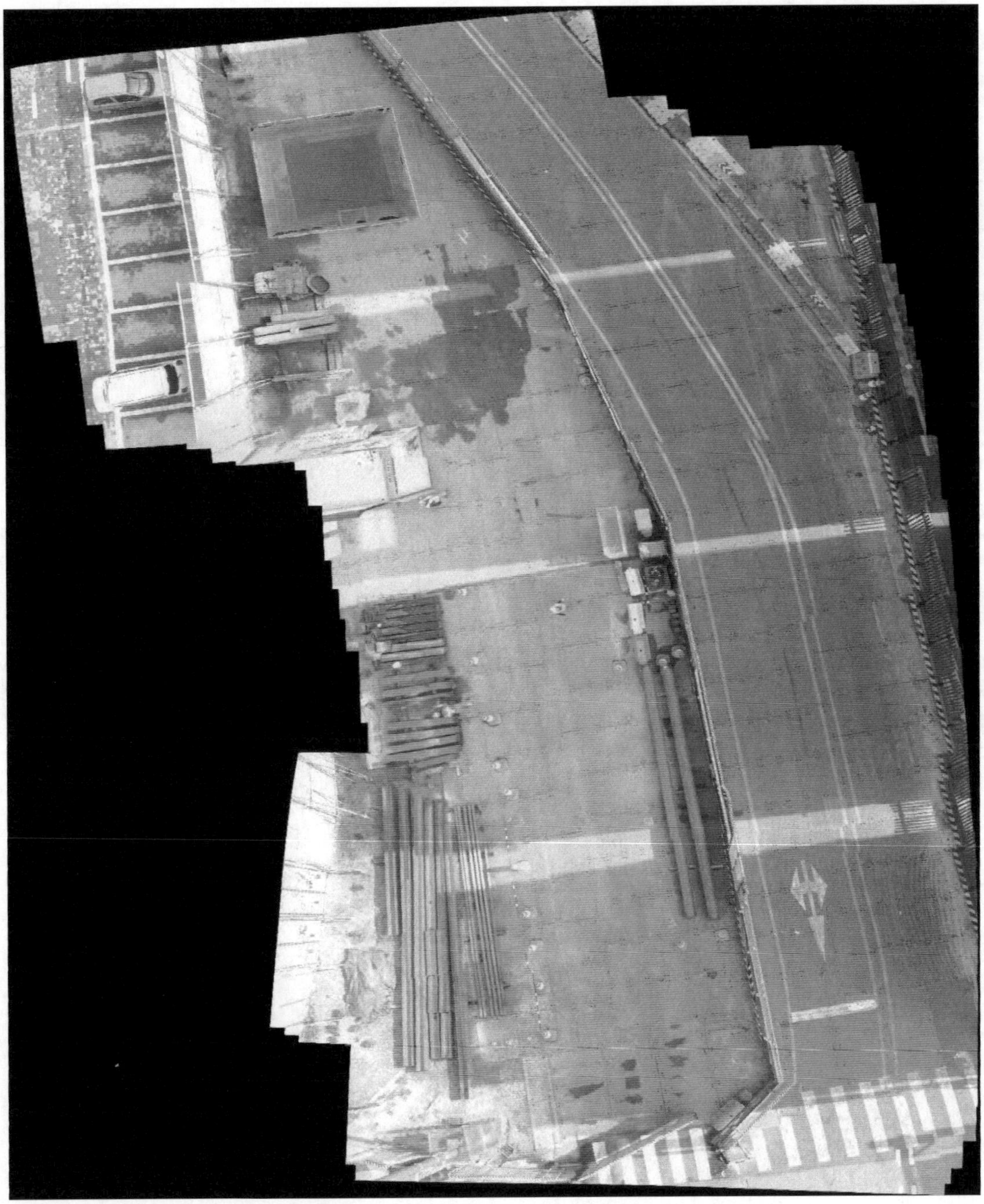

Figure 2. The result of 2D high-resolution map using the UAV video of the road construction site

Investigating the accuracy of 3D models created using SfM

Kosei Ishida[1]

[1] School of Architecture, Kogakuin University, Tokyo, Japan
E-mail : k_ishida@cc.kogakuin.ac.jp

Abstract

Technological innovations in photogrammetry have progressed over time, and civil engineers in Japan have employed this technology for earthworks. Photogrammetry has also been introduced to building construction work, and in particular, it is used in the surveying of buildings and the production of architectural drawings in repair work.

With this motivation, the authors evaluate the accuracy of Three-dimensional (3D) model extraction using structure from motion (SfM) photogrammetry. In particular, the authors evaluate the accuracy of the height measurements (Z axis).

The authors survey the accuracy of a 3D model using a cube with a side length of 5 cm. Our surveying procedure consists of the following four steps:

1. **Place the cube on a turntable, rotate the turntable in 5° increment, and capture 72 photographs.**
2. **Select 12 photographs considering six control points. Edit the control points given their 3D positions in each of the 12 photographs.**
3. **Extract the 3D model using SfM software.**
4. **Import the 3D model and measure the height of the cube at 121 points.**

In this paper, we discuss the differences in the measurement accuracy resulting from the differences in the camera position.

Keywords –
Photogrammetry; SfM; Accuracy; Height

1 Introduction

Technological innovations in photogrammetry have progressed over time, and structure from motion (SfM) is a rapidly advancing method for extracting Three-dimensional (3D) models from multiple overlapping photographs. The Ministry of Land, Infrastructure, Transport and Tourism recommends the introduction of ICT technology to civil engineering work, which is referred to as "i-Construction." As a result, civil engineers have employed this technology in surveying tasks such as earthworks management of the amount of soil in drilling and filling work.

Photogrammetry has also been introduced to building construction work, and in particular, it is used in the surveying of buildings and the production of architectural drawings in repair work. However, it is important to evaluate the accuracy of the three-dimensional positions during the 3D modeling of buildings. Therefore, we evaluate the accuracy of 3D model extraction using SfM photogrammetry. Several researchers have investigated the accuracy of 3D models created using SfM photogrammetry [1] [2]. Several researchers also have investigated the accuracy of Unmanned Aerial Vehicle (UAV) photogrammetry [3]. Terrestrial laser scanning (TLS) is also important technology of 3D modelling. According to D. Skarlatos et al. [4], Bundler-PMVS seems to have an advantage in terms of methodology and accuracy in small and medium size objects and distances. On the other hands, TLS is better in terms of quality and processing time in large scale objects. Here, we evaluate the accuracy of the height measurements (Z axis).

In order to extract a 3D model using SfM, photographs of a building exterior are often captured using a drone. The typical procedure of UAV photogrammetry consists of the following steps (see Figure.1):

1. Measure the control points using a total station. The control points are used to extract the 3D model using SfM.
2. Capture photographs of the entire building exterior using a UAV.
3. Produce a 3D surface model from multiple overlapping photographs using SfM software.
4. Draw a 3D solid model or BIM model from the 3D surface model.

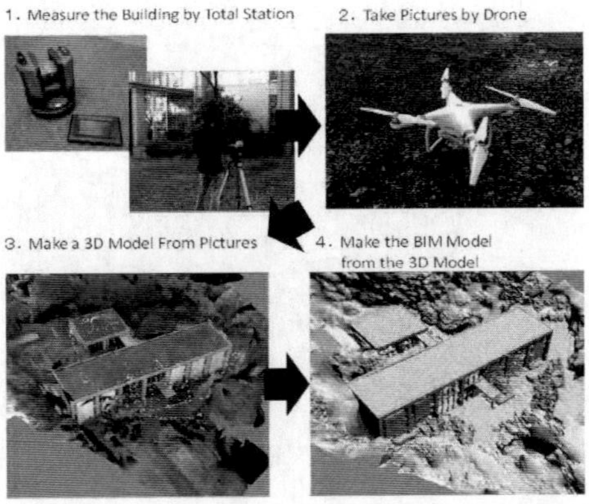

Figure.1 Typical procedure of UAV photogrammetry

In order to create a 3D model with a high accuracy and precision, it is necessary to survey the appropriate flight altitude and turning radius.

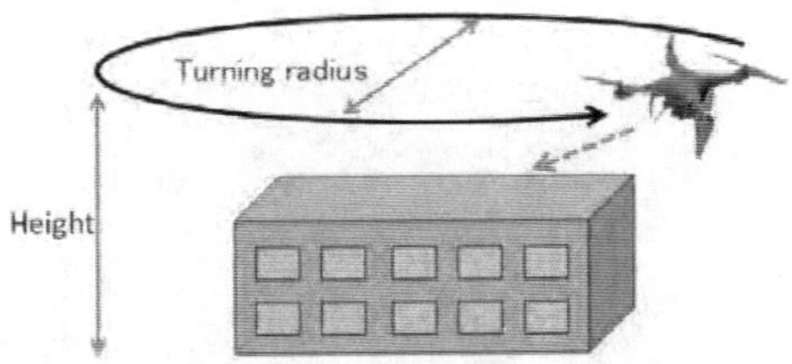

Figure.2 Flight altitude and turning radius

2 Evaluation of the accuracy of 3D model extraction

We evaluate the accuracy of 3D model extraction using SfM photogrammetry. We survey the accuracy of a 3D model using a cube with a side length of 5 cm. We evaluate the differences in the measurement accuracy resulting from the differences in the camera angle and the distance between the subject and the camera.

2.1 Surveying procedure for 3D model accuracy

The distance between the subject and the camera influences the accuracy. In order to investigate this influence, it is necessary to develop a method for capturing the entire exterior of a subject at the same distance and angle. For this, we used a cube with a side length of 5 cm and a turntable. We rotated the turntable in 5° increments and captured 72 photographs.

Our surveying procedure consists of the following four steps:

1. Place the cube on the turntable, rotate the turntable in 5° increments, and capture 72 photographs.
2. Select 12 photographs considering six control points. Edit the control points given their 3D positions in the 12 photographs.
3. Extract the 3D model using SfM software.
4. Import the 3D model and measure the height of the cube at 121 points.

2.2 Photographs of the cube acquired from multiple directions

We acquired photographs to evaluate a 3D model created using photogrammetry. As the subject of the photographs, we created a cube with a known side length of 5 cm. Our camera specifications are listed in Table.1. Figure.3 shows the setup of the camera in front of the cube on a turntable used in this experiment.

Table.1 Camera Specifications

Type	
Camera Name	Nikon 1 J1
Sensor Size	13.2 mm × 8.8 mm
Focal Length	10 mm
Image size	3,872 × 2,592

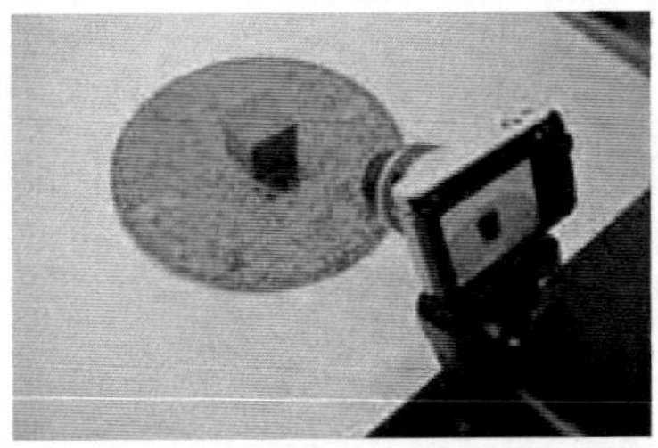

Figure.3 Camera setup in front of a cube on a turntable

We rotated the turntable in 5° increments and captured 72 photographs. Figure.4 shows the positions of the camera and cube used in the experiment. We acquired 24 sets of photographs captured at various camera heights and distances. An example of some photographs is shown in Figure.5.

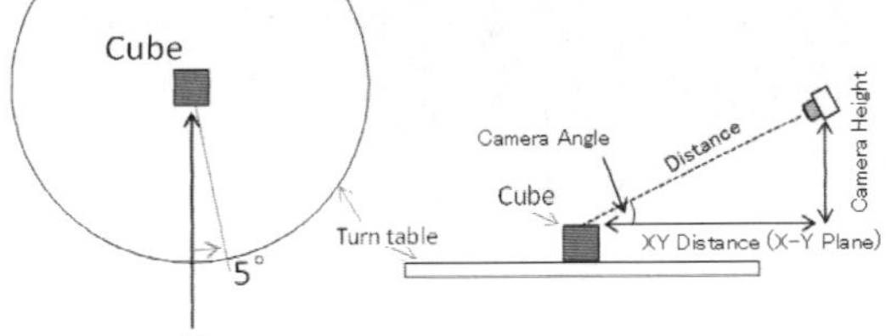

Figure.4 Positions of the camera and cube used in this experiment

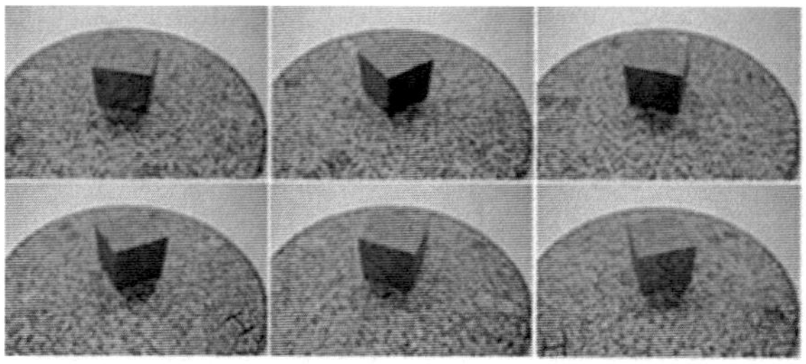

Figure.5 Photographs of the cube from multiple directions

2.3 Selection of control points and checking automatic tie points

SfM software can automatically generate a large number of tie points. However, in order to determine the actual size, we need to define the physical point at which a 3D position is prescribed in the real world. This physical point in the photographs is called the "control point." We determined the control points in different photographs, which correspond to the projections of the same physical point in the scene.

In order to determine the size of the 3D model, we decided to use six control points, as shown in Figure.6. We edited these six control points given their three-dimensional positions in each of the 12 photographs, which consist of three photographs each for four directions. Figure.7 shows an example of the 3D model extracted using SfM software. The blue and red points indicate the camera positions for the 72 photographs, where the red points represent the camera positions of the twelve selected photographs given the six control points.

Figure.8 shows the automatic tie points. We confirmed that a sufficient number of tie points were automatically extracted.

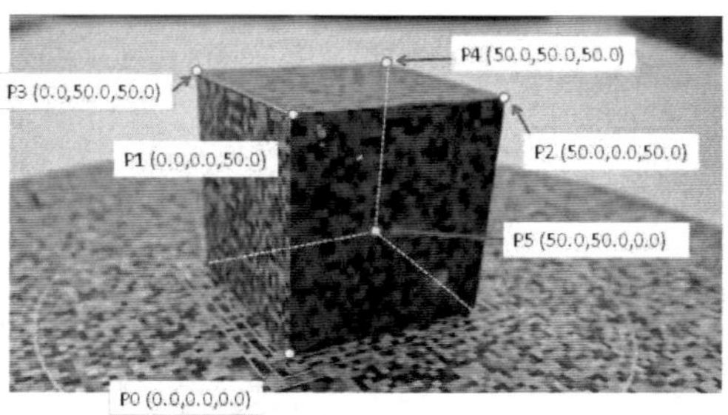

Figure.6 Six control points in this experiment

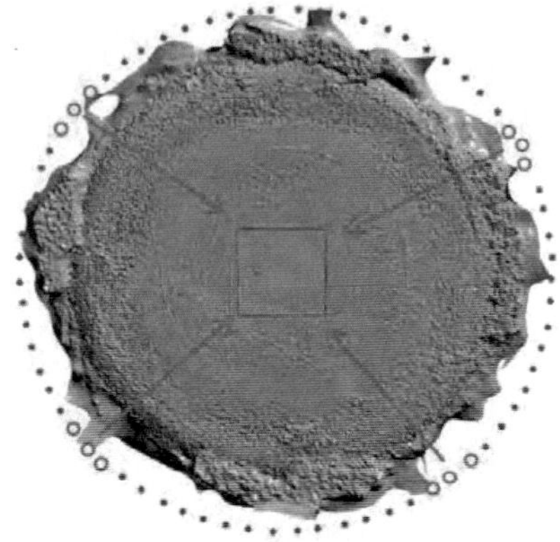

Figure.7 Camera positions for the twelve photographs selected given the six control points

Figure.8 Automatic tie points

2.4 Extraction of the 3D model using SfM software

We then used SfM software to extract the 3D model from the photographs. We used "ContextCapture" to create the 3D models. An example of a 3D model created from the photographs is shown in Figure.9. Figure.10 presents a comparison of the 3D model with a photograph.

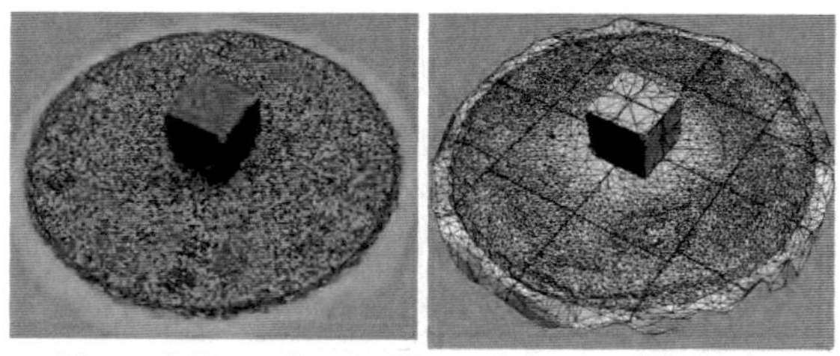

Figure.9 Example of a 3D model created from the photographs

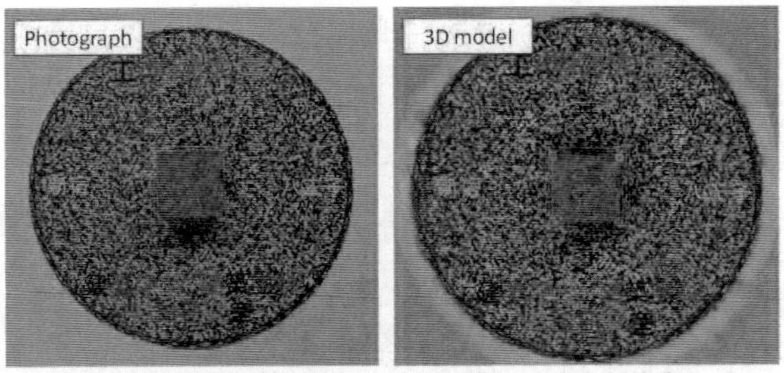

Figure.10 Comparison of the 3D model and the photograph

2.5 Importing a 3D surface model to 3D-CAD

The 3D model was output as STL data and then imported into 3D-CAD. In addition, the camera positions were output as XYZ data and imported into 3D-CAD. Figure.11 shows the 3D model and camera positions in 3D-CAD. The blue points represent the camera positions for the 72 photographs.

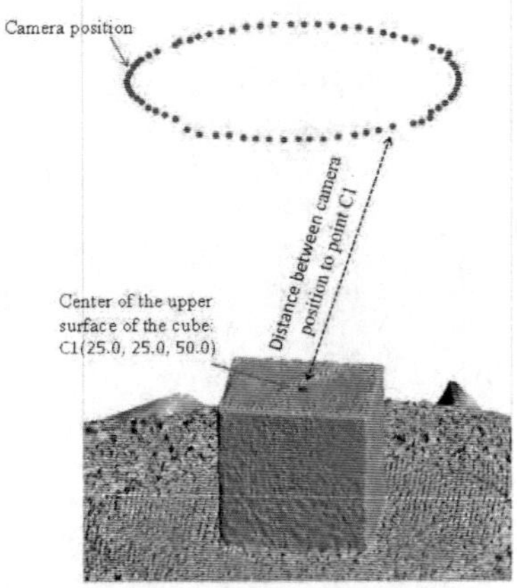

Figure.11 3D model and camera positions in 3D-CAD

We then sliced the cube shown in Figure.11 in the X–Z plane and measured its dimensions. The results are shown in Figure.12. We calculated the distance from each camera position to the center of the top face of the cube (point C1). The coordinates of the point C1 are (25.0, 25.0, 50.0). Figure.13 shows the set of camera positions for the 72 photographs.

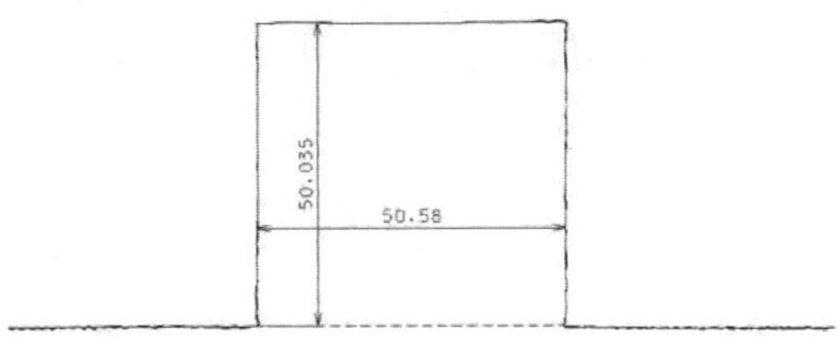

Figure.12 Cube dimensions in the X–Z plane

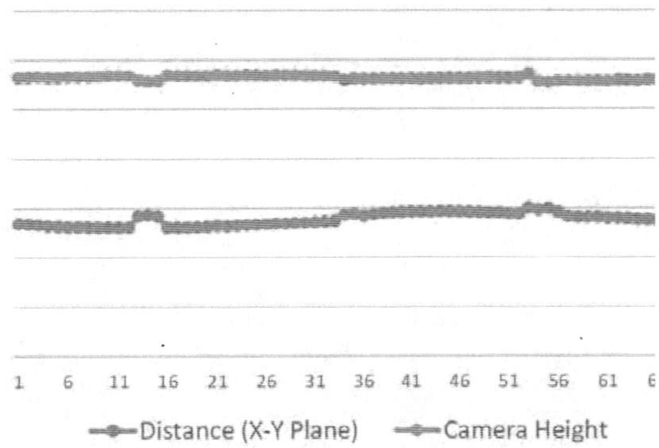

Figure.13 Camera positions for the 72 photographs

2.6 Measurement of the height of the cube

As shown in Figure.14, the 3D model extracted by SfM consists of polygon models. Moreover, the surfaces of the 3D models are not smooth and contain minute irregularities. These minute irregularities lead to variabilities in the cube dimensions. Therefore, in order to eliminate these variabilities in cube dimensions, we decided to measure many points on the cube surface.

Figure.15 shows the measurement positions of the height and width dimensions. The method for calculating the point at which the reference line and surface model intersect consists of the following steps:

1. Select all of the polygons that make up a cube.
2. Remove one polygon and calculate the equation of the plane using three points defining that polygon. (The equation defining a plane is "ax + by + cz + d = 0." We calculated the variables a, b, c, and d.)
3. Compute an intersection point. Substitute the values X and Y into the equation of the plane "ax + by + cz + d = 0" and then obtain the value for Z.
4. Calculate whether the intersection point is inside the polygon.
5. Repeat Steps 2–4 until all polygons are checked.

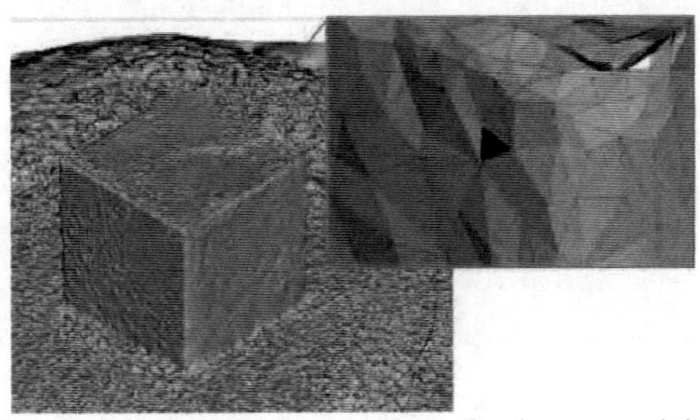

Figure.14 3D model consisting of polygon models

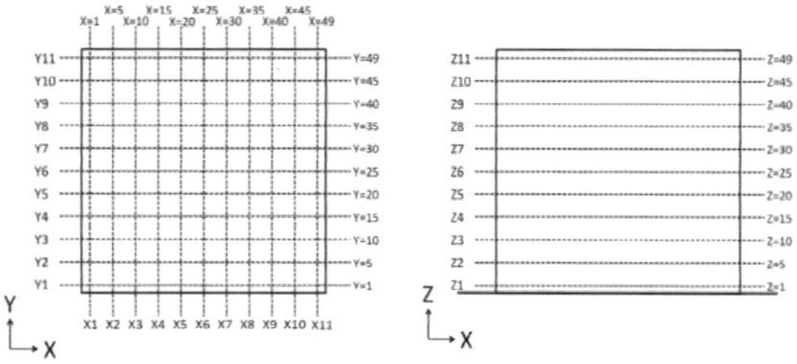

Figure.15 Measurement positions of the height and width dimensions

Figure.16 shows the calculated results for the points at which the reference line and surface model intersect. Table.2 presents an example of the results of the calculation. These 121 points represent the calculation results for the intersection points on the top face of the cube. The camera height for this data is 104.9 mm, the camera angle is 33.8°, and the distance between the camera and the center of the top face of the cube is 188.4 mm. The average value of the height dimension is 50.029 mm, and the sample standard deviation is 0.109 mm.

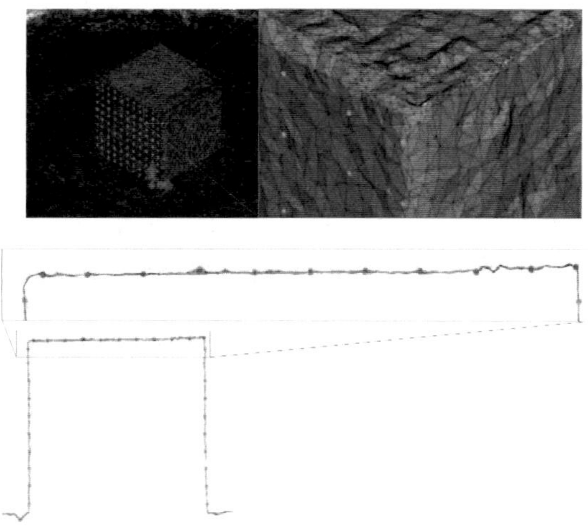

Figure.16 Calculated results for the points at which the reference line and surface model intersect

Table.2 Calculation results for the intersection points on the top face of the cube.

	Y1	Y2	Y3	Y4	Y5	Y6	Y7	Y8	Y9	Y10	Y11
X1	49.830	49.986	49.948	49.925	49.931	50.162	50.353	50.113	49.957	50.053	50.037
X2	49.847	50.059	50.109	50.135	50.101	50.130	50.124	50.102	50.131	50.129	50.126
X3	49.824	50.038	49.959	49.922	49.887	49.897	49.887	49.944	50.018	50.172	50.072
X4	49.901	50.070	49.956	49.899	49.862	49.843	49.861	49.880	49.926	50.128	50.099
X5	49.953	50.107	49.989	49.932	49.918	49.947	49.944	49.964	50.035	50.137	50.074
X6	49.973	50.124	49.979	49.913	49.886	49.906	49.929	49.991	49.963	50.083	50.026
X7	49.982	50.146	50.019	49.926	49.912	49.915	49.917	49.909	49.927	50.023	50.064
X8	49.995	50.177	50.108	49.984	49.953	49.912	49.944	49.931	50.080	50.085	50.103
X9	50.083	50.227	50.118	50.041	49.964	49.950	49.924	50.019	50.028	50.063	50.062
X10	50.199	50.235	50.182	50.132	50.085	50.144	50.109	50.042	50.064	50.078	50.137
X11	50.303	50.248	50.239	50.191	50.150	50.087	50.050	50.021	50.019	50.038	50.180

3 Discussion of measurement accuracy

We extracted 3D models from 26 data sets, each of which consisted of 72 photographs, and then measured the height of each 3D model. The measurement results for the height dimension and standard deviation are presented in Table.3.

Figure.17 shows the accuracy of the polygon model for cube No. 3 and No. 11. The red color indicates a part that is larger than its original size. On the other hand, the blue color indicates a part that is smaller than its original size.

As shown in Figure.18, the 3D models for data sets 24, 25, and 26 consist of cubes without a top face. Therefore, they constitute error data. Table.4 summarizes the ground pixel size for each distance between the camera and the subject.

Table.3 Results for the height dimensions and standard deviation

No.	Dimension		Camera position			
	Height dimension (mm)	Sample standard deviation	XY Distance (mm)	Camera height (mm)	Camera angle (°)	Distance (mm)
1	49.98	0.15	417.2	21.1	2.9	417.7
2	50.11	0.28	224.0	16.5	4.2	224.6
3	50.01	0.13	356.5	35.9	5.7	358.3
4	49.84	0.29	242.9	24.5	5.8	244.1
5	50.01	0.17	203.8	30.4	8.5	206.1
6	50.03	0.12	147.6	22.0	8.5	149.3
7	50.07	0.11	349.4	76.1	12.3	357.6
8	50.13	0.12	347.2	129.6	20.5	370.6
9	50.00	0.13	159.7	64.5	22.0	172.3
10	50.37	0.11	377.7	243.4	32.8	449.3
11	50.03	0.11	156.5	104.9	33.8	188.4
12	50.03	0.13	357.5	292.9	39.3	462.1
13	50.14	0.10	700.3	628.6	41.9	941.1
14	50.28	0.10	153.5	152.8	44.9	216.5
15	50.24	0.10	373.6	397.4	46.8	545.4
16	50.50	0.15	186.2	202.1	47.3	274.8
17	50.09	0.19	608.7	806.4	53.0	1010.3
18	50.81	0.40	59.0	87.2	55.9	105.3
19	50.34	0.10	182.7	291.7	57.9	344.2
20	51.57	0.15	71.9	137.2	62.3	154.9
21	51.47	0.40	55.4	112.7	63.8	125.5
22	51.22	0.18	85.0	183.2	65.1	202.0
23	49.70	0.30	189.8	561.5	71.3	592.7
24	Error		189.4	−22.6	−6.8	190.7
25	Error		182.0	−8.2	−2.6	182.1
26	Error		246.7	16.3	3.8	247.3

Table.4 Ground pixel size in this experiment

Distance between camera and object (mm)	200	300	400	500	600	700	800	900	1000
Ground pixel size (mm/pixel)	0.06	0.10	0.13	0.17	0.20	0.24	0.27	0.30	0.34

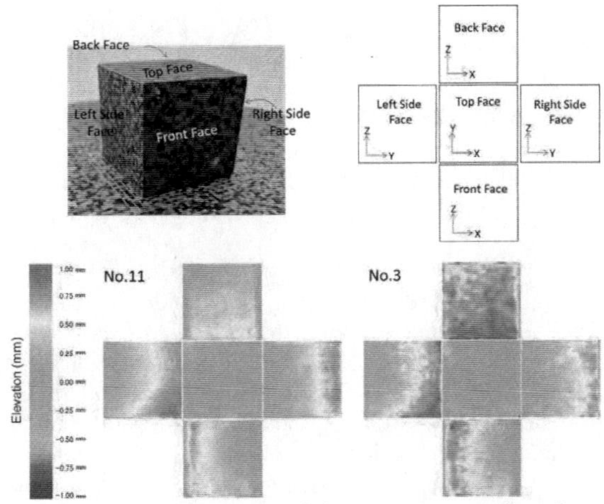

Figure.17 Accuracy of the measurement data for cube No. 3 and No. 11

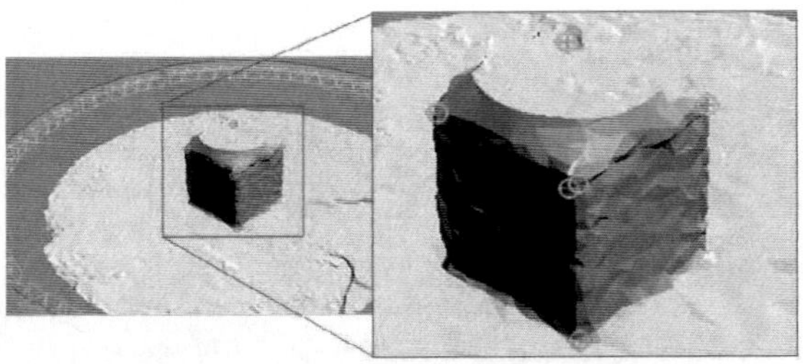

Figure.18 3D model (No. 25) without a top face.

The scatterplot in Figure.19 shows the relationship between the camera angle and the measurement accuracy. The error bars in the scatterplot indicate the sample standard deviation of the height dimension. This scatterplot shows that as the camera angle approaches 90°, the measurement accuracy deteriorates.

The scatterplot in Figure.20 shows the relationship between the measurement accuracy and the distance between the camera position and the center of the top face of the cube.

The camera angle has a stronger influence on the accuracy than the distance to the subject.

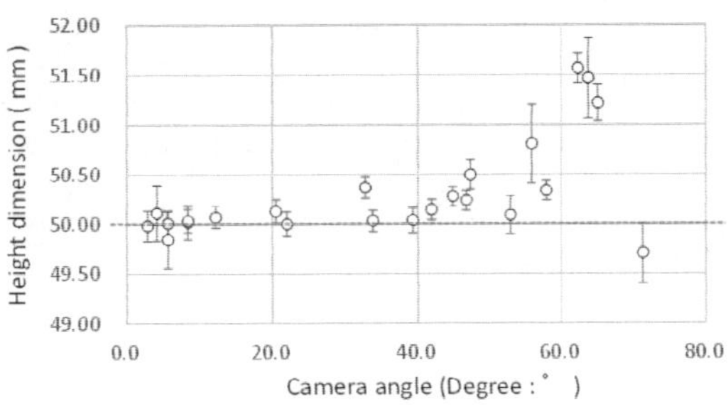

Figure.19 Relationship between the camera angle and the measurement accuracy

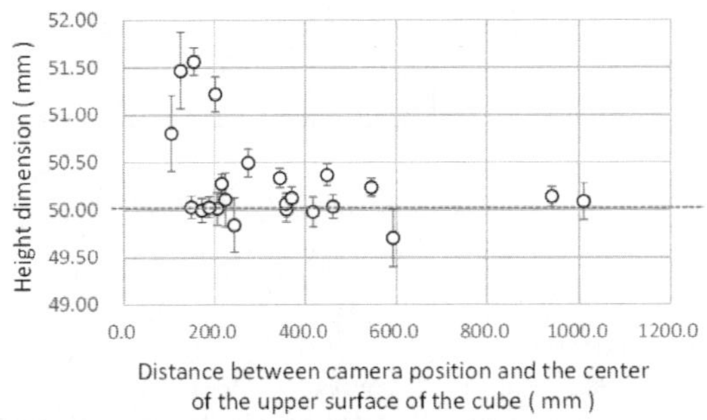

Figure.20 Relationship between the camera distance and the measurement accuracy

4 Conclusion

In this study, we have evaluated the height measurement accuracy of 3D model extraction using SfM photogrammetry. We have surveyed the accuracy of a 3D model using a cube with a side length of 5 cm. We extracted 3D models from 26 data sets, each of which consisted of 72 photographs, and then measured the height of each 3D model. As a result, we have determined that the accuracy of the height dimension is influenced by the camera angle.

References

[1] Micheletti N., Chandler J H., and Lane S N., 2014. Investigating the geomorphological potential of freely available and accessible structure-from-motion photogrammetry using a smartphone, Earth Surface Processes and Landforms, 40 (4), pp.473-486,

[2] Caroti G., Martínez-Espejo Zaragoza I., and Piemonte A., 2015. Accuracy Assessment in Structure from Motion 3D Reconstruction from UAV-Born Images: the Influence of the Data Processing Methods, ISPRS - International Archives of the Photogrammetry, Remote Sensing and Spatial Information Sciences, Volume XL-1/W4, pp.103-109

[3] Marinus Axel Boon, Richard Greenfield, Solomon Tesfamichael, 2016, Unmanned Aerial Vehicle (UAV) Photogrammetry Produces Accurate High-resolution Orthophotos, Point Clouds and Surface Models for Mapping Wetlands, South African Journal of Geomatics, Vol. 5. No. 2, pp.186-200

[4] D. Skarlatos, S. Kiparissi, 2012. COMPARISON OF LASER SCANNING, PHOTOGRAMMETRY AND SFM-MVS PIPELINE APPLIED IN STRUCTURES AND ARTIFICIAL SURFACES, ISPRS Annals of the Photogrammetry, Remote Sensing and Spatial Information Sciences, Volume I-3, Commission III, Melbourne, Australia, pp.299-304

Construction Inspection through Spatial Database

Ahmad Hasan[a], Ashraf Qadir[a], Ian Nordeng[a] and Jeremiah Neubert[a]

[a]Department of Mechanical Engineering, University of North Dakota, United States of America
E-mail: ahmadjarjis.hasan@und.edu, ashraf.qadir@und.edu, ian.nordeng@und.edu, jeremiah.neubert@und.edu

Abstract –

This paper presents a novel pipeline for development of an efficient set of tools for extracting information from the video of a structure, captured by an Unmanned Aircraft System (UAS) to produce as-built documentation to aid inspection of large multi-storied building during construction. Our system uses the output from a Simultaneous Localization and Mapping system and a 3D CAD model of the structure in order to construct a spatial database to store images into the 3D CAD model space. This allows the user to perform a spatial query for images through spatial indexing into the 3D CAD model space. The image returned by the spatial query is used to extract metric information. The spatial database is also used to generate a 3D textured model which provides a visual as-built documentation.

Keywords –

Computer Vision; Construction Inspection; Spatial Query; Metric Data Analysis

1 Introduction

The objective of our system is to reduce the amount of labor and time required to collect important construction site information, and enable engineers to generate as-built documentation of building elements. Generation of as-built documentation requires engineers to search through large amounts of video of the construction site for a specific region of interest. This is time-consuming and an inefficient method. The proposed system provides a credible solution to that problem by aligning a spatial database to the 3D CAD model in order to facilitate a spatial query for the images desired. This is done by allowing the user to click on the region of interest on the 3D CAD model.

The system takes a previously generated 3D CAD model and the images collected with the UAS, which may be processed by a variety of keyframe based Simultaneous Localization and Mapping (SLAM) [1,2,3] systems, as input. The SLAM system is used to estimate camera poses and generate a 3D point cloud map of the structure, which forms the spatial database. Each map point in the point cloud stores the 3D pose in world coordinate m^w, its index, and a reference to the index of the source keyframe where it was first detected. Tagging the index of the 3D point to the index of the keyframe provides the spatial index of the database. Our system takes the point cloud map and aligns it to the 3D CAD model to update the geometry of the spatial database. This allows the user to perform a spatial query through the 3D CAD model by use of mouse clicks on the model to search for the desired images. Our system also provides a metric data analysis tool to analyze the queried image. Finally, a textured 3D model of the structure is generated to serve as an overall visual as-built documentation of the structure.

Building inspection for quality assurance often involves analyzing images and detecting any anomaly or defect in the construction. For a multi-storied large building, it is necessary that the images are collected from a close range for proper visualization and inspection. Using a UAS allows us to collect high resolution building images from close range, and perform detailed image processing and analysis for anomaly detection. This was necessary as ground based images would not be able to provide the required fidelity to perform accurate building inspections.

The tools that have been developed to generate as-built documentation are:

Spatial Query for Images: Our contribution is to allow a user to perform an efficient spatial query for images by aligning the database to the 3D CAD model. The user may easily visualize the 3D CAD model, making our implementation convenient for anyone to search for images of a specific region of interest without having to search through time intensive videos.

Metric Data Analysis: Our system accurately calculates the metric distance between any two places of the as built structure. A 3D CAD model-image correspondence is used to ensure the robustness of the calculation. To facilitate a better experience, our system has a magnifier which works on mouse over to ensure a higher precision for selecting the desired pixel.

Visualization through 3D textured model: The proposed system also generates a 3D textured model with high-resolution images to create a virtual reality of the scene. The virtual reality is visualized with an OpenGL window implemented with a moving camera to allow the user to move around the scene. The 3D textured model enables engineers or facility mangers to visually assess the building and track construction progress.

2 Related Research

There have been more than a million small UAS sold in the United States over the past few years according to news reports [4, 5]. As the use of small UAS grows the need for cost-effective methods for accessing and processing data will grow.

Contractors are already using small UAS to gather information about their worksite and inspect structures. In [6, 7] small UASs are used to construct a detailed 3D map of work sites. Others have used small UAS to inspect existing structures [7]. In [8] they outline the potential applications of UAS in the new construction such as monitoring the build process, creating "as-built" documentation and automated defect detection. Researchers are currently working on creating the algorithms needed to exploit this potential. One such system presented in [9] was used to aid in the creation of "as-built" documentation. Their work generated reasonable output, but the dimensions had more than 5% error.

The D4AR modeling [10] uses an unordered collection of images of the structure to generate the underlying geometric model by using a Structure-from-Motion (SfM). They solve the similarity transform between the model and 3D point cloud found from the SfM using minimum user inputs to transform the SfM coordinate system to the 3D CAD model's coordinate system to allow the aligning the SLAM photos to the CAD model. There are popular methods being used to create 3D mesh models by using images. Such methods have been used in [11, 12, 13, 14, and 15].

A different approach was taken by [16]. Instead of creating a 3D mesh model using images, they rely on an existing semantic 3D CAD model known as Building Information Model (BIM). This modeling is widely available nowadays to facilitate easier construction as it provides prior detailed information about the building or structure to be constructed.

Methods described in [10, 11, 12, 13, 14, 15] use images to create underlying geometry. Such methods are not useful for construction sites because it only provides visual information of an already built structure. Construction sites require a pre-designed CAD model or blueprints to be able to have ground truth for the construction and compare it to the as-built structure to detect anomalies. Considering this, our method uses a 3D CAD model designed with AutoCAD Revit as [16].

In this work, out design considerations are governed by the need of engineers on a construction site. PCL construction provided us with a checklist which specifically requested the need for an accurate metric data analysis as a priority. Another important tool required in that checklist was the need for a user-friendly tool to easily search for images through the 3D CAD model.

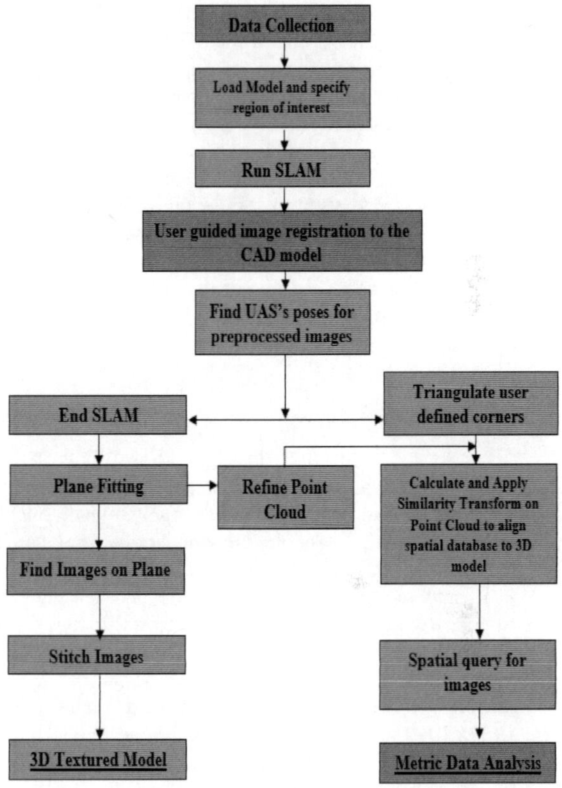

Figure 1. The process flow chart for system overview. Blue boxes indicate processes that require no user interaction. Red boxes indicate steps where user input is required.

3 System Overview

Figure 1 outlines our system. First, the user specifies the region of interest for inspection from the model (Section 5.1). The system leverages user input to register the first keyframe to the 3D CAD model. Our system prompts the user to specify a four-point correspondence between the model and the 1st keyframe at the beginning to aid the registration process (Section 5.1). This registration explores the pose of the UAS in model coordinates using the p3p algorithm (Section 5.1). The system takes a 3D point cloud map from the SLAM as input. The point cloud is refined with a plane-fitting algorithm (Section 5.3). A similarity transform is applied to the point cloud to transform the point cloud to the 3D CAD model coordinate system (Section 5.2) thus updating the geometry of the spatial database. The updated spatial database uses the spatial index to run a query for images of the desired region specified by the user (Section 5.4). The image returned from the spatial query may further be used to evaluate distance in the as-built structure. To generate the 3D textured model, our system retrieves images on the planes and stitches them together to texture map those images to the specified zone of the textured model on a different window (Sec 5.6). The metric information is mainly intended to find dimensions of various entities in the as-built structure (Section 5.5). The user must click on different corners of the model to inspect the size of that specific entity i.e. windows, doors, the length of a column etc. After clicking on the model, the corresponding image of that area will be seen, and the user will specify the desired dimension with mouse clicks.

4 Data Collection

Our system utilized the DJI Spreading Wings S1000+ as our UAS. One of the most important features of the S1000+ is a low gimbal mounting bracket which enables a wide range of possible viewing angles and camera motions [17]. It was fully compatible with the Zenmuse Z15 camera gimbals from DJI which stabilizes the camera to the desired orientation during flight. A Canon EOS 5D Mark III camera has been used to capture video of the scene. The camera has a 22.3 Megapixel CMOS sensor. High-Definition video with a resolution of 1920×1040 was collected at 30 frames per second.

The SLAM requires a calibrated camera system. Calibration images were taken before every flight. The 3D CAD model used was loaded in the system using Open Asset Import Library (Assimp) and OpenGL.

5 Methodology

Our system requires inputs from the user for registering images to the 3D CAD model. In this section, the methodology of the system has been described. The image registration process, alignment of 3D CAD model to the spatial database, and spatial query has been explained in detail.

With the registration of the images, plane fitting is performed which facilitates further processing such as 3D textured reconstruction, and metric information extraction.

5.1 Image Registration to 3D CAD Model

The user must provide four 3D-2D point correspondences between the 3D CAD model and the first keyframe. The Perspective-Three-Point problem determines the position and orientation of the camera while capturing the first keyframe in the model coordinate system. The p3p algorithm provides up to four solutions which are disambiguated by using a fourth point [18]. The user clicks on four corners of an entity, such as a window, from the model. That entity must be seen in the first keyframe so that the user can then click on corresponding four corners of the same entity from the image in a sequential order.

5.2 Alignment of Spatial Database to the 3D CAD Model

Initially the 3D point cloud, and the camera poses generated by the SLAM are expressed in the camera coordinate system with the first camera pose set as the identity. In order to align the spatial database to the 3D CAD model, the 3D point cloud along with the camera poses need to be transformed to the CAD model coordinate system from the camera coordinate system. Since monocular SLAM suffers from scale ambiguity, it is required to find the metric scale for the SLAM generated map that matches the scale of the 3D CAD model that has been provided.

Transformation between the CAD model coordinate system and the camera coordinate system is performed by first creating a set of 3D-3D points correspondences between the two coordinate systems, and then by computing the similarity transforms between the 3D point correspondences. That similarity transform is applied to the point cloud and camera poses to transform them to the 3D CAD model coordinate system.

To create the 3D-3D correspondences, the points selected by the user as described in section 5.1 were used. The selected points on the CAD model constitute the 3D points set in the CAD model coordinate system. Now we need their corresponding 3D points in the camera coordinate system. This is done by using the 2D image points selected by the user in the first keyframe. Since it is not guaranteed that the selected 2D points correspond to any 3D point in the SLAM generated map, their 3D positions are estimated by finding their 2D correspondences in the second keyframe. As our system takes an ordered photo collection, the specified entity should appear in the second keyframe too. Our system creates small 11×11 patches around the selected points of the first keyframe. Then a normalized cross-correlation based matcher finds the corresponding four 2D points in the second keyframe. Once a set of 2D-2D points correspondences are found, they are triangulated to estimate their 3D positions in the camera coordinate system.

Once the set of 3D-3D correspondences are found between the CAD model coordinate and camera coordinate systems, they are used to compute the similarity transform,

$$T = \begin{bmatrix} R & \vec{t} \\ \vec{0} & s \end{bmatrix} \quad (1)$$

between the CAD model and the camera coordinate system. The similarity transform is computed using our implementation of the Horn's method [19]. The similarity transform is then used to transform the point cloud and the camera poses to the CAD model coordinate system.

5.3 Plane Fitting

Assuming the structure to be piecewise planar, our system implemented a plane fitting algorithm for both assigning points to planes as well as reject outliers. A RANSAC [20] based plane fitting method was implemented that uses a voting scheme for assigning points in the 3D point cloud to individual planes. For a set of 3D data points $\{P_i(x_i, y_i, z_i); i = 1, \ldots.., N\}$, where N is total number of 3D points in the point cloud, the plane equation has been defined as,

$$ax + by + cz + d = 0 \quad (2)$$

where, a, b, c are slope parameters and d is the of the plane from the origin.

Planes are extracted by randomly constructing different planes from point cloud data. Three random points are sampled from the point cloud and checked for collinearity. If collinear or coincident, new points are considered. A plane hypothesis is created, and this process is repeated for a predefined number of times. The resulting candidate planes are scored against all points in the cloud to validate the candidate plane. In a candidate plane, the points that falls onto that plane, votes for the plane. The total vote for a candidate plane is the score of that plane. After a predefined number of trials, the candidate plane having the highest score is validated as a plane. Points voting for the valid plane are tagged to that plane and removed from the plane fitting consideration. The procedure is then repeated on the remainder of the point cloud to find subsequent planes. The planes are finally made robust using least-square constrains. All the points not included in any plane are considered as outliers and removed from the cloud. Figure 2(a) shows the region selection for inspection and 2(b) shows the point cloud generated with SLAM after being refined with plane fitting.

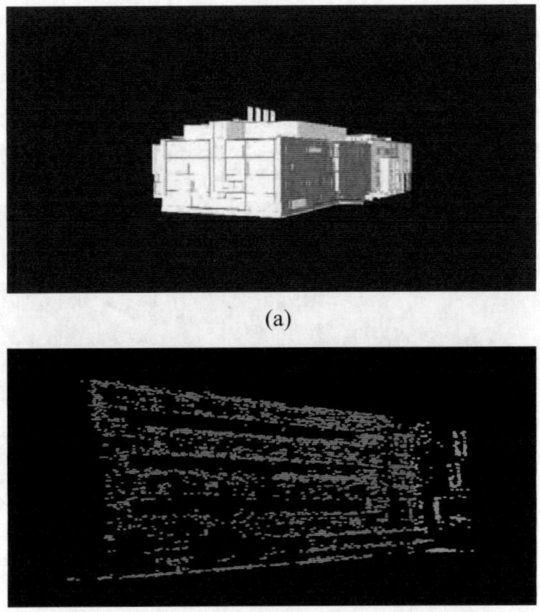

(a)

(b)

Figure 2. (a) The blue area indicates the region selected for inspection (b) Point cloud generated by SLAM and refined with plane fitting.

5.4 Spatial Query for Images

As the source keyframes are tagged along with each point in the cloud, a spatial index of images is used for efficient spatial image query. The entries in the spatial index depend on the vertices location in the model coordinate system.

The user must click on the area or entity on the visualization window for inspection. A ray casting method was used to select the 3D location of the mouse click in the model. The selected 3D point of the model was snapped to the nearest vertex, so that the user does not have to click precisely. The nearest point of the point cloud to the selected vertex is selected as the entry to spatial index for query and the associated source keyframe to that point is returned.

5.5 Metric Data Analysis

Our system assumes the user desires distance from one vertex to another to check the dimensions of different entities. By using the spatial query for an image, the user finds the image of the entity to be inspected. A binary image of the queried image is created and contours are found by using [21]. A Doughlas-Peucker [22] algorithm has been used to find rectangular shapes from the contours found. The contour area is calculated to make sure the area is larse enough to be considered as a window.

The actual height of the window is taken during alignment of the spatial database to the 3D CAD model. The pixel distance of the height of the window then sets a scale factor. The user clicks on two points of the image to find the 3D distance and the nearest windows scale factor is used to find the 3D distance. A magnifier has been implemented to aid the accuracy of clicking on the image. Figure 3 shows the detected windows by using this system.

Figure 3. Automatically detected windows as described in section 5.5.

5.6 3D Textured Model Generation

All keyframes for each plane are then stored in a vector to be stitched together as [23]. Due to a high number of key-points our system takes in every tenth image from the keyframe vector and stitches them together. The user must specify a required area to be inspected during image registration to the 3D mesh model. That process takes in the boundaries for each plane in the specified region. That boundary is used to texture map the stitched image to its corresponding plane. A photorealistic 3D texture model is created in this manner.

6 Experimental Result

In this section, we compare our proposed algorithm against the latest version of the state of the art Pix4D mapper (version 3.1) [24]. The primary focus of this research is to incorporate a BIM to access the point cloud which provides us a prior knowledge on how the building was intended to be constructed. The CAD model provides entry to the spatial database eliminating the need to perform a dense reconstruction as Pix4D. We will show that not having a dense reconstruction does not affect our metric data calculations while and outperforming Pix4D.

The point cloud generated by the SLAM have been aligned to the 3D CAD model by a similarity transform. Each point has a keyframe associated with it, which creates the spatial index of images. The user clicks on the desired entity. The nearest point from the point cloud to the vertex is selected, working as an entry to the spatial index for query and efficiently searches for the spatial image data. Figure 4 shows some of the outputs using the spatial query.

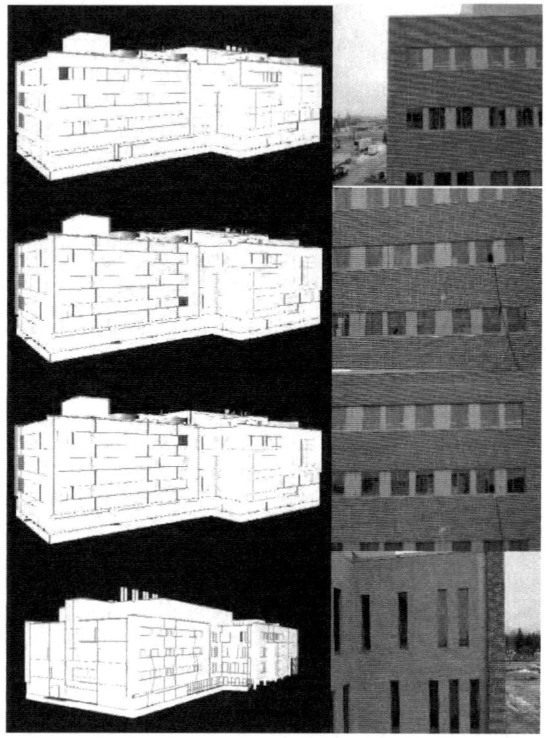

Figure 4. Spatial query: The window to be inspected in the model is marked with blue square and the image of that region found through the spatial query.

After a query the user may choose to check dimensions of various entities which is the metric data analysis in our system. We have compared our dimensional calculations to Pix4D's calculations. Our system showed significant improvements over Pix4D in Metric data extraction. However, Pix4D requires geotagged images to assign scale and orientation. It is not always possible to have geotag information with images, so our algorithm was delepoed keeping this in mind and geotags was not included in the proposed algorithm. As our system does not require geotagged images, we provided the scale manually to Pix4D. To apply a scale constraint into Pix4D the recommended process is to click on both vertices from the dense point cloud that Pix4D generates and provide the accurate metric distance. Moreover, it is recommended by Pix4D to correct the vertices in at least two of the corresponding images. Our system applies scale by using two consecutive images. As a direct comparison, we applied scale to the Pix4D model using the dense point cloud and two images for scale correction.

Pix4D does recommend more scale constraints as it will provide better accuracy. This is true for both Pix4D, as well as the proposed algorithm. Figure 5 below provides the process to apply scale using Pix4D mapper.

Figure 5. Applying scale constraint on Pix4D mapper.

After applying scale constraints, we have tested the height and width of 15 windows providing 60 total distances for each method. For verification, these same distances were manually measured. The actual width and height were 2.01168 meters(m) (6.6 feet) and 1.8288 meters(m) (6 feet). The CAD model provides these distances as 1.8288 meters for both width and height. The scale was applied based on the height of the windows, as provided by the CAD model, on both the proposed model and Pix4D. We will show that the proposed system and Pix4D both determine the building was not constructed per the CAD model dimensions.

The proposed model resulted in a mean squared error (MSE) of 31.9 cm^2 whereas Pix4D mapper's MSE was of 45.6 cm^2. For Pix4D's width calculation, it has a standard deviation of 4.92 cm, as opposed to the proposed system's standard deviation of 4.28 cm. For height calculation, Pix4D's standard deviation resulted in 4.17 cm, where the proposed algorithm provided a standard deviation of 3.27 cm. Pix4D had a combined standard deviation of 6.45 cm where the proposed algorithms combined standard deviation was 5.39 cm. Figure 6 provides the width and height calculations along with the actual.

A t-test was performed on the width and height averages to verify the statistical significance of the calculated errors. Tables 1 and 2 show the resulting t-tests from the two-sample width and height data assuming unequal variances. In both cases, an $\propto$ of 0.01 was used. This shows we are 99% confident that the proposed method is statistically different from Pix4D. The null hypotheses for both cases were set as there are no significant difference.

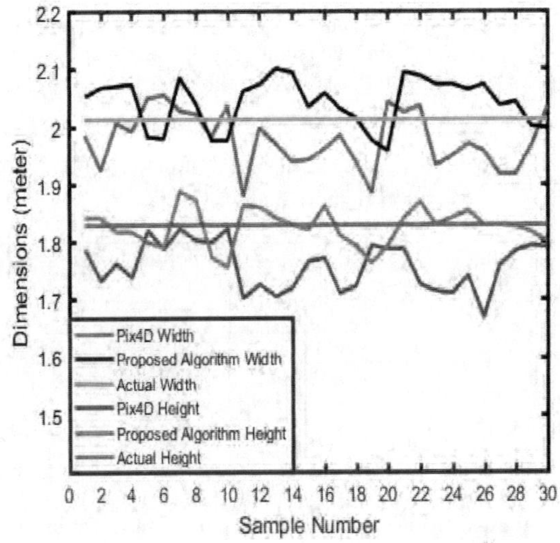

Figure 6. Dimensions calculated with Pix4D and Proposed Algorithm along with theit actual values.

Table 1. t-test results on two sample widths assuming unequal variances

	Pix4D Width	Proposed Algorithm Width
Mean	1.97673	2.04084
t Stat	-5.38338	
t Critical 2 tail	2.66487	
P two-tail	1.4E-06	

Table 2. t-test results on two sample heights assuming unequal variances

	Pix4D Height	Proposed Algorithm Height
Mean	1.75758	1.82494
t Stat	-6.95956	
t Critical 2 tail	2.66822	
P two-tail	4.4E-09	

Due to the p_{value} for both resulted in a much lower value than the $\propto$ value, the null hypothesis can be rejected. We can therefore state that the mean average in the proposed algorithm were significantly closer to the actual values than Pix4D.

An analysis of variance (ANOVA) was performed on the variances of the data analyses to verify the statistical significance of the sample variables. Table 3 provides the ANOVA output.

Table 3. Two-factor ANOVA with replication results on both width and height with $\propto = 0.01$

Source of Variation	F	P-value	F Critical
Sample	73.39466	5.25E-14	6.858521
Width/Height	803.691	5.6E-54	6.858521
Interaction	0.044844	0.832588	6.858521

Pix4D and the proposed algorithm are represented as the two samples. The p_{value} $(sample)$ was lower than $\propto$. The $F_{Crirtical}$ $(sample)$ is lower than $F(sample)$ showing both samples have significant differences. The second row in Table 3 represents the effect of both samples on width and height calculations and the $p_{value}(width/height)$ is much less than $\propto$. Therefore, we can reject the null hypothesis and conclude both systems are significantly different on the width and height calculations. The p_{value} $(interaction)$ calculated a value greater than $\propto$, showing we can conclude the interaction between width and height have no significant difference, meaning the effect of width or height does not depend on one another.

We can therefore assume the proposed algorithm provides significantly more accurate results than Pix4D data analysis.

Another tool developed was for visualization through 3D textured model. To visualize the 3D texture model, the OpenGL library was used with a moving camera implementation. This allows the user to roam through the 3D textured model to look for any visual anomalies. We have compared the proposed system to Pix4D mapper's textured reconstruction. For a fair comparison, we have selected the high resolution texture mapping option for Pix4D. Figure 7 shows some of the images taken from the proposed systems 3D textured modelling.

Figure 7. 3D texture model generated with the proposed system from various camera locations

Figure 8 shows the differences between the proposed system and Pix4D's 3D texture models.

Figure 8. Left Column: Pix4D's textured model
Right Column: Proposed Algorithm's textured model

Pix4D's textured reconstruction results in clearly visible holes and artifacts. Windows are not realistic and provide visible anomalies. Straight lines are distorted resulting in wavy lines, unusable for detailed visual inspections. The proposed method does not have any of these artifacts resulting in a photorealistic rendition.

7 Conclusion

In this system, we have successfully demonstrated spatial query into the spatial database through the provided 3D CAD model. We have performed metric data analysis along with visualization through a 3D textured model. A comparison with the state of the art Pix4D mapper was given where our proposed system has been proven to significantly improve the metric data analysis and provided better and photorealistic 3D textured model. Our system is easy to use and does not require the user to have any previous knowledge of visualization, rendering or CAD software. . Interesting steps towards further research could be window detection with deep convolutional neural networks, automatic scheduling monitoring and temporal navigation using the 3D CAD model, and optimizing the system to implement on an on-board environment or mobile devices.

References

[1] Qadir, A. "A Large Scale Inertial Aided Visual Simultaneous Localization and Mapping (SLAM) System for Small Mobile Platforms," Ph.D. dissertation, Dept. Mech. Eng., Univ. of North Dakota, Grand Forks, ND, 2016.

[2] Klein, G., & Murray, D. (2007, November). Parallel tracking and mapping for small AR workspaces. In *Mixed and Augmented Reality, 2007. ISMAR 2007. 6th IEEE and ACM International Symposium on* (pp. 225-234). IEEE.

[3] Mur-Artal, R., Montiel, J. M. M., & Tardos, J. D. (2015). ORB-SLAM: a versatile and accurate monocular SLAM system. *IEEE Transactions on Robotics*, *31*(5), 1147-1163.

[4] Lowy, J. "Drone sightings up dramatically (update)," 2014. [Online]. Available: http://phys.org/news/2014-11-drone-sightings.html. dramatically/SrrcKzjnphejy4Zvf3OaEI/story.html.

[5] Jansen, B. "FAA: Drone sightings on pace to quadruple this year," USA TODAY, 2015. [Online]. Available: http://www.usatoday.com/story/news/2015/08/13/drone-sightings-faa-newark/31614151/.

[6] Siebert, S., & Teizer, J. (2014). Mobile 3D mapping for surveying earthwork projects using an Unmanned Aerial Vehicle (UAV) system. *Automation in Construction*, *41*, 1-14.

[7] Eschmann, C., Kuo, C. M., Kuo, C. H., & Boller, C. (2012, July). Unmanned aircraft systems for remote building inspection and monitoring. In *6th European workshop on structural health monitoring* (pp. 1-8).

[8] Yue, K., Huber, D., Akinci, B., & Krishnamurti, R. (2006, June). The ASDMCon project: The challenge of detecting defects on construction sites. In *3D Data Processing, Visualization, and Transmission, Third International Symposium on* (pp. 1048-1055). IEEE.

[9] Klein, L., Li, N., & Becerik-Gerber, B. (2012). Imaged-based verification of as-built documentation of operational buildings. *Automation in Construction*, *21*, 161-171.

[10] Golparvar-Fard, M., Peña-Mora, F., & Savarese, S. (2011). Integrated sequential as-built and as-planned representation with D 4 AR tools in support of decision-making tasks in the AEC/FM industry. *Journal of Construction Engineering and Management*, *137*(12), 1099-1116.

[11] Snavely, N., Seitz, S. M., & Szeliski, R. (2006, July). Photo tourism: exploring photo collections in 3D. In *ACM transactions on graphics (TOG)* (Vol. 25, No. 3, pp. 835-846). ACM.

[12] Debevec, P. E., Taylor, C. J., & Malik, J. (1996, August). Modeling and rendering architecture from photographs: A hybrid geometry-and image-based approach. In *Proceedings of the 23rd annual conference on Computer graphics and interactive techniques* (pp. 11-20). ACM.

[13] Sinha, S. N., Steedly, D., Szeliski, R., Agrawala, M., & Pollefeys, M. (2008, December). Interactive 3D architectural modeling from unordered photo collections. In *ACM Transactions on Graphics (TOG)* (Vol. 27, No. 5, p. 159). ACM.

[14] Xu, K., Zheng, H., Zhang, H., Cohen-Or, D., Liu, L., & Xiong, Y. (2011, August). Photo-inspired model-driven 3D object modeling. In *ACM Transactions on Graphics (TOG)* (Vol. 30, No. 4, p. 80). ACM.

[15] Colburn, A., Agarwala, A., Hertzmann, A., Curless, B., & Cohen, M. F. (2013). Image-based remodeling. *IEEE transactions on visualization and computer graphics*, *19*(1), 56-66.

[16] Karsch, K., Golparvar-Fard, M., & Forsyth, D. (2014). ConstructAide: analyzing and visualizing construction sites through photographs and building models. *ACM Transactions on Graphics (TOG)*, *33*(6), 176.

[17] D. A. R. Reserved, "Spreading wings S1000+," 2016. [Online]. Available: http://www.dji.com/spreading-wings-s1000-plus.

[18] Kneip, L., Scaramuzza, D., & Siegwart, R. (2011, June). A novel parametrization of the perspective-three-point problem for a direct computation of absolute camera position and orientation. In *Computer Vision and Pattern Recognition (CVPR), 2011 IEEE Conference on* (pp. 2969-2976). IEEE.

[19] Horn, B. K. (1987). Closed-form solution of absolute orientation using unit quaternions. *JOSA A*, *4*(4), 629-642.

[20] Fischler, M. A., & Bolles, R. C. (1981). Random sample consensus: a paradigm for model fitting with applications to image analysis and automated cartography. *Communications of the ACM*, *24*(6), 381-395.

[21] Suzuki, S. (1985). Topological structural analysis of digitized binary images by border following. *Computer vision, graphics, and image processing*, *30*(1), 32-46.

[22] Douglas, D. H., & Peucker, T. K. (2011). Algorithms for the Reduction of the Number of Points Required to Represent a Digitized Line or its Caricature. *Classics in Cartography: Reflections on Influential Articles from Cartographica*, 15-28.

[23] Brown, M., & Lowe, D. G. (2007). Automatic panoramic image stitching using invariant features. *International journal of computer vision*, *74*(1), 59-73.

[24] "Generate 2D and 3D Information, Purely from Images with Pix4D." Pix4D. N.p., n.d. Web. 09 Apr. 2017 retrived from. https://pix4d.com/

34th International Symposium
on Automation and Robotics
in Construction

ISARC 2017

June 28 – July 1
TAIPEI, TAIWAN

LASER SCANNING / BIG DATA

Developing a Pattern Model of Damage Types on Bridge Elements Using Big Data Analytics

S. Lim[a*], S. Chung[a], S. Chi[a]

[a]Department of Civil and Environmental Engineering, Seoul National University, Republic of Korea

E-mail: sorami@snu.ac.kr*, hwani751@snu.ac.kr, shchi@snu.ac.kr

Abstract

The number of over 30–year-old bridge structures has increased rapidly in Korea. Due to the lack of maintenance budget and professional inspectors, the demands for more effective and cost efficient bridge condition monitoring solutions have increased. The primary purpose of this study is to develop a model using big data analytics to recognize bridge damage patterns that show the relationships between bridge-related variables and damage types on different bridge elements. This research covered the total of 6,773 bridges in Korea and analyzed Bridge Management System (BMS) data with weather and contractor-related variables brought from the outside of the BMS database. After preprocessing, key predictors (i.e., independent variables) were selected by the association rule discovery algorithm and then damage patterns were extracted by decision tree. The pilot study results with the data originated from three cities in Korea, Ulju-gun, Inje-gun, and Mungyeong-si, showed that different predictors derived by region, and the extracted patterns implied geographical characteristics such as heavy snow and different construction capacities of contractors. The derived patterns were expected to give bridge inspectors prior information about the primary inspection area.

Keywords

Bridge Management, Infrastructure Maintenance, Damage Pattern, Big Data analytics, O&M

1 Introduction

Bridge plays a significant role in public transportation networks but its damage can threat public safety and hinder economic activity. Timely bridge management is needed not only to keep operating services but also to ensure traffic safety. Since a number of bridges were built with rapid economic growth all over the world, nowadays economically developed countries such as the U.S. and Korea are suffering from the substantial number of over 30–year-old bridge structures, the aged structures. The average age of the U.S.'s 607,380 bridges was 42 years old in 2016 [1], and the aged bridges in Korea is expected to increase up to over three times in 2025 from 3,094 in 2015 [2].

The Korean government legislated on the Special Act on Safety Control for Infrastructure in 1995 after the collapse of Seongsu Bridge in Seoul in 1994. The aim of the Act is to enforce periodic inspection of major infrastructure facilities to provide on-time repairs according to the infrastructures' conditions graded by "A" (excellent) to "E" (poor). The grades of "C", "D" and "E" indicate that the structure needs to get repaired. Such visual periodic inspection is manually performed every six month, and further detailed inspection and precise safety diagnosis including structural analysis and safety assessment are performed every two to six years according to the safety grades. However, due to the limited number of budget, time, and professional inspectors, the quality of such periodic inspection becomes less objective and not guaranteed.

To support the limitation of the manual inspections, researchers have developed condition monitoring approaches of bridge structures. One of traditional real-time monitoring methods is sensor-based structural health monitoring (SHM). This method keeps monitoring sensing data obtained from the attached sensors of the bridge and detects outliers such as displacement captured from strain gage sensors while explaining abnormal conditions. The Golden Gate Bridge in 2006 and Jindo Bridge in Korea are widely known as application examples of such sensor-based approaches [1]. Other researchers investigated a bridge condition analysis model by analyzing bridge inspection data. They normally focused on the development of bridge condition deterioration models by applying regression and Markov models [3-4].

Although the previous research showed potential benefits for preventive maintenance, they failed to explain detailed mechanism for damage causation of individual bridge elements. Moreover, the sensor-based monitoring approaches have difficulties from detecting damages that do not cause structural movement changes such as concrete delamination and clogged drains. Even though a main element of the bridge is broken, it sometimes cannot be detected since the structural is normally designed with safety margins. Additionally, regression and Markov models are normally hard to handle multidimensional data [4] and thus bridge deterioration models utilize a limited number of both independent and dependent variables.

Thus, this study aims to develop a model to explore bridge damage patterns for specific bridge elements. A range of factors containing both structural (e.g., thickness of bridge deck) and non-structural information (e.g., traffic volume, weather, and constructor) were considered as independent variables in the model, and damages that do not cause abnormal structural movements were also considered as dependent variables. Once inspectors do their inspection, they can recognize the primary inspection area based on the derived patterns by matching independent variables in the targeted bridge information with those in the extracted patterns.

The model analyzed data stored in Bridge Management System (BMS) developed by the Korea Institute of Civil Engineering and Building Technology (KICT), and big data algorithms including association rules discovery and decision trees in classification were applied to extract patterns of bridge damages.

This paper introduces the framework to develop the pattern recognition model and pilot test results.

2 Research Methodology

2.1 Research Framework

Figure 1 illustrates the model development framework using big data analytics.

After collecting data, data passed through the preprocessing steps including data reduction, data cleaning, data transformation, and data integration. The preprocessed data was then divided by predominant features such as cities. Due to a large number of variables, the data dimensionality was reduced with the execution of feature selection and the damage patterns were finally extracted by the selected features. Cross validation was applied to validate the findings. The framework was developed and implemented by R software version 3.3.2 for statistical computing and graphics.

2.2 Data Collection

In addition to inspection data obtained from the total of 6,773 bridges in Korea, stored in BMS, weather data and contractor-related information were collected from the websites of the Korean Meteorological Administrat-

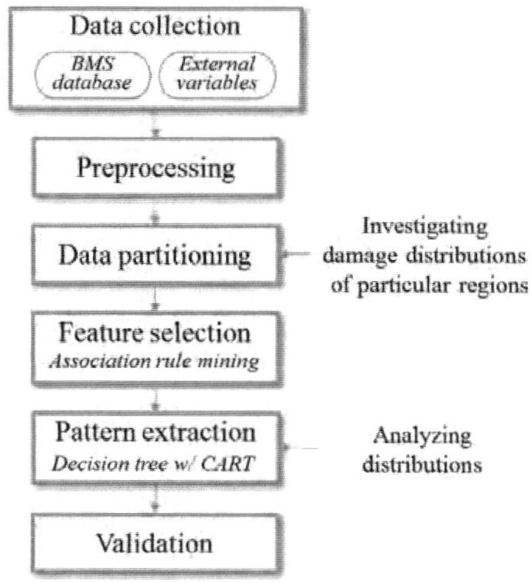

Figure 1. Model Development Framework

ration (KMA) and the Construction Association of Korea (CAK) respectively. KMA published annual weather reports from 1992 to 2015 and the reports covered only 78 regional measurements. Such weather information of the 78 cities was synchronized to the same cities declared in the BMS database. The weather information for the remaining cities in BMS except 78 was filled with the nearest neighbor's information. In the latter case, the distance to determine the nearest was measured by the simple Euclidean method.

The contractor-related information was explained by the level of construction capability of Korean contractors evaluated by CAK. The level of construction capability was calculated by sum of construction cost for recent three years, management performance, engineering capabilities, and company's credit rating. Such company's constructability level was captured from the CAK database and matched to the contractor information stored in BMS as the constructor of the bridge.

2.3 Data Characteristics

The collected data consisted of general, structural, and inspection information of bridges, weather information, and contractor-related variables. The general information bridge information included bridge classes, locations, competent authorities, offsets, detours, traffic volumes, lengths, widths, the number of lanes, the number of spans, main structure types, substructure types, design live loads, attached facilities, and others. The structural information included span lengths, decks, girders, diaphragms, ribs for spans and support types, abutments, piers, expansion joints, shoes, and stopper factors for supports.

Such general and structural information could not be changed by time since they represent innate characteristics confirmed after construction. However, condition grades in inspection information are updated after the inspection. The weather information including temperature, relative humidity, precipitation (i.e., rainfall), sunshine, wind, and specific atmospheric phenomena (e.g., freezing and snow cover) was selected based on previous studies and time series variables [5-8].

The BMS database covered 6,773 bridges built from 1966 to 2016 in Korea and structural data included 19,625 records for spans and 32,805 rows for supports. The inspection data stored 9,775 detailed inspection records and 900 precise safety diagnoses results from 1994 to 2015, and consequently, the inspection data included 834,815 records in a level of different damage types of bridge elements.

The distribution of condition grade of the whole bridge was different from that of damages on bridge elements. The condition grades of the whole bridge were distributed as grades "A" (29.0%), "B" (67.1%), "C" (3.8%), "D" (0.1%), "E" (0%) (Figure 2). For the

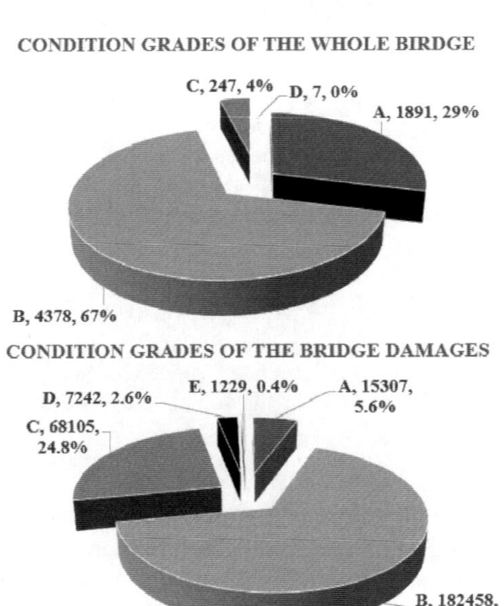

Figure 2 Distributions of condition grades: (a) Condition grades of the whole bridge; (b) Condition grades of the bridge elements

bridge damage level, grades "A" (5.6%), "B" (66.5%), "C" (24.8%), "D" (2.6%), "E" (0.4%), were shown in Figure 2. The number of damaged grades from "C" to "E" of the whole bridges took a substantially smaller portion than normal grades from "A" to "B". The proportion of the damaged grades of the bridge damage level, however, was larger than the other grades.

2.4 Data Preprocessing

Data preprocessing including data reduction, data cleaning, data transformation, and data integration was conducted to improve the quality of the data and thus to obtain more accurate analysis results consequently [9]. At first, data reduction was performed to reduce the number of variables as follows: text variables including descriptive variables (e.g., bridge name) which can be replaced with the codes (e.g., bridge code), data entry variables (e.g., input date), and high-rate missing and redundant values.

Next data cleaning was conducted to remove noise and inconsistent data and to fill in missing values. A specific type of bridge data was deleted since the number of data was too small to find a pattern: nine cable-stayed bridges and two suspension bridges. Randomly and frequently distributed blank cells were filled in a global constant, "9999", which could still indicate missing value by support running the process. The reason why a global constant was used was to maintain the current distribution of data.

After downsizing the data, data transformation was performed to generate hierarchy of variables and to discretize the data. Nominal variables had the relationship between subordinates and superiors, for example, cities and provinces. By investigating such hierarchical relationships, subordinate variables were remained and superiors were removed. Data discretization was processed to replace numeric variables (e.g., bridge length) by interval categories (e.g., 11.7–23.6, 23.6-35.2, and others).

Before categorizing the continuous numerical values, the number of bins (k) was decided by Sturge's rule (see Equation (1)) using size of data (n) [10].

$$k = 1 + 3.322(log_{10}\text{n}) \qquad (1)$$

The computed k values were 18 for 170,609 records of the span data and 17 for 69,858 rows of the support data. K-means clustering was then utilized to distribute one continuous variable into k clusters. This technique was used for data discretization [5, 9] based on a rule that minimizes intra-cluster distances and maximizes inter-cluster distances [11]. More specifically, k points, or *centroids*, were first initiated to the target data space. Next, each data point was assigned to the closest centroid to make k clusters. Euclidean distance was used to calculate the closeness between the data point and the centroid. The positions of the k centroids were then recalculated with newly assigned data points to the centroids. The second and third steps were repeated until all the centroid positions did not change further [5, 9]. The "discretize" function in the "arules" package in R software was utilized.

The last preprocessing step was data integration to generate data subsets for analyses. The general and structural variables were combined according to the representative bridge number and the inspection variables were added based on the span or support numbers of the bridge having the same bridge number. The weather variables were included to the datasets according to the regional information and the contractor-related variables were added by contractor names. The final dataset consisted of 70 variables from the BMS database, 20 variables from the annual weather data, and four contractor-related variables. Since the structural variables of spans and supports were in different data formats, two datasets were eventually developed separately: the span dataset with 170,609 records and the support dataset with 69,858 records.

2.5 Data Partitioning

The region (i.e., cities) was chosen as a data partitioning criterion. Among 137 cities, three cities were selected for a pilot test based on three criteria: comparatively large amount of data, exposure to harsh weather condition (e.g., heat wave and heavy snow), and geographical conditions of the locations (e.g., mountain regions, seaboards, islands, and inland areas).

2.6 Feature Selection

When a dataset is composed of numerous predictors (i.e., independent variables), feature selection, also known as dimensionality reduction, is normally conducted to improve the learning performance of a classification model such as decision trees [12]. The feature selection increases the accuracy of results and the speed of learning by excluding less correlated features and giving higher weights for more correlated features [13], and it can consequently prevent overfitting [9].

The feature selection mainly focuses on choosing the best subset of attributes [6], so that association rule mining was implemented in this study [13]. The "association rule" is extracted based on co-occurrence of attributes, and therefore, the rule does not mean causality. This method was originated from basket analysis which aimed to find frequent buying patterns by analyzing market transaction data [13]. In this study, attributes (i.e., columns) and inspections (i.e., rows) of the span and the support datasets corresponded to items and transactions respectively.

Let $I = \{i_1, i_2, i_3, \cdots, i_d\}$ be the set of all items in market basket data and $T = \{\boldsymbol{t_1}, \boldsymbol{t_2}, \boldsymbol{t_3}, \cdots, \boldsymbol{t_n}\}$ be the set of all transactions. Each transaction $\boldsymbol{t_i}$ has a combination of items from I. An association rule is a set of items which contains an antecedent (i.e., "if") and a consequent (i.e., "then") with the implication expression of the form $A \rightarrow B$, where A and B are disjoint (i.e., $A \cap B \neq \emptyset$) [9, 13, 14].

To find frequent item sets, two measurements, *support* and *confidence*, act as determinants. Above all, *support count*, which is distinguished from just *support*, is defined as the number of transactions that contain a particular item set X. *Support count*, σ(X), can be mathematically expressed as follows:

$$\sigma(X) = n(\{t_i \mid X \subseteq t_i, t_i \in T\}) \quad (2)$$

Minimum support and *confidence* to detect the interesting rules are predefined by an analyst. *Support* counts the number of transactions which covers all items in A and B together. *Confidence* indicates how frequently items in B appear in transactions that include A. The concept of *confidence* resembles conditional probability, so that the expressions of both formulas are similar. The metrics are shown below [13]:

$$Support(A \rightarrow B) = \sigma(A\ and\ B) \quad (3)$$
$$Confidence(A \rightarrow B) = \sigma(A\ and\ B)/\sigma(A) \quad (4)$$

Apriori algorithm, proposed by Agrawal and Srikant in 1994 [15], was utilized in this research due to its efficiency on big data analysis [9]. The "apriori" function in the "arules" package on R software was used in this study.

2.7 Pattern Extraction

From the sets of selected features, damage patterns were able to be determined by decision tree analysis. A path of the tree from the top to the bottom can be a pattern.

CART (Classification And Regression Tree) algorithm was utilized in this study. Introduced by Breiman, CART was named with the emphasis on the two types of trees: the classification tree for the continuous target variables and the regression tree for discrete variables [16]. CART generates binary trees which have a branch with only two leaves, dissimilar to other decision tree algorithms such as ID3, C4.5, and CHAID [9, 17]. The Gini index is used for the attribute selection of discrete attributes and for the decrement of variance of continuous attributes. This study applied

CART because it can handle both continuous and discrete attributes simultaneously and has benefits on the treatment of missing values [18-19]. The "rpart" function in the "rpart" package on R software was utilized for the model building.

2.8 Validation

Once damage patterns were extracted, the developed model was validated by cross validation: K-fold validation. The whole dataset was divided into k subsets, and then the algorithm selected one subset as a testing set and remainders as a training set. This step was repeated k times and the model parameters were updated until the fluctuation of performance is reduced [9].

3 Pilot Results and Discussions

In this study, a pilot test using the span dataset was conducted to show the validity of the developed research framework. Through data partitioning, Ulju-gun, Inje-gun, and Mungyeong-si in Korea were selected (Figure 3). The primary criterion was the size of data and the second was specific geographical conditions of the locations: seaboards, mountain regions, and inland areas. 59 (Ulju-gun), 64 (Inje-gun), 118 (Mungyeong-si) rules were found by the association rule mining, and approximately 30 attributes appeared within the extracted rules. Table 1 summarizes data properties and the results of the association rule mining for three cities. Using the extracted rules, three decision trees were developed and patterns were produced when they met the target class of the condition level "C". A pattern for the Ulju-gun dataset explained a regional characteristic, such as heavy snow in winter, but the derived patterns were difficult to represent general regional characteristics. Nevertheless, the fact that extracted attributes were differentiated from each city represents the significance of regional partitioning.

The extracted attributes from the Ulju-gun dataset contained thickness of decks and deck elements which were distinct from the other two cities. The only one pattern was extracted as below:

- {Thickness of rib bottom flange < 24mm, thickness of web < 16mm, thickness of deck < 27.5cm, height of rail = 108cm, span length = 50m}.

The major damages found in Ulju-gun were cracks in reinforced concrete (RC) decks and clogged drainages.

Inje-gun was represented with the attributes related to decks and adjacent elements to decks. The selected patterns are found as below:

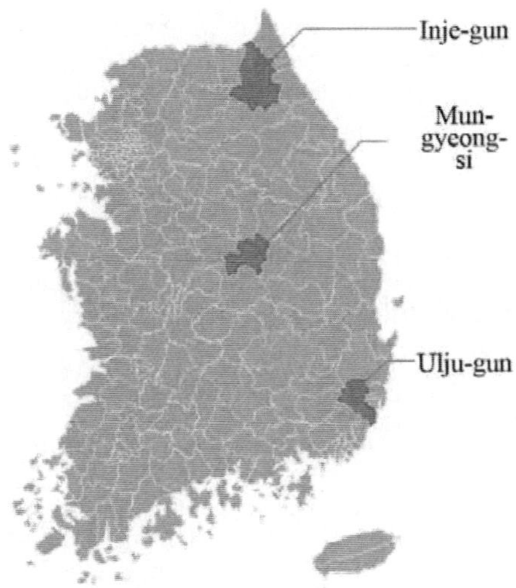

Figure 3. The selected regions in the pilot study

Table 1. The properties and the results of association rule mining for three cities

Region	Data properties			Results of association rule mining		
	No. of bridges	No. of inspec-tions	No. of dama-ges	No. of rules	Sup-port	No. of attribu-tes
Ulju-gun	45	187	3324	59	.05	29
Inje-gun	71	196	2853	64	.10	27
Mun-gyeong-si	22	116	2806	118	.12	26

* confidence = 0.5, the maximum length of rule = 4

1) {Thickness of deck pavement >= 6.5cm, depth of water < 2.15m, constructor = (L company, D1 company, D2 company, I company), height of the median >= 39cm, number of the up lines ≠ (1, 2, 3), depth of water < 0.75m},
2) {Thickness of deck pavement < 6.5cm, type of the median = (concrete, the others), rail material = (steel, aluminum alloy, the others), constructor = (D3 company, H company)},
3) {Thickness of deck pavement < 6.5cm, type of the median = (concrete, the others), rail material = concrete}.

Contractor information, thickness of decks, and the structural information of medial and rail appeared frequently from the patterns.

Damages which took large proportions of the whole Inje-gun data were (1) cracks in RC decks, (2) cracks in concrete rails and curbs, (3) breakages of concrete rails and curbs, (4) leakages and efflorescence in RC decks, and (5) clogged drainages.

The second pattern indicated that efflorescence and corrosion in RC decks frequently occurred due to chlorides [20]. Ulju-gun has a lot of snow in winter so that de-icing chemicals containing chlorides are widely used.

In addition, water-related damages including clogged drainages, leakages, efflorescence, and corrosion in RC decks took relatively large proportion of damages.

In the Mungyeong-si dataset, plane shape and pavement related attributes were found to be used as predictors for the decision tree. Three patterns were determined as below:

1) {Pavement area < $5872m^2$, constructor = (G company, D4 company), plane shape = (straight bridge with skew, curved bridge without skew), number of the down lines = 0, maximum span length >= 42.5m},
2) {Pavement area < $5872m^2$, constructor = (G company, D4 company), plane shape = (straight bridge with skew, curved bridge without skew), number of the down lines = 2, depth of water >= 0.35m, pavement area >= $4620m^2$},
3) {Pavement area >= $5872m^2$}.

The main damage types were breakages in RC decks and breakages in steel rails and curbs. The first pattern appeared to be linked with the characteristics of breakages in RC decks, streel rails, and curbs. The first and second pattern indicated breakages in steel rails and curbs were critical. However, bridges built by Contractor G in other regions also showed 68% of breakages in steel rails and curbs.

4 Conclusions

This study developed a model to find patterns of damages on bridge elements through big data analyses. Weather and contractor-related variables were added on the BMS database, and the preprocessed dataset was divided into three cities for the pilot study. Association rule mining performed feature selection and the patterns were extracted by decision trees.

More specifically, the pilot test was conducted with the span data subsets of Ulju-gun, Inje-gun, and Mungyeong-si. The extracted patterns were different by different cities with the implication of geographical characteristics such as heavy snow and different construction capacities of contractors. The research findings showed potential to support decision makers for strategic preventive bridge maintenance.

In further study, data partitioning with various criteria will be performed using data clustering and other association rule mining and decision tree algorithms will be applied and compared to develop the best fit model.

Acknowledgments

This work was funded by the Promising-Pioneering Researcher Program through Seoul National University (SNU) in 2015.

References

[1] Liang Y., Wu D., Liu G., Li Y., Gao C., Ma Z. J., and Wu W. Big data-enabled multiscale serviceability analysis for aging bridges. *Digital Communications and Networks*, 2(3): 97-107, 2016.

[2] Ministry of Land, Infrastructure, and Transport of Korea *Yearbook of road bridges and tunnels*, 2015.

[3] Adarkwa O. and Attoh-Okine N. Prediction of Structural Deficiency Ratio of Bridges Based on Multiway Data Factorization. *ASCE-ASME Journal of Risk and Uncertainty in Engineering Systems, Part A: Civil Engineering*, F4016002, 2016.

[4] Bu G., Lee J., Guan H., and Loo Y. Prediction of Long-Term Bridge Performance: Integrated Deterioration Approach with Case Studies. *Journal of Performance of Constructed Facilities*, 29(3): 04014089, 2015.

[5] Huang R., Mao I. S., and Lee H. Exploring the deterioration factors of RC bridge decks: a rough set approach. *Computer-Aided Civil and Infrastructure Engineering*, 25(7):517-529, 2010.

[6] Melhem H. G., Cheng Y., Kossler D., and Scherschligt D. Wrapper Methods for Inductive Learning: Example Application to Bridge Decks. *Journal of Computing in Civil Engineering*, 17(1):46-57, 2003.

[7] Huang R. and Hsu W. Reliability-based component deterioration model for bridge life-cycle cost analysis. *Journal of Chinese Institute of Civil and Hydraulic Engineering*, 17(4):679-691, 2005.

[8] Creary P. A. and Fang F. C. The Data Mining Approach for Analyzing Infrastructure Operating Conditions. *Procedia - Social and Behavioral Sciences*, 96:2835-2845, 2013.

[9] Han J., Kamber M., and Pei J. *Data mining: concepts and techniques*, Third Edition. Elsevier, Waltham, United States, 2012.

[10] Nie N. H., Hull C. H., Jenkins J. G., Steinbrenner K., and Bent D. H. *SPSS statistical package for the social sciences*, Second Edition. McGraw-Hill, New York, United States, 1975.

[11] Morcous G., Rivard H., and Hanna A. Modeling bridge deterioration using case-based reasoning. *Journal of Infrastructure Systems*, 8(3):86-95, 2002.

[12] Kohavi R. and John G. H. Wrappers for feature subset selection. *Artificial intelligence*, 97(1-2):273-324, 1997.

[13] Rajeswari K. Feature Selection by Mining Optimized Association Rules based on Apriori Algorithm. *International Journal of Computer Applications*, 119(20), 2015.

[14] Slimani T. and Lazzez A. Efficient analysis of pattern and association rule mining approaches. *ArXiv preprint arXiv*: *1402.2892*, 2014.

[15] Agrawal R. and Srikant R. Fast algorithms for mining association rules. Proceedings of 20th International Conference on Very Large Data Bases, 1215:487-499, 1994.

[16] Breiman L., Friedman J. H., Olshen R. A., and Stone C. J. *Classification and regression trees*. Wadsworth & Brooks. Monterey, CA, 1984.

[17] Huang R. and Chen P. F. Analysis of influential factors and association rules for bridge deck deterioration with utilization of national bridge inventory. *Journal of Marine Science and Technology*, 20(3):336-344, 2012.

[18] Patidar P. and Tiwari A. Handling Missing Value in Decision Tree Algorithm. *International Journal of Computer Applications*, 70(13):31-36, 2013.

[19] Ture M., Tokatli F., and Kurt I. Using Kaplan–Meier analysis together with decision tree methods (C&RT, CHAID, QUEST, C4. 5 and ID3) in determining recurrence-free survival of breast cancer patients. *Expert Systems with Applications*, 36(2):2017-2026, 2009.

[20] Cusson D., Lounis Z., and Daigle L. Durability Monitoring for Improved Service Life Predictions of Concrete Bridge Decks in Corrosive Environments. *Computer-Aided Civil and Infrastructure Engineering*, 26(7):524-541, 2011.

Prediction of Concrete Compressive Strength from Early Age Test Result Using an Advanced Metaheuristic-Based Machine Learning Technique

D. Prayogo1[a], M.-Y. Cheng2[b], J. Widjaja3[a], H. Ongkowijoyo4[a], and H. Prayogo5[a]

[a]Department of Civil Engineering, Petra Christian University, Indonesia
[b]Department of Civil and Construction Engineering, National Taiwan University of Science and Technology, Taiwan
E-mail: prayogo@petra.ac.id, myc@mail.ntust.edu.tw, m21414116@john.petra.ac.id, m21414084@john.petra.ac.id, m21415097@john.petra.ac.id

Abstract –

Determining accurate concrete strength is a major civil engineering problem. Test results of 28-day concrete cylinder represent the characteristic strength of the concrete that has been prepared and cast to form the concrete work. It is important to wait 28 days to ensure the quality control of the process, although it is very time consuming. Machine learning techniques are progressively used to simulate the characteristic of concrete materials and have developed into an important research area. This study proposed a comprehensive study using an advanced machine learning technique to predict the compressive strength of concrete from early age test results. In this case, early age test data are being used to get reliable values of the two constants which are required for the prediction. A total of 28 historical cases were used to establish the intelligence prediction model. Obtained results show the performance of the advanced hybrid machine learning technique in predicting the concrete strength with a relatively high accuracy measured by four error indicators. Therefore, the proposed study can offer a high benefit for construction project managers in decision-making processes based on early strength test results.

Keywords –
Concrete; Compressive Strength; Early Strength; Machine Learning; Prediction

1 Introduction

Concrete is a material used in construction that has great versatility and which is used across the globe. Concrete has several advantages, including good compressive strength, durability, workability, construction availability, and low cost. Nevertheless, these benefits have a serious reliance on curing, placing, and the appropriate mix. Strength, within the construction industry, is a major criterion in the selection of a concrete to be used in a specific application. Construction concrete will gain strength throughout an extended period after it has been poured. Concrete's nominal strength has a definition of a sample's compressive strength that is 28 days old.

If there is a faulty mix preparation or mix design on site, the results of the test may indicate failure to reach the required strength, triggering a mandatory repeating of the whole process, and this can be time consuming and costly. For all failures, another 28 days must be awaited, so there has long been a requirement for an ability to estimate, at an early age, the concrete's final strength. Therefore, an appropriate, quick method of predicting concrete strength would be a significant advancement for the industry [1]. An ability to predict the compressive strength of concrete early allows constructers to quickly understand the concrete's probable weaknesses and make a decision to manage a destruction process or continue with construction. Further, to the benefit of both user (or purchaser) and producer, reliably and rapidly predicting the results of a 28-day test would benefit all stakeholders as opposed to waiting the full, conventional, 28 days.

Researchers are eager to explore the behavior of the concrete, therefore predicting concrete strength, and this has been targeted as an active area to be researched. There are multiple recent studies exploring concrete's behavior and the possibility of improving characteristic strength predictions. The studies have revealed that many tests have concentrated on the way that the strength of the concrete is affected by the mix, but only a small number of studies have focused on the relation between early testing (for example, 7 and 14 days) and the full 28-day compressive strength test. Moreover, most of the studies have some limitations; for example, an absence of advanced method for measuring accuracy, no validating techniques involved, or using only a conventional approach.

Machine learning methods are proven to outperform conventional methods because of their excellent learning functions [2-6]. The conventional approaches, such as linear regression and decision tree, are not sufficient to build a satisfactory model in terms of accuracy and computational time. Therefore, this primary research aim is to construct an advanced accurate model to improve early age predictions of concrete strength.

This research proposes a new advanced machine learning technique called Symbiotic Organisms Search-based Support Vector Regression (SOS-SVR). SOS-SVR fuses an accurate prediction technique, Least Squares Support Vector Regression (LS-SVR) [7], and a very promising metaheuristic, Symbiotic Organisms Search (SOS) [8]. The proposed model will be investigated alongside other prediction methods in building an accurate prediction model of concrete strength in correspondence to the early age strength test results.

2 The Proposed Metaheuristic-Based Machine Learning Technique

2.1 Least Squares Support Vector Regression

LS-SVR is a modified version of the support vector regression (SVR) [9]. The LS-SVR is a statistical learning theory that adopts a least squares linear system as a loss function instead of the quadratic program in the original support vector machine (SVM) [10]. The optimization problem and the constraints for LS-SVR can be stated by the following formulation.

Minimize: $J_p(w, e) = \frac{1}{2}w^T w + \gamma \frac{1}{2}\sum_{k=1}^{N} e_k^2$ (1)

Subjected to: $y_k = w^T\phi(x_k) + b + e_k, k = 1, \ldots, N$ (2)

where $w \in R$ is an undetermined parameter vector; $\phi(\cdot)$ is the nonlinear function introduced by a kernel function for mapping the input space to a high-dimensional feature space; $e_k \in R$ is error variable; γ denotes a regularization constant.

It is worth noting that the objective function includes a sum of squared fitting error and a regularization term. However, when w becomes infinite dimensional, one cannot solve this primal problem. Therefore, it is necessary to construct the Lagrangian and derive the dual problem. The resulting LS-SVR model for function estimation is expressed as:

$$y(x) = \sum_{k=1}^{N} \alpha_k K(x_k, x_l) + b \quad (3)$$

where α_k and b are the solution to the linear system. The kernel function that is often utilized is the radial basis function (RBF) kernel. Description of RBF kernel is given as follows:

$$K(x_k, x_l) = exp\left(\frac{\|x_k - x_l\|^2}{2\sigma^2}\right) \quad (4)$$

where σ is the kernel function parameter.

There has been an increased used of the LS-SVR in many engineering fields. However, the LS-SVR presents a problem in prediction accuracy whenever its parameters are not fine-tuned. The regularization parameter (γ) is a positive cost parameter similar to the C in SVM/SVR, while the kernel parameter (σ) is an additional parameter since LS-SVR utilized the RBF kernel function. These two parameters need to be specified in order to find the best prediction model. Thus, there is a need of metaheuristic algorithm in order to find the most effective combination of LS-SVR parameters. It has been shown that the self-tuned parameter approaches, involving the metaheuristic as the optimizer, have been studied extensively and has produced an increased prediction accuracy [11, 12].

2.2 Symbiotic Organisms Search

In recent years, many researchers proposed a new metaheuristic algorithm to search for optimality in various optimization problems. It has been proven that metaheuristic algorithm can be an effective tool in dealing with challenging problems. Some notably, examples of recent metaheuristic algorithms applications in tackling hard and complex problems can be seen in [13-18].

Another new and powerful metaheuristic algorithm namely symbiotic organisms search (SOS) has been proposed by Cheng and Prayogo [8] to solve the continuous based optimization problems. The experimental results have been shown in the high efficiency of SOS performance in comparison with the conventional metaheuristic techniques including genetic algorithm (GA), particle swarm optimization (PSO), and differential evolution (DE).

The SOS algorithm adopts the interaction behaviour among organisms which are living together in one ecosystem. It gradually guides a population of candidate solutions, denoted as organisms, towards promising regions in the solution space. Each candidate solution, denoted as organism, has a certain fitness value which corresponds to the objective function or goal of the problem. The main steps of the SOS algorithm can be seen as follows.

1: Initialize random population
2: **Do**

3: Mutualism phase
4: Commensalism phase
5: Parasitism phase
6: Preserve the current best solution
7: **Until** termination criteria are met

For more details, the reader can refer to [8]. Since the first publication in 2014, the use of SOS in solving different problems has been significantly increased in various research topics [19-27]. It shows the potential of SOS algorithm as an attractive tool to search for the optimality.

2.3 Symbiotic Organisms Search-Based Support Vector Regression

The symbiotic organisms search-based support vector regression (SOS-SVR) model is a hybrid machine learning system that combines the two different techniques of SOS and LS-SVR. In this system, the LS-SVR acts as a supervised-learning-based predictor to build the accurate input-output relationship of the dataset; and the SOS works to optimize the LS-SVR parameters, the σ and γ parameters.

The SOS-SVR involves eight major steps which are categorized into two phases, beginning with a training phase followed by the testing phase. The whole procedure of SOS-SVR is shown in Figure 1. An explanation of the major steps involved in SOS-SVR is given below:

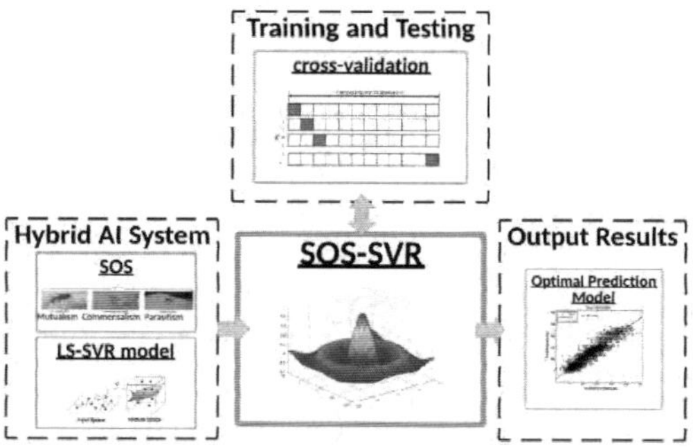

Figure 1. SOS-SVR framework.

1. Hybrid AI System:
   LS-SVR addresses the complex relationship between input and output variables. LS-SVR requires two parameters, σ and γ, to finish the learning process. LS-SVR will gradually obtain the tuning parameters from the optimizer – the SOS algorithm.
   In this hybrid system, SOS is utilized to explore the various combinations of σ and γ parameters to look for the best set of them. SOS utilizes mutualism, commensalism, and parasitism phases to gradually improve the fitness value of each solution.
   The process terminates when the termination criterion is satisfied. While still unsatisfied, the model will proceed to the next iteration. As SOS-SVR uses SOS, the termination criterion used in this study was the SOS iteration number.

2. Training and testing:
   The data for training are obtained from data collection. A fitness function is now developed to evaluate accuracy of the learning system. This function has a correlation with the accuracy of the prediction model. The most accurate prediction model is represented by the combination of σ and γ parameters that produce the best fitness value.
   Instead of randomly splitting the data, it was partitioned into two subsets: learning subset and validation subset. To avoid the sampling bias, the 5-fold cross-validation technique is used in splitting the training data into learning and validation subsets. In this study, the fitness function utilizes mean square error (MSE) of the validation dataset. When the termination criterion is fulfilled, the loop stops. This condition means that the prediction model has identified the input/output mapping relationship with the optimal σ and γ parameters.
   Meanwhile, the data for testing is obtained from data collection. The optimal σ and γ parameters obtained from the training phase are used to establish the prediction model for predicting the testing data.

3. Output Results:
   The optimal σ and γ parameters obtained from the training phase are used to establish the prediction model for predicting the testing data. The accuracy measurement methods are used to measure the performance of the SVR-based prediction model.

3 Experimental Settings

3.1 Data Preparation

The historical database for experiment was obtained from previous literature [28]. In this research, the database is assigned into two group of dataset: (1) the mix proportions with 7-day strength test result and (2) the mix proportions with 14-day strength test result.

Each dataset has a total of 28 records of concrete mix proportion and were used to analyse the behaviour of concrete mixture with strength at the various early ages. Every dataset has 6 input variables and 1 output variable, including the coarse aggregate to cement ratio (CA/C), fine aggregate to cement ratio (FA/C), water to cement ratio (W/C), water (W), coarse aggregate size ratio of 10 mm:20 mm (CAS), 7-day or 14-day strength test result depending on each group (fc-7 or fc-14), and the 28-day strength test result (fc-28), respectively. The characteristics of the variables of each dataset can be seen in Table 1.

Table 1 Statistical description of concrete mix proportion

Variables	Unit	Group	Min	Max	Avg.	Std. dev.
CA/C	-	1, 2	1.4	2.24	1.811	0.256
FA/C	-	1, 2	2.22	2.97	2.591	0.236
W/C	-	1, 2	0.4	0.52	0.460	0.041
W	kg/m3	1, 2	185	190	187.5	2.546
CAS	-	1, 2	1	2	1.464	0.508
fc-7	MPa	1	13.84	26.02	19.025	3.957
fc-14	MPa	2	17.8	35.35	23.659	4.673
fc-28	MPa	1, 2	19.53	37.4	26.972	4.143

3.2 Accuracy Measurement and Parameter Settings

Machine learning is recently employed to predict behavior in many research areas. For the comparison purpose to our proposed method, this research applies five widely used machine learning methods including classification and regression tree (C&R tree), neural network (neural net), regression, support vector machine (SVM), chi-squared automatic interaction detector (CHAID) to model the prediction behavior of concrete strength at various early ages. All aforementioned methods were performed using the IBM SPSS Modeler.

The parameters of the machine learning techniques were set to default for fair comparison. Through trial-and-error, a suitable parameter setting for SOS-SVR is determined as follows: the maximum number of iteration was set to 50; population size was set to 25; and search range for the σ and γ parameters starts from 10^{-5} to 10^{5}.

The effectiveness of state-of-the art machine learning methods can be determined using the accuracy measurement methods. In this research, three measurement methods were adopted – coefficient of correlation (R); root mean squared error (RMSE); mean absolute error (MAE); and mean absolute percentage error (MAPE). The lowest RMSE, MAE, and MAPE values alongside with a highest R value indicate the best model outcome.

3.3 Cross-validation

To obtain the prediction result from each machine learning technique, the dataset must be labelled into two groups of training data and testing data. At first, the training data is used by one method to generate the prediction model. Then, the prediction model is applied to validate the testing data. Once the predicted output is generated, the accuracy measurement is performed to check the deviation between the predicted and actual output.

During the partitioning of training and testing subset, there is a possibility for the bias associated with the random sampling. To minimize the sampling bias, cross-validation technique is utilized in this research. The main goal of the cross-validation technique is to ensure the partitioning process generating the independent and unbiased data subset. The data is then labelled into 5 folds and the performance of each fold is recorded and validated in accordance to each accuracy measurement.

4 Results and Discussions

The training and testing processes are performed for each predictive model. The complete experimental results based on 7-day and 14-day test results can be seen in Table 2 and Table 3, respectively. It can be seen that SOS-SVR has performed better in all measurement categories in the testing dataset.

For training dataset, the SOS-SVR prediction model has produced a decent training accuracy as shown in Table 2. It is worth noting that C&R tree prediction model yielded the highest accuracy on training dataset but the model performs poorly on the testing set. The over-fitting had occurred during the training process of C&R tree. As a result, the generalization ability of C&R tree model to predict the future data is declined drastically. Although the SOS-SVR model did not have the highest accuracy on the training dataset, still it has the highest accuracy on the testing dataset. The metaheuristic-tuned framework helps the SOS-SVR to determine the right balance between learning and generalization paradigm.

Table 2 Comparative experimental results among predictive methods for 7-day strength test result.

Machine Learning Techniques	Average Result			
	R	RMSE (MPa)	MAE (MPa)	MAPE (%)
Training				
C&R tree	0.9965	0.33	0.15	0.55
Neutral net	0.8744	2.06	1.55	5.84
Regression	0.9050	1.71	1.39	5.21
SVM	0.8709	2.18	1.42	5.26
CHAID	0.8618	2.06	1.42	5.14
SOS-SVR	0.8853	1.99	1.30	5.11
Testing				
C&R tree	0.7567	3.55	2.92	10.75
Neutral net	0.8708	2.58	2.19	8.08
Regression	0.8457	3.52	2.92	10.88
SVM	0.8460	2.71	2.26	8.35
CHAID	0.6788	3.77	2.96	10.79
SOS-SVR	0.8793	2.49	2.01	6.96

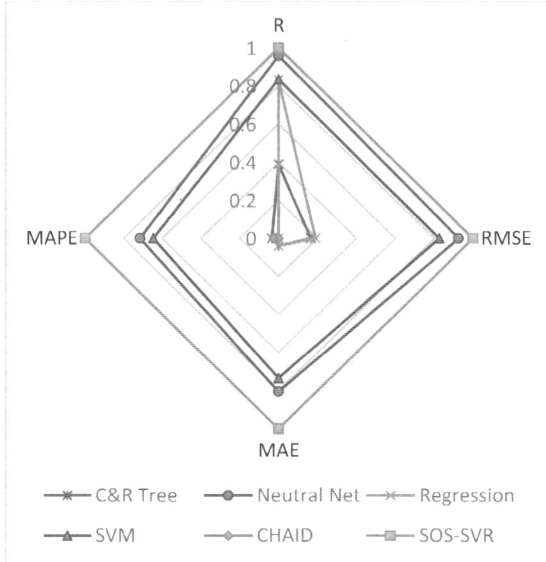

Figure 2. Radar plot of normalized accuracy measures (7-day strength test result)

Table 3 Comparative experimental results among predictive methods for 14-day strength test result.

Machine Learning Techniques	Average Result			
	R	RMSE (MPa)	MAE (MPa)	MAPE (%)
Training				
C&R tree	0.9983	0.22	0.12	0.45
Neutral net	0.8467	2.20	1.58	5.81
Regression	0.9180	1.58	1.21	4.53
SVM	0.8962	2.00	1.21	4.41
CHAID	0.8946	1.78	1.03	3.53
SOS-SVR	0.8765	1.85	1.14	5.00
Testing				
C&R tree	0.7516	4.19	3.35	12.87
Neutral net	0.8215	2.86	2.43	9.69
Regression	0.8289	2.86	2.21	8.63
SVM	0.8499	2.64	2.13	8.43
CHAID	0.8029	3.39	2.82	11.45
SOS-SVR	0.8563	2.51	2.07	7.26

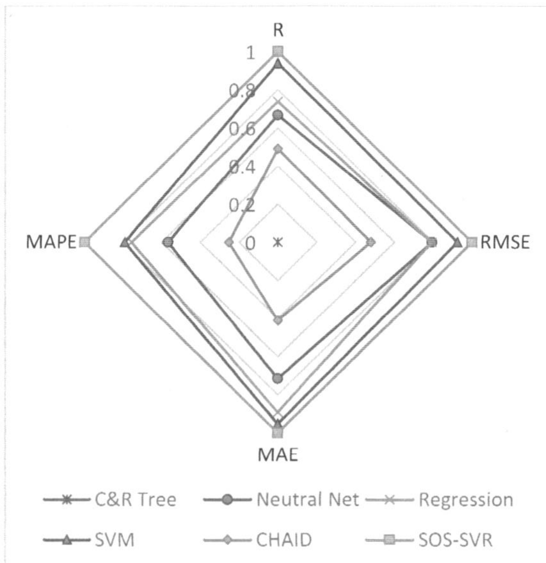

Figure 3. Radar plot of normalized accuracy measures (14-day strength test result)

Figures 2 and 3 further illustrate the comparative performance from each measurement category. Each value from average testing result is normalized into 0 to 1 in which 0 denotes the current worst performance and

1 denotes the current best performance. The outermost line from the figures denotes the best performer. It can be seen that the SOS-SVR prediction model has delivered the best performance in all categories in comparison with other machine learning techniques.

Figures 4 and 5 present a comparison of the actual and predicted values of the concrete strength at 7-day and 14-day test result, respectively. It shows the fit curve between the target and predicted output at training and testing stages of one fold. This further shows the effectiveness of SOS-SVR as the most reliable method for establishing the prediction model in this experiment.

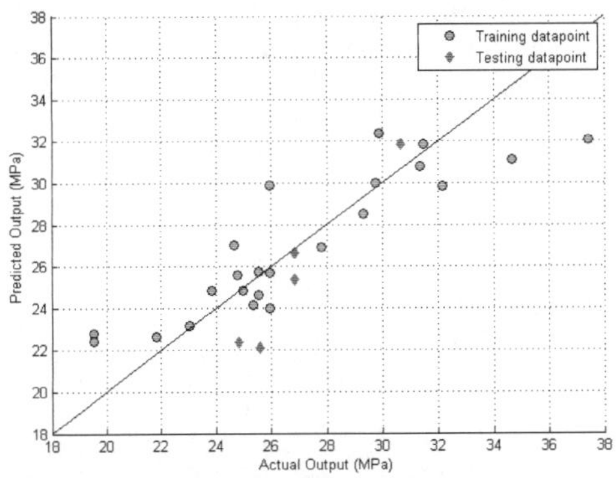

Figure 4. Actual and predicted concrete strength by SOS-SVR prediction model at training and testing stages for one fold (7-day strength test result).

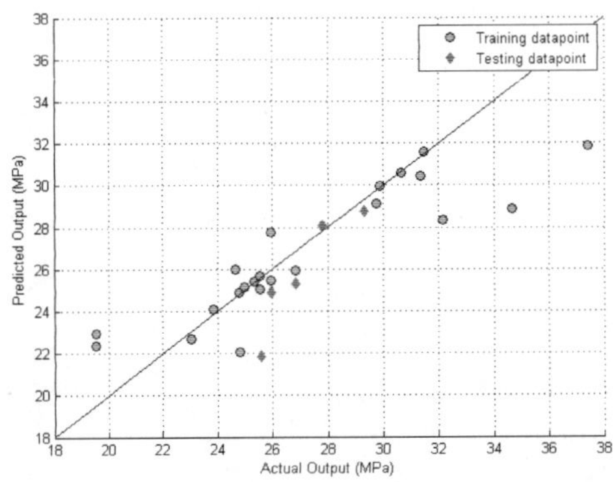

Figure 5. Actual and predicted concrete strength by SOS-SVR prediction model at training and testing stages for one fold (14-day strength test result).

5 Conclusion

This study developed a new prediction method called symbiotic organisms search-based support vector regression (SOS-SVR) to predict the strength of concrete mixtures from the early age test result. To investigate the accuracy of the proposed method, five machine learning techniques were used as benchmarks for the SOS-SVR. The experimental data set was acquired from laboratory test of 28 samples. In this investigation, the proposed predictive techniques were applied to the prepared training and testing datasets generated by 5-fold cross-validation.

The proposed SOS-SVR was further compared for performance outcomes by using four different performance measures (R, RMSE, MAE, and MAPE) to obtain a comprehensive comparison of the applied predictive techniques. The findings showed that the proposed SOS-SVR achieved the best accuracy for all performance measures.

This study presents a significant contribution to address the importance of the problem of early ages prediction of concrete strength. By accurately predicting the concrete behavior, the SOS-SVR assists the users and concrete designers in decision-making processes based on early strength test results. Analytical results indicate that SOS-SVR is the most reliable model for building accurate prediction behavior of concrete mix proportion at various early ages of strength.

Acknowledgement

The authors would express appreciation to the Ministry of Research, Technology, and Higher Education of the Republic of Indonesia for the financial support through the "Bantuan Seminar Luar Negeri 2017" funding program.

References

[1] Kheder G. F., Gabban A. M. A., and Abid S. M. Mathematical model for the prediction of cement compressive strength at the ages of 7 and 28 days within 24 hours, *Materials and Structures*, 36(10): 693, 2003.

[2] Cheng M.-Y., Firdausi P. M., and Prayogo D. High-performance concrete compressive strength prediction using Genetic Weighted Pyramid Operation Tree (GWPOT), *Engineering Applications of Artificial Intelligence*, 29 104-113, 2014.

[3] Cheng M.-Y., Prayogo D., and Wu Y.-W. Novel Genetic Algorithm-Based Evolutionary Support Vector Machine for Optimizing High-Performance Concrete Mixture, *Journal of Computing in Civil Engineering*, 28(4), 06014003, 2014.

[4] Cheng M.-Y., Wibowo D. K., Prayogo D., and Roy A. F. V. Predicting productivity loss

caused by change orders using the evolutionary fuzzy support vector machine inference model, *Journal of Civil Engineering and Management*, 21(7): 881-892, 2015.

[5] Hoang N.-D. and Pham A.-D. Hybrid artificial intelligence approach based on metaheuristic and machine learning for slope stability assessment: A multinational data analysis, *Expert Systems with Applications*, 46 60-68, 2016.

[6] Cheng M.-Y., Prayogo D., Ju Y.-H., Wu Y.-W., and Sutanto S. Optimizing mixture properties of biodiesel production using genetic algorithm-based evolutionary support vector machine, *International Journal of Green Energy*, 13(15): 1599-1607, 2016.

[7] Suykens J. A. K. and Vandewalle J. Least Squares Support Vector Machine Classifiers, *Neural Process. Lett.*, 9(3): 293-300, 1999.

[8] Cheng M.-Y. and Prayogo D. Symbiotic Organisms Search: A new metaheuristic optimization algorithm, *Computers & Structures*, 139 98-112, 2014.

[9] Suykens J. A. K. and Vandewalle J. Least squares support vector machine classifiers, *Neural Processing Letters*, 9(3): 293-300, 1999.

[10] S. K. K., Kim D., S.L.Sekar, and Samui P. Model of least square support vector machine (LSSVM) for prediction of fracture parameters of concrete, *International Journal of Concrete Structures and Materials*, 5(1): 29-33, 2011.

[11] Cheng M.-Y. and Prayogo D. Modeling the permanent deformation behavior of asphalt mixtures using a novel hybrid computational intelligence. In *ISARC 2016 - 33rd International Symposium on Automation and Robotics in Construction*, pages 1009-1015, Auburn, USA, International Association for Automation and Robotics in Construction, 2016.

[12] Cheng M.-Y. and Hoang N.-D. Estimating construction duration of diaphragm wall using firefly-tuned least squares support vector machine, *Neural Computing and Applications*, 1-9, 2017.

[13] Mirjalili S. The Ant Lion Optimizer, *Advances in Engineering Software*, 83 80-98, 2015.

[14] Mirjalili S. Moth-flame optimization algorithm: A novel nature-inspired heuristic paradigm, *Knowledge-Based Systems*, 89 228-249, 2015.

[15] Civicioglu P. Backtracking Search Optimization Algorithm for numerical optimization problems, *Applied Mathematics and Computation*, 219(15): 8121-8144, 2013.

[16] Cheng M. Y. and Prayogo D. Fuzzy adaptive teaching–learning-based optimization for global numerical optimization, *Neural Computing and Applications*, 1-19, 2016.

[17] Cheng M.-Y. and Prayogo D. A novel fuzzy adaptive teaching–learning-based optimization (FATLBO) for solving structural optimization problems, *Engineering with Computers*, 33(1): 55-69, 2017.

[18] Cheng M.-Y., Prayogo D., Wu Y.-W., and Lukito M. M. A Hybrid Harmony Search algorithm for discrete sizing optimization of truss structure, *Automation in Construction*, 69 21-33, 2016.

[19] Tejani G. G., Savsani V. J., and Patel V. K. Adaptive symbiotic organisms search (SOS) algorithm for structural design optimization, *Journal of Computational Design and Engineering*, 2016.

[20] Tran D.-H., Cheng M.-Y., and Prayogo D. A novel Multiple Objective Symbiotic Organisms Search (MOSOS) for time–cost–labor utilization tradeoff problem, *Knowledge-Based Systems*, 94 132-145, 2016.

[21] Cheng M.-Y., Chiu C.-K., Chiu Y.-F., Wu Y.-W., Syu Z.-L., Prayogo D., and Lin C.-H. SOS optimization model for bridge life cycle risk evaluation and maintenance strategies, *Journal of the Chinese Institute of Civil and Hydraulic Engineering*, 26(4): 293-308, 2014.

[22] Kamankesh H., Agelidis V. G., and Kavousi-Fard A. Optimal scheduling of renewable micro-grids considering plug-in hybrid electric vehicle charging demand, *Energy*, 100 285-297, 2016.

[23] Ayala H., Klein C., Mariani V., and Coelho L. Multi-objective Symbiotic Search Algorithm Approaches for Electromagnetic Optimization, *IEEE Transactions on Magnetics*, PP(99): 1-1, 2017.

[24] Panda A. and Pani S. A Symbiotic Organisms Search algorithm with adaptive penalty function to solve multi-objective constrained optimization problems, *Applied Soft Computing*, 46 344-360, 2016.

[25] Yu V. F., Redi A. A. N. P., Yang C.-L., Ruskartina E., and Santosa B. Symbiotic organisms search and two solution representations for solving the capacitated vehicle routing problem, *Applied Soft Computing*, 52 657-672, 2017.

[26] Cheng M. Y., Prayogo D., and Tran D. H. Optimizing Multiple-Resources Leveling in Multiple Projects Using Discrete Symbiotic Organisms Search, *Journal of Computing in Civil Engineering*, 30(3), 04015036, 2016.

[27] Prayogo D., Cheng M.-Y., and Prayogo H. A Novel Implementation of Nature-inspired Optimization for Civil Engineering: A Comparative Study of Symbiotic Organisms Search, *Civil Engineering Dimension*, 19(1): 36-43, 2017.

[28] Alavi A. H., Ameri M., Gandomi A. H., and Mirzahosseini M. R. Formulation of flow number of asphalt mixes using a hybrid computational method, *Construction and Building Materials*, 25(3): 1338-1355, 2011.

An Investigation of Google Tango® Tablet for Low Cost 3D Scanning

Mark Froehlich[a], Salman Azhar[b], and Matthew Vanture[c]

[a]McCarthy Holdings Inc. Company, Dallas, Texas, United States
[b]McWhorter School of Building Science, Auburn University, Auburn, Alabama, United States
[c]The Whiting-Turner Contracting Company, Tampa, Florida, United States
E-mail: mfroehlich@mccarthy.com, salman@auburn.edu, Matt.Vanture@whiting-turner.com

Abstract –

Google Tango® is an augmented reality based Android tablet that contains customized hardware, an array of sensors, and software designed to track the full 3D motion of the device while simultaneously creating a map of the environment. The embedded sensors allow the device to make over a quarter million 3D measurements every second, updating its position and orientation in real-time, and combining that data into a single 3D model of the space around it. This arrangement allows users to create experiences such as indoor navigation, 3D mapping, augmented reality, physical space measurement, and environment recognition in a virtual world. The aim of this research study is to investigate the scope and accuracy of Google Tango® tablet for generating 3D point cloud based scans that could be used for schematic designing, space management, and for renovations and retrofitting planning. In addition, other functionalities of Tango® tablet that could be useful to designers, constructors, and facility managers (FM) are investigated. A mixed-methods approach was adopted for this study. Four testing scenarios were designed to collect the necessary quantitative data. In addition, interviews with BIM/laser scanning professionals were conducted to get their insight. The data analysis indicated that the accuracy of 3D scans in most cases is within ±5% of actual measurements which is fairly reasonable for schematic planning. Hence this technology could provide significant cost and time savings for the design, construction and FM industries. The paper first discusses the main concept and applications of Tango® tablet. After that, its possible applications for design, construction, and facility management are illustrated. This is followed by a brief discussion on the methodology and data collection methods. Based on the presented results, major findings and conclusions are outlined and recommendations for researchers and practitioners are presented.

Keywords –

Project Tango®; Simultaneous Localization and Mapping (SLAM) devices; 3D scanning; Point cloud; Augmented reality

1 Introduction

Tango® is an augmented reality computing platform, developed and authored by Google [1]. It uses computer vision to enable mobile devices, such as tablets and smartphones, to detect their position relative to the world around them without using GPS or other external signals. This arrangement allows developers and users to create experiences that include 3D mapping, indoor navigation, augmented reality, physical space measurement, and environment recognition in a virtual world [2]. Google has produced two devices to demonstrate the Tango® technology: the discontinued Peanut® phone and a 7-inch tablet. More than 3,000 of these devices had been sold chiefly to researchers and software developers interested in building applications for the Tango® platform [2]. This research study tested the 7-inch Tango® tablet which is shown in Figure 1.

Figure 1 The Google® Tango Tablet [1]

The Tango® platform is developed based on the concepts of motion tracking, depth perception, and area learning [3]. Tango® devices contain customized hardware sensors and software designed to track the full 3D motion of the device, while simultaneously creating a map of the environment. The sensors allow the device to make over a quarter million 3D measurements every second, updating its position and orientation in real-time, and combining that data into a single 3D model of the space around the user [3]. The main specifications of Tango® tablet are shown in Table 1.

Table 1. Specifications of Google® Tango Tablet [2]

Screen	7.02" 1920x1200 HD IPS display (323 ppi) Scratch-resistant Corning glass
Dimensions	119.77 x 196.33 x 15.36mm
Weight	0.82 lbs (370g)
Cameras	4 MP 2mm RGB-IR pixel sensor, 1 MP front facing, fixed focus
OS	Android 4.4 KitKat or higher
Wireless	4G LTE Dual-band Wi-Fi (2.4GHz/5GHz) WiFi a/b/g/n NFC chip
Audio Output	Dual stereo speakers 3.5mm audio connector
Memory	128 GB Storage; 4 GB RAM
Ports	Micro HDMI; USB 3.0; Micro SD card; Nano SIM slot
Processor	NVIDIA® Tegra K1 w/ 192 CUDA cores
Sensors	Motion tracking camera; 3D depth sensing; Accelerometer; Ambient Light; Barometer; Compass; GPS; Gyroscope

The applications of Tango® tablet are much broader than design, construction and facility management, as it could be used in other sectors such as retail, automotive, fashion, and service industries [3]. Tango® tablet does not offer laser-quality scanning but it allows users to perform rapid 3D acquisition of the surrounding space. It could create maps or floor layouts of large facilities. Once a building project is occupied, users can populate the building with more interior details. A Tango-enabled device could eventually access other analytics, such as temperature, vibration or noise, to create a rich map of the facility [4].

2 Background and Rationale

Over the last twenty years or so, laser-scanning technologies have seen a significant shift in applications across multiple disciplines [5]. Currently, there are a large number of commercial 3D scanners available for many different markets and applications covering art, automotive industry, archaeology, product design, architecture, fashion design, body measurement, civil engineering, metrology, forensics, and national security [6]. There are benefits and limitations of each type of scanner in terms of resolution, scan area, mobility, data capturing speed, tolerances and cost [7].

Clark and Liu conducted a field testing for drywall installation using laser scanning and BIM. They determined that field scanning requirements for simple rooms such as offices are minimal requiring only one medium resolution scan with an image capture for reference - a 20 minute time investment for each office space. The accuracy of scans proved that the integration of laser scanning and BIM technologies can be used in a specific area without having to make any (traditional) field measurements. The study additionally concluded that on most construction projects, the window for conducting scans is relatively short and suggested the need for a faster scanning process [5].

Becerik-Gerber et al. performed research on data acquisition errors during laser scanning caused by target setup, acquisition, and reorientation. They explored how different target types and target layouts affect the registration accuracy [8]. Their research suggested that laser scanning is not a straight forward process and a great care is needed to achieve reliable results. 3D laser scanning requires stillness in the scene that is captured. Thus, ongoing production activities can prove troublesome. If the geometries in the scene are numerous and complex, more data captures are required to get comprehensive spatial data of the entire scene [9]. Hence laser scanning requires both time and effort which in turn increases its overall cost.

The Google Tango® tablet provides a new avenue for conducting fast and low-cost scanning. Combining Infrared (IR) depth sensors and cameras paired with Simultaneous Localization And Mapping (SLAM) technology allows for scanning on the jobsite quickly with minimal training and preparation time by any construction personnel or trade worker. The purpose of this research study is to test the feasibility of Google Tango® tablet and its applications for use in the design, construction and facility management practice. This research specifically focused on preparation time, scanning time, and accuracy of the scans.

3 Google Tango® Applications (Apps) for Design, Construction, and Facility Management

Following is a brief description of Google Tango® applications for design, construction, and facility management industries. Readers interested in more details are advised to read the detailed study by Froehlich [10] or visit Tango® website [2].

3.1 Tango® Constructor

The most important Tango® app is "Constructor" which allows users to scan and view 3D models (or meshes) of the surrounding space. The users may export 3D scan files to various BIM software or other VR/AR tools (e.g. Unity®) [11].

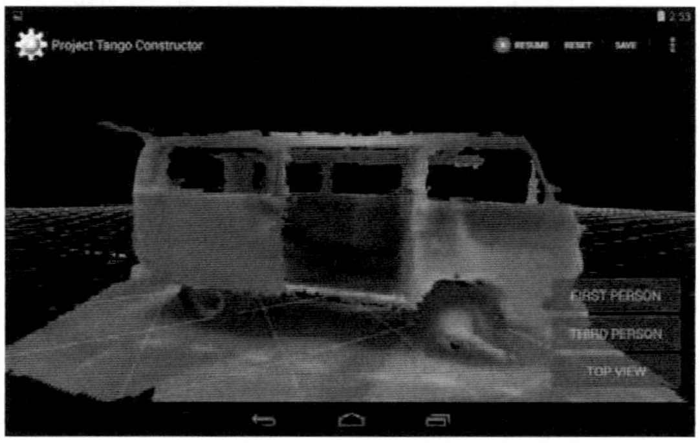

Figure 2 Google Tango® Constructor Application [11]

3.2 Tango® MeasureIt

"MeasureIt" is a simple application that allows users to take measurements of the surrounding space with a Tango® device. Users can simply point the device at what they would like to measure and use the cursor to place points. The measurements are displayed in an Augmented Reality view [12].

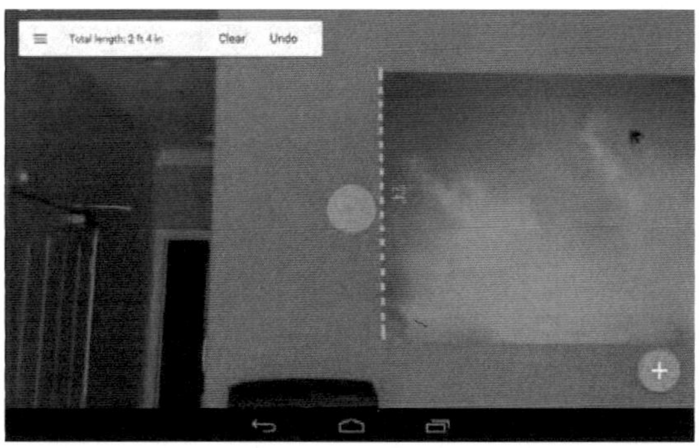

Figure 3 Google Tango® MeasureIt Application [12]

3.3 MagicPlan

MagicPlan creates floor plans. It measures existing rooms and surrounding spcaes, and draws floor plans just by taking pictures. Users have to simply add objects, annotations, and attributes to create the complete plan of a property. These plans can be saved in PDF, JPG, PNG, SVG, DXF, and CSV formats [2].

Figure 4 MagicPlan Screenshot [2]

3.4 Matterport Scenes

This app allows users to capture real-life scenes such as small spaces or objects in digital 3D. They can view, measure and share, from anywhere. In addition, they can measure any point-to-point dimension or use 3D trim to crop objects of interest. These objects can be shared via a Cloud-based library [2].

3.5 Wayfair™ View

WayfairView uses Augmented Reality (AR) technology to place real (furniture and related) products from Wayfair's (www.wayfair.com) extensive catalogue into any room on a 1:1 scale. WayfairView allows users to check the scale of products, see how items match their current décor, rearrange furniture in the space, and save or share photos [13].

Figure 5 Wayfair View Application [13]

3.6 Lowe's Tango® App

For the past two years, building products retailer Lowe's has been developing visualization applications using augmented and virtual reality tools, and Tango® made this technology accessible for everyday customers, right in their hands. Lowe's considered it a big step which can allows customers to better visualize and select products for their homes [2].

Figure 6 A screenshot of Lowe's Vision Tango App [2]

In addition to these six apps which were extensively tested in this research, there are several other available apps (most of them are currently in their beta versions) which are worth testing but were considered outside the scope of this study.

4 Research Aim, Objectives, and Scope

The aim of this research study is to investigate the scope and accuracy of Google Tango® tablet for generating low-cost point cloud based 3D scans that could be used for conceptual designing, space management, and for renovations and retrofitting planning. The research objectives includes the following:

1. Investigation of the state-of-the-art handheld Simultaneous Localization And Mapping (SLAM) enabled devices for onsite scanning and BIM model development;
2. Demonstration of their applications in design, construction, and facility management; and
3. Development of relevant examples for classroom and industry demonstrations.

This research was conducted using Google Tango® tablet which at the time of this research was the only SLAM enabled device available in the market.

5 Methodology and Data Collection

5.1 Methodology

A mixed-methods approach was adopted for data collection due to its strength for including both quantitative and qualitative data streams [14]. The major steps in this study were as follows:

1. The preliminary data was collected via extensive literature search. Related case studies published in various technical journals, magazines and conference proceedings were collected, reviewed and summarized.
2. Initial testing of the tablet was carried out though scanning as-built conditions. Three as-built area scans, a small conference room, a mechanical room, and a graduate students office in the Miller Gorrie Center on the campus of Auburn University, were generated. The produced scans were then compared to the as-built drawings. Time to prepare, conduct, and process the scans was assessed for each scenario.
3. Next, 3D scans of a mailroom located in the Whiting-Turner Contracting Company office, were created using terrestrial laser scanning and Tango® tablet and compared. The accuracy of both scans was verified using an already developed BIM model of the facility.
4. The results were presented to a focus group of 4 experts with experience in BIM, laser scanning, and advanced visualization technologies. Their feedback was collected using a series of questions and open discussion.
5. At the end, both quantitative and qualitative data were compared to draw necessary conclusions and recommendations.

5.2 Data Collection

5.2.1 Quantitative Data

The following procedure was adopted for scanning and/or creating point cloud using the Tango® tablet:

Due to the SLAM technology in the Tango® tablet and its ability to compute and recognize its location independently, without the aid of GPS require minimal preparation time (typically 3-5 minutes). Additionally, there was no need for markers to be placed as required by laser scanners. Unlike a fixed scanner, (e.g. laser scanners) the Tango® tablet utilizes built-in motion tracking which allows the user to sweep the area from different angles scanning untethered within a space.

The Tango® Constructor app was used to conduct the scans. Upon starting the tablet it calibrates itself requiring the user to extend the tablet out in front of him/her in a steady position for a few seconds. Once the tablet is calibrated, the user has full range to scan spaces within 0.5 to 4 meters (1.5 to 13 feet) away from the tablets sensors. Currently the tablet is designed to work best indoors at moderate distances [3]. The application supports functions allowing the user to pause, resume, or reset a scan in progress. This allows scanning to stop or pause and resume scanning of the same area. Additional options allow for the scan to be saved in the native file type for the application, as well as an .OBJ file type.

The results of a scan produce a 3D mesh which can be saved in the .OBJ file format and transferred to a computer for further processing. The file was then imported into Autodesk® Memento software and exported as an Autodesk® Recap Point Cloud file. The converted scan was then imported into Autodesk® Revit as a point cloud file thereby providing the same functionality as a laser scan produced point cloud.

5.2.2 Qualitative Data

The quantitative data analysis results were presented to a focus group of 4 experts who have 8 years or more experience in the field of BIM, laser scanning, and other advanced visualization technologies. The discussion started with the following questions: (1) Is Google Tango® a viable alternate to laser scanning? (2) What are its best applications in design, construction, and facility management? (3) What are its biggest benefits and limitations? and (4) What is the future of Tango® and other similar (SLAM) devices in the AEC industry? The experts provided their opinions on these four questions and then open-discussion was held to reach a final consensus. The major findings of the focus group are reported in section 6.5.

6 Results

6.1 Scenario 1: Conference Room Scan

The size of the chosen conference room was approximately 13ft x 18ft with a conference table and chairs centrally located in the room. To test the accuracy of the produced 3D models from the scan, two starting points, point A and point B were determined in the conference room, as shown in Figure 7.

The scanning path followed the perimeter of the room, clockwise from each starting point. A total of ten scans were conducted, five scans with a start at point A, and five scans with a start at point B. A live view of the 3D Mesh could be seen on the tablet screen as it scans the area. The view shows the mesh model simultaneously being constructed and refining itself as the tablet sweeps the area. The user has the following options for viewing scan on the screen: *first person*, *third person* or *top view*. First person view was chosen as it displays scanning in real-time as the user moves the tablet over areas to be scanned. Each scan was completed within the time range of 5-8 minutes. The scanning times and results are dictated by the user and their movements through the space.

The duration of a scan using the Tango® tablet is constrained by both hardware and software limitations. Hardware limitations include the battery life (appox. 80 minutes) and overheating of the device after 30 minutes of use which considerably affects the accuracy of the sensors. Main software limitation is the position loss of the Tango® tablet (drifting) during a scan due to motion tracking failures that results in noisy 3D models [3, 10].

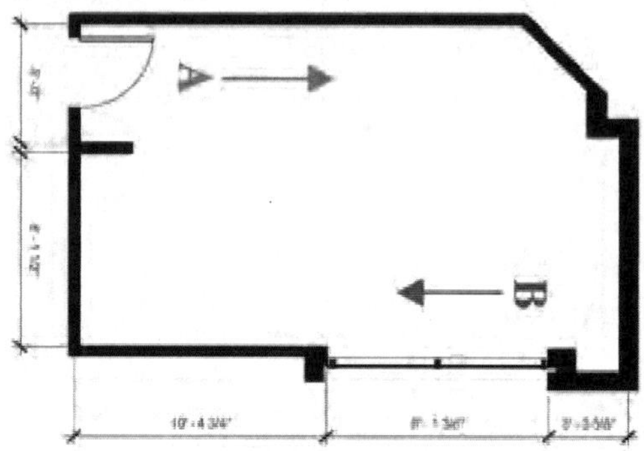

Figure 7 Starting Points of Scan

Figure 8 illustrates the results of a scan initiated at starting point A, scanning clockwise around the room.

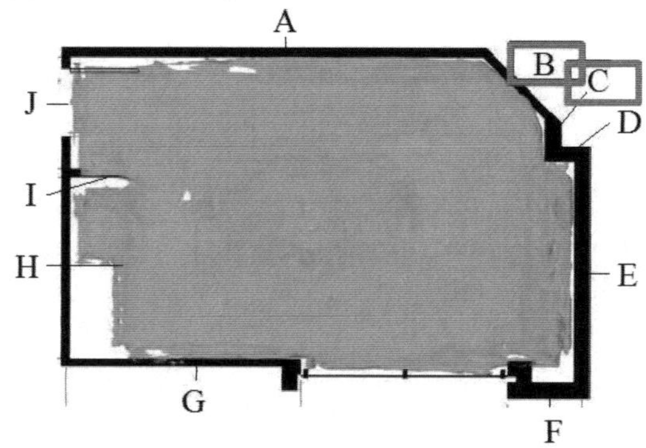

Figure 8 Tango® Scan of Small Conference Room

Table 2 shows a comparison of the results with the as-built model measurements.

Table 2 Small conference room scan analyses of As-built measurements and Tango® scanned measurements

Wall	As-Built Model m	As-Built Model ft	Tablet Scan Average m	Tablet Scan Average ft	Variance %
A	5.54	18.16	5.65	18.53	2.01
B	1.19	3.90	0.96	3.14	19.44
C	0.63	2.07	0.53	1.75	15.38
D	0.39	1.29	0.39	1.28	0.94
E	2.65	8.71	2.66	8.74	0.34
F	0.93	3.04	0.90	2.95	2.92
G	2.59	8.50	2.64	8.65	1.77
H	2.47	8.11	2.48	8.14	0.27
I	0.66	2.17	0.69	2.27	4.99
J	1.52	5.00	1.55	5.07	1.49

As can be noted that the percentage variance for most measurements is within ±5 percent except for walls B and C. Both walls were unevenly illuminated by the natural light coming from the front window and that may be the cause of this large error. This assumption was further supported by the data collected in scenario 2. It is important to note that the scanning accuracy of the overall model deteriorated with time and its movement along the path [10]. This was due to drifting of the tablet and its ability to maintain its known location in the space. The portions affected by the drift maintained a greater level accuracy based of its new coordinate location within the space.

6.2 Scenario 2: Mechanical Room Scan

Next the mechanical room within the Gorrie Center was chosen for scanning. The room is filled with mechanical equipment, piping, and ducts. The small space would be a challenge for laser scanning, and multiple views were required to develop a 3D scan of the space. Figure 9 shows a point cloud of the Mechanical Room produced from the tablet. The scan of the Mechanical Room gave the tablet more surfaces and objects to reference itself to as compared to scenario 1. Overall there was only minor drifting which occurred along the bottom wall as shown in Figure 9. The location where the drifting happened was poorly lit as well as having only flat surfaces, both of which are the likely cause of the drifting. The entire scan of the room was completed in 19 minutes. The complexity and tight spaces of the room required additional attention in order to get a usable model. The existing Revit model did not include the modelling of the pumps, pipes, ducts and equipment. The point cloud produced from the tablet scan shows the location of those items. Table 3 shows a comparison of measurements using As-built model and Tango scans. It is important to note that the variances for all walls are within ±5% except Walls D, E, and G due to either uneven illumination or poor lighting conditions.

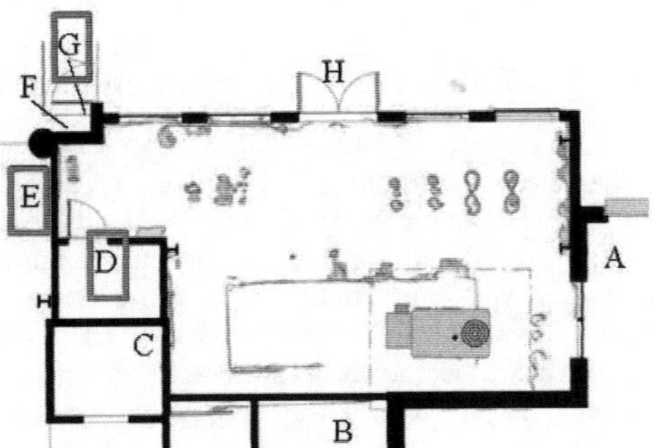

Figure 9 Mechanical Room Plan and Superimposed Scan

Table 3 Mechanical room scan analyses of As-built measurements and Tablet scanned measurements

Wall	As-Built Model		Tablet Scan Average		Variance
	m	ft	m	ft	%
A	6.28	20.60	6.49	21.29	3.3
B	9.22	30.25	9.65	31.66	4.7
C	3.62	11.88	3.81	12.50	5.2
D	2.50	8.20	2.11	6.92	15.6
E	2.22	7.28	3.15	10.33	41.9
F	1.04	3.41	1.07	3.51	2.9
G	0.46	1.51	0.41	1.35	10.9
H	10.68	35.04	10.90	35.76	2.1

6.3 Scenario 3: Graduate Students Office

In this scenario, the accuracy of the Google Tango's MeasureIt® app was tested by taking measurements of the Graduate Students Office located in the Gorrie center. Table 4 shows the variance percentage between the results of a laser scan and Tango measurements. The results of this scenario showed greater improvement in accuracy in comparison to the 3D space scanning performed in scenario 1 and 2. Overall time to prepare, conduct, and process the data was almost one-third as compared to scenarios 1 and 2.

Table 4 Graduate students' office scan analyses of As-built measurements and Tablet scanned measurements through MeasureIt App

Wall	Laser Scan based Measurement		Tango based Measurement		Variance
	(Meters)	(Feet)	(Meters)	(Feet)	(Percent)
A	1.63	5.34	1.60	5.24	1.86%
B	3.26	10.71	3.24	10.64	0.61%
C	0.59	1.93	0.60	1.95	-1.19%
D	3.78	12.41	3.76	12.33	0.61%
E	12.11	39.73	11.97	39.26	1.18%

It is important to note that the measurements of the spaces produced varying results based on where the points were selected within the physical space. Measurements were taken both at the bottom and at the top of walls. This arrangement reduced discrepancies within the data sets utilizing the straight edge of the floor and ceiling to get a perfect straight line. Additionally, measuring a space within a fixed point of view is proved to be more accurate than those requiring walking or moving through the space with the tablet.

6.4 Scenario 4: Mailroom Scan

In the final scenario, the Tango® tablet was used to scan the mailroom of the Whiting-Turner Contracting Company office located in Tampa, Florida. The intent of this testing was twofold: (1) to compare measurements taken by laser and Tango® scans; and (2) to visually assess the level of detail for both scans and to their determine their viability for schematic planning.

Figures 10-12 (all images courtesy of the Whiting-Turner Contracting Company) compares the level of details using an actual photo, a laser scan, and a Tango® scan of the same area.

Figure 10 Mailroom photo taken by iPhone 5S

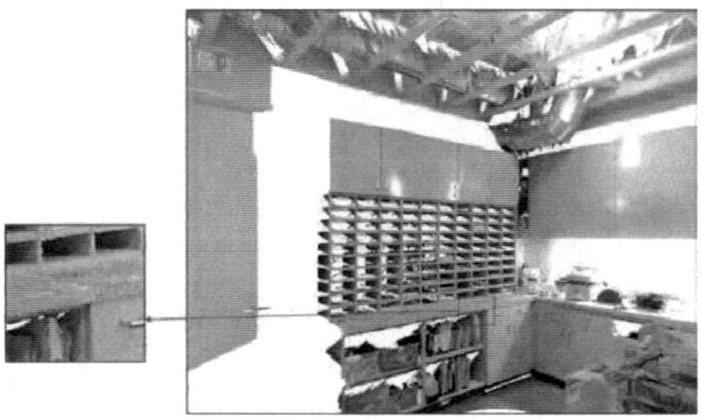

Figure 11 Mailroom image through a Laser scan

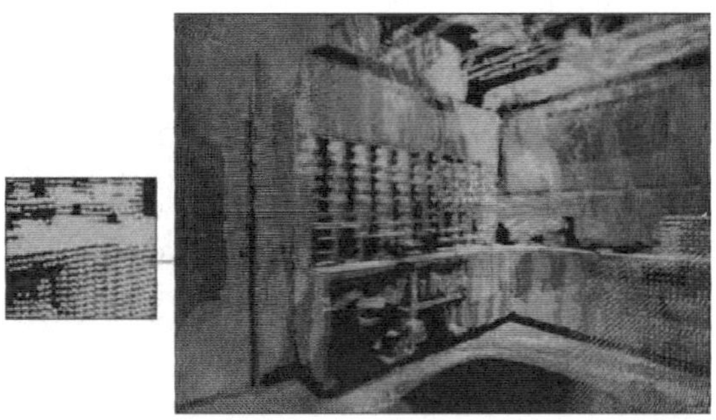

Figure 12 Mailroom image through a Tango® scan

Although not shown here but all measurements of laser and Tango® scans were very close to each other and within a variance of ±5 percent. In this case, all scans were created under artificial lighting and every effort was made to minimize the illumination effect that was the major error source in scenarios 1 and 2. By comparing Figures 10-12, one can determine that the level of detail of Tango® is the poorest but it can still recognize objects fairly well and can be reasonably used for schematic planning.

6.5 Main Findings of the Focus Group

The focus group members agreed on the following conclusions:

- Google Tango® is an innovative piece of technology that has a lot of potential in the design, construction an FM industries. At this point, it is neither very accurate not very reliable to be considered as an alternate of laser scanning however it can be reasonably used in the initial planning process as a substitute of laser scanning especially for maintenance and retrofitting projects.
- It can be used for schematic design comparisons, jobsite layout, measuring the height of overhead obstructions for safety planning, and/or planning for confined spaces.
- Its biggest advantage is its cost which is around $500 and can be easily afforded by most design and construction firms. Since the tablet is at its infancy stage and most of the apps are in their beta versions hence it is difficult to fully rely on the results they produce.
- The tablet offers several promising applications for design, construction and facility management. It is hoped that the software developers will further develop and refine their apps.
- The focus group members were in agreement that a time period of at least 3-5 years would be needed before one can see its commercial use in the design, construction, and FM industries.

7 Conclusions and Recommendations

The Google Tango® tablet is an innovative piece of technology allowing the user to quickly scan and/or measure surrounding spaces. While the technology is still in its infancy it shows a promising future for use in the design, construction, and FM industries. The cost paired with the Android OS provide a versatile tool to aid any user regardless of the industry. With many retailers developing applications for consumer shopping and home use, SLAM and AR are not going away.

While scans generated through Tango® tablet are not of the same quality or high accuracy as of terrestrial laser scans, they can serve as a beneficial aid during design and construction. Time and skill required to complete scans on site are minimal and can prove to be a beneficial tool. Additionally, there are significant costs and time savings in comparison to the cost of traditional laser scanners.

This research determined that there is a wide range of factors that influence the integrity of a produced point cloud. For any area in order to be scanned must have surfaces capable of being registered by the tablet sensors. Translucent or glass material cannot be read by the tablet, as well as shiny or surfaces with a gloss finish. The tablet sensors are not able to get an accurate depth reading due to the scattering of the transmitted laser and IR signals. Uniform lighting is very important to fully scan an area without noise or distortion. Too much light, or not enough light caused drifts in the tablet's location within the space as well as holes in the produced model or point cloud. Direct sunlight on surfaces, specifically cause holes, as they are not able to be scanned, similar to glass or windows. As this research was conducted indoors, lighting was found to be detrimental to the overall quality of a scan. The users' method of scanning a space, in regards to how they hold the tablet, their speed through the space, and ability to thoroughly scan around the space from multiple perspectives can affect the scan quality. Further research to determine the best practices for scanning using the Tango® tablet would be beneficial. Additionally, through this research it can be determined that the quantity and close proximity of surfaces to one another can produce a more accurate model. The more surfaces the tablet is able to identify and register, the more accurate it can identify its own location within an area to be scanned.

The use of the Google Tango® allows users to relay information of which it is spatial aware of, thereby improving the coordination and communication of trades on site. Deploying a downloadable application of the Model allows for simple exchange of information. Not only can the Google Tango® tablet serve for trade coordination but it can serve as a digital manual for building handover. Providing an owner with SLAM device with the spatially aware information within a BIM model can serve as a beneficial information tool.

This research tested the applications of area scanning for design, construction and FM industries and determined that Tango® tablet has the great potential for fast area scanning. While scanning of large spaces is not nearly as accurate as scanning smaller ones, they both offer advantages. Future research will build upon this study, continuing research with the Tango® Tablet, as well as the Microsoft® HoloLens. The HoloLens® is a mixed reality headset which utilizes SLAM technology to overlay information within a space. The headset allows users a hands-free experience and like the Tango® tablet, it scans the space as the user moves around. Finally, the SLAM is an innovative technology that has yet to be fully realized but offers great potential to innovate the design, construction and FM industries.

8 References

[1] Wikipedia. (2016). Tango Platform. Retrieved from https://en.wikipedia.org/wiki/Tango_(platform).

[2] Google Tango. (2017). Project Tango. Retrieved from https://get.google.com/tango/.

[3] Diakite, A.A., & Zlatanova, S. (2016). First experiments with the Tango tablet for indoor scanning, ISPRS Annals of the Photogrammetry, Remote Sensing & Spatial Information Sciences, Vol. III-4.

[4] Wilkinson, P. (2014). BIM+ - Google's Project Tango? Retrieved from: http://www.bimplus.co.uk/technology/what-googles-project-tango/

[5] Caitlin T. C. & Liu, J. (2014). A case study of integrating 3D high definition laser scanning technology and BIM into drywall installation. 50th ASC Annual International Conference Proceedings.

[6] Wulf, O., and Bernardo, W. (2003). Fast 3D Scanning Methods for Laser Measurement System. Retrieved from: http://citeseerx.ist.psu.edu/viewdoc/download?doi=10.1.1.576.9832&rep=rep1&type=pdf

[7] Aniwaa. (2016). 3D Scanners Comparison. Retrieved from: http://www.aniwaa.com/comparison/3d-scanners/.

[8] Becerik-Gerber B, Jazizadeh F, Kavulya G, Calis G. (2011). Assessment of target types and layouts in 3D laser scanning for registration accuracy. Automation in Construction 2011;20:649–58. doi:10.1016/j.autcon.2010.12.008.

[9] Kabir, H., Biswasa D, Boschéa F, & Suna M. (2015). Planning for Scanning Using Building Information Model: A Novel Approach with Occlusion Handling. Symposium on Automation and Robotics in Construction and Mining (ISARC 2015), vol. 15, 2015, p. 18.

[10] Froehlich, M. (2016). Handheld Simultaneous Localization and Mapping Devices for BIM Visualization. Graduate Capstone, Auburn University, Auburn, Alabama, USA.

[11] Google. (2016a). Tango Constructor Project Tango. Google Play. Retrieved from: https://play.google.com/store/apps/collection/promotion_3001310_project_tango_featured.

[12] Google. (2016b). Tango MeasureIt. Google Play. Retrieved from: https://play.google.com/store/apps/collection/promotion_3001310_project_tango_featured.

[13] WayfairView. (2016). Google Play. Retrieved from: https://play.google.com/store/apps/details?id=com.wayfair.wayfairview

[14] Research Rundowns. (2016). Mixed Methods Research Designs, https://researchrundowns.com/mixed/mixed-methods-research-designs

Building element recognition with thermal-mapped point clouds

Pileun Kim[a], Jingdao Chen[b], and Yong Kwon Cho[a]

[a] School of Civil and Environmental Engineering, Georgia Institute of Technology
[b] School of Electrical and Computer Engineering, Robotics Institute, Georgia Institute of Technology
E-mail: pkim45@gatech.edu, jchen490@gatech.edu, yong.cho@ce.gatech.edu

Abstract –

Automated recognition of building elements convey vital information for inspection, monitoring and maintenance operations in indoor environments. However, existing object recognition methods from point clouds suffer from problems due to sensor noise, occlusion and clutter, which are prevalent in indoor environments. This paper proposes an object recognition method based on thermal-mapped point clouds for building elements consisting of electrical systems and heating, ventilation, and air-conditioning (HVAC) components. The proposed processing pipeline involves data collection from a mobile robot using both laser scanners and a thermal camera where temperature mapping can be performed from thermal images to point cloud. Next, the ceiling region containing the building elements of interest is identified and extracted from the point cloud. Segmentation of peak and valley thermal intensity regions is carried out based on absolute and relative temperature threshold values. The identified point cloud clusters can be each associated with a building element and localized based on the cluster center. The proposed building element recognition method was validated with two sets of laser scan data collected in an indoor laboratory. Experimental results for detection of lighting elements and cooling elements showed that the method achieved an average of 100% precision, 90% recall, and 0.25m root mean squared error (RMSE).

Keywords –

Laser scanning; Point cloud; Object recognition; Thermal image

1 Introduction

Laser scanned data in the form of 3D point clouds has emerged in recent years as one of the primary methods of data collection from built environments as well as construction sites. Due to the large number of possible applications, laser scanning technology has rapidly improved in terms of scanning speed and accuracy [1]. 3D point clouds are advantageous because they contain valuable geometric information at a high resolution that can be used to perform semantic modelling [2], quality assessment [3], defect analysis [4], and asset management [5].

Researchers in the field have explored various ways to automatically recognize building elements from unstructured point cloud data in order to reduce the manual effort required to annotate and label the point clouds. Examples of building elements that can be recognized are walls, ceilings, floors, doors, windows, and roofs [2][6]. In particular, recognition of planar structures such as walls and floors is a key component of robotic mapping of indoor environments [7]. Recognition of building elements is also highly relevant for Building Information Modelling (BIM) applications, which involve the creation of a BIM model representing the shape, identity, and relationships of scanned objects [8][9][10]. However, point cloud data is often collected at a high volume which imposes high memory and processing requirements for automated point cloud processing algorithms due to the large number of points involved [11]. In addition, raw point cloud data collected from indoor environments is also known to suffer from sensor noise, occlusion, and clutter [12].

In short, the requirement for automatic object recognition methods is increasing by using robots deployed in field operations. To address the need for more sophisticated building element recognition methods from point clouds, this study proposes a method using a combination of thermal images and laser-scanned point cloud to detect and localize electrical systems and heating, ventilation, and air-conditioning (HVAC) components. In general, the thermal information provides many advantages because people can be detected easily both indoor and outdoor environments. Also, the point cloud offers 3D geometric data. Therefore, the 3d point cloud with thermal data can observe a sort of concealed objects such as heating pipes as well as structural properties such as defects. The method relies on the heat signature emitted by these building elements to accurately extract and identify the building elements from a thermal-mapped point cloud. This is because regions in the point cloud with high thermal intensity can be attributed to lighting elements whereas regions with low thermal intensity can be attributed to cooling elements. The following sections will present an overview of the related work, methodology, results, discussion, and conclusion.

2 Related Work

Recognition of building elements from a job site can be performed through various different mediums such as visual images [13][14][15][16], thermal images [17], [18], and laser-scanned point clouds [2][6]. Each sensory medium has its own advantages and disadvantages when applied to the task of object recognition. For example, 2D-based sensing is vulnerable to changes in view angle, lighting conditions, as well as cases with multiple occluding objects [19]. On the other hand, 3D-based sensing requires a computationally intensive registration step to combine scan data [20] and is also subject to clutter and point density variations [21]. The following subsections summarize the common techniques that are used for object recognition from point clouds and thermal images respectively.

2.1 Object recognition from point clouds

Object recognition methods from point clouds rely on geometric information from point samples to infer semantic properties of the unknown object. For objects that exhibit primitive shapes such as walls and floors, simple geometric reasoning can be used to identify these components in a point cloud [2], [6]. For objects that exhibit complex geometry, one possible method is to use a lower dimensional feature representation of the point cloud in a vector form known as a descriptor [22]. A 3D descriptor can be calculated based on histograms of orientations [23], curvatures [24], or length and area statistics [25]. Additionally, when a computer-aided design (CAD) model of the target object is available, object recognition can be performed by directly registering the CAD model with the point cloud object and examining the registration error [11]. Alternatively, the 3D data can be projected into a 2D form to take advantage of existing 2D recognition algorithms [26].

2.2 Object recognition from thermal images

There are various methods for recognizing objects from a thermal image. Treptow et al. [27] used an elliptical model with particle filter to identify the contours of a person from the thermal images. This method shows that accurate detection at high frame rates, but it only works for one person with the highest measurement probability. Similarly, Socolinsky et al. [28] revealed that performing facial recognition by using thermal image is more accurate than visual image because of high invariance to illumination condition. However, these studies focused on 2D thermal imaging for the small objects and did not consider global perception at the entire space. To solve the problem of detecting multiple objects from a single image, Davis et al. [29] used the technique of contour enhancement. This technology successfully detects multiple objects on a thermal image over a wide range of environmental conditions. It is also robust with respect to object shape and can extract silhouettes such as people, dogs and vehicles. However, it only generates an overview of the detected objects and does not explicitly indicate the meaning of each object. To detect building elements, Balaras and Argiriou [30] used thermal infrared imaging for inspecting of building elements. Also, Freitas et al. [31] used infrared thermography for a rapid non-destructive diagnosis of building envelopes. However, their work only 2D thermal images so that it might be easy to detect where energy leaks and perform diagnotics, but difficult to identify the shape for object recognition.

3 Methodology

The task of recognizing building elements from thermal-mapped point clouds can be divided into three steps: (i) collect data from a mobile robot, (ii) perform thermal mapping for the point cloud data, and (iii) perform recognition of building elements. Each step will be described in detail in the subsections below:

3.1 Data collection from a mobile robot

A robotic system, as shown in Figure 1, is used to acquire map information from the environment in the form of point cloud data as well as thermal images. The robotic system consists of a hybrid scanning framework with 4 units of SICK laser scanners [32] and one thermal camera (640 x 480 pixels) [33] mounted on a rotating body. The laser scanners used are line laser scanners with a horizontal resolution of 0.072 degrees and a vertical resolution of 0.167 degrees. It performed scans 150° horizontally and 190° vertically. Therefore, the total number of collected points was around 1,000,000. The infrared camera is able to capture 25 degrees in the horizontal direction and 18 degrees in the vertical direction at a time; therefore, at least six panning movements are required to cover 150 degrees. In this experiment, three tilting movement are applied to cover 54 degrees vertically.

Figure 1: Sensor setup for the robotic system

3.2 Thermal mapping for point cloud data

The following subsection describes how to combine the acquired point cloud data with thermal images. The process works by mapping each 3D point in the point cloud, (x, y, z) to 2D homogenous image coordinates (u, v, w) in thermal images and extracting the temperature information from that pixel location (Equation 1). The camera extrinsic matrix, which contains a rotation term, R and a translation term, T, maps 3D points from the world coordinate frame to the camera coordinate frame. This matrix can be determined by the position and orientation of the thermal camera when each thermal image is captured. On the other hand, the camera intrinsic matrix, K, describes the mapping between 3D points in camera coordinates to 2D points in image coordinates. This matrix is shown in Equation 2, where f_x and f_y are the focal lengths in the x and y axes whereas (c_x, c_y) is the image center. The values of these parameters can be obtained by a camera calibration process.

$$\begin{bmatrix} u \\ v \\ w \end{bmatrix} = K\,[R\mid t]\begin{bmatrix} x \\ y \\ z \\ 1 \end{bmatrix} \qquad (1)$$

$$K = \begin{bmatrix} f_x & 0 & c_x \\ 0 & f_y & c_y \\ 0 & 0 & 1 \end{bmatrix} \qquad (2)$$

The panoramic view of the laser scanned point cloud can be created directly using the angular coordinates of each point. Equation 3 parameterizes the mapping from the points of 3D camera coordinates (u, v, w) to its corresponding points on the panoramic view (x, y) where x, y are the image coordinates of the projected point cloud.

$$\begin{aligned} x &= \frac{1}{\Delta\theta}\tan^{-1}\frac{v}{w} \\ y &= \frac{1}{\Delta\theta}\tan^{-1}\frac{w}{\sqrt{u^2+v^2}} \end{aligned} \qquad (3)$$

3.3 Recognition of building elements

After a thermal-mapped point cloud is acquired, building elements of interest can be detected and localized from the point cloud. First, the point cloud is subdivided into different subsections based on the z-value, which is the height. The subsection with the largest z-value, which represents the ceiling region, is extracted from the point cloud. An example of such an extracted subsection is shown in Figure 2.

Figure 2: Subsection of a building point cloud consisting of the ceiling region

Next, the thermal information contained in the point cloud is used to identify heat-emitting objects, which in this case are lighting elements, as well as heat-absorbing objects, which in this case are cooling elements. The main idea is to identify regions in the point cloud which constitutes peaks (high intensity) or valleys (low intensity) in terms of the temperature. This is achieved by using a combination of absolute thresholds and relative thresholds. An absolute threshold is a fixed value that applies to the entire point cloud. On the other hand, a relative threshold is applied by comparing the ratio of the temperature of a point to the temperatures of the neighbouring points. This helps to overcome the problem of non-uniform distribution of temperature across the point cloud. Then, points which have temperatures greater than a predetermined upper threshold will be considered as occupying peak regions whereas points which have temperatures less than a predetermined lower threshold will be considered as occupying valley regions. This is illustrated in Figure 3, where thermal intensity peak and valley regions are highlighted with respect to the original point cloud.

(a)

(b)

Figure 3: Thermal intensity (a) peak and (b) valley regions highlighted in red.

Finally, the peak and valley regions are subdivided into individual elements using a clustering algorithm. The clustering algorithm works by randomly sampling seed points and expanding individual clusters by incrementally assigning points which are geometrically close to each other to be part of the same cluster. This procedure is repeated until all points are assigned to a unique cluster. However, clusters with less than 50 points are filtered out since it is assumed that clusters which are too small do not originate from building elements but are a result of noise in the data. A visualization of this step is shown in Figure 4, where individual point cloud clusters are highlighted in different colours. Each point cloud cluster is considered a prediction of a building element and the centroid of the point cloud cluster is taken to be the centroid of the predicted building element.

(a)

(b)

Figure 4: Point cloud clusters of detected building elements: (a) lighting elements and (b) cooling elements

4 Results

The proposed method is evaluated with laser-scanned point clouds of an indoor laboratory in Georgia Institute of Technology. Two different laser scans were collected from the site. The thermal mapping process described in Section 3.2 was used to map the temperature information from thermal images to the point cloud. Figure 5 shows the resulting thermal-mapped point cloud in an indoor lab setting.

Figure 5: Thermal-mapped point cloud in an indoor lab setting

Next, the method proposed in Section 3.3 was used to detect and localize lighting elements and cooling elements. The accuracy of the proposed building element recognition method is evaluated with three different metrics. The precision metric measures the number of correct predictions of building elements compared to the total number of predictions made. The recall metric measure the number of correct predictions of building elements compared to the actual number of building elements present in the original scene. The root mean squared error (RMSE) metric measures the distance error between the centroid of a predicted building element and the ground truth centroid which is obtained manually. Table 1 shows the precision, recall, and RMSE results for lighting and cooling elements in two different laser scans (labelled A and B respectively).

Table 1: Precision, recall and RMSE for each building element

Building element	Precision	Recall	RMSE (m)
Lighting A	1.00	0.86	0.146
Cooling A	1.00	1.00	0.098
Lighting B	1.00	0.75	0.277
Cooling B	1.00	1.00	0.489

5 Discussion

The performance of the proposed recognition system can be analysed with respect to the precision, recall, and RMSE metrics. Table 1 shows that the proposed method is able to achieve perfect precision in terms of the building elements that are recognized. This is because the combination of thermal information and geometric information helps to eliminate false positives and the resulting detections can be labelled as building elements with high confidence. However, in the case of recall rates, the method only achieves around 80% for lighting elements. This is due to the fact that the laser scan data collected is non-uniform in resolution. For regions in the point cloud that are far away from the laser scan origin, the point cloud resolution is low and the thermal mapping process is error-prone. Thus, building elements that are located far away from the laser scan origin tend to result in missed detections. Finally, the results for RMSE showed that the proposed method can localize building elements to within about 0.25m of the ground truth. The high RMSE value for the "Cooling B" test set is due to an outlier where the low temperature region is mistakenly assigned to an adjacent patch so the predicted centroid is inconsistent with the ground truth centroid.

6 Conclusion

This research proposes a method for recognizing building elements from thermal-mapped point clouds based on regions of high and low thermal intensity. The proposed processing pipeline involves data collection from a mobile robot using both laser scanners and a thermal camera where temperature mapping can be performed from thermal images to point cloud. Next, the ceiling region containing the building elements of interest is identified and extracted from the point cloud. Segmentation of peak and valley thermal intensity regions is carried out based on absolute and relative temperature threshold values. The identified point cloud clusters can be each associated with a building element and localized based on the cluster center. The proposed building element recognition method was validated with two sets of laser scan data collected in an indoor laboratory using the metrics of precision, recall and RMSE. Experimental results showed that the method achieved perfect precision, which means that there are no false positive results. In addition, a promising 90% recall rate and 0.25m RMSE was achieved as well. Future work in this area would involve increasing the types of building elements that can be recognized as well as quantifying the accuracy on a larger dataset.

7 Acknowledgement

This material is based upon work supported by the National Science Foundation (Award #: CMMI-

1358176). Any opinions, findings, and conclusions or recommendations expressed on this material are those of the authors and do not necessarily reflect the views of the NSF.

References

[1] S. El-Omari and O. Moselhi, "Integrating automated data acquisition technologies for progress reporting of construction projects," *Autom. Constr.*, vol. 20, no. 6, pp. 699–705, 2011.

[2] C. Wang, Y. K. Cho, and C. Kim, "Automatic BIM component extraction from point clouds of existing buildings for sustainability applications," *Autom. Constr.*, vol. 56, pp. 1–13, 2015.

[3] L. Bu and Z. Zhang, "Application of Point Clouds From Terrestrial 3D Laser Scanner for Deformation Measurements," *Int. Arch. Photogramm. Remote Sens. Spat. Inf. Sci.*, vol. XXXVII, pp. 545–548, 2008.

[4] E. B. Anil, P. Tang, B. Akinci, and D. Huber, "Deviation analysis method for the assessment of the quality of the as-is Building Information Models generated from point cloud data," *Autom. Constr.*, vol. 35, pp. 507–516, 2013.

[5] H. Son, F. Bosché, and C. Kim, "As-built data acquisition and its use in production monitoring and automated layout of civil infrastructure: A survey," *Adv. Eng. Informatics*, vol. 29, no. 2, pp. 172–183, 2015.

[6] X. Xiong, A. Adan, B. Akinci, and D. Huber, "Automatic creation of semantically rich 3D building models from laser scanner data," *Autom. Constr.*, vol. 31, pp. 325–337, 2013.

[7] S. Thrun, C. Martin, Y. Liu, and D. Hahnel, "A real-time expectation maximization algorithms for acquiring multi-planar maps of indoor environments with mobile robots," *IEEE Trans. Robot.*, vol. 20, no. 3, pp. 433–443, 2004.

[8] S. Azhar, "Building Information Modeling (BIM): Trends, Benefits, Risks, and Challenges for the AEC Industry," *Leadersh. Manag. Eng.*, vol. 11, no. 3, pp. 241–252, 2011.

[9] P. Tang, D. Huber, B. Akinci, R. Lipman, and A. Lytle, "Automatic reconstruction of as-built building information models from laser-scanned point clouds: A review of related techniques," *Autom. Constr.*, vol. 19, no. 7, pp. 829–843, 2010.

[10] V. Pătrăucean, I. Armeni, M. Nahangi, J. Yeung, I. Brilakis, and C. Haas, "State of research in automatic as-built modelling," *Adv. Eng. Informatics*, vol. 29, pp. 162–171, 2015.

[11] F. Bosché, "Automated recognition of 3D CAD model objects in laser scans and calculation of as-built dimensions for dimensional compliance control in construction," *Adv. Eng. Informatics*, vol. 24, no. 1, pp. 107–118, 2010.

[12] A. Dimitrov and M. Golparvar-fard, "Segmentation of building point cloud models including detailed architectural / structural features and MEP systems," *Autom. Constr.*, vol. 51, pp. 32–45, 2015.

[13] J. Seo, S. Han, S. Lee, and H. Kim, "Computer vision techniques for construction safety and health monitoring," *Adv. Eng. Informatics*, vol. 29, no. 2, pp. 239–251, 2015.

[14] M. Memarzadeh, M. Golparvar-Fard, and J. C. Niebles, "Automated 2D detection of construction equipment and workers from site video streams using histograms of oriented gradients and colors," *Autom. Constr.*, vol. 32, pp. 24–37, 2013.

[15] S. Han and S. Lee, "A vision-based motion capture and recognition framework for behavior-based safety management," *Autom. Constr.*, vol. 35, pp. 131–141, 2013.

[16] S. Chi and C. H. Caldas, "Automated Object Identification Using Optical Video Cameras on Construction Sites," *Comput. Civ. Infrastruct. Eng.*, vol. 26, no. 5, pp. 368–380, 2011.

[17] C. Wang and Y. K. Cho, "Automatic 3D Thermal Zones Creation for Building Energy Simulation of Existing Residential Buildings," *Constr. Res. Congr. 2014*, pp. 1014–1022, 2014.

[18] K. Zheng, Y. K. Cho, C. Wang, and H. Li, "Noninvasive Residential Building Envelope R - Value Measurement Method Based on Interfacial Thermal Resistance," vol. 22, no. 4, pp. 1–9, 2016.

[19] J. Gong and C. H. Caldas, "An object recognition, tracking, and contextual reasoning-based video interpretation method for rapid productivity analysis of construction operations," *Autom. Constr.*, vol. 20, no. 8, pp. 1211–1226, 2011.

[20] Y. Kim, P., Chen, J., Cho, "Automatic registration of laser scanned color point clouds based on common feature extraction," in *International Symposium on Automation and Robotics in Construction (ISARC)*, 2016.

[21] Y. Guo, M. Bennamoun, F. Sohel, M. Lu, and J. Wan, "3D object recognition in cluttered scenes with local surface features: A survey," *IEEE Trans. Pattern Anal. Mach. Intell.*, vol. 36, no. 11, pp. 2270–2287, 2014.

[22] J. Chen, Y. Fang, Y. K. Cho, and C. Kim, "Principal Axes Descriptor for Automated Construction-Equipment Classification from Point Clouds," *J. Comput. Civ. Eng.*, pp. 1–12,

2016.

[23] F. Tombari, S. Salti, and L. Di Stefano, "Unique signatures of histograms for local surface description," *Lect. Notes Comput. Sci. (including Subser. Lect. Notes Artif. Intell. Lect. Notes Bioinformatics)*, vol. 6313 LNCS, no. PART 3, pp. 356–369, 2010.

[24] H. Tanaka, M. Ikeda, and H. Chiaki, "Curvature-based Face Surface Recognition Using Spherical Correlation Principal Directions for Curved Object Recognition," *Proc. Third IEEE Int. Conf. Autom. Face Gesture Recognit.*, pp. 372–377, 1998.

[25] W. Wohlkinger and M. Vincze, "Ensemble of shape functions for 3D object classification," *2011 IEEE Int. Conf. Robot. Biomimetics*, pp. 2987–2992, 2011.

[26] Y. K. Cho and M. Gai, "Projection-Recognition-Projection Method for Automatic Object Recognition and Registration for Dynamic Heavy Equipment Operations," *J. Comput. Civ. Eng.*, vol. 28, no. 5, p. A4014002, Sep. 2014.

[27] A. Treptow, G. Cielniak, T. Duckett, "Active people recognition using thermal and grey images on a mobile security robot", 2005 IEEE/RSJ Int. Conf. Intell. Robot. Syst. IROS. (2005) 3610–3615. doi:10.1109/IROS.2005.1545530.

[28] D. Socolinsky, L. Wolff, J. Neuheisel, "Illumination invariant face recognition using thermal infrared imagery", IEEE Comput. Vis. Pattern Recognit. 1 (2001) 527–534. doi:10.1109/CVPR.2001.990519.

[29] J.W. Davis, V. Sharma, "Background-subtraction in thermal imagery using contour saliency", Int. J. Comput. Vis. 71 (2007) 161–181. doi:10.1007/s11263-006-4121-7.

[30] Balaras, C. and Argiriou, A. "Infrared thermography for building diagnostics". Energy and Buildings 34, pp. 171–183.

[31] Freitas, SS. Freitas VP., and Barreira, E. "Detection of façade plaster detachments using infrared thermography – A nondestructive technique", Construction and Building Materials, Vol. 70, pp. 80-87.

[32] "SICK." [Online]. Available: https://www.sick.com/us/en/.

[33] "FLIR" [Online] Available: http://www.flir.com/automation/display/?id=41330

Optimal Placement of Precast Bridge Deck Slabs with respect to Precast Girders using LiDAR

S.H. Yoon[a], Q. Wang[a,b] and H. Sohn[a]

[a]Department of Civil and Environmental Engineering,
Korea Advanced Institute of Science and Technology, South Korea
[b]Department of Civil and Environmental Engineering,
The Hong Kong University of Science and Technology, Hong Kong
E-mail: sh.yoon@kaist.ac.kr, qwangau@ust.hk, hoonsohn@kaist.ac.kr

Abstract –

Many bridges have been constructed using precast components because they offer faster production, lower cost, and more efficient construction compared to conventional in-situ construction. For coupled behavior of bridges slabs and girders, shear pockets on the deck slabs and shear connectors on the girders need to be properly connected. However, precast girders can be easily deformed once they are placed on sites because of their heavy weights, time dependent effects of creep and shrinkage, pre or post tensioning, etc. Once the girders are deformed, shear pockets and shear connectors may no longer match properly. This study proposes a technique which can automatically identify mismatches between shear connectors and shear pockets using LiDAR and identify the optimal placement of precast deck slabs with respect to precast bridge connectors. First, precast bridge deck slabs and precast girders are scanned using a LiDAR. Then, unnecessary noise data are removed using a DBSCAN-based algorithm. Afterwards, the dimensional features such as locations and sizes of shear pockets and shear connectors are estimated from the point cloud data, respectively. Finally, the optimal placement of deck slabs is identified to minimize the mismatches between shear pockets and shear connectors using the Levenberg-Marquardt (LM) method. To validate the effectiveness of the proposed technique, an experiment was conducted on a real precast girder. The experimental results demonstrate that the proposed technique can effectively and efficiently estimate the optimal placement of deck slabs with respect to precast girders.

Keywords –

LiDAR; Precast girder; Precast bridge deck slab; Precast bridge connector; Optimal placement

1 Introduction

Precast bridges are constructed by connecting precast components, such as precast girders and precast bridge deck slabs, on construction sites. For coupled behavior of precast bridge deck slabs and precast girders, the shear pockets on the deck slabs and the shear connectors on the girders need to be properly connected [1]. However, three reasons may lead to the mismatch between the shear pockets and the shear connectors including (1) dimensional errors of precast girders and deck slabs occurred in fabrication process, (2) deformation of precast girders due to their weights, time dependent creep and shrinkage, and pre or post tensioning [2], and (3) incorrect orientation and location of shear connectors. Therefore, it is of great significance to identify the optimal placement of precast bridge deck slabs with respect to precast girders to minimize the mismatches between shear pockets and shear connectors. However, precast components are currently manually inspected by certified inspectors using traditional tools such as measuring tapes, which is time-consuming, expensive, and subjective [3]. Hence, it is necessary to develop techniques that can automatically and accurately assess the dimensional quality of precast components such as precast bridge deck slabs and precast girders for estimating the optimal placement of precast bridge deck slabs.

This study proposes an automated optimal placement estimation technique for precast bridge deck slabs with respect to precast girders using LiDAR. After obtaining scan data using LiDAR, the proposed technique firstly removes noise components using a DBSCAN-based algorithm. Then, a RANSAC-based plane detection algorithm is used to extract scan data associated with shear connectors. Next, a two-class classification algorithm is applied to extract scan data which only belong to shear connector heads. Afterwards, the orientation and location of each shear connector are estimated. Finally, the optimal placement of precast bridge deck slabs is estimated by solving a nonlinear minimization problem based on the previously estimated orientation and location of shear connectors and the shear pocket location obtained in a previous study [4]. This paper is organized as follows. Section 2 provides the research background on precast bridge construction and LiDAR. The developed optimal placement estimation technique for precast bridge deck slabs are described in Section 3. Section 4 validates the proposed technique using a field experiment. Finally, Section 5 provides a summary and conclusions of this study.

2 Research background

2.1 Precast bridge connections

For the safety and serviceability of precast bridges, it is critical to ensure a tight connection between precast bridge deck slabs and precast girders [1]. The deck slabs and the girders are connected by the shear pockets on the deck slabs and the shear connectors on the girders. A shear pocket is usually a rectangular hole throughout a precast bridge deck slab as shown in Figure 1 (a). On the other hand, a stud type shear connector, which is the type of shear connectors considered in this study, is a steel rod welded on the top flange of a precast girder and is composed of a connector head and a connector body as shown in Figure 1 (b). Multiple shear connectors are placed inside a single shear pocket as shown in Figure 1 (c), and the gap is filled by cement for coupled behavior. The main function of the coupled shear pockets and shear connecters is to prevent vertical separation of precast bridge deck slabs from precast girders and to minimize longitudinal movement between the deck slabs and the girders [5]. From time to time, shear pockets and shear connectors are mismatched due to dimensional errors of precast bridge deck slabs and precast girders, time dependent deformations of precast girders, and faulty shear connectors to name a few. However, such mismatches are hardly discovered until precast bridge deck slabs are actually placed on precast girders, resulting in construction delay and cost overrun. Therefore, it is necessary to identify the foreseeable mismatches at the fabrication stage of precast components.

2.2 Light Detection And Ranging (LiDAR)

LiDAR measures the distance to a target by illuminating the target with laser beams and analyzing the reflected signals. Due to the rapid rotation of the scanner head in both horizontal and vertical directions, a laser scanner can measure the distances at different directions at a high speed within a region of interest (ROI). The obtained scanned points called 'point cloud data' consist of three-dimensional X, Y, and Z coordinates of each scan point with respect to the scanner.

The distance can be measured by two different principles, time-of-flight and phase-shift. The time-of-flight system measures the round trip time between the emitted light pulse and the returned pulse. Since the velocity of the laser pulse is known, the distance can be derived from the round trip time. On the other hand, the phase-shift system emits an amplitude modulated continuous wave and estimates the distance by measuring the phase difference between the emitted and reflected waves. The time-of-flight system is usually adopted for a long range measurement while the phase-shift system is adopted for a short distance sensing with a higher accuracy [6].

Although LiDAR has been used for the quality assessment of precast deck slabs [7,8,9,10], no study has been conducted on the quality assessment of precast girders or the optimal placement of precast deck slabs.

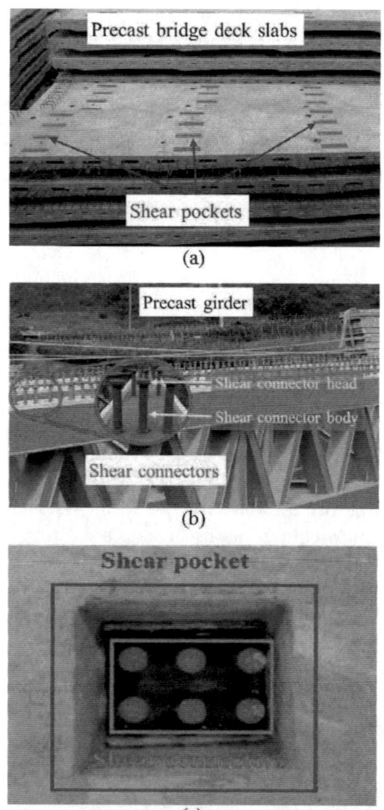

(c)

Figure 1. Coupling between shear pockets and shear connectors. (a) Precast bridge deck slabs with shear pockets. (b)Precast girders with shear connectors. (c) A connection between shear pockets and shear connectors.

3 Optimal placement of precast bridge deck slabs with respect to precast girders

This Section illustrates the developed technique for estimating the optimal placement of precast bridge deck slabs. Before the optimal placement estimation, the dimensions of precast bridge deck slabs and precast girders are assessed, which are illustrated in Sections 3.1 and 3.2, respectively. Afterwards, based on the dimension estimation, the optimal placement of precast bridge deck slabs is estimated with respect to precast girders by minimizing mismatches between shear pockets on the deck slabs and shear connectors on the girders, which is illustrated in Section 3.3.

3.1 Dimension estimation of precast bridge deck slabs

The dimensions of precast bridge deck slabs are assessed during fabrication in the precast concrete factory. The technique developed in the authors' previous study [11] is used, which has the following four steps. Firstly, the scan data of precast bridge deck slabs are acquired using LiDAR. Then, coordinate transformation and noise data removal are undertaken. Thirdly, key dimensional features, i.e. edges and corners of shear pockets, are extracted from the scan data. Finally, the dimensional properties of the shear pockets are computed from the extracted corners.

3.2 Dimension estimation of precast girders

The dimensions of precast girders are estimated at the construction site since deformations can occur after they are placed on site. LiDAR is located on a crane to acquire the scan data of the target girder. The acquired scan data are processed through four steps to estimate the dimensions of precast girder, (1) data preprocessing, (2) shear connector identification, (3) shear connector head extraction, and (4) estimation of shear connector orientation and location, as follows.

3.2.1 Data preprocessing

This step aims to remove noise data from raw scan data of precast girders and align the coordinates of the scan data with respect to the local coordinates of the precast girders. Scan data usually contain three categories of data including valid points, background points, and mixed pixels, as shown in Figure 2. A mixed pixel occurs when the laser beam is split by edges of target objects and can be located anywhere along the line of the laser beam [12]. Both mixed pixels and background points are considered as noise data and need to be removed.

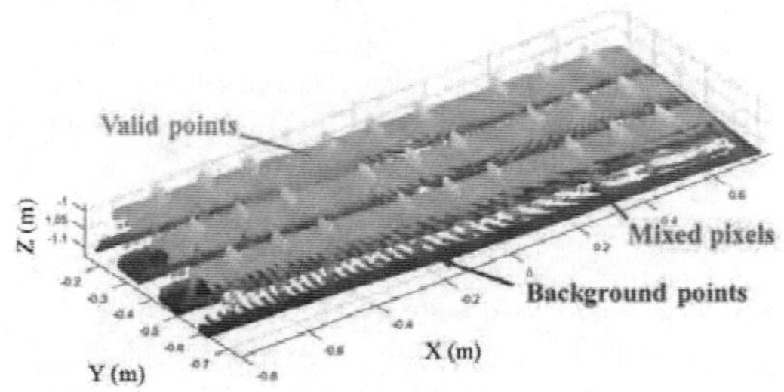

Figure 2. Three categories of scan data: valid points, background points, and mixed pixels.

Compared to valid points and background points, mixed pixels have larger distances to their neighbors, resulting in a lower spatial density. Thus, a DBSCAN-based data classification algorithm developed by Wang et al. [13] is applied for data classification. This DBSCAN-based algorithm performs classification based on spatial densities such that mixed pixels are classified as noise data due to low density while valid points and background points become two clusters. Furthermore, valid points are distinguished from background points as they have shorter distances to the LiDAR. After noise removal, valid points representing the as-built object (the precast girder) are remained.

Once the as-built object is extracted, the 3D coordinates of the as-built object are transformed into a new Cartesian coordinate system so that the as-built object best matches the as-design object. The same coordinate transformation method as used in the previous study [14] is applied to the as-built object as follows. (1) Extract the fitted planes of the as-built and as-design objects. (2) Transform the as-built object so that the two fitted planes are overlapped. (3) Find the transformation that minimizes the root mean square distances (RMSD) between the as-built and as-design object boundaries by iterative search. (4) Transform the as-built object with the previously obtained transformation to match the as-design object.

3.2.2 Shear connector identification

A precast girder mainly consists of two parts, a girder plane and shear connectors. This step aims to distinguish the girder plane and shear connectors, and extract shear connectors from the scan data. The girder plane and the shear connectors are distinguished using a plane extraction algorithm, namely Random Sample Consensus (RANSAC) [15]. The girder plane is estimated using RANSAC with a reference vector of [0

0 1], as shown in Figure 3 (a). To implement RANSAC, three parameters (*S*, *N*, and δ) are decided as follows. (1) The number of random selection (*S*) should be set such that the mathematical model can be estimated. Since a plane is extracted from the scan data, *S* is set as three. (2) The maximum number of random trials (*N*) is determined based on a probabilistic approach. The probability (*P*) that at least one out of *N* random trials can successfully select three scan points belonging to the girder plane is represented as follows:

$$P = 1 - (1 - \alpha^3)^N \quad (1)$$

where α is the ratio between the number of girder plane data and the number of the total scan data. The ratio α is estimated as the ratio between the girder plane area and the total girder area and approximately equals to 0.90, and *P* is set as 0.999. Hence, *N* is set as six according to Equation (2).

$$N = \frac{\log(1 - 0.999)}{\log(1 - 0.90^3)} \approx 5.29 \quad (2)$$

(3) The tolerance (δ) refers to the maximum orthogonal distance from an inlier to the plane. Since girders require cambering in order to compensate for dead load deflections, δ is set as 1.2 × maximum camber, where 1.2 is a safety factor. Given the above three parameters, the girder plane is estimated using RANSAC. Scan data which do not belong to the estimated girder plane are regarded and extracted as shear connectors, as shown in Figure 3 (b).

3.2.3 Shear connector head extraction

The extracted shear connector data contain three parts including (1) shear connector heads, (2) shear connector bodies, and (3) noise data. Due to the LiDAR position, some parts of shear connector bodies are occluded by their own shear connector heads. Therefore, only the shear connector heads are extracted in this study, using the K-means clustering method [16].

To partition the shear connector heads from the other scan data, three features are selected for K-means clustering as follows. Firstly, the neighboring scan points of each scan point within a radius of 2 cm are extracted (Figure 4 (a)). Then, the eigenvalues ($\lambda_1 \geq \lambda_2 \geq \lambda_3 \geq 0$) of the covariance matrix of its neighbors are calculated [17]. Based on these eigenvalues, two geometric features, namely planarity and variance, are defined and calculated as $((\lambda_2 - \lambda_3) / \lambda_1)$ and $(\lambda_1 / \lambda_1 + \lambda_2 + \lambda_3)$, respectively. Furthermore, a third geometric feature, namely average height, is defined and calculated as the average height of neighboring scan points. As shown in Figure 4 (b), shear connector heads, shear connector bodies, and noise data have substantially different values for these three geometric features. Using the three features, K-means clustering algorithm is applied to extract the shear connector heads.

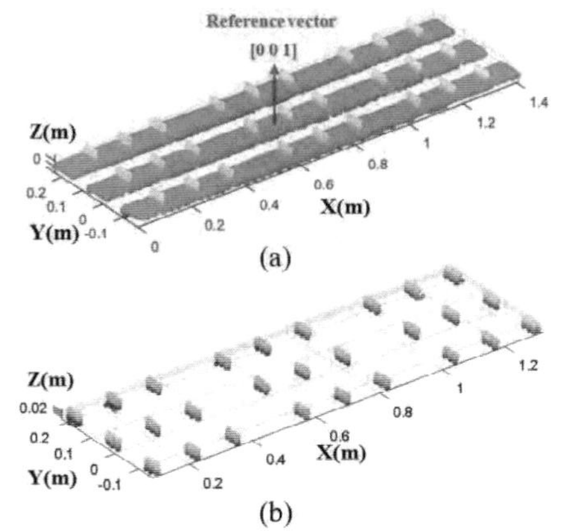

Figure 3. Identification of shear connectors using RANSAC. (a) Scan data of the as-built precast girder. (b) Identified shear connectors.

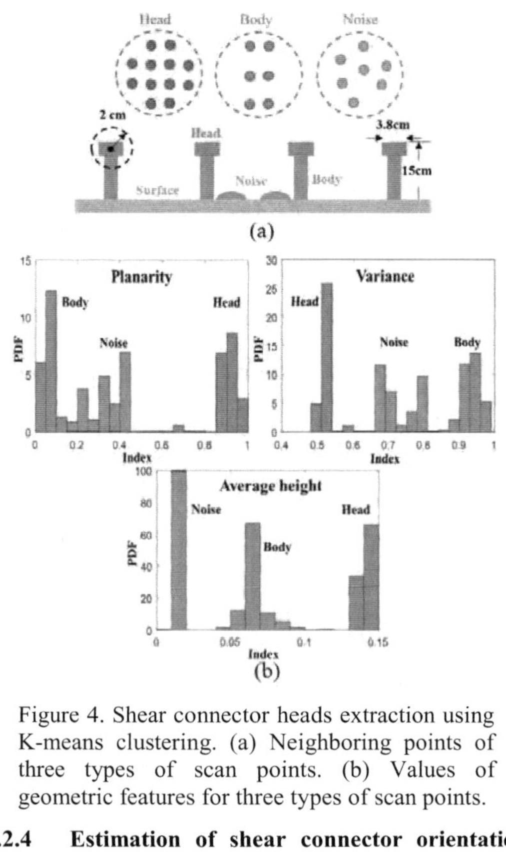

Figure 4. Shear connector heads extraction using K-means clustering. (a) Neighboring points of three types of scan points. (b) Values of geometric features for three types of scan points.

3.2.4 Estimation of shear connector orientation and location

The center position of each shear connector head is

computed as the mean value of all the scan points located on the shear connector head and considered as the location of the shear connector head (L_Head). Afterwards, the orientation (θ) of each shear connector is estimated by calculating the angle between the normal vector of each shear connector head and a reference vector [0 0 1] using dot production, as illustrated in Figure 5 (a).

$$\theta = \cos^{-1}\left(\frac{n_1 \cdot n_2}{\|n_1\|\|n_2\|}\right) \tag{3}$$

where n_1 and n_2 are the normal vector of shear connector head and the reference vector at the center position of L_Head, respectively. The position of each shear connector on the girder plane (L) is estimated based on the L_Head, θ, and the length of the shear connector body provided from the as-design model, as shown in Figure 5 (b).

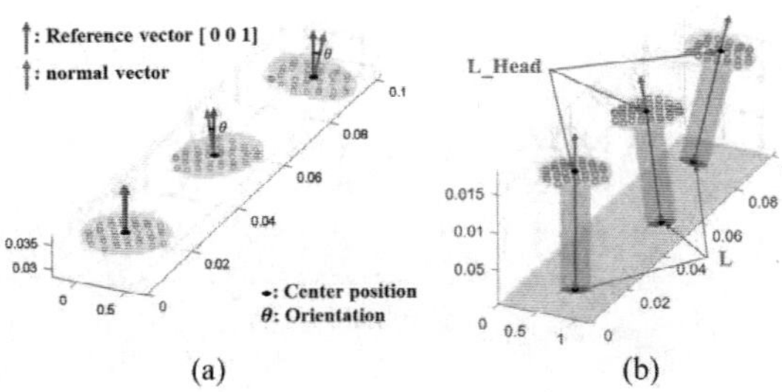

Figure 5. Orientation and location estimation of shear connector. (a) Estimation of L_Head and orientation (θ) of the shear connector head. (b) Estimation of the shear connector position (L).

3.3 Optimal placement of precast bridge deck slabs

The optimal placement of as-built precast bridge deck slabs suggests the location of each deck slab with respect to the deformed as-built precast girders such that mismatches between shear pockets and shear connectors can be minimized. First, three terminologies are defined before proceeding to the optimal placement estimation: *integrated deck slab*, *bundle area* (S_i), and *inner area* (A_i), as illustrated in Figure 6. All the precast bridge deck slabs that will be placed within a single span is defined as one *integrated deck slab* as shown in Figure 6 (a). Then, S_i is defined by the following three steps: (1) All the shear connectors that are supposed to be placed within a single shear pocket are identified; (2) the location and the area of each shear connector's head and body (red and blue circles in Figure 6 (b)) are estimated based on L_Head, L, and as-design diameters of the shear connector's head and body; (3) A rectangular area including all the areas of the shear connector's heads and bodies is defined as S_i. Finally, the fraction of S_i that falls within the shear pocket, is defined as A_i (Figure 6 (c)). The optimal placement is estimated through two steps, (1) coarse matching, and (2) fine matching, as described below.

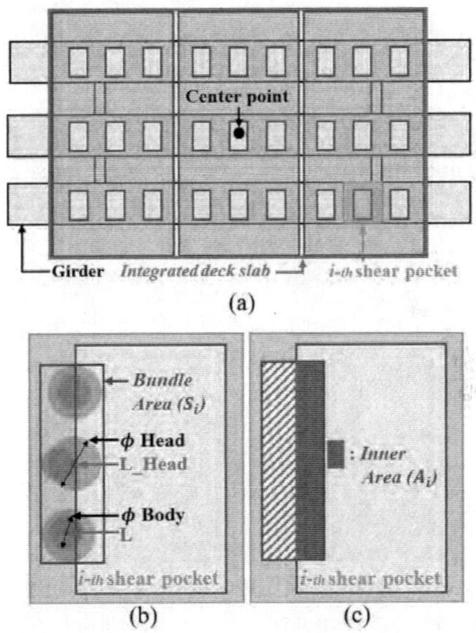

Figure 6. Fundamental concepts for the optimal placement estimation. (a) *Integrated deck slab*. (b) *Bundle area* (S_i), and (c) *inner area* (A_i).

3.3.1 Coarse matching

In the coarse matching step, an integrated deck slab is shifted by an initial translation vector T_0 so that its center point aligns with the center point of the precast girder. Then, all the *inner areas* within the integrated deck slab are calculated and their summation (ΣA_i) is computed. Next, the integrated deck slab is shifted again to maximize ΣA_i. If all the bundle of shear connectors are placed inside the boundaries of the shear pocket where they are supposed to be in, ΣA_i is maximized and becomes equal to the summation of all *bundle areas* (ΣS_i). Hence, the optimal placement estimation becomes the problem of minimizing the squared sum of the discrepancies between S_i and A_i as follows.

$$\mathbf{T}^c = \underset{\mathbf{T}^c}{\operatorname{argmin}} \sum_{i=1}^{n} (S_i - A_i)^2 \tag{4}$$

where n is the number of shear pockets within the *integrated deck slab* and $\mathbf{T}^c = [T_x^c, T_y^c]^T$, where T_x^c and T_y^c are the coarse translation values of the *integrated deck slab* in X and Y axes, respectively. This nonlinear optimization problem is solved using the Levenberg-Marquardt (LM) method [18]. S_i and A_i at the *i-th* shear pocket are obtained by Equations (5) and (6), respectively.

$$S_i = (a_{x,i}^2 - a_{x,i}^1) \times (a_{y,i}^2 - a_{y,i}^1) \quad (5)$$

$$A_i = [\min(c_{x,i}^2, a_{x,i}^2) - \max(c_{x,i}^1, a_{x,i}^1)] \times [\min(c_{y,i}^2, a_{y,i}^2) - \max(c_{y,i}^1, a_{y,i}^1)] \quad (6)$$

where $a_{x,i}^1$, $a_{x,i}^2$, $a_{y,i}^1$, and $a_{y,i}^2$ are the X and Y coordinates of corner points of *i-th bundle area* as shown in Figure 7 and $c_{x,i}^1$, $c_{x,i}^2$, $c_{y,i}^1$, and $c_{y,i}^2$ are the X and Y coordinates of the four corner points of the coarsely translated i-th shear pocket and are determined by the translation vector $\mathbf{T^c}$ as follows (Figure 7).

$$c_{x,i}^1 = b_{x,i}^1 + T_x^c,\ c_{x,i}^2 = b_{x,i}^2 + T_x^c$$
$$c_{y,i}^1 = b_{y,i}^1 + T_y^c,\ c_{y,i}^2 = b_{y,i}^2 + T_y^c \quad (7)$$

where $b_{x,i}^1$, $b_{x,i}^2$, $b_{y,i}^1$, and $b_{y,i}^2$ are the X and Y coordinates of the four corner points of the *i-th* as-design shear pocket location in X and Y axes, respectively, (Figure 7). The translation vector $\mathbf{T^c}$ is iteratively estimated by solving the nonlinear minimization problem described in Equation 4.

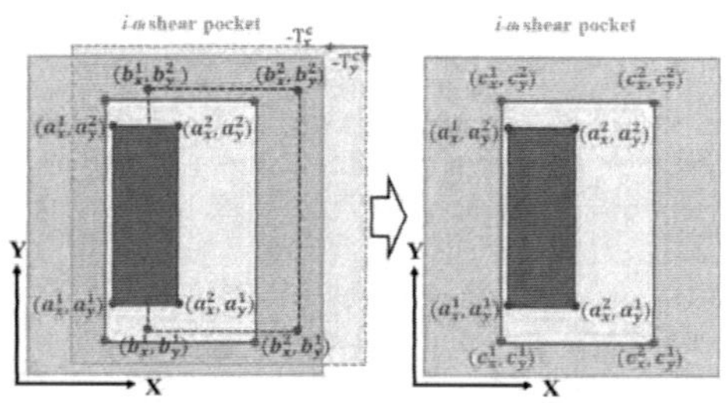

Figure 7. Change of the overlap area between the *bundle area* (S_i) and the *inner area* (A_i) after coarse matching by $\mathbf{T^c}$.

3.3.2 Fine matching

Once the integrated deck slab is shifted as a whole, now each deck slab within the integrated deck slab is individually shifted for fine matching. Fine matching starts with the deck slab which has the largest number of remaining mismatches. Then, fine matching operates on the next slab with the second largest number of mismatches, and repeats until all the mismatches disappear or until the last slab is shifted. For the *k-th* deck slab, the fine translation vector is found as follows.

$$\mathbf{T}_k^F = \underset{\mathbf{T}_k^F}{argmin} \sum_{j=1}^{m} (S_j - A_j)^2 \quad (8)$$

where $\mathbf{T}_k^F$ is the fine translation vector of the *k-th* deck slab, and m is the number of shear pockets within the *k-th* deck slab. Note that there are the maximum allowable horizontal and vertical relative movements between two adjacent deck slabs, denoted as σ_x and σ_y in Figure 8, respectively. Therefore, the values of $\mathbf{T}_{k,x}^F$ and $\mathbf{T}_{k,y}^F$ in $\mathbf{T}_k^F$ are limited within the range of $[-\sigma_x/2, \sigma_x/2]$ and $[-\sigma_y/2, \sigma_y/2]$, respectively. Similar to Eq. (4), this optimization problem is again solved using the LM method. After coarse matching and fine matching, the total translation vector $\mathbf{T}_k$ of the k-*th* deck slab becomes:

$$\mathbf{T}_k = \mathbf{T_0} + \mathbf{T^c} + \mathbf{T}_k^F \quad (9)$$

Finally, any remaining mismatched shear connectors are visualized in Figure 8 so that they can be adjusted or repaired if necessary.

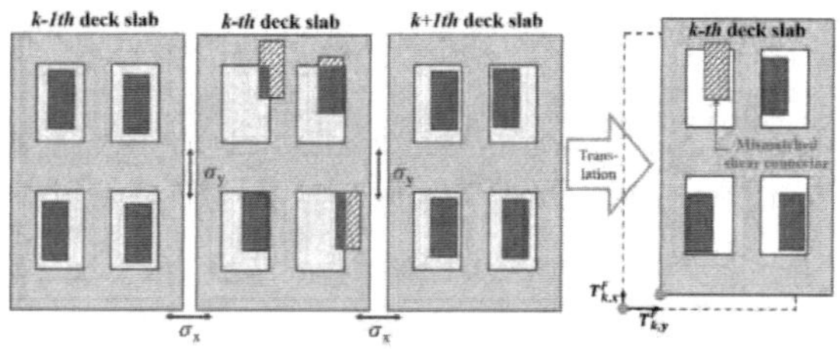

Figure 8. Fine matching of the *k-th* deck slab and visualization of the mismatched shear connector.

4 Field experiment

4.1 Experimental set-up

To validate the proposed technique for optimal placement estimation of precast bridge deck slabs, a field test was conducted on a real precast bridge girder. The field test was conducted at a precast concrete factory located in Jincheon-gun, Republic of Korea. Scan data of the precast girder were acquired using a FARO Focus 3D laser scanner, as shown in Figure 9 (a). The precast girder had dimensions of 13,100 mm × 3,200 mm × 1,503 mm with 234 shear connectors as shown in Figure 9 (b). Each shear connector had a height, head diameter, and body diameter of 150 mm, 38 mm, and 25 mm, respectively. The scan distance from the precast girder to the laser scanner was 13 m. Due to the large scale of the precast girder and the long scan distance, the highest angular resolution, i.e., 0.009°, was selected. The data acquisition and data processing took about 20 minutes and 30 minutes on average, respectively. As described in Section 3.1, the dimensions of the deck slabs were estimated using the technique developed in Kim et al. [11], and the developed dimensions estimation technique was applied for the girders.

(a)

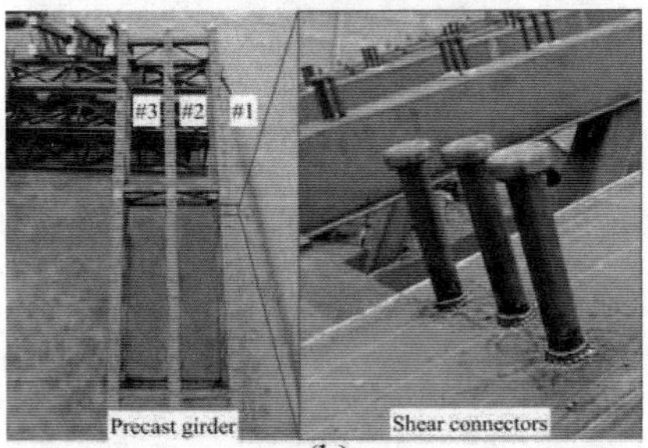

(b)

Figure 9. Test configuration of the full-scale girder scanning. (a) Test set-up, (b) A full-scale precast girder.

4.2 Experimental results for optimal placement estimation

The proposed optimal placement estimation technique was applied to the precast girder used in the experiment. When the deck slabs were placed based on the as-design placement, there were a total 54 mismatches between the shear pockets and shear connectors, as shown in Figure 10 (a). After the proposed coarse matching step, the number of mismatches was reduced to 4, as shown in Figure 10 (b). Finally, all the mismatches were eliminated after the fine matching step, as shown in Figure 10 (c). The amount of horizontal and vertical translations for the optimal placement of all the six deck slabs were presented in Table 1.

Table 1 The amount of horizontal and vertical translations of all the six deck slabs for optimal deck slab placement.

Translation (mm)	X axis	Y axis	X axis	Y axis	X axis	Y axis
Slab No.	#1		#2		#3	
Coarse	26.5	-25.0	26.5	-25.0	26.5	-25.0
Fine	0	4.6	0	2.9	0	-3.9
Total	26.5	-20.4	26.5	-22.1	26.5	-28.9
Slab No.	#4		#5		#6	
Coarse	26.5	-25.0	26.5	-25.0	26.5	-25.0
Fine	0	-5.9	0	0	0	0
Total	26.5	-30.9	26.5	-25.0	26.5	-25.0

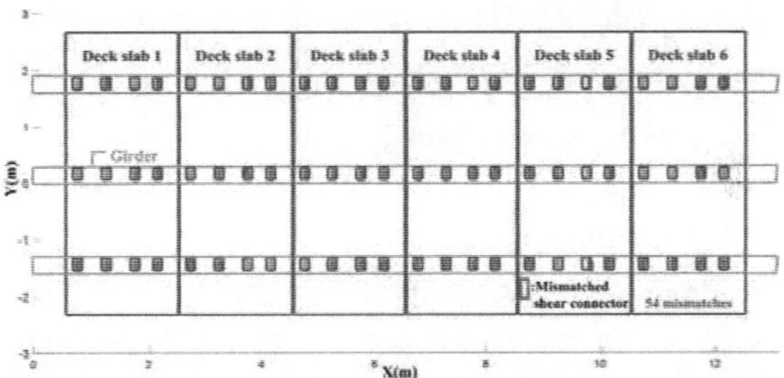

(a)

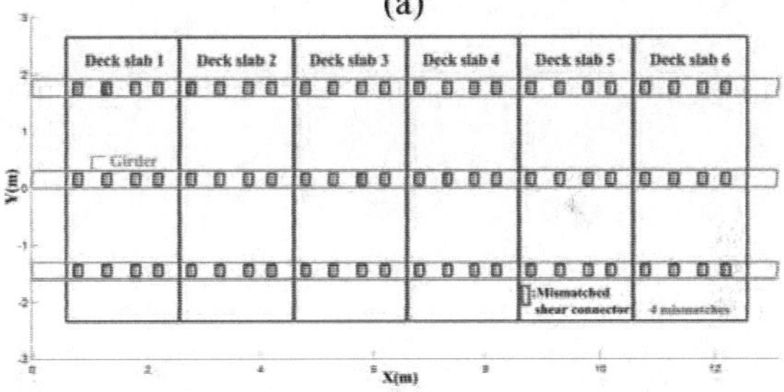

(b)

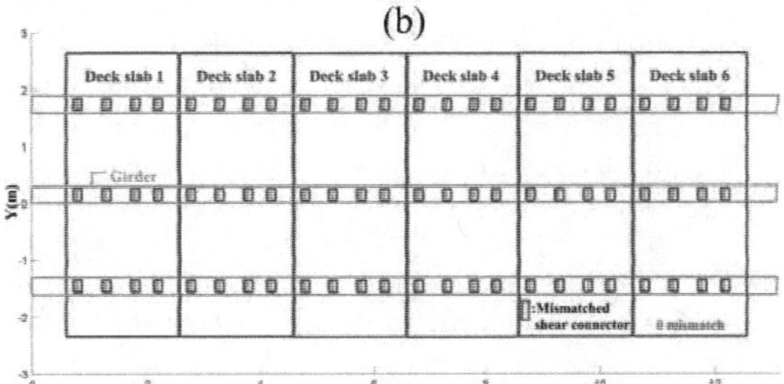

(c)

Figure 10. Result of optimal placement. (a) As-design placement. (b) Coarse matching. (c) Fine matching.

5 Conclusions

This study presents an optimal placement estimation technique for precast bridge deck slabs with respect to precast girders to minimize mismatches between the shear pockets on the deck slabs and the shear connectors on the girders using LiDAR. The scan data from a girder are acquired using LiDAR and processed through five steps, (1) data preprocessing, (2) shear connector identification, (3) shear connector head extraction, (4) estimation of shear connector orientation and location, and (5) optimal placement estimation. The proposed technique was validated through a full-scale precast girder experiment at an actual construction site. During the optimal placement estimation, 54 mismatches were reduced to 4 mismatches after coarse matching, and no mismatch existed after fine matching.

One major limitation of this study is that LiDAR must be located on the top of precast girders to acquire laser scan data. However, it can be sometimes difficult to impossible to place the LiDAR in actual construction sites. One possible solution is to mount the LiDAR on an unmanned aerial vehicle (UAV) for scanning.

Acknowledgement

This research was supported by a grant (13SCIPA01) from Smart Civil Infrastructure Research Program funded by Ministry of Land, Infrastructure and Transport (MOLIT) of Korea Government and Korea Agency for Infrastructure Technology Advancement (KAIA).

References

[1] M.A. Issa, T.A Patton, H.A. Abdalla, A.A. YOUSIF, and M.A. Issa, Composite behavior of shear connections in full-depth precast concrete bridge deck panels on steel stringers, PCI journal, 48(5):76-89, 2003.

[2] K.A. Byle, N.H. Burns, and R.L. Carrasquillo, Time-dependent deformation behavior of prestressed high performance concrete bridge beams, *Work*, 580(6), 1997.

[3] B.M. Phares, G.A. Washer, D.D. Rolander, B.A. Graybeal, and M. Moore, Routine highway bridge inspection condition documentation accuracy and reliability, *Journal of Bridge Engineering*, 9(4):403-413, 2004.

[4] M.-K. Kim, H. Sohn, and C.-C. Chang, Automated dimensional quality assessment of precast concrete panels using terrestrial laser scanning, *Automation in Construction*, 45:163-177, 2014.

[5] J. Qureshi, D. Lam, and J. Ye, Effect of shear connector spacing and layout on the shear connector capacity in composite beams, *Journal of Constructional Steel Research*, 67(4):706-719, 2011.

[6] J. Shan, and C.K. Toth, Terrestrial laser scanners, Topographic laser ranging and scanning: principles and processing, *CRC press*, 2008.

[7] M.-K. Kim, H. Sohn, and C.-C. Chang, Localization and quantification of concrete spalling defects using terrestrial laser scanning, *Journal of Computing in Civil Engineering*, 29(6), 2014.

[8] M-K Kim, J.C. Cheng, H Sohn, and C.-C. Chang, A framework for dimensional and surface quality assessment of precast concrete elements using BIM and 3D laser scanning, *Automation in Construction*, 49:225-238, 2015

[9] Q. Wang, M-K Kim, H Sohn, and J.C. Cheng, Surface flatness and distortion inspection of precast concrete elements using laser scanning technology, *Smart Structures and Systems*, 18(3):601-623, 2016.

[10] Q. Wang, H Sohn, and J.C. Cheng, Development of a mixed pixel filter for improved dimension estimation using AMCW laser scanner, *Journal of Photogrammetry and Remote Sensing*, 119:246-258, 2016.

[11] M.-K. Kim, Q. Wang, J.-W. Park, J.C. Cheng, H. Sohn, and C.-C. Chang, Automated dimensional quality assurance of full-scale precast concrete elements using laser scanning and BIM, *Automation in Construction*, 72:102-114, 2016.

[12] M. Hebert, and E. Krotkov, 3D measurements from imaging laser radars: how good are they?, *Image and Vision Computing*, 10(3):170-178, 1992.

[13] A.A. Siddiqui, A New Inspection Method Based on RGB-D Profiling, *Virginia Tech*, 2015.

[14] Q. Wang, M.-K. Kim, J.C. Cheng, and H. Sohn, Automated quality assessment of precast concrete elements with geometry irregularities using terrestrial laser scanning, *Automation in Construction*, 68:170-182, 2016.

[15] M.A. Fischler, and R.C. Bolles, Random sample consensus: a paradigm for model fitting with applications to image analysis and automated cartography, *Communications of the ACM*, 24(6):381-395, 1981.

[16] S. Lloyd, Least squares quantization in PCM, *IEEE transactions on information theory*, 28(2):129-137, 1982.

[17] R. Blomley, M. Weinmann, J. Leitloff, and B. Jutzi, Shape distribution features for point cloud analysis-a geometric histogram approach on multiple scales, ISPRS Annals of the Photogrammetry, *Remote Sensing and Spatial Information Sciences*, 2(3):9, 2014.

[18] D.W. Marquardt, An algorithm for least-squares estimation of nonlinear parameters, *Journal of the society for Industrial and Applied Mathematics*, 11(2):431-441, 1963.

Using SLAM-based Handheld Laser Scanning to Gain Information on Difficult-to-access Areas for Use in Maintenance Model

T. Makkonen[a], R. Heikkilä[a], P. Tölli[b], F. Fedorik[a]

[a]Construction Technology Research Center, University of Oulu, Finland
[b]First Quantum Minerals, Pyhäsalmi Mine, Finland

E-mail: Tomi.Makkonen@oulu.fi, Rauno.Heikkila@oulu.fi, Filip.Fedorik@oulu.fi
Pekka.Toelli@fqml.com

Abstract

Laser scanning is the de-facto system to collect large amounts of geometrical information, especially indoors. Total stations are usually used for accuracy and information-rich results. However, in some cases the target area is not suitable for using a terrestrial laser scanner. For instance, too cramped or difficult-to-access space can be a limiting factor. In complex cases, standard scanning methods are usually time-consuming, leading to unexpected costs and sometimes scanning can be omitted for time and cost reasons.

Simultaneous location and mapping (SLAM)-based rapid laser scanning could be considered as a practical method to perform scanning in poorly accessible spaces. In this paper, we present results obtained using a ZEB1 handheld SLAM-based laser scanner to ~1440 m depth in a mine shaft in Finland while standing on the roof of the moving elevator, for very fast data collection in an extreme environment.

In a scanning time of approximately one hour, we collected and archived 120 million data points. Compared with standard tripod lasers, the accuracy was low, but could be improved using positioned targets. The method also gave an imperfect model which required manual correction. Considering the rapid maturation of SLAM-based systems, the method can be seen a promising tool to collect laser point data in difficult-to-access areas in the near future for maintenance model applications. These systems already work well enough when speed is required over accuracy and in cases where the environment is too difficult to use standard total stations, as in the case presented in this paper.

Keywords –
mine shaft; BIM; laser scanning; SLAM; maintenance model

1 Introduction

This study is an extension of previous work in which we studied the practicality of using simultaneous location and mapping (SLAM)-based rapid laser scanning for mine inventory model purposes [1]. The case presented in this paper is probably one of the most difficult possible: a very long narrow space (elevator shaft), rain due to water leaking from rocks, moisture on the walls and a relatively fast moving platform. The latter made it impossible to scan the full 360°, and instead about 270° were scanned while going down and another 270° while going up. The expectations for perfect results were very low; the intention was rather to test the limits of current performance of rapid laser scanning systems. However, the results were better than expected and using the same elevator shaft later should give us a great baseline to compare the advances in SLAM rapid laser scanning in upcoming years.

In Finland, YIV2015 [2] and COBIM2012 [3] specifications are used in infrastructure and building specifications, but the mining sector applies specific regulations. Mine shafts must be regularly inspected, as required by national standards, for instance the U.S. Code of Federal Regulations for Mineral Resources (U.S. Government Publishing Office). At Pyhäsalmi Mine in Finland, a mine elevator shaft used for personnel transport is inspected at least once every three months. At least four inspectors must use a specific inspection platform under or on the roof of the car, with one person in the car with radio connection to an operator of the elevator at surface level. Hence, five people are needed in the elevator and one on the surface. The radio operator and inspectors must have a clear connection, causing no misunderstanding that can lead to critical situations. The inspection takes about one hour and during this time the elevator is disabled for other uses.

Thus improving the inspection method and procedure in Pyhäsalmi Mine consists of: reducing the man hours needed, shortening the inspection time to free up the elevator for other uses and finding and responding to changes and faults more frequently than every three months. This can be achieved by finding ways to generate a maintenance model.

2 Laser scanning with SLAM

One of the promising candidates for performing specialist tasks such as shaft monitoring is applying laser scanners with SLAM capabilities. SLAM-based laser scanners usually include, in addition to the laser scanner part, an inertia measurement unit (IMU) that records angular rate and specific forces. In addition, they can use other positional data in sensor fusion, e.g. GNSS location or odometer information, if available, in mathematical calculation of simultaneous location and scanning.

The commercial name of the SLAM-based laser scanner tested in this study is ZEB1 (Figure 1), but the entire system, developed by CSIRO, is called Zebedee. Tutorials and video presentations on the device can be obtained from CSIRO [4], Geoslam [5] and 3D Laser Mapping [6]. The handheld laser scanner generates a 3D map automatically while measurement is being performed. Thus for instance a whole tunnel network (or whole building) can be scanned with only one continuous measurement. The main parts of the ZEB1 are a spring-mounted 2D laser, data logger and inertia measurement unit (IMU). The range typically used for scanning is 15-20 m, scan rate 43200 points per second, noise ± 30 mm and field of view horizontally 270°. The moved-around vertical field of view is approximately 120°. Post analysis of data is performed at dedicated servers in the cloud. Details of the construction and design parameters are available elsewhere [7]. The accuracy of ZEB1 in a mine environment has previously been tested at Kemi Mine [8]. Some differences between total station and SLAM-based rapid laser scanning are listed in Table 1.

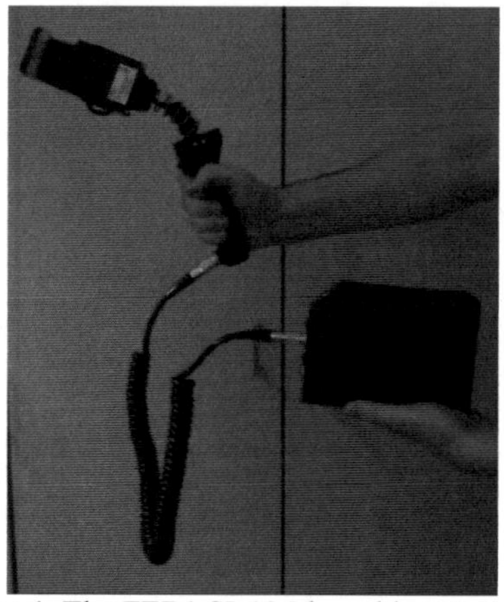

Figure 1. The ZEB1 SLAM-based laser scanner.

Table 1 Differences between total station and SLAM-based handheld rapid laser scanning

Total station	SLAM-based rapid laser scanner
Needs setting up and referencing per measurement location	Is ready to go from the start. However, requires closed loop path, so that measurement is started and stopped at the same location
Is required to stand in one location without distributions	Measurement device is moved around, and location is auto-generated relative to the environment
Accuracy and noise are good	Accuracy and noise are low compared with total station
No drift error	Drift error, which is halved by closing the measurement loop
Scanning speed slow in cramped space because it needs setting up per scan location. For narrow spaces, many measurement locations are needed	Very fast, because measuring device can move around while continuously measuring
Not economical in general in cramped space, because needs long time to set up to measure from multiple places with referencing. Also needs skilled workforce	More economical in cramped space, because is faster to perform and skill cap is low
With working reference networks combining point clouds gives great results, without references needs referencing first	Currently does not use reference points 'inside' the measurement. This is also a disadvantage if accuracy is the main goal, over speed

3 Rapid laser scanning experiment in Pyhäsalmi Mine shaft

A laser scanning experiment using ZEB1 was performed at Pyhäsalmi Mine, Finland, in an elevator shaft called Timo's Shaft. The length of the shaft is approximately 1440 m and the diameter is 5 m, and it had never been laser-scanned previously. The laser scanning measurements were performed during a regular inspection. While the inspectors were standing on the roof of the car, laser scanning was performed at the highest point of the elevator system. This gave a relatively obstruction-free view to the wall of the shaft (Figures 2 and 3). The operator waved the ZEB1 up and down in one position at about 1 Hz frequency, to scan well over half (270°) of the tunnel when going down. After the elevator started to ascend, the position was rotated 180° to face the opposite wall, resulting in overlapping point clouds (approximately 90° overlap both sides) covering the entire shaft wall.

The elevator speed was regulated to 1 m/s, giving a time from ground level to bottom of about 25 minutes and another 25 minutes back up. The total measurement time was about an hour, including time needed for preparation.

The conditions were demanding for laser scanning, since water penetrating through the rocks resulted in a constant rain in the shaft for the first 1000 m. There was no rain cover in use and therefore the rain wet the glass in the measuring glass head, but also the rock walls, resulting in shiny surfaces. Furthermore, rain itself is not suitable for laser measurement, as it obscures the path of the laser rays [9].

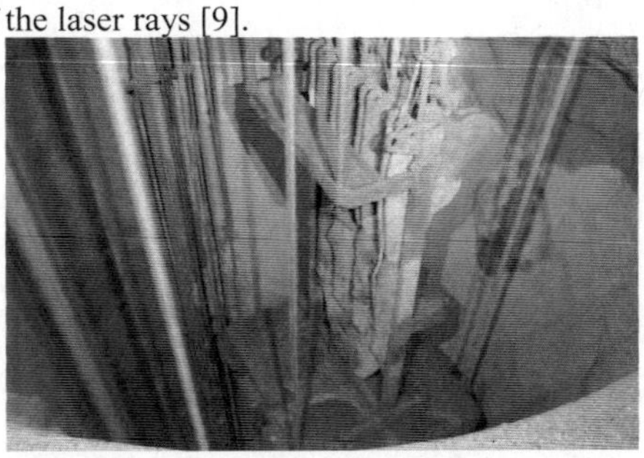

Figure 2. Measurement position, the highest platform in the image.

Figure 3. Overall image of the measurement position in relation to the upper part of the elevator.

4 Results of measurements in the mine shaft

A graphical expression of the point cloud obtained from the measurements is presented in Figure 4. It shows the point cloud of the whole shaft on the left and detailed shaft from the highest point of the elevator to the first level. The point cloud was made from 120 million points in *ply* format. The calculated approximation for point density during movement of 1 m/s was 2778 points per square metre, meaning that points were separated by 2 cm. Because the measurement method involves hand movements during scanning, this value can vary significantly, depending on the orientation of the measurement head.

Compatibility of the measured point cloud with the mining software Surpac was achieved using a Matlab script to convert *ply* format to native Surpac *str*. An example of data for *str* covering the first three measured points is given in Table 2.

After measurement, the collected data were sent for post processing in Geoslam, where SLAM-based data fitting failed completely twice, but gave results at the third attempt. The best result obtained is presented in Figure 4. Complete success was not achieved, as part of the point cloud failed to match measurements between 252 and 434 m, as shown in Figure 5. As can be seen, the results, represented by two half-cylindrical rings, did not overlap correctly. A correction procedure was applied for part of the shaft to test the accuracy and possibility of improving the results obtained. The two overlapping rings were first classified, resulting in two separate point clouds, by using time stamps as shown in Figure 6. Given the vague shape and asymmetry of the clouds, automatic alignment was not possible. Registering points manually was also too difficult to achieve suitable accuracy.

Table 2 Example of *str* format for the first three measured points

0.000,	0.000,	0.000,	0.000,	0.000,	0.000
1,	3.622,	-1.643,	-75.120,		
1,	3.723,	-1.430,	-75.187,		
1,	3.739,	0.820,	-75.469,		
0,	0.000,	0.000,	0.000,		
0,	0.000,	0.000,	0.000,	END	

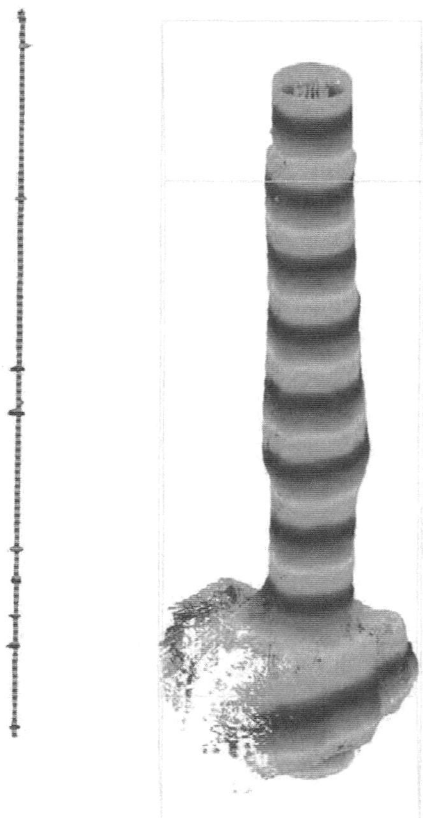

Figure 4. Left: The whole scan of the 1440-m Timo's shaft. Right: Upper part of the scan.

To solve the problem, five cables were segmented out of both clouds and automatically aligned (Figure 7). The transformation matrix was then applied to the point clouds, resulting in the new point cloud in Figure 8. The quality of the results was confirmed by comparing computed distance of overlapping parts (Figures 9 and 10), with mean error 37 mm and standard deviation 19 mm.

The quality of SLAM-generated results was studied by using part of the scan, from 130 m depth. Here the rock surface was characteristically more similar than for most of the tunnel's uneven bare rock surface. Although the time difference between up and down scans was among the greatest, measurement in this location was still systematically similar to the basic procedure. Repeated time classification was used to generate two separate point clouds and the differences in the overlapping area were studied. The resulting mean distance error was 27 mm and the standard deviation 17 mm (Figures 11 and 12).

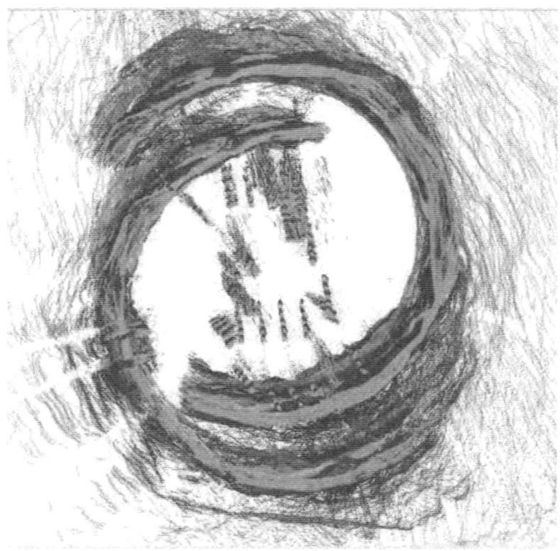

Figure 5. Poorly shaped SLAM-generated point cloud between 252 and 434 m.

Figure 6. Two separate point clouds, classified using time stamps.

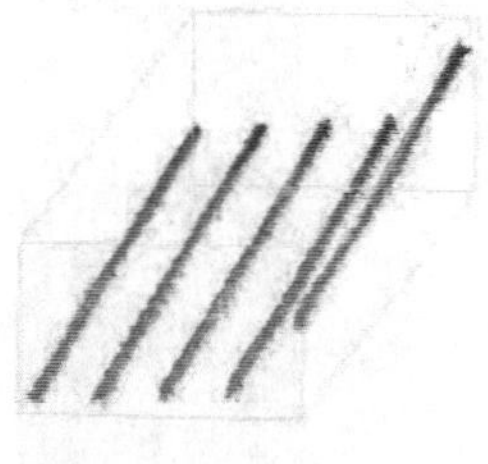

Figure 7. Aligned cables to generate the transformation matrix.

Figure 8. Semi-manually merged point clouds.

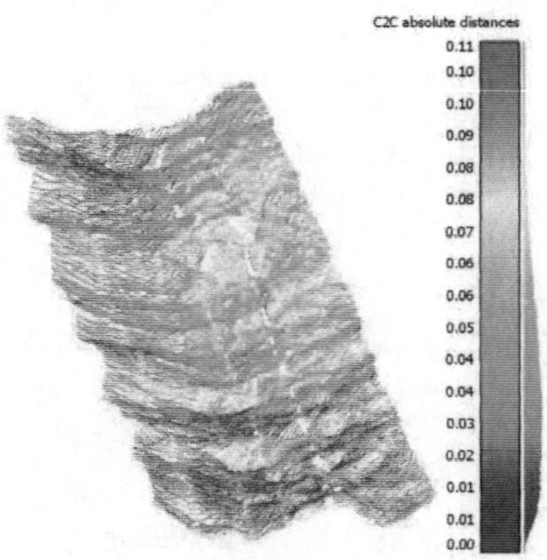

Figure 9. Distance error (m) in merged point clouds after transformation.

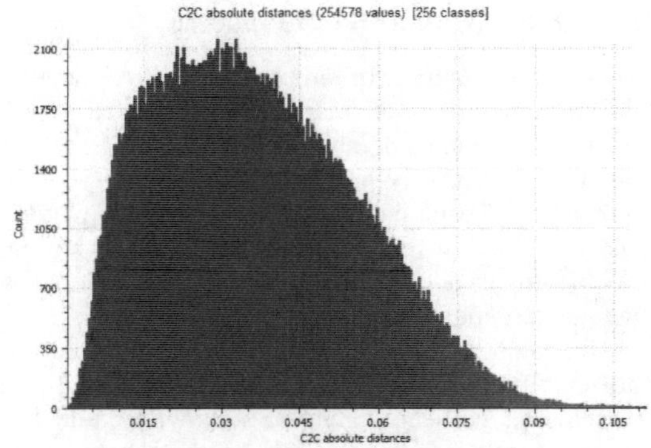

Figure 10. Histogram of distance error (m) for manually corrected point cloud.

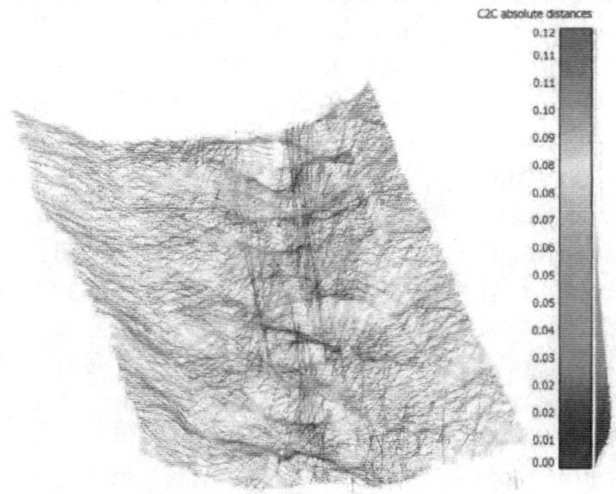

Figure 11. Distance error (m) in SLAM algorithm-generated point cloud, comparing time stamp-classified clouds.

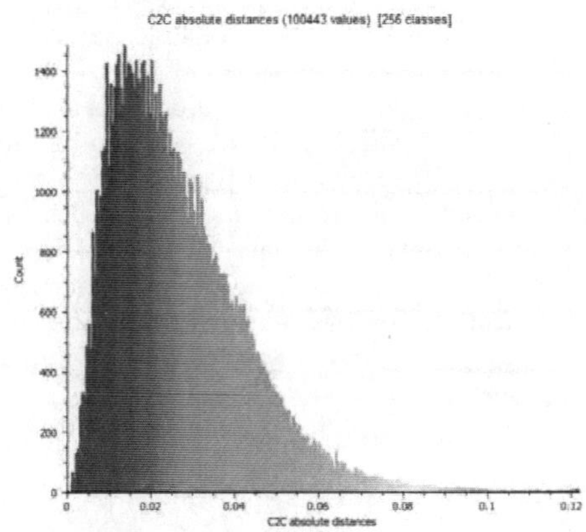

Figure 12. Histogram of SLAM algorithm-generated point clouds, comparison of time-classified up and down scans (m).

5 Discussion and conclusions

This measurement experiment in extreme conditions demonstrated the potential of SLAM-based scanning systems to generate point clouds. The SLAM generation of laser scanners is not fully mature yet, but shows promise for future automated elevator-shaft monitoring and maintenance model generation. In its current state, the system is not ready yet for the demanding task it was given in this paper.

It should be noted that the scanning can most likely be improved greatly, even using the same equipment. The most significant change results-wise could be the scanning pattern, as only 270° of the shaft wall were scanned, while a whole 360° scan would have been preferable. This could be achieved using automated movements of the scanner, or designing better, more systematic movement in the handheld method, whereby more of the cylindrical shape is covered. The most efficient way would probably be a laser scanner system fixed to the elevator with a uniform circular moving mechanism.

Nevertheless, the results suggest that applying the laser scanner for mine shaft point cloud generation has much potential, e.g. the method is extremely fast and does not require wide spaces to perform. Noise value for mean distance error of 27 mm and a standard deviation of 17 mm in an automatically generated point cloud is already acceptable for some usages. The results obtained seemed to be unaffected too much by rain and humidity condensing on the device, but it is probably advisable to protect the scanner from falling rocks and/or other elements and to prevent rain and humidity from influencing the measured results. Results from our earlier work [1] also support use of rapid handheld laser scanning in difficult-to-access areas. In that study scans were performed in an old, cramped hoisting tower (Figure 13), a steep stairway to an evacuation shelter and even places where the operator had to crawl through a small hole.

In future work, analysing positional measurement errors and finding more efficient methods to obtain point clouds will be the main goals. On achieving accurate scanned results to monitor a shaft, a statistical method and a suitable number of scans should be determined. Hence, continuous scanning and applied statistical techniques might allow development of SLAM-based point cloud measurements of sufficient quality for maintenance monitoring.

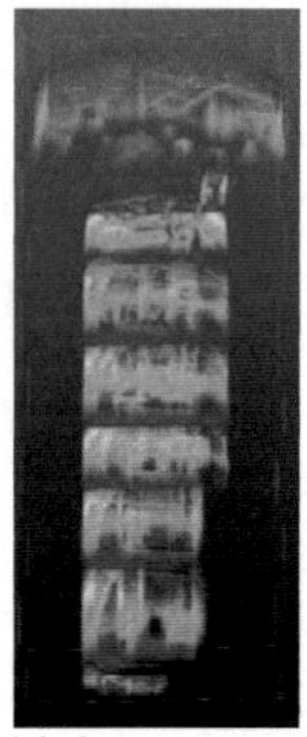

Figure 13. Rapid laser scanning results of a hoisting tower – cross-section of the whole tower [1].

References

[1] T. Makkonen, R. Heikkilä, J. Jylänki, and S. J. Fraser, "Reopening an Abandoned Underground Mine-3D Digital Mine Inventory Model from Historical Data and Rapid Laser Scanning," in *ISARC. Proceedings of the International Symposium on Automation and Robotics in Construction*, 2015, vol. 32, p. 1.

[2] "Yleiset inframallivaatimukset 2014 – YIV 2014 | InfraBIM." .

[3] "COBIM 2012 | buildingSMART Finland." [Online]. Available: http://www.en.buildingsmart.kotisivukone.com/3. [Accessed: 31-Mar-2016].

[4] "Autonomous Systems - Autonomous Systems." [Online]. Available: https://wiki.csiro.au/display/ASL/Autonomous+Systems. [Accessed: 22-Mar-2015].

[5] "Geoslam." [Online]. Available: http://www.geoslam.com/. [Accessed: 22-Mar-2017].

[6] "3D Laser Mapping." [Online]. Available: http://www.3dlasermapping.com/. [Accessed: 22-Mar-2017].

[7] M. Bosse, R. Zlot, and P. Flick, "Zebedee: Design of a Spring-Mounted 3-D Range Sensor with Application to Mobile Mapping," *IEEE Trans. Robot.*, vol. 28, no. 5, pp. 1104–1119, 2012.

[8] T. Makkonen, R. Heikkilä, A. Kaaranka, and M. Naatsaari, "The Applicability of the Rapid Handheld Laser Scanner to Underground Tunnel Surveying," in *ISARC. Proceedings of the International Symposium on Automation and Robotics in Construction* 2015

[9] [G. Heritage and A. Large, *Laser Scanning for the Environmental Sciences*. John Wiley & Sons, 2009.

34th International Symposium on Automation and Robotics in Construction

ISARC 2017

June 28 – July 1 TAIPEI, TAIWAN

ENERGY & SUSTAINABILITY

Hybrid Ventilation Potential Investigation for Small and Medium Office Buildings in Different US Climates Under Uncertainties

Jianli Chen[a], Godfried Augenbroe[b] and Xinyi Song[a]

[a]School of Building Construction, Georgia Institute of Technology, USA
[b]School of Architecture, Georgia Institute of Technology, USA
E-mail: jianlichen@gatech.edu, godfried.augenbroe@coa.gatech.edu, xinyi.song@coa.gatech.edu

Abstract

Hybrid ventilation, or mixed mode ventilation, is a technique that uses a mixture of natural and mechanical ventilation for maintaining a thermally comfortable indoor environment for building occupants. With its potential for improving both the energy efficiency and occupants comfort at the same time, its real application benefits across different US climate zones are still not clear, especially considering the possible variations of building context and building properties in real scenarios. In this paper, by taking these uncertainties into account, we have investigated the potential of utilizing hybrid ventilation on small to medium size commercial buildings across different US climate zones. One prototypical small to medium size office building was selected first. Uncertainties related to meteorology, building microclimate, building properties and operation have been applied for a thorough investigation of hybrid ventilation potential based on our baseline building. A rule based control has been specified based on the adaptive thermal comfort model in the thermal comfort standard. The results showed that the mean energy saving varies from 17% to 43.8% in different climate zones in US.

Keywords –

Hybrid Ventilation; Potential Investigation; Uncertainty Analysis;

1 Introduction

Air conditioning has long been recognized as one of the major sources of energy consumption during the operation of commercial buildings. Accompanying with the benefits such as stable and relatively comfortable indoor environment, the air conditioning system could also possibly bring some healthy problems such as sick building syndrome due to insufficient ventilation. On the other hand, naturally ventilated building, which is able to provide sufficiently large amount of fresh air, has become an increasingly popular option that is recognized by owners for moving towards greener building these years. However, it also suffers from constraints such as uncertain performance during the building operation due to its susceptibility to outdoor environment. As a compromise of both types of buildings, hybrid ventilation building has the capability of providing a comfortable indoor environment with sufficient ventilation in an energy-efficient way compared to air conditioned buildings. Meanwhile, it's more reliable than naturally ventilated building when maintaining indoor environment comfortable since it has the capability to transfer into air conditioned mode when the outdoor climate is not favorable.

Recognizing its benefits, some hybrid ventilation buildings were already established in Europe as pioneer studies [1]. For example, in Belgium, a small office building with 1500 m^2 area utilized large louvers on different façades for summer cooling while a small mechanical cooling unit was also installed to reduce the supply air temperature when the outdoor air temperature is too high. Based on the feedbacks, the ventilation system achieved good performance except in one office that got severest exposure to the sun. Also, in Italy, a four stories open plan office buildings with approximately 45 occupants were equipped with both natural ventilation system and a fan coil system for both summer cooling and winter heating. Similar buildings also exist in Denmark, Sweden, Norway etc. However, in United States, hybrid ventilation buildings are still rarely seen. One example of this type of building is San Francisco Federal building. Air conditioning is utilized at the lower floors of the building while upper floors are cross ventilated. Nevertheless, despite its excellent expected performance, the building suffers from large amount of complains ranging from thermal comfort to acoustics [2].

As the first step to popularize hybrid ventilation building in US, its application potential in different US climate zones for saving energy and improving indoor environment should be thoroughly investigated. As one of the most widely used techniques for assessing building performance, building simulation is an important tool in potential investigation of hybrid ventilation due to its reliability. Nonetheless, currently almost all the potential investigations of hybrid ventilation are still based on deterministic building simulation results, which could be insufficient to account for possible variations spring from simulation scenarios [3]. Thus, in this paper, different level of uncertainties in building simulation, ranging from meteorological, microclimate, building properties and operation scenarios, are quantified and discussed first. The hybrid ventilation potentials are then investigated based on our prototypical small to medium office building using the uncertainties we have quantified. The results of potential investigation will finally be presented. The following sections will be structured as follows:

2 Literature Review

Uncertainties in building simulation have been investigated by researchers for years. According to [4], the uncertainties could be categorized as structural uncertainty and parameter uncertainty. In the building simulation field, structural uncertainty represents the simulation inadequacy that springs from the simplified physical modeling process. A good example of this category of uncertainty is shown by Han et al [5]. In the paper, different methodologies such as computational fluid dynamics and default modeling in EnergyPlus have been compared. The results showed that the accuracy of energy simulation could increase up to 11% in certain cases. Also, Yuming et al [6] have quantified the solar diffuse irradiation uncertainty of Perez Model on the inclined surface based on real measurement data. A two-phase regression model was developed and the validation result showed that the model bias error could be reduced by this newly introduced model. In addition to the structural uncertainty, parametric uncertainty, which is caused by insufficiency of building information in the design phase, is also an important uncertainty source to consider. For example, the properties of construction material are one of the critical inputs for building simulation. However, they are rarely known before the construction process, which might cause large deviations of results in the modeling process. In order to address this deficiency, Domínguez-Munoz et al [7] have done an investigation about the variations of thermal conductivity and density for most of the mainstream construction materials. Similarly, to fully account for the inaccuracy of convection heat transfer, Palyvos [8] have summarized and provided a critical discussion on all the available correlations of external wind coefficients. Also, Thevenard and Haddad [9] have compared results from two building simulation models, in which one of them used typical reflectance for meteorological years while the other use actual recorded data. The results show that the difference could be up to 10.9% on yearly basis while 23.3% on a monthly basis.

However, despite of the existence of uncertainties that could have non-negligible impacts on the simulation outcomes, almost all the hybrid ventilation potential investigations are based on deterministic simulation results. Ezzeldin and Rees [10] have done a systematic investigation of hybrid ventilation potential in arid climates. In addition to basic hybrid ventilation strategy, the authors have also combined hybrid ventilation with other potential energy saving techniques such as evaporative, radiant or ground cooling. Deterministic simulations were implemented for representative cities in the arid climates. The conclusion showed that the energy saving could be up to 90% after combining the technologies. Also, Emmerich [11] has utilized a coupled multi-zone airflow simulation tool – CONTAMR to model a 5-story building in five representative cities in U.S. The results confirmed that the hybrid ventilation system could reduce energy consumption compared with mechanical system while maintaining the satisfactory indoor environment. Thus, considering the possible deviations that could be caused by building simulation uncertainties mentioned above, a more thorough investigation of hybrid ventilation potential taking fully account of all the possible variations of meteorology, building microclimates, properties and operations is needed to provide more reliable information of actual hybrid ventilation across different climates.

3 Methodology

3.1 Research Objective and Summary

Overall, the main objective of this research is to investigate the hybrid ventilation potential across different climate zones of U.S. with the different uncertainties presented thus to provide a more reliable and informative reference for the potential saving of hybrid ventilation. In order to achieve the goal, a representative building model will be selected first. The uncertainties from different levels will then be quantified and applied in our building simulation. The GURA workbench, which is specially designed for uncertainty analysis for building simulation, will be used to automate the process. The results and discussions will finally be presented.

3.2 Building Model

Considering the diverse types of zones including lobbies, office rooms etc., a campus building, whose layout is shown in Figure 1 below, has been selected as our baseline commercial building in the hybrid ventilation potential investigation. The whole building is composed of 21 zones with the detailed construction information listed in the Table 1 below. As to building operation related parameters, the lighting, electric equipment and occupancy are set to 8.5 W/m2, 8 W/m2 and 0.05 person/m2 respectively based on recommendations by ASHRAE [12]. And their schedules are also based on typical settings of commercial buildings in the ASHRAE guideline [13]. In addition, each zone has at least one window (2.1m * 2.4m) attached to meet the requirements of ASHRAE Standard 62.1 [14] for ventilation purpose in natural ventilation mode. Lastly, the HVAC system of the building was configured according to the recommendation of DOE reference commercial building [15]. Thus, VAV (Variable Air Volume) system was attached. The COP (coefficient of performance) for cooling was set to 3.4 while the efficiency of heating was set as 0.8. The supply air temperature from air conditioning unit was constantly 15℃ throughout the year. Air economizer was also used to adjust the outdoor air inflow rate based on the outdoor dry bulb temperature.

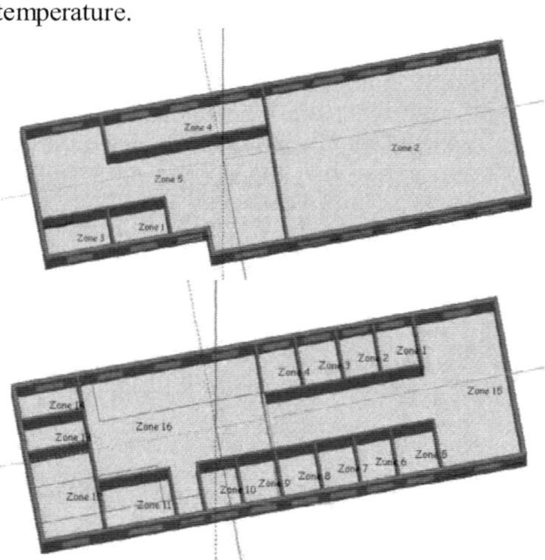

Figure 1. Baseline Building Layout

Table 1. Building Construction Details

Exterior Wall	U Value: 0.55 W/m^2-K Brick: 5 cm, Concrete Block: 10 cm, Rigid Insulation: 5cm, Gypsum Plasterboard: 1.3cm
Ground Floor	U Value: 1.96 W/m^2-K Concrete Slab: 15 cm
Roof	U Value: 0.73 W/m^2-K Bitumen Layer: 1 cm, Rigid Insulation: 4 cm, Cast Concrete: 12 cm, Plasterboard: 1.3 cm
Window	U Value: 2 W/m^2-K SHGC: 0.55

3.3 Uncertainty Quantification

3.3.1 Overall Setting

To efficiently uncover the potential of hybrid ventilation in different climates in US, a robust and thorough uncertainty quantification is essential. Since the purpose of this paper is for general investigation of hybrid ventilation potential, it is assumed that no further building context information is available such that a general version of uncertainty analysis is applied. More specifically, the uncertainties could be divided into 5 scales, i.e. meteorological uncertainty, urban uncertainty, building uncertainty, system uncertainty and occupant uncertainty. In our analysis, all the uncertainties from different levels are more or less considered. The detailed uncertainty quantification processes will be provided from Section 3.3.2 to Section 3.3.5 below. All the works are based on model discrepancy of EnergyPlus, which is the building simulation software we use for simulation.

3.3.2 Meteorological Uncertainty

Meteorological uncertainty depicts the uncertainties cause by variability of overall weather condition on one area or possible long term climate changes. Weather file is doubtlessly reported as one of the most critical inputs for the building simulation. However, currently in the building simulation practice, almost all the building modelers would only use TMY (typical meteorological year) or TRY (test reference year), which leads to the negligence of influence cause by possible weather fluctuations in that area. Accordingly, in our work, instead of using TMY, historical meteorological years are used as replacements in order to thoroughly investigate the potential and performance of hybrid ventilation under different weather conditions. 10 years have been randomly selected from 30 years historical meteorological years data and used as inputs for our simulation.

3.3.3 Urban Uncertainty

The urban uncertainty, or microclimate uncertainty, represents the possiblily different influence of building microclimate on the building simulation. Obviously, the microclimate around one building could be very different from the other (e.g. suburb vs. downtown) in real scenarios. However, most of these effects are still neglected in the building simulation. In our study, the influence of urban heat island effect (UHI), local wind speed and ground reflectance have been quantified and applied to generate better simulation results compared to deterministic simulation.

3.3.3.1 Uncertainty Quantification for Urban Heat Island Effect

Urban heat island effect, which commonly exists in most of large cities around the world, is one of the most important phenomenon to consider to achieve more accurate hybrid ventilation investigation results due to its influential impacts on the surround outdoor temperature of a building. In our works for urban heat island effect quantification, the high-fidelity TEB (tower energy budget) model was used to compute the temperature difference Δ_T between urban and rural area at the first place. Then, a statistical model is developed for prediction of Δ_T based on the weather condition at certain time steps. The equation is shown below:

$$\Delta_T = \beta_0 + \sum_{i=1}^{6} \beta_i w_i + \beta_7 T_r$$

in which $w = (w_1, .., w_6)$ are short wave and long wave solar radiation, air temperature, wind speed, pressure and specific humidity respectively and T_r is rural temperature. These parameters can all be generated from the weather data at the simulation time step. Since the microclimate of a building is uncertain, our inputs for TEB model vary from case to case to generate different sets of possible Δ_T under different scenarios. All the inputs including canyon height, canyon ratio, pervious road fraction and building roof area fraction for TEB model are based on a global database for urban geometric parameters [16]. Finally, a Gaussian process [17] emulator $\hat{\beta}(z)$ was established for the prediction of β under different scenarios.

3.3.3.2 Uncertainty Quantification for Local Wind

Local wind speed around the building is one of the most unpredictable variables in the building simulation. To quantify the uncertainty of local wind speed, the high fidelity model Community Land Model (CLM) [] was employed. Similar to the uncertainty quantification for urban heat island effect, the difference between the results of CLM and EnergyPlus prediction were calculated under different settings shown in Table 2 below. Piecewise linear regression was used as regression model,

$$\Delta_W = \begin{cases} \beta_{01} + \beta_{11} z + \xi & \xi \in N(0,\sigma^2) \quad when\ z < \tau \\ \beta_{02} + \beta_{12} z + \xi & \xi \in N(0,\sigma^2) \quad when\ z \geq \tau \end{cases}$$

in which z is the central height of building surface in order to further account for the variation of wind at different heights. All the coefficients in piecewise linear regression were estimated using maximum likelihood method. In our uncertainty quantification process, for all the experiment settings, the R square of regression was above 95%, which means that the uncertainties of wind speed were sufficiently quantified.

Table 2. Experiment Settings for Local Wind Uncertainty Quantification

	H		H/W		B_L/B_W	
	Min	Max	Min	Max	Min	Max
Country	3	3	0	0.3	0.25	1
Urban	99 cases from database				0.25	1
Cities	26 cases from database				0.25	1

3.3.3.3 Uncertainty Quantification for Ground Reflectance

Lastly, as one of the aspects of microclimate uncertainty, the uncertainty of ground reflectance was quantified in city, urban/suburban and open country areas. The equation below shows the calculation of ground reflectance from $f_{pervious}$ (the road area fraction of pervious road) and $\rho_{pervious}$ (pervious road solar reflectance).

$$\rho_{ground} = \rho_{pervious} f_{pervious} + \rho_{impervious}(1 - f_{pervious})$$

Monte Carlo method was utilized in the sampling process based on the data from [18].

3.3.4 Building Uncertainty

In addition to the meteorological and urban level uncertainty, building uncertainty is another significant source of uncertainties that could impose non-negligible impacts on the simulation outcomes. In our uncertainty analysis, for the building level uncertainty, mainly the material uncertainty and convection uncertainty were taken into account.

Building material uncertainty could be caused by many reasons such as different manufacturing and construction process, possible change of material property in specific environments etc. With respect to this uncertainty, one of the most thorough works was done by Domínguez-Munoz et al [7], in which all the uncertainties of thermal conductivity and density for mainstream construction material were quantified. In our uncertainty analysis, a relative normal distribution with 10% variance was applied to conductivity of window, density of normal construction material and solar transmittance of the window etc. Meanwhile, a relative normal distribution with 5% variance was imposed for material properties such as conductivity of normal construction material, front and back side visible light transmittance etc.

Except for the material uncertainty, uncertainties exist in building convective heat transfer also plays significant roles in the building simulation.

To sufficiently quantify the possible variations of convective heat transfer coefficient, firstly, literature surveys have been done to collect the results from different experiments. Using the following equation to calculate the external convective heat transfer coefficient,

$$h_c = aV_z + b$$

bivariate kernel density estimator was employed to model and estimate the density and covariance of a and b based on collected experiment results (V_z is the local wind speed). Lastly, optimization technique was utilized to determine the bandwidth H matrix for the kernel.

3.3.5 Building Operation Uncertainty

Last but not least, the uncertainties exist in the building operation, including uncertainty of lighting consumption, electric equipment consumption and occupancy consumption, are also significant to consider for an accurate prediction of the building energy consumption and indoor thermal environment. These related parameters are also the hardest parameters to accurately determine in the design phase of a building. However, due to the lack of related data, their uncertainties are usually not quantified for the simulation practice. In our uncertainty analysis, relative normal distribution with 30% variance were applied for both lighting usage density and occupancy density respectively for illustration purposes. Also, according to the ASHRAE fundamental handbook [19] for electric equipment consumption in the office with light and heavy office works, the possible variation of electric equipment consumption was set to uniformly distributed from 5.4 W/m2 to 16.1 W/m2 in our study.

3.3.6 Uncertainty Distribution

To efficiently distribute the uncertain parameter, the GURA (Georgia Tech Uncertainty and Risk Analysis) workbench was developed to automate this process. The process of conducting uncertainty analysis using GURA is shown in Figure 2 below. Firstly, a XML file that defines the distribution of uncertain parameters should be established as UQ Repository. Sampling plugin will then generate samples according to the UQ Repository and then distribute them into building and weather module for further process. After the processing, the EnergyPlus built in the simulation engine module will run the IDF file and generate the results for final processing. 200 runs were performed for our analysis since no obvious change of results was observed if we further increased the number of runs of uncertainty analysis.

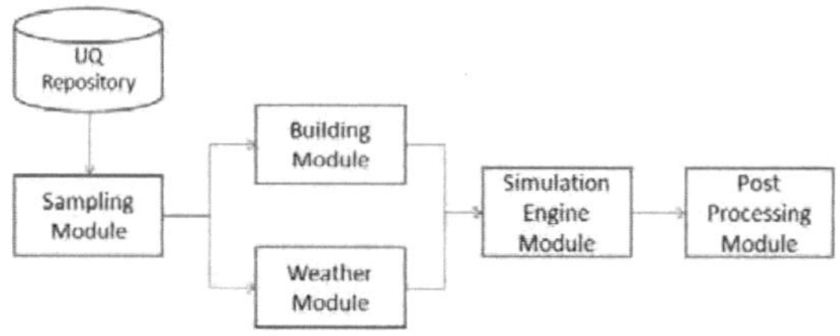

Figure 2. GURA workflow

3.3.7 Thermal Comfort Criteria

The argument for which thermal comfort criteria should be applied for hybrid ventilation building exists for a long period of time. Nowadays, in the European standard (EN 15251 [20]), the adaptive model could be applied for the hybrid ventilation building when the building is natural ventilation mode as long as (1) the building has operable window (2) no clothing protocol is forced. On the other hand, in US standard, the hybrid ventilation building is classified as air conditioned building since it has air conditioned presented. Thus, the graphical zone method, which is based on Fanger's PMV/PPD model [21], should be applied. However, recently, there is an increasing argument that the adaptive model is more suitable for thermal comfort evaluation of hybrid ventilation building compared to the PMV/PPD method. Thus, in our analysis, the adaptive model from ASHRAE 55 is applied for evaluation of thermal comfort of hybrid ventilation building. As shown in Figure 3 below, the thermal comfort zone (80% satisfactory) is determined by mean outdoor temperature for 15 consecutive days.

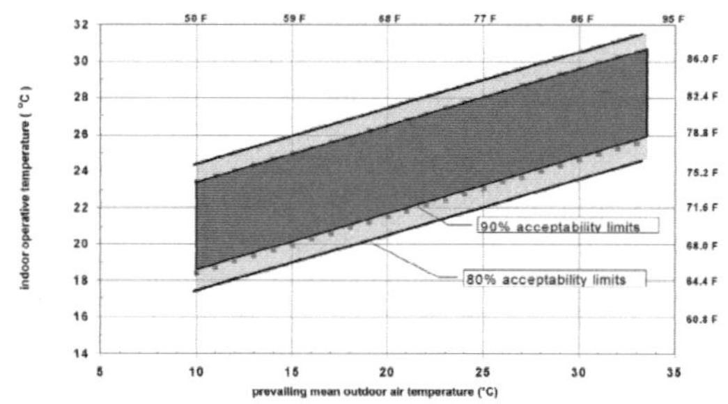

Figure 3. Adaptive Thermal Comfort Model

3.3.8 Building Control Strategy

As to control of the building, rule based control is used in our analysis for hybrid ventilation operation. It is the most widely used control methods for hybrid ventilation control due to its simplicity of implementation and easiness of design. In the rule based control strategy, the window openings or louver openings of a building are usually determined by the outdoor air temperature and wind speed.

In our analysis, for all the climates, the windows for the whole building were set to open when the outdoor temperature is between 18 ℃ and 24 ℃ based on the adaptive thermal comfort model mentioned above. Since the operative temperature for a building (especially for the second floor of our baseline building with roof exposure to the sun) could usually be several Celsius higher than the air temperature, this is a very reasonable range. No humidity limit was applied since the adaptive thermal comfort model has already taken it into account. The window is set to be off when the outdoor local wind speed is larger than 10 m/s.

4 Discussion and Result

After applying the uncertainty quantified for our analysis, both the results of potential energy saving and airflow volume increase in natural ventilation hours in four different representative cities that have most hybrid ventilation potential were presented below. Meanwhile, during natural ventilation hours for all cities, the thermally uncomfortable time are all within the 3% limits out of office hours.

As an example, Figure 4 below showed the result for Atlanta, which include the histogram of energy saving percentage and opening hour percentage (out of total office hour). For climate of Atlanta (3A mixed-humid), the energy saving is approximately 17% and has the standard deviation of 4.1%. Also, the windows could be opened for 12% of office hour throughout the year. Similarly, Figure 5 to 7 showed the results for Los Angeles (3B Coast), San Francisco (3C Marine) and Seattle (4C Marine). Overall, Table 3 summarized these results, in which the average and standard deviation for all the outputs were listed.

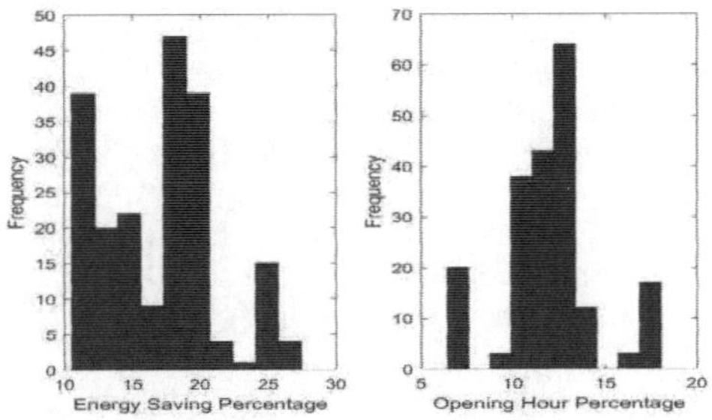

Figure 4. Atlanta result

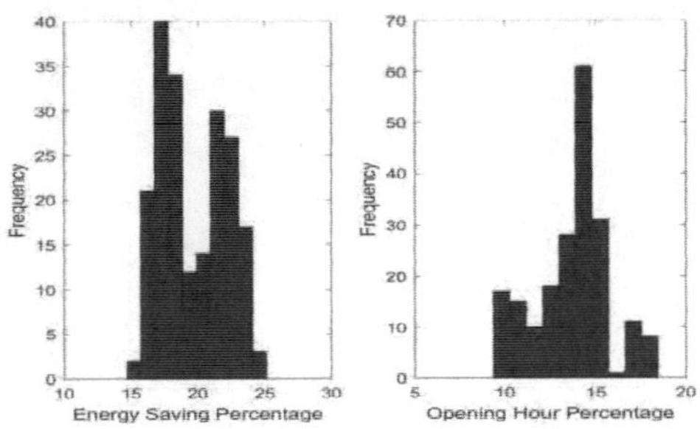

Figure 5. Seattle result

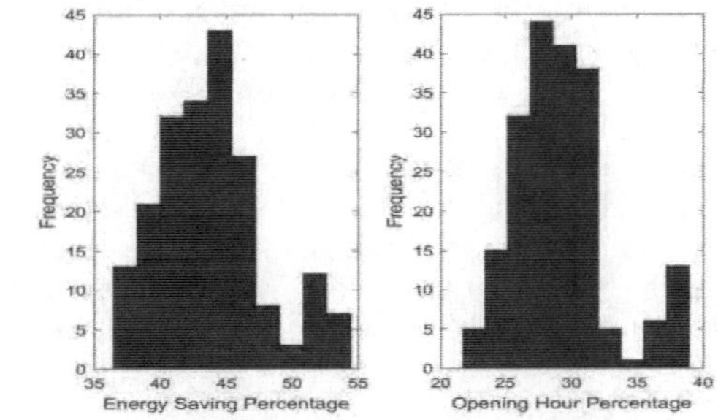

Figure 6. San Francisco result

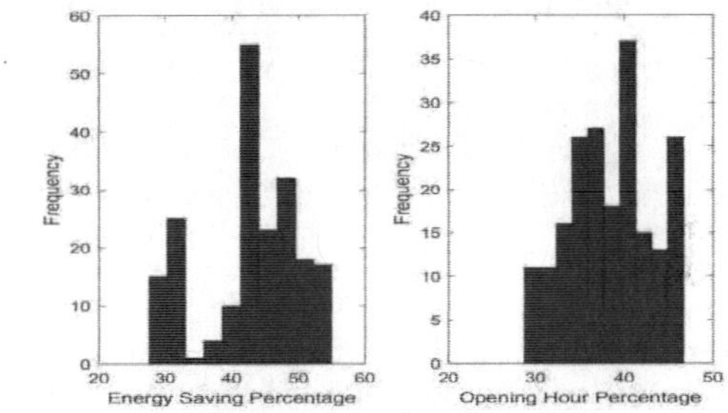

Figure 7. Los Angeles result

Table 3. Summary of Hybrid Potential Investigation Result Using Rule Based Control

City	Energy Saving		Window Opening Hour	
	Mean	Std	Mean	Std
Atlanta	17%	4.1%	12%	2.5%
Seattle	19.7%	2.5%	13.7%	2.1%
San Francisco	43.8%	4%	29%	3.6%
Los Angeles	42.9%	7.4%	38.5%	4.7%

5 Conclusion and Limitation

In this study, an uncertainty analysis was performed to evaluate the hybrid ventilation potential for possibly the most suitable climates for natural ventilation in US. Different levels of uncertainty were quantified first and then applied on our prototypical office building for simulation. The results showed that the energy saving on average could range from approximately 15% to 45% with an increase of airflow rate from 13 to 23 times. And the most suitable climates for hybrid ventilation are climates in Los Angeles (3B) and San Francisco (3C). As to limitation of the research, firstly, one important assumption of our analysis is that the occupants should operate the window in exactly the same way we have defined (open window as soon as the outdoor temperature is between 18℃ and 24℃). Thus, in the study, the uncertainties spring from occupant behavior are not accounted, which leads to the fact that our results tend to be more optimistic. With the advance of control technology, more complicated and effective control (e.g. model predictive control) could be applied for hybrid ventilation building as well. Consequently, in our future study, the uncertainties of occupants will further be quantified and more advance control technologies will be tested to improve our results of hybrid potential investigation.

6 Reference

[1] Brager, Gail, Sam Borgeson, and Yoonsu Lee. "Summary report: control strategies for mixed-mode buildings." (2007).

[2] Fowler, K., et al. "Re-assessing green building performance: A post occupancy evaluation of 22 GSA buildings." Pacific Northwest National Laboratory: Richland, WA, USA (2011).

[3] De Wit, Sten, and Godfried Augenbroe. "Analysis of uncertainty in building design evaluations and its implications." Energy and Buildings 34.9 (2002): 951-958.

[4] Draper, David. "Assessment and propagation of model uncertainty." Journal of the Royal Statistical Society. Series B (Methodological) (1995): 45-97.

[5] Han, Guiyuan, Jelena Srebric, and Elena Enache-Pommer. "Different modeling strategies of infiltration rates for an office building to improve accuracy of building energy simulations." Energy and Buildings 86 (2015): 288-295.

[6] Sun, Yuming, et al. "Quantification of model form uncertainty in the calculation of solar diffuse irradiation on inclined surfaces for building energy simulation." Journal of Building Performance Simulation 8.4 (2015): 253-265.

[7] Fernando Domínguez-Muñoz, Brian Anderson, José M. Cejudo-López, Antonio Carrillo-Andrés, Uncertainty in the thermal conductivity of insulation materials, Energy and Buildings, Volume 42, Issue 11, November 2010, Pages 2159-2168, ISSN 0378-7788, http://dx.doi.org/10.1016/j.enbuild.2010.07.006

[8] Palyvos, J. A. "A survey of wind convection coefficient correlations for building envelope energy systems' modeling." Applied Thermal Engineering 28.8 (2008): 801-808.

[9] Thevenard, D., and K. Haddad. "Ground reflectivity in the context of building energy simulation." Energy and Buildings 38.8 (2006): 972-980.

[10] Ezzeldin, Sherif, and Simon J. Rees. "The potential for office buildings with mixed-mode ventilation and low energy cooling systems in arid climates." Energy and Buildings 65 (2013): 368-381.

[11] Emmerich, Steven J. "Simulated performance of natural and hybrid ventilation systems in an office building." Hvac&R Research 12.4 (2006): 975-1004.

[12] Handbook, A. S. H. R. A. E. "Fundamentals." American Society of Heating, Refrigerating and Air Conditioning Engineers, Atlanta 111 (2001).

[13] Standard, A. S. H. R. A. E. "Energy standard for buildings except low-rise residential buildings." ASHRAE/IESNA Standard 90.1 (1999).

[14] Standard, A. S. H. R. A. E. "62.1. 2007, Ventilation for Acceptable Indoor Air Quality." American Society of Heating, Refrigerating and Air-Conditioning Engineers, Inc., Atlanta, GA (2007).

[15] Deru, Michael, et al. "US Department of Energy commercial reference building models of the national building stock." (2011): 1.

[16] Jackson, Trisha L., et al. "Parameterization of urban characteristics for global climate modeling." Annals of the Association of American Geographers 100.4 (2010): 848-865.

[17] Rasmussen, Carl Edward. "Gaussian processes for machine learning." (2006).

[18] Ahrens, C. Donald. Meteorology today: an introduction to weather, climate, and the environment. Cengage Learning, (2012).

[19] Handbook, A. S. H. R. A. E. "Fundamentals." American Society of Heating, Refrigerating and Air Conditioning Engineers, Atlanta 111 (2001).

[20] CEN, EN15251. "15251, Indoor environmental input parameters for design and assessment of energy performance of buildings addressing indoor air quality, thermal environment, lighting and acoustics." European Committee for Standardization, Brussels, Belgium (2007).

[21] Fanger, Poul O. "Thermal comfort. Analysis and applications in environmental engineering." Thermal comfort. Analysis and applications in environmental engineering. (1970).

Assessment of Utilizing Mobile App in Delivering Eco-Feedback Information for Building Energy Conservation

Jing Lin [a], Guofeng Ma [b], Nan Li [a,*], and Junjie Zhou [b]

[a] Department of Construction Management, Tsinghua University, Beijing 100084, China, China
[b] Department of Construction Management and Real Estate, Tongji University, Shanghai 200092
E-mail: linjing_tju@163.com, guofengma@tongji.edu.cn, nanli@tsinghua.edu.cn, 1432701@tongji.edu.cn

Abstract –

Researchers have developed various eco-feedback systems that aim at reshaping energy consumption behaviors of building occupants for energy conservation. Different technologies have been used to build eco-feedback systems through which occupants receive and respond to eco-feedback information. One emerging technology that has been envisioned for this purpose but rarely tested in practice is mobile app. This paper aims to examine the possibility of utilizing mobile app in eco-feedback system, and assess its effectiveness, benchmarked against web technology, a technology that has been proven effective in prior research for delivering eco-feedback information. Two eco-feedback systems were developed in this study and implemented in three student dormitories on a university campus. These systems used web technology and mobile app, respectively, to deliver the same eco-feedback information, and they differed significantly in their level of accessibility. Participants of a 15-week experiment were divided into two groups, each provided with access to one of the two systems. Experiment results showed that mobile app-based eco-feedback system led to relatively more intensive engagement of the participants with the system, and was significantly more effective in promoting building energy conservation.

Keywords –

Mobile app; Website; Behavior; Eco-feedback technology; Energy consumption

1 Introduction

Energy consumed in buildings has been considered a prime target for energy conservation, as buildings are responsible for 39.78% of whole energy consumption in the U.S. in 2015. Eco-feedback, which provides occupants with their energy consumption information, is a promising approach to gradually reshaping building occupants' energy consumption behavior and hence save energy [1]. The effectiveness of eco-feedback systems in energy conservation varies from 5% to 55% [2,3]. There is a number of factors impacting the effectiveness of eco-feedback systems that have been examined in prior study. Examples of such factors include culture [1,4], information representation [5] and interface design [6]. Technology used for developing eco-feedback systems, in particular, has been proven an important factor. Eco-feedback technologies can take on various forms such as utility bills, messages, emails, home-displays and websites. Website is one of the most effective and widely researched technology in prior studies [6,7]. Prior research also discussed the potential of utilizing mobile app in delivering eco-feedback information, as mobile phones provide a powerful and ubiquitous computing platform [8]. However, the effectiveness of app-based eco-feedback system is rarely experimentally examined, and its relative performance compared to more widely used eco-feedback technologies is unknown. To fill this gap, this paper aims to assess the effectiveness of app-based eco-feedback system, benchmarked with web-based eco-feedback system, and discuss possible causes of the discrepancies between them if any. The remainder of this paper is organized as follows: section 2 presents a brief overview of related research, followed by section 3 that introduces the settings of two eco-feedback systems developed in this study. Section 4 describes the design and implementation of an eco-feedback experiment. Section 5 then presents findings of this study. Section 6 concludes the paper.

2 Related Research

Various forms of eco-feedback systems developed with different technologies have been reported in prior research and used for reshaping occupants' energy consumption behavior in buildings [9]. Traditionally, energy consumption information was delivered by bills and posters. However, the frequency of updating information in these traditional methods is too low to provide real-time eco-feedback information, and is therefore a barrier in improving the awareness of habitual energy consumption behavior. Messages, emails, websites and in-home displays are then introduced, thanks to fast development of information and computing technologies in the last decades. A comparative study that involved website, in-home displays and improved electricity bill reported that occupants utilizing website-based eco-feedback system saved more energy than those utilizing the other two [7]. More recently, the possibility of utilizing mobile app to eco-feedback information has drawn considerable attention, due to the ubiquitous computing capabilities mobile app provides [10].

Eco-feedback system is a type of information system, whose accessibility can vary when different technologies are used to deliver information. Accessibility is an important dimension of system quality, and it is a major challenge in eco-feedback systems [8]. Accessibility refers to "the degree to which the system and the information it contains can be accessed with relatively low effort" [11]. According to information system success model [12], accessibility influences system success through impacting user satisfaction, attitude and user engagement with a system. For eco-feedback system, in particular, its effectiveness is influenced by attitude [13] and engagement [14] on which accessibility can be impactful [15]. Thus, the different effectiveness of eco-feedback systems built with different technologies can be possibly explained from the accessibility point of view.

3 Eco-Feedback Systems

Two eco-feedback systems, which differ in the way eco-feedback information is delivered, are developed in this study, as shown in Figure 1. One is a widely researched web-based eco-feedback system built with web technology. The other one is an app-based eco-feedback system built with emerging mobile computing technology. These two systems share the same data capture and processing component, including electric meters for each residential unit, data concentrators for each building, cables and a server where information is stored, processed, and prepared for delivery. The server updates daily energy consumption data for each unit and prepares it for user request. Web-based eco-feedback system delivers eco-feedback information through website interface. Information provided includes personal daily energy consumption data, historical personal data from the last seven days and last thirty days, and normative comparison with the user's peers. App-based eco-feedback system provides the same information as web-based eco-feedback system, but through mobile app. Furthermore, mobile app provides a gadget in the notification panel of mobile phones, where energy consumption overview of the previous day and the last seven days is displayed to the user. The two systems also share an email portal, which is responsible for sending emails to users to remind them of using eco-feedback systems to access eco-feedback information. For web-based eco-feedback system, the email portal sends reminders every Monday morning. For app-based eco-feedback system, the email portal sends reminders whenever a user remains inactive for over 24 hours, which is considered an alert that the mobile app has been killed or malfunctioning for some reason.

A fundamental difference of the two systems is their level of accessibility. The accessibility of online systems can be assessed in three dimensions, including physical accessibility, interface accessibility and information accessibility [16]. Physical accessibility assesses users' access to terminal and access to online system [16]. Interface accessibility assesses users' use of command language in online system [17]. Information accessibility assesses the extent to which users can employ a system to reach information [17]. The app-based eco-feedback technology has higher accessibility than the web-based eco-feedback technology due to the following reasons: 1) mobile app users do not need to remember or type in account information as website users have to do, leading to better access to online system and hence high physical accessibility; 2) the app-based system includes a gadget that provides an additional pathway to the full interface, which improves the interface accessibility of the system; and 3) the use of gadget improves the information accessibility of the system, by providing users with a quick overview of key eco-feedback information in the notification bar without having to access the full interface.

4 Eco-Feedback Experiment

4.1 Test-bed Buildings and Participants

An experiment was designed and carried out in this study, to empirically compare the effectiveness of eco-feedback systems with different levels of accessibility in changing the energy consumption behaviors of occupants. Test-bed buildings used in this study were three six-story student dormitories on the campus of Tongji University. Each residential unit in these test-bed buildings had a bathroom, a bedroom and a balcony, and could accommodate one or two occupants depending on its area that varied between $8m^2$ and $13.77m^2$. Given that each unit had a designated electric meter and hence the energy consumption was measured at the unit level, only single units were included in the experiment so that the eco-feedback information could be processed and provided at the individual occupant level. Typical appliances in these units included lights, air conditioners, phone chargers and computers, according to site visits and informal interviews. Students living in single units in the test-bed buildings were face-to-face interviewed, during which they were explained the purpose of the experiment, and informed that they would be provided with an individual login account to the website or mobile app for accessing their respective electricity usage data and the averaged data of their peers. They were also informed that they would receive emails reminding them of checking the energy consumption information. A total of 33 single-unit occupants in the test-bed buildings signed up to participate in the experiment.

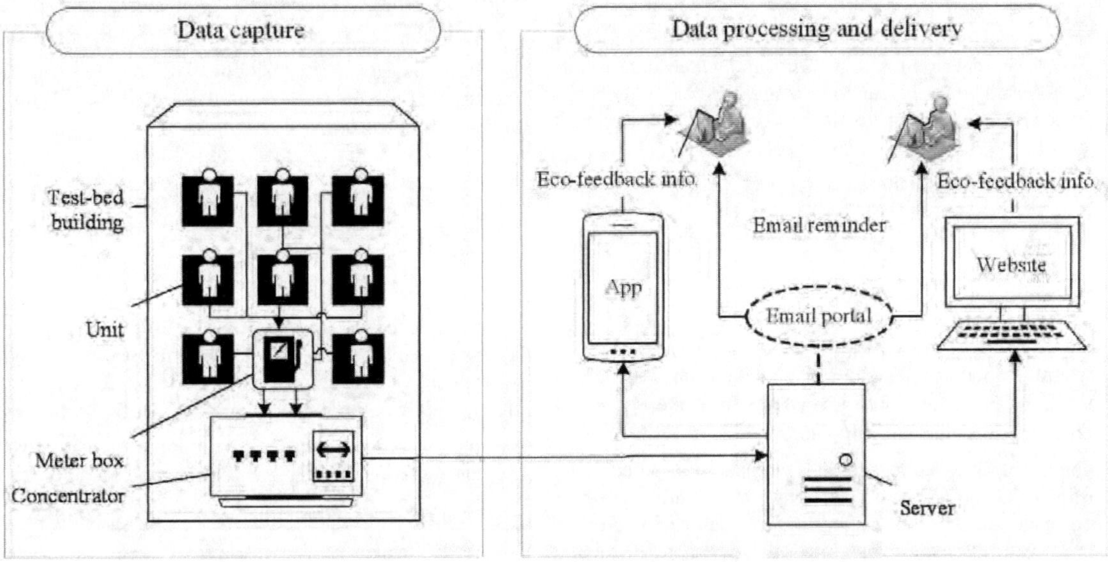

Figure 1: Architecture of two eco-feedback systems

4.2 Experiment Design

The experiment lasted for 15 weeks, from October 17, 2015 to January 29, 2016. It included a pre-study phase and a study phase. The pre-study period (phase 0) lasted for 6 weeks from October 17 to November 27, 2015. Daily energy consumption data were collected during this period and used for offsetting the effect of inherent difference between two SGs. Participants received no eco-feedback information or behavior intervention of any kind during this period. Then, participants received an email containing their individual account information of the eco-feedback system, and a link to the login page of the website or download page of the app on November 28. The study period (phase 1) lasted for 9 weeks from November 28, 2015 to January 29, 2016. During phase 1, participants using the web-based system received emails every Monday morning reminding them of logging into the website to check their eco-feedback information, and participants using the app-based system received emails once their app remained inactive for over 24 hours. Daily energy consumption data in all single units were collected during phases 0 and 1 for comparing the effectiveness of two eco-feedback systems in reshaping the participants' energy consumption behaviors. All single-unit occupants in the test-bed buildings were divided into the following three groups:

• Study Group A (SG A) – Experiment participants provided with web-based access to eco-feedback information.

• Study Group B (SG B) – Experiment participants provided with app-based access to eco-feedback information.

• Control Group (CG) – Occupants not participating individually in the experiment and not provided with access to any eco-feedback information.

There were 14 and 19 participants in SG A and SG B, respectively, and 257 occupants in CG. They were all PhD students between 21 and 30 years old. Among participants in SGs, 54.55% were male and 45.45% were female. A total of 54.55% participants indicated in a questionnaire that normally they would check emails immediately when they were alerted by mobile app of incoming new emails, and 96.97% of them would check emails at least once per day. A total of 76.76%

participants indicated they would keep their mobile phones connected to Internet at least 8 hours a day, and 39.39% of them would keep their mobile phones online all the time.

5 Findings and Discussions

After missing or corrupted data points were cleaned, a total of 25,858 valid data points of daily energy usages in single occupancy units were collected during the entire experiment. These included 22,617 data points for CG, 1,375 data points for SG A, and 1,866 data points for SG B. The statistical percentage difference in energy consumption resulting from the use of the eco-feedback system (Δ_b (%)), a measure of behavior change of occupants, was calculated as follows ([5]):

$$\delta_{pre-study\_b}(\%) = \frac{\sum_{i=1}^{n}((P-C)/C)}{n} \quad (1)$$

$$\delta_{study\_b}(\%) = (P-C)/C \quad (2)$$

$$\Delta_b(\%) = \delta_{study\_b} - \delta_{pre-study\_b} \quad (3)$$

where P denotes the energy consumption of a given room on a given day, C denotes the average daily consumption per unit in CG, Δ_b denotes the energy consumption behavior changes, measured in percentage, of a SG in energy consumption, relative to the corresponding CG, between the pre-study period and the study period, $\delta_{pre-study\_b}$ denotes the average percentage difference between P and C for a given day of the week in the pre-study period, δ_{study_b} denotes the percentage difference between P and C for a given day of the week in the study period, and n denotes the number of the days of the pre-study period.

The statistical absolute difference in energy consumption resulting from the eco-feedback information (Δ_{con} (kWh)), a measure of energy consumption savings of occupants, was calculated based on the following equations ([5]):

$$\delta_{pre-study\_con}(kWh) = \frac{\sum_{i=1}^{n}(P-C)}{n} \quad (4)$$

$$\delta_{study\_con}(kWh) = P - C \quad (5)$$

$$\Delta_{con}(kWh) = \delta_{study\_con} - \delta_{pre-study\_con} \quad (6)$$

where Δ_{con} denotes the absolute energy savings, measured in kWh, of a SG in energy consumption, relative to the corresponding CG, between the pre-study period and the study period, $\delta_{pre-study\_con}$ denotes the average absolute difference between P and C for a given day of the week in the pre-study period, δ_{study_con} denotes the absolute difference between P and C for a given day of the week in the study period, n denotes the number of days of the pre-study period. It needs to be noted that Δ_b and Δ_{con} differ mainly in the way they factor in variable C , which varies depending on external factors such as temperature. It needs to be noted that it is mathematically possible, and was actually observed in the experiment, that trend of energy consumption savings (Δ_{con}), i.e. increase or decrease, may not always agree with the trend of energy consumption behavior change (Δ_b).

Cumulative energy savings (Δ_{cum} (kWh)), a measure of cumulative energy savings of a SG since the beginning of phase 1, was calculated based on the following equation:

$$\Delta_{cum}(kWh) = \sum_{i=1}^{d}\Delta_{con} \quad (7)$$

where Δ_{cum} denotes the cumulative energy savings in kWh for a SG in energy consumption, relative to the corresponding CG, and d denotes any given day in the study phases.

Both energy consumption behavior changes and energy savings of the participants in SG A and SG B, were analyzed based on data collected in the experiment. Table 1 shows the mean Δ_b and Δ_{con} of SG A and SG B in phase 1. Negative values in the table indicate positive behavior changes or energy savings, and vice versa.

Table 2: Behavior change of SG A and SG B

	Δ_b (%)	Δ_{con} (kWh)
SG A	-0.26	0.37
SG B	-24.27	-0.78
Sig.	0.000	0.000

To compare the results between two groups and between two phases, independent-samples T test was conducted using SPSS 20. The results showed that, at a confidence level of 95%, participants using web-based eco-feedback system barely changed their energy consumption behavior, with negligible positive behavior change by 0.26% (p-value=0.968>0.05). Meanwhile, they significantly increased their energy consumption by 0.37 kWh per participant per day on average (p-value=0.004<0.05). Similar observation where eco-feedback led to more energy consumption was also reported in prior studies. For instance, Jain et al. [5] found that unsuitable information representation in eco-feedback system could cause increase of energy consumption. Another possible reason for this observation is that some participants, by receiving eco-feedback information, may realize that they overestimated their daily energy consumption before and they were actually using less energy. Consequently, they may become less concerned about energy conservation and start increasing their energy consumption. While this hypothesis requires further validation, it suggests the possibility that eco-feedback may be counterproductive for users who are already over concerned about energy conservation. In addition, the participants came from different regions in China with different cultural and social background. Culture [1,4] and social norm [18] are influential factors on the effectiveness of eco-feedback. They could be another reason why the participants reacted differently to eco-feedback, with some of them showing significant negative behavior changes.

To the contrary, participants using app-based eco-feedback system showed significant positive behavior change of 24.27% (p-value=0.000<0.05), and significant energy savings of 0.78 kWh per participant per day on average (p-value=0.000<0.05). At a 95% confidence level, both Δ_b and Δ_{con} were statistically different between the two SGs in phase 1 (p-value=0.000<0.05 for both variables). In other words, the difference in eco-feedback technology resulted in different behavior-changing and energy-saving effects in the experiments. For behavior change (Δ_b), participants utilizing app-based eco-feedback system saved 23.01% more than participants utilizing web-based system on average, which indicated that app-based eco-feedback system was more effective than web-based eco-feedback system in reshaping occupants' energy consumption behavior. For energy consumption change (Δ_{con}) , participants utilizing app-based eco-feedback system saved 1.15 kWh more electricity per participant per day on average than participants utilizing web-based eco-feedback system, which indicated that app-based eco-feedback system saved more energy than web-based system.

Cumulative energy savings (Δ_{cum}) of two groups were also analyzed to assess the cumulative energy-saving effectiveness of the two systems over the entire experiment. The results are showed in Figure 2. As can be seen in the figure, when participants were not affected by eco-feedback during phase 0, their energy consumption behavior generally remained stable, and the two SGs were alike. Once participants started receiving eco-feedback information, the results showed that those in SG B who used app-based eco-feedback system changed their behaviors towards energy conservation, and their cumulative energy savings was positive and steadily increasing. Participants in SG A who used web-based eco-feedback system, however, reported negative cumulative energy savings of up to 62.31 kWh over the entire phase 1.

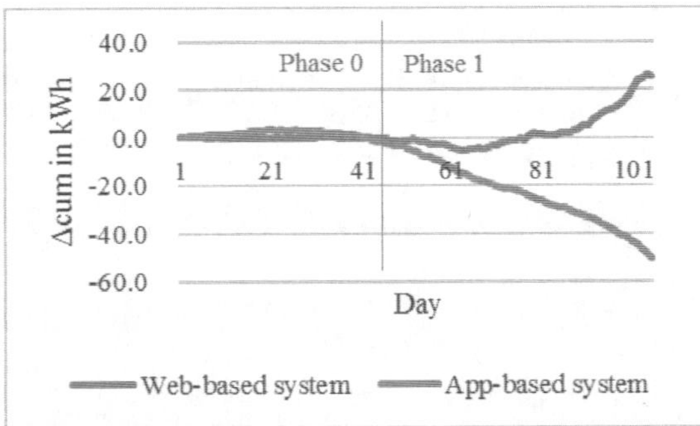

Figure 2: Cumulative energy savings among phases

To sum up, during phase 1, the app-based system resulted in positive energy consumption behavior change (Δ_b=-24.27%), whereas the web-based system did not yield any significant behavior changes. Such difference could probably be attributed to the different levels of accessibility between the two systems. As discussed in Section 3, the app-based system had higher accessibility than the web-based system. Drawing on the information system success model, success of information systems, in this case positive change of energy consumption behavior and positive energy savings achieved by eco-feedback systems, could be impacted by accessibility through participants' engagement [19]. Prior research has found that participants' engagement was correlated with the behavior-changing effect of eco-feedback systems [6]. The participants frequently engaging in eco-feedback were more liked to be affected by eco-feedback and would tend to save more energy. To test this hypothesis in this study, participants' engagement with eco-feedback systems recorded in the server was further analyzed, and the results are shown in Figure 3. The results showed that participants in SG A were much less active in engaging with the system to retrieve eco-feedback information than participants in SG B, with an average of 1.15 and 54.90 activities (logins plus data views) per week per participant, respectively. It is therefore hypothesized that accessibility might have positively affected occupants' engagement with eco-feedback systems which, in turn, positively affected the effectiveness of the eco-feedback systems in reshaping participants energy consumption behaviors.

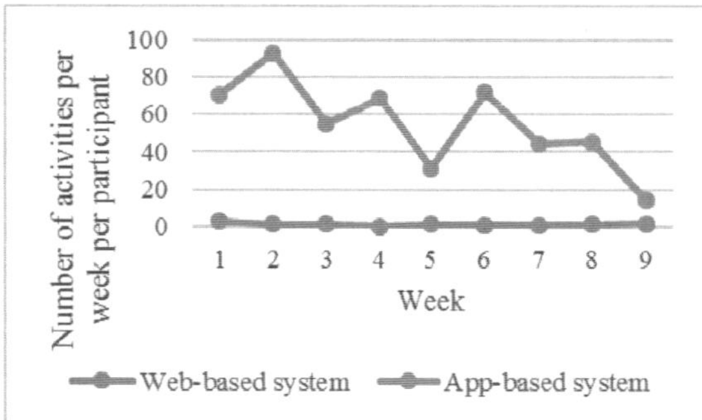

Figure 3: Engagement of participants with two eco-feedback systems

6 Conclusions

This study assesses the effectiveness of utilizing eco-feedback systems, built with two technologies including web technology and mobile app, in reshaping the energy consumption behavior of building occupants for building energy conservation. Two eco-feedback systems were developed in this study, using web and mobile app technologies, respectively, and deployed in three test-bed buildings, where an eco-feedback experiment was conducted over 15 weeks. By analyzing the behavior changes and energy consumption changes of the participants utilizing two eco-feedback systems, the results showed that the two eco-feedback systems exhibited noticeably different behavior-changing and energy-saving effectiveness. Overall speaking, mobile app-based system achieved significantly better performance, and a probable reason for such difference is the higher accessibility of app-based system. Practically speaking, mobile app provides more effective delivery of eco-feedback information and should be considered as an effective alternative technology in eco-feedback systems. It suggests that technologies and system designs that can improve system accessibility and increase user engagement should be encouraged in the development of eco-feedback systems in practice.

References

[1] Ma G, Lin J, Li N, Zhou J. Cross-Cultural Assessment of the Effectiveness of Eco-Feedback in Building Energy Conservation. *Energy Build*, 134:329–38, 2017.

[2] Darby S. The Effectiveness of Feedback on Energy Consumption. *A Rev DEFRA Lit Metering* 2006.

[3] Petersen JE, Shunturov V, Janda K, Platt G, Weinberger K. Dormitory residents reduce electricity consumption when exposed to real-time visual feedback and incentives. *Int J Sustain High Educ*, 8:16–33, 2007.

[4] Lin J, Nan L, Guofeng M, Zhou J. The Impact of Eco-Feedback on Energy Consumption Behavior: A Cross-Cultural Study. *In Proceedings of 33rd Int. Symp. Autom. Robot. Constr. Min. (ISARC 2016)* , pages 274-181, Auburn, USA, 2016.

[5] Jain RK, Taylor JE, Peschiera G. Assessing eco-feedback interface usage and design to drive energy efficiency in buildings. *Energy Build*, 48:8–17, 2012. doi:10.1016/j.enbuild.2011.12.033.

[6] Vassileva I, Odlare M, Wallin F, Dahlquist E. The impact of consumers' feedback preferences on domestic electricity consumption. *Appl Energy*, 93:575–82, 2012.

[7] Bartram L. Design challenges and opportunities for eco-feedback in the home. *IEEE Comput Graph Appl*, 35:52–62, 2015.

[8] Hermsen S, Frost J, Renes RJ, Kerkhof P. Using feedback through digital technology to disrupt and change habitual behavior: A critical review of current literature. *Comput Human Behav*, 57:61–74, 2016.

[9] Ehrhardt-Martinez K, Donnelly KA, Laitner S, others. Advanced metering initiatives and residential feedback programs: a meta-review for household electricity-saving opportunities, 2010.

[10] Zheng Y, Zhao K, Stylianou A. The impacts of information quality and system quality on users' continuance intention in information-exchange virtual communities: An empirical investigation. *Decis Support Syst*, 56:513–24, 2013.

[11] Delone WH, Mclean ER. The DeLone and McLean Model of Information Systems Success: A Ten-Year Update. *J Manag Inf Syst*, 19:9–30, 2003.

[12] Kim YH, Dan JK, Wachter K. A study of mobile user engagement (MoEN): Engagement motivations, perceived value, satisfaction, and continued engagement intention. *Decis Support Syst*, 56:361–70, 2013.

[13] Mohammadmoradi H, Gnawali O, Moss D, Boelzle R, Wang G. The impact of user engagement in the effectiveness of energy saving programs: poster abstract. *In Proceedings of 15th Int. Conf. Inf. Process. Sens. Networks*, pages 47, Vienna, Austria, 2016.

[14] Lee Y, Kozar KA. Understanding of website usability: Specifying and measuring constructs and

their relationships. *Decis Support Syst*, 52:450–63, 2012.

[15] Culnan MJ. The dimensions of accessibility to online information: implications for implementing office information systems. *Acm Trans Inf Syst*, 2:141–50, 1984.

[16] Loiacono ET, Djamasbi S, Kiryazov T. Factors that Affect Visually Impaired Users' Acceptance of Audio and Music Websites. *Int J Hum Comput Stud*, 71:321–34, 2013.

[17] Jain RK, Taylor JE, Culligan PJ. Investigating the impact eco-feedback information representation has on building occupant energy consumption behavior and savings. *Energy Build*, 64:408–14, 2013. doi:10.1016/j.enbuild.2013.05.011.

[18] Anderson K, Song K, Lee SH, Krupka E, Lee H, Park M. Longitudinal Analysis of Normative Energy Use Feedback on Dormitory Occupants. *Appl Energy* 189:623–39, 2017. doi:10.1016/j.apenergy.2016.12.086.

[19] Karahanna E, Limayem M. E-Mail and V-Mail Usage: Generalizing Across Technologies. *J Organ Comput Electron Commer*, 10:49–66, 2000.

Integration of control policies for comfort improvement in a large public building

M. Vaccarini, A. Piccinini, L. D'Angelo, A. Carbonari

[a]Department of Civil and Building Engineering and Architecture (DICEA), Università Politecnica delle Marche, via Brecce Bianche, 60131, Ancona, Italy
E-mail: m.vaccarini@staff.univpm.it, alessandro.piccinini@outlook.com, dangelo.letizia@gmail.com, alessandro.carbonari@staff.univpm.it

ABSTRACT

Some public buildings, which date back to the 1980-1990s, often miss to comply with comfort basic requirements. Also, their energy consumption is often above thresholds posed by regulations presently in force. Hence, low-cost technology improvements for renovation with high return on investment is a good opportunity in order to limit the required budget. To this purpose, the integration of advanced control policies may optimize energy use of buildings with minimum hardware enhancement.

In this paper, we evaluated the performances of an existing building that belongs to the campus of the Università Politecnica delle Marche (Ancona, Italy), and that is made of offices, teaching rooms, laboratories, a library and other public spaces. Thanks to an extensive monitoring of indoor comfort conditions, the main inefficiencies in terms of comfort were analyzed and a Dymola model of the actual building was worked out. Then, some energy improvement actions were assumed and tested by means of numerical simulation in the Dymola environment, and the benefits in terms of comfort and energy saving were estimated. Meanwhile, a BIM model of the building was built, regarding both the present status and each improved scenario, that allowed to perform a detailed cost estimation of each scenario. As a result, a cost-benefit analyses of each renovation scenario was performed and all the options were compared. The analyses showed that some renovation actions could be supported by energy savings with reasonable return on investment.

Keywords –
comfort improvement; control systems; energy analysis; cost-benefit analysis; BIM.

1 Introduction

Large buildings are the cause of a remarkable part of worldwide energy consumption, and keeping in mind that a huge part of large buildings is represented by public buildings, the development of customized refurbishment approaches is critical for the benefit of the whole society [1]. Evaluating refurbishment actions requires that an accurate model of the current and design scenarios of the building are setup. The development of the current scenario means setting the baseline, which must be validated against measured consumption data. Also, the collection of physical and geometrical data through on-site surveys may be necessary. What is more important, the simulation tool must be very flexible, in order to mirror both the current scenario and several possible design scenarios, in terms of performance and convenience; also, it must be supported by dynamic simulations [2]. The use of dynamic simulation tools will make the evaluation of the benefits deriving from customized – even if minor – renovation actions more accurate and reliable [3]. In this paper, a case study relative to the evaluation of the convenience of energy renovation actions applied to a large public building located in Italy is presented. Both energy analyses, indoor comfort assessment and cost analyses were performed in order to evaluate the best renovation actions. These analyses took advantage of BIM modelling for information generation and management about the current and future scenarios of the building undergoing renovation.

2 The case study

2.1 General description

Our case study is a building called "Eustachio", that hosts the rooms of the Faculty of Medicine and Surgery of the Università Politecnica delle Marche, located in

Ancona, Italy. Eustachio is made of two blocks, the northern and southern ones, which are connected by two transverse wings inside which the staircases and elevators are placed (Figure 1). For the purpose of this paper, only the southern block of the building was analysed, because users reported that the majority of drawbacks related to environmental discomfort in the summer season were experienced inside this block. The sourthern block of Eustachio accommodates lecture halls, educational laboratories and staff offices. Its southern façade is made of a curtain wall, that extends over the whole length and height of the building block.

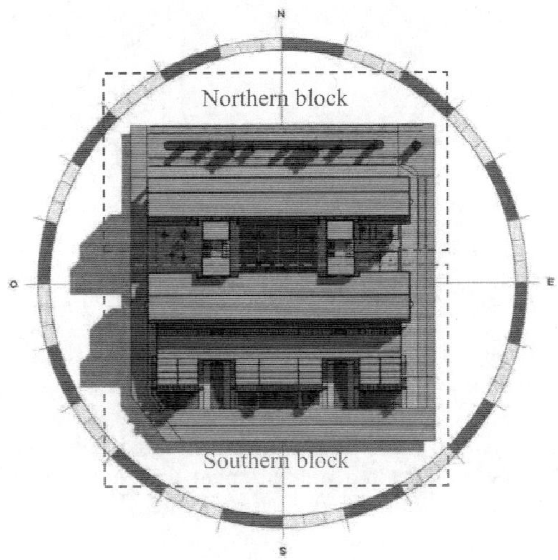

Figure 1. Plan and orientation of the University building considered as a case study.

According to what is depicted on Fig. 2, the air conditioning system of the southern block was split into several sections. The educational laboratories and staff offices are supplied with fresh or hot air (depending on the season) by three AHUs placed on the roof. The lecture halls are supplied by seventeen AHUs placed on the first level of the building. Furthermore, VAVs for local air-conditioning were installed inside every laboratory and fan coils are installed in the remaining rooms.

More specifically, fan coils are located above the entrance door of each room and staff office, which face the south orientation and whose external wall is made of a curtain wall. Instead, the VAVs were arranged in the corridors and in the educational laboratories, which face the internal courtyard of the building (they are north facing rooms). All the AHUs, fan coils and VAVs are served by a two pipe system working both in summer and winter, and connected to the central plant that includes four boilers placed in technical room at the basement level, and two chillers placed on the roof, i.e. on the fifth level.

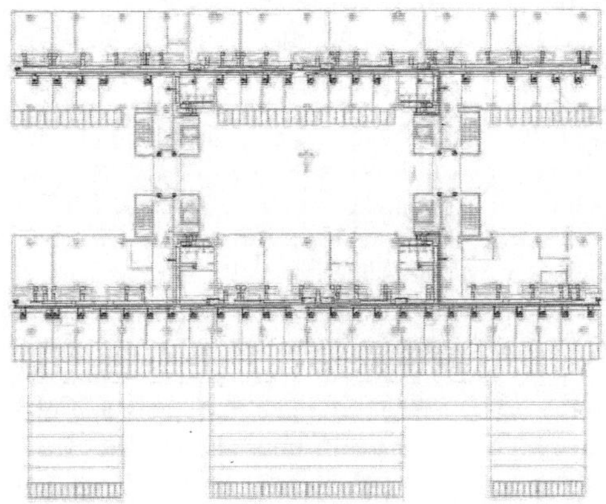

Figure 2. Plan of the fourth level

2.2 Breakdown of energy consumption

As a first step, energy consumption bills were examined and users were interviewed, who reported that the most relevant discomfort was experienced in summer, particularly from June to September. However, even spring and fall days - when a high level of irradiation occurs – can too often cause indoor overheating.

Table 1. Monthly consumption of electric energy.

	Electric energy (kWh)			
Month	**2011**	**2012**	**2013**	**2014**
Jan	150242	148034	153013	145977
Feb	151019	150231	148877	139205
Mar	165008	155370	158433	156506
Apr	133405	123241	131138	129658
May	136429	133284	135521	124810
Jun	236866	203897	186091	156443
Jul	242928	283445	259033	221287
Aug	191435	167258	196223	165546
Sep	228419	161058	203745	156150
Oct	150696	145358	151167	145758
Nov	145377	148630	156828	144243
Dec	132017	130551	137588	127120
TOT.	2063841	1950357	2017657	1812703

Table 2. Overall electric consumptions of the chiller in the summer season due to Eustachio.

Month	Jun	Jul	Aug	Sep	TOT
kWh	18392	37940	12894	15424	84652

The consumption of the chiller was estimated by

analysing monthly electric bills charged by the utility company all over the past four years, which are detailed in Table 1. Keeping in mind that electric bills include every type of electric consumption, that is not limited to just cooling, and that the cooling system in the Eustachio building works from from June to September, the estimation of electric consumption caused by cooling was worked out as the difference between the average consumption from June to September and the average in the remaining months of the year. This approach was considered valid because the lighting system is always on, irrespective of the season (indeed, no dimming control system is installed).

Then, the monthly electric energy required by the chiller was averaged over the four years and the means listed in Table 2 were obtained. These figures were used to validate the numerical Dymola model reported in the next Section.

3 Simulation and assessments

The numerical simulation model that is described in this Section, was aimed first at the diagnosis of the main drwabacks of the building in terms of indoor summer comfort; secondly, at the assessment of low-cost energy enhancement actions, that can improve its behaviour. Hence, the whole model was developed by means of the Modelica-Dymola simulation tool (sub-Section 3.1). This model was validated with respect to the invoiced cooling consumption estimated as reported in sub-Section 2.2. Finally, the comfort improvement and energy savings than can derive from requalification strategies and control policies will be presented in sub-section 3.2.

3.1 The simulation model

The model that we want to showcase in this paper was developed by means of the Dymola™ development platform, which is based on the Modelica simulation language, whose reliability and benefits were demonstrated in previous papers [4] [5].

The whole building's model was developed thanks to the exploitation of two Dymola libraries: the first one is the Modelica Standard library and the second one is the Modelica "Buildings Library" v3.0 [6]. When necessary, those libraries were integrated with new components developed by the research team and customized for the Eustachio building.

The whole model is managed in Dymola as a layered representation of several connected sub-components. Figure 3 depicts a graphical schematic and logic representation of the top layer of the Dymola model for our case study. It is made up of five components that are integrated at the model's top layer, namely: the thermal zones of the building (A), the Air Handling Units sub-system (B), the model of the air duct and hydronic sub-system (C), the chiller (D) and lastly, the model of the fan coil, that was managed as an underlying layer (i.e. new class component). Due to length restrictions of this paper, not all the components will be described in detail, rather the methodology that was used to develop the class model of the fan coil will be reported in detail. The remaining sub-systems were developed similarly.

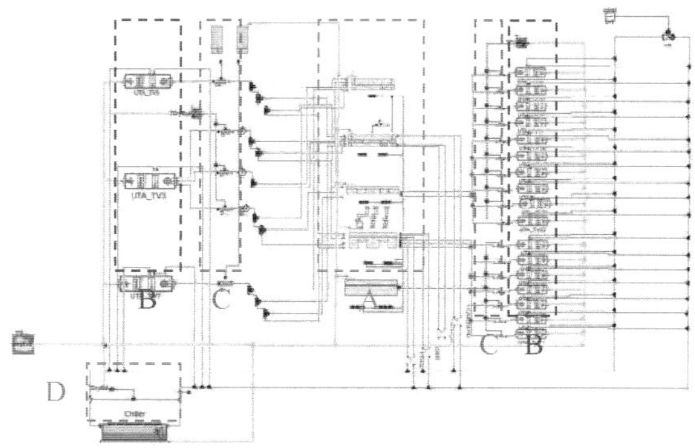

Figure 3. The sub-model of the cooling system.

The first phase related to the fan coil sub-model development was data collection. These data were taken from design drawings and technical reports provided by the office in charge of Eustahio's management, from surveys and from a measurement campaign carried out over two months by means of a Hobo™ sensor setup, including temperature and humidity tracking.

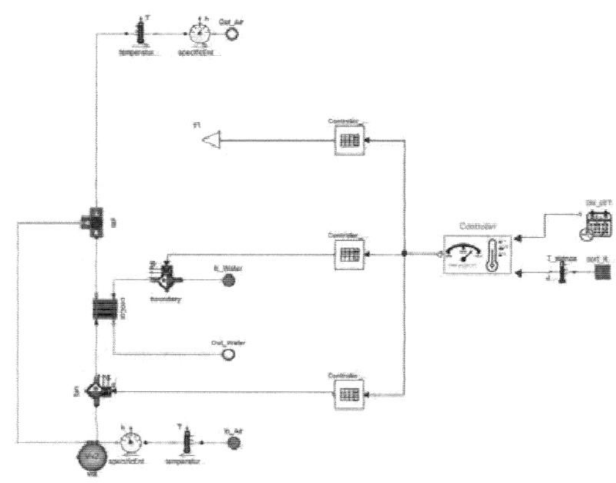

Figure 4. The sub-model of the fan coil.

Every component of the sub-system was progressively created and integrated in the model, namely: the heat exchange battery, the fan, the controller that is in charge of mass flow regulation through switching the heat exchange battery and adjusting the fan coil speed over three levels, a mathematical component representing the ventilation's effectiveness and, finally, all the connection devices.

The model of the fan coil (ref. Fig. 4) was connected to the thermal zones of the building (A), each grouping several rooms of the building and whose boundaries were chosen according to the use, orientation and reciprocal positions of the rooms. Every thermal zone was modelled through the "Room" component from the Modelica Building's library.

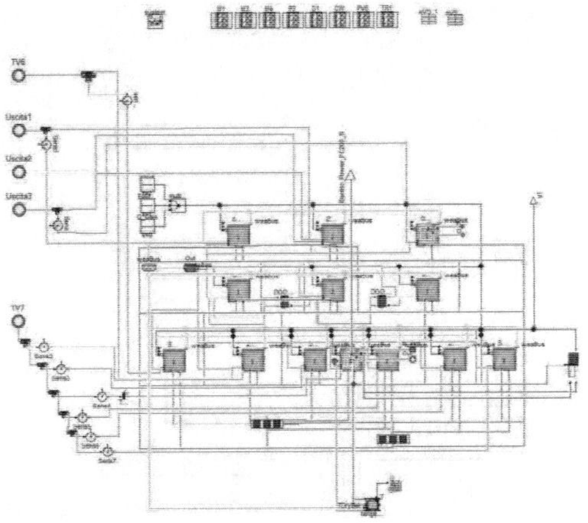

Figure 5. The layer with the thermal zones of the third floor

In Fig. 5 the thirteen thermal zones of the third floor are shown. Every thermal zone of the Dymola model was assigned the corresponding thermal characteristics of the Eustachio envelope, floor, ceiling, internal gains and windows.

Validation of the class component of the fan coil was carried out by means of comparison with the data collected during a dedicated experimental campaign. Most of the sensors were installed inside room no. 90, and the remaining ones were installed in the rooms located at the boundaries. Hence, a parametric study made of a number of simulations was carried out, while varying several parameters (e.g. heat exchange battery efficiency, ventilation efficiency, air flow figures), until simulated outputs matched with temperature and humidity measures. The model was refined until there was a good matching between simulation and experimental results.

All the remaing components mentioned at the beginning of this Section was validated against experimental data, similarly to what done with the fan coil's sub-model. As a result, the whole building model depicted on Fig. 3 was worked out and validated.

The time step used to perform simulations was equal to 300 seconds. The simulation time window was restricted to the the months from June to September, that is when the cooling system was on. The indoor comfort can be evaluated by means of many parameters. Among them, the Eustachio model was assessed in terms of energy consumption of the chiller unit, comfort level measured as PMV and PPD.

These values were plotted against the simulation month (x-axis), in such a way that for every month the average, confidence interval and standard deviation of the relevant parameters are provided on the y-axis. In Figs. 6 and 7 the electrical energy consumption and the thermal comfort of the benchmark (e.g. current state of the building's south block) are depicted.

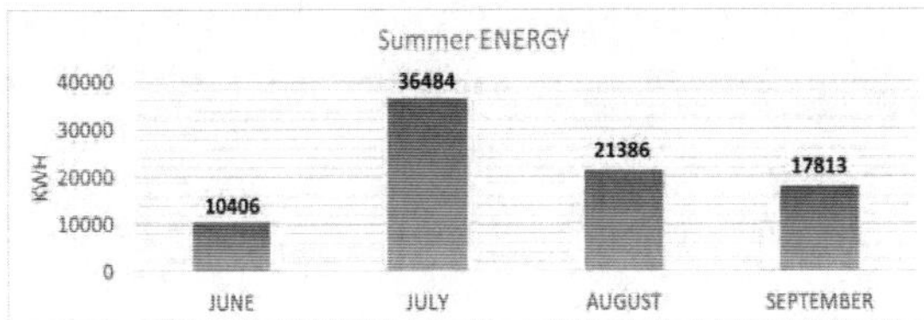

Figure 6. Electrical energy consumption during the summer.

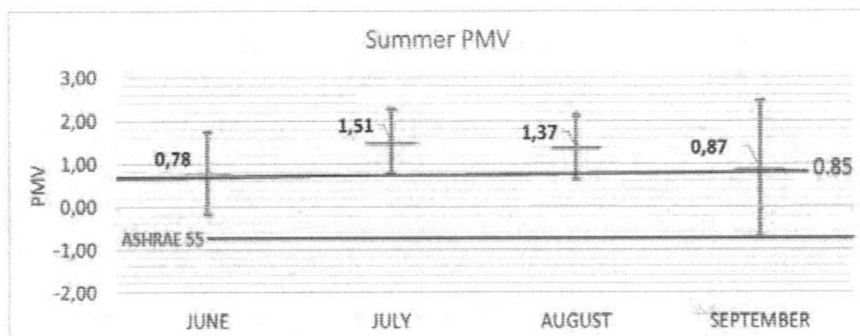

Figure 7. Thermal comfort during the summer and comparison with ASHRAE thresholds.

The overall electrical consumption within the considered time window exceeds 86.000 kWh. This value is in accordance with the estimateion carried out in sub-Section 2.2. Furthermore, every month an environmental discomfort with average index higher than ASHRAE 55's threshold is experienced [7], that is in accordance with complaints and witnesses reported by the interviewed users.

For that reason, some energy enhancement action strategies customized to the building and having the objective to improve the environmental comfort and decrease energy consumption were analyzed.

3.2 Energy enhancement scenarios

Four enhancement scenarios were analyzed to improve the building's comfort, each of them was compared with the current scenario in order to evaluate the benefits that could be had in terms of environmental comfort and energy saving. Per each of these scenarios, a Dymola model was developed, following the approach described in sub-Secton 3.1.

3.2.1 Enhanced ventilation effectiveness (A)

Ventilation effectiveness is an index representing the ability of a system to mix efficiently supply air with the indoor air of the building. According to literature [8], ventilation effectiveness in all the rooms of the southern block is particularly low in the current scenario. Ventilation effectiveness was estimated by means of the index sugeìgested by the ASHRAE's table [9] and approved by the guidelines provided by Rehva asscociation [8], that considers the adoption of a tracer gas as the standard technical approach.

From the technical point of view, this enhancement enatils moving the supply air inlet of the fan coil from the current position, that is above the entrance door, towards the centre of the room, through an extension of the air supply duct. Thanks to this enhancement, ventilation effectiveness would rise from 0.65 up to 1. As reported in Table 3, the adoption of this scenario would save 3563 kWh per year, that is equal to 4% of the overall current electric consumption for cooling, while the average PMV would be increased as of 6%.

3.2.2 VRF cooling system (B)

The second scenario asks for the installation of Variable Refrigerant Flow (VRF) system, that can replace the AHU currently serving the fan coils. In Dymola, just the cooling behaviour of this system was simulated, hence it was simulated as a two pipe VRF without heat recovery to serve the fan coil situated in the southern zone of the Eustachio's building. This enhancement would cause 8780 kWh of energy saving, but with no improvement in terms of environmental comfort.

3.2.3 Solar shading system (C)

The third scenario is a solar shading system installed on the southern façade of the Eustachio southern block.

Dymola allows users to simulate the shading feature through re-setting some parameters on the glazed surfaces: change of two coefficients of the glass component (i.e. solar transmission coefficient τ and solar reflection coefficient ρ). The new parameters were computed by means of a software that is based on th technical standard UNI 13363 [10]. The solar shading system giving the lowest solar transmission value G was selected. Furthermore, a control system to adjust the fan coil operation according to the solar shading system's performances was simulated. A fixed set point temperature regulation was assumed.

Thanks to the described improvement, 10612 kWh of energy can be saved, corresponding to 12% of current consumption. PMV would increase up to 58%, and its mean would often fall within ASHRAE 55 recommended values.

3.2.4 Exchange of RTD on the AHU (D)

Finally, the replacement of the resistance temperature detector sensors in order to regulate precisely the air supply temperature by the AHUs was simulated. This scenario would allow to adjust air supply temperature according to Italian standards, and would determine energy savings as high as 11240 kWh, that is equivalent to 13% when air supply temperature of the AHU is set at 26°C in the summer period. Conversely, there is a worsening of the environmental comfort, because air is currently kept at a lower temperature, due to inaccurate controlling devices.

Table 3. Comfort and energy improvements expected from the four enhancement scenarios.

Scenario	Energy saving (kWhel)	Energy saving (%)	Δ average PMV	Δ average PMV (%)
A	3563,94	4.13	0.08	6.19
B	5261,83	6.11	0	0
C	10612,22	12.33	0.47	58.28
D	11240,83	13.06	-0.09	-8.18

3.2.5 Combination of enhancement scenarios

Table 4 lists all the combined energy enhancement strategies that were simulated and compared with the current scenario. The marginal improvements generated by the various combinations were exploited in the following cost-benefit analyses and labelled as energy and comfort benefits. Eight combinations were considered on the overall.

Table 4. Comfort and energy benefits determined by the combinations of enhancement strategies.

Code	Energy saving (kWhel)	Energy saving (%)	Δ average PMV	Δ average PMV (%)
C+D	24656,94	29	0,43	38
A+D	15984,72	19	0,01	1
B+D	19994,78	23	-0,09	-8
A+C	11815,00	14	0,47	42
A+C+D	26756,94	31	0,44	39
A+B+C+D	27613,67	32	0,44	39
A+C+D+fc	21556,94	25	0,75	66
A+B+C+D+fc	23018,48	27	0,76	67

The last two combinations in Table 4 are similar to the 6th one, except that the indoor temperature was set at 24°C.

4 The BIM Model

Cost-benefits analyses requires an accurate cost estimation. Traditionally, the bill of quantities was prepared based on detailed drawings and specifications [11]. Bill of quantities is a very important document which provides a structured assessments for cost estimation of construction projects. Hence, a Building Information Model (BIM) of our case study was developed, because it can provide detailed estimates, 3D visualization, cost estimation in an electronic format and can easily be extended to further scenarios or be updated with as-built information.

4.1 BIM Model development

Every enhancement scenario described in Section 3 and the current scenario of the building were developed in the form of BIM models. This process required a parametric modelling of the whole building and the definition of a hierarchical structure of the relevant information. All the data collected through surveys and from desing drawings were transferred into the BIM. In our test case, we developed first the model of the bearing structure, secondly the architectural model and, finally, the mechanical system (i.e. MEP).

Autodesk Revit™ was used as the development platform, where the main components were included: terrain, walls, roofs, floors, ceilings, doors, windows, curtain walls, stairs and railings based on the project's drawings and specifications. Additionally, we modelled all the components relevant to describe the project geometry and its use.

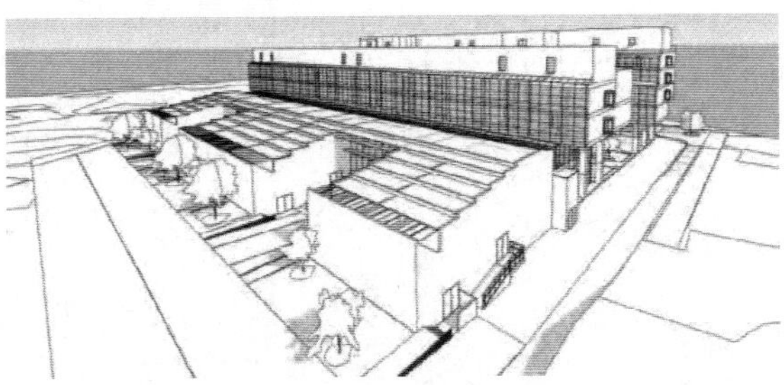

Figure 8. Structural and architectural BIM models of Eustachio building.

Since the enhancement strategies concerned specifically the mechanical system, a deep concern was posed in the modelling of the mechanical system.

To this purpose, Autodesk Revit MEP™ was adopted, which allowed us to integrate both plumbing, hydronic and air supply system. More specifically, we modelled ducts, flexible ducts and their joints, the air vents and the mechanical equipment like the Air Handling Units and the VAVs; similarly, we modelled the chillers, the boilers and the fan coils.

Thanks to this model, the construction cost was estimated for each scenario, and a bill of quantities was worked out, along with a management plan of the construction work and layouts for site organization over the execution phases.

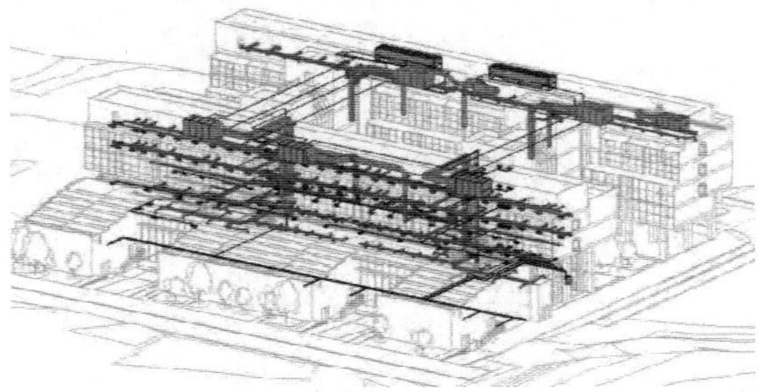

Figure 9. BIM-MEP model of Eustachio.

5 Cost-benefit analyses

Thanks to the definition of construciton phases in the BIM models, the bill of quantities relative to every demolition and construction phase were automatically generated by means of Revit™ tools. These documents were then used to estimate the total cost starting from unist costs of all the components. The relative grouping of all these quantities led to the cost estimation of every enhancement scenarios, as listed in Section 3.

5.1 Assessment of requalification strategies

As a result, this procedure allowed us to estimate the cost relative to every enhancement strategy. Economic benefits related to electric savings were estimated based on the unit cost worked out from bills, that resulted equal to 0.23€/kWh_{el}. Then, one cash-flow was developed for each of the considered scenarios, where positive cash-flows were determined by energy cost savings, and negative cash-flows were determined by the cost caused by technology enhancement. First, a simple comparison between cost and benefits (B/R ratio) was provided, like in Table 5.

This approach was used to perform a first evaluation, limited to the economic point of view. Checking Table 5 and given the renovation cost provided therein, the fourth scenario is ranked as the best one. In fact, installing new sensors in the AHUs is quite cheap, much cheaper than the other scenarios, because it costs only 2'410.86 € (Table 5) and would provide 2'585.32€ saving in annual electric bills (in fact it would be paied back in one year).

If the strategy of the user is just maximizing energy savings, then the combined scenario (A+B+C+D) in row number 10 would work best. It includes installing new sensors in the AHUs, the solar shading system on the south facing façade and the VRF cooling system. It would provide rather high saving, as high as 6'351.14€ per year, but the installation is very expensive.

Table 5. List of enhancement strategies, with costs and benefits and estimation of the cost/benefit ratio.

Scenario's Scenario	Requalification (€)	Annual bill saving (€)	Δ B/R (%)
A	8'700.86	819.72	9,42
B	134'952.74	2'019.46	1,50
C	232'086.08	2'440.76	1,05
D	2'410.86	2'585.32	107,24
C+D	234'496.94	5'561.10	2,37
A+D	11'111.72	3'676.49	33,09
B+D	137'363.60	4'598.80	3,35
A+C	240'786.94	2'717.45	1,13
A+C+D	243'197.80	6'154.10	2,53
A+B+C+D	378'150.54	6'351.14	1,68
A+C+D+fc	243'197.80	4'958.10	2,04
A+B+C+D+fc	378'150.54	5'294.25	1,40

Once the data provided in Table 5 are integrated with those ones that are provided in Table 4 about thermal comfort improvement, if facing thermal discomfort becomes of primary importance, we must select one between the last two combination scenarios, because they could provide respectively 66% and 67% improvement in the average PMV values.

As a further step, a ten year long financial evaluation about the convenience of every enhancement strategy was worked out, simulating the situation of a contractor that is committed to renovate the building in the first year and to manage the building over the 10 year period. Hence, the contractor would benefit from renovation because it lowers management costs. Table 6 reports a sample analysis about the costs involved in the third scenario, that is the solar shading system.

The technology enhancement costs reported in Table 6 were estimated by means of the BIM Revit models (Fig. 10); benefits were derived from simulations carried out in the Modelica/Dymola™ models (one example is provided in Fig. 11, where the solar shading component of the Dymola model is depicted).

Every scenario considered in this step was assessed by means of the Net Present Value (NPV) approach, which evaluates the financial convenience that a contractor would have at the end of a 10 year long period, if committed to manage the building. The 10 year long horizon was justified by suggestion provided in guidelines for EPC contracting. As a result, Table 7 splits two relevant controbutions. In the second column, that part of the the overall investment that would be re-paid back as a result of energy savings, that can be defned as the private contribution. In the third column that other part of the investment that must be covered by the owner of the building, which is complementary to the first part, as compared to the overall investment cost.

Table 6. Summary of the management costs before and after the enhancement due to the the third scenario.

Cost incurred in the "CURRENT SITUATION" scenario	
Operation cost	62'314.79 €
Gas consumption cost	192'955.98 €
Electric energy consumption cost	418'755.58 €
Total bill consumption cost	*674'026.35 €*
Cost incurred in the "ENHANCEMENT YEAR" scenario	
Technology enhancement cost	232'086.08€
Operation cost	62'314.79€
Gas consumption cost	192'955.98€
Electric energy consumption cost	396'788.74€
Total bill consumption cost	*884'145.59€*
Cost incurred in the"post-ENHANCEMENT" scenario	
Operation cost	62'314.79€
Gas consumption cost	192'955.98€
Electric energy consumption cost	415'777.85€
Total bill consumption cost	*670'848.62€*

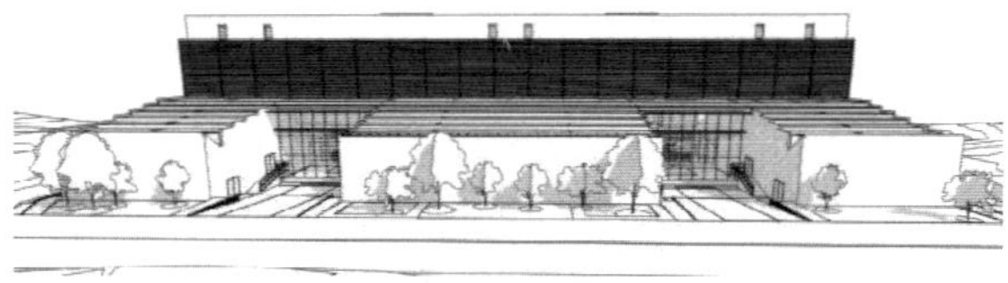

Figure 10. BIM model of the solar shading system.

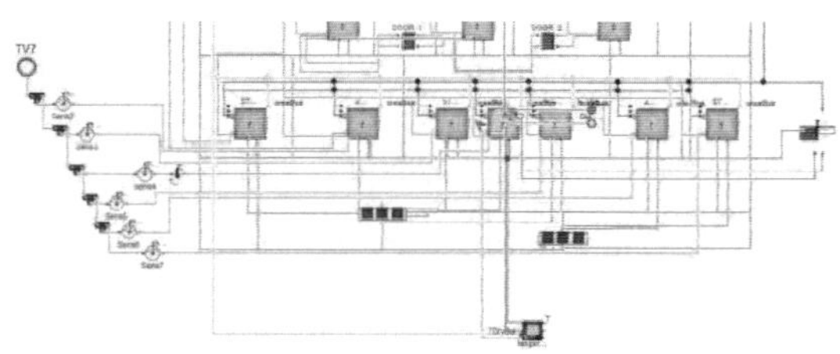

Figure 11. Dymola™ model of the solar shading system.

Apparently, the fourth scenario is again the best one, because benefits pay back the whole investment cost in far less than ten years.

However, this scenario is very limited and does not provide relevant comfort improvement. When energy and comfort evaluations are integrated, the last two scenarios are more advantageous, being the first one able to save 38'604.57€ out of the overall investment, and the second one to save 41'221.61€ out of the overall investment. These figures represent the amount of money saved by the owner in case the investment is carried out by the contractor that manages and operates the building.

Table 7. Estimation of costs that would be covered by energy savings over a 10-year-long investment.

Scenario's Scenario	Private financing (€)	Owner contribution (€)
A	6'382.43	2'318.43
B	15'723.77	119'228.97
C	19'004.02	213'08205
D	2'410.96	0
C+D	44'155.77	189'421.47
A+D	10'399.35	712.47
B+D	35'806.75	101'556.95
A+C	21'158.36	218'708.78
A+C+D	47'916.46	194'361.64
A+B+C+D	49'450.70	327'780.14
A+C+D+fc	38'604.57	203'673.83
A+B+C+D+fc	41'221.61	336'009.23

6 Conclusion

The case study described in this paper is representative of many large public buildings, which are often outdated. As a consequence, they are high energy consumers and do not provide optimal comfort for users.

In this paper we showed that Modelica/Dymola™ models are accurate enough to simulate even customized and partial energy enhancement scenarios. In addition, BIM models can facilitate and make more accurate the process of cost estimation. Once those data are available, informed cost-benefit analyses can be done.

Cost-benefit cannot be the only approach to perform evaluations. Rather, user satisfaction should be considered as another more important parameter.

The approach presented in this paper can be useful both to owners of large outdated public buildings, who are willing to evaluate several enhancement scenarios or who want to control sub-contractors, in charge of the management of those buildings, such as in EPC contracting.

Also, this paper showed that even cheap technology enhancement can cause dramatic improvement in terms of comfort perfromances and energy savings. The development of a BIM model can act as a hierarchical information database to track the status of the building and to check whether contracts were applied as agreed.

References

[1] Short C.A., Cook M., Cropper P.C., Al-Maiyah S. Low energy refurbishment strategies for health buildings. *Journal of Building Performance Simulation*, 3(3): 197-216, 2010.

[2] Felgner F., Ag ustina R., Cladera Bohigas R., Merz R., Litz L. Simulation of Thermal Building Behaviour in Modelica. In: *Proceedings of the 2nd International Modelica Conference*, 2002, March 18-19, DLR in Oberpfaffenhofen, Germany.

[3] Saidur R., Hasanuzzaman M., Yogeswaran S., Mohammed H.A., Hossan M. S. An end-use energy analysis in a Malaysian public hospital. *Energy*, 35:4780-4785, 2010.

[4] Pontoni R., Giuliani M., Carbonari A., Vaccarini M. A Lumped Parameter Model for Dynamic Simulation of Energy Control Policies Applied to Small Public Buildings. In *Proceedings of the 32nd ISARC*, 2015, June 15-18, Oulu, Finland.

[5] Carbonari A., Meschini S., Benedettelli M. Design and simulation of a fuzzy supervisory control system integrated in a small public building. In *Proceedings of the 33rd ISARC*, 2016, June 15-18, Oulu, Finland.

[6] Wetter M. Modelica Library for Building Energy and Control Systems: Buildings 1.0_build2, available at:http://simulationresearch.lbl.gov/publications/recent-developments-modelica, March 2016.

[7] ASHRAE, ASHRAE Standard 55: Thermal Environmental Conditions for Human Occupancy, 2013.

[8] Mundt E., Mathisen H.M., Nielsen P.V., Moser A., "Ventilation Effectiveness", Guidebook no.2, REHVA guidebook 2004.

[9] ASHRAE, ASHRAE Standard 62.1: Ventilation for Accetable Indoor Air Quality, 2013

[10] Standard UNI EN 13363-1:2004, Dispositivi di protezione solare in combinazione con vetrate; calcolo della trasmittanza totale e luminosa, metodo calcolo semplificato, 2004

[11] Nadeem, A., Wong, A.K.D. & Wong, F.K.W. Bill of Quantities with 3D Views Using Building Information Modeling, *Arab J Sci Eng* (2015) 40: 2465.

Mitigating High Energy Consumption for Residential Buildings in The Gambia.

Saikou Sonko Author1[a] and Tsung-Juang Wang, PhD Author2[b]
Department of Architecture and Urban Design

National Taipei University of Technology Taipei, Taiwan.

Sonkobaa3889@yahoo.com

Abstract

As the price of energy goes up, more and more concern has been focused on the sustainable development of residential houses. One of the best solutions would be construction of low energy consuming "passive" houses. A study of residential buildings in The Gambia shows that most occupants of the buildings have persistent and growing problems with the indoor environment. Most buildings are characterized by poor design in relation to the climate, which requires a great deal of energy for cooling during extreme climate. Other problems are poor natural ventilation, inadequate surface volume ratio and poor building orientation. The main question analyzed was whether comfort is achievable solely by the use of passive design strategies, or if energy consuming mechanical systems are required. However, there is no official residential passive housing standard at the present time in The Gambia. This paper focus on the experimental and numerical investigations to determine how the building construction mass, the climatic parameters and the management of air changes rates behave on the effectiveness of natural ventilation and on indoor comfort conditions. Using portable thermometers monitoring instruments data were acquired during a period of a hottest summer phase, in parallel with occupancy survey. Passive cooling techniques can reduce the peak cooling load in buildings, thus reducing the size of the air conditioning equipment and the period for which it is generally required. This paper reviews and critically analyzes various passive cooling techniques and their role in providing thermal comfort and its significance in energy conservation.

Keywords-

Climate; Environment; Innovation; Passive House; Ventilation; Residential Building; The Gambia.

1. Introduction

Technologies and measures which are aimed at reducing the use of energy in buildings could have several advantages, such as lower energy bills, increasing comfort of living or working, and reduced impact on the environment, including reduction of CO2 emissions. [1] the buildings sector has the greatest potential for climate change mitigation The options considered for energy savings particularly leading to CO_2 emission reductions include the following: Use of renewables for heating, cooling and electricity; improvements to the building envelope, including materials, natural ventilation and daylighting De Dear, Richard J., et al. [2]. The hot dry climate with its extremely high temperatures and intense solar radiation is becoming hotter and drier in this era of climate change and global warming. Thus providing for indoor thermal comfort and reducing energy use in residential buildings is become in increasingly difficult. This has called for new ways of thinking and re-evaluation of the existing methods of tackling this problem.

Almost all the residential buildings in The Gambia depend on artificial lighting and ventilation. However, frequent power disruption and load shedding in The Gambia, sometimes over six hours or more a day, amid hot and dry conditions, have made the life of urban dwellers miserable Sanneh, Edward Saja, et al.[3]. As the load-shedding situation continues to worsen, the excessive heat drives people to use more electricity at their homes and offices. The situation becomes worse during the peak dry season from mid-March to early May when the ambient temperature becomes very high. During this period, the demand for electricity use goes up to its highest level because of hot weather, as well as a huge need for cooling energy.

Reduction of energy consumption in buildings can be achieved by simple methods and techniques using an appropriate building design and energy-efficient system and technology, such as passive cooling system. The passive air cooling system provides cooling through the use of passive process without using customary mechanical units such as fan, compressor, etc. in regions where cooling is a dominant problem. Passive cooling strategy maximizes the efficiency of the building envelope by minimizing heat gain from the external sources and assisting heat loss to the natural sources of cooling such as air movement, cooling breezes, evaporation and earth coupling. The principle of passive cooling is to prevent heat from getting into a building during a warm day and bringing in cool air from outside when the temperature drops.

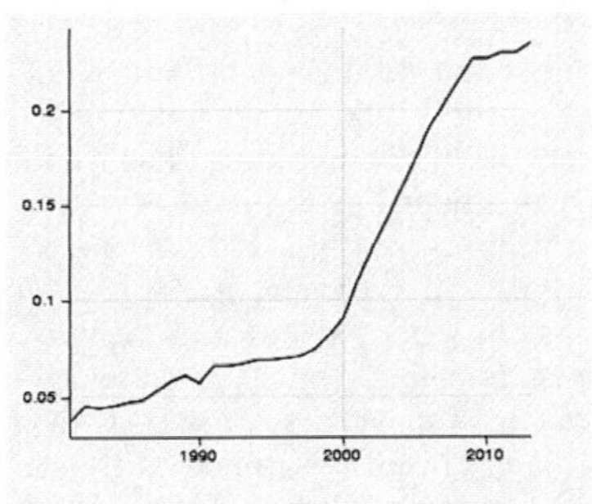

Figure 1. net energy consumption

Particularly energy-efficient buildings offer economic benefits in the face of rising energy prices, help combat climate change and provide a very pleasant room temperature. In The Gambia buildings that meet the low-energy house standard has not been officially been pronounce. Africa is said to be the hottest continent as well as the world's largest consumer of biomass energy, which accounts for two-third of total energy consumption, contributing about 3.7% of total world energy-related carbon emissions. According to UNEP [4] annual report, "per capita levels of commercial energy consumption and energy related carbon emissions tend to be lower in Africa than in developed countries, with the levels projected to remain roughly flat through 2020." [5] Give a clue to future impacts of climate change. The net energy consumption of the Gambia, as of 2011, is 0.2 TWh. See figure 1. For detail

2. Research Problem Statement

In order to achieve the goals two research objectives were laid. The key research questions articulated to facilitate the inquiry are:

- The potential for energy saving: the comparison of Design and operational effects
- The possibility of house design for mixed-mode Operation.

In order to identify innovative concrete technology that are sustainable, an intensive review of literature was done to identify the various innovation in concrete technology being used worldwide that are environmentally friendly. The purpose of this paper is to study the feasibility of passive house in The Gambia. In order to achieve this goal, several passive house projects in hot humid regions in Africa have been studied to gain valuable experience and mature technical skills. This will be of great help for the just started passive house field in The Gambia. However, economic aspect and technology aspect will be analyzed for developing passive house. It is not only getting a better understands of passive house from both technical and economic aspect, but also the paper will include the information and technique use on passive houses based on the examples in Europe.

3. Research Methods

3.1 Characteristic of local climate

The Greater Banjul area is about 93sq km and a population size of 357,000 (26% of the total country's population). Greater Banjul Area, which includes the City of Banjul and the Kanifing Municipal Council. Is located in the west-eastern part of The Gambia at latitude: 13° 27' 9.86" N and Longitude: 16° 34' 40.91" W 9° 49'E [6]. The Gambia's climate is a sub-tropical climate with two variations of distinct dry and rainy seasons. The dry season is between November and May and a rainy season from June to mid-October, due to the African monsoon. The dry season weather in the Gambia is influence by the northeasterly dry winds call Hamattan from the Sahara desert [7]. See figure 2, 3 below for detail. The raining season it's hot during the day, with highs around 32°C to 34°C on the coast and 34°C to 36 °C in the inland areas, but with peaks of 38°C to 40°C even on the coast. The wet season is in the summer beginning June until October.

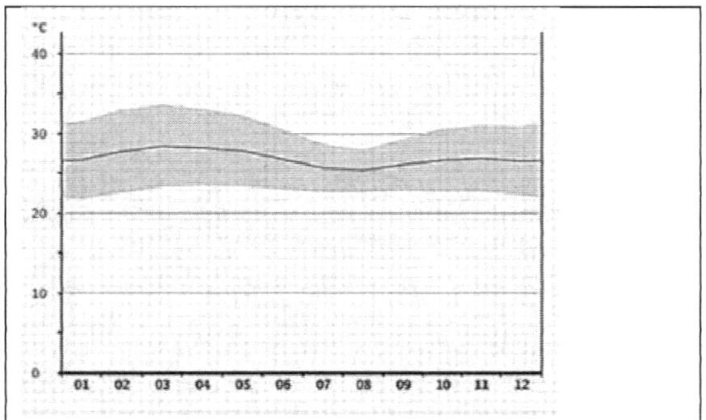

Figure 2. Shows The temperature in March averages 28.4 °C. The coldest month, August with temperatures averaging 25.3 °C.

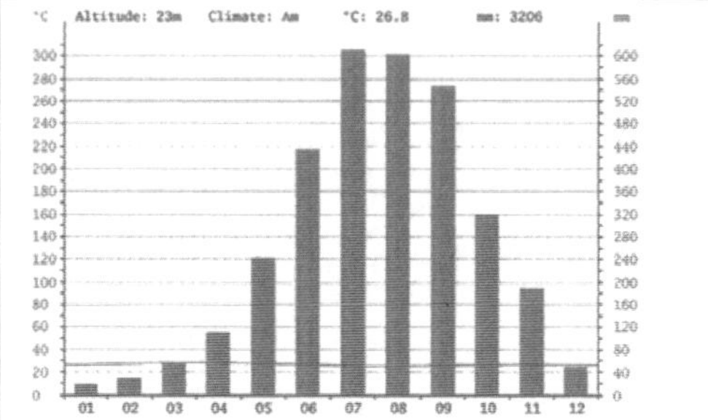

Figure 3. The driest month is January, with 18 mm of rain. Most of the precipitation here falls in July, averaging 608 mm.

3.2 Outline of survey

In this study, 100 structured questionnaires were distributed to samples of the target population in eight wards in Kanifing municipality. The questionnaire, which had the aim to get information on the user-satisfaction and on the building use, covers the following issues. Temperature and humidity (comfort) from the user point of view. The questionnaire were distributed to

Each household depending on the number of adults in the family, using a short term monitoring with portable instruments thermometers was applied with the aim to assess the main thermal performance of the buildings. Thermometers were fixed onto the wall at window level, one for the living room, bedroom and the other for outside under a shaded wall. Residents were required to fill the questionnaires while the research personnel record the environmental measurement three times in a day 7-10am, 12-3pm, and 5-8pm. The contents of the questionnaires mainly include resident's demographic and socio economics characteristics, building characteristics, thermal comfort perception of the living spaces etc. The temperature measurements in the buildings shows that the conditions are bad since the indoor temperature rises to extremely high value of 37°C in the afternoon see figure 4.1,4.2

3.3 Outcome of investigation

Cement sand Crete structure plays an important role in the building enclosure with the proportion of 48%, while 20% of the buildings are built with mud bricks and another 15% with stone structures, 6% with burnt clay bricks and compressed earth bricks while just 5% are built of thatch. The mean size of sleeping room per-household floor areas is from 10.28m 2 while the average room height is 2.7m.The average window area is 0.89m 2. The windows play a dominant role in inducing indoor ventilation due to wind forces. See figure 5 below. From the environmental measurement carried out, the temperature measurements in the buildings shows that the conditions are bad since the indoor temperature rises to extremely high value of 38°C in the afternoon as shown in figure 4.3, 4.4

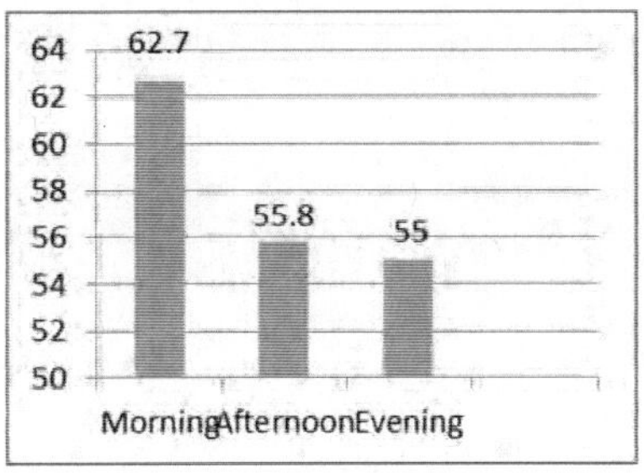

Figure 4.3 indoor relative humidity

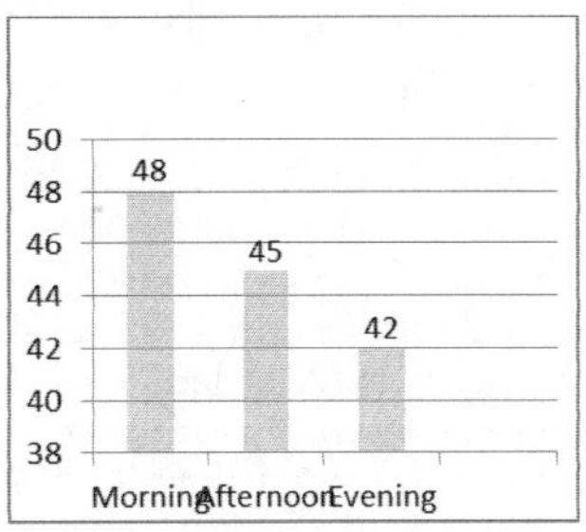

Figure 4.4 Outdoor relative humidity

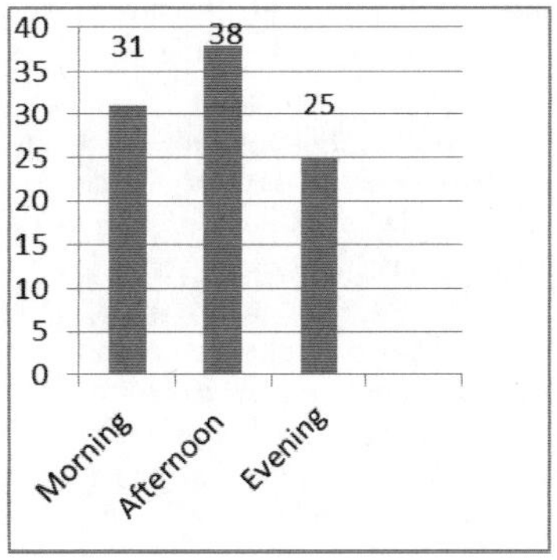

Figure 4.1 Mean indoor temperature

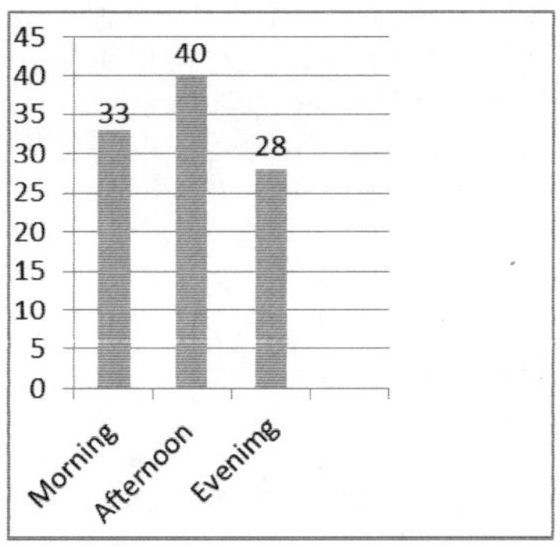

Figure 4.2 mean outdoor temperature

4. The need and benefits of passive design approach in residential buildings

To be comfortable, buildings in all Gambian climates require some form of cooling at some time of the year. There are many ways you can design or modify your home to achieve comfort through passive (non-mechanical) cooling, as well as hybrid approaches which utilize mechanical cooling systems.

The quality of lives of human beings to a very large extent depends on the quality of their indoor environments because people spend most of their time indoors. Human beings partake in various activities within building enclosures and these activities can only be performed best when the environmental conditions are favorable. The hot-dry climate poses environmental challenges of high temperature and this extreme temperature has to be catered for at the design stage to avoid absolute dependence on active energy systems for indoor comfort. In the Gambia, where power supply is epileptic and erratic, most residential buildings depend on energy excessively to attain indoor comfort particularly in buildings that are not passively design. Passive design is a low energy-intensive method of keeping a building cool by relying on architectural design Knowles, Ralph L. [8]. Heat avoidance techniques, natural lighting and natural cooling methods are incorporated in the structure to minimize energy consumption while improving the indoor comfort level Santamouris, Matheos, and Dimosthénis [9]. The benefits of passive design are obvious; considerable peak load reduction for the utility company, improved comfort, lower utility bills and little additional cost to the builder.

4.1 Ecology and Sustainability

Passive Houses are eco-friendly by definition [10].They use extremely little primary energy, leaving sufficient energy resources for all future generations without causing any environmental damage. This seems so obvious that there is no immediate need for additional illustrations. It is rather worth mentioning though, that the Passive House standard provides this level of sustainability for anyone wishing to build a new construction or renovating an older one at an affordable price contribution to protecting the environment [11].

4.2 Heat gain prevention techniques

The interaction of solar radiation by the building is the source of maximum heat gain inside the building space. The natural way to cool a building, therefore, is to minimize the incident solar radiation, proper orientation of the building, adequate layout with respect to the neighboring buildings and by using proper shading devices to help control the incident solar radiation on a building effectively Akbari et al. [12] Good shading strategies help to save 10%-20% of energy for cooling. Properly designed roof overhangs can provide adequate sun protection, especially for south facing surfaces. Vertical shading devices such as trees, trellises, trellised vines, shutters, shading screens awnings and exterior roll blinds are also effective. [13]. these options are recommended for east-facing and west-facing windows and walls. If ambient temperatures are higher than the room temperature, heat enters into the building by convection due to undesirable ventilation, which needs to be reduced to the minimum possible level. Adequate wind shelter and sealing of windows reduces the air infiltration and this requires proper planning and landscaping. See figuer 5,5.1 for detail.

4.3 Window opening type and opening percentages or angles

Several studies have been carried out in recent years regarding air change rates, indoor air quality and window opening habits in residential buildings [14]. There is a large variety of window opening types available. Different window opening types have different properties regarding weather protection, maximum achievable ventilation rates, adjustability of opening sizes and possible interferences of opened windows with furniture. In warm climates, the effective opening size and the adjustability might be more important than protection from weather or rain. Generally, the choice of window opening types can be essential for the resulting air exchange rates as well as for the resulting user behavior. Some window opening types can be adjusted either steeples or in several steps like for example sliding, pivoted, side or top hung windows. Others, e.g. bottom hung windows in contrast are typically either opened to a fixed angle or closed. Figure 5 provides an overview on the properties of different window opening types. Therefore it can be assumed that opening properties predefine the related window opening behavior. [15] Indicate that as long as window design allows, occupants take advantage of continuous window opening adjustment possibilities.

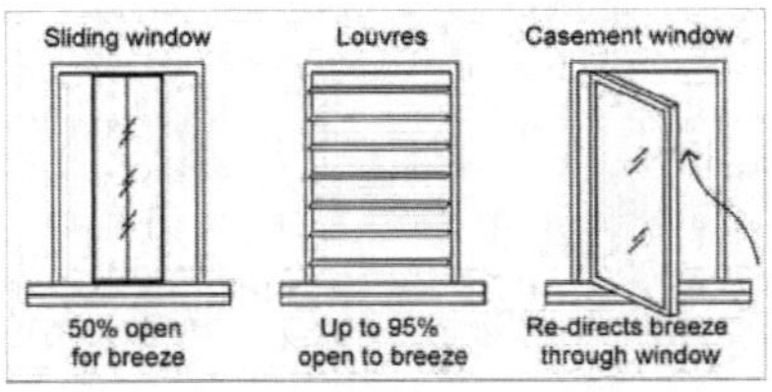

Figure 5. diagram shows three different types of windows:

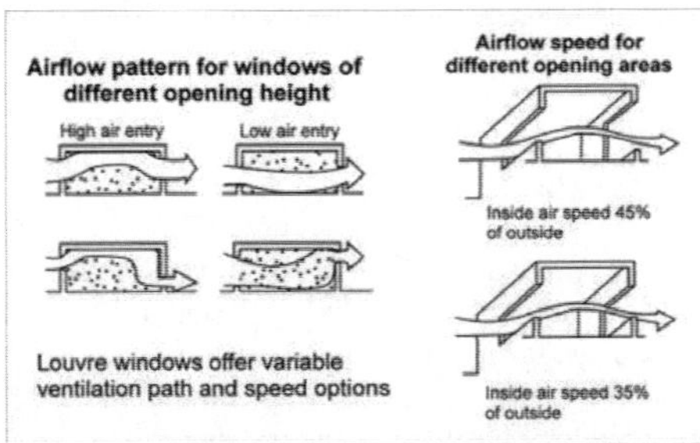

Figure 5.1 How different openings can affect the design

4.4 Proper Orientation of Building

In the Gambia, proper orientation of most houses are not given due considerations. According to Knowles, et al. [16,17] the longer axis of the building should laid along east-west direction for minimum solar heat gain by the building envelope [18,19,20] Proper orientation of buildings to reduce the impact of unfavorable weather conditions like solar radiation, driving rain and thunderstorm. In the house, the rooms should be located in such a way that the ones frequently used should be elongated along the east- west dimensions to mitigate heat gain in summer and also making efficient use of sun. The passive design feature on orienting the longer axis of the building towards east-west direction, as suggested [21] may not always be possible, especially due to actual orientation of the site, that is, when the site itself is longer on the west and east sides. In such a case, the west facade needs more attention because it heats up in the afternoon and important rooms such as bedrooms are generally used later during the day when residents return from office. The east side is less problematic as it warm only in the morning when only few households occupy the major rooms.

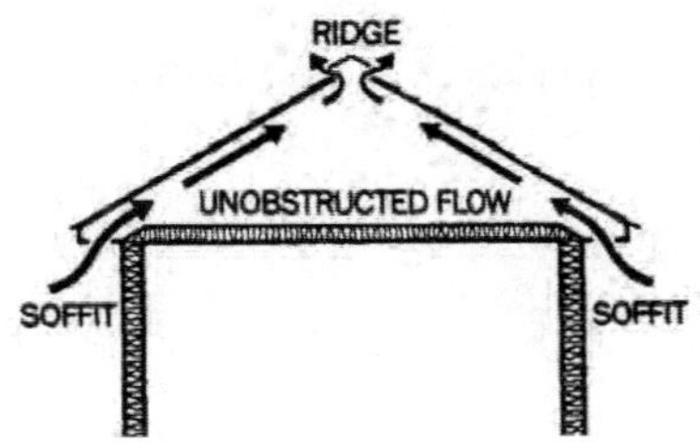

Figure 5.2 This picture shows the typical attic ventilation requirements as well as how ridge vents work. Click image to enlarge

4.5 Utilizing natural resources as building materials

Natural Building includes a variety of building techniques that focus on creating sustainable buildings which minimize their negative ecological impact [22]. Natural Buildings often rely on non-industrial, minimally processed, locally available, and renewable materials and can also utilize recycled or salvaged materials. Natural Building ideally incorporates sustainable design practices to integrate the building into its environment. It may also integrate electricity production, water catchment, passive heating and cooling, and alternative waste-treatment.

In the Gambia, buildings are usually built with cement blocks but in recent times, in the construction industry the use of local materials is not being promoted and a lot of research should be done to look into its efficiency and sustainability for achieving good indoor climate. In almost all localities, nature has provided us with some good quality materials to build with and some of these materials require little processing or transporting [23]. Some of these materials are renewable resources like trees and straw, and some may be so abundant that their supply seems almost inexhaustible like rocks, clay and sand. One of the beauties of building with local materials is that they seem to fit well with the feeling of the place naturally [24].

5. Improving thermal comfort and natural cooling in residential buildings

Some common design elements that directly or indirectly affect thermal comfort conditions and thereby the energy consumption in a building includes:

5.1 Vegetation, landscaping and greenery

Landscaping which can be used to enhance a passive cooling system. The use of vegetation to shade a house is a cooling technique that can be very effective [25]. A well-placed plant can deliver effective, cool shade, as well as add to the aesthetic value of a building. Trees control climate by moderating the effects of the sun, rain and wind. Leaves absorb and filter the sun's radiant energy, keeping things cool in summer. Trees also preserve warmth by providing a screen from harsh wind [26, 27]. In addition to influencing wind speed and direction, they shield us from the downfall of rain, sleet and hail. Trees also lower the air temperature and reduce the heat intensity of the greenhouse effect by maintaining low levels of carbon dioxide.

5.2 Courtyard Basics

A courtyard is also an open-air space. It typically offers little to no cover from the elements. Buildings and homes are built up around an outdoor area, which is then outfitted with benches, paved areas, plants and other elements to create an aesthetically pleasing and functional design.

5.3 Interior layouts

A good interior layout will facilitate many of the passive strategies recommended in this paper. Following is shown the ideal location for the different rooms of the houses.

- Kitchens should ideally be located within the building in such a way as to avoid over-heating, either the kitchen itself or the rest of the building. One way to ensure this is to avoid placing kitchens on the western elevation. In most instances, this will cause overheating in the warm summer months, so an ideal location for a kitchen is on the eastern side of the building. This catches the morning sun but not the warmer, late afternoon sun.
- Rooms that are occupied predominantly in the evening should be located on the western side of the building, in order to take advantage of the evening sun.
- Frequently used rooms, should be located on the southern side where they can be warmed but sunlight throughout the day.
- Bedrooms generally require less heat. Decisions for the location of bedrooms can largely be based on aesthetics and occupant or designer preferences in addition to thermal comfort considerations.

6. Conclusion

Gambia is the only country in the world that pays the highest electricity bills according to European Union (EU) programed manager. Siylvain Lequere. However, this research work has shown that the thermal discomfort in the study area is aggravated by the epileptic power supply, high cost of air conditioning systems, absence or scanty vegetation around houses, improper positioning of windows for cross ventilation, lack or poor insulation of floors, walls and roofs, activities of non-professionals in the industry, use of good conducting materials on buildings, and the use of incandescent bulbs and kerosene lamps as lighting devices. By adopting the general rule for energy efficient building in The Gambia as follows:

- Building openings should be of suitable size and should be orientated to enable natural airflow from the windward to the leeward side. A building should not have too deep a plan and should be relatively free of major obstructions within the interior.
- Retractable awnings or fixed overhangs over north-facing windows can provide complete shading from the direct sun during summer, and still enable solar penetration in winter. By adapting to these design strategies it will limit the energy demand for cooling and will also result to an adapted architecture to the climatic environment which will encourage innovation design solutions for building professionals in hot-dry climate.

Reference

[1] Climate Change 2007: Mitigation of Climate Change. Working Group III contribution to the Fourth Assessment Report of Contribution of the Intergovernmental panel on climate Change. Figure SPM.6. Cambridge University Press

[2] De Dear, Richard J., et al. "Developing an adaptive model of thermal comfort and preference / discussion." *ASHRAE transactions* 104 (1998): 145.

[3] Sanneh, Edward Saja, and Allen H. Hu. "Lighting rural and peri-urban homes of the gambia using solar photo-voltaics (PV)." *The Open Renewable Energy Journal* 2 (2009): 00-00.

[4] Cheng, Chia-Chin, et al. "The Kyoto Protocol, The clean development mechanism and the building and construction sector: A report for the UNEP Sustainable Buildings and Construction Initiative." (2008).

[5] Meehl, Gerald A., et al. "The WCRP CMIP3 multimodel dataset: A new era in climate change research." *Bulletin of the American Meteorological Society*88.9 (2007): 1383-1394.

[6] Ampomah, Gifty, et al. "climate change vulnerability assessment and adaptation in greater Banjul-Gambia." (2012).

[7] Giglioli, Marco EC, and I. Thornton. "The mangrove swamps of Keneba, Lower Gambia River Basin. I. Descriptive notes on the climate, the mangrove swamps and the physical composition of their soils." *Journal of Applied Ecology* (1965): 81-103.

[8] Knowles, Ralph L. "The solar envelope: its meaning for energy and buildings." *Energy and buildings* 35.1 (2003): 15-25.

[9] Santamouris, Matheos, and Dimosthénis Asimakopoulos. *Passive cooling of buildings*. Earthscan, 1996.

[10] Knudstrup, Mary-Ann, Hanne Tine Ring Hansen, and Camilla Brunsgaard. "Approaches to the design of sustainable housing with low CO 2 emission in Denmark." *Renewable Energy* 34.9 (2009): 2007-2015.

[11] Späth, Philipp. "District heating and passive houses—interfering strategies towards sustainable energy systems." *Proceedings of the 4S/EASST Conference 2004*. 2005.

[12] Santamouris, Mattheos, and Dionysia Kolokotsa. "Passive cooling dissipation techniques for buildings and other structures: The state of the art." *Energy and Buildings* 57 (2013): 74-94.

[13] Akbari, Hashem, Melvin Pomerantz, and Haider Taha. "Cool surfaces and shade trees to reduce energy use and improve air quality in urban areas." *Solar energy* 70.3 (2001): 295-310.

[14] Lundqvist GR. Indoor air quality and air exchange in bedrooms. Proceedings of the 6th AIVC Conference 1985, A I X Coventry Great Britain, p. 5.1e5.8.

[15] Yun, Geun Young, Paul Tuohy, and Koen Steemers. "Thermal performance of a naturally ventilated building using a combined algorithm of probabilistic occupant behaviour and deterministic heat and mass balance models." Energy and buildings 41.5 (2009): 489-499.

[16] Tantasavasdi, Chalermwat, Jelena Srebric, and Qingyan Chen. "Natural ventilation design for houses in Thailand." *Energy and Buildings* 33.8 (2001): 815-824.

[17] Rabah K. Development of energy-efficient passive solar building design in Nicosia Cyprus. Renewable Energy 2005;30(6):937e56.

[18] Knowles, Ralph L. "The solar envelope: its meaning for energy and buildings." *Energy and buildings* 35.1 (2003): 15-25.

[19] Aksoy UT, Inalli M. Impacts of some building passive design parameters on heating demand for a cold region. Building and Environment 2006;41 (12):1742e54.

[20] Andersson B, et al. The impact of building orientation on residential heating and cooling. Energy and Buildings 1985;8(3):205e24

[21] Bekkouche, S. M. A., et al. "Introduction to control of solar gain and internal temperatures by thermal insulation, proper orientation and eaves." *Energy and buildings* 43.9 (2011): 2414-2421.

[22] Shashua-Bar, Limor, and Milo E. Hoffman. "Geometry and orientation aspects in passive cooling of canyon streets with trees." *Energy and Buildings* 35.1 (2003): 61-68.

[23] Kibert, Charles J., and Robert R. Ries. "Green building education and research at the University of Florida." *International Proceedings of the 45th ASC Annual Conference, Gainesville, FL*. 2009.

[24] Moavenzadeh, Fred. "Construction industry in developing countries." *World Development* 6.1 (1978): 97-116.

[25] Akande, O. K. "Passive design strategies for

residential buildings in a hot dry climate in Nigeria." *WIT Transactions on Ecology and the Environment* 128 (2010): 61-71.

[26] Santamouris, Matheos, and Dimosthénis Asimakopoulos. *Passive cooling of buildings*. Earthscan, 1996.

[27] Lindzen, Richard S. "Some coolness concerning global warming." *Bulletin of the American Meteorological Society* 71.3 (1990): 288-299.

An Empirical Investigation on Construction Companies' Readiness for Adopting Sustainable Technology

M. Foroozanfar[a], S. M.E. Sepasgozar[b] and H. Arbabi[c]

[a] Department of Construction Project Management, University of Tarbiyat Modares, Iran
[b] Faculty of Built Environment, University of UNSW, Australia
[c] Department of Construction Project Management, University of Tarbiyat Modares, Iran

Abstract

New digital technology has the potential to decrease the environmental footprint of construction activities, while they increase the productivity and safety involved in major construction projects. Digital technologies such as web-based building information modelling (BIM) and sensing technology are emerging technology that can be widely employed in construction projects in developing countries. However, the construction industry tends to be slow in adopting new digital technology to meet their objectives in developing countries. This paper aims to investigate construction technology readiness for utilization of digital technology in construction projects and to identify the factors that can potentially facilitate construction stockholders' application of digital technology.

A questionnaire is designed, and an empirical investigation is conducted to collect data from construction companies including industrial and infrastructure construction companies. Regression analysis was utilized to analyze the data to measure the relations of latent factors through significance values using SPSS. The findings show the importance of a series of factors influencing construction stakeholders' intention to use a new digital technology. Based on the extensive review on the relevant literature, few empirical studies have been conducted to examine the proposed constructs including innovativeness and optimism for emerging digital technology. The results are also expected to provide a guidance to broaden understanding of adoption behavior within this context and thereby increasing the chances for successful adoption of sustainable technology and develop activity-level.

Keywords

Construction Industry; Digital technology; Sustainability; Technology Adoption.

1 Introduction

The construction industry has been under pressure to move toward the sustainability because different governments have legislated regulations for protecting the environment [1]. A large number of organizations tend to increase their sustainable practices to save their resources and competitive advantage [1]. The intention for implementing sustainability concepts in construction projects has increased uncertainty and complexity and created many challenges for construction stakeholders [2]. On the other hand, organizations need to make decision about the utilization of new technologies to be competitive in the industry. That is why construction stakeholders tends to apply different technological tools to meet the requirements sustainability in all phases of projects [1].

One of the most effective information technology toward sustainable construction is digital technology which its influence on construction industry has become greater. The digital technique cover the entire process of building program from design to construction [3]. In today's digital world, innovative solutions for projects managers to solve problems of coordination and delivering projects on-time, on-budget and according to its scope and sustainability specifications are presented by digital technology such as web-based technique; however, there is no comprehensive understanding on how this technology is implemented in construction industry [4]. In addition, the construction industry is slow in up taking digital technologies and there are challenges during the implementation phase [5].

To address this challenge, the aim of this study is to empirically examine the factors affecting construction stakeholders' adoption of digital technology and to investigate the research question in this study: what are the effects of the organizational, social, technical, and individual factors on adoption behaviors as related to digital technology by construction stakeholders?

The paper first systematically reviews the literature. It follows with presenting how the literature of digital technology applications is growing while the number of publications which presents green technologies has been constant in the last three years. Second, the research design and approach are presented. Finally, the results of the study are discussed following with an agenda for future research.

2 Literature review

This section systematically reviews the most recent literature in both green and digital technologies. Figure 1 presents the process of the technology transfer in the construction industry from technology development to implementation. This paper focuses on the adoption phase referring to the process which a company make decision to utilize a modern technology.

In order to systematically analyses the most recent literature three journals were selected such as Journal of Information Technology in Construction, Building Research & Information and Journal of Construction Engineering and Management then their papers were reviewed. Each of journal is devoted to a different special topic, so by choosing them a wide range of articles were reviewed. Regarding green technology, we classify papers in to four non-green specific non-technology, green specific non-technology, non-green specific technology and green-based technology categories. Figure 2 illustrates the number of papers in two non-green specific technology and green-based technology categories in three recent years. It shows rapid growth in non-green specific technology papers in 2016, accordingly it can be predictable that by improving of the technology, this increasing trend will continue in the coming years. However, the figure shows despite of the increased sense of responsibility to sustainable development, the researchers neglected focusing on green technology adoption. Figure 3 demonstrates that green specific technologies are reviewed less than non-green specific technologies, and it shows that researchers should be encouraged to pay more attention to the gap of green specific technology adoption in construction.

Based on the current review, four main themes were identified: 1) Off-line digital technologies; 2) BIM based technologies; and 3) web-based technologies; 4) Green technologies. Each theme is reviewed in the following sections. The most recent technologies intend to address sustainable concerns and therefore tend to identify new applications for decreasing the environmental footprint of construction activities.

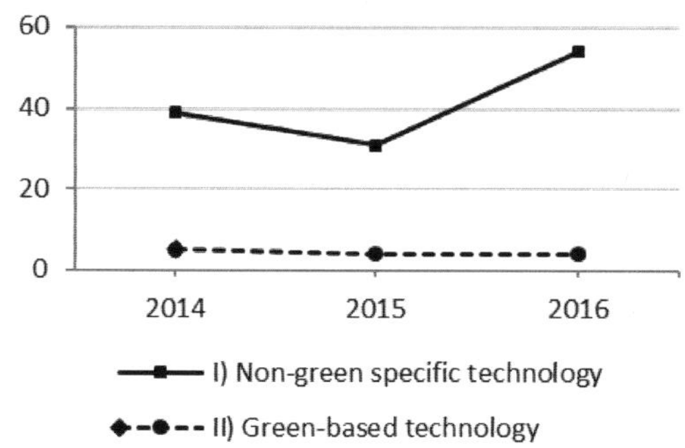

Figure 2 total technology related publications in three mentioned journals in tree recent years

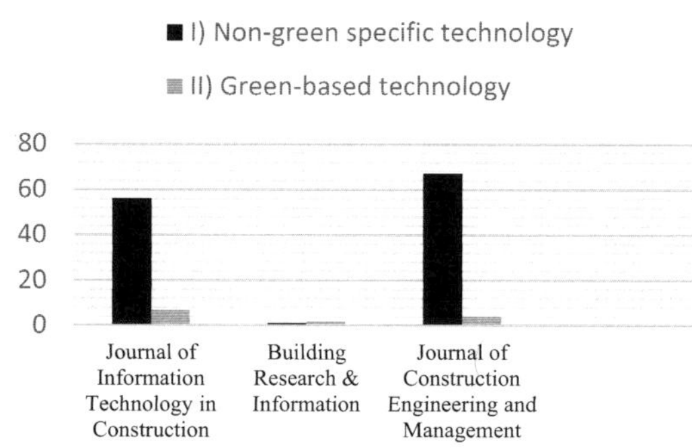

Figure 3 technology related publications in each of the three mentioned journals

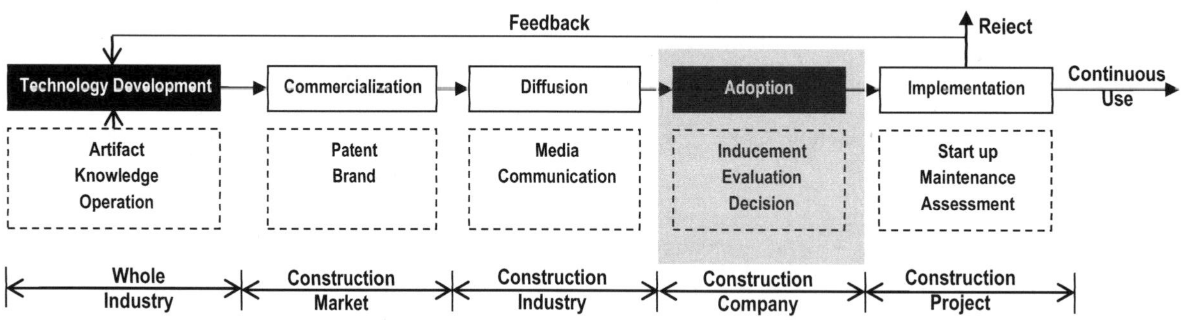

Figure 1 The process of technology transfer in the construction industry

2.1 Theme 1: Off-line program

Construction digital technologies such as digital leveling devices, building information modeling, 3D printers, smart roads, transparent solar panels, smart helmet and anti-collision software are increasingly being used in the construction industry as a new trend. For example, drone is an aerial quadcopter that is armed with video cameras, in construction industry drone enables project managers to monitor the site, maintain the asset and capture images and videos from different positions throughout the worksite [6, 7], and virtual reality contains real life images which helps us to analyze building structures in different environmental status or to promote construction safety [8, 9]. This paper refers to a Construction digital technology as any type of electronic device that is used for visualization, process, monitoring, tracking and etc.

Construction digital technology in construction industry use for different purposes such as management and delivery of projects [10]. Construction digital technology is used in all of the phases of the life cycle of the construction projects from project design to construction for different purposes [3]. Such as building lifecycle management, digital representation of information using different visualization and modelling techniques and software applications or coordination and management issues around digital delivery of projects. Although offered definition focuses on the technical aspects of construction technology, it contains practices and processes that support technology [10]. These technology include tools and technology which are independent, integrated and web-based and apply to capture, store, process, display and communicate data and information in different phases of construction [11].

2.1 Theme 2: BIM based technologies

Building information modeling (BIM) has been developed recently in the construction industry. By using BIM we can increase project quality, schedule timetables accurately, and reduce total costs of the project [12]. BIM is a technology which is used in construction industry to generate and manage a parametric model of a building to decrease document errors and rework, and reduce the time of the design process Son, et al. [13]. Using BIM enable project team to create and use digital models, to design construction and to manage operation of the construction process and finally to achieve the project life management [14].

2.2 Theme 3: Web-based technology

The character of the construction industry's organization is unique because the nature of its projects is one-off in planning and operation. Each project has its own challenges, risks and problems which are different from others but experiential knowledge and information data are important to solve these problems and challenges. Besides, each construction project is a temporary organization which is the combined of several organizations, it has a multi-participant feature; there for, exchanging information is critical for it. In each phase of project and for solving the problems, relevant organizations should collaborate with each other to obtain solutions and determine appropriate concessions [5].

One of the construction innovations is in the area of IT which is developing rapidly, such as several tools software with different usages including, quantity calculation software, cost evaluation software and quota management software. One of this applications that has an upward trend growth is in the field of communication networks, which enables the involved stakeholders to exchange information and data. These web-based communication networks can enhance the coordination and cooperation among team members [5]. Web-based platform solution apply latest progression in computing technologies, using it enable us to access to sustainability resources in the form of interactive, dynamic, and user-oriented services seamlessly (Petri, Beach, Rezgui [1]. There are several applications of information and communication technology in construction industry as are shown in table 1. Table 1 shows different types of technologies that are introduced or implemented for construction purposes. In addition, it shows that the recent information and communication technology products are more complicated and may need specific considerations for utilizing in construction projects.

2.3 Theme 4: Green-based technology

One of the most critical factors in reaching the goal of sustainable construction, is using the new technology in this field, but most of these technologies have their own complex nature and operational challenges that prevent the successful use of them or reduce their application. Therefore, to achieve sustainability objectives, relevant technical considerations must be investigated. For example, alongside the development of new sustainable construction technology, the efficiency and effectiveness of existing technology must be improved, so their weaknesses should be known; besides, access to technical standard procedures, suitable guides and adequate and reliable data about the performance of sustainable construction technology can encourage participants to use these technology [15].

One of the modern technology with increasing usage for sustainable construction is digital technology. Some studies indicate that IT and digital technology and software are so significant in reaching environmentally friendly features into construction projects helping in analyzing the sustainable building performance such as daylighting, energy efficiency and sustainable materials [16]. Digital technology enable stakeholders to cooperate and communicate with each other effectively to equilibrate the artistry and function of buildings, it provides the condition for related participants to carry out various analysis about energy consumption, carbon balance, lighting and noise. On the other hand it enables experts to simulate physical environment which will be helpful to design and built a building with low energy consumption and carbon balance [3].

Table 1 Example of technology applications for all identified themes

Theme	Non-Green Specific Technology	Theme 4: Green-Based Technology
Theme 1: Off-Line Program	Computer Aided Architectural Design (CAAD) to integrate different applications of Architecture Science, Artificial Intelligence and Graphics [3]	physical environment simulation software to analyze energy consumption [3]
	Analyzing information of the construction and managing the documents in digital form [17]	Sustainable Construction Modelling evaluate sustainability of the construction by the system [18]
Theme 2: BIM Based Platform	Integrating the project management and cost estimating with three-dimensional (3D) model [12]	Determination the orientation of the building, studying daylight to determine buildings position on the site and Calculation of the carbon emissions and usage operational energy [12]
	preparing a more simplified business process and provide solutions to decrease costs and increase productivity and efficiency through effective communication in remote construction projects [19]	Advanced simulation of energy, realization of the design problem regarding to energy, such as thermal comfort [14]
Theme 3: Web-Based Tchnology	Web-based Enterprise Resource Planning to manage construction supply chain [11]	BMS system for doing energy management initiatives [20]
	Using video cameras to control and monitor construction sites remotely [21]	web-based self-assessment prototype tool to manage sustainability at the strategic level of corporate [2]

Most of the construction stakeholders are familiar to web-based services such as e- mail, network news, remote login, and file transfer, these new opportunities help them to move toward improved sustainable construction processes because by using them construction stakeholders can achieve accurate project information in needed formats, anywhere and anytime without delay across the world. Even they can have a live vision of site location happenings if they aren't able to be there which is really critical for sustainable construction goals. By these web-based applications, virtual online design studios are available that enable stakeholders to interact and discuss applying auditory and pictorial conferencing, online electronic bidding systems helps us upload, download and update the documents online and reduce the paper usage, time and cost of bidding, using on-line project administration systems can help us to access the on-time information project status and specifications, control reports and submit on-line change orders. Web-based catalogues are building product information sources which include numerous information about products such as their performance data, standards, and instructions of installation which enable us to order on line; therefore, we can save time and energy. Moreover, web-based technology provide the opportunity to monitor and control the project via web cams [21].

2.4 DCT concepts and adoption barriers

The literature shows that current DCT application and practices can be divided into three types of studies: a) technology development, b) technology adoption, and c) technology implementation.

The first group of studies are based on field experimentations to develop new prototypes to introduce it to the construction industry or improve a current technology in terms of accuracy and quality.

The second studies present or use models that are drawn from the established body of knowledge predominately such as socioeconomic or psychological models. For example, Son, et al. [13] showed that two distinct constructs (usefulness and ease of use) are positively associated with an individual's behavioral intention to utilize a new technology in construction. Some other studies tend to understand the adoption process through an exploratory process which is mainly case based study. For example, Mitropoulos and Tatum [22] investigated factors affecting the adoption of information systems in construction, and the results of such exploratory studies are not necessarily tailored or adjusted with classic models in information systems.

The third group of studies focus on the way of operating an innovative technology to resolve a specific problem in construction. For example, Wang and Cao [23]

review main project-level factors which influence IT implementation and analyze how these factors effect IT implementation in construction industry. In other paper the implementation of user-based innovative construction technologies in Australia is investigated and it's studied if the construction industry environment is supportive for innovation or not [24].

One of the important points of interest in the investigation of influencing factors on construction stakeholders' decisions for application of digital technology is recognizing the challenges of them. These challenges can be divided to the challenges of the own technology and the problems of the digital technology' application. The following table is a collection of well-known own technology challenges.

Table 2 The application of digital technologies in two major groups: a) green-based; and b) non-green specific technologies

	Technology Theme	Focus and Current Direction	Limitation and Suggested Improvement
Green-Based Technology	Theme 1: Off-line program	To develop a simulation tool for predicating carbon emission by virtual prototyping technologies [25]	The current tools cannot monitor construction mechanisms and evaluate the sensitivity of planning schedules
	Theme 2: BIM based platform	To generate documentation necessary required for green building certification [26]	Limited to only a group of the BREEAM materials and compatible with ArchiCAD® software
	Theme 3: Web-based technology	To develop BIM-based web platform for green building services such as energy simulation and code checking [27]	Technically limited functions are delivered and cannot cover critical qualitative factors such as human comfort
Non-Green Specific Technology	Theme 1: Off-line program	Computer Aided Architectural Design (CAAD) to integrate different applications of Architecture Science, Artificial Intelligence and Graphics [3]	The function of the technology is limited to design and architecture. The technology should be compatible with other information available in construction sites
	Theme 2: BIM based platform	The design process is improved, because values are generated through waste control and management [28]	The linkage between the project design information and the project support information should be improved for sustainability purposes
	Theme 3: Web-based technology	To develop a web-based performance management system for supplier selection and monitoring their performance [29]	The system should be integrated with the current systems (e.g. project databases, tender documents) considering their green practices

The main reasons of the slow technology uptake are cost, too many functionalities, rework problems due to data entry, difficulty in learning and complexity in usage, requiring computer knowledge, less user-friendly characteristics and low efficiencies. Lack of ability to use some of these technology in the early stages of design reduces its performance in this field and lack of integrity of some of the analytical tools with key software, uncertainty and distrust in data sources, persistence to making change in lifestyle of the project, industry and organizational realized barriers, can prevent its widespread application [1, 3, 30].

It is proposed that a construction company should seek for possible solutions for a need or to improve the current situation. For example, to boost the productivity of construction in restricted areas, mini-equipment as new technologies used extensively [31]. On the other hand, vendors correspondingly support the potential adopters in several stages. For example, the vendor support the customer by know-how knowledge Hommels, et al. [32] and awareness Frambach and Schillewaert [33] as crucial stages to the success of a technology innovation.

Several innovations are used in construction industry for different purposes such as increasing speed or reducing the time but some of them are related to sustainability, for instance for suit sustainable construction and energy efficiency we can use improved design, for reducing energy consumption in old buildings we can retrofit solar passive principles to them, new building materials can be used for insulating or improving environmental efficiency. Thorpe, et al. [34], argue that these innovations are used because they are 'good ideas' and key drivers for business which increase the competitiveness as the point of difference and using them causes the improvement of reputation, prospects, productivity and profit.

3 Research Method

3.1 Participants profile

Construction stakeholders who use digital technology in their projects with the aim of sustainability are studied in this paper. A total number of 60 construction project participants from different construction companies were selected for the survey. In these companies considering the principles of sustainable construction is a priority. The survey was conducted using a questionnaire for data collection. Sufficient description is provided to prevent the misinterpretation by respondents. All of the respondents had experience using sustainable construction digital technology. Of the respondents in the survey, about 40% were men and about 60% were women. 100 %of the respondents had university degree. They were aged between 27–47 years. The sample demographics are listed in Table 3.

3.2 Measurements

The survey is divided into two parts. The first section solicited demographic information including gender, age, education level, job title, and tenure with the company. The second section asked the respondents to indicate their level of agreement with 42 different statements (the measurement items) on a five-point Likert type scale, with 1 displaying "strongly disagree" and 5 displaying "strongly agree." the measurement items were adapted from previous studies and reworded to render the items relevant to digital technology usage. Measurement items are grouped according to the six main constructs (organizational facilitating conditions, expected performance, expected efforts, innovativeness, optimism, and user performance), for each construct 5-8 questions were designed to assess the impact of 4 independent variable (Table4) on dependent variables (perceived usefulness and perceived ease). Regression analysis was utilized to analyze the data to measure the relations of structures through significance values using SPSS. Finally, respondents were asked to express their opinions in the form of a number of open ended questions.

Table 3 Demographic attributes of the respondents

Characteristics	Number	Percent
Gender		
Male	24	40
Female	36	60
Education		
University	19	31
Graduate school	41	69
Age		
20-30	12	20
30-40	42	70
≥40	6	10
Job title		
Consultant	12	20
Designer	14	24
Supervisor	9	15
Contractor	10	16
Client	3	5
Workshop manager	6	10
Project manager	6	10

Table 4 Definition of independent variables

Abbreviation	Independent variables	Definition
UP	User Performance	The user's efficiency for performing a duty, which refers to the conducted act's accuracy and race in the period of a task [35].
OFC	Organizational Facilitating Condition	Availability of different technological and organizational resources, such as opportunities, science, and resources which can enable employees to eliminate the obstacle of applying a system [36].
O	Optimism	A positive belief that by technology application and people can have more controllable, flexible, and proficient life which cause people have a positive opinion about technology [37].
I	Innovativeness	Innovation is carrying out new or current operations by a new method which has been defined in respect to new production, new operations, new materials, new organizational structures [38].

4 Results

In this research, to recognize the key factors that facilitate the digital technology implementation in construction industry an analysis was carried out. To achieve this aim, the relevant literature was reviewed to identify the associated factors with the adoption of digital technologies in the construction industry. Accordingly, researchers such as Davis [39] proposed that two internal factors -perceived usefulness and perceived ease of use determine the intention to use of the technology and some external factors influence on these two internal factors. In this paper, the effects of the four independent variables such as organizational facilitating conditions, innovativeness, optimism, and user performance (which are described in Table 4) on the internal factors are studied.

Firstly, for testing the internal consistency as a measure of reliability for each sub-scale, the Cronbach's alpha was assessed. According Nunnally [38], alphas above .70 for group analyses are acceptable. In this research, the Cronbach's alpha is .706 which indicates it is acceptable and no item shouldn't be excluded. Regression analysis was conducted to estimate the relationships among dependent and independent variables, in model summary, R factor shows the intensity of the correlation between two variables while the conditional expectation of the dependent variable given the independent variables is determined through R^2. The results of ANOVA illustrate if Regression model can significantly predict dependent variable changes, if Sig. in Coefficient is less than 0.05, the correlation is meaningful. The Beta analysis shows the amount of dependent variable impact on independent valuable.

Table 5 Results of the Regression analysis of Independent variables and perceived ease of use

Model Summary	
R	R Square
.668a	.447

ANOVA					
Model	Sum of Squares	df	Mean Square	F	Sig.
Regression	6.010	4	1.502	9.890	.000b
a. Dependent Variable: perceived ease of use					
b. Predictors: (Constant), UP, OFC, I, O					

Coefficients					
Independent variables	Unstandardized Coefficients		Standardized Coefficients	t	Sig.
	B	Std. Error	Beta		
OFC	.255	.129	.247	1.982	.004
I	.282	.074	.508	3.821	.000
O	-.011	.141	-.012	-.082	.935
UP	.131	.097	.161	1.348	.184

The results indicate that there is a significant correlation between the four mentioned variables and both internal factors although the correlation intensity for perceived ease of use is greater; also, the results demonstrate that the four independent variables can explain perceived usefulness for 33% and perceived ease of use for 44%.

Table 6 Results of the Regression analysis of Independent variables and perceived usefulness

Model Summary	
R	R Square
.577a	.333

ANOVA					
Model	Sum of Squares	df	Mean Square	F	Sig.
Regression	10.261	4	2.565	6.110	.000b
a. Dependent Variable: perceived usefulness					
b. Predictors: (Constant), UP, OFC, I, O					

Coefficients					
Independent variables	Unstandardized Coefficients		Standardized Coefficients	t	Sig.
	B	Std. Error	Beta		
OFC	-.356	.214	-.228	2.200	.005
I	-.099	.123	-.118	-.806	.424
O	.501	.234	.346	2.140	.004
UP	.601	.162	.485	3.708	.001

Table 5 and 6 show that these 4 structures can significantly and appropriately predict the two dependent variables and among the independent variables, user performance is the most significant variable for perceived usefulness that can explain it is up to 48% and innovativeness is the most effective variable for perceived ease of use that can explain it is up to 50%.

Following tables illustrate the result of Regression analysis.

According to the results, positive relationship between user performance and two dependent variables shows that if the efficiency of a user is improved by using a technology, he or she recognizes it more useful and if it could enhance his or her efficiency it states that using the technology is easy.

The positive relationship between optimism and two dependent variables indicates that optimistic individuals toward the technology find it more beneficial and easy to use.

The relationship between organizational facilitating conditions and two dependent variables significantly positive which implies that if fewer obstacles in the use of technology prevent users for adopting technology and needed technological and organizational resources are available, they will recognize it more useful and easier to use.

The positive relationship between innovation and two dependent variables suggests that if a technology causes innovation creation in the process of company which results in competitive advantage, individual finds it easy to use a technology and has a positive opinion about the efficiency of it.

5 Conclusion

The paper was aimed to identify key factors influencing the companies' decisions in terms of digital technology implementation. The current literature is paid enough attention to the process of technology adoption in construction; however, none of them evaluates the readiness of construction companies regarding to the process of digital technology considering sustainable approaches. This paper contributes to the body of knowledge by developing an initial framework for measuring readiness index for predicting the possibilities of successful implementation process in a construction company.

The results show that the participants prefer to use a technology that links to the context of innovation and creativity in the company and create a competitive advantage for their company. The results also show that 'easy of learn' and 'ease of use' are two main factors influencing the companies' decision to use a new technology. In addition, the results show that the high cost of technology, lack of financial support, lack of executive specialist, resistance of traditional systems, the uncertainty of the return of investment, lack of encouragement of the government, unfamiliarity with equipment performance would prevent the use of the technology.

Respondents believe that by holding training courses,

promoting the corporate culture, informing about the advantages of sustainable construction technologies usage and legalizing it, adopting motivational tools and using penalties, the rate of green technology adoption can be increased. Furthermore, major environmental impacts of construction projects should be identified and the culture of sustainable development at all levels of the company should be created to increase using green technology. Future studies should apply the proposed model in different contexts to examine and identify how a new green technology can be utilized in a shorter time.

6 References

[1] I. Petri, T. Beach, Y. Rezgui, I. E. Wilson, and H. Li, "Engaging Construction Stakeholders With Sustainability Through a Knowledge Harvesting Platform," *Computers in Industry,* vol. 65, pp. 4449-469, 2014.

[2] I. O. Adetunji, "Sustainable Construction: A Web-Based Performance Assessment Tool," Doctor of Engineering, Civil & Building Engineering, Loughborough, 2005.

[3] L. Yang, "Some Thoughts on the Development of Digital Technology in the Construction Industry," *Advanced Materials Research,* vol. 243-249, pp. 6637-6643, 2011.

[4] M. Ibrahim and O. Moselhi, "Inertial measurement unit based indoor localization for construction applications," *Automation in Construction,* 2016.

[5] Y. P. Feng, "Application of Information Technology in Construction Management " in *The CRIOCM 2006 International Symposium on "Advancement of Construction Management and Real Estate"*, Beijing, China, 2006.

[6] J. Irizarry, M. Gheisari, and B. N. Walker, "Usability Assessment of Drone Technology As Safety Inspection Tools," *Journal of Information Technology in Construction,* vol. 17, pp. 194-212, 2012.

[7] C. Snow, "The Truth About Drones in Construction and Infrastructure Inspection," Skylogic Research2016.

[8] M. Setareh, D. A. Bowman, and A. Kalita, "Development of a Virtual Reality Structural Analysis System " *Journal of Architectural Engineering,* vol. 11, no. 4, pp. 156-164, 2005.

[9] D. Zhao and J. Lucas, "Virtual Reality Simulation for Construction Safety Promotion," *International Journal of Injury Control and Safety Promotion,* vol. 22, no. 1, pp. 57-67, 2015.

[10] A. Shibeika and C. Harty, "Diffusion of Digital Innovation in Construction: A Case Study of a UK Engineering Firm," *Construction Management and Economics,* vol. 33, no. 5-6, pp. 453-466, 2016.

[11] E. O. Ibem and S. Laryea, "Survey of Digital Technologies in Procurement of Construction Projects," *Automation in Construction,* vol. 46, pp. 11-21, 2014.

[12] P. Bynum, R. R. A. Issa, and S. Olbina, "Building Information Modeling in Support of Sustainable Design and Construction," *JOURNAL OF CONSTRUCTION ENGINEERING AND MANAGEMENT,* vol. 139, no. 1, pp. 24-34, 2013.

[13] H. Son, S. Lee, and C. Kim, "What Drives the Adoption of Building Information Modeling in Design Organizations? An Empirical Investigation of the Antecedents Affecting Architects' Behavioral Intentions," *Automation in Construction,* vol. 49, Part A, no. 0, pp. 92-99, 1// 2015.

[14] S. Jiang and W. Lei, "The Application of BIM in Green Building Energy Saving: Take Helsinki Music Center as an Example," *Advanced Materials Research,* vol. 935, pp. 3-7, 2014.

[15] A. P. C. Chan, A. Darko, E. E. Ameyaw, and D.-G. Owusu-Manu, "Barriers Affecting the Adoption of Green Building Technologies," *Journal of Management in Engineering,* 2016.

[16] N. H. Ibrahim, "Reviewing the Evidence: Use of Digital Collaboration Technologies in Major Building and Infrastructure Projects," *Journal of Information Technology in Construction,* vol. 18, pp. 40-63, 2013.

[17] A. P. Chassiakos, "The Use of Information and Communication Technologies in Construction," in *Civil Engineering Computations: Tools and Techniques*, B. H. V. Topping, Ed. St. Julians, Malta: Saxe-Coburg Publications, 2007, pp. 79-104.

[18] M. Matar, H. Osman, M. Georgy, and A. Abou-Zeid, "Sustainable construction modelling: a systems engineering approach," in *The 19th CIB World Building Congress*, Queensland, Australia, 2013, pp. 1-14: Queensland University of Technology.

[19] B. Sidawia and A. A. Al-Sudairia, "The Use ofAdvanced Computer Based Management Systems by Large Saudi CompaniesforManaging Remote Construction Projects," *Procedia Engineering,* vol. 77, pp. 161 – 169, 2014.

[20] A. Kazim, "Construction of Green Buildings in Dubai International Academic City," in *4th IEEE International Conference on Digital*

Ecosystems and Technologies: Conference Proceedings of IEEE-DEST 2010, Dubai, United Arab Emirates, 2010: IEEE.

[21] I. Ahmad, S. Azhar, and S. M. Ahmed, "Web-based Construction Project Management: Current State, Trends and Potentials " in *International Conference" Construction Innovation and Global Competitiveness"*, Cincinnati, Ohio, 2002, pp. 599-610.

[22] P. Mitropoulos and C. Tatum, "Technology Adoption Decisions in Construction Organizations," *Journal of Construction Engineering and Management,* vol. 125, no. 5, pp. 330-338, 1999.

[23] G. Wang and D. Cao, "Research on the project-level influencing factors on information technology implementation in construction industry," in *International Conference on Management and Service Science (MASS)*, TBD Wuhan, China, 2010.

[24] J. Shelton, I. Martek, and C. Chen, "Implementation of innovative technologies in small-scale construction firms: Five Australian case studies," *Engineering, Construction and Architectural Management,* vol. 23, no. 2, pp. 177-191, 2016.

[25] J. K. Wong, H. Li, H. Wang, T. Huang, E. Luo, and V. Li, "Toward low-carbon construction processes: the visualisation of predicted emission via virtual prototyping technology," *Automation in Construction,* vol. 33, pp. 72-78, 2013.

[26] B. Ilhan and H. Yaman, "Green building assessment tool (GBAT) for integrated BIM-based design decisions," *Automation in Construction,* vol. 70, pp. 26-37, 2016.

[27] J. C. Cheng and M. Das, "A BIM-based web service framework for green building energy simulation and code checking," *Journal of Information Technology in Construction (ITcon),* vol. 19, no. 8, pp. 150-168, 2014.

[28] Y. Arayici, P. Coates, L. Koskela, M. Kagioglou, C. Usher, and K. O'Reilly, "Technology Adoption in the BIM Implementation for Lean Architectural Practice," *Automation in Construction,* vol. 20, no. 2, pp. 189-195, 2011.

[29] Z. Q. Li, H. C. Tan, C. Anumba, and F. C. Chia, "Development of a web-based system for managing suppliers' performance and knowledge sharing in construction project," *Built Environment Project and Asset Management,* vol. 7, no. 2, 2017.

[30] Y.-C. Lin, "Development of Web-based Teams Management System in Construction " *World Academy of Science, Engineering and Technology,* vol. 4, no. 5, 2010.

[31] G. D. Holt and D. J. Edwards, "Analysis of United Kingdom Off-Highway Construction Machinery Market and Its Consumers Using New-Sales Data," *Journal of Construction Engineering and Management,* vol. 139, no. 5, pp. 529-537, 2012.

[32] A. Hommels, P. Peters, and W. E. Bijker, "Techno therapy or nurtured niches? Technology studies and the evaluation of radical innovations," *Research Policy,* vol. 36, no. 7, pp. 1088-1099, 9// 2007.

[33] R. T. Frambach and N. Schillewaert, "Organizational Innovation Adoption: A Multi-level Framework of Determinants and Opportunities for Future Research," *Journal of Business Research,* vol. 55, no. 2, pp. 163-176, 2002.

[34] D. Thorpe, N. Ryan, and M. B. Charles, "Environmental Sustainability - A Driver for Innovation in Construction SMES?," in *Third International Conference of the Cooperative Research Centre (CRC) for Construction Innovation*, Gold Coast, Australia, 2008, pp. 12-14.

[35] F. D. Davis, R. P. Bagozzi, and P. R. Warshaw, "User Acceptance of Computer Technology: A Comparison of Two Theoretical Models," *Management Science,* vol. 35, no. 8, pp. 982-1003, 1989.

[36] M. Nicosia, A. Oulasvirta, and P. O. Kristensson, "Modeling the Perception of User Performance," in *SIGCHI Conference on Human Factors in Computing Systems*, Toronto, Ontario, Canada, 2014, pp. 1747-1756.

[37] V. Venkatesh, M. G. Morris, G. B. Davis, and F. D. Davis, "User Acceptance of Information Technology: Toward a Unified View," *MIS Quarterly,* vol. 27, no. 3, pp. 425-478, 2003. Management Information Systems Research Center, University of Minnesota

[38] P. Godoe and T. S. Johansen, "Understanding Adoption of New Technologies: Technology Readiness and Technology Acceptance as An Integrated Concept," *Journal of European Psychology Students,* vol. 3, 2012.

[1] I. Petri, T. Beach, Y. Rezgui, I. E. Wilson, and H. Li, "Engaging Construction Stakeholders With Sustainability Through a Knowledge Harvesting Platform," *Computers in Industry,* vol. 65, pp. 4449-469, 2014.

[2] I. O. Adetunji, "Sustainable Construction: A Web-Based Performance Assessment Tool,"

Doctor of Engineering, Civil & Building Engineering, Loughborough, 2005.

[3] L. Yang, "Some Thoughts on the Development of Digital Technology in the Construction Industry," *Advanced Materials Research,* vol. 243-249, pp. 6637-6643, 2011.

[4] M. Ibrahim and O. Moselhi, "Inertial measurement unit based indoor localization for construction applications," *Automation in Construction,* 2016.

[5] Y. P. FENG, "APPLICATION OF INFORMATION TECHNOLOGY IN CONSTRUCTION MANAGEMENT," in *The CRIOCM 2006 International Symposium on "Advancement of Construction Management and Real Estate"*, Beijing, China, 2006.

[6] J. Irizarry, M. Gheisari, and B. N. Walker, "Usability Assessment of Drone Technology As Safety Inspection Tools," *Journal of Information Technology in Construction,* vol. 17, pp. 194-212, 2012.

[7] C. Snow, "The Truth About Drones in Construction and Infrastructure Inspection," Skylogic Research2016.

[8] M. Setareh, D. A. Bowman, and A. Kalita, "Development of a Virtual Reality Structural Analysis System " *Journal of Architectural Engineering,* vol. 11, no. 4, pp. 156-164, 2005.

[9] D. Zhao and J. Lucas, "Virtual Reality Simulation for Construction Safety Promotion," *International Journal of Injury Control and Safety Promotion,* vol. 22, no. 1, pp. 57-67, 2015.

[10] A. Shibeika and C. Harty, "Diffusion of Digital Innovation in Construction: A Case Study of a UK Engineering Firm," *Construction Management and Economics,* vol. 33, no. 5-6, pp. 453-466, 2016.

[11] E. O. Ibem and S. Laryea, "Survey of Digital Technologies in Procurement of Construction Projects," *Automation in Construction,* vol. 46, pp. 11-21, 2014.

[12] P. Bynum, R. R. A. Issa, and S. Olbina, "Building Information Modeling in Support of Sustainable Design and Construction," *JOURNAL OF CONSTRUCTION ENGINEERING AND MANAGEMENT,* vol. 139, no. 1, pp. 24-34, 2013.

[13] H. Son, S. Lee, and C. Kim, "What Drives the Adoption of Building Information Modeling in Design Organizations? An Empirical Investigation of the Antecedents Affecting Architects' Behavioral Intentions," *Automation in Construction,* vol. 49, Part A, no. 0, pp. 92-99, 1// 2015.

[14] S. Jiang and W. Lei, "The Application of BIM in Green Building Energy Saving: Take Helsinki Music Center as an Example," *Advanced Materials Research,* vol. 935, pp. 3-7, 2014.

[15] A. P. C. Chan, A. Darko, E. E. Ameyaw, and D.-G. Owusu-Manu, "Barriers Affecting the Adoption of Green Building Technologies," *Journal of Management in Engineering,* 2016.

[16] N. H. Ibrahim, "Reviewing the Evidence: Use of Digital Collaboration Technologies in Major Building and Infrastructure Projects," *Journal of Information Technology in Construction,* vol. 18, pp. 40-63, 2013.

[17] A. P. Chassiakos, "The Use of Information and Communication Technologies in Construction," in *Civil Engineering Computations: Tools and Techniques*, B. H. V. Topping, Ed. St. Julians, Malta: Saxe-Coburg Publications, 2007, pp. 79-104.

[18] M. Matar, H. Osman, M. Georgy, and A. Abou-Zeid, "Sustainable construction modelling: a systems engineering approach," in *The 19th CIB World Building Congress*, Queensland, Australia, 2013, pp. 1-14: Queensland University of Technology.

[19] B. Sidawia and A. A. Al-Sudairia, "The Use ofAdvanced Computer Based Management Systems by Large Saudi CompaniesforManaging Remote Construction Projects," *Procedia Engineering,* vol. 77, pp. 161 – 169, 2014.

[20] A. Kazim, "Construction of Green Buildings in Dubai International Academic City," in *4th IEEE International Conference on Digital Ecosystems and Technologies: Conference Proceedings of IEEE-DEST 2010*, Dubai, United Arab Emirates, 2010: IEEE.

[21] I. Ahmad, S. Azhar, and S. M. Ahmed, "Web-based Construction Project Management: Current State, Trends and Potentials " in *International Conference" Construction Innovation and Global Competitiveness"*, Cincinnati, Ohio, 2002, pp. 599-610.

[22] P. Mitropoulos and C. Tatum, "Technology Adoption Decisions in Construction Organizations," *Journal of Construction Engineering and Management,* vol. 125, no. 5, pp. 330-338, 1999.

[23] G. Wang and D. Cao, "Research on the project-level influencing factors on information technology implementation in construction industry," in *International Conference on Management and Service Science (MASS)*, TBD Wuhan, China, 2010.

[24] J. Shelton, I. Martek, and C. Chen, "Implementation of innovative technologies in small-scale construction firms: Five Australian case studies," *Engineering, Construction and Architectural Management,* vol. 23, no. 2, pp. 177-191, 2016.

[25] J. K. Wong, H. Li, H. Wang, T. Huang, E. Luo, and V. Li, "Toward low-carbon construction processes: the visualisation of predicted emission via virtual prototyping technology," *Automation in Construction,* vol. 33, pp. 72-78, 2013.

[26] B. Ilhan and H. Yaman, "Green building assessment tool (GBAT) for integrated BIM-based design decisions," *Automation in Construction,* vol. 70, pp. 26-37, 2016.

[27] J. C. Cheng and M. Das, "A BIM-based web service framework for green building energy simulation and code checking," *Journal of Information Technology in Construction (ITcon),* vol. 19, no. 8, pp. 150-168, 2014.

[28] Y. Arayici, P. Coates, L. Koskela, M. Kagioglou, C. Usher, and K. O'Reilly, "Technology Adoption in the BIM Implementation for Lean Architectural Practice," *Automation in Construction,* vol. 20, no. 2, pp. 189-195, 2011.

[29] Z. Q. Li, H. C. Tan, C. Anumba, and F. C. Chia, "Development of a web-based system for managing suppliers' performance and knowledge sharing in construction project," *Built Environment Project and Asset Management,* vol. 7, no. 2, 2017.

[30] Y.-C. Lin, "Development of Web-based Teams Management System in Construction " *World Academy of Science, Engineering and Technology,* vol. 4, no. 5, 2010.

[31] G. D. Holt and D. J. Edwards, "Analysis of United Kingdom Off-Highway Construction Machinery Market and Its Consumers Using New-Sales Data," *Journal of Construction Engineering and Management,* vol. 139, no. 5, pp. 529-537, 2012.

[32] A. Hommels, P. Peters, and W. E. Bijker, "Techno therapy or nurtured niches? Technology studies and the evaluation of radical innovations," *Research Policy,* vol. 36, no. 7, pp. 1088-1099, 9// 2007.

[33] R. T. Frambach and N. Schillewaert, "Organizational Innovation Adoption: A Multi-level Framework of Determinants and Opportunities for Future Research," *Journal of Business Research,* vol. 55, no. 2, pp. 163-176, 2002.

[34] D. Thorpe, N. Ryan, and M. B. Charles, "Environmental Sustainability - A Driver for Innovation in Construction SMES?," in *Third International Conference of the Cooperative Research Centre (CRC) for Construction Innovation*, Gold Coast, Australia, 2008, pp. 12-14.

[35] M. Nicosia, A. Oulasvirta, and P. O. Kristensson, "Modeling the Perception of User Performance," in *SIGCHI Conference on Human Factors in Computing Systems*, Toronto, Ontario, Canada, 2014, pp. 1747-1756.

[36] V. Venkatesh, M. G. Morris, G. B. Davis, and F. D. Davis, "User Acceptance of Information Technology: Toward a Unified View," *MIS Quarterly,* vol. 27, no. 3, pp. 425-478, 2003. Management Information Systems Research Center, University of Minnesota

[37] P. Godoe and T. S. Johansen, "Understanding Adoption of New Technologies: Technology Readiness and Technology Acceptance as An Integrated Concept," *Journal of European Psychology Students,* vol. 3, 2012.

[38] J. C. Nunnally, *Psychometric theory*, 2nd ed. New York:: McGraw-Hill, 1978.

[39] F. D. Davis, R. P. Bagozzi, and P. R. Warshaw, "User Acceptance of Computer Technology: A Comparison of Two Theoretical Models," *Management Science,* vol. 35, no. 8, pp. 982-1003, 1989.

[40] J. C. Nunnally, *Psychometric theory*, 2nd ed. New York:: McGraw-Hill, 1978.

7 Appendix I:

The following selected questions were asked from the participants.

Table 7 Selected questions

Identifier	Proposed Factors Influencing the Successful Implementation of Sustainable Construction Technology By a Company
OFC12	Organizational culture affects the decision to use the technology.
OFC12	Motivational mechanisms of the senior managers affect my decision in the use of technology.
EP5	Creating competitive advantage resulting from the use of technology affects my company decision in the use of technology.
EE3	Feasibility of applying this technology without prior experience affects my company decision in the use of technology.
EE5	Adoption of sustainable practices or procedures (e.g. waste management) affects the decision to use the technology.
EOU1	Ease of learning how to use technology affect my decision in the use of technology.
EE11	Creating the innovation in the process of company by using the technology, affects the decision to use the technology.
O5	Improving the quality of provided services by using the technology, affects the decision to use the technology.
UE1	Enhancing job performance using technology, affects the decision to use the technology.
UE3	Increasing individual motivation using technology, affects the decision to use the technology.
	Open Ended Questions
OP1	How is the decision-making process regarding the utilization of a new sustainable technology in your company?
OP2	What prevents the successful use of sustainable construction technology?
OP3	What are the challenges of sustainable construction technologies usage?
OP4	What are your solutions to facilitate and accelerate the use of sustainable construction technology?

34th International Symposium
on Automation and Robotics
in Construction

ISARC 2017

June 28 – July 1
TAIPEI, TAIWAN

Challenges to the Utilization of BIM in the Palestinian Construction Industry

A. Enshassi1[a] and L. AbUHamra2[b]

[a]Department of Civil Engineering, IUG, Palestine
[b]Department of Civil Engineering, IUG, Palestine
E-mail: aenshassi@gmail.com, Lina@gmail.com

Abstract –

This Building Information Modelling (BIM) is considered an innovative method to design, construct and manage a building. BIM can reduce time, cost and enhance project quality. The study reported in this paper was aimed at identifying and evaluation challenges that are facing the application of BIM in the Palestinian construction industry. A total of 270 professionals were surveyed to elicit their views regarding challenges to the utilization of BIM in the construction industry. Factors were ranked according to their relative importance index through SPSS version 22. The results of this study revealed that the main challenges to the utilization of BIM were: lack of awareness of BIM by stakeholders, lack of knowledge about BIM software, lack of awareness of BIM benefits, lack of engineers' skills in BIM, lack of education and training on the use of BIM, lack of demand from clients, and lack of governmental regulations. It is recommended to conduct awareness program for stockholders regarding MIM benefits. Education and training for engineers at universities is strongly recommended so that they can apply BIM in the practice.

Keywords –

Construction; BIM; Challenges; Implementation; Performance

1 Introduction

There are several problems when implementing BIM in the fragmented construction industry and which connected with many different barriers hindering effective adoption of BIM [1, 2]. Some of these barriers are quite simple to be removed, while others could be considered impossible to mitigate [3]. Yan and Damian [4] indicated, that the barriers to implement BIM in UK and USA are: people refuse to learn and think current design technology is enough for them to design the projects; people think that (BIM) is unsuitable for the projects; about (40%) of respondents from USA and about (20%) respondents from UK believe that BIM wastes time and human resources and their companies have to allocate lots of time and human resources to the training process; in addition to the cost of copyright and training.

Howard and Björk [5] found many obstacles to implement BIM in Denmark, Hong Kong, Holland, Norway, Sweden, UK and USA. The obstacles were: need of education; need of sharing information; lack of standards; and lack of legal issues to implement BIM. Arayici et al. [6] stated that the reasons to implement BIM in UK and Finland were: firms are not familiar enough with BIM use; reluctance to initiate new workflows, or train staff; firms do not have enough opportunity for BIM implementation; benefits from BIM implementation do not outweigh the costs to implement it; benefits are not tangible enough to warrant its use; and BIM does not offer enough of a financial gain to warrant its use.

Construction projects in Gaza strip suffer from many complex issues due to the fragmented nature of the construction industry and the lack of knowledge sharing as well as lack of communication between different professionals and stakeholders. In addition, the rising cost of construction projects remains the greatest problem the construction industry is facing now in Gaza Strip. Hence, the adoption of new and developing technology by construction firms would be pivotal in the overall development and success of the industry. The objective of this study is to identify and evaluate challenges that are facing the application of BIM in the Palestinian construction industry.

2 Literature review

Participants in the building process are constantly challenged to deliver successful projects despite tight budgets, limited manpower, accelerated schedules, and waste [7]. The Architecture, Engineering and Construction (AEC) industry has long sought to adopt techniques to decrease project cost, increase productivity and quality, reduce project delivery time, and eliminate waste [8]. One of these techniques is Building Information Modeling (BIM). Azhar et al. [9] said that BIM has recently attained widespread attention in the AEC industry.

Keegan [10] identified several barriers to the utilization of BIM in USA, namely: lack of knowledge about BIM by the owner; lack of t knowledge about the software; and cost of implementing and updating the system. Becerik-Gerber et al. [11] reported two main groups of challenges to implement BIM: technology and process challenges; and organizational challenges. Lahdou and Zetterman [12] stated that the challenges for adoption of BIM in the construction project process in Sweden were: personal opinions towards BIM; difficulties in implementation of BIM software; and lack of knowledge.

Kjartansdóttir [12] reported that the reasons for not applying BIM in Iceland were: BIM is lacking features or flexibility to create building model/drawing; clients are not requiring BIM; BIM is too expensive; and the lack of training in BIM software. Khosrowshahi and Arayici [14] identified the most significant reasons to failure to implement BIM in the UK and Finland as: firms are not familiar enough with BIM use; reluctance to initiate new workflows or train staff; benefits from BIM implementation do not outweigh the costs to implement it; benefits of BIM are not tangible enough to warrant its use; lacks the capital to invest in having started with hardware and software; resistance to culture change; and no demand for BIM use. Kassem et al. [15] investigated the barriers to adopt BIM and 4D in the UK civil and building industry. They found that the most of the barriers were shortage of experience within the workforce; and lack of awareness by stakeholders.

Elmualim and Gilder [16] investigated the challenges that are facing the construction industry in the installation of BIM in the UK, Europe, USA, India, Ghana, China, Russia, South Africa, Australia, Canada, Malaysia and UAE. Findings from their study showed that: (20.4%) of the respondents stated that they lack the capital to invest in getting started with the hardware and software; about (2%) stated that BIM is too risky from a liability standpoint to warrant its use, (15.3%) stated that the benefits of BIM do not outweigh the cost to implement it; (15.3%) stated that the benefits are not tangible enough to warrant its use. Thurairajah and Goucher [17] mentioned several obstacles for implementing BIM: overall lack of knowledge and understanding of what BIM is; a strong training requirement associated with BIM implementation to gain the full advantages from it; and the need for detailed understanding of cost consultants' challenges during the implementation of 5D BIM in construction projects.

Crowley [18] revealed that the potential barriers to BIM implementation are: lack of training/ education; lack of client demand; lack of government lead/direction; and lack of standards. Aibinu and Venkatesh (28) reported the barriers to the adoption of BIM were: cost of implementation; lack of awareness of the benefits from cost benefit analysis perspective; lack of demand by clients; lack of trust in the integrity of BIM; and adaptation issues; and technology change and ability of firms to adapt to the change from cultural perspective and financial perspective.

Fischer and Kunz [19] reported two main groups of obstacles: technical obstacles; and managerial obstacles. Arayici et al. [20] said that the (BIM) barriers can be grouped into four categories: the legal issues; the cultural issues; the technological issues; and the fragmented nature of the construction industry industry. Becerik-Gerber et al. [11] reported two main groups of challenges to implement BIM: technology and process challenges; and organizational challenges. Both and Kindsvater [21] grouped BIM barriers into the following four categories: technological issues; normative issues; general issues; and education and training. Gu et al. [22] categorized barriers to adopt BIM in the AEC industry into three groups: product; process; and people.

3 Methodology

A quantitative survey approach involving professionals (Architects, Civil engineers, Mechanical engineers, Electrical engineers, and any other and any other related specialization) in the construction industry in Gaza strip, Palestine has been adopted. Convenience sample was chosen as the type of sample. A total of 275 copies of the questionnaire were distributed and 270 copies of the questionnaire were returned from the respondents with respondents' rate of 97.8 %. Personal delivery for the whole sample helped to increase the rate of response and thus the representation of the sample.

A total of 36 challenges to BIM utilization in the construction industry were assembled from extensive literature review which was further refined based on a pilot study. As a result of the pilot study, some items have been selected; other items have been modified, while others have been merged, as well as some items have been added. As a result 18 challenges were selected for this study. Analysis of the data was undertaken using IBM SPSS Statistics (Statistical Package for the Social Sciences) Version 22(IBM).

The relative importance index method (RII) was used to determine the ranks of BIM barriers as perceived by the respondents. Respondents were invited to rate their response on a five- point Likert Scale. The relative importance index was computed as [23, 24, 25]:

$$RII=\Sigma W/ (A*N)$$

Where:

W = the weighting given to each factor by the respondents (ranging from 1 to 5)

A = the highest weight (i.e. 5 in this case)

N = the total number of respondents

The RII value had a range from 0 to 1 (0 not inclusive), the higher the value of RII, the more impact of the attribute. However, RII doesn't reflect the relationship between the various items.

4 Results and Discussion

Out of 36 challenges that were assembled from literature review, 18 potential challenges to BIM implementation were selected after modifications according to the pilot study. The descriptive statistics, i.e. means, standard deviations (SD), t-value (two-tailed), probabilities (P-value), relative importance indices (RII), and ranks were established and presented in Table (1). Items/Variables were categorized with ratings from 77.33 % to 66 % (Figure 1).

The findings indicated that "Lack of awareness of BIM by stakeholders" (BA 2) is the most influential barrier to adopt BIM in the construction industry in Gaza Strip. It has been ranked as the first position with (RII = 77.33%) and (P-value = 0.00*). This result indicates that a significant proportion of respondents have little or no understanding of BIM concept. This is consistent with the result which has been found by Kassem et al. [15]. Also, this result is in line with Thurairajah and Goucher [17] who found that, there is a lack of knowledge and understanding of BIM among professionals in the construction industry. According to the study of Löf and Kojadinovic [26] in Sweden, the reason for the lack of knowledge of BIM is the lack of guidelines on how to use and align BIM in the production phase of construction projects.

"Lack of knowledge of how to apply BIM software" (BA 3) (RII = 76.80 %; P-value = 0.00*) was ranked as the second BIM barrier for adopting BIM in the construction industry in Gaza Strip. Due to the complexity of gathering all the relevant information when working with BIM on a building project some companies have developed software designed specifically to work in a BIM framework. This result is consistent with Tse et al. [27] results, who found that a large part of the architects stated that BIM is "not easy to use". Also, this result is in line with which has been found in Sweden by Lahdou and Zetterman [11], who found that project managers in construction projects, claimed that the implementation of BIM is not always as easy as software developers suggest. A usual problem is getting different file formats to function properly when creating a combined building information model.

"Lack of awareness of the benefits that BIM can bring to engineering offices, companies and projects" (BA 5) was ranked in the third position with (RII of 76.22 %; P-value = 0.00*). There is a knowledge gap regarding BIM application in the Gaza Strip. The result is agreed with those reported about barriers for BIM adoption in the UK by Arayici et al. [6], and Kassem et al. [15]. Also, this outcome corroborates the findings of the studies of Khosrowshahi and Arayici [14], Aibinu and Venkatesh [28], and Elmualim and Gilder [16], whose research studies determined lack of awareness of BIM benefits as one of the main barriers associated with BIM implementation in the AEC industries in (UK and Finland), Australia, and (UK, Europe, USA, India, Ghana, China, Russia, South Africa, Australia, Canada, Malaysia and UAE), respectively. In addition, professionals in Australia displayed a degree of hesitancy in implementing BIM on a project because of the lack of knowledge about BIM and its distinctive capabilities in the field of construction industry [29]. In Hong Kong, Tse et al. [27] revealed that a large part of the architects did not find the tools in BIM to satisfy their needs. Thus, BIM benefits are still often misunderstood or not known to those not use it in their works [26].

"Necessary high costs to buy BIM software and costs of the necessary hardware updates" (BA 1) (BIM barrier) was ranked as the lowest barrier for BIM adoption in the 18th position with (RII = 66 %; P-value = 0.00*). This view has more than one interpretation, based on the point of view from some respondents; they do not really know how much costs need to adopt BIM. While there was another point of view from other respondents who are working in consulting offices, which is that the initial costs that must be spend in the beginning will not affect the organization for a long time as long as there are great benefits to be gained from BIM adoption in the long run. There are several examples of the high costs that are needed to implement BIM, such as: software licensing; the costs to improve server capacity to suit having such a high IT requirements; ongoing maintenance fee; the cost of the proper creation of a building model; and the costs of training.

Regarding results for all BIM barriers, the mean for all those items equals 3.59 and the total RII equals 71.80 %, which is greater than 60% (the neutral value of RII (3/5)*100 = 60%. The value of t test equals 14.54, which is greater than the critical value of t that equals 1.97. As well as, the total p-value of the all items equals 0.00 and it is less than the significance level of 0.05. Based on all previous results, BIM barriers are greatly affecting the adoption of BIM in the construction industry in the Gaza Strip.

Table (1): Challenges to BIM utilization

No.	Challenges statements	Mean	SD	RII (%)	t-value (two-tailed)	P-value (Sig.)	Rank
BA 2	Lack of awareness of BIM by stakeholders	3.87	0.99	77.33	14.34	0.00*	1
BA 3	Lack of knowledge of how to apply BIM software	3.84	0.95	76.80	14.50	0.00*	2
BA 5	Lack of awareness of the benefits that BIM can bring to engineering offices, companies and projects	3.81	0.98	76.22	13.63	0.00*	3
BA 14	Lack of interest in Gaza strip to pursue the condition of the building over the life after completion of implementation stage	3.75	1.10	75.04	11.29	0.00*	4
BA 15	Lack of architects/engineers skilled in the use of BIM programs	3.71	1.11	74.15	10.47	0.00*	5
BA 16	Lack of education or training on the use of BIM, whether in the university or any governmental or private training centers	3.69	1.02	73.78	11.14	0.00*	6
BA 12	Lack of demand and disinterest from clients regarding with using BIM technology in design and construction of the project	3.69	1.11	73.78	10.22	0.00*	7
BA 11	Lack of governmental regulations to fully support implementation of BIM	3.68	1.14	73.68	9.82	0.00*	8
BA 4	Professionals think that the current CAD system and other conventional programs satisfy the need of designing and performing the work and complete the project efficiently	3.67	1.01	73.46	10.97	0.00*	9
BA 13	Lack of real cases in Gaza strip or other nearby areas in the region that have been implemented by using BIM and have proved positive return of investment	3.67	1.12	73.46	9.74	0.00*	10
BA 6	Lack of effective collaboration among project stakeholders to exchange necessary information for BIM application, due to the fragmented nature of the AEC industry in Gaza strip	3.57	0.98	71.41	9.57	0.00*	11
BA 18	Reluctance to train architects/engineers due to costly training requirements in terms of time and money	3.51	1.08	70.30	7.81	0.00*	12
BA 8	Lack of the financial ability for the small firms to start a new workflow that is necessary for the adoption of BIM effectively	3.42	1.17	68.43	5.90	0.00*	13
BA 9	Companies prefer focusing on projects (under working/ construction) rather than considering, evaluating and implementing BIM	3.40	1.07	67.93	6.08	0.00*	14
BA 10	Difficulty of finding project stakeholders with the required competence to participate in BIM	3.36	1.03	67.21	5.75	0.00*	15
BA 7	Resistance by companies and institutions for any change can occur on the workflow system and refusal of adopting a new technology	3.36	1.08	67.21	5.41	0.00*	16
BA 17	Unwillingness of architects /engineers to learn new applications because of their educational culture and their bias toward programs that are familiar to them	3.33	1.10	66.54	4.90	0.00*	17

Table (1): Challenges to BIM utilization

No.	Challenges statements	Mean	SD	RII (%)	*t*-value (two-tailed)	*P*-value (Sig.)	Rank
BA 1	Necessary high costs to buy BIM software and costs of the necessary hardware updates	3.30	1.12	66	4.41	0.00*	18
	All barriers	3.59	0.67	71.80	14.54	0.00*	

Critical value of t: at *degree of freedom (df)* = [*N-1*] = [270-1] = 269 and *significance (Probability) level* 0.05 equals "1.97"

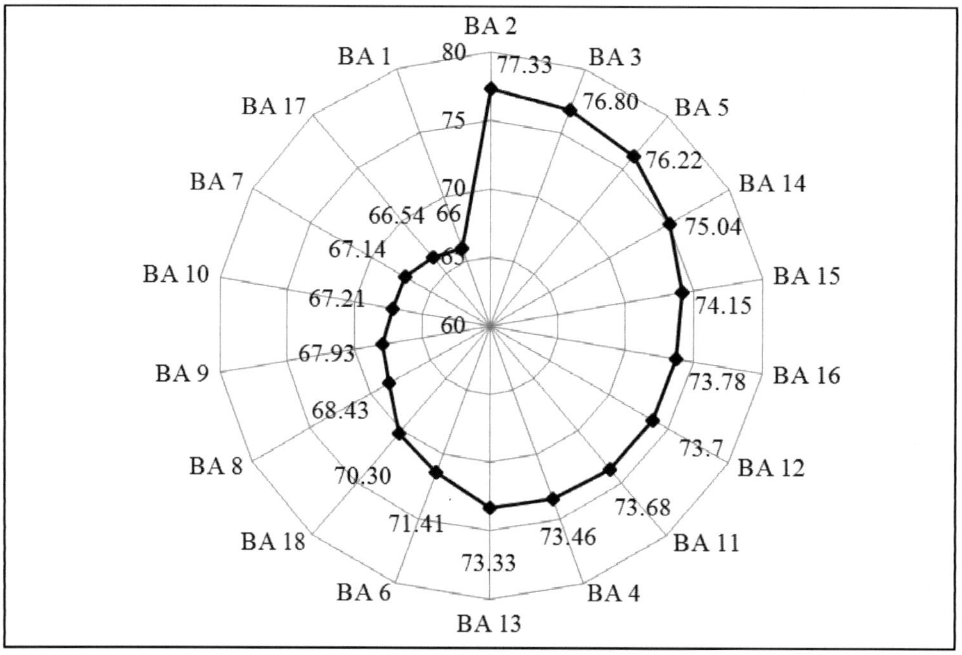

Figure (1): RII of BIM challenges (BA 1 to BA 18)

4 Conclusion

A research study has been carried out to identify and explore the challenges to BIM utilization in the construction industry. The results of this study revealed that the main challenges to the utilization of BIM were: lack of awareness of BIM by stakeholders, lack of knowledge about BIM software, lack of awareness of BIM benefits, lack of engineers' skills in BIM, lack of education and training on the use of BIM, lack of demand from clients, and lack of governmental regulations.

5 Recommendation

It is recommended to conduct awareness program for stockholders regarding the benefits from BIM utilization as BIM can reduce time, cost and enhance project quality. Education and training for engineers at universities is strongly recommended so that they can apply BIM in the practice. This study provides useful information for stockholders regarding challenges to BIM adoption so that they can overcome their obstacles. It is suggested to conduct further research in BIM application and benefits by applying BIM in a real case study, so that the benefits can be clear to contract parties.

References

[1] Lindblad, H. *Study of the implementation process of BIM in construction projects/ analysis of the barriers limiting BIM adoption in the AEC-industry*. Master thesis, Department of Real Estate and Construction Management, Architecture and Build Environment, ABE, Architectural Design and Construction Project Management, KTH Architecture and Build Environment, Stockholm. 2013.

[2] Mandhar, M., & Mandhar, M. Biming the architectural curricula – integrating building information modelling (BIM) in architectural education. *International Journal of Architecture* (IJA), 1(1): 1-20, 2013.

[3] Gökstorp, M. (2012). *BIM implementation and potential benefits for the facility managers*. Master thesis, Department of Civil and Environmental Engineering, Chalmers University of Technology, Göteborg, Sweden, 2012.

[4] Yan, H., & Damian, P. Benefits and Barriers of Building Information Modelling. *In proceedings of the 12th International conference on Computing in Civil and Building* (INCITE 2008) pages 1-5, Beijing, China, 2008.

[5] Howard, R., & Bjork, B. C. Building information modeling-Experts' views on standardization and industry deployment. *Advanced Engineering Informatics*, 22(2): 271–280, 2008.

[6] Arayici, Y., Khosrowshahi, F., Ponting, A. M., & Mihindu, S. Towards implementation of building information modelling in the construction industry. *In Proceedings of the Fifth International Conference on Construction in the 21st Century "collaboration and integration in Engineering, Management and Technology"*. Pages 1342-1351, Istanbul, Turkey: Middle East Technical University and Florida International University, 2009.

[7] NBS. (2013). *NBS international BIM* report. UK: The national BIM library.

[8] Azhar, S., Nadeem, A., Mok, J. Y., & Leung, B. H. Building information modeling (BIM): a new paradigm for visual interactive modeling and simulation for construction projects. *In proceedings of the first International Conference on Construction in Developing Countries (ICCIDC–I)*, pages 435-446. Karachi, Pakistan: Advancing and Integrating Construction Education, Research & Practice, 2008b.

[9] Azhar, S., Hein, M., & Sketo, B. (2008a). Building information modeling (BIM): benefits, risks and challenges. *Proceedings of the 43rd ASC National Annual Conference*, pages 1-11, Auburn, Alabama, 2008a.

[10] Keegan, C. J. *Building information modeling in support of space planning and renovation in colleges and universities*. Master thesis, Construction Project Management, Faculty of Worcester Polytechnic Institute (WPI), USA, 2010.

[11] Becerik-Gerber, B., Jazizadeh, F., Li, N., & Calis, G. Application areas and data requirements for BIM-enabled facilities management. *Journal of Construction Engineering and Management*, 138(3): 431-442, 2011.

[12] Lahdou, R., & Zetterman, D. *BIM for project managers; how project managers can utilize BIM in construction projects*. Master thesis, Division of Construction Management, Visualization Technology, Department of Civil and Environmental Engineering, Chalmers University of Technology, Göteborg, Sweden, 2011.

[13] Kjartansdóttir, I. B. *BIM adoption in Iceland and its relation to lean construction*. Master thesis, Construction Management, The School of Science and Engineering, Reykjavík University, Iceland, 2011.

[14] Khosrowshahi, F., & Arayici, Y. Roadmap for implementation of BIM in the UK construction industry. Engineering, *Construction and Architectural Management,* 19(6): 610-635, 2012.

[15] Kassem, M., Brogden, T., & Dawood, N. BIM and 4D planning: a holistic study of the barriers and drivers to widespread adoption. *KICEM Journal of Construction Engineering and Project Management*, 2(4): 1-10, 2012.

[16] Elmualim, A., & Gilder, J. BIM: innovation in design management, influence and challenges of implementation. *Architectural Engineering and Design Management*, 10(1080):1745-2007, 2013.

[17] Thurairajah, N., & Goucher, D. Advantages and challenges of using BIM: a cost consultant's perspective. In proceedings of the 49th ASC Annual International Conference, pages 1-8, Birmingham, United Kingdom: 2013.

[18] Crowley, C. Identifying opportunities for quantity surveyors to enhance and expand the traditional quantity surveying role by adopting building information modelling. In proceedings of the CITA BIM Gathering. Pages 71-77, Ireland: The Construction IT Alliance, 2013.

[19] Fischer, M., & Kunz, J. *The scope and role of information technology in construction*. CIFE technical report No.156 Stanford: CIFE center for integrated facility engineering/ Stanford University. 2004.

[20] Arayici, Y., Aouad, G., & Ahmed, V. Requirements engineering for innovative integrated ICT systems for the construction industry. *Construction Innovation*, 5(3):179 – 200, 2005.

[21] Both, P., & Kindsvater, A. Potentials and barriers for implementing BIM in the German AEC market: results of a current market analysis, *In proceedings of the 30th eCAADe Conference*, pages 151-158, Moscow, Russia, 2012.

[22] Gu, N., Singh, V., London, K., & Brankovic, L. BIM: expectations and a reality check. *In proceedings of 12th International Conference on Computing in Civil and Building Engineering & 2008 International Conference on Information Technology in Construction*, pages1-6, Beijing, China: QUT Digital Repository, Icon.Net Pty Ltd, 2008.

[23] Enshassi A., Mohamed S., Mustafa Z., and Mayer P. Factors affecting labour productivity in building projects in the Gaza Strip, *Journal of Civil Engineering and Management*, 13(4):245-254, 2007.

[24] Enshassi, A., Al-Najjar, J., and kumaraswamy, M. Delays and cost overruns in construction projects in the Gaza Strip, *Journal of Financial Management of Property and Construction*, 14(2): pp.126-251, 2009.

[25] Field, A. *Discovering statistics using SPSS*, 3rd Edition. London: SAGE Publications Ltd, 2009.

[26] Löf, M. B., & Kojadionovic, I. *Possible utilization of BIM in the production phase of construction projects*. Master thesis, Department of Real Estate and Construction Management, Architecture and Build Environment, ABE, Architectural Design and Construction Project Management, Skanska, Sweden, 2012.

[27] Tse, T. C. K., Wong, K. D. A., & Wong, K. W. F. The utilization of building information models in modeling: a study of data interfacing and adoption barriers. *ITcon*, 10: 85-110, 2005.

[28] Aibinu, A., & Venkatesh, S. Status of BIM adoption and the bim experience of cost consultants in Australia. *American Society of Civil Engineers* (ASCE), 140(3): 1-10, 2013.

[29] Mitchell, D., & Lambert, S. BIM: rules of engagement. *In proceedings of the CIB World Building Congress*, pages 1-5, Brisbane, 2013.

A Methodology of Building Code Checking System for Building Permission based on openBIM

Jungsik Choi[a] and Inhan Kim[b]

[a]Research center, buildingSMART Korea, Republic of Korea
[b]Department of Architecture, Kyung Hee University, Republic of Korea
E-mail: jungsikchoi@gmail.com, ihkim@khu.ac.kr

Abstract

The purpose of this study is to suggest methodology of building code checking system for building permission based on open BIM[1]. To achieve this purpose, the authors have suggested an open BIM-based building code checking process and developed a building code checking system called KBim Assess-Lite. According to this process, prevent various problem which may occur in developing the automated design quality checking system.

Keywords –

Automated Code Checking System; Building Code; Building Information Modeling (BIM); Design Quality; Industry Foundation Classes (IFC)

1 Introduction

Architectural information is managed throughout the life cycle of a building. Building information modeling (BIM) serves as a shared knowledge resource for information about a facility, forming a reliable basis for decisions during its life cycle from inception onward [1]. BIM technology has been utilized increasingly in quantitative ways in the architecture, engineering, and construction disciplines. However, owing to increasing requirements for an improvement in qualitative factors in BIM-based design projects, it has become necessary to develop a checking and evaluation process for improvement of BIM-based design quality [2]. The delivery of the BIM data is mandatory in advanced countries, and these countries are promoting the automated checking for BIM quality, including compliance with the building code. For example, Singapore has developed a BIM-based automated regulation checking process through the software FORNAX and has built a construction administration system, CORENET. The SMARTcodes project in the USA has structuralized the regulations of the International Code Council (ICC) and developed automatic code compliance checking systems [3]. In particular, code checks through and automated code checking system can reduce errors, time, and the inefficient use of human resources through objective verification. The purpose of this study is to suggest methodology of building code checking system for design quality checking based on BIM. In this study, the scope of applied design quality checking is confined to Korean building codes. The methodology of this study to achieve this purpose is as follows.

1. The authors have suggested an open BIM-based design quality checking process, called in KBIM e-Submission process. The architects create open BIM data using guideline and BIM library that was developed this study. The BIM data is checked through building code checking system called KBim Assess-Lite.
2. The authors developed a logic rule-based management system called KBim Logic. This system translates Korea building code (natural language) into computer-executable rule set files. Following the logic-rule based mechanism, the building code sentences are converted into intermediate script language called KBim Code.
3. The authors developed a building code checking system called KBim Assess-Lite. This system can assess quality of the IFC-based open BIM data. Major aspect of the BIM model quality is conformance of the building code (KBim code).

According to this process, prevent various problem which may occur in developing the automated design quality checking system. Consequently, the architects can achieve the set up goals much easier and faster through the developed automated design quality checking system.

[1] Open BIM is a universal approach to the collaborative design, realization and operation of buildings based on open standards and workflows. Open BIM is an initiative of buildingSMART and several leading software vendors use the open buildingSMART data model.

2 Background

Building code checking is the most actively applied quality checking research. It is reviewed through national laws and is based on each country's regulations. It is processed through building code checking system developed using the IFC related information. Research of last four years were analyzed into the vertex. In the research of João Poças Martins et al., the system LicA was developed to examine the plumbing facilities and plumbing equipment of the provisions of Portugal. This paper presented a development plan, developed the LicAXML format to save the Pset and developed additional attribute information using the IFC format to another LicAXML for IFC compatible BIM software such as Revit MEP [4]. In the research of Sijie Zhang, a review was conducted for the safety regulations applicable to BIM model of construction process. Review of the potential risks arising from the construction process does not take into account of the design stage. Scaffolding was installed through automation and safety devices in order to solve this problem [5]. In the research of Sager Malsane et al., study of the IFC schema was performed to propose measures to increase the potential of the UK regulatory review through the localization of specific terms in the United Kingdom. This research attempted to improve interoperability by the reconstruction of IFC structure and applied model [6]. Inhan Kim et al. conducted a building code checking research of safety regulations targeting super-tall buildings. Using Application Program Interface (API) of Solibri Model Checker (SMC), the development of the ruleset in code checking system and the development of an evacuation elevator and safety evacuation zone sector utilizing the IFC Pset [7]. Junho Choi derived entries based on the building permission regulation check-list and proceeded with building code checking using Pset of the IFC. Case of determining whether or not to use entries in the living room space were randomly checked by defining the Pset [8]. Hyeonsu Lee et al. researched on analysing the human-readable language sentence in form of building codes objects and properties of building regulations, the associated functional conversion classification as well as the classification items as pseudo code of the previous stage of the system functions. This paper shows the possibility of utilizing building code checking from a variety of programs by establishing a web-based database based on the derived the entries and transformed codes [9].

The delivery of BIM data is mandatory in advanced countries, and these countries promote automated building code checking. For example, Singapore has developed a BIM-based automated code checking process through the software FORNAX and has built a construction administration system, CORENET [10]. The SMARTcodes project in the USA has structuralized the regulations of the International Code Council (ICC) and developed an automatic building code checking system [11]. In particular, legality checking through an automated legality checking system can reduce errors, time, and inefficient use of human resources. Therefore, it is necessary to develop a building code checking system for design quality improvement based on open BIM in Korea, as an improvement over the conventional 2D drawings and documents.

3 Open BIM-based building code checking process

The Construction industry involves compiling a multitude of data from various sources working within different disciplines in a process that includes a variety of design stages. This process support collaboration between numerous participants. Architect can BIM data checking before Seumter (building administration permission services in Korea) submission through KBim Assess-Lite. Architect can pre-checking for Seumter building permission through BIM quality center verification process. Approver can Korean building code checking through KBim Assess in the Seumter environment [12].

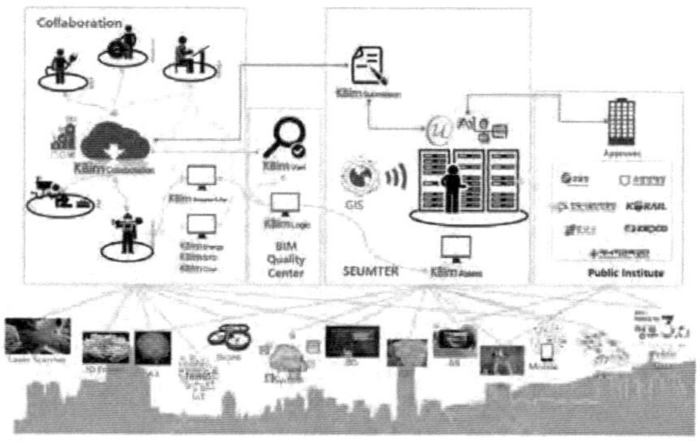

Figure 1. Open BIM-based design quality checking process

3.1 Design phase

3.1.1 Design Collaboration system

KBim Collaboration was developed to assist building design project management and communication in BIM and web-based environments. The purpose is to facilitate information transfers and seamless communication between different stakeholders and to prompt data sharing in an open BIM environment. This system thrives on optimizing essential features for BIM-based design collaboration by providing desirable industry features: [Project Information Management],

[Design Management], [Online Meeting], and [Project File Management]. These management tool functions were derived through a comprehensive analysis of the industry's needs (through a Delphi survey and expert interviews) to understand and overcome the limits of existing BIM collaboration systems.

In utilizing KBim Collaboration, the project participants are free to work across different BIM applications with the supported international open standard formats, Industry Foundation Classes (IFC), and BIM Collaboration Format (BCF). It is possible to extract required project management information from the IFC model and to communicate in BCF regarding design coordination.

Furthermore, KBim Collaboration serves as a platform for integrating other KBim-applications, such as KBim Submission, KBim D-Generator, and KBim Assess-Lite. The system automatically delivers all necessary and suited data for each application, including project information, BIM model files, drawings, and other project documents.

Figure 2. KBim Collaboration

3.1.2 BIM quality self-assessment

KBim Assess-Lite is a public-access version of KBim Assess software that supports BIM model checking for individual users. It operates as a rule-based model quality checker for IFC-based BIM models. KBim Assess-Lite aims to provide a convenient tool for building project participants to review their BIM models for building code compatibilities and general errors from the early design phase until submissions for project approvals.

Key functions include:

- IFC-based BIM model quality assessment by the rule-based checking methodology
- Provides text description of relevant building clauses from the Korean Building Act, its enforcement decree, and associated regulations.
- Execution criteria are defined and managed with code checklists of existing Korean building regulations and codes
- Checked results visualized in real-time 3D graphics
- Review reports can be outputted as MS Excel or .pdf files
- Modifiable execution criteria can be defined with user-customizable scripts

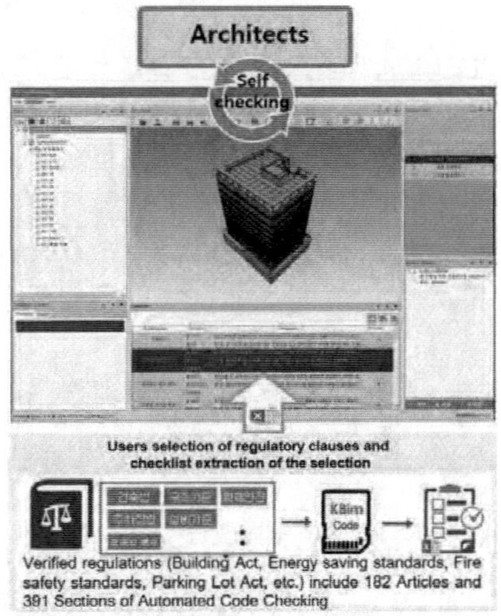

Figure 3. KBim Assess-Lite

3.1.3 Energy saving and automatic checking

KBim Energy is an automation program that generates energy performance evaluations from open BIM-based models. It is used in reference to energy-saving standards and the energy efficiency rating certification system of Korea for buildings with a total floor area of more than 3000m^2. The energy ratings from the automated analysis process are calculated according to ISO13790 to ensure transparency and reliability.

It provides a means for the rapid energy analysis of design alternatives; thus, it enables users to better determine energy requirement-compatible designs, enhances design workflow, and effectively reduces the amount of repetitive work. Energy Performance Index (EPI) output is also available for users from the web-based interface.

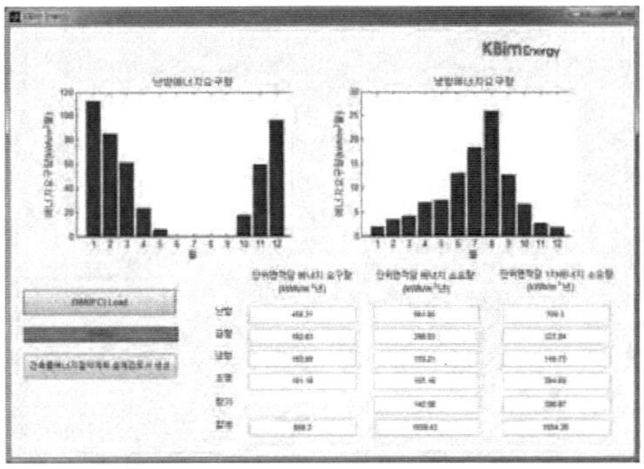

Figure 4. KBim Energy

3.2 Verification phase

3.2.1 BIM quality verification

KBim Veri is a BIM model quality assessment module for verifying KBim Assess regulatory compliance and before submitting the BIM model data to SEUMTER for project approval.

KBim Veri functions as a part of the BIM Quality Center service, which is a proposed web-based service for authenticating verified BIM models for SEUMTER submissions.

Key functions include:

- BIM model evaluation for file errors, such as invalid IFC version or wrong format.
- Preliminary checking of BIM models prior to KBim Assess on proper input of model objects and attributes necessary for automated code checking.
- Basic quality assessment of model objects' dimensions, location, type, and object relations.
- Reports generation of checked results

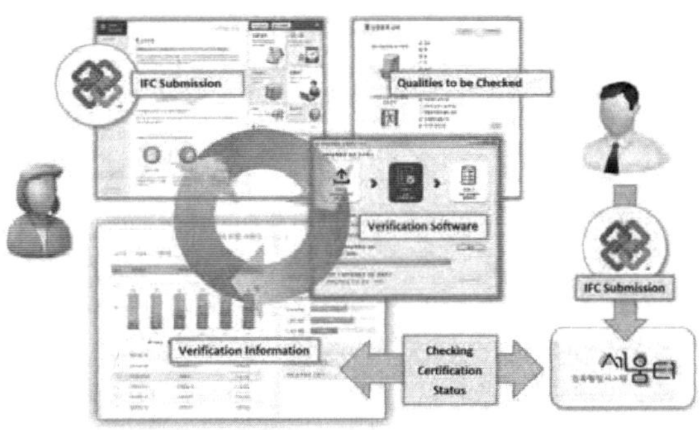

Figure 5. KBim Veri

3.2.2 Logical rule for building code

The KBim Logic application is a logic rule-based management system that governs the process of digitalizing the National Building Regulations and Codes of Korea (natural language) into KBim Code. It is comprised of 1) a logic rule-based meta database of Korean building regulations and codes; 2) the KBim Code authoring tool; 3) and the KBim Code database, a comprehensive management tool, and system databases.

The logic rule-based meta database provides the classification of regulatory-specific objects and properties, sentence relations etc. Following the logic rule-based mechanism, the KBim Code authoring tool accesses the meta database for the generation and management of intermediate script language, KBim Code.

KBim Code is the intermediate script language codes of Korea Building Code sentences. It was established as a database to be reused and executed in accordance with the Building Act. KBim Code is generated and managed by KBim Logic, the logic rule-based management application. An exported series of KBim Code can be used in KBim Assess, an open BIM-based model checker for building permits in Korea.

Distinguishing features of KBim Code include the following: it is a software-independent and standardized approach to the rule-making process and it is officially maintained and managed by government authorities, guarantying reliability for public use as well as ensuring an up-to-date database.

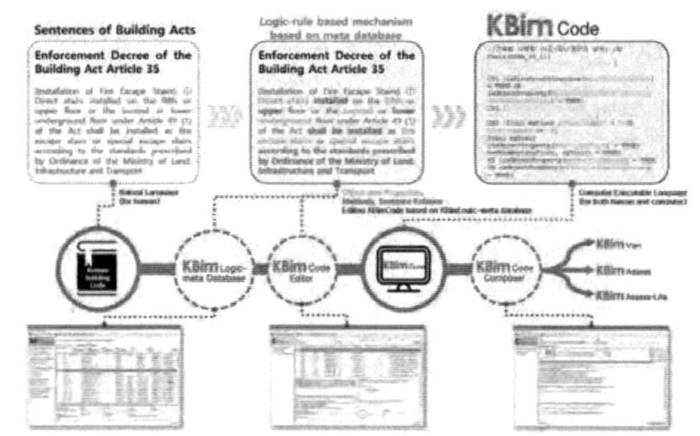

Figure 6. KBim Logic and Code

3.3 Submission phase

3.3.1 Automated submission system

The KBim Submission program facilitates the organization and management of building permits. It enables direct data linkage to SEUMTER (a Korean national e-submission system), therefore minimizing manual administrative work on project information throughout the design process. It revolutionizes the submission process by automating submission data entries in the SEUMTER system, providing an efficient management platform. Stepping away from traditional processes, 280 different categories of information (based on new builds) required for building permission can now be automated, avoiding inaccuracy from manual input and the inefficiency of manual data collection.

KBim Submission enables the extraction and evaluation of the data necessary for building permits directly from BIM models and fully supports open BIM. With IFC-based BIM models, users can manage the design-to-submission processes with ease due to the highly automated functionality of KBim Submission. The program also includes a standardized library of 399 types of materials for the systematic management of design drawings, specifications, and documentations applicable to building permit applications. Furthermore, the data submitted is automatically converted and linked to the SEUMTER internal database, eliminating the need for information re-entry.

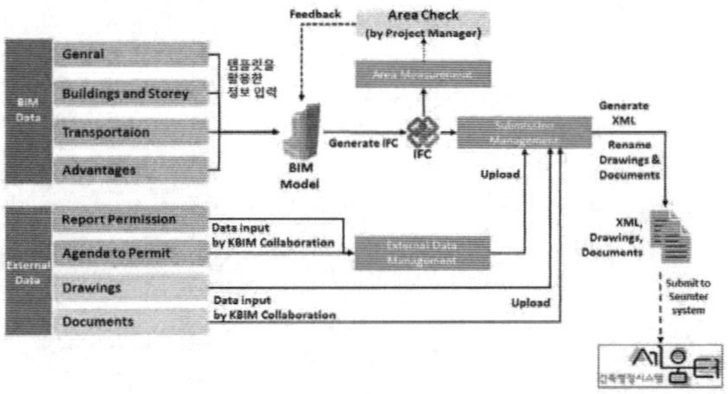

Figure 7. KBim Submission

3.3.2 2D drawing creation from IFC

The KBim D-Generator program automatically generates building permit submissions applicable to 2D drawings, such as plans, elevations, and sections, directly from the 3D IFC model. The generated drawings are interactive 2D drawings connected to the 3D model. Through the KBim D-Generator program, users can view both the 3D model and the generated drawings in parallel and access specific objects and information on the interactive drawings according to their needs.

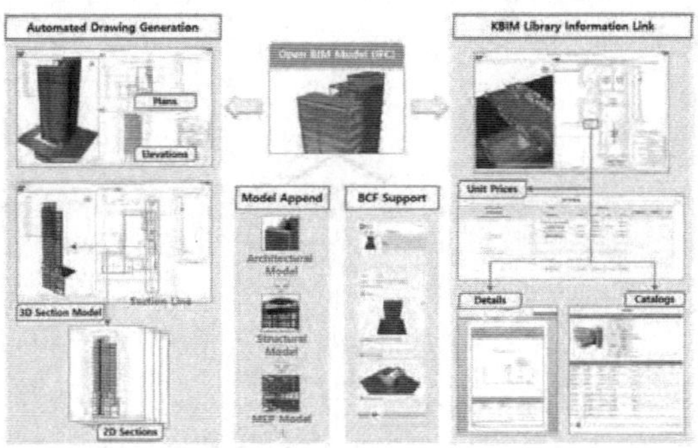

Figure 8. KBim D-Generator

3.4 Permission phase

KBim Assess is a rule-based BIM model that checks software for screening eligible BIM-based projects prior to project approvals from SEUMTER. SEUMTER is the Korea national construction administrative online service, which includes e-submissions.

KBim Assess rejects any faulty or defective models that contain errors or insufficient data for the SEUMTER approval process. This pre-checking procedure saves time and manual efforts effectively for governmental officers. Unlike KBim Assess-Lite, which can be executed as a stand-alone application, KBim Assess operations depend on the SEUMTER system and execute checking only for the models provided by the system. It performs checking with the latest versions of the Korean building regulations, codes a logic rule-based meta database, and delivers results directly back to the SEUMTER system.

Key functions include:

- IFC-based BIM model quality assessment by a rule-based checking methodology
- Checking results visualized in real-time 3D graphics
- Interoperability with the SEUMTER system on BIM model downloads, results uploads to the SEUMTER server, and project data transfer.
- Connects to the Korean building regulations and codes database and executes the latest versions.

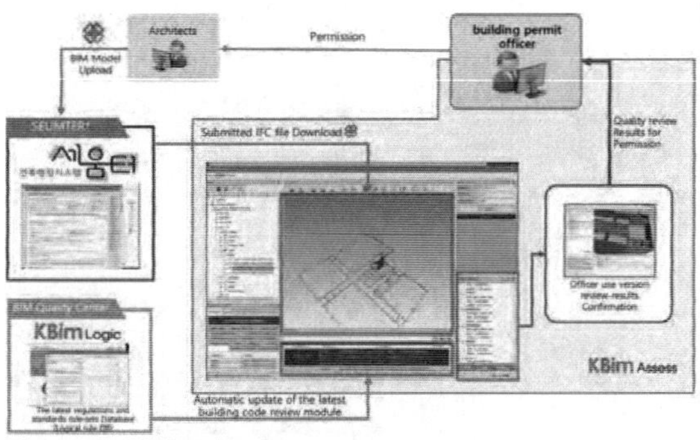

Figure 9. KBim Assess

4 Conclusion

OpenBIM-based automated code checking is recognized as the prominent future of BIM worldwide. It is expected to bring changes to the architecture and construction industry for the next 20 years. In Korea, in order to advance on the architectural and construction technology of the nation and strengthen the competitiveness of small and medium-sized design firms, government funded the BIM-related project [13]. This project develop applied technology for building permission documentation through standardized data format, submission format, criteria and requirements for the benefit of BIM data exchange; and to develop an integrated collaborative project administrative system.

Prior to the KBIM e-Submission development, building permission process of Seumter relied greatly on 2D drawings with around 150 pieces of documents required in total from architectural, structural, MEP and etc. KBIM e-Submission process reduces 1/3 of the time required to generate required documentations for submissions by automating the drawing generating process, pre-checking BIM models with the latest building codes, allowing a sustainable platform for collaboration, and automating the code compliance assessment processes.

Acknowledgements

This research was supported by a grant (17AUDP-B127891-01) from Architecture & Urban Development Research Program funded by Ministry of Land, Infrastructure and Transport of Korean government.

References

[1] Eastman, C., Teicholz, P., Sacks, R. and Liston, K. *BIM Handbook*, John Wiley & Sons, Hoboken, NJ, 2008.

[2] Choi, J., Choi, J. and Kim, I. Development of BIM-based evacuation regulation checking system for high-rise and complex buildings. *Automation in Construction*, 46:38-49, 2014.

[3] Choi, J. and Kim, I. Development of an Open BIM-based Legality System for Building Administration Permission Services. Journal of Asian Architecture and Building Engineering, 14(3):577-584, 2015.

[4] João P. and André M. Lica: A BIM based automated code-checking application for water distribution systems. *Automation in Construction*, 29:12-23, 2013.

[5] Sijie Z., Kristiina, S., Markku, K., Ilkka, E., Eastman, C. and Jochen, T. BIM-based fall hazard identification and prevention in construction safety planning, *Safety Science*, 72:31-45, 2015.

[6] Malsane, S. Matthews, J., Lockley, S., Love, P. and Greenwood, D. Development of an object model for automated compliance checking. *Automation in Construction*, 49:51-58, 2015.

[7] Kim, I., Choi, J. and Cho, G. Development of Rule-based Checking Modules for the Evacuation Regulations of Super-tall Buildings in Open BIM Environments. *Transactions of the Society of CAD/CAM Engineers*, 18(2):83-92, 2013.

[8] Choi, J. A Study on the Development of Code Checking System for Building Administration Process Applying Open BIM-based Process. *Master dissertation*, Kyung Hee University, 2014.

[9] Lee, H., Park, S., Kim, I. and Lee, J. A Logical Rule-based Approach to the Korea Architecture Code Sentences for BIM-enabled Design Assessment Systems. *Journal of Korea Design Knowledge*, 34:101-110, 2015.

[10] Liebich, T., Wix, J. and Qi, Z. Speeding-up the submission process, The Singapore e-Plan Checking project offers automatic plan checking based on IFC. In *INCITE2004*, pages 245-252, Malaysia, 2004.

[11] Conover, D. and Lee, E. SMARTcodes. *Journal of Building Information Modeling (JBIM)*, 22-23, 2008.

[12] Choi, J. and Lee, M. Findings from the Korea Government-funded BIM R&D Project. *The BIM*, 16:12-14, 2016.

[13] Kyung Hee University Consortium. Open BIM based Technical Environment for Building Design Quality Enhancement. On-line: http://dq.kbims.or.kr, Accessed: 30/03/2017.

Contribution to the study of collaborative working in BIM-based projects in the context of public works

Miceli Junior, G.,[a] Teixeira, A.C., [a] Nascimento, A.F. [a] and Pellanda, P.C. [a]

[a] Defense Engineering Graduate Program, Military Institute of Engineering
E-mail: gmicelijr@uol.com.br, acruz75@gmail.com, alexandrefitzner@gmail.com, pcpellanda@ieee.org

Abstract –

The great development of Information and Communication Technology (ICT) in recent years, especially network-based and cloud-based technologies, has led to a significant increase in efficiency of Building Information Modelling (BIM) platforms, which motivated a radical change in project managing in Architecture, Engineering and Construction (AEC) companies through implementation of collaborative working environments. However, the methodology used for implementing all the involved concepts has a direct influence on the results in terms of quality of AEC projects and efficiency in their development process. This paper focuses on the similarities and differences between two collaborative arrangements for BIM adoption in AEC projects, with a view to choosing best practices for civil construction works management and inspection offices in the context of public works in the Brazilian federal government. The first arrangement uses software from the same developer and proprietary formats, while the second uses software from different developers and formats suitable for interoperability between them. Based on the results, this work points out the limitations, advantages and disadvantages of using both arrangements, indicating research perspectives for industry and academia.

Keywords – Collaborative Design; Building Information Modelling; Public Works; Interoperability.

1 Introduction

Construction projects have become larger and more complicated in the first decade of this century. The amount of information required has been increasing, leading to greater complexity in the collaboration between the designers themselves and between them and various possible stakeholders.

Among several technologies appeared in the Architecture, Engineering and Construction (AEC) sector, certainly Building Information Modelling (BIM) is the most important. BIM is an emerging technology focused methodology that can be used to improve the performance and productivity of an asset´s design, construction, operation and maintenance process [1].

According to Succar, BIM is described as a digital representation of physical and functional characteristics of a facility [2]. Succar also describes BIM as a set of technologies, processes and policies enabling multiple stakeholders to collaboratively design, construct and operate facility in virtual space [3]. Eastman et al. describe BIM as a modelling technology and an associated set of processes to produce, communicate and analyse construction models [4].

This study focuses on the similarities and differences between two collaborative arrangements for the adoption of BIM in AEC projects, to choose best practices for civil construction works management and inspection office in the context of public works in the Brazilian federal government, notably Brazilian Army Regional Works Offices. Based on the results from two case studies, we aim to discuss about limitations, advantages and disadvantages of each arrangement, providing useful information for future implementation of collaborative BIM-based working environments.

Section 2 presents research background. Section 3 presents possible BIM collaborative arrangements used in two case studies. In Section 4, the results obtained in these case studies are discussed. Section 5 proposes possible applications in the context of public works in the Brazilian federal government. Finally, Section 6 concludes the paper.

2 Research Background

In this section, the main characteristics of collaboration, interoperability and application of BIM in the public sector are presented.

2.1 BIM in the public sector

Public sector plays an important role in leading the industry towards BIM adoption. BIM implementations continue to increase intensively as more and more government bodies and non-profit organizations of various countries worldwide implemented BIM in their projects and provided different BIM standards and solutions [5].

It becomes important in public sector to review and evaluate the current performance of all processes, to ensure the public sector obtaining a greater value for the money in their construction projects [6].

Many countries around the world have adopted BIM technology. The United States is believed to be one of the pioneering countries for BIM adoption. Many public-sector bodies at different levels in the United States have established BIM programs and set up BIM goals, implementation roadmaps and BIM standards [5].

Apart from the United States, many countries in Europe have embarked on significant BIM implementations, like Netherlands, Finland, Norway and United Kingdom. Although BIM adoption in the public sector came later in Asia, BIM has now developed rapidly in Asian regions, for example, in Singapore, Hong Kong, China and South Korea [4;5].

Succar [3] listed the publicly-available guides, reports and visions related to BIM in Australia, Denmark, Finland, Netherlands, Norway, the United States and a consortium of organizations in Europe.

Based on the review and comparison of BIM variables and implementations in different countries, six major roles of the public sector regarding BIM adoption were analyzed in [5], which are initiators and drivers, regulators, educators, funding agencies, demonstrators, and researchers, as illustrated in Figure 1.

Figure 1. Roles of the public sector for BIM adoption. (adapted from [5])

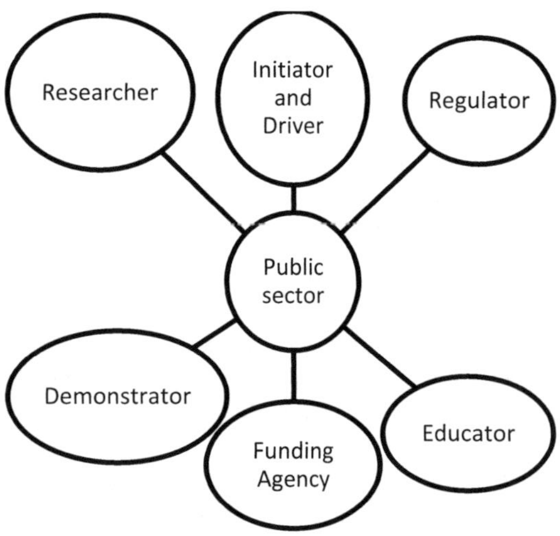

Unfortunately, BIM experiences in Brazil related to these roles of public sector have been sparse and not coordinated. Possibly the first state action with public results with respect to BIM projects took place in 2010, when an initial version of a BIM component library for state program "Minha Casa Minha Vida" was developed. Since its completion, this library has been largely distributed in Internet, being a reference for popular housing projects.

The first bidding process that refers to BIM in Brazil was also launched in 2010. Several constructions for the 2014 World Cup in Brazil and the 2016 Olympic and Paralympic Games in Rio de Janeiro were developed with BIM processes; however, it was an option of contractors themselves, and not a state imposition.

Perhaps the most successful BIM experience is that of the State of Santa Catarina, where the developers of two health institute projects were selected to be contracted in 2014/2015 by using technical evaluation of proposals with help of Solibri Model Checker [7].

In other public institutes, BIM has generally been used in common tasks like quantification take-off and geometric modelling of civil engineering projects [8].

In another point of view, Brazilian Army has improved real estate management quality by using the Unified System of Works Process (from the Portuguese "Sistema Unificado do Processo de Obras", whose acronym is OPUS), an information and communication technology system developed by its Directorate of Civil Works. Nascimento at al. describe OPUS as an integration of specialized Enterprise Resource Planning software with construction building information models, created for supporting the built environment lifecycle management including buildings and assets related to AEC [8,9].

OPUS was internationally recognized as unique in the world in BIM technology (BIM online collaboration, standards and protocols) for infrastructure and public sector [8].

2.2 Collaboration

The notion that building design is a multi-disciplinary process involving contributions from an increasingly broad range of specialists is well understood and generally accepted. Building design could be described as a cooperative process with brief episodes of collaboration where team members come together to resolve issues through negotiation and evaluation. Collaboration arrangement is traditionally used in AEC. It has revolved around the exchange of 2D drawings and documents. As Eastman et al. explained in their book [4], BIM technology makes easier the simultaneous work by multiple design disciplines. Although collaboration with drawing is possible and used, it is inherently more difficult and slower than working with one or more coordinated 3D models, in which change control can be well managed [10,11,12].

BIM-based collaboration project should be run with programs that provide functions necessary to complete the tasks in each of its phases [13].

Several countries around the world have established or are establishing rules and guidelines for the collaborative process. In the United Kingdom, BS 1192: 2007 establishes the methodology for managing the production, distribution and quality of information in construction, including those generated by CAD systems, using a disciplined process for collaboration in a specified policy [14].

To provide an efficient collaboration environment, the interoperability is an important point that must be solved in early design. This concept will be well described in the next topic.

2.3 Interoperability

No application can support alone all the tasks associated with building design. The interoperability represents the necessity of data exchange by the multiples types of applications, allowing multiples professionals to collaborate with the design. Data exchange between two applications is made by many ways. The principals are: direct link, used when the connection between two applications called from one or both application user´s interface; proprietary file exchange formats, when it is developed by a commercial organization for interfacing with that company´s application, like DXF (Data eXchange Format), by Autodesk Company; public product data model exchange formats, when an open-standard building model is used, e.g. IFC (Industry Foundation Classes) - developed model schema that is able to support a semantically rich representation of a building for use during the life cycle design and management of a project. Despite buildingSMART efforts, exchange of data via IFC still continues with data loss of geometry and information, even in its current version IFC4 [4,11,17].

The first collaborative arrangement presented in this paper is based on the first two ways of data exchange. The second one used IFC to exchange the data between the members of the team.

3 Collaborative arrangements

Two collaboration arrangements are presented for developing projects, with the aim of applying them to a management execution office of public works in the Brazilian Army: the first, using software from the same developer; the second, using software from different developers.

Each one was employed in a different educational project. While the former involved the application of BIM in the construction of a sports school, the latter involved the construction of an elementary sustainable school. Each case study involved the development of its own interoperability, collaboration, and production solutions within a team, which is presented in the next sections.

3.1 Case study 1: sports school

The first arrangement was based on Autodesk solutions contained in Design Suite 2016, which brings together the two most used tools in Brazil: AutoCAD and Revit. A summary description of the used suite products is described in Table 1.

Table 1. Summary description of the software used in case study 1 [4]

Software	Summary description
Autodesk AutoCAD	CAD-based software, used mainly for the elaboration of 2D/3D technical drawings. Autocad is considered as a support tool, not as a BIM solution itself.
Autodesk Revit	Solution developed specifically for BIM, allowing the development of models with features for modeling, quantitative surveying, camera generation, views, captions, tables and interactive tours.
Autodesk Navisworks	Solution that enables AEC professionals to review integrated models and data for better control over project results through interference checking, quantitative extraction, and creation of 4D analysis of the work.
Autodesk Robot Structural Analysis	Robot Structural Analysis Professional provides structural engineers with advanced analysis for large and complex structures.

The sports school project was developed by a group of eight professionals. There were one architect, four civil engineers - including the first author of this paper - and three senior students - an architecture and two civil engineering students. The tasks were divided among the architects and the engineers so that the architecture, structure and installations projects were then developed.

Each program had a specific purpose and usage by the task group, whether or not being part of Design Suite 2016, as shown in Table 2.

For the project development, the federated model was adopted, where several local or central files were created so that they could be linked to each other. In this way, all team members were able to work concurrently the different components of the same or different disciplines, after an initial meeting to discuss the concept of the project and the development of an initial model in Level of Development (LOD)100.

Table 2. Description of software used in case study 1.

Software used	Objective
Autodesk AutoCAD 2016	2D Specific details
Autodesk Revit 2016	3D BIM models
Autodesk Navisworks 2016	Clash detection, 4D BIM Model and quantification take-off
Autodesk Robot Structural Analysis 2016	Structural model analysis
Lumion 3D	Images and 3D videos creation
Adobe Photoshop	Image edition
Microsoft Project 2013	Schedule development
Microsoft Excel 2013	Auxiliary worksheet development

Within each central file, worksets were created in order to make ease the interface and compatibility between all discipline models, as shown in Figure 2.

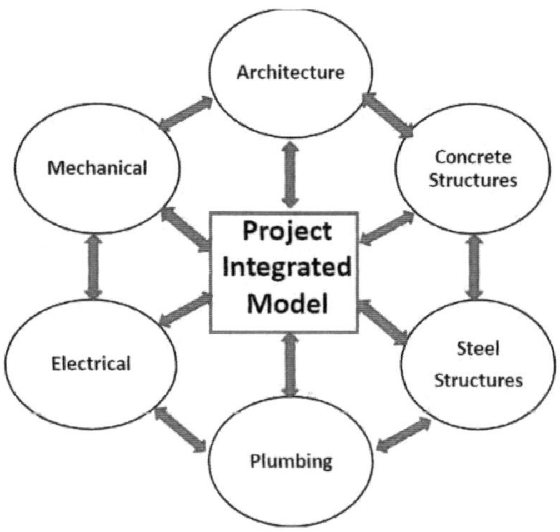

Figure 2. Information exchange between all project disciplines

Each central file was related to a design discipline: architecture, structural (steel and concrete), mechanical, electrical or building installations. As a draft project, all BIM components were modelled in LOD 200, except for a few MEP components available only on higher LOD at pages such as BIMObject.com and Autodesk Seek.

Other specific components developed specially for this work, like those related to gym and bodybuilding apparatus, were stored in a library located in a virtual work environment.

For specific situations, Robot structural analysis files and Navisworks clash detection reports were separated from other folders. All related files to AutoCAD drafts, Revit BIM models, Navisworks reports and Robot structural analysis were deposited in a Dropbox cloud environment, managed by the group coordinator.

Communication between individuals and software was done by direct links or proprietary file exchange formats whenever possible. Priority was given to direct links between Autodesk software that were already available, rather than using the IFC interoperability format, as shown in Figure 3.

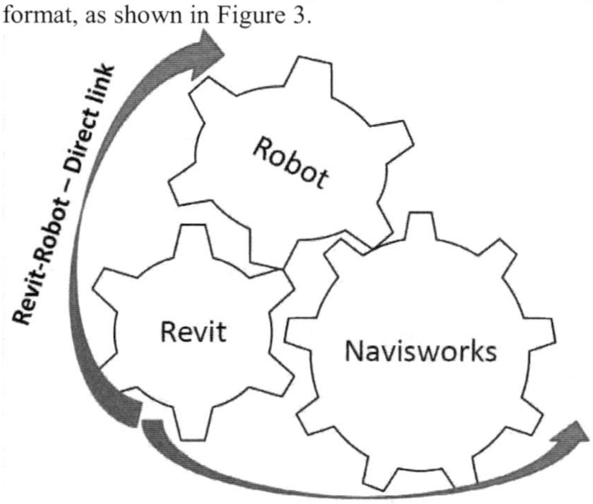

Figure 3. Information exchange inside structure discipline

Work meetings were held weekly and virtual. GoToMeeting application was successfully used for group discussion, project analysis, and action. For written communications about general project information and other questions, Trello application allowed tracking of all project discussions, by means of practitioners' computers or smart phones.

Visual interference control was done throughout the work inside or outside the meetings, either by the manager or by each member of the group.

Clash detection routines through Autodesk Navisworks were performed in rounds. In each round, all models passed through the tool in pairs, and clash reports were received by MEP engineer. According to the orientation of the group coordinator, the models were corrected by the corresponding members responsible for the involved disciplines, starting then a new round of clash detection routines, until full completion of all clashes.

3.2 Case study 2: sustainable school

The second arrangement - where the second author of this paper has worked - was made using several software solutions for AEC design, communication management, model validation and compatibilization. In contrast to Case Study 1, Case Study 2 used interoperability solutions, like IFC format. Table 3 shows all software used in Case Study 2 while Table 4 shows their purpose in this research.

Table 3 Software Use

Software	Use
Graphisoft ArchiCAD 19	Architecture Design
Autodesk Revit 2016	MEP and Structure Design
Autodesk Robot Structural Analysis 2016	Structure Analysis
Solibri Model Checker(SMC)	Model Validation and Compatibilization
Graphisoft BIMx	3D videos creation and presentation

Table 4 Software Feature

Software	Description
Graphisoft ArchiCAD 19	Oldest BIM tool on the market. Intuitive interface and the native library is wide
Solibri Model Checker (SMC)	The software is an IFC reader that performs interference checks and preloaded engineering rules. These rules can be customized by the user with logical instruction. It also has the functionality to coordinate projects, creating a communication log so that the project team can analyze and correct the interferences and inconsistencies found in the model.
Graphisoft BIMx	Software used for virtual walkthrough and video presentation of BIM models in Android operational system.

The aim was to design an Elementary Sustainable School collaboratively by a multidisciplinary team, according to BIM-process premises. The data exchange format used in this project was IFC while the communication between team members was the BIM Communication Format (BCF).

The workflow, shown in Figure 4, begun with a meeting to discuss the concept of the project, resulting in a necessity program. After this, the architecture team developed an initial model. When it achieved a level of maturity that allowed the work of other disciplines, the IFC file was made available. A cloud storage service was used to be the collaborative files repository.

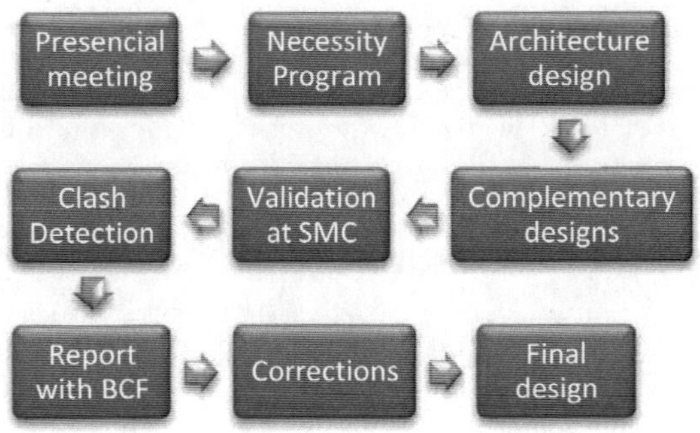

Figure 4 - Design workflow

From this point, the other disciplines developed their projects in their preferred software on the architectural basis while the architecture continued to evolve its LOD to perform compatibility checks.

By the time the complementary designs reached a degree of maturity that could be checked, they were submitted to a validation evaluation at SMC. This validation was related to interferences between objects of the same model, dimensions of environments and considerations about the necessity program and it was done by the designer himself.

Having in hands the designs validated by SMC, a meeting was scheduled to compatible them. All the models were assembled in a federated central model to perform a Clash Detection, as shown in Figure 5.

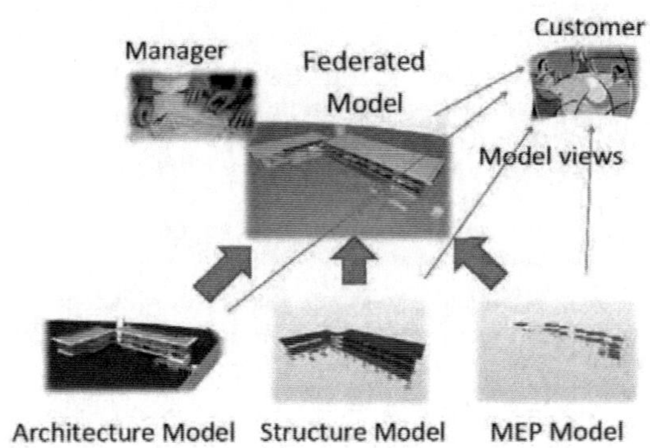

Figure 5 – Federated Model

The clashes were then analyzed by all members and

the responsibilities for their correction were defined, if necessary. To report what should be done and who should do it, BCF communication format was used. The BCF is part of SMC and can be viewed by anyone who has the BCF viewer installed. Once the corrections have been finalized, the central model is updated in SMC, as well as the information of the interferences in it.

The used model to work collaboratively is called Federate Model (Figure 5). In a simplified way, it means that there are distinct models, logically linked by the SMC. This implies that a change in one model does not automatically change the other models.

4 Results discussion

In both case studies, the design process has become more intense because of typical collaborative and interoperability features of BIM projects. The solutions in all disciplines were quickly discussed among the task group, during meetings or using network solutions like Trello or Netmeeting.

Designers and their experiences are highly valued in BIM processes because other technical and relational skills and abilities are required. Human relations between professionals have become even more important, requiring the project manager and team members to develop relationship skills.

Weaknesses in design and technical requirements have arisen in both case studies during the components modelling in Autodesk Revit or Graphisoft ArchiCAD, instead of appearing during a theoretical bidding or construction phase of the design.

Weekly, design clash detection rounds were performed by using SMC or Autodesk Navisworks, with the goal of further increasing project maturity, reducing the clashes to the lowest possible level in its final delivery, and enabling a feasible and consistent result.

Unfortunately, interoperability problems have become an important issue in project development, either with Autodesk direct links and proprietary file exchange formats or with IFC interoperability format; three of which are highlight in the following:

- Sometimes ArchiCAD IFC translator did not perform actual representation of some instances, losing some important information;
- Some components like lighting devices, mechanical equipment and ceiling elements were not recognized by modeling software. Modelers were usually asked to launch absent component again and thus continue the project immediately;
- In structural analysis, some nodes and elements were not recognized after having been transferred from Autodesk Revit to Autodesk Robot Structural Analysis. Structural engineers were then asked to review structural model and correct it again, element by element.

All team members have had their productivity increased, despite the problems described above. This may be related to the fact that most people involved have had little or no prior experience with any collaborative arrangement that uses interoperability or network solutions.

5 Application

5.1 Description

The aim of this section is to discuss applicability of one of the case studies presented in this work in civil construction works management and inspection offices of public works in Brazilian federal government.

Brazilian Army, through its Directorate of Civil Works (DCW), a technical and normative support agency, developed the OPUS system – a BIM cloud based platform already described in [9] – to overcome difficulties related to the lack of integrating tools and the existence of several databases that could not be updated systematically and equitably.

The arrival of OPUS enabled DCW to a better work and asset management. Military constructions being built in the Amazon or anywhere in Brazil can now be easily managed by DCW director or by his staff despite the diversity of imposed constraints related to legal, normative, technical, military, cultural and environmental aspects, many of them imposed by Brazil public procurement laws.

All military constructions are managed by 12 Regional Works Offices (RWO), where AEC works are projected and supervised by engineers and architects. However, in contrast to the successfully strategies for developing and adopting of such a city information modeling (or a building information management) tool like OPUS, most RWO are in pre-BIM level, if considered Succar BIM adoption matrix, whose levels of adoption are illustrated in Figure 6 [3]. This is evident because most projects are developed by using CAD-based software with DXF interoperability pattern. BIM proper usage is seen only in a few projects. Only two of twelve RWO have just reached Succar Level 1 of BIM adoption – even if it is used only on architectural models in an *ad hoc* way.

Additionally, there has been only one architectural template in development by DCW; however, it's rarely been used by all RWO. Structural and MEP models, clash detection routines and 4D modelling are still rare and uncommon.

Surveys performed by the first author of this paper in his PhD thesis pointed out the following complaints among engineers and architects working on RWO about

BIM programs:

- Use of BIM solutions – notably Autodesk Revit Architecture – is relatively limited in all RWO;
- There is few incentive in learning of BIM programs;
- Few AEC professionals can discern the difference between Revit and BIM or why it can be so important;
- Only a few can apply the benefits of BIM to RWO activities like design and construction supervision.

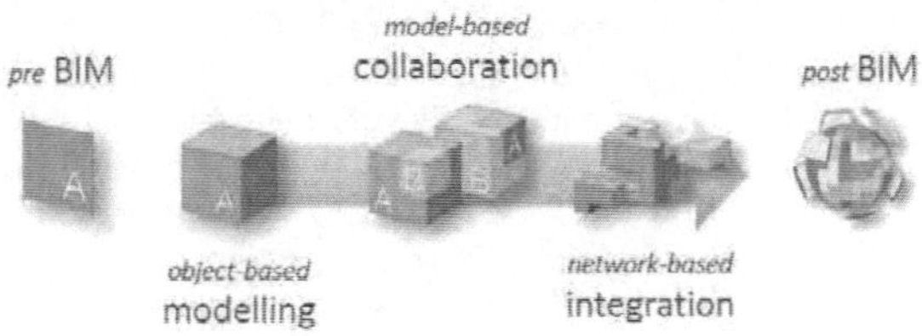

Figure 6 - BIM stages of adoption (adapted from [3])

Most of the constructions developed by DCW have got functional use and no more than three floors, with exception of construction of some residential buildings for military personnel. Their structure is generally of reinforced concrete and its MEP installations do not have great complexity.

Given the BIM conditions in RWO, applicability of case studies presented in Sections 3 and 4 will be discussed bellow.

5.2 Applicability of case studies

First, a solution or set of BIM solutions must be chosen to meet all needs. The two case studies have shown that any of them is suitable to meet their needs. However, it is emphasized that it is difficult to find solutions from only one software factory to meet all requirements, as described in Case Study 1, given the diversity of RWO activities.

Once the BIM software solution has been chosen, a set of defined workflows must be adopted, as also shown in the two case studies, both for the elaboration of the projects – internal standards for modeling – and for clash detection.

Modeling rules should be complemented by the definition of a BIM Execution Plan (BEP) that will be used as a subsidy for the preparation of projects in all RWO and for the contracting of BIM complementary projects.

About BEP, English standard PAS 1192: 2-2013 defines and splits it into two phases, where the first one refers to the implementation, goals and design milestones to be reached, while the second part refers to information about management, planning, BIM solutions and standard methods of BIM adoption [15,16].

BEP must contain an initial set of building component library for model execution and definition of interoperability protocols. It must also enforce all work groups – who work in project design, procurement or construction management – to follow its defined protocols.

The involvement, incentive and encouragement of all stakeholders are of central importance in this process, so that all RWOs have a homogeneous BIM adoption evolution.

Another important factor is the training program of professionals, which may involve basic and intermediate level training of BIM tools defined in BEP and/or management training within each workflow defined in the BEP, focusing on collaborative work practices.

Practical exercises should be done for training project teams. In this way, Eastman argues that it can be done first with a short pilot project, with qualified groups and with a clear goal aligned with each previously presented goal. This long-term training program would aim the achievement of Level 2 in terms of Succar index of BIM adoption [3].

Work can still be expanded to larger projects after having been reached BIM Level 2. More complex network and cloud environments can be used after diversified and trained project teams would be created. Given the unique nature of all project developed – Brazilian Army barracks and headquarters – proprietary cloud solutions like Collaboration for Revit (C4R) or Autodesk A360 must be avoided.

Quantity takeoff engineers will be able to participate in 4D/5D modeling with BIM software e.g. Autodesk Navisworks, VICO Office and/or a proper budgeting software. It is also important the definition of modeling guidelines for quantity takeoff-oriented BIM design, such as proposed in [18]. Hence, budgets and schedule activities would be made more accordingly to the services to be executed in each construction.

6 Conclusions

This study has focused on the similarities and differences between two collaborative arrangements for adoption of BIM in AEC projects. Two case studies have been presented with different interoperability arrangements. While in the first, Autodesk direct links or proprietary file exchange formats have been used, IFC format has been used in the second one.

The aim was to conclude about a future application of one of the case studies in inspection offices of Brazilian federal government, notably the Brazilian Army Directorate of Civil Works and all its Regional

Works Offices located throughout the Brazilian territory.

Although the initial comparison between the given case studies concluded that both could be used in any government office, it must be emphasized that any proprietary solution, with software from a specific developer, can meet all the needs from DCW. At some moment in the near future, IFC interoperability format will have to be used, despite all its limitations.

From performed surveys, one can conclude that the main difficulty for BIM adoption is not the proper BIM solution that will be employed. In general, professionals have got a limited point of view of what BIM really is or what it can do to make their job easier.

Then, the key word of BIM adoption is proper training and an appropriate definition of methods that will make professionals more confident about the outcome everyone wants and expects.

Given all these conditions, a training program and a BIM execution plan – as defined in Section 5 – must be the start point of an integrated BIM adoption in all RWO, instead of an *ad hoc* implementation under the initiative of a few professionals in one or two work offices.

The proficiency in usage of BIM software and the creation of proper network solutions can boost the development of computational tools to improve the existing OPUS system, integrating them with current management tools that already exist.

For that, a definition of a conceptual framework for the public management of military works, both for the management and design of projects, is highly recommended to achieve a better level of BIM adoption.

References

[1] Love, P.E.D. et al. From justification to evaluation: building information modelling for asset owners. *Automation in Construction* 35:208-216, 2013.

[2] Liu, H et al. BIM-based integrated approach for detailed construction scheduling under resource constraints. *Automation in Construction* 53:29-43, 2015.

[3] Succar, B. Building information modelling framework: A research and delivery foundation for industry stakeholders. *Automation in Construction* 18:257-275, 2009.

[4] Eastman, C. Teicholtz, P. Sacks, R. Liston, K. *BIM Handbook: A guide to building information modeling for owners, managers, designers, engineers, and contractors*. John Willey & Sons Inc. , New Jersey, 2008.

[5] Cheng, J.C.P. and Lu, Q. A review of the efforts and roles of the public sector for BIM adoption worldwide. *Journal of Information Technology in Construction*. 20: 442-478, 2015.

[6] Porwal, A. and Hewage, K.H. Building Information Modeling (BIM) partnering framework for public construction projects. *Automation in Construction* 31:204-214. 2013

[7] Cunha, J.M.F. and Silva, R.F.T. The experience of the State of Santa Catarina, Brazil, in bidding a BIM project. In *4th BIM International Conference*, pages 67-68, São Paulo, Brazil, 2016.

[8] Kassem, M. Mohamad Kassem: "Strategy for the diffusion of BIM in Brazil". On-line: http://www.makebim.com/2016/08/31/mohamad-kassem-strategy-for-the-diffusion-of-bim-in-brazil/?lang=en, Accessed: 27/03/2017.

[9] Nascimento, A.F. Miceli Jr, G. and Pellanda, P.C. Built environment lifecycle management by using large-scale BIM: a Brazilian Army case study. In *4th BIM International Conference*, pages 49-50, São Paulo, Brazil, 2016.

[10] Kvan, T., Collaborative design: what is it? *Automation in Construction* 9:409–415, 2000.

[11] Plume, J. and Mitchell, J. Collaborative design using a shared IFC building model—Learning from experience. *Automation in Construction* 16:28-36, 2007.

[12] Singh, V., Gu, N. and Wang, X. A theoretical framework of a BIM-based multi-disciplinary collaboration platform. *Automation in Construction* 20:134-144, 2011.

[13] Minho, O., Lee, J., Hong, S.W. and Jeong, Y. Integrated system for BIM-based collaborative design. *Automation in Construction* 58: 196-206, 2015.

[14] BS 1192:2007. *Collaborative production of architectural, engineering and construction information - Code of practice*. British Standards Institution, London, 2007

[15] PAS 1192-2:2013. *Specification for information management for the capital & delivery phase of construction projects using BIM.* British Standards Institution, London, 2013.

[16] PAS 1192-3:2014. *Specification management for the operational phase of assets using building information modelling. British Standards Institution,* London, 2014.

[17] Ryu, J., Lee, J. and Choi, J. Development of process for interoperability improvement of BIM data for free-form buildings design using the IFC standard. *International Journal of Software Engineering and Its Applications*, 10 (2):127-138, 2016.

[18] Monteiro, A. and Martins, J.P. A survey on modeling guidelines for quantity takeoff-oriented BIM-based design. *Automation in Construction* 35:238-253, 2013.

Improve Indoor Acoustics Performance by Using Building Information Modeling

Y. Tan[a], Y. Fang [b], T. Zhou [b], Q. Wang[a,c], and J.C.P. Cheng[a*]

[a]Department of Civil and Environmental Engineering, The Hong Kong University of Science and Technology, Hong Kong
[b]Department of Mechanical and Aerospace Engineering, The Hong Kong University of Science and Technology, Hong Kong
[c]Department of Civil and Environmental Engineering, Korea Advanced Institute of Science and Technology, South Korea

E-mail: ytanai@connect.ust.hk, yfangah@connect.ust.hk, tzhouag@connect.ust.hk, qwangau@connect.ust.hk, cejcheng@ust.hk

Abstract

Indoor acoustics is usually an important criterion of evaluating the performance of building rooms such as concert halls, lecture rooms, and wards. Indoor acoustic simulation during architecture design and before room decoration can improve room acoustic performance. However, 3D room modeling usually takes much time before acoustic simulation. Building information modeling (BIM) is currently widely applied through life cycle of a building because it dramatically improves modeling efficiency by providing both geometric and semantic information. This study aims at integrating BIM technology with acoustic simulation to improve indoor acoustic performance. A BIM model containing information of three main factors affecting indoor acoustic, namely architecture geometry, speaker position, and decoration materials, is built and slightly modified so that it can be used to conduct acoustic simulation. Acoustic simulations with different architecture geometries, speaker locations, and decoration materials were performed to study how indoor acoustic performance was affected by examining variation of reverberation time, which is an important indoor acoustic evaluation criteria. Sensitivity analysis of the three main factors that affect indoor acoustics was also conducted. Finally, the acoustic simulation results with BIM and without BIM were compared. It was found that the simulation results were the same and the acoustic simulation time was reduced by using BIM.

Keywords –
Architecture design, BIM, Decoration Materials, Indoor Acoustic Simulation, Speaker Position

1 Introduction

Acoustic performance is an important aspect that has to be carefully considered in the design and construction process of buildings, especially for buildings with specific acoustic requirements, such as lecture rooms or concert halls. Once a building is constructed, it would be very costly in both finance and manpower to fix inappropriate acoustic performance. Therefore, it is necessary to conduct acoustic simulation and prediction to help designers and engineers to analyze the acoustic performance of a building in the design stage.

To evaluate the acoustic performance of a building, a couple of parameters are commonly used, including reverberation time (RT), sound intensity level, noise, sound uniformity and intelligibility [41]. This study focuses on the reverberation time since it is one of the most important indicators and can be calculated by relatively simple equations. Reverberation time is defined as the time required for reflections of a direct sound to decay 60dB. The preferred reverberation time is dependent on the purpose of buildings. For classrooms, the recommended reverberation time is tenths of seconds [33], whereas for larger buildings such as concert halls, this value usually varies from 1 to 2 seconds [26,27].

Although many simulation methods are available, prediction of building acoustic performance is still a very complicated work. In real situations, reverberation time of a room is affected by many factors, including room size [4,23], enclosure shape [13], materials of walls and ceilings [32], absorption of seats and audiences [11,12,25]. In the acoustic design process, all those factors require careful considerations. In most cases, it is extremely hard to find proper values for all those parameters in a short time. Plenty of time is required in the process of trials and errors.

Building information modeling (BIM) has been commonly used in the architecture, engineering, and construction (AEC) industry. With BIM technology, an accurate virtual model of a building is constructed digitally [18]. A building information model contains precise geometry and relevant data needed to support through the lifecycle of the buildings. Many studies on BIM have been conducted at different building life cycle stages such as design [39], construction [24,29,30], operation and maintenance [6,7,36], and demolition [14,31], which have shown that BIM can provide geometric and semantic information of buildings accurately and effectively at any stage. Therefore, in this study, BIM is integrated with acoustic simulation tools to evaluate and improve the acoustic performance of concert halls.

This study aims at improving the acoustics performance of concert halls by integrating BIM with existing acoustic simulation software. The rest of this paper is organized as follows. Section 2 presents the literature review. The integration of BIM and acoustic simulation software is illustrated in Section 3. Section 4 uses an example to illustrate the proposed methodology. Discussions and conclusions are included in Section 5.

2 Literature Review

2.1 Reverberation Time Prediction

Prediction of acoustic performance of a building has been studied for over a hundred years. The most commonly used reverberation time prediction formulas were proposed by Sabine [34] and Eyring [19], respectively. Their equations were derived from different considerations but the same assumptions. In the following decades, based on their works, a series of improved methodologies were developed to solve problems in different cases [5,9,20]. After 1960s, introduction of computer modelling greatly increased the efficiency and availability of building acoustic simulation [35,38].

A couple of computer models have been developed for acoustic simulation, such as particle tracing models [40], rays tracing model [37], beam tracing model [21], and hybrid model [38]. Based on those simulation models, effects of different factors were investigated. Using the acoustic rays and particle tracing technology, Passero and Zannin [32] evaluated the acoustic quality of real open-plan office and improved the acoustic condition of the office by performing modifications on the office room. Their study concluded that divider panels between the work stations and ceilings with high sound-absorbing materials could ensure the acoustic quality. With ray-based programs, Bistafa and Bradley [9] predicted the reverberation time of a classroom model and analysed the effects of different construction materials. Vorländer [40] considered the uncertainties of computer acoustic simulations.

2.2 Industry Foundation Classes (IFC)

Industry Foundation Classes (IFC), which is defined in ISO/PAS 16739 [2], is currently the main neutral file format for data exchange among BIM software [1]. IFC is designed to represent as much building information as possible through the life cycle and to facilitate the data transfer among various BIM-supported software, thereby is used in this study. The latest version of IFC is IFC 4, which includes 766 entities [3]. Related information is stored in terms of entities such as the *IfcProject*, *IfcSite*, *IfcWall*, *IfcSpace*, etc. *IfcSpace* is usually used to represent an area or volume that provides a certain function within a building. *IfcSpace* has been studied to extract spaces from buildings to conduct indoor navigation [17], indoor path planning [28], emergency response [10], and indoor localization [15]. Since acoustic simulation requires the volume of the concert, *IfcSpace* is also studied in this paper.

2.3 BIM Applications on Acoustic Simulation

BIM has been commonly applied through the lifecycle of buildings. However, as for acoustic simulation, only a few studies are reported. For example, Deng et al. [16] integrated BIM and 3D geographic information system (GIS) to combine traffic noise evaluation in both outdoor and indoor environments in a single platform. A high level of detailed information from BIM such as interior walls and interior rooms was used in a 3D GIS platform. Wu and Clayton [41] presented a BIM-based software prototype that can speed the prediction of the acoustic performance of an indoor space at the building design stage. In this study, BIM was used to provide related building information to calculate the current acoustic performance of the indoor environment based on RT and sound intensity level (SIL). However, no comparisons among different designs of architecture geometry, decorated materials, and speaker locations have been conducted. Besides, the presented software prototype has to refer to multiple platforms. Therefore, BIM-based acoustic simulation studies to improve the indoor acoustic performance are still limited.

3 Methodology

In this study, a BIM-based methodology is proposed (see Figure 1) to improve indoor acoustic performance. First, an IFC file is exported from the concert BIM model. Then, information including architecture geometry, speaker location, and decorated materials is extracted from the IFC file and serves as the input to set up acoustic simulation environment. Finally, acoustic simulations with different architecture geometries, speaker locations, and decorated materials are conducted according to the RT of concert hall to improve its acoustic performance. More details of the methodology are introduced in the following sections.

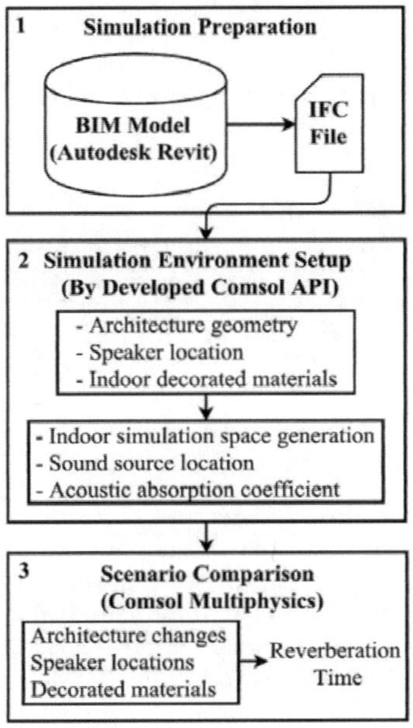

Figure 1. The proposed methodology.

3.1 Simulation Preparation

The BIM model of a building project is usually available at the early stage of its lifecycle in the AEC industry. Acoustic performance therefore can be evaluated and improved in an efficient manner by directly using the building information from BIM model. In this study, the 3D architectural model is built by Autodesk Revit, which is commonly used BIM software in the AEC industry. Through Revit, the building information can be easily extracted and exported into IFC format.

The IFC file, which is based on EXPRESS data structure, contains the geometric information, object location, and the types of all elements in a building. *IfcSapce* entity, as introduced in Section 2.2, is used to obtain the geometry of the concert hall. The local placement property of the speaker in the IFC file can provide its location with a coordinate (*x, y, z*). As for decorated materials, whose sound absorption coefficients are vital for acoustic simulation, they can be determined by the types of the wall and seats.

3.2 Simulation Environment Setup

Acoustic analysis is conducted by a commercial finite element solver COMSOL Multiphysics. To run an acoustic analysis, three parameters are needed to setup the simulation environment including indoor simulation space, sound source location, and acoustic absorption coefficient.

Firstly, necessary information for acoustic analysis is extracted from the IFC file through a developed COMSOL application programming interface (API), including architecture geometry, speaker location, and indoor decorated materials. The COMSOL API file is programmed based on Java. Through the API, all operations and manipulations performed in COMSOL Desktop can be reflected in programming lines. To run COMSOL simulation case, a Java class file (.class) is built through Jave IDE. Java Development Kit (JDK) and COMSOL server are used to compile the Jave-file that COMSOL generates.

Secondly, all the three parameters for acoustic analysis are generated from the information extracted from the IFC file. The architecture geometry and speaker location can be directly used in COMSOL as the indoor simulation space and sound source location, respectively. Furthermore, indoor decorated materials are mapped to the corresponding acoustic absorption coefficients through a developed plug-in.

3.3 Acoustic Simulation

A typical acoustic simulation can be divided into three parts: a sound source which is used to mimic realistic sound generation, the propagation media (air in most cases) with boundary settings in the simulated space and the 'microphone' which receives sound response information. The interested parameters and results can be obtained through post-processing of the three parts. Since RT is one of the critical factors for evaluating performance of a concert hall, it is studied in this project.

In this project, the reverberation time is calculated by the Eyring-Norris's equation (in SI units):

$$\mathrm{RT}_{60} = 0.161 \frac{V}{-S\log_{10}(1-\frac{A}{S})}, \qquad (1)$$

where V is the room volume, S is the total area of all interior surfaces in the room, A is the total sound absorption in the room, which can be expressed as:

$$\mathrm{A} = \sum a_i S_i \qquad (2)$$

where S_i is the area of each surface in the room, including the walls, doors, windows, floor and so on, and a_i is the corresponding absorption coefficient of each surface. For the same material, the absorption coefficient is dependent on sound wave frequency (as listed in Table 1). Hence, the reverberation time of a room will differ within different frequency ranges.

Table 1 Absorption coefficient of several building materials and sectional areas at different frequencies [8,22].

Materials	Absorption Coefficients					
	125 Hz	250 Hz	500 Hz	1000 Hz	2000 Hz	4000 Hz
Brick, unglazed, painted	0.01	0.01	0.02	0.02	0.02	0.03
Carpet, heavy, on concrete	0.02	0.063	0.14	0.37	0.60	0.65
Concrete block, painted	0.10	0.05	0.06	0.07	0.09	0.08
Wood floors	0.15	0.11	0.10	0.07	0.06	0.07
Seats area (Mitaka Hall)	0.44	0.52	0.52	0.41	0.41	0.37
Seats area (TOC Concert Hall)	0.41	0.36	0.37	0.33	0.25	0.21

Based on the three parts in acoustic simulation, the settings and study cases are listed as follows. Two point sources are set in simulations to mimic loudspeakers arranged on both lateral sides inside the hall. The sources will be fixed on the ceiling and moved from the first row to the last row of seating area so that the influence of sound source positions on RT can be evaluated.

As for the propagation media and boundary conditions, the whole space will be filled with the air which matches with realistic situations of all the existing concert halls and boundaries will be set based on the characteristics of interested cases, such as sound absorption coefficient settings of walls, floors, seats and so on. In this study, RT of concert hall with decoration materials with varied sound absorption capabilities will be studied and associating regularity will be concluded. It should be mentioned that the acoustic response will vary at different frequency ranges. Hence, the sound absorption coefficient of each area will be considered in a reasonable range which can cover acoustic response over broadband.

At the last, to evaluate the effects of geometry of concert hall, other similar halls with the same volume but different shapes will be simulated for comparison.

4 Illustrative Example

4.1 Introduction to Concert Model

In order to validate the proposed BIM-based methodology, a real concert hall in the Hong Kong University of Science and Technology (HKUST), which is a space with volume of about 2230 m^3, was used (see Figure 2). After necessary parameters were collected through a site visit, the BIM model of the concert hall was created in Autodesk Revit based on the geometric information of seating area, walls, floors, etc. (see Figure 2).

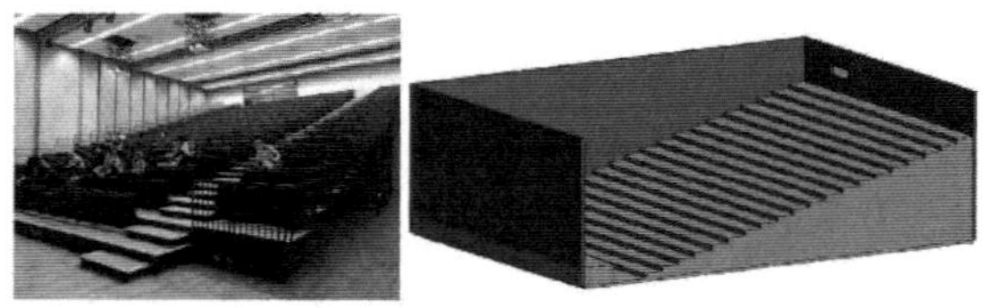

Figure 2. Example concert hall and its BIM model in Revit.

The initial model used in COMSOL simulations was a simplified prototype of the above-mentioned concert hall. The main characteristics including walls, floors, seating area and stage were extracted for improving acoustic performance based on the proposed methodology. To evaluate the influence of speaker locations and properties of decoration materials, a series of positions of sound source and varied reasonable materials of each component were considered. Besides, to obtain the effect of concert hall geometry on acoustic property, the original model was modified into fan-shaped with the same relative placements of stage and seating area, identical height in z direction and the same volume, as shown in Figure 3.

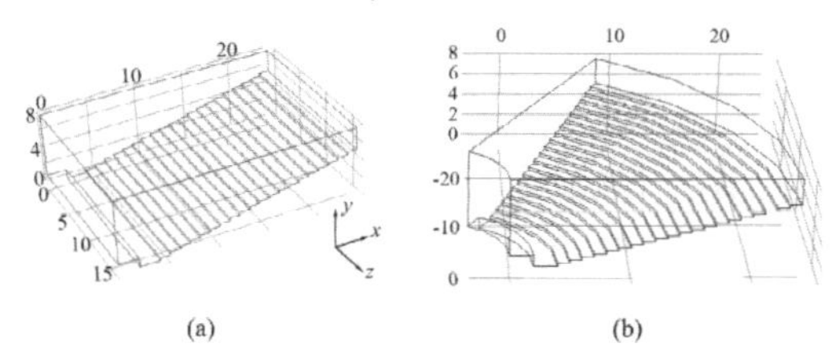

Figure 3. COMSOL models of concert halls. (a) Initial model from concert BIM model; (b) Modified model for study the effect of geometry

4.2 Simulation Results

4.2.1 Effects of Position of Sound Source

To study the influence of positions of sound source on RT, the sound absorption coefficients of the interested regions, i.e., floor, walls and seating area, are set as an initial case ($\alpha_{floor}=\alpha_{wall}=0.05$, $\alpha_{seats}=0.3$). It is assumed that the two loudspeakers hanged on the ceiling are moved backward synchronously. Figure 4 shows the simulation results, where the *x* axis indicates the distance between the front wall and the sound sources and the *y* axis represents the RT. The results show that the RT is almost the same with various positions of sound sources.

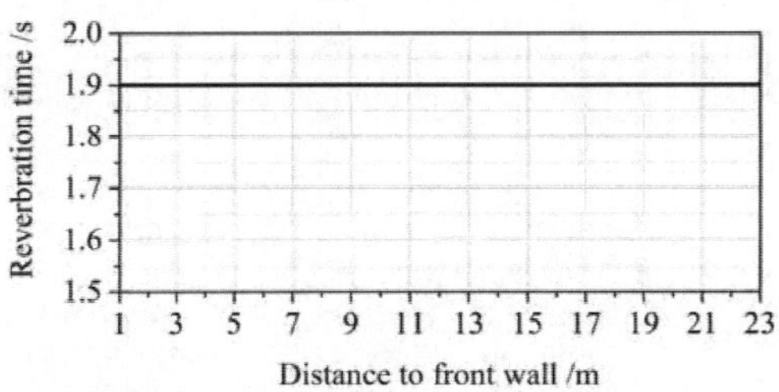

Figure 4. Reverberation time response of varied position of sound source.

4.2.2 Effects of Geometry of Concert Hall

To evaluate the effects of geometry of concert hall on RT, the original model is modified. The size of the cross-section in *x-y* plane remains the same and it is revolved so that the new concert hall has the same internal volume as the original one (2230 m^3), which is shown in Figure 3. Here, the sound absorption coefficients setting are the same as the initial case mentioned in Section 4.2.1 and the two speakers are hanged at the initial position (1 m from the front wall). For the original concert hall, the RT based on the above-mentioned settings is 1.9 s, however, this value changes to 1.61 s. It indicates that the change of concert hall geometry, which results in variation of the total sound absorption coefficient, also exerts important influence on RT and it should be designed carefully after determining the whole volume.

4.2.3 Effects of Sound Absorption Coefficients

The acoustic properties of decoration materials play an important role on performance of a concert hall. The sound absorption coefficients of materials have deciding effects on RT. In this study, the walls and seating area occupy the most area inside the concert hall. Hence, they are considered as two vital factors and studied in detail on the premise that the $\alpha_{floor}=0.05$ (initial case). The walls are always made of concrete or woods with varied thicknesses. Their sound absorption coefficients usually lie between almost 0.01 and 0.1. In terms of seats, they are usually made of porous material, which is soft and comfortable. The sound absorption coefficients can locate in a wider range, from almost 0.1 to about 0.5. Considering these two variables, the reverberation time of the simulation model shown in Figure 3(a) is evaluated. Figure 5-6 give the dependence of reverberation time on sound absorption coefficients of seating area and walls, respectively. It can be seen that as sound absorption coefficients of the walls and seating area increase, the reverberation times reduce.

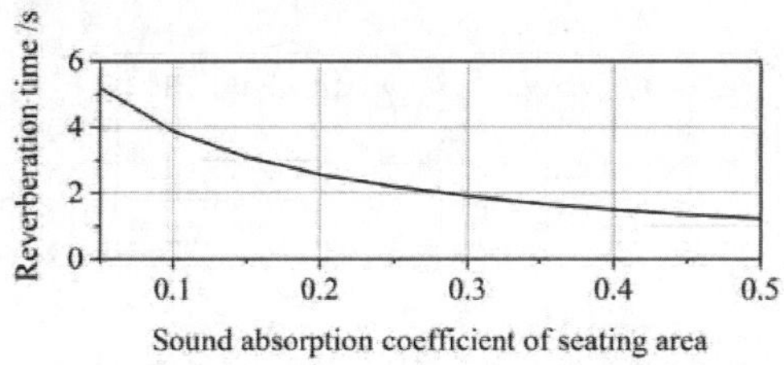

Figure 5. Reverberation time response of various sound absorption coefficients of seating area ($\alpha_{wall}=0.05$).

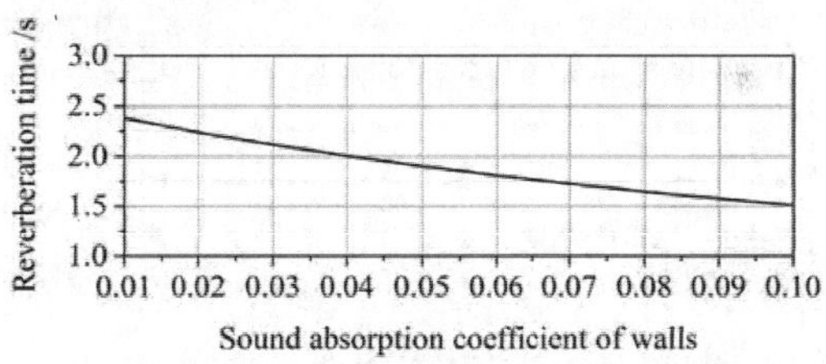

Figure 6. Reverberation time response of various sound absorption coefficients of walls ($\alpha_{seats}=0.3$).

Furthermore, the combination of materials of walls and seating area is simulated and optimized. As shown in Figure 6, the sound absorption coefficients of both walls and seating area are considered simultaneously. Given an optimization objective, such as a RT of 1.8 s, proper combinations of materials of walls and seating area can be easily obtained (dashed line shown in Figure 7).

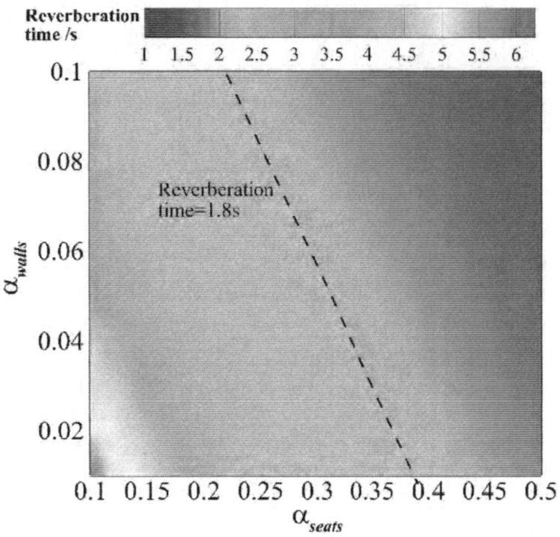

Figure 7. Reverberation time response of various sound absorption coefficients of walls and seating area.

In addition to conducting the acoustic simulation of concert hall by using BIM, the same acoustic simulations has also been conducted manually. It shows that the simulation results obtained from both methodologies are the same. However, acoustic simulation with BIM takes less time at the simulation environment setup stage. Therefore, using BIM to conduct acoustic simulation can help evaluate and improve the acoustic performance of concert hall in an efficient manner. Besides, with the proposed BIM-based methodology, the more complex the concert hall is, the more time can be saved for the acoustic simulation.

5 Discussions and Conclusions

This paper integrates BIM with existing acoustics simulation software to improve acoustic performance of concert halls by utilizing the rich building information from concert BIM model. API of the simulation software COMSOL Multiphysics is used to develop algorithm that can extract building information directly from IFC file, which is exported from Autodesk Revit, and to set up the acoustic simulation environment automatically. This developed methodology reduces the time of simulation environment setup. Scenario comparisons has also been conducted to provide guidance for the decisions on architecture geometry, speaker location, and decorated materials. To validate the proposed BIM-based methodology, the example concert model was created to set up the simulation environment. Simulation comparisons show that the same test results can be obtained, while the simulation environment setup time was increased. In addition, the simulation environment setup time saved by the proposed BIM-based methodology increases with the complexity of concert halls.

However, there are still some limitations in this study. The building information extraction algorithm works well only on regular architecture geometry. To make the proposed methodology more general, extraction algorithm should be furtherly improved. In addition, the developed function of decorated materials-sound absorption coefficient mapping only contains a few commonly used decorated materials. More potential materials should be included in the future work.

References

[1] Industry Foundation Classes buildingSMART International Ltd., 2016.

[2] Industry Foundation Classes Release 4, Model Support Group of buildingSMART International Ltd.

[3] ISO/PAS 16739:2005, ISO, 2005, pp. Industry Foundation Classes, Release 2x, Platform Specification (IFC2x Platform).

[4] C. Andree, The effect of position on the absorption of materials for the case of a cubical room, J Acoust Soc Am 3 (4) (1932) 535-551.

[5] H. Arau-Puchades, An improved reverberation formula, Acta Acustica united with Acustica 65 (4) (1988) 163-180.

[6] Y. Arayici, T. Onyenobi, C. Egbu, Building information modelling (BIM) for facilities management (FM): The MediaCity case study approach, International Journal of 3D Information Modelling 1 (1) (2012) 55-73.

[7] B. Becerik-Gerber, F. Jazizadeh, N. Li, G. Calis, Application areas and data requirements for BIM-enabled facilities management, Journal of construction engineering and management 138 (3) (2011) 431-442.

[8] L.L. Beranek, T. Hidaka, Sound absorption in concert halls by seats, occupied and unoccupied, and by the hall's interior surfaces, J Acoust Soc Am 104 (6) (1998) 3169-3177.

[9] S.R. Bistafa, J.S. Bradley, Predicting reverberation times in a simulated classroom, J Acoust Soc Am 108 (4) (2000) 1721-1731.

[10] P. Boguslawski, L. Mahdjoubi, V. Zverovich, F. Fadli, H. Barki, BIM-GIS modelling in support of emergency response applications, Building Information Modelling (BIM) in Design, Construction and Operations 149 (2015) 381.

[11] J. Bradley, The sound absorption of occupied auditorium seating, J Acoust Soc Am 99 (2) (1996) 990-995.

[12] J.S. Bradley, Predicting theater chair absorption from reverberation chamber measurements, J Acoust Soc Am 91 (3) (1992) 1514-1524.

[13] M. Carroll, C. Chien, Decay of reverberant sound in a spherical enclosure, J Acoust Soc Am 62 (6) (1977) 1442-1446.

[14] J.C.P. Cheng, L.Y.H. Ma, A BIM-based system for demolition and renovation waste estimation and planning, Waste Management 33 (6) (2013) 1539-1551.

[15] A. Costin, J. Teizer, Utilizing bim for real-time visualization and indoor localization of resources, Computing in Civil and Building Engineering (2014), 2014, pp. 649-656.

[16] Y. Deng, J.C.P. Cheng, C. Anumba, A framework for 3D traffic noise mapping using data from BIM and GIS integration, Structure and Infrastructure Engineering 12 (10) (2016) 1267-1280.

[17] A.A. Diakité, S. Zlatanova, Extraction of the 3D Free Space from Building Models for Indoor Navigation, ISPRS Annals of Photogrammetry, Remote Sensing and Spatial Information Sciences (2016) 241-248.

[18] C. Eastman, C.M. Eastman, P. Teicholz, R. Sacks, K. Liston, BIM handbook: A guide to building information modeling for owners, managers, designers, engineers and contractors, John Wiley & Sons, 2011.

[19] C.F. Eyring, Methods of calculating the average coefficient of sound absorption, J Acoust Soc Am 4 (3) (1933) 178-192.

[20] D. Fitzroy, Reverberation formula which seems to be more accurate with nonuniform distribution of absorption, J Acoust Soc Am 31 (7) (1959) 893-897.

[21] T. Funkhouser, I. Carlbom, G. Elko, G. Pingali, M. Sondhi, J. West, A beam tracing approach to acoustic modeling for interactive virtual environments, Proceedings of the 25th annual conference on Computer graphics and interactive techniques, ACM, 1998, pp. 21-32.

[22] R.A. Hedeen, Compendium of materials for noise control, Department of Health, Education, and Welfare, Public Health Service, Center for Disease Control, National Institute for Occupational Safety and Health, Robert A. Taft Laboratories, 1980.

[23] W. Joyce, Exact effect of surface roughness on the reverberation time of a uniformly absorbing spherical enclosure, J Acoust Soc Am 64 (5) (1978) 1429-1436.

[24] M. Kassem, N. Dawood, R. Chavada, Construction workspace management within an Industry Foundation Class-Compliant 4D tool, Automation in construction 52 (2015) 42-58.

[25] D.L. Kirkegaard, Sound absorption of occupied chairs as a function of chair design and audience clothing, J Acoust Soc Am 99 (4) (1996) 2458-2500.

[26] Z. Kui-sheng, Heritage and development of architectural acoustics design for concert hall, Technical Acoustics 1 (2011) 006.

[27] H. Kuttruff, Room acoustics, Crc Press, 2016.

[28] Y.-H. Lin, Y.-S. Liu, G. Gao, X.-G. Han, C.-Y. Lai, M. Gu, The IFC-based path planning for 3D indoor spaces, Advanced Engineering Informatics 27 (2) (2013) 189-205.

[29] Q. Lu, J. Won, J.C.P. Cheng, A financial decision making framework for construction projects based on 5D Building Information Modeling (BIM), International Journal of Project Management 34 (1) (2016) 3-21.

[30] D. Mitchell, 5D BIM: Creating cost certainty and better buildings, Proceedings of 2012 RICS Cobra Conference, 2012.

[31] J.W. Park, G.W. Cha, W.H. Hong, H.C. Seo, A Study on the Establishment of Demolition Waste DB System by BIM-Based Building Materials, Applied Mechanics and Materials, Vol. 522, Trans Tech Publ, 2014, pp. 806-810.

[32] C.R.M. Passero, P.H.T. Zannin, Acoustic evaluation and adjustment of an open-plan office through architectural design and noise control, Applied Ergonomics 43 (6) (2012) 1066-1071.

[33] R. Plomp, H. Steeneken, T. Houtgast, Predicting speech intelligibility in rooms from the modulation transfer function. II. Mirror image computer model applied to rectangular rooms, Acta Acustica united with Acustica 46 (1) (1980) 73-81.

[34] W.C. Sabine, M.D. Egan, Collected papers on acoustics, J Acoust Soc Am 95 (6) (1994) 3679-3680.

[35] M. Schroeder, B. Atal, C. Bird, Digital computers in room acoustics, Proc. 4th ICA, Copenhagen M 21 (1962).

[36] P. Teicholz, BIM for facility managers, John Wiley & Sons, 2013.

[37] J.-P. Vian, D. Van Maercke, Calculation of the room impulse response using a ray-tracing method, Proc. ICA symposium on acoustics and theatre planning for the performing arts, Vancouver, Canada, 1986, pp. 74-78.

[38] M. Völander, Performance of computer simulations for architectural acoustics, Proceedings of 20 th International Congress on Acoustics, ICA, 2010.

[39] R. Volk, J. Stengel, F. Schultmann, Building Information Modeling (BIM) for existing buildings — Literature review and future needs, Automation in construction 38 (0) (2014) 109-127.

[40] M. Vorländer, Simulation of the transient and steady - state sound propagation in rooms using a new combined ray - tracing/image - source algorithm, J Acoust Soc Am 86 (1) (1989) 172-178.

[41] C. Wu, M. Clayton, BIM-based acoustic simulation Framework, 30th CIB W78 International Conference, 2013.

Smoothing Process of Developing the Construction MEP BIM Model - A Case Study of the Fire-Fighting System

C.W. Feng [a] and W.T. Lin [b]

Department of Civil Engineering, National Cheng Kung University, Taiwan
E-mail: [a] cfeng@mail.ncku.edu.tw, [b] N66021464@ncku.edu.tw

Abstract –

Conflict analysis is one of the most recognized benefits of employing Building Information Modeling (BIM) to the mechanical, electrical and plumbing (MEP) system, especially at the design stage. However, at the construction stage, the MEP professional still strives to install the MEP system only according to the BIM model due to insufficient information. This is because the BIM model of the MEP system is not well developed from the information requirements of the construction professional. Furthermore, it may also take too much effort to develop an information- sufficient BIM model because of the complexity involved. Therefore, there is a need of taking the more effective and efficient approach that can quickly develop the MEP BIM model from the construction perspective.

This proposed study takes the fire-fighting system as the example to smooth the processes of developing the constructive MEP BIM model. First, a review of the construction specification and an interview with fire-fighting construction professionals are conducted to identify the information requirements. Then an ontology that employs the above result to determine the rules of building the information-sufficient BIM model is developed. Furthermore, a series of Dynamo modules are established to smooth the process of developing the BIM model for the fire-fighting system.

Keywords –

BIM; Dynamo; Ontology; Material Management

1 Introduction

In order to meet the safety and requirement of modern buildings, various functions of the equipment and the pipeline need to be installed in limited space. Due to the space constraints, it's not easy to integrate the MEP items. In addition, the traditional lofting method is not accurate in piping engineering, so the pipeline installation often needs to modify the pipe's length. Conflict analysis is one of the most recognized benefits of employing Building Information Modeling (BIM) to the mechanical, electrical and plumbing (MEP) system, especially at the design stage.

At the construction site, even well-experienced MEP professionals are still unable to avoid mistakes, which results in secondary construction in piping engineering. This is not only time-consuming and laborious, but also difficult to control quality of piping installation. In order to avoid the lack of information, the BIM model of MEP system needs to be developed from the requirements of construction professional. Wu et al. [1] proposed the BIM-based material management framework. He built a database to manage informationized BIM elements, so the material management, inventory, and replenishment management can be implemented in an efficient way. Wu et al. [2] proposed the information flow on various types of components in MEP system. The message providers and receivers were identified in each information flow chart.

This paper takes the fire-fighting system as an example, and it's organized as follows. First, a review of the studies on fire-fighting system, material management and ontology is presented. Next, the methods and the tools which are used in this research are explained. Then, the automatic modeling module is developed to create MEP elements in the model by Dynamo. Furthermore, the advantage and disadvantage of traditional method and using Dynamo were discussed. Finally, the limitations and the contribution of this paper is presented.

2 Literature Review

At present, there is not a detailed schedule in mechanical and electrical construction. MEP subcontractors usually know their work when they arrive at the site. Another problem is that MEP subcontractors could have hard time to keep up with the schedule if other structural-related activities are delayed. It means the waste on time and cost, and the poor quality. Moreover, wages account for 40% and material costs account for 20% of the total cost in the electrical and mechanical engineering. Therefore, proper planning of material management can effectively improve the duration, cost, and quality. Also, the supplier can offer the materials timely, which makes the construction process smoothly [3].

2.1 Fire-Fighting System

Fire system can be divided into four categories— alarm, fire, smoke, and evacuation. Alarm category includes fire automatic alarm system, and fire detection system. Fire category includes automatic sprinkler system, foam extinguishing system, and fire hydrant system. Smoke category includes smoke protection equipment. As for evacuation category, it includes fire emergency illuminating light, emergency fire broadcast system and fire extinguishers [4].

In general construction, automatic sprinkler system, foam extinguishing system, and fire hydrant system are classified into water supply and drainage engineering. Automatic fire alarm system, fire detection system, fire emergency illuminating light and emergency fire broadcast system are classified into electrical engineering [5].

2.2 Application of BIM in MEP Material Management

Lee et al. [6] proposed the significance of enterprise management— the purchasing department of the company is responsible for purchasing quality and price control, and the construction site is responsible for the acceptance of quality and quantity control. Wu et al. [1] proposed four factors of material management— material trends, storage management, stock control and replenishment management. He also used IDEF0 to analyze the information needs and information flow at each stage.

In order for the BIM model component attributes to meet the material management information requirements, Lee et al. [7] divided mechanical and electrical system into three levels— system, circuit, component. He also established BIM extension properties that meet the construction requirements (Estimation, Procurement, Allocation, Priority, and Testing) in pipe, wire, fitting, support, equipment and panel board. So, an information framework of MEP model was proposed.

2.3 Ontology

Ontology is a philosophical term. It's a research that explores the nature of life or the existence of biology [8]. Ontology defines a collection of representational primitives, and establishes a field of knowledge or a model of the paper [9]. The concept of ontology is to convey knowledge and experience in a simple but systematic and logical way. By the cause of fault analysis and semantic decomposition, the smallest unit in the lowest level of the class can be obtained. In this study, Facility Product Model and Maintenance Model were created, object attributes and associations were clarified by semantic analysis. As a result, different categories can be connected according to attribute and association [10].

Following are the steps of ontology building [11]:

1. Determine the domain and scope of established ontology.
2. Consider using existing ontologies.
3. List important words in ontology.
4. Define classes and levels of the classes.
5. Define the property of category (Property or Slot).
6. Defines the direction of attribute
7. Create instance(Individual).

2.4 Literature Summary

In the past, material management which applied to BIM was developed to manage the building materials. With the improvement of MEP modeling, more and more people are engaged in MEP material management which applied to BIM. According to MEP engineering characteristics, Lee et al. [7] classified the components and defined the attributes required for MEP construction. Wu et al. [1] developed the material management database framework as a reference for another field of material management database which applied to BIM. Noy et al. [11] provided the steps to create an ontology model. Wang et al. [10] defined ontology concept and constructed an ontology model based on facility maintenance.

Although, a lot of model information needed to be expanded or customized in the past literature, the information is not easy to be created in the model. In other words, if there is no effective way to handle these processes, it's hard to match model information with practice.

3 The Framework of Developing the Construction MEP BIM Model

In this paper, a framework was established to smooth the process of developing the construction MEP model. First of all, fire-fighting engineers are interviewed to identify the construction method of fire-fighting equipment and materials supply in the construction site. In addition, 2D document such as CAD drawings, was used to identify the material requirements. Apart from this, the specification of materials needs to be confirmed on the basis of

"Standard for Installation of Fire Safety Equipment".

In order to interpret the knowledge of material management obtained from the interview with fire-fighting engineer, material supply process was analyzed and developed by IDEF0 method. Thus, the material management requirement was identified in each phase.

Then, a fire-fighting engineering ontology model was created to integrate the material management requirement and BIM elements. So the rule for establishing BIM model can be defined from the ontology. Moreover, the knowledge-based model can be modified and used.

Finally, Autodesk Revit 2017 and plug-in software, Dynamo, were used for modeling and giving information automatically. As a result, the list of BIM elements can be imported into Access to create a project database which can provide construction information. Figure 1 shows the framework of this paper.

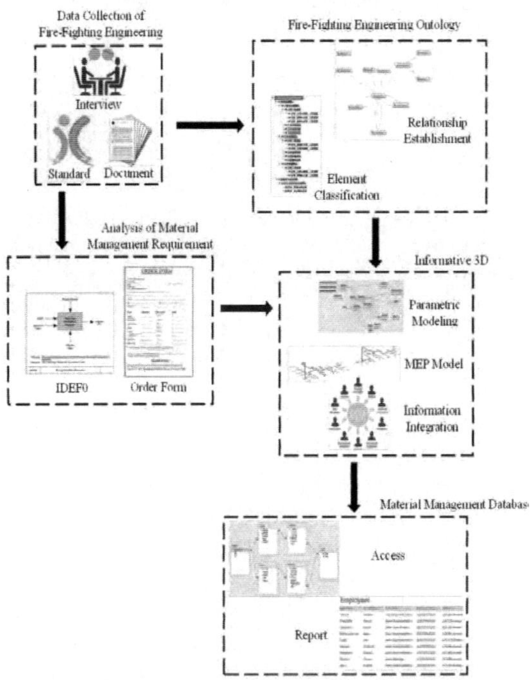

Figure 1. Framework of material management

3.1 Analyze the Process of Material Management and Information Requirements Applied to BIM

In order for the BIM model to meet material management requirements, it's important that material information need to be considered in modeling process. Material management process is divided into five stages— procurement, delivery, warehouse, construction and track, as shown in Table 1.

Table 1. Work contents in material management

Stage of Material Management	Contents
Modeling	In modeling step, BIM model was created with correct information according to the usage purpose (Component name, specification, type, function, delivery date, warehouse location, construction date, construction location, etc.). Also, these systems and circuits in the MEP model need to be defined according to their function.
Procurement	The component database and material requirement plan (MRP) were created on the basis of the information from the model. Then, the material can be ordered with the correct specifications, types and quantity of the equipment components on a correct date.
Delivery	Check the material quality and equipment function according to specification, and reject failed items.
Warehouse	Determine the most suitable storage location and supply path, considering the restrictions of construction path, material timeliness and storage method.
Construction	Supply the correct construction materials timely according to construction sequence and method.
Track	Material management system was established to control stock and replenishment. Then the results were feedbacked to the procurement stage to improve material requirements plan.

This study proposes material management process and information requirements which applied to BIM. Further, IDEF0 method was used to analyze the items and participants of each stage from the procurement stage. Then, material information which needs to be given or expanded on BIM model can be obtained. Figure 2 shows the material management process.

3.1.1 Model Informationization Stage

Model informationization process is divided into two steps: drawing modeling and informationization of model components. The input from the left side is 2D drawing, a 3D model can be created through the modeling program with full discussion and drawing explanation by modeling staff, designers, and MEP professional contractor. According to the purposes and requirements, parameters and information were given to the elements which were built in BIM model. As a result, informative elements list which was exported from the model at A1 stage was generated.

3.1.2 Purchase Informationization Stage

Procurement informationization stage can be divided into the establishment of project component database and the material requirements plan. The database which includes material information from the BIM elements list was created. In addition, the supplier's ability to provide enough materials needs to be confirmed for the project according to the order form and vendor catalog. Then, material requirement plan was developed according to the construction sequence, progress, and logic. Lastly, project order information can be generated.

3.1.3 In-Site Material Information Management Stage

After the materials were sent to the construction site, the material manager needs to confirm the quantity, specifications, and quality of the material. They need to reject the failed product according to the inspection process and purchase records in delivery stage. Then MEP professional contractor follows the piping logic and equipment functions to sort the materials according to different systems or circuits. Material manager and warehouse manager have to coordinate and determine whether it is necessary to store based on material properties and space constraints. If the materials need to be stored, the storage location should be determined according to the material timeliness and construction sequence. Storage methods and construction path restrictions also need to be considered. Finally, the material configuration information can be obtained from requisition for materials after construction. After material manager and MEP professional contractor confirm the correctness of material configuration information according to the construction sequence and drawings, the material usage information can be generated.

3.1.4 Stock Control and Replenishment Management Stage

In A4 stage, it's divided into database update, materials quantity check, and the stock analysis. First, the construction progress was updated in the database from the material usage information and material stock in the construction site. Additionally, the material stock need to be inspected at the construction site by comparing materials that have already arrived and the ones that have been used. After warehouse staff confirm the inventory, the progress and construction sequence can be used to analyze the current stock and determine whether the order needs to be added or reduced.

3.1.5 Information Feedback Stage

The productivity of MEP professional contractor can be analyzed according to the comparison between scheduled progress and actual progress in information feedback stage. Then, the material requirement plan can be improved by the productivity of contractor and replenishment requirement. As a result, the optimal order which can meet the real productivity was generated.

3.1.6 Information Requirements

Through the analysis of material management process above, the information requirements are summarized in Table 2.

3.2 Ontology for Material Management of Fire-Fighting System in BIM

In order to confirm BIM model meet the requirements of the order, in this study, a fire-fighting engineering ontology was not only used to integrate the material management requirement and BIM elements, but also used to establish the interrelationships among various elements.

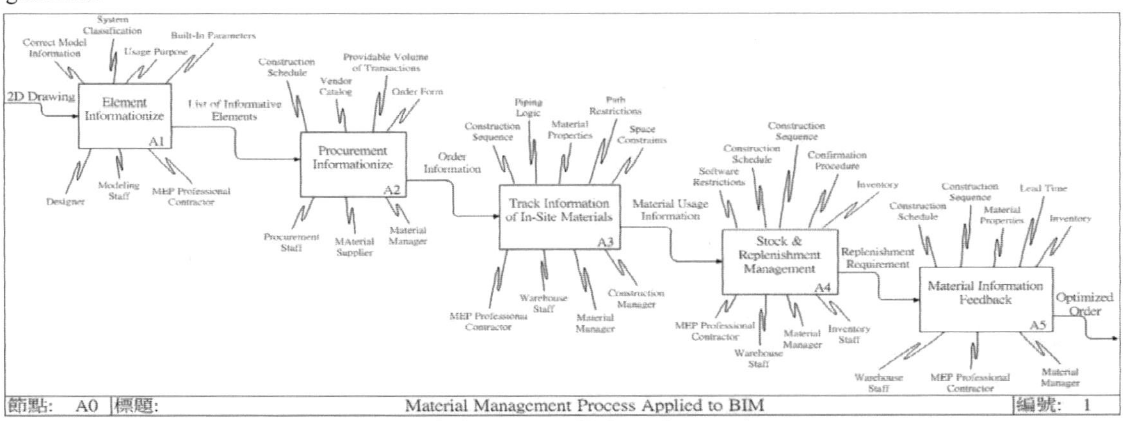

Figure 2. Analysis process of information requirement

Table 2. Information requirement in each item

Stage	Items	Information Requirements
Procurement	Confirm the material information	Material name, quantity, size, type
	Confirm the form of material order	Order form
	Identify supplier capability	Identify capacity of supplier
	Build the component database	Integrate the materials and required information for the project
	Set up a material requirements plan	Expected construction schedule, material lead time
Delivery	Confirm material entry	Entry sequence, space planning, transportation equipment and methods
	Inspect material	Inspection process, methods and standards
	Build purchase records	Integrate the information of delivered material
Warehouse	Classify system and circuit	Construction sequence, equipment function
	Analyze the path	Restrictions of the storage area
	Analyze the storage space	Material type, volume, stock quantity, storage method
	Integrate material configuration information	Material storage location
Construction	Establish the information requirement for construction	Pre-item, construction drawing and sequence, material requirement
	Analyze the supply path	Path restrictions of storage area and construction area
	Create material usage information	List of requisition for materials
	Update the component database	Construction schedule, material stock
Track	Check the quantity of storage materials	Material inventory list
	Analyze replenish requirements	Future construction schedule, material stock
	Analyze the productivity	Scheduled construction progress and actual construction progress
	Optimize future orders	Construction sequence and schedule, lead time, material stock, professional contractor's actual productivity

3.2.1 BIM Elements

To fully describe the relationship between the components of fire-fighting system, in this paper, the fire-fighting system is divided into five categories—"Pipe", "Equipment", "Machine", "Fitting" and "Valve".

1. "Pipe" includes standpipe (riser), main-pipe, and branch.
2. "Equipment" means the elements which are connected to terminal pipes, such as sprinklers.
3. "Machine" means the elements which can provide water or electricity in the system, such as pump.
4. "Fitting" includes various types of connectors, such as straight joints, elbow joints, and T-joints.
5. "Valve" includes various water valve devices which are placed in the pipe or machine to control water.

3.2.2 Establish Relationship between Classes

In language, "connect" is an only word to describe the connection of the pipeline, but we can't use a word to describe all connections between the elements in the ontology. For example, there are three different types of connection which has a similar meaning in T-joint on the main-pipe — "main-pipe to the T-joint", "T- joints to the main-pipe", "T- joints to the branch". Although they are all relationship of connection, the objects which were described in these three connections are all different in the ontology model. So, if there isn't a clear division among their relationship, the reasoning process will be confused in each class.

In this paper, the fire-fighting engineering ontology was developed with four main classes — BIM elements, fire-fighting system, construction information requirement and storage information requirement. "Fire-fighting system" will be connected with "BIM elements" by relationship "HasInstances". Apart from this, "Fire-fighting system" is linked to "construction information requirement" by using "HasWorkitemInformation". "BIM elements" is linked to "storage information requirement" by using "HasElementInformation". Thus, the requirements of material management and construction can be described by these relationships. Figure 3 shows the ontology schema used in this study.

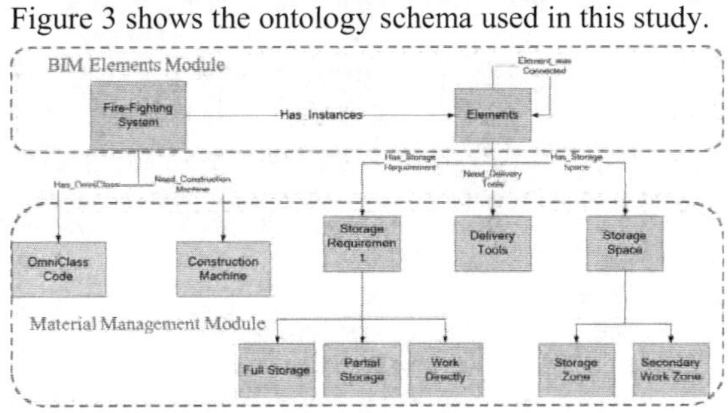

Figure 3. The schema of the ontology model

3.2.3 Reasoning Result

Due to the expanse of knowledge, it's not easy that integrate all requirements and BIM elements. In order to solve this problem, SWRL language was used to encode the rule. The semantic of fire-fighting engineering knowledge was translated into computer language to establish the attributes and relationships. Thus, the BIM elements can be classified into different systems. Besides, the requirements of material management can be identified. Then, the modeling rules can be followed according to the information which was obtained from ontology.

For example, omniclass number which is in "construction information requirement" class is used to classify the location of the work item. But it's not easy to identify all materials in ontology, SWRL language can help us to classify the materials into a different omniclass number according to material's attribute, as shown in Figure 4.

Figure 4. Reasoning results of SWRL rules

3.3 Automated Modeling with Dynamo Parametric Design Software

In recent years, with the vigorous development of BIM software, various BIM software developers also commit to reduce the obstacles of model drawing. The modeling software used in this study was Revit which was developed by Autodesk Company. Even if the method of MEP modeling is increasing, traditional MEP modeling methods still cost a lot of time to create and check the model to solve the more complex system circuit. So the plug-in software, Dynamo, which was developed by Autodesk Company was used. Comparing to traditional manual modeling method, Dynamo is able to place the components in correct position and reset the component attributes in a parametric way. In MEP modeling stage, the modeling rules were followed and Python language was used for automatic modeling.

This paper takes automatic sprinkler system as an example. MEP automatic modeling is divided into six main processes, and these six processes were integrated as a prototype of automated modeling in the future. The Dynamo module was shown in Figure 5.

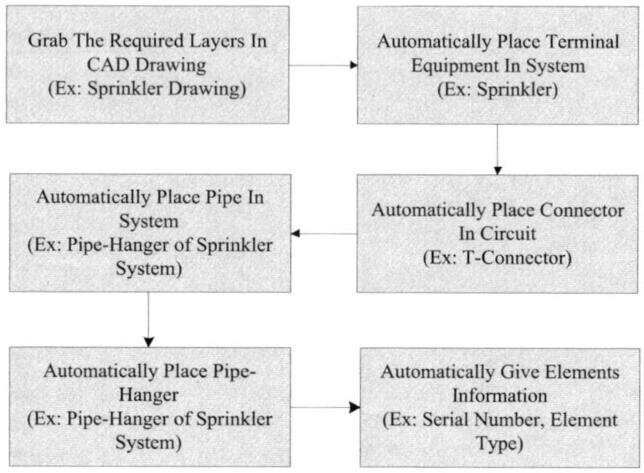

Figure 5. Automated modeling module

3.3.1 Grab the Required Layers in CAD Drawing

Because CAD layer is complex, if we want to build a 3D model, we need to remove the useless information and layers. Then, the required layers were imported into the BIM modeling software. Due to the complicatation of the process, Dynamo function node was used for layer selection. Further, the target layer can be captured according to usage requirement.

3.3.2 Automatically Place Terminal Equipment in System

Because the coordinate of the equipment is a single point in the drawing, the deleted function can be used to delete information other than "point". Therefore, Dynamo built-in function node was used to create elements on these points. In addition, parameter modification function node was used to modify elements' parameters such as displacement.

3.3.3 Automatically Place Connector on Circuit

The purpose of this stage is to establish the connector elements in the system and circuit. We focus on the intersection of each line. First, the deleted function was used to delete information other than "Line" and "Poly Curve". Then, they were divided into two lists, line and poly curve, and made them intersect with each other to find out the intersection point. Finally, these points were used to place the connectors.

3.3.4 Automatically Place Pipe in System

Due to complications of the MEP piping path, it's not easy to create the piping model in the traditional way. Therefore, Python script was edited to establish a rule which can create pipes in the model. In addition, both the system and the type of pipe need to be defined in the model at first.

3.3.5 Automatically Place Pipe-Hanger

First, the pipes which need to generate the hanger was selected. Then, the pipe-hangers are created with the appropriate height and location by the input of parameters. Finally, parameter modification function node was used to modify elements' parameters such as displacement and angle.

3.3.6 Automatically Give Elements Information

In order to meet the material management requirement, the information of BIM elements needs to be expanded, such as serial number and manufacturer. However, it's not easy to choose the target in hundreds of elements in MEP model. Dynamo function node was used to select the elements of a specified parameter, so that the process can be smooth. Figure 6 shows the process of automation.

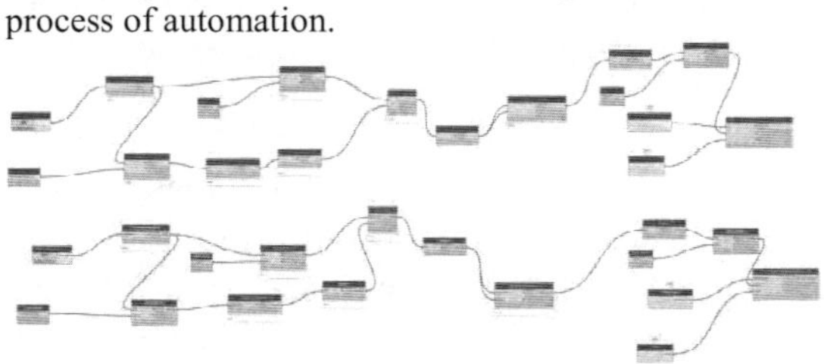

Figure 6. The Dynamo nodes of information giving

3.4 Comparison of Common Practice and Automated Modeling Method

In this study, the Dynamo module is used to change the traditional modeling method. MEP model was created by automated modeling module, as shown in Figure 7. Parameter design not only makes the modeling process faster and more efficient, but also makes the model change or update more convenient. Table 3 shows the advantages of modeling and informing with Dynamo.

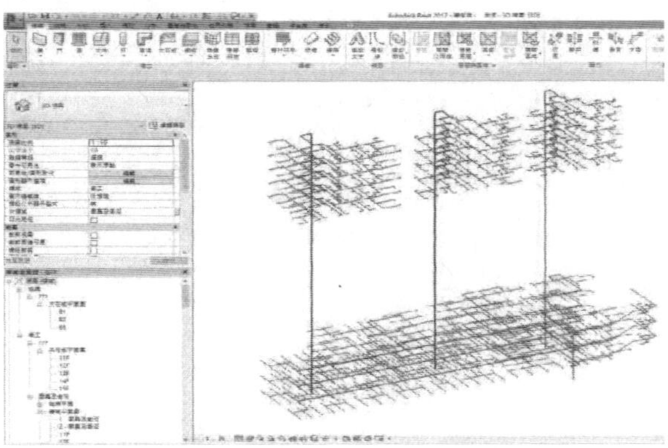

Figure 7. MEP model of the case study in BIM

Table 3. Comparison of traditional method and using Dynamo

Phase	Method	Process	Advantages & Disadvantages
Modeling Phase	Traditional Method	Placing the equipment before establishing the piping system or creating pipe element directly.	Advantage: The modeling process is simple and easy. Disadvantage: The process is time-consuming and the accuracy is low.
	Dynamo	Using Dynamo modules to place the equipment, fittings, and pipes.	Advantage: Quick modeling and quick modification. Disadvantage: The modeling procedure is complex.
Information Given Phase	Traditional Method	In addition to the model built-in information, components of different attribute need to be distinguished manually.	Advantage: n.a. Disadvantage: The process is laborious and inaccurate.
	Dynamo	Filtering different attributes and properties of the components by Dynamo module, and give the element information automatically.	Advantage: Quickly select the target component and edit the information. Disadvantage: Parametric method may cause the error, so the input of the parameters should be cautious.

4 Conclusions

As the discussion and analysis above, this paper proposes a MEP material management which applied to BIM. We hope to provide more and more effective information for MEP professional contractors by means of technology-based and informatization. The information not only enhance the efficiency of MEP professional contractor, but also achieve the purpose of cost control.

In this study, the method of MEP construction was recognized to know the MEP engineering sequence. Further, the sequence and requirement of material supply were identified to establish a framework of material management. In order to establish a modeling rule which is based on MEP material management, the fire-fighting ontology model was developed. Then the BIM model was used to be an information integration tool to provide required information to the elements. In order to smooth the process of developing the MEP BIM model, Dynamo parametric plug-in software was used to develop the modeling automation module. It can quickly establish the MEP model according to the drawing of complex system or circuit. Thus, the accuracy and time required for MEP model can be improved.

As a result, the elements data sheet which contains the required information was used to be the basis for material supply. Apart from this, the inventory could be maintained in the best range to achieve the material requirement plan by analysis of contractor's productivity and material stock. Through the above procedures, the model can provide construction workers not only the material, but also allow MEP professional contractors to clearly understand the construction location and time course of these materials.

Due to the current constraints, there are some issues which supposed to be solved are not completed:

1. Other categories of fire-fighting system in the building.
2. The common materials were used in this study, but in practice, there are various types of fitting.
3. Dynamo module still has room to improve its function and effect.

Acknowledgement

This work was supported by the Ministry of Science and Technology, Taiwan under Grant MOST105-2221-E-006-047-MY2.

References

[1] Wu K. C. and Feng C. W., *Developing the BIM-based Framework For Construction Engineering Material Management*, Thesis of National Cheng Kung University, Taiwan, 2013.

[2] Wu M. R. and Feng C. W., *Developing BIM For Material Management - a case study of MEP System*, Thesis of National Cheng Kung University, Taiwan, 2015.

[3] Wu F. C. and Wang W. C., *Factors Affecting the Cost Estimations of Mechanical/Electrical/ Plumping Construction*, Thesis of National Chiao Tung University, Taiwan, 2006.

[4] Chen T. L., *Practice of Construction and Supervision of MEP Engineering*, 175-206, CHAN'S Arch-Publishing co., Taipei, Taiwan 2013.

[5] Public Construction Commission, *Tutorial of MEP quality management*, Executive Yuan, Taipei, Taiwan, 2013.

[6] Lee J. X. and Lin Z. D., *Construction Industry Materials-and-Supplies Management System*, Thesis of National Central University, Taiwan, 2002.

[7] Lee K. W. and Feng C. W., *The Study of Developing 3D Drawing Objects Model According to the Needs of the MEP Sub-Contractors*, Thesis of National Cheng Kung University, Taiwan, 2009.

[8] Gruber, T., Ontology, Online: http://tomgruber.org/writing/ontology-definition-2007.htm, Accessed: 22/03/2017.

[9] Ting K. P. and Chou C. C., *A Reasoning Mechanism Using Ontology, Protégé and SWRL Tools for Building Information Model Data*, Thesis of National Central University, Taiwan, 2013.

[10] Wang P. R. and Feng C. W., *Applying Ontology and BIM to MEP Facility Management - A Case Study of Replacing Large-size Equipment*, Thesis of National Cheng Kung University, Taiwan, 2016.

[11] Noy N. F. and McGuinness D. L., Ontology Development 101: A Guide to Creating Your First Ontology. Online: http://liris.cnrs.fr/alain.mille/enseignements/Ecole_Centrale/What%20is%20an%20ontology%20and%20why%20we%20need%20it.htm, Accessed: 22/03/2017.

Changes in the BIM Service Provision of Top BIM Companies: A Five-Year Longitudinal Study

Namcheol Jung[a] and Ghang Lee[b]

[a,b]Department of Architecture and Architectural Engineering, Yonsei University, South Korea.
E-mail: [a]chuliee@nate.com. Corresponding author: [b]glee@yonsei.ac.kr.

Abstract –

This paper presents the results of a five-year longitudinal study conducted to analyze the changes in the building information modeling (BIM) services provided by top contractors from 2011 to 2015. Many previous studies analyzed the trend of BIM adoption through questionnaires, case studies, or interviews, all methods that are subject to respondents' perspectives. To objectively analyze the changes in BIM services throughout the years, the research team tracked the BIM services provided by top contractors, which were selected using the 2011 to 2015 Giants 300 reports. The BIM services that each company provided were collected in 2011, 2013, and 2015 from the company's websites, brochures, and/or reports about its BIM projects. The collected BIM services were regrouped using the BIM use classification suggested by Pennsylvania State University. The frequency of BIM services and the correlations between the BIM service types were analyzed. The results showed that the average number of BIM services provided by one company increased from 5.16 in 2011 to 10.91 in 2015. The top five BIM services provided by the top BIM companies include phase planning, 3D coordination, design reviews, cost estimation, and site utilization. While about 70% of the top BIM companies provided these services in 2011, these services were provided by most companies (approximately 85%) in 2015. It is expected that the results of this study will help the industry to overview trends in the BIM market and decide which technologies they need to adopt in order to keep pace with major BIM companies.

Keywords –

Building information modeling (BIM); BIM service; BIM use; Longitudinal study

1 Introduction

For over a decade, many companies in the architecture, engineering, and construction industry have been adopting building information modeling (BIM). Many studies have been conducted to understand the current status of BIM adoption as well as its future implementation and direction. This study aims to identify the ways in which leading construction companies' utilization of BIM has changed since 2011 through a five-year longitudinal study.

Previous studies depended on interviews or surveys, both of which are methods that study the perceptions of participants. To minimize the subjective judgment in the analysis, this study tracked the changes in the BIM services of the top BIM contractors by analyzing their websites, brochures, reports about their BIM projects in 2011, 2013, and 2015, respectively. The top BIM contractors were selected from Giants 300 reports, which are annually published reports on the rankings of design firms and contractors that conduct BIM projects [1]. A content analysis was conducted to analyze the changes in the BIM services provided. By analyzing the BIM services that each company provides rather than analyzing surveys, this study attempted to capture the companies' commitment to BIM rather than BIM users' personal views. The eventual goal is to understand the changes in the practical uses of BIM rather than those in the perceived uses of BIM. Also the methodology utilized in this study will improves time-consuming works to investigate the BIM trend.

This paper is divided into six sections. The next section reviews previous studies related to BIM trends. The third section describes this study's research questions and the data collection and analysis methods used. Section 4 reports the analytical results, and the final section concludes the paper.

2 Previous Studies on BIM Adoption

Many studies have been conducted to research the statuses of BIM adoption and implementation around the world. Yalcinkaya and Singh conducted a research pattern analysis and derived three main research areas from 220 publications directly relevant to BIM implementation and adoption [2]: 1) adoption and implementation processes, 2) project delivery methods, and 3) benefit realization. Azhar and Cochran also conducted a survey to determine the adoption status of

BIM in the electrical construction industry in the United States [3]. They conducted a survey using questionnaires to determine the popular BIM uses of and obstacles involved in BIM adoption. In addition, Kiziltas and Akinci conducted interviews and surveys with BIM experts to understand the lessons learned from BIM projects [4]. They summarized the results using three phases of construction—before construction, during construction, and after construction— and categorized BIM services according to these three phases.

Further, Becerik-Gerber et al. investigated the connection between BIM and facility management [5], and Alin et al. conducted a study on the inter-firm processes involved in implementing BIM [6]. Won et al. conducted an international survey to discover the critical success factors involved in the successful adoption of BIM within an organization, BIM project selection, BIM service selection, and BIM software selection [7]. Samuelson and Björk conducted a longitudinal study to investigate the adoption status of building-related information technologies (IT) in Sweden [8]. The surveys were conducted from 1998 to 2011 and utilized the IT barometer. Furthermore, Lee and Jung proposed a simplified model to evaluate and represent the level of BIM implementation in multiple countries according to four factors: 1) the depth of BIM implementation, 2) years using BIM, 3) the level of BIM proficiency, and 4) BIM adoption rate [9].

The SmartMarket Report series published by McGraw Hill Construction Research and Analytics (currently, Dodge Analytics) provides quantitative survey data related to the status of BIM adoption around the world starting from 2007 [10]. In addition, the National Building Specification (NBS) for the United Kingdom has published national BIM reports related to BIM adoption since 2011 [11].

Despite the strengths of interviews and surveys as research methods, they also have limitations in that interviewees may answer perceived facts rather than actual facts [12]. For this reason, this study longitudinally tracked the BIM services that each company declared it provided as a basis to analyze the status of BIM implementation in top BIM contractors.

3 Research Design

3.1 Research Questions

The primary aim of this research was to study the changes in the use of BIM from 2011 to 2015 by top global contractors. Based on this, the following research questions were developed.

- Q1: How has the number of BIM services provided by each company changed over time?
- Q2: Were there any changes in the main BIM services provided each year?
- Q3: What BIM services were provided for each project phase?
- Q4: What are the most popular BIM services provided by leading BIM companies?
- Q5: What BIM services were newly introduced in each year?
- Q6: Is there any correlation between BIM services?
- Q7: What are the BIM-related characteristics of the contractors that stayed on the Giants 300 reports from 2011 to 2015?

3.2 Classification of BIM Services and Data Coding

To classify the collected BIM services, the BIM use list developed by Pennsylvania State University (Penn State) [13] was used. The Penn State classification is composed of 25 BIM services throughout four project phases. The 25 BIM services are divided into two types: primary and secondary BIM uses. Each item is listed in Table 1. The Penn State classification was selected for the following reasons:

1. The BIM services are clearly categorized into primary and secondary services and by project phase.
2. The classification is simple and easy to understand.
3. The classification is already well known to the BIM community and has been used in other studies [14, 15].

Among the 25 BIM services listed, several services related to engineering analyses, such as structural analyses, lighting analyses, energy analyses, and mechanical analyses, were merged with engineering services to prevent ambiguity in their definitions. Disaster planning was also excluded from the analysis because all the contractors claimed that they did not support disaster planning. Therefore, a total of 20 BIM services from the original 25 BIM services specified by Penn State were used in this analysis.

3.3 Data Collection

We gathered texts from the webpages, brochures, and case study reports of the top BIM contractors to conduct a content analysis. The provision of a BIM service field was checked yes (1) if a company provided the BIM service and no (0) otherwise.

Table 1 BIM uses [13]

Item	Construction Phase	Type	Note
Existing Conditions Modeling	Plan, Design, Construct, Operate	Primary	
Cost Estimation	Plan, Design, Construct, Operate	Primary	
Phase Planning	Plan, Design, Construct	Primary	
Programming	Plan, Design	Primary	
Site Analysis	Plan, Design	Primary	
Design Reviews	Plan, Design	Secondary	
Design Authoring	Design	Primary	
Structural Analysis	Design	Primary	Merged
Lighting Analysis	Design	Primary	Merged
Energy Analysis	Design	Primary	Merged
Mechanical Analysis	Design	Secondary	Merged
Other Engineering Analysis	Design	Secondary	Merged
LEED Evaluation	Design	Secondary	
Code Validation	Design	Secondary	
3D Coordination	Design, Construct	Primary	
Site Utilization Planning	Construct	Primary	
Construction System Design	Construct	Secondary	
Digital Fabrication	Construct	Secondary	
3DControl and Planning	Construct	Primary	
Record Model	Construct, Operate	Primary	
Maintenance Scheduling	Operate	Primary	
Building System Analysis	Operate	Primary	
Asset Management	Operate	Secondary	
Space Management/Tracking	Operate	Secondary	
Disaster Planning	Operate	Secondary	Not used

The contractors used in the content analysis were selected from the top 100 BIM contractors listed in the Giants 300 reports published by Building Design and Construction. Giants 300 reports have been published every year since 2010. Each year, the report ranks the leading companies in the architecture, engineering, and construction industry by sector based on their annual revenues.

The 2011 Giants 300 report listed leading BIM companies in two sectors: design firms and contractors. The 2012 Giants 300 report listed leading BIM companies in three sectors: architecture firms, engineering firms, and construction firms. The following reports used the same categorization. This study conducted an analysis focusing on leading contractors using information from their webpages, brochures, and reports. A list of the companies included in the analysis is given in Table 2.

For data collection from 2011, we used the Giants 300 report published on December 16, 2010 that investigated revenue of the top companies in 2009 [16]. From the report, 32 companies were selected as target companies based on their provision of the proper materials. For 2013, we used the Giants 300 report published on September 20, 2012 to collect data on 34 companies [17]. Similarly, in 2015, we used the Giants 300 report published on August 4, 2014 to collect data on 65 companies [18].

3.4 Data Analysis

To answer the seven sub-questions, we conducted a frequency analysis and a cross-tabulation analysis.

To answer Q1 (how has the number of BIM services provided by each company changed over time?), we tracked and analyzed the average amount of BIM services provided by each company in 2011, 2013, and 2015, respectively.

To answer Q2 (were there any changes in the main BIM services provided each year?), we analyzed variation in 2013 and variation in 2015. Variation in 2013 represents the subtracted frequency of BIM services provided in 2011 from that of BIM services provided in 2013, while variation in 2015 represents the subtracted frequency of BIM services in 2013 from that of BIM services in 2015. The calculated ratio of the two values was utilized to show the changes.

Table 2 Source of analysis for each year

Year of Analysis	Number of Samples	Source
2011	32	2010 Giants 300 report
2013	34	2012 Giants 300 report
2015	65	2014 Giants 300 report

To answer Q3 (what BIM services were provided for each project phase?), we classified the 20 BIM services according to the four project phases (planning, design, construction, or operation) defined by Penn State. The percentage of companies that provided BIM services for each project phase was measured.

To answer Q4 (what are the most popular BIM services provided by leading BIM companies?), the top 5 BIM services, which were those most frequently used in each year, were selected.

To answer Q5 (what BIM services were newly introduced in each year?), the top 5 BIM services that had the highest variation values in 2013 and 2015 were selected.

To answer Q6 (is there any correlation between BIM services?), we conducted Fisher's exact test. The test measured whether there was a significant relationship between BIM service pairs. For example, the first BIM service, existing condition modeling, was compared with the other 19 BIM services. This was done for each service.

To answer Q7 (what are the BIM-related characteristics of the contractors that stayed on the Giants 300 reports from 2011 to 2015), we conducted a t-test to compare the average amount of BIM services provided by companies that stayed on the Giants 300 reports from 2011 to 2015 and those that did not.

4 Results

Through the analysis described above, the collected data were analyzed. The frequency analyses conducted to answer Q1, Q2, and Q3 derived the following results. Regarding Q1, the average number of BIM services provided by companies increased from 5.16 in 2011 to 10.91 in 2015. The adoption rates of some BIM services increased evenly for five years, while some BIM services were adopted in the same year.

Regarding Q2, changes in the ranking of each BIM service are presented in Table 3.

Regarding Q3, in the four phases of the building life cycle, there were steady increases in the BIM services from 2011 to 2015.

- Plan: 40.6% in 2011 to 62.6% in 2015 (+22.0%)
- Design: 31.3% in 2011 to 59.4% in 2015 (+28.1%)
- Construct: 45.8% in 2011 to 59.3% in 2015 (+13.5%)
- Operate: 17.4% in 2011 to 50.6% in 2015 (+33.2%)

In total, the BIM services provided increased from 25.8% in 2011 to 54.5% in 2015.

Regarding Q4, the top 5 BIM services were phase planning, design reviews, site utilization, 3D coordination, and cost estimation, and they did not change over 5 years. The average adoption rate of the top 5 BIM services was approximately 70% in 2011. In 2015, this rate increased to 85%.

Regarding Q5, an analysis of the top BIM services with the highest variations was conducted to find the newly introduced BIM services in 2013 and 2015. In 2013, the adoption rates of services related to building system analysis, asset management, space management tracking, maintenance scheduling, and design authoring significantly increased. In 2015, many companies adopted services related to design authoring, code validation, programming, site analysis, and engineering analysis. Few companies used operation and management services in 2011; however, in 2013, more than half of the companies adopted these services. Similarly, less than 10% of companies used design-related BIM services in 2011; however, approximately 40% of the companies adopted these services in 2015. Therefore, there was a significant increase in the services adopted related to the operation phase in 2013 and those related to the design phase in 2015.

Regarding Q6, the relationships between each BIM service were analyzed using Fisher's exact test. The analyzed BIM services were grouped into five. The following list explains the five groups of related BIM services.

- Group 1 is comprised of services related to existing condition modeling and record model. It is expected that existing condition modeling, including laser scanning, affects record model.
- Group 2 is comprised of services related to cost estimation and phase planning. Group 2 consists of the BIM services considered to be fundamental for leading construction companies. More than 90% of the top companies have provided BIM services related to phase planning since 2011. BIM services related to cost estimation were also adopted by many companies in 2013 and 2015.

Table 3 Rankings of BIM services each year

BIM Services	2011		2013		2015	
	Change	Ranking	Change	Ranking	Change	Ranking
Existing Conditions Modeling	.	10	▼4	14	▼5	19
Cost Estimation	.	4	.	4	▼1	5
Phase Planning	.	1	.	1	.	1
Programming	.	16	.	16	▲2	14
Site Analysis	.	16	.	16	▲2	14
Design Reviews	.	3	▲1	2	▲1	1
Design Authoring	.	16	▲2	14	▲7	7
Engineering Analysis	.	16	.	16	▲2	14
LEED Evaluation	.	9	▼4	13	▲4	9
Code Validation	.	16	▼4	20	▲6	14
3D Coordination	.	2	.	2	▼1	3
Site Utilization Planning	.	5	.	5	▲1	4
Construction System Design	.	6	.	6	.	6
Digital Fabrication	.	8	▼4	12	▼2	14
3D Control and Planning	.	11	▼8	19	▼1	20
Record Model	.	7	▼4	11	▲3	8
Maintenance Scheduling	.	11	▲3	8	▼5	13
Building System Analysis	.	11	▲4	7	▼3	10
Asset Management	.	11	▲3	8	▼3	11
Space Management/Tracking	.	11	▲3	8	▼3	11

Table 4 Answers to the research questions

Question	Answer	
Q1	The average number of BIM services provided by leading companies was 5.16 in 2011, 8.35 in 2013, and 10.91 in 2015.	
Q2	Some BIM services increased evenly, while some were adopted at the same time.	
Q3	Plan	: 40.6% → 62.6%
	Design	: 31.3% → 59.4%
	Construct	: 45.8% → 59.3%
	Operate	: 17.4% → 50.6%
Q4	Phase planning, design reviews, site utilization, 3D coordination, cost estimation	
Q5	2013:	Building system analysis, asset management, space management tracking, maintenance scheduling, design authoring
	2015:	Design authoring, code validation, programming, site analysis, engineering analysis
Q6	Group 1:	Existing condition modeling, record model
	Group 2:	Cost estimation, phase planning
	Group 3:	3D Coordination, design reviews, site analysis, construction system design
	Group 4:	Programming, design authoring, engineering analysis, code validation, site analysis
	Group 5:	Maintenance scheduling, building system analysis, asset management, space management/tracking
Q7	No significant differences exist between companies that stayed on the Giants 300 reports from 2011 to 2015 and those that did not.	

- Group 3 is comprised of services related to 3D coordination, design reviews, site analysis, and construction system design. In 2011, only 3D coordination and design reviews services were provided. In 2013, site utilization planning and construction system design were also added, but they were not related to 3D coordination or design reviews. In 2015, there was a correlation between all four BIM services included in Group 3.
- Group 4 is comprised of services related to programming, design authoring, engineering analysis, code validation, site analysis, and LEED evaluation. In 2011 and 2013, few companies utilized these services. However, by 2015, more than one third of the leading companies adopted and provided these services.
- Group 5 is comprised of services related to maintenance scheduling, building system analysis, asset management, and space management/tracking. It is expected that the development of facility management and building operation led to the use of these four services.

Regarding Q7, the t-test, which was conducted to compare the average amount of BIM services that were provided by the companies that stayed on the list from 2011 to 2015 and those that did not, showed no significant statistical difference between these two groups.

5 Conclusion

The major findings of this study are listed in Table 4, and their implications for contractors are as follows.

In 2015, more than 40% of the companies adopted all the primary BIM services defined by Penn State except for 3D control and planning. In addition, the adoption rate of newly introduced BIM services increased. In 2015, most companies (94%) provided BIM services related to phase planning and design reviews.

In 2011, construction-related functions (site utilization, phase planning, etc.) tended to be utilized more than other BIM services by contractors. In 2013, Engineering analysis, code validation, and those for building operation are increased. BIM uses for communication (design authoring, programming, etc.) between stakeholders are also rapidly increased in 2015.

In this study, a cross-tabulation analysis showed that BIM services with similar functions were adopted together. BIM services could be grouped into five. Group 1 was related to existing condition or record model. Group 2 was related to cost and schedule management, and Group 3 was related to the design reviews. Group 4 was related to design and engineering, and Group 5 was related to facility management and operation.

The top 5 BIM services provided include phase planning, design reviews, site utilization, 3D coordination, and cost estimation. The adoption rate of design authoring, site utilization, building system analysis, asset management, space management/tracking, and maintenance scheduling increased more than 40% from 2011 to 2015. These services became common among construction companies. In addition, design authoring, asset management, building system analysis, maintenance scheduling, and space management/tracking were newly introduced in 2013 and design authoring, site analysis, engineering analysis, code validation, and programming were newly introduced in 2015. BIM, which is defined as "process of generating and managing building information in an interoperable and reusable way [19]", is a tool that emphasis co-work between project participants. This study contributes to offering longitudinal BIM trend derived from websites, brochures, and/or reports generated by top BIM contractors. The trend will notice what BIM functions companies have to complement to become a real BIM partner in AEC industry.

Though this study aimed to be objective, this study has a limitation in that an ambiguous representation of BIM functions in the text may be interpreted differently by different researchers.

An analysis using 2017 data is still too early to be conducted. The study will be continued to continuously track the changes in the BIM service provisions.

References

[1] Building design + Construction. Giants 300 Report. Online: https://www.bdcnetwork.com/giants-300, Accessed: 02/20/2017.

[2] Yalcinkaya M. and Singh V. Patterns and trends in Building Information Modeling (BIM) research: A Latent Semantic Analysis. *Automation in Construction*, 59(2015):68-80, 2015.

[3] Azhar S. and Cochran S. Current status of Building Information Modeling (BIM) adoptability in the U.S. electrical construction industry. *Proceedings of the 5th International Conference on Construction in the 21st Century (CITC-V)*, Istanbul, Turkey, 2009

[4] Kiziltas S. and Akinci B. Lessons learned from utilizing building information modeling for construction management tasks. *Proceedings of the Construction Research Congress 2010: Innovation for Reshaping Construction Practice*, 318-327, Banff, Canada, 2010

[5] Becerik-Gerber B., Jazizadeh F., Li N. and Calis G. Application areas and data requirements for BIM-enabled facilities management. *Journal of construction engineering and management*,

138(3):431-442, 2011.

[6] Alin P., Maunula A. O., Taylor J. E. and Smeds R. Aligning Misaligned Systemic Innovations: Probing Inter - Firm Effects Development in Project Networks. *Project Management Journal*, 44(1):77-93, 2013.

[7] Won J., Lee G., Dossick C. and Messner J. Where to focus for successful adoption of building information modeling within organization. *Journal of Construction Engineering and Management*, 139(11):04013014, 2013.

[8] Samuelson O. and Björk B.-C. A longitudinal study of the adoption of IT technology in the Swedish building sector. *Automation in Construction*, 37(2014):182-190, 2014.

[9] Lee G. and Jung W. Slim BIM Charts for Rapidly Visualizing and Quantifying Levels of BIM Adoption and Implementation. *Journal of Computing in Civil Engineering*, 30(4):Online publication, 2016.

[10] McGraw Hill Construction Research & Analytics. Smart Market Report. Online: http://analyticsstore.construction.com/smartmarket-reports.html, Accessed: 02/16/2017.

[11] National Building Specification. NBS National BIM Report Knowledge Articles | NBS. Online: https://www.thenbs.com/knowledge/nbs-national-bim-report, Accessed: 02/02/2017.

[12] Ketokivi M. A. and Schroeder R. G. Perceptual measures of performance: fact or fiction? *Journal of Operations Management*, 22(3):247-264, 2004.

[13] Messner J., Anumba C., Dubler C., Goodman S., Kasprzak C., Kreider R., Leicht R., Saluja C. and Zikic N. *BIM Project Execution Planning Guide*, The Computer Integrated Construction Research Program, Pennsylvania State University, United States, 2010

[14] Kreider R., Messner J. and Dubler C. Determining the frequency and impact of applying BIM for different purposes on projects. *Proceedings of the 6th International Conference on Innovation in Architecture, Engineering and Construction (AEC)*, PennState, United States, 2010

[15] Kreider R. G. and Messner J. I., *The Uses of BIM: Classifying and Selecting BIM Uses Version 0.9*, Penn State University, United States, 2013

[16] Building design + Construction. 2010 Giants 300 BIM Report. Online: https://www.bdcnetwork.com/sites/default/files/Top Firms by BIM Project Revenue.pdf, Accessed: 02/20/2017.

[17] Building design + Construction. 2012 Giants 300 Report. Online: https://www.bdcnetwork.com/bim-finally-starting-pay-aec-firms, Accessed: 02/03/2017.

[18] Building design + Construction. 2014 Giants 300 Report. Online: https://www.bdcnetwork.com/bim-giants-firms-enhance-bimvdc-advanced-collaboration-tools-2014-giants-300-report, Accessed: 02/02/2017.

[19] Lee G., Sacks R. and Eastman C. M. Specifying parametric building object behavior (BOB) for a building information modeling system. *Automation in Construction*, 15(2006):758-776, 2006.

34th International Symposium on Automation and Robotics in Construction

ISARC 2017

June 28 – July 1

TAIPEI, TAIWAN

Cartesian-Space Motion Planning for Autonomous Construction Machines

M. Ragaglia[a], A. Argiolas[a] and M. Niccolini[a]

[a]Yanmar Research & Development Europe, Viale Galileo 3/A, 50125, Firenze, Italy
E-mail: matteo_ragaglia@yanmar.com, alfredo_argiolas@yanmar.com, marta_niccolini@yanmar.com

Abstract -

Nowadays, the construction industry is probably the least productive and most dangerous among the various industry sectors. Given this scenario, it is quite clear that the introduction of Autonomous Construction Machines (ACMs) could represent a great opportunity to improve both productivity and safety. To this purpose, a fundamental problem that has to be tackled is trajectory planning. In the last 15 years, several sample-based algorithms have been proposed, that relies on Joint-Space sampling. Unfortunately, this feature often results in trajectories that are quite counterintuitive from the point of view of a human being. In this work we propose "cart-RRT", a Cartesian-Space randomized algorithm that improves the intuitiveness of the output trajectory, while ensuring both its safety (in terms of collision avoidance) and its feasibility.

Keywords -

Robotics; Motion Planning; Construction Machines;

1 Introduction

Nowadays, in the construction industry, several operations that require both high power and high accuracy (such as panel positioning, plumbing, material handling) are still manually performed by human workers, in very inefficient and dangerous ways. After a thorough field investigation, we believe that the construction industry could definitely benefit from the introduction of Autonomous Construction Machines (ACMs), in terms of increased productivity and human workers' safety. Among the huge number of functionalities that are required to develop an ACM, motion planning plays a fundamental role. As a matter of fact, the planner has the responsibility to compute a trajectory that allows the ACM to travel from its original configuration to a desired one, while satisfying kino-dynamic constraints and avoiding collision with obstacles.

Given the fact that robots are complex systems, often characterized by nonlinear dynamics and actuation constraints, the problem of trajectory planning can easily become computationally intractable. For this reason, randomized sample-based planning algorithms [7] have emerged as a reliable and appealing alternative to search-based planning techniques [9] and model predictive control approaches [14]. Randomized sample-based planners like Rapidly-exploring Random Trees (RRT) [8] have experienced a growing popularity since their introduction in the late '90s. The main reason behind this success relies in a rather simple, yet effective idea: sample multiple robot configurations that do not entail collisions with obstacles and connect them to build either a graph or a tree of feasible trajectories. Then, a solution is extracted from this tree (or graph), in terms of a sequence of edges that connect couples of nodes. Furthermore, it has been demonstrated in [1] that sample-based planning algorithms are probabilistically complete. Even though they have been originally introduced to solve the trajectory planning problem for holonomic robots, RRT-based algorithms have been extended to tackle more complex problems, like for instance the optimal and constrained trajectory planning. Several solutions have been proposed in the last few years, including optimal and non-linear RRTs [2], Rapidly-exploring Random Graphs (RRG), and RRT* [5, 4]. Several approaches have also been proposed to solve the optimal and constrained trajectory planning problem for arbitrary kino-dynamic systems, like for instance [3, 11, 12, 13].

A typical feature of all these randomized planning algorithms consists in sampling the nodes directly in the robot Configuration-Space, also known as Joint-Space. Unfortunately, Joint-Space randomized algorithms tend to output quite counterintuitive trajectories from the point of view of a human being. Given the fact that in construction sites humans and machines usually work side-by-side, and considering the results presented in [15], it is clear that the intuitiveness of the planned trajectories can have a strong impact on the comfort, the productivity and the level of safety perceived by human workers sharing their workspace with ACMs. In order to avoid these drawbacks, a possible solution is represented by addressing the motion planning problem directly at the Cartesian-Space level. We here propose "cart-RRT", a Cartesian-Space randomized algorithm that allows the tree to be extended directly towards the goal, as it has been originally proposed in [6]. In cart-RRT edges are computed as linear segments in the Cartesian-Space and the corresponding joint positions profiles are determined using Inverse Kinematics. Once the tree reaches the goal, a path shortening procedure is performed to reduce the length of the output trajectory while keeping it collision-free. Then, a spline-based interpolation is used to eliminate Joint-Space velocity discontinuities. Finally, kinematic scaling is applied to guarantee that joint velocities do not exceed their lower and upper bounds. The proposed algorithm has been tested and validated in multiple simulated scenarios, using the model of a UR10 manipulator as test bench.

2 Proposed Algorithm

One of the fundamental features of cart-RRT is the fact that nodes are computed as linear segments in the Cartesian-Space. In order to draw linear edges in the Cartesian-Space, while computing their counterpart in Joint-Space, the algorithm relies on the following data-structures:

- **node**: the basic element of the random tree. It is defined by: a set of joint positions $\boldsymbol{q}$, the corresponding Cartesian pose $\boldsymbol{p}$, and a reference to its parent node within the tree n_{par};
- **edge**: the element that connects two distinct nodes. It is defined by two matrices: $\boldsymbol{Q}$ and $\boldsymbol{P}$ that contain, respectively, the series of intermediate Joint-Space and Cartesian-Space poses visited by the robot while traveling from the starting node to the destination one;
- **Tree**: the random tree built by the algorithm. It contains two main fields: a list of nodes named **Nodes**, and a list of edges named **Edges**.

Algorithm 1 explains how the cart-RRT planner works. initially, the tree only contains the starting node n_0. Then, following the approach originally proposed in [6], a random selector is sampled to select between the two possible extension modalities: random and goal-oriented. To this purpose, the value of the parametric threshold "rand$_{th}$" plays a fundamental role, since it affects the balance between exploration (random expansion) and exploitation (goal-oriented expansion). Once the tree has been extended, the algorithms checks if the goal region "cart$_g$" has been reached. If the tree reaches the goal region before its cardinality overcomes the value of the "maxNodes" parameter, the algorithm extracts a trajectory from the tree (function "GetTrajectory"). Then, it applies a post-processing technique (procedure "PostProcess", whose details are given in Section 2.4) and, finally, it returns the output trajectory. Otherwise, the algorithm returns a failure (i.e. a "null" value).

2.1 Tree Extension

The two procedures that allow to extend the random tree are detailed in Algorithms 2 and 3. The first procedure implements the standard RRRT extension strategy, while the latter tries to establish a series of nodes connecting the tree directly with the goal region.

Algorithm 1 cart-RRT Algorithm pseudo-code

```
Global Parameters: rand_th, maxNodes
procedure CART-RRT(n_0, cart_g)
    Tree.Nodes ← n_0
    Tree.Edges ← ∅
    while (|Tree.Nodes| ≤ maxNodes)
        rand ← RAND (0, 1)
        if (rand ≤ rand_th)
            Tree ← ExtendRandomly (Tree)
        else
            Tree ← ExtendTowardsGoal (Tree, cart_g)
        end if
        if (∃ n ∈ Tree.Nodes ⟹ FKine (n.q) ∈ cart_g)
            trajectory ← GetTrajectory (Tree)
            trajectory ← PostProcess (trajectory)
            return trajectory
        end if
    end while
    return null
end procedure
```

Algorithm 2 Random extension procedure

```
Parameters: maxLength
procedure EXTENDRANDOMLY(Tree)
    n_rnd ← RandomSampling ()
    n_near ← NearestNode (Tree.nodes, n_rnd)
    n_new ← GetNewNode (n_near, n_rnd, maxLength)
    edge ← EdgeCalc (n_near, n_new)
    if (edge ≠ null)
        n_new.n_par ← n_near
        Tree.Nodes ← Tree.Nodes ∪ n_new
        Tree.Edges ← Tree.Edges ∪ edge
    end if
    return Tree
end procedure
```

2.2 Edge Calculation

One of the fundamental features of cart-RRT is the fact that nodes are computed as linear segments in the Cartesian-Space. Given two distinct nodes in the tree, the connecting edge is computed according to Algorithm 4. Starting from the first node, the procedure iteratively applies differential inverse kinematics in order to reach the destination node. Intermediate Joint-Space and Cartesian-Space positions are stored in the output variable "edge". A threshold distance (parameter "edgeThresh") is set in order to exit the procedure once a given neighborhood of the destination node is reached. The procedure returns a failure in two distinct cases: whenever a kinematic singularity is encountered and in case the inverse kinematics does not converge after a given number of iterations (parameter "edgeCountMax").

Algorithm 3 Goal-oriented extension procedure

```
Parameters: maxLength
procedure ExtendTowardsGoal(Tree, cart_g)
    n_rnd ← SampleFromGoal (cart_g)
    n_old ← NearestNode (Tree.nodes, n_rnd)
    while (1)
        n_new ← GetNewNode (n_old, n_rnd, maxLength)
        edge ← EdgeCalc (n_old, n_new)
        if (edge ≠ null)
            n_new.n_par ← n_old
            Tree.Nodes ← Tree.Nodes ∪ n_new
            Tree.Edges ← Tree.Edges ∪ edge
            if (FKine (n_new.q) ∈ cart_g)
                return Tree
            else
                n_old ← n_new
            end if
        else
            return Tree
        end if
    end while
end procedure
```

Algorithm 4 Edge calculation procedure pseudo-code

```
Parameters: edgeThresh, edgeCountMax
procedure EdgeCalc(n_0, n_f)
    edge.Q ← ∅
    edge.P ← ∅
    reached ← 0
    counter ← 0
    q = n_0.q
    while (reached = 0)
        if (counter > edgeCountMax)
            return null
        end if
        edge.Q ← edge.Q ∪ q
        edge.P ← edge.P ∪ p
        p = FKine (q)
        Δp = n_f.p − p
        if (||Δp|| < edgeThresh)
            reached ← 1
        else
            if (det (J (q)) ≈ 0)
                return null
            else
                q = q + (J (q)^-1 Δp)
            end if
        end if
        counter ← counter + 1
    end while
    return edge
end procedure
```

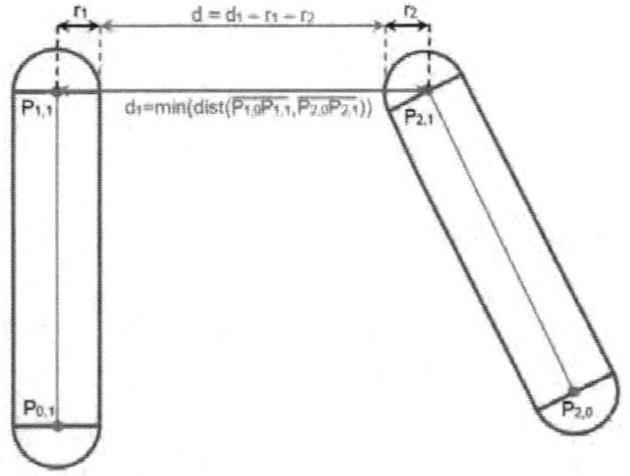

Figure 1. Minimum distance between two capsules.

2.3 Collision Checking Strategy

One of the fundamental element of every sample-based planning algorithm (and also one of their major bottlenecks) is represented by the collision-checking strategy. As a matter of fact, directly taking into account the complex geometry of a robot and of generic obstacles would almost surely result in enormous computation time spent on checking possible collisions. For this reason, a capsule-based geometric model of both the robot and the obstacles has been adopted. We chose capsules since they represent a quite simple (yet effective) tool to model complex geometries without being too conservative. To define a capsule only two points in the Cartesian Space and a radius are needed. The main computational advantage provided by capsules is represented by the fact that to compute the minimum distance between two capsules it is sufficient to compute the minimum distance between the two segments P_0-P_1 (that can be determined analytically) and then subtract the radius of each capsule, as it is shown in Figure 1. Moving back to our algorithm, every time a new edge is computed the "**CollisionCheck**" (see Algorithm 5) procedure is invoked to verify if the motion of the entire robot along the edge may cause collisions with the obstacles. For the sake of clarity, the "RobotCapsules" function is responsible for computing the capsule-based geometry model of the manipulator, see for instance Figures 3(a)-3(c).

2.4 Trajectory Post Processing

As soon as the tree reaches the goal region, the sequence of nodes connecting the starting configuration to the final one is extracted from the tree and a three-stage post-processing procedure is performed:

1. **Path Shortening**: according to the algorithm proposed in [10], two distinct configuration are sampled along the trajectory. Then, the edge connecting the two sampled configurations is computed and checked for collisions. In case the edge is collision-free, the

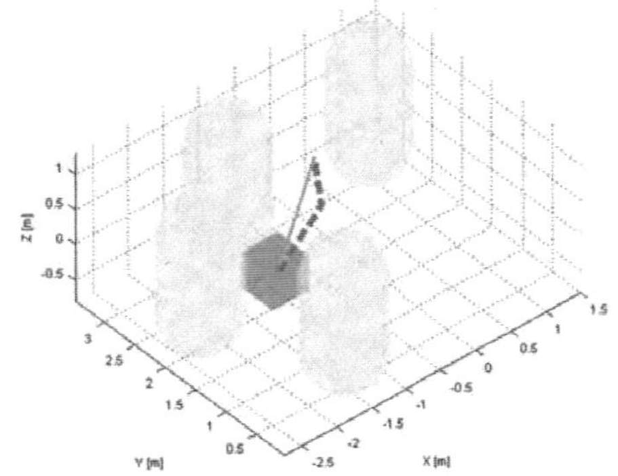

(a) Path Shortening

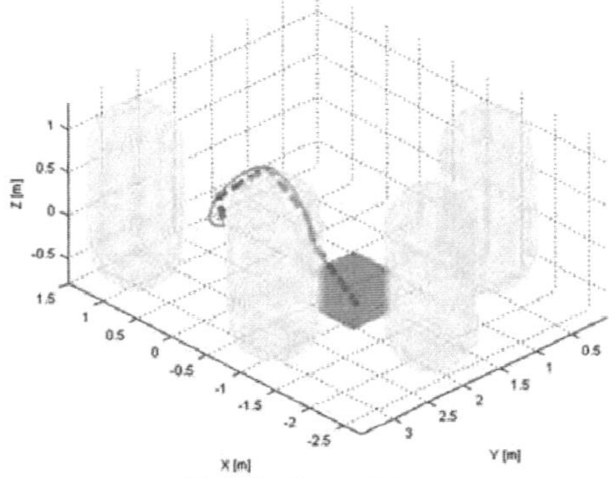

(b) Spline Interpolation

Figure 2. Trajectory post-processing stages: original trajectory (dashed red), post-processed trajectory (solid green), obstacles (yellow capsules) and gaol region (red cube).

Algorithm 5 Collision checking procedure pseudo-code

```
procedure CollisionCheck(edge, obstCaps)
    collFree ← 1
    for all q ∈ edge.Q
        robCaps ← RobotCapsules(q)
        for all rCap ∈ robCaps
            for all oCap ∈ obstCaps
                minDist ← MinDistance(rCap, oCap)
                if (minDist < 0)
                    collFree ← 0
                end if
            end for
        end for
    end for
    return collFree
end procedure
```

trajectory is modified by inserting the new edge between the two sampled configuration. This procedure is repeated until a given number of iterations is reached. As a matter of fact, this path shortening procedure results in an a-posteriori optimization of the trajectory in the Cartesian Space. This approach can be thought as a convenient alternative with respect to tree optimization strategies normally exploited by optimal algorithms. An example of path shortening is shown in Figure 2(a);

2. **Spline Interpolation**: a series of evenly spaced configurations are selected along the trajectory to be used as via points for a cubic spline interpolation. In this way, sharp corners due to the connection of linear Cartesian-Space edges can be removed, ensuring that the trajectory belongs to the C^2 class of functions. An example of spline interpolation is shown in Figure 2(b);

3. **Kinematic Scaling**: up to this point the trajectory is computed taking into consideration a virtual time coordinate τ_v, ranging from 0 to 1. In this stage, joint positions are differentiated to find joint velocities and, in turn, joint accelerations. Then, a time-scaling parameter k_τ is computed to ensure that both joint velocities and accelerations do not exceed the imposed upper and lower bounds. Consequently, a real time coordinate τ_r can be computed as: $\tau_r = k_\tau \tau_v$.

3 Simulation Results

In order to validate the cart-RRT algorithm, a MATLAB implementation has been developed and a simulated UR10 manipulator has been considered as test bench. As far as actuation limits are concerned, we took into consideration the maximum and minimum joint velocities and accelerations displayed in Table 1. In order to resemble the operations that an ACM could possibly perform in an actual construction yard, a pick-and-place task involving a panel has been chosen. Then, the three simulation scenarios pictured in Figure 3 have been designed. More in depth:

- Scenario #1 (grasping panel "easy"), where the robot has to reach the goal region, while avoiding three obstacles. See Figure 3(a);
- Scenario #2 (grasping panel "difficult"), similar to Scenario #1, with an additional obstacle. See Figure 3(b);
- Scenario #3 (depositing panel), where the robot has grasped the panel and has to deposit it in the goal region. See Figure 3(c);

Table 1. UR10 Actuation Limits

Joint	$\dot{q}_{min}$ [°/s]	$\dot{q}_{max}$ [°/s]	$\ddot{q}_{min}$ [°/s^2]	$\ddot{q}_{max}$ [°/s^2]
1	-120	+120	-20	+20
2	-120	+120	-20	+20
3	-180	+180	-20	+20
4	-180	+180	-20	+20
5	-180	+180	-20	+20
6	-180	+180	-20	+20

For the sake of clarity, in each scenario the robot is represented in its starting configuration with links pictured with solid black lines and the corresponding capsules drawn in blue. On the other hand, capsules representing the obstacles are shown in yellow. The goal region is highlighted with a red cube and the panel is pictured with green lines and capsules. Finally, the Cartesian reference frames corresponding to the robot base, the robot end-effector and the Cartesian goal are pictured with red (X), green (Y) and blue (Z) dashed lines. Examples of possible solutions are shown in Figures 3(d)-3(f), thus demonstrating the effectiveness of the proposed algorithm. Furthermore, Figure 4(a) and 4(b) shows the joint velocity and acceleration profiles, respectively. These data clearly prove that the kinematic scaling procedure correctly scales the trajectory in order to always satisfy the aforementioned actuation bounds.

3.1 Comparative Evaluation

To properly assess the cart-RRT algorithm in terms of effectiveness, efficiency, and intuitiveness of the output, a comparative evaluation against state-of-the-art alternatives has been performed. More in depth, we took into account the following planning algorithms: RRT [7], RRT* [5] and SJRRT [6]. Similarly to cart-RRT, the selected state-of-the-art algorithms have been modified to return a failure if the maximum cardinality value (set to 500 nodes) is reached without finding a solution. For the sake of completeness, notice that SJRRT has been modified by removing the resolution of the kinematic redundancy, since the UR10 is not kinematically redundant. Five thousand simulations have been performed for each algorithm in each scenario in order to collect statistically significant datasets. In the following, the results of these simulations are presented and discussed. More in detail, Figures 5(a)-5(c) contain histograms that represent the average success rate of the different algorithms, with respect to each scenario. On the other hand, Figures 5(d)-5(f), Figures 5(g)-5(i), and Figures 5(j)-5(l) show box plots representing execution time, tree cardinality, and linear length of the output trajectory, respectively.

As regards the success rate, it is clear that cart-RRT and SJRRT perform consistently better than the others algorithms in all the considered scenarios. This situation is almost surely determined by the possibility to extend the random tree directly towards the goal. Moreover, it is worth noticing that RRT* was not able to provide solutions in scenario #3. Moving to the execution time, in Scenarios #1 and #3 cart-RRT and SJRRT are the fastest algorithms able to find a solution. On the other hand, in Scenario #2, the execution time of cart-RRT is significantly greater than the one of SJRRT and it is also greater than the one of RRT. Not surprisingly, RRT* is the slowest algorithm to converge in all the scenarios, since it optimizes the random tree at each iteration. Moving to tree cardinality, it is quite clear that cart-RRT and SJRRT are able to find solutions without having to reach the maximum allowed tree cardinality, as RRT normally does. As far as RRT* is concerned, the tree cardinality is always equal to the maximum, since the algorithm keeps adding nodes to the tree in order to optimize the output trajectory. Finally, if we take into consideration trajectory length, we can state that in Scenario #1 the outputs of cart-RRT, RRT* and SJRRT can be considered similar, while RRT finds longer trajectories. In Scenario #2, RRT* still provides very short trajectories, while cart-RRT performs worse than RRT, but better than SJRRT. In Scenario #3, finally, cart-RRT finds significantly shorter trajectories with respect to RRT and SJRRT.

Finally, in order to evaluate and compare the intuitiveness of the trajectories computed by the different algorithms, a questionnaire has been designed and submitted to 15 workers. For each scenario, the interviewed workers were presented one example of output trajectory for each algorithm. Then, they were asked to rate the intuitiveness of each trajectory, i.e. its similarity with respect to the trajectory they would have followed. Interviewed workers were asked to rate each trajectory, by assigning a score between "0" (meaning very counterintuitive trajectory) and "5" (i.e. very intuitive trajectory). Results were aggregated and they are displayed in Figures 6, once again in the form of box plots. The comparison clearly demonstrate that the outputs of cart-RRT and RRT* (if present) are considered much more intuitive by the interviewed workers with respect to the outputs of RRT and SJRRT. To sum up, cart-RRT guarantees much better performances than RRT* in terms of success rate and execution time, while RRT* is able to provide shorter trajectories only in Scenario #2. Consequently, we can state that our algorithm represents a promising solution to the motion planning problem for ACMs sharing their workspace with human workers in actual construction yards.

4 Conclusions and Future Developments

In this work, an innovative Cartesian-Space randomized algorithm, named cart-RRT, is proposed. It consists in a modified version of RRT and it is able to provide more intu-

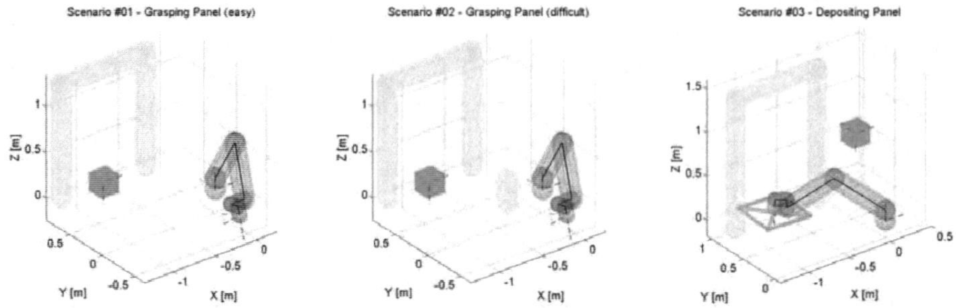

(a) Scenario #1 - grasping panel "easy" (b) Scenario #2 - grasping panel "difficult" (c) Scenario #3 - depositing panel

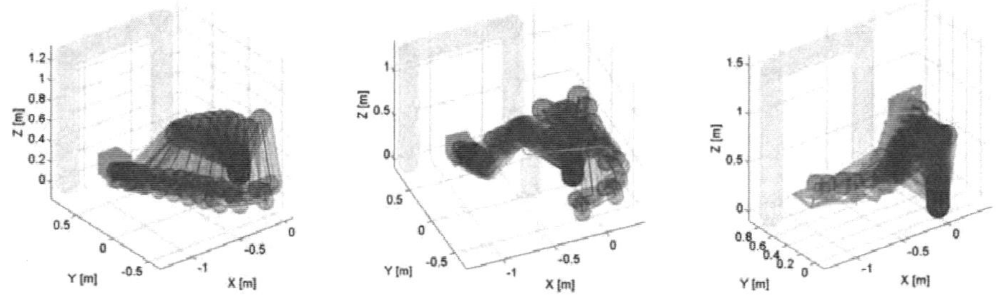

(d) Scenario #1 - grasping panel "easy" (e) Scenario #2 - grasping panel "difficult" (f) Scenario #3 - depositing panel

Figure 3. Simulation scenarios and examples of solutions.

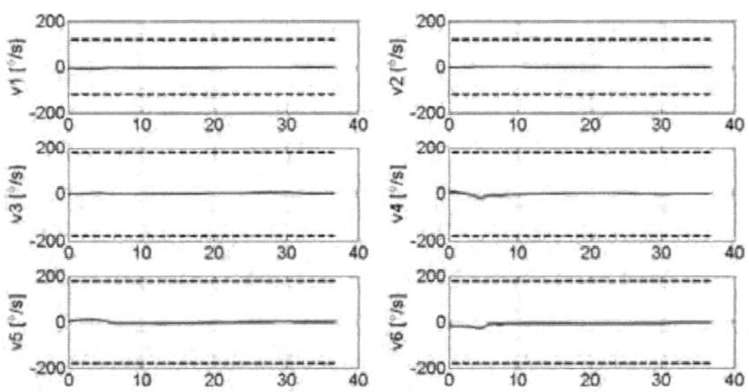

(a) Joint velocity profiles

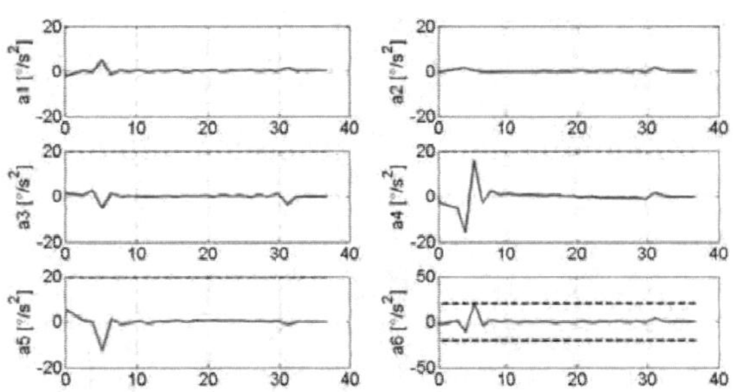

(b) Joint acceleration profiles

Figure 4. Example of solution for Scenario #1.

itive output trajectories thanks to two distinct features: the fact that nodes are sampled directly in the Cartesian-space, and the possibility to extend the tree directly towards the goal. Furthermore, the algorithm ensures that the output trajectories are collision-free and consistent with respect to actuation constraints. The effectiveness of the proposed approach is demonstrated in multiple simulated scenarios. Moreover, a comparative analysis of the results obtained by our algorithm with respect to the ones produced by other state-of-the-art planners is shown. In the end, not only we demonstrate that the efficiency of our algorithm is comparable with respect to state-of-the-art alternative, but we can also state that our algorithm is able to provide more intuitive trajectories with respect to state of the art planners, from the point of view of a human construction worker. At the moment, the most promising future development is represented by the possibility to extend this algorithm to the case of kinematically redundant ACMs. A study will be conducted to investigate how extra-DoFs can be used in order to improve optimality and safety (in terms of distance with respect to the obstacles) of the resulting trajectory.

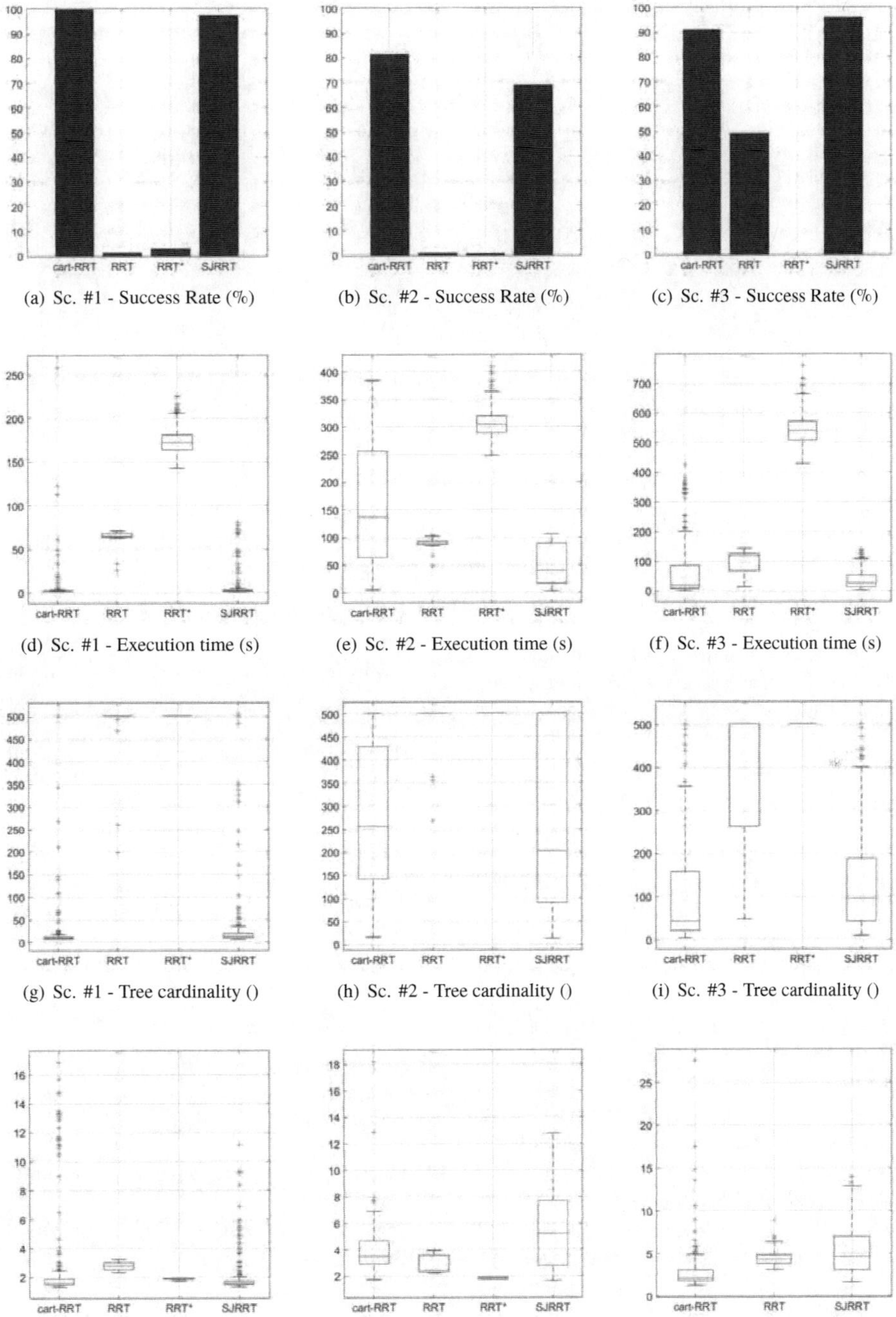

(a) Sc. #1 - Success Rate (%)
(b) Sc. #2 - Success Rate (%)
(c) Sc. #3 - Success Rate (%)

(d) Sc. #1 - Execution time (s)
(e) Sc. #2 - Execution time (s)
(f) Sc. #3 - Execution time (s)

(g) Sc. #1 - Tree cardinality ()
(h) Sc. #2 - Tree cardinality ()
(i) Sc. #3 - Tree cardinality ()

(j) Sc. #1 - Trajectory lenght (m)
(k) Sc. #2 - Trajectory lenght (m)
(l) Sc. #3 - Trajectory lenght (m)

Figure 5. Comparative evaluation results

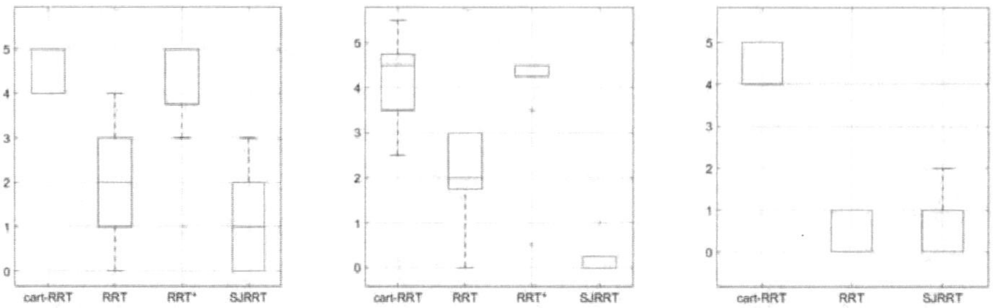

(a) Scenario #1 (b) Scenario #2 (c) Scenario #3

Figure 6. Evaluated intuitiveness of the output trajectory

References

[1] Jerome Barraquand, Lydia Kavraki, Jean-Claude Latombe, Rajeev Motwani, Tsai-Yen Li, and Prabhakar Raghavan. A random sampling scheme for path planning. The International Journal of Robotics Research, 16(6):759–774, 1997.

[2] M. S. Branicky, M. M. Curtiss, J. Levine, and S. Morgan. Sampling-based planning, control and verification of hybrid systems. IEE Proceedings - Control Theory and Applications, 153(5):575–590, Sept 2006.

[3] J. h. Jeon, S. Karaman, and E. Frazzoli. Anytime computation of time-optimal off-road vehicle maneuvers using the rrt*. In Decision and Control and European Control Conference (CDC-ECC), 2011 50th IEEE Conference on, pages 3276–3282, Dec 2011.

[4] S. Karaman and E. Frazzoli. Sampling-based optimal motion planning for non-holonomic dynamical systems. In Robotics and Automation (ICRA), 2013 IEEE International Conference on, pages 5041–5047, May 2013.

[5] Sertac Karaman and Emilio Frazzoli. Sampling-based algorithms for optimal motion planning. Int. J. Rob. Res., 30(7):846–894, June 2011.

[6] Bakir Lacevic and Paolo Rocco. Safety-oriented path planning for articulated robots. Robotica, 31(6):861âĂŞ874, Feb 2013.

[7] S. M. LaValle and J. J. Kuffner. Randomized kinodynamic planning. In Robotics and Automation, 1999. Proceedings. 1999 IEEE International Conference on, volume 1, pages 473–479 vol.1, 1999.

[8] Steven M. LaValle and Jr. James J. Kuffner. Randomized kinodynamic planning. The International Journal of Robotics Research, 20(5):378–400, 2001.

[9] Maxim Likhachev and Dave Ferguson. Planning long dynamically feasible maneuvers for autonomous vehicles. The International Journal of Robotics Research, 28(8):933–945, 2009.

[10] College of Engineering Control Systems Laboratory University of Illinois at Urbana-Champaign. Rapidly exploring random tree (rrt) path planning.

[11] A. Perez, R. Platt, G. Konidaris, L. Kaelbling, and T. Lozano-Perez. Lqr-rrt*: Optimal sampling-based motion planning with automatically derived extension heuristics. In 2012 IEEE International Conference on Robotics and Automation, pages 2537–2542, May 2012.

[12] M. Ragaglia, M. Prandini, and L. Bascetta. Poli-rrt*: Optimal rrt-based planning for constrained and feedback linearisable vehicle dynamics. In Control Conference (ECC), 2015 European, pages 2521–2526, July 2015.

[13] M. Ragaglia, M. Prandini, and L. Bascetta. Multi-Agent Poli-RRT*, pages 261–270. Springer International Publishing, 2016.

[14] A. Tahirovic and G. Magnani. General framework for mobile robot navigation using passivity-based mpc. IEEE Transactions on Automatic Control, 56(1):184–190, Jan 2011.

[15] A. M. Zanchettin, L. Bascetta, and P. Rocco. Achieving humanlike motion: Resolving redundancy for anthropomorphic industrial manipulators. IEEE Robotics Automation Magazine, 20(4):131–138, Dec 2013.

Pose and Position Estimation of Dozer Blade in 3-dimensional by Integration of IMU with Two RTK GPSs

D.I. Sun[a] and S.H. Kim[b], Y.S. Lee[b], S.K. Lee[a], C.S. Han[c]

[a]Mechatronics Engineering of Hanyang University of Republic of Korea
[b]Graduate School of Hanyang University, Republic of Korea
[c]Robot Engineering of Hanyang University of Republic of Korea
E-mail: jeniussdi@hanyang.ac.kr, cshan@hanyang.ac.kr

Abstract –

Automation system in construction machine such as excavator, loader and other things has been developed to improve the performance and reduce the cost, time and manpower. Dozer is no exception.

Dozer is a crawler equipped with a substantial metal plate known as blade used to push large quantities of soil, sand, or other materials during leveling, digging or conversion work. Dozer can be found on a wide range of sites, which are mines, quarries and road. Therefore, the cost and time consuming in dozer working are highly significant matter.

However, in order to implement the automation system in construction equipment, guidance system for construction working has to be done before. Because the automation system can perform the operation and work better with information such as velocity, pose and posture etc. thorough the sensor. These system is called machine guidance system because guidance system is applied to construction machine. Shortly, if guidance system is done very well, the achievement of automation system could be better. Dozer is also no exception.

This paper will describe the method for estimating the pose and position of dozer blade, namely guidance of dozer, by sensor fusion or integration of inertial measurement unit device with two RTK GPSs. And then the proposed method is verified by experimental result.

Keywords –

Machine Guidance; Dozer; Sensor Fusion

1 Introduction

Dozer is a crawler equipped with a substantial metal plate known as blade used to push large quantities of soil, sand, or other such material during construction such as grading, leveling, digging or conversion work. [1] Dozer can be found on a wide range of sites which are mines, quarries, road and express construction etc. Therefore, the cost and time consuming in dozer working are highly significant matters.

The automation system of dozer is assistance system for improving the performance such as accuracy and efficiency. A driver drives and commands the construction works through the system interface. After that, system do work according to order by using the information from sensor systems installed in dozer and field. Operator just pay attention to only driving. Blade motions such as lifting, tilting, angling known as pitch, roll and yaw are operated automatically during driving. As mentioned above, consequently, automated dozer brings the enhancement of productivity and efficiency rather than manual. Figure 1 includes the these contents. Komatsu, representative Japanese company wrote a technical paper about dozer automation system which is including the efficiency. [2]

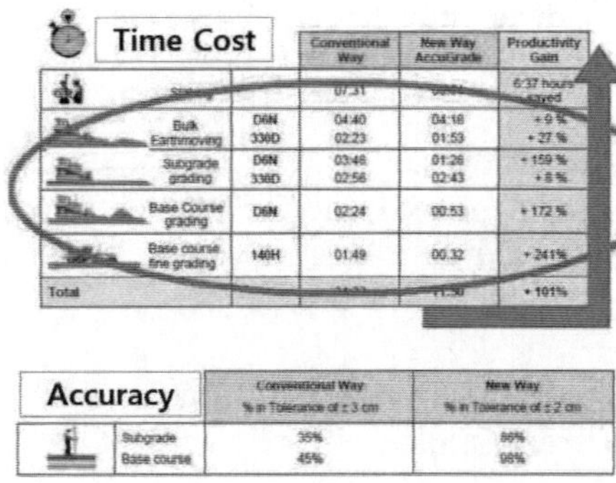

Figure 1. The table is presented in Komatsu, Japanese development company: Advanced performance in both cost and accuracy part

The automation system in civil engineering is called machine control. And the system which incorporates the position sensors, motion measuring units and other devices is to provide on-board with information about the movement of machine such as dozer and excavator. And this system is called machine guidance. In this paper, these systems are applied to dozer, so we can name it as a dozer machine control and guidance.

As mentioned above, dozer machine control is a factor that can bring a lot of benefits. In order to implement the dozer machine control, however, machine guidance must be done before. Because, the control can be performed by receiving the movement information like position, orientation, pose and velocity etc. by using the sensor systems. To put it briefly, if machine guidance is performed greatly, the achievement of dozer machine control system could be better. For this reason, machine guidance is also significant part, and the sensor system should be installed well too.

This paper will describe the method about pose and position estimation of blade by integration of IMU with two RTK GPSs. And then proposed algorithm is verified by experimental results.

2 Well Used Sensor Fusion Method Examples about Pose Estimation

Many researcher have worked about sensor fusion method for estimating the pose and position of vehicle. The UAV(Unmanned Aerial Vehicle), ship and flight has to install the navigation system for driving as well.

Keong, [3], proposed the method about determining heading and pitch by using a single difference GPS. Kumar, [4], used the extended Kalman filter for doing integration of inertial navigation system and global positioning system. Godha, [5], suggested the integrated GPS and INS system for pedestrian navigation in a single degrade environment. This paper has a novelty value in approach to non-vehicle system. In domestic, Ko, [6], performed the indoor localization of a mobile robot by using external sensor. Yeom, [7], used the GPS and accelerometer for estimating the position of moving object.

To implement the dozer machine control, Xue, [8], analyzed the kinematics of dozer by simulation, and Sun, [9], me, studied on sensor fusion algorithm for estimating the attitude of dozer blade.

3 Estimating the Position and Pose of Dozer Blade

3.1 System configuration and coordinates

In machine guidance of dozer, sensor system may have to be installed to dozer. Firstly, two IMU are used to get the pose, namely, roll, pitch and yaw of blade and cabin at period of 100Hz. Secondly, GPS antennas are used to get the position. The one is base antenna that will be installed on the outside of dozer, the others are attached on the blade. The rover will be in the cabin. Figure 2 illustrates a system configuration in detail.

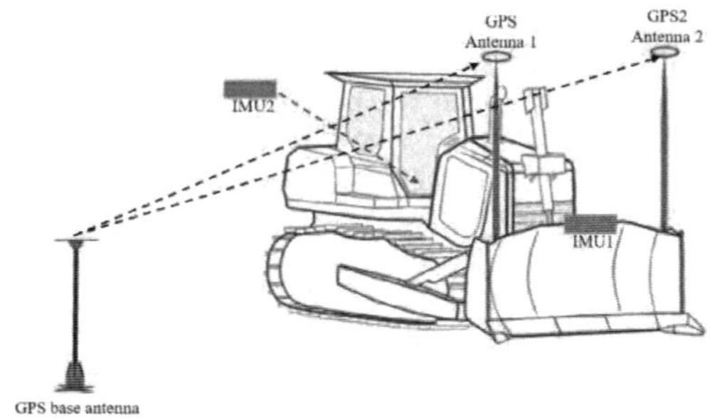

Figure 2. System configuration: IMU and GPS

To generate the Cartesian coordinates system, instead of GPS which has longitude, latitude and height, UTM (Universal Transverse Mercator) coordinates is used. Based on this, the frame of dozer, sensor and reference frame can be formed. Reference frame is naturally set to be the origin of UTM coordinate system. And then frame with respect to cabin of dozer and blade are set by convention, i.e., roll is defined by rotation about axis that pointing forward and yaw is rotation that done by upward axis. Pitch is done by right hand rule.

We installed the GPS and IMU for receiving the pose and position of blade and cabin. Especially, pose of blade is estimated by calculation of GPS data for compensating the unstable in IMU device, which is expressed with respect to origin of frame in UTM coordinates system. This is why frame of sensor system is necessary.

3.2 Pose Estimation

The process of linear Kalman filter is used for estimating the pose of blade reliably. Figure 3 is algorithm about linear Kalman filter that is used in.

$$(N \le k \le 2N-1)\ \hat{x}_k^- = A\hat{x}_{k-1} + u_k$$

$$\hat{x}_N^- = A\hat{x}_{N-[\,]},\quad [\,] = \left[\frac{IMU\_data\_sampling\_frequency}{GPS\_data\_sampling\_frequency}\right]$$

$$P_N^- = AP_{N-[\,]}A^T + Q$$

$$K_N = P_N^- H^T (HP_N^- H^T + R)^{-1}$$

$$\hat{x}_N = \hat{x}_N^- + K_N (Z_N - H\hat{x}_N^-)$$

$$P_N = (I - K_N H) P_N^-$$

$$u_N \sim u_{2N-1} = K_N (Z_N - H\hat{x}_N^-)$$

Figure 3. Modified Kalman filter algorithm for estimating the pose

3.2.1 Prediction and Measurement Modeling

There are two process modellings in the linear Kalman filter. One is prediction and another is measurement. Prediction modelling is defined as follows, equation (1).

$$x_{k+1} = Ax_k, \tag{1}$$

where

$$x = \begin{bmatrix} 1 \\ Roll_{IMU} \\ Pitch_{IMU} \\ yaw_{IMU} \end{bmatrix}, A(N) = \begin{bmatrix} 1 & \\ Roll_{IMU} & \\ Pitch_{IMU} & 0_{4\times3} \\ Yaw_{IMU} & \end{bmatrix}$$

N is the step parameter for GPS data, and state of x will be compared with pose derived from measurement modelling.

Measurement modelling is defined as follows, equation (2), (3), (4) and (5).

$$z_N = \begin{bmatrix} Roll_{GPS}(N) \\ Pitch_{GPS}(N) \\ Yaw_{GPS}(N) \end{bmatrix} \tag{2}$$

$$Roll_{GPS} = \arcsin((z_1 - z_2)/Baseline) \tag{3}$$

where z_1 and z_2 is height data of GPS

$$Yaw_{GPS} = \arctan((y_1 - y_2)/(x_1 - x_2)) + \pi/2 \tag{4}$$

Where, $x_1 > x_2$, $\pi/2 \leq \arctan((y_1 - y_2)/(x_1 - x_2)) \leq \pi/2$

$$Yaw_{GPS} = \arctan((y_1 - y_2)/(x_1 - x_2)) - \pi/2 \tag{5}$$

Where, $x_1 < x_2$, $\pi/2 \leq \arctan((y_1 - y_2)/(x_1 - x_2)) \leq \pi/2$

Each function is used in linear Kalman filter so that the attitude of blade can be estimated more precisely.

3.3 Estimation of Edge Position

Most important things in dozer machine guidance in construction working is to know the bottom position of blade, namely edge. Because, landform have a considerable effect on working process. However, edge position cannot be estimated directly because of sensor installation on the top of the blade. Therefore, in order to estimate the edge position by using the sensor system, kinematics has to be used as alternative solution. Figure 4 is explaining overall frame of body, link and joint. Each Table 1 and 2 is Denavit-Hartenberg parameters table about figure 4.

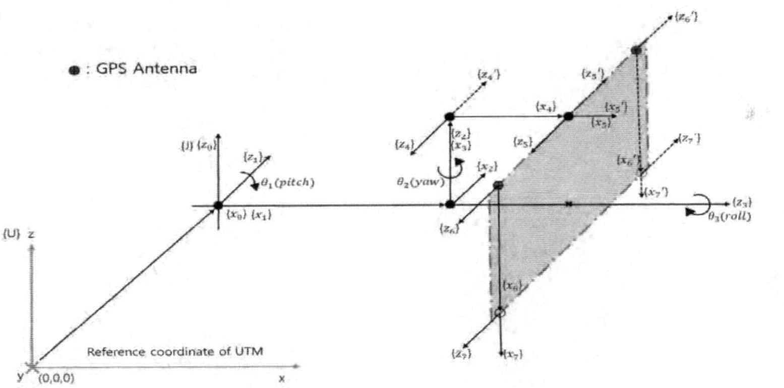

Figure 4. The assignment of link frames in dozer blade having 3 degrees of freedom

Table 1. D-H parameters table concerned with antenna 1

i	α_{i-1}(deg)	$a_{i-1}(m)$	$d_i(m)$	θ_i(deg)
1	-90	0	0	θ_1
2	90	a_1	0	$90+\theta_2$
3	90	0	0	$90+\theta_3$
4'	**90**	a_3	**0**	**90**
5'	**0**	a_4	**0**	**0**
6'	**0**	**0**	**Baseline/2**	**90**
7'	**0**	a_6	**0**	**0**

Table 2. D-H parameters table concerned with antenna 2

i	α_{i-1}(deg)	$a_{i-1}(m)$	$d_i(m)$	θ_i(deg)
4	**-90**	a_3	**0**	**-90**
5	**0**	a_4	**0**	**0**
6	**0**	**0**	**Baseline/2**	**-90**
7	**0**	a_6	**0**	**0**

The reference frame is denoted by {U}. {J} or {0} is frame of pitch joint and origin of this is on the middle of two pitch joint. Frame 7 is end effector and 6 is used for estimating the origin position of {0}. Theta 1, 2 and 3 is angle of pitch, yaw and roll. The transformation matrix of body expressed with respect to UTM coordinates system can be derived by IMU data which is installed in cabin. By doing the forward kinematics overall, edge position of blade will be estimated.

4 Experiment and Evaluation

4.1 Experimental Setup

Figure 5 is showing the dozer and sensor system and where experiment was done.

Figure 5. Experimental setup: Dozer, sensors and location that is Hanyang Univ. at Ansan

The sampling rate of GPS and IMU are set to be different. GPS is 20Hz and IMU is 100Hz. Link parameters and joint parameters of D-H parameter were measured. Baseline is 2.78m, a_1 is 1.69m, a_3 is 0.51m, a_4 is 0.23m, a_6 is 0.85m and d_6 is Baseline/2. Each parameter Q, R, H and P_0 of sensor integration part are given by equation (6).

$$Q = \begin{bmatrix} 0 & 0 & 0 & 0 \\ 0 & 1 & 0 & 0 \\ 0 & 0 & 0 & 0 \\ 0 & 0 & 0 & 1 \end{bmatrix}, R = \begin{bmatrix} 0 & 0 & 0 \\ 0 & 1 & 0 \\ 0 & 0 & 0 \end{bmatrix} \tag{6}$$

$$H = \begin{bmatrix} 0 & 1 & 0 & 0 \\ 0 & 0 & 1 & 0 \\ 0 & 0 & 0 & 1 \end{bmatrix}, P(0) = O_{4\times4}$$

4.2 Result and Evaluation

Experiment was done in roll, pitch and yaw motion of blade with non-driving. In this paper, experimental result will be mentioned about roll motion with non-driving representatively.

The results of estimating the pose in roll motion at period of 100Hz through the sensor integration are in Figure 6. The results of graph which represent the estimated edge position with regard to x, y and z axis in roll motion can be founded in Figure 7, 8, and 9.

In Figure 6, pose, in other words, rotation data from IMU and estimations derived from sensor integration are matched well. If more anti-vibration system to sensor is applied well, the result will be better. Additionally, Figure 7, 8 and 9 are modified little for analysing the proposed method. In the figure, UTM zone position value is removed. Consequently, edge position of blade is estimated as expected.

5 Conclusion

In this paper, estimation method about pose and position of dozer blade was proposed for machine control. To do that, IMU and RTK GPS were installed in dozer system.

The estimation results about edge position were done as expected. However, sensor error such as black out, satellite factor and inaccurate measurement of link parameter were existed. This is a problem awaiting to be solved with software and hardware.

The estimated position was not compared with real edge position. Therefore, more reliable evaluation is needed and should be done later.

Pose and position of dozer, not blade, was not handled. But for doing machine control, guidance system about cabin and blade of dozer are necessary. So, this will have to be done later also.

As mentioned in introduction, machine guidance is necessary part in implementation of machine control. Final goal of this research is to do the machine control. When machine guidance is completed, machine control of dozer will be progressed based on this research.

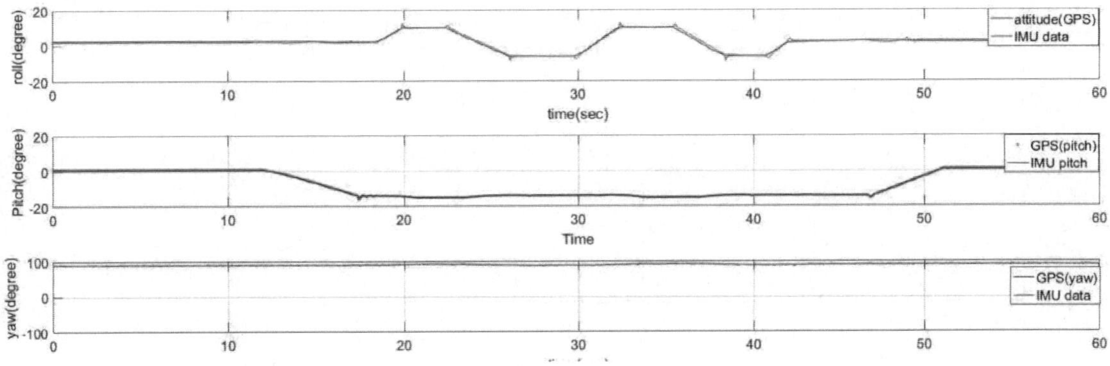

Figure 6. The graph of test results about estimated pose of blade in roll motion: IMU data is compensated by GPS

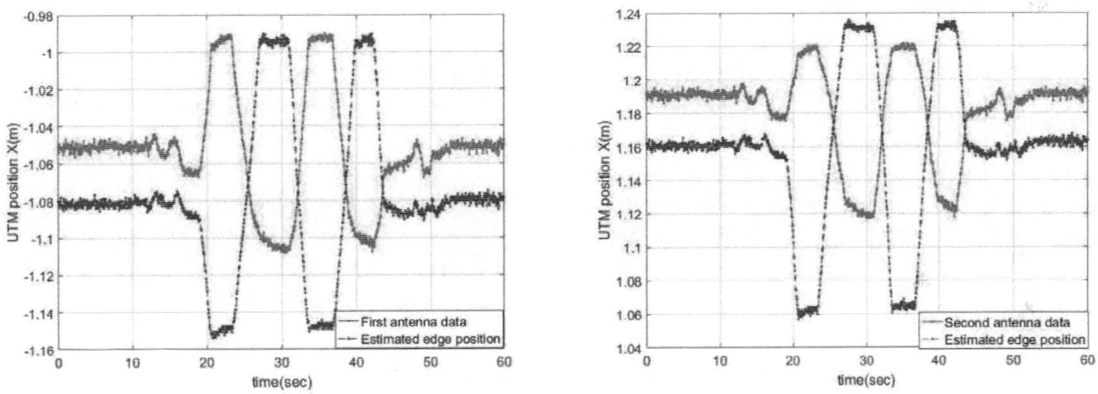

Figure 7. The result graph of test about estimating the edge position with regard to x axis in roll motion

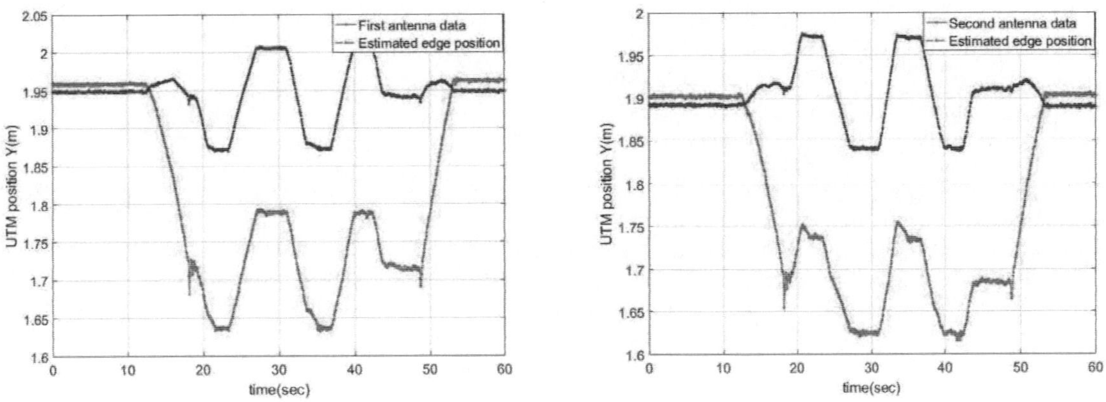

Figure 8. The result graph of test about estimating the edge position with regard to y axis in roll motion

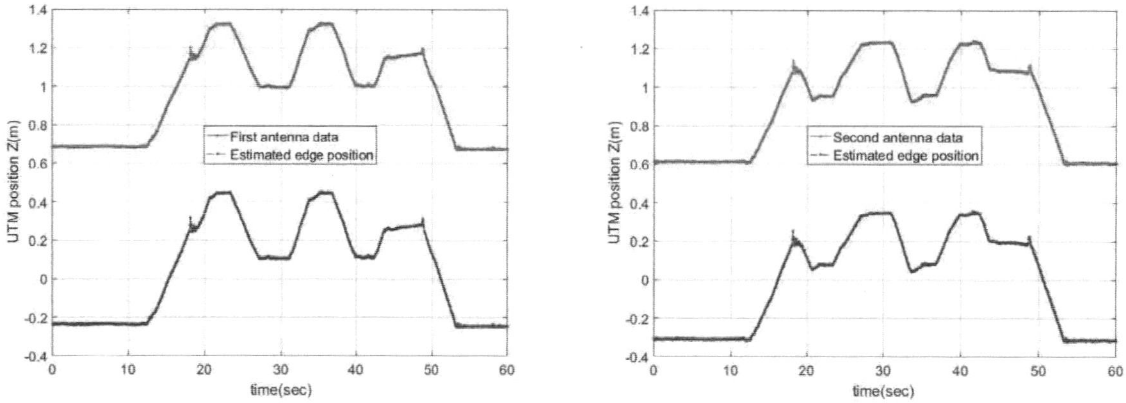

Figure 9. The result graph of test about estimating the edge position with regard to z axis in roll motion

6 Acknowledgements

This research was supported by the Industrial Strategic technology development program (No.10052965) funded by the Ministry of Trade, Industry and Energy (MOTIE), KOREA and was supported by a grant(14SCIP-B079344-01) from Smart Civil Infrastructure Research Program funded by Ministry of Land, Infrastructure and Transport(MOLIT) of Korea government and Korea Agency for Infrastructure Technology Advancement(KAIA)

References

[1] Wikipedia, http://en.wikipedia.org/Bulldozer.

[2] Kazuhiko Hayashi, et al. " Development of D61EXi_PXi-23 Bulldozer with automatic control system of work equipment." Komatsu technical report, 59. 166 (2013): 9-16.

[3] Keong, Jiun Han., " Determining heading and pitch using a single difference GPS/GLONASS approach." National Library of Canada= Biblioth èque nationale du Canada, 2001.

[4] Kumar, Vikas. Integration of inertial navigation system and global positioning system using Kalman filtering. Diss. INDIAN INSTITUTE OF TECHNOLOGY, BOMBAY MUMBAI, 2004.

[5] Godha, Saurabh, G é rard Lachapelle, and M. Elizabeth Cannon. "Integrated GPS/INS system for pedestrian navigation in a signal degraded environment." ION GNSS. Vol. 2006. 2006.

[6] Ko, Nak-Yong, and Tae-Gyun Kim. "Indoor Localization of a Mobile Robot Using External Sensor." Journal of Institute of Control, Robotics and Systems 16.5 (2010): 420-427.

[7] Yeom, Jeong-Nam, et al. "Position Estimation System of Moving Object using GPS and Accelerometer." Journal of Korea Multimedia Society 12.4 (2009): 600-607.

[8] Gang Xue, C. S. Han, Y. S. Lee, C.U.Ji, M. S. Kang. "A study on the kinematic analysis and dynamic simulation for automatic 3D control of the dozer blade." Conference paper of KSPE, (2015.5): 419-420.

[9] D. I. Sun, C. S. Han, Y. S. Lee, S. H. Kim, S. H. Lee. "A Study on Sensor Fusion Algorithm for Dozer' s Blade Position and Attitude Estimation." Conference paper of KSPE, (2016.5): 47-48.

Optimal Variable Speed Limit Control under Connected Work Zone and Connected Vehicle Environment

S. Du[a], S. Razavi[b] and W. Genders[a]

[a]Department of Civil Engineering, McMaster University, Canada
[b]Chair in Heavy Construction, Department of Civil Engineering, McMaster University, Canada
E-mail: dus5@mcmaster.ca, razavi@mcmaster.ca, genderwt@mcmaster.ca

Abstract

Despite the long-term benefits gained through road maintenance projects, the work zone involved can cause serious traffic congestion and greatly increase traffic delay. Variable speed limit (VSL) enforcement provides a means to reduce the travel time and alleviate the traffic congestion in work zones. This paper presents a novel proactive VSL control algorithm for freeway work zone traffic management under connected work zone and connected vehicle environment. In a connected work zone environment, the estimated work zone capacity can be easily shared considering the work zone configuration factors such as number of lane closure, signal distance, presence of work activity, etc. After incorporating the estimated capacity which contributes to a better modeling of the traffic disturbance caused by work zone, the traffic flow predictive model was established to obtain preventive measures and avoid traffic breakdown. In addition, based on the trajectories of individual vehicles recorded and shared by connected vehicles, the objective function was developed for the VSL control algorithm to minimize the Total Travel Time which can be used to measure level of the traffic congestion. The VSL control was implemented and tested in a traffic simulator SUMO. The results showed that the designed VSL algorithm effectively decreased the total travel time with the reduction of the speed variance which potentially improved the traffic safety at work zones.

Keywords

Variable Speed Limit; Connected Work Zone; Connected Vehicles; Total Travel Time

1 Introduction

Traffic congestion frequently occurred in freeway has become a critical issue in modern societies. To improve the traffic condition, $66 billion are estimated to be invested into maintaining and repairing road infrastructure in Canada between 2013 and 2023 [1]. However, these projects such as resurfacing, lane widening, drainage realignment, etc. create numerous construction work zones in freeways and other urban arterial roads. Such work zones, which involve lane closure, will have serious impacts on traffic mobility. According to [2], in 2014 the estimated 888 million hours delay and the fuel loss of over 310 million gallon are attributed to work zones in the US. Therefore it is essential to develop an effective control strategy on freeway traffic operation to mitigate the traffic congestion caused by work zones.

Variable speed limit, one of the Intelligent Transportation System (ITS) control strategies, can dynamically change the posted speed limits along the upstream segments of a bottleneck. As such, VSL is able to reduce the traffic inflow or homogenize the traffic flow to avoid traffic breakdown, reduce the total travel time, achieve uniform distribution of traffic flow and improve the safety. However, the majority of the research developed the VSL control strategies based on macroscopic speed-limit model, which extracts the aggregate traffic states such as flow, density and average speed. This makes it difficult to accurately acquire and analyze the microscopic behavior (acceleration, merge location, etc.) of individual vehicles given the work zone disturbance.

Connected vehicle (CV) environment enables vehicle-to-vehicle and vehicle-to-infrastructure wireless communication to better share the traffic states of each CV-enabled vehicle. Few studies have considered making the work zone connected. Connected work zone enables a more accurate and timely assessment of the traffic impacts caused by the configuration factors of work zone such as work zone ahead warning signal distance, number of lane closure, work activity inside work zone, etc.

Since the microscopic vehicle behavior and work zone configuration have great impacts on traffic flow, the VSL control strategy should be developed given due consideration to such impacts. Therefore a microscopic traffic speed-limit model, capable of obtaining more accurate traffic states as the input of the VSL controller, is in need. Meanwhile, the work zone as the disturbance to the traffic also needs to be modeled in order to consider the work zone configuration.

This paper introduces a novel proactive VSL control strategy to optimize the VSL, minimize the impact on travel time and mitigate the traffic congestion under connected work zone and connected vehicle environment. The rest of this paper is organized as follows. Recent studies are reviewed in Section 2. The research objectives are described in Section 3. The development of a model predictive controller to optimize the VSL given the constraints of posted speed limit on the freeway is presented in Section 4.1. The connected work zone is analyzed in Section 4.2. After the introduction of experiment setup, the effectiveness of the designed VSL control was tested in the traffic simulator SUMO [3] as described in Section 5. Section 6 concludes this paper.

2 Literature Review

There are a number of studies on developing VSL control strategies to reduce the traffic congestion and improve traffic condition. The Model Predictive Control (MPC) was utilized by [4] to proactively coordinate VSL to eliminate the traffic shock waves. The macroscopic model METANET was modified to predict the traffic flow, density, average speed and queue length so that VSL controller would be able to reduce the effects of moving jams. The SPECIALIST algorithm was designed based on the shock wave theory by using the fundamental diagram (FD) [5]. The shock wave detection, resolve and assessment were proposed by analysing the traffic states on the FD. The FD was also employed in cell transmission model design to optimize the VSL control [6]. The modified FDs were proposed at the active bottleneck on the freeway to model the capacity drop and variable speeds. The optimization of VSL proved to improve the traffic flow and capacity. The VSL optimization was also proposed and developed to improve traffic safety by homogenizing the traffic flow given different driver compliance rates [7].

The control strategies adopted in the above-mentioned studies were able to optimize the VSL proactively to suppress the shock wave, improve the traffic flow capacity and enhance the traffic safety. However, the control method was developed under the macroscopic simulation environment. Therefore, the behaviour and trajectories of microscopic individual vehicles, which are vitally important to analyze the disturbance of traffic flow, were neglected. Inappropriate lane changing behavior can easily cause the shock wave. The vehicle sudden acceleration and deceleration behaviors due to velocity changes of the leading or following vehicles also greatly affect the traffic flow. To accurately analyze the traffic flow on the freeway and develop optimal VSL control strategy, it is essential to take the behavior of individual vehicle into consideration.

A number of VSL control strategies were also developed in work zone applications to mitigate the negative effects such as traffic congestion, green house emission, safety issue, etc. A multi-objective function was established to optimize the travel time, improve safety and reduce gas emission under microscopic simulation [8]. In this study, a MPC was implemented to predict the microscopic vehicle behaviors considered a hypothetical incident with speed limit on the freeway. In another study, a macroscopic traffic flow model was established to develop the VSL controller for the freeway work zone operations [9]. In [10], to study the effectiveness of VSL, a control logic was designed to determine the posted speed limits on the highway with a work zone.

The work zone applications of VSL have been explored to study the interaction between the traffic flow and work zone so as to optimize the VSL and reduce the negative impacts due to the presence of work zone. However, the details of work zone such as the number of lane closure and warning signal distance have not yet considered. Instead, the work zone disturbance was simplified as a static speed limit on the freeway. In reality, the different configuration factors of work zone will result in different capacity and have different impacts on traffic flow. Thus having a better understanding of work zone as the disturbance is of great importance to develop an effective VSL control strategy.

Though extensive studies have been conducted to develop VSL control strategies and explore the work zone applications, the lack of the behavior of each vehicle reacted to the work zone and understanding of work zone disturbance makes it difficult to obtain optimal VSL. Therefore, a proactive VSL control which is capable of incorporating the microscopic vehicle behavior and work zone configuration needs to be developed.

3 Research Objectives

Developing the optimal VSL control for freeway work zone traffic management to mitigate the traffic congestion and reduce travel time is the major objective of this study. The following objectives are included: (1) developing a predictive VSL control under connected vehicle environments to incorporate the behavior of individual vehicle, (2) analyzing the effects of connected work zone on VSL controller design and (3) evaluating the effectiveness of the designed control strategy in SUMO.

4 Methodology

To reduce the traffic congestion caused by work zone, a MPC is proposed to achieve optimal VSL control on the freeway under the connected work zone and connected vehicle environment. To obtain the traffic states and predict the behavior of each vehicle, the modified Krauß car-following model and 4-layer motivation lane changing model are adopted [11][12]. Since the freeway work zone can easily lead to non-recurring congestion and has significant influence on traffic flow, the work zone configuration factors are analyzed to establish the vehicle merge location model, estimate the work zone capacity and evaluate the effects of work zone disturbance. The comprehensive traffic data such as speed and position of individual vehicle and the work zone factors are available under connected environment.

To make full use of the traffic states and work zone information, the MPC is developed and connected work zone effects are analyzed in the following sections.

4.1 Model Predictive Control

To develop VSL control for traffic flow, MPC is proposed to follow the controller design procedures showed in Figure 1.

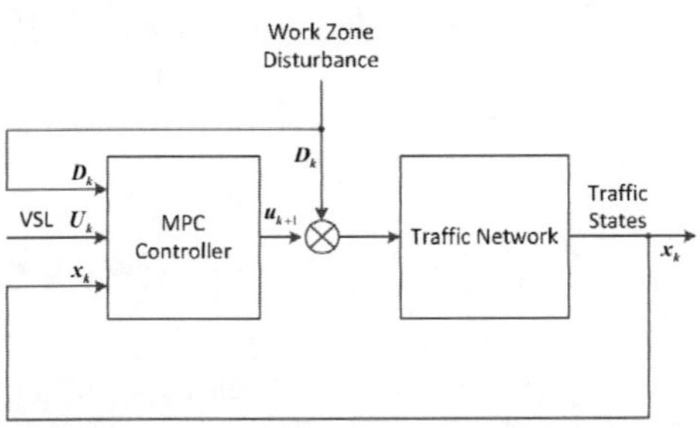

Figure 1. MPC Control for Traffic Network

The MPC controller is designed to control the traffic network through the control signal VSL $\boldsymbol{u}_{k+1}$ at next time step k+1. Three signals are input to MPC controller to optimize the VSL $\boldsymbol{u}_{k+1}$. The work zone on the freeway is treated as $\boldsymbol{D}_k$ which consists of N_c controller time steps of disturbance from work zone. The VSL control signal $\boldsymbol{U}_k$ as the second input has the VSL solutions for N_c controller time steps. The traffic states $\boldsymbol{x}_k$ such as the velocity, position of each vehicle at time step k collected from the connected vehicle environment are fed back to the MPC controller for each simulation step as the third input signal. By taking the feedback traffic states $\boldsymbol{x}_k$, together with the VSL $\boldsymbol{U}_k$ and work zone disturbance $\boldsymbol{D}_k$, MPC controller is developed to optimize the VSL.

The components inside MPC controller are illustrated in Figure 2. The objective function is designed to optimize the next time step VSL control $\boldsymbol{u}_{k+1}$ by minimizing the total travel time. The travel time is calculated based on the predicted N_p time steps of traffic states. Specifically, the velocity and position of each vehicle are predicted for N_p time steps by using the car-following model. Thus in the designed MPC control, the traffic states are predicted in N_p prediction horizon while the control signals are predicted in N_c controller horizon. However, only the first optimal VSL control $\boldsymbol{u}_{k+1}$ is applied in traffic network. Then the prediction horizon shift one time step when the new traffic states are obtained. The VSL control action is optimized in this rolling horizon scheme considering the constraints of VSL in real world.

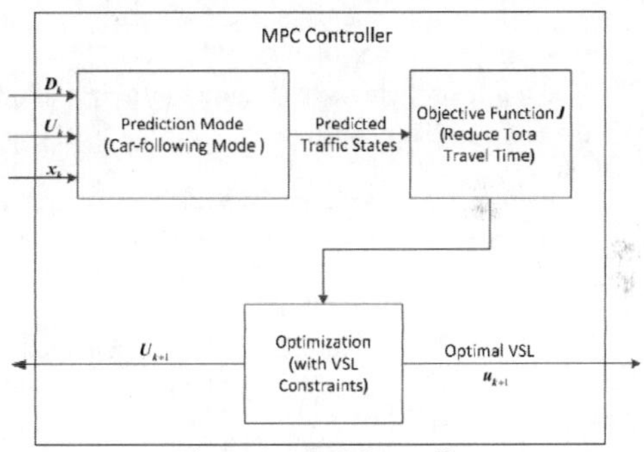

Figure 2. MPC Controller

4.1.1 Car-following Model

The car-following model is developed based on the modified Krauß model which has low computation complexity and was proved to have high validity [11].

There are three speed constraints needed to be considered for this car-following model. The first one is safe speed. The speed v of the vehicle which follows the leader vehicle whose speed is v_{leader} should not exceed the safe speed v_{safe} meaning the gap d_{gap} between the vehicle and its leader should be larger than the brake distance d_{brake}. This prevents the vehicle having the rear-end collision. The brake distance d_{brake} and safe speed are calculated as:

$$d_{brake} = \frac{v_{safe}^2}{2a_{\max dec}} \quad (1)$$

$$v_{safe} = \sqrt{(\tau * a_{\max dec})^2 + v_{leader}(t-1)^2 + 2 * a_{\max dec} * d_{gap}(t-1)} - \tau * a_{\max dec} \quad (2)$$

where the $a_{\max dec}$ is the maximum deceleration ability, τ is the driver reaction time, $v_{leader}(t-1)$ is the velocity of leader vehicle at previous time step t-1 and $d_{gap}(t-1)$ is the gap between the vehicle and its leader at previous time step t-1.

In addition to the constraint of vehicle deceleration ability, the speed of vehicle also needs to satisfy the acceleration ability showed in Equation (3).

$$v_{acc} = v(t-1) + a_{\max acc} \quad (3)$$

where $a_{\max acc}$ is the maximum deceleration ability and $v(t-1)$ is the speed of the vehicle at previous time step t-1. However, in reality when the speed is close to the maximum speed, the acceleration tends to be smaller. Therefore, a decay factor was applied to the acceleration in Equation (4).

$$a_{\max adjusted} = a_{\max acc}(1 - \frac{v}{v_{\max}}) \quad (4)$$

where $v_{\max}$ is the speed limit posted.

The third constraint is maximum speed $v_{\max}$ allowed on the freeway. This constraint reflects the VSL control action.

Therefore the vehicle speed which satisfied these three constraints is the desired speed showed in Equation (5).

$$v_{desire} = \min\{v_{safe}, v_{acc}, v_{\max}\} \quad (5)$$

When the driver's imperfection is considered, the desired speed will be reduced by a stochastic deceleration expressed by

$$v = \max\{0, v_{desire} - \lambda * a_{\max adjusted} * \delta\} \quad (6)$$

where λ is the imperfection factor of the driver and δ is a random number between 0 and 1.

4.1.2 Lane Changing Model

The lane changing model which has a low wrong-lane teleport rate in SUMO was adopted [12] in this study. This model analyzes four motivations to perform the lane changing behavior: (1) strategic change, (2) cooperative change, (3) tactical change and (4) regulatory change. In this study, the focus is on performing lane changing behavior due to the lane closure at the work zone area. Therefore, strategic change, which happens when the dead lane is in front, is mainly discussed.

Before analyzing the speed changes of vehicles, the some terms for the lane and vehicles will be briefly introduced in Figure 3. As it is showed, the ego vehicle is the vehicle which is performing the lane changing behavior to the target lane. The leader and follower vehicles potentially block the ego vehicle.

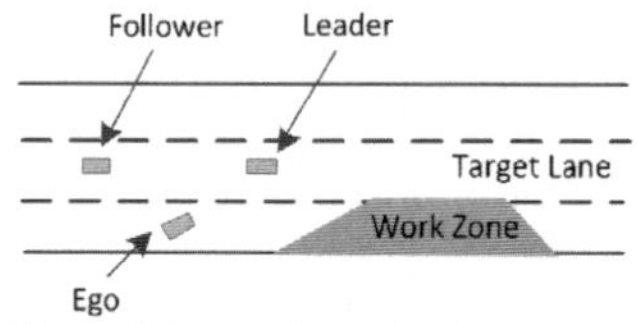

Figure 3. Lane Changing Behaviour

To successfully change to the target lane, the speed adjustment should be made based on the flow diagrams illustrated in Figure 4 and Figure 5.

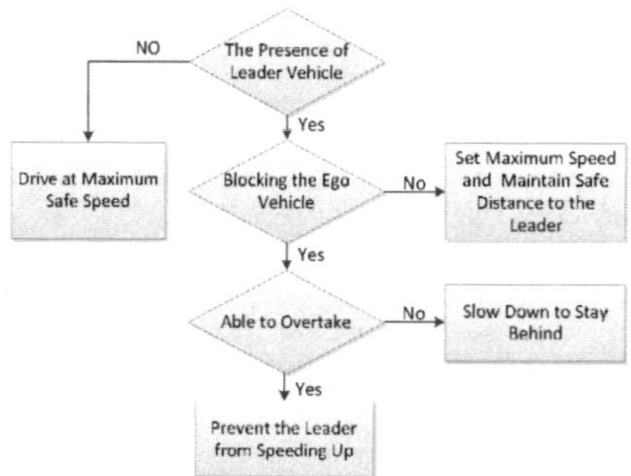

Figure 4. Speed Adjustment with the Presence of Leader

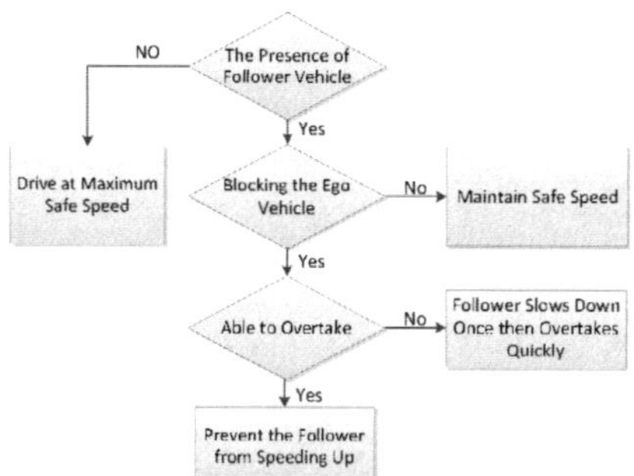

Figure 5. Speed Adjustment with the Presence of Follower

The speed of ego, leader, and follower vehicles should be adjusted to be complied with the basic rules set in the above flow diagrams when ego vehicle performs the lane changing behaviour.

4.1.3 Objective Function

The objective function is designed to optimize VSL by minimizing the total travel time spent in the upstream and work zone area. To calculate the total travel time, the velocity and position of each vehicle are predicted first. The velocity is updated based on the maximum safe speed discussed in car-following model. The VSL will affect the safe speed by setting the maximum speed limit v_{max}. Once the speed of vehicle i is updated at time step t+1, the position of vehicle i is updated as:

$$x_i(t)=x_i(t-1)+\frac{v_i(t)^2-v_i(t-1)^2}{2a_i}=x_i(t-1)+\frac{v_i(t)+v_i(t-1)}{2} \quad (7)$$

The total travel time J_{travel} is obtained using Equation (8). It can be seen that total travel time is calculated by taking the sum of vehicles in the upstream of work zone. Because one simulation time step in SUMO represents one second, then the N_p prediction horizon is N_p seconds. When N_p time steps of vehicle position are predicted, the total number of vehicle stays in the upstream segments will determine the total travel time.

$$J_{travel}=T\sum_{k=t}^{N_p} n_k \quad (8)$$

where T equals to one simulation time step which is one second and n_k is the total number of vehicle in the upstream work zone and work zone area at time step t.

The genetic algorithm was utilized to optimize the VSL. By changing the VSL, the maximum speed of each vehicle is changed. Correspondingly the predicted the position of each vehicle is changed. Then Equation (8) is utilized to find the optimal VSL with minimum total travel time.

4.1.4 Constraints

The three constraints showed in Equation (9) are considered during the optimization of VSL.

$$\begin{cases} v \in \{50,60,70,80,90,100\} \\ |v_j(t)-v_{j+1}(t)| \le 10 \\ |v_j(t)-v_j(t+1)| \le 10 \end{cases} \quad (9)$$

where the $v_j(t)$ represents the posted speed limit in the position j at time step t.

The VSL should be in the set of speed limit from 50km/h to 100km/h with the increment of 10km/h. The 50km/h is set to ensure the travel efficiency of freeway and 100km/h is the maximum speed limit in free-flow status. Such numbers may differ for different jurisdictions. To avoid the sudden acceleration and deceleration, the difference between the posted speed limits on two consecutive VMS should be less than 10km/h and the speed limit at the same position for two consecutive controller time steps should be less than 10km/h too.

4.2 Connected Work Zone

The work zone is considered under connected environment in this study. By making the work zone connected, more information of work zone such as the approaching distance to work zone, number of lane closure, activity inside work zone, etc. will be shared with the vehicles in the upstream segment. This will help drivers make better decision before they approach the work zone and will reduce the possibility of traffic congestion. The merge location, capacity estimation and disturbance caused by work zone will be discussed under the connected environment.

4.2.1 Merge Location

The early merge location determines when the vehicle in the lane which will be closed due to work zone should change to the target lane earlier. The distance is more than 550 meters, where 50% of vehicles merge, away from work zone [13]. It will greatly affect the lane changing behavior when the lane changing model in 4.1.2 is utilized.

The early merging location model was adopted to consider the density and speed in the lane of work zone and target lane [14]. The merge location is calculated using Equation (10) and the obtained location will act as one input signal to the lane changing model.

$$L_{merge}=\exp(\boldsymbol{\alpha}\cdot\boldsymbol{\omega}+\varepsilon) \quad (10)$$

where L_{merge} is the merge location, $\boldsymbol{\alpha}$ is the parameter vector, $\boldsymbol{\omega}$ consists of the variables of the speed and density and ε is the random error follows the normal distribution.

Under connected environment, the distance to the work zone and traffic states will be accessible to perform better lane changing behavior.

4.2.2 Capacity

The work zone capacity which is the traffic breakdown flow is an essential element to analyze the traffic congestion. However, the capacity estimation does not only depend on work zone layout and other geometry factors. It also correlates with the states of traffic flow near the work zone.

To estimate the capacity of the 3-to-2 type of work zone, the estimation model was established in [15] as:

$$C_{wz}=\boldsymbol{\beta}\cdot\boldsymbol{\chi} \quad (11)$$

where $\boldsymbol{\beta}$ is the parameter vector and $\boldsymbol{\chi}$ consists of the variable of the distance of warning sign, the percentage of traffic on each lane, work activity in work zone, heavy vehicle percentage, etc.

Then the estimated work zone capacity under connect environment is utilized to analyze the disturbance of work zone in the following section.

4.2.3 Disturbance

The work zone acts as the disturbance in VSL control showed in Figure 1. Since the presence of work zone will greatly affect the traffic flow and cause traffic congestion in the upstream segments of work zone. It is essential to model this disturbance in microscopic simulation.

Before modeling the work zone disturbance, the traffic stages in which VSL will be applied will be discussed first. There are three stages (free flow, transition and congestion) to describe the traffic flow [16]. In free flow stage, there is a low traffic density and the traffic will flow stably. In transition stage, the small disturbance in the traffic flow will be disappeared. However, the large disturbance will cause the traffic breakdown. While in congestion stage, the traffic congestion has already happened or a small disturbance will lead to traffic congestion.

VSL control will only apply to the transit stage. In free flow stage, VSL control will restrict the traffic inflow and increase the time delays. While it will not have effective impacts on traffic control when the traffic is in congestion. However, in transit stage, VSL control will have the possibility to absorb the small disturbance and reduce or eliminate the effects caused by the disturbance.

Therefore the speed-flow relationships in transit stage will be utilized to analyze the work zone disturbance. This relationship is expressed as [16]:

$$v_{limit} = v_{max} - (v_{max} - 18.7) * (\frac{Q - 900}{2182.2 - 4.3 * v_{max}})^{3.6} \quad (12)$$

where v_{max} is the maximum speed limit and Q is the work zone capacity.

The disturbance is modelled as the speed limit in the work zone area. Under connected environment, the speed limit will be obtained based on the calculated work zone capacity (section 4.2.2).

5 Experiment and Results Analysis

5.1 Experimental Setup

The effectiveness of the proposed VSL control was assessed in traffic simulator SUMO. A hypothetical three lanes of freeway traffic network was established under connected work zone and connected vehicle environment as showed in Figure 6. The upstream road of the work zone was divided into 5 segments. Each segment had 800 meters. There was a 3-to-2 type of work zone set in the 6th segment with the length of 100m. In addition, the 7th segment represented the downstream road of the work zone with 800 meters. It was assumed that work zone will not be removed during the simulation.

There were six VMSs set up to post the speed limits as presented in Figure 6. The first 5 VMSs along the upstream road will show the speed limits acquired from the VSL control and the 6th VMS will post the speed limit result from the work zone disturbance. The initial speed limit was set to be 100km/h for all the 6 VMSs before the VSL was applied. Once the vehicle travels to a new segment, it will set the posted speed limit in the new segment as its maximum speed limit.

The work zone in 6th segment was set up with the following configuration settings: heavy vehicle percentage was 5%; the warning signal distance was 3.2km; lateral clearance was 8ft; lane width was 12ft; there was no work activity presented inside the work zone.

The vehicle parameters in SUMO were set with maximum acceleration 2.6 m/s^2; maximum deceleration 4.5 m/s^2; the driver's imperfection 0.9. The traffic demand was 3500 vehicles/h which is close to the estimated work zone capacity 4000 vehicles/h by HCM 2000 [17]. The parameter vectors of merge distance and capacity estimation were adopted from [14] and [15]. The prediction horizon N_p was selected as 3.5 minutes which is an approximate to the time to travel the whole road network under free flow and the controller horizon N_c was selected as 2 minutes.

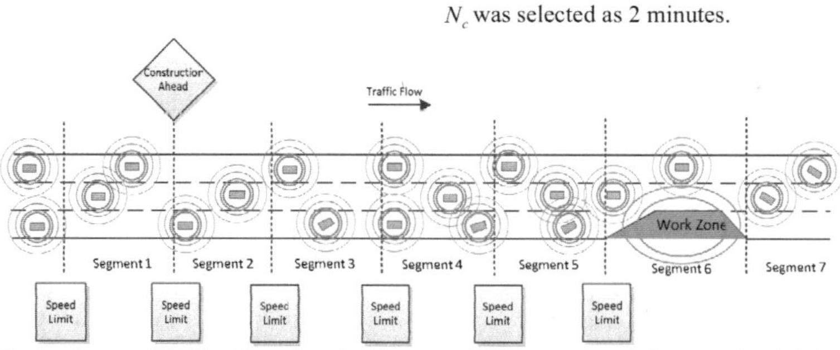

Figure 6. VSL for Freeway Traffic Control under Connected Work Zone and Connected Vehicle Environment

5.2 Results Analysis

To assess the impacts of connected environment and the effectiveness of VSL control, 4 scenarios are considered in the simulation: (1) VSL with connected environment; (2) VSL without connected environment; (3) uniform speed limit with connected environment and (4) uniform speed limit without connected environment.

The simulation was run for 20 minutes with the assumption of 100% market penetration rate of the connected vehicle. Since it takes time for the traffic to arrive at the work zone area and have a stable traffic condition, the simulation data of the first 10 minutes was discarded.

The estimated average travel time of each segment are showed in Table 1.

Table 1 Average Travel Time for Each Segment

Travel Time (s)	Scenario			
	VSL & Connected	VSL	Connected	Normal
Segment 1	30.2	30.2	30.2	30.2
Segment 2	33.7	35.3	35.8	37.2
Segment 3	89.5	101.5	89.1	97.6
Segment 4	92.6	117.2	130.5	151.4
Segment 5	68.4	182.2	209.1	217.2
Segment 6	66.4	177.2	181.0	193.7

As we can see from Table 1, VSL with connected environment has the shortest travel time in all the six segments. Though scenario 2 has similar travel time to scenario 1 in the first 2 segments, the congestion happened in scenario 2 before the work zone delays the traffic significantly. Because the information of work zone is not shared with the vehicles earlier before they arrive at the work zone, the vehicle tends to perform inappropriate lane changing behavior which will easily lead to traffic congestion. The effects of earlier information are also showed in the uniform speed limit control between connected and non-connected environment. The outperformance of VSL is illustrated in both connected and non-connected environment as the average travel time of VSL is much less than that of uniform speed limit. Although this outperformance is not significant in the first 3 segments, it saves considerable time when the traffic is closed to work zone area. The improvement of homogenization of the traffic flow because of VSL control accounts for the less travel time compared with traffic network with uniform speed limit. This argument was also demonstrated in the speed profile of each segment in Figure 7.

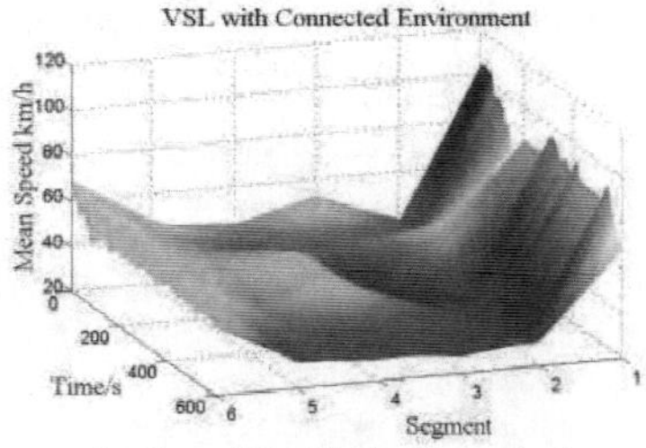

(a) Speed Profile in Scenario 1

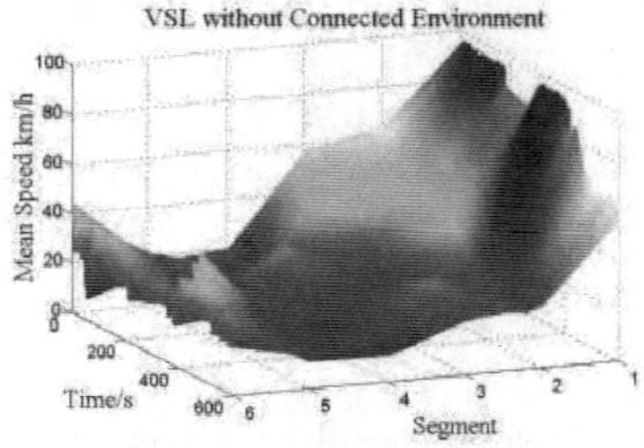

(b) Speed Profile in Scenario 2

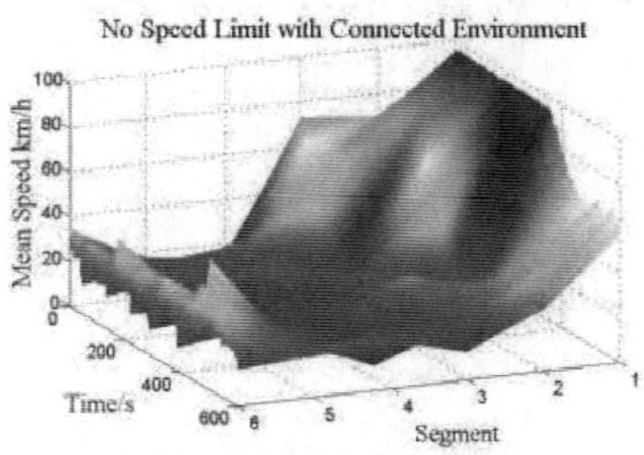

(c) Speed Profile in Scenario 3

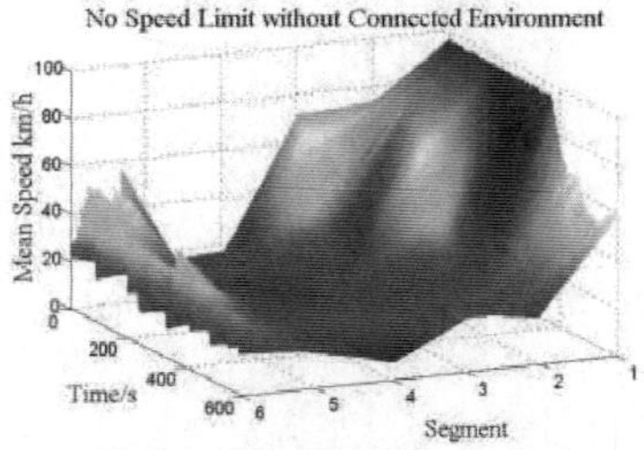

(d) Speed Profile in Scenario 4

Figure 7. Speed Profile in Different Scenarios

As it can be seen from Figure 7, the vehicles in upstream segment maintain higher speed (40-60km/h) and experience less speed variance under VSL control with connected environment. Without connected environment, only VSL can achieve relatively small variance. However, the travel speed (20-40km/h) in segment 3, 4, 5 and 6 are much slower. A large speed variance exists at segment 6 which is the work area for VSL control without connected environment due to the lack of work zone information, which leads to more stop-to-go vehicle statuses and higher safety risk. The homogenization effects are also demonstrated between speed profiles of the VSL and uniform speed limit.

It can be seen from Table 1 and Figure 7, VSL control significantly decreases the travel time, enhances the speed homogeneity and reduces the speed variance which potentially improves the safety.

6 Conclusion and Future Work

To mitigate the traffic congestion and reduce the travel time, a VSL control strategy was proposed under connected work zone and connected vehicle environment. By optimizing the VSL using MPC and connected work zone, the proposed approach was able to significantly reduce the total travel time and decrease the speed variance which has the potential to improve the safety around the freeway work zone area.

In this study, the 100% market penetration of connected vehicle and perfect wireless communication were assumed. The driver compliance rate was not considered. In future work, the effects of low market penetration will be explored and the driver reaction to VSL will be incorporated to achieve optimal VSL control.

References

[1] McKinsey Global Institute. Infrastructure Productivity : How to save $ 1 trillion a year, Insight and Publications Report. Online: http://www.mckinsey.com/industries/capital-projects-and-infrastructure/our-insights/infrastructure-productivity, Accessed: 01/03/2017.

[2] Federal Highway Administration. Facts and Statistics – Work Zone Mobility. Online: http://www.ops.fhwa.dot.gov/wz/resources/facts_stats/mobility.htm, Accessed:01/03/2017

[3] Krajzewicz D., Erdmann J, Behrisch M. and Bieker L. Recent development and applications of SUMO - Simulation of Urban Mobility. *International Journal On Advances in Systems and Measurements*, 5(3&4): 128–138, 2012.

[4] Hegyi A., DeSchutter B., and Hellendoorn J. Optimal Coordination Of Variable Speed Limits To Suppress Shock Waves. *IEEE Transactions on Intelligent Transportation Systems*, 6(1):102-112, 2005.

[5] Hegyi A., Hoogendoorn S.P., Schreuder M., Stoelhorst H. and Viti F. SPECIALIST: A dynamic speed limit control algorithm based on shock wave theory. In *Proceedings of the 11th International IEEE Conference on Intelligent Transportation Systems*, page 827-832, Beijing, China, 2008.

[6] Hadiuzzaman Md. and Qiu T.Z. Cell Transmission Model Based Variable Speed Limit Control For Freeways. *Canadian Journal of Civil Engineering*, 40(1): 46-56, 2013.

[7] Yu R. and Abdel-Aty M.. An Optimal Variable Speed Limits System to Ameliorate Traffic Safety Risk. *Transportation Research Part C: Emerging Technologies* 46: 235-246, 2014.

[8] Khondaker B. and Kattan L.. Variable Speed Limit: A Microscopic Analysis In A Connected Vehicle Environment. *Transportation Research Part C: Emerging Technologies*, 58:146-159, 2015.

[9] Yang X., Lu Y. and Lin Y. Optimal Variable Speed Limit Control System For Freeway Work Zone Operations. *Journal of Computing in Civil Engineering*, 31(1): 04016044, 1-9, 2017.

[10] Fudala N. and Fontaine M. Interaction Between System Design And Operations Of Variable Speed Limit Systems In Work Zones. *Transportation Research Record: Journal of the Transportation Research Board* 2169: 1-10, 2010.

[11] Krajzewicz D. *Traffic Simulation with SUMO – Simulation of Urban Mobility. In: Fundamentals of Traffic Simulation International Series in Operations Research and Management Science.* volume 145. Springer New York, 2010.

[12] Erdmann J. *SUMO's Lane-Changing Model. In: Behrisch M., Weber M. (eds) Modeling Mobility with Open Data. Lecture Notes in Mobility.* Springer, Cham, 2015.

[13] McCoy P. and Pesti G. Dynamic Late Merge–Control Concept for Work Zones on Rural Interstate Highways. *Transportation Research Record* 1745: 20-26, 2010.

[14] Weng J. and Meng Q. Modeling speed-flow relationship and merging behavior in work zone merging areas. *Transportation Research Part C: Emerging Technologies*,19(6): 985-996, 2011.

[15] Kevin H., Kondyli A., Arguea D., Elefteriadou L. and Sullivan F. Estimation Of Freeway Work Zone Capacity Through Simulation And Field Data. *Transportation Research Record: Journal of the Transportation Research Board* 2130:16-24, 2009.

[16] Avrenli K., Benekohal R. and Hani Ramezani. Four-Regime Speed-Flow Relationships For Work Zones With Police Patrol And Automated Speed Enforcement. *Transportation Research Record: Journal of the Transportation Research Board* 2272: 35-43, 2012.

[17] TRB. *Highway Capacity Manual*. TRB, National Research Council, Washington, D.C., 2000

A Behavior-based Architecture for Excavation Tasks

T. Groll, S. Hemer, T. Ropertz, K. Berns

Robotics Research Lab, Technical University of Kaiserslautern, Germany
E-mail: groll@cs.uni-kl.de, s_hemer@cs.uni-kl.de, ropertz@cs.uni-kl.de, berns@cs.uni-kl.de

Abstract -

The paper describes a software architecture for autonomous excavation tasks. For this, it uses a behavior-based approach which allows adaptation to the changing environment and unexpected events. As part of the presented architecture, basic primitives are defined which solve spatially limited motions. These can be activated as and when required. It is possible to have multiple of them running in parallel. Normally, an excavation task can be split up into several different phases. During these phases, the primitives will be activated as needed. To be reusable, the primitives are designed in a very general way. To show the performance of the presented architecture, an example application is built up which can be used to dig a trench. This application is implemented on a system running on a real backhoe loader.

Keywords -
primitives; behavior-based control architecture; excavator

1 Introduction

Excavators are commonly used machines on construction sites and in mining environments. One of the main reasons for that is their multi-functionality. They can be deployed for transport or loading of material for example and as well for all kinds of digging scenarios. Targeting this, excavators are available from small machines with weights around 1 t to huge machines with over 200 t. Additionally, a huge set of different attachments is available for the use with excavators. Besides a lot of different kinds of buckets, it is also possible to use tools like drillers with such machines (compare [6]).

Because of the kinematic structure of a common excavator arm, it is necessary to control multiple degrees of freedom at the same time to solve a task properly. This requires some expertise. The basic kinematic structure is comparable to the one of common industrial robots but most of the excavator arms are controlled by hydraulic actuators. So it is comprehensible that the proceeding automation does not stop in the area of construction machines. In the past years, some projects in this field were executed. For example, there are Thor [15] and Lucie [1].

According to the wide range of possible tasks which can be solved by an excavator, an architecture for an autonomous control system for such a machine should be adaptable. Another big issue while working on construction sites is that here the environment is very unstructured and can even heavily change while solving the task. In most cases, the changing of the environment is the goal of the excavation work. To deal with this situation the behavior of a control system should also be very adaptive.

Such an architecture for the control software of an autonomous excavator is described in this paper. With the presented approach the task of digging a trench as an example application can be executed autonomously. This typical excavation problem can be divided into four subtasks: *moving to trench start, digging, moving to dumping position, dumping*. So the implemented solution should address these phases as part of its architecture. Additionally, it should be possible to exchange the implementation for adaptability. If this is done properly, the same solution can be used to solve different tasks with only exchanging, for example, a single phase implementation. This could be the exchanging of the digging strategy between the excavation of big masses and a solution for creating a more precise shape of the pit. Another problem which is addressed by the architecture is the switching between the subtasks while system operation. This should happen according to the system state measured by the sensor system. In previous research presented in [2], it was shown that expert human operators overlap the solving of the subtasks a little bit. So the architecture presented here also should allow such overlapping. On the lower level of the software stack, primitive motion control modules are applied which can be reused for different tasks or subtasks. Another goal for the architecture design is to be scalable. This means that in future works the same solution could be used to extend the local trenching to an implementation for building up a whole construction pit. To show the performance of the presented architecture, an example application is built up which digs a trench. This is implemented on a system running on a *John Deere 410J TMC* backhoe loader.

Followed by this introduction, similar projects and other architecture techniques are presented. Then in section 3, the developed architecture is explained in a general way. In section 4, the implementation of the architecture for the trenching task is presented. Following this, the experiments and their results are shown. At the end, a conclusion is given which also includes a short explanation of possible future work.

2 Related Work

For every system which implements an autonomous excavation task, it is necessary to implement subsystems for solving smaller parts of the work cycle. Additionally, a strategy is needed which controls the execution of the sub-implementations. For solving this, different approaches can be found in the literature.

2.1 Autonomous Excavators

A lot of projects in which autonomous excavation was implemented use finite state machines (FSM) to fulfill the task. For example in [3] a solution is presented which uses a FSM in a software system which allows a wheel-loader to load a pile of soil on a truck. In the implementation of mass excavation including truck loading for the *Automated Loading Systems* of *Carnegie Mellon University* parameterized scripts are used. These scripts implement the control algorithms for the different joints of the used excavator and are used in parallel. They are synchronized by event-triggered state changes, which allows interpreting the single scripts also as finite state machines. This solution is described in [12] and [13]. Schmidt describes in [15] a control system for the automated excavator THOR which also uses a FSM. Here, the state machine provides the input for a behavior-based system which generates the moving trajectory by activating primitives. Each of this primitives calculates control signals for every possible moving direction. In this works also some algorithms for environmental perception and interpretation are described. For example, he explains a gridmap-based solution to find the best starting point for the next excavation cycle.

On the low level of the control stack for excavation automation operates the solution presented by Maeda in [8] and [9]. Here, an adaptive impedance control implementation is shown which can control the bucket of an excavator along a given trajectory while moving through soil with a lot of unknown disturbances. Targeting the problems occurring because of this disturbances, the system can adopt the control algorithm while executing the excavation cycles and improve the precision of following the trajectory with every new iteration of the digging process. Besides these projects which all implement software system to drive real excavators, Du et al. describe in [2] a system that should improve simulation by including a virtual operator into the tested machine. For this, the developed algorithm should act as similar as possible to the movement a human operator would control. To find out how a human really solves the excavation task, they did detailed research with expert machine operators. Based on this, a detailed description of the excavation cycle while trenching is given. Starting from this knowledge a virtual controller for the excavation simulation was implemented. In the implementation also a state machine with hard state changes is used. However, one conclusion of the tests with the experts was that very experienced human operators let the states overlap a bit.

Also in the work presented in [16] no real excavator is used. There, a high-level task planner is presented which can be used to divide the earthwork of a whole construction site into small portions. The resulting segmentation delivers positions which should be used for local digging by an excavator. An algorithm to reach this location in a good manner is then given by Kim et al. in [5].

2.2 Motion Primitives

Using primitives is a good strategy to develop modules that can be reused for different tasks. If the primitives are additionally adaptable in their behavior, the range of possible usage is extended a lot. Addressing this, Ewerton describes in [4] an approach which uses movement primitives with parameters. In this work, he teaches a robot to play golf by demonstrating the basic trajectory to it. With the implemented algorithm the robot was then able to adapt this trajectory to vary the velocity of the golf club when it hits the ball. This is possible because they approximate the given trajectory with parameterized Gaussian functions. To adapt the velocity, and with it the hitting strength, the parameters of the Gaussian kernels are changed.

A similar solution is used for the *Dynamic Motion Primitives* (DMP) introduced by Schaal in [14]. Here, parameterized functions are used to approximate a trajectory which also include Gaussian basis functions.

Another approach for implementing movement primitives is given by Luksch in [7]. There the primitives are described as small systems that generate local control commands. For example, it exist a primitive to control the velocity of an object according to one criterion. The resulting modules can run in parallel and are triggered by stimulation which is given by a hierarchical behavior-based network. This uses an architecture earlier introduced in [10] which is called *integrated behavior-based control* (IB2C).

2.3 Robot Architectures

To design robot control software, it is necessary to use an arbitrary software architecture with structure modules and define rules for the execution of those. Weidauer has done so in [17] for a standard industrial manipulator. To provide the possibility of reuse, system manipulation primitives are introduced which implement parts of the system each. To give the execution a appropriate order it is organized by a petri net.

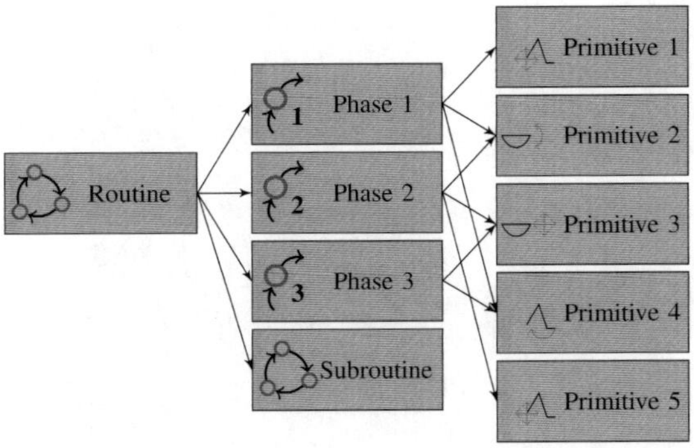

Figure 1. The structure of the behavior-based control architecture

3 Architecture

With the presented architecture the previously explained subdivision of the trenching process into subtasks is targeted. Additionally, the implementation of the subtasks should be exchangeable and on the lower level primitives are used to provide a good re-usability for control laws in the different phases and for other tasks.

3.1 Structure

In general, the architecture structures the implementation hierarchically. So the overall control system solving the task is called a routine. A routine consists of phases that represent the subtasks a routine can be split into. In every phase a set of primitives will be activated. It is possible that a primitive is actually used by multiple phases. This structure is graphical shown in figure 1.

To adapt the implementation of a subtask, for example to switch from mass excavation to precise trench shaping, it is only necessary to exchange one single phase while the remaining implementation can stay the same. Besides the option of using a set of primitives directly, it is possible to use a subroutine as implementation of a phase.

3.2 Phase

Subtasks, that a routine can be split into, are represented as phases. This structure element is used for arbitration; a phase should become active if specific criteria of the system state are fulfilled. These can include measured data from the environment like the actual trench shape as well as the machine state, for example the actual angle of the bucket according to the machine frame. During the development process for the implementation of the trenching task, it appeared that a phase should become active if some conditions are fulfilled and should be held active while some other criteria are in the appropriate range. Targeting this, the developed architecture defines a set of preconditions that has to be fulfilled at the moment of phase activation. Additionally, an invariant, represented as a set of conditions, has to be fulfilled during the phase is running.

For the trenching process, four phases are defined according to the subtasks introduced before: *move to start, digging, transport to dumping and dumping*. In the *move to start* phase the bucket should be controlled to the start position of the next digging cycle. This is the position at which the bucket should begin to retrieve the next portion of soil out of the trench. To determine this position, rating functions are used which operate on grid maps of the desired and actual surface. These functions are introduced in [15]. Additionally to reaching the position, this first phase should curl the bucket to an angle which allows to break into the soil in an appropriate manner. During the *digging* phase, the bucket should move through the soil to scoop up as much soil as possible. This leads to a trajectory which moves the bucket from an open angle to a closed one. Also, the bucket should move into the direction of the excavator base. When the bucket is filled, the soil has to be brought to the dumping position. This can be a target pile or the loading area of a truck, for example. The transfer to this position is done in the *transport to dumping* phase. Arrived there, the material has to be dumped out of the bucket. For this, the *dumping* phase is responsible. When the bucket is empty again, the digging cycle should start with the next iteration.

To formulate the conditions, mainly the pose of the bucket given in Cartesian space Π_{bucket} is used. This includes the bucket position P_{bucket} and its orientation Ω_{bucket}.

$$\Pi_{bucket} = (P_{bucket}, \Omega_{bucket}) \tag{1}$$

While digging and dumping, the main movement should be a curling of the bucket to gather soil into it or drop it. Addressing that, the angle of the bucket according to the basis frame of the arm θ_{bucket} is used here to measure the progress. Additionally, some target or threshold values are defined to formulate criteria for fulfillment of the conditions. These are Π_{start} for the desired start pose of the next digging process, θ_{lift} for the angle that should be used while lifting the bucket out of the trench, P_{dump} for the desired position of the bucket while dumping and θ_{dump} for the bucket angle in which it is fully open. In table 1 the used phases with their preconditions and invariants are listed. With a clever definition of the preconditions and invariants, it is possible to let the execution of the phases overlap appropriately.

phase	precondition	invariant
move to digging	$\theta_{bucket} \leq \theta_{dump}$	$\Pi_{bucket} \not\approx \Pi_{start}$
digging	$\Pi_{bucket} \approx \Pi_{start}$	$\theta_{bucket} \leq \theta_{lift}$
move to dumping	$\theta_{bucket} \approx \theta_{lift}$	$P_{bucket} \not\approx P_{dump}$
dumping	$P_{bucket} \approx P_{dump}$	$\theta_{bucket} > \theta_{dump}$

Table 1. The phases of the digging cycle with their preconditions and invariants

3.3 Motion Primitive

On the lowest level of the presented architecture, primitives are used. They implement local control laws which generate velocity vectors for the TCP. Finally, these local velocity vectors are combined to an overall control vector. This will be given to the low-level system which calculates the control commands for the machine. A primitive can generate a fixed velocity as well as a dependent one which is calculated based on a sensor data input vector.

Primitives will be activated during the phases. It is possible that one phase uses multiple primitives and one primitive can be used by multiple phases. This improves the re-usability of the structure elements of the system. The standard case of every phase is that a set of primitives is active in parallel.

To solve the trenching task the following primitives are used:

Rotate: This primitive is used to rotate the swing to a given target angle. To do this, it generates a rotational movement around the z-axis. The desired velocity is proportional to the angular distance to the target angle. Here, the maximum angular velocity in the experiments on the machine were $0.25\ \frac{\text{rad}}{\text{s}}$.

Curl bucket: Similar to the first primitive, it generates an angular velocity which curls the bucket to a given target angle. This is a rotation around the y-axis of the backhoe plane coordinate system (maximum velocity: $0.5\ \frac{\text{rad}}{\text{s}}$).

Move to position: The target of this primitive is to move the bucket to a given position in the two-dimensional backhoe plane. To reach this, a 2D velocity vector is generated which is directed to the target point and has a length proportional to the remaining distance. In the experiments, this movement is done with a maximum velocity of $1\ \frac{\text{m}}{\text{s}}$.

Close bucket: This one curls the bucket with a configured fixed angular velocity in the closing direction (velocity: $1\ \frac{\text{rad}}{\text{s}}$).

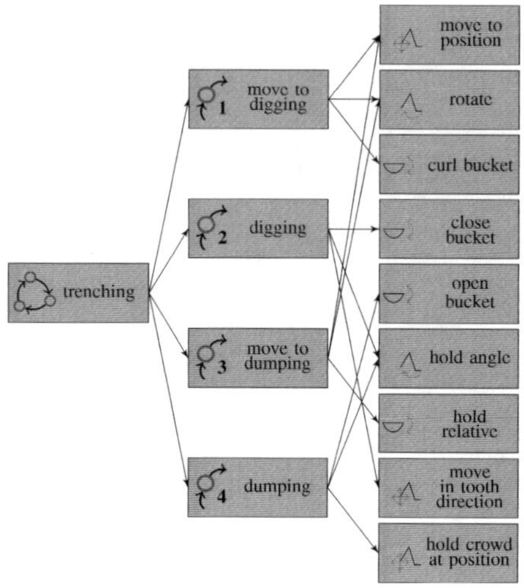

Figure 2. The architecture of the trenching process

Open bucket: Similar to the previous primitive, this curls the bucket in the opening direction (velocity: $1\ \frac{\text{rad}}{\text{s}}$).

Hold angle: With this, the backhoe plane should be held at a fixed angle. This is done by generating an angular movement around the z-axis of the arm base coordinate system if the actual angle differs from the configured one. Here, the value of the generated velocity is proportional to the angular error.

Hold crowd at position: By activating this primitive, the crowd joint will be held at a given position on the backhoe plane. This is done in a similar way as in the primitive explained before but here a linear velocity will be generated according to the error instead of an angular one.

Move in tooth direction: From this, a velocity vector for the TCP with a fixed length is generated which targets into the direction of the tooths of the bucket. To be able to generate a vector which targets into the right direction, the mounting angle of the tooths has to be configured. Additionally, the actual bucket angle according to the machine coordinate system is used to generate the movement (velocity: $1\ \frac{\text{m}}{\text{s}}$).

Hold relative: This primitive generates an angular velocity which holds the bucket in a configured angle relative to a given reference plane.

In figure 2, the whole architecture for the trenching process is shown.

4 Implementation

For the implementation of the architecture, a framework is used which defines special modules that can be used together as a behavior-based control network. This is introduced in [10] as IB2c. A short introduction to this is also given in the next section.

4.1 IB2C

In IB2c, a basic module type is defined that can be used to compose a net which makes a control system for robots. Such a node has a set of well-defined inputs and outputs, figure 3 shows a schematic representation. An IB2c module is defined with the three-tuple $B = (f_a, f_r, F)$ where $r = f_r(\iota)$ is the target rating function, $a = f_a(r, \iota)$ the activity function and $\vec{u} = F(\vec{e})$ the transfer function. Additionally, each module has a set of input and output signals which include the meta data: Activity $a \in [0; 1]$, target rating $r \in [0; 1]$, stimulation $s \in [0; 1]$ and inhibition $i \in [0; 1]$ as well as the input vector $\vec{e} \in \mathbb{R}^m$ and the output vector $\vec{u} \in \mathbb{R}^n$. With the transfer function F, the output vector will be calculated from the input vector according to the internal control law. Using the meta data, the influence of the module in the whole system is calculated and coded in a. This is done by calculating the activation ι of the module from the meta data inputs. In difference to the original introduction of IB2c, in the presented approach the following equation is used to calculate the activation.

$$\iota = \min(s, 1 - i) \tag{2}$$

With this value, the target rating r can be calculated by the user-defined function f_r. It should express how much the calculated output vector $\vec{u}$ should influence the system if the module is fully active. From this values, the activity a of the module can be calculated.

$$a = \min(\iota, r) \tag{3}$$

With this equation, it can be ensured that the activity has in average case the value of the target rating r and at maximum the value of the stimulation s. To combine the control vectors generated by different IB2c nodes, a fusion module is used. For the combination, it uses the meta data of the corresponding modules. From this, the fusion module calculates a new output vector in which every input vector is included according to its corresponding activity and target rating. To perform this, different fusion methods can be used; for example a maximum function or a weighted sum. If the maximum function is used, the output of the module with the highest activity is directly used as the output of the fusion module. By using the weighted sum solution the input vectors of the fusion are multiplied with their corresponding activities and summed up.

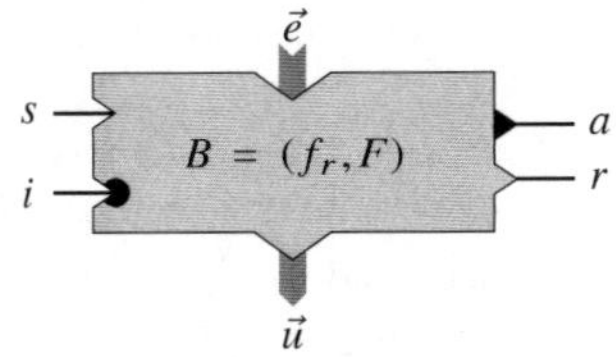

Figure 3. The basis IB2c module

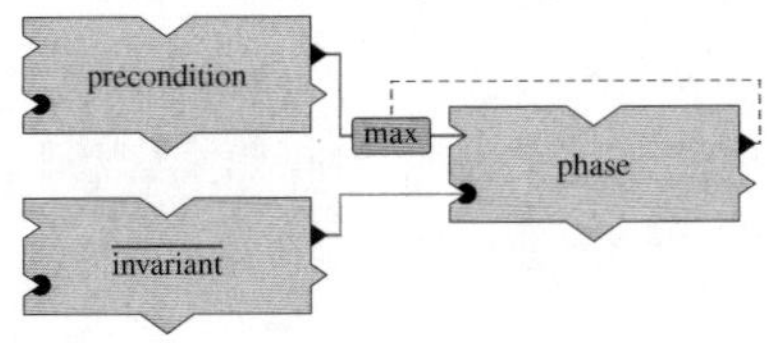

Figure 4. Phase arbitration with IB2c

4.2 Phase Arbitration

For the implementation of the conditions used for the arbitration of the phases, IB2c modules are used. Also, the phase itself is implemented as such a behavior-based node. It is not possible to use the condition modules as simple stimulation for the phase, because it should become active if the precondition is fulfilled, but stay active until the invariant is not fulfilled anymore. So it should not become active if the invariant is fulfilled but the precondition is not and it should stay active if the criteria for the precondition were in the right range but are not anymore. Addressing this problem, the pattern shown in figure 4 is used for the arbitration. To hold the maximum activation given by the precondition, a self-stimulation of the phase module is used. Without inhibition, this leads to an activation for the phase $\iota_{phase_{i=0}}$ which is the maximum of the previous activities given from the precondition. The set of the previous activities of the precondition is defined as $A_{pre} = \{a_{pre_0}, a_{pre_1}, ..., a_{pre_t}\}$ where t is the actual time step.

$$\iota_{phase_{i=0}} = \max(A_{pre}) \tag{4}$$

When the work is done, the phase has again to be deactivated. For this the invariant is used which is connected to the inhibition input of the phase module. With this implementation, the activity of the invariant lowers again the activation of the phase. So the invariant condition has to be implemented in a way that it becomes active if the invariant is violated. This pattern allows also to implement the condition with fuzzy activity functions which lead to a partial activation of the phase.

4.3 Conditions

In the trenching implementation, two kinds of conditions are used. One is to proof if an input angle is in a

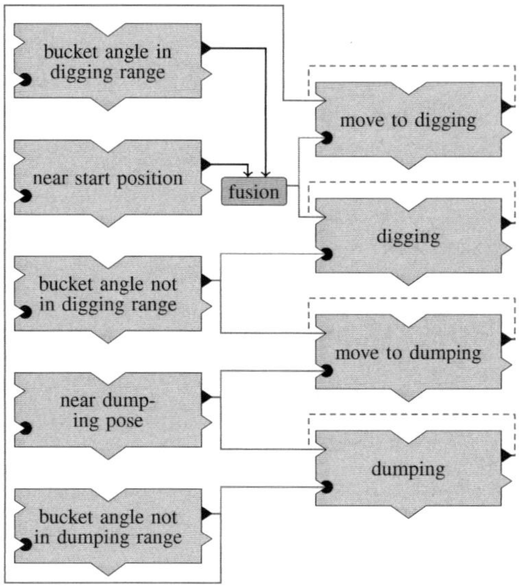

Figure 5. The phase arbitration of the trenching process

configured range and the other one is to test if a given position is in the near or behind a plane defined in Cartesian space. For the first condition type, three parameters are defined. With the start angle θ_s and the end angle θ_e an angular area is defined in which the condition should be fully active. The third parameter θ_a defines an angular distance to the start angle in which the condition starts to become active. This leads to the following activity function for the condition. Here θ_{in} is the measured input angle which should be tested.

$$r_{cond_a} = \begin{cases} 1, & \text{if } \theta_s < \theta_{in} < \theta_e \\ \frac{\theta_s - \theta_{in}}{\theta_a}, & \text{if } (\theta_s - \theta_a) < \theta_{in} \leq \theta_s \\ 0, & \text{otherwise} \end{cases} \quad (5)$$

With the second condition, a given input position P_{in} is tested if it is in the near or behind a plane. It starts to become active if the distance d_p of P_{in} to the plane is smaller than the activity distance d_a and fully active if the input position moves behind the plane. From this the activity function of this condition r_{cond_p} can be formulated as given in the following equation.

$$r_{cond_p} = \begin{cases} 1, & \text{if } d_p < 0 \\ \frac{d_p}{d_a}, & \text{if } 0 \leq d_p < d_a \\ 0, & \text{otherwise} \end{cases} \quad (6)$$

To test if the bucket has reached the start position or the dumping position respectively, the second condition type is used. The target plane is defined with the target position as supporting point and a normal which directs into

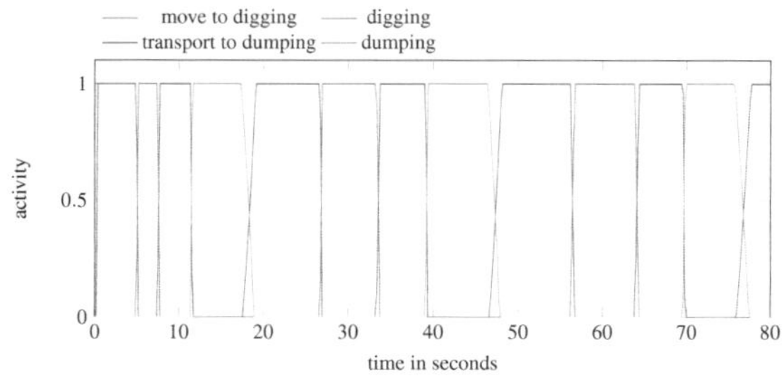

Figure 6. The activity of the different phases while trenching

the starting position of the movement. Using the angle checking condition, three modules are implemented. One for checking if the bucket angle is in the right angle range for digging, another one if it is not in this range anymore and a third one which checks if the angle has passed the right angle for dumping. In figure 5 the network for the phase arbitration of the trenching process is shown.

4.4 Adaption Layers

With the described method, the trajectories are generated in a very free way. To ensure given constraints, the generated velocity vector has to be adapted according to the environment and the desired target trench. This is done by two additional behavior-based subnets which implement an adaption of the velocity vector.

To prevent the bucket from hitting an obstacle, which also includes the remaining soil of the ground, an IB2C network is developed which adjust the velocity vector of the TCP if the distance to the next object is too small. For this a grid map is used which represents the measured environment by containing a height information in every cell, representing the surrounding. If the cell in the configured safety distance in the desired moving direction contains a value that is higher as the current z-position of the bucket, the net blocks the moving in this direction.

Similar to the obstacle avoidance there is a net implemented which prevents the bucket to hit the borders of the desired target trench. It works in the same way as the obstacle avoidance but the gridmap which is used for the adaption does not contain measured obstacles, it contains a representation of the desired trench shape.

5 Experiments

To proof the performance of the developed approach, an experiment was executed. For this, the presented approach for autonomous trenching was implemented on a *John Deere 410J TMC* backhoe loader. This is a machine with an overall weight of around 8 t. Additionally to the

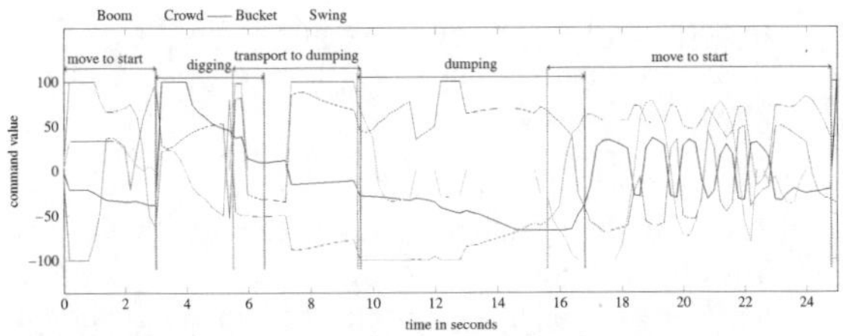

Figure 7. The generated joystick commands while trenching

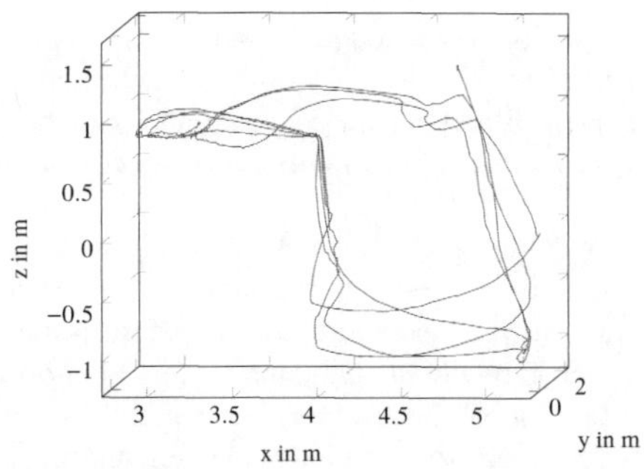

Figure 8. The trajectory of the bucket while digging

real machine, a simulation environment was used for testing. To do this, a realistic simulation model of the *John Deere* backhoe loader was built up. With the usage of the simulation platform V-Rep (see [11]), it was possible to represent the dynamics of the vehicle. For the experiment and the results presented in the following part the real machine was used. In the implemented solution a target trench could be configured with length, depth and starting position. Here, a length of 1 m and a depth of 1 m are used. The starting position was given to $(5\,\text{m}, 0\,\text{m})$ which represents a point in the local coordinate system of the used machine. The origin of this coordinate system is located in the swing joint which is basically the origin of the backhoe arm. It could be shown that the implemented control system was able to dig the configured trench. In figure 8 the trajectory of the bucket recorded while this process is plotted. Here, it can be seen that the resulting trajectory is not as smooth as it would be if a well adjusted industrial robot would execute it with a closed loop control. But these uncertainties are originated from the constant adaption of the behavior-based system. For example at the boundaries of the desired trench dimensions a behavior becomes active which adapts the control vector to keep the bucket inside this boundaries. Another adaption behavior prevents the bucket from colliding with the ground and other obstacles. Furthermore, for the trenching process the small uncertainties are not relevant because they are not detectable in the end result. Additionally, the activity of the phases implemented with the presented approach was recorded. This data is visualized in figure 6. There, it can be seen that one excavation cycle lasts around 20 s to 30 s, which is approximately in the same range as Du et al. have measured with human experts in [2]. Additionally, with the architecture used here, it is even possible to overlap the different phases. This can also be seen in the data showed in figure 6.

The machine used for the experiments can be controlled with joystick commands which are directly translated into flow rates for the hydraulic cylinders. These commands are in a range between -100 and 100. Using the solution presented here, the calculated control velocity vector is converted to appropriate joystick commands with the help of the inverse kinematic. It is possible to calculate them directly because the machine which is used here has only four joints which leads to a uniquely defined system. So four command values are generated, one for each joint. Figure 7 shows the commands generated by the control system during the trenching process. In this figure additionally, the active phases are marked.

6 Conclusion and Outlook

With the presented approach, an architecture is defined which can organize a control system in phases. For the arbitration of the phases, conditions are used. Because of this structure the system is similar to a finite state machine. But with the help of the used behavior-based implementation the conditions have fuzzy like activity functions. So it is possible that multiple phases are active at the same time. This leads to a smoother state change as in a standard FSM. Additional this adds a property to the control system which let it more behave like expert human machine operators.

Although with the used behavior-based method, it is possible to build up reactive systems which can adapt in a proper way to high variety in the environment. To target new requirements, it is very easy to add additional primitives which can implement local control systems to address additional problems. Overall the usage of this primitives provides the possibility of re-using the same control law to solve similar subproblems in different tasks.

In the future the system will be extended to solve a more global task like building up a larger excavation pit. This will also include a transfer of the whole machine to have a larger range. A possibility to do this is also given with the presented architecture with the definition of subroutines which can be used as phase implementation. So the presented local trenching process can be such a subroutine in a larger control system. To be able to give the trench a better shape it should be possible to switch to a more precise digging technique which is more optimized for shaping than for mass excavation. Then there are at least two possibilities for digging available which makes it necessary to choose the appropriate one for the given requirements. Targeting this, a planning system should be designed in future work. This should be as autonomous as possible. So, the first step towards this can be to add some properties to the primitives and phase implementations which can help the planning system to choose the right solution for solving the task. For example, one possibility for such a property can be the desired digging shape which is expected as a result of the appropriate module.

References

[1] David A. Bradley and Derek W. Seward. The development, control and operation of an autonomous robotic excavator. Journal of Intelligent and Robotic Systems, 21(1):73–97, January 1998.

[2] Yu Du, Michael C Dorneich, and Brian Steward. Virtual operator modeling method for excavator trenching. Automation in Construction, 70:14–25, 2016.

[3] Ahmed Elezaby, Mohamed Abdelaziz, and Sabri Cetinkunt. Operator model for construction equipment. In Proceedings of the International Conference on Mechtronic and Embedded Systems and Applications (MESA), pages 582–585. IEEE, 2008.

[4] Marco Ewerton, Guilherme Maeda, Gerhard Neumann, Viktor Kisner, Gerrit Kollegger, Josef Wiemeyer, and Jan Peters. Movement primitives with multiple phase parameters. In Robotics and Automation (ICRA), 2016 IEEE International Conference on, pages 201–206. IEEE, 2016.

[5] Sung-Keun Kim, Jongwon Seo, and Jeffrey S Russell. Intelligent navigation strategies for an automated earthwork system. Automation in Construction, 21:132–147, 2012.

[6] Horst König. Maschinen im Baubetrieb. Springer, 4. edition, 2014.

[7] Tobias Luksch, Michael Gienger, Manuel Mühlig, and Takahide Yoshiike. A dynamical systems approach to adaptive sequencing of movement primitives. In Proceedings of the 7th German Conference on Robotics (ROBOTIK), Munich, Germany, May 21-22 2012.

[8] Guilherme J Maeda and David C Rye. Learning disturbances in autonomous excavation. In Intelligent Robots and Systems (IROS), 2012 IEEE/RSJ International Conference on, pages 2599–2605. IEEE, 2012.

[9] Guilherme J. Maeda, David C. Rye, and Surya P. N. Singh. Iterative autonomous excavation. In Kazuya Yoshida and Satoshi Tadokoro, editors, Field and Service Robotics: Results of the 8th International Conference, pages 369–382, Berlin, Heidelberg, 2014. Springer Berlin Heidelberg.

[10] Martin Proetzsch, Tobias Luksch, and Karsten Berns. The behaviour-based control architecture iB2C for complex robotic systems. In Proceedings of the 30th Annual German Conference on Artificial Intelligence (KI), pages 494–497, Osnabrück, Germany, September 10-13 2007.

[11] E. Rohmer, S. P. N. Singh, and M. Freese. V-rep: a versatile and scalable robot simulation framework. In Proc. of The International Conference on Intelligent Robots and Systems (IROS), 2013.

[12] Patrick Rowe and Anthony T. Stentz. Parameterized scripts for motion planning. In Proceedings of the IEEE/RSJ International Conference on Intelligent Robotic Systems, volume 2, pages 1119–1124, Grenoble , France, September 7-11 1997.

[13] P.S. Rowe. Adaptive motion planning for autonomous mass excavation. PhD thesis, Carnegie Mellon University, 1999.

[14] Stefan Schaal. Dynamic movement primitives-a framework for motor control in humans and humanoid robotics. In Adaptive motion of animals and machines, pages 261–280. Springer, 2006.

[15] Daniel Schmidt. Shaping the Future - A Control Architecture for Autonomous Landscaping with an Excavator. RRLab Dissertations. Verlag Dr. Hut, München, 2016. `http://www.dr.hut-verlag.de/978-3-8439-2816-8.html` ISBN-13: 978-3-8439-2816-8.

[16] Jongwon Seo, Seungsoo Lee, Jeonghwan Kim, and Sung-Keun Kim. Task planner design for an automated excavation system. Automation in Construction, 20(7):954–966, 2011.

[17] Ingo Weidauer, Daniel Kubus, and Friedrich M Wahl. A hierarchical extension of manipulation primitives and its integration into a robot control architecture. In Robotics and Automation (ICRA), 2014 IEEE International Conference on, pages 5401–5407. IEEE, 2014.

Modelling of industrial robotic brick system

Ing. Vyacheslav Usmanov
Department of Construction Technology (K122), Faculty of Civil Engineering in Prague Thákurova 7/2077, 166 29 Prague 6 - Dejvice, Czech Republic
vyacheslav.usmanov@fsv.cvut.cz

Ing. Michal Bruzl
Department of Construction Technology (K122), Faculty of Civil Engineering in Prague Thákurova 7/2077, 166 29 Prague 6 - Dejvice, Czech Republic
michal.bruzl@fsv.cvut.cz

doc. Ing. Pavel Svoboda, CSc.
Department of Construction Technology (K122), Faculty of Civil Engineering in Prague Thákurova 7/2077, 166 29 Prague 6 - Dejvice, Czech Republic
pavel.svoboda@fsv.cvut.cz

Ing. Rostislav Šulc, Ph.D.
Department of Construction Technology (K122), Faculty of Civil Engineering in Prague Thákurova 7/2077, 166 29 Prague 6 - Dejvice, Czech Republic
rostislav.sulc@fsv.cvut.cz

Abstract – The paper will describe a design of functional model of a bricklaying industrial robot that is able to technically implement the complete process of bricklaying. The system will be created at a reduced scale and will be able to brick object. The robotic system for bricklaying will use special developed software which will convert the necessary data of mathematical model from BIM environment to format KUKA for industrial robots KRC4 and make optimal design of movements of the robotic arm. As a result will be developed: a functional model of the robotic bricklaying system including technology (de-palletizing, motion control, cutting and gluing bricks on the bond); Software for data migration (IFC -> KRC4), technical equipment (grip, applying mortar / glue, cutting bricks, control systems); control system including optimization processes (energy, time, material consumption).

Keywords –

Robotic; brick; system; BIM; IFC; robot; KUKA

1 Introduction

Industrial revolution 4.0 predicts that in 10 years the factories will be run by themselves and the productivity will increase by one third. Currently construction industry drops behind in using of robotic systems compared to other industrial branches, because of the difficulty of implementing various and dynamic work activities into robotic systems. Manual and mechanical work is usually being used in construction industry. Because of large number of injuries suffered and high extent of burden relocation by human strength during constructions, implementing of robotic bricklaying system is very topical. An attempt of implementing robotic systems into construction industry is very pressing, because of total absence of these systems in construction industry.

Currently there is no mobile industrial bricklaying robotic system available on the market. Up until now he have known mainly stationary robots used in industry for specific actions. There are robotic construction systems on moving machine basis, which can brick up liner objects (fence, wall). Due to large extents and weight they need a large workspace and because of that they are not universally usable for building spatial objects.

Focusing on development of industrial robotic systems means to put an end to unnecessary delays, ineffective use of material and energy, unnecessary health risks caused by hard working conditions. The aim of this research was to lower these problems to minimum or eliminate them completely by implementing robotic systems to construction raw and finishing processes. A functional model of bricklaying robot with mathematical modeling of optimal trajectory was the main and original part of the research. Some of the aforementioned parts (optimization procedures, video logs) were transferred into an internet application.

Created robotic bricklaying system with the help of uniquely developed software can convert needed data of mathematical model from BIM into format of industrial robots KUKA KRC4 and perform optimal suggestion of movements of the robotic arm. We also researched other factors influencing construction process:

- optimization of material consumption.
- optimization of costs and energy.
- efficient time management and optimal movement trajectory of robotic units.

2 Mathematical model and system concept

The Listed problems are partially removed by industrial bricklaying robotic system, which is the subject of the research. It is a system consisting of three main elements: robotic end device (grip, glue nozzle and brick cutting system), robotic arm and computer operating system (see Figure 1). All parts of the system are connected through data collector, which allows input and output information flow with following analysis. Robotic bricklaying model has the ability of adaptive and interactive system and it can react to changing environment and human factor. Bricklaying robot can be used for automatic walling of vertical constructions (supporting and non-bearing) based on model BIM. Because of high investment costs, the robotic system was created in lowered scale as a functional model.

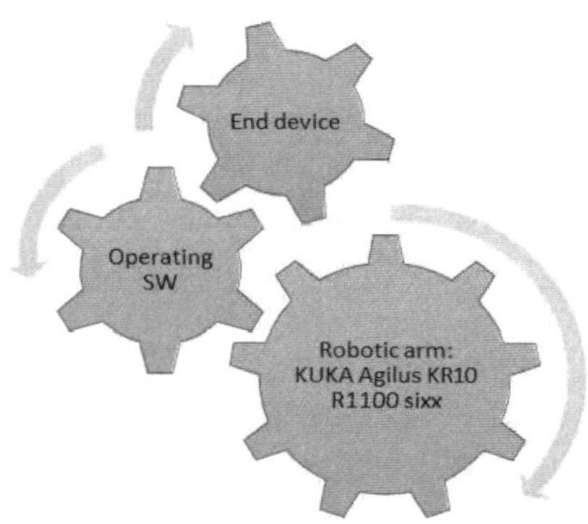

Figure 1 - Schematics of robotic bricklaying system

We will formulate basic steps of development of robotic bricklaying system (see Figure 2):

1. Choice of concepts and materials for modeling of bricklaying robot.
2. Specification of technological and economic aspects and demands during the process.
3. Development of SW for migration of data from BIM to KUKA KRC4.
4. Computer simulation of robotic system.
5. Design and manufacture of robotic end device (grip, mortar/glue fetching).
6. Design and manufacture of brick cutting device.
7. Design and programming of optimal procedures and robot movements.
8. Verification on functional model, quality control.
9. Elaboration of research report with user manual, internet application a video logs.

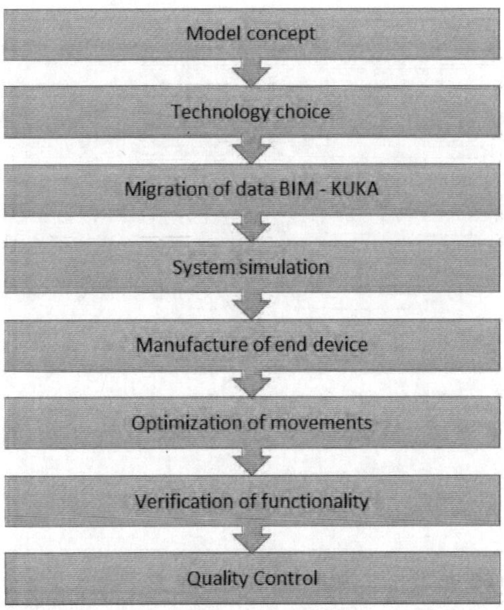

Figure 2 - Flow diagram of system development

Mathematical model and simulation were conducted in a special simulation SW Matlab Simulink [8] and subsequently verified by a simple example. Mathematical model is made from 4 parts: inputs, mathematical core, outputs and external influences [3], [4] (Figure 3). The system takes inputs from BIM, where construction objects are conclusively described (technical parameters, drawing documentation) and construction elements (size and weight of objects for walling). Subsequently developed converter (see further description) converts data from BIM into industrial robot KUKA KRL [9] and optimization SW performs a suggestion of efficient movements of robotic arm. Also, critical points and technical restrictions will be evaluated including possible collisions. The output of the system will be a list of optimized robotic movements according to following aspects: speed of object construction, energy consumption and environmental impact, quality of work [5].

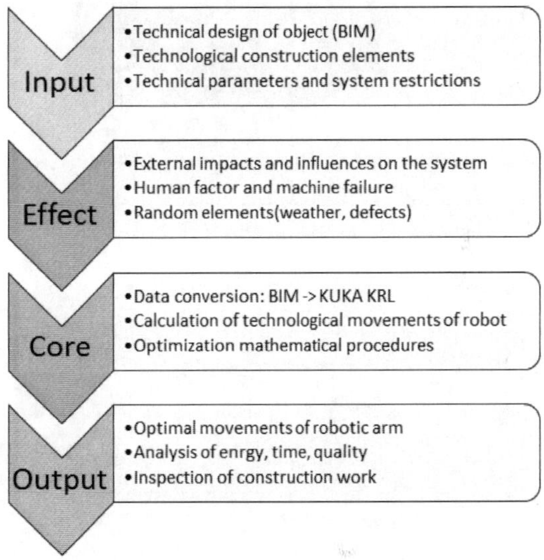

Figure 3 - Mathematical model of robotic bricklaying system

3 Converter BIM (IFC) -> KUKA (KRC4)

Building Information Modeling *(BIM)* is a modern intelligent process for creation and management of projects based on model [10]. It eases the transfer of information in process of designing of the project, construction and usage of building. It allows the creation and managing of projects of ground and engineering infrastructure buildings - faster, more economically and with lesser environment impact.

It was decided to use technical drawings from IBM as a pattern for input system, which definitively determine the need for functioning of the information system. Following chart (see Table 1) describes the concept of developed converter in 6 steps. For between-step format standard IFC was used.

Step 1: Model creation (BIM)

AUTODESK® REVIT® 2016

Step 2: Model export to IFC format

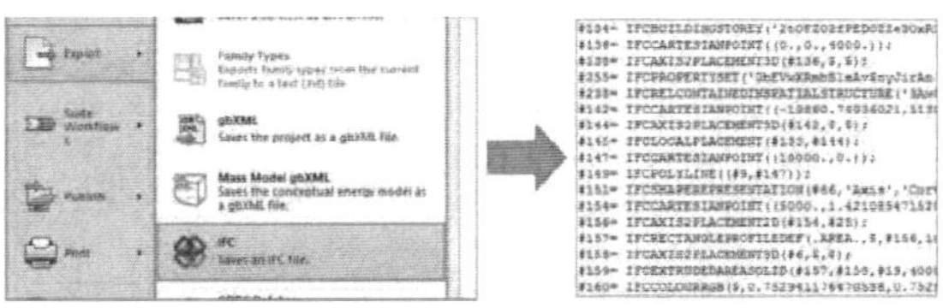

Step 3: Parsing IFC file

```
#257= IFCRELDEFINESBYPROPERTIES('1aSfmB5efCDA_Q
#222= IFCCARTESIANPOINT((19860.74036021,-4980.8
#132= IFCAXIS2PLACEMENT3D(#6,$,$);
#133= IFCLOCALPLACEMENT(#32,#132);
#134= IFCBUILDINGSTOREY('2tOFZOZfPEDOZZe3OxR3jb
#136= IFCCARTESIANPOINT((0.,0.,4000.));
#138= IFCAXIS2PLACEMENT3D(#136,$,$);
#255= IFCPROPERTYSET('0bEVwXRmbBleAv$nyJirAn',#
#238= IFCRELCONTAINEDINSPATIALSTRUCTURE('3Aw$FV
#142= IFCCARTESIANPOINT((-19860.74036021,5130.8
#144= IFCAXIS2PLACEMENT3D(#142,$,$);
#145= IFCLOCALPLACEMENT(#133,#144);
#147= IFCCARTESIANPOINT((10000.,0.));
#149= IFCPOLYLINE((#9,#147));
#151= IFCSHAPEREPRESENTATION(#86,'Axis','Curve2
#154= IFCCARTESIANPOINT((5000.,1.42108547152020
```

- coordinate system (clockwise Cartesian coordinate system) ($WORD)
- matrix of coordinate points [X, Y, Z] ($BASE)

Step 4: Processing of points

- robot placement ($ROBROOT);
- verification of robot reach ($TOOL -> $BASE);
- setting of priority points (movement direction) ($BASE);
- task assignment.

Step 5: Files creation .DAT/.SRC (KUKA KRL)

$ROBROOT, $TOOL, $BASE

.DAT

```
;ENDFOLD (USER EXT)
;ENDFOLD (EXTERNAL DECLARATIONS)
DECL BASIS_SUGG_T LAST_BASIS={POINT1[] "p5                    ",
"PDAT1                ",CONT[] "                   ",CP_VE
",SPL_NAME[] "S0                    ",A_PARAMS[] "ADAT0
DECL E6POS XP1={X 146.914566,Y 135.745819,Z -12.2283487,A 83.62067
DECL FDAT FP1={TOOL_NO 3,BASE_NO 3,IPO_FRAME #BASE,POINT2[] " ",TQ
DECL PDAT PPDAT1={VEL 100.000,ACC 100.000,APO_DIST 100.000,APO_MOD
DECL E6POS XP2={X 58.7224655,Y 23.5809097,Z -12.4140739,A 62.92398
DECL FDAT FP2={TOOL_NO 3,BASE_NO 3,IPO_FRAME #BASE,POINT2[] " ",TQ
DECL LDAT LCPDAT1={VEL 2.00000,ACC 100.000,APO_DIST 100.000,APO_FA
50.0000,EXAX_IGN 0}
DECL E6POS XP3={X 23.6995068,Y 81.2394409,Z -11.8915625,A 82.92398
DECL FDAT FP3={TOOL_NO 3,BASE_NO 3,IPO_FRAME #BASE,POINT2[] " ",TQ
DECL LDAT LCPDAT2={VEL 2.00000,ACC 100.000,APO_DIST 100.000,APO_FA
50.0000,EXAX_IGN 0}
DECL LDAT LCPDAT3={VEL 2.00000,ACC 100.000,APO_DIST 100.000,APO_FA
```

movements, tasks

.SRC

```
;FOLD PTP HOME  Vel= 100 % DEFAULT;%{PE}%MKUKATPBASIS,%CMOVE,%VPTP,%
$BWDSTART = FALSE
PDAT_ACT=PDEFAULT
FDAT_ACT=FHOME
BAS (#PTP_PARAMS,100 )
$H_POS=XHOME
PTP  XHOME
;ENDFOLD
;FOLD LIN P1 Vel=2 m/s CPDAT4 Tool[3]:fix1 Base[3]:hne_lev;%{PE}%R 8
$BWDSTART=FALSE
LDAT_ACT=LCPDAT4
FDAT_ACT=FP1
BAS(#CP_PARAMS,2)
LIN XP1
;ENDFOLD
;FOLD LIN P2 Vel=2 m/s CPDAT1 Tool[3]:fix1 Base[3]:hne_lev;%{PE}%R 8
```

Step 6: Import of files .DAT/.SRC to robot KUKA

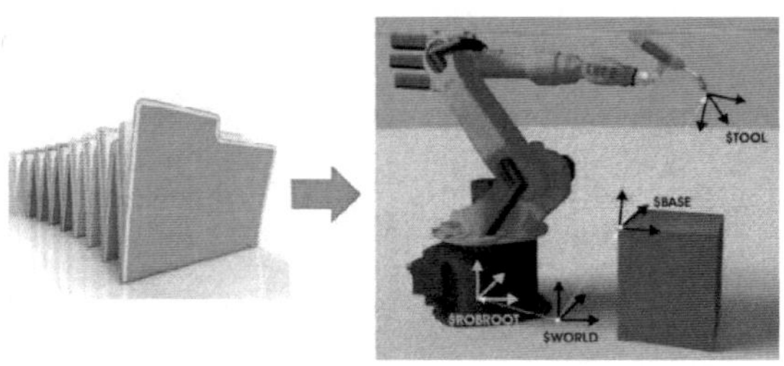

Table 1 - Concept of converter BIM -> KRL

IFC (The Industry Foundation Classes) is an open neutral file format supporting sharing of data on BIM principle, which allows communication between individual participants of construction process and their software BIM tools. Specification IFC is registered as an international standard ISO 16739:2013. Current formats are IFC 2x2, IFC 2x3 (2006), IFC 4 (2013) [10].

Industrial robot system KUKA uses its own programming language KRL, which is derived from group of basic languages C++. Because the programming language KRL [9] has limited options, the system core of the converter was designed in language C++, which transfers format IFC into format KUKA KRL, which is comprehensible for the robot. We are interested in the level of support bearing construction elements (drawing documentation, critical coordinates and intersections of surfaces) and technical parameters of construction elements (size, weight, strength characteristics). BIM system saves needed information and data into IFC format. Then the converter takes and processes the data and saves it into .SRC and .DAT files accessible also for the robot [9]. Industrial robot, which will be in external automated mode or awaiting input parameters, will begin to conduct the technological process. We presuppose that the sequence of movements of technological process will be optimized according to pre-determined optimization parameters.

4 Optimization of robotic movements and trajectory

Very important part of the system is the optimal design of movement for obtaining needed technological process. Optimization is made through converting outputs of BIM system (technical drawings and description of construction elements) based on the following parameters:

- speed of object construction.
- energy consumption.
- environmental impact.
- quality of work.
- malfunctions quantity and influence of random factors.

Technological process must be strictly respected with the lowest possible deviation from required quality and technical instruction of construction work. Optimizing parameters may be evaluated also based on a multiple-criteria decision.

With the help of simulation in program Matlab Simulink [8] and based on outputs of optimizing SW we will receive several possible solutions, which will fulfill given task. Not all of the permissible solutions will lead to optimal and efficient solution. We have to assign optimization criteria in advance and find the most efficient solution. All permissible solutions may create in graphic form an optimal curve, where global, respectively local extremes will be the sought solution of the problem.

There are many fast converging optimization methods [7]. After evaluating several methods it was decided to use the theory of convex polyhedron for a group of optimal points of linear optimization task. Each evaluating parameter can be assigned importance by using coefficients from 0 to 1. The sum of all optimization coefficients equals exactly „1“. If we have a group of permissible points, we can draw a polyhedron in space, where the sides or apexes will create the optimal solution. In mathematical expression the maximum value of parameter will definitively determine the result of the problem. If we include several evaluating parameters into the task, this method allows the search for a complex solution of the problem:

$$\max F(x) = \sum_{i=1}^{k} p_i\, x_i, \text{ [-]}$$

where p – weight coefficient, [-];
k – number of evaluating system parameters, [-].

In the next chapters of the article a detailed description of aims, methods and results of the research work will be provided. Also, the way of realization of modeling and simulation of technological construction processes of bricklaying will be explained.

5 Technical and technological aspects of the bricklaying process

5.1 Bricklaying done by a man

During the bricklaying the first layer of bricks is put on a perfectly horizontal and coherent layer of base mortar of minimal depth of 10 mm. The first important step is height location of baseplate (ceiling) in places where the walls are going to be. The correct location is made after the melting of isolation strips to the underlay in the place of the wall. During the leveling, the highest point is set and from this it is determined how to lay the first layer of bricks. After the setting of height the mortar bed is applied and straightened, where it is needed to pay attention to the correct consistence of the base mortar. After the application the mortar is straightened by aluminum lath.

The first layer of bricks is set directly into the mortar bed. The exact straightening of bricks will be achieved by percussion of masonry hammer or rubber hammer. The level will be checked by water-level and masonry cord which holds the level. This way we will brick up the whole wall. If a brick does not fit exactly, we can break it on demand by percussion of masonry hammer.

Approximately 1 centimeter wide gap is left between the bricks. These vertical connective gaps are filled by mortar. The surplus mortar from connective and storing gaps is let to flow downwards on the wall and is removed by a trowel. To achieve the correct adhesiveness of mortar to bricks we will also wet the upper part of bricked wall. We will apply mortar on the damp surface to create horizontal storing gap and the bricks are placed into it from the corners using the masonry cord. If we are adding a new partition to the current supporting structure, it will be tied by chiseled pockets, or by putting a reinforcing iron into the gap in the new wall and by drilling it to the main wall.

5.2 Bricklaying done by a robot

5.2.1 De-palletization

Automated de-palletization can be made with robotic arm or with portal robots. It is the opposite of palletization process, where the products are removed from palette and placed in their spot (in our case a robotic arm for walling is used). The information was obtained with the help of video files and photo documentation from actual de-palletization performed by a man on a construction site, which led to creating the trajectory for unloading the material. It was necessary to create a trajectory for precise removal of the walling elements from the palette to avoid collision with neighboring material on the palette. An electromagnet was used as the gripper in our model (see Figure 4), which is in on-off mode depending on the movement and action it is currently doing. Moreover, all the bricks were adjusted by creating a groove on the surface and inserting of a small metal plate. The development of the gripper for real brick blocks will be a part of the future research.

Figure 4 – Gripper of the walling robot (electromagnet)

In our model, two kinds of palettes were used. The first type simulates whole walling elements (100mm x 50mm) and they are placed on the palette in the way they are distributed by the manufacturer. On second palette there are halves of walling elements (50mm x 50mm), which are used for beginnings or ending of the wall, to ensure proper bonding of bricks (see Figure 5).

Figure 5 – De-palletizing by the robot (two pallet types – on the left)

Each walling element must be calibrated after the extraction from the palette. This is made via calibration model, into which the brick is put after de-palletization, straightened and again gripped to be put into the wall. The main reason is to eliminate inaccuracies and small deviations that are created by incorrect holding of the brick by the gripper. These inaccuracies would have a negative impact on the flatness of the wall.

5.2.2 Robotic bricklaying

After the removal and calibration of the walling element from the palette, it is allocated to the exact spot where it should be by the instructions of the 3D. Robot abides by the same principles as a man, it starts bricklaying from the corner of the wall, it chooses half a brick for correct linking, it does not put bricks where there should be an open space.

The walling elements must be suitably arranged. They have to be tied up in layers of the wall or a pillar, so that they act as a one monolithic supporting element. To ensure proper binding of the wall, bricks in height not over 25 mm should be bound to length equal to higher of value 0, 4 times height of walling element or 4 mm and bricks higher than 25 mm to length equal to higher of value 0,2 times height of walling element or 10 mm. In corners or places of wall connections, the length of binding of walling element should not be lower than its width. If the walls cross, one is let to go through and the second will be bricked up. In the next layer it is vice versa, second is let to go through and the first is bricked up (see Figure 6).

Figure 6 – Intertwined masonry in the corner of the wall

Advantages:

- Time saving.
- Faster construction (only necessary technological breaks).
- Higher quality (geometric precision, complexity, accuracy, flatness)
- Lower construction costs.
- Protection of human health in risky locations.
- Allocating of large and heavy elements.

Disadvantages:

- Manipulation space.
- Ability to adjust.
- Purchase price.

6 Analysis of energy and time consumption

Based on the example case, which contains a cut of construction work (part of wall construction in corner and a round chimney), an analysis of time and energy consumption was created. Also we analyzed the strain of six motors/transmissions of the robot.

The comparison was made for two states:

- without smoothing and without optimization of movements.
- with smoothing and optimization protocol in place.

Environment of WorkVisual offers a graphic comparison of different states in the form of graph (see Figure 7) and it can compare time and energy consumption or the strain of the robot.

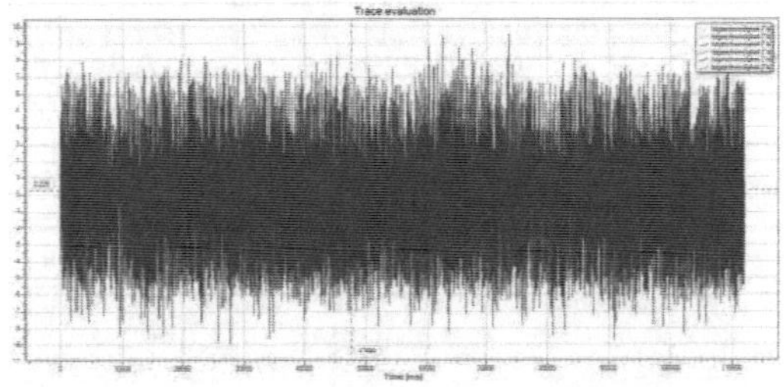

Figure 7 – Observation and analysis of robot movements in environment Oscilloscope in WorkVisual

Subsequently a calculation and a comparison of states were made. As it is obvious from Table 2 and Figure 8, the developed optimization algorithm enabled a more efficient robot deployment on construction site.

Parameter	Before	After
I. Time [s]	1250	1130
II. Consumption energy [Wh]	945	869
III. Max angular velocity [′/s]	1050	941
IV. Average angular velocity [′/s]	380	354

Table 2 – Comparison of states before and after optimization

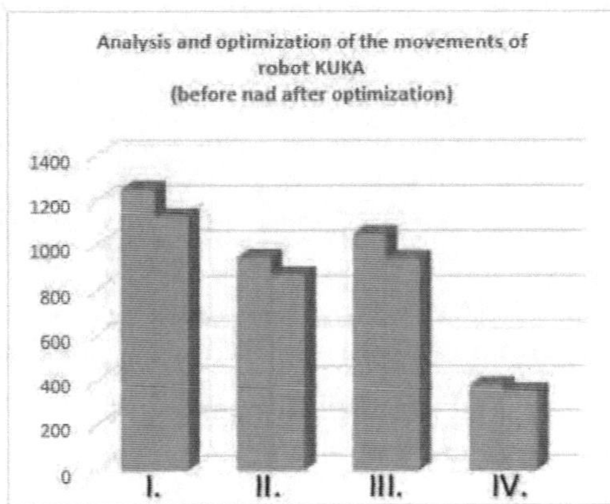

Figure 8 – Comparison of states before and after optimization

Using of optimization algorithm showed efficiency 7-9%, which has a very positive effect for deployment of robots on construction sites. Wear and tear of robot transmissions will be lowered, time of task completion will be lowered.

7 Conclusion

Development of walling robot should mostly contribute to implementing of robotization and automatization to construction industry, which is completely different from industrial production due to constantly changing work space. Robot can work nearly without a break, it is only forced to follow technological breaks. Human can only place 300 – 500 bricks a day, robot can place 800 – 1200 bricks a day, which increases the productivity. The primary aim is not to entirely replace a human mason by a robot, but to increase the productivity of bricklaying in combination human – robot. Robot eases the work by lifting heavy burdens, frequency of repeating individual processes, higher quality etc., but it still cannot react and evaluate crisis situations on construction site the same way a human can.

In the next phase of development of the bricklaying robot we want to focus on a bonding material, in which it seems to be more economic to use foam instead of mortar. For this to work it will be necessary to develop or upgrade the gripper in a way that it is compatible with the walling gripper.

Furthermore, we want to examine wall cutting. When a non-standard situation on the walling occurs, robot can locate, cut and place the brick into the free space. This will be made by a modified circular saw, which will be a part of the robot.

References

[1] Robots Lay Three Times as Many Bricks as Construction Workers. MIT Technology Review [online]. Australia, 2015 [cit. 2017-03-13]. Dostupné z: https://www.technologyreview.com/s/540916/robots-lay-three-times-as-many-bricks-as-construction-workers/

[2] Příručka pro provádění zdiva. Heluz [online]. Praha, 2015 [cit. 2017-03-13]. Dostupné z: http://www.heluz.cz/files/Prirucka-pro-provadeni

[3] SOKOLOWSKI, J. A., BANKS, C. M.: Modeling and simulation fundamentals. New Jersey: John Wiley & Sons, 2010. 427 s. ISBN 978-0-470-48674-0.

[4] KLVAŇA, J. Modelování 20: Operační výzkum. 3. vyd. 246 s. Praha: ČVUT, 2005. ISBN 80-01-03263-9.

[5] SVÍTEK, M., BORKA, J., VLČEK M.: Modelování systémů a procesů. Praha: ČVUT v Praze, 2001. 135 s. ISBN 80-01-02361-3.

[6] NOSKIEVIČ, P.: Modelování a identifikace systémů. Ostrava: Montanex a.s. Ostrava, 1999. 275 s. ISBN 80-7225-030-2.

[7] REKTORYS, K.: Přehled užité matematiky, Prometheus, Praha 2000, ISBN 9531107

[8] Simulační SW Matlab Simulink v. R2010b.

[9] KUKA System Software 8.3, Verze: KSS 8.3 END V3

[10] Eastman, C., Tiecholz, P., Sacks, R., Liston, K.: BIM Handbook: A Guide to Building Information Modeling for Owners, Managers, Designers, Engineers and Contractors (2nd ed.). Hoboken, New Jersey: John Wiley, 2011

34th International Symposium
on Automation and Robotics
in Construction

ISARC 2017

June 28 – July 1
TAIPEI, TAIWAN

COMPUTER VISION / SENING & MONITORING

Efficient Face Recognition Using FPGA and Semantic Features for Security Controls

Ying-Hao Yu [a*], Yu-Chen Xu [b], Yi-Siang Ting [c] and Ngaiming Kwok

[a]Department of Electrical Engineering, National Chung Cheng University,
No. 168, University Rd., Min-Hsiung Township, Chiayi County 62102, Taiwan
[b]School of Mechanical and Manufacturing Engineering,
The University of New South Wales Sydney, NSW 2052, Australia
E-mail: yinhaun@gmail.com, p1239688@gmail.com, tingyisiang@gmail.com, nmkwok@unsw.edu.au

Abstract –

Facial features are essential to biometric authentication. Nowadays an effective face recognition system is mostly composed of feature's transformation, geometric analysis, and recursive training to resist illegal intrusion. However, the design challenge always arises from the dilemma of lower computing resource usages, power consumption, and real-time performance under rigorous operations of construction sites. Here a chip-based face recognition is proposed by using semantic features to achieve a minimal dataset and high processing speed without any complicated formulization processes. Our experimental results on single FPGA chip demonstrate the feasibility to realize a miniature and efficient security control system for construction facilities in the future.

Keywords –

Face recognition; Semantic features; FPGA; Security control; Biometrics

1 Introduction

In a construction project, safety and security controls are indispensable for loss reduction from accidents and crimes. The main issue of safety is about the labor working with hazard materials and danger zones. Such problems can be fundamentally precluded by the management to carry out training and operating discipline. However, an impervious security control system relies on high techs to take over manpower for 24 hours a day. Unlike the key and traditional sensors which have been consider insufficient for authentication, the biometric technologies recently has been reported as an efficient choice to resist illegal access by unique features [1].

Biometric authentication is an identification technology based on the unique physiological and behavioral traits, which involves face, fingerprints, ears, keystroke dynamics, voiceprint, gait, and the shape of body [1]. Among these biometric technologies, the face recognition is the favorite one due to advantages of lower cost, non-invasive sensing, and portable dimension [2], [3]. This kind of methodology not only enables security control system to identify but also tracks people with a longer distance at public or domestic places [3], [4]. Currently, the biometric design with digital camera can be also found on embedded system (e.g., smart phones and tablets) as a new trend to check user's access identity [5].

According to the discussion in [1], the way to design a biometric system with lower dataset, computing load, memory space, and time delay on mobile devices still remains a challenge. Although the biometrics designed with external server and sophisticated face recognition algorithms can achieve a good result [6], the security functionality will be totally collapsed if the network is disconnected [1]. For the face detection on embedded systems, the algorithm named Adaboost with Haar-like features is broadly adopted and improved [7]. However, the Haar-like features is too rough and inaccurate for face features' classification. Currently, the most of face recognition systems employ principal component analysis (PCA) and support vector machine (SVM) for better classification, but the processing speed is usually deteriorated by complicated algorithms [8].

In this paper, we propose a real-time face recognition by using feature extraction of semantics-based vague image representation (SVIR) on single field-programmable gate-array (FPGA) chip, using hardware circuit designs [9]. The SVIR is a rule-based feature extraction without operations of trigonometric functions and matrices, and it also has no need to analyze a pattern via geometry. Moreover, the concise algorithm enables feature's evolutions and recognition to be performed during image pixels scanning without any external memory space. Thus the advantages of miniature dimension, real-time performance, economical resource usages, and power saving by lower system clock speed are entirely sufficient for construction sites, warehouse, and engineering vehicles.

This paper is organized as the follows. In the section 2, the methods of skin color detection using HSV color space, applied SVIR algorithm, and the mechanism of face recognition will be detailed. The experimental results on single FPGA chip is in the section 3 followed by discussion in section 4. Finally, the conclusion is drawn in section 5.

2 Methods

2.1 Skin Color Detection by HSV

The skin color in a picture is firstly transferred into binary image as the input of SVIR. By considering the work of Yang et al. [10], we sampled faces from Asians and defined new HSV skin color threshold for $160 < H < 184$, $V > 900$, and $S < 50$ with the scale of 0 to 255. At the output of binary image, a real-time noise filter from [11] was implemented for:

$$I_{(i,j)}\langle 1 \mid 0 \rangle = I_{(i-1,j)} \cap I_{(i-2,j)} \cap I_{(i-3,j)} \quad (1)$$

where (i, j) denotes the location of pixel in picture and I is the output binary image. The true logic will be confirmed with three successive "1" outputs while the random noise produces a false output logic "0", as shown in Figure 1.

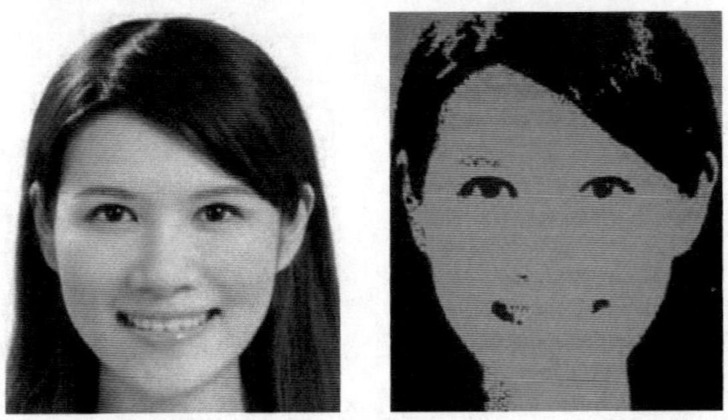

(a) Original image (b) After noise filtering

Figure 1. Binary image with noise filtering

2.2 SVIR Algorithm

The method of SVIR is based on our previous work in [9], which is consisted of bipolar image encoding, vertical evolution of sub-patterns, and lateral combination with sub-windows to perform feature's extraction and evolutions. Due to the small dimension of facial features, the algorithm of lateral combination was unemployed in this design and a couple of SVIR rules were also amended for face recognition.

2.2.1 Bipolar Image Encoding

Unlike the Adaboost sampling a picture pixel-by-pixel for sub-images matching [8], the SVIR samples a picture by stationary sub-windows $S_{(i,j)}$. Each sub-window is also split up into ten sub-sections s_y in order to integrate pixels number c during image pixels scanning as:

for $2 \le y \le 9$,

$$s_y(P,N) = \begin{cases} (1,0), & if \quad |c_y - c_{y-1}| > |c_{y-1} - c_{y-2}| \\ (0,1), & if \quad |c_y - c_{y-1}| < |c_{y-1} - c_{y-2}| \\ (0,0), & if \quad |c_y - c_{y-1}| = |c_{y-1} - c_{y-2}| \end{cases} \quad (2)$$

where y is sub-section's order by image pixels scanning direction from the left- to right-hand side. The P and N are sub-pattern working as polarities of bipolar encoding to represent pixels' variation. Meanwhile, the s_1 is an exception of (2) determined by the variation between c_0 and c_1, and s_0 is undefined in SVIR.

In addition to P and N annotations, the sampled sub-pattern can be also described with a concave or convex trend. Thus the superscript "+" is for $c_y > c_{y-1}$ (concave) and the superscript "–" is defined for $c_y < c_{y-1}$ (convex). When c_y is equal to c_{y-1}, the trend of sub-pattern is determined by c_{y-1} and c_{y-2} instead. Besides, s_y will be denoted as Z_e for a rectangle pattern, and the notation Z_0 is for empty sub-sections from c_{y-2} to c_y. Furthermore, a Z_u is defined for positive slope and Z_d is for the negative condition of Z_u.

The sketched sampling result of bipolar encoding is shown in Figure 2. Nine sub-patterns from s_1 to s_9 compose of a bigger sub-feature. It can be seen that SVIR does not express a feature by geometric formulization or abstract patterns (e.g., Fourier and Haar-like features) but with a series of comprehensible contours. Such result provides a convenience to represent a pattern in semantic way, e.g., a pattern in concave, convex, rectangular, or protrudent shape.

2.2.2 Vertical Evolution

Pattern recognition with irregular contour usually dramatically increases the difficulty of classification and training. For this, SVIR provides rule-based evolutions during image pixels scanning in order to approximate a sampled feature [9]. The proposed evolution rules for two sub-features stacking up are as follows:

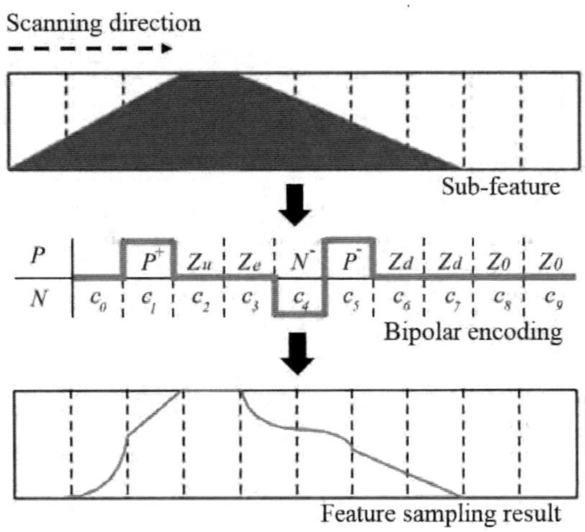

Figure 2. Sub-feature sampling by bipolar encoding

1. Proposition 1: Invariability

Two homogeneous sub-patterns stacking up will result in a similar pattern to original one as:

$$P^{(+,-)}\left(N^{(+,-)}\right)+P^{(+,-)}\left(N^{(+,-)}\right)=P^{(+,-)}\left(N^{(+,-)}\right) \quad (3)$$

and

$$Z_{(u,d,e)}+Z_{(u,d,e)}=Z_{(u,d,e)} \quad (4)$$

2. Proposition 2: Cancellation

Two different sub-patterns stacking up with homogeneous superscript will result in invariant conditions of $Z_{(u,d)}$.

$$P^{(+,-)}\left(N^{(+,-)}\right)+N^{(+,-)}\left(P^{(+,-)}\right)=Z_{(u,d)} \quad (5)$$

3. Proposition 3: Complementation

If both sub-patterns have homogeneous polarity and different superscripts, the evolution result will be the same as the sub-pattern at the bottom.

$$P^{(+,-)}\left(N^{(+,-)}\right)+P^{(-,+)}\left(N^{(-,+)}\right)=P^{(-,+)}\left(N^{(-,+)}\right) \quad (6)$$

4. Proposition 4: Mirroring

The evolution result becomes an invariant state (rectangle) while both stacking sub-patterns are in a mirroring relationship to each other,

$$P^{(+,-)}\left(N^{(+,-)}\right)+N^{(-,+)}\left(P^{(-,+)}\right)=Z_e \quad (7)$$

and

$$Z_{(u,d)}+Z_{(d,u)}=Z_e \quad (8)$$

5. Proposition 5: Carry

SVIR's evolution is not only occurring at individual sub-section but also involves with vicinities by carrying mechanisms as addition. With this mechanism, evolution results after stacking up sub-features will be gradually dominated by the bottom (recent) features. The outmoded (top) sub-features will be continually phased out in two directions of s_6 to s_9 and s_5 to s_2. There are five cases for SVIR's carry as:

Case 1: $P^{(+,-)}$ and $N^{(+,-)}$ stacking with $Z_{(u,d,e)}$

This case is applied to sub-sections s_2 to s_9 and will produce a carry as original one ($P^{(+,-)}$ or $N^{(+,-)}$).

$$P^{(+,-)}\left(N^{(+,-)}\right)+Z_{(u,d,e)}=Z_{(u,d,e)} \quad (9)$$

and

$$Z_{(u,d,e)}+P^{(+,-)}\left(N^{(+,-)}\right)=Z_{(u,d,e)}$$

$$carry\ P^{(+,-)}\left(N^{(+,-)}\right) \quad (10)$$

Case 2: $Z_{(u,d)}$ stacking with Z_e

Once $Z_{(u,d)}$ stacks with Z_e among sub-sections s_2 to s_9, the carry is similar to *Case 1* as:

for $s_* \neq Z_e$,

$$Z_{(u,d)}\left(Z_e\right)+Z_e\left(Z_{(u,d)}\right)=Z_e$$

$$carry\ Z_{(u,d)} \quad (11)$$

where s_* denotes the next sub-section. This case will not be triggered if the evolution result of s_* is still at Z_e.

Case 3: Stacking at Mirroring and Invariability

The carry operation should follow propositions of invariability or mirroring and is inactive as:

$$Z_{(u,d)} + Z_{(d,u)} = Z_e \text{, } mirroring \quad (12)$$

$$Z_{(u,d,e)} + Z_{(u,d,e)} = Z_{(u,d,e)}, \; invariability \quad (13)$$

Case 4: Stacking with Z_0

The carry operation is also inactive once sub-patterns stack with Z_0.

$$Z_0 + P^{(+,-)}\left(N^{(+,-)}\right) = P^{(+,-)}\left(N^{(+,-)}\right) \quad (14)$$

$$P^{(+,-)}\left(N^{(+,-)}\right) + Z_0 = P^{(+,-)}\left(N^{(+,-)}\right) \quad (15)$$

$$Z_0(Z_{(u,d,e)}) + Z_{(u,d,e)}(Z_0) = Z_{(u,d,e)} \quad (16)$$

Case 5: Carry at s_1

The carry at s_1 is also disable and the evolution rules at this sub-section are defined as:

$$\left(P^{(+,-)}, N^{(+,-)}\right)_1 + Z_0 = \left(P^{(+,-)}, N^{(+,-)}\right)_1 \quad (17)$$

and

$$Z_0 + \left(P^{(+,-)}, N^{(+,-)}\right)_1 = \left(P^{(+,-)}, N^{(+,-)}\right)_1 \quad (18)$$

$$Z_e + \left(P^{(+,-)}, N^{(+,-)}\right)_1 = \left(P^{(+,-)}, N^{(+,-)}\right)_1 \quad (19)$$

$$\left(P^{(+,-)}, N^{(+,-)}\right)_1 + Z_e = \begin{cases} \left(P^{(+,-)}, N^{(+,-)}\right), & if \;\; entire \; [c_2, c_9] \neq Z_e \\ Z_e, & if \;\; entire \; [c_2, c_9] = Z_e \end{cases} \quad (20)$$

In the end, an evolution result for two sub-features' stacking up is shown in Figure 3. Two different sub-features are located at successive sub-windows and combine into a larger pattern (feature). Generally, with a larger pattern, the system also needs more memory spaces to keep dataset at the same time. However, after applying vertical evolution, only a set of bipolar encoding is required to describe the larger pattern with the same budget of memory space. This leads significant improvements in computing load, and the memory consumption and access time are also less than the other traditional algorithms.

2.3 Face Sampling and Recognition

Face recognition is usually based on the classification of facial features for eyes, nose, mouth, and ears. In our design, we classified different faces by eyes and nose after considering the stability of image resources. As shown in Figure 4, the testee aligned his

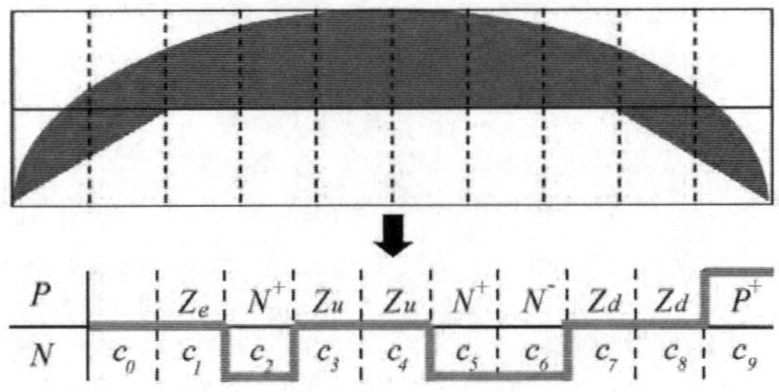

Figure 3. Vertical evolutions for two sub-features

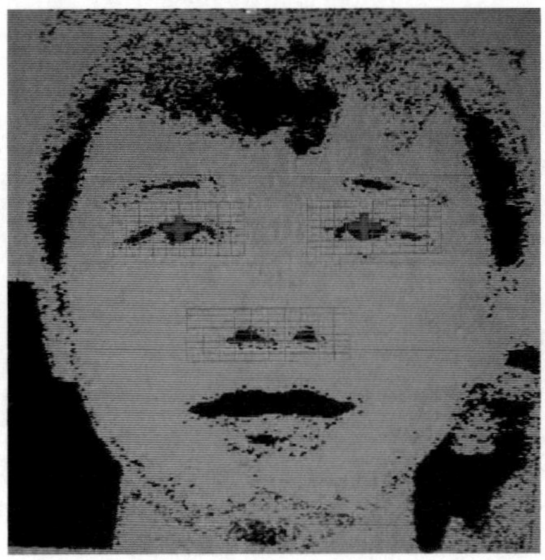

Figure 4. Face features sampling by sub-windows

face by overlapping pupils on the cross marks. Meanwhile, there were eight sub-windows at every facial feature, and each feature was also split into two parts in order to perform vertical evolution by four sub-windows. Thus we totally obtained six features from the locations of eyes and nose.

For the dataset collection, we firstly provided forty SVIR eye-templates for eyes and the other ten nose-templates for nose. Next the system sampled a face for thirty times within one second and then determined testee's face type by the highest similarity to feature templates. The face, thus, can be easily represented by a set of codes and recorded in chip's registers without the use of external memory, as shown in the Table 1.

Finally, we could achieve face recognition by checking the similarity of sub-features between testee and dataset. Here we empirically adjusted different weights for each sub-section at upper eyes (5 points), lower eyes (3 points), and nose (2 points). Accordingly, there is a total of 162 points with six sub-features, and the similarity between testee and dataset can be determined by:

$$\text{Similarity}(\%) = \frac{\text{Final score}}{162} \times 100 \qquad (21)$$

3 Experimental Results

The proposed face recognition system was implemented on a developing platform with Cyclone II EP2C70F896C6 FPGA chip. Our digital camera module's resolution was set for 1280×1024 pixels. The video frame rate was at 34 fps with 77 MHz pixel clock. This pixel clock was also employed to synchronize with SVIR and recognition circuits.

The definition for sub-patterns and corresponding codes are list in Table 1. Each sub-pattern discussed in Section 2 can be simply translated into a number. Thus the representation of a sub-feature before/ after evolutions only requires a set of nine numbers. This design concept is very important to chip and embedded system designs in that miniature systems usually do not equip with sufficient onboard and inbound memories.

As the example in Figure 5, forty-six Asian faces were collected as strangers to compete with our testee. It can be seen that similarity between testee and recorded dataset is noticeably higher to the other strangers. This result can help us to set a threshold for discriminating testee from strangers. In our design, three testees with different face types had been invited to compete with these Asian faces. The collected test results then became a useful basis for threshold determination.

Table 1. The list of SVIR coding with corresponding sub-patterns

Code Name	Code	Pattern	Code Name	Code	Pattern
P^+	11	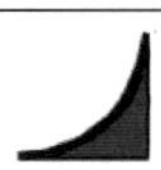	Z_0	00	
P^-	12		Z_u	01	
N^+	21	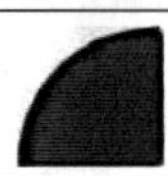	Z_d	02	
N^-	22		Z_e	03	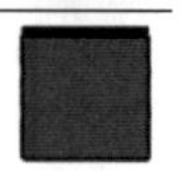

According to our tests with three testees, we derived false rejection rates (FRRs) and false acceptance rates (FARs) based on Bayes' theorem. Here we picked out highest FAR and FRR from tests by different similarity thresholds, as shown in Figure 6. The equal error rate (EER) with FAR and FRR then could be found about 85% similarity, which is the balanced point for FAR and FRR adjustments. This threshold makes a clear separation of our testees from strangers for 7% at least.

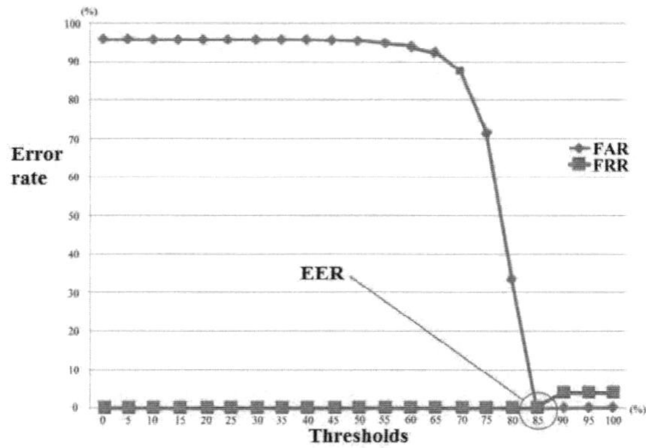

Figure 6. Face recognition error rates with different recognition thresholds

Table 2. Hardware resource usages of FPGA

Designs	Logic elements (%)
Image capture/ Demosaicking/VGA	1,700(2.4%)
HSV colour space/ Binary image / noise filter	1,300(1.9%)
SVIR×6	5,898(8.6%)
Template Comparison×6	5,497(8%)
Dataset & Recognition	698(1%)
Test circuits (LCD/LED)	736(1%)
Total	15,829(23%)
Inbound RAM	82,864 bits (7%)

4 Discussion

Face recognition based on SVIR can attain a significant improvement in hardware resource usages. As the logic elements (LEs) consumption in Table 2, we initially duplicated our SVIR circuits for upper eyes, lower eyes, and nose with higher LEs usage. Nevertheless, our LEs consumption for feature extraction (5,898 LEs) is still much lower than the chip-based principle component analysis (PCA) (24,958 LEs) and Fourier Transform (12,014 LEs) designs [12], [13].

Besides, another salience of the proposed system is real-time performance. Since SVIR can perform feature extraction during image pixels scanning at ears and nose [9], the similarity comparison between testee and datasets can be easily performed in the rest of picture. This advantage allows us to check testee's face image quality for 30 times per second and then pick out the best face image for recognition processes.

Finally, the use of 90 SVIR templates for defining testee's face type can reach 1.6 million combinations, which is sufficient for real-life applications. Here testees' face features are only stored in chip's registers with concise data format. Such advantage is very important to reduce the power consumption and installation dimension for construction sites or vehicles.

5 Conclusion

Face recognition is usually impeded by image's resolution and system clock speed. Many designs, therefore, adopt the strategy to degrade picture's quality or dramatically increase system clock speed in order to attain real-time performance. Contrarily, the rule-based SVIR can avoid such problem from feature's extraction, evolutions, and recognition during image pixels scanning. Additionally, the concise features format also allows us to leave out the requirement of external memory. The use of SVIR evolutions and templates can avoid tedious training processes and designs. These advantages imply a possibility to install the proposed system at a narrow and rigorous environment without the supports from external server and networks. It can be seen that the proposed face recognition system is promising for the application in construction sites. The future work will focus on the recognition with different face features.

Acknowledgement

Supports from Ministry of Science and Technology funded by the government of Taiwan under Grant: MOST 104-2221-E-194-049-MY2 is gratefully acknowledged.

References

[1] Mitra S., Wen B., and Gofman M. *Biometrics in a data driven world*, CRC Press, 6000 Broken Sound Parkway NW, Suite 300, Boca Raton, FL 33487-2742, 2016.

[2] Shams N., Hosseini I., Sadri M S., and Azarnasab E. Low cost fpga-based highly accurate face recognition system using combined wavelets with subspace methods. In *Proceedings of the International Conference in Image Processing*, pages 2077–2080, Atlanta, USA, 2006.

[3] Zuo F. and Peter H N. Real-time embedded face recognition for smart home. *IEEE Trans on Consumer Electronics*, 51(1), 183-190, 2005.

[4] Jain A K., Duin R P W., and Mao J. Statistical pattern recognition: a review. *IEEE Trans on Pattern Analysis and Machine Intelligence*, 22(1), 4-37, 2004.

[5] Chimote M G R. and Tarbani N M., A survey paper on authentication system in android phones. *The International Journal of Scientific & Engineering Research*, 7(2), 238-240, 2016.

[6] Kremic E., Subasi A., Hajdarevic K. Face recognition implementation for client server mobile application using pca. In *Proceedings of the International Conference on Information Technology Interfaces (ITI)*, Dubrovnik, Croatia, 2012.

[7] Cho J., Mirzaei S., Oberg J., and Kastner R. Fpga-based face detection system using haar classifiers. *In Proceedings of International Conference on Field Programmable Gate Arrays (ACM/SIGDA)*, pages 103-112, 2009.

[8] Jeon D., Dong Q., Kim Y., Wang X., Chen S., Yu H., Blaauw D., and Sylvester D. A 23-mw face recognition processor with mostly-read 5t Memory in 40-nm cmos. *IEEE Journal of Solid-state Circuits*. (To appear)

[9] Yu Y H., Lee T T., Chen P Y., and Kwok N. On-chip real-time feature extraction using semantic annotations for object recognition," *Real-Time Image Processing*, DOI 10.1007/s11554-014-0474-2, 2014. (6 pages)

[10] Yang L., Li H., Wu X Y., and Zhao D. An algorithm of skin detection based on texture. *In Proceedings of IEEE International Congress on Image and Signal Processing (CISP 2011)*, Shanghai, China, 2011.

[11] Yu Y H., Kwok N M., and Ha Q. Color tracking for multiple robot control using a system-on-programmable-chip. Automation in Construction, 20, 669-676, 2011.

[12] Zhong F., Capson D W., and Schuurman D C. Parallel architecture for pca image feature detection using fpga. *In Proceedings of IEEE Conference on Electrical and Computer Engineering*, Canada, pages 1341-1344, 2008.

[13] Bahoura M. and Ezzaidi H. Fpga implementation of a feature extraction technique based on fourier transform. *In Proceedings of IEEE International Conference on Microelectronics*, Algiers, Algeria, pages 1-4, 2012.

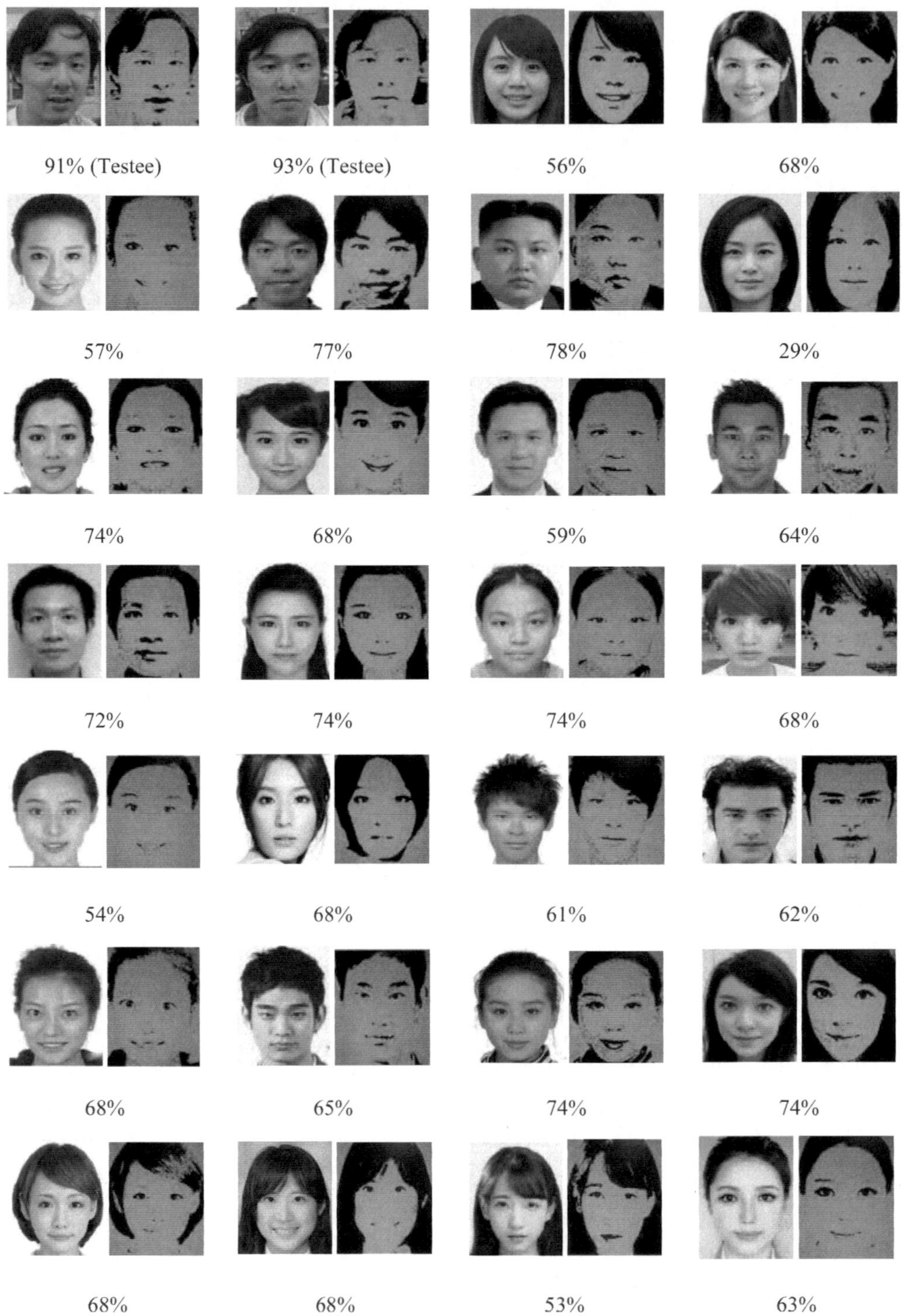
91% (Testee)
93% (Testee)
56%
68%
57%
77%
78%
29%
74%
68%
59%
64%
72%
74%
74%
68%
54%
68%
61%
62%
68%
65%
74%
74%
68%
68%
53%
63%

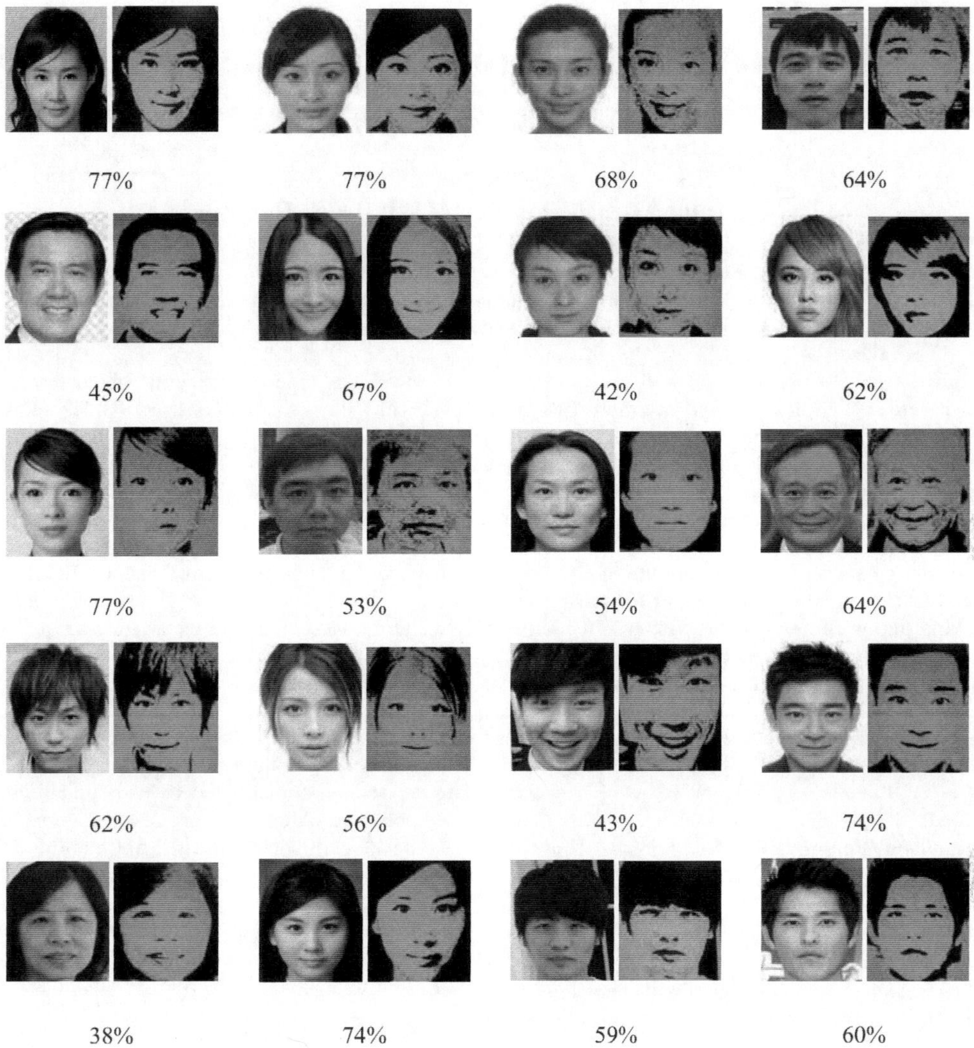

Figure 5. Face recognitions by competing similarity between testee and strangers

BIM-based Indoor Positioning Technology Using a Monocular Camera

Yichuan Deng[a], Hao Hong[b], Hui Deng[c], Han Luo[d]

[a,b,c,d] School of Civil Engineering and Transportation, South China University of Technology, China
[a]State Key Laboratory of Sub-tropical Building Science, South China University of Technology, China
E-mail: ctycdeng@scut.edu.cn

Abstract

In recent years, increasing attention has been paid to indoor positioning. This paper presents a BIM-based Indoor positioning technology using a monocular camera. This technology utilizes a monocular camera to obtain on-site image. An improved corner detection algorithm is used to identify the corners of the ground tile in the on-site image. Combining the geometric information from BIM, this paper completes the conversion of image pixel coordinates to the actual coordinates of the indoor ground. Based on the location information of camera in BIM model, this paper achieves the precise positioning of actual indoor target in BIM model.

Keywords

Indoor positioning; BIM; Image recognition; Computer Vision

1 Introduction

The research and application of real-time acquisition of 3-D position information of people and objects has become the hotspot in recent years. Real-time positioning technology is of enormous benefit in people's lives, promoting the progress of transportation and surveying[1]. Also, it plays a significant role in the AEC industry. The 3-D position information of construction personnel and machinery has wide application prospect in construction management, including schedule management, quality management, safety management and operation analysis[2].The Manuscript should not exceed 8 pages in length, including text, figures, tables, and references. Electronic file should not exceed 5 megabytes.

Previously, some scholars have developed several positioning and tracking methods for positioning personnel or machinery on the construction site. Yang et al identified people and specific machinery and realized action tracking by using trained classifier. Combining image recognition and pattern recognition, this method showed good performance in object recognition and tracking in on-site construction[3]. However, this method could not acquire the position information of object. Fang et al acquired real-time position information of target through 3-D point cloud technology[4]. This method used UAV equipped with high-definition camera to obtain image information from different angles in the air and developed a point cloud model. Then, the real-time position information of the target was obtained by matching the real-time image with the feature point. Although this method achieved the position tracking of on-site construction vehicle, it still had defects such as inapplicability to track interior objects of building, inaccurate position information, and sensitivity to complex environments. Compared to outdoor environment, the interior of building is more complicated. After literature review, a conclusion could be developed that the identification ability of computer vision complement with massive information (geometry, material, coordinates) in BIM model. In order to realize real-time positioning of indoor target, this paper presents a BIM-based real-time calibration and positioning method using monocular camera.

2 Literature Review

GPS (Global Positioning System) positioning, the most popular positioning technology nowadays, calculates 3-D position information of GPS receiving devices by using not only signal transmission between different satellites and GPS receiving devices, but also geometric relations. On the one hand, GPS has various advantages like wide range application, low cost, mature technology. On the other hand, GPS has some shortcomings. For example, civilian GPS has low accuracy of 10m[5]. The signals of GPS are susceptible to building disturbances. The accuracy of GPS could not meet with the requirement of indoor positioning. Aaron Costin at el combined BIM and RFID technology, in which RFID tags are pre-implanted into the target device and on-site RFID reader, as well as BIM reader, is used for reading position information. This realizes the positioning, tracking and visualization of target device[6]. UWB (Ultra Wide-Band) is another technology for indoor high-accuracy positioning. UWB needs to set base station with known coordinates. It uses receiving device to read the tag carried by target object and then calculates the relative position. The accuracy of this technology could be up to centimeters. Even though both RFID-based and UWB based positioning technology perform well in indoor positioning, they still strongly depend on pre-planted tags in target. In addition, UWB has defects like short signal coverage radius of only about 10 meters and expensive cost of $140 per square meter[7]. Aiming to solve above problems and realize indoor positioning without pre-implanted tags, Huang el al utilized a device equipped with two cameras to acquire image information. After conversion from RGB image to HSV space, they set threshold to generate binary image and identified the plate in the image. According to the principles of binocular calibration and stereo imaging, the authors calculated the 3-D coordinates of feature points in the plate edge[8]. This method could realize random target positioning without pre-implanted tags on the basis of binocular calibration and image identification. But it also greatly depended on the success matching of feature points. While complex indoor environment easily resulted in difficulties of extracting and matching the target feature points.

3 Methodology

In order to avoid the influence of collision and realize indoor positioning of target without pre-implanted tags, this paper presents a new BIM-based real-time calibration and positioning method using monocular camera.

Camera calibration is a process for solving parameters of camera imaging model. Here, imaging model represents the correspondence between a point in a spatial object and corresponding point in the image. When calibrating, it is necessary to obtain the real world coordinates corresponding to several pixel points in advance and the true distance between these points. BIM is not only a digital representation of the physical and functional characteristics of facilities, but also a shared knowledge resource[9]. Besides a 3-D visual architectural geometric model, BIM is an information model including semantic information such as construction project progress, cost. BIM could provide geometric information, i.e. accurate distance between specific points in the camera calibration phase. When camera calibration is finished, BIM could also provide position information to accurate indoor positioning. Based on reasons above, the methodology of positioning technology proposed in this paper is shown in Fig.1.

3.1 Camera Calibration

The function of the camera calibration is to find the correspondence between the pixel coordinates of the points in the image and the real coordinates of the corresponding points in the real scene. The correspondence could be expressed in matrix form as follows:

$$\begin{bmatrix} x \\ y \\ 1 \end{bmatrix} = sMW \begin{bmatrix} X \\ Y \\ Z \\ 1 \end{bmatrix} = sM[R \;\; T] \begin{bmatrix} X \\ Y \\ Z \\ 1 \end{bmatrix} \quad (1)$$

where, x and y represent lateral and vertical coordinates of the point in the image respectively. X, Y and Z represent the real coordinates of the corresponding points in the real scene. M is a 3*3 matrix, standing for the intrinsic matrix of the camera. W is a 3*4 matrix, standing for the extrinsic matrix of the camera. R, a 3*3 matrix, represents the rotation matrix. T, a 3*1 vector, represents the translation vector. And s is a ratio with random scale.

From Eq. (1), the process of camera calibration is actually a calculation process of solving the intrinsic and extrinsic matrixes of camera. It is worth mentioning that the intrinsic matrix makes up of camera's inherent parameters such as focal distance, which would not change with the external influence of image shooting angle, and degree of definition. If the intrinsic matrix of camera is known, the calculation will convert from finding correspondence between pixel coordinate and real coordinates to solving the extrinsic matrix of the camera. The extrinsic matrix of camera represents the relationship between the coordinates of camera itself and world coordinate system. It includes the rotation matrix R and translation vector T. The rotation matrix stands for the angle relation between camera coordinate system and world coordinate system. And translation vector represents the spatial distant relationship between camera and target point.

Nowadays, monocular camera calibration mostly uses the calibration method proposed by Zhengyou Zhang in 1988[10]. This calibration method uses a chessboard with known fixed-size grids as the calibration target. By recognizing the corners of the chessboard's grids at multiple angles, the camera calculates the intrinsic matrix of camera and the extrinsic matrix of camera in each image. During calibration, it is important to input the number of corners and the size of grids, as well as strictly define the arrangement of the plane corners. Zhang's method has the advantages of high calculation efficiency and high calibration accuracy. Yet when it comes to the calibration requirements in this paper, the size of chessboard is too small as compared to the whole indoor

Error! No topic specified.

Figure 1. Methodology

space to be used as standard calibration reference. With the observation that here is obvious corner feature in the joints intersect of ceramic tiles, this paper improves the calibration method based on theoretical basis of Zhang's calibration method and utilizes the characteristic corner of indoor ceramic tiles to realize calibration without a calibration plate.

3.2 Corner Recognition

As referred, the basis of camera calibration is the correct identification of corner in the target object. The existing corner recognition algorithm, based on Zhang's calibration method, only performs well when recognizing the corner of the black and white chessboard, which shows in Fig.2. As for the indoor condition, the algorithm could not identify the general indoor corner. In order to solve those problems, this paper has developed a more general algorithm for indoor ceramic tile corner recognition. The process of this algorithm shows in Fig.3.

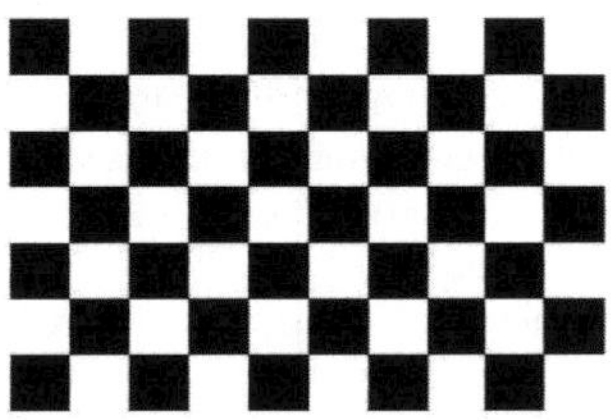

Figure 2. Chessboard

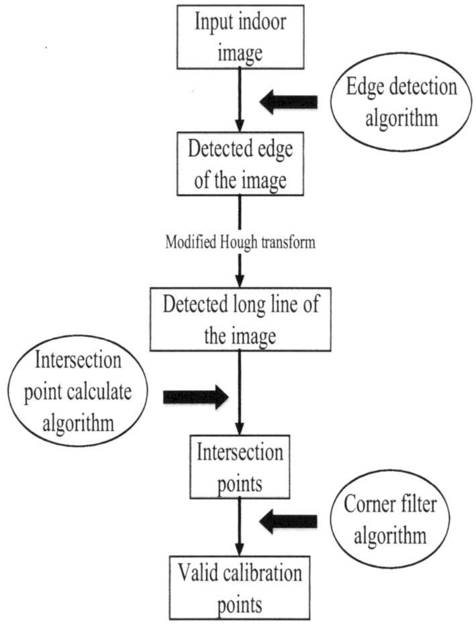

Figure 3. Calibration point extraction process

As Fig.4 (a), indoor ceramic tiles have obvious line feature in joints. For this property, this paper first uses the line detection algorithm to extract all the long straight lines in the image. And corner mostly exists in the intersection of two straight lines, shown in fig. 4(b). Therefore, the algorithm of calculating the intersection of two straight lines would be used in all the extracted straight lines from last steps. Inevitably, on-site construction images are influenced by factors like lights and sundries. This tends to produce disturbed straight lines and extra corners when extracting correct and useful ceramic tile corners. Moreover, because of the problems from shooting angles, it could not be guaranteed that the correctly recognized corners of ceramic tiles belong to the same piece of ceramic tile. Also, it is hard to arrange the order of the corners in the same way as a chessboard. During actual calibration, the real vertical coordinate, Z, of the calibrated point is usually set to 0. With Eq. (1), it just needs 4 calibrated points for finding the exact solution of extrinsic matrix. Thus, this paper used screening algorithm for searching for four corners in the same ceramic tiles. Then, these four corners are used as the final calibrated points, illustrated as Fig. 4(c).

Figure 4(a). Original indoor image

Figure 4(b). Detected lines and intersection points

Figure 4(c). Valid calibration corners

3.3 BIM Information Extraction

An important part of realizing camera calibration is assigning a predefined real coordinate to the calibration point of target. In Zhang's calibration method, predefined real coordinates are determined manually by pre-input according to the known size of chessboard. Based on the correct extraction of four corners in the same ceramic tile, this paper refers to the geometric information from BIM model directly without manually inputting coordinates.

BIM model contains the geometry and position information for all components of the building. Through automatic reading of the geometric information of the corresponding tile in BIM model, the technology proposed by this paper could achieve a fast automatic monocular camera calibration. When calibration is completed, the 3-D real coordinates of positioning target in the image will be determined by the solved coordinate transformation matrix. Strikingly, the 3-D coordinates, solved at this time, refer to the coordinates system established by four calibration points. For the sake of obtaining the actual coordinates of the target relative to the whole building, it's a key to read the camera's position information that predefined in the BIM model. After simple coordinates transformation, the target will be accurately positioning in the end.

4 Case Study

The proposed indoor positioning technology is verified in an academic building. This academic building has 6 floors and hundreds of sub-space including conference rooms, labs, classrooms, office rooms and so on. Then, an image shooting device is fixed on a lab in the 4th floor. And the intrinsic matrix of camera has been acquired by general camera calibration method.

Fig.5(a) shows the random image shooting by the fixed camera. Now, the blued stool is used as the positioning target for the verification of feasibility of the developed method. First of all, after the detection using the straight line detection algorithm, there are two problems. One is that the number of straight lines in the joints of ceramic tiles is too small. Another one is that a large number of disturbing invalid lines present in the rest of the image. With studying the original image, three reasons are found for these phenomena: (1) since the light is too strong in the left side of the image and makes serious reflection in the left side of the ground, there are a lot of small pieces of straight lines after initial detection; (2) the target itself has the linear characteristic; and (3) due to poor construction quality and long-term use, the width of ceramic tiles' joints are not uniform and sometimes the joints are too small to show the colors in the image. This results in difficulty of forming the straight line feature in lab's ceramic tiles. Fig.5(b) presents the intersection of straight line determined on the basic of straight line detection. From this, it is easy to know that invalid corners are occurred in disturbing straight lines. As expected, the four valid corners are still selected after doing the corner screening algorithm, shown in Fig.5(c). That indicates the algorithm for recognizing valid calibration points has the ability to resist indoor light changes and object interference.

Figure 5(a). Image with positioning object

Figure 5(b). Line and intersection points detect

Figure 5(c). Valid calibration corners

After the extraction of valid calibration points, the geometric information of ceramic tile is automatic obtained from BIM model. Fig.6 illustrates the geometric information of the ceramic tile. In this case, the size of ceramic tile is 39.5cm*39.5cm. The initial real coordinates (unit in mm) of the four valid points are (0, 0, 0), (0, 395, 0), (395, 0, 0) and (395, 395, 0), respectively. And the corresponding pixel coordinates of them are (415, 1945), (1162, 1375), (1283, 2401) and (1944, 1648).

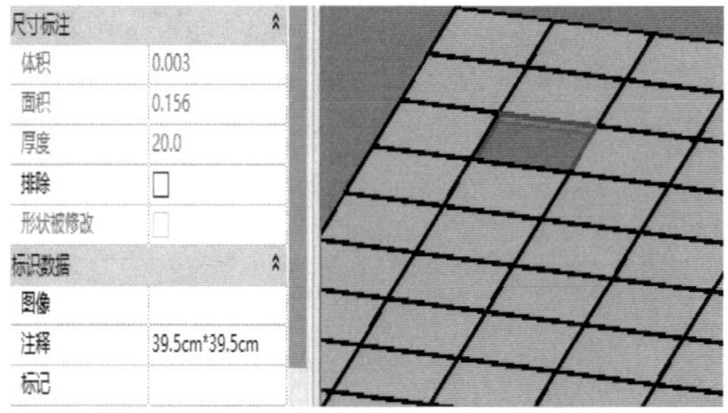

Figure 6. Geometry information of calibration object in BIM

According to the known intrinsic matrix of camera and the coordinates of valid calibration points, authors get the extrinsic matrix by using the function of solving extrinsic matrix. That function is from OpenCV, the open source library in computer vision domain. Then, authors obtain the coordinate transformation matrix of the image pixel and its corresponding real point. Now, in order to positioning the blue stool and find its real 3-D coordinates, a simple algorithm is used to select the positioning point determined by the triangle centroid of three stool feet. The coordinates of the positioning point are (2277, 1137) in the image, shown in Fig. 7. Then, a coordinate system is built with an origin in the lower left corner in Fig. 5(c). Based on this coordinate system, the real 3-D coordinates of the positioning point are (256.3, 590.5, 0), calculated by the matrix transformation algorithm. For the aim of positioning target object in the whole building, the position information of camera is read from BIM model. In this case, the camera is located 75cm above the 4th floor and the lower left corner of the whole building is treated as the origin of coordinate system. Accordingly, the camera coordinates are (4220, 2377.5 11550) and the coordinates of the lower left corner in Fig. 5(c) are (-945, 790, -750) relative to the camera. Finally, the real coordinates of target are (3531.3, 3758, 10800). The comparison between the coordinates of the target and the on-site actual measurement results is shown in Table 1.

Figure 7. Positioning point of the object

Table 1 Comparsion between calculated coordinate and real coordinate

Direction	X	Y	Z
Calculated coordinate (mm)	3531.3	3758	10800
Real coordinate (mm)	3580	3917.5	10800
Accurancy (%)	98.64	95.93	100

From Table 1, the proposed method in this paper has high accuracy. And with the use of BIM, the height information could guarantee 100% accuracy. However, the errors of X and Y directions could be up to 10cm. Through analysis, two reasons are found: (1) the number of valid calibration points chosen for calibration is not enough so that the resulting transformation matrix may not be the optimal solution; and (2) there is a major error in selecting positioning point of target object.

5 Conclusion and Future Work

5.1 Conclusion

After analyzing the shortcomings of various positioning technology like GPS and UWB, this paper proposed a new indoor positioning method combining monocular camera calibration and BIM technology. The main results are as follows:

- Proposed the camera calibration method implemented by using indoor ceramic tiles and geometric information from BIM model without the use of chessboard-like calibration plate;
- Developed a new corner recognition algorithm by combining straight line detection algorithm and line intersection algorithm;
- Developed a corner screening algorithm for the extraction of valid calibration point;
- Established a complete indoor positioning process by using the camera position in BIM and the coordinates transformation relation of camera calibration;
- Realized the positioning of target object in one lab of a 6-floor academic building with 100% accuracy in Z direction, 95% accuracy in X and Y directions.

5.2 Future Work

There are some limitations and shortcomings in the proposed indoor positioning method. For instance, since there are a lot of pixel points around each stool feet in the image, using triangle centroid of three stool feet as positioning point in the case study has a negative influence on the accuracy of positioning point. This brings major error to the selection of target positioning point. Furthermore, the deformation of the stool due to imaging principle also causes difficulties in the selection positioning point.

References

[1] Zhai hongyan. Analysis and Comparison of Several Positioning Techniques in Internet of Things. Software Engineering[J],2014(10): 32-34.

[2] Liu Wenping. Research on Early Warning Mechanism of Construction Accident Based on BIM and Location Technology [D]. Tsinghua University,2015.

[3] Yang, J., et al., Construction performance monitoring via still images, time-lapse photos, and video streams: Now, tomorrow, and the future. Advanced Engineering Informatics, 2015. 29(2): p. 211-224.

[4] Fang, Y., et al. A point cloud-vision hybrid approach for 3D location tracking of mobile construction assets. 2016. Auburn, AL, United states: International Association for Automation and Robotics in Construction I.A.A.R.C).

[5] Yang Jie and Zhang Fan. A Comparative Study of High Precision GPS Differential Positioning Technique. Mobile Communications,2014(02): 54-58+64.

[6] Costin, A.M., J. Teizer and B. Schoner, RFID and bim-enabled worker location tracking to support real-time building protocol control and data visualization. Journal of Information Technology in Construction, 2015. 20: p. 495-517.

[7] Zhang Jie. Comparison of Ultra - Wideband Location and Other Wireless Location

Technology.The 22nd National Conference on Coal Mine Automation and Informatization: Shaanxi Xi'an China.4.

[8] Huang, G. and W. Zhang. Recognizing and locating of objects using binocular vision system. 2013. Tainan, Taiwan: IEEE Computer Society.

[9] He Qinghua et al. Research on the Present Situation and Obstacles of BIM Application at Home and Abroad. Journal of Engineering Management,2012(01): 12-16.

[10] Zhang Z. A Flexible New Technique for Camera Calibration[J]. IEEE Transactions on Pattern Analysis & Machine Intelligence, 2000, 22(11):1330-1334.

RTLS systems as a Lean Management tool for productivity improvement

Piotr Nowotarski[a] Jerzy Pasławski[a] Maciej Skrzypczak[b] Radosław Krygier[b]

[a] Institute of Structural Engineering, Poznan University of Technology, Poland
[b] Student at Institute of Structural Engineering, Poznan University of Technology, Poland

E-mail:piotr.nowotarski@put.poznan.pl, jerzy.paslawski@put.poznan.pl, maciej.r.szkypczak@student.put.poznan.pl, radoslaw.krygier@student.put.poznan.pl

Abstract – There exists many types of different RTLS systems that can be used in the construction sector for tracking people and assets. Previous research showed that such system can be used with success especially during the construction stage of the object. Lean Management is a methodology based on waste reduction in production processes focused on productivity improvement. This methodology is widely used in different sectors, including construction, where it can be effective and results can be visible for all interested involved parties. A goal of this paper is to present RTLS systems used for tracking construction process and with the focus on Bluetooth technology for object tracking. Authors presents the idea of using locating systems for efficiency improvement as a part of Lean Management strategies implementation on the construction site. Discussion is made in the field of construction process showing where they can be used and at what stage in introduction LM strategies RTLS system is most useful. Authors illustrate this idea with the results of the research performed with Bluetooth RTLS system used for tracking objects in the building planned for renovation.

Keywords –
Indoor positioning systems; Sensor technologies; Automated data acquisition; RTLS; Lean Management; Beacons; Bluetooth; Lean Construction

1 Introduction

In recent years, many advanced wireless tracking technology solutions have been developed by scientists and companies interested in implementation of those solutions into constructions sector. Numerous research studies have developed approaches for applying Real Time Locating Systems technologies to construction projects and facility/infrastructure management, and in particular, construction site assets tracking [1]. The development and extensive use of technologies: Radio Frequency Identification (RFID), Ultra-Wideband (UWB) and Bluetooth Low Energy (BLE) has improved location tracking for allocating labor, materials and equipment resources as well as improved health and safety [2 ,3 , 4] in many sectors of industry.

New technologies are important help for the Lean Management methodologies, since having possibility of storing real data from construction site it is easier to analyze it and later use it for process improvement that results in time, money, and resources savings in the processes. So far many times it was not possible for introduction of improvements, because it was difficult to track and save data to realize where the problems lies, and which kind of actions can be taken to resolve them. RTLS systems can serve as a tools that can effectively find the waste associated primarily with the movement of workers and resources

Authors in this study decided to present the possibility of using Bluetooth based RTLS system for assets tracking checking its accuracy and performing root cause analysis for the results that was obtained, to check why results are not as precise as expected, and how were they influenced during the research.

2 RTLS systems

There is no universal definition of RTLS system, but the main idea is to combine the appropriate hardware and software to automatically determine the coordinates of the tracked object in real time, within a given area. Data collected by RTLS can be used not only for control at the given moment, but also for further analysis, what, if there is a need to eliminate unwanted actions and behavior, can be very helpful [5].

There are many types of systems RTLS consisting of different types of location sensors and operating using many location technologies available on the market. The idea of the system lies in communication with the tag by the signal receivers, and its location is calculated using a variety of software algorithms to analyze the received data. Depending on the type of technology, the accuracy of the tracking system varies. Table 1 shows the most popular localization technologies used in construction, along with information on their accuracy.

Table 1. Accuracy of location technologies used in construction sector [5]

Localization Technology:	Accuracy:
RFID	0,86-2,6 m
Satellite (GPS)	2,15-4,36 m
UWB	0,3 m
Vision analysis	0,658 m
WLAN	1,5-4,57 m
Ultrasounds	0,04 m
Infrared	--

As shown in Table 1, the best accuracy is achieved using a system based on ultrasound signals and UWB. The least accurate system is the satellite GPS system. It should be noted that in December 2016 a satellite system GALILEO was launched [6], which accuracy can be closed to UWB levels, but there is a need to wait for attempts to use this type of system in the construction industry.

Attempting on location of the object during the process of construction is also associated with certain limitations in terms of the technologies used. Some of them (GPS) will not work indoors - as necessary in this case is the visibility of the satellite, others (UWB, WLAN) will require the installation of special additional network and receiving equipment on the construction site. There is currently no universal solution that would be good for all applications in the construction industry, which is why the selection of the system should be carried out carefully taking into account the objectives and purpose of its implementation and also the type of tracked objects (tools, material or people).

2.1 Bluetooth in construction

The idea of use of Bluetooth technology as RTLS for the construction industry purposes is not widely available in the literature. In the Thomson Reuters Web of Science Core Collection database, which comprises of seven different citation databases including different information collected from journals, conferences, reports, books and book series [7], there are only 4 articles related to RTLS and Bluetooth keywords. One of them is associated with checking the accuracy of the system for the health care purposes [8]. The results showed that for the given boundary conditions, the locations become unreliable as they change too much and only under specific requirements. Authors indicated the need for more precise studies in this field. Other research was performed in the same area and it was a pilot study prepared to serve as a guide for developers of applications (for software as part of localization) and as help to identify potential research problems and future products [9]. Bluetooth location systems were also studied in terms of big data process environment [10] and was taken as a reference point for comparison with ZigBee solutions, which are similar to Bluetooth technology [11].

Majority of authors of analyzed works indicated that the use of positioning technologies such as Bluetooth, infrared, audible sound, and magnetic signals have yet to be studied (or reported) in the construction industry [5] and the location tracking system needs to be improved further to make them more reliable [8], which is why results presented in this article fill the existing knowledge gap on accuracy research of this technology.

3 Lean Management

Lean Management is an idea identified with the management, which has its foundations in MIT. It was first used by Johnny Krafcik in publication that discusses the results of an international research project on the automotive industry, related with the look for the reasons for predominance of the Japanese producers delivering cars using traditional management techniques over different countries [12, 13].

Lean thinking can be summarized in five fundamental goals [14]:

- eliminate waste,
- establish value stream,
- encourage flow,
- pull production,
- pursue perfection.

The reduction of waste in places where it is possible, and orientation to creation of added value for the end customer [15] is one of the most important, which is implemented through a variety of methods introduced at the stage of planning and production of the product.

The implementation of lean production in construction is known as Lean Construction [16]. Wasteful practices can be described as those which increase time and cost and reduce reliability without adding value to the final outcomes of the processes [17]. In the literature [18, 19] 8 main types of waste are indicated which with help of the Lean Management tools can be successfully eliminated:

- overproduction,
- waiting,
- over-processing,
- transportation,

- storage,
- unnecessary movement,
- defects,
- non-utilized talents.

All of these waste can be observed also during analysis of the construction processes, so Lean approach is nowadays more and more widely used in construction [20]. At the same time more often those methodologies are applied to improve the implementation of other measures in areas such as health and safety [21] and BIM [22], using the assessment of, among others, elements of fuzzy logic known from decision-making in the production management [23].

4 Case study

For the purpose of the article authors decided to perform research connected with checking the possibility of locating objects in existing building. The idea was to use the Bluetooth system for improving renovation processes by enabling direct location of the objects like equipment or materials.

Figure 1. Faculty of Civil and Environmental Engineering Building – front view [photo by Poznan University of Technology]

4.1 System description

Authors performed test on the devices received from one of the European Beacon producers. Beacons are based on Bluetooth Low Energy (BLE) which is a medium ranged transmission standard aiming for cheaper devices with longer lasting batteries [24]. Tested Beacons were designed initially for advertisement industry.

During the research there were totally 15 devices used. Each Beacon has dimensions about 60 mm in diameter and about 15 mm thickness. Devices were equipped with the CR2032 battery and had range, according to the producer, of 70 m. Devices were design to work inside of the building. Producer together with the equipment provides online CMS platform, where it is possible to:

- Manage the Beacons (also of other manufacturers) - adding devices and assigning them to different campaigns, locating navigation units in a construction plan integrated with Open Street Maps and defining area labels.
- Monitor and modify navigation parameters such as frequency, signal strength, battery status.
- Read data and traffic statistics or the number of people present in the area of interest and their visual analysis.

Except from described platform there was also mobile application available which is required for communication between CMS system and Beacons. It is responsible for updates sent to the content management system. The update occurs when the device (mobile phone) with the application enabled is within range of the Beacon.

There was also another mobile application which was specially design to search and locate for individual items that, upon discovery, appear on the on-line building plan in CMS. The application acts as a scanner - it is necessary to appear within the Beacon range in order to read the location of individual asset. These data are then sent to the web platform, which displays the location. Tested application was in beta version, since the producer was still working on improving the localization system at the moment of performing the research.

4.2 Research procedure

Research was performed on the left wing of 3rd floor of Faculty if Civil and Environmental Engineering building located in Poznan, Poland. Building is planned for renovation within 24 months. The three-store building with basement was made of demolition brick, with DMS prefabricated ceilings with reinforced concrete was constructed between 1953-55y on the old military baronets of fort Rauch in Poznan (Figure 1).

Authors prepared digital plan on the basis of a manual inventory drawing of the object. Then it was entered into the platform integrated with Open Street Maps, as shown in the Figure 2.

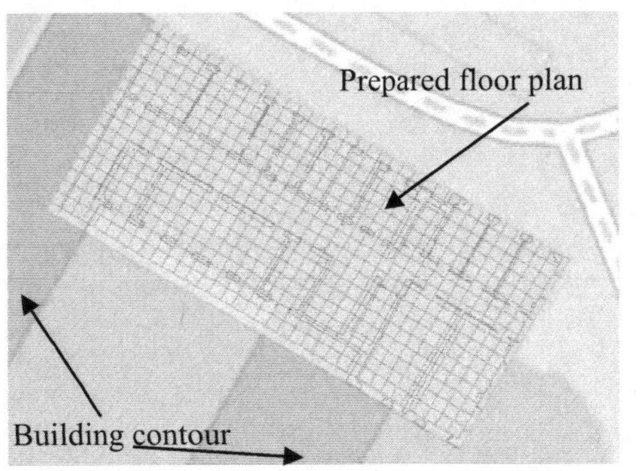

Figure 2. Floor plan in the on-line tracking system integrated with Open Street Maps

The Beacon producer makes it possible to change the frequency, wavelength and signal strength of each Beacon on the internet platform. Sensors were programmed to -23dB for the whole test procedure, which was the smallest possible value that the system can provide. Signal strength has been selected based on manufacturer's recommendations.

It has to be noted that in the online system user has to assign role for each Beacon. Some of the Beacons acts like a grid for location objects, others acts as searched assets.

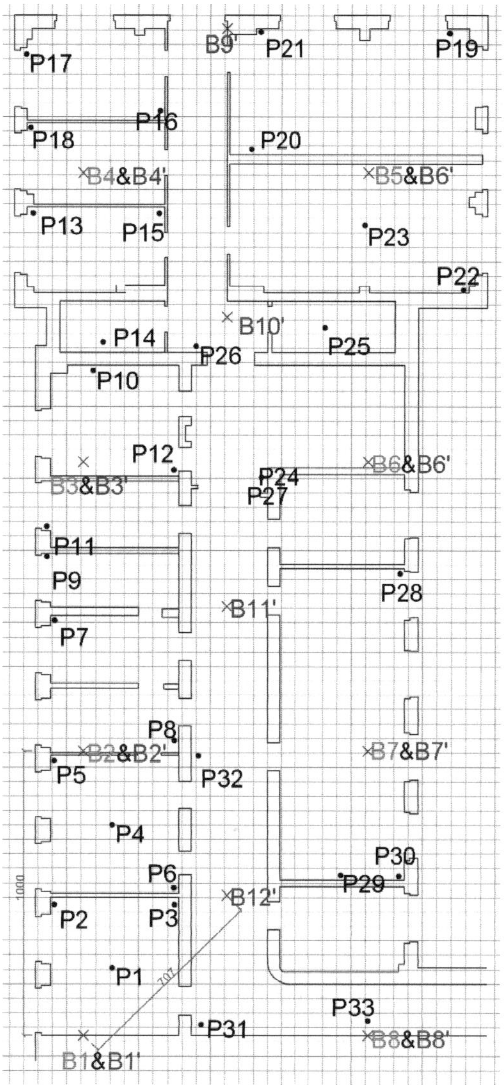

Figure 3. Grid scheme (green- 1st with 10,0m spacing; red 2nd with 7,07m) with location of 33 measuring locations (P1-P33)

In the next step, the Beacon grid was depicted as shown in Figure 3 and the first series of measurements was made. Next, the grid was consolidated by adding three additional sensors, which reduced the distance between grid points.

The distance was (A) – 10 m and (B) – 7.07 m, respectively. Each Beacon was placed at the same height - 1m, and at appropriate distances from the walls, according to the drawings. There were in total 66 measurements in two series for 33 each. Tracked objects were placed at different distances from the mesh sensors as shown in Figure 3, where positions were marked as (P). During conducting measurements in the second grid, the layout of the searched objects was reconstructed and the impact on the mesh density was investigated. By changing the placement of asset's Beacons, a mobile device was approached to communicate with the system which was necessary to save the searched location sensors into online system.

The final test results were calculated using screenshots from mobile phone (Android system), on-line CMS accessed via desktop computer and the actual location of the sensors.

After receiving results there was also performed root-cause-analysis to find out what caused such results and to assess the possibilities of use tested equipment in construction sector.

Performed procedures and action are illustrated in the Figure 4 for better understanding of what steps have been taken by the researchers.

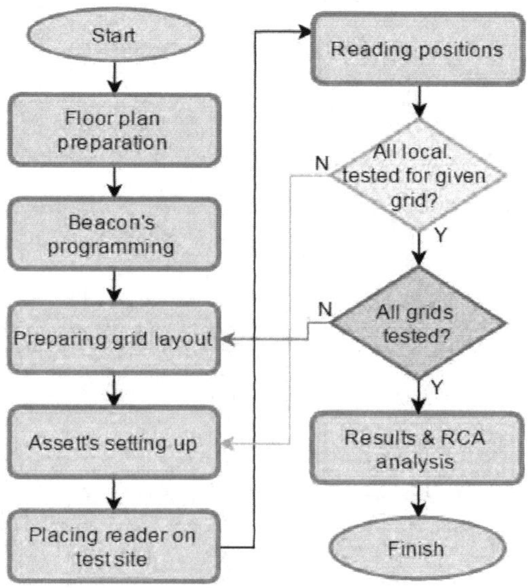

Figure 4. Research procedure for asset localization

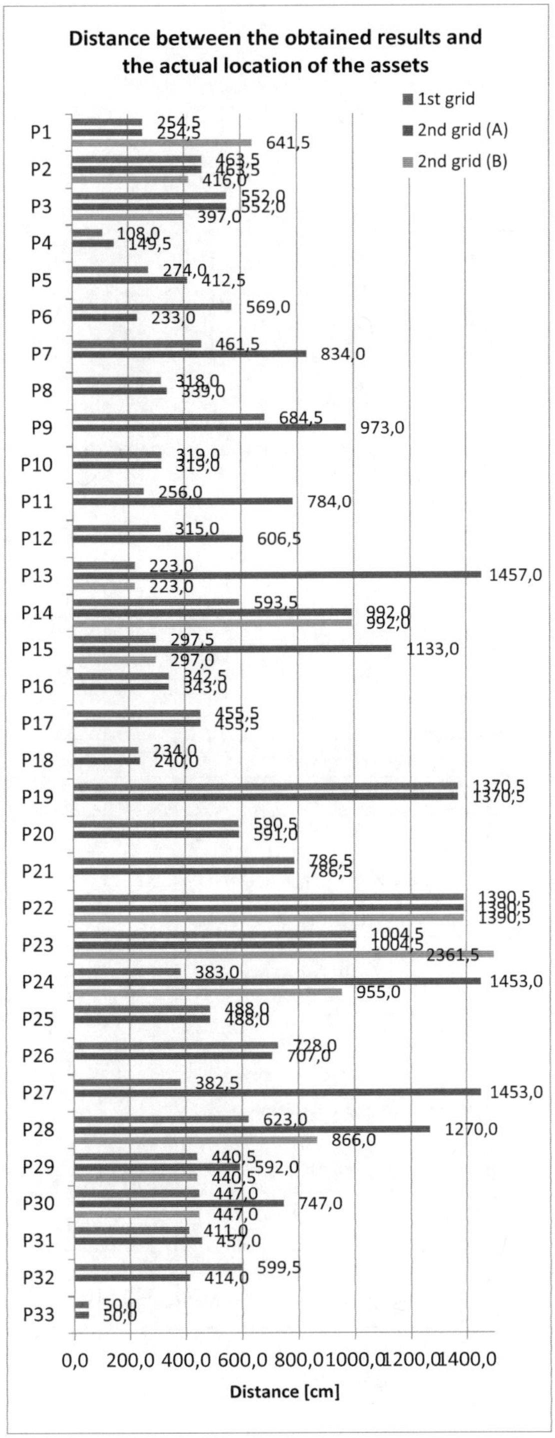

Figure 5. Distances between the results obtained in the tracking application and the actual location

4.3 Study results and discussion

Results of the performed analysis showed that during the measurements, there were four situations in which, for one placement of assets, the application returned two different results. Situation happened only at the second (concentrated) grid arrangement. Both results were analyzed and those obtained at the second grid arrangement for the first indications of the application, will be called Case A in the further discussion. Similarly, the set of results obtained at the second grid arrangement, taking the second indications of the application, will be called Case B.

In Figure 5 the distances between the actual location of the assets and those specified by the application was compared to determine the effect of change of the Beacons grid on the accuracy of the measurements.

The average measurement error from the performed study is:

- For the first grid laying - 497.5 cm
- For case A - 706.5 cm
- For Case B - 649.5 cm
- Total average - 618.0 cm

Authors analyzed results in detail to check if changing the grid (making it more dense) from 10,0m to 7,07m had any influence on the final results. It occurred that for the A case, the mesh density had a positive effect on the accuracy of three measurements, eighteen negative, while the twelve results remained unchanged. In case B the mesh thickening has had a positive effect on the accuracy of the five measurements, negative - seventeen, while eleven results remained unchanged. Results are presented in Figure 6.

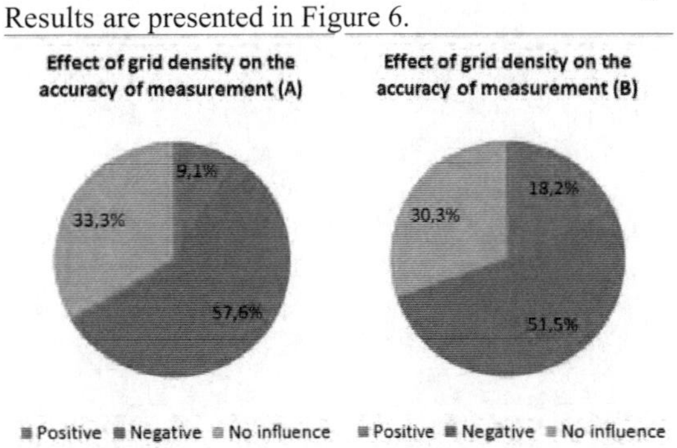

Figure 6. Change in grid density on the final results for Case A and Case B

It can be easily seen that changing the density(make it more dense) of the Beacon grid had negative effect on the measurement which means that the grid of 10,0 m was better that 7,07 m grid in terms of accuracy measurements. This clearly shows that analyzed system is not suitable for small areas, since it cannot provide enough precision in localization of the assets. It has to be also noted that even for first grid measurement error was at the level of 497,5 cm, what make it impossible to precisely indicate the asset position and limits possible application in process improvement methods.

After interviewing the Beacon producer it was also expected that the location of the search in the application would overlap with the Beacon location closest to the grid. The actual indications differed from the expected results depending on which grid (1st or 2nd) was used and analyzed. Figure 7 graphically presents this phenomena. Three criteria were used, which are below for better understanding how graph was prepared.

- Overlap/matching - the searched object was in the application at the point of the nearest grid Beacon.
- Does not overlap - the searched object was in the application at a point other than the nearest grid Beacon.
- The object is located between grid Beacons - the searched object was in the application at a point between two grid Beacons (not always closest).

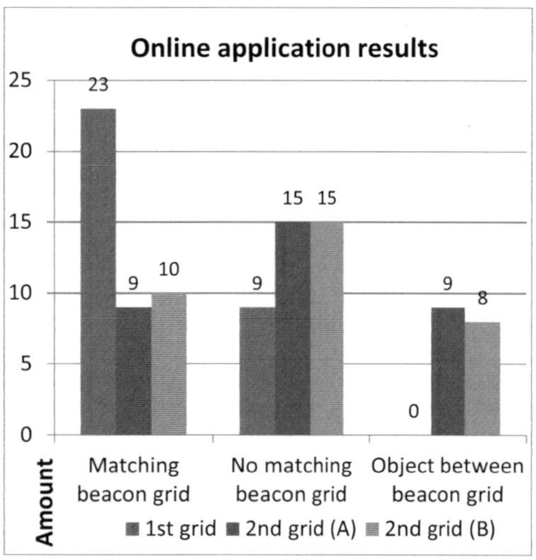

Figure 7. Beacons localization in the application compared to the grid position A and B

Figure 7 confirms that 1st grid had better precision that 2nd, since in 23 measurements Beacons indicated position of the closest grid point, while in the 2nd grid (denser) it was only 9 positions. Even the situation that for the 2nd grid 2 different cases (A and B) were received indicates, that the precision problem increased in this set up.

4.4 Root cause analysis

After reviewing obtained results it was decided to

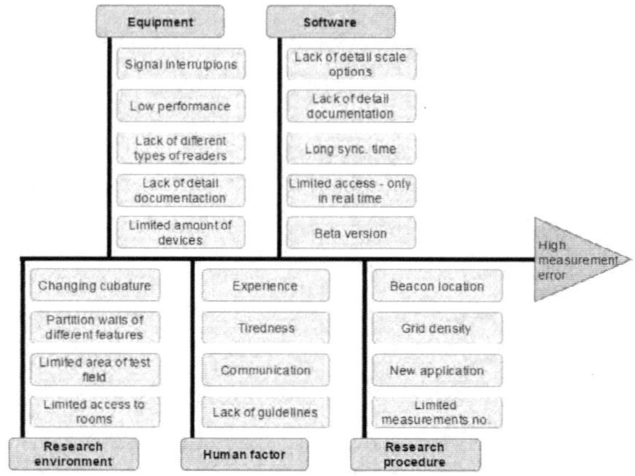

Figure 8. Ishikawa diagram prepared for high measurement error causes

	Problem	O	D	S	RPN
Equipment	Low performance	3	6	10	180
	Lack of different reader	5	1	3	15
	Signal interruptions	5	8	7	280
	Lack of detail documentation	9	1	2	18
	Lack of different types of readers	10	1	3	30
	Limited amount of devices	8	3	3	72
Environment	Changing cubature	8	1	4	32
	Partition walls of different features	4	3	4	48
	Limited area of test fields	5	2	4	40
	Limited access to rooms	3	2	4	24
Human f.	Tiredness	3	5	5	75
	Experience	7	2	4	56
	Communication	3	4	3	36
	Lack of guidelines	10	1	7	70
Software	Lack of detail documentation	5	1	6	30
	Long sync. Time	5	7	4	140
	Lack of detail scale options	9	7	3	189
	Limited access - only in real time	10	1	5	50
	Beta version	10	1	10	100
Procedure	Beacon location	5	4	2	40
	Limited meas. no.	2	2	2	8
	Grid density	6	5	8	240
	New application	10	1	8	80

Figure 9. FMEA analysis on the high measurement error

perform root cause analysis to find out what are the reasons for that situation and how they can be resolved in future studies. For this purpose Ishikawa diagram (Figure 8) was prepared with groups of couses and detail problems that were observed during the research procedure. Five main categories were defined: Equipment, Software, Research environment, Human factor and Research procedure. In each category precise problems were indicated.

After preparing the Ishikawa graph, FMEA analysis was performed to asses and decide which of the detailed problems had the highest influence on the received results (Figure 9). For this purpose 3 categories were selected: Occurrence, Detection and Severity, for which scale from 1-10 was adopted. For each category "1" indicated positive effect (respectively: never happened, easy to detect, no influence on the process), and "10" meant negative effect (always, hard to detect, huge influence on the process). Risk Priority Number (RPN) was then calculated by multiplying numbers in each category. In this way 3 the most important problems were defined, which are

- signal interruptions (RPN=280),
- grid density(RPN=240),
- lack of scale options in the software. (RPN=189).

Signal interruptions may be connected with the other devices using Bluetooth technology in the range of Beacons as well as walls and other physical obstacles in the range of Beacons. Grid density caused problem, since due to the building size it was difficult to introduce less dense grid. Lack of scale options is related to the precision of the measurements, since it should be able to precisely set the scale in the software before performing actual tests.

It has to be noted that tested system was never previously used in this kind of application thus observed problems may be corrected in the final version of the software and further research should be then performed with the analyzed Beacon technology.

5 Conclusions

Conducted analysis of the available literature with the focus on different RTLS technologies and performed study on the possibility of use BLE Beacons – Bluetooth devices as a locating systems allows to formulate the following conclusions:

- The construction industry is struggling with problems that can be resolved through the implementation of the Lean Management philosophy, resulting in resources savings, speeding up the construction process and waste removal.
- The Bluetooth based RTLS systems can be useful device in construction sector even taking into account its limitations.
- The measurement error for the grid of 10m was 497.5 cm, which significantly limits the usefulness of the system for precise localization, but opens possibilities for location on the bigger area.
- Denser Beacon grid results in less accurate measurement results for analyzed cases.
- 3 the most important problems have influenced such results were defined: signal interruptions, grid density, and lack of scale options in the software.
- Most suitable stage for using RTLS systems in productivity improvement is measure step, where they can be used as a tool for data acquisition.

The purpose of the study was also to check the suitability of the system in construction industry. Based on the obtained results, it is proposed to use this type of the systems to locate large scale objects where the exact location of the object is not required, but only the approximate location, such as the area in which it is located. The potential for using the system to track employee traffic has also been noted, assuming that in the future system will be equipped with the additional options based on the access to historical data.

It has to be also noted that one features of Beacons is the possibility of sending specific messages to smartphones within its reach. Authors are of the opinion that this feature can be useful in construction industry, especially in the area of alerting people on site of dangerous areas, transmitting health and safety instructions, etc. Therefore, it is also suggested to test equipment in this respect.

Authors plan to perform further research in the area for the possibility of use of RTLS in construction sector, for the purpose of process improvement, since those systems are response to the need of data acquisition necessary for performing Lean Management methods and improvements. What is further tests should be performed on final version of the analyzed Beacon software, as improvements might influence in positive manner on the final results of further tests.

Acknowledgements

The publication was created with the support of statutory funds of Institute of Structural Engineering at Poznan University of Technology.

References

[1] Jiang S., Jang W. S. and Skibniewski M. J. Selection of wireless technology for tracking construction materials using a fuzzy decision model.

Journal of Civil Engineering and Management. 18(1): 43-59, 2012.

[2] Li H., Yang X., Wang F., Rose T., Chan G., and Dong S. Stochastic state sequence model to predict construction site safety states through Real-Time Location Systems. *Safety science.* 84: 78-87, 2016.

[3] Isaac S. and Edrei T. A statistical model for dynamic safety risk control on construction sites. *Automation in Construction.* 63: 66-78, 2016.

[4] Kim H., Lee H. S., Park M., Chung B. and Hwang, S. Automated hazardous area identification using laborers' actual and optimal routes. *Automation in Construction.* 65: 21-32, 2016.

[5] Li H., Chan G., Wong J. K. W. and Skitmore M. Real-time locating systems applications in construction. *Automation in Construction.* 63: 37-47, 2016.

[6] Galileo Declares: Open for Business!, *GPS World.* 28(1): 10-12, 2017.

[7] Chadeganil A. A., Salehi H., Yunus M. M., Farhadi H., Fooladi M., Farhadi M. and Ebrahim N. A. A Comparison between Two Main Academic Literature Collections: Web of Science and Scopus Databases. *Asian Social Science.* 9(5): 1911-2025, 2013.

[8] Han G., Klinker G. J., Ostler D. and Schneider A. Testing a proximity-based location tracking system with Bluetooth Low Energy tags for future use in the OR. In E-health Networking, Application and Services (HealthCom). *2015 17th International Conference on.* 17-21, 2015.

[9] Arisaka N., Mamorita N., Isonaka R., Kawakami T. and Takeuchi A. Trial of real-time locating and messaging system with Bluetooth low energy. *Technology and Health Care.* 24(5): 689-699, 2016.

[10] Kim N. J. and Choi E. I. A Stream Data Processing Framework for Location-Based Service Using NoSQL Technology. *Applied Mechanics and Materials.* 763 :159-163, 2015.

[11] Cui Y. and Zhao J. Real-Time Location System and Applied Research Report. *In International Conference on Internet of Vehicles.* 49-57, 2015.

[12] Krafcik J. F. Triumph of the lean production system MIT. *Sloan Management Review,* 30(1): 41, 1998.

[13] Lodgaard E., Ingvaldsen J. A., Gamme I. and Aschehoug S. Barriers to Lean Implementation: Perceptions of Top Managers, Middle Managers and Workers. *Procedia CIRP.* 57: 595-600, 2016.

[14] Womack J. P. and Jones D. T. *Lean thinking: banish waste and create wealth in your corporation.* Simon and Schuster. 2010.

[15] Brioso X. Integrating ISO 21500 Guidance on Project Management, Lean Construction and PMBOK. *Procedia Engineering.* 123: 76-84, 2015.

[16] Golzarpoor H., González V., Shahbazpour M. and O'Sullivan M. An input-output simulation model for assessing production and environmental waste in construction. *Journal of Cleaner Production.* 143: 1094-1104, 2017.

[17] Alves T. D. C., Pestana A. C. V., Gilbert E. and Hamzeh F. Lean Principles for the Management of Construction Submittals and RFIs. *Journal of Professional Issues in Engineering Education and Practice.* 142(4): 05016004, 2016.

[18] Kadarova J. and Demecko M. New Approaches in Lean Management. *Procedia Economics and Finance.* 39: 11-16, 2016.

[19] Plebankiewicz E. and Mitera E. Sources of waste at construction site[in Polish]. *Materiały Budowlane.* 6/2016:184, 2016.

[20] Shakeri I., Boroujeni K. A. and Hassani H. Lean Construction: From theory to practice. International. *Journal of Academic Research.* 7(1):129, 2015.

[21] Pasquire C. L. and Gibb A. G. A lean and agile construction system as a set of countermeasures to improve health, safety and productivity in mechanical and electrical construction. *Lean construction Journal.* 61-76, 2009.

[22] Sacks R., Koskela L., Dave B. A. and Owen R. Interaction of lean and building information modeling in construction *Journal of construction engineering and management.* 136(9): 968-980. 2010.

[23] Vinodh S. and Aravindraj S. Evaluation of leagility in supply chains using fuzzy logic approach. International. *Journal of Production Research.* 51(4): 1186-1195, 2013.

[24] Posdorfer W. and Maalej W. Towards Context-aware Surveys Using Bluetooth Beacons. *Procedia Computer Science.* 83: 42-49, 2016.

Sliding behavior monitoring of a deep seated landslide with differential interferometric synthetic aperture radar, mems tiltmeter and unmanned vehicle images

K.-L. Wang[a], Y.-M. Hsieh[b], J.-T. Lin[a] and M.-H. Hsi[a]

[a]Department of Civil Engineering, Nationl Chi Nan University, Taiwan
[b]Department of Construction Engineering, National Taiwan University of Science and Technology, Taiwan
E-mail: klwang@ncnu.edu.tw, ymhsieh@mail.ntust.edu.tw, nokilin@gmail.com, ilovevachen@gmail.com

Abstract

Landslide is a common issue in Taiwan with more than 70% area of this island is mountainous land. Typhoons and earthquakes attack Taiwan every year even by month. People lost their lives, properties and wounded when landslides happened during typhoons, heavy rainfall and earthquake. Landslide hazard mitigation is very important in hazard prevention chain. Two deep-seated landslides selected for monitoring in this research. A visual approach calls red relief image is applied in the beginning to search scars, geological structure on the ground surface. Synthetic aperture radar images acquired by JAXA (Japan) are used in this research to monitor landslide displacement in large area. The electromagnetic waves from temporal images can use to derive differential between time period. Thus the displacement can be coverted from phase differences. Mems monitoring stations are set up after differential interferometric results. The memes sensors are cost effective to monitoring landslide tilt behavior 24 hours. Moreover, unmanned vehicle system is adopted in this research to see optical differences on ground surface. The results show good coincidence with each other monitoring methods.

Keywords –

Landslide; SAR; Mems; UAS

1 Introduction

Taiwan locates in the collision zone of sea plate and continental plate, which inducing earthquake and orogeny. Moreover, this island is surrounded with warm and cool sea water. Typhoons and heavy rainfall attacked this island frequently, especially after year 2000. Several typhoons and heavy rainfall with unexpected large rainfall attacked Taiwan in past 10 years. Typhoon Morakot in 2009 brings maximum 3,000 mm accumulated rainfall and the rainfall is 3/4 of average annual precipitation.

Landslide potential map is the first work for landslide hazard mitigation. There are several methods to produce landslide potential map. The logistic regression method combines potential factors with landslides and give landslide potential prediction. Two major categories including bivariate analysis and multivariate analysis are used in logistic regression analysis. Gupta and Joshi (1990) used Landslide Nominal Risk Factor (LNRF) to derive dimensionless factors and calculate the potential risk. Jade and Sarkar (1993) adopted Information Theory and classified the potential risk into three classes. Keefer (2000) proposed a landslide concentration (LC) factor to establish the relationship between landslide and earthquake. The landslide susceptibility mapping using bivariate analysis without seismic condition was conducted by different researchers. (Çevik et al., 2003; Chau et al., 2004; Lee, 2004; Lee et al., 2004; Lee, 2005; Ohlmacher and Davis, 2003). However, the bivariate analysis does not take into account the independency of factors. Therefore, multivariate analysis has been developed for that purpose. The independency of factors must be checked and the matrix of factors is thus established. Two classification algorithms – logistic regression and discriminating analysis are the most well-known methods to identify the susceptibility of landslide. Lin and Tung (2004) used structural equation model to establish a measuring matrix for landslide evaluation for the cases of Chi-Chi earthquake. Süzen and Doyuran (2004) compared the landslide susceptibilities using bivariate and multivariate analysis (logistic regression) based on northwest Turkey study area.

The neural network analysis uses factors related to landslides to perform training and evaluation of landslide susceptibility. The accuracy can be as high as 90% in the same area and using same event (Lee et al., 2003). However, the training process needs to restart when encountering different condition.

Semi-logistic regression methods which combined

factor of safety to produce probability of landslide can be used to establish the susceptibility map. Pack et al. (1998, 1998a, 2001) used the stability index derived from factor of safety to classify susceptibility into several groups: stable, moderately stable, quasi-stable, lower threshold, upper threshold, and defended. Lan et al. (2004) used the module to analyze the landslide hazard in Yunnan, China. Another semi-logistic method adopts displacement as the index of landslide susceptibility. Wang et al. (2010) proposed a semi-mechanical and semi-regression method to produce landslide potential map.

The study is trying to propose a method to produce landslide potential map from differential interferometric synthetic aperture radar (DInSAR). Potential landslide zones are mapped after fringes and displacements are generated from ALOS/PALSAR radar image processing. Meanwhile, a deep-seated landslide site has been selected for instrumentation. Monitoring data from Mems has been collected and analyzed to find relationship between landslide displacement and rainfall, and groundwater. Mems accelerometer is designed for active or quick displacement area. The design is considered tilt and dynamic data collection. Moreover, unmanned vehicle system is adopted to produce high resolution, high precision digital elevation to compare previous results.

2 Methodology

The study area is located at central Taiwan and the elevation ranges from 1000m to 2500m. A landslide potential map was mapped by expert from LiDAR data, which means the scars were generated by some events with unknown years and unknown active status. The scars in study area are as shown in Figure 1. Most aspects of scars are facing north or south-north owing to geological condition. The geological condition in this area is quite unique, which is slate of Lushan Formation, Miocene. Lushan Formation is slate interlayered with thin metamorphosed sandstone and slate. The dip angles of slate in this area ranges from 20-70 degrees owing to gravity and tectonic forces.

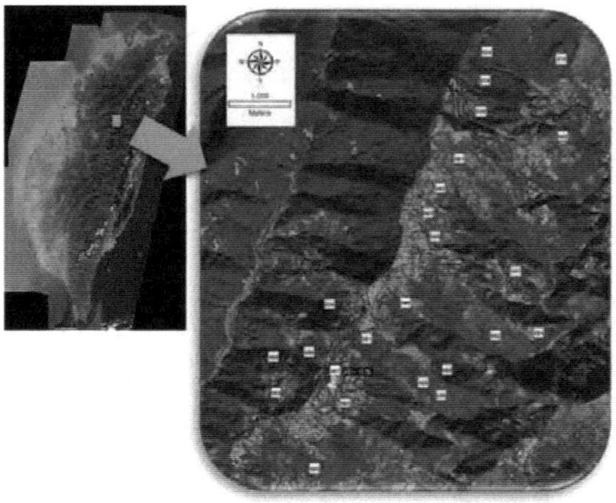

Figure 1 Landslide mapping with LiDAR data in study area.

2.1 Benchmark survey

Benchmark survey was executed by local government from plain area to mountain area. The differences in elevation is as shown in Figure 2 The differences are almost increasing except some benchmarks. The increasing of elevation is undoubtedly the effect orogeny. If rank the elevation difference and select a threshold for stable zone, the others can be eliminated the average value of stable zone. The benchmarks locate in landslide scars can be identified and the true landslide displacement can be calculated based on this assumption as shown in Figure 3. There are four benchmarks show elevation decreasing, which means locate in landslide scar and moving between 2002~2008. The displacement has been calculated as shown in Tab. 1. The values have been transferred to annual velocity for DinSAR comparison.

Figure 2 The elevation change from benchmark survey in 8 years (2002-2009)

Figure 3 Elevation differences adjusted for landslide displacement

Table 1 Elevation differences derived for landslide velocity

Landslide Number	Relative decreasing velocity (mm/year)
C050	8.02
C047	12.03
C045	9.90
C044	22.06

2.2 Potential landslide mapping with DInSAR

SAR measures the distance between satellite and ground surface by sending and receiving radar signals. Measurement of ground topography using SAR represents two locations of antenna that sensing the surface and are separated by a baseline. If the viewing geometry is controllable or known with sufficient accuracy, then the topography h can be derived from the phase measurement of the two sensing radar waves. The topography h can be obtained by geometry of satellite and observation point. Two important conditions that need to be understood for detecting and measuring surface change with SAR are:

(1) The changes between two successive images must not be too large.

(2) The radar-scattering characteristics within each pixel must remain similar.

The deformation of ground surface derived from DInSAR could be less than centimeter with high accuracy digital elevation model. When doing interferometry analysis by temporal SAR data, the phase information generated is including topography, change of ground characteristics, ground movement and atmosphere effect, etc. (eq.1) Phase difference of topography can be removed by high accuracy digital elevation model and short base line. Atmosphere effect can be reduced by long term analysis. The deformation can be trusted to high precision and after removed previous described errors and simply left ground deformation and noise.

$$\emptyset_{Int} = 4\pi\frac{R_1-R_2}{\lambda} = \Phi_{Topography} + \Phi_{Change} + \Phi_{Movement} + \Phi_{Atmosphere}$$

$$= \frac{4\pi B_n}{\lambda R\sin\theta}h + \Phi_{change} + \frac{4\pi}{\lambda}\Delta R_{Movement} + \Phi_{Atmosphere}$$

$$= \frac{4\pi B_n}{\lambda R\sin\theta}h + \frac{4\pi}{\lambda}\Delta R_{Movement} + \Phi_{Noise} \quad [1]$$

in which,

R: distance from satellite to observation position

B: Baseline length

h: elevation

λ: wave length of radar

θ: side looking angle

DInSAR, PS, and SBAS has been well developed in recent years thus the monitoring of landslide displacement is becoming more possible. (Pieraccini, Casagli et al. 2003, Tarchi, Casagli et al. 2003, Guzzetti, Manunta et al. 2009, Cal, x00F et al. 2013, Liu, Wang et al. 2013, Lowry, Gomez et al. 2013, Jebur, Pradhan et al. 2015, Tang, Chen et al. 2015, Casagli, Cigna et al. 2016, Uhlemann, Smith et al. 2016)

The SAR images used in this research is Advanced Land Observing Satellite (ALOS) Phased Array type L-band Synthetic Aperture Radar (PALSAR) L band image. The polarization of radar electromagnetic wave is Horizontal - Horizontal (HH) mode and incident angle is 34.3 degree. The range resolution from this area is approximately 10 meter. There are several SAR images can be used for differential interferometric SAR. However, ALOS PALSAR with L band has longer wave length, which is 23cm and possible could be eliminated the effect of vegetation.

The results of no significant rainfall event and typhoon event are as shown in Figure 4 and Figure 5, separately. Nine large landslide scars were mapped comparing with event and no event data. Comparing with Figure 4 and Figure 5, there are some higher displacement points located in the middle of slope. This might be due to landslide moving and the scar comes out from middle of slope.

DInSAR method is fine to observe landslide scar at two images with close dates. However, the displacement data shows too much noise to identify accuracy and comparison with benchmark data. Thus another approach small baseline subset (SBAS) is adopted for higher accuracy. This method searches points with the same radar signal strength through observing years and keep tracking locations in each scene. Figure 6 shows the result of SBAS method with larger zone of study area. The RMS error directly from radar shows 10mm with 95% confidence.

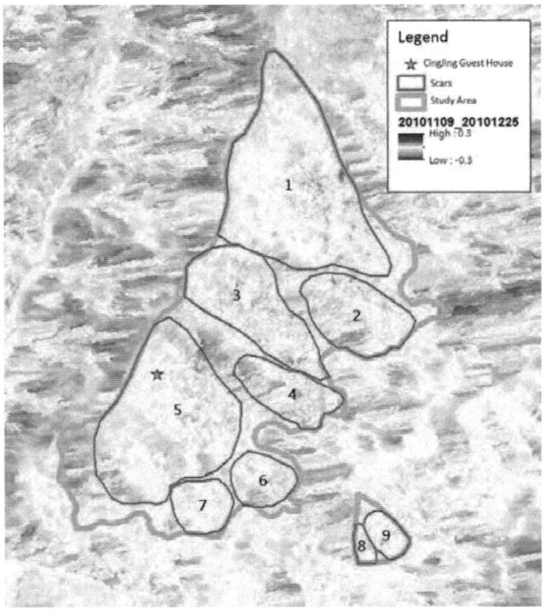

Figure 4 The displacement generated by DInSAR from multi-temporal data (2006-2010; no events)

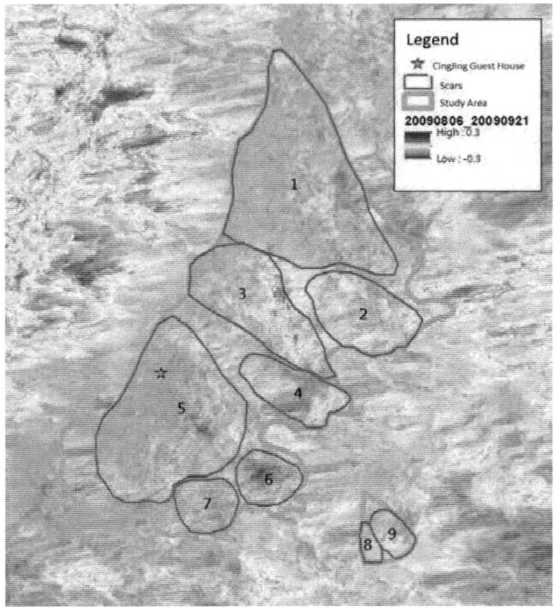

Figure 5 The displacement generated by DInSAR from multi-temporal data (2006-2010; typhoon event)

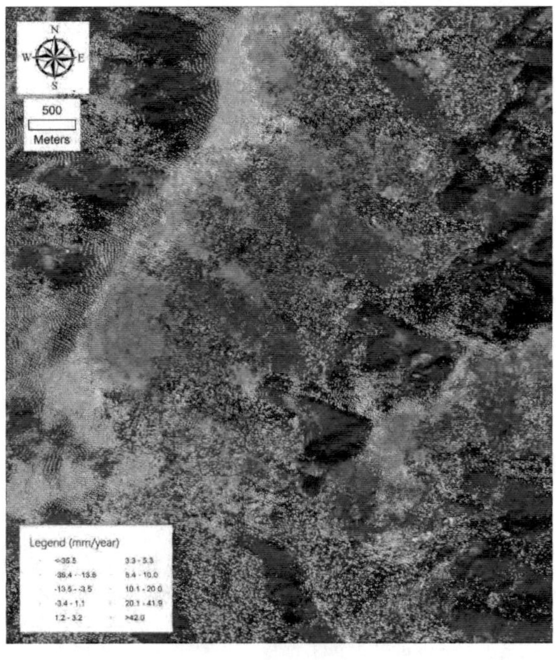

Figure 6 Displacement velocity map produced with SBAS from 2006-2016

2.3 Mems instrumentation

Owing to the area of slide number 8 and 9 in Fig. 4 is about 34 hectares, total landslide instrumentation is too expensive and unnecessary. In this work, we present a low-cost slope-monitoring device. The device is constructed using components purchased online at reasonable costs. The device is capable of recording triaxial accelerations at 200Hz onto standard SD (secure digital) memory cards. In addition, the sensed triaxial accelerations can be converted to tilt, which can then be sent via onboard GPRS module to a cloud server. Thus, the constructed device can be used for both static measurements of the slope surface and dynamic measurements of ground-surface acceleration. This information can assist assess slope condition regularly and dynamic force experienced by the slope after earthquake. The recorded data has been transferred into tilt as shown in Figure 7.

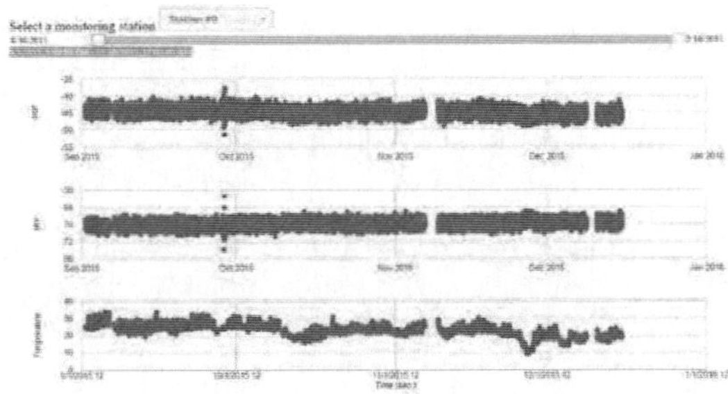

Figure 7 Tilt measurement data within three months

2.4 Unmanned vechicle system

Field investigation is time consuming, high budget, limited data collected and dangerous. Investigation with satellite images has disadvantages such as less of the actual situation and poor resolution. Thus the possibility for slope investigation with UAV will be proposed and discussed in this research. Hazard investigation and monitoring is adopted UAV in recent years. UAV has advantages such as light weight, small volume, high mobility, safe, easy maintenance and low cost. Investigation can be executed in high risk area. Use the mature aero photogrammetry , combines aero photos with control point. Digital surface model (DSM) and Ortho photos can be produced with control points aligned. The resolution can be less than 5cm thus can be used as temporal creeping monitoring before landslide happens. Figure 8 shows the red relief image map from UAS result. The scars can be identified from that and have nice coincidence with previous results.

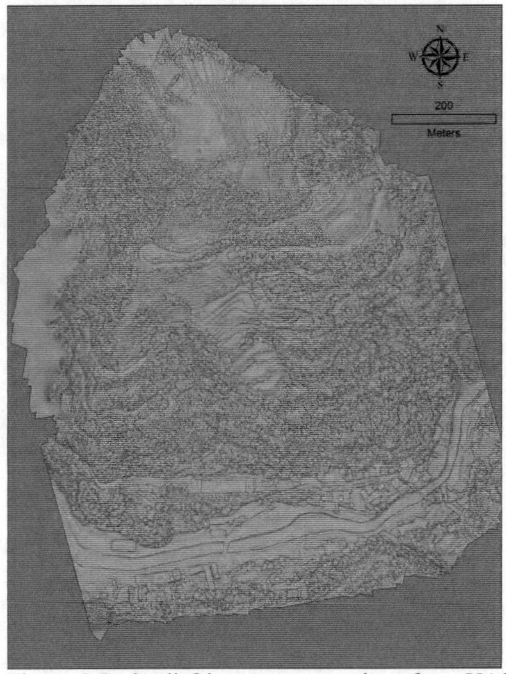

Figure 8 Red relief image map produce from UAS result

3 Conclusions

Landslide potential map is typically generated by factors by statistic method or mechanic equilibrium approach. In this work we propose a monitoring methodology from DInSAR and SBAS data processing. This method can identify active landslide scars in the same rainfall condition area, which means should pay more attention to monitoring. DInSAR can map landside scars from fringe and displacement map. The SBAS has more precise displacement monitoring with linear modeling.

A deep seated landslide was selected for instrumentation to compare with DInSAR and SBAS method.

The underground instrumentations show that landslide displacement has good coincidence with accumulative rainfall. However, landslide instrumentation is too expensive when there are many potential scars in study area. Thus a low-cost mems monitoring sensor was developed in this work. In this work, we share our experience on building a low-cost slope-monitoring device that senses temperature and dynamic accelerations at 200Hz, provides 6-month long-term storage with a 16GB micro SD card, and sends calculated tilt to cloud servers via GPRS module. The low-cost system works well and provides more quick response to landslide warning much earlier than DInSAR and SBAS methods.

References

[1] Cal, F., F. Ardizzone, R. Castaldo, P. Lollino, P. Tizzani, F. Guzzetti, R. Lanari and M. Manunta (2013) Landslide analysis through the multi-sensor SBAS-DInSAR approach: The case study of Assisi, Central Italy. 2013 IEEE International Geoscience and Remote Sensing Symposium - IGARSS. 2916-2919

[2] Casagli, N., F. Cigna, S. Bianchini, D. Hölbling, P. Füreder, G. Righini, S. Del Conte, B. Friedl, S. Schneiderbauer, C. Iasio, J. Vlcko, V. Greif, H. Proske, K. Granica, S. Falco, S. Lozzi, O. Mora, A. Arnaud, F. Novali and M. Bianchi (2016) Landslide mapping and monitoring by using radar and optical remote sensing: Examples from the EC-FP7 project SAFER. Remote Sensing Applications: Society and Environment 4: 92-108.

[3] Chau, K.T., Y.L. Sze, M.K. Fung, W.Y. Wong, E.L. Fong and L.C.P. Chan (2004) Landslide hazard analysis for Hong Kong using landslide inventory and GIS. Computers and Geosciences. 30:429-443

[4] Gupta, R.P and B.C. Joshi (1990) Landslide Hazard Zoning Using the GIS Approach – A Case Study from the Ramganga Catchment, Himalayas. Engineering Geology. 28:119-131

[5] Guzzetti, F., M. Manunta, F. Ardizzone, A. Pepe, M. Cardinali, G. Zeni, P. Reichenbach and R. Lanari (2009). Analysis of Ground Deformation Detected Using the SBAS-DInSAR Technique in Umbria, Central Italy. Pure and Applied Geophysics. 166(8): 1425-1459.

[6] Jade S. and S. Sarkar (1993) Statistical models for slope instability classification. Engineering Geology., 36:91-98

[7] Jebur, M. N., B. Pradhan and M. S. Tehrany (2015) Using ALOS PALSAR derived high-resolution DInSAR to detect slow-moving landslides in tropical forest: Cameron Highlands, Malaysia. Geomatics, Natural Hazards and Risk 6(8): 741-759.

[8] Keefer D.K. (2000) Statistical analysis of an earthquake-induced landslide distribution – the 1989 Loma Prieta, California event. Engineering Geology. 58:231-249

[9] Lan H.X., C.H. Zhou, L.J. Wang, H.Y. Zhang and R.H. Li (2004) Landslide hazard spatial analysis and prediction using GIS in the Xiaojiang watershed, Yunan, China. Engineering Geology. 76:109-128

[10] Lee, C. N. (2001) Preliminary study on the Tsao-Ling landslide area under earthquake. Master's Thesis. National Taiwan University

[11] Lee S. (2004) Application of likelihood ratio and logistic regression models to landslide susceptibility mapping using GIS. Environmental Management. 34(2):223-232

[12] Lee S. (2005) Application of logistic regression model and its validation for landslide susceptibility mapping using GIS and remote sensing data. International Journal of Remote Sensing. 26(7):1477-1491

[13] Lin, M. L. and C.C. Tung (2004), A GIS-based potential analysis of the landslides induced by the Chi-Chi earthquake. Engineering Geology. 71:63-77

[14] Liu, G., R. Wang, Y. K. Deng, R. Chen, Y. Shao, W. Xu and D. Xiao (2013). Monitoring of ground deformation in Beijing using SBAS-DInSAR technique. 2013 Asia-Pacific Conference on Synthetic Aperture Radar (APSAR). 300-303

[15] Lowry, B., F. Gomez, W. Zhou, M. A. Mooney, B. Held and J. Grasmick (2013) High resolution displacement monitoring of a slow velocity landslide using ground based radar interferometry. Engineering Geology. 166: 160-169.

[16] Ohlmacher G. C. and J.C. Davis (2003) Using multiple logistic regression and GIS technology to predict landslide hazard in northeast Kansas, USA. Engineering Geology. 69:331-343

[17] Pack, R. T., D.G. Tarboton and C. N. Goodwin (1998) The SINMAP approach to terrain stability mapping. The 8th Congress of the International Association of Engineering Geology, Vancouver, Canada

[18] Pack, R. T., D.G. Tarboton and C.N. Goodwin (1998a) Terrain stability mapping with SINMAP, technical description and user guide for version 1.00. Report Number 4114-0, Terratech Consulting Ltd., Salmon Arm, B.C., Canada

[19] Pack, R. T., D.G. Tarboton and C.N. Goodwin (2001) Assessing terrain stability in a GIS using SINMAP. The 15th annual GIS conference, Vancouver, Canada

[20] Pieraccini, M., N. Casagli, G. Luzi, D. Tarchi, D. Mecatti, L. Noferini and C. Atzeni (2003) Landslide monitoring by ground-based radar interferometry: A field test in Valdarno (Italy). International Journal of Remote Sensing. 24(6): 1385-1391.

[21] Süzen, M.L. and V. Doyuran (2004) A comparison of the GIS based landslide susceptibility assessment methods: multivariate versus bivariate. Environmental Geology. 45:665-679

[22] Tang, P., F. Chen, H. Guo, B. Tian, X. Wang and

N. Ishwaran (2015) Large-Area Landslides Monitoring Using Advanced Multi-Temporal InSAR Technique over the Giant Panda Habitat, Sichuan, China. Remote Sensing. 7(7): 8925.

[23] Tarchi, D., N. Casagli, S. Moretti, D. Leva and A. J. Sieber (2003) Monitoring landslide displacements by using ground-based synthetic aperture radar interferometry: Application to the Ruinon landslide in the Italian Alps. Journal of Geophysical Research: Solid Earth. 108(B8): n/a-n/a.

[24] Uhlemann, S., A. Smith, J. Chambers, N. Dixon, T. Dijkstra, E. Haslam, P. Meldrum, A. Merritt, D. Gunn and J. Mackay (2016) Assessment of ground-based monitoring techniques applied to landslide investigations. Geomorphology. 253: 438-451.

[25] Wang, K.-L. and M.-L. Lin (2010) Development of Shallow Seismic Landslide Potential Map Based on Newmark's Displacement: The Case Study of Chi-Chi Earthquake, Taiwan. Environmental Earth Sciences. 60(4), 775-785

An Integrated GIS and Wi-Fi Based Locating System for Improving Construction Labor Communications

S. Shirowzhan[a] , S.M.E.Sepasgozar[b] , I.Zaini[c] and Cynthia Wang[d]
[a,b,,c,d] Faculty of Built Environment, University of New South Wales, Australia

E-mail: s.shirowzhan@unsw.edu.au, sepas@unsw.edu.au , Izmilzaini@googlemail.com, and cynthia.wang@unsw.edu.au

Abstract – Locating construction assets such as equipment and workforce during construction are critical for improving safety and productivity. The current indoor locating practices utilize real time locating system by alerting staffs when entering hazardous, high risk, or off-limits areas. This is a risk based study by means of probability analysis based on the distance of the worker from hazard. Not much attention has been given on communication especially on improving two-way communication between workers and supervisors in conjunction with tracking them in a 3D environment for forensic purposes and performance analysis. The aim of this study is to develop a Wi-Fi-based positioning and communicating system for indoor positioning with radio communication systems which is called the Voice Communication and Locating System (VCLS) and integrate it with other applications namely BIM and GIS. The BIM and GIS are used for displaying the positions of mobile devices for tracking the workers in a proposed area. The system development process includes determination of the functionalities of VCLS components and the interrelationships between them presenting a 3D environment. A series of experiments were carried out at a current facility at the University of NSW. The initial results of this on-going study shows that VCLS has the capability to track the approximate locations of workers in 3D environment. The integration of VCLS and GIS is extremely valuable to construction users, because it increases the communication quality and efficiency of their interaction on construction site.

Keywords – Voice Communication and Locating System; Construction Site; Real-Time System; BIM; GIS.

1 Introduction

Real-time locating system (RTLS) is a part of contemporary topic as it is one of the fastest growing technology markets. RTLS has been considered as a highly effective tool for tracking and monitoring the location of materials and vehicles in different businesses such as healthcare and transportation. For example, the Veterans Affairs Department of the US Government planned $550 million for utilizing RTLS to improve efficiency, track equipment and monitor employees in hospitals. This technology is advancing and the new generation of Global Navigation Satellite Systems will provide coded signals, which would offer cheaper receivers with millimeter in precision applications in the future. Receiver devices cost will drop while the complexity of the technology will rise. It means that the availability of the technology, wider application and consequently the market penetration will increase.

One of the main challenges of construction operation is the communication and the ability of operators to identify obstacles around them. Any deficiency in communication with workers and plant operators on site may cause accident and damage. Figure 1a shows a dump truck crashes on a security patrol car on site. On the other hand, there are many advance technologies such as machine control system which can be used on construction site. For example, Figure 1b shows some of RTLS standard technologies such as global positioning system (GPS) and other sensors, which are applicable for earthmoving vehicles. Some pioneers currently utilise only the standard navigation technologies, which is a one-way communication system for single equipment. A wider range of new RTLS technologies including novel sensors and devices enable two-way communication have practically not been investigated nor widely tested in construction operation management. Application of

RTLS technology in construction industry is still in the early stages due to the characteristics of the construction site.

The locating and tracking systems have gradually become important and can be widely used for monitoring workers and vehicles to enhance productivity and safety if the obstacles be identified and resolved [1]. This study aims to develop and evaluate the applications of RTLS for construction operation purposes by testing the selected RTLS in a laboratory and on the field.

a) Dump truck without RTLS

Figure 1: A dump truck without RTLS and a sensor enabled dozer in earthmoving

Exploring the application of complex technologies including new RTLS technologies is an increasingly important area of research from government and construction industry perspective. From government perspectives, national policy makers make efforts to encourage the construction industry main players to increase the adoption of modern technologies in order to enhance safety, productivity and accuracy. For example, Australian Construction Vision 2020 encourages the construction industry to move to partner with other industries to facilitate digital technology diffusion to the industry [2] The U.S. General Service Administration recently started encouraging their projects to use modern technologies as a prominent vehicle for acquiring building digital data [3]. Therefore, there is a rising desire among policy makers and academics to promote rapid digital technology implementation. RTLS is one of the most efficient tools to generate digital data for tracking and positioning in construction.

From the construction industry perspective, there are many evidences show that utilization of RTLS as a digital navigation and time technology potentially improves productivity and safety in construction, and so its new applications should be explored. For example, Grau et al. [4] reported that the average work time spent per component for a boiler on a construction site is 36 min in traditional tracking process, and they improve the amount of time to 4 min by using a RTLS-based tracking system and this potentially saves $120,000 for the task and save $12.5 per tag. Talavera et al. [5] estimated an annual saving of US$412, 057 for a steel production company by utilizing a RTLS technology (i.e. Radio Frequency Identification) for shared inventory allowing forklift operators to place the finished product anywhere on the production site in order to maximize the utilization of the storage area. The potential benefit of RTLS justifies the necessity of future research to determine the customized solutions in construction operations.

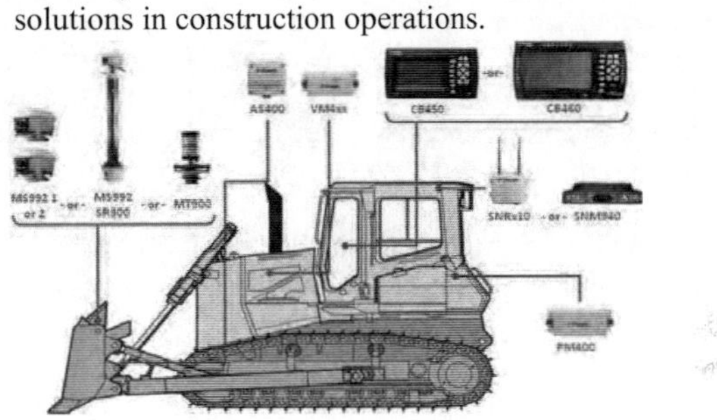

b) Dual GPS with laser augmentation and Universal Total Station suitable for Dozer

2 Real-time Location-Based Systems

The current literature mainly focuses on a wide range of the common geospatial technologies such as radio frequency identification (RFID), ultra-wide band (UWB) tags, geographic information systems (GIS), and global positioning systems (GPS). This study investigates among others; sorting information capability, receiving the stored information, communication between the workers and devices, modifying errors and transferring information to wireless local area networks (WLANs). This section reviews the relevant literature, and mainly focuses on the current practices of workers' communications.

In construction, some papers discuss the importance of the local area network and global positioning system (GPS) using Wi-Fi and ultra-wide band technologies [6]. For example, Khoury and Kamat [6] demonstrated the ability of Indoor GPS to estimate users' location with a low uncertainty about two centimeters. However, many other locating sensors have not been tested on construction sites. UWB also has known as an effective real time locating system for construction purposes. For example, Cheng et al. [7] use the UWB tools for a 3D material tracking in construction projects. However, in their discussion, other sensing technology such as laser scanners can easily identify open space areas better than UWB for creating spatial models. They also discussed the need to test other tracking technologies to complement the current limitations of UWB technology in terms of mobility and range. Zhang et al. [8] study tests on GPS data logger attached to the construction hard-hat of a concreter who was pouring cast-in-place concrete columns, show that the data logger has a comparatively low accuracy in an indoor area, and it is suitable for outdoor environment. There is a need to test a wider range of sensor technology for indoor construction application. In addition, GPS based system is more suitable for larger assets tracking and monitoring due to economic factor as the system is more expensive. Therefore, there is a need to replace these expensive technologies with other technologies for construction projects. Table 1 summaries key characteristics of these technologies in construction.

Table 1: Comparing three main technologies for construction purposes

Technology	Data acquisition	Dependency of the technology	Way
RFID	Accurate and timely information about shipping and receiving [9]	Needs other tools to be accurate [10]	One way
UWB	Communicating receivers and tags over a large bandwidth [11]	Independently provide an accurate location	One way
GPS	Low accuracy in indoor area [8]	Should be attached to objects	disconnections [12]

3 Methodology

This paper attempts to implement a model to improve workers' communications in construction sites using real-time locating systems. Based on the literature, the overall process for real-time locating is proposed consisting of using tag, collecting data, and process. In the data collection step, an educational building under renovation at the Kensington Campus of the University of New South Wales has been chosen. In order to implement the proposed system for construction operations, the voice communication and locating system (VCLS) is developed in the first place. The research scope is limited to explore the potential of the locating systems for improving the worker communication in the construction projects to enhance safety or productivity. In order to accomplish this objective, the VCLS is developed as a guide to implement the sensors at UNSW campus both in the laboratory (as the pilot study) and the filed. VCLS consists of following steps, which all have to be performed to obtain system integrity generating useful data:

Step 1: Setting the Goals and System – the goals and functionality of the system is determined at the first step.

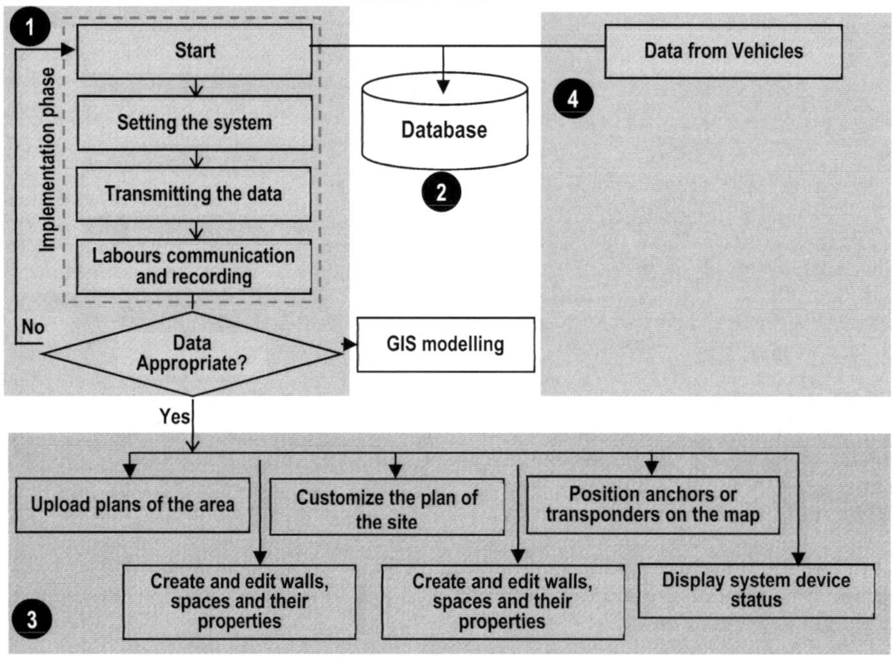

Figure 2: The flowchart of the system including data acquisition and applications

Step 2: System Segmentation – the mission to employ RTLS in construction operation is broken down into segments which will be executed and tested.

Step 3: Determine Performance Factors – the key performance areas of RTLS is defined such as efficiency (e.g. connect to Wi-Fi, standard loading time, work time), reliability (e.g. lost time, duration of defect, route reliability), productivity (e.g. orders picked per time) and visibility (e.g. location of objects, status of the objects, display the current order and progress in order).

Step 4: Development of Prototype Application – the prototype will be designed that can cover different factors such as metric specification, graphical representation, data collection plan, and utilization. The prototype can include: base station (Anchor), tag, sentry, data transmission channel (Data Channel), the positioning

. The data acquisition sub-system includes intercom, which is equipped with a built-in tag. The VCLS is configured via web-client application. The web-client application enables the user to browse locations of workers and/ or machineries on the map, device information and additional data provided by the system. Table 2 shows the details of the instruments are used in our experimentations such as access points, sentry cradle, built-in tags and analyzer.

Table 2: A list of VCLS components and instruments

	Instruments	Specification	Units
1	Indoor access point	5V, 1A, connector 5.5/2.1 ("+" inside)	6
3	Sentry Cradle	12V, 0.5A connector 3.5/1.35 ("+" inside)	3
4	Intercom	Oval - Micro USB – 5V, 1A,	3
	RIO	Square - 5V, 1A, connector 5.5/2.1 ("+" inside)	
5	Analyzer	12V, 0.5A connector 5.5/2.1 ("+" inside)	1
6	Portable screen	21" (at office) and 16" (in the field)	3
7	Smart phone	Samsung Galaxy	1

engine server (Location Engine), RIO Intercom. The RIO intercom will be used to determine the location of the object (i.e. vehicle or employee) and allow two-way communication between the person and his supervisor anytime when he is moving. This device also allows the person to connect the telephone network.

Step 5: System Implementation – the system will be used in the lab and/or field and the observations will be reported.

4 VCLS Architecture

The voice communication and locating system (VCLS) includes three main phases: data acquisition sub-system and transmission, data storage and processing, real-time visualization and application as shown in

5 VCLS Implementation and Results

The developed VCLS was tested four times in the laboratory and two times in the field. Figure 3 shows the field test area in the sample building and the drawing of the selected area on the 3rd and 4th floors. The area includes windows, doors and stairs area and it is complex enough to explore the accuracy of the work in different architectural environment. Figure 3 shows the experimentation area including anchors and other devices.

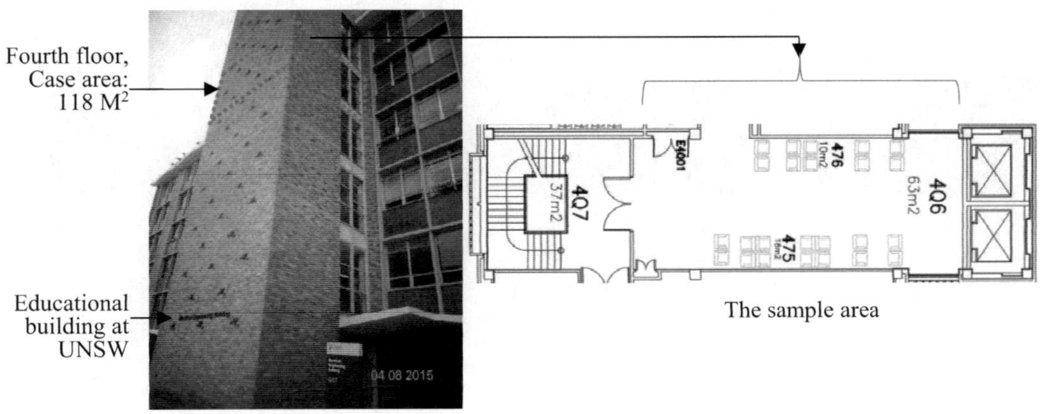

Figure 3: The sample location and layout plan

The experiment includes collecting data from tags on level 4.

Figure 4 shows the setup details of the proposed system. Six access points (1 to 6) were installed in different places in the selected area in level 4. Sample of Figure 4.

The intercoms with built-in tag allow workers to communicate with other and enable us to identify the current location of the workers as signals are transmitted via the anchors, intercoms and tags. The workers' location is visible to all colleagues and the site manager via the system's monitor.

It takes five minutes to lock in the devices after the connection has been established. All hardware devices will be online once the setup process is completed and VCLS is launched. Each device will automatically lock-in and displayed in the web-application side bars. The intercoms on display can be manually selected by setting filters on the left sidebar, which helps supervisors to monitor specific tags.

Error! Reference source not found. shows the web-application's interface includes the following key intercoms with built-in tag are shown in elements: Top toolbar, Sidebar, Map area.

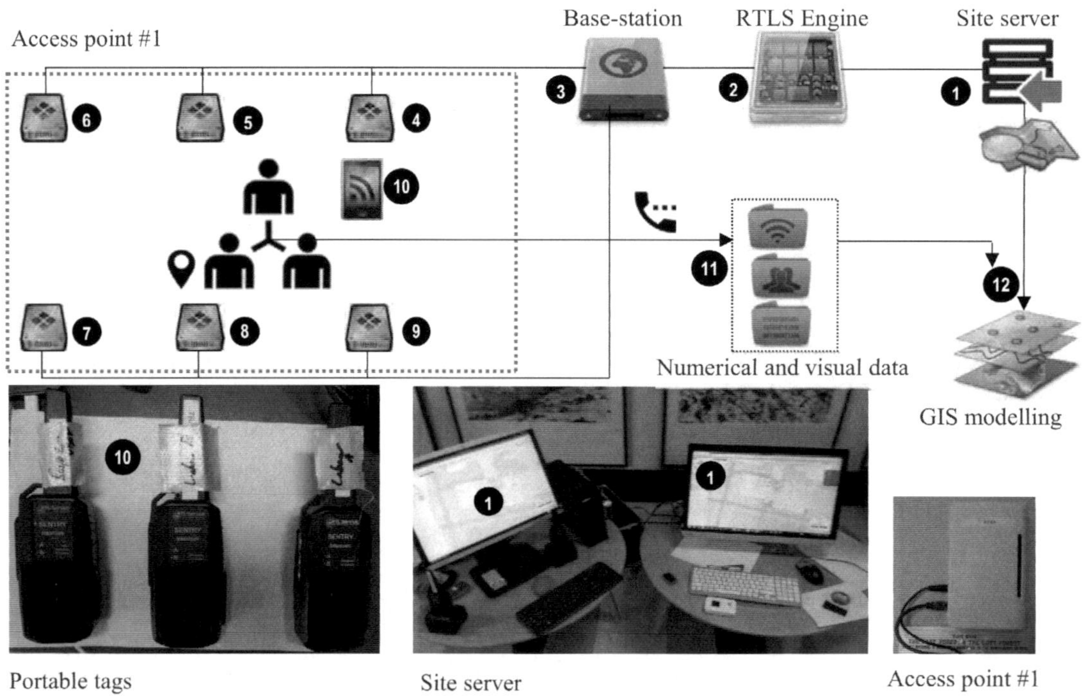

Figure 4: The VCLS architecture including all devices utilized in the field experimentation

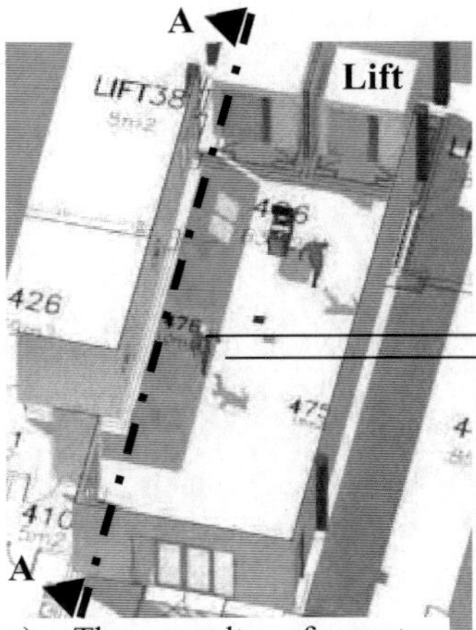

a) The result of system implementation

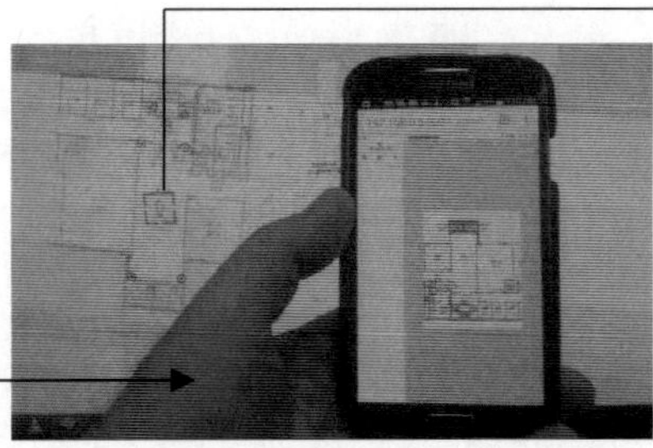

b) The online application launch with smart phone

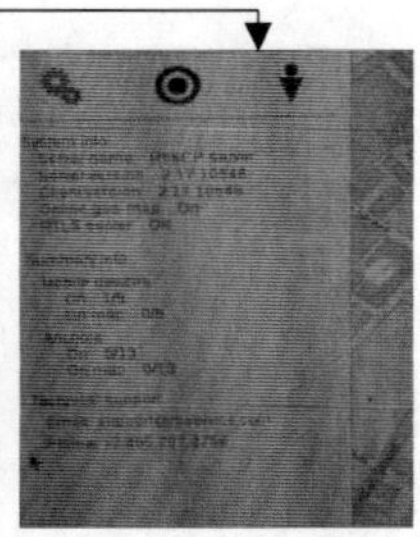

c) Web application system interface. All active and inactive tags are shown on the left sidebar, where details can be viewed by expanding the menu.

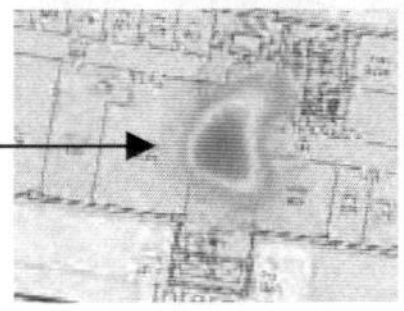

d) Heat map of the intensive area of workers

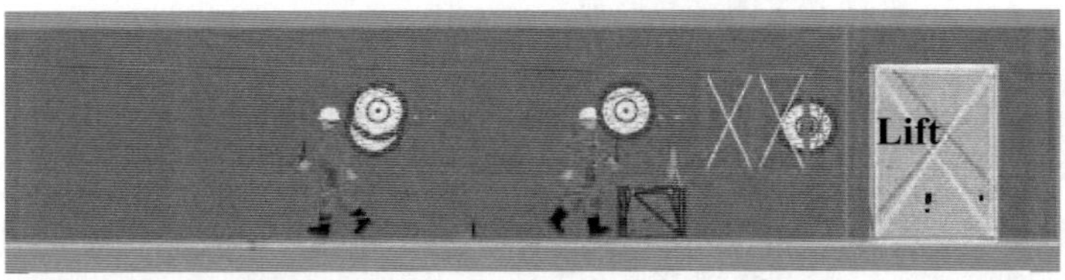

e) Section A-A showing access points and the workers

Figure 5: The visual representation of the results in the field

Figure 6 shows the result of tracking three labors in GIS environment. The figure shows 10 spot for each labor with 10 seconds interval. The total time for each labor is 90 seconds. The distance analysis also shows when the labors are working closer to each other in time intervals 3 (after 20 seconds) and 5 (after 40 seconds), and further distance in time intervals 1, 4 (after 30 seconds), 10 (after 90 seconds). The result shows three sets of data for each labor. Data recorded were the communication between labors and the location data in conjunction with the verbal data. This is important because all activities on site can be recorded and used for forensic purposes. The implemented measurement system will assist project managers to obtain detailed knowledge of workers' conditions on construction site periodically. Jiang et al [12] intend to measure the labor consumption in a dam monolith project using GPS and GIS. However, they reported communication failures in various positions in the project due to obstacles interrupting communication between the receivers and satellites.

We did not experience such interruptions because the system doesn't need to communicate with satellites. **Error! Reference source not found.** briefly shows the contribution of the paper in the body of knowledge in construction.

The field experimentation also shows that there are some limitations that should be addressed in next experiment. The data can be stored in the GIS environment but it cannot be dynamically tracked in GIS. The verbal communication also cannot be stored or analyzed in GIS and there is a need to investigate any other environment to store and analyze this type of data.

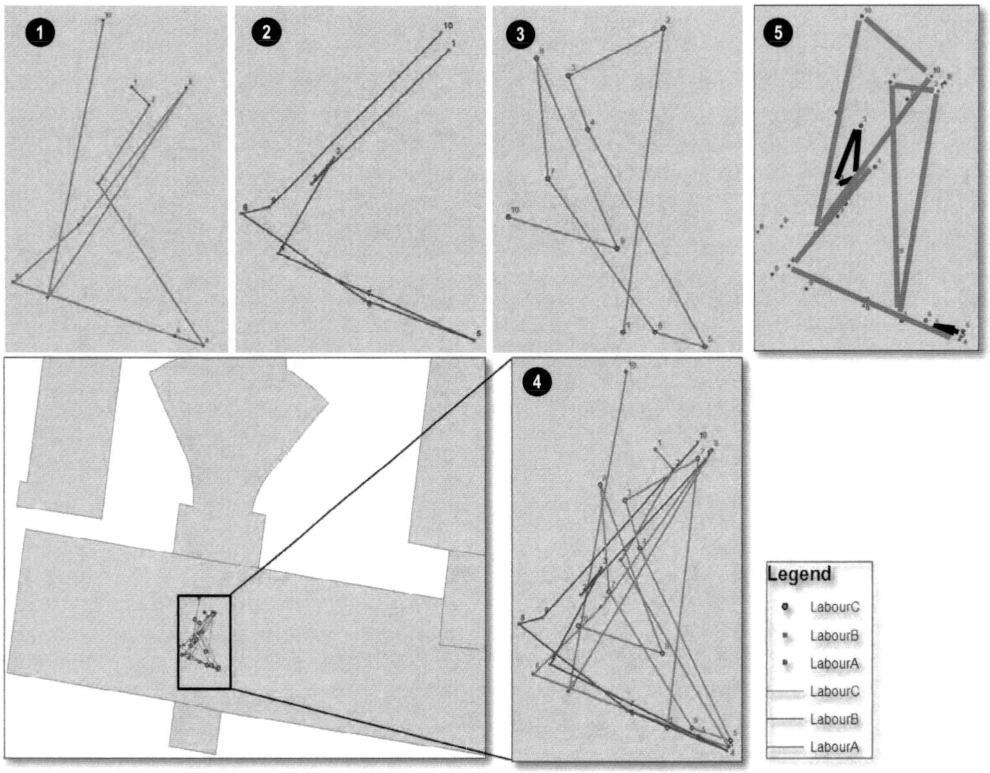

Figure 6: The results of tracking 3 labors (A, B and C) in ArcGIS with 10 seconds time interval, 90 seconds altogether

Table 3: Advantages and limitations of VCLS comparing the current practices

Our VCLS advantage or limitation	Example of current practices	Limitation or advantage of current practices
Wi-Fi-based locating system and do not need the satellite connections	GPS is used in a dam in china [12]	Communication fails reported, because of obstacles interrupting communication between satellite and receivers [12]
Supervisors and labours can communicate (two-way) while the system shows their location	Locating systems are used for alerting and messaging which is an one-way communication [12]	
The system shows who is in the hazardous or off-limits areas.	Chirp Spread Spectrum (CSS) is developed to alert labors [13]	Li et al. [13] reports the system is practically cannot be used in construction sites and needs to be further developed
The system is Linux based and can be customized for any projects	The systems are not open source	
VCLS require less than $300 to equip each person	UWB and Indoor GPS are used for locating users in sites [6]	Using both UWB and Indoor GPS are expensive and require significant time and effort to provide sufficient coverage in the site, for example individual receivers are $2195 each [6]

6 Conclusion

The purpose of this paper was to implement the proposed framework of the VCLS that can be utilized for two-way communication i.e. between labors and supervisors and in conjunction with tracking them. This study intends to improve the communication quality on construction sites for safety and forensic purposes. A series of laboratory and field experimentations were utilized to test VCLS in different situation. The result shows that VCLS is a useful platform for construction sites, since it provides two-way communication between workers while at the same time, the location of workers can be tracked by the main server. The system has adjustable time interval of voice and track recording varying from 2 seconds to 15 minutes and equipped with heat map analysis showing the most concentrated areas of workers' activities in the designated area. Furthermore, the study also presented the initial 3D models in GIS which needs further work to evaluate the feasibility of a semi-automated process. The finding is important because there is a need to offer solutions in which voice and visual data can be integrated for improving the overall construction performance.

7 Acknowledgements

The authors acknowledge Brian Rowland for providing the instruments for the field experimentations. The authors also thank Nabeel Tahir to assist us for producing 3D figures.

References

[1] Lee, H.-S., Lee, K.-P., Park, M., Baek, Y. and Lee, S. (2012) "Rfid-Based Real-Time Locating System for Construction Safety Management", *Journal of Computing in Civil Engineering,* 26(3), 366-377.

[2] Hampson, K. D. and Brandon, P. *Construction 2020 - a Vision for Australia's Property and Construction Industry,* Brisbane, Australia: CRC Construction Innovation.

[3] GSA (2013) ' Gsa Building Information Modeling Guide Series', [online], available: http://www.gsa.gov/portal/content/102282 [accessed September 2016].

[4] Grau, D., Caldas, C. H., Haas, C. T., Goodrum, P. M. and Gong, J. (2009) *Impact of Fast Automated Tracking of Construction Components on Labor Productivity,* translated by.

[5] Talavera, H. E., Banks, J., Smith, N. R. and Cárdenas-Barrón, L. E. (2015) "Enhancing the Management of Shared Inventory in the Steel Industry Using Rfid: An Alternative to Bar Codes", *International Journal of Machine Learning and Cybernetics,* 6(5), 733-745.

[6] Khoury, H. M. and Kamat, V. R. (2009) "Evaluation of Position Tracking Technologies for User Localization in Indoor Construction Environments", *Automation in Construction,* 18(4), 444-457.

[7] Cheng, T., Mantripragada, U., Teizer, J. and Vela, P. A. (2011) "Automated Trajectory and Path Planning Analysis Based on Ultra Wideband Data", *Journal of Computing in Civil Engineering,* 26(2), 151-160.

[8] Zhang, S., Teizer, J., Pradhananga, N. and Eastman, C. M. (2015) "Workforce Location Tracking to Model, Visualize and Analyze Workspace Requirements in Building Information Models for Construction Safety Planning", *Automation in Construction,* 60, 74-86.

[9] Song, J., Haas, C. T., Caldas, C., Ergen, E. and Akinci, B. (2006) "Automating the Task of Tracking the Delivery and Receipt of Fabricated Pipe Spools in Industrial Projects", *Automation in Construction,* 15(2), 166-177.

[10] Shahi, A., Aryan, A., West, J. S., Haas, C. T. and Haas, R. C. (2012) "Deterioration of UWB Positioning During Construction", *Automation in Construction,* 24, 72-80.

[11] Omar, T. and Nehdi, M. L. (2016) "Data Acquisition Technologies for Construction Progress Tracking", *Automation in Construction,* 70, 143-155.

[12] Jiang, H., Lin, P., Qiang, M. and Fan, Q. (2015) "A Labor Consumption Measurement System Based on Real-Time Tracking Technology for Dam Construction Site", *Automation in Construction,* 52, 1-15.

[13] Li, H., Chan, G., Huang, T., Skitmore, M., Tao, T. Y., Luo, E., Chung, J., Chan, X. and Li, Y. (2015) "Chirp-Spread-Spectrum-Based Real Time Location System for Construction Safety Management: A Case Study", *Automation in Construction,* 55, 58-65.

34th International Symposium
on Automation and Robotics
in Construction

ISARC 2017

June 28 – July 1
TAIPEI, TAIWAN

PROJECT MANAGEMENT

WEB-BASED INTEGRATED PROJECT CONTROLS SYSTEM

A. Montaser[a] and A. Montaser[b]

[a] Department of Structural Engineering, University of Ain Shams, Egypt
[b] Department of Structural Engineering, University of Ain Shams, Egypt
E-mail: ahmed_mohsen1313@yahoo.ecom, structural_ali@yahoo.com

Abstract –

The early detection of overruns in construction industry is essential for effective project management. Inefficient and wasteful project reporting systems contribute in project failure as it deprives project management team from taking immediate corrective actions. Reporting and dashboard tools allow multiple data source extraction and processing, which facilitate producing integrated web-based visualized report in near-real-time to be available for project parties. This paper presents a web-based platform that supports real estate developers through integrating cost data with visualized progress information. The platform development tool is ORACLE Business Intelligence Enterprise (BIE). A specially designed relational database was developed to store progress and cost data and to produce the visualized progress layout using mapping techniques and dashboard design. Cost data is imported from a customized cost reporting system with a distinct database. The developed platform creates a practical integration between cost reporting and facilitates data reporting in visualized and graphical representation. The platform has been tested and applied to actual case study to demonstrate and verify its ease of use and capabilities. The case study project is a residential community composed of villas, building and supporting infrastructure works in addition to community building in the Kingdom of Saudi Arabia with 53,800 m2 land area and a 39,526 m2 built up area. The developed model supports project controls more thoroughly to make better business decision.

Keywords –

Web-based reporting, integrated reporting, cost reporting, project control and ORACLE Business Intelligence.

1 Introduction

Project managers could utilize earned value management methodologies the most as it keeps them up-to-date with the early indicators and progress of a project's execution plans and alerts them to take corrective action when needed [17]. Many research studies have been dedicated to this area to enhance and to show its capabilities [3] [8]. Early finding of actual or forecast cost overrun in the field of construction projects is vital to assist project managers [6]. It supports and gives the opportunity to develop remedial actions and rises the chance of controlling such issues or lessening their overall impacts on the project. Hence cost overruns reduce revenues of projects, it is very clear to understand why both project managers and project executives are sensitive to any data deviations. This necessitates the need of project manager's ability to plan, implement, and sustain an efficient and comprehensive methodology for progress monitoring to promptly recognize, process and communicate variations between actual and planned performances as early as possible [5] [6]. The main challenge for project managers and project controls managers alike is the essential desire to combine and integrate the information gathered and maintained by these various entities within the organization [10] [11] [16]. Project controls usually pursue to integrate variant tools and applications together so that they can get reliable information across those tools and applications. One main method in which this characteristic of effective project controls process may be achieved is through integrated reporting systems. Recording systems are a significant concept for integrated reporting module which means that all information distributed in the form of reports or interactive dashboards need to be reconciled to a single information's source [1].

Cost data is mainly the criteria which most companies evaluate the success or inadequacy of their project controls techniques. Project financial information has many unique features which aren't limited to simple budgeted amount presentation but also, the desire to track much more discreet information. This information is regularly contained within the system of record but this data should be addressed in the overall reporting of projects [9] [12]. Currently, most of organizations have a financial system of record is often a large Enterprise Resource Planning (ERP) application such as SAP or ORACLE. Integration of information from these systems of record involves that the required information to be exchanged easily between of these unique systems. ORACLE BIE applications include Extract Transform Load (ETL) routines to extract, transform, and load data into the ORACLE Business Analytics Ware-house. It connects to multiple data sources such as ORACLE DB & Exadata, QL Server,

DB/2, Teradata Essbase, ORACLE OLAP and MSAS and Integral part of ORACLE Fusion Middleware [13], [14] [15].

The developed model facilitates cost reporting based on Cost Break-down Structure (CBS) or Work Break-down Structure (WBS) hierarchy. Additionally, it visualizes the imported progress from scheduling tools to progress curves and bar charts. Colored progress layouts are used as a major part of the system to show the progress of each project component in a different color matching with its progress. The developed model facilitates data reporting in visualized and graphical representation. The developed model is very flexible to modification of data presentation and manipulation method and can show any data extracted from the supporting systems. Stakeholders can take immediate corrective actions based on the web based system output. A warehouse database is specially designed to gather information from scheduling and cost tools to produce the required visualized web based reports.

2 Proposed Methodology

The reporting system is developed to provide project stakeholders with integrated project reporting through logging in any internet browser. Cost information is provided to show budget of work and payments details and status. Visualization capabilities is a major consideration during developing the model consequently the model contains interactive mapping layouts which represents graphical representation for work progress based on color scheme. The web-based application is providing user with visualization capabilities through interactive dashboards. The dashboard allows user to draw interactive graphs and interactive progress layouts through transferring progress % to color schemes for any WBS/CBS level. The model allows users to understand tabularly, graphically, visually and textually the reporting information and project status in easy manner as soon as the data is updated in supporting systems.

The main steps of the proposed methodology are following the regular reporting steps, as shown in Figure 1. cost tool is used to develop the cost monitoring system and for following payments. Budget data is the baseline or the cost system and is used as the main guideline during project life cycle. Actual cost and payments incurred is gathered and stored in the cost system. After updating the system with the latest status, it stores all the cost information in cost tool database to be transferred to the ware-house database. Progress information will be imported directly to the databased through exporting it from the used scheduling tool. Integrated report is issued automatically after updating any data from cost, and importing the required data from scheduling tool. Web-based application is utilized to issue the integrated report after collecting all information. The application should provide ease of usage in both creation and viewing project progress dashboards. The dashboard contains all imported data in addition to charts, curves and progress layouts through manipulating the imported information. Finally, the interactive dashboards and report are generated through the system and can be viewed by all users. Reports are saved inside the ware-house database to be used as a historical information for all project data to be used later if needed.

As shown in figure 2, First stage in developing the progress report is deciding which levels is going to be reported (EPS-WBS-activities) in addition to dashboard and report design that should consider Stakeholders requirements. Graphical representation of data is obtained through the report through progress status which is a bar chart showing planned and actual % while time status is a pie chart representing elapsed and remaining duration. Resource assignments which are exported from scheduling tool into ware-house database is utilized inside the reporting application to produce S-curve to show both cumulative and periodically planned and actual progress graphical presentation. The S cueve curve is plotting work progress against time. Work progress may be expressed in terms of units of work, budget, man-hours, percent complete. It is a result of scheduling sequence and dates represented by units of work allocation through project period.

Budget data in the developed model is limited to budget at completion for the work which represents the cumulative value of budget cost work scheduled also the cumulative value of the performance measurement baseline MB. While the actual status of the project is reported through three numbers which are certified gross of work done, certified net amount and net paid to date. Certified gross to paid represents the amount that already certified without any deductions such as advance payment recovery and retention while the net certified amount represents the work amount after the contractual deductions. Net paid to date is the amount of money already spent to date. It is important to show both paid to date and net certified amount to show the variances between the two numbers which highlights any delayed amount due to any financial reason such as payments still not released by finance department or still under process. Cost reporting information is selected by project responsible to satisfy stakeholders requirement. The data is imported to all project and portfolios levels to be part of each level integrated report. The model could also be extended easily to include any cost data such as cost variance, cost performance index and estimate at completion, which enriches the process of cost reporting.

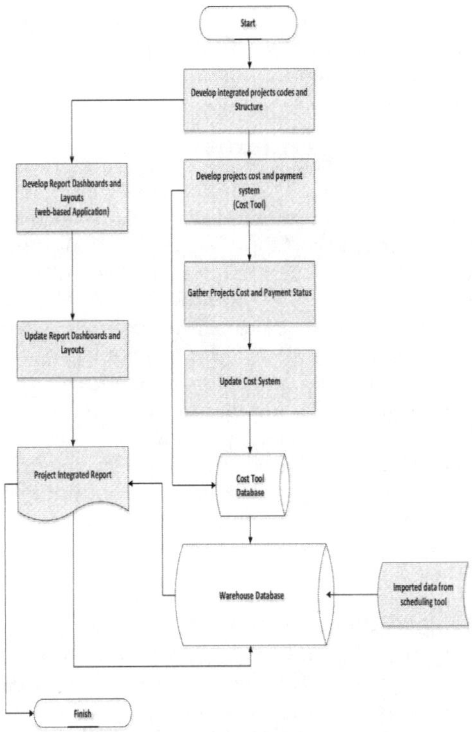

Figure 1- Proposed methodology

Progress Visualization is a requirement for reporting complex information for stakeholders in easy manner. Through colored progress status or highlighting a specific part of the project gives the report viewer the summarization of project status in a simple way. To prepare progress layouts, layers of progress should be defined which are matching with WBS and activities to ease the progress interpretation. Each layer has a specific code similar to the code used in scheduling tool. Programming logic is written to color the required components based on the imported progress from the scheduling tool. The color scheme is based on conditional formatting function to show similar progress percent in same color while lower in a different color to report progress %. Also, the selected WBS is highlighted in the overall project layout upon selection. Counting the WBS or activities based on their color gives the viewer overall progress status such as how many WBS started, completed or in progress beside the total number of WBS which contains a selected activity. As shown in figure 3, the coloring report is generated as a final step and stored inside the ware-house database to be used as a part integrated report. The used application for reporting should allow the mapping and presenting coloring schemes to be utilized in visual representation. Finally, the integrated report is saved in ware-house database after updating all reports to be kept for recording purposes. Stakeholders can view report immediately and a customized notification is sent to them once an update occurs.

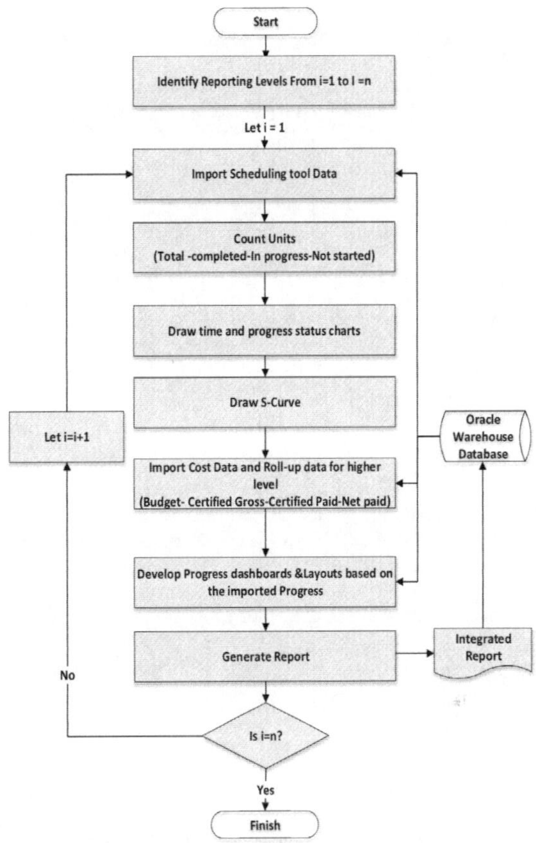

Figure 2- Report Development Process

Ware-house database is the central storage area for the reporting model. It is basically developed to integrate data from scheduling tool and cost tool to satisfy reporting requirements through providing requests to retrieve the information needed by reporting software beside being the main storage for reporting historical information. Activities ID along with their features (name, type, status, float, dates, duration, WBS, codes, etc.) are the main attributes which is exported from scheduling tool for the activity entity. The activity entity is the lowest level to measure project performance. WBS levels is the main source for reporting cost data beside scheduling and progress photos. Every WBS has a unique identifier based on its level on the project and type of work done to facilitate date transfer to the integrated report from other reporting tools. WBS has many attributes such as name, ID, dates, scheduled % and performance % which are imported from scheduling tool while budget, certified gross value, certified net amount and paid to date is imported from scheduling tool. Each WBS has its reporting photos extracted from web-based photos software.

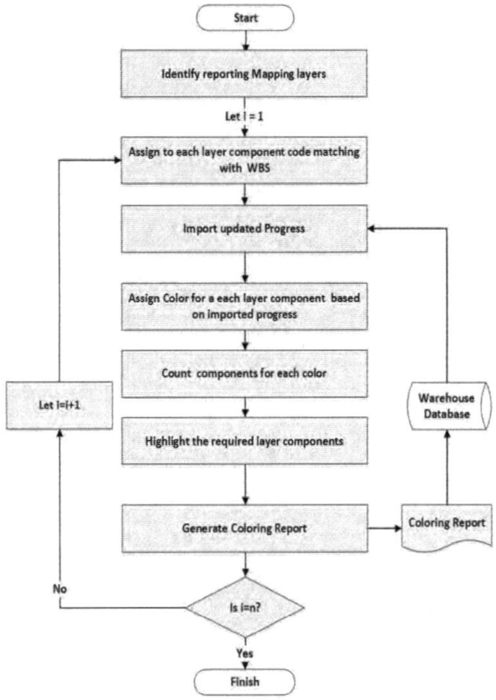

Figure 3- coloring layouts process

All imported attributes are presented in integrated way based on WBS ID matching. Activity spreadsheet and resources assignment are another entity that is extracted from scheduling tool to produce S-curves through manipulating the resources distribution for each period. It contains periodically and cumulative planned and actual activity resources assignments distribution. EPS is the highest level for reporting which divides the projects in portfolios and programs to summarize information to executive directors. Each tool has different tables for EPS, WBS, Activities to report the output for each level and the integration occurs inside the reporting module through requesting the data from the database based on the required reporting level and type of data required. Database retrieves the stored data based on the level ID which is unified for all tools. Portfolio, program, project, WBS and activities history entities is created inside the database to record all historical information for the reported data and to save the reports when created. Due to space limitation, the entity relationship diagram of the designed database will not be presented. Sample of relationship between entities are the relation (1:M) between WBS and activities, EPS and WBS and WBS and photos while the relation M:N Between activity and resources and activity and activity codes.

3 Model Verification

The developed model is implemented in a web-based software, which has similar structure to the illustrated methodology model. In-House web-based cost reporting module has been used as a cost tool and it depend on ORACLE database which facilitates data transfer. A separate customized web-based application is developed to allow project reporting and viewing. Visualization and integration of reports data are a major advantage of the developed reporting module. The application main coding language is SQL (Structured Query Language) while the main development tool is ORACLE Business Intelligence Enterprise which is used to design reports and dashboards. It uses ORACLE data integrator to support Extract Load and Transform (E-LT) technology that improves performance and maintain data integration. It utilizes ORACLE ware-house database to store tabular data and for integration with cost and progress information to gather progress data. It is a central repository of integrated data from one or more disparate sources. It stores current and historical data and is used for creating report dashboards.

Three-tier standalone architecture was chosen for the developed software and other reporting systems as shown in figure 4. Tier one represents the presentation tier, which is mainly a web-based interface for the system. The user interacts with that interface through predesigned quires regarding project development and updates. There are two-way data flow between UI and event handler to get the required data and apply cost calculations. Tier two is cost logic where all cost algorithms and procedures in the data analysis and processing are implemented. Cost system uses ORACLE E-T-L tool built in the system to process data to processing tier to implement cost calculations and return it to user interface. Ware-house database is the main database for photos developed application to be stored and used during report generating. ORACLE Business Intelligence Enterprise Edition 11g(OBIEE) is the main system for developing the integrated report. It was chosen to be compatible with other used application in addition to mapping and reporting capabilities which produce interactive dashboards. User interface is the first tier of the system it allows user to select the required dashboard to be viewed through multi selection pane that enables him to filter report results. Reporting processing is the main logic for the application which resides in tier two that processing utilizes the stored data in the ware-house database to generate reports through data manipulation, filtration and graphical presentation.

Data integration process is another part of tier 2 which integrate data from cost database and transfer it to ware-house database to be processed during report generation. Ware-house databased stores all dispatched reports inside it to be used for recording purposes.

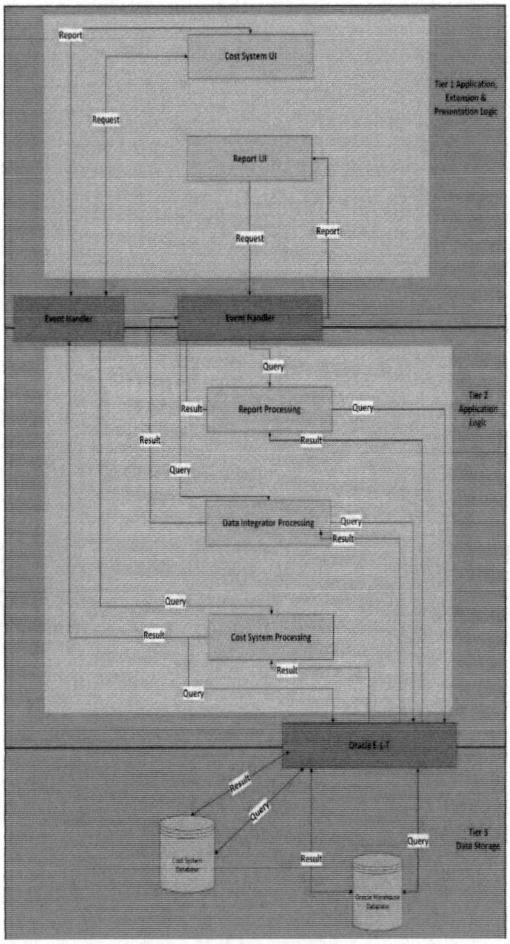

Figure 4- Reporting systems architecture

Use-case diagram is an illustration of a user's communication with the system. It describes the relationship between the user and the different users or systems. Figure 5 represents the use-case diagram for all the integrated components. There are four actors to generate the integrated report. First actor is cost engineer who is responsible for developing cost system, collecting cost updates and updating the system. Both planning engineers and cost engineers should use the same hierarchy during developing both systems beside developing the report and dashboard design which satisfies all stakeholders requirements. The concerned stakeholders, 2nd actor, including cost engineer are able to log-in and view reporting system. The 3rd actor is the cost system which executes cost calculations to cost system in addition to importing all the data (planned, budget and Actual) to each system database. The last actor is the OBIEE which is the main reporting application which supports dashboard and report design, colored layouts based on visualized maps and support simple reporting calculations. Data is imported from ware-house database after manipulation and filtration to generate the integrated web-based report.

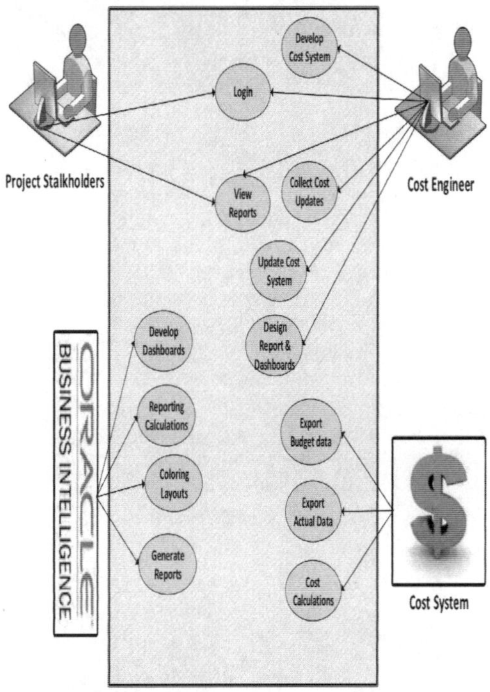

Figure 5- Use-Case diagram

The developed web-based system has been applied to real case study project to test and verify its capabilities. The project is LIMAS/ Jubail project at Alhamra district, Jubail City, Kingdom of Saudi Arabia. LIMAS project is a gated housing project (110 villa + 20 apartments) is being built on a 53,800 m2 land area and a 39,526 m2 built up area. The land is surrounded with plenty of houses and villas served with complete infrastructure as shown in Figure 6. The project is composed of four different types of villas type A, B, C, and type D, Amenities Building (recreation center) and infrastructure works. The project is composed of 26 villas type A with BUA=393 m2/ villa, 54 villa type B with BUA= 319 m2/ villa, 30 Villa type C with BUA= 266 m2/ villa ,1 building (20 apartments) with BUA= 2070 m2 and recreation center with BUA=2032 m.

Figure 6- Project perspective

Cost process starts with developing CBS which should be matching with WBS levels. After budget decomposition and contractor creation. BOQ items were loaded in the system for each contract coded with its cost code based on cost hierarchy as shown in figure 7. For each item, description, unit of measurements quantity and rate were inserted while amount was calculated automatically inside the application. These parameters can be inserted manually or through exporting from excel or any other similar application. Allocation of all project budget is necessary in this stage to assure accuracy of cost report. Budget on the integrated report is extracted during this stage and stored in the ware-house database.

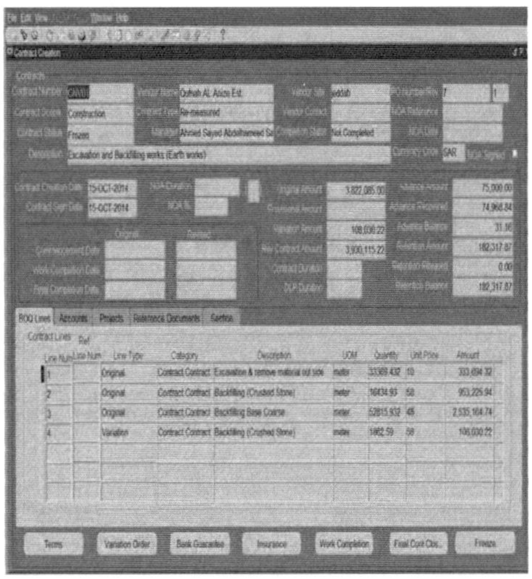

Figure 7- Contract Creation Process window

Regular updates for cost data occurs in case of any approved variation or change order. Figure 8 is illustrating the process of updating cost data based on progress on site to issue payment certificates. Contractors submit progress per each BOQ item to consultant for review. After consultant review to check that the payment is matching with actual approved works on site, cost engineer update the system accordingly. Payment certificate are issued after approval form all concerned people as shown in figure 9. The payment certificate shows the gross value of work done and net amount of work done which are used as a part of the integrated report. payment issuance from finance department is part of the cost system. Based on finance approval and transferring amount to contractor's accounts, net paid amount is calculated inside the system and transferred to ware-house database for reporting purposes. ORACLE Cost systems contains many parameters and automated reports which are generated automatically inside the system upon sending queries.

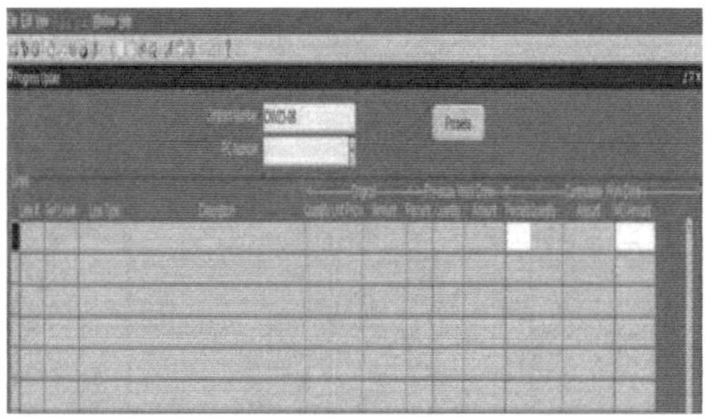

Figure 8- BOQ items Update Window

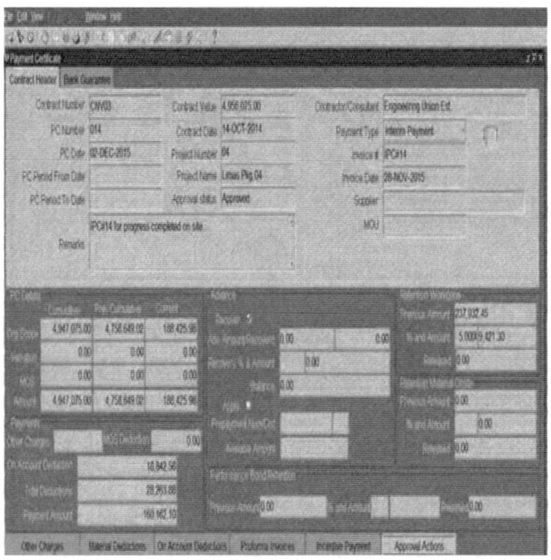

Figure 9- Payment Certificate Window

Dashboards design is the first stage for issuing the progress report. The executives have visualization requirement to view the whole progress on colored layout based on units' progress for each activity and to highlight the selected area in the master plan based on selection. Master plan AutoCAD file was modified and layers were assigned to match with WBS hierarchy and conditional formatting was used to differentiate between unit's progress. Moreover, the selected part of reporting is shown in red color to add visualization effect to the report. Report dashboards were similar to the manual reporting which was reported before developing the new system. S-curves were created inside the reporting systems based on the extracted data from scheduling tool to simplify the numbers by using graphical presentations.

First dashboards were designed to show portfolio progress to assess the overall situation of company projects in summarized way. It contains map location of each project, progress status and time status charts for each project which were created using OBIEE capabilities and overall time and cost status for the projects as shown in figure 10. The second portfolio dashboard shows S-curve for contractors and projects based on multi selection pane to choose period, project or zone as shown in figure 11. This dashboard enables user to evaluate the overall situation of project or contractor during selected time period.

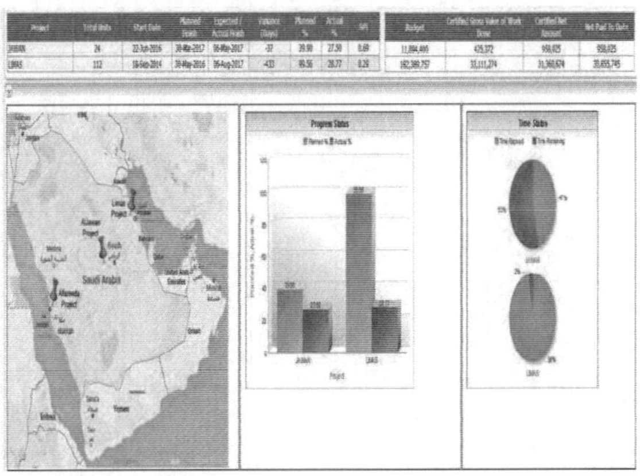

Figure 10- Portfolio dashboard

Project dashboards contain interactive map show units' color based on the actual progress for each unit as shown in figure 12 which is showing phase dashboard. This interactive colored layout is developed inside the OBIEE automatically to allow progress visualization. Activities progress layout is the last dashboard in the report showing the status of each activity in colored map in addition to legend showing the count of completed, in progress and not started units for that activity as shown in figure 13. Same dashboard was developed for infrastructure works with some modifications based on the nature of infrastructure works. OBIEE provide flexibility in modifying dashboard contents and information presented based on user requirements.

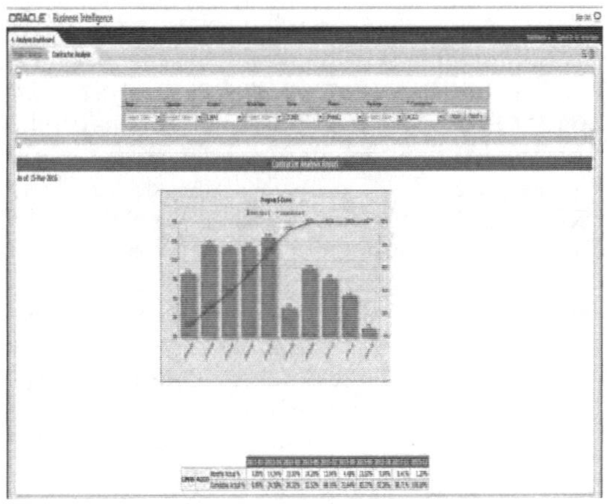

Figure 11- Portfolio S-curve dashboard

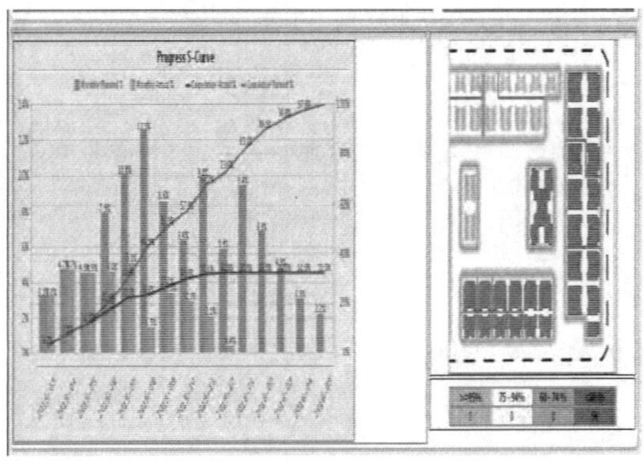

Figure 12- Phase dashboard

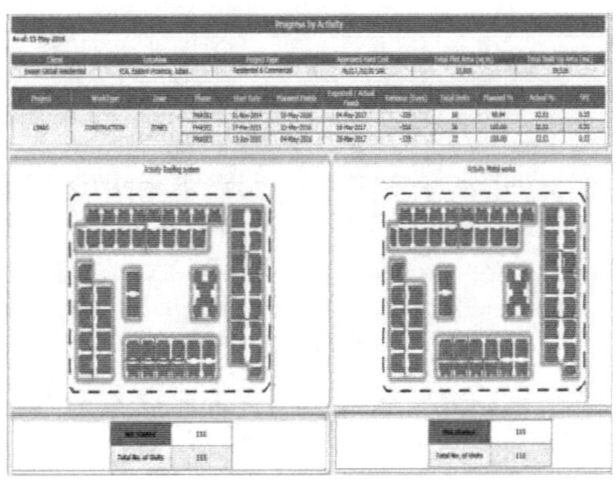

Figure 13- Activity dashboard

4 Conclusion

This paper represents design and implementation of automated integrated web-based reporting system for construction projects. The developed model integrates project time and cost data in addition to visualization capabilities through utilization of mapping techniques. The report goes through 3 main stages for achieving the final output, which is web-based integrated report. Cost reporting is the first stage of the process in which budget is created and actual costs and payments are recorded. Cost engineer is responsible for this process using customized ORACLE cost system. Web-based application was developed to store progress updates and store it to be used in creation of the developed reporting model. Web-based application was developed using OBIEE to support visualization and integration of data in addition to designing interactive dashboards. At the core of the developed model lies a data analysis and processing framework that stores, retrieves and processes the captured data from cost system. The developed model facilitates data reporting in visualized and graphical representation. The customized integrated web based module constitutes a step ahead of current market applications solution for integrated web-based reporting. The developed model supports monitoring performance more thoroughly and make better business decision through generating one simple integrated reports to assist in speedy assessment, analysis and to take suitable actions in timely manner. The updated progress information is available immediately to all concerned parties with high quality and reliable timing. The developed model is very flexible to modification of data presentation and manipulation method and can show any data extracted from the supporting systems.

References

[1] Blodgett, J. and Criss, B. Integrated Project Reporting Using Dashboards: Harnessing the Power of Primavera P6. *Cost Engineering Journal*. 57(1):4-11, 2015.

[2] Bosche, F. Automated recognition of 3D CAD model objects in laser scans and calculation of as-built dimensions for dimensional compliance control in construction. *Advanced Engineering Informatics*, 24(1):107–118, 2010.

[3] Cioffi, D. Designing project management: A scientific notation and an improved formalism for earned value calculations. *Intonational Journal of Project Management*, 24(1):136-144, 2006.

[4] Fleming, Q.W. and Koppelmam, J.M. Earned Value of Project Management. *Cost Engineering, AACE*, 39(2):13 – 15, 1997.

[5] Fleming, Q.W. and Koppelmam, J. M. *Earned Value Project Management*. 2nd edition, Project Management Institute, Inc., USA, 2000.

[6] Halpin D. *Construction Management,* 3rd Edition, John Wiley and Sons, New York, USA, 2006.

[7] Hassanein, A. Planning and Scheduling Highway Construction Using GIS and Dynamic Programming. *PhD's dissertation*, Civil, Building, and Environmental Engineering, Concordia University, Montreal, Canada, 2002.

[8] Iranmanesh, S.H.; Kimiagari, S.; Mojir, N. A new formula to estimate at completion of a project's time to improve earned value management system. *In Proceedings of the IEEE International Conference on Industrial Engineering and Engineering Management (IEEM)*, Singapore: 1014–1020, 2007.

[9] Kerzner, H. *Project Management -A Systems Approach To Planning, Scheduling And Controlling*. 10th edition. John Wiley and Sons, New York, USA, 2009.

[10] Koo B. and Fischer M. Feasibility Study of 4D in Commercial. *Construction. Journal of Construction Engineering and Management*, 126(4): 251-260, 2000.

[11] Kymell W. B*uilding information modeling: planning and managing construction projects with 4D CAD and simulations*. McGraw-Hill. New York. USA, 2008.

[12] Mubarak, S. *construction project scheduling and contro*l. 3rd edition. John Wiley and Sons, New York, USA, 2015.

[13] ORACLE. ORACLE® Fusion Middleware Getting Started with ORACLE Business Intelligence Publisher, 2013.

[14] ORACLE .ORACLE BUSINESS INTELLIGENCEFOUNDATION SUITE 11g DATA SHEET, 2013.

[15] ORACLE. Powerful Visual Analytics for the Entire Organization ORACLE Business Intelligence 12c, 2015.

[16] Poku S. and Arditi D. Construction scheduling and progress control using geographical information systems. *Journal of Computing in Civil Engineering*. 20(5): 351-360, 2006.

[17] Vanhoucke, M. 2012. Measuring the efficiency of project control using fictitious and empirical project data. *Intonational Journal of Project Management*, 30(1): 252–263, 2012.

Dynamic Scheduling Framework to Overcome Deficiency of Skilled Workers

Pawan Pandey1[a], J. Uma Maheswari 2[b], Ratnesh Kumar3[c]

*a*Corresponding Author, *b*Assistant Professor, *c*Research Scholar,
Department of Civil Engineering, IIT Delhi, New Delhi, India
E-mail: pvnengr@gmail.com, uma.iit@gmail.com, ratnesh.iitd@outlook.com

Abstract –

Substantial investments in infrastructure world over and rapid advancement in construction technology has created acute shortage of skilled workers. The skill deficiency gets aggravated further due to absenteeism, migration, etc. In practice, it was observed that there exist several partially trained semi-skilled workers having expertise in one or more trade. Using these semi-skilled workers as substitute can address the shortage of skilled workers to some extent. On the contrary, skilled workers alone are considered while planning and scheduling in most of the construction projects. Here, the potential capability of these partially trained workers remains unexploited because of their default grouping with unskilled workers.

The root cause of this skill treatment can be the absence of standardised system of screening and classification of workers into different skill category as highly skilled, skilled, semi-skilled, or unskilled and the ambiguity in updating or maintaining the worker's database. Hence, the objective of the present study is to review the "skill" levels and the worker's skill utilization to the maximum extent possible. To achieve this, a framework has been proposed to generate worker's multi skill database (WMSD) and dynamic scheduling to address the uncertainty caused by deficit of skilledworkers.

Keywords –

Dynamic Scheduling Frame Work, Workers Multi Skilled Database (WMSD), Productivity Matrix, Absenteeism

1 Introduction

Construction industry remains one of the most labour intensive industries despite mechanization. This implies although mechanisation reduces the overall requirement of workers in a project, yet demand of skilled workers has been increasing with regard to emerging tools, techniques, new construction material, and technology [1].

Quality construction typically demands a certain number of skilled workers in a crew [2]. The formation of skilled workers crew becomes restricted due to fragmented construction industry, system of subcontracting and short employment duration. Also, few workers acquire proficiency in more than one trade/craft (i.e., multi-skilled) due to frequent changes in work assignment [3, 4]

As each construction project is unique and requires minimum number of skilled workers to meet quality and contract conditions, only skilled workers are generally considered while scheduling and the partially trained workers (i.e semi-skilled or multi-skilled workers) are ignored due to their default clubbing with the unskilled workers. Low wages, poor working condition and overall poor image of the construction industry forces the workers to frequently absent, migrate to other countries, etc. [5]. This short term deficiency of skilled workers reduces crew productivity at activity level and the impact of the same gets magnified at project level resulting into time-cost overrun.

The present situation offers an opportunity to utilise the partially skilled workers efficiently to overcome short-term scarcity of skilled workers. This can be achieved by developing an appropriate dynamic scheduling framework that is reliable. In the long term, the offer of longer employment, higher compensation and variety in assignment to semi-skilled and multi-skilled workers are likely to motivate huge pool of unskilled workers towards early skill upgradation and thereby reduce absenteeism.

Hence, the objective of this study is to propose a framework which utilises semi-skilled and multi-skilled workers as substitute to limit crew productivity loss caused by absence of skilled workers.

2 Background

2.1 Defining Skill

Defining skill itself is complex, highly subjective,

and means differently to different researchers. Workers perceive skills as competence acquired through education, training and experience which may not have been put to use in the past (i.e effective skill), the employer considers only a portion of the effective skill suitable for impending task (nominal skill) and sociologist considers that skill is an evolving processes emerging from social interaction between workers and employer [6]. Generally skill is considered as "a set of knowledge, attributes and capacities that enable individual workers to successfully and consistently perform an activity or task and that skill can be built upon and extended through training, learning and experience" [7]. Irrespective of industry, International Standard Classification of Occupations (ISCO-08) divides workers into four skill levels based on education and job specific training and experience (i.e., 1st level- primary education, 2nd level-high school, 3rd level- graduation and 4th level- post graduation) [8]. The groups relevant to construction industry are 1) construction professional (i.e., engineers, project managers and technical associates) to have graduation and post-graduation and training; 2) skilled construction workers, plant/ machine operators, assemblers require high school education, craft certification like ITI/Diploma and experience of 2 years or more; and 3) Other construction workers or unskilled workers.

2.2 Construction Workers at Site

CII in 1998 proposed multiskilling as "worker utilization strategy where workers in possession of a range of skills appropriate for more than one work process and these can be used flexibly on a project or within an organization, not limited to traditional job boundaries", to reduce overall number of skilled workers[9]. However, it may be difficult for a worker to acquire 'skilled' status in more than one trade due to evolving work processes.

In India, there exist three distinct career paths for construction worker as 1) traditional craftsmen (who consider the work as family tradition and learn trade expertise from their parents); 2) higher secondary qualified student (who joins ITI or other structured skill development program to get trained in the trade of their interest); and 3) migrant workers (who move as nomads for livelihood) [10]. The traditional construction workers are small in number and focused on specialist work like repairs and maintenance of heritage structures, monumental stone work and carvings etc. The ITI qualified workers form the major chunk of skilled construction workers; whereas, third category comprises all other construction workers with many of them partially skilled but clubbed with unskilled workers while project scheduling [11]. These partially skilled workers may not be able to execute quality work in isolation; but will have decent productivity when employed under supervision of skilled worker in a crew [12].

2.3 Scheduling Techniques

2.3.1 MRCPSP (Multi-Mode Resource Constrained Project Scheduling Problem)

RCPSP [13] assumes that an activity can be executed only in one mode (i.e. specific type of resource required for fixed duration to execute an activity), contrary to the fact that most of the activities can be executed in more than one mode. Here, the schedule which considers all possible modes offers more efficient utilization of the resource and is referred as MRCPSP. Invariably researchers make four basic assumptions - any worker can do any job; mode of execution cannot be changed during execution; availability of manpower is uniform; no breaks allowed once an activity is started [14]. However, these assumption makes such schedules unrealistic.

2.3.2 MSPSP (Multi-Skilled Project Scheduling Problem)

Kadrou (2006) highlighted the complexities and issues associated with modelling of multi skilled (MS) worker with varying degree of productivity and did not test the proposed solution with Project data [15]. There will be additional modes if MS worker capable to perform more than one job are considered making scheduling process complex. However, employment of MS workers brings in additional flexibility, encourages innovation [16], saving of 5-20% labour cost, about 35% reduction in required workforce [17]. However benefits are marginal beyond 20% MS workforce, and little benefits beyond 2-3 trades [18].

Most of the MSPSP studies focused on optimum utilization of skilled workers possessing equal expertise in more than one trade which is an unrealistic assumption. Hegazy et al [12] suggested storing individual workers multi skill data (i.e semi-skilled workers with expertise in more than one trade) and substitution (i.e 2 semi-skilled workers for each skilled worker) to overcome skill deficiency. But, several issues were left for further research (i.e method to identify multi skilled workers in project site, substitution rules assuming non uniform skill level and productivity, possible deviations in quality of work by substituted workers, constant resource pools over the activity duration, decision process of selecting important activities to which resources are to be diverted). Implications of short term skill deficiency while scheduling have not been attempted in these studies although a common phenomenon in view of high demand, poor working conditions and wide spread absenteeism exist.

In view of above there is a need to define skill level amongst construction workers and formulate a suitable frame work which allows substitution of partially trained workers to overcome short term deficiency of skilled workers and provide realistic project duration considering variation in crew productivity due to substitution, if any.

3 Skill levels in Construction Workers

In the present study five category of construction workers have been considered as described below;

1. Highly Skilled (HS) – Worker can visualize the entire scope of work, impart training to crew, ability to coordinate work of other crews/craft and ultimately can produce the best quality work.
2. Skilled (SK) - Can read engineering drawing, lead the crew to execute all kind of works and overall an ideal quality/ productivity levels.
3. Semi-Skilled (SS) - Will have expertise in any one work process and can work as team member in a crew and on the whole expected to have desirable productivity levels
4. Multi Skilled (MS) – Semi skilled worker who have capabilities in more than one work processes/trade.
5. Unskilled (US) - Untrained form of worker who can be used as helper in any crew.

4 Dynamic Scheduling Frame Work

4.1 Assumptions

The present study uses simplified yet realistic assumptions reflecting actual site conditions and are listed below;

1. A project can be divided into fixed number of activities with predefined logic constraints.
2. The skill screening process is expected to be done fairly by site in-charge/ experts as per project requirement.
3. A crew can be worked out for each activity using work study giving output equal to baseline productivity (BP) at economical cost.
4. Due to higher demand of SK and HS in any trade and presence of very few SK and HS in more than one trade, secondary and tertiary trade for SK and HS were neither captured during screening nor considered while scheduling.
5. A productivity table can be generated indicating productivity of various feasible combination of crew using work study. The productivity table can be further extended to reflect fixed number of likely working conditions.
6. Once known in the beginning, desired number of skilled, semiskilled and multi-skilled workers can be arranged on desired date with minimum fluctuation.
7. Workers availability for any date can be simulated to represent the real presence of workers during project execution.

4.2 The Framework

The initial schedule generated based on network scheduling technique can be improved in successive stages to; 1) reduce project duration by reducing duration of critical activity and efficient utilisation of critical plant/ machineries/ workers; 2) reduce cost of hiring/ firing of workers and mobilisation/ demonization cost of plants/ machineries by resource levelling; 3) Limit time-cost overrun by identifying minimum requirement of workers including reserve workers to counter absenteeism and factoring loss in crew productivity despite substitution; 4) Realistic schedule utilising actual availability of workers from dynamically updated skill database. The proposed dynamic scheduling framework is represented as a schematic diagram in figure 1.

4.3 Worker's Skill Database

The skill level of an individual worker can at best be quantified based on nominal screening process which involves;1) Possession of ITI diploma or equivalent certification; 2) Performance on test sample task; and 3) unstructured interview formulated by site engineer and foreman. Workers are designated as 'skilled' if they possess requisite educational qualification, job specific training/experience and exhibit productivity equal to benchmark productivity. Currently, partial skills possessed by the workers are completely neglected at the time of screening.

It was observed during data collection that even this nominal practice of skill screening was seldom followed due to adhoc recruitment, work pressure and lack of concern at higher management level, and so on. And no structured worker's skill data base was available except the workers certificate. The survey also revealed that existence of several multi-skilled workers.

Therefore to generate reliable database all workers joining the project are to be tested on their primary, secondary and tertiary test because there is limited benefit of multi-skilling beyond three work processes [18]) The flow diagram shown in figure 2 shows how the workers can be categorized into HS, SK, SS, MS and US and details collected in the database.

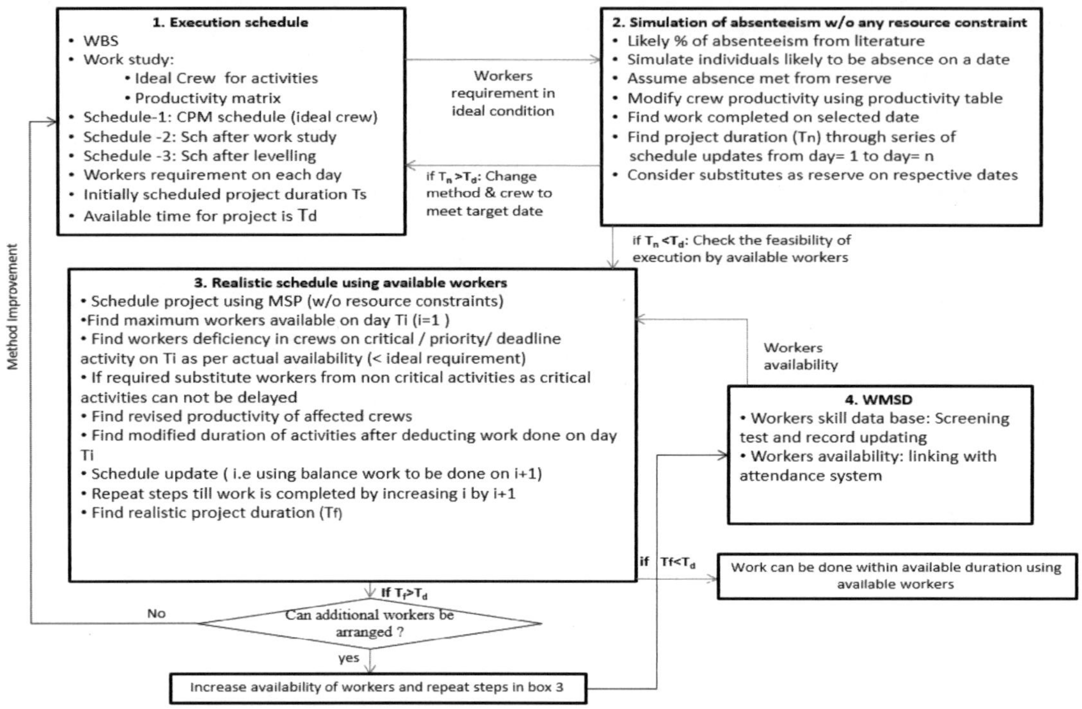

Figure 1. Dynamic Scheduling Frame Work

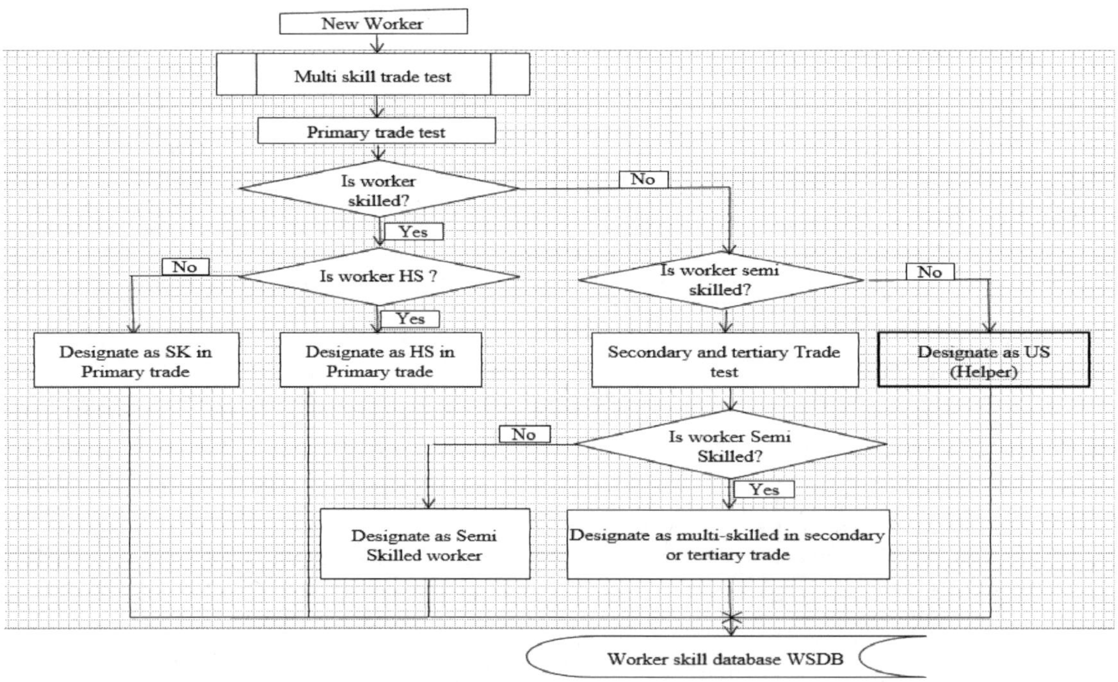

Figure 2. Flow Chart for Screening and Categorisation of Worker

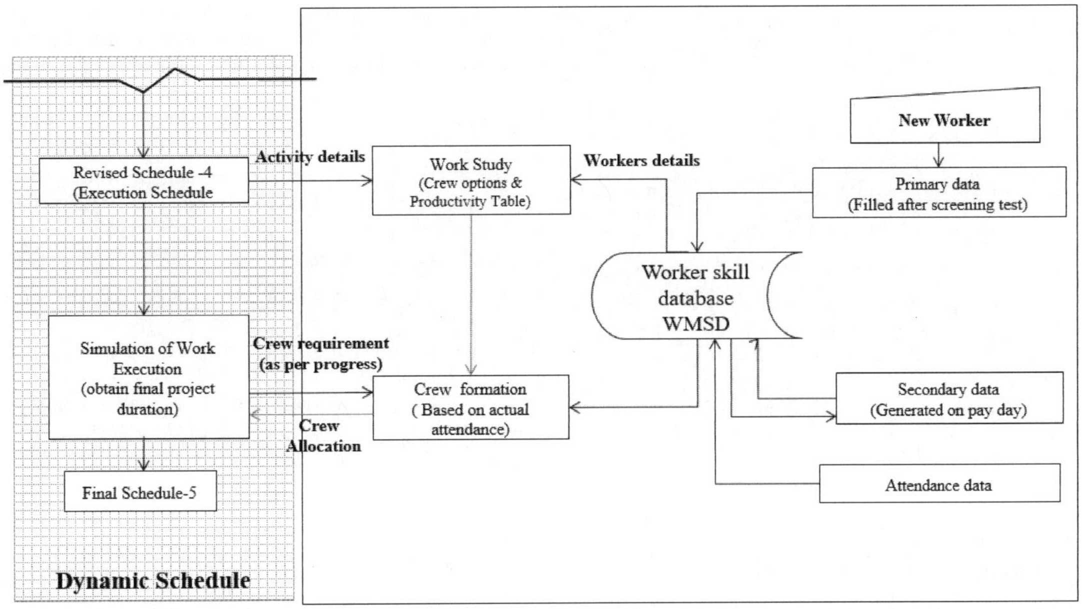

Figure 3. Flowchart for WMSD

This database can be linked to attendance module to obtain real time availability of workers and their respective skill levels as seen in Figure 3. This database can be also further extended to multi-project environment by linking the workers profile with their social identification number and managed through a third party as seen in figure 3.

4.3.1 Crew Productivity Matrix

The actual productivity of a crew for specific activity can be measured using work study [19]. The absence of a member will have adverse effect on crew productivity because rate of activity execution is a function of task complexity, desired quality in task, and skill level of workers, etc. Although it may be feasible for the crew to progress the work with less number of workers, the productivity will be lower even if deficiency is compensated by substitution due to lack of co-ordination between existing member and new member. The loss in productivity will also depend on skill of the absent worker and skill level of substitute worker [12]. In addition there may be limiting condition such as minimum number of SK workers or maximum acceptable crew productivity loss without critically affecting quality, etc. [2]. To cater all these conditions, a productivity matrix can be generated for each activity using work study.

A hypothetical productivity matrix has been shown in Table 1 to explain the concept. The matrix has three parts – 1, 2 & 3. The combination of these parts can be used to revise crew productivity. The process has been explained for an activity which ideally requires 3x SK, 6 x SS/MS, 6xUS as given in Table 1.

Example: 2 x SK workers are absent.

Considering substitution rule (Part 1 of Table 1) work can be executed even if two skilled workers are absent because 1x SK worker from original crew is still available and maximum productivity loss is only 20%. Following substitution options are feasible when 2 xSK workers are absent.

1. Case 1: If no substitution, crew productivity loss will be 20% (see Part 2 of Table 1).
2. Case 2: If 2xSK workers are substituted from reserve pool - productivity gain will be 10% and crew productivity loss will reduce to 10% (i.e 20%-10%).
3. Case 3- If void is filled by SS/MS workers from reserve pool of workers, the productivity loss will be;

- 10% (i.e 20% - 4x2.5%) if 4 Nos of SS/MS are used as substitute;
- 12.5% (i.e 20% - 3x2.5%) if 3 Nos of SS/MS are used as substitute;
- 15% (i.e 20% - 2x2.5%) if 2 Nos of SS/MS are used as substitute;
- 17.5% (i.e 20% - 1x2.5%) if 1 Nos of SS/MS are used as substitute

4.3.2 Execution Schedule

As shown in Box 1 of figure 1, process includes finding scope of work for each activity utilising work breakdown structure (WBS), ideal crew composition for these activities utilising work study and subsequent series of schedules to find execution schedule. The initial schedule will provide project duration (T_s) and workers requirement considering nil workers deficiency.

4.3.3 Simulation of Project Execution

As shown in Box 2 of figure 1, the exported report has three Excel worksheets, namely Task Table (includes activity details), Resource Table (includes resource details) and Assignment Table (includes details on resource allocation to activities). From Assignment Table trade wise resource requirement for all activities can be summed up day wise.

Table 1: Productivity Matrix

Part 1: Substitution Rules						
1. Minimum 1 SK to be present in a crew						
2. Crew not employed if Productivity loss > 33%						
Part 2: Productivity loss due to absence in %						
Worker	Ideal Nos	Min	Absence			
			-1	-2	-3	-4
SK	3	1	-10	-20	NA	NA
SS/MS	6	3	-5	-10	-15	NA
US	6	5	-5	NA	NA	NA
Part 3: Productivity gain due to substitution						
Worker	Ideal Nos	Max Subtn	Substitution			
			1	2	3	4
SK	3	1	5	NA	NA	NA
SS/MS	6	4	2.5	5	7.5	10
US	6	4	1.25	2.5	3.75	5
Note: The productivity loss and gain to be found after work study.						

For given % of absenteeism, likely absence of specific workers can be simulated. Assuming availability of adequate reserve, modified productivity of affected crew using productivity matrix (Table 1) and revised project duration can be calculated. There are two possible outcomes after considering productivity loss due to absenteeism;

1. Likely project duration considering absenteeism is more than available time ($T_n > T_d$). In this case duration of activities on critical path would require reduction through method study and identification of ideal crew combination (see arrow from box 2 to box 1). The revised crew composition may not give best individual worker productivity but revision will be unavoidable to meet project dead line (T_d). The simulation will have to be attempted again till likely project duration (T_n) is less than time available (T_d).
2. Likely duration is less than available duration ($T_n < T_d$). In this situation, attempt can be made to find project duration based on available workers (see arrow from box 4 to box 3, figure 1). There could be two possibilities;

- Calculated project duration is more than available time ($T_f > T_d$);
  - In case additional workers can be arranged, steps given in box 3 to be repeated.
  - If additional workers are not available, schedule can be improved through method improvement and flow process design (Box 1, figure 1).
  - In case it is not possible to meet the project deadline despite method revision, the project delay would have to be accepted.
  - If calculated duration $T_f < T_d$, the schedule can be used for execution of workers utilising available workers.

5 Methodology

5.1 Execution schedule

Scheduling based on logic constraints is simple and adequate for monitoring by executives at project level. However, project manager has to face daily challenge of managing crucial resource. Thus resource constrained project scheduling (RCPSP) become a necessity. RCPSP requires crew composition for each activity and corresponding activity duration. Activity duration and ideal crew composition can be obtained using WBS and work study. In traditional RCPSP problems, the resource allocated for any activity is assumed to be uninterrupted during the execution and activity execution is considered by a unique crew taking fixed duration. In case of MRCPS multiple modes are considered for execution of activities. This makes scheduling very complex without any surety that mode selected through MRCPSP will be available during execution.

To overcome the challenge two pronged strategy needs to be adopted;

- Find all possible crew combination for execution of each activities and likely rate of execution in the form of productivity matrix (Table 1) using work study.

- Identify important activities {i.e activities on the critical path, activities having completion deadlines, and activities having high priority (not on critical path but engaging high rental equipment)} and allocate resources on priority before attempting other activities

The proposed steps to find execution schedule are as follows (Box 1 of figure 1);

- Identify ideal crew composition for work packages or activities identified from WBS.
- Schedule the project assuming that desired crews can be arranged whenever required using MSP (Micro Soft Project software).
- Find scheduled project duration t_s, required number of ideal crews, crew employment period and requirement of workers on each day.
- Compare desired duration (t_d) with scheduled project duration (t_s).
  - If $t_s > t_d$: It infers that despite abundance of workers, it is not possible to execute the work in time. Hence, need to improve technology to change the logic constraint and/or improve the crew productivity (i.e method change) to shorten the critical path.
  - If $t_s < t_d$: It infers that execution is feasible within available duration when workers are in abundance but may not be feasible under resource constrain and productivity loss due to absenteeism etc. Hence there is need for simulation of productivity loss due to short term deficiency and finding requirement of reserve pool of workers to limit time cost overrun.

5.2 Simulation of Project Execution

The schedule can be further improved by simulating workers absenteeism to find reserve pool of workers and subsequent simulation utilising available workers less reserve to cater likely absenteeism etc. The information about availability of workers can be obtained from dynamically updated WMSD. The steps are as follows;

- On a selected date, workers likely to be absent can be identified using simulation for given % of absenteeism or industry average of absenteeism over duration t_d.
- On day one (t=1) identify activities in progress and affected by absent individuals. Modify their productivity on t=1 assuming that absence met from reserve pool of workers (see section 5.2.2). Critical and important activities to be given priority.
- Calculate quantum of work executed by modified crew in affected activities on t=1 and calculate additional time required to complete those activities assuming no absence afterwards.
- Modify activity duration of all affected activities at the end of day t=1 by finding % of work balance after t=1 and find duration of remaining part of the activities, assuming that all workers are present thereafter.
- On t=2, use modified activity duration to reschedule project using any network scheduling software (i.e MSP etc).
- Through series of schedule updates, find project duration (T_n) when no balance work is left. This will be project duration with abundance of workers and productivity loss due to absenteeism.
- If $T_n > T_d$: It infers that absenteeism will delay the project despite availability of ideal crew & requisite reserves. Hence, there will be a need to improve technology to change the logic constraint and/or improve the crew productivity (i.e method change) to shorten the critical path.
- If $T_n < T_d$: It infers that work can be completed with ideal crew combination and in presence adequate reserve.
  - Available total buffer is equal to T_d-T_n days.
  - As per the steps given in Box 3, project duration can be found considering available worker from WMSD less required reserve.
  - Find realistic project duration (T_f).
  - If $T_f > T_d$
    - Arrange additional resources and follow steps given in Box 3, or
    - Change activity duration through method improvement and repeat steps given in Box 1, 2 and 3 of figure 1 in a sequence.

6 Discussion

Categorisation of workers based on their education, proficiency in work process, experience and existing norms in Indian proposed (see section 3) will yield desired result when the WMSD is built over a period of time. The initial category will evolve with progress of time when workers employment duration, productivity and sincerity (i.e., less absence, less sick leaves, etc.) are mapped (see section 4.3). The productivity matrix proposed in section 5 will require rigorous investigation using work study tools and form the first set of standardisation, presently lacking in construction industry. The recognition of crew productivity loss due to sudden change in crew composition will motivate various stake holders in evolving suitable approach to limit the short term absence. Concept of creating reserve pool of workers by simulating absenteeism will guide contractors regarding number of workers to be kept in their permanent roster. This reserve can be utilised by main contractor to assist subcontractors when they face short term deficiency. The realistic schedule will result in lesser variation in activity durations and higher productivity as well as limited time-cost overrun. Once automated, the final schedule will facilitate project managers to select best possible crew composition for activity in progress on selected date considering the work progress up to previous day. The awareness about likely completion date on any day will facilitate all stakeholders to take appropriate measures.

7 Limitations

The proposed framework is being applied on actual project data to validate concept and hence minor changes in the framework cannot be ruled out. The concept focused on short term deficiency of skilled worker only. In addition, assumption given in section 4.1 may itself become the first set of challenge. Since, entire process is complex, the automation with different modules catering different types of project will be essential in creating of robust WMSD. Although it may be easy to create database for workers engaged in one project yet database involving input from multiple subcontractors located in different geographical locations would be a challenge but essential to get real benefit.

8 References

[1] Dobbs, R., Madgavkar, A., Barton, D., Labaye, E., Manyika, J., Roxburgh, C., Lund, S., &Madhav, S. (2012). "The world at work: Jobs, pay, and skills for 3.5 billion people". McKinsey Global Institute Report.

[2] Tenth Five Year Plan (2002-2007), Construction, Planning Commission, New Delhi, India, Govt. of India, 2 (2002) 847-885.

[3] R. C. Burleson, C. T. Haas, R. L. Tucker and A. Stanley, Multiskilled labor utilization strategies in construction, Journal of Construction Engineering and Management, ASCE, 124 (1998) 480–489.

[4] Construction Industry Institute (CII), An analysis of multiskilled labor strategy in construction, Multiskilled craft capabilities in construction research team, Research summary, CII, 137 (1) (1998) 01–30.

[5] Wells, Jill. The Construction Industry in the Twenty-first Century: Its Image, Employment Prospects and Skill Requirements. International Labour Organization, 2001.

[6] Wood, Stephen. "Redundancy and female employment." The Sociological Review 29.4 (1981): 649-683.

[7] Baum, Tom. "Skills and training for the hospitality sector: a review of issues."Journal of Vocational Education and Training 54.3 (2002): 343-364.

[8] Greenwood, Adriana Mata. Updating the International Standard Classification of Occupations, ISCO-08. UN, 2004.

[9] L. Florez, D. Castro and A. Medaglia, Maximizing labor stability as a sustainability performance indicator in project scheduling, Construction Research Congress 2012, ASCE, (2012) 465–474

[10] Saboia, João. (1997. "Contract labour in the Brazilian construction industry", in Contract labour: Looking at issues (Geneva, ILO), Labour Education, 1997/1-2, No. 106/107.

[11] Vaid, K.N. (1999). "Contract worker in the construction industry in India", in D.P.A. Naidu (ed.): Contract worker in South Asia. Geneva, ILO, Bureau for Workers' Activities.

[12] T. Hegazy, A. K. Shabeeb, E. Elbeltagi and T. Cheema, Algorithm for scheduling with multiskilled constrained resources, Journal of construction Engineering and management, ASCE, 126 (6) (Nov. – Dec.) (2000) 414 – 421.

[13] P. Brucker, A. Drexel, R. Mohring, K. Neumann and E. Pesch, Resource-constrained project scheduling: notation, classification, models and methods, European Journal of Operations Research, Science Direct, 112 (1999) 3–41.

[14] S. Hartmann and R. Kolisch, Experimental evaluation of state-of-the-art heuristics for resource constrained project scheduling problem, European Journal of Operational Research, Science Direct, 127 (2001), 394–407.

[15] Y. Kadrou, and N. M. Najid, Tabu search for the multi-mode resource constrained project scheduling problem with resource flexibility, Proceedings of the 19th International Conference on Production Research, Valparaiso, Chile, July 29 – 31, 2007.

[16] R. C. Burleson, C. T. Haas, R. L. Tucker and A. Stanley, "Multiskilled labour utilization strategies in construction", ASCE Journal of Construction Engineering and Management, 124, 480-489, 1998. Burleson et al. 1998

[17] Clough R. H.; Sears, G. (1994). "Construction contracting" 6th Ed., John Wiley and Sons, New York.

[18] Gomar, J. E.; Hass, C. T.; Morton D. P. (2002). "Assignment and allocation optimization of partially multiskilled workforce" Journal of construction engineering and management, ASCE, 128(2), 103-109.

[19] Drewin, F., (1982). "Construction Productivity: Measurement and Improvement Through Work Study".

Sharing roles and responsibilities of construction project delivery methods in Taiwan

Hsiang-Wen Li 1[a] and Tsung-Chieh Tsai 2[b]

[a] Doctorate Candidate, Graduate School of Engineering Science and Technology, National Yunlin University of Science &Technology, Taiwan
[b] Associate Professor, Department of Construction Engineering, National Yunlin University of Science &Technology, Taiwan
E-mail: hedy416@gmail.com, tsaitc@yuntech.edu.tw

Abstract –

This study explains the procurement of construction projects for public and private in Taiwan firstly. And introduce a complicate project delivery system of PCM (Professional Construction Management) which combine pure CM and Design-Build and defined by Government Procurement Act of Taiwan. This project delivery method widely used for large engineering projects and clients who have little experience of construction projects. Then the relationship of the participations was used to illustrate the sharing roles and responsibilities of project delivery methods in Taiwan. The framework of management in execution level is examined completely in case study. Finally, the deficiencies of this system will be discussed in conclusion for the reference to improvement of other systems.

Keywords –

delivery method; PCM; procurement; roles and responsibility

1 Introduction

This study explains the procurement of construction projects for public and private in Taiwan firstly. And introduce a complicate project delivery system of PCM (Professional Construction Management) which combine pure CM and Design-Build and defined by Government Procurement Act of Taiwan. This project delivery method widely used for large engineering projects and clients who have little experience of construction projects. Then the relationship of the participations was used to illustrate the sharing roles and responsibilities of project delivery methods in Taiwan. The framework of management in execution level is examined completely in case study. Finally, the deficiencies of this system will be discussed in conclusion for the reference to improvement of other systems.

In Taiwan, the governmental agencies that are responsible for making policies and systems to the construction industry can be roughly classified into two groups. The first group plays as owners, who perform as the clients for asking engineering professionals to conduct planning, design, and supervision jobs, constructors to construct facilities, or professional construction management (PCM) to act as a technical substitute or surrogate of owners in ascertaining their interests during the project delivery process. The ministries of Transportation and Communication, Economic Affairs, Education belong to this group. Every ministry may have specified necessary technical codes (such as specifications of materials and scale magnitude, composition of materials, or tested methods and criteria) for the construction companies to deliver the facilities they need.

The next group plays as the policy maker, or designer and regulator of the systems for the construction industry. The Public Construction Committee (PCC) of Executive Yuan and the Construction and Planning Agency (CPA), Ministry of Interior fall into this group. PCC is primarily in chare of the regulations of engineering professionals (including professional engineers, professional engineer firms, and engineering consultant companies), Government Procurement Law, and Promotion of Private Participation. The regulations of constructors and architects, and buildings are conducted by CPA.

2 Budget introduce process of Public project

On the national budgeting basis, PCC will review all the proposals of public investment of construction facilities, either buildings or civil works, by checking the necessity and budget amount of the required facilities. As shown in Figure1, basically it is very similar to most of projects. Feasibility study is necessary in plan stage, especially economic efficiency is enough to project is going or not going. The fundamental design and detail design should have complete information to revise the preliminary budget of plan stage, and to set up the contract budget. One another important job in design stage is to get the construction permit by regulation. The major issue in construction stage is to make sure every activity in site is fully the requirement of contract document. Finally, the supervisor should assist the client to inspect and survey the project is acceptable, for client can complete the project.

Basically, in public procurement, there are three ways to issue the bid.

1) Open tendering procedures: the procedures under which a public notice is given to invite all interested suppliers to submit their tenders.

2) Selective tendering procedures: the procedures under which a public notice is given to invite all interested suppliers to submit their qualification documents for pre-qualification evaluation basing upon specific qualification requirements and, after such evaluation, the qualified suppliers are invited to tender.

3) Limited tendering procedures: the procedures under which, where no public notice is given, two or more suppliers are invited to compete or only one supplier is invited for tendering.

And, there is little different to select the contractor in Public Procurement and Private Procurement.

The principles of awarding contract in Public Procurement are as bellows.

1) The Lowest Bid with government estimate: the lowest tender within the government estimate.

2) The Lowest Bid without government estimate: the lowest tender within the budget amount.

3) The most advantageous tender: may or may not consider tendering price when evaluating, tenders are allowed to submit tenders for construction work, property, and services with different qualities.

4) Multiple award: awarded to different renderers by different items or different quantities

The lowest bid is the major type of awarding contract.

Normally there is only one type of awarding contract in Private Procurement. The tendering procedure is invitation bid in generally, and principle of awarding contract is the most advantageous tender for client.

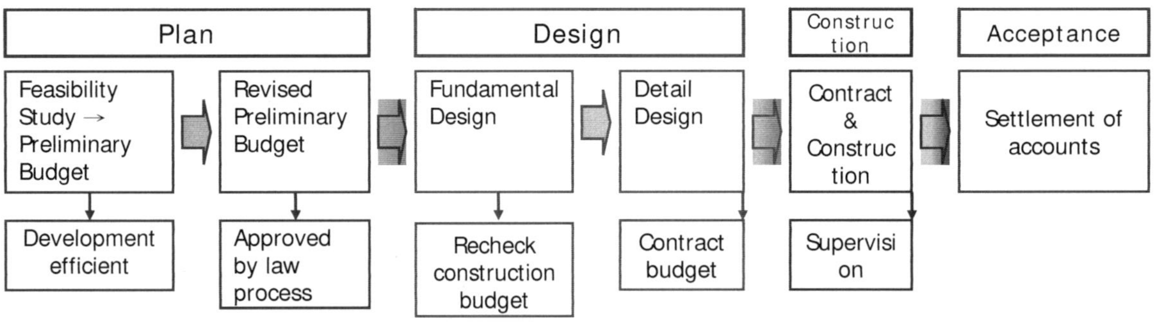

Figure 1. Budget introduce process of Public project

3 Sharing roles and responsibility in PCM

The major construction companies in Taiwan include: architect firm, professional engineer (PE) firm, engineering consultant company, construction company (contractor), PCM firm, and developer company, as listed in Figure 2. An architect firm shall have at least one registered architect. Several architects can jointly run an associate architect firm. If necessary, an architect firm can hire in-house professional engineers for conducting specified technical works such as structural engineering works. Otherwise, the structural engineering works have to subcontract to other PE firm or engineering consultant company. PE firm shall have at least one registered professional engineer, while engineering consultant company shall have many registered professional engineers with different professional expertise so that the engineering consultant company can provide services that are matched with the expertise areas of the professional engineers. PCM firm shall hire engineering professionals who have same qualifications as that required for architects or registered professional engineers.

Construction company (contractor) shall have at least one registered professional engineer in the areas of civil engineering, structural engineering, hydraulic engineering, survey engineering, geotechnical engineering, environmental engineering, or architect. In order to operate its business, a constructor company has to hire registered professional engineer who has practical experiences of more than 2 years. At construction sites, constructor companies must hire a chief site manager, who has related engineering background and taken a training course designed by CPA, is responsible for all the activities. Quality control engineer is also required at the construction sites to make sure the desired quality is fulfilled. The registered professional engineer of a constructor company is primarily responsible for the issues of technical and safety at the sites. All the professional engineers at the sites must follow the process requested by government, conduct tests on materials specified in construction codes, and fill out necessary documents for recording construction details. The main professional service is interface management, including the interfaces of cost, schedule, quality, etc. The main technical services may have building, civil work, mechanic, electric, pluming, elevator, lighting, security, landscape, Eco. Certification, etc.

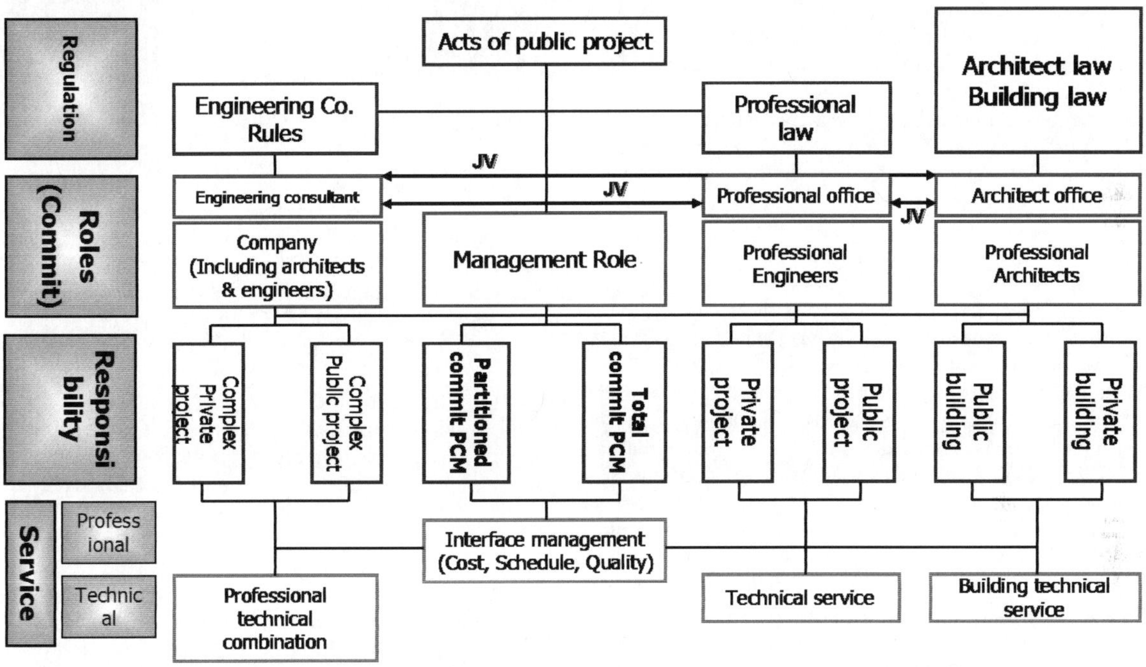

Figure 2. Sharing roles and responsibility in PCM

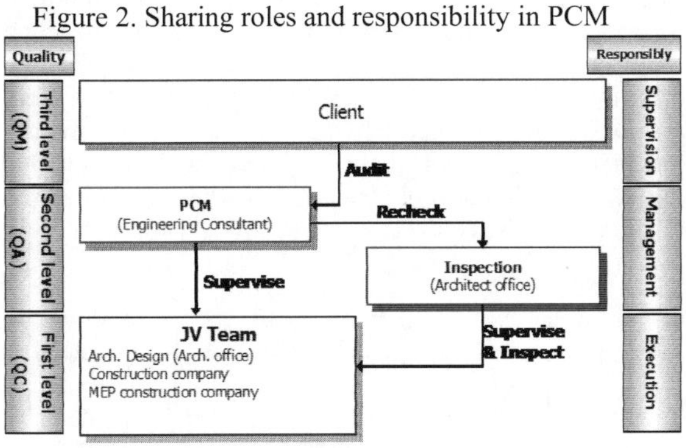

Figure 3. Quality management in PCM

For example the quality management in PCM delivery method is implementation of 3-level quality control systems aims at upgrading the public construction quality. Effective quality control system should be established for motivating, remedying, preventing the defects of the construction company's (contractor's) quality control. Three levels of public project quality control systems are established to enhance public construction quality control. The content of the establishment are all parties respectively as the authority, the client supervision party, PCM, architect inspection and JV team (design architect, construction company, and MEP construction company). Their construction quality controls are as shown in the Figure 3. The construction quality of public projects must be independently completed and guaranteed by the JV team, i.e., the JV team itself must be capable of processing the quality control system through self inspection, quality assurance, quality directing and quality audit. The target of contractual construction quality shall be completed under such a system.

For confirming the execution result of the construction quality management, the construction authority should apply construction quality evaluation through a very objective manner. The construction quality's degree of good or bad should be determined according to the appropriate quality evaluation standards. The result of the evaluation can be applied as basis for evaluating the authority and it also can be used as reference for improving the contractor's quality control operation and selection of excellent bidder. Moreover, by convincing the client (or supervisory party) and contractor's substantial practice of quality control, it is intended to achieve the target of upgrading construction quality. Project inspection operation should be established with "Construction Evaluation Team's Operation Regulations". The regulations were established in accordance with the Government's Procurement Laws Clause 70, Sub-clause 4 and "Project construction evaluation team" shall be stipulated according the laws. In their evaluation, the said team should process in compliance with the construction quality control system announced by the Public Construction Commission, Executive Yuan, relevant laws & regulations and the requirement of the contract conditions. Further, by referring to the construction evaluation operation reference standards in their evaluation on the items such as construction quality and progress etc. The major evaluating items to be performed by the Evaluation Team are as follows:

1) Quality directing mechanism of the authority, the record of the reviewed supervision plan, construction progress management measures, handling of construction defect and the defect improvement tracing.

2) Supervision organization of the supervisory party, the review procedure of construction plan & quality plan, the evaluation procedures and standards of material/equipment random inspection and construction evaluation, quality audit, document record management system etc from the supervision plan content and executed condition; the executed condition of defect remedy tracing and construction progress supervision etc.

3) Contractor's quality control organization, construction outlines, quality control standard, material & construction inspection procedure, self inspection checklist, control of unqualified product or work, remedy & prevention measures, internal quality audit, document record management system etc. in the quality plan content and execution condition; the execution condition of construction progress management, catch up plan, safety/hygiene & environmental protection measures etc.

4) Construction planning & designing, ecological environmental protection, material & equipment, significant defects of drawings & specifications, necessity of design change, whether the architect, contractor's professional engineer and quality control personnel perform their obligations in compliance with relevant laws & regulations and contractual requirement etc.

4 Case study

As shown in Figure 4, the case study is a complexion building with 11 blocks of public project. The site area has 77,667 M2, and the budget is 5.7 billion NTD. The JV (joint venture) team should complete geotechnical, structure (RC), finish for exterior wall and interior partition, basic mechanical/electrical/pluming, and facilities of site landscape.

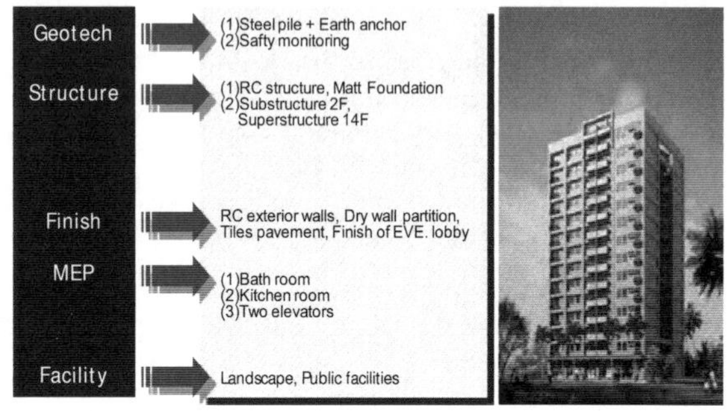

Figure 4. Introduction of case study

The schedule, shown as figure 5, from land acquirement, realize project requirement, boring investigation for earth condition, basic design for construction permit, detail design for coordination with electric/fire/communication/water/sewage system , approval for issue the contract to commence the project, inspection for project completion, and to handover is 1170 calendar days. The main project participations, shown as figure 6, are client, PCM, supervisor and contractor (JV team). As shown in Figure 2, this delivery method has very complicate relationship for all participations. But the major roles in the framework of management are PCM and supervisor. Because the PCM involve in the project on very early stage. And settle the management system for project which will be executed through the project. The framework of management is also developed by this management system. The supervisor plays as the role of inspection architect, whose rights and obligation are stipulated by Building Code and Architect Law, but not by Government Procurement Law. The conflicts of these regulation systems also make the defects of sharing roles and responsibilities of this delivery system. The framework of management and these two major roles will be discussed in detail to illustrate this delivery method. And, the deficiencies will be discussed in conclusions.

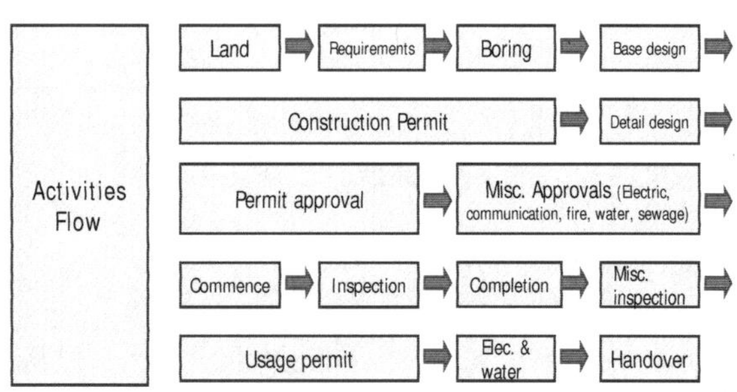

Figure 5. The main activities of case study

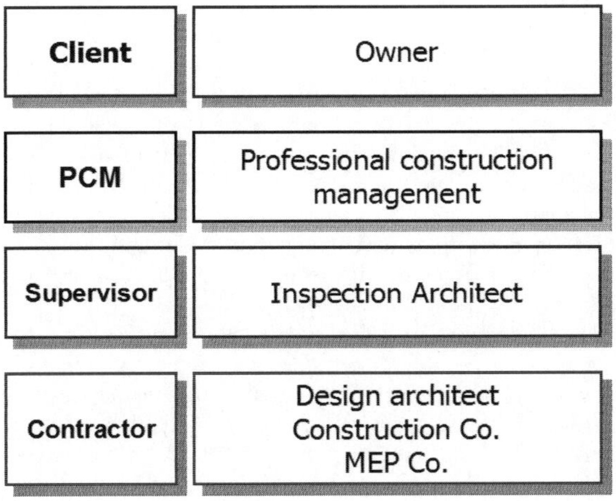

Figure 6. The main project participations of case study

4.1 The main responsibility of PCM

The main responsibilities of PCM in case study, shown as figure 7, are to get the charter from client, to settle the management system for project, to assist the client selecting contractor, to manage the construction of contractor, and to coordinate the building handover. From the view of time series, PCM should settle management system for project to review design drawings, specification & budget, inspection plan of supervisor, approval flow of material & equipment, etc. And coordinate with client to make the decision whether issue the contract or not. In contract stage, PCM should review the proposals from contractors, and decide which contractor is proper to client, then coordinate with client to sign the contract. A lot of activities which are supervision, approval, execution, review will be processed concurrently to reach the objectives of project. Which objectives are concerning with cost, schedule, quality, and safety. The most important responsibility of PCM in construction stage is to coordinate designer and contractor to produce the completion drawings, then to get the usage permit. Finally, in handover stage, PCM should coordinate other participations to handover the project and complete records for client.

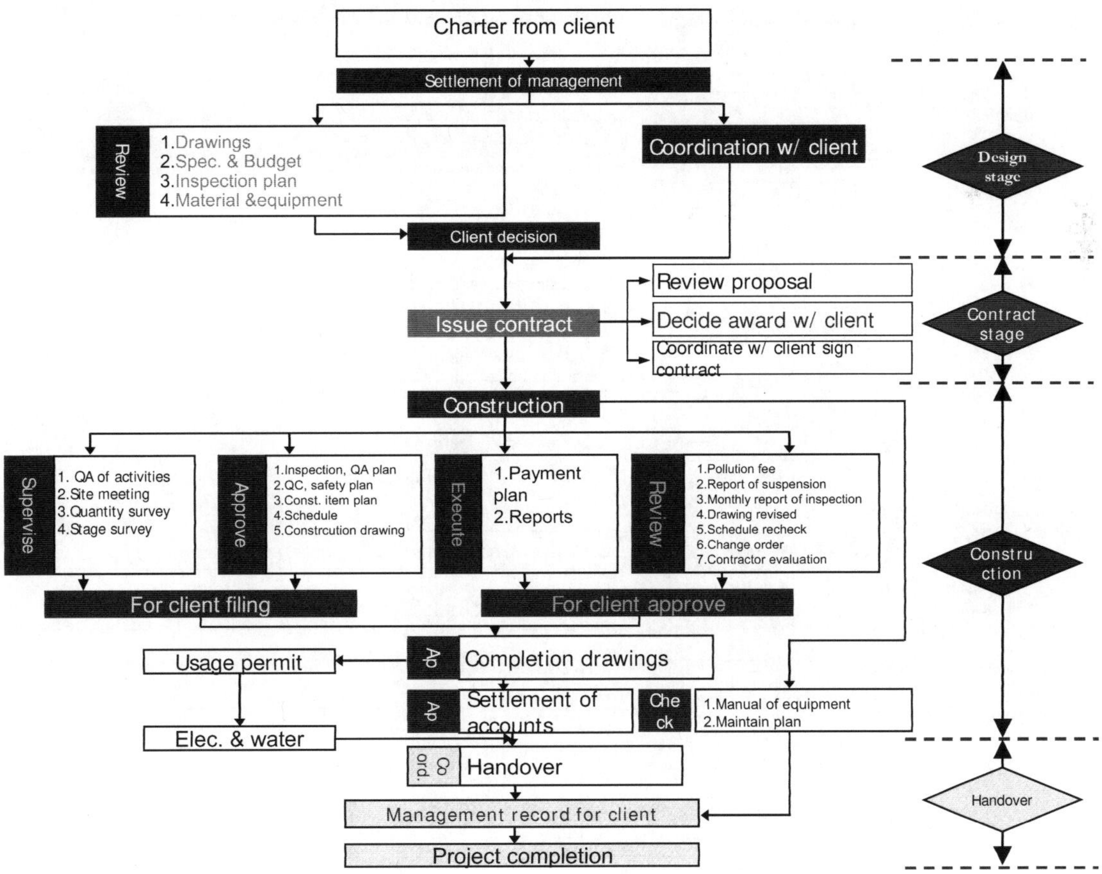

Figure 7. The main responsibilities of PCM

4.2 The main responsibilities of inspection architect

The main responsibilities of supervisor in case study as shown in Figure 8.

1) Providing inspection plan and bid document to explain quality assurance and information to estimation for contractors. And client can issue the bid and sign the contract in order to commence construction project and inspect construction activity.

2) Setting up the regular meetings which explain the details of project in order to build mechanism of project control through Q&A of specification and drawings, and coordination of interfaces among the contractors or subcontractors. In the same time, reviewing the construction method, construction plan, quality management plan, etc.

3) Structuring the document review system with other project participations on the principles of review then inspection, efficient and effective.

4) Making the conscience with client, PCM, user, contractor and inspector to realize the work standards for contractor, subcontractors, suppliers, specialists. And expedite the materials and equipment can be inspected then used in the job site.

5) Drawing out the major working items to discuss the construction method for the details of construction processes & checkpoints, quantified management for quality control and tracing & control base on concurrent management.

6) Inspecting the deficiency of activities and recoding on construction & inspection reports to require improvement.

7) Collecting the record document and rechecking the site condition to assist the client survey the real quantity of working items then asking client to sign the payment.

8) Constructing the working flow of change order to deal with conflicts between client and contractor rationally, completely and effectively. And making effort to tell the clear story of whole project for every party, in order to make the completion inspection smoothly.

9) Assisting the client to accept the project according to the specification and drawing, and handover to settle the account and make sure this project can be operated immediately.

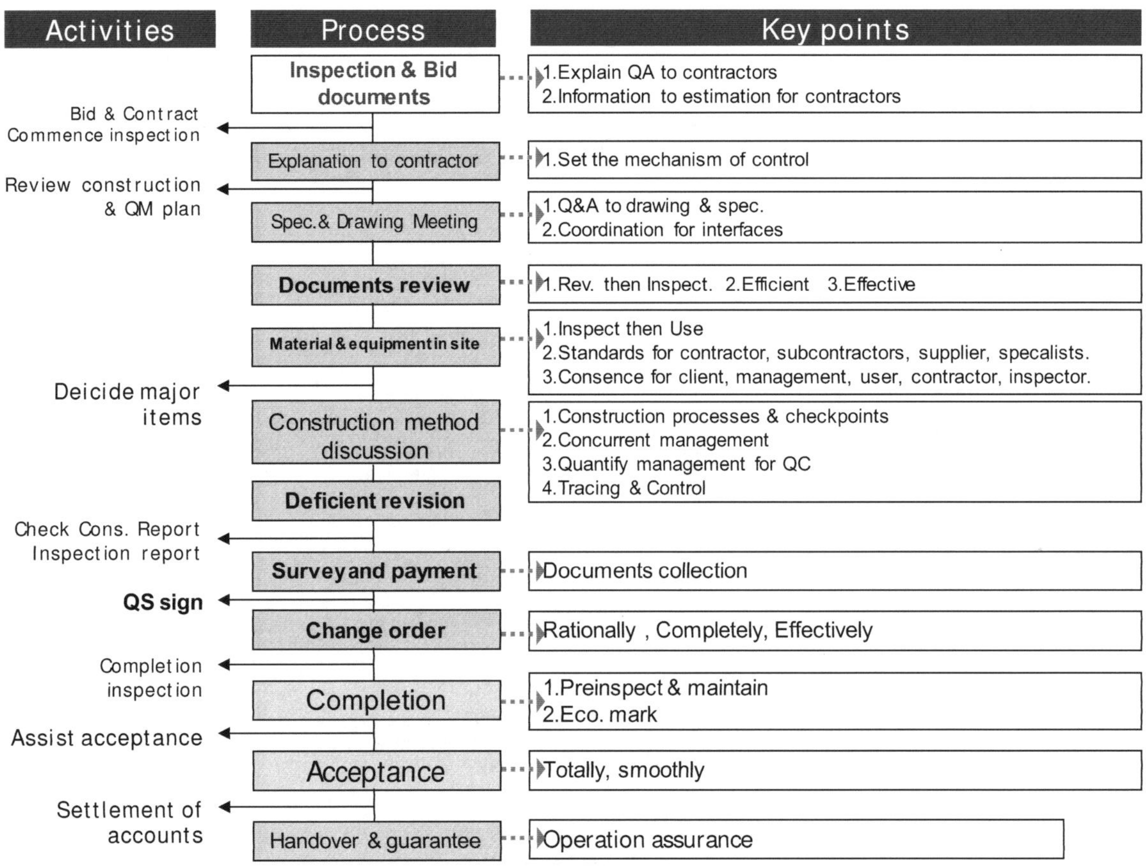

Figure 8. The main responsibilities of inspection architect

4.3 The framework of management

The framework of management for construction, shown as figure 9, basically is for quality assurance, quality control, surveying, and payment. The processes consist with review then inspect, inspect then use, construction inspect, and revise for every activity, material, equipment etc. used in site.

1) The main purposes of review then inspect are to make sure of the qualification of contractors, subcontractors, suppliers, etc., and review the construction drawings & plans. The tools may have review list, activities & QA process & list, weekly examination and monthly control. The frequency will be depended on the schedule of production and inspection of materials or equipment. Normally, the proposals or samples should be submitted two months before construction in site.

2) The main purpose of inspect then use is to realize the construction team, including contractors, subcontractors, suppliers and specialists. The tools may examine the ability of production, supply, construction method, etc. to build up the standards of inspection. The frequency will be different in working items, including civil works, architecture works, MEP works etc.

3) The main purpose of construction inspection is to do quality assurance of supervisor and contractor. The tools may have quantified management using spread sheets with quantity and weight division of working items, inspection mode whether complete or sampling, management mode from mobile to trust.

4) The main purposes of revise are to stipulate the time of improvement and to prevent the repeat deficiency. The tools may have control improvement from payment, engineer in chart for improvement, household control in finish stage and record deficiency & difference in handover for guarantee.

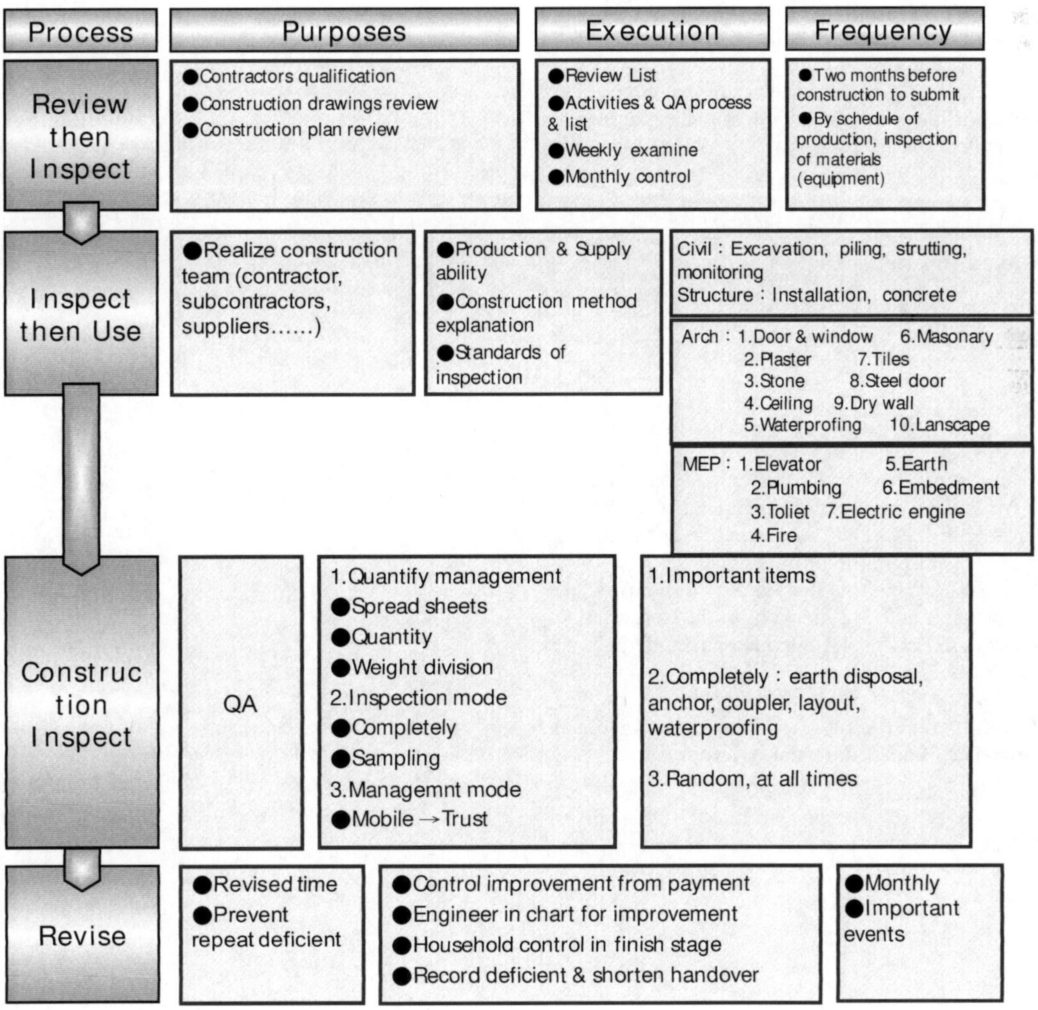

Figure 9. The framework of management for construction

5. Conclusions

As described before, the PCM project delivery method is widely used large engineering projects and clients who have little experience of construction projects. But the complicate and huge relationship of project participations and conflicts of professional law and Government Procurement Act make the confusion and ambiguous of sharing roles and responsibilities in this project delivery method. And this paper may mark some conclusions as below.

1) The combination of pure CM and Design-Build defined by Government Procurement Act of Taiwan.

2) The providing the three levels management, especially for quality management to large engineering projects and clients who have little experience of construction projects.

3) But, extending the management process of project and increasing the interfaces of roles/responsibilities of project participations.

4) PCM may positively take over the role of inspection & supervision, but on the contrary, PCM may reluctant to take the responsibilities because the role conflicts with architects.

5) PCM may only response the views of client, and neglect reasonable processes, professional judgments, dispute treatment, etc.

Acknowledgment

Thank Sun, W.Y. Architect & Assoc. provides the data of case study for this paper to analysis.

References

[1] Government Procurement Act. (1999) Public Construction Commission, Executive Yuan, Taiwan.
[2] Constructor Law. (2003) Construction and Planning Agency, Ministry of Interior, Taiwan.
[3] M. T. Wang (2008) A Study on the Systems of the Professionals and Organizations of the Construction Industry of Taiwan, 1st international conference of quality assurance, Kyoto.
[4] T.C. Tsai (2008) A Study on the Quality Assurance of Construction Project in Taiwan, 1st international conference of quality assurance, Kyoto.

Activity Based Prediction of On-site CO_2 Emissions Containing Uncertainty

Chulu Nam[a], Dongyoun Lee[a], Goune Kang[a], Hunhee Cho[a] and Kyung-In Kang[a]

[a]School of Civil, Environmental, and Architectural Engineering, Korea University, Republic of Korea
E-mail: glock9@korea.ac.kr, dy_lee@korea.ac.kr, gkang17@korea.ac.kr, hhcho@korea.ac.kr, kikang@korea.ac.kr

Abstract –

Recently, the importance of reducing embodied carbon has become clear. The construction stages, is the stage where the building production takes place, and a large quantity of embodied carbon is emitted. However, because this stage presents a variety of sources of uncertainty at building sites, it is difficult to compute and predict precise CO_2 emissions. To solve this problem, existing research has estimated emissions amounts by considering the variability of the main materials' carbon emission factor, as well as the variability of the equipment's activity conditions. However, these approaches are unable to reflect uncertainty at activity level, leading to an underestimation of CO_2 emissions. In this research, we perform an analysis by considering the uncertainty of CO_2 emissions in the construction stage at activity level. In addition, from the results, we recognize the relevance of considering uncertainty for each activity. Therefore, we present a CO_2 emission prediction method using a Monte Carlo simulation and confirm its effectiveness. We believe that the outcome of this research advocates for the necessity of considering the uncertainty in each activity and contributes to the prediction and management of on-site emissions.

Keywords –

CO_2; activity based; uncertainty; Monte Carlo simulation

1 Introduction

Recently, as operational carbon emissions (OC) have been reduced, it has gradually become clear that it is important to reduce the embodied carbon emissions (EC), which occur latently during the production, construction, maintenance and discarding of building materials [1]. The construction stage is a good example of the seriousness of EC. In this stage, 10 to 30% of the materials' lifetime CO_2 are emitted within a relatively short time [2]. In addition, the construction stage is the stage in which the usage of the building materials is completed, and thus, mechanical equipment and materials are used, causing a large amount of CO_2 emissions.

However, the CO_2 emissions that occur in the building construction stage are difficult to compute and predict accurately due to various sources of uncertainty that are present at a building site [3]. Choosing and managing goals of CO_2 emission reduction based on uncertain prediction values cause problems in meeting these goals [4]. To solve this problem, existing research has estimated emissions amounts by considering the variability of the main materials' carbon emission factors, as well as the variability of the equipment's activity conditions. However, these approaches are based on resources, not based on activity, so it is possible that the CO_2 emissions will be underestimated. Furthermore, another limitation is the suggestion of only piecemeal reduction methods such as a reduction or replacement of input resources.

To address those limitations, in this research, we aim to analyze the construction stage CO_2 emissions' uncertainty at activity level. We also aim to present a method that can use the analyzed uncertainty data to probabilistically predict CO_2 emissions.

We selected concrete construction as our research target, as it uses a large amount of material and has a high material carbon emission factor. For the uncertainty that can occur in concrete placement, measured equipment operating times of both the concrete pump and the concrete mixer truck were considered. These operating times were converted to CO_2 emissions, and the uncertainty was analyzed by comparing the actual value and the planned emissions based on the equipment operation time. Finally, the analyzed data was used to derive a method to predict CO_2 emissions, which is based on a Monte Carlo simulation, and their effects were verified.

2 Computing the construction stage CO_2

A deterministic emission calculation obtained by multiplying the CO_2 emission factor and the amount of input material can cause problems due to uncertainty. Therefore, existing research has evaluated the uncertainty when estimating carbon emissions and then presented a prediction method that considers this. We reviewed several researches in terms of materials and equipment.

Regarding materials, there is a research that considers the most used building materials to perform an analysis of statistical properties, and based on this, it probabilistically predicts the materials' construction stage emissions [4]. Another research analyzes the variability of emission factor of the most used materials based on accumulated greenhouse gas emissions, as well as energy emissions factor, in order to evaluate the uncertainty when estimating the emissions of apartment housing construction stage [5].

Regarding equipment, there is a research on a simulation technique, which is developed to calculate CO_2 emissions in a road paving work while considering the variability in the amount of fuel consumption according to engine load [6]. Similarly, another research uses a CYCLONE simulation to predict emissions probabilistically while considering the variability of fuel consumption according to the activity of earthwork equipment [7].

However, the above-mentioned research has the limitation of being mainly focused on materials and earthwork equipment. We perceived a lack of research on building construction equipment, which is a direct source of emissions at the building site. Also, we analyzed existing research is not considering uncertainty at activity level. Therefore, in this research, we evaluate uncertainty in CO_2 emissions focusing on building construction equipment at activity level, and then present a prediction method that considers our analysis.

3 Evaluating and analyzing uncertainty in CO_2 emissions

3.1 Measuring and analyzing equipment operation time

To analyze the uncertainty sources in carbon emissions, we estimated fuel consumption according to measured equipment operation time and planned equipment operation time to obtain CO_2 emissions, and then compared these values. To analyze at activity level, we measured on-site data about equipment operation times (Table 1). For the information of used equipment, we referred to the report from the Korea Specialty Contractors Association [8], which contains the standard measurements for construction machines in Korea(Table 2).

Table 1. Overview of measuring equipment operation time

Components	Contents
Facility type and scale	1 education research facility, 2 stories underground/6 stories aboveground
Total area	22,910 m^2
Location	Seongbuk-gu, Seoul, Korea
Measurement target process	Concrete slab placement work
Data gathering date	2017-3-10

Table 2. Information of used equipment

Used equipment	Concrete mixer truck	Concrete pump
Type	6 m^3	36 m
Activity amount	6 m^3	80-95 m^3/h
Fuel efficiency(l/hr)	13	17.7
Fuel type	Diesel	Diesel

We analyzed the measured equipment operation time in the following way. A single concrete mixer truck entering the site and leaving was set as one activity cycle. Our analysis did not include the start and end of the depositing work or the periods of time when the activity was suspended. In this research, we analyzed a total of 21 activity cycles. We then set the detailed activities for each piece of equipment. For the concrete mixer truck, we considered four activities; stopping after entering the site, waiting after stopping, pouring concrete, and leaving. For the concrete pump, only one activity was considered; pumping the concrete. Following this, we measured the equipment operation time for each activity, and obtained a total of 21 data sets with the five activities.

However, the planned equipment operation time is not calculated for the defined activity units; therefore, in this research, we used several processes to estimate the values. For concrete pump cars, the time spent to digest the concrete ($6m^3$) of a concrete mixer truck was calculated based on the average value of the pump amount per hour ($87.5m^3/hr$) which isprovided in [8]. In the case of the concrete mixer truck, we did not present any specific activity time information, and assumed the average equipment operation time to be the planned equipment operation time (Table 3).

3.2 Calculating emissions from equipment operation times

This section describes the process for calculating CO_2 emissions by estimating fuel consumption from equipment operation times. The fuel consumption for each activity was calculated via Equation (1). For the fuel efficiency, the equipment data presented in Table 2 was used. The load factor reflects the equipment's fuel consumption according to the activity. We used the load factors analyzed in [7], and the load factors for each activity were Low (25%) for stopping after entering and leaving, Idle (10%) for waiting after stopping, and Accelerated (100%) for pouring concrete and pumping concrete.

Fuel consumption (l) = Equipment operation time (s) × Fuel efficiency (l/hr) × Load factor (%) × 1/3600 (hr/s) (1)

The fuel consumption calculated in Equation (1) was used along with Equation (2) to calculate the CO_2 emissions for each activity. In this research, we used the data provided in [9]; the oil conversion factor is 0.000845, the fuel carbon emission factor is 0.837. Here, the constant of 44/12 is the ratio of CO_2's molecular weight to carbon's atomic weight, and the constant of 10^6 converts the emission unit from tons to grams.

CO_2 emissions (gCO_2) = Fuel consumption (l) × Oil conversion factor (toe/T) × Fuel carbon emission factor (T×C/toe) × 44/12 (Carbon conversion) × 10^6(g/T) (2)

3.3 Comparative analysis of emissions

Through the process above, emissions based on measured times and emissions based on planned times were calculated. The results for the calculated emissions are shown in Table 4. The emissions based on measured time are shown along with descriptive statistic values for the emissions of each activity, whereas the emissions based on planned time are shown as specific values.

From the data based on activity cycles, the calculated value present emissions based on planned time were around 3.0 $kgCO_2$ more than the minimum and were 4.4 $kgCO_2$ less than the maximum, respectively. In terms of error rate, the differences were -42.7% and 63.1% respectively. When we compare the average values for the emissions based on planned time and the emissions based on measured time, the error is around 5%, which is smaller than the difference between maximum and minimum.

If emissions are predicted as deterministic values, a relatively accurate calculation with less than 5% error is obtained from the average values of the equipment data and measured data. However, this approach does not create acceptable predictions. As can be seen from the comparative results in Table 4, this is because the emissions obtained from measurements present a broad range of deviation. This can cause problems in managing emissions related to underestimation or overestimation of the CO_2 emissions in the planning stage. Therefore, in order to make realistic emission predictions, we must consider the uncertainty of each activity.

4 Predicting CO_2 emissions using a Monte Carlo simulation

4.1 Monte Carlo Simulation

This section presents a method of predicting CO_2 emissions probabilistically by using a Monte Carlo simulation. The Monte Carlo simulation is one of the main techniques used in probabilistic analysis. It creates a probabilistic model of variables' uncertainty, and it

Table 3. Measured equipment operation time and planned equipment operation time

Activity	Stopping after entering	Waiting after stopping	Pouring concrete	Leaving	Pumping concrete	Cycle
Average measured equipment operation time (Standard deviation)	36.3 (27.3)	190.5 (243.2)	343.8 (95.5)	9.3 (3.5)	299.6 (83.2)	879.4 (279.35)
Planned equipment operation time	36.3	190.5	343.8	9.3	246.9	826.8

* The unit of all items is sec.

presents statistical results from simulated experiments [10]. These statistical results allow for more effective decision-making in situations where the effects of uncertainty are clear.

In this research, we used the Crystal Ball program, which is an effective tool to execute Monte Carlo simulations. Crystal Ball supports probabilistic model analysis and simulation tests of the data needed for a Monte Carlo simulation. A method for making predictions using Crystal Ball follows the next steps. First, use Crystal Ball to perform a goodness of fit test on the data for the emissions based on measured time. To take the uncertainty at activity level into account, perform the goodness of fit test for each activity. Next, use the analyzed test result values to construct a model of the probability distribution for each activity. In addition, analyze the correlations between emissions for each activity in order to perceive the relation between the different activities. Finally, perform simulation tests that reflect the correlations in the constructed probability model, and analyze the results.

4.2 Goodness of fit test

For the goodness of fit test on the emission data obtained from the measured times, we used the Kolmogorov-Smirnov (K-S) test. When we compared the K-S test to the Anderson-Darling test, which is also a goodness of fit test, the former showed a more sensitive tendency in the median area of the distribution than the tail areas [11]. In this research, the K-S test was used to focus on the general situation where the number of observations is larger than the specific situation where the equipment operation time is measured long.In the test, we considered 14 types of probability distributions supported by the Crystal Ball program.

The results of the test for each activity are shown in Table 5. Test values (D) were below the rejection value of 0.287 at a significance level of 0.05 when N was 21. Therefore, we accepted the null hypothesis of the K-S test, which means the emission data fit well in specific distributions.

Analysis showed that the distributions for stopping after entering, waiting after stopping, and leaving activities follow a log normal distribution, whereas pouring concrete follows a logistic distribution, and pumping concrete follows an extreme value distribution.

4.3 Correlation analysis

To consider the relationship between activities' emissions in the simulation test, we calculated their Pearson correlation factors. The emission correlation factors are shown in Table 6. Through the analysis, we found 4 significant relationships.

First, the activities with the highest correlation were concrete pouring and concrete pumping, with an emission correlation coefficient of 0.854, which is a relatively high value. This correlation is present because the concrete is being poured and pumped at the same time. Second, leaving and waiting after stopping had a significant correlation value of 0.617. This is caused by the delays in the concrete pouring activity and delays in moving vehicles due to crowding around the concrete pump. Stopping after entering and concrete pumping had a correlation of 0.387, while stopping after entering

Table 4. Calculation results for CO_2 emissions according to equipment operation time

Activity		Stopping after entering	Waiting after stopping	Pouring concrete	Leaving	Pumping concrete	Cycle
Emissions for measured equipment operation time	Average	85.1	178.4	3219.7	21.7	3819.7	7324.4
	Median	63.2	112.4	3118.4	18.7	3664.9	7220.7
	Standard Deviation	64.0	227.8	894.2	8.1	1061.4	1877.2
	Kurtosis	12.4	5.7	1.6	1.8	-0.2	0.4
	Skewness	3.3	2.5	1.0	1.4	0.4	0.6
	Min. Value	44.5	9.4	1947.9	14.0	1879.6	3984.3
	Max. Value	334.8	889.6	5637.5	44.5	5903.4	11333.1
Emissions for planned equipment operation time		85.1	178.4	3219.7	21.7	3442.6	6947.5

* The unit of all items except kurtosis and skewness is gCO_2.

and concrete pouring had a correlation of 0.339, which are similar values. The reasons for these correlations are that parking and concrete pouring can be slowed down by the skill of the mixer truck driver, which also influences pumping.

4.4 Simulation test results and analysis

The simulation test was performed considering the probability distribution models obtained via the goodness of fit tests and the calculated correlation coefficients. In the simulation test, the results were obtained probabilistically by generating random values that consider the correlations for the probability distribution of a specific activity. The upper and lower bounds for the emissions can be determined by adjusting the confidence level. In this research, we executed the simulation test 10,000 times at a 95% confidence level.

From the test results, Figure 1 shows the probabilities of the emissions that occur during a single activity cycle. The outcomes suggest that a log normal function is the most fitting probability distributions. An analysis of the results shows that the probability of emitting less CO_2 than the planned emissions amount is 46.2%, which means that the probability of obtaining more emissions than planned is higher. Therefore, we can see that for the planned emissions, the CO_2 emissions that occur during the activity are underestimated.

Table 5. Goodness of fit test results by activity

Goodness of fit test (Kolmogorov-Smirnov test)					
Activity	Stopping after entering	Waiting after stopping	Pouring concrete	Leaving	Pumping concrete
Fit distribution	Log normal	Log normal	Logistic	Log normal	Extreme value
D	0.108	0.108	0.137	0.158	0.076

* At a significance level of 0.05 when N = 21, the rejection value is 0.287

Table 6. Results of analysis of correlation between activities

Activity	Stopping after entering	Waiting after stopping	Pouring concrete	Leaving	Pumping concrete
Stopping after entering	1				
Waiting after stopping	-0.040	1			
Pouring concrete	0.339*	-0.079	1		
Leaving	0.284	0.617**	0.140	1	
Pumping concrete	0.387*	-0.298	0.854**	0.110	1

* Significant at 0.05 level of significance
** Significant at 0.01 level of significance

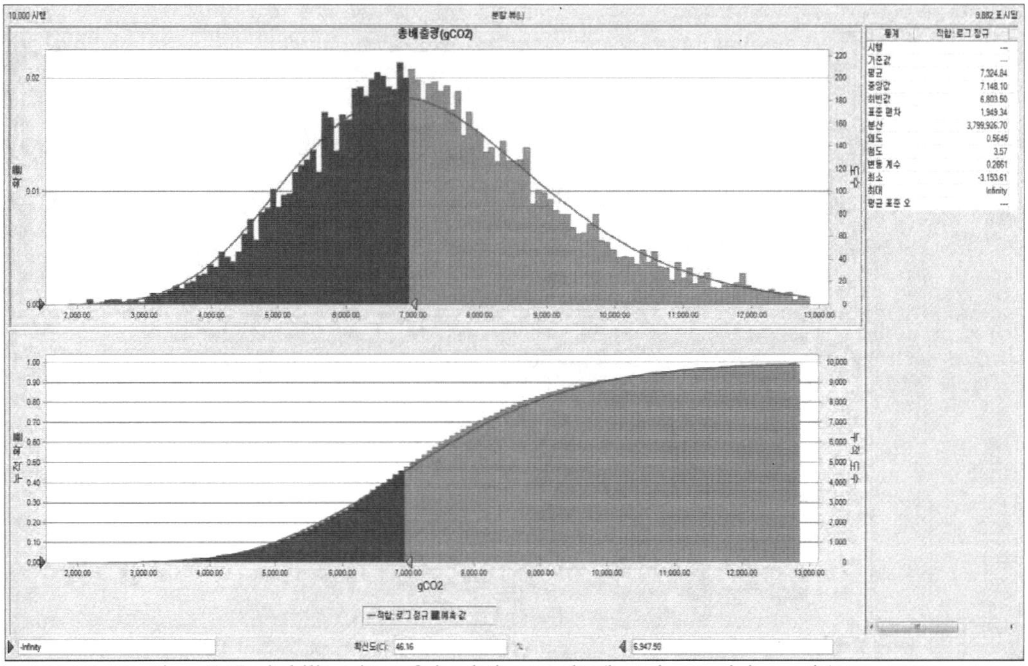

Figure1. Probability view of simulation results (based on activity cycle)

Based on an activity cycle (Table 7), the lower bound for emissions was 4039.8 gCO_2, and the upper bound was 11833.9 gCO_2 at a confidence level of 95%. For each value, when the emissions were compared to those obtained from the planned equipment operation time, there was a 70.3% to - 41.9% difference. From this, it is possible to predict with 95% of confidence that the actual emissions will be within 70.3% to - 41.9% of the planned emissions. In addition, the average value of the simulation was 5.5% higher than the planned value, at 7324.8 gCO_2. This gives a smaller error than the comparison with the bound values.

From the results by activity (Table 7), concrete pouring and concrete pumping have lower kurtosis and skewness than other activities so they follow a gradual distribution, but they present a high emission ratio. In both activities, the equipment is practically working, so the load factor is high and the activity time is long and it causes high amount of emissions. In addition, we think that the activity timespan is determined by the equipment that present relatively uniform work speeds; therefore, the distribution is relatively gradual.

On the other hand, in the case of stopping after entering, waiting after stopping, and leaving, the emission ratio is low but the kurtosis and skewness are both large. We consider that this result occurs because the activities load factor is low and the activity timespan is short. The reason for the kurtosis of these activities to be high is that they are relatively simple activities so they are often completed within a fixed amount of time. Sometimes, however, due to the delay in former activity and preparations for tasks, the activity timespan is extremely long. These extreme value causes an increase in the degree of skewness.

Based on this analysis, reducing activity deviation and improving activity efficiency of equipment can be good methods to both increase prediction accuracy and reduce emissions. However, in order to consider uncertainty effectively, the activities must be considered from a management perspective. Looking at waiting after stopping, the emissions show a difference of more than five times with respect to the average. In addition, the difference between the upper and lower bound amounts to 87 times. That is, as the activities are being performed, certain activities have extreme time increases that can cause an increase in emissions and prediction error. Therefore, we believe that to increase prediction accuracy and reduce emissions, such activities must be well identified and managed.

Table 6. Simulated test results

Activity		Stopping after entering	Waiting after stopping	Pouring concrete	Leaving	Pumping concrete	Activity cycle
Simulated test result emissions	Average	85.4	187.7	3699.1	22.5	3864.5	7325.8
	Median	64.2	99.5	3152.8	18.7	3663.0	7112.0
	Standard Deviation	75.7	286.1	866.5	13.4	1186.7	1973.8
	Kurtosis	155.0	65.8	4.3	142.7	6.0	0.8
	Skewness	9.0	5.79	0.2	7.7	1.3	5.1
	Lower**	45.0	10.9	1445.1	13.8	2158.7	4039.8
	Upper**	257.2	945.6	4940.0	55.0	6711.7	11833.9
Emissions based on planned equipment operation time		85.1	178.4	3219.7	21.7	3442.6	6947.5

* The unit of all items except kurtosis and skewness is gCO_2.

** 95% confidence level

5 Conclusion

In this research, to analyze the uncertainty in CO_2 emissions in the construction stage at activity level, we gathered data about equipment operation time by activity and used this to analyze the related uncertainty. The results were used to show the need for considering uncertainty at activity level, as well as to propose a CO_2 emission prediction method using a Monte Carlo simulation and verify its outcomes.

The conclusions obtained from this research are the following. First, we propose that in order to make a realistic prediction of CO_2 emissions in the building construction stage, it is important to consider uncertainty at activity level. Second, we obtained a realistic emission prediction by presenting a probabilistic method of predicting CO_2 emissions using activity data. Third, by analyzing emissions by activity, we presented a plan to increase the accuracy of emissions predictions and reduce emissions.

We believe that the results of this research will emphasize the need for considering uncertainty at activity level and assist in the prediction and management of on-site emissions. In the future, we will consider plans at activity level and gather additional data at activity level in order to increase the accuracy of predictions and expand the use of the proposed method.

Acknowledgement

This work was supported by the National Research Foundation of Korea grant funded by the Korea government(2016R1A2B3015348). The contribution of the Ministry of Science, ICT and Future Planning is gratefully acknowledged.

References

[1] Ibn-Mohammed T. Greenough R. Taylor S. Ozawa-Meida L. and Acquaye A. Operational vs. embodied emissions in buildings—A review of current trends. *Energy and Buildings*, 66:232-245, 2013.

[2] Daewoo E&C. 2014 *Daewoo E&C sustainability report*, volume 1. Daewoo E&C, 75, Saemunan-ro, Jongno-gu, Seoul, Republic of Korea, 2014.

[3] González V. and Echaveguren T. Exploring the environmental modeling of road construction operations using discrete-event simulation. *Automation in Construction*, 24:100-110, 2012.

[4] Monahan J. and Powell JC. An embodied carbon and energy analysis of modern methods of construction in housing: A case study using a lifecycle assessment framework. *Energy and Buildings*, 43(1):179-188, 2011.

[5] Kang G. Kim T. Kim YW. Cho H. and Kang KI. Statistical analysis of embodied carbon emission

for building construction. *Energy and Buildings*, 105:326-333, 2015.
[6] Zhang H. Simulation-based estimation of fuel consumption and emissions of asphalt paving operations. *Journal of Computing in Civil Engineering*, 29(2): 04014039, 2014.
[7] Yi CY. Gwak HS. and Lee DE. Stochastic carbon emission estimation method for construction operation. *Journal of Civil Engineering and Management*, 1-13, 2016.
[8] Korea Specialty Contractors Association. *Machine cost report of construction machinery 2017*, volume 1. Korea Specialty Contractors Association, 15, Boramae-ro 5-gil, Dongjak-gu, Seoul, Republic of Korea, 2017.
[9] IPCC. *2006 IPCC guidelines for national greenhouse gas inventories*, volume 2. IGES, 2108 –11, Kamiyamaguchi, Hayama, Kanagawa, Japan, 2006.
[10] Touran A. and Wiser EP. Monte Carlo technique with correlated random variables. *Journal of Construction Engineering and Management*, 118(2):258–272, 1992.
[11] Huynh VN. Nakamori Y. Lawry J. and Inuiguchi M. *Integrated uncertainty management and applications*, volume 68. Springer Science & Business Media, Heidelberger Platz 3, 14197, Berlin, Germany, 2010.

Effectiveness of On-Site Communication in Residential Housing Projects

M.L. Djajalaksana[a] and P.R. Zekavat[b] and S. Moon[c]

[a]Department of Construction Management, Curtin University, Australia
[b]Discipline of Building, Victoria University, Australia
[c]Department of Civil and Construction Engineering, Swinburne University of Technology, Australia
E-mail: maya.djaja@gmail.com, payam.rahnamayiezekavat@vu.edu.au, sungkon.moon@gmail.com

Abstract –

Ineffective communication at work-front is frequently reported to hinder the successful delivery of construction projects. However, the efforts to address this shortcoming mainly revolve around adopting corporate communication strategies at filed-level. But, temporary nature of projects inhibits implementation of strategies developed based on permanent communication arrangement of established organizations. To bridge this gap, this study aims at understanding how effective communication is defined by construction professionals who manage construction processes on-site. Such definition determines requirements of an ideal communication system on-site and accordingly provides a measure to identify inherent characteristics of construction processes that compromise robustness of work-front communication. Finally, the input from selected construction professionals will be analyzed to map existing solutions to identified communication barriers. The mapping is essential to a) modify existing solutions to better overcome the associated communication obstacle and b) craft new solutions to tackle overlooked communication flaws.

As residential projects are known for their relatively simpler structure, they have been selected as the starting point of the attempt to modify project communication based on the constraints imposed by construction process. Semi-structured interviews were carried out with 10 construction professionals selected through judgmental sampling. It means that the research team picked one representative per residential building project based on overview of construction site, available personnel, and the likelihood of sharing the information sought in the study. Careful content analysis resulted in extraction of similarities, differences and trends from interview manuscripts that were required to deliver research objectives. The results surfaced nine attributes of effective communication on jobsite. Interestingly, fitness of the communication medium for the purpose of transferred message ranked the most significant factor assuring integrity of the communication. Accordingly, our results show that although linguistics means is the most problematic communication medium on an active jobsite, dialogue within a crew is less vulnerable to failure. Lastly, analysis of data corpus revealed a trend towards creation of an informal communication protocol as a natural response to the lack of officially defined communication conduits.

Keywords –

Communication; Construction Site; Residential Housing

1 Introduction

Although the concept of communication has been studied for centuries, there is still prevailing confusion regarding the definition of communication [1]. Bubshait et al. [2] refer to communication as "a process by which parties assign and convey meaning in an attempt to create understanding." Communication as a process comprises of several key elements: senders, encoding, the message itself, a transmission channel/medium, decoding, receivers, noise and feedback [3]. Senders encodes (creates) the message that will be sent through transmission channel/medium (e.g. email or text) to the receivers. The receivers then decodes (translates) the message received from the senders. The communication process is usually completed with feedback that is sent back to the senders.

Communication is an essential element of project management as it helps to ensure smooth and successful delivery of the project [4]. The several aspects of communication that need to be considered are namely a) Project communication level [5], b) Lines of communication in project management [6], and c) Methods of communication [7].

Ineffective communication during project execution

causes detrimental problems in terms of cost, project success and health and safety [8]. If problem caused by communication arises in the beginning of the project and is not fixed immediately, the cost and opportunity to fix it will become higher and harder as the project progresses through its life cycle. Although the importance of communication has been acknowledged, many studies report that ineffective communication is still one of the top reasons contributing to the failure of construction projects [9]. There are studies such as [10] that highlights the negative consequences of poor communication in terms of increased number of disputes in residential building. Similarly, [11] discuss ineffective communication in residential projects resulting in cost and schedule overruns. But, In particular, the communication mechanisms in domestic construction have been overlooked. Accordingly, this study aims to have a closer look at the communication architecture at the work-front of residential projects and evaluate it effectiveness to enhance project performance. To achieve these targets, 10 appropriate projects were selected by the research team and semi-structured interviews with project representatives were carried out. Judgemental sampling was preferred over random selection of projects as the research team anticipated that randomly selected participants may not provide honest answers to the sensitive questions regarding their communication and information sharing strategies. In other words, judgmental sampling was deployed to avoid unnecessary bias associated with people's inherent reluctance to share sensitive information with strangers.

2 Towards Effective On-Site Commuincation

This section reports the results obtained from the interviews. Firstly, to understand the importance of effective communication in projects in the housing industry, participants' perspective on the impact of effective communication on project performance was examined. Collectively, 19 different reasons, classified in six categoeirs, were elaborated by the interviewees in support of the statement that effective communication is essential to project success. The most common category of reponse with six repetitions was the need for a mechanism to reliably convey correct and complete information among many stakeholders:

> Participant 6: "*there is also chain of communication or instructions that need to be passed on from say the project designer to the project manager to the one who actually builds it on site. Of course there is also external consultants that need to be involved and hence effective, efficient and very smooth communication is essential.*"

Other response categories highlighting that project prosperity rests on effective communication include:

- Communication is key to problem solving (frequency: 3).
- Conflict avoidance through enhanced communication (frequency: 3).
- Essential role of communication to foresee life-cycle requirements of the project (frequency: 3).
- Multi-agent nature of construction projects urges to improve communication (frequency: 3).
- Interruption to project is inevitable in case of communication breakdown (frequency: 1).

However, despite the stressed importance of communication, its reach, particularly on-site, is significantly tuned by the act of project management team. Accordingly, our participants were asked about their opinion on relationship between progress of on-site processes and degree of communication exhibited by the management team. In one way or another, all respondents mentioned that project management's communication skill consequently affects the "*chain of works*" at field. For example, Participant 2 explains how communication skills are implemented by the management team to fulfil client's expectations based on resources available at the work-front.

> Participant 2: "*It [site management's communication skills] will affect the success. So the way the project manager handles clients and workers so that the project can run is very important.*"

The results indicate that three attributes of project manager role, mainly determine the extent of the communication taking place at the construction process level. Those three are:

1. Being the focal point to funnel information flow in- and out-wards the site boundaries. In the context of residential construction where referral projects are vital, satisfied clients will result in more opportunities. Therefore, in capacity of a mediator, project managers are to reflect client's message on site activities while reporting field constraints back to the client.
2. Offering a range of skills solidified by effective communication: Knowledge of and work experience in residential sector provide the foundation to the project managers to justify what they say.

3. Leadership provision and influencing project personnel by means of communication: Given their higher position in the project hierarchy, project managers' strategy in regard to utilisation of existing communication capacity determines the accessibility to the critical project information. It goes without saying that in case of smaller scale residential projects, the responsibility to keep workforce motivated rest solely on the shoulders of the project manager.

2.1 Elements of Effcetive Communication in Residential Construction

Careful analysis of participants' insight in respect to attributes of effective communication resulted in nine pillar attributes listed in Table 1. The criterion to consider an element impacts quality of communication was the requirement to be mentioned by at least two of the participants. It is seen from Table 1 that there often exist competing forces such as completeness versus succinctness which eventually hinder effective communication. In other words, an effort to assure a message is complete may unintentionally results in less succinct statement. Another example is difficulty in utilizing the full capacity of existing communication routes if they are partially reserved to deliver frequently updated information. Therefore, depending on the occasion, elements of effective communications should be prioritized in an order that warrants the highest performance of the construction process in terms of productivity, quality and safety.

Table 1. Attributes of effective communication for residential projects

Attribute	Frequency	Remarks
Communication medium fits the designated purpose	7	a) If a solution made on site, write it down and sign off b) Samples are provided for visual interaction c) Meet your client directly rather than sending email d) Table meeting are the most effective tool for problem solving
Clarity of message	4	a) Instructions need to be clear b) Unclear message creates confusion and delays work
Predefined routes	4	a) Structured chain of communication prohibits shortcuts b) Exclusive points of contact for urgent matters
Customized to the recipient	3	a) Deploy a style understandable to your target person b) Instructions need to be objective in a sense that everybody knows what it is saying
Completeness of the message	2	a) Drawings need to be complete
Succinctness of the message	2	a) Use less amount of words that explain the most amount of meaning b) Be to the point
Openness of the communicators	2	a) Be open to opposite opinions b) Be open to question if you've understood properly
Regularly updated message	2	a) Constantly report progress of project to every stakeholder b) Update everyone about the problems encountered on site
Responsive two-way interaction	2	a) Two-way means there is a response b) Feedback might be collected through other means such as monitoring requests for information or defects trend

2.2 Tackling Communication Deficiency in Housing Construction

To better understand the root causes of communication deficiency on a jobsite, the research team provided participants the deconstructed communication model presented in Table 2. Each respondent was then asked to rank the most problematic component of each construct. Subjects highlighted in Table 2 show the items determined by the interviewees to be the dominant form of communication failures in their daily job. Interestingly, problems with informal transactions demonstrate the highly fragmented communication environment of smaller building projects. In fact, the tendency to deploy informal avenues even while formal arrangements are in place has been mentioned as main concern area in residential projects:

> Participant 4: "*our client in this instance, really like to visit the project site every day without telling us. In my opinion, if they hire us or any other agencies to manage their project, they should have trust us to manage the project and they should not visit the site to give instruction to the workers themselves.*"

On the other hand, associated with less organisational development of builders involved in residential housing, management of external affairs surfaced to be more challenging compared to handling of site issues. The fundamental differences between clients of housing projects compared to commercial or civil construction were expressed to be the root cause of failed external communications. In particular, clients' lack of understanding about construction was stated to hinder effective communication in many occasions.

Another key finding discussed in this section is the struggle of housing projects to adopt the most basic form of communication i.e. spoken language. This observation, however, doesn't mean that residential builders have mastered use of other communication media but it implies the fact that implementation of other methods is less vulnerable to error in the context of residential construction. In case of drawings, for example, the complexity of design is far less complicated compared to high-rise buildings which results in accurate interpretation of drawings when reviewed on-site.

Table 2. Bottlenecks of on-site communication

Construct	1st order Metric	2nd order Metric	Example
Line of Communication	Formal	-	✓ Method statement
	Informal	-	✓ Casual quotes for the best price
Communication Reach	Internal	-	✓ Negotiating sequence of work among trades
	External	-	✓ Change orders from the client ✓ Acquiring legal permits
Medium	Linguistic	**Spoken**	✓ Oral communication
		Non-Spoken	✓ Writing progress report
	Non-Linguistic	Spoken	✓ Altering tone to express dissatisfaction ✓ Whistling to grab attention ✓ Laughter
		Non-Spoken	✓ Drawings ✓ Pointing ✓ Body language

Definitions:
Formal: Structured channels officially enforced by the management
Casual: Unstructured personal links to selected partners
Internal: Matter of interacting with people present on jobsite
External: Sourcing information beyond site boundaries
Linguistic: Formation of statement using words and grammar consolidated under one particular language
Spoken: Interacting by creating voice uttered through the mouth

2.2.1 Barriers to Effective Communication

The research team were able to categorise seven most prevalent impediments to effective communication in a domestic construction setting into three main classes of a) Physical interruption, b) Language barrier, and c) Self-interest. The disruptive architecture of intra-project communication was notably the most detrimental aspect of physical barriers by counting for 27% of cases where interviewees mentioned about barriers to effective communication. Collected data suggests that a robust communication platform is an overarching one that is centrally developed and managed. An established system with explicitly defined links enables multiple users to draft their message in a standard fashion which is immune to noise, break down or information overload. Such a backbone system facilitates expansion of the communication matrix in a standard way when the project is growing or immediate responses are required to address dynamics of the jobsite.

Probably attributed to the more routine construction processes carried out in residential developments the dialogue between crew members is less vulnerable to language barriers. Our interview data reveals that even newcomers get used to jargons and slangs to refer to particular types of work in relatively short period of time. It was noticed that despite having different mother-tongues, the simplicity of the construction processes offsets the risk of communication failure even if broken common language (i.e. English) is used to exchange information at the work-front. With total weight of 13%, items of language barriers including jargon and different languages were the least concerning barriers among three recognized classes.

Self-interest, on the other hand was elaborated to count for the remaining 60% weight of significant barriers on site. Self-interest was found out to distort a message by compromising one's capacity as either receiver or sender. Selective perception, weighted 27%, deteriorates effective communication as the receiver tends to pick those parts of the message that are aligned with her/his self-interest. For example, labours interpret the verbal instruction in a way that results in less physical work usually by taking shortcuts. The same scenario applies to incomplete drawings allowing tradespersons make assumptions in their own favour. Oppositely, site personnel avoid initiation of a work-related conversation with people positioned higher than themselves in the hierarchy. This phenomenon, referred to as status block in the literature [12], identified to negatively impact site communication in 20% of occasions. However, status block isn't the only factor stoping field workers to send a message, in particular questing for information. To maintain a good reputation, rather than concentrating on the quality of outcome, sometimes workers lean to promise "unrealistic goals" or to be the "smart alec". Diverting the focus from construction process to self-image has been mentioned to reflect on accuracy of the exchanged information and consequently cause miscommunication. With similar 6.5% to the self-image barrier, site personnel's tendency to refine a message when acting as a mediator was recognised as the seventh common barrier to communication in demotic construction. Predominantly, if there is interest in hiding the information, refinement may be replaced by filtering. In either case, the spread of an altered message is a risk to communication.

2.2.2 Strategies to Alleviate the Barriers

Three tiers of preventive actions were extracted from responses provided by the interviewees. In fact, actions were grouped in different tiers based on frequency of their appearance as a practical solution to the issue of failed communication. The most demanding solution, sitting alone in tier 1 with 30% weight in terms of impact, is design of a robust architecture for site communication. The second tier incorporates regular update and proper documentation each with designated weight of 17.5%. The remaining five suggested solutions are grouped in the third tier given that they received the equal weight of 7% each. These five solutions include a) Maintaining clarity of information source, b) Confirming essence of the message with receiver, c) Provision of communication training, d) On-going discussion, and e) Development of ability to adapt. Figure 1 depicts how suggested solutions are mapped to the identified barriers.

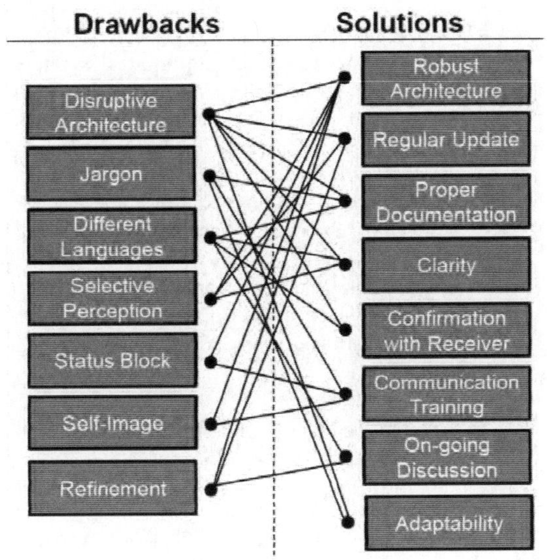

Figure 1. Potentail impact areas of suggest solutions

3 Conclusion

Project management literature refers to communication as the key cooperative element for success. However, pattern of communication at construction projects specifically at field-level is not profoundly examined yet. Accordingly, this study evaluated the existing communication architecture that governs information flow at residential construction jobsites. For this purpose, 10 ongoing projects were targeted by the research team and a representative from each site was interviewed. The analysis of collected data depicted the critical impact of project managers on effectiveness of communication in housing projects. Consistent with the project management literature, participants emphasized the substantial role of communication in successful delivery of residential projects. But, to some extent, in contrast to large-scale projects where official conduits are in existence to connect different teams, shortcuts were found to dominantly portray the information links between stakeholders attending the site. Finally, eight recommended actions were extracted from the interview data to address communication deficiency in domestic construction projects. Through deeper understanding of characteristics of and barriers to effective communication, this research persuades practicing professionals in the industry to optimise communication in their projects.

References

[1] Trenholm S. *Human Communication Theory*, 2nd ed. Englewood Cliffs, N.J.: Englewood Cliffs, N.J. Prentice Hall, 1991.

[2] Bubshait A. and Siddiqui M.K. and Al-Buali A.L. Role of Communication and Coordination in Project Success: Case Study. *Journal of Performance of Constructed Facilities*, 2014.

[3] Zulch B. Communication: The Foundation of Project Management. *Procedia Technology*, 2014.

[4] Emmitt S. and Gorse A.C. Investigating Interpersonal Communication During Construction Progress Meetings: Challenges and Opportunities. *Engineering, Construction and Architectural Management*, 10(4):234-244, 2003.

[5] Verčič T. and Verčič A.D. and Sriramesh K. Internal Communication: Definition, Parameters, and the Future. *Public Relations Review,* 38(2):223-230, 2012.

[6] Fisk E.R. and Reynolds W.D. *Construction Project Administration*, Prentice Hall, 2010.

[7] Jones S.E. and LeBaron C.D. Research on the relationship between verbal and nonverbal communication: Emerging integrations. *Journal of Communication*, 52(3)499-521, 2002.

[8] Long D.N. and Ogunlana S.O. and Lan D.T.X. A study on project success factors in large construction projects in Vietnam. *Engineering, Construction and Architectural Management*, 11(6):404-413, 2004.

[9] El-Saboni M. and Aouad G. and Sabouni A. Electronic Communication Systems Effects on the Success of Construction Projects in United Arab Emirates. *Advanced Engineering Informatics*, 23(1):130-138, 2008.

[10] Ayudhya B.I.N. Appraisal of Common Dispute Problems on Public Works Projects in Thailand. *Journal of Civil Engineering and Architecture*, 5(10), 2011.

[11] Koushki P.A. and Al-Rashid K. and Kartam N. Delays and cost increases in the construction of private residential projects in Kuwait. *Construction Management and Economics*, 23(3): 285-294, 2005.

[12] Rai U. *Effective Communication,* Himalaya Pub. House, Mumbai, India, 2009.

Interaction Matters? Exploring Interactive Music as a Reminder to Break Sedentary Office Time

Xipei Ren, Yuan Lu, Vincent Visser, P.D.Huy Le and Reijn van den Burg

Department of Industrial Design, Eindhoven University of Technology, the Netherlands
E-mail: {x.ren, y.lu}@tue.nl, {v.j.j.visser, p.d.h.le, r.s.v.d.burg}@student.tue.nl

Abstract –

This paper presents a within-subject field test (n=24) with Flow platform, a smart cushion that uses interactive music to motivate office workers to break excessive sedentary time. In this study, we compared continuous music and interactive music as reminders to inform sedentary time by every 15-min. After each treatment, a questionnaire adapted from Flow State Scale was used to measure participants' experiences and thus motivations with interactive and non-interactive music. Moreover, a post-interview was conducted with every participant individually to gather insights on the challenges and opportunities for current design. This study examines if the interactivity as an additional factor could influence the music as a motivator to prevent sedentary behaviours among office workers. Our findings suggest a further design iteration to appropriate the concept more to the context of office.

Keywords –

Sedentary behaviour; Interactive music; Flow experience; Motivational design

1 Introduction

Nowadays, the rapid growth of office automation is largely associated with sedentary behaviours, which has been proven to threaten cardio metabolic health among office workers and thus to contribute to various preventable causes of death in later life stage [1]. Evidence shows that the effect of too much sitting displaces the benefit from partial exercise [2]. To overcome this issue, Owen *et al.* [3] suggested to break excessive sedentary time and to encourage light intensity exercises[1] in repeated sessions throughout the day, such as the frequent changes of sitting postures from one to another. Based on this observation, we previously designed Flow platform [4], a novel technology that learns and logs sitting behaviour from users and in return uses interactive music to encourage users to change sitting postures with different dynamics.

From the earlier study [4], we learned that music in general motivates physically more active performance. Yet, it is unclear: (i) if the current interactivity of music in Flow platform can leverage the motivation to aid sedentary behaviours among office workers; (ii) how to improve the current interaction for a better motivation in the future.

To investigate this further, we conducted a within-subject field-test (n=24) with Flow platform, where we compared continuous music and interactive music respectively as reminders to inform sedentary time by 15-min intervals. We carried each treatment out by an hour per participant and recorded their sitting posture data by our technology. After each treatment, participants were asked to fill in the questionnaire adapted from Flow State Scale [5], a 36-item five Likert Scale questionnaire explains why people engage in certain activities intrinsically, to understand their experiences using Flow platform and thus motivations to sit actively with non-interactive/interactive music. Moreover, we also conducted a post hoc interview with every participant individually to gain some insights for further design.

By comparing quantitative data, we learned the current interaction could not make too much impact to prevent sedentary behaviours for target users. From the qualitative results, however, we found a few insights suggested us to implement the interactive music more appropriately for the office workers. This paper makes two contributions. First, it examines interactivity as another factor to influence motivational music in aiding sedentary office people. Second, it identifies extra insights to account for when designing interactive music as an interface to support behaviour change.

This paper proceeds as follows. Next section we provide a brief overview of Flow platform [4] and the review on current ambient technologies that reduce sedentary office time, which motivate us to conduct this study. In section 3, we describe on how we designed our experiment, which is followed by the results from this study in section 4. Conclusion is given in the end.

---

[1] **light intensity exercise:** the exercise in the energy expenditure range of 1.9 to 2.9 METs (multiples of the basal metabolic rate)

2 Related Work

2.1 Flow Cushion

Earlier to this study, we designed Flow platform [4], a smart cushion that can be placed on every normal chair to track and archive sitting behaviours from the user. Consequently, if the user remains one sitting posture for an extended period of time (15 min in our case), the system would use the interactive music to stimulate the user to shift to the different postures according to the musical dynamics (see Figure 1).

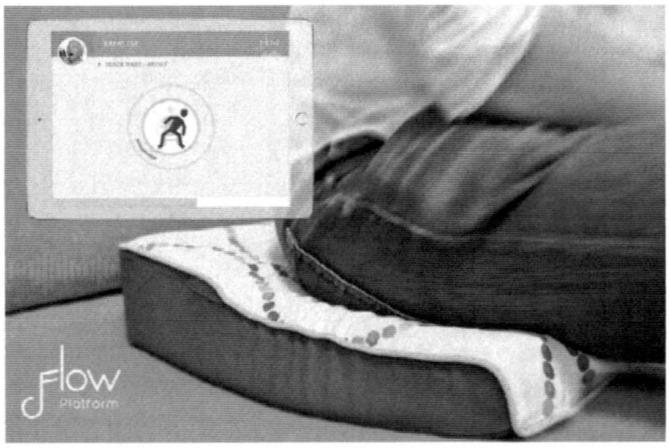

Figure 1. Flow platform is designed to reduce sedentary time by enabling the user to interact with different musical dynamics.

Flow platform was developed in three steps, consisting sensor implementation, interaction design, and experience design. Since in this study we have only examined on the interactive music part, in the following we elaborate on the first two steps specifically to appropriate how we implement the system technically and how the music works in the system.

2.1.1 Sensor Implementation

Inspired by previous design research [6] and products [7], we built a fabric pad (see Figure 2) that could be embedded into the seat cushion. Six square type force resistor sensors (FRS) were placed according to sedentary pressure map for efficient detection area coverage. We connected the sensor pad with Arduino Yun, a small micro-controller prototyping board, which connected to a Bluetooth module. In this way, Flow platform was able to synchronise data periodically (once every second) over Bluetooth to computer or smartphone.

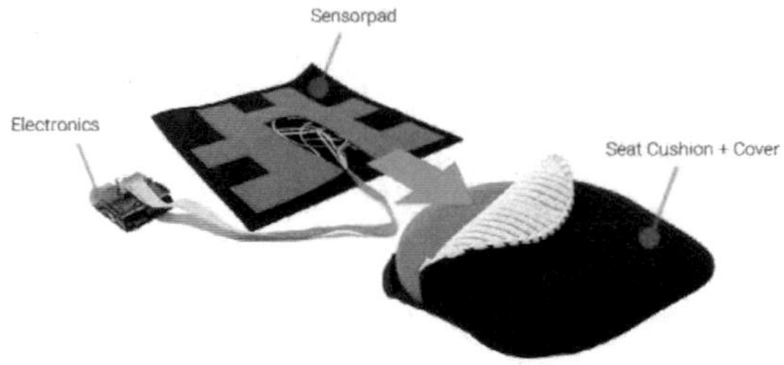

Figure 2. The sensor-pad of Flow platform, which contains six square FRS configured with the reference of [8].

To determine the user's sitting posture, we applied a specialized Artificial Neuron Network (ANN) [4] in the software (see Figure 3), by which the system can calibrate movement range according to different person. With the Wi-Fi connection, therefore, Flow platform can measure and analyse the user's sitting behaviour over a distance and accordingly customise the sitting exercise.

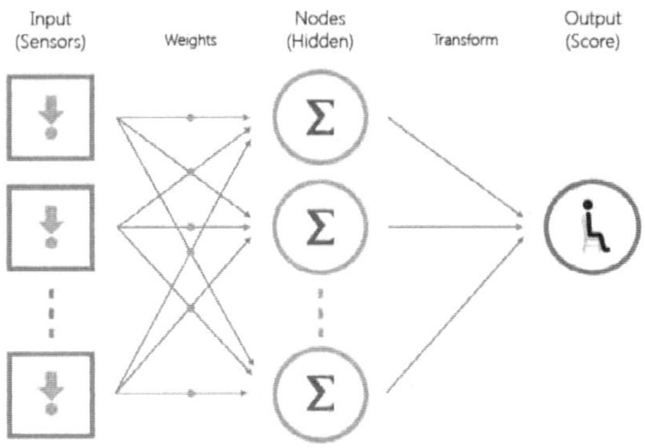

Figure 3. The ANN used in Flow platform.

2.1.2 Interaction Design

The musical interaction in Flow platform was designed based on observation [4] in an elderly care centre as well as several ideation sessions in the university. The instrumental music was chosen in the end aiming to improve the exercise adherence [9]. We designed the mechanism as follows (see Figure 4): *If the user sits stationary for 15 minutes, the volume of the music will decrease gradually. Then the user needs to shift the sitting postures actively to literally interact with the music for 2 minutes. When the goal has been achieved, the volume goes back to the normal range.*

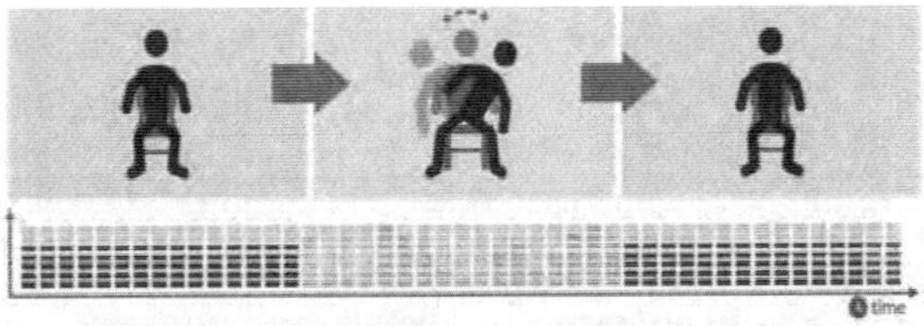

Figure 4. the logic of how the interactive music work in Flow platform to support the user to avoid sedentariness.

This system was originally designed for the elderly people. During the user test and expert consultation, however, both the target users and physiotherapists suggested the musical intervention we used here could also be used for the office workers to aid the prolonged sedentary working time. They gave this advice as they thought the interactive music reminds people to be aware of the sedentary behaviour in a non-obtrusive way. In addition, as far as we know, music can sometimes accompany office workers during work hours to increase the work efficiency [10]. Therefore, we decide to examine our design with the office workers as well to explore the opportunities and challenges of our design.

In fact, designing ambient technology [11] to remind users on the sedentary behaviour in the office settings was not new in HCI, as in such context people wish to focus on the task at hand rather than bothered by the technology [12]. Many prior studies have been investigated this area using different interaction techniques [13]. In the following, we review the earlier works in detail.

2.2 Ambient Display to Aid Sedentary Time

As a growing design and research focus in HCI, ambient technology [11] is featured by presenting information to users requiring the smallest amount of attention and allows users to interact with the technology in a non-obtrusive way. In the scope of this study, ambient technology offers a trade-off between raising office workers' awareness on sedentary behaviours and the demand to focus on work at hand. As one of the earliest study in this area, Jafarinaimi *et al.* [14] identified design guidelines for ambient display to promote physically active behaviours, consisting ***abstract***, ***non-intrusive***, ***public***, and ***aesthetic***. They also developed a tangible prototype called Breakaway using biomimetic metaphor to further explain their design goals. Similar to Breakaway, Hong *et al.* [15] designed a flower-shaped avatar to provide an ambient feedback about different sitting behaviours. In addition, other display techniques, such as ambient lighting [16] [17] and haptic feedback [18], have also been widely employed in relevant design.

In a more recent PhD project, Probst [19] explored this area from a task-based perspective, where different forms of peripheral interaction [20] were embedded in different objects (e.g. [21], [22]) in the office environment. As such, the proposed technologies enable office workers to use diverse body movements to execute different office tasks in order to enhance physical activeness during work hours. Probst's works clearly broaden ambient technology for sedentary behaviour from only stimulating the periphery of perception to stimulating the periphery of physical interaction [20]. In line with this direction, Flow platform uses interactive music to not only remind people in a way fits for their working routine, but also provide a channel for them to do physical exercise while working without any disturbance.

To examine further, we set out a within-subject experiment (n=24) to explore the potential role of interactive music in Flow platform to motivate more activeness among office workers during work hours.

3 Experiment Design

3.1 Scenario

To better explain how Flow Cushion should work in the workplace environment, we apply it to a use case and propose the following scenario:

Jeff is a junior computer programmer, who has four tasks to accomplish this week. Therefore, he spent eight hours per day sitting at the desk to write codes for two ongoing projects, look up the algorithms for a new project, and write a report for a past project. He starts to feel pain in his back and neck recently and wants to reduce the prolonged sitting. Yet, he always forgets how long he has been sitting when focusing on the task at hand. Plus, he does not want to use any system that literally takes him away from the work. He looks up the products that match his needs and then finds Flow platform. He really appreciates the idea of using music to remind the inappropriate sitting behaviours, as he always listens to the music while working. So he thinks the musical reminder will fit his working routine in a non-obtrusive way. After installed Flow cushion on his chair and Flow app on his computer, he wears headset, switches on his favourite music album, and carries on working. 15 minutes later, he finds the music becomes lower. Then he changes his posture one to another randomly within his preferred movement range and keeps his main focus on current task. After several attempts on different movements, the volume of the music backs to the normal range. Jeff then stops shifting postures and continue on working without any pause. At the end of the week, Jeff finishes all the four tasks in time and still feels energetic for the weekend. Jeff thinks that Flow platform can help him to maintain physical active during the work without any interruption. He feels refreshed after the sitting exercise, rewarded when the music comes back, and more focused on the work.

This scenario illustrates how the interactive music system works in the workplace settings, following the design guideline from [14], which abstracts the information and makes the presentation non-obtrusively. Our system also provides office workers a channel to interact with the music in a pleasant way. To eliminate the disturbance on co-workers in the same place, Flow platform in this case is advised to use headset to restrict the publicity of the music.

3.2 Participants

We recruited the participants by spreading information via word-of-mouth through a snowball sampling approach: We began by asking people we knew that exhibited similar characteristics as our target users. We then asked them to pass along the information about our study to their social contacts. Eventually, there were 24 volunteers participating in the study. This closes to the minimum number of people required as to obtain statistically sound results. The participants were all university students or junior office workers. They were aged between 18 and 25 (mean age: 21.46, standard deviation: 1.67).

3.3 Procedure

Each participant was subjected to two tests – one test in which they used the FLOW cushion with interactive music, and one test in which they used the FLOW cushion with non-interactive music. 50% of the participants was subjected to the interactive music first, the other 50% to the non-interactive music.

Before each test, participants were asked to sign a consent form and were briefly introduced to the concept of the FLOW platform - that the cushion registers their sitting behaviour and that the music was part of the platform. Moreover, participants were instructed that they were allowed to do any activity as long as they remained seated on the cushion. Participants were furthermore not instructed to achieve any particular goal.

The sitting positions of the participants were logged during the tests. Each of these tests had a duration of one hour. We pre-set the volume interruption of the music at 15-min interval persisting 2-min slot for interaction. At the end of each test, participants were immediately instructed to fill in a questionnaire and to answer various open questions regarding their experience with the FLOW cushion and the music.

3.4 Data Collection

During each test, three forms of data were collected: the registrations of shifts in postures, questionnaire results (quantitative data) and interview results (qualitative data).

3.4.1 Registration of the Shift in Postures

The software automatically registers and collects the date and time of each shift in posture and the specific sitting position (represented as a number) during each test. This data is collected during both the tests with continuous music and interactive music, and will be compared with each other in order to conclude whether there are any significant differences.

3.4.2 Questionnaire

The questionnaire is based on the FLOW state scale. It measures nine aspects experienced by the participant during the tests: challenge-skill balance, clear goals, paradox of control, autotelic experience, transformation of time, loss of self-consciousness, concentration on task at hand, unambiguous feedback, and action-awareness merging. These categorized questions give us a better understanding of the challenges and feelings experienced by the participants during the tests [5].

Challenge-Skill Balance (continuous music (C): $\alpha = .590$; interactive music (I): $\alpha = .362$): "*In flow, the person perceives a balance between the challenges of a situation and one's skills, with both operating at a personally high level.*"

Action-Awareness Merging (C: $\alpha = .565$; I: $\alpha = .728$): "*Involvement in the flow activity is so deep that it becomes spontaneous or automatic. There is no awareness of self as separate from the actions one is performing.*"

Clear Goals (C: $\alpha = .912$; I: $\alpha = .858$): "*Goals in the activity are clearly defined (either set in advance or developed out of involvement in the activity), giving the person in flow a strong sense of what he or she is going to do.*"

Unambiguous Feedback (C: $\alpha = .823$; I: $\alpha = .795$): "*Immediate and clear feedback is received, usually from the activity itself, allowing the person to know he or she is succeeding in the set goal.*"

Concentration on Task at Hand (C: $\alpha = .876$; I: $\alpha = .844$): "*Total concentration on the task at hand occurs when in flow.*"

Paradox of Control (C: $\alpha = .725$; I: $\alpha = .798$): "*A sense of exercising control is experienced, without the person actively trying to exert control.*"

Loss of Self-Consciousness (C: $\alpha = .395$; I: $\alpha = .832$): "*Concern for the self disappears during flow as the person becomes one with the activity.*"

Transformation of Time (C: $\alpha = .797$; I: $\alpha = .839$): "*Time alters perceptibly, either slowing down or* *speeding up. Alternatively, time may simply become irrelevant and out of one's awareness."*

Autotelic Experience (C: α= .900; I: α= .824): "*An autotelic experience is an intrinsically rewarding experience.*"

3.4.3 Interview

At the end of each session with the FLOW cushion, participants were asked a few open questions regarding their experience with the flow cushion and the interactive/non-interactive music. All the interviews were audio recorded upon the approval of the participants. Although not all interviews were completely similar, they generally followed the script below:

After non-interactive music treatment:

- What do you feel with FLOW?
- How did you feel about doing exercises, such as shifting postures, with the FLOW cushion?
- What do you think about music as a reminder to break your sedentary time?
- What kind of music would you like to hear with FLOW?

After interactive music treatment:

- What do you feel with FLOW?
- How did you feel about doing exercises, such as shifting postures, with the FLOW cushion?
- What do you think of interactive music as a reminder for shifting postures?
- What kind of music would you like to hear with FLOW?
- Would you like to use the FLOW cushion more often?
- What would you like to change to the design?

3.5 Data Analysis

3.5.1 Registered Posture Data Analysis

We initially planned to compare the registered posture data between the treatment with continuous music and another with interactive music, to check if interactive music contributes to more posture shifts at a significant level. However, due to a logical error in our program in the beginning, we did not successfully archive the posture data in our cloud. We only realised this error at the very last stage of the experiment. Given this issue, we did not collect the complete sitting data from each test for any participant. In this study, therefore, we are not able to analyse if there was an impact of interactive music for more dynamic sitting posture shifts for the office workers. Yet, it was a meaningful lesson for us to learn that before the experiment every single detail needs to be double-checked in case any error occurs in the experiment. Moreover, this mistake also helped us to fine-tune our program technically for future design and research.

3.5.2 Questionnaire Data Analysis

To compare the questionnaire data between two treatments, the normality test of distribution was conducted at first using Kolmogorov-Smirnov test. Given most of the categories between two treatments were not normally distributed, we then conducted related-sample nonparametric analysis using Wilcoxon test to compare the difference between two tests.

3.5.3 Interview Data Analysis

We clustered qualitative data based on logical closeness on different feedbacks following the inductive approach [23]. Here we aimed to inquire any valuable insights that emerged from our study to confirm our current design concept to some extent as well as to enlighten our future design.

4 Findings

In this section, we report on the result from both our questionnaire and our post hoc interviews.

4.1 Quantitative Findings

Table 1. Mean score and comparison of flow experience between continuous and interactive music

Continuous Music (n=24)	Interactive Music (n=24)	Wilcoxon Test
Challenge-Skill Balance		
M 3.24 SD 0.70	M 3.29 SD 0.62	*Z* -.607 *P* .544
Action-Awareness Merging		
M 3.67 SD 0.63	M 3.69 SD 0.77	*Z* -.677 *P* .498
Clear Goals		
M 3.77 SD 1.05	M 3.46 SD 1.09	*Z* -1.569 *P* .117
Unambiguous Feedback		
M 3.20 SD 0.73	M 3.18 SD 0.90	*Z* -.264 *P* .792
Concentration on Task at Hand		
M 3.17 SD 1.03	M 2.84 SD 0.99	*Z* -1.412 *P* .158
Paradox of Control		
M 3.82 SD 0.62	M 3.64 SD 0.92	*Z* -1.523 *P* .128
Loss of Self-Consciousness		
M 4.14 SD 0.63	M 4.17 SD 0.86	*Z* -.328 *P* .743
Transformation of Time		
M 2.54 SD 1.00	M 2.53 SD 1.10	*Z* -.019 *P* .985
Autotelic Experience		
M 2.85 SD 0.78	M 2.99 SD 0.82	*Z* -1.073 *P* .283

As can be seen from Table 1, the Wilcoxon test

suggests none of the considered categories between two treatments are different at the significant level below 0.05. From this study, we learned that the current interactivity of the music in Flow platform made no different flow state to the normal music. For the office workers, we therefore conclude that using both forms of music to break sedentary time causes the engaging user experiences at the same level.

4.2 Qualitative Findings

Our findings from the interviews compensate the quantitative results from the questionnaire. In what follows, we present the general experience with Flow platform as well as the opportunities for future design.

4.2.1 General Experience

From the interview, the majority of participants indicated that Flow platform increased their awareness about their sedentariness and posture behaviour. More specifically, both two musical treatments were considered to be appropriate motivators for the participants to get sitting posture shifts. We also learned that the future use of Flow platform was generally considered plausible by most of our participants. However, participants, who already moved often during work, did not see the clear benefit from Flow to make them more active on the working routine. Which means that our design is supposed to be more appropriate for the real sedentary office workers. This finding suggests a thorough study on profiling the target users should be done here to better positioning our work for the context.

What is more, a few participants added the comments that the final execution of the concept would influence their willing to use the system to some extent. Therefore, we need to do more works to finalise the detail design of Flow platform. In the part of design opportunities, we will discuss this further.

4.2.2 Design Opportunities

During the interviews, we also found some issues that made negative impacts on our test. However, we believe these findings will help us to refine the current design largely and in turn appropriate Flow platform more to our target users. Below we discuss on what participants found about our design. In response, we propose the design strategies we are going to use to optimise Flow platform.

First of all, although the Flow cushion itself is a prototype, some participants left the remarks that the current design was not being very physically comfort enough to support long-term use. These comments were basically related to the wires were exposed on the seat. Since our prototype needs an external power supply for long-term usage, we connected the cushion with a power plug during our experiments. This setting, however, restricted participants' freedom to move during the study. For the future step, we consider to embed a rechargeable battery in the cushion, which has enough capacity for longer time use (e.g., one week per charge). As such, we are confident that the problem in regard with the ergonomics of Flow platform will be addressed.

Second, most of the concerns we received from the participants were pointed to the interactivity of the music we implemented in the test. On the one hand, some participants claimed that the volume of the interactive music changed too quickly for activities that require more serious concentration, and thus study and office work became irritating in the end. This suggests us to lower the speed of volume alternation. On the other hand, one participant stated that the intended effect (the interactivity of the volume) should be stronger to increase the awareness on sitting postures. All these comments suggest we should cater the changing pace for every individual. A more viable way would be enabling user to customise the speed themselves.

Third, although participants found the alternation of volume was interesting and motivational for them to break prolonged sitting, they also encouraged us to investigate on the applicability of other psycho-acoustic parameters to leverage active behaviours. For example, it is also possible to suggest an active break by slowing down the rhythm of the music. It is hard to say which parameter of music is the best motivator in our case. Following the prior suggestion, our design could be made to be more customisable to let users to select their own favourite parameter(s) and even more to pre-set diverse combinations of parameter(s) according to the context differences.

Fourth, some participants commented that different working tasks require different levels of concentrations. Thus not all types of music were appropriate in this context. Instrumental music is perhaps beneficial for physical activities, but it is too energetic for moderate exercise when the major focus was on the working task. Most of the participants suggested us to use ambient sound and peripheral music for future design.

5 Conclusion

In this paper, we present a within-subject user test (n=24) on our design of Flow platform with office people to explore how to implement interactive music into our technology to break sedentary office time more appropriately. Our findings suggested in general the music was considered to be an appropriate motivator for the office workers to aid their sedentary behaviours. Yet there is room for our interaction design to be more appreciated by the target users. On the one hand, specifying sportive people and sedentary people in our case and accordingly using different design strategies would be helpful to narrow down the scope and improve the effectiveness of our design. On the other hand, the threshold of interactivity and the interactive parameters of music could be designed customisable for the end users. Moreover, the type of the music and the ergonomics of the cushion should be more comfortable for the given context. Although a lot of effort needs to be spent for further development and validation of our design, we still believe the results from this study can contribute to relevant works on the interactive music for motivational design and thus design for health and wellbeing in the HCI community.

Acknowledgement

The first author of this paper is being sponsored by China Scholarship Council (CSC). We thank all the participants who volunteered to take part in this study.

References

[1] Owen N, Bauman A, Brown W. Too much sitting: a novel and important predictor of chronic disease risk?. *British Journal of Sports Medicine*, 43(2): 81-83, 2008.

[2] Healy GN, Dunstan DW, Salmon J, Cerin E, Shaw JE, Zimmet PZ, Owen N. Breaks in sedentary time. *Diabetes care*,31(4):661-666, 2008.

[3] Owen N, Healy GN, Matthews CE, Dunstan DW. Too much sitting: the population-health science of sedentary behavior. *Exercise and sport sciences reviews*, 38(3):105-113, 2010.

[4] Ren X, Visser V, Lu Y, Brankaert R, Offermans S, Nagtzaam H. FLOW pillow: exploring sitting experience towards active ageing. In *Proceedings of the 18th ACM International Conference on Human-Computer Interaction with Mobile Devices and Services Adjunct*, page 706-713, Florence, Italy, 2016.

[5] Jackson SA, Marsh HW. Development and validation of a scale to measure optimal experience: The Flow State Scale. *Journal of sport and exercise psychology*, 18(1):17-35, 1996.

[6] Min DA, Kim Y, Jang SA, Kim KY, Jung SE, Lee JH. Pretty Pelvis: A Virtual Pet Application That Breaks Sedentary Time by Promoting Gestural Interaction. In *Proceedings of the 33rd Annual ACM Conference Extended Abstracts on Human Factors in Computing Systems*, page 1259-1264, Soul, Korea, 2015.

[7] Darma co. Darma Sit Smart. On-line: http://darma.co/index.html, Accessed: 28/03/2017.

[8] Brienza D, Kelsey S, Karg P, Allegretti A, Olson M, Schmeler M, Zanca J, Geyer MJ, Kusturiss M, Holm M. A randomized clinical trial on preventing pressure ulcers with wheelchair seat cushions. *Journal of the American Geriatrics Society*, 58(12):2308-2314, 2010.

[9] Schutzer KA, Graves BS. Barriers and motivations to exercise in older adults. *Preventive medicine*, 39(5):1056-1061, 2004.

[10] Lesiuk T. The effect of music listening on work performance. *Psychology of music*, 33(2):173-191, 2005.

[11] Weiser M, Brown JS. The coming age of calm technology. In Beyond calculation, page 75-85, New York, USA, 1997.

[12] Cook DJ, Augusto JC, Jakkula VR. Ambient intelligence: Technologies, applications, and opportunities. *Pervasive and Mobile Computing*, 5(4):277-298, 2009.

[13] Haller M, Richter C, Brandl P, Gross S, Schossleitner G, Schrempf A, Nii H, Sugimoto M, Inami M. Finding the right way for interrupting people improving their sitting posture. In *IFIP Conference on Human-Computer Interaction 2011*, page 1-17, Lisbon, Portugal, 2011.

[14] Jafarinaimi N, Forlizzi J, Hurst A, Zimmerman J. Breakaway: an ambient display designed to change human behavior. In *Proceedings of the 23rd Annual ACM Conference Extended Abstracts on Human Factors in Computing Systems*, page 1945-1948, Portland, USA, 2005.

[15] Hong JK, Koo BC, Ban SR, Cho JD, Bianchi A. BeuPo: a digital plant that you can raise and customize with your current posture. In Adjunct Proceedings of *the 2015 ACM International Joint Conference on Pervasive and Ubiquitous Computing*, page 1015-1020, Osaka, Japan, 2015.

[16] Fortmann J, Stratmann TC, Boll S, Poppinga B, Heuten W. Make me move at work! An ambient light display to increase physical activity. In *Proceedings of the 7th International Conference on Pervasive Computing Technologies for Healthcare,* page 274-277, Venice, Italy, 2013.

[17] Mateevitsi V, Reda K, Leigh J, Johnson A. The health bar: a persuasive ambient display to improve the office worker's well being. In *Proceedings of the 5th Augmented Human International Conference*, page 21-22, Kobe, Japan, 2014.

[18] Zheng Y, Morrell JB. A vibrotactile feedback approach to posture guidance. In *IEEE Haptics Symposium*, page 351-358, Waltham, USA, 2010.

[19] Probst K. Active office: designing for physical activity in digital workplaces. In *Proceedings of the 14th International Conference on Mobile and*

Ubiquitous Multimedia, page 433-438, Linz, Austria, 2015.

[20] Bakker S, van den Hoven E, Eggen B. Peripheral interaction: characteristics and considerations. *Personal and Ubiquitous Computing*, 19(1):239-254, 2015.

[21] Probst K, Lindlbauer D, Haller M, Schwartz B, Schrempf A. A chair as ubiquitous input device: Exploring semaphoric chair gestures for focused and peripheral interaction. In *Proceedings of the 32nd Annual ACM Conference on Human Factors in Computing Systems*, page 4097-4106, Toronto, Canada, 2014.

[22] Probst K, Lindlbauer D, Perteneder F, Haller M, Schwartz B, Schrempf A. Exploring the use of distributed multiple monitors within an activity-promoting sit-and-stand office workspace. In *IFIP Conference on Human-Computer Interaction 2013*, page 476-493, Cape Town, South Africa, 2013.

[23] Thomas DR. A general inductive approach for analyzing qualitative evaluation data. *American journal of evaluation*, 27(2):237-246, 2006.

AUTHOR INDEX

AUTHOR INDEX

AUTHOR INDEX

AUTHOR INDEX

AUTHOR INDEX

International Association for Automation
and Robotics in Construction (IAARC)
Slovak University of Technology in Bratislava
Faculty of Engineering
Radlinského 11
813 68 Bratislava
SK - Slovakia

ISBN 978-1-5108-4473-5

CURRAN ASSOCIATES INC.
proceedings.com